Lecture Notes in Artificial Intelligence 4304

Edited by J. G. Carbonell and J. Siekmann

Subseries of Lecture Notes in Computer Science

RECEIVED

FEB 2 1 2007

UTD LIBRARY
ACQUISITION

Abdul Sattar   Byeong-Ho Kang (Eds.)

AI 2006: Advances in Artificial Intelligence

19th Australian Joint Conference on Artificial Intelligence
Hobart, Australia, December 4-8, 2006
Proceedings

Springer

Series Editors

Jaime G. Carbonell, Carnegie Mellon University, Pittsburgh, PA, USA
Jörg Siekmann, University of Saarland, Saarbrücken, Germany

Volume Editors

Abdul Sattar
Griffith University, Institute for Integrated and Intelligent Systems
Parklands Drive, Southport, QLD, Australia
E-mail: a.sattar@griffith.edu.au

Byeong-Ho Kang
University of Tasmania, School of Computing
Hobart Campus, Centenary Building, Hobart, TAS 7001, Australia
E-mail: bhkang@utas.edu.au

Library of Congress Control Number: 2006936897

CR Subject Classification (1998): I.2, F.4.1, H.3, H.2.8, F.1

LNCS Sublibrary: SL 7 – Artificial Intelligence

ISSN   0302-9743
ISBN-10   3-540-49787-0 Springer Berlin Heidelberg New York
ISBN-13   978-3-540-49787-5 Springer Berlin Heidelberg New York

This work is subject to copyright. All rights are reserved, whether the whole or part of the material is
concerned, specifically the rights of translation, reprinting, re-use of illustrations, recitation, broadcasting,
reproduction on microfilms or in any other way, and storage in data banks. Duplication of this publication
or parts thereof is permitted only under the provisions of the German Copyright Law of September 9, 1965,
in its current version, and permission for use must always be obtained from Springer. Violations are liable
to prosecution under the German Copyright Law.

Springer is a part of Springer Science+Business Media

springer.com

© Springer-Verlag Berlin Heidelberg 2006
Printed in Germany

Typesetting: Camera-ready by author, data conversion by Scientific Publishing Services, Chennai, India
Printed on acid-free paper   SPIN: 11941439   06/3142   5 4 3 2 1 0

Organization

AI 2006 was hosted and organized by the University of Tasmania, Australia. The conference was held at the Wrest Point Hotel, Hobart, from December 4 to 8, 2006.

Conference Committee

Conference Co-chairs
    John Lloyd (Australian National University)
    Christopher Lueg (University of Tasmania)
Program Committee Co-chairs
    Abdul Sattar (Griffith University)
    Byeong Ho Kang (University of Tasmania)
Local Organizing Co-chairs
    Ray Williams (University of Tasmania)
    Mark Hepburn (University of Tasmania)
Workshop Chair
    Peter Vamplew (University of Ballarat)

Senior Program Committee

Paul Compton (University of New South Wales)
Dan Corbett (Science Applications International Corporation (SAIC), USA)
John Debenham (University of Technology, Sydney)
Aditya K. Ghose (University of Wollongong)
Vladmir Estvill-Castro (Griffith University)
Achim Hoffmann (University of New South Wales)
Ray Jarvis (Monash University)
John Lloyd (Australian National University)
Hiroshi Motoda (Asian Office of Aerospace Research & Development, US (AOARD), Japan)
Dickson Lukose (DL Informatique)
Mehmet Orgun (Macquarie University)
Markus Stumptner (University of South Australia)
Toby Walsh (National ICT Australia)
Geoff West (Curtin University)
Wayne Wobcke (University of New South Wales)
Chengqi Zhang (University of Technology, Sydney)

Program Committee

Hussein Abbass, Australia
Anbulagan, Australia
Douglas Aberdeen, Australia
Tae-Chon Ahn, Korea
David Albrecht, Australia
Sansanee Auephanwiriyakul, Thailand
James Bailey, Australia
Mike Bain, Australia
Nick Barnes, Australia
Peter Baumgartner, Australia
David Benn, Australia
Mohammed Bennamoun, Australia
David Billington, Australia
Michael Blumenstein, Australia
Adi Botea, Canada
Mike Brooks, Australia
Olivier Buffet, Australia
Gulcin Buyukozkan, Turkey
Mike Cameron-Jones, Australia
Longbing Cao, Australia
Lawrence Cavedon, Australia
Qingfeng Chen, Australia
Yi-Ping Phoebe Chen, Australia
William Cheung, Hong Kong
Sung-Bae Cho, Korea
Vic Ciesielski, Australia
Bob Colomb, Australia
David Cornforth, Australia
Khanh Hoa Dam, Australia
Sandy Dance, Australia
Richard Dazeley, Australia
David DOWE, Australia
Hongjian Fan, Australia
Jesualdo Tomás Fernandez-Breis, Spain
Robert Fitch, Australia
Alfredo Gadaldon, Australia
Xiaoying Gao, New Zealand
Raj Gopalan, Australia
Guido Governatori, Australia
Alban Grastien, Australia
Jong-Eun Ha, Korea
Jacky Hartnett, Australia
Ray Hashemi, USA

Patrik Haslum, Australia
Tim Hendtlass, Australia
Mark Hepburn, Australia
Nicole Herbert, Australia
He Huang, China
Xiaodi Huang, Australia
Heath James, New Zealand
Andrew Jennings, Australia
Huidong Jin, Australia
Zhi Jin, China
Sung-Hae Jun, Korea
Waleed Kaduos, Australia
Byeong Ho Kang, Australia
Seung-Shik Kang, Korea
Masayoshi Kanoh, Japan
Andrei Kelarev, Australia
Graham Kendall, UK
Paul Kennedy, Australia
ByungJoo Kim, Korea
Mihye Kim, Korea
Taek-Hun Kim, Korea
Irwin King, Hong Kong
Kevin Korb, Australia
Lara Kornienko, Australia
Ryszard Kowalczyk, Australia
Bor-Chen Kuo, Taiwan
Mu-Hsing Kuo, Taiwan
Mihai Lazarescu, Australia
Chris Leckie, Australia
Chang-Hwan Lee, Korea
Jungbae Lee, Korea
Kyung Ho Lee, Korea
Kang Hyuk Lee, Korea
Ian Lewis, Australia
Jiaming Li, Australia
Li Li, Australia
Wei Li, Australia
Xiaodong Li, Australia
Chunsheng Li, China
Yuefeng Li, Australia
Gang Li, Australia
Li Lin, Australia
Wolfram-M Lippe, Germany

Wei Liu, Australia
Jingli Lu, New Zealand
Xudong Luo, UK
Michael Maher, Australia
Ashesh Mahidadia, Australia
Frederic Maire, Australia
Vishv Malhotra, Australia
Graham Mann, Australia
Eric Martin, Australia
Rodrigo Martínez-Béjar, Spain
Barry McCollum, UK
Thomas Meyer, Australia
Kyoungho Min, New Zealand
Azah Mohamed, Malaysia
Sanjeev Naguleswaran, Australia
David Newth, Australia
Minh Le Nguyen, Japan
Ann Nicholson, Australia
Kozo Ohara, Japan
Robert Ollington, Australia
Kok-Leong Ong, Australia
Maurice Pagnucco, Australia
Adrian Pearce, Australia
Sanja Petrovic, UK
Tuan Pham, Australia
Zhenxing Qin, Australia
Marcus Randall, Australia
Jochen Renz, Australia
Debbie Richards, Australia
Jussi Rintanen, Australia
Daniel Rolf, Australia
Bernard Rolfe, Australia
Malcolm Ryan, Australia
Seiichiro Sakurai, Japan
Arthur Sale, Australia
Claude Sammut, Australia
Conrad Sanderson, Australia
Ruhul Sarker, Australia
Abdul Sattar, Australia
Rolf Schwitter, Australia
Zhiping Shi, China
Simeon J. Simoff, Australia
Andrew Slater, Australia
Liz Sonenberg, Australia
Andy Song, Australia

Marcilio Carlos Pereira de Souto, Brazil
David Squire, Australia
Andrew Stranieri, Australia
Kaile Su, Australia
Ramasubramanian Sundararajan, India
John Thangarajah, Australia
John Thornton, Australia
Shusaku Tsumoto, Japan
Rafael Valencia-García, Spain
Peter Vamplew, Australia
Hans van Ditmarsch, New Zealand
David Vengerov, USA
Brijesh Verma, Australia
Bao Vo, Australia
Vilem Vychodil, Czech Republic
Kewen Wang, Australia
Jiaqi Wang, Australia
Dianhui Wang, Australia
Chao Wang, Australia
Lipo Wang, Singapore
Mary-Anne Williams, Australia
Ray Williams, Australia
Bill Wilson, Australia
Michael Winikoff, Australia
Michal Wozniak, Poland
Fengjie Wu, Australia
Shuxiang Xu, Australia
Zhuoming Xu, China
Takahira Yamagucti, Japan
Ying Yang, Australia
Jianhua Yang, Australia
John Yearwood, Australia
Kenichi Yoshida, Japan
Dingrong Yuan, China
Debbie Zhang, Australia
Minjie Zhang, Australia
Zili Zhang, Australia
Mengjie Zhang, New Zealand
Yan Zhang, Australia
Jilian Zhang, China
Weicun Zhang, China
Guangquan Zhang, Australia
Shichao Zhang, Australia
Yanchang Zhao, Australia
Tatjana Zrimec, Australia

Additional Reviewers

Abramov, Vyacheslav
Achuthan, N. R.
Adbun, Mahnood
Albrecht, David
Allison, Lloyd
Atul, Sajjanhar
Bae, Eric
Bai, Quan
Banerjee, Arun
Barakat, Nahla
Becket, Ralph
Berg, George
Bindoff, Ivan
Blair, Alan
Booth, Richard
Botea, Adi
Brzostowski, Jakub
Cassidy, Steve
Chan, Jeffrey
Chen, Feng
Chen, Jie
Chen, Qingliang
Chen, Wanli
Chojnacki, Wojciech
Chomnphuwiset, Phattanaphong
Clarke, Bertrand
Cregan, Anne
De Voir, John
Elfeky, Ehab
Fogelman, Shoshana
Geng, Xin
Gomes, Eduardo
Green, Steve
Guo, Ying
Harland, James
Hasan, Kamrul
Haslum, Patrik
He, Hongxing
He, Minghua
Hebden, Peter
Hengel, Anton van den
Hernandez Orallo, Jose
Hong, Jin-Hyuk
Huang, Henry
Huang, Jianye

Huda, Shamsul
Hwang, Keum-Sung
Jeff , Riley
Jianye, Huang
Jingyu , Hou
Johnson, David
Johnston, Benjamin
Joselito, Chua
Kaufhold, John
Keeratipranon, Narongdech
Kilby, Philip
Kim, Kyung-Joong
Kim, Yang Sok
Kirley, Michael
Kulkarni, Sidhivinayak
Kwok, Rex Bing Hung
Lamont, Owen
Law, Terence
Li, Wenyuan
Li, Jiaming
Li, Jifang
Li, Qing
Li, Ron
Li, Zeng
Lin, Han
Lin, Weiqiang
Loewenich, Frank
Luo, Suhuai
Luo, Xiangyu
Ma, Jun
Mahmood, Adbun
Marom, Yuval
Mayer, Wolfgang
McCarthy, Chris
Mian, Ajmal S.
Milton, John
Mingxuan , Huang
Mitchell-Wong, Juliana
Moratori, Patrick
Muecke , Nial
Mueller, Ingo
Muthukkumarasamy, Vallipuram
Ng, Kee Siong
Nguyen, Xuan Thang
Oboshi, Tamon

Ofoghi, Bahadorreza
Overett, Gary
Pan, Jeff Z.
Park, Laurence
Peng, Tao
Petersson, Lars
Polpinij, Jantima
Pullan, Wayne
Qi, Guilin
Qiao, Rong-Yu
Qingyong, Li
Qiu, Bin
Rahman, Masudur
Rajaratnam, David
Rasmussen, Rune
Rifeng, Wang
Rintanen, Jussi
Roos, Teemu
Sabu, John
Scanlan, Joel
Schlegel, Tino
Shaw, David
Shen, Weicheng
Shui, Yu
Simon, Moncreiff
Singh, Kalvinder
Song, Xueyan
Stuckey, Peter
Sucahyo, Yudho Giri
Sulan, Zhang
Suter, David
Sutharshan, Rajasegarar
Tahaghoghi, Saied
Tan, Tele
Tele, Tan
Thormaehlen, Thorsten
Tischer, Peter
Tu, Yiqing
Ubaudi, Franco
Ullah, Barkat
Uren, Philip
Ursani, Ziauddin
Uther, Will
Wang, Mei
Werner, Felix
Wong, Angela
Wu, Boris
Wu, Henry
Xiaofeng, Zhu
Xiurong, Zhao
Yang, Fengzhao
Yang, Howard Hua
Yang, Hui
Yang, Xiaowei
Yanping, Guo
Yi, Guo
Yu, Donggang
Yue, Weiya
Yuk-Hei , Lam
Yuval , Marom
Zelniker, Emanuel
Zeman, Astrid
Zhang, Jian Feng
Zhang, Jian Ying
Zhang, Lin
Zheng, Zheng
Zhenghao, Shi
Zhiwei, Shi
Zhou, Chenfeng
Zhuang, Ling
Zrimec, Tatjana

Table of Contents

Invited Talks

PART I: Regular Papers

Foundations and Knowledge Based System

Knowledge Representation and Reasoning

Machine Learning

Connectionist AI

Data Mining

Intelligent Agents

Cognition and User Interface

Vision and Image Processing

Natural Language Processing and Web Intelligence

Neural Networks

Search and Planning

Robotics

AI Applications

PART II: Regular Papers (5–7 Pages)

What Can We Do with Graph-Structured Data?
– A Data Mining Perspective

Hiroshi Motoda[*]

Institute of Scientific and Industrial Research,
Osaka University
8-1, Mihogaoka, Ibaraki, Osaka 567-0047, Japan
`motoda@ar.sanken.osaka-u.ac.jp`

Recent advancement of data mining techniques has made it possible to mine from complex structured data. Since structure is represented by proper relations and a graph can easily represent relations, knowledge discovery from graph-structured data (graph mining) poses a general problem for mining from structured data. Some examples amenable to graph mining are finding functional components from their behavior, finding typical web browsing patterns, identifying typical substructures of chemical compounds, finding typical subsequences of DNA and discovering diagnostic rules from patient history records. These are based on finding some typicality from a vast amount of graph-structured data. What makes it typical depends on each domain and each task. Most often frequency which has a good property of anti-monotonicity is used to discover typical patterns. The problem of graph mining is that it faces with subgraph isomorphism which is known to be NP-complete. In this talk, I will introduce two contrasting approaches for extracting frequent subgraphs, one using heuristic search (GBI) and the other using complete search (AGM). Both uses canonical labelling to deal with subgraph isomorphism. GBI [6,4] employs a notion of chunking, which recursively chunks two adjoining nodes, thus generating fairly large subgraphs at an early stage of search. It does not use the anti-monotonicity of frequency. The recent improved version extends it to employ pseudo-chunking which is called chunkingless chunking, enabling to extract overlapping subgraphs [5]. It can impose two kinds of constraints to accelerate search, one to include one or more of the designated subgraphs and the other to exclude all of the designated subgraphs. It has been extended to extract unordered trees from a graph data by placing a restriction on pseudo-chunking operations. GBI can further be used as a feature constructor in decision tree building [1]. AGM represents a graph by its adjacency matrix and employs an Apriori-like bottom up search algorithm using anti-monotonicity of frequency [2]. It can handle both connected and disconnected graphs. It has been extended to handle a tree data and a sequential data by incorporating to each a different bias in joining operators [3]. It has also been extended to incorporate taxonomy in labels to extract generalized

---

[*] Current address: AFOSR/AOARD, 7-23-17 Roppongi, Minato-ku, Tokyo 105-0032, Japan, e-mail: hiroshi.motoda@aoard.af.mil

A. Sattar and B.H. Kang (Eds.): AI 2006, LNAI 4304, pp. 1–2, 2006.
© Springer-Verlag Berlin Heidelberg 2006

subgraphs. I will show how both GBI and AGM with their extended versions can be applied to solve various data mining problems which are difficult to solve by other methods.

References

1. W. Geamsakul, T. Yoshida, K. Ohara, H. Motoda, H. Yokoi, and K. Takabayashi. Constructing a decision tree for graph-structured data and its applications. *Fundamenta Informaticae*, 66(1-2):131–160, 2005.
2. A. Inokuchi, T. Washio, and H. Motoda. Complete mining of frequent patterns from graphs: Mining graph data. *Machine Learning*, 50(3):321–354, 2003.
3. A. Inokuchi, T. Washio, and H. Motoda. General framework for mining frequent subgraphs from labeled graphs. *Fundamenta Informaticae*, 66(1-2):53–82, 2005.
4. T. Matsuda, H. Motoda, and T. Washio. Graph-based induction and its applications. *Advanced Engineering Informatics*, 16(2):135–143, 2002.
5. P. C. Nguyen, K. Ohara, H. Motoda, and T. Washio. Cl-gbi: A novel approach for extracting typical patterns from graph-structured data. In *Proceedings of the 9th Pacific-Asia Conference on Knowledge Discovery and Data Mining*, pages 639–649, 2005.
6. K. Yoshida and H. Motoda. Clip : Concept learning from inference pattern. *Journal of Artificial Intelligence*, 75(1):63–92, 1995.

Morphological Computation – Connecting Brain, Body, and Environment

Rolf Pfeifer

Artificial Intelligence Laboratory,
University of Zurich, Switzerland

Abstract. Traditionally, in robotics, artificial intelligence, and neuroscience, there has been a focus on the study of the control or the neural system itself. Recently there has been an increasing interest into the notion of embodiment – and consequently intelligent agents as complex dynamical systems – in all disciplines dealing with intelligent behavior, including psychology, cognitive science and philosophy. In this talk, we explore the far-reaching and often surprising implications of this concept. While embodiment has often been used in its trivial meaning, i.e. „intelligence requires a body", there are deeper and more important consequences, concerned with connecting brain, body, and environment, or more generally with the relation between physical and information (neural, control) processes. Often, morphology and materials can take over some of the functions normally attributed to control, a phenomenon called "morphological computation". It can be shown that through the embodied interaction with the environment, in particular through sensory-motor coordination, information structure is induced in the sensory data, thus facilitating perception and learning. An attempt at quantifying the amount of structure thus generated will be introduced using measures from information theory. In this view, "information structure" and "dynamics" are complementary perspectives rather than mutually exclusive aspects of a dynamical system. A number of case studies are presented to illustrate the concepts introduced. Extensions of the notion of morphological computation to self-assembling, and self-reconfigurable systems (and other areas) will be briefly discussed. The talk will end with some speculations about potential lessons for robotics, artificial intelligence, and cognitive science.

Bio

Rolf Pfeifer received his master's degree in physics and mathematics and his Ph.D. in computer science from the Swiss Federal Institute of Technology (ETH) in Zurich, Switzerland. He spent three years as a post-doctoral fellow at Carnegie-Mellon University and at Yale University in the US. Since 1987 he has been a professor of computer science at the Department of Informatics, University of Zurich, and director of the Artificial Intelligence Laboratory. Having worked as a visiting professor and research fellow at the Free University of Brussels, the MIT Artificial Intelligence Laboratory, the Neurosciences Institute (NSI) in San Diego, and the Sony Computer Science Laboratory in Paris, he was elected "21st Century COE Professor,

A. Sattar and B.H. Kang (Eds.): AI 2006, LNAI 4304, pp. 3–4, 2006.
© Springer-Verlag Berlin Heidelberg 2006

Information Science and Technology" at the University of Tokyo for 2003/2004, from where he held the first global, fully interactive, videoconferencing-based lecture series "The AI Lectures from Tokyo" (including Tokyo, Beijing, Jeddah, Warsaw, Munich, and Zurich). His research interests are in the areas of embodiment, biorobotics, artificial evolution and morphogenesis, self-reconfiguration and self-repair, and educational technology. He is the author of the book "Understanding Intelligence", MIT Press, 1999 (with C. Scheier). His new popular science book entitled "How the body shapes the way we think: a new view of intelligence," MIT Press, 2006 (with Josh Bongard) is scheduled to appear this fall.

Interaction-Oriented Programming: Concepts, Theories, and Results on Commitment Protocols

Munindar P. Singh

North Carolina State University
http://www.csc.ncsu.edu/faculty/mpsingh/

Abstract. Unlike traditional information systems, modern systems are _open_, consisting of autonomous, heterogeneous parties interacting dynamically. Yet prevalent software techniques make few accommodations for this fundamental change. Multiagent systems are conceptualized for open environments. They give prominence to flexible reasoning and arms-length interactions captured via communications. On the backdrop of multiagent systems, Interaction-Oriented Programming is the idea of programming with interactions as first-class entities instead of, e.g., objects. Protocols are to interactions as classes are to objects: because of their key nature, protocols have obtained a lot of research attention. Modeling protocols suitably for open environments meant modeling their content, not just the surface communications. In a number of important cases, such as business processes and organizations, the content is best understood using the notion of commitments of an agent to another agent in an appropriate context. Our theory of protocols supports flexible enactment of protocols, a treatment of refinement and composition of protocols, and their relationship with organizations and contracts, thus reducing the gap between agents and conventional computer science. This talk will review the key concepts, theories, and results on commitment protocols, and some important challenges that remain..

Bio

Dr. Munindar P. Singh is a professor in the department of computer science at North Carolina State University. From 1989 through 1995, he was with the Microelectronics and Computer Technology Corporation (MCC). Munindar's research interests include multiagent systems and service-oriented computing, wherein he specifically addresses the challenges of trust, service discovery, and business processes and protocols in large-scale open environments. Munindar is widely published and cited. Munindar's 1994 book Multiagent Systems, was published by Springer-Verlag. He coedited Readings in Agents, which was published by Morgan Kaufmann in 1998. Munindar edited the Practical Handbook of Internet Computing published by Chapman & Hall / CRC Press in October 2004 and coauthored a new text, Service-Oriented Computing published by Wiley in January 2005. Munindar cochaired the 2005 edition of AAMAS, the International Joint Conference of Autonomous Agents and MultiAgent Systems. He serves on the Board of Directors of the newly synthesized IFAAMAS, the International Foundation of Autonomous

A. Sattar and B.H. Kang (Eds.): AI 2006, LNAI 4304, pp. 5 – 6, 2006.
© Springer-Verlag Berlin Heidelberg 2006

Agents and MultiAgent Systems. Munindar was the editor-in-chief of IEEE Internet Computing from 1999 to 2002 and continues to serve on its editorial board. He is also a founding member of the editorial boards of the Journal of Autonomous Agents and Multiagent Systems, the Journal of Web Semantics, the International Journal of Agent-Oriented Software Engineering, and the Journal of Service-Oriented Computing and Applications. Munindar's research has been recognized with awards and sponsorship by the National Science Foundation, DARPA, IBM, Cisco Systems, and Ericsson. Munindar obtained a B.Tech. in Computer Science and Engineering from the Indian Institute of Technology, Delhi in 1986 and a Ph.D. in Computer Sciences from the University of Texas at Austin in 1993.

Symmetry Breaking

Toby Walsh[*]

National ICT Australia and School of CSE, University of New South Wales, Sydney,
Australia
tw@cse.unsw.edu.au

Symmetry occurs in many problems in aritifical intelligence. For example, in the n-queens problem, the chessboard can be rotated 90°. As a second example, several machines in a factory might have the same capacity. In a production schedule, we might therefore be able to swap the jobs on machines with the same capacity. As a third example, two people within a company might have the same skills. Given a staff roster, we might therefore be able to interchange these two people. And as a fourth example, when configuring a computer, two memory boards might be identical. We might therefore be able to swap them around without changing the performance of the computer.

Symmetries come in many different forms. For instance, there are rotation symmetries (e.g. the 90° rotations of the chessboard in the n-queens problem), reflection symmetries (e.g. the reflection of the chessboard along one of its diagonals), and permutation symmetries (e.g. the interchange of equally skilled personnel in a staff roster). Problems may have many symmetries at once (e.g. the n-queens problem simultaneously has several rotation and reflection symmetries). We can describe such symmetries using group theory. The symmetries of a problem form a group. Their actions are to map solutions (a n-queens solution, a production schedule, a staff roster, a configuration, etc.) onto solutions. We must deal with such symmetry or we will waste much time visiting symmetric solutions. In addition, we must deal with symmetry during our search for solutions otherwise we will visit many search states that are symmetric to those that we have already visited.

One well known mechanism to deal with symmetry is to add constraints which eliminate symmetric solutions [1]. I will describe a general method for breaking any type of symmetry [2,3]. The basic idea is very simple. We pick an ordering on the variables, and then post symmetry breaking constraints to ensure that the final solution is lexicographically less than any symmetric re-ordering of the variables. That is, we select the "lex leader" assignment. In theory, this solves the problem of symmetries, eliminating all symmetric solutions and pruning many symmetric states. Unfortunately, the set of symmetries might be exponentially large (for example, there are $n!$ symmetries if we have n identical machines in a factory).

---

[*] Thanks to my co-authors, especially Christian Bessiere, Alan Frisch, Emmanuel Hebrard, Brahim Hnich, Zeynep Kiziltan and Ian Miguel. NICTA is funded through the Australian Government's *Backing Australia's Ability* initiative, in part through the Australian Research Council.

A. Sattar and B.H. Kang (Eds.): AI 2006, LNAI 4304, pp. 7–8, 2006.
© Springer-Verlag Berlin Heidelberg 2006

By exploting properties of the set of symmetries we can sometimes overcome this problem. I will describe two special cases where we can deal with an exponential number of symmetries in polynomial time. These two cases frequently occur in practice. In the first case, we have a matrix (or array) of decisions variables in which the rows and/or columns are symmetric and can be permuted. In this case, we can lexicographical order the rows and/or columns [4,5]. In the second case, we have values for our decision variables which are symmetric and can be interchanged. I show how so called "value precedence" can be used to break all such symmetry [6,7].

Finally, I will end with a discussion of open research questions in this area. Are they efficient ways to combine together these symmetry breaking constraints for particular types of symmetries? Are there useful subsets of these symmetry breaking constraints when there are too many to post individually? Are there problems where the symmetry breaking constraints can be simplified? Can these symmetry breaking methods be used outside of constraint programming?

References

1. Puget, J.F.: On the satisfiability of symmetrical constrained satisfaction problems. In Komorowski, J., Ras, Z., eds.: Proceedings of ISMIS'93. LNAI 689, Springer-Verlag (1993) 350–361
2. Crawford, J., Luks, G., Ginsberg, M., Roy, A.: Symmetry breaking predicates for search problems. In: Proceedings of the 5th International Conference on Knowledge Representation and Reasoning, (KR '96). (1996) 148–159
3. Walsh, T.: General symmetry breaking constraints. In: 12th International Conference on Principles and Practices of Constraint Programming (CP-2006), Springer-Verlag (2006)
4. Shlyakhter, I.: Generating effective symmetry-breaking predicates for search problems. In: Proceedings of LICS workshop on Theory and Applications of Satisfiability Testing (SAT 2001). (2001)
5. Flener, P., Frisch, A., Hnich, B., Kiziltan, Z., Miguel, I., Pearson, J., Walsh, T.: Breaking row and column symmetry in matrix models. In: 8th International Conference on Principles and Practices of Constraint Programming (CP-2002), Springer (2002)
6. Law, Y., Lee, J.: Global constraints for integer and set value precedence. In: Proceedings of 10th International Conference on Principles and Practice of Constraint Programming (CP2004), Springer (2004) 362–376
7. Walsh, T.: Symmetry breaking using value precedence. In: Proceedings of the 17th ECAI, European Conference on Artificial Intelligence, IOS Press (2006)

Heyting Domains for Constraint Abduction

Michael Maher

National ICT Australia and University of NSW
Sydney, Australia
Michael.Maher@nicta.com.au

Abstract. We investigate constraint domains in which answers to constraint abduction problems can be represented compactly by a most general answer. We demonstrate several classes of domains which have this property, but show that the property is not compositional.

1 Introduction

Abduction is the inference rule that derives A from B and C, such that $A, B \vdash C$. It was considered by Peirce [12] to be – along with deduction and induction – one of the fundamental forms of reasoning. Mostly, abduction has been addressed in a setting of partially determined predicates or propositions. *Constraint abduction* refers to abduction in a setting where A, B and C come from a set of pre-defined, completely-defined relations (constraints, in the sense of constraint logic programming [6]) and A, the answer, ensures $A \wedge B \to C$.

Traditionally, abduction has been used as a model of scientific hypothesis formation and diagnosis, and as a component of belief revision and machine learning, but the applications of constraint abduction are in query evaluation using views [18], which is used for integrating multiple sources of information [9], and in logic program analysis [3,4] and type inference [16].

In previous work it has been established that some natural constraint domains, such as linear arithmetic and finite terms, do not in general admit compact explicit representations of all answers to an abduction problem [11,10]. In this paper we investigate constraint domains in which all answers can be represented by a single most general answer. We call these *Heyting domains*. It turns out that many of the simpler, commonly-used constraint domains have this property although even in these simple cases there are provisos. Furthermore, we show that even a simple combination of Heyting domains loses the Heyting property. This presents a substantial difficulty for the use of constraint abduction, but we demonstrate one case in which the composition of Heyting domains is Heyting. We leave for future work a detailed study of this issue.

2 Background

The syntax and semantics of constraints are defined by a constraint domain [6]. Given a signature Σ, and a set of variables $Vars$ (which we assume is infinite), a *constraint domain* is a pair $(\mathcal{D}, \mathcal{L})$ where $\mathcal{D}$ is a Σ-structure and $\mathcal{L}$ (the language of constraints)

A. Sattar and B.H. Kang (Eds.): AI 2006, LNAI 4304, pp. 9–18, 2006.

© Springer-Verlag Berlin Heidelberg 2006

is a set of Σ-formulas including constraints equivalent to *true* and *false*, and closed under conjunction and renaming of free variables. If, for some set of constraints $S \subseteq \mathcal{L}$, every constraint is equivalent to a conjunction of renamed S-constraints then we refer to S as a set of *primitive constraints*. We say the constraint domain is *unary* if $\mathcal{L}$ is generated by a set of unary primitive constraints. We say the constraint domain is *finite* if the set of values in $\mathcal{D}$ is finite.

When the constraint domain $\mathcal{D}$ is clear from the context we write $C \rightarrow C'$ as an abbreviation for $\mathcal{D} \models C \rightarrow C'$. We say C is *more general than* C' if $C' \rightarrow C$. Two constraints C and C' are *equivalent* if $C \rightarrow C'$ and $C' \rightarrow C$.

The constraints (modulo equivalence) form a partially ordered set (poset) where $C_1 \leq C_2$ iff $C_1 \rightarrow C_2$. *true* and *false* are, respectively, the top and bottom elements of the poset. The greatest lower bound $C_1 \sqcap C_2$ of two constraints C_1 and C_2 is their conjunction $C_1 \wedge C_2$. The least upper bound $C_1 \sqcup C_2$ of two constraints may not exist.

When every pair of elements in a poset has a least upper bound and a greatest lower bound, the poset is called a *lattice*. If, in addition, top and bottom elements exist it is called a *bounded lattice*. A lattice is said to be *distributive* if, for all elements x, y and z, $x \sqcap (y \sqcup z) = (x \sqcap y) \sqcup (x \sqcap z)$ and $x \sqcup (y \sqcap z) = (x \sqcup y) \sqcap (x \sqcup z)$. The complement of an element x in a bounded lattice is an element $\overline{x}$ such that $x \sqcup \overline{x}$ is the top element and $x \sqcap \overline{x}$ is the bottom element. A distributive lattice where every element has a unique complement is called a *Boolean lattice*. A Boolean lattice that provides the operations $\sqcap$, $\sqcup$ and $^-$ is called a *Boolean algebra*. The *relative pseudo-complement* of an element x with respect to an element y is the unique (if it exists) greatest element z such that $z \sqcap x \leq y$. A bounded lattice where every element has a relative pseudo-complement with respect to each element in the lattice is called a *Heyting lattice*. Every Boolean lattice is a Heyting lattice. If the dual of a Heyting lattice $\mathcal{H}$ (the "upside-down" lattice, where $\leq$ and $\geq$ are interchanged) also forms a Heyting lattice then $\mathcal{H}$ is a *bi-Heyting lattice*. Heyting lattices and algebras originated as models for intuitionistic logic and bi-Heyting algebras have been proposed as models for spatial reasoning calculi [15], but these uses seem largely unconnected to their use in this paper.

3   Simple Constraint Abduction

We can now formally define the simple constraint abduction problem. Extensions of this problem are discussed in Section 5. We write $\tilde{\exists}$ for the existential quantification of all variables.

Definition 1. *The* Simple Constraint Abduction Problem *is as follows:*
Given a constraint domain $(\mathcal{D}, \mathcal{L})$, *and given two constraints* $B, C \in \mathcal{L}$ *such that* $\mathcal{D} \models \tilde{\exists} B \wedge C$, *for what constraints* $A \in \mathcal{L}$ *does*

$$\mathcal{D} \models (A \wedge B) \rightarrow C$$

and

$$\mathcal{D} \models \tilde{\exists} (A \wedge B)$$

An instance *of the problem has a fixed constraint domain and fixed constraints* B *and* C. *We call* A *an* answer *to the problem instance.*

Throughout this paper, A, B and C refer to the constraints in a simple constraint abduction problem. In an abuse of terminology, we often will refer to an instance as a SCA problem.

Traditional abduction might be thought of as a kind of higher-order constraint abduction where the variables range over relations and the constraints state tuples that are in/out of those relations. Abductive logic programming [7] is somewhat different because the deductive relation $\vdash$ is different from material implication. ACLP [8] extends abductive logic programming with constraints. It uses a version of SCA (where C is $false$) in its algorithm but it does not perform abduction on constraints, only on formulas that may involve constraints.

It is easy to see that there is always an answer to a SCA problem by taking A to be C. However, in general there may be infinitely many answers, and we will need a finite representation. For example, if B is $x \geq 5$ and C is $x \geq 7$ then C is an answer, but so is every constraint $x \geq k$ where $k \geq 7$. In this case, C can be considered to be a representation of all answers since, for all answers A', A' is the conjunction of some constraint with C (that is, $A' \rightarrow C$), and every A' that is obtained in this way and is consistent with B is an answer. We say an answer A to a SCA problem is the *most general answer* if, for every answer A' of the problem, $A' \rightarrow A$.

Unfortunately, not all SCA problems have a most general answer, which leads us to consider *maximally general answers*. A maximally general answer is an answer A such that there is no answer strictly more general than A. These answers are singled out by the parsimony principle of abductive inference, which suggests choosing explanations with weakest explanatory power. They also have the potential to compactly represent many answers, since they can be taken to represent all stronger answers.

However, in some constraint domains such as linear arithmetic constraints, the maximally general answers do not represent all answers [11]. Furthermore, even in constraint domains where maximally general answers represent all answers, there may be infinitely many maximally general answers [10,11]. It is clear from [10,11] that only rarely do SCA problems have most general answers in the constraint domains that those papers address (linear arithmetic constraints/finite term equalities).

In this paper, we will focus on identifying constraint domains for which every SCA problem has a most general answer. This property ensures that the difficulties mentioned above are avoided, though at the cost of limiting the constraint domains over which we can abduce. In the type-inference setting, it corresponds to the principal-types property. We say $\mathcal{D}$ is a *Heyting domain* when this property holds. This terminology is justified by the following result.

Proposition 1. *Let $\mathcal{D}$ be a constraint domain. $\mathcal{D}$ is a Heyting domain iff the poset of constraints has the property that every element has a relative pseudo-complement with respect to each element in the poset.*

This proposition essentially reformulates the property of being a Heyting domain in order-theoretic terms. The relative pseudo-complement is precisely the most general answer. Note that a Heyting domain might not form a Heyting lattice because it might not be a lattice. However, every finite unary constraint domain forms a lattice, and thus every finite unary Heyting domain is a Heyting lattice.

4   Heyting Constraint Domains

We have seen that constraint domains involving arithmetic over the reals or integers, or equality over finite terms, are not Heyting domains. Of the remaining widely-used constraint domains, most are unary. In particular, bound and interval constraints are used for solving integer problems [5] and non-linear continuous problems [1], as well as problems over finite sets [2,13] or multisets [17], while finite domain constraints are the basis of CSPs. Hence we focus on establishing if/when these and similar domains are Heyting, and presenting a constructive characterization of their most general answers, when it exists.

4.1   Boolean Constraint Domains

Proposition 1 characterizes the Heyting domains, but it does not provide a way to compute a most general answer. However, in Boolean algebras we can compute the most general answer directly, using the algebraic operations.

Proposition 2. *Let $\mathcal{D}$ be a constraint domain where the constraints form a Boolean lattice. Consider the SCA problem over $\mathcal{D}$ and define A to be $\overline{B} \sqcup C$.*
   Then A is the most general answer.

The above result applies to finite domains, as used in CSPs, where each variable ranges over a finite set of values D and the only primitive constraints restrict variables to a subset of D. The complement of a constraint $x \in S$ is $x \in (D \backslash S)$. The least upper bound is obtained by union: $(x \in S_1) \sqcup (x \in S_2)$ is $x \in (S_1 \cup S_2)$. The greatest lower bound is expressed by conjunction and can be obtained by intersection in a similar way. Similarly, the constraint domain where constraints are unions of floating point-bounded intervals over the real numbers, as used in Echidna [14], is a Heyting domain by this result.

Note that the least upper bound operation $\sqcup$ is, in general, different from disjunction $\vee$. However, when the constraint language is closed under negation the constraint domain forms a Boolean lattice and, in that case, the least upper bound is exactly disjunction.

So far we have been using the lattice structure of the *constraints*. In the remainder of this section we consider constraint domains where the partial order of *values* is important.

4.2   Bounds on a Total Order

Interval/bounds constraints on a total order cannot be addressed by Proposition 2 because these constraints do not form a distributive lattice and a complement is not defined.

Example 1. Consider bounds constraints on integers. Let C_1 be $x \in 1..5$, C_2 be $x \in 7..9$ and C_3 be $x \in 5..6$. Then $C_1 \sqcup C_2$ is $x \in 1..9$. and $(C_1 \sqcup C_2) \wedge C_3$ is $x \in 5..6$. However, $C_1 \wedge C_3$ is $x \in 5..5$ and $C_2 \wedge C_3$ is *false* so $(C_1 \wedge C_3) \sqcup (C_2 \wedge C_3)$ is $x \in 5..5$. Thus the lattice of constraints is not distributive.

The complement of C_1 must be an interval disjoint from 1..5 and hence contains only numbers less than 1 or greater than 5. But then the least upper bound of these two intervals cannot be the full interval of integers. Thus complement is not defined.

Consider any constraint domain $\mathcal{D}$ with a total ordering $\leq$. If we consider bounds, primitive constraints have the form $c \leq x$ or $x \leq c$, where c is a constant and x is a variable. If we consider intervals, primitive constraints have the form $x \in l..u$, meaning $l \leq x \leq u$, where l and u are constants and x is a variable. In both cases we might also consider strict inequalities; to simplify the exposition we will omit this possibility, which complicates the description but does not raise any new issues.

For both bounds and intervals, a solved form can be computed by essentially intersecting all intervals (or bounds) concerning each variable. Thus, in a solved form there is at most one interval or two bounds (one lower, one upper) on each variable. If there is no lower (upper) bound on a variable in a constraint we can act as though there is an infinitely small (large) bound.

Given B and C, which we can assume without loss of generality are in solved form, we construct A to satisfy $A \wedge B \to C$ as follows:

For each variable x in B or C, let l_B^x and l_C^x (u_B^x and u_C^x) be the lower (respectively, upper) bounds for x in B and C. Define A as follows. For each variable x:

$d \leq x$ is in A iff $d = l_C^x$, $d \neq -\infty$ and $l_B^x < l_C^x$

$x \leq d$ is in A iff $d = u_C^x$, $d \neq \infty$ and $u_C^x < u_B^x$

That is, A contains those bounds of C that are strictly tighter than those of B.

The following result follows directly from the construction of A.

Theorem 1. *Consider the SCA problem over a constraint domain of bounds on a total order. Let A be as defined above. Then A is the unique most general answer.*

For example, let B be $2 \leq x \wedge x \leq 8$ and C be $1 \leq x \wedge x \leq 5$. Then $l_B^x \not< l_C^x$ ($2 \not< 1$) and so A contains no lower bound for x. But $u_C^x < u_B^x$ ($5 < 8$) so A contains $x \leq 5$.

Since intervals are essentially the conjunction of two bounds, we can achieve a similar result in that case, but only if there is minimum possible lower bound and maximum possible upper bound. If there is no minimum possible lower bound and maximum possible upper bound (for example, if bounds of intervals can be any integer, but not an infinite bound) then SCA problems can have an infinite ascending chain of answers, and thus not have a maximally general answer.

Example 2. Consider a constraint domain of finite intervals over the integers. If B is $x \in 0..5$ and C is $x \in 1..10$ then any constraint A of the form $x \in 1..n$ is an answer, but the set of all such constraints forms an ascending chain with no least upper bound.

4.3 Bounds on a Quasi Order

If the values of a constraint domain $\mathcal{D}$ are quasi-ordered by a relation $\sqsubseteq$ then we can consider the induced equivalence relation $\sim$ and the induced partial order over the quotient $D/\sim$. Any $\sqsubseteq$-bound on D induces a bound on $D/\sim$, and vice versa. Thus

Proposition 3. *Let $\mathcal{D}$ be the constraint domain of a quasi-order over a set D and let $\mathcal{D}'$ be the corresponding quotient constraint domain of the induced partial order.*
Then $\mathcal{D}$ is a Heyting domain iff $\mathcal{D}'$ is a Heyting domain.

Constraint domains with a quasi-ordering can arise when truncation or rounding of numbers is performed. For example, the set of real numbers is quasi-ordered by the ordering on integers via truncation: $x \sqsubseteq y$ iff $\lfloor x \rfloor \leq \lfloor y \rfloor$. Thus the above result implies that limitations of machine representability of numbers will not affect the Heyting domain property for the constraint domains discussed in the previous subsection. Later we will see an example involving finite sets.

4.4 Bounds on a Partial Order

We can obtain similar results for some partially ordered constraint domains with bounds constraints. Unlike a total order, a partially ordered constraint domain with bounds constraints forms a Heyting domain only when the partial order has a certain structure: $\mathcal{D}$ must be a bi-Heyting lattice. However, the construction of this answer cannot be given in such simple terms, in general.

Theorem 2. *Consider the SCA problem over a constraint domain $\mathcal{D}$ of bounds on a partial order that forms a bi-Heyting lattice. Then $\mathcal{D}$ is a Heyting domain.*

In this result the relative pseudo-complement is needed for the upper bounds, while its dual is needed for the lower bounds.

We can construct most general answers if the partial order of values forms a Boolean lattice – for example, bounds on set variables [2] where we assume the values that a set variable can take are bounded above by a finite set. Suppose B and C are in solved form. We define A by, for each variable x,
$$d \leq x \text{ is in } A \text{ iff } d = (l_C^x \sqcap \overline{l_B^x}) \text{ and } l_B^x \not\sqsupseteq l_C^x$$
$$x \leq d \text{ is in } A \text{ iff } d = (u_C^x \sqcup \overline{u_B^x}) \text{ and } u_C^x \not\sqsupseteq u_B^x$$
where $\sqcup$ and $\sqcap$ are the lattice operations and $\overline{p}$ denotes the complement of p. Then A is the most general answer.

Theorem 3. *Consider the SCA problem over a constraint domain $\mathcal{D}$ of bounds on a partial order that forms a Boolean lattice. Suppose that the set of values of variable-free terms in $\mathcal{D}$ (that can be used as bounds in constraints) is closed under the operations of the lattice ($\sqcup$, $\sqcap$, and complement). Let A be as defined above.*
Then A is the most general answer.

Theorem 3 applies directly to the constraint domain of bounds on finite sets.

Example 3. Consider interval constraints on finite sets as in Conjunto [2]. Here the lattice operations on the values of the domain are the familiar union, intersection and set complement operations. Suppose that x has a lower bound in B of $l_B^x = \{1, 2, 3\}$ and a lower bound in C of $l_C^x = \{1, 2, 4\}$. Then $l_B^x \not\sqsupseteq l_C^x$ and $l_A^x = \{1, 2, 4\} \sqcap \overline{\{1, 2, 3\}} = \{4\}$.

If the values of this constraint domain are the finite subsets of an infinite set and constraints are intervals then, like Example 2, there may not be a most general answer. But if the values are the subsets of a finite set then there is a most general answer, even when the constraints are intervals. However, this only applies to finite sets where subset bounds are the only constraints; if there are more constraints then there might not be a most general answer.

4.5   Finite Set Constraints

In [13], the conventional containment partial ordering on finite sets is used in combination with a quasi-ordering based on the cardinalities of sets and a lexicographic ordering on sets. The quasi-ordering $\sqsubseteq$ is defined by $X \sqsubseteq Y$ iff $|X| \leq |Y|$, where $|X|$ denotes the cardinality of X. The lexicographic ordering is defined recursively as follows:

$X \preceq Y$ iff $X = \emptyset$ or $x < y$ or $x = y \wedge X \backslash \{x\} \preceq Y \backslash \{y\}$

where $x = max(X)$, the largest element in X, and $y = max(Y)$ and $<$ is a total order over all elements that may appear in a set. For example, suppose elements of the sets are natural numbers and $<$ is the usual ordering on natural numbers. Then[1] $\{3, 2, 1\} \preceq \{4, 2\}$ since $max(\{3, 2, 1\}) = 3 < 4 = max(\{4, 2\})$. The characteristic vector representation of a set S is a list χ_S of 0/1 values such that the i'th value is in S iff the i'th element of χ_S is 1, where the values are ordered from $<$-largest to smallest. The *longest common prefix* of two sets U and V is the longest common prefix of χ_U and χ_V.

By Theorem 1, the constraint domain of bounds over sets under the lexicographic ordering is a Heyting domain. Similarly, by Proposition 3 and Theorem 1, the constraint domain of bounds on cardinalities of sets is a Heyting domain. As we saw above, from Theorem 3, the constraint domain of bounds over sets under the containment ordering is a Heyting domain. However, it does not immediately follow that a constraint domain combining these kinds of constraints has unique most general answers.

Indeed, consider the SCA problem where B is $|x| = 1$ and C is $x = \{1\}$. Then both $x \subseteq \{1\}$ and $\{1\} \subseteq x$ are maximally general answers. Thus the constraint domain which uses all three kinds of constraints (as in [13]) is not a Heyting domain (and neither are the constraint domains using only cardinality constraints and either of the other two orderings). That is, the property of being a Heyting domain is not compositional, even in this relatively simple case.

Definition 2. *The* composition by language *of constraint domains* $(\mathcal{D}, \mathcal{L}_1)$ *and* $(\mathcal{D}, \mathcal{L}_2)$ *having the same structure* $\mathcal{D}$ *is* $(\mathcal{D}, \mathcal{L})$ *where* $\mathcal{L}$ *is the conjunctive closure of* $\mathcal{L}_1 \cup \mathcal{L}_2$.

Proposition 4. *The Heyting property is not closed under composition by language.*

Let us now consider the constraint domain involving only containment and lexicographic bounds constraints. A constraint in this constraint language is in *simplified form* if the constraints on any variable x have the form $d_l \subseteq x \subseteq d_u, f_l \preceq x \preceq f_u$ where

- $d_l \subseteq f_u \subseteq d_u$ and $d_l \subseteq f_l \subseteq d_u$,
- $max(d_u) = max(f_u)$ and $max(\overline{d_l}) = max(\overline{f_l})$, where $\overline{x}$ denotes the complement of x,
- the longest common prefix of f_l and f_u is a prefix of d_l and d_u.

This simplified form is similar to the normal form of rewrite rules given in [13], though formulated differently, and every constraint can be simplified by similar rules.

Let B be in simplified form and let BC be the simplified form of $B \wedge C$. We define A' using the methods defined earlier for Theorem 1 (in the case of $\preceq$ constraints) and

---

[1] Elements are written in descending order in this example to emphasize the lexicographic nature of $\preceq$.

for Theorem 3 (in the case of $\subseteq$ constraints). Thus A' contains $\emptyset \preceq x$ if $l_{BC} \preceq l_B$ and $l_{BC} \preceq x$ otherwise, where l_B and l_{BC} are the corresponding $\preceq$-lower bounds for x in B and BC respectively. A' contains $\emptyset \subseteq x$ if $l_{BC} \subseteq l_B$ and $l_{BC} \cap \overline{l_B} \subseteq x$ otherwise, where here l_B and l_{BC} are the corresponding $\subseteq$-lower bounds for x in B and BC respectively. Upper bounds are similar.

Given A', we expand the bounds, if possible as follows. Let A'_{-l} and A'_{-u} be A' except that the $\preceq$-lower bound (respectively $\preceq$-upper bound) of x is omitted. If $(A'_{-l} \wedge B) \rightarrow \neg(l_B \preceq x \prec l_{A'})$ holds then the $\preceq$-lower bound of x in A' is replaced by the $\subseteq$-lower bound. Similarly, if $A'_{-u} \wedge B \rightarrow \neg(u_{A'} \prec x \preceq u_B)$ holds then the $\preceq$-upper bound of x in A' is replaced by the $\subseteq$-upper bound. A similar expansion on the $\subseteq$-bounds is unnecessary. The resulting constraint is A.

Theorem 4. *Consider an SCA problem over the constraint domain of finite sets with both containment and lexicographic orderings. Let A be as defined above.*
Then A is the most general answer.

Example 4. Consider the SCA over the combined domain on subsets of $\{4, 3, 2, 1\}$. Let B be $\{3, 1\} \preceq x$, or $\emptyset \subseteq x \subseteq \{4, 3, 2, 1\}$, $\{3, 1\} \preceq x \preceq \{4, 3, 2, 1\}$ in simplified form, and let C be $\{2\} \subseteq x \subseteq \{4, 3, 2\}$, $\{3, 2\} \preceq x \preceq \{4, 3, 2\}$. Then BC is identical to C and A' is $\{2\} \subseteq x \subseteq \{4, 3, 2\}$, $\{3, 2\} \preceq x \preceq \{4, 3, 2\}$. Expanding the lower bound since $A'_{-l} \wedge B \rightarrow x \neq \{3, 1\}$, we find A is $\{2\} \subseteq x \subseteq \{4, 3, 2\}$, $\{2\} \preceq x \preceq \{4, 3, 2\}$.

4.6   Bounds on Finite Multisets

When the constraint domain consists of bounds over multiset variables [17] the above results do not apply. First we define this constraint domain. A multiset is a function in $S \rightarrow \mathbb{N}$, where S is a finite set. Thus a multiset $m = \{\{a, b, b\}\}$ is formalized as a function where $m(a) = 1$, $m(b) = 2$, and $m(c) = 0$. Multisets are ordered by the containment ordering: $m_1 \leq m_2$ iff $\forall s \in S\ m_1(s) \leq m_2(s)$. We say m_1 is contained in m_2 or m_1 is a submultiset of m_2. Under the containment ordering there is a least multiset (the empty multiset, which is the zero function), and we assume there is a greatest multiset M. The least upper bound and greatest lower bound under this ordering can be defined as follows: $(m_1 \sqcup m_2)(s) = max\{m_1(s), m_2(s)\}$ and $(m_1 \sqcap m_2)(s) = min\{m_1(s), m_2(s)\}$.

The primitive constraints in a multiset constraint domain have the form $x \leq m$ or $m \leq x$, specifying that the value of the variable x is contained in (respectively, contains) the multiset m. For any constraint C, we can express C as $\bigwedge_{x \in Vars}(l_C^x \leq x \wedge x \leq u_C^x)$. We assume there is a greatest multiset M in the constraint domain, the multisets under consideration are exactly the submultisets of M, and that each submultiset can be represented by a constant expression. Thus, if M is finite then multiset variables range over a finite collection of multisets.

The containment ordering on multisets does form a distributive lattice, but it does not have a complement. Thus the lattice of values is not a Boolean lattice, and thus Theorem 3 does not apply. Similarly, Proposition 2 does not apply because the lattice of constraints is not a Boolean lattice.

Nevertheless, we can construct the most general answer to any SCA problem over such multiset constraints. For a given SCA problem, define A as $\bigwedge_{x \in Vars}(l_A^x \leq x \wedge x \leq u_A^x)$ where

$$l_A^x(s) = \begin{cases} 0 & l_B^x(s) \geq l_C^x(s) \\ l_C^x(s) & \text{otherwise} \end{cases}$$

and

$$u_A^x(s) = \begin{cases} M(s) & u_B^x(s) \leq u_C^x(s) \\ u_C^x(s) & \text{otherwise} \end{cases}$$

Then A defines the most general answer for the SCA problem over multiset constraints.

Theorem 5. *Consider the SCA problem over a constraint domain of bounds on multiset variables where there is a greatest multiset M. Let A be as defined above.*
Then A is the unique most general answer.

For example, if B is $\{\{a, b, b\}\}$ (that is, $m_B(a) = 1$, $m_B(b) = 2$, $m_B(c) = 0$) and C is $\{\{a, b, c\}\}$ ($m_C(a) = 1$, $m_C(b) = 1$, $m_C(c) = 1$) then A is $\{\{c\}\}$ ($m_A(a) = 0$, $m_A(b) = 0$, $m_A(c) = 1$).

5   Generalized Constraint Abduction

Simple constraint abduction is the basis for other abduction problems [10]. In joint constraint abduction, several simple problems must be solved simultaneously by an answer A. In variable-restricted constraint abduction, A may use only a limited set V of variables[2]. For Heyting domains, joint problems can be decomposed into simple abduction problems, and the conjunction of all most general answers provides a most general answer to the joint problem iff the joint problem has an answer. See [10].

Corresponding to every variable restricted problem is the *unrestricted* problem – the simple problem where the variable restriction is ignored. We can use the unrestricted problem to address the variable-restricted problem.

Theorem 6. *Let $(\mathcal{D}, \mathcal{L})$ be a constraint domain where constraints are generated from unary primitive constraints. Then the variable-restricted constraint abduction problem wrt V has an answer iff $\mathcal{D} \models B \rightarrow \exists_V C$.*
Furthermore, assuming the variable-restricted problem has an answer, if the unrestricted problem has a most general answer then that is also the most general answer to the restricted problem.

6   Conclusions

We have investigated the class of Heyting domains: those domains where every simple constraint abduction problem that has an answer has a most general answer. We have

---

[2] This set corresponds to the notion of *abducibles* in traditional abduction.

demonstrated classes of domains that have this property, which include many of the simpler commonly-used constraint domains. We have seen that the Heyting property is not preserved under constraint language composition, but that it does extend to joint constraint abduction and, for unary domains, to variable-restricted abduction.

References

1. F. Benhamou & W.J. Older, Applying Interval Arithmetic to Real, Integer, and Boolean Constraints, *J. Logic Programming* 32(1): 1–24, 1997.
2. C. Gervet, Interval Propagation to Reason about Sets: Definition and Implementation of a Practical Language, *Constraints* 1(3): 191–244, 1997.
3. R. Giacobazzi, Abductive Analysis of Modular Logic Programs, *Journal of Logic and Computation* 8(4):457-484, 1998.
4. J. M. Howe, A. King, & L. Lu, Analysing Logic Programs by Reasoning Backwards in: M. Bruynooghe and K.-K. Lau (Eds), Program Development in Computational Logic, LNCS 3049, 152–188. 2004.
5. P. Van Hentenryck, *Constraint Satisfaction in Logic Programming*, MIT Press, 1989.
6. J. Jaffar & M.J. Maher, Constraint Logic Programming: A Survey, *Journal of Logic Programming 19 & 20*, 503–581, 1994.
7. A.C. Kakas, R.A. Kowalski & F. Toni, Abductive Logic Programming, *J. Logic and Computation* 2(6): 719–770, 1992.
8. A.C. Kakas, A. Michael & C. Mourlas, ACLP: Abductive Constraint Logic Programming, *J. Logic Programming* 44(1-3): 129–177, 2000.
9. A.Y. Levy, A. Rajaraman & J.J. Ordille, Query-Answering Algorithms for Information Agents, *Proc. AAAI/IAAI*, Vol. 1, 40–47, 1996.
10. M.J. Maher, Herbrand Constraint Abduction, *Proc. Logic in Computer Science Conf.*, 397–406, 2005.
11. M. Maher, Abduction of Linear Arithmetic Constraints, *Proc. International Conference on Logic Programming*, LNCS 3668, Springer, 174–188, 2005.
12. C.S. Peirce, Collected Papers of Charles Saunders Peirce, C. Hartshorne & P. Weiss (Eds), Belknap Press of Harvard University Press, 1965.
13. A. Sadler & C. Gervet, Hybrid Set Domains to Strengthen Constraint Propagation and Reduce Symmetries *Proc. CP*, 604–618, LNCS 3258, 2004.
14. G. Sidebottom & W.S. Havens, Hierarchical Arc Consistency for Disjoint Real Intervals in Constraint Logic Programming, *Computational Intelligence* 8: 601–623, 1992.
15. J.G. Stell & M.F. Worboys, The Algebraic Structure of Sets of Regions, *Proceedings Spatial Information Theory, COSIT'97*, 163–174, 1997.
16. M. Sulzmann, T. Schrijvers & P.J. Stuckey, Type Inference for GADTs via Herbrand Constraint Abduction, draft paper, 2006.
17. T. Walsh, Consistency and Propagation with Multiset Constraints: A Formal Viewpoint, *Proc. CP*, 724–738, LNCS 2833, 2003.
18. J. Wang, M. Maher & R. Topor, Rewriting Unions of General Conjunctive Queries Using Views, *Proc. EDBT*, LNCS 2287, 52–69, 2002.

Feedback in Multimodal Self-organizing Networks Enhances Perception of Corrupted Stimuli

Andrew P. Papliński[1] and Lennart Gustafsson[2]

[1] Clayton School of Information Technology,
Monash University, Victoria 3800, Australia
`app@csse.monash.edu.au`
[2] Computer Science and Electrical Engineering,
Luleå University of Technology, S-971 87 Luleå, Sweden
`Lennart.Gustafsson@ltu.se`

Abstract. It is known from psychology and neuroscience that multimodal integration of sensory information enhances the perception of stimuli that are corrupted in one or more modalities. A prominent example of this is that auditory perception of speech is enhanced when speech is bimodal, i.e. when it also has a visual modality. The function of the cortical network processing speech in auditory and visual cortices and in multimodal association areas, is modeled with a Multimodal Self-Organizing Network (MuSON), consisting of several Kohonen Self-Organizing Maps (SOM) with both feedforward and feedback connections. Simulations with heavily corrupted phonemes and uncorrupted letters as inputs to the MuSON demonstrate a strongly enhanced auditory perception. This is explained by feedback from the bimodal area into the auditory stream, as in cortical processing.

1 Introduction

Bimodal integration of sensory information is advantageous when phenomena have qualities in two modalities. Audiovisual speech, i.e. speech both heard and seen by lip reading is a case where such sensory integration occurs. An important advantage that this integration yields is that audiovisual speech is more robust against noise, see e.g. [1]. This advantage, robustness of identification against noise in stimuli, that are sensed by more than one sensory modality is a general property of bimodal integration. Bimodal and multimodal integration has been studied extensively, for reviews, see [2].

There are areas in cortex that have long been recognized as multimodal association areas, such as the superior temporal polysensory area (STP), see e.g. [3,4], but more recently it has been established that multimodal convergence also occurs earlier in cortical sensory processing, in unimodal (which thus are not exclusively unimodal) sensory cortices [5,6].

Convergence of signals conveying auditory and visual information onto a neuron or a neural structure can be mediated in different ways. Feedforward or

A. Sattar and B.H. Kang (Eds.): AI 2006, LNAI 4304, pp. 19–28, 2006.
© Springer-Verlag Berlin Heidelberg 2006

bottom up connections from lower levels to higher levels in the neural hierarchy and feedback or top down connections going in the opposite direction both serve to integrate information from different sensory modalities, see e.g. [7,8].

The functionality of feedforward and feedback connections has been extensively studied in vision. When a visual stimulus is presented there follows a rapid forward sweep of activity in visual cortex, with a delay of only about 10 msec for each hierarchical level [9]. The initial activity is thus determined mainly by feedforward connections. Feedback will then dynamically change the tuning of neurons even in the lowest levels.

Feedback in bimodal sensory processing has been found to be important in the bimodal processing of audiovisual speech. Speech is processed in several cortical regions, see [10] for a review. Sensory specific cortices provide one of the stages in the human language system [11,5,12]. Auditory processing for phoneme perception takes place in the left posterior Superior Temporal Sulcus (STSp) see e.g. [13,14]. Bimodal integration of audiovisual speech takes place in the multimodal association area in the Superior Temporal Sulcus (STS) and the Superior Temporal Gyrus (STG) [12], located between the sensory-specific auditory and visual areas.

In audiovisual speech there is a perceptual gain in the sensory-specific auditory cortex as compared to purely auditory speech. This corresponds to increased activity in unimodal cortices when bimodal stimuli are presented and is believed to be the result of top-down processing from the STS, modulating the activity in unimodal cortices [15]. It has been found, see [4], that while sensory convergence in higher-order bi- and multisensory regions of the superior temporal sulcus (STS) is mediated by feedforward connections, the visual input into auditory cortex is mediated by feedback connections. This enhances the perception in auditory cortex. As a comparatively recent development in human evolution language also exists in written form, i.e. language has yet another set of visual properties. Simultaneous presentation of written text and auditory speech is not as "natural" an occurrence as lip reading and auditory speech, yet, if the written text and the auditory speech are congruent, speech perception is improved, see [16,17]. Neural resources for processing of letters exist in or close to the left fusiform gyrus, see [18,19,20]. Bimodal integration of phonemes and letters takes place in the STS [21,22].

In this paper, which is a direct continuation of work presented in [23,24], we model the processing of phonemes and letters both in sensory-specific areas and in sensory integration. The network we use for this study is a multimodal self-organizing network (MuSON) which consists of unimodal maps with phonetic and graphic inputs respectively, and an integrating bimodal map. These maps correspond to the cortical architecture for phonetic processing in the STSp, processing of letters in the fusiform area and the bimodal integration in the STS. In [22] it is argued that there is feedback from the bimodal integrating area down into the sensory-specific auditory area. This feedback is also part of our model, in that we introduce a second auditory map which accepts feedforward inputs from the first auditory map as well as feedback inputs from the bimodal map.

We have earlier shown [24] in a study without feedback how templates for phonemes and letters result from self-organization of the unimodal maps and integrate into templates for the bimodal percepts in the bimodal map. We have also shown that the bimodal percepts are robust against additive noise in the letters and phonemes. The purpose of this study is to show how this robustness of the bimodal percepts can be "transferred" down in the processing stream by feedback. In essence we want to show that we hear a noisy phoneme better when we see the corresponding uncorrupted letter.

2   The Multimodal Self-Organizing Networks

Self-organizing neural networks have been inspired by the possibility of achieving information processing in ways that resemble those of biological neural systems.

In particular, pattern associators based on Hebbian learning [25] and self-organizing maps [26] show similarities with biological neural systems. Pattern associators have been employed to simulate the multimodal sensory processing in cortex [27]. Kohonen Self-Organizing Maps (SOMs) are well-recognized and intensively researched tools for mapping multidimensional stimuli onto a low dimensionality (typically 2) neuronal lattice, for an introduction and a review,

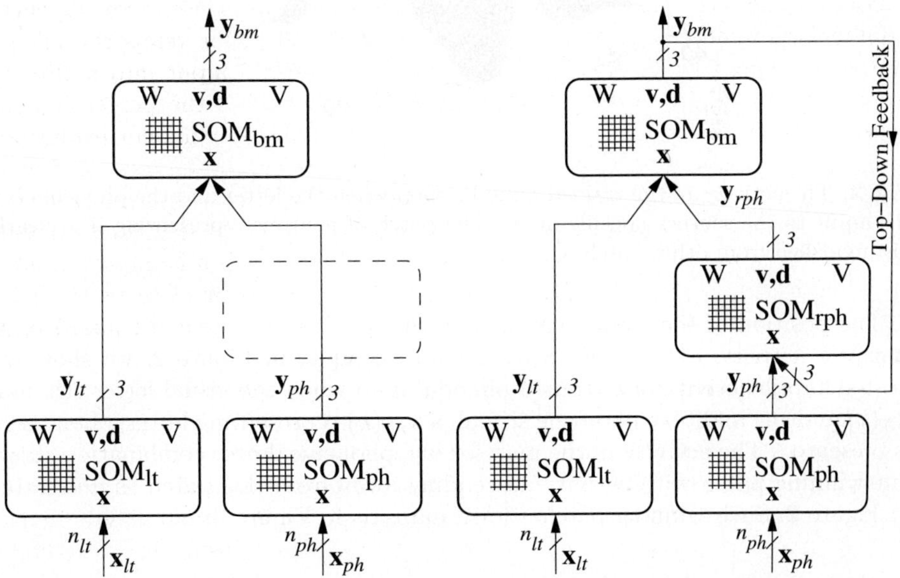

Fig. 1. Left: A two-level feedfoward-only Multimodal Self-Organizing Network (MuSON) processing auditory and visual stimuli. The auditory stimuli are processed in SOM_{ph}, and the visual stimuli in SOM_{lt}. Bimodal integration then takes place in SOM_{bm}. Right: A three-level Multimodal Self-Organizing Network (MuSON) with a feedback connection from the bimodal level to the auditory stream.

see [26]. In this paper we will employ a network of interconnected SOMs, referred to as Multimodal Self-Organizing Networks (MuSONs), see [23,24].

We consider first a feedforward Multimodal Self-Organizing Network (MuSON) as presented in the left part of Figure 1. The pre-processed sensory stimuli, $\mathbf{x}_{lt}$ and $\mathbf{x}_{ph}$ form the inputs to their respective unisensory maps, SOM_{lt} and SOM_{ph}. Three-dimensional outputs from these maps, $\mathbf{y}_{lt}$ and $\mathbf{y}_{ph}$, are combined together to form a six-dimensional stimulus for the higher-level bimodal map, SOM_{bm}. The learning process takes place concurrently for each SOM, according to the well-known Kohonen learning law, see [23,24] for details. After self-organizations each map performs the mapping of the form: $\mathbf{y}(k) = g(\mathbf{x}(k); W, V)$, where $\mathbf{x}(k)$ represents the kth stimulus for a given map, W is the weight map, and V describes the structure of the neuronal grid. The 3-D output signal $\mathbf{y}(k)$ combines the 2-D position of the winner with its 1-D activity.

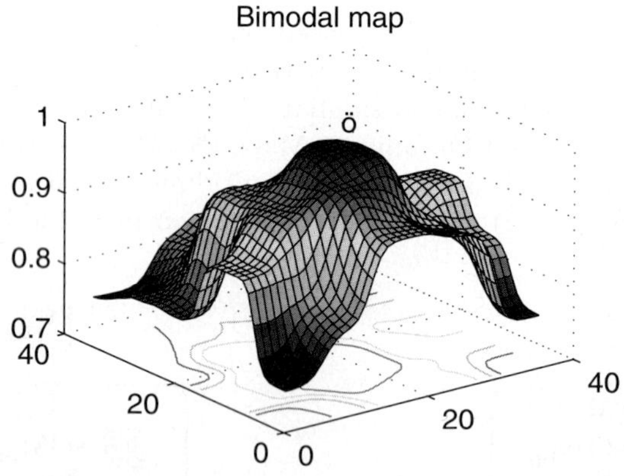

Fig. 2. The activity in the trained bimodal map when the letter and the phoneme **ö** is the input to the sensory-specific maps. The patch of neurons representing **ö** is clearly distinguished from other patches.

The position of the winner can be determined from the network (map) post-synaptic activity, $d(k) = W \cdot \mathbf{x}(k)$. As an example, in Figure 2, we show the post-synaptic activity of a trained bimodal map when the visual letter stimulus $\mathbf{x}_{lt}(k)$ and the auditory phoneme stimulus $\mathbf{x}_{ph}(k)$ representing letter/phoneme **ö** is presented. The activity in the map for one phoneme/letter combination shows one winning patch with activity descending away from this patch as illustrated in Figure 2. Such winning patches form maps as in Figure 3.

3 The Unimodal Visual Map for Letters and Auditory Map for Phonemes

The stimulus $\mathbf{x}_{lt}$ to the visual letter map is a 22-element vector obtained by a principal component analysis of scanned (21×25)-element binary character

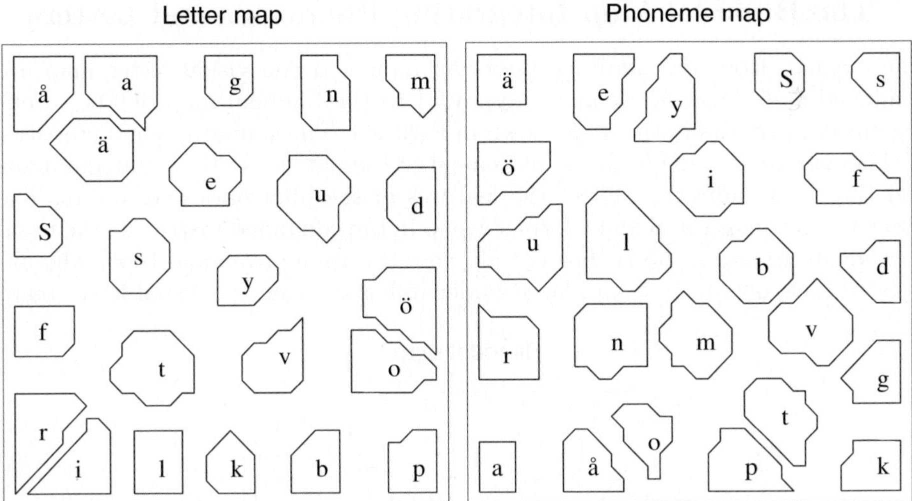

Fig. 3. Patches of highest activity for labeled letters and phonemes after self-organization on a map of 36×36 neurons

images. After self-organization is complete, the properties of the visual letter map SOM_{lt} may be best summarized as in the left part of Figure 3. The map shows the expected similarity properties — symbols that look alike are placed close to each other in the map. Note, for example, the cluster of characters **f**, **t**, **r**, **i**, **l**, **k**, **b** and **p** based predominantly on a vertical stroke. The patches in Figure 3 cover populations of neurons which show the highest activity for their respective stimuli. The neuronal populations within the patches of the letter map constitute the detectors of the respective letters.

In 1988 Kohonen presented the "phonetic typewriter", a phonotopic SOM that learned to identify Finnish phonemes [28,26]. In our study the auditory material consists of twenty-three phonemes as spoken by ten native Swedish speakers. Thirty-six melcepstral coefficients [29] represent each phoneme spoken by each speaker. These feature vectors were averaged over the speakers, yielding one thirty-six element feature vector for each phoneme. This averaged set of vectors constitutes the inputs $\mathbf{x}_{ph}$ to the auditory map SOM_{ph}. After the learning process we obtain a phoneme map as presented in the right part of Figure 3. As for the letter map the patches of neuronal populations constitute the detectors of the respective phonemes.

Note that the plosives **g**, **k**, **t** and **p**, the fricatives **s**, **S** (this is our symbol for the sh-sound as in English she) and **f** and the nasal consonants **m** and **n** form three close groups on the map. Vowels with similar spectral properties are placed close to each other. The back vowels **a**, **å** and **o** are in one group, the front vowels **u ö**, **ä**, **e**, **y** and **i** in another group with the tremulant **r** in-between. The exact placing of the groups vary from one self-organization to another, but the existence of these groups is certain.

4 The Bimodal Map Integrating Phonemes and Letters

The outputs from the auditory phoneme map and the visual letter map are combined as 6-dimensional inputs $[\mathbf{y}_{lt}\ \mathbf{y}_{ph}]$ to the bimodal map SOM_{bm}. Self-organization results in the map shown in Figure 4. The similarity characteristics of this map are derived from the placement of the patches in the unimodal maps and thus only indirectly reflect the features of the phonemes and letters. The fricative consonants **s**, **S** and **f** form a group in the combined map as do the nasal consonants **m** and **n**. Most, but not all, vowels form a group and those who are isolated have obviously been placed under influence from the visual letter map.

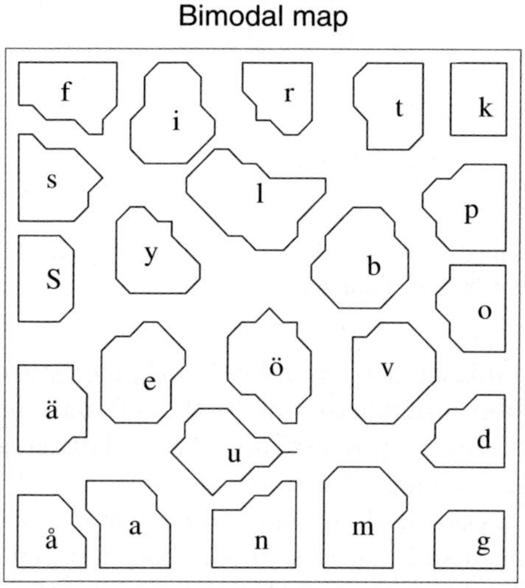

Fig. 4. Bimodal SOM_{bm} map. Patches of highest activity for labeled letter/phoneme combinations after self-organization on a map of 36×36 neurons.

5 Robustness of the Bimodal Percepts Against Unimodal Disturbances

An important advantage of integration of stimuli from sensory-specific cortices into multimodal percepts in multimodal association cortices is that even large disturbances in the stimuli may be eliminated in the multimodal percepts. Our model has the same advantage, as can easily be demonstrated.

We choose to study the processing of the three letters **i**, **å** and **m** which are all uncorrupted. The corresponding phonemes i, å and m are heavily corrupted however, and these corrupted phonemes cause the activity on the phoneme map to move as shown in the left part of Figure 5. In the bimodal map, the right part of Figure 5, the activities have moved very little. The recognition of these bimodal percepts is much less influenced by the auditory corruptions than the recognition in the phoneme map. This holds for all other letter/phoneme combinations as well.

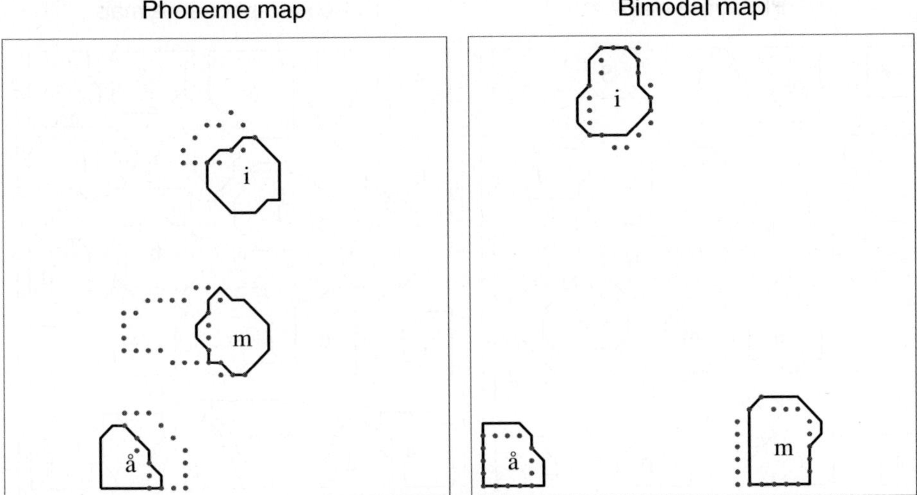

Fig. 5. The maximum activities, shown by solid lines in the maps when the inputs consist of different uncorrupted phonemes and by dotted lines when the inputs consist of heavily corrupted phonemes. Notice the difference in changes of activities in the phoneme map and in the bimodal map.

6 Introduction of Feedback and Its Significance for Auditory Perception

The robustness of the bimodal percepts, demonstrated in Figure 5, can be employed to benefit through feedback to enhance auditory perception, as is the case in cortex [15,22].

We introduce feedback in our MuSON through the re-coded phoneme map SOM_{rph}, see the right part of Figure 1. The 6-dimensional input stimuli to the re-coded phoneme map $[\mathbf{y}_{ph}\ \mathbf{y}_{bm}]$ is formed from the feedforward connection from the sensory phoneme map SOM_{ph} and the top-down feedback connection from the bimodal map SOM_{bm}.

In the second phase of self-organization the re-coded phoneme map SOM_{rph} is initialized to have the 23 winners in the same positions as in the phoneme map SOM_{ph}. The weight vectors are then trained by the Kohonen rule. Relaxation is included between two maps in the feedback loop. After self-organization the re-coded phoneme map SOM_{rph} is seen to be similar to the phoneme map SOM_{ph} as illustrated in Figure 6. However, the re-coded phoneme map through its feedback connections has a dramatically different property when phonemes are corrupted, as seen in Figure 7. The map shows post-synaptic activities for the three letters and phonemes, **i**, **å** and **m**, when the auditory and visual inputs to the sensory-specific maps are perfect (solid lines), and when the auditory inputs to the unisensory phoneme map are heavily corrupted (dotted lines). Note that when phonemes are corrupted the activities in the re-coded phoneme map change

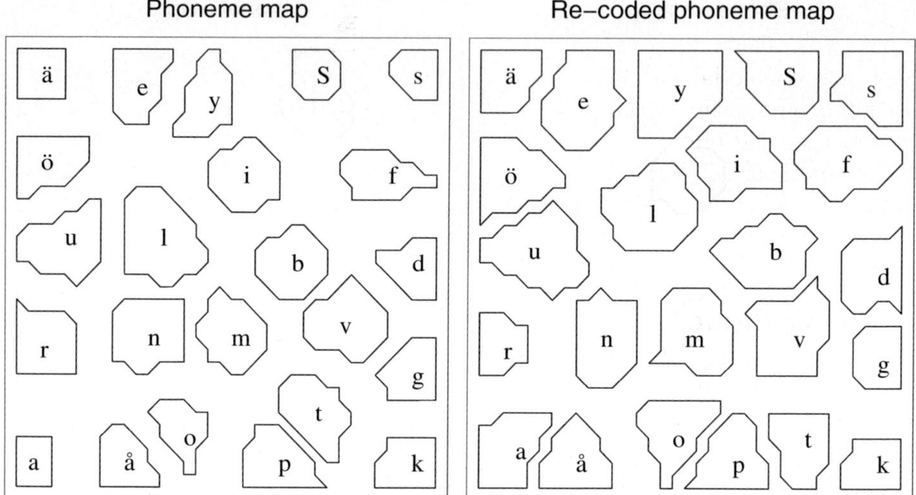

Fig. 6. The phoneme SOM_{ph} map and the corresponding re-coded phoneme SOM_{rph} map

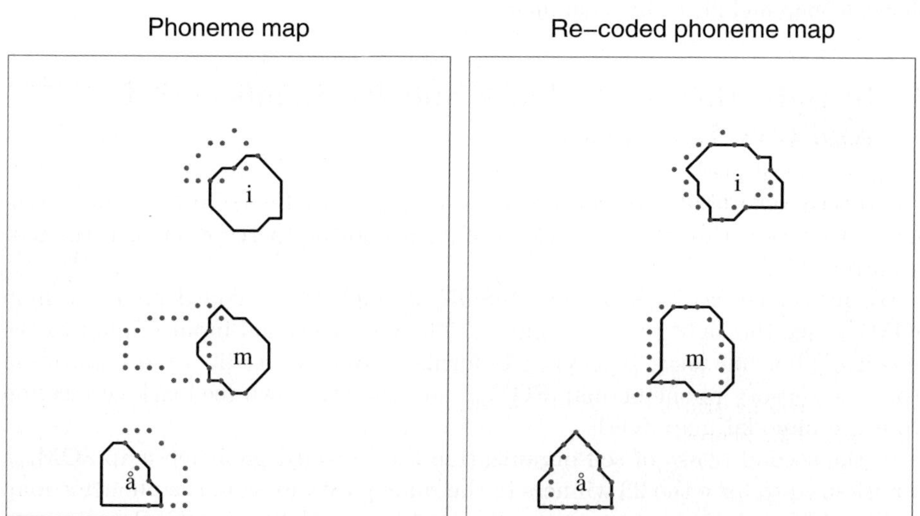

Fig. 7. The maximum activities for the three letters and phonemes **i**, **å** and **m**, shown by solid lines when the auditory and visual inputs to the sensory-specific maps are perfect and by dotted lines when the auditory inputs to the unisensory phoneme map are heavily corrupted

insignificantly compared to the activities in the phoneme map. This holds for all other letter/phoneme combinations as well.

7   Conclusion

With a Multimodal Self-Organizing Network we have simulated bimodal integration of audiovisual speech elements, phonemes and letters. We have demonstrated that the bimodal percepts are robust against corrupted phonemes and that when these robust bimodal percepts are fed back to the auditory stream the auditory perception is greatly enhanced. These results agree with known results from psychology and neuroscience.

Acknowledgment

The authors acknowledge financial support from the 2006 Monash University Small Grant Scheme.

References

1. Sumby, W., Pollack, I.: Visual contribution to speech intelligibility in noise. J. Acoust. Soc. Am. **26** (1954) 212–215
2. Calvert, E.G., Spence, C., Stein, B.E.: The handbook of multisensory processes. 1st edn. MIT Press, Cambridge, MA (2004)
3. Schroeder, C., Foxe, J.: The timing and laminar profile of converging inputs to multisensory areas of the macaque neocortex. Cognitive Brain Research **14** (2002) 187–198
4. Schroeder, C., Smiley, J., Fu, K., McGinnis, T., O'Connell, M., Hackett, T.: Anatomical mechanisms and functional implications of multisensory convergence in early cortical processing. Int. J. Psychophysiology **50** (2003) 5–17
5. Calvert, G.A., Bullmore, E.T., Brammer, M.J., Campbell, R., Williams, S.C., McGuire, P., Woodruff, P.W., Iversen, S.D., David, A.S.: Activation of auditory cortex during silent lipreading. Science **276** (1997) 593–596
6. Driver, J., Spence, C.: Crossmodal attention. Curr. Opin. Neurobiol. **8** (1998) 245–253
7. Calvert, G.A., Thesen, T.: Multisensory integration: methodological approaches and emerging principles in the human brain. J. Physiology Paris **98** (2004) 191–205
8. Foxe, J.J., Schroeder, C.E.: The case for feedforward multisensory convergence during early cortical processing. Neuroreport **16**(5) (2005) 419–423
9. Lamme, V., Roelfsema, P.: The distinct modes of vision offered by feedforward and recurrent processing. Trends Neuroscience **23** (2000) 571–579
10. Price, C.J.: The anatomy of language: contributions from functional neuroimaging. J. Anat. **197** (2000) 335–359
11. Binder, J.R., Frost, J.A., Hammeke, T.A., Bellgowan, P.S.F., Springer, J.A., Kaufman, J.N., Possing, E.T.: Human temporal lobe activation by speech and nonspeech sounds. Cerebral Cortex **10** (2000) 512–528
12. Calvert, G.A., Campbell, R.: Reading speech from still and moving faces: The neural substrates of visual speech. J. Cognitive Neuroscience **15**(1) (2003) 57–70
13. Dehaene-Lambetrz, G., Pallier, C., Serniclaes, W., Sprenger-Charolles, L., Jobert, A., Dehaene, S.: Neural correlates of switching from auditory to speech perception. NeuroImage **24** (2005) 21–33

14. Möttönen, R., Calvert, G.A., Jääskeläinen, I., Matthews, P.M., Thesen, T., Tuominen, J., Sams, M.: Perceiving identical sounds as speech or non-speech modulates activity in the left posterior superior temporal sulcus. NeuroImage **19** (2005)
15. Calvert, G., Campbell, R., Brammer, M.: Evidence from functional magnetic resonance imaging of crossmodal binding in the human heteromodal cortex. Current Biology **10** (2000) 649–657
16. Frost, R., Repp, B., Katz, L.: Can speech perception be influenced by simultaneous presentation of print? J. Mem. Lang. **27** (1988) 741–755
17. Dijkstra, T., Frauenfelder, U.H., Schreuder, R.: Bidirectional grapheme-phoneme activation in a bimodal detection task. J. Physiology Paris **98**(3) (2004) 191–205
18. Gauthier, I., Tarr, M.J., Moylan, J., Skudlarski, P., Gore, J.C., Anderson, A.W.: The fusiform "face area" is part of a network that processes faces at the individual level. J. Cognitive Neuroscience **12**(3) (2000) 495–504
19. Polk, T.A., Farah, M.J.: The neural development and organization of letter recognition: Evidence from functional neuroimaging, computational modeling, and behavioral studies. PNAS **98** (1998) 847–852
20. Polk, T.A., Stallcup, M., Aguire, G.K., Alsop, D.C., D'Esposito, M., Detre, J.A., Farah, M.J.: Neural specialization for letter recognition. J. Cognitive Neuroscience **14**(2) (2002) 145–159
21. Raij, T., Uutela, K., Hari, R.: Audiovisual integration of letters in the human brain. Neuron **28** (2000) 617–625
22. van Atteveldt, N., Formisano, E., Goebel, R., Blomert, L.: Integration of letters and speech sounds in the human brain. Neuron **43** (2004) 271–282
23. Papliński, A.P., Gustafsson, L.: Multimodal feedforward self-organizing maps. In: Lecture Notes in Computer Science. Volume 3801., Springer (2005) 81–88
24. Gustafsson, L., Papliński, A.P.: Bimodal integration of phonemes and letters: an application of multimodal self-organizing networks. In: Proc. Int. Joint Conf. Neural Networks, Vancouver, Canada (2006) 704–710
25. Hopfield, J.: Neural networks and physical systems with emergent collective computational properties. PNAS USA **79** (1982) 2554–2588
26. Kohonen, T.: Self-Organising Maps. 3rd edn. Springer-Verlag, Berlin (2001)
27. Rolls, E.T.: Multisensory neuronal convergence of taste, somatosentory, visual, and auditory inputs. In Calvert, G., Spencer, C., Stein, B.E., eds.: The Handbook of multisensory processes. MIT Press (2004) 311–331
28. Kohonen, T.: The "neural" phonetic typewriter. Computer (1988) 11–22
29. Gold, B., Morgan, N.: Speech and audio signal processing. John Wiley & Sons, Inc., New York (2000)

Identification of Fuzzy Relation Model Using HFC-Based Parallel Genetic Algorithms and Information Data Granulation

Jeoung-Nae Choi, Sung-Kwun Oh, and Hyun-Ki Kim

Department of Electrical Engineering, The University of Suwon, San 2-2 Wau-ri,
Bongdam-eup, Hwaseong-si, Gyeonggi-do, 445-743, South Korea
ohsk@suwon.ac.kr

Abstract. The paper concerns the hybrid optimization of fuzzy inference systems that is based on Hierarchical Fair Competition-based Parallel Genetic Algorithms (HFCGA) and information data granulation. The granulation is realized with the aid of the Hard C-means clustering and HFCGA is a kind of multi-populations of Parallel Genetic Algorithms (PGA), and it is used for structure optimization and parameter identification of fuzzy model. It concerns the fuzzy model-related parameters such as the number of input variables, a collection of specific subset of input variables, the number of membership functions, and the apexes of the membership function. In the hybrid optimization process, two general optimization mechanisms are explored. The structural optimization is realized via HFCGA and HCM method whereas in case of the parametric optimization we proceed with a standard least square method as well as HFCGA method as well. A comparative analysis demonstrates that the proposed algorithm is superior to the conventional methods.

Keywords: fuzzy relation model, information granulation, genetic algorithms, hierarchical fair competition (HFC), HCM, multi-population.

1 Introduction

Fuzzy modeling has been a focal point of the technology of fuzzy sets from its very inception. Fuzzy modeling has been studied to deal with complex, ill-defined, and uncertain systems in many other avenues. In the early 1980s, linguistic modeling [1] and fuzzy relation equation-based approach [2] were proposed as primordial identification methods for fuzzy models. The general class of Sugeno-Takagi models [3] gave rise to more sophisticated rule-based systems where the rules come with conclusions forming local regression models. While appealing with respect to the basic topology (a modular fuzzy model composed of a series of rules) [4], these models still await formal solutions as far as the structure optimization of the model is concerned, say a construction of the underlying fuzzy sets—information granules being viewed as basic building blocks of any fuzzy model.

Some enhancements to the model have been proposed by Oh and Pedrycz [5,6]. As one of the enhanced fuzzy model, information granulation based fuzzy relation fuzzy

A. Sattar and B.H. Kang (Eds.): AI 2006, LNAI 4304, pp. 29–38, 2006.

© Springer-Verlag Berlin Heidelberg 2006

model was introduced. Over there, binary-coded genetic algorithm was used to optimize structure and premise parameters of fuzzy model, yet the problem of finding "good" initial parameters of the fuzzy sets in the rules remains open.

This study concentrates on optimization of information granulation-oriented fuzzy model. Also, we propose to use hierarchical fair competition-based parallel genetic algorithm (HFCGA) for optimization of fuzzy model. GAs is well known as an optimization algorithm which can be searched global solution. It has been shown to be very successful in many applications and in very different domains. However it may get trapped in a sub-optimal region of the search space thus becoming unable to find better quality solutions, especially for very large search space. The parallel genetic algorithm (PGA) is developed with the aid of global search and retard premature convergence. In particular, as one of the PGA model, HFCGA has an effect on a problem having very large search space [9].

In the sequel, the design methodology emerges as two phases of structural optimization (based on Hard C-Means (HCM) clustering and HFCGA) and parametric identification (based on least square method (LSM), as well as HCM clustering and HFCGA). Information granulation with the aid of HCM clustering helps determine the initial parameters of fuzzy model such as the initial apexes of the membership functions and the initial values of polynomial function being used in the premise and consequence part of the fuzzy rules. And the initial parameters are adjusted effectively with the aid of the HFCGA and the LSM.

2   Information Granulation (IG)

Usually, information granules [7] are viewed as related collections of objects (data point, in particular) drawn together by the criteria of proximity, similarity, or functionality. Granulation of information is an inherent and omnipresent activity of human beings carried out with intent of gaining a better insight into a problem under consideration and arriving at its efficient solution. In particular, granulation of information is aimed at transforming the problem at hand into several smaller and therefore manageable tasks. In this way, we partition this problem into a series of well-defined subproblems (modules) of a far lower computational complexity than the original one. The form of information granulation (IG) themselves becomes an important design feature of the fuzzy model, which are geared toward capturing relationships between information granules.

It is worth emphasizing that the HCM clustering has been used extensively not only to organize and categorize data, but it becomes useful in data compression and model identification [8]. For the sake of completeness of the entire discussion, let us briefly recall the essence of the HCM algorithm.

We obtain the matrix representation for hard c-partition, defined as follows.

$$M_C = \left\{ U \mid u_{gi} \in \{0,1\}, \ \sum_{g=1}^{c} u_{gi} = 1, \ 0 < \sum_{i=1}^{m} u_{gi} < m \right\} \tag{1}$$

[**Step 1**]. Fix the number of clusters $c(2 \le c < m)$ and initialize the partition matrix $\mathbf{U}^{(0)} \in M_C$

[Step 2]. Calculate the center vectors $\mathbf{v}_g$ of each cluster:

$$v_g^{(r)} = \{v_{g1}, v_{g2}, \cdots, v_{gk}, \cdots, v_{gl}\} \tag{2}$$

$$v_{gk}^{(r)} = \sum_{i=1}^{m} u_{gi}^{(r)} \cdot x_{ik} \left/ \sum_{i=1}^{m} u_{gi}^{(r)} \right. \tag{3}$$

Where, $[u_{gi}] = \mathbf{U}^{(r)}$, $g = 1, 2, \ldots, c$, $k=1, 2, \ldots, l$.

[Step 3]. Update the partition matrix $\mathbf{U}^{(r)}$; these modifications are based on the standard Euclidean distance function between the data points and the prototypes,

$$d_{gi} = d(\mathbf{x}_i - \mathbf{v}_g) = \|\mathbf{x}_i - \mathbf{v}_g\| = \left[\sum_{k=1}^{l}(x_{ik} - v_{gk})^2\right]^{1/2} \tag{4}$$

$$u_{gi}^{(r+1)} = \begin{cases} 1 & d_{gi}^{(r)} = \min\{d_{ki}^{(r)}\} \ \text{ for all } \ k \in c \\ 0 & \text{otherwise} \end{cases} \tag{5}$$

[Step 4]. Check a termination criterion. If

$$\| \mathbf{U}^{(r+1)} - \mathbf{U}^{(r)} \| \leq \varepsilon \ \ \text{(tolerance level)} \tag{6}$$

Stop ; otherwise set $r = r + 1$ and return to **[Step 2]**

3   Design of Fuzzy Model with the Aid of IG

The identification procedure for fuzzy models is usually split into the identification activities dealing with the premise and consequence parts of the rules. The identification completed at the premise level consists of two main steps. First, we select the input variables $x_1, x_2, \ldots, x_k$ of the rules. Second, we form fuzzy partitions of the spaces over which these individual variables are defined. The identification of the consequence part of the rules embraces two phases, namely 1) a selection of the consequence variables of the fuzzy rules, and 2) determination of the parameters of the consequence (conclusion part). And the least square error (LSE) method used at the parametric optimization of the consequence parts of the successive rules.

In this study, we use the isolated input space of each input variable and carry out the modeling using characteristics of input-output data set. Therefore, it is important to understand the nature of data. The HCM clustering addresses this issue. Subsequently, we design the fuzzy model by considering the centers of clusters. In this manner the clustering help us determining the initial parameters of fuzzy model such as the initial apexes of the membership functions and the order of polynomial function being used in the premise and consequence part of the fuzzy rules.

3.1   Premise Identification

In the premise part of the rules, we confine ourselves to a triangular type of membership functions whose parameters are subject to some optimization. The HCM clustering helps us organize the data into cluster so in this way we capture the characteristics of the experimental data. In the regions where some clusters of data have been identified, we end up with some fuzzy sets that help reflect the specificity of the data set

The identification of the premise part is completed in the following manner. Given is a set of data $\mathbf{U}=\{\mathbf{x}_1, \mathbf{x}_2, \ldots, \mathbf{x}_l ; \mathbf{y}\}$, where $\mathbf{x}_k =[x_{1k}, \ldots, x_{mk}]^T$, $\mathbf{y} =[y_1, \ldots, y_m]^T$, l is the number of variables and , m is the number of data.

[Step 1]. Arrange a set of data $\mathbf{U}$ into data set $\mathbf{X}_k$ composed of respective input data and output data.

$$\mathbf{X}_k=[\mathbf{x}_k ; \mathbf{y}] \tag{7}$$

$\mathbf{X}_k$ is data set of k-th input data and output data, where, $\mathbf{x}_k =[x_{1k}, \ldots, x_{mk}]^T$, $\mathbf{y} =[y_1, \ldots, y_m]^T$, and $k=1, 2, \ldots, l$.

[Step 2]. Complete the HCM clustering to determine the centers (prototypes) $\mathbf{v}_{kg}$ with data set $\mathbf{X}_k$.

> **[Step 2-1].** Classify data set $\mathbf{X}_k$ into c-clusters, which in essence leads to the granulation of information.
> We can find the data pairs included in each cluster based on the partition matrix u_{gi} by (5) and use these to identify the structure in conclusion part.

> **[Step 2-2].** Calculate the center vectors $\mathbf{v}_{kg}$ of each cluster.

$$\mathbf{v}_{kg} = \{v_{k1}, v_{k2}, \ldots, v_{kc}\} \tag{8}$$

Where, $k=1, 2, \ldots, l$, $g = 1, 2, \ldots, c$.

[Step 3]. Partition the corresponding isolated input space using the prototypes of the clusters $\mathbf{v}_{kg}$. Associate each clusters with some meaning (semantics), say Small, Big, etc.

[Step 4]. Set the initial apexes of the membership functions using the prototypes $\mathbf{v}_{kg}$.

3.2 Consequence Identification

We identify the structure considering the initial values of polynomial functions based upon the information granulation realized for the consequence and antecedents parts.

[Step 1]. Find a set of data included in the fuzzy space of the j-th rule.

> **[Step 1-1].** Find the input data included in each cluster (information granule) from the partition matrix u_{gi} of each input variable by (5).

> **[Step 1-2].** Find the input data pairs included in the fuzzy space of the j-th rule
> **[Step 1-3].** Determine the corresponding output data from above input data pairs.

[Step 2]. Compute the prototypes $\mathbf{V}_j$ of the data set by taking the arithmetic mean of each rule.

$$\mathbf{V}_j = \{V_{1j}, V_{2j}, \ldots, V_{kj}; M_j\} \tag{9}$$

Where, $k=1, 2, \ldots, l$. $j=1, 2, \ldots, n$. V_{kj} and M_j are prototypes of input and output data, respectively.

[Step 3]. Set the initial values of polynomial functions with the center vectors $\mathbf{V}_j$.

The identification of the conclusion parts of the rules deals with a selection of their structure that is followed by the determination of the respective parameters of the local functions occurring there.

In Case of Type 2: Linear Inference (linear conclusion)
The conclusion is expressed in the form of a linear relationship between inputs and output variable. This gives rise to the rules in the form

$$R^j : \text{If } x_1 \text{ is } A_{1c} \text{ and } \cdots \text{ and } x_k \text{ is } A_{kc} \text{ then } y_j - M_j = f_j(x_1,\cdots,x_k) \tag{10}$$

The calculations of the numeric output of the model, based on the activation (matching) levels of the rules there, rely on the expression

$$y^* = \frac{\displaystyle\sum_{j=1}^{n} w_{ji} y_i}{\displaystyle\sum_{j=1}^{n} w_{ji}} = \frac{\displaystyle\sum_{j=1}^{n} w_{ji}(f_j(x_1,\cdots,x_k)+M_j)}{\displaystyle\sum_{j=1}^{n} w_{ji}} = \sum_{j=1}^{n} \hat{w}_{ji}(a_{j0}+a_{j1}(x_1-V_{j1})+\cdots+a_{jk}(x_k-V_{jk})+M_j) \tag{11}$$

Here, as the normalized value of w_{ji}, we use an abbreviated notation to describe an activation level of rule R^j to be in the form

$$\hat{w}_{ji} = \frac{w_{ji}}{\displaystyle\sum_{j=1}^{n} w_{ji}} \tag{12}$$

where R^j is the j-th fuzzy rule, x_k represents the input variables, A_{kc} is a membership function of fuzzy sets, a_{j0} is a constant, M_j is a center value of output data, n is the number of fuzzy rules, y^* is the inferred output value, w_{ji} is the premise fitness matching R^j (activation level).

Once the input variables of the premise and parameters have been already specified, the optimal consequence parameters that minimize the assumed performance index can be determined. In what follows, we define the performance index as the mean squared error (MSE).

$$\text{PI} = \frac{1}{m}\sum_{i=1}^{m}(y_i - y_i^*)^2 \tag{13}$$

where y^* is the output of the fuzzy model, m is the total number of data, and i is the data number. The minimal value produced by the least-squares method is governed by the following expression:

$$\hat{\mathbf{a}} = (\mathbf{X}^T \mathbf{X})^{-1}\mathbf{X}^T \mathbf{Y} \tag{14}$$

where

$$\mathbf{x}_i^T = [\,\hat{w}_{1i} \cdots \hat{w}_{ni}\ (x_{1i}-V_{11})\hat{w}_{1i} \cdots (x_{1i}-V_{1n})\hat{w}_{ni} \cdots (x_{ki}-V_{k1})\hat{w}_{1i} \cdots (x_{ki}-V_{kn})\hat{w}_{ni}\,],$$

$$\hat{\mathbf{a}} = [a_{10} \cdots a_{n0}\ a_{11} \cdots a_{n1} \cdots a_{1k} \cdots a_{nk}]^T,$$

$$\mathbf{Y} = \left[y_1 - \left(\sum_{j=1}^{n} M_j w_{j1}\right)\ \ y_2 - \left(\sum_{j=1}^{n} M_j w_{j2}\right)\ \ \cdots\ \ y_m - \left(\sum_{j=1}^{n} M_j w_{jm}\right) \right]^T$$

$$\mathbf{X} = [\mathbf{x}_1\ \ \mathbf{x}_2\ \ \cdots\ \ \mathbf{x}_i\ \ \cdots\ \ \mathbf{x}_m]^T .$$

4 Optimization by Means of HFCGA

The premature convergence of genetic algorithms is a problem to be overcome. The convergence is desirable, but must be controlled in order that the population does not get trapped in local optima. Even in dynamic-sized populations, the high-fitness individuals supplant the low-fitness or are favorites to be selected, dominating the evolutionary process. Fuzzy model has many parameters to be optimized, and it has very large search space. So HFCGA may find out a solution better than GAs.ï

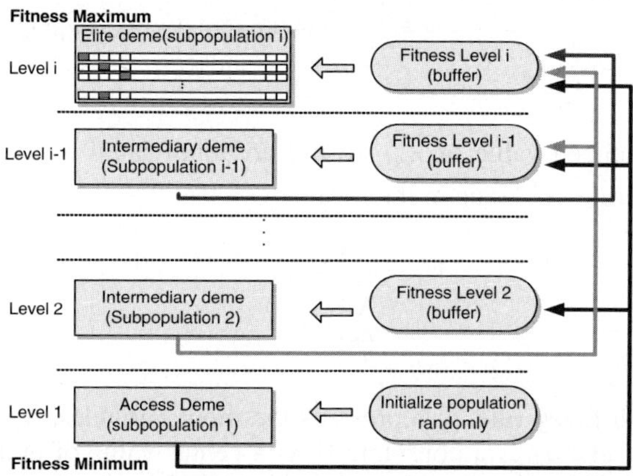

Fig. 1. HFC based migration topology

In HFCGA, multiple demes (subpopulation) are organized in a hierarchy, in which each deme can only accommodate individuals within a specified range of fitness. The universe of fitness values must have a deme correspondence. Each deme has an admission threshold that determines the profile of the fitness in each deme. Individuals are moved from low-fitness to higher-fitness subpopulations if and only if they exceed the fitness-based admission threshold of the receiving subpopulations. Thus, one can note that HFCGA adopts a unidirectional migration operator, where individuals can move to superior levels, but not to inferior ones. The figure 1 illustrates the migration topology of HFCGA. The arrows indicate the moving direction possibilities. The access deme (primary level) can send individuals to all other demes and the elite deme only can receive individuals from the others. One can note that, with respect to topology, HFCGA is a specific case of island model, where only some moves are allowed.

Fig. 2 depict flowchart of implemented HFCGA, it is real-coded type, we can choice the number of demes, size of demes, and operators(selection, crossover, mutation algorithms) for each deme. Where, each deme can evolve with different operators. In this study, we use five demes (subpopulation), Size of demes is 100, 80, 80, 80, and 60 respectively, where elite deme is given as the least size. And we use same operators as such linear ranking based selection, modified simple crossover, and dynamic mutation algorithm for each deme.

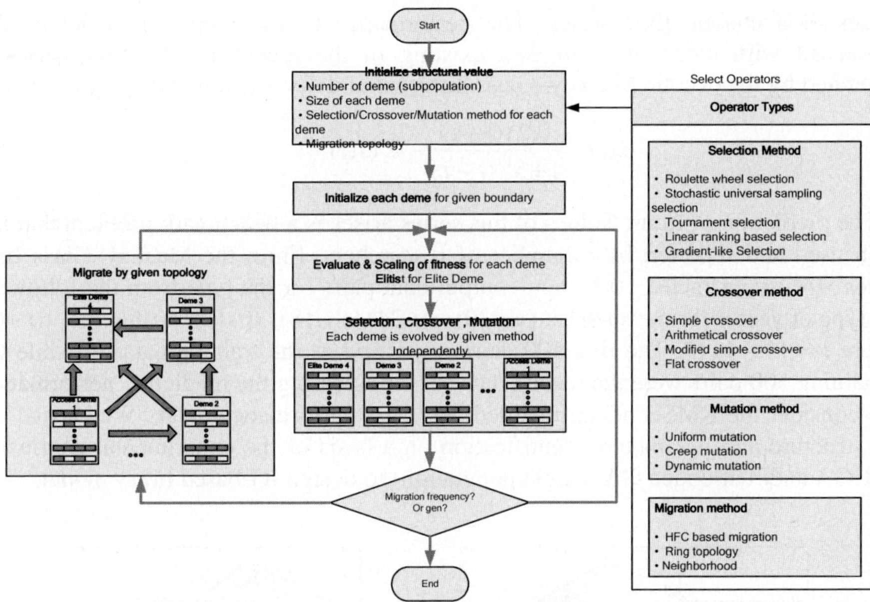

Fig. 2. Flowchart for HFCGA

Identification procedure of fuzzy model consists of two phase, structural identification and parametric identification. HFCGA is used in each phase. At first, in structural identification, we find the number of input variables, input variables being selected and the number of membership functions standing in the premise and the type of polynomial in conclusion. And then, in parametric identification, we adjust apexes of the membership functions of premise part of fuzzy rules. Figure 3 shows an arrangement of chromosomes to be used in HFCGA.

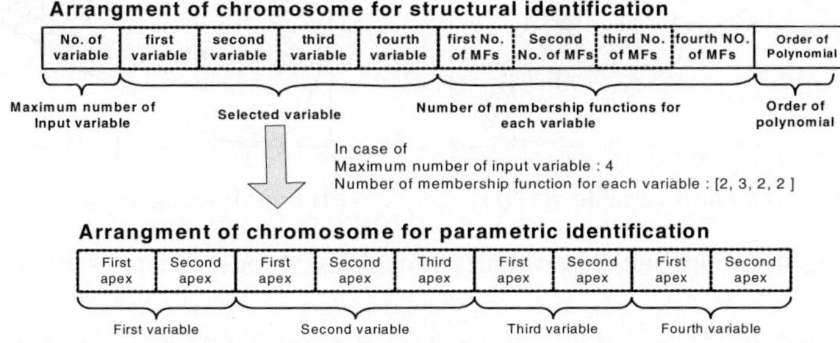

Fig. 3. Arrangement of chromosomes for identification of structure and parameter identification

5   Experimental Studies

In this section we consider comprehensive numeric studies illustrating the design of the fuzzy model. We demonstrate how IG-based FIS can be utilized to predict future values of a chaotic time series. The performance of the proposed model is also contrasted with some other models existing in the literature. The time series is generated by the chaotic Mackey–Glass differential delay equation [10] of the form:

$$\dot{x}(t) = \frac{0.2x(t-\tau)}{1+x^{10}(t-\tau)} - 0.1x(t) \tag{15}$$

The prediction of future values of this series arises is a benchmark problem that has been used and reported by a number of researchers. From the Mackey–Glass time series $x(t)$, we extracted 1000 input–output data pairs for the type from the following the type of vector format such as: $[x(t\text{-}30), x(t\text{-}24), x(t\text{-}18), x(t\text{-}12), x(t\text{-}6), x(t); x(t+6)]$ where $t = 118\text{–}1117$. The first 500 pairs were used as the training data set while the remaining 500 pairs were the testing data set for assessing the predictive performance. We consider the RMSE being regarded here as a performance index. We carried out the structure and parameters identification on a basis of the experimental data using HFCGA and real-coded GA (single population) to design IG-based fuzzy model.

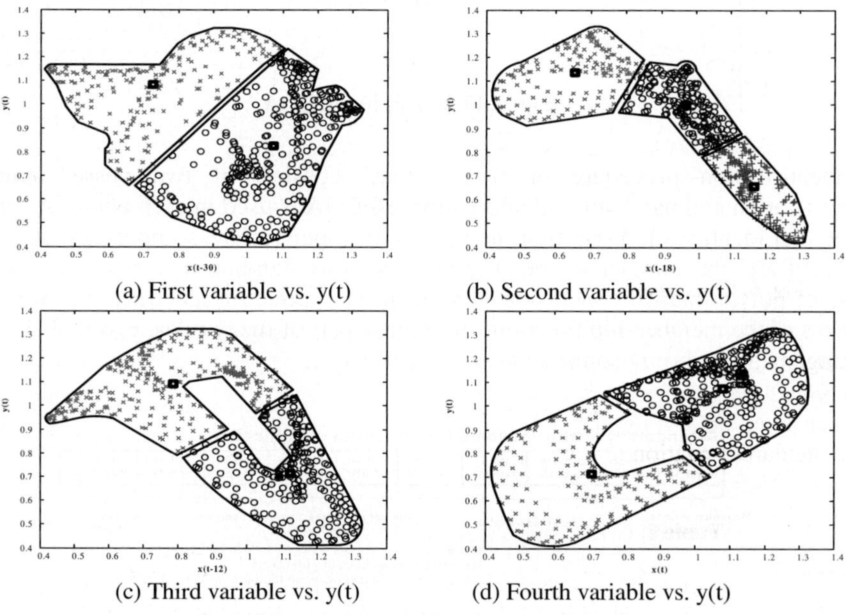

(a) First variable vs. y(t)   (b) Second variable vs. y(t)

(c) Third variable vs. y(t)   (d) Fourth variable vs. y(t)

Fig. 4. Groups and central values through HCM for each input variable

Figure 4 depicts groups and central values through HCM for each input variable. Where, the number of input variables and number of groups (membership function) to be divided are obtained from structural optimization procedure. Clustering results are used for information granulation.

Table 1 summarizes the performance index for real-coded GA and HFCGA. It shows that the performance of the HFCGA based fuzzy model is better than real-coded GA based one for premise identification. However, for structure identification, same structure is selected. To compare real-coded GA with HFCGA, show the performance index for the Type 2 (linear inference). Figure 5 show variation of the performance index for real-coded GA and HFCGA in premise identification phase.

Table 1. Performance index of IG-based fuzzy model by means of HFCGA

Evolutionary algorithm	Structure Identification					Parameter Iden.	
	Input variables	No. of MFs	Type	PI	E_PI	PI	E_PI
HFCGA	x(t-30)	2	Type 3 (Quadratic)	0.00008	0.00021	0.00006	0.00007
	x(t-18)	3					
	x(t-12)	2	Type 2 (Linear)	0.00140	0.00136	0.00070	0.00062
	x(t)	2					

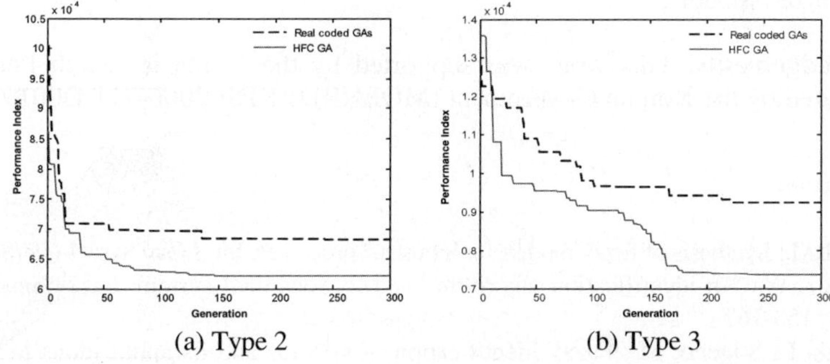

(a) Type 2   (b) Type 3

Fig. 5. Convergence process of performance index for real-coded GA and HFCGA

The identification error (performance index) of the proposed model is also compared with the performance of some other models; refer to Table 2. Here the non-dimensional error index (NDEI) is defined as the root mean square errors divided by the standard deviation of the target series.

Table 2. Comparison of identification error with previous models

Model	No. of rules	PI_t	PI	E_PI	NDEI
Wang's model [10]	7	0.004			
	23	0.013			
	31	0.010			
Cascaded-correlation NN [11]					0.06
Backpropagation MLP [11]					0.02
6th-order polynomial [11]					0.04
ANFIS [12]	16		0.0016	0.0015	0.007
FNN model [13]			0.014	0.009	
Our model	24		0.00006	0.00007	0.00032

6 Conclusions

In this paper, we have developed a comprehensive hybrid identification framework for information granulation-oriented fuzzy model using hierarchical fair competition–based parallel genetic algorithm. The underlying idea deals with an optimization of information granules by exploiting techniques of clustering and genetic algorithms. We used the isolated input space for each input variable and defined the fuzzy space by information granule. Information granulation with the aid of HCM clustering help determine the initial parameters of fuzzy model such as the initial apexes of the membership functions and the initial values of polynomial function being used in the premise and consequence part of the fuzzy rules. The initial parameters are fine-tuned (adjusted) effectively with the aid of HFCGA and the least square method. The experimental studies showed that the model is compact (realized through a small number of rules), and its performance is better than some other previous models. The proposed model is effective for nonlinear complex systems, so we can construct a well-organized model.

Acknowledgements. This work was supported by the Korea Research Foundation Grant funded by the Korean Government (MOEHRD)(KRF-2006-311-D00194).

References

1. Tong RM.: Synthesis of fuzzy models for industrial processes. Int. J Gen Syst. 4 (1978) 143-162
2. Pedrycz, W.: An identification algorithm in fuzzy relational system. Fuzzy Sets Syst. 13 (1984) 153-167
3. Takagi, T., Sugeno, M.: Fuzzy identification of systems and its applications to modeling and control. IEEE Trans Syst, Cybern. SMC-15(1) (1985) 116-132
4. Sugeno, M., Yasukawa, T.: Linguistic modeling based on numerical data. In: IFSA'91 Brussels, Computer, Management & System Science. (1991) 264-267
5. Oh, S.K., Pedrycz, W.: Identification of fuzzy systems by means of an auto-tuning algorithm and its application to nonlinear systems. Fuzzy Sets and Syst. 115(2) (2000) 205-230
6. Park, H.S., Oh, S.K.: Fuzzy Relation-based Fuzzy Neural-Networks Using a Hybrid Identification Algorithm. International Journal of Control Automation and Systems. 1(3) (2003) 289-300
7. Pderycz, W., Vukovich, G.: Granular neural networks. Neurocomputing. 36 (2001) 205-224
8. Krishnaiah, P.R., Kanal, L.N., editors.: Classification, pattern recognition, and reduction of dimensionality, volume 2 of Handbook of Statistics. North-Holland, Amsterdam. (1982)
9. Hu, J.J., Goodman, E.: The Hierarchical Fair Competition (HFC) Model for Parallel Evolutionary Algorithms. Proceedings of the 2002 Congress on Evolutionary Computation: CEC2002. IEEE. Honolulu. Hawaii. (2002)
10. Wang, L.X., Mendel, J.M.: Generating fuzzy rules from numerical data with applications. IEEE Trans. Systems, Man, Cybern. **22(6)** (1992) 1414-1427
11. Crowder, R.S. III.: Predicting the Mackey-Glass time series with cascade-correlation learning. In D. Touretzky, G. Hinton, and T. Sejnowski, editors, Proceedings of the 1990 Connectionist Models Summer School. Carnegic Mellon University. (1990) 117-123
12. Jang J.R.: ANFIS: Adaptive-Network-Based Fuzzy Inference System. IEEE Trans. System, Man, and Cybern. **23(3)** (1993) 665-685
13. Maguire, L.P., Roche, B., McGinnity, T.M., McDaid, L.J.: Predicting a chaotic time series using a fuzzy neural network. Information Sciences. **112** (1998) 125-136

Evaluation of Incremental Knowledge Acquisition with Simulated Experts[*]

Paul Compton and Tri M. Cao

School of Computer Science and Engineering
University of New South Wales
Sydney 2052 , Australia
{compton, tmc}@cse.unsw.edu.au

Abstract. Evaluation of knowledge acquisition (KA) is difficult in general. In recent times, incremental knowledge acquisition that emphasises direct communication between human experts and systems has been increasingly widely used. However, evaluating incremental KA techniques, like KA in general, has been difficult because of the costs of using human expertise in experimental studies. In this paper, we use a general simulation framework to evaluate Ripple Down Rules (RDR), a successful incremental KA method. We focus on two fundamental aspects of incremental KA: the importance of acquiring domain ontological structures and the usage of cornerstone cases.

1   Introduction

Knowledge acquisition is widely considered as a modelling activity [14,19]. Most of the KA approaches for building knowledge based systems support a domain analysis (including problem analysis and/or problem solving analysis) by the domain experts and the knowledge engineers. This process possibly involves steps like developing a conceptual model of the domain knowledge, distinguishing subtasks to be solved, differentiating types of knowledge to be used in the reasoning process, etc. Eventually, this knowledge engineering approach results in a model of the domain knowledge that can be turn into an operational system manually or automatically.

On the other hand, the incremental KA approach aims to skip the time consuming process of analysing expertise and domain problem by a knowledge engineer [4,6]. It rather allows the experts themselves to communicate more directly their expertise to the system. This communication is usually triggered by real data that the experts encounter in their normal workflow. Over time, a complex system will be gradually developed. As many knowledge based systems are classification systems, from this point on, we focus on classification based systems, even though most of our arguments are easy to adapt to other types of knowledge based systems. We also use the term production systems as a synonym for classification systems.

---

[*] A shorter version of this paper has been accepted to the conference EKAW 2006.

A. Sattar and B.H. Kang (Eds.): AI 2006, LNAI 4304, pp. 39–48, 2006.

© Springer-Verlag Berlin Heidelberg 2006

The following algorithm describes a general incremental knowledge acquisition process.

Algorithm 1. General Incremental KA

1. Start with an empty KB
2. Accept a new data case
3. Evaluate the case against the knowledge base (KB).
4. If the result is not correct, an expert is consulted to refine the KB
5. If the overall performance of the KB is satisfactory, then terminate, otherwise go to Step 2

It is important to note that this scheme largely corresponds to the maintenance phase of any knowledge based system (KBS). The KB processes cases; when its performance is judged unsatisfactory or inadequate, changes are made and the performance of the new KB is validated. RDR is an extreme example of this maintenance model as it starts the maintenance cycle immediately with an empty KB. The first industrial demonstration of this was the PEIRS system, which provided clinical interpretations for reports of pathology testing and had almost 2000 rules built by pathologists [5,13]. Since then, RDR has been applied successfully to a wide range of tasks: control [16], heuristic search [1] and document management [10]. The level of evaluation in these studies varies, but overall they clearly demonstrate very simple and highly efficient knowledge acquisition. There is now significant commercial experience of RDR confirming the efficiency of the approach.

Following the PEIRS example, one company, Pacific Knowledge Systems, supplies tools for pathologist to build systems to provide interpretative comments for medical Chemical Pathology reports. One of their customers now processes up to 14,000 patient reports per day through their 23 RDR knowledge bases with a total of about 16,000 rules, giving very highly patient-specific comments. They have a high level of satisfaction from their general practitioner clients and from the pathologists who keep on building more rules or rather who keep on identifying distinguishing features to provide subtle and clinically valuable comments. A pathologist generally requires less than one day's training and rule addition is a minor addition to their normal duties of checking reports; it takes an average of 77 seconds per rule [8].

Given the success of the knowledge representation scheme and the knowledge revision procedure, it is of interest to investigate the properties of RDR to account for its success and shape its future developments. In this paper, we use a general simulation framework proposed in [7] and developed in [2] to evaluate two interesting features of incremental knowledge acquisition: the usage of supporting data (aka cornerstone cases) in interactions with human experts and the importance of domain ontology acquisition.

The structure of the paper is as follows: in section 2, we sketch a brief description of the simulation framework that is used in the paper. Section 3 describes

Flat Ripple Down Rules (FRDR), the incremental knowledge acquisition method being investigated. In section 4, we experiment with the usage of cornerstone cases in FRDR. The importance of domain ontology acquisition is investigated in section 5, and we conclude in section 6.

2   Simulation Framework

Evaluation of KA tools and methodologies is difficult [11,15]. The essential problem is the cost of human expertise to build a KBS. A solution to this is the use of simulated experts in evaluation studies. A simulated expert is not as creative or wise as a human expert, but it readily allows for repeated control experiments. The simulation framework we use in this paper is described in [2]. In this section, we outline the main features of this framework.

We characterise an expert by two parameters: overspecialisation and overgeneralisation. Overspecialisation is the probability that a definition excludes data which it should cover. Overgeneralisation, on the other hand, is the probability that a definition includes data which it should not cover. This is depicted in Figure 1. In this figure, the human expert tries to capture a target concept by providing the system a rule or rules; however as the expert is not perfect, the defined concept deviates from target concept. The deviation can be quantified through two parameters: overspecilisation and overgeneralisation.

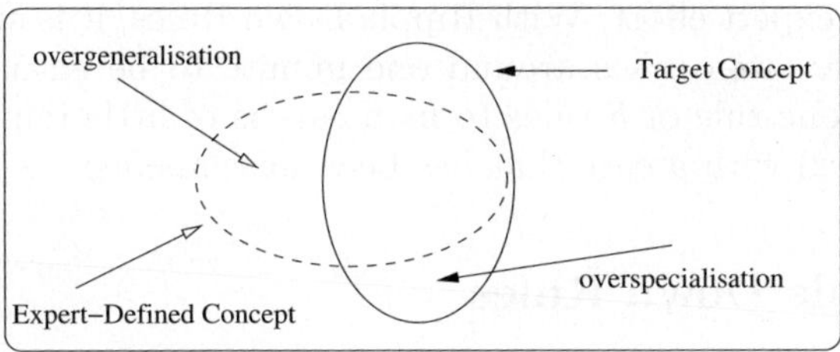

Fig. 1. Overspecialisation and overgeneralisation

In classification based systems, errors of overspecialisation and overgeneralisation are often called false negative and false positive, respectively. These errors not only apply to individual classification rules, but to complex classifiers too. Moreover, they also apply to other aspects of knowledge based system. With a planning system, the KBS has error components that cause an incorrect plan to be produced for the data provided. That is, the data was covered inappropriately; there was overgeneralisation. However, the system also failed to cover the data correctly, and that was overspecialisation. In a similar manner, these errors also apply to ontology acquisition. The definitions of concepts, or the relations between them result in objects failing to be covered or being covered inappropriately. If an expert provides too many repeated low level definitions rather than developing abstractions, there is an overspecialisation error.

In this study, we simulate obtaining rules from the expert and so apply these errors at the rule level. A given rule may cover data for which the conclusion is not appropriate; that is, it is too general. Or the rule is too specific and so excludes some data that should be covered. The intuitive response to an overgeneralised rule is to add conditions and for an overspecialised rule to remove conditions. However, whether one does this or corrects the system in some other way depends on the KA tool or method being used.

These characterisations can be used to describe different levels of expertise (for example, experienced experts and trainees). These errors also increase with the difficulty of the domain. Trainees will be associated with higher overgeneralisation and/or overspecialisation errors than experienced experts in the domain. One major problem with previous work that used simulated experts is how to model levels of expertise. For example in [6], levels of expertise are represented by picking various subsets of the full condition. There is no such difficulty in our approach as we model the effects of different levels of expertise by using different combinations of overgeneralisation and overspecialisation.

As mentioned above, the simulation here is restricted to classification. Secondly the domain is assumed to be made up of non-overlapping regions. The minimum number of rules required is therefore the number of regions in the domain. This assumption is made for the sake of simplicity and can easily be relaxed to allow for more complex domains.

Expert effort is often measured by the number of rules created in the knowledge base. We suggest that the number of knowledge acquisition sessions is a better metric for expert effort. With Ripple Down Rules, it is shown in [8] that, in practice, a rule often takes around one minute to be actually encoded. So whether it takes one rule or 5 rules to fix a case is of little importance; the key issue is how to deal with a case that has been misclassified.

3   Flat Ripple Down Rules

The RDR variant we use in this experiment is the flat rule version. The reason behind this choice is that Flat Rule is a simplified version of multiple classification RDR which is used in practical systems, e.g. from Pacific Knowledge Systems. Flat RDR can be considered as a *n-ary* tree of depth two. Each node of the tree is labelled with a primary rule with the following properties:

- The root is a default rule which gives a default dummy conclusion (for example **unknown**).
- The rules in the nodes of depth 1 give a classification to a data case
- The rules in the nodes of depth 2 are called deletion rule and work as refinements to the the classification rules.

Figure 2 shows an example of Flat RDR. Flat RDR works as follow: a data case is passed to root. As the root always fires, a dummy conclusion is recorded. After that, the case is passed to *all* the classification rules (the rules of depth 1). A conclusion of a rule is recorded (and overrides the dummy conclusion) if and only

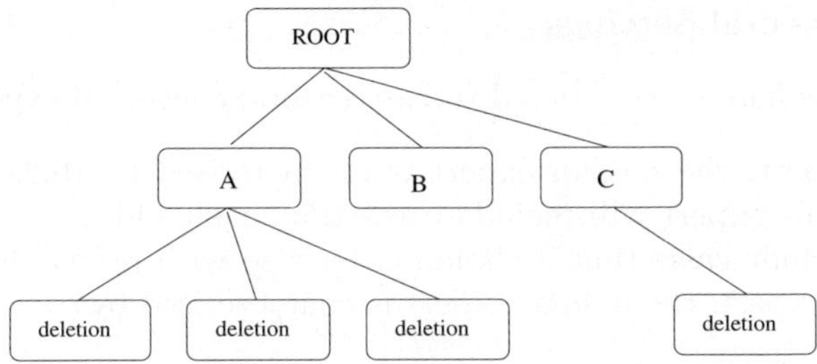

Fig. 2. Flat RDR

if the condition of this rule is satisfied and *none* of its children (its deletion rules) fires. The final classification given to the case is all the conclusions recorded from the classification rules. If there is an undesired conclusion, the human expert will be asked to provide a deletion rule to remove it. The new deletion rule is added as a child to the classification rule that misfires the case. On the other hand, if the expert decides that a classification should be given to this data case, a classification rule (rule of depth 1) will be added.

FlatRDR is capable of handling the Multiple Classification Problem, i.e, a data case can be classified with more than one label. In this simulation framework, however, we just apply Flat RDR to the single classification problem.

Although we have described Flat RDR with the RDR tree and refinement structure, it also corresponds to the general case of simple classification, where new rules are added or rules are refined when incorrect.

4   Cornerstone Cases

Cornerstone cases are data cases that trigger the creation of new rules. One of the hallmark features of RDR is the employment of cornerstone cases. They serve two purposes:

- as a means of maintaining past performance by imposing consistency
- as a guide to help the experts make up the new rules.

The cornerstone cases are used in the following manner: when a data case is misclassified by the system, an expert is consulted and asked to provide a new rule (or rules) to deal with this case. The new rule then is evaluated against all the cornerstone cases stored in the system. If any of the cornerstone cases is affected by the new rule, the expert is asked to refine it. Only when the system confirms that the new rule does not affect any of the cornerstone cases then it is added to the knowledge base, and the current data case becomes the new rule's cornerstone. In practice, the expert might decide to allow the rule to apply an existing cornerstone case, but this evaluation excludes this.

The first question for the evaluation is the importance of cornerstone cases. Or more generally, what is the importance of validating performance against test data after modifying a KBS.

4.1   Experimental Settings

The simulations here are restricted to two artibrary levels of expertise:

- 'Good' Expert: the human expert is characterised by $(0.2, 0.2)$, i.e a rule made by this expert will include cases that it should not with probability 0.2 and exclude cases that it should cover also with probability 0.2.
- 'Average' Expert: the human expert is characterised by $(0.3, 0.3)$.

Our naming of these levels of expertise is arbitrary; our intention is simply to distinguish higher and lower levels of expertise. With each level of expertise, we run the simulation with two options: with or without cornerstone cases. The simulation is run with 100000 data cases from a domain of 20 regions, and the number of required KA sessions is recorded.

4.2   Result and Discussion

The following figures show the number of KA acquisition sessions as a function of data cases presented to the system. As a KA is required each time a data case is misclassified, the slope of this graph can also be considered as the error rate for the acquired system.

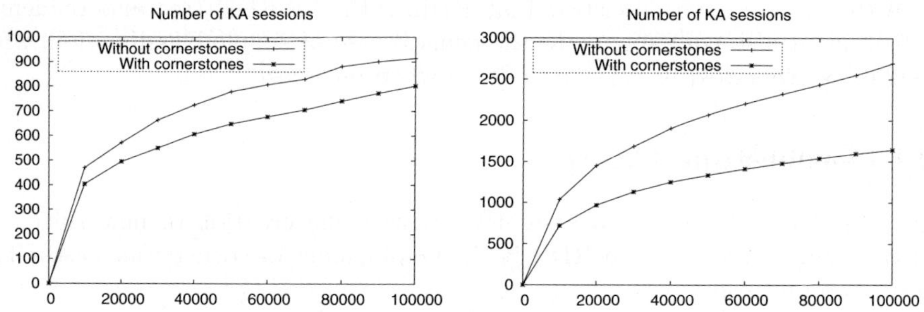

Fig. 3. Good and Average Expertise Levels

It can be seen from the graphs that when a good level of expertise is available, there is not much difference in the performance of the acquired knowledge base whether or not cornerstone cases are employed. However, when the available expertise is average, the system with cornerstone cases clearly outperforms the one without, in terms of the number of KA sessions (or error rate). In a KA session with the system that uses cornerstone cases, the expert is usually asked to create more primary rules. However, this is perfectly acceptable since the number of KA sessions is a better measure of human experts' time than the number of primary rules.

5   Domain Ontology Acquisition

In recent years, the use of explicit ontologies in knowledge based systems has
been intensively investigated [18,9,12,17]. Heuristic classification was first introduced by [3] and remains a popular problem solving method (PSM). It can be
understood as a PSM using a very simple ontological structure of intermediate
conclusions. It is comprised of three main phases:

- abstraction from a concrete, particular problem description to a problem
  class definition that applies to
- heuristic match of a principal solution to the problem class
- refinement of the principal solution to a concrete solution for the concrete
  problem

This process can be seen in the following figure

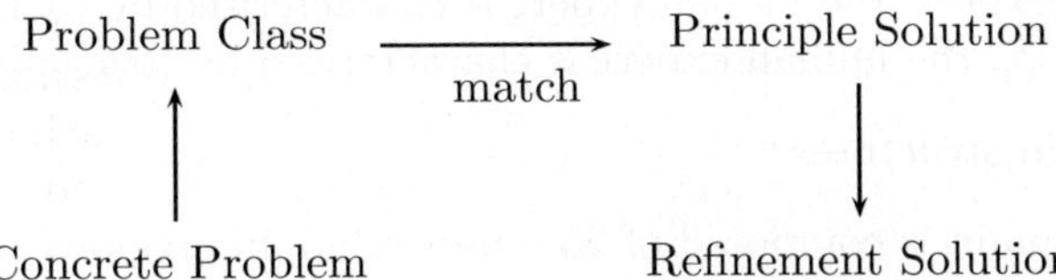

In practice, it is not always the case that all three phases of heuristic classification
are employed. The example we look at in the next subsection will show how a
simple taxonomy is used with classification systems.

5.1   Example

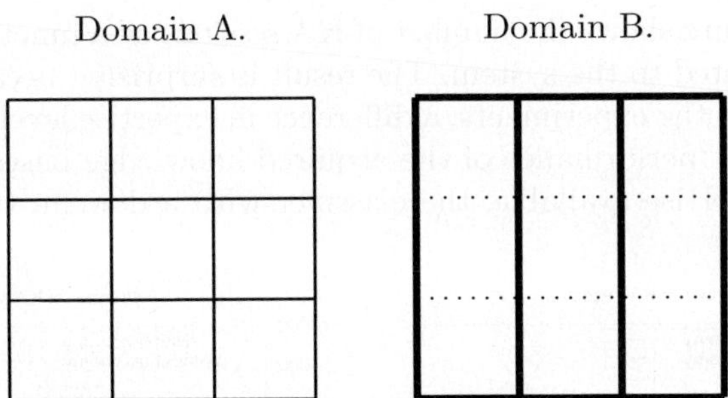

We look at two domain structures as in the picture above. The task here is to
acquire a classifier for this domain from human experts. There are nine elementary classifications as shown in case A. In case B, however, we assume that there
is a known taxonomy of classifications: the domain is divided into three general
classes and each general class contains three elementary classifications. This taxonomy can be considered as a very simple ontology. We now describe how this
explicit taxonomy of classification is used in a classification system and how we
evaluate its usage.

In case A, the classifier produces one of the nine classifications. Revision of the knowledge base when a data case is misclassified is done similarly as in Section 3. On the other hand, in case B, classification is done in a two-step process. First, the classifier assigns a general class (from three classes in this particular example) to the input data. After that, the data is passed to a second sub-classifier which (based on the general class assigned) gives the sub-classification associated with this case. When there is a misclassification, the classifier (or classifiers) will be revised. As a consequence, one can argue that, revision in this case is likely to be more complex than that in case A. However, in our experiments, we still count each revision to deal with a case as a KA session.

5.2   Experiment Settings

The simulations here are restricted to two arbitrary levels of expertise:

- 'Average' Expert: the human expert is characterised by $(0.3, 0.3)$,
- 'Bad' Expert: the human expert is characterised by $(0.4, 0.4)$.

and two domain structures

- (A) the domain is composed of 25 non-overlapping regions
- (B) the domain is composed of 5 non-overlapping regions, and each region, in turn, is composed of 5 sub-regions.

Again, the naming of the levels of expertise is arbitratry. The simulation is run with 100000 data cases and the number of required KA sessions is recorded.

5.3   Result and Discussion

The following figure shows the number of KA sessions as a function of number of data cases presented to the system. The result is surprising because even with a fixed taxonomy in the experiments, a difference in expertise level can lead to such a difference in the performance of the acquired knowledge bases. While there is a reasonable expertise available, the classifier with a domain taxonomy clearly

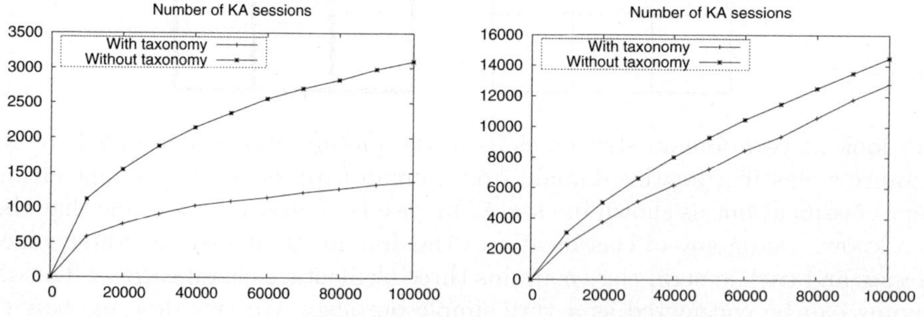

Fig. 4. Average and Bad Expertise Levels

outperform the one without. However, when the level of available expertise is poor, performance is similar so it might be better not to use the domain ontology because knowledge acquisition is simpler.

6   Conclusion

In this paper, we use the simulation framework developed in [2] to investigate two interesting aspects of incremental knowledge acquisition, namely, the usage of supporting data cases and explicit domain ontology. We do not claim that our model accurately reflects the real life situation, or our results quantitatively apply to the a real knowledge based system, the simulation still shows interesting observations.

We observe that the use of cornerstone cases in Ripple Down Rules system shows a real improvement of the knowledge base performance. While the expert has work a bit more at each knowledge acquisition session, the number of KA sessions will be less over time. In particular, when a high level of expertise is not available, the use of cornerstone cases significantly improves the experts' performance.

The second observation is that explicit domain ontology brings significant improvement in the resulting system's performance if high levels of expertise are available. However, explicit ontologies do not have as much positive effect when the domain is dynamic (due to its changing nature, or unestablished tacit knowledge).

Aspects of these conclusions are entirely obvious and be accepted by all: that validation and ontologies are both useful. However, the methodology also raises the question that as we move into less well defined area relating to personal and business preferences, validation becomes more critical while perhaps ontologies are less valuable.

In the future, we would like to investigate other aspects of evaluating KA: more complex domain structure or in multiple experts settings.

Acknowledgements

This work is supported by CRC for Smart Internet Technology, Australia.

References

1. G. Beydoun and A. Hoffmann. Incremental acquisition of search knowledge. *Journal of Human-Computer Studies*, 52:493–530, 2000.
2. T. Cao and P. Compton. A simulation framework for knowledge acquisition evaluation. In *Proceedings of 28th Australasian Computer Science Conference*, pages 353–361, 2005.
3. W. J. Clancey. Heuristic classification. *Artificial Intelligence*, 27:289–350, 1985.
4. P. Compton, G. Edwards, A. Srinivasan, P. Malor, P. Preston, B. Kang, and L. Lazarus. Ripple-down-rules: Turning knowledge acquisition into knowledge maintenance. *Artificial Intelligence in Medicine*, 4:47–59, 1992.

5. P. Compton and R. Jansen. A philosophical basis for knowledge acquisition. *Knowledge Acquisition*, 2:241–257, 1990.

6. P. Compton, P. Preston, and B. Kang. The use of simulated experts in evaluating knowledge acquisition. In B. Gaines and M. Musen, editors, *9th Banff KAW Proceeding*, pages 1–12, 1995.

7. Paul Compton. Simulating expertise. In *PKAW*, pages 51–70, 2000.

8. Paul Compton, Lindsay Peters, Glenn Edwards, and Tim Lavers. Experience with ripple-down rules. In *Applications and Innovations in Intelligent Systems*, pages 109–121, 2005.

9. Asuncion Gomez-Perez. Evaluation of ontologies. *Int. J. Intelligent Systems*, 16:391–409, 2001.

10. B. Kang, K. Yoshida, H. Motoda, and P. Compton. A help desk system with intelligence interface. *Applied Artificial Intelligence*, 11:611–631, 1997.

11. T. Menzies and F. Van Hamelen. Editorial: Evaluating knowledge engineering techniques. *Journal of Human-Computer Studies*, 51(4):715–727, 1999.

12. Natalya Fridman Noy, Michael Sintek, Stefan Decker, Monica Crubézy, Ray W. Fergerson, and Mark A. Musen. Creating semantic web contents with protégé-2000. *IEEE Intelligent Systems*, 16(2):60–71, 2001.

13. P. Preston, G. Edwards, and P. Compton. A 2000 rule expert system without a knowledge engineer. In B. Gaines and M. Musen, editors, *8th Banff KAW Proceeding*, 1994.

14. Guus Schreiber, Bob J. Wielinga, Robert de Hoog, Hans Akkermans, and Walter Van de Velde. Commonkads: A comprehensive methodology for kbs development. *IEEE Expert*, 9(6):28–37, 1994.

15. N. Shadbolt and K. O'Hara. The experimental evaluation of knowledge acquisition techniques and methods: history, problem and new directions. *Journal of Human-Computer Studies*, 51(4):729–775, 1999.

16. G. Shiraz and C. Sammut. Combining knowledge acquisition and machine learning to control dynamic systems. In *Proceedings of the 15th International Joint Conference in Artificial Intelligence (IJCAI'97)*, pages 908–913. Morgan Kaufmann, 1997.

17. York Sure, Asunción Gómez-Pérez, Walter Daelemans, Marie-Laure Reinberger, Nicola Guarino, and Natalya Fridman Noy. Why evaluate ontology technologies? because it works!. *IEEE Intelligent Systems*, 19(4):74–81, 2004.

18. G. van Heijst, A. Th. Schreiber, and B. J. Wielinga. Using explicit ontologies in kbs development. *Journal of Human-Computer Studies*, 45:183–292, 1997.

19. Gertjan van Heijst, Peter Terpstra, Bob J. Wielinga, and Nigel Shadbolt. Using generalised directive models in knowledge acquisition. In *EKAW*, pages 112–132, 1992.

Finite Domain Bounds Consistency Revisited

C.W. Choi[1], W. Harvey[2], J.H.M. Lee[1], and P.J. Stuckey[3]

[1] Department of Computer Science and Engineering, The Chinese University of Hong Kong, Shatin, N.T., Hong Kong SAR, China
{cwchoi, jlee}@cse.cuhk.edu.hk
[2] CrossCore Optimization Ltd, London, United Kingdom
warwick.harvey@crosscoreop.com
[3] NICTA Victoria Laboratory, Department of Computer Science & Software Engineering, University of Melbourne, 3010, Australia
pjs@cs.mu.oz.au

Abstract. A widely adopted approach to solving constraint satisfaction problems combines systematic tree search with constraint propagation for pruning the search space. Constraint propagation is performed by propagators implementing a certain notion of consistency. Bounds consistency is the method of choice for building propagators for arithmetic constraints and several global constraints in the finite integer domain. However, there has been some confusion in the definition of bounds consistency and of bounds propagators. We clarify the differences among the three commonly used notions of bounds consistency in the literature. This serves as a reference for implementations of bounds propagators by defining (for the first time) the *a priori* behavior of bounds propagators on arbitrary constraints.

1   Introduction

One widely-adopted approach to solving CSPs combines backtracking tree search with constraint propagation. This framework is realized in constraint programming systems, such as ECL^iPS^e [4], SICStus Prolog [23] and ILOG Solver [10], which have been successfully applied to many real-life industrial applications.

Constraint propagation, based on local consistency algorithms, removes infeasible values from the domains of variables to reduce the search space. The most successful consistency technique was *arc consistency* [15], which ensures that for each binary constraint, every value in the domain of one variable has a supporting value in the domain of the other variable that satisfies the constraint.

A natural extension of arc consistency for constraints of more than two variables is *domain consistency* [24] (also known as *generalized arc consistency* and *hyper-arc consistency*). Checking domain consistency is NP-hard even for linear equations, an important kind of constraints.

To avoid this problem weaker forms of consistency were introduced for handling constraints with large numbers of variables. The most successful one for linear arithmetic constraints has been *bounds consistency* (sometimes called *interval consistency*). Unfortunately there are *three* commonly used but *incompatible* definitions of bounds consistency in the literature. This is confusing to

A. Sattar and B.H. Kang (Eds.): AI 2006, LNAI 4304, pp. 49–58, 2006.
© Springer-Verlag Berlin Heidelberg 2006

practitioners in the design and implementation of efficient bounds consistency algorithms, as well as for users of constraint programming systems claiming to support bounds consistency. We clarify the three existing definitions of bounds consistency and the differences between them.

Propagators are functions implementing certain notions of consistency to perform constraint propagation. For simplicity, we refer to propagators implementing bounds consistency as *bounds propagators*. We aim to formalize (for the first time) precisely the *operational semantics* of various kinds of bounds propagators for arbitrary constraints. The precise semantics would serve as the basis for all implementations of bounds propagators. We also study how bounds propagators are used in practice.

2   Propagation Based Constraint Solving

We consider integer constraint solving using constraint propagation. Let $\mathcal{Z}$ denote the integers, and $\mathcal{R}$ denote the reals.

We consider a given (finite) set of integer variables $\mathcal{V}$, which we shall sometimes interpret as real variables. Each variable is associated with a finite set of possible values, defined by the domain. A *domain* D is a complete mapping from a set of variables $\mathcal{V}$ to finite sets of integers. The *intersection* of two domains D and D', denoted $D \sqcap D'$, is defined by the domain $D''(v) = D(v) \cap D'(v)$ for all $v \in \mathcal{V}$. A domain D is *stronger* than a domain D', written $D \sqsubseteq D'$, iff $D(v) \subseteq D'(v)$ for all variables $v \in \mathcal{V}$. A domain D is equal to a domain D', denoted $D = D'$, iff $D(v) = D'(v)$ for all variables $v \in \mathcal{V}$. A domain D is stronger than (equal to) a domain D' w.r.t. variables V, denoted $D \sqsubseteq_V D'$ (resp. $D =_V D'$), iff $D(v) \subseteq D'(v)$ (resp. $D(v) = D'(v)$) for all $v \in V$.

Let *vars* be the function that returns the set of variables appearing in an expression or constraint. A *valuation* θ is a mapping of variables to values (integers or reals), written $\{x_1 \mapsto d_1, \ldots, x_n \mapsto d_n\}$. Define $vars(\theta) = \{x_1, \ldots, x_n\}$. In an abuse of notation, we define a valuation θ to be an element of a domain D, written $\theta \in D$, if $\theta(v) \in D(v)$ for all $v \in vars(\theta)$. Given an expression e, $\theta(e)$ is obtained by replacing each $v \in vars(e)$ by $\theta(v)$ and calculating the value of the resulting variable free expression.

We are interested in determining the infimums and supremums of expressions with respect to some domain D. Define the *infimum* and *supremum* of an expression e with respect to a domain D as $\inf_D e = \inf\{\theta(e) | \theta \in D\}$ and $\sup_D e = \sup\{\theta(e) | \theta \in D\}$ respectively. A *range* is a contiguous set of integers, and we use *range notation*: $[l..u]$ to denote the range $\{d \in \mathcal{Z} \mid l \leq d \leq u\}$ when l and u are integers. A domain is a *range domain* if $D(x)$ is a range for all x. Let $D' = \mathrm{range}(D)$ be the smallest range domain containing D, i.e. domain $D'(x) = [\inf_D x .. \sup_D x]$ for all $x \in \mathcal{V}$.

A constraint places restriction on the allowable values for a set of variables and is usually written in well understood mathematical syntax. More formally, a *constraint* c is a relation expressed using available function and relation symbols in a specific constraint language. For the purpose of this paper, we assume the usual integer interpretation of arithmetic constraints and logical operators such as $\neg$, $\wedge$, $\vee$, $\Rightarrow$, and $\Leftrightarrow$. We call valuation θ an *integer (resp. real) solution* of c iff $vars(\theta) = vars(c)$ and $\mathcal{Z} \models \theta(c)$ ($\mathcal{R} \models \theta(c)$). We denote by $solns(c)$ all the *integer solutions* of c.

We can understand a domain D as a constraint: $D \leftrightarrow \bigwedge_{v \in \mathcal{V}} \bigvee_{d \in D(v)} v = d$. A *constraint satisfaction problem* (CSP) consists of a set of constraints read as conjunction.

A *propagator* f is a monotonically decreasing function from domains to domains, i.e. $D \sqsubseteq D'$ implies that $f(D) \sqsubseteq f(D')$, and $f(D) \sqsubseteq D$. A propagator f is *correct* for constraint c iff for all domains D

$$\{\theta \mid \theta \in D\} \cap solns(c) = \{\theta \mid \theta \in f(D)\} \cap solns(c)$$

This is a weak restriction since, for example, the identity propagator is correct for all constraints c.

Typically we model a CSP as a conjunction of constraints $\wedge_{i=1}^{n} c_i$. We provide a propagator f_i for each constraint c_i where f_i is correct and checking for c_i. The propagation solver is then applied on the set $F = \{f_i \mid 1 \leq i \leq n\}$. The consideration of individual constraints is *crucial*. By design (of what constraints are supported in a constraint language), we can provide efficient implementations of propagators for particular constraints, but not for arbitrary ones. A human modeler should exploit this fact in constructing efficient formulations of problems.

A *propagation solver* for a set of propagators F and current domain D, $solv(F, D)$, repeatedly applies all the propagators in F starting from domain D until there is no further change in the resulting domain. In other words, $solv(F, D)$ returns a new domain defined by $solv(F, D) = \mathrm{gfp}(iter_F)(D)$ where $iter_F(D) = \bigsqcap_{f \in F} f(D)$, and gfp denotes the greatest fixpoint w.r.t. $\sqsubseteq$ lifted to functions.

Propagators are often linked to some notion of implementing some consistency for a particular constraint. A propagator f *implements* a consistency $\mathcal{C}$ for a constraint c, if $D' = solv(\{f\}, D)$ ensures that D' is the greatest domain $D' \sqsubseteq D$ that is $\mathcal{C}$ consistent for c.[1] Note that f only needs to satisfy this property at its fixpoints.

Definition 1. *A domain D is* domain consistent *for a constraint c where $vars(c) = \{x_1, \ldots, x_n\}$, if for each variable $x_i, 1 \leq i \leq n$ and for each $d_i \in D(x_i)$ there exist integers d_j with $d_j \in D(x_j)$, $1 \leq j \leq n, j \neq i$ such that $\theta = \{x_1 \mapsto d_1, \ldots, x_n \mapsto d_n\}$ is an integer solution of c, i.e. $\mathcal{Z} \models_\theta c$.*

We can now define a *domain propagator*, $dom(c)$, for a constraint c as:

$$dom(c)(D)(v) = \begin{cases} \{\theta(v) \mid \theta \in D \wedge \theta \in solns(c)\} & \text{where } v \in vars(c) \\ D(v) & \text{otherwise} \end{cases}$$

---

[1] This assumes that such a greatest domain exists. This holds for all sensible notions of consistency, including all those in this paper.

From the definition, it is clear that the domain propagator $dom(c)$ implements domain consistency for the constraint c. The domain propagator $dom(c)$ is clearly idempotent.

3   Different Notions of Bounds Consistency

The basis of bounds consistency is to relax the consistency requirement to apply only to the lower and upper bounds of the domain of each variable x. There are three incompatible definitions of bounds consistency used in the literature, all for constraints with finite integer domains. For bounds($\mathcal{D}$) consistency each bound of the domain of a variable has integer support among the values of the domain of each other variable occurring in the same constraint. For bounds($\mathcal{Z}$) consistency the integer supporting values need only be within the range from the lower to upper bounds of the other variables. For bounds($\mathcal{R}$) consistency the supports can be real values within the range from the lower to upper bounds of the other variables.

Definition 2. *A domain D is* bounds($\mathcal{D}$) consistent *for a constraint c where* $vars(c) = \{x_1, \ldots, x_n\}$, *if for each variable $x_i, 1 \leq i \leq n$ and for each $d_i \in \{\inf_D x_i, \sup_D x_i\}$ there exist **integers d_j** with $d_j \in D(x_j)$, $1 \leq j \leq n, j \neq i$ such that $\theta = \{x_1 \mapsto d_1, \ldots, x_n \mapsto d_n\}$ is an **integer solution** of c.*

Definition 3. *A domain D is* bounds($\mathcal{Z}$) consistent *for a constraint c where* $vars(c) = \{x_1, \ldots, x_n\}$, *if for each variable $x_i, 1 \leq i \leq n$ and for each $d_i \in \{\inf_D x_i, \sup_D x_i\}$ there exist **integers d_j** with $\inf_D x_j \leq d_j \leq \sup_D x_j$, $1 \leq j \leq n, j \neq i$ such that $\theta = \{x_1 \mapsto d_1, \ldots, x_n \mapsto d_n\}$ is an **integer solution** of c.*

Definition 4. *A domain D is* bounds($\mathcal{R}$) consistent *for a constraint c where* $vars(c) = \{x_1, \ldots, x_n\}$, *if for each variable $x_i, 1 \leq i \leq n$ and for each $d_i \in \{\inf_D x_i, \sup_D x_i\}$ there exist **real numbers d_j** with $\inf_D x_j \leq d_j \leq \sup_D x_j$, $1 \leq j \leq n, j \neq i$ such that $\theta = \{x_1 \mapsto d_1, \ldots, x_n \mapsto d_n\}$ is a **real solution** of c.*

Definition 2 is used in for example for the two definitions in Dechter [6, pages 73 & 435]; Frisch *et al.* [7]; and implicitly in Lallouet *et al.* [12]. Definition 3 is far more widely used appearing in for example Van Hentenryck, Saraswat, & Deville [24]; Puget [19]; Régin & Rueher [21]; Quimper *et al.* [20]; and SICStus Prolog [23]. Definition 4 appears in for example Marriott & Stuckey [17]; Schulte & Stuckey [22]; Harvey & Schimpf [9]; and Zhang & Yap [27]. Apt [1] gives both Definitions 3 (called interval consistency) and 4 (called bounds consistency).

3.1   General Relationship and Properties

Let us now examine the differences of the definitions. The following relationship between the three notions of bounds consistency is clear from the definition.

Proposition 1. *If D is bounds($\mathcal{D}$) consistent for c it is bounds($\mathcal{Z}$) consistent for c. If D is bounds($\mathcal{Z}$) consistent for c it is bounds($\mathcal{R}$) consistent for c.* $\square$

Example 1. Consider the constraint $c_{lin} \equiv x_1 = 3x_2 + 5x_3$. The domain D_2 defined by $D_2(x_1) = \{2, 3, 4, 6, 7\}$, $D_2(x_2) = [0..2]$, and $D_2(x_3) = [0..1]$ is bounds($\mathcal{R}$) consistent (but not bounds($\mathcal{D}$) consistent or bounds($\mathcal{Z}$) consistent) w.r.t. c_{lin}.

The domain D_3 defined by $D_3(x_1) = \{3, 4, 6\}$, $D_3(x_2) = [0..2]$, and $D_3(x_3) = [0..1]$ is bounds($\mathcal{Z}$) and bounds($\mathcal{R}$) consistent (but not bounds($\mathcal{D}$) consistent) w.r.t. c_{lin}.

The domain D_4 defined by $D_4(x_1) = \{3, 4, 6\}$, $D_4(x_2) = [1..2]$, and $D_4(x_3) = \{0\}$ is bounds($\mathcal{D}$), bounds($\mathcal{Z}$) and bounds($\mathcal{R}$) consistent w.r.t. c_{lin}.

The relationship between the bounds($\mathcal{Z}$) and bounds($\mathcal{D}$) consistency is straightforward to explain.

Proposition 2. *D is bounds($\mathcal{Z}$) consistent with c iff* range(D) *is bounds($\mathcal{D}$) consistent with c.*   □

The second definition of bounds consistency in Dechter [6, page 435] works only with range domains. By Proposition 2, the definition coincides with both bounds($\mathcal{Z}$) and bounds($\mathcal{D}$) consistency. Similarly, Apt's [1] interval consistency is also equivalent to bounds($\mathcal{D}$) consistency. Finite domain constraint solvers do not always operate on range domains, but rather they use a mix of propagators implementing different kinds of consistencies, both domain and bounds consistency.

Example 2. Consider the same setting from Example 1. Now range(D_3) is both bounds($\mathcal{D}$) and bounds($\mathcal{Z}$) consistent with c_{lin}. As noted in Example 1, D_3 is only bounds($\mathcal{Z}$) consistent but *not* bounds($\mathcal{D}$) consistent with c_{lin}.

Both bounds($\mathcal{R}$) and bounds($\mathcal{Z}$) consistency depend only on the upper and lower bounds of the domains of the variables under consideration.

Proposition 3. *For $\alpha = \mathcal{R}$ or $\alpha = \mathcal{Z}$ and constraint c, D is bounds(α) consistent for c iff* range(D) *is bounds(α) consistent for c.*   □

This is not the case for bounds($\mathcal{D}$) consistency, which suggests that, strictly, it is not really a form of *bounds consistency*. Indeed, most existing implementations of bounds propagators make use of Proposition 3 to avoid re-executing a bounds propagator unless the lower or upper bound of a variable involved in the propagator changes.

Example 3. Consider the same setting from Example 1 again. Both D_3 and range(D_3) are bounds($\mathcal{Z}$) and bounds($\mathcal{R}$) consistency with c_{lin}, but only range(D_3) is bounds($\mathcal{D}$) consistent with c_{lin}.

There are significant problems with the stronger bounds($\mathcal{Z}$) (and bounds($\mathcal{D}$)) consistency. In particular, for linear equations it is NP-complete to check bounds($\mathcal{Z}$) (and bounds($\mathcal{D}$)) consistency, while for bounds($\mathcal{R}$) consistency it is only linear time (e.g. see Schulte & Stuckey [22]).

Proposition 4. *Checking bounds($\mathcal{Z}$), bounds($\mathcal{D}$), or domain consistency of a domain D with a linear equality $a_1x_1 + \cdots a_nx_n = a_0$ is NP-complete, where $\{a_0, \ldots, a_n\}$ are integer constants and $\{x_1, \ldots, x_n\}$ are integer variables.* $\square$

There are other constraints where bounds($\mathcal{R}$) consistency is less meaningful. A problem of bounds($\mathcal{R}$) consistency is that it may not be clear how to interpret an integer constraint in the reals.

3.2   Conditions for Equivalence

Why has the confusion between the various definitions of bounds consistency not been noted before? In fact, for many constraints, the definitions are *equivalent*.

Following the work of Zhang & Yap [27] we define n-ary monotonic constraints as a generalization of linear inequalities $\sum_{i=1}^{n} a_ix_i \leq a_0$. Let $\theta \in_{\mathcal{R}} D$ denote that $\theta(v) \in \mathcal{R}$ and $\inf_D v \leq \theta(v) \leq \sup_D v$ for all $v \in vars(\theta)$.

Definition 5. *An n-ary constraint c is* monotonic with respect to variable $x_i \in vars(c)$ *iff there exists a total ordering $\prec_i$ on $D(x_i)$ such that if $\theta \in_{\mathcal{R}} D$ is a real solution of c, then so is any $\theta' \in_{\mathcal{R}} D$ where $\theta'(x_j) = \theta(x_j)$ for $j \neq i$ and $\theta'(x_i) \preceq_i \theta(x_i)$. An n-ary constraint c is* monotonic *iff c is monotonic with respect to all variables in $vars(c)$.*

The above definition of monotonic constraints is equivalent to but simpler than that of Zhang & Yap [27], see Choi *et al.* [5] for justification and explanation. Examples of monotonic constraints are: all linear inequalities, and $x_1 \times x_2 \leq x_3$ with non-negative domains, i.e. $\inf_D(x_i) \geq 0$. For this class of constraints, bounds($\mathcal{R}$), bounds($\mathcal{Z}$) and bounds($\mathcal{D}$) consistency are equivalent to domain consistency.

Proposition 5. *Let c be an n-ary monotonic constraint. Then bounds($\mathcal{R}$), bounds($\mathcal{Z}$), bounds($\mathcal{D}$) and domain consistency for c are all equivalent.* $\square$

Although linear disequality constraints are not monotonic, they are equivalent for all the forms of bounds consistency because they prune so weakly.

Proposition 6. *Let $c \equiv \sum_{i=1}^{n} a_ix_i \neq a_0$. Then bounds($\mathcal{R}$), bounds($\mathcal{Z}$) and bounds($\mathcal{D}$) consistency for c are equivalent.* $\square$

All forms of bounds consistency are also equivalent for binary monotonic functional constraints, such as $a_1x_1 + a_2x_2 = a_0$, $x_1 = ax_2^2 \wedge x_2 \geq 0$, or $x_1 = 1 + x_2 + x_2^2 + x_2^3 \wedge x_2 \geq 0$.

Proposition 7. *Let c be a constraint with $vars(c) = \{x_1, x_2\}$, where $c \equiv x_1 = g(x_2)$ and g is a bijective and monotonic function. Then bounds($\mathcal{R}$), bounds($\mathcal{Z}$) and bounds($\mathcal{D}$) consistency for c are equivalent.* $\square$

For linear equations with at most one non-unit coefficient, we can show that bounds($\mathcal{R}$) and bounds($\mathcal{Z}$) consistency are equivalent.

Proposition 8. *Let $c \equiv \Sigma_{i=1}^{n} a_i x_i = a_0$, where $|a_i| = 1, 2 \leq i \leq n$, a_0 and a_1 integral. Then bounds($\mathcal{R}$) and bounds($\mathcal{Z}$) consistency for c are equivalent.* $\square$

Even for linear equations with all unit coefficients, bounds($\mathcal{D}$) consistency is different from bounds($\mathcal{Z}$) and bounds($\mathcal{R}$) consistency.

In summary, for many of the commonly used constraints, the notions of bounds consistency are equivalent, but clearly not for all, for example c_{lin}.

4   Different Types of Bounds Propagators

In practice, propagators implementing bounds($\mathcal{Z}$) and/or bounds($\mathcal{R}$) consistency for individual kinds of constraints are well known. Propagators implementing bounds($\mathcal{Z}$) consistency (and not bounds($\mathcal{D}$) or bounds($\mathcal{R}$) consistency) are defined for the `alldifferent` constraint in e.g. Puget [19], Mehlhorn & Thiel [18], and López-Ortiz *et al.* [14]; for the global cardinality constraint in e.g. Quimper *et al.* [20], and Katriel & Thiel [11]; and for the global constraint combining the sum and difference constraints in Régin & Rueher [21]. Propagators implementing bounds($\mathcal{R}$) consistency for common arithmetic constraints are defined in e.g. Schulte & Stuckey [22].

On the other hand, propagators implementing bounds($\mathcal{D}$) consistency explicitly are rare. The `case` constraint in SICStus Prolog [23] allows compact representation of an arbitrary constraint as a directed acyclic graph. The `case` constraint implements domain consistency by default, but there are options to make the constraint prune only the bounds of variables using bounds($\mathcal{D}$) consistency. We can also enforce the multibound-consistency operators of Lallouet *et al.* [12] to implement bounds($\mathcal{D}$) consistency by using only a single cluster.

In the following, we give *a priori* definitions of the different types of bounds propagators for an arbitrary constraint. Although these definitions are straightforward to explain, we are not aware of any previous definition.

Bounds($\mathcal{D}$) Propagator. We can define bounds($\mathcal{D}$) propagators straightforwardly. Let c be an arbitrary constraint, then a bounds($\mathcal{D}$) propagator for c, $dbnd(c)$, can defined as $dbnd(c)(D) = D \sqcap \mathrm{range}(dom(c)(D))$. This definition is also given in Lallouet *et al.* [12]. There they implicitly define bounds consistency as the result of applying this propagator.

The bounds($\mathcal{D}$) propagator $dbnd(c)$ implements bounds($\mathcal{D}$) consistency for the constraint c.

Theorem 1. *Given a constraint c, if $D' = dbnd(c)(D)$, then D' is the greatest domain $D' \sqsubseteq D$ that is bounds($\mathcal{D}$) consistent for c.* $\square$

Like the domain propagator $dom(c)$, the bounds($\mathcal{D}$) propagator $dbnd(c)$ is clearly idempotent (as a result of Theorem 1).

Bounds($\mathcal{Z}$) Propagator. We can also define bounds($\mathcal{Z}$) propagators straightforwardly. Let c be an arbitrary constraint, then a bounds($\mathcal{Z}$) propagator for c, $zbnd(c)$, can be defined as $zbnd(c)(D) = D \sqcap \mathrm{range}(dom(c)(\mathrm{range}(D)))$.

Unlike the bounds($\mathcal{D}$) propagator $dbnd(c)$, the bounds($\mathcal{Z}$) propagator $zbnd(c)$ is *not* idempotent.

Theorem 2. *$zbnd(c)$ implements bounds($\mathcal{Z}$) consistency for constraint c.*   □

Bounds($\mathcal{R}$) Propagator. The basis of a bounds($\mathcal{R}$) propagator is to relax the integral requirement to reals. We define a *real domain* $\hat{D}$ as a mapping from the set of variables $\mathcal{V}$ to sets of reals. We also define a valuation θ to be an element of a real domain $\hat{D}$, written $\theta \in \hat{D}$, if $\theta(v_i) \in \hat{D}(v_i)$ for all $v_i \in vars(\theta)$. We can similarly extend the notions of *infimum* $\inf_{\hat{D}} e$ and *supremum* $\sup_{\hat{D}} e$ of an expression e with respect to real domain $\hat{D}$.

We will define the behavior of bounds($\mathcal{R}$) propagators by extending the definition of domain propagators to reals. Given a constraint c and a real domain $\hat{D}$, define the *real domain propagator, $rdom(c)$,* as

$$rdom(c)(\hat{D})(v) = \begin{cases} \{\theta(v) \mid \theta \in \hat{D} \wedge \mathcal{R} \models_\theta c\} & \text{where } v \in vars(c) \\ \hat{D}(v) & \text{otherwise} \end{cases}$$

We use *interval* notation, $[l\text{---}u]$, to denote the set $\{d \in \mathcal{R} \mid l \leq d \leq u\}$ where l and u are reals. Let $\hat{D} = \text{real}(D)$ be the real domain $\hat{D}(v) = [\inf_D(v)\text{---}\sup_D(v)]$ for all $v \in \mathcal{V}$. Let $D' = \text{integral}(\hat{D})$ be the domain $D'(v) = [\lceil\inf_{\hat{D}}(v)\rceil .. \lfloor\sup_{\hat{D}}(v)\rfloor]$ for all $v \in \mathcal{V}$.

We can now define bounds($\mathcal{R}$) propagators straightforwardly. This is the first time (that we are aware of) that this has been formalized. Let c be an arbitrary constraint, then the bounds($\mathcal{R}$) propagator for c, $rbnd(c)$ is defined as

$$rbnd(c)(D) = D \sqcap \text{integral}(rdom(c)(\text{real}(D))).$$

Similar to bounds($\mathcal{Z}$) propagators, the previous example clearly shows that bounds($\mathcal{R}$) propagators are *not* idempotent. The bounds($\mathcal{R}$) propagator does not guarantee bounds($\mathcal{R}$) consistency except at its fixpoints.

Theorem 3. *$rbnd(c)$ implements bounds($\mathcal{R}$) consistency for constraint c.*   □

5   Related Work

In this paper we consider *integer* constraint solving. Definitions of bounds consistency for real constraints are also numerous, but their similarities and differences have been noted and explained by e.g. Benhamou *et al.* [2]. Indeed, we can always interpret integers as reals and apply bounds consistency for real constraints plus appropriate rounding, e.g. CLP(BNR) [3]. However, as we have pointed out in Section 3.1, there exist integer constraints for which propagation is less meaningful when interpreted as reals.

Lhomme [13] defines *arc B-consistency*, which formalizes bounds propagation for both integer and real constraints. He proposes an efficient propagation algorithm implementing arc B-consistency with complexity analysis and experimental results. However, his study focuses on constraints defined by numeric relations (i.e. numeric CSPs).

Walsh [25] introduces several new forms of bounds consistency that extend the notion of (i, j)-consistency and relational consistency. He gives theoretical analysis comparing the propagation strength of these new consistency forms.

Maher [16] introduces the notion of propagation completeness with a general framework to unify a wide range of consistency. These include hull consistency of real constraints and bounds($\mathcal{Z}$) consistency of integer constraints. Propagation completeness aims to capture the timeliness property of propagation.

The application of bounds consistency is not limited to integer and real constraints. Bounds consistency has been formalized for solving set constraints [8], and more recently, multiset constraints [26].

6   Conclusion

The contributions of this paper are two-fold. First, we point out that the three commonly used definitions of bounds consistency are incompatible. We clarify their differences and study what is actually implemented in existing systems. We show that for several types of constraints, bounds($\mathcal{R}$), bounds($\mathcal{Z}$) and bounds($\mathcal{D}$) consistency are equivalent. This explains partly why the discrepancies among the definitions were not noticed earlier. Second, we give *a priori* definitions of propagators that implement the three notions of bounds consistency, which can serve as the basis for verifying all implementations of bounds propagators.

Acknowledgements

We thank the anonymous referees for their constructive comments. The work described in this paper was substantially supported by grant number CUHK4219/04E from the Research Grants Council of Hong Kong SAR.

References

1. K. Apt. *Principles of Constraint Programming*. Cambridge University Press, 2003.
2. F. Benhamou, D. McAllester, and P. Van Hentenryck. CLP(Intervals) revisited. In *ILPS 1994*, pages 124–138, 1994.
3. F. Benhamou and W. J. Older. Applying interval arithmetic to real, integer, and boolean constraints. *JLP*, 32(1):1–24, 1997.
4. A. Cheadle, W. Harvey, A. Sadler, J. Schimpf, K. Shen, and M. Wallace. ECL^{i}PS^e: An introduction. Technical Report IC-Parc-03-1, IC-Parc, Imperial College London, 2003.
5. C. W. Choi, W. Harvey, J. H. M. Lee, and P. J. Stuckey. A note on the definition of constraint monotonicity. Available from http://www.cse.cuhk.edu.hk/~cwchoi/monotonicity.pdf, 2004.
6. R. Dechter. *Constraint Processing*. Morgan Kaufmann, 2003.

7. A. Frisch, B. Hnich, Z. Kiziltan, I. Miguel, and T. Walsh. Global constraints for lexicographic orderings. In *CP2002*, pages 93–108. Springer-Verlag, 2002.
8. C. Gervet. Interval propagation to reason about sets: Definition and implementation of a practical language. *Constraints*, 1(3):191–244, 1997.
9. W. Harvey and J. Schimpf. Bounds consistency techniques for long linear constraints. In *Proceedings of TRICS: Techniques foR Implementing Constraint programming Systems*, pages 39–46, 2002.
10. ILOG. *ILOG Solver 5.2: User's Manual*, 2001.
11. I. Katriel and S. Thiel. Fast bound consistency for the global cardinality constraint. In *CP 2003*, pages 437–451, 2003.
12. A. Lallouet, A. Legtchenko, T. Dao, and A. Ed-Dbali. Intermediate (learned) consistencies. Research Report RR-LIFO-2003-04, Laboratoire d'Informatique Fondamentale d'Orléans, 2003.
13. O. Lhomme. Consistency techniques for numeric CSPs. In *IJCAI 93*, pages 232–238, 1993.
14. A. López-Ortiz, C.-G. Quimper, J. Tromp, and P. van Beek. A fast and simple algorithm for bounds consistency of the alldifferent constraint. In *Proceedings of the 18th International Joint Conference on Artificial Intelligence (IJCAI 2003)*, pages 245–250, 2003.
15. A. K. Mackworth. Consistency in networks of relations. *Artificial Intelligence*, 8(1):99–118, 1977.
16. M. Maher. Propagation completeness of reactive constraints. In *ICLP 2002*, pages 148–162, 2002.
17. K. Marriott and P. J. Stuckey. *Programming with Constraints: an Introduction*. The MIT Press, 1998.
18. K. Mehlhorn and S. Thiel. Faster algorithms for bound-consistency of the sortedness and the alldifferent constraint. In *CP2000*, pages 306–319, 2000.
19. J.-F. Puget. A fast algorithm for the bound consistency of alldiff constraints. In *Proceedings of the 15th National Conference on Artificial Intelligence (AAAI 98)*, pages 359–366, 1998.
20. C.-G. Quimper, P. van Beek, A. López-Ortiz, A. Golynski, and S. B. Sadjad. An efficient bounds consistency algorithm for the global cardinality constraint. In *CP 2003*, pages 600–614, 2003.
21. J.-C. Régin and M. Rueher. A global constraint combining a sum constraint and difference constraints. In *CP 2000*, pages 384–395, 2000.
22. C. Schulte and P. J. Stuckey. When do bounds and domain propagation lead to the same search space. In *Proceedings of the 3rd International Conference on Principles and Practice of Declarative Programming (PPDP 2001)*, pages 115–126, 2001.
23. SICStus Prolog. *SICStus Prolog User's Manual, Release 3.10.1*, 2003.
24. P. Van Hentenryck, V. Saraswat, and Y. Deville. Design, implementation and evaluation of the constraint language cc(FD). *Journal of Logic Programming*, 37(1–3):139–164, 1998.
25. T. Walsh. Relational consistencies. Research Report APES-28-2001, APES Research Group, 2001.
26. T. Walsh. Consistency and propagation with multiset constraints: A formal viewpoint. In *Proceedings of the 9th International Conference on Principles and Practice of Constraint Programming (CP 2003)*, pages 724–738, 2003.
27. Y. Zhang and R. H. C. Yap. Arc consistency on n-ary monotonic and linear constraints. In *CP 2000*, pages 470–483, 2000.

Speeding Up Weighted Constraint Satisfaction Using Redundant Modeling*

Y.C. Law, J.H.M. Lee, and M.H.C. Woo

Department of Computer Science and Engineering
The Chinese University of Hong Kong, Shatin, N.T., Hong Kong
{yclaw, jlee, hcwoo}@cse.cuhk.edu.hk

Abstract. In classical constraint satisfaction, combining mutually redundant models using channeling constraints is effective in increasing constraint propagation and reducing search space for many problems. In this paper, we investigate how to benefit the same for *weighted constraint satisfaction problems* (WCSPs), a common soft constraint framework for modeling optimization and over-constrained problems. First, we show how to generate a redundant WCSP model from an existing WCSP using *generalized model induction*. We then uncover why naively combining two WCSPs by posting channeling constraints as hard constraints and relying on the standard NC* and AC* propagation algorithms does not work well. Based on these observations, we propose m-NC_c^* and m-AC_c^* and their associated algorithms for effectively enforcing node and arc consistencies on a combined model with m sub-models. The two notions are strictly stronger than NC* and AC* respectively. Experimental results confirm that applying the 2-NC_c^* and 2-AC_c^* algorithms on combined models reduces more search space and runtime than applying the state-of-the-art AC*, FDAC*, and EDAC* algorithms on single models.

1 Introduction

Redundant modeling [1] has been shown effective in increasing constraint propagation and solving efficiency for *constraint satisfaction problems* (CSPs). While the technique, which is to combine two different CSP models of a problem using *channeling constraints*, is applied successfully to classical CSPs, we study in this paper how to benefit the same for *weighted CSPs* (WCSPs) [2,3], a common soft constraint framework for modeling optimization and over-constrained problems.

Unlike classical CSPs, obtaining mutually redundant WCSP models for a problem is more difficult in general, since each problem solution is associated with a cost. Besides the problem requirements, we need to ensure the same cost distribution on the solutions of the two models. While model induction [4] can automatically generate a redundant classical CSP from a given one, we *generalize* the notion so that redundant permutation WCSPs can also be generated.

* We thank the anonymous referees for their constructive comments. The work described in this paper was substantially supported by grants (CUHK4358/02E and CUHK4219/04E) from the Research Grants Council of Hong Kong SAR.

A. Sattar and B.H. Kang (Eds.): AI 2006, LNAI 4304, pp. 59–68, 2006.

© Springer-Verlag Berlin Heidelberg 2006

For classical CSPs, we can rely on the standard propagation algorithms of the channeling constraints to transmit pruning information between sub-models to achieve stronger propagation. We can also do the same to combine WCSPs by posting the channeling constraints as hard constraints. However, we discover that applying the standard AC* algorithm [3] to the combined model results in weaker constraint propagation and worse performance. Some prunings that are available when propagating a single model alone would even be missed in a combined model. Based on these observations, we *generalize* the notions of node and arc consistencies and *propose* m-NC$_c^*$ and m-AC$_c^*$ for a combined model with m sub-models. The m-NC$_c^*$ (*resp.* m-AC$_c^*$) has strictly stronger propagation behavior than NC* (*resp.* AC*) and degenerates to NC* (*resp.* AC*) when $m = 1$. While m-NC$_c^*$ and m-AC$_c^*$ are applicable to general WCSPs, we focus our discussions on permutation WCSPs. Experimental results confirm that 2-AC$_c^*$ is particularly useful to solve hard problem instances. Enforcing 2-AC$_c^*$ on combined models reduces significantly more search space and runtime than enforcing the state-of-the-art local consistencies AC* [3], FDAC* [5], and EDAC* [6] on single models.

2   Background

WCSPs associate *costs* to tuples [7]. The costs are specified by a *valuation structure* $S(k) = ([0, \ldots, k], \oplus, \geq)$, where $k \in \{1, \ldots, \infty\}$, $\oplus$ is defined as $a \oplus b = \min\{k, a + b\}$, and $\geq$ is the standard order among naturals. The minimum and maximum costs are denoted by $\bot = 0$ and $\top = k$ respectively. A binary WCSP is a quadruplet $\mathcal{P} = (k, \mathcal{X}, \mathcal{D}, \mathcal{C})$ with the valuation structure $S(k)$. $\mathcal{X} = \{x_1, \ldots, x_n\}$ is a finite set of *variables* and $\mathcal{D} = \{D_{x_1}, \ldots, D_{x_n}\}$ is a set of finite *domains* for each $x_i \in \mathcal{X}$. An *assignment* $x \mapsto a$ in $\mathcal{P}$ is a mapping from variable x to value $a \in D_x$. A *tuple* is a set of assignments in $\mathcal{P}$. It is *complete* if it contains assignments of all variables in $\mathcal{P}$. $\mathcal{C}$ is a set of unary and binary *constraints*. A unary constraint involving variable x is a cost function $C_x : D_x \to \{0, \ldots, k\}$. A binary constraint involving variables x and y is a cost function $C_{x,y} : D_x \times D_y \to \{0, \ldots, k\}$. We also assume a *zero-arity* constraint $C_\emptyset$ in $\mathcal{P}$ which is a constant denoting the global lower bound of costs in $\mathcal{P}$. Fig. 1(a) shows a WCSP with variables $\{x_1, x_2, x_3\}$ and domains $\{1, 2, 3\}$. We depict the unary costs as labeled nodes and binary costs as labeled edges connecting two assignments. Unlabeled edges have $\top$ cost; $\bot$ costs are not shown for clarity.

The cost of a complete tuple $\theta = \{x_i \mapsto v_i \mid 1 \leq i \leq n\}$ in $\mathcal{P}$ is $\mathcal{V}(\theta) = C_\emptyset \oplus \sum_i C_{x_i}(v_i) \oplus \sum_{i<j} C_{x_i, x_j}(v_i, v_j)$. If $\mathcal{V}(\theta) < \top$, θ is a *solution* of $\mathcal{P}$. Solving a WCSP is to find a solution θ with minimized $\mathcal{V}(\theta)$, which is NP-hard. The WCSP in Fig. 1(a) has a minimum cost 2 with the solution $\{x_1 \mapsto 3, x_2 \mapsto 1, x_3 \mapsto 2\}$ ($C_\emptyset \oplus C_{x_1}(3) \oplus C_{x_2}(1) \oplus C_{x_3}(2) \oplus C_{x_1, x_2}(3, 1) \oplus C_{x_1, x_3}(3, 2) \oplus C_{x_2, x_3}(1, 2) = 1 \oplus 0 \oplus 1 \oplus 0 \oplus 0 \oplus 0 \oplus 0 = 2$). A WCSP is equivalent to a classical CSP if each cost in the WCSP is either $\bot$ or $\top$. Two WCSPs are *equivalent* if they have the same variables and for every complete tuple θ, $\mathcal{V}(\theta)$ is the same for both WCSPs.

Following Schiex [2] and Larrosa [3], we define node and arc consistencies for WCSPs as follows. They degenerate to their standard counterparts for CSPs.

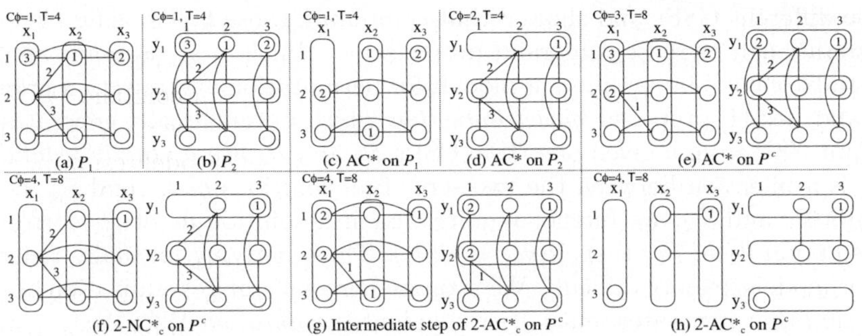

Fig. 1. Enforcing node and arc consistencies on WCSPs $\mathcal{P}_1$, $\mathcal{P}_2$, and $\mathcal{P}^c$

Definition 1. *Let* $\mathcal{P} = (k, \mathcal{X}, \mathcal{D}, \mathcal{C})$ *be a binary WCSP.*

Node consistency. *An assignment* $x \mapsto a$ *in* $\mathcal{P}$ *is* star node consistent *(NC*) if* $C_\emptyset \oplus C_x(a) < \top$. *A variable* x *in* $\mathcal{P}$ *is NC* if (1) all assignments of* x *are NC* and (2) there exists an assignment* $x \mapsto a$ *of* x *such that* $C_x(a) = \bot$. *Value* a *is a* support *for* x. $\mathcal{P}$ *is NC* if every variable in* $\mathcal{P}$ *is NC*.*

Arc consistency. *An assignment* $x \mapsto a$ *in* $\mathcal{P}$ *is* arc consistent *(AC) with respect to a constraint* $C_{x,y}$ *if there exists an assignment* $y \mapsto b$ *of* y *such that* $C_{x,y}(a,b) = \bot$. *Value* b *is a* support *for* $x \mapsto a$. *A variable* x *in* $\mathcal{P}$ *is AC if all assignments of* x *are AC with respect to all binary constraints involving* x. $\mathcal{P}$ *is* star arc consistent *(AC*) if every variable in* $\mathcal{P}$ *is NC* and AC.*

NC* and AC* are enforced by forcing supports for variables and pruning node inconsistent values. Supports can be forced by *projections* of unary (*resp.* binary) constraints over $C_\emptyset$ (*resp.* unary constraints) [2,3]. Let $0 \leq b \leq a \leq k$, we define *subtraction* as $a \ominus b = a - b$ if $a \neq k$; and $a \ominus b = k$ otherwise. Let $\alpha = \min_{a \in D_x}\{C_x(a)\}$ for a variable x. *Projection* of C_x over $C_\emptyset$ [3] is defined such that $C_\emptyset := C_\emptyset \oplus \alpha$ and for each $a \in D_x$, $C_x(a) := C_x(a) \ominus \alpha$. After forcing supports for all variables, all assignments $x \mapsto a$ with $C_\emptyset \oplus C_x(a) = \top$ can be removed. NC* can be enforced on a WCSP with n variables and maximum domain size d in $O(nd)$ time [3]. The WCSP in Fig. 1(a) is not NC*. Removing value 1 from D_{x_1} makes it NC*. Similarly, given variables x and y, for each $a \in D_x$, let $\alpha_a = \min_{b \in D_y}\{C_{x,y}(a,b)\}$. *Projection* of $C_{x,y}$ over C_x [2,3] is defined such that for each $a \in D_x$, $C_x(a) := C_x(a) \oplus \alpha_a$ and for each $a \in D_x$ and $b \in D_y$, $C_{x,y}(a,b) := C_{x,y}(a,b) \ominus \alpha_a$. Consider the assignment $x_1 \mapsto 2$ and the constraint C_{x_1,x_2} in Fig. 1(a), all the costs $C_{x_1,x_2}(2,1)$, $C_{x_1,x_2}(2,2)$, and $C_{x_1,x_2}(2,3)$ are non-$\bot$. We can subtract 2 from each of these costs and add 2 to $C_{x_1}(2)$ to force a support for $x_1 \mapsto 2$. AC* can be enforced using a fix point algorithm in $O(n^2d^3)$ time [3]. The WCSP in Fig. 1(c) is AC* and equivalent to the one in Fig. 1(a).

3 Generating Redundant WCSP Models

Deriving multiple classical CSP models for the same problem is common, although not trivial. Cheng et al. [1] modeled a real-life nurse rostering problem as two different CSPs and showed that combining those CSPs using *channeling constraints* can increase constraint propagation. Hnich et al. [8] made an extensive study of combining different models for permutation and injection problems. Law and Lee [4] proposed *model induction* which automatically generates a redundant CSP from a given one. Two CSPs $\mathcal{P}_1$ and $\mathcal{P}_2$ are *mutually redundant* if there is a bijection between the two sets of all solutions of $\mathcal{P}_1$ and $\mathcal{P}_2$. For two WCSPs $\mathcal{P}_1$ and $\mathcal{P}_2$, we further require that if a solution θ_1 of $\mathcal{P}_1$ corresponds to a solution θ_2 of $\mathcal{P}_2$, then $\mathcal{V}(\theta_1) = \mathcal{V}(\theta_2)$. Thus, deriving mutually redundant WCSP models is more difficult. We propose here a slight generalization of model induction that generates mutually redundant *permutation* WCSPs.

A *permutation WCSP* (PermWCSP) is a WCSP in which all variables have the same domain, the number of variables equals the domain size, and every solution assigns a permutation of the domain values to the variables, i.e., we can set the costs $C_{x_i,x_j}(a,a) = \top$ for all variables x_i and x_j and domain value a. Given a PermWCSP $\mathcal{P} = (k, \mathcal{X}, \mathcal{D}_\mathcal{X}, \mathcal{C}_\mathcal{X})$, we can always interchange the roles of its variables and values to give a *dual* PermWCSP. If $\mathcal{X} = \{x_1, \ldots, x_n\}$ and $D_{x_i} = \{1, \ldots, n\}$, a dual model is $\mathcal{P}' = (k, \mathcal{Y}, \mathcal{D}_\mathcal{Y}, \mathcal{C}_\mathcal{Y})$ with $\mathcal{Y} = \{y_1, \ldots, y_n\}$ and $D_{y_j} = \{1, \ldots, n\}$. The relationship between the variables in $\mathcal{X}$ and $\mathcal{Y}$ can be expressed using the channeling constraints $x_i = j \Leftrightarrow y_j = i$ for $1 \leq i,j \leq n$.

Generalized model induction of a WCSP $\mathcal{P}$ requires a *channel function* that maps assignments in $\mathcal{P}$ to those in another set of variables. If $\mathcal{P}$ is a PermWCSP, we always have the bijective channel function $f(x_i \mapsto j) = y_j \mapsto i$. The constraints $\mathcal{C}_\mathcal{Y}$ in the *induced model* are defined such that for $1 \leq a,i \leq n$, $C_{y_a}(i) = C_{x_i}(a)$, and for $1 \leq a,b,i,j \leq n$, $C_{y_a,y_b}(i,j) = C_{x_i,x_j}(a,b)$ if $i \neq j$; and $C_{y_a,y_b}(i,j) = \top$ otherwise. Note that the induced model must be a PermWCSP, since $C_{y_a,y_b}(i,i) = \top$ for all $1 \leq a,b,i \leq n$. Fig. 1(b) shows the induced model $\mathcal{P}_2$ of $\mathcal{P}_1$ in Fig. 1(a). In the example, we have, say, the unary cost $C_{y_1}(2) = C_{x_2}(1) = 1$ and the binary cost $C_{y_2,y_3}(1,2) = C_{x_1,x_2}(2,3) = 3$.

4   Combining Mutually Redundant WCSPs

Following the redundant modeling [1] technique, we can also combine two mutually redundant WCSPs $\mathcal{P}_1 = (k_1, \mathcal{X}, \mathcal{D}_\mathcal{X}, \mathcal{C}_\mathcal{X})$ and $\mathcal{P}_2 = (k_2, \mathcal{Y}, \mathcal{D}_\mathcal{Y}, \mathcal{C}_\mathcal{Y})$ using a set of channeling constraints $\mathcal{C}^c$ to give the *combined model* $\mathcal{P}^c = (k_1 + k_2, \mathcal{X} \cup \mathcal{Y}, \mathcal{D}_\mathcal{X} \cup \mathcal{D}_\mathcal{Y}, \mathcal{C}_\mathcal{X} \cup \mathcal{C}_\mathcal{Y} \cup \mathcal{C}^c)$. In $\mathcal{P}^c$, $\mathcal{C}^c$ contains hard constraints. For example, the channeling constraint $x_1 = 2 \Leftrightarrow y_2 = 1$ has the cost function $C_{x_1,y_2}(a,b) = \bot$ if $a = 2 \Leftrightarrow b = 1$; and $C_{x_1,y_2}(a,b) = \top$ otherwise. We perform some preliminary experiments in ToolBar,[1] a branch and bound WCSP solver maintaining local consistencies at each search node, using Langford's problem (prob024 in CSPLib[2]) to evaluate the performance of the single and combined models. The single model $\mathcal{P}$ used is based on a model by Hnich et al. [8] Since the problem is over-constrained for many instances, we soften $\mathcal{P}$ so that the

---

¹ Available at http://carlit.toulouse.inra.fr/cgi-bin/awki.cgi/ToolBarIntro.
² Available at http://www.csplib.org.

constraints except the all-different constraint can have random non-$\top$ costs. The combined model $\mathcal{P}^c$ contains $\mathcal{P}$ and its induced model as sub-models.

Unlike redundant modeling in classical CSPs, enforcing AC* on $\mathcal{P}^c$ is even weaker than on $\mathcal{P}$. The former requires more backtracks than the latter in the search, and solving $\mathcal{P}^c$ is about three times slower than solving $\mathcal{P}$. This is mainly due to three reasons. First, there are more variables and constraints in $\mathcal{P}^c$ than in $\mathcal{P}$. It takes longer time to propagate the constraints. Second, we need a large number of channeling constraints to connect two models. For CSPs, efficient global constraints exist for propagating the channeling constraints, but there are no such counterparts for WCSPs. Third, we discover that enforcing AC* on $\mathcal{P}^c$ can miss some prunings which are available even in $\mathcal{P}$. For example, Figs. 1(a) and 1(b) show two mutually redundant WCSPs $\mathcal{P}_1$ and $\mathcal{P}_2$ respectively, which can be combined using $\mathcal{C}^c = \{x_i = j \Leftrightarrow y_j = i \,|\, 1 \leq i, j \leq 3\}$ to give $\mathcal{P}^c$ with the valuation structure $S(4+4) = S(8)$ and $C_\emptyset = 1 \oplus 1 = 2$. Figs. 1(c), 1(d), and 1(e) respectively show the AC* equivalent of $\mathcal{P}_1$, $\mathcal{P}_2$, and $\mathcal{P}^c$. Note that $x_1 \mapsto 1$ and $y_1 \mapsto 1$ are removed in Figs. 1(c) and 1(d) respectively. However, in Fig. 1(e), the assignments are still NC* and AC*, since $C_\emptyset \oplus C_{x_1}(1) = 3 \oplus 3 < \top = 8$ and $C_\emptyset \oplus C_{y_1}(1) = 3 \oplus 2 < \top = 8$, and thus they cannot be removed. This example shows the undesirable behavior that enforcing AC* on a combined model results in weaker constraint propagation than on its sub-models individually.

To remedy the drawbacks, in the following subsections, we propose m-NC$_c^*$ and m-AC$_c^*$ and their associated algorithms for effectively improving propagation in a combined model with m sub-models. We also reveal that the propagation of pruning information among sub-models can be done by enforcing m-NC$_c^*$. This means that redundant modeling can be done with no channeling constraints.

4.1   Node Consistency Revisited

We observe in the above example that given any solution θ of $\mathcal{P}^c$, if an assignment $x_i \mapsto j$ in $\mathcal{P}_1$ is in θ, then according to the channeling constraints, the corresponding assignment $y_j \mapsto i$ in $\mathcal{P}_2$ must be also in θ, and vice versa. Therefore, we can check the node consistencies of $x_i \mapsto j$ and $y_j \mapsto i$ simultaneously. If $C_\emptyset \oplus C_{x_i}(j) \oplus C_{y_j}(i) = \top$, then both $x_i \mapsto j$ and $y_j \mapsto i$ cannot be in any solution of $\mathcal{P}^c$ and can be pruned. Consider the assignments $x_1 \mapsto 1$ and $y_1 \mapsto 1$ in $\mathcal{P}^c$ in Fig. 1(e), since $C_\emptyset \oplus C_{x_1}(1) \oplus C_{y_1}(1) = 3 \oplus 3 \oplus 2 = 8 = \top$, both assignments should be pruned, thus restoring the available prunings in the single models.

Furthermore, a complete tuple of $\mathcal{P}_2$ must contain exactly one assignment of, say, y_1. The set of assignments $\{y_1 \mapsto 1, y_1 \mapsto 2, y_1 \mapsto 3\}$ in $\mathcal{P}_2$ corresponds to $\theta = \{x_1 \mapsto 1, x_2 \mapsto 1, x_3 \mapsto 1\}$ in $\mathcal{P}_1$. Therefore, a solution of $\mathcal{P}^c$ must contain exactly one assignment in θ. In Fig. 1(e), the minimum cost among $C_{x_1}(1)$, $C_{x_2}(1)$, and $C_{x_3}(1)$ is $1 > \bot$, we can use such information to tighten $C_\emptyset$ of $\mathcal{P}^c$.

By capturing the previously described ideas, we propose a new notion of node consistency m-NC$_c^*$ for combined WCSP models with m sub-models. Note that m-NC$_c^*$ is a general notion; it is not restricted to PermWCSPs only. In the following, we assume $\mathcal{P}_s = (k_s, \mathcal{X}_s, \mathcal{D}_s, \mathcal{C}_s)$ for $1 \leq s \leq m$ are m mutually redundant WCSPs, where $\mathcal{X}_s = \{x_{s,i} \,|\, 1 \leq i \leq n_s\}$ and $\mathcal{D}_s = \{D_{x_{s,i}} \,|\, 1 \leq i \leq n_s\}$

$(n_s = |\mathcal{X}_s|)$. $\mathcal{C}_{s,t}$ is the set of channeling constraints connecting $\mathcal{P}_s$ and $\mathcal{P}_t$, and $\mathcal{C}^c = \bigcup_{s<t} \mathcal{C}_{s,t}$. $\mathcal{P}^c = (k, \mathcal{X}, \mathcal{D}, \mathcal{C})$ is a combined model of all m sub-models $\mathcal{P}_s$, where $k = \sum_s k_s$, $\mathcal{X} = \bigcup_s \mathcal{X}_s$, $\mathcal{D} = \bigcup_s \mathcal{D}_s$, and $\mathcal{C} = \bigcup_s \mathcal{C}_s \cup \mathcal{C}^c$. Function $f_{s,t}$ is a bijective channel function from assignments in $\mathcal{P}_s$ to those in $\mathcal{P}_t$. By definition, $f_{t,s} = f_{s,t}^{-1}$ and $f_{s,s}$ is the identity function. $\vartheta_t(x_{s,i}) = \{f_{s,t}(x_{s,i} \mapsto a) \mid a \in D_{x_{s,i}}\}$ is a set of all the corresponding assignments of $x_{s,i}$ in $\mathcal{P}_t$.

Definition 2. *Let $\mathcal{P}^c$ be a combined model of m sub-models $\mathcal{P}_s$ for $1 \leq s \leq m$.*

- *An assignment $x_{s,i} \mapsto a$ is m-channeling node consistent (m-NC_c^*) if $C_\emptyset \oplus \sum_t C_{x_{t,j}}(b_t) < \top$, where $f_{s,t}(x_{s,i} \mapsto a) = x_{t,j} \mapsto b_t$ for $1 \leq t \leq m$.*
- *A variable $x_{s,i} \in \mathcal{X}$ is m-NC_c^* if (1) all assignments of $x_{s,i}$ are m-NC_c^* and (2) for $1 \leq t \leq m$, there exists an assignment $(x_{t,j} \mapsto b) \in \vartheta_t(x_{s,i})$ such that $C_{x_{t,j}}(b) = \bot$. The assignment $x_{t,j} \mapsto b$ is a c-support for $\vartheta_t(x_{s,i})$.*
- *$\mathcal{P}^c$ is m-NC_c^* if every variable in $\mathcal{X}$ is m-NC_c^*.*

For example, $\mathcal{P}^c$ in Fig. 1(e) is NC* (and AC*) but *not* 2-NC_c^*, since (1) $C_\emptyset \oplus C_{x_1}(1) \oplus C_{y_1}(1) = \top$ and (2) there are no c-supports for the tuple $\{x_1 \mapsto 1, x_2 \mapsto 1, x_3 \mapsto 1\}$. Fig. 1(f) shows its 2-$NC_c^*$ equivalent. Note that 1-NC_c^* is equivalent to NC*, while m-NC_c^* is a stronger notion of consistency than NC*.

Theorem 1. *Let $\mathcal{P}^c$ be a combined model of m sub-models $\mathcal{P}_s$ for $1 \leq s \leq m$. Enforcing m-NC_c^* on $\mathcal{P}^c$ is strictly stronger than enforcing NC* on $\mathcal{P}^c$.*

To enforce m-NC_c^* on $\mathcal{P}^c$, we propose a new form of projection which forces c-supports for tuples. Given a tuple θ, let $\alpha = \min_{(x \mapsto a) \in \theta}\{C_x(a)\}$. *C-projection* of a tuple θ over $C_\emptyset$ is a flow of α cost units such that $C_\emptyset := C_\emptyset \oplus \alpha$ and for each $(x \mapsto a) \in \theta$, $C_x(a) := C_x(a) \ominus \alpha$. C-projection is a generalization of ordinary projection. The former allows the assignments in θ from *different* variables, while the latter is equivalent to c-projection of *all* assignments of a *single* variable. Clearly, after c-projection of a tuple θ, there must exist an assignment $(x \mapsto a) \in \theta$ such that $C_x(a) = \bot$. Given a variable $x_{s,i}$ in $\mathcal{P}^c$, c-projection of $\vartheta_t(x_{s,i})$ over $C_\emptyset$ transforms $\mathcal{P}^c$ into an equivalent WCSP.

In the above example, $\theta = \{x_1 \mapsto 1, x_2 \mapsto 1, x_3 \mapsto 1\}$ is the set of assignments in $\mathcal{P}_1$ corresponding to the set of all assignments of y_1 in $\mathcal{P}_2$. Hence, c-projection of θ over $C_\emptyset$ maintains the same cost distribution on complete tuples. In the original $\mathcal{P}^c$, $C_{x_1}(1) = 3$, $C_{x_2}(1) = 1$, and $C_{x_3}(1) = 2$, c-projection of θ deducts 1 from each of the costs and increases $C_\emptyset$ by 1, forcing a c-support $x_2 \mapsto 1$ for θ.

Fig. 2(a) shows an algorithm for enforcing m-NC_c^* on a combined model $\mathcal{P}^c$. The algorithm first forces a c-support for each $\vartheta_t(x_{s,i})$ by c-projecting each $\vartheta_t(x_{s,i})$ over $C_\emptyset$. Next, for each $x_{s,i} \in \mathcal{X}$, pruneVar$_c(x_{s,i})$ is called to prune any non-m-NC_c^* assignments. By using table lookup, a channel function can be implemented in $O(1)$ time. Therefore, pruneVar$_c()$ and NC$_c^*()$ runs in $O(md)$ and $O(mnd)$ time respectively, where d is the maximum domain size and $n = |\mathcal{X}|$.

In $\mathcal{P}^c$, when an assignment $x_{s,i} \mapsto a$ is not m-NC_c^*, all $f_{s,t}(x_{s,i} \mapsto a)$ for $1 \leq t \leq m$ are also not m-NC_c^* and can be pruned. Therefore, enforcing m-NC_c^* on $\mathcal{P}^c$ has already done all the propagation of the channeling constraints, and

procedure $NC_c^*(k, \mathcal{X}, \mathcal{D}, \mathcal{C})$
1. **for each** $x_{s,i} \in \mathcal{X}$ each $1 \leq t \leq m$ **do**
2. $\alpha := \min_{(x_{t,j} \mapsto b) \in \vartheta_t(x_{s,i})} \{C_{x_{t,j}}(b)\}$;
3. $C_\emptyset := C_\emptyset \oplus \alpha$;
4.   **for each** $(x_{t,j} \mapsto b) \in \vartheta_t(x_{s,i})$ **do**
5. $C_{x_{t,j}}(b) := C_{x_{t,j}}(b) \ominus \alpha$;
6. **for each** $x_{s,i} \in \mathcal{X}$ **do**
7.   pruneVar$_c(x_{s,i})$;
endprocedure
procedure pruneVar$_c(x_{s,i})$
8. **for each** $a \in D_{x_{s,i}}$ **do**
9.   **let** $x_{t,j} \mapsto b_t \equiv f_{s,t}(x_{s,i} \mapsto a)$ for $1 \leq t \leq m$
10.   **if** $C_\emptyset \oplus \sum_t C_{x_{t,j}}(b_t) = \top$ **then**
11.   **for each** $1 \leq t \leq m$ **do**
12. $D_{x_{t,j}} := D_{x_{t,j}} \setminus \{b_t\}$;
endprocedure

(a) $m\text{-}NC_c^*$ algorithm

procedure findCSupport$(x_{s,i}, x_{s,j})$
1. **for each** $1 \leq t \leq m$ **do**
2. $supported := true$;
3.   **for each** $(x_{t,i'} \mapsto a') \in \vartheta_t(x_{s,i})$ **do**
4.   **if** $S(x_{t,i'} \mapsto a', x_{s,j}, t) \notin \vartheta_t(x_{s,j})$ **then**
5.   **let** $x_{t,j*} \mapsto b^* \equiv argmin_{(x_{t,j'} \mapsto b') \in \vartheta_t(x_{s,j})} \{C_{x_{t,i'},x_{t,j'}}(a',b')\}$;
6. $S(x_{t,i'} \mapsto a', x_{s,j}, t) := (x_{t,j*} \mapsto b^*)$;
7. $\alpha := C_{x_{t,i'},x_{t,j*}}(a',b^*)$; $C_{x_{t,i'}}(a') := C_{x_{t,i'}}(a') \oplus \alpha$;
8.   **for each** $(x_{t,j'} \mapsto b') \in \vartheta_t(x_{t,j})$ **do**
9. $C_{x_{t,i'},x_{t,j'}}(a',b') := C_{x_{t,i'},x_{t,j'}}(a',b') \ominus \alpha$;
10.   **if** $C_{x_{t,i'}}(a') = \bot \wedge \alpha > \bot$ **then** $supported := false$;
11.   **if** $\neg supported$ **then**
12.   **let** $x_{t,i*} \mapsto a^* \equiv argmin_{(x_{t,i'} \mapsto a') \in \vartheta_t(x_{s,i})} \{C_{x_{t,i'}}(a')\}$;
13. $S(x_{s,i}, t) := (x_{t,i*} \mapsto a^*)$;
14. $\alpha := C_{x_{t,i*}}(a^*)$; $C_\emptyset := C_\emptyset \oplus \alpha$;
15.   **for each** $(x_{t,i'} \mapsto a') \in \vartheta_t(x_{s,i})$ **do**
16. $C_{x_{t,i'}}(a') := C_{x_{t,i'}}(a') \ominus \alpha$;
endprocedure

(b) $m\text{-}AC_c^*$ algorithm

Fig. 2. Algorithms for enforcing $m\text{-}NC_c^*$ and $m\text{-}AC_c^*$

we can skip posting them in $\mathcal{P}^c$ to save propagation overhead. In subsequent discussions, we assume no channeling constraints exist in a combined model if $m\text{-}NC_c^*$ is enforced.

4.2   Arc Consistency Revisited

Consider variable x_2 in $\mathcal{P}_1$, the set of assignments $\{x_2 \mapsto 1, x_2 \mapsto 2, x_2 \mapsto 3\}$ in $\mathcal{P}_1$ corresponds to $\theta = \{y_1 \mapsto 2, y_2 \mapsto 2, y_3 \mapsto 2\}$ in $\mathcal{P}_2$. Now for the assignment $y_2 \mapsto 1$ in Fig. 1(e), there is a binary cost $C_{y_2,y_j}(1,2)$ incurred between $y_2 \mapsto 1$ and $(y_j \mapsto 2) \in \theta$. (When $j = 2$, there are actually no such cost, but we assume without losing generality that such "cost" is $\top$.) Since the minimum cost among $C_{y_2,y_1}(1,2) = 2$, "$C_{y_2,y_2}(1,2)$" $= \top$, and $C_{y_2,y_3}(1,2) = 3$ is $2 > \bot$, we can use such information to tighten the bound on the cost $C_{y_2}(1)$. Thus, we can propose a new arc consistency $m\text{-}AC_c^*$ for combined models with m sub-models.

Definition 3. *Let $\mathcal{P}^c$ be a combined model of m sub-models $\mathcal{P}_s$ for $1 \leq s \leq m$.*

- *An assignment $x_{s,i} \mapsto a$ in $\mathcal{P}^c$ is m-channeling arc consistent ($m\text{-}AC_c^*$) with respect to constraint $C_{x_{s,i},x_{s,j}}$ if for $1 \leq t \leq m$, there exists an assignment $(x_{t,j'} \mapsto b') \in \vartheta_t(x_{s,j})$ such that $C_{x_{t,i'},x_{t,j'}}(a',b') = \bot$, where $x_{t,i'} \mapsto a' = f_{s,t}(x_{s,i} \mapsto a)$. The assignment $x_{t,j'} \mapsto b'$ is a c-support for $x_{t,i'} \mapsto a'$.*
- *A variable $x_{s,i} \in \mathcal{X}$ is $m\text{-}AC_c^*$ if all assignments of $x_{s,i}$ are $m\text{-}AC_c^*$ with respect to all constraints involving $x_{s,i}$.*
- *$\mathcal{P}^c$ is $m\text{-}AC_c^*$ if each $x_{s,i} \in \mathcal{X}$ is $m\text{-}NC_c^*$ and $m\text{-}AC_c^*$.*

The combined model $\mathcal{P}^c$ in Fig. 1(e) is AC* but *not* 2-AC$_c^*$, since there are no c-supports among $\{y_1 \mapsto 2, y_2 \mapsto 2, y_3 \mapsto 2\}$ for $y_2 \mapsto 1$. Fig. 1(h) shows an equivalent 2-AC$_c^*$ WCSP. Again, 1-AC$_c^*$ is equivalent to AC*, while $m\text{-}AC_c^*$ is a stronger notion of consistency than AC*.

Theorem 2. *Let $\mathcal{P}^c$ be a combined model of m sub-models $\mathcal{P}_s$ for $1 \leq s \leq m$. Enforcing $m\text{-}AC_c^*$ on $\mathcal{P}^c$ is strictly stronger than enforcing AC* on $\mathcal{P}^c$.*

To enforce m-AC^*_c on a combined model, we extend the definition of c-projections which can force c-supports for assignments. Given a tuple θ and an assignment $(x \mapsto a) \notin \theta$, let $\alpha = \min_{(y \mapsto b) \in \theta}\{C_{x,y}(a,b)\}$. *C-projection* of θ over $x \mapsto a$ is a flow of α cost units such that $C_x(a) := C_x(a) \oplus \alpha$ and for each $(y \mapsto b) \in \theta$, $C_{x,y}(a,b) := C_{x,y}(a,b) \ominus \alpha$. C-projection of the set of all assignments of a variable y over $x \mapsto a$ is equivalent to ordinary binary projection. Given an assignment $(x_{t,j} \mapsto a) \notin \vartheta_t(x_{s,i})$ in a combined model $\mathcal{P}^c$, c-projection of $\vartheta_t(x_{s,i})$ over $x_{t,j} \mapsto a$ transforms $\mathcal{P}^c$ into an equivalent WCSP.

Recall the combined model $\mathcal{P}^c$ in Fig. 1(e), $y_2 \mapsto 1$ has no c-supports in $\theta = \{y_1 \mapsto 2, y_2 \mapsto 2, y_3 \mapsto 2\}$, which is the corresponding set of all assignments of x_2 in $\mathcal{P}_1$. C-projections of θ over $y_2 \mapsto 1$ yields $C_{y_2,y_1}(1,2) = \bot$, $C_{y_2,y_3}(1,2) = 1$, and $C_{y_2}(1) = 2$, as shown in Fig. 1(g). In the figure, $\{x_1 \mapsto 1, x_2 \mapsto 1, x_3 \mapsto 1\}$ is also c-projected over $C_\emptyset$ such that $C_{x_1}(1) = 2$, $C_{x_2}(1) = \bot$, $C_{x_3}(1) = 1$, and $C_\emptyset = 4$. The assignments $x_1 \mapsto 1$, $x_1 \mapsto 2$, $y_1 \mapsto 1$, and $y_2 \mapsto 1$ are consequently not 2-NC^*_c, since $C_\emptyset \oplus C_{x_1}(1) \oplus C_{y_1}(1) = C_\emptyset \oplus C_{x_1}(2) \oplus C_{y_2}(1) = 4 \oplus 2 \oplus 2 = \top$, and are thus pruned. Variable x_1 is now bound and further propagation yields the 2-AC^*_c WCSP in Fig. 1(h). The optimal solution is $\{x_1 \mapsto 3, x_2 \mapsto 1, x_3 \mapsto 2, y_1 \mapsto 2, y_2 \mapsto 3, y_3 \mapsto 1\}$, which has aggregate cost 4.

Fig. 2(b) shows the core algorithm for enforcing m-AC^*_c. It uses two data structures $S(x_{t,i'} \mapsto a', x_{s,j}, t)$ and $S(x_{s,i}, t)$. They store the current c-support for the assignment $x_{t,i'} \mapsto a'$ among $\vartheta_t(x_{s,j})$ and for $\vartheta_t(x_{s,i})$ respectively. The algorithm generalizes the procedure findSupport() by Larrosa [3] so that for each sub-model $\mathcal{P}_t$, it forces a c-support among $\vartheta_t(x_{s,j})$ for each assignment $(x_{t,i'} \mapsto a') \in \vartheta_t(x_{s,i})$ (lines 3–10). C-projections over $C_\emptyset$ are done when necessary (lines 11–16). After finding c-supports, any assignments that are not m-NC^*_c are pruned using pruneVar$_\mathrm{c}$() in Fig. 2(a). The findCSupport() algorithm runs in $O(md^2)$ time, hence the overall m-AC^*_c algorithm runs in $O(mn^2d^3)$ time.

5   Experiments

We implement the 2-NC^*_c and 2-AC^*_c algorithms in ToolBar to evaluate their efficiency on combined models, using Langford's problem and Latin square problem, both of which can be modeled as PermWCSPs. Enforcing AC* on combined models is highly inefficient, so comparisons are made among AC* [3], FDAC* [5], and EDAC* [6] on a single model $\mathcal{P}$ and 2-AC^*_c on a combined model $\mathcal{P}^c$, which contains $\mathcal{P}$ and its induced model as sub-models. Experiments are run on a 1.6GHz US-IIIi CPU with 2GB memory. We use the *dom/deg* variable ordering heuristic [6] and the smallest-cost-first value ordering heuristic [6]. The initial $\top$ provided to ToolBar is n^2, where n is the number of variables in a single model. We report in the tables the average number of fails (i.e., the number of backtracks occurred in solving a model) and CPU time in seconds to find an optimal solution. The first column shows the problem instances; those marked with "*" have a $\bot$ optimal cost. The subsequent columns show the results of enforcing various local consistencies on either $\mathcal{P}$ or $\mathcal{P}^c$. A cell labeled with "-" denotes a timeout after 2 hours.

Table 1. Experimental results on solving Langford's problem and Latin square problem

(a) Classical Langford's problem

(m,n)	AC* on $\mathcal{P}$ fail	time	FDAC* on $\mathcal{P}$ fail	time	EDAC* on $\mathcal{P}$ fail	time	2-AC$_c^*$ on $\mathcal{P}^c$ fail	time
$(2,6)$	80	0.01	80	**0**	80	0.01	**57**	0.01
$(2,7)$*	**1**	**0**	**1**	0.01	**1**	0.01	2	0.01
$(2,8)$*	22	0.01	18	**0**	18	**0**	**0**	0.01
$(2,9)$	8576	0.85	8576	0.8	8576	1	**4209**	**0.79**
$(2,10)$	48048	5.03	48048	4.91	48048	6.07	**22378**	**4.48**
$(2,11)$*	11	**0.01**	8	**0.01**	11	**0.01**	**3**	0.02
$(2,12)$*	7	**0.01**	7	**0.01**	7	**0.01**	**0**	0.02
$(2,13)$	14.73M	1769.8	14.73M	1691.98	14.73M	2164.02	**5.72M**	**1325.92**
$(3,5)$	6	**0**	6	**0**	6	**0**	**4**	0.01
$(3,6)$	20	**0.01**	20	**0.01**	20	**0.01**	**14**	0.02
$(3,7)$	62	0.04	62	**0.03**	62	0.04	**29**	0.04
$(3,8)$	238	0.13	238	**0.1**	238	0.12	**96**	0.13
$(3,9)$*	195	0.12	193	**0.09**	192	0.11	**53**	0.11
$(3,10)$*	590	0.42	560	0.29	560	0.39	**97**	**0.23**
$(3,11)$	14512	10.01	14512	7.45	14512	10.09	**3029**	**6.15**
$(3,12)$	62016	46.13	62016	35.19	62016	46.69	**12252**	**25.57**
$(3,13)$	300800	247.45	300800	191.56	300800	257.59	**45274**	**108.71**
$(3,14)$	1.37M	1185.08	1.37M	933.22	1.37M	1229.83	**190153**	**525.88**
$(3,15)$	7.52M	6992.09	7.52M	5419.17	-	-	**851968**	**2750.05**

(b) Soft Langford's problem

(m,n)	AC* on $\mathcal{P}$ fail	time	FDAC* on $\mathcal{P}$ fail	time	EDAC* on $\mathcal{P}$ fail	time	2-AC$_c^*$ on $\mathcal{P}^c$ fail	time
$(2,6)$	180	**0.01**	177	0.02	176	0.02	**130**	0.02
$(2,7)$*	91	**0.01**	**79**	**0.01**	80	**0.01**	90	**0.01**
$(2,8)$*	162	**0.01**	146	**0.01**	145	0.02	**137**	0.03
$(2,9)$	10850	1.09	10823	1.05	10823	1.05	**5074**	**1**
$(2,10)$	59681	6.38	59654	6.14	59655	6.13	**25718**	**5.42**
$(2,11)$*	595	**0.07**	547	**0.07**	546	0.08	**405**	0.1
$(2,12)$*	795	**0.1**	708	**0.1**	700	0.12	**576**	0.17
$(2,13)$	18.16M	2232.01	18.16M	2142.04	18.16M	2160.93	**6.46M**	**1552.65**
$(3,5)$	368	**0.04**	285	**0.04**	**279**	0.05	284	0.06
$(3,6)$	901	0.14	681	**0.12**	668	0.15	**600**	0.19
$(3,7)$	2286	0.51	1687	**0.42**	1687	0.53	**1559**	0.67
$(3,8)$	2735	0.81	1976	**0.61**	**1962**	0.75	2398	1.23
$(3,9)$*	2665	0.53	**1169**	**0.31**	1360	0.44	3893	1.94
$(3,10)$*	6612	1.74	4026	**1.34**	**3903**	1.6	8715	5.2
$(3,11)$	77507	43.8	69356	33.89	69455	42.3	**27992**	**33.27**
$(3,12)$	275643	178.25	252527	136.27	252830	171.14	**82804**	**119.44**
$(3,13)$	949361	736.13	920007	577.57	919749	720.18	**174881**	**354.39**
$(3,14)$	4.41M	3704.64	4.32M	2886.43	4.32M	3624	**596201**	**1448.54**
$(3,15)$	-	-	-	-	-	-	**2.31M**	**6742.25**

(c) Classical Latin square problem

n	AC* on $\mathcal{P}$ fail	time	FDAC* on $\mathcal{P}$ fail	time	EDAC* on $\mathcal{P}$ fail	time	2-AC$_c^*$ on $\mathcal{P}^c$ fail	time
5*	**0**	**0**	**0**	**0**	**0**	0.01	**0**	0.01
10*	1	**0.13**	1	0.14	1	0.16	**0**	0.23
15*	14	**1.44**	14	1.55	14	1.99	**0**	2.62
20*	712	**11.5**	546	13.79	747	24.33	**0**	18.49
25*	4	65.54	18	**36.88**	173	41.79	**0**	130.66

(d) Soft Latin square problem

n	AC* on $\mathcal{P}$ fail	time	FDAC* on $\mathcal{P}$ fail	time	EDAC* on $\mathcal{P}$ fail	time	2-AC$_c^*$ on $\mathcal{P}^c$ fail	time
3	5	**0**	**4**	**0**	**4**	**0**	5	**0**
4	172	**0.01**	100	0.02	**98**	0.03	125	0.02
5	25016	**2.61**	10038	2.85	**7325**	5.08	14950	2.78
6	751412	153.82	111545	105.95	**76353**	144.6	203839	**93.93**
7	-	-	1.87M	3558.49	**950480**	3632.47	3.62M	**2933.09**

Table 1(a) shows the results on the (m,n) instances of Langford's problem solved using a classical CSP model [8] containing $m \times n$ variables. In the problem, only a few instances are satisfiable (marked with "*"). Therefore, we soften the problem as described in Section 4. For each soft instance, we generate 10 instances and report the average results in Table 1(b). For both classical and soft cases, 2-AC$_c^*$ achieves the fewest number of fails among all four local consistencies in most instances. This shows that it does more prunings than AC*, FDAC*, and EDAC*, reduces more search space, and transmits both pruning and cost projection information better. The 2-AC$_c^*$ is clearly the most efficient for the larger and more difficult instances, which require more search efforts to either prove unsatisfiability or find an optimal solution. We improve the number of fails and runtime of, say, $(3,14)$, by factors of 7.2 and 2.2 respectively. There are even instances that enforcing AC*, FDAC*, and EDAC* on $\mathcal{P}$ cannot be solved before timeout but enforcing 2-AC$_c^*$ on $\mathcal{P}^c$ can. The reduction rates of number of fails and runtime of $\mathcal{P}^c$ to $\mathcal{P}$ increase with the problem size. Exceptions are the "*" instances, in which once an $\perp$ cost solution is found, we need not prove its optimality and can terminate immediately. Such instances require relatively fewer search efforts, and the overhead of an extra model may not be counteracted.

Table 1(c) and 1(d) show the experimental results of classical and soft Latin square problem (prob003 in CSPLib) respectively. Contrary to Langford's problem, Latin square problem has many solutions. We assert preferences among the solutions by assigning to each allowed binary tuple a random cost from $\perp$ to n inclusive. We can see from Table 1(c) that classical Latin square problem is easy to solve up to $n = 25$. The amount of search is small even using a single model. Enforcing 2-AC$_c^*$ on $\mathcal{P}^c$ can still reduce the search space to achieve no backtracks. However, the runtime is not the fastest due to the overhead of an extra model. The soft Latin square problem is more difficult than the classical one since we are searching for the most preferred solution. We can solve only the smaller instances within 2 hours. Although 2-AC_c^* does not achieve the smallest number of fails among other local consistencies, its runtime is highly competitive as shown in Table 1(d). Like in the case of the classical and soft Langford's problem, 2-AC_c^* is the most efficient for the larger instances. In fact, even with two models, the time complexity of 2-AC_c^* is only a constant order higher than AC^*, while $FDAC^*$ and $EDAC^*$ have higher time complexities $O(n^3 d^3)$ and $O(n^2 d^2 \max\{nd, \top\})$ respectively. In this problem, 2-AC_c^* strikes a balance between the amount of prunings and the time spent on discovering these prunings. This explains why 2-AC_c^* has more fails than $EDAC^*$ but the fastest runtime.

6   Conclusion

While we can rely on the standard propagation algorithms of the channeling constraints for classical CSPs to transmit pruning information between sub-models, we have shown that this approach does not work well when combining WCSPs. Instead, we have generalized NC^* and AC^* and proposed the strictly stronger $m\text{-NC}_c^*$ and $m\text{-AC}_c^*$ respectively for combined models containing m sub-models. Experiments on our implementations of 2-NC_c^* and 2-AC_c^* have shown the benefits of extra prunings, which lead to a greatly reduced search space and better runtime than the state-of-the-art AC^*, $FDAC^*$, and $EDAC^*$ algorithms on both classical and soft benchmark problems, especially for *hard* instances.

Redundant modeling for WCSPs is a new concept and has plenty of scope for future work. We can investigate how generalized model induction can generate induced models of non-PermWCSPs, and to apply $m\text{-AC}_c^*$ to the resultant combined models. It would be also interesting to incorporate c-supports and c-projections to $FDAC^*$ and $EDAC^*$ to obtain $m\text{-FDAC}_c^*$ and $m\text{-EDAC}_c^*$.

References

1. Cheng, B., Choi, K., Lee, J., Wu, J.: Increasing constraint propagation by redundant modeling: an experience report. Constraints **4**(2) (1999) 167–192
2. Schiex, T.: Arc consistency for soft constraints. In: Proc. of CP'00. (2000) 411–424
3. Larrosa, J.: Node and arc consistency in weighted CSP. In: Proc. of AAAI'02. (2002) 48–53
4. Law, Y., Lee, J.: Model induction: a new source of CSP model redundancy. In: Proc. of AAAI'02. (2002) 54–60
5. Larrosa, J., Schiex, T.: In the quest of the best form of local consistency for weighted CSP. In: Proc. of IJCAI'03. (2003) 239–244
6. de Givry, S., Heras, F., Zytnicki, M., Larrosa, J.: Existential arc consistency: Getting closer to full arc consistency in weighted CSPs. In: Proc. of IJCAI'05. (2005) 84–89
7. Schiex, T., Fargier, H., Verfaillie, G.: Valued constraint satisfaction problems: hard and easy problems. In: Proc. of IJCAI'95. (1995) 631–637
8. Hnich, B., Smith, B., Walsh, T.: Dual modelling of permutation and injection problems. JAIR **21** (2004) 357–391

Verification of Multi-agent Systems
Via Bounded Model Checking*

Xiangyu Luo[1], Kaile Su[2,4,**], Abdul Sattar[2], and Mark Reynolds[3]

[1] Department of Computer Science, Guilin University of Electronic Technology, Guilin, China
shiangyuluo@gmail.com
[2] Institute for Integrated and Intelligent Systems, Griffith University, Brisbane, Australia
{k.su, a.sattar}@griffith.edu.au
[3] School of CSSE, The University of Western Australia, Perth, Australia
mark@csse.uwa.edu.au
[4] Department of Computer Science, Sun Yat-sen University, Guangzhou, China

Abstract. We present a bounded model checking (BMC) approach to the verification of temporal epistemic properties of multi-agent systems. We extend the temporal logic CTL^* by incorporating epistemic modalities and obtain a temporal epistemic logic that we call CTL^*K. CTL^*K logic is interpreted under the semantics of synchronous interpreted systems. Though CTL^*K is of great expressive power in both temporal and epistemic dimensions, we show that BMC method is still applicable for the universal fragment of CTL^*K. We present in some detail a BMC algorithm and prove its correctness. In our approach, agents' knowledge interpreted in synchronous semantics can be skillfully attained by the state position function, which avoids extending the encoding of the states and the transition relations of the plain temporal epistemic model for time domain.

Keywords: bounded model checking, multi-agent systems, temporal epistemic logic, bounded semantics.

1 Introduction

Model checking is a technique for automatic formal verification of finite state systems. There are many practical applications of the technique for hardware and software verification. Recently, verification of multi-agent systems (MAS) has become an active field of research. Verification of MAS has mainly focused on extending the existing model checking techniques usually used for verification of reactive systems. In the multi-agent paradigm, particular emphasis is given to the formal representation of the mental attitudes of agents, such as agents' knowledge, beliefs, desires, intentions and so on. However, the formal specifications used in the traditional model checking are most commonly expressed as formulas of temporal logics [1] such as CTL and LTL.

* This work was partially supported by the Australian Research Council grant DP0452628, National Basic Research 973 Program of China under grant 2005CB321902, National Natural Science Foundation of China grants 60496327, 10410638 and 60473004, and Guangdong Provincial Natural Science Foundation grants 04205407 and 06023195.
** Corresponding author.

A. Sattar and B.H. Kang (Eds.): AI 2006, LNAI 4304, pp. 69–78, 2006.

© Springer-Verlag Berlin Heidelberg 2006

So, the research of MAS verification has focused on the extension of traditional model checking techniques to incorporate epistemic modalities for describing information and motivation attitudes of agents [2].

The application of model checking within the context of the logic of knowledge was first proposed by Halpern and Vardi [3] in 1991. In [4,5], van der Hoek and Wooldridge analyzed the application of SPIN and MOCHA respectively to model checking of LTL and ATL extended by epistemic modalities. In 2004 Meyden and Su took a promising step towards model checking of anonymity properties in formulas involving knowledge [6], and then based on the semantics of interpreted systems with local propositions, Su developed an approach to the BDD-based symbolic model checking for CKL_n [7].

Unfortunately, the main bottleneck of BDD-based symbolic model checking is the state explosion problem. One of complementary techniques to BDD-based model checking is *Bounded Model Checking* (BMC). The basic idea of BMC is to explore a part of the model sufficient to check a particular formula and translate the existential model checking problem (the problem that decides whether there is a path in a system satisfying a given formula) over the part of the model into a test of propositional satisfiability.

BMC for LTL was first introduced by Biere *et al.* [8]. Then, Penczek *et al.* developed a BMC method for $ACTL$ (the universal fragment of CTL) [9]. In addition, Penczek and Lomuscio presented a BMC method for $ACTLK$ in [10], which incorporates epistemic modalities to $ACTL$ logic. Further, in [11], Woźna proposed a BMC method for $ACTL^*$, the universal fragment of CTL^*, which subsumes both $ACTL$ and LTL. Naturally, it is meaningful to develop a BMC method for those languages that incorporate epistemic modalities into $ACTL^*$.

In computer science, many protocols are designed so that their actions take place in rounds or steps (where no agent starts round $m + 1$ before all agents finish round m), and agents know what round it is now at all times. Therefore, we restrict MAS to a *synchronous* one, in which agents have access to a shared clock and run in synchrony.

This paper is the extension version of our previous work [12]. The aim of this paper is to develop a BMC method for an expressive branching time logic of knowledge that we call $ACTL^*K$, which incorporates epistemic modalities into $ACTL^*$. We adopt the *synchronous* interpreted systems semantics [13]. Moreover, in order to avoid extending the encoding of the states and the transition relations of the *plain* temporal epistemic model for the time domain of synchronous interpreted systems, we introduce a *state position function* to attain agents' knowledge.

The significance of $ACTL^*K$ is that the temporal expressive power of $ACTL^*K$ is greater than that of $ACTLK$ extended from $ACTL$. For example, we permit the subformula of an epistemic formula (its main operator is an epistemic one) to be a *state* or *path* formula, while $ACTLK$ only subsumes *state* formulas. It is convenient to use $ACTL^*K$ to specify and verify dynamic knowledge of agents in the dynamic environment of a MAS.

The rest of this paper is organized as follows. Section 2 introduces synchronous semantics of multi-agent interpreted systems. Section 3 defines the syntax and the synchronous semantics of CTL^*K, and Section 4 defines the bounded semantics of $ECTL^*K$. In Section 5, the equivalence between the synchronous semantics and the bounded one is to be proven. Section 6 describes the BMC method for $ECTL^*K$. Finally, we conclude this paper in Section 7.

2 Synchronous Temporal Epistemic Model

We begin with a short discussion about a set of agents $A = \{1, \ldots, n\}$ and their environment e. Let L_e be a set of possible states for the environment and L_i a set of possible local states for each agent $i \in A$. We take $G = L_e \times L_1 \times \ldots \times L_n$ to be the set of global states and define function $l_i \colon G \longrightarrow L_i$, which returns the local state of agent i from a global state $s \in G$. A *run* over G is a function from the time domain (natural numbers) to G, it can be identified with a sequence of global states in G. A *point* is a pair (r, m) consisting of a run r and time m. The global state at the point (r, m) is defined as $r(m) = (s_e, s_1, \ldots, s_n)$, where $s_e \in L_e$ and $s_i \in L_i$ for all $i \in A$. So $r(0)$ is the initial state. Further we define $r_e(m) = s_e$ and $r_i(m) = s_i$ for all $i \in A$. Thus, $r_i(m)$ is agent i's local state at the point (r, m). We define a *system $\mathcal{R}$ over G* as a set of runs over G. (r, m) is a point in system $\mathcal{R}$ if $r \in \mathcal{R}$. An *interpreted system $\mathcal{I}$* is a structure $(\mathcal{R}, \mathcal{V})$, where $\mathcal{R}$ is a system over G and $\mathcal{V}$ assigns truth values to the primitive propositions at the global states in G.

We introduce the indistinguishability relation $\sim_i$ for each agent $i \in A$ as follows. Let s and s' be two global states in G, $s \sim_i s'$ denotes that s and s' are indistinguishable to agent i, i.e., i has the same local state in both s and s'. If $r(n) = s$ and $r'(n') = s'$, we use $(r, n) \sim_i (r', n')$ to denote $s \sim_i s'$. Let $i \in A$ and $\Gamma \subseteq A$, we introduce four epistemic modalities $\mathbf{K}_i$(knows), $\mathbf{D}_\Gamma$(distributed knowledge), $\mathbf{E}_\Gamma$(everyone knows) and $\mathbf{C}_\Gamma$(common knowledge) to our logic CTL^*K. Their epistemic relations are defined as $\sim_i$, $\sim {}^D_\Gamma = \bigcap_{i \in \Gamma} \sim_i$, $\sim {}^E_\Gamma = \bigcup_{i \in \Gamma} \sim_i$ and $\sim {}^C_\Gamma$, respectively, where $\sim {}^C_\Gamma$ is the transitive closure of $\sim {}^E_\Gamma$ [13]. By modifying the epistemic accessibility relation in Definition 2.1 in [10] to a synchronous one, we get the following definition:

Definition 1. *Given an interpreted system $\mathcal{I} = (\mathcal{R}, \mathcal{V})$ and a set of agents $A = \{1, \ldots, n\}$, we associate with $\mathcal{I}$ a* Synchronous Temporal Epistemic Model *$M = (S, s_0, T, \overset{T}{\sim}_1, \ldots, \overset{T}{\sim}_n, \mathcal{V})$, where (1) S is a finite set of global states of $\mathcal{I}$; (2) s_0 is the initial state of $\mathcal{I}$; (3) $T \subseteq S \times S$ is a total binary (successor) relation on S such that $r(m)Tr(m+1)$ for each $r \in \mathcal{R}$ and natural number $m \geq 0$; (4) $\overset{T}{\sim}_i$ is a synchronous epistemic accessibility relation on the points of $\mathcal{I}$ for agent $i \in A$. Let (r, m) and (r', m') be two points of $\mathcal{I}$, then $(r, m) \overset{T}{\sim}_i (r', m')$ iff $r_i(m) = r'_i(m')$ and $m = m'$; (5) $\mathcal{V} \colon S \times \mathcal{PV} \longrightarrow \{true, false\}$ is a truth assignment function for a set of propositional variables $\mathcal{PV}$ such that $\mathcal{V}(s)(p) \in \{true, false\}$ for all $s \in S$ and $p \in \mathcal{PV}$.*

Thus, given a synchronous temporal epistemic model M, each run of M can be obtained by infinitely unfolding T of M from the initial state s_0. In addition, let $i \in A$ and $\Gamma \subseteq A$, the epistemic relations used by $\mathbf{K}_i$, $\mathbf{D}_\Gamma$, $\mathbf{E}_\Gamma$ and $\mathbf{C}_\Gamma$ are defined as $\overset{T}{\sim}_i$, $\overset{T}{\sim} {}^D_\Gamma = \bigcap_{i \in \Gamma} \overset{T}{\sim}_i$, $\overset{T}{\sim} {}^E_\Gamma = \bigcup_{i \in \Gamma} \overset{T}{\sim}_i$, and $\overset{T}{\sim} {}^C_\Gamma$, respectively, where $\overset{T}{\sim} {}^C_\Gamma$ is the transitive closure of $\overset{T}{\sim} {}^E_\Gamma$. It is easy to prove that for any points (r, n) and (r', n') of a synchronous system, $(r, n) \overset{T}{\sim} {}^Y_\Gamma (r', n')$ iff $(r, n) \sim {}^Y_\Gamma (r', n')$ and $n = n'$, where $Y \in \{\mathbf{D}, \mathbf{E}, \mathbf{C}\}$. Hereafter, we use *"model"* to denote *synchronous temporal epistemic model*.

3 CTL^*K Logic and Its Subsets

Here we extend the temporal logic CTL^* [1] by incorporating epistemic modalities, which include $\mathbf{K}_i$, $\mathbf{D}_\Gamma$, $\mathbf{E}_\Gamma$ and $\mathbf{C}_\Gamma$, where $i \in A$ and $\Gamma \subseteq A$. In order to solve the existential model checking problem, we add four dual epistemic modalities related to the modalities mentioned above. If $\mathbf{Y} \in \{\mathbf{K}_i, \mathbf{D}_\Gamma, \mathbf{E}_\Gamma, \mathbf{C}_\Gamma\}$ and φ is a formula, then $\overline{\mathbf{Y}}$ is the dual modalities of $\mathbf{Y}$ and $\mathbf{Y}\,\varphi \equiv \neg\overline{\mathbf{Y}}\,\neg\varphi$. We call the resulting logic CTL^*K. Assume the familiarity with the syntax of CTL^*, we refer to [1] for more details. The only thing about CTL^*K mentioned here is that if α is a *path* formula, then epistemic formula $\overline{\mathbf{Y}}\alpha$ is a *state* formula. Note that any state formula is also a path formula.

We define the $ECTL^*K$ logic as the restriction of CTL^*K such that the negation can be applied only to propositions and the operators are restricted to $\mathbf{E}$ (in some path), $\overline{\mathbf{K}}_i$, $\overline{\mathbf{D}}_\Gamma$, $\overline{\mathbf{E}}_\Gamma$ and $\overline{\mathbf{C}}_\Gamma$. The $ACTL^*K$ logic is also the restriction of CTL^*K such that its language is defined as $\{\neg\varphi | \varphi \in ECTL^*K\}$, in which the path quantifier $\mathbf{A}$ (in all paths) is defined as $\mathbf{A}\varphi \equiv \neg\mathbf{E}\neg\varphi$.

Definition 2 (Synchronous Semantics of CTL^*K). *Let M be a model, (r, n) a point of M and α, β be CTL^*K formulas. $(M, r, n) \models \alpha$ denotes that α is true at point (r, n). M is omitted if it is implicitly understood. The relation $\models$ is defined as follows:*

$(r, n) \models p$ *iff* $\mathcal{V}(r(n))(p) = true,$ $(r, n) \models \neg p$ *iff* $(r, n) \not\models p,$

$(r, n) \models \alpha \wedge \beta \mid \alpha \vee \beta$ *iff* $(r, n) \models \alpha$ *and (or)* $(r, n) \models \beta$, $(r, n) \models \mathbf{X}\alpha$ *iff* $(r, n + 1) \models \alpha$,

$(r, n) \models \mathbf{F}\alpha$ *iff* $\exists_{n' \geq n}(r, n') \models \alpha$, $(r, n) \models \mathbf{G}\alpha$ *iff* $\forall_{n' \geq n}(r, n') \models \alpha$,

$(r, n) \models \alpha \mathbf{U} \beta$ *iff* $\exists_{n' \geq n}\big((r, n') \models \beta$ *and* $\forall_{n \leq i < n'}(r, i) \models \alpha\big)$,

$(r, n) \models \mathbf{E}\alpha$ *iff there is a run r' and time n' with $r(n) = r'(n')$ such that $(r', n') \models \alpha$*,

$(r, n) \models \mathbf{Y}\alpha$ *iff there is a run r' and time n' with $(r, n) \overset{Y}{\sim} (r', n')$ and $n = n'$ such that $(r', n') \models \alpha$, where $(\mathbf{Y}, \overset{Y}{\sim}) \in \{(\overline{\mathbf{K}}_i, \sim_i), (\overline{\mathbf{D}}_\Gamma, \sim \overset{D}{\Gamma}), (\overline{\mathbf{E}}_\Gamma, \sim \overset{E}{\Gamma}), (\overline{\mathbf{C}}_\Gamma, \sim \overset{C}{\Gamma})\}$.*

As for modality $\mathbf{K}_i$, here we capture the intuition that agent i knows formula φ at point (r, n) of M exactly if at all points with the same time n that i considers possible at (r, n), φ is true. For the modality $\overline{\mathbf{K}}_i$, conversely, $(M, r, n) \models \overline{\mathbf{K}}_i\varphi$ if and only if φ is true at some point with the same time n that i considers possible at (r, n). So with such a synchronous semantics, only one point with time n should be considered in a run.

Definition 3 (Validity). *A CTL^*K formula φ is valid in M (denoted $M \models \varphi$) iff $(M, r, 0) \models \varphi$ for all run r in M, i.e., φ is true in the initial state of M.*

4 Bounded Semantics of $ECTL^*K$

In this section we combine the bounded semantics for $ECTL^*$ [11] with epistemic modalities so that the BMC problem for $ECTL^*K$ can be translated into a propositional satisfiability problem.

Let M be a model and k a positive natural number. A *k-path* is a path of length k, i.e. *k-path* is a finite sequence $\pi_k = \{s_0, \ldots, s_k\}$ of states such that $(s_i, s_{i+1}) \in T$ for all $0 \leq i < k$, and state s_i of π_k can be denoted by $\pi_k(i)$. A finite *k-path* can also represent an infinite path if there is a *back loop* from the last state to a certain previous state of the *k-path*. So, a *k-path* π_k is a *(k,l)-loop* if $(\pi_k(k), \pi_k(l)) \in T$ for some $0 \leq l \leq k$.

Given a model M, in order to translate the existential model checking into bounded model checking and the SAT problem, we only consider a part of the model M, i.e., the compact model consists of all the *k-paths* of M and is defined as follows.

Definition 4 (k-model). *Let* $M = (S, s_0, T, \overset{T}{\sim}_1, \ldots, \overset{T}{\sim}_n, \mathcal{V})$ *be a model and* k *a positive natural number. A k-model of* M *is a tuple* $M_k = (S, s_0, P_k, \overset{T}{\sim}_1, \ldots, \overset{T}{\sim}_n), \mathcal{V}$, *where* s_0 *is the initial state of* M *and* P_k *is the set of all the k-paths of* M.

Since the bounded semantics for temporal operators depends on whether the considered k-path π_k of the k-model M_k is a loop or not, we define a function $loop(\pi_k) = \{l | 0 \leq l \leq k$ and $(\pi_k(k), \pi_k(l)) \in T\}$ so as to distinguish whether π_k is a loop.

A k-path π_k in M_k can be viewed as a part of a *run* r in M. So, we can project a partial r onto a k-path π_k. Let $r(n) = \pi_k(m)$ for some $n \geq 0$ and $0 \leq m \leq k$, by Definition 5, we can calculate the position $i \in \{0, \ldots, k\}$ of the state of π_k such that $\pi_k(i) = r(c)$ for some *time* $c \geq n$, i.e., state $\pi_k(i)$ of M_k represents state $r(c)$ of M.

Definition 5 (State Position Function). *Let* M *be a model and* M_k *a k-model of* M. *Assume that* r *is a* run *in* M *and the corresponding* π_k *is a k-path of* M_k *such that* $r(n) = \pi_k(m)$ *for some* $n \geq 0$ *and* $m \leq k$, *then, for time* $c \geq n$ *and* $l \leq k$, *function*
$$pos(n, m, k, l, c) := \begin{cases} m + c - n, & if\ c \leq n + k - m; \\ l + (c - n - l + m)\%(k - l + 1), & else\ if\ l \in loop(\pi_k). \end{cases}$$
returns the position of the state of π_k *such that* $\pi_k(pos(n, m, k, l, c)) = r(c)$, *where* $\%$ *is modular arithmetic. Further define* State Position Function $f_I(k, l, c) := pos(0, 0, k, l, c)$ *for epistemic operators.*

Thus, a part of an *infinite* run of M may be represented by a (k, l)-loop of M_k by means of the state position function. Next, we extend Definition 4.3 in [11] to Definition 6, the bounded semantics of $ECTL^*K$, by incorporating the time domain and epistemic operators.

Definition 6 (Bounded Semantics of $ECTL^*K$). *Let* M_k *be a k-model and* α, β $ECTL^*K$ *formulas. Given a natural number* $l \in \{0, \ldots, k\}$, *if* α *is a state formula, then* $[M_k, \pi_k, l, m, c] \models \alpha$ *denotes that* α *is true at the state* $\pi_k(m)$ *of* M_k *and time* c. *If* α *is a path formula, then* $[M_k, \pi_k, l, m, c] \models \alpha$ *denotes that* α *is true along the suffix of the k-path* π_k *of* M_k, *which starts at the position* m *and time* c. M_k *is omitted if it is implicitly understood. The relation* $\models$ *is defined inductively as follows:*

$[\pi_k, l, m, c] \models p\ iff\ \mathcal{V}(\pi_k(m))(p) = true.$ $[\pi_k, l, m, c] \models \neg p\ iff\ \mathcal{V}(\pi_k(m))(p) = false.$

$[\pi_k, l, m, c] \models \alpha \wedge \beta \mid \alpha \vee \beta\ iff\ [\pi_k, l, m, c] \models \alpha\ and\ (or)\ [\pi_k, l, m, c] \models \beta.$

$[\pi_k, l, m, c] \models \boldsymbol{E}\alpha\ iff\ \exists \pi'_k \in P_k\ (\pi'_k(0) = \pi_k(m)\ and\ \exists_{0 \leq l' \leq k}\ [\pi'_k, l', 0, c] \models \alpha).$

$[\pi_k, l, m, c] \models \boldsymbol{X}\alpha\ iff$
$$\begin{cases} if\ m \geq k\ then\ false\ else\ [\pi_k, l, m+1, c+1] \models \alpha, & if\ l \notin loop(\pi_k); \\ if\ m \geq k\ then\ [\pi_k, l, l, c+1] \models \alpha\ else\ [\pi_k, l, m+1, c+1] \models \alpha, & otherwise. \end{cases}$$

$[\pi_k, l, m, c] \models \boldsymbol{F}\alpha\ iff\ \begin{cases} \exists_{m \leq i \leq k}\ [\pi_k, l, i, c+i-m] \models \alpha, & if\ l \notin loop(\pi_k); \\ \exists_{m \leq i \leq k}\ [\pi_k, l, i, c+i-m] \models \alpha\ or \\ \exists_{l \leq i < m}\ [\pi_k, l, i, c+k-m+1+i-l] \models \alpha, & otherwise. \end{cases}$

$[\pi_k, l, m, c] \models \boldsymbol{G}\alpha\ iff\ \begin{cases} false, & if\ l \notin loop(\pi_k); \\ \forall_{m \leq i \leq k}\ [\pi_k, l, i, c+i-m] \models \alpha, & if\ l \in loop(\pi_k)\ and\ l \geq m; \\ \forall_{l \leq i < m}\ [\pi_k, l, i, c+k-m+1+i-l] \models \alpha\ and \\ \forall_{m \leq i \leq k}\ [\pi_k, l, i, c+i-m] \models \alpha, & if\ l \in loop(\pi_k)\ and\ l < m. \end{cases}$

$[\pi_k, l, m, c] \models \alpha \mathbf{U} \beta$ iff

$$\begin{cases} \exists_{m \le i \le k} \left([\pi_k, l, i, c + i - m] \models \beta \text{ and } \forall_{m \le j < i} [\pi_k, l, j, c + j - m] \models \alpha\right), \text{ if } l \notin loop(\pi_k); \\ \exists_{m \le i \le k} \left([\pi_k, l, i, c + i - m] \models \beta \text{ and } \forall_{m \le j < i} [\pi_k, l, j, c + j - m] \models \alpha\right) \text{ or} \\ \exists_{l \le i < m} \left([\pi_k, l, i, c + k - m + 1 + i - l] \models \beta \text{ and} \forall_{m \le j \le k} [\pi_k, l, j, c + j - m] \models \alpha \text{ and} \\ \forall_{l \le j < i} [\pi_k, l, j, c + k - m + 1 + j - l] \models \alpha\right), \qquad\qquad otherwise. \end{cases}$$

$[\pi_k, l, m, c] \models \mathbf{Y}\alpha$ iff $\exists \pi'_k \in P_k$ such that $\pi'_k(0) = s_0$ and, if $c \le k$, then $\pi_k(m) \overset{Y}{\sim} \pi'_k(c)$ and $\exists_{0 \le l' \le k} [\pi'_k, l', c, c] \models \alpha$; else $\exists_{0 \le l' \le k} \left(l' \in loop(\pi'_k) \text{ and } \pi_k(m) \overset{Y}{\sim} \pi'_k(f_I(k, l', c)) \text{ and } [\pi'_k, l', f_I(k, l', c), c] \models \alpha\right)$, where $(\mathbf{Y}, \overset{Y}{\sim}) \in \left\{(\overline{\mathbf{K}}_i, \sim_i), (\overline{\mathbf{D}}_\Gamma, \sim \overset{D}{\Gamma}), (\overline{\mathbf{E}}_\Gamma, \sim \overset{E}{\Gamma})\right\}$.

$[\pi_k, l, m, c] \models \overline{\mathbf{C}}_\Gamma \alpha \Leftrightarrow [\pi_k, l, m, c] \models \bigvee_{i=1}^{k} (\overline{\mathbf{E}}_\Gamma)^i \alpha$.

From the above bounded semantics, we can see that when checking a temporal formula at the state $\pi_k(m)$ and time c, the time domain c' corresponding to the i-th state of π_k can be calculated as follows: $c' := c + i - m$ if $i \ge m$, or else if $l \in loop(\pi_k)$, then $c' := c + k - m + 1 + i - l$, where i is the position of the current state under consideration in π_k. It assures that the time c' always increases.

As for the knowledge modality $\mathbf{K}_i$, we consider whether or not there is a k-path π'_k from the *initial* state (the first state $\pi_k(0)$ of π_k is equal to the initial state of M) that results in a state s' that agent i consider possible in $\pi_k(m)$ and the current time in s' is equal to c. Note that the position of s' should be calculated by the state position function $f_I(k, l, c)$ only if the time $c > k$.

Definition 7 (Validity for Bounded Semantics). *An $ECTL^*K$ formula φ is valid in a k-model M_k (denoted $M \models_k \varphi$) iff $[M_k, \pi_k, l, 0, 0] \models \varphi$ for some $0 \le l \le k$, where $\pi_k(0)$ is the initial state of M_k.*

5 Correctness of the Bounded Semantics

In this section we will prove that the $ECTL^*K$ model checking problem ($M \models \varphi$) can be reduced to the $ECTL^*K$ BMC problem ($M \models_k \varphi$). Some proofs similar to [10] or [11] are omitted for the limited space. Note that the bounded semantics of $ECTL^*K$ differs from those of other papers because we add time to it. We first define the length of a formula φ (denoted by $|\varphi|$) as the number of operators (exclude $\neg$) in φ. Then, by Lemma 4.3 of [11], Lemma 1 and 2, we can determine the maximum bound k (called Diameter of M) that an $ECTL^*K$ formula should be checked with to guarantee that the property holds.

Lemma 1. *Let M be a model, α an LTL formula and $Y \in \{E, \overline{K}_i, \overline{D}_\Gamma\}$, then, if $(M, r, 0) \models \mathbf{Y}\alpha$ then there exists $k \le |M| \cdot |\alpha| \cdot 2^{|\alpha|}$ with $[M_k, \pi_k, l, 0, 0] \models \mathbf{Y}\alpha$ for some $0 \le l \le k$, where $r(0)$ and $\pi_k(0)$ are equal to the initial state of M.*

Proof. (1) Let $\mathbf{Y} = \mathbf{E}$. The case can be easily proven by Lemma 4.3 of [11] when time domain is added to it. (2) Let $\mathbf{Y} = \overline{\mathbf{K}}_i$ and $(r, 0)$ be an initial point of M. By Definition 2, $(M, r, 0) \models \overline{\mathbf{K}}_i\alpha$ iff there is a run r' and time c' with $(r, 0) \sim_i (r', c')$ and $c' = 0$ such that $(M, r', c') \models \alpha$. Since $r'(0)$ is also the initial state of M, it can be viewed as an existential model checking problem [8] for the LTL formula α in M. So the case holds by Lemma 4.3 of [11]. (3) Let $\mathbf{Y} = \overline{\mathbf{D}}_\Gamma$. Straightforward by definition from the case $\mathbf{Y} = \overline{\mathbf{K}}_i$. $\qquad\square$

Lemma 2. *Let M be a model, φ be an $ECTL^*K$ formula and (r, c) be a point of M. If $(M, r, c) \models \varphi$, then there exists $k \leq |M| \cdot |\varphi| \cdot 2^{|\varphi|}$ and a k-path π_k of M_k with $r(c) = \pi_k(m)$ such that $[M_k, \pi_k, l, m, c] \models \varphi$ for some $0 \leq l \leq k$.*

Proof. By induction on the length of φ. The lemma follows directly when φ is a propositional variable or its negation. Assume that the hypothesis holds for all the proper subformulas of φ. The lemma is easily proven when $\varphi = \alpha \vee \beta | \alpha \wedge \beta$. Consider case 1) and 2):

1) Assume that α is an LTL formula. Let $\varphi = \overline{\mathbf{E}}_\Gamma \alpha$. Since $\overline{\mathbf{E}}_\Gamma \alpha = \bigvee_{i \in A} \overline{\mathbf{K}}_i \alpha$, by induction the lemma follows from the case of $Y = \overline{\mathbf{K}}_i$ in Lemma 1 for some $i \in A$ and the case for the boolean connectives. Let $\varphi = \overline{\mathbf{C}}_\Gamma \alpha$. Because $(M, r, c) \models \overline{\mathbf{C}}_\Gamma \alpha$ iff $(M, r, c) \models \bigvee_{i \leq |S|} (\overline{\mathbf{E}}_\Gamma)^i \alpha$, where $|S|$ is the number of reachable states in M, the lemma holds by induction on the former cases. Because the state under consideration may be the subsequent one of the initial state, the existential model checking problem here is not harder than that of Lemma 1, so Lemma 1 can be applied in the former cases.

2) Assume that α is not a *"pure"* LTL formula, we extend the state labelling technique of [1] to epistemic operators. Let $\varphi = \mathbf{Y}\alpha$ and $\mathbf{Y}, \mathbf{Y}_i, \mathbf{Z} \in \{\mathbf{E}, \overline{\mathbf{K}}_i, \overline{\mathbf{D}}_\Gamma, \overline{\mathbf{E}}_\Gamma, \overline{\mathbf{C}}_\Gamma\}$ for $i = 1, \ldots, n$. Let $\mathbf{Y}_1\alpha_1, \ldots, \mathbf{Y}_n\alpha_n$ be the list of all maximal subformulas of α (i.e., each $\mathbf{Y}_i\alpha_i$ is a state subformula of $\mathbf{Y}\alpha$, but not a subformula of any subformula $\mathbf{Z}\beta$ of $\mathbf{Y}\alpha$, where $\mathbf{Z}\beta$ is different from $\mathbf{Y}\alpha$ and $\mathbf{Y}_i\alpha_i$ for $i = 1, \ldots, n$).

Next, we introduce for each $\mathbf{Y}_i\alpha_i$ a fresh atomic proposition p_i, where $1 \leq i \leq n$, and augment the labelling of *each* state s of M_k with p_i iff $\mathbf{Y}_i\alpha_i$ holds at state s, then we replace each subformula $\mathbf{Y}_i\alpha_i$ by p_i and obtains a new formula α', which is a *"pure"* LTL formula. Furthermore, by the definition of synchronous semantics, we have that verifying $(M, r, c) \models \overline{\mathbf{E}}_\Gamma \alpha' \mid \overline{\mathbf{C}}_\Gamma \alpha'$ can also be viewed as an existential model checking problem for the LTL formula α' in M. The proof is similar to that in the case of $\mathbf{Y} = \overline{\mathbf{K}}_i$ in Lemma 1. Hence, by Lemma 1 and the cases above, we have that $(M, r, c) \models \mathbf{Y}\alpha$ implies $[M_k, \pi_k, l, m, c] \models \mathbf{Y}\alpha$. $\square$

The condition $r(c) = \pi_k(m)$ in Lemma 2 expresses that the state $\pi_k(m)$ of M_k is the same as the state $r(c)$ of M. We have the following theorem based on Lemma 2.

Theorem 1. *Let M be a model and φ an $ECTL^*K$ formula. Then $M \models \varphi$ iff $M \models_k \varphi$ for some $k \leq |M| \cdot |\varphi| \cdot 2^{|\varphi|}$.*

Theorem 1 shows that the satisfiability of an $ECTL^*K$ formula φ in the bounded semantics is equivalent to the unbounded one with the diameter $|M| \cdot |\varphi| \cdot 2^{|\varphi|}$.

6   Bounded Model Checking for $ECTL^*K$

In this section we present a BMC method for $ECTL^*K$ in synchronous interpreted systems. It is an extension of the method presented in [11].

The main idea of the BMC method is that the validity of an $ECTL^*K$ formula φ can be determined by checking the satisfiability of a propositional formula $[M, \varphi]_k := [M^{\varphi, s_0}]_k \wedge [\varphi]_{M_k}$, where $[\varphi]_{M_k}$ is a number of constraints that must be satisfied on M_k for φ to be satisfied, and $[M^{\varphi, s_0}]_k$ represents the (partial) k-model M_k under consideration (the *submodel* of M_k), which consists of a part of valid k-paths in M_k. Definition

8 determines the number of those k-paths that is sufficient for checking formula φ, such that the validity of φ in M_k is equivalent to the validity of φ in the part of M_k. The bound $f_k(\varphi)$ can be obtained by structural induction on the translation of φ in Definition 10.

Definition 8. *Define a function f_k from the set of formulas to natural number:*
$f_k(p) = f_k(\neg p) = 0$, *where* $p \in \mathcal{PV}$, $f_k(\alpha \vee \beta) = max(f_k(\alpha), f_k(\beta))$,
$f_k(\alpha \wedge \beta) = f_k(\alpha) + f_k(\beta)$, $\qquad f_k(\mathbf{X}\alpha) = f_k(\mathbf{F}\alpha) = f_k(\alpha)$,
$f_k(\mathbf{G}\alpha) = (k+1) \cdot f_k(\alpha)$, $\qquad f_k(\alpha \mathbf{U}\beta) = k \cdot f_k(\alpha) + f_k(\beta)$,
$f_k(\overline{\mathbf{C}}_\Gamma \alpha) = f_k(\alpha) + k$, $\qquad f_k(\mathbf{Y}\alpha) = f_k(\alpha) + 1$, *where* $\mathbf{Y} \in \{\mathbf{E}, \overline{\mathbf{K}}_i, \overline{\mathbf{D}}_\Gamma, \overline{\mathbf{E}}_\Gamma\}$.

Once $[M, \varphi]_k$ is constructed, the validity of formula φ over M_k can be determined by checking the satisfiability of $[M, \varphi]_k$ via a SAT solver. We give the BMC algorithm for $ACTL^*K$ as follows: Let φ be an $ACTL^*K$ formula. Then, start with $k := 1$, test the satisfiability of $[M, \neg\varphi]_k$ via a SAT solver, and increase k by one either until $[M, \neg\varphi]_k$ becomes satisfiable or k reaches $|M| \cdot |\neg\varphi| \cdot 2^{|\neg\varphi|}$. If $[M, \neg\varphi]_k$ is satisfiable with some k, then returns "$M \not\models \varphi$", returns "$M \models \varphi$" otherwise.

We now give details of the translations for the propositional formulas $[M^{\varphi,s_0}]_k$ and $[\varphi]_{M_k}$. Because the synchronous model is identical with the model in [10] except that the synchronous epistemic accessibility relation, the technical details of translation are similar to that of [10], from which we only introduce some propositional formulas. Let w, v be two global state variables, s and s' two states encoded by w and v respectively. $I_s(w)$ encodes state s of the model by global state variable w; $T(w, v)$ encodes $(s, s') \in T$; $p(w)$ represents $p \in \mathcal{V}(s)$; $H(w, v)$ represents the fact that w, v represent the same state; $H_i(w, v)$ represents that the i-local state in s and s' is the same; $L_{k,j}(l) := T(w_{k,j}, w_{l,j})$ represents that the j-th k-path is a (k, l)-loop. See [10] for more details.

The propositional formula $[M^{\varphi,s_0}]_k$ represents the transitions in the k-model M_k, its definition is given as below.

Definition 9 (Unfolding of Transition Relation). *Let* $M_k = (S, s_0, P_k, \overset{T}{\sim}_1, \ldots, \overset{T}{\sim}_n, V)$ *be a k-model of M, $s \in S$, and φ an $ECTL^*K$ formula. Define formula*
$$[M^{\varphi,s}]_k := I_s(w_{0,0}) \wedge \bigwedge_{1 \leq j \leq f_k(\varphi)} \bigwedge_{0 \leq i \leq k-1} T(w_{i,j}, w_{i+1,j}),$$
where $w_{0,0}$ and $w_{i,j}$ are global state variables for $i = 0, \ldots, k$ and $j = 1, \ldots, f_k(\varphi)$. $w_{i,j}$ also represents the ith state of the jth symbolic k-path.

According to the bounded semantics of $ECTL^*K$, we make the following definition for translating an $ECTL^*K$ formula φ into a propositional formula.

Definition 10. *Given a k-model M_k and $ECTL^*K$ formulas α, β and let $L_{k,i} := \bigvee_{l'=0}^{k} L_{k,i}(l')$, $x \in \{k, (k, l)\}$, we use $[\alpha]_k^{[m,n,c]}$ to denote the translation of α at state $w_{m,n}$ and time c into a propositional formula based on the bounded semantics of a non-loop k-path, whereas the translation of $[\alpha]_{k,l}^{[m,n,c]}$ depends on the bounded semantics of a (k, l)-loop. The translation of φ is defined inductively as follows:*

$[p]_x^{[m,n,c]} := p(w_{m,n})$, $\quad [\neg p]_x^{[m,n,c]} := \neg p(w_{m,n})$, $\quad [\alpha \wedge \beta]_x^{[m,n,c]} := [\alpha]_x^{[m,n,c]} \wedge [\beta]_x^{[m,n,c]}$,
$[\mathbf{X}\alpha]_k^{[m,n,c]} := if\ m < k\ then\,[\alpha]_k^{[m+1,n,c+1]}\ else\,false$, $[\alpha \vee \beta]_x^{[m,n,c]} := [\alpha]_x^{[m,n,c]} \vee [\beta]_x^{[m,n,c]}$,
$[\mathbf{X}\alpha]_{k,l}^{[m,n,c]} := if\ m < k\ then\,[\alpha]_{k,l}^{[m+1,n,c+1]}\ else\,[\alpha]_{k,l}^{[l,n,c+1]}$, $[\mathbf{F}\alpha]_k^{[m,n,c]} := \bigvee_{i=m}^{k}[\alpha]_k^{[i,n,c+i-m]}$,
$[\mathbf{F}\alpha]_{k,l}^{[m,n,c]} := \bigvee_{i=m}^{k}[\alpha]_{k,l}^{[i,n,c+i-m]} \vee \bigvee_{i=l}^{m-1}[\alpha]_{k,l}^{[i,n,c+k-m+1+i-l]}$, $\quad [\mathbf{G}\alpha]_k^{[m,n,c]} := false$,

$$[\boldsymbol{G}\alpha]_{k,l}^{[m,n,c]} := \textit{if } l \geq m \textit{ then } \bigwedge_{i=m}^{k}[\alpha]_{k,l}^{[i,n,c+i-m]}$$
$$\textit{else } \bigwedge_{i=m}^{k}[\alpha]_{k,l}^{[i,n,c+i-m]} \wedge \bigwedge_{i=l}^{m-1}[\alpha]_{k,l}^{[i,n,c+k-m+1+i-l]},$$
$$[\alpha\boldsymbol{U}\beta]_{k}^{[m,n,c]} := \bigvee_{i=m}^{k}([\beta]_{k}^{[i,n,c+i-m]} \wedge \bigwedge_{j=m}^{i-1}[\alpha]_{k}^{[j,n,c+j-m]}),$$
$$[\alpha\boldsymbol{U}\beta]_{k,l}^{[m,n,c]} := \bigvee_{i=m}^{k}([\beta]_{k,l}^{[i,n,c+i-m]} \wedge \bigwedge_{j=m}^{i-1}[\alpha]_{k,l}^{[j,n,c+j-m]}$$
$$\vee \bigvee_{i=l}^{m-1}([\beta]_{k,l}^{[i,n,c+k-m+1+i-l]} \wedge \bigwedge_{j=m}^{k}[\alpha]_{k,l}^{[j,n,c+j-m]} \wedge \bigwedge_{j=l}^{i-1}[\alpha]_{k,l}^{[j,n,c+k-m+1+j-l]}),$$
$$[\boldsymbol{E}\alpha]_{x}^{[m,n,c]} := \bigvee_{i=1}^{f_k(\varphi)}(H(w_{m,n},w_{0,i})\wedge((\neg L_{k,i} \wedge [\alpha]_{k}^{[0,i,c]}) \vee \bigvee_{l'=0}^{k}(L_{k,i}(l') \wedge [\alpha]_{k,l'}^{[0,i,c]}))),$$
$$[\boldsymbol{Y}\,\alpha]_{x}^{[m,n,c]} :=$$
$$\begin{cases} \textit{if } c \leq k \textit{ then } \bigvee_{i=1}^{f_k(\varphi)}(I_{s_0}(w_{0,i}) \wedge \boldsymbol{Z}(H_a(w_{m,n},w_{c,i})) \\ \qquad\qquad \wedge((\neg L_{k,i} \wedge [\alpha]_{k}^{[c,i,c]}) \vee \bigvee_{l'=0}^{k}(L_{k,i}(l') \wedge [\alpha]_{k,l'}^{[c,i,c]}))) \\ \textit{else } \bigvee_{i=1}^{f_k(\varphi)}(I_{s_0}(w_{0,i}) \wedge \bigvee_{l'=0}^{k}(L_{k,i}(l') \wedge \boldsymbol{Z}(H_a(w_{m,n},w_{f_I(k,l',c),i})) \wedge [\alpha]_{k,l'}^{[f_I(k,l',c),i,c]})), \end{cases}$$
$$\textit{where } (\boldsymbol{Y},\boldsymbol{Z}) \in \{(\overline{\boldsymbol{K}}_a,\epsilon),(\overline{\boldsymbol{D}}_\Gamma,\textstyle\bigwedge_{a\in\Gamma}),(\overline{\boldsymbol{E}}_\Gamma,\textstyle\bigvee_{a\in\Gamma})\} \textit{ and } \epsilon \textit{ denotes that } \boldsymbol{Z} \textit{ is empty.}$$
$$[\overline{\boldsymbol{C}}_\Gamma\alpha]_{x}^{[m,n,c]} := [\textstyle\bigvee_{1\leq i\leq k}(\overline{\boldsymbol{E}}_\Gamma)^{i}\,\alpha]_{x}^{[m,n,c]}.$$

Lemma 3. *Let M_k be a k-model, φ be an $ECTL^*K$ formula, $m,l \leq k$ and $c \geq 0$. Then, $[M_k,\pi_k,l,m,c] \models \varphi$ iff there is a submodel M_k' of M_k with $|P_k'| \leq f_k(\varphi)$ such that $[M_k',\pi_k,l,m,c] \models \varphi$.*

Lemma 4. *Let M_k be a k-model of M, φ an $ECTL^*K$ formula, and $l \leq k$. For each point (r,c) of M, if there is a submodel M_k' of M_k with $|P_k'| \leq f_k(\varphi)$, where P_k' is the set of all the k-paths of M_k', and there is a k-path $\pi_k \in P_k'$ with $\pi_k(m) = r(c)$ $(m \leq k)$, then $M_k', [(\pi_k,l),m,c] \models \varphi$ if and only if the following conditions holds:*
1. *$[M^{\varphi,\pi_k(m)}]_k \wedge [\varphi]_k^{[0,0,c]}$ is satisfiable, if φ is a state formula;*
2. *$[M^{\varphi,\pi_k(m)}]_k \wedge [\varphi]_k^{[m,0,c]}$ is satisfiable, if φ is a path formula and π_k is not a (k,l)-loop;*
3. *$[M^{\varphi,\pi_k(m)}]_k \wedge [\varphi]_{k,l}^{[m,0,c]}$ is satisfiable, if φ is a path formula and π_k is a (k,l)-loop.*

Lemma 3 and 4 can be proven by structural induction on the length of φ. These proofs are omitted here for the limited space. Then, from the two Lemmas we immediately have Theorem 2, which guarantee the correctness of the translation.

Theorem 2. *Let M be a model, φ be an $ECTL^*K$ formula, and k be a bound. Then, $M \models_k \varphi$ iff $[M^{\varphi,s_0}]_k \wedge [\varphi]_{M_k}$ is satisfiable, where $[\varphi]_{M_k} = [\varphi]_k^{[0,0,0]}$.*

7 Conclusions

Formal verification methods in multi-agent systems have traditionally been associated with specifications. In this paper we present the BMC method for branching time logic of knowledge $ACTL^*K$ in synchronous multi-agent systems. We define synchronous interpreted system semantics via revising epistemic accessibility relations of interpreted system semantics of [10] to synchronous ones. On the other hand, in order to obtain more expressive power in temporal dimension, we adopt $ACTL^*$ [11] as the underlying temporal logic and incorporate epistemic modalities to it. So, the resulting logic $ACTL^*K$ is different from $ACTLK$.

After construction of the model of a MAS, we can use the proposed BMC algorithm for $ACTL^*K$ to check (1) agents' knowledge about the dynamic world, (2) agents' knowledge about other agents' knowledge, (3) the temporal evolution of the above knowledge, and (4) any combination of (1), (2), and (3). For example, consider a typical synchronous multi-agent system, the well-known Muddy Children Puzzle with n children [13], the $ACTL^*K$ formula $AGC_\Gamma(\bigwedge_{i=1}^{n}(m_i \Rightarrow FK_i\ m_i))$ says that in any situation, all children have the common knowledge that the child whose forehead is muddy will eventually know his/her forehead has mud, where $m_i = true$ represents that child i's forehead is muddy. Note that the subformula $\bigwedge_{i=1}^{n}(m_i \Rightarrow FK_i\ m_i)$ is a *path* subformula. Obviously, the $ACTL^*K$ formula cannot be expressed in $ACTLK$ [10]. In addition, if an $ACTL^*K(ECTL^*K)$ specification is $false(true)$, then we can get a counterexample (witness) of *minimal length*. This feature helps the users to analyze a MAS for faults more easily.

We are also keen to explore how to develop a security protocol verifier using our BMC method, which can automatically construct the synchronous temporal epistemic model of security protocols and check various security properties such as privacy, anonymity and nonrepudiation. In addition, in order to overcome the intrinsic limitation of BMC techniques, we will extend our BMC method to *Unbounded Model Checking* (UMC) [14] for the full CTL^*K language.

References

1. Clarke, E.M., Grumberg, O., Peled, D.A.: Model Checking. The MIT Press, Cambridge, MA (2000)
2. Bordini, R.H., Fisher, M., Pardavila, C., Wooldridge, M.: Model checking agentspeak. In: Proceedings of the second international joint conference on Autonomous agents and multiagent systems, ACM Press (2003) 409–416
3. Halpern, J.Y., Vardi, M.Y.: Model checking vs. theorem proving: A manifesto. In: KR. (1991) 325–334
4. van der Hoek, W., Wooldridge, M.: Model checking knowledge and time. In: Proc. 19th Workshop on SPIN, Grenoble (2002)
5. van der Hoek, W., Wooldridge, M.: Tractable multiagent planning for epistemic goals. In: Proc. of the 1st Int. Conf. on Autonomous Agents and Multi-Agent Systems (AAMAS02). Volume III., ACM (2002) 1167–1174
6. van der Meyden, R., Su, K.: Symbolic model checking the knowledge of the dining cryptographers. In: Proc of 17th IEEE Computer Security Foundations Workshop. (2004) 280–291
7. Su, K.: Model checking temporal logics of knowledge in distributed systems. In McGuinness, D.L., Ferguson, G., eds.: AAAI, AAAI Press / The MIT Press (2004) 98–103
8. Biere, A., Cimatti, A., Clarke, E.M., Zhu, Y.: Symbolic model checking without BDDs. In: Proc. of TACAS99. Volume 1579 of LNCS., Springer-Verlag (1999) 193–207
9. Penczek, W., Woźna, B., Zbrzezny, A.: Bounded model checking for the universal fragment of CTL. Fundamenta Informaticae **51** (2002) 135–156
10. Penczek, W., Lomuscio, A.: Verifying epistemic properties of multi-agent systems via bounded model checking. Fundamenta Informaticae **55** (2003)
11. Woźna, B.: ACTL* properties and Bounded Model Checking. Fundamenta Informaticae **63**(1) (2004) 65–87
12. Luo, X., Su, K., Sattar, A., Chen, Q., Lv, G.: Bounded model checking knowledge and branching time in synchronous multi-agent systems. In Dignum, F., Dignum, V., Koenig, S., Kraus, S., Singh, M.P., Wooldridge, M., eds.: AAMAS, ACM (2005) 1129–1130
13. Fagin, R., Halpern, J., Moses, Y., Vardi, M.: Reasoning about knowledge. MIT Press, Cambridge, MA (1995)
14. Kacprzak, M., Lomuscio, A., Penczek, W.: From bounded to unbounded model checking for temporal epistemic logic. Fundamenta Informaticae **62** (2004) 1–20

Logical Properties of Belief-Revision-Based Bargaining Solution*

Dongmo Zhang and Yan Zhang

Intelligent Systems Laboratory
School of Computing and Mathematics
University of Western Sydney, Australia
{dongmo, yan}@scm.uws.edu.au

Abstract. This paper explores logical properties of belief-revision-based bargaining solution. We first present a syntax-independent construction of bargaining solution based on prioritized belief revision. With the construction, the computation of bargaining solution can be converted to the calculation of maximal consistent hierarchy of prioritized belief sets. We prove that the syntax-independent solution of bargaining satisfies a set of desired logical properties for agreement function and negotiation function. Finally we show that the computational complexity of belief-revision-based bargaining can be reduced to $\Delta_2^P[\mathcal{O}(\log n)]$.

Keywords: belief revision, bargaining theory, automated negotiation.

1   Introduction

Much recent research has shown that belief revision is a successful tool in modeling logical reasoning in bargaining and negotiation[2,4,5,14,15,16]. These studies have established a qualitative solution to bargaining problem, which differentiates them from the traditional game-theoretic solution[6,11]. In [16], Zhang *et al.* proposed an axiomatic system to specify the logical reasoning behind negotiation processes by introducing a set of AGM-like postulates[1]. In [4,5], Meyer *et al.* further explored the logical properties of these postulates and their relationships. In [16], Zhang and Zhang proposed a computational model of negotiation based on Nebel's syntax-dependent belief base revision operation[8] and discussed its game-theoretic properties and computational properties. It was shown that the computational complexity of belief-revision-based negotiation can be Π_2^P-hard. However, it is left unknown that how this computational model related to the axiomatic approach to negotiation. In this paper, we shall establish a relationship between these two approaches and reassess the computational complexity of belief-revision-based negotiation. In order to make two different modeling approaches comparable, we shall redefine the bargaining solution given in [16] based

---

* This work is partially supported by UWS Research Grant Scheme.

A. Sattar and B.H. Kang (Eds.): AI 2006, LNAI 4304, pp. 79–89, 2006.
© Springer-Verlag Berlin Heidelberg 2006

on the assumption that bargaining inputs are logically closed[1]. We then shown that the computation of agreements can be reduced to the construction of maximal consistent hierarchies of negotiation items. Based on this result, we show that the new definition of bargaining solution satisfies most of desired logical properties of negotiation. Finally we present a completeness result on computational complexity of belief-revision-based bargaining solution, which shows that the decision problem of bargaining solution is $\Delta_2^P[\mathcal{O}(\log n)]$-complete. This result significantly improves the result presented in [16].

Similar to the work in [16], we will restrict us to the bargaining situations within which only two agents are involved. We assume that each agent has a set of negotiation items, referred to as *demand set*, which is describable in a finite propositional language $\mathcal{L}$. The language is that of classical propositional logic with an associated consequence operation Cn in the sense that $Cn(X) = \{\varphi : X \vdash \varphi\}$, where X is a set of sentences. A set K of sentences is *logically closed* or called a *belief set* when $K = Cn(K)$. If X, Y are two sets of sentences, $X + Y$ denotes $Cn(X \cup Y)$.

Suppose that X_1 and X_2 are the demand sets of two agents. To simplify exploration, we will use X_{-i} to represent the other set among X_1 and X_2 if X_i is one of them.

2   Prioritized Belief Revision

Suppose that K is a belief set and $\preceq$ a pre-order[2]. We define recursively a hierarchy, $\{K^k\}_{k=1}^{+\infty}$, of K with respect to the ordering $\preceq$ as follows:

1. $K^1 = \{\varphi \in K : \neg\exists\psi \in K(\varphi \prec \psi)\}$; $T^1 = K\backslash K^1$.
2. $K^{k+1} = \{\varphi \in T^k : \neg\exists\psi \in T^k(\varphi \prec \psi)\}$; $T^{k+1} = T^k\backslash K^{k+1}$.

where $\varphi \prec \psi$ denotes $\varphi \preceq \psi$ and $\psi \npreceq \varphi$. The intuition behind the construction is that each time collects all maximal elements and remove them from the current set.

We will write $K^{\leq l}$ to denote $\bigcup_{k=1}^{l} K^k$. The following lemma shows that the hierarchy can only be finite if $\preceq$ satisfies the following *logical constraint*:

(LC) If $\varphi_1, \cdots, \varphi_n \vdash \psi$, $\min\{\varphi_1, \cdots, \varphi_n\} \preceq \psi$.

It is easy to see that such an order can induce an AGM epistemic entrenchment and vice versa[3]. Therefore such an ordering will be referred to as a *epistemic entrenchment(EE) ordering*. The following lemma is easy to verify and will be used intensively throughout the paper.

---

[1] This assumption is essential because Zhang and Zhang's construction of bargaining solution is syntax-dependent(logically equivalent inputs could result different outcomes). Without the assumption, even the most fundamental postulates, such as *Extensionality*, cannot be satisfied.

[2] A pre-order over a set we mean in this paper is a complete ordering over the set which is transitive and reflexive.

Lemma 1. *Let K be a belief set and $\preceq$ a pre-order over K which satisfies (LC), then*

1. *for any l, $K^{\leq l}$ is a belief set.*

2. *There exists a number N such that $K = \bigcup_{k=1}^{N} K^k$.*

Let O be any set of sentences in $\mathcal{L}$, we define the *degree of coverage* of O over K, denoted by $\rho_K(O)$, to be the greatest number l that satisfies $K^{\leq l} \subseteq O$.

It is well-known that an AGM belief revision operator can be uniquely determined by an epistemic entrenchment ordering. Similar to [16], we will define a belief revision function based on the idea of maximizing retainment of most entrenched beliefs.

By following the convention introduced by Nebel[8], for any belief set K, a set of sentences F, and an EE ordering $\preceq$ over K, we define $K \Downarrow F$ as follows: for any $H \in K \Downarrow F$,

1. $H \subseteq K$;
2. for all $k = 1, 2, \cdots$, $H \cap K^k$ is set-inclusion maximal among the subsets of K^k such that $\bigcup_{j=1}^{k} (H \cap K^j) \cup F$ is consistent.

We call $\otimes$ a *prioritized revision function* over $(K, \preceq)$ if it is defined as follows:

$$K \otimes F \overset{def}{=} \bigcap_{H \in K \Downarrow F} Cn(H) + F.$$

Lemma 2. [Nebel 1992] $\otimes$ *satisfies all AGM postulates.*

3   Belief-Revision-Based Bargaining Solution

Now we redefine the bargaining solution given in [16]. Different from their work, we will define a bargaining game as a pair of prioritized belief sets rather than a pair of prioritized belief bases. We will see that this change results a significant differences in logical properties.

Definition 1. *A bargaining game is a pair of prioritized belief sets $((K_1, \preceq_1), (k_2, \preceq_2))$, where K_i is a belief set in $\mathcal{L}$ and $\preceq_i$ $(i = 1, 2)$ is an EE ordering over K_i.*

The definition of deals remains the same as in [16].

Definition 2. *Let $B = ((K_1, \preceq_1), (K_2, \preceq_2))$ be a bargaining game. A deal of B is a pair (D_1, D_2) satisfying the following two conditions: for each $i = 1, 2$,*

1. $D_i \subseteq K_i$;
2. *for each $k = 1, 2, \cdots$, $D_i \cap K_i^k$ is set-inclusion maximal among the subsets of K_i^k such that $\bigcup_{j=1}^{k} (D_i \cap K_i^j) \cup D_{-i}$ is consistent.*

The set of all deals of B is denoted by $\Omega(B)$.

We remark that since K_1 and K_2 are belief sets, it is easy to prove that for any deal (D_1, D_2), both D_1 and D_2 are logically closed. This property will play a key role in the proofs of theorems in Section 4.

Definition 3. *For any bargaining game $B = ((K_1, \preceq_1), (K_2, \preceq_2))$, we call $\Phi = (\Phi_1, \Phi_2)$ the core of the game if*

$$\Phi_1 \stackrel{def}{=} \bigcap_{(D_1, D_2) \in \gamma(B)} D_1, \quad \Phi_2 \stackrel{def}{=} \bigcap_{(D_1, D_2) \in \gamma(B)} D_2$$

where

$$\gamma(B) = \{D \in \Omega(B) : \rho_B(D) = \rho_B\}$$
$$\rho_B(D) \stackrel{def}{=} \min\{\rho_{K_1}(D_1), \rho_{K_2}(D_2)\}$$
$$\rho_B \stackrel{def}{=} \max\{\rho_B(D) : D \in \Omega(B)\}$$

It is easy to see that $\gamma(B)$ represents the subset of $\Omega(B)$ that contains the deals with the highest degree of coverage over all deals in $\Omega(B)$. The min-max construction of the core captures the idea that the final agreement should maximally and evenly satisfy both agents's demands(see [16]).

Now we can finalize the reconstruction of bargaining solution.

Definition 4. *A bargaining solution is a function $\mathbf{A}$ which maps a bargaining game to a set of sentences (agreement), defined as follows. For each bargaining game $B = ((K_1, \preceq_1), (K_2, \preceq_2))$*

$$\mathbf{A}(B) \stackrel{def}{=} (K_1 \otimes_1 \Phi_2) \cap (K_2 \otimes_2 \Phi_1) \tag{1}$$

where (Φ_1, Φ_2) is the core of B and $\otimes_i$ is the prioritized revision function over $(K_i, \preceq_i)$.

We call $\mathbf{A}(B)$ (sometimes we write it as $\mathbf{A}(K_1, K_2)$) an *agreement function*. It is easy to see that the outcomes of an agreement function do not depend on the syntax of its inputs. However, if the bargaining solution defined in [16] takes belief sets as inputs, it will give exactly the same outcomes as the above definition. In such a sense, the logical properties we discuss in the following section can be viewed as the idealized properties of the bargaining solution defined in [16].

4   Logical Properties of Bargaining Solution

In this section, we will present a set of logical properties of the bargaining solution we introduced in the previous section. We will show that the solution satisfies most desired properties for agreement functions and negotiation functions. To establish these properties, we need a few technical lemmas. Note that none of the lemmas holds without the assumption of the logical closeness of belief sets.

Lemma 3. *Given a bargaining game $B = ((K_1, \preceq_1), (K_2, \preceq_2))$, let $\pi_{max} = \max\{k : K_1^{\leq k} \cup K_2^{\leq k}$ is consistent$\}$ and $(\Psi_1, \Psi_2) = (K_1^{\leq \pi_{max}}, K_2^{\leq \pi_{max}})$. Then $(\Phi_1, \Phi_2) = (K_1 \cap (\Psi_1 + \Psi_2), K_2 \cap (\Psi_1 + \Psi_2))$.*

Proof. Before we prove the main result of the lemma, we first show that $\rho_B = \pi_{max}$. For any deal $D = (D_1, D_2)$ such that $K_1^{\leq \pi_{max}} \subseteq D_1$ and $K_2^{\leq \pi_{max}} \subseteq D_2$, we have $\rho_B(D) \geq \pi_{max}$. Thus $\rho_B \geq \pi_{max}$. On the other hand, for any $D \in \gamma(B)$, $\rho_B(D) = \rho_B$, which means that $\rho_{K_1}(D_1) \geq \rho_B$ and $\rho_{K_2}(D_2) \geq \rho_B$. According to the definition of degree of coverage of a deal, $K_1^{\leq \rho_{K_1}(D_1)} \subseteq D_1$ and $K_2^{\leq \rho_{K_2}(D_2)} \subseteq D_2$. So either $\rho_{K_1}(D_1) \leq \pi_{max}$ or $\rho_{K_2}(D_2) \leq \pi_{max}$ as $D_1 \cup D_2$ is consistent. Therefore $\rho_B \leq \min\{\rho_{K_1}(D_1), \rho_{K_2}(D_2)\} \leq \pi_{max}$. We have proved that $\rho_B = \pi_{max}$.

Obviously if $K_1 \cup K_2$ is consistent, then $(\Phi_1, \Phi_2) = (K_1, K_2)$. Therefore we will assume that $K_1 \cup K_2$ is inconsistent. We only prove $\Phi_1 = K_1 \cap (\Psi_1 + \Psi_2)$. The second component is similar.

For any $(D_1, D_2) \in \gamma(B)$, we have $\Psi_1 \subseteq D_1$ and $\Psi_2 \subseteq D_2$. In fact, we can prove that $K_1 \cap (\Psi_1 + \Psi_2) \subseteq D_1$. If it is not the case, there exists a sentence $\varphi \in K_1 \cap (\Psi_1 + \Psi_2)$ but $\varphi \notin D_1$. On one hand, $\varphi \in K_1 \cap (\Psi_1 + \Psi_2)$ implies that $D_1 \cup D_2 \vdash \varphi$. On the other hand, $\varphi \notin D_1$ implies that $\{\varphi\} \cup D_1 \cup D_2$ is inconsistent (otherwise D_1 will include φ). It follows that $D_1 \cup D_2 \vdash \neg\varphi$. Therefore $D_1 \cup D_2$ is inconsistent, a contradiction. We have proved that for any deal $(D_1, D_2)\gamma(B)$, $K_1 \cap (\Psi_1 + \Psi_2)$. Thus $K_1 \cap (\Psi_1 + \Psi_2) \subseteq \bigcap_{(D_1, D_2) \in \gamma(B)} D_1 = \Phi_1$.

Now we prove that $\Phi_1 \subseteq K_1 \cap (\Psi_1 + \Psi_2)$. To this end, we assume that $\varphi \in \Phi_1$. If $\varphi \notin \Psi_1 + \Psi_2$, we have $\{\neg\varphi\} \cup \Psi_1 \cup \Psi_2$ is consistent. On the other hand, since $K_1^{\leq \pi_{max}+1} \cup K_2^{\leq \pi_{max}+1}$ is inconsistent, there exists a sentence $\psi \in K_1^{\leq \pi_{max}+1}$ such that $\neg\psi \in K_2^{\leq \pi_{max}+1}$ (because both $K_1^{\leq \pi_{max}+1}$ and $K_2^{\leq \pi_{max}+1}$ are logically closed). Since $\{\neg\varphi\} \cup \Psi_1 \cup \Psi_2$ is consistent, there is a deal $(D_1, D_2) \in \gamma(B)$ such that $\{\neg\varphi \vee \psi\} \cup \Psi_1 \subseteq D_1$ and $\{\neg\varphi \vee \neg\psi\} \cup \Psi_2 \subseteq D_2$. We know that $\varphi \in \Phi_1$, so $\varphi \in D_1 + D_2$. Thus $\psi \wedge \neg\psi \in D_1 + D_2$, a contradiction.   ¶

Lemma 4. *Assume that Ψ_1, Ψ_2 and π_{max} are defined as Lemma 3. Then*

$$K_1 \otimes_1 \Phi_2 = K_1 \otimes_1 \Psi_2 \text{ and } K_2 \otimes_2 \Phi_1 = K_2 \otimes_2 \Psi_1$$

where $\otimes_i$ is the prioritized revision function over $(K_i, \preceq_i)$.

Proof. We only present the proof for the first statement. The second one is similar.

First it is easy to prove that $\Psi_1 + \Psi_2 \subseteq K_1 \otimes_1 \Phi_2$. It follows that $K_1 \otimes_1 \Phi_2 = K_1 \otimes_1 \Phi_2 + (\Psi_1 + \Phi_2)$. On the other hand, according to Lemma 3, we have $\Phi_2 \subseteq \Psi_1 + \Psi_2$. Since $\otimes_1$ satisfies the AGM postulates, we then have $K_1 \otimes_1 \Phi_2 + (\Psi_1 + \Phi_2) = K_1 \otimes_1 (\Psi_1 + \Psi_2)$. Therefore $K_1 \otimes_2 \Phi_2 = K_1 \otimes_1 (\Psi_1 + \Psi_2)$. In addition, it is easy to prove that $\Psi_1 \subseteq K_1 \otimes_1 \Psi_2$. By AGM postulates again, we have $K_1 \otimes_1 \Psi_2 = K_1 \otimes_1 \Psi_2 + \Psi_1 = K_1 \otimes_1 (\Psi_1 + \Psi_2)$. Therefore $K_1 \otimes_1 \Phi_2 = K_1 \otimes_1 \Psi_2$.¶

Lemma 5. *Assume that Ψ_1, Ψ_2 and π_{max} are defined as Lemma 3. Let*

$\Psi'_1 = K_1^{\leq \pi^1_{max}}$, where $\pi^1_{max} = \max\{k : K_1^{\leq k} \cup K_2^{\leq \pi_{max}} \text{ is consistent}\}$,
$\Psi'_2 = K_2^{\leq \pi^2_{max}}$, where $\pi^2_{max} = \max\{k : K_1^{\leq \pi_{max}} \cup K_2^{\leq k} \text{ is consistent}\}$.
Then

$$K_1 \otimes_1 \Phi_2 = \Psi'_1 + \Psi_2 \text{ and } K_2 \otimes_2 \Phi_1 = \Psi_1 + \Psi'_2.$$

where $\otimes_i$ is the prioritized selection revision over $(K_i, \preceq_i)$.

Proof. We only present the proof for the first statement. The second one is similar. According to Lemma 4, $K_1 \otimes_1 \Phi_2 = K_1 \otimes_1 \Psi_2$. Therefore we only need to prove that $K_1 \otimes_1 \Psi_2 = \Psi'_1 + \Psi_2$.

If $\Psi'_1 = K_1$, $K_1 \cup \Psi_2$ is consistent. Then $K_1 \otimes_1 \Psi_2 = K_1 + \Psi_2$. We have $K_1 \otimes_1 \Psi_2 = K_1 + \Psi_2 = \Psi'_1 + \Psi_2$, as desired. If $\Psi'_1 \neq K_1$, according to the definition of π^1_{max}, we have $K_1^{\leq \pi^1_{max}+1} \cup K_2^{\leq \pi_{max}}$ is inconsistent. It follows that there exists a sentence $\varphi \in K_1^{\leq \pi^1_{max}+1}$ such that $\neg\varphi \in K_2^{\leq \pi_{max}} = \Psi_2$ (note that both $K_1^{\leq \pi^1_{max}+1}$ and $K_2^{\leq \pi_{max}}$ are logically closed). Now we can come to the conclusion that $\Psi'_1 + \Psi_2 = K_1 \otimes_1 \Psi_2$. In fact, by the construction of prioritized belief revision, we can easily verify that $\Psi'_1 + \Psi_2 \subseteq K_1 \otimes_1 \Psi_2$. To prove the other direction of inclusion, we assume that $\psi \in K_1 \otimes_1 \Psi_2$. If $\psi \notin \Psi'_1 + \Psi_2$, then $\{\neg\varphi\} \cup \Psi'_1 \cup \Psi_2$ is consistent. So is $\{\neg\psi \vee \varphi\} \cup \Psi'_1 \cup \Psi_2$. Notice that $\neg\psi \vee \varphi \in K_1^{\leq \pi^1_{max}+1}$. There exists $H \in K_1 \Downarrow \Psi_2$ such that $\{\neg\psi \vee \varphi\} \cup \Psi'_1 \subseteq H$. Since $\psi \in K_1 \otimes_1 \Psi_2$ and H is logically closed, we have $\varphi \in H$, which contradicts the consistency of $H \cup \Psi_2$. Therefore $K_1 \otimes_1 \Psi_2 \subseteq \Psi'_1 + \Psi_2$. ¶

Having the above lemmas, the verification of the following theorems becomes much easier.

The first theorem shows that the calculation of agreement function can be transferred to the calculation of maximal consistent hierarchies of two belief sets.

Theorem 1. *For any bargaining game* $B = ((K_1, \preceq_1), (K_2, \preceq_2))$,

$$\mathbf{A}(B) = (\Psi'_1 + \Psi_2) \cap (\Psi_1 + \Psi'_2) \tag{2}$$

where Ψ_1, Ψ'_1, Ψ_2 *and* Ψ'_2 *are defined as Lemma 5.*

Proof. Straightforward from Lemma 5. ¶

We remark that the computation of Ψ_1, Ψ'_1, Ψ_2 and Ψ'_2 has much less cost than the calculation of the core of a game. This explains why we could reduce the computational complexity of bargaining solution significantly (see Section 5).

The following theorem shows the basic logical properties of agreement function.

Theorem 2. *Let* **A** *is a bargaining solution. For any bargaining game* $B = ((K_1, \preceq_1), (K_2, \preceq_2))$, *the following properties hold:*

(A1) $\mathbf{A}(K_1, K_2) = Cn(\mathbf{A}(K_1, K_2))$.
(A2) $\mathbf{A}(K_1, K_2)$ *is consistent.*

(A3) $\mathbf{A}(K_1, K_2) \subseteq K_1 + K_2$.
(A4) *If $K_1 \cup K_2$ is consistent, then $K_1 + K_2 \subseteq \mathbf{A}(K_1, K_2)$.*

Given Theorem 1, the verification of the above properties is trivial. One may notice that the above properties are similar to the postulates introduced in [4] for negotiation outcomes. In fact, our agreement function satisfies all the postulates for negotiation outcomes except

(O4) $K_1 \cap K_2 \subseteq A(K_1, K_2)$ or $A(K_1, K_2) \cup (K_1 \cap K_2) \models \bot$.

The following is a counterexample.

Example 1. *Let $K_1 = Cn(\{p, \neg q, r\})$ with the hierarchy $K_1^1 = Cn(\{p\})$ and $K_1^2 = K_1 \backslash K_1^1$. Let $K_2 = Cn(\{q, \neg p, r\})$ with the hierarchy $K_2^1 = Cn(\{q\})$ and $K_2^2 = K_2 \backslash K_2^1$. Then $\Psi_1 = \Psi_1' = Cn(\{p\})$ and $\Psi_2 = \Psi_2' = Cn(\{q\})$. It follows that $A(K_1, K_2) = Cn(\{p, q\})$. Obviously **(O4)** does not hold for this example.*

The reason that both agents give up their common demand r is the following. If an agent demands $\neg p$ or $\neg q$, the agent needs to commit herself to accept its logical consequence $\neg p \vee \neg r$ or $\neg q \vee \neg r$, respectively. Since r is less entrenched than the commitment by both agents, they have to give up r in order to reach the agreement $Cn(\{p, q\})$. If the agents do not mean that, it should be explicitly expressed in the initial demands.

The following theorem shows that with our construction of bargaining solution, we can define a negotiation function which is similar to the negotiation function introduced by Zhang *et al.* in [14].

Theorem 3. Let N be a function defined as follows:

$$N(K_1, K_2) = (K_1 \otimes_1 \Phi_2, K_2 \otimes_2 \Phi_1)$$

where (Φ_1, Φ_2) is the core of the bargaining game $B = ((K_1, \preceq_1), (K_2, \preceq_2))$. Then N has the following properties (N_i is the i-th component of N):

(N1) $N_i(K_1, K_2) = Cn(N_i(K_1, K_2))$. *(Closure)*
(N2) $N_i(K_1, K_2) \subseteq K_1 + K_2$. *(Inclusion)*
(N3) If $K_1 \cup K_2$ is consistent, then $K_1 + K_2 \subseteq N_i(K_1, K_2)$. *(Vacuity)*
(N4) If $K_i \cup N_i(K_1, K_2)$ is consistent, then $K_i \subseteq N_i(K_1, K_2)$. *(Consistent Expansion)*
(N5) If $K_i \cup N_{-i}(K_1, K_2)$ is consistent, then $N_{-i}(K_1, K_2) \subseteq N_i(K_1, K_2)$. *(Safe Expansion)*

Proof. The proof of (N1)-(N4) is trivial by Theorem 1. For (N5), since $K_i \cup (\Psi_{-i}' + \Psi_i)$ is consistent, we have $\pi_{max}^{-i} = \pi_{max}$. It follows that $N_{-i}(K_1, K_2) = \Psi_1 + \Psi_2$. Therefore $N_{-i}(K_1, K_2) \subseteq N_i(K_1, K_2)$. ¶

We can see that the properties of *Closure, Inclusion, Vacuity and Consistent* are exactly the same as the corresponding postulates in [14]. The postulate *Extensionality* is trivially true for our definition. The postulates *Inconsistency and Iteration* are not applicable in our case because we do not consider inconsistent inputs and iteration operations. The following example illustrates that the postulate *No Recantation* is invalid for our definition of negotiation function.

Example 2. *Let $K_1 = Cn(\{p, \neg q, r\})$ with the hierarchy $K_1^1 = Cn(\{p\})$, $K_1^2 = Cn(\{p, \neg q\}) \backslash K_1^1$ and $K_1^3 = K_1 \backslash K_1^{\leq 2}$. Let $K_2 = Cn(\{q, r\})$ with the hierarchy $K_2^1 = Cn(\{q\})$ and $K_2^2 = K_2 \backslash K_2^1$. Then $\Psi_1 = \Psi_1' = Cn(\{p\})$, $\Psi_2 = Cn(\{q\})$ and $\Psi_2' = Cn(\{q, r\})$. It follows that $N_1(K_1, K_2) = Cn(\{p, q\})$ and $N_2(K_1, K_2) = Cn(\{p, q, r\})$. Therefore $r \in K_1 \cap N_2(K_1, K_2)$ but $r \notin N_1(K_1, K_2)$.*

As a weak version of *No Recantation* the property of *Safe Expansion* says that if an agent initiatively gives up all the demands which conflicts to the other agent, the other agent should accept all the consistent demands from the first agent.

5 Computational Complexity

In [16], Zhang and Zhang shows that the complexity of belief-revision-based bargaining solutions is Π_2^P-hard. This result is based on the syntax-dependent construction of bargaining solution.

Theorem 4. [Zhang and Zhang 2006] *Let B be a bargaining game and φ a formula. Deciding whether $\mathbf{A}(B) \vdash \varphi$ is Π_2^P-hard, where $\mathbf{A}(B)$ is the agreement function defined in [16].*

In this section, we will show that the complexity can be reduced if we use the syntax-independent construction of agreement function.

We assume that readers are familiar with the complexity classes of P, NP, coNP, $\Delta_2^P[\mathcal{O}(\log n)]$, Δ_2^P and Π_2^P. It is well known that $P \subseteq NP \subseteq \Delta_2^P[\mathcal{O}(\log n)] \subseteq \Delta_2^P \subseteq \Pi_2^P$, and these inclusions are generally believed to be proper (readers may refer to [10] for further details).

Given a bargaining game $B = ((K_1, \preceq_1), (K_2, \preceq_2))$, since K_1 and K_2 are logically closed, they are infinite sets even though the language we consider is finite. To make computation possible, we assume that equivalent statements are represented by only one sentence, so a belief set can be finite[3]. In such a sense, we will refer a bargaining game B to a pair of prioritized belief sets, $((X_1, \preceq_1), (X_2, \preceq_2))$, where X_i is finite sets of sentences and $\preceq_i$ is a pre-order over X_i which satisfies logic constraint (LC). According to Lemma 1, for each $i = 1, 2$, we can always write $X_i = X_i^1 \cup \cdots \cup X_i^m$, where $X_i^k \cap X_i^l = \emptyset$ for any $k \neq l$. Also for each $k < m$, if a formula $\varphi \in X_i^k$, then there does not exist a $\psi \in X_i^l$ ($k < l$) such that $\varphi \prec_i \psi$. Therefore, for the convenience of our complexity analysis, in the rest of this section, we will specify a bargaining game as $B = (X_1, X_2)$, where $X_1 = \bigcup_{i=1}^m X_1^i$ and $X_2 = \bigcup_{j=1}^n X_2^j$, and $X_1^1, \cdots, X_1^m$, and $X_2^1, \cdots, X_2^n$ are the partitions of X_1 and X_2 respectively and satisfy the property mentioned above. According to Theorem 1, we can define an agreement function by Equation (2) as

$$\mathbf{A}(B) = (\Psi_1' + \Psi_2) \cap (\Psi_1 + \Psi_2') \tag{3}$$

---

[3] In fact, the complexity results presented in this section do not require the belief sets in a bargaining game to be logically closed. We can view Equation (2) as the approximation of bargaining solution.

where

$$(\Psi_1, \Psi_2) = (\bigcup_{i=1}^{\pi} X_1^i, \bigcup_{i=1}^{\pi} X_2^i);$$

$$(\Psi_1', \Psi_2') = (\bigcup_{i=1}^{\pi_1} X_1^i, \bigcup_{i=1}^{\pi_2} X_2^i);$$

$$\pi = \max\{k : (\bigcup_{i=1}^{k} X_1^i) \cup (\bigcup_{i=1}^{k} X_2^i) \text{ is consistent}\};$$

$$\pi_1 = \max\{k : (\bigcup_{i=1}^{k} X_1^i) \cup (\bigcup_{i=1}^{\pi} X_2^i) \text{ is consistent}\};$$

$$\pi_2 = \max\{k : (\bigcup_{i=1}^{\pi} X_1^i) \cup (\bigcup_{i=1}^{k} X_2^i) \text{ is consistent}\}.$$

Theorem 5. *Let $B = (X_1, X_2)$ be a bargaining game and φ a formula. Deciding whether $\mathbf{A}(B) \vdash \varphi$ is $\Delta_2^P[\mathcal{O}(\log n)]$-complete.*

Proof. Membership proof. Let $B = (X_1, X_2)$, where $X_1 = \bigcup_{i=1}^{m} X_1^i$ and $X_2 = \bigcup_{i=1}^{n} X_2^j$. Firstly, we can find a maximal k with a binary search such that if $\bigcup_{i=1}^{k} X_1^k \cup \bigcup_{i=1}^{k} X_2^k$ is consistent but $\bigcup_{i=1}^{k+1} X_1^i \cup \bigcup_{i=1}^{k+1} X_2^i$ is no longer consistent. Obviously, we will need $\mathcal{O}(\log n)$ times search. Secondly, we fix $\bigcup_{i=1}^{k} X_1^i$, and find a maximal p such that $\bigcup_{i=1}^{k} X_1^i \cup \bigcup_{j=1}^{p} X_2^j$ is consistent but $\bigcup_{i=1}^{k} X_1^k \cup \bigcup_{j=1}^{p+1} X_2^j$ is not inconsistent. Note that we should have $k \leq p$. Also, this can be done with a binary search in time $\mathcal{O}(\log n)$. In a similar way, we can fix $\bigcup_{i=1}^{k} X_2^i$ and find a maximal q satisfying that $\bigcup_{i=1}^{q} X_1^i \cup \bigcup_{j=1}^{k} X_2^j$ is consistent and $\bigcup_{i=1}^{q+1} X_1^i \cup \bigcup_{j=1}^{k} X_2^i$ is not consistent. So we can compute $\mathbf{A}(B)$ using a deterministic Turing machine with $\mathcal{O}(\log n)$ queries to an NP oracle. Finally, we check the consistency of $\mathbf{A}(B) \cup \{\neg\varphi\}$ with one query to an NP oracle. So the problem is in $\Delta_2^P[\mathcal{O}(\log n)]$.

Hardness proof. By restrict $B = (X_1, X_2)$ where $X_1 = \{\varphi_1\} \cup \cdots \cup \{\varphi_n\}$ and $X_2 = \{\psi\}$. Then it is easy to see that our bargaining problem is identical to the cut base revision, which implies that deciding whether $\mathcal{A}(B) \vdash \varphi$ is $\Delta_2^P[\mathcal{O}(\log n)]$-hard [7]. ¶

The following result shows that if we restricts the language to be Horn clauses, the decision problem of bargaining solution is tractable.

Theorem 6. *Let B be a bargaining game and φ a formula where all formulas occurring in B and φ are Horn clauses. Deciding whether $\mathbf{A}(B) \vdash \varphi$ is in P.*

6 Conclusion and Related Work

In this paper, we have presented a set of logical properties of belief-revision-based bargaining solution. By representing bargaining game as a pair of prioritized belief sets, the computation of bargaining solution can be converted to the construction of maximal consistent hierarchy of two agents' belief sets. Based on the result, we have shown that the agreement function and negotiation function defined by the bargaining solution satisfies most of postulates introduced in the literature. Our complexity analysis indicates that in general the computation of belief-revision-based bargaining solution can be reduced or be approximated to $\Delta_2^P[\mathcal{O}(\log n)]$-complete.

This work is closed related to [16]. In fact, we can view the syntax-independent bargaining solution is a special case of syntax-dependent bargaining solution when bargaining games consists of belief sets. Although the assumption of logical closeness is just an idealized case, it is essential to disclose the logical properties behind the negotiation reasoning. We have shown that our solution to negotiation problem is different from the axiomatic approaches [4,5,14]. Since these formalisms are all based on the assumption of logical closeness, it is possible to apply our approach to develop a concrete construction for their negotiation functions.

References

1. E. Alchourrón,P. Gärdenfors and Makinson, On the logic of theory change: partial meet contraction and revision functions, *The Journal of Symbolic Logic*, 50(2)(1985), 510-530.
2. R. Booth,A negotiation-style framework for non-prioritised revision. *TARK: Theoretical Aspects of Reasoning about Knowledge*, 137-150, 2001.
3. P. Gärdenfors and D. Makinson, Revisions of knowledge systems using epistemic entrenchment, In *Proceedings of TARK*, 83-95, 1988.
4. T. Meyer, N. Foo, R. Kwok and D. Zhang, Logical foundations of negotiation: outcome, concession and adaptation, *Proceedings of the 19th National Conference on Artificial Intelligence (AAAI-04)*, 293-298, 2004.
5. T. Meyer, N. Foo, R. Kwok and D. Zhang, Logical foundations of negotiation: strategies and preferences, *Proceedings of the 9th International Conference on the Principles of Knowledge Representation and Reasoing(KR'04)*, 311-318, 2004.
6. A. Muthoo, 1999. *Bargaining Theory with Applications*, Cambridge University Press, 1999.
7. B. Nebel, How Hard is it to Revise a Belief Base?, in D. Dubois and H. Prade (eds.), *Handbook of Defeasible Reasoning and Uncertainty Management Systems*, Vol. 3: Belief Change , Kluwer Academic, 77-145, 1998.
8. B. Nebel, 1992. Syntax-Based Approaches to Belief Revision, in: P. Grdenfors (eds.), *Belief Revision*, Cambridge University Press, 52-88, 1992.
9. M. J. Osborne and A. Rubinstein, 1990. *Bargaining and Markets*, Academic Press, 1990.
10. C.H. Papadimitriou, 1994. *Computational Complexity*. Addison Wesley, 1994.
11. A. Roth, 1979. *Axiomatic Models of Bargaining*, Springer-Verlag, 1979.
12. A. Rubinstein, 2000. *Economics and Language: Five Essays*, Cambridge University Press, 2000.
13. D. Zhang and N. Foo, 2001. Infinitary belief revision,*Journal of Philosophical Logic*, 30(6):525-570, 2001.

14. D. Zhang, N. Foo, T. Meyer and R. Kwok, 2004. Negotiation as mutual belief revision, *Proceedings of the 19th National Conference on Artificial Intelligence (AAAI-04)*, 317-322, 2004.
15. D. Zhang, 2005. A logical model for Nash bargaining solution, *Proceedings of the Nineteenth International Joint Conference on Artificial Intelligence (IJCAI-05)*, 983-988, 2005.
16. D. Zhang and Y. Zhang, A computational model of logic-based negotiation, *Proceedings of the 21st National Conference on Artificial Intelligence (AAAI-06)*, 728-733, 2006.

Knowledge Compilation for Belief Change

Maurice Pagnucco

ARC Centre of Excellence for Autonomous Systems and National ICT Australia, School of
Computer Science and Engineering, The University of New South Wales, Sydney, NSW 2052,
Australia
morri@cse.unsw.edu.au
http://www.cse.unsw.edu.au/~morri/

Abstract. Techniques for knowledge compilation like prime implicates and binary decision diagrams (BDDs) are effective methods for improving the practical
efficiency of reasoning tasks. In this paper we provide a construction for a belief
contraction operator using prime implicates. We also briefly indicate how this
technique can be used for belief expansion, belief revision and also iterated belief change. This simple yet novel technique has two significant features: (a) the
contraction operator constructed satisfies all the AGM postulates for belief contraction; (b) when compilation has been effected only syntactic manipulation is
required in order to contract the reasoner's belief state.

Keywords: knowledge representation and reasoning, belief revision and update,
common-sense reasoning.

Transforming the sentences of a knowledge base KB into an alternate, regular form
KB' has proven to be an effective technique in improving the efficiency of reasoning
tasks. This idea is referred to as *knowledge compilation* and has been widely applied
to problems in truth maintenance, constraint satisfaction, satisfiability, planning and
scheduling, etc. Some of the more commonly used transformations for knowledge compilation include prime implicates and implicants, binary decision diagrams (BDDs) and
decomposable negation normal form (see [2] for an introduction and analysis of these
methods).

We shall concern ourselves with prime implicates and apply this compilation method
to constructing an operator for belief change where the idea is to maintain the beliefs
of a reasoner as new information is acquired. In this paper we show how to construct
a belief contraction operator using prime implicates and also indicate how this can be
used for belief expansion and belief revision as well. The elegance of the technique
that we describe has two significant advantages: (i) once compilation has been effected
only syntactic manipulation is required in order to contract the reasoner's belief state,
and; (ii) we do not sacrifice theory for the sake of implementation like many syntactic
methods but remain loyal to the AGM postulates for belief contraction. Furthermore,
we briefly indicate how this technique can be used to implement iterated belief change.

Knowledge compilation techniques, including prime implicates have been used in
implementing belief maintenance systems in the past. The closest work to our own
is that of Gorogiannis and Ryan [8] who implement AGM belief change using BDDs,
analysing the complexity of these operators in specific instances. However, their method

A. Sattar and B.H. Kang (Eds.): AI 2006, LNAI 4304, pp. 90–99, 2006.

© Springer-Verlag Berlin Heidelberg 2006

does not give a natural way to achieve iterated belief change. Prime implicates have been suggested for truth maintenance in the work of Reiter and de Kleer [13] as well as that of Kean [11]. Due to the lack of preferential information, it is difficult to implement general forms of belief change and iterated belief change in the frameworks they develop. Implementations of AGM belief change such as those by Dixon and Wobcke [4] and Williams [18] rely on theorem proving which we try to do away with once the reasoner's epistemic stated has been compiled.

The rest of this paper is set out as follows. The next section covers the requisite logical background for prime implicates and belief change. Section 2 explains how prime implicates are utilised to represent the reasoner's epistemic state. The construction of belief contraction operators is described in Sections 3 and 4. Our conclusions together with a brief discussion of how this construction can be used for belief expansion, belief revision and iterated belief change are presented in Section 5.

1   Background

We assume a fixed finite object language $\mathcal{L}$ with standard connectives $\neg$, $\wedge$, $\vee$, $\rightarrow$, $\equiv$ as well as the logical constants $\top$ (truth) and $\perp$ (falsum). Cn represents the classical consequence operator while $\vdash$ represents the classical consequence relation (i.e., $\Gamma \vdash \phi$ iff $\phi \in Cn(\Gamma)$). We also adopt the following linguistic conventions to simplify the presentation. Lower case Greek letters ϕ, ψ, χ, ... denote sentences of $\mathcal{L}$. Upper case Greek letters Δ, Γ, ... denote sets of formulas. Theories (closed under logical consequence) are used to represent reasoners' belief states in AGM belief change and will be denoted by upper case Roman characters H, K, ... (i.e., $K = Cn(K)$). The inconsistent belief set is denoted $K_\perp$. Lower case Roman characters l, k, ... denote literals (both positive and negative). Upper case Roman characters C, D, ... denote clauses.

1.1   Prime Implicates

Prime implicates are minimal length clauses (in terms of set inclusion) implied by a knowledge base. Transforming a knowledge base into a set of prime implicates gives a uniform way of equivalently expressing the sentences in the knowledge base that can be exploited to enhance computational efficiency.

Definition 1. *A clause C is an* implicate *of a set of clauses Γ iff $\Gamma \vdash C$. A non-tautologous clause C is a* prime implicate *of a set of clauses Γ iff C is an implicate of Γ, and $\Gamma \not\vdash D$ for any $D \vdash C$ and $D \neq C$.*

When Γ is a single clause, representing clauses as sets of literals allows us to simplify this definition by replacing $\vdash$ by $\subset$ (where $\subset$ denotes proper set inclusion). This will simplify the exposition below. We shall denote the set of prime implicates of an arbitrary set of sentences Γ by $\Pi(\Gamma)$ and note that these sets are logically equivalent.

Proposition 1. *Given a set of clauses Γ, $\Gamma \vdash \phi$ iff $\Pi(\Gamma) \vdash \phi$.*

When Γ is a set of clauses, prime implicates are easily computed by repeated application of resolution and removing subsumed (i.e., implied) clauses [17]. This basic strategy can be improved upon by using incremental methods [12,10] that do not require the implicates to be re-computed when new sentences are added to the knowledge base and through the use of efficient data structures [13] for subsumption checking. Compiling a knowledge base into prime implicate form can lead to an exponential number of implicates in the number of atoms (see [15]).

1.2   AGM Belief Change

Belief change takes the following model as its departure point. A reasoner is in some state of belief when it acquires new information. Belief change studies the manner in which this newly acquired information can be assimilated into its current state of belief in a rational way. We base our account here on the popular framework developed by Alchourrón, Gärdenfors and Makinson (AGM) [1,5,7]. In the AGM the reasoner's belief state is modelled as a deductively closed set of sentences referred to as a *belief set*. That is, for a belief set K, $K = Cn(K)$. In this way, the reasoner has one of three *attitudes* towards a sentence $\phi \in \mathcal{L}$: ϕ is *believed* when $\phi \in K$; ϕ is *disbelieved* when $\neg\phi \in K$; or the reasoner is *indifferent* to ϕ when $\phi, \neg\phi \notin K$. A belief set can be viewed as representing the sentences the reasoner is committed to believing regardless of whether that commitment can be feasibly realised. Shortly, when we discuss epistemic entrenchment, we will consider belief sets with additional preferential information about the reasoner's beliefs which we will refer to as the reasoner's *epistemic state*. These preferences over beliefs can be seen as offering a greater set of epistemic attitudes to the three outlined above.

The AGM considers three types of transformations on beliefs sets as new information is acquired: *belief expansion* by ϕ in which this new information is added to the initial belief set K without removal of any existing beliefs ($K + \phi$); *belief contraction* by ϕ where belief in ϕ is suspended ($K \dot{-} \phi$); and, *belief revision* by ϕ where ϕ is assimilated into K while existing beliefs may need to be suspended in order to maintain consistency ($K * \phi$). Here we are only concerned with belief contraction and so we will present and motivate the AGM rationality postulates for this operation.

(K$\dot{-}$1) For any sentence ϕ and any belief set K,
  $K \dot{-} \phi$ is a belief set
(K$\dot{-}$2) $K \dot{-} \phi \subseteq K$
(K$\dot{-}$3) If $\phi \notin K$, then $K \dot{-} \phi = K$
(K$\dot{-}$4) If $\nvdash \phi$ then $\phi \notin K \dot{-} \phi$
(K$\dot{-}$5) If $\phi \in K$, $K \subseteq (K \dot{-} \phi) + \phi$
(K$\dot{-}$6) If $\vdash \phi \equiv \psi$, then $K \dot{-} \phi = K \dot{-} \psi$
(K$\dot{-}$7) $K \dot{-} \phi \cap K \dot{-} \psi \subseteq K \dot{-} (\phi \wedge \psi)$
(K$\dot{-}$8) If $\phi \notin K \dot{-} (\phi \wedge \psi)$, then $K \dot{-} (\phi \wedge \psi) \subseteq K \dot{-} \phi$

Postulate (K$\dot{-}$1) stipulates that contraction of a belief set should return a new belief set. (K$\dot{-}$2) states that no beliefs should be added during contraction while (K$\dot{-}$3) says that the belief state is unaltered when there is no need to remove anything. (K$\dot{-}$4) ensures that the new information is removed where possible and (K$\dot{-}$5) that removal followed by addition of the same information results in no change. Postulate (K$\dot{-}$6) says that the syntax of the information is irrelevant. (K$\dot{-}$7) states that any beliefs removed in the contraction of $\phi \wedge \psi$ should be removed when contracting by ϕ alone or contracting by ψ

alone (or possibly both) while (K$\dot{-}$8) that if ϕ is removed when given a choice between removing ϕ or ψ, then whatever beliefs are removed along with ϕ are also removed along with $\phi \wedge \psi$. These fairly intuitive conditions are capable of characterising quite powerful forms of contraction.

1.3   Epistemic Entrenchment

It can be easily verified that the rationality postulates for AGM belief contraction describe a class of functions rather than prescribing a particular method. If we wish to focus on a particular function within this class of rational contractions we need to apply extralogical information (which is the whole point of the study of belief change). The two main constructions are a semantic one based on possible worlds due to Grove [9] and a syntactic one providing preferences over beliefs known as *epistemic entrenchment* due to Gärdenfors and Makinson [6]. Intuitively, epistemic entrenchment orders the reasoner's beliefs with tautologies being most strongly believed as these cannot be given up. An entrenchment relation $\phi \leq \psi$ says that ψ is at least as entrenched as ϕ and expresses the notion that when faced with the choice between giving up ϕ and giving up ψ, the reasoner would prefer to relinquish belief in ϕ. The formal properties of this ordering are:

(EE1)   If $\phi \leq \psi$ and $\psi \leq \gamma$ then $\phi \leq \gamma$
(EE2)   If $\{\phi\} \vdash \psi$ then $\phi \leq \psi$
(EE3)   For any ϕ and ψ, $\phi \leq \phi \wedge \psi$ or $\psi \leq \phi \wedge \psi$
(EE4)   When $K \neq K_\perp$, $\phi \notin K$ iff $\phi \leq \psi$ for all ψ
(EE5)   If $\phi \leq \psi$ for all ϕ then $\vdash \psi$

(EE1) tells us that the relation is transitive. (EE2) states that consequences of belief are at least as entrenched since it is not necessary to give up (all) consequences in order to suspend judgement in the belief itself. Taken together with (EE1) and (EE2), property (EE3) enforces that a conjunction be entrenched at the same level as the least preferred of its conjuncts (i.e., $\phi \wedge \psi = min(\phi, \psi)$). These properties also allow us to state an important property regarding disjunctions—these are at least as entrenched as the most preferred of the two disjuncts (i.e., $\phi \vee \psi \geq max(\phi, \psi)$). This second property will be very important in what follows. Finally, (EE4) says that non-beliefs are least preferred and hence minimally entrenched while (EE5) says that tautologies are most preferred and therefore maximally entrenched.

Putting all of these properties together we see that an entrenchment ordering is a total pre-order—or ranking—of beliefs in which tautologies are most preferred and non-beliefs least preferred. Since the non-beliefs are clumped together as the minimal elements of an epistemic entrenchment ordering, this gives a convenient way of obtaining the belief state of the reasoner. We can simply obtain the belief state as the set of all non-minimal elements of the epistemic entrenchment: $K_\leq = \{\phi : \phi > \perp\}$. It is therefore common, particularly in iterated belief change [3], to refer to an epistemic entrenchment relation $\leq$ as the reasoner's *epistemic state* and we shall adopt this convention here. To be more specific, the technique we introduce in this paper is centred around compiling the reasoner's epistemic state (epistemic entrenchment relation) using prime implicates.

It remains to describe how we can effect belief contraction given an epistemic entrenchment ordering and vice versa. The first part is given by the condition [6]

$$(C \dot{-}) \quad \psi \in K \dot{-} \phi \text{ iff } \psi \in K \text{ and either } \phi < \phi \vee \psi \text{ or } \vdash \phi$$

It states that a belief is retained after contraction when there is independent evidence for retaining it or it is not possible to effect the contraction[1]. There is independent evidence for retaining a belief ψ whenever the disjunction $\phi \vee \psi$ is strictly more entrenched than ϕ (the sentence we are trying to remove)[2]. Moving in the opposite direction and determining an epistemic entrenchment relation from a contraction function is specified by the following condition [6]:

$$(C \leq) \quad \phi \leq \psi \text{ iff } \phi \notin K \dot{-} (\phi \wedge \psi) \text{ or } \vdash \phi \wedge \psi$$

That is, we prefer ψ to ϕ whenever given the choice between giving up one or the other, we choose to give up ϕ.

The correctness of the foregoing properties and conditions is given by the following two results due to Gärdenfors and Makinson.

Theorem 1. *[6]*
If an ordering $\leq$ satisfies (EE1) – (EE5), then the contraction function which is uniquely determined by $(C \dot{-})$ satisfies $(K \dot{-} 1) – (K \dot{-} 8)$ as well as condition $(C \leq)$.

Theorem 2. *[6]*
If a contraction function $\dot{-}$ satisfies $(K \dot{-} 1) – (K \dot{-} 8)$, then the ordering that is uniquely determined by $(C \leq)$ satisfies (EE1) – (EE8) as well as condition $(C \dot{-})$.

2 Compiling Epistemic Entrenchment

We now begin to describe our simple yet elegant technique for compiling an epistemic entrenchment ordering and using it for belief contraction. We proceed as follows. Firstly we describe how to compile the epistemic entrenchment relation using prime implicates. In the next section we use this compiled representation to effect belief contraction using single clauses. We then extend these results to arbitrary formulas by converting them into Conjunctive Normal Form (CNF).

In specifying how to compile the reasoner's epistemic state, represented by an epistemic entrenchment relation, we take as our point of departure an important observation on epistemic entrenchment relations due to Rott.

Proposition 2. *[14] Given an arbitrary sentence $\psi \in \mathcal{L}$, $\{\phi : \psi \leq \phi\} = Cn(\{\phi : \psi \leq \phi\})$.*

Put simply, if we were to "cut" the epistemic entrenchment relation at any level, the beliefs that are at least this entrenched would form a set that is deductively closed (i.e., a belief set)[3]. This gives us yet another way to view the epistemic entrenchment relation. Starting from the tautologies which are maximally entrenched we can see that the cuts form ever expanding theories nested one within the other. This should not be surprising

---

[1] When the new information ϕ is a tautology.
[2] Recall from above that $\phi \leq (\phi \vee \psi)$.
[3] The term "cut" is due to Rott [14] and we shall adopt it here.

given the relationship established by Gärdenfors and Makinson [6] between epistemic entrenchment and Grove's [9] possible worlds semantics for AGM belief change.

It is this fact which is the key to our representation. Ignoring the tautologies which are always maximally entrenched, we take each successive cut of the epistemic entrenchment relation and compile the resulting theory. In so doing we end up with ordered sets of prime implicates. To be a little more precise, we start with the sentences in the most entrenched cut less than the tautologies and convert this theory into its equivalent set of prime implicates. We then proceed to the next most entrenched cut and use an incremental prime implicate algorithm to compute the additional prime implicates to be added at this level of entrenchment. These prime implicates will be one of two types: (i) prime implicates that subsume more entrenched prime implicates; and, (ii) "new" prime implicates. In either case, we add these additional prime implicates to an epistemic entrenchment relation over prime implicates only.

In this way we take an epistemic entrenchment relation $\leq$ and transform it into a new epistemic entrenchment relation $\leq_\Pi$ equivalent to the first but that orders clauses (i.e., prime implicates) only. Clauses occur in the ordering $\leq_\Pi$ only when they are prime implicates of some cut of the original epistemic entrenchment relation $\leq$ (i.e., $C \in \Pi(\{\phi : \phi \leq \psi\})$ for some $\psi \in \mathcal{L}$). This new epistemic entrenchment relation is formally specified as follows.

Definition 2. *($\leq_\Pi$)*
Given an epistemic entrenchment ordering $\leq$ satisfying properties (EE1)–(EE5) we define a compiled epistemic entrenchment ordering $\leq_\Pi$ as follows. For any two clauses $C, D, C \leq_\Pi D$ iff all of the following hold:

1. $C \leq D$;
2. $C \in \Pi(\{\phi : \phi \leq \psi\})$ for some $\psi \in \mathcal{L}$; and,
3. $D \in \Pi(\{\phi : \phi \leq \chi\})$ for some $\chi \in \mathcal{L}$

Note that the empty clause is less entrenched than all clauses which we denote $\bot \leq C$ (for all clauses C).

This definition simply specifies that one clause is at least as epistemically entrenched as another precisely when it is at least as epistemically entrenched as the other in the original entrenchment ordering $\leq$ and both clauses are prime implicates of some cut of $\leq$. Importantly, clauses that are not prime implicates for any cut do not appear in the ordering at all and are not required for effecting belief contraction. In this way, the epistemic entrenchment relation only orders a minimal set of required sentences to effect belief contraction rather than ordering all sentences in the language. Furthermore, we can see whether a sentence ϕ is believed ($\phi \in K_{\leq_\Pi}$) with respect to the compiled ordering $\leq_\Pi$ by converting it into prime implicate form and ensuring that each of the implicates in $\Pi(\phi)$ is subsumed by some non-minimal clause in $\leq_\Pi$. That this ordering is correct as far as our intended purpose is concerned can be seen from the following result.

Lemma 1. *Let $\phi \in \mathcal{L}$, $\leq$ satisfy properties (EE1)–(EE5) and $\leq_\Pi$ be obtained from $\leq$ via Definition 2. Then for each clause $C \in \Pi(\phi)$ there is some clause D such that $D \subseteq C$ and $\bot <_\Pi D$ iff $\phi \in K_{\leq_\Pi}$*

3 Belief Change Using the Compiled Epistemic Entrenchment: Contraction by a Single Clause

Using the (C$\dot{-}$) condition we now provide a way of using the compiled entrenchment relation to effect belief contraction. Since this condition is specified in terms of disjunction, prime implicates are an ideal candidate for knowledge compilation and we start by considering contraction of single clauses only. Simply put, our technique works by applying the (C$\dot{-}$) condition directly to the prime implicates for each cut, with one twist. Instead of simply removing clauses as determined using (C$\dot{-}$) we consider whether they should be replaced by weaker clauses otherwise we risk removing too many clauses and not remaining faithful to the AGM postulates (K$\dot{-}$1)–(K$\dot{-}$8).

One concept we require is the level of entrenchment of a clause.

Definition 3. *Let C be a clause, $max_{\leq_\Pi}(C) = D$ such that*

1. *D is a clause where $D \subseteq C$, and*
2. *there is no other clause $D' \subset C$ and $D \leq_\Pi D'$.*

A clause's level of entrenchment is at the same level as the maximally entrenched clause that subsumes it under $\leq_\Pi$ (which is obviously the same level at which it would appear in $\leq$). Using this definition and condition (C$\dot{-}$), when contracting by a clause C we automatically retain all clauses D strictly more entrenched than C (i.e., $max_{\leq_\Pi}(C) < D$). The remaining clauses—those less or equally as entrenched as C—need to be considered in turn.

Again using condition (C$\dot{-}$) for any clauses D, where $D \leq_\Pi max_{\leq_\Pi}(C)$ and $max_{\leq_\Pi}(C) < max_{\leq_\Pi}(C \cup D)$ we can retain D. However, clauses D failing to meet these conditions will be removed from the ordering. As indicated at the start of this section, simply removing these clauses will result in too many beliefs being given up to satisfy all the AGM postulates. In some cases we need to consider how to replace these clauses with weaker versions. For these clauses, $max_{\leq_\Pi}(C) = max_{\leq_\Pi}(C \cup D)$ and we replace D in $\leq_\Pi$ by all $(D \cup E) \backslash C$ (here $\backslash$ is set substraction) where $max_{\leq_\Pi}(C) < max_{\leq_\Pi}(C \cup D \cup E)$.[4] The contracted entrenchment $\leq_\Pi^C$ is now formally defined.

Definition 4. *Given an epistemic entrenchment relation $\leq$ and a clause C we define the contraction $\leq_\Pi^C$ of a compiled epistemic entrenchment $\leq_\Pi$ by C as $D \leq_\Pi^C E$ iff either*

1. *$D \leq E$ and $max_{\leq_\Pi}(C) < D, E$, or*
2. *$max_{\leq_\Pi}(D) \leq max_{\leq_\Pi}(C)$, $max_{\leq_\Pi}(C) < E$, $D = C \cup F \cup G > max_{\leq_\Pi}(C)$ and $F \leq max_{\leq_\Pi}(C)$ for clauses F, G, or*
3. *$max_{\leq_\Pi}(D) \leq max_{\leq_\Pi}(C)$, $D = C \cup F \cup G > max_{\leq_\Pi}(C)$, $F \leq max_{\leq_\Pi}(C)$ for clauses F, G and $max_{\leq_\Pi}(E) \leq max_{\leq_\Pi}(C)$, $E = C \cup A \cup B > max_{\leq_\Pi}(C)$, $A \leq max_{\leq_\Pi}(C)$ for clauses A, B*

And we can see that this definition is correct as far as single clauses are concerned.

Theorem 3. *Let K be a belief set, $\dot{-}$ an AGM contraction function satisfying (K$\dot{-}$1)–(K$\dot{-}$8) and $\leq$ an epistemic entrenchment relation defined from $\dot{-}$ by condition (C$\leq$). Furthermore, let C be a clause. Then $K \dot{-} C = Cn(\{D : \bot <_\Pi^C D\})$.*

---

[4] Including where $C \cup D \cup E$ may be a tautology.

4 Belief Change Using the Compiled Epistemic Entrenchment: Belief Contraction by an Arbitrary Sentence

The results of the previous section can now be easily extended to arbitrary sentences ϕ by converting these sentences into CNF and considering some results relating to contraction of conjunctions and epistemic entrenchment. As indicated earlier, belief contraction by an arbitrary sentence ϕ is achieved by contracting by $CNF(\phi)$.

The first result we give states that in contraction by conjunctions we only need to consider the removal of the least entrenched conjuncts.

Proposition 3. *Given a belief set K, an AGM contraction function $\dot{-}$ and epistemic entrenchment relation $\leq$ related by conditions (C$\dot{-}$) and (C$\leq$) and $\phi, \psi \in \mathcal{L}$ such that $\phi < \psi$, then*

1. $K \dot{-} (\phi \wedge \psi) = K \dot{-} \phi$
2. $\psi \in K \dot{-} (\phi \wedge \psi)$

Furthermore, for equally entrenched conjuncts, we can simply take what is common to the contractions of each of the conjuncts.

Proposition 4. *Given a belief set K, an AGM contraction function $\dot{-}$ and epistemic entrenchment relation $\leq$ related by conditions (C$\dot{-}$) and (C$\leq$) and $\phi, \psi \in \mathcal{L}$, then*
$$\text{If } \phi = \psi, \text{ then } K \dot{-} (\phi \wedge \psi) = K \dot{-} \phi \cap K \dot{-} \psi$$

These two results together with (K$\dot{-}$6) mean that contracting by an arbitrary sentence ϕ can be achieved by considering only the minimally entrenched clauses in $CNF(\phi)$, contracting by each of these clauses individually and taking what is common to all. We can specify this formally as follows.

Definition 5. *Let $\phi \in \mathcal{L}$, the set of minimal conjuncts of ϕ is defined as follows:*
$min_{\leq_\Pi}(CNF(\phi)) = \{C_i \in CNF(\phi) : max_{\leq_\Pi}(C_i) \leq_\Pi max_{\leq_\Pi}(C_j) \text{ for all } C_j \in CNF(\phi)\}$.

And extend our previous result to show that this is indeed correct.

Theorem 4. *Let K be a belief set, $\dot{-}$ an AGM contraction function satisfying (K$\dot{-}$1)–(K$\dot{-}$8) and $\leq$ an epistemic entrenchment relation defined from $\dot{-}$ by condition (C$\leq$). Furthermore, let ϕ be a sentence. Then $K \dot{-} \phi = \bigcap_{C_i \in min_{\leq_\Pi}(CNF(\phi))} Cn(\{D : \bot <_\Pi^{C_i} D\})$.*

5 Conclusions

In this paper we have introduced a simple yet elegant technique for compiling an epistemic entrenchment ordering into prime implicates and showing how it can be used to effect AGM belief contraction. Our technique satisfies all the standard AGM postulates and, furthermore, once compilation has been performed only syntactic manipulation and subsumption checking is required.

Belief expansion by a sentence ϕ with respect to a compiled entrenchment ordering is not difficult to achieve but requires additional information to be specified in a similar way to Spohn's [16] framework for belief change. This additional piece of information takes the form of a level of entrenchment where each of the clauses in $CNF(\phi)$ will be entrenched after expansion. If any of these clauses is more entrenched, they can be ignored. Also, prime implicates for each cut lower in the entrenchment ordering may need to be recalculated. Using the Levi Identity: $K * \phi = (K \dot{-} \neg\phi) + \phi$ [5] we can achieve belief revision. Again, we need to specify a level of entrenchment for the clauses in $CNF(\phi)$ that will be required during the expansion phase. It may however be more profitable to attempt to implement belief revision more directly through a modification of the technique presented here. Iterated belief contraction on the other hand is already within our grasp. The definitions above indicate how to modify the compiled entrenchment ordering after contraction by ϕ. The resulting, contracted epistemic entrenchment ordering, can be used for subsequent contractions.

Note that in the worst case [15] since there are potentially an exponential number of prime implicates in the number of atoms we may need to consider exponentially many clauses when contracting by highly entrenched clauses. In practice, knowledge compilation has proven to be more effective in most cases.

Future work will concentrate on the various possibilities for iterated belief change as well as alternate forms of belief contraction. Extending these results to first-order belief change in non-finite domains is also of interest.

Acknowledgments

We would like to thank the anonymous referees whose comments were highly useful in improving this paper.

References

1. Carlos E. Alchourrón, Peter Gärdenfors, and David Makinson. On the logic of theory change: Partial meet contraction and revision functions. *Journal of Symbolic Logic*, 50:510–530, 1985.
2. Adnan Darwiche and Pierre Marquis. A knowledge compilation map. *Journal of Artificial Intelligence Research*, 17:229–264, 2002.
3. Adnan Darwiche and Judea Pearl. The logic of iterated belief revision. *Artificial Intelligence*, 89:1–29, 1997.
4. Simon Edmund Dixon and Wayne Wobcke. The implementation of a first-order logic AGM belief revision system. In *Proc. of the Fifth IEEE Int. Conf. on Tools in Artificial Intelligence*, 1993.
5. Peter Gärdenfors. *Knowledge in Flux: Modeling the Dynamics of Epistemic States*. Bradford Books, MIT Press, Cambridge Massachusetts, 1988.
6. Peter Gärdenfors and David Makinson. Revisions of knowledge systems using epistemic entrenchment. In *Proceedings of the Second Conference on Theoretical Aspect of Reasoning About Knowledge*, pages 83–96, 1988.
7. Peter Gärdenfors and Hans Rott. Belief revision. In Dov M. Gabbay, Christopher J. Hogger, and John A. Robinson, editors, *Handbook of Logic in Artificial Intelligence and Logic Programming Volume IV: Epistemic and Temporal Reasoning*, pages 35–132. Oxford University Press, 1995.

8. Nikos Gorogiannis and Mark D. Ryan. Implementation of belief change operators using BDDs. *Studia Logica*, 70(1):131–156, 2002.

9. Adam Grove. Two modellings for theory change. *Journal of Philosophical Logic*, 17:157–170, 1988.

10. Peter Jackson. Computing prime implicates incrementally. In *Proceedings of the Eleventh Conference on Automated Deduction*, June 1992.

11. Alex Kean. A formal characterisation of a domain independent abductive reasoning system. Technical Report HKUST-CS93-4, Dept. of Computer Science, The Hong Kong University of Science and Technology, 1993.

12. Alex Kean and George Tsiknis. An incremental method for generating prime implicants/implicates. *Journal of Symbolic Computation*, 9:185–206, 1990.

13. Raymond Reiter and Johan de Kleer. Foundations of assumption-based truth maintenance systems: Preliminary report. In *Proc. of the Nat. Conf. in Artificial Intelligence*, pages 183–188, 1987.

14. Hans Rott. Preferential belief change using generalized epistemic entrenchment. *Journal of Logic, Language and Information*, 1:45–78, 1992.

15. Robert Schrag and James M. Crawford. Implicates and prime implicates in random 3-SAT. *Artificial Intelligence*, 81(1-2):199–222, 1996.

16. Wolfgang Spohn. Ordinal conditional functions: A dynamic theory of epistemic states. In William L. Harper and Brian Skryms, editors, *Causation in Decision, Belief Change, and Statistics, II*, pages 105–134. Kluwer Academic Publishers, 1988.

17. Pierre Tison. Generalization of consensus theory and application to the minimization of boolean functions. *IEEE Trans. on Elec. Computers*, 4:446–456, 1967.

18. Mary-Anne Williams. Iterated theory change: A computational model. In *Proceedings of the Fourteenth International Joint Conference on Artificial Intelligence*, pages 1541–1550, San Francisco, 1995.

Design Methodologies of Fuzzy Set-Based Fuzzy Model Based on GAs and Information Granulation

Sung-Kwun Oh[1], Keon-Jun Park[1], and Witold Pedrycz[2]

[1] Department of Electrical Engineering, The University of Suwon, San 2-2 Wau-ri, Bong-dam-eup, Hwaseong-si, Gyeonggi-do, 445-743, South Korea
ohsk@suwon.ac.kr
[2] Department of Electrical and Computer Engineering, University of Alberta, Edmonton, AB T6G 2G6, Canada and Systems Research Institute, Polish Academy of Sciences, Warsaw, Poland

Abstract. This paper concerns a fuzzy set-based fuzzy system formed by using isolated fuzzy spaces (fuzzy set) and its related two methodologies of fuzzy identification. This model implements system structure and parameter identification by means of information granulation and genetic algorithms. Information granules are sought as associated collections of objects (data, in particular) drawn together by the criteria of proximity, similarity, or functionality. Information granulation realized with HCM clustering help determine the initial parameters of fuzzy model such as the initial apexes of the membership functions in the premise and the initial values of coefficients of polynomial function located in the consequence. And the initial parameters are tuned by means of the genetic algorithms and the least square method. To optimally identify the structure and parameters of fuzzy model we exploit two design methodologies such as a separative and a consecutive identification for tuning of the fuzzy model using genetic algorithms. The proposed model is contrasted with the performance of the conventional fuzzy models presented previously in the literature.

1 Introduction

Fuzzy modeling has been studied to deal with complex, ill-defined, and uncertain systems in many other avenues. The researches on the process have been exploited for a long time. Linguistic modeling [1] and fuzzy relation equation-based approach [2] were proposed as primordial identification methods for fuzzy models. The general class of Sugeno-Takagi models [3] gave rise to more sophisticated rule-based systems where the rules come with conclusions forming local regression models. While appealing with respect to the basic topology (a modular fuzzy model composed of a series of rules) [4], these models still await formal solutions as far as the structure optimization of the model is concerned, say a construction of the underlying fuzzy sets—information granules being viewed as basic building blocks of any fuzzy model. Some enhancements to the model have been proposed by Oh and Pedrycz [5], yet the problem of finding "good" initial parameters of the fuzzy sets in the rules remains open.

In this study we consider the problem of fuzzy set-based fuzzy model that is a development of information granules-fuzzy sets. The design methodology emerges as

A. Sattar and B.H. Kang (Eds.): AI 2006, LNAI 4304, pp. 100–109, 2006.

© Springer-Verlag Berlin Heidelberg 2006

a hybrid structural and parametric optimization. We exploit a separative and a consecutive optimization of fuzzy model that is formed by using isolated fuzzy spaces to optimally identify the structure and parameters. The proposed model is evaluated through numeric experimentation.

2   Information Granules (IG)

Roughly speaking, information granules (IG) [6], [7] are viewed as related collections of objects (data point, in particular) drawn together by the criteria of proximity, similarity, or functionality. Granulation of information is an inherent and omnipresent activity of human beings carried out with intent of gaining a better insight into a problem under consideration and arriving at its efficient solution. In particular, granulation of information is aimed at transforming the problem at hand into several smaller and therefore manageable tasks. In this way, we partition this problem into a series of well-defined subproblems (modules) of a far lower computational complexity than the original one. The form of information granulation themselves becomes an important design feature of the fuzzy model, which are geared toward capturing relationships between information granules.

It is worth emphasizing that the HCM clustering has been used extensively not only to organize and categorize data, but it becomes useful in data compression and model identification. For the sake of completeness of the entire discussion, let us briefly recall the essence of the HCM algorithm [8].

We obtain the matrix representation for hard c-partition, defined as follows.

$$M_C = \left\{ U \mid u_{gi} \in \{0,1\}, \ \sum_{g=1}^{c} u_{gi} = 1, \ 0 < \sum_{i=1}^{m} u_{gi} < m \right\} \tag{1}$$

[Step 1]. Fix the number of clusters $c(2 \leq c < m)$ and initialize the partition matrix $\mathbf{U}^{(0)} \in M_C$

[Step 2]. Calculate the center vectors $\mathbf{v}_g$ of each cluster:

$$\mathbf{v}_g^{(r)} = \{v_{g1}, \ v_{g2}, \cdots, \ v_{gk}, \cdots, \ v_{gl}\} \tag{2}$$

$$v_{gk}^{(r)} = \sum_{i=1}^{m} u_{gi}^{(r)} \cdot x_{ik} \left/ \sum_{i=1}^{m} u_{gi}^{(r)} \right. \tag{3}$$

Where, $[u_{gi}] = \mathbf{U}^{(r)}$, $g = 1, 2, \ldots, c$, $k = 1, 2, \ldots, l$.

[Step 3]. Update the partition matrix $\mathbf{U}^{(r)}$; these modifications are based on the standard Euclidean distance function between the data points and the prototypes,

$$d_{gi} = d(\mathbf{x}_i - \mathbf{v}_g) = \left\| \mathbf{x}_i - \mathbf{v}_g \right\| = \left[\sum_{k=1}^{l} (x_{ik} - v_{gk})^2 \right]^{1/2} \tag{4}$$

$$u_{gi}^{(r+1)} = \begin{cases} 1 & d_{gi}^{(r)} = \min\{d_{ki}^{(r)}\} \ \text{for all} \ k \in c \\ 0 & \text{otherwise} \end{cases} \tag{5}$$

[Step 4]. Check a termination criterion. If

$$\| \mathbf{U}^{(r+1)} - \mathbf{U}^{(r)} \| \leq \varepsilon \ \text{(tolerance level)} \tag{6}$$

Stop ; otherwise set $r = r + 1$ and return to **[Step 2]**

3 Fuzzy Set-Based Fuzzy Systems with IG

The identification procedure for fuzzy models is usually split into the identification activities dealing with the premise and consequence parts of the rules. The identification completed at the premise level consists of two main steps. First, we select the input variables x_1, x_2, …, x_k of the rules. Second, we form fuzzy partitions of the spaces over which these individual variables are defined. The identification of the consequence part of the rules embraces two phases, namely 1) a selection of the consequence variables of the fuzzy rules, and 2) determination of the parameters of the consequence (conclusion part). And the least square error method used at the parametric optimization of the consequence parts of the successive rules. In this study, we carry out the modeling using characteristics of experimental data. The HCM clustering addresses this issue. Subsequently we design the fuzzy model by considering the centers (prototypes) of clusters. In this manner the clustering help us determine the initial parameters of fuzzy model such as the initial apexes of the membership functions in the premise and the initial values of polynomial function in the consequence part of the fuzzy rules. The design process of fuzzy model based on information granules for two-input single-output system is visualized in figure 1. Here, V_{jk} and M_j is a center value of the input and output data, respectively.

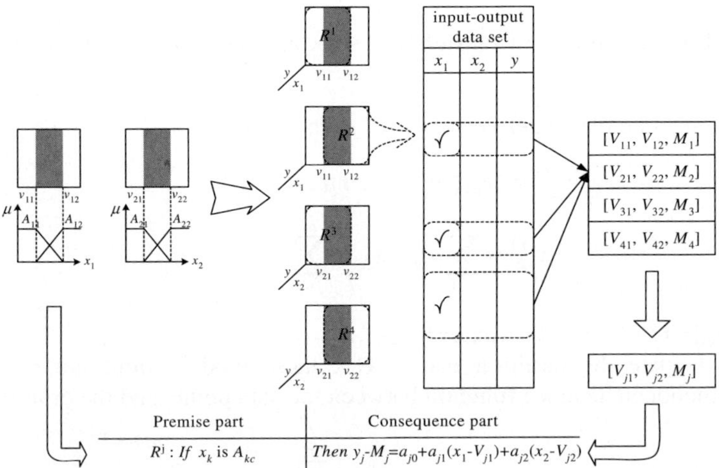

Fig. 1. Fuzzy set-based fuzzy systems with IG; illustrate is a case of the two-input single-output system

3.1 Premise Identification

In the premise part of the rules, we confine ourselves to a triangular type of membership functions whose parameters are subject to some optimization. The HCM clustering helps us organize the data into cluster so in this way we capture the characteristics of the experimental data. In the regions where some clusters of data have been identified, we end up with some fuzzy sets that help reflect the specificity of the data set. In the sequel, the modal values of the clusters are refined (optimized) using genetic optimization, and genetic algorithms (GAs), in particular.

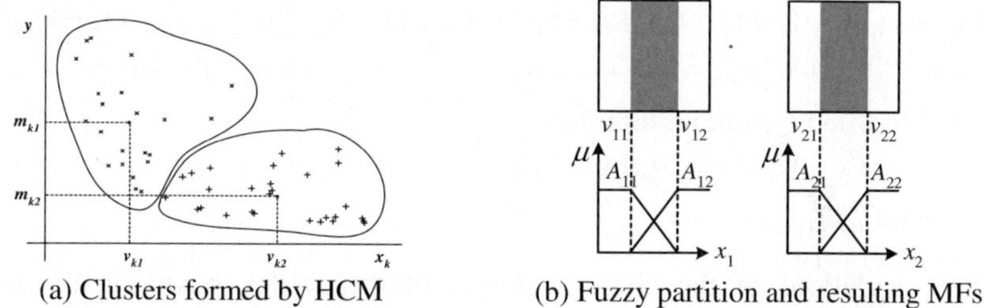

(a) Clusters formed by HCM   (b) Fuzzy partition and resulting MFs

Fig. 2. Identification of the premise part of the rules of the system

Given is a set of data $\mathbf{U}=\{\mathbf{x}_1, \mathbf{x}_2, ..., \mathbf{x}_l ; \mathbf{y}\}$, where $\mathbf{x}_k =[\mathrm{x}_{1k}, ..., \mathrm{x}_{mk}]^T$, $\mathbf{y} =[\mathrm{y}_1, ..., \mathrm{y}_m]^T$, l is the number of variables and , m is the number of data.

[Step 1]. Arrange a set of data $\mathbf{U}$ into data set $\mathbf{X}_k$.

$$\mathbf{X}_k=[\mathbf{x}_k ; \mathbf{y}] \tag{7}$$

[Step 2]. Complete the HCM clustering to determine the centers (prototypes) $\mathbf{v}_{kg}$ with data set $\mathbf{X}_k$. and calculate the center vectors $\mathbf{v}_{kg}$ of each cluster.

$$\mathrm{v}_{kg} = \{v_{k1}, \ v_{k2}, ..., v_{kc}\} \tag{8}$$

[Step 3]. Partition the corresponding isolated input space using the prototypes of the clusters $\mathbf{v}_{kg}$. Associate each clusters with some meaning, say Small, Big, etc.
[Step 4]. Set the initial apexes of the membership functions using the prototypes $\mathbf{v}_{kg}$.

3.2 Consequence Identification

We identify the structure considering the initial values of polynomial functions based upon the information granulation realized for the consequence and antecedents parts.
[Step 1]. Find a set of data included in the fuzzy space of the j-th rule.
[Step 2]. Compute the prototypes $\mathbf{V}_j$ of the data set by taking the arithmetic mean of each rule.

$$\mathrm{V}_j = \{V_{1j}, V_{2j}, ..., V_{kj}; M_j\} \tag{9}$$

[Step 3]. Set the initial values of polynomial functions with the center vectors $\mathbf{V}_j$.

The identification of the conclusion parts of the rules deals with a selection of their structure that is followed by the determination of the respective parameters of the local functions occurring there.
The conclusion is expressed as follows.

$$R^j : If \ x_k \ is \ A_{kc} \ then \ y_j - M_j = f_j(x_1, \cdots, x_k) \tag{10}$$

Type 1 (Simplified Inference): $f_j = a_{j0}$

Type 2 (Linear Inference): $f_j = a_{j0} + a_{j1}(x_1 - V_{j1}) + \cdots + a_{jk}(x_k - V_{jk})$

Type 3 (Quadratic Inference):

$$f_j = a_{j0} + a_{j1}(x_1 - V_{1j}) + \cdots + a_{jk}(x_k - V_{kj}) + a_{j(k+1)}(x_1 - V_{1j})^2 + \cdots + a_{j(2k)}(x_k - V_{kj})^2$$
$$+ a_{j(2k+1)}(x_1 - V_{1j})(x_2 - V_{2j}) + \cdots + a_{j((k+2)(k+1)/2)}(x_{k-1} - V_{(k-1)j})(x_k - V_{kj}) \tag{11}$$

Type 4 (Modified Quadratic Inference):

$$f_j = a_{j0} + a_{j1}(x_1 - V_{1j}) + \cdots + a_{jk}(x_k - V_{kj}) + a_{j(k+1)}(x_1 - V_{1j})(x_2 - V_{2j})$$
$$+ \cdots + a_{j(k(k+1)/2)}(x_{k-1} - V_{(k-1)j})(x_k - V_{kj})$$

The calculations of the numeric output of the model, based on the activation (matching) levels of the rules there, rely on the expression

$$y^* = \frac{\sum_{j=1}^{n} w_{ji} y_i}{\sum_{j=1}^{n} w_{ji}} = \frac{\sum_{j=1}^{n} w_{ji}(f_j(x_1, \cdots, x_k) + M_j)}{\sum_{j=1}^{n} w_{ji}} = \sum_{j=1}^{n} \hat{w}_{ji}(f_j(x_1, \cdots, x_k) + M_j) \tag{12}$$

Here, as the normalized value of w_{ji}, we use an abbreviated notation to describe an activation level of rule R^j to be in the form

$$\hat{w}_{ji} = \frac{w_{ji}}{\sum_{j=1}^{n} w_{ji}} \tag{13}$$

If the input variables of the premise and parameters are given in consequence parameter identification, the optimal consequence parameters that minimize the assumed performance index can be determined. In what follows, we define the performance index as the mean squared error (MSE).

$$PI = \frac{1}{m} \sum_{i=1}^{m} (y_i - y_i^*)^2 \tag{14}$$

The minimal value produced by the least-squares method is governed by the following expression:

$$\hat{a} = (X^T X)^{-1} X^T Y \tag{15}$$

4 Two Methodologies of Optimization Using GAs

The need to solve optimization problems arises in many fields and is especially dominant in the engineering environment. There are several analytic and numerical optimization techniques, but there are still large classes of functions that are fully addressed by these techniques. Especially, the standard gradient-based optimization techniques that are being used mostly at the present time are augmented by a differential method of solving search problems for optimization processes. Therefore, the optimization of fuzzy models may not be fully supported by the standard gradient-based optimization techniques, because of the nonlinearity of fuzzy models represented by rules based on linguistic levels. This forces us to explore other optimization techniques such as genetic algorithms. It has been demonstrated that genetic algorithms [9] are useful in a global optimization of such problems given their ability to efficiently use historical information to produce new improved solutions with enhanced performance.

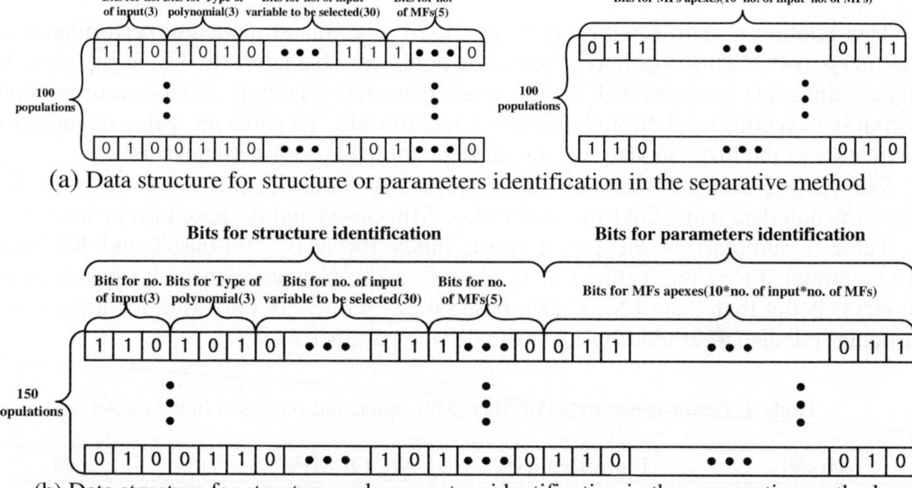

(a) Data structure for structure or parameters identification in the separative method

(b) Data structure for structure and parameters identification in the consecutive method

Fig. 3. Data structure of genetic algorithms used for the optimization of the fuzzy model

In this study, in order to identify the fuzzy model we determine such a structure as the number of input variables, input variables being selected and the number of the membership functions standing in the premise part and the order of polynomial (Type) in conclusion. The membership parameters of the premise are genetically optimized. For the identification of the system we conduct two methods as a separative and a consecutive identification. The former is that the structure of the system is identified first and then the parameters are identified later. And the latter is that the structure and parameters is simultaneously identified in the consecutive chromosomes arrangement. Figure 3 shows an arrangement of the content of the string to be used in genetic optimization. Here, parentheses denote the number of chromosomes allocated to each parameter.

For the optimization of the fuzzy model, genetic algorithms use the serial method of binary type, roulette-wheel in the selection operator, one-point crossover in the crossover operator, and invert in the mutation operator. In the case of separative method we use 150 generations and run the GA of a size of 100 individuals for structure identification and GA was run for 300 generations and the population was of size 100 for parameters identification. In the other case we use 300 generations and run the GA of a size of 150 individuals. We set up the crossover rate and mutation probability to be equal to 0.65, and 0.1, respectively (the choice of these specific values of the parameters is a result of intensive experimentation; as a matter of fact, those are quite similar to the values reported in the literature).

5   Experimental Studies

This section includes comprehensive numeric study illustrating the design of the proposed fuzzy model. We consider a nonlinear static system.

$$y = (1 + x_1^{-2} + x_2^{-1.5})^2, \quad 1 \le x_1, x_2 \le 5 \tag{16}$$

This nonlinear static equation is widely used to evaluate modeling performance of the fuzzy model. This system represents the nonlinear characteristic. Using Eq. (16), 50 Input-output data are generated: the inputs are generated randomly and the corresponding output is then computed through the above relationship. To come up with a quantitative evaluation of the fuzzy model, we use the standard MSE performance index.

We carried out the structure and parameters identification on a basis of the experimental data using GAs to design Max_Min-based and IG-based fuzzy model.

Table 1 summarizes the performance index for Max_Min-based and IG-based fuzzy model. From the Table 1 it is clear that the performance of IG-based fuzzy model is better than that of Max_Min-based fuzzy model not only after identifying the structure but also after identifying optimally the parameters.

Table 1. Performance index of Max_Min-based and IG-based fuzzy model

Model	Identification	No. Of MFs	Type	PI
Max/Min_FIS	separative	s p	5+5	Type 3	0.0014 $5.072e^{-20}$
IG_FIS		s p	5+5	Type 3	$1.152e^{-23}$ $1.444e^{-26}$
Max/Min_FIS	consecutive	s+p	5+5	Type 3	$5.625e^{-19}$
IG_FIS		s+p	5+5	Type 3	$4.612e^{-25}$

s : structure, p : parameter

Figure 4 shows the partition of the spaces and their optimization for the IG-based fuzzy model with MFs 5+5 and Type 3 by means of a separative and a consecutive identification. Figure 5 depicts the values of the performance index produced in successive generations of the genetic optimization. We note that the performance of the IG-based fuzzy model is good starting from some initial generations; this could have been caused by the characteristics of the experimental data at hand.

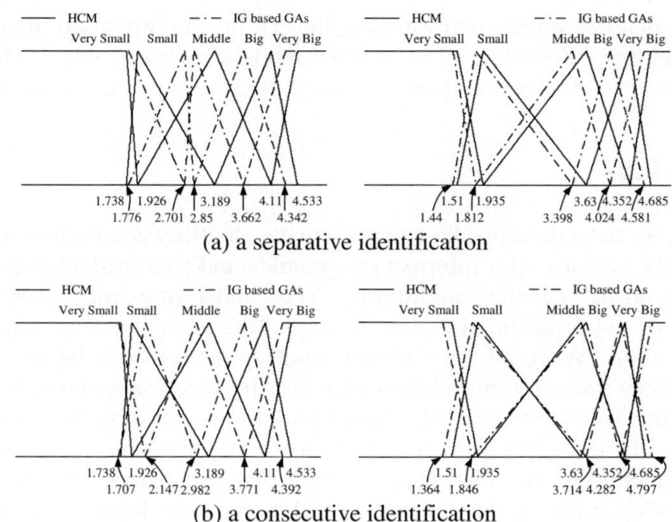

(a) a separative identification

(b) a consecutive identification

Fig. 4. Initial and optimized membership functions for the IG-based fuzzy model

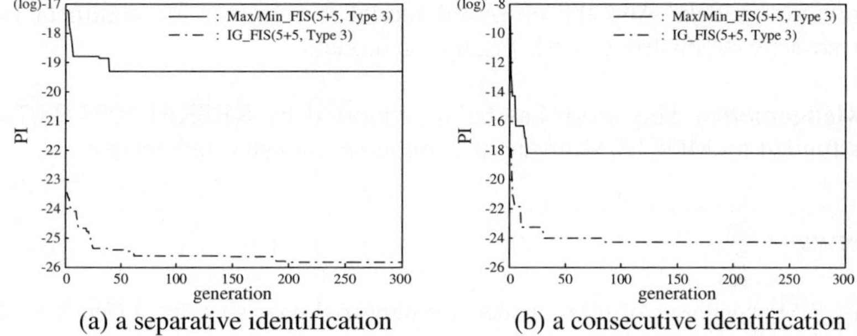

(a) a separative identification   (b) a consecutive identification

Fig. 5. Optimal convergence process of performance index for Max_Min-based and IG-based fuzzy model

Table 2. Comparison of identification error with previous models

Model		No. of rules	PI
Sugeno and Yasukawa[10]		6	0.079
Gomez-Skarmeta et al[11]		5	0.070
Kim et al.[12]		3	0.019
Kim et al.[13]		3	0.0089
Oh et al.[14]	Basic PNN		0.0212
	Modified PNN		0.0041
Park et al.[15]	BFPNN	9	0.0033
	MFPNN	9	0.0023
Our Model	separative	10	$1.444e^{-26}$
	consecutive	10	$4.612e^{-25}$

The identification error (performance index) of the proposed model is also compared with the performance of some other models; refer to Table 2. The proposed fuzzy model outperforms several previous fuzzy models known in the literature.

6 Conclusions

In this paper, we have developed a comprehensive identification framework for fuzzy set-based fuzzy systems with information granules and two methodologies of fuzzy identification using genetic algorithms. The underlying idea deals with an optimization of information granules by exploiting techniques of clustering and genetic algorithms. We used the isolated input space for each input variable and defined the fuzzy space by information granules. Information granules realized with HCM clustering help determines the initial parameters of fuzzy model such as the initial apexes of the membership functions in the premise and the initial values of polynomial function in the consequence. The initial parameters are tuned (adjusted) effectively with the aid of the genetic algorithms and the least square method. To optimally identify the structure and parameters of fuzzy model we exploited two methodologies of a separative and a consecutive identification using genetic algorithms. The experimental studies showed the performance is better than some other previous models. And the proposed model is effective for nonlinear complex systems, so we can construct a well-organized model.

Acknowledgements. This work has been supported by KESRI(I-2004-0-074-0-00), which is funded by MOCIE(Ministry of commerce, industry and energy).

References

1. Tong, R.M.: Synthesis of fuzzy models for industrial processes. Int. J Gen Syst. **4** (1978) 143-162
2. Pedrycz, W.: An identification algorithm in fuzzy relational system. Fuzzy Sets Syst. **13** (1984) 153-167
3. Takagi, T, Sugeno, M.: Fuzzy identification of systems and its applications to modeling and control. IEEE Trans Syst, Cybern. **SMC-15**(1) (1985) 116-132
4. Sugeno, M, Yasukawa, T.: Linguistic modeling based on numerical data. In: IFSA'91 Brussels, Computer, Management & System Science. (1991) 264-267
5. Oh, S.K., Pedrycz, W.: Identification of fuzzy systems by means of an auto-tuning algorithm and its application to nonlinear systems. Fuzzy Sets and Syst. **115**(2) (2000) 205-230
6. Zadeh, L.A.: Toward a theory of fuzzy information granulation and its centrality in human reasoning and fuzzy logic. Fuzzy Sets and Syst. **90** (1997) 111-117
7. Pderycz, W., Vukovich, G.: Granular neural networks. Neurocomputing. **36** (2001) 205-224
8. Krishnaiah, P.R., Kanal, L.N., editors.: Classification, pattern recognition, and reduction of dimensionality. volume 2 of Handbook of Statistics. North-Holland Amsterdam (1982)
9. Golderg, D.E.: Genetic Algorithm in search, Optimization & Machine Learning. Addison Wesley. (1989)

10. Sugeno, M., Yasukawa, T.: A Fuzzy-Logic-Based Approach to Qualitative Modeling. IEEE Trans. on Fuzzy systems. **1**(1) (1993) 7-13
11. Gomez Skarmeta, A. F., Delgado, M., Vila, M. A.: About the use of fuzzy clustering techniques for fuzzy model identification. Fuzzy Sets and Systems. **106** (1999) 179-188
12. Kim, E. T., Park, M. K., Ji, S. H., Park, M. N.: A new approach to fuzzy modeling. IEEE Trans. on Fuzzy systems. **5**(3) (1997) 328-337
13. Kim, E. T, Lee, H. J., Park, M. K., Park, M. N.: a simply identified Sugeno-type fuzzy model via double clustering. Information Sciences. **110** (1998) 25-39
14. Oh, S. K., Pedrycz, W., Park, B. J.: Polynomial Neural Networks Architecture: Analysis and Design. Computers and Electrical Engineering. **29**(6) (2003) 703-725
15. Park, B. J., Pedrycz, W., Oh, S. K.: Fuzzy Polynomial Neural Networks: Hybrid Architectures of Fuzzy Modeling. IEEE Trans. on Fuzzy Systems. **10**(5) (2002) 607-621
16. Oh, S.K., Park, B.J., Kim, H.K. : Genetically Optimized Hybrid Fuzzy Neural Networks Based on Linear Fuzzy Inference Rules. International Journal of Control Automation and Systems. **3**(2) (2005) 183-194

$\mathcal{ALE}$ Defeasible Description Logic

Pakornpong Pothipruk[1] and Guido Governatori[2]

School of ITEE, University of Queensland, Australia
[1]pkp@itee.uq.edu.au,
[2]guido@itee.uq.edu.au

Abstract. One of Semantic Web strengths is the ability to address incomplete knowledge. However, at present, it cannot handle incomplete knowledge directly. Also, it cannot handle non-monotonic reasoning. In this paper, we extend $\mathcal{ALC}^-$ Defeasible Description Logic with existential quantifier, i.e., $\mathcal{ALE}$ Defeasible Description Logic. Also, we modify some parts of the logic, resulting in an increasing efficiency in its reasoning.

1 Introduction

Description logics, which is a language based on first order logic, cannot handle incomplete and inconsistent knowledge well. In fact, it is often that complete and consistent information or knowledge are very hard to obtain. Consequently, there is an urgent need to extend description logics with a non-monotonic part which can handle incomplete and inconsistent knowledge better than themselves. Defeasible logic, developed by Nute [6], is a well-established nonmonotonic reasoning system. Its outstanding properties are low computational complexity [4,5] and ease of implementation. Thus, this logic is the nonmonotonic part of our choice. Governatori [3] proposes a formalism to extend $\mathcal{ALC}^-$ with defeasibility. He combines the description logic formalism with the defeasible logic formalism, creating a logic that can handle both monotonic and non-monotonic information coherently. We extend this work to be able to handle existential quantification.

2 Introduction to Defeasible Logic

Defeasible logic handles non-monotonicity via five types of knowledge: Facts, Strict Rules (monotonic rules, e.g., $LOGISTIC\_MANAGER(x) \rightarrow EMPLOYEE(x)$), Defeasible Rules (rules that can be defeated, e.g., $MANAGE\_DELIVER(x) \Rightarrow EMPLOYEE(x)$), Defeaters (preventing conclusions from defeasible rules, e.g., $TRUCK\_DRIVER(x) \rightsquigarrow \neg EMPLOYEE(x)$), and a Superiority Relation (defining priorities among rules). Note that We consider only propositional rules. Rule with free variables will be propositionalized, i.e., it is interpreted as the set of its grounded instances. Due to space limitation, see [6,2,1] for detailed explanation of defeasible logic.

Defeasible logic is proved to have well behavior: 1) coherence and 2) consistence [3,2]. Furthermore, the consequences of a propositional defeasible theory D can be

A. Sattar and B.H. Kang (Eds.): AI 2006, LNAI 4304, pp. 110–119, 2006.

© Springer-Verlag Berlin Heidelberg 2006

derived in $O(N)$ time, where N is the number of propositional literals in D [4]. In fact, the linear algorithm for derivation of a consequence from a defeasible logic knowledge base exploits a transformation algorithm, presented in [1], to transform an arbitrary defeasible theory D into a *basic defeasible theory* D_b.Given a defeasible theory D, the tranformation procedure $Basic(D)$ consists of three steps: 1) normalize the defeasible theory D, 2) eliminate defeaters: simulating them using strict rules and defeasible rules, and 3) eliminate the superiority relation: simulating it using strict rules and defeasible rules. The procedure then returns a transformed defeasible theory D_b. Given a basic defeasible theory D_b, defeasible proof conditions in [3] are simplified as follows:

$+\partial$: If $P(i+1) = +\partial q$ then either
    1. $+\Delta q \in P(1..i)$ or
    2. (a) $\exists r \in R[q], \forall a \in A(r) : +\partial a \in P(1..i)$ and
       (b) $\forall s \in R[\sim q], \exists a \in A(s) : -\partial a \in P(1..i)$

$-\partial$: If $P(i+1) = -\partial q$ then
    1. $-\Delta q \in P(1..i)$ and either
    2. (a) $\forall r \in R_{sd}[q], \exists a \in A(r) : -\partial a \in P(1..i)$ or
       (b) $\exists s \in R[\sim q], \forall a \in A(s) : +\partial a \in P(1..i)$

3 $\mathcal{ALC}^-$ Defeasible Description Logic

In this section, we introduce an $\mathcal{ALC}^-$ defeasible description logic [3], i.e., an extension of $\mathcal{ALC}^-$ description logic with defeasible logic. Also, we introduce several adjustments for the logic.

    Like $\mathcal{ALC}$ description logic, $\mathcal{ALC}^-$ knowledge base consists of a Tbox and Abox(es), i.e., $\Sigma = \langle \mathcal{T}, \mathcal{A} \rangle$. The Abox $\mathcal{A}$, containing a finite set of concept membership assertions ($a : C$ or $C(a)$) and role membership assertions ($(a,b) : R$ or $R(a,b)$), corresponds to the set of facts F in defeasible logic. The Tbox $\mathcal{T}$ contains a finite set of axioms of the form: $C \sqsubseteq D \mid C \doteq D$, where C and D are concept expressions. Concept expressions are of the form: $A \mid \top \mid \bot \mid \neg (atomiconly)C \mid C \sqcap D \mid \forall R.C$ where A is an atomic concept or concept name, C and D are concept expressions, R is a simple role name, $\top$ (top or full domain) represents the most general concept, and $\bot$ (bottom or empty set) represents the least general concept. Their semantics are similar to ones in $\mathcal{ALC}$ description logic. In fact, the Tbox $\mathcal{T}$ corresponds to the set of strict rules in defeasible logic. Governatori [3] shows how can we transform $\mathcal{ALC}^-$ Tbox to a set of strict rules, using a set of axiom pairs and a procedure $ExtractRules$. Basically, an inclusion axiom: $C_1 \sqcap C_2 \sqcap ... \sqcap C_m \sqsubseteq D_1 \sqcap D_2 \sqcap ... \sqcap D_n$ in the Tbox $\mathcal{T}$ is transformed into an axiom pair (ap): $\langle \{C_1(x), C_2(x), ..., C_m(x)\}, \{D_1(x), D_2(x), ..., D_n(x)\} \rangle$. Eventually, for the Tbox $\mathcal{T}$, we get the set AP of all axiom pairs. Then, the procedure $ExtractRules$ in Algorithm 1 is used to transform AP into a set of strict rules in defeasible logic.

    This procedure progresses recursively until all axiom pairs in AP are transformed. Here is an example: let $\mathcal{T} = \{C_1 \sqcap \forall R_1.C_2 \sqsubseteq C_3 \sqcap \forall R_2.(C_4 \sqcap \forall R_3.C_5)\}$. Here are the steps for strict rules generation.

Algorithm 1. *ExtractRules(AP)*:

if $ap = \langle\{C(x)\}, \{D(x)\}\rangle \in AP$ **then**
    $C(x) \to D(x) \in R_s$
    $\neg D(x) \to \neg C(x) \in R_s$
    $AP = AP - \{ap\}$
else if $ap = \langle\{C_1(x_{C1}), ..., C_{Cn}(x_n), R_1(x_{R1}, y_{R1}), ...,$
$R_m(x_{Rm}, y_{Rm})\}, \{D_1(x_{D1}), ..., D_k(x_{Dk}), \forall S_1.D_{S1}(x_{S1})$
$..., \forall S_k.D_{Sk}(x_{Sk})\}\rangle \in AP$ **then**
    for all $I \in \{D_1, ..., D_k\}$ **do**
        $C_1(x_{C1}), ..., C_n(x_{Cn}), R_1(x_{R1}, y_{R1}), ..., R_m(x_{Rm}, y_{Rm})$
        $\to I(x_I) \in R_s$
    end for
    for $1 \leq i \leq k$ **do**
        $AP = AP - \{ap\} \cup$
$$\left\{ \begin{array}{l} \langle\{C_1(x_{C1}), ..., C_n(x_{Cn}), R_1(x_{R1}, y_{R1}), \\ ..., R_m(x_{Rm}, y_{Rm}), S_i(x_{Si}, y_{Si})\}, \{D_1^{Si}(y_{Si}), \\ ..., D_l^{Si}(y_{Si}), \forall T_1^{Si}.D_1^{Si}(y_{Si}), ..., \forall T_p^{Si}.D_p^{Si}(y_{Si})\}\rangle \end{array} \right\},$$
        where
            y_{Si} is a new variable (not in ap), and
            $D_i = D_1^{Si} \sqcap ... \sqcap D_l^{Si} \sqcap \forall T_1^{Si}.D_1^{Si} \sqcap ... \sqcap \forall T_p^{Si}.D_p^{Si}$
    end for
end if
if AP is not an empty set **then**
    ExtractRules(AP)
end if
return R_s

Step 1: $AP = \{\langle\{C_1(x), \forall R_1.C_2(x)\}, \{C_3(x), \forall R_2.(C_4 \sqcap \forall R_3.C_5)(x)\}\rangle\}$.
Step 2: *ExtractRules(AP)* : $C_1(x), \forall R_1.C_2(x) \to C_3(x) \in R_s$.
Step 3: $AP = \{\langle\{C_1(x), \forall R_1.C_2(x), R_2(x, y)\},$
$\{C_4(y), \forall R_3.C_5(y)\}\rangle\}$.
Step 4: *ExtractRules(AP)* : $C_1(x), \forall R_1.C_2(x), R_2(x, y) \to C_4(y) \in R_s$.
Step 5: $AP = \{\langle\{C_1(x), \forall R_1.C_2(x), R_2(x, y), R_3(y, z)\}$
$, \{C_5(z)\}\rangle\}$.
Step 6: *ExtractRules(AP)* : $C_1(x), \forall R_1.C_2(x), R_2(x, y),$
$R_3(y, z) \to C_5(z) \in R_s$.
Step 7: $AP = \{\}$, the procedure ends here.

Notice that the *ExtractRules* procedure takes care of description logic's universal quantified concepts, occurring on the RHS of the Tbox axioms, by transforming those axioms into first order logic rules. However, description logic's universal quantified concepts occurring on the LHS of the Tbox axioms still remain in those rules. At this point, we need additional inference rules to deal with the remaining universal quantified concepts. However, universal quantified concepts' semantics take into account all individuals in the description logic knowledge base, in particular the Abox. Consequently, the proof conditions for universal quantified concepts will incorporate the domain of Abox $\mathcal{A}$, i.e., $\Delta_{\mathcal{A}}^{\mathcal{I}}$ (see [7]), in themselves as follows:

$+\Delta\forall R.C$: If $P(i+1) = +\Delta\forall R.C(a)$ then $\forall b \in \Delta_{\mathcal{A}}^{\mathcal{I}}$,
    1) $-\Delta R(a,b)$ or 2) $+\Delta C(b)$.
$-\Delta\forall R.C$: If $P(i+1) = -\Delta\forall R.C(a)$ then $\exists b \in \Delta_{\mathcal{A}}^{\mathcal{I}}$,
    1) $+\Delta R(a,b)$ and 2) $-\Delta C(b)$.
$+\partial\forall R.C$: If $P(i+1) = +\partial\forall R.C(a)$ then $\forall b \in \Delta_{\mathcal{A}}^{\mathcal{I}}$,
    1) $-\partial R(a,b)$ or 2) $+\partial C(b)$.
$-\partial\forall R.C$: If $P(i+1) = -\partial\forall R.C(a)$ then $\exists b \in \Delta_{\mathcal{A}}^{\mathcal{I}}$,
    1) $+\partial R(a,b)$ and 2) $-\partial C(b)$.

Now, we have a complete formalism to deal with $\mathcal{ALC}^-$ literals. Since a rule, however, can consists literal(s) with variable(s), there can be *vague rules* such as: $C(x), D(y) \rightarrow E(z)$. In this rule, variables in one literal are not bounded/connected with variables in any other literals. Given a rule, we say that the rule is a *variable connected rule* if, for every literal in the rule, there exists another literal which has the same variables as the literal, and a *vague rule* otherwise . Here are examples of variable connected rules: $C(x), D(x) \rightarrow E(x)$ and $C(x), R(x,y) \Rightarrow P(y,z)$. In this thesis, we only allow variable connected rules in the knowledge base.

Governatori [3] also shows that $\mathcal{ALC}^-$ defeasible description logic is coherent and consistent. He also proves that the complexity of $\mathcal{ALC}^-$ defeasible description logic w.r.t. a defeasible description theory D is $O(n^4)$, where n is the number of symbols in D. This proof is based on two steps: 1) propositionalization the theory, and 2) analyze proof conditions for universal quantified concepts. For the first step, in short, since the logic allows roles (binary predicates), the size of resulting theory is $O((\Delta_{\mathcal{A}}^{\mathcal{I}})^2)$, assuming the number of rules is much less than the size of $\Delta_{\mathcal{A}}^{\mathcal{I}}$. For the second step, the proof conditions for universal quantified concepts are embedded in the propositionalization procedure. Let n be the size of $\Delta_{\mathcal{A}}^{\mathcal{I}}$. For each universal quantified literal $\forall R.C(x)$ in an antecedent of a rule, i.e., U, create the following propositional auxiliary rules:

$$\text{for every } a_i \text{ in } \Delta_{\mathcal{A}}^{\mathcal{I}},$$
$$RC(a_i, a_1), ..., RC(a_i, a_n) \rightarrow \forall R.C(a_i)$$
$$\text{for every } a_j \text{ in } \Delta_{\mathcal{A}}^{\mathcal{I}},$$
$$\sim R(a_i, a_j) \rightarrow RC(a_i, a_j)$$
$$C(a_j) \rightarrow RC(a_i, a_j).$$

using the *AuxiliaryRules(U)* procedure. In fact, each universal quantified literal in antecedent adds at most $(\Delta_{\mathcal{A}}^{\mathcal{I}})^2$ auxiliary rules to the theory. Thus, the size of the new propositionalized theory is $O((\Delta_{\mathcal{A}}^{\mathcal{I}})^4)$ in size of the original theory D. As mentioned before that a consequence of a propositional defeasible theory D can be derived in $O(N)$ time, where N is the number of propositional literals in D [4], thus, a consequence of a propositional $\mathcal{ALC}^-$ defeasible description logic theory can be derived in polynomial time, i.e., $O(N(\Delta_{\mathcal{A}}^{\mathcal{I}})^4)$ time.

In summary, to reason with an $\mathcal{ALC}^-$ defeasible description logic, we must do the following steps:

Step 1: Transform $\mathcal{T}$ to R_s using the *ExtractRules* procedure, and transform $\mathcal{A}$ to F, hence, $D_{DDL} = \langle \mathcal{T}, \mathcal{A}, R, > \rangle$ is reduced to $D = \langle F, R, > \rangle$. Given a theory D_{DDL} and a conclusion T (i.e., a query), this step is designed such that $D_{DDL} \vdash T$ iff $D \vdash T$ [3].

Step 2: Propositionalize D to D_{prop}, including propositional auxiliary rules for universal quantified literals ($\forall R.C(x)$) in antecedents of rules. Given a theory D and a conclusion T (i.e., a query), this step guarantees that $D \vdash T$ iff $D_{prop} \vdash T$. It is easy to verify that this statement follows immediately from the first step statement plus the nature of propositionalization.

Step 3: Apply *LinearProve(Basic(D_{prop}))*, the linear algorithm. Note that $\mathcal{A}$ and $\mathcal{T}$ are normalized in this step.

Even the complexity of the logic is polynomial in time, in practice, the reasoning process on $\mathcal{ALC}^-$ defeasible description logic still suffers a huge number of additional propositionalized rules, regardless of the linear time complexity reasoning for a propositionalized theory. Consequently, it is essential to optimize the reasoning, especially the propositionalization step (Step 2).

Before we proceed with the optimization, we show that the propositionalization step is not trivial. First, we re-consider (variable-connected) simple strict rules:

$$
\begin{aligned}
r1 &: C(x), D(x) &&\rightarrow E(x) \\
r2 &: C(x), R(x,y) &&\rightarrow E(x) \\
r3 &: C(x), D(x) &&\rightarrow R(x,y) \\
r4 &: R(x,y), P(y,z) &&\rightarrow C(x)
\end{aligned}
$$

As you can notice that there is no variable quantifier in the rules. In fact, the above four rules are equal to the following rules:

$$
\begin{aligned}
r1 &: \forall x && C(x), D(x) &&\rightarrow E(x) \\
r2 &: \forall x \forall y && C(x), R(x,y) &&\rightarrow E(x) \\
r3 &: \forall x \forall y && C(x), D(x) &&\rightarrow R(x,y) \\
r4 &: \forall x \forall y \forall z && R(x,y), P(y,z) &&\rightarrow C(x)
\end{aligned}
$$

Let n be the number of individuals in the domain $\Delta_{\mathcal{A}}^{\mathcal{I}}$. In rule $r1$, there is only one variable, thus it is propositionalized to n propositional rules. In rule $r2$ and $r3$, there are two variables in each rule, thus each rule is propositionalized to n^2 propositional rules. In rule $r4$ there are three variables. In fact, the rule $r4$ is equivalent to $\forall x \forall y \forall z RP(x,y,z) \rightarrow C(x)$, where $\forall x \forall y \forall z RP(x,y,z) \equiv \forall x \forall y \forall z R(x,y) \wedge P(y,z)$. Consequently, the rule $r4$ is propositionalized to n^3 propositional rules, which will make the complexity of a conclusion derivation in $\mathcal{ALC}^-$ defeasible description logic increase to $O(n^5)$. In fact, the size of propositionalized theory is $O(n^{n_v})$ of the size of the original theory, where n_v is the maximum number of variables in rules.

Second, we re-consider (variable-connected) strict rules with universal quantified concepts (e.g., $\forall R.C(x)$) in their consequents. For example, the rule $r1 : \forall x C(x) \rightarrow \forall R.\forall P.D(x)$ is transformed to $r1 : \forall x \forall y \forall z C(x), R(x,y), P(y,z) \rightarrow D(z)$, using the *ExtractRule* procedure, which is, in turn, propositionalized to n^3 propositional rules. In fact, given a rule of this kind, propositionalization will generate additional $n^{d_\forall + 1}$ rules, where $d_\forall$ is depth of the nested universal quantified consequent, e.g., $d_\forall = 2$ for $\forall R.\forall P.D(x)$.

Third, we re-consider (variable-connected) strict rules with universal quantified concepts (e.g., $\forall R.C(x)$) in their antecedents. For example, the rule $r1 : \forall x \forall R.\forall P.D(x) \rightarrow C(x)$ is propositionalized to n propositional rules, plus the following propositional auxiliary rules:

$$
\begin{aligned}
&\text{for every } a_i \text{ in } \Delta_{\mathcal{A}}^{\mathcal{I}}, \\
&\quad R\forall P.C(a_i,a_1),...,R\forall P.C(a_i,a_n) \rightarrow \forall R.\forall P.C(a_i) \\
&\quad \text{for every } a_j \text{ in } \Delta_{\mathcal{A}}^{\mathcal{I}}, \\
&\qquad \sim R(a_i,a_j) \rightarrow R\forall P.C(a_i,a_j) \\
&\qquad \forall P.C(a_j) \rightarrow RC(a_i,a_j) \\
&\qquad PC(a_j,a_k),...,PC(a_j,a_k) \rightarrow \forall P.C(a_j) \\
&\qquad \text{for every } a_k \text{ in } \Delta_{\mathcal{A}}^{\mathcal{I}}, \\
&\qquad\quad \sim P(a_j,a_k) \rightarrow PC(a_j,a_k) \\
&\qquad\quad C(a_k) \rightarrow PC(a_j,a_k).
\end{aligned}
$$

It is easy to verify that the number of propositional auxiliary rules generated is $n(2n+1)^2$, or about n^3. In fact, given a rule of this kind, propositionalization will generate additional $n(2n+1)^{d_\forall}$ auxiliary rules for each universal quantified antecedent, where $d_\forall$ is depth of the nested universal quantified antecedent, e.g., $d_\forall = 2$ for $\forall R.\forall P.D(x)$.

At this point, the size of the new propositionalized theory D_{prop} can exceed $O((\Delta_{\mathcal{A}}^{\mathcal{I}})^4)$ in size of the original theory D. In fact, the size of the new propositionalized theory D_{prop} is $O((\Delta_{\mathcal{A}}^{\mathcal{I}})^{n_{max}})$ in size of the original theory D, where n_{max} is the maximum of $(n_v + d_\forall^{RHS}, d_\forall^{LHS} + 3)$, n_v is the maximum number of variables in each rule, $d_\forall^{LHS}$ is depth of the biggest nested universal quantified antecedent, and $d_\forall^{RHS}$ is depth of the biggest nested universal quantified consequent. However, we can modify the *ExtractRules* and *AuxiliaryRules* procedures such that size of the resulting propositionalized theory is $O(n^3)$ and $O(n^4)$ the size of the original theory respectively.

Since we allow only variable-connected rules with at most two variables, the *ExtractRules* procedure is transformed to *ExtractRules₂*. In the algorithm 2, we introduce the notion of *intermediate literals*. The intermediate literal correctness follows immediately from the semantics of universal quantified concept in description logic. Here is a simple example demonstrating a usage of the algorithm 2: let $\mathcal{T} = \{C_1 \sqcap \forall R_1.C_2 \sqsubseteq C_3 \sqcap \forall R_2.(C_4 \sqcap \forall R_3.C_5)\}$. Here are the steps for strict rules generation.

Step 1: $AP = \{\langle\{C_1(x), \forall R_1.C_2(x)\}, \{C_3(x), \forall R_2.(C_4 \sqcap \forall R_3.C_5)(x)\}\rangle\}$.

Step 2: $ExtractRules(AP) : \forall x, C_1(x), \forall R_1.C_2(x) \rightarrow C_3(x) \in R_s$.

Step 3: $AP = \{\langle\{C_1(x), \forall R_1.C_2(x), R_2(x,y)\}, \{C_4(y),$
$R_3C_5(y)\}\rangle, \langle\{R_3C_5(x)\}, \{\forall R_3.C_5(x)\}\rangle\}$.

Step 4: $ExtractRules(AP) :$
  - $\forall x \forall y, C_1(x), \forall R_1.C_2(x), R_2(x,y) \rightarrow C_4(y) \in R_s$.
  - $\forall x \forall y, C_1(x), \forall R_1.C_2(x), R_2(x,y) \rightarrow R_3C_5(y) \in R_s$.

Step 5: $AP = \{\langle\{R_3C_5(x)\}, \{\forall R_3.C_5(x)\}\rangle\}$.

Step 6: $ExtractRules(AP)$: get no rule.

Step 7: $AP = \{\langle\{R_3C_5(x), R_3(x,y)\}, \{C_5(y)\}\rangle\}$

Step 8: $ExtractRules(AP)$: $\forall x \forall y, R_3C_5(x), R_3(x,y) \rightarrow C_5(y) \in R_s$

Step 9: $AP = \{\}$, the procedure ends here.

Algorithm 2. *ExtractRules$_2$(AP)*:

---

if $ap = \langle \{C(x)\}, \{D(x)\} \rangle \in AP$ **then**

     $\forall x, C(x) \rightarrow D(x) \in R_s$

     $\forall x, \neg D(x) \rightarrow \neg C(x) \in R_s$

     $AP = AP - \{ap\}$

else if $ap = \langle \{C_1(x), ..., C_{Cn}(x), R_1(x,y), ..., R_m(x,y)\},$

$\{D_1(x), ..., D_k(x)\} \rangle \in AP$ **then**

     for all $I \in \{D_1, ..., D_k\}$ **do**

         $\forall x \forall y, C_1(x), ..., C_{Cn}(x), R_1(x,y), ..., R_m(x,y) \rightarrow I(x)$

     end for

     $AP = AP - \{ap\}$

else if $ap = \langle \{C_1(x), ..., C_{Cn}(x), R_1(x,y), ..., R_m(x,y)\},$

$\{D_1(x), ..., D_k(x), \forall S_1.D_{S1}(x), ..., \forall S_k.D_{Sk}(x)\} \rangle \in AP$ **then**

     for all $I \in \{D_1, ..., D_k\}$ **do**

         $\forall x \forall y, C_1(x), ..., C_n(x), R_1(x,y), ..., R_m(x,y)$

         $\rightarrow I(x) \in R_s$

     end for

     for $1 \leq i \leq k$ **do**

         $AP = AP - \{ap\} \cup$

$$\left\{ \begin{array}{l} \langle \{C_1(x), ..., C_n(x), R_1(x,y), ..., R_m(x,y), \\ S_i(x,z)\}, \{D_1^{Si}(z), ..., D_l^{Si}(z), T_1^{Si}D_1^{Si}(z), ..., \\ T_p^{Si}D_p^{Si}(z)\} \rangle \langle \{T_1^{Si}D_1^{Si}(x)\}, \\ \{\forall T_1^{Si}D_1^{Si}(x)\} \rangle ... \langle \{T_p^{Si}D_p^{Si}(x)\}, \\ \{\forall T_p^{Si}D_p^{Si}(x)\} \rangle \end{array} \right\},$$

         where

             y and z are new variables (not in ap),

             $D_i = D_1^{Si} \sqcap ... \sqcap D_l^{Si} \sqcap \forall T_1^{Si}.D_1^{Si} \sqcap ... \sqcap \forall T_p^{Si}.D_p^{Si}$,

             and $T_1^{Si}D_1^{Si}(x), ..., T_p^{Si}D_p^{Si}(x)$ are *intermediate*

             literals.

     end for

end if

if AP is not an empty set **then**

     ExtractRules$_2$(AP)

end if

return R_s

It is easy to verify that the *ExtractRules$_2$* procedure in algorithm 2 generates the resulting propositionalized theory of the size $O(n^3)$ the size of the original theory.

Regarding the *AuxiliaryRules* procedure, since we allow only variable-connected rules with at most two variables, for each universal quantified literal $U = \forall R.C(x)$ in an antecedent of a rule, call the procedure *AuxiliaryRules$_2$(U)*:

Algorithm 3. $AuxiliaryRules_2(U)$:

$\forall x, RC(x,a_1),...,RC(x,a_n) -> \forall R.C(x) \in R_s$
$\forall x \forall y, \sim R(x,y) -> RC(x,y) \in R_s$
$\forall x \forall y, C(y) -> RC(x,y) \in R_s$.
if $C(y)$ is a universal quantified literal, then
$\qquad AuxiliaryRules_2(C(x))$

---

where $\Delta_{\mathcal{A}}^{\mathcal{I}}$ is $\{a_1,...,a_n\}$, and n is the size of $\Delta_{\mathcal{A}}^{\mathcal{I}}$. After this, we can propositionalize the additional auxiliary rules as usual. It is easy to verify that the size of of the propositionalized theory is $O(n^4)$ the size of the original theory.

In summary, if we limit the number of variable $n_v = 2$, and use the above specifications for universal quantified literals (both antecedent and consequent), we will regain complexity of the modified $\mathcal{ALC}^-$ defeasible description logic to be $O(n^4)$ again.

Fourth, we show how a defeater can be propositionalized. A defeater is only used to prevent a conclusion, thus its consequent is a negative literal. In $\mathcal{ALC}^-$ defeasible description logic, only atomic negation is allowed. In addition, we only allow variable-connected rules with at most two free variables in each rule. Consequently, a defeater can be of the form $\forall x(\forall y), LHS \rightsquigarrow \neg C(x) \mid \forall x \forall y, LHS \rightsquigarrow \neg R(x,y)$. In propositionalization, a defeater is treated as a rule. For example, let $\Delta_{\mathcal{A}}^{\mathcal{I}} = \{a,b,c\}$, a defeater $\forall x, C(x) \rightsquigarrow \neg D(x)$ is propositionalized to: $C(a) \rightsquigarrow \neg D(a) \in R_{dft}, C(b) \rightsquigarrow \neg D(b) \in R_{dft}, C(c) \rightsquigarrow \neg D(c) \in R_{dft}$. A propositionalized defeater will prevent conclusion by not firing defeasible rule(s) whose head(s) has(have) literal(s) which is(are) negativity of the defeater head if all literals in the defeater body are provable. Since, a defeater is propositionalized in the same way as a rule is, it will prevent correct propositionalized defeasible rule(s) from firing. Hence, the size of propositionalized R_{dft} is $O(n^2)$ the size of original R_{dft}. Thus, complexity of the $\mathcal{ALC}^-$ defeasible description logic is still $O(n^4)$.

Lastly, we show how can we extend the superiority relation to cover the propositionalized rules. Since the superiority relation is defined over pairs of rules, in particular defeasible rules, which have contradictory heads, we only need to extend the superiority relation to cover the corresponding pairs of propositionalized rules. We illustrate this fact by a simple example. Let $\Delta_{\mathcal{A}}^{\mathcal{I}} = \{a,b,c\}$, we have a set of rules: $r1 : \forall x, C(x) \Rightarrow E(x), r2 : \forall x, D(x) \Rightarrow \neg E(x)$, and the superiority relation $> = \{\langle r1, r2 \rangle\}$. The set of rules are propositionalized to: $r1a : C(a) \Rightarrow E(a), r1b : C(b) \Rightarrow E(b), r1c : C(c) \Rightarrow E(c), r2a : D(a) \Rightarrow \neg E(a), r2b : D(b) \Rightarrow \neg E(b), r2c : D(c) \Rightarrow \neg E(c)$, and the extended superiority relation $> = \{\langle r1a, r2a \rangle, \langle r1b, r2b \rangle, \langle r1c, r2c \rangle\}$. Hence, the size of extended superiority relation is $O(n^2)$ the size of original superiority relation. Thus, complexity of the $\mathcal{ALC}^-$ defeasible description logic is still $O(n^4)$. Note that the propositionalized R_{dft} and the extended $>$ will be absorbed into R_{sd} in the linear algorithm for a conclusion derivation.

4 $\mathcal{ALE}$ Defeasible Description Logic

In this section, we introduce an $\mathcal{ALE}$ defeasible description logic, i.e., an extension of $\mathcal{ALC}^-$ defeasible description logic with existential quantification constructor.

$\mathcal{ALE}$ knowledge base consists of a Tbox and Abox(es), i.e., $\Sigma = \langle \mathcal{T}, \mathcal{A} \rangle$. The Abox $\mathcal{A}$, containing a finite set of concept membership assertions ($a : C$ or $C(a)$) and role membership assertions ((a,b) $: R$ or $R(a,b)$), corresponds to the set of facts F in defeasible logic. The Tbox $\mathcal{T}$ contains a finite set of axioms of the form: $C \sqsubseteq D \mid C \doteq D$, where C and D are concept expressions. Concept expressions are of the form: $A \mid \top \mid \bot \mid \neg(atomiconly)C \mid C \sqcap D \mid \forall R.C \mid \exists R.C$ where A is an atomic concept or concept name, C and D are concept expressions, R is a simple role name, $\top$ (top or full domain) represents the most general concept, and $\bot$ (bottom or empty set) represents the least general concept. Their semantics are similar to ones in $\mathcal{ALC}$ description logic. Tbox $\mathcal{T}$ corresponds to the set of strict rules in defeasible logic. The algorithm 4 shows how can we transform $\mathcal{ALE}$ Tbox to a set of strict rules, using a set of axiom pairs and a procedure $ExtractRules_3$. Basically, an inclusion axiom: $C_1 \sqcap C_2 \sqcap ... \sqcap C_m \sqsubseteq D_1 \sqcap D_2 \sqcap ... \sqcap D_n$ in the Tbox $\mathcal{T}$ is transformed into an axiom pair (ap): $\langle \{C_1(x), C_2(x), ..., C_m(x)\}, \{D_1(x), D_2(x), ..., D_n(x)\} \rangle$. Eventually, for the Tbox $\mathcal{T}$, we get the set AP of all axiom pairs. Then, the procedure $ExtractRules_3$ is used to transform AP into a set of strict rules in defeasible logic.

Algorithm 4. $ExtractRules_3(AP)$:

---

if $ap = \langle \{C_1(x), ..., C_{Cn}(x), R_1(x,y), ..., R_m(x,y)\},$
$\{D_1(x), ..., D_k(x), \exists S_1.D_{S1}(x), ..., \exists S_k.D_{Sk}(x)\} \rangle \in AP$ **then**
    for all $I \in \{D_1, ..., D_k\}$ **do**
        $\forall x \forall y, C_1(x), ..., C_n(x), R_1(x,y), ..., R_m(x,y)$
        $\rightarrow I(x) \in R_s$
    end for
    $AP = AP - \{ap\} \cup$
    $\left\{ \begin{array}{l} \langle \{C_1(x), ..., C_n(x), R_1(x,y), ..., R_m(x,y)\}, \\ \{S_1(x,y_1), D_{S1}(y_1), ..., S_k(x,y_k), D_{Sk}(y_k)\} \rangle \end{array} \right\}$,
else if $ap = \langle \{\exists S_1.D_{S1}(x), ..., \exists S_k.D_{Sk}(x)\},$
$\{C_1(x), ..., C_n(x)\} \rangle \in AP$ **then**
    $AP = AP - \{ap\} \cup$
    $\left\{ \begin{array}{l} \langle \{S_1 D_{S1}(x), ..., S_k D_{Sk}(x)\}, \{C_1(x), ..., C_n(x)\} \rangle \\ \langle \{S_1(x,y), D_{S1}(x)\}, \{S_1 D_{S1}(x)\} \rangle, ... \\ \langle \{S_k(x,y), D_{Sk}(x)\}, \{S_k D_{Sk}(x)\} \rangle \end{array} \right\}$,
else $ExtractRules_2(AP)$
end if
if AP is not an empty set **then**
    $ExtractRules_3(AP)$
end if
return R_s

It is easy to verify that the $ExtractRules_3$ procedure in algorithm 4 generates the resulting propositionalized theory of the size $O(n^4)$ the size of the original theory. Consequently, we get a defeasible description logic with very efficient derivation process for a conclusion.

5   Discussion and Future Works

This paper introduces several modifications to the existing $\mathcal{ALC}^-$ defeasible description logic such that its derivation can be accomplished in $O(n^4)$. This makes the language more useful in practice. Further, we extend the logic with existential quantification constructor, resulting in a new logic, i.e., $\mathcal{ALE}$ defeasible description logic, which can handle more expressive knowledge base in the context of non-monotonic reasoning. Our work is significant because it is a foundation to be extended to higher expressive non-monotonic description logic. In the history of Description Logics, increasing-in-expressiveness description logics have been studied chronologically, in order to find the highest expressive description logic that are still decidable. Also, the maximum bound of tractable logic has been found. However, those are the cases for *monotonic* description logics, not for *nonmonotonic* description logics. This work presents a new result showing a nonmonotonic description logic that is still tractable, i.e., $\mathcal{ALE}$ defeasible description logic. In the near future, we will study how we can add the full negation constructor to the logic, resulting in $\mathcal{ALC}$ defeasible description logic. However, it is still arguable whether nonmonotonic logic can be extended to $\mathcal{SHOIN}(\mathcal{D})$, which is equal to $\mathcal{ALC}^{\mathcal{R}^+}\mathcal{HOIN}(\mathcal{D})$. Transitive roles, role inclusions, one-of operators, inverse roles, qualified number restriction, and concrete domain must be added to the $\mathcal{ALC}$ defeasible description logic, in order to achieve the $\mathcal{SHOIN}(\mathcal{D})$ defeasible description logic. These additions are still open issues.

Acknowledgements

This work was supported by Australia Research Council under Discovery Project No. DP0558854.

References

1. G. Antoniou, D. Billington, G. Governatori, and M. Maher, Representation Results for Defeasible Logic, *ACM Transactions on Computational Logic*, 2(2), pp. 255-287, 2001.
2. D. Billington, Defeasible Logic is Stable, *Journal of Logic and Computation*, 3, pp. 370-400, 1993.
3. G. Governatori, Defeasible Description Logics, In G. Antoniou and H.Boley, editors, *Proceedings of Rules and Rule Markup Languages for the Semantic Web: Third International Workshop (RuleML 2004)*, pp. 98-112, 2004.
4. M. J. Maher, Propositional Defeasible Logic has Linear Complexity, *Theory and Practice of Logic Programming*, 1(6), pp. 691-711, 2001.
5. M. J. Maher, A. Rock, G. Antoniou, D. Billignton, and T. Miller, Efficient Defeasible Reasoning Systems, *International Journal of Artificial Intelligence Tools*, 10(4), pp. 483-501, 2001.
6. D. Nute, Defeasible Logic, In *Handbook of Logic in Artificial Intelligence and Logic Programming*, vol. 3, pp. 353-395, Oxford University Press, 1987.
7. P. Pothipruk and G. Governatori, A Formal Ontology Reasoning with Individual Optimization: A Realization of the Semantic Web, In M. Kitsuregawa, E. Neuhold and A. Ngu, editors, *Web Information Systems Engineering (WISE 2005)*, pp. 119-132, 2005.

Representation and Reasoning for Recursive Probability Models

Catherine Howard[1] and Markus Stumptner[2]

[1] Electronic Warfare and Radar Division, Defence Science and Technology Organisation,
PO Box 1500, Edinburgh, South Australia, 5111
[2] Advanced Computing Research Center, University of South Australia, Adelaide, 5095

Abstract. This paper applies the Object Oriented Probabilistic Relational Modelling Language to recursive probability models. We present two novel anytime inference algorithms for recursive probability models expressed using this language. We discuss the strengths and limitations of these algorithms and compare their performance against the Iterative Structured Variable Elimination algorithm proposed for Probabilistic Relational Modelling Language using three different non-linear genetic recursive probability models.

1   Introduction

Recursive probability models (RPMs) are described as models where a variable may probabilistically depend on a potentially infinite, but finitely describable, set of variables [1,2]. These types of models are natural in genetic or language applications. For example, the propagation of genetic material through a family tree, shown in Fig 1, is an example of a RPM. In a RPM, a variable associated with a particular model entity may probabilistically depend on the same variable associated with a different model entity: in Fig 1, a person's MChromosome attribute is dependent on their mother's MChromosome attribute.

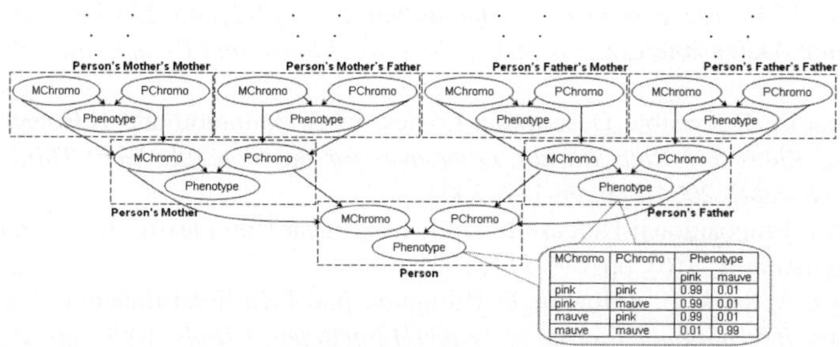

MChromo	PChromo	Phenotype	
		pink	mauve
pink	pink	0.99	0.01
pink	mauve	0.99	0.01
mauve	pink	0.99	0.01
mauve	mauve	0.01	0.99

Fig. 1. An example of a Recursive Probability Model

Recursive models present two main challenges: how to compactly represent the model and how to perform inference. This paper addresses these challenges by applying the Object Oriented Probabilistic Relation Modelling Language (OPRML) to

A. Sattar and B.H. Kang (Eds.): AI 2006, LNAI 4304, pp. 120–130, 2006.

RPMs. In Section 2, we illustrate the use of the OPRML to represent RPMs in using three simple genetic models, demonstrating that this language can represent recursive models in a compact and structured fashion. In Section 3, we present the Iterative Junction Tree Construction Algorithms which are novel interruptible inference algorithms for RPMs expressed using OPRML. In Section 4, as the OPRML is similar to the Probabilistic Relational Modelling Language (PRML) [2], we experimentally evaluate the IJC algorithm against the Iterative Structured Variable Elimination (ISVE) algorithm [1] which was proposed for RPMs expressed using the PRML. The evaluation is carried out using the three genetic models developed in Section 2 and as a part of this evaluation, we have implemented the ISVE algorithm, which to our knowledge, has not been previously implemented.

2   Representing Recursive Probability Models

The OPRML is a typed relational language. Models produced using this language are called Object-Oriented Probabilistic Relational Models (OPRMs) [3,4].

Definition: An OPRM, Ξ, is a pair (RC, PC) of a relational component RC and a probabilistic component PC. The RC consists of:

- A set of classes, $\mathbf{C} = \{C_1, C_2,\ldots, C_n\}$
- A partial ordering over $\mathbf{C}$ which defines the class hierarchy
- A set of named instances, $\mathbf{I} = \{I_1, I_2,\ldots, I_n\}$, which represent instantiations of the classes
- A set of descriptive attributes, $\Delta_{\mathbf{C}} = \{\delta_1, \delta_2,\ldots, \delta_n\}$, for each class C in $\mathbf{C}$. Attribute δ_x of class C_1 is denoted $C_1.\delta_x$
- A set of complex attributes, $\Phi_{\mathbf{C}} = \{\phi_1, \phi_2,\ldots, \phi_n\}$, for each class C in $\mathbf{C}$. Attribute ϕ_x of class C_1 is denoted $C_1.\phi_x$

The PC consists of:

- A conditional probability model for each descriptive attribute δ_x, $P(\delta_x|\mathbf{Pa}[\delta_x])$ where $\mathbf{Pa}[\delta_x] = \{Pa_1, Pa_2,\ldots, Pa_n\}$ is the set of parents of δ_x

OPRMs specify a template for the probability distribution over a knowledge base (KB), where a knowledge base is defined as consisting of a set of OPRM classes and instances, a set of inverse statements and a set of instance statements. OPRMS are discussed further in [3,4].

When representing and reasoning with RPMs, we place the following restrictions on the knowledge base: there are no inverse statements in the KB and there are a finite number of named instances in the KB. The first restriction is used to simplify the discussion in this paper and can easily be lifted (see [1]). The second restriction cannot, because, as will be seen in Section 3, in order to reason with RPMs, we approximate the model using a *finite* network.

In the remainder of this section, we outline the three non-linear genetic recursive probability models which are later used to compare algorithm performance. We consider multiple models in this paper in order to illustrate firstly how RPMs can be compactly represented using OPRMs and secondly that the relative performance of the inference algorithms discussed in Sections 3 and 4 is model dependant. The models in Fig 2 are presented in order of their complexity. The simplest model, Fig. 2a and d, models a person's happiness as being dependant only on their mother's happiness. In this model, $C=\{Person\}$, $\Delta_{Person}=\{Happy\}$ and $\Phi_{Person}=\{Mother, Father\}$. The query is performed on the Person's Happy attribute.

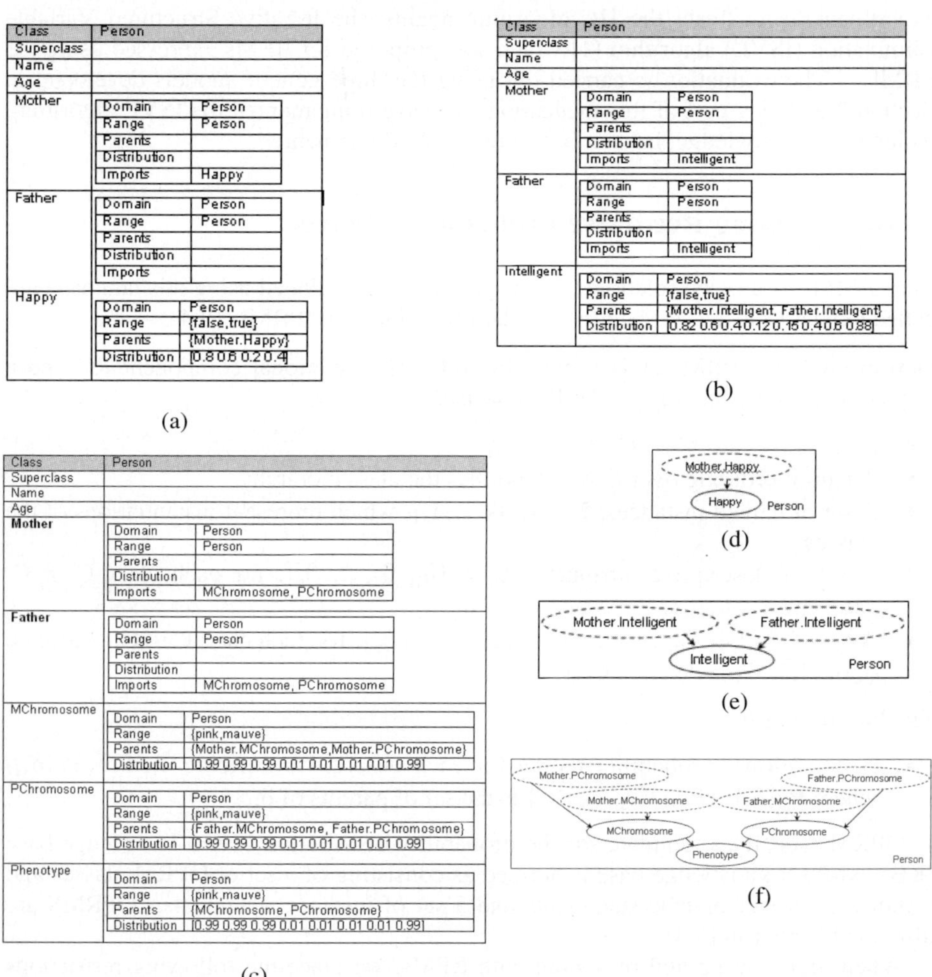

(a)

Class	Person	
Superclass		
Name		
Age		
Mother	Domain	Person
	Range	Person
	Parents	
	Distribution	
	Imports	Happy
Father	Domain	Person
	Range	Person
	Parents	
	Distribution	
	Imports	
Happy	Domain	Person
	Range	{false,true}
	Parents	{Mother.Happy}
	Distribution	[0.8 0.6 0.2 0.4]

(b)

Class	Person	
Superclass		
Name		
Age		
Mother	Domain	Person
	Range	Person
	Parents	
	Distribution	
	Imports	Intelligent
Father	Domain	Person
	Range	Person
	Parents	
	Distribution	
	Imports	Intelligent
Intelligent	Domain	Person
	Range	{false,true}
	Parents	{Mother.Intelligent, Father.Intelligent}
	Distribution	[0.82 0.6 0.4 0.12 0.15 0.4 0.6 0.88]

(c)

Class	Person	
Superclass		
Name		
Age		
Mother	Domain	Person
	Range	Person
	Parents	
	Distribution	
	Imports	MChromosome, PChromosome
Father	Domain	Person
	Range	Person
	Parents	
	Distribution	
	Imports	MChromosome, PChromosome
MChromosome	Domain	Person
	Range	{pink,mauve}
	Parents	{Mother.MChromosome,Mother.PChromosome}
	Distribution	[0.99 0.99 0.99 0.01 0.01 0.01 0.01 0.99]
PChromosome	Domain	Person
	Range	{pink,mauve}
	Parents	{Father.MChromosome, Father.PChromosome}
	Distribution	[0.99 0.99 0.99 0.01 0.01 0.01 0.01 0.99]
Phenotype	Domain	Person
	Range	{pink,mauve}
	Parents	{MChromosome, PChromosome}
	Distribution	[0.99 0.99 0.99 0.01 0.01 0.01 0.01 0.99]

(d)

(e)

(f)

Fig. 2. (a-c) OPRM class specifications and (d-f) equivalent networks for the recursive models

The second model, Figs. 2b and e, describes the relationship between a person's intelligence and that of both their parents. In this model, $C=\{Person\}$ where $\Delta_{Person}=\{Intelligent\}$ and $\Phi_{Person}=\{Mother, Father\}$. The query is performed on the Person's Intelligent attribute. The most complex model, Figs. 2c and f, is the OPRM equivalent of the RPM shown in Fig 1. It is a model of the transmission of a single eye colour gene through a family tree and was and used by [1] in their treatment of

RPMs. In this model, $\mathbf{C}$={Person}, $\Delta_{\mathbf{Person}}$={MChromo, PChromo, Phenotype} and $\Phi_{\mathbf{Person}}$={Mother, Father}. The query is performed on the Person's Phenotype. By comparing Fig 1 with Fig 2f, it can be seen how compactly the OPRML can represent RPMs. Each of our example RPMs can be represented using one OPRM class.

3   Reasoning with Recursive Probability Models

In this section we present the Iterative Junction Tree Construction (IJC) Algorithms which are novel inference algorithms for RPMs expressed using OPRML. These algorithms are iterative extensions of the Junction Tree model Construction (JC) algorithm for OPRMs presented in [4]. It was necessary to develop model construction and inference algorithms specifically for recursive models because the previously published model construction algorithms for OPRMs, namely the Knowledge Based Model Construction (KBMC) and Junction Tree Construction (JC) algorithms, are unsuitable for RPMs. Because the set of variables in the recursive model is potentially infinite, the KBMC and JC model construction algorithms may never terminate.

The IJC algorithms are an iterative Anytime Algorithms (AA). AAs offer a trade off between solution quality and execution time. The IJC algorithms embody this tradeoff by using the OPRM class specifications for a RPM (e.g. those shown in Figs 2a-c) to automatically and incrementally construct a larger, more complex finite model which approximates the infinite network described by the RPM. That is, each iteration of the algorithms makes the model a better approximation to the infinite network. The nth iteration of the IJC algorithms produces a model which is the nth order approximation to the infinite network. As the approximate model constructed by the IJC algorithms describe the probability distribution over a finite number of variables, the model construction process will terminate and exact inference can be performed in a finite time.

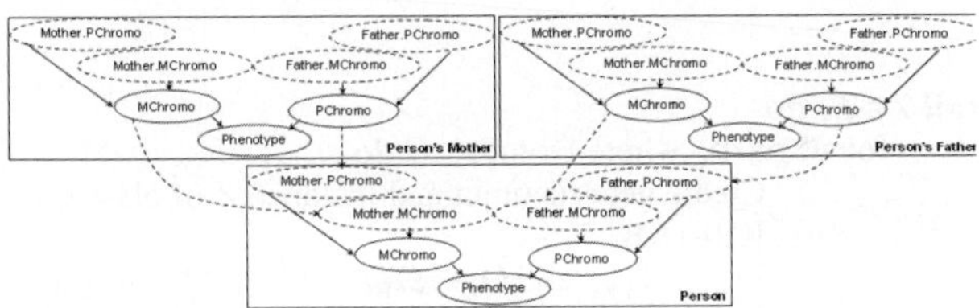

Fig. 3. The second order approximation to the infinite BN shown in Fig. 1

In our example models, the order of approximation, n, equates to the number of generations of people included in the model (e.g. Fig 3). As n (and hence execution time) increases, the algorithm includes more and more generations in the approximate model and returns a better and better approximate solution to the query. IJC is an interruptible anytime algorithm: it can be interrupted at any time and will return the best computed approximation so far. The meta-level control paradigm employed by the IJC algorithm is to allocate the desired order of approximation and allow the algorithm to construct a model of this order, regardless of how long it takes.

At design time, the OPRM class specifications are used to create the equivalent object oriented probabilistic network for each class in the KB. This network is then translated directly into a 'local' junction tree. An interface clique is created which consists of all nodes in the class's interface, κ. This interface clique is connected into the local junction tree and any loops created during this process are removed. Thus at design time, each class is 'precompiled' into a local junction tree and these local junction trees are stored in a cache.

When the IJC algorithms are executed, the user specifies their desired order of approximation, n_D. Then, starting with n=1 and increasing n by one in each iteration, the algorithm builds on the model from the previous iteration by instantiating the required number of generic unnamed instances of the Person class and adding them to the model. The number of instances that are required to be added to the model will be specified by the probabilistic dependencies within the class. For example, in the Phenome and Intelligence models, a person's Phenome and Intelligence attributes depend on Phenome and Intelligence attributes of *both* parents, so 2^n generic instances will be added to the model in each iteration, while for the Happiness model, only 1^n instances will be added.

For the IJC Version 1 algorithm (IJCV1), whenever a generic unnamed instance, Z, of the Person class is created, the appropriate local rooted junction tree, JT_{RZ}, is instantiated. A root clique for the model is created which contains all the nodes in all the generic unnamed instances' interfaces, κ_Z. Each local instance junction tree is then connected to the root clique, ω_R, to create the 'global junction tree'. The IJC Version 2 algorithm (IJCV2), varies from IJCV1 in how the global junction tree is created. Instead of creating a root clique and connecting each local instance junction tree to it, IJCV2 connects the interface cliques of the individual local instance junction trees as appropriate to form the global junction tree. In both algorithms, exact inference is performed on the global junction tree using standard message passing techniques.

```
Set n_D
n=1
Ω_KB←Ω_C
while n ≤ n_D
        if n > 1
                forall Z∈ Z_T do
                        forall φ∈ Φ_C where Dom[φ]=C do
                                Create generic unnamed instance Z of class C
                                KB←KB∪Z
                                JT_RZ←JT_RC and Ω_Z←Ω_RC
                                Ω_KB←Ω_KB∪(Ω_Z -ω_κ)
                                set the reference relationships
                                ω_R← ω_R∪κ_Z
                        end forall
                end forall
        end if
        Ω_KB←(Ω_KB∪ω_R)
        Create ψ_KB from ψ_Z
        perform reasoning
        n←n+1
end while
```

Algorithm 1. The IJCV1 algorithm

In the pseudocode presented in this section, we denote the rooted junction tree for class C by JT_{RC}, the set of cliques for the class C by $\Omega_C=\{\omega_1, \omega_2,..., \omega_n\}$, and the set of cliques in the KB by $\Omega_{KB}=\{\Omega_{Z1}, \Omega_{Z1},..., \Omega_{Zn}\}$. The interface clique for the class is denoted by ω_κ and the root clique for the KB by ω_R. ψ_{KB} is the separator matrix for the KB while ψ_Z is the separator matrix for generic instance Z. Z_T denotes the set of generic instances at the root of the model. For our three example models, C will be the Person class. The Design Time phase of the IJC algorithms is the same as the Design Time Phase of the JC presented in [4], so we do not duplicate it here. The run time phase of the algorithm, however, is very different:

```
Set n_D
n=1
Ω_KB←Ω_C
while n ≤ n_D
        if n > 1
                forall Z∈ Z_T do
                        forall φ∈ Φ_C where Dom[φ]=C do
                                Create generic unnamed instance Z of class C
                                KB←KB∪Z
                                JT_RZ←JT_RC and Ω_Z←Ω_RC
                                Ω_KB←Ω_KB∪Ω_Z
                                set the reference relationships
                        end forall
                end forall
        end if
        Create ψ_KB from ψ_Z
        perform reasoning
        n←n+1
end while
```

Algorithm 2. The IJCV2 algorithm

4   Experimentation and Discussion

In this section, we experimentally evaluate the IJC algorithms described above against the ISVE algorithm which was proposed for RPMs expressed using the PRML. All three algorithms were implemented in MATLAB augmented with the BN Toolbox (BNT) and executed using a Pentium III, 1.2GHz computer. The performance of the ISVE algorithm was measured using PRMs equivalent to the OPRMs described in Section 2.

Both ISVE and the IJC algorithms approximate the infinite BN by a finite network which is incrementally built up to the desired order of approximation. The algorithms then use exact reasoning techniques on these resulting model approximations to obtain a probability distribution over the query variables. ISVE uses a structured form of variable elimination while the IJC algorithms use a message passing. During our experiments, we measured the 'wall clock' execution time as well as the total time taken by the algorithm to construct the model and the time taken to perform inference. In Figs 4-6, we present comparisons the model construction, inference and execution times for each algorithm. These plots allow us to gain an insight into which phase each algorithm spends the majority of its time and whether this behaviour is consistent from one model to the next.

We are unable to present a comparison of inference times for the IJCV1 algorithm for the Intelligence and Phenome models. From the pseudocode in Section 3, it is clear that root clique contains all the interface nodes of all the generic unnamed instances currently in the model. This means that, depending on the model, the root clique can become very large very quickly. This is an important limitation of the technique. During our experiments, we found that as a result of the way probability distributions are represented and manipulated in BNT, MATLAB's maximum variable size was quickly exceeded when attempting to perform inference with the Phenome and Intelligence models.

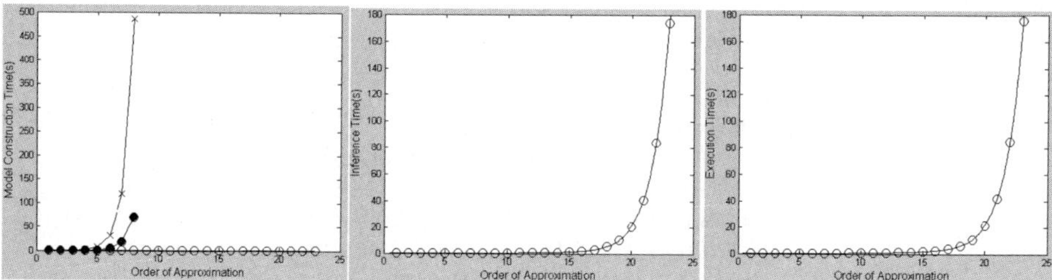

Fig. 4. The IJCV1 model construction, inference and execution times for the Happiness (hollow circles), Intelligence (filled circles) and Phenome (crosses) models

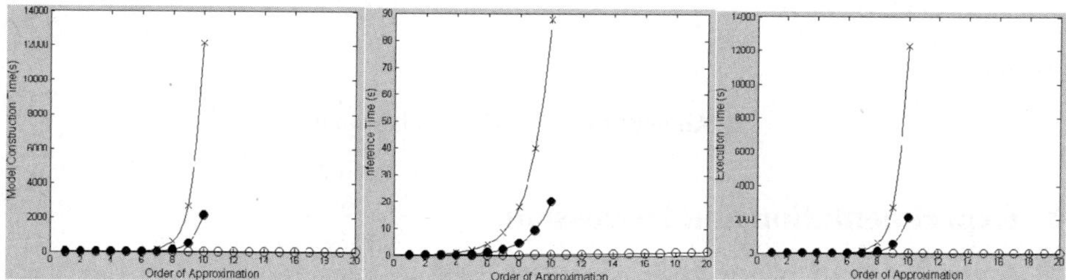

Fig. 5. The IJCV2 model construction, inference and execution times for the Happiness (hollow circles), Intelligence (filled circles) and Phenome (crosses) models

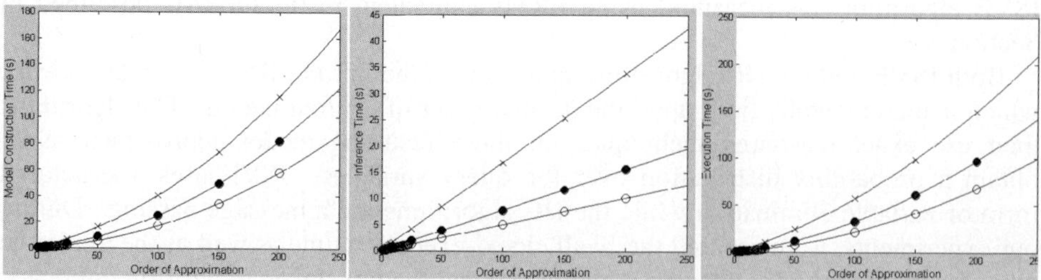

Fig. 6. The ISVE model construction, inference and execution times for the Happiness (hollow circles), Intelligence (filled circles) and Phenome (crosses) models

Figs 4a, 5a and 6a illustrate that, as expected, the time taken by each algorithm to automatically construct the model is dependent on the model. For example, the model construction times for the more complex Phenome model are considerably longer than those of the simpler Happiness model. This is expected because during each iteration, the Phenome model adds $2^{(n-2)}\times10$ nodes, the Intelligence model adds 2^n nodes, and the Happiness model adds 1 node to the model. Figs 4b, 5b and 6b show that, as expected, the time taken by each algorithm to perform inference is also model dependant.

Comparing Figs 4a and b shows that, for the Happiness model, the IJCV1 algorithm spends the majority of its time in the inference phase. The ratio of model construction time to inference time decreases as the order of approximation increases. For the IJCV2 algorithm, the model construction phase is the dominant phase for all three models (reflected in Figs 5a and b). We observed that for all three models, the ratio of model construction time to inference time steadily increases as the order of approximation increases. The rate of this increase was found to be dependant on the model. For the ISVE algorithm, the model construction phase is the dominant phase for all three models (reflected in Figs 6a and b). The ratio of model construction time to inference time increases linearly for each of the three models.

In Fig 6c, we observe what was theorized in [1]: that under certain restrictions, which our models satisfy, the amount of work done in each iteration of ISVE is a constant. The observed behaviour of the ISVE execution time is quadratic, but the coefficient of the squared term is approximately 1% of the coefficient of the linear term. So the behaviour of this algorithm is clearly dominated by the linear term.

Figs 7-9 present an analysis of the relative performance of the algorithms for each model in turn. Comparing these figures allows us to draw conclusion that, for each phase, the relative performance of the algorithms is dependant on the model used.

For the Phenome and Intelligence models, the performance of the three algorithms is comparable for $n \leq 7$. At this point, the IJCV1 and IJCV2 algorithms begin to diverge significantly. For both these models, the IJCV2 experiences a blowout in the model construction time which can be seen clearly in Figs 8a and 9a.

For the Happiness model, Fig 7 shows that the performance of the IJCV1 and ISVE algorithms is comparable for $n \leq 17$. However, after this, IJCV1 experiences a blowout in inference time, which can be seen clearly in Fig 7b. Figs 7a and c show that while the IJCV2 algorithm experiences a marked increase in model construction time, it overall performance remains comparable to ISVE for $n \leq 40$. Fig 7 also shows that IJCV2 outperforms IJCV1 for all orders of approximation.

Thus for small orders of approximation, the three algorithms are comparable. However, for larger orders of approximation, the ISVE algorithm is clearly superior. Being a query based approach, the ISVE algorithm produces a query specific network, while the IJC, having a data based approach, produces a situation specific model approximation. As discussed in [4], a query based approach has some disadvantages in data driven domains. As the recursive modeling examples considered in this paper are not data driven, the IJC algorithms have little or no advantage.

For the models considered so far in this paper, both the IJC algorithms and ISVE take advantage of the structure of the domain by encapsulating information within individual objects so that only a small amount of information needs to be passed between objects. And both types algorithms take advantage of reuse. The IJC algorithm takes advantage of reuse by building on the model from the previous iteration and reusing class models and junction trees. ISVE takes advantage of reuse by maintaining a persistent cache of results of queries executed on the generic instances. By maintaining such a cache, computations can be reused for different generic instances of the same class. This is very beneficial when there are many reoccurrences of generic unnamed instances of the same type of class in the model, as occurs in the examples we have considered. Another major advantage of this caching is that solutions can be reused from one iteration to the next. However, this caching is mainly useful for generic instances. In cases where the knowledge base consists of a finite set of nameod instances or a mixture of generic unnamed instances and named instances, we would expect ISVE to loose some of the advantages afforded by caching because ISVE has no mechanism for reuse for named instances.

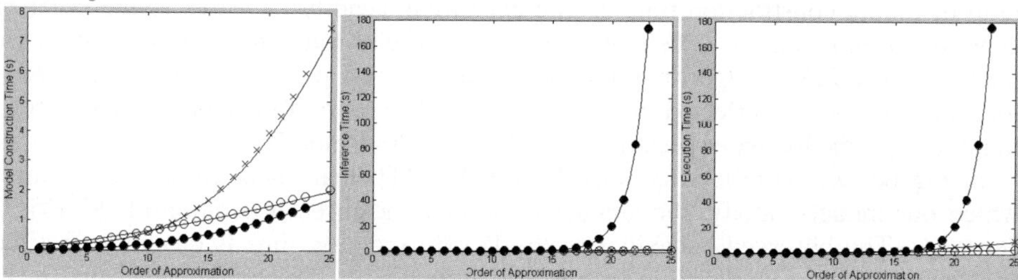

Fig. 7. The ISVE (hollow circles) IJCV1 (filled circles) and IJCV2 (crosses) model construction, inference and execution times for the Happiness model

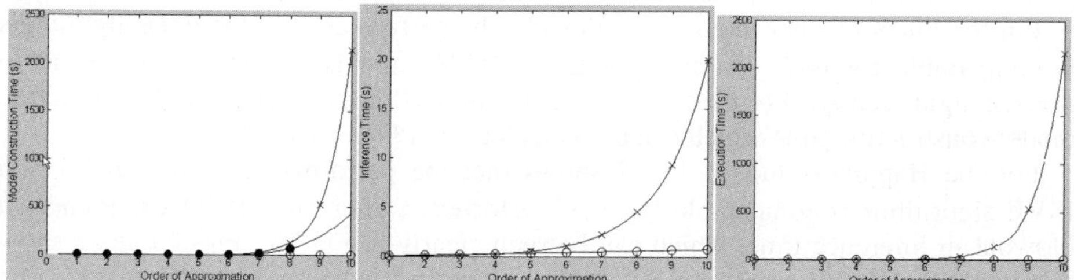

Fig. 8. The ISVE (hollow circles) IJCV1 (filled circles) and IJCV2 (crosses) model construction, inference and execution times for the Intelligence model

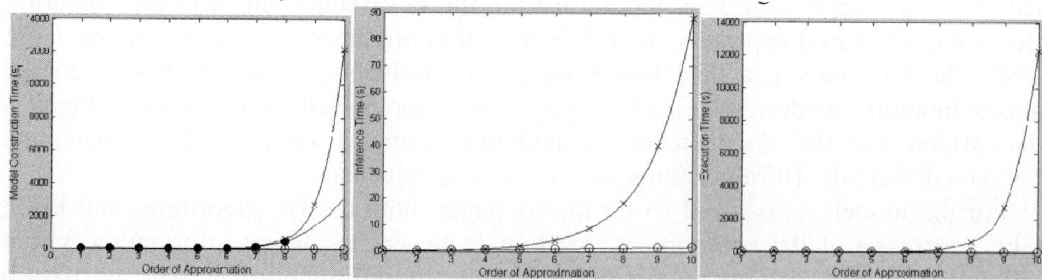

Fig. 9. The ISVE (hollow circles) IJCV1 (filled circles) and IJCV2 (crosses) model construction, inference and execution times for the Phenome model

In order to compare the performance of the IJC algorithms and ISVE, we need to be able to define objective metrics for measuring the quality of the solution provided by the algorithm at any given time. A traditional measure for the quality of an approximate algorithm is the Kullback-Leibler Entropy. We cannot use this measure for algorithms applied to RPMS: as these models can potentially include an infinite number of variables, the true probability distribution over the query variables can not be calculated.

The behaviour of an anytime algorithm is often described using a performance profile (PP) which describes how the quality of the algorithm output varies as a function of some varying resource such as execution time, inference accuracy or representational precision. The quality of the output can be any characteristic of the output which is deemed to be significant. The output of the ISVE and IJC algorithms is an approximation of the probability distribution over the query variables. So, in order to compare the performance of the algorithms, a metric for measuring the quality of the approximation of the probability distribution over the query variables at any given time is required. Fig 10 shows the PP of the three algorithms for the three models where the quality is defined as one minus the normalized rate of change of the probability distribution over the query variable. This figure shows that the quality measure of all three algorithms is a non-decreasing function of execution time and that the rate of change of the quality of the solution is large at the early stages of execution and decreases over time. As can be seen from Fig 10, for all models, the IJC algorithms reach a steady state in a shorter time. However, because these plots are a representation of the quality/execution time trade-off, they do not reflect the blowouts experienced by the IJC algorithms for larger values of n.

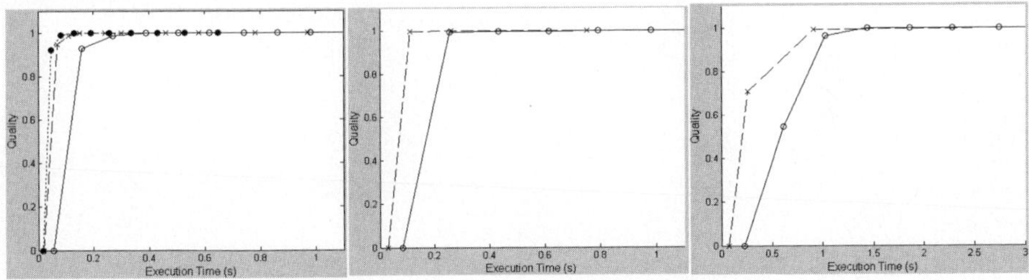

Fig. 10. The quality of the ISVE (hollow circles), IJCV1 (filled circles) and IJCV2 (crosses) algorithms for the Happiness, Intelligence and Phenome models

5 Conclusions

In this paper we have applied the OPRML to RPMs. We developed two iterative extensions to the JC algorithm. These algorithms make it possible to compute an approximate solution to the infinite BN described by a RPM in a finite time. We compared the IJC algorithms against the ISVE algorithm, highlighting the strengths and limitations of each for the recursive domain. We found that, as you would expect, the comparative performance of the algorithms is model dependant.

References

1. Pfeffer, A. and D. Koller. Semantics and Inference for Recursive Probability Models. in National Conference on Artificial Intelligence (AAAI). 2000.
2. Pfeffer, A.J., Probabilistic Reasoning for Complex Systems, PhD thesis in Department of Computer Science. Stanford University. 1999.
3. Howard, C. and M. Stumpter. Situation Assessments Using Object Oriented Probabilistic Relational Models. in Proc.8th Int'l Conf on Information Fusion. Philadelphia. 2005.
4. Howard, C. and M. Stumptner. Model Construction Algorithms for Object Oriented Probabilistic Relational Models. in 19th Int'l Florida Artificial Intelligence Research Society's Conf. Florida. 2006.

Forgetting and Knowledge Update

Abhaya Nayak[1], Yin Chen[2], and Fangzhen Lin[3]

[1] Intelligent Systems Group, Department of Computing
Division of ICS, Macquare University, Sydney, NSW 2109, Australia
abhaya@ics.mq.edu.au
[2] Department of Comp. Sc., South China Normal University
Guangzhou, P.R. China
ychen@cs.ust.hk
[3] Department of Comp. Sc., Hong Kong U of Sc. & Tech.
Clear Water Bay, Hong Kong
flin@cs.ust.hk

Abstract. Knowledge Update (respectively Erasure) and Forgetting are two very different concepts, with very different underlying motivation. Both are tools for knowledge management; however while the former is meant for accommodating new knowledge into a knowledge corpus, the latter is meant for modifying – in fact reducing the expressivity – of the underlying language. In this paper we show that there is an intimate connection between these two concepts: a particular form of knowledge update, namely the KM update using Dalal Distance, and literal forgetting are inter-definable. This connection is exploited to enhance both our understanding of update as well as forgetting in this paper.

Keywords: Knowledge Update, Forgetting, Dalal Distance.

Knowledge management involves removal of dated information as well as incorporation of new information. Relevant literature provides many different approaches to accomplishing this tasks, including belief change (revision and contraction) [1,2], knowledge update (update and erasure) [3], belief merging [4,5] and forgetting (forgetting and remembering) [6,7,8]. While these are related but different concepts, there is a crucial difference between the first three on one hand and forgetting on the other. The former three manipulate knowledge – they deal with the stuff deemed known; the beliefs. Forgetting on the other hand manipulates the language in which knowledge is expressed. Consider erasure and forgetting, for instance. We may dramatize the difference between them as follows: while belief erasure purports to answer the question "What should I believe if I can no longer support the belief that the cook killed Cock Robin?", forgetting purports to answer the question "What should I believe if *Killing* was a concept not afforded in my language?".[1]

On the face of it, it would appear that forgetting on the one hand, and belief change (or update) on the other are completely different concepts, and the relation, if any, between the language manipulating function forgetting and the belief manipulating functions such as belief change and belief update, would be a tenuous one. The difference becomes very explicit from the way these functions are constructed. In a certain sense,

---

[1] It's a case of Relational forgetting, of which, literal forgetting that we will largely deal with in this paper is a special case – propositions are 0-ary relations.

A. Sattar and B.H. Kang (Eds.): AI 2006, LNAI 4304, pp. 131–140, 2006.
© Springer-Verlag Berlin Heidelberg 2006

the result of forgetting a primitive relation (including 0-ary relations or atomic sentences) is computed by projecting the original knowledge corpus into a less expressive language, one in which the relation in question has been eliminated. On the other hand, the result of discarding a belief is computed by removing the belief in question as well as a judicious selection of its supporting beliefs – a process for which, an extra-logical choice mechanism (such as a distance function between the possible worlds or entrenchment relation among beliefs) is utilised. Perhaps surprisingly, it turns out that belief update (respectively erasure) and forgetting, in a qualified way, are inter-translatable. This paper focuses on this inter-translatability.

While we believe that our results can be generalised to a broad spectrum of belief manipulating functions, we largely restrict out attention to belief update and belief erasure, based on Hamming Distance (or Dalal measure) [9] between worlds on the one hand, and literal forgetting as applied to a finite representation of the knowledge base on the other [8]. We believe this connection will have bearing upon many interesting related issues.

1   Background

Belief manipulation such as updating and erasing, as well as language manipulation such as forgetting involve representation of an agent's current stock of beliefs in a specified language as well as a manipulator operator. In this section we detail crucial parts of the notation used, as well as how the notions of update, erasure and forgetting are formally captured. We adopt a few other conventions that are not explicated here due to lack of space, but we hope they will be obvious from the context.

1.1   Notation

We shall consider a framework based on a finitary propositional language $\mathcal{L}$, over a set of atoms, or propositional letters, $\mathcal{A} = \{a, b, c, \ldots\}$, and truth-functional connectives $\neg, \wedge, \vee, \rightarrow$, and $\leftrightarrow$. $\mathcal{L}$ also includes the truth-functional constants $\top$ and $\perp$. To clarify the presentation we shall use the following notational conventions. Upper-case Roman characters (A, B, $\ldots$) denote sets of sentences in $\mathcal{L}$. Lower-case Roman characters (a, b, $\ldots$) denote arbitrary sentences of $\mathcal{L}$.

An *interpretation* of $\mathcal{L}$ is a function from $\mathcal{A}$ to $\{T, F\}$; Ω is the set of interpretations of $\mathcal{L}$. A *model* of a sentence x is an interpretation that makes x true, according to the usual definition of truth. A model can be equated with its defining set of literals, or alternatively, with the corresponding **truth-vector**. $[x]_{\mathcal{L}'}$ denotes the set of models of sentence x over any language $\mathcal{L}'$, while by $[x]$ we will denote $[x]_{\mathcal{L}}$. For interpretation ω we write $\omega \models x$ for x is true in ω. For $x \in \mathcal{L}$, we will define $\mathcal{L}(x)$, the sub-language of $\mathcal{L}$ in which x is expressed, as comprising the minimum set of atoms required to express x, as follows, where x_q^p is the result of substituting atom q everywhere for p in x:

$$\mathcal{L}(x) = \{p \in \mathcal{A} \mid x_\top^p \not\equiv x_\perp^p\} \cup \{\top, \perp\}$$

This set of atoms is unique. Thus $\mathcal{L}(p \wedge (q \vee \neg q)) = \mathcal{L}(p) = \{p, \top, \perp\}$. This can be extended to sets of sentences in the obvious way. It follows trivially that if $\models x \leftrightarrow y$ then $\mathcal{L}(x) = \mathcal{L}(y)$. Also note that if $\models x$ then $\mathcal{L}(x) = \{\top, \perp\}$.

By a theory or belief set we will mean a subset of $\mathcal{L}$ that is closed under Cn. We denote the set of all theories by $\mathcal{T}$. Since $\mathcal{L}$ is finitary, any theory T can be represented by a finite set $T' \subseteq \mathcal{L}$ such that $Cn(T') = T$, and consequently by a single sentence $\bigwedge_{x_i \in T'} x_i$. Traditionally, in the case of knowledge update, erasure and forgetting, a belief corpus is represented by a single sentence, where as in accounts of belief change, knowledge corpus is represented as a theory. Both update and erasures are defined as functions $\oplus, \ominus : \mathcal{L} \times \mathcal{L} \to \mathcal{L}$ where as forgetting is defined as a function $\odot : \mathcal{L} \times \mathcal{A} \to \mathcal{L}$. Thus, for instance, $k \oplus e$ and $k \ominus e$ are respectively the sentences representing the update and erasure of sentence e from the knowledge base represented by sentence k. On the other hand, $k \odot a$ is the theory that results from forgetting atom a from knowledge k.

1.2 Update and Erasure

Let $\leq_\omega$ for any world ω be a preorder (reflexive and transitive relation) over Ω that is *faithful* with respect to ω, ie., for every $\omega' \in \Omega$, we have both $\omega \leq_\omega \omega'$ and if $\omega' \leq_\omega \omega$ then $\omega = \omega'$. If, in addition, either $\omega' \leq_\omega \omega''$ or $\omega'' \leq_\omega \omega'$, for every pair ω', ω'', then it is a *total* preorder. By $\omega_1 \leq_\omega \omega_2$ we intuitively mean that ω_2 is at least as similar to ω as ω_1 is. As usual, the strict part of $\leq_\omega$ is denoted by $<_\omega$ This relation compares worlds with respect to their similarity or proximity to ω. By $\omega' \sim_\omega \omega''$ we will denote that ω' and ω'' are not comparable under $\leq_\omega$; and by $\omega' \equiv_\omega \omega''$ we will understand that both $\omega' \leq_\omega \omega''$ and $\omega'' \leq_\omega \omega'$. Given a set Δ of worlds, by $min_{\leq \omega}(\Delta)$ we denote the set $\{\omega' \in \Delta \mid \omega'' \not<_\omega \omega'$ for any $\omega'' \in \Delta\}$ of worlds in Δ that are closest or most similar to ω.

Definition 1 (Update). $[k \oplus e] = \bigcup_{\omega \in [k]} min_{\leq \omega}([e])$

Intuitively, when we learn that e has been effected, with respect to each world allowed by our current knowledge, we compute what the scenario would be if e was effected in that world; and collating all those worlds results in the models of our updated knowledge. Obtaining a knowledge base k' from these models is not difficult.

Definition 2 (from Update to Erasure). *The erasure operation is defined by reduction to update using the Harper Identity.* $k \ominus e = k \vee (k \oplus \neg e)$

Alternatively, Erasure can be directly defined, and update can be defined from erasure using the Levi Identity.

Definition 3 (Erasure). *Erasure can be directly defined as:*
$[k \ominus e] = [k] \cup \bigcup_{\omega \in [k]} min_{\leq \omega}([\neg e])$

Definition 4 (from Erasure to Update). $k \oplus e = (k \ominus \neg e) \wedge e$

The definitions we have provided of Update and Erasure are constructive. Alternatively these operations can be defined as those (with appropriate signature) that satisfy the respective rationality postulates of update and erasure. The postulates can be found in [3].

It is easily noticed that $[k \oplus e] \subseteq [e]$, equivalently, $k \oplus e \vdash e$, i.e., a successful update by a sentence e leads to e being believed. Furthermore, since we assume that a world closer it itself than any other world, and the worlds in $[e]$ that are $\leq_\omega$-minimal with respect to *any* world ω in $[k]$ constitute $[k \oplus e]$, it is clear that $[k] \cap [e] \subseteq [k \oplus e]$. Since $[k] = ([k] \cap [e]) \cup ([k] \cap [\neg e])$, we get the following as an immediate consequence:

Observation 1. $[k] \subseteq ([k \oplus e] \cup [k \oplus \neg e])$ *wherefore* $k \vdash (k \oplus e) \vee (k \oplus \neg e)$.

Now, it follows from Definition 2 that $(k \ominus e) \vee (k \ominus \neg e) = k \vee (k \oplus e) \vee (k \oplus \neg e)$. However, as we just noticed, $k \vdash (k \oplus e) \vee (k \oplus \neg e)$. This leads to:

Observation 2. $(k \oplus e) \vee (k \oplus \neg e) \equiv (k \ominus e) \vee (k \ominus \neg e)$.

We will call $(k \oplus e) \vee (k \oplus \neg e)$ the *symmetric update*, and $(k \ominus e) \vee (k \ominus \neg e)$ the *symmetric erasure*, of k by e. [2] Thus Observation 2 shows the equivalence between symmetric update and symmetric erasure.

1.3   Literal Forgetting

Forgetting, as we mentioned earlier, is more about language manipulation than about belief manipulation. While belief erasure (say by e) involves traversing to a judicious belief state where e is not believed (while allowing the possibility of $\neg e$ still believed), forgetting of e involves moving to a state where no information regarding e is still retained. Forgetting has a clear meaning when e in the above case is an atomic sentence – forgetting of an atom a means moving to a state where neither a nor $\neg a$ is believed. Hence it is often called *literal forgetting*.

Let us look at forgetting from an information theoretic point of view. We are considering the forgetting of atom a from knowledge base k. For an interpretation ω of k, we will call ω' the a-dual of ω if it differs from ω exactly on the truth assignment to a. For instance, if the first bit in $\omega = 1011$ is the assignment to a, then the a-dual of ω is 0011. A belief base k has information regarding atom a just in case $[k]$ contains some interpretation ω but not its a-dual.

Definition 5 (a-**dual closure**). *Given a set of interpretations Δ and atomic sentence a, Δ is closed under a-dual iff the a-dual of every interpretation $\omega \in \Delta$ is also in Δ.*

Clearly, if Δ is closed under a-dual, then it contains no information regarding atom a. Furthermore, if Δ is closed under a-dual for every atom $a \in \mathcal{A}$, then Δ has no information at all. Since forgetting by a from k is meant to removing all information pertaining to a from k, it suggests the operation of closing the set of interpretations $[k]$ under a-dualship. Accordingly we define the dual closure operation $\uplus : 2^{\Omega} \times \mathcal{A} \to 2^{\Omega}$ as follows:

Definition 6 (dual-closure operation). *Given a set of interpretations $\Delta \subseteq \Omega$ and atomic sentence $a \in \mathcal{A}$, the dual-closure $\uplus(\Delta, a)$ of Δ under a is the smallest subset of Ω that includes Δ and is closed under a-dual. In other words, $\uplus(\Delta, a) = \Delta' \subseteq \Omega$ such that (1) $\Delta \subseteq \Delta'$, (2) Δ' is closed under a-dual, and (3) if any Δ'' such that $\Delta \subseteq \Delta'' \subseteq \Delta'$ is closed under a-dual, then $\Delta' = \Delta''$.*

This definition of $\uplus$ is non-constructive. It also presumes that $\uplus(\Delta, a)$ is unique. Theorem 1 below guides its construction, and justifies the presumption in question. We need one more definition before producing this result.

---

[2] Our terminology is somewhat at variance with that used by Katsuno and Mendelzon [3] who refer to $(k \oplus e) \vee (k \oplus \neg e)$ as the symmetric erasure of e from k, and mention that Winslett, in an unpublished manuscript, calls it the forgetting of e from k.

Definition 7 (Model Counterparts). *For language $L \subseteq \mathcal{L}$, by $[S]_{\Omega_L}$ we denote the set of "model counterparts" of S in Ω_L. It can be represented as the set of interpretations ω_2 in Ω_L that are sub-model of some interpretation or other, $\omega_1 \in [S]$ as: $\{\omega_2 \in \Omega_L \mid \exists_{\omega_1 \in [S]}\omega_2 \subseteq \omega_1$. Alternatively, it is the set of interpretations in Ω that are super models of any such model ω_2 in Ω_L: $\{\omega \in \Omega \mid \exists_{\omega_1 \in [S],\ \omega_2 \in \Omega_L}\omega_2 \subseteq \omega_1$ and $\omega_1 \subseteq \omega\}$.*

Example 1. Consider for example $\mathcal{A} = \{a, b, c\}$, $L = \mathcal{L}(\{b, c\})$ and $S = \{a \leftrightarrow b\}$. Now, $[S] = \{11\star, 00\star\}$, and its submodels in Ω_L are $\{-\star\star\} = \Omega_L$.[3] Alternatively, the desired set is $\{\star\star\star\} = \Omega$.

Theorem 1. [4] *Given a set of interpretations $\Delta \subseteq \Omega$, an atomic sentence $a \in \mathcal{A}$ and a sentence d such that $[d] = \Delta$ (a canonical construction of d in DNF[5] can be easily carried out taking the composition of Δ as a guide), $\uplus(\Delta,\ a) = [d]_{\Omega_{\mathcal{L}}(\mathcal{A}\backslash\{a\})}$.*

We introduced the concept of dual-closure being motivated by an intuitive notion of literal forgetting. Now we are in a position to offer a semantic definition of literal forgetting.

Definition 8 (Semantics of literal forgetting). *The forgetting operation $\odot: \mathcal{L} \times \mathcal{A} \to \mathcal{L}$ should be such that $[\odot(k,\ a)] = \uplus([k],\ a)$.*

The following example illustrates that this definition captures the semantic intuition behind forgetting. It also substantiates our generic claim in the introductory note that forgetting is more about language manipulation than about belief manipulation.

Example 2. Let $\mathcal{A} = \{a, b, c\}$ and $k = (a \wedge \neg b) \vee c$. Thus $[k] = \{001, 101, 100, 111, 011\}$. We tabulate two distinct but equivalent constructions of $[\odot(k,\ a)]$ in accordance with the two definitions of $[S]_{\Omega_L}$. In some truth vectors '$-$' is used to indicate the absence of relevant atoms.

$[k]$	$[\odot_1(k,\ a)]$	$[\odot_2(k,\ a)]$
001	-01	001
101		101
100	-00	100
		000
111	-11	111
011		011

Thus $\odot_1(k,\ a)$ can be expressed as $\neg(b \wedge \neg c)$ where as $\odot_2(k,\ a)$ can be expressed as $\neg(a \wedge b \wedge \neg c) \wedge \neg(\neg a \wedge b \wedge \neg c)$. While the latter is expressed in $\mathcal{L}$, the former is couched in a language devoid of the atom a. Nonetheless, both are equivalent, and are equivalent to $(b \to c)$ as expected.

---

[3] From here onwards we use $\star$ to indicate atoms with "wild" truth-values, and $-$ to indicate location of "non-applicable atoms" by a given interpretation.

[4] Proofs of results provided in this paper can be obtained from the first named author upon request.

[5] A formula in **D**isjunctive **N**ormal **F**orm is a disjunction of conjunctive clauses, where a *conjunctive clause* is a conjunction of literals.

Now that the semantics of forgetting is in place, we consider its syntactic characterisation. For this purpose, we borrow the definition of forgetting from [8].

Definition 9 (Syntactic representation of literal forgetting). *For any formula x and atom a, denote by $x|_{1\leftarrow a}$ the result of replacing every occurrence of a in x by $\top$, and by $x|_{0\leftarrow a}$ the result of replacing every occurrence of a in x by $\bot$. Forgetting an atom a from a knowledge base represented as sentence k is defined as: $k \odot a = k|_{1\leftarrow a} \vee k|_{0\leftarrow a}$.*

Continuing with the Example 2, we find that $\odot((a \wedge \neg b) \vee c, a) \equiv ((\top \wedge \neg b) \vee c) \vee ((\bot \wedge \neg b) \vee c) \equiv \neg b \vee c \equiv (b \to c)$ matching the result from semantic analysis. The following result shows that this match is not due to a lucky choice of example.

Theorem 2. *For any sentence $k \in \mathcal{L}$, atom $a \in \mathcal{A}$ and world $\omega \in \Omega$, it holds that $\omega \in [\odot(k, a)]$ (as defined in Definition 9) if and only if $\omega \in \uplus([k], a)$.*

The Definition 9 also makes it clear that $\oplus$ is very syntax-sensitive. Consider for instance four knowledge bases that are equivalent when conjunctively interpreted: $B_1 = \{a, b\}$, $B_2 = \{a \wedge b\}$, $B_3 = \{a \to b, a \vee b\}$, and $B_4 = \{a, \neg a \vee b\}$, and we are to forget a. If we were to understand forgetting from a set as piece-wise forgetting from the member-formulas in the set, then we get as result: $B_1' = \{\top, b\}$, $B_2' = \{b\}$, $B_3' = \{\top\}$ and $B_4' = \{\top\}$. This explains why we represent the knowledge base as a single sentence. If, however, piece-meal processing is desirable, it can be done by representing the knowledge base as a set of conjunction of literals, disjunctively interpreted (corresponding to DNF), and process the set-members individually. This would closely correspond the semantic account given in Example 2. In contrast, if a base is represented as a set of clauses, it will give the wrong result, as in the case of B_4 above.

2 Connecting Update with Forgetting

We have noticed that update and erasure are belief manipulating operations whereas forgetting is a language manipulating operation. Nonetheless, it is clear that both in case of erasure and forgetting, beliefs are lost. We now explore whether, and if so, how they can be inter-defined.

The first obstacle to this is the fact that where as in case of erasure, a particular relational measure is assumed to exist over Ω, no such measure is assumed in case of forgetting. Inter-translatability between these two concepts therefore will therefore push us to either impose such a external measure on Γ and accordingly generalize the definition of forgetting, or restrict us to measures on Γ that are implicitly given by the knowledge base k itself and look at more restricted forms of erasure. Lang and Marqis [6] choose the former option. In this paper we choose the latter. This brings us to specific approaches to erasure (and updating) – in particular one based on Hamming Distance – between interpretations.

2.1 A Concrete Update Operator Via Hamming-Distance

Belief update operators presume a preorder over interpretations, as do belief revision operators. Two such classes preorders have drawn much attention from the researchers in the area, one based on the Hamming Distance between the sets of literals representing two interpretations, and the other based on the symmetric difference between them. The former was introduced for belief update by Forbus [10], and by Dalal [9] for belief revision. On the other hand, Winslett [11] introduced the latter for belief update, and Satoh [12] for belief revision. Here we restrict our attention to updates only, and that too based on Hamming Distance. As popularly called in the area, refer to Hamming distance as "Dalal Distance".

Definition 10 (Dalal Distance). *Given ω, $\omega' \in \Omega_L$ for some language **L**, the Dalal Distance $dist_D(\omega, \omega')$ between ω and ω' is the number of atoms that they assign different values: $dist_D(\omega, \omega') = |\{a \in Atoms(\mathbf{L}) \mid \omega \models a \text{ iff } \omega' \vdash \neg a\}|$.*

Thus, $dist_D(101, 001) = 1$, $dist_D(101, 011) = 2$, and $dist_D(101, 110) = 2$.

Definition 11 (Dalal Preorder $\leq^D$). *Given three interpretations ω, ω' and $\omega'' \in \Omega_L$ for some language **L**, $\omega' \leq^D_\omega \omega''$ iff $dist_D(\omega, \omega') \leq dist_D(\omega, \omega'')$.*

Thus, we get $001 <^D_{101} 011$, $001 <^D_{101} 110$, and $011 \equiv^D_{101} 110$. It is easily verified that Given a language **L** and a world $\omega \in \Omega_L$, the Dalal Preorder $\leq^D_\omega$ is a faithful, total preorder over Ω_L. The definitions of update and erasure we provided earlier (see Definitions 1 – 4) directly or indirectly employ a preorder. We can thus obtain specific update and erasure operators simply by plugging in Dalal and Winslett preorders.

Definition 12 (Dalal Update and Dalal Erasure). *By $\oplus^D$ and $\ominus^D$ we will denote the Dalal Update and Erase operations obtained by plugging in the Dalal Preorder in the definitions of Update and Erasure provided in Definitions 1 and 2 (or alternatively in Definitions 3 and 4).*

Example 3. Let $\mathcal{A} = \{a, b, c\}$ and $k = (a \wedge \neg b) \vee c$ as in Example 2. Thus $[k] = \{001, 101, 100, 111, 011\}$ and $[a] = \{100, 101, 110, 111\}$. We consider updates and erasures of k by a.   We need to compute $\bigcup_{\omega \in [k]} min_{\leq\omega}([a])$ for update and $\bigcup_{\omega \in [k]} min_{\leq\omega}([\neg a])$ for erasure with respect to $\leq^D$. Since $\leq^D$ is faithful, for any sentence e,

$$\bigcup_{\omega \in [k]} min_{\leq\omega}([e]) = ([k] \cap [e]) \cup \bigcup_{\omega \in [k]\setminus[e]} min_{\leq\omega}([e]).$$

In the table below, the first two columns give the $\leq_\omega$-ordering of $[a]$ for $\omega \in ([k] \setminus [a])$, and the last three columns give the $\leq_\omega$-ordering of $[\neg a]$ for $\omega \in ([k] \setminus [\neg a])$. The comma's are to be interpreted as $\equiv_D$. The minimal elements of interest are in bold face.

$[a]^{\leq^D_{001}}$	$[a]^{\leq^D_{011}}$	$[\neg a]^{\leq^D_{100}}$	$[\neg a]^{\leq^D_{101}}$	$[\neg a]^{\leq^D_{111}}$
110	100	011	010	000
100, 111	101, 110	001, 010	000, 011	001, 010
101	**111**	**000**	**001**	**011**

Accordingly we get:

$[k \oplus^D a] = \{100, 101, 111\}$ and $[k \ominus^D a] = \{001, 011, 100, 101, 111, 000\}$, whereby, $(k \oplus^D a) \equiv a \wedge (b \rightarrow c)$ and $(k \ominus^D a) \equiv b \rightarrow c$. Furthermore, using Definitions 2 and 4 (or going through semantics again) we can also show that $(k \oplus^D \neg a) \equiv \neg a \wedge (b \rightarrow c)$ and $(k \ominus^D \neg a) \equiv (a \wedge \neg b) \vee c \equiv k$.

2.2   Updating, Erasing and Forgetting – Literally

We discussed updating and erasing with Dalal and Winslett preorders so that these accounts of belief manipulation and the account of language manipulation via forgetting are on the same footing, with the hope that we can easily explore the connection between them. Since the account of forgetting involves forgetting of literals, we will also assume in this section that the update and erase in question are those by literals. Hence it is prudent to start with exploring the connection between them, and later on to convert that connection with respect to updates.

Now, both forgetting and erase are operations that *remove* information. In fact, while erase removes some information involving the sentence a, forgetting removes *all* information regarding the atom a – it semantically removes a itself. Hence it would appear that perhaps forgetting of a would involve erasure of both a and $\neg a$. The following result shows that it indeed is the case – forgetting of atom a indeed results in that part of the knowledge that is left untouched by both erasure of a and erasure of $\neg a$ under Dalal Preorder!

Theorem 3 (Dalal Erasure to Forget). *Given k a knowledge base, $k \odot a \equiv (k \ominus^D a) \vee (k \ominus^D \neg a)$ for every atom a.*

Appealing to Observation 2 and Theorems 3, we right away obtain:

Corollary 1 (Dalal Update to Forget). $k \odot a \equiv (k \oplus^D a) \vee (k \oplus^D \neg a)$ *for any knowledge base k and any atom a.*

Putting the results Theorems 3 and Corollary 1 we obtain:

Theorem 4 (Dalal Update/Erasure to Forget). *For any knowledge base k and any atom a, the formula $k \odot a$ is equivalent to each of the following two formulas:*

1. $(k \oplus^D a) \vee (k \oplus^D \neg a)$
2. $(k \ominus^D a) \vee (k \ominus^D \neg a)$

Theorem 4 shows that forgetting of a knowledge base k by atom a can be identified, individually, with both the symmetric update and the symmetric erasure of k by a where the update and erasure operations in question are based on Dalal preorder. The following example illustrates this result.

Example 4. Let's revisit Example 3. We know that

1. $(k \oplus^D a) \equiv a \wedge (b \rightarrow c)$
2. $(k \ominus^D a) \equiv b \rightarrow c.$
3. $(k \oplus^D \neg a) \equiv \neg a \wedge (b \rightarrow c).$
4. $(k \ominus^D \neg a) \equiv (a \wedge \neg b) \vee c.$

Plugging in these values, we get

– $(k \oplus^D a) \vee (k \oplus^D \neg a) \equiv$
  $(a \wedge (b \rightarrow c)) \vee (\neg a \wedge (b \rightarrow c)) \equiv b \rightarrow c.$
– $(k \ominus^D a) \vee (k \ominus^D \neg a) \equiv$
  $(b \rightarrow c) \vee ((a \wedge \neg b) \vee c) \equiv b \rightarrow c.$

Thus in either case the value obtained for $k \odot a$, reduced to $\oplus^D$ or $\ominus^D$, is $b \rightarrow c$, matching the value of $k \odot a$ independently obtained in Example 2.

Now that we know how Literal Forgetting can be reduced to update (or erase) of a special nature, we consider the question whether the reduction can be done in the reverse direction. Theorem 5 below shows that Dalal literal updates can be defined via forgetting. The similarity between this result, and the definition of updating via erasure (Definition 4) is rather striking.

Theorem 5 (Forget to Update). *Given any knowledge base k and any atom a, the following equivalences hold:*

1. $(k \oplus^D a) \equiv (k \odot a) \wedge a$
2. $(k \oplus^D \neg a)\, \neg a) \equiv (k \odot a) \wedge \neg a$

This result shows how the update of a knowledge base by a literal, modulo Dalal preorder, can be computed purely by syntactic manipulation in an efficient manner. Syntactic characterisation of most belief update and belief revision operators have been done by del Val [13,14]. del Val's approach assumes conversion of the relevant portion of the knowledge base k, and the evidence e into DNF. However, as pointed out by Delgrande and Schaub [15], this may require exponential time step and exponential space. Clearly update and erasure operations via forgetting are a lot more frugal.

Applying Definition 2, that $k \ominus e = k \vee (k \oplus \neg e)$, to the definition of updates in Theorem 5, and noting that $[k] \subseteq [k \odot a]$, we reduce Dalal erasure to forgetting as well:

Theorem 6 (Forget to Erase). *Given any knowledge base k and any atom a, the following equivalences hold:*

1. $(k \ominus^D a) \equiv (k \odot a) \wedge (a \rightarrow k)$
2. $(k \ominus^D \neg a) \equiv (k \odot a) \wedge (\neg a \rightarrow k)$

In Example 5 below we complete the circle in the sense that the results of updates and erasures computed in accordance with Theorems 5 and 6 are shown to match to those independently computed in Example 3.

Example 5. We know from Example 2 that $(k \odot a) \equiv (b \rightarrow c)$. Applying Theorems 5 and 6 we get

1. $(k \oplus^D a) \equiv (b \rightarrow c) \wedge a$
2. $(k \oplus^D \neg a) \equiv (b \rightarrow c) \wedge \neg a$
3. $(k \ominus^D a)$
   $\equiv (b \rightarrow c) \wedge a \rightarrow ((a \wedge \neg b) \vee c)$
   $\equiv (b \rightarrow c) \wedge (a \rightarrow (b \rightarrow c)) \equiv (b \rightarrow c).$
4. $(k \ominus^D \neg a)$
   $\equiv (b \rightarrow c) \wedge \neg a \rightarrow ((a \wedge \neg b) \vee c)$
   $\equiv (b \rightarrow c) \wedge ((a \wedge \neg b) \vee (a \vee c))$
   $\equiv (b \rightarrow c) \wedge (a \vee c) \equiv (a \wedge \neg b) \vee c.$

matching the values of updates and erasures independently computed in the Example 3.

3 Conclusion

In this paper, we started with intention of showing that belief-manipulation mechanisms such as updating and language manipulation mechanisms like forgetting are interconnected. In Section 1, after introducing the notation, we provided the background information on both knowledge update and erasure, restricted to Dalal measure. In the next section we showed how update and erasure, given Dalal measure and a literal input on the one hand, and forgetting on the other, are inter-definable.

Our preliminary results show that these results can be extended to more general measures such as Winslett measure. We also believe interesting accounts of forgetting, where the input is an arbitrary sentence, can be given based on the interconnections provided here. We will address these issues in our future work.

References

1. Alchourrón, C.E., Gärdenfors, P., Makinson, D.: On the logic of theory change: Partial meet contraction and revision functions. Journal of Symbolic Logic **50** (1985) 510–530
2. Gärdenfors, P.: Knowledge in Flux: Modeling the Dynamics of Epistemic States. Bradford Books, MIT Press, Cambridge Massachusetts (1988)
3. Katsuno, H., Mendelzon, A.O.: On the difference between updating a knowledge base and revising it. In Gärdenfors, P., ed.: Belief Revision. Cambridge (1992) 183–203
4. Everaere, P., Konieczny, S., Marquis, P.: Introduuota and gmin merging operators. In: International Joint Conference on Artificial Intelligence. (2005) 424–429
5. Konieczny, S., Pinoperez, R.: Merging information under constraints: a qualitative framework. Journal of Logic and Computation **12(5)** (2002) 773–808
6. Lang, J., Marquis, P.: Resolving inconsistencies by variable forgetting. In: Proceedings of the 8th KR. (2002) 239–250
7. Lang, J., Liberatore, P., Marquis, P.: Propositional independence: Formula-variable independence anf forgetting. J. of Artificial intelligence Research **18** (2003) 391–443
8. Lin, F., Reiter, R.: Forget it! In: Procs. of the AAAI Symp. of Relevance. (1994) 154–159
9. Dalal, M.: Investigations into a theory of knowledge base revision: Preliminary report. In: Proceedings of the Seventh National Conference on Artificial Intelligence. (1988)
10. Forbus, K.: Introducing actions into qualitative simulation. In: International Joint Conference on Artificial Intelligence. (1989) 1273–1278
11. Winslett, M.: Reasoning about action using a possible models approach. In: Proceedings of the 7th National Conference in Artificial Intelligence. (1988) 88–93
12. Satoh, K.: Nonmonotonic reasoning by minimal belief revision. In: Proceedings of the International Conference of Fifth Generation Computer Systems. (1988) 455–462
13. del Val, A.: Computing knowledge base updates. In: Third Conference on Pronciple of Knowledge (KR-92). (1992) 740–750
14. del Val, A.: Syntactic characterizations of belief change operators. In: IJCAI. (1993) 540–547.
15. Delgrande, J., Schaub, T.: A consistency-based approach for belief change. Artificial Intelligence **151(1-2)** (2003) 1–41

Enhanced Temporal Difference Learning Using Compiled Eligibility Traces

Peter Vamplew[1], Robert Ollington[2], and Mark Hepburn[2]

[1] School of Information Technology and Mathematical Sciences, University of Ballarat,
PO Box 663, Ballarat, Victoria 3353, Australia
`p.vamplew@ballarat.edu.au`
[2] School of Computing, University of Tasmania, Private Bag 100, Hobart,
Tasmania 7001, Australia
`{Robert.Ollington, Mark.Hepburn}@utas.edu.au`

Abstract. Eligibility traces have been shown to substantially improve the convergence speed of temporal difference learning algorithms, by maintaining a record of recently experienced states. This paper presents an extension of conventional eligibility traces (compiled traces) which retain additional information about the agent's experience within the environment. Empirical results show that compiled traces outperform conventional traces when applied to policy evaluation tasks using a tabular representation of the state values.

1 Introduction

Reinforcement learning addresses the problem of an agent interacting with an environment. At each step the agent observes the current state and selects an action. The action is executed and the agent receives a scalar reward. The agent has to learn a mapping from state-to-action to maximise the long-term reward. One way to do this is to learn the expected return, either per state or per state-action pair. Many algorithms for learning these values are based on the use of temporal differences (TD) [1] where the value of the current state at each step is used to update the estimated value of previous states.

It is well-established that the use of eligibility traces can substantially improve the performance of TD algorithms [2, 3]. These traces maintain a record of states and actions experienced earlier in the current episode, allowing the value of those states and actions to be updated based on the states and rewards encountered later in the episode. However conventional eligibility traces exhibit a fundamental limitation, in that they only record information about the current episode – all information about the previous episode is discarded from the traces whenever a new episode commences. (Watkins' $Q(\lambda)$ algorithm [4] is even more restricted in that it clears all traces whenever a non-greedy action is performed).

In this paper we propose the use of an extended form of eligibility trace which we will refer to as a compiled trace. Compiled traces differ from conventional traces in two important ways: First, they contain information about states experienced in all previous episodes, not just those from the current episode. Second, compiled traces identify and support learning for sequences of states which are possible within the environment but which have yet to actually be observed (Note: for reasons of

© Springer-Verlag Berlin Heidelberg 2006

simplicity we will primarily discuss compiled traces with regards to the task of learning state values under a fixed policy (policy evaluation). However these methods can readily be extended to learning action values, as required in policy-iteration algorithms such as Q-Learning).

2 Motivation for Compiled Traces

Figure 1 illustrates a simple, probabilistic environment designed to illustrate the difference between conventional eligibility traces and compiled traces. Consider a situation in which the first two episodes observed by the agent consisted of the sequences A-C-D-F and B-C-D-G. Assuming that the estimated value of all states starts at 0, then following the first episode the values for states A, C and D will all have been trained towards the reward of -1 received on the final transition. After the second episode, the values for states B, C, and D will have been altered in the direction of the reward of +10. Notice that using conventional traces states A and B will have quite different values at this point, although it can be seen from Figure 1 that in the long run they should have identical values. This illustrates a weakness of conventional eligibility traces — by including only information gathered during a single episode they are sensitive to any stochasticity in the environment.

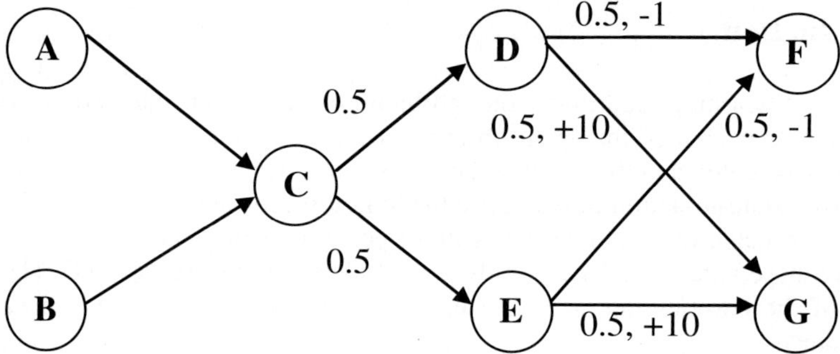

Fig. 1. An environment consisting of 7 states. A and B are the possible starting states for each episode, whilst F and G are terminal states. Arcs are labelled with their transition probabilities and associated rewards; probabilities of 1.0 and rewards of 0 have been omitted for clarity.

This issue can be addressed by extending the eligibility traces to include the values from multiple prior episodes. One relatively simple approach would be to calculate a single-episode trace as per normal, but in addition to calculate and store at each state a mean of the values of the single-episode trace at each time that state was visited. These mean-traces would then be used within the TD updates. In the previous example, this would result in state A also being updated during the second episode as state D would have 0.5 as its mean-trace value for both states A and C.

Whilst this mean-trace approach should offer some benefits compared to conventional single-episode traces in alleviating the effects of under-sampling within probabilistic environments, it still fails to fully utilise all information available from previous episodes. Consider again the environment from Figure 1, and assume that the following three episodes have been observed: A-C-D-F, B-C-E-F and B-C-E-G. Under these circumstances the mean-trace algorithm will not update the value of state A during the third episode because the mean-trace stored for A at E is 0, as no sequence with A as a predecessor of E has ever been observed.

However from the two earlier episodes it is possible to infer that whilst the sequence A-C-E has not yet been observed, it must be possible for it to occur as the sub-sequences A-C and C-E have been observed (assuming the environment to be Markov, which is a standard assumption for many applications of TD methods). Therefore it should be possible to include state A amongst the states to be updated as a result of the third episode. A possible means by which such updates could be achieved is to use the traces stored at each state to indicate not just the extent to which other states have previously been observed as predecessors to this state (as is the case for mean-traces) but more generally the extent to which those other states could occur as predecessors to this state. We will refer to any trace algorithm which works in this manner as a compiled trace, as the trace stored at each state is essentially a compilation of all knowledge about that state's predecessors.

Construction of compiled traces can be achieved through a relatively simple boot-strapping process. Whenever a transition is observed from state s to state s' then the compiled traces stored at s' are updated to reflect a combination of their current values and the values of the compiled traces stored at s.

In the previous example, following the first episode state C will have a non-zero compiled trace for state A, whilst state D will have non-zero traces for both of its observed predecessors (states A and C). Following the second episode, state C will have non-zero traces for both A and B, whilst state E will have non-zero traces for both these states as well (having 'learnt' about state A via state C). Therefore when the transition from E to G occurs during the third episode, the value for state A will be updated (in addition to the values for B and C which would have been updated using conventional or mean traces).

3   Comparison to Other Learning Algorithms

3.1   Conventional Eligibility Traces

The example given in the previous section demonstrates the primary advantages of compiled traces over conventional eligibility traces. By storing information gleaned from all previous episodes rather than just the current episode, compiled traces allow a larger number of states to be updated after each step within the environment which should enable faster learning. This would be expected to be of particular benefit in environments containing a stochastic element in either the state transitions or the rewards, as under-sampling in such environments can skew the estimated values of states. By sharing updates across all possible predecessor states compiled traces reduce the effects of such stochastic variations.

3.2 Model-Based Methods

The ability to learn about sequences of states which are possible within an environment but yet to be observed can also be provided by model-based learning methods such as the Dyna methods [5]. These learning algorithms build a model of the environment by learning the successors of each state on the basis of observed transitions. This model can then be used to generate simulated episodes which can be learnt from in the same manner as episodes experienced in the actual environment.

There are some similarities between model-based methods and compiled trace methods. Both construct values reflecting the likelihood of particular sequences of states being observed on the basis of the actual sequences observed. However the model-based approach is 'forward-looking' and limited in scope in that each state learns the likelihood of every state occurring as an immediate successor, whereas compiled traces are 'backward-looking' and unlimited in scope, as each state learns about all of its predecessor states, no matter how distant.

3.3 Goal-Independent Reinforcement Learning

Some similarities also exist between compiled traces and goal-independent reinforcement learning algorithms such as DG-Learning [6] and Concurrent Q-Learning [7]. These goal-independent algorithms also update the values of states and actions which were not directly involved in the current episode. This is achieved by learning an explicit distance from every state to each other state, and using graph relaxation techniques to dynamically update these values if a shorter path is experienced.

These goal-independent algorithms have similar storage costs as compiled traces, but potentially are more computationally expensive. In addition the relaxation techniques are only applicable in the context of learning distance values, whilst compiled traces can be applied to arbitrary reward functions.

4 Implementing Compiled Traces

The main issue to be resolved in implementing compiled traces is the manner in which the current values of the traces at state s' are combined with the values stored at s when a transition from s to s' is observed (as outlined in Section 2). We believe there are three main approaches which could be utilised.

4.1 All-Visits-Mean (AVM) Traces

This algorithm sets the compiled traces at each state equal to the mean of the traces of its immediate predecessor states, weighted by the frequency of occurrence of each predecessor. Each state s in S (the set of all states in the environment) stores a compiled trace $T_{s,s'}$ for all states s' in S. Each state s also stores N_s which records the number of times that state has been visited, and $E_{s,s'}$ which sums the values of the traces of the state immediately preceding s over all episodes. All of these variables are initialised to zero. Whenever a transition from state s to s' is observed $T_{s,s}$ is set to 1, and the values stored at s' are updated as follows (where λ is the trace decay parameter as used in conventional eligibility traces, and γ is the discounting factor):

$$N_{s'} = N_{s'} + 1 \tag{1}$$

$$E_{s',s''} = E_{s',s''} + \lambda\gamma T_{s,s''} \; \forall \; s'' \text{ in } S \tag{2}$$

$$T_{s',s''} = E_{s',s''}/N_{s'} \; \forall \; s'' \text{ in } S \tag{3}$$

If state s' is the first state in an episode, then step 2 is omitted (effectively combining the current compiled trace with an all-zero predecessor trace).

AVM traces reflect the best approximation to the predecessor probabilities observed so far, and so should promote rapid, accurate learning. However they weight all episodes equally regardless of their currency, which means the traces will adapt slowly (and never fully) to dynamic environments in which the transition probabilities change over time. This is a significant issue for the use of compiled traces for policy iteration, as in Q-learning [4]. Even if the environment itself is static, the policy followed will change over time, so retaining information from previous episodes may result in traces which no longer reflect the current policy. This is an issue with any compiled trace algorithm, but particularly for the AVM form due to its infinitely long memory of previous episodes.

4.2 Recent-Visit-Mean (RVM) Traces

To adapt to dynamic environments, the traces must forget about episodes which are no longer relevant. One means to achieve this is to calculate the mean of the predecessor traces over a fixed number of recent visits to this state. This can be implemented using a queue at each state to store the predecessor traces for the most recent visits. The sensitivity of the agent to environmental changes can be altered by varying the length of the queue (which we will denote as v). Smaller values for v will place a greater emphasis on more recent learning episodes.

4.3 Blended Traces

This approach aims to be better suited to dynamic tasks than AVM traces by placing a larger weighting on more recent experience. In this algorithm the traces at s' are updated following a transition from s, as follows (where β is a blending parameter with $0 \le \beta \le 1$):

$$T_{s',s''} = \beta T_{s',s''} + (1-\beta)\lambda\gamma T_{s,s''} \; \forall \; s'' \text{ in } S \tag{4}$$

If s' is an initial state then the update is:

$$T_{s',s''} = \beta T_{s',s''} \; \forall \; s'' \text{ in } S \tag{5}$$

The emphasis which the system places on recent experience is controlled via the β parameter – larger values of β will place more weight on accumulated past experience (as stored in $T_{s',s''}$), whilst smaller values will make the system more responsive to recent experience ($T_{s,s''}$).

One possible problem with using a static value for β is that it will take several visits to a state before the traces grow significantly from their initial zero values, which will slow early learning. A potential solution is suggested by viewing AVM traces as a specialised form of a blending trace where the value of β is determined dynamically at each state to be equal to $(N_{s'} - 1)/N_{s'}$. It may be possible to combine the early learning speed of this approach with the on-going flexibility of blending by setting $\beta = ((N_{s'})^n-1)/ (N_{s'})^n$ where n is a fixed value just smaller than 1, or alternatively $\beta = \min((N_{s'} - 1)/N_{s'}, \beta_{max})$ where β_{max} is an upper-bound on the value of β. In this paper we have used the latter approach.

4.4 Comparison of Compiled Trace Algorithms and TD(λ)

Both RVM and blended traces can be seen as generalizations of the AVM form of compiled traces. Behaviour identical to AVM can be produced by setting the algorithm's parameter appropriately (using queue length $v=\infty$ for RVM or $\beta_{max}=1$ for blended traces). Similarly both algorithms are a generalization of conventional TD(λ), which can be obtained by using $v=1$ or $\beta_{max}=0$. Therefore previous proofs of the convergence properties of TD(λ) (such as [8]) apply to these compiled trace algorithms, at least for these specific choices of parameter. Extension of these proofs to the more general form of these algorithms is an area for future work.

5   Experimental Method

5.1 The Layered-DAG Test Suite

The experiments reported in this paper are based on a suite of policy evaluation test-beds (the Layered-DAG tests). These test environments are defined by a directed acyclic graph, in which each node represents a state. The nodes have been grouped into layers with arcs from each node in a layer to each node in the next layer. Each arc is assigned a random probability of occurrence (with the probabilities on the outgoing arcs from each node normalised so as to sum to 1). Each episode commences at a randomly selected node in the first layer, and ends when the environment moves to a terminal state in the final layer. Each terminal node is assigned a unique fixed reward in the range 0..n-1, where n is the number of nodes in the terminal layer.

This test suite has two advantages for this research. First the true value of each state in the environment can readily be calculated. Secondly a range of problems of varying difficulty can be generated by altering the number of layers in the graph or the number of nodes in each layer. This allows an investigation of the relative ability of compiled and conventional traces as the problem difficulty is increased.

For each variant of the Layered-DAG problem, various combinations of parameters were trialed for both the conventional and compiled traces. For each set of parameters, 20 trials were run using different random number generators to assign the transition probabilities within the graph. Each trial consisted of a number of episodes equal to 50 multiplied by the number of nodes per layer, to ensure that on average each state would experience the same number of visits regardless of the size of the test problem.

After each learning episode, the root mean squared error of the table values compared to the actual state values was calculated. This was averaged over all episodes of each trial, and then over each trial for a given set of parameters. This measure was chosen as it requires the learning algorithm to converge both rapidly and to a good approximation of the actual values — both of these properties are important if the algorithm is to be applied in an on-line context.

5.2 Choice of Compiled Trace Algorithm

Both the RVM and blended forms of the compiled trace algorithm were investigated. Very little variation was noted between the two algorithms, and hence for space reasons only the blended trace results will be reported in this paper. Given the similar performance of these two variants of compiled traces the blended form is to be preferred due to the much greater space requirements of the RVM form.

6 Results and Discussion

The graphs in Figures 2–4 summarise the performance of the blended trace algorithm across all variants of the Layered-DAG problem (each variant is identified as NxM where N is the number of nodes per layer, and M is the number of layers). The results have been normalized relative to the lowest mean RMSE error obtained on each variant of the problem, to facilitate comparison across these variants. Note that the first data-point in each graph (corresponding to $\beta_{max}=0$) indicates the performance of conventional TD(λ).

It can immediately be seen that non-zero values of β_{max} significantly improve on the results achieved using standard TD(λ). The TD(λ) error rate is on the order of 15-45% higher than that obtained using the optimal value of β_{max}. Importantly the trend is that the improvement due to using compiled traces increases as the complexity of the problem (either the number of states per layer or the number of layers) is increased. The performance of the compiled trace algorithm is relatively insensitive to the choice of β_{max}, with good results obtained across a wide range of values. As the episode length increases, the optimal choice of β_{max} gradually decreases.

An examination of the best values found for the α (learning rate) and λ parameters gives some insight into the reasons for the improved performance achieved using compiled traces. Whilst the α values were similar for both algorithms, the best results for compiled traces were achieved using far higher values for λ than were beneficial for TD(λ). Figure 5 illustrates this for the 8x20 Layered-DAG problem, showing the mean RMSE curves achieved during learning for both algorithms using their optimal parameter settings and for TD(λ) using the compiled trace algorithm's optimal settings. A higher λ value produces rapid early learning when used in conjunction with either conventional or compiled traces, by maintaining higher traces for (and therefore making larger changes to the values of) states encountered early in each episode. However TD(λ) fails to converge to a suitably low final error when using this λ value, as these large trace values cause the values of early states to be overly reactive to the reward received in the most recent episode. In contrast using compiled traces, the change due to most recent reward are 'spread' across many possible predecessor states, thereby moderating the impact of the stochasticity in the environment.

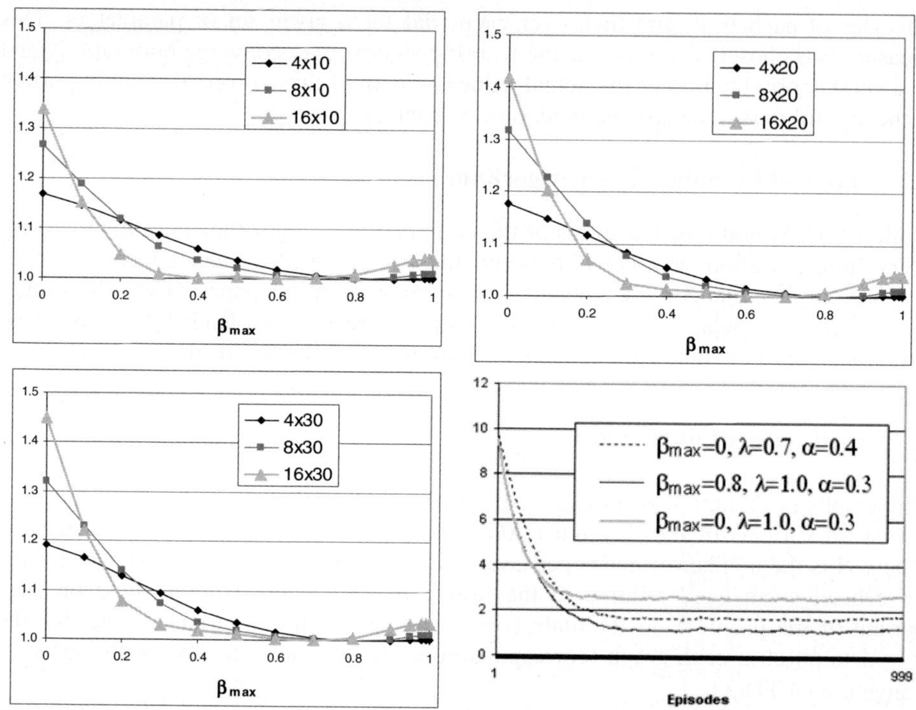

Figs. 2–5. Graphs of the normalised root-mean-squared-error (averaged over all episodes of all trials) against the β_{max} parameter for the 10 layer (top-left), 20 layer (top-right) and 30 layer (bottom-left) variants of the Layered-DAG task. The bottom-right graph plots RMSE (averaged over all trials) against learning episodes on the 8x20 task.

7   Conclusion and Future Work

7.1   Comparison of Conventional and Compiled Traces

This paper has demonstrated both by example and through empirical results that the use of compiled eligibility traces can substantially improve the performance of an agent using temporal difference learning methods, particularly in highly probabilistic environments. On the Layered-DAG tasks compiled traces gave a 15-45% reduction in error compared to conventional traces (implemented by setting the β_{max} parameter to 0).

Thus far these benefits have been demonstrated only in the simplest of TD learning tasks (evaluation of a static policy using a tabular representation of the state values). Further experiments will be required to determine the effectiveness of compiled traces on more difficult problems, particularly those involving dynamic environments or policy iteration tasks. It would also be beneficial to formally analyse the convergence properties of TD algorithms using compiled traces.

The primary disadvantage of compiled traces compared to conventional traces is the additional memory and computational overhead — a naïve implementation of

TD(λ) is order $O(|S|)$, whereas a similarly naïve compiled traces implementation is $O(|S|^2)$. It has been shown that the costs of TD(λ) can be reduced by ignoring near-zero traces — this technique may be less applicable to compiled traces however as the proportion of near-zero traces is likely to be significantly lower. However it may still be possible to discover more efficient implementation techniques similar to those used for TD(λ) by [9].

A further issue which needs to be addressed in order to scale up compiled traces to larger problems is integrating it into function approximation algorithms, as has previously been achieved for conventional traces. We intend to explore the use of compiled traces with localized function approximators such as tile-partitioning [10] and resource-allocating networks [11].

7.2 Using Compiled Traces for Planning and Exploration

Whilst the primary motivation behind compiled traces is to maximise the amount of learning possible from each interaction with the environment, it may also prove possible to utilise the additional information stored in these traces for other purposes. One possible application would be within an intelligent exploration algorithm. Such an algorithm might identify a particular state worthy of further exploration (for example a region of state space which has been under-explored, or not visited recently) and set it as a sub-goal. With conventional eligibility traces there is no knowledge within the system of how to move from the current state towards this sub-goal state, and hence no way to select actions to carry out the desired exploration.

Compiled traces also do not directly encode this information in the way that a goal-independent learning system does. However the values of the traces at the sub-goal state may give some guidance as to the correct actions. If the system is currently at state s and the sub-goal is at state g, we would select the action a which gives the highest value for $T_{g,s,a}$ (the trace stored at g for the combination of s and a). If $T_{g,s,a}$ is zero for all values of a, then we could either select an action randomly, or select greedily with respect to the overall goal. This process would be repeated for all states encountered until state g is reached. Whilst this approach is unlikely to result in the optimal path from s to g, it is likely to be significantly more efficient than random exploration in search of g, and incurs a relatively small computational cost.

This may be an area in which compiled traces provide benefits relative to model-based approaches, as the 'backward-looking' nature of compiled traces may allow more efficient planning of exploration than the 'forward-looking' models can achieve.

References

1. Sutton, R.S. (1988). Learning to predict by the methods of temporal differences, Machine Learning, 3:9-44.
2. Singh, S.P. and Sutton, R.S. (1996), Reinforcement learning with replacing eligibility traces, Machine Learning, 22:123-158.
3. Rummery, G. and M. Niranjan (1994). On-line Q-Learning Using Connectionist Systems. Cambridge, Cambridge University Engineering Department.
4. Watkins, C.J.C.H. (1989), Learning from Delayed Rewards, PhD Thesis, Cambridge University.

5. Sutton, R.S, (1991), Dyna, an integrated architecture for learning, planning and reacting. SIGART Bulletin, 2:160-163.
6. Kaelbling, L. P. (1993). Learning to Achieve Goals, Proceedings of the Thirteenth International Joint Conference on Artificial Intelligence, Chambéry, France.
7. Ollington, R. and Vamplew, P. (2003), Concurrent Q-Learning for Autonomous Mapping and Navigation, The 2nd International Conference on Computational Intelligence, Robotics and Autonomous Systems, Singapore.
8. Jaakkola, T., Jordan, M.I. and Singh, S.P. (1994), On the convergence of stochastic iterative dynamic programming algorithms, Neural Computation, 6:1185-1201
9. Cichosz, P. (1995), Truncating temporal differences: On the efficient implementation of TD(λ) for reinforcement learning, Journal of Artificial Intelligence Research, 2:287-318.
10. 10.Sutton R.S. (1996). Generalisation in reinforcement learning: Successful examples using sparse coarse coding. In Touretzky D.S., Mozer M.C., & Hasselmo M.E. (Eds.). Advances in Neural Information Processing Systems: Proceedings of the 1995 Conference (1038-1044). Cambridge, MA: The MIT Press.
11. Kretchmar, R. M. and Anderson, C.W. (1997). Comparison of CMACs and RBFs for local function approximators in reinforcement learning, IEEE International Conference on Neural Networks.

A Simple Artificial Immune System (SAIS) for Generating Classifier Systems

Kevin Leung and France Cheong

School of Business IT, RMIT University, Melbourne, Australia
{kevin.leung, france.cheong}@rmit.edu.au

Abstract. Current artificial immune system (AIS) classifiers have two major problems: (1) their populations of B-cells can grow to huge proportions and (2) optimizing one B-cell (part of the classifier) at a time does not necessarily guarantee that the B-cell pool (the whole classifier) will be optimized. In this paper, we describe the design of a new AIS algorithm and classifier system called simple AIS (SAIS). It is different from traditional AIS classifiers in that it takes only one B-cell, instead of a B-cell pool, to represent the classifier. This approach ensures global optimization of the whole system and in addition no population control mechanism is needed. We have tested our classifier on four benchmark datasets using different classification techniques and found it to be very competitive when compared to other classifiers.

Keywords: artificial immune systems, classification, instance-based learning.

1   Introduction

Classification is a commonly encountered decision making task. Categorizing an object into a predefined group or class based on a number of observed attributes related to an object is a typical classification problem [1]. There have been many studies which examined artificial intelligence (AI) techniques such as genetic algorithms (GAs) and artificial neural networks (ANNs) as classifier systems [2,3].

A more recent form of AI technique known as the artificial immune system (AIS) is rapidly emerging. It is based on the natural immune system principles and it can offer strong and robust information processing capabilities for solving complex problems. Just like the ANNs, AIS can learn new information, recall previously learned information, and perform pattern recognition in a highly decentralised way [4]. Applications of AIS are various and include data analysis, scheduling, machine learning, classification, and security of information systems [5,6,7,8,9,10].

However, there are two fundamental problems with current AIS classifier systems. The first problem is related to population control where it has been found that the number of B-cells, which match some antigens, increases, through cloning and mutation, to such an amount that it usually overtakes the entire population of B-cells [11]. Also, most AIS classifier systems make use of populations of B-cell pools and the problem which we have identified is that optimizing one B-cell, which is only part of the classifier, at a time does not guarantee the optimization of the B-cell pool, which is the complete classifier.

A. Sattar and B.H. Kang (Eds.): AI 2006, LNAI 4304, pp. 151–160, 2006.
© Springer-Verlag Berlin Heidelberg 2006

This study will introduce a new AIS algorithm and classifier system which will resolve the two problems mentioned above without sacrificing the classification performance of the system. Section 2 provides some related work, whilst in section 3 we explain the algorithm and development of our new classifier system. Section 4 provides details of the tests performed and results obtained, and section 5 concludes our paper.

2 Related Work

2.1 Resource Limited AIS (RLAIS)/Artificial Immune Network (AINE)

One of the first AIS created by Timmis, Neal and Hunt [7] was an effective data analysis tool. It could achieve good results by classifying benchmark data into specific clusters. However, a number of problems were clear: (1) the population control mechanism was not efficient in preventing an exponential population explosion with respect to the network size and (2) the resultant networks became so large that it was difficult to interpret the data and identify the clusters [12]. To address the issues raised by their first AIS, the authors proposed a new system called the resource limited AIS (RLAIS), which was later renamed to artificial immune network (AINE).

RLAIS makes use of artificial recognition balls (ARBs), which was inspired by Farmer, Packard and Perelson [13] in describing antigenic interaction within an immune network. The ARBs represent a number of identical B-cells and must compete for these B-cells based on their stimulation level. They reduce complexity and redundant information in the network [14]. However even though ARBs are essentially a compression mechanism that takes B-cells to a higher granularity level, the population of B-cells still grows rapidly to huge proportions [11]. A limited number of B-cells were predefined in RLAIS in order to solve this problem and to have an effective control of the population expansion.

2.2 Self-Stabilising AIS (SSAIS)

SSAIS was developed in an attempt to solve the problems of RLAIS. Neal [9] identified the two fundamental problems in (1) the nature of the resource allocation mechanism and (2) the non-continuous nature of the synchronous update mechanism. The second problem is of minor importance to this study and it was easily resolved by reviewing the RLAIS algorithm and making sure that all the functionalities of the algorithm were fully operational after the presentation of every data items.

The first problem was resolved by making changes to the way the ARBs work. In RLAIS, a predefined number of resources were allocated to ARBs by order of and in proportion to their stimulation level. On the other hand, in SSAIS, each ARB can increase its own resource allocation by registering the highest stimulation level for any incoming data item and increment its resource holding by adding its current stimulation level [9]. Since there are no longer a limited number of resources in SSAIS, it makes use of a mortality constant and a decay rate to control the network population size.

2.3 Artificial Immune Recognition System (AIRS)

The main study which regards AIS as a classifier system was done by Watkins [8]. The classifier system was named artificial immune recognition system (AIRS) and it was based on the principle of RLAIS and made use of ARBs. AIRS has proved to be a very powerful classification tool and when compared with the 10-30 best classifiers on publicly available classification problem sets, it was found to be among the top 5-8 classifiers for every problem set, except one in which it ranked second [15].

AIRS is considered to be a one-shot algorithm in that the training data is presented to the system only once [15]. One major problem that we have identified with the AIRS learning algorithm is that it does not guarantee the generation of an optimal classifier. AIRS also controls its population by removing the least stimulated cells from its B-cell pools; however, these cells, being the least stimulated for one specific antigen, might have very high affinity with other antigens, leading to a loss of good B-cells. Thus, optimizing one part (B-cell) of the system might have a negative effect on the immune system as a whole.

3 Simple Artificial Immune System (SAIS)

This section presents an overview of our proposed algorithm and classifier system which was named simple artificial immune system (SAIS). As its name implies, SAIS is very simple in that it adopts only the concept of affinity maturation which deals with stimulation, cloning and mutation as opposed to currently available AIS which tend to focus on many particular subsets of the features found in the natural immune system. It is also a compact classifier using only a predefined number of exemplars per class. This will be further discussed in the next section.

3.1 The Algorithm

In a conventional AIS algorithm (such as [15]), a classifier system is constructed as a set of exemplars that can be used to classify a wide range of data and in the context of immunology, the exemplars are known as B-cells and the data to be classified as antigens. A typical AIS algorithm operates as follows:

1. First, a set of training data (antigens) is loaded and an initial classifier system is created as a pool of B-cells with attributes either initialised from random values or values taken from random samples of antigens.
2. Next, for each antigen in the training set, the B-cells in the cell pool are stimulated. The most highly stimulated B-cell is cloned and mutated, and the best mutant is inserted in the cell pool. To prevent the cell pool from growing to huge proportions, B-cells that are similar to each other and those with the least stimulation levels are removed from the cell pool.
3. The final B-cell pool represents the classifier.

The conventional AIS algorithm is shown in Figure 1. From the description of the algorithm, three problems are apparent with conventional AIS algorithms:

1. Only one pass through the training data does not guarantee the generation of an optimal classifier.
2. Finding optimal exemplars does not guarantee the generation of an optimal classifier as local optimizations at the B-cell level does not necessarily imply global optimization at the B-cell pool level.

```
Conventional AIS algorithm
1 Load antigen population (training data)
2 Generate pool of B-cells with random values
      or values from random antigens
3 For all antigens in population
  3.1 Present antigen to B-cell pool
  3.2 Calculate stimulation level of B-cells
  3.3 Select most highly stimulated B-cell
  3.4 If stimulation level > threshold
        Clone and mutate most highly stimulated B-cell
        Select best mutants and insert into B-cell pool
      Endif
  3.5 Delete similar and least stimulated B-cells from B-cell pool
  EndFor
4 Classifier = B-cell pool
```

Fig. 1. Conventional AIS Algorithm

3. The simple population control mechanism of removing duplicates cannot guarantee a compact B-cell pool size. Many of the early AIS classifiers reported in the literature [14,11] suffer from the problem of huge size. Good exemplars may be lost during the removal process. We have experimented with a conventional AIS classifier and found the size of the cell pool to grow to astronomical proportions when using such a simple population control mechanism.

In order to address these issues, we have designed the SAIS algorithm to operate as follows:

1. First, a set of training data (antigens) is loaded and an initial classifier system is created as single B-cell containing a predefined number of exemplars initialized from random values. The purpose and content of our B-cell is different from the one used in conventional AIS algorithms. Our B-cell represents the complete classifier and it contains one or more exemplars per class to classify data. A B-cell in a conventional AIS algorithm, however, represents exactly one exemplar and the complete classifier is made up of a pool of B-cells.
2. Next, an evolution process is performed and iterated until the best possible classifier is obtained. The current B-cell is cloned and mutants are generated using the hypermutation process found in natural immune systems. If the classification performance of the best mutant is better than that of the current B-cell, then the best mutant is taken as the current B-cell. Our measure of stimulation is different from one used in conventional systems in that we make use of classification performance as a measure of stimulation of the complete classifier on all the training data rather than the distance (or affinity) between part of the classifier (a B-cell) and part of the data (an antigen).
3. The current B-cell represents the classifier.

The SAIS algorithm is shown in Figure 2.

```
SAIS algorithm
1 Load antigen population (training data)
2 Current B-cell = randomly initialized B-cell
3 Termination condition = B-cell with maximum performance
4 Do while NOT termination condition
   4.1 Clone and mutate current B-cell
   4.3 Calculate stimulation levels of mutated B-cells
          (classification performance of B-cell)
   4.2 New B-cell = mutated B-cell with best performance
   4.3 If performance of new B-cell > current B-cell
          Current B-cell = new B-cell
   EndDo
5 Classifier = current B-cell
```

Fig. 2. SAIS Algorithm

Using a B-cell to represent the whole classifier rather than part of the classifier has several advantages:

1. Optimizations are performed globally rather than locally and nothing gets lost in the evolution process.
2. There is no need for any population control mechanism as the classifier consists of small predefined number of exemplars. So far in our experiments we have used only one exemplar per class to be classified. This ensures the generation of the most compact classifier that is possible.

3.2 Model Development

SAIS was implemented in Java using the Repast[1] agent-based modelling framework. We used three different types of classification methods. One of the methods is exemplar-based while the two others are function-based. In the exemplar-based method, the attributes of a single exemplar per class are stored in the classifier. If there are two classes, for instance, the complete classifier will consist of two exemplars and their attributes. However, in the function-based method, the whole classifier will consist of only one set of parameters for the function in question whether or not there are one or multiple classes.

Minimum Distance Classification Method. In this exemplar-based method, we make use of a distance measure to classify the data. This approach is adapted from instance-based learning (IBL) [16] which is a paradigm of learning in which algorithms store the training data and use a distance function to classify the testing data. The distance function used in our classifier is the heterogeneous Euclidean-overlap metric (HEOM)

---

[1] Available from http://repast.sourceforge.net

[17] which can handle both categorical and continuous attributes. Categorical attributes are handled by the overlap part:

$$overlap(x, y) = \begin{cases} 0, & \text{if } x = y \\ 1, & \text{otherwise} \end{cases} \qquad \text{where } x = \text{exemplar}; y = \text{antigen}$$

while continuous ones are handled by the Euclidean part which is calculated as:

$$euclidean(x, y) = \sqrt{\sum_{i=1}^{n}(x_i - y_i)^2} \qquad \text{where } n = \text{number of attributes}$$

The overall HEOM distance function is thus defined as:

$$distance(x, y) = \begin{cases} overlap(x, y), & \text{if categorical attributes} \\ euclidean(x, y), & \text{if continuous attributes} \end{cases}$$

The minimum distance is then chosen to determine to which class to classify each antigen (see Figure 3). The percentage of correctly classified data can thus be generated. The B-cell undergoes cloning and mutation and the process described above is repeated until an optimal classifier is obtained. It should also be noted that our model normalize all the training data before the training process starts. This avoids the problem of overpowering the other attributes if one of them has a relatively large range.

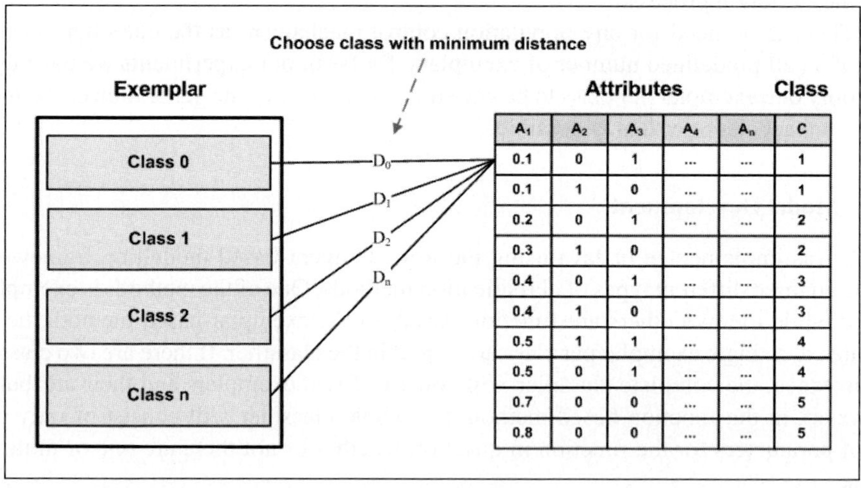

A₁	A₂	A₃	A₄	Aₙ	C
0.1	0	1	...	...	1
0.1	1	0	...	...	1
0.2	0	1	...	...	2
0.3	1	0	...	...	2
0.4	0	1	...	...	3
0.4	1	0	...	...	3
0.5	1	1	...	...	4
0.5	0	1	...	...	4
0.7	0	0	...	...	5
0.8	0	0	...	...	5

Fig. 3. Minimum Distance Classification Method of SAIS

Discriminant Analysis (DA) Classification Method. A second classification method that we use is an equivalent of the discriminant analysis (DA) statistical technique. DA is usually used to classify observations into two or more mutually exclusive groups by using the information provided by a set of predictor attributes. It was first proposed by

Fisher [18] as a classification tool and has been reported as the most commonly used data mining techniques for classification problems. It takes the form of:

$$y = c + \sum_{i=1}^{n} w_i x_i \qquad \text{where } n = \text{number of attributes} \qquad \text{(Normal DA)}$$

DA multiplies each independent attribute (x) by its corresponding weight (w) and adds these products and the constant (c) together, resulting in a single discriminant score (y). In our model, we did not make use of the constant and our DA function-based classification method is thus defined as:

$$y = \sum_{i=1}^{n} w_i x_i \qquad \text{(SAIS DA)}$$

The classifier optimizes the classification score (y) by evolving the weights (w) of all the different attributes (x).

Polynomial Classification Method. A third method that we use as part of our classifier is polynomial, a non-linear statistical technique. A polynomial is a mathematical expression involving the sum of a constant and powers in one or more attributes multiplied by the coefficients:

$$y = c + \sum_{i=1}^{n} w_i x_i^{p_i} \qquad \text{(Normal Polynomial)}$$

Just like our DA technique, the polynomial technique used in SAIS is function-based and does not make use of the constant. It is defined as:

$$y = \sum_{i=1}^{n} w_i x_i^{p_i} \qquad \text{(SAIS Polynomial)}$$

In this classification method, the function is being optimized by evolving the weights (w) and powers (p) of all the different attributes (x).

4   Experimentation and Results

The classification performance of SAIS was tested on four benchmark datasets. These publicly available datasets were selected from the machine learning repository of the University of California Irvine [19] and they are the (1) Breast cancer dataset with 9 categorical attributes and 2 classes; (2) Pima diabetes dataset with 8 continuous attributes and 2 classes; (3) Iris plant dataset with 4 continuous attributes and 3 classes; and (4) Wine recognition dataset with 13 continuous attributes and 3 classes.

To remain comparable with other classifiers reported in the literature [20], a 10-fold cross-validation technique was used to partition each dataset into a training and a testing set. SAIS was run 600 times on the 10 training sets of each dataset and results show that, on average, the classifier becomes optimal after 200 iterations (see Figure 4). The classifier was then run on the 10 testing sets of each dataset and the results were averaged. Table 1 shows the classification performance of SAIS when using the different classification techniques while Table 2 shows the classification performance of SAIS when compared to some other classifiers obtained from [20,15].

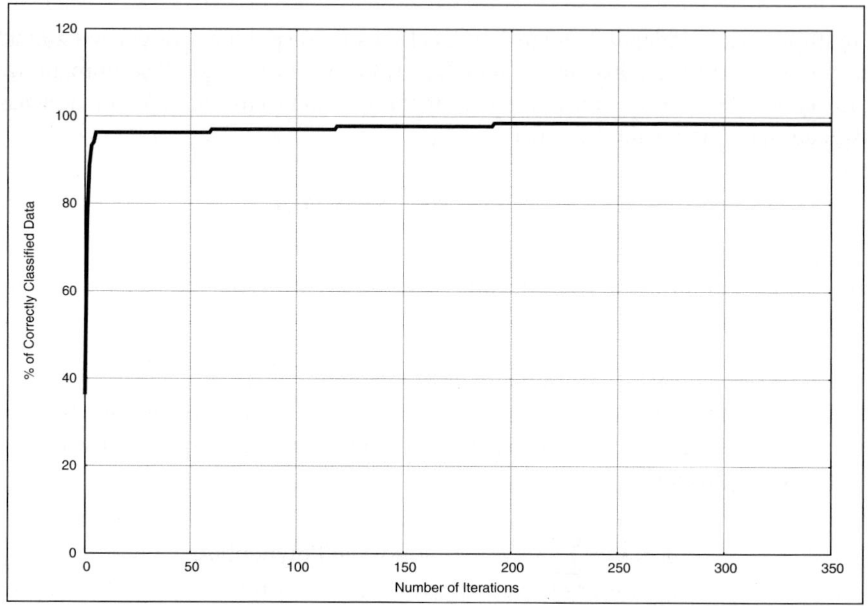

Fig. 4. Training Classification Performance V/S Number of Iterations

Table 1. Comparison of SAIS performance of different classification techniques

Classification Method	Breast	Diabetes	Iris	Wine
Minimum Distance	94.4%	**77.4%**	**97.3%**	**97.1%**
Discriminant Analyis	**96.3%**	77.0%	53.3%	33.1%
Polynomial	95.5%	69.8%	62.0%	33.1%

Based on our experiments, we have found that the minimum distance method is best for datasets with linear attributes while DA and polynomial techniques show very poor performances. When categorical attributes are involved, the difference in the classification performances of the three different techniques is not big. We also found that, on average, SAIS classifies well for datasets with few attributes (<10) and classes (<4). The difference in performance between the best classifier of each dataset and SAIS is small as shown in Table 2. SAIS even outperforms AIRS in the Diabetes and Iris datasets. We also found that the classification accuracy of SAIS tends to decrease as the number of attributes and classes increase. There is clearly a need to improve this particular aspect of our classifier.

Table 2. Comparison of SAIS and other classifiers classification performance

Rank	Breast Cancer		Diabetes		Iris		Wine	
1	NB + kernel est	97.5%	Logdisc	77.7%	Gronian	100.0%	IncNet	98.9%
2	SVM	97.2%	IncNet	77.6%	SSV	98.0%	SSV, opt prune	98.3%
3	Naive MFT	97.1%	DIPOL92	77.6%	C-MLP2LN	98.0%	kNN	97.8%
4	kNN	97.1%	LDA	77.5%	PVM	98.0%	SSV opt node	97.2%
5	GM	97.0%	**SAIS**	**77.4%**	**SAIS**	**97.3%**	**SAIS**	**97.1%**
6	VSS	96.9%	SMART	76.8%	AIRS	96.7%	FSM	96.1%
7	FSM	96.9%	GTO DT	76.8%	Fune-I	96.7%		
8	...		...					
12	**SAIS**	**96.3 %**	LVQ	75.8%				
13	...		...					
23	MML tree	94.8%	AIRS	74.1%				
24	ASR	94.7%	C4.5 DT	73.0%				

5   Conclusion and Future Work

In this study, we implemented a new and simple AIS algorithm and classifier to remove some of the drawbacks (population control, local and one-shot optimization) of other AIS classifiers without sacrificing the classification performance of the system. We tested SAIS on four benchmark datasets using different classification techniques and found it to be a very competitive classifier.

Future work lies in improving the classification performance of SAIS for datasets with large number of attributes (>10) and classes (>4). We intend to use multiple exemplars per class and hope to get better classification performances. We also have the intention of using an average and a k-Nearest Neighbour (kNN) distance method instead of the minimum distance method. Replacing our HEOM with an heterogeneous value difference Metric (HDVM) distance function is also envisaged since HDVM tends to produce better results than HEOM [17]. This is to be used on datasets with a mixture of categorical and continuous attributes.

References

1. Zhang, G.P.: Neural networks for classification: A survey. In: Proc. IEEE Transactions on Systems, Man, and Cybernetics - Part C: / Applications and Reviews. Volume 30. (2000) 451–462
2. Desai, V.S., Convay, D.G., Crook, J.N., Overstreet, G.A.: Credit scoring models in the credit union environment using neural networks and genetic algorithms. IMA Journal of Mathematics Applied in Business and Industry **8** (1997) 323–346
3. Malhotra, R., Malhotra, D.K.: Evaluating consumer loans using neural networks. The International Journal of Management Science **31** (2003) 83–96

4. Tarakanov, A., Dasgupta, D.: A formal model of an artificial immune system. BioSystems **55** (2000) 155–158

5. Hunt, J.E., Cooke, D.E.: An adaptive, distributed learning system based on the immune system. In: Proc. IEEE International Conference on Systems, Man and Cybernetics. (1995) 2494–2499

6. Dasgupta, D., Forrest, S.: Novelty detection in time series data using ideas from immunology. In: Proc. ICSA 5th International Conference on Intelligent Systems, Reno, Nevada (1996) 87–92

7. Timmis, J., Neal, M., Hunt, J.: An artificial immune system for data analysis. BioSystems **55** (2000) 143–150

8. Watkins, A.: AIRS: A Resource Limited Artificial Immune Classifier. Master's thesis, Mississippi State University (2001) Available at http://www.cse.msstate.edu/~andrew/research/publications.html.

9. Neal, M.: An artificial immune system for continuous analysis of time-varying data. In: Proc. First International Conference on Artificial Immune Systems (ICARIS), Canterbury, UK (2002) 76–85

10. Engin, O., Doyen, A.: A new approach to solve hybrid shop scheduling problems by an artificial immune system. Future Generation Computer Systems **20** (2004) 1083–1095

11. Nasraoui, O., Dasgupta, D., Gonzlez, F.: A novel artificial immune system approach to robust data mining. In: Proc. Genetic and Evolutionary Computation Conference (GECCO), New York (2002) 356–363

12. Timmis, J., Knight, T.: Artificial Immune Systems: Using the Immune System as Inspiration for Data Mining. In: Data Mining: A Heuristic Approach. Idea Group Publishing (2002) 209–230

13. Farmer, J.D., Packard, N.H., Perelson, A.S.: The immune system, adaptation and machine learning. Physica D **20** (1986) 187–204

14. Timmis, J., Neal, M.: A resource limited artificial immune system for data analysis. Knowledge-Based Systems **14** (2001) 121–130

15. Watkins, A., Timmis, J., Boggess, L.: Artificial immune recognition system (AIRS): An immune-inspired supervised learning algorithm. Genetic Programming and Evolvable Machines **5** (2004) 291–317

16. Aha, D.W., Kibler, D., Albert, M.K.: Instance-based learning algorithms. Machine Learning **6** (1991) 37–66

17. Wilson, D.R., Martinez, T.R.: Improved heterogeneous distance functions. Journal of Artificial Intelligence Research **6** (1997) 1–34

18. Fisher, R.A.: The use of multiple measurements in taxonomic problems. Ann. Eugenics **7** (1936) 179–188

19. Blake, C.L., Merz, C.J.: UCI Repository of Machine Learning Databases (1998) http://www.ics.uci.edu/~mlearn/MLRepository.html.

20. Duch, W.: Datasets used for Classification: Comparison of Results (2000) http://www.phys.uni.torun.pl/kmk/projects/datasets.html.

Selection for Feature Gene Subset in Microarray Expression Profiles Based on an Improved Genetic Algorithm

Chen Zhang, Yanchun Liang[*], Wei Xiong, and Hongwei Ge

College of Computer Science and Technology, Jilin University, Key Laboratory of Symbol
Computation and Knowledge Engineering of
Ministry of Education, Changchun 130012, China
ycliang@jlu.edu.cn

Abstract. It is an important subject to extract feature genes from microarray expression profiles in the study of computational biology. Based on an improved genetic algorithm (IGA), a feature selection method is proposed in this paper to find a feature gene subset so that the genes related to diseases could be kept and the redundant genes could be eliminated more effectively. In the proposed method, the information entropy is used as a separate criterion, and the crossover and mutation operators in the genetic algorithm are improved to control the number of the feature genes in the subset. After analyzing the microarray expression data, the artificial neural network (ANN) is used to evaluate the feature gene subsets obtained in different parameter conditions. Simulation results show that the proposed method can be used to find the optimal or quasi-optimal feature gene subset with more useful and less redundant information.

Keywords: microarray, feature selection, genetic algorithm, information entropy.

1 Introduction

Currently, the genetics research concerns about the relationship of different genes more than the function of a certain gene with the development of microarray technology, which is one of the most recent and important experimental breakthroughs in molecular biology. We can look at many genes at once and determine which are expressed in a particular cell type by microarray. Further, we expect to find some genes that relate directly to a particular disease. So we need to exploit effective methods of feature selection to eliminate redundant genes as many as possible from a large amount of genes [1,2].

Two key problems have to be solved for feature selection: one is the criterion of best features, and the other is the search method to find best features.

In machine learning, there are two approaches for feature selection: filter and wrapper [3]. Some quantificational criterions are usually used to evaluate feature subset in filter approach instead of accurate rate of classification in wrapper.

---

[*] Corresponding author.

A. Sattar and B.H. Kang (Eds.): AI 2006, LNAI 4304, pp. 161–169, 2006.
© Springer-Verlag Berlin Heidelberg 2006

Based on an improved genetic algorithm (IGA), a feature selection method is proposed in this paper to find a feature gene subset. The separate criterion based on the information entropy is used as the fitness evaluation in the IGA. The crossover and mutation operators are improved to meet the need of feature selection. The IGA is used to select the feature gene subset which contains the information of classification as much as possible from microarray expression profiles. The number of feature genes is controlled during the process of genetic operation to obtain subsets with different quantity of features. Then the subsets are evaluated by running the ANN classifier for different generations and deferent numbers of feature genes. In the end, we analyze the microarray data with 62 tissue samples of human gene expressions. The simulation results show that a better subset of feature genes can be found effectively using the proposed method.

2   Mathematics Descriptions and Model of Feature Selection

Microarray data consists of p probes and n DNA samples [4], and it can be described as a $p \times n$ matrix, $X = [x_{ij}]$, where x_{ij} represents the expression data of the ith gene g_i on the jth sample X_j.

A vector can be used to represent a sample

$$X_i = \{x_{1i}, x_{2i}, ..., x_{pi}\} \tag{1}$$

Suppose that the DNA samples belong to K categories,

$$\omega_1, \omega_2, ..., \omega_K , \tag{2}$$

The categories classified according to the ith feature gene subset are denoted as

$$\omega_1^i, \omega_2^i, ..., \omega_K^i . \tag{3}$$

Let n_k be the number of samples which belong to ω_k and n_k^i be the number of samples which belong to ω_k^i according to the ith feature gene subset, and let $n_{k,j}^i$ be the number of samples which belong to both ω_j and ω_k^i .

Since it has be supposed that there are p probes in microarray, the length of genetic string should be p. The string is denoted by a binary vector G

$$G = \{g_1, g_2, ..., g_p\} \tag{4}$$

Each position in the string represents whether the corresponding gene would be involved in the subset. It means that if the ith gene is selected, $g_i = 1$; otherwise $g_i = 0$. For example, a five-featured set is $G = \{g_1, g_2, g_3, g_4, g_5\}$, a string $\{1\ 1\ 0\ 1\ 0\}$ means that the subset is $G = \{g_1, g_2, g_4\}$.

3 Improved Genetic Algorithm

A wrapper approach and an improved genetic algorithm is used in this paper to select the feature gene subset.

After encoding the individual, the initial population is created randomly: for each individual, according to the number of assigned feature genes FN, we randomly select FN positions and set them as 1. The initial population size is taken as between 50 and 100.

With the purpose of selecting feature gene subset in which the genes contribute to classifying more greatly for microarray data, the evaluation process of feature gene subset is made up of four steps:

(1) Create an initial population with some individuals selected randomly. Every individual denotes a feature gene subset.
(2) Take the genes that are not in the current subset out of the initial training set which is classified by k-means clustering to obtain the compressed training set.
(3) Reclassify the compressed training set, and calculate its fitness value.
(4) The fitness here corresponds to the separate criterion in feature extraction. Gini impurity index based on the information entropy is adopted in this paper, which is obtained through the Shannon entropy [5]

$$H\left(P(\omega_1 \mid x), P(\omega_2 \mid x), ..., P(\omega_n \mid x)\right) = -\sum_{i=1}^{n} P(\omega_i \mid x)\log P(\omega_i \mid x) \tag{5}$$

Square it under some condition:

$$H^2\left(P(\omega_1 \mid x), P(\omega_2 \mid x), ..., P(\omega_n \mid x)\right) = 2\left(1 - \sum_{i=1}^{n} P^2(\omega_i \mid x)\right) \tag{6}$$

Where $P(\omega_i \mid x)$ represents the conditional probability of the sample belonging to the category ω_i under the condition x, and must satisfy the following condition: $\sum_{j=1}^{n} P(\omega_j \mid x) = 1$

The Gini index is calculated for each category which is classified according to the current genetic individual. The fitness would be the reciprocal of the sum of all indexes. It can be written as

$$f(G_i) = \frac{C}{\sum_{k=1}^{K} Gini(\omega_k^i)} \tag{7}$$

Where C is an adjustment factor, and G_i represents the ith genetic individual after encoding. $\sum_{k=1}^{K} Gini(\omega_k^i)$ is the sum of the Gini index, in which

$$Gini(\omega_k^i) = \begin{cases} \dfrac{1}{2} H^2\left[P(\omega_1 \mid \omega_k^i), P(\omega_2 \mid \omega_k^i), ..., P(\omega_K \mid \omega_k^i)\right] & n_k^i \neq 0 \\ 1 & n_k^i = 0 \end{cases} \tag{8}$$

When the number n_k^i of the samples which belong to ω_k according to the individual G_i, is equal to 0, the diversity of categories is supposed to be the maximum, with the value of 1, otherwise, it is equal to the Gini index, in which

$$P\left(\omega_j \mid \omega_k^i\right) = \frac{n_{k,j}^i}{n_k^i}.$$

(9)

It represents the posterior probability that the samples which belong to ω_k^i after reclassifying according to G_i belonged to ω_j. Apparently the following equation is satisfied

$$\sum_{j=1}^{K} P\left(\omega_j \mid \omega_k^i\right) = 1$$

(10)

After proposing the fitness function, the genetic operations are considered. The operators in the IGA are fitness proportional selection operator, the improved one-point crossover operator, and the improved basic mutation operator.

(1) Selection operator
The fitness proportional selection operator is adopted for the selection operation.
(2) Improved crossover operator
It is required that the number of feature genes must keep fixed during the process of the genetic operation, so an improved one-point crossover operator is developed here. In the crossover operation, it would be ensured that the positions of 0 and the positions of 1 which change by crossover operator are in the same quantity. First, for either of two individuals, calculate the number n_0 of the positions of 0 and n_1 of the positions of 1 when the corresponding positions in the two individuals are not the same. Then let n be equal to the smaller one. Choose a value of an integer k randomly in the range of $[0,1,\ldots,n]$ and cross over the first k positions of 0 and the first k positions of 1 that have been considered with the other individual. An example for the improved one-point crossover operator for $n=n_1=n_0=5$ and $k=3$ is shown in Figure 1.

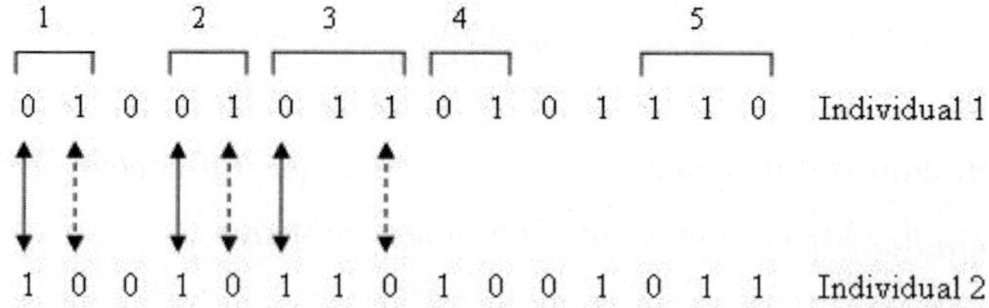

Fig. 1. An example of the improved one-point crossover operator

(3) Improved mutation operator

For the same reason, an improved basic mutation operator is also developed. A position of 0 and a position of 1 are selected randomly to turn over at the same time.

The number of generations is determined by the simulated experiment.

After the genetic process, the individual whose fitness value is the largest one in the last generation is decoded. Then we have got the best feature gene subset.

A classifier constructed by ANN is used to evaluate the results [6] using the test set. The index for evaluation is the accuracy of classification (Acc for short here), which is calculated by comparing with the initial classifying result.

In addition, the fitness value based on the Gini index given in the algorithm can be another index for evaluation. It will be used to compare with the Acc value.

4   Simulated Experimental Results and Analysis

The results are reported from analyzing the colon set from Affymetrix which is arrayed with 62 tissue samples of 65000 oligonucleotide probes (40 tumors and 22 normal tissues). The data set actually used here is the expressions with 2000 human genes picked by Alon et al. [7].

50 samples (4/5 of all) are taken out of the data set as the training set and the rest samples are used as test set to evaluate the results.

The evaluation process consists of two steps. First, the number of generations is increased to determine which one is the best according to the evaluated result. Then the feature subsets of different numbers of genes are evaluated to examine the effectiveness of the algorithm.

I. Decision of the number of generations

Two kinds of feature gene subsets whose FNs are 40 and 41 are evaluated with different numbers of generations (GNO for short) which are in the range of [0, 1200]. The Acc values are recorded every 100 generations. The results are shown in Table 1:

Table 1. The results of different generations

FN=40				FN=41			
GNO	Acc.	GNO	Acc.	GNO	Acc.	GNO	Acc.
100	0.3843	700	0.3962	100	0.4536	700	0.5833
200	0.4692	800	0.5971	200	0.5968	800	0.5293
300	0.5126	900	0.4167	300	0.3900	900	0.6920
400	0.5852	1000	0.6667	400	0.4230	1000	0.7821
500	0.5833	1100	0.5157	500	0.5833	1100	0.7209
600	0.4729	1200	0.5320	600	0.5833	1200	0.6853

It can be seen from the table that there is a highest Acc value when GNO=1000. The intuitional exhibition is illustrated in Figure 2:

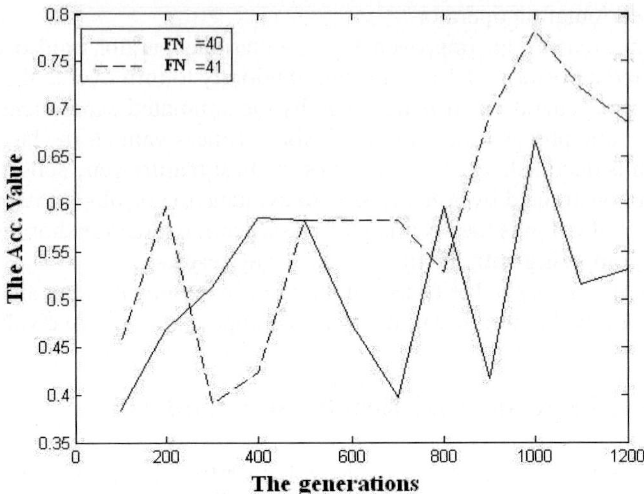

Fig. 2. The trend of the results of different generations

Table 2. Evaluation results of subsets with different sizes

The fitness based on the Gini index				The classified accuracy			
FN	Fitness	FN	Fitness	FN	Acc.	FN	Acc.
42	2.1552	21	0.468395	42	0.5833	21	0.5000
41	2.30992	20	0.449494	41	0.7812	20	0.4167
40	1.92105	19	0.526317	40	0.6667	19	0.4342
39	1.1568	18	0.416938	39	0.5000	18	0.4722
38	1.00237	17	0.465198	38	0.6732	17	0.5278
37	1.165	16	0.41551	37	0.7084	16	0.4588
36	0.977234	15	0.375939	36	0.5382	15	0.3889
35	0.985222	14	0.377843	35	0.6235	14	0.3889
34	0.958801	13	0.378833	34	0.4167	13	0.5556
33	0.900999	12	0.381596	33	0.5625	12	0.6944
32	0.810726	11	0.349861	32	0.5000	11	0.4722
31	0.784092	10	0.377626	31	0.8351	10	0.3541
30	0.599155	9	0.34971	30	0.4791	9	0.3750
29	0.768351	8	0.347612	29	0.5278	8	0.3056
28	0.783772	7	0.353035	28	0.5208	7	0.6042
27	0.684251	6	0.324282	27	0.5417	6	0.4444
26	0.569178	5	0.322767	26	0.2778	5	0.4167
25	0.570255	4	0.336273	25	0.5833	4	0.3889
24	0.500108	3	0.3103	24	0.5556	3	0.2916
23	0.517073	2	0.28773	23	0.5278	2	0.3611
22	0.531320	1	0.26843	22	0.3895	1	0.1255

II. Evaluation of the subsets with different sizes

Next, 42 feature gene subsets whose GNOs are all 1000 and FNs are between 42 and 1 are analyzed to find which subset is the best. The fitness value and the Acc value are recorded for each subset. Furthermore, the effect of the algorithm will be shown by comparing the values and the trends of the two indexes.

From Table 2, it can be seen that there is a highest fitness value when FN=41, while the Acc value is the highest when FN=31. There are high Acc values when FN=38, 41, and the classification accuracy of the three subsets are all more than 80%.

The trend charts are shown in Figure 3 and Figure 4.

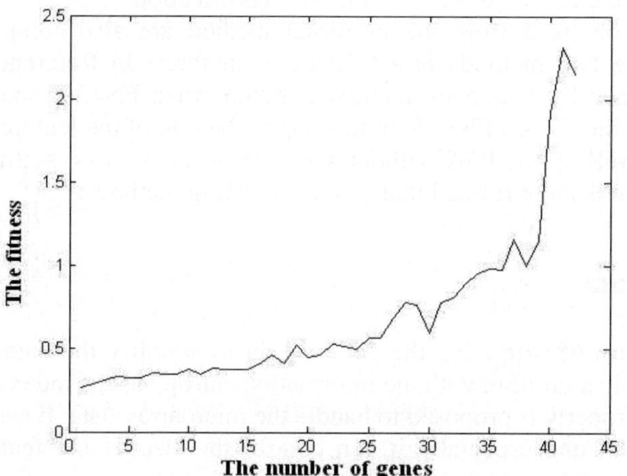

Fig. 3. The trend of fitness value

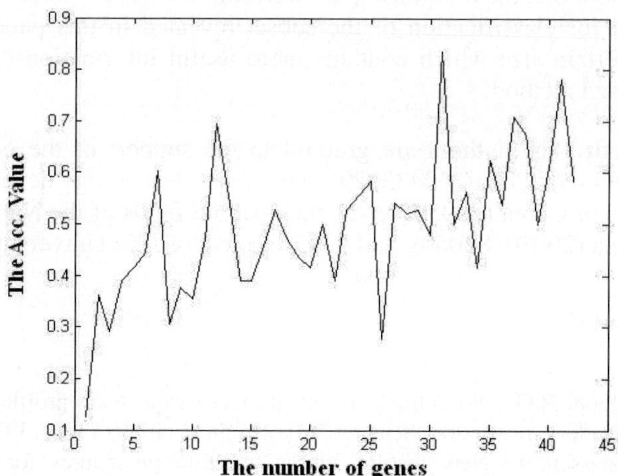

Fig. 4. The trend of Acc value

In Figure 3, it can be seen that the fitness value increases calmly with the size of subsets, and increases sharply when FN=40. There is a highest value when FN=41.

In Figure 4, it can be seen that there is a trend of fluctuant increasing with the increase of the size of the subsets.

Comparing the two charts, it is found that there is always a wave crest in the Acc value trend when the fitness value is a local extremum at the same point. For example, the fitness value is an extremum of 0.768351 when FN=29, while the Acc value is 0.8351 when FN=31; the fitness is 1.165 when FN=37 while the Acc is 0.7084; the fitness is 2.30992 when FN=41 while the Acc is 0.7821.

It means that there is something related between the two indexes that reflect the contribution of the feature gene subset to the classification.

The results obtained from the proposed method are also compared with those obtained by the two methods based on decision forest in Reference [8]. The best accuracy obtained by frequency method is 0.865 when FN=39, and it is 0.8767 by permutation method when FN=26. In this paper, the size of the feature gene subset can be controlled well by the IGA without assisting of other ways, so the process of the feature selection is more refined that in some existing methods.

5   Conclusions

With the purpose of extracting the gene subset to simplify the data for analysis, an improved genetic algorithm with the information entropy as the index of evaluation for classification property is proposed to handle the microarray data. It can eliminate more redundant genes, and especially it can control the size of the feature gene subset accurately for the future research of the attribute of genes.

It can be seen from the simulated experimental results that the classified accuracy of the feature gene subset with the best size and the best number of generations reaches more than 80%. It is shown that the subset obtained using the proposed algorithm is satisfying. Furthermore, the relationship between the size of the feature gene subset and the contribution for classification of the subset revealed in this paper shows that a subset with a certain size which contains more useful information can be found by using the proposed method.

Acknowledgment. The authors are grateful to the support of the National Natural Science Foundation of China (60433020), the science-technology development project of Jilin Province of China (20050705-2), the doctoral funds of the National Education Ministry of China (20030183060), and "985" project of Jilin University of China.

References

1. Lee, C.K., Klopp, R.G.., Weindruch, R., et al.: Gene expression profile of aging and its retardation by caloric restriction. Science. Vol. 285 5432 (1999) 1390-1393
2. Bo, T.H., Jonassen, I.: New feature subset selection procedures for classification of expression profiles. Genome Biology. Vol.3 4 (2002) 0017.1-0017.11
3. Yang, J., Honavar, V.: Feature subset selection using a genetic algorithm. IEEE Intelligent Systems & Their Applications. Vol.13 2 (1998) 44-49

4. Li, X., Rao, S.Q., Wang, Y.D., et al.: Gene mining: a novel and powerful ensemble decision approach to hunting for disease genes using microarray expression profiling. Nucleic Acids Research. Vol. 32. 9 (2004) 2685-2694
5. Mayoral, M.M.: Renyi's entropy as an index of diversity in simple-stage cluster sampling. Information Sciences. Vol.105 3 (1998) 101-114
6. Khan, J., Wei, J.S., Ringner, M., et al.: Classification and diagnostic prediction of cancers using gene expression profiling and artificial neural networks. Nature Medicine. Vol.7 6 (2001) 673-679
7. Alon, U., Barkai, N., Notterman, D.A., et al.: Broad patterns of gene expression revealed by clustering analysis of tumor and normal colon tissues probed by oligonucleotide arrays. Cell Biology. Vol.96 12 (1999) 6745-6750
8. Lv, S.L., Wang, Q.H., Li, X., et al.: Two feature gene recognition methods based on decision forest. China Journal of Bioinformatics. Vol.3 2 (2004) 19-22

An Efficient Alternative to SVM Based Recursive Feature Elimination with Applications in Natural Language Processing and Bioinformatics

Justin Bedo[1,2], Conrad Sanderson[1,2], and Adam Kowalczyk[1,2,3]

[1] Australian National University, ACT 0200, Australia
[2] National ICT Australia (NICTA), Locked Bag 8001, ACT 2601, Australia
[3] Dept. Electrical & Electronic Eng., University of Melbourne, VIC 3010, Australia

Abstract. The SVM based Recursive Feature Elimination (RFE-SVM) algorithm is a popular technique for feature selection, used in natural language processing and bioinformatics. Recently it was demonstrated that a small regularisation constant C can considerably improve the performance of RFE-SVM on microarray datasets. In this paper we show that further improvements are possible if the explicitly computable limit $C \to 0$ is used. We prove that in this limit most forms of SVM and ridge regression classifiers scaled by the factor $\frac{1}{C}$ converge to a centroid classifier. As this classifier can be used directly for feature ranking, in the limit we can avoid the computationally demanding recursion and convex optimisation in RFE-SVM. Comparisons on two text based author verification tasks and on three genomic microarray classification tasks indicate that this straightforward method can surprisingly obtain comparable (at times superior) performance and is about an order of magnitude faster.

1 Introduction

The *Support Vector Machine* based *Recursive Feature Elimination* (RFE-SVM) approach [1] is a popular technique for feature selection and subsequent classification, especially in the bioinformatics area. At each iteration a linear SVM is trained, followed by removing one or more "bad" features from further consideration. The goodness of the features is determined by the absolute value of the corresponding weights used in the SVM. The features remaining after a number of iterations are deemed to be the most useful for discrimination, and can be used to provide insights into the given data. A similar feature selection strategy was used in the *author unmasking* approach, proposed for the task of authorship verification [2] (a sub-area within the natural language processing field). However, rather than removing the worst features, the best features were iteratively dropped.

Recently it has been observed experimentally on two microarray datasets that using very low values for the regularisation constant C can improve the performance of RFE-SVM [3]. In this paper we take the next step and rather than pursuing the elaborate scheme of recursively generating SVMs, we use the limit $C \to 0$. We show that this limit can be explicitly calculated and results in a centroid based classifier. Furthermore, unlike RFE-SVM, in this limit the removal of one or more

A. Sattar and B.H. Kang (Eds.): AI 2006, LNAI 4304, pp. 170–180, 2006.

© Springer-Verlag Berlin Heidelberg 2006

features does not affect the solution of the classifier for the remaining features. The need for multiple recursion is hence obviated, resulting in considerable computational savings.

This paper is structured as follows. In Section 2 we calculate the limit $C \to 0$. In Sections 3 and 4 we provide empirical evaluations on two authorship attribution tasks and on three bioinformatics tasks, respectively, showing that the centroid based approach can obtain comparable (at times superior) performance. A concluding discussion is given in Section 5.

2   Theory

Empirical risk minimisation is a popular machine learning technique in which one optimises the performance of an estimator (often called a predictor or hypothesis) on a "training set" subject to constraints on the hypothesis class. The predictor f, which is constrained to belong to a Hilbert space $\mathcal{H}$, is selected as the minimiser of a regularised risk which is the sum of two terms: a regulariser that is typically the squared norm of the hypothesis, and a loss function capturing its error on the training set, weighted by a factor $C > 0$.

The well studied limit of $C \to \infty$ results in a hard margin Support Vector Machine (SVM). In this paper, we focus on the other extreme, the limit when $C \to 0$. This is a heavy regularisation scheme as the regularised risk is dominated by the regulariser. We formally prove that if the classifier is scaled by the factor $1/C$, then it converges in most cases to a well defined and explicitly calculable limit: the centroid classifier.

We start with an introduction of the notation necessary to precisely state our formal results. Given a training set $\mathbb{Z} = \{(\mathbf{x}_i, y_i)\}_{i=1,\ldots,m} \in \mathbb{R}^n \times \{\pm 1\}$ a Mercer kernel $k : \mathbb{R}^n \times \mathbb{R}^n \to \mathbb{R}$ with the associated reproducing kernel Hilbert space $\mathcal{H}$, and with a feature map $\Phi : \mathbb{R}^n \to \mathcal{H}$, $\mathbf{x} \mapsto k(\mathbf{x}, .)$ [4,5], the regularised risk on the training set is defined as

$$R^{reg}[f, b] := \|f\|_{\mathcal{H}}^2 + \sum_{y=\pm 1} C_y \sum_{i, y_i = y} l(1 - y_i(f(\mathbf{x}_i) + b)) \tag{1}$$

for any $(f, b) \in \mathcal{H} \times \mathbb{R}$. Here $C_{+1}, C_{-1} \geq 0$ are two class dependent regularisation constants. These constants can be always written in the form

$$C_y = C \frac{1 + yB}{2m_y}, \tag{2}$$

where $C > 0$, $0 \leq B \leq 1$ and $m_y := |\{i \ ; \ y_i = y\}|$ is the number of instances with label $y = \pm 1$. Further, $l(\xi)$ is the *loss function* of one of the following three forms: $l(\xi) := \xi^2$ for the *ridge regression* (RREGR), the *hinge* loss $l(\xi) := \max(0, \xi)$ for the *support vector machine* (SVM) with the *linear* loss, and $l(\xi) := \max(0, \xi)^2$ for the SVM^2 with *quadratic* loss [4,5,6].

We are concerned with *kernel machines* defined as

$$f := f_H := \arg\min_{f \in \mathcal{H}} R^{reg}[f, 0], \tag{3}$$

in the *homogeneous case* (no bias b), or, in general, as the sum

$$f = f_H + b, \qquad (f_H, b) := \arg\min_{(f, b) \in \mathcal{H} \times \mathbb{R}} R^{reg}[f, b]. \tag{4}$$

2.1 Centroid Based Classification and Feature Selection

The *centroid classifier*, is defined for every $-1 \leq B \leq 1$ as

$$g_B(x) = \frac{1+B}{2} \sum_{i,y_i=+1} \frac{k(x,x_i)}{m_{+1}} - \frac{1-B}{2} \sum_{i,y_i=-1} \frac{k(x,x_i)}{m_{-1}} + b, \tag{5}$$

where b is a bias term. In terms of the feature space, it is the projection onto the vector connecting (weighted) arithmetic means of samples from both class labels, degenerating to the mean of a single class for $B = \pm 1$, respectively. For a suitable kernel k it implements the difference between Parzen window estimates of probability densities for both labels. For the linear kernel it simplifies to

$$g_B(\mathbf{x}) = \mathbf{w}_B \cdot \mathbf{x} + b := \left(\frac{1+B}{2} \bar{\mathbf{x}}_{+1} - \frac{1-B}{2} \bar{\mathbf{x}}_{-1} \right) \cdot \mathbf{x} + b \tag{6}$$

where $\bar{\mathbf{x}}_y$ is the centroid of the samples with label $y = \pm 1$ and $\mathbf{w}_B \in \mathbb{R}^n$ is the *centroid weight vector*. In this work we have used $B = 0$ and $b = \mathbf{w}_B \cdot (\bar{\mathbf{x}}_{+1} + \bar{\mathbf{x}}_{-1})/2$, such that the decision hyperplane is exactly halfway between $\bar{\mathbf{x}}_{+1}$ and $\bar{\mathbf{x}}_{-1}$. The weight vector can be used for feature selection in the same manner as in RFE-SVM, i.e. the features are selected based on a ranking given by $r_i = |\mathbf{w}_B^{(i)}|$, where $\mathbf{w}_B^{(i)}$ is i-th dimension of $\mathbf{w}_B$. However, unlike RFE-SVM, the removal of a feature does not affect the solution for the remaining features, thus the need for recursion is obviated. This form of combined feature selection and subsequent classification will be referred to as the *centroid approach*.

For the case of the linear kernel, we note that the centroid classifier can be interpreted as a form of *Linear Discriminant Analysis (LDA)* [7], where the two classes share a diagonal covariance matrix with equal diagonal entries.

2.2 Formal Results

In the following theorem we prove that using either form of *quadratic* loss specified above, the machine converges to the centroid classifier as $C \to 0$; the case of linear loss is covered by Theorem 2.

Theorem 1. *Let $|B| < 1$, the loss be either $l(\xi) = \xi^2$ or $l(\xi) = \max(0,\xi)^2$ and*

$$f_C := f_{H,C} + b_C \quad \text{and} \quad (f_{H,C}, b_C) := \arg\min_{f,b} R^{reg}[f,b]$$

for every $C > 0$. Then $\lim_{C \to 0+} |\frac{f_C}{C}| = \infty$ if $B \neq 0$ but $\lim_{C \to 0+} \frac{f_{H,C}}{C} = \frac{1-B^2}{2} g_0$.

Thus although the whole predictor f_C scaled by C^{-1} diverges with $C \to 0$, its homogeneous part $f_{H,C}$ converges to an explicitly calculable limit.

Proof outline. The kernel trick reduces the proof to the linear kernel $k(\mathbf{x}, \mathbf{x}') := \mathbf{x} \cdot \mathbf{x}'$ on $\mathbb{R}^n$ and $\mathcal{H}$ isomorphic to $\mathbb{R}^n$. In such a case $f(\mathbf{x}) = f_H(\mathbf{x}) + b = \mathbf{x} \cdot \mathbf{w} + b$, where $\mathbf{w} \in \mathbb{R}^n$ and $\|f_H\|_{\mathcal{H}}^2 = \|\mathbf{w}\|^2$. The regularised risk (1-2) takes the form

$$R^{reg}[\mathbf{w}, b] = \|\mathbf{w}\|^2 + C \sum_i \frac{y_i + B}{2m_{y_i}} l(1 - y_i(\mathbf{w} \cdot \mathbf{x}_i + b))$$

and $f_C(\mathbf{x}) = \mathbf{w}_C \cdot \mathbf{x} + b_C$ for $\mathbf{x} \in \mathbb{R}^n$, where $(\mathbf{w}_C, b_C) = \arg\min_{\mathbf{w},b} R^{reg}[\mathbf{w}, b]$. This implies

$$\|\mathbf{w}_C\|^2 \leq \min_{\mathbf{w},b} R^{reg}[\mathbf{w}, b] \leq R^{reg}[0, 0] = C \tag{7}$$

and due to continuous differentiability of the loss functions in our case:

$$\frac{\partial R^{reg}}{\partial b}[\mathbf{w}_C, b_C] = 0 \quad \text{and} \quad \frac{\partial R^{reg}}{\partial \mathbf{w}}[\mathbf{w}_C, b_C] = 0. \tag{8}$$

Solving the first of these equations we obtain

$$b_C = B - \sum_{i \in SV} \frac{1 + y_i B}{2m_{y_i}} \mathbf{w}_C \cdot \mathbf{x}_i$$

where $SV := \{i : 1 - y_i \mathbf{x}_i \mathbf{w}_C - y_i b_C > 0\}$ is the set of the *support vectors* in the case of the hinge loss and the whole training set in the case of regression. Furthermore,

$$\left| \sum_{i \in SV} \frac{1 + y_i B}{2m_{y_i}} \mathbf{w}_C \cdot \mathbf{x}_i \right| \leq \|\mathbf{w}_C\| \max_i \|\mathbf{x}_i\| \leq \sqrt{C} \max_i \|\mathbf{x}_i\|,$$

by (7), hence, denoting by $O(\xi)$ a term such that $|O(\xi)/\xi|$ is bounded on for $\xi > 0$, we get

$$b_C = B + O(\sqrt{C}). \tag{9}$$

Case $l(\xi) = \xi^2$. The first equation in (8) takes the form

$$2\mathbf{w}_C - 2C \sum_i \frac{y_i + B}{2m_{y_i}} (1 - y_i(\mathbf{w}_C \cdot \mathbf{x}_i + b_C))\mathbf{x}_i = 0$$

which upon consideration of (7) and (9) implies

$$\frac{f_{H,C}(\mathbf{x})}{C} = \sum_i y_i \frac{(1 + y_i B)(1 - y_i B + O(\sqrt{C}))}{2m_{y_i}} \mathbf{x}_i \cdot \mathbf{x} = \frac{1 - B^2}{2} g_0(\mathbf{x}) + O(\sqrt{C}),$$

$$\frac{f_C(\mathbf{x})}{C} = \frac{f_{H,C}(\mathbf{x}) + b_C}{C} = \frac{1 - B^2}{2} g_0(\mathbf{x}) + O(\sqrt{C}) + \frac{B + O(\sqrt{C})}{C}.$$

This immediately proves the current case of the theorem.

Case $l(\xi) = \max(0, \xi)^2$. This case reduces to the previous one, once we observe that every training sample becomes a support vector for sufficiently small C. Indeed, Eqns. (7) and (9) imply

$$1 - y_i \mathbf{w}_C \cdot \mathbf{x}_i - y_i b_C = 1 - y_i B + O(\sqrt{C}) > 0$$

if C is sufficiently small, since $|B| < 1$. $\qquad\square$

Note that the above result does not cover the most popular case of SVM, namely the linear loss $l(\xi) := \max(0, \xi)$. The example in Figure 1 shows that in such a case the limit depends on the particular data configuration in a complicated way and cannot be simply expressed in terms of the centroid classifier.

The extension of Theorem 1 to the case of homogeneous machines follows. A proof, similar to the above one, is not included.

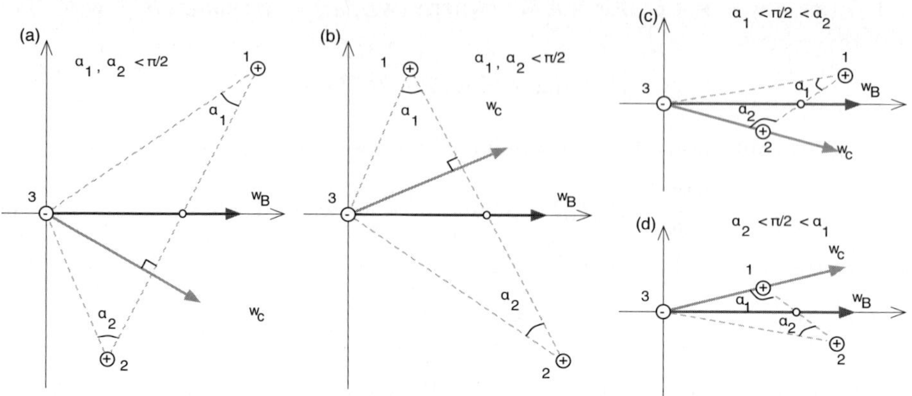

Fig. 1. Four subsets of $\mathbb{R}^2$ that share the same centroid classifier (the blue line), but which have different vectors $\mathbf{w}_C$ (the red line) for the SVM[1] solutions $\mathbf{x} \to \mathbf{w}_C \cdot \mathbf{x} + b_C$

Theorem 2. *Let $|B| \leq 1$, the loss be either $l(\xi) = \xi^2$ or $l(\xi) = \max(0, \xi)^p$ for $p = 1, 2$ and $f_C := \arg\min_f R^{reg}[f, 0]$ for every $C > 0$. Then $\lim_{C \to 0+} f_C/C = g_B(\boldsymbol{x})$.*

Using the kernel $k(\mathbf{x}, \mathbf{x}') + 1$ rather than $k(\mathbf{x}, \mathbf{x}')$ in the last theorem we obtain:

Corollary 1. *Let $f_C := f_{H,C} + b_C$ for every $C > 0$, where $(f_{H,C}, b_C) := \arg\min_{f,b} R^{reg}[f, 0] + b^2$. Then $\lim_{C \to 0+} f_C/C = g_B(\boldsymbol{x}) + B$.*

Note the differences between the above of results. Theorem 2 and Corollary 1 hold for $B = \pm 1$, which is virtually the case of one-class-SVM, while in Theorem 1 this is excluded. The limit $\lim_{C \to 0} f_C/C$ exists in the case of Corollary 1 but does not exist in the case of Theorem 1. Finally, in Theorem 1 the convergence is to a scaled version of the balanced centroid g_0 for all $|B| < 1$, while in the latter two results to g_B or $g_B + B$, respectively.

3　Authorship Verification Experiments

Recently an RFE-SVM based technique was proposed for the task of authorship verification [2], a particular problem that spans the fields of natural language processing and humanities [8,9]. Given a reference text from a purported author and a text of unknown authorship, an *author unmasking* curve is built as follows. Both texts are divided into chunks, with each chunk represented by counts of pre-selected words. The chunks are partitioned into training and test sets. Each point in the author unmasking curve is the accuracy of discriminating (using a linear SVM) between the test chunks from the two texts. At the end of each iteration several of the most discriminative words are removed from further consideration. A vector representing the essential features of the curve is then classified as representing either the "same author" or a "different author" [2].

The underlying hypothesis is that if the two given texts have been written by the same author, the differences between them will be reflected in a relatively

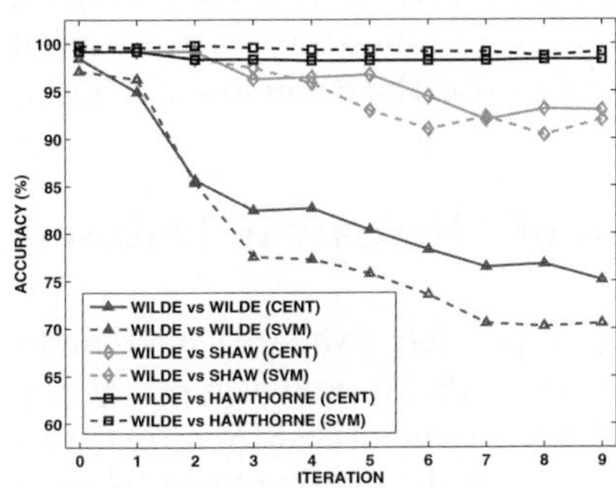

Fig. 2. Demonstration of the unmasking effect for Wilde's *An Ideal Husband* using Wilde's *Woman of No Importance* as well as the works of other authors as reference texts.

Table 1. Classification accuracies for same-author (SA) and different-author (DA) curves on the Gutenberg dataset, using centroid and RFE-SVM (R/S) based unmasking. The second term in the method indicates the secondary classifier type.

Method	SA Acc.	DA Acc.	Time
Cent. + Cent.	92.3%	95.4%	14 mins
Cent. + SVM	92.3%	95.2%	14 mins
R/S + Cent.	69.2%	97.2%	124 mins
R/S + SVM	92.3%	94.9%	124 mins

Table 2. As per Table 1, but for the Columnists dataset

Method	SA Acc.	DA Acc.	Time
Cent. + Cent.	82%	89.6%	99 mins
Cent. + SVM	86%	90.3%	104 mins
R/S + Cent.	84%	88%	709 mins
R/S + SVM	84%	88.5%	714 mins

small number of features. In [2] it was observed that for texts authored by the same person, the extent of the accuracy degradation is much larger than for texts written by different authors.

We used two datasets in our experiments: Gutenberg and Columnists. The former is comprised of 21 books[1] from 10 authors (as used in [2]), while the latter (used in [10]) is comprised of 50 journalists (each covering several topics) with two texts per journalist; each text has approximately 5000 words.

The setup of experiments was similar to that in [2], with the main difference being that books by the same author were not combined into one large text. For the Gutenberg dataset we used chunks with a size of 500 words, 10 iterations of removing 6 features, and using 250 words with the highest average frequency in both texts as the set of pre-selected words. For each pair of texts, 10 fold cross-validation was used to obtain 10 curves, which were then represented by a mean curve prior to further processing. For the Columnist dataset we used a chunk size of 200 words and 100 pre-selected words (based on preliminary experiments).

A leave-one-text-out approach was used, where the remaining texts were used for generation of same-author (SA) and different-author (DA) curves, which in turn were used for training a secondary classifier. The left-out text was then unmasked against the remaining texts and the resulting curves classified. For the Gutenberg dataset, there was a total of 24 SA and 394 DA classifications, while for Columnists there were 100 SA and 9800 DA classifications. Both the SVM and the centroid were also used as secondary classifiers.

---

[1] Available from Project Gutenberg (www.gutenberg.org).

As can be observed in Figure 2 (which closely resembles Fig. 2 in [2]) the unmasking effect is also well present for the centroid based approach. Tables 1 and 2 indicate that the RFE-SVM and the centroid based unmasking approaches are comparable in terms of performance, but differ greatly in terms of wall clock time[2]. The centroid based approach is about seven to nine times faster even if we neglect the significant time required for searching for the optimal C for SVM.

4   Experiments with Classification of Microarray Datasets

In this section we perform experiments on three publicly available microarray datasets: *colon* cancer [1,3,11], *lymphoma*[3] cancer [3,12,13], and *Cancer of Unknown Primary (CUP)* [14]. All these datasets are characterised by a relatively small number of high dimensional samples. The colon dataset contains 62 samples (22 normal and 40 cancerous), with each sample containing the expression values for 2000 genes. The lymphoma dataset contains three subtypes, with a total of 62 samples. Each sample contains measurements of 4026 expressed genes. We used a subset of the CUP dataset, containing samples for 226 primary and metastatic tumours, representing 12 tumour types[4], with each sample containing expressions for approximately 10500 genes measured on a cDNA microarray platform. The number of samples per class widely varied, form 8 to 50.

In addition to comparing the performance of the RFE-SVM and centroid approaches, we also evaluated the multi-class *Shrunken Centroid (SC)* approach [15], proposed specifically for processing microarray datasets. Briefly, in SC the class centroids are shrunk towards the overall centroid, involving a form of t-statistic which standardises each feature for each class by its within-class standard deviation. The degree of shrinking for all features is controlled by single tunable parameter. A feature is not used for prediction of a class when it has been shrunk "past" the overall centroid. We note that one of the main differences between SC and the centroid approach (described in Section 2.1) is that in the latter standard deviation estimates are not used.

For classification of multiclass datasets (lymphoma and CUP) with the centroid and SVM approaches we have used the popular one-vs-all architecture [16]. Feature selection for these datasets was done follows. Each class had its own ranking of features, of which the top T were selected. The final set of features was the union of the selected features from all classes. Note that since the class-specific feature subsets can intersect, the final set can have less than KT features, where K is the number of classes.

The experiment setup for the colon and lymphoma datasets followed [3]. Testing was based on 50 random splits of data into disjoint training and test sets. In each split approximately 66% of the samples from each class were assigned to

---

[2] Using LibSVM 2.71 (www.csie.ntu.edu.tw/~cjlin/libsvm) on a Pentium 4, 3.2 GHz.
[3] Available from http://llmpp.nih.gov/lymphoma/data.shtml
[4] The full CUP dataset contains 13 tumour types, however we have omitted the smallest class ('testicular') which was represented by only three samples.

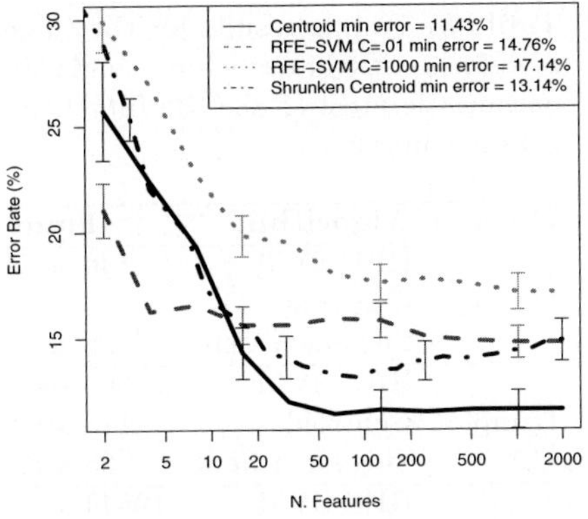
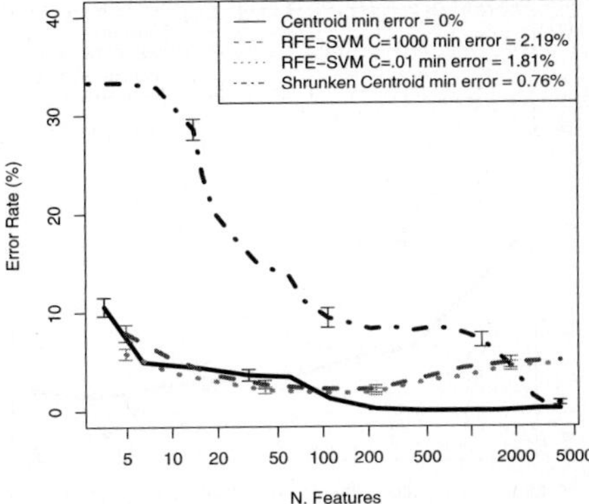

Fig. 3. Results for the colon dataset   **Fig. 4.** Results for the lymphoma dataset

the training set, with the remainder assigned to the test set. Results are shown[5] in Figures 3 and 4.

As the value of C decreases, the performance of RFE-SVM improves, achieving a minimum error rate of 14.76% and 1.82% with $C = 0.01$ on the colon and lymphoma datasets, respectively. This agrees with results in [3]. When using the $C \to 0$ limit (i.e. the centroid approach), the minimum error rate drops to 11.43% and 0% on the colon and lymphoma datasets, respectively. The SC approach obtained a comparable minimum of 13.14% and 0.76% on the respective datasets. However, on the lymphoma dataset its error rate increased rapidly when fewer features were used. Specifically, the centroid approach selects about 200 genes that achieve near perfect discrimination, while the SC approach requires the full set of 4026 genes to achieve its minimum error rate.

The CUP dataset was compiled for building a clinical test for determination of the origin of a cancer from the tissue of a secondary tumour. In [14] it was demonstrated (for a subset of five cancers), that in order for CUP classification to be practical for use in pathology tests, it is necessary to select a relatively small subset of marker genes which are then used to build a considerably cheaper and also more robust test, i.e., via a Polymerase Chain Reaction based microfluidic card, where the maximum number of gene probes is 384 [14]. Results in Figure 5 indicate that the centroid method outperforms both RFE-SVM and SC in the practically important region of a few hundred genes.

Finally, wall clock timing results shown in Table 3 indicate that the centroid approach is between four to 17 times faster than RFE-SVM. For for the two-class dataset (colon), the SC and centroid approaches have comparable time requirements. For multi-class datasets (lymphoma and CUP), the SC approach is roughly twice as fast as the centroid approach, due to the use of the one-vs-all architecture for the latter.

---

[5] We have evaluated a large set of C values for RFE-SVM, however for purposes of clarity only a subset of the results is shown in Figures 3, 4 and 5.

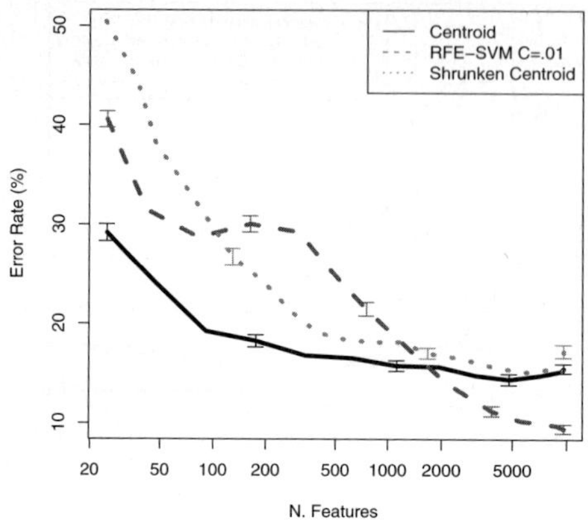

Fig. 5. Results for the CUP dataset

Table 3. Timing results for the three bioinformatics datasets for a single 50-permutation test (1.83 GHz Intel Core Duo machine)

Dataset	Algorithm	Time
	RFE-SVM	146 sec.
Colon	Centroid	20 sec.
	Shrunken Cent.	28 sec.
	RFE-SVM	505 sec.
Lymp.	Centroid	113 sec.
	Shrunken Cent.	66 sec.
	RFE-SVM	19844 sec.
CUP	Centroid	1135 sec.
	Shrunken Cent.	478 sec.

5   Discussion and Conclusions

In this paper we have shown that the limit $C \to 0$ of most forms of SVMs is explicitly computable and converges to a centroid based classifier. As this classifier can be used directly for feature ranking, in the limit we can avoid the computationally demanding recursion and convex optimisation in RFE-SVM. We have also shown that on some real life datasets the use of the centroid approach can be advantageous.

The results may at first be surprising, raising the question: *Why is the centroid approach a competitor to its more sophisticated counterparts?* One explanation is that its relative simplicity makes it well matched for classification and feature selection on datasets comprised of a small number of samples from a noisy distribution. On such datasets, even the simple estimates of variance which are employed by, say, the Shrunken Centroid approach [15], are so unreliable that their usage brings more harm than good. Indeed, the results presented in Section 4 support this view.

Another independent argument can be built on the theoretical link established in Section 2, relating the centroid classifier to the extreme case of regularisation of SVMs. Once we accept the well supported perspective of the SVM community that regularisation is the way to generate robust empirical estimators and classifiers from noisy data [4,6], it naturally follows that we should get robust classifiers in the limit of extreme regularisation. It is worth noting that the same argument can be extended to the theoretical and empirical comparison with LDA [7], which due to space restrictions is not included here.

We note that the centroid approach has been applied and used sporadically in the past: we have already linked it to the "ancient" Parzen window; its kernel form can be found in [4] and its application to classification of breast cancer microarrays is given in a well cited paper [17]. However, what this paper brings is a formal link to SVMs and RFE-SVM in particular. Furthermore, to our knowledge, this type of centroid based feature selection and subsequent classification has not been explored previously and its application to genomic microarray classification as well as natural language processing (author unmasking) is novel.

To conclude, we have shown that the centroid approach is a good baseline alternative to RFE-SVM, especially when dealing with datasets comprised of a low number of samples from a very high dimensional space. It it very fast and straightforward to implement, especially in the linear case. It has a nice theoretical link to SVMs and regularisation networks, and given the right circumstances it can deliver surprisingly good performance. We also recommend it as a baseline classifier, facilitating quick sanity cross checks of more involved supervised learning techniques.

Acknowledgements

We thank Simon Guenter for useful discussions. National ICT Australia (NICTA) is funded by the Australian Government's *Backing Australia's Ability* initiative, in part through the Australian Research Council.

References

1. Guyon, I., Weston, J., Barnhill, S., Vapnik, V.: Gene selection for cancer classification using support vector machines. Machine Learning **46** (2002) 389–422
2. Koppel, M., Schler, J.: Authorship verification as a one-class classification problem. In: Proc. 21st Int. Conf. Machine Learning (ICML), Banff, Canada (2004)
3. Huang, T.M., Kecman, V.: Gene extraction for cancer diagnosis by support vector machines - an improvement. Artificial Intelligence in Medicine **35** (2005) 185–194
4. Schölkopf, B., Smola, A.: Learning with Kernels. MIT Press, Cambridge (2002)
5. Cristianini, N., Shawe-Taylor, J.: An Introduction to Support Vector Machines. Cambridge University Press, Cambridge, UK (2000)
6. Vapnik, V.: Statistical Learning Theory. John Wiley and Sons, New York (1998)
7. Duda, R., Hart, P., Stork, D.: Pattern Classification. John Wiley & Sons (2001)
8. Gamon, M.: Linguistic correlates of style: authorship classification with deep linguistic analysis features. In: Proc. 20th Int. Conf. Computational Linguistics (COLING), Geneva (2004) 611–617
9. Love, H.: Attributing Authorship: An Introduction. Cambridge University Press, U.K. (2002)
10. Sanderson, C., Guenter, S.: Short text authorship attribution via sequence kernels, Markov chains and author unmasking: An investigation. In: Proc. 2006 Conf. Empirical Methods in Natural Language Processing (EMNLP), Sydney (2006) 482–491
11. Ambroise, C., McLachlan, G.: Selection bias in gene extraction on the basis of microarray gene-expression data. Proc. National Acad. Sci. **99** (2002) 6562–6566
12. Alizadeh, A., Eisen, M., Davis, R., et al.: Distinct types of diffuse large B-cell lymphoma identified by gene expression profiling. Nature **403** (2000) 503–511
13. Chu, F., Wang, L.: Gene expression data analysis using support vector machines. In: Proc. Intl. Joint Conf. Neural Networks. (2003) 2268–2271

14. Tothill, R., Kowalczyk, A., Rischin, D., Bousioutas, A., Haviv, I., et al.: An expression-based site of origin diagnostic method designed for clinical application to cancer of unknown origin. Cancer Research **65** (2005) 4031–4040
15. Tibshirani, R., Hastie, T., et al.: Class prediction by nearest shrunken centroids, with applications to DNA microarrays. Statistical Science **18** (2003) 104–117
16. Rifkin, R., Klautau, A.: In defense of one-vs-all classification. Journal of Machine Learning Research **5** (2004) 101–141
17. van't Veer, L., Dai, H., van de Vijver, M., He, Y., Hart, A., et al.: Gene expression profiling predicts clinical outcome of breast cancer. Nature **415** (2002) 530–536

Efficient AUC Learning Curve Calculation

Remco R. Bouckaert

Computer Science Department, University of Waikato, New Zealand
remco@cs.waikato.ac.nz

Abstract. A learning curve of a performance measure provides a graphical method with many benefits for judging classifier properties. The area under the ROC curve (AUC) is a useful and increasingly popular performance measure. In this paper, we consider the computational aspects of calculating AUC learning curves. A new method is provided for incrementally updating exact AUC curves and for calculating approximate AUC curves for datasets with millions of instances. Both theoretical and empirical justifications are given for the approximation. Variants for incremental exact and approximate AUC curves are provided as well.

1   Introduction

Receiver operator characteristics (ROC) graphs allow for easily visualizing performance of classifiers. ROC graphs stem from electronics [3], but has been widely adopted by the medical decision making community [2,8]. Recently, the machine learning community has taken an interest as well [4,5][1].

Probabilistic classifiers, such as naive Bayes, are often selected based on accuracy, where the class with the highest probability is taken to be the class prediction. However, classification performance can be increased by optimally choosing the threshold for selecting a class [6], so that for instance a binary class the positive class is selected when the classifier predicts at least a 0.6 probability for this class, and otherwise selects the negative class. One of the benefits of ROC curves is that it allows selection of classifiers that make a probabilistic prediction without committing to a threshold. ROC has further desirable properties for situations where misclassification has different costs for the classes and where the classes are unbalanced [4,5]. The area under the ROC curve (AUC) is a summary statistic that can be used to select such a classifier.

Another graphical tool for visualizing classifier performance is the learning curve where the x-axis represents the number of training instances and the y-axis the classifiers performance. Learning curves show when enough data is obtained, and no further learning takes place. Also, it can show relative performance of different classifiers at various data set sizes and crossing points. The advantages of AUC as a measure of classification performance especially for probabilistic classifiers make it a good candidate for the y-axis of a learning curve. However,

---

[1] See for example Workshops on ROC Analysis in AI and ML http://www.dsic.upv.es/~flip/ROCAI2004, ROCML2005, ROCML2006.

A. Sattar and B.H. Kang (Eds.): AI 2006, LNAI 4304, pp. 181–191, 2006.
© Springer-Verlag Berlin Heidelberg 2006

calculating the AUC exactly requires sorting over all instances in the dataset, which may become cumbersome when there are many AUCs to be calculated or when there the size of the dataset becomes very large.

In this paper, we consider the computational problems of calculating the AUC for learning curves. Creating a learning curve can be done by incrementally testing the classifier with an instance in the data set, and then training it with the same instance. In this situation, it would be computationally desirable to efficiently update the AUC incrementally with the prediction made by the classifier. We address this issue in Section 3. For datasets with millions of instances, it may become computationally infeasible to calculate the AUC exactly, so we look at a way to approximate AUC in Section 4 and justify its performance. Finally, we consider incremental updating of approximate AUC in Section 5.

2   Exact Area Under ROC Curve

ROC curves deal with situations with a binary class where the class outcomes are called positive and negative. The classifier assigns a probability that the outcome is positive or negative. For a given threshold t, if the probability of the positive class is higher than t, the prediction is positive, otherwise negative. So, by varying t, the number of positive and negative prediction changes. The *true positives* are instances where the prediction is positive while the outcome is indeed positive, and the *false positives* are instances where the prediction is erroneously positive. The same definitions hold for *true negatives* and *false negatives* but for the negative outcomes. A ROC curve is a graph where the x-axis represents the number of true negatives and the y-axis the number of true positives.

The area under the ROC curve is a summary statistic that indicates higher overall performance with higher AUC. AUC is typically calculated [4] by

- Order the sequence of outcomes giving the sequence $o_1, \ldots, o_n$ such that $o_i \leq o_{i+1}$ and let c_i be the class value corresponding to outcome o_i.
- Determine points on ROC curve as follows; sequentially processing the class values $c_1, \ldots, c_n$ the curve starts in the origin and goes one unit up, for every negative outcome the curve goes one unit to the right. Units on the x-axis are $\frac{1}{\#TN}$ and on the y-axis $\frac{1}{\#TP}$ where $\#TP$ ($\#TN$) is the total number of true positives (true negatives). This gives the points on the ROC curve $(0,0), (x_1, y_1), \ldots, (x_n, y_n), (1,1)$.
- Calculate the area under the ROC curve, as the sum over all trapezoids with base x_{i+1} to x_i, that is, $AUC = \sum_{i=0}^{n} (y_{i+1} + y_i)/2(x_{i+1} - x_i)$.

The first step requires sorting the n outcomes, which is $O(n \log n)$ in computational cost. This may not be desirable, for instance, when n is large (say in the millions) and one is interested in the learning curve in terms of AUC. Then the number of times the AUC needs to be calculated is proportional to n, giving $O(n^2 \log n)$ computational cost.

3   Incremental Exact AUC

We start with an ordered sequence of outcomes o_1, o_2 with $o_1 = 0$ and $o_2 = 1$. For each outcome o_{new} with class c_{new}, we record the number of preceding true positives TP_i^n and true negatives TN_i^n and the total number of true positives $\#TP^n$ and negatives $\#TN^n$. So, we have the points (x_i, y_i) on the ROC curve with $x_i = TN_i^n/\#TN^n$ and $y_i = TP_i^n/\#TP^n$. Suppose we have the $n-1$ outcomes ordered $o_1, \ldots, o_{n-1}$ for which we have calculated the AUC as AUC_{n-1}. Further, we have the number of true positives TP_i^{n-1} and the number of true negatives TN_i^{n-1} ($1 \leq i \leq n-1$). If we receive a new outcome o_{new}, it should be inserted in the sequence of outcomes. Let j be the index of the outcome such that $o_j \leq o_{new} \leq o_{j+1}$. Then we have by definition $AUC_n = \sum_{i=0}^{n} (y_{i+1} + y_i)/2 \times (x_{i+1} - x_i)$ which equals $\sum_{i=0}^{n} (TP_{i+1}^n/\#TP^n + TP_i^n/\#TP^n)/2 \times (TN_{i+1}^n/\#TN^n - TN_i^n/\#TN^n)$ $= \sum_{i=0}^{n} (TP_{i+1}^n + TP_i^n)/2 \times (TN_{i+1}^n - TN_i^n)\frac{1}{\#TN^n\#TP^n}$. We have to distinguish between c_{new} being positive or negative.

If c_{new} is negative, then TP_i^{n-1} equals TP_i^n for any $0 \leq i <\leq n$ and TN_i^{n-1} equals TN_i^n for $i < j$ but $TN_i^{n-1} + 1 = TN_{i+1}^n$ for $i > j$.

So, we can write the above sum as $\sum_{i=0}^{j-1} \frac{(TP_{i+1}^{n-1} + TP_i^{n-1})/2 \times (TN_{i+1}^{n-1} - TN_i^{n-1})}{\#TN^n\#TP^n} +$ $\sum_{i=j}^{n-1} \frac{(TP_{i+1}^{n-1} + TP_i^{n-1})/2 \times (TN_{i+1}^{n-1} - TN_i^{n-1})}{\#TN^n\#TP^n} + \frac{(TP_{j+1}^{n-1} + TP_i^{n-1})/2 \times (TN_{i+1}^{n-1} - TN_i^{n-1})}{\#TN^n\#TP^n} -$ $\frac{(TP_{j+1}^{n-1} + TP_i^{n-1})/2 \times (TN_{i+1}^{n-1} - TN_i^{n-1})}{\#TN^n\#TP^n} + \frac{(TP_j^n + TP_{j-1}^n)/2 \times (TN_j^n - TN_{j-1}^n)}{\#TN^n\#TP^n} +$ $\frac{(TP_{j+1}^n + TP_i^n)/2 \times (TN_{j+1}^n - TN_j^n)}{\#TN^n\#TP^n}$. Now, the first three terms are equal to AUC_{n-1} weighed with $\frac{\#TN^{n-1}}{\#TN^n}\frac{\#TP^{n-1}}{\#TP^n}$. So, we have $AUC_n = AUC_{n-1}\frac{\#TN^{n-1}}{\#TN^n}\frac{\#TP^{n-1}}{\#TP^n} -$ $\frac{(TP_{j+1}^{n-1} + TP_i^{n-1})/1 \times (TN_{i+1}^{n-1} - TN_i^{n-1})}{\#TN^n\#TP^n} + \frac{(TP_j^n + TP_{j-1}^n)/2 \times (TN_j^n - TN_{j-1}^n)}{\#TN^n\#TP^n} +$ $\frac{(TP_{j+1}^n + TP_i^n)/2 \times (TN_{j+1}^n - TN_j^n)}{\#TN^n\#TP^n}$.

If c_{new} is positive $TP_i^{n-1} + 1 = TP_{i+1}^n$ and $TN_i^{n-1} = TN_{i+1}^n$ $TP_i^{n-1} = TP_i^n$ and $TN_i^{n-1} = TN_i^n$ for $i < j$ and TN_i^{n-1} equals TN_i^n but TP_i^{n-1} equals $TP_{i+1}^n - 1$ for $i > j$. So, we get an extra term $+ \sum_{i=j}^{n} \frac{1/2 \times (TN_j^n - TN_{j-1}^n)}{\#TN^n\#TP^n} = \frac{1/2 \times (\#TN^n - TN_{j-1}^n)}{\#TN^n\#TP^n}$ to add.

In short, AUC_n can be calculated in constant time, provided we know the terms in the formula. Effectively, we need to maintain correct administration of the number of preceding true positives and negatives at each point. Wlog, suppose the outcome c_j is positive, then $TP_i^n = TP_i^{n-1}$ for $i < j$ and $TP_i^n = TP_i^{n-1} + 1$ for $j \geq i$, so updating the true positives is linear in the number of instances. The total complexity of calculating AUC_n given AUC_{n-1} thus is $O(\log n)$ for finding the location of o_{new} in the outcomes and $O(n)$ for maintaining the cumulative true positive and negative counts. So, instead of spending $O(n \log n)$ for calculating a complete AUC, this can be achieved in $O(n + \log n)$. Therefore, calculating a learning curve becomes $O(n^2 + n \log n)$ instead of $O(n^2 \log n)$ for the non-incremental method. Note that most of the work goes into maintaining the true positive and negative counts (TP_i^n and TN_i^n). Section 5 provides a technique that can be adapted to reduce the amount of work of updating for every new outcome from $O(n)$ to $O(\sqrt{n})$.

4   Approximate AUC

Instead of calculating the AUC exactly, we can approximate it by taking m bins, each bin containing the number of positive and negative outcomes in an interval. The procedure then goes as follows;

- For every result r_i representing the probability of the positive class, identify the bin $j = \lfloor r_i \cdot m \rfloor$ and update bin j with outcome o_i. To prevent many bins remaining empty (see experiments later in this section), apply a log-log transform first ($r'_i = \frac{1}{2} - \alpha \log(-log(r_i))$ if $r_i \leq \frac{1}{2}$ and $r'_i = \frac{1}{2} + \alpha \log(-log(1 - r_i))$) if $r_i > \frac{1}{2}$ where α is an appropriate chosen constant which can be determined from a small sample of the data)
- Approximate ROC curve. Let b_i^+ be the number of positive outcomes in bin i, and b_i^- be the number of negative outcomes. The curve starts in the origin and goes with a straight line to (b_0^+, b_0^-). Then, from (b_0^+, b_0^-) to $(b_0^+ + b_1^+, b_0^- + b_1^-)$. From $(b_0^+ + b_1^+, b_0^- + b_1^-)$ to $(b_0^+ + b_1^+ + b_2^+, b_0^- + b_1^- + b_2^-)$, etc. So, we build the ROC curve from line pieces from $(\sum_{j=0}^{i} b_j^+, \sum_{j=0}^{i} b_j^-)$ to $(\sum_{i=0}^{j+1} b_j^+, \sum_{j=0}^{i+1} b_j^-)$.
- Let $y_i = \sum_{i=0}^{j+1} b_j^+$ and $x_i = \sum_{j=0}^{i+1} b_j^-$ the cumulative number of true positives and negatives, then we can calculate the area under the approximate ROC curve, as before, by taking the sum over the trapezoids $AUC = \sum_{i=0}^{m}(y_i + y_{y-1})/2 \times (x_i - x_{i-1})$, where by convention $x_{-1} = y_{-1} = 0$.

The complexity of this procedure with m bins is $O(n + m)$. Note that we need to keep track of the cumulative number of true positives y_i, but not necessarily the cumulative number of true negatives x_i, since $AUC = \sum_{i=1}^{m}(y_i + y_{y-1})/2 \times (x_i - x_{i-1}) = \sum_{i=1}^{m}(y_i + y_{y-1})/2 \times b_i^-$. Calculating a learning curve then becomes $O(n^2 + n \cdot m)$ instead of $O(n^2 \log n)$.

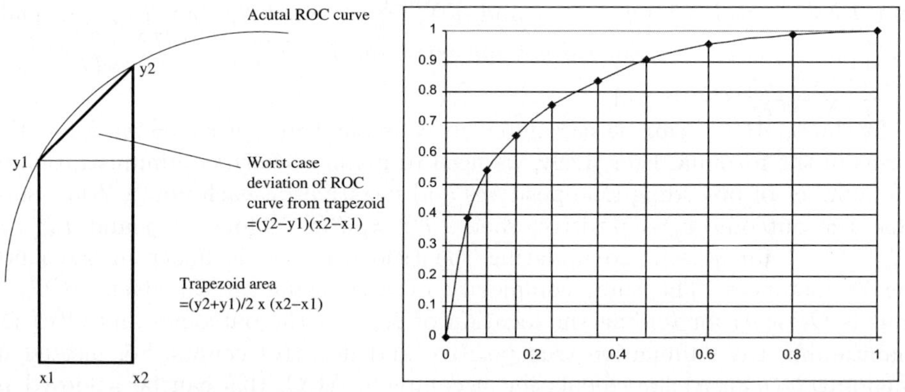

Fig. 1. Trapezoid approximation of ROC curve (left). Bins can have different widths (right). Here 10 bins are shown, the smallest is 5% the widest 20%.

4.1   Maximum Error of Approximation

A trapezoid is used to approximate the ROC curve in various steps where the beginning and ending of the upward part of the trapezoid coincide with the ROC curve. Since the ROC curve is a monotonely rising function, the worst possible deviation of the ROC curve from the trapezoid is if it goes straight up and then straight sidewards (or vise versa). In this case, the difference between the ROC curve and its approximation is the area in the triangle of the trapezoid (see left of Figure 1).

Figure 1 (right) illustrates that the bins need not be of equal width in terms of the ROC curve. This is caused by the fact that the class probability does not need to be uniformly distributed, and in fact it often is not. Therefore, the number of negatives represented on the x-axis need not be uniformly distributed over the bins. As a consequence, the maximum error cannot be estimated before the data has been seen. However, once the bins are filled, the error cannot exceed the triangle above the trapezoid of the bin. which can be calculated as $max\_error = \sum_{i=0}^{m}(y_i - y_{i-1})/2 \times (x_i - x_{i-1})$ (taking the same form as the approximate AUC, but with a plus sign replaced by a minus) which equals $\sum_{i=0}^{m} \frac{1}{2}b_i^+ \times b_i^-$.

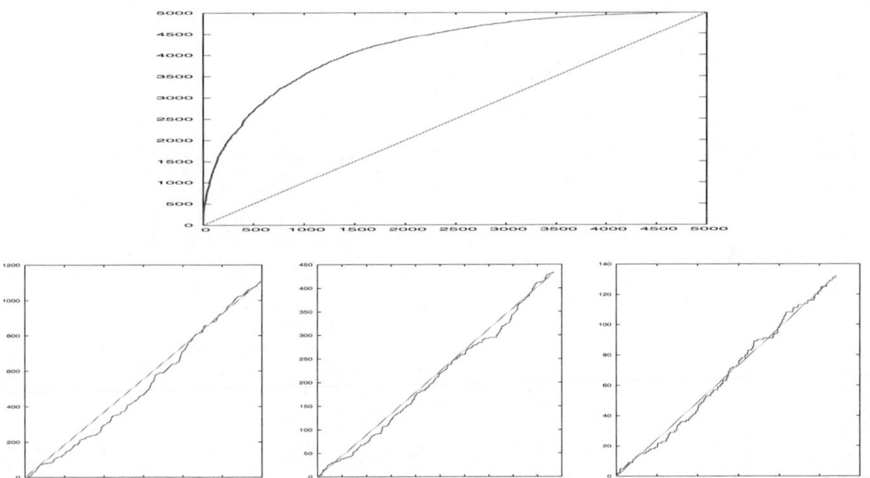

Fig. 2. ROC-plots of extended simple model. Top, full curve, below zoomed into true negative intervals starting at 0.1 (left), 0.5 (middle) and 0.9 (right) and extending 0.001.

To verify that the ROC curve does not deviate too far from the straight line, we created a simple probabilistic model $P(y, x1, x2, x3)$ taking the decomposition $P(y)P(x1|y)P(x2|y)P(x3|y)$ with $P(y = 0) = P(y = 1) = \frac{1}{2}$, $P(x1|y) = P(x2|y) = N(y, 1)$ where $N(m, \sigma)$ is the normal distribution with mean m and variance σ. The distribution over the binary attribute $x3$ is defined by $P(x3 = 0|y = 0) = P(x3 = 1|y = 1) = \frac{1}{4}$. The full ROC curve was generated using 10 thousand instances and is shown in Figure 2. The snapshots at the bottom of the figure show the ROC curve from 0.1 to 0.1001 (left), from 0.5 to 0.5001 (middle), and from 0.9 to 0.9001 (right), generated with 10 million samples (with x and y axis scaled to make them equally sized). A line starting in the left lower corner to the right upper corner is shown as well. Similar results hold when the discrete attribute $x3$ is removed from the model (not shown).

4.2   Error Estimate

To get a better estimate of the error in the approximation, consider Figure 2 bottom middle. Judging from Figure 2 it appears to be a reasonable assumption that within a sufficiently small bin, the probability of getting a positive or negative class is constant. With every positive class, the curve goes up, with every negative class the curve goes sideways. So, the ROC curve within a bin can be considered a one dimensional random walk with probability p of going up, and $1 - p$ of going down once we rotate the curve 45% (see Figure 3). Furthermore, we know the start and end point of the random walk, namely starting at the origin and ending for bin j at (b_j^+, b_j^-).

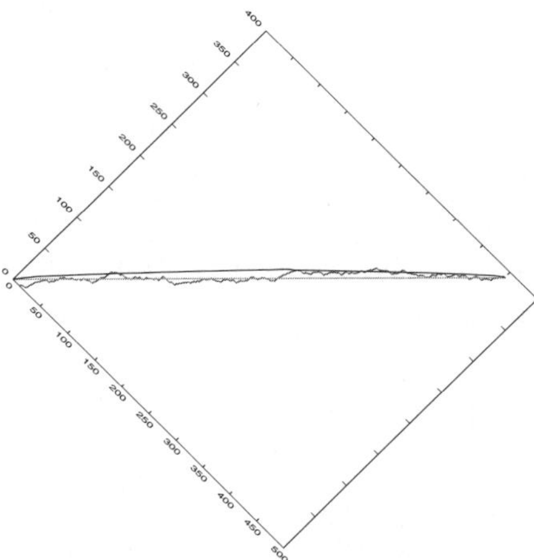

Fig. 3. Figure 2 bottom middle rotated, showing random walk. Expected deviation from line is drawn as well.

The expected value of a one dimensional random walk with unit steps and equal probability of positive and negative steps after k steps is $\sqrt{k}$. So, the expected area (representing the error in the AUC estimate) after k steps is bounded by $\sum_{i=0}^{k} \sqrt{i} \approx \int_0^k \sqrt{x}dx = \frac{2}{3}k\sqrt{k}$. However, the steps in the ROC curve are not of unit length, but only of size $\sqrt{\frac{1}{2}}$ due to the rotation, leaving an expected area of $\sqrt{\frac{1}{2}} \times \frac{2}{3}k\sqrt{k} = \frac{1}{3}k\sqrt{2k}$ after k steps. Now, due to symmetry, the expected area due to the first half number of steps contained in a bin equals the expected area in the second half. So a bin containing $q = b^+ + b^-$ ($b^+ = b^-$) outcomes has an expected error of $2 \times \frac{1}{3}\frac{q}{2}\sqrt{2\frac{q}{2}} = \frac{1}{3}q\sqrt{q}$ for a single bin. Furthermore, though the probability of a positive or negative outcome remains approximately constant within a bin, it changes from bin to bin. A bin with b^+ positive classes and b^- negative classes can be rescaled so that every positive step is $b^-/(b^+ + b^-)$ and every negative step $b^+/(b^+ + b^-)$. Instead of a unit length random walk, we get a $\sqrt{(\frac{b^+}{(b^++b^-)})^2 + (\frac{b^-}{(b^++b^-)})^2} = \sqrt{\frac{(b^+)^2+(b^-)^2}{(b^++b^-)^2}}$. So, a bin with $q = b^+ + b^-$ outcomes can expect to have its error bounded by $2 \times \sqrt{\frac{(b^+)^2+(b^-)^2}{(b^++b^-)^2}}\frac{1}{3}\frac{q}{2}\sqrt{2\frac{q}{2}} = \frac{1}{3}\sqrt{(b^+)^2 + (b^-)^2}\sqrt{q}$ (using $\sqrt{\frac{1}{(b^++b^-)^2}} = \frac{1}{q}$). The complete set of bins thus can be calculated as $\sum_{i=1}^{m} \frac{1}{3}\frac{\sqrt{(b_i^+)^2+(b_i^-)^2}}{\#TP+\#TN}\sqrt{\frac{b_i^++b_i^-}{\#TP+\#TN}}$. Like for the maximum error, we see that the bin containing the most outcomes contributes most to the expected error.

4.3 Empirical Evaluation of Approximation

Figure 4 shows the mean error of AUC estimates for 100 samples drawn from the model $P(y, x1, x2, x3)$ described above. The exact AUC was calculated as described in Section 2. The top line shows the AUC estimate based on a sample from the outcomes drawn from the model [1]. Though this approach does not require visiting all outcomes, it does require sorting. The bottom line is the average error obtained with the bin based approach. The maximum error found is indicated with an upwards error bar. The minimum absolute error turned out to be zero for the bin based approach and just above the mean bin based error for the sample based approach (not shown to prevent cluttering the graphs). The middle line is the maximum error estimate for the bin based approach.

Clearly, the bin based approach is superior over sampling in its error; the maximum error is consistently below the error of the sampling based approach.

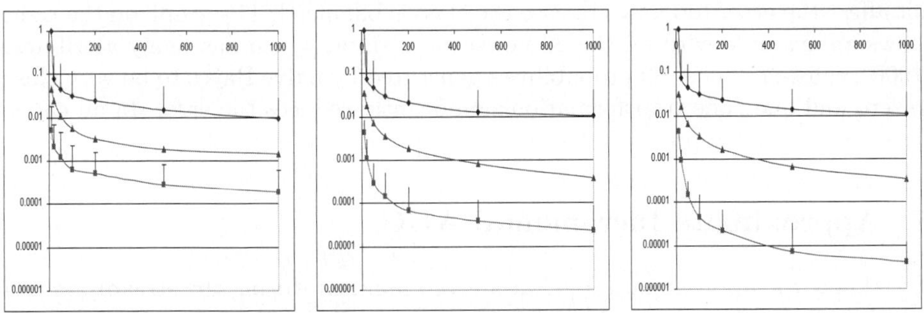

Fig. 4. Absolute error in AUC estimate (on y-axis) for different numbers of bins/samples (on x-axis). Top line is sampling based, middle line is maximum estimate of error for bin based estimate, bottom line is bin based. Left, dataset with 100 instances, middle with 1000 instances, and right 10000 instances.

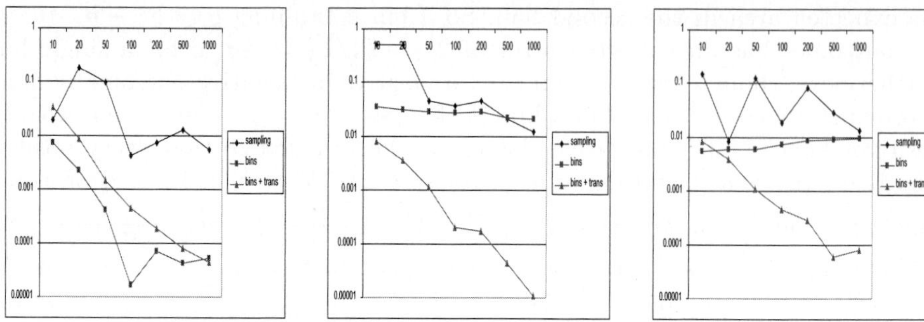

Fig. 5. Error in AUC on letter and newsgroup data. Axis same as Figure 5. Left class is A to O in letter, middle class is A in letter, right newsgroup.

Surprisingly, even with as little as ten bins, the error remains mostly below 1%, quickly dropping off with increasing number of bins. For sufficiently large data sets and 1000 bins, the estimate can become accurate up to the fifth digit. Increasing the number of outcomes decreases the error for the bin based approach, but not for the sampling based approach. The maximum error is clearly a large upper bound on the actual error with a factor of 10.

Unbalanced classes can cause many bins to be empty, and hence the error to become larger. Figure 5 shows this effect with the UCI letter dataset trained with naive Bayes. Left, the class is distinguishing between the letters A to O and the remaining letters, which gives a reasonably balanced class. The error of the approximate approach behaves as for the simple data source. In the middle, the class is recognizing A against all other letters, which gives an unbalanced class. The approximate approach does not increase performance with increasing number of bins because most bins remain empty. To remedy this, the bin can be chosen based on a log-log transform. The line ending lowest in the middle of Figure 5 represents the error with bins chosen this way. The left figure shows that the transformation does not affect the error too much in case the class is balanced. The graph on the right shows an example with on the 20 newsgroups data, which has many attributes (1000), causing almost all probabilities generated by naive Bayes to be very close to zero and one. The transformation significantly reduces the error. In all cases, the sample based approach is outperformed by the transformation.

5   Approximate Incremental AUC

If we know the approximate AUC at a given moment during the stream, we can calculate the approximate AUC for the next point in the curve. Suppose the next outcome is positive. Figure 6 shows how we need to adept the curve.

Let y_{j-1} (x_{j-1}) be the number of true positives (negative) before the target bin, and y_j (x_j) be the number of true positives (negative) before and including the target bin. So, we have $y_j = y_{j-1} + b_j^+$ and $x_j = x_{j-1} + b_j^-$ where b_j^+ (b_j^-) is the number of positives (negatives) in the target bin. Then, after a similar derivation as in Section 3, we get

$$AUC_n = AUC_{n-1}\frac{\#TP^n - 1}{\#TP^n} + \frac{x_m - x_j}{\#TP^n\#TN^n} + \frac{(y_j + y_{j-1} + 1)b_j^-}{2\#TP^n\#TN^n} - \frac{(y_j + y_{j-1})b_j^-}{2(\#TP^n - 1)\#TN^n}$$

where $b_j^- = x_j^n - x_{j-1}^n = x_j^{n-1} - x_{j-1}^{n-1}$. And if c_j is negative, we get

$$AUC_n = AUC_{n-1}\frac{\#TN^n - 1}{\#TN^n} + \frac{(y_j + y_{j-1} + 1)b_j^-}{2\#TP^n\#TN^n} - \frac{(y_j + y_{j-1})(b_j^- - 1)}{2\#TP^n(\#TN^n - 1)}$$

where $b_j^- = x_j^n - x_{j-1}^n$ is the number of true negatives in bin j after the update. To calculate AUC_n we need to know y_j and x_j. This can be calculated by $y_j = \sum_{i=0}^{j} b_i^+$ and $x_j = \sum_{i=0}^{j} b_i^-$ however, this is linear in m. We can improve this at the cost of some memory as follows. Suppose that we evenly distribute a anchor points over the m bins, and each anchor point we track the number of positives and negative for between the anchor points. So, when a positive item is added, we update the target bin, and also the anchor point containing the target bin. When $j = \lfloor \frac{r_i}{n} \rfloor$ is the index of the target bin, then $k = \lfloor \frac{j}{a} \rfloor$ is the corresponding target anchor point.

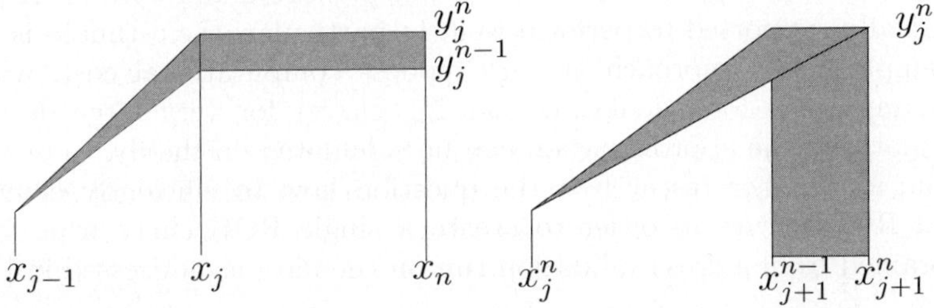

Fig. 6. Areas affected by updating the ROC curve with a positive class (left) and a negative class (right)

Now, to calculate y_j (and x_j), we sum all b_i to the nearest lower anchor, then proceed adding the positives (negatives) in the previous anchors. Let a_i^+ and a_i^j be the number of positives and negatives kept in anchor a_i and let i_a be the index of the bin associated with the anchor. Then we have

$$y_j = \sum_{i=i_a}^{j} b_i^+ + \sum_{j=1}^{k} a_i^+, \quad x_j = \sum_{i=i_a}^{j} b_i^- + \sum_{j=1}^{k} a_i^-$$

The expected number of summations is the expected number of $(j - i_a)$ for the bins and the expected number of k for the anchors. Assuming both are uniformly distributed, we have $E\{j - i_a\} = \frac{1}{2}\frac{m}{a}$ and $E\{k\} = \frac{1}{2}a$. So, together we expect $f(a) = \frac{1}{2}\frac{m}{a} + \frac{1}{2}a$ summations instead of $\frac{1}{2}m$. To minimize $f(a)$ we take the derivative and get $-\frac{1}{2}\frac{m}{a^2} + \frac{1}{2} = 0$ which gives $a = \sqrt{m}$. Then the expected number of summations is $f(\sqrt{(m)}) = \frac{1}{2}\sqrt{m} + \frac{1}{2}\sqrt{m} = \sqrt{m}$. So, calculating the next AUC becomes $O(\sqrt{m})$ instead of $O(m)$.

Two simple optimizations are possible;

- Instead of summing over bins to the first preceding anchor point, we can sum to the *closest* anchor point. This halves the number of expected summations over the bins. If the closest anchor point is a higher bin than the current bin, we calculate y_j as $-\sum_{i=j+1}^{i_a} jb_i^+ - \sum_{j=1}^{k+1} a_i^+$.
- We can keep track of the total number of positives and negatives and for every anchor points over half the number of bins we sum over all following anchors (instead of all preceding ones). This halves the expected number of anchor points. To calculate the contribution at the anchor point, just subtract the obtained sum from the total number of positives or negatives. Note that the technique with the anchor points can be adapted to the exact AUC calculation in Section 3.

6   Conclusions

This article presents a computationally efficient approach for calculating AUC learning curves by incrementally updating the AUC with each new sample point. Further, a method for approximating AUC was presented and both theoretically and empirically supported to perform well. In particular, the estimate is better than a sample based approach at very modest computational cost, which is especially important for calculating learning curves for very large data sets. Error estimates of the approximation can be calculated on the fly.

An issue for further research is the question how to efficiently combine a number of ROC curves in order to create a single ROC curve from various curves obtained from a cross validation run on the data as addressed in [7].

Acknowledgements

I thank all Waikato University Machine Learning group members for his stimulating discussions on this topic, especially Richard Kirkby for making his Moa system for experimenting on large data streams available.

References

1. Andriy Bandos, Howard E. Rockette, D. Gur. Resampling Methods for the Area Under the ROC Curve. ROCML'06.
2. J. Beck. and E. Schultz. The use of ROC curves in test performance evaluation. Archives of Pathology and Laboratory Medicine, 110:13-20. 1986
3. J. P. Egan. Signal Detection Theory and ROC analysis. Academic Press, New York, 1975.
4. Tom Fawcett. ROC Graphs: Notes and Practical Considerations for Data Mining Researchers Technical Report HPL-2003-4, HP Labs
5. P.A. Flach. The geometry of ROC space: understanding machine learning metrics through ROC isometrics. ICML, 2003.

6. N. Lachiche and P.A. Flach. Improving accuracy and cost of two-class and multi-class probabilistic classifiers using ROC curves. ICML, 2003.
7. Sofus A. Macskassy, Foster Provost and Saharon Rosset. Pointwise ROC Confidence Bounds: An Empirical Evaluation. ROCML'05.
8. J.A. Swets. Measuring the accuracy of diagnostic systems. Science 240, 1285-93, 1988.
9. I.H. Witten and E. Frank. Data mining: Practical machine learning tools and techniques with Java implementations. Morgan Kaufmann, San Francisco, 2000.

Learning Hybrid Bayesian Networks by MML

Rodney T. O'Donnell, Lloyd Allison, and Kevin B. Korb

School of Information Technology
Monash University
Clayton, Victoria 3800, Australia

Abstract. We use a Markov Chain Monte Carlo (MCMC) MML algorithm to learn hybrid Bayesian networks from observational data. Hybrid networks represent local structure, using conditional probability tables (CPT), logit models, decision trees or hybrid models, i.e., combinations of the three. We compare this method with alternative local structure learning algorithms using the MDL and BDe metrics. Results are presented for both real and artificial data sets. Hybrid models compare favourably to other local structure learners, allowing simple representations given limited data combined with richer representations given massive data.

1   Introduction

There is a large literature on methods of learning Bayesian networks from observed data. Much of that work has focused solely on learning network structure, treating network parameterization as a separate process. However, some work has been done on learning network structure and parameters simultaneously and many algorithms exist for performing this task. Most techniques involve a heuristic search through network space to find the optimal combination of directed acyclic graph (DAG) and the set of associated conditional probability tables (CPTs).

For discrete networks, CPTs are the most powerful representation of a child node's probability distribution. Any variety of interaction between parent states may be expressed (where the parameters are entirely independent of each other); likewise, any variety of functional dependence between parameters may be expressed, such as noisy-OR models. When, as in this last case, some parameters are highly dependent upon others, this is described as *local structure*. The expressive power, and complexity, of CPTs is wasted in such cases, and so there is value in finding simpler representations, such as modest-sized decision trees.

For example, consider Figure 1, which shows a CPT requiring 8 continuous parameters; expressed in a decision tree form it requires only four. The benefit of non-CPT models quickly becomes apparent as more parent variables are added. The advantage of using local structures that are more economical than CPTs in Bayesian networks has been clearly shown in [1,2] and elsewhere.

Here we apply CaMML (Causal discovery via MML) [3,4,5] to the learning of local structure in Bayesian networks in an especially flexible way, using either full CPTs, logit models or decision trees, or any combination of these determined

A. Sattar and B.H. Kang (Eds.): AI 2006, LNAI 4304, pp. 192–203, 2006.

© Springer-Verlag Berlin Heidelberg 2006

on a node-by-node basis (hybrid models). We compare our approach with Nir Friedman's implementation[1] of the Bayesian BDe metric[6] and the MDL[7] metric, which were also applied to learning local structure using decision trees, although without hybrid model learning.

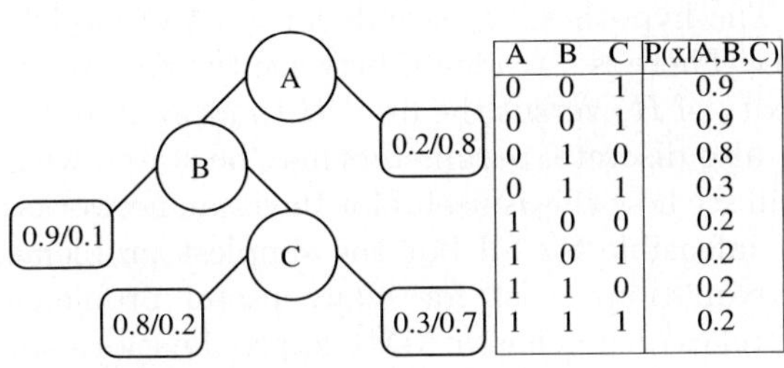

A	B	C	P(x|A,B,C)
0	0	1	0.9
0	0	1	0.9
0	1	0	0.8
0	1	1	0.3
1	0	0	0.2
1	0	1	0.2
1	1	0	0.2
1	1	1	0.2

Fig. 1. Decision Tree and CPT example

Previous work in learning local structure with MML used logit models[2]. Here we extend CaMML to decision trees, making it possible to compare the results effectively with local structure learning elsewhere. This requires a new coding technique, distinct from that of [8], since the decision trees involved in local structure have unique constraints. The hybrid learning likewise extends earlier work, allowing all varieties of local structure to be represented in forms suitable to their complexity. This provides an effective means of automatically adapting representational complexity to the amount of data available.

2   Metrics

Minimum Message Length (MML) inference is a method of estimating a fully parameterized model, using Bayes's Theorem: [9]

$$P(H\&D) = P(H).P(D|H) = P(D).P(H|D)$$

for a hypothesis (e.g., a Bayesian network), H, and data, D, and on Shannon's law for optimal codes [10]

$$msgLen(E) = -\log P(E)$$

requiring an event, E, to have a code of length $-\log P(E)$. Lengths can be measured in bits (log base 2) or, if mathematically convenient, in nits (natural logs). In any case,

$$msgLen(H\&D) = msgLen(H) + msgLen(D|H)$$

Being Bayesian, MML requires an explicit prior distribution, $P(H)$, on hypotheses. If the hypothesis space is discrete, a hypothesis has a non-zero probability and computing a message length is in principle straightforward, involving a negative log probability and a negative log likelihood. If however a hypothesis has one or more continuous parameters, it has no probability as such, rather a probability density. MML requires that continuous parameters be stated to optimal, *finite* precision. The latter point is subtle, allowing Bayes's rule to be applied to continuous hypothesis spaces. MML inference is consistent and is invariant under monotonic transformations of parameters.

MML uses an enumerable code-book of hypotheses previously agreed to by a *transmitter* and a *receiver*. The transmitter sends data to the receiver in a two-part message, sending a hypothesis, H, and then data coded on the assumption that the hypothesis is true, $D|H$. The hypothesis H stands for a set of models and thus gets a non-zero probability. There is a trade-off between the specificity of the set (equivalently the complexity of H) versus the fit of H to *expected* data. Note that not only continuous, but also discrete, parameters may be stated with less than maximum precision; we will see how this is useful for Bayesian networks.

Strict MML is computationally infeasible for all but the simplest problems [11], but practical, efficient approximations exist for many useful problems [12,13,14]. The work described here relies on stochastic MML approximations [3].

Minimum description length (MDL) inference was developed as an alternative to MML [15] and uses the same message length paradigm. MDL, however, favours universal priors and the selection of a model class rather than a parameterized model. A detailed comparison of MML and MDL has been given elsewhere by Baxter and Oliver[16].

BDe has its roots in Bayesian statistics. Like MDL, BDe attempts to find a model class rather than a parameterized model. BDe integrates over its prior on continuous parameters, whereas MML tries to segment continuous parameters into optimally sized regions and returns (a representative of) this region as an estimate.

3   Bayesian Networks

A Bayesian network is a directed acyclic graph (DAG) over a set of variables. Each node of the DAG represents a single variable. An arc commonly represents a direct causal relationship between a parent and a child. A node specifies the relationship between the node's variable and its parents, if any; we use conditional probability tables (CPTs), logit models and decision trees [8] for this.

An important concept when dealing with Bayesian nets is the 'statistical equivalence class' (SEC) [17]. Two DAGs in the same equivalence class can be parameterized to give an identical joint probability distribution; there is no way to distinguish between the two using only observational data over the given variables, although they may be distinguished given experimental data. Another important concept is the 'totally ordered model' (TOM). A TOM consists of a set of connections and a total ordering (permutation) of variables. Just as several DAGs may be in one SEC, several TOMs may realize a single DAG. TOMs are discussed further when introducing our MML coding scheme in §4.3.

4   Learning Global Structure

We briefly discuss the coding scheme for MDL and BDe and then build on these to present our MML coding scheme.

4.1   MDL Coding Scheme

We now summarize how Friedman encodes Bayesian networks using MDL. First the network structure is encoded, then the parameters of the network, and finally the data given the parameterized network. To encode the network structure, send the number of parents each node possesses, followed by the selection of parents out of all possible selections:

$$MDL_H = \sum_i \left(\log k + \log \binom{k}{|\pi(i)|} \right)$$

where k is the number of nodes and $\pi(i)$ is the parent set for node i. Friedman's code requires $\frac{1}{2} \log N$ nits per parameter to state a CPT (N is the sample size). The data requires

$$MDL_D = -\sum_{i=0}^{N} \log P(D_i)$$

Using this coding scheme and a heuristic search the DAG with the shortest description length is accepted as the best model.

4.2   BDe Scheme

Heckerman et al.'s BDe metric [6], based on a previous Bayesian metric of Cooper and Herskovits [18], has been augmented with decision trees by Friedman [1] whose implementation is used for comparison in section 7. The suggested prior is based on an edit distance from an expert supplied network. Friedman uses a prior based on the MDL prior outlined above with $P(H) \propto 2^{MDL_H}$.

Once the network structure has been stated, we must integrate

$$P(D|H) = \int P(D|\theta, H)P(\theta|H)d\theta$$

where $P(\theta|H)$ is the prior parameter density and $P(D|\theta, H)$ is the probability of the data given the parameterized network. Using a Dirichlet prior, the closed form solution is:

$$P(D|H) = \prod_i \prod_{pa_i} \frac{\Gamma(\sum_{x_i} N'_{x_i|pa_i})}{\Gamma(\sum_{x_i} N'_{x_i|pa_i} + N(pa_i))} \times \prod_{x_i} \frac{\Gamma(N'_{x_i|pa_i} + N(x_i, pa_i))}{\Gamma(N'_{x_i|pa_i})}$$

where x_i is a node instantiation, pa_i an instantiation of $\pi(i)$, $\Gamma(x)$ is the Gamma function and $N(\cdot)$ counts the number of sample cases matching an instantiation. Heckerman's default "equivalent sample size", $N' = 5$, is used.

4.3   MML Coding Scheme

Whereas many methods use a uniform prior over SECs or DAGs, CaMML uses uniform priors over totally ordered models (TOMs). In essence, TOMs are DAGs with a consistent total order selected. A TOM can be thought of as a way of realizing its DAG; in effect, TOMs specify distinct possible worlds in which the DAG is true. We should prefer to employ non-informed, uniform (maximum entropy) priors only as a last resort — i.e., when we arrive at the most primitive level of description, in this case TOMs rather than DAGs or SECs. (For more discussion of this approach see Korb and Nicholson, Chapter 8. [4].)

CaMML's stochastic search algorithm (section 4.4) samples TOMs, but it counts DAGs, so that each TOM contributes probability mass to its DAG. The algorithm has the following stages. A TOM is sampled in the MCMC process. The corresponding DAG is "cleaned" by deleting weak arcs whenever this reduces the total message length. The clean DAG is counted, as is its SEC. Repeated counting allows us to estimate the posterior distribution over the DAG and SEC space.

When the sampling phase is over, SECs are also grouped in case the data does not justify choosing between them, using a Kullback-Leibler (KL) divergence test. As with lower level groupings, such a group may gain enough probability mass to be preferred even when the single SEC might not.

The encoding of a single TOM has two parts: a list of arcs and a total ordering. The arcs can be encoded in $m \times \log P_a + \left(\frac{k(k-1)}{2} - m \right) \times \log(1 - P_a)$ nits. Where k is the number of nodes in the network, m is the number of arcs present and P_a is the prior probability of arc presence (default 0.5). The cost to state the total ordering is simply $\log k!$ nits. In addition to the TOM we must also use $\log p(data|TOM)$ nits to express the data — see section 5. This scheme forms the basis of an efficient code employing our prior beliefs about network structure.

4.4   The MCMC Search

An MCMC search algorithm allows us to approximate a posterior distribution over DAGs and SECs by sampling TOM space. The algorithm used follows:

1. Simulated Annealing to find the best single TOM. This optimal TOM is used to estimate P_a and provide a starting position for our sampling.
2. Attempt a mutation on the current TOM M transforming it into M'
   (a) **Temporal:** Swap the order of two neighbouring nodes in the total ordering. If an arc exists, reverse its direction.
   (b) **Skeletal:** Select two nodes at random and toggle the existence of an arc between them.
   (c) **DoubleSkeletal:** Select three nodes at random, toggle the arcs from the first two to the final (in the total ordering).
   (d) **ParentSwap:** Select three nodes such that $a \rightarrow c$ but not $b \rightarrow c$ [1] and toggle the arcs ac and bc. This effectively removes one parent and replaces it with another.

   Mutations c and d are not strictly necessary since they are compositions of the other mutations, but speed up the sampling process.

---

[1] If not possible, choose a different mutation.

3. Accept M' as the new sampled M if

$$\frac{\log P_{MML}(M) - \log P_{MML}(M')}{temperature} > \log U[0,1]$$

else retain M. Sampling is conducted at $temperature = 1.8$. [2]

4. Add a weight of $\frac{\exp^{P_{MML}} - \exp^{P_{BestMML}}}{temperature}$ to the current TOM's clean representative DAG and SECs accumulated weight. $P_{BestMML}$ is a common factor to avoid underflow.

5. Loop to 2 until a set number of steps are complete.

Step 4 above refers to a TOM's clean representative DAG and SEC; mutation continues from the current "unclean" TOM at the next iteration.

5 Learning Local Structure

The coding scheme and the MCMC algorithm place few requirements on the method used to represent local structure (i.e., parent-child relationships). The present work uses CPTs, decision trees, logit and hybrid models. The later is able to choose between other model types on a node-by-node basis. Here we describe the addition of local structure to our MML metric; see Friedman [1] for more detail on MDL and BDe.

5.1 CPTs

The CPT is the standard building block of Bayesian networks. If a child (variable) takes one of S possible values (states), a multi-state distribution having $S - 1$ parameters must be specified for each combination of parent values. Wallace and Boulton [12] gave the MML calculations for the multi-state distribution. The message length was shown to be equivalent to using an adaptive code for the data with an extra "penalty" of a fraction of a nit, $\frac{1}{2}\log\frac{\pi e}{6} \approx 0.176$ per parameter; the consequence of using an estimate with optimum precision [13]:

$$mesglen = \frac{|pa| \times (|x| - 1)}{2}\log\frac{\pi e}{6} + \sum_{pa_i}^{|pa|}\log\left(\frac{(N(pa_i) + |x| - 1)!}{(|x| - 1)! \times \prod_{x_i}^{|x|}(N(pa_i, x_i)!)}\right)$$

where $|pa|$ and $|x|$ are the number of parent and child states respectively.

5.2 Decision Trees

To code decision trees we begin with Wallace and Patrick's code [8]. Briefly, each leaf or split node has an initial cost of 1 bit. Also stated for each split

---

[2] A high temperature causes TOM space to be more widely traversed, low temperature makes the sampling more likely to stay near the original model. A temperature of 1.8 was used throughout this work.

is the variable being split; an attribute cannot be "reused" and so this costs $\log k - depth$ nits. Each leaf must state a model for that node and also the data given that model. Multinomial models are the natural choice here, as with CPTs.

In general, decision trees can ignore parents that are irrelevant by never splitting on them. However, in the context of local structure in a Bayesian network, we require that all parents should be used, since the coding of the network structure implies it. To achieve this, we correct the message length and force our trees to split on each parent at least once, other trees being disallowed. Selecting parents is thus the responsibility of network topology discovery, rather than the local node encoding.

Our strategy for reclaiming lost probability (from the disallowed trees) is based on the number of trees with n split nodes, i.e., on the Catalan numbers defined as $cat(n) = \begin{pmatrix} 2n \\ n \end{pmatrix} \div (n + 1)$. The prior probability of a given tree structure with n splits can be calculated as $p_n = 2^{-(2n+1)}$; by multiplying these numbers we calculate the total proportion of prior allocated to models with n splits. As can be seen in Table 1, this prior is skewed towards models with few splits, as one would expect.

Table 1. Catalan numbers

n	cat(n)	$2^{-(2n+1)}$	$p_n cat(n)$	$\sum_{i=0}^{n} p_i cat(i)$
0	1	0.50000	0.50000	0.50000
1	1	0.12500	0.12500	0.62500
2	2	0.03125	0.06250	0.68750
3	5	0.00781	0.03906	0.72656
4	14	0.00195	0.02734	0.75391
5	42	0.00048	0.02051	0.77441
6	132	0.00012	0.01611	0.79053
7	429	0.00012	0.01309	0.80362
∞	∞	0		1

Table 2. Savings made(in bits): where n = number of parents, min = minimum number of splits, max = maximum number of splits, s = number of splits made

n	min/max	$P_{s<min}$	$P_{s\geq max}$	$P_{invalid}$	saving
0	0...0	0.000	0.500	0.500	1.00
1	1...1	0.500	0.375	0.125	3.00
2	2...3	0.625	0.273	0.102	3.30
3	3...7	0.688	0.196	0.116	3.11
4	4...15	0.727	0.140	0.133	2.91
5	5...31	0.754	0.099	0.147	2.77

Taking the simplest example of a leaf with no parents, it is obvious that our tree structure will be that of a single leaf. However, under our original prior 1 *bit* is still required to express this structure. For a (binary) tree with 1 parent the original structure cost would be 3 *bits*, one for each split and one for each leaf. In general, it would be reasonable to pay this cost as a decision to split is actually made, but in the present context the modified priors tell us that these splits are always required, so the penalty should be removed. A slightly more complex case arises for two or more parents where our priors allow between N and 2^N splits where N is the number of parents. An approximation of this saving is used for n-ary variables.

To further illustrate, we take a tree with three binary parents. We are constrained to have at least three splits, and no more than seven splits. From Table 1 we see that trees with less than three splits use 0.688 and trees with more than seven use $1 - 0.804 = 0.196$ of our prior hypothesis space. So our original prior has 0.884 wasted on impossible hypotheses! By subtracting $-\log 1 - 0.884 = 3.11$ *bits* from our message length we effectively redistribute the probability mass from trees with an invalid number of splits to those with appropriate splits.

By examining the table of Catalans, Table 1, it is possible to calculate the saving, as seen in Table 2. Using our true prior to calculate Decision Tree costs is obviously a good thing to do, but it becomes especially important when dealing with hybrid models so that CPTs and trees can compete fairly.

5.3   Logit

We follow Neil et al.[2] in supporting MML logit models of local structure. A CPT treats each parameter independently. It is common, however, for there to be local structure, with some parameters dependent upon others. A first order logit model allows us to exploit this.

$$P(X = x | Z_1 = z_1, Z_2 = z_2, \ldots Z_n = z_n) = \frac{e^{a_i + b_{iz_1} + c_{iz_2} \cdots}}{\sum_{j=1}^{|X|} e^{a_j + b_{jz_1} + c_{jz_2} \cdots}}$$

When the effect of each parent is independent of the effects of other parents (so joint parameters are not independent), we expect our logit model to give a better representation of a distribution than a CPT would; given interacting parents a CPT or DTree would be expected to perform better. To get the best of both worlds, hybrid models are useful.

5.4   Hybrid Models

Hybrid models, as the name suggests, are combinations of two or more models of local structure; in this case we have combined CPTs, decision trees and logit models. Previous work [2] combined CPTs and logit models with some success.

We define a hybrid model as a model which can choose between competing local models to give the best result. The search costs a decision tree, a logit and a CPT, choosing the one with the lower MML cost. It is also necessary to add $\log 3$ *nits* to the node's cost — treating models as being equally likely apriori. This extra cost is not required for models with zero or one parent, as the CPT, DTree and logit have equivalent expressive power, so a CPT is used.

6   Evaluation

Our results are defined in terms of KL divergence from a true model (when known) or log probability on a test set (otherwise). Tests were also run on data generated from artificial networks having varying degrees of "decision treeness" in their local structure to show cases where CPTs should be favoured and where decision trees should be favoured; hybrid models should perform well across the whole spectrum. This degree is quantified by P_l: the number of leaves a decision tree will possess is at least $P_l \times |pa|$. Low values correspond to more local structure. Our simple tree generation algorithm follows:

1. Begin with an empty tree (single leaf)
2. Choose a leaf at random and split using any unused variables
3. If less than $P_l \times |pa|$ leaves exist, goto 2
4. For each variable has not been split on, choose a leaf and split on it.

In addition to the random networks we use the "insurance" network, which consists of 27 variables with arity ranging from 2 to 5, and also the real datasets listed in Table 3.

Error bars are shown at 1.96 standard errors (SE) from the mean. Ideally, pairwise comparisons over all searches would have been used, but this becomes unmanageable when comparing 8 unique metrics.

7 Results

Figure 2 shows results for artificial networks with varying degrees of local structure, i.e., varying P_l. The KL divergence of the inferred network from the true network is plotted. Results are averaged over 100 trials. CaMML performs much better than BDe and MDL, both with CPTs and DTrees. As expected, CaMML DTrees do best when P_l is low and CPTs do better when P_l is high. The CPT-DTree hybrid model does well across the range.

Logit models were excluded in this test, as their assumption of a non-interactive distribution is not met here and as such would perform poorly. An experiment similar to this (although varying levels of first and second order effect strength) compares CPTs, logits and CPT-logit hybrid models is found in Neil et al[2]. That paper shows logit models outperforming CPTs when their assumptions are warranted and CPTs winning when they are not. Again, hybrid models perform well throughout.

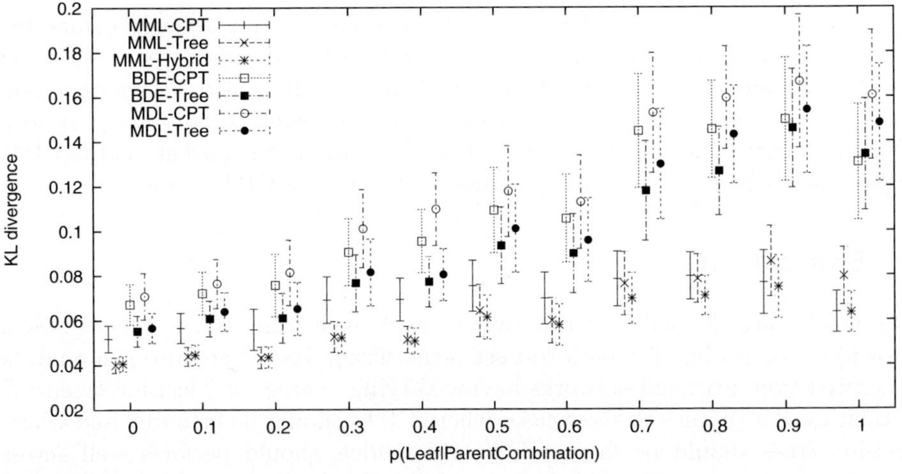

Fig. 2. Varying P_l, $P_a = 0.25$, N = 1000, 7 nodes, arity = [2,5,10,7,3,3,2] 100 folds. Error bars at 1.96 SE.

Figure 3 shows KL divergence results for the insurance network while varying the training sample size. Plots were adjusted by multiplying KL values by $N/\log N$, as suggested in [1], keeping values approximately constant across the graph by compensating for KL asymptoting to zero as N grows large. As in Figure 2, MDL performs worse than MML and BDe across the board and as such has been removed to reduce clutter.

It is clear that for both MML and BDe (and unshown MDL results) that tree based learners often outperform CPT learners for this example. The difference is especially evident for large datasets. This result confirms Friedman's work.

Of more interest, however, is the comparison to logit based learners. Our MML logit learner significantly outperforms all out non-logit learners for small datasets ($N < 1000$) but performs much worse for large datasets. This is due to logit models being able to approximate first order effects better than CPTs or trees for small datasets, but being unable to express second (and higher) order effects when enough data is given to reveal them. Our MML hybrid learner (using CPT, tree and logit models) has the best of both worlds. It performs as well as the logit model for small sample sizes, then roughly as well as the tree based model for larger data sizes.

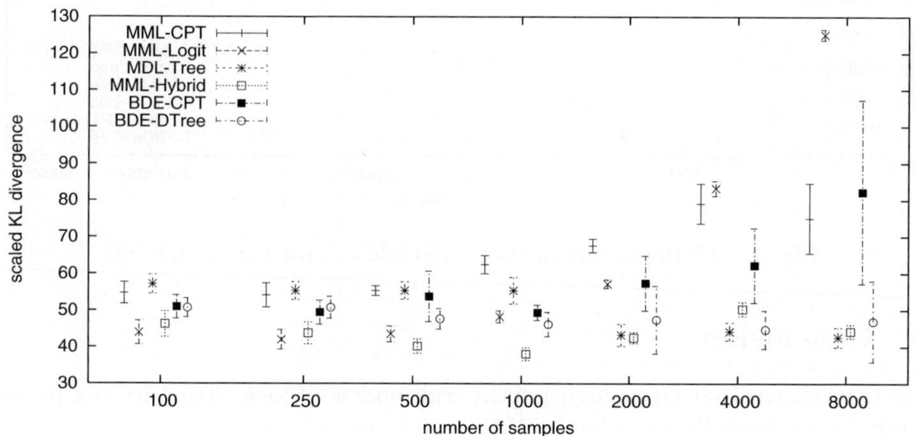

Fig. 3. Insurance Network, 10 folds. Error bars at 1.96 SE

Figure 4 shows results from several real datasets summarized in Table 3. Once again MDL performed badly and is not shown. Eight datasets are examined including the six used by Neil [2] and two larger datasets where we would expect decision trees to do well.

Table 3. Real Datasets from UCI repository

name	description	k	arity	N
Zoo	Animal attributes	17	2-7	101
ICU	Intensive care unit	17	2-3	200
Flare	Solar flares	13	2-7	323
Voting	US congress votes	17	2-3	435
Popularity	Childhood popularity.	11	2-9	478
kr-vs-kp	Chess end games	37	2-3	3196
Mushroom	Poison mushrooms	23	2-12	8124
Nursery	Child care.	9	2-5	12960

Comparing CPT and tree based learners (for MML and BDe) we see comparable results for small datasets, with tree learners performing significantly better on larger datasets. To visually clarify results, all *logP* values have been normalized by the worst scoring learner for each dataset.

In all but one of our datasets there is a clear winner between MML Tree and logit models, with logit winning on small datasets, but loosing badly on large datasets. Our hybrid model does well throughout having several significant wins against each rival MML learner, but no significant losses.

In results not shown, all hybrid combinations of models were examined. That is a "CPT-Logit", "CPT-Tree" and "Tree-Logit", in addition to the "CPT-Tree-Logit" shown. It was evident that when our hybrid model contained a logit component, it did well on small datasets and when it contained a tree component it did well on large datasets. Hybrid learners with both options did well on small and large datasets. Removing CPTs from the "CPT-Tree-Logit" learner has a much smaller effect than removing either the tree or logit component.

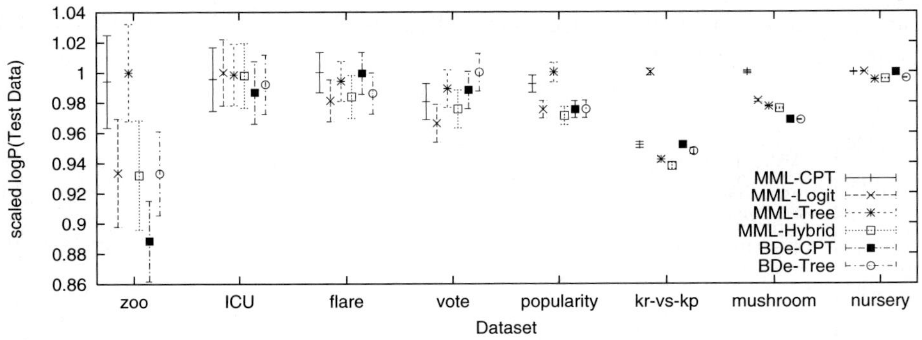

Fig. 4. UCI Repository datasets, 100 folds. Error bars at 1.96 SE

8   Conclusion

We have shown that trees, logit and hybrid models of local structure can be introduced successfully into the CaMML search procedure. When jointly incorporated in hybrid model discovery, the result is a flexible learning procedure which automatically accommodates data set sizes by preferring simple local structure representations (e.g., logit models) given small data sets and by finding richer representations (decision trees and/or CPTs) given large data sets. We also compared these MML metrics with BDe and MDL metrics, finding that generally BDe and CaMML do well, with MDL performing poorly.

Acknowledgments

We would like to acknowledge the late Chris Wallace whose work on MML, Bayesian networks and decision trees was pioneering.

References

1. Friedman, N., Goldszmidt, M.: Learning Bayesian networks with local structure. In: Uncertainty in Artificial Intelligence. (1996)
2. Neil, J.R., Wallace, C.S., Korb, K.B.: Learning Bayesian networks with restricted causal interactions. In: Uncertainty in Artificial Intelligence. (1999)
3. Wallace, C.S., Korb, K.B.: Learning linear causal models by MML samplling. In Gammerman, A., ed.: Causal Models and Intelligent Data Management. Springer-Verlag (1999)
4. Korb, K., Nicholson, A.: Bayesian Artificial Intelligence. CRC Press (2003)
5. O'Donnell, R.T., Nicholson, A.E., Han, B., Korb, K.B., Alam, M.J., Hope, L.R.: Causal discovery with prior information. 19th Australian Joint Conf on AI (2006)
6. Heckerman, D., Geiger, D., Chickering, D.: Learning bayesian networks: The combination of knowledge and statistical data. Machine Learning **20**(197-243) (1995)
7. Lam, W., Bacchus, F.: Learning Bayesian belief networks. Computational Intelligence **10** (1994)
8. Wallace, C., Patrick, J.: Coding decision trees. Machine Learning **11** (1993) 7
9. Bayes, T.: An essay towards solving a problem in the doctrine of chances. Philosophical Transactions of the Royal Soc. of London (1764/1958) reprinted in Biometrika 45(3/4) 293-315 Dec 1958.
10. Shannon, C.E.: A mathematical theory of communication. Bell System Technical Journal **27**(3) (1948) 379–423
11. Farr, G.E., Wallace, C.S.: The complexity of strict minimum message length inference. Computer Journal **45**(3) (2002) 285–292
12. Wallace, C.S., Boulton, D.M.: An information measure for classification. The Computer Journal **11** (1968) 185–194
13. Wallace, C.S.: Statistical and Inductive Inference by Minimum Message Length. Springer, Berlin, Germany (2005)
14. Allison, L.: Models for machine learning and data mining in functional programming. Journal of Functional Programming (2005)
15. Rissanen, J.: Modeling by shortest data description. Automatica **14** (1978)
16. Baxter, R., Oliver, J.: MDL and MML: similarities and differences. Technical Report 207, Dept of Computer Science, Monash University (1994)
17. Chickering, D.: A transformational characterization of equivalent Bayesian network structures. In: Uncertainty in Artificial Intelligence. (1995)
18. Cooper, G., Herskovits, E.: A Bayesian method for the induction of probabilistic networks from data. Machine Learning **9** (1992) 309–347

This maximization is equivalent to find eigen-pairs (λ_i, v_i) , $i=1,2,\cdots,d$, $\lambda_1 \geq \lambda_2 \geq \cdots \geq \lambda_d$ for the generalized eigenvalue problem

$$S_b^{NW} v = \lambda S_w^{NW} v \tag{4}$$

3 *K*-Nearest-Neighbor Classifier Based on Adaptive NWFE Separability

In this section, a *k*-nearest-neighbor classifier based on adaptive nonparametric separability (*k*NN-ANS) is proposed. This technique is based on the exploitation of the results of the nonparametric weighted feature extraction (NWFE) performed on a set of K_m patterns in a nearest neighborhood of unclassified pattern x^*. Only using these K_m patterns can emphasize the influence of the separability around x^*.

Suppose $R(x^*, K_m)$ is the set of these K_m patterns around x^*. The between-class scatter matrix $S_b^{ANW}(x^*)$ and the within-class scatter matrix $S_w^{ANW}(x^*)$ of NWFE are defined as

$$S_b^{ANW}(x^*) = \sum_{i=1}^{L} P_i \sum_{\substack{j=1 \\ j \neq i}}^{L} \sum_{x_k^{(i)} \in R(x^*, K_m)} \frac{\lambda_k^{(i,j)}}{n_i} (x_k^{(i)} - M_j(x_k^{(i)}))(x_k^{(i)} - M_j(x_k^{(i)}))^T , \tag{5}$$

$$S_w^{ANW}(x^*) = \sum_{i=1}^{L} P_i \sum_{x_k^{(i)} \in R(x^*, K_m)} \frac{\lambda_k^{(i,i)}}{n_i} (x_k^{(i)} - M_i(x_k^{(i)}))(x_k^{(i)} - M_i(x_k^{(i)}))^T , \tag{6}$$

where $M_j(x_k^{(i)}) = \sum_{l=1}^{n_j} w_{kl}^{(i,j)} x_l^{(j)}$, $\lambda_k^{(i,j)} = \dfrac{\mathrm{dist}(x_k^{(i)}, M_j(x_k^{(i)}))^{-1}}{\sum_{l=1}^{n_i} \mathrm{dist}(x_l^{(i)}, M_j(x_l^{(i)}))^{-1}}$, $w_{kl}^{(i,j)} = \dfrac{\mathrm{dist}(x_k^{(i)}, x_l^{(j)})^{-1}}{\sum_{l=1}^{n_i} \mathrm{dist}(x_k^{(i)}, x_l^{(j)})^{-1}}$, n_i

denotes the training sample size of class i, $M_j(x_k^{(i)})$ denotes the weighted mean of $x_k^{(i)}$ in class j, L is the number of classes and $\mathrm{dist}(x, z)$ be the Euclidean distance from x to z .

Let $\Lambda = \lambda_1 + \lambda_2 + \cdots + \lambda_d$ (λ_i is the *i*-th large eigenvalue of $(S_W^{ANW}(x^*))^{-1} S_b^{ANW}(x^*)$, $i=1,2,...,d$). Define a new weighted metric

$$\Sigma = \frac{\lambda_1}{\Lambda} v_1 v_1^T + \frac{\lambda_2}{\Lambda} v_2 v_2^T + \cdots + \frac{\lambda_d}{\Lambda} v_d v_d^T . \tag{7}$$

Note that

$$\Sigma v_i = \frac{\lambda_i}{\Lambda} v_i, \forall i = 1, 2, \cdots d.$$

Thus $(\frac{\lambda_i}{\Lambda}, v_i), \forall i = 1, 2, \cdots, d$ are eigen-pairs of Σ .

Let $x^T \Sigma x = 1, x \in \Re^d$. Since $\{v_1, v_2, \cdots, v_d\}$ forms a new basis for $\Re^d$, thus $x = \bar{x}_1 v_1 + \bar{x}_2 v_2 + \cdots + \bar{x}_d v_d$. So we have

$$x^T \Sigma x = (\bar{x}_1 v_1 + \bar{x}_2 v_2 + \cdots + \bar{x}_d v_d)^T \Sigma (\bar{x}_1 v_1 + \bar{x}_2 v_2 + \cdots + \bar{x}_d v_d)$$

$$= \sum_{i=1}^{d} \sum_{j=1}^{d} (\bar{x}_i v_i^T) \Sigma (\bar{x}_j v_j).$$

Since

$$(\bar{x}_i v_i^T) \Sigma (\bar{x}_j v_j) = (\bar{x}_i \bar{x}_j) v_i^T \Sigma v_j$$

$$= (\bar{x}_i \bar{x}_j) v_i^T (\frac{\lambda_j}{\Lambda} v_j)$$

$$= \begin{cases} \dfrac{\lambda_i}{\Lambda} \bar{x}_i^2 & \text{if } i = j \\ 0 & \text{if } i \neq j \end{cases}, \forall i, j = 1, 2, \cdots d,$$

thus

$$1 = x^T \Sigma x = \frac{\lambda_1}{\Lambda} \bar{x}_1^2 + \frac{\lambda_2}{\Lambda} \bar{x}_2^2 + \cdots + \frac{\lambda_d}{\Lambda} \bar{x}_d^2$$

$$= \frac{\bar{x}_1^2}{(\frac{\Lambda}{\lambda_1})} + \frac{\bar{x}_2^2}{(\frac{\Lambda}{\lambda_2})} + \cdots + \frac{\bar{x}_d^2}{(\frac{\Lambda}{\lambda_d})},$$

i.e., $\{x \mid x^T \Sigma x = 1\}$ forms an ellipse and the lengths of axes are $2\sqrt{\dfrac{\Lambda}{\lambda_1}}, 2\sqrt{\dfrac{\Lambda}{\lambda_2}}, \cdots, 2\sqrt{\dfrac{\Lambda}{\lambda_d}}$.

From this result it is easy to find that the length of major axis with respect to v_1 (and v_d) is the shortest (and longest respectively).

This distance measure between the unknown x^* and the training sample $x_j^{(i)}$ is

$$d(x^*, x_j^{(i)}) = (x^* - x_j^{(i)})^T \Sigma (x^* - x_j^{(i)}). \tag{8}$$

The modification shrinks the neighborhood according to the degree of the separability around x^*. We use this to find k nearest points within the points in the training set from the unknown observation x^* and assign the label of the unknown observation using the majority vote.

Fig. 1 shows an example of our metric. There are two classes in two dimensions. The target point (shown as **x**) is chosen to be near the class boundary. The first panel shows the 5 nearest neighbors of **x** using the Euclidean distance. The second panel shows the same size neighborhood using our new weighted metric. Notice how the modified neighborhood shrinks in the direction (approximately orthogonal to the decision boundary which is around **x**) with high separability and extends further in the other direction.

The value of K_m must be reasonably large corresponding to the problem's dimension d, since the initial neighborhood $R(x^*, K_m)$ is used to estimate the scatter matrices. Often a smaller number k of neighbors is preferable for the final classification rule.

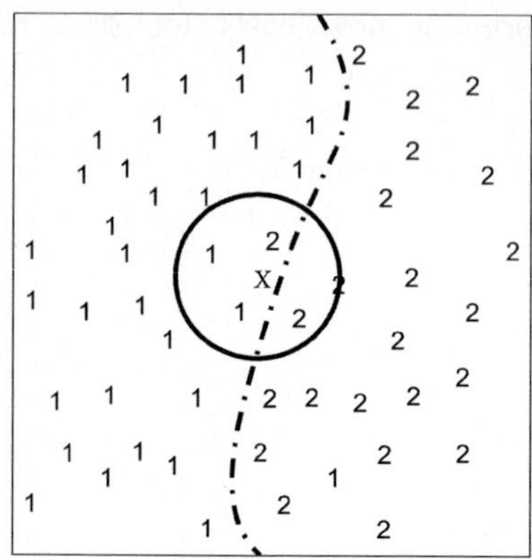 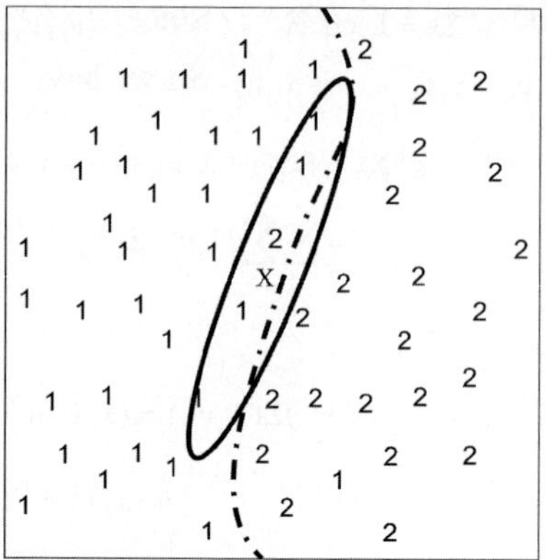

Fig. 1. The first panel shows the spherical neighborhood containing 5 points. The second panel shows the ellipsoidal neighborhood found by the 5NN-ANS procedure, also containing 5 points. The latter is elongated along the direction (approximately parallel to the decision boundary which is around **x**) of less separability and flattened orthogonal to it.

These procedures are summarized in kNN-ANS Algorithm.

kNN-ANS Algorithm

Input:
    d-   the problem's dimension.

    N-   the number of training samples.

    L-   the number of pattern classes.

    (x_i, j_i)- N ordered pairs, where x_i is the i-th training sample and j_i is its class number ($1 \le j_i \le L$ for all i).

    k-   the order of k-NN classifier.

    x^*-   an incoming pattern.

    K_m-   the number of $R(x^*, K_m)$.

Output:
    l-   the number of class into which x^* is classified.

Step 1.   Set $X^T = [x_1, \cdots, x_N]$ and $S = \{(x_i, j_i)\}_{i=1}^N$.

Step 2.   Compute $S_b^{ANW}(x^*)$ and $S_w^{ANW}(x^*)$ to find eigen-pairs (λ_i, v_i), $i=1,2,\ldots,d$ of $(S_W^{ANW}(x^*))^{-1} S_b^{ANW}(x^*)$, with $\lambda_1 \ge \lambda_2 \ge \cdots \ge \lambda_d$.

Step 3.   Establish $\Sigma = \dfrac{\lambda_1}{\Lambda} v_1 v_1^T + \dfrac{\lambda_2}{\Lambda} v_2 v_2^T + \cdots + \dfrac{\lambda_d}{\Lambda} v_d v_d^T$, $\Lambda = \lambda_1 + \lambda_2 + \cdots + \lambda_d$.

Step 4.   Find $(y, j_0) \in S$ which satisfies $y = \underset{(z,j) \in S}{\arg\min}(z^T)\Sigma(x^*)$.

Step 5.   If $k = 1$ set $l = j_0$ and stop; else initialize a L- dimensional vector IC: $IC(i') = 0, i' \ne j_0; IC(j_0) = 1$ and set $S = S - \{(y, j_0)\}$.

Step 6.   For $i_0 = 1,2,\cdots,k-1$ do steps 7-8.

Step 7.   Find $(y, j_0) \in S$ such that $y = \underset{(z,j) \in S}{\arg\min}(z^T)\Sigma(x^*)$.

Step 8.   Set $IC(j_0) = IC(j_0) + 1$ and $S = S - \{(y, j_0)\}$.

Step 9.   Set $l = \max(IC(i)), 1 \le i \le L$ and stop.

4   A Simplified Version of the *KNN-ANS*

So far our technique outperforms enormously original k-NN but the shortcoming is "time consuming," in that for every unclassified pattern x^*, it is necessary to be recomputed $S_b^{ANW}(x^*)$, $S_w^{ANW}(x^*)$, and all eigen-pairs of $(S_W^{ANW}(x^*))^{-1}S_b^{ANW}(x^*)$.

Sometimes using the separability of NWFE with whole training samples (i.e., the eigen-pairs of $(S_W^{NW})^{-1}S_b^{NW}$) still has not bad performance. In particular, the ratio of the number of training samples to the dimension d is small. This method is denoted by kNN-NS, a special case of kNN-ANS with K_m which is equal to the number of training samples. However, in kNN-NS, the weighted metric is the same for all unclassified patterns and so it can get over the problem of "time consuming."

The kNN-NS can be summarized in the following steps:

1.   Compute the eigen-pairs (λ_i, v_i), $i=1,2,...,d$ of $(S_W^{NW})^{-1}S_b^{NW}$, where S_b^{NW} and S_w^{NW} are the scatter matrices of NWFE with whole training samples.

2.   Establish $\Sigma = \dfrac{\lambda_1}{\Lambda}v_1v_1^T + \dfrac{\lambda_2}{\Lambda}v_2v_2^T + \cdots + \dfrac{\lambda_d}{\Lambda}v_dv_d^T$, where $\Lambda = \lambda_1 + \lambda_2 + \cdots + \lambda_d$.

3.   Do steps 4-9 in kNN-ANS Algorithm.

Fig. 2 shows the main difference of kNN-ANS and kNN-NS. There are two classes, one of which almost surrounds the other in two dimensions. The first panel shows the "unit sphere," an ellipsoid using kNN-ANS metric around the unclassified samples which are chosen to be near the class boundary. The second panel shows the "unit sphere," an ellipsoid using kNN-NS metric on the same samples. The red lines in the first panel are the axes using the method kNN-ANS. We can image that the axes will change when the different unclassified sample is coming. The blue lines in the second panel are the axes using the method kNN-NS. It is easy to point that the axes are the same for all unclassified samples. Notice that the local class posterior probabilities around unclassified patterns in the first panel are approximately constant. As we will see in our experiments, kNN-ANS can often provide improvement in classification performance.

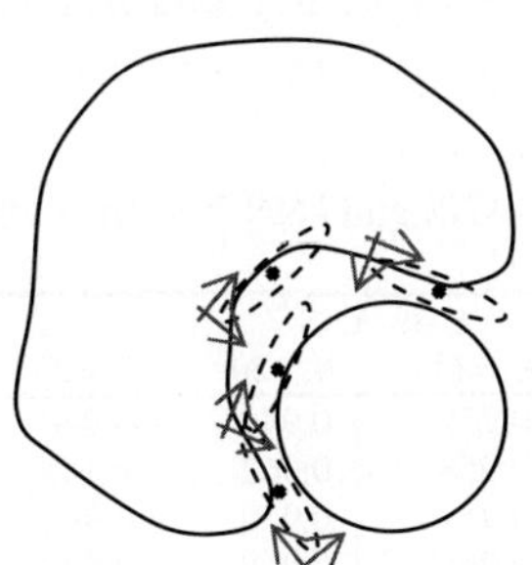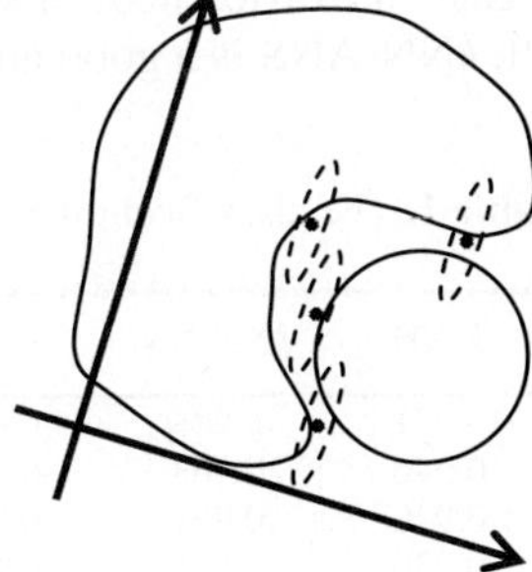

Fig. 2. The first panel shows the "unit sphere," an ellipsoid using kNN-ANS metric around the unclassified samples which are chosen to be near the class boundary. The second panel shows the "unit sphere," an ellipsoid using kNN-NS metric on the same samples. The red lines in the first panel are the axes using the method kNN-ANS. The blue lines in the second panel are the axes using the method kNN-NS.

5 Experiment Design and Results

The design of Experiment is to compare the multiclass classification performances of four classifiers: kNN-ANS, kNN-NS, k-NN, support vector classifier using cross-validation (SVC-CV), and Parzen [4]. Two real data experiment results are displayed. One is the Fisher's Iris data published by Fisher [5] and the other is the Washington, DC Mall image data [6]. At each experiment, 10 training and testing data sets are selected for computing the accuracies of algorithms.

5.1 Fisher's Iris Data

The Iris flower data were originally published by Fisher [5], for examples in discriminant analysis and cluster analysis. Four parameters, including sepal length, sepal width, petal length, and petal width, were measured in millimeters on fifty iris specimens from each of three species, Iris setosa, Iris versicolor, and Iris virginica. One class (Iris setosa) is linearly separable from the two other; the latter (Iris versicolor and Iris virginica) are not linear separable from each other. So the set of data contains 140 examples with 4 dimensions and 3 classes. In every data set, we randomly choose 10 samples from every class to form the training data and the other 40 samples in every class are assigned to the testing data.

In this experiment, the grid method is used to determine K_m. Six parameters, 5, 10, 15, 20, 25, and 30, are used. Similarly, the k, the order of k-NN classifier, are set by 1, 3, 5, 7, 9, 11, 13, 15, and 17. The results are displayed in Table 1. The red parts mean that the performances of our methods are better than original k-NN. The comparisons of our proposed methods and other classifiers are summarized in Table 2, where the results k-NN, kNN-ANS, and kNN-NS are chosen from Table 1 with the best accuracies, respectively.

Tables 1 and 2 show the following.

1. The results using modified metric are better than using the Euclidean distance.
2. For decreasing degree of accuracies, our methods outperform k-NN when k is increasing.
3. In this case, the differences of kNN-ANS and kNN-NS are not remarkable.
4. Overall, kNN-ANS is a good and robust choice.

Table 1. The classification accuracies using kNN-ANS and kNN-NS (Iris data)

k	k-NN	kNN-NS	kNN-ANS					
			K_m=5	K_m=10	K_m=15	K_m=20	K_m=25	K_m=30
1	0.931	**0.945**	0.931	0.942	0.938	0.941	0.940	0.945
3	0.946	0.963	0.939	0.954	0.958	0.962	**0.963**	0.963
5	0.940	0.961	0.946	**0.963**	0.957	0.960	0.961	0.961
7	0.930	0.968	0.942	0.962	0.965	**0.968**	0.968	0.968
9	0.926	0.966	0.938	0.960	0.960	**0.968**	0.967	0.966
11	0.913	0.960	0.942	0.963	0.963	0.963	**0.963**	0.960
13	0.904	0.960	0.943	0.964	0.962	0.963	**0.964**	0.960
15	0.883	0.959	0.941	**0.962**	0.959	0.961	0.960	0.959
17	0.868	0.958	0.929	0.958	0.955	**0.961**	0.959	0.958

Table 2. The classification accuracies using kNN-ANS, kNN-NS, k-NN, SVC-CV, and Parzen (Iris data)

kNN-ANS	kNN-NS	k-NN	SVC-CV	Parzen
0.968	**0.968**	0.946	0.951	0.941
k=7, K_m=20	k=7	k=3		

5.2 Washington, DC Mall Image Data

The Washington, DC Mall image data [6] is used for this experiment. There are 7 classes, 191 spectral bands in it, and 20, 40, 100 training samples in each class in this experiment. For investigating the influences of the ratio of the number of training samples to the dimension using our methods, every 5-th band, which begins from the first one, are selected for the 39 bands case. Table 3 shows classification accuracies of testing data using 20, 40, and 100 training samples with 39 bands in each class. This result indicates that the 1NN-ANS has the best performances. In addition, using 1NN-ANS has more outstanding effect than using 1NN-NS, i.e., the use of an adaptive metric allows improving the accuracy of a fixed metric, thus confirming the influence of the shape and size of the neighborhood around unclassified pattern.

Table 3. The classification accuracies using kNN-ANS, kNN-NS, k-NN, SVC-CV, and Parzen (DC Mall, 39 bands)

# of training Samples	1NN-ANS	1NN-NS	1-NN	SVC-CV	Parzen
20	**0.878** K_m=45	0.825	0.708	0.771	0.722
40	**0.908** K_m=50	0.860	0.761	0.828	0.777
100	**0.942** K_m=105	0.895	0.818	0.893	0.832

Table 4. Maximum, medium, minimum and standard deviation of the accuracy on the testing set using 1NN-ANS for varying values of the parameter K_m (DC Mall, 39 bands)

# of training samples	Max	Medium	Min	Std dev
20	0.878 K_m=45	0.856 K_m=80	0.825 K_m=140	0.017
40	0.908 K_m=50	0.885 K_m=160	0.860 K_m=280	0.017
100	0.942 K_m=105	0.917 K_m=345	0.892 K_m=675	0.016

In order to evaluate the results with respect to the choice of K_m, Table 4 shows the maximum, medium, minimum, and standard derivation of the accuracies attained in a number of experiments using 1NN-ANS. As it would be expected, when the value of

k is fixed, the choice of K_m is very important. By the way, the maximum accuracies always occur at small K_m. Oppositely, the minimum accuracies always occur at larger K_m, i.e., it approximates kNN-NS.

Fig. 3. A color IR image of a portion of the DC data set

Fig. 4. The classified result of Fig. 3 with 39 bands using 1NN-ANS (K_m=10) classifier

Fig. 5. The classified result of Fig. 3 with 39 bands using 1NN-NS classifier

Fig. 6. The classified result of Fig. 3 with 39 bands using 1NN classifier

Fig. 7. The classified result of Fig. 3 with 39 bands using SVC-CV classifier

Fig. 8. The classified result of Fig. 3 with 39 bands using Parzen classifier

Fig. 3 shows a color IR image of a portion of the DC Mall area for reference. Fig. 4-8 are the classification results of the portion of DC Mall image with 39 bands using 1NN-ANS (K_m=45), 1NN-NS, 1NN, SVC-CV, and Parzen classifiers with 20 training samples. Comparing Figure 10 and 11, we see that the performance of 1NN-ANS is similar to that of 1NN-NS, although the accuracy of 1NN-ANS is higher than that of 1NN-NS. Obviously, the results of1NN-ANS and 1NN-NS are better than those of other classifiers.

6 Concluding Comments

In this paper we proposed a new approach using k-nearest neighbor classification scheme based on adaptive separability metric of NWFE (kNN-ANS) and its special case kNN-NS. Real hyperspectral images and Iris data show that the average classification accuracy of applying kNN-ANS is better than those of applying other classifiers. Some findings are summarized in the following:

1. All performances of kNN-ANS are the best.
2. kNN-ANS and kNN-NS are more robust than traditional k-NN.
3. The thematic maps of 1NN-ANS outperform those of other classifiers.
4. From above experiences, it is valid that using the adaptive separability of NWFE estimates an effective metric for computing a new neighborhood. The modified neighborhood shrinks in the direction with high separability and extends further in the other direction. Then the class conditional probabilities tend to be more homogeneous.

From the above findings, we may say that the use of kNN-ANS is more beneficial and yielding better results than other classifiers.

References

1. T. Hastie and R. Tibshirani: Discriminant Adaptive Nearest Neighbor Classification. IEEE Trans. on Pattern Analysis and Machine Intelligence, Vol. 18, No. 6, (1996) 607–615
2. Bor-Chen Kuo and David A. Landgrebe: Nonparametric Weighted Feature Extraction for Classification. IEEE Trans. Geosci. Remote Sens., Vol. 42, No. 5, (2004) 1096–1105
3. K. Fukunaga: Introduction to Sataistical Pattern Recognition. San Diego, CA: Academic (1990)
4. R. P. W. Duin: PRTools, a Matlab Toolbox for Pattern Recognition. [Online]. Available: http://www.ph.tn.tudelft.nl/prtools/ (2002)
5. Fisher R.A.: The Use of Multiple Measurements in Taxonomic Problems. Annual Eugenics, 7, Part II, (1936) 179-188
6. Landgrebe, D. A.: Signal Theory Methods in Multispectral Remote Sensing. John Wiley and Sons, Hoboken, NJ: Chichester (2003)
7. J. Friedman: Flexible Metric Nearest Neighbour Classification. Tech. Rep., Stanford University, November (1994)
8. R. Short and K. Fukanaga: A New Nearest Neighbor Distance Measure. In Proc. 5th IEEE Int. Conf. on Pattern Recognition (1980) 81-86
9. Menahem Friedman and Abraham Kandel: Introduction to Pattern Recognition: Statistical, Structural, Neural, and Fuzzy Logic Approaches. World Scientific, Singapore (1999)
10. G. Baudat and F. Anouar: Generalized Discriminant Analysis Using a Kernel Approach. Neural computation, 12 (2000) 2385-2404

Virtual Attribute Subsetting

Michael Horton, Mike Cameron-Jones, and Ray Williams

School of Computing, University of Tasmania, TAS 7250, Australia
`{Michael.Horton, Michael.CameronJones, R.Williams}@utas.edu.au`

Abstract. Attribute subsetting is a meta-classification technique, based on learning multiple base-level classifiers on projections of the training data. In prior work with nearest-neighbour base classifiers, attribute subsetting was modified to learn only one classifier, then to selectively ignore attributes at classification time to generate multiple predictions. In this paper, the approach is generalized to any type of base classifier. This 'virtual attribute subsetting' requires a fast subset choice algorithm; one such algorithm is found and described. In tests with three different base classifier types, virtual attribute subsetting is shown to yield some or all of the benefits of standard attribute subsetting while reducing training time and storage requirements.

1 Introduction

A large portion of machine learning research covers the field of 'classifier learning'. Given a list of instances with known classes as training data, a classifier learning algorithm creates a classifier which, given an instance with unknown class, attempts to predict that class. More recent classifier research includes 'meta-classification' algorithms, which learn multiple classifiers and combine their output to yield classifications that are often more accurate than those of any of the individual classifiers used. Some common meta-classification techniques are bagging [1], boosting [2] and stacking [3]. Attribute subsetting [4, 5] is another meta-classification technique, based on learning multiple base classifiers on different training sets. Each training set contains all the original training instances, but only a subset of the attributes.

In [4], for the specific case of nearest-neighbour base classifiers, one copy of the training data was kept; at classification time, attributes in the instance to classify were selectively ignored. In this paper, the same approach is generalized: regardless of classifier type, only one base classifier is learnt, although multiple attribute subsets are still generated. At classification time the instance to be classified is copied, with one copy for each subset. The values of the attributes not appearing in the corresponding subset are set to 'unknown', and the instance is passed to the single base classifier. The predictions from each instance copy are then combined for a final prediction. This form of attribute subsetting generates multiple classifications without learning multiple classifiers, so it is called 'virtual attribute subsetting'.

This paper describes existing work on standard attribute subsetting and classification given instances with unknown attribute values, followed by an outline of the

A. Sattar and B.H. Kang (Eds.): AI 2006, LNAI 4304, pp. 214–223, 2006.

© Springer-Verlag Berlin Heidelberg 2006

virtual attribute subsetting algorithm. The choice of attribute subsets is significant, so four different attribute subset choice algorithms are considered. Results are then reported from tests using virtual attribute subsetting with three different base classifier types: Naïve Bayes, decision trees and rule sets. It is shown to gain some or all of the improved accuracy of standard attribute subsetting, while reducing training time and storage requirements. The degree of benefit depends on the base classifier type.

2 Previous Work

The most relevant previous work, that on standard attribute subsetting, will now be discussed, after a description of the related and better known technique of bagging. Virtual attribute subsetting depends on its base classifier being able to classify instances with missing values, so some methods used to achieve this are also covered.

2.1 Bagging

'Bagging', or 'bootstrap aggregating' [1], is an ensemble classification technique. Multiple training datasets, each with as many instances as the original training data, are built by sampling with replacement. A classifier is then learnt on each training dataset. When a new instance is classified, each classifier makes a prediction and the predictions are combined to give a final classification. (The training sets are known as 'bootstrap' samples, hence 'bootstrap aggregating').

2.2 Standard Attribute Subsetting

Attribute subsetting is an ensemble classification technique with some similarities to bagging. Multiple training datasets are generated from the initial training data; each training set contains all the instances from the original data, but some attributes are removed. A classifier is then learnt on each training set. To classify a new instance, predictions from each classifier are combined to give a final classification.

If the training data are stored in a table with each row representing an instance and each column containing attribute values, bagged training samples are built by selecting table rows, while attribute subsetted training samples are built by selecting table columns. Because of this similarity, the technique has been referred to as 'attribute bagging' [6] and 'feature bagging' [7], terms which are only technically accurate if the attributes are sampled with replacement. This has been tested [4], but did not significantly affect accuracy.

Attribute subsetting has been effective when the base classifiers are neural networks [8], decision trees [5, 6], nearest-neighbour classifiers [4] or conditional random fields [7].

2.3 Unknown Attributes in Classification

Sometimes the instance provided to a classifier does not have a known value for every attribute. Many types of classifiers can handle such 'missing values', but their means of doing so differ. such classifiers are considered here.

Naïve Bayesian classifiers can easily handle missing values by omitting the term for that attribute when the probabilities are multiplied [9].

Decision tree classifiers can deal with missing values at classification time in several different ways [10]. In the specific case of C4.5, if a node in the tree depends on an attribute whose value is unknown, all subtrees of that node are checked and their class distribution summed [11].

Many rule induction algorithms can handle missing values, but some, such as RIPPER, use rules which fail if they need a missing value [12]. Early tests showed that virtual attribute subsetting performs poorly with base classifiers of this type. A more promising rule set inductor is the PART algorithm, which trains decision nodes with the C4.5 decision tree learning algorithm [13].

3   The Virtual Attribute Subsetting Algorithm

Bay [4] applied attribute subsetting to a single nearest-neighbour base classifiers. The only training step for nearest-neighbour classifiers is reading the training data. At classification time, multiple nearest-neighbour measurements were made, each measuring a subset of attributes. This saved training time and storage space, while returning identical results to standard attribute subsetting.

This approach can be generalized to other base classifiers, to create 'virtual attribute subsetting'. At training time, a single base classifier is learnt on all of the training data and multiple subsets are generated. To classify an instance, one copy of the instance is created for each subset. Each instance copy has the attributes missing from its subset set to value 'unknown'. The instance copies are then passed to the base classifier, and its predictions are combined to give an overall prediction.

For most base classifiers, virtual attribute subsetting may be less accurate than standard attribute subsetting, but will have used less training time and storage. Specifically, it only needs the time and space required by the single base classifier and the subsets; even the subset generation can be left to classification time if necessary. A side benefit is that, if a classifier has already been learnt, virtual attribute subsetting can be applied to it, if it can classify instances with missing attributes.

Three parameters of a virtual attribute subsetting classifier may be varied: the subsets, the type of base classifier, and the means of combining predictions.

3.1   Subset Choice

There are many ways to select multiple subsets of attributes. Some subsets were chosen by hand, based on domain knowledge [7, 8]. Pseudorandom subsets have also been used, as in [5], where each subset contained 50% of the attributes. Pseudorandom subsets may be optimized by introspective subset choice: a portion of the training data is set aside for evaluation, and only those subsets which lead to accurate classifiers on the evaluation data are used. Introspective subset choice may be based on learning the optimal proportion of attributes for each dataset [4], or by generating many subsets and only using those which lead to accurate classifiers [6].

Virtual attribute subsetting is intended to be a fast, generic technique. The subsets should not be based on domain knowledge (which is not generic) or introspective subset choice (which is time-consuming). For this experiment, four types of pseudorandom subset choice were tested: random, classifier balanced, attribute balanced and both balanced. Each subset generation algorithm receives three parameters:

a, the number of attributes,
s, the number of subsets to generate and
p, the desired proportion of attributes per subset.

3.1.1   Random Subsets

This is the simplest algorithm. It iterates through the $a{\times}s$ attribute/subset pairs, randomly selecting $a{\times}s{\times}p$ attributes to include in the subsets.

3.1.2   Classifier Balanced Subsets

This algorithm chooses subsets such that each subset contains close to $s{\times}p$ attributes. Since $s{\times}p$ may not be an integer, it rounds some subsets up and some down to build an overall proportion as close to p as possible.

3.1.3   Attribute Balanced Subsets

This algorithm chooses subsets such that each attribute appears in close to $a{\times}p$ subsets. Since $a{\times}p$ may not be an integer, it rounds some counts up and some down to build an overall proportion as close to p as possible.

3.1.4   Both Balanced Subsets

Creating subsets that are balanced both ways is slightly more difficult. The algorithm is illustrated by associating a line segment with each attribute. The length of each line segment is the number of further times that attribute needs to be added to a subset. It is described in pseudocode below and the process is illustrated in Figure 1. Note that this algorithm may create duplicated subsets; a version that created non-duplicated subsets was tested and had an insignificant effect on accuracy.

```
Determine the number of times each attribute should ap-
    pear; if achieving the correct proportion requires
    varied attribute counts (some must be rounded up and
    some rounded down), randomly distribute the counts
For each subset:
    Randomly arrange the attributes in line
    selectorLength ← number of subsets remaining
    selectorPos ← random(0..selectorLength-1)
    While selectorPos lies adjacent to an attribute:
        Add that attribute to the subset
        selectorPos ← selectorPos + selectorLength
    Reduce the lengths of the chosen attributes by 1
```

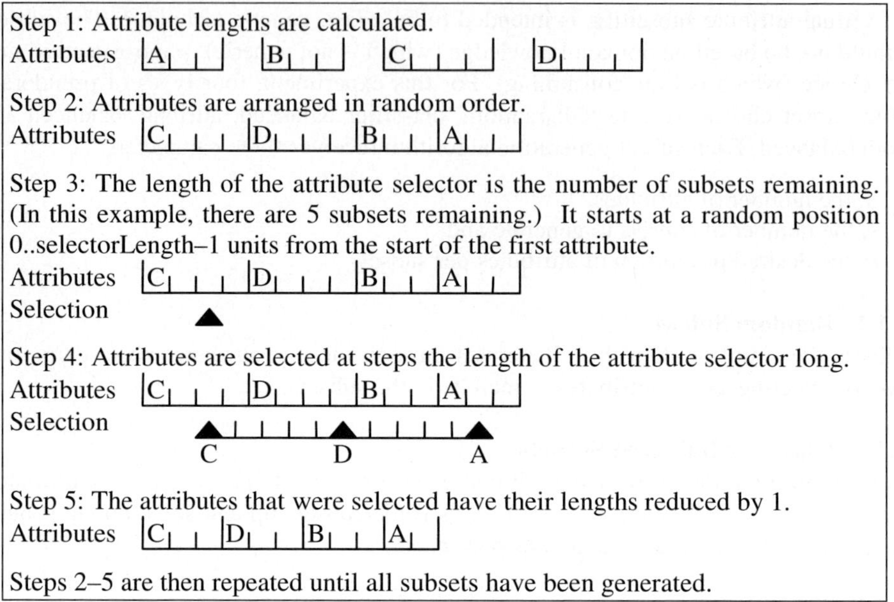

Fig. 1. Visualization of steps in both balanced attribute subset choice

Sample subsets generated by all four subset choice algorithms are shown in Table 1.

Table 1. Examples of subset generation algorithm output with $a = 4$, $s = 5$ and $p = 0.7$

No balancing

Subsets	Attributes				Ttl
	0	1	1	1	3
	0	1	1	1	3
	1	1	0	1	3
	0	1	0	0	1
	1	1	1	1	4
Ttl	2	5	3	4	14

Classifier balanced

Subsets	Attributes				Ttl
	0	1	1	1	3
	1	1	0	1	3
	0	1	0	1	2
	1	1	1	0	3
	1	1	1	0	3
Ttl	3	5	3	3	14

Attribute balanced

Subsets	Attributes				Ttl
	1	0	1	1	3
	0	1	0	1	2
	1	1	1	0	3
	0	0	1	1	2
	1	1	1	1	4
Ttl	3	3	4	4	14

Both balanced

Subsets	Attributes				Ttl
	1	0	1	1	3
	1	1	1	0	3
	1	1	0	1	3
	0	1	1	1	3
	0	0	1	1	2
Ttl	3	3	4	4	14

3.2 Base Classifiers

The effect of virtual attribute subsetting will vary, depending upon the base classifier used. Three types of base classifiers were tested in this experiment: Naïve Bayes, C4.5 decision trees, and PART rule sets.

Making an attribute value unknown makes Naïve Bayes ignore it as if it were not in the training data, so there should be no difference in output between standard and virtual attribute subsetting if both use Naïve Bayes as the base classifier.

For C4.5 and PART base classifiers, virtual attribute subsetting is unlikely to match the accuracy of standard attribute subsetting, as the classifiers used for each subset will not be appropriately independent. However, virtual attribute subsetting may still be more accurate than a single base classifier.

3.3   Combining Predictions

There are many ways to combine the predictions of multiple classifiers. The method chosen for virtual attribute subsetting was to sum and normalize the class probability distributions of the base classifiers to give an overall class probability distribution.

4   Method

All experiments were carried out in the Waikato Environment for Knowledge Analysis (WEKA), a generic machine learning environment [14]. WEKA conversions of 31 datasets from the UCI repository [15, 16] were selected for testing. The dataset names are listed in Table 8. Each test of a classifier on a dataset involved 10 repetitions of 10-fold cross-validation.

Both standard and virtual attribute subsetting have three adjustable settings: subset choice algorithm ('no balancing', 'classifier balanced', 'attribute balanced' or 'both balanced'), attribute proportion (a floating point number from 0.0 to 1.0) and number of subsets (any positive integer). Preliminary tests showed that reasonable default settings are balancing='both balanced', attribute proportion=0.8 and subsets=10.

The decision tree classifier used was 'J4.8', the WEKA implementation of C4.5. The rule set classifier was the WEKA implementation of PART. All base classifier settings were left at their defaults. For both J4.8 and PART, this meant pruning with a confidence factor of 0.25 and requiring a minimum of two instances per leaf.

5   Results

A virtual attribute subsetting classifier with a given base classifier may be considered to succeed if it is more accurate than a single classifier with the same type and settings: it has made an accuracy gain for no significant increase in storage space or training time. This is the main comparison made here; standard attribute subsetting results are also shown, as they provide a probable upper bound to accuracy.

5.1   Naïve Bayes

The ability of virtual attribute subsetting to yield exactly the same results as standard attribute subsetting when Naïve Bayes is the base classifier was verified, so the results shown apply to both standard and virtual attribute subsetting. Attribute subsetting only significantly improved the accuracy of a Naïve Bayesian classifier when attributes were balanced (Table 2). The best attribute proportion was 0.9 (Table 3).

Table 2. Win/draw/loss analysis for standard/virtual attribute subsetting with varying subset choice algorithms compared with a single Naïve Bayesian classifier

Subset algorithm	No balancing	Classifier balanced	Attribute balanced	Both balanced
Wins	16	17	20	21
Draws	0	1	1	0
Losses	15	13	10	10
Wins-Losses	1	4	10	11

Table 3. Win/draw/loss analysis standard/virtual attribute subsetting with varying proportion compared with a single Naïve Bayesian classifier

Proportion	0.1	0.2	0.3	0.4	0.5	0.6	0.7	0.8	0.9
Wins	6	9	15	15	17	18	19	21	22
Draws	0	0	0	0	1	0	0	0	0
Losses	25	22	16	16	13	13	12	10	9
Wins-Losses	-19	-13	-1	-1	4	5	7	11	13

5.2 Decision Trees

With J4.8 as the base classifier, standard attribute subsetting performed well with all subset algorithms, but virtual attribute subsetting was only effective when attributes were balanced and improved when both attributes and classifiers were balanced (Table 4). Both attribute subsetting classifiers were most accurate with attribute proportions of 0.8 or 0.9 (Table 5). The accuracies achieved with optimal settings are listed in Table 8.

Three additional experiments with J4.8 were undertaken but not reported in detail. The number of subsets was varied from 2 to 40, with no significant improvement in virtual attribute subsetting once the number of subsets was at least 6. Unpruned decision trees were also tested; virtual attribute subsetting using unpruned J4.8 as the base classifier performed well against a single unpruned J4.8 classifier but poorly against a single pruned J4.8 classifier. Finally, measurement of decision tree size showed that neither method learnt consistently larger trees.

Table 4. Wins–losses for standard/virtual attribute subsetting with varying subset choice algorithms compared with a single J4.8 classifier

Subset algorithm	No balancing	Classifier balanced	Attribute balanced	Both balanced
Standard attribute subsetting	19	22	23	20
Virtual attribute subsetting	3	3	8	12

Table 5. Wins–losses for standard/virtual attribute subsetting with varying proportion compared with a single J4.8 classifier

Proportion	0.1	0.2	0.3	0.4	0.5	0.6	0.7	0.8	0.9
Standard attribute subsetting	-17	-11	1	10	18	20	21	20	23
Virtual attribute subsetting	-25	-25	-17	-9	-5	1	7	12	4

Table 6. Wins–losses for standard/virtual attribute subsetting with varying subset choice algorithms compared with a single PART classifier

Subset algorithm	No balancing	Classifier balanced	Attribute balanced	Both balanced
Standard attribute subsetting	26	23	29	29
Virtual attribute subsetting	0	4	15	13

5.3   Rule Learning

Results for the PART algorithm are similar to those for J4.8. Standard attribute subsetting gave good results, while virtual attribute subsetting was only effective when the

Table 7. Wins–losses for standard/virtual attribute subsetting with varying proportion compared with a single PART classifier

Proportion	0.1	0.2	0.3	0.4	0.5	0.6	0.7	0.8	0.9
Standard attribute subsetting	-25	-5	3	14	18	22	20	29	28
Virtual attribute subsetting	-29	-23	-17	-9	-5	-3	6	13	15

Table 8. Percentage accuracy over the 31 datasets for J4.8, PART and standard/virtual attribute subsetting using PART and J4.8 base classifiers; wins and losses are based on a simple accuracy comparison and were measured before these figures were truncated for presentation

	J4.8				PART			
Dataset	Single classifier	Virtual attribute subsetting		Standard attribute subsetting		Single classifier	Virtual attribute subsetting		Standard attribute subsetting	
anneal	98.6	98.6	D	98.6	W	98.3	98.3	D	98.5	W
audiology	77.3	77.4	W	79.2	W	79.4	79.3	L	82.2	W
autos	81.8	81.9	W	83.4	W	75.1	75.1	W	81.7	W
balance-scale	77.8	78.3	W	78.2	W	83.2	83.5	W	86.1	W
horse-colic	85.2	85.2	W	85.2	D	84.4	84.7	W	85.4	W
credit-rating	85.6	85.6	W	85.7	W	84.4	84.5	W	86.7	W
german_credit	71.3	71.3	W	72.1	W	70.5	70.7	W	74.6	W
pima_diabetes	74.5	74.8	W	74.7	W	73.4	74.1	W	74.5	W
Glass	67.6	68.0	W	68.8	W	68.7	68.8	W	74.0	W
cleveland-heart	76.9	76.7	L	77.4	W	78.0	78.1	W	81.8	W
hungarian-heart	80.2	80.2	L	80.3	W	81.1	81.3	W	81.9	W
heart-statlog	78.1	78.4	W	78.9	W	77.3	77.3	D	81.2	W
hepatitis	79.2	79.3	W	79.7	W	79.8	79.9	W	82.7	W
hypothyroid	99.5	99.5	D	99.6	W	99.5	99.5	W	99.6	W
ionosphere	89.7	89.8	W	90.6	W	90.8	90.8	D	92.9	W
iris	94.7	94.7	D	94.7	D	94.2	94.2	D	94.6	W
kr-vs-kp	99.4	99.4	D	99.4	W	99.2	99.2	D	99.4	W
labor	78.6	78.6	D	78.6	D	77.7	77.7	D	79.1	W
lymphography	75.8	75.4	L	75.9	W	76.4	76.4	W	81.0	W
mushroom	100.0	100.0	L	100.0	D	100.0	100.0	D	100.0	D
primary-tumor	41.4	42.0	W	43.1	W	40.9	41.6	W	45.2	W
segment	96.8	96.8	W	97.0	W	96.6	96.6	L	97.9	W
sick	98.7	98.5	L	98.7	L	98.6	98.6	L	98.6	W
sonar	73.6	73.7	W	74.3	W	77.4	77.4	D	81.2	W
soybean	91.8	91.6	L	92.3	W	91.4	91.4	W	93.7	W
tic-tac-toe	85.6	86.4	W	87.3	W	93.6	93.6	W	95.8	W
vehicle	72.3	72.4	W	73.6	W	72.2	72.6	W	74.5	W
vote	96.6	96.6	D	96.6	D	96.0	96.0	D	96.6	W
vowel	80.2	80.3	W	82.2	W	77.7	77.7	W	91.6	W
waveform	75.3	76.8	W	78.1	W	77.6	78.9	W	84.2	W
zoo	92.6	92.6	D	92.6	D	93.4	93.4	D	93.4	D
Win/Draw/Loss		18/7/6		24/6/1			18/10/3		29/2/0	
Wins-Losses		12		23			15		29	

attributes were balanced. Balancing both attributes and classifiers did not cause further improvements (Table 6). Of the attribute proportions tested, attribute subsetting using PART was most accurate with proportions of 0.8 or 0.9 (Table 7). The accuracies achieved with optimal settings are listed in Table 8.

5.4 Training Time

Since the training for virtual attribute subsetting is limited to building subsets and training one classifier on the original training data, it was expected to have similar training time to a single classifier. The standard attribute subsetting algorithm builds a classifier for each subset. Since the training data for each classifier have some attributes removed, the individual classifiers may take less time to learn than a single classifier (as there are less attributes to consider) or more time (as more steps may be needed to build an accurate classifier).

To test this, the ratios of attribute subsetting training time to single classifier training time were measured. As expected, virtual attribute subsetting took comparable training time to a single classifier, while standard attribute subsetting needed at least six times longer than both a single classifier and virtual attribute subsetting.

5.5 Accuracy Table

The accuracies returned by the most accurate attribute subsetting and virtual attribute subsetting classifiers using J4.8 and PART base classifiers are shown in Table 8, along with their win/draw/loss totals.

6 Conclusions and Further Work

Compared to standard attribute subsetting using Naïve Bayes as the base classifier, virtual attribute subsetting produces the same classifications with reduced training time and storage. Virtual attribute subsetting also frequently increases the accuracy of C4.5 decision trees and PART rule sets for no significant increase in training time.

The subset choice was found to be significant. Balancing attributes consistently improved accuracy; balancing attributes and subsets usually increased it further. More sophisticated subset choice algorithms may be better still.

Virtual attribute subsetting may also help other base-level classifiers. Whether it does so depends upon how they handle unknown attribute values. It may also be possible to combine virtual attribute subsetting with other meta-classification techniques.

Acknowledgements

We would like to thank the authors of WEKA, [14], the compilers of the UCI repository [15, 16] and Mr Neville Holmes and the anonymous reviewers for their comments, some of which we were unable to implement due to time constraints.

This research was supported by a Tasmanian Graduate Research Scholarship.

References

1. Breiman, L.: Bagging Predictors. In: Machine Learning **24** (1996) 123-140
2. Freund, Y., Schapire, R.E.: A Short Introduction to Boosting. In: Journal of the Japanese Society for Artificial Intelligence **14** (1999) 771-780
3. Wolpert, D.H.: Stacked Generalization. In: Neural Networks **5** (1992) 241-259
4. Bay, S.D.: Combining Nearest Neighbor Classifiers Through Multiple Feature Subsets. In: Proceedings of the 15th International Conference on Machine Learning, Morgan Kaufmann (1998) 37-45
5. Ho, T.K.: The random subspace method for constructing decision forests. In: IEEE Transactions on Pattern Analysis and Machine Intelligence **20** (1998) 832-844
6. Bryll, R., Gutierrez-Osuna, R., Quek, F.: Attribute bagging: improving accuracy of classifier ensembles by using random feature subsets. In: Pattern Recognition **36** (2003) 1291-1302
7. Sutton, C., Sindelar, M., McCallum, A.: Feature Bagging: Preventing Weight Undertraining in Structured Discriminative Learning. In: CIIR Technical Report IR-402 (2005)
8. Cherkauer, K.J.: Human expert-level performance on a scientific image analysis task by a system using combined artificial neural networks. In: Chan, P., (ed.): Working Notes of the AAAI Workshop on Integrating Multiple Learned Models, AAAI Press (1996) 15-21
9. Kohavi, R., Becker, B., Sommerfield, D.: Improving Simple Bayes. In: Someren, M.v., Widmer, G., (eds.): 9th European Conference on Machine Learning, Springer-Verlag (1997)
10. Quinlan, J.R.: Unknown Attribute Values in Induction. In: Segre, A.M., (ed.): Proceedings of the 6th International Workshop on Machine Learning, Morgan Kaufmann (1989) 164-168
11. Quinlan, J.R.: Unknown Attribute Values. In: C4.5: Programs for Machine Learning. Morgan Kaufmann (1993) 30
12. Cohen, W.W.: Fast Effective Rule Induction. In: Prieditis, A., Russell, S.J., (eds.): Proceedings of the 12th International Conference on Machine Learning, Morgan Kaufmann (1995) 115-123
13. Frank, E., Witten, I.H.: Generating Accurate Rule Sets Without Global Optimization. In: Proceedings of the 15th International Conference on Machine Learning, Morgan Kaufmann (1998) 144-151
14. Witten, I.H., Frank, E.: Data Mining: Practical machine learning tools and techniques. 2nd edn. Morgan Kaufmann (2005)
15. Newman, D.J., Hettich, S., Blake, C.L., Merz, C.J.: UCI Repository of machine learning databases. http://www.ics.uci.edu/~mlearn/MLRepository.html (1998)
16. Weka 3 - Data Mining with Open Source Machine Learning Software in Java - Collections of datasets. http://www.cs.waikato.ac.nz/~ml/weka/index_datasets.html

Parallel Chaos Immune Evolutionary Programming

Cheng Bo, Guo Zhenyu, Bai Zhifeng, and Cao Binggang

Research & Development Center of Electric Vehicle Xi'an Jiaotong
University, Xi'an Jiaotong University, Xi'an 710049, China
Chengbo7681@163.com

Abstract. Based on Clonal Selection Theory, Parallel Chaos Immune Evolutionary Programming (PCIEP) is presented. On the grounds of antigen-antibody affinity, the original antibody population can be divided into two subgroups. Correspondingly, two new immune operators, Chaotic Clonal Operator (CCO) and Super Mutation Operator (SMO), are proposed. The former has strong search ability in the small space while the latter has better search ability in large space. Thus, combination searching local optimum with maintaining population diversity can be actualized by concurrently operating CCO and SMO. Compared with the Classical Evolutionary Programming (CEP) and Evolutionary Algorithms with Chaotic Mutations (EACM), experimental results show that PCIEP is of high efficiency and can effectively prevent premature convergence. Therefore, it can be employed to solve complicated machine learning problems.

Keywords: Immune algorithm, chaos search, evolution programming, parallel evolution.

1 Introduction

Evolutionary Algorithm (EA) has been widely applied to solving NP complete problems, but its poor local search ability, low convergent speed and premature convergence limit the functions of algorithm. However, the immunology has become a source of inspiration to improve EA, and many immune operators have been presented on the basis of various immune mechanisms to better the performance of EA [1][2]. In the field of machine learning, multi-modal function optimization is very complicated because of frequent variable coupling. As a result, searching mechanism of the traditional Artificial Immune System Algorithm (AISA) is not perfect. It is of poor local search capacity and insufficient parallelism inherent, which restrains the improvement of searching efficiency [3][4][5].

In order to overcome the weakness of AISA, a novel immune algorithm, PCIEP, is put forward based on the clonal selection theory. According to difference of anti-body-antigen affinity, original antibody population is divided into two subgroups, low-affinity one and high-affinity one. Correspondingly, two operators, Chaotic Clonal Operator (CCO) and Super Mutation Operator (SMO), are proposed. The former has strong search ability in the small space while the latter has better search ability in large space. Thus, combination searching local optimum with maintaining

A. Sattar and B.H. Kang (Eds.): AI 2006, LNAI 4304, pp. 224–232, 2006.
© Springer-Verlag Berlin Heidelberg 2006

population diversity can be actualized by concurrently operating CCO and SMO, which can enhance search efficiency of the algorithm.

The remainder of this paper is organized as follows: in section 2, based on clonal selection theory, PCIEP is depicted; in section 3, experimental results of multi-modal function optimization are demonstrated; in section 4, conclusions are drawn.

2 Parallel Chaos Immune Evolutionary Programming

2.1 The Clonal Selection Theory

Clonal Selection Theory was put forward by Burnet in 1958. Its main points are as follows: when biological immune system is exposed to invading antigen, B cells of high affinity with antigen-antibody are selected to proliferate, clone and hyper-mutate so that B cells with higher affinity in local scope can be searched; At the same time, phenomenon of receptor editing takes place among the B Cells with low antigen-antibody affinity, so they can be mutated into some points far away from the original one in the shape space, which is conducive to search B cells with high affinity. Besides, some B Cells are dead and replaced by the new ones generated from bone marrow to maintain the population diversity. After many generations of evolution and selection, B cells with very high affinity are produced finally, which are further differentiated into plasma cells, generating a lot of antibodies having the same shape with receptors to annihilate the antigens [6][7].

2.2 Chaotic Clonal Operator

In the study of parallel immune evolutionary algorithm, the commonly used method of population partition and manipulation of evolutionary operators is that several original populations are simultaneously adopted, or one original population is divided into several subgroups with the same scale, then evolutionary operators with similar functions are involved, and the optimal antibodies are produced by competition and communication among the subgroups [5][8][9].

In this paper, according to the antigen antibody affinity, a new method of population partition is proposed. The original population is divided into two subgroups, of which, one has high affinity above the average, and the other has sub-average affinity. These two immune operators are complementary and adopted to obtain the optimal antibodies.

The parallel operation mechanism of PCIEP is that Chaotic Clone Operator (CCO) is designed according to the phenomena of B Cell clonal expansion and hyper-mutation, whereas Super Mutation Operator (SMO) is designed according to the phenomena of Receptor editing phenomenon.

The current population P_k is a N dimension vector, and $P_K = \{a_1, a_2, ..., a_N\}$. Real coding method is adopted here, and antibody code length is L. After computing antibody affinity, antibody population is divided into two subgroups, A_k and B_k, in accordance with CCO and SMO. Evolutionary process from P_k to P_{k+1} will be shown in Fig. 1:

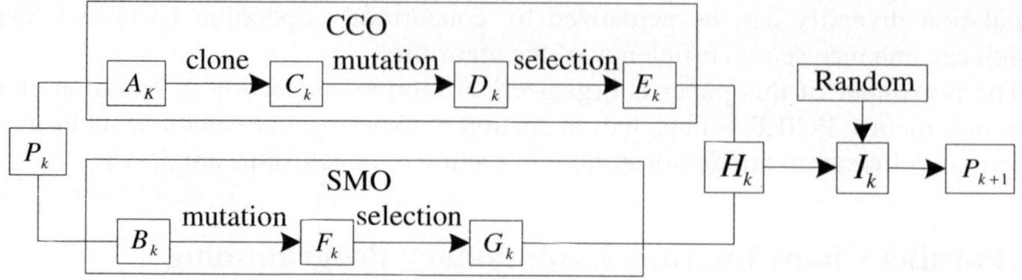

Fig. 1. Population evolution chart

In the population A_k, according to the affinity function $f(*)$, an individual a_i will be divided into q_i antibodies, $a_{i1}, a_{i2}, ..., a_{iq_i}$, then new population C_k will be produced. The steps of ECO are described as follows:

Clone: A_k is a M dimension vector and the population size M is decided by the original population size N together with λ [10]. Generally speaking, M is:

$$M = Int(\lambda N), 0 \leq \lambda \leq 1.\tag{1}$$

In this paper, M is decided by a new method. A_k is defined as follows:

$$A_k = \left\{ a_i \middle| f(a_i) \geq \bar{f}, i \in N \right\}\tag{2}$$

Where

$$\bar{f} = \frac{1}{N}\sum_{l=1}^{N} f_l, l = 1,2,\cdots,N.\tag{3}$$

According to the average affinity, the original population is divided into two subgroups by the formulas (2) and (3).

In population A_k, antibody a_i will be cloned into q_i antibodies. q_i is defined as:

$$q_i = Int(C * P_i), i = 1,2,...,M.\tag{4}$$

q_i is adjusted to being adaptable according to C and p_i. $C > M$, $C = \sum_{i=1}^{M} q_i$. The constant C is a given integer related to the clonal size. $Int(*)$ rounds the elements of X to the integers nearest towards infinity and $Int(X)$ returns to be the smallest integer bigger than X. Here, p_i, the probability of antibody a_i, produces new antibodies, it is as:

$$P_i = f(i) \Big/ \sum_{j=1}^{M} f(j), i = 1,2,...,M.\tag{5}$$

After population clone, C_k replaces the population A_k:

$$C_k = \{A_1', A_2', ..., A_M'\}. \tag{6}$$

Where $\quad A_j' = \{a_{j1}, a_{j2}, ..., a_{jq_j}\}, j = 1, 2, ..., M$.

Mutation: Three main rules of chaotic theory are randomness, exhaustive search and regularity of the chaotic sequence. Different from some stochastic optimization algorithms, chaos search has better search ability in small space [11]. Therefore, instead of the commonly used mutation operator of conventional evolutionary Programming (CEP), namely Gaussian mutation, chaotic mutation operator is adopted to seek the optimal solutions.

In the population C_k , the corresponding updating equation of positional parameters is given by:

$$x'_i(j) = x_i(j) + \sigma_i(j)K_i(0,1) . \tag{7}$$

Where $K(0,1)$ is a chaos sequence located in [-2,2], whose mutation size is similar to that of standard Gauss operator. Here, one dimension Logistic map is adopted, whose equation is shown as follows:

$$r_{n+1} = \lambda r_n(1 - r_n), \ r_n \in [0,1] . \tag{8}$$

Where λ is a control parameter located in $[0,4]$, Logistic map is an irreversible map in $[0,1]$. It can be proved, if λ is equal to 4, the system is in the state of chaos, r_n is ergodic in $[0,1]$, $K(0,1)$ can be obtained with r_n enlarging and translating in $[0,1]$.

Clonal selection: After mutation, D_k replaces C_k . In the population D_k , if there exist the following antibodies:

$$d_i = \{a'_{ij} | \max f(a'_{ij}), j = 1, 2, ...q_i\} . \tag{9}$$

Then a new population D_k' replaces D_k , and $D_k' = \{ d_i \}, i = 1, 2, ..., M$ .Then D_k' is incorporated into the original population A_k to build the parent population. Q is set, which is the parameter of tournament scale. After the competition and exchanging information among the antibodies, a new offspring population is produced, then E_k replaces D_k .

The Chaotic Clonal Operator (CCO) combines chaos search method with immune clonal operation in order to merge their advantages. CCO produces a series of antibodies similar to the original one. An optimal antibody is obtained through the competition and exchanging information among them. The nature of CCO is to search a neighborhood of the single antibody and then find an optimum in this neighborhood to replace the original single antibody. It is useful to improve the local search ability of CCO.

2.3 Super Mutation Operator

The fatter tail of the Cauchy distribution generates a greater probability of taking large steps, which could help in offered as a possible escaping local optima, and this has been explanation for its enhanced performance [12]. In this paper, Super Mutation Operator will utilize improved Cauchy mutation to maintain the population diversity.

All the antibodies of B_k have sub-average affinity, B_k is shown as follows:

$$B_k = \{a_{ij}\}, \quad i = 1,2,...,S, \ S = N - M, \ j = 1,2,...,L. \tag{10}$$

S represents the population size of B_k. Real coding method is adopted and L stands for the antibody encoding length. Here, a_{ij} is the jth component of the antibody a_i, the corresponding positional parameter updating equations are given by:

$$a'_i(j) = a_i(j) + \sigma'_i(j) \cdot \lambda \cdot C_i(0,1). \tag{11}$$

$$\sigma'_i(j) = \sigma_i(j) \exp(\tau' C(0,1) + \tau C_i(0,1)). \tag{12}$$

Where $a'_i(j)$, $a_i(j)$, $\sigma'_i(j)$ and $\sigma_i(j)$ denote the jth component of the vector a'_i, a_i, σ'_i and σ_i, respectively. $C(0,1)$ denotes a Cauchy random number centered at zero with a scale parameter of 1, $C_j(0,1)$ indicates that the random number is generated anew for each value of j. The factors τ and τ' are set to 1/sqrt (2sqrt(n)) and l/(2sqrt(n)), respectively. λ is a self-adaptive parameter, it is defined as follows:

$$\lambda = \exp\left(-\frac{f_i}{f_{max}}\right). \tag{13}$$

Where f_i denotes the affinity function of antibody a_i, f_{max} denotes the maximum affinity function of all the antibodies. It is obvious that the lower an antibody's affinity, the greater it's mutation degree, then a further solution space can be searched, which is helpful to maintain the population diversity.

After the mutation and selection operation, G_k replaces the population B_k. G_k is incorporated into the population E_k, then H_k is produced. After sorting on the grounds of the population affinity, I_k replaces H_k. In the population I_k, new members produced randomly replace the antibodies with poor affinity. The number of new members is $Int(\eta N)$ in which η is in 0.1~0.15 and N stands for the population size. Accordingly, P_{k+1} replaces P_k.

2.4 Parallel Immune Evolutionary Programming

Based on the Chaotic Clone Operator and the Super Mutation Operator, parallel chaos immune evolutionary programming algorithm is put forward.

$f: R^M \to R$ is the optimized object function. Without loss of generality, we consider the problem of the minimum of affinity function. The affinity function of antibody and antigen $\Phi: I \to R$, where $I = R^m$ is the individual space. N stands for the population size and M indicates the number of the optimized variables, so

$a = \{x_1, x_2, \cdots, x_m\}$ represents the details of the overall algorithm, which is shown as follows:

$$K = 0$$

Initial: setting the parameters of CCO, such as mutation probability, meanwhile, setting the clonal size C; setting the parameters of SMO, such as the death rate of the antibodies. At the same time, the Termination Criterion of the algorithm is given. Then the original antibody population is randomly generated, and the antibody $P_0 = \{a_1(0), a_2(0), ..., a_N(0)\}$ is in the interval $[0,1]$.

Domain Transform: $x_{ij} \in a_i$ can be scaled to $\overline{x}_{ij} \in [d_j, u_j]$ as follows:

$$\overline{x}_{ij} = (u_j - d_j) \times x_{ij} + d_j \quad j = 1, 2, \cdots, M \quad x_{ij} \in a_i. \tag{14}$$

Then
$$\overline{a}_i = \{\overline{x}_{i1}, \overline{x}_{i2}, \cdots, \overline{x}_{im}\} \tag{15}$$

Calculating the affinity: $p(0) : \{\Phi(p(0))\} = \{\Phi(\overline{a}_1(0)), \Phi(\overline{a}_2(0)), \cdots, \Phi(\overline{a}_n(0))\}$

If the Termination Criterion cannot be satisfied, the original Antibody Population can be updated. Then next circulation begins.

> **While** CCO is running, namely, $A_k \rightarrow C_k \rightarrow D_k \rightarrow E_k$
>
> **and** SMO is running, namely, $B_k \rightarrow F_k \rightarrow G_k$
>
> **Incorporating the subgroup** E_k **with** G_k, and then H_k is obtained.
>
> **Calculating the affinity:** H_k
>
> **Producing new antibodies:** some random numbers are added to H_k
>
> $k = k + 1$
>
> **End**

In fact, the optimum solution of the antibody population could not be bettered by applying restricted iterative number, constant iterative of several generations, or the compromise of these two methods. The definition of the Termination Criterion is different from the reference [13].

$$\left| f^* - f^{best} \right| < \varepsilon. \tag{16}$$

Where, f^* is the optimum value of the object function, f^{best} is the best value of the in the current generation. When $0 < \left| f^* \right| < 1$,

$$\left| f^* - f^{best} \right| < \varepsilon \left| f^* \right|. \tag{17}$$

3 Experimental Results

In order to analysis the performance of the PCIEP, five standard testing functions are involved. The results are compared with that of Classical Evolutionary Programming (CEP) and Evolutionary Algorithms with Chaotic Mutations (EACM) [14].

(1) $F_5 = 0.002 + \sum_{j=1}^{25} [1 / (j + \sum_{i=1}^{2} (x_i - a_{ij})^6)] , |x_i| \le 65.56$.

(2) $F_6 = 0.5 + (\sin^2 \sqrt{x_1^2 + x_2^2}) / (1 + 0.001(x_1^2 + x_2^2))^2 , -100 \le x_1, x_2 \le 100$.

(3) $F_7 = (x_1^2 + x_2^2)^{0.25} [\sin^2 (50(x_1^2 + x_2^2)^{0.1}) + 1], -10 \le x_1, x_2 \le 10$.

(4) $F_8 = \sum_{i=1}^{N} (x_i^2 - 10\cos(2\pi x_i) + 10) , -5.12 \le x_i \le 5.12$.

(5) $F_9 = -\sum_{i=1}^{N} \left(\sum_{j=1}^{i} x_j \right)^2 , -100 \le x_i \le 100$.

In order to compare the performance of these three algorithms, they are run 30 times independently using the same initial populations. The initial populations are randomly produced and the real coding method is adopted. The population size of these three algorithms is 100. The tournament size, Q , is 10. When F_8 and F_9 are tested, N is 10 and the maximum generation is 15000.

The performance comparison of CEP, EACM and PCIEP is shown in Table 1. 'Max Gen.' denotes the maximal evolutionary generation. 'Stuck' denotes the number that the algorithm is stuck at a local optimum within 30 times. 'Gen' denotes the average generation. 'Thresh' denotes the threshold values of these three functions, F_5 , F_6 and F_7 .

It is obvious to see from Table 1 that PCIEP evaluates function value in the less generations compared with CEP and EACM, which indicates that the convergent speed of PCIEP is quicker. The reason is that parallel operation makes PCIEP search more large space within fewer generations. At the same time, it is clear to see that PCIEP rarely traps into the local optimal values. It means that the global searching ability of PCIEP is stronger than that of CEP and EACM.

The optimization results of the functions are given in Table 2, including three items: average value, average error and standard deviation. When the multi-modal functions with serious coupling variables are optimized, the search performance of PCIEP is better than CEP and EACM. Experiment shows that the parallel operation of CCO and CMO is successful, and it is efficient enough to improve the local search ability of PCIEP and enlarge its searching space to maintain the population diversity.

Table 1. The performance comparison of CEP, EACM and PCIEP

Function	Max Gen.	Stuck. CEP	EACM	PCIEP	Gen. CEP	EACM	PCIEP	Thresh.
F5	100	6	0	0	69.35	38.73	14.36	1.000
F6	200	6	3	0	165.90	96.66	25.57	0.001
F7	500	8	4	0	389.90	216.66	24.96	0.005
F8	15000	8	4	0	231.54	203.15	26.43	/
F9	15000	5	3	0	373.90	166.56	28.27	/

Table 2. The optimization results of CEP, EACM and PCIEP

Func-tion	Global optimum	Average value			Average error			Standard deviation		
		CEP	EACM	PCIEP	CEP	EACM	PCIEP	CEP	EACM	PCIEP
F5	1.001	1.001	1.001	1.001	0	0	0	0	0	0
F6	0	1.563e-5	1.233e-5	1.163e-6	1.563e-5	1.233e-5	1.053e-6	1.163e-5	1.233e-5	1.5792e-6
F7	0	1.565e-3	1.135e-3	1.045e-4	1.565e-3	1.135e-3	1.023e-4	1.125e-5	1.245e-5	3.909e-7
F8	100	97.83	98.85	100	2.226e-2	1.486e-2	1.036e-3	-1.113e-2	-1.106e-2	6.536e-4
F9	/	-5.323e-2	-3.323e-5	-4.356e-9	-5.323e-3	-3.323e-5	-2.235e-9	-1.146e-2	-1.123e-2	5.643e-4

Fig.2 and Fig.3 are the optimization results of F_8 and F_9, which are the curves of theaverage function value versus evolutionary generations. As generations increase, a series of local optimums are obtained. Experiment shows that the number of local

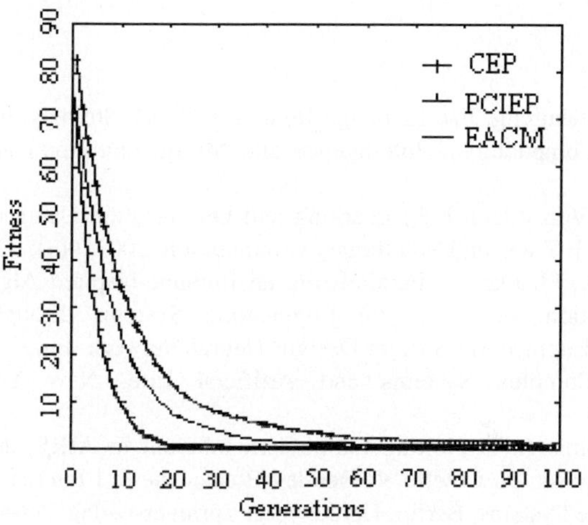

Fig. 2. The optimization results of F_8

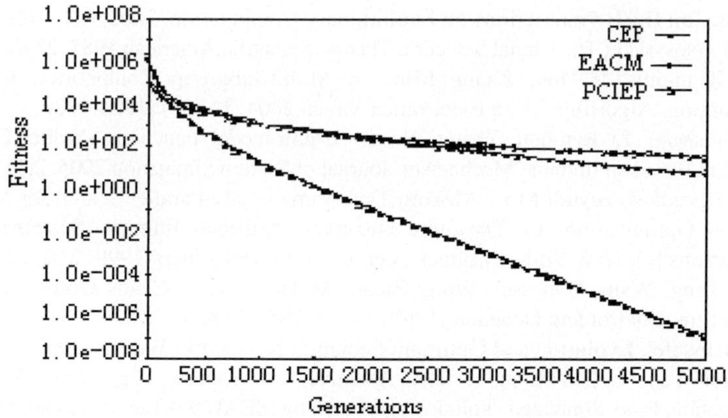

Fig. 3. The optimization results of F_9

optimums of PCIEP is more than that of CEP and EACM. It means that the population diversity of PCIEP is more effectively maintained, and also, the global search ability is improved.

4 Conclusion

Based on the clonal selection theory, two new immune operators, Chaotic Clone Operator and Super Mutation Operator are designed. Experiment shows that combination the excellent local search ability of CCO with the diversity preservation ability of SMO can improve the search efficiency. It is proved that parallel operation mechanism of CCO and SMO is successful. Parallel chaos immune evolutionary programming can effectively prevent premature convergence, and also, it can be adopted to solve complicated machine learning problems.

References

1. Liu Ruochen, Du Haifeng, Jiao Licheng.: Immunity Clonal Strategy. In: Fifth International Conference on Computational Intelligence and Multimedia Applications (ICCIMA'03). 2003:290-295.
2. De Castro L N, Von Zebun F J.: Learning and Optimization Using the Clonal Selection Principle. In: IEEE Trans on Evolutionary Computation, 2002, 6(3): 239-251.
3. Watkins A, Bi X, Phadke A.: Parallelizing an Immune-inspired Algorithm for Efficient Pattern Recognition. In: Intelligent Engineering Systems through Artificial Neural Network: Smart Engineering System Design: Neural Network, Fuzzy Logic, Evolutionary Programming, Complex Systems and Artificial Life, New York: ASME Press, 2003.13:225-230.
4. Watkins A, Timmis J.: Exploiting Parallelism Inherent in AIRS, an Artificial Immune Classifier. In: Nicosia G, Cutello V, Bentley P eds. The 3rd International Conference on Artificial Immune Systems, Berlin, Heidel berg: Springer-verlag, 2004:427-438.
5. Yang KongYu, Wang XiuFeng.: Research and Implement of Adaptive Multimodal Immune Evolution Algorithm. Control and Decision.2005, 20(6): 717-720.
6. De Castro L N, Von Zuben F J.: Learning and Optimization Using the Clonal Selection Principle. In: IEEE Transactions on Evolutionary Computation. 2002, 6(3): 239-251.
7. Ada G L, Nossal G.: The Clonal Selection Theory. Scientific American,1987, 257(2): 50-57.
8. Wang Xiangjun, Ji Dou, Zhang Min.: A Multi-Subgroup Competition Evolutionary Programming Algorithm. Acta Electronica Sinica.2004, 11 (32): 1824-1828.
9. Luo Yin-sheng, Li Ren-hou, Zhang Weixi. : Multi-modal Functions Parallel Optimization Algorithm Based on Immune Mechanism. Journal of System Simulation.2005, 2(11): 319-322.
10. Toyoo Fukuda, kazuyudi Mori, Makoto Tsukiyama, et al.: Parallel Search for Multi-modal Function Optimization. In: Dasgupta Dipankar. Artificial Immune Systems and Their Applications [c]: New York: Springer -Verlag Berlin Heidelberg.1999: 210- 220.
11. Zhang Tong, Wang Hongwei, Wang Zicai.: Mutative Scale Chaos Optimization and its Application. Control and Decision .1999, 14 (3): 285 - 288.
12. H.P.: Schwefel, Evolution and Optimum Seeking. New York: Wiley, 1995.
13. Zhengjun Yan, Lishan Kang. : The Adaptive Evolutionary Algorithms for Numerical Optimization. Proc. Simulated evolution and Learning (SEAL'96),Taejon, Korea, 53-60, 1996.
14. Luo Chenzhong, Shao Huihe.: Evolutionary Algorithms with Chaotic Mutations. Control and Decision. 2000, 15(5): 557-560.

Classification of Lung Disease Pattern Using Seeded Region Growing

James S.J. Wong[1] and Tatjana Zrimec[2]

[1] School of Computer Science and Engineering,
University of New South Wales, Sydney, NSW 2052, Australia
jamesjw@cse.unsw.edu.au
[2] Centre for Health Informatics,
University of New South Wales, Sydney, NSW 2052, Australia
tatjana@cse.unsw.edu.au

Abstract. Honeycombing is a disease pattern seen in High-Resolution Computed Tomography which allows a confident diagnosis of a number of diseases involving fibrosis of the lung. An accurate quantification of honeycombing allows radiologists to determine the progress of the disease process. Previous techniques commonly applied a classifier over the whole lung image to detect lung pathologies. This resulted in spurious classifications of honeycombing in regions where the presence of honeycombing was highly improbable. In this paper, we present a novel technique which uses a seeded region growing algorithm to guide the classifier to regions with potential honeycombing. We show that the proposed technique improves the accuracy of the honeycombing detection. The technique was tested using ten-fold cross validation on forty two images over eight different patients. The proposed technique classified regions of interests with an accuracy of 89.7%, sensitivity of 96.6% and a specificity of 88.6%.

1   Introduction

High-Resolution Computed Tomography (HRCT) of the lung is a technique which allows radiologists to view cross-sectional slices of the lung. HRCT scanning involves a radiation source and a set of detectors. The radiation source circles around the body and the detectors measure the attenuation level of the radiation at each possible angle. These attenuation levels are measured in Hounsfield Units (HU). These attenuation levels allow radiologists to recognise abnormalities in the lung parenchyma.

Our aim is to build a system to automatically detect lung disease processes in HRCT images. In this paper, we focus on the detection of honeycombing. Pathologically, honeycombing is defined by the presence of small air-containing cystic spaces with thickened walls composed of dense fibrous tissue [1]. In HRCT, the low-density cystic airspaces appear as roughly circular, dark patches; the denser walls which surround these airspaces appear as white borders (see Fig. 1). The characteristic "honeycombing" appearance is the result of these cystic airspaces

A. Sattar and B.H. Kang (Eds.): AI 2006, LNAI 4304, pp. 233–242, 2006.
© Springer-Verlag Berlin Heidelberg 2006

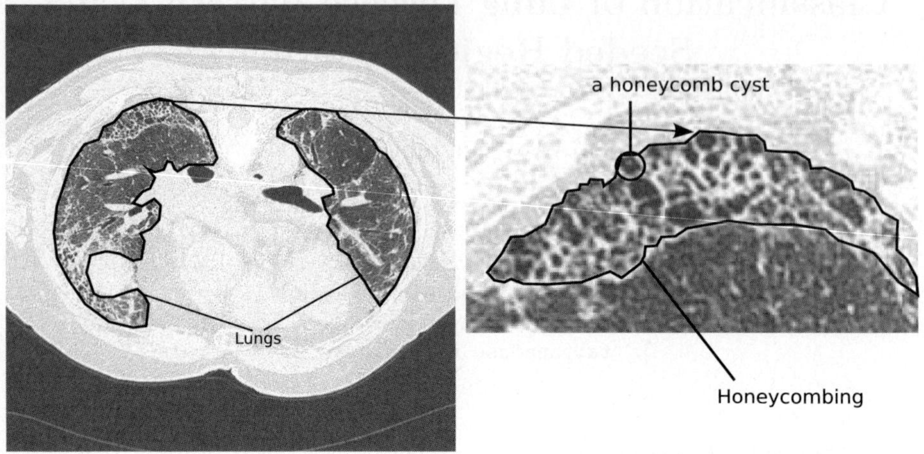

Fig. 1. (Left) Lungs with honeycombing. Lungs have been outlined. (Right) Close-up of region with honeycombing. Honeycombing region is outlined.

occurring in clusters. Honeycombing cysts occur predominantly in the periphery of the lung (see Fig. 2). The periphery of the lung is defined as the region approximately 1 cm–2 cm from the surface of the lung.

Various automated detection algorithms have been developed to detect honeycombing. Uppaluri *et al.* [2] divided the lung into overlapping, square regions. An adaptive multiple-feature method was used to assess twenty-two independent

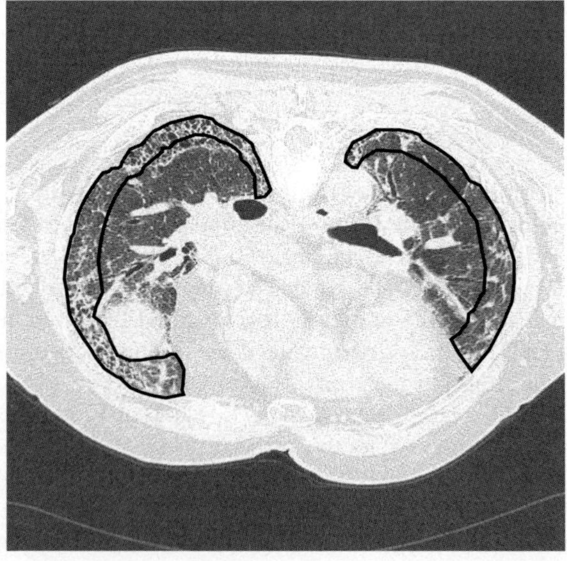

Fig. 2. Approximate lung periphery outlined

texture features to classify a tissue pattern in these square regions. The features with the greatest ability to discriminate between the different patterns were chosen and a Bayesian classifier is trained using the selected features. Similarly, Uchiyama *et al.* [3] proposed a scheme in which the lung is segmented and divided into many contiguous regions of interest (ROIs) with a size of 32×32. Twelve features were calculated per ROI; six from the 32×32 matrix and six from a 96×96 region. An Artificial Neural Network was used to classify the ROI.

In this paper, we present a novel technique which incorporates a seeded region growing (SRG) algorithm to guide the system to ROIs which potentially have honeycombing. We show that this classification scheme improves the accuracy of the honeycombing detection.

In the next section we give a description of the proposed system. Experiment results and discussions are given in Section 3, followed by the concluding remarks in Section 4.

2   Method

In order to recognise honeycombing patterns, we adopt a machine learning approach in which the system builds a classifier from training examples labelled by the domain expert. A diagram of the process is shown in Fig. 3. In this case, we supply example HRCT images that have particular regions marked by a radiologist as being either honeycombing or non-honeycombing. From these regions, we extract ROIs that are represented by a vector of features. We reduce the dimensionality of the feature space by selecting a subset of these features, using a feature selection scheme to choose the most relevant features. In the final stage, a classifier is built by a machine learning algorithm which uses the input feature vectors to learn the decision boundary in feature space. New test samples are classified using this classifier.

The novelty of the system is in the selective classification of ROIs. In many systems [2,3], all ROIs within the lung region are classified. However, not all regions have the same likelihood of showing honeycombing. In particular, the central regions of the lung are less likely to have honeycombing. Honeycombing

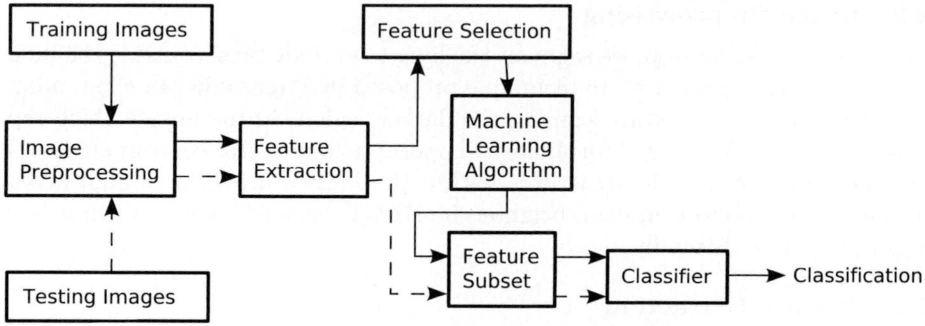

Fig. 3. Process flow diagram for honeycombing detection

cysts in the central region of the lung are predominantly adjacent to other honeycombing cysts in the peripheral lung region. In Section 2.6, we describe how we use this information to improve the performance of the honeycombing detection.

The following sections explain each of the processes to detect honeycombing in detail.

2.1   Data Set

For the study, we collected scans from 8 different patients (7 of whom showed honeycombing in the HRCT scan, and 1 without). We chose 42 images from these scans that showed patterns representative of honeycombing and non-honeycombing tissue. A radiologist was asked to mark some areas where honeycombing appears on those images. Additional honeycombing regions were marked by the first author. Non-honeycombing regions were also marked. From the marked regions, ROIs of sizes, 7×7 pixels and 15×15 were extracted (see Fig. 4). Adjacent ROIs overlapped such that the centres of adjacent ROIs are three pixels apart. The final set contained 18407 labelled ROIs: 9467 ROIs were honeycombing and 8940 ROIs were non-honeycombing. These ROIs represent only a fraction of all the ROIs in the lung to limit the number of regions. Using all the ROIs in the lung would bias the non-honeycombing regions as there are generally more non-honeycombing regions than honeycombing regions.

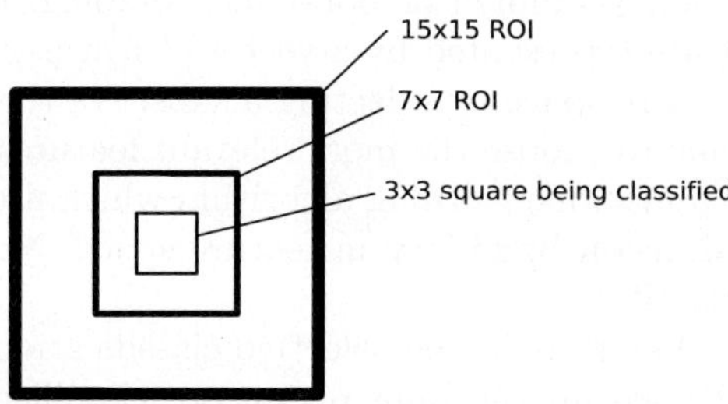

Fig. 4. Features are extracted from the 7×7 and 15×15 ROI. The classification is applied to the 3×3 square.

2.2   Image Preprocessing

In our preprocessing step, we segment the lung to exclude pixels outside the lung. We use the lung segmentation technique proposed by Papasoulis [4]. First, adaptive thresholding is used to segment the darker regions in the image which represents the air-filled lung. Morphological operators and active contour snakes [4] are then used to include structures within the lung which have a high attenuation (and therefore appear brighter in HRCT images). The resulting lung segmentation is shown in Fig. 5.

2.3   Feature Extraction

Having segmented the lung, we proceed to extract features from ROIs within the lung image. In our feature extraction step, we use ROIs of multiple sizes:

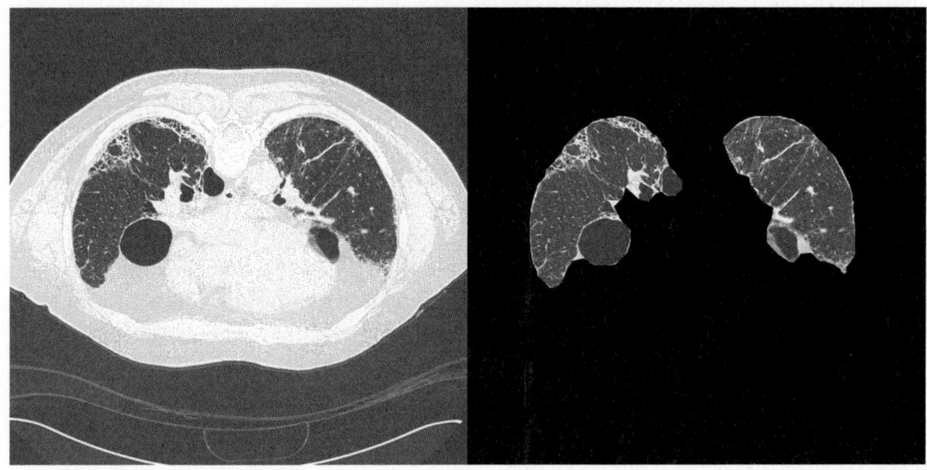

Fig. 5. (Left) Original image. (Right) Segmented lungs.

7×7 and 15×15 (see Fig. 4). This enables us to capture the characteristics of small and larger honeycombing cysts. Our system utilises texture-based features to discriminate between ROIs containing honeycombing and non-honeycombing. We calculate both first order, second order texture features, and gray-level difference features for each ROI. The first order texture features measure the gray-level distribution within the ROI. The specific features calculated are: mean HU, variance, skewness, kurtosis, energy, and entropy.

The second order features describe the spatial distribution of the gray-levels within these ROIs. To do this, a co-occurrence matrix is calculated, which specifies the frequency of a particular gray-level occurring near another gray-level. The characteristics of this co-occurrence matrix are measured by the following features: energy, entropy, contrast, homogeneity, correlation and inverse difference moment. The co-occurrences of the gray-levels for four different directions are measured: $0°$, $45°$, $90°$, $135°$.

The gray-level difference features measure the distribution of the difference of pairs of gray-levels within the ROI. A table of frequencies of gray level differences are calculated. The features extracted are the descriptions of the gray-level difference distribution in the ROI: mean, variance, contrast, entropy, angular second moment. The gray-level difference features for four different directions are also measured.

In total, 63 features are extracted per ROI size: since we have two different ROI sizes, our feature vector has 126 dimensions.

2.4   Feature Selection

With a feature vector of 126 dimensions, the classifier generated would be computationally intractable. We reduce the dimensionality by selecting a subset of features which are the most discriminative in identifying honeycombing and non-honeycombing regions.

In this study, we use two different feature selection schemes. The first feature selection scheme is called, Correlation-based Feature Selection (CFS) [5]. CFS uses a correlation-based heuristic to evaluate the worth of a subset of features. It selects subsets of features that are highly correlated with the class, while having low inter-correlation. A high correlation between attributes is not desirable as the information provided by the attributes are redundant.

The second feature selection scheme is called Consistency Feature Selection (ConsistencyFS) [6]. Ideally, instances with features of similar values should have the same class label. This feature selection scheme evaluates the worth of a subset of features by the level of consistency in the class values in the training set.

2.5   Classifier

We use the reduced vectors of features from the feature subset selection step to build a classifier. Two classifiers were evaluated: Naive Bayes and the Weka J48 decision tree classifier [7].

The Naive Bayesian classifier is based on a probability model. The probability of a class given a feature vector is determined using Bayes' rule:

$$P(c|F) = \frac{P(F|c)P(c)}{P(F)} \qquad (1)$$

where c is the class and F is the vector of features.

The class with the highest probability is assigned to the ROI. The Naive Bayes classifier is known to be optimal when the features are independent, however, in reality the classifier still works well without this assumption.

The Weka J48 decision tree classifier [7] is built by selecting features to partition examples at each node in the tree. A feature is chosen if it results in the highest information gain at that particular node. The information gain represents the expected reduction in entropy from using a particular feature in the node.

2.6   Seeded Region Growing

In many systems (e.g. [2,3]), all ROIs within the lung are classified. However, for diseases that show honeycombing, we observe that the process spreads from the periphery of the lung. This information allows us to contain the area that requires classification to be performed.

We incorporated a SRG algorithm in the system. In conventional SRG [8], a region is added to a set if it is sufficiently similar to other regions in the set. In our system, we use a SRG algorithm to guide a classifier to ROIs with potential honeycombing. The regions that are grown by the SRG algorithm includes ROIs that have already been classified, and ROIs that are yet to be classified. ROIs in the periphery of the lung are set as the "seed points" for the SRG algorithm. To determine the periphery of the lung, we use the lung region segmentation method proposed by Zrimec et al. [9]. The ROIs are then classified by a classifier built in the step outlined in Section 2.5. When ROIs are classified as Honeycombing, nearby unclassified ROIs are included in SRG regions for classification. The algorithm stops, when there are no more ROIs available for classification. In summary, the SRG algorithm guides the system to classify all ROIs that are either in the periphery of the lung or in close proximity to other ROIs classified as Honeycombing. The classification scheme incorporating the SRG algorithm is shown in Fig. 6.

Guiding a Classifier Using Seeded Region Growing

   Data: *remaining_rois*
1 initialise all ROIs in the lung to non-honeycombing;
2 initialise *remaining_rois* with all the ROIs in the periphery of the lung;
3 **while** *remaining_rois* $\neq \emptyset$ **do**
4   Choose a ROI, *roi*, from *remaining_rois*;
5   *classification* ← *classify(roi)*;
6   **if** *classification=Honeycombing* **then**
7   apply classification to *roi*;
8   insert all unclassified ROIs within a threshold distance of *roi* in *remaining_rois*;
9   **end**
10   remove *roi* from *remaining_rois*;
11 **end**

Fig. 6. Guiding a Classifier Using Seeded Region Growing

In Fig. 7, we show the steps of the proposed classification scheme. The classifier was not required to classify the whole lung region.

3   Results and Discussions

In order to test the clinical viability of the system, we evaluate the performance of our system over all the ROIs in lung region. We divided the forty-two HRCT slices into ten groups. The marked ROIs (see Section 2.1) of nine of these groups are used as training for our classifier. In the remaining group, we use all the ROIs in the lung regions for testing. We ensure that all ROIs with honeycombing have been marked; the unmarked ROIs are assumed to be non-honeycombing. This process is repeated for all ten combinations of training and testing sets.

The number of ROIs used for testing varies for each fold, as the size of the lung in each slice affects the number of ROIs that we extract. On average 9337 ROIs were used for testing (876 ROIs for honeycombing and 8461 ROIs for non-honeycombing on average). The number of ROIs containing non-honeycombing is significantly larger as most of the lung region does not show honeycombing. The average of these results are shown in Tables 1 and 2 (of the Weka J48 and Naive Bayes classifiers respectively). The accuracy measures the percentage of ROIs that were classified correctly; sensitivity measures the proportion of ROIs

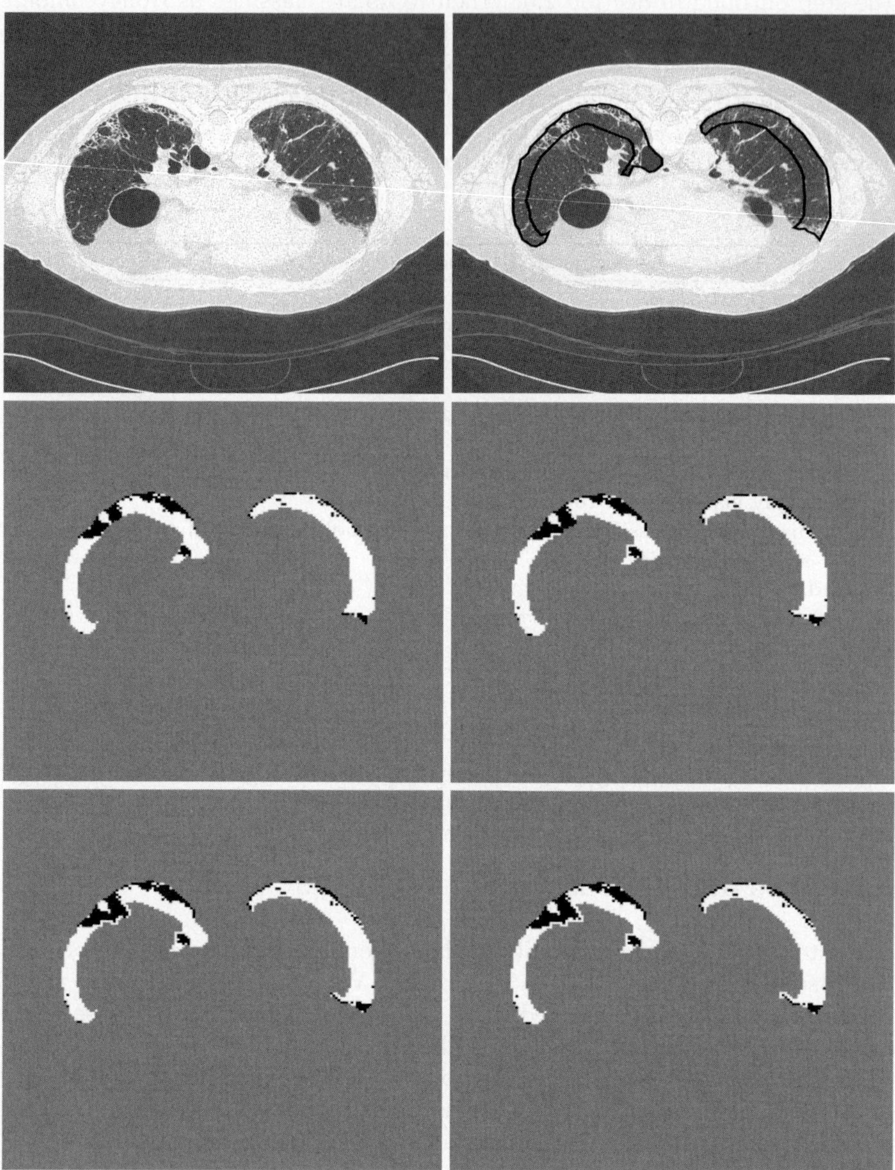

Fig. 7. The seeded region growing algorithm used to guide the classifier in honey-combing detection. Top row: (Left) Original image. (Right) Lung periphery outlined. 2nd and 3rd row: ROIs classified as honeycombing are in black. ROIs classified as non-honeycombing are in white. Unclassified regions are in gray. Initially only ROIs in the lung periphery were classified. Eventually, some ROIs in the non-peripheral lung region were classified if the ROI was near other ROIs which have already been classified as Honeycombing. In the final image (bottom right), large portions of the image were still not required to be classified by the classifier.

Table 1. Results of the Weka J48 classifier

Feature Selection Scheme	CFS		ConsistencyFS	
Classifier Scheme	Normal	SRG-Guided	Normal	SRG-Guided
Accuracy	88.2%	89.7%	88.0%	89.6%
Sensitivity	96.7%	96.6%	97.3%	97.1%
Specificity	86.8%	88.6%	86.5%	88.4%

Table 2. Results of the Naive Bayesian classifier

Feature Selection Scheme	CFS		ConsistencyFS	
Classifier Scheme	Normal	SRG-Guided	Normal	SRG-Guided
Accuracy	85.5%	87.2%	83.3%	85.1%
Sensitivity	97.5%	97.4%	97.0%	96.9%
Specificity	83.5%	85.5%	80.8%	83.1%

with honeycombing that were correctly classified; while specificity measures the proportion of ROIs with non-honeycombing that were correctly classified.

In all four combinations of feature-selection and classifier, the use of a SRG algorithm to guide the classifier improved the accuracy of the honeycombing detection. The improvement in accuracy is attributed to a decrease in false positive classifications (sometimes by over 2%). The increase in accuracy shows that the technique is well suited for honeycombing detection. A paired t-test, shows the improvement in accuracy to be statistically significant ($p < 0.0001$).

4   Conclusion

Honeycombing is an important disease pattern in HRCT which allows radiologists to make a diagnosis. The accurate quantification of honeycombing allows radiologists to track the progress of the disease. Many reported systems apply a classifier over the whole lung image. In contrast, we have proposed a novel technique, which used a seeded region growing algorithm to guide a system to regions with potential honeycombing. We implemented this classification scheme in a system using a machine learning approach to automatically detect honeycombing. Our system was able to detect honeycombing with a high degree of accuracy, sensitivity and specificity. The improvement was evident using both the Weka J48 decision tree classifier and the Naive Bayes classifier. We tested the system with eight different patients over forty-two images with a high degree of accuracy (89.7%) and sensitivity (96.6%). This empirically verifies that our classification scheme is well-suited for the purpose of honeycombing detection.

Acknowledgements

We thank Min Sub Kim for his valuable comments on drafts of this paper. This research was partially supported by the Australian Research Council through a Linkage grant (2002-2004), with I-Med Network Ltd. and Philips Medical Systems as partners.

References

1. Webb, W.R., Müller, N.L., Naidich, D.P.: High-Resolution CT of the Lung. 3rd ed edn. Lippincott Williams & Wilkins, Philadelphia (2001)
2. Uppaluri, R., Hoffman, E.A., Sonka, M., Hartley, P.G.: Computer Recognition of Regional Lung Disease Patterns. American Journal of Respiratory and Critical Care Medicine **160**(2) (1999) 648–654
3. Uchiyama, Y., Katsuragawa, S., Abe, H., Shiraishi, J., Li, F., Li, Q., Zhang, C.T., Suzuki, K., Doi, K.: Quantitative computerized analysis of diffuse lung disease in high-resolution computed tomography. Medical Physics **30**(9) (2003) 2440–2454
4. Papasoulis, J.: LMIK — Anatomy and Lung Measurements using Active Contour Snakes. Undergraduate thesis, Computer Science and Engineering, University of New South Wales, Sydney, Australia (2003)
5. Hall, M.A.: Correlation-based feature selection for discrete and numeric class machine learning. In: ICML '00: Proceedings of the Seventeenth International Conference on Machine Learning, San Francisco, CA, USA, Morgan Kaufmann Publishers Inc. (2000) 359–366
6. Liu, H., Setiono, R.: A Probabilistic Approach to Feature Selection – A Filter Solution. In: Proceedings of the 13th Intl. Conf. Machine Learning. (1996) 319–327
7. Witten, I.H., Frank, E.: Data Mining: Practical machine learning tools and techniques. 2nd edition edn. Morgan Kaufmann, San Francisco (2005)
8. Adams, R., Bischof, L.: Seeded region growing. Pattern Analysis and Machine Intelligence, IEEE Transactions on **16**(6) (1994) 641–647
9. Zrimec, T., Busayarat, S., Wilson, P.: A 3D model of the human lung with lung regions characterization. ICIP 2004 Proc. IEEE Int. Conf. on Image Processing **2** (2004) 1149–1152

Voting Massive Collections of Bayesian Network Classifiers for Data Streams

Remco R. Bouckaert

Computer Science Department, University of Waikato, New Zealand
`remco@cs.waikato.ac.nz`

Abstract. We present a new method for voting exponential (in the number of attributes) size sets of Bayesian classifiers in polynomial time with polynomial memory requirements. Training is linear in the number of instances in the dataset and can be performed incrementally. This allows the collection to learn from massive data streams. The method allows for flexibility in balancing computational complexity, memory requirements and classification performance. Unlike many other incremental Bayesian methods, all statistics kept in memory are directly used in classification.

Experimental results show that the classifiers perform well on both small and very large data sets, and that classification performance can be weighed against computational and memory costs.

1 Introduction

Bayesian classifiers, such as naive Bayes [5] can deal with large amounts of data with minimal memory requirements and often perform well. The assumption underlying naive Bayes that all attributes are independent given the class value often does not hold and by incorporating more dependencies, its performance can be improved. Tree augmented naive Bayes (TAN) [3], super parent TAN [6] and forest augmented naive Bayes [7] provide such methods. However, these methods are not well suited for incremental learning since the structure of the network cannot be fixed.

Averaged One-Dependence Estimators (AODE) [8] can be considered as voting of various Bayesian classifiers, one for each attribute in the data set. It performs well on both small and large data sets and can be trained incrementally. This raises the expectation that voting more Bayesian classifiers can improve performance. In this article, we analyze how to vote a collection of forest augmented naive Bayes (FAN) under some mild conditions. The number of FANs is exponential in the number of attributes making it seemingly impossible to use in a practical way. However, we show how to calculate the vote in quadratic time instead of exponential. We also consider some other collections. The goal of this paper is to design a classifier that is able to learn and classify from massive amounts of data, is able to incrementally update the classifier, show good classification performance on both massive and small data sets, has a reasonable memory requirement, and can be able to incorporate external information about the model.

A. Sattar and B.H. Kang (Eds.): AI 2006, LNAI 4304, pp. 243–252, 2006.
© Springer-Verlag Berlin Heidelberg 2006

The following section introduces Bayesian classifiers formally. In Section 3, we present Bayesian network classifiers for streams and explains how to perform calculation efficiently. Section 4 presents some experimental results and discussion and we conclude with some remarks and directions for further research.

2   Bayesian Network Classifiers

Given a set of *attributes* $\{X_1, \ldots, X_n\}$ with values $\{x_1, \ldots, x_n\}$ we want to predict the value of a *class* variable Y taking one of k class values $\mathcal{C} = \{y_1, \ldots, y_k\}$. We assume attributes and classes to be discrete. In order to do so, we train a classifier using a data set D, which is a set of m tuples $\{x_{1j}, \ldots, x_{nj}, y_j\}$ $(1 \leq j \leq m)$ of values of attributes and a value of the class.

A *Bayesian network* B over a set of variables V consists of a network structure B_S, which is a directed acyclic graph over V and a set of conditional probability table B_P, one table $P(X|pa_X)$ for each variable in $X \in V$ conditioned on the variables that form the parent set pa_X in B_S. A Bayesian network defines a probability distribution over V by the product of the attributes $P(V) = \prod_{X \in V} P(X|pa_X)$. A Bayesian network can be used as a classifier by training the network over the attributes and class variables on the data and use the represented distribution using the following decision rule: $argmax_{y \in \mathcal{C}} P(Y|X_1, \ldots, X_n)$ which, using Bayes rule, is the same as taking $argmax_{y \in \mathcal{C}} P(Y, X_1, \ldots, X_n)$ which equals $argmax_{y \in \mathcal{C}} \prod_{X \in \{Y, X_1, \ldots, X_n\}} P(X|pa_X)$. So, an instance $\{x_1, \ldots, x_n\}$ is classified by plugging in the values in the Bayesian network and taking the class value with the highest probability represented by the network.

In general, there are two learning tasks when training a Bayesian network: learning the network structure, and learning the conditional probability tables. Network structures can be fixed, as for naive Bayes, or learned based on dependencies in the data, as for tree augmented naive Bayes. The conditional probability tables are typically learned using an estimator based on the counts in the data D. Throughout, we use the Laplace estimator which estimates the probability of X taking value x given its parents taking values pa_x to be

$$P(x|pa_x) = \frac{\#(x, pa_x) + 1}{\#(pa_x) + |X|}$$

where $\#(x, pa_x)$ is the count of instances in D where X takes value x and its parents take values pa_x. Likewise $\#(pa_x)$ is the count of parents taking value pa_x. We use $|X|$ to denote the number of values that X can take. Note that if the parent set of X is empty, we have by convention that $\#(pa_x) = 0$ and $\#(x, pa_x)$ is the count of X taking value x. One of the benefits of this estimator is that we can keep the counts $\#(x, pa_x)$ and $\#(pa_x)$ in memory and easily update them with each new incoming instance. So, updating the probability tables incrementally is trivial.

In the remainder of this section, we consider some popular Bayesian classifiers and explain how they relate to the Bayesian network classifier.

Naive Bayes. The naive Bayes classifier [5] can be considered a Bayesian network with a fixed network structure where every attribute X_i has the class as its parent and there are no other edges in the network. Therefore, naive Bayes has the benefit that no effort is required to learn the structure and the probability tables can be updated online by keeping counts in memory.

Tree Augmented Naive Bayes. The disadvantage of the naive Bayes classifier is that it assumes that all attributes are conditionally independent given the class, while this often is not a realistic assumption. One way of introducing more dependencies is to allow each (except one) attribute to have an extra parent from among the other attributes. Thus, the graph among the attributes forms a tree, hence the name tree augmented naive Bayes (TAN) [3]. The network structure is typically learned by determining the maximum weight spanning tree where the weight of links is calculated using information gain. This algorithm is quadratic in the number of attributes. Also, it requires gathering of the counts $\#(x, pa_x)$ and $\#pa_x$, where pa_x takes values for each class and attribute Z with $Z \in \{X_1, \ldots, X_n\} \backslash X$. So, we need to keep track of a lot of counts not ultimately used in the classifier. Further, incremental updating requires retraining the network structure, which is $O(n^2)$ in computational complexity.

Forest Augmented Naive Bayes. The TAN cannot distinguish between relevant links in the final tree and irrelevant links. The forest augmented naive Bayes (FAN) [7] is a network structure where each attribute has at most one extra attribute (apart from the class) as parent, but it can as well have no other parent. The benefit is that such a structure can be learned using a greedy approach with a scoring based metric like Bayesian posterior probability, minimum description length, Akaike information criterion, etc which provide a natural stopping criterion for adding edges to the network. Still, since all possible edges need to be considered, like for TANs many counts need to be kept in memory which are not utilized in the probability tables.

Super Parent Classifier. The super parent classifier [6] selects one of the attributes and calls this the super parent. Each of the other attributes get this attribute and the class as its parent set while the super parent only has the class as its parent. Determining which parent becomes the super parent requires collecting many counts that are discarded as for TANs.

Averaged One Dependence Classifiers. AODE classifiers [8] uses the super parent idea to build a collection of n Bayesian classifiers by letting each of the attributes be super parent in one of the networks. The class is predicted by taking the sum of the probabilistic votes of each of the networks. Clearly, the benefit is that all counts are utilized and no effort is spent in learning network structures.

3   Bayesian Network Classifiers for Data Streams

In the previous section, we shortly listed a number of popular Bayesian network architectures. In this section, we set out to design a Bayesian classifier that can deal with the following requirements;

- be able to learn from massive amounts of data,
- be able to classify massive amounts of data,
- be able to incrementally update the classifier when more data becomes available,
- show good classification performance,
- have a reasonable memory requirement,
- be able to incorporate external information about the model.

Naive Bayes is a fast learner and classifier, but lacks in performance. TANs, FANs and super parents require quadratic time in learning a network structure, which makes incremental updating cumbersome. AODEs perform well in most areas though classification time is quadratic in the number of attributes and it lacks the ability of incorporating prior information. An algorithm that performs well in terms of the first four requirements is the Hoeffding tree algorithm [2]. However, because it grows trees incrementally, it can require large amounts of memory. Also, incorporating knowledge obtained from experts or prior experience is awkward. The Hoeffding criterion can be applied to Bayesian network learning [4] thereby incrementally learning the network structure. Still, many statistics need to be kept in memory thereby slowing down the number of instances that can be processed per second. Also, single Bayesian networks tend to be less accurate classifiers than ensembles like AODE.

This last general observation leads us to try to use a collection of a large number of Bayesian networks. Consider the collections of FANs where the order of attributes $\{X_1, \ldots, X_n\}$ is fixed in such a way that an edge between two attributes always goes from the lower ordered to the higher ordered, that is the network can only contain edges $X_i \rightarrow X_j$ if $i < j$. Now consider the classifier that sums over all such FANs

$$P(Y, X_1, \ldots, X_n) = C \sum_{FAN} P_{FAN}(Y, X_1, \ldots, X_n) \tag{1}$$

where C is a normalizing constant. An attribute X_i can have i different parent sets, namely only the class variable $\{Y\}$ or the class with one of the previous attributes $\{Y, X_1\}, \ldots \{Y, X_{i-1}\}$. So, in total there are $n!$ such FANs to sum over. Obviously, this is not a practical classifier for even a moderate number of attributes. However, we show using induction that the sum (1) can be rewritten as

$$CP(Y) \prod_{i=1}^{n} \left(P(X_i|Y) + \sum_{j=1}^{i-1} P(X_i|C, X_j) \right) \tag{2}$$

The induction is by the number of attributes. For a single attribute $n = 1$, there is only one network structure and trivially Equations (1) and (2) are equal. For $n-1 \geq 1$ attributes, suppose that (1) and (2) are equal and we have n attributes. Now observe that

$$\sum_{FAN} P_{FAN}(Y, X_1, \ldots, X_n) = \sum_{FAN} P(Y) \prod_{i=1}^{n} P(X_i|pa_i^{FAN})$$

where pa_i^{FAN} is the parent set of X_i in the FAN. X_n can be taken outside the product, giving

$$\sum_{FAN} P(Y) \prod_{i=1}^{n-1} P(X_i|pa_i^{FAN})P(X_n|pa_n^{FAN}) \tag{3}$$

Now, observe that for every sub-forest over the first $n-1$ attributes there is one network in the set FAN where X_n has no parents but Y, one where X_n has Y and X_1 as parent, one where Y and X_2 are parents, etc, and one where Y and X_n are parents. Let FAN' represent the set of all sub-forests over the first $n-1$ attributes. Then, we can rewrite Equation 3 as

$$\Big(\sum_{FAN'} P(Y) \prod_{i=1}^{n-1} P(X_i|pa_i^{FAN'}) \Big) \cdot \Big(P(X_n|Y) + \sum_{j=1}^{n-1} P(X_n|C,X_j) \Big)$$

and by our induction hypothesis $\sum_{FAN'} P(Y) \prod_{i=1}^{n-1} P(X_i|pa_i^{FAN'})$ is equal to $P(Y) \prod_{i=1}^{n-1} \Big(P(X_i|Y) + \sum_{j=1}^{i-1} P(X_i|C,X_j) \Big)$, so the above equals

$$P(Y) \prod_{i=1}^{n-1} \Big(P(X_i|Y) + \sum_{j=1}^{i-1} P(X_i|C,X_j) \Big) \Big(P(X_n|Y) + \sum_{j=1}^{n-1} P(X_n|C,X_j) \Big)$$

which by taking the last part inside the product equals Equation (2).

Note that the calculation of (2) is only quadratic in the number of attributes. So, we can sum over $n!$ FANs in $O(n^2)$ time, which means we can perform classification efficiently. No network structure learning is required, just maintaining counts $\#(X_i, X_j, C)$ and $\#(X_j, C)$ to calculate the probabilities in (2) is needed. So, only a fixed amount of memory is quadratic in n required.

3.1   Collection Architectures

So far, we only considered the collection of all FANs for a given node ordering. However, the same derivation as above can be done for any collection of parent sets, giving rise to a choice of collection architectures. Let PA_i be a set of i_n parent sets $PA_i = \{pa_{i,1}, \ldots, pa_{i,i_n}\}$ for node X_i such that all nodes in any parent set $X_j \in pa \in PA_i$ is lower ordered than X_i (that is, $j < i$). Then, if we have such a parent set for each attribute, we can sum over the collection of all Bayesian networks with any combination of these parent sets. The distribution (corresponding to Equation 2) is

$$P(Y, X_1, \ldots, X_n) = CP(Y) \prod_{i=1}^{n} \sum_{pa_i \in PA_i} P(X_i|C, pa_i)$$

Note that inserting parent sets that do not contain the class node is not useful, since for the purpose of classification the terms $P(X_i|pa_i \setminus \{C\})$ are constant, so these do not impact the class probabilities.

Which collection architectures can be expected to result in well performing classifiers? The collection of all FANs we considered is essential the set of all networks where each attribute has the class as parent and at most one other (lower ordered) attribute. We abbreviate such FAN collection as FANC. A natural extension is to consider all networks with the class plus *at most* two other lower ordered attributes, and call this a two-parent collection or TC for short. The number of parent sets grows quadratically with the number of attributes, and the memory requirements for storing the conditional probability tables for $P(X_i|C, X_j, X_k)$ grows cubic in the number of attributes. So, it may be quite costly when the number of attributes is large. A reasonable alternative is to identity one attribute as super parent and consider for every attribute of the class, possibly the super parent and possibly one other lower ordered attribute. We call this the super parent collection, or SPC for short.

3.2   Practical Miscellany

We shortly discuss some pragmatic issues.

Missing values. When there are attributes with missing values we can easily take these attributes out of the network by just ignoring the terms containing the attributes. So, terms $P(X_i|Y, X_j)$ and $P(X_i|Y)$ where X_i or X_j are missing in the instance are simply removed from Equation 2.

Non-discrete attributes. Though the above technique works for non-discrete attributes as well, efficient representation of such attributes requires some care. For the sake of simplicity, we concentrate on discrete variables.

Prior knowledge. In the FANC implementation of Equation 2 each term has the same weight. When there is strong evidence that some attributes, say X_i and X_j $(j > i)$ are highly related given Y, the term $P(X_j|Y, X_i)$ can be given a higher weight in (2) than the other terms. Alternatively, the parent set collection can be tuned such that all parent sets of X_i contains X_j.

Attribute ordering. In preliminary experiments, we found that the order of attributes had moderate impact on the performance of the FANC classifier. This issue is open for further research.

Super parent selection. In our experiments we found that the choice of the super parent has considerable impact on the performance. Selecting the super parent based on classification accuracy on the training set would give many draws. In our experiments, taking the sum of class probabilities (i.e. $\sum_{i=1}^{m} P(y_i|x_{1i}, \ldots x_{ni})$) as selecting criteria led to few draws and well performing classifiers. We call this SPC variant SPCr. This appears to work well for small datasets in batch learning, but is harder to implement incrementally for large datasets.

The following table gives an overview of the memory requirements, training time on a data set and test time of an instance with a representing the maximum number of values of attributes.

	NB	TAN	AODE	FANC	SPC	SPCr	TC
Memory	$O(ma)$	$O(ma^2)$	$O(m^2a^2)$	$O(m^2a^2)$	$O(m^2a^3)$	$O(m^2a^3)$	$O(m^3a^3)$
Train time	$O(nm)$	$O(nm^2)$	$O(nm^2)$	$O(nm^2)$	$O(nm^2)$	$O(nm^3)$	$O(nm^3)$
Test time	$O(m)$	$O(m)$	$O(m^2)$	$O(m^2)$	$O(m^2)$	$O(m^2)$	$O(m^3)$

4   Empirical Results

We evaluate the Bayesian classifiers both for small and large datasets to get an
impression of its performance in a range of circumstances, both in batch mode
and in incremental mode. All experiments were performed using Weka [9] on a
PC with a 2.8GHz CPU and 1 GB memory.

4.1   Small Data Sets

Batch mode experiments were performed with naive Bayes, TAN, AODE, FANC,
SPC, SPCr and TC using ten times repeated ten fold cross validation on 28
UCI data sets[1]. All datasets are discretized using three bins, but missing values
were not filled in, except for TAN, which uses the most common value instead
of missing values. We judged the various algorithms on classification accuracy,
train time, test time and memory requirements. Due to space limitations, we
cannot present the results of these experiments extensively, but give a summary
of our findings.

We confirmed results from [3] and [8] that classification performance of TAN
and AODE is better than naive Bayes and that AODE tends to perform similar to TAN but with lower train time. Further, of the collections (FANC, SPC,
SPCr and TC) classified all consistently better than naive Bayes, except for the
labor dataset, which is due to the small number of instances (just 57 in total).
Better performance comes from being able to capture more dependencies, while
diminished performance for very small datasets is caused by the increased variance in parameter estimation since for the collections a much larger number of
parameters need to be established. Training and testing times as well as memory
requirements are consistently larger than naive Bayes. Compared with TAN and
AODE, FANC classification performs is about the same, SPC and TC slightly
better and SPCr considerably better overall. This is illustrated in Table 1 on the
right, which shows the classification performance of AODE (x-axis) versus SPCr
(y-axis). Points over the x=y-line favor SPCr and below AODE. The significant
outlier under the line is labor, but the most other sets show a difference in favor
of SPCr. Table 1 on the left gives a summary of our findings and allows for
selecting a Bayesian classifier that fits the requirements of a particular learning
situation.

---

[1] anneal, balance scale, breast cancer, wisconsin breast cancer, horse colic, credit-
rating, german credit, pima diabetes, glass, cleveland heart-disease, hungarian
heart-disease, heart-statlog, hepatitis, hypothyroid, ionosphere, iris, kr-vs-kp, labor,
lymphography, primary-tumor, segment, sonar, soybean, vehicle, vote, vowel, waveform, zoo.

In short, though SPCr classification performance is the best overall, this comes at a cost of training time. If memory is not a restriction, TC performs best; but if it is, SPC tends to perform quite well.

Table 1. A quick rough overview of some relevant properties of the various algorithms giving a guideline on which algorithm is appropriate for a particular situation

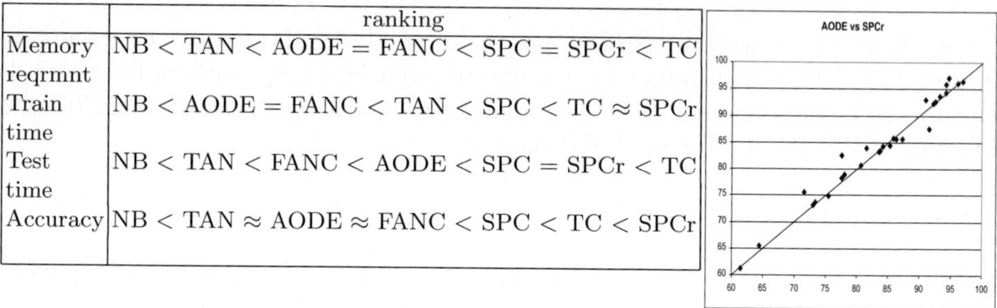

	ranking
Memory reqrmnt	NB < TAN < AODE = FANC < SPC = SPCr < TC
Train time	NB < AODE = FANC < TAN < SPC < TC ≈ SPCr
Test time	NB < TAN < FANC < AODE < SPC = SPCr < TC
Accuracy	NB < TAN ≈ AODE ≈ FANC < SPC < TC < SPCr

4.2   Large Data Sets

We randomly generated models over 20 binary variables (a class and 19 binary attributes), a naive Bayes (19 edges), a TAN (37 edges) and Bayesian net augmented naive Bayes (BAN) model (50 edges) representing increasingly complex concepts. From each model a dataset of 10 million instances was generated. Figure 1 shows the learning curve for each of the models for various learning algorithms. Since the curve is rather flat fairly quickly, Figure 2 shows the curve for the first hundred thousand instances as well.

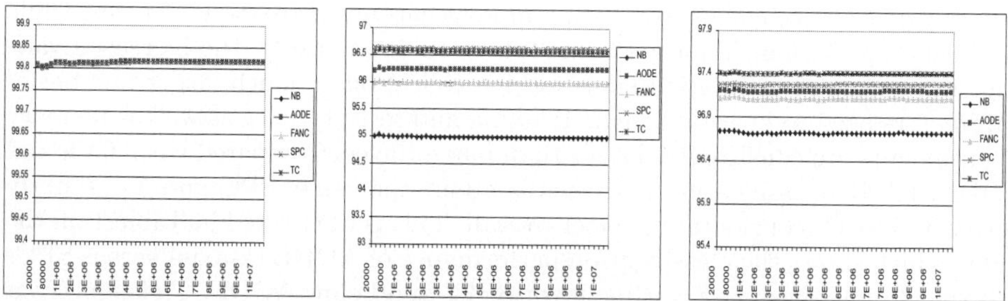

Fig. 1. Mean accuracy of various variants on streams of synthetic data with 10 million instances. The x-axis shows the number of instances used for training, the y-axis is accuracy. Left, sourced from naive Bayes, in the middle sourced from a TAN and right sourced from a BAN.

For the naive Bayes data source, all algorithms perform very similar. For the TAN data source, naive Bayes is consistently outperformed by the other algorithms and SPC and TC outperforms the other algorithms. For the BAN data

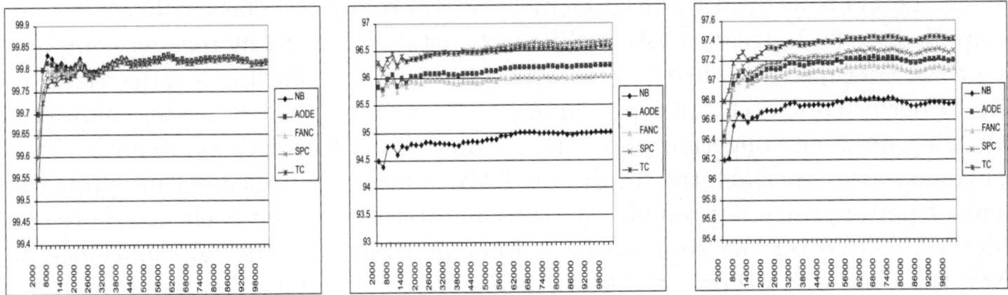

Fig. 2. Mean accuracy of various variants on streams of synthetic data. The x-axis shows the number of instances used for training, the y-axis is accuracy. Left, sourced from naive Bayes, sourced from TAN in the middle and right, sourced from BAN.

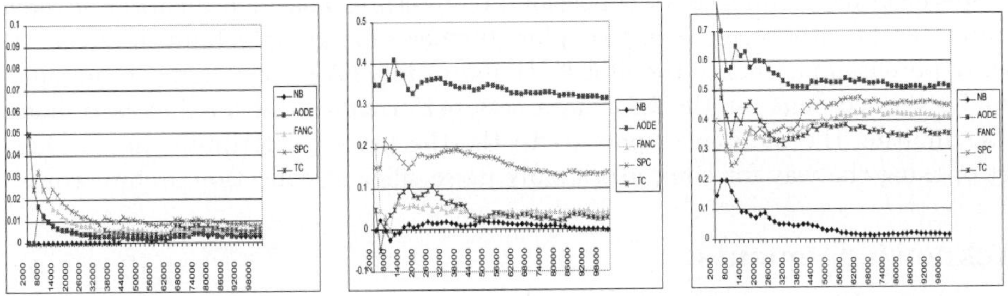

Fig. 3. Difference of accuracy when ten noisy attributes are added

source results are similar to the TAN data source, except that TC outperforms SPC. This leads to two conclusions; the collections perform well on large data sets and with increasing complexity of the concept to be learned increasingly complex collections help in performance.

We investigated sensitivity to noise by adding 10 binary attributes with no dependencies to any of the other attributes. Figure 3 shows the degradation in performance for the various algorithms. Naive Bayes is least affected, but the other do not degrade that much leaving naive Bayes still the worst performer. AODE is most affected while FANC, TC and SPC are in between. It appears that working with large collections of classifiers decreases sensitivity to irrelevant attributes. We also investigated the performance degradation due to adding 5% noise to the attributes and inserting 5% missing values (not shown due to space limitations) but found that all algorithms were affected similarly.

5   Conclusions

Voting with massive collections of Bayesian networks was shown to be feasible in polynomial time and these collections perform well both on small and large data sets. Since the training time is linear in the data set size and learning of

UTD LIBRARY

its parameters is naturally suited to incremental updating, these collections can be applied to data streams with millions of instances. With increasing complexity of the concept to be learned increasingly complex collections were found to help in performance but cost in memory and classification time. This allows for balancing between collection performance and computational complexity.

Preliminary experimental results on FAN collections indicate that improvement of performance is possible by ordering attributes. For each attribute the order was based on the sum over all other attributes of the information gain given the class and other attribute. Also, some preliminary results of selection of a proper super parent in a super parent collection (for instance through training set performance) can improve the classification performance of a collection. However, in both these situations, it is not clear how to adapt the algorithm without sacrificing much memory or computational cost. Both these are issues we hope to address more comprehensively in the future.

Specification of alternative collection architectures and identification of their properties are interesting areas to explore further. For example, limiting the number of potential parents to the first k attributes in a FAN collection. Preliminary results showed that for the UCI data sets performance did not increase much after limiting the number of parents to the the first six attributes in the data set opening the way for computationally more efficient collection architectures.

Acknowledgements

I thank all Waikato University Machine Learning group members for stimulating discussions on this topic, especially Richard Kirkby for making his Moa system for experimenting on large data streams available.

References

1. C.L. Blake and C.J. Merz. UCI Repository of machine learning databases. Irvine, CA: University of California, 1998.
2. P. Domingos and G. Hulten. Mining High-Speed Data Streams. SIGKDD 71–80, 2000.
3. N. Friedman, D. Geiger, and M. Goldszmidt. Bayesian Network Classifiers. In Machine Learning 29:131–163, 1997
4. G. Hulten and P. Domingos. Mining complex models from arbitrarily large databases in constant time. SIGKDD 525–531, 2002.
5. G.H. John and Pat Langley. Estimating Continuous Distributions in Bayesian Classifiers. Uncertainty in Artificial Intelligence, 338–345, 1995.
6. E. Keogh and M. Pazzani. Learning augmented Bayesian classi- fiers: A comparison of distribution-based and classification-based approaches. AIStats 225 230, 1999.
7. J.P. Sacha. New synthesis of Bayesian network classifiers and interpretation of cardiac SPECT images, Ph.D. Dissertation, University of Toledo, 1999.
8. Geoffrey I. Webb , Janice R. Boughton , Zhihai Wang. Not so naive Bayes: aggregating one-dependence estimators, Machine Learning, 58(1) 5-24, 2005.
9. I.H. Witten and E. Frank. Data mining: Practical machine learning tools and techniques with Java implementations. Morgan Kaufmann, San Francisco, 2000.

UTD LIBRARY

Feature Weighted Minimum Distance Classifier with Multi-class Confidence Estimation

Mamatha Rudrapatna[1] and Arcot Sowmya[1,2]

[1] School of Computer Science and Engineering, University of New South Wales, Sydney,
NSW 2052, Australia
[2] Division of Engineering, Science and Technology, UNSW Asia, Singapore
{mamathar, sowmya}@cse.unsw.edu.au

Abstract. In many recognition tasks, a simple discrete class label is not sufficient and ranking of the classes is desirable; in others, a numeric score that represents the confidence of class membership for multiple classes is also required. Differential diagnosis in medical domains and terrain classification in surveying are prime examples. The Minimum Distance Classifier is a well-known, simple and efficient scheme for producing multi-class probabilities. However, when features contribute unequally to the classification, noisy and irrelevant features can distort the distance function. We enhance the minimum distance classifier with feature weights leading to the Feature Weighted Minimum Distance classifier. We empirically compare minimum distance classifier and its enhanced feature weighted version with a number of standard classifiers. We also present preliminary results on medical images with acceptable performance and better interpretability.

1 Introduction

Machine vision is a challenging domain for the application of machine learning. Vision systems are increasingly used in the fields of navigation, medicine and defence applications. The core tasks of such systems include detection, recognition and classification of future observations, based on a learned model. To be useful in critical applications (where automated systems often serve as a second opinion), the recognition rates must have high accuracy. In such situations, it is not sufficient to obtain a single decision label or a mere ranking of decisions. The experts require information on the likelihood of other labels, represented as a numerical score, in order to reconsider options. We may denote this score as an estimate of class membership probability or as a measure of confidence in the prediction, and use these terms synonymously.

There exist many other situations where a confidence measure is likely to be useful. One such example is in multistage processing. As outputs of one level are fed as inputs to another, a hard decision made at each level may result in early rejection of potential candidates. Instead, soft decisions at each level would facilitate a final decision based on all such confidence measures. Confidence measures also appear in the field of information fusion, where information from multiple sources is combined after being weighted by their reliability. They are also used to model uncertainty and imprecision, and are useful where cost-sensitive decisions are needed.

A. Sattar and B.H. Kang (Eds.): AI 2006, LNAI 4304, pp. 253–263, 2006.

© Springer-Verlag Berlin Heidelberg 2006

Our focus is on lung diseases detection using High Resolution Computed Tomography (HRCT) images. Here, pixels represent body tissue and may display the appearance of a number of abnormal overlapping patterns. As in any biological process, there is no clear-cut separation of regions, and huge variations are usually observed. This makes automated recognition a very difficult task. We are building a multistage hierarchical vision system, where soft decisions are made at each level.

In our application, there are thousands of pixels in the training dataset, and while in use, millions of pixels must be classified within a reasonable time. Therefore, a simple, computationally efficient classifier capable of producing confidence values is required. The Minimum distance classifier (MDC), also known as nearest centroid method, meets this requirement [1]. Class centroids are the only parameters learned during the training phase. During the testing phase, the distance from a test sample to the class centroids is all that is needed to produce both a class label and a confidence. This classifier has been successfully used on high dimensional and complex gene expression micro-array data [2]. When features contribute unequally towards classification, it is well known that noisy and irrelevant features can distort the distance function. Feature weights can be used to improve the contribution of dominant features and reduce the disproportionate effect of weaker ones [3]. We call this the *feature weighted Minimum Distance classifier* (fwMDC).

In this paper, we empirically compare a number of standard classifiers with both the MDC and the enhanced fwMDC. We evaluate them on prediction accuracy, performance on ranking and spread of probabilities, for their suitability in vision systems. In section 2 we review the classifiers that we have considered for comparison and in section 3, we discuss the MDC. Section 4 explains the experimental setup, and results on benchmark datasets are presented in section 5. Section 6 has preliminary results on HRCT data. We conclude with a summary and future directions in section 7.

2   Related Work

Theoretically, any classifier that produces a numerical output may be used to obtain a confidence measure. However, it is well known that while classifiers do produce accurate ranking based on such scores, these do not serve well as probability estimates. Recently there has been increasing interest in augmenting many popular classification schemes to improve their probability estimates. However many of these are designed to improve the prediction accuracy or ranking, rather than the true probability estimates.

We selected some popular classifiers from four categories namely, Bayesian, Decision Trees, Lazy Learners and Ensembles in our study.

Classifiers based on Bayesian decision theory have proven to be very effective. Naïve Bayes is a simple and efficient classifier and has excellent performance. Even though it produces poor probability estimates [4], it is found to perform well in ranking [5]. Naïve Bayes is a confident classifier and the posterior probabilities tend to clump towards the two extremes of 0 and 1, especially when the attributes are positively correlated. Zadrozny and Elkan [6] have proposed binning of examples based on their scores and estimating probability as the fraction of training examples in a bin that belongs to the predicted class. The authors show that even though this increases the accuracy of prediction, resolution is reduced. Others have tried to accommodate attribute dependencies, leading to augmented naïve Bayesian networks [7]. However learning an optimal structure for such networks is a difficult task.

Decision tree (C4.5) offers a simple, intuitive and effective learning method and is widely used in classification tasks [8]. However, they perform poorly with regard to probability estimates and ranking [9]. Many improvements have been made to enhance the class probability estimates. Provost and Domingos [10] recommend turning off error based pruning and smoothing the estimation with Laplace correction and call this method C4.4. However, this unpruned tree can become very large and complex. Kohavi created a hybrid decision tree by embedding a naïve Bayes classifier at each leaf node [11]. Margineantu and Dietterich [12] propose bagged lazy option trees. They grow the tree upon seeing a test example much like the nearest neighbour classifier. This dynamic behaviour makes them computationally expensive during the testing phase. Ling and Yan [13] use the weighted average of probability estimates from all leaves of the tree and show that this method outperforms C4.4 for ranking.

Instance based learners such as k-nearest neighbour classifier (kNN) simply store the training examples and predict the class label and probability estimates based on the training examples nearest to the test example. Delany et al [14] have used the distances to nearest like and unlike neighbours as the confidence metric. However kNN is computationally very expensive.

In ensemble learning, a voted classifier is built by combining decisions of a group of classifiers [15]. We considered Bagging and Boosting with C4.4 as the base classifier, which are probably the most well known ensemble learners. Bagged trees have performed well over many other probability estimation trees for ranking [13]. Classifiers such as support vector machines and neural nets also produce numerical outputs, however they are not easily interpretable.

Poor probability estimates provided by classifiers that perform well in ranking can be improved with calibration. Zardonzy and Elkan [16] and Niculescu-Mizil and Caruana [17] have suggested some calibration techniques. The effectiveness of such measures is still debated and their extension to multi-class problems is not trivial.

3   Minimum Distance Classifier

The Bayesian decision rule for minimum-error-rate classification can be written as

$$C_l = \arg\max_k \{\ln p(x|c_k) + \ln p(c_k)\} \tag{1}$$

which assigns class label C_l to be class k that has maximum posterior probability for a feature vector x [1]. If the class-conditional densities are assumed to be multivariate normal $N(\mu_k, \Sigma_k)$ with equal prior probabilities, equation (1) can be written as

$$C_l = \arg\min_k \{(x - \mu_k)^T \Sigma^{-1} (x - \mu_k)\} \tag{2}$$

where μ is the class mean of the feature vector, Σ is the covariance matrix and T is the transpose operator on a matrix. The term $(x - \mu_k)^T \Sigma^{-1} (x - \mu_k)$ in equation (2) is the squared Mahalanobis distance between x and μ.

When Σ is an identity matrix (implying mutual independence among variables), the squared Mahalanobis distance is simply the squared Euclidean distance given by

$$d_k^2 = \sum_{i=1}^{d} (x_i - \mu_{ki})^2 \tag{3}$$

where d is the dimension of the feature vector, and x_i is the ith feature. The decision rule in equation (2) can now be written as

$$C_l = \arg\max_k \{\frac{1}{d_k}\} \tag{4}$$

The resulting decision rule classifies an instance based on its Euclidean distance to class centroids in feature space. This yields the very simple and computationally effective classifier known as the *Minimum Distance* classifier. To moderate the contribution of less-relevant features to the distance, feature weights can be introduced into the distance function in equation (3). The feature weighted distance term is given by

$$d_k^2 = \sum_{i=1}^{d} w_i (x_i - \mu_{ki})^2 \tag{5}$$

where w_i is the feature weight for the ith feature. The decision rule in equation (4) with the modified distance term in equation (5) yields the *feature weighted minimum distance classifier* (fwMDC). Various feature weighting methods are found in the literature [3, 18] and are usually chosen such that

$$w_i \geq 0 \text{ and } \sum_{i=1}^{d} w_i = 1$$

In our implementation, we use the normalised scores produced by ReliefF [19] algorithm for feature weights. The feature-weighted distance as defined above can be used as confidence measures.

4 Experimental Setup

We used 11 datasets summarised in Table 1 from the UCI repository [20] and the 10 classifiers listed in Table 2. We implemented MDC and fwMDC in Weka [21]. We used the Weka implementation for the other learners. Since the two minimum distance classifiers currently handle only numerical attributes, datasets with numerical attributes alone were chosen after removing any attributes that were unique for every instance (eg. the attribute 'ID' in Wisconsin Diagnostic Breast Cancer (Wis_BC_Diag dataset). The results of the experiments are discussed in the next section.

5 Results

We evaluated the classifiers on the datasets by conducting 10-fold cross-validation that was repeated 10 times. We also conducted paired t-test at 95% confidence level with Naïve Bayes as the base classifier. The prediction accuracies are shown in Table 3 along with the outputs of the paired t-test, where '^' denotes a statistically better performance and 'v' denotes worse performance. The classifiers are ordered left to right in the decreasing order, based on their average prediction accuracy.

Most training datasets only contain class labels; information on the rank or the class membership probabilities is usually unknown. Area under the ROC (Receiver

Table 1. UCI Datasets

Dataset	# of Attributes	# of Classes	# of Instances	Missing Attrib
Pima-Diabetes	9	2	768	No
Ecoli	8	7	336	No
Glass	10	6	214	No
Ionosphere	35	2	351	No
Iris	4	3	150	No
Liver	7	2	345	No
Segmentation	20	7	1500	No
Wine	14	3	178	No
Wis_BC_Diag	31	2	569	No
Wis_BC_Prog	34	2	198	Yes
Yeast	9	10	1484	No

Table 2. Classifiers

Classifier	Comment
Naïve Bayes	
C4.5	
kNN	Inverse Distance Weighted
MDC	
fwMDC	
C4.4	C4.5 no pruning, with Laplace smoothing
NBTree	
Boosted C4.4	C4.5 no pruning, with Laplace smoothing
Bagged C4.4	C4.5 no pruning, with Laplace smoothing

Table 3. Prediction Accuracy

Data sets	Bagged C4.4	Boosted C4.4	NBTree	C4.5	C4.4	kNN	Naïve Bayes	MDC	fw-MDC
ecoli	84.27 (0.94)v	83.33 (1.20)v	82.26 (0.97)v	82.33 (0.7)v	81.20 (1.05)v	80.62 (0.57)v	85.77 (0.88)	84.12 (0.36)v	81.99 (0.54)v
glass	74.34 (2.29)^	75.89 (1.92)^	71.19 (2.18)^	67.56 (2.06)^	67.64 (2.45)^	70.30 (1.25)^	47.75 (1.79)	43.41 (1.97)v	36.48 (2.22)v
iris	94.4 (0.64)v	93.87 (0.61)v	93.8 (0.77)v	94.73 (0.8)v	93.93 (0.58)v	95.40 (0.38)	95.53 (0.45)	92.33 (0.96)v	96.00 (0.00)
ionosphere	92.19 (0.66)^	92.42 (1.26)^	89.12 (1.14)^	89.74 (1.44)^	89.88 (1.15)^	87.10 (0.49)^	82.17 (0.49)	74.07 (0.57)v	75.16 (0.64)v
liver	72.15 (1.32)^	67.96 (1.91)^	65.12 (1.31)^	65.84 (1.71)^	65.63 (2.05)^	62.22 (1.15)^	54.89 (1.14)	58.18 (1.35)^	56.24 (1.20)
pima	75.71 (1.06)	73.03 (1.39)v	75.24 (0.98)	74.49 (0.91)v	73.98 (1.04)v	70.62 (0.84)v	75.75 (0.44)	72.95 (0.29)v	73.57 (0.37)v
wis-bc_d	95.24 (0.52)^	95.5 (0.89)^	93.89 (1.09)	93.27 (0.95)	93.15 (1.02)	95.64 (0.28)^	93.31 (0.17)	93.75 (0.19)^	94.87 (0.18)^
wis-bc_p	78.55 (1.48)^	75.98 (2.32)^	73.55 (2.28)^	72.59 (2.73)^	72.08 (2.86)^	70.68 (1.75)^	67.29 (1.14)	65.04 (0.85)v	63.86 (0.87)v
wine	95.28 (1.19)v	93.77 (1.07)v	96.62 (1.17)	93.2 (1.38)v	92.70 (1.35)v	95.12 (0.39)v	97.46 (0.47)	96.12 (0.60)v	92.85 (0.37)v
yeast	59.69 (0.93)^	57.27 (1.28)	57.49 (0.63)	56.61 (0.89)v	53.00 (1.00)v	52.61 (0.31)v	57.99 (0.44)	50.62 (0.23)v	49.05 (0.30)v
segment	96.73 (0.21)^	97.63 (0.14)^	94.17 (0.38)^	95.71 (0.37)^	96.06 (0.25)^	96.33 (0.23)^	81.12 (0.17)	83.91 (0.16)^	84.75 (0.15)^
Average	83.5 (1.02)	82.42 (1.27)	81.13 (1.17)	80.55 (1.27)	79.93 (1.35)	79.69 (0.69)	76.28 (0.69)	74.05 (0.68)	73.17 (0.62)

Operating Characteristics) curve, abbreviated as AUC, can be used to compare the quality of ranking [22]. A classifier with AUC score of 1 is said to rank the instances perfectly (i.e. the positive instances are consistently ranked above negative instances). For calculating multi-class AUC, we have adopted the simpler method proposed by Domingos and Provost [23]. Here, per-class AUC is calculated by considering instances belonging to one class at a time as positive examples and all other classes as negative examples. The per-class AUCs, denoted as $AUC(c_i)$ for class c_i, are then combined by weighting each by its prior class probability $p(c_i)$ as shown below:

$$AUC_{total} = \sum_{c_i \in C} AUC(c_i) \times p(c_i)$$

For a more rigorous method to derive multi-class AUC, the M-Measure proposed by Hand and Till [24] may be used. The ranking performance is shown in Table 4 and is generated by averaging AUC scores from 10 x 10-fold cross-validation runs.

A classifier that produces excellent results for ranking need not produce good estimates of class membership probabilities. For example, estimates of 0.49 and 0.51 respectively for two classes can produce the correct ranking even if the actual probabilities are 0.30 and 0.70. Since class membership probabilities are usually unknown, it is difficult to compare probability estimates of classifiers. One of the commonly used metrics is the Mean Square Error (MSE) also known as the Brier Score [25] given by:

Table 4. Area under ROC curve

Data set	Bagged C4.4	Boosted 4.4	Naïve Bayes	NB Tree	fw-MDC	MDC	C4.4	IBk	C4.5
ecoli	0.95 (0.00)	0.94 (0.00)	0.96 (0.00)	0.94 (0.00)	0.95 (0.00)	0.95 (0.00)	0.93 (0.01)	0.88 (0.00)	0.90 (0.01)
glass	1.00 (0.02)	1.00 (0.02)	0.91 (0.01)	1.00 (0.02)	0.90 (0.00)	0.92 (0.01)	0.98 (0.02)	0.96 (0.01)	0.95 (0.02)
iris	0.94 (0.01)	0.93 (0.01)	0.95 (0.00)	0.95 (0.01)	0.96 (0.00)	0.95 (0.00)	0.93 (0.00)	0.95 (0.00)	0.94 (0.00)
iono-sphere	0.97 (0.00)	0.95 (0.01)	0.93 (0.00)	0.92 (0.01)	0.86 (0.00)	0.86 (0.00)	0.94 (0.01)	0.97 (0.00)	0.90 (0.02)
pima	0.82 (0.01)	0.77 (0.02)	0.81 (0.00)	0.80 (0.01)	0.80 (0.00)	0.79 (0.00)	0.79 (0.01)	0.72 (0.01)	0.76 (0.01)
liver	0.74 (0.02)	0.72 (0.01)	0.63 (0.01)	0.62 (0.02)	0.61 (0.01)	0.63 (0.01)	0.67 (0.02)	0.65 (0.01)	0.66 (0.02)
wis-bc_d	0.98 (0.00)	0.98 (0.00)	0.98 (0.00)	0.95 (0.00)	0.99 (0.00)	0.98 (0.00)	0.96 (0.00)	0.95 (0.00)	0.93 (0.01)
wis-bc_p	0.69 (0.03)	0.69 (0.02)	0.65 (0.01)	0.61 (0.04)	0.73 (0.00)	0.67 (0.01)	0.60 (0.05)	0.55 (0.02)	0.56 (0.05)
wine	0.99 (0.00)	0.98 (0.01)	1.00 (0.00)	1.00 (0.00)	0.99 (0.00)	1.00 (0.00)	0.95 (0.01)	1.00 (0.00)	0.95 (0.01)
yeast	0.80 (0.00)	0.78 (0.00)	0.80 (0.00)	0.79 (0.00)	0.75 (0.00)	0.77 (0.00)	0.74 (0.01)	0.70 (0.00)	0.72 (0.01)
seg-ment	1.00 (0.00)	1.02 (0.00)	0.99 (0.00)	1.00 (0.00)	0.98 (0.00)	0.99 (0.00)	1.00 (0.00)	1.00 (0.00)	1.00 (0.00)
Average	0.90 (0.01)	0.89 (0.01)	0.87 (0.00)	0.87 (0.01)	0.87 (0.00)	0.86 (0.00)	0.86 (0.01)	0.85 (0.01)	0.84 (0.02)

$$MSE = \frac{1}{n}\sum_{1}^{n}\sum_{c_i \in C}(p(c_i|x) - \hat{p}(c_i|x))^2$$

where, n = number of instances, $p(c_i|x)$ = 1 if the nth instance has a class label of c_i, 0 otherwise and $\hat{p}(c_i|x)$ is the predicted membership probability for class c_i.

The probability estimates can be subjectively compared using the histogram of the estimated scores. Reliability diagrams [26] can be used in binary class problems. Table 5 gives the MSE scores and Figure 1 shows the histograms for Bagged C4.4 which had the best average performance in Table 4, and also for naïve Bayes, fwMDC and MDC for comparison.

5.1 Discussion

From the prediction accuracy measures, it can be seen that all the other classifiers outperform fwMDC. This is to be expected, given the unrealistic assumptions made by fwMDC about the independence of attributes and the distribution of class conditional densities. The AUC scores however show that fwMDC is outperformed only by more sophisticated meta and hybrid learners apart from Naïve Bayes. MSE scores are poor for fwMDC, however, the calculation of MSE is itself biased towards classifiers that produce unrealistic probability estimates of 0 and 1 due to assumptions made on actual probabilities. This is evident from the performance on the Iris dataset on which fwMDC outperforms every other classifier in predicting the class labels and still has the highest MSE. The histograms in Fig. 1 show that fwMDC produces probability estimates that have a better spread.

Despite these shortcomings, fwMDC has certain advantages, especially for vision systems.

1. Simplicity and Computational Efficiency: The learned model is simply a set of feature vectors representing the centroid of each class, whereas learners such as C4.4 produce a tree with 800 nodes on the yeast dataset. The classification is based on a distance measure, which is efficient to compute.
2. Intuitiveness: The distance measure can be seen as a measure of similarity. This property can be very useful in medical domains where the experts are familiar with 'case based reasoning' to explain the decisions.
3. No irrelevant features: No feature selection scheme is needed before the classification stage, as the effects of less relevant features are compensated for by their lower feature weights. Unlike the machine learning field, where correlated features are removed because they do not aid in classification and tend to produce complicated models, medical experts see the presence of one or more correlated features as adding further confidence to the decision. This feature is mimicked in the fwMDC classification scheme, which increases the confidence of the decision in the presence of more dominant features. Similarly, the presence of rare features, which are usually seen as irrelevant, are not ignored.
4. Usable confidence measures for multi-stage processing: As seen from the probability histograms (Fig 1), unlike most other learners, fwMDC histograms show better spread. This is useful in multi-stage processing, since confidence scores that are close to 0 or 1 tend to produce results that are equivalent to hard classification.

Table 5. Mean Squared Error

Dataset	Bagged C4.4	NB Tree	Boosted C4.4	C4.4	C4.5	Naïve Bayes	kNN	MDC	fW-MDC
ecoli	0.33	0.35	0.41	0.39	0.38	0.31	0.48	0.40	0.60
glass	0.51	0.54	0.58	0.69	0.70	0.89	0.70	0.69	0.74
iris	0.09	0.10	0.11	0.11	0.11	0.07	0.09	0.13	0.15
ionosphere	0.13	0.20	0.16	0.18	0.20	0.33	0.28	0.36	0.42
liver	0.40	0.47	0.56	0.49	0.54	0.52	0.74	0.49	0.50
pima	0.33	0.35	0.51	0.37	0.39	0.36	0.59	0.39	0.41
diag	0.09	0.13	0.10	0.13	0.14	0.14	0.11	0.15	0.20
prog	0.32	0.37	0.45	0.43	0.48	0.56	0.59	0.46	0.46
wine	0.13	0.10	0.15	0.17	0.18	0.09	0.14	0.23	0.36
yeast	0.65	0.58	0.82	0.86	0.74	0.59	0.97	0.72	0.79
segment	0.10	0.13	0.08	0.12	0.11	0.38	0.11	0.42	0.54
Average	0.28	0.30	0.36	0.36	0.36	0.39	0.44	0.40	0.47

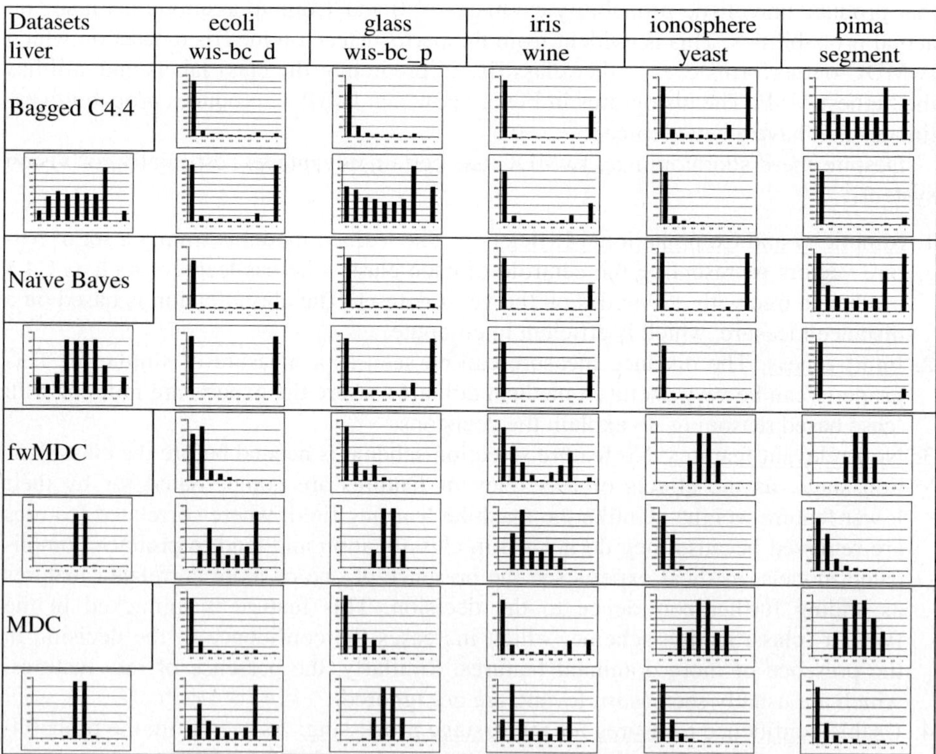

Fig. 1. Histogram of probability estimates

6 Application to Vision Systems

We applied the fwMDC on HRCT images to distinguish between five patterns namely Emphysema, Normal, Ground glass opacity (GGO), Honeycombing and Vasculature. We extracted 7 features - mean value of pixel intensity, and 6 levels of the wavelet transform in a 3 x 3 neighbourhood. The training set had 140,000 instances. We evaluated the effectiveness of fwMDC by observing the confidence assigned to pixels with known class label (Table 6). The confidence values reflect the similarity and overlap among patterns well known to radiologists. To visually assess the results, we assigned the label with maximum confidence to each pixel. Fig 2 shows each pixel classified into one of 5 classes and the results agree with expert marking. We ran 10x10 randomised cross-validation experiments on a subsampled (without replacement) dataset of 2500 instances (500 per class) derived from the above training set. The results are summarised in Table 7. Even though the prediction accuracy of MDC and fwMDC are poorer compared to that of Naïve Bayes, the AUC scores are very close. However, the average time taken to classify an instance is an order-of-magnitude better than the Naïve Bayes, and the histograms again show that the class probabilities are not clumped towards zero and one. These latter qualities indicate the suitability of minimum distance classifiers for multi-level vision tasks.

Table 6. Confidence values; Pixels of interest – central pixel in the 3x3 neighbourhood

Class	Pixel	Confidence Values (rounded)
GGO		Honeycombing = 0.18, Emphysema = 0.16, Normal = 0.18, GGO = 0.27, Vasculature = 0.20
Vascula-ture		Honeycombing = 0.21, Emphysema = 0.16, Normal = 0.17, GGO = 0.21, Vasculature = 0.24
Emphy-sema		Honeycombing = 0.20, Emphysema = 0.24, Normal = 0.21, GGO = 0.17, Vasculature = 0.18
Normal		Honeycombing = 0.14, Emphysema = 0.24, Normal = 0.37, GGO = 0.14, Vasculature = 0.11
Honey-combing		Honeycombing = 0.23, Emphysema = 0.18, Normal = 0.21, GGO = 0.17, Vasculature = 0.19

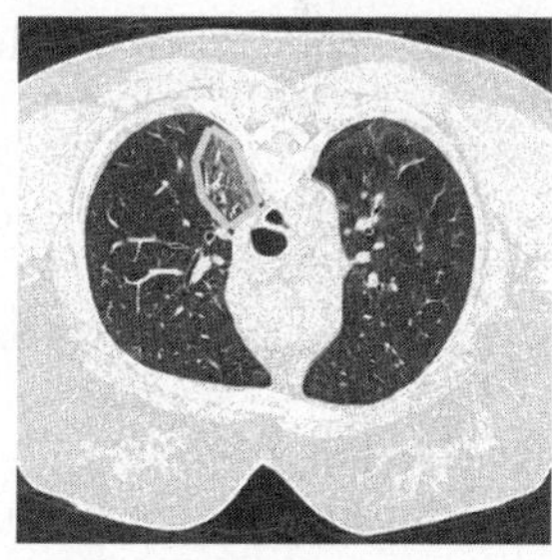

(a) Original Image (dominant abnormal region marked)

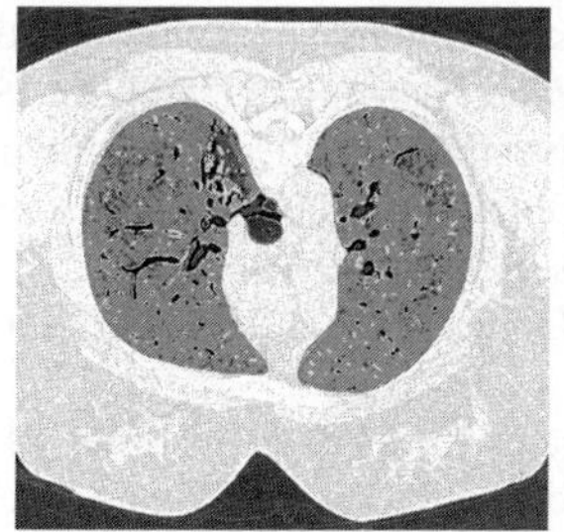

(b) Predicted Labels

Fig. 2. HRCT Data

Table 7. Performance on HRCT data

	Bagged C4.4	Naïve Bayes	fwMDC	MDC
Accuracy %	75.02(0.31)	71.64(0.11)	69.20(0.11)	68.96(0.16)
Time per Instance (ms)	0.11	0.14	0.02	0.03
AUC	0.93 (0.00)	0.91 (0.00)	0.90 (0.00)	0.90 (0.00)
MSE	0.33	0.44	0.47	0.64
Histogram				

7   Concluding Remarks

In this paper we have empirically evaluated the feature weighted minimum distance classifier on a number of measures, in order to benchmark its performance against some of the well-known classifiers. Experiments have shown that the fwMDC has acceptable performance and is suitable for computationally heavy multi-stage vision tasks due to its simplicity and intuitiveness. We have used fwMDC to classify lung pixels into 5 classes and the results are consistent with expert opinion.

Acknowledgement. This research was partially supported by the Australian Research Council through a Linkage grant (2002-2004), with I-Med Network Ltd. and Philips Medical Systems as partners. We thank Alena Shamsheyeva for providing wavelet features for HRCT data.

References

1. R. Duda, P. Hart, D. Stork, Pattern Classification, 2nd ed, *Wiley Interscience*, 2001
2. A.R. Dabney, Classification of microarrays to nearest centroids, *Bioinformatics*, 21, 4148-54, 2005
3. D. Wettschereck, D.W. Aha, T. Mohri, A Review and Empirical Evaluation of Feature Weighting Methods for a Class of Lazy Learning Algorithms, *Artificial Intelligence Review*, 11, 273-314, 1997
4. Domingos, Pazzani M, Beyond Independence: Conditions for the Optimality of the Simple Bayesian Classifier, *Machine Learning*, 29, 103-30, 1997
5. H. Zhang, J.Su, Naive Bayesian Classifiers for Ranking, ECML, *LNAI* 3201, 501-12, 2004
6. B. Zadrozny, C. Elkan, Obtaining calibrated probability estimates from decision trees and naive Bayesian classifiers, *ICML*, 609-16, 2001
7. Friedman N, Greiger D, Goldszmidt M, Bayesian Network Classifiers, *Machine Learning*, 29, 103-30, 1997
8. J.R. Quinlan, C4.5: programs for machine learning, *Machine Learning*, 1993
9. Provost F, Fawcett T, Kohavi R, The case against accuracy estimation for comparing induction algorithms, *ICML*, 445-53, 1998
10. Provost F and Domingos P, Tree induction for probability-based ranking, *Machine Learning*, 52, 199-215, 2003

11. Kohavi R, Scaling up the accuracy of naïve bayes classifiers: A decision-tree hybrid, *KDD*, 202-07, 1996

12. Margineantu, D. D. and Dietterich, T. G, Improved class probability estimates from decision tree models, *LNS,* 171, 169-84, 2002

13. C. Ling, J. Yan, "Decision tree with better ranking", *ICML*, 480-87, 2003

14. S. Delany, P. Cunningham, D. Doyle, A. Zamolotskikh, Generating Estimates of Classification Confidence for a Case-Based Spam Filter, *LNCS*, 3620, 177-90, 2000

15. Dietterich T. G, Ensemble Learning Methods, in *Handbook of Brain Theory and Neural Networks*, 2nd ed, MIT Press, 2001

16. B. Zadrozny, C.Elkan, Transforming Classifier Scores into Accurate Multiclass Probability Estimates, *KDD*, 259-68, 2002

17. A. Niculescu-Mizil, R. Caruana, Predicting Good Probabilities With Supervised Learning, *ICML*, 625-32, 2005

18. H. Lin, A. Venetsanopoulos, A Weighted Minimum Distance Classifier for Pattern Recognition, *Canadian Conference on Electrical and Computer Engineering*, 904-07, 14-17, 1993

19. Kononenko. I, Estimating attributes: analysis and extensions of Relief, *ECML*, 171–182, 1994

20. Newman D.J, Hettich S, Blake C.L, Merz C.J. UCI Repository of machine learning databases, University of California, Irvine, http://www.ics.uci.edu/~mlearn/MLRepository.html.

21. I.H. Witten, E.Frank, Data Mining: Practical machine learning tools and techniques, 2nd ed, 2005

22. Bradley A. P, The use of the area under the ROC curve in the evaluation of machine learning algorithms, *Pattern Recognition*, 30, 1145-59, 1997

23. P. Domingos and F. Provost, Well-trained PETs: Improving probability estimation trees, CDER *Working Paper #00-04-IS, Stern School of Business*, NYU, NY 10012, 2000

24. Hand D. J, Till R. J, A simple generalization of the area under the ROC curve to multiple class classification problems, *Machine Learning*, 45, 171-86, 2001

25. Brier G. W, Verification of forecasts expressed in terms of probability, *Monthly Weather Review*, 78, 1-3, 1950

26. DeGroot M, Fienberg S, The comparison and evaluation of forecasters, *Statistician*, 32, 12-22, 1982

z-SVM: An SVM for Improved Classification of Imbalanced Data

Tasadduq Imam, Kai Ming Ting, and Joarder Kamruzzaman

Gippsland School of Information Technology, Monash University, Australia
{tasadduq, kaiming, joarder}@infotech.monash.edu.au

Abstract. Recent literature has revealed that the decision boundary of a Support Vector Machine (SVM) classifier skews towards the minority class for imbalanced data, resulting in high misclassification rate for minority samples. In this paper, we present a novel strategy for SVM in class imbalanced scenario. In particular, we focus on orienting the trained decision boundary of SVM so that a good margin between the decision boundary and each of the classes is maintained, and also classification performance is improved for imbalanced data. In contrast to existing strategies that introduce additional parameters, the values of which are determined through empirical search involving multiple SVM training, our strategy corrects the skew of the learned SVM model automatically irrespective of the choice of learning parameters without multiple SVM training. We compare our strategy with SVM and SMOTE, a widely accepted strategy for imbalanced data, applied to SVM on five well known imbalanced datasets. Our strategy demonstrates improved classification performance for imbalanced data and is less sensitive to the selection of SVM learning parameters.

Keywords: class imbalance, support vector machine, SMOTE, z-SVM.

1   Introduction

Support Vector Machine (SVM) classifier [1,2] has found popularity in a wide range of classification tasks due to its improved performance in binary classification scenario [3,4,5,6]. Given a dataset, SVM aims at finding the discriminating hyperplane that maintains an optimal margin from the boundary examples called support vectors. An SVM, thus, focusses on improving generalization on training data. A number of recent works, however, have highlighted that the orientation of the decision boundary for an SVM trained with imbalanced data, is skewed towards the minority class, and as such, the prediction accuracy of minority class is low compared to that of the majority ones. Strategies like SVM ensemble trained at varying sampling rate [7,8], SVM with different cost [9] and SMOTE (Synthetic Minority Oversampling Technique) [10,11] have, therefore, been investigated to improve the minority classification accuracy for imbalanced data. A concern regarding the use of these strategies in practical applications is the necessity to pre-select a good value of the parameters that are introduced in

A. Sattar and B.H. Kang (Eds.): AI 2006, LNAI 4304, pp. 264–273, 2006.

© Springer-Verlag Berlin Heidelberg 2006

the standard SVM learning scheme. Due to the lack of defined guideline on how to select these parameters for imbalanced data, users are required to perform empirical search for the best value of the parameters through multiple training of SVM. Also, it is not much clear as to how a particular value of these parameters affect the SVM hyperplane and the generalization capability of the learned model.

We present in this paper, a novel strategy that addresses the aforementioned issue. In particular, we focus on post adjusting the learned decision boundary of an SVM model so that it maintains a good margin from the data of both the classes, and thereby, improves classification performance in imbalanced data domain. In contrast to the existing approaches, our strategy does not require any parameter pre-selection and auto-adjusts the decision hyperplane based on training data. The rest of this paper is organized as follows. In Section 2, we briefly focus on the theory of support vector learning scheme. Then in Section 3, we discuss the effect of class imbalance and existing techniques to address class imbalance for SVM. Section 4, introduces our proposed scheme followed by some experimental results and comparative discussion in Section 5.

2   Support Vector Machine (SVM)

SVM [1,12] is a discriminant based classifier that focusses on finding the optimal separating hyperplane between class samples by finding the solution for the following quadratic optimization problem:

$$min_{\mathbf{w},b} J(\mathbf{w}, b) = \frac{1}{2}\|\mathbf{w}\|^2 + C\sum \xi_i \tag{1}$$

$$s.t.$$

$$y_i(\mathbf{w}.\mathbf{x_i} + b) \geq 1 - \xi_i$$

$$\xi_i \geq 0$$

where, $(X = \{\mathbf{x_i}\}, Y = \{y_i\})$ denotes the set of feature vectors and set of class labels, respectively for the training dataset. $\mathbf{w}$ is the weight vector for learned decision hyperplane and b is the model bias. ξ_i are the slack variables that, in geometric perspective, indicates how far a particular instance is from its correct side of the decision boundary and is non-zero for examples violating the constraint $y_i(\mathbf{w}.\mathbf{x_i} + b) \geq 1$. The parameter C is a trade-off between maximization of margin and minimization of error, with higher value of C focussing more on minimizing error. With concern that data could be linearly inseparable, SVM exploits the use of kernel functions [1,12] to compute dot product in a mapped high dimensional space and the learning task comprises finding the solution for the following dual problem of (1):

$$max_\alpha L(\alpha) = \sum \alpha_i - \frac{1}{2}\sum \alpha_i\alpha_j y_i y_j K(\mathbf{x_i}, \mathbf{x_j}) \tag{2}$$

$$s.t.$$

$$\sum \alpha_i y_i = 0$$

$$0 \leq \alpha_i \leq C$$

where $K(.,.)$ is the kernel function, α_i are the Lagrangian multiplicative constants associated with each training data point. At the optimal point for (2), either $\alpha_i = 0$ or $0 < \alpha_i < C$ or $\alpha_i = C$. The input vectors for which $\alpha_i > 0$, are termed as support vectors. These are the only important information from the perspective of classification, as they define the decision boundary, while the rest of the inputs may be ignored. The optimal decision boundary is expressed as,

$$\mathbf{w} = \sum \alpha_i y_i \phi(\mathbf{x}_i) \tag{3}$$

where ϕ is a mapping function such that $K(\mathbf{x_i}, \mathbf{x_j}) = \phi(\mathbf{x}_i).\phi(\mathbf{x}_j)$.

For classification of a given test instance $\mathbf{x}$, SVM uses the following decision function:

$$f(\mathbf{x}) = \sum_{\mathbf{x_i} \varepsilon SV} \alpha_i y_i K(\mathbf{x}, \mathbf{x_i}) + b \tag{4}$$

where, SV is the set of support vectors. The value of $f(\mathbf{x})$ denotes the distance of the test instance from the discriminating hyperplane, while the sign indicates the class label (either positive or negative).

3   Class Imbalance and Support Vector Machine

Even though the problem of class imbalance has been well known to machine learning research community for a number of years, the study of its effect on SVM learning is relatively recent. It has been pointed out that heavily skewed class distribution causes the solution to (1) or (2), be dominated by the majority class [9,7,11]. We regard in the rest of the paper, the positive class to be the minority class for a binary classification task. As outlined in (1), SVM focuses on maximizing the margin between examples of opposite classes with a penalty for errors. For imbalanced training data, the penalty introduced in the objective function (1) for the relatively small number of positive samples is outweighed by that introduced by the large number of negative samples. As a consequence, the minimization problem in (1) focusses more on maximizing margin from the majority samples, resulting in a decision hyper-plane more skewed towards the minority class.

To cope with this skew, Veropoulos et. al. [9] suggested setting different penalties for misclassification of the classes. But there is no defined indication, as to what values should be preselected as the penalty parameters and the choice is totally empirical. The alternative to this strategy, as has been investigated in literature, is SMOTE (Synthetic Minority Oversampling Technique) [10,11]. The strategy comprises of oversampling the minority class by introducing artificial minority samples based on interpolation between a given minority sample and its nearest minority class neighbours. SMOTE has gained popularity in solving imbalance problem due to its performance. However, this strategy is also dependent on the proper determination of the user dependent parameter, which in this case is the percentage of oversampling. Also, SMOTE generates noisy artificial data causing high rate of false alarms [8]. Some of the other interesting approaches, that have been suggested, are alignment of kernel boundary by conformal transformation (KBA) [13] and SVM ensembles [7,8]. KBA requires several iterative retraining of an SVM and therefore is excessively expensive in terms of computation time. SVM ensemble combines the output of a set of SVM trained at different sampling ratios of positive and negative examples. But performance of this strategy is also dependent on sampling parameters.

4   Proposed Approach

We present in this section a strategy, that focuses on improving classification performance for imbalanced data by auto-adjusting the hyperplane and reducing skew towards the minority class.

4.1   Mathematical Formulation

The learned weight vector equation of (3) can be re-written as:

$$\mathbf{w} = \sum \alpha_p y_p \phi(\mathbf{x}_p) + \alpha_n y_n \phi(\mathbf{x}_n) \tag{5}$$

and the classification decision equation of (4) can be re-written as:

$$f(\mathbf{x}) = \sum_{\mathbf{x_p} \varepsilon SV; y_p > 0} \alpha_p y_p K(\mathbf{x}, \mathbf{x_p}) + \sum_{\mathbf{x_n} \varepsilon SV; y_n < 0} \alpha_n y_n K(\mathbf{x}, \mathbf{x_n}) + b \tag{6}$$

where, $\mathbf{x_p}$ and $\mathbf{x_n}$ are positive and negative support vectors in a learned SVM model. α_p, y_p and α_n, y_n are the associated Lagrange multipliers and corresponding labels respectively. As equation (6) illustrates, classification depends on a number of factors: the Lagrange multiplication constants associated with positive and negative support vectors, the kernel function and also the number of positive and negative support vectors. For imbalanced training data, these factors are more biased toward negative (majority) class, causing the tendency to classify an unknown instance as negative.

To reduce the bias of a trained SVM to majority class for imbalanced data, we introduce a multiplicative weight, z, associated with each of the positive class support vectors. We thus reformulate equation (6) as:

$$f(\mathbf{x}, z) = z \sum_{\mathbf{x_p} \varepsilon SV; y_p > 0} \alpha_p y_p K(\mathbf{x}, \mathbf{x_p}) + \sum_{\mathbf{x_n} \varepsilon SV; y_n < 0} \alpha_n y_n K(\mathbf{x}, \mathbf{x_n}) + b \tag{7}$$

This may be viewed as weighting the α_p-s such that minority classification is improved. Also, as equation (5) illustrates, the weight vector of a learned SVM model may be expressed as a linear combination of the positive and negative support vectors. Weighting the α_p-s shifts the weight vector from learned position to another position that is further away from the positive class, thereby reducing the skew towards minority class. Thus the introduced multiplicative weight z is, in effect, a correction of the originally learned boundary of SVM to cope with class imbalance. Technique to determine the appropriate value of z automatically, is presented in next section.

4.2   The Gmean Measure and Determination of z

We focus on setting z to the value that shifts the hyperplane to a position such that the geometric mean (gmean) of the accuracy of positive and negative samples, is maximized for the training data.

The gmean measure [14] is defined as: $gmean = \sqrt{acc_+ * acc_-}$, where acc_+ and acc_- are the classification rate for positive class (sensitivity) and negative class samples respectively. The gmean measure has already been used in a number of research works [11,13] as a performance measure for imbalanced data. Our use of gmean in this context, however, is focussed on shifting trained decision boundary to a well-balanced position to improve classification in imbalanced data scenario. As gmean penalizes heavily the misclassification of small class (eg.,failing to recognize a few minority examples reduces acc_+ drastically) , adjusting the SVM learning according to this measure, in effect, auto-incorporates the class imbalance information for training an SVM.

Returning to our discussion on setting a proper value of z, if we gradually increase the value of z from 0 to some positive value, M, in (7), and use the new model to classify data, the new model will classify everything as negative for $z = 0$ and classify everything as positive for $z = M$. Since gmean is a function of accuracy of both classes, it's value will increase from 0 (at some point $z = z^l$) to some maximum value (at $z = z^*$) and drop to 0 again (at $z = z^h$). The problem is an unconstrained optimization problem of single variable z and is stated in (8).

$$max_z J(z) = \sqrt{\frac{\sum_{\mathbf{x_u} \epsilon X; y_u > 0} I(y_u f(\mathbf{x_u}, z))}{P} \cdot \frac{\sum_{\mathbf{x_v} \epsilon X; y_v < 0} I(y_v f(\mathbf{x_v}, z))}{N}} \qquad (8)$$

where, (X, Y) denotes the set of of training vectors. P and N are the total number of positive and negative training vectors respectively. $I(y f(\mathbf{x}, z))$ is a function such that it has value 1 if $y f(\mathbf{x}, z) \geq 0$ and value 0, otherwise. $(\mathbf{x_u}, y_u)$ and $(\mathbf{x_v}, y_v)$ are the set of positive and negative training vectors respectively. The term $\sum_{\mathbf{x_u} \epsilon X; y_u > 0} I(y_u f(\mathbf{x_u}, z))$ calculates the number of positive class training samples correctly classified. Similarly the term $\sum_{\mathbf{x_v} \epsilon X; y_v < 0} I(y_v f(\mathbf{x_v}, z))$ indicates the same of negative class training samples. Overall, the function $J(z)$ formulates the gmean measure on the training set.

A univariate unconstrained optimization technique is used to determine the optimal $z = z^*$. For our strategy, we have applied Golden section search algorithm [15] for the optimization process. The derived z^* value is used to make classification decision according to (7). We call our approach z-SVM.

4.3 Runtime Complexity

The runtime complexity of our approach is computed as follows. The runtime complexity of an SVM algorithm depends on the the number of training points and the kernel computation involved in quadratic optimization. Hence for training set of size N, the complexity of SVM learning is $O(N^3)$, with some practical software approximating it close to $O(N^2)$ [16]. For the extra processing involved in our approach, the overhead is only due to the number of gmean evaluations. For each of these evaluations, the time complexity involved is $O(N^2)$ due to kernel evaluation for each of the data points. For the search strategy, the total number of gmean evaluations with respect to training size is $O(1)$. Hence the total runtime complexity of z-SVM, derived from an SVM trained using $O(N^2)$ complexity algorithm, is $O(N^2)$.

5 Experiments and Results

5.1 Experimental Settings

We investigated the effect of incorporating the z-SVM strategy on a learned SVM model using five well known UCI datasets, as presented in Table 1. The original dataset files have been processed as in [13]. A 10-fold cross validation was used for each of the datasets.

As performance measure, we have used gmean and sensitivity (minority class recognition accuracy). We have not considered Area under ROC (AUC) [17]. Our experiments show that for highly imbalanced datasets, even when the sensitivity is 0, AUC value is considerably high. For instance, an SVM trained on highly imbalanced dataset as abalone, yeast and car, results in AUC values 0.6765, 0.8573 and 0.9889 respectively, while the corresponding sensitivity values are 0. This is because AUC is more focussed on how a classifier relatively ranks the positive and negative class, rather than what label is predicted for a particular test data. Hence for deterministic classifier like SVM, AUC calculated using the decision score of SVM model [17] can be high even when the actual prediction performance is low, especially for minority class. As such we focussed on the use of gmean and sensitivity as performance measure.

5.2 Effect of z-SVM on Decision Boundary

To investigate the effect our algorithm on decision boundary, we trained SVM model for each of the five datasets.We then applied z-SVM on these models. Fig. 1 indicates average distance of the positive (filled up bars) and negative (blank bars) training samples from the learned boundary. It is evident that, for highly imbalanced data (abalone, car, yeast), the trained positive examples fall on average on the wrong side of the hyper-plane, implying that in test environment, that model is more likely to misclassify positive samples. Also, the average distance of decision hyperplane from positive examples is relatively much less than that of negative examples, implying the skew of hyperplane towards the

Table 1. Five UCI datasets with extreme imbalance to moderate imbalance

DATA SET	TOTAL	POSITIVE (%POS.)	NEGATIVE (%NEG.)
ABALONE	4177	32 (0.77%)	4145 (99.23%)
YEAST	1484	51 (3.44%)	1433 (96.56%)
CAR	1728	69 (3.99%)	1659 (96.01%)
EUTHYROID	2000	238 (11.90%)	1762 (88.10%)
SEGMENTATION	210	30 (14.29%)	180 (85.71%)

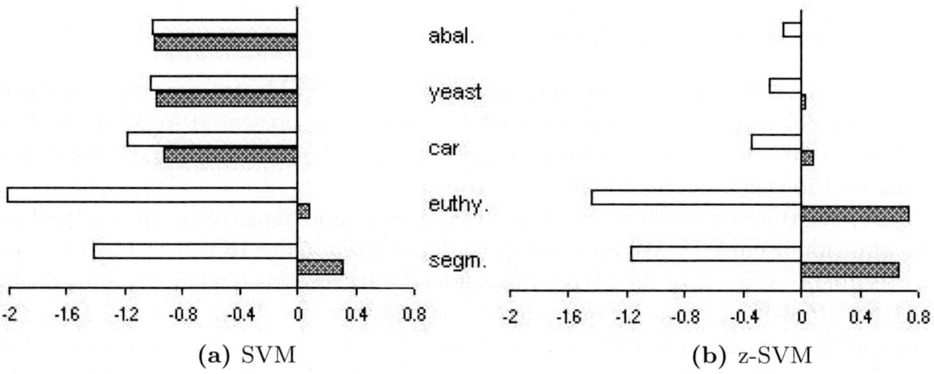

Fig. 1. Average distance of examples from the trained hyperplane for (a) SVM and (b) z-SVM derived from SVM. White bar and shaded bar indicate the distance for negative and positive classes respectively, with positive class being the minority.

positive (minority) class. The z-SVM algorithm attempts to reduce the skew in learned decision boundary by keeping a relatively balanced margin between the decision hyperplane and class points. Comparison of Fig. 1(a) and (b) illustrates that z-SVM has been successful in overcoming the error in training by shifting the decision boundary away from positive class and thus increasing the margin of hyperplane from positive class data. The better positioning of the hyperplane in z-SVM means a test sample is more likely to be classified correctly than that in SVM.

5.3   Performance of z-SVM

In Table 2, we present the effect of our strategy on the performance of SVM trained with with Radial Basis kernel and $\gamma = 1$ ($K(x_1, x_2) = exp^{(-\gamma||x_1 - x_2||^2)}$).

Table 2 indicates that z-SVM improves the prediction capability of a standard SVM in terms of gmean and sensitivity, indicating a good classification performance on imbalanced data. Incorporation of z-SVM allows the model to recognize minority samples, even when the original model fails to recognize any minority sample (as is the case for abalone, yeast and car dataset). A comparison with SMOTE employing 100% oversampling rate (Table 2) shows that z-SVM, derived from standard SVM, performs much better than SMOTE in terms of gmean and sensitivity when the dataset is highly imbalanced, while showing slight improvement or comparable results on moderately imbalanced data. The results imply that, for any degree of imbalance in dataset, z-SVM performs better or at least comparable to SMOTE.

Table 2. Performance comparison for different SVM based techniques with Radial Basis kernel and $\gamma=1$. Oversampling rate,$N=100\%$ for SMOTE.

DATA	SCHEME	GMEAN	SENS.
ABALONE	SVM	0.0000	0.0000
	SMOTE	0.0000	0.0000
	z-SVM	0.6202	0.6267
YEAST	SVM	0.0000	0.0000
	SMOTE	0.0000	0.0000
	z-SVM	0.7281	0.6667
CAR	SVM	0.0000	0.0000
	SMOTE	0.3486	0.1833
	z-SVM	0.9361	0.9167
EUTHYROID	SVM	0.7259	0.5417
	SMOTE	0.8930	0.8306
	z-SVM	0.9042	0.8686
SEGMENTATION	SVM	0.9266	0.8667
	SMOTE	0.9759	0.9667
	z-SVM	0.9759	0.9667

We also investigated the effect of varying oversampling rate ($N=100\%$ to 1000%) on SMOTE and z-SVM derived from SMOTE's model. Fig.2 presents the values of gmean and sensitivity, averaged over all five datasets, corresponding to each oversampling rate. Results show that z-SVM outperforms SMOTE over the whole range of oversampling rate.

We also investigated the effect of varying kernel parameters on the performance of SVM. Fig. 3 shows the performance of SVM with kernel parameters at $\gamma = 0.1$, 0.2, 0.4, 0.6, 0.8 and 1.0, averaged over all five datasets, along with z-SVM applied to SVM's models. As illustrated, while there is considerable variation in the performance of SVM, the performance of z-SVM is less drastic and remains better over the full range. This shows that z-SVM is relatively less sensitive to user specified kernel parameters.

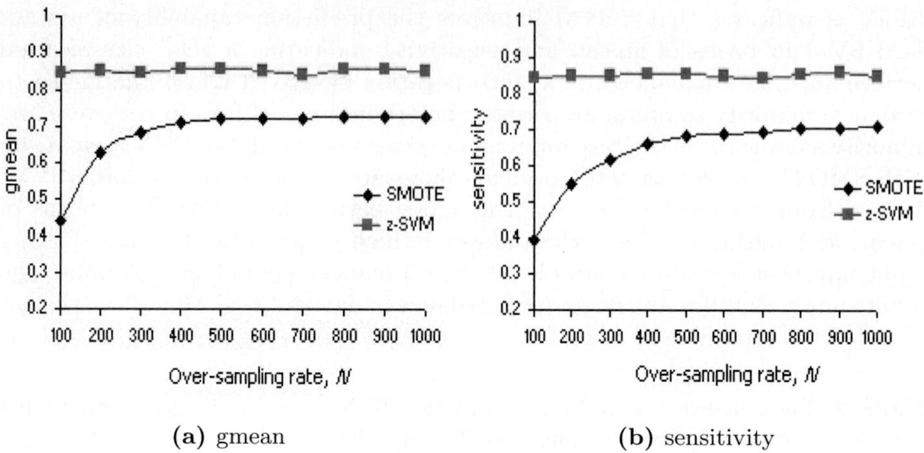

(a) gmean

(b) sensitivity

Fig. 2. Average (a) gmean and (b) sensitivity, for change of N for SMOTE and z-SVM over five UCI datasets

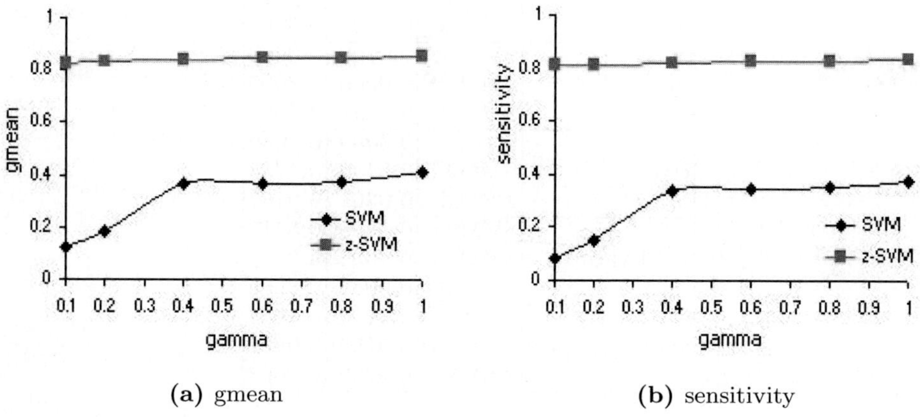

(a) gmean

(b) sensitivity

Fig. 3. Average (a) gmean and (b) sensitivity, for change of γ for SVM over five UCI datasets

6   Conclusion

In this paper we presented a new strategy to improve prediction accuracy of SVM for imbalanced data. The proposed strategy reduces the skewness of the decision boundary towards the minority class and automatically orients the boundary so that a good margin is maintained for each class, which yields better recognition performance for imbalanced data. In contrast to existing schemes like SMOTE which requires multiple training of SVM to select appropriate parameters for good performance, our scheme eliminates such need and is less sensitive to the selection of learning parameters, no matter the base models are derived from

SVM or SMOTE. Future research will be directed towards further alignment of SVM boundary considering individual support vector and spatial distribution of its neighborhood.

References

1. Vapnik, N.V.: The Nature of Statistical Learning Theory. Springer-Verlag, New York. (2000)
2. Schlkopf, B., Smola, A.J.: Learning with Kernels: Support Vector Machines, Regularization, Optimization, and Beyond. The MIT Press (2001)
3. Begg, R., Palaniswami, M., Owen, B.: Support vector machines for automated gait classification. IEEE Trans. Biomedical Engineering **52**(5) (2005) 828–838
4. Mukkamala, S., Janoski, G., Sung, A.: Intrusion detection using neural networks and support vector machines. In: International Joint Conference on Neural Networks. Volume 2. (2002) 1702–1707
5. Joachims, T.: Text categorization with support vector machines: learning with many relevant features. In: ECML. (1998) 137–142
6. Drucker, H., Wu, D., Vapnik, N.V.: Support vector machines for spam categorization. IEEE Trans. Neural Networks **10**(5) (1999) 1048–1054
7. Yan, R., Liu, Y., Jin, R., Hauptmann, A.: On predicting rare classes with svm ensembles in scene classification. In: Proc. 2003 IEEE International Conference on Acoustics, Speech, and Signal Processing (ICASSP '03). Volume 3. (2003) III–21–4
8. Liu, Y., An, A., Huang, X.: Boosting prediction accuracy on imbalanced datasets with svm ensembles. In: PAKDD. (2006) 107–118
9. Veropoulos, K., Campbell, C., Cristianini, N.: Controlling the sensitivity of support vector machines. In: International Joint Conference on Artificial Intelligence(IJCAI99). (1999) 55–60
10. Chawla, N.V., Bowyer, K.W., Hall, L.O., Kegelmeyer, W.P.: Smote: Synthetic minority over-sampling technique. J. Artif. Intell. Res. (JAIR) **16** (2002) 321–357
11. Akbani, R., Kwek, S., Japkowicz, N.: Applying support vector machines to imbalanced datasets. In: ECML. (2004) 39–50
12. Cristianini, N., Shawe-Taylor, J.: An Introduction to Support Vector Machines. Cambridge University Press (2000)
13. Wu, G., Chang, E.Y.: Kba: Kernel boundary alignment considering imbalanced data distribution. IEEE Trans. Knowl. Data Eng. **17**(6) (2005) 786–795
14. Kubat, M., Matwin, S.: Addressing the curse of imbalanced training sets: One-sided selection. In: ICML. (1997) 179–186
15. Gill, P.E., Murray, W., Wright, M.H.: Practical Optimization. Academic Press (1981)
16. Collobert, R., Bengio, S., Bengio, Y.: A parallel mixture of svms for very large scale problems. Neural Computation, **14**(5) (2002) 1105–1114
17. Fawcett, T.: Roc graphs: Notes and practical considerations for researchers (2004)

GP for Object Classification: Brood Size in Brood Recombination Crossover

Mengjie Zhang, Xiaoying Gao, and Weijun Lou

School of Mathematics, Statistics and Computer Science
Victoria University of Wellington, P.O. Box 600, Wellington, New Zealand
{mengjie, xgao, norman}@mcs.vuw.ac.nz

Abstract. The brood size plays an important role in the brood recombination crossover method in genetic programming. However, there has not been any thorough investigation on the brood size and the methods for setting this size have not been effectively examined. This paper investigates a number of new developments of brood size in the brood recombination crossover method in GP. We first investigate the effect of different fixed brood sizes, then construct three dynamic models for setting the brood size. These developments are examined and compared with the standard crossover operator on three object classification problems of increasing difficulty. The results suggest that the brood recombination methods with all the new developments outperforms the standard crossover operator for all the problems. As the brood size increases, the system effective performance can be improved. When it exceeds a certain point, however, the effective performance will not be improved and the system will become less efficient. Investigation of three dynamic models for the brood size reveals that a good variable brood size which is dynamically set with the number of generations can further improve the system performance over the fixed brood sizes.

1 Introduction

Classification tasks arise in a very wide range of applications, such as detecting faces from video images, recognising words in streams of speech, diagnosing medical conditions from the output of medical tests, and detecting fraudulent credit card fraud transactions [1,2,3,4]. In many cases, people (possibly highly trained experts) are able to perform the classification task well, but either there is a shortage of such experts, or the cost of people is too high. Given the amount of data that needs to be classified, automatic computer based classification programs/systems are of immense social and economic value.

A classification program must correctly map an input vector describing an instance (e.g. an object) to one of a small set of class labels. Writing classification programs that have sufficient accuracy and reliability is usually very difficult and often infeasible: human programmers often cannot identify all the subtle conditions needed to distinguish between all instances of different classes.

Derived from genetic algorithms and evolutionary programming [5,6], Genetic programming (GP) is a relatively recent and fast developing approach to automatic programming [7,8,9]. In GP, solutions to a problem can be represented

A. Sattar and B.H. Kang (Eds.): AI 2006, LNAI 4304, pp. 274–284, 2006.

© Springer-Verlag Berlin Heidelberg 2006

in different forms but are usually interpreted as computer programs. Darwinian principles of natural selection and recombination are used to evolve a population of programs towards an effective solution to specific problems. The flexibility and expressiveness of computer program representation, combined with the powerful capabilities of evolutionary search, make GP an exciting new method to solve a great variety of problems. A strength of this approach is that evolved programs can be much more flexible than the highly constrained, parameterised models used in other techniques such as neural networks and support vector machines.

Since the early 1990s, there has been a number of reports on applying GP techniques to a range of object recognition problems such as shape classification, face identification, and medical diagnosis [10,11,12,13,14,15,16,17]. While showing promise, current GP techniques are limited and frequently do not give satisfactory results on difficult classification tasks. While there are many problems in the current GP techniques that often lead to unacceptable programs in a reasonable time frame, one main problem is that the crossover operator is not sufficiently powerful to generate good solutions.

The crossover genetic operator has been considered a centre storm of GP [18]. In the current crossover operator, two sub-programs (crossover points) are randomly chosen from two parent programs, and two new programs are generated by simply swapping them. However, the totally random choice is clearly unable to guarantee the best choice. Furthermore, this often destroys the good "building blocks" (sub-programs which are good for that task) in evolved programs.

To improve the standard crossover operator and preserve potential good building blocks, Tackett [19] introduced the "brood recombination" method. In this method, a "brood" N was created for each crossover operation and the operation was repeated N times to produce $2N$ child programs, but only the best two children were kept and all others were killed. In this way, better child programs will survive and be put into the next generation while the population size will remain constant.

Clearly, the brood size plays an important role in the brood recombination crossover method in genetic programming. While a brood size of four is commonly used in the method as in [19], a larger size might result in better performance but if the size is too large, there might be some negative effects. This size might also be task dependent and related to the evolutionary process parameters such as the number of generations and program size. However, there has not been any clear investigation on effects of the brood size, and the methods for setting this size have not been effectively examined.

1.1   Goals

The goal of this paper is to further analyse the brood size for the brood recombination crossover method in terms of effect of a fixed brood size and a variable size. To do this, the brood recombination crossover with different brood sizes will be compared with the standard crossover operator in GP on three object classification problems of increasing difficulty to measure the effectiveness and efficiency. Specifically, we are interested in the following issues:

- whether increasing the brood size can result in better performance,
- whether there exists a diversity point for the brood size in the situation that a larger size does not improve the system performance, and
- whether a variable size is better than a fixed size, and
- to find a effective heuristic for setting the brood size.

1.2   Organisation

The remainder of this paper is organised as follows. Section 2 gives a brief overview of the brood recombination crossover method and describes the experiment configurations. Section 3 investigates the use and effect of fixed brood sizes. Section 4 describes the analysis of a variable brood size compared with the fixed size and the basic GP approach. Finally, we draws conclusions in section 5.

2   Brood Recombination and Experiment Configuration

2.1   Brood Recombination Overview

The main objective of the brood recombination crossover method is to reduce the destructive effect of crossover and preserve good potential building blocks. The rationale behind this method is that many animal species produce far more offspring than are expected to live. Although there are many different mechanisms, the excess offspring die. So should GP crossover. A simple illustration of the method is shown in figure 1.

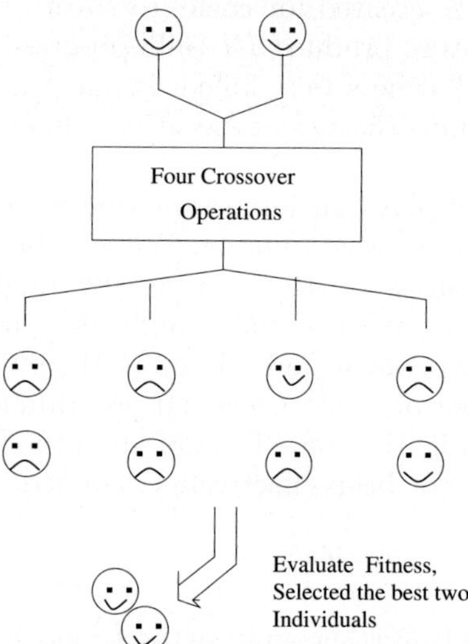

Fig. 1. The brood recombination crossover method

In this method, a "brood" N was created for each crossover operation. The standard crossover operation was repeated N times on the two same parent programs selected from the population and $2N$ child programs are generated. These child programs are then evaluated and their fitness ranked. The two programs with the best fitness are considered the "real" children of the parents and retained, but other children are discarded.

From the effectiveness point of view, this method improves the standard crossover from two-fold. Firstly, it reduces the effect of disrupting potential building blocks of the standard crossover operator through multiple trials of searching for good crossover points. Secondly, it actually adds a kind of hill-climbing search into the genetic beam search in GP.

Clearly, the brood size is a key parameter in this approach. However, Tackett [19] did not make further investigation and analysis on this parameter, which is the main focus of this paper.

2.2   Experiment Configuration

Image Data Sets. Experiments were conducted on three different image data sets providing object classification problems of increasing difficulty. Sample images for each data set are shown in figure 2.

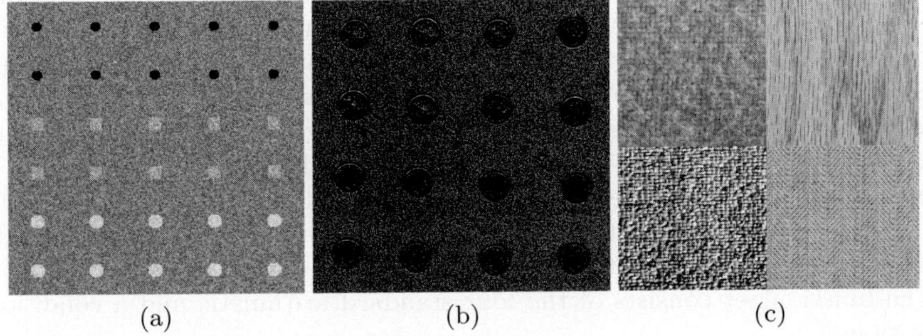

(a)   (b)   (c)

Fig. 2. Sample images in datasets: (a) Shape; (b) Coins; (c) Texture

The first data set (*shape*, figure 2a) was generated to give well defined objects against a reasonably noisy background. The pixels of the objects were produced using a Gaussian generator with different means and variances for each class. Four classes of 600 small objects (150 for each class) were cut out from the images and used to form the classification data set. The four classes are: *dark circles, grey squares, light circles* and *noisy background*.

The second set of images (*coin*, figure 2b) contains scanned 10 cent New Zealand coins. The coins were located in different places with different orientations and appeared in different sides (head and tail). The background was also cluttered. Three classes of 500 objects were cut out from the large images to form the data set. The three classes are: *head, tail* and *background*. Among the 500 cutouts, there are 160 cutouts for *head*, 160 cutouts for *tail* and 180 cutouts for *background* respectively. Compared with the *shape* data set, the classification problem in this data set is harder. Although these objects are still regular, the problem is quite hard due to the noisy background and the low resolution.

The third set of images (figure 2c) contains four different kinds of texture images, which are taken by a camera under the natural light. The images are taken from a web-based image database held by SIPI of USC [20]. The four texture classes are named *woollen cloth*, *wood grain*, *raffia* and *herringbone weave* respectively. Because they are quite similar in many aspects, this classification task is expected to be more difficult than that in the *coin* data set. There are 900 sample cutouts from four large images, and each class has 225 samples. This dataset is referred to as *texture*.

For all the three data sets, the objects were equally split into three separate data sets: one third for the training set used directly for learning the genetic program classifiers, one third for the validation set for controlling overfitting, and one third for the test set for measuring the performance of the learned program classifiers.

Terminal Set and Function Set. For image recognition tasks, terminals correspond to image features. In this approach, we use four simple pixel statistics extracted from each data set as terminals. Given an object cutout image, the four pixel statistics, *mean, standard deviation, skewness*, and *kurtosis*, are calculated as features. Since the ranges of these four feature terminal values are quite different, we scale them into the range [-1, 1] based on all object image examples to be classified. While these features are not the best for these particular problems, our goal is to investigate the brood size rather than finding good features for a particular task, which is beyond the scope of this paper.

In addition to the feature terminals, we also used a constant terminal for all the three tasks. To be consistent with the feature terminals, we also set the range of these constant terminals to [-1, 1].

The function set consists of the four standard arithmetic and a conditional operations:

$$FuncSet = \{+, -, *, /, if\} \tag{1}$$

The $+$, $-$, and $*$ operators have their usual meanings — addition, subtraction and multiplication, while $/$ represents "protected" division which is the usual division operator except that a divide by zero gives a result of zero. Each of these functions takes two arguments. The *if* function takes three arguments. The first argument, which can be any expression, constitutes the condition. If the first argument is negative, the *if* function returns its second argument; otherwise, it returns the third argument.

Fitness Function. We used classification accuracy on the training set of object images as the fitness function. The classification accuracy of a genetic program classifier refers to the number of object images that are correctly classified by the genetic program classifier as a proportion of the total number of object images in the training set. According to this design, the best fitness is 100%, meaning that all object images have been correctly recognised.

The output of a genetic program in the GP system is a floating point number. In this approach, we used a variant version of the *program classification map*

[17] to translate the single output value of a genetic program into a set of class labels.

Parameters and Termination Criteria. The parameter values used in this approach are shown in table 1. These values are determined based on empirical search. The evolutionary process is terminated when the number of generations reaches the pre-defined number, *max_generations*, or when the classification problem has been solved on the training set or the accuracy on the validation set starts falling down, in which case the evolution was terminated earlier.

Table 1. Main parameter values for the three data sets

Parameter	Shape	Coin	Texture	Parameter	Shape	Coin	Texture
Pop. Size	300	500	500	Crossover rate:	50%	50%	50%
Initial Max. Depth	5	5	5	Mutation rate:	30%	30%	30%
Max. Depth	5	6	8	Reproduction rate	20%	20%	20%
Max_generations	50	50	50	Brood size	see later		

In the approach, we used the tree-structure to represent genetic programs [8]. The ramped half-and-half method was used for generating programs in the initial population and for the mutation operator [7]. The proportional selection mechanism and the reproduction, crossover and mutation operators [9] were used in the learning and evolutionary process.

To compare the results with different brood sizes and the standard crossover operator, we use the classification accuracy, training time and the number of generations to measure the performances of these methods. For each experiment, we run 80 times and the average results are presented in the following sections.

3   Investigation of Fixed Brood Sizes

To investigate the effect of the brood size in the brood recombination crossover method, we did experiments on the three data sets using different fixed brood sizes with 2, 4, 6, 8, and 10 respectively. The average results on the *test set* of the GP system with these brood sizes together with the standard crossover operator (N = 1) are presented in table 2. The first line of the table shows that for the shape data set on the 80 runs, the GP system with the standard crossover operator (N = 1) used 8.59 generations and 0.09 second for the evolutionary process on average and achieved an average classification accuracy of 96.16% on the test set.

As shown in table 2, for all brood sizes investigated here, the brood recombination crossover method achieved better classification accuracy than the standard crossover operator for all the data sets. Although the number of generations used in the evolutionary process for the brood recombination method was smaller than the standard crossover operator, the actual training time was increased. This is mainly because the number of real evaluations in each generation in the brood

Table 2. Results of brood recombination crossover with fixed different brood sizes

Dataset	Brood size N	Generation	Time(s)	Accuracy(%)
Shape	1	8.59	0.09	96.16%
	2	5.26	0.10	98.25%
	3	3.48	0.10	98.25%
	6	3.01	0.11	98.12%
	8	2.66	0.12	98.44%
	10	2.30	0.13	98.03%
Coin	1	28.64	1.78	90.37%
	2	21.88	2.50	92.42%
	4	19.70	3.07	93.08%
	6	17.59	3.53	92.82%
	8	15.85	3.89	93.08%
	10	15.94	4.49	92.80%
Texture	1	29.99	1.83	72.45%
	2	26.01	3.31	76.68%
	4	21.82	4.23	76.46%
	6	23.69	6.09	79.82%
	8	20.00	6.12	80.71%
	10	17.80	6.50	78.13%

recombination method was increased. These results further confirmed Tackett's conclusions [19].

The results also show that different brood sizes resulted in different results. For the object classification problems investigated here, it seems that a brood size of 4–8 could be a good starting point.

Further Analysis. Further inspection of the results reveals that as the brood size increases at a certain number (4 or 8 for different data sets), the classification accuracy is increased. When the brood size exceeds this number, the accuracy achieved starts falling down. These results suggests that there exists such a brood size that could lead to the best performance for a particular task. We refer to this number as *the brood-diversity point*.

In the biological world, the chromosomes for a particular species are usually quite long and the crossover can occur in multiple genes in different positions. Accordingly, a huge number of crossover points can be provided, which allows a large size of brood to produce *distinguished* child chromosomes.

In most GP systems, however, the program size is limited to the parameter, *maximum program size*. In addition, the GP crossover only choose a single point and swap the subtrees in the parent programs. Accordingly, when the brood size increases to a certain number, the probability of the crossover operation on the same two parent programs to produce redundant programs will be extremely high. In other words, when the brood size exceeds the brood-diversity point, the brood recombination crossover operator will not only be unable to produce *distinguished* child programs, but also have to take longer time for more evaluations. This will result in a longer training time with non-improved even slightly worse performance in effectiveness due to possibility of pre-mature convergence.

Clearly, the brood-diversity point is related to the maximum program size. In general, the larger the program size, the bigger the brood size effectively allowed. When setting the brood size, one should consider the *threshold* number of crossover points that the two parent programs with the maximum program size can provide. We expect that this heuristic can help users to choose the brood size when using the brood recombination method in GP.

4   Investigation of Variable Brood Sizes

Since the genetic programs can have a different size and the program size varies with the number of generations in the evolutionary process, this section investigate a different approach from the last section — a variable brood size with respect to the number of generations.

According to the GP building block analysis [7], the GP crossover operator will generally preserve small building blocks and construct larger building blocks in the first stage of evolution, but is very likely to tear the larger building blocks apart in the later stage. Accordingly, we can allow the brood size to grow as evolution proceeds in order to protect the larger building blocks.

To meet these requirements, we proposed three ways to dynamically grow the brood size, as shown in equations 2, 3 and 4, respectively, where N is the brood size and gen is the number of generations.

$$N_1(gen) = round(0.14 \times gen) + 1 \tag{2}$$

$$N_2(gen) = round(0.0028 \times gen^2) + 1 \tag{3}$$

$$N_3(gen) = round(0.98 \times \sqrt{gen}) + 1 \tag{4}$$

The main consideration for the three formulas is as follows. In the previous experiment, the results suggest that the maximum brood-diversity point for the three data set is 8. Since the maximum number of generations in this approach is 50, we allow the brood size gradually increases to this number as evolution proceeds closely to generation 50. Notice that the growing speed of the three methods for the brood size is different.

To investigate whether dynamic variable brood size is better than the fixed size and the standard crossover approach, we examined them on the three data sets. The average results are shown in table 3.

According to tables 3, the classification accuracies achieved by brood recombination method with the three variable models for the brood size were better than the standard crossover. Inspection of table 2 reveals that these results were also better than that with almost all of the fixed brood sizes. In addition, the evolutionary training times for the variable brood sizes were also shorter than those for the fixed sizes. These results suggest that in the brood recombination method, a variable brood size which is dynamically set with the increase of the number of generations can further improve the system performance.

Table 3. Results of brood recombination with variable brood sizes and the basic approach. (A1: the basic approach with standard crossover; A2: the linear variable model — equation 2; A3: the squared variable model — equation 3; A4: the squared root model — equation 4).

Data set	Shape				Coin				Texture			
Approach	A1	A2	A3	A4	A1	A2	A3	A4	A1	A2	A3	A4
Generation	8.59	4.40	4.54	4.99	28.64	20.59	20.59	15.85	29.99	21.93	21.31	23.24
Time (s)	0.09	0.10	0.11	0.10	0.09	2.50	3.14	3.24	1.83	3.31	4.25	4.12
Accuracy(%)	96.16	98.62	99.03	98.63	90.37	93.08	93.54	93.21	72.45	80.05	81.34	80.16

Inspection of the three variable models for the brood size reveals that the squared variable model (equation 3) achieved the best results for all the three data sets. The other two models achieved similar results but it is not clear which one is better than the other on these data sets. This might be because the squared variable model is closer to the building block disruption trend in evolution, but further investigation is needed in the future.

5   Conclusions

The goal of this paper was to investigate the effect of brood size in the brood recombination crossover method in GP for object classification problems. The goal was successfully achieved by testing five different fixed brood sizes and designing three variable models for the brood size. The brood recombination methods with these new developments were examined and compared with the standard crossover operator on three object classification problems of increasing difficulty. The experiment results suggest that the brood recombination method with all these new developments achieved better classification performance than the standard crossover operator while the evolutionary training time was increased.

The results also suggest that different brood sizes usually result in different performances. As the brood size increases by the *brood-diversity point*, the system effective performance can be improved. When the brood size exceeds this point, however, the effective performance will not be improved and the system will become more inefficient. Our research suggests that the brood size is related to the number of generations, the program size and specific tasks.

Investigation of three dynamic models for setting the brood size reveals that a variable brood size which is dynamically set with the number of generations can further improve the system performance over the use of fixed brood sizes. In particular, the squared variable model achieved the best performance for the three data sets examined here.

Although developed for object classification problems, this approach could be expected to be applied to even more general problems.

This work reveals that the brood size is closely related to the maximum program size parameter. We will investigate the relationship between the brood size and the program size together with the number of generations in the future.

Acknowledgement

This work was supported in part by the national Marsden Fund at Royal Society of New Zealand (05-VUW-017) and the University Research Fund at Victoria University of Wellington.

References

1. Eggermont, J., Eiben, A.E., van Hemert, J.I.: A comparison of genetic programming variants for data classification. In: Proceedings of the Third Symposium on Intelligent Data Analysis, LNCS 1642, Springer-Verlag (1999)
2. Howard, D., Roberts, S.C., Ryan, C.: The boru data crawler for object detection tasks in machine vision. In: Applications of Evolutionary Computing, Proceedings of EvoWorkshops2002. Kinsale, Ireland, Springer (2002) 220–230
3. Valentin, D., Abdi, H., O'Toole: Categorization and identification of human face images by neural networks: A review of linear auto-associator and principal component approaches. Journal of Biological Systems **2**(3) (1994) 413–429
4. Poli, R.: Genetic programming for image analysis. In Koza, J.R., Goldberg, D.E., Fogel, D.B., Riolo, R.L., eds.: Genetic Programming 1996: Proceedings of the First Annual Conference, Stanford University, CA, USA, MIT Press (1996) 363–368
5. Holland, J.H.: Adaptation in Natural and Artificial Systems: An Introductory Analysis with Applications to Biology, Control, and Artificial Intelligence. Ann Arbor : University of Michigan Press; Cambridge, Mass. : MIT Press (1975)
6. Michalewicz, Z.: Genetic algorithms + data structures = evolution programs (3rd ed.). Springer-Verlag, London, UK (1996)
7. Banzhaf, W., Nordin, P., Keller, R.E., Francone, F.D.: Genetic Programming: An Introduction on the Automatic Evolution of computer programs and its Applications. Morgan Kaufmann Publishers (1998)
8. Koza, J.R.: Genetic programming : on the programming of computers by means of natural selection. Cambridge, Mass. : MIT Press, London, England (1992)
9. Koza, J.R.: Genetic Programming II: Automatic Discovery of Reusable Programs. Cambridge, Mass. : MIT Press, London, England (1994)
10. Howard, D., Roberts, S.C., Brankin, R.: Target detection in SAR imagery by genetic programming. Advances in Engineering Software **30** (1999) 303–311
11. Loveard, T., Ciesielski, V.: Representing classification problems in genetic programming. In: Proceedings of the Congress on Evolutionary Computation. Volume 2., Seoul, Korea, IEEE Press (2001) 1070–1077
12. Song, A., Ciesielski, V., Williams, H.: Texture classifiers generated by genetic programming. In: Proceedings of the 2002 Congress on Evolutionary Computation CEC2002, IEEE Press (2002) 243–248
13. Tackett, W.A.: Genetic programming for feature discovery and image discrimination. In: Proceedings of the 5th International Conference on Genetic Algorithms, Morgan Kaufmann (1993) 303–309
14. Winkeler, J.F., Manjunath, B.S.: Genetic programming for object detection. In: Proceedings of the Second Annual Conference, Morgan Kaufmann (1997) 330–335
15. Zhang, M., Ciesielski, V.: Genetic programming for multiple class object detection. In: Proceedings of the 12th Australian Joint Conference on Artificial Intelligence, Sydney, Australia, Springer(1999) 180–192

16. Zhang, M., Andreae, P., Pritchard, M.: Pixel statistics and false alarm area in genetic programming for object detection. In: Applications of Evolutionary Computing, Lecture Notes in Computer Science, Vol. 2611, Springer (2003) 455–466
17. Zhang, M., Ciesielski, V., Andreae, P.: A domain independent window-approach to multiclass object detection using genetic programming. EURASIP Journal on Signal Processing, **2003**(8) (2003) 841–859
18. Banzhaf, W., Nordin, P., Keller, R.E., Francone, F.D.: Genetic Programming - An Introduction. Morgan Kaufmann Publishers (1998)
19. Tackett, W.A.: Recombination, Selection and the Genetic Construction of Computer Programs. PhD thesis, University of Souithern California, Department of Electrical Engineering Systems (1994)
20. Webpage: `http://sipi.usc.edu/services/database/database.cgi?volume=textures`. (by Signal & Image Processing Institute of University of Southern California. *accessed on 22 July, 2004*)

IPSOM:
A Self-organizing Map Spatial Model of How Humans Complete Interlocking Puzzles

Spyridon Revithis, William H. Wilson, and Nadine Marcus

Artificial Intelligence Group, School of Computer Science and Engineering,
University of New South Wales, UNSW-Sydney, NSW 2052, Australia
`revithiss@cse.unsw.edu.au`, `billw@cse.unsw.edu.au`,
`nadinem@cse.unsw.edu.au`

Abstract. Cognitive modeling methodologies form the groundwork of significant studies in cognitive science. In this work a prototype computational model, called IPSOM, is introduced, which charts the spatial cognitive human behaviour in completing interlocking puzzles. IPSOM is a neural network of the class of self-organizing maps, and has been implemented using an artificial data set that consists of synthesized patterns of puzzle completion. The results show that the model is particularly successful in depicting valid cognitive behavioural patterns with a very high degree of confidence. Based on IPSOM's performance and structure, it is argued that a scaled-up version of this model could readily be used in representing real-life puzzle-completion patterns.

Keywords: Self-Organizing Map, Kohonen Map, Cognitive Modeling of Human Behaviour, Neural Network Applications.

1   Introduction

The domain of modelling cognitive behaviour encompasses a wide range of applications, and has significant implications in related fields of study including artificial intelligence, computer science, cognitive science, and computational neuroscience [1, 2, 3, 4]. Relative to previous research [5], the incentive for this work was to introduce a computational tool that could contribute towards cataloguing and classifying the vast repertoire of human cognitive behaviour in a particular family of activities. The focus here is on tasks, which contain a significant spatial-reasoning component; specifically, the completion of *interlocking puzzles* (Jigsaw puzzles).

A major challenge in modeling the way that humans complete interlocking puzzles is managing the large human behavioural space; in this case, the set of all possible ways, in which a human would complete a given puzzle. An important consideration here is finding a method that could process such an immense space at reasonable speed. Such a requirement can be met in using a class of artificial neural networks known as Kohonen maps, self-organizing feature maps, or, simply, *self-organizing maps* (SOM) [6, 7]. The latter biologically inspired computational method is capable of data compression, and although resists formal analysis it has been empirically shown to be reliable and efficient; Cottrell [8], and Haykin [9] offer a detailed account on SOM's technical aspects and features.

A. Sattar and B.H. Kang (Eds.): AI 2006, LNAI 4304, pp. 285–294, 2006.
© Springer-Verlag Berlin Heidelberg 2006

The proposed SOM-based model provides a way of revealing existing spatial behavioural patterns in completing an interlocking puzzle via unsupervised learning. In all but the trivial cases the behavioural space that characterizes certain groups of individuals for the given cognitive task is generally much less than the vast space of all possibilities (*n!* for an *n*-piece puzzle) but still prohibitively large. Effectively, it can be compressed via neural self-organization into an easily manageable set of behavioural patterns in the form of a cognitive map that represents statistically dominant cognitive strategies in dealing with the given task. It is expected that the resulting map would be a close approximation of the behavioural space (Fig. 1).

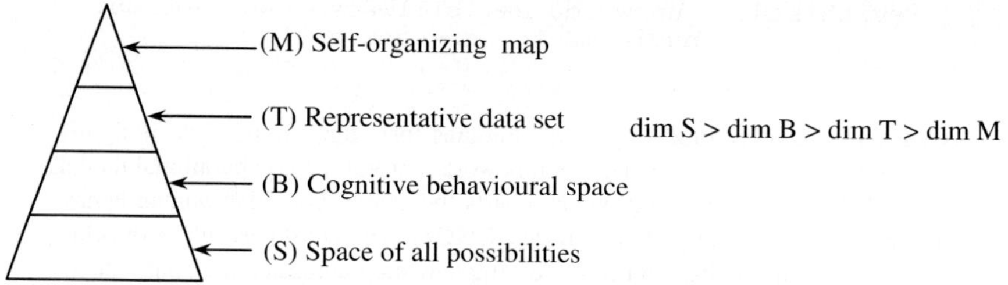

Fig. 1. Significant space reduction can be achieved via training a SOM network using a representative data set drawn from the behavioural space

A prototype model, called *IPSOM* (Interlocking Puzzle Self-Organizing Map), has been constructed and evaluated using synthetic data (i.e. artificial data). The latter was a critical step since synthetic data enabled the process of determining the model's effectiveness and measuring its efficiency.

1.1 Paper Overview

In sections 2 and 3 the architecture of IPSOM and the structure of the synthetic data set are presented, respectively. Section 4 discusses the implementation and mathematical aspects of the model. In section 5 the results obtained from the model are presented, and IPSOM's performance is analyzed. The concluding section 6 is devoted to certain considerations that need to be taken into account when exposing IPSOM to real-life data.

2 Structure and Configuration of IPSOM

IPSOM is configured to model the case of a person being presented with a 20-piece interlocking puzzle on a 4-by-5 puzzle board. Each location on the puzzle board is encoded with a unique *position value* (Fig. 2.a). The constant difference between consecutive position values is 0.050, which is the *position matrix resolution*. Given a certain value, the maximum tolerance in clearly determining a position on the puzzle board is less than 0.025, which is the *discrimination threshold*.

IPSOM is principally accommodated in a 6-by-6 lattice that hosts the resulting map of the model after the SOM training (Fig. 2.b). Conceptually, this map is a two-dimensional lattice that contains a number of valid cognitive behavioural *patterns*, each of which is represented as a 20-unit vector that holds position values. Each vector in the map potentially represents a statistically dominant way (i.e. ways that humans frequently use) of completing the given 20-piece puzzle, where each puzzle piece in the vector is denoted by its position value on the puzzle board. The order of the puzzle completion is the order that the puzzle pieces are arranged in the vector (e.g., the first piece placed on the puzzle board has its position value placed in the first slot of the vector.) The size of IPSOM's two-dimensional lattice is independent of the size of the position matrix (and, consequently, of the puzzle board), and depends on the number of cognitive behavioural patterns that IPSOM is expected to represent.

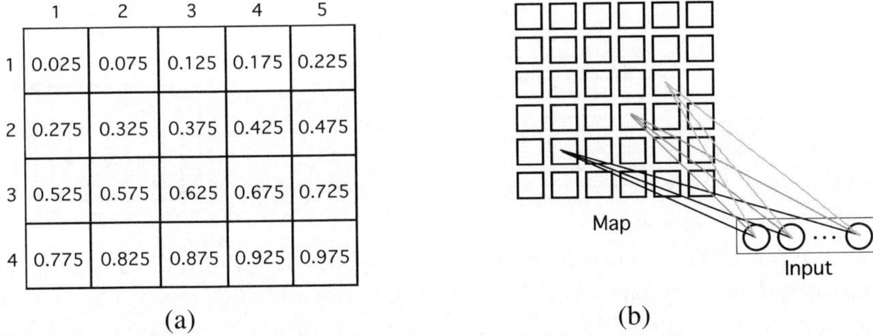

	1	2	3	4	5
1	0.025	0.075	0.125	0.175	0.225
2	0.275	0.325	0.375	0.425	0.475
3	0.525	0.575	0.625	0.675	0.725
4	0.775	0.825	0.875	0.925	0.975

(a)   (b)

Fig. 2. (a) The position matrix contains position values for all the puzzle pieces on the puzzle board. For example, the top left corner puzzle piece *(puzzle coordinates (1, 1))* has a position value of *0.025*, and the puzzle piece with position value *0.775* belongs to the bottom left corner of the puzzle *(puzzle coordinates (4, 1))*. **(b)** IPSOM's map structure is according to the Kohonen SOM model (note: the 20-unit *input* connects to all the neurons in the map).

3   The Synthetic Data Set

The IPSOM neural network was trained using a data set that consisted of synthetic data; four patterns, namely *H*, *V*, *PH*, and *PV*, were designed (Fig. 3) and presented to the network in a pseudo-random fashion repeatedly in order for the latter to converge to a stable state.

The design principle behind the selected patterns is based on simplicity, clustering, and periphery control. The working guidelines were to construct a small number of patterns that are straightforward, to utilize the concept of topological clustering (i.e. avoid continuous random jumping from one place to another when completing a puzzle), and occasionally to give priority to the basic puzzle-solving strategy of determining the puzzle board periphery during the puzzle completion task. The selected patterns were favoured over random patterns because they are easily recognizable, and relate to real-life strategies.

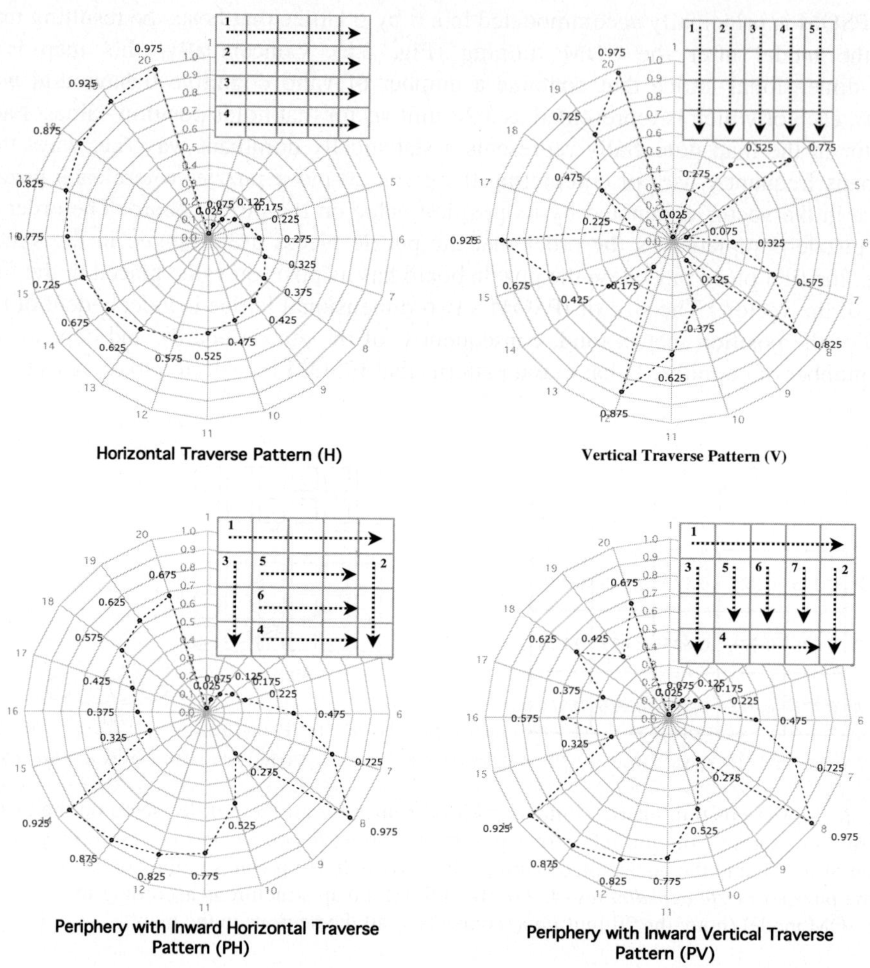

Horizontal Traverse Pattern (H)

Vertical Traverse Pattern (V)

Periphery with Inward Horizontal Traverse Pattern (PH)

Periphery with Inward Vertical Traverse Pattern (PV)

Fig. 3. An illustration of the data set synthetic patterns. For each pattern a radar-graph depicts the order of puzzle completion: the radial axis shows position value, and the angular axis shows the (discrete) filling sequence numbers. E.g., H vector pattern: [0.025, 0.075, 0125, 0175, 0225, 0275, ..., 0.975]. By connecting the points on the graph a distinct visual pattern is formed. Attached to each graph, a puzzle board depicts the puzzle completion in a conventional way.

4 Implementation

The IPSOM model has been implemented in ANSI C, and it is based on a static data structure framework that primarily consists of arrays and global variables. Specifically, the self-organizing map is a three-dimensional lattice that holds the synaptic weight vectors of the neural network. The first two dimensions represent the coordinates needed to refer to a specific neuron in the lattice. The third dimension is used for holding the synaptic weight vector of that neuron.

The data generator uses the position matrix as a base for creating instances (i.e. input vectors) of the training set during the SOM training. Each input vector spawned by the generator is a one-dimensional numerical array that contains position values in accordance with one of the synthetic training set patterns. The training set is balanced and exhibits no bias towards any of the patterns.

The mathematical and algorithmic form of the SOM neural network used is according to Haykin [9], and is outlined here for the sake of comprehensibility together with IPSOM network parameter customization. The IPSOM training process consists of four conceptual parts: lattice *initialization*, neuron *competition*, neuron *cooperation*, and synaptic *adaptation*. During initialization the position matrix is created, and the neuron weights within the lattice are set to a value that ensures a non-biased initial state for the SOM; this is achieved by using a pseudo-random number generator that spawns values within the interval of the position matrix value set.

Competition, cooperation, and synaptic adaptation are sequenced within a loop that constitutes the SOM main training loop. Competition among neurons within the network lattice occurs every time a new feature (i.e. input vector) is introduced to the neural network. The winning neuron i with lattice index $i(x)$ of each competition satisfies the Euclidean minimization equation

$$i(x) = \arg \min_{j} \| x - w_j \|, \quad j = 1, 2, \ldots, l \tag{1}$$

where x is the feature vector, w is the weight vector of the winning neuron, and l is the number of neurons in the lattice.

Each time a winning neuron has been determined the training goes through the cooperation part, in which the neighbourhood of that neuron is specified and all neurons within it are considered excited. The topological neighbourhood of the winning neuron i is specified by the translation invariant Gaussian function

$$h_{j,i(x)}(n) = \exp\left(-\frac{d^2_{j,i(x)}}{2 \cdot \sigma^2(n)}\right), \quad n = 0, 1, 2, \ldots, t \tag{2}$$

where j denotes an excited neuron, and n refers to the current iteration of the main training loop. The value t was set to 250000, which –due to the training set size– in effect translates to 62500 epochs. The neighbourhood function h is based on the lateral distance d between the winning neuron i and an excited neuron j, which is expressed by the equation

$$d^2_{j,i} = \| r_j - r_i \|^2 \tag{3}$$

where vector r defines the position of j and i neuron, respectively.

The width σ of the neighbourhood function h is a temporal function with exponential decay, which is specified by the equation

$$\sigma(n) = \sigma_0 \cdot \exp\left(-\frac{n}{\tau_1}\right), \quad n = 0, 1, 2, \ldots, t \tag{4}$$

where σ_0 is the initial width of h, and τ_1 is a temporal constant. The temporal constant was set according to the formula

$$\tau_1 = \frac{t}{\log \sigma_0} \; . \tag{5}$$

The value of σ_0 was initially set to be equal to the radius of the lattice but the results were not satisfactory; it was finally set to 1.2.

In the last part of the main training loop the network goes through a synaptic adaptation process, in which the synaptic weights of the neurons within the neighbourhood of the winning neuron are updated in relation to the feature vector. The updated weight vector $w\,(n + 1)$ of each excited neuron j is computed by the equation

$$w_j(n+1) = w_j(n) + \eta(n) \cdot h_{j,i(x)}(n) \cdot (x - w_j(n)) \tag{6}$$

where x is the feature vector, and $w(n)$ denotes the old weight vector. IPSOM uses stochastic approximation, which is realized by the time-varying learning rate η with exponential decay as depicted in the equation

$$\eta(n) = \eta_o \cdot \exp(-\frac{n}{\tau_2}), \quad n = 0, 1, 2, \ldots, t \tag{7}$$

where η_0 denotes the initial learning rate, and τ_2 is a temporal constant. The initial learning rate was set to 1 since smaller values rendered the network unable to reach a small approximation error.

The main training loop is executed until the network converges to a stable state at a rate largely dictated by the parameter values presented in this section. After the SOM training has been completed a numerical map is produced that presents the synaptic weights of the whole network in the topology that the neurons exhibit within the network lattice.

5   Discussion and Evaluation

IPSOM's resulting numerical map (Fig. 4.a) hosts 36 topologically ordered neurons, each of which holds a synaptic weight vector. The map successfully represents all four patterns of the training set with variable level of accuracy. Optimal representation of a pattern resides in a *primary core neuron*, and to a slightly lesser degree in a *core neuron*. Each vector in the map was visualized and matched against the training data set visual patterns; the best matching pattern was used as a reference in order to calculate the *mean absolute error* (MAE), based on the L1-Norm, of the current vector as follows

$$\mathrm{MAE} = \left(\sum_{k=1}^{c} |v_{jk} - p_{ik}| \right) \div d, \quad d = \dim w = \dim p \tag{8}$$

where the j_{th} vector v is matched against the i_{th} pattern p, and d denotes the dimensionality of the vector and, thus of the pattern (since the input vectors have the same structure and size with the synaptic weight vectors). The MAE measure shows the deviation of the neuron from its reference pattern, and indicates the strength of the representation (Fig 4.b). It is a measure that produces values compatible with the position value range. *Standard neurons* achieve an effective representation of a pattern although their MAE is significantly higher than a primary core neuron's. *Weak neurons* have approximately a ten times higher MAE than standard neurons but still match a data set synthetic pattern visually. As MAE approaches 0.1, which is four times the value of the discrimination threshold, the neuron's representational acuity degrades to one of an *undecided neuron.*

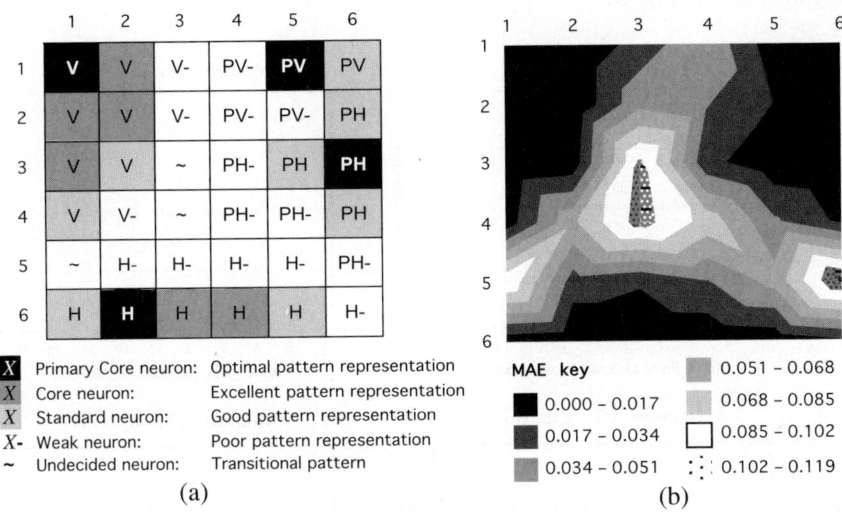

Fig. 4. (a) IPSOM model's resulting map using visual-numerical pattern matching. Each neuron in the map has been visually matched with one of the data set patterns (*X*: V, PV, H, or PH; as in Fig. 3), and a numerical follow-up determined its representational strength. (b) A graphical illustration of the map using the representational mean absolute error of each neuron.

The map produced by the IPSOM neural network using the synthetic data set exhibits all the properties of self-organizing maps [9]. Specifically, the map's set of synaptic weights provides an excellent representation of the input space since it contains the complete set of input patterns from the training set. It also exhibits a valid topological ordering of the synthetic patterns with well-defined regions for each pattern in the map. In addition, the map reflects the statistical distribution idiosyncrasies (density) on the input space since all four patterns are equally well represented. Finally, the map satisfies the feature selection property since it holds the complete set of features that are needed in order to approximate the input space.

It might be thought that the trained SOM does not satisfy the density matching property, based on the observation that although the PV synthetic pattern was emphasized during training just as much as H, V, and PH it occupies a relatively small area compared to the rest of the patterns in the map (Fig. 4.a). A closer look on the PV visual pattern (Fig. 3) reveals that PV is a pattern that, unlike the rest, assumes a noncurved shape in the last 5 positions. Furthermore, and most importantly, it differs from its sister pattern PH in only four positions (16 through 19), and substantially in only three (16, 17, 19). Based solely on visual matching the neurons with coordinates (3, 4) and (2, 6) match the PV pattern more than the PH one; however, at these locations the neuron's MAE is lower when representing PH than representing PV, thus the latter two neurons were assigned to the PH pattern instead (Fig 4.a). As a result, PV seems to be slightly underrepresented in the map.

A quantitative analysis of the primary core neurons reveals the high level of confidence with which the map depicts the synthetic patterns, and consequently illustrates the validity of the initial assumption regarding the function, effectiveness and efficiency of IPSOM (Fig. 5).

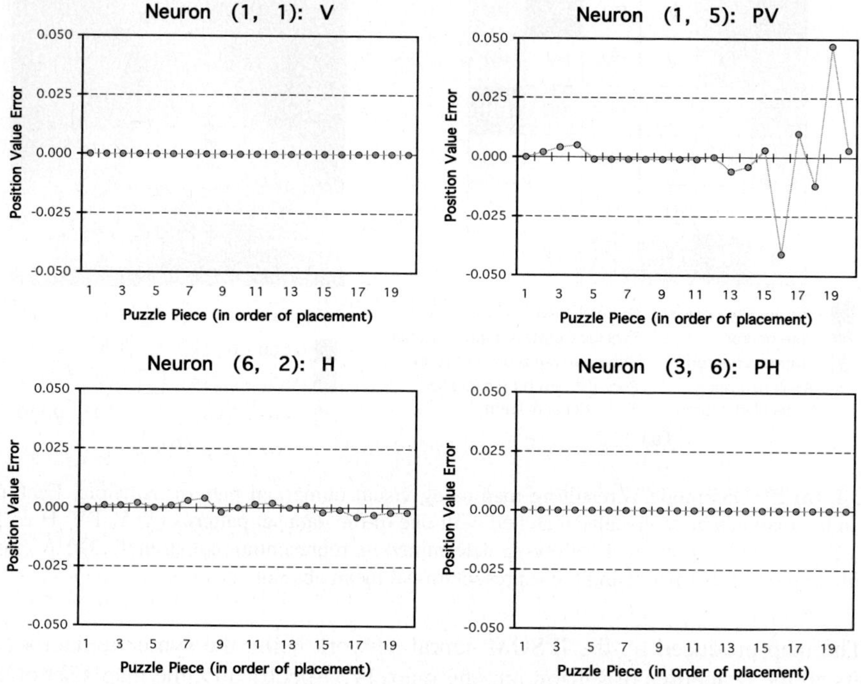

Fig. 5. The relative position value error in representing the four patterns of the training set at the primary core neurons

Specifically, the mean absolute error is very well suppressed and reaches only 28% of the discrimination threshold. Pattern V, and pattern PH are represented with 0 MAE. Pattern PV is represented with 0.007 MAE, whereas two position values in the synaptic weight vector of the neuron exceed the discrimination threshold but are located well within the position matrix resolution zone. Finally, pattern H is represented with 0.002 MAE.

6 Concluding Remarks

For the case of synthetic data IPSOM demonstrates high performance, and ease of use in determining the pre-designed patterns of the training set. Extending the model to handle real data poses certain considerations towards ensuring equally reliable results.

Unlike synthetic data sets, empirical training sets, which consist of real data, contain patterns that reflect the numerous subtle variations of human behaviour. It is expected that the size of the SOM lattice will, thus, need to be expanded to accommodate all statistically dominant patterns and create enough room for smooth transitions between primary core nodes; the presence of undecided neurons that host transitional patterns needs to be facilitated. Accordingly, the training algorithm could be adjusted to use a slightly larger initial width of the topological neighbourhood function in order to encourage the formation of more transitional neurons in the resulting map (Fig. 6).

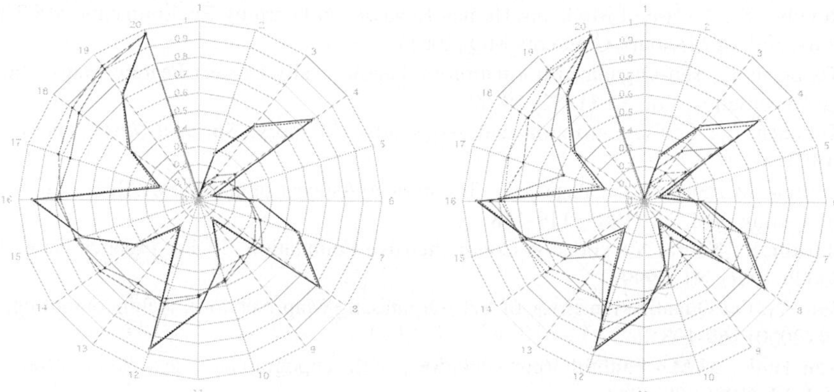

Fig. 6. Combined illustration of map's first-row neurons after two separate SOM training sessions of IPSOM for different initial neighbourhood widths (all other network parameters identical). A larger initial neighbourhood width, $\sigma_0=1.8$ (right diagram), facilitates the representation of transition between patterns unlike a smaller one, $\sigma_0=1.0$ (left diagram), where the neurons are tightly grouped into two patterns with no transitional members.

The idea is that transitional neurons can be used both as a guide to determine valid patterns of the input space, and to exclude outliers. However, the use of a very large topological neighbourhood is not prudent since it can severely degrade the approximation of the input space [10]. Similarly, setting the initial neighbourhood width to be equal to the radius of the lattice [9] did not yield satisfactory results during the early stages of IPSOM training.

Furthermore, the map's representational sensitivity must be maintained to an optimal level. Although it was not observed with IPSOM, it has been established that standard SOM formation can suffer from degraded density matching [9, 11], thus SOM alternatives or modifications could be considered in order to ensure a statistically faithful distribution of patterns without bias as in the case of the SOM with conscience [12]. Finally, alternative SOM algorithms with dynamic map dimensionality as in Marsland et al. [13] can be used to accommodate input spaces that are characterized by a widely varying number of patterns; fast forming SOMs [14] could be employed to tackle exceedingly large data sets; and more careful weight initialization [15] might be considered to improve performance even further.

References

1. Sun, R., Coward, L. A., Zenzen, M. J.: On Levels of Cognitive Modeling. Philosophical Psychology, Vol. 18. (2005) 613-637
2. Sun, R., Ling, C.: Computational Cognitive Modeling, the Source of Power and Other Related Issues. AI Magazine, Vol. 19. (1997) 113-120
3. Shultz, T. R.: Computational Developmental Psychology. The MIT Press, Cambridge, Massachusetts (2003)
4. Polk T. A., Seifert, C. M. (ed.): Cognitive Modeling. The MIT Press, Cambridge, Massachusetts (2002)
5. Revithis, S.: A Case of Modeling Human Behavior in Learning Environments. MS Thesis. University of Missouri, Columbia MO (2001)
6. Kohonen, T.: Self-Organized Formation of Topologically Correct Feature Maps. Biological Cybernetics, Vol. 53. (1982) 59-69
7. Kohonen, T.: Self-Organization and Associative Memory. Springer-Verlag, New York (1984)
8. Cottrell, M., Fort, J. C., Pagès, G.: Theoretical Aspects of the SOM Algorithm. Neurocomputing, Vol. 21. (1998) 119-138
9. Haykin, S.: Neural Networks: A Comprehensive Foundation. 2nd Ed. Prentice Hall, Upper Saddle River NJ (1999)
10. Sun, Y.: On Quantization Error of Self-Organizing Map Network. Neurocomputing, Vol. 34 (2000) 169-193
11. Van Hulle, M.M.: Faithful Representations with Topographic Maps. Neural Networks, Vol. 12. (1999) 803-823
12. DeSieno, D.: Adding a conscience to competitive learning. IEEE International Conference, Neural Networks, Vol. 1. San Diego CA (1988) 117-124
13. Marsland, S., Shapiro, J., Nehmzow, U.: A Self-Organising Network that Grows when Required. Neural Networks, Vol. 15. (2002) 1041-1058
14. Su, M., Chang, H.: Fast Self-Organizing Feature Map Algorithm. IEEE Transactions on Neural Networks, Vol. 11. (2000) 721-733
15. Han, L.: Initial Weight Selection Methods for Self-Organizing Training. Proc. IEEE International Conference on Intelligent Processing Systems, Vol. 1, (1997) 404-406

Mining Generalised Emerging Patterns

Xiaoyuan Qian, James Bailey, and Christopher Leckie

Department of Computer Science and Software Engineering
University of Melbourne, Australia
{jbailey, caleckie}@csse.unimelb.edu.au

Abstract. Emerging Patterns (EPs) are a data mining model that is useful as a means of discovering distinctions inherently present amongst a collection of datasets. However, current EP mining algorithms do not handle attributes whose values are asscociated with taxonomies (*is-a* hierarchies). Current EP mining techniques are restricted to using only the leaf-level attribute-values in a taxonomy. In this paper, we formally introduce the problem of mining generalised emerging patterns. Given a large data set, where some attributes are hierarchical, we find emerging patterns that consist of items at any level of the taxonomies. Generalised EPs are more concise and interpretable when used to describe some distinctive characteristics of a class of data. They are also considered to be more expressive because they include items at higher levels of the hierarchies, which have larger supports than items at the leaf level. We formulate the problem of mining generalised EPs, and present an algorithm for this task. We demonstrate that the discovered generalised patterns, which contain items at higher levels in the hierarchies, have greater support than traditional leaf-level EPs according to our experimental results based on ten benchmark datasets.

1   Introduction

An important problem in data mining is how to characterise the differences between two data sets. A widely used approach to this problem is to find *emerging patterns* (EPs), which can be used to describe significant changes between two data sets [4]. EPs are conjunctions of simple conditions representing a particular class of records. Emerging patterns have strong discriminating power and are thus very useful for describing the contrasts that exist between two classes of data. A key advantage of emerging patterns is that since they are basically conjunctions of simple conditions, they are very easy to understand.

In many application domains, taxonomies (is-a hierarchies) are available for attribute values. For example, an attribute corresponding to a geographical location can be represented at many levels of detail, such as city, state or country. Until now, emerging patterns have only been capable of representing contrasts between datasets whose attributes are non-hierarchical. The patterns discovered have been at the lowest level of representation. This can lead to a large number of similar patterns being discovered, involving attribute values that are distinct, yet semantically related, e.g., cities that are all part of the same state. This

A. Sattar and B.H. Kang (Eds.): AI 2006, LNAI 4304, pp. 295–304, 2006.

© Springer-Verlag Berlin Heidelberg 2006

creates two problems. First the support of each such pattern in isolation will be less than the support of the group of related patterns. Second, a large group of similar patterns is more difficult for an expert to understand and verify, compared to a single, more general pattern. Therefore, our motivation has been to develop an algorithm to mine *generalised* emerging patterns, which has the capability of dealing with data sets whose attribute values are associated with *is-a* taxonomies, such that the contrasts discovered will contain information not only from the lowest-level but also from any other levels in the hierarchy.

In this paper, we introduce a new knowledge pattern, called *generalised emerging patterns*. With the introduction of the concept of generalised EPs, we now allow a much larger and semantically extended pattern base by associating membership hierarchies with some attributes in a dataset. Quantitative data such as age or income can have meaningful attribute hierarchies achieved by aggregating base values into increasing, non-overlapping ranges. For example, dates of birth can be generalised to month and year of birth.

A critical challenge in this problem is that attribute hierarchies greatly expand the size of the space of potential EPs that could be mined. Patterns can now express a mixed level of concept aggregation realized by using combinations of values across different levels in the hierarchy. These patterns have improved expressiveness, potential usefulness and understandability for decision makers over traditional EPs. The resultant multi-dimensional generalised EPs are more concise and interpretable. If we are given a large number of leaf-level EPs, we might prefer having fewer, but more expressive EPs that have stronger discriminating power when used to describe distinctive characteristics of a class of data objects. However, the pattern space is greatly expanded after including attribute values from higher levels in the hierarchy. This makes the efficient computation of generalised EPs an important research challenge.

There are three main contributions of this research. First, we introduce the concept of generalised emerging patterns and formulate the problem of mining generalised emerging patterns. Second, we develop an algorithm for mining these patterns, and analyse the behaviour of the mining method over a variety of real-life datasets. Third, we analyse the compactness and expressiveness of the discovered generalised EPs in comparison with traditional non-hierarchical EPs.

2   Background and Related Work

In this section we present basic definitions which are used throughout this paper. All definitions in this section are adapted from [7]. A dataset is a collection of data objects of the form $(a_1, a_2, ..., a_n)$ following the schema $(A_1, A_2, ..., A_n)$ where each of the objects is called an **instance** and $A_1, A_2, ..., A_n$ are called **attributes**. Each data object in the dataset is also labelled by a class label $C \in \{C_1, C_2, ..., C_k\}$ which indicates the class to which the data object belongs.

An item is a pair of the form (attribute-name, attribute-value). A set of items is called an **itemset** (or a **pattern**). We say any instance S contains an itemset

X, if $X \subseteq S$. The **support** of an itemset X in a dataset D, denoted as $sup_D(X)$, is $Count_D(X)/|D|$, where $Count_D(X)$ is the number of instances in D containing X, and $|D|$ is the total number of instances in D.

We follow the definition of Emerging Patterns used in [4],[3],[2],[6], also known as Jumping Emerging Patterns. An EP is an itemset whose support increases abruptly from zero in one data set (known as the negative data set), to non-zero in another data set (known as the positive data set) - the ratio of support-increase being infinite. An itemset X is said to be an **Emerging Pattern (EP)** from D_1 to D_2 if $sup_{D_1}(X) = 0$ and $sup_{D_2}(X) > 0$. An EP X is said to be **minimal** if there does not exist another EP Y such that $Y \subset X$.

A related field is the problem of mining association rules. The aim of mining association rules is to find all rules that satisfy a user-specified minimum *support* and minimum *confidence* [9],[11],[12]. The concept of *generalised* association rules first appeared in [10]. The problem of mining *generalised* association rules was defined informally as – given a set of transactions and a taxonomy, find association rules where the items may be from **any level** of the taxonomy. The approach taken by [10] was to first add all ancestors of each item in a trans-action T, so that an extended transaction T' is obtained. After extending all transactions, any non-generalised association rule mining algorithm could then be applied to the extended transactions. Work presented in [5] further improved efficiency by imposing a lexicographic order on the itemsets such that rightmost Depth-First-Search could be used. In more recent work [8], "more-general-than" hierarchies were associated with *each* attribute in a large dataset. This is more closely related to our research, since in the context of mining emerging patterns, attributes are often multivalued and are independent from any other attributes in the data set. Their algorithm, **GenTree** [8], uses a tree structure which rep-resents the multi-dimensional generalisation relations among all data tuples in a relational dataset over a set of hierarchical attributes.

Current techniques of mining EPs do *not* handle hierarchical attributes. More-over, the techniques of dealing with taxonomy structures in the context of as-sociation rule mining cannot be applied directly to EP mining. A key difference is that while association rule learning requires search for high support rules in a single data set, EPs must satisfy the additional constraint of low support in the negative data set. This additional constraint creates the need for an efficient algorithm to prune candidate generalisations that match the negative data set. Therefore, the aim of our research is to address the problem of mining emerg-ing patterns for data sets whose attributes are associated with hierarchies. This problem is formally described in the next section.

3   Generalised Emerging Patterns

The type of datasets we consider have a set of attributes, one or more of which is associated with a taxonomy which represents *is-a* relations on its attribute-values. Figure 1 shows an example of two such attribute taxonomies.

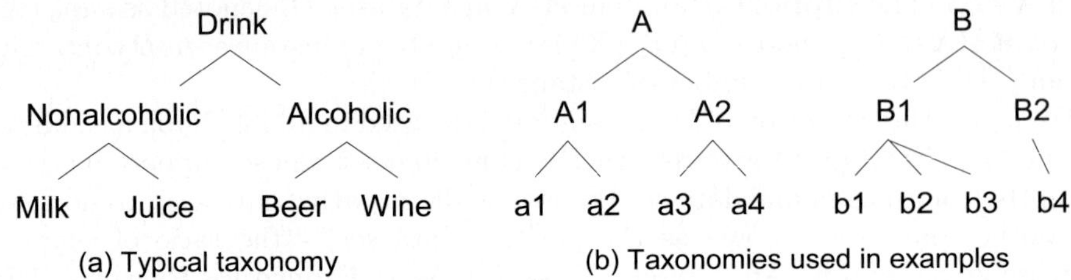

Fig. 1. Examples of taxonomic hierarchies

A **generalised emerging pattern** is an itemset whose support increases from zero in one class of data, to non-zero in another class, and consists of items from any level of the taxonomy associated with the corresponding attributes.

Given a data set D_1 of positive instances and data set D_2 of negative instances, there are many different EPs that can be found. An **EP space** is defined as the set of all EPs with repsect to D_1 and D_2 [9]. It has been shown that the EP space can be concisely represented by two bounds $\langle \mathcal{L}, \mathcal{R} \rangle$ [9]. The left bound $\mathcal{L}$ contains the *most general* EPs, and the right bound $\mathcal{R}$ contains the *most specific* EPs, where an itemset X is said to be more general than another itemset Y if $X \subset Y$.

Note that in the case of generalised EPs, the pattern space has been greatly expanded since items are not only restricted to leaves of the taxonomies. Our aim is to find the most general EPs from a class of data. The traditional approach is to take the collection from the left bound of the border representation because the concept of generality is equivalent to the concept of minimality (or most expressive). However, with the introduction of non-leaf-level items in the patterns, the word *general* here no longer implies *minimal* only, it also implies the *highest-level* possible in the hierarchy. Therefore an extended generalisation measure must be introduced to allow comparison of two patterns and decide whether or not one is considered more general than the other in both dimensions.

The existence of hierarchies on attribute-values now enables us to compare one item with another within the same attribute domain. We say that one item X is *more general* than another item Y if and only if $X \in ancestors(Y)$. In our small example above, item $A1$ is considered more general than $a2$. Considering the fact that hierarchies exist on individual attributes, only items from the same attribute are comparable. Therefore, only EPs consisting of items from the same combinations of attributes are are allowed to be compared with each other. For example, $\{A_1, c1\}$ and $\{b2, c1\}$ are not comparable because they have items from different attributes. However, $\{A_1, c1\}$ and $\{a3, c2\}$ *are* considered to be comparable. We refer to pattern $X = \{x_1, x_2, ..., x_m\}$ as a **generalisation** of pattern $Y = \{y_1, y_2, ..., y_m\}$ if $x_i \in ancestors(y_i)$ for at least one i ($i = 1$ to m); and $x_k = y_k$ for $k = 1$ to m where $k \neq i$. X is also said to be **more general** than Y. Y is referred to as a **specialisation** of (or **more specific** than) X.

3.1 Problem Description

Given a set of instances $\mathcal{D}$, our aim is to mine the most general set of generalised EPs which contain items from any level of the taxonomies *and* have infinite growth rate from one class to all others.

Considering a generalised pattern X, if any of its specialisations is supported by the positive instances, X itself is then supported by the positive instances. Likewise, if any of its specialisatons is supported by the negative instances, X itself is then supported by the negative instances. Therefore, all legal generalised emerging patterns must have the following properties: (1) At least **one** of its specialisations must have non-zero support in the *positive* class; (2) **None** of its specialisations should have non-zero support in the *negative* class.

The set of most general EPs that could be mined with respect to a set of positive instances and a set of negative instances should satisfiy the properties of **completeness** and **conciseness**.

- **Completeness**: The set of most general EPs mined, S, can be used as the left bound of the border representation such that $< S, \mathcal{R} >$ represents all generalised EPs in a set.
- **Conciseness**: The set S is in its most concise form because any pattern which is either a subset or a generalisation of some EP in the set is not an EP anymore.

4 Algorithms for Mining Generalised EPs

Let us consider the problem of mining generalised EPs in terms of the problem of mining traditional EPs. A set of EPs mined from a dataset can be represented using a border representation $< \mathcal{L}, \mathcal{R} >$, where the left bound $\mathcal{L}$ is the set of minimal EPs, while the right bound $\mathcal{R}$ is the set of most specific EPs. Since we intend to discover the most general set of generalised EPs, we can therefore post-process on the left bound to obtain a more general set $\mathcal{L}'$ such that the new border $< \mathcal{L}', \mathcal{R} >$ represents the complete set of generalised emerging patterns in a dataset. In this generalised case, a pattern is considered an EP if:

- it is a superset *or* a specialisation *or* a specialisation of a superset of a certain EP in $\mathcal{L}'$ *and*
- it has a leaf-level specialisation that is a subset of an EP in $\mathcal{R}$

Hence, the problem of discovering the most general generalised EPs can be decomposed into three subproblems:

1. Mine the set of non-hierarchical EPs from the dataset using an existing mining algorithm
2. Enumerate all possible generalised patterns based on the set $\mathcal{L}$ of minimal EPs (the left bound)
3. Prune all generalised patterns that are not legal emerging patterns

Since existing EP mining techniques can be used for subproblem 1, our algorithm has been developed to address subproblems 2 and 3.

4.1   Algorithm GTree

Let us motivate our algorithm by first considering a brute-force approach to mining generalised EPs. The brute-force approach is to take each EP from the collection and enumerate all possible generalised patterns using ancestors of the items this EP contains. For each generalised pattern, check through the negative instances to see whether it is supported in the negative class and delete it if it has non-zero support in the negative class. All generalised patterns that do not have support in the negative class are considered as legal generalised EPs. Before each generalised EP is added to the final set, it must be checked against the EPs that are already in the set to make sure that it is removed if it is a specialisation of any existing patterns *or* the specialisations of it are removed if it is considered more general than any existing patterns.

We have added several important optimisations to the Brute-Force algorithm to develop an algorithm called **GTree**. In this algorithm, the enumeration of generalised patterns is now achieved through building a set enumeration tree for each input non-hierarchical EP. There are several important optimisations introduced by this approach. First, fewer passes are needed over the negative class data, since only one pass over the negative instances is required for each tree, instead of for each generalised pattern. In addition, taxonomic information can improve pruning efficiency. When an item is pruned, its corresponding generalisations are also pruned, since all generalisations will also be supported by the negative class. Let us now examine the GTree algorithm in detail.

4.2   GTree Construction

A GTree is a set-enumeration tree constructed from a set of leaf-level EPs in the left bound $\mathcal{L}$. The tree represents all possible generalised patterns that can be enumerated from the leaf-level EPs. The tree has a root, which does not contain any value. The height of the tree is the length of the EP. Each path from root to leaf represents one possible generalised pattern. There is an imposed ordering on the tree, such that paths from left to right are patterns from the most specific to the most general.

Before constructing the tree, we first divide the set of non-hierarchical EPs into a number of **groups** (G_1 to G_k) according to the combination of attributes in each pattern, such that each **group** should only contain patterns with items from the *same* combination of attributes. For each group, we then construct one tree for all EPs in that group. For example, consider a dataset where $\{a1, b2\}$ and $\{a1, b3\}$ are both non-hierarchical EPs. Each of these EPs will result in the same generalisation tree. Consequently, these patterns should be grouped together, so that a single tree can be constructed for both patterns, and then pruning on the negative patterns need occur only once.

The formal algorithm for constructing a tree is outlined as follows:

```
Given E = {input non-hierarchical minimal EPs}
for each group of EPs G_k ∈ L do
   Tree r ← buildTree( G_k )

buildTree( group G )
   initialise empty tree with root node r
   for each EP p_j ∈ G do
      insertPattern( r, p_j, 1 )
   return tree r;

Let p[l] denote the L^{th} item in pattern p
   e.g., if p = (a_1, b_1), p[1] = a_1, p[2] = b_1
assign ordering of values in attribute l and their generalisations,
most specific to most general

insertPattern( node n, pattern p, l )
   for each item i ∈ p[l] ∪ generalisations(p[l])
      if ( i ∉ n.children ) then
          insert i in n.children according to ordering of attribute l
      find node n_i ∈ n.children that corresponds to item i
      insertPattern( n_i, p, ++l );
```

As an example, consider a group of EPs from $\{a1, b1\}$, $\{a2, b2\}$, $\{a3, b3\}$, $\{a3, b4\}$, $\{a4, b3\}$. The resulting GTree is shown in Figure 2. We can see that items along each path from the root to the leaf forms a generalised itemset and paths from left to right are from the most specific to the most general.

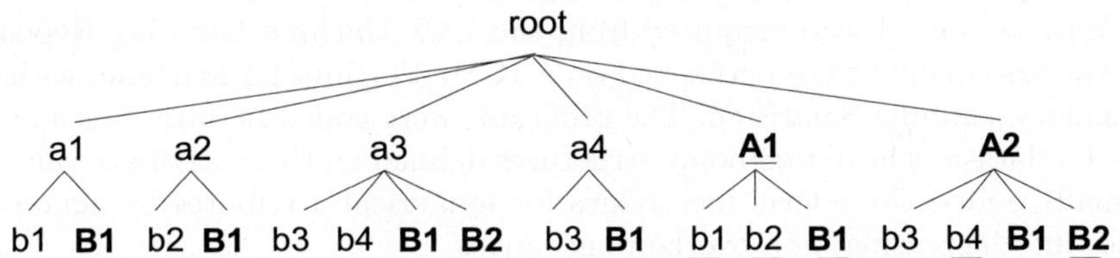

Fig. 2. A *GTree* built based on the example in Figure 1. Note that underlined nodes correspond to nodes that are pruned in the discussion in Section 4.3

4.3   Pruning from GTree

Pruning invalid EPs from the tree requires testing all negative instances one by one.

```
For each pattern p from the negative dataset
   pruneTree( root r, p, 1 )

pruneTree( node n, pattern p )
   for each item i ∈ pattern p
      if ( attribute type of item i ≠ attribute type of n.children ) then
         skip to next item i
      else
         for each node n_c ∈ n_c.children
            if ( n_c.children ≠ null ) then
               pruneTree( n_c, p )
            else delete n_c from n_c.children
```

For example, if the training dataset contains the negative instances $\{a1, b2\}$, $\{a1, b3\}$, $\{a2, b1\}$, $\{a4, b4\}$, then the underlined nodes are pruned from the tree in Figure 2.

4.4   Extracting the Most General EPs

After pruning, only complete paths (from root to leaf) that are left in the tree are legal generalised EPs. However, these EPs do not necessarily represent the most general set since there might still exist some patterns that are specialisations of others. Since the paths in the tree have an imposed order as introduced previously, we can take out patterns from the tree in order from *right to left*, such that more general ones are always taken out first. In this way, when each EP is extracted it can be compared with those that were extracted earlier to see whether it is a specialisation of those.

For example in Figure 2, once $\{A2, B1\}$ has been extracted, $\{A2, b3\}$ is redundant, since $b3$ is a specialisation of $B1$. Finally, the set of generalised EPs obtained from each group of input leaf-level EPs are unioned together, in order to ensure that only the most general EPs are kept.

5   Evaluation

We now present experimental results to demonstrate the performance of our **GTree** algorithm on various data sets and with different taxonomy structures. The data sets used were acquired from the UCI Machine Learning Repository [1]. All experiments were performed on a 1GHz Pertium III machine, with 2GB of memory, running Solaris 86. The programs were coded in Java. Since none of the UCI data sets have taxonomy structures defined on their attribute values, we manually added three-level hierarchies for numerical attributes by aggregating values into increasing, non-overlapping ranges.

5.1   Dimensionality and Scalability

Our goal is to evaluate the scalability of our algorithm by measuring its runtime performance on data sets with different dimensionalities. As a basis for comparison, we have implemented a **Brute-Force** algorithm, which enumerates all generalised patterns for each EP, and compares each generalised pattern against the negative class data. In terms of implementation, both algorithms used the **Border-Diff** algorithm for mining non-hierarchical EPs, although any other existing EP mining technique can be used. Therefore the total execution time recorded did *not* include the time taken for the mining of non-hierarchical EPs. The following table shows the experimental results of using the the two algorithms. Note that **GTree** is able to find all generalised EPs for all datasets, and in substantially less time than the brute force approach.

Table 1. Experimental results on 10 data sets

Dataset	No. Attr	No. Hierarchical Attributes	Input No. EPs	No. Groups	Output No. G.EPs	Execution Times(s) Brute -Force	GTree
iris	4	4	9	4	6	0.16	**0.10**
glass	9	9	84	63	96	15.3	**1.0**
autos	25	15	662	658	675	108.6	**8.4**
breast-w	9	9	781	243	860	188.9	**7.1**
breast-cancer	9	3	949	302	933	7.4	**1.2**
diabetes	8	8	1015	224	1038	455.8	**12.0**
credit-a	15	6	3015	1375	2982	65.7	**13.8**
heart-statlog	13	6	4032	1792	4016	295.7	**16.8**
vehicle	18	18	34463	14253	36902	>10000	**994.1**
ionosphere	34	20	126330	79060	125434	>10000	**5035.2**

5.2 Compactness and Support Evaluation of Generalised EPs

Having examined the efficiency of our algorithm, we now want to evaluate its effectiveness, in terms of the quantity and quality of the generalised EPs.

There are generally three classes of generalised EPs that occur, namely:

- **Class 1**: patterns consisting of leaf-level items only; e.g.$\{a1, b2\}$
- **Class 2**: patterns consisting of at least *one* non-leaf item and at least *one* leaf item; e.g. $\{a1, B2\}$, and
- **Class 3**: patterns consisting of non-leaf items only; e.g. $\{A1, B2\}$

Note that when we talk about non-leaf and leaf items, we refer to those attributes that are associated with hierarchies.

We now observe the quantity and quality of the patterns that belong to each of the three classes. The term *quality* refers to the expressiveness of each pattern. A pattern is considered powerful or expressive if it is sharply discriminative or strongly representative of the instances of a particular class, and this can be measured in terms of its support. The results shown in Table 2 are obtained from the same ten data sets which were used for producing Table 1. *Average support of each pattern* in the table reflects the average support of each pattern from a particular category, which is a direct indication of the expressiveness of a pattern. The results obtained are averaged over the ten data sets.

Table 2. Experimental results showing the proportion of generalised EPs in each class

Property	Class 1	Class 2	**Class 3**
Proportion of the complete set	23.5%	51.4%	**25.1%**
Average support of each pattern	1.7%	3.9%	**10.7%**

After post-processing, more than 76% of the output set are patterns that include items from higher levels in the taxonomy. Only less than 24% still remain as leaf-level EPs, without being generalised. Table 2 shows that the average support of **Class 3** patterns is significantly higher than that of the patterns belonging to the other two classes. **Class-1** patterns, which are a subset of the original EPs, have the smallest average support. It is clear that **Class 3**, which corresponds to generalised EPs, has the highest average support. Overall, we find that the generalised EPs that are mined using our algorithm have much higher support than the traditional non-hierarchical EPs. This improvement means that the generalised EPs are representative of a larger proportion of the positive dataset, and are thus of higher quality than the original leaf-level EPs.

6   Conclusion and Future Work

We have introduced the concept of *Generalised Emerging Patterns*. We have presented an algorithm for generating generalised EPs in data sets whose attributes are associated with membership hierarchies. Our experimental results based on ten real-life data sets show that our **GTree** algorithm is capable of mining generalised EPs from high-dimensional data sets within a small period of time. The complete set of most general generalised EPs mined from a data set has approximately the same size as its minimal non-hierarchical EP set. Moreover, the generalised EPs have higher support than traditional leaf-level EPs.

Acknowledgements. This research was supported in part by the Australian Research Council.

References

1. C. L. Blake and P. M. Murphy. UCI Machine Learning Repository.
2. H. Fan and K. Ramamohanarao. Efficient mining of emerging patterns: Discovering trends and differences. *PAKDD-2002*, 2002.
3. J. Li G. Dong and X. Zhang. Discovering jumping emerging patterns and experiments on real datasets. *IDC99*, 1999.
4. G.Dong and J. Li. Efficient mining of emerging patterns: Discovering trends and differences. *KDD-99*, pages 43–52, 1999.
5. R. Wirth J. Hipp, A. Myka and U. Guntzer. A new algorithm for faster mining of generalised association rules. *PKDD-98*, pages 74–82, 1998.
6. T.Manoukian J.Bailey and K.Ramamohanarao. Fast algorithms for mining emerging patterns. *PKDD-2002*, 2002.
7. J. Li. Mining emerging patterns to construct accurate and efficient classifiers. *PhD Thesis, The University of Melbourne*, 2001.
8. Y. Li and L. Sweeney. Learning robust rules from data. *Carnegie Mellon University, Computer Science Tech Report CMU ISRI 04-107, CMU-CALD-04-100*, Feb 2004.
9. T. Imielinski R. Agrawal and A. Swami. Mining association rules between sets of items in large databases. *ACM SIGMOD-93*, pages 207–216, 1993.
10. R. Agrawal R.Srikant. Mining generalised association rules. *VLDB-95*, 1995.
11. G. Webb. Efficient search for association rules. *KDD-2000*, pages 99–107, 2000.
12. S. Zhang X. Wu, C. Zhang. Efficient mining of both positive and negative association rules. *ACM Trans. Inf. Syst. 22(3)*, pages 381–405, 2004.

Using Attack-Specific Feature Subsets
for Network Intrusion Detection

Sung Woo Shin[1] and Chi Hoon Lee[2,*]

[1] School of Business Administration,
110-745, Seoul, Korea
shinswoo@{skku.edu, kaist.ac.kr}
[2] School of Information and Communication Engineering,
440-746, Suwon, Korea
chihoon@samsung.com

Abstract. One of the essential tasks for building a network intrusion detection system might be to differentiate a salient feature subset from noisy and/or redundant features. Especially, in real-time environment too many features to be monitored deteriorate the system performance. In this paper, we focus on extracting robust feature subsets that maximizes inter-classes seperability with minimized subset size based on a genetic algorithm-based optimization, reducing both false positive and false negative errors by learning class-specific feature subsets. Experimental results show that the proposed approach is especially effective in detecting totally unknown attack patterns compared with single feature-subset model.

Keywords: Network intrusion detection, feature selection, class-specific learning, genetic algorithm, data mining.

1 Introduction

In this study, we present a genetic algorithm-based feature selection method considering attack or normal class-specific behaviors based on k nearest neighbor matching rule and empirically demonstrate its robustness using the feature-sensitive naïve Bayesian network classifier.

Recently, intrusion detection system (IDS) has been receiving great attention as number of novel cyber attacks is rapidly increased. The primary goal of IDS lies in building a predictive model capable of distinguishing between "bad" connections, called intrusions or attacks, and "good" normal connections [12].

Any computers connected to the Internet are always under threat from viruses, worms and attacks from hackers [5]. Moreover, because most attacks tend to utilize unknown weaknesses of the computer system, it is very difficult to achieve a security-level integrity.

An intrusion is defined to be a violation of the security policy of the system: an intrusion detection system (IDS) thus 1) refers to the mechanism that are developed to

---

* Corresponding author.

A. Sattar and B.H. Kang (Eds.): AI 2006, LNAI 4304, pp. 305–311, 2006.
© Springer-Verlag Berlin Heidelberg 2006
"

detect violations of system security policy [1], and 2) act as a final defense, looking for known or potential threats recorded in network traffics and/or audit data [3].

Recent advances in data mining algorithms make it possible for IDS to be more robust (e.g., [4], [10], [14]), and therefore many actual IDSs have shown to be successful in detecting familiar known attacks. However, for many unknown attacks IDSs still suffer from false negative alarms (ie., predict an attack as a normal pattern), even though, as noted by Stolfo et al. [12], some intrusion experts believe that most novel attacks can be detected by using a signature of known attacks.

In this context, we investigate an effectiveness of class-specific feature selection approach within a framework of genetic optimization, expecting that the mixture of anomaly and misuse detection approaches can decrease both false positive and false negative errors simultaneously.

The rest of this paper is organized as follows. In Section 2, the proposed genetic feature selection algorithm is presented. In Section 3, the experimental results are summarized, and in the final section future research direction is discussed.

2 Proposed Approach

In this paper, we propose the mixture model of anomaly detection and misuse detection approaches. The objective of our approach is 1) to detect novel attacks using an optimized feature subset representing normal behaviors; and 2) at the same time, maintaining high true-positive detection rate for known attacks. To achieve this, we develop a genetic algorithm-based feature selection procedure using an enhanced fitness function that can find multi class-specific feature subsets in a non-redundant fashion.

In the genetic feature selection, it is critical to design an appropriate fitness function to avoid local mina. Traditionally, the fitness function used for a GA is simply the hit rate (HR). While this function gives reasonable performance, it does not generate a set of features which yields the maximum class separation. A better fitness metric might be using the concept of a k nearest neighbor classification rule that considers the number of neighbors that are not used in the majority decision [11], [6].

When a domain problem has many class labels such as intrusion detection, we should additionally consider a feature subset that can maximize inter-class separability. For this reason, the *cMiC* criterion as the cardinality of the minority classes within the group of k neighbors might be helpful.

In addition, the *cMaP* criterion reflecting the cardinality of the near-neighbor majority patterns can be combined in a synergistic fashion. Hence, more complicated but probably more powerful fitness function can be defined in the following way:

$$Fitness = \alpha\, HR + \beta\, \{\, (cMaP - cMiC) \,/\, (k \times p)\, \} - \gamma(s\,/\,t) \tag{1}$$

Here α, β, γ are tunable parameters, k the number of neighbors, p the total number of training patterns, and the last term denotes a *cost* in which s is the number of selected features, t is the number of total features.

3   Experimental Investigation

The data used for this study is the KDDCUP99 data set. It was originally provided by DARPA/MIT Lincoln Labs and later preprocessed for KDD competition (see http://www.ics.uci.edu/~kdd/databases/kddcup99/kddcup99.html for detail) by the Columbia IDS group [13]. Despite of some critiques[1] against this data set including procedural or reality problems of generated data [9] and an overoptimistic tendency in intrusion detection [8], we believe this data set is still the most useful test bed for intrusion detection evaluations, making it possible to compare the published results of wide range of state-of-the-art data mining approaches. Moreover, as noted by Mahoney and Chan [8] it might be the most comprehensive test set available today, supporting both host and network based methods.

3.1   Data

The KDDCUP99 data set includes a wide variety of intrusions simulated in a military network environment, in which 41 input features represent basic TCP connections, packet contents within a connection suggested by domain knowledge, and traffic information computed using a two-second time window. Each connection is labeled as either normal, or as one of the following four attack types [12]:

> *Probe*: surveillance and other probing, eg., port scanning;
> *DOS*: denial-of-service, eg. syn flood;
> *U2R*: unauthorized access from a user to root privilege,
>   e.g., various ``buffer overflow" attacks;
> *R2L*: unauthorized access from remote to local machine,
>   e.g. guessing password;

It should be noted that the test data is not from the same probability distribution as the training data. Moreover, the test data includes novel attack types that were not appeared in the training data. This formation makes IDSs difficult to correctly predict and thus one could evaluate a true impact of newly developed algorithms in real-world situation. In this study we used "10% data sets" which were randomly selected from 4,940,000 connections.

Synthesis of the data distributions by attack types used in experiments is summarized in Table 1. Table 1 displays a total of 23 training attack types, with additional 17 unknown types. Note that *warezclient* attack belonging to R2L is the majority patterns in the training set, however in the test set *guess_passwd* and *warezmaster* comprises most patterns of R2L. This reflects real situation that the frequency of majority attack patterns will vary time over time.

To test these combined scenarios separately, we build two kinds of test sets from 311,029 records; first set consists of 3, 000 normal patterns plus 3,039 known attacks with different distribution, and the second set consists of another 3,000 normal patterns plus 3,031 totally unknown attack patterns. For each of probe, DOS, and R2L attacks, 1,000 patterns are randomly selected for both test sets, while 39 and 31 U2R patterns are included into the known and unknown test set, respectively.

---

[1]   The authors thank an anonymous reviewer for pointing out this issue.

Table 1. Data sets used for this study

Class	Attack type	Tr	TeK	TeU
Normal	—	563	3,000	3,000
Probe	ipsweep	141	208	
	nmap	122	54	
	portsweep	142	248	
	satan	158	490	
	saint			451
	mscan			549
DOS	back	84	148	
	land	9	5	
	neptune	123	253	
	pod	60	51	
	smurf	163	536	
	teardrop	124	7	
	apache2			268
	processtable			257
	udpstorm			2
	mailbomb			473
U2R	buffer_overflow	12	22	
	loadmodule	6	2	
	perl	2	2	
	rootkit	6	13	
	xterm			13
	ps			16
	sqlattack			2
R2L	ftp_write	5	2	
	guess_passwd	27	716	
	imap	9	0	
	multihop	6	13	
	phf	3	0	
	spy	1	0	
	warezclient	500	0	
	warezmaster	12	269	
	snmpgetattack			664
	named			11
	xlock			4
	xsnoop			3
	sendmail			16
	httptunnel			90
	worm			1
	snmpguess			211
Total		**2,278**	**6,039**	**6,031**

Note. Tr, TeK, TeU denote training set, test set of known attacks, and test set of unknown attacks, respectively.

The training set comprises 2,278 patterns (selected from 494,020 records) in which 563 normal patterns and equal-sized three attacks including probe, DOS, and R2L comprises 2,252 patterns. Because the number of records of U2R and R2L types in the original training set is relatively very small—52 and 1,126 respectively, we randomly selected 50% records in order to avoid sample bias. Based on the class of R2L, for the other three classes—normal, probe, and DOS—563 patterns are equally selected.

3.2 Experimental Setup

For both GA optimization and intrusion detection, the hit rate was measured by selective naïve Bayesian network (SNBN) classifier [7] since SNBN is very sensitive according to the features selected. The discretization method used for the Bayesian classifier was the minimal entropy heuristic [2]. The average probability of crossover was set to 0.6; that of mutation to 0.05; population size to 30; maximum generation to 50; parameters k, α, β, and γ to 7, 1.0, 0.0009, and 0.0005, respectively.

3.3 Results

Table 2 (a) shows single feature subset model that was simultaneously optimized for 5 classes, while Table 2 (b) displays five models optimized for each of 5 classes including 4 attack types plus the normal class.

Not surprisingly, as can be seen in the interpretation of Table 2 (b), class-specific five models clearly reflect an inherent characteristic of five classes. For Probe, DOS, and R2L attacks, GA selected only 25% features at most, while U2R and Normal models share similar features due to its similar behavior.

Table 2. Optimized feature subsets

(a) Single model

Features selected (feature number)
duration(1), service(3), flag(4), src_bytes(5), wrong_fragment(8), urgent(9), hot(10), num_failed_logins(11), num_compromised(13), root_shell(14), is_guest_login(22), same_srv_rate(29), srv_diff_host_rate(30), dst_host_count(32), dst_host_diff_srv_rate(35), dst_host_same_src_port_rate(36), dst_host_serror_rate(38), dst_host_srv_serror_rate(39)

(b) Five class-specific models

Class	Features selected (feature number)	Interpretation
Probe (5)	service(3), src_bytes(5), dst_bytes(6), logged_in(12), diff_srv_rate(30)	3, 30: inherent port (service) scanning related features 5, 6: interactive traffic related features of port scanning
DOS (8)	service(3), src_bytes(5), land(7), wrong_fragment(8), hot(10), dst_host_diff_srv_rate(35), dst_host_same_src_port_rate(36), dst_host_serror_rate(38)	7, 8: inherent DOS-specific features 3: related to a denial of specific service 35, 36, 38: related to an multi-target attack
U2R (14)	service(3), src_bytes(5), land(7), hot(10), logged_in(12), num_root(16), serror_rate(25), srv_rerror_rate(27), srv_diff_host_rate(31), dst_host_count(32), dst_host_srv_count(33), dst_host_same_src_port_rate(36), dst_host_srv_diff_host_rate(37), dst_host_rerror_rate(40)	10, 12, 16: related to a login to a specific service, buffer overflow Overall, wide-range of features were selected since its behavior is similar to Normal class
R2L (9)	protocol_type(2), service(3), flag(4), hot(10), num_access_files(19), count(23), diff_srv_rate(30), dst_host_count(32), dst_host_srv_serror_rate(39)	4: connection status indicating various trials to obtain a legitimate user privilege 19, 23, 32: indicating repeated trial-and-errors
Normal (14)	duration(1), service(3), flag(4), rc_bytes(5), wrong_fragment(8), hot(10), is_guest_login(22), rerror_rate(27), same_srv_rate(29), diff_srv_rate(30), dst_host_count(32), dst_host_same_src_port_rate(36), dst_host_srv_diff_host_rate(37), dst_host_serror_rate(38)	1: representing legitimate continued operations 10, 22: related to login procedure 3, 29, 30, 37: related to user service Overall, very similar to U2R

In Table 3, the detail classification results produced by SNBN using five class-specific feature subsets are presented. As expected, the performance gap measured by a "true positive" alarming rate between known and unknown attacks was above 20% (88.4% versus 67.2%, see overall performance in Table 4).

Table 3. Confusion matrix of five attack-specific SNBN models

(a) Known attack

Class	Normal	Probe	Dos	U2R	R2L	TP %
Normal	**2946**	24	8	10	12	**98.2**
Probe	1	**996**	3	0	0	**99.6**
DOS	7	6	**984**	2	1	**98.4**
U2R	5	0	2	**25**	7	**64.1**
R2L	601	11	0	1	**387**	**38.7**

(b) Unknown attack

Class	Normal	Probe	Dos	U2R	R2L	TP %
Normal	**2954**	18	5	9	14	**98.5**
Probe	83	**807**	1	109	0	**80.7**
DOS	497	89	**256**	0	158	**25.6**
U2R	2	1	0	**22**	6	**71.0**
R2L	29	89	866	0	**16**	**1.6**

More specifically, for the known attack test set, Normal, Probe, and DOS models predicted almost perfectly, however, for U2R and R2L models its accuracy was remarkably deteriorated mainly due to the different distribution of subclasses. For unknown data set, the detection rates for DOS and R2L classes were disappointing. Especially, the detection rate for R2L is just 1.6%, predicting most attack patterns as DOS class. In contrast, Probe and U2R models were relatively robust by showing 80.7% and 71.0% detection rate, respectively.

Table 4. Performance summary of three approaches

Class	Known attack			Unknown attack		
	Full set (41)	SM (18)	CSM	Full set (41)	SM (18)	CSM
Normal	97.5	98.5	98.2	97.4	98.7	98.5
Probe	99.3	99.6	99.6	62.3	68.2	80.7
DOS	82.0	98.7	98.4	11.9	15.1	25.6
U2R	71.8	56.4	64.1	61.3	51.6	71.0
R2L	30.2	33.8	38.7	0.4	0.5	1.6
Mean	**83.9**	**87.8**	**88.4**	**61.1**	**63.2**	**67.2**

Note: SM, CSM denote a single model, class-specific model, respectively.

As a summary, Table 4 contrasts the performance of the SNBN using class-specific feature subsets against both NBN using full features and SNBN using single feature subset. As can be seen in the last row, the overall accuracy of the class-specific model was the best in both test sets, especially for unknown test set it significantly outperformed single feature model.

4 Conclusion

The five feature subsets found by this study showed the promising results in terms of detecting unknown attack patterns.

It is important to note that currently our research purpose is not to discover *oracle* feature subsets that are working on diverse data sets, but to develop a competitive intrusion detection model based on the methodological or technical enhancement. Accordingly, feature subsets found by this study should be further investigated and refined through additional real world evaluations.

References

1. Chebrolu, S., A. Abraham, J. Thomas, 2005, Feature deduction and ensemble design of intrusion detection systems, *Computers & Security*, 24, 295-307
2. Fayyad, U. M., K. B. Irani, 1993, Multi-interval discretization of continuous-valued attributes for classification learning, *Proc. of 13th International Joint Conf. on Artificial Intelligence*, 1022-1027.
3. Giacinto, G., F. Roli, L. Didaci, 2003, Fusion of multiple classifiers for intrusion detection in computer networks, *Pattern Recognition Letters*, 24, 1795-1803.
4. Han, S. J., S-B., Cho, 2003, Detecting intrusion with rule-based integration of multiple models, *Computers & Security*, 22, 613-623.
5. Hansman, S. Hunt, R., 2005, A Taxonomy of network and computer attacks, *Computers & Security*, 24, 31-43.
6. Kim, S. H., Shin, S. W., 2000, Identifying the impact of decision variables for nonlinear classification, *Expert Systems With Applications*, 18, 201-214.
7. Langley, P. and Sage. S., 1994, Induction of Selective Bayesian classifiers, *Proc. of the 10th Conference on Uncertainty in Artificial Intelligence*, 399-406.
8. Mahoney, M. V., P. K. Chan, 2003, An analysis of the 1999 DARPA/Lincoln Laboratory evaluation data for network anomaly detection. Proc. 6th Intl. Symposium on Recent Advances in Intrusion Detection (RAID 2003) 53, *LNCS 2820*, 220-237
9. McHugh, J. 2000. Testing intrusion detection systems: A critique of the 1998 and 1999 DARPA intrusion detection system evaluations as performed by Lincoln Laboratory. ACM Trans. Information System Security 3 (4), 262–294.
10. Mukkamala, S., Andrew H. Sunga, A. Abraham, 2005, Intrusion detection using an ensemble of intelligent paradigms, *Journal of Network and Computer Applications*, 28, 167–182.
11. Punch, W. F., Goodman, E. D., et al., 1993, Further research on feature selection and classification using genetic algorithms, *Int. Conf. on Genetic Algorithms*, 557-564.
12. Stolfo, S., W. Fan, W. Lee, A. Prodromidis, P. K. Chan. 2000, Cost-based Modeling for Fraud and Intrusion Dectection: Results from the JAM Project, *DARPA Information Survivability Conference*.
13. Stolfo, S., W. Lee, P. K. Chan, W. Fan, and E. Eskin. 2001, Data mining-based intrusion detectors: An overview of the Columbia IDS project, *ACM SIGMOD Record*, 30 (4), 5-14.
14. Zhang, C., J. Jiang, M. Kamel, 2005, A Intrusion detection using hierarchical neural networks, *Pattern Recognition Letters*, 779–791.

Time Series Analysis Using Fractal Theory and Online Ensemble Classifiers

Dalton Lunga and Tshilidzi Marwala

University of the Witwatersrand
School of Electrical and Information Engineering
Private Bag 3 Wits 2050,
Johannesburg, South Africa
{d.lunga, t.marwala}@ee.wits.ac.za
http://www.ee.wits.ac.za/~marwala

Abstract. Fractal analysis is proposed as a concept to establish the degree of persistence and self-similarity within the stock market data. This concept is carried out using the rescaled range analysis (R/S) method. The R/S analysis outcome is applied to an online incremental algorithm (Learn++) that is built to classify the direction of movement of the stock market. The use of fractal geometry in this study provides a way of determining quantitatively the extent to which time series data can be predicted. In an extensive test, it is demonstrated that the R/S analysis provides a very sensitive method to reveal hidden long run and short run memory trends within the sample data. The time series data that is measured to be persistent is used in training the neural network. The results from Learn++ algorithm show a very high level of confidence of the neural network in classifying sample data accurately.

1   Introduction

The financial markets are regarded as complex, evolutionary, and non-linear dynamical systems [1]. Advanced neural techniques are required to model the non-linearity and complex behavior within the time series data [2]. In this paper we apply a non-parametric technique to select only those regions of the data that are observed to be persistent. The selected data intervals are used as inputs to the incremental algorithm that predicts the future behavior of the time series data. In the following sections we give a brief discussion on the proposed framework which implements a fractal theory technique and an online incremental Learn++ algorithm. The findings from this investigation are also discussed.

2   Fractal Analysis

The fractal dimension of an object indicates something about the extent to which the object fills space. On the other hand, the fractal dimension of a time series shows how turbulent the time series is and also measures the degree to which the time series is scale-invariant. The method used to estimate the fractal dimension

A. Sattar and B.H. Kang (Eds.): AI 2006, LNAI 4304, pp. 312–321, 2006.
© Springer-Verlag Berlin Heidelberg 2006

using the Hurst exponent for a time series is called the rescaled range (R/S) analysis, which was invented by Hurst [3] when studying the Nile River in order to describe the long-term dependence of the water level in rivers and reservoirs. The estimation of the fractal dimension given the approximated Hurst exponent will be explained in the following sections. The simulated fractal Brownian motion time series in Fig. 1 was done for different Hurst exponents. In Fig. 1 (1a) we have an anti-persistent signal with a Hurst exponent of $H = 0.3$ and the resulting fractal dimension is $D_f = 1.7$. In Fig. 1 (1b) the figure shows a random walk signal with $H = 0.5$ and the corresponding fractal dimension is $D_f = 1.5$. In Fig. 1(1c) a persistent time series signal with $H = 0.7$ and a corresponding fractal dimension of $D_f = 1.5$ is also shown. From Fig. 1 we can easily note the degree to which the data contain jagged features that needs to be removed before the signal is passed over to the neural network model for further analysis in uncovering the required information.

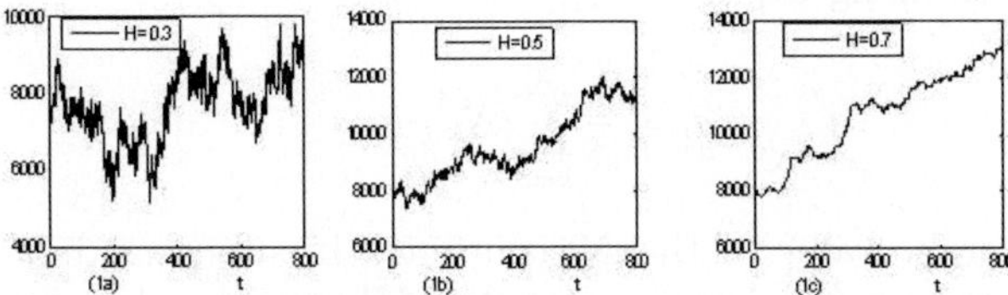

Fig. 1. Fractal Brownian motion simulation: (1a) anti-persistent signal, (1b) random walk signal, and (1c) persistent signal

2.1 Re-scaled Range Analysis

In this section we describe a technique for estimating the quality of a time series signal to establish the intervals that are of importance to use in our Neural Network. The rescaled range (R/S) analysis is a technique that was developed by Hurst, a hydrologist, who worked on the problem of reservoir control on the Nile River dam project at around 1907. His problem was to determine the ideal design of a reservoir based upon the given record of observed river discharges. An ideal reservoir never empties or overflows. In constructing the model, it was common to assume that the uncontrollable process of the system, which at the time was the influx due to the rainfall, followed a random walk due to the many degrees of freedom in the weather. When Hurst examined this assumption he gave a new statistical measure, the Hurst exponent (H).

2.2 The R/S Methodology

The main idea behind using the (R/S) analysis for our investigation is that we establish the scaling behavior of the rescaled cumulative deviations from the mean, or the distance that the system travels as a function of time relative to the mean. The intervals that display a high degree of persistence are selected and grouped in rebuilding the signal to be used as an input to the Neural Network model. A statistical correlation approach is used in recombining the intervals. From the discovery that was made by Hurst, the distance covered an independent system increases, on average, by the square root of time. If the system covers a larger distance than this, it cannot be independent by this definition; the changes must be influencing each other and therefore have to be correlated [4]. In our approach we first start with a time series in prices of length M. This time series is then converted into a time series of logarithmic ratios or returns of length $N = M - 1$ as shown by Eq. 1

$$N_i = \log(\frac{M_{i+1}}{M_i}), i = 1, 2, ..., (M-1) \tag{1}$$

We divide this time period into T contiguous sub periods of length j, such that $T * j = N$. Each sub period is labeled I_t, with $t = 1, 2...T$. Then, each element in I_t is labeled $N_{k,t}$ such that $k = 1, 2, , j$. For each sub period I_t of length j the average is calculated using Eq. 2

$$e_t = \frac{1}{j} \sum_{k=1}^{j} N_{k,t} \tag{2}$$

Thus, e_t is the average value of the N_i contained in sub-period I_t of length j. We then calculate the time series of accumulated departures $X_{k,t}$ from the mean for each sub period I_t, defined as Eq. 3

$$X_{k,t} = \sum_{i=1}^{k} N_{i,t} - e_t; k = 1, 2, ..., j \tag{3}$$

Now, the range that the time series covers relative to the mean within each sub period is defined by Eq. 4

$$R_{I_t} = max(X_{k,t}) - min(X_{k,t}); 1 < k < j \tag{4}$$

Next we calculate the standard deviation of each sub-period using Eq. 5

$$X_{k,t} = \sqrt{\frac{1}{j} \sum_{i=1}^{k} (N_{i,t} - e_t)^2} \tag{5}$$

Then, the range of each sub period R_{I_t} is rescaled/normalized by the corresponding standard deviation S_{I_t}. This approach is done for all the T sub intervals we have for the series. As a result the average R/S value for length j is given by Eq. 6

$$e_t = \frac{1}{T} \sum_{t=1}^{T} (\frac{R_{I_t}}{S_{I_t}}) \tag{6}$$

Now, the calculations from the above equations are repeated for different time horizons. This is achieved by successively increasing j and repeating the calculations until we covered all j integers. After having calculated the R/S values for a large range of different time horizons j, we plot $log(R/S)_j$ against $log(n)$. By performing a least squares linear regression with $log(R/S)_j$ as the dependent variable and $log(n)$ as the independent one, we find the slope of the regression which is the estimate of the Hurst exponent (H). The relationship between the fractal dimension and the Hurst exponent is modeled by Eq. 7

$$D_f = 2 - H \tag{7}$$

2.3   The Hurst Interpretation

If, $H \in (0.5; 1]$ it implies that the time series is persistent which is characterized by long memory effects on all time scales [5]. This also implies that all hourly prices are correlated with all future hourly price changes; all daily price changes are correlated with all future daily prices changes, all weekly price changes are correlated with all future weekly price changes and so on. This is one of the key characteristics of fractal time series as discussed earlier. The persistence implies that if the series has been up or down in the last period then the chances are that it will continue to be up and down, respectively, in the next period. The strength of the trend reinforcing behavior, or persistence, increases as H approaches 1. This impact of the present on the future can be expressed as a correlation function G as shown by Eq. 8

$$G = 2^{2H-1} - 1 \tag{8}$$

In the case of $H = 0.5$ the correlation $G = 0$, and the time series is uncorrelated. However, if $H = 1$ we see that that $G = 1$, indicating a perfect positive correlation. On the other hand, when $H \in [0; 0.5)$ we have an antipersistent time series signal (interval). This means that whenever the time series has been up in the last period, it is more likely to be down in the next period. Thus, an antipersistent time series will be more jagged than a pure random walk as shown in Fig. 1. The intervals that showed a positive correlation coefficient were selected as inputs to the Neural Network.

3   Incremental Online Learning Algorithm

An incremental learning algorithm is defined as an algorithm that learns new information from unseen data, without necessitating access to previously used data. The algorithm must also be able to learn new information from new data and still retain knowledge from the original data. Lastly, the algorithm must be able to teach new classes that may be introduced by new data.

In this paper, we propose a framework that implements fractal theory and online incremental learning algorithm with application to time series data. This proposed approach is implemented and tested on the classification of stock options movement direction with the Dow Jones data used as the sample set for the experiment. We make use of the Learn++ incremental algorithm in this study. Learn++ is an incremental learning algorithm that uses an ensemble of classifiers that are combined using weighted majority voting. Learn++ was developed by Polikar [6]. Each classifier is trained using a training subset that is drawn according to a distribution. The classifiers are trained using a weakLearn algorithm. The requirement for the weakLearn algorithm is that it must be able to give a classification rate of at least 50% initially and then the capability of Learn++ is applied to improve the short fall of the weak MLP. For each database D_k that contains training sequence, S, where S contains learning examples and their corresponding classes, Learn++ starts by initializing the weights, w, according to the distribution D_T, where T is the number of hypothesis. Initially the weights are initialized to be uniform, which gives equal probability for all instances to be selected to the first training subset and the distribution is given by Eq. 9

$$D = \frac{1}{m} \tag{9}$$

Where m represents the number of training examples in database S_k. The training data are then divided into training subset T_R and testing subset T_E to ensure weakLearn capability. The distribution is then used to select the training subset T_R and testing subset T_E from S_k. After the training and testing subset have been selected, the weakLearn algorithm is implemented. The weakLearner is trained using subset, T_R. A hypothesis, h_t obtained from weakLearner is tested using both the training and testing subsets to obtain an error,ϵ_t:

$$\epsilon_t = \sum_{t:h_t(x_i) \neq y_i} D_t(i) \tag{10}$$

The error is required to be less than $\frac{1}{2}$; a normalized error β_t is computed using:

$$\beta_t = \frac{\epsilon_t}{1 - \epsilon_t} \tag{11}$$

If the error is greater than $\frac{1}{2}$, the hypothesis is discarded and new training and testing subsets are selected according to D_T and another hypothesis is computed. All classifiers generated so far, are combined using weighted majority voting to obtain composite hypothesis, H_t

$$H_t = \arg\max_{y \in Y} \sum_{t:h_t(x)=y} \log \frac{1}{\beta_t} \tag{12}$$

Weighted majority voting gives higher voting weights to a hypothesis that performs well on its training and testing subsets. The error of the composite hypothesis is computed as in Eq.13

$$E_t = \sum_{t:H_t(x_i) \neq y_i} D_t(i) \tag{13}$$

If the error is greater than $\frac{1}{2}$, the current composite hypothesis is discarded and the new training and testing data are selected according to the distribution

D_T. Otherwise, if the error is less than $\frac{1}{2}$, the normalized error of the composite hypothesis is computed as:

$$B_t = \frac{E_t}{1 - E_t} \tag{14}$$

The error is used in the distribution update rule, where the weights of the correctly classified instances are reduced, consequently increasing the weights of the misclassified instances. This ensures that instances that were misclassified by the current hypothesis have a higher probability of being selected for the subsequent training set. The distribution update rule is given by Eq. 15

$$w_{t+1} = w_t(i) \cdot B_t^{[|H_t(x_i) \neq y_i|]} \tag{15}$$

Once the T hypotheses are created for each database, the final hypothesis is computed by combining the composite hypothesis using weighted majority voting developed by Littlestone [7]. Eq. 16 summaries the weighted majority voting.

$$H_t = \arg\max_{y \in Y} \sum_{k=1}^{K} \sum_{t:H_t(x)=y} \log \frac{1}{\beta_t} \tag{16}$$

3.1   Confidence Measurement

An intimately relevant issue is the confidence of the classifier in its decision, with particular interest on whether the confidence of the algorithm improves as new data become available. The voting mechanism inherent in Learn++ hints to a practical approach for estimating confidence: decisions made with a vast majority of votes have better confidence than those made by a slight majority [8]. We have implemented McIver and Friedl's weighted exponential voting based confidence metric [9] with Learn++ as in Eq. 17

$$C_i(x) = P(y = i|x) = \frac{\exp^{F_i(x)}}{\sum_{k=1}^{N} \exp^{F_k(x)}}, 0 \leq C_i(x) \leq 1 \tag{17}$$

Where $C_i(x)$ is the confidence assigned to instance x when classified as class i , $F_i(x)$ is the total vote associated with the ith class for the instance x and N is the number of classes. The total vote $F_i(x)$ class received for any given instances is computed as in Eq. 18

$$F_i(x) = \sum_{t=1}^{N} \left(\begin{matrix} \log \frac{1}{\beta_t}, \; if \; h_t(x) = i \\ 0, \; otherwise \end{matrix} \right) \tag{18}$$

The confidence of winning class is then considered as the confidence of the algorithm in making the decision with respect to the winning class [10]. Since $C_i(x)$ is between 0 and 1, the confidences can be translated into linguistic indicators as shown in Table 1. These indicators are adopted and used in interpreting our experimental results. Equations 17 and 18 allow Learn++ to determine its own

Table 1. Confidence estimation representation

Confidence range (%)	Confidence level
$90 \leq C \leq 100$	Very High (VH)
$80 \leq C < 90$	High (H)
$70 \leq C < 80$	Medium (M)
$60 \leq C < 70$	Low (L)
$C < 60$	Very Low (VL)

confidence in any classification it makes. The desired outcome of the confidence analysis is to observe a high confidence on correctly classified instances, and a low confidence on misclassified instances, so that the low confidence can be used to flag those instances that are being misclassified by the algorithm. A second desired outcome is to observe improved con-fidences on correctly classified instances and reduced confidence on misclassified instances, as new data become available, so that the incremental learning ability of the algorithm can be further confirmed.

4   Forecasting Framework

4.1   Experimental Data

The database used consisted of 800 instances of the Dow Jones average consisting of the Open, High, Low and Close values during the period of January 2003 to December 2005; 400 instances were used for training and all the remaining instances were used for validation.

4.2   Model Input Selection

The R/S analysis function makes it easy for determining regions of persistence and non-persistence within the time series. using the estimation parameters discussed above. The processed time series data from the R/S analysis function is combined with extra four data points (Fig. 2) to form a signal that is used as an input to the Neural Network model. The proposed framework is to classify the eighth day's directional movement given the selected persistent signal from the previous seven days data. Within each of the seven previous days, four data points were collected in the form of the *Open, High, Low* and *Close* values.

A combination of all data processed in Fig. 2 points resulted in a signal with twenty-eight points (Fig. 3(a)) All highly correlated data intervals were combined to construct a signal (Fig. 3(b)) that was later mixed with the previous day's raw data and the resulting signal was used as an input to the Learn++ algorithm.

4.3   Experimental Results

The training dataset of 400 instances were divided into three subsets S_1 to S_3, each with 100 instances containing both classes to be used in four training sessions. In each training session, only one of these datasets was used. For each

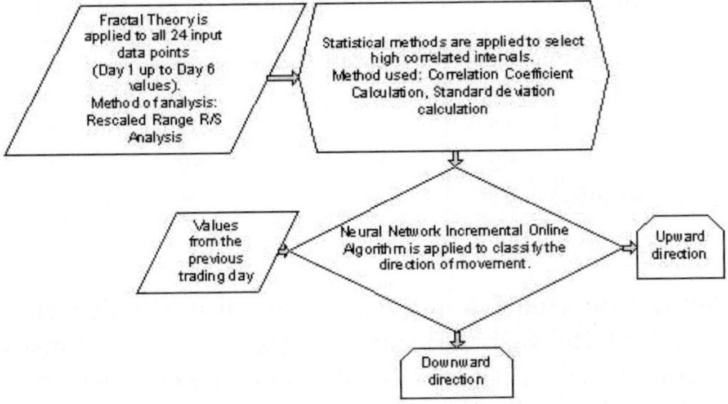

Fig. 2. Model framework for fractal R/S technique and online neural network time series analysis

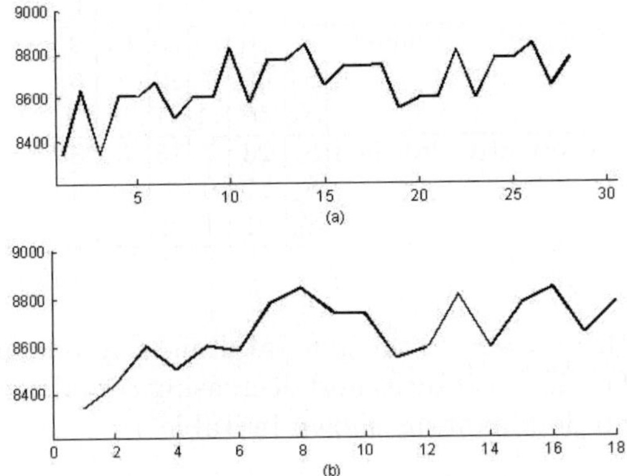

Fig. 3. A simulated fractal analysis signal. (a) Represent the original time 7-day time series with H=0.563 (b) Represents the reconstructed signal from the optimal intervals that consist of persistent intervals from the original 7-day series.

training session $k, (k = 1, 2, 3)$ three weak hypotheses were generated by Learn ++. Each hypothesis h_1, h_2 and h_3 of the kth training session was generated using a training subset T_{R_t} and a testing subset T_{E_t}. The results shows an improvement from coupling the MLP with a fractal R/S analysis algorithm; MLP hypothesis (weakLearner) preformed over 59% compared to the 50% without fractal analysis. We also observed a major classification accuracy improvement from 73% (without fractal R/S algorithm) to over 83% (with fractal R/S algorithm implemented). This result demonstrates the more performance improvement property of Fractal-Learn++ (as inherited from Learn++) on a given database. The last row of Table 2 shows the classification performance on the validation dataset, which gradually improved indicating that the confidence of framework increases as hard examples are introduced. The improvement is modest, however, as

Table 2. Training and generalisation performance of Fractal-Learn++ model

Database	Class(1)	Class(-1)	Test Performance (%)
S_1	68	32	74
S_2	55	45	77
S_3	62	48	82
Validate	58	52	–

majority of the new information is already learned in the first training session. Table 3 indicates that the vast majority of correctly classified instances tend to have very high confidences, with continually improved confidences at consecutive training sessions. While a considerable portion of misclassified instances also had

Table 3. Algorithm confidence results on time series analysis

		VH	H	M	VL	L
Correctly classified	S_1	26	24	25	15	9
	S_2	60	7	22	8	5
	S_3	55	11	21	3	11
Incorrectly classified	S_1	23	7	13	3	8
	S_2	27	0	1	3	4
	S_3	21	1	2	4	2

high confidence for this database, the general desired trends of increased confidence on correctly classified instances and decreasing confidence on misclassified ones were notable and dominant, as shown in Table 4.

Table 4. Confidence trends for the time series

	Increasing Steady	Decreasing
Correctly classified	123	3
Misclassified	9	31

5   Conclusion

In this paper, we propose the use of a fractal theory technique in conjunction with a neural network incremental algorithm to predict financial markets movement direction. As demonstrated in our empirical analysis, fractal re-scaled range (R/S) analysis algorithm is observed to establish clear evidence of persistence and long memory effects on the time series data. For any $7-day$ interval analysis we find the Hurst exponent of $H = 0,563$ for the range $7 < n < 28$, where n is the number of data points. A high degree of self-similarity is noted in the daily time series data. These long-term memory effects may be caused by the rate at which information is shared amongst the investors. Thus, the Hurst exponent can be said to be a measure of the impact of market sentiment, generated by past events, upon future returns in the stock market. The importance of using fractal R/S algorithm is noted in the improved accuracy and confidence measures of the Learn++ algorithm. This is a very comforting outcome, which further indicates that the proposed methodology does not only process and analyze time series data but it can also incrementally acquire new and novel information from additional data. As a recommendation future work should be conducted to implement techniques that are able to provide stable estimates of the Hurst exponent for small data sets as this was noted to be one of the major challenges posed by time series data.

References

1. Freund, Y., Schapire, R.: Stock market prices do not follow random walks: evidence from a simple specification test. Review of Financial Studies (1988) 41–66
2. Mandelbrot, B.: Fractals and scaling in finance: Discontinuity. Springer (1997)
3. Hurst, H.E.: Long-term storage of reservoirs. Transactions of the American Society **116** (1951) 770–778
4. Skjeltorp, J.: Scaling in the norwergian stock market. Physica A **283** (2000) 486
5. Gammel, B.: Hurst's rescaled range statistical analysis for pseudorandom number generators used in physical simulations. The American Physical Society **58** (1998) 2586
6. Byorick, J., Moreton, M., Marino, A., Polikar, R.: Learn++: A classifier independent incremental learning algorithm. Joint Conference on Neural Networks **31** (2002) 1742–1747
7. Littlestone, N., Warmuth, M.: Weighted majority voting algorithm. information and computer science **108** (1994) 212–216
8. Polikar, R., Udpa, L., Udpa, S., Honavar, V.: An incremental learning algorithm with confidence estimation for automated identification of nde signals. Transactions on Ul-trasonic Ferroelectrics, and Frequency Control **51** (2004) 990–1001
9. McIver, D., Friedl, M.: Estimating pixel-scale land cover classification confidence using nonparametric machine learning methods. Transactions on Geoscience and Remote Sensing **39** (2001)
10. S.Krause, Polikar, R.: An ensemble of classifiers approach for the missing feature proble. (2003) 553

MML Mixture Models of Heterogeneous Poisson Processes with Uniform Outliers for Bridge Deterioration

T. Maheswaran[1], J.G. Sanjayan[2], David L. Dowe[3], and Peter J. Tan[3]

[1] VicRoads, Metro South East Region, 12 Lakeside Drive, Burwood East, Vic 3151,
(Previously Department of Civil Engineering, Monash University when the research presented
in this paper was carried out)
Mahes.Maheswaran@roads.vic.gov.au
[2] Department of Civil Engineering, Monash University, Clayton, Vic 3800, Australia
jay.sanjayan@eng.monash.edu.au
[3] School of Computer Science and Software Engineering, Monash University, Clayton, Vic
3800, Australia
dld@bruce.csse.monash.edu.au,
Peter.Jing.Tan@infotech.monash.edu.au

Abstract. Effectiveness of maintenance programs of existing concrete bridges is highly dependent on the accuracy of the deterioration parameters utilised in the asset management models of the bridge assets. In this paper, bridge deterioration is modelled using non-homogenous Poisson processes, since deterioration of reinforced concrete bridges involves multiple processes. Minimum Message Length (MML) is used to infer the parameters for the model. MML is a statistically invariant Bayesian point estimation technique that is statistically consistent and efficient. In this paper, a method is demonstrated estimate the decay-rates in non-homogeneous Poisson processes using MML inference. The application of methodology is illustrated using bridge inspection data from road authorities. Bridge inspection data are well known for their high level of scatter. An effective and rational MML-based methodology to weed out the outliers is presented as part of the inference.

1 Introduction

Bridge asset management is an emerging concept in road authorities. Bridge management is a systematic process of maintaining, upgrading and operating bridge assets cost-effectively. It combines engineering principles with sound business practices and economic theory, and it provides tools to facilitate a more logical approach to decision-making. Thus, bridge asset management provides a framework for handling both short-and long-term planning. As defined by the American Public Works Association Asset Management Task Force, asset management is "... a methodology needed by those who are responsible for efficiently allocating generally insufficient funds amongst valid and competing needs."

Asset management of bridges has come of age because of (1) an increase in allowable truck loads of bridges, (2) changes in public expectations, and more importantly

A. Sattar and B.H. Kang (Eds.): AI 2006, LNAI 4304, pp. 322–331, 2006.
© Springer-Verlag Berlin Heidelberg 2006

(3) extraordinary advances in information technology and data-mining. Currently, bridge investment and maintenance decisions are based on tradition, intuition, personal experience, resource availability, and political considerations, with systematic application of objective analytical techniques applied to a lesser degree. Many road authorities limit application of their management systems to monitoring conditions and then plan and program their projects on a "worst first" basis.

The deterioration of a reinforced concrete element is not a homogeneous process. It involves chloride ingress, corrosion initiation, crack initiation and crack propagation stages. Therefore, a multi-stage, non-homogeneous Poisson process with multiple deterioration rates to capture the entire phenomenon of concrete bridge deterioration is adopted. The application of the methodology is illustrated using real-life bridge inspection data from VicRoads. VicRoads is a state government owned agency responsible for the maintenance and management of bridges on state highways and main roads in Victoria, Australia. In Victoria alone, more than \$50 million per year is spent on the maintenance and upgrade of bridges valued at more than \$6 billion.

Bridge inspection data used in this paper for modelling deterioration is based on Level 2 inspections according to VicRoads [9]. Level 2 inspections are managed on a state-wide basis to assess the condition state of each structure and its main components. The frequency of inspection varies between 2 and 5 years depending on bridge rating. The bridge element condition state is described on a scale of 1 to 4, where 1 stands for "excellent condition" and 4 stands for "serious deterioration". The inspector records the condition states of the bridge element and the percentage of that element in a particular condition state.

2   Non-homogeneous Poisson Process

The non-homogeneous or non-stationary Poisson process is a process where the arrival rate, $r(t)$ at time t, is a function of t. The counting process $\{N(t), t \geq 0\}$ is said to be a non-homogeneous Poisson process with intensity function $r(t)$, $t \geq 0$, if

(i) $N(0) = 0$;   (ii)   The   process   has   independent   increments;
(iii) $P\{N(t+h) - N(t) \geq 2\} = o(h)$; (iv) $P\{N(t+h) - N(t) = 1\} = r(t)h + o(h)$

Let $m(t) = \int_0^t r(s)\,ds$. Then it can be shown that the probability of n parts moving from condition state i to state $i+1$ can be expressed by Equation (1).

$$P\{N(t+s) - N(s) = n\} = e^{-[m(t+s)-m(t)]} \frac{[m(t+s) - m(t)]^n}{n!} \quad n \geq 0 \tag{1}$$

3   Minimum Message Length

The Minimum Message Length (MML) principle [12][16][14][3][2][11] is widely used for model selection in various machine learning, statistical and econometric problems [12][16][14][13][15][10][1][2][5][8][11] and references therein. The principle is that the best theory for a body of data is the one that minimises the size of the theory plus the amount of information necessary to specify the exceptions relative to the theory.

A Bayesian interpretation of the MML principle is that it variously states that the best conclusion to draw from the data is the theory with the highest posterior probability or, equivalently, that theory which maximises the product of the prior probability of the theory with the probability of the data occurring in light of that theory. For a hypothesis (or theory), H, with prior probability $\Pr(H)$ and data, D, the relationship between the probabilities can be written [4][15] as shown in the Equation (2) by application of Bayes's Theorem.

$$\Pr(H \ \& \ D) = \Pr(H) \cdot \Pr(D \mid H) = \Pr(D) \cdot \Pr(H \mid D) \ . \tag{2}$$

Equation (3) can be derived by re-arranging Equation (2).

$$\Pr(H \mid D) = \Pr(H) \cdot \Pr(D \mid H) / \Pr(D) \ . \tag{3}$$

Since D and $\Pr(D)$ are given and H needs to be inferred, the problem of maximising the posterior probability, $\Pr(H \mid D)$, can be regarded as the one of choosing H so as to maximise $\Pr(H) \cdot \Pr(D \mid H)$. Elementary coding theory tells us that an event of probability, p, can be coded by a message length $l = -\log(p)$. So, the length of a two-part message $(MessLen)$ conveying the parameter estimates based on some prior and the data encoded based on these estimates can be given as in Equation (4). In this paper, natural logarithms are used and the message lengths are in nits.

$$MessLen(H \ \& \ D) = -\log(\Pr(H)) - \log(\Pr(D \mid H)) \ . \tag{4}$$

Since $-\log(\Pr(H) \cdot \Pr(D \mid H)) = -\log(\Pr(H)) - \log(\Pr(D \mid H))$, maximising the posterior probability, $\Pr(H \mid D)$, is equivalent to minimising $MessLen(H \ \& \ D)$ given in Equation (4). The receiver of such a hypothetical message must be able to decode the data without using any other knowledge. The model with the shortest two-part message length is considered to give the best explanation of the data. For a discussion of the relationship of the works of Solomonoff, Kolmogorov and Chaitin with MML and the subsequent Minimum Description Length (MDL) principle [7], see [14] and [2].

The Poisson distribution is used in this paper for modelling bridge element deterioration. Let r be the rate at which the number of parts in a bridge element moving from condition state i to $i+1$, t_i be the length of the time interval and c_i be the number of parts moved in that time interval. In order to infer the rate of the process [13][15], first a Bayesian prior density on r is required. Let this prior be $h(r) = \left(e^{-r/\alpha}\right)/\alpha$ for some α. The message length can be expressed as in Equation (5), where L is the (negative) log-likelihood and F is the Fisher information [15][16][14][13].

$$MessLen = -\log(h) + L + \frac{1}{2}\log(F) + \frac{1}{2}\left(1 - \log(12)\right)$$

$$= \log(\alpha) + \frac{r}{\alpha} + r\sum_{i=1}^{N} t_i - \log(r)\sum_{i=1}^{N} c_i - \sum_{i=1}^{N} c_i \log(t_i) + \sum_{i=1}^{N}\log(c_i!) \tag{5}$$

$$-\frac{1}{2}\log(r) + \frac{1}{2}\log\left(\sum_{i=1}^{N} t_i\right) + \frac{1}{2}\left(1 - \log(12)\right)$$

We then estimate r by minimizing the message length [15] (Equation (6)).

$$\hat{r}_{MML} = \frac{C + \dfrac{1}{2}}{T + \dfrac{1}{\alpha}} \quad \text{where } C = \sum_{i=1}^{N} c_i \text{ and } T = \sum_{i=1}^{N} t_i \,. \tag{6}$$

Cut-points can be found in many machine learning problems. They arise where there is a need to partition data into groups which are to be modelled distinctly [4][10]. A piece-wise function is used for partitioning of data and the rate of the Poisson process, r, is assumed to be constant in between the cut-points. The message length including the penalties for cutting the data into groups assuming a uniform prior can be roughly given as a first draft (see, e.g. [10]) by Equation (7):

$$MessLenCp = \log(n+1) + \log\left(\frac{n!}{(n-ncp)!\,ncp!}\right) + \sum_{i=1}^{ncp+1} MessLen(i) \tag{7}$$

where $MessLenCp$ = Message length including penalties for cutting data; $MessLen(i)$ = Message length for data in between cut-point $(i-1)$ and cut-point (i) calculated using Equation (5); n = Maximum possible number of cut-points; and ncp = number of cut-points.

Note that it costs $\log_e(n+1)$ nits to specify ncp because $0 \leq ncp \leq n$. Letting cut-point $ncp + 1$ refer to the end of the data, $MessLen(ncp + 1)$ refers to the data after the ncp^{th} (i.e., last) cut.

A separate code-word of some length can be set aside for missing data. The transmission of the missing data will be of constant length regardless of the hypothesis classification, and as such will affect neither the minimisation of the message nor the (statistical) inference (Section 2.5 of [15] and Section 5, p.42 of [13]).

3.1 Models of Outliers and Multi-state Distribution

The bridge inspection data is from visual inspections, and the inspectors employed to gather data varied in their experience. Factors such as visibility at the time of inspection and resources available for the inspectors to carry out the bridge inspections, etc. may also affect the reliability of the data. A comprehensive study [6] on visual bridge inspection data concluded that significant measurement errors exist. Further, the measurement errors may show a seeming improvement in bridge element conditions (which is a physical impossibility) or give rather unusual data. Therefore, the data has to be screened or have an explicit model of outliers in order to take account of those errors before modelling the data.

If there are $n(t)$ members of class t $(t = 1,2,...,T)$ then the label used in the description of a thing to say that it belongs to class t will occur $n(t)$ times in the total message. If the relative frequency of class t is estimated as $p(t)$ (where $p(1)+p(2)+...+p(T) = 1$) then the information needed to quote the class membership of all things is given in Equation (8) [12]:

$$-\sum_{t=1}^{T}\left(n(t)+\frac{1}{2}\right)\log p(t) \tag{8}$$

It was decided to have two classes in our bridge deterioration modelling problem, known as outliers and non-outliers. An outlier is a datum which is considered to be erroneous and therefore not used in the estimation of Poisson rate. However, there is a cost to classifying a datum as an outlier. The message length for the outlier data points is given by

$$MessLenOl = N_{ol}\left(-\log(q)+\log(H+1)\right) \tag{9}$$

where $MessLenOl$ = Message length from (uniform) outlier data; N_{ol} = Number of

data in outlier class; q = Frequency of outlier class = $\dfrac{N_{ol}+\dfrac{1}{2}}{N+1}$ (see Equation (12));

$H+1$ = Maximum possible values a data point can take (100+1= 101);
N = Total number of data. These two terms in Equation (9) are so because each outlier must be encoded as an outlier and then its value is encoded as being uniformly equally likely from (H+1) different values.

The message length for non-outlier data points is given by

$$MessLenNol = N_{Nol}\left(-\log(1-q)\right)+MessLenCp \tag{10}$$

where $MessLenNol$ = Message length from non-outlier data; N_{Nol} = Number of data in non-outlier class = $N - N_{ol}$; and $MessLenCp$ is from Equation (7)

The message length for all data points is then given by

$$MessLenMo = MessLenNol + MessLenOl \tag{11}$$

Multi-state Distribution
For a multi-state distribution with M states, a uniform prior, $h(p) = (M-1)!$ is assumed over the $(M-1)$-dimensional region of hyper-volume $1/(M-1)!$ given by $p_1 + p_2 + ... + p_M = 1$; $p_i \geq 0$. Letting n_m be the number of things in state m and $N = n_1 + n_2 + ... + n_M$, minimising the message length Equation (5) gives that the MML estimate $\hat{p}_m$ of p_m is given by [13][15][12][16]:

$$\hat{p}_m = (n_m + 1/2)/(N + M/2) \tag{12}$$

Substituting Equation (12) into the message length Equation (5) gives rise to a (minimum) message length shown in the Equation (13) for both stating the parameter estimates and then encoding the things in light of these parameter estimates [12],[15].

$$MessLenMs = \frac{M-1}{2}\left(\log\left(\frac{N}{12}\right)+1\right) - \log(M-1)! - \sum_{i=1}^{M}\left(n_i + \frac{1}{2}\right)\log(p_i) \tag{13}$$

In this case, the distribution is binomial with M=2 in Equation (13). The total message length, including the cost of cut-points, modelling outliers and multi-state variables, is given by Equation (14).

$$MessLenT = MessLenMo + MessLenMs \tag{14}$$

4 Application of Methodology

4.1 Bridge Inspection Data

There are four condition states defined in the bridge inspection data. The deterioration process of a bridge element can be treated as three separate Poisson processes as defined below (there can be deterioration observed, but we assume not improvement; and we also assume Cs13=Cs14=Cs24=0):

– <u>Process Cs12:</u> parts of element deteriorate from condition state 1 to state 2
– <u>Process Cs23:</u> parts of element deteriorate from condition state 2 to state 3
– <u>Process Cs34:</u> parts of element deteriorate from condition state 3 to state 4

It is assumed that the bridge element was new when constructed, and the initial condition state is assumed to be $pc1i = 100$, $pc2i = 0$, $pc3i = 0$, and $pc4i = 0$ at time 0. Since the bridge inspections are recorded in percentages, it is assumed that there are 100 parts in an element. $Cs12$, $Cs23$ and $Cs34$ during the period between the ith and $(i+1)^{th}$ inspections can be calculated from Equations (15) to (17).

$$Cs12 = pc1\,i - pc1f \tag{15}$$

$$Cs34 = pc4f - pc4\,i \tag{16}$$

$$\begin{aligned} Cs23 &= (pc3f - pc3\,i) + (pc4f - pc4\,i) \\ &= (pc1\,i - pc1f) - (pc2f - pc2\,i) \end{aligned} \tag{17}$$

where $pc1i$, $pc2i$, $pc3i$, $pc4i$ = number of parts in condition state 1, 2, 3 and 4 respectively at the ith inspection; $pc1f$, $pc2f$, $pc3f$, $pc4f$ = number of parts in condition state 1, 2, 3 and 4 respectively at the $(i+1)^{th}$ inspection.

The bridge element considered in this study is precast concrete deck/slab (element number: 8P) in the most aggressive environment [9]. This element includes all precast concrete deck slabs and superstructure units forming the span and the deck of a bridge. Bridge inspection records (from 1996 to 2001) of 22 bridges were selected from the VicRoads database for the analysis. Fig. 1 shows the condition states of these selected bridges versus the ages (ranging from 0 to 39) of the bridges in years.

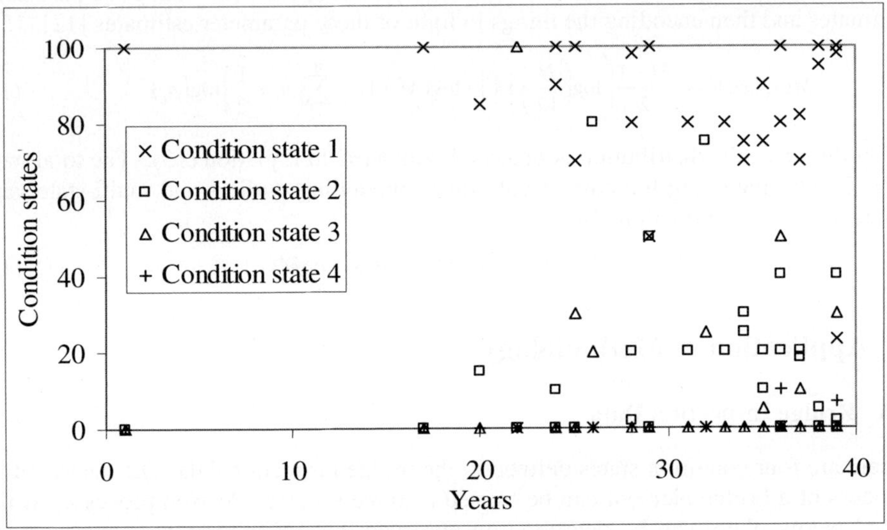

Fig. 1. Bridge element 8P in aggressive environment – condition states

4.2   Number and Location of Cut-Points and Estimation of Parameters

The minimum size of the time interval for the estimation of cut-points is assumed to be less than or equal to two years because the minimum time period for Level 2 inspections is in two-year cycles [9] and the deterioration expected for a bridge element within this time is relatively small.

The cut-points and the rates of the Poisson processes are estimated for each process by minimising the Message Length for each process separately. The multi-state (binary) distribution is used with both the Outlier model and the non-homogenous Poisson process model.

Fig. 2 shows the message lengths for various cut-points for Poisson process Cs12. The minimum message length of 165.24 nits was found to be with two cut-points (ncp=2) - hence three Poisson rates are estimated. Table 1 gives the cut-points followed by the Poisson rates of the processes. A closer examination of Poisson rates r_1, r_2 and r_3 reveals that $r_3 =0.041$ could not occur unless there is an improvement in the condition states of the bridge element. An improvement in condition states of bridge element can occur by carrying out repair works or by measurement errors. A closer look at the bridge inspection data revealed that this is most probably the improvement works carried out to bridge elements (after 37 years of service) rather than measurement errors. But, this conclusion cannot be confirmed, since the data available is for 39 years only. We therefore decided to exclude the Poisson rate r_3 from further calculations.

Table 1. Poisson rates and cut-points for Cs12

Cut number (i)	Start	1	2	End
Cut location cp_i (years)	0	29	37	39
Range between cut-points		<1	1 - 2	>2
Rates r_i		0.017	0.664	0.041
No. of Data in Outlier		8	4	1

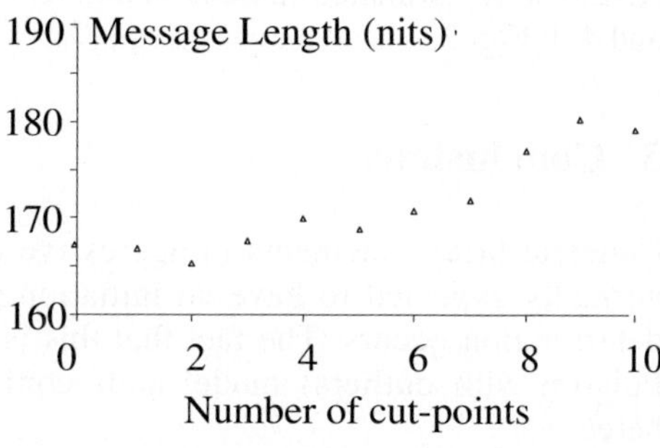

Fig. 2. Variation of message length for Poisson process Cs12

The message lengths for the Poisson process Cs23 for various cut-points were estimated in similar manner. The minimum message length of 85.66 nits was estimated for this Poisson process with no-cut point and a (very small) rate of 0.0006.

Similarly, the message lengths for Poisson process Cs34 for various cut-points resulted in the minimum message length of 36.42 nits for no-cut point and an even smaller Poisson process rate of 0.0005.

The estimated Cs12, Cs23 and Cs34 values are used to calculate the distribution of condition states of the bridge element. Fig. 3 shows the deterioration model for the bridge element estimated from the bridge inspection data shown in Figure 1. The

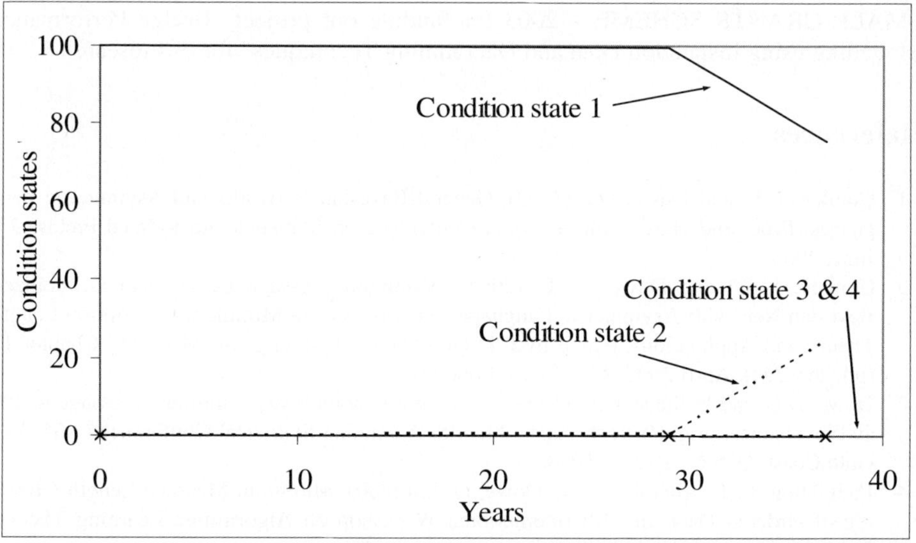

Fig. 3. Deterioration model for the bridge element 8P

number of parts moved from condition state 3 to state 4 (Cs34) and state 2 to state 3 (Cs23) were estimated to be zero and therefore there are no parts in condition states 3 and 4 in Fig. 3.

5   Conclusions

Concrete bridge elements in aggressive environments considered in this paper are normally expected to have an initiation period of about 30 years during which no deterioration occurs. The fact that this is accurately inferred by our (heterogeneous Poisson with outliers) model adds confidence to this modelling process adopted here.

Deterioration models for predicting the distribution of future condition states of bridge elements are an essential part of a bridge asset management system. It has been shown in this paper that concrete deterioration can be modelled using a non-homogeneous Poisson process together with an application of MML inference to estimate the Poisson rates. The estimated cut-points and Poisson rates in turn are used for predicting the distribution of the future condition states of bridge elements.

Bridge inspection data contain measurement errors or highly erroneous data due to a range of reasons including inspector subjectivity. Past attempts to model the data indicated that finding a structure in this type of data is very difficult. The methodology illustrated using bridge inspection data in this paper gives an objective and reasonably accurate way to identify and exclude measurement errors in the data.

Acknowledgement

The authors acknowledge Monash University's FACULTY OF ENGINEERING SMALL GRANTS SCHEME – 2003 for funding our project "Bridge Performance Modelling using Inspection Data and Data mining Techniques" for this research.

References

1. Comley, J.W. and Dowe, D.L.(2003), General Bayesian Networks and Asymmetric Languages, Proc. 2nd Hawaii International Conference on Statistics and Related Fields, 5-8 June, 2003.
2. Comley, J. W. and Dowe, D. L. (2005), Minimum Message Length and Generalized Bayesian Nets with Asymmetric Languages, in Advances in Minimum Description Length Theory and Applications, edited by P D Grunwald, I J Myung and M A Pitt, Chapter 11 (pp. 265-294), April 2005 (MIT Press: London).
3. Dowe, D.L. and Wallace, C.S. (1998), Kolmogorov complexity, minimum message length and inverse learning, abstract, page 144, 14th Australian Statistical Conference (ASC-14), Gold Coast, Qld, 6 - 10 July 1998.
4. Fitzgibbon, L. J., Allison, L. and Dowe, D. L. (2000), Minimum Message Length Grouping of Ordered Data, in 11th International Workshop on Algorithmic Learning Theory, 2000, LNAI 1968, Springer, Sydney, Australia, pp. 56 - 70.

5. Fitzgibbon, L.J., D. L. Dowe and F. Vahid (2004). Minimum Message Length Autoregressive Model Order Selection. In M. Palanaswami, C. Chandra Sekhar, G. Kumar Venayagamoorthy, S. Mohan and M. K. Ghantasala (eds.), Interna-tional Conference on Intelligent Sensing and Information Processing (ICISIP), Chennai, India, 4-7 January 2004. ISBN: 0-7803-8243-9, IEEE, pp. 439-444.

6. Moore, M., Phares, B., Graybeal, B., Rolander, D. and Washer, G. (2001), Reliability of Visual Inspection for Highway Bridges, Volume: 1 and 2, Final Report, Report No: FHWA-RD-01-020, NDE Validation Center, Office of Infrastructure Research and Development, Federal Highway Administration, McLean, VA, USA.

7. Rissanen, J. J. (1978), Modeling by Shortest Data Description, Automatica, 14, pp. 465 - 471.

8. Tan, P.J., and Dowe, D.L. (2004). MML Inference of Oblique Decision Trees, Proc. 17th Australian Joint Conference on Artificial Intelligence (AI'04), Cairns, Qld., Australia, Dec. 2004, Lecture Notes in Artificial Intelligence (LNAI) 3339, Springer-Verlag, pp1082-1088.

9. VicRoads (1995), VicRoads Bridge Inspection Manual, Melbourne, Australia.

10. Viswanathan, M., Wallace, C. S., Dowe, D. L. and Korb, K. B. (1999), Finding Cutpoints in Noisy Binary Sequences - A Revised Empirical Evaluation, Proc. 12th Australian Joint Conference on Artificial Intelligence, Lecture Notes in Artificial Intelligence (LNAI) 1747, Springer, Sydney, Australia, pp. 405 - 416.

11. Wallace, C. S. (2005), Statistical and inductive inference by minimum message length, Springer, Berlin, New York, ISBN 0-387-23795-X, 2005.

12. Wallace, C. S. and Boulton, D. M. (1968), An Information Measure for Classification, Computer Journal, 11, pp. 185 - 194.

13. Wallace, C. S. and Dowe, D. L. (1994), Intrinsic Classification by MML - the Snob Program, Proc. 7th Australian Joint Conference on Artificial Intelligence, UNE, World Scientific, Armidale, Australia, pp. 37 - 44.

14. Wallace, C. S. and Dowe, D. L. (1999), Minimum Message Length and Kolmogorov Complexity, Computer Journal (Special issue on Kolmogorov Complexity), 42(4), pp. 270 - 283.

15. Wallace, C. S. and Dowe, D. L. (2000), MML Clustering of Multi-State, Poisson, von Mises Circular and Gaussian Distributions, Statistics and Computing, 10, pp. 73 - 83.

16. Wallace, C. S. and Freeman, P. R. (1987), Estimation and Inference by Compact Coding, J. Royal Statistical Society (Series B), 49, pp. 240 - 252.

Extracting Structural Features Among Words from Document Data Streams

Kumiko Ishida[1], Tomoyuki Uchida[2], and Kayo Kawamoto[2]

[1] Depart. of Computer and Media Tech., Hiroshima City University, Japan
`k_ishida@toc.cs.hiroshima-cu.ac.jp`
[2] Faculty of Information Sciences, Hiroshima City University, Japan
`{uchida@cs, kayo@im}.hiroshima-cu.ac.jp`

Abstract. We consider the online data mining problem of continuously extracting all structural features among words from an infinite sequence of tree structured documents. In order to represent structural features among words appearing in tree structured documents, firstly, we introduce a consecutive path pattern (CPP, for short) on a list of words. A CPP is a sequence of consecutive paths from leaves to leaves. Then, we give a matching function over CPPs with respect to the recent frequency of a CPP, the recency of a CPP and the viewing time of tree structured document in which a CPP appears. Secondly, we present an online algorithm based on a sliding window strategy for extracting continuously all maximal CPPs as characteristic structural features from an infinite sequence of tree structured documents. Finally, by reporting experimental results on our algorithm, we show the good performance of our algorithm.

1 Introduction

Recently, XML and HTML documents have been rapidly increasing due to the rapid growth of Information Technologies. Since such documents have tree structures but no rigid structure, they are called tree-structured documents. A tree-structured document d can be represented by a rooted tree T_d, called the *OEM tree* of d. T_d has strings such as tags or texts in node labels and all internal nodes of T_d have ordered children. For example, an XML document *xml_sample* in Fig. 1, which is a tree-structured document, is represented by the tree T given in Fig. 2, that is, T is the OEM tree of *xml_sample*.

In the fields of data mining and knowledge discovery, many researchers [1,10,12] have developed techniques for extracting frequent tree patterns from a set of tree-structured documents such as XML documents, Web documents. Moreover, for data streams such as transaction records or Web visit logs which may be endless and arrive continuously in a time-varying way, mining techniques [2,5,11] are proposed. In order to discover characteristic schema, many researchers focused on tags and their structured relations. But in this paper, we are more interested in words and their structured relations rather than tags. The aim of this paper is to present an online algorithm for continuously extracting recent structural features among words commonly appearing in tree-structured documents which a user views recently.

A. Sattar and B.H. Kang (Eds.): AI 2006, LNAI 4304, pp. 332–341, 2006.

© Springer-Verlag Berlin Heidelberg 2006

```
<REUTERS>
    <DATE>26-FEB-1987</DATE>
    <TOPICS><D>cocoa</D></TOPICS>
    <TITLE>BAHIA.COCOA.REVIEW</TITLE>
    <DATELINE>SALVADOR,Feb 26</DATELINE>
    <BODY>
      Showers continued throughout the week in the Bahia cocoa zone,···
    </BODY>
</REUTERS>
```

Fig. 1. An XML document *xml_sample*

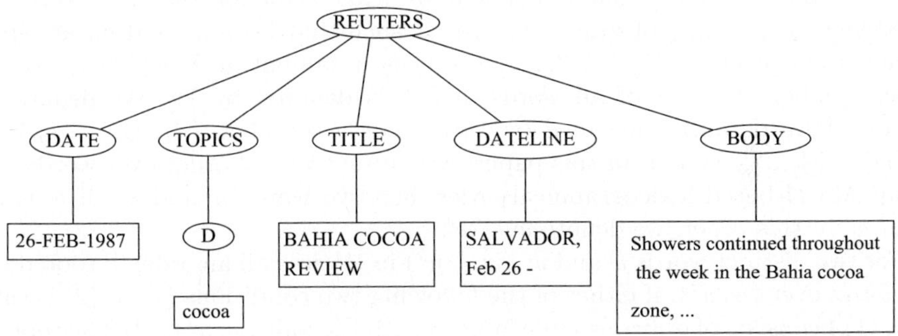

Fig. 2. The OEM tree T of an XML document *xml_sample*

Uchida et al. [12] presented a *consecutive path pattern* (*CPP*, for short) for representing structural features among words in tree-structured documents. We consider the problem for continuously extracting characteristic CPPs with respect to the recent frequency of a CPP, the viewing time and the recency of a document in which the CPP appears from an infinite sequence of documents. In this paper, by using the algorithm given in [12] for extracting all characteristic CPPs from a finite set of tree-structured documents, we present an online algorithm based on sliding window strategy for solving this problem. Our algorithm uses compact data structures such as a trie [6] and an FPTree [8] for maintaining all extracted characteristic CPPs. Next, we evaluate the performance of our online algorithm by implementing it on a PC and reporting some experimental results. Our algorithm for data streams gathers the information of the recent frequency, the viewing time and the recency of a document. Hence, we can design a new keyword search engine over XML/HTML documents by applying the gathered information to ranking pages over XML/HTML documents such as XRANK [7] and XSEarch [4].

2   Preliminaries

In this paper, we simply call a tree-structured document a *document*. For a set A, we denote by A^∞ the set of all infinite sequences on A. For a set or list S, the number of elements of S is denoted by $|S|$. Let Max_TIME be a nonnegative integer. Let $\mathcal{D}$ be the set of all documents and $\mathcal{N}$ the set of all nonnegative integers less than or equal to Max_TIME. A pair (d,n) of a document $d \in \mathcal{D}$ and an integer $n \in \mathcal{N}$ is said to be *document data*. Then, we model a *document data stream* as an infinite sequence of document data $S = ((d_1,n_1),\ (d_2,n_2),\ldots) \in (\mathcal{D} \times \mathcal{N})^\infty$. For any $i \geq 1$, n_i indicates a time needed for viewing the document d_i. For two integers i,j $(1 \leq i \leq j)$, $S[i,j]$ denotes the subsequence of S from i-th up to j-th document data.

An *alphabet* is a set of finite symbols and is denoted by Σ. We assume that Σ includes the space symbol " ". A finite sequence $(a_1, a_2, \ldots, a_n)$ of symbols in Σ is called a *string* and is denoted by $a_1 a_2 \cdots a_n$ for short. A *word* is a substring $a_2 a_3 \cdots a_{n-1}$ of $a_1 a_2 \cdots a_n$ over Σ such that both a_1 and a_n are space symbols and each a_i $(i = 2, 3, \ldots, n-1)$ is a symbol in Σ which is not the space symbol. The set of all words in Σ^+ is denoted by $\mathcal{W}$. We denote the set of all words appearing in d by $\mathcal{W}(d)$. For a set $D = \{d_1, d_2, \ldots, d_m\}$, let $\mathcal{W}(D) = \bigcup_{1 \leq i \leq m} \mathcal{W}(d_i)$. In this paper, we assume a total order over words such as an ASCII-based lexicographical order. For two words w and w', if w is less than w' in this order, we denote $w < w'$.

For two distinct words w and w' $(w < w')$ in $\mathcal{W}$, we call an ordered rooted tree t a *2-tree* over (w, w'), if either of the following two conditions (1) or (2) is satisfied. (1) t consists of only one node labeled with the pair (w, w'). (2) The root of t has just two children. The two leaves u and v of t have the words w and w' as node labels, respectively. All internal nodes including the root of t have strings over Σ as nodes label. The lefthand (resp. righthand) leaf of t is denoted by $ll(t)$ (resp. $rl(t)$). If t consists of only one node, we assume that $ll(t)=rl(t)$. For example, in Fig. 3, t_1, t_2, t_3 and t_4 are 2-trees over (BAHIA, SALVADOR), (SALVADOR, cocoa), (cocoa, week) and (cocoa, week). $ll(t_1)$ and $rl(t_1)$ are the leaves of t_1 which are labeled with the words "BAHIA" and "SALVADOR", respectively. Moreover, $ll(t_3) = rl(t_3)$ is the only one node which t_3 consists of.

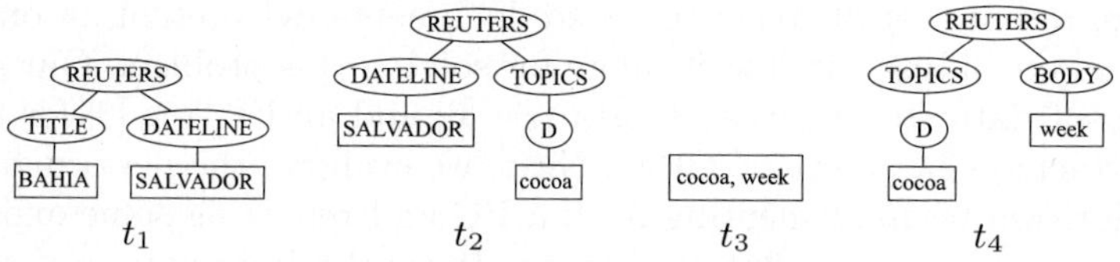

Fig. 3. 2-trees t_1, t_2, t_3 and t_4

Let $(w_1, w_2, \ldots, w_k)$ be a list of k words such that $w_i \leq w_{i+1}$ for $1 \leq i \leq k-1$. A list $\langle t_1, t_2, \ldots, t_{k-1} \rangle$ of $k-1$ consecutive 2-trees is called a *consecutive path pattern* (a *CPP*, for short) over $(w_1, w_2, \ldots, w_k)$ if for any $1 \leq i \leq k-1$, the node labels of $ll(t_i)$ and $rl(t_i)$ are w_i and w_{i+1}, respectively. Notice that, for any $1 \leq i \leq k-2$, both $rl(t_i)$ of t_i and $ll(t_i + 1)$ of t_{i+1} have the same label. For example, in Fig. 4, we give a CPP $\alpha = \langle t_1, t_2, t_4 \rangle$ over (BAHIA, SALVADOR, cocoa, week), where t_1, t_2, t_4 are given in Fig. 3.

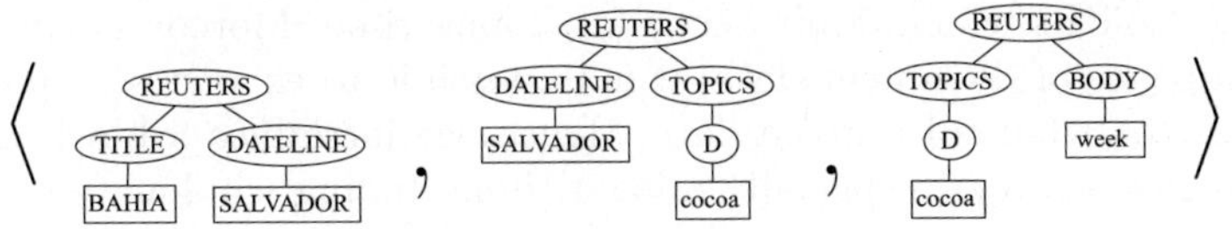

Fig. 4. A CPP α over $(\texttt{BAHIA}, \texttt{SALVADOR}, \texttt{cocoa}, \texttt{week})$

We define a *matching function of t for d* $\pi : V_t \to V_{T_d}$ as follows, where V_t and V_{T_d} are the node sets of t and T_d, respectively.

(1) π is a one-to-one mapping. That is, for any $v_1, v_2 \in V_t$, if $v_1 \neq v_2$ then $\pi(v_1) \neq \pi(v_2)$.

(2) π preserves the parent-child relation. That is, $\{v_1, v_2\}$ is an edge of t if and only if $\{\pi(v_1), \pi(v_2)\}$ is an edge of T_d.

(3) If t consists of only one node v, $\pi(v)$ is a leaf of T_d and the two words in the node label of v appear in the node label of $\pi(v)$ in V_{T_d}. Otherwise, for each leaf $v \in V_t$, $\pi(v)$ is a leaf of T_d and the word in the node label of v appears in the node label of the leaf $\pi(v)$ as a word and for each internal node $u \in V_t$, $\pi(u)$ is an internal node of T_d which has the same node label of u.

Two 2-trees t and t' are *isomorphic*, denoted by $t \cong t'$, if both a matching function of t for t' and a matching function of t' for t exist. For a CPP $\alpha = \langle t_1, t_2, \ldots, t_k \rangle$, we can also define a *matching function* $\pi : \bigcup_{1 \leq i \leq k} V_i \to V_{T_d}$ as follows, where V_i is the node set of the 2-tree t_i for each $1 \leq i \leq k$.

(1) For any $1 \leq i \leq k$, there exists a matching function $\pi_i : V_i \to V_{T_d}$ such that for any node v in t_i, $\pi(v) = \pi_i(v)$.

(2) For any $1 \leq i \leq k-1$, $\pi(rl(t_i)) = \pi(ll(t_{i+1}))$.

Two CPPs $\alpha = \langle s_1, \ldots, s_k \rangle$ and $\beta = \langle t_1, \ldots, t_k \rangle$ are said to be *isomorphic*, denoted by $\alpha \cong \beta$, if for any $1 \leq i \leq k$, $s_i \cong t_i$. Moreover, a CPP $\alpha = \langle s_1, \ldots, s_k \rangle$ is said to be a *sub-CPP* of $\gamma = \langle t_1, \ldots, t_\ell \rangle$ if there exists i $(1 \leq i \leq \ell - k + 1)$ such that for any $1 \leq j \leq k$, $t_{i+j-1} \cong s_j$. Especially, if $i > 1$ or $i + k - 1 < \ell$ then α is said to be a *proper* sub-CPP of γ. For a CPP $\alpha = \langle t_1, t_2, \ldots, t_k \rangle$, $\mathcal{CPS}_d(\alpha)$ denotes the set $\{(\pi(ll(t_1)), \pi(rl(t_1)), \ldots, \pi(rl(t_k))) \mid \pi$ is a matching function of α for $d\}$. If $|\mathcal{CPS}_d(\alpha)| \geq 1$, we say that α *appears* in d. For example, the a CPP $\langle t_1, t_2, t_4 \rangle$ appears in *xml_sample* given in Fig. 1, but the a CPP $\langle t_1, t_2, t_3 \rangle$ does not appear in *xml_sample* because there exists no matching function π such that $\pi(rl(t_2)) = \pi(ll(t_3)) = \pi(rl(t_3))$.

3   Online Mining from Document Data Streams

We consider the online data mining problem of extracting all characteristic CPPs in a given document data stream $\mathcal{S}$ at time of user's request. We naturally adopt the concepts that the following CPPs (a)-(c) are not important for users. (a) infrequent CPPs, (b) CPPs which only appear in documents having little viewing times, or (c) CPPs which don't occur for a long time. Moreover, since $\mathcal{S}$ is an infinite sequence of document data, it is impossible to extract all characteristic CPPs from the whole of a given data. Hence, we introduce a block B and the sliding window strategy, especially a logarithmic sliding window (see [5]). Let k and sw be integers which are given by users as a *block size* and a *sliding window's size*, respectively. In order to simplify our discussion, we assume $sw = 2^p - 1$ for some $p > 0$. For an integer j $(1 \leq j)$, the subsequence $B_j = \mathcal{S}[(j-1)*k+1, j*k]$ is said to be the *j-th block*, which is simply called a *block* if clear from context. We regard $\mathcal{S}$ as an infinite sequence of blocks. For two integer i, j $(1 \leq i \leq j)$, $\mathcal{B}[i, j]$ denotes a sequence of blocks from i-th to j-th block. For an integer j $(j \geq sw)$, we define a list of blocks $BL = \mathcal{B}[j - sw + 1, j] = \mathcal{S}[(j - sw) * k + 1, j * k]$.

Our final aim is to extract all characteristic CPPs from BL. Now, we indicate some definitions. $Occ_B(\alpha) = |\{h \mid (j-1)*|B|+1 \leq h \leq j*|B|, \mathcal{CPS}_{d_h}(\alpha) \neq \emptyset\}|$ and $VT_B(\alpha) = \max\{n \mid (d, n) \in B, \mathcal{CPS}_d(\alpha) \neq \emptyset\}$. $f_{\mathcal{B}[i,j]}(\alpha) = Occ_{\mathcal{B}[i,j]}(\alpha)/|\mathcal{B}[i,j]|$ denotes the frequency of α between i-th and j-th block.

$$F_{BL}(\alpha) = \frac{f_{\mathcal{B}[j,j]}(\alpha) + f_{\mathcal{B}[j-2,j-1]}(\alpha) + \cdots + f_{\mathcal{B}[j-sw+1,,j-(sw-1)/2]}(\alpha)}{p}$$

For example, when $p = 4$ and $sw = 15$, for an integer j $(j \geq 15)$,

$$F_{BL}(\alpha) = \frac{f_{\mathcal{B}[j,j]}(\alpha) + f_{\mathcal{B}[j-2,j-1]}(\alpha) + \cdots + f_{\mathcal{B}[j-14,j-7]}(\alpha)}{4}$$

$F_{BL}(\alpha)$ denotes the frequency of α in BL.

In order to extract all characteristic CPPs from B, we introduce a *matching function* G' defined as follows. Let (ρ_1, ρ_2) be a pair of real numbers in $(0, 1]$ which is called a *minimum support* (*minsup*, for short) *pair* of G'.

$$G'(\alpha, B, (\rho_1, \rho_2)) = \begin{cases} 1 & \text{if } \dfrac{Occ_B(\alpha)}{|B|} \geq \rho_1 \wedge \dfrac{VT_B(\alpha)}{Max\_TIME} \geq \rho_2, \\ 0 & \text{otherwise.} \end{cases}$$

Also, in order to extract all characteristic CPPs from BL, we define a *matching function* G as follows. Let $PT_{BL}(\alpha) = \sum_{h=m}^{j*k} n_h$, where $m = \max\{i \mid (j - sw) * k + 1 \leq i \leq j * k, \mathcal{CPS}_{d_i}(\alpha) \neq \emptyset\}$, and (ℓ_1, ℓ_2) a pair of two real numbers in $(0, 1]$, which is called a *minsup pair* on G.

$$G(\alpha, BL, (\ell_1, \ell_2)) = \begin{cases} 1 & \text{if } F_{BL}(\alpha) \geq \ell_1 \wedge \dfrac{1}{PT_{BL}(\alpha)} \geq \ell_2, \\ 0 & \text{otherwise,} \end{cases}$$

Moreover, α is said to be an (ℓ, ρ)-*maximal CPP on BL w.r.t.* (G, G') if the following two conditions hold. (1) there exists no CPP α' such that $G(\alpha', BL, \ell) = 1$ and α is a proper sub-CPP of α'. (2) there exists j $(q - sw \leq j \leq q)$ such that $G'(\alpha, B_j, \rho) = 1$, where $q = \lfloor i/k \rfloor$. At this time, α is also said to be an (ℓ, ρ)-*maximal CPP on $\mathcal{S}$ w.r.t.* (G, G') *at the stage i* $(i \geq sw * k)$.

Then, the online data mining problem for a document data stream is defined as follows.

Online Maximal CPPs Problem on (G, G')

Given: A document data stream $\mathcal{S} \in (\mathcal{D} \times \mathcal{N})^\infty$, a minsup pair ℓ of G and a minsup pair ρ of G'.

Problem: Find all (ℓ, ρ)-maximal CPPs on $\mathcal{S}$ w.r.t. (G, G') at time of a user's request.

Since the data streams may arrive in time-varying way, uncharacteristic CPPs are needed to account for CPPs becoming from uncharacteristic to characteristic. Hence, for a minsup pair (ℓ_1, ℓ_2), we define a pair (ℓ'_1, ℓ'_2) with $\ell'_1 < \ell_1$ and $\ell'_2 < \ell_2$ which is called by an *approximate minsup pair* for (ℓ_1, ℓ_2). We keep all characteristic CPPs α such that $G(\alpha, BL, \ell') = 1$ for each step $j \geq sw$. In Fig. 5, we present an online algorithm *Find_Maximal_CPP* and its procedures *Maintain_StorageTree* and *Pruning_StorageTree* for solving Online Maximal CPPs Problem on (G, G'). We illustrate an overview of the process of i-th stage of *Find_Maximal_CPP* in Fig. 6. In *Find_Maximal_CPP*, We adopt a compact data structure such as trie [6] and FPTree [8] to manage efficiently the information of all extracted characteristic CPPs. This data structure, called a *CPP Tree*, is given by an ordered rooted tree T satisfying the following conditions.

(1) For any node v except the root v_0 of T, v has a 2-tree $t(v)$ and the information being used to calculate G or G' for α_v as its label, where $\alpha_v = \langle t(v_1), t(v_2), \ldots, t(v_j), t(v)\rangle$ is a CPP lying on the path $(v_0, v_1, v_2, \ldots, v_j, v)$ from v_0 to v. In case of G, v has the information of the list $\mathbf{Occ} = (Occ_{\mathcal{B}[i,i]}(\alpha_v), Occ_{\mathcal{B}[i-1,i-1]}(\alpha_v), \ldots, Occ_{\mathcal{B}[i-sw+1,i-sw+1]}(\alpha_v))$ of integers and $PT(v) = PT_{BL}(\alpha_v)$. In case of G', the list $\mathbf{Occ}$ and $VT(v) = VT_{BL}(\alpha_v)$.

(2) Any child u of the root has a list of all nodes v except u such that $t(v) \cong t(u)$ holds. Such a list is denoted by $L(t(u))$ and used to determine whether or not a CPP $\alpha_v = \langle t(v_1), t(v_2), \ldots, t(v_j), t(v)\rangle$ is (ℓ, ρ)-maximal.

We present a CPP Tree T in Fig. 7. For v_9 of T, the CPP $\alpha = \langle t_1, t_2, t_3\rangle$ lies on the path (v_1, v_2, v_6, v_9) from the root v_1 to v_9. Moreover, since $t(v_9) = t_3$ and v_4 is the child of v_1 such that $t(v_4) \simeq t(v_9)$, $L(t(v_4)) = L(t_3)$ contains v_9. In Fig. 7, we show it by the broken arrow from v_4 to v_9. By using $L(t_3)$, we can see that two CPPs $\beta_1 = \langle t_2, t_3\rangle$ and $\beta_2 = \langle t_3\rangle$ which are lie on the path (v_1, v_3, v_7) and (v_1, v_4) respectively are proper sub-CPPs of α. Therefore α is (ℓ, ρ)-maximal, but β_1 and β_2 are not. We call CPP Trees based on G and G' a *Storage Tree* and a *Block Tree*, respectively.

Using Fig. 6, we explain *Find_Maximal_CPP*. Firstly, given a block B_i and a minsup pair ρ, *Find_CPP_for_Block* at line 5 of *Find_Maximal_CPP* returns a Block Tree BT_i storing all CPPs α such that $G'(\alpha, B_i, \rho) = 1$ holds. We can present *Find_CPP_for_Block* by slightly modifying the algorithm *Find_Freq_CPP* given in [12]. Next, *Maintain_StorageTree* removes the oldest Block Tree BT_{i-7} from the list of Block Trees SS, that is, moves the sliding window to the right by one block. Then, the current Storage Tree ST is updated to the updated Storage Tree ST' by removing the information about BT_{i-7} from ST and merging the information of BT_i to ST. In Fig. 6, since both BT_i and ST have the CPP $\alpha = \langle t_1, t_2, t_3\rangle$, the information of each node on the path expressing α is revised by merging BT_i to ST. On the other hand, since the CPP $\beta = \langle t_1, t_3\rangle$ exists in BT_i, but not in ST, we expand a new node which has t_3 as its label and the information of β, from the node in ST which has t_1 as its label. *Pruning_StorageTree* removes all infrequent CPPs from ST'. When the CPP $\gamma = \langle t_2, t_5\rangle$ becomes to be infrequent (i.e., $G(\gamma, BL, \ell') = 0$), γ is pruned from ST'. Finally, in order to gather all (ℓ, ρ)-maximal CPPs from ST' at i-th stage, *Mining_Charac_CPP*

Algorithm Find_Maximal_CPP with (G, G')
Input: A document data stream $((d_1, n_1), (d_2, n_2), \cdots, (d_i, n_i), \cdots) \in (\mathcal{D} \times \mathcal{N})^\infty$,
         a minsup pair ℓ of G, an approximate minsup pair ℓ' for ℓ of G,
         a minsup pair ρ of G', a sliding window's size sw and a block size k.
Output: All $(\ell,\ \rho)$-maximal CPPs w.r.t. (G, G') if a user requests.

begin
1. $ST := \texttt{NULL};\ SS := ();\ B := \emptyset;$
2.   **for** $i = 1$ **to** ∞ **do begin**   // i-th stage
3.   add the document (d_i, n_i) to B;
4.   **if** $|B| = k$ **then begin**
5. $(ST, SS):=Maintain\_StorageTree(ST, SS, sw, Find\_CPP\_for\_Block(B,\rho),\ \ell')$;
6. $B := \emptyset$
7.   **end**;
8.   **if** a user requests **then return** $Mining\_Charac\_CPP(ST,\ell)$;
9.   **end**
end.

Procedure Maintain_StorageTree
Input: A Storage Tree ST, a list of Block Trees SS, a slide window's size sw,
              a Block Tree BT and an approximate minsup pair ℓ'.
Output: The updated Storage Tree ST and the updated list of Block Trees SS.

begin
1.   **if** $|SS| < sw$ **then** $BT' := \texttt{NULL}$ **else** let BT' be the oldest element in SS;
2.   update ST by removing the information about BT' from ST;
3.   **if** $BT' \neq \texttt{NULL}$ **then** remove BT' from SS;
4.   append BT to SS and merge the information which BT keeps to ST;
5.   **if** $|SS| = sw$ **then** $ST :=Pruning\_StorageTree(ST, SS, \ell')$;
6.   **for** each node u of ST which is not a child of the root of ST **do**
7.   add u to the node list $L(w)$ of the child w of the root of ST such that
        $2tree(w) \cong 2tree(u)$;
8.   **return** ST and SS
end;

Procedure Pruning_StorageTree
Input: A Storage Tree ST, a list of blocks SS and an approximate minsup pair ℓ'.
Output: The updated Storage Tree ST.

begin
1.   **for** each node v of ST in pre-order **do begin**
2.   let α_v be a CPP which lies on the path from the root of ST to v;
3.   **if** $G(\alpha_v, SS, \ell') = 0$ **then begin**
4.   **for** each descendant u of v such that u is not a child of the root of ST **do**
5.   remove u from the node list of the child w of the root of ST
            such that $2tree(w) \cong 2tree(u)$;
6.   prune the root v and all descendants of v from ST
7.   **end**
8.   **end**
9.   **return** ST
end;

Fig. 5. Algorithm **Find_Maximal_CPPs** with $(G,\ G')$ and Procedures **Maintain_StorageTree** and **Pruning_StorageTree**

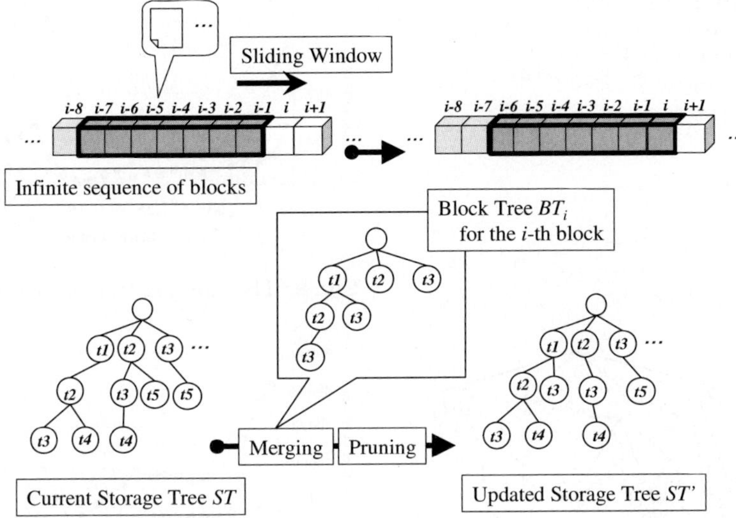

Fig. 6. Illustration of the process of $Find\_Maximal\_CPP$ at i-th stage

visits each leaf v in ST' in pre-order and calculate $G(\alpha_v, BL, \ell)$, for $\alpha_v = \langle t_1, ..., t_k \rangle$ lying on the path from the root to v. If $G(\alpha_v, BL, \ell) = 1$, Min-ing_Charac_CPP checks whether or not α_v is (ℓ, ρ)-maximal by using $L(t_k)$. If α_v is (ℓ, ρ)-maximal, $Mining\_Charac\_CPP$ returns α_v. In Fig. 6, $Mining\_Charac\_$ CPP returns the CPPs such as $\langle t_1, t_2, t_3 \rangle$ and $\langle t_1, t_2, t_4 \rangle$.

4   Experimental Results

In this section, we explain our experimental environments. Then, we discuss the performance of our algorithm $Find\_Maximal\_CPP$ given in previous section by reporting our experimental results.

We implemented our algorithm in C++ on a PC running Red Hat Linux8.0 with a 3.4GHz Pentium 4 processor and a 1GB of main memory. A document data stream used in experiments was generated from Reuters-21578 text categorization collection in [9], which has 21,578 SGML documents, and its size is about 28MB. Each document is assigned a randomly produced integer up to 600 from 1 as a viewing time of it. We set Max_TIME for 600. Let $PT = (500*600*3)$. We experimented in cases that a minsup pair for G is $(0.02, 1/PT)$ or $(0.07, 1/PT)$, a minsup pair for G' is $(0.1, 0.5)$ and an approximate minsup pair for G is $(0.016, 0.8/PT)$ or $(0.056, 0.8/PT)$. We set a block size for 500 and a sliding window's size for 15. In each experiment, we measured the running time taken for executing operations from line 4 to line 7 of $Find\_Maximal\_CPP$ and the number of nodes in the resulting Storage Tree after executing $Maintain\_StorageTree$. These experimental results are shown in Fig. 8 and Fig. 9, respectively. Since $Pruning\_StorageTree$ executes after the 15th block, both the running time and

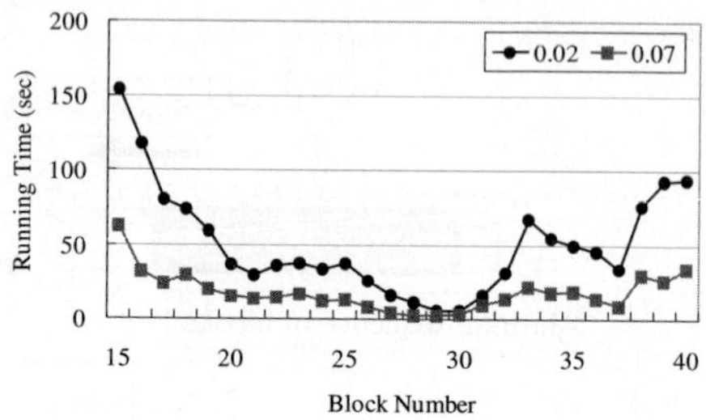

Fig. 8. The running time for each block

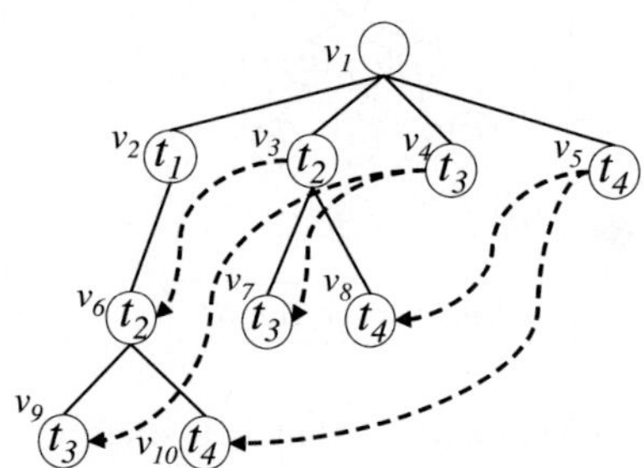

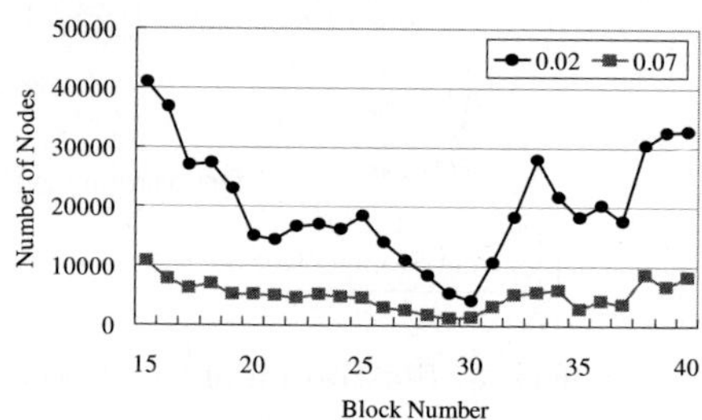

Fig. 7. A CPP Tree T. A directed edge $e = (u, v)$, which is drawn a broken arrow, indicates that the node list of the node u contains the node v.

Fig. 9. The number of nodes of a Storage Tree for each block

the number of nodes monotonously increase before the 15th block and decrease drastically after the 15th block. Hence, Fig. 8 and Fig. 9 show the results from the 15th block. From two experimental results, we can find a correlation between these measures. We can see that the number of nodes in case of 0.07 is stable around 5000 in Fig.9. This reason is that almost all CPPs extracted at each stage are similar to previous extracted CPPs due to the high frequency. On the other hand, consider the extreme fluctuation of the number of nodes in Fig. 9 in case of 0.02. It's not stable. This reason is that a lot of CPPs are extracted but almost those CPPs will be pruned later due to the low frequency. Moreover, in Fig.8, we can see that the running time of analyzing each block is at most 150 sec even if the frequency is 0.02. From these results, we can see that *Find_Maximal_CPPs* have good performance for extracting structural features among words from an infinite sequence of documents.

5   Concluding Remarks

In this paper, we have considered the problem of extracting structural features among words from a document data stream. In order to represent structural features among words, we have introduced a consecutive path pattern (CPP, for short) presented in [12]. Moreover, we give a matching function over CPPs with respect to the recent frequency of a CPP, the recency of a CPP and the viewing time of tree structured document in which a CPP appears. Then, we have given an online algorithm based on a sliding window strategy for finding all maximal CPPs from a document data stream. As a future task, by modifying the ranking mechanisms such as XRANK [7] or XSEarch [4] to a page ranking mechanism over CPPs, we will construct an efficient search engine for XML documents as an application of our online algorithm for a document data stream. Moreover, we will apply the techniques presented in this paper to implementation of logic wrappers using XSLT transformation language [3].

References

1. T. Asai, K. Abe, S. Kawasoe, H. Arimura, H. Sakamoto, and S. Arikawa. Efficient substructure discovery from large semi-structured data. *Proc. 2nd SIAM Int. Conf. Data Mining (SDM-2002)*, pages 158–174, 2002.
2. T. Asai, H. Arimura, K. Abe, S. Kawasoe, and S. Arikawa. Online algorithms for mining semi-structured data stream. *Proc. IEEE International Conference on Data Mining (ICDM'02)*, pages 27–34, 2002.
3. A. Badica, C. Badica, and E. Popescu. Implementing logic wrappers using XSLT stylesheets. *International Multi-Conference on Computing in the Global Information Technology, ICCGI 2006 (to appear)*, 2006.
4. S. Cohen, J. Mamou, Y. Kanza, and Y. Sagiv. XSEarch: A semantic search engine for XML. *Proc. 29th VLDB Conference*, 2003.
5. C. Giannella, J. Han, J. Pei, X. Yan, and P.S. Yu. *Next Generation Data Mining*, chapter Mining Frequent Patterns in Data Streams at Multiple Time Granularities, pages 191–212. AAAI/MIT, 2003.
6. G. Gonnet and R. Baeza-Yates. *Handbook of Algorithms and Data Structures*. Addison-Wesley, 1991.
7. L. Guo, F. Shao, C. Batev, and J. Shanmugasundaram. XRANK: ranking keyword search over XML documents. *ACM SIGMOD*, pages 16–27, 2003.
8. J. Han and M. Kamber. *Data Mining: Concepts and Techniques*. Morgan Kaufmann Publishers, 2001.
9. D. Lewis. Reuters-21578 text categorization test collection. *UCI KDD Archive, http://kdd.ics.uci.edu/databases/reuters21578/reuters21578.html*, 1997.
10. T. Miyahara, Y. Suzuki, T. Shoudai, T. Uchida, K. Takahashi, and H. Ueda. Discovery of frequent tag tree patterns in semistructured web documents. *Proc. PAKDD-2002, Springer-Verlag, LNAI 2336*, pages 341–355, 2002.
11. R. Rastogi. Single-path algorithms for querying and mining data streams. *Proc. HDM'02*, pages 43–48, 2002.
12. T. Uchida, T. Mogawa, and Y. Nakamura. Finding frequent structural features among words in tree-structured documents. *Proc. PAKDD-2004, Springer-Verlag, LNAI 3056*, pages 351–360, 2004.

Clustering Similarity Comparison Using Density Profiles

Eric Bae[1], James Bailey[1], and Guozhu Dong[2]

[1] NICTA Victoria Laboratory, Department of Computer Science and Software
Engineering, University of Melbourne, Australia
[2] Department of Computer Science and Engineering, Wright State University, USA

Abstract. The unsupervised nature of cluster analysis means that objects can be clustered in many ways, allowing different clustering algorithms to generate vastly different results. To address this, clustering comparison methods have traditionally been used to quantify the degree of similarity between alternative clusterings. However, existing techniques utilize only the point memberships to calculate the similarity, which can lead to unintuitive results. They also cannot be applied to analyze clusterings which only partially share points, which can be the case in stream clustering. In this paper we introduce a new measure named ADCO, which takes into account density profiles for each attribute and aims to address these problems. We provide experiments to demonstrate this new measure can often provide a more reasonable similarity comparison between different clusterings than existing methods.

1 Introduction

Cluster analysis is a fundamental machine learning task which discovers patterns, relationships and structures in an unsupervised manner. It has been used in a variety of fields, including biomedicine, information retrieval and financial institutions, to discover hidden knowledge and information. However, clustering is naturally an ill-posed problem, where the act of grouping similar data objects is a subjective notion and highly dependent on the clustering criterion. For this reason, a vast number of algorithms have been developed, each aiming to address different aspects of the problem, yet such algorithms often provide very different results. Moreover, even when a single algorithm is used, different alternative clusterings[1] can easily be generated, simply by changing the initial conditions of the algorithm. Therefore, in order to provide a measure of comparison between clusterings, cluster analysis has been often accompanied by a comparison method. Formally called external validation [16], this provides a quantitative measure of the degree to which two different clusterings are similar/different.

However, the current comparison measures suffer from a fundamental problem of judging the clustering similarity/difference purely on the membership of points to clusters. While these point-to-cluster assignments can be an important

---

[1] A clustering is a set of clusters.

A. Sattar and B.H. Kang (Eds.): AI 2006, LNAI 4304, pp. 342–351, 2006.
© Springer-Verlag Berlin Heidelberg 2006

determining factor in defining clusterings, they completely neglect other important aspects of data, which can seriously affect the outcome. These measures also suffer from the limitation that they are not applicable for comparing clusterings which may partially or not at all share points.

2   Problems and Motivations

We illustrate the problem in Fig. 1. Here we have three clusterings, each with three clusters. Figure 1(a) is a pre-defined clustering which is compared against 1(b) and 1(c). Both clusterings 1(b) and 1(c) have five points clustered differently compared to 1(a).

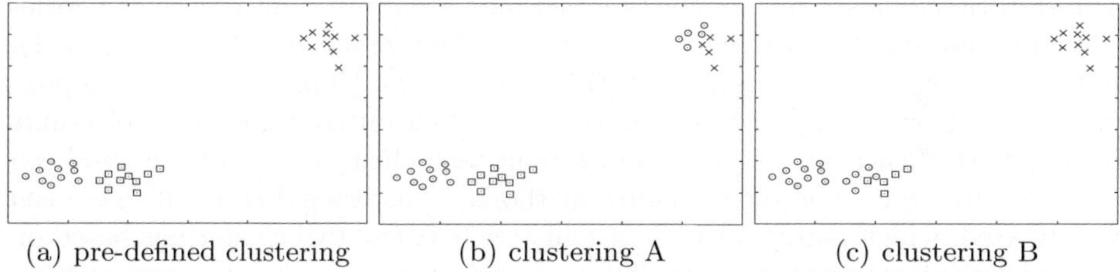

(a) pre-defined clustering   (b) clustering A   (c) clustering B

Fig. 1. Pre-defined clustering 1(a) containing 3 clusters, compared to two clusterings 1(b) and 1(c). Membership based measures give the same values for both comparisons.

Let clustering comparison X be between 1(a) and 1(c), while comparison Y is between 1(a) and 1(b). When comparing in terms of either cluster representatives (i.e. centroids), shapes or point distributions of clusters it seems intuitive that the degree of similarity for comparison X should not be the same as the degree of similarity for comparison Y. For example, suppose a new point were added to the dataset and it was merged with the closest cluster. It seems more probable that in both 1(a) and 1(c) it would join the same cluster. However, for 1(b), it is more likely that it might join a different cluster than 1(a). This is because 1(a) and 1(c) share a higher structural similarity, than 1(a) and 1(b). However, the available comparison measures are not able to recognize this difference. For example, a popular pair-counting measure, the Rand Index [13], gives a similarity value of 0.44 for both comparisons[2]. In fact, it is easily possible to generate arbitrary clusterings, which give the exactly same Rand index value when compared to 1(a), provided it has just five points clustered differently. Therefore treating point-to-cluster assignments as a primary (if not only) measure of comparison has limitations and does not necessarily correspond with intuition.

In this paper, we address this problem by developing a new clustering similarity measure we term ADCO[3]. The contribution of ADCO is to address two main limitations of existing methods.:

---

[2] Section 3 and 5 provide descriptions of these measures and experimental results.
[3] **Attribute Distribution Clustering Orthogonality.**

- **Addressing Non intuitive Behaviour:** ADCO incorporates distribution information of data points along each attribute, considering the shapes or density profiles of the clusters. This provides more intuitive comparisons than membership based techniques like Rand [13] or Jaccard [7] indices.
- **Applicability to stream data clustering:** ADCO can compare clusterings that may be built upon entirely different point sets and thus support the post-analysis phase in stream data clustering, where clusterings using different stream windows are compared. Comparison of clusterings using different sets of points is impossible for membership based techniques.

3   Related Work

The traditional clustering comparison methods are divided into three categories : 1) pair counting, 2) set matching and 3) variation of information (see table 1). In the pair-counting category, Rand [13] and Jaccard [7] indices have been popular for their simplicity. These methods are based on counting pairs of points belonging to the same or different clusters in each clustering and have also been extended in [8,2]. For set-matching methods, Clustering Error [10] has been widely used, which matches the 'best' clusters between two clusterings based on the number of points they share. The comparison is given by the total number of points shared between pairs of matching clusters over all the points. Other set-matching methods are described in [10]. Finally, Variation of Information introduced in [11], is based on information theory, measuring the amount of mutual information between two clusterings via the number of points they share. More recently, authors in [17] applied Mallows distance function to cluster representatives to calculate a comparison. Its method, although addresses a similar problem to ours, is nevertheless still more similar to the membership-based approaches, supplementing them with additional information about cluster centroids.

Clustering comparison methods have also been applied within the context of *ensemble clustering*, which merges several clusterings to form a consensus clustering. A popular technique for merging is called 'majority voting' [3], which is a pair-counting method extended over multiple clusterings. Using a co-association matrix of data points, pairs of points are given a score if they appear in the same cluster over all available clusterings. The pairs with a score higher than

Table 1. Definitions of Rand index (RI), Jaccard index (JI), Clustering Error (CE) and Variation of Information (VI). For RI and JI, N_{11} and N_{00} refer to the 'agreement' while N_{10} and N_{01} are 'disagreement' values between two clusterings. For CE, n is the number of objects and K is the number of clusters in each clustering. $n_{k,\sigma(k)}$ finds the 'best match' between pairwise clusters. For VI, $H(C)$ refers to the entropy of the clustering C, while $H(C,C')$ is the joint entropy of two clusterings.

RI	$RI(C,C') = \frac{N_{11}+N_{00}}{N}$	JI	$JI(C,C') = \frac{N_{11}}{N_{11}+N_{10}+N_{01}}$
CE	$CE(C,C') = 1 - \frac{1}{n}max \sum_{k=1}^{K} n_{k,\sigma(k)}$	VI	$VI(C,C') = 2H(C,C') - H(C) - H(C')$

pre-defined threshold are then 'voted' to be in the same cluster. In [15], clusterings are represented as a set of connected hypergraphs. Here, vertices connected by edges are objects in the same cluster over all clusterings. HyperGraph Partitioning algorithm [9] is then applied to find the consensus clustering by cutting a minimum number of hyper-edges. Although the approach is different, its underlying idea is to find dense intersections between clusterings and the method is considered as a variant to membership based methods.

Comparison methods are also used is in stream data clustering [1,2]. This raises an interesting analysis task, as clusterings can evolve over time and studying this evolution can uncover valuable information. In [1], the comparison of evolving clusterings is described, where clusterings at different periods are compared by observing any newly formed, removed or modified clusters. The technique used is membership-based and assumes that clusterings have at least some non-empty overlaps of data points, meaning windows for which clusterings do not share any points cannot be compared.

4   The ADCO Similarity Measure

4.1   Background and Terminology

Before presenting our new measure for comparing (dis)similarity between clusterings, we provide some necessary definitions here. Let $D = \{d_1, d_2, .., d_n\}$ be a dataset of n objects, described by r attributes $\{a_1, \ldots, a_r\}$. Let $d_i[a_j]$ refer to the value of object d_i on attribute a_j. A *clustering* C, is partition of d into a set of clusters. i.e. $C = \{c_1, \ldots, c_k\}$, where each c_i is a *cluster* (set of points).

Let $C_1 = \{c_1, \ldots, c_k\}$ and $C_2 = \{c'_1, \ldots, c'_k\}$ be the two clusterings which will be compared. Note that we assume the number of clusters in each clustering must be the same (an assumption also shared by existing measures). The *similarity* between two clusterings, $Sim(C_1, C_2)$, is a function which computes their similarity, with higher values of the measure indicating higher dissimilarity (less similarity). Various similarity measures have been defined in existing work, e.g. Rand index (Sim_{Rand}) [13], Jaccard index ($Sim_{Jaccard}$ [7]). Our measure will be referred to as ADCO, or Sim_{ADCO}.

4.2   Computing the ADCO Similarity Measure

Our ADCO similarity measure aims to determine the similarity between two clusterings based on their density profiles along each attribute. Essentially, r-dimensional space is chopped up into a "hyper grid". Points from the dataset occupy exactly one of the cells in this grid. The similarity between two clusters corresponds to how similarly the point sets from each cluster are distributed across the grid. The similarity between two clusterings then corresponds to the amount of similarity between their component clusters.

Suppose (the range of) each attribute a_i is divided into q bins, using some discretisation method. Let a_i^j refer to the set of values of attribute a_i within

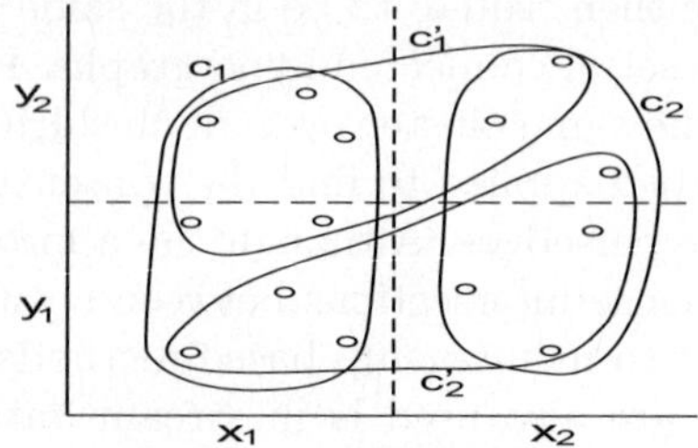

Fig. 2. Two Clusterings C and C' with attributes X and Y binned into q number of intervals where $C = \{c_1, c_2\}$ and $C' = \{c_1', c_2'\}$

bin j. Figure 2 shows two clusterings, each with two clusters ($k = 2$) and each attribute has been divided into two bins ($q = 2$).

To compute Sim_{ADCO}, we start by computing the density of cluster c for each bin j along each attribute a_i.

$$dens_c(a_i, j) = |\{d \in c \mid d[a_i] \in a_i^j\}| \tag{1}$$

where $dens_{c_i}(a_i, j)$ is the density (number of points) of cluster c for an attribute-bin pair (a_i, j). For a given cluster c, we can compute the $dens_c(a_i, j)$ measure for all bins (r of them), of all attributes (q of them), giving $r \times q$ measures in total. Using some arbitrary ordering scheme, we can form a vector of length $r \times q$ containing these measures, $dens_c = \{dens_c(a_1, 1), dens_c(a_1, 2), .., dens_c(a_r, q)\}$.

Example: For Fig. 2 where two clusterings C and C' are present, let $(X, x_1), (X, x_2), (Y, y_1), (Y, y_2)$ be the ordering of the attribute bin pairs. Then $dens_{c_1} = (8, 0, 5, 3)$, $dens_{c_2} = (0, 6, 3, 3), dens_{c_1'} = (5, 2, 2, 5), dens_{c_2'} = (3, 4, 6, 1)$.

It is now possible to compare two clusterings, by measuring the similarity between their component clusters with a dot product operation as follows:

$$C \cdot C' = \sum_{i=1}^{k} dens_{c_i} \cdot dens_{c_i'}. \tag{2}$$

where the $\cdot$ refers to the dot product between two vectors. A zero value indicates that the two clusterings are dissimilar and a large value indicates similarity. Equation 2 compares clusters on a pairwise basis, (c_1 versus c_1', c_2 versus c_2', etc). However, it is important that our similarity measure be independent of the actual cluster names assigned. We thus need to be able to permute the second clustering to calculate all possible pairings of clusters from C_1 and C_2. From these permutations, we select the pairing which gives a maximum value. i.e.

$$PairwiseSim(C, C') = max_P[C \cdot P(C')] \tag{3}$$

where P ranges over all permutations of C'. For Fig. 2 with a pairing $C = (c_1, c_2)$ and $C' = (c_1', c_2')$, we get $C \cdot C' = 110$. In contrast, if C' is permuted such that c_2' is renamed to c_1' and c_1' is renamed to c_2', we get $C \cdot P(C') = 90$. This shows that the first pairing of c_1 and c_1' and c_2 and c_2' gives the largest value.

The final step is then to normalize this pairwise similarity value, with respect to the maximum possible similarity. This is then subtracted from 1, so that 0 indicates highly similar and 1 indicates highly dissimilar.

$$Sim_{ADCO}(C, C') = 1 - \frac{PairwiseSim_D(C, C')}{MaxSim(C, C')} \tag{4}$$

where $MaxSim(C, C') = max(C \cdot C, C' \cdot C')$. Note that $MaxSim(C, C') = max(C \cdot C, C' \cdot C')$ is the upper bound on the dot product values involving at least one of C and C'. For the example of Fig. 2, $Sim_{ADCO}(C, C') = 0.276$.

ADCO Properties: Finally, we briefly describe properties ADCO.

- **Symmetry:** $ADCO(C, C') = ADCO(C', C)$
- **Triangle inequality:** For any three clusterings C, C' and C'',

$$ADCO(C, C') + ADCO(C', C'') \geq ADCO(C, C''). \tag{5}$$

The advantages of these properties are well understood and described in [10].

5 Experiments

We compared the behaviour of ADCO with four existing measures described in table 1. Clusterings were generated using k-means, EM, CURE, FarthestFirst (FF), Average-Linkage (AL), Complete-Linkage (CL) and Single-Linkage (SL). All initial parameters of algorithms were kept constant throughout the experiment and values are measured between 0 and 1, where a high value indicates a high dissimilarity. For all experiments, we set the number of bins q to 10, which is a commonly used choice for discretising data [14].

Synthetic Datasets: Two synthetic datasets are shown in Fig. 1 and 3. As already mentioned in section 2, in Fig. 1 clusterings in Fig. 1(a) and 1(c) are more similar than 1(a) and 1(b). Similarly in Fig. 3, Fig. 3(a) and 3(c) are closer

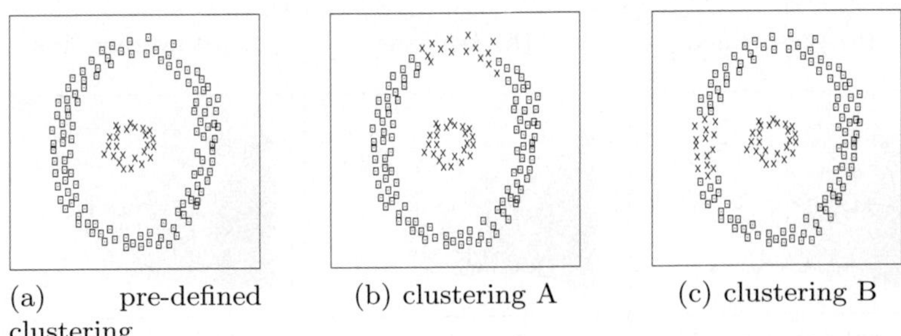

(a)   pre-defined clustering   (b) clustering A   (c) clustering B

Fig. 3. Pre-defined clustering of two clusters compared to two clusterings 3(b) and 3(c). Membership based measures give exactly same values for both comparisons.

Table 2. Comparing clusterings in 1 and 3. For both datasets, ADCO is the only measure that detects the structural difference.

	ADCO	RI	JI	CEM	VI
figures 1(a) vs. 1(c)	**0.16**	0.17	0.41	0.17	0.16
figures 1(a) vs. 1(b)	**0.47**	0.17	0.41	0.17	0.16
figures 3(a) vs. 3(c)	**0.16**	0.44	0.45	0.32	0.15
figures 3(a) vs. 3(b)	**0.33**	0.44	0.45	0.32	0.15

in regards to point distributions than 3(a) and 3(b). Table 2 shows how ADCO can recognise this distinction, while other measures fail to do so, giving the same value for all comparisons.

Real Datasets: We looked at two real world datasets, 'diabetes' and 'credit'. Each dataset comes with a pre-defined clustering (or class labels), which we then compared against clusterings generated by each of the clustering algorithms. The dataset 'diabetes' in Fig. 4(a), contains two natural clusters. In Fig. 4(b) and 4(c), we show the clusterings of k-means and AL, projected onto two attributes, to assist in visualisation. Similar to previous examples, we can see that Fig. 4(c) is more dissimilar to the Fig. 4(a) than the clustering in Fig. 4(b). However, when we observe the comparison measures in 3, this is not reflected by four membership based measures. In fact, ADCO is the only measure that can correctly describe this increase in dissimilarity from k-means to AL. For dataset 'credit', we also see a similar trend. The comparison between 4(d) and 4(f) is more dissimilar than the one between 4(d) and 4(e). Table 3 shows that ADCO is the only measure that can recognise this correctly.

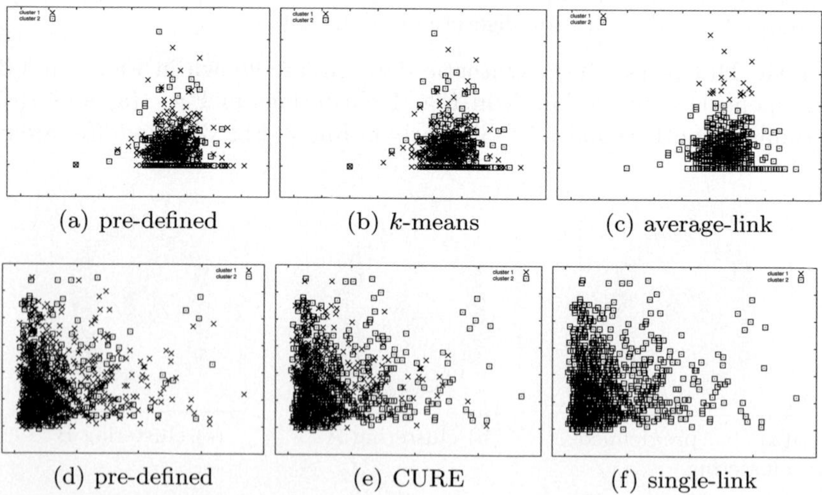

(a) pre-defined   (b) k-means   (c) average-link

(d) pre-defined   (e) CURE   (f) single-link

Fig. 4. Three clusterings of diabetes (4(a),4(b),4(c)) and credit (4(d),4(e),4(f)) datasets projected onto two attributes

We can connect these results to the problems of traditional methods mentioned in the section 2. It is clear that from the Fig. 4, the clusterings differ in membership of the points, as well as their point distributions. In particular, 4(c) and 4(f) show higher dissimilarity when compared to the pre-defined clusterings 4(a) and 4(d), than clusterings of other algorithms. This dissimilarity is correctly displayed through the ADCO measure. However, all other measures incorrectly capture these comparisons by actually giving a smaller value, implying that 4(c) and 4(c) are actually more similar to the pre-defined clusterings than others. This is because these methods consider only the intersected groups of points between two clusterings, regardless of the overall structures of clusterings. For other real world datasets tested, we have observed the same problem and ADCO was the only measure able to highlight the differences accurately.

5.1   Using ADCO for Evolution Analysis in Data Stream Clustering

As mentioned in section 2, clustering comparison is important in stream data for evaluating cluster evolution over time. [1]. However, the traditional measures require clusterings to have the same data points. Hence clusterings with a large time gap between them in the stream (thus using totally different points) cannot be compared. In contrast, ADCO's use of density profiles for each attribute makes such a comparison possible, as we now illustrate.

In Fig. 5 we have divided the dataset 'IRIS' into five subsets, where subset 1 and 2 share 50% of the points while the rest are different, though similar in their values. Therefore, when comparing them, ADCO should return a low dissimilarity value. Subsets 2 and 3 are similar to 1 and 2, but non-overlapping points are different, hence ADCO value should be higher. Subsets 3 and 4 are independent, but their values are similar and subsets 4 and 5 are independent and their values are highly different. Hence, ADCO values should be low for the comparison of 3 and 4 and high for when comparing 4 and 5. These comparisons can be seen as evaluating clustering evolution at different time periods.

Table 3. Dissimilarity values when comparing seven clusterings using five dissimilarity measures for dataset diabetes and credit

dataset	measures	k-means	EM	FF	CURE	AL	CL	SL
	ADCO	0.02	0.03	0.03	0.07	0.25	0.07	0.3
	RI	0.5	0.5	0.49	0.5	0.44	0.5	0.42
diabetes	JI	0.61	0.61	0.61	0.62	0.46	0.62	0.42
	CE	0.44	0.43	0.42	0.45	0.33	0.46	0.3
	VI	0.18	0.18	0.18	0.19	0.12	0.19	0.09
	ADCO	0.17	0.1	0.2	0.16	0.33	0.26	0.26
	RI	0.5	0.5	0.49	0.5	0.45	0.5	0.5
credit	JI	0.63	0.64	0.57	0.62	0.47	0.63	0.64
	CE	0.45	0.48	0.42	0.45	0.35	0.45	0.45
	VI	0.2	0.2	0.17	0.19	0.11	0.2	0.2

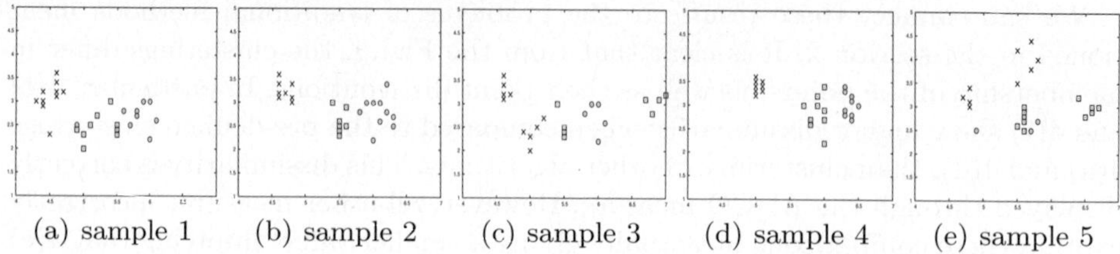

(a) sample 1	(b) sample 2	(c) sample 3	(d) sample 4	(e) sample 5

Fig. 5. 5 samples IRIS dataset with varying degrees of overlapping regions between them. ADCO provides intuitive comparisons even when clusterings only partially overlap or share no points.

Table 4. Dissimilarity values between different samples of the dataset IRIS

sample pairs	ADCO
sample 1 & sample 2	0.15
sample 2 & sample 3	0.37
sample 3 & sample 4	0.31
sample 3 & sample 5	0.41

We can see from the table 4 that ADCO corresponds with expectation. This means that ADCO can help speed up the stream analysis process, by allowing users to compare clusterings from any time windows, and they can investigate further if ADCO value indicates there is significant dissimilarity.

5.2   ADCO Limitations

Although providing more accurate measures by corresponding to the intuition, the current state of ADCO is limited to only comparing clusterings with equal number of clusters. Moreover, we have not discussed methods to handle soft clusterings or subspace clusterings. ADCO would need to be generalized to handle these cases. Finally we have used $q = 10$ as a best-practice value for the number of bins, but the real impact of this variable can be studied further.

6   Future Work and Conclusion

Clustering comparison is an important task in the overall cluster analysis process. In this paper we have discussed the limitations of existing methods, which consider only point-to-cluster assignments as the determining factor for dissimilarity between clusterings. This ignores other important feature-related information and may mislead users with inaccurate or even incorrect evaluations.

We have presented a new measure, ADCO, which takes a different approach to determining clustering dissimilarity, by using the distribution information for each attribute. We have experimentally shown that ADCO can indeed lead to more reasonable measures, than existing methods. Furthermore, we have identified an important application of ADCO for stream data clustering, where it is able to compare clusterings drawn from non-overlapping time windows. This kind of analysis is impossible with the other clustering comparison measures.

For future work, we would like to investigate generalizing ADCO to handle clusterings with different numbers of clusters. This could facilitate the tighter integration of ADCO into areas such as ensemble clustering. We would also like examine other cluster features for comparison, such as boundaries and centroids.

References

1. Aggarwal, C., Han, J., Wang, J., Yu, P.: A framework for clustering evolving data streams. 29th VLDB Conference. (2003)
2. Aggarwal, C.: A framework for diagnosing changes in evolving data streams. Intern. Conf. on Management of Data (2003) 575–586
3. Fred, A., Jain A.: Combining Multiple Clusterings Using Evidence Accumulation. Transac. on Pattern Analysis and Machine Intelligence. **27** (2005) 835–850
4. Fred, A., Jain, A.: Robust data clustering. Comp. Soc. Conf. on Computer Vision and Pattern Recognition. (2003) 128–133
5. Grossman, S.: Elementary Linear Algebra. Saunders College Publishing. (1994)
6. Halkidi, M., Batistakis, Y., Vazirgiannis, M.: On Clustering Validation Techniques. Journ. of Intelligent Info. Sys. **17** (2001) 107–145
7. Hamers, L., Hemeryck, Y.: Similarity measures in scientometric research: the Jaccard index vs. Salton's cosine formula. Info. Process. and Manage. **25** (1989) 315–318
8. Hubert, L.: Comparing partitions. Journ. of classification. (1985) 193–218
9. Karypis, G., Aggarwal, R., Kumar, V.: Multilevel hypergraph partitioning: application in VLSI domain. Ann. Conf. on Design Automation. (1997) 526–529
10. Meila, M.: Comparing Clusterings. Statistics Technical Report. http://www.stat.washington.edu/www/research/reports/2002/ (2005)
11. Meila, M.: Comparing Clusterings - An Axiomatic View. 22nd International Conference on Machine Learning. (2005)
12. O'Callaghan, L., Mishra, N., Meyerson, A.: Streaming-Data Algorithms for High-Quality Clustering Intern. Conf. on Data Engineering. (2002)
13. Rand, W.: Objective criteria for the evaluation of clustering methods. Journ. of the American Statistical Association. **66** (1971) 846–850
14. Ratanamahatana, C.: CloNI : Clustering of square root of N interval discretization. Data Mining IV, Info. and Comm. Tech. **29** (2003)
15. Strehl, A., Ghosh, J.: Cluster Ensembles - A Knowledge Reuse Framework for Combining Multiple Partitions. Jour. on Machine Learning. **3** (2002) 583–617
16. Theodoridis, S., Koutroumbas, K.: Pattern Recognition. Academic Press. (1999)
17. Zhou, D., Li, J., Zha, H.: A new Mallows distance based metric for comparing clusterings. Intern. Conf. on Machine Learning. (2005)

Efficient Mining of Frequent Itemsets in Distorted Databases

Jinlong Wang and Congfu Xu*

Institute of Artificial Intelligence, Zhejiang University
Hangzhou, 310027, China
zjupaper@yahoo.com, xucongfu@cs.zju.edu.cn

Abstract. Recently, the data perturbation approach has been applied to data mining, where original data values are modified such that the reconstruction of the values for any individual transaction is difficult. However, this mining in distorted databases brings enormous overheads as compared to normal data sets. This paper presents an algorithm GrC-FIM, which introduces granular computing (GrC), to address the efficiency problem of frequent itemset mining in distorted databases. Using the *key granule* concept and granule inference, support counts of candidate non-key frequent itemsets can be inferred with the counts of their frequent sub-itemsets obtained during an earlier mining. This eliminates the tedious support reconstruction for these itemsets. And the accuracy is improved in dense data sets while that in sparse ones is the same.

1   Introduction

In recent years, the wide availability of personal data has been regularly collected and mined for the requirements of decision making processes. This analyzing activity opens new threats to the privacy of the individual data and has resulted in the development of privacy-preserving data mining (PPDM) [1]. As an important field, PPDM lets business derive the understanding they need without collecting accurate personal information.

One of the methods for PPDM is based on the data perturbation technique, in which random noises are added to the original data, and only the disguised data are shared for data mining [1,2,3]. Through mining distorted databases, frequent itemset models can be built with the reconstruction method [2,4]. This brings expensive overheads as compared to directly mining original data sets. Such as, with the probabilistic distortion of user data mentioned in [2,4], reconstructing the support count of a k-itemset needs to keep track of the counts of the 2^k combinations produced from it. In [4], the authors made use of the basic formula from set theory to eliminate these counting overheads. With the formula, the counts of all these 2^k combinations generated from an itemset can be calculated from the its counts and the counts of its subsets which are available from previous passes over the distorted database. Although this method shows

---

* Corresponding author.

A. Sattar and B.H. Kang (Eds.): AI 2006, LNAI 4304, pp. 352–361, 2006.
© Springer-Verlag Berlin Heidelberg 2006

good performance with sparse data sets, with dense data sets, such as census data, in which a number of long frequent itemsets exist, the method needs the tedious reconstruction operations to compute support counts of all the itemsets. This makes its performance degrade incredibly.

To further improve the efficiency in mining frequent itemsets under the perturbation framework in EMASK [4], we introduce granular computing (GrC) into the mining process, and present an effective algorithm GrC-FIM (**G**ranular **C**omputing based **F**requent **I**temset **M**ining in distorted databases). In the algorithm, the *key granule* concept is presented for granule inference. With the method, support counts of candidate non-key frequent itemsets can be inferred with the counts of their frequent sub-itemsets generated during an earlier mining. This eliminates the tedious support reconstruction for these itemsets. And the accuracy is improved in dense data sets while that in sparse ones is the same. Based on elementary granules represented with bitmap, support counts can be computed quickly with the bit AND and bit COUNT operations. The extensive experiments confirm that GrC-FIM provides significant improvement over the EMASK algorithm [4], for dense as well as sparse data sets.

The rest of the paper is organized as follows. Section 2 introduces GrC and its related content. In section 3, granule inference is described based on the concept *key granule*, and support counts can be computed with the inference method. Section 4 presents the algorithm GrC-FIM step by step. The experiment results are showed in section 5. Finally we summarize our work in section 6.

2   Granular Computing

The term "Granular computing (GrC)" was first introduced in 1997 by T. Y. Lin [5] to label studies on information granulation and computations [6]. As an important technique in the process of problem solving, the basic study issues of GrC constitute two related aspects, granulation and computation. The former deals with the construction, interpretation, and representation of granules, and the latter deals with the computing and reasoning with granules. Granules usually can be constructed through the use of information tables.

Formally, an information table is a quadruple [7,8]: $S = (U, At, \{V_a | a \in At\}, \{I_a | a \in At\})$. Where, U is a finite nonempty set of objects, At is a finite nonempty set of attributes, V_a is a nonempty set of values for $a \in At$, and $I_a{:}U \to V_a$ is an information function. Each information function I_a is a total function that maps an object of U to exactly one value in V_a.

In general, the inverse image of the attribute a with a fixed value $c_o \in V_a$ composes a granule [8]: $G(a, c_o) = \{x \in U | I_a(x) = c_o\}$. The aggregate consists of all objects which take only the value c_o from the domain V_a of a. It is a subset of the universe U. The granule is also called elementary granule [8]. The method of constructing elementary granules can be easily extended to cases more than one attribute. For a pair of attribute $a, b \in At$ and two values $v_a \in V_a$, $v_b \in V_b$, one can obtain the granule: $G((a, v_a) \wedge (b, v_b)) = G(a, v_a) \cap G(b, v_b)$. [8]

used the granular-bitmap technique to mine frequent itemsets and association rules.

In this paper, we simplify the expression of information table into transaction table, where the domain V_a of attribute a is $\{0,1\}$. For simplicity, the value 0 is omitted.

3   Support Counts Computation

In this section, we first describe the granule inference, wherein the *key granule* concept is introduced. Then, we show how the support counts computation can be implemented with granule inference.

3.1   Granule Inference

In the description of granule inference, the attribute independent concept of rough set in [9] is adopted. Let $B \subseteq At$ denote an itemset, and let $a \in B$, a is dispensable in B if $G(B) = G(B - \{a\})$, otherwise a is indispensable. The itemset B is independent if all its attribute items are indispensable.

Definition 1. *If attribute set (itemset) B ($B \subseteq At$) is independent, the granule $G(B)$ is called a key granule; otherwise it is a non-key granule.*

Lemma 1. *For $B \subseteq At$:*

1. *$G(B)$ is a key granule iff $\forall a \in B$, $G(B) \neq G(B - \{a\})$.*
2. *$G(B)$ is a non-key granule iff $\exists a \in B$, $G(B) = G(B - \{a\})$.*

Proof. Refer to the Definition 1. $\qquad\qquad\square$

Lemma 2. *For $B \subseteq C \subseteq At$, $G(C) \subseteq G(B)$.*

Proof. $G(C) = G(C + B - B) = G(B) \cap G(C - B) \subseteq G(B)$. $\qquad\qquad\square$

Let $|\cdot|$ denote the number of elements in the aggregate.

Theorem 1. *If $G(C)$ is a non-key granule, then $G(C) = MIN\{G(B)|\ B \subseteq C \subseteq At$, and $|B| = |C| - 1\}$. The notion $MIN\{\cdot\}$ denotes the aggregate, in which the number of elements is the least.*

Proof. This theorem can be proven by contradiction. Assume $G(C) \neq MIN\ \{G(B)\}$, since $B \subseteq C$, according to Lemma 2, we can obtain $G(C) \subset G(B)$ and $G(C) \neq G(B)$. Due to $|B| = |C|-1$, so $\nexists a \in C$, which make $G(C) = G(C-\{a\})$. Thus $G(C)$ is a key granule. This contradicts with the condition. $\qquad\square$

Corollary 1. *If $G(C)$ is a key granule, then $G(C) \neq MIN\{G(B)|\ B \subseteq C \subseteq At$, and $|B| = |C| - 1\}$.*

Theorem 2. *Let $B \subseteq C \subseteq At$, if $a \in B$ is dispensable in B, then a is dispensable in C, $G(C - \{a\}) = G(C)$.*

Proof. Since a is dispensable in B, $G(B - \{a\}) = G(B)$. $G(C) = G(C - \{a\} + \{a\}) = G(C - \{a\}) \cap G(\{a\})$. Since $\{a\} \subseteq B$ and a is dispensable in B, $G(\{a\}) \supseteq G(B) = G(B - \{a\})$, then $G(C) \supseteq G(C - \{a\}) \cap G(B - \{a\})$. Because $(B - \{a\}) \subseteq (C - \{a\}) \Rightarrow G(C - \{a\}) \subseteq G(B - \{a\})$, then $G(C) \supseteq G(C - \{a\})$. The opposite inequality is obvious. So, $G(C - \{a\}) = G(C)$. $\qquad\qquad\square$

Then, from Theorem 2 and the Definition 1, we can obtain the following corollary.

Corollary 2. *Let $B \subseteq C$: if $G(B)$ is a non-key granule, then $G(C)$ is a non-key granule; if $G(C)$ is a key granule, then $G(B)$ is a key granule.*

The Corollary 2 shows that the key granule satisfies the anti-monotone property. Thus the non-key granule can be identified quickly.

3.2   Support Counts Computation Based on Granule Inference

From the granule concept described in GrC, the support count of an itemset can be calculated through the cardinal number of the corresponding granule. For an itemset B, $sup(B) = |G(B)|$. The support $sup(B)$ used is absolute occurrence frequency, not the relative ones as in some literatures.

Theorem 3. *For $A, B \subseteq At$, if $G(A) = G(B)$, then $|G(A)| = |G(B)|$.*

Proof. Refer to the definition and properties of granule. $\qquad\qquad\square$

Corollary 3. *$|G(C)| = min\{|G(B)| \mid B \subseteq C \subseteq At$ and $|B| = |C| - 1\}$ iff $G(C)$ is a non-key granule, where $min\{| \cdot |\}$ denotes the least one among the element number in $| \cdot |$.*

Proof. Refer to Theorem 1 and Theorem 3, the 'if' part is obvious. The 'only if' part can be obtained with Corollary 1. $\qquad\qquad\square$

In frequent itemset mining, if $G(B)$ is a key granule, the itemset B is called a key itemset; otherwise, it is called a non-key one. The key itemset notion was also proposed by [10], however we use the granule inference in GrC to introduce and interpret the concept, which are different from the literature [10].

In this paper, we have the following equivalence notions. The itemset B is independent $\leftrightarrow$ $G(B)$ is a *key granule* $\leftrightarrow$ B is a *key itemset*. So the properties of granule can be used in frequent itemset mining. For convenience, we use the support counts computation of itemsets to replace the one of granules in the following paper. With a minimum support threshold s defined by user, for a frequent itemset B, if it is a key itemset, it is also called a key frequent one. With the candidate set generation-and-test approach [11], if all the proper sub-itemsets of a candidate frequent itemset are key frequent itemsets, it is called a candidate key frequent itemset; otherwise it is a cardidate non-key one.

4   The GrC-FIM Algorithm

In this section, we will develop the GrC-FIM algorithm. First, we describe the frequent itemset mining in distorted databases with GrC, where the granule inference method mentioned above is adopted. Then, the pseudo-code of the algorithm is given.

4.1   Frequent Itemset Mining in Distorted Databases

Under the framework of symbol-specific distortion process in [4], '1' and '0' in the original database are respectively flipped with $(1-p)$ and $(1-q)$. Through scanning the distorted database, support counts of itemsets in the original database can be reconstructed and frequent itemsets can be obtained. For the sake of clarity, both the candidate frequent itemsets and frequent itemsets in follows refer to the results in the original database.

Based on the candidate set generation-and-test approach, candidate frequent k-itemsets (C_k) are generated from frequent $(k\text{-}1)$-itemsets (Fp_{k-1}). For a candidate k-itemset, if it is a candidate non-key frequent itemset, its support count can be inferred using the minimum support counts of its $(k\text{-}1)$-itemsets obtained during previous processes; otherwise, its support count needs to be reconstructed through scanning the distorted database. When reconstructing the support count of this candidate key frequent itemset, the basic formula from set theory in [4] are introduced.

With the inference method, if a frequent itemset is non-key, according to Corollary 2, all its supersets are non-key itemsets, and their support counts in the original database can be inferred without reconstructing. Thus, only for the key frequent itemsets, their support counts in the distorted database need to be recorded (These support counts are simplified as distorted support counts). This is different from EMASK [4], and reduces the memory.

In the process of computing support counts in the distorted database, the bit AND and bit COUNT operations in granular-bitmap method [8] are used. In order to overcome memory restrictions, our approach horizontally partitions the transaction data sets into disjointed subsets. Thus each partition represented by bitmap can be loaded into main memory for efficient processing. The partitioning method makes it suit the parallel and distributed processing.

4.2   The Process of Algorithm

Algorithm 1 describes the process of GrC-FIM. In the algorithm, the distorted data file D^*, the distortion parameters p, q (as [4]) and minimum support s are input. After mining, the frequent itemsets Fp can be obtained. In the process, for the key frequent itemsets in Fp, their support counts in the distorted database are saved. And a boolean variable is used to indicate whether or not an itemset is a (candidate) key one.

Algorithm 1. GrC-FIM Algorithm

Data: D^*, distortion parameters p and q, minimum support s.
Result: Frequent itemsets Fp in the original database D.
begin
    Step 1. Compute Fp_1 and transform D^*;
    for *each* $a \in At$ **do**
        $sup(a, D^*) = 0$;
        $sup(a, d_i{}^*) = 0 \ (i \in \{1, 2, \cdots, w\})$;
    for $i = 1; i \leq w; i + + $ **do**
        $Tr\_GrBit(d_i{}^*)$;
        for *each* $a \in At$ **do**
            $sup(a, D^*) + = sup(a, d_i{}^*)$;
        Save GrBit($d_i{}^*$) into hard disk;
    for $\forall a \in At$ **do**
        Reconstruct its support count $sup(a, D)$ with p, q;
    Generate Fp_1;
    Step 2. Create Fp_1' and determine the key itemsets.
    for $\forall fp \in Fp_1$ **do**
        if $fp.sup == 100\%$ **then**
            $fp.key = false$
        else
            $fp.key = true$;
            Save the distorted support count $fp.disup \ (sup(fp, D^*))$;
    Step 3. Compute frequent itemsets $Fp_k (k \geq 2)$;
    for $k = 2; Fp_{k-1} \neq \Theta; k + + $ **do**
        Step 3.1 Generate candidate sets C_k
        $C_k = gen\_candidate(Fp_{k-1})$;
        Step 3.2 Compute the support of candidate frequent itemsets
        for $\forall c \in C_k$ **do**
            if $c.key == false$ **then**
                $c.sup = c.predsup$
            if $c.key == true$ **then**
                Reconstruct the support $c.sup$ of itemset c;
                if $sup(c, D) == c.predsup$ **then**
                    $c.key = flase$
        Generate Fp_k;
        Save the distorted support counts of the key itemsets in Fp_k;
end

In the algorithm, when scanning the distorted database for the first time, step 1 obtains the frequent itemsets Fp_1 (satisfying minimum support s) through computing and reconstructing the support counts of 1-itemsets in At. At the same time, the distorted database is transformed into the elementary granule bitmap representation partition by partition through the function $Tr\_GrBit()$, and each of them $\mathrm{GrBit}(d_i{}^*)$ is saved into hard disk. Suppose D^* is horizontally partitioned into w partitions $D^* = d_1{}^* d_2{}^* \cdots d_w{}^*$, and w is determined with the constraints of memory amount. The support count of the itemset B in the distorted database D^* is $sup(B, D^*) = \sum_{i=1}^{w} sup(B, d_i{}^*)$. In step 2, if the support counts of the frequent itemsets are equal to the number of transactions $|D|(|D^*|)$, that is $p.sup == 100\%$, they are marked as non-key itemsets. For key frequent 1-itemsets, their distorted support counts need to be saved.

Step 3 iterates and generates Fp_k $(k > 2)$. First, step 3.1 generates candidate k-itemsets with frequent $(k\text{-}1)$-itemsets. In the candidate frequent k-itemset C_k, predicted support counts are recorded, and candidate frequent non-key k-itemsets are determined with the Corollary 3. Then step 3.2 computes the support counts of candidate key k-itemsets through scanning the distorted database D^*. According to the minimum support, the frequent itemsets Fp_k can be obtained. Finally, if the reconstructed support counts equal to the predicted ones, they are marked as non-key itemsets. And the distorted support counts of key frequent itemsets in Fp_k are saved.

5   Performance Evaluation

For performance evaluation of the algorithm GrC-FIM, we used the original source of EMASK [4], provided by the authors in the website[1]. And for the better performance evaluation, we also implemented the algorithm GB-FIM (Only based on granular-bitmap (GB) without granule inference of this paper, same as the algorithm in [12]). All the algorithms are coded in C++. All the experiments were conducted on a 3GHz Intel Pentium PC with 512MB memory and Windows XP Service Pack 2 installed.

Our experiments were performed on the synthetic and real data sets. The synthetic databases T25.I4.D1M.N1K (T25) and T40.I10.D1M.N1K (T40), using the IBM synthetic market-basket data generator [11], mimic the transactions in a retailing environment. The pumsb* data contains census data. This is a real data set, and is very dense, i.e., it produces many long frequent itemsets even for very high support values. Whereas, the synthetic data sets are sparser when compared to the real one. Table 1 shows the characteristics of these data sets. In the following experiments, these data sets were distorted with parameters $p = 0.5$ and $q = 0.97$ as EMASK [4]. In the experiments, for a fair comparison on algorithms and scalable requirements, GB-FIM and GrC-FIM were run where only 5K transactions were loaded into the main memory one time.

5.1   Efficiency Analysis

The Fig.1 and Fig.2 show the execution time of these three algorithms on different data sets. In each graph, the horizontal axis is the size of minimum support threshold, and the vertical axis is the CPU execution time (sec).

---

[1] http://dsl.serc.iisc.ernet.in/projects/software/software.html.

Table 1. Transaction database characteristics

Database	#Transactions	Items	Max. Length	Avg. Length
T25.I4.D1M.N1K	1000000	972	53	24.9
T40.I10.D1M.N1K	1000000	991	81	39.6
pumsb*	49046	7117	50	50

As shown in the figures, GrC-FIM and GB-FIM outperform EMASK, and the margin grows as the minimum support decreases. For sparse datasets (T25 and T40), the execution time of GrC-FIM and GB-FIM increase slowly as the minimum support decreases. And because support counts of few itemsets can be obtained through inference, both the time costs of the algorithms approximate, but GrC-FIM also outperforms GB-FIM. For real datasets pumsb*, as the minimum support decreases, the time required by all the algorithms increases. However, at the low support, the predominance of performance in the algorithms GrC-FIM and GB-FIM is evident due to the use of granular-bitmap. And GrC-FIM outperforms GB-FIM. This is because that there are a number of frequent itemsets produced in the situation, especially with the long itemsets, the effect of granule inference is evident and the time cost of computation is reduced.

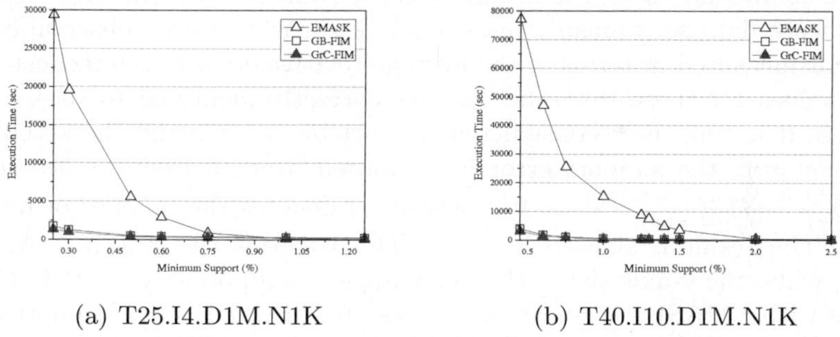

(a) T25.I4.D1M.N1K   (b) T40.I10.D1M.N1K

Fig. 1. Execution time on synthetic data sets

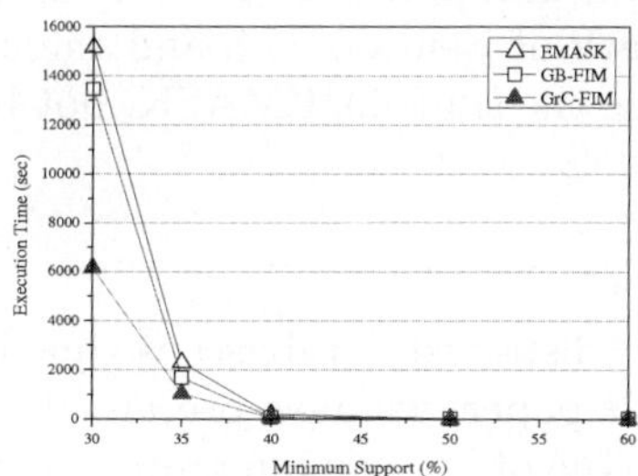

Fig. 2. Execution time on real data set pumsb*

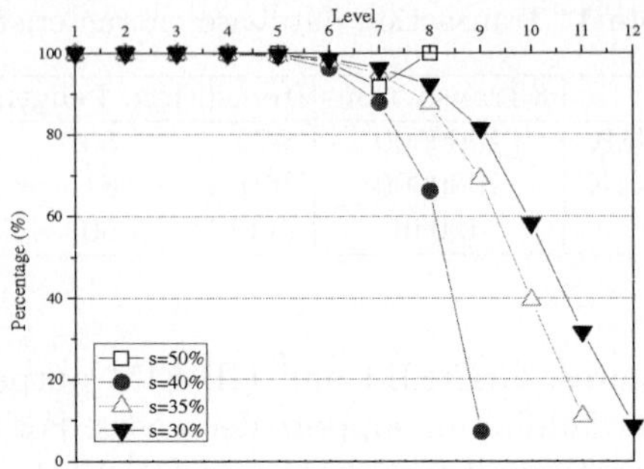

Fig. 3. Accuracy comparison on real data sets pumsb*

5.2 Accuracy Analysis

In this section, the experiments are conducted to evaluate the accuracy of results. With the error metrics proposed in [2], two kind of mining errors, the support error and identify error are evaluated. Our main improvement is on the support error. And at the identify error, our algorithm approximates to the algorithm EMASK [4] and GB-FIM without granule inference. Since the results of EMASK and GB-FIM are the same, we only compare GrC-FIM with EMASK for convenience. In sparse data sets, the accuracy is the same in both the algorithms. In the real dense data set pumsb*, the support error is improved observably. The accuracy is measured in terms of the average relative error in the reconstructed support values for those itemsets that are correctly identified to be frequent. As in [2,4], denoting the reconstructed support by rec_sup and the actual support by act_sup, the support error is computed over all frequent itemsets as $\rho = \frac{1}{|f|} \sum_f \frac{|rec\_sup_f - act\_sup_f|}{act\_sup_f} \times 100$, where $|f|$ denotes the number of frequent itemsets. The result is showed in Fig.3. The x-axis is the length of frequent itemsets, while the y-axis shows the percentage of support error that GrC-FIM compares with EMASK. As shown in the results, the accuracy of support in our algorithm GrC-FIM is not below the EMASK algorithm. When the itemsets are long, at high level, the improvement of accuracy is evident. Even though when renewing reconstruction error and percolating error can counteract, for example in the Fig.3, when the length of itemsets is 8 and the support is $s = 50\%$, the accuracy of our algorithm is the same as EMASK, not below it.

6   Conclusions

Mining frequent itemsets in distorted databases is time-consuming as compared to normal databases. In this paper, we have presented an effective algorithm to address the problem. By virtue of granule inference, support counts of candidate non-key frequent itemsets can be inferred from frequent *key* itemsets obtained in previous mining processes, which improves the efficiency. Since the support count's error of a non-key frequent itemset only comes from the one of its sub-itemsets, renewing reconstruction error is avoided and the accuracy is improved, especially for long frequent itemsets in dense data sets. The experiments show that our algorithm is more efficient than EMASK. In particular the granule inference plays a more important role in dense data sets than sparse data sets.

Acknowledgements

We gratefully acknowledge assistance from domain experts T. Y. Lin. And thank Ren'ai Zhang for his experiment works. This paper was supported by the Natural Science Foundation of China (No. 60402010), Zhejiang Provincial Natural Science Foundation of China (Y105250).

References

1. Agrawal, R., and Srikant, R.: Privacy-preserving data mining. In: Proc. ACM-SIGMOD Int. Conference on Management of Data (SIGMOD'00). 439-450
2. Rizvi, S., and Haritsa, J.: Maintaining data privacy in association rule mining. In: Proc. VLDB Int. Conference on Very large data bases (VLDB'02). 682-693
3. Du,W., and Zhan, Z.: Using randomized response techniques for privacy-preserving data mining. In: Proc. ACM-SIGKDD Int. Conference on Knowldge discovery and data mining (SIGKDD'03). 505-510
4. Agrawal, S., Krishnan, V., and Haritsa, J.: On addressing efficiency concerns in privacy-preserving mining. In: Proc. DASFAA Int. Conference on Database systems for advances applications (DASFAA'04). 113-124
5. Lin, T. Y.: Granular computing. In: Announcement of the BISC Special Interest Group on Granular Computing (1997).
6. Zadeh, L. A.: Towards a theory of fuzzy information granulation and its centrality in human reasoning and fuzzy logic. Fuzzy Sets and Systems, v.90 n.2, 111-127, Sept. 1. 1997.
7. Yao, Y. Y., and Zhong, N.: Granular computing using information tabules. In: Data Mining, Rough Sets and Granular Computing, Physica-Verlag, 2002. 102-124
8. Lin, T. Y., and Louie, E.: Data mining using granular computing: fast algorithms for finding association rules. In: Data Mining, Rough Sets and Granular Computing, Physica-Verlag, 2002. 23-45
9. Pawlak, Z.: Some issues on rough sets. Transactions on Rough Sets I. 2004. 1-58
10. Bastide, Y., Taouil, R., Pasquier, N., Stumme, G., and Lakhal, L.: Mining frequent patterns with counting inference. SIGKDD Explorations. 2(2), 2000. 66-75
11. Agrawal, R., and Srikant, R.: Fast algorithms for mining association rules. In: Proc. VLDB Int. Conference on Very large data bases (VLDB'94). 487-499
12. Xu, C., Wang, J., Dan, H., and Pan, Y.: An improved EMASK algorithm for privacy-preserving frequent pattern mining. In: Proc. Computational Intelligence and Security International Conference (CIS'05). Springer, 752-757

Improved Support Vector Machine Generalization Using Normalized Input Space

Shawkat Ali[1] and Kate A. Smith-Miles[2]

[1] School of Information Systems, Central Queensland University, QLD 4702, Australia
[2] School of Engineering and Information Technology, Deakin University, VIC 3125, Australia
s.ali@cqu.edu.au, katesm@deakin.edu.au

Abstract. Data pre-processing always plays a key role in learning algorithm performance. In this research we consider data pre-processing by normalization for Support Vector Machines (SVMs). We examine the normalization affect across 112 classification problems with SVM using the rbf kernel. We observe a significant classification improvement due to normalization. Finally we suggest a rule based method to find when normalization is necessary for a specific classification problem. The best normalization method is also automatically selected by SVM itself.

Keywords: Normalization, Classification, Support Vector Machines.

1 Introduction

Support Vector Machines (SVMs) [1-3] are a comparatively new machine learning tool, which adopted statistical learning theory. From the beginning SVMs are solving classification, regression and novelty detection tasks with better generalization compared with traditional learning algorithms. Let us consider, a linear SVM over a training set $\{(\vec{x}_i, y_i), i = 1, \cdots, n\}, x_i \in \Re^m, y_i \in \{-1, 1\}$ where the task is to estimate a weight vector $\vec{\omega}$ and a scalar bias factor b to construct an optimal hyperplane defined by $\vec{\omega} \cdot \vec{x} + b$ as follows:

$$\min_{\vec{\omega}, b} \frac{1}{2} \langle \vec{\omega} \cdot \vec{\omega} \rangle, subject\ to\ y_i \left(\langle \vec{\omega} \cdot \vec{x}_i \rangle + b \right) \geq 1 \tag{1}$$

which leads to the dual optimisation problems such as:

$$\max_{\alpha}\ W(\alpha) = \sum_i \alpha_i - \frac{1}{2} \sum_{i,j} \alpha_i \alpha_j y_i y_j \langle \vec{x}_i \cdot \vec{x}_j \rangle$$
$$subject\ to\ \sum_i y_i \alpha_i = 0, \quad \alpha_i \geq 0 \tag{2}$$

Now the goal is to estimate the optimisation parameter properly. The generalization performance of SVMs and many other function estimation algorithms depend on appropriate parameter estimation. Data normalization has already shown significant improvement of generalization performance with SVM [4-6]. Herbrich and Graepel

A. Sattar and B.H. Kang (Eds.): AI 2006, LNAI 4304, pp. 362–371, 2006.

© Springer-Verlag Berlin Heidelberg 2006

[7] have shown in terms of image classification that normalization is a data pre-processing method which plays an important role in SVM classification with real world problems. In this research we examine the issue of data normalization across a wide range of classification problems and propose a simple rule based methodology to find where normalization is beneficial. Finally we select an appropriate normalization method with the help of popular learning algorithms with priority given to SVM itself for a particular classification problem.

Our methodology seeks to understand the characteristics of the data (for particular classification problems), and explain why a normalized input space might offer better generalization. First we classify a wide range of natural classification problems with the most popular learning algorithms including SVM with radial basis function (rbf) kernel. Automatic rbf kernel parameter selection is described in [8]. After that we use all these problems with SVM rbf with modified input space with different normalized methods. Then we identify the dataset characteristics matrix by statistical measures to generate rules where normalization is necessary or not. Finally we use a set of popular learning algorithms to predict the appropriate normalization method selection for a particular classification problem with SVM. All 112 classification problems are considered from the UCI Repository [9] and Knowledge Discovery Central [10] database. Over all the experiments we consider 10 Fold Cross Validation (10FCV) performances.

Our paper is organized as follows: In Section 2, we provide some theoretical frameworks regarding SVM input space normalization. Section 3 analyses the experimental results. All statistical measures to identify the dataset characteristics matrix are summarized in Section 4. A priori normalization/non normalization method selection will be explained in 5. The most suitable normalization method selection is described in section 6. Finally we conclude our research towards the end in Section 7.

2 SVMs Input Space

One specialty of SVM is that it transforms the data by adopting kernel. During the transformation some kernels essentially normalize the data points automatically. For example, rbf kernel [3] as follows:

$$K\left(\vec{x}_i, \vec{x}_j\right) = \exp\left(-\frac{\left\|\vec{x}_i - \vec{x}_j\right\|^2}{2\sigma^2}\right) \quad \text{where } \sigma > 0 \tag{3}$$

We propose some additional normalization for SVM before we transform the data points in the kernel space. The normalized data is used in the SVM kernel space rather than natural data input. We provide the preference on global attributes values normalization rather than single attribute normalization. Two types of normalization [11] are examined in this research as described in the following sections.

2.1 Min-Max Normalization

The formulation of the Min-Max normalization is:

$$D'(i) = \frac{D(i) - \min(D)}{\max(D) - \min(D)} * (U - L) + L \tag{4}$$

where D' is the normalized data matrix, D is the natural data matrix and U and L are the upper and lower normalization bound.

This type of normalization method is used to normalize the data matrix into a desired bound. The most popular bound is between 0 and 1. We also change the bound values to between 0 and –1, as well as between 1 and –1.

2.2 Zero-Mean Normalization

The formulation of the Zero-Mean normalization is as follows:

$$D' = \left(\frac{D - \overline{D}}{\sigma} \right) \tag{5}$$

where $\overline{D}$ is the mean of the natural data matrix D and σ is the standard deviation of the same data matrix. In this normalization method, the mean of the normalized data points is reduced to zero. Due to this, the mean and standard deviation of the natural data matrix is required.

The normalization performance of UCI dataset 'xab' is shown in Figure 1. The natural data distribution is highly positive skewed. But the normalized 'xab' dataset is shown balanced skewed.

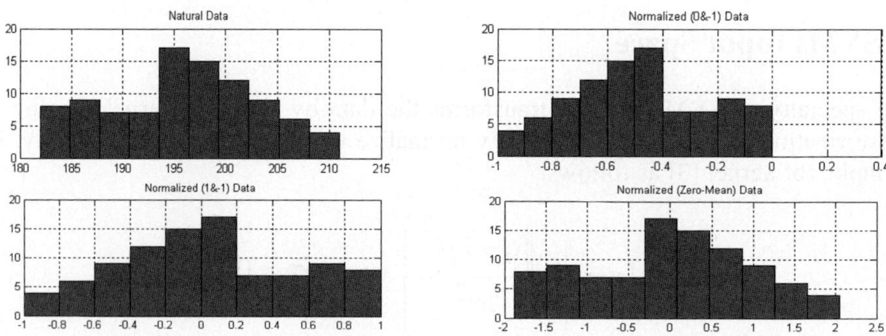

Fig. 1. Graphically represents the natural and different normalized distribution of 'xab' UCI dataset. The data distribution scale is completely different after getting normalization.

The hyperplane construction procedure of SVM with rbf kernel for UCI data set 'wpbc' is shown in Figure 2. During the classification of SVM using natural data points, several optimal decision boundaries are constructed, but only a single boundary is constructed for normalized data points. The misclassification error is higher for natural data points than normalized data points.

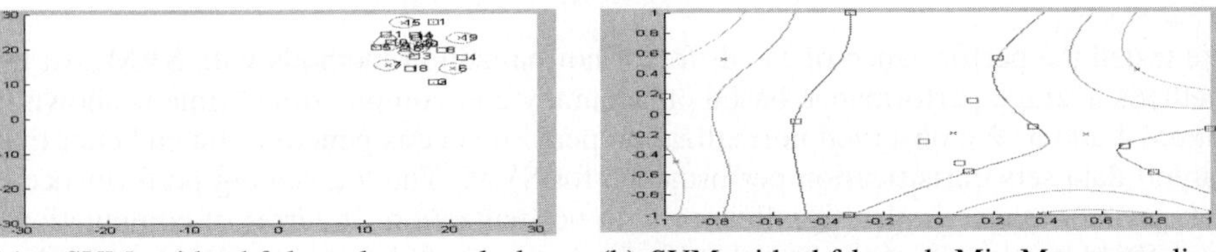

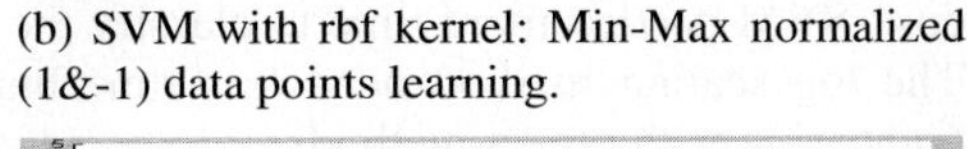

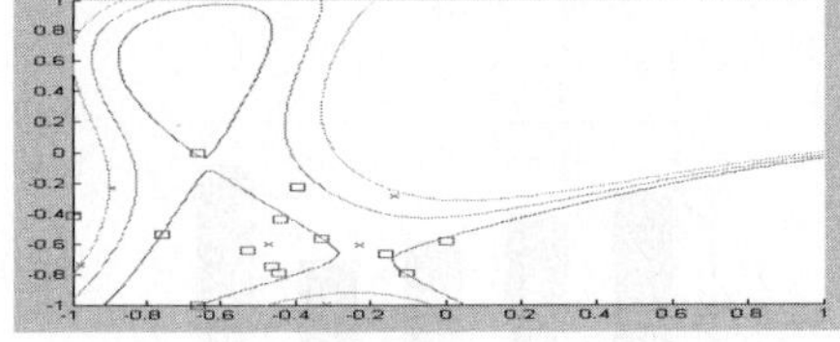

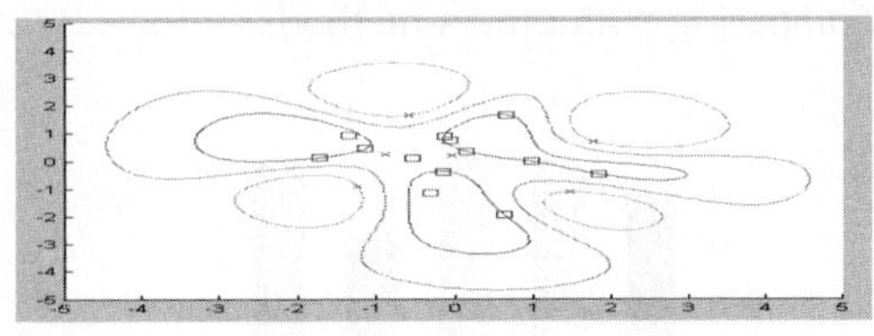

(a) SVM with rbf kernel: natural data points learning.

(b) SVM with rbf kernel: Min-Max normalized (1&-1) data points learning.

(c) SVM with rbf kernel: Min-Max normalized (0&-1) data points learning.

(d) SVM with rbf kernel: Zero-Mean normalized data points learning.

Fig. 2. The hyperplane construction procedure for natural and normalized data points of UCI 'wpbc' data set. SVM considers the rbf kernel with width 1.

We now compare the scaling method with normalization performance. This method also transforms the data points within a certain range.

2.3 Log Scaling

The formulation of the log scaling [12] method of the data matrix is as follows:

$$D'(i) = \log(D(i)) \tag{6}$$

This type of scaling transforms the data points within a logarithmic scale as shown in Figure 3.

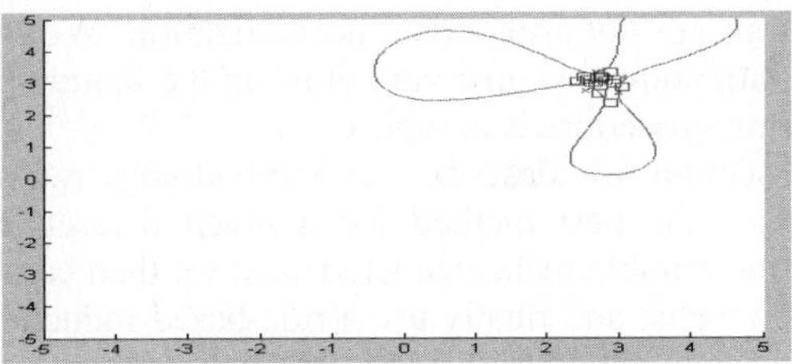

Fig. 3. The hyperplane construction procedure for log scaled data points of UCI 'wpbc' data set. SVM considers the rbf kernel width 1.

It is observed that log scaled data points are very difficult to classify by SVM. The classification error is higher than normalized data classification performance.

3   Experimental Results

3.1   Classification and Computational Performance

We tested the performance of the different normalization methods with SVM. All the methods average performance based on accuracy and computational time is shown in Figure 4 and 5. We observed normalization performance is generally much better than natural data sets classification performance for SVM. The log scaling performance is not carried out the similar significance with normalization. In terms of computational cost, SVM needed more time to classify the normalized data points than natural data. The log scaling method need less time but the accuracy performance is the worst comparing with others methods.

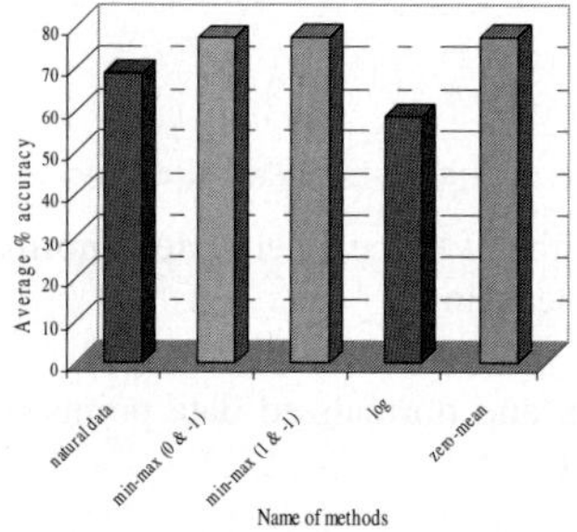

Fig. 4. Average accuracy performance of different normalization and scaling methods compared to natural data points classification performance

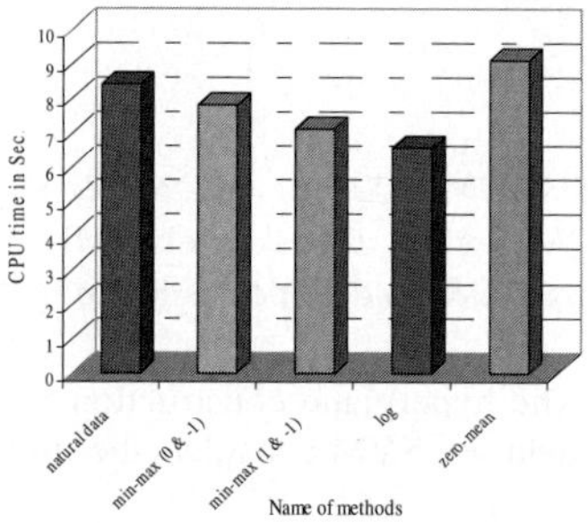

Fig. 5. Average computational performance of different normalization and scaling method with comparing natural data points classification performance

We observed among the different methods of normalization 84% of the 112 data sets performed better with some kind of normalization. The rest of the data sets showed better classification performance without normalizing or scaling. We found some visual reason why normalization is not always effective. First, those data sets with a combination of negative continuous and discrete attributes values, second a combination of categorical and continuous and finally those holding categorical attributes values that all are not aspect data normalization. We observed after getting normalization these data points become very close in the feature space, where optimal hyperplane construction procedure is complex.

In the following section we describe the methodology we use to assist in the appropriate selection of the best method for a given dataset. First each dataset is described by a set of measurable meta characteristics; we then combine this information with the performance results; and finally use a rule-based induction method to provide rules describing when each method for SVM is likely to perform well.

4   Datasets Characteristics Measurement

Each dataset can be described by simple, distance and distribution based statistical measures [13,14]. These three sets of measures characterized the datasets in different ways. First, the simple classical statistical measures identify the data characteristics based on variable to variable comparisons. Then, the distance based measures identify the data characteristics based on sample to sample comparisons. Finally, the density based measures consider the single data point from a matrix to identify the datasets characteristics. We average most of statistical measures over all the variables and take these as global measures of the dataset characteristics.

4.1 Simple Statistical Measures

Descriptive statistics can be used to summarise any large dataset into a few numbers that contain most of the relevant characteristics of that dataset. The following table lists the statistical measures used in this work as provided by the Matlab Statistics Toolbox [12] and some other different sources [15, 16] as follows: Geometric mean, Harmonic mean, Trim mean, Standard deviation, Interquartile Range, Max. and Min. eigenvalue, Skewness, Kurtosis, Correlation Coefficient and Prctile.

4.2 Distance Based Measures

Distance based measures calculate the dissimilarity between samples. We measure the euclidean, city block and mahalanobis distance between each pair of observations for each dataset as follows: Euclidean distance, City Block distance and Mahalanobis distance.

4.3 Distribution Based Measures

The probability distribution of a random variable describes how the probabilities are distributed over the various values that the random variable can take on. We measure the probability density function (pdf) and cumulative distribution function (cdf) for all datasets by considering different types of distributions as follows: Chi-square pdf, Normal pdf, Binomial pdf, Exponential pdf, Gamma pdf, Lognormal pdf, Rayleigh pdf, Chi-square cdf, Normal cdf, Discrete uniform cdf, F pdf, Poisson pdf, Student's t pdf, and Noncentral T cdf (nctcdf).

These measures are all calculated for each of the datasets to produce a dataset characteristics matrix. Finally by combining this matrix with the performance results we can derive rules to suggest when certain methods are appropriate.

5 Rule Generations

The trial-and-error approach could be a very common procedure to select the normalization or non normalization method for SVM classification. It is a computationally complex task to find an appropriate method by following this procedure. If we are interested in applying a specific method to a particular problem we have to consider which method is more suitable for which problem. The suitability test can be done from rules developed with the help of the data characteristics properties.

Rule based learning algorithms, especially decision trees (also called classification trees or hierarchical classifiers), are a divide-and-conquer approach or a top-down induction method, that have been studied with interest in the machine learning community. Quinlan [17] introduced C4.5 and then C5.0 algorithms to solve classification problems. C5.0 works in three main steps. First, the root node at the top node of the tree considers all samples and passes them through to the second node called 'branch node'. The branch node generates rules for a group of samples based on an entropy measure. In this stage C5.0 constructs a very big tree by considering all attribute values and finalizes the decision rule by pruning. It uses a heuristic approach for pruning based on statistical significance of splits. After fixing the best rule, the branch nodes send the final class value in the last node called the 'leaf node' [17,18]. C5.0 has two parameters: the first one is called the pruning confidence factor (c) and the second one represents the minimum number of branches at each split (m). The pruning factor has an effect on error estimation and hence the severity of pruning the decision tree. The smaller value of c produces more pruning of the generated tree and a higher value results in less pruning. The minimum branches m indicates the degree to which the initial tree can fit the data. Every branch point in the tree should contain at least two branches (so a minimum number of $m = 2$). For detail formulations see [17].

Now that the characteristics of each dataset can be quantitatively measured, we can combine this information with the empirical evaluation of normalization/natural classification performance and construct the dataset characteristics matrix. Thus, the result of the jth method on the ith dataset is calculated as:

$$R_{ij} = 1 - \frac{e_{ij} - \max(e_i)}{\min(e_i) - \max(e_i)} \tag{7}$$

where e_{ij} is the percentage of correct classification for the jth method on dataset i, and e_i is a vector of accuracy for dataset i. The class values in the matrix are assigned based on the performance. If the normalization method is showed better performance than natural data set classification, then class value is 1, otherwise 2. Based on the 112 classification problems we can then train a rule-based classifier (C5.0) to learn the relationship between dataset characteristics and normalization/natural method performance. We split the matrix 90% to construct the model tree. The process is then repeated using a 10 fold cross validation approach so that 10 trees are constructed. From these 10 trees, the best rules are found for normalization/non normalization method selection based on the best test set results. The generalization of these rules is then tested by applying each of the randomly extracted test sets and calculating the average accuracy of the rules as discussed below in Table 1.

5.1 Rules for Normalization/Non Normalization Selection

The rules for normalization method selection are generated with c = 85% and m = 4 as follows:

Rule 1: IF (nctcdf > 0) OR (median > 68.9) THEN do normalization.
Rule 2: IF (nctcdf < 0) OR (median < 68.9) THEN don't do normalization.

Include principle rules are data driven was to determine if data will benefit from normalization, or is it already normalized enough. For example, we can explain the nature of the rules by following noncentral T cumulative distribution function (nctcdf) value p as shown in Figure 6.

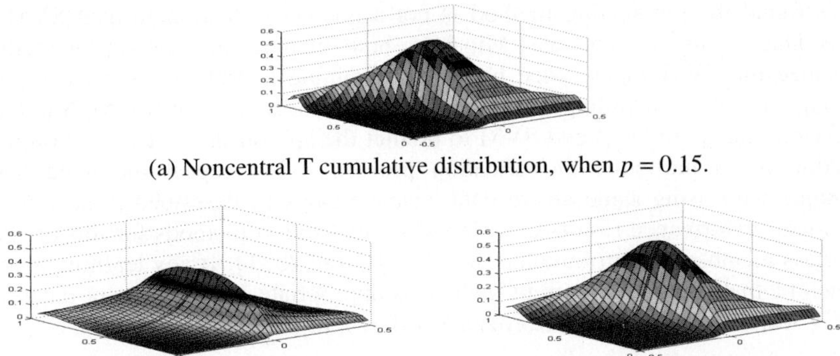

(a) Noncentral T cumulative distribution, when $p = 0.15$.

(b) Noncentral T cumulative distribution, when $p < 0$.

(c) Noncentral T cumulative distribution, when $p > 0$.

Fig. 6. Noncentral T cumulative distribution with different p values that explains the nature of data distribution

Noncentral T cumulative distribution [20-21] is highly positively skewed with respect to normal distribution as shown in Figure 6(a). We observed from the experiment, when p value of the noncentral T cumulative distribution function is less than 0, then the distribution nature of the data is closely normal. In that case, data normalization is not required for SVM classification. On the other hand, when p value is greater than 0 that means data is closely distributed as like noncentral T cumulative distribution. These types of data are needed to normalization for SVM classification.The 10 fold cross validation performance of the rule generation process is summarized in Table 1.

Table 1. Confusion matrix based on 10FCV results for the normalization/nonnormalization method selection rule (Accuracy = 89.40%)

		Normalization Method Best	
		Y	N
Data Condition Satisfied	Y	88	6
	N	6	12

We summarized the best rules from 10FCV performance. Average 10FCV performance is more than 89%. However, which method is best for individual datasets has been shown to be quite data dependent. These rules might be useful to determine where normalization is most appropriate for which problem.

Now, we can decide either normalization is necessary or not. Then we need to find which normalization method is suitable for a particular problem. We analyzed the performance of each normalization and scaling method with comparing natural datasets classification performance.

We found the log scaling method is not a good way to transformed SVM input space. Due to this we consider Min-Max both method and zero-mean method to normalize the SVM input space before start mining a problem. We used a set of learning algorithms including neural network (NN), decision tree (C4.5), Naive Bayes (NB) (for details see [19]) and SVM to predict the appropriate normalization method selection for a particular classification problem. We repeat the meta learning technique with using same above data characteristics as described in section 4 and individual performance results of different normalization methods for this prediction. The class membership attribute has designed by {+1 & -1}. If any learning algorithm predict +1 that means the current method is appropriate for the present dataset. The 10FCV learning algorithms performance is shown in Table 2.

Table 2. Normalization method selection performances with different learning algorithms

Test set classification performance (% Accuracy)			
	MLP	NB	C4.5	SVM
Max-Min method [0&-1]	75	66.67	66.67	83.33
Max-Min method [1&-1]	75	50	91.67	100
Zero-Mean method	91.67	50	83.33	100

SVM has shown best performance for a specific normalization method selection. So, SVM can help itself when this algorithm needs proper normalization method.

6 Discussions

We investigate the normalization affect across a large scale classification problem with one of the popular classification algorithm SVM. The normalization of SVM input space can significantly influence the higher accuracy performance of the classification procedure. We find out the reason why some problems are not required normalization method. Some normalization method required less computational time to classify the normalized data sets rather than natural data sets. We proposed a priori rule based method when normalization is necessary for SVM with a specific problem. SVM is also used by itself to find out the most suitable normalization methods for a specific problem. The limitation of this research is we have not considered different types of data, for instance gene expression data. This methodology could be examined with bioinformatics problem. This research could also be re-examining with other popular SVM kernels.

References

1. Vapnik, V.: The Nature of The Statistical Learning Theory, Springer-Verlag, New York. (1995).
2. Vapnik, V.N.: Statistical Learning Theory, John Wiley & Sons, Inc. (1998).
3. Vapnik, V.N.: An Overview of Statistical Learning Theory, IEEE Transaction on Neural Networks, 10(5) (1999) 988-999.

4. Graf, A. and Borer, S.: Normalization in Support Vector Machines, in Proc. DAGM Pattern Recognition. Berlin, Germany: Springer-Verlag. (2001).

5. Pontil M. and Verri, A.: Support Vector Machines for 3-D Object Recognition, IEEE Trans.Pattern Anal. Machine Intell. 20 (1998) 637–646.

6. Graf, A.B.A., Smola, A.J. and Borer, S.: Classification in a Normalized Feature Space Using Support Vector Machines, IEEE Transactions on Neural Networks, 14(3) (2003) 597-605.

7. Herbrich R. and Graepel, T.: A PAC-bayesian margin bound for linear classifiers: Why SVM's work. Advances in Neural Information Processing Systems. 13 (2001).

8. Ali, S. and Smith, K.A.: Kernel Width Selection for SVM Classification-A Meta-Learning Approach, International Journal of Data Warehousing and Mining, Idea Publishers, USA. (2005) 78-97.

9. Blake, C. and Merz, C.J.: UCI Repository of Machine Learning Databases. http://www.ics.uci.edu /~mlearn/MLRepository.html, Irvine, CA: University of California. (2002).

10. Lim, T.-S.: Knowledge Discovery Central, Datasets, http://www.KDCentral.com/. (2002).

11. Kennedy, R.L., Lee, Y., Roy, B.V., Reed, C.D. and Lippman, R.P.: Solving Data Mining Problems Through Pattern Recognition, Prentice-Hall, NJ. (1997).

12. Statistics toolbox user's guide, Version 3, The MathWorks, Inc. USA. (2001).

13. Smith, K.A., Woo, F., Ciesielski, V. and Ibrahim, R.: Modelling The Relationship Between Problem Characteristics and Data Mining Algorithm Performance Using Neural Networks, C. Dagli et al. (eds.), Smart Engineering System Design: Neural Networks, Fuzzy Logic, Evolutionary Programming, Data Mining, and Complex Systems, ASME Press, 11 (2001) 357-362.

14. Smith, K.A., Woo, F., Ciesielski, V. and Ibrahim, R.: Matching Data Mining Algorithm Suitability to Data Characteristics Using a Self-Organising Map, in A. Abraham and M. Koppen (eds.), Hybrid Information Systems, Physica-Verlag, Heidelberg, (2002) 169-180.

15. Mandenhall, W. and Sincich, T.: Statistics for Engineering and The Sciences, 4th eds. Prentice Hall. (1995).

16. Tamhane, A.C. and Dunlop, D.D.: Statistics and Data Analysis, Prentice Hall. (2000).

17. Quinlan, R.: C4.5: Programs for Machine Learning, Morgan Kaufman Publishers, San Mateo, CA. (1993).

18. Duin, R.P.W.: A note on comparing classifier, Pattern Recognition Letters, 1 (1996) 529-536.

19. Witten, I.H. and Frank, E.: Data Mining: practical machine learning tool and technique with Java implementation, Morgan Kaufmann, San Francisco, (2000).

20. Evans, M., Hastings, N. and Peacock, B.: Statistical Distributions, Second Edition, John Wiley and Sons. (1993).

21. Johnson, N. and Kotz, S.: Distributions in Statistics: Continuous Univariate Distributions-2, John Wiley and Sons. (1970).

An Efficient Similarity Measure for Clustering of Categorical Sequences*

Sang-Kyun Noh[1], Yong-Min Kim[2], DongKook Kim[3], and Bong-Nam Noh[3],**

[1] Interdisciplinary Program of Information Security,
Chonnam National University, Korea
`guru@lsrc.chonnam.ac.kr`
[2] Dept. of Electronic Commerce,
Chonnam National University, Korea
`ymkim@chonnam.ac.kr`
[3] Div. of Electronics Computer Engineering,
Chonnam National University, Korea
`{dkim, bbong}@chonnam.ac.kr`

Abstract. In this paper, we propose an efficient similarity measure as pre-processing method for clustering of categorical and sequential attributes. The similarity measure is based on a new dynamic programming algorithm, which computes sequence comparison scoring from the gap penalty matrix. This is presented by normalizing sequence comparison scoring. Self-evaluation of the proposed similarity measure is conducted by experimental results of clustering, which is an unsupervised learning algorithm greatly influenced by similarity measure between clusters. In the experiment, Tcpdump Data from DARPA 1999 Intrusion Detection Evaluation Data Sets are used. These transmission data are composed of sequential packet data in a network. Finally, the results of comparison experiments are discussed.

Keywords: Similarity measure, Dynamic programming, Sequence clustering.

1 Introduction

We propose an efficient similarity measure for clustering data with categorical and sequential attributes. It is based on the dynamic programming (DP) algorithm, which computes the sequence comparison scoring from the gap penalty matrix. The similarity measure is presented by normalizing the sequence comparison scoring. The method that uses similarity measure is more efficient than distance measure, because the sequential data is categorical. We propose a new DP algorithm as an optimal alternative for comparing the sequential data. DP is an algorithm which can compare them when each pattern length is different.

Dynamic programming, such as the divide-and-conquer method, solves problems by combining the solutions to subproblems. Divide-and-conquer algorithms partition

---

* This research was supported by the MIC(Ministry of Information and Communication), Korea, under the ITRC(Information Technology Research Center) support program supervised by the IITA(Institute of Information Technology Assessment) (IITA-2006-C1090-0603-0027).
** Correspondent author.

A. Sattar and B.H. Kang (Eds.): AI 2006, LNAI 4304, pp. 372–382, 2006.
© Springer-Verlag Berlin Heidelberg 2006

the problem into independent subproblems, solve the subproblems recursively, and then combine the resulting solutions to solve the original problem. In contrast, DP is applicable when the subproblems are not independent, that is, when subproblems share sub-subproblems [1]. DP is one of classic programming paradigm, introduced even before the term Computer Science was firmly established.

We take the place of self-evaluation of the proposed similarity measure by experimental results of clustering, which is an unsupervised learning algorithm greatly influenced by similarity measure between clusters. For the clustering, we use the ROCK algorithm [7] that clustering of categorical attributes is possible in the data mining algorithms. Tcpdump Data of DARPA 1999 Intrusion Detection Evaluation Data Sets are used in the experiment [8]. These data consist of sequential packets which are transmitted for conducting normal or attack activities. In the experiment, the effectiveness of the proposed similarity measure is shown through comparison with traditional methods.

In the remainder of this paper, we proceed as follows: In Sect. 2, methods of sequence alignment using dynamic programming techniques are presented. In Sect. 3, the new dynamic programming algorithm is proposed. In Sect. 3.1, the transition direction of sequences is introduced. In Sect. 3.2, computation of the sequence comparison scoring using the gap penalty matrix is proposed. In Sect. 4, our similarity measure is proposed. In Sect. 4.1, the normalization method of similarity is presented. In Sect. 4.2, the sequence similarity algorithm is presented. In Sect. 5, we analyze the effectiveness of the proposed method on several experiment results. In Sect. 5.1, the ROCK clustering algorithm we use is introduced. In Sect. 5.2, we compare and analyze the experimental results. Finally, this paper concludes in Sect. 6.

2 Related Works

The basic dynamic programming method for sequence alignment forms the core of the pair-wise and multiple sequence and structure comparisons, as well as of sequence database searching [2]. A protein is defined as an indexed string of elements at each level in the hierarchy of protein structure: sequence, secondary structure, super-secondary structure, and so on. The elements, for example, residues or secondary structure segments such as helices or beta-strands, are associated with a series of properties and can be involved in a number of relationships with other elements. Element-by-element dissimilarity matrices are then computed and used in the alignment procedure based on the sequence alignment algorithm of Needleman & Wunsch, expanded by the simulated annealing technique to take into account relationships as well as properties.

In a speech recognition system, dynamic time warp (DTW) [4] calculations are reduced by use of a search strategy for an optimal non-linear search warp function path: signals predictive of the similarity are generated responsive to the preceding and current correspondence signals of the input word to reference warped words. It has the advantage of conceptual simplicity and robust performance, and has been used in the matching of human movement patterns recently. As far as DTW is concerned, even if the time scale between a test pattern and a reference pattern may be inconsistent, it can still successfully establish matching as long as time ordering constraints hold. However, cross alignment is impossible.

Edit distance [5, 6] between two strings is defined as the minimum number of edit operations (insertions, deletions and substitutions) needed to transform the first string into the other. Each operation is given a score (weight), usually, insert and delete operations are given the same score, and using alignment algorithms, we search for the minimal scoring (or the maximum negative scoring), representing the minimal number of edit operations. Since most of the changes to a DNA during evolution are due to the three common local mutations: insertion, deletion and substitution, the edit distance can be used as a way to roughly measure the number of DNA replications that occurred between two DNA sequences. However, each operation has a cost associated with an optimal edit transcript that describes the transformation.

3 Dynamic Programming

We describe the gap penalty matrix based on the DP algorithm. For example, let the two sequences S_i={A,B,G,C,C,B,A,E,C,F} and S_j={C,B,C,D,D,E,E,B,D,C,F,F} be given. A *sequence* is a multiple and sequential set.

3.1 The Transition Direction

To explicitly describe the word order difference between source and target language, [3] applied the regular alignment concept (in which a source position i is mapped to a target position j, such as Fig. 1) and the inverted alignment concept (in which a target position j is mapped to a source position i, such as Fig. 2), as follows:

$$\text{Regular alignment:} \quad i \rightarrow j = s_i \tag{1}$$

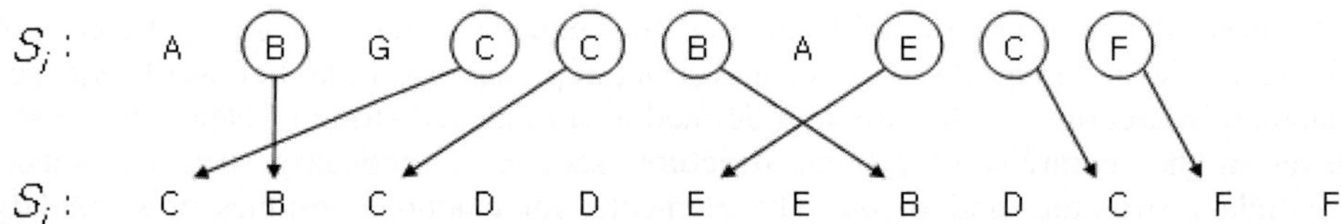

Fig. 1. Regular alignment for the transition direction S_i to S_j

$$\text{Inverted alignment:} \quad j \rightarrow i = s_j \tag{2}$$

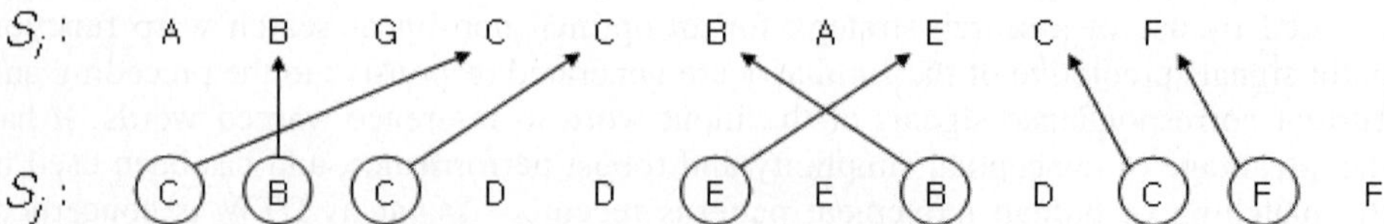

Fig. 2. Inverted alignment for the transition direction S_j to S_i

On the other hand, we propose the combined alignment because for each source there can not be exactly one target on the alignment path. And the alignment method is symmetrical. It is shown in Fig. 3. The arrow linking the two B^1s has double head, because of a symmetrical matching.

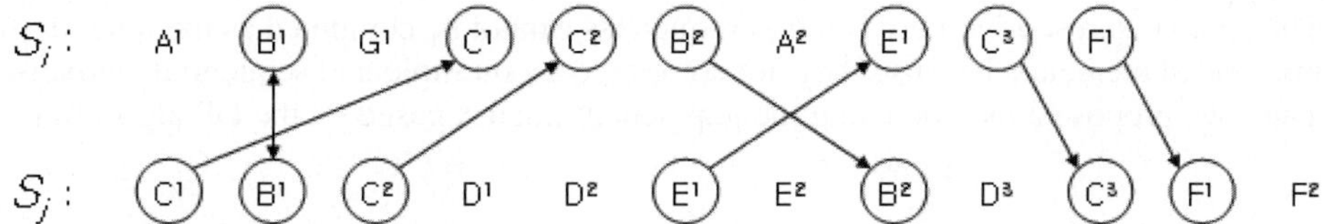

Fig. 3. Combined alignment for the transition direction of S_i and S_j. These are mapped in order, beginning with the heads of the two sequences.

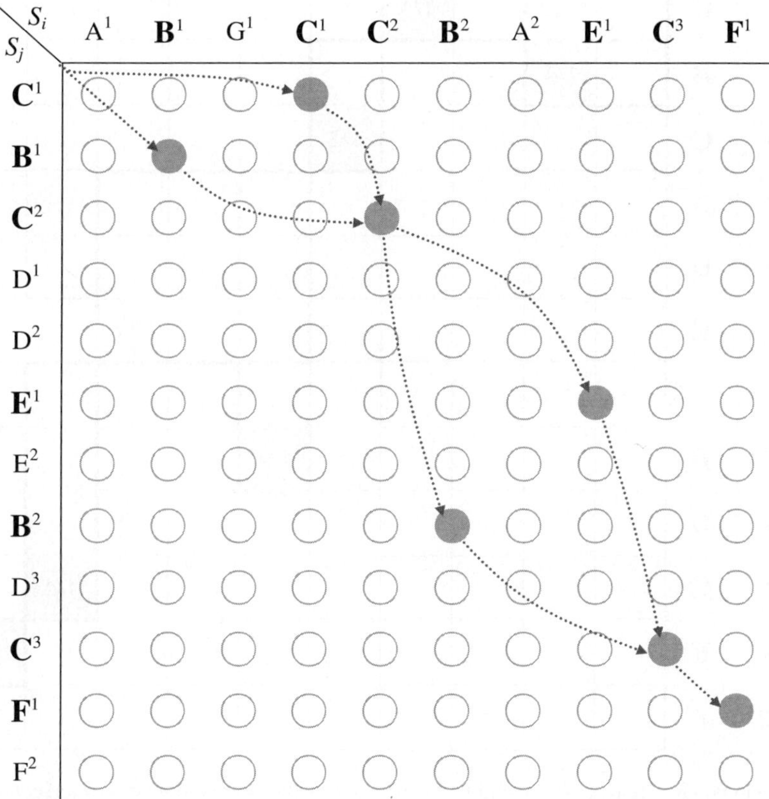

Fig. 4. The transition direction of the two sequences S_i and S_j by combined alignment

In the combined alignment, the cross elements (which are $B^1|C^1$ and $B^2|E^1$) are in need of cross alignment for independent partial orders. Thus, the following elements (which are C^2 and C^3) have a duplicated previous-path as in Fig. 4. This creates difficulty in sequential scoring for comparing similarity between two sequences. We solve the problem in Sect. 3.2.

3.2 Sequence Comparison Scoring (SCS)

The comparison scoring between two sequences cannot be computed as the number of intersected elements, because they are composed by multiple and sequential elements. Thus, we propose a method using the gap penalty matrix based on the DP algorithm.

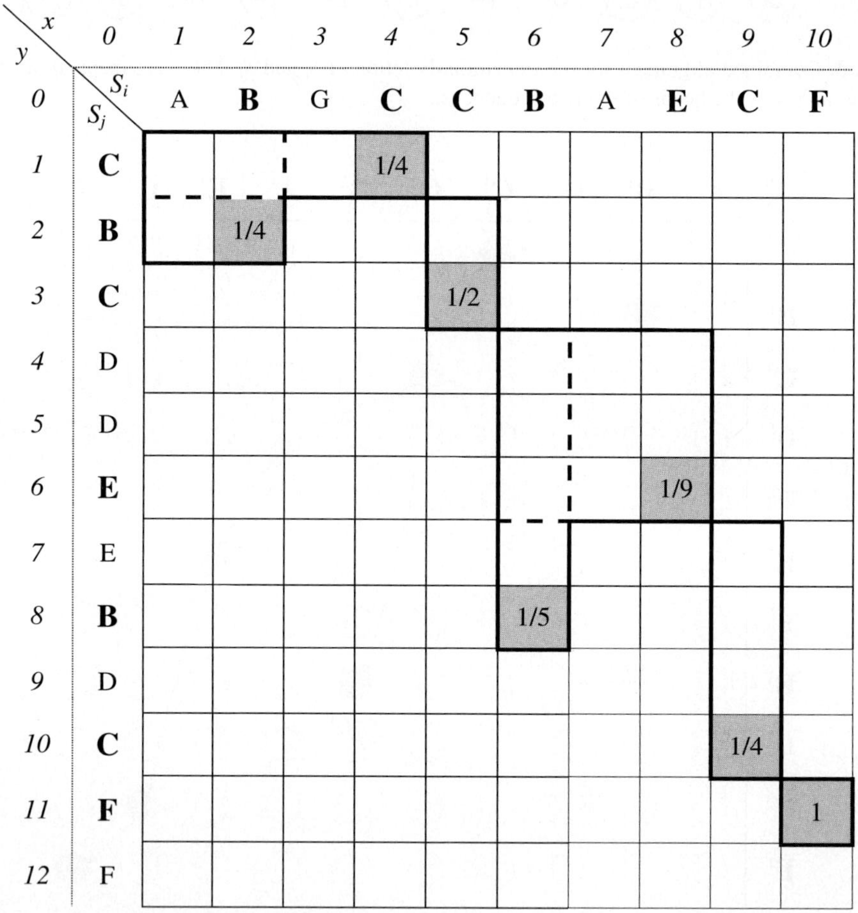

x / y	S_i →	0	1 A	2 B	3 G	4 C	5 C	6 B	7 A	8 E	9 C	10 F
0	S_j ↓											
1	C					1/4						
2	B			1/4								
3	C						1/2					
4	D											
5	D											
6	E									1/9		
7	E											
8	B							1/5				
9	D											
10	C										1/4	
11	F											1
12	F											

Fig. 5. Gap penalty matrix. The weights are affected by the nearest previous-intersected blocks.

Scorings on the intersected blocks are shown in Fig. 5. When (x_{prev}, y_{prev}) is a previous-intersected block of (x,y), $(x_{prev} < x)$ and $(y_{prev} < y)$ must be satisfied. The scoring is the reciprocal of the distance from the previous-intersected block. We use the conception of the gap penalty for computing the distance.

$$BS_{(x,y)} = \frac{1}{D_{(x,y)}}, \quad D_{(x,y)} = (x-x_{prev})(y-y_{prev}) \geq 1 \tag{3}$$

$BS_{(x,y)}$ is the *Block Scoring (BS)* of the intersected block (x,y). $D_{(x,y)}$ is the distance from the nearest previous-intersected block (x_{prev},y_{prev}) to the block (x,y). The distance means a number of blocks between (x_{prev},y_{prev}) and (x,y).

We depend on the three rules for selecting the nearest previous-intersected block (x_{prev},y_{prev}) of the intersected block (x,y) as follows:

Rule 1.   A previous-intersected block which has a maximum of $(x_{prev} \times y_{prev})$ about an arbitrary (x_{prev},y_{prev}) is selected. If it does not exist, the initial intersected block $(0,0)$ is selected.

Rule 2.   If the blocks that satisfy *Rule 1* are duplicated, a block which has a maximum value of $(x_{prev} + y_{prev})$ is selected.

Rule 3.   If the blocks that satisfy *Rule 2* are duplicated, the selecting is depended upon the following two additional rules.

Rule 3-1.   If x and y of block (x,y) are equal, the weight is the same even though nothing is selected between them.

Rule 3-2.   If x and y of block (x,y) are not equal, a block which has less value between minimum values of $(x - x_{prev})$ and $(y - y_{prev})$ is selected.

In Fig. 5, the initial intersected block is $(0,0)$ and the *BS* is zero. The nearest previous-intersected block of intersected block $(5,3)$ becomes the block $(4,1)$ to be based on *Rule 2*. In addition, the nearest previous-intersected block of intersected block $(9,10)$ becomes the block $(8,6)$ to be based on *Rule 3-2*. The intersected block $(10,11)$ has 1 for the maximum *BS*, because F follows on C in all the two sequences. The comparison scoring measure simply depends on the hierarchical transition from previous orders. The sequence comparison scoring is the sum of the *BS*s, as follows:

$$SCS = \sum_{(x,y)\in I} BS_{(x,y)} \tag{4}$$

where I is a set of the intersected blocks between sequences. The *SCS* of the two sequences S_i and S_j is 2.56, as $(1/4 + 1/4 + 1/2 + 1/5 + 1/9 + 1/4 + 1)$.

4 Similarity Measure

Our similarity measure between the two sequences S_i and S_j is defined as follows:

$$Sim(S_i,S_j) = \frac{\{S_i \cap S_j\}}{\{S_i \cup S_j\}} \Rightarrow \frac{SCS}{\{S_i \cup S_j\}} \tag{5}$$

The more elements that the two sequences S_i and S_j have in common, that is, the larger intersection $\{S_i \cap S_j\}$ is, the more similar the sequences. Dividing by union $\{S_i \cup S_j\}$ is the scaling factor ensuring similarity between 0 and 1. $\{S_i \cap S_j\}$ is replaced as the sequence comparison scoring.

4.1 Normalization

The union is used for the normalization of the similarity measure. The union value is simply the number of all elements such as Fig. 6. Thus, the union value of S_i and S_j is 15 as bag {A,G,A,B,C,C,B,E,C,F,D,D,E,D,F}. A bag is a set which allows multiple occurrences of the elements.

Fig. 6. The union bag of the two sequences S_i and S_j. The union value is the number of all elements.

4.2 Sequence Similarity

Ultimately, the similarity of the two sequences S_i={A,B,G,C,C,B,A,E,C,F} and S_j={C,B,C,D,D,E,E,B,D,C,F,F} is computed as 0.17 (= 2.56 / 15). The similarity function is symmetrical. The sequence similarity algorithm is presented in Fig. 7.

```
procedure SeqSim(S_i, S_j)
begin
    set I[x,y] for every intersected blocks between S_i and S_j
    for each (x,y) ∈ I do {
      (x_prev, y_prev) := select_nearest_prev-intersected_block(x,y)
      D := compute_distance(x, y, x_prev, y_prev)
      BS := reciprocal(D)
      intersection := intersection + BS
    }
    union := size of bag { S_i ∪ S_j }
    similarity := intersection / union
End
```

Fig. 7. An algorithm of the sequence similarity measure

5 Experimental Results

Self-evaluation of the proposed similarity measure is conducted by the evaluation of experimental results of clustering, which is an unsupervised learning algorithm greatly affected by the similarity measure between clusters. For clustering, we use the ROCK algorithm, that clustering of categorical attributes is possible in the data mining algorithms. We experiment with Tcpdump Data of DARPA 1999 Intrusion Detection Evaluation Data Sets. These are sets of network packets generated over five weeks of simulated network traffic in a hypothetical military Local-Area-Network. These data consist of packets transmitted for achieving normal and attack activities. The set of packets transmitted for the one activity is called session, which is a sequential transmission unit of a series of packets from connection opening to closing. We show the results of clustering achieved by comparison of the similarities of the normal and attack sessions.

5.1   The ROCK Clustering Algorithm

ROCK is an agglomerative hierarchical clustering algorithm based on the link relationship of neighbors. We introduce the similarity measure at the stage of prioritizing neighborhood computations (that is pre-processing stage of ROCK) in priority. See [7] for details of the clustering process. A point's neighbors are those points that are considerably similar. Let $Sim(p_i, p_j)$ be a similarity measure between the pair of points p_i and p_j. It is assumed that Sim takes values between 0 and 1, with larger values indicating that the points are more similar.

$$Sim(p_i, p_j) \geq \theta \tag{6}$$

In the above equation, θ is a user-defined parameter. Assuming that Sim is 1 for identical points and 0 for totally dissimilar points, a value of 1 for θ constrains a point to be a neighbor only to other identical points. On the other hand, a value of 0 for θ permits any arbitrary pair of points to be neighbors.

Let $Sim(T_1, T_2)$ be a similarity measure between the two transactions T_1 and T_2. Also, let T_1 be $\{1,2,3\}$ and T_2 be $\{3,4,5\}$. Sim is the following:

$$Sim(T_1, T_2) = \frac{|T_1 \cap T_2|}{|T_1 \cup T_2|} \tag{7}$$

where $|T_i|$ is the number of items in T_i. The intersection $|T_1 \cap T_2|$ is 1, that is $|\{3\}|$, and the union $|T_1 \cup T_2|$ is 5, that is $|\{1,2,3,4,5\}|$. Therefore, the similarity of T_1 and T_2 is 0.2 (= 1 / 5). The neighbor(the notion of adjacency) of T_1 and T_2 can be determined as 1(link) or 0(non-link) according to whether Sim is greater than θ. Such similarity measures influence the accuracy of the clustering.

5.2   Comparative Analysis

First of all, we show the results of original clustering, in which non-sequential similarity measure is applied. Table 1 describes the results of clustering using Jaccard algorithm that is the similarity measure introduced in Sect. 5.1. And then, Table 2 describes the clustering results using DTW and edit distance as sequential similarity measure. Finally, Table 3 describes the clustering results from our similarity measure.

Table 1. Results of clustering using Jaccard. This is the non-sequential similarity measure.

Attack Class	Data Sets			Clustering Results					
				Normal Cluster			Attack Cluster		
	Total	Normal	Attack	Total	True	False	Total	True	False
guesspop	59	29	30	29	29	0	30	30	0
ftpwrite	56	53	3	56	53	3	0	0	0
guessftp	133	53	80	117	42	75	16	5	11
ncftp	56	53	3	56	53	3	0	0	0
mailbomb	5115	2749	2366	616	116	500	4499	1866	2633
netbus	2751	2749	2	2749	2749	0	2	2	0
sendmail	2750	2749	1	2750	2749	1	0	0	0
phf	104	100	4	89	88	1	15	3	12

Clustering using Jaccard Similarity Measure

Table 2. Results of clustering using DTW and edit distance. These are the sequential similarity measures.

Clustering using DTW

Attack Class	Data Sets			Clustering Results					
				Normal Cluster			Attack Cluster		
	Total	Normal	Attack	Total	True	False	Total	True	False
guesspop	59	29	30	29	29	0	30	30	0
ftpwrite	56	53	3	54	53	1	2	2	0
guessftp	133	53	80	36	36	0	97	80	17
ncftp	56	53	3	53	53	0	3	3	0
mailbomb	5115	2749	2366	4426	2749	1677	689	689	0
netbus	2751	2749	2	2749	2749	0	2	2	0
sendmail	2750	2749	1	2750	2749	1	0	0	0
phf	104	100	4	103	100	3	1	1	0

Clustering using Edit Distance

Attack Class	Data Sets			Clustering Results					
				Normal Cluster			Attack Cluster		
	Total	Normal	Attack	Total	True	False	Total	True	False
guesspop	59	29	30	29	29	0	30	30	0
ftpwrite	56	53	3	56	53	3	0	0	0
guessftp	133	53	80	8	8	0	125	80	45
ncftp	56	53	3	53	53	0	3	3	0
mailbomb	5115	2749	2366	68	68	0	5047	2366	2681
netbus	2751	2749	2	2749	2749	0	2	2	0
sendmail	2750	2749	1	2750	2749	1	0	0	0
phf	104	100	4	104	100	4	0	0	0

As the table illustrates, two clusters are identified, one containing normal sessions and the other containing attack sessions. The labels of the clusters are set classified as *Normal Cluster* or *Attack Cluster*. These are results of clustering of network sessions about normal and attack traffic.

In the Tables, *False* members of *Normal Cluster* indicate attack, and *False* members of *Attack Cluster* indicate normal as well. The unit of articles is the number.

Table 3. Results of clustering using the proposed similarity measure

Attack Class	Data Sets			Clustering Results					
				Normal Cluster			Attack Cluster		
	Total	Normal	Attack	Total	True	False	Total	True	False
guesspop	59	29	30	29	29	0	30	30	0
ftpwrite	56	53	3	53	53	0	3	3	0
guessftp	133	53	80	53	53	0	80	80	0
ncftp	56	53	3	53	53	0	3	3	0
mailbomb	5115	2749	2366	2761	2749	12	2354	2354	0
netbus	2751	2749	2	2749	2749	0	2	2	0
sendmail	2750	2749	1	2749	2749	0	1	1	0
phf	104	100	4	100	99	1	4	3	1

As expected, the quality of the clusters generated by the Jaccard algorithm is very poor. For instance, in the cluster for the *mailbomb* class found by the traditional method, as many as around 81% of the members are attacks. The proposed method only has false members of approximately 0.4%. This is because the similarity function does not consider the sequence of attributes. The traditional distance measures, i.e., DTW and edit distance, were not useful to the categorical and sequential attributes in need of cross alignment. In the Table 2, they show the *False* results of similar distribution. Ultimately, the results show that the effectiveness of the proposed similarity measure is better in the experiment.

6 Conclusion

In this paper, we proposed a new similarity measure for clustering of data with categorical and sequential attributes in need of cross alignment. It is based on a dynamic programming algorithm, which computes the sequence comparison scoring from the gap penalty matrix. The similarity measure was presented by normalizing the sequence comparison scoring. Self-evaluation of the proposed similarity measure was taken by the evaluation of experimental results of clustering, which is unsupervised learning algorithm greatly influenced by similarity measure between clusters. We experimented with Tcpdump Data of DARPA 1999 Intrusion Detection Evaluation Data Sets. The results of the experimental study with the data sets were very encouraging. The proposed method can be used, not only for intrusion detection of network activities with abnormal sequences, but also protein homology search, RNA structure prediction, gene finding, and so forth.

References

1. T. H. Cormen, C. E. Leiserson, and R. L. Rivest, *Introduction to algorithms*, MIT Press and McGraw-Hill Book Company, 14th edition, 1994.
2. A. Sali and T. L. Blundell, "Definition of general topological equivalence in protein structures: A procedure involving comparison of properties and relationships through simulated annealing and dynamic programming," J. Mol. Biol. 212, 403-428, 1990.

3. Christoph Tillmann and Hermann Ney: "Word Reordering and a Dynamic Programming Beam Search Algorithm for Statistical Machine Translation," Computational Linguistics 29(1): 97-133, 2003.
4. Cory Myers, et. al., "Performance Tradeoffs in Dynamic Time Warping Algorithms for Isolated Word Recognition," IEEE Trans. on acoustics, speech, and signal processing, vol. ASSP-28, No.6, Dec 1980.
5. Mikhail J. Atallah, *Algorithms and Theory of Computation Handbook*, CRC Press, 2000 N.W. Corporate Blvd., Boca Raton, FL 33431-9868, USA, 1999.
6. L. Allison, *Dynamic programming algorithm (DPA) for edit-distance*, In Algorithms and Data Structures Research & Reference Material, School of Computer Science and Software Engineering, Monash University, Australia, 1999.
7. Sudipto Guha, Rajeev Rastogi, and Kyuseok Shim, "ROCK: A Robust Clustering Algorithm for Categorical Attributes," Proceeding of the IEEE International Conference on Data Engineering, Sydney, March 1999.
8. MIT Lincoln Laboratory, DARPA Intrusion Detection Evaluation Data Sets, http://www.ll.mit.edu/IST/ideval/data/data_index.html.

SDI: Shape Distribution Indicator and Its Application to Find Interrelationships Between Physical Activity Tests and Other Medical Measures

Ashkan Sami[1], Ryoichi Nagatomi[2], Makoto Takahashi[1], and Takeshi Tokuyama[3]

[1] Graduate School of Engineering, Tohoku University, Japan
[2] Department of Medicine and Science in Sports and Exercise, Graduate School of Medicine, Tohoku University, Japan
[3] Graduate School of Information Sciences, Tohoku University; Japan
ashkan.sami@most.tohoku.ac.jp

Abstract. Comprehensibility is driving force in medical data mining results since doctors utilize the outputs and give the final decision. Another important issue specific to some data sets, like physical activity, is their uniform distribution due to tile analysis that was performed on them In this paper, we propose a novel data mining tool named SDI (Shape Distribution Indicator) to give a comprehensive view of co-relations of attributes together with an index named ISDI to show the robustness of SDI outputs. We apply SDI to explore the relationship of the Physical Activity data and symptoms in medical test dataset for which popular data mining methods fail to give an appropriate output to help doctors decisions. In our experiment, SDI found several useful relationships.

1 Introduction

Medical data are very unique, therefore medical data mining is very valuable. Even though knowledge discovery from medical data is very rewarding in a lot of aspects, great challenges face the data miners who would like to work in this field. Since specific challenges of medical data issues have been discussed in [1-2], we will not go into detail of specific characteristics that medical knowledge discovery projects have. However as you will see in this paper, comprehensibility is a driving force in overall theme of this work.

Physical Activity tests are mainly designed for elderly people to find the status of their Physical Activity and risk to fall. These measures are very good indicators of overall 'healthiness' of elderly people. What makes majority of these measurements different from regular tests is the way they are interpreted.

Interpretation of Physical Activity tests are based on tile analysis which is a quite popular in medical studies on regular people (in contrast to patients) [3] which is called social or community study in medical terms. In other words, the data is sorted based on results of a test that we want to make tiles. Based on total frequency the data is divided to the specific number of tiles, in most cases four or three which is called quartile and tertile. In contrast to Blood analysis or Urine analysis where clear thresholds separating healthiness or suspect of illness exist, in Physical Activity Quartiles or Tertiles provide interpretations.

A. Sattar and B.H. Kang (Eds.): AI 2006, LNAI 4304, pp. 383 – 392, 2006.

© Springer-Verlag Berlin Heidelberg 2006

Interrelationship of Physical Activity test and symptoms is very important for medical doctors in Sports Medicine. These relationships can lead them to design or suggest activities that can improve or stop the symptoms or at least slow the progress of the symptom. Since data came from a community study number of cases having symptoms is low. Even though in Statistics information hidden in low frequent cases will not be significant, in data mining information hidden in cases with low frequency is very important. A typical two dimensional distribution of fitness test and LUT symptoms are presented in Table 1 and 2.

Table 1. An example of strong relationship with SDI's values

	Quartile 1	Quartile 2	Quartile 3	Quartile 4	SDI_i's
Value = 0	73	65	53	43	-51
Value = 1	32	30	27	25	-12
Value = 2	1	5	5	4	4.5
Value = 3	2	4	6	9	11.5
Value = 4	1	4	9	10	16
Value = 5	1	2	10	19	31
Sum	110	110	110	110	

Table 2. An example of no relationship

	Quartile 1	Quartile 2	Quartile 3	Quartile 4	SDI_i's
Value = 0	59	54	50	50	-15.5
Value = 1	30	32	32	33	4.5
Value = 2	1	2	8	10	16.5
Value = 3	5	5	7	2	-3.5
Value = 4	2	9	5	11	11.5
Value = 5	10	8	8	4	-9
Sum	110	110	110	110	

Even though extensive number of well established statistical algorithms [4] exist that can find correlations between the two values, results of these tests will obey the major distributions that are concentrated around low values of the symptoms. The high value symptoms that have much less frequency and may contain a lot of knowledge will have no significant contribution to the results. Although in Statistics we do not consider small values, in data mining these values with small frequencies are highly regarded.

Decision tree analysis to find rules sharing both categories can be performed by deploying some objective measures like Quality of Life [5] as the consequent of the rules. However these analyses will lead to results that are very hard for experts to interpret if we use the data that is not tiled. In contrast, using tiles will lead to severe loss of accuracy.

Clustering of data in high dimension like [6] is hard to interpret and as it will be shown in the experimental results any form of subspace clustering like Frequent Itemsets [7], although provides results, no knowledge can be gained from it. Other methods of driving rules like association rule classifiers [8-10] may indeed lead to no answer.

The main purpose of this paper is to present a general methodology named Shape Distribution Indicator (*SDI*) to find relationship between two factors when one of the distributions has changed to uniform due to tile analysis and the other factor has highly uneven distribution. *SDI* calculates set of indicators that comparison of each indicator with prior or next one present the type of change that occurred from one set to the other. The values are indicators in a sense that only comparisons among them convey meaning. Each specific indicator of *SDI* is a real number that its magnitude presents how far the distribution is relatively tilted and sign of the Indicator presents which side. To be more precise a larger value compare to a smaller value indicates a shift toward higher values, where a smaller *SDI* value presents frequencies that are tilted toward lower tile values. Each value may not convey all information concerning distribution by itself. Anyhow they are only indicators and comparison conveys meaning.

To have an appreciation of the method, the method is used on Physical Activity test which is an excellent example of these kinds of problems. The use of regular data mining algorithms even though provide answers, the answers are not comprehensive in a way to convey knowledge concerning the data. Thus, in this paper, first we will talk about Physical Activity dataset and its pre-processing. Section 3 presents the Shape Distribution Indicator (*SDI*). Section 4 presents *ISDI* which is an index to assess the quality of the *SDI*. Section 5 describes the experimental results and summary comes in section 6.

2   Physical Activity Dataset and Its Pre-processing

In this section we will talk about the data that we used in the analysis and the pre-processing we performed on it. The file Physical Activity dataset contained results of different Physical Activity tests on 960 Japanese aged over 70 years. We used this dataset based on two reasons. First, the data quality was much higher than regular datasets available since various quality control routines were performed on the dataset. Just to show an example, the missed or suspected values were measured again by asking the volunteers to return to the medical center and repeat the measurements. Secondly, other teams performed other tests on the same people. Thus results of examinations from other departments were also available so that interrelationship finding can be performed.

Pre-processing of Physical Activity Dataset was mainly eliminating unrelated fields. The Physical Activity Dataset which originally had 64 fields was reduced based on detailed definition of the items from literature review and interviews. The data consisted of some overall physical condition of the volunteer, the tests that were performed on him/her and, some extra information like the physician(s) who performed the tests and etc.

Other items related to illnesses and "pain conditions at the test date" were also deleted. These questions were asked in order to help the Physical Activity examiners to grab the physical status and eligibility of the subjects for Physical Activity measurement and not intended to be analyzed in the context of geriatric assessment. Of course some items related to the condition of measurement may be used to compensate or to adjust the measured values.

The remaining data consisted of the results of tests that were performed on the volunteers. All these deletions were based on expert's approval. Due to space limitation, we will not go through details of the data. The main characteristics of the items in file are as follows:

1.   The length that volunteers could reach in different positions.
2.   The time or number steps that it took them to perform a task.

All parameters are numeric. Except 'Risk to Fall' which is presented by digits. Since all of items like Timed Up-and-Go Test (TUGT) were measured twice, average, maximum or minimum was used based on verification of the expert.

3   Shape Distribution Indicator (SDI)

Now, we present a novel quantitative approach that can explain the overall relationship between the values. Due to quartile analysis of the each of the Physical Activity measures, the base distribution is a uniform distribution. If a pattern exists among any of the two measures of the two categories, the 'overall' distribution of the Physical Activity measure with respect to desired symptom values should shift. This *constant* movement from one quartile to another quartile, if exists, presents the pattern we are looking for. To have a better understanding, refer to Tables 1 and 2.

For different desired symptom values, this distribution spreads over different values. In Tables 1 and 2, a Physical Activity test that contains four quartiles is considered against a symptom. All quartiles exactly have the same overall frequency. In other words, the overall distribution is uniform for the Physical Activity test. To consider relationship with a symptom, respective frequencies of Physical Activity quartiles will spread among different values of the symptom, (0 to 5). In case there is a relationship (like in Table 1), as we move from the symptom values of 0 to five the distribution shifts from lower quartile values smoothly to higher values. The shift indicates that there is dependency between the two and increase of one factor leads to increase or decrease of the other. In contrast, in Table 2, there is no specific pattern among the distribution of Physical Activity quartiles with respect to symptom. Thus no relationship exists.

By SDI values, we would like to find an indicator to express how the shape of distribution is tilted. To do so, we consider mean as a baseline. Then we find the difference of each frequency from the mean. Summation of weighted differences by the numerical values of its corresponding tile gives the number that represents how much and to which side the overall shape is tilted. Since we are only looking for indicators and indicators are only used for comparison no need for normalization exists.

To provide a clearer definition, consider A is a desired symptom and B is a result of a specific Physical Activity test after quartile. When there is no relationship between A and B, distribution of B with respect to each specific value of A should not vary in a consistent way or the shape of distribution may stay the same. In contrast, when a relationship exists, the variation of distribution should follow a pattern. To find the pattern, we should provide a quantitative way of explaining the distribution. Then based on the change in the Indicator (constantly increasing or decreasing), we can say a pattern exists. If no consistent change or no change at all exists, then no relationship between the two exists.

In other words, consider in general A can have values A_0, A_1, ..., A_n and B is distributed among B_1, B_2, ..., B_m. At the same time f_{ij} is the number of samples for a specific i and j, where $0 \leq i \leq n$, $1 \leq j \leq m$, A is A_i and B is B_j. In case of a relationship, the Indicator for distribution of f_{ij}'s for a specific value i when i changes from 0 to n should constantly increase or decrease. To present the Indicator we have to consider for each specific A_i, in this study $\sum_{j=1}^{m} f_{ij}$ is not only same, but also changes drastically.

Now the problem is to find a Indicator to quantitatively present the frequency distribution for each specific A_i. Since we are only interested on the change in the Indicator, we do not find a formula for the whole Indicator but what the Indicator depends on.

First, we will simply assign 1 to m to B_1, B_2,..., B_m respectively. This assignment is performed so that contribution of each f_{ij} will be appointed to a specific B_j. This assignment can be done anyway that we want as long as B's are strictly increasing integers and the larger assignments represent higher tiles. Since the calculated values of the Indicator for different i's are only compared together and the effect of the assignment affects all the SDI_i's equally, no need for normalization exists.

Secondly, the Indicator should not depend on frequencies value of distribution but on weighted difference from mean.

The contribution of each f_{ij} to the Indicator is related to distance form mean of the distribution weighted by the value of respected B. Thus if we define

$$\mu_{f_i} = \sum_{j=1}^{m} f_{ij} / m \tag{1}$$

the shape distribution Indicator for a specific i or SDI_i is simply related to the product of the two. Thus for each f_{ij}, we can write:

$$SDI_i \propto (f_{ij} - \mu_{f_i}) B_j \tag{2}$$

Summation of the contributions will provide the SDI_i. Thus,

$$SDI_i = \sum_{j=1}^{m} (f_{ij} - \mu_{f_i}) B_j \tag{3}$$

The general form of the equation becomes:

$$
\begin{bmatrix} SDI_0 \\ SDI_1 \\ \vdots \\ SDI_n \end{bmatrix}
=
\begin{bmatrix} f_{01} & f_{01} & \cdots & f_{0m} \\ f_{11} & f_{12} & \cdots & f_{1m} \\ \vdots & \vdots & \ddots & \vdots \\ f_{n1} & f_{n2} & \cdots & f_{nm} \end{bmatrix}
\begin{bmatrix} B_1 \\ B_2 \\ \vdots \\ B_m \end{bmatrix}
-
\begin{bmatrix} \mu_{f_0} & \mu_{f_0} & \cdots & \mu_{f_0} \\ \mu_{f_1} & \mu_{f_1} & \cdots & \mu_{f_1} \\ \vdots & \vdots & \ddots & \vdots \\ \mu_{f_n} & \mu_{f_n} & \cdots & \mu_{f_n} \end{bmatrix}
\begin{bmatrix} B_1 \\ B_2 \\ \vdots \\ B_m \end{bmatrix}
\tag{4}
$$

It just happens that SDI_i can be also written as:

$$SDI_i = \sum_{j=1}^{m} (B_j - \mu_B) f_{ij} \tag{5}$$

where

$$\mu_B = \sum_{j=1}^{m} B_j \, / \, m \tag{6}$$

Even though there is no physical interpretation for it. The proof is written in Appendix.

Since the method is not related to frequencies of each set of A_i, it can be used in situations where the distributions are not even, like ours. Each SDI_i gives a quantitative Indicator for the distribution. Thus a pattern is nothing but a strict change through all the values of SDI_i's. In other words, we should obtain all the values of SDI_i for different i's. If one of the below mentioned equalities hold, a relationship exists.

$$SDI_0 < SDI_1 < ... < SDI_n \tag{7}$$

$$SDI_0 > SDI_1 > ... > SDI_n \tag{8}$$

Inequality (7) indicates a direct relationship. That is, by increase in i, the distribution shifts from lower values of B_j to higher values of B_j. As an example, in our case by increase in desired symptom values the distribution shifts to higher quartiles. Or as symptoms get worse, the indicated Physical Activity becomes worse. This kind of relation is of importance for Physical Activities that are measured in seconds. Thus by increase in those we see an improvement. Stated differently, for the cases like Max Speed that is measured in Seconds, any direct relationship with any desired symptom value will indicate that category of Physical Activity can improve the symptoms. Thus by assigning that type of exercise we can reduce the symptoms. For the measures that count steps, direct relationship may provide insight or it may be only an observation.

In contrast, Inequality (8) presents a reverse relationship which by increase of i the distribution has a shift to lower values. These kinds of relationships, for the measures that count steps, will provide an indicator that exercises of that category will reduce the symptoms. The Physical Activity tests that are measured in seconds like TUGT will not give insight by this measure and is just an indication of status if lower time measures are considered better values.

4   ISDI an Index for SDI

An index will present the strength of the measure. $ISDI_i$ or an Index for SDI_i is used to measure the strength of relationships presented by SDI methodology in each layer (or each specific i). It will be shown that for each layer $ISDI_i$,

$$ISDI_i = \frac{\left(\sum_{j=1}^{m} B_j f_{ij} - m\mu_B \mu_{f_i} \right)^2}{\left(\sum_{J=1}^{m} B_j^2 - m\mu_{B_i} \right)\left(\sum_{J=1}^{m} f_{ij}^2 - m\mu_{f_i} \right)} \tag{9}$$

And total index $ISDI_i$,

$$ISDI = \sum_{i=0}^{n} \left(ISDI_i \times \sum_{j=1}^{m} f_{ij} \right) \tag{10}$$

We use the sum of square errors in each layer of i and based on that, we define the index. As can be seen SDI_i is not independent of type of distribution B_j's have. However, SDI_i's are compared together. Thus, no needs to include factors that are purely depend on B_j's distribution or values exist. But to present distribution based on SDI_i, we should multiply it by specific values that are dependent on distribution and specific values of B_j's which would be a function purely based on B_j's. Considering two degrees of freedom to present the data distribution, we find SSE for f_{ij}'s of each i as follows:

$$SSE_i = \sum_{j=1}^{m} \left(SDI_i \times f(B_1, B_2, \ldots, B_m) \times B_j - C_i - f_{ij} - SDI_i \times f(B_1, B_2, \ldots, B_m) \times \mu_B - \mu_{f_i} \right)^2 \tag{11}$$

Since we did not define a hard set of rules for enumerating B_j's, $f(B_1, B_2, \ldots, B_m)$ is given to compensate the flexibility in B_j's assignments. C_i is the second degree of freedom. Expansion of the above given equation and substituting C_i with the value of:

$$C_i = \frac{\sum_{j=1}^{m} f_{ij} - SDI_i f(B_1, B_2, \ldots, B_m) \sum_{j=1}^{m} B_j}{m} = \mu_{f_i} - SDI_i \times f(B_1, B_2, \ldots, B_m) \times \mu_B \tag{12}$$

we will obtain SSE_i as Equation (13) where $ISDI_i$ is given in Equation (9):

$$SSE_i = \left(\sum_{j=1}^{m} f_{ij}^2 - \mu_{f_i} \sum_{j=1}^{m} f_{ij} \right) (1 - ISDI_i) \tag{13}$$

Thus, the higher the index, the better the assumption. It is important to note since assigning $f(B_1, B_2, \ldots, B_m)$ were done arbitrary (and is same for all i in the experiments), the function does not play an important role in index and it is affecting all the analysis the same way. A lot of ways exit that we can combine $ISDI_i$ for different i's and come up with a more concrete index. The method we used in our analysis was to consider the frequency as a weight for the index. Thus, Equation (10) is the $ISDI$. Definitely the higher the value of $ISDI$ the more concrete the analysis was performed.

5   Experimental Results

Finding interrelationships between Physical Activity and other symptoms is in its infancy stage. Since majority of Physical Activity tests were performed independent of other measurements or tests and the main objective of those tests were to assess the risk to fall of elderly, results of this kind of analysis are not abundant to make a cross-checking analysis.

To evaluate our method we have tested our approach to find interrelationship between Physical Activity and IPSS (International Prostate Symptom Score) [11] symptoms in male volunteers. IPSS is a symptom index of benign prostatic hyperplasia (BPH). BPH is fundamentally a disease that causes morbidity through the urinary symptoms with which it is associated. Due to popularity of BPH among old male population and costs associated with this disease, it is very important to prolong the increase of symptoms and if possible improve the symptoms. The index includes seven questions covering frequency, nocturia, weak urinary stream, hesitancy, intermittence, incomplete emptying and urgency with answers ranging from zero to five, where zero means no symptom and five indicates severe problem with the urinary symptom.

Before presenting the results with our approach, we will give more descriptive reasons why Frequent Itemsets (FI) fail to provide appropriate answers. We present the results of FI's because they were the only set of algorithms that provided results. In this part, we only present FI that exist between Risk to fall, presented by R and Nocturia, represented by N. By setting Sup=10%, the FI are presented in Table 3. Table 4 presents the results when Sup=3%. Even though several FI were found and all of them are not statistically explainable, no specific knowledge is drawn from these Itemsets. Thus now we explain our findings based on SDI.

It is important to note, since the dataset we had was based on a social study on old Japanese aged over 70 years old who had no major illness, the number of volunteers with acute symptoms was rare.

In order to overcome the problem of frequency and make the algorithm more robust, we ran the algorithm on frequencies of A_0, A_1, summation of A_2 and A_3 and summation of A_4 and A_5. In other words, frequencies of IPSS values of 2 and 3 were grouped and considered as one category and frequencies of values of IPSS equal to 4 and 5 were grouped and were considered as one category. Even after this aggregation, most of the summed categories were still much less than 10% of the population. It makes sense to consider more populated samples than solely relay on few numbers that existed in the high values of IPSS since these aggregates will dampen the huge effect of small variations in few data. It is important to emphasis that the proposed method is not dependant on the percentage of the distribution among values; however, the summation will make the analysis more robust.

Table 3. Part of Frequent Itemsets that have Nocturia and Risk to fall, Sup=10%

R2=31.6%	N1=37.4%	N2,R4=12.8%	N1,R2=15%
R3=30%	N2=32.1%	N1,R3=13.5%	
R4= 37%	N3=14.3%	N2,R2=10.1%	

Table 4. Part of Frequent Itemsets that have Nocturia and Risk to fall, Sup=3%

R2=31.6%	N1=37.4%	N2,R4=12.8%	N1,R2=15%
R3=30%	N2=32.1%	N1,R3=13.5%	N0,R3=3.6%
R4= 37%	N3=14.3%	N2,R2=10.1%	N3,R3=3.4%
N0,R4=3.6%	N0=9.7%	N0,R4=3.6%	N2,R3=9.2%
N3,R4=8.0%	N0,R3=3.6%	N0,R4=3.6%	

In the male population, a relationship was found between Incontinence and Physical Activity tests that measure brisk walking. Moreover, the algorithm showed males suffering form incontinent were walking more slowly at shorter steps than regular people. However, we found no major interrelationship between Physical Activity and nocturia. Even though a pattern between Urgency and Physical Activity was observed, the pattern was not interesting just descriptive. In other words, the people who suffered from Urgency showed they are faster walkers. A slight relationship between Functional reach category and hesitancy was observed. Since the relationship was not very strong it was not regarded highly. Weak stream had relationship with stretching, so it can be concluded that exercises that improve dynamic body balance can improve stream problems. Frequency symptoms also become worse for people who are slow walkers. Emptying symptom affects extension and Max Speed.

The relation between incontinent and Physical Activity was very important for physicians. Since, urine incontinence or unintended leak of urine makes patients embarrassed, and occasionally results in home-bound without going out of home. This relationship shows, appropriate amount of exercise not only keeps them fit but also can prevent the embarrassment they encounter due to incontinent.

6 Summary

This paper presents *SDI* (Shape Distribution Indicator) that can be deployed in situations where relationship between two variables, one having uniform distribution due to tile analysis and the other is distributed very unevenly is under investigation. *SDI* value quantitatively represents the distribution. We presented some experimental results based on LUT and Physical Activity tests and found very useful relationships. Moreover, the robustness of the measure was provided by the *ISDI* index.

Acknowledgement

The authors are very grateful to the Tsurugaya Project Team members at Graduate School of Medicine; Tohoku University. We would like to express our deepest appreciation for help, support and data that they provided.

This research was mainly funded by Ministry of Education, Science, sports, and culture of Japan under a Ph.D. fellowship to the first author.

References

1. Krzysztof J. Cios,and G. William Moore, 'Uniqueness of medical data mining', Artificial Intelligence in Medicine 26(1-2), 1–24. (2002)
2. Sami, A., "Obstacles and Misunderstandings Facing Medical Data Mining", In: Li, Xue; Zaiane, Osmar R.; Li, Zhanhuai (Eds.) Advanced Data Mining Applications, Lecture Notes in Artificial Intelligence, Vol. 4093. Springer-Verlag, pages 856-863, (2006).
3. American Society of Consultant Pharmacologists, Health Trends, http://www.ascp.com/public/pubs/tcp/1998/jun/trends.shtml, last visted 3/2006. (1998).

4. StatSoft, Inc. (2006). Electronic Statistics Textbook. Tulsa, OK: StatSoft. WEB: http://www.statsoft.com/textbook/stathome.html.
5. Okamura K, Usami T, Nagahama K, Maruyama S., Mizuta E., '"Quality of life" assessment of urination in elderly Japanese men and women with some medical problems using International Prostate Symptom Score and King's Health Questionnaire', European Urology, Apr;41(4):411-9, (2002)
6. Agrawal, R. Gehrke, J. Gunopulos, D. and Raghavan, P., "Automatic Subspace Clustering of High Dimensional Data for Data Mining Applications," SIGMOD RECORD, VOL 27; NUMBER 2, pages 94-105. (1998)
7. J. Han, J. Pei, and Y. Yin, "Mining frequent patterns without candidate generation.", *Proc. 2000 ACM SIGMOD Int. Conf. Management of Data (SIGMOD'00)*, pages 1-12, Dallas, TX, May 2000.
8. Liu, B., Hsu, W., Ma, Y.: Integrating classification and association rule mining. Proc. of Fourth International Conference on Knowledge Discovery and Data Mining (1998)
9. Li, W., Han, J., Pei, J.: Cmar: Accurate and efficient classification based on multiple class-association rules. Proc. of First IEEE International Conference on Data Mining (2001) 369–376
10. Dong, G., Zhang, X., Wong, L., Li, J.: Caep: Classification by aggregating emerging patterns. Proc. of Second International Conference on Discovery Science, Lecture Notes in Computer Science **1721** (1999) 30–42
11. Barry, M. J. , Fowler, F. J. Jr, O'Leary, M. P., Bruskewitz R. C. , Holtgrewe, H. L., Mebust, W. K., Cockett AT, 'The American Urological Association symptom index for benign prostatic hyperplasia,' Journal of Urology, Nov;148(5):1549-57, (1992).

Appendix

In this appendix we prove that cumulative sum of weighted difference of each frequency with respect to mean of frequencies is equal to summation of difference between assigned weights from their mean times frequencies.

$$SDI_i = \sum_{j=1}^{m}(f_{ij} - \mu_{f_i})B_j = \sum_{j=1}^{m}(f_{ij} - \sum_{j=1}^{m} f_{ij}/m)B_j = \sum_{j=1}^{m}(f_{ij}B_j) - \sum_{j=1}^{m}(f_{ij}/m)\sum_{j=1}^{m}B_j$$

$$= \sum_{j=1}^{m}(f_{ij}B_j) - \sum_{j=1}^{m}(f_{ij})\sum_{j=1}^{m}B_j/m = \sum_{j=1}^{m}(f_{ij}B_j) - \sum_{j=1}^{m}(f_{ij})\mu_B = \sum_{j=1}^{m}(f_{ij}B_j) - \sum_{j=1}^{m}(f_{ij})\mu_B$$

$$= \sum_{j=1}^{m}(f_{ij}B_j) - \sum_{j=1}^{m}(f_{ij}\mu_B) = \sum_{j=1}^{m}(f_{ij}B_j - \mu_B f_{ij}) = \sum_{j=1}^{m}(B_j - \mu_B)f_{ij}$$

A Flexible Framework for SharedPlans

Minh Hoai Nguyen and Wayne Wobcke

School of Computer Science and Engineering
University of New South Wales
Sydney NSW 2052, Australia
{minhn, wobcke}@cse.unsw.edu.au

Abstract. SharedPlans is an agent teamwork model that provides a formalization of the conditions under which a group of agents has a collaborative plan. This paper describes a general framework for implementing SharedPlans theory that addresses the computational issues of team formation, group plan elaboration and plan execution, involving coordination, communication and monitoring. The framework includes a team-oriented programming language for specifying recipes for SharedPlans, and an extension to a BDI architecture with several metaplans for interpreting the plan language. We indicate how the formal requirements for the establishment of SharedPlans are fulfilled within the framework.

1   Introduction

Agent teamwork models have been shown to be suitable in a range of applications, such as simulating air combat in a military training environment, Tidhar, Heinze and Selvestrel [19], and investigating Robot-Agent-Person teams in rescue domains, Scerri *et al.* [16]. Teamwork is characterized by a high degree of communication, collaboration and cooperation. A team of agents not only work together to achieve common goals, but maintain an ongoing commitment to the team, helping one another when necessary, keeping others informed of relevant information, etc. In addition, agents working in teams must maintain mutual beliefs about the world (beliefs not only about the world but about other agents' beliefs), joint goals (goals the team has collectively agreed to adopt), and joint intentions (commitments not only to the agent's own actions but to those of the team). Underpinning this team behaviour must be mechanisms to support team formation, communication between agents, synchronization and monitoring of the execution of joint plans, and the fulfilment of obligations to inform the team of relevant facts or notify the team when it is appropriate to abandon a team plan. Thus teamwork applications are complex, both theoretically and computationally.

There is a substantial "gap" between theory and practice for computational models of teamwork. For example, the theory of joint intentions developed by Cohen and Levesque [4] presents a formalization of joint persistent goals, goals such that all team members have that goal, and in addition, a commitment that on dropping the goal, to notify the other team members that the goal is no longer mutually held. This obligation to inform others on abandoning the team goal captures only one basic computational aspect of teamwork. The more complex theory of SharedPlans developed by Grosz and Kraus [7] shows how structured team plans can be modelled. SharedPlans theory aims

A. Sattar and B.H. Kang (Eds.): AI 2006, LNAI 4304, pp. 393–402, 2006.

© Springer-Verlag Berlin Heidelberg 2006

to formalize the conditions on the mental attitudes an agent must have to engage in collaborative activities, similar to the approach of Bratman [2] in which cooperative activity results from an interlocking set of intentions held by multiple agents. Shared-Plans theory captures some complex constraints on the beliefs of agents participating in a team, but does not address computational issues surrounding the formation and execution of SharedPlans.

In contrast, existing computational approaches to teamwork, e.g. Tidhar [18], Kinny *et al.* [11], Tambe [17], Pynadath *et al.* [13] and the JACKTeams model [10], are focused on the need for efficient architectures, languages and platforms for developing applications, and are not well-motivated theoretically. This makes team-based applications built using those approaches difficult to understand. Hence we believe there is a need for a more theoretically sound implementation of a teamwork model that also addresses computational concerns.

In this paper, we present a general, flexible, approach to implementing the Shared-Plans theory of Grosz and Kraus [7,8] using as a basis a PRS-type architecture, Rao and Georgeff [14]. The approach, implemented using JACK Intelligent Agents^{TM}, allows programmers to specify team plans that are executed using the standard BDI interpreter. The approach addresses the computational concerns of team-oriented programming, such as how agents in a team agree on the form and structure of the team, how they synchronize their actions with one another, how and when they communicate with one another, how they monitor the execution of their joint plans and activities, and how they agree to abandon infeasible joint activities – in a manner consistent with SharedPlans theory. Our framework thus goes much further than the work of Grosz and Kraus [8], who describe an implementation of SharedPlans in a "Truckworld" environment, and that of Grosz and Hunsberger [5], who propose a SharedPlans extension (unimplemented) to the IRMA architecture of Bratman, Israel and Pollack [3], as both these approaches present general-purpose algorithms, but which are domain independent only for the formation of SharedPlans. The main contribution of our approach is a team-oriented programming language for specifying team plans, and domain-independent mechanisms, inherited from JACK, for SharedPlan execution and monitoring.

It should be noted that there are also several other implementations of SharedPlans. These include an electronic commerce system, Hadad and Kraus [9], a collaborative interface for distance learning (DIAL), Ortiz and Grosz [12], and a multi-agent system for collaboration of heterogeneous groups of people and computer systems (GigAgent), described in Grosz, Hunsberger and Kraus [6]. SharedPlans theory is also used as the basis of the Collagen dialogue system, Rich and Sidner [15]. However, all these systems are special-purpose implementations of the theory for specific problems and do not provide a general implementation of SharedPlans, whereas our framework provides an architecture for implementing SharedPlans independent of any specific application.

The remainder of this paper is organized as follows. Section 2 contains a summary of the main definitions of SharedPlans theory. In Section 3, we present MIST, our framework for implementing SharedPlans, describing the language for team-oriented programming and the internal architecture of MIST agents. In Section 4, we indicate how the basic definitions of SharedPlans theory are satisfied in MIST. Finally, we compare MIST to other general team-oriented programming platforms.

2   SharedPlans Theory

SharedPlans theory is a formalization of the mental attitudes of agents engaging in group activities. In SharedPlans theory, a group of agents have a collaborative plan when they each hold certain beliefs, desires and intentions. Thus the formalization attempts to define some complex concepts, such as full SharedPlans and partial SharedPlans, based on these basic mental attitudes. The formalization is given in first-order logic enhanced with several primitive predicates, modal operators, meta-predicates and action functions. Some axioms also govern the commitments and behaviour of agents. This section provides a brief summary of the main definitions of SharedPlans theory.

SharedPlans theory distinguishes between two kinds of intentions: intentions to perform an action (IntTo) and intentions that a proposition holds (IntThat). An agent intending to do an action must commit to doing that action, and must hold appropriate beliefs about its ability to perform the action, Grosz, Hunsberger and Kraus [6]. IntThat is used to represent an agent's expectation that some proposition will hold or some actions will be performed (possibly by other agents). An agent intending that a proposition holds must be committed to doing what it can to help make the proposition hold. However, unlike IntTo, with IntThat, it is not necessary for the agent to do or to be able to do anything.

A group of agents are said to have a SharedPlan for doing an action if they mutually believe that all members of the group are committed to having the action done. In addition, there exists a recipe such that the group mutually believe the need to perform all subactions in the recipe. Furthermore, the group must mutually believe that every subaction is catered for by a capable agent or subgroup of agents.

A more formal definition of a full SharedPlan, adapted from Grosz, Hunsberger and Kraus [6], is as follows. Let $FSP(Gr, \alpha, R_\alpha)$ denote that a group Gr has a full Shared-Plan to do action α using recipe R_α. $FSP(Gr, \alpha, R_\alpha)$ holds if and only if the following conditions are satisfied:

1. Gr has a mutual belief that each member of Gr intend that Gr do α.
2. Gr has a mutual belief that Gr has a full recipe R_α for doing α.
3. For each subaction β in R_α, (i) there is an agent A_β in Gr having an individual plan or a subgroup Gr' of Gr having a full SharedPlan to do β, (ii) Gr has a mutual belief that A_β/Gr' has an individual plan/full SharedPlan to do β and is able to do β – note that agents who are not members of Gr' are not required to know the recipe involved, and (iii) Gr has a mutual belief that each agent in Gr intends that A_β/Gr' be able to do β.

Note that the theory is typically understood as providing conditions on the attribution of SharedPlans to a group of agents *at the time of plan formation*. It is unclear what SharedPlans a group of agents has during execution (e.g. whether they continue to have the whole SharedPlan or only the part remaining to be executed). This is because the notion of individual intention used in SharedPlans theory is not precisely defined (one might go further, in that if the theory of Bratman [1] is followed, an agent would no longer intend to do an action already completed; in such a case, the SharedPlans held by the agents are constantly changing as execution proceeds).

SharedPlans theory also provides a definition of a partial SharedPlan. Partial Shared-Plans are plans in which the recipes for the actions might be incomplete or in which some subactions have not been assigned to any agent or any subgroup of agents. In the case of a partial SharedPlan, the group must have a full plan for elaboration of the partial plan into a full plan.

A more formal definition of a partial SharedPlan (with the amendments to the full SharedPlan definition highlighted) is as follows. Let $PSP(Gr, \alpha, R_\alpha)$ denote that a group Gr has a partial SharedPlan to do action α using recipe R_α. $PSP(Gr, \alpha, R_\alpha)$ holds if and only if the following conditions are satisfied:

1. Gr has a mutual belief that each member of Gr intend that Gr do α.
2. Gr has a mutual belief that Gr has a full recipe R_α for doing α *or that Gr has a partial recipe R_α that may be extended into a full recipe and a full SharedPlan for selecting such an extended recipe.*
3. For each subaction β in R_α, *either* (i) there is an agent A_β in Gr having an individual plan or a subgroup Gr' of Gr having a partial SharedPlan to do β, (ii) Gr has a mutual belief that A_β/Gr' has an individual plan/partial SharedPlan to do β and is able to do β, and (iii) Gr has a mutual belief that each agent in Gr intends that A_β/Gr' be able to do β, *or (iv) Gr has a mutual belief that there is some agent in Gr or subgroup Gr'' of Gr that can do β and that there is a full SharedPlan to select such an agent/subgroup.*

The formalism of Grosz and Kraus [8] makes clearer some subtle points in the definition, e.g. with "Gr' is able to do β" each agent in Gr' must know a recipe that the group can use to do β (but the notion of group ability is not analysed any further), hence with the corresponding mutual belief of Gr, each agent of Gr must believe, for some candidate subgroup Gr', that that subgroup can do β, but Gr need not know (at the time of forming the partial SharedPlan) which subgroup Gr' will be selected. Now since SharedPlans theory is unclear about the SharedPlans held during plan execution, it is not clear whether, at the time that Gr' is selected, all agents in Gr are required to know the identity of Gr'. However, we believe that this condition would need to be satisfied in any practical implementation.

3   MIST: Minimal Infrastructure for SharedPlans Theory

This section describes MIST, our general implementation of SharedPlans theory using the JACK agents platform. First, MIST provides a specification language for recipes; recipes, once adopted by a group of agents that form the relevant beliefs and intentions then become SharedPlans held by the group. Second, MIST extends the JACK architecture with particular JACK plans (from now on called *processes* to avoid confusion with SharedPlans or domain plans) that embody the mechanisms for a group of agents to form teams, settle on a team plan, then execute a team plan (each agent synchronizing with and communicating relevant information to other agents in the team, while monitoring events in the environment that affect the success or failure of the team plan). MIST provides a generic platform for SharedPlans theory in that recipes are specified in a general "team-oriented programming" language independent of the JACK plans used to form and execute SharedPlans (these JACK plans act more like meta-plans in taking SharedPlans as arguments). In MIST, the JACK platform is used for the implementation of the individual agents, for event processing and inter-agent communication, and to implement the MIST infrastructure processes for team plan formation and execution. MIST agents also use the JACK plan library for representing individual agent plans.

3.1 Team-Oriented Programming in MIST

Each agent system contains a set of agents and their capabilities (actions that can be executed by an agent). The main part of the agent is the recipe library. Each recipe is a hierarchical plan, containing a list of subactions and a list of actions it supports (can be used to fulfil). In addition, the recipe includes the interdependencies between subactions. As in the approach of Kinny *et al.* [11], each recipe also contains information about the roles in the recipe. Finally, each recipe contains a definition of its success and failure conditions. Each recipe is of the following form.

$$
\begin{aligned}
&\texttt{recipe => supports-action}_1\texttt{, ..., supports-action}_n &&(1)\\
&\texttt{SUBACTIONS : subaction}_1\texttt{, ..., subaction}_m &&(2)\\
&\texttt{SUCCEEDS\_WHEN : Dependency-Expression} &&(3)\\
&\texttt{role :: subaction}_1\texttt{, ..., subaction}_l &&(4)\\
&\texttt{subaction := Dependency-Expression} &&(5)
\end{aligned}
$$

The first three lines are followed by any number of lines of the format (4) or (5). A line in format (4) gives a role name followed by the subactions carried out by agents in filling that role. This use of roles provides a convenient way to express constraints that some subactions must be performed by the same agent. A line in format (5) describes the start condition for a subaction. The formula means that the subaction on the left hand side should start when the conditions in the right hand side are met. The dependency expressions on the right hand side can be any boolean combinations of atomic conditions of the form `Condition+Time`, where an optional `+Time` is a numeric offset and `Condition` is of the form `Event` (meaning successful termination of an action or some condition in the world) or `Event@FAILURE` (meaning termination of an action execution with failure). MIST uses communication between agents, where possible, to synchronize execution using these conditions, so as to minimize the amount of monitoring required of the individual agents (this communication is also minimized).

As an example, consider a scenario involving a team of three scouting helicopters and an infantry platoon. The mission is to get the majority of the infantry platoon to the battlefield; it is considered successful even if some scouting helicopters or individual soldiers are shot down. The team can use a `ScoutMoveRec` recipe, defined as follows, with `Scouting2` as an alternative plan to be used when `Scouting` fails.

```
ScoutMoveRec => MoveToBattleField
SUBACTIONS : Scouting, Scouting2, Move, BuildBridge, PumpFuel
SUCCEEDS_WHEN : Move
MainTroopRole :: Move, BuildBridge
ScoutingRole :: PumpFuel, Scouting
Move := BuildBridge+5 AND (Scouting OR Scouting2)
Scouting := PumpFuel AND Sunrise
Scouting2 := Scouting@FAILURE
```

3.2   MIST Agent Architecture

The basis of MIST is an extension to JACK providing processes for team formation, group plan elaboration and SharedPlan execution. There are four types of processes: group-related elaborator processes (GREPs), group-related intention processes (GRIPs), single-agent processes (SAPs) and permanent monitor processes (PMPs). GREPs and GRIPs are responsible for coordinating group activities based on SharedPlans theory, and are similar to the processes used in Grosz and Kraus [8] (in turn similar to the algorithms of Kinny *et al.* [11]), while SAPs and PMPs are responsible for executing domain specific actions. Figure 1 depicts the internal architecture of a MIST agent in terms of the messages sent and received by the agent's processes.

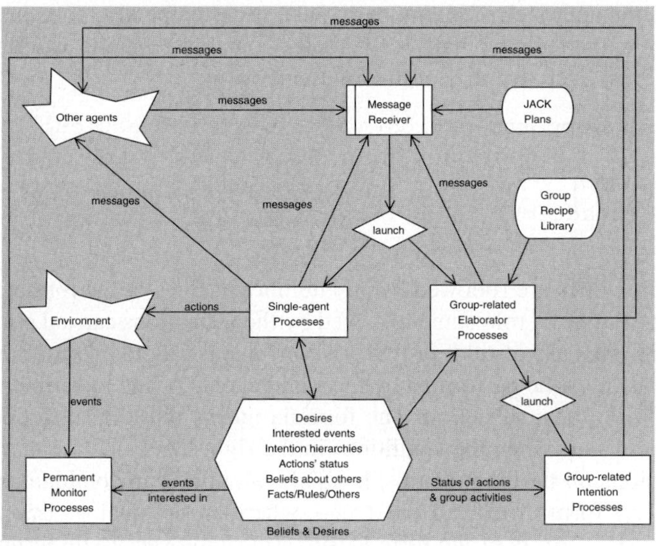

Fig. 1. MIST Internal Agent Architecture

Team Formation: When a message is sent to an agent requesting it to execute a group action, the agent invokes a GREP, which computes a list of all groups possibly able to perform the action (based only on a predefined list of agent capabilities). From this list, one group and a group leader are randomly chosen, and the GREP sends a message to every agent in the chosen group asking them to execute the group action. From this time onwards, the group leader is responsible for coordinating the group activities. If some agents refuse to participate or fail to respond before the timeout, the leader broadcasts termination messages. If the leader receives messages from every agent confirming their commitment to the action, it broadcasts this information to every member and the team is considered formed. Intuitively, the group now mutually believes that it is working towards an intention to do the group action, though it is yet to commit to that intention until a recipe for doing the group action is determined.

Group Plan Elaboration: GREPs are used for forming and elaborating group plans. After a team is formed to execute a group action, GREPs are used to identify a recipe for the action. The team leader sends messages to each team member requesting them to propose a recipe for the group action. The plan to execute the action fails if no replies are received, otherwise the leader selects one of the proposed recipes and requests each team member to commit to the chosen recipe. The recipe will only be adopted if all agents respond with such a commitment, and in this case, the team leader informs each team member of their commitment to the selected team plan.

The team leader now initiates a role assignment phase. The leader requests bids from each team member concerning the roles in the recipe they are prepared to fill, then computes a role assignment consistent with the bids. This does not guarantee the SharedPlan is executable, so next the team leader proposes the role assignment to each team member, requesting confirmation of the assigned role(s). The agent confirms only if it believes the whole role assignment is feasible. Again, the group action fails if there is no agreement on the role assignment. But if agreement is reached, intuitively the team now has a mutual belief that they have an intention to do the group action and a recipe for executing it (conditions (1) and (2) in the definition of a partial SharedPlan).

SharedPlan Execution: After a SharedPlan has been established, agents must fulfil their allocated roles by executing their part of the team plan. For this purpose, GRIPs are launched for each subaction in the plan. Moreover, as there might be interdependencies between these subactions, the group action and environmental conditions, agents need to keep track of these dependencies in order to execute their subactions at the appropriate time. The GRIP that corresponds to the intention to execute α first waits for the start condition of the super-action of α to be satisfied, then waits for the start condition of α to be satisfied, then executes α, and finally notifies all agents that need to know about the success or failure of α. Let us examine these steps in more detail. First, since each GRIP must wait for the start condition of its super-action, rather than having the GRIP itself monitor this condition, in MIST an agent relies on the team leader to notify it when this condition is satisfied. The team leader therefore needs to monitor this condition. Second, agents know the start conditions of their subactions by looking at the containing recipe. Third, actions are executed by launching either a SAP, in the case of an individual plan, or a GREP, in the case of a group action. Finally, the agents needing to know about the status of α can be determined from the recipe and the assignment of roles to agents. An agent only reports the execution result of an action to the responsible agents of dependent actions. The responsible agent of a group action is the team leader, and for an individual action is the agent executing the action. By using this mechanism, agents in the team are able to synchronize their execution of the team plan through communication, minimizing the individual monitoring done by each agent.

Permanent monitoring processes (PMPs) are domain-specific processes invoked only once when the agent is created. PMPs constantly monitor the environment for events the agent is currently interested in (determined from the agent's belief set). If a PMP detects such an event, it sends appropriate messages to the message receiver which is responsible for invoking the relevant processes to handle the event, for example, to execute an action. PMPs, therefore, can be used to initiate the whole process of team formation, group plan elaboration and SharedPlan execution.

4   Satisfying SharedPlans Theory

In this section, we discuss informally how MIST satisfies the requirements of Shared-Plans theory in attributing a partial SharedPlan to a group of MIST agents – focusing on the time at which a team plan is formed by the group as the standard case.

First note that though the basis of SharedPlans theory is mutual belief, the theory provides no indication of how mutual beliefs can be attained in practice. In MIST, agents have beliefs annotated by a set of agent names, so each agent can explicitly represent beliefs about other agents. The basic mechanism for a set of agents to reach mutual belief is communication. Here we make some standard assumptions about the truthfulness of agents and the reliability of the communication channel that guarantee that mutual belief can be attained. More precisely, it is assumed that an agent will only send a message that it believes α if it does believe α, and that any message sent will eventually be received (correctly) by its recipients within a known finite amount of time.

The beliefs required by SharedPlans theory are not explicitly represented in MIST agents, but instead are derived from their intention structures. In particular, a belief in the intention to do α is attributed to a MIST agent if it contains GRIPs to execute its part of a recipe R_α for doing α. The team formation and group elaboration procedures described above result in each agent in the team instantiating appropriate GRIPs, and moreover, each agent also knows that all agents in the team instantiate their appropriate GRIPs. So the mutual beliefs for conditions (1) and (2) of the partial SharedPlans definition obtain. For condition (2), it is also required that there be a mutual belief that the recipe R_α, if partial, can be extended into a full recipe for α, and that there is a full SharedPlan for selecting such an extended recipe. Here again, it is assumed that the GREP for group elaboration provides a mutually known mechanism for extending and selecting a full recipe for α extending R_α. However, note that MIST agents (like JACK agents) do no computation to determine whether R_α can in fact be extended to a full SharedPlan for α (doing so would require predicting the state of the world when R_α needs to be extended), but simply accept this, which is sufficient for condition (2), since SharedPlans theory requires only the *belief* that R_α can be extended to a full recipe.

Consider now condition (3) on partial SharedPlans. For individual actions, clauses (i)–(iii) are satisfied, whereas for group actions, clause (iv) is satisfied. First, for an individual action, the existence of a plan for the subaction follows from the role assignment phase of the group plan elaboration procedure. Each agent is explicitly required to check its assigned subactions against its capabilities, and only commit to the team plan if there is no conflict (however, MIST agents, again as in JACK, do not look for potential conflicts between different team plans). This procedure also establishes clause (ii), mutual belief in the relevant agent having an appropriate recipe to fulfil its role(s) in the team plan, and, in the case of group actions, clause (iv), however, only on the understanding that agents assume that for any group subaction β, a subsequently invoked GREP (the mutually known mechanism) will succeed in elaborating that action into a full SharedPlan for β. Clause (iii) involves IntThat. Here it is unclear what SharedPlans theory formally requires (see the discussion in Grosz and Kraus [8]), but we take it that any agent intends that the agent or group assigned to an action do that action. Appropriate communicative actions are included in the GRIPs for executing the SharedPlan to enable synchronization of action execution and abandonment of the SharedPlan.

Finally, note that although we have focused on the definitions of full and partial SharedPlans, SharedPlans theory also includes "rationality axioms" about the agents, such as that agents do not adopt conflicting intentions, Grosz and Kraus [7]. We have made no attempt to satisfy these axioms, as this would require highly complex reasoning by the agents that would at best undermine the efficiency of the computational model of teamwork, or at worst be impossible to compute, especially as agents are typically operating under a high degree of uncertainty in a dynamically changing environment.

5   Related Work

In this section, we discuss related general computational approaches to modelling teamwork and supporting team-based applications. First, Kinny *et al.* [11] presented a BDI-style approach to representing and executing team plans, which introduced the use of roles needed to be filled by the agents in a team. Synchronization amongst team members occurs through rewriting the team plan to include communicative actions between agents informing them when a synchronization point has been reached.

In the STEAM architecture, Tambe [17], each agent has a copy of a plan in which certain steps are designated as team plans (requiring coordinated execution). For each team plan, the team leader sends a synchronizing message to establish a joint persistent goal. Once messages from each team member have been broadcast confirming the establishment of the joint goal, execution can commence. Having the team plan as a joint persistent goal places obligations on team members to inform the team when that goal is dropped. The work of Pynadath *et al.* [13] extended STEAM to a more general framework for team-oriented programming using TEAMCORE.

Finally, the JACKTeams model released as part of the JACK platform [10] presents an extension of this approach, based on the work of Tidhar [18]. In the JACKTeams framework, (software) team agents are treated on a par with individual agents in having explicit team beliefs, goals and intentions. Team agent beliefs are both derived from, and propagate to, the beliefs of the individual agents in the team, and the team agent mediates the interaction between team members and acts as a central point of control. Thus the agents in the team do not need to be aware of one another, whereas SharedPlans theory requires teams with more autonomous agents which must know one another in order to negotiate and participate in team activities.

6   Conclusion

In this paper, we presented a general framework for implementing the SharedPlans theory of collaborative action, addressing computational issues such as team formation, group plan elaboration, and SharedPlan execution, which involves communication, coordination, synchronization and monitoring. The framework includes a team-oriented programming language for the specification of recipes for SharedPlans, and uses the JACK platform and BDI architecture for implementing the framework and interpreting SharedPlans expressed in the recipe language. Our overall aim has been to develop a generic computational teamwork model that is theoretically well-motivated and more directly related to the supporting theory.

References

1. Bratman, M.E. (1987) *Intention, Plans, and Practical Reason.* Harvard University Press, Cambridge, MA.
2. Bratman, M.E. (1992) 'Shared Cooperative Activity.' *The Philosophical Review*, **101**, 327–341.
3. Bratman, M.E., Israel, D.J. & Pollack, M.E. (1988) 'Plans and Resource-Bounded Practical Reasoning.' *Computational Intelligence*, **4**, 349–355.
4. Cohen, P.R. & Levesque, H.J. (1991) 'Teamwork.' *Noûs*, **25**, 487–512.
5. Grosz, B.J. & Hunsberger, L. (2004) 'The Dynamics of Intention in Collaborative Activity.' Paper presented at the Conference on Collective Intentionalty IV, Siena, Oct, 2004.
6. Grosz, B.J., Hunsberger, L. & Kraus, S. (1999) 'Planning and Acting Together.' *AI Magazine*, **20**(4), 23–34.
7. Grosz, B.J. & Kraus, S. (1996) 'Collaborative Plans for Complex Group Action.' *Artificial Intelligence*, **86**(2), 269–357.
8. Grosz, B.J. & Kraus, S. (1999) 'The Evolution of SharedPlans.' in Wooldridge, M. & Rao, A.S. (Eds) *Foundations of Rational Agency.* Kluwer Academic Publishers, Dordrecht.
9. Hadad, M. & Kraus, S. (1999) 'SharedPlans in Electronic Commerce.' in Klusch, M. (Ed.) *Intelligent Information Agents.* Springer-Verlag, Berlin.
10. Howden, N., Rönnquist, R., Hodgson, A. & Lucas, A. (2001) 'JACK Intelligent Agents™ – Summary of an Agent Infrastructure.' Paper presented at the Second International Workshop on Infrastructure for Agents, MAS, and Scalable MAS, Montreal, May, 2001.
11. Kinny, D.N., Ljungberg, M., Rao, A.S., Sonenberg, E.A., Tidhar, G. & Werner, E. (1994) 'Planned Team Activity.' in Castelfranchi, C. & Werner, E. (Eds) *Artificial Social Systems.* Springer-Verlag, Berlin.
12. Ortiz, C.L. & Grosz, B.J. (2002) 'Interpreting Information Requests in Context: A Collaborative Web Interface for Distance Learning.' *Autonomous Agents and Multi-Agent Systems*, **5**, 429–465.
13. Pynadath, D.V., Tambe, M., Chauvat, N. & Cavedon, L. (1999) 'Toward Team-Oriented Programming.' in Jennings, N.R. & Lespérance, Y. (Eds) *Intelligent Agents VI.* Springer-Verlag, Berlin.
14. Rao, A.S. & Georgeff, M.P. (1992) 'An Abstract Architecture for Rational Agents.' *Proceedings of the Third International Conference on Principles of Knowledge Representation and Reasoning (KR'92)*, 439–449.
15. Rich, C. & Sidner, C.L. (1998) 'COLLAGEN: A Collaboration Manager for Software Interface Agents.' *User Modeling and User-Adapted Interaction*, **8**, 315–350.
16. Scerri, P., Pynadath, D., Johnson, L., Rosenbloom, P., Si, M., Schurr, N. & Tambe, M. (2003) 'A Prototype Infrastructure for Distributed Robot-Agent-Person Teams.' *Proceedings of the Second International Joint Conference on Autonomous Agents and Multiagent Systems*, 433–440.
17. Tambe, M. (1997) 'Agent Architectures for Flexible, Practical Teamwork.' *Proceedings of the Fourteenth National Conference on Artificial Intelligence (AAAI-97)*, 22–28.
18. Tidhar, G. (1993) 'Team-Oriented Programming: Preliminary Report.' Technical Note 41, Australian Artificial Intelligence Institute, Apr, 1993.
19. Tidhar, G., Heinze, C. & Selvestrel, M. (1998) 'Flying Together: Modelling Air Mission Teams.' *Applied Intelligence*, **8**, 195–218.

An Analysis of Three Puzzles in the Logic of Intention

Wayne Wobcke

School of Computer Science and Engineering
University of New South Wales
Sydney NSW 2052, Australia
wobcke@cse.unsw.edu.au

Abstract. In this paper, we generalize our formal approach to modelling PRS agents away from PRS-specific assumptions to more general theories of rationality, while not losing the concreteness of the connection between an agent's mental states and formal models, using three "puzzles" in the logic of intention to motivate our extended approach. We show how the theory can be used to represent solutions to the puzzles and draw out insights into how agent architectures may be extended to handle these more complex scenarios.

1 Introduction

In previous work, we developed formal models of a class of BDI agent architectures based on PRS, Georgeff and Lansky [7]. We have developed an operational semantics aiming to succinctly abstract the main properties of this family of architectures [16], models of the "mental states" of agents based on this architecture [17], a formal derivation of some "rationality postulates" of those agents [18], extended the account to incorporate belief update and attempts [19], and examined model generation algorithms for use in model checking for agents of this class [20]. The theory is based on a logic, Agent Dynamic Logic (ADL), that combines elements from Computation Tree Logic [6], Propositional Dynamic Logic [13], and BDI Logic [14]. The motivation of all this work is to develop a logical framework that is at once rigorous in providing formal notions of belief, desire and intention, yet which is also closely enough aligned to the operational behaviour of this architecture to enable formal reasoning about the behaviour of agents in this class.

Some aspects of this prior work are specific to the PRS architecture, and do not extend to more general theories of rational agency. This has been deliberate, in order to capture the behaviour of the PRS family in as concise a manner as possible. Two major assumptions peculiar to PRS are (i) that the agent attempts to execute only one action at a time, and (ii) that the agent selects a plan to fulfil an achievement goal only at the time of executing that plan (at the latest possible moment). These assumptions reflect more the simplicity of PRS and/or the characteristics of the environments in which PRS agents are expected to operate, rather than general principles of rationality.

The main objective of this paper is to generalize the formal theory away from these PRS-specific assumptions, while not losing the concreteness of the connection between an agent's mental states and formal models. To do this, we use as motivation three puzzles in the logic of intention and action, two discussed extensively by Bratman [2]

A. Sattar and B.H. Kang (Eds.): AI 2006, LNAI 4304, pp. 403–412, 2006.
© Springer-Verlag Berlin Heidelberg 2006

(a "Video Game" example and a "Strategic Bomber" example), and one presented here derived from the Lottery Paradox. These examples highlight some of the complexities of Bratman's account which have not been captured in previous formal models of rational agency. Our work is a step towards capturing these more complex properties.

One lesson from developing the formal models used in ADL is that the theory of action modelling the dynamics of the environment must be clearly distinguished from the beliefs of the agent, whereas in approaches such as Cohen and Levesque [4], the theory of action is part of the agent's beliefs. For PRS, these must be kept distinct because the PRS agent does no reasoning about actions and plans except insofar as it selects a plan from a given plan library to achieve some goal. Thus the theory of action *describes* the agent's behaviour, but is not part of the agent's explicit beliefs. Bratman, however, considers highly sophisticated rational agents who know their own theory of action; nevertheless, this theory still needs to be distinguished from the beliefs of an agent about the environment during the execution of some plan. Thus ADL uses two modal operators to capture this distinction, a modal operator B for the agent's beliefs, and a modal operator A to capture the theory of action implicit in the formal models.

In ADL, the intentions of the agent are those actions the agent chooses from its plans that eventually it will successfully perform, represented using a formula Iπ for a program term π, and are relative to a *situation*, a state in a branching-time execution structure that includes accessibility relations for beliefs, alternative actual situations, and a subworld relation denoting the successful executions of an action. The language of programs in ADL extends that for PDL in including special actions of the form *achieve* γ where γ is a PC formula, and *attempt* π where π is a program term. The language also includes test statements α? and program terms for conditional and iterative constructs, defined as follows:

$$\textbf{if } \alpha \textbf{ then } \pi \textbf{ else } \psi \equiv (\alpha?; \pi) \cup (\neg\alpha?; \psi)$$
$$\textbf{while } \alpha \textbf{ do } \pi \equiv (\alpha?; \pi)^*; \neg\alpha?$$

The semantics of ADL is given in terms of computation trees, extending the approach of Harel [8]. Union is defined in terms of "tree merging", sequencing in terms of "tree adjoining" and iteration in terms of the transitive closure of the sequencing operation.

The organization of this paper is as follows. We first describe the three "puzzles" in the logic of intention that are addressed in this work, and then provide a critique of earlier formal approaches to these problems (newly identifying shortcomings in that earlier work). We then present definitions of ADL that extend the theory towards more general models of rational agency, and indicate how the three puzzles are handled in the framework. The exercise yields insights into how agent architectures may be extended to handle these more complex scenarios.

2   Three Puzzles in the Logic of Intention

2.1   The "Package Deal" Problem: Strategic Bomber

The "package deal" problem, Bratman [2, Ch. 10], concerns the relationship between choice and intention and the question of whether an agent intends the consequences of its intended actions. The example discussed by Bratman compares a "Strategic Bomber" with a "Terror Bomber", both engaged in a war with the enemy.

Terror Bomber plans to bomb a school in enemy territory, thereby killing the children, and so to terrorize the enemy population. Strategic Bomber plans to bomb a munitions plant, thereby undermining the enemy's war effort. However, Strategic Bomber knows that the munitions plant is next to a school, and by bombing the munitions plant, he will also kill the children. Strategic Bomber has taken this into consideration in choosing his course of action. Does Strategic Bomber intend to kill the children?

The problem arises because agents, in their reasoning about what to do, choose between alternative "scenarios", sets of actions and their consequences taken into account in their reasoning. So in particular, Strategic Bomber considers and deliberately chooses to kill the children (as part of a larger scenario involving bombing the munitions plant). So shouldn't this mean that Strategic Bomber intends to kill the children? Bratman argues that it does not, essentially because "killing the children" does not play the three characteristic functional roles of intention, e.g. the agent does not pursue means towards killing the children (so would not plan to kill the children, for example, if they were evacuated from the school). The question is how to capture these complex properties of intentions in a formal model.

2.2 Motivational Potential: The "Video Game" Example

Bratman [1] uses the example of a video game to argue against what he calls the "Simple View", the view that if I do A intentionally then I intend to A. The argument hinges on Bratman's requirement for intentions to be strongly consistent, relative to the agent's beliefs, and hence the example provides support for this requirement.

Consider a video game, played with two hands, whose aim is to hit one of two targets. A missile shot using the left hand heads towards one target; one shot using the right towards another. The game ends when one of the targets is hit, and it is impossible (and the agent knows this) to hit both targets simultaneously (in this case, the game shuts down just prior to this happening). A natural strategy is for the agent to continually guide two missiles, one with each hand, each towards the appropriate target, each trying to hit that target. Does the agent intend to hit the targets?

The *strong consistency* requirement on beliefs and intentions is never defined precisely by Bratman; the closest statement approaching a definition is that an agent's intentions are strongly consistent relative to its beliefs when it is 'possible for [its] plans taken together to be successfully executed in a world in which [its] beliefs are true', Bratman [3, p. 19], though Bratman realizes that this is in need of further elaboration, Bratman [2, p. 179, note 3].

Now given this requirement, in the video game example, the agent does not intend to hit either target. Let the targets be target 1 and target 2. Then the two intentions, the intention to hit target 1 and the intention to hit target 2, are not strongly consistent with the belief that it is impossible to hit both target 1 and target 2. The question is that if the agent does not intend to hit target 1 nor intend to hit target 2, what does the agent intend? Bratman [1] suggests a number of possibilities, such as the agent intends to shoot at the targets, to try to hit the targets, to hit each target if it can, or to hit one of the two targets. The questions for formal modelling are which of these is right and how can the differences between them be captured.

2.3 Rational Belief and Intention: The Lottery Problem

The Lottery Paradox, usually attributed to Kyburg [10], is a problem about the rationality of belief under uncertainty.

> Consider a fair lottery with a million tickets and only one winning ticket. An agent with one ticket has a very small chance of winning, hence it is rational for the agent to believe that it will have a losing ticket. However, by similar reasoning, it is rational for the agent to believe that any given ticket is a losing ticket. But the agent also believes that there will be a winning ticket. So it seems that the agent's beliefs, each one of which is rational, are together inconsistent, a clear violation of rationality.

The Lottery Paradox has been extensively discussed in the literature, and it is not possible in this paper to survey all the proposed approaches. One reasonable way out of this paradox, proposed by Pollock [12], is to deny that the agent is rational in believing it has a losing ticket (Pollock treats this as a case of "collective defeat", i.e. the collection of equally supported beliefs that each ticket will not win, together with the belief that one ticket will win, defeats the conclusion that any particular ticket will not win).

The *Lottery Problem* that I want to raise here is: is it possible for the agent to intend to win the lottery, and further, if so, is the agent rational to so intend? After all, if the agent is irrational in believing it will lose, the strong-consistency requirement on intentions and beliefs does not block the (rational) agent's intention to win the lottery. But now the agent will start planning to win the lottery, planning what to do with the money, etc., and this seems to be irrational behaviour. The problem is that Bratman's theory seems to leave no room for the agent to intend and plan on the basis of the likelihoods of achieving the intended outcomes, including extremely improbable outcomes such as winning the lottery, and this, at root, is because beliefs are "flat out", not graded with any degree of certainty. The issue for formal modelling is how to model the agent's beliefs and intentions while respecting the answers given to the above questions.

3 Critique of Previous Approaches

Despite the strong influence of Bratman's work in the BDI agents literature, there have been surprisingly few attempts to develop logics of belief, desire and intention. There are, broadly speaking, two approaches to modelling intention, corresponding roughly to Bratman's "two faces" of intention. According to Bratman [1], intentions are Janus-faced, looking to the present (in controlling action) and to the future (in planning). In their future-directed aspect, actions of the agent are modelled as attempts with no guarantee of success; viewed from the aspect of the (ongoing) present, the world is modelled as a sequence of states that actually occur, and intentions are incorporated into such models. The approaches of Cohen and Levesque [4] and Wooldridge [21] are of the latter type, taking as a starting point models of actions inspired by computational logic; those of Rao and Georgeff [14,15] and Wobcke [17,19] are more oriented towards the future-directed aspects (so notions of the agent's possible actions and the failure of an agent's attempts are more pertinent in these approaches). However, the main difficulty is to capture both present-directed and future-directed aspects of intention in one formalism.

Perhaps the best known work is that of Cohen and Levesque [4], in which, rather than formalizing intentions directly, intentions are reduced to persistent goals. Persistence is related to Bratman's notion of stability [2], but is much simpler, in referring only to a temporal property. More precisely, an intention towards an action a to achieve a goal p, relative to some condition q, is characterized, with respect to a state in a sequence of states, as a persistent goal that the agent has that a be done and then p hold immediately after believing that a will happen and then p hold. For a goal to be a *persistent goal* (P-GOAL) relative to some condition q, the agent must believe the goal does not currently hold, and have the goal and continue to have the goal until either believing it to be achieved, believing it to be impossible or believing q to be false.

However, because the intentions of the agent are relative to a sequence of states (world), the agent's intentions can only be determined once it is known which world the agent inhabits, as this will determine which goals persist. While in one sense, whether a goal persists should depend on the world (e.g. a goal should be dropped if some adverse circumstance in the world arises), in another sense the intentions of the agent should be able to be related to the agent's mental state at any point in time without reference to a particular sequence of states, which is impossible with Cohen and Levesque's theory as intentions vary from world to world.

The problem can be illustrated with a simple counterexample. Suppose on Monday at noon I decide on the goal of having lunch at the cafe, and I persist in having this goal until I later have lunch at the cafe. Now on Tuesday at noon I decide exactly the same thing, except on my way to the cafe I hear music and instead go to a nearby concert, and have my lunch later at home. On Cohen and Levesque's account, on Monday at noon I had the P-GOAL (hence the intention) to have lunch at the cafe, while on Tuesday at noon, I did not have this P-GOAL and intention (at no stage did I ever think it achieved or infeasible to have lunch at the cafe, so the P-GOAL was only relative to the fact that there was no concert). On the temporal understanding of "persistent goal" this is right: as my behaviour showed, I was not really "committed" on Tuesday to having lunch at the cafe, preferring instead the concert, but the problem is that, intuitively, there is no difference between my mental states at noon on Monday and at noon on Tuesday (recall I do not know about the concert on Tuesday until later), so why should I be considered to have an intention (to have lunch at a cafe) on Monday that I don't have on Tuesday? The problem stems from the fact that the sequences of states in Cohen and Levesque's models represent courses of events of the actual world and the agent's beliefs and goals in the actual world, and not the possible actions the agent could or would have performed in a range of "possible worlds".

Finally, in terms of addressing the logical properties of intentions, as Cohen and Levesque themselves note, the approach of reducing intention to persistent goals is problematic in that the logic of P-GOAL possesses the right properties for capturing the side-effect free principle, that a rational agent need not intend the believed necessary consequences (side-effects) of an intention, but 'because of what [they] believe to be the wrong reasons,' Cohen and Levesque [4, p. 238].

Properly capturing the future-directed aspect of intention requires taking into account the nondeterminism of actions and the possibility of the agent's attempts to fail. An alternative approach more consistent with this aspect of intention was presented by Rao and Georgeff [14,15], whose formalism is based on branching-time structures, in which a situation (a state in a branching-time structure) can have multiple successor situations determined by the different choices available to the agent. One other difference between this approach and Cohen and Levesque's is that intention is treated as a primitive concept.

Rao and Georgeff [15] explicitly address the side-effect problems using the Strategic Bomber example, however in this formalism, "bombing the plant" and "killing the children" are modelled as propositions, not actions. Thus it is open to Rao and Georgeff to use standard devices in modal logic to ensure the desired conclusions, such as allowing a possible world with a situation that satisfies "bombing the plant" but not "killing the children", and making such a situation goal and intention-accessible to a situation modelling the state of the world, but not belief-accessible. However, the problem now is how to relate those situations to the agent; such situations relate to the agent only insofar as they satisfy (or not) the agents beliefs, goals and intentions, so this does not (without further elaboration) provide any explanation of any underlying properties of those beliefs, desires and intentions. The difficulty is that such a situation is not one that can actually occur (that is, is not even a possible state the agent will consider or encounter), so it is difficult to independently motivate the inclusion of such a situation in a model of the agent's reasoning or behaviour.

The formalism of Rao and Georgeff [14] is more complicated, despite appearing earlier. In this model, "transitions" between situations in branching-time structures are labelled with events, for each transition at most one event that either succeeds or "fails". The aim is to capture the attempts of an agent, which may succeed or fail; note that a failed event labelling a transition refers to the action the agent was trying to do, not to what actually happened as the result of that failed attempt. In any case, it is not clear under this approach how the side-effect free principles *for actions*, in particular for the Strategic Bomber example, should be modelled. Presumably "bombing the plant" and "killing the children" are events, but the restriction in the models to at most one successful or failed event per transition means that the relation between actions of killing the children *by* bombing the plant cannot be modelled, as properly this requires modelling two actions using the one transition (treating "killing the children" as a proposition holding at situations would not work in general). Moreover, as Strategic Bomber succeeds in killing the children (though that was not his intention), even allowing multiple successful events per transition does not solve the problem, because there would still need to be a way to differentiate those successful events the agent intended from those it did not (c.f. the motivational potential of the intention, Bratman [2, Ch. 8]).

In earlier work, Wobcke [17,19], motivated by these problems, we developed a logic called Agent Dynamic Logic (**ADL**), in which possible worlds corresponded to branching-time execution structures of an agent program, and in which intentions referred to actions successfully performed by the agent. In Wobcke [18], we showed how to automatically derive various logical properties of intention. However, the definitions there were too strong, for, while intention satisfied the side-effect free principle, in that an agent intending to bomb the plant did not intend to kill the children, it was not possible to model an agent that did intend actions that are consequences of its intended actions, so the agent which intended both to bomb the plant *and* kill the children could not be modelled. More formally, the formula $I\pi \wedge I(\pi \cup \psi)$ was not satisfied unless $\pi \cup \psi$ was the same action as π (when π is an action, $\pi \cup \psi$ subsumes π and can be viewed as a consequence of π). The basic reason for this is that the definitions assumed, as in Rao and Georgeff [14], that only one primitive action could be attempted by the agent in any one situation (as is the case for typical BDI agents, though see Pollack [11]). That this assumption is too strong can be seen by looking at an example of action specialization. Suppose there are two equally desirable plans with the postcondition of being in San Francisco, the "Buridan" case discussed by Bratman [2, p. 11]. On the earlier definitions, an agent intending to execute one of the plans could not also intend to achieve being in San Francisco. This works for PRS agents, which select a plan to achieve a goal only at the time of execution. However, this will not work for planning agents in general, and so won't adequately cope with the Strategic Bomber example where reasoning and plan selection precede execution.

4   Modelling the Examples

The original semantics of action in **ADL** was based on execution structures, which provided, for each program term π in the language, a computation tree $\mathcal{R}_\pi$ with each node corresponding to a state of the environment together with some associated set of belief-alternative states, defined by an accessibility relation $\mathcal{B}$. Branching in execution structures corresponded to nondeterminism in the execution of actions. With each mental state of the agent was associated a PRS-interpretation, also an execution structure (modelling the possible executions of the current set of plans of the PRS agent), with in addition, a subworld relation $\mathcal{I}$ defining the successful action executions (those achieving the postcondition) and an accessibility relation $\mathcal{A}$ giving the "actual alternative" situations, those situations that the agent could be in given its past history, starting from some initial state of the environment and partially executing plans and making observations. An agent was defined to have an intention to do an action π in a situation σ in an execution structure if for any $\mathcal{A}$-related situation σ' of σ, some initial part of the execution structure emanating from σ' was isomorphic to the execution structure in $\mathcal{R}_\pi$ whose root had the same environment state and set of belief-alternatives as σ'.

However, as discussed above, this definition is too strong, as it embodies an assumption, acceptable for PRS, that only one atomic action is attempted at any point in time. To properly model the examples, in particular Strategic Bomber where two actions ("bombing the plant" and "killing the children") are performed with the same behaviour, as also studied by Israel, Perry and Tutiya [9], we need to modify the definitions as follows. An agent is now defined to have an intention to do an action π in a situation σ in an execution structure if for any $\mathcal{A}$-related situation σ' of σ, some initial part of the execution structure emanating from σ' is isomorphic to *a subworld of* the execution structure in $\mathcal{R}_\pi$ whose root has the same environment state and set of belief-alternatives as σ'. This definition makes valid the formula $I\pi \Rightarrow I(\pi \cup \psi)$, so care must be taken to interpret formulae such as $I(\pi \cup \psi)$, which no longer mean an agent has a choice between π and ψ (it being satisfied, for example, if the agent already intends π).

4.1   Video Game

In the Video Game example, the agent cannot both intend to hit target 1 and intend to hit target 2. However, what the agent does intend depends, we believe, on the exact strategy adopted. The program language in **ADL** allows the expression of a number of strategies. Perhaps the most natural strategy is that the agent intends to repeatedly guide the relevant missiles (m_1 and m_2) towards targets 1 and 2 (t_1 and t_2) until either is hit, expressed as the program '**repeat** *attempt* $(guide(m_1, t_1) \cup guide(m_2, t_2))$ **until** $(hit(t_1)$ $\vee hit(t_2))$'. Here "guiding a missile towards a target' is understood as an action that succeeds iff the missile is directed towards the target, but which may also fail (otherwise); in both cases, the agent continues executing the plan. The expression $guide(m_1, t_1) \cup$ $guide(m_2, t_2)$ here means that the agent has a choice between guiding m_1 to t_1 and m_2 to t_2, which is needed because **ADL** does not model the execution of two actions simultaneously. Thus what differentiates this example from the normal cases of intention is that the intended action is a complex program, consisting of a variety of constituent actions (e.g. to guide a missile to target 1 or target 2), and the intention is directed towards an attempt to perform the complex program, not towards the component actions. Thus, returning to Bratman's suggestions, under this strategy, the agent (i) does not intend to shoot at the targets (if this is equivalent to $guide(m_1, t_1) \cup guide(m_2, t_2)$), since the agent only intends to attempt this (though repeatedly), (ii) does not intend (merely) to try to hit the targets (*attempt achieve*$(hit(t_1) \vee hit(t_2))$), since (iii) does intend to hit one of the targets (*achieve* $(hit(t_1) \vee hit(t_2))$), assuming the complex program is a plan for its achievement. The language of **ADL** provides no way to express conditional intentions of the form to "hit each target if it can".

4.2   Strategic Bomber

To correctly model the Strategic Bomber, and to handle the side-effect free principle for actions, the action of "bombing the plant" cannot be subsumed by "killing the children". But there is no transition on states of the environment for "bombing the plant" in which the children are not also killed. To model this scenario, we therefore make use of the idea that agents "keep track" of the intended effects of their actions, as discussed in Bratman [2, p. 159]. Tracking is handled in **ADL** through modelling the agent's changes of belief as the result of observations associated with the agent's chosen actions. In the case of Strategic Bomber, we can define the observation associated with "bombing the plant" to include the issue of whether the plant is destroyed, but not to include the issue of whether the children are killed. In this way, there will be a transition *on situations* for "bombing the plant", corresponding to a successful execution (hence is intended), that is not also a transition for "killing the children" (a transition resulting in the same state of the environment but differing in the agent's beliefs), so that "bombing the plant" is not subsumed by "killing the children", and the agent intending the former is not forced to intend the latter. However, the malicious Strategic Bomber, who does intend to kill the children, can also be modelled, as the observation associated with "bombing the plant" for him will include the issue of whether the children are killed, so for him, the action of "bombing the plant" is subsumed by "killing the children", and the intention to bomb the plant implies the intention to kill the children.

4.3   Lottery Problem

As far as the logic of intention in ADL is concerned, there is no reason to prefer a modelling where the agent can intend, or cannot intend, to win the lottery, since ADL does not deal in probabilities and utilities. However, for the formal semantics to intuitively represent the agent's reasoning and behaviour, there is strong reason to prefer a model in which the agent *cannot* intend to win the lottery. Moreover, this is preferred if we also understand Bratman's role of intention in "controlling the conduct" of the agent to mean, not just that an intention to A leads to the agent trying to A, but that normally the agent is expected to succeed in that execution, as is required for the intention to be used reliably in coordinating the agent's activities. Hence the simple solution to the problem is to model the agent as intending to *attempt* to win the lottery, by choosing a plan, such as "buy ticket", whose postcondition is having a ticket, with the possible side-effect of winning the lottery.

This solution suggests a simple extension to standard BDI architectures, which typically include plans whose postconditions are *both* the success conditions on their execution and the motivation for their consideration. The example suggests differentiating these two functions, reserving the postcondition for the success condition (here having a ticket), and providing the plan with an additional *goal* (here winning the lottery). As in the Video Game example, the unusual aspect of this situation is that the agent's intention is merely an attempt whose success falls outside the agent's control; thus the situation is similar to that of speech acts, as discussed by Cohen and Levesque [5], where a speech act such as a 'request' is modelled as an attempt (to get the hearer to perform some action). Again, the agent needs to adopt a plan whose postcondition is the successful execution of the attempt and whose goal is the further perlocutionary effect. So this mechanism would enable a plan-based approach to communication via speech acts to be incorporated into a BDI architecture.

5   Conclusion

Motivated by three examples from the literature on rational agency and intention, we generalized our earlier approach to modelling BDI agents away from PRS-specific assumptions, whilst maintaining the close correspondence between an agent's internal states and the formal models. We showed how a modification to the formal theory enabled the examples to be modelled, in particular by (i) allowing more than one action to be performed with the same behaviour, and (ii) including a notion of goals which may be indirect effects of plans in BDI agent architectures.

Acknowledgement

This work is funded by an Australian Research Council Discovery Project Grant.

References

1. Bratman, M.E. (1984) 'Two Faces of Intention.' *The Philosophical Review*, **93**, 375–405.
2. Bratman, M.E. (1987) *Intention, Plans, and Practical Reason.* Harvard University Press, Cambridge, MA.
3. Bratman, M.E. (1990) 'What is Intention?' in Cohen, P.R., Morgan, J. & Pollack, M.E. (Eds) *Intentions in Communication.* MIT Press, Cambridge, MA.
4. Cohen, P.R. & Levesque, H.J. (1990) 'Intention is Choice with Commitment.' *Artificial Intelligence*, **42**, 213–261.
5. Cohen, P.R. & Levesque, H.J. (1990) 'Rational Interaction as the Basis for Communication.' in Cohen, P.R., Morgan, J. & Pollack, M.E. (Eds) *Intentions in Communication.* MIT Press, Cambridge, MA.
6. Emerson, E.A. & Clarke, E.M. (1982) 'Using Branching Time Temporal Logic to Synthesize Synchronization Skeletons.' *Science of Computer Programming*, **2**, 241–266.
7. Georgeff, M.P. & Lansky, A.L. (1987) 'Reactive Reasoning and Planning.' *Proceedings of the Sixth National Conference on Artificial Intelligence (AAAI-87)*, 677–682.
8. Harel, D. (1979) *First-Order Dynamic Logic.* Springer-Verlag, Berlin.
9. Israel, D.J., Perry, J.R. & Tutiya, S. (1991) 'Actions and Movements.' *Proceedings of the Twelfth International Joint Conference on Artificial Intelligence*, 1060–1065.
10. Kyburg, H.E., Jr. (1961) *Probability and the Logic of Rational Belief.* Wesleyan University Press, Middletown, CT.
11. Pollack, M.E. (1991) 'Overloading Intentions for Efficient Practical Reasoning.' *Noûs*, **25**, 513–536.
12. Pollock, J.L. (1995) *Cognitive Carpentry.* MIT Press, Cambridge, MA.
13. Pratt, V.R. (1976) 'Semantical Considerations on Floyd-Hoare Logic.' *Proceedings of the Seventeenth IEEE Symposium on Foundations of Computer Science*, 109–121.
14. Rao, A.S. & Georgeff, M.P. (1991) 'Modeling Rational Agents within a BDI-Architecture.' *Proceedings of the Second International Conference on Principles of Knowledge Representation and Reasoning (KR'91)*, 473–484.
15. Rao, A.S. & Georgeff, M.P. (1991) 'Asymmetry Thesis and Side-Effect Problems in Linear-Time and Branching-Time Intention Logics.' *Proceedings of the Twelfth International Joint Conference on Artificial Intelligence*, 498–504.
16. Wobcke, W.R. (2001) 'An Operational Semantics for a PRS-Like Agent Architecture.' in Stumptner, M., Corbett, D. & Brooks, M. (Eds) *AI 2001: Advances in Artificial Intelligence.* Springer-Verlag, Berlin.
17. Wobcke, W.R. (2002) 'Modelling PRS-Like Agents' Mental States.' in Ishizuka, M. & Sattar, A. (Eds) *PRICAI 2002: Trends in Artificial Intelligence.* Springer-Verlag, Berlin.
18. Wobcke, W.R. (2002) 'Intention and Rationality for PRS-Like Agents.' in McKay, B. & Slaney, J. (Eds) *AI 2002: Advances in Artificial Intelligence.* Springer-Verlag, Berlin.
19. Wobcke, W.R. (2004) 'Model Theory for PRS-Like Agents: Modelling Belief Update and Action Attempts.' in Zhang, C., Guesgen, H.W. & Yeap, W.K. (Eds) *PRICAI 2004: Trends in Artificial Intelligence.* Springer-Verlag, Berlin.
20. Wobcke, W.R., Chee, M. & Ji, K. (2005) 'Model Checking for PRS-Like Agents.' in Zhang, S. & Jarvis, R. (Eds) *AI 2005: Advances in Artificial Intelligence.* Springer-Verlag, Berlin.
21. Wooldridge, M.J. (2000) *Reasoning About Rational Agents.* MIT Press, Cambridge, MA.

Building Intelligent Negotiating Agents

John Debenham and Simeon Simoff

Faculty of IT, University of Technology, Sydney, Australia
{debenham, simeon}@it.uts.edu.au
http://e-markets.org.au/

Abstract. We propose that the key to building intelligent negotiating agents is to take an agent's historic observations as primitive, to model that agent's changing uncertainty in that information, and to use that model as the foundation for the agent's reasoning. We describe an agent architecture, with an attendant theory, that is based on that model. In this approach, the utility of contracts, and the trust and reliability of a trading partner are intermediate concepts that an agent may estimate from its information model. This enables us to describe intelligent agents that are not necessarily utility optimisers, that value information as a commodity, and that build relationships with other agents through the trusted exchange of information as well as contracts.

1   Introduction

The potential value of the e-business market — including e-procurement — is enormous. Given that motivation and the current state of technical development it is surprising that a comparatively small amount of automated negotiation is presently deployed.[1] Technologies that support automated negotiation include multiagent systems [1] and virtual institutions [2], game theory and decision theory.

Game theory tells an agent what to do, and what outcome to expect, in many well-known negotiation situations [3], but these strategies and expectations are derived from assumptions about the agent's preferences and about the preferences of the opponent. One-to-one negotiation is generally known as *bargaining* [4] — it is the natural negotiation form when the negotiation object comprises a number of issues. For example, in bargaining over the supply of steel issues could include: the quantity of the steel, the quality of the steel, the delivery schedule, the settlement schedule and, of course, the price. Beyond bargaining there is a wealth of material on the theory of *auctions* [5] for one-to-many negotiation, and *exchanges* for many-to-many negotiation. Fundamental to this analysis is the central role of the utility function, and the notion of rational behaviour by which an agent aims to optimise its utility, when it is able to do so, and to optimise its *expected* utility otherwise.

---

[1] Auction bots such as those on eBay, and automated auction houses do a useful job, but do not automate negotiation in the sense described here.

A. Sattar and B.H. Kang (Eds.): AI 2006, LNAI 4304, pp. 413–422, 2006.
© Springer-Verlag Berlin Heidelberg 2006

We propose that utility functions, or preference orderings, are often not known with certainty; further, the uncertainty that underpins them is typically in a state of flux. We propose that the key to building intelligent negotiating agents is to take an agent's historic observations as primitive, to model that agent's changing uncertainty in that information, and to use that model as the foundation for the agent's reasoning. We call such agents *information-based agents*. In Sec. 2 we describe these agents. Sec. 3 relates commitments made to their eventual execution, this leads to a formalisation of trust. Strategies for information-based agents are discussed in Sec. 4.

2   The Architecture of Information-Based Agents

The architecture of our information-based agent is shown in Fig. 1. The agent begins to function in response to a percept (received message) that expresses a need $N \in \mathcal{N}$. A *need* can be either exogenous (typically, the agent receives a message from another agent $\{\Omega_1, \ldots, \Omega_o\}$), or endogenous. The agent has a set of pre-specified *goals* (or *desires*), $G \in \mathcal{G}$, from which one or more is selected to satisfy its perceived needs. Each of these goals is associated with one or more plans, $s \in S$. This is consistent with the BDI model [1], and we do not detail these aspects here. The agent in Fig. 1 also interacts with information sources $\{\theta_1, \ldots, \theta_t\}$ that in our experiments[2] include unstructured data mining and text mining 'bots' that retrieve information from the agent market-place and from general news sources.

Finally the agent in Fig. 1 interacts with an 'Institution Agent', ξ, that reports honestly and promptly on the fulfilment of contracts. The Institution Agent is a conceptual device to prevent the requirement for agents to have 'eyes' and effectors. For example, if agent Π wishes to give an object Y to agent Ω_k then this is achieved by Π sending a message to ξ requesting the transfer of the ownership of Y from Π to Ω_k, once this is done, ξ sends a message to Ω_k advising him that he now owns Y. Given such an Institution Agent, agents can negotiate and evaluate trade by simply sending and receiving messages.

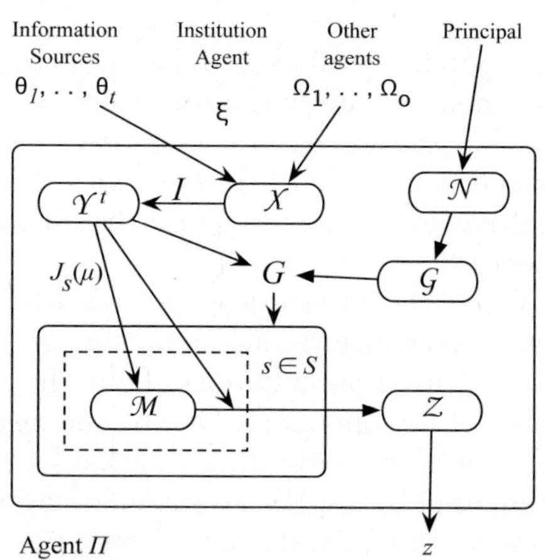

Fig. 1. Agent architecture

Π has two languages: $\mathcal{C}$ and $\mathcal{L}$. $\mathcal{L}$ is a first-order language for internal representation — precisely, it is a first-order language with sentence probabilities optionally attached to each sentence representing Π's epistemic belief in the validity of that sentence. $\mathcal{C}$ is an illocutionary-based language for communication [6]. Messages

---

[2] `http://e-markets.org.au`

expressed in $\mathcal{C}$ from $\{\theta_i\}$ and $\{\Omega_i\}$ are received, time-stamped, source-stamped and placed in an *in-box* $\mathcal{X}$. The illocutionary particles in $\mathcal{C}$ are:

– Offer(Π, Ω_k, δ). Agent Π offers agent Ω_k a contract $\delta = (\pi, \varphi)$ with action commitments $\pi \in \mathcal{L}$ for Π and $\varphi \in \mathcal{L}$ for Ω_k.
– Accept(Π, Ω_k, δ). Agent Π accepts agent Ω_k's previously offered contract δ.
– Reject$(\Pi, \Omega_k, \delta[, \mathit{info}])$. Agent Π rejects agent Ω_k's previously offered contract δ. Optionally, information $\mathit{info} \in \mathcal{L}$ explaining the reason for the rejection can be given.
– Withdraw$(\Pi, \Omega_k[, \mathit{info}])$. Agent Π breaks down negotiation with Ω_k. Extra $\mathit{info} \in \mathcal{L}$ justifying the withdrawal may be given.
– Inform$(\Pi, \Omega_k, \mathit{info})$. Agent Π informs Ω_k about $\mathit{info} \in \mathcal{L}$ and commits to the truth of info.
– Reward$(\Pi, \Omega_k, \delta, \phi[, \mathit{info}])$. Intended to make the opponent accept a proposal with the promise of a future compensation. Agent Π offers agent Ω_k a contract δ. In case Ω_k accepts the proposal, Π commits to make $\phi \in \mathcal{L}$ true. The intended meaning is that Π believes that worlds in which ϕ is true are somehow desired by Ω_k. Optionally, additional information in support of the contract can be given.
– Threat$(\Pi, \Omega_k, \delta, \phi, [\mathit{info}])$ Intended to make the opponent accept a proposal with the menace of some sort of retaliation. Agent Π offers agent Ω_k a contract δ. In case Ω_k does not accept the proposal, Π commits to make $\phi \in \mathcal{L}$ true. The intended meaning is that Π believes that worlds in which ϕ is true are somehow *not* desired by Ω_k. Optionally, additional information in support of the contract can be given.
– Appeal$(\Pi, \Omega_k, \delta, \mathit{info})$ Intended to make the opponent accept a proposal as a consequence of the belief update that the accompanying information might bring about. Agent Π offers agent Ω_k a contract δ, and passes information in support of the contract.

The accompanying information, *info*, can be of two basic types: (i) referring to the process (plan) used by an agent to solve a problem, or (ii) data (beliefs) of the agent including preferences. When building relationships, agents will therefore try to influence the opponent by changing their processes (plans) or by providing new data.

2.1   World Model

Everything that Π has at its disposal is derived from the messages in the in-box $\mathcal{X}$. As messages age, the degree of belief that Π associates with them will decrease. We call this *information integrity decay*. A factor in the integrity of a message will be the reliability of the source. This subjective decay is a feature of the agent, and agents will differ in their subjective estimates.

Each plan is driven by its expectations of the state of the world, and by the states of the other agents. These states will generally be quite numerous, and so we assume that at any time the agent's active plans will form expectations of certain *features* only, where each feature will be in one of a finite number of states[3]. Suppose that there are m such features, introduce m random variables, $\{X_i\}_{i=1}^m$. Each value, $x_{i,j}$, of the i'th random variable, X_i, denotes that the i'th feature is in the j'th perceivable state, or *possible world*, of that feature.

The messages in $\mathcal{X}$ are then translated using an *import function* I into sentences expressed in $\mathcal{L}$ that have integrity decay functions (usually of time) attached to each sentence, they are stored in a *repository* $\mathcal{Y}^t$. And that is all that happens until Π triggers a goal.

In general Π will be uncertain of the current state of each feature. Π's *world model*, $M \in \mathcal{M}$, consists of probability distributions over each of these random variables. If these m features are independent then the overall uncertainty, or entropy, of Π's world model is: $\mathbb{H}^t(M) = -\sum_{i=1}^m \mathbb{E}(\ln \mathbb{P}^t(X_i))$. The general idea is that if Π receives new information then the overall uncertainty of the world model is expected to decrease, but if Π receives no new information then it is expected to increase.

2.2 Reasoning

Consider first what happens if Π receives no new information. Each distribution, $\mathbb{P}^t(X_i)$, is associated with a *decay limit distribution*, $\mathbb{D}(X_i)$, that represents the expected limit state of the i'th feature in the absence of any observations of the state of that feature: $\lim_{t \to \infty} \mathbb{P}^t(X_i) = \mathbb{D}(X_i)$. For example, if the i'th feature is whether it is raining in Sydney, and $x_{i,1}$ means "it is raining in Sydney" and $x_{i,2}$ means "it is not raining in Sydney" — then if Π believes that it rains in Sydney 5% of the time: $\mathbb{D}(X_i) = (0.05, 0.95)$. If Π has no background knowledge about $\mathbb{D}(X_i)$ then the decay limit distribution is the maximum entropy, "flat", distribution. In the absence of incoming information, $\mathbb{P}(X_i)$ decays by:

$$\mathbb{P}^{t+1}(X_i) = \Delta_i(\mathbb{D}(X_i), \mathbb{P}^t(X_i))$$

where Δ_i is the *decay function* for the i'th feature satisfying the property that $\lim_{t \to \infty} \mathbb{P}^t(X_i) = \mathbb{D}(X_i)$. For example, Δ_i could be linear:

$$\mathbb{P}^{t+1}(X_i) = (1 - \nu_i)\mathbb{D}(X_i) + \nu_i \times \mathbb{P}^t(X_i) \tag{1}$$

where $\nu_i < 1$ is the decay rate for the i'th feature. Either the decay function or the decay limit distribution could also be a function of time: Δ_i^t and $\mathbb{D}^t(X_i)$.

If Π receives a message expressed in $\mathcal{C}$ then it will be transformed by inference rules into statements expressed in $\mathcal{L}$. We introduce this procedure with an example. Preference information is a statement by an agent that it prefers one class of contracts to another where contracts may be multi-issue. Preference illocutions may refer to particular issues within contracts — e.g. "I prefer red to yellow", or to combinations of issues — e.g. "I prefer a car with a five year warranty to the same car than costs 15% less with a two year warranty". An agent will find it useful to estimate which contract under consideration is favoured most by the opponent. Preference information can assist with this estimation as the

---

[3] We thus exclude the possibility of continuous variables.

following example shows. Suppose Π receives preference information from Ω_k through an Inform($\Omega_k, \Pi, info$) illocution: $info =$ "for contracts with property Q_1 or property Q_2, the probability that the contract Ω_k prefers most will have property Q_1 is z" — the ontology in $\mathcal{C}$ is assumed to contain an illocutionary particle that can express this statement. What happens next will depend on Π's plans. Suppose that Π has an active plan $s \in S$ that calls for the probability distribution $\mathbb{P}^t(\text{Favour}(\Omega_k, \Pi, \delta)) \in M$ over all δ, where $\text{Favour}(\Omega_k, \Pi, \delta)$ means that "δ is the contract that Ω_k prefers most from Π". Suppose Π has a prior distribution $\boldsymbol{q} = (q_1, \dots)$ for $\mathbb{P}^t(\text{Favour}(\cdot))$. Then s will require an inference rule: $J_s^{\text{Favour}}(info)$ that is the following linear constraint on the posterior $\mathbb{P}^t(\text{Favour}(\Omega_k, \Pi, \delta))$ distribution:

$$
z = \frac{\sum_{\delta:Q_1(\delta)} p_\delta}{\left(\sum_{\delta:Q_1(\delta)} p_\delta\right) + \left(\sum_{\delta:Q_2(\delta)} p_\delta\right) - \left(\sum_{\delta:Q_1 \wedge Q_2(\delta)} p_\delta\right)} \tag{2}
$$

and is determined by the *principle of minimum relative entropy* — a form of Bayesian inference that is convenient when the data is sparse [7] — as described generally below. The inference rule $J_s^{Favour}(\cdot)$ infers a constraint on a distribution in M from an illocution expressed in $\mathcal{C}$. Inferences of this sort are necessary for Π to operate, but their validity is a personal matter for Π to assume.

Now, more generally, suppose that Π receives a percept μ from agent Ω_k at time t. Suppose that this percept states that something is so with probability z, and suppose that Π attaches an epistemic belief probability $\mathbb{R}^t(\Pi, \Omega_k, \mu)$ to μ. Π's set of active plans will have a set of model building functions, $J_s(\cdot)$, such that $J_s^{X_i}(\mu)$ is a set of linear constraints on the posterior distribution for X_i where the prior distribution is $\mathbb{P}^t(X_i) = \boldsymbol{q}$. Let $\boldsymbol{p} = (p_1, \dots)$ be the distribution with minimum relative entropy with respect to $\boldsymbol{q}$: $\boldsymbol{p} = \arg\min_{\boldsymbol{p}} \sum_j p_j \log \frac{p_j}{q_j}$ that satisfies the constraints $J_s^{X_i}(\mu)$. Then let $\boldsymbol{r}$ be the distribution:

$$
\boldsymbol{r} = \mathbb{R}^t(\Pi, \Omega_k, \mu) \times \boldsymbol{p} + (1 - \mathbb{R}^t(\Pi, \Omega_k, \mu)) \times \boldsymbol{q}
$$

and then for a small time step δt let:

$$
\mathbb{P}^{t+\delta t}(X_i) = \begin{cases} \boldsymbol{r} & \text{if } \mathbb{K}(\boldsymbol{r} \| \mathbb{D}(X_i)) > \mathbb{K}(\mathbb{P}^t(X_i) \| \mathbb{D}(X_i)) \\ \boldsymbol{q} & \text{otherwise} \end{cases} \tag{3}
$$

where $\mathbb{K}(\boldsymbol{x} \| \boldsymbol{y}) = \sum_j x_j \ln \frac{x_j}{y_j}$ is the Kullback-Leibler distance between two probability distributions $\boldsymbol{x}$ and $\boldsymbol{y}$. The idea in Eqn. 3 is that the vector $\boldsymbol{r}$ will only update $\mathbb{P}^t(X_i)$ if it contains more information with respect to the decay limit distribution than the prior $\boldsymbol{q}$. Then combining Eqn. 3 with Eqn. 1 let:

$$
\mathbb{P}^{t+1}(X_i) = (1 - \nu_i)\mathbb{D}(X_i) + \nu_i \times \mathbb{P}^{t+\delta t}(X_i) \tag{4}
$$

and note that this procedure has dealt with integrity decay, and with two probabilities: first, the probability z in the percept μ, and second the epistemic belief probability $\mathbb{R}^t(\Pi, \Omega_k, \mu)$ that Π attached to μ. Given a probability distribution

q, the *minimum relative entropy distribution* $\boldsymbol{p} = (p_1, \ldots, p_I)$ subject to a set of J linear constraints $\boldsymbol{g} = \{g_j(\boldsymbol{p}) = \boldsymbol{a}_j \cdot \boldsymbol{p} - c_j = 0\}, j = 1, \ldots, J$ (that must include the constraint $\sum_i p_i - 1 = 0$) is: $\boldsymbol{p} = \arg\min_{\boldsymbol{p}} \sum_j p_j \log \frac{p_j}{q_j}$. This may be calculated by introducing Lagrange multipliers $\boldsymbol{\lambda}$: $L(\boldsymbol{p}, \boldsymbol{\lambda}) = \sum_j p_j \log \frac{p_j}{q_j} + \boldsymbol{\lambda} \cdot \boldsymbol{g}$. Minimising L, $\{\frac{\partial L}{\partial \lambda_j} = g_j(\boldsymbol{p}) = 0\}, j = 1, \ldots, J$ is the set of given constraints $\boldsymbol{g}$, and a solution to $\frac{\partial L}{\partial p_i} = 0, i = 1, \ldots, I$ leads eventually to $\boldsymbol{p}$.

2.3   Estimating $\mathbb{R}^t(\Pi, \Omega_k, \mu)$

Π attaches an epistemic belief probability $\mathbb{R}^t(\Pi, \Omega_k, \mu)$ to each message μ. A historic estimate of $\mathbb{R}^t(\Pi, \Omega_k, \mu)$ may be obtained by measuring the 'difference' between commitment and execution. Π's plans will have constructed a set of distributions. We measure this 'difference' as the error in the effect that μ has on each of Π's distributions. Suppose that μ is received from agent Ω_k at time u and is verified at some later time t. For example, μ could be a chunk of information: "the interest rate will rise by 0.5% next week", and suppose that the interest rate actually rises by 0.25% — represent what the message should have been μ'. What does all this tell agent Π about agent Ω_k's reliability? Consider one of Π's distributions for X that is $\boldsymbol{q}^u$ at time u. Let $\boldsymbol{p}^u_\mu$ be the posterior minimum relative entropy distribution subject to the constraint $J^X_s(\mu)$, and let $\boldsymbol{p}^u_{\mu'}$ be that distribution subject to $J^X_s(\mu')$. We now estimate what $\mathbb{R}^u(\Pi, \Omega_k, \mu)$ should have been in the light of knowing *now*, at time t, that μ should have been μ'.

The idea of Eqn. 3, is that the current value of $\mathbb{R}^t(\Pi, \Omega_k, \mu)$ should be such that, *on average*, $\boldsymbol{p}^u_\mu$ will be "close to" $\boldsymbol{p}^u_{\mu'}$ when we eventually discover μ' — no matter whether or not μ was used to update the distribution for X, as determined by the acceptability test in Eqn. 3 at time u. The *observed reliability* for μ and distribution X, $\mathbb{R}^t_X(\Pi, \Omega_k, \mu) | \mu'$, on the basis of the verification of μ with μ', is the value of r that minimises the Kullback-Leibler distance:

$$\mathbb{R}^t_X(\Pi, \Omega_k, \mu) | \mu' = \arg\min_r \mathbb{K}(r \cdot \boldsymbol{p}^u_\mu + (1 - r) \cdot \boldsymbol{q}^u \| \boldsymbol{p}^u_{\mu'})$$

If $\mathbf{X}(\mu)$ is the set of distributions that μ affects, then the overall *observed reliability* on the basis of the verification of μ with μ' is:

$$\mathbb{R}^t(\Pi, \Omega_k, \mu) | \mu' = 1 - (\max_{X \in \mathbf{X}(\mu)} |1 - \mathbb{R}^t_X(\Pi, \Omega_k, \mu) | \mu'|)$$

Then for each ontological context o_j, at time t when μ has been verified with μ':

$$\mathbb{R}^{t+1}(\Pi, \Omega_k, o_j) = (1 - \nu) \times \mathbb{R}^t(\Pi, \Omega_k, o_j) + \nu \times \mathbb{R}^t(\Pi, \Omega_k, \mu) | \mu' \times \mathrm{Sim}(o_j, O(\mu))$$

where Sim measures the semantic distance between two sections of the ontology, and ν is the learning rate. Over time, Π notes the ontological context of the various μ received from Ω_k, and over the various ontological contexts calculates the relative frequency, $\mathbb{P}^t(o_j)$, of these contexts, $o_j = O(\mu)$. This leads to an overall expectation of the *reliability* that agent Π has for agent Ω_k:

$$\mathbb{R}^t(\Pi, \Omega_k) = \sum_j \mathbb{P}^t(o_j) \times \mathbb{R}^t(\Pi, \Omega_k, o_j)$$

3   Commitment and Execution

The interaction between agents Π and Ω_k will eventually lead to some sort of *contract*: $\delta = (\pi, \varphi)$ where π is Π's commitment and φ is Ω_k's commitment. No matter what these commitments are, Π will be interested in any variation between Ω_k's commitment, φ, and what actually happens, the execution, φ'. The form of this commitment could be a promise to deliver goods, or abide by certain trading terms that extend over a period of time, or that some information that may, or may not, prove to be correct. We denote the relationship between commitment and execution, $\mathbb{P}^t(\text{Execute}(\varphi')|\text{Commit}(\varphi))$ simply as $\mathbb{P}^t(\varphi'|\varphi)$. In general we assume that such commitment and execution takes place in the context of a *relationship* ρ between Π and Ω_k.

Beliefs 'evaporate' as time goes by. If we don't keep an ongoing relationship, we become unsure how *trustworthy* a trading partner is. This decay is what justifies a continuous relationship between agents. The conditional probabilities, $\mathbb{P}^t(\varphi'|\varphi)$, should tend to ignorance as represented by the *decay limit distribution* $\boldsymbol{d} = \{d_i\}$. If we have the set of observations $\Phi = \{\varphi_1, \varphi_2, \ldots, \varphi_n\}$ then complete ignorance of the opponent's expected behaviour means that given the opponent commits to φ, the conditional probability for each observable outcome φ' becomes $d_i = \frac{1}{n}$, but Π may have background beliefs about Ω_k's decay limit distribution. This natural decay of belief is offset by new observations. We define the evolution of the probability distribution as: $\mathbb{P}^{t+1}(\varphi'|\varphi) = \big((1 - \nu) \cdot \boldsymbol{d} + \nu \cdot \mathbb{P}^t_+(\varphi'|\varphi)\big)$, where $\nu \in [0, 1]$ is the learning rate, and $\mathbb{P}^t_+(\varphi'|\varphi)$ represents the posterior distribution for $(\varphi'|\varphi)$ given an observed contract execution as the following shows.

Suppose that Π has a business relationship ρ with agent Ω_k, that Ω_k commits to φ, and this commitment is sound. The material value of φ to ρ will depend on the future use that Π makes of it, and that is unlikely to be known. So Π estimates the value of φ to the relationship ρ he has with Ω_k using a probability distribution $(p_1, \ldots, p_n)$ over a *relationship evaluation space* $E = (e_1, \ldots, e_n)$ that could range from "that is what I expect from the perfect trading partner" to "it is totally useless" — E may contain hard or fuzzy values. $p_i = w_i(\rho, \varphi)$ is the probability that e_i is the correct evaluation of the enactment φ in the context of relationship ρ, and $\boldsymbol{w} : \mathcal{L} \times \mathcal{L} \to [0, 1]^n$ is the *evaluation function*.

Let $(\varphi_1, \ldots, \varphi_m)$ be the set of possible contract executions in some order. Then for a given φ_k, $(\mathbb{P}^t(\varphi_1|\varphi_k), \ldots, \mathbb{P}^t(\varphi_m|\varphi_k))$ is the prior distribution of Π's estimate of what will actually occur if Ω_k committed to φ_k occurring and $\boldsymbol{w}(\rho, \varphi_k) = (w_1(\rho, \varphi_k), \ldots, w_n(\rho, \varphi_k))$ is Π's evaluation over E with respect to the relationship ρ of Ω_k's commitment φ_k. Π's expected evaluation of what will occur given that Ω_k has committed to φ_k occurring is:

$$\boldsymbol{w}^{\exp}(\rho, \varphi_k) = \left(\sum_{j=1}^{m} \mathbb{P}^t(\varphi_j|\varphi_k) \cdot w_1(\rho, \varphi_j), \ldots, \sum_{j=1}^{m} \mathbb{P}^t(\varphi_j|\varphi_k) \cdot w_n(\rho, \varphi_j) \right).$$

Now suppose that Π observes the event $(\phi'|\phi)$ in another relationship ρ' also with agent Ω_k. Eg: Π may buy wine and cheese from the same supplier.

Π may wish to revise the prior estimate $\boldsymbol{w}^{\mathrm{exp}}(\rho, \varphi_k)$ in the light of the observation $(\phi'|\phi)$ to:

$$(\boldsymbol{w}^{\mathrm{rev}}(\rho, \varphi_k) \mid (\varphi'|\varphi)) = \boldsymbol{g}(\boldsymbol{w}^{\mathrm{exp}}(\rho, \varphi_k), \boldsymbol{w}(\rho', \varphi), \boldsymbol{w}(\rho', \varphi'), \rho, \rho', \varphi, \varphi, \varphi'),$$

for some function $\boldsymbol{g}$ — the idea being, for example, that if the commitment, φ, concerning the purchase of cheese, ρ', was not kept then Π's expectation that the commitment, φ, concerning the purchase of wine, ρ, will not be kept should increase. We estimate the posterior $\mathbb{P}^t_+(\varphi'|\varphi)$ by applying the principle of minimum relative entropy: $\left(\mathbb{P}^t_+(\varphi_j|\varphi)\right)^m_{j=1} = \arg\min_{\boldsymbol{p}} \sum^m_{i=1} p_i \log \frac{p_i}{\mathbb{P}^t(\varphi_i|\varphi)}$ where $\boldsymbol{p} = (p_j)^m_{j=1}$, satisfies the n constraints:

$$\sum^m_{j=1} p_j \cdot w_i(\rho, \varphi_j) = g_i(\boldsymbol{w}^{\mathrm{exp}}(\rho, \varphi_k), \boldsymbol{w}(\rho', \varphi), \boldsymbol{w}(\rho', \varphi'), \rho, \rho', \phi, \varphi, \varphi')$$

for $i = 1, \ldots, n$. This is a set of n linear equations in m unknowns, and so the calculation of the minimum relative entropy distribution may be impossible if $n > m$. In this case, we take only the m equations for which the change from the prior to the posterior value is greatest. That is, we attempt to select the most significant factors.

Consider a distribution of expected fulfilment of commitments that represent Π's "ideal" for a relationship with Ω_k, in the sense that it is the best that Π could reasonably expect Ω_k to do. This distribution will be a function of Ω_k, Π's history with Ω_k, anything else that Π believes about Ω_k, and general environmental information including time — denote all of this by e, then we have $\mathbb{P}^t_I(\varphi'|\varphi, e)$. For example, if Π considers that it is unacceptable for the execution φ' to be less preferred than the commitment φ then $\mathbb{P}^t_I(\varphi'|\varphi, e)$ will only be non-zero for those φ' that Π prefers to φ. The distribution $\mathbb{P}^t_I(\cdot)$ represents what Π expects, or hopes, Ω_k will do. *Trust* is the relative entropy between this ideal distribution, $\mathbb{P}^t_I(\varphi'|\varphi, e)$, and the distribution of the observation of fulfilled commitments, $\mathbb{P}^t(\varphi'|\varphi)$. That is:

$$\mathrm{Trust}(\Pi, \Omega_k, \varphi) = 1 - \sum_{\varphi'} \mathbb{P}^t_I(\varphi'|\varphi, e) \log \frac{\mathbb{P}^t_I(\varphi'|\varphi, e)}{\mathbb{P}^t(\varphi'|\varphi)} \tag{5}$$

where the "1" is an arbitrarily chosen constant being the maximum value that trust may have. This equation defines trust for one, single commitment φ — for example, my trust in my butcher if he commits to provide me with a 10% discount for the rest of the year. It makes sense to aggregate these values over a class of commitments, say over those φ that are subtypes of a particular relationship ρ, that is $\varphi \leq \rho$. In this way we measure the trust that I have in my butcher in relation to the commitments he makes for red meat generally:

$$\mathrm{Trust}(\Pi, \Omega_k, \rho) = 1 - \frac{\sum_{\varphi:\varphi\leq\rho} \mathbb{P}^t(\varphi) \left[\sum_{\varphi'} \mathbb{P}^t_I(\varphi'|\varphi, e) \log \frac{\mathbb{P}^t_I(\varphi'|\varphi,e)}{\mathbb{P}^t(\varphi'|\varphi)} \right]}{\sum_{\varphi:\varphi\leq\rho} \mathbb{P}^t(\varphi)}$$

where $\mathbb{P}^t(\varphi)$ is a probability distribution over the space of commitments that the next commitment Ω_k will make to Π is φ. Similarly, for an overall estimate of Π's trust in Ω_k:

$$\text{Trust}(\Pi, \Omega_k) = 1 - \sum_\varphi \mathbb{P}^t(\varphi) \left[\sum_{\varphi'} \mathbb{P}^t_I(\varphi'|\varphi, e) \log \frac{\mathbb{P}^t_I(\varphi'|\varphi, e)}{\mathbb{P}^t(\varphi'|\varphi)} \right]$$

4   Strategies

An agent requires *strategies* for deciding who to interact with, and for deciding how to manage interaction using the language $\mathcal{C}$. In $\mathcal{C}$ as defined in Sec. 2, contracts may be for a single trade, or may encapsulate an on-going trading relationship. Interaction is normally bound by an *interaction protocol* that moderates the interaction sequence, and so may limit the range of model building functions, $J_s(\cdot)$. Consider the protocol in which statements in $\mathcal{C}$ are exchanged between pairs of agents, and Offer$(\cdot)$ statements are binding until countered, or until one of the pair issues a Quit$(\cdot)$. That is, an agent would only enter into a negotiation — ie: offer exchange — if it were prepared to commit. To manage this protocol, agent Π requires the following probability estimates in M where Π is bargaining with opponent Ω_k, in satisfaction of some need $N \in \mathcal{N}$:

1. $\mathbb{P}^t(val(\Pi, \Omega_k, N, \delta) = v_i)$ — for any deal, δ, the probability distribution over some valuation space $\{v_i\}$ that measures how "good" the deal δ is to Π.
2. $\mathbb{P}^t(acc(\Pi, \Omega_k, \delta))$ — for any deal, δ, the probability that Ω_k would accept δ.
3. $\mathbb{P}^t(conv(\Pi, \Omega_k, \Delta))$ — for any sequence of offer-exchanges, Δ, the probability that that sequence will converge to an acceptable deal.
4. $\mathbb{P}^t(trade(\Pi, \Omega_k, o_j) = u_i)$ — for an ontological context o_j, the probability distribution over some valuation space $\{u_i\}$ that measures how "good" Ω_k is as a trading partner to Π for deals in ontological context o_j.

The estimation of these distributions has been described previously [8]. Π's strategy determines how it uses these distributions. An approach to issue-tradeoffs is described in [9]. That strategy attempts to make an acceptable offer by "walking round" the iso-curve of Π's previous offer (that has, say, an acceptability of α) towards Ω_k's subsequent counter offer. In terms of the machinery described here: $\arg\max_\delta\{\ \mathbb{P}^t(acc(\Pi, \Omega_k, \delta)) \mid \mathbb{E}^t(val(\Pi, \Omega_k, N, \delta)) \approx \alpha\ \}$. By including the "information dimension" Π can implement strategies that go beyond utilitarian thinking. Π evaluates every illocution for its utilitarian value, and for its value as information. For example, the *equitable information revelation* strategy [8] responds to a message μ with a message that gives the recipient expected information gain similar to that which μ gave to Π; these responses are also "reasonable" from a utilitarian point of view. An information-based agent evaluates all exchanges in terms of both their estimated utilitarian value, and their information value.

Estimations of trust — Sec. 3 — may be used to select a trading partner. One interesting question is to determine a set of partners to maintain for deals from a particular section of the ontology — this is a question of risk management. Having identified such a set, the agent then has to decide which one of these partners to use for the next negotiation. A nice strategy is to choose the partner with a probability equal to the probability that they are the best choice — as determined by trust, or some other means.

An information-based agent additionally requires strategies to manage the exchange of information, and to be strategic in their information acquisition. This includes strategies for dealing with the information sources $\{\theta_1, \ldots, \theta_t\}$, which becomes interesting if those sources are not always available, charge a fee, or take some time to deliver. This also includes strategies for the acquisition of information by both covert and overt strategic interaction with other agents $\{\Omega_1, \ldots, \Omega_o\}$. These information strategies are the subject of current research.

5   Conclusion

We do not claim that this is the end of the matter in deploying automated negotiators, and the approach described here has yet to be trialed extensively. But we do maintain the strategic apparatus of intelligent negotiating agents should include the intelligent use of information. We have proposed a theoretical basis for managing information in the context of competitive interaction, and have shown how that theory may be computed by an intelligent agent. Information theory provides the theoretical underpinning that enables such an informed agent to value, manage and exchange her information intelligently.

References

1. Wooldridge, M.: Multiagent Systems. Wiley (2002)
2. Arcos, J.L., Esteva, M., Noriega, P., Rodríguez, J.A., Sierra, C.: Environment engineering for multiagent systems. Journal on Engineering Applications of Artificial Intelligence **18** (2005)
3. Rosenschein, J., Zlotkin, G.: Rules of Encounter. MIT Press (1998)
4. Muthoo, A.: Bargaining Theory with Applications. Cambridge UP (1999)
5. Klemperer, P.: The Economic Theory of Auctions : Vols I and II. Edward Elgar (2000)
6. Sierra, C., Jennings, N., Noriega, P., Parsons, S.: A framework for argumentation-based negotiation. In: Intelligent Agents IV: Agent Theories, Architectures, and Languages (ATAL-97), Springer-Verlag: Heidelberg, Germany (1998) 177–192
7. Cheeseman, P., Stutz, J.: On The Relationship between Bayesian and Maximum Entropy Inference. In: Bayesian Inference and Maximum Entropy Methods in Science and Engineering. American Institute of Physics, Melville, NY, USA (2004) 445 – 461
8. Debenham, J., Simoff, S.: An agent establishes trust with equitable information revelation. In Subrahmanian, V., Regli, W., eds.: Proceedings of the 2005 IEEE 2nd Symposium on Multi-Agent Security and Survivability, Drexel University, Philadelphia, USA, IEEE (2005) 66 – 74
9. Faratin, P., Sierra, C., Jennings, N.: Using similarity criteria to make issue trade-offs in automated negotiation. Journal of Artificial Intelligence **142** (2003) 205–237

Towards Goals in Informed Agent Negotiation

Paul Bogg

University of Technology Sydney, Broadway NSW 2007, Australia
plbogg@it.uts.edu.au

Abstract. Negotiation is typically the way in which real world multi-issue commercial contracts are resolved and signed. The automation of negotiation in the B2B context has the potential to revolutionise trade. Intelligent agents have been proposed as the software architecture for automating negotiation. Goals are important based on the notion from [1] that an agent which focuses on negotiating by *interests* or *goals* rather than *positions* may increase the quality of agreement, and the speed of reaching the agreement. This paper presents work towards goal based negotiation that extends an information theoretical framework for automating negotiation using maximum entropy inference with linear information constraints. We place goals in terms of the information inference model, with a view towards goal based argumentation.

1 Introduction

Negotiation is defined as a process whereby two (or more) individuals with conflicting interests reach a mutually beneficial agreement on a set of issues. The parties may attempt to cooperate in order to achieve this agreement despite having conflicting interests. The way they do this is to make proposals, offer concessions, and trade views [2]. Up to a certain point determined by individual preferences, each individual is willing to give up ground on particular issues to reach this agreement.

E-commerce trade is one situation where intelligent agents have been applied to automate negotiation. Agents act on behalf of humans in order to conduct commerce trade automatically over the Internet. In game theoretical approaches to automated negotiation, preferences and interests are represented in the form of a utility function. Utility encapsulates an agents preference over the set of possible agreements. However, there are two observed difficulties with respect to a utilitarian approach to automating negotiation with agents on behalf of humans. It is these two difficulties that inspire further research.

The first observed difficulty that has been frequently identified is that instilling an agent with a utility function is difficult and unrealistic compared to what happens in the real world [3]. Humans have great trouble knowing their utility, and consequently, instilling an agent with this utility function is also difficult. Additionally, in maximising expected utility, an agent estimates a probability distribution inference model over outcomes based on information received from the environment. Information however, may be volatile and in a continual state

A. Sattar and B.H. Kang (Eds.): AI 2006, LNAI 4304, pp. 423–432, 2006.
© Springer-Verlag Berlin Heidelberg 2006

of decay, changing the shape of the probability distribution. It has been argued by [4][5] that catering for volatility of information and changing distribution models is crucial to being able to negotiate successfully.

The second observed difficulty is based on the general observation which follows the argument from [1]: by focusing on interests or goals rather than positions, it may be possible to reach agreements faster, and with a higher quality. However, there is currently no model for automated negotiation which represents an agent's interests and accounts for an agent's information distribution model.

An information theoretical model for negotiation[4] aims to estimate for a set of possible negotiation outcomes, the likelihood of each outcome being true. The way this is done is with maximum entropy inference with linear constraints. The linear constraints are specified in terms of decaying integrity estimates for information the agent believes at a particular point in time. By using maximum entropy, the distribution model is maximally non-committal with regards to missing information.

The two above mentioned difficulties inspire this paper towards developing a model for integrating goals into an information theoretical model for automated negotiation. We proposed a goal based model which extends the information theoretical model for negotiation. An agent estimates goal probabilities which are sensitive to decaying information. Additionally, the model may be extended towards goal *sharing* with argumentation.

This paper is organised as follows. Section 2 defines goals in terms of negotiating intelligent agents. Section 3 outlines an approach for incorporating goals into an information theoretical framework for negotiation from [5][4].

2   Arguing for Goals in Negotiation

The main premise of this paper lies in the observation that negotiation, for instance in e-commerce trade, is primarily a goal oriented process. A proponent identifies a set of needs that requires fulfilment by way of the trade of a particular *item*. The proponent identifies a set of goals that will fulfill the needs. These goals are used to both guild the negotiation process and determine how successful the proponent was in fulfilling his/her needs. These goals differ in type, and somehow guide the proponent's actions.

We argue that intelligent agents that automatically negotiate on behalf of humans for the trade of *items* over the Internet should also adopt some form of goal based approach for a number of reasons. Firstly, an explicit representation of goals in negotiation allows an agent to identify what a successful negotiation is. For instance, an agent that satisfies all of its goals by the conclusion of a negotiation may be determined as being *completely successful*. It also may not necessarily be important to the agent that all the goals are achieved, so long as a few key goals are satisfied ("I really need to buy this fridge within my budget, with high energy efficiency being very important, but not essential"). In this case, it is *partially successful*. Secondly, by explicitly representing goals it becomes possible to negotiate by focusing on interests instead of positions[1].

Exchanging goals during the course of a negotiation enables the possibility to reach agreements quicker and with outcomes closer to what is needed. Furthermore, argumentation about goals may also be incorporated into the negotiation.

We make the distinction in this paper between three types of goals. The first is the *primary trading goal*. This goal type is a principle reason that the agent initiates a negotiation. The trading goal(s) prompts the agent to negotiate for trade such that by the conclusion of the negotiation, the trading goal(s) are fulfilled. For example, an agent may adopt the primary goal to buy a fridge. Additionally, the agent may adopt a trading goal to improve a business relationship.

The second type of goal is the *resource-based goal*. Resource based goals determine to what extent the agent is able to concede in terms of resources. An agent has a set of resources to negotiate with, and may attempt to make other traders commit to giving a certain amount of resources in return. Traditionally, an agent has a preference structure over resources which identifies to a certain degree what quantity/quality of a resource is preferred over other quantities/qualities. In this respect it is possible to view resource based goals as a strong type of preference relation. For instance, an agent has a preference relation over money where it prefers spending less money than more but may spend more if circumstances allow, and additionally, has a budget of only X to spend, which is the upper limit for the spending. Throughout this paper we make the distinction between preferences and resource goals and identify a distinct boundary.

Finally, the third type of goal is the *attribute goal*. Attribute goals are goals on the quantity/quality of the attributes of the item which is to be traded. In certain situations, an agent knows what type of item it wants, but does not know precisely which item. It also knows particular attributes of the type of item which are important to it. For instance, an agent is unsure of what fridge to purchase, though it maintains that the fridge must be highly energy efficient and be of a particular size. Attribute goals are the quantity/quality of an attribute that the item must possess if the agent is agree to a trade. In a sense, attribute goals generalise the type of item to be traded. Attribute goals differ from resource based goals since the former is limited to the objective availability of types of items satisfying the attributes quantity/quality, where as the latter may be a subjective formulation over resources available to trade for a specific item.

3   Goals and Information in Negotiation

We formalise our approach to incorporating goals in the negotiation. An information theoretical framework for agent based negotiation[4] forms the base within which goals are defined. The reader is referred to [4][5] for a proper treatment.

Let α and β be two agents willing to negotiate with each other for the exchange of some item, *item*. The agents each have a knowledge base $\mathcal{K}$ and a belief set $\mathcal{B}$ containing statements in first order predicate logic form. $\mathcal{K}$ contains logic statements which are true that represent all that the agent knows about the world. $\mathcal{B}$ contains logic statements $b \in \mathcal{B}$ which are given a sentence probability $\mathbf{B}(b) \in [0, 1]$ that represents the degree of belief in the truth of that statement.

β has an *item* and α wishes to acquire one somehow. α wants to negotiate a contract $\delta = (\Gamma_\alpha, \Gamma_\beta)$ whereby if α agrees to a set of terms Γ_α and β agrees to some other set of terms, Γ_β, α will acquire *item*. The set of all possible contracts is the *deal space*, $\mathcal{V}$. The agent assumes $\mathcal{V}$ is finite. An *issue* in a negotiation is an attribute of the contract. The *domain* of an issue is the set of possible values that an attribute may have. The tuple $((j_{\alpha 1}, j_{\alpha 2}, \ldots, j_{\alpha m}), (j_{\beta 1}, j_{\beta 2}, \ldots, j_{\beta n}))$ represents the issues that α and β are responsible for. Let D_j be the domain of issue j. For example, an agent may determine for an issue *price* a domain set $D_{price} = \{\$5.6, \$5.7, \$5.8\}$, and so forth. For all domains of each issue, the set of possible contracts is defined as $\mathcal{V} \subseteq \{(x,y) | x \in D_{j_{\alpha 1}} \times D_{j_{\alpha 2}} \times \ldots \times D_{j_{\alpha m}}, y \in D_{j_{\beta 1}} \times D_{j_{\beta 2}} \times \ldots \times D_{j_{\beta n}}\}$. A possible contract is some $\delta \in \mathcal{V}$. When an offer is accepted, $\{acc(\delta)\} \cup \mathcal{K}$. When negotiation breaks down, $\{bd\} \cup \mathcal{K}$.

An agent describes a probability distribution(s) over the set of possible outcomes. This distribution model(s) represents the likelihoods of particular outcomes being true by the conclusion of the negotiation. An agent may know its utility function, in which case it may estimate *expected utility* as a combination of the utility and probability of each outcome being true. The concern of the information theoretic approach is the role that information plays in shaping the probability distribution. The approach that it takes is to use *maximum entropy* with *random worlds* described here. A *random world* for $\mathcal{K}$ is a probability distribution $\mathcal{W}_a = \{p_i\}$ over $\mathcal{V}$, which expresses the agents belief that a particular world is the actual world. $\mathcal{M} = \{\mathcal{W}_a\}$ is the agent's *world model*.

Information is used to constrain the construction of the probability distributions in $\mathcal{M}$. For some information type a, the agent uses information of this type to constrain the construction of a distribution $\mathcal{W}_a$. The agent determines what type of information it is interested in. For instance, it may be interested in information about a market price, and not in the weather tomorrow. We use two types of information here — self-valuation information, and opponent acceptability information. Self-valuation information is information the agent believes about what is likely to be the most acceptable contract for itself. For α, we will refer to information of this type with the predicate symbol 'Accept$_\alpha$'. Opponent acceptability information is information the agent receives about what is most likely to be the most acceptable contract for the opponent. For α, we will refer to information of this type with the predicate symbol 'Accept$_\beta$'.

Information that is sent to the agent is stored as a belief in $\mathcal{B}$ with an associated sentence probability $\mathbf{B}(b)$. For example, if α receives an offer of δ from β and α estimates β tells the truth 90% of the time, α may store the belief 'Accept$_\beta(\delta)$' with $\mathbf{B}(\text{Accept}_\beta(\delta)) = 0.9$. Given information that the agent believes, the agent may construct the distribution $\mathcal{W}_a$ that is the least biased distribution with regards to information it doesn't know. The way this is done, is with maximum entropy. Traditionally, the entropy of a discrete random variable X with probability mass function $\{p_i\}$ is defined:

$$H(X) = -\sum_n p_n \log p_n$$

where $p_n \geq 0$ and $\sum_n p_n = 1$. Let $\mathcal{W}_a$ be the "maximum entropy probability distribution over $\mathcal{V}$ that is consistent with $\mathcal{B}$". That is, given some beliefs of a particular information type a, the maximum entropy distribution is the construction of the random world probability distribution which is "maximally non-committal with regards to missing information". The distribution is constructed using the same technique as for the maximum likelihood problem for the Gibbs distribution – the reader is referred to [5][4] for a proper treatment.

In the following section, we extend this information theoretic framework to include goals as a kind of strong preference statement, which reflects what an agent needs to achieve out of a negotiation, and reflects on negotiation as a goal oriented process.

3.1   Representing Goal Types

The goals for an agent at a given point in time determine what they need to achieve in the future. We limit the discussion to goals which are directly related to a negotiation situation. Let $\mathcal{G}$ be the set of negotiation goals that an agent manages. $\mathcal{G} \subseteq GOALS$, the set of all goals that an agent has at time t. We refer to negotiation goals here as just 'goals'. An agent has a partial ordering of goals. This represents the notion that an agent views that some goals are more important than other goals, though may not necessarily be the case for all goals. The ordering of goals is represented by the ordering relation $>_\mathcal{G}$ over $\mathcal{G}$ which is both irreflexive and transitive. The ordering of goals generalises the notion of motivation — an agent is said to be more *motivated* to satisfy $g \in \mathcal{G}$ than $g' \in \mathcal{G}$ iff $g >_\mathcal{G} g'$. This does not mean that an agent in a negotiation wants to satisfy g instead of g', but rather, the agent wishes to satisfy both, with g taking on a marginally greater importance to the agent. When g and g' are equally motivated, it is possible to say $g \sim_\mathcal{G} g'$, where $\sim_\mathcal{G}$ is reflexive, symmetric, and transitive. For the time being, we can assume that all goals are equally important — that is, $\forall_{g,g' \in \mathcal{G}}(g \sim_\mathcal{G} g')$.

A goal takes the form of a tuple (p, f, C, r) where p is a predicate statement of the first order type, f is a goal statement evaluation which determines how the goal is evaluated, C provides the set of conditions which determine when the goal is evaluated, and r is the present state of the goal. $\mathcal{G}$ contains goals of this type. An agent with a goal of this form constantly evaluates its actions against f to maintain goal consistency such that it does not end up violating the goal when the conditions C occur. The occurrence of C is taken here to mean a certain state of $\mathcal{K}$ or $\mathcal{B}$, however can be extended to involve a more temporal conditions. By default until the event C occurs, r is $UNKN$. When C occurs, the evaluation on f results in the update of r where r is $SUCC$ iff f is *true* , $FAIL$ iff f is *false*. In the discussion of goals, we refer to the $p, f, C,$ or r of a goal g with the symbol $\triangleright$ (for instance, $f \triangleright g$ reads "the statement evaluation f of goal g", $r \triangleright g$ reads "the present state r of goal g").

An example of a goal might be agent α has a budget of \$250 to spend. By the conclusion of a negotiation whereby a trade was agreed, α must satisfy not having agreed to spend greater than \$250. This is represented by the tuple

$(budget(250), (j_{price} \in \delta, j_{price} < \$250), \{\exists_{acc(\delta)\in\mathcal{K}} \vee \exists_{bd\in\mathcal{K}}\}, UNKN)$. The goal is evaluated when a contract is accepted or a breakdown occurs. If the negotiation ends in breakdown, then this goal ultimately fails.

We now define three differing types of goals as identified earlier. Let $G_t \subseteq \mathcal{G}$ be the set of *primary trading goals*. A *primary trading goal* is one which exists prior to the negotiation and motivates an agent to negotiate for trade. The *default* primary trading goal here is the e-commerce goal to buy or sell an item (depending on the agent's perspective) — $buy(item)$ or $sell(item)$. For instance, α's default set of primary trading goals is $G_{t,\alpha} = \{(buy(item), (\exists_{acc(\delta)\in\mathcal{K}}), \{\exists_{acc(\delta)\in\mathcal{K}} \vee \exists_{bd\in\mathcal{K}}\}, UNKN)\}$. α is successful for $buy(item)$ when a deal δ is agreed upon at the conclusion of a negotiation (i.e., $\exists_{acc(\delta)\in K}$ holds true). It is important to note that the success of each $g \in G_t$ may not depend on an agreement of δ. For example, suppose that an agent comes to the negotiation table with the intention of discovering or confirming information about competitors as well as striking a deal. The agent may achieve the negotiation goal by the conclusion, regardless of agreement, in which case the negotiation may still be viewed as partially successful. Argumentation in negotiation [6] certainly lends itself to these type of goals. For the time being however, we limit this discussion to $buy(item)$ and $sell(item)$ as possible primary trading goals — which is to say, only goals that are evaluated on the existence of a deal or breakdown.

Let $G_r \subset \mathcal{G}$ be the set of resource goals. An agent maintains knowledge about a set of resources in $\mathcal{K}$, however, for the purpose of the negotiation, may have a set of goals which limit the amount of resources the agent is able to negotiate with or wants to receive from the other trading agent. Resource goals directly relate to issues in negotiation — that is, an issue may be a resource to be negotiated for. The agent specifies resource goals with the aim of coming to an agreement on issues in δ which stay inline with the resource goal specifications. For instance, the goal to maintain a budget of spending no more than \$250 identified earlier is a resource goal when the issue is money. Another example is for the case α wants to get a warrantee equal to or over 2 years for the purchase of *item*: $(warrantee(2), (j_{warr} \in \delta, j_{warr} \geq 2), \{\exists_{acc(\delta)\in\mathcal{K}} \vee \exists_{bd\in\mathcal{K}}\}, UNKN)$.

Let $G_m \subset \mathcal{G}$ be the set of attribute goals. An attribute goal is defined as the quality/quantity of an attribute of an *item-type* which is required by the agent. In a negotiation, an agent can negotiate for the purchase of a general type of item by negotiating across item attributes. For instance, α may negotiate with β for the purchase of a fridge. α does not know much about fridges, but knows that the fridge must be a minimum 4 star energy efficient and of a particular size. The options of fridges available which suit the attribute goals are limited by what stock β has. Precisely how α proceeds to select the fridge to purchase whilst keeping in line with the attribute goals depends on the negotiation protocol ([2] for more details).

Issue goals are goals that an agent wishes to achieve for the issues in the negotiation. The set of issue goals that an agent has is equivalent to $G_r \cup G_m$. An agent may have k issues, however the cardinality of $|G_r \cup G_m| \leq k$. Generally, an agent requires that $(\forall_{g \in G_r \cup G_m}, f \triangleright g$ is true for $j \in \delta)$ prior to an opposing trader agreeing to δ.

The sets of goals G_t, G_r, and G_m are partitions of $\mathcal{G}$. That is, $G_t \cap G_r \cap G_m = \{\varnothing\}$. Additionally, all goals identified here are *internal* to a trading agent. They are *internal* from the perspective that they represent only what the agent needs to achieve in the negotiation and directly restrict and guide the agents behaviour. An agent only knows its own goals, and makes no assumptions about what other agent's goals are. Argumentation makes goal sharing possible, in which case an agent not only knows its own set $\mathcal{G}$, but may also know some of all of another agent's goal set, e.g. $\mathcal{G}_\beta$. This is not explored here, and for now we assume the agent only knows its internal goals.

3.2   Goals and Acceptability

The internal goals that an agent has affects what an agent finds acceptable in a negotiation. Given that goals are a kind of hard preference, we integrate goals in the form of hard linear constraints upon which the maximum entropy distributions are constructed.

An agents possible deal set, $\mathcal{V}$, is *internally restricted* in some sense by goals. We define this restriction by $\mathcal{G}$ on the set of random worlds $\mathcal{W}_a$. Because $\mathcal{G}$ represents an agents own goals, the restriction on $\mathcal{W}_a$ occurs for only information that are *self valuation* information types. Here, we only consider the self valuation that α has on itself, 'Accept$_\alpha$'. Thus, α attempts to estimate the random world distribution $\mathcal{W}_{Accept_\alpha}$ given both $\mathcal{G}$ and $\mathcal{B}$ constraints.

A $g \in \mathcal{G}$ enforces a linear constraint if and only if $f \rhd g$ is an *issue-based* evaluation statement. In this model, the set $G_r \cup G_m$ satisfies this feature. The set $\mathcal{W}_a$ is internally restricted by a goal constraint for $g \in G_r \cup G_m$ of the form:

$$\forall_{v \in \mathcal{V}} \sum_n \{p_n \in \mathcal{W}_a : f \rhd g \text{ is false for } v_n\} = 0$$

That is to say, the random worlds distribution for a consistent with $G_r \cup G_m$ is the set of worlds where $\mathbf{P}_{\mathcal{W}_a}(v) = 0$. Why are the probabilities zero? In the case of goals, an agent that knows its goals does not want to violate the goals. Goals represent what an agent needs to achieve from the negotiation as a kind of minimal mark. The maximum entropy inference operates over the set of possible deals which do not violate the goals.

An example illustrates the above. α intends to purchase a seconds goods fridge from β. In this situation, α knows precisely the fridge model wanted, however does not know what price nor how many days it will take to be delivered — these are the two negotiation issues. α has a budget of \$200 which it does not want to reach, so anything under this is ok. α also knows that tomorrow is not a good day to deliver the fridge (no one will be there to receive it), but any other day is fine up until 5 days time. α represents its goals as such: $\mathcal{G}_\alpha = \{$

$(buy(fridge), \exists_{acc(\delta) \in \mathcal{K}}, \{\exists_{\delta \in \mathcal{K}} \vee \exists_{bd \in \mathcal{K}}\}, \mathit{UNKN}),$
$(budget(), (j_{cost} \in \delta, j_{cost} < \$200), \{\exists_{acc(\delta) \in \mathcal{K}} \vee \exists_{bd \in \mathcal{K}}\}, \mathit{UNKN}),$
$(daysdel(), (j_{day} \in \delta, 2 \leq j_{day} \leq 5), \{\exists_{acc(\delta) \in \mathcal{K}} \vee \exists_{bd \in \mathcal{K}}\}, \mathit{UNKN})\}.$

α attempts to estimate $\mathcal{W}_a$ for 'Accept$_\alpha$' based on the preferential relations $\forall_{x,y\in j_{price}}((x>y)\rightarrow(\mathrm{Accept}_\alpha(x,z)\rightarrow\mathrm{Accept}_\alpha(y,z)))$ and $\forall_{x,y\in j_{days}}((x>y)\rightarrow(\mathrm{Accept}_\alpha(z,x)\rightarrow\mathrm{Accept}_\alpha(z,y)))$. The probability estimates for the likelihood that an offer is the highest offer α is willing to accept, that maximises entropy given $\mathcal{G}$ constraints, looks like:

	1	2	3	4	5
\$160	0.0	0.9408	0.7056	0.4704	0.235
\$170	0.0	0.7056	0.5292	0.3529	0.1765
\$180	0.0	0.4704	0.3529	0.2359	0.1176
\$190	0.0	0.2359	0.1765	0.1176	0.0588
\$200	0.0	0.0	0.0	0.0	0.0

3.3 Opponent and Towards Goal Based Strategy

An agent aims to estimate the probabilities for the highest offer that an opposing trading agent will make. The agent then negotiates according to its estimates for its own acceptability, the opposing traders acceptability, and the goals that it knows it must achieve. For any strategy to be successful, a definition of a successful negotiation is needed.

In the negotiation an agent aims to trade successfully by the conclusion of the negotiation. In the literature for automated negotiation, what it means to be successful is defined as traditionally being the agreement of a contract which maximises utility (or for instance reaches an *pareto optimal* solution). However, utility may be unknown in real world negotiation, and imparting such information on an agent difficult. Here, we seek to define a successful negotiation in terms of the achievement of goals.

Definitions of a successful negotiation. An agent is *completely successful* in their negotiation trade iff $\forall_{g\in\mathcal{G}}, r \rhd g \equiv SUCC$. An agent has a *partial success* iff $\exists_{g,g'\in\mathcal{G}_t}, r \rhd g \equiv SUCC, r' \rhd g' \equiv FAIL$. An agent has a *compromised success* iff $\forall_{g\in\mathcal{G}_t}\exists_{g'\in G_r\cup G_m}, r \rhd g \equiv SUCC, r' \rhd g' \equiv FAIL$. An agent has a *partial and compromised success* iff $\exists_{\delta\in\mathcal{K}}\exists_{g'\in G_r\cup G_m}\exists_{g''\in\mathcal{G}_t}, r \rhd g \equiv SUCC, r' \rhd g' \equiv FAIL, r'' \rhd g'' \equiv FAIL$. An agent is *completely unsuccessful* iff $\forall_{g\in\mathcal{G}}, r \rhd g \equiv FAIL$.

A partial success is described for the situation where at least one of the primary trading goals has been satisfied. A compromised success is the case where although all of the primary trading goals are satisfied, at least one of the issue goals has not been satisfied. In this case, for this to occur, an agent must have compromised the requirement of the satisfaction of one or more goals. A partially successful and compromised negotiation outcome is one where a deal was agreed, however at least one primary trading goal was not satisfied, and at least one issue goal was not satisfied.

The agent negotiating towards a successful negotiation conclusion may do so by defining what degree of success is required. A completely successful negotiation may be important to an agent. On the other hand, a compromised success might be still be acceptable. For now, we look at the case where an agent is satisfied at the conclusion iff it is completely successful in terms of goals. We limit the agents scope to either a completely successful or completely unsuccessful negotiation.

A goal oriented strategy for an agent is one which aims to negotiate towards achieving the agents goals. A default trading goal oriented strategy is one which aims to achieve the agents issue goals (and henceforth, satisfy the default trading goal). In order for the agent to strategically reason about whether or not information views concur with its goals for the negotiation, the agent seeks to estimate the likelihood of goal success. The way this is modeled here is with *possible goal worlds.*

A possible goal world is a type of possible world where, for all goals, it is known whether they succeed or fail (for this we refer to $r \rhd g \equiv SUCC$ as just $g\top$, and $r \rhd g \equiv FAIL$ as just $g\bot$). The set of possible goal worlds $\mathcal{R}$ is the set of possible worlds where each is a permutation of a goal set. For example, if $G_r \cup G_m = \{g, g'\}$ then $\mathcal{R}_{G_r \cup G_m} = \{g\top \wedge g'\top, g\top \wedge g'\bot, g\bot \wedge g'\top, g\bot \wedge g'\bot\}$. An agent considers the possibility that any of the goal permutations might be the actual permutation.

An agent attempts to define the distribution $\mathcal{W}_a$ constrained by information of type a. The distribution is created in similar way as for α's acceptability, with an alternative assumption over the preferential relation, and no assumption on the opposing agents goals $\mathcal{G}_\beta{}^1$.

For issue goals in negotiation, the agent estimates possible goal worlds for $\mathcal{V}$. The set of worlds in $\mathcal{V}$ consistent with some $s \in \mathcal{R}_{G_r \cup G_m}$ is the set:

$$\mathcal{D}(s) = \{v \in \mathcal{V} | ((s \equiv g\top) \to f \rhd s \text{ for } v) \vee ((s \equiv g\bot) \to \neg f \rhd s \text{ for } v)\}$$

The probability that an opposing agent will accept a possible goal world is:

$$\mathbf{P}(s \in \mathcal{R}_{G_r \cup G_m}) = \sum_n \{p_n \in \mathcal{W}_a : v_n \in \mathcal{D}(s)\}$$

where $\forall_{s \in \mathcal{R}_{G_r \cup G_m}}, \sum \mathbf{P}(s) = 1$. This says that the probability of each possible goal world associated with an information view a is equal to the sum of the random world probability distribution estimates consistent with the issue goals.

For example, α estimates β's acceptability distribution $\mathcal{W}_{Accept_\beta}$ for selling a fridge where the issues are *price* and *days delivery.* α receives an offer from β to sell the fridge for \$190 with 4 days delivery. Based only on this information, the probability distribution which maximises entropy looks like:

	1	2	3	4	5
\$160	0.0563	0.0563	0.0563	0.0563	0.01
\$170	0.0563	0.0563	0.0563	0.0563	0.01
\$180	0.0563	0.0563	0.0563	0.0563	0.01
\$190	0.0563	0.0563	0.0563	0.0563	0.01
\$200	0.01	0.01	0.01	0.01	0.01

---

[1] $\mathcal{G}_\beta \neq \{\varnothing\}$ in the case of goal exchange via argumentation, discussed separately.

There exists a single belief $\mathbf{B}(Accept_\beta(4, \$190)) = 0.9$. α retains its goals $\mathcal{G}_\alpha$. Thus, there are four goal permutations: $\mathcal{R} = \{budget()\top \wedge daysdel()\top, budget()\top \wedge daysdel()\bot, budget()\bot \wedge daysdel()\top, budget()\bot \wedge daysdel()\bot\}$. Consequently, based on $\mathcal{W}_{Accept_\beta}$, α estimates the likelihood of β reaching each possible goal world $s \in \mathcal{R}$ as $\mathcal{D}_0$: $\mathbf{P}(s_1) = 0.7156, \mathbf{P}(s_2) = 0.2252, \mathbf{P}(s_3) = 0.04, \mathbf{P}(s_4) = 0.02$. The entropy is measure as $\mathbf{H}(\mathcal{D}_0) = 0.33969$.

The agent can now represent with certainty the least biased estimates on goal success that 'maximises entropy' based only on known information. Due to the size limitations of this paper, we do not consider strategy in this paper. We leave as 'future work' the development of goal based strategies, extension towards argumentation, and the evaluation of the framework.

4   Conclusion

Goals and interests are important in negotiating for e-commerce trade. In automating negotiation, by using goals and interests in negotiation it may be possible to reach faster outcomes, with a higher quality. There has been little research towards negotiating with goals and interests.

We place a goal oriented framework in terms of the information theoretical approach. The certainty of the goals are an extension of information that the agent knows at a particular point in time. The estimate of negotiation success in terms of goals is the maximally non-committal estimate with regards to missing information. Goals are aimed towards constructing a negotiation framework where interests may be shared in negotiation via argumentation.

References

1. Rahwan, I., Sonenberg, L., Dignum, F.: Towards interest-based negotiation. In: AAMAS '03: Proceedings of the second international joint conference on Autonomous agents and multiagent systems, New York, NY, USA, ACM Press (2003) 773–780
2. Jennings, N., Faratin, P., Lomuscio, A., Parsons, S., Sierra, C., Wooldridge, M.: Automated negotiation: prospects, methods and challenges. Int. J. of Group Decision and Negotiation **10** (2001) 199–215
3. Henrich, J.: Does culture matter in economic behavior? ultimatum game bargaining among the machiguenga of the peruvian amazon. American Economic Review **90** (2000) 973–979
4. Debenham, J.: Bargaining with information. In: AAMAS '04: Proceedings of the Third International Joint Conference on Autonomous Agents and Multiagent Systems, IEEE Computer Society (2004) 663–670
5. Debenham, J.: Auctions and bidding with information. In: AAMAS '04: Agent Mediated Electronic Commerce Workshop AMEC within Third International Conference on Autonomous Agents and Multi Agent Systems, IEEE Computer Society (2004) 15–28
6. Rahwan, I., Ramchurn, S.D., Jennings, N.R., McBurney, P., Parsons, S., Sonenberg, L.: Argumentation-based negotiation. The Knowledge Engineering Review 18 (4) (2003) 343–375

Application of OWA Based Classifier Fusion in Diagnosis and Treatment offering for Female Urinary Incontinence

Behzad Moshiri[1], Parisa Memar Moshrefi[2], Maryam Emami[3], and Majid Kazemian[1]

[1] Control and Intelligent Processing Center of Excellence, Electrical and Computer Eng.
Department, University of Tehran, Tehran, Iran
[2] Department of Computer Engineering, Islamic Azad University Science and Research Branch,
Tehran, Iran
[3] Urology Department, Iran University of Medical Science, Tehran, Iran
moshiri@ut.ac.ir, parisa.moshrefi@gmail.com,
am.kazemian@ece.ut.ac.ir, emami59658@yahoo.com

Abstract. Classifier fusion is a process that combines a set of outputs from multiple classifiers in order to achieve a more reliable and complete decision. In this work, the application of Ordered Weighted Averaging (OWA) operator as a classifier fusion approach, for diagnosing and offering the treatment of female urinary incontinence has been investigated. In this study, a classifier combination system has been constructed on four underlying individual classifiers, with different approaches including two multi-layer perceptrons, a generalized feed forward and a support vector machine. The system combines the decisions of these classifiers and is considered as a medical council based on only clinical patients data. Instead of choosing very accurate and expensive data sources like urodynamic, cystoscopy and voiding cystourethrogeram as paraclinical tests, we can nominate a small group of experts and use not so costly clinical measurements and then take experts' judgments and weight them by the level of expertise they have. Considering only clinical patient data which gathered from Iran urology medical center, the accuracy of OWA based classifier fusion system in diagnosis of urinary incontinence types improved 2.02%, 4.11% and 8.27% comparing the accuracy obtained by best individual underlying classifier, simple averaging and majority voting respectively.

Keywords: Urinary incontinence, Classifiers combination, OWA, Confusion matrix.

1 Introduction

Classifier fusion systems are considered in decision level information fusion. Data may be combined, or fused, at a variety of levels like raw data level, feature level and decision level. In combining classifiers instead of fusing raw data, the final outputs of the classifiers should be combined. The advantage of such classifier fusion system is that each individual underlying classifier may represent different useful aspect of the problem, or of the input data, while none of them represent all useful aspects [1]. Less

A. Sattar and B.H. Kang (Eds.): AI 2006, LNAI 4304, pp. 433–442, 2006.

© Springer-Verlag Berlin Heidelberg 2006

costly information is the other important advantage of such system where it could utilize less accurate sources.

Urinary incontinence is uncontrolled leakage of urine causing hygienic and social problems. This is one of the most common chronic medical conditions seen in primary care practice. Incontinence is an expensive problem, generating more costs each year. Women have higher rates of urinary incontinence than men. Prevalence increases with age; one third of women older than 65 years have some degree of incontinence, and 12 percent have daily incontinence [16]. Evaluation of the disease includes clinical and paraclinical tests [18]. The clinical tests consist of patient history, physical and manual measurements which are relatively cheap but the paraclinical tests including some examinations such as urodynamic, cystoscopy and voiding cystourethrogeram (VCUG) are very expensive.

With clinical assessments, women could be categorized into four groups of diagnostic requirements: incontinence on physical activity (stress incontinence), incontinence with mixed symptom, incontinence with urgency and incontinence with complex history [11]. Complex history such as infection, radical pelvic surgery, and fistula are not considered in our work. The differential diagnosis of three other groups of female urinary incontinence is a quite difficult for physician because of the patient's reporting symptoms which are not always present truly and also because of the overlapping class boundaries. Consequently, prescribing appropriate treatment can be difficult. A variety of methods have been used to classify a patient on the basis of patient information [3-5].

In this study, we classify different types of female incontinence and select treatment options based on only clinical patient data. In most patients with stress incontinence, surgery is the final way of treatment. The request of the patient for the way of treatment is important and this is due to the inconvenience that incontinence makes for the patient. In mixed type, regarding to the severity of stress, treatment has been led to surgery or medication. The urge incontinence can be treated with medication.

As individual underlying classifiers, we utilize four neural networks which fed with processed clinical patient data. Classifiers' expertises in diagnosis were calculated via their confusion matrices. The accuracy of the classifier fusion system with OWA has been explored and compared to some other combining methods like simple averaging and majority voting. The results demonstrate a considerable improvement of disease diagnosis, even using not expensive clinical patient data. The rest of this study is organized as follows: Section 2 describes the combining classifiers. The mathematics of optimistic and pessimistic OWA is described in section 3. Different sources of information and also the data pre-processing phase are described in section 4. Section 5 demonstrates the architecture of classifier fusion system. The experimental results are shown in section 6 and finally the conclusion is drawn in section 7.

2 Combining of Classifiers

Recently, deriving more enriched decision from multiple sources using different methods of evidence combination has been widely investigated and applied in statistic, management, science, pattern recognition and bioinformatics [6].

Each classifier simply maps the input feature vector X to c-dimensional output vector, where c is the number of class labels. The classifier output could be arranged in a vector with supports to each class as follows:

$$D_i(x) = [d_{i,1}(x), ..., d_{i,c}(x)]^T \tag{1}$$

Without loss of generality, we could restrict all $d_{i,j}(x)$ between 0 and 1. If a crisp class label of x is needed, we can use the maximum membership rule as argument of maximum value in vector $D_i(x)$:

$$class = ArgMax(D_i(x)) \tag{2}$$

In our case there are four class labels including: stress-surgery, mixed-surgery, mixed-medication and urge-medication, which are labeled as c_1, c_2, c_3 and c_4 respectively.

In the context of classifier fusion with L classifiers, which aims at a higher accuracy than each individual member, the output of classifiers could be set into the decision profile matrix which represents the soft labels. Here, we utilized four underlying classifiers so the decision profile is configured as:

Output of classifier $D_2(x)$

$$DP(x) = \begin{bmatrix} d_{1,1}(x) & d_{1,2}(x) & d_{1,3}(x) & d_{1,4}(x) \\ d_{2,1}(x) & d_{2,2}(x) & d_{2,3}(x) & d_{2,4}(x) \\ d_{3,1}(x) & d_{3,2}(x) & d_{3,3}(x) & d_{3,4}(x) \\ d_{4,1}(x) & d_{4,2}(x) & d_{4,3}(x) & d_{4,4}(x) \end{bmatrix} \tag{3}$$

Support from classifiers $D_1,...,D_4$ for class c_3

Here, we applied OWA aggregation operator for fusion of class supports obtained by underlying classifiers. This fusion operator just uses the jth column of DP(x), regardless of what the support for the other classes is. These types of methods are known as class-conscious methods [20].

3 Ordered Weighted Averaging (OWA)

An OWA operator was first introduced by Yager [8], as a universal operator which is, in some reasonable sense, the best to use [9]. The OWA of dimension c is a mapping $F: \mathfrak{R}^c \to \mathfrak{R}$, and provides a means for aggregating objects support from classifiers $D_1, D_2, ..., D_L$ for class c_j. This operator for combining classifiers works as follows[20]:

> 1. Find vector $W = [w_1, w_2, ..., w_L]$ considering the following constraint:
>
> $$\sum_{i=1}^{L} w_i = 1, \quad w_i \in [0,1]$$
>
> 2. For $k = 1,...,c$
>
>   a) $\underset{i=1,...,L}{Sort} \; d_{i,k}(x)$ in descending order, so that:
>
> $$a_1 = \max d_{i,k}(x), \text{ and } a_L = \min d_{i,k}(x)$$
>
>   b) Calculate the support for class c_k
>
> $$d_k(x) = \sum_{i=1}^{L} w_i a_i$$

What is considerable in OWA operator is the ordering step of elements. Particular w_i is not associated with a_i but rather associated with the particular ordered position of the arguments.

OWA is a very flexible operator in which by selecting appropriate weights, a variety of arithmetic operators could be extracted. For example when the first weight in W is set to one and others to "zero" then OWA acts as a pure "or", vice versa when the last weight is "one" while others are zero then OWA operates as "and". It is proved that the aggregated done by an OWA operator is always between maximum and the minimum. A degree of orness associated with OWA with weigh vector W was defined as [10]:

$$orness(W) = \frac{1}{n-1} \sum_{i=1}^{n} (n-i)w_i . \tag{4}$$

To obtain weightening vector, a class of OWA named exponential OWA operators is used which generates a set of OWA weights satisfying a given degree of orness. For solving this problem there exist a simple relationship between the orness and the elements of weight vector of OWA [10]:

- Optimistic weights:

$$w_1 = \alpha \; ; w_2 = \alpha(1-\alpha); w_3 = \alpha(1-\alpha)^2; ...; w_{n-1} = \alpha(1-\alpha)^{n-2}; w_n = (1-\alpha)^{n-1} . \tag{5}$$

- Pessimistic weights:

$$w_1 = \alpha^{n-1}; w_2 = (1-\alpha)\alpha^{n-2}, w_3 = (1-\alpha)\alpha^{n-3}; w_{n-1} = (1-\alpha)\alpha; w_n = (1-\alpha) \tag{6}$$

Where parameter α belong to the unit interval, $0 \le \alpha \le 1$ and is related to *orness* value and number of data sources [10].

4   Data Sources to Be Used

For evaluating and offering the treatment of the disease, some different diagnostic clinical parameters including subjective and objective data were gathered from several data sources. The data sources consist of symptoms, which are the history of the patients, and signs, which are qualitative or quantitative measurements of physical examinations [18].

Subjective data consist of irritating and obstructive symptoms. These are some questions about having frequency, dysuria, urgency, urge symptom, stress symptom, nocturnal, incomplete evacuation, force and calibre, hesitancy and post void

Table 1. Clinical parameters for evaluation of common types of urinary incontinence

Parameter			Value	
Subjective	Irritating	Frequency	Yes, No	
		Dysuria	Yes, No	
		Urgency	Yes, No	
		Urge Symptom	Yes, No	
		Nocturia	Yes, No	
		Stress Symptom	Yes, No	
	Obstructive	Incomplete evacuation	Yes, No	
		Force	Yes, No	
		Hesitancy	Yes, No	
		Post void dribbling	Yes, No	
Age			Numerical	Young (< 55), old (≥ 55)
Body mass index			Numerical	Thin (< 18.5),Normal (18.5-25), Fat (25-40), Very fat (≥40)
Pregnancy	Cesarean		Numerical	
	Normal, vaginal delivery		Numerical	
Past medical History(PMH)	Discopathy		Yes, No	
	Radical hysterecty		Yes, No	
	Ap repair		Yes, No	
	Diabetes mellitus		Yes, No	
Medication			-	
objective	Neurological exam	Perineal sensation	Normal, Abnormal	
		Bulbocavernous reflex	Normal, Abnormal	
	Vaginal exam	Rectocel	Yes, No	
		Entrocel	Yes, No	
		Uterine prolapsed	Yes, No	
		Cystocel	Yes, No	
		Cystocel grade	Numerical	Weak (1-2) Strong (3-4)

dribbling. The age of patient causes the loss of estrogens and neurological changes and also reduces the elasticity. Pregnancy and childbirth affects hormones, causing the pressure of uterus and contents and mechanical disruption of muscles and sphincters. Body mass index is included in the parameters and it is calculated considering the weight and height of the patient [17].

$$\text{Body Mass Index (BMI)} = \text{weight} / (\text{height}^2). \tag{7}$$

Objective data is obtained from general examinations such as observation of urine loss with cough, neurological and vaginal exams including pelvic and rectal exams.

The diagnostic parameters, gathered from different sources of information, are mixed. There are both binary and continuous parameters. Continuous variables become discrete asking the experts (Table 1).

For evaluating the disease, we find out that more than one parameter might have the same behavior in the system. We simplify our problem by replacing a group of parameters with a single new parameter. For reducing the parameters, we have consulted with the physicians. Table 2 shows the reduced parameters for evaluating the disease. For example parameters: Incomplete evacuation, force, hesitancy and post void dribbling is replaced by new parameter: 'difficulty with voiding'.

Table 2. Reduced parameters for evaluation of urinary incontinence with clinical data

Parameter			Value	
Subjective	Irritating	Frequency	Yes, No	
		Dysuria	Yes, No	
		Urgency	Yes, No	
		Urge Symptom	Mild, Severe	
		Nocturia	Yes, No	
		Stress Symptom	Mild, Severe	
	Obstructive	Difficulty with voiding	Yes, No	No(Incomplete evacuation =No & Force=No & Hesitancy=No & Post void dribbling=No) Yes(otherwise)
Age			Numerical	Young (< 55), old (≥ 55)
Objective	Physical exam	Prolapse	Yes, No	No (Rectocel=No & Entrocel =No & Uterine prolapsed=No & Cystocel=weak) Yes(otherwise)
		Post void residue	Numerical	Normal (0–100ml), Abnormal (>100ml)

Data used in this study was gathered from Iran urology medical center (Hasheminejad hospital). Data of 120 patients was used for training the system and the 48 remaining cases for testing its performance. Table 3 shows the frequency of each type in both training and testing datasets.

The patients' data, like the other real-world information, are incomplete. There are a lot of missing data. Several approaches have been explored to act with missing data [12]. As all of parameters could be considered in yes or no category, so we simply replaced yes with 1, no with -1, and missing data with 0.

Table 3. Statistics of training and testing datasets

Dataset \ Diagnosis class	Stress- medication surgery	Mixed – surgery	Mixed – medication	Urge – medication
Train	61	39	11	9
Test	20	16	9	3

5   Underlying Classifiers

Neural networks are the best known types of classifiers. Selecting the appropriate weights for a neural network is an optimization problem. Combining multiple decisions generated by different networks improves the final decision. The main idea is to develop L independent neural networks, separately train them with relevant features, classify a given input pattern by each of them, and then apply an appropriate fusion method to derive the final class [13].

As individual underlying classifiers, we explored different neural networks and investigated their results and finally choose four neural networks with best exclusive results. Figure 1 shows the architecture of our classifier fusion system.

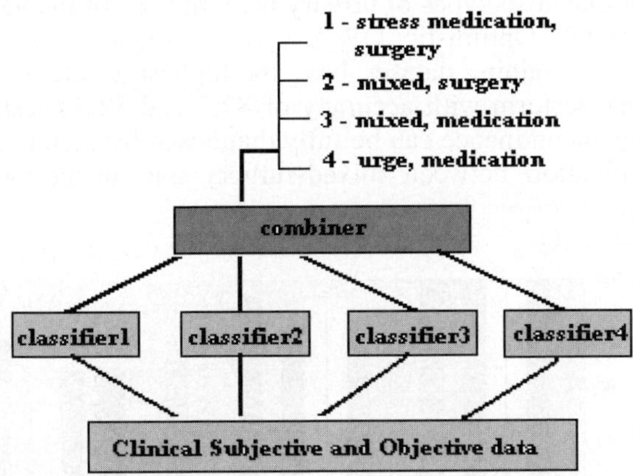

Fig. 1. The classifier fusion system architecture for diagnosis of urinary incontinence

Selected Networks as classifiers in this study are: (1) A Multi Layer Perceptron (MLP) with 3 hidden layer and 3 neurons in each layer, (2) MLP with 1 hidden layer and 6 neurons in its hidden layer, (3) A general feed forward network with 6 neurons on its hidden layer and (4) a support vector machine.

The results of the classifiers were summarized in the form of the confusion matrix. A confusion matrix is a matrix in which the actual class under test is represented by the matrix rows, and the class estimated by the classifiers represented by the matrix columns [14]. Table 4 exemplifies the confusion matrix of General Feed Forward classifier.

Table 4. Confusion matrix of General Feed Forward classifier in diagnosis of urinary incontinence types

Predicted class Actual Class	stress-medication or surgery	Mixed-surgery	mixed-medication	urge-medication
stress- medication or surgery	20	0	0	0
mixed-surgery	1	15	0	0
mixed-medication	1	3	5	0
urge-medication	0	0	0	3

The element M[i][j] repreosents the number of times that class i was assigned to class j, and indicates the correct classification. If the matrix is not normalized the sum of rows i is total number of cases in class i and the sum of elements of column j is the total num of predicted cases in class j.

6 Experimental Results

Figure 2 represents the accuracy of the individual classifiers as well as different fusion methods for diagnosis types of urinary incontinence of the gathered cases from Hasheminejad hospital. Optimistic OWA and pessimistic OWA operators, with optimized alpha in the training dataset, have the highest accuracy. Simple averaging and majority voting perform with accuracy of 87.5 and 83.34 respectively. Patients with stress and urge incontinence can be fully diagnosed by each utilized combination method. Misclassification between mixed-surgery and mixed-medication has the

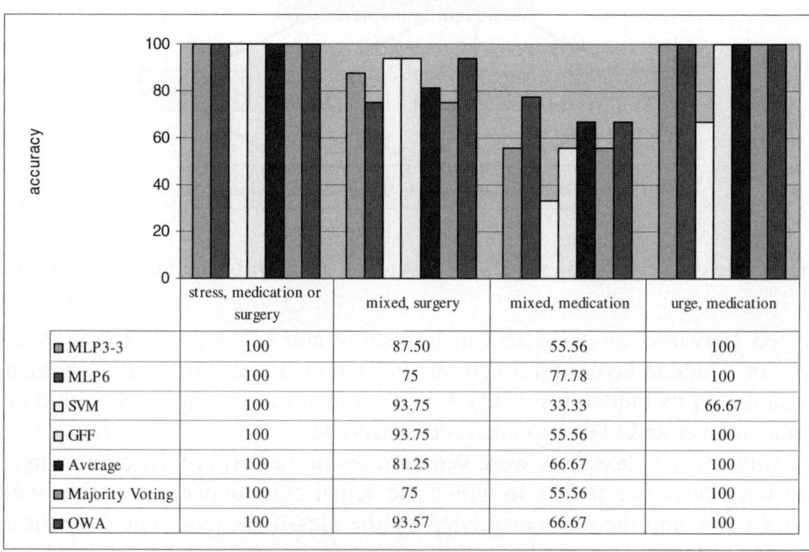

	stress, medication or surgery	mixed, surgery	mixed, medication	urge, medication
MLP3-3	100	87.50	55.56	100
MLP6	100	75	77.78	100
SVM	100	93.75	33.33	66.67
GFF	100	93.75	55.56	100
Average	100	81.25	66.67	100
Majority Voting	100	75	55.56	100
OWA	100	93.57	66.67	100

Fig. 2. Accuracy of individual underlying classifiers and Fusion Methods in diagnosis of female urinary incontinence types

highest rate. Discrimination between prescribing medication and performing surgical operation on a patient with mixed incontinence is difficult. In some cases physician intended to perform surgical operation on patient with mixed type of incontinence while they are not sure.

The most complex class, mixed-medication, can be classified with OWA and simple averaging methods better than majority voting. Although this class can be diagnosed with one hidden layer MLP (second classifier) better than the other used approaches, but the overall accuracy of OWA in diagnosis is the highest among the underlying individual classifiers and other fusion methods. The accuracy of OWA considering only clinical patient data reaches to 91.61%. This accuracy is higher than the total accuracy of each individual classifier: MLP3-3, MLP6, SVM and GFF which are 87.5, 87.5, 83.33 and 89.58 respectively.

It is clear that the more the numbers of the cases in each class are, the better the accuracy of that class becomes. In our work, the number of patients in mixed-medication group was almost restricted.

7 Conclusions and Future Research

Referring to every year growing cost of evaluation and treatment of urinary incontinence, using the proposed classifier system could be considerably useful. We achieved to noticeable diagnosis accuracy, using only clinical patient data. The system does not need expensive parameters evaluation procedure for making decision about the type of incontinence. Further more, we can utilize this system in medical centers which are not equipped with expensive paraclinical instruments.

To further improve the system, using individual classifiers with more accuracy and diversity could reduce the overall rate of misclassification. We are also exploring the system with paraclinical patient data such as urodynamic measurements on the basis of classifiers combination. Closing the eyes to the cost of the urodynamic test, where using it is inevitable, it is clear that the results will improve. As a further research, a supplementary system could be designed and added to urodynamic software for interpreting the measured signals during the test and supporting the decision about the type of the disease.

References

1. T.H. Ho, J. J. Hull, S.N. Srihari, "Decision Combination in Multiple Classifier System", IEEE Transaction on pattern Analysis and Machine Intelligence, vol. 16, pt. 1, pp. 66-75, 1994.
2. M.A. Abidi, R.C. Gonzalez, "Data fusion in robotics and machine intelligence", ISBN: 0120421208, Publisher: Academic Pr, 1992.
3. J. Laurikkala, M. Juhola, "Learning diagnostic rules from a urological database using a genetic algorithm". In: Alander JT, eds. Proceedings of the third Nordic workshop on genetic Algorithms and their applications. Helsinki: Finnish Artificial Intelligence Society, pp. 233-44, 1997.
4. J. Laurikkala, M. Juhola, "A genetic-based machine learning system to discover the diagnostic rules for female urinary incontinence", Computer Methods and Program in Biomedicine 55, pp. 217-228, 1998.

5. L. Breiman, JH. Friedman, RA. Olshen, CJ. Stone, "Classification and regression Tree", Belmont: Wadsworth Internatinal Group, 1984.
6. L. XU, A. Krzyzak, C.Y. Suen, "Methods of combining Multiple classifier and their applications to handwriting recognition", IEEE Trans. SMC, vol. 22, No. 3, pp. 418-435, 1992.
7. D. Ruta, B. Gabrys, "An Overview of classifier fusion methods", Computing and information system, 7, pp.1-10, 2000.
8. R. R. Yager, "On ordered weighted averaging aggregation operators in multi-criteria decision making", IEEE Transaction on System, Man and Cybernetics 18, pp.183-190, 1988.
9. R.R. Yager, V. Kreinovich, "Main Idea behind OWA Lead to Universal and Optimal Approximation Scheme", pp. 428- 433, 2002.
10. D. P. Filev, R. R. Yager, "Learning OWA operator weights from data", Proceedings of the Third IEEE International Conference on Fuzzy Systems, Orlando, pp. 468-473, 1994.
11. J.W. Thueroff, P. Abrams, W. Aartibani, F. Haab, S. Khoury, H. Madersbacher, R. Nijman, P. Norton, "Clinical Guidelines for the Management of Incontinence", health publications Ltd., Plymounth, 1999, pp. 933-943
12. J. Laurikala, M. Juhola, S. Lammi, J. Penttinen, P. Aukee, "Analysis of the imputed female urinary incontinence data for the evaluation of expert system parameters", Computers in biology and medicine 31, pp. 239-257, 2001.
13. S. Cho, J. H. Kim, "Combining Multiple Neural Networks by Fuzzy Integral for Robust Classification", IEEE Transaction on Systems, Man and Cybernetics, vol. 25, No. 2, 1995.
14. J. R. Parker, "Rank and response combination from confusion matrix data", Information Fusion 2(2), pp. 113-120, 2001.
15. M. Kazemian, B. Moshiri, H. Nikbakht, C. Lucas, "Protein Secondary Structure Classifiers Fusion using OWA", Lecture Notes in Computer Science, LNCS Springer-Verlag Berlin Heidelberg 3745 - pp. 338 -345, 2005.
16. D. W. Barry, "Selecting Medications for the Treatment of Urinary Incontinence", American Family Physician, Volume 71, Number 2, 2005.
17. G. Herold und Mitarbeiter, "Innere Medizine: Eine vorlesungsorientierte Darstellung", 2003
18. S. Raz, "Female Urology", ISBN: 0-7216-6723-6, Second Edition, Publisher: W.B. Saunders Company, 1996.
19. L. I. Kuncheva, "Combining Classifiers: Soft Computing Solutions", In S.K. Pal and A. Pal, editors, Pattern Recognition, From Classical to Modern Approaches, chapter 15, pp. 427-452. World Scientific, 2001.

Automatic Generation of Funny Cartoons Diary for Everyday Mobile Life

Injee Song, Myung-Chul Jung, and Sung-Bae Cho

Computer Science Department, Yonsei University
134 Shinchon-dong, Sudaemoon-ku, Seoul 120-749, South Korea
{schunya, mcjung, sbcho}@sclab.yonsei.ac.kr

Abstract. The notable developments in pervasive and wireless technology enable us to collect enormous sensor data from each individual. With context-aware technologies, these data can be summarized into context data which support each individual's reflection process of one's own memory and communication process between the individuals. To improve reflection and communication, this paper proposes an automatic cartoon generation method for fun. Cartoon is a suitable medium for the reflection and the communication of one's own memory, especially for the emotional part. By considering the fun when generating cartoons, the advantage of the cartoon can be boosted. For the funnier cartoon, diversity and consistency are considered during the cartoon generation. For the automated generation of diverse and consistent cartoon, context data which represent the user's behavioral and mental status are exploited. From these context information and predefined user profile, the similarity between context and cartoon image is calculated. The cartoon image with high similarity is selected to be merged into cartoon cuts. Selected cartoon cuts are arranged with the constraints for the consistency of cartoon story. To evaluate the diversity and consistency of the proposed method, several operational examples are employed.

1   Introduction

Personal computing and World Wide Web fundamentally change each individual's communication environments at home and the office. People in the world present their thought to a large number of anonymous people by posting articles, images and videos on their personal web page and web log. Pervasive computing environments will not only change and enhance our office environment, but also change our communication pattern in the relationship with family members, friends and business partners in the everyday life. Through the sensors such as GPS, visual and audio sensors in the pervasive computing environment, we can collect log data from each individual's daily life. These logged sensor data can be employed to infer user's behavior, emotion, and surrounding by applying context-aware techniques [1]. Such inferred context data are useful in summarizing an individual's daily life. Like a personal diary, summarized context is helpful in reflecting one's own memory and in strengthening the communication process. Even though multimedia representation of the context like images and videos could be more ambiguous than simple text, multimedia is

A. Sattar and B.H. Kang (Eds.): AI 2006, LNAI 4304, pp. 443–452, 2006.
© Springer-Verlag Berlin Heidelberg 2006

more effective to represent and share impressions about personal experiences [2]. Particularly, cartoon can effectively express personal emotion. Especially, funny cartoon enhances the impression of personal experiences. To make funnier contents, contents designers make various patterns in the contents while considering users' cognitive acceptability [3].

For the automated generation of funny cartoon, this paper focuses on the diversity of cartoon images and the consistency of cartoon story. Various patterns in the cartoon can be made by diverse cartoon images, and the readers' cognitive acceptance is considered using consistency constraints. For the automatic cartoon generation, we compose character and background images from the user profile and landmark context events. To find appropriate character and background images to the landmark events, semantic similarities between landmarks and images are computed. After composing, generated cartoon cuts are reorganized using consistency constraints to keep the consistency of cartoon story. The formal description of the problem is the following. Given a tuple L of n landmark events, a cartoon story CS_k is a tuple $CS_k = \{CC_{k1}, CC_{k2}, ..., CC_{kn}\}$ where

- $L = \{event_1, event_2, ...,event_n\}$ is a tuple of landmark events in which $event_x$ is the xth event that is extracted from daily context data;
- $CC_{km} = \{CartoonImage_{km1}, CartoonImage_{km2}, ..., CartoonImage_{kmh}\}$ is a tuple of cartoon images that describe the cartoon cut which is the most feasible to the mth event and consists of h different types of cartoon images, i.e., character image, background image, etc.

2   Related Works

With the development of pervasive and mobile technology, many context aware researches are going on. R. Miikkulainen proposed the episodic memory that categorizes stories and recovers script-based memories by hierarchal SOM (Self-organizing Feature Maps) [4]. E. Horvitz made a Bayesian network representation of human cognitive behavior [5]. He also designed a learning model that found out critical events from online calendar software. Together with context-aware researches, many researchers have been studying about how to represent these context data effectively. B. Viégas of MIT media laboratory created two visualizations from email archives. One highlighted social networks and the other depicted the temporal rhythms of interactions with individuals [6]. ATR institute in Japan developed ComicDiary [7] as subsystem of C-MAP project which was visitor guidance system for exhibition. ComicDiary organizes scenario using story stream method and composes the cartoon image and text using template. Because our research focuses on the diversity of cartoon composition, methods using predefined templates will generate only limited set of cartoons.

When we compose character and background images with user's context data, there is a process that finds relevant images for context data. For this process, we can utilize image retrieval researches [8]. Generally, there are two kinds of retrieval methods. One is QBK (query-by-keywords) and the other is QBE (query-by-example). When we compose a cartoon image, QBK approach is more applicable. M. Inoue suggested image retrieval methods for lightly annotated images by using image similarities [9]. In the retrieval process, annotations in the images can be used as query keywords. In the sense that QBK requires to compute semantic similarities between keywords and the images, it can be compared to compose various web services with user's service request. To compose web services, Cardoso matches service object (SO) and service template (ST) in the web service searching process [10]. SO defines the real web service component that system wants to find, and ST defines the functions of web service that user requests. Matching process consists of two steps. At first, system looks for SO that is the most similar to ST using structural and functional similarities. After that, to match actual functions of SO and ST, semantic similarity between functions of SO and ST is utilized. Semantic similarity is computed by employing the numerical formula similar to Jaccard coefficient like in the following equation.

$$S_{ij} = \frac{p}{p+q+r} .$$
(1)

S_{ij} = similarities between object i and object j
p = number of attributes that object i and object j all have
q = number of attributes that object i has, but object j does not
r = number of attributes that object j has, but object i does not

3   Cartoon Generation Method

Entire process of cartoon generation is shown in Fig. 1. It selects the character and background images in the similarity of the landmark events from context-aware module with user profile. In selection process, semantic similarity between images and landmark event is calculated based on the predefined annotations for the images. After selecting the cartoon cuts for each context data, they are organized into a story stream to make a plausible cartoon story.

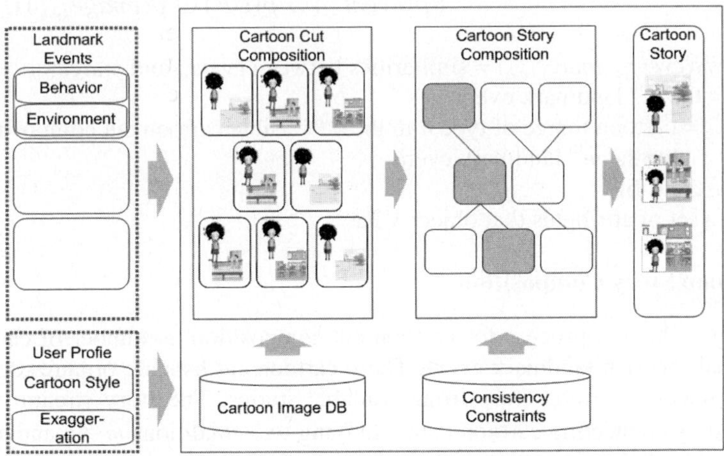

Fig. 1. Cartoon generation process

3.1 Landmark Events and User Profile

For the automatic composition of cartoon images, landmark events and user profile are used as input data. A landmark event is a set $event_x = (\textit{<Behavior>}, \textit{<Environment>})$ is the xth landmark event in chronological order which describes the important event happened during the day consists of behavioral and environmental landmarks. Behavioral landmark $\textit{<Behavior>}$ is a set of keywords that describe user's action and emotional status. Environmental landmark $\textit{<Environment>}$ is a set of keywords that describe the place where the event happens. User profile is a set of values that reflect the user's preference about the cartoon image. For example, preferred character style and preference about exaggerated cartoon image can be used as attribute.

3.2 Cartoon Cut Composition

A cartoon cut $CC_{km} = \{\textit{CharacterImage}_{km},\ \textit{BackgroundImage}_{km}\}$ is a tuple of cartoon images that describe the cartoon cut which is the kth feasible cartoon composition to the mth landmark event. To compose diverse cartoon cuts, we match landmark events and images in a flexible way. After defining images with the keywords from the same ontology which define landmark events, we match cartoon image and landmark event by the semantic similarity between them. By means of semantic similarity, we cannot only acquire various cartoon cuts from limited images but also easily expand the annotation of landmark events and cartoon images. At first, semantic similarity between the cartoon images and landmark events are calculated using the equation (2). From the images whose similarities are higher than certain threshold, the candidate list of cartoon cuts is generated. The threshold for image selection is defined in consideration of cartoon diversity and computational cost. Lower threshold will compose more diverse cartoon expression, but will require more computational cost for the consistency constraint violation check.

$$Similarity\,(event_m,\, image_{kmh}) = \frac{|\,(p(event_m) \cup p(UP)) \cap p(image_{kmh})\,|}{|\,p(event_m) \cup p(UP) \cup p(image_{kmh})\,|} \tag{2}$$

$Similarity(event_m,\, image_{kmh})$ = similarities between $event_m$ and $image_{kmh}$
$event_m$ = the mth landmark event
$image_{kmh}$ = cartoon image of type h in the kth feasible cartoon cut composition
         for the mth landmark event
UP = user profile
$p(X)$ = a set of attributes that object X has

3.3 Cartoon Story Composition

After image selection process for cartoon cut composition, a number of cartoon cuts are acquired for each landmark event. These cartoon cut lists are organized into story stream to select a cartoon story from available stories. The story stream is a graph constructed by connecting cartoon cuts satisfying two conditions $m \neq n$ and a tuple $\{x, y\} \notin \textit{ConsitencyConstraints}$ where

- CC_{km} is the kth feasible cartoon cut composition for the mth landmark event;
- CC_{jn} is the jth feasible cartoon cut composition for the nth landmark event;
- *ConsistencyConstraints* is a set of tuples that consist of the preceding relationship between behaviors or emotions. For the preceding relationship in consistency constraints, the consistency of the character's emotion and the character image style are considered. For an example, ("Very happy", "Very sad");
- $K(X) = p(image_{X1}) \cup p(image_{X2}) \cup ... \cup p(image_{Xl})$ is a union of keyword sets that describe images ($image_{X1}$, $image_{X2}$,..., $image_{Xl}$) in cartoon cut X;
- $x \in K(CC_{km})$ and $y \in K(CC_{jn})$.

After constructing cartoon story stream, available cartoon stories are searched through the story stream graph using the process described in Fig. 2. One cartoon story among the candidates is randomly selected to be presented as the final cartoon.

```
CartoonCutList, two dimensional array of cartoon cuts
ConsistencyConstraints, the constraints defines the
preceding relationship between two behaviors and emo-
tions
ComicStripList, a list of available cartoon stories

Func Init()
  For C1 in CartoonCutList[1]
    Scenario.add(C1)
    Search(C1, Scenario, 1)
    Scenario.remove(C1)

Func Search(CartoonCut Cn, Scenario, Level)
  Scenario.add(Cn)
  IF (Level is CartoonCutList.Length)
    ComicStripList.add(Scenario)
    Scenario.remove(Cn)
    return
  For dc in CartoonCutList[Level+1]
    Bn ← BehaviorOf(Cn)
    Bn+1 ← BehaviorOf(Cn+1)
    If (ConsistencyConstraints.has(Bn,Bn+1) is false)
      Search(Cn+1, Scenario, Level+1)
  Scenario.remove(Cn)
  return
```

Fig. 2. Cartoon story search process

4 Operational Examples

4.1 Scenario Description

Because the fun of the composed cartoon story can be evaluated only in subjective measure, we evaluate the fun of the cartoon by evaluating the diversity and the consistency of the cartoon. To evaluate the generated cartoon story, we create sample

Table 1. Landmark context examples

ID	Behavioral landmark	Environmental landmark
1	Movement, MP3, Stand	Bus, Indoor
2	Study, Take a class, Sit	University, Classroom, Indoor
3	Eat, Korean Food, Sit	University, Dinning Room, Indoor
4	Talk, Phone, Stand	University, Outdoor
5	Cheer, Watch, Stand	Stage, Outdoor
6	Drink, Beer, Sit, Cheer	Drinking House, Bar, Indoor

Table 2. Category of keywords to describe landmark events and cartoon images

Category of keywords			Examples (Number of total keywords)
Behavior	Action	Basic	Stand, Sit, Phone, See, Kiss, …(16)
		Movement	Work, Run, Car, Bus, Subway(5)
		At home	Cook, Bath, Shower, Makeup, …(12)
		In work	Study, Paperwork, Presentation, …(11)
		Entertainment	Dance, Sing, Game, …(8)
		Sports	Tennis, Golf, Ski, Swim, Yoga, …(9)
		Others	Umbrella, Shopping back, …(7)
	Emotion		Happy, Sad, Angry, Surprised, …(30)
Environment	Place	Indoors	Classroom, Cafeteria, Office, … (45)
		Midtown	Sidewalk, Shopping quarters, … (21)
		Nature	Woods, Mountain, Beach, …(38)
		Imaginary	Heaven, Hell, Space, …(35)
	Weather		Sunny, Cloudy, Rainy, …(21)
	Time	Season	Spring, Summer, Fall, Winter (4)
		Day/Night	Day, Night (2)
User ence	Prefer-	Cartoon Style	Oriental, Western (2)
		Exaggeration	Exaggerated, Non-exaggerated (2)

scenario landmark events described in Table 1. Description of landmark events and cartoon images are annotated using the keywords in Table 2. For each landmark, similarities between cartoon character and background images are computed with user profile using the cartoon cut composition method. Computed similarities between a landmark event $event_4$ and some of cartoon character images are shown in Fig. 3. The cartoon images with the similarity over 50% are selected to be composed into cartoon cuts. These cartoon cuts are arranged into story stream graph as depicted in Fig. 4. Using story stream graph, final cartoon story is composed with cartoon cuts which do not violate the consistency constraints as shown in Table 3. With composed cartoon

Landmark Event		
ID	**Behavior**	**Environment**
4	Talk, Phone, Stand	University, Outdoor

User Profile	
Character Style	**Exaggeration Preference**
Western	Exaggeration

		Character Image Annotation			
ID	**Behavior**	**Style**	**Exaggeration**	**Sim.**	**Image**
1000	Talk, Phone, Stand	Oriental	Non-exaggerated	42.8%	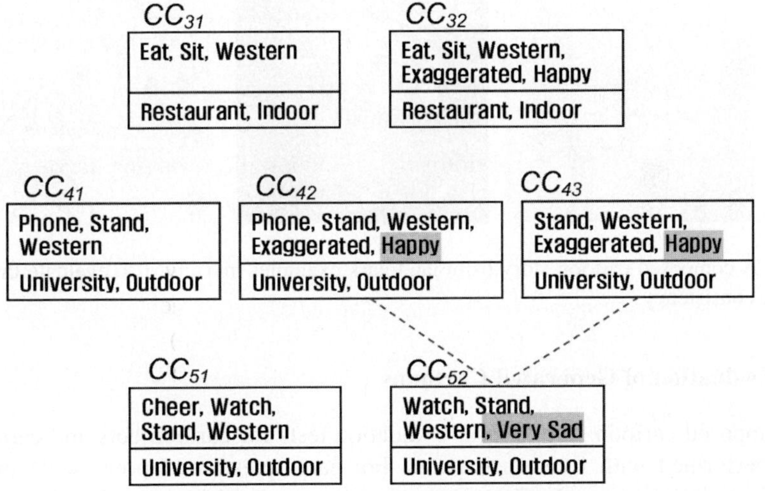
1001	Talk, Phone, Stand, Mad	Oriental	Exaggerated	50%	
1100	Talk, Phone, Stand	Western	Non-exaggerated	57%	
1101	Talk, Phone, Stand, Happy	Western	Exaggerated	62.5%	
2000	Inspect, See	Oriental	Non-exaggerated	0%	

Fig. 3. Semantic similarities between character images and the landmark event (ID =4)

CC_{31}

Eat, Sit, Western
Restaurant, Indoor

CC_{32}

Eat, Sit, Western, Exaggerated, Happy
Restaurant, Indoor

CC_{41}

Phone, Stand, Western
University, Outdoor

CC_{42}

Phone, Stand, Western, Exaggerated, Happy
University, Outdoor

CC_{43}

Stand, Western, Exaggerated, Happy
University, Outdoor

CC_{51}

Cheer, Watch, Stand, Western
University, Outdoor

CC_{52}

Watch, Stand, Western, Very Sad
University, Outdoor

Fig. 4. A part of story streams generated by a landmark context example. Each box represents a cartoon cut which composed of annotations of the character image and the background image. Dashed lines represent the relation which violates the consistency constraints.

story, image representation of cartoon story is generated. Fig. 5 shows one of the composed cartoon story images. To evaluate the method, four composed cartoon story images were shown to participants.

Table 3. A part of consistency constraints

ID	Preceding Behavior	Irrelevant Following Behavior
1	Very Happy	Very Sad
2	Happy	Very Sad
3	Very Sad	Very Happy
4	Very Sad	Happy
5	Sad	Very Happy

Fig. 5. A composed cartoon story from landmark examples in Table 1 to evaluate the diversity and the consistency

4.2 Evaluation of Generated Cartoons

By composed cartoon stories, user evaluation tests for cartoon cuts and cartoon story were performed with five participants. For each question, 5-point scale measure is used to grade an item. At first, cartoon cuts generated by a landmark example are evaluated in the criteria of the diversity and the descriptiveness. Fig. 6 shows the evaluation results. After cartoon cut evaluation, diversity and consistency of four cartoon stories were asked to the participants. The diversity of cartoon stories results in 4.0 of average and 1.0 of standard deviation, and the consistency of cartoon stories is depicted in Fig. 7. These data are not enough for the statistical significance, but the average score for each criterion shows affirmative tendencies.

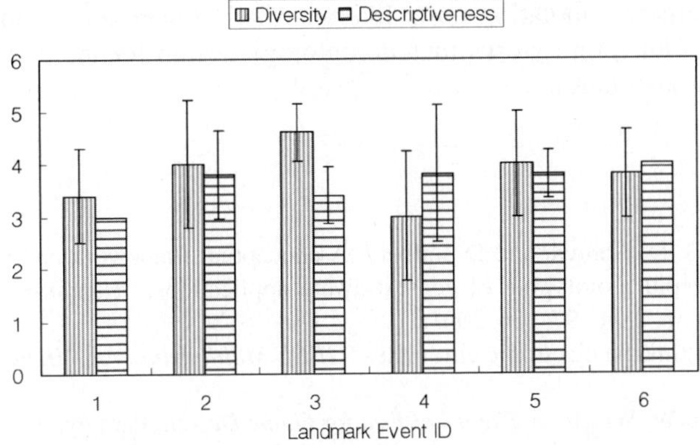

Fig. 6. Diversity and descriptiveness of cartoon cuts for each landmark event (n=5)

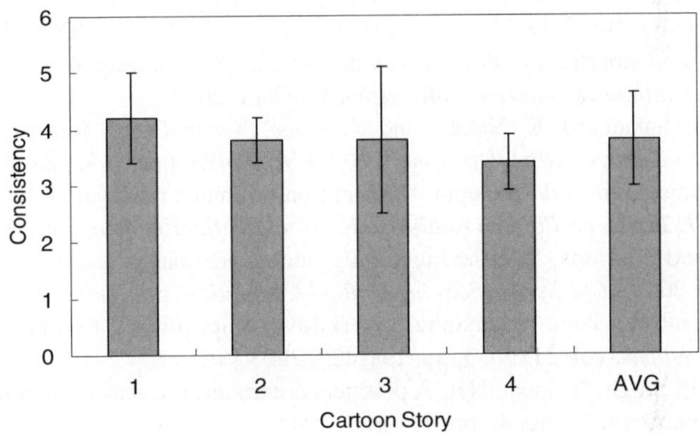

Fig. 7. Consistency Scores of four generated cartoon stories (n=5)

5   Concluding Remarks

To strengthen the effectiveness of the communication and the reflection of everyday mobile life, we generate funny cartoons with landmark events. From the landmark events and the user profile, we generate the cartoon cuts by applying semantic similarity between the context events and the images. Theses cartoon cuts are arranged into a cartoon story by using consistency constraints. By these methods, we can acquire a variety of funny cartoon stories with consistency. Because we use rather simple composition method, it can be smoothly used to compose the cartoons even in the limited mobile environments. But in this method, to exactly describe cartoon images, too much annotation costs are required to define cartoon images with a set of keywords. To decrease the annotation cost, cartoon image should be described with less keywords. By utilizing ontology like ConceptNet [11] or LifeNet [12], the annotation cost can be decreased. The approach with ontology can be also useful in selecting appropriate cartoon images and cartoon stories with landmark events. More fundamental approach to decrease annotation costs like automatic image annotation [13] will be more plausible. Our future works include ontology design for the cartoon image and landmark events annotation.

References

1. A. K. Dey, G. D. Abowd, and D. Salber, "A conceptual framework and a toolkit for supporting the rapid prototyping of context-aware applications," *Human-Computer Interaction*, vol. 16, no. 2, pp. 97-166, 2001.
2. R. Jain, "Multimedia electronic chronicles," *IEEE Multimedia*, vol. 10, no. 3, pp. 111-112, 2003.
3. R. Koster and W. Wright, *A Theory of Fun for Game Design*, Paraglyph Press, 2004.
4. R. Miikkulainen, "Script recognition with hierarchical feature maps," *Connection Science*, vol. 2, pp. 83-101, 1990.
5. E. Horvitz, S. Dumais, and P. Koch, "Learning predictive models of memory landmarks," *CogSci 2004: 26th Annual Meeting of the Cognitive Science Society*, 2004.
6. F. B. Viégas, D. Boyd, D. H. Nguyen, J. Potter, and J. Donath, "Digital artifacts for remembering and storytelling: Post history and social network fragments," *Proc. of the 37th Hawaii International Conference on System Sciences*, 2004.
7. Y. Sumi, R. Sakamoto, K. Nakao, and K. Mase, "ComicDiary: Representing individual experience in a comic style," *UbiComp 2002, LNCS 2498*, pp. 16-32, 2002.
8. A. W. M. Smeulders and A. Gupta, "Content-based image retrieval at the end of the early years," *IEEE Trans. on Pattern Analysis and Machine Intelligence*, vol. 22, no. 12, 2000.
9. M. Inoue and N. Ueda, "Retrieving lightly annotated images using image similarities," *Proc. of the 2005 ACM Symposium on Applied Computing*, pp. 1031-1037, 2005.
10. J. Cardoso and A. Sheth, "Semantic e-workflow composition," *Journal of Intelligent Information Systems*, vol. 21, no. 3, pp. 191-225, 2003.
11. H. Liu and P. Singh, "ConceptNet: A practical commonsense reasoning tool-kit," *BT Technology Journal*, vol. 22, no. 4, pp. 211-226, 2004.
12. P. Singh and W. Williams, "LifeNet: A propositional model of ordinary human activity," *Distributed and Collaborative Knowledge Capture Workshop*, 2003.
13. K. Barnard, P. Duygulu, D. Forsyth, N. de Freitas, D. M. Blei, and M. I. Jordan, "Matching words and pictures," *Journal of Machine Learning Research*, vol. 3, pp. 1107-1135, 2003.

Welfare Interface Using Multiple Facial Features Tracking

Yunhee Shin and Eun Yi Kim*

Department of Internet and Multimedia Engineering, Konkuk Univ., Korea
(ninharsa, eykim)@konkuk.ac.kr

Abstract. We propose a welfare interface using multiple facial features tracking, which can efficiently implement various mouse operations. The proposed system consist of five modules: face detection, eye detection, mouth detection, facial features tracking, and mouse control. The facial region is first obtained using skin-color model and connected-component analysis (CCs). Thereafter the eye regions are localized using neural network (NN)-based texture classifier that discriminates the facial region into eye class and non-eye class, and then mouth region is localized using edge detector. Once eye and mouth regions are localized, they are continuously and correctly tracking by mean-shift algorithm and template matching, respectively. Based on the tracking results, mouse operations such as movement or click are implemented. To assess the validity of the proposed system, it was applied to the interface system for web browser and was tested on a group of 25 users. The results show that our system have the accuracy of 99% and process more than 12 frames/sec on PC for the 320×240 size input image, as such it can supply a user-friendly and convenient access to a computer in real-time operation.

1 Introduction

During the last decades, considerable interest has been shown in video-based interfaces that use human gestures to convey information or to control devices and applications [1, 2]. Users can create gestures by a static hand or body pose, or by a physical motion, including eye blinks or head movements. Among these, facial features movements, such as a gaze direction, eye blinks, and mouth opening/closing, have a high communicative value that allows for the extraction of human intention or orders. Moreover these interfaces can support the people who can not use the keyboard or mouse due to severe disabilities, Due to these, such an interface has gained many attractions, so far many systems have been developed [3-6].

Most of them use eye movements or mouth movements to control some devices. The systems using eye movements were developed in [3, 4]. These systems can supply the mouse movement and click event through moving the eye and gazing a specific area during one more seconds. As well as eye movements, the systems proposed in [5, 6] were developed using mouth movement. These systems were applied as musical controller, digital painting and fatigue detection of driver by detecting various mouth movements.

---

* Corresponding author. Tel.: +82-2-450-4135; Fax: +82-2-450-4135

A. Sattar and B.H. Kang (Eds.): AI 2006, LNAI 4304, pp. 453–462, 2006.

© Springer-Verlag Berlin Heidelberg 2006

However, the above mentioned systems are difficult to operate user's various intentions in the systems. For instance, in case of only using eye movement, we can easily catch user's detail intentions such as moving cursor all the time with no difficulty in display, but such systems need the waiting time to select the menu. On the contrary, it is difficult to present user' detail intentions using mouth movement, while it is possible to imply various meaning using mouth shape such as mouth opening/closing. As a result, the systems using one more facial feature should be developed to support various operations in real-time.

In this paper, we develop the PC-based HCI system using multiple facial features such as mouth and eye, then the major advantage of our system is to provide accurate features tracking and to express user's various intentions. Our system consists of 5 modules, as shown in Fig. 1. It receives and displays a live video of the user sitting in front of the computer. The user moves the mouse pointer by moving his (her) eyes, and clicks icons and menus by opening and closing his (her) mouth.

In our system, the facial region is first obtained using skin-color model and connected-component analysis. Thereafter the eye regions are localized using NN-based texture classifier that discriminates the facial region into eye class and non-eye class, and then mouth regions localized using edge detector. Once the eye and mouth regions are localized, they are continuously and correctly tracking by mean-shift algorithm and template matching, respectively. This enables us to accurately detect user's facial feature regions even if they put on the glasses in the cluttered background. Once facial feature regions are detected in the first frame, the respective features are continuously tracked by a mean-shift and template matching. Based on the tracking results, mouse operations such as movement or click are implemented.

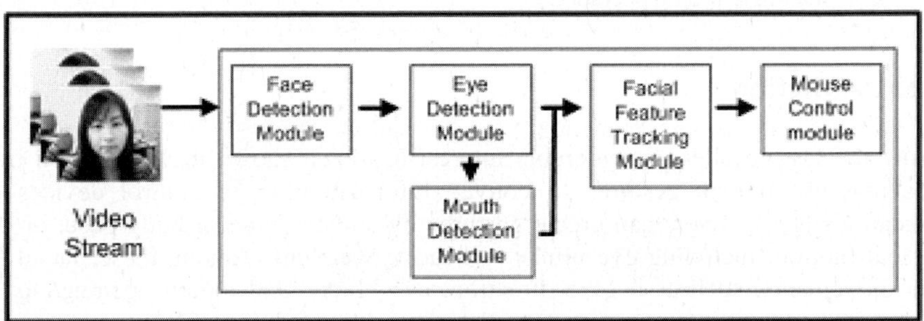

Fig. 1. The overall configuration of the system

To evaluate the proposed method, the system is used as the interface of web browser to convey the user's command via his (her) eye and mouth movements. The proposed system is tested on 25 peoples, then the results shows that our system has the accuracy of 99% and confirms that our system is robust to the time-varying illumination and less sensitive to the specula reflection of eyeglasses. Also, the results show that it can be efficiently and effectively used as the interface to provide a user-friendly and convenient communication device.

This paper is structured as follows. We describe face detection in section 2, eye detection in section 3 and mouth detection in Section 4. And then section 5 describes facial features tracking in section 5. Based on the tracking results, mouse operations implemented in section 6. Section 7 contains results of using this system, and then section 8 concludes with general observations.

2   Face Detection Module

Detecting pixels with a skin-color offers a reliable method for detecting face part. In this work, the skin color is represented in the chromatic color space, where it can be represented by 2D Gaussian distribution, $G(m, \Sigma^2)$, as follows.

$$m = (\bar{r}, \bar{g}) , \quad \bar{r} = \frac{1}{N}\sum_{i=1}^{N} r_i , \quad \bar{g} = \frac{1}{N}\sum_{i=1}^{N} g_i , \quad \Sigma = \begin{bmatrix} \sigma_{rr} & \sigma_{rg} \\ \sigma_{gr} & \sigma_{gg} \end{bmatrix} \tag{1}$$

where $\bar{r}$ and $\bar{g}$ represent Gaussian means of r and g color distribution respectively, and Σ^2 represent covariance matrix of each distribution. These parameters were provided in our previous work [7].

Once the skin-color model is created, the most straightforward way to locate the face is to match the skin-color model with the input image to identify facial regions. As such, each pixel in the original image is converted into chromatic color space, and then compared with the distribution of the skin-color model. By thresholding the matched results, we obtain a binary image (see Fig. 2(b)). For the binary image, the connected-component analysis is performed to remove the noise and small region. The generated connected-components (CC) are filtered by their attributes, such as size, area, and location. Then the facial regions are obtained by selecting the largest components with skin-colors, which is shown rectangular box in Fig. 2(b).

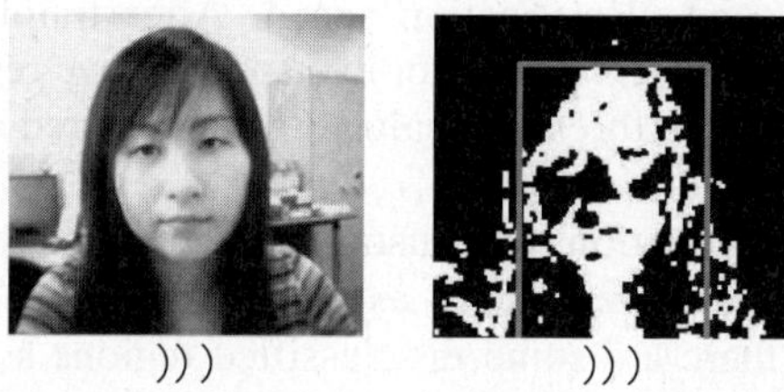

Fig. 2. Face detection results. (a) original color images, (b) extracted facial regions.(the facial region is surrounded by a rectangle)

3   Eye Detection Module

After detecting face regions, we try to detect eye locations. Generally, the eye region has the following properties: 1) it has the high brightness contrast between white eye sclera and dark iris and pupil, along the texture of the eyelid; 2) it has place in the upper of the facial region. These properties help reduce the complexity of the problem, and facilitate the discrimination between the eye regions from the whole face.

Here we use an MLP to automatically discriminate between eye regions and non-eye ones in various environments. The input layer of the network has 57 nodes, the hidden layer has 33 nodes, and the output layer has 1 node. Adjacent layers are fully connected. Fig. 3 describes the architecture of the MLP-based texture classifier.

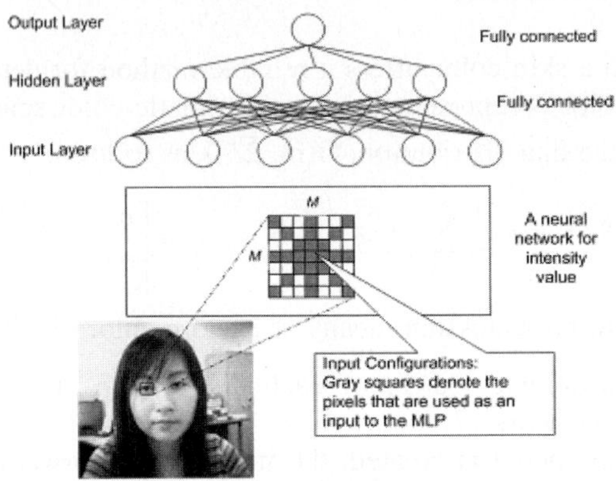

Fig. 3. The architecture of the MLP-based texture classifier

We assume that the eye has a different texture from the facial region and is detectable. The simplest way to characterize the variability in a texture pattern is by noting the gray-level values of the raw pixels. However, since the main disadvantage is that the size of the feature vector is large, we use a configuration from autoregressive features (only the shaded pixels from the input window in Fig. 3) [8]. This reduces the size of the feature vector from M^2 to ($4M$-3), thereby resulting in an improved generalization performance and classification speed. After training, the neural network outputs the real value between 0 and 1 for eye and non-eye respectively. If a pixel has a larger value than the given threshold value, it is considered as an eye; otherwise it is labeled as non-eye.

Fig. 4 shows the classification result using the neural network. In Fig. 4(a), the pixels to be classified as eyes are marked as blue. We can see that all of the eyes are labeled correctly, but there are some misclassified regions as eye. To eliminate false alarms, we use the CCs result posterior to the texture classification. We have then applied three-stage filtering on the CCs:

Stage 1: Heuristics using features of CCs such as area, fill factor, and horizontal and vertical extents; The width of the eye region must be larger than the predefined minimum width (Min_width), the area of the component should be larger than Min_area and smaller than Max_area, and the fill factor should be larger than Min_fillfactor.

Stage 2: Heuristics using the geometric alignment of eye components; we check the rotation angles of two nearby eye candidates. They have to be smaller than the pre-defined rotation angle.

Using these two heuristics, the classified images are filtered, then the resulting image is shown in Fig. 4(b). In Fig. 4(b), the extracted eye region is filled green for the better viewing.

(a)   (b)

Fig. 4. An example of eye detection. (a) the classified image by the neural network, (b) the detected eye region after post-processing.

4 Mouth Detection Module

In this section, we try to detect mouth region. To reduce the complexity of mouth detection, it is detected based on the position of eye regions. Then two following properties are used: 1) it has place within distance between two eye positions, and in the lower facial region than the eye region; 2) it has high contrast in comparison with surroundings. Thus, mouth region is localized using the edge detector within the search region estimated using several heuristic rules based on eye position.

The search window for mouth detection is shown in Fig. 5. Let denote D_H and D_V be the distance between eyes and distance between search window and the center of two eyes, respectively. Then, the search window, $S=(S_H, S_W)$ is defined as follows: the width of search window, S_W is equal to the D_H, and the height of search window, S_H is experimentally determined to the 40px.

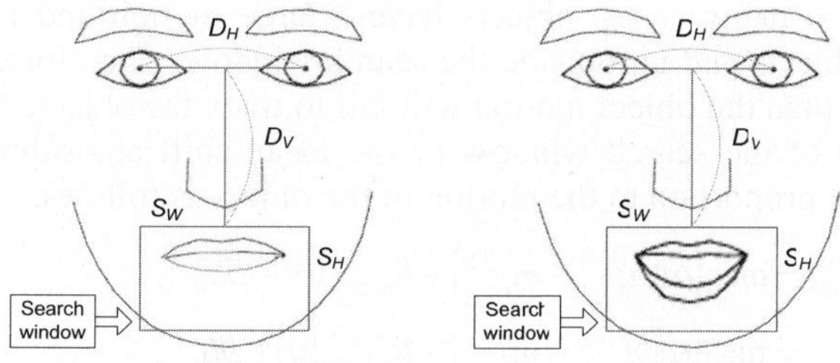

Fig. 5. Search window to detect mouth region

The detection result within estimated region includes narrow edges and noises. For discrimination of these noises, we use CCs result posterior to the facial boundary. Through this post-processing, the resulting image is shown in Fig. 6. Facial boundary was detected as well as real mouth region by edge detector, as can be seen in Fig. 6(a) – 6(c). Then the final mouth region after post-processing is shown in Figs. 6(b) - 6(d).

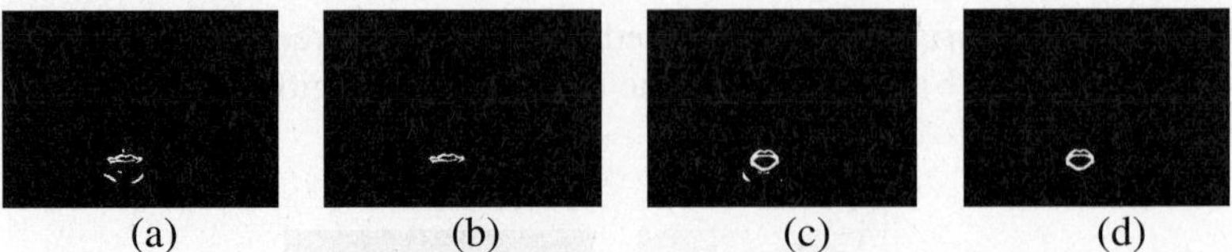

(a)   (b)   (c)   (d)

Fig. 6. An example of mouth detection

5 Facial Features Tracking Module

5.1 Mean-Shift Based Eye Tracking

To track the detected eyes, a mean shift algorithm is used, which finds the object by seeking the mode of the object score distribution [9]. In the present work, the color distribution of detected pupil, $P_m(g_s)=-(2\pi\sigma)^{-1/2}\exp\{(g_s-\mu)^2\sigma^1)\}$, where the μ and σ are set to 40 and 4 respectively. The distribution is used as the object score distribution at site s, which represents the probability of belonging to an eye.

A mean shift algorithm is a nonparametric technique that climbs the gradient of a probability distribution to find the nearest dominant mode. The algorithm iteratively shifts the center of the search window to the weighted mean until the difference between the means of successive iterations is less than a threshold. The search window is moved in the direction of the object by shifting the center of the window to the weighted mean.

The weighted mean, i.e. the search window center at iteration $n+1$, m_{n+1} is computed using the following equation,

$$m_{n+1} = \sum_{s\in W} P_m(g_s)\cdot s \Big/ \sum_{s\in W} P_m(g_s) \tag{2}$$

The search window size for a mean shift algorithm is generally determined according to the object size, which is efficient when tracking an object with only a small motion. However, in many cases, objects have a large motion and low frame rate, which means the objects end up outside the search window. Therefore, a search window that is smaller than the object motion will fail to track the object. Accordingly, in this paper, the size of the search window of the mean shift algorithm is adaptively determined in direct proportion to the motion of the object as follows:

$$W_{width}^{(t)} = \max\left(\alpha\left(\left|m_x^{(t-1)} - m_x^{(t-2)}\right| - B_{width)}\right),0\right) + \beta B_{width}$$
$$W_{height}^{(t)} = \max\left(\alpha\left(\left|m_x^{(t-1)} - m_x^{(t-2)}\right| - B_{height)}\right),0\right) + \beta B_{height} \quad (t>2) \tag{3}$$

where α and β are constant and t is the frame index. This adaptation of the window size allows for accurate tracking of highly active objects.

5.2 Template Matching Based Mouth Tracking

When the mouse is detected in the first frame, it is used as a template to determine the position of the mouse in the next frame [10]. Using this template, the tracking module searches user's mouse in a search window, where is centered at the position of the feature in the previous frame. The template is iteratively shifted within this search window and compared with the underlying sub-images.

The result of the comparison of a template and a sub-image is represented by a matching score. The matching score between a component and a template is calculated using Hamming distance. The Hamming distance between two binary strings is defined as the number of digits in which they differ. We calculate the matching score for all templates and pick the templates with the best matching score.

Fig. 7 shows the results of the facial features tracking, where the facial features are drawn out white for the better viewing. As can be seen in Fig. 7, the facial features are accurately tracking.

Fig. 7. Tracking result of facial features

6   Mouse Control Module

The computer translates the user's eye movements into the mouse movements by processing the images received from the PC camera. Then the processing of a video sequence is performed by the proposed facial features tracking system. The system determines the center of eyes in the first frame as the initial coordinates of mouse, and then computes it automatically in subsequent frames. The coordinates of mouth region and eye regions in each frame are sent to the operating system through Window functions. Mouse point is moved according to eye movement. If a user opens one's mouth, we consider that a user intends to click menus or icons on the corresponding screen region.

7   Experimental Results

To show the effectiveness of the proposed method, it was used as the interface of web browser to convey users' command via an eye and mouth movements. Fig. 8 shows the interface using our method, which consists of a PC camera and a computer. The PC camera, which is connected to the computer through the USB port, supplies 30 color images of size 320×240 per second. The computer is a PentiumIV−3GHz with the Window XP operating system.

Fig. 8. The interface system using our facial features tracking method

To assess the effectiveness of the proposed method, the interface system was tested on 25-users under the various environments. Then, the tracking results are shown in Fig. 9. The extracted facial features have been drawn white for better viewing. The features are tracked throughout the 100 frames and not lost once.

Fig. 9. Tracking result

To quantitatively evaluate the performance of the interface system, it was tested through web browsers. Each user was given an introduction to how the system worked and then allowed to practice moving the cursor for five minutes. The five-minute practice period was perceived as sufficient training time by the users. In the web browser, each user was asked to click the item (ex. Icon, menu, etc.) thirty times with the mouse using eyes and mouth movement. For each test, the user's time to click the item was recorded.

Table 1. Processing time

Stage	Time (*ms*)
Face Detection	60
Eye Detection	110
Mouth Detection	47
Tracking	62
Average Time	78

Table 1 shows the time taken to detect the face and facial features, and track them. In the system, the detections of facial features and face are once executed for the initial frame and only tracking is executed for the subsequent frame. Therefore, the average time to process a frame is about 78 *ms*, and then our system process more than 12 frames/sec.

Table 2. Test result

	Click	Non-Clicked
When the user intends to click the menu	747 (99%)	3 (1%)
Otherwise	0 (0%)	750 (100%)

Table 2 presents the accuracy to be taken to click the item in the web browser. We can see that the system is mostly operated the mouse operation correctly.

Table 3. Timing comparison between Eye Mouse [4] and proposed method

Method	Measure	Time/sec
Eye Mouse [4]	Mean	0.66s
	Deviation	0.02s
Proposed method	Mean	0.49s
	Deviation	0.06s

Table 3 presents the average time to be taken to click the item in the web browser, when playing with Eye Mouse [4] and the proposed method. The latter is about 1.5-times faster than the former and can reduce tiredness and inconvenience.

Consequently, the proposed system can supply a user-friendly and convenient access to a computer in real-time operation. Since it does not need any additional hardware except a general PC and an inexpensive PC camera, the proposed system is very efficient to realize many applications using real-time interactive information between users and computer systems.

8 Conclusions

In this paper, we implemented the welfare interface using multiple facial features detection and tracking as PC-based HCI system. The proposed system has worked very well in a test database and cluttered environments. It is tested with 25 numbers of people, and then the result shows that our system has the accuracy of 99% and is robust to the time-varying illumination and less sensitive to the specula reflection of eyeglasses. And the experiment with proposed interface showed that it has a potential to be used as generalized user interface in many applications, i.e., game systems and home entertainments.

Acknowledgments. This work was supported by Korea Research Foundation Grant (KRF-2004-041-D00643).

References

1. Sharma, R., Pavlovic, V.I., Huang, T.S.: Toward multimodal human-computer interface. Proceedings of the IEEE , volume 86 , pp. 853 – 869 (1998)
2. Scassellati, Brian.: Eye finding via face detection for a foveated, active vision system. American Association for Artificial Intelligence. (1998)
3. Anthony Hornof, Anna Cavender, and Rob Hoselton.: EyeDraw: A System for Drawing Pictures with Eye Movements. ACM SIGACCESS Accessibility and Computing, Issue 77-78 (2003)
4. Eun Yi Kim, Sin Kuk Kang.: Eye Tracking using Neural Network and Mean-shift. LNCS, volume 3982, pp. 1200-1209 (2006)
5. Michael J. Lyons.: Facial Gesture Interfaces for Expression and Communication. IEEE International Conference on Systems, Man, Cybernetics, volume 1, pp. 598-603. (2004)
6. Yang Jie, Yin DaQuan, Wan WeiNa, Xu XiaoXia, Wan Hui.: Real-time detecting system of the driver's fatigue. ICCE International Conference on Consumer Electronics, 2006 Digest of Technical Papers. pp. 233 - 234 (2006)
7. Schiele, Bernet., Waibel, Alex.: Gaze Tracking Based on Face-Color. School of Computer Science. Carnegie Mello University (1995)
8. Chan a. d. c., Englehart K., Hudgins B., and Lovely D. F.: Hidden markov model classification of myoeletric signals in speech. IEEE Engineering in Medicine and Biology Magazine, volume 21, no. 4, pp. 143-146 (2002)
9. D. Comaniciu, V. Ramesh, and P. Meer.: Kernel-Based Object Tracking. IEEE Trans. Pattern Analysis and Machine Intelligence, volume 25, no. 5, pp. 564-577 (2003)
10. C.F. Olson.: Maximum-Likelihood Template Matching. Proc. IEEE Conf. Computer Vision and Pattern Recognition, volume 2, pp. 52-57 (2000)

Towards an Efficient Implementation of a Video-Based Gesture Interface

Jong-Seung Park, Jong-Hyun Yoon, and Chungkyue Kim

Department of Computer Science & Engineering, University of Incheon,
177 Dohwa-dong, Nam-gu, Incheon, 402-749, Republic of Korea
{jong, jhyoon, ckkim}@incheon.ac.kr

Abstract. Human-computer interactions in augmented reality games are generally based on human gestures. For each input video frame captured from a live video camera, image analysis technologies are used to infer the human intension. The development of augmented reality user interfaces is a difficult task due to the instability of the gesture analysis. The user interfaces cannot be efficiently developed with traditional development techniques. In this paper, we investigate an effective development methodology for gesture-based augmented reality interfaces by means of three different approaches. The implementation requires a real-time tracking of bare hands or real rackets to allow fast movements and interactions without delay. We also verify the applicability of the prototyping mechanism by implementing and demonstrating an augmented reality game played with either bare hands or real rackets.

1 Introduction

Nowadays, many 3D games are readily accessible and they are played widely. Some 3D games are extremely realistic and sophisticated. Recently, several augmented reality (AR) games have appeared, which provide immersive 3D gaming environments [1,2,3]. Testbeds for AR tabletop gaming have also been proposed to explore the possibilities of realistic and interactive game interface [4]. In recent years, vision-based gestural interfaces have been proposed [5,6]. However, in the field of gesture analysis and recognition, it is very difficult to understand hand gestures, mainly due to the large degrees of freedom (DoF) of the hand configuration. Most of the previous work has been focused on the recognition of static hand postures. The fusion of the dynamic characteristics of gestures has only recently been taken much interest [7].

Recently, a use of AR technologies in computer gaming has been reported in various sports computer game genres. One of the most strong points of AR games is that games built in AR can blend physical aspects and mental aspects to enhance existing classical game styles [8]. A virtual tennis game was developed on a virtual environment connecting an egocentric head mounted display system with an exocentric fishtank system [2]. Their system uses a head mounted display and a motion tracker without using sports instruments. A tennis game is a kind of a paddle game which is a game involving the use of a paddle. Some paddle

A. Sattar and B.H. Kang (Eds.): AI 2006, LNAI 4304, pp. 463–472, 2006.
© Springer-Verlag Berlin Heidelberg 2006

game applications use real balls and real rackets [3][9]. On the other hand, some applications use real rackets without a real ball [10].

Real world sports games are strongly physical and players are free to use their whole physical bodies. AR games could also be physical as real world games. Moreover, the same content can be injected seamlessly into the real world [8]. However, current AR devices are indisposed and are not ready to such AR games. Physical interactions are significantly limited by input devices, such as paddle trackers in table tennis games.

For the physical interactions at such gaming environments, the eventual goal is to use means of human-to-human interactions. The purpose of the natural interface is to develop a conversational interface based on natural human-to-human interactions. Gestures are typical natural expressions and they could be integrated with other expressions, such as voice, body movements, and facial expressions. In the gestural interface, hand poses and specific gestures are used as commands. Naturalness of the interface requires that any gesture should be interpretable [7]. In a vision-based gesture interface, it is required to discriminate gestures given purposefully from unintended gestures and to use the intended gestures as instructions to the interface system. Computer vision algorithms for interactive graphics applications need to be robust and fast. They should be reliable, work for different people, and work against unpredictable backgrounds.

Unfortunately, the current vision-based gesture interfaces do not provide a satisfactory solution due to the complexity of the gesture analysis. Vision-based gesture interfaces have suffered from enormous variety of environments. It is hard to deal with uncontrolled light conditions, complex background of users, and movements of human hands based on articulation and deformation. These reasons increase difficulty both in the hand-tracking step and in the gesture recognition step. A compromise would be possible using some sensory devices, using simple shaped markers, using marked gloves, using simple backgrounds, or restricting the poses of hands to frontal directions. Those approaches might not be appropriate for the gesture interface since they restrict natural human gestures or environments. For the practical implementation, one should determine the level of knowledge that should be extracted from the image analysis. The knowledge extraction is difficult and time consuming, and it makes the gesture interface not applicable to game applications.

An important property of AR applications is that the application context restricts the possible visual interpretations and the user exploits the immediate visual feedback to adjust their gestures.Such consideration makes the problem much easier. Moreover, the high-level knowledge is seldom required for most sports games. The minimum level of knowledge could be extracted by a simple and fast method not discarding the naturalness of the interface. These cues motivate us to investigate a development methodology of a versatile gesture interface for AR games. Our goal is to develop gesture-based natural interfaces that are fast and stable without any strict restrictions on human motion or environments.

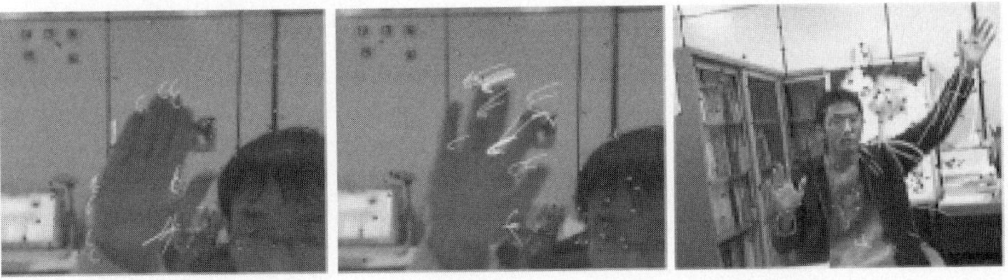

Fig. 1. Feature point tracking for gesture recognition

Fig. 2. Identifying hand gestures by filtering skin color regions and tracking feature points

In the remainder of this paper we first describe the interaction method with bare hands in section 2. We then describe the interaction method with a real paddle in section 3. Finally we show experimental results in section 4 and provide conclusions in section 5.

2   Interactions with Bare Hands

A considerable amount of research has been conducted on computer-related gesture-recognition technology for various reasons. For example, a gesture interface would speedup the player's reaction time in playing games. It can also be used to help physically disabled persons to interact with computers [11]. Though there is a lot of potential uses in niche applications, the gesture recognition technology faces several critical problems. As an interface-related issue, there are many different hand motions of making the same gesture and there is not a common rule in interpreting the detected hand motions. The robust

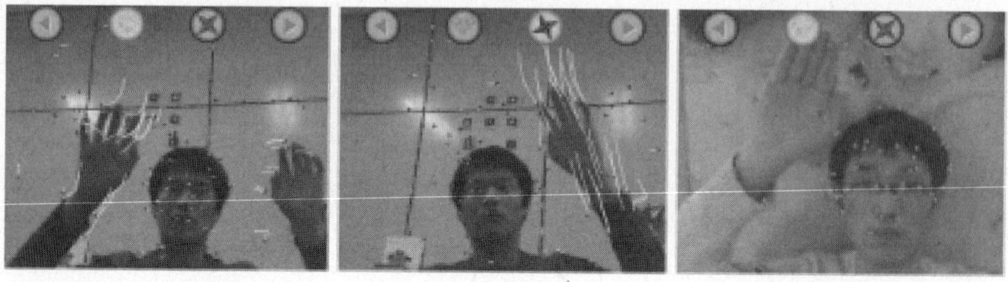

Fig. 3. Hand gestures to click virtual menu buttons

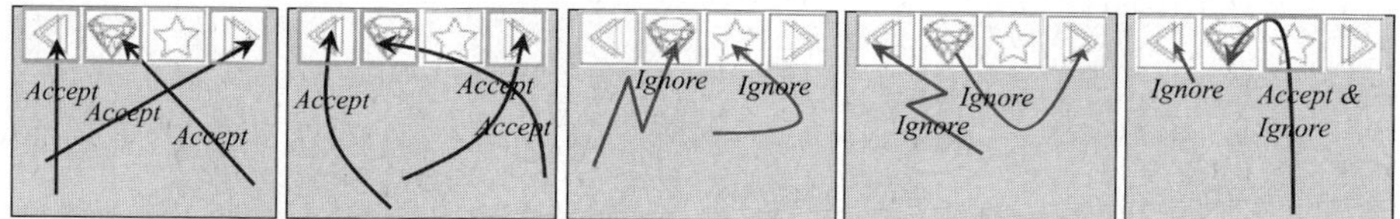

Fig. 4. Identifying intended gestures for menu selection

recognition of hand motions in an arbitrary environment is nearly impossible since there are too many varying factors to be considered such as varying illumination conditions, confusing backgrounds, and irregular speed of hand motions.

Feature point tracking is a typical approach for many motion recognition problems. First, many good features are extracted and tracked as shown in Fig. 1. Then, the point trajectories are analyzed to infer the motion information. There are huge amount of ambiguity in the inference and it is hard to assure the recognition result is correct.

A promising way to reduce the ambiguity is to use the color information. The feature points in skin regions are more important than other feature points. We could put more weights on skin points in hand gesture recognition. The hand motion can be inferred from the consistent movements skin points. Fig. 2 shows feature tracking and skin region detection for the hand gesture recognition.

We have investigated a fast hand motion detection and gesture classification method. Great emphasis has been placed on the processing speed and the stability of the responses. The screen shots of our interface prototype is shown in Fig. 3. From the tracked point features, motions of moving objects are analyzed. The moving patterns of the motion trajectories are used to determine whether the motion is intended gestures. We first filter skin color regions and determine the candidate image locations of hands. Then, we detect and track the skin regions. The intended hand motion is inferred based on the trajectories of the moving regions in a short time interval.

For the proximity touch to a menu icon, we locate skin color regions and analyze gestures occurred in the regions. The hue-saturation-value(HSV) model is used to intuitively specify skin color properties. *Hue* is a 360 degree value through color families and *saturation* is the degree of strength of a color. The important property of hue is the invariancy to the various lighting environment.

For the values r, g, and b in the RGB color space, each range from 0 to 1, hue h ranges from 0 to 360° and saturation s ranges from 0 to 1. For general digital color images, r, g, and b have discrete values range from 0 to 255 to store each color value in one byte. The HSV color space also should be quantized to fit into the byte range.

The shape of the skin color subspace in the HSV space can be estimated by successive adjustments according to skin regions of test images[12]. Based on the statistics, we filter skin color regions. We only use h and s to eliminate the lighting effects. However, under the very dark illumination, the skin color filtering is unstable and we exclude the case when the intensity is too small. The skin color subspace is bounded by the following constraint:

$$(h \in [0, 22] \text{ or } h \in [175, 180]) \text{ and } s \in [58, 173] \text{ and } v \in [76, 255]$$

The skin color boundary also works well for different human races and different lighting conditions.

We determine the proximity of hand parts to action positions and infer the intended user click action from the moving direction of hand parts. We detect enough number of feature points and track them. Fig. 2 shows skin color regions in the green color and feature points in the red color. It is common that some feature points disappear due to the point matching failure. When a feature point is failed in matching, a new one is detected and added to the feature point set. The trajectories of all feature points are kept during a fixed time interval. When one of the skin color regions overlaps with one of action regions, the trajectories of feature points inside the skin color regions are inspected. If the trajectory shape agrees the user intended menu selection, the corresponding action is activated. Unintended gestures have inconsistently moving trajectories such as an abrupt change of moving direction, small motion trajectories where the starting positions are near an action region, and confusing trajectories crossing multiple action regions. Fig. 4 shows how we can discriminate intended gestures from accidental gestures.

3 Interactions with a Real Paddle

More stable interaction is possible if we use a paddle instead of using human hands. Contrary to human hands, paddles are rigid objects and they have less degree of freedoms in their shapes. A table tennis game is a typical example of the paddle-based game. In an augmented reality version of the table tennis computer game, the camera detects the spatial movement of the player's real paddle and moves the virtual paddle accordingly. It can be played as a single player playing against a computerized opponent, or against another player through the network.

A table tennis racket is made from a wooden blade, which is flat and rigid. The sides of the blade used to hit the ball are covered in a rubber. If both sides of the bat are used to hit the ball, one rubber must be bright red, and the other rubber black. If only one side of the bat is used, the rubber on that side can be red or black, but the other side without rubber must be the other color. We

Fig. 5. Interactions with a real paddle: A user is stroking the virtual ball while watching a screen monitor (left) or a HMD (right) on which the virtual ball is displayed

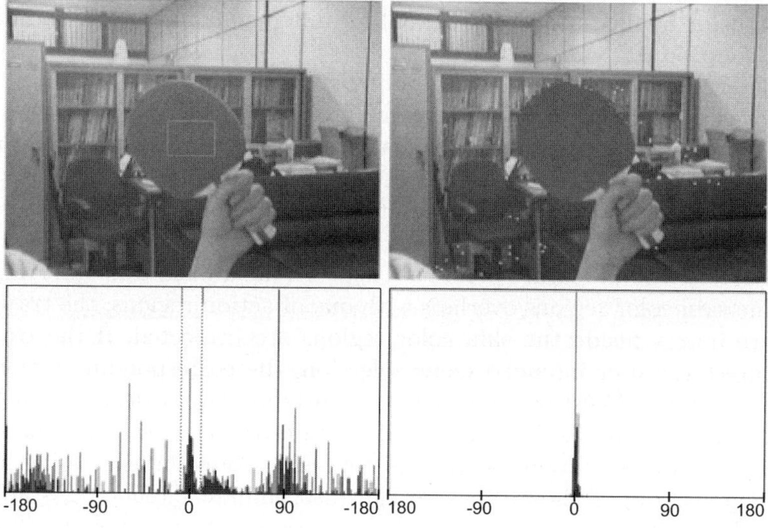

Fig. 6. Color histogram analysis to detect the racket region

assume a single side of the racket is used in playing and that side is covered in a red-colored rubber.

To locate the racket position, we filter racket-colored pixels and count the number of pixels in two directions of the image coordinate system. The hue value of the true red color is $0°$. The negative range is changed to the positive range by adding $360°$ to it. Since we are interested in the red colored racket we set the range of hue to $[-180°, 180°]$ instead of $[0°, 360°]$ to preserve the continuity around $0°$. By default, pixels whose hue value is in $[-10°, 10°]$ are regarded as the candidate pixels of the racket surface. The range depends on the lighting conditions and physical environments. For example, at the playing environment shown in Fig. 5, most pixels in the racket falls in the range $[-15°, 5°]$. We discard too dark pixels by enforcing the third component at the HSV space must be greater than 0.5.

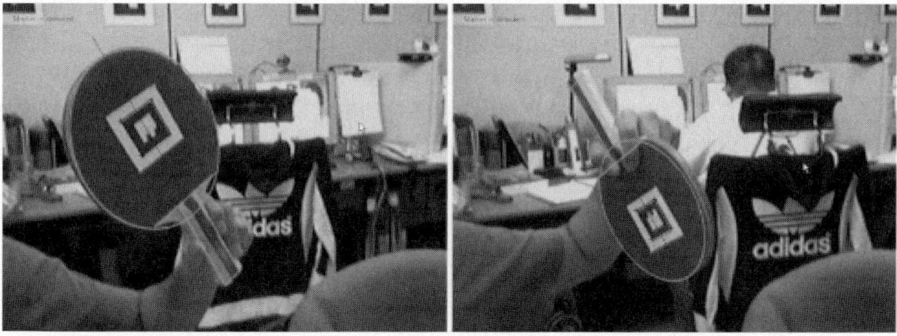

Fig. 7. Racket pose estimation using the attached marker

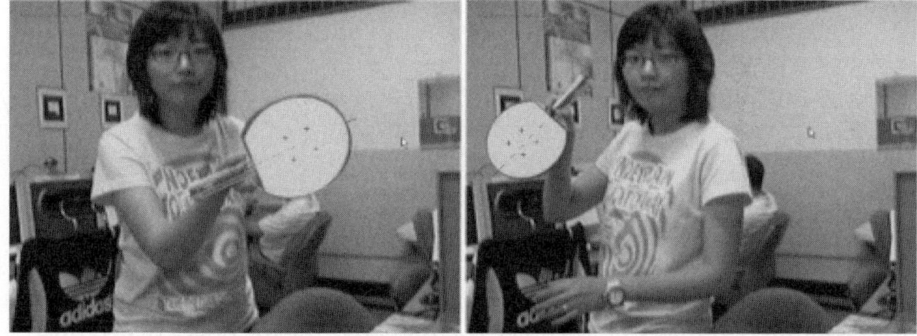

Fig. 8. Tracking a real racket using a marker attached to the racket

To adaptively determine the racket pixels, the hue range should be computed according to the current lighting condition. Fig. 6 shows the hue range decision process. For a racket image, we locate the racket positioned at the image center and compute the average value and the variation of the hue component. In the figure, the computed hue range is [-8°, 12°]. Then, using the range, pixels are filtered and the racket pixels are determined, as shown in the right figure, where the pink colored pixels are in the hue range. The histogram in the left figure is for the distribution of hue values for all pixels and the histogram in the right figure is for the pixels in the centered rectangular region. The hue values for the racket pixels are densely located around the mean value.

The number of pixels in the racket color range is counted along both the horizontal and the vertical directions and the counts along the two directions are stored in two separate arrays. The image location of the racket is given by seeking the peak points at the arrays. This technique is robust even for the case when some racket-like regions exist around the real racket.

Many augmented reality applications use black and white fiducial markers for both the fast and easy detection of features and the direct computation of camera poses. Using a marker-attached racket, we can utilize further information for the racket poses such as the position and orientation of the racket relative

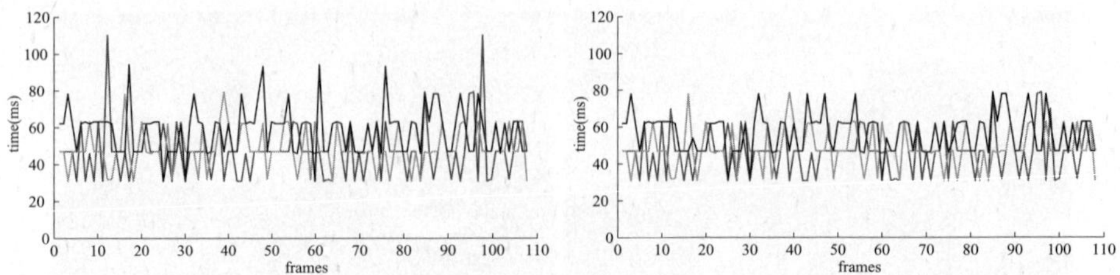

Fig. 9. The frame processing time for three untrained users

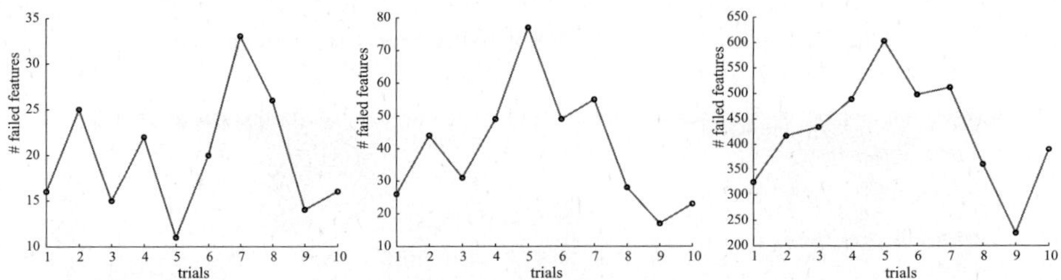

Fig. 10. Feature tracking errors for three types of hand gestures

to the camera. We attach a marker on the racket surface and detect the marker during the game playing. Fig. 7 shows the detected racket poses overlapped with the marker-attached racket images. Fig. 8 shows the racket tracking results. The identified racket face is marked in yellow color together with the marker corners and the marker direction.

4   Experimental Results

We implemented three different types of interaction: interactions using bare hands, using a real racket, and using a marker-attached real racket. To demonstrate the interactions with bare hands, a gesture-controlled music player is implemented. In the music player, several menu icons are displayed on the top of the screen and a user can activate a menu by hand gestures as shown in Fig. 3. The music player has four menu icons to play, stop, forward, and backward a music. Each menu item is shown as an icon located on the top side of the camera view.

For the paddle-based interactions, we implemented an augmented reality table tennis game. In the virtual table tennis game, a user can hit the virtual ball with a table tennis racket. The game player manipulates the virtual ball in the 3D space by means of a red colored table tennis racket in front of the camera field of view. A player playing the virtual table tennis game is shown in Fig. 5. The player can see her current pose in the virtual space on a monitor screen or on a stereoscopic head-mounted display.

The experiments were performed on a 2.6GHz Pentium 4-based system. The frame size is fixed to 320×240. Color space conversion to HSV space and skin color region filtering requires 15 milliseconds on average per frame. Feature tracking for 500 feature points requires 42 milliseconds on average per frame. The overall processing speed for the interface is about 15 frames per second. The frame processing time for input video streams are shown in Fig. 9.

The interface using a real racket is much faster than the feature-based interface. Red color pixel filtering and histogram accumulation provides candidate positions of the racket. Though much complicated and accurate analysis is possible, it is just a trade-off between time and cost. The overall physics and rendering speed for the game is 75 frames per second. Due to the limit of our camera capability, the frame rate is limited by at most 30 frames per second. Thus, the actual paddle movement is reflected to the virtual paddle position at the frame rate.

Our experimental results showed that using only feature tracking for the gesture recognition is significantly unstable. Fig. 10 shows the number of failed features during the tracking. We have tried to recognize ten different gestures for three different motions: normal motions, fast motions, and abrupt fast motions. For each trial we find 100 feature points at the first frame and track them during next 100 frames. If some feature points are failed in tracking they are immediately excluded and new feature points are inserted. For each frame we count the number of failed features and accumulate it during the 100 frames. For each of the three cases, the accumulated number of failed features in tracking is shown in Fig. 10. For the normal motion the number of failed features are about 20 on average. For the fast motion many features points get out of the field of view and the number of failed features are about 40 on average. The abrupt motion makes the worst result. The number of failed features are about 425 on average or, equivalently, 4 failed features per frame.

We found a much stable interaction is possible when using both feature tracking and color filtering. To observe the effectiveness of the hybrid approach, we measured the menu activation accuracy for the gestures of three untrained users. Each user tried 100 menu selection gestures in front of the camera watching the monitor. One of them is requested to do them with fast hand motion. For normal speed gestures of two users, 93 trials and 95 trials were succeeded. For fast gestures of the third user, 84 trials were succeeded.

To validate the superiority of the gesture interface over the classical interface, we also implemented a conventional mouse and keyboard based interface in the virtual table tennis game. Untrained users have tried both interfaces. All the users confirmed that the gesture interface using the paddle is true to nature and easier to operate than the conventional interface.

5   Conclusion

We described gesture-based interaction methods applicable to augmented reality applications. Three different approaches were described: using human hand gestures, using a red colored racket, using a marker-attached racket. For the discrimination of intended gestures from accidental hand movements, we use a smooth trajectory toward one of virtual objects or menus. We have implemented and tested two simple augmented reality applications: a gesture-based music player and a virtual table tennis game. Though the accuracy of interactions depends on many factors regarding the given gestures, a considerable improvement was possible by using both the fiducial tracking and the color filtering of the human hands or the real racket surface.

We have still lots of room for improvement in accuracy. Obvious directions for future work for the improvement include applying the unified framework of various user contexts such as velocity of motion, patterns of hand gestures, current human postures, and human body silhouette.

Acknowledgements. This work was supported in part by grant No. RTI05-03-01 from the Regional Technology Innovation Program of the Ministry of Commerce, Industry and Energy(MOCIE) and in part by the Brain Korea 21 Project in 2006.

References

1. Torre, R., Balcisoy, S., Fua, P., Ponder, M., Thalmann, D.: Interaction between real and virtual humans : Playing checkers. In: Eurographics Workshop on Virtual Environments. (2000)
2. Asutay, A., Indugula, A., Borst, C.: Virtual tennis: a hybrid distributed virtual reality environment with fishtank vs. hmd. In: IEEE International Symposium on Distributed Simulation and Real-Time Applications. (2005) 213–220
3. Govil, A., You, S., Neumann, U.: A video-based augmented reality golf simulator. In: ACM Multimedia 2000. (2000) 489–490
4. Nilsen, T., Looser, J.: Tankwar tabletop war gaming in augmented reality. In: Proceedings of 2nd International Workshop on Pervasive Gaming Applications. (2005)
5. O'Hagan, R., Zelinsky, A.: Finger track - a robust and real-time gesture interface. In: Australian Joint Conference on Artificial Intelligence. (1997) 475–484
6. von Hardenberg, C., Brard, F.: Bare-hand humancomputer interaction. In: Proceedings of Perceptual User Interfaces. (2001)
7. Pavlovic, V., Sharma, R., Huang, T.S.: Visual interpretation of hand gestures for human-computer interaction: A review. IEEE Transactions on Pattern Analysis and Machine Intelligence **19**(7) (1997) 677–695
8. Nilsen, T., Linton, S., Looser, J.: Motivations for augmented reality gaming. In: Proceedings Fuse 04, New Zealand Game Developers Conference. (2004) 86–93
9. Ishii, H., Wisneski, C., Orbanes, J., Chun, B., Paradiso, J.: Pingpongplus: Design of an athletic-tangible interface for computer supported cooperative play. In: Proceedings of Conference on Human Factors in Computing Systems. (1999) 394–401
10. Woodward, C., Honkamaa, P., Jappinen, J., Pyokkimies, E.: Camball - augmented virtual table tennis with real rackets. In: Proc. ACE 2004. (2004) 275–276
11. Geer, D.: Will gesture-recognition technology point the way? IEEE Computer **37**(10) (2004) 20–23
12. Garcia, C., Tziritas, G.: Face detection using quantized skin color regions merging andwavelet packet analysis. IEEE Transactions on Multimedia **1**(3) (1999) 264–277

Turkish Fingerspelling Recognition System Using Axis of Least Inertia Based Fast Alignment

Oğuz Altun, Songül Albayrak, Ali Ekinci, and Behzat Bükün

Yıldız Technical University, Computer Engineering Department, Yıldız, İstanbul, Türkiye
{oguz, songul}@ce.yildiz.edu.tr, behzat.bukun@gmail.com,
ali_ekinci@yahoo.com

Abstract. Fingerspelling is used in sign language to spell out names of people and places for which there is no sign or for which the sign is not known. In this work we describe a Turkish fingerspelling recognition system that recognizes all 29 letters of the Turkish alphabet. A single representative frame is extracted from the sign video, since that frame is enough for recognition purposes of the letters mentioned. Processing a single frame, instead of the whole video, increases speed considerably. The skin regions in the representative frame are extracted by color segmentation in YCrCb space before clearing noise regions by morphological opening. A novel fast alignment method that uses the angle of orientation between the axis of least inertia and y axis is applied to hand regions. This method compensates small orientation differences but increases big ones. This is desirable when differentiating the fingerspelling signs, some of which are close in shape but different in orientation. Also the use of minimum bounding *square* is advised, which helps in resizing without breaking the alignment. Binary values of this minimum bounding square are directly used as feature values, and that allowed experimenting with different classification schemes. Features like mean radial distance and circularity are also used for increasing success rate. Classifiers like kNN, SVM, Naïve Bayes, and RBF Network are experimented with, and 1NN and SVM are found to be the best two of them. The video database was created by 3 different signers, a set of 290 training videos, and a separate set of 174 testing videos are used in experiments. The best classifiers 1NN and SVM achieved a success rate of 99.43% and 98.83% respectively.

Keywords: Turkish Fingerspelling Recognition, Fast Alignment, Angle of orientation, Axis of Least Inertia, Minimum Bounding Square, Classification.

1 Introduction

Sign Language is a visual means of communication using gestures, facial expression, and body language. Sign Language is used mainly by deaf people and people with hearing difficulties. There are two major types of communication in sign language. The first one has word based sign vocabulary, where gestures, facial expression, and body language are used for the most common words. The second one has letter based vocabulary, and is called fingerspelling, which is a method of spelling words using hand movements. Fingerspelling is used in sign language to spell out names of people

A. Sattar and B.H. Kang (Eds.): AI 2006, LNAI 4304, pp. 473–481, 2006.
© Springer-Verlag Berlin Heidelberg 2006

and places for which there is no sign and can also be used to spell words for signs that the signer does not know the sign for, or to clarify a sign that is not known by the person reading the signer [1].

Sign languages develop specific to their communities and are not universal. For example, ASL (American Sign Language) is totally different from British Sign Language even though both countries speak English [2]. In the automatic sign language recognition, there are successful systems for American Sign Language (SL) [3], Australian SL [4], and Chinese SL [5] .

Previous approaches to word level sign recognition rely heavily on statistical models such as Hidden Markov Models (HMMs). A real-time ASL recognition system developed by Starner and Pentland [3] used colored gloves to track and identify left and right hands. They extracted global features that represent positions, angle of axis of least inertia, and eccentricity of the bounding ellipse of two hands. Using an HMM recognizer with a known grammar, they achieved a 99.2% accuracy at the word level for 99 test sequences. For TSL (Turkish Sign Language) Haberdar and Albayrak [6], developed a TSL recognition system from video using HMMs for trajectories of hands. The system achieved a word accuracy of 95.7% by concentrating only on the global features of the generated signs. The developed system is the first comprehensive study on TSL and recognizes 50 isolated signs. This study is improved with local features and performs person dependent recognition of 172 isolated signs in two stages with an accuracy of 93.31% [7].

For fingerspelling recognition, most successful approaches are based on instrumented gloves, which provide information about finger positions. Lamar and Bhuiyant [8] achieved letter recognition rates ranging from 70% to 93%, using colored gloves and neural networks. More recently, Rebollar et al. [9] used a more sophisticated glove to classify 21 out of 26 letters with 100% accuracy. The worst case, letter 'U', achieved 78% accuracy. Isaacs and Foo [10] developed a two layer feed-forward neural network that recognizes the 24 static letters in the American Sign Language (ASL) alphabet using still input images. ASL fingerspelling recognition system is with 99.9% accuracy with an SNR as low as 2. Feris, Turk and others [11] used a multi-flash camera with flashes strategically positioned to cast shadows along depth discontinuities in the scene, allowing efficient and accurate hand shape extraction. Altun et al. [12] increased the effect of fingers in Turkish fingerspelling shapes by thick edge detection and correlation with penalization. They achieved 99% accuracy out of 203 sign videos of 29 the Turkish alphabet letters.

In this work, we have developed a signer independent fingerspelling recognition system for Turkish Sign Language (TSL). The representative frames are extracted from sign videos. Hand objects in these frames are segmented out by skin color in YCrCb space. These hand objects are aligned using the novel angle of orientation based fast alignment method. Then, the aligned object is moved into the center of a minimum bounding square, and resized. The binary values of the minimum bounding square are used as classification features, in addition to the binary object features like mean radial distance and circularity. We experimented with different classification schemes and reported their success rate.

The remaining of this paper is organized as follows: In Section 2 we describe the representative frame extraction, our fast alignment method, and extraction of additional object features. Section 3 covers the video database we use. We listed the classification schemes we used in Section 4. Finally, conclusions and future work are addressed in Section 5.

2 Feature Extraction

Contrary to Turkish Sign Language word signs, Turkish fingerspelling signs, because of their static structure, can be discriminated by shape alone by use of a representative frame. To take advantage of this and to increase processing speed, these representative frames are extracted and used for recognition. Fig. 1 shows representative frames for all 29 Turkish Alphabet letters.

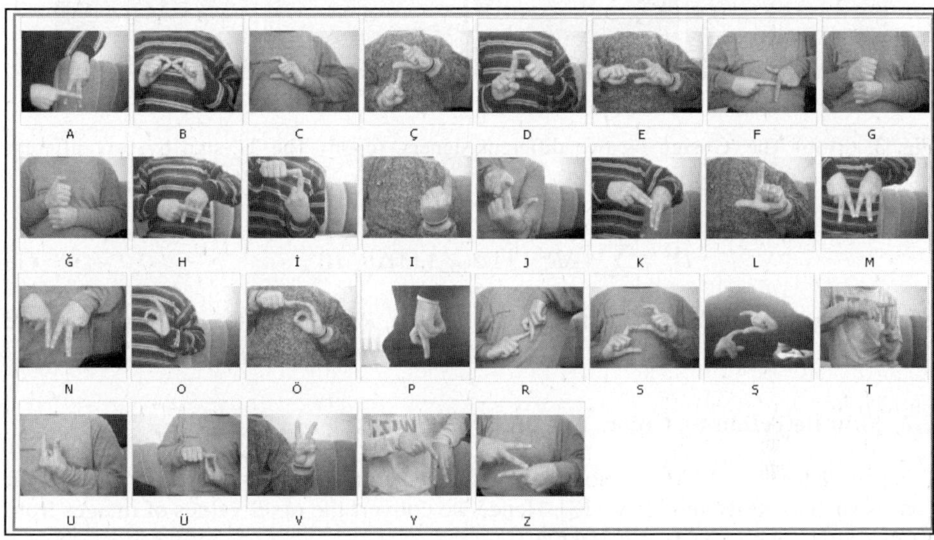

Fig. 1. Representative frames for all 29 letters in Turkish Alphabet

In each representative frame, hand regions are determined by skin color. From the binary images that show hand and background pixels, the regions we are interested in are extracted, aligned and resized. In addition to aligned binary pixel values, binary object features are extracted to support maximum correlation based matching.

Each process is summarized below:

2.1 Representative Frame Extraction

In a Turkish fingerspelling video, representative frames are the ones with least hand movement. Hence, the frame whose distance to its successor is minimum can be chosen as a representative frame. Distance between successive frames f and $f+1$ is given by the sum of the city block distance between corresponding pixels:

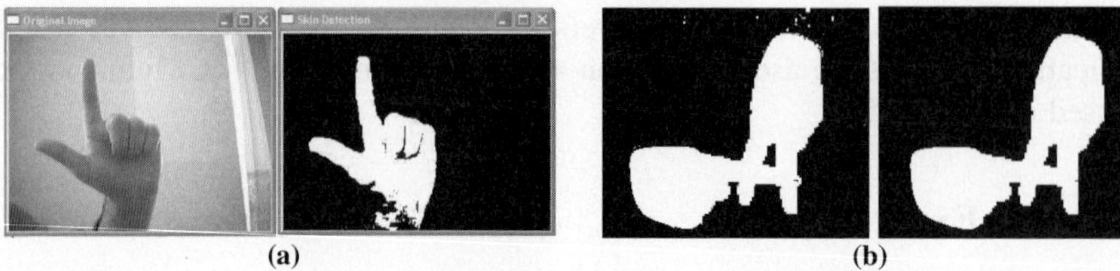

Fig. 2. (a) Original image and detected skin regions after pixel classification, (b) result of the morphological opening.

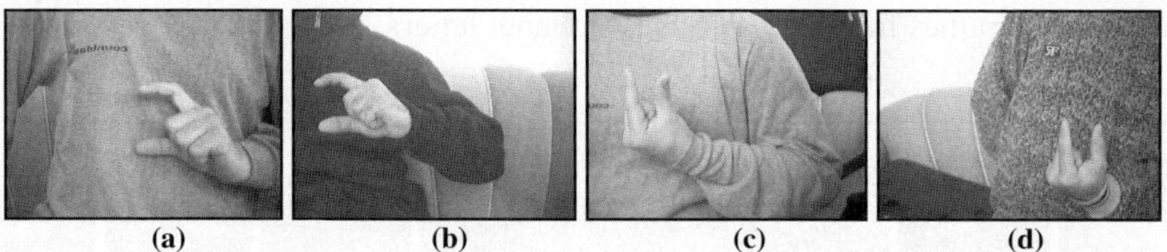

Fig. 3. (a)-(b) The 'C' sign by two different signers. (c)-(d) The 'U' sign by two different signers.

$$D^f = \sum_n |\Delta R_n^f| + |\Delta G_n^f| + |\Delta B_n^f|, \tag{1}$$

where f iterates over frames, n iterates over pixels, R, G, B are the components of the pixel color, $\Delta R_n^f = R_n^{f+1} - R_n^f$, $\Delta G_n^f = G_n^{f+1} - G_n^f$, and $\Delta B_n^f = B_n^{f+1} - B_n^f$.

2.2 Skin Detection by Color

For skin detection, YCrCb color-space has been found to be superior to other color spaces such as RGB and HSV [13]. Hence we convert the pixel values of images from RGB color space to YCrCb using (2).

In order to decrease noise, each of the Y, Cr and Cb components of the image are smoothed with the 2D Gaussian filter given by (3), where σ is its standard deviation

$$Y = 0.299R + 0.587G + 0.114B, \; C_r = B - Y, \; C_b = R - Y \tag{2}$$

$$F(x, y) = \frac{1}{2\pi\sigma^2} \exp(-\frac{x^2 + y^2}{2\sigma^2}), \tag{3}$$

Chai and Bouzerdom [14] report that pixels that belong to the skin region have similar Cr and Cb values, and give a distribution of the pixel color in Cr-Cb plane. Consequently, we classified a pixel as skin if the Y, Cr, Cb values of it falls inside the ranges $135 < Cr < 180$, $85 < Cb < 135$ and $Y > 80$ (Fig. 2.a).

After clearing small skin colored regions by morphological opening (Fig. 2.b), skin detection is completed.

2.3   Fast Alignment for Maximum Correlation Based Template Matching

Template matching is very sensitive to size and orientation changes. A scheme that can compensate size and orientation changes is needed. Eliminating orientation information totally is not appropriate however, as depicted in Fig. 3. Fig. 3a-b show two 'C' signs that we must be able to match each other, so we must compensate the small orientation difference. In Fig. 3c-d we see two 'U' signs that we need to differentiate from 'C' signs. 'U' signs and 'C' signs are quite similar to each other in shape, luckily orientation is a major differentiator. As a result we need a scheme that not only can compensate small orientation differences of hand regions, but also is responsive to large ones.

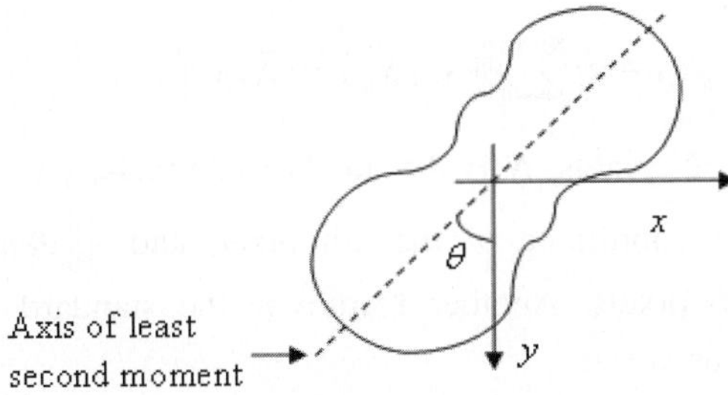

Fig. 4. Axis of least second moment and the angle of orientation

We propose a fast alignment method that works by making the *angle of orientation* (θ) zero. Angle of orientation, given by (4), is the angle between y axis and the *axis of least moment* (shown in Fig. 4).

$$2\theta = \arctan\left(\frac{2M_{11}}{M_{20} - M_{02}}\right) \tag{4}$$

where ($I(x,y) = 1$ for pixels on the object, 0 otherwise), $M_{11} = \sum_x \sum_y xyI(x, y)$, $M_{20} = \sum_x \sum_y x^2 I(x, y)$, and $M_{02} = \sum_x \sum_y y^2 I(x, y)$.

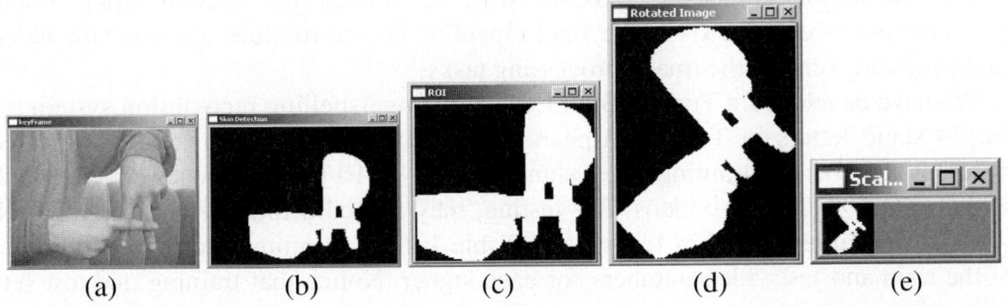

Fig. 5. Stages of fast alignment. (a) Original frame. (b) Detected skin regions. (c) Region of Interest (ROI). (d) Rotated ROI. (e) Resized bounding square with the object in the center.

Let's define *bounding square* as the smallest square that can completely enclose all the pixels of the object. After putting images in the center of a bounding square, and than resizing the bounding square to a fixed, smaller resolution, the fast alignment process ends (Fig. 5).

2.4 Additional Binary Object Features

Instead of using only pixel values in the bounding square, additional binary object features are extracted to support decision process.

These features include area, center of area, perimeter [15], angle of orientation (defined above), and circularity (defined as *perimeter^2/area*). In addition, mean radial distance μ_R is extracted:

$$\mu_R = \tfrac{1}{N} \sum_n \left\| (x_n, y_n) - (\bar{x}, \bar{y}) \right\| \tag{5}$$

where n iterates over all pixels, N is the number of pixels, $(\bar{x}, \bar{y})$ is the center of area, (x_n, y_n) is the coordinate of the nth pixel, and $\left\| \cdot \right\|$ denotes the Euclidean distance between two pixels. Another feature is the standard deviation of radial distance σ_R, defined as

$$\sigma_R = \left(\tfrac{1}{N} \sum_n \left[\left\| (x_n, y_n) - (\bar{x}, \bar{y}) \right\| - \mu_R \right]^2 \right)^{1/2}. \tag{6}$$

As the last binary object feature, a second circularity measure C is computed by

$$C = \mu_R / \sigma_R . \tag{7}$$

To summarize, 9 binary object features are added to the 30x30 binary values of the minimum boundary square.

3 Video Database

The training and test videos are acquired by a Philips PCVC840K CCD webcam. The capture resolution is set to 320x240 with 15 frames per second (fps). While programming is done in C++, the Intel OpenCV library routines are used for video capturing and some of the image processing tasks.

We have developed a Turkish Sign Language fingerspelling recognition system for the 29 static letters in Turkish Alphabet. The training set was created using three different signers. For training, they signed a total of 10 times for each letter, which sums up to 290 training videos. For testing, they signed a total of 6 times for each letter, which sums up to 174 test videos. Table 1 gives a summary of the distribution of the train and test video numbers for each signer. Notice that training and test sets are totally separated.

Table 1. Distribution of train and test video numbers for each signer

	Signer 1		Signer 2		Signer 3		Total	
	Train	Test	Train	Test	Train	Test	Train	Test
A	4	2	4	2	2	2	10	6
⋮	⋮	⋮	⋮	⋮	⋮	⋮	⋮	⋮
Z	4	2	4	2	2	2	10	6
Total							290	174

Table 2. Success rates of most successful classifiers on fingerspelling data

Classifier	Success Rate (%)
1NN [16]	99.43
SVM [17]	98.85
Random Forest [18]	97.13
RBF Network [19]	96.55
Multinomial Naive Bayes [20]	93.68
Naive Bayes [21]	88.51
J48 [22]	85.63

4 Classification Comparison

A set of different classification algorithms are applied to the features extracted as explained in Section 2 and obtained results are sorted according to their success rates. These classification results are summarized in Table 2.

The most successful classifiers are one nearest neighbor (1NN) and support vector machine (SVM). These methods classified more than 98% successfully, as seen in Table 2. The biggest problem is in the classification of the letter 'S', which is confused by 'Ş'. A second problem letter was 'G', which is confused by 'Ğ'. The confused characters are very similar to each other in shape, as seen in Fig. 6.

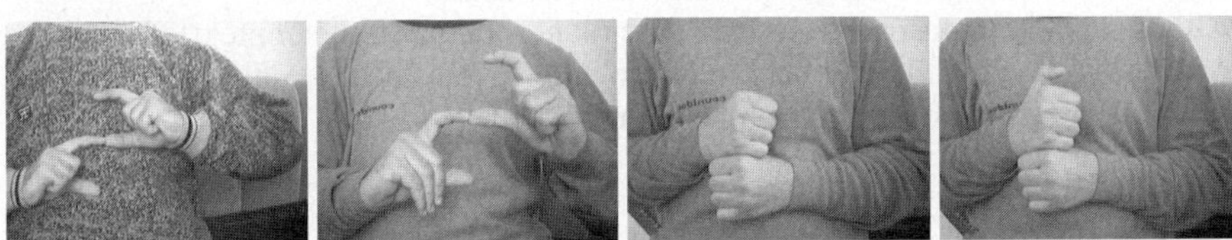

Fig. 6. Two difficult cases where our methods may fail. Left to right: 'S' and 'Ş', 'G' and 'Ğ'.

5 Conclusions and Future Work

A Turkish fingerspelling recognition system is tested and found to have more than 99% accuracy. Testing and training sets is created by multiple signers, as a consequence the developed system is signer independent. Accuracy is the result of the fast alignment process we applied. This process brings objects with similar orientation into same alignment, while bringing objects with high orientation difference into different alignment. This is a desired result, because for fingerspelling recognition, shapes that belong to different letters can be very similar, and the orientation can be the main differentiator. After the alignment, to resize without breaking the alignment, the object is moved into the center of a *minimum bounding square*. The binary values in minimum bounding square are used as the features. In addition, we used binary object features like circularity and mean radial distance, which helped increasing success rate.

Our method is robust to the problem of occlusion of the hands, because the fast alignment method allows us to process an ensemble of one or more connected components in the same way.

The system is fast due to representing the sign video by a single frame, the speed of fast alignment process, and resizing the bounding square to a smaller resolution. The amount of resizing can be arranged for different applications.

Since we used binary pixel values as ordinary features, we are able to experiment with different classification algorithms, amongst which are kNN, SVM, RBF Network, Naïve Bayes, Random Forest, and J48 tree. The 1NN and SVM give the best success rates, 99.43% and 98.85% respectively.

Not all letters in Turkish alphabet are representable by one single frame, 'Ş' being an example. The sign of letter involves some movement that differentiates it from 'S'. In fact, this letter is the one that prevented us achieving a 100% success rate. Still, representing the whole sign by one single frame is acceptable since this work is actually a step towards making a full blown Turkish Sign Language recognition system that can also recognize word signs. That system will incorporate not only shape but also the movement, and the research on it is continuing.

The importance of successful segmentation of the skin and background regions can not be overstated. In this work we assumed that there is no skin colored background regions and used color based segmentation in YCrCb space. The systems' success depends on that assumption, and research on better skin segmentation is invaluable.

The fast alignment and classification schemes presented would work equally well in the existence of a face in the frame, even though in this study we used only hand regions when creating the fingerspelling video database.

Although we demonstrated the fast alignment method in the context of hand shape recognition, it is equally applicable to other problems where shape recognition is required, for example to the problem of shape retrieval.

References

1. http://www.british-sign.co.uk/learnbslsignlanguage/whatisfingerspelling.htm.
2. http://www.deaflibrary.org/asl.html.
3. Starner, T., Weaver, J., Pentland, A.: Real-time American sign language recognition using desk and wearable computer based video. Ieee Transactions on Pattern Analysis and Machine Intelligence **20** (1998) 1371-1375
4. Holden, E.J., Lee, G., Owens, R.: Australian sign language recognition. Machine Vision and Applications **16** (2005) 312-320

5. Gao, W., Fang, G.L., Zhao, D.B., Chen, Y.Q.: A Chinese sign language recognition system based on SOFM/SRN/HMM. Pattern Recognition **37** (2004) 2389-2402
6. Haberdar, H., Albayrak, S.: Real Time Isolated Turkish Sign Language Recognition From Video Using Hidden Markov Models With Global Features. Lecture Notes in Computer Science **LNCS 3733** (2005) 677
7. Haberdar, H., Albayrak, S.: Vision Based Real Time Isolated Turkish Sign Language Recognition. International Symposium on Methodologies for Intelligent Systems, Bari, Italy (2006)
8. Lamar, M., Bhuiyant, M.: Hand Alphabet Recognition Using Morphological PCA and Neural Networks. International Joint Conference on Neural Networks, Washington, USA (1999) 2839-2844
9. Rebollar, J., Lindeman, R., Kyriakopoulos, N.: A Multi-Class Pattern Recognition System for Practical Fingerspelling Translation. International Conference on Multimodel Interfaces, Pittsburgh, USA (200)
10. Isaacs, J., Foo, S.: Hand Pose Estimation for American Sign Language Recognition. Thirty-Sixth Southeastern Symposium on, IEEE System Theory (2004) 132-136
11. Feris, R., Turk, M., Raskar, R., Tan, K.: Exploiting Depth Discontinuities for Vision-Based Fingerspelling Recognition. 2004 IEEE Computer Society Conference on Computer Vision and Pattern Recognition Workshops(CVPRW'04) (2004)
12. Altun, O., Albayrak, S., Ekinci, A., Bükün, B.: Increasing the Effect of Fingers in Fingerspelling Hand Shapes by Thick Edge Detection and Correlation with Penalization. PSIVT 2006 (2006)
13. Sazonov, V., Vezhnevetsi, V., Andreeva, A.: A survey on pixel vased skin color detection techniques. Graphicon-2003 (2003) 85-92
14. Chai, D., Bouzerdom, A.: A Bayesian Approach To Skin Colour Classification. TENCON-2000 (2000)
15. Umbaugh, S.E.: Computer Vision and Image Processing: A Practical Approach Using CVIPtools. Prentice Hall (1998)
16. Aha, D.W., Kibler, D., Albert, M.K.: Instance-Based Learning Algorithms. Machine Learning **6** (1991) 37-66
17. Keerthi, S.S., Shevade, S.K., Bhattacharyya, C., Murthy, K.R.K.: Improvements to Platt's SMO algorithm for SVM classifier design. Neural Computation **13** (2001) 637-649
18. Breiman, L.: Random forests. Machine Learning **45** (2001) 5-32
19. Fritzke, B.: Fast Learning with Incremental Rbf Networks. Neural Processing Letters **1** (1994) 2-5
20. McCallum, A., Nigam, K.: A Comparison of Event Models for Naive Bayes Text Classification. AAAI-98, Workshop on Learning for Text Categorization (1998)
21. John, G.H., Langley, P.: Estimating Continuous Distributions in Bayesian Classifiers. Eleventh Conference on Uncertainty in Artificial Intelligence. Morgan Kaufmann, San Mateo (1995) 338-345
22. Quinlan, R.: C4.5: Programs for Machine Learning. Morgan Kaufmann Publishers, San Mateo, CA (1993)

Vehicle Tracking and Traffic Parameter Extraction Based on Discrete Wavelet Transform*

Jun Kong[1,2], Huijie Xin[1,2], Yinghua Lu[1], Bingbing Li[1,2], and Yanwen Li[1,2,*]

[1] Computer School, Northeast Normal University, Changchun, Jilin Province, China
[2] Key Laboratory for Applied Statistics of MOE, China
{kongjun, xinhj490, luyh, libb613, liyw085}@nenu.edu.cn

Abstract. Real time traffic monitoring is one of the most challenging problems in machine vision. In this paper, we describe a method for multiple vehicles tracking and traffic parameters extraction based on Discrete Wavelet Transform. Occlusion is effectively resolved by our method. In order to describe the traffic characteristics, some important traffic parameters are extracted, such as vehicle trajectory, vehicle speed and vehicle count. Discrete wavelet transform is used to prevent the disturbance of fake motions. Experimental results are presented to demonstrate the effective tracking of vehicles on an urban artery.

Keywords: Vehicle tracking; Traffic parameter extraction; ITS; DWT.

1 Introduction

Animals can rapidly detect changes and take corresponding actions in their normal environment. This particular ability to be aware of environmental changes is very important for survival, for example, to find food or avoid predators. The sense of vision plays an important part in animal behavior either for tracking movements or identifying objects of interest. If such an ability can implemented artificially, these applications will be used in automated inspection and surveillance tasks.

The Intelligent Transportation System (ITS) technology has significantly enhanced the ability to provide relevant information to road users. Traffic parameters can be extracted to describe the characteristics of the vehicles and their movement. Various kinds of traffic control systems, such as automatic tolls, congestion detection, and automatic routing, can be implemented with these traffic parameters.

In spite of these advantages, the vehicle detection error due to occlusion makes serious difficulties to extract traffic parameters. In our method the occlusion is effectively resolved. The experiment results show that our proposed framework is effective in vehicle tracking and parameter extraction.

This paper is organized as follows: In Section 2, we give a brief review of related work. In Section 3, the motion detection method based on background subtraction is presented. In Section 4, the vehicle tracking and parameter extraction is described. Experimental results and conclusion are given in Section 5.

---

* This work is supported by science foundation for young teachers of Northeast Normal University, No. 20061002, China.
* Corresponding author.

A. Sattar and B.H. Kang (Eds.): AI 2006, LNAI 4304, pp. 482–490, 2006.

© Springer-Verlag Berlin Heidelberg 2006

2 Related Work

Recently, several motion detection algorithms have been proposed. Generally, they can be classified into three methodologies: the background subtraction based, the probability based, and the temporal difference based approach.

There were some systems that achieved the vehicle detection through background subtraction based method [1, 2]. Background subtraction based on standard intensities may work well if the illumination for the processed images is constant. Any change in illumination will significantly affect intensities, because the traditional background subtraction method uses a fixed reference background image. Therefore, quality and illumination-tolerance provided by background subtraction are the basic criteria to be concerned for moving object detection and image segmentation. Hence, the traditional background subtraction based on standard intensities is inherently sensitive to illumination. So background updating is needed due to the change of ambient lighting, shadow, weather, etc.

The method proposed by Paragios and Deriche in [3] considered a probability and statistics problem, and used the observed information to obtain a classification equation of probability and statistics problem, and used the observed information to obtain a classification equation of probability to segment image. These detection models can detect and track moving objects perfectly, but they suffer from the high computing cost, so it is hard to be used in real-time surveillance application.

Temporal difference method computes difference between two successive frames so as to remove static part and get moving part within the image [4][5]. It is a simple method for detecting moving objects in a static environment, but unable to detect static vehicle. Moreover, the results of inter-frame subtraction are also influenced by speed of vehicles. Too low or too high speed may cause errors in vehicle detection.

3 Motion Detecting

In the following sub-sections, we will present the motion detection approach. Firstly, the background is extracted, then the background subtraction method based on wavelet transform is proposed, and at last the bounding boxes of the moving vehicles are located.

3.1 Background Extraction

In image sequences of road traffic, it might be impossible to acquire a background image. Hence, it is desirable that the initial background image is automatically extracted from a sequence of road traffic images. Therefore, we use the background extraction method described in [6]. The approach is summarized as follows:

Step 1: Subtract the successive frames to get the motion difference image,
Step 2: Segment the difference images to get foreground and background region,
Step 3: If one pixel never appears in the background image, but now in the background region, then we use its corresponding value to update the background image,
Step 4: Repeat steps 1, 2 and 3, until the percentage of extracted background pixels reaches to a predefined value, end.

The detail procedure of background extraction is shown in Fig. 1. The result of the inter-frame subtraction method is influenced by the speed of the vehicles, and if the speed is very slow, the difference images will have little moving pixels. In order to reduce the processing time and increase the calculating accuracy, we use two pairs of images like frame t and frame t-4, frame t and frame t+4, but not every frame.

Fig. 1. The sequential procedure of background subtraction: (a)-(c) are the initial input images, (d)-(f) are the first, second, and third iteration results

3.2 Background Subtraction Based on Discrete Wavelet Transform

When digital images are to be viewed or processed at multiple resolutions, the discrete wavelet transform (DWT) is the mathematical tool of choice [7]. In addition to being an efficient, highly intuitive framework for the representation and storage of multi-resolution images, the DWT provides powerful insight into an image's spatial and frequency characteristics. The Fourier transform, on the other hand, reveals only an image's frequency attributions.

Two dimensional discrete wavelet transform (DWT) can be used to decompose an image into four sub-images (LL, HL, LH, HH). An example of two-scale DWT is shown in Fig .2. We can see that the high frequency sub-images, such as LH and HL, contain the detailed information like the edge of leaves. Because the moving vehicles belong to low frequency part, and we don't take more attention to the detailed information of the frame, so our background subtraction is performed in the low frequency sub-image of the two-scale DWT (the upper left corner in Fig. 3.(b)) due to the consideration of low computing cost and noise reduction issue.

After the background generated, the foreground pixels can be extracted by the following equation:

$$DB_{i,j}^{t} = \begin{cases} 1 & \text{if } \left| C_{LL2}^{t}(i, j) - B_{LL2}^{t}(i, j) \right| \geq T \\ 0 & \text{otherwise} \end{cases} . \tag{1}$$

where $C_{LL2}^{t}(i, j)$ and $B_{LL2}^{t}(i, j)$ are the low frequency sub-image of the two-scale DWT corresponding to the current input image and background image, t is the frame number, and T is a predefined threshold.

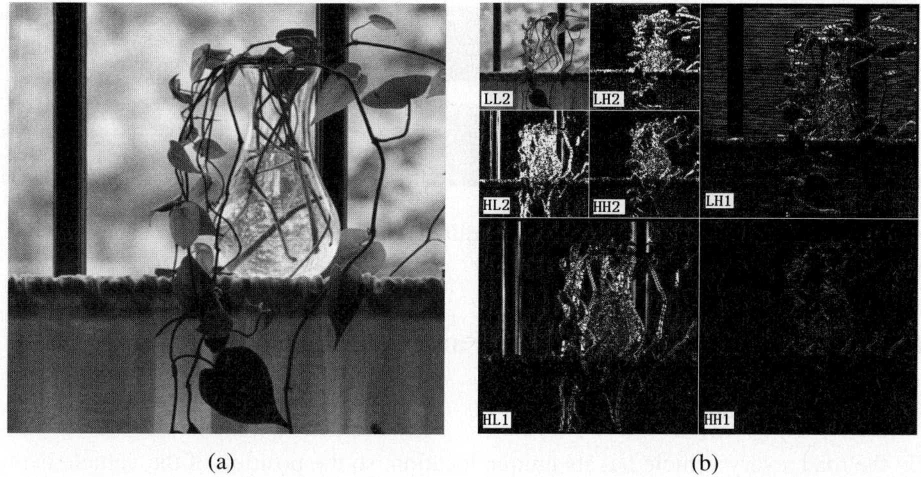

(a)   (b)

Fig. 2. Two-scale DWT: (a) Original image, (b) The result of two-scale DWT

3.3 Background Updating

In background subtraction method, the result of object region extraction strongly depends on the quality of background image. Thus, it is necessary to develop a robust background image updating approach along with the illumination variance.

The current input image also contains foreground objects, so we have to classify the pixels as foreground and background, and only use the background pixels to update the current background. The binary inter-frame difference image $DB_{i,j}$ is used to classify the pixels.

$$B = \begin{cases} B_{new} & \text{if } DB_{i,j} = 0 \\ B_{old} & \text{otherwise} \end{cases}. \tag{2}$$

The formulation shows that in the current input image, if pixel (x, y) belongs to the foreground, then we keep the former corresponding background pixel unchanged, otherwise we use the value of this pixel to update the former background image. Because the background subtraction is performed on the low frequency sub-image, and in the low frequency sub-image some details like the trees' movement are disappeared, so the updated cycle can be longer than the method without DWT.

3.4 Bounding Box of the Moving Objects

After the background subtraction procedure, we extract the connected component of the foreground images. No connected component means that there is no object moving in the frame. If there are more than one connected components, we can conclude that there are moving objects in the frame. The bounding box can be located by computing the maximum and minimum value of x and y coordinates of connected components. The extracted bounding boxes are shown in Fig. 3. (c).

Fig. 3. (a) Image of current frame, (b) The difference image of background subtraction, (c) Moving objects bounded by bounding boxes

4 Object Tracking and Traffic Parameter Extraction

4.1 Feature Extraction

On the road, every vehicle has its unique location, so the position of the vehicle is the most important feature for vehicle tracking in our method. The centroid coordinates of connected components are used to represent the vehicle's location. The bounding box's width and height are also used as another two important features. So the feature vectors of object i can be represented as follow:

$$F^i = (width^i, height^i, x^i, y^i) , \tag{3}$$

where $width^i$, $height^i$, x^i and y^i are bounding box's width, bounding box's height, object centroid's x-axis coordination and object centroid's y-axis coordination.

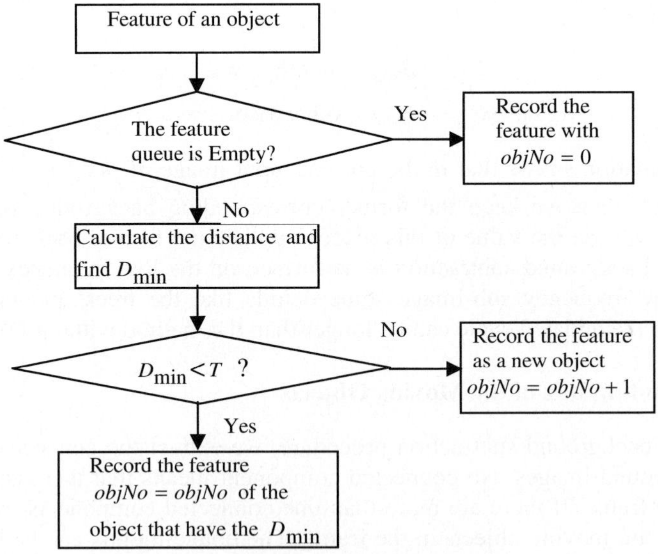

Fig. 4. The flow of the object identification

4.2 Object Identification

A feature queue is created to save the features of the moving objects. When a new object enters the system, it will be tracked, and the features of the object are extracted and recorded into the queue. Once a moving object i is detected, the system will extract the feature F^i and identify it from the identified objects in the queue by computing the distance $D(F^i, F^j)$ as follow:

$$D(i, j) = \left\| F^i - F^j \right\| , \qquad (4)$$

where j is one of the n identified objects, and the flow chart of the object identification is described in Fig. 4.

4.3 Occlusion Removal

In order to resolve the occlusion problem, firstly we must detect the occurrence of the occlusion. This problem is resolved by examining the following conditions:

a. Inspect the size of the bounding box. If the width and the height of a region are more than certain thresholds, that region includes more than one object.
b. Inspect the ratio of the width to the height of the bounding box. If the ratio is out of a certain threshold, that region includes more than one object.
c. Inspect the location of the region. If a new object appears in the middle part of the scene, this region must be an occluded one.

There are mainly two cases that cause the occurrence of the object merge as follows:

a. Horizontal merging. A vehicle merges with the one that is in its right side or left side.
b. Vertical merging. A vehicle merges with the one located at its back or front.

Taking the vertical merging as an example, we describe the procedure of splitting the occluded vehicles as follow:

Step 1: Extract the connected component of the occluded region,
Step 2: For every row, calculate the horizontal distance D_i ($i = 1, 2, \dots, n$, where n is the number of the rows) of the occluded region,
Step 3: Subtract the neighboring D_i as follow:

$$G_i = D_{i+1} - D_i \quad \text{if } D_i \neq 0 , \qquad (5)$$

where i is the corresponding row number,
Step 4: Find the maximum G_i and the row number i is the split row position, end.

Because the bottom and the top of the object in the bounding box may have the maximum G_i, but they are not the split place we want, so we divide the region into three parts: upper part, middle part and lower part, and G_i is calculated only in the middle part. The ratio of the three parts in our study can be defined as shown in Fig.5. (a). The final result is shown in Fig. 5. (b).

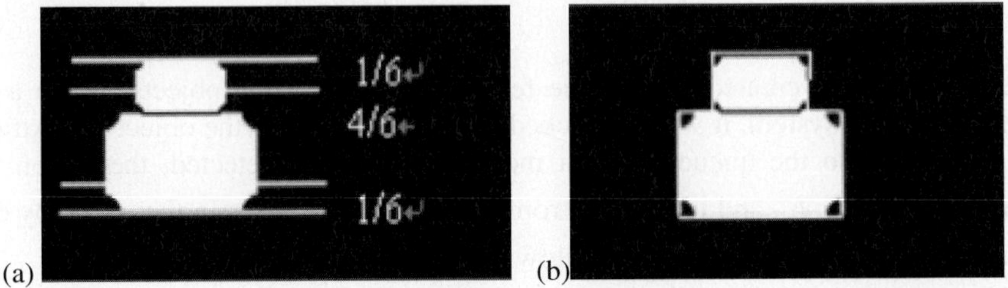

(a)   (b)

Fig. 5. (a) The ratio of the three parts, (b) The final result of the split

4.4 Traffic Parameter Extraction Via Vehicle Tracking

Traffic parameters, such as vehicle count, vehicle speed, and vehicle trajectory are extracted via vehicle detection and tracking method. When a vehicle is detected, it is added to the linked-list that includes the information about it, such as position and frame count. The information is updated as the vehicle object moving within the tracking area. Vehicle trajectory is drawn based on the position in the linked-list. Two kinds of vehicle count are calculated: the vehicle count moving left and right. Vehicle speed is computed by the equation as below:

$$Speed = \frac{D \times F}{F_t} \, ,$$

(6)

where D is the physical distance between entrance and exit of the tracking area, F and F_t are frame rate and number of frames taken for the vehicle to pass the tracking area, respectively.

5 Experimental Results and Conclusions

5.1 Experimental Results

In order to test our algorithm, we used image sequences of about 400 frames of a video tape captured by a common CCD camera. This camera was installed on the side-view of the road.

The result of the tracking without occlusion removal is shown in Fig. 6. (a) and (b). The vehicle count is shown on the upper left corner of the image, and classify into left and right moving direction. For example, the number of left direction vehicles in Fig. 6. (a) is 4, and the number of right direction vehicles is 1. The vehicle direction is described using the right and left arrow on the vehicle. We can see that the bounding box in the lower left corner of Fig. 6. (a) contains two vehicles which merged together. In some papers like [8], they can only separate overlapping vehicles with the similar size, but with our proposed method, the occlusion is successfully resolved as shown in Fig. 6. (c). The bounding box in the middle right of Fig. 6. (b) also contain two vehicles. The two vehicles merge together because of the shadow. They are separated, and the final result is shown in Fig. 6. (d).

The background subtraction in the low frequency sub-image can successfully prevent the disturbance of fake motions. We can see that there are trees in the background image, and they swing with the wind. When the vehicles move nearby the trees, the vehicles and trees may be merged together, and the bounding boxes will be larger than vehicles. In this situation, when the bounding boxes' properties satisfy the occlusion removal conditions, then these vehicles will be split as shown in Fig. 7. (a), but use the DWT, the fake motions are successfully removed, and the result is shown in Fig. 7. (b). The wavelet used in our paper is Haar wavelet.

(a)   (b)

(c)   (d)

Fig. 6. (a) The 17th frame without occlusion removal, (b) The 120th frame without occlusion removal, (c) The 17th frame with occlusion removal, (d) The 120th frame with occlusion removal

(a)   (b)

Fig. 7. (a) The result of 27th frame without DWT, (b) The result of 27th frame with DWT

6 Conclusions

A method for vehicle tracking and traffic parameter extracting is proposed in this paper. The occlusion is successfully removal. The vehicle speed, vehicle count, and vehicle trajectory are given to describe the traffic characteristic. The fake motions always belong to high frequency, and our background subtraction is operated on the low frequency sub-images of current frames and corresponding backgrounds, so the fake motions like the movement of the trees are successfully removed.

References

1 N. Hoose: Computer Image Processing in Traffic Engineering. Taunton Research Studies Press,UK,1991

2 Roya Rad, Mansour Jamzad: Real time classification and tracking of multiple vehicles in highways. Pattern Recognition Letters (2005) 1597-1607

3 N. Paragios, R. Deriche: Geodesic active contours and level sets for the detection and tracking of moving object. IEEE Trans. Pattern Anal. Mach. Intell. ,22 (2000) 266–280

4 H. Zhang, Y. Gong, D. Patterson, A. Kankanhalli: Moving object detection, tracking and recognition. The Third International Conference on Automation, Robotics and Computer Vision (1994) 1990–1994

5 Fang-Hsuan Cheng, Yu-Liang Chen: Real time multiple objects tracking and identification based on discrete wavelet transform. Pattern Recognition, 39 (2006) 1126-1139

6 D.M..Ha, J.-M.Lee,Y.-D.Kim: Neural-edge- based vehicle detection and traffic parameter extraction. image and vision computing (2004) 899-907

7 Rafael C. Gonzalez, Richard E. Woods, Steven L.Eddins: Digital Image Processing Using MATLAB (2004)

8 S.-C. Chen, M.-L. Shyu, C. Zhang, and R. L. Kashyap: Object Tracking and Augmented Transition Network for Video Indexing and Modeling. 12th IEEE International Conference on Tools with Artificial Intelligence (2000) 428-435

Neural Net Based Division of an Image Blob of People into Parts of Constituting Individuals

Yongtae Do

School of Electronic Engineering, Daegu University
Gyeongsan-City, Gyeongbuk, 712-714, Korea
`ytdo@daegu.ac.kr`

Abstract. This paper presents an example-based learning approach to divide a foreground blob of people into its constituents on a surveillance video camera image. As people tend to walk and interact in groups with other people, occlusions frequently happen in camera images. They are detected in the same foreground image blob and dividing it into image parts of constituting individuals is a prerequisite for high-level vision processing like people tracking and activity understanding. The division is easy for a human observer but difficult in computer vision especially when the image resolution is low. We treat this task as a pattern classification problem by identifying partial outline shape patterns of a foreground blob, which can characterize the position where the blob can be well divided. When a probabilistic neural network was employed to identify the pattern, the network showed over 80% correct recognition rates in experiments.

1 Introduction

Video surveillance (VS) is one of most active subjects in current computer vision research [1]. Humans and vehicles are two important targets in VS. Processing human images is generally with more difficulties than processing vehicle images mainly due to the non-rigid nature of a human. Various approaches have been proposed for processing human images for various applications. Some analyzed body parts and shapes in relatively high-resolution images [2,3], while others tracked multiple people in low-resolution images of a camera installed in a far distance [4]. If the size of a human is small, it is difficult to build and maintain the statistics in his or her image features such as colour and shape. This is the case we consider in this paper.

A frequent problem that is encountered when processing people images of a VS camera is occlusion. People tent to walk and interact in groups with other people, thereby increasing the chances that a person being targeted will be occluded by other person moving together completely or partially [5]. When a camera is fixed at a far distance from the scene, people can have relatively slow non-linear motions. Then occlusions tend to last for long period of time [4].

The simplest method tackling the occlusion problem is avoiding it by placing a camera looking vertically downwards [6]. However, this misses most information of people monitored such as colour and silhouette. In [7], an occlusion is detected when multiple people are predicted to be located within the same region in the new video image frame based on their velocities. No attempt was made to segment a blob of

A. Sattar and B.H. Kang (Eds.): AI 2006, LNAI 4304, pp. 491–499, 2006.

© Springer-Verlag Berlin Heidelberg 2006

multiple humans occluded each other into constituents but just waiting de-occlusion. Analyzing the outline of a blob of multiple humans, for example using an active contour (snakes), is another way of segmentation [8]. However, when people are in low-resolution, it cannot be effective as small error in foreground extraction causes significant change in the contour. Practically, outdoor surveillance cameras are often installed at a far distance for wider field-of-view.

In this paper, a foreground blob of multiple humans is segmented into individual constituents using a probabilistic neural network (PNN) [9], which is a neural network implementation of the Bayesian classifier. Fig. 1(b) shows an example of segmentation, which the PNN employed in this paper does. We regard detecting the segmentation position separating two people (occluding and occluded) vertically in a foreground blob as a two-class pattern recognition problem; positions need to be segmented constitute a class and others constitute the other class. The use of a PNN is motivated by two factors. First, it is hard to characterize occlusion by some feature variables. Different clothes and postures will cause different variable values. Secondly, if images are in low-resolution, it is difficult to build a reliable statistical model because the number of pixels is not enough. Although a human observer can separate people in occlusion without difficulty, the method implicitly used by human intelligence is hard to formulize. In this situation, while most existing techniques cannot provide an effective solution, a PNN is expected to determine where a blob of people needs to be segmented by learning based on shape information of upper body parts of human targets.

(a)   (b)

Fig. 1. Two approaches to tackle occlusion problem when processing human images: (a) shows an example image of a camera fixed to look downward. The occlusion problem can be virtually avoided but significant information loss is resulted, (b) shows a segmentation example of occluded humans. In this paper we design a PNN to get a result like this.

2   Extracting Data Patterns for PNN

The goal of research work described in this paper is finding the best segmentation position for a foreground image blob of multiple people. After various attempts, we arrived at a conclusion that using a PNN is an effective method. Fig. 2 shows the procedures from input image processing to PNN-based segmentation. As shown in the figure, two important steps are required to extract data for a neural network employed: extracting humans as foreground and obtaining their shape information for segmentation.

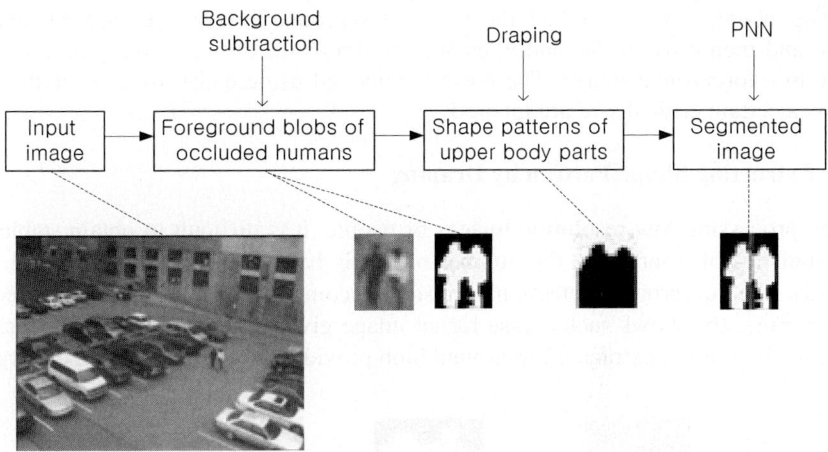

Fig. 2. Procedures to segment an image of occluded humans. Occlusion occurs not only when a person is occluded by another person but also when people are in close distance each other.

2.1 Extracting Foreground Blobs of Humans

The initial stage of VS is moving target detection. There are three conventional methods to detect moving targets in video sequences: temporal differencing [10], optical flow [11], and background subtraction [12]. Temporal differencing is very adaptive to dynamic environments, but generally does a poor job of extracting all relevant feature pixels. Optical flows can be used to detect independently moving targets in the presence of camera motion. However, most optical flow computation methods are very complex and are inapplicable to real-time algorithms if there is no specialized hardware like [13]. The background subtraction technique generally provides the most complete feature data. One disadvantage is its high sensitivity to dynamic scene change. In this paper, an adaptive background subtraction method proposed by Lipton et al [12] is used. For each pixel value p_n in the n'th image frame, a running average $\bar{p}_n$ and a form of standard deviation σ_{p_n} are maintained by temporal filtering. Due to the filtering process, these statistics change over time reflecting dynamism in the environment. The filter is of the form

$$F(t) = e^{t/\tau}.$$
(1)

where τ is a time constant which can be configured to refine the behaviour of the system. The filter is implemented like

$$\bar{p}_{n+1} = (1-\alpha)\bar{p}_n + \alpha p_{n+1}.$$
(2)
$$\bar{\sigma}_{n+1} = (1-\alpha)\bar{\sigma}_n + \alpha \mid p_{n+1} - \bar{p}_{n+1} \mid.$$

where $\alpha = \tau \times f$, and f is the frame rate. If a pixel has a value that is more than 2σ from $\bar{p}_n$, then it is considered a foreground pixel.

To obtain better foreground images, two error types are considered. First, similar colours of background and human clothing often result in incomplete extraction of a moving people. We processed the binary foreground image extracted by dilation twice and then erosion. Secondly, erroneous extraction of non-moving pixels brings noise in a foreground image. The noise is removed using a size filter whereby blobs below a certain critical size are ignored.

2.2   Extracting Shape Pattern by Draping

When processing low-resolution images of people, it is difficult to obtain stable colour statistics of a target as the number of pixels belong to the target is small. Frequently, the foreground detection is also not complete and the problem becomes worse. Fig. 3(b) shows such a case for an image given as Fig. 3(a). In this situation, the boundary of the extracted foreground blob provides useful information on targets.

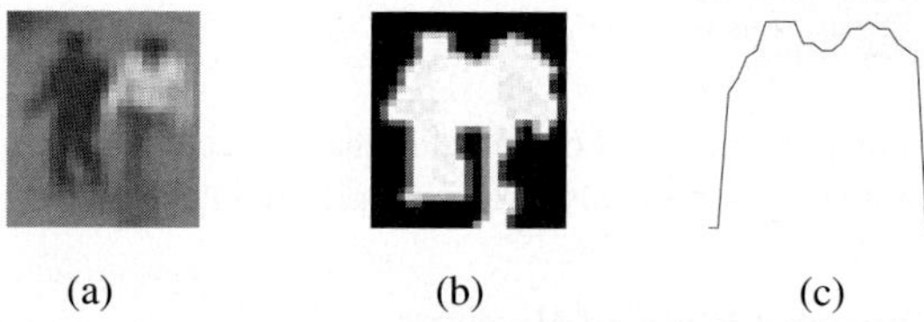

(a)   (b)   (c)

Fig. 3. Example of draping: (a) original image, (b) foreground blob extracted, (c) draping line

We propose to use a line connecting top pixel of each column in an image blob (we call it *'draping line'*) as a shape descriptor. As illustrated in Fig. 4, draping line is different from contour, which follows the boundary of a blob. Actually, draping is much simpler than boundary following and output data length is shorter. Although it misses most shape information of the side and lower parts of a human blob (e.g., neck and legs), it becomes hardly a problem in low resolution images. Instead, using a draping line is robust against the error of foreground extraction. Incomplete foreground extraction inside a blob does not change the line shape at all. Fig. 3(c) shows such a case. Top pixels of columns in an image usually represent upper parts of human bodies and the upper parts are relatively stable than lower parts in extracted images when a person moves [14].

Partial slope of a draping line is used as a clue to detect vertical segmentation position for an image blob of occluded humans. The slope of a pixel p_n on a draping line can be determined as one of three cases; $+1$ *if* $p_n < p_{n+1}$, 0 *if* $p_n = p_{n+1}$, and -1 *if* $p_n > p_{n+1}$. The draping line of Fig. 4(a), for example, can be represented by slope code like $\{+1, +1, +1, +1, -1, -1, +1, +1, 0, 0, -1, -1, -1, 0\}$. This representation is simpler than those used for boundary following such as chain code. Assuming Fig. 4 shows a real foreground image of two humans, a good segmentation position is at the 7th (or 8th) column. This position corresponds to slope pattern $\{-1, +1, +1\}$. We use a part of slope code as the input of a PNN employed to determine a segmentation position.

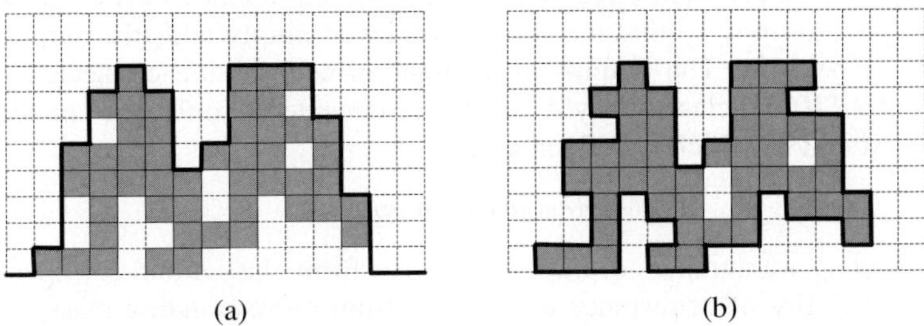

(a)   (b)

Fig. 4. Draping: (a) shows an example of draping line compared to boundary following illustrated in (b)

3 Probabilistic Neural Network

In this paper, we employ a PNN to find a good segmentation position. The PNN introduced by Specht [9] is essentially based on the well-known Bayesian classifier technique commonly used in many classical pattern recognition problems. When using a Bayesian classifier, a key step is estimating the probability density functions (pdf) from the training patterns. However, in some cases like the problem of this paper, estimating the pdf is difficult. In the PNN, a nonparametric estimation technique known as Parzen window [15] is used to construct the class-dependent pdf for each classification category required by Bayes' theorem. This allows determination of the chance a given vector pattern lies within a given category. By the following equation, we can estimate the conditional pdf

$$f_j(x) = ((2\pi)^{M/2}\sigma^M)^{-1}(PT_j)^{-1}\sum_{i=1}^{PT_j}\exp[-(x-x_i^j)^T(x-x_i^j)/2\sigma^2]. \tag{3}$$

where M stands for the dimensionality of input patterns, PT_j denotes the number of training patterns belong to class j, x_i^j represents the i'th input training pattern from class j, and σ is a smoothing parameter (Although we used σ as standard deviation in eq. (2), smoothing parameter is usually denoted as σ and we redefine the symbol here). Combining this with the relative frequency of each category, the PNN selects the most likely category for the given pattern vector.

A PNN consists of four layers: input layer, pattern layer, summation layer, and output layer. The input layer represents the m input variables $\{x_1, x_2, ..., x_m\}$. The input neurons distribute all of the variables x to all the neurons in the pattern layer. The pattern layer is fully connected to the input layer, with one neuron for each pattern in the training set. The weight values of the neurons in pattern layer are set equal to the different training patterns. Each pattern-layer neuron j performs a dot product on the input pattern vector **x** with the pattern stored as a weight vector such that $Z_j = xw_j$.

Then a nonlinear transfer function $\exp[(Z_j-1)/\sigma^2]$ is done before outputting to the summation neuron. Both **x** and $\mathbf{w}_j$ are normalized to unit length. The summation of the exponential terms is carried out by the summation-layer neurons. There is one summation neuron for each class. The weights on the connections to the summation layer are fixed to unity so that the summation layer simply adds the outputs from the pattern-layer neurons. The outputs of summation neurons are connected to a single output neuron. If two classes, say A and B, are considered, only one connection has a non-unity weight C, which is defined as

$$C = -(h_B l_B n_A)/(h_A l_A n_B). \tag{4}$$

where n_A and n_B are numbers of training patterns from class A and B respectively, h is a priori probability of occurrence of patterns from corresponding class, and l is the loss function associated with a decision.

4 PNN Design for Segmenting an Image Blob of Occluded People

We regard the segmentation problem as a classification problem. Segmentable positions constitute a class (say class S), and other positions constitute the other class (say class N). As there are only one or few good segmentation positions, the size of class S is much smaller than that of class N (i.e., much more training data for class N).

The input vector of the PNN designed has a part of slope code from a draping line as its elements. This means that the PNN estimates a good segmentation position from the local shape pattern of upper body parts of humans in occlusion. For a foreground

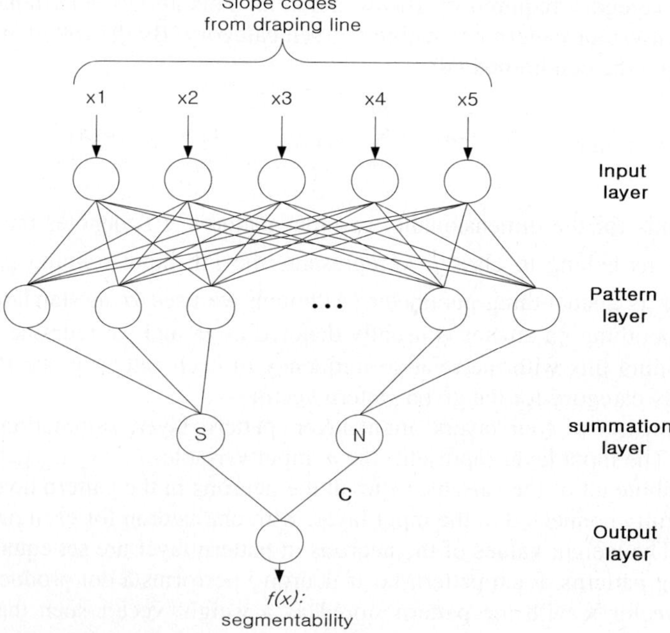

Fig. 5. PNN implementation to segment a foreground image blob of occluded humans. In our experiment relatively good results were obtained by a reasonable network size when five consequent elements of slope code were used as network input.

image blob in the width of c pixels, c-1 elements of slope code are first obtained from a draping line. The network output represents the goodness of segmentation at the centre position of a part of slope code inputted. In a typical pattern recognition problem using a PNN, the output is binary. But, in this paper, we get real output values for input vectors, and the position of the maximum output value is selected assuming an occluding person and an occluded person are in a foreground image blob. For image blobs to be used for training the PNN, the best segmentation positions are set manually and used for true output values. However, since the position at which a foreground image blob can be best segmented is difficult to determine but it can be within some range, three left and right positions of user determined segmentation position are also set to class S. In this way, the size of class S becomes larger.

5   Results

A PNN designed to segment image blobs of occluded people is tested. Various occlusion cases are synthesized randomly with five people sampled from real images as shown in Fig. 6. Training data for the PNN are obtained from 30 occlusion cases synthesized using Person 1 and Person 2, in addition to other 30 cases using Person1 and Person 3. Sixty testing data are generated by overlapping images of Person 2 and Person 3, and images of Person 4 and Person 5. Testing data from Person 4 and Person 5 are interesting because both images are not used for training. Furthermore extracting foreground pixels of Person 4 is quite incomplete particularly around his neck and shoulder, which are upper body parts and important in determining a draping line.

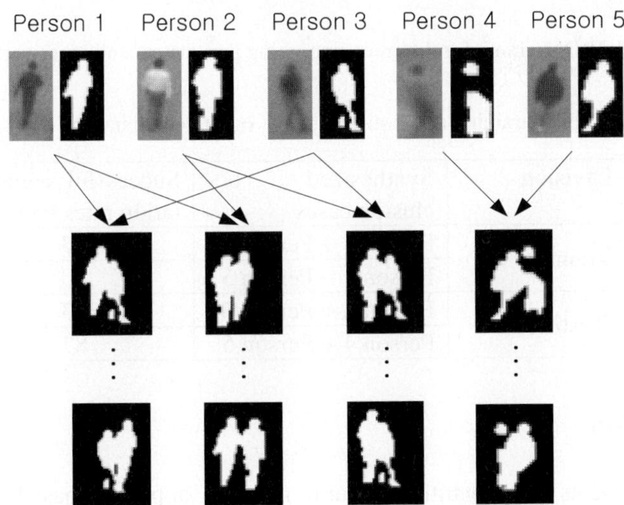

Fig. 6. Synthesized occlusions for training (Person 1 + Person 2 and Person 1 + Person 3) and for testing (Person 2 + Person 3 and Person 4 + Person 5) the PNN designed

Unlike multilayer perceptron networks (MLPNs), weights of connections in a PNN are fixed. So, training a PNN can be regarded as finding the best smoothing parameter σ for a set of training input vectors that maximizes the classification accuracy. Fig. 7 shows the training result. We used five elements of slope code for the network input with $\sigma = 1.2$. Although better training can be done with a larger network input vector, the differences are not significant while the training performance of using three input elements is notably bad. Table 1 shows the training and testing results. Notice that network showed still good result for the occlusions between Person 4 and Person 5.

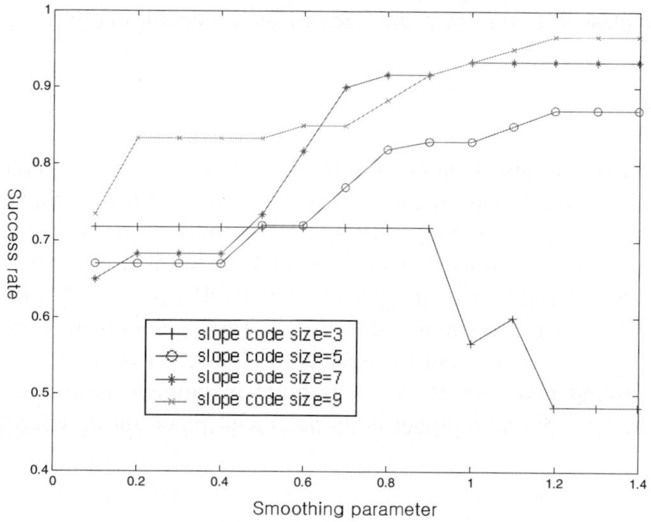

Fig. 7. PNN training for different smoothing parameters and input vectors

Table 1. PNN training and testing results (input vector size=5, $\sigma = 1.2$)

Division	Synthesized oc-clusion cases	Successful segmen-tation rates [%]
Training	Person 1 + Person 2	87
	Person 1 + Person 3	80
Testing	Person 2 + Person 3	83
	Person 4 + Person 5	87

6   Conclusion

This paper presents a probabilistic neural network approach based on the well-established Bayesian classifier method to segment a foreground image blob of occluded people for video surveillance applications. The network design technique and procedures to extract data to train the network are described. It is discussed that a PNN is useful to tackle occlusions particularly in low-resolution images, where existing contour-based technique is difficult to be used. Upper body parts of humans are relatively stable during human motion and patterns of partial draping line are used for the network learning. Unlike multilayer perceptrons, which need ad hoc design procedures and long training time, implementation of a PNN is straightforward and simple. In experiments using various synthesized occlusion cases of people, the PNN designed showed over 80% success rates.

References

1. Collins, R.T., Lipton, A.J., Kanade, T.: Introduction to the Special Section on Video Surveillance. IEEE Trans. Pattern Analysis and Machine Intelligence, Vol. 22, No. 8 (2000) 745-746
2. Haritaoglu, I, Harwood, D., Davis, L.S.: Ghost: A Human Body Part Labeling System Using Siluettes. In: Proc. Int. Conf. Pattern Recognition (1998) 77-82
3. Wren, C., Azarbayejani, A., Darrell, T., Pentland, A.: Pfinder: Real-Time Tracking of the Human Body. IEEE Trans. Pattern Analysis and Machine Intelligence, Vol. 22, No. 8 (1997) 780-785
4. KaewTrakulPong, P., Bowden, R.: A Real Time Adaptive Visual Surveillance System for Tracking Low-Resolution Colour Targets in Dynamically Changing Scenes. Image and Vision Computing 21 (2003) 913-929
5. Khan, S., Shah, M.: Tracking People in Presence of Occlusion. In: Proc. Asian Conf. Computer Vision (2000)
6. Grimson, W.E.L., Stauffer, C., Romano, R., Lee, L.: Using Adaptive Tracking to Classify and Monitor Activities in a Site. In: Proc. IEEE Computer Vision and Pattern Recognition (1998) 22-29
7. Sindhu, A.J., Morris, T.: A Region Based Approach to Tracking People before, during, and after Occlusions. In: Proc. IASTED Int. Conf. Visualization, Imaging, and Image Processing (2004)
8. Heikkila, J., Silven, O.: A Real-Time System for Monitoring Cyclists and Pedestrians. In: Proc. IEEE Workshop on Visual Surveillance (1999) 74-81
9. Specht, D.F.: Probabilistic Neural Networks for Classification, Mapping, or Associate Memory. In: Proc. IEEE Int. Conf. Neural Networks, Vol. I (1988) 525-532
10. Anderson, C., Burt, P., Wal, G. van der: Change Detection and Tracking Using Pyramid Transformation Techniques. In: Proc. SPIE, Vol. 579 (1985) 72-78
11. Barron, J., Fleet, D., Beauchemin, S.: Performance of Optical Flow Techniques. Int. Journal of Computer Vision, Vol. 12, No. 1 (1994) 42-77
12. Lipton, A.J., Fujiyoshi, H., Patil, R.S.: Moving Target Classification and Tracking from Real-Time Video. In: Proc. IEEE Workshop on Application of Computer Vision (1998) 8-14
13. Becanovic, V., Hosseiny, R. Indiveri, G.: Object Tracking Using Multiple Neuromorphic Vision Sensors. In: Proc. RoboCup (2004) 426-433
14. Haritaoglu, I, Harwood, D., Davis, L.S.: W4: Who? When? Where? What? A Real Time System for Detecting and Tracking People. In: Proc. Int. Conf. Face and Gesture Recognition (1998) 222-227
15. Parzen, E.: On Estimation of a Probability Density Function and Mode. Annals of Mathematical Statistics, Vol. 33 (1962) 1065-1076

Applying Learning Vector Quantization Neural Network for Fingerprint Matching

Ju Cheng Yang[1], Sook Yoon[2], and Dong Sun Park[1]

[1] Dept. of Infor.& Comm.Eng., Chonbuk National University,
Jeonju, Jeonbuk, 561-756, Korea
`yangjucheng@hotmail.com`
[2] Major in Multimedia Eng., Dept. of Infor. Eng., Mokpo National University,
Jeonnam, 534-729, Korea

Abstract. A novel method for fingerprint matching using Learning Vector Quantization (LVQ) Neural Network (NN) is proposed. A fingerprint image is preprocessed to remove the background and to enhance the image by eliminating the LL4 sub-band component of a hierarchical Discrete Wavelet Transform (DWT). Seven invariant moment features, called as a fingerCode, are extracted from only a certain region of interest (ROI) of the enhanced fingerprint. Then an LVQ NN is trained with the feature vectors for matching. Experimental results show the proposed method has better performance with faster speed and higher accuracy comparing to the Gabor feature-based fingerCode method.

1 Introduction

Because of durability and uniqueness, fingerprints have been widely applied in several fields of personal identification including criminal identification and access control. However, a reliable and efficient fingerprint recognition system is difficult achieved. Contrary to popular belief with extensive research efforts, automated fingerprint recognition is not a closed problem. Reliable matching of fingerprints is still a challenging problem. No two different impressions of a fingerprint are identical, even when they come from the same individual. The matching algorithm has to be therefore invariant to changes in orientation, displacement, elastic deformation, occlusion and missing features. Varying skin conditions may produce poor quality images that can lead to feature extraction errors and subsequently error during recognition.

Usually, there are two main methods for fingerprint matching: minutiae-based matching methods [1-2] and texture-based matching methods [3-4]. The more popular and widely used techniques, minutiae-based matching methods, use a feature vector extracted from the fingerprints and stored as sets of points in the multi-dimensional plane. The feature vector may contain minutiae's positions, orientations or both of them, etc. It essentially consists of finding the best alignment between the template and the input minutiae sets. However, minutiae-based matching methods may not utilize the rich discriminatory information available in the fingerprints and very time consuming [5]. The texture-based matching methods use different types of features from the fingerprint ridge patterns such as local orientation and frequency, ridge shape and texture information. The features may be extracted more reliably than

A. Sattar and B.H. Kang (Eds.): AI 2006, LNAI 4304, pp. 500–509, 2006.
© Springer-Verlag Berlin Heidelberg 2006

minutiae. Among various texture-based matching methods, Gabor feature-based fingerCode method [3] and its improved version [4] show relatively high performance comparing to previous works. These approaches use a fixed length representation, called as a fingerCode, to represent each fingerprint.

A good fingerprint recognition system should be robust to rotation, since we can not insure all the input fingerprints have alignments with the template fingerprints stored in the database. Inaccurate or wrong results may result in if the input fingerprints are rotated. To achieve approximate rotation invariance, the Gabor feature-based fingerCode method requires five templates per fingerprint in database during enrollment, and each corresponds to a rotated version of the original template. Furthermore, five more templates are generated by rotating the image by 11.25^0 during the matching procedure for rotation invariance. Therefore, each fingerprint is represented with ten associated templates stored in the database. The template with the minimum score is considered as the rotated version of the input fingerprint image. With this approach, the matching requires a larger storage space and a significantly high enroll time for rotation invariant. Therefore, new approaches need to explore to resolve the demerits of this method.

Invariant moments are one of the principal approaches used in image processing to describe the texture of a region. The set of moments [7] can be invariant to translation, rotation, and scale changes [6], and be used for fingerprint matching to solve the rotation problem.

To employ the merits of the invariant moments and to resolve the demerits of the Gabor feature-based fingerCode method, a new matching method is proposed in this paper. Instead of using multiple rotation templates for rotation invariant as in Gabor feature-based fingerCode method, we used seven invariant moments from a cropped enhanced fingerprint image (or a ROI) based on a reference point. In our approach, a vector with these seven invariant moments, called as a fingerCode as in ref [3], is used to represent a fingerprint. Then the matching itself is realized by an LVQ NN, which is a supervised pattern classification method with each output unit representing a particular class or category. LVQ NN learning has the advantage of very fast convergence [11] and has favorable classification ability. In addition, as one of the texture-based matching methods, the invariant moment fingerCode based method also takes of rich discriminatory information available in the fingerprints, so it is able to match with high accuracy.

The paper is organized as follows: The theory of invariant moments and LVQ NN are briefly reviewed in section 2 and 3 respectively. In section 4, the fingerprint enhancement is introduced. The proposed matching method is explained in details in section 5 and experimental results are shown in section 6. Finally, conclusion remarks are given in section 7.

2 Invariant Moments

In order to extract moment features, which may be invariant to translation, rotation and scale changes, from the ROI of the enhanced image, we used the moments defined in ref [6, 7].

For a 2-D continuous function $f(x,y)$, the moment of order (p +q) is defined as

$$m_{pq} = \int\limits_{-\infty}^{\infty} \int\limits_{-\infty}^{\infty} x^p y^q f(x, y) dx dy \qquad \text{for p, q= 0, 1, 2,...} \tag{1}$$

A uniqueness theorem states that if $f(x,y)$ is piecewise continuous and has nonzero values only in a finite part of the xy-plane, moment of all orders exist, and the moment sequence (m_{pq}) is uniquely determined by $f(x,y)$. Conversely, (m_{pq}) is uniquely determined by $f(x,y)$.

The central moments are defined as

$$\mu_{pq} = \int\limits_{-\infty}^{\infty} \int\limits_{-\infty}^{\infty} (x-\bar{x})^p (y-\bar{y})^q f(x, y) dx dy \tag{2}$$

Where $\qquad \bar{x} = \dfrac{m_{10}}{m_{00}} \qquad and \qquad \bar{y} = \dfrac{m_{01}}{m_{00}}$

If $f(x,y)$ is a digital image, then Eq.(2) becomes

$$\mu_{pq} = \sum_x \sum_y (x-\bar{x})^p (y-\bar{y})^q f(x, y) \tag{3}$$

and the normalized central moments, denoted η_{pq} , are defined as

$$\eta_{pq} = \frac{\mu_{pq}}{\mu_{00}^{\gamma}}, \quad \text{where} \quad \gamma = \frac{p+q}{2} + 1 \text{ for p+q = 2,3,....} \tag{4}$$

A set of seven invariant moments can be derived from the second and the third moments by Hu [7]. As shown below, Hu derived the expressions from algebraic invariants applied to the moment generating function under a rotation transformation. They consist of groups of nonlinear centralized moment expressions. The result is a set of absolute orthogonal moment invariants, which can be used for scale, position, and rotation invariant pattern identification.

$$\phi_1 = \eta_{20} + \eta_{02} \tag{5}$$

$$\phi_2 = (\eta_{20} - \eta_{02})^2 + 4\eta_{11}^2 \tag{6}$$

$$\phi_3 = (\eta_{30} - 3\eta_{12})^2 + (3\eta_{21} - 3\eta_{03})^2 \tag{7}$$

$$\phi_4 = (\eta_{30} + \eta_{12})^2 + (\eta_{21} + \eta_{03})^2 \tag{8}$$

$$\phi_5 = (\eta_{30} - 3\eta_{12})(\eta_{30} + \eta_{12})[(\eta_{30} + \eta_{12})^2 - 3(\eta_{21} + \eta_{03})^2]$$
$$+ (3\eta_{21} - \eta_{03})(\eta_{21} + \eta_{03})[3(\eta_{30} + \eta_{12})^2 - (\eta_{21} + \eta_{03})^2] \tag{9}$$

$$\phi_6 = (\eta_{20} - \eta_{02})(\eta_{30} + \eta_{12})[(\eta_{30} + \eta_{12})^2 - (\eta_{21} + \eta_{03})^2]$$
$$+ 4\eta_{11}(\eta_{30} + \eta_{12})(\eta_{21} + \eta_{03}) \tag{10}$$

$$\phi_7 = (3\eta_{21} - \eta_{03})(\eta_{30} + \eta_{12})[(\eta_{30} + \eta_{12})^2 - 3(\eta_{21} + \eta_{03})^2]$$
$$+ (3\eta_{12} - \eta_{30})(\eta_{21} + \eta_{03})[3(\eta_{30} + \eta_{12})^2 - (\eta_{21} + \eta_{03})^2] \qquad (11)$$

3 LVQ NN

The topology of LVQ NN was originally suggested by Tuevo Kohonen in the mid 80's, well after his original work in self-organizing maps. Both this network and self-organizing maps are based on the Kohonen layer, which is capable of sorting items into appropriate categories of similar objects. Specifically, LVQ NN is an artificial NN model used both for classification and image segmentation problems.

Topologically, the network contains an input layer, a single Kohonen layer and an output layer. An example network is shown in Fig.1. The output layer has as many processing elements as there are distinct categories, or classes. The Kohonen layer has a number of processing elements grouped for each of these classes. The number of processing elements per class depends upon the complexity of the input-output relationship. Usually, each class will have the same number of elements throughout the layer. It is the Kohonen layer that learns and performs relational classifications with the aid of a training set. This network uses supervised learning rules.

LVQ NN classifies its input data into groupings that it determines. Essentially, it maps an n-dimensional space into an m-dimensional space. That is it takes n inputs and produces m outputs. The networks can be trained to classify inputs while preserving the inherent topology of the training set. Topology preserving maps preserve nearest neighbor relationships in the training set such that input patterns which have not been previously learned will be categorized by their nearest neighbors in the training data.

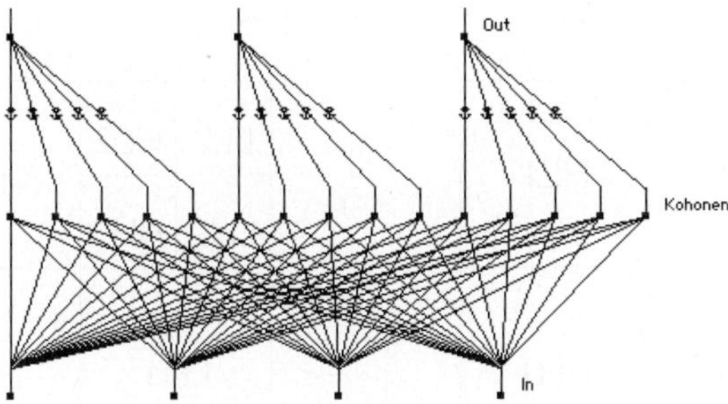

Fig. 1. The architecture of an LVQ NN

In the training mode, this supervised network uses the Kohonen layer such that the distance of a training vector to each processing element is computed and the nearest processing element is declared the winner. There is only one winner for the whole layer. The winner will enable only one output processing element to fire, announcing the class or category the input vector belonged to. If the winning element is in the expected class of the training vector, it is reinforced toward the training vector. If the winning element is not in the class of the training vector, the connection weights entering the processing element are moved away from the training vector. This later operation is referred to as repulsion. During this training process, individual processing elements assigned to a particular class migrate to the region associated with their specific class.

4   Fingerprint Image Enhancement

A fingerprint image is preprocessed to remove the background and to enhance the image by eliminating the LL4 sub-band component of a DWT. Since the details of images keep no background part of an original image, the values in these sub-band images, their combinations, or the derived features from these bands uniquely characterize features better than those from an original image. So we use the enhanced fingerprint image for processing instead of the original image. The theory of multi-resolution analysis with DWT is briefly introduced below.

In multi-resolution analysis, the DWT of a given signal may be interpreted as the decomposition of the signal into a set of time frequency functions by the use of translated and dilated basis functions of a mother wavelet. An image can be decomposed by DWT with divided into four sub-bands and critically sub-sampled by applying DWT as shown in Fig.2 (a). These sub-bands labeled LH1, HL1, and HH1 represent the finer scale wavelet coefficients or details of image, while the sub-band LL1 corresponds to coarse level coefficients or approximation image. To obtain the next coarse level of wavelet coefficients, the sub-band LL1 alone is further decomposed and critically sampled. This results in two-level wavelet decomposition as shown in Fig.2 (b). Similarly, to obtain further decomposition, LL2 can be used and this process continues until some final scale is reached. Also, we can reconstruct the original image by combining any combination of sub-band images from each level of decomposition.

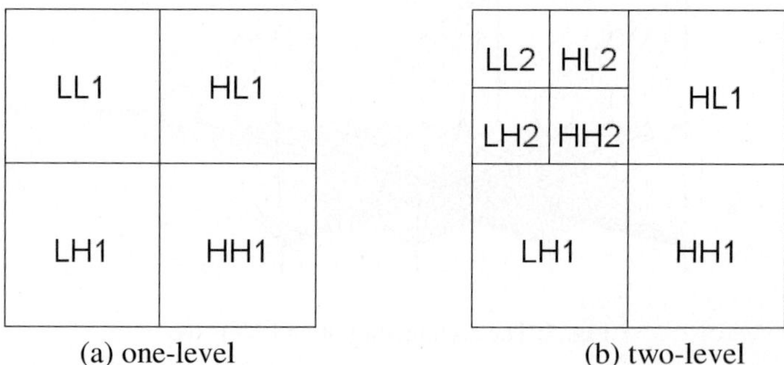

Fig. 2. Image decomposition by DWT (a) one-level (b) two-level

It is known that to subtract the reconstructed images with higher level decomposition compoments usually keeps more information than lower level wavelet decomposition. In this paper, wavelet decomposition is used to decompose the input fingerprint image into four different scales. The flow diagram of enhancement algorithm is shown in Fig.3. A four-level DWT is performed on an original fingerprint image (X) and reconstruct an image (X_4) using only LL4 sub-band component by the inverse DWT (IDWT). Then we subtract the reconstructed image(X_4) from the original image (X), and the residua (X - X_4) represents the enhanced version of the fingerprint image. Daubechies wavelet filters are reasonable tools for decomposing images [8], here for simplicity, Db4 is chosen as the wavelet basis.

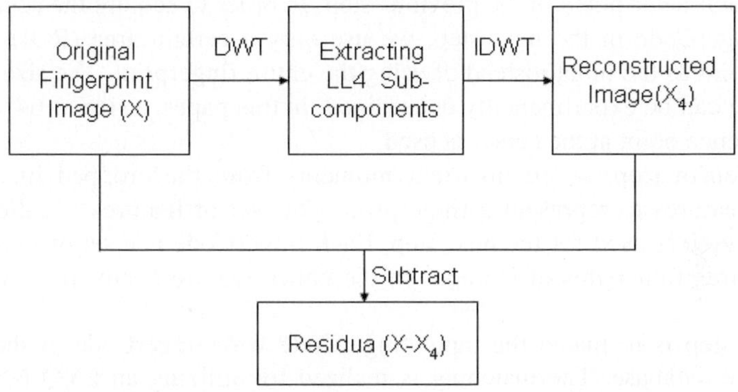

Fig. 3. The flow diagram of enhancement algorithm

5 Proposed Matching Approach

The proposed matching approach contains five main steps as shown in Fig.4.

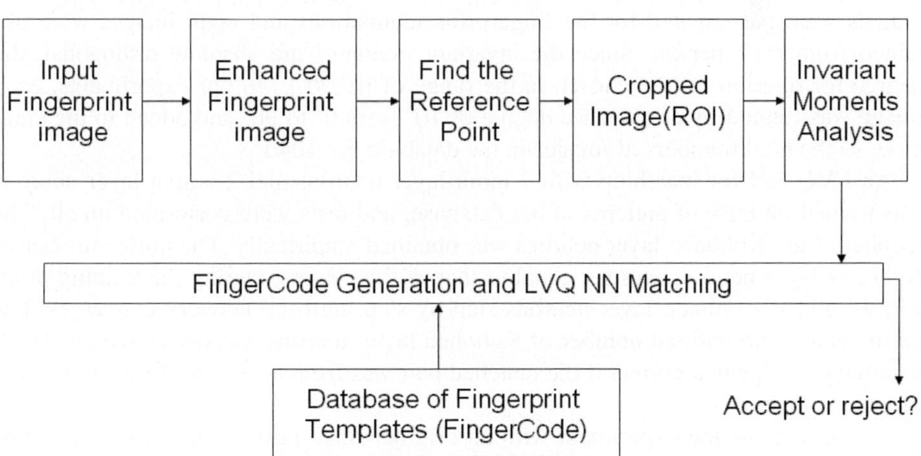

Fig. 4. The flow diagram of our matching algorithm

The first step is to enhance the original fingerprint image using DWT as descried in Section 4.

In the second step, we determine the reference point from the enhanced image of a fingerprint. The exact reference point is determined by using a complex filters [9]. Complex filters, applied to the orientation field in multiple resolution scales, can be used to detect the symmetry and the types of symmetry very well. The maximum response point of the complex filter can be considered as the reference point.

The third step is to segment fingerprint image by cropping the image based on the reference point determined in the previous step. In order to acquire the acute invariant moment fingerCode in the next step, we use only a certain area (ROI) around the reference point as the input instead of using the entire fingerprint. The size of the area for cropping can be experimentally determined, in this paper, a size of 64×64 image with a reference point at the center is used.

At the fourth step, seven invariant moments from the cropped image are extracted as features to represent a fingerprint. This set of features is called as a fingerCode, which is used for the next step. Each fingerCode is a set of long type data which requires four bytes of storage, so the entire feature vector requires 28 bytes of storage.

The last step is to match the input fingerCode with fingerCode of the templates stored in the database. The matching is realized by utilizing an LVQ NN to distinguish between the two corresponding fingerCodes of the test fingerprint image and temple fingerprint image in the database into a match and non-match.

6 Experimental Results

The fingerprint image database used in this experiment is the database [10] containing 104 fingerprint images of 256×256 pixels in 256 gray scale levels. Thirteen individuals were participated for the fingerprint acquisitions and eight images were obtained from each person. Since the invariant moments are absolute orthogonal, the degree for rotation can be chosen in the range of $[0^0, 90^0]$. In our experiments, each image was rotated by an increase degree of 10^0 from 0^0 to 90^0 and added to the database, so the total numbers of images in the database are 1040.

An LVQ NN for matching with 7 input layer neurons and 2 output layer neurons was trained on 80% of patterns in the database, and tests were performed on all. The number of the Kohonen layer neurons was obtained empirically. The initial number of Kohonen layer neurons was set equal to that of the classes. During the training process, we added Kohonen layer neurons step by step until the network converges. Experimentally, the optimal number of Kohonen layer neurons was determined to 16. A matching was labeled correct if the matched pair was from an identical fingerprint and incorrect otherwise.

An example of the experiment with an original fingerprint is shown in Fig.5. The experimental results of the enhanced image, the determined reference point and cropped images are shown in Fig.5 (b), (c), (d) respectively.

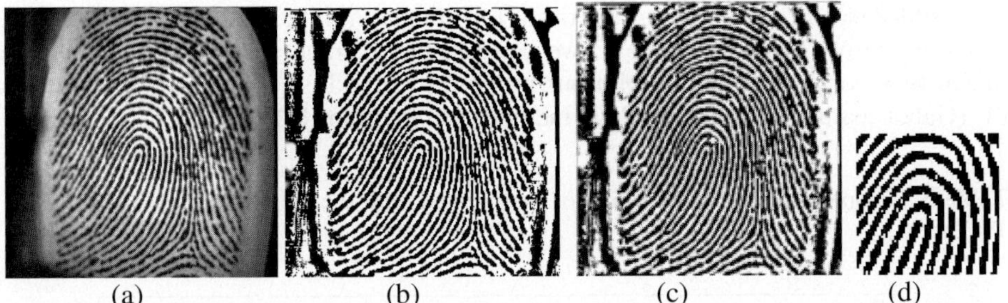

(a)   (b)   (c)   (d)

Fig. 5. (a) the original fingerprint of 2_5.tif (size 256 X 256) (b) the enhanced image (c) the determined reference point (d) the cropped image (size 64 X 64)

The rotation of the input fingerprint is the primary effect during the recognition process. The invariant moments of the rotated fingerprints are not the same with the original one [6]. Table 1 shows the seven invariant moments with different rotations of fingerprint 2_5.tif. We can see that the invariant moments of the rotated fingerprint are slight difference. However, severe difference exist between the different fingerprints, the seven invariant moments of different fingerprints (2_5.tif, 3_3.tif, 4_6.tif, 5_3.tif, 6_5.tif, 7_8.tif) are shown in table 2.

Table 1. The seven invariant moments with the different rotations of fingerprint 2_5.tif. (IM= Invariant Moments, RA= Rotation Angle)

IM / RA	ϕ_1	ϕ_2	ϕ_3	ϕ_4	ϕ_5	ϕ_6	ϕ_7
0	3.638998	5.790981	9.406715	8.109718	16.013728	11.257869	17.634734
10^0	3.165232	5.475452	3.185487	8.969868	16.608253	5.958084	11.207698
20^0	3.507695	6.609280	9.055225	5.727607	9.554618	11.127802	17.477927
30^0	3.363911	4.191439	4.382081	6.942658	16.439649	11.299093	15.889649
50^0	3.141671	4.813770	7.192613	5.396760	13.713264	7.988141	11.647312
70^0	3.156682	5.737732	8.328947	7.820797	5.588973	10.464465	16.922432

Table 2. The seven invariant moments with the different fingerprints 2_5.tif,3_3.tif,4_6.tif, 5_3.tif,6_5.tif,7_8.tif. (IM= Invariant Moments, DF= Different Image)

IM / DF	ϕ_1	ϕ_2	ϕ_3	ϕ_4	ϕ_5	ϕ_6	ϕ_7
2_5.tif	3.638998	5.790981	9.406715	8.109718	16.013728	11.257869	17.634734
3_3.tif	0.685542	0.363004	3.816905	3.349534	2.839905	3.399062	0.322383
4_6.tif	2.211359	5.202222	9.802340	9.135755	20.986924	11.606508	15.757113
5_3.tif	0.814568	1.872518	2.392312	0.499981	0.001734	0.257572	3.171350
6_5.tif	0.985913	0.836690	4.885397	5.320638	11.794020	5.639698	7.574177
7_8.tif	10.048416	19.105554	30.044806	30.02917	84.018507	39.706427	50.869092

In addition, the comparing of the time consuming with the methods is shown in table 3 as below. From the table 3, we can see the average enroll time and average match time for one matching with our proposed method is less than the method sha [4] (Gabor feature-based fingerCode method) (on Pentium 1.2 GHZ PC).

Table 3. Comparing the time consuing of our method with the method of Sha (M=Method , T= Time)

T / M	Average enroll time	Average match time
Method of Sha	0.54 seconds	0.18 seconds
Our proposed	0.24 seconds	0.09 seconds

For evaluating the recognition rate performance of the proposed method, a Receiver Operating Characteristic (ROC) is used, which is a plot of Genuine Acceptance Rate (GAR) against False Acceptance Rate (FAR). Fig.6. compares the ROCs of the method of Sha with our proposed matching method. Since the ROC curve of our proposed (solid line) is above the method of Sha (dashed line), we consider our proposed method performs better of accuracy than the method of Sha on this database. For example, at a 0.6% FAR, the GAR of our proposed matching method is 96.1%, while method of Sha is 92.3% respectively.

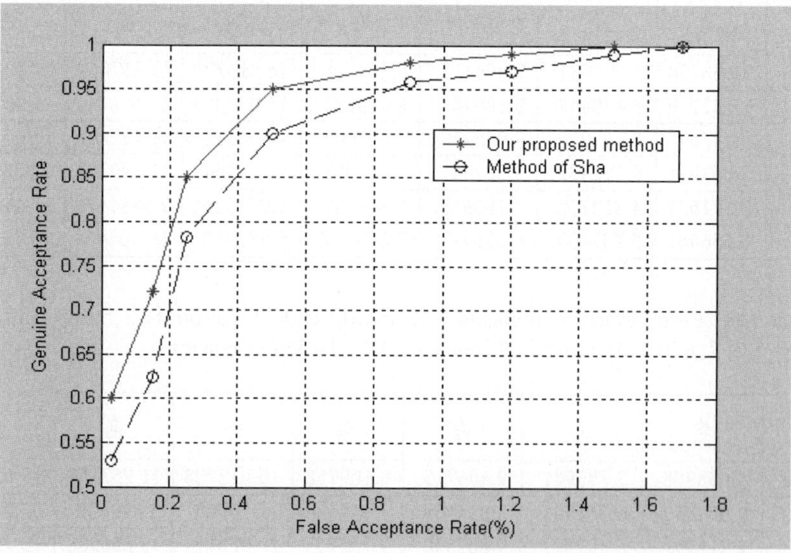

Fig. 6. The ROC curve comparing the recognition rate performance of the proposed method with method of Sha

7 Conclusion

In this paper, a novel method for fingerprint matching using an LVQ NN is proposed. A fingerprint image is firstly enhanced using DWT, and a set of invariant moment features are extracted from a ROI of the fingerprint enhanced images for matching with five main steps. Then fingerprint matching is realized by using an LVQ NN. Comparing with the method of Sha, the experimental results show that our proposed method has better performance of processing speed and accuracy.

Similarly to most other methods, our work is based on the reference point detection. When the reference point cannot be reliably detected, or it is close to the border of the fingerprint area, the feature extraction may be incomplete or incompatible with respect to the template. Further work should be proceeded to improve the accuracy and the reliability of our proposed method.

Acknowledgement

This work was supported by the second stage of Brain Korea 21 Project in 0000.

References

1. Jang X. and Yau W.Y.: Fingerprint Minutiae Matching Based on the Local and Global Structures, in Proc. Int. Conf. on Pattern Recognition (15th) Vol.2, pp. 1024-1045, 2000.
2. Liu J., Huang Z., and Chan K.: Direct Minutiae Extraction from Gray-Level Fingerprint Image by Relationship Examination, in Int. Conf. on Image Processing, Vol. 2, pp 427-430, 2000.
3. Jain A.K., Prabhakar S., Hong L., Pankanti, S.: Filterbank-based Fingerprint Matching, IEEE Transactions on Image Processing, vol9, no.5, pp 846-859,2000.
4. Sha L.F., Zhao F, Tang X.O.: Improved Fingercode for Filterbank-based Fingerprint Matching, International Conference on Image Processing, Vol. 2, pp: II-895-8, vol.3, 2003.
5. Maltoni Davide, Maio D. el.: Handbook of Fingerprint Recognition, Springer, 2003.
6. R. C. Gonzalez, R E. woods: Digital Image Processing (second edition), Prentice Hall, 2002.
7. Hu M-K.: Visual Pattern Recognition by Moment Invariants, IRE Trans. on Information Theory, IT-8:pp. 179-187, 1962.
8. Mallat S.: A Wavelet Tour of Signal Processing, Academic Press. 1998.
9. Kenneth Nilsson and Josef Bigun: Localization of Corresponding Points in Fingerprints by Complex Filtering, Pattern Recognition Letters, 24, pp2135-2144, 2003.
10. Biometric Systems Lab., University of Bologna, Cesena-Italy, (www.csr.unibo.it /research/biolab/).
11. Kohonen T. etc., LVQ_PAK: The Learning Vector Quantization Program Package Version 3.1, 1995, Espoo, Finland: Helsinki University of Technology. http://www.cis.hut.fi/ research/som_lvq_pak.shtml (Accessed 15 Oct 2001).

Moving Object Detecting Using Gradient Information, Three-Frame-Differencing and Connectivity Testing

Shuguang Zhao[1,2], Jun Zhao[2], Yuan Wang[2], and Xinlin Fu[2]

[1] College of Information Sciences and Technology, Donghua University
Shanghai 201620, P.R. China
[2] School of Electronic Engineering, Xidian University
Xi'an 710071, P.R. China
Sg.zhao@126.com, jun_zhao@263.net

Abstract. The motivation and the current status of moving object detecting research were reviewed firstly, and a novel method was proposed and verified in this paper. The method works mainly by using gradient information, three-frame-differencing and connectivity-testing-based noise reduction. The results of theoretical analyses and computer simulation show that the method has some advantages over its competitors, e.g., a wider application range, a less computation complexity and a faster processing speed. Thus, it is able to work rather robust in a noisy environment.

Keywords: Digital image identifying, video sequence processing, moving object detecting, three-frame-differencing, connectivity testing.

1   Introduction

Effectively detecting and extracting objects of interest from video sequences are vital in some AI field relevant to image, e.g., computer vision, robotics. They are prerequisite to further processing such as object tracking and object identifying. Processing speed (or computation complexity) and reliability are two primary problems to be solved for moving object detecting (MOD), which give rise to two important indexes to evaluate the relevant algorithms [1].

There are mainly three categories of conventional approaches to MOD: temporal differencing, background subtraction and optical flow. Temporal differencing techniques are based on the subtraction of two consecutive frames followed by threshold detection [2]. They can adapt well to a changing environment, but they are usually incapable of extracting the complete contours of moving objects [3]. Optical flow techniques detect moving objects by estimating the motion field and merging the motion vectors with similarities. They can work rather well in the presence of camera motion. However, most of them are computationally complex and sensitive to noise. If without help of specialized hardware, they will be incompetent to real-time processing of full-frame video streams [4]. Background-subtraction techniques detect moving objects by calculating the differences between the current frame and the background image for each pixel and applying threshold detection to them [5]. While they are capable of identifying almost all pixels involved with the motion, they are

A. Sattar and B.H. Kang (Eds.): AI 2006, LNAI 4304, pp. 510–518, 2006.

© Springer-Verlag Berlin Heidelberg 2006

very sensitive to dynamic changes of the environment, e.g., changes of lighting and extraneous events [3].

In this paper, we proposed a novel method for automated extraction of moving objects. The method proposed mainly consists of four ordered procedures: the image enhancement based on gradient information, the three-frame-differencing to identify the contours of moving objects, the threshold detection to extract the B&W contours from the background, and the connectivity testing to reduce the remaining background noise. Theoretical analyses and experimental results were presented below to validate it, which argue that it has some advantages over its competitors, including a wider application range, less computation amount, higher real-time performance and being more robust to interferences.

2   The Improved Method for MOD

2.1   Gradient Information Extracting

The key of temporal differencing is to ascertain exactly the difference between the objects and the noises so as to obtain as accurate as possible a threshold value for segmentation. As the signal noise ratio (SNR) varies with the video frames, the threshold value should be adaptable. In general a fixed threshold value is inappropriate for a robust autonomous vision system, and it should be calculated dynamically according to the image content. Besides the quantization noise inevitably introduced during the image acquisition process, automatic adjustments of aperture, which are probably required for monitoring complex scenes, make it more difficult to extract moving objects by direct image differencing. Fortunately, (in theory) image gradient information is much less affected the quantization noise and abrupt change of illumination, and consequently it could be used to effectively improve the robustness of MOD methods [6]. Inspired by this observation, gradient information extracting is chosen as the first procedure of our algorithm.

Sobel operators have better performance of edge detecting as compared with Roberts operators and Prewitt operators, and they are generally less computation expensive and more suitable for hardware realizations and real time tasks as compared with LOG operators and Canny operators [7, 8, 9]. Therefore, as indicated by Equation (1) and Equation (2), two Sobel operators are used to calculate the horizontal gradient and the vertical gradient for each pixel, respectively, and the magnitude of the gradient vector is estimated.

$$f_E(x, y, t) = \mid f(x, y, t) * H_X \mid + \mid f(x, y, t) * H_Y \mid \tag{1}$$

$$H_X = \frac{1}{4} \begin{bmatrix} -1 & 0 & 1 \\ -2 & 0 & 2 \\ -1 & 0 & 1 \end{bmatrix} \quad H_Y = \frac{1}{4} \begin{bmatrix} -1 & -2 & -1 \\ 0 & 0 & 0 \\ 1 & 2 & 1 \end{bmatrix} \tag{2}$$

Where, $f(x, y, t)$ denotes a pixel of a gray-scale image, $f_E(x, y, t)$ denotes the gradient of the pixel, H_X and H_Y denote the horizontal and the vertical transform matrix, respectively. For a color image to be processed, it should be converted beforehand into a gray image according to Equation (3)

$$f(x,y,t) = 0.587 \cdot f(x,y,t)_G + 0.114 \cdot f(x,y,t)_B + 0.29 \cdot f(x,y,t)_R \tag{3}$$

Using the above Equations, gradient information can be extracted one pixel by one pixel, forming the gradient-based images of a video sequence.

2.2 Three-Frame Differencing

Two-frame differencing that avails MOD of the difference of two consecutive frames is commonly used. Theoretical results [10] show that the difference of two consecutive frames (of a gray-scale image sequence) regarding a static (background) pixel, d(x, y, Δt), can be modeled with a zero-mean Normal distribution, N(0, σ^2). Its probability density function can be expressed as

$$p(d_k \mid H_0) = \exp[-d_k^2/(2\sigma^2)]/[\sqrt{2\pi}\sigma] \tag{4}$$

Where, d_k denotes a pixel of a gray-scale image, H_0 denotes the assumption that d_k is a static pixel (i.e., it belongs to the background), σ is the unknown variance. Equation (4) indicates that $p(d_k \mid H_0)$ depends on $(d_k/\sigma)^2$. The parameter σ can be estimated instantly from a fixed region defined in advance, or alternatively, estimated in advance and applied in a real-time system afterwards. On these bases, using the well-known "three sigma's" rule of mathematical statistics for hypothesis testing, a threshold value, $T=3\sigma$, can be properly set to subtract the background. As a result, two-frame differencing has some advantages such as being straightforward, easy to realize, and applicable to real-time tasks; meanwhile, it has some important problems in practice [11]. As shown in Fig. 1(d) and Fig. 1(e), its outcome usually contains both the moving objects' contours of interest and the unwanted contours of screened areas and the background texture.

To solve these problems, it is reasonable to ascertain the moving objects' contours in a frame by finding out the contours whose counterparts in the two relevant two-frame-differenced images coincide with each other. The operation is detailed in Equation (5). Where, $f_E(x, y, t\text{-}1)$, $f_E(x, y, t)$ and $f_E(x, y, t\text{+}1)$ denote a pixel of the preceding frame, the current frame and the sequent frame of a gradient-based image sequence, respectively. Such a processing, called three-frame-differencing, is inspired by the observation that the contours of moving objects of an original image highly correlate with that of moving regions of a temporal differenced image. As demonstrated in Fig. 1(f), the problem of moving object occlusion and texture reappearing could be solved rather effectively in this way [12].

$$D(x,y,\Delta t) = \mid f_E(x,y,t) - f_E(x,y,t-1)\mid \cdot \mid f_E(x,y,t) - f_E(x,y,t+1)\mid \tag{5}$$

The background contour in an image resulted from such a gradient-based three-frame differencing can be largely eliminated, whereas the contours of moving objects can be remarkably enhanced. Thus, by transforming the image into its binary (B&W)

counterpart (refer to Equation (6)) with a properly chosen threshold value, T, the moving objects can be extracted form the background automatically.

$$BD(x, y, \Delta t) = \begin{cases} 255 & D(x, y, \Delta t) > T \\ 0 & D(x, y, \Delta t) < T \end{cases} \tag{6}$$

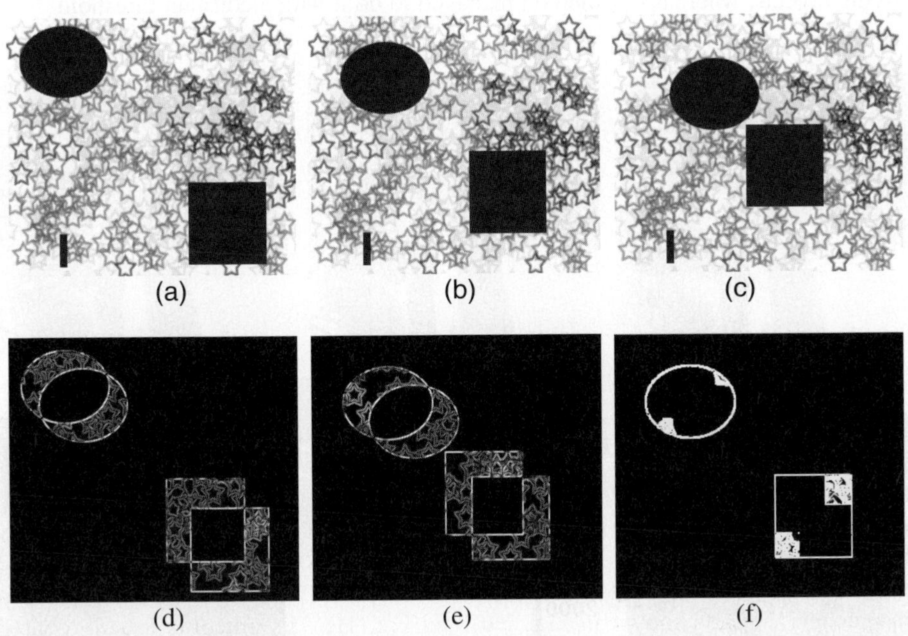

Fig. 1. Experimental comparison between two-frame differencing and three-frame differencing. The original images: (a) the 1st frame, $f(t\text{-}1)$, (b) the 2nd frame, $f(t)$, and (c) the 3rd frame $f(t\text{+}1)$. The results of two-frame differencing: (d) $|f_E(t)\text{-} f_E(t\text{-}1)|$ and (e) $|f_E(t)\text{-} f_E(t\text{+}1)|$. (f) The result of three-frame differencing: $|f_E(t)\text{-} f_E(t\text{+}1)|\cdot|f_E(t)\text{-} f_E(t\text{-}1)|$.

The method proposed above was validated by the experiments on actual image sequences. For example, the images shown in Fig. 2(a) are three consecutive frames that differ from each other in the overall brightness ($|\Delta|=4$). Of these images, the processing result of gradient-based three-frame differencing was shown in Fig. 2(b), while the result of traditional three-frame differencing was shown in Fig. 2(c). Obviously, the former exceed the latter in clearness of the contours.

With the original images that remain the same except a larger brightness variation, i.e., $|\Delta|=10$, further experiments were performed to verify the effects of utilizing gradient information. From the results shown in Fig. 3, it is observed that after being processed with our method, most pixels of the background have lower values in the brightness histogram, while most pixels of the moving objects have higher values. This fact argues that in conjunction with our method, a constant threshold, T, is usable to subtract the background. In contrast, after the traditional three-frame differencing, pixels of the background were not separated well from that of the moving objects. Consequently, an adaptive threshold is required to subtract the background, which will unfortunately bring biggish background noise.

Moreover, as shown in Fig. 4, further simulations shown that when brightness variations is no less than 14 (e.g., $|\Delta I|=14$), the traditional method failed to extract the moving objects, whereas our method managed to do it with a constant threshold.

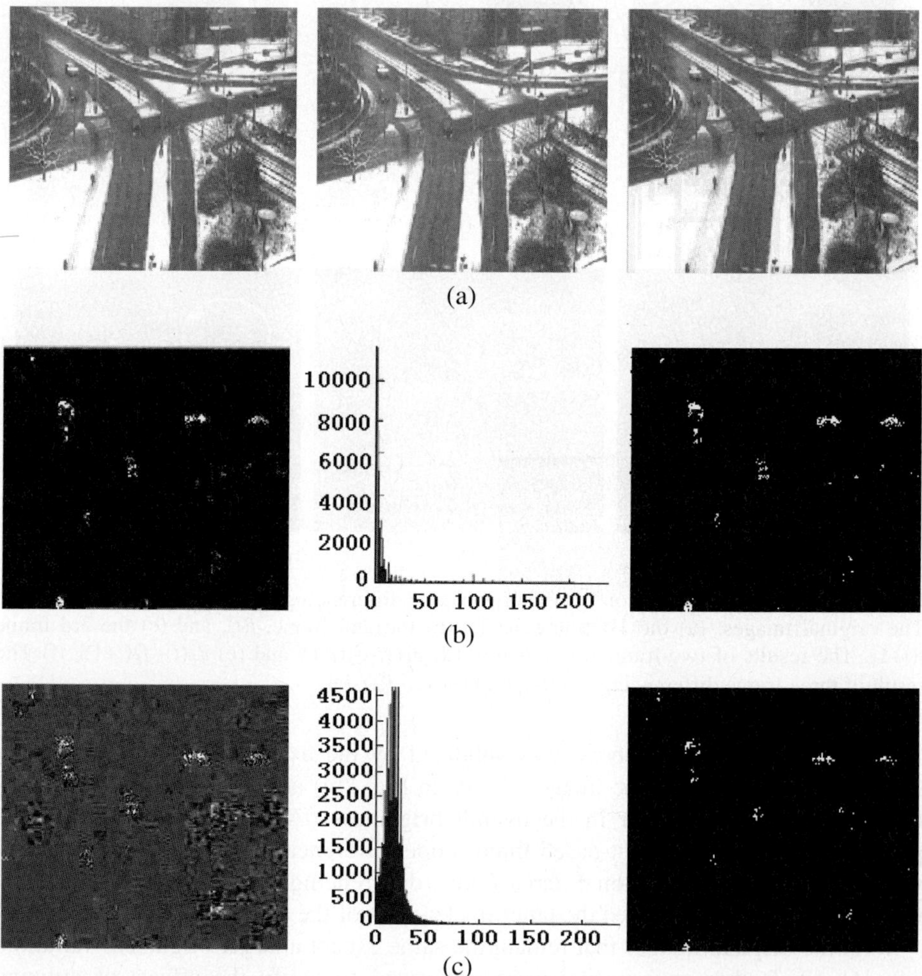

Fig. 2. Experimental comparison between our method and the traditional three-frame-differencing. (a) Three original consecutive frames differ in overall brightness (extracted from "dtnew winter.mpg", $|\Delta I|=4$). The differenced image, the brightness histogram, and the binary (B&W) differenced image of: (b) our method, (c) traditional three-frame-differencing.

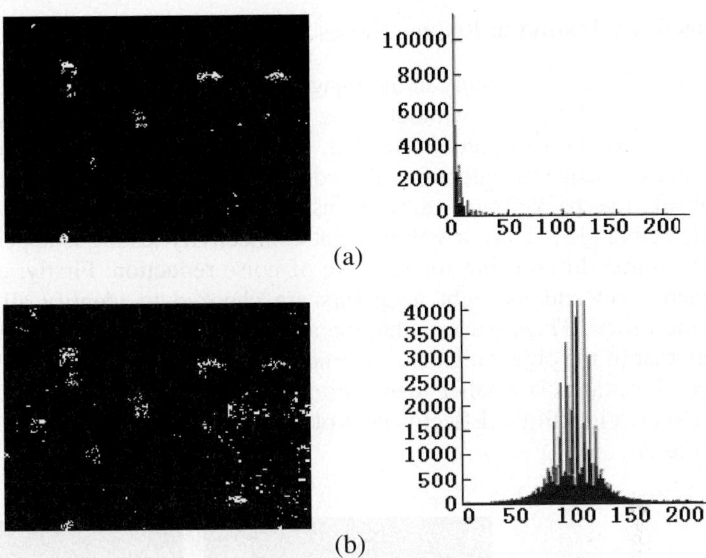

Fig. 3. Further comparison with an increased brightness variation, |Δ|=10. The differenced image and brightness histogram of: (b) our method, (c) the traditional three-frame differencing.

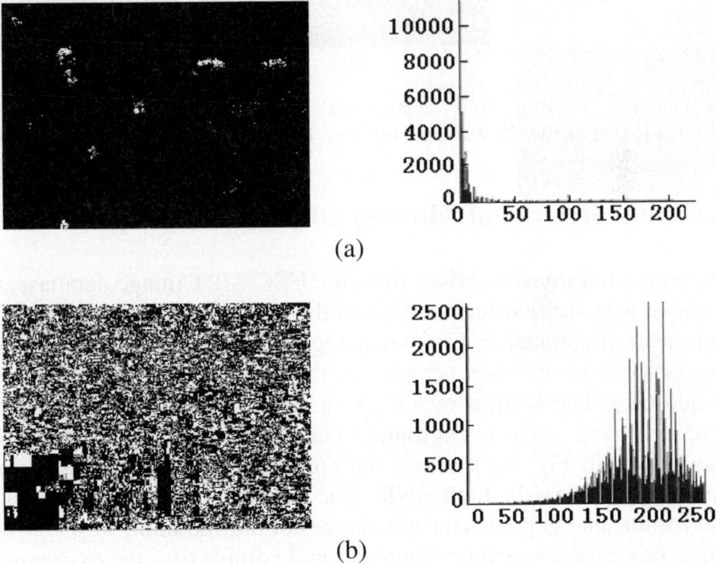

Fig. 4. Further comparison with an increased brightness variation, |Δ|=14. The differenced image and brightness histogram of: (a) our method, (b) the traditional three-frame differencing.

2.3 Connectivity Testing to Reduce Noises

In practice, noise exists inevitably in the images captured by a surveillance system. If the noise pixels are left alone without being reduced, they will definitely affect the effectiveness of MOD. Connectivity testing, which checks the connections between the pixels and consequently identifies the connected regions, is useful to reduce or even eliminate the background noise if used in conjunction with a subsequent threshold detecting [13]. Thus, a step of eight-connectivity testing is appended to the step of three-frame differencing for the sake of noise reduction. Firstly, connections between each pixel and its eight neighbors are checked to identify all connected regions in the image. Then, because that the areas of noise regions are usually much smaller than that of the object regions, most noise regions can be removed by clearing the regions whose area is smaller than a properly chosen threshold, say, 4% of the largest region area identified. Effectiveness of such a processing was demonstrated in Fig. 5 and Fig. 6.

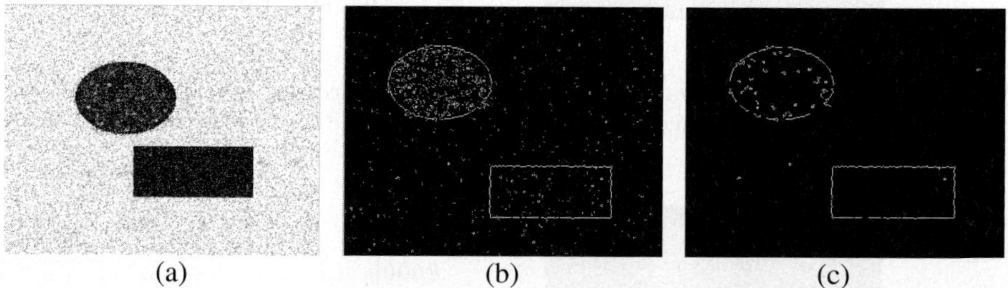

(a)   (b)   (c)

Fig. 5. An example of connectivity testing: (a) Noised image. (b) Result of three-frame-differencing. (c) Effect of the subsequent connectivity testing

3 Experiment Results of Moving Object Detecting

With some sequential images chosen from the USC-SIPT image database, the method proposed above was verified further. Due to the length limit, just two examples of the experiments were illustrated in Fig. 6 and Fig. 7, respectively. For the first example, the moving objects are two toy vehicles in the front of the scene, as shown in Fig. 6(a), 6(b) and 6(c). The interim result before connectivity-testing was shown in Fig. 6(d), where still exist some background noises. The final result after connectivity-testing was shown in Fig. 6(e), where the complete contours of the moving objects were obtained with a fairly high SNR. The second example as well as the other experiment results also validates the advantages of our method.

Regarding processing speed or computational complexity, the experimental results are also encouraging. The algorithm embodied by a VC6.0 application demonstrated an average processing speed of 15ms per frame, with some real video sequences (320×240 pixels, 24 bits color) and a typical desktop PC (Pentium®4 2.8G, 512M DDR). In contrast with it, the Mixed Gauss Model (MGM) [14] frequently adopted by background subtraction methods working with no shadow restraining processing just attained a processing speed of 38ms per frame under the same experimental condition.

Another disadvantage of MGM is that its detection performance tends to be largely affected by shadows, while our method does not.

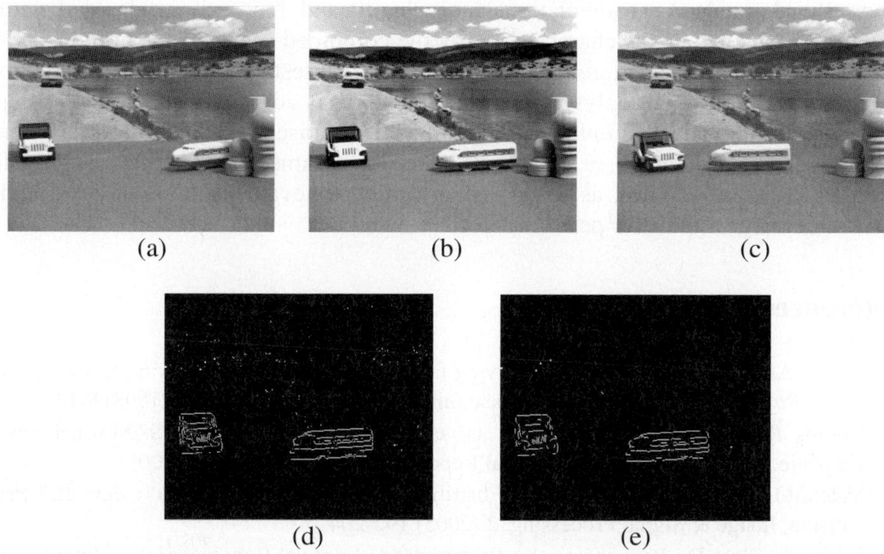

(a)   (b)   (c)

(d)   (e)

Fig. 6. Experimental results with a real image sequence (Toy Vehicle). The original images: (a) the 1st frame, (b) the 2nd frame and (c) the 3rd frame. The MOD results of: (d) three-frame-differencing, (e) three-frame-differencing and subsequent connectivity-testing.

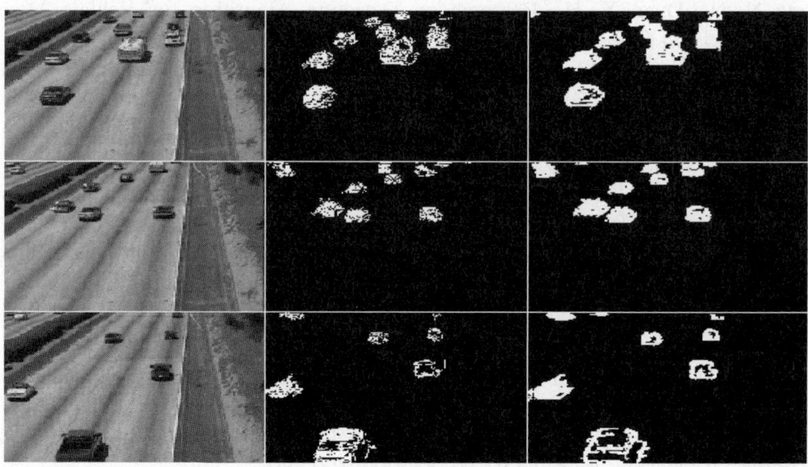

Fig. 7. Experimental results with another real image sequence (highwayII_raw.mpg). These images are captured at frame #35, frame #65, and frame #241. The first column shows the original image, the second column shows the result of the traditional three-frame differencing method, and the third column shows the result of our method.

4 Conclusions

A novel method for automatic extraction of moving objects was proposed in this paper. Besides a lower computational complexity and fewer demands of dynamic parameter adjustment, it is characterized by the combined use of gradient information extracting, three-frame-differencing and connectivity-testing-based noise reduction. Shown by the results of analyses and experiments, it is adaptable to random changes of the environmental illumination, quantization noise and aperture variations. Moreover, its main steps (e.g., Sobel operation, frame differencing) are fit for hardware realization. Thus, as compared with other relevant methods, it has a wider application range and better performances.

References

1. Lipton A., Fujiyoshi H., Patil R.: Moving target classification and tracking from real-time video. Proc IEEE Workshop on Applications of Computer Vision, NJ. (1998) 8-14
2. Collins Retal.: A system for video surveillance and monitoring: VSAM final report. Carnegie Mellon University, Technical Report: CMU-RI-TR-00-12. (2000)
3. Valera M., Velastin S. A.: Intelligent distributed surveillance systems: a review. IEE Proc. –Vision, Image & Signal Processing. 2 (2005) 192-204
4. Barron J., Fleet D., Beauchemin S.: Performance of optical flow techniques. International Journal of Computer Vision. 12 (1994) 42-77
5. Arseneau S., Cooperstock J.: Real-time image segmentation for action recognition. In: Proc IEEE Pacific Rim Conference on Communications, Computers and Signal Processing, Victoria, Canada. (1999) 86-89.
6. Tong N. N., Duan X. H.: Research on Detection Algorithm for Vehicle Monitoring System. The first Chinese Conference on Intelligent Surveillance, Beijing. (2002)
7. I.Abdou.: Quantitative Methods of Edge Detection. SCIPI Report 830, Image Processing, University of Southern California, Los Angeles. (1973).
8. Nevatia R., Babu K.R.: Linear Feature Extraction and Description. Computer Graphics and Image Processing. 13 (1980) 257-269
9. Abdou I.E., Pratt W.K.: Quantitative Design and Evaluation of Enhancement/Thresholding Edge Detectors. IEEE Proceedings. 5(1979) 753-763
10. Neri A., Colonnese S., Russo G., Talone P.: Automatic moving object and background separation. IEEE Transactions on Signal Processing. 10 (1998) 219-232
11. Yang L., Li Y. S.: Detection Contours of Multiple Moving Objects with Complex Background. Journal of Electronics and Information Technology. 2 (2005) 306-309
12. Dubuisson M. P., Jain A. K.: Contour extraction of moving objects in complex outdoor scenes. International Journal of Computer Vision. 14 (1995) 83-105
13. Haritaoglu I., Harwood D., Davis L. S.: W4: Real-Time Surveillance of People and Their Activities. IEEE Transactions on PAMI. 8 (2000) 809-830
14. Stauffer C., Grimson W. E. L.: Adaptive background mixture models for real-time tracking. In: Proceedings 1999 IEEE Computer Society Conference on Computer Vision and Pattern Recognition, Fort Collins CO, USA. 2 (1999) 252-259

3D Morphable Model Parameter Estimation

Nathan Faggian[1], Andrew P. Paplinski[1], and Jamie Sherrah[2]

[1] Monash University, Australia, Faculty of Information Technology, Clayton
[2] Clarity Visual Intelligence, Australia
{nathan.faggain, andrew.paplinski}@infotech.monash.edu.au,
jsherrah@clarityvi.com

Abstract. Estimating the structure of the human face is a long studied
and difficult task. In this paper we present a new method for estimating
facial structure from only a minimal number of salient feature points
across a video sequence. The presented method uses both an Extended
Kalman Filter (EKF) and a Kalman Filter (KF) to regress 3D Morphable
Model (3DMM) shape parameters and solve 3D pose using a simplified
camera model. A linear method for initializing the recursive pose filter is
provided. The convergence properties of the method are then evaluated
using synthetic data. Finally, using the same synthetic data the method
is demonstrated for both single image shape recovery and shape recovery
across a sequence.

1   Introduction

To accurately determine the fine three-dimensional shape of an object using
spatio-temporal information is a difficult and well studied task. Reffered to as
"structure from motion" there are many methods both in the linear and nonlinear domain. This paper presents a recursive filter approach to the problem using
Kalman filters. Through the use of two KFs we are able to track across variation
in pose and identity. The requirement for real-time operation is a complication
that is also addressed by our method since it is limited to extrapolation from a
small number of salient features.

There are a number of popular methods to compute the motion and structure
of the human face either from a single view or from a series of them. Recently
there has been much development of methods that combine salient feature tracking and 3D structure estimation as one step [11,6,1]. Generally these methods
lead to a search for the full set of parameters of a model and its camera matrices.
Given this trait such methods are dubbed parametric methods. Such methods
are well suited to tracking if the object is known. They restrict the object to a
certain domain and can produce highly accurate results. This paper contributes
another parametric fitting method that makes use of the popular KF as a way to
fit the high dimensional 3DMM to images using a subset of salient feature points.

As it stands there are already a number of parametric methods. Baker and
Matthews focus specifically on the extension of the popular Active Appearance
Model (AAM) of Cootes et al [5]. They called the extension the 2D+3D AAM

A. Sattar and B.H. Kang (Eds.): AI 2006, LNAI 4304, pp. 519–528, 2006.

© Springer-Verlag Berlin Heidelberg 2006

[11], where the key difference is a computationally efficient fitting strategy called Inverse Compositional Image Alignment [2]. In their method they first track a person specific AAM and then build a 3D Morphable Model using non-rigid factorization. The method requires a two step process of tracking and model construction before it can perform 3D tracking.

Dornaika and Ahlberk [6] also presented a modified AAM fitting algorithm. This was applied to the well known CANDIDE animation model using weak perspective projection. The solution demonstrated that a small generic model was capable of tracking accross pose end expression using their modified directed search. However the CANDIDE model is not a true statistical model of shape. Specifically it is not constructed from labelled training data.

In what could be considered a continuation of the AAM Anisetti et al. [1], extended the forwards additive approach of Lucas et al. [10] to fit the CANDIDE model to images. For the purpose of robust tracking they build a template from a single view, using a small number of points that correspond to vertices in their model. Using the defined template they are able to track across pose, illumination and expression. However in order for tracking to occur a 3D template needs to be defined.

In yet another development, Mittrapiyanuruk et al. [12] also demonstrated an accurate method for tracking rigid objects accross pose. Using a modified Gauss-Newton optimization to accommodate M-estimation they were able to demonstrate rotational accuracy within 5 degrees for estimates of yaw, pitch and roll. Similar to [1] a user labeled 3D template is required.

Our solution is similar to the mentioned methods but does not require hand labeling of a 3D template nor does it require a complex optimization framework. It only requires a somewhat accurate tracking of the feature points, which could be provided by an AAM or other tracking algorithm. Tracking can contain missing or inaccurate results because we make use of two Kalman filters, which allows for missing or bad data to be factored out of the fitting process. Our solution also exploits the relationship between the 3DMM and the subset of 2D feature points.

2   3D Morphable Models

A 3D Morphable Model (3DMM) is a statistical representation of both the 3D shape and texture of an object in a certain domain. 3DMMs are popular in the face domain [15,13] and are used for this task in our work. A 3DMM is built from 3D laser scans of human faces, which are then put into dense correspondence [3]. The 3D scans are then packed to form $M \times N$ shape and texture matrices. The dimensionality of the shape and texture matrices are very high, in our case we have 75 3D heads and 150,000 shape and texture points per head. Using the aligned heads PCA is used to construct a 3DMM and provide equations for shape and texture variation:

$$\hat{s} = \bar{s} + S \cdot \mathrm{diag}(\sigma_s) \cdot c_s \quad \hat{t} = \bar{t} + T \cdot \mathrm{diag}(\sigma_t) \cdot c_t \tag{1}$$

where $\hat{s}$ and $\hat{t}$ are novel $3N \times 1$ shape and texture vectors, $\bar{s}$ and $\bar{t}$ are the $3N \times 1$ mean shape and texture vectors, S and T are the $3N \times M$ column (eigenvectors) spaces of the shapes and textures, σ are the corresponding eigenvalues, c_s and c_t are shape and texture coefficients. These linear equations describe the variation of shape and texture within the span of 3D training data. The coefficients c_s, c_t are scaled by the corresponding eigenvalue σ_s, σ_t of the eigenvector S, T. In comparison 3DMMs are similar to the 2D Active Appearance Model (AAM) [5], although 3DMM model sizes are larger and pose and illumination changes can be addressed in the 3DMM fitting process.

2.1 3DMM Fitting

One of the most accurate fitting methods for 3DMMs is the gradient descent approach [15]. This is analysis by synthesis where the coefficients are inferred from a difference between a rendered head (image) and the input image. Effectively it is the minimization of the cost function:

$$\epsilon = ||F(c_s, c_t) - I||^2 \tag{2}$$

where F is a rendering function that when provided with shape and texture coefficients produces an image that is aligned with the input image, I. Speed is an issue for such a method and generally model coefficients are determined in a number of minutes.

2.2 Regularized 3DMM Point Based Fitting

In our work we would like to avoid optimization as much as possible to determine the shape coefficients for a 3DMM. Instead of fitting a 3DMM using texture information we use only 2D point features. Such features could be provided by any reliable face tracking algorithm, such as the AAM. We use the recently proposed method by Blanz et al [4] which is a concise and mathematically optimal method to reconstruct a 3DMM from a sparse set of either 2D or 3D feature points. The method has two advantages: 1) it relies on only linear operators and 2) operates in real-time (at the expense of model accuracy.) Using the assumption that only a small set of corresponding 2D feature points are available the method minimizes the cost function:

$$\epsilon = ||L \cdot V \cdot S \cdot \text{diag}(\sigma_s)c_s - r||^2 \tag{3}$$

where L is a camera matrix containing the projection parameters, V is a $2P \times 3N$ subset selection matrix, S is the $(3N \times M)$ column (eigenvectors) space of the training shapes, σ are the corresponding eigenvalues, c_s are shape coefficients and r is a $2P \times 1$ set of feature points. Solving for the coefficients is done in a statistically optimal way using the pseudo-inverse operation and a regularization term [4]:

$$c_s = U^T \frac{s_i}{s_i^2 + \eta} V y \tag{4}$$

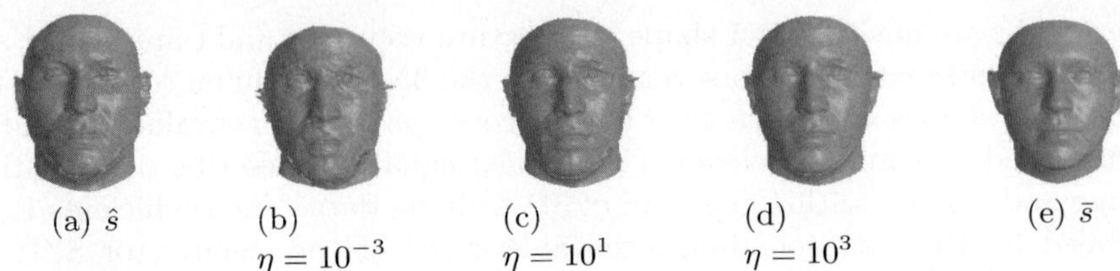

(a) $\hat{s}$ (b) $\eta = 10^{-3}$ (c) $\eta = 10^{1}$ (d) $\eta = 10^{3}$ (e) $\bar{s}$

Fig. 1. Variation of η, using only 46 Farkas [7] points, on a random 3D head, $\hat{s}$. Showing that as η increases the estimate approaches the mean, $\bar{s}$.

where both U, V are orthogonal matrices provided by Singular Value Decomposition [4]. However, when using this approach, the regularization term (η) is not intuitively set. For our selection of η we performed an experiment (in section 4), which shows that for our 3D data $\eta = 1$ was optimal for tracking. Figure 1 demonstrates the effect of changing η, as it increases the shape estimate approaches the mean of the 3DMM.

3   KF 3DMM Point-Based Fitting

In this paper we extend Blanz' method to work on a video sequence using two Kalman filters (KFs). Specifically we combine both an EKF and a KF. Using an EKF we first solve the non-linear problem of exterior orientation and then with a linear KF we determine the model coefficients assuming that the Blanz shape estimate is optimal given a reduced set of salient features (this is shown in [4]). With this assumption in hand our use of the KF means that we infer the state parameters, which are the 3DMM shape coefficients, directly from feature measurements. This also relies on the assumption that identity should not vary across pose changes. During our work we found the assumption that identity should remain constant is highly dependant on two key factors: 1) accurate measurements of salient features and 2) accurate estimates of 3D to 2D model alignment (exterior orientation), both of which are hard to guarantee. Using KFs we can effectively factor out these inaccuracies as process noise. Using KFs our technique models the human face using a 3DMM under the assumption that there is one true identity state (plus some process noise) across a video sequence.

Figure 2 shows the filter series for one image in the sequence. Given a measurement, r, a prior estimate of the camera matrix, and and a prior estimate of the identity parameters, the pose filter, Ω_P, estimates the current camera matrix, L_n. Using r and L_n, the identity filter, Ω_I, estimates the current identity parameters. Its output are the coefficients, c_s, estimated using the Blanz et al. method, Ω_V, for a given regularization term, η. From this Ω_I estimates the shape coefficients assuming it is measuring the state directly. The estimated camera matrix and shape coefficients then become the prior for the next image in the sequence.

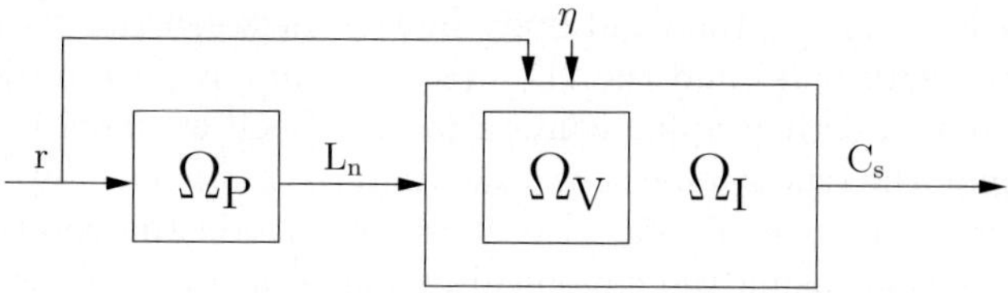

Fig. 2. DUAL KF Framework: Ω_P - pose filter, Ω_V - 3DMM fitting, Ω_I - identity filter

3.1 Kalman Filtering

Since its introduction by Rudolph Kalman in 1960 [8], the Kalman filter has been applied to a great many parameter estimation taks. This work demonstrates another application for the discrete filter. The Kalman filter is composed of two components, where the first is a set of time update equations:

$$\hat{x}_n = \hat{x}_{n-1} + Bu_{n-1} \tag{5}$$

$$P_n = AP_{n-1}A^T + Q \tag{6}$$

The time update equations project forward the current state $\hat{x}_n$ and error co-variances P_n. This leads to the estimate of the system state which is used in the measurement update equations:

$$K_n = P_n H^T (HP_n H^T + R)^{-1} \tag{7}$$

$$\hat{x}_n = \hat{x}_n K_n(z_n - H\hat{x}_n) \tag{8}$$

$$P_n = I - K_n HP_n \tag{9}$$

The measurement update equations are responsible for feedback. The equations change the apriori error covariance estimate (P_n) and obtain an improved posteriori state estimate ($\hat{x}_n$) of the system. The Kalman gain (K_n) is also calculated, a variable that could be considered as how much to "trust" the estimate versus the measurement.

For the task of 3D face tracking the relationship between the model and projects are non-linear in measurement. Both the projection and rotations of the 3DMM introduce the non-linearity. It is for this reason that the Extended Kalman Filter (EKF) is used. The EKF is a linearisation of the measurement and time update equations for the Kalman filter [9]. It is applied when a non linearity is present in the update or measurement equations. The difference in equations is subtle and well covered in the literature.

3.2 Pose Estimation

Solving for three-dimensional pose and structure is accomplished in our framework using two KFs. The first (pose, Ω_p) filter is used to solve the well-known exterior-orientation problem, minimizing the cost function:

$$\epsilon = ||(s \cdot R \cdot V \cdot (\bar{s} + S \cdot \text{diag}(\sigma) \cdot f(c_s)) + t) - r||^2 \tag{10}$$

This is the task of solving the rigid body motion between the 3D model (given a set of model coefficients) and the 2D measurements, r. As a projection model we chose the scaled orthographic camera model. Such a model is a reasonable approximation of the true perspective case when the distance between the object and the camera is not small. We also chose to encode the rotation as a first order approximation, using the canonical exponential form, shown here as the Rodriguez equation:

$$R = I_3 + \hat{w} \sin \theta + \hat{w}^2 (\cos\theta - 1) \tag{11}$$

where $\hat{w}$ is a skew symmetric matrix. If it is assumed that rotations are relatively small then the first order approximation is a good estimate for rotation. This in combination with the scaled projection forms the measurement equation for the EKF:

$$\begin{bmatrix} x_r \\ y_r \end{bmatrix} = \begin{bmatrix} s\ 0\ 0 \\ 0\ s\ 0 \end{bmatrix} \begin{bmatrix} 1 & -w_z & w_y \\ w_z & 1 & -w_x \\ -w_y & w_x & 1 \end{bmatrix} \begin{bmatrix} X_m \\ Y_m \\ Z_m \end{bmatrix} + \begin{bmatrix} t_x \\ t_y \\ t_z \end{bmatrix} \tag{12}$$

where (x_r, y_r) are a measurement point which correspond to the current iterations 3D model point (X_m, Y_m, Z_m), s is scale, (t_x, t_y, t_z) is translation and (w_x, w_y, w_z) encodes the incremental rotation of the model. After estimation, the parameters are packed into the camera matrix, L.

3.3   Structure Estimation

Structure is determined using the second identity KF to estimate the shape parameters for the 3DMM. These shape parameters span the facial structure variation that is present in our database of 3D heads. This is achieved in two steps. Firstly we use Blanz's [4] method to effectively minimize the cost function shown in (3), using the camera matrix L provided by the first pose EKF. We then use this estimate of shape coefficients, c_s, as input to a the second identity KF. Assuming that we are measuring the state directly and that the process noise will encode the variation due to bad estimates of pose and error of feature extractions. After applying the KF to the measurement the new estimate of the shape coefficients, c_s, is produced.

3.4   The Dual KF Algorithm

Our dual KF tracking is a combination of steps. Firstly a measurement is extracted from the video frame. The measurement, r, and the current estimate of 3D shape (for the first frame of a sequence this is the mean 3DMM shape) is used as an input to the the (pose) EKF. The pose EKF provides an estimate of the rigid body motion and projection parameters that align the projection of the estimated 3D shape to the measurement from the frame. This alignment information is then used to construct the camera matrix, L.

The rotation estimate of the EKF pose filter is then used to update a global estimate of rotation. It encodes the incremental change in object rotation. This is required because we assume that the feature tracking is happening in real-time and measurement differences between frames are small. A first order estimate for pose is good enough to model this situation. After this update L contains the global estimate of rotation.

Secondly we use the camera matrix L and extracted measurement, r, as input to the second identity filter (Ω_I). This provides an estimate of the 3D structure for the object, c_s. This is an estimate that is restricted to the span of the 3DMM and in the form of 3DMM coefficients. As the last step we use the new estimate of 3D shape to update the current estimate of 3D shape. The camera matrix, L and the estimate of 3D shape are ready for use in the next frame.

4   Experiments and Results

Using 75 laser scanned heads from the USF database [14], we constructed a 3DMM that retained 95% variance in shape and texture. This 3DMM was then used as the input to our DUAL KF fitting algorithm to test pose and identity estimates. We also determined the required 3D mapping for 47 Farkas [7] points and their projection to 2D, to generate the ground truth measurements for all experiments.

4.1   Optimizing η

To determine an optimal regularization term, η, we performed a simple optimization. Our experiment consisted of generating 100 random heads (using the 3DMM) and optimizing for the parameter η, minimizing the error function:

$$\epsilon = ||\Omega_I(\eta, r, L) - R_s||^2 \tag{13}$$

where $\Omega_I(r, L, \eta)$ is the Kalman Filter estimate of identity, using the regularization term η, the camera matrix L and salient features r, and R_s is the set of ground truth shape coefficients for the individual random head. Across the sample of randomly generated heads we found consistently that the minimum error was attained at $\eta = 1$. This was surprising since unity did not appear to present the most perceptually pleasing shape estimate, visualy.

For another experiment and to demonstrate the influence of η on the tracking task we used a fixed identity rotating in a fixed motion and then varied η. This was repeated for 100 randomly generated identities. As expected the tracking performance improves as the value of η is accurately set to unity, this is with respect to the minimum Euclidean distance to the ground truth (figure 3). During this experiment we also found empirically that the DUAL KF identity estimates converged to a solution 95% of the time within as few as 25 frames.

4.2   Pose Estimate Accuracy

Using our 3DMM it was possible to accurately test pose; the 3DMM provides ground truth data. Thus using a specified set of yaw, pitch and roll functions

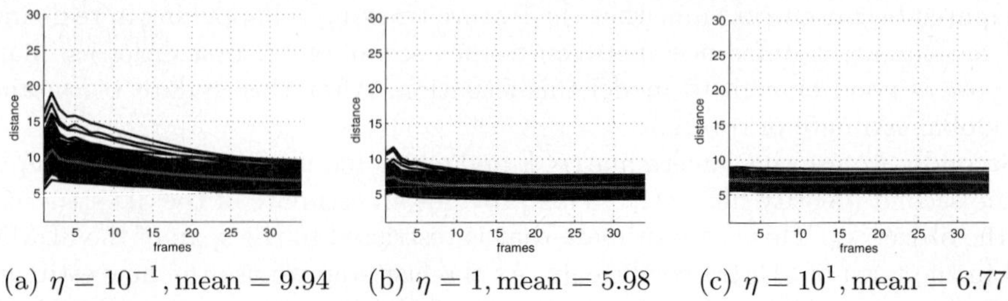

(a) $\eta = 10^{-1}$, mean = 9.94   (b) $\eta = 1$, mean = 5.98   (c) $\eta = 10^{1}$, mean = 6.77

Fig. 3. Identity KF convergence, distance versus measurements for different η

the method was examined by generating 100 random heads and using the dual KF approach to track the head motion. The accuracy was measured as the RMS error between the dual KF prediction and the actual measurement. When this error was below 1 pixel then the model was deemed to have accurately converged to the correct pose/identity.

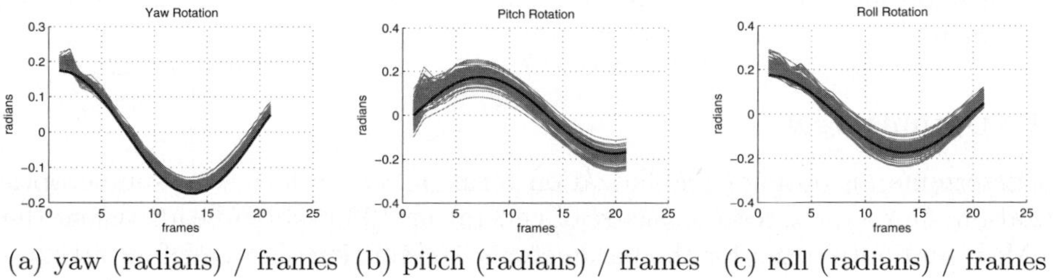

(a) yaw (radians) / frames (b) pitch (radians) / frames (c) roll (radians) / frames

Fig. 4. Pose estimation convergence

The results in figure 4 show that the dual KF approach is capable of tracking the motion quiet successfully. In the case of the 100 random heads, the mean error in yaw pitch and roll was 2.21, -2.87 and -3.34 degrees respectively. We found empirically that the dual KF approaches pose estimates converged to a solution 95% of the time within as few as 4 frames.

4.3   Structure Estimate Accuracy

To test the structure estimates we used the 3DMM to generate a random head and then attempted to fit the dual KF in two situations, 1) given only a single measurement and 2) as the random head was rotated in a similar fashion to the pose experiment. We used the Euclidean distance (as per first experiment) between the known and estimated heads as a measure of fitting error. In both cases we used only 46 Farkas [7] feature points for estimation.

The first experiment demonstrated that the DUAL KF could be used to reconstruct the dense 3D shape given only one frame. The fitting was achieved by repeating the single frame measurement. In this case the pose of the model

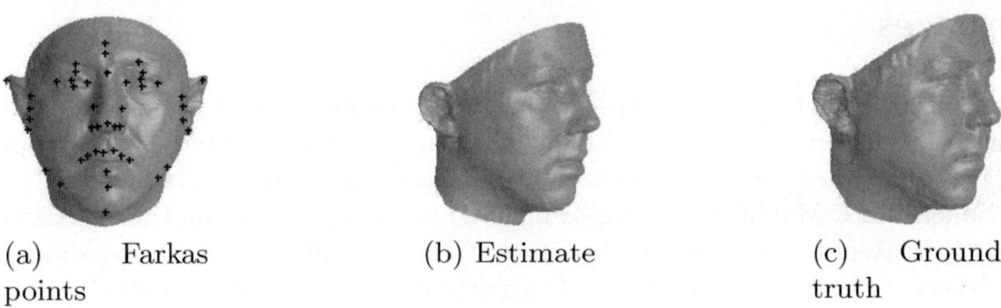

(a)   Farkas points   (b) Estimate   (c)   Ground truth

Fig. 5. Example of shape estimation from a single frame, distance from ground truth (8.76)

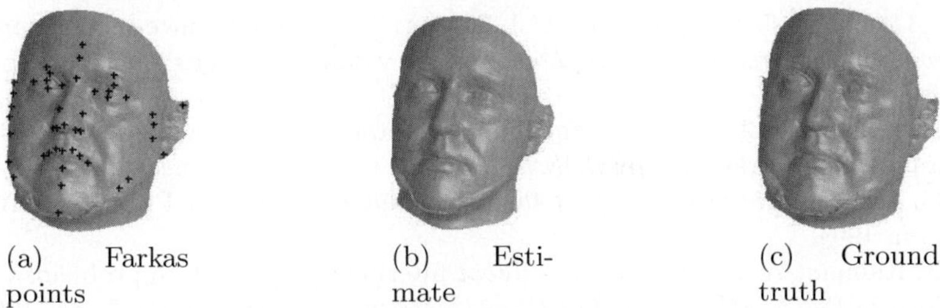

(a)   Farkas points   (b)   Estimate   (c)   Ground truth

Fig. 6. Example of shape estimation from a sequence, distance from ground truth (5.39)

was estimated in only 3 frames of measurements. An example fitting is shown in figure 5. This fitting achieved a distance from the ground truth, 8.76 units

The second experiment showed that the DUAL KF could refine an identity estimate across pose changes. Figure 6 shows the converged identity after a successful tracking. Note the distance from the ground truth, 5.39 units, smaller than the single image case.

5   Conclusion

In this paper we have proposed a new method to estimate the parameters for a 3DMM from a sparse set of salient features tracked across a video sequence. We have combined the benefits of Blanz's optimal 3D parameter estimation and the Kalman filter to achieve exactly this. Our method shows that the EKF estimates 3DMM pose parameters to an (average) accuracy of below 3 degrees, in as few as 4 measurements. Our method also shows that a single identity is inferred using a linear KF in as few as 25 images.

Acknowledgments

The authors would like to thank Sami Romdhani from the Faculty of Computer Science at UniBas for putting the USF data [14] into correspondence and the Australian Research Council for its continued funding.

References

1. M. Anisetti, V. Bellandi, L. Arnone, and F. Bevernia. Face tracking algorithm robust to pose, illumination and face expression changes: a 3D parametric approach. In *Proceedings of Computer Vision Theory and Applications*, 2006.
2. S. Baker and I. Matthews. Lucas-kanade 20 years on; a unifying framework: Part1. Technical Report 16, Robotics Institute, Carnegie Mellon University, 2002.
3. C. Basso, T. Vetter, and V. Blanz. Regularized 3D morphable models. In *Workshop on: Higher-Level Knowledge in 3D Modeling and Motion Analysis*, 2003.
4. V. Blanz, A. Mehl, T. Vetter, and H. Seidel. A statistical method for robust 3D surface reconstruction from sparse data. In *Second International Symposium on 3D Data Processing, Visualization and Transmission*, pages 293–300, 2004.
5. T.F. Cootes, G.J. Edwards, and C.J. Taylor. Active appearance models. In *Proc. European Conference on Computer Vision*, volume 2, pages 484–498. Springer, 1998.
6. F. Dornaika and J. Ahlberg. Face model adaptation for tracking and active appearance model training. In *British Machine Vision Conference*, 2003.
7. L. G. Farkas. *Anthropometry of the Head and Face*. Raven Press, New York, 2 edition, 1994.
8. R. E. Kalman. A new approach to linear filtering and prediction problems. *Transactions of the ASME–Journal of Basic Engineering*, 82(Series D):35–45, 1960.
9. T. Lefebvre, H. Bruyninckx, and J. De Schutter. Kalman filters for non-linear systems: a comparison of performance. *Control*, 77(7):639–653, May 2004.
10. B. Lucas and T. Kanade. An iterative image registration technique with an application to stereo vision. In *Proceedings of the International Joint Conference on Artificial Intelligence*, pages 674–679, 1981.
11. I. Matthews, J. Xiao, S. Baker, and T. Kanade. Real-time comined 2D+3D active appearance models. In *Computer Vision and Pattern Recognition*, volume 2, pages 535–542, 2004.
12. P. Mittrapiyanuruk, G. DeSouza, and A. Kak. Accurate 3D tracking of rigid objects with occlusion using active appearance models. In *Proceedings of the IEEE Workshop on Motion and Video Computing*, 2005.
13. S. Romdhani and T. Vetter. Estimating 3D shape and texture using pixel intensity, edges, specular highlights, texture constraints and a prior. In *IEEE Conference on Computer Vision and Pattern Recognition*, 2005.
14. S. Sarkar. USF DARPA humanID 3D face database. University of South Florida, Tampa, FL.
15. T. Vetter and V. Blanz. A morphable model for the synthesis of 3D faces. In *Siggraph 1999, Computer Graphics Proceedings*, pages 187–194, 1999.

Reconstructing Illumination Environment by Omnidirectional Camera Calibration

Yong-Ho Hwang and Hyun-Ki Hong

Dept. of Image Eng., Graduate School of Advanced Imaging Science, Multimedia and Film,
Chung-Ang Univ., 221 Huksuk-dong, Dongjak-ku, Seoul, 156-756, Korea
hwangyh@wm.cau.ac.kr, honghk@cau.ac.kr

Abstract. This paper presents a novel approach to reconstruct illumination environment by omnidirectional camera calibration. The camera positions are estimated by our method considering the inlier distribution. The light positions are computed with the intersection points of the rays starting from the camera positions toward the corresponding points between two images. In addition, our method can integrate various synthetic objects in the real photographs by using the distributed ray tracing and HDR (High Dynamic Range) radiance map. Simulation results showed that we can generate photo-realistic image synthesis in the reconstructed illumination environment.

1   Introduction

The seamless integration of synthetic objects with real photographs or video images has long been one of the central topics in computer vision and computer graphics [1,2]. In special, it should be maintained both the geometric property and illumination effects. More specifically, in order to generate a high quality synthesized image, we have to make an appropriate view of the synthetic object in accordance with the real camera movements, and then shade the synthetic objects so that they appear to be illuminated by the same lights as another object in the background image. However, in general, the integration of synthetic objects in a realistic and believable way is labor intensive process and not always successful due to the enormous complexities of real-world. Therefore, an automatic tool to reconstruct 3D structure and illumination environment of the scene allows users to alleviate much effort for realistic composition.

An omnidirectional camera system is given increasing interest by researchers working in computer vision, because it can capture large part of a surrounding scene. Therefore, wide angle of view often makes it possible to establish many spacious point correspondences which lead to more complete 3D reconstruction from few images. In addition, it is widely used to capture the scene and illumination from all directions from far less number of images.

This paper presents a novel method for obtaining the information of the light source in the real image to illuminate the computer generated images. At the first, we capture the scene structure and the illumination from all directions by using omnidirectional images taken by a digital camera with a fisheye lens. In the second step, we

A. Sattar and B.H. Kang (Eds.): AI 2006, LNAI 4304, pp. 529–535, 2006.
© Springer-Verlag Berlin Heidelberg 2006

estimate the projection model and the extrinsic parameters—the translation and the rotation—of the camera by considering a distribution of the inlier set. After calibrating the camera, we reconstruct the scene structure and the positions of the light sources automatically from the correspondences over images. In final, our method can generate photo-realistic scenes illuminated by distant natural light sources in 3D illumination environment.

The remainder of this paper is structured as follows: Sec. 2 reviews previous studies on calibration of omnidirectional camera and image synthesis, and then the calibration method of the omnidirectional camera is discussed in Sec. 3. Sec. 4 presents 3D reconstruction of the scene and generating the illuminated scene images with the identified lights. Finally, the conclusion is described in Sec. 5.

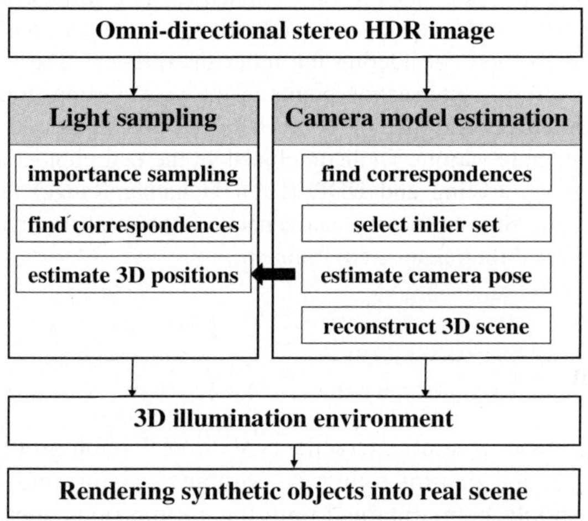

Fig. 1. Block diagram for the proposed method

2 Previous Studies

Many image synthesis researches for illuminating scenes and objects with images of light from the real world have been proposed up to now. The first part of the table 1 shows previous studies may be classified according to three viewpoints: how to construct a geometric model of the scene, how to capture illuminations of the scene, and how to render the synthetic scene.

Founier et al cannot consider illumination from outside of the input image without user's specification owing to using 2D photographs [1]. Debevec's, which is one of the most representative works, constructs light-based model by mapping reflections on a spherical mirror onto a geometric model of the scene. Therefore, a degree of realism of the rendering image depends on selecting viewpoints for observing the mirror so that the reflections on the mirror can cover the entire geometric model of the scene [2]. Gibson et al proposed integration algorithms at interactive rates for augmented reality applications by using a single graphics pipeline [3].

Table 1. Previous studies on image synthesis and calibration of omnidirectional images

Work	Geometry	Capturing illumination	Rendering
image synthesis	user's specification for 3D construction of the scene [1]	Photographs	radiosity
	user's input and light-based model: distant and local scene, synthetic objects [2]	HDRI of scene reflections on a spherical mirror	RADIANCE system
	user's specification [3]	HDRI by sphere mapping	OpenGL
	Image Acquisition	**Limitations**	
calibration of omnidirectional images	self-calibration of fisheye lens and capturing the spherical panorama with 3~4 images [4]	- restricted camera setup: by rotating the camera 90 degree - based on equi-distance camera model	
	using omnidirectional pairs for scene modeling & scene radiance computing [6]	- 3D reconstruction and lighting - strong pre-calibration and scene constraints	
	automatic reconstruction of un-calibrated omnidirectional images [9]	- applications problems in image sequence: correspondence, frame grouping	

In addition, the second part of table 1 shows the calibration methods of the omnidirectional camera. Xiong et al register four fisheye lens images to create the spherical panorama, while self-calibrating its distortion and field of view [4]. However, camera setting is required, and the calibration results may be incorrect according to lens because it is based on equi-distance camera model. Sato et al simplify user's direct specification of a geometric model of the scene by using an omnidirectional stereo algorithm, and measure the radiance distribution. However, it is required in advance a strong camera calibration for capturing positions and internal parameters, which is complex and difficult process [6]. Pajdla et al use the epipolar constraint to estimate both the camera model and the extrinsic parameters simultaneously [7]. The method may be affected by sampling corresponding points between a pair of the omindirectional images [9]. However, it requires further consideration to select a proper inlier set.

3   Omnidirectional Camera Estimation

The camera projection model describes how 3D scene is transformed into 2D image. The light rays are emanated from the camera center and determined by a rotationally symmetric mapping function with respect to the the radius of the point and the angle between a ray and the optical axis. We derive an one-parametric non-linear model for Nikon FC-E8 fisheye converter [10].

One of the main problems is that estimating the essential matrix is sensitive to the point location errors. In order to cope with the unavoidable outliers inherent in the correspondence matches, our method is based on 9-points RANSAC that calculates the point distribution for each essential matrix. Since the essential matrix contains

2   Gaussian-Cylindroid Color Model

Although any camera system is used under the same illuminators, the single-colored objects have different color distributions due to irregular brightness and the viewing geometry of the camera system. Since chrominance tends to be less affected by brightness changes, such chromatic color spaces as the UV plane of the YUV space, the HS plane of the HSI space, and the normalized RG plane of the RGB space have often been used in color based vision applications. However, the chromatic color distribution ignoring the luminance component is not invariant enough to disregard the the effect of brightness changes completely. Some adaptive methods using the shape of the color histogram were proposed to cope with the brightness changes, but they still have difficulties in being adapted to the abrupt changes in brightness and the initial detection(segmentation) problem under various brightness conditions.

We present a new parametric color model to fit all chromatic distributions on the effective brightness range.

2.1   Elliptical Analysis of Chromatic Color Distribution

The assumption that a single-colored object has a Gaussian distribution on any 2D chromatic color plane gives

$$p(\mathbf{c}) = \frac{1}{2\pi\sqrt{|\mathbf{\Sigma}|}} \exp\left(\frac{(\mathbf{c} - \bar{\mathbf{c}})^T \mathbf{\Sigma}^{-1}(\mathbf{c} - \bar{\mathbf{c}})}{-2}\right) \tag{1}$$

where $\mathbf{c}$ is a 2D vector of chromatic color components, and $\bar{\mathbf{c}}$ and $\mathbf{\Sigma}$ are its mean and covariance, respectively. Letting a^2 and b^2 are the eigenvalues of $\mathbf{\Sigma}$, from the elliptical geometry we can get the ellipse with the center $\bar{\mathbf{c}}$, the major axis length a, the minor axis length b and the orientation of the major axis ϕ:

$$(\mathbf{c} - \bar{\mathbf{c}})^T \left(\mathbf{R}(\phi)\mathbf{L}(a,b)\mathbf{R}(\phi)^T\right)^{-1} (\mathbf{c} - \bar{\mathbf{c}}) = 1 \tag{2}$$

where

$$\mathbf{R}(\phi) = \begin{pmatrix} \cos\phi & -\sin\phi \\ \sin\phi & \cos\phi \end{pmatrix}, \quad \mathbf{L}(a,b) = \begin{pmatrix} a^2 & 0 \\ 0 & b^2 \end{pmatrix}, \quad \bar{\mathbf{c}} = (\bar{c}_x \ \bar{c}_y)^T. \tag{3}$$

2.2   Gaussian-Cylindroid Color Model

One ellipse represented by $\chi = (\bar{c}_x \ \bar{c}_y \ a \ b \ \phi)^T$ from (2) corresponds to one Gaussian distribution of the chromatic color components as shown in Figure 1. This section gives the method to obtain the Gaussian-cylindroid color model representing all chromatic color distributions with respect to the effective brightness range.

Fist of all, the effective brightness range needs to be uniformly divided with any fixed interval. Then the training samples(pixels) are grouped by each divided

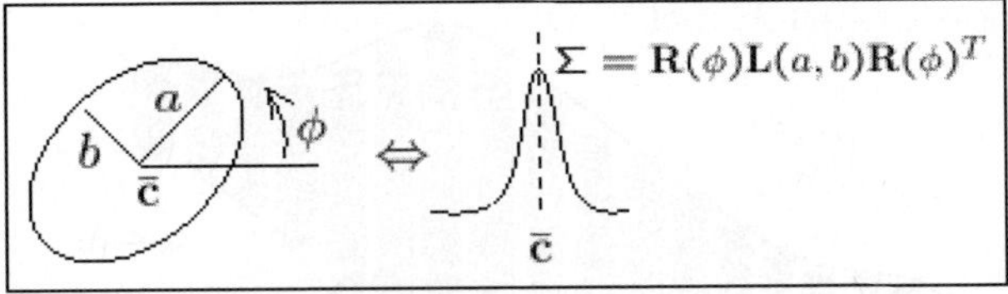

Fig. 1. Ellipse corresponding to Gaussian Distribution

brightness range, which is the quantization of the brightness. Applying the elliptical analysis to the mean and the covariance of the chromatic components at each group, we obtain

$$\chi_q^{(f)} = (\bar{c}_x^{(f)}\ \bar{c}_y^{(f)}\ a^{(f)}\ b^{(f)}\ \phi^{(f)})_q^T, \quad q = 1, 2, \cdots, M \tag{4}$$

where q is a quantized brightness.

B-spline functions have been used widely in tracking of objects represented by contours [1][7]. Now B-spline functions are introduced to serve as a tool for parametric construction of all chromatic color distributions over the effective brightness range.

B-spline function can be shown as the matrix product of the B-spline basis vector and the control vector [1]:

$$x(s) = \mathbf{B(s)}^T \mathbf{Q^x} \tag{5}$$

where

$$\mathbf{B(s)} = (B_0(s)\ B_1(s)\ \cdots\ B_{N_B-1}(s))^T$$
$$\mathbf{Q^x} = (x_0\ x_1\ \cdots\ x_{N_B-1})^T.$$

Likewise, the B-spline fitted elliptical parameters are shown as

$$\chi(s) = (\bar{c}_x(s)\ \bar{c}_y(s)\ a(s)\ b(s)\ \phi(s))^T$$
$$= (\mathbf{Q^{\bar{c}_x}}\ \mathbf{Q^{\bar{c}_y}}\ \mathbf{Q^a}\ \mathbf{Q^b}\ \mathbf{Q^\phi})^T \mathbf{B(s)} \tag{6}$$

and geometrically form a cylindroid in the 3-D color space(chrominance 2-D+brightness 1-D) as described in Figure 2.

The surface-displacement measure $\mathcal{D}\left(\chi^{(f)}, \chi(s)\right)$ is defined as

$$\mathcal{D}\left(\chi^{(f)}, \chi(s)\right) = \frac{1}{M}\Sigma_{q=1}^{M} d\left(\chi_q^{(f)}, \chi(s_q)\right) \tag{7}$$

$$= \frac{1}{MN}\Sigma_{q=1}^{M}\Sigma_{i=1}^{N} \parallel \mathbf{R}_{\phi_q}^{(f)}\mathbf{P}_{\theta_i}\mathbf{r}_q^{(f)} - \mathbf{R}_{\phi(s_q)}\mathbf{P}_{\theta_i}\mathbf{r}(s_q) + \bar{\mathbf{c}}_q^{(f)} - \bar{\mathbf{c}}(s_q) \parallel^2$$

where

$$\mathbf{r} = (a\ b)^T$$
$$\mathbf{P}_{\theta_i} = \begin{pmatrix} \cos\theta_i & -\sin\theta_i \\ \sin\theta_i & \cos\theta_i \end{pmatrix}, \quad \theta_{i+1} = \theta_i + \Delta,\ 0 \leq \theta_i \leq 2\pi.$$

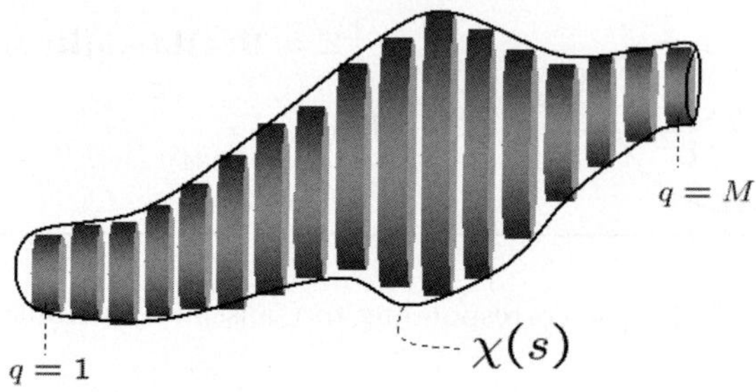

Fig. 2. Gaussian-Cylindroid Color model

To obtain the control vectors of $\mathbf{Q}^{\bar{c}_x}$, $\mathbf{Q}^{\bar{c}_y}$, $\mathbf{Q}^a$, $\mathbf{Q}^b$ and $\mathbf{Q}^\phi$ minimizing the surface-displacement measure of (7), we used Powell's nonlinear optimization method [14].

We obtained the Gaussian-cylindroid color model of skin using the skin patch images collected from various brightness conditions. Its parameters are shown separately in Figure 3-5 where the dotted lines denote the estimations of (4) from the training samples and the solid lines shows the resulting parameters by Powell's nonlinear optimization of (7).

The preliminary test shows the robustness of the proposed color model against the irregular brightness changes as shown in Figure 6.

3   Tracking with Adaptive Mean Shift

Bandwidth selection is crucial to the performance of mean-shift iterations. In this section we introduce how to obtain an optimal bandwidth and show the adaptive mean-shift tracking with the optimal bandwidth and the Gaussian-cylindroid color model.

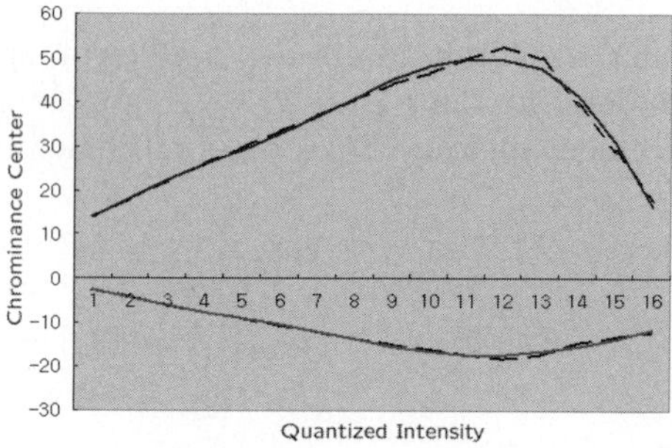

Fig. 3. Chrominance Center. The upper and the lower lines are for U and V components(chorminance), respectively.

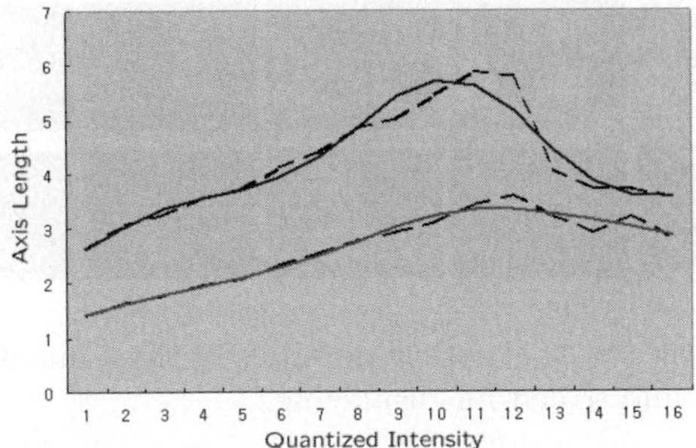

Fig. 4. Axis Length. The upper and the lower lines are for the lengths of the major axis length a and the minor axis length b, respectively.

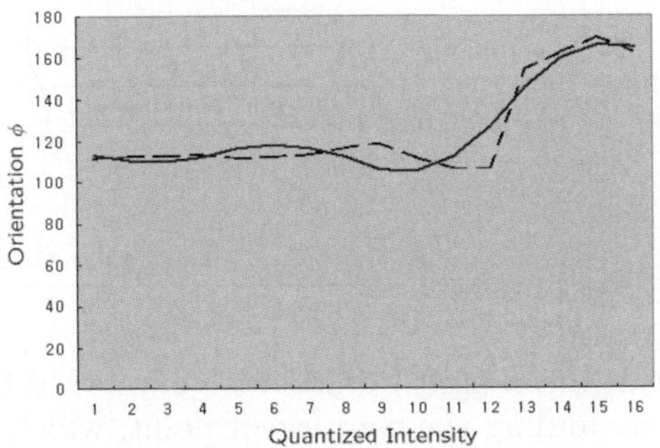

Fig. 5. Orientation of Major Axis

3.1   Kernel Density of Skin Region and Mean Shift

We assume that a tracking target of a single color such as the skin is maintained within its Gaussian-cylindroid color space while the target has a bounded region on the image plane. To represent the target model including its spatial domain and color domain, we introduce the method of the sample kernel density estimation that is the most popular density estimation method. Employing the multivariate normal kernel gives at the point $\mathbf{x}$,

$$p(\mathbf{x}) = C \sum_{i=1}^{N_{\mathbf{H}}} |\mathbf{H}|^{-\frac{1}{2}} \exp\left(\frac{(\mathbf{x} - \mathbf{x}_i)^T \mathbf{H}^{-1}(\mathbf{x} - \mathbf{x}_i)}{-2} \right)$$

$$\times |\mathbf{\Sigma}_{q_i}|^{-\frac{1}{2}} \exp\left(\frac{(\mathbf{c}_i - \bar{\mathbf{c}}_{q_i})^T \mathbf{\Sigma}_{\mathbf{q_i}}^{-1}(\mathbf{c}_i - \bar{\mathbf{c}}_{q_i})}{-2} \right) \tag{8}$$

where C is the normalizing constant and q_i is a quantized intensity of pixel i.

Since the mean-shift vector points toward the direction of maximum increase in the sample kernel density and converges the local maxima of the density [2],

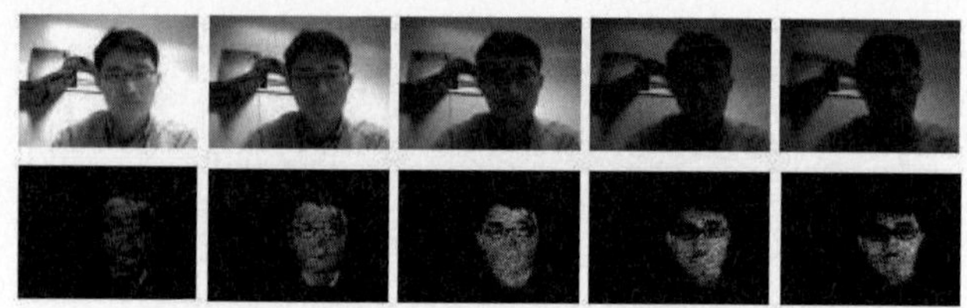

Fig. 6. Skin Color Area. The probability densities of the trained skin color model are calculated at all pixels and scaled for displaying.

the mean-shift iteration procedure has often been used in the areas of clustering, detection and tracking.

The mean-shift vector over the distribution shown in (8) is given as

$$\mathbf{m}(\mathbf{x}) = \frac{\sum_{i=1}^{N_H} \mathbf{x}_i \psi_i \exp\left(\frac{(\mathbf{x}-\mathbf{x}_i)^T \mathbf{H}^{-1}(\mathbf{x}-\mathbf{x}_i)}{-2}\right)}{\sum_{i=1}^{N_H} \psi_i \exp\left(\frac{(\mathbf{x}-\mathbf{x}_i)^T \mathbf{H}^{-1}(\mathbf{x}-\mathbf{x}_i)}{-2}\right)} - \mathbf{x} \tag{9}$$

where

$$\psi_i = |\mathbf{\Sigma}_{q_i}|^{-\frac{1}{2}} \exp\left(\frac{(\mathbf{c}_i - \bar{\mathbf{c}}_{q_i})^T \mathbf{\Sigma}_{q_i}^{-1}(\mathbf{c}_i - \bar{\mathbf{c}}_{q_i})}{-2}\right).$$

The convergence of this mean-shift iteration is guaranteed by the use of normal kernels [2]. In our case, finding the convergent point, which is the peak point of the distribution (8), corresponds to tracking of the single-colored target.

3.2   Maximum Likelihood Estimation of Bandwidth

If the multivariate normal kernel is considered, the bandwidth denotes its covariance matrix. If the data statistics is homogeneous, then one global bandwidth suffices for the mean shift iteration. However, if the data statistics are changing across the feature space of the spatial and the color domain, the local bandwidths should be computed. We have only to estimate the spatial bandwidth since the variable bandwidth in the color domain was predetermined with respect to intensities using the Gaussian-cylinder color model described in the previous section.

The local log-likelihood of the target is from (8) as(up to scale)

$$\mathcal{L} = 2 \ln \sum_{i=1}^{N_\mathbf{H}} |\mathbf{H}|^{-\frac{1}{2}} \exp\left(\frac{(\mathbf{x} - \mathbf{x}_i)^T \mathbf{H}^{-1}(\mathbf{x} - \mathbf{x}_i)}{-2}\right) \psi_i. \tag{10}$$

By Jensen's inequality [12] the lower bound of $\mathcal{L}$, $\hat{\mathcal{L}}$, is obtained as

$$\mathcal{L} \geq \hat{\mathcal{L}} = \sum_{i=1}^{N_\mathbf{H}} \psi_i \left(\ln |\mathbf{H}|^{-1} - (\mathbf{x} - \mathbf{x}_i)^T \mathbf{H}^{-1}(\mathbf{x} - \mathbf{x}_i)\right). \tag{11}$$

We obtain the spatial bandwidth maximizing the lower bound of the local log-likelihood $\mathcal{L}$ as follows:

$$\mathbf{H} = \frac{\sum_{i=1}^{N_{\mathbf{H}}} \psi_i (\mathbf{x} - \mathbf{x}_i)(\mathbf{x} - \mathbf{x}_i)^T}{\sum_{i=1}^{N_{\mathbf{H}}} \psi_i}. \tag{12}$$

3.3 Adaptive Tracking

After learning of the Gaussian-cylindroid color model of the tracking target, our adaptive mean-shift algorithm is performed as follows:

1. Initialize the position and the spatial bandwidth of the target.
   (a) Set the spatial bandwidth $\mathbf{H}^{(0)}$.
   (b) Find $\mathbf{x}^{(0)}$ by scanning or random search where $\mathcal{L}(\mathbf{x}^{(0)}, \mathbf{H}^{(0)})$ is larger than the specified threshold.
2. Iterate for $t = 0, 1, \cdots$.
   (a) Update the position.

$$\mathbf{x}^{(t+1)} = \frac{\sum_{i=1}^{N_{H^{(t)}}} \mathbf{x}_i \psi_i \exp\left(\frac{(\mathbf{x}^{(t)} - \mathbf{x}_i)^T \mathbf{H}^{(t)-1}(\mathbf{x}^{(t)} - \mathbf{x}_i)}{-2} \right)}{\sum_{i=1}^{N_{H^{(t)}}} \psi_i \exp\left(\frac{(\mathbf{x}^{(t)} - \mathbf{x}_i)^T \mathbf{H}^{(t)-1}(\mathbf{x}^{(t)} - \mathbf{x}_i)}{-2} \right)}$$

   (b) Update the spatial bandwidth.

$$\mathbf{H}^{(t+1)} = \frac{\sum_{i=1}^{N_{\mathbf{H}^{(t)}}} \psi_i (\mathbf{x}^{(t+1)} - \mathbf{x}_i)(\mathbf{x}^{(t+1)} - \mathbf{x}_i)^T}{\sum_{i=1}^{N_{\mathbf{H}^{(t)}}} \psi_i}$$

In the preliminary experiment of the algorithm, although the tracking started at the distracted position and the poor bandwidth, the target position and bandwidth were recovered by at most 3-5 iterations as shown in Figure 7.

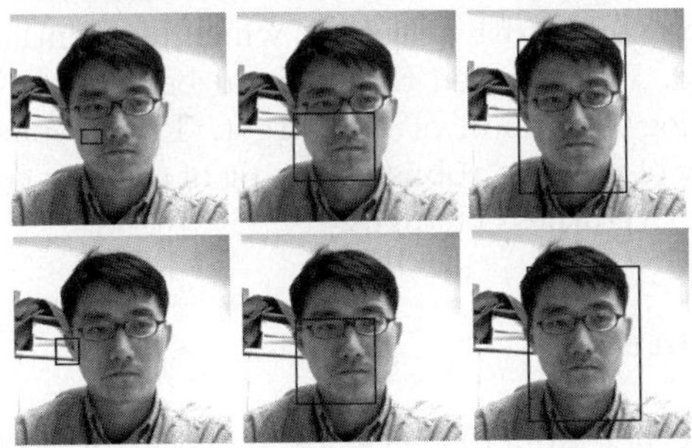

Fig. 7. Mean-Shift with Adaptive Bandwidth: In the top row the initial mean-shift tracker is within the face region. In the bottom row the tacker initially half deviates from the face region.

4 Testing and Performance

We applied the proposed algorithm to the skin region tracking under irregular illumination changes. Through several experiments, our tracking system demonstrated real-time applicability where its frame rate is more than 15 fps with Pentium IV PC(3GHz). Figure 8 shows the effective and robust tracking even in abrupt changes in illumination.

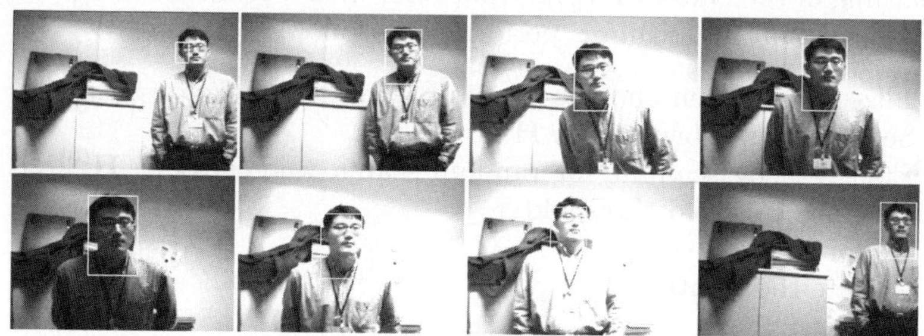

Fig. 8. Robust Tracking Against Irregular Illumination Changes. Frame goes from the left-top to the right down.

5 Conclusions

We proposed a new method for robust and real-time object tracking using a Gaussian-cylindroid color model and an adaptive mean shift. In the Gaussian-cylindroid color model all chromatic distributions within the brightness range are parameterized as a compact form with the B-spline functions.

A target for tracking is expressed by a kernel density function that incorporates the Gaussian-cylindroid color distribution into the positional domain in image lattice. Tracking is performed using the mean-shift algorithm where a simple method for the bandwidth selection, which is essential to tracking performance, was proposed. We achieved the optimal bandwidth that maximizes the lower bound of the log-likelihood of the target. The proposed method demonstrated the capability of fast and robust tracking of a single-colored target against irregular and abrupt brightness changes.

Acknowledgments

This research has been conducted as a part of the project, "Development of Cooperative Network-based Humanoids' Technology", funded by the Ministry of Information and Communication (MIC) of Korea.

References

1. A. Blake and Isard,M., "Active contour", Springer-Verlag, London, 1998.
2. Y. Cheng, "Mean Shift, Mode Seeking and Clustering", IEEE Trans. Pattern Analysis and Machine Intelligence, Vol. 17, No. 8, 1995.
3. D. Comaniciu, V. Ramesh, and P. Meer, "Real-time tracking of non-rigid objects using mean shift", IEEE Conference on Computer Vision and Pattern Recognition, vol. II, Hilton Head, SC, June 2000, pp. 142-149.
4. D. Comaniciu, "An algorithm for Data-Driven Bandwidth Selection", IEEE Trans. Pattern Analysis and Machine Intelligence, Vol. 25, No. 2, May 2003.
5. S. Feyrer and A. Zell, "Detection, Tracking, and Pursuit of Humans with an Autonomous Mobile Robot", Proc. of IEEE/RSJ International Conference on Intelligent Robots and Systems, 1999, pp. 864-869.
6. G. Jang and I. Kweon, "Robust Object Tracking using an Adaptive Color Model", Proc. of IEEE International Conference on Robotics and Automation, Seoul, Korea, May 21-26, 2001, pp. 1677-1682.
7. Jeong,Mun-Ho, Kuno,Y., Shimada,N., Shirai,Y., "Recognition of Shape-Changing Hand Gestures", IEICE TRANSACTIONS on Information and Systems, Vol.E85-D, No.10, 2002, pp.1678-1687.
8. M. Jones, J. Marron, and S. Sheather, "A Brief Survey of Bandwidth Selection for Density Estimation", J. Am. Statistical Assoc., Vol. 91, 1996, pp. 401-407.
9. C. H. Kim, B. J. You, H. Kim, and S..R. Oh, "A Robust Color Segmentation Technique for Tangible Space", Proceedings of XVth Triennial Congress of International Ergonomics Association, Seoul, Korea, 2003.
10. J. Yang and A. Waibel, "A Real-time Face Tracker", Proc. of IEEE Workshop on Application of Computer Vision, 1996, pp. 142-147.
11. Y. Leung, J. Zhang, and Z. Xu, "Clustering by Scale-Space Filtering", IEEE Trans. Pattern Analysis Machine Intelligence, Vol. 22, No. 12, 2000, pp. 1396-1410.
12. S. Loyka and A. Kouki, "On the use of Jensen's inequality for MIMO channel capacity estimation", Electrical and Computer Engineering, Canadian Conference on, Vol. 1, 2001, pp. 475-480.
13. E. J. Pauwels and G. Frederix, "Finding Salient Regions in Images", Computer Vision and Image Understanding, Vol. 75, 1999, pp. 73-85.
14. W. H. Press, S. A. Teukolsky, W. T. Vetterling, and B. P. Flannery, "Numerical Recipes in C++", Cambridge Univ. Press, 2002.
15. S. Sheather and M. Jones, "A Reliable Data-Based Bandwidth Selection Method for Kernel Density Estimation", J. Royal Statistical Soc. B, Vol. 53, 1991, pp. 683-690.

A3:

If you are able to see the Internet then it sounds like it is working, you may want to get in touch with your IT department to see if you need to make any changes to your settings to get it to work. *Try performing a soft reset, by pressing the stylus pen in the small hole on the bottom left hand side of the Ipaq and then release.*

A4:

I would recommend doing a soft reset by pressing the stylus pen in the small hole on the left hand side of the Ipaq and then release. Then charge the unit overnight to make sure it has been long enough and then see what happens. If the battery is not charging then the unit will need to be sent in for repair.

Fig. 2. Sample answers that share a sentence

The methods developed on the basis of these two dimensions are: *Retrieve Answer, Predict Answer, Predict Sentences, Retrieve Sentences* and *Hybrid Predict-Retrieve Sentences.* The first four methods represent the possible combinations of information-gathering technique and level of granularity; the fifth method is a hybrid where the two information-gathering techniques are applied at the sentence level.

Our aim in this paper is to investigate when the different methods are applicable, and whether individual methods are uniquely successful in certain situations. For this purpose, we decided to assign a level of success not only to complete responses, but also to partial ones (obtained with the sentence-based methods). The rationale for this is that we believe that a partial high-precision response is better than no response, and better than a complete response that contains incorrect information. We plan to test these assumptions in future user studies.

The rest of this paper is organized as follows. In the next section, we describe our five methods, followed by the evaluation of their results. In Section 4, we discuss related research, and then present our conclusions and plans for future work.

2   Information-Gathering Methods

In the following sub-sections we present the implementation details of the various methods. Note that some of the methods were implemented independently of each other at different stages of our project. Consequently, there are minor implementational variations, such as choice of machine learning algorithms and some discrepancy regarding features. We plan to bridge these differences in the near future, but are confident that they do not impede the aim of the current study: exploring the performance of our system along the two dimensions mentioned in the previous section (information-gathering approach, and level of granularity).

2.1   Retrieve a Complete Answer

This method retrieves a complete document (answer) on the basis of request lemmas. We use cosine similarity to determine a retrieval score, and use a minimal retrieval threshold that must be surpassed for a response to be accepted.

Initially, we retain the answer sentences whose recall exceeds a threshold.[6] Once we have the set of candidate answer sentences, we attempt to remove redundant sentences. This requires the identification of sentences that are similar to each other — a task for which we use the sentence clusters described in Section 2.3. Given a group of answer sentences that belong to the same cohesive cluster, we retain the sentence with the highest recall (in our current trials, a cluster is sufficiently cohesive for this purpose if its cohesion ≥ 0.7). In addition, we retain all the answer sentences that do not belong to a cohesive cluster. All the retained sentences will appear in the answer.

2.5   Hybrid Predict-Retrieve Sentences

It is possible that the Sentence Prediction method predicts a sentence cluster that is not sufficiently cohesive for a confident selection of a representative sentence. However, a sentence can still be selected by using clues from the request. For example, selecting between a group of sentences concerning the installation of different drivers might be possible if the request mentions a specific driver. Thus the Sentence Prediction method is complemented with the Sentence Retrieval method to form a hybrid, as follows.

- For highly cohesive clusters predicted with high confidence, we select a representative sentence as before.
- For clusters with medium cohesion predicted with high confidence, we attempt to match the sentences with the request sentences, using the Sentence Retrieval method but with a lower recall threshold. This reduction takes place because the high prediction confidence provides a guarantee that the sentences in the cluster are suitable for the request, so there is no need for a conservative recall threshold. The role of retrieval is now to select the sentence whose content lemmas best match the request, regardless of how good this match is.
- For uncohesive clusters or clusters predicted with low confidence, we have to resort to word matches, which means reverting to the higher, more conservative recall threshold, because we no longer have the prediction confidence.

3   Evaluation

As mentioned in Section 1, our corpus was divided into topic-based datasets. We have observed that the different datasets lend themselves differently to the various information-gathering methods described in the previous section. In this section, we examine the overall performance of the five methods across the corpus, as well as their performance for different datasets.

3.1   Measures

We are interested in two performance indicators: *coverage* and *quality*.

Coverage is the proportion of requests for which a response can be generated. The various information gathering methods presented in the previous section have accep

---

[6] To assess the goodness of a sentence, we experimented with *f-scores* that had different weights for recall and precision. Our results were insensitive to these variations.

tance criteria that indicate that there is some level of confidence in generating a response. A request for which a planned response fails to meet these criteria is not covered, or addressed, by the system. We are interested in seeing if the different methods are applicable in different situations, that is, how exclusively they address different requests. Note that the sentence-based methods generate partial responses, which are considered acceptable so long as they contain at least one sentence generated with high confidence. In many cases these methods produce obvious and non-informative sentences such as "Thank you for contacting HP", which would be deemed an acceptable response. We have manually excluded such sentences from the calculation of coverage, in order to have a more informative comparison between the different methods.

Ideally, the **quality** of the generated responses should be measured through a user study, where people judge the correctness and appropriateness of answers generated by the different methods. However, we intend to refine our methods further before we conduct such a study. Hence, at present we rely on a text-based quantitative measure. Our experimental setup involves a standard 10-fold validation procedure, where we repeatedly train on 90% of a dataset and test on the remaining 10%. We then evaluate the quality of the answers generated for the requests in each test split, by comparing them with the actual responses given by the help-desk operator for these requests.

We are interested in two quality measures: (1) the precision of a generated response, and (2) its overall similarity to the actual response. The reason for this distinction is that the former does not penalize for a low recall — it simply measures how correct the generated text is. As stated in Section 1, a partial but correct response may be better than a complete response that contains incorrect units of information. However, more complete responses are generally better than partial ones, and so we use the second measure to get an overall indication of how correct and complete a response is. We use the traditional Information Retrieval precision and f-score measures [1], employed on a word-by-word basis, to evaluate the quality of the generated responses.[7]

3.2   Results

Table 1 shows the overall results obtained using the different methods. We see that combined the different methods can address 72% of the requests. That is, at least one of these methods can produce some non-empty response to 72% of the requests. Looking at the individual coverages of the different methods we see that they must be applicable in different situations, because the highest individual coverage is 43%.

The Answer Retrieval method addresses 43% of the requests, and in fact, about half of these (22%) are uniquely addressed by this method. However, in terms of the quality of the generated response, we see that the performance is very poor (both precision and f-score have very low averages). Nevertheless, there are some cases where this method uniquely addresses requests quite well.

The Answer Prediction method can address 29% of the requests. Only about a tenth of these are uniquely addressed by this method, but the generated responses are of a fairly high quality, with an average precision and f-score of 0.82. Notice the large standard deviation of these averages, suggesting a somewhat inconsistent behaviour. This

---

[7] We have also employed sequence-based measures using the ROUGE tool set [7], with similar results to those obtained with the word-by-word measure.

Table 1. Performance of the different methods, measured as coverage, precision and f-score

Method	Coverage	Precision Ave (stdev)	F-score Ave (stdev)
Answer Retrieval	43%	0.37 (0.34)	0.35 (0.33)
Answer Prediction	29%	0.82 (0.21)	0.82 (0.24)
Sentence Prediction	34%	0.94 (0.13)	0.78 (0.18)
Sentence Retrieval	9%	0.19 (0.19)	0.12 (0.11)
Sentence Hybrid	43%	0.81 (0.29)	0.66 (0.25)
Combined	72%	0.80 (0.25)	0.50 (0.33)

is due to the fact that this method gives good results only when complete generic responses are found. In this case, any re-used response will have a high similarity to the actual response. However, when this is not the case, the performance degrades substantially, resulting in inconsistent behaviour. This means that Answer Prediction is suitable when requests that share some regularity receive a complete template answer.

The Sentence Prediction method can find regularities at the sub-document level, and therefore deal with cases where partial responses can be generated. It produces responses for 34% of the requests, and does so with a consistently high precision (average 0.94, standard deviation 0.13). Only an overall 1% of the requests are uniquely addressed by this method. However, for the cases that are shared between this method and other ones, it is useful to compare the actual quality of the generated response. In 5% of the cases, the Sentence Prediction method either uniquely addresses requests, or jointly addresses requests together with other methods but has a higher f-score. This means that in some cases a partial response has a higher quality than a complete one.

Like the document-level Answer Retrieval method, the Sentence Retrieval method performs poorly. It is difficult to find an answer sentence that closely matches a request sentence, and even when this is possible, the selected sentences tend to be different to the ones used by the help-desk operators, hence the low precision and f-score. This is discussed further below in the context of the Sentence Hybrid method.

The Sentence Hybrid method extends the Sentence Prediction method by employing sentence retrieval as well, and thus has a higher coverage (45%). In fact, the retrieval component serves to disambiguate between groups of candidate sentences, thus enabling more sentences to be included in a generated response. This, however, is at the expense of precision, as we also saw for the pure Sentence Retrieval method. Although retrieval selects sentences that match closely a given request, this selection can differ from the "selections" made by the operator in the actual response. Precision (and hence f-score) penalizes such sentences, even when they are more appropriate than those in the model response. For example, consider request-answer pair **RA5**. The answer is quite generic, and is used almost identically for several other requests. The Hybrid method almost reproduces this answer, replacing the first sentence with **A6**. This sentence, which matches more request words than the first sentence in the model answer, was selected from a sentence cluster that is not highly cohesive, and contains sentences that describe different reasons for setting up a repair (the matching word in **A6** is "screen"). Overall, the Hybrid method outperforms the other methods in about 10% of the cases, where it either uniquely addresses requests, or addresses them jointly with other methods but produces responses with a higher f-score.

> **RA5:**
> *My screen is coming up reversed (mirrored). There must be something loose electronically because if I put the stylus in it's hole and move it back and forth, I can get the screen to display properly momentarily. Please advise where to send for repairs.*
>   To get the iPAQ serviced, you can call `1-800-phone-number`, options 3, 1 (enter a 10 digit phone number), 2. Enter your phone number twice and then wait for the routing center to put you through to a technician with Technical Support. They can get the unit picked up and brought to our service center.

> **A6:**
>   To get the iPAQ repaired (battery, stylus lock and screen), please call `1-800-phone-number`, options 3, 1 (enter a 10 digit phone number), 2.

3.3   Summary

In summary, our results show that, with the exception of the Sentence Retrieval method, the different methods are applicable in different situations, all occurring significantly in the corpus. The Answer Retrieval method uniquely addresses a large portion of the requests, but many of its attempts are spurious, thus lowering the combined overall quality shown at the bottom of Table 1 (average f-score 0.50), calculated by using the best performing method for each request. The Answer Prediction method is good at addressing situations that warrant complete template responses. However, its confidence criteria might need refining to lower the variability in quality. The combined contribution of the sentence-based methods is substantial (about 15%), suggesting that partial responses of high precision may be better than complete responses with a lower precision.

4   Related Research

There are very few reported attempts at corpus-based automation of help-desk responses. The retrieval system *eResponder* [8] retrieves a list of request-response pairs and presents a ranked list of responses to the user. In contrast, our system can re-use a single representative response, or collate sentences from multiple responses to generate a single (possibly partial) response. Lapalme and Kosseim [9] employ a rule-based, template-based system that extracts patterns from a corpus of request-response emails in order to personalize a response (for example substituting names of individuals and companies). They also employ a retrieval component that finds answers from a corpus of technical documents (not the request-response corpus). Our work differs from these two examples in that it applies a purely data-driven approach that uses a corpus of request-response pairs to generate a single response. Similar corpus-based approaches have been implemented in FAQ domains [10,11]. However, this kind of corpus is significantly different to ours in that it lacks repetition and redundancy and the responses are not personalized.

5   Conclusion and Future Work

We have presented four basic methods and one hybrid method for addressing help-desk requests. The basic methods represent the four ways of combining level of granularity (sentence and document) with information-gathering technique (prediction and retrieval). The hybrid method applies prediction possibly followed by retrieval to information at the sentence level. Our results show that with the exception of Sentence Retrieval, the different methods can address a significant portion of the requests. A future avenue of research is thus to characterize situations where different methods are applicable, in order to derive decision procedures that determine the best method automatically. We have also started to investigate an intermediate level of granularity: paragraphs.

Our results suggest that the automatic evaluation method requires further consideration. As seen in Section 3, our f-score penalizes the Sentence Prediction and Hybrid methods when they produce good answers that are more informative than the model answer. As mentioned previously, a user study would provide a more conclusive evaluation of the system, and could be used to determine preferences regarding partial responses.

Finally, we propose the following extensions to our current implementation. First, we would like to improve the representation used for clustering, prediction and retrieval by using features that incorporate word-based similarity metrics [12]. Secondly, we intend to investigate a more focused sentence retrieval approach that utilizes syntactic matching of sentences. For example, if a sentence cluster is strongly predicted by a request, but the cluster is uncohesive because of a low verb agreement, then the retrieval should favour the sentences whose verbs match those in the request.

Acknowledgments

This research was supported in part by grant LP0347470 from the Australian Research Council and by an endowment from Hewlett-Packard. The authors also thank Hewlett-Packard for the extensive help-desk data.

References

1. Salton, G., McGill, M.: An Introduction to Modern Information Retrieval. McGraw Hill (1983)
2. Filatova, E., Hatzivassiloglou, V.: A formal model for information selection in multi-sentence text extraction. In: COLING'04 – Proceedings of the 20th International Conference on Computational Linguistics, Geneva, Switzerland (2004) 397–403
3. Wallace, C., Boulton, D.: An information measure for classification. The Computer Journal **11**(2) (1968) 185–194
4. Oliver, J.: Decision graphs – an extension of decision trees. In: Proceedings of the Fourth International Workshop on Artificial Intelligence and Statistics, Fort Lauderdale, Florida (1993) 343–350
5. Banerjee, S., Pedersen, T.: The design, implementation, and use of the Ngram Statistics Package. In: Proceedings of the Fourth International Conference on Intelligent Text Processing and Computational Linguistics, Mexico City (2003)
6. Marom, Y., Zukerman, I.: Towards a framework for collating help-desk responses from multiple documents. In: Proceedings of the IJCAI05 Workshop on Knowledge and Reasoning for Answering Questions, Edinburgh, Scotland (2005) 32–39
7. Lin, C., Hovy, E.: Automatic evaluation of summaries using n-gram co-occurrence statistics. In: Proceedings of the 2003 Language Technology Conference (HLT-NAACL 2003), Edmonton, Canada (2003)

```
Page  ::= HTML Tree
HTML Tree  ::= <Start tag> Child * <End tag>
Child  ::= HTML Tree | Leaf
Leaf  ::= Token*
```

A tree consists of many sub-trees or leaves. Each node represents the start tag and the end tag of a (sub) tree. The (sub)tree rooted at a node in an upper level represents a parent tree and a tree rooted at a lower level node represents a child tree. For example, we call the TD tree a child tree of the TR tree. Figure 1 shows an example page represented in HTML and its tree structure.

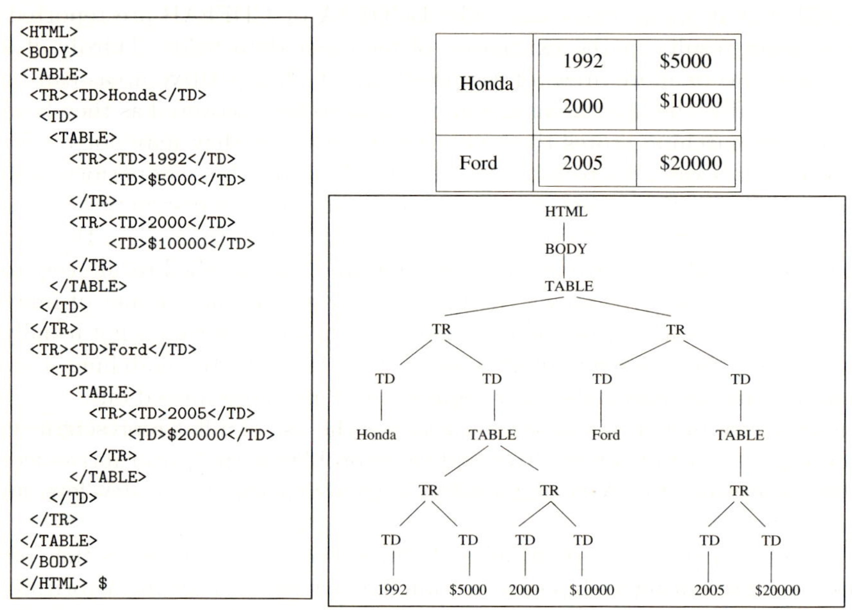

Fig. 1. An example page, HTML source, and its tree representation

The trees transformed from HTML pages expose several interesting properties for the detection of repetitive regions. Firstly, all elements of a tree (including its decedents) are consecutive in the original HTML page. They are in fact all elements between the starting and ending tags of the tree. Therefore, a repetitive region remains repetitive in the tree that contains the region. Secondly, all repetitive units (rows) in a regular region often have the same mark-up tags, and thus each region usually comprises of one or more sub-trees of the tree containing the region. The regularity detection problem is then transformed to finding the repetitive sub-trees in a tree.

Detecting regular data areas by finding similar sub-trees overcomes the four limitations mentioned in the introduction. Firstly, using a tree-like structure, the Smith-Waterman algorithm can be applied to each sub-tree separately rather than a long list of tokens, so the efficiency is greatly improved. Secondly, a row (repetitive unit) of repeated data will start at the beginning of a sub tree, and end at the leaves of a sub-tree. This will eliminate the confusion of deciding the start and end of rows in repetitive detection. Thirdly, different regular regions often have different mark-up tags or are located in different areas on the page, and they will be in different (sub)trees. As we detect repetitive structure on each (sub)tree separately, the detection will not affect each other. Finally, if the page has nested repetitive structure, it will be reflected in repetition in more than one level of the tree-like structure. If we discover the repetition in the child tree (inner structure, sub-tree) first, the repetitive property of the parent tree (outer structure) still remains. In this way, the nested structure can be detected by applying the detection algorithm on the tree level by level in a bottom-up fashion.

2.2   Building the Tree

Tokenising and Preprocessing. The HTML pages are tokenised into a list of tokens, each of which is either a tag token or a text token. Each token contains features of the token such as open or close type of tag tokens and the associated attributes. The data elements are usually contained between the open and close BODY tags, thus to reduce complexity, we only need to extract the list from BODY and /BODY tags. In other words, the tree we are interested in will have the root at BODY and /BODY tags.

Parsing the Token List to a Tree. Most browsers nowadays allow a flexible HTML scheme, such as missing tags. The /TR and /TD and /P tags, among other examples, can be omitted in many web pages. However, their absence can be detected easily if the context is examined. For example, a TD tree will have its closing tag just before a new TD tag is introduced, or when the end of the table or table row is reached. Theoretically, on a page with consistent display, all missing closing tags can be reconstructed, otherwise data will be positioned in a non-deterministic manner.

In the HTML syntax, there are tags that must have a corresponding closing tag such as TABLE, which we called *strict tags*. Other tags like P or TD are called *lax tags* as they do not require a closing tag, in fact they will be automatically closed when encountering some tags. However, not all encountered tags can close a lax tag. For example, a TR can close a TD tag but cannot be closed by a TD because TD is considered a sub component of TR. The TR tag cannot be closed by a TABLE tag either though TR is a sub component of TABLE. The reason is that a new table may be contained within the current TR row. Fortunately, if such a case happens, the super component is a strict tag.

Looking for closing tags of strict tags is normally trivial as they are usually present in the page. In other words, if a closing tag is found, it will be matched with the nearest currently opening tag of the same type. However, matching for lax tags is rather complicated. A network of relationships between those tags is built to represent how each tag can be closed by other tags. Generally, a lax tag can close a tag of the same type, or a sub component of the lax tag. For the current stage, only relationships of tags defining positioning (TR, TD, P, Hx, LI etc) are processed. Other tags like FONT and B have not been processed yet. The parsing algorithm is described as follows.

- Create a tree, with the root corresponding to the opening BODY tag at the start of the token list. Maintain a tree pointer initialised to the root tree.
- Iterate through the token list:
  - If a token is a strict tag and it is a open tag, a new tree is created as a child of the current tree (the tree that the pointer points to). The pointer is changed to point to the newly created tree.
  - If the token is an open tag of a lax tag, the pointer traverses up to close any can-be-closed opening trees. A new tree of the tag is then created as the next sibling of the last closed tree.
  - If a token is a close tag, the pointer will traverse up to look for the opening tree of that tag. If it is found, all of its descendents will be automatically closed. The tree is then closed and the pointer will point to its parent as the current working tree.
  - If the token is not one of any above categories, it will be add to the children list of the current tree as a leaf.

3   Detecting Regular Regions by Finding Similar Sub-trees

This section first introduces the language used to represent an information extraction pattern, then describes the algorithm that learns patterns by detecting similar sub-trees.

3.1   Pattern Representation

We represent Web pages with regular data areas as a tree. A pattern is what we use to represent the regularity so that data can be extracted from the regular data areas. A language is designed to represent an information extraction pattern for representing the repeated sub-trees. As we are only interested in trees containing repetitive structure, trees without any regular regions will be considered as leaves. The syntax of the pattern language is defined as follows.

```
Pattern ::= Repeated Tree
Repeated Tree ::= <Start tag> Child* <End tag>
Child ::= Repeated Region | Leaf
Repeated Region ::= Repeated Unit +
Repeated Unit ::= Genetic token + | Repeated Tree
Genetic token ::= tag | dataN | Optional token | Alternative token
Leaf ::= Token*
```

Where the * and + notations denote zero or more and one or more respectively. A Repeated Region represents a regular data area. Each regular area consists of one or more *Repeated Units*. A Repeated Unit consists of a sequence of tag and data tokens and is created by merging similar sub-trees. We introduce *optional* and *alternative* tokens to merge approximately repetitive sub-trees.

This pattern language is very expressive. A repeated tree can have multiple Repeated Regions as its children, which allows multiple regular data areas in a page. A Repeated Unit itself can be another repeated tree and this multi-layer nature of the tree allows nested repetition. Thus, this pattern language will be able to express nested and multiple occurrences of regular regions.

3.2 Learning Patterns by Finding Similar Sub-trees

In our learning algorithm, the tree-like Web page structure will be visited in the post-order, that is, all children will be visited before doing so for the parent. In each visit to a tree, the Smith-Waterman algorithm is applied to detect the repetitive sub trees of the tree. When a sub-tree applies the Smith-Waterman algorithm, it is flattened into list of tokens and the algorithm is used in a similar way to that of AutoWrapper [7].

If similar sub-trees (child trees) are discovered, the tree (parent tree) will be marked as repeated tree and all the similar sub-trees are generalised/merged to repeated units and a sequence of repeated units form a repeated region. The way to generalise/merge the sub-trees is the same as that in AutoWrapper which is based on a standard machine learning algorithm Find-S[10]. A tree is also marked as a repeated tree if one of its children is a repeated tree. Once all regular areas are detected, the (sub)trees not marked as repeated are recorded as list of leaves.

For the example page shown in figure 1, the pattern learned using this learning algorithm and its tree representation are shown in figure 2.

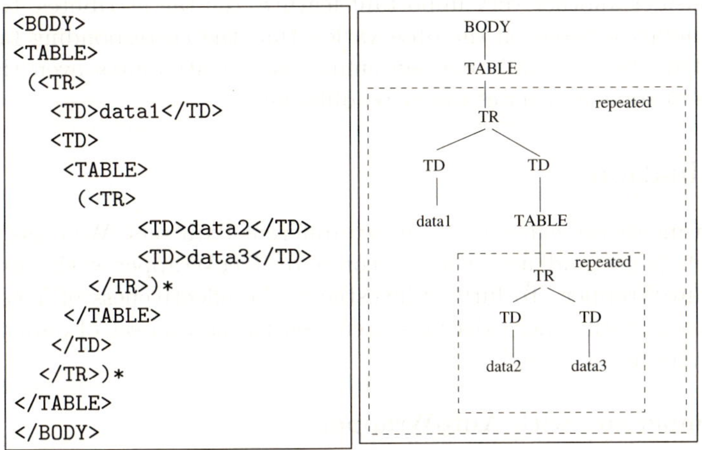

Fig. 2. Pattern for the example page shown in figure 1

But how do we judge whether the user is interested in a page or not? Although in usage-based recommendation systems we can't expect users to express likes or dislikes explicitly, it is quite reasonable to suppose that if a user is not interested in a page, he/she won't spend much time viewing it and vice versa. As pageview duration can be calculated from web logs, it is a good choice for inferring user interest.

In this paper we use the time length of a pageview to estimate its importance in a transaction, in order to capture the user's interest more precisely. We assign a weight to each pageview in a transaction according to its duration, and perform *Weighted Association Rule (WAR)* mining to discover significant page sets. In the on-line recommendation phase, we also take pageview duration into account to better capture the current user's interest and generate more relevant recommendations.

The rest of the paper is organized as follows: Section 2 talks about related work. In Section 3 we introduce our approach of *Weighted Association Rule* mining to discover significant page sets. Section 4 introduces the online recommendation method using pageview duration as an indicator of user interest. Section 5 introduces our experiment and results. Section 6 concludes this paper and provides ideas for future work.

2 Related Work

The overall process of web personalization generally consists of three phases: data preparation and transformation, pattern discovery, and recommendation [1]. Data preparation and pattern discovery are performed offline. The data preparation phase transforms raw web log files into transaction data ready for data mining tasks [3]. Here a transaction contains a list of pageviews along with their weights. The weights can be binary, representing the existence or non-existence of a pageview in the transaction, or they can be some function of the pageview duration. In the pattern discovery phase, various data mining techniques can be applied to the transaction data, such as clustering, association rule mining, and sequential pattern discovery. Only the recommendation phase is performed online. In the recommendation phase, the recommendation engine considers the active user session in conjunction with the discovered patterns to recommend pages that the user is most likely to visit. [1]

Some studies have considered the approach of using pages interesting to the user for the recommendation process. In [2], Mobasher et al use statistical significance testing to judge whether a page is interesting to a user. Its main idea is: A duration threshold is calculated for each page using the average duration and standard deviation of the visits to the page; if the duration of a pageview is longer than the threshold, that pageview is considered interesting to the user and vice versa. The drawback of such an approach is that it simply divides pageviews into interesting and uninteresting groups, and neglects the difference in the degrees of interest. For one thing, there isn't a clear division between interesting and uninteresting pages; for another, the degrees of interest are probably not the same for all the interesting (and uninteresting) pages. In this work we try to assign a quantitative weight to each pageview, taking into account the degree of interest.

Clustering and collaborative filtering approaches are ready to incorporate both binary and non-binary weights of pageviews, although binary weights are usually used for computing efficiency [1] [2]. *Association Rule (AR)* mining and *Sequential Pattern (SP)* mining [4] can lead to higher recommendation precision [1], and are easy to scale to large datasets, but how to incorporate pageview weight into the *AR* and the *SP* models has not been explored in previous studies.

Weighted Association Rule (WAR) mining allows different weights to be assigned to different items, and is a possible approach to improving the *AR* model in the web personalization process. Cai et al. [5] proposed assigning different weights to items to reflect their different importance. In their framework, two ways are proposed to calculate itemset weight: total weight and average weight. Weighted support of an itemset is defined as the product of the itemset support and the itemset weight. Tao et al. [6] also proposed assigning different weights to items, the itemset/transaction weight is defined as the average weight of the items in the set/transaction, and weighted support of an itemset is the fraction of the weight of the transactions containing the itemset relative to the weight of all transactions. Both models attempt to give greater weights to more important items, facilitating the discovery of important but less frequent itemsets and association rules. However, both models assume a fixed weight for each item, while in the context of web usage mining a pageview might have different importance in different sessions.

3 Incorporating Pageview Weight

3.1 How to Determine Pageview Weight

In this paper we simply use the duration of a pageview as its weight in the transaction. Using more complex functions to determine pageview weights is left for future work. Several reasons validate the simple approach of using pageview duration as its weight. First, it reflects the relative importance of each pageview, because a user will generally spend more time on a more useful page. Second, the rates of most human beings getting information from web pages shouldn't differ greatly. If we assume a similar rate of acquiring information from pages for each user, the time a user spends on a page is proportional to the volume of information useful to him/her, and is a quantitative measure of the usefulness of that pageview to the user.

3.2 Weighted Support of an Itemset

We use the preprocessing techniques discussed in [3] to extract transaction data from web log files. After the preprocessing phase, we get a set of n pageviews, $P = \{p_1, p_2, ..., p_n\}$, and a set of m user transactions, $T = \{t_1, t_2, ..., t_m\}$, each transaction t is an l-length sequence of ordered pairs: $t = < (p_1, w(p_1)), (p_2, w(p_2)), ..., (p_l, w(p_l)) >$, where each $p_i \in P$, and the weight $w(p_i)$ associated with p_i is the duration (in seconds) of pageview p_i in transaction t.

For example, $t = \{(A, 60), (B, 20), (C, 70), (D, 90)\}$ is a transaction which records that the user spent 60 seconds on page A, 20 seconds on page B, 70 seconds on page C and 90 seconds on page D.

Definition-1 Weighted support of an itemset by a transaction: Weighted support *WSP* of an itemset X by a transaction t is the quantity of X contained in t, denoted as $wsp(X, t)$:

$$wsp(X, t) = min\{w(p_{x1}), w(p_{x2}), \dots , w(p_{xk})\} , \tag{1}$$

where $w(p_{xi})$ is the weight of pageview p_{xi} in transaction t. In simple words, weighted support of an itemset by a transaction implies how much of the itemset is contained in the transaction. For example, $wsp(A, t) = 60$, $wsp(B, t) = 20$, $wsp(AB, t) = 20$, $wsp(ACD, t) = 60$, $wsp(ABCD, t) = 20$.

Definition-2 Weighted support of an itemset (across all transactions): Weighted support $wsp(X)$ of an itemset X across all transactions is defined as follows:

$$wsp(X) = \frac{\sum_{t_i \in T} wsp(X, t_i)}{|T| \cdot \overline{w}} , \tag{2}$$

where T is the set of all transactions, $\overline{w}$ is the average weight of all the items across all transactions. The numerator is the sum of the weighted support of X over all transactions; the denominator is just a normalizing factor making most *wsp* values fall between 0 and 1.

3.3 Weighted Association Rules

In our framework, a weighted association rule has the form of $X => Y$, where X and Y are two itemsets, $X \cap Y = \varnothing$.

Definition-3: Weighted support of the association rule

$$wsp(X => Y) = wsp(X \cup Y) . \tag{3}$$

Definition-4: Confidence of the weighted association rule

$$conf(X => Y) = \frac{wsp(X \cup Y)}{wsp(X)} . \tag{4}$$

3.4 Mining Significant Itemsets

In the traditional association rule mining framework, an itemset is denoted *large* if its *support* is above a predefined minimum support threshold. In our framework, we say an itemset is *significant* if its *weighted support* is above a predefined weighted support threshold. Our approach to mining significant itemsets is based on the Apriori [7] algorithm. Apriori algorithm is an efficient algorithm utilizing the *downward closure property*, that is, any subset of a large itemset is also large. By our definition of weighted support and significant itemsets, there is a property that any subset of a significant itemset is also significant, here called a *weighted downward closure property*. The property always holds because for any itemset X and Y, if $X \subset Y$, by definition we have $wsp(X, t) \geq wsp(Y, t)$, hence $wsp(X) \geq wsp(Y)$.

4   Recommendation Engine

4.1   Significant Itemset Graph

We use a *Significant Itemset Graph* to improve the recommendation efficiency. Fig. 1 gives an example of the *Significant Itemset Graph*. The idea comes from [8], in which the data structure is called the *Frequent Itemset Graph* because the itemsets stored in it are *frequent* itemsets.

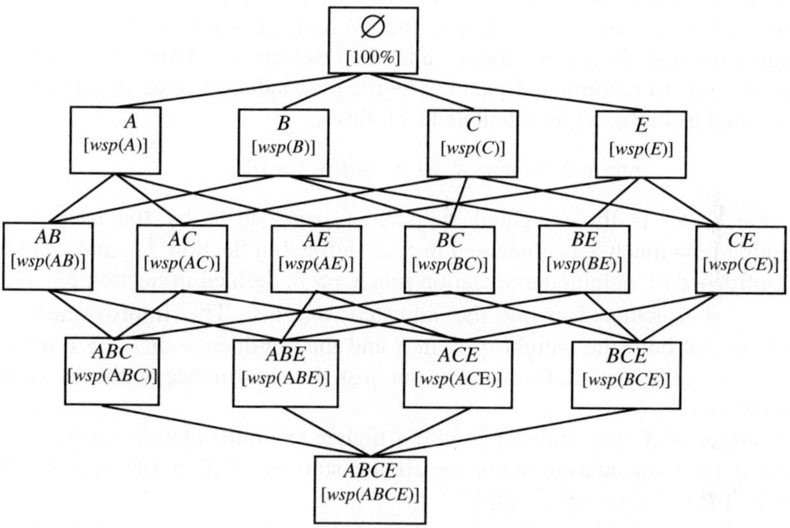

Fig. 1. *Significant Itemset Graph* - An Example. Each node stores a significant itemset along with its weighted support. For a node N containing itemset X, each child node of N corresponds to a significant itemset $X \cup \{p\}$.

4.2   Capture the User's Current Interest More Precisely

Previous works [1] [4] [8] go the right way to use only the last visited pages to generate the recommendations because *earlier* portions of a user session are not likely to represent the user's *current* information need. However, it still has another shortcoming: It doesn't eliminate the uninteresting pages. We propose that both page freshness and interestingness should be considered to capture a user's current interest. To give more recently visited pages more consideration, we attenuate the weight of each pageview as it gets obsolete. Pageviews with the greatest adjusted weights in the current user session are used to generate the recommendations. In this paper we use a most simple method to attenuate the weight of each pageview: linear attenuation. Given an active user session $S = \, < (p_1, w_1), (p_2, w_2), \dots , (p_k, w_k) >$, the attenuation factors and adjusted weights are calculated as follows:

Table 1. Attenuation factors and adjusted weights

Pageview	(p_1, w_1)	(p_2, w_2)	...	(p_{k-1}, w_{k-1})	(p_k, w_k)
Attenuation factor	$\frac{1}{k}$	$\frac{2}{k}$	...	$\frac{k-1}{k}$	$\frac{k}{k} = 1$
Adjusted weight	$\frac{1}{k} w_1$	$\frac{2}{k} w_2$	...	$\frac{k-1}{k} w_{k-1}$	w_k

4.3 Recommendation Process

The recommendation process is as follows: Given an active user session S, the recommendation engine first adjusts the weight of each pageview, using the attenuation method discussed above, and then selects the pages with the highest adjusted weights to recommend pages. The recommendation score of page p by page set X, denoted as $rec(p, X)$, is calculated as follows:

$$rec(p, X) = wsp(X, S) \cdot conf(X => p) , \tag{5}$$

where $wsp(X, S)$ is the weighted support of page set X by the user session S, representing how much X is contained in S, as defined in Section 3.2, and $conf(X => p)$ is the confidence of weighted association rule $X => p$, defined in Section 3.3. Note that $wsp(X, S)$ is calculated using the adjusted weights. The improvement of this approach is that both the weight of X in S and the confidence of $X => p$ are used to determine the recommendation score, not just the confidence value as is used in previous works.

As a subset of X may support p with a higher recommendation score, we choose the highest recommendation score by all the subsets of X as the recommendation score of p. Table 2 gives an example.

The confidence of $\varnothing => p$ is simply the wsp of p. The weight of $\varnothing$ is set to be the average duration of all pageviews across all transactions (here assumed to be 50 seconds in this example). For the last pageview whose duration could not be calculated from the access log, we also use the average duration to estimate it.

Table 2. Calculation of recommendation scores: Given association rules, and the pageview set $X = \{(A, 100), (B, 200)\}$ selected to generate the recommendation, the highest score of 140 is stored as the recommendation score of page p

Weighted Association Rule	$\varnothing$ $=> p$	A $=> p$	B $=> p$	AB $=> p$
Confidence	40%	60%	70%	90%
Recommendation Score	50 · 40% = 20	100 · 60% = 60	200 · 70% = 140	100 · 90% = 90

For a k-item page set X, we use all the 2^k subsets of X to generate the candidate recommendations, and the candidates with the highest recommendation scores are recommended to the user. For computing efficiency, an upper bound of 5 is set for the number of pages used to generate recommendations, as the number of subsets increases exponentially with k.

5   Experimental Evaluation

The recommendation approach based on *Weighted Association Rule* (*WAR*) proposed in this work is compared with the traditional *Association Rule* (*AR*) based approach.

5.1   Test Data Set

The comparative test is performed on the preprocessed and filtered sessionized data from the main DePaul CTI web server (http://www.cs.depaul.edu). The data is based on a random sample of users visiting this site during a 2 week period in April of 2002. The original (unfiltered) data contains a total of 20950 sessions from 5446 users. The filtered data files are produced by filtering out low support pageviews, and eliminating sessions of size 1. The filtered data contains 13745 sessions and 683 pageviews. We treat each session as a transaction. Each transaction contains a sequence of pageviews along with their weights (durations).

We perform 10-fold cross-validation on the CTI data set. In each of the 10 iterations, the data set is divided into training (90%) and evaluation (10%) data sets. The training set is used to generate the *AR* (*frequent* itemsets) and *WAR* (*significant* itemsets) models, and the evaluation set is used to test the recommendation effectiveness of the *AR* and *WAR* based approach.

5.2   Evaluation Methodology

To compare the *AR* based model and the *WAR* based model fairly, we select the same number of frequent/significant k-itemsets for the *AR* and *WAR* based models. Our evaluation methodology is as follows: Each transaction t in the evaluation set is divided into two halves; pages in the first half are used to generate the recommendation set R_t, and pages in the second half, denoted as $Eval_t$, are used to evaluate the generated recommendations. For the *AR* based recommendation approach, we adopt the *active session window*, i.e. the last visited $|win|$ pages [1] [4] [8], to generate recommendations. For the *WAR* based model, we use pages which are both fresh and significant in the first half to generate recommendations, as described in Section 4.2. To guarantee an appropriate minimum session size, we only use transactions containing at least 4 pageviews.

We propose two evaluation metrics: *Weighted Coverage* (*WC*) and *Weighted Precision* (*WP*), to evaluate the recommendation effectiveness. For a transaction t in the evaluation data set, *Weighted Coverage* and *Weighted Precision* are defined as:

$$WC_t = \frac{\sum_{R_t \cap Eval_t} w(p_i)}{\sum_{Eval_t} w(p_i)} , \tag{6}$$

$$WP_t = \frac{\sum_{R_t \cap Eval_t} w(p_i)}{|R_t| \cdot \overline{w}} , \tag{7}$$

where $w(p_i)$ is the weight of pageview p_i in transaction t; $|R_t|$ is the size of the recommendation set, and $\overline{w}$ is the average pageview weight across all transactions.

For example, if the evaluation half of transaction t is $Eval_t = \{(A, 80), (B, 20)\}$, it is evident that recommending page A captures more of the user's information need than recommending page B. In previous studies, however, recommending either page will get a coverage score of 50%. In our framework, weighted coverage of recommending page A is 80%, and that of recommending page B is 20%.

Still with the evaluation half $Eval_t = \{(A, 80), (B, 20)\}$, a recommendation set $R_1 = \{A, C, D, E\}$ is more precise than $R_2 = \{B, C, D, E\}$, although both have only 1 page needed by the user. Assuming the average pageview weight $\overline{w} = 50$, weighted precision of R_1 is $\dfrac{80}{4 \cdot 50} = 40\%$, and weighted precision of R_2 is $\dfrac{20}{4 \cdot 50} = 10\%$.

The overall weighted coverage and weighted precision scores are calculated as the *weighted* average of WC_t and WP_t scores for each transaction, where $w_t = \sum_{Eval_t} w(p_i)$ is the weight of $Eval_t$, and $Eval$ is the evaluation (10%) data set:

$$Weighted\ Coverage = \frac{\sum_{Eval} WC_t \cdot w_t}{\sum_{Eval} w_t}, \tag{8}$$

$$Weighted\ Precision = \frac{\sum_{Eval} WP_t \cdot w_t}{\sum_{Eval} w_t}, \tag{9}$$

The reason to use *weighted* average is that we should give greater weight to a score associated with a more important session. For example, a weighted coverage score of 80% may be meaningless, if $Eval_t = \{(A, 8), (B, 2)\}$ with page A being recommended, as the user is probably not interested in either page, spending less than 10 seconds on each. Another weighted coverage score of 60% may be more meaningful, if $Eval_t = \{(A, 60), (B, 40)\}$ with page A being recommended, because this evaluation set contains pages more useful to the user.

The overall weighted coverage and weighted precision scores for each of the 10 rounds are averaged to calculate the final weighted coverage and weighted precision scores.

5.3 Experimental Results

We vary the number of recommended pages from 4 to 14. Experimental results show that the *WAR* based model increases weighted coverage by 10% ~ 20% and weighted precision by 20% ~ 30%. We also vary $|win|$ from 3 to 5 and do not observe significant impact on the weighted coverage and weighted precision scores.

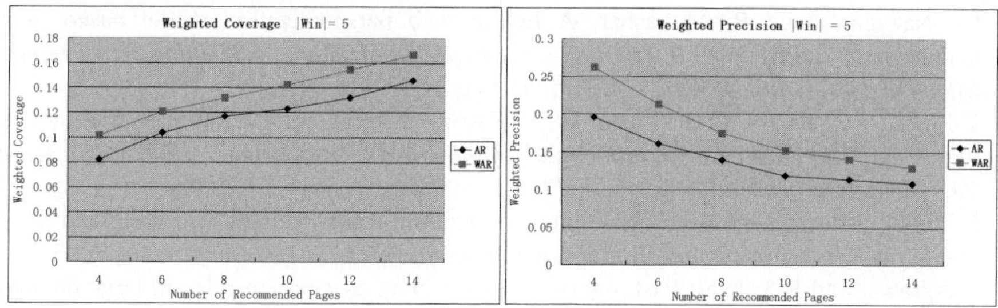

Fig. 2. Weighted coverage and weighted precision of the *AR* and *WAR* based models

6   Conclusion and Future Work

Most usage based page recommendation systems proposed in previous studies do not distinguish the importance of different pageviews in a transaction, and pages interesting to the user and those otherwise are equally used to capture user behavior patterns and generate recommendations. In this work, we use pageview duration to judge the importance of a pageview to a user, and try to give more consideration to pages which are more useful to the user, in order to capture the user's information need more precisely and recommend pages more useful to the user.

We have made a preliminary attempt to incorporate pageview weight into the association rule based recommendation system, and proposed a *Weighted Association Rule* model as an improvement to the traditional *Association Rule* model. Comparative experiments were performed to test the recommendation effectiveness of the traditional *AR* based model and the proposed *WAR* based model. Experimental results have shown that the *WAR* based model could significantly improve the recommendation effectiveness. The reason behind the improvement is that by taking the difference of the interestingness of the visited pages into consideration, our approach can discover and predict user interest more precisely.

The *WAR* based model can easily be extended to the (*contiguous*) *sequential pattern* model, which is a direction for future work. What is the optimal approach to determine pageview weight also needs further study.

Acknowledgment. This work was supported by Chinese 973 Research Project under grant No. 2004CB719401.

References

[1] B. Mobasher: Web Usage Mining and Personalization, In *Practical Handbook of Internet Computing*. Munindar P. Singh (ed.), CRC Press, 2005.

[2] B. Mobasher, H. Dai, T. Luo, and M. Nakagawa: Improving the Effectiveness of Collaborative Filtering on Anonymous Web Usage Data. In *Proceedings of the IJCAI 2001 Workshop on Intelligent Techniques for Web Personalization (ITWP01)*, Seattle, WA, August 2001.

[3] R. Cooley, B. Mobasher, and J. Srivastava: Data Preparation for Mining World Wide Web Browsing Patterns. In *Journal of Knowledge and Information Systems*, 1(1): 5–32, 1999.

[4] M. Nakagawa and B. Mobasher: A hybrid web personalization model based on site connectivity. In *The Fifth International WEBKDD Workshop: Webmining as a Premise to Effective and Intelligent Web Applications*, pages 59 ~ 70, 2003.

[5] C.H. Cai, A.W.C. Fu, C.H. Cheng, W.W. Kwong: Mining association rules with weighted items. In *Database Engineering and Applications Symposium, 1998. Proceedings. IDEAS'98. International*, pages 68 ~ 77, July 1998.

[6] F. Tao, F. Murtagh, M. Farid: Weighted Association Rule Mining using Weighted Support and Significance Framework. In *Proceedings of the 9th SIGKDD Conference*, 2003.

[7] R.Agrawal and R.Srikant: Fast algorithms for mining association rules in large databases. In *Proceedings of the 20th International Conference on Very Large Data Bases (VLDB'94)*, Santiago, Chile, 1994, pages 487-499.

[8] B. Mobasher, H. Dai, T. Luo, and M. Nakagawa: Effective personalization based on association rule discovery from web usage data. In *Proceedings of the 3rd ACM Workshop on Web Information and Data Management (WIDM01)*, Atlanta, Georgia, November 2001.

Neural Networks Fusion to Overlay Control System for Lithography Process

Jihyun Kim[1], Sanghyeok Seo[2], and Sung-Shick Kim[3]

[1] Research Institute for Information and Communication Technology,
Korea University, Seoul, 1, 5-ka, Anam-dong, Sungbuk-ku, 136-701, Korea
[2] Research Institute for Information and Communication Technology,
Korea University, Seoul, 1, 5-ka, Anam-dong, Sungbuk-ku, 136-701, Korea
charmi@korea.ac.kr
[3] Department of Industrial Systems and Information Engineering, Korea University,
Seoul, 1, 5-ka, Anam-dong, Sungbuk-ku, 136-701, Korea

Abstract. This paper presents a neural network based overlay control system for lithography process. The control system is structured to be compatible with the existing control system. The two main components of the control system are neural network prediction module for metrology prediction and a control module for various control methods. The prediction module utilizes various overlay metrologies and other process related parameters to assess the process conditions and make accurate predictions of the output metrologies. Based on the prediction results, control module calculates the appropriate control parameter settings. The prediction module is constructed using the Levenberg-Marquardt method to compensate for the small to medium size neural network and the demand for speed. The control module incorporates both popular control methods and specific engineering process control (EPC). Evaluation results are presented to illustrate the control system performance.

Keywords: Neural Networks, Lithography, Run-to-Run Control.

1 Introduction

Semiconductor manufacturing is regarded as one of the highly advanced and fast evolving industry. Strict production control limits of the semiconductor manufacturing industry are becoming tighter as the design patterns grow smaller and more complicated. With the increase in both the wafer size and the complexity of fabrication process, the accuracy of the process setting is becoming more critical. Furthermore, uncertainty (pre-run sampling) and misjudgement (rework) of process condition that lead to non-product wafer usage and repeated processes are very costly. Thus, the assurance of optimal process control for each lot (or run) has become vital to the industry. In order to acquire the optimal process settings for each lot, the control system must possess the ability to accurately assess the current process condition and compute the optimal process settings (recipes). Such requirements led to the development and implementation of Advanced Process Control (APC).

A. Sattar and B.H. Kang (Eds.): AI 2006, LNAI 4304, pp. 587–596, 2006.
© Springer-Verlag Berlin Heidelberg 2006

APC is evaluated as a viable solution to meet the increased efficiency and product quality demands of semiconductor manufacturing facilities. Three primary focus areas for the application of APC techniques are 1) statistical process control (SPC), 2) run-to-run (R2R) control, and 3) real-time process control (RTPC) [1]. Application of RTPC is often difficult due to limitations of integration with tool software and the reliability and long term reproducibility of real-time sensors [2]. Although efficient in detecting aberrations, traditional SPC based system does not provide solution to controlling them. Recently, R2R feedback control method has become popular in the semiconductor manufacturing industry. R2R is used to fine tune the recipes 'in between runs' using a collection of both statistical and engineering process control techniques. Progressing from SPC, exponentially weighted moving average (EWMA) is one of the most commonly used control method used in R2R and vigorous research has been conducted [4],[5],[6],[7]. With linear relationship assumption between control parameters and output metrologies, the gain is fixed and the process condition change is compensated by adjusting the intercept. The linearity assumption holds for many cases and has yielded successful application to semiconductor fabrication, especially for overlay control of the lithography process. However, the increasing requirements for smaller process margin necessitate tighter control of the process.

Typical structures of the control models are either a single-input single-output (SISO) or a multiple-input single-output (MISO) with small number of input parameters. For example, one parameter is generally used to control one overlay metrology in lithography overlay process. If additional process related parameters are introduced to the control model, it may lead to reduction of the disturbance factor resulting in an improved control parameter setting. However, identifying the relationship between the new parameters with the existent parameters is very demanding. Furthermore, uncontrollable parameters and qualitative variables are difficult to incorporate in control models.

An alternative is to improve the evaluation of the process condition. If the present status can be evaluated more accurately, it will greatly benefit the precision of the control model which uses this information. For this purpose, a process predictor is proposed. The intricate nature of the process is yet to be fully understood and therefore very difficult to model. Therefore, the use of neural networks for prediction purposes is employed for it does not require assumptions over input/output relationships and learns arbitrary mappings of complex nonlinear data. Output prediction is made using the initial recipe setting obtained from the control model. The control parameter and the predicted metrology acts as the information of a 'pseudo lot' that is fabricated with the current process conditions. Next, the predicted information is fed back to the control model. With this new information that better represents current process condition, an improved control setting can be obtained.

2   Neural Network for Process Assessment

Neural network gains process information from raw data without the need for algorithm or rule development. The innate ability of neural network to approximate complex nonlinear relationships and outperforming analytical model make neural network a perfect solution for process modeling [8],[9]. Neural network learns the patterns of systems through the training process and then predicts the system's performance without exact knowledge of the system dynamics. Such characteristics of neural network led to its application in manufacturing process modeling [10],[11]. Nonetheless, field of application of neural network in semiconductor fabrication is narrow. It has been mainly used to obtain estimates of the weight factor for EWMA control [12],[13]. And neural network has also been used for prediction purposes alone [14],[15]. R2R controller is predominantly a two-step procedure. The first step is to estimate the process condition. The control parameter settings and the resulting metrologies of the previously fabricated lots are used for this purpose. The second step is to calculate the control parameter setting. Using the process condition with the metrology target value, the recipe setting is obtained. Therefore, the outcome of the actual fabrication is dependent on the accuracy of the control model and in turn heavily influenced by the quality of the condition assessment. Process state evaluation, in general, is performed by using only a small portion of the information from the previously fabricated lots. In some cases, the evaluation outcome is adjusted by the engineers before being used to obtain the control parameters. With such limited information and possible outside interference, process condition assessment may be erroneous. To ascertain the proper process condition of the present, the use of metrology prediction to generate a 'pseudo lot' using neural network is suggested.

The pseudo lot contains the predicted outcome for the given input, i.e. the likely process result for a specific control setting. When calculating the prediction, the neural network utilizes various process related parameters that are excluded from commonly employed control systems. Along with the ability to model complex relationship, the resulting prediction is more accurate. Next, this information is used in calculating the control parameter as if it is a previously fabricated lot. With better reflection of the process condition, improvement in obtaining the control setting for the current 'actual lot' is possible. And because the pseudo lot data is exactly the same as any other previous lots, implementation to existing control system is straightforward.

3   System Architecture

The proposed neural network embedded control scheme is mainly divided into two components, Process controller and Process Predictor. Initial control parameter calculation, control parameter recalculation, and parameter search are performed in the process controller. In the process predictor, metrology prediction and evaluation is performed. Initial control parameters for the proposed heuristic controller are obtained by using the existing R2R controller, e.g. EWMA or engineering process control (EPC).

Next, the proposed parameter values along with additional process parameters are used as input for the neural network prediction. If the predicted outcome meets the predefined conditions, the lot is fabricated using the calculated recipe

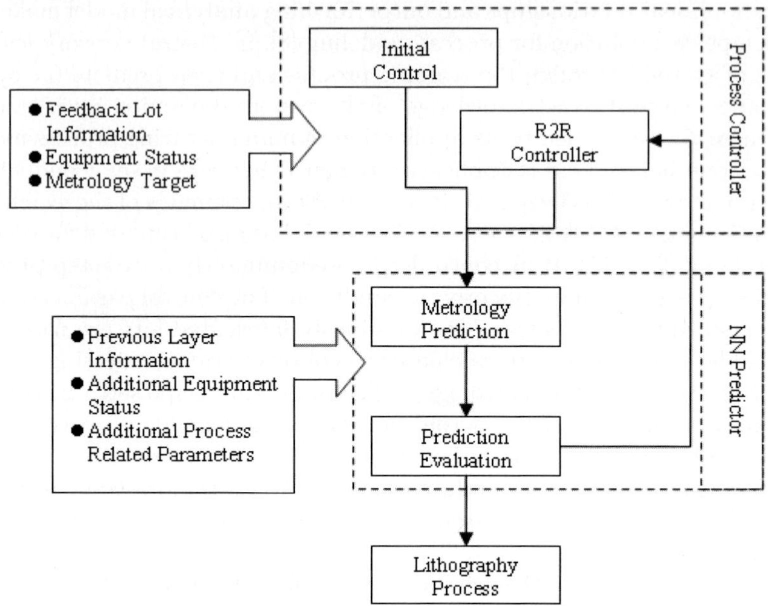

Fig. 1. The diagram of the control scheme

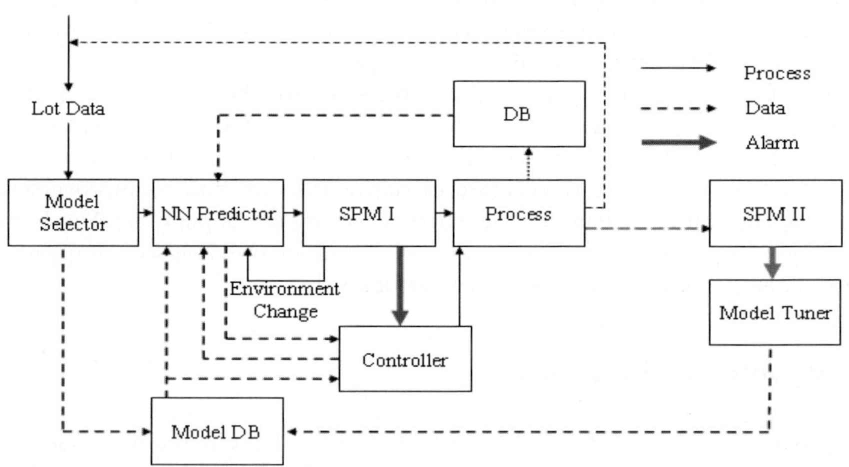

Fig. 2. The Overall Controller Structure

setting. If not, the suggested control setting with the predicted result is returned to the process controller. The pseudo lot information is used to recalculate the recipe setting. The schematic of the proposed methodology is shown in figure 1.

One of the major concerns when using neural network for actual process control and modeling is the requirement for obtaining enough data for training. Lack of a process controller is unacceptable for manufacturing process. Therefore, a model DB is included in the system. It contains some of the basic control models, such as moving average (MA) or weighted moving average (WMA), to be used in the absence of a trained network. Statistical Process Monitoring II (SPM II) module along with the Model Tuner module monitors and evaluates the trained network. Once proven to be sufficient, process prediction is performed using the neural network predictor. Such structure is useful for updating the neural network as well. Because back-propagation algorithm, specifically the Levenberg-Marquardt method, is adopted for neural network learning, batch learning is unavoidable. By way of SPM I module, the performance of the process prediction is monitored. A series of large errors can be a criterion for model update. The overall system is depicted in figure 2.

4   Neural Network Based Prediction Model

In this study, the input layer of units consists of control parameters and wafer state variables which are varied at the pre-process' equipments. Input variables are normalized to a range of $[-\alpha, \alpha]$ where α is a value between 0 and 1. Such approach is taken to maximize the use of the steepest portion of the sigmoid function. The learning begins and the error for each output unit is computed and summed over all training patterns. The residual error is the difference between measured states' change and predicted states' change multiplied by the derivative of the output unit's activation function. The hyper tangent function is used for hidden layer's activation function and linear function is used for outputs to take on any value.

Levenberg-Marquardt method is adopted to minimize residual error. In general, on function approximation problems, for networks that contain up to a few hundred weights, the Levenberg-Marquardt algorithm has the fastest convergence. This advantage is especially noticeable if very accurate training is required like in FAB modeling. However, as the number of weights in the network increases, the advantage of the Levenberg-Marquardt decreases. The conjugate gradient algorithms seem to perform well particularly for networks with a large number of weights. The models usually made on FAB processes are relatively small size and the Levenberg-Marquardt are enough for most cases.

Since the system is nonlinear, values of training data sets should be varying for different operational conditions. The trained process predictor model will be more accurate in the range of the training data. In order to model each process up-to-dately, data from the new run is used to train neural network periodically or on demand.

5   Application

5.1   Lithography Process in Semiconductor Fabrication

Lithography is one of the key processes in semiconductor fabrication. It is the first step where patterns of each layers are masked on photoresist (PR). The patterned

PR is used to etch desired patterns on the deposited layers. Typical procedures of the lithography process are depicted in figure 3. The highlighted steps are the three key steps in lithography. With the aligned mask (or reticle) positioned in the desired location, PR is irradiated using radiation of appropriate wavelength (e.g. deep ultraviolet) during exposure. During development, the exposed PR is dissolved leaving the desired pattern behind. Major control parameters of the stepper/scanner are associated with mask alignment and exposure.

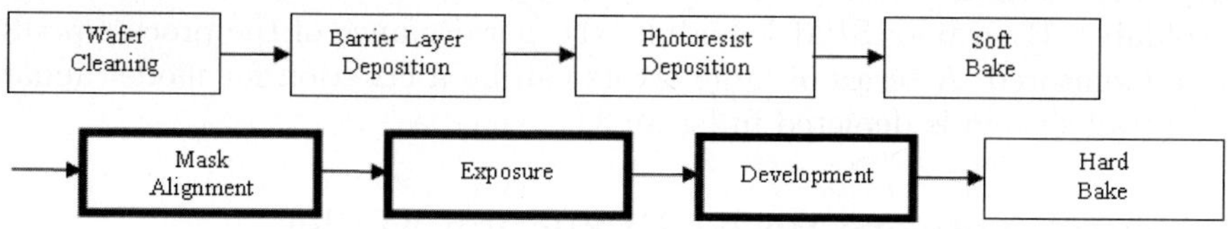

Fig. 3. Lithography procedures

There are two major metrologies related to the lithography process and they are critical dimensions (CD) and Overlays. Critical dimension is the dimension of the smallest geometrical features (width or space between interconnect lines, contacts, trenches, etc.). Overlay is the superposition of the current pattern to the previously created. The overlay errors are divided into inter-field and intra-field and shown in figure 4. The offset x and offset y represent the degrees of deviation for the patterns formed on a wafer, horizontally and vertically. The wafer-rotation represents the rotation degree of alignment axes of the patterns with respect to the base alignment axes.

The scale x and scale y represent the degrees of expansion for the patterns. The orthogonality represents the deviation degree of two base alignment axes. The reticle rotation represents the rotation degree of alignment axes for the patterns with respect to the base alignment axes caused by incorrect settings of reticles and the reticle magnification represents the degree of reduction for the

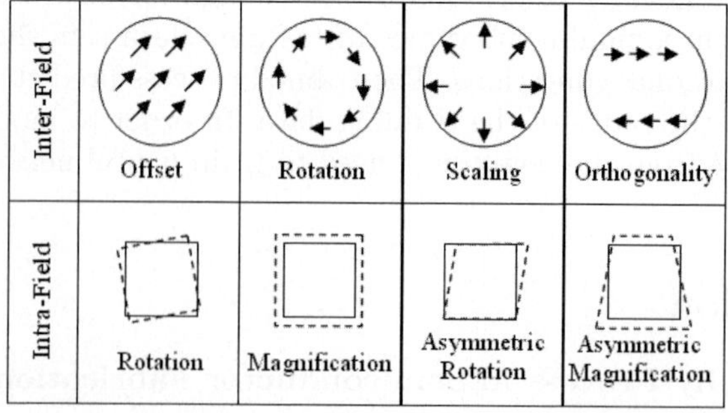

Fig. 4. Overlay errors classification

patterns. Each of these overlay parameters are adjusted using dedicated control parameters in the stepper/scanner [16].

Two layers are used for adjusting and measuring overlays - align layer and overlap target layer. First, align layer is used as a basis to align the stepper/scanner. Next the current layer is targeted to be masked exactly on top of the overlap layer. Offset is the distance between the overlap layer and the current layer. The ideal offset control setting, with no noise factor, and the actual offset error is depicted in figure 5.

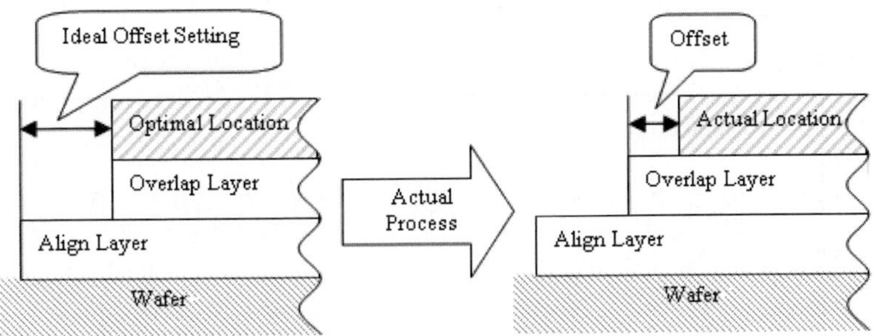

Fig. 5. Diagram of offset control parameter and metrology

5.2 Implementation to Overlay Lithography Process

The proposed control scheme was evaluated on a 256M DRAM lithography process. The offset overlay parameters were collected from a bit line generation step. The target plant uses an SISO EPC control model that assumes linear relationship between the control parameter and the associated overlay metrology. The control model utilizes the control parameter setting and the resulting overlay metrology to calculate the recipe setting for the new lot.

In addition to the previous lot information, previous layer metrology information, current equipment status as well as other qualitative parameters were employed to construct the neural network prediction model. Generally, the overlay control parameters are assumed to be independent. And although some correlation may exist, it is of minor scale and not included in the model. The data set, collected in the span of three months, was divided into three groups. Forty percent of the data set was used for training the network and thirty percent was used for validation to prevent overfitting when training the network. The remaining thirty percent was used for testing.

Building the neural network's architecture starts from the simplest network with one hidden layer with one hidden neuron. The network is trained with the training data set and the squared error of network with validation data set is monitored to determine the iteration length and adequacy of the learning. The correlation coefficient (R) between the real metrology and the forecasted metrology is also monitored. If a predetermined value of R ($= 0.75$) is not achieved, another hidden node is added. If the predetermined value is reached, no more nodes are added and network architecture is fixed. Then, this trained network is used as forecasting model.

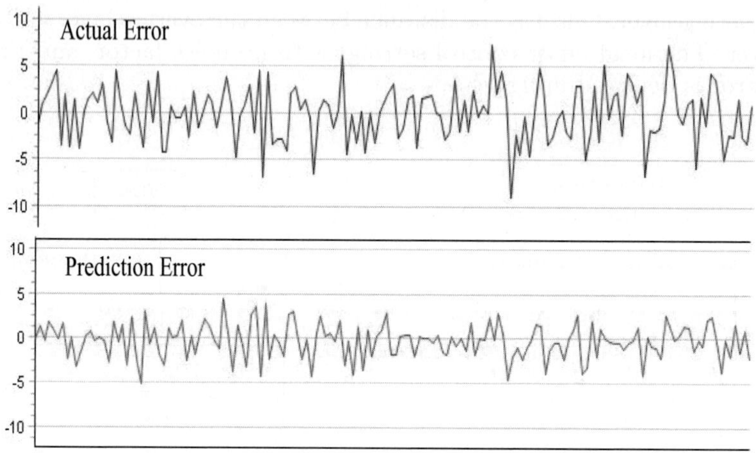

Fig. 6. Plot of actual Error versus Prediction Error

The results show that, on average, the correlation between the observation and prediction for the testing data set is 0.82. Next, linear regression was performed on the same set of data. Although the target factory also uses a linear model EPC, it is based on a SISO system. In order to make a proper comparison, multivariate regression is performed with the same input parameters (excluding qualitative parameters). The test data set yielded 0.75 correlation. This shows that although linearity may be assumed for the data set, the prediction performance is weaker compared to neural network. The predicted error from the proposed control system and the actual error from the existing process for offset x overlay parameter are depicted in figure 6. Based on existing control system, the predicted error is zero; it is assumed that the optimal metrology, i.e. zero deviation, can be obtained by using the calculated control parameter. Therefore, such comparison can shed light on the performance of each control system. Based on the adjusted scales between -10 to 10, the standard deviations are 2.96 and 1.92. The sum of squared error shows that the introduction of additional parameters lead to 24% reduction of prediction errors for linear regression and 35% reduction for neural network.

6   Conclusion

Control system for improving the performance of the semiconductor fabrication control by introducing neural network prediction into the overall control scheme have been proposed in this paper. By generating a pseudo lot that reflects the current process condition, evaluation of the current process condition is distinguishably improved. Using this information, the performance of the control model is enhanced. With the library of control models available, the control system can be implemented even for a newly beginning products. And the structure of the control system is such that it can be implemented on a existing control system with ease. In addition, improvement in process evaluation can help in reducing outside interruptions rising from uncertainty regarding the existing control system. Evaluation of the proposed control system shows promising results.

Acknowledgement. This work is financially supported by the Ministry of Education and Human Resources Development(MOE), the Ministry of Commerce, Industry and Energy(MOCIE) and the Ministry of Labor(MOLAB) through the fostering project of the Lab of Excellency and also supported by the Brain Korea 21 Project in 2006. The authors would like to thank J.W. Ju and J.Y. Moon of Hynix Semiconductor Inc. and W.K. Choi of Bistel Inc. for their valuable advices throughout this study.

References

1. Han, S., May, G.: Using neural network process models to perform PECVD silicon dioxide recipe synthesis via genetic algorithms. IEEE Transactions on Semiconductor Manufacturing. **10** (1997) 279–287
2. Kim, B., May, G.: An optimal neural network process model for plasma etching. IEEE Transactions on Semiconductor Manufacturing. **7** (1994) 12–21
3. Moyne, J., Del Castillo, E., Hurwitz, A. M.: Run-to-Run Control in Semiconductor Manufacturing. CRC Press, Boca Raton FL (2000)
4. Ingolfsson, A., Sachs, E.: Stability and Sensitivity of an EWMA Controller. Journal of Quality Technology. **25** (1993) 271–287
5. Butler, S., Stefani, J.: Supervisory run-to-run control of polysilicon gate etch using In Situ ellipsometry. IEEE Transactions on Semiconductor Manufacturing. **7** (1994) 193–201
6. Del Castillo, E., Rajagopal, R.: A multivariate double EWMA process adjustment scheme for drifting processes. IIE Transactions. /bf 34 (2002) 1055–1068
7. Tseng, S-T, Chou, R-J, Lee, S-P: A study on a multivariate EWMA controller. IIE Transactions. **34** (2002) 541–549
8. Holter, T., Yao, X., Rabelo, L. C., Jones, A., Yih, Y.: Integration of Neural Networks and Genetic Algorithms for an Intelligent Manufacturing Controller. Computers & Industrial Engineering. **29** (1995) 211–215
9. Meng, H., Russell, P. C., Lisboa, P. J. G., Jones, G. R.: Modelling and Control of Plasma Etching Processes in the Semiconductor Industry. Computers & Industrial Engineering. **37** (1999) 367–370
10. Haley, P., Soloway, D., Gold, B.: Real-time Adaptive Control Using Neural Generalized Predictive Control. Proceedings of the American Control Conference. San Diego CA (1999)
11. Soloway, D., Haley, P.: Neural Generalized Predictive Control: A Newton-Raphson Implementation. Proceedings of the IEEE CCA/ISIC/CACSD. IEEE Paper No. ISIAC-TA5.2. (1996)

financial metrics [18-21]. This is not surprising since these models utilise more information such as inter market indicators, fundamental indicators and technical indicators. Furthermore, neural networks are capable of describing the dynamics of nonstationary time series due to their non-parametric, adaptive and noise tolerant properties [17].

However, despite the encouraging results of using artificial neural networks (ANN) for financial time series prediction compared to linear statistical models, the robustness of these findings has been questioned [6], due to a number of well known problems with neural models such as:

1. Different neural networks algorithms can produce different results when trained and tested on the same data set. This is because there are different classes of decision boundaries that different ANN prefers.
2. For any given type of neural network, the network is sensitive to the network size and the size of the data set. Neural networks suffer from overfitting and as a result network architecture, learning parameters and training data have to be selected carefully in order to achieve good generalisation, which is critical when using the network for financial time series prediction.
3. The inherent nonlinearity of financial time series can prevent a single neural network from being able to accurately forecast an extended trading period even if it could forecast changes in the testing data.

In this paper we introduce a novel architecture, named higher order polynomial pipelined neural networks. The networks consist of pi-sigma neural networks [7] and recurrent pi-sigma networks [8] concatenated with each other. Each of these networks forms a single unit in the proposed network.

The proposed model is applied to three exchange rate time series namely the Euro to USA dollar, Japanese Yen to USA dollar and Sterling Pound to USA dollar during the period from 17th October 1994 to 6th July 2001 totalling 1730 trading days.

The remainder of the paper is organised as follows: section 2 presents the structure of the pi-sigma and recurrent pi-sigma networks. Section 3 shows the structure of the novel polynomial pipelined neural network and how it can be adaptively trained. In section 4 we discuss the simulation results. Several alternative neural network architectures benchmark the performance of the proposed network, these include the multilayer perceptron, the functional link network and the recurrent pi-sigma network [8].

2 Pi-sigma and Recurrent Pi-sigma Network

The pi-sigma network is a multiple-layer, higher-order neural network. It was introduced by Gosh and Shin [7] to perform function approximation and classification. The network has two layers with product and the summing units. The input patterns are presented to the network, which calculates the sums of the weighted inputs at the summing unit layer. The outputs of the summing units layer are directly multiplied with each other, since the weights between the summing and the product units layers are fixed to unity, and the result is forwarded to usually a nonlinear transfer function. The number of summing units corresponds to the order of the network, i.e., a second- order network contains two summing units, while a third order network contains three summing units and so on. This means that the network has a parsimonious structure, in contrast to higher-order networks, which have irregular structure since increasing their order will excessively increase the number of interconnected weights.

The recurrent pi-sigma network (RPSN) was introduced by Hussain and Liatsis [8] to perform function approximation and image compression. It is obtained from the standard model by adding to the input models a delayed feedback unit from the network output.

Consider a RPSN with M external inputs and one output. The total number of inputs is M+2 (M external inputs, one input for the bias, and one input for the recurrent link). Let the number of summing units be k with W a weight matrix of size $k \times (M+2)$.

If $x_m(t)$ represents the mth external input and y(t) represents the output of the network at time t, then the input vector of the network is a concatenation defined by z(t):

$$z_j(t) = \begin{cases} x_j(t), & \text{if } 1 \le j \le M \\ 1, & \text{if } j = M+1 \\ y(t), & \text{if } j = M+2 \end{cases} \tag{1}$$

The processing equations of the RPSN are given as follows:

$$h_L(t+1) = \sum_{m=1}^{M+2} w_{Lm} z_m(t)$$

$$y(t+1) = f(\prod_{L=1}^{k} h_L(t+1)) \tag{2}$$

where $h_L(t+1)$ represents the net sum of the L unit at time t+1, and the output of this network is y(t+1). The unit's activation function f is a nonlinear transfer function and is usually taken to be the logistic sigmoid.

The network is trained using dynamic backpropagation [9], which is a gradient descent learning algorithm based on the assumption that the initial state of the network is independent of the initial weights and

$$\Delta w_{ij}(t+1) = -\eta \frac{\partial J(t+1)}{\partial w_{ij}} + \alpha \Delta w_{ij}(t) \tag{3}$$

where $\Delta w_{ij}(t+1)$ is the change of change of w_{ij} at time t+1, η is a positive real number representing the learning rate, α is the momentum and J(t+1) is the cost function.

3   Polynomial Pipelined Neural Network

This section introduces a new type of pipelined neural network, called polynomial pipelined neural network (PPNN). The network consists from a number of pi-sigma networks and recurrent pi-sigma networks linked together in sequence. It is designed to adaptively predict highly nonlinear and nonstationary signals such as the exchange rate time series in a similar way to the fully recurrent pipelined neural network [10].

Let q represent the total number of feedforward and recurrent pi-sigma neural networks, which are concatenated with each other. Each neural network is called a unit (or a module) and consists of M external inputs and one output. All the units of the PPNN are pi-sigma neural networks, except the last module in the pipeline, which is a recurrent pi-sigma network where its output is fedback to the inputs. For the recurrent pi-sigma network, the output is fedback to the inputs and it is forwarded to the next unit as input. The bias is included into the structure of the unit by adding an extra input line of value 1.

The total number of trainable weights in each unit is $(M+1) \times k$ where k is the order of the network and each module holds a copy of the same weight matrix W. The output of the PPNN is the output of the first unit in the pipeline. The detailed structure of the proposed network is shown in Figure 1.

In what follows, the inputs and the processing equations of the polynomial pipelined neural networks are presented. The input vectors to unit i comprise time shifted tapped delay lines with windows of size M, given by:

$$X_i(t) = [x_{i,1}(t), \ x_{i,2}(t),\ldots\ldots\ldots, x_{i,M}(t)]^T$$
$$= [S(t-i), \ S(t-(i+1)),\ldots\ldots, S(t-(i+M-1))]^T \tag{4}$$

where $\{S(t)\}$ is the source signal consisting of the time series to be modelled. Let $y_{i,}$ represent the output vector of module i.

The inputs to the last module is augmented by a recurrence connection:

$$Z_q(t) = [x_{q,1}(t) \ x_{q,2}(t),\ldots\ldots\ldots, x_{qi,M}(t),1, y_q(t-1)]^T \tag{5}$$

For all other units of the network, the input vector is augmented with the output from the next module in the pipeline:

$$Z_i(t) = [x_{i,1}(t) \ x_{i,2}(t),\ldots\ldots\ldots, x_{i,M}(t),1, y_{i+1}(t)]^T \tag{6}$$

The processing equations of the PPNN are given by:

$$y_i(t) = f(v_i(t)) \tag{7}$$

For the last module of the polynomial pipelined network:

$$v_q(t) = \prod_{L=1}^{k} h_{q,L}(t+1) \tag{8}$$

$$h_{a,L}(t+1) = \sum_{m=1}^{M} w_{q,Lm} \cdot x_{q,m}(t) + w_{q,L(M+1)} \cdot 1 + w_{q,L(M+2)} \cdot y_q(t-1)$$

where k is the order of the recurrent pi-sigma unit.

For all other modules:

$$v_i(t) = \prod_{L=1}^{k} h_{i,L}(t+1) \tag{9}$$

$$h_{i,L}(t+1) = \sum_{m=1}^{M} w_{i,Lm} \cdot x_{i,m}(t) + w_{i,L(M+1)} \cdot 1 + w_{i,L(M+2)} \cdot y_{i+1}(t)$$

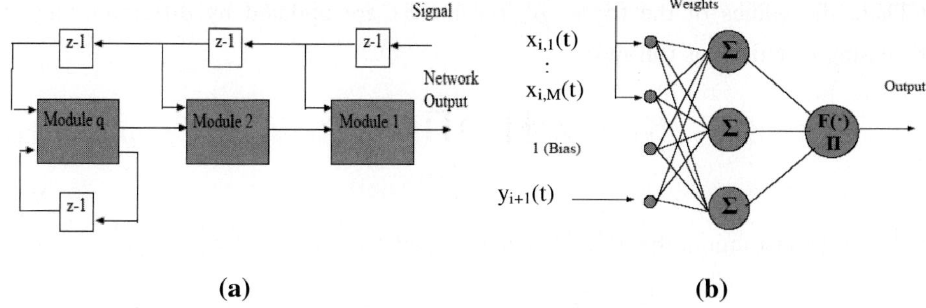

Fig. 1. The structure of the polynomial pipelined neural network; (b) the structure of unit i of the network

4   Parameters Estimation for the Polynomial Pipelined Neural Network

The proposed network is trained using the real-time learning algorithm developed by Williams and Zipser [11]. The advantage of this learning algorithm is that the epoch boundaries are no longer required, making the implementation of the algorithm simpler and letting the network be trained for indefinite period of time. The error from each module contributes to an overall cost function:

$$\varepsilon(t) = \sum_{i=1}^{q} \lambda^{i-1} e_i^2(t),$$ (10)

where λ is an exponential forgetting factor selected in the range (0, 1) and $e_i(t)$ is the error of module i at time t calculated by taking the difference between the actual and the predicted values. The weights are iteratively updated by gradient descent:

$$\Delta w_{ml}(t) = -\eta \frac{\partial \varepsilon(t)}{\partial w_{ml}},$$ (11)

where $\square$ is the manually adjusted gain and

$$\frac{\partial \varepsilon(t)}{\partial w_{ml}} = 2 \sum_{i=1}^{q} \lambda^{i-1} e_i(t) \frac{\partial e_i(t)}{\partial w_{ml}},$$
$$= -2 \sum_{i=1}^{q} \lambda^{i-1} e_i(t) \frac{\partial y_i(t)}{\partial w_{ml}}$$ (12)

Define

$$p_{ml}^i(t) = \frac{\partial y_i(t)}{\partial w_{ml}}.$$

Then, the values of the triple $p^i_{ml}(t)$ matrix are updated by differentiating the processing equations as follows:

$$P^i_{ml}(t) = f'(\prod_{L=1}^{k} h_L) \prod_{\substack{L=1 \\ L \neq m}}^{k} h_L Z_l \tag{13}$$

while for the last unit q, the $p^q_{ml}(t)$ matrix is updated recursively:

$$P^q_{ml}(t) = f'(\prod_{L=1}^{k} h_L) \prod_{\substack{L=1 \\ L \neq m}}^{k} h_L \left(Z_l(t) + w_{m(M+2)} P^q_{ml}(t-1) \right) \tag{14}$$

5 Simulation Results

Three exchange rates time series have been used for our experiments, namely the Euro to USA dollar (EUR/USD), Japanese Yen to USA dollar (YEN/USD) and Sterling Pound to USA dollar (POUND/USD) in the period between 17th of October 1994 to 6th of July 2001. This period amounts to 1730 trading days. Since the Euro was not traded until the 4th January 1999, the earlier samples are a retropolated synthetic series using the USA Dollar to German Mark (USD/DEM) daily rate combined with the fixed EUR/DEM conversion rate agreed in 1998. The performance of the network is evaluated using the signal processing and trading metrics defined in Table 1.

The input data are pre-processed between 0 and 1 and passed to the neural networks as nonstationary data. The network is trained to predict the next incoming time point.

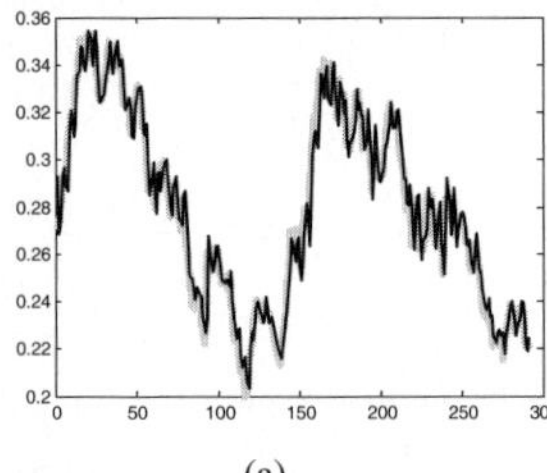
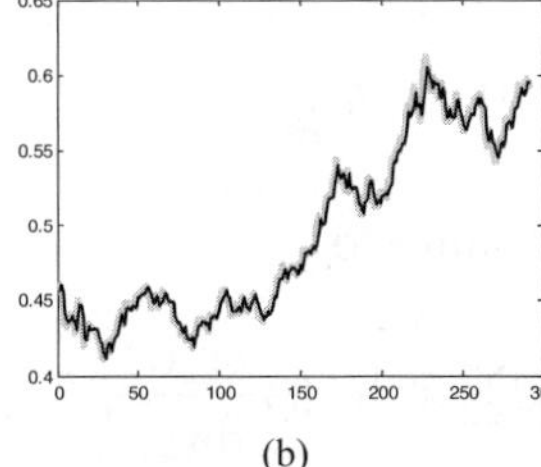
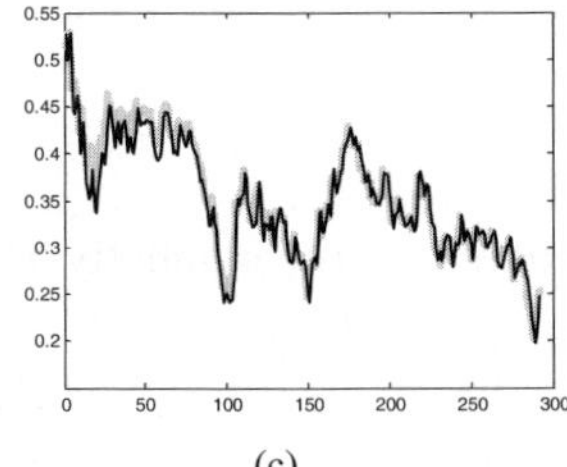

(a)   (b)   (c)

Fig. 2. Part of the prediction of the daily exchange rate using the PPNN in the period between 17th of October 1994 to 6th of July 2001 between the (a) Euro; (b) Yen; (c) Pound and the USA. The dashed line indicates the predicted signal, while the grey line shows the original signal.

When using the polynomial pipelined neural network to predict the daily exchange rates between the three currencies and the USA dollar in the trading period, the logistic sigmoid transfer function was utilised at the output layer. The weights of one unit of the network were initialised between -0.2 and 0.2 and trained for 2000 epochs using 300 data points. The learning rate and the momentum term were set to 0.4 and 0.2, respectively. The weights are then copied for the whole units of the polynomial pipelined network, which can then be trained adaptively as discussed in section 5. This procedure is necessary to initialise the weights of the network rather than using random weight values. The number of input units and the order of the networks, which are required to obtain the necessary mapping accuracy, are determined experimentally by trial and error.

Table 1. Signal processing and trading simulation performance measures

Mean square of the error	$$MSE = \frac{1}{N}\sum_{i=1}^{N} e_i(t)^2$$ where e(t) is the error and N is the number of data points
Signal to noise ratio	$$SNR = 10 \cdot \log_{10}\left(\frac{Max(y)^2 \cdot N}{SSE}\right)$$ where SSE is the sum square of the error and Max(y) is the maximum value of the signal.
Annualised return	$$R^A = 252 \cdot \frac{1}{N}\sum_{t=1}^{N} R_t$$ $$R_t(t) = \frac{\hat{y}(t)}{\hat{y}(t-1)} - 1$$
Maximum drawdown	$$MD = Min\left(\sum_{i=1}^{N} CP_{i-1} - Max(CP_1,......CP_i)\right)$$ where $$CP = \sum_{i=1}^{N} \hat{R}_i(t)$$

Figure 2 shows part of the prediction of the exchange rate time series using the polynomial pipelined network on out-of-sample data with the best result achieved. The performance of the PPNN network was benchmarked with the MLP network, the functional link neural network (FLN) and the recurrent pi-sigma network. For the MLP, nine different network architectures were used. The weights of the networks were set randomly between 0 and 1. The networks were trained for 20 times to give an average value using the resilient backpropagation learning algorithm and early stopping strategy. The learning rate was set to 0.05. For the FLN, five network architectures were used. Similar to the MLPs, each network was reinitialized and retrained 20 times to give an average value. The networks were trained using the resilient backpropagation with early stopping. The learning rate was set to 0.05. For the recurrent pi-sigma networks, a total of 16 different network configurations were used with 4, 5, 6 and 7 inputs, the order of the network was changed between 2 and 5. The networks were trained using the dynamic backpropagation-learning algorithm.

The initial weights were set between -0.2 and 0.2. Each network was retrained 20 times in order to provide an average value. The learning rate was set to 0.4 and the momentum rate was set to 0.2.

Table 2. Benchmark performance averaging each model over 20 runs

Network	EUR/USD	YEN/USD	POUND/USD	
MLP	0.09481	0.25005	0.33859	MSE
	6.47	32.37	17.33	SNR (dB)
	-0.98	10.90	-5.49	Annualized return (%)
	-10.05	-6.68	-3.49	Maximum Drawdown (%)
FLN	0.23827	0.24329	0.35170	MSE
	5.53	26.87	13.43	SNR (dB)
	1.09	5.49	-7.04	Annualized return (%)
	-8.11	-8.80	-3.87	Maximum Drawdown (%)
Recurrent Pi-sigma	3.372560	0.06286	1.64110	MSE
	9.94	33.65	21.86	SNR (dB)
	14.30	-3.85	6.91	Annualized return (%)
	-9.66	-11.44	-8.13	Maximum Drawdown (%)
PPNN	0.07532	0.01413	0.10604	MSE
	26.91	38.91	29.08	SNR (dB)
	5.05	4.53	7.42	Annualized return (%)
	-13.05	-6.24	-6.59	Maximum Drawdown (%)

As it can be noticed from Tables 2, the PPNN achieves significant improvement using the signal to noise ratio over the MLP, the FLN, and the recurrent pi-sigma network. For the annualised return trading measure, the network showed profit for all signals and produced better values for the prediction of the POUND/USD and EUR/USD in comparison to the MLP network.

6 Conclusion and Further Work

In this paper, we have presented a novel type of pipelined neural network, the polynomial pipelined neural network. The proposed network was used to predict the exchange rate time series. Three FX time series were used, EUR/USD, YEN/USD and POUND/USD exchange rates. The aim of the new proposed network is to predict highly nonlinear and nonstationary signals using the concept of divide and conquer. Meaning that if the problem is big enough, then this problem can be divided into a number of smaller problems. The simulation results have shown that using the signal processing measures such as the signal to noise ratio, the new polynomial pipelined network showed the best results when benchmarked with the MLP, the FLN and the recurrent Pi-sigma network.

Further work will involve the use of genetic algorithms to determine the best choice of the network architecture, the number of units concatenated in the pipelined structure and the number of inputs. Furthermore, this research has focused on one-step ahead prediction. It would be interesting to determine the capabilities of higher-order pipelined neural networks to predict multi-step ahead prediction.

References

1. WILSON, R. and SHARDA, R., 1994. Bankruptcy prediction using neural networks. *Decision Support Systems* **11**, pp. 545–557.
2. A.N. REFENES, M. AZEMA-BARAC, L. CHEN & S.A. KAROUSSOS "Currency exchange rate and neural networks design strategies". In *Neural Computing and Application*. Vol. 1, No. 1. Pages 46-58. 1993
3. R.G BROWN. "Smoothing, forecasting and prediction of discrete time series". New Jersey, USA Prentice Hall. 1963
4. J.E HANKE & A.G REITSCH. "Business forecasting". London, UK. Allyn and Bacon. 1989.
5. D.T PHAM "Neural networks for identification, prediction and control". London, UK Springer-Verlag. 1995
6. M. Versace, R. Bhatt, O. Hinds and M. Shiffer, "Predicting the exchange traded fund DIA with a combination of genetic algorithms and neural networks", Expert systems with applications, 27, 2004, pp. 417-425.
7. J. GHOSH, & Y. SHIN. "Efficient higher-order neural networks for classification and function approximation". In *International Journal of Neural Systems*. Vol. 3, No. 4. pages 323-350. 1992.
8. A. HUSSAIN and P. LIATSIS, "Recurrent Pi-Sigma neural network for DPCM image coding," In *Neurocomputing*, Vol. 55. Pages 363-382. 2003
9. C.M KUAN.: "Estimation of neural network models", PhD thesis, University of California, San Diego, USA. 1989.
10. S. HAYKIN & L. LI. "Nonlinear adaptive prediction of nonstationary signals", In *IEEE Transactions on Signal Processing*, Vol. 43, No. 2. Pages 526-535. 1995.
11. R.J. WILLIAMS & D. ZIPSER. "A learning algorithm for continually running fully recurrent neural networks". In *Neural Computation*. No. 1. Pages 270-280. 1989.
12. T. MASTERS. "Practical neural network recipes in C++". Morgan Kaufmann, San Francisco, USA, 1993. ISBN: 0-12-479040-2
13. M. RIEDMILLER, & H. BRAUN. "A direct adaptive method or faster backpropagation learning" In *Proc. of the IEEE Intl. Conf. on Neural Networks*, San Francisco, USA, Pages 586 -591. 1993.
14. ODAM, M.D. and SHARDA, R, 1990. *A neural network model for bankruptcy predictionIn: Proceedings of the IEEE International Joint Conference on Neural Networks. San Diego, CA* **vol. 2** pp. 163–168.
15. GINZBURG, I. and HORN, D., 1994. Combined neural networks for time series analysis. *Advances in Neural Information Processing Systems Sci- Systems* **6**, pp. 224–231.
16. C. DUNIS and M. WILLIAMS, "Modelling and trading the EUR/USD exchange rate: Do neural network models perform better?", *Derivatives Use, Trading and Regulation*, **8**, 3, pp. 211-239.

17. L. CAO and F. E. H. TAY, "Financial Forecasting Using Vector Machines", In *Neural Computing and Application.* Vol. 10, 2001, pp. 184-192.
18. S. M. ABECASIS and E. S. LAPENTA, " Modeling multivariate time series with neural networks: comparison with regression analysis", Proceeding of the INFONOR'96: IX International Symposium in Informatic Applications, Antofagasta, Chile, 1996, pp. 18-22.
19. W. CHENG, L. WANGER, and C.H. LIN, "Forecasting the 30-year US treasury bond with a system of neural networks", *J Computational Intelligence in Finance*, 4, 1996, pp.10-16.
20. R. SHARDA and R. B. PATIL, "A connectionist approach to time series prediction: an empirical test", *Neural Networks in Finance Investing*, 1993, pp. 451-464.
21. E. VAN and J. ROBERT, "The application of neural networks in forecasting of share prices", Finance and Technology Publishing, Haymarket, VA, USA, 1997.

Using Neural Networks to Tune the Fluctuation of Daily Financial Condition Indicator for Financial Crisis Forecasting

Kyong Joo Oh[1], Tae Yoon Kim[2], Chiho Kim[3], and Suk Jun Lee[4]

[1] Department of Information and Industrial Engineering, Yonsei University, Seoul, Korea
johanoh@yonsei.ac.kr
[2] Department of Statistics, Keimyung University, Daegu, Korea
tykim@kmu.ac.kr
[3] Korea Deposit Insurance Corporation, Seoul, Korea
chihokim@kdic.or.kr
[4] Department of Information and Industrial Engineering, Yonsei University, Seoul, Korea
lsj77@yonsei.ac.kr

Abstract. Recently, Oh et al. [11, 12] developed a daily financial condition indicator (DFCI) which issues an early warning signal based on the daily monitoring of financial market volatility. The major strength of DFCI is that it is expected to serve as a quite useful early warning system (EWS) for the new type of crisis which starts as an instability of the financial markets and then develops into a major crisis (e.g., 1997 Asian crises). One of the problems with DFCI is that it may show a high degree of fluctuation because it handles daily variable, and this may harm its reliability as an EWS. The main purpose of this article is to propose and discuss a way of smoothing DFCI, i.e., it will be tuned using long-term (monthly or quarterly) fundamental economic variables. It turns out that such a tuning procedure could reveal influential macroeconomic variables on financial markets. Since tuning DFCI is done by the method of fitting various types of data simultaneously, neural networks are employed. Tuning the DFCI for the Korean financial market is given as an empirical example.

1 Introduction

Recently, Oh et al. [11, 12] developed a daily financial condition indicator (DFCI) which issues an early warning signal for possible crisis by monitoring the daily volatility of the financial market. The major motivation behind their work is a recent trend for financial crises to start out as financial market instability and then develop very quickly into major crises without any significant long-term deterioration of economic fundamentals (e.g., 1997 Asian crisis). Since this runs counter to the existing hypothesis that an financial crisis is an outcome of the deterioration of economic fundamentals over a long period of time, most of the existing early warning systems (EWS's) fail to function for the new type of crisis. Recall that the commonly used EWS's concentrate on long-term (monthly or quarterly, in some cases yearly) movements of macroeconomic variables and issue a warning signal when current

A. Sattar and B.H. Kang (Eds.): AI 2006, LNAI 4304, pp. 607–616, 2006.

© Springer-Verlag Berlin Heidelberg 2006

financial conditions are approaching a pre-defined financial crisis in terms of the movements of such variables.

Since DFCI by Oh et al. [11, 12] often exhibits a high degree of day-to-day fluctuations, which is natural because it handles daily variables, it seems necessary to smooth out DFCI in order to detect a reliable signal (otherwise such a high degree of fluctuations is likely to cause an observer of the DFCI to be confused between true warning signals and noise). The main purpose of this study is to propose and discuss a way of smoothing DFCI, i.e., tuning DFCI with long-term (monthly or quarterly) fundamental economic variables. The tuned DFCI will be termed t-DFCI hereafter. The tuning method is discussed as a procedure associating DFCI as a daily response variable with monthly or quarterly predictor variables. In addition, the *new scoring variables* that relate to economic fundamental weakness in terms of time length are introduced to improve the tuning effect. Since t-DFCI is to be constructed to handle various types of data, neural networks (NN) are employed as a key tool because of their well-known technical advantage, i.e., flexibility in handling various types of data [2, 14].

This study consists of four sections. Section 2 presents a t-DFCI construction procedure (tuning procedure) of associating long-term variables with daily variables. Section 3 gives a specific illustration of t-DFCI construction by establishing t-DFCI for the Korean financial market (i.e., tuning the DFCI using the method of Oh et al. [11, 12]). The concluding remarks are given in Section 4.

2 t-DFCI Construction Procedure

In this section, we present a t-DFCI construction procedure, as a smoothing procedure. Actually two phases precedes this tuning procedure [11, 12]: (1) construction of sub-DFCI's based on daily volatility of financial variables such as the stock price index (SPI), the foreign exchange rate (FER) and interest rates (INT), (2) combining individual sub-DFCI's into one i-DFCI (integrated DFCI). These two phases are discussed in detail by Oh et al. [11, 12] and hence we mainly focus on the third phase here, i.e. tuning i-DFCI by long term variables. See Fig. 1 for t-DFCI construction architecture.

The main function of DFCI proposed by Oh et al. [11, 12] is to classify the current conditions of the financial markets according to their volatility level: (i) low volatility level period (LP), (ii) transition period (TP), and (iii) high volatility level period (HP). It is TP that gives the unique feature to DFCI, i.e., DFCI is designed to activate (i.e., issue warning) whenever the financial market enters TP. As TP usually occurs just prior to a crisis it can be interpreted as a phase through which the financial market makes a transition from LP to HP or vice versa. Often it is called a grey zone where the self-correcting mechanism of the financial market deteriorates and it is characterized by a sudden change of volatility level and rapid swings in market sentiment. Note that the financial market in TP may either proceed to a crisis or return to a stable condition.

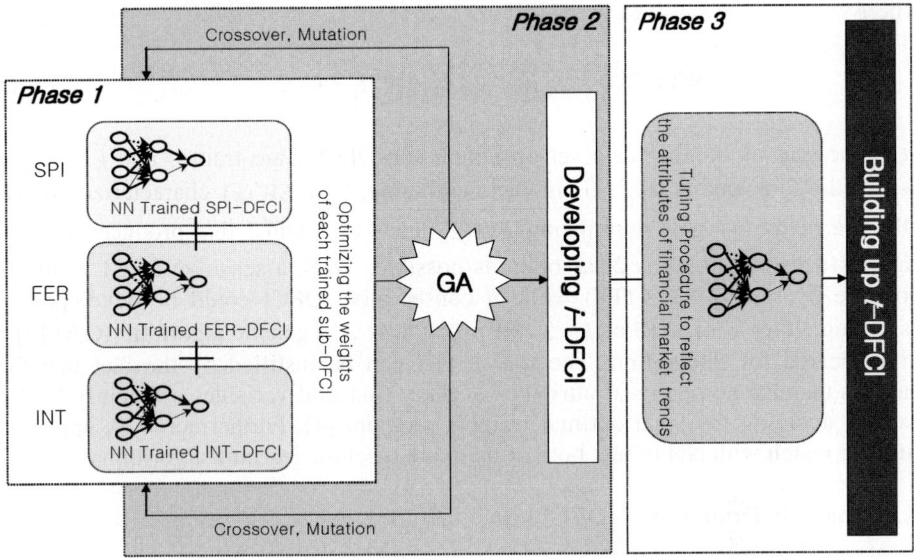

Fig. 1. DFCI construction architecture

2.1   A Brief Review of Phases 1 and 2

Phase 1 consists of two critical steps for successful construction of DFCI, i.e. (i) selecting financial variables and then establishing training data for each financial variable with appropriate input and output variables, (ii) training an efficient classifier (sub-DFCI) for each financial variable. To make our discussion simple, it is assumed here and below that SPI, FER and INT are selected as appropriate financial variables. A key issue to the first step is that training data are to be established such that each period (LP, TP and HP) has its own distinctive feature. At the second step, the NN is designed and trained as a sub-DFCI for each financial variable. The usefulness of the NN in establishing sub-DFCI is discussed in detail by Kim et al. [8, 9, 10].

At Phase 2, the individually trained sub-DFCI's are combined into one integrated DFCI (i -DFCI) as follows:

$$i - DECI_t = w_1\,S_t + w_2\,F_t + w_3\,I_t\,, \tag{1}$$

where $w_1, w_2, w_3 \geq 0$ are weights with $w_1 + w_2 + w_3 = 1$, and S_t, F_t and I_t are the outputs of the trained sub-DFCI's of SPI, FER and INT at given time t. Note that S_t, F_t or I_t takes on values 1, 2, or 3 representing LP, TP and HP, respectively. One key issue in Phase 2 is finding the optimal weights in equation (**1**). For this, Oh et al. [11, 12] introduces a fitness function $EW(k)$ and finds its minimizer (w_1, w_2, w_3), i.e.,

$$\arg\min_{w_1, w_2, w_3} EW(k) \tag{2}$$

where

$$EW(k) = \sum_{i=1}^{n} (w_1 \, S_i + w_2 \, F_i + w_3 \, I_i - k)^2, \tag{3}$$

n is the size of training data set on which sub-DFCI's are trained and k is a pre-assigned value among 1, 2, 3. In fact, assigning k in $EW(k)$ characterizes the i-DFCI, e.g., the weights minimizing $EW(2)$ leads to i-DFCI that produces warning signals (days classified as 2) as often as possible. Thus, a sensitive DFCI would be possible by setting $k = 2$ (TP) while a conservative DFCI could be developed by assigning either $k = 1$ (LP) or $k = 3$ (HP). At Phase 2, a genetic algorithm (GA) [5] is implemented for integration. The use of GA can be justified by the fact that GA handles the chaotic nonlinear attributes of daily financial variables efficiently [13] as well as avoiding the local optimal weights problem [4]. Furthermore GA appears a suitable match with NN in that both of them are machine learning algorithms.

2.2 Phase 3: Tuning the i-DFCI into t-DFCI

The main difficulty in tuning the i-DFCI with long-term (monthly or quarterly) variables is that daily response variable (i-DFCI) and monthly or quarterly predictor variables do not match in their frequency. There are two ways of resolving the frequency mismatches. One way is to transform the high frequency response to low frequency by aggregating daily responses. The other way is to transform low frequency predictors to high frequency by making more frequent observations of predictors. The first approach is not appropriate for our case since our objective is to build a "daily" financial condition indicator, while the second approach is desirable but not available since most macroeconomic variables can be recorded only on a monthly, quarterly, or in some cases yearly basis.

The approach we follow here is that the long-term predictor observed at the given point of time serves as the daily predictor for the rest of the period (i.e., between the given point and the next point of observation of the long term predictor). Intuitively one may naturally expect that this approach could smooth out DFCI sufficiently since it assumes that the effect of macroeconomic variables on DFCI lasts for the month or quarter without change. Recall that the financial markets more often than not react sharply to announcements of past macroeconomic performances. Technically this approach causes the "repeat measurements" problem [3] since then the macroeconomic variables are to be treated as predictor variables assumed constant throughout month or quarter for a varying response variable (i.e., the volatility level of financial markets). An empirical study in Section 3 demonstrates this.

Selection of long-term or macroeconomic predictor variables for tuning is of course important. One reasonable selection might be the variables that have been considered for other traditional EWS's based on long-term macroeconomic variables [6, 7]. It is well known that such macroeconomic variables measure the weakness of the financial market quantitatively. As alternatives, here we introduce the *new scoring variables* that relate to the fundamental economic weakness somewhat qualitatively. Indeed they are designed to measure how long financial market is exposed to a potentially dangerous situation that may bring a possible financial collapse. For example, a successive run of large trade deficits may put the financial market on the brink of breakdown. Our empirical example in Section 3 shows that such *scoring variables* have significant influence upon financial market volatility. At this phase, NN may be recommended as the main training tool since t-DFCI is to be constructed to handle various types of data including quantitative as well as qualitative data.

Table 1. Input variables considered

Variable name	Numerical formula	Description
IND	x_t	Index or Rate
DRF	$p_t = \dfrac{x_t - x_{t-1}}{x_{t-1}}$	Daily rise and fall rate
MA(m)	$\overline{p}_{m,t} = \displaystyle\sum_{i=t-(m-1)}^{t} p_i$	m-day moving average
MV(m)	$s_{m,t}^2 = \displaystyle\sum_{i=t-(m-1)}^{t} (p_i - \overline{p}_{m,t})^2$	m-day moving variance

3 t-DFCI for the Korean Financial Market

The Korean financial crisis that broke out in late 1997 and persisted for over a year brought a wave of large-scale insolvencies throughout the financial and industrial sectors. At that time it was a quite new experience for Korea, which had become accustomed to steady economic growth. This unprecedented crisis brought much attention to the study of EWS. In particular, a new EWS such as DFCI was strongly called for since there was a widespread suspicion that the 1997 crisis was a result of abrupt and unexpected movements of international capital, not a result of the long-term deterioration of economic fundamentals. In response to the strong call, Oh et al. [11, 12] provided DFCI whose main function is to monitor the daily volatility level of financial markets and issue a warning signal whenever it reaches TP. In fact, they chose SPI, FER, and INT, and proposed a method for measuring their volatility levels. See Tables 1 – 2, Fig. 2, Fig. 3 – 4 (see Fig. 3 – 4 of [11]), and Fig. 5 for the input variables employed by them for measuring volatility. With these input variables a training dataset is established for each financial variable (Table 3). Since the year 1997 when financial crisis broke out shows very obvious clear changes of volatility levels in every financial market, it is used as base period for the training data. More precisely, Oh et al. [11, 12] define TP around the volatility level change point (Fig. 4 (see Fig. 4 of [11]) and then LP and HP before and after TP, respectively. After successful training of sub-DFCI's on the training set, the three sub-DFCI's are integrated into i-DFCI by genetic algorithms (GAs) which select fitness function EW(3) where $k = 3$ is chosen with respect to its reproduction fidelity to the Korean financial market (i.e. we measure how frequently it measures to the events of the Korean financial market). The resulting i-DFCI is given

Table 2. List of input and output variables for each sub-DFCI

Variable name	Input variables	Output variables
DFCI (SPI)	IND, DRF, MA(5), VA(5)	LP: 1
DFCI (FER)	DRF, MA(5), VA(5), MA(20), VA(20), MA(60), VA(60)	TP: 2
DFCI (INT)	DRF, MA(5), VA(5), MA(20), VA(20), MA(60), VA(60)	HP: 3

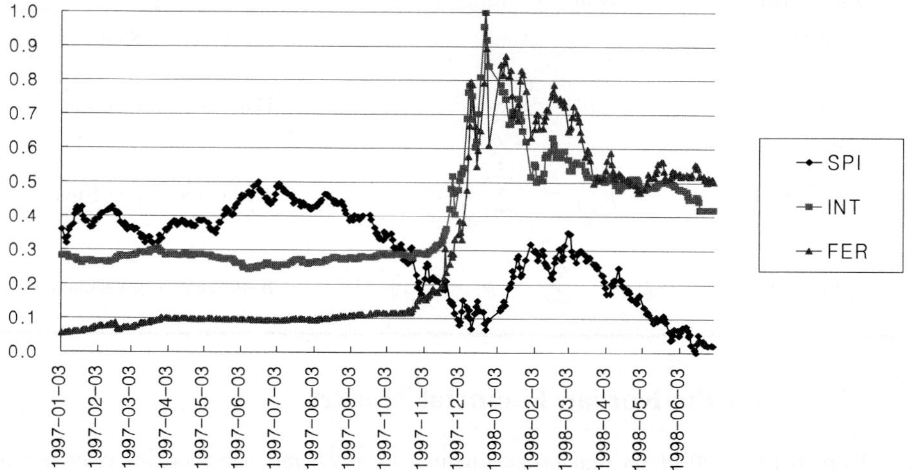

Fig. 2. Korea stock price index, foreign exchange rates and interest rates of 1997, which are scaled from 0 to 1

in Fig. 5, which is established by Phases 1 and 2 of Section 2 [11, 12]. One may easily note the high daily fluctuations of i-DFCI from Fig. 5.

To tune i-DFCI into t-DFCI, it remains to complete Phase 3. In order to tune i-DFCI with macroeconomic variables, those used for the other traditional EWS's are considered first. Then among them, some variables that could affect financial market significantly are carefully selected. In addition, the *new scoring variables* (derivatives of the selected macroeconomic variables) are included in order to measure how long the Korean financial market is exposed to a potentially dangerous situation. Such types of variables considered in our empirical study are $X2$, $X9$, $X13$, $X14$, $X17$ and $X18$ which count the number of successive months during which dangerous situations continue in Table 4 (see Appendix). For example, note that $X18$ counts the number of successive months of trade deficits.

As discussed in Section 2.2, we assume that observation of the long-term predictor variables at the given point remains constant for the rest of the month or quarter. With these predictors and the responses (i-DFCI on the training period of Fig. 5) it is easy to establish the training data for Phase 3. The training algorithms used are the well-known backpropagation neural networks (BPN) where the number of hidden layers

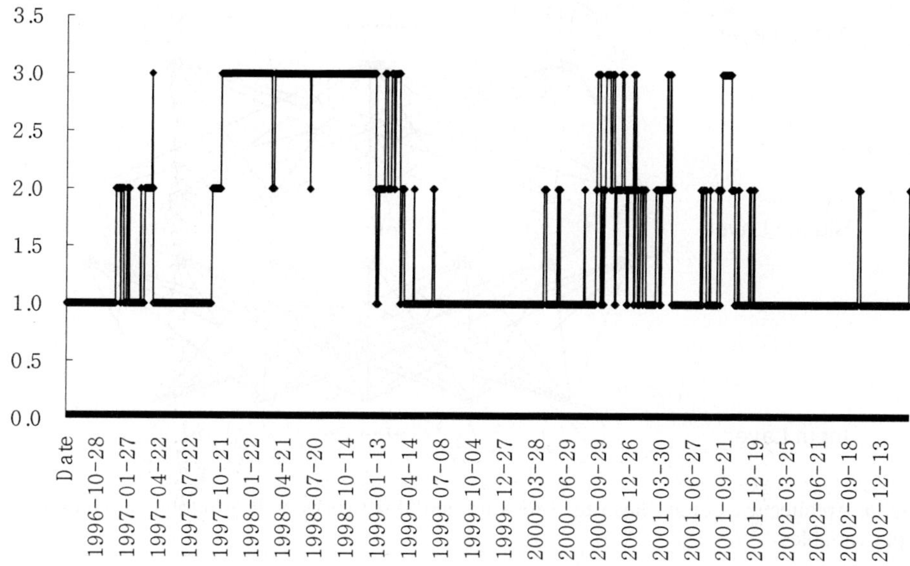

Fig. 5. Classification result of i - DFCI with $k = 3$ from May 1996 to March 2003

Table 3. Specific dates of LP, TP and HP for training data

Index	LP	TP	HP
SPI	Apr. 22, 97 – Sep. 18, 97	Sep. 19, 97 – Oct. 21, 97	Oct. 22, 97 – Mar. 10, 98
FER	Jul. 1, 97 – Oct. 26, 97	Oct. 27, 97 – Nov. 30, 97	Dec. 1, 97 – Mar. 20, 98
INT	Aug. 21, 97 – Nov. 12, 97	Nov. 13, 97 – Dec. 12, 97	Dec. 13, 97 – Mar. 16, 98

ranges from 2 to 6 and as an activation function, the logistic function is utilized with learning rate, momentum and initial weight given by 0.1, 0.1 and 0.3, respectively. Refer Fig. 6 for the BPN architecture employed.

After training is completed, t-DFCI is applied to test the data as was done with i-DFCI in Fig. 5, which produces Fig. 7. Compared to i-DFCI of Fig. 5, t-DFCI smoothes out i-DFCI sufficiently, which confirms the tuning (or smoothing) effect. Finally the significance of *scoring variables* is examined by decision tree analysis [1], which shows that $X2$ (continual increasing of note default rate) and $X9$ (continual deterioration of industrial production) are particularly significant in tuning DFCI. Decision tree analysis also reveals that the note default rate ($X1$) and the rate of change of M3 growth ($X4$, total liquidity) have significant impact. Detailed analysis results from the decision tree are omitted. Furthermore, Oh et al. [12] evaluates usefulness of proposed model based on monthly chronicle of Korean financial market.

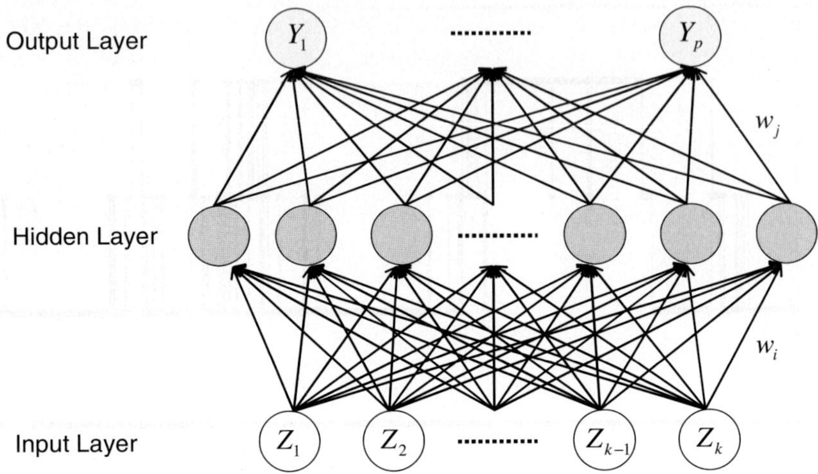

Fig. 6. Architecture of multilayer feed-forward neural networks with p output variables and k input variables

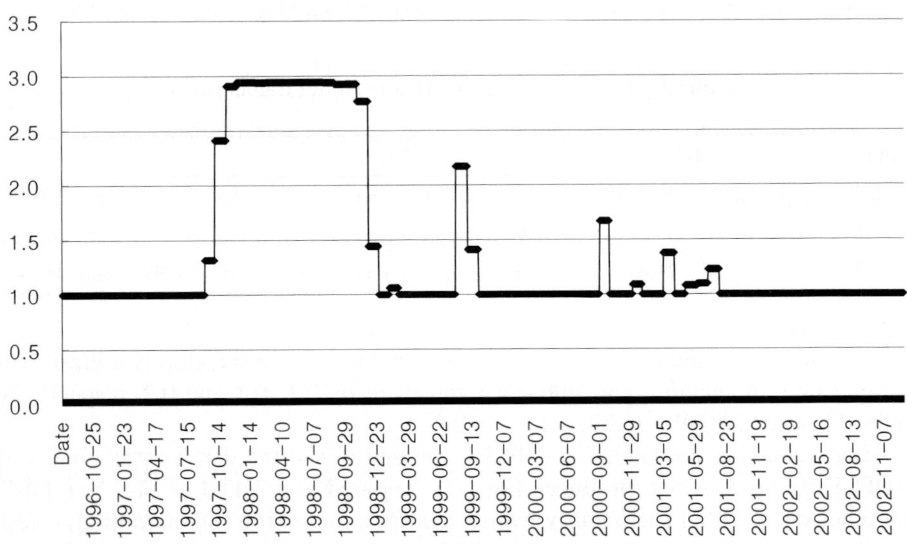

Fig. 7. t-DFCI from May, 1995 to March, 2003 including the training period 1997

4 Concluding Remarks

The DFCI proposed by Oh et al. [11, 12] stressed the importance of monitoring daily volatility levels of financial markets in issuing warnings against the new type of financial crisis. However, such DFCI tends to suffer from noise contamination because it handles high frequency daily variables. In this study a tuning procedure with long term macroeconomic variables is proposed and discussed as a method of removing the noise and an empirical example demonstrates tuning or smoothing effect. In addition, the *new scoring variables* introduced to measure how long the financial market has been exposed to a dangerous situation appear to play effective role in issuing warning signals. Our empirical study confirms that such variables could be those that have a great impact on financial markets. Further studies would be worthy to think about how proposed model can be done for other markets to test its practical relevance.

Acknowledgment

This work was supported by KRF 2003-070-C00008.

References

1. Adei, H., Hung, S.: Machine Learning: Neural Networks, Genetic Algorithsms, and Fuzzy Systems. Wiely New York (1995).
2. Rumelhart, D.E., Hinton, G.E., Williams, R.J.: Learning internal representations by back propagation. In: D.E. Rumelhart, J.L. McClelland, and PDP research group (Eds.); Parallel Distributed Processing. MA: MIT Press, Cambridge (1986).
3. Draper, N.R., Smith, H.: Applied Regression Analysis. Wiley, New York (1981).
4. Goldberg, D.E.: Genetic Algorithms in Search, Optimization and Machine Learning. Addison-Wesley, New York (1989).
5. Holland, J.H.: Adaptation in natural and artificial systems: an introductory analysis with applications to biology, control and artificial intelligence. University of Michigan Press, Michigan (1975).
6. Khalid, A.M., Kawai, M.: Was financial market contagion the source of economic crisis in Asia? Evidence using a multivariate VAR model. Journal of Asian Economics **14** (2003) 131-156.
7. Kaminsky, G.L., Reinhart, C.M.: The twin crises: The causes of banking and balance-of-payments problems. American Economic Review **89** (1999) 473-500.
8. Kim, T.Y., Hwang, C., Lee, J.: Korea financial condition indicator using a neural network trained on the 1997 crisis. Journal of Data Science **2** (2004a) 371-381.
9. Kim, T.Y., Oh, K.J., Sohn, I., Hwang, C.: Usefulness of artificial neural networks for early warning system of economic crisis. Expert Systems with Applications **26** (2004b) 585-592.
10. Kim, T.Y., Oh, K.J. Kim, C., Do, J.D.: Artificial Neural Networks for Non-Stationary Time Series. Neurocomputing **61** (2004c) 439-447.
11. Oh, K.J., Kim, T.Y., Lee, H.Y., Lee, H.: Using Neural Networks to support Early Warning System for Financial Crisis Forecasting. Lecture Notes in Artificial Intelligence **3809** (2005) 284-296.
12. Oh, K.J., Kim, T.Y., Kim, C.: An Early Warning System for detection of Financial Crisis using Financial Market Volatility. Expert Systems **23** (2006) 83-98.
13. Peters, E.E.: Chaos and Order in the Capital Markets. Wiley, New York (1991).
14. Rosenblatt, F.: Principles of Neurodynamics. Spartan, New York (1962).

Appendix

Table 4. The final input variables including the derived ones

	Selected Variable	Explanation
X1	Note Default Rate (NDR)	Using raw data
X2	Size of the run of increasing X1 during the latest 12 months	$\sum_{i=t-11}^{t}(Z1)_i$, $Z1 = 1$ if $(X1)_t > (X1)_{t-1}$; 0, otherwise
X3	Change rate of foreign exchange holdings (FEH)	Ratio of the current month to the same month of the last year
X4	Change rate of money stock	Ratio of the current month to the same month of the last year
X5	Change rate of producer price index	Ratio of the current month to the same month of the last year
X6	Change rate of consumer price index	Ratio of the current month to the same month of the last year
X7	Change rate of balance of trade	Ratio of the current month to the same month of the last year
X8	Change rate of index of industrial production	Ratio of the current month to the same month of the last year
X9	Size of the run of decreasing X8 during the latest 12 months	$\sum_{i=t-11}^{t}(Z8)_i$, $Z8 = 1$ if $(X8)_t < (X8)_{t-1}$; 0, otherwise
X10	Change rate of index of producer shipment	Ratio of the current month to the same month of the last year
X11	Change rate of index of equipment investment	Ratio of the current month to the same month of the last year
X12	FEH per gross domestic products	FEH/GDP
X13	Size of the run of decreasing X12 during the latest 12 months	$\sum_{i=t-11}^{t}(Z8)_i$, $Z12 = 1$ if; 0, otherwise
X14	Size of the run of decreasing monthly change of X12 during the latest 12 months	$\sum_{i=t-11}^{t}(Z12)_i$ $Z12 = 1$ if $(X12)_t < (X12)_{t-1}$; 0, otherwise
X15	Change rate of FEH per GDP	Ratio of the current month to the same month of the last year
X16	Balance of trade per GDP	BOT/GDP
X17	Size of the run of increasing X16 during the latest 16 months	$\sum_{i=t-11}^{t}(Z16)_i$, $Z16 = 1$ if $(X16)_t < (X16)_{t-1}$; 0, otherwise
X18	Size of the run of negative X16 during the latest 16 months	$\sum_{i=t-11}^{t}(Z16)_i$ $Z16 = 1$ if $(X16)_t < 0$; 0, otherwise
X19	Balance of payments (direct Investment)	Using raw data
X20	Balance of payments (securities investment)	Using raw data
X21	Other balance of payments	Using raw data
X22	Amount of foreigners' investment in the stock market	Using raw data

Predicting Stock Market Time Series Using Evolutionary Artificial Neural Networks with Hurst Exponent Input Windows

Somesh Selvaratnam and Michael Kirley

Department of Computer Science and Software Engineering
The University of Melbourne
Parkville, Victoria 3010, Australia
{sase, mkirley}@csse.unimelb.edu.au

Abstract. Predicting stock market time series is a challenging problem due to their random nature, non-stationarity and noise. In this study, we introduce an enhanced evolutionary artificial neural network (EANN) model to meet this challenge. Here, fractal analyses based on Hurst exponent calculations are used to characterize the time series and to identify appropriate input windows for the EANN. We investigate the efficacy of the model using closing price time series for a suite of stocks listed on the SPI index on the Australian Stock Exchange. The results show that Hurst exponent configured models out-perform basic EANN models in terms of average trading profit found using a simple trading strategy.

1   Introduction

Randomness, non-stationarity and noise characterize stock market time series. These factors make prediction a challenging task. Consequently, a great deal of effort has been devoted to developing robust time series forecasting models. Well-established techniques reported in the literature include: (a) linear models, such as the autoregressive integrated moving average (ARIMA); (b) non-linear models, such as artificial neural networks (ANN) and fuzzy system models (see [1,4,7,8,14,15,16])and (c) a combination of linear and non-linear models. The common characteristic of all approaches is that they attempt to predict a future value based on the output of the predictive model constructed from observed past data.

Stock market time series often exhibit high degrees of non-linear variability, and frequently have fractal properties [2,9,15]. The fractal dimension and *Hurst exponent* [6] provide unique insights into the characteristics of the time series and provide information about the presence of long-range correlations in the time series. The Hurst exponent measures the level of persistency (or anti-persistency) in the given signal and thus may be used as an indicator as to whether it is possible to forecast the time series with a reasonable degree of accuracy.

Recently, there has been increased interest in combining the information obtained from fractal analysis with ANNs for time series prediction

A. Sattar and B.H. Kang (Eds.): AI 2006, LNAI 4304, pp. 617–626, 2006.
© Springer-Verlag Berlin Heidelberg 2006

[3,9,11,12,13,14,15,16]. In this study, we extend this idea and introduce a novel method for determining the input time window to be used in evolutionary artificial neural networks (EANN) for stock market forecasting. Here, the size of the input time window is determined by the largest Hurst exponent of the time series, calculated using *rescaled range analysis* (R/S) [6,10]. It could be argued that the Hurst exponent cannot be representative for the entire time series. However, the underlying hypothesis on which our model is based is that the largest Hurst exponent does contain useful information about the time series and thus provides a framework for identifying an appropriate input time window for the EANN. Thus, the major contribution of this work is in determining how the prediction accuracy of alternative EANN models is affected by varying the input window size based on the fractal properties of a time series.

The remainder of the paper is organized as follows: In Section 2, we describe the fractal properties of time series data in more detail. Section 3 introduces the Hurst exponent based EANN model. In section 4, experiments and results are described in detail using a number of stocks listed on the Australian Stock Exchange. Section 5 concludes the paper and identifies potential future directions of research.

2   Hurst Exponent and Fractal Properties

Modelling time series using the inherent fractal properties provides valuable information and increases our understanding of the nature of movement and trends within the time series data. The fractal dimension, D, of a stock market time series measures the degree of fluctuation in price changes [9]. The value of D is typically stated in terms of the Hurst exponent (H), described by the equation: $D = 2 - H$. In the remainder of this section, we describe the steps necessary to calculate the value of H for a given time series.

The Hurst exponent is represented by a value in the range $0 < H < 1$. As such, the magnitude of H provides a measure of persistence and "fractality" of a time series and can be used to characterize the underlying stochastic processes, long-range correlations and periodicities in the series [6,9,10,11]. A time series with $H \neq 0.5$ can be theoretically forecasted because of its persistent or anti-persistent nature [6]. Subsequently, a three-phase classification scheme may be used:

$H = 0.5$ indicates a random walk (purely Brownian change)
$H > 0.5$ persistent changes, characterized by long memory effects
$H < 0.5$ anti-persistent, indicating no clear trends

The value of H is estimated by calculating the average rescaled range over multiple regions of the time series, $X_1, X_2, ...X_n$. To do this, we have implemented a version of the *rescaled range analysis* (R/S) technique as described by Peters [9]. The rescaled range is calculated by finding the range between the maximum and minimum distances that the cumulative sum of a stochastic random variable has moved from its mean and then dividing this by its standard deviation. We briefly describe the steps used to estimated H below.

1. Mean $\mu = 1/n \sum_{i=1}^{n} X_i$
2. Mean adjusted series $Y_i = X_i - \mu$
3. Cumulative deviate series $Z_i = \sum_{i=1}^{i} Y_i$
4. Range series $R_t = max(Z_1, Z_2, .., Z_t) - min(Z_1, Z_2, .., Z_t)$
5. Standard deviation series $S_t = \sqrt{1/t \sum_{i=1}^{t}(X_i - \mu)^2}$
6. Rescaled range series $(R/S)_t = R_t/S_t$
7. Hurst exponent $H = \frac{log_2(R/S)_t}{log_2 t}$

where: $(R/S)_t$ is averaged over $[X_1 - X_t], [X_{t+1} - X_{2t}], \ldots, [X_{(m-1)t+1} - X_{mt}]$; t is the index of periodicities $(t = 1, 2, \ldots, n)$; n is the number of time instants of t per scale; and $m = \lfloor n/t \rfloor$.

3   The Model

3.1   Artificial Neural Network Architecture

There is no standard architectural model of ANNs that is considered best for predicting stock market time series. Subsequently, we have elected to use two alternative models – *Elman Recurrent Neural Network* (ERNN) and a *Multi Layer Perceptron* (MLP) – as the base networks to be examined here. The rationale for this choice was based on the need to keep the ANN predictive models simple in architecture and size, and the fact that the ERNN architecture (Fig 1) is very popular in predicting time series [5,14].

ERNN are modelled on the Markov assumption that the next series can be predicted based on the previous state of the time series. The ERNN may be described as follows:

$$h_i(t) = \sigma(\sum_{j=1}^{m} \sigma(V_{ij}x_j(t)) + \sigma(W_{ij}x_j(t)))$$

$$o(t) = \sum_{j=1}^{n} \sigma(Z_j h_j(t))$$

where: σ is the sigmoid function; m is the number of input nodes; n is the number of hidden nodes not in the Elman layer; h the number of hidden nodes in the Elman layer; V, W and Z are real valued weights and o the output node. However, the structure of the above formalism will change because parameters m, n and h are expected to vary as the network evolves. (see Section 3.3).

The MLPs used in this study, have single hidden layers, whilst the ERNNs have two hidden layers (including one recurrent layer). The number of nodes in the hidden layer is kept equal to the number of input nodes. The hidden nodes used sigmoid activation functions with the output node a linear activation function. Here, the base architectures have 5 inputs. The input values are normalized log values of the daily returns for each stock.

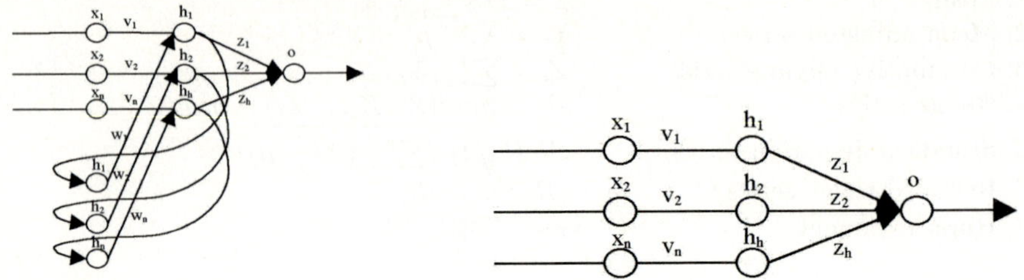

Fig. 1. Base artificial neural networks. *Left:* ERNN, and *Right:* MLP.

3.2   Determining the Input Window

Many papers have dealt with input selection to ANNs when it comes to mapping financial indexes and stocks (see [1] for an overview). Generally, the input window size is selected using a heuristic or a trial and error approach. Our model extends the work introduced by (a) Chen et al. [3], who used a sliding input windows to predict stock prices, (b) Tekbas [13], who introduced a variable-length windowing approach for chaotic time series modelling, and (c) Yakuwa et al. [15], who have used fractal properties of the stock market indices to configure ANNs used to forecast prices.

The premise in this study is that specific time windows of time series data carry more persistent or anti-persistent information than some arbitrary window. Consequently, by identifying appropriate input time windows, the predictive accuracy of the EANN should be improved. Here, we use fractal properties to configure sizes of input windows of the EANN. The Hurst exponent based input window was derived using the R/S analysis method. We calculate H for each scale (t) in the range $t = 3$ to $t = 50$ for each time series. We then selected the scale with the maximum H value and used its size as the size of the input window to each EANN.

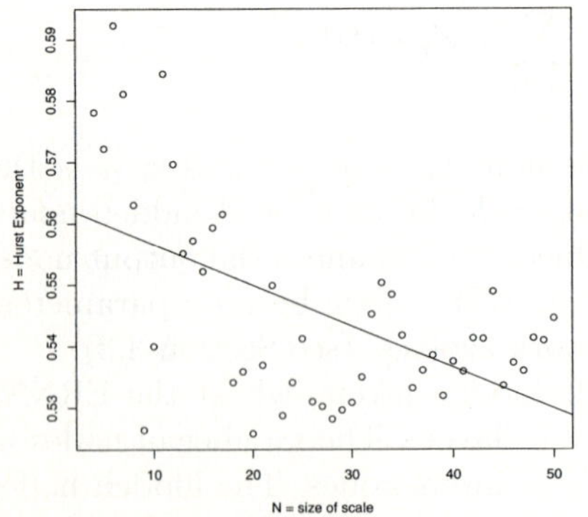
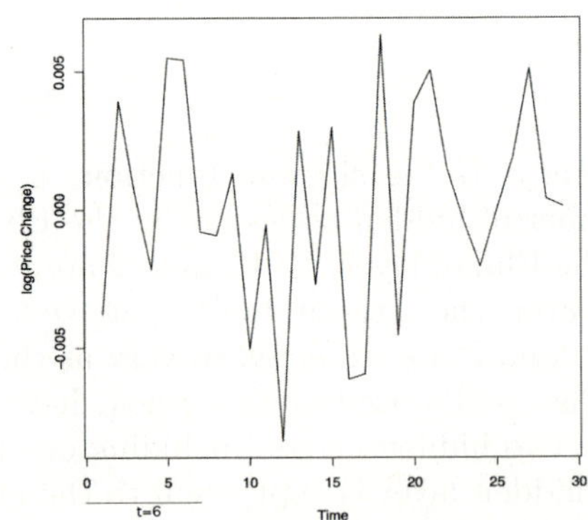

Fig. 2. *Left:* H at scales $t = 3$ to $t = 50$ for NAB stock, and *Right:* Sliding time window for NAB stock. Here, the window is fixed to 6.

Fig 2 provides a description of the output of the calculations using NAB, a stock listed on the Australia Stock Exchange (see data sets in Section 4). The plot on the left shows that for the NAB time series, the strongest persistency is when H is 0.59 at scale 6. Therefore, the input window size to the EANN used to forecast the time series was set to 6. The plot also shows that as the size of the scale increases, H approaches 0.50, which means that as the scale increases, the information relating to the nature of the time series diminishes. However, this behaviour varies for each time series. In Fig 2, the plot on the right shows the NAB time series over 30 trading days with a representation of the sliding time window of $t = 6$, used as the input window. As this window traverses the time series, its values are used as the inputs to the ANN.

3.3 Evolving the Artificial Neural Network

To improve the prediction accuracy of the ANN models, each model was trained and evolved using a version of EPNet [17,18]. After each generation, the networks go through further training epochs to optimize performance. The algorithm utilizes four evolutionary operators, which are applied to each individual (single network) at each time step. Here, we have implemented an elitist version of the algorithm. The specific steps are as follows:

```
Run for 20 Generations
    Clone net to tempNet
    Evolve net
    (select operation that improves fitness)
        Delete connection
        Add connection
        Delete node
        Add node
    Train net for y epochs
    Validate (net.fitness < tempNet.fitness)
        Replace net with tempNet
```

Algorithm 1. Basic evolutionary algorithm for EPNet

4 Experiments and Results

The aim of the following experiments was to test the hypothesis that a Hurst exponent based input window configuration would improve the prediction accuracy of EANN. We were particularly interested in both the accuracy of the predictive models and the performance of a simple trading strategy when $H > 0.5$ or $H < 0.5$. In addition, we wished to identify whether there was a correlation between the value of H and the results of the predictive models.

4.1 Data Sets

Fifteen stocks from the SPI index on the ASX (Australian Stock Exchange) were selected to evaluate the Hurst exponent-based EANN models. The time series contained closing prices and approximately 600 days of trading history between

Table 1. RMSE Performance. On the left, are the associated H values for each stock. On the right, are the RMSE results for each of the four models considered.

Stock	Hurst			Model			
	H	*Max H*	*Scale H*	*ERNN*	*H-ERNN*	*MLP*	*H-MLP*
AMP	0.55	0.61	3	0.01	0.04	0.01	0.04
ANZ	0.50	0.52	29	0.02	0.05	0.03	0.21
BHP	0.50	0.56	3	0.02	0.06	0.02	0.06
CBA	0.54	0.56	5	0.04	0.12	0.02	0.14
CML	0.62	0.70	10	0.01	0.04	0.02	0.04
FGL	0.33	0.43	3	0.01	0.11	0.01	0.11
NAB	0.55	0.59	6	0.02	0.04	0.01	0.03
QBE	0.47	0.51	6	0.03	0.08	0.02	0.13
SGB	0.52	0.53	34	0.01	0.05	0.01	0.05
SUN	0.60	0.61	29	0.02	0.11	0.01	0.11
TLS	0.52	0.53	37	0.02	0.03	0.02	0.05
WBC	0.56	0.57	48	0.03	0.09	0.05	0.14
WES	0.33	0.38	3	0.01	0.04	0.01	0.07
WOW	0.54	0.54	11	0.02	0.04	0.01	0.05
WPL	0.38	0.42	13	0.02	0.05	0.02	0.10
			Mean	0.02	0.06	0.02	0.09
			Std	0.01	0.03	0.01	0.05

the years 2002 – 2004. The fractal and statistical properties were calculated for each time series by taking the log differences of daily returns. Table 1 lists the stock (column 1), H value (column 2), the maximum H value (column 3) and the resulting scale (column 4). The scale value is used to set the size of the input window for each of the Hurst-exponent based models (H-ERNN and H-MLP). In the data sets used, the Hurst exponents for the stocks range from 0.33 – 0.62, with approximately half the stocks having values in the range 0.45 – 0.55.

4.2   Model Parameters

The data sets were divided into 2 components, where 500 days were used to train the EANN and the remaining 100 days were used to validate and test the model. The EANN were trained using the Back Propagation algorithm over 200 epochs. The training of each network was combined with 20 generations of evolution (using a version of EPNet). The learning rate was kept at a static rate of 0.7. The *root mean squared error* (RMSE) was used as the fitness function during training and prediction.

4.3   Results

Table 1 lists the accuracy in terms of the RMSE value of each the EANN models for each of the stocks considered. The average accuracy of the Hurst exponent input window based models (H-ERNN and H-MLP) across all stocks is comparatively inferior to the base architectures (ERNN and MLP). The simple models with an input window size of 5 performed better on average. However, it should

Table 2. Returns for the predictive models using a simple trading strategy. Here, stocks from Table 1 with $0.48 < H < 0.52$ were excluded.

Stock	Model			
	ERNN	*H-ERNN*	*MLP*	*H-MLP*
AMP	7 %	37 %	33 %	28 %
CBA	-66 %	-11 %	-62 %	-20 %
CML	-18 %	5 %	-22 %	5 %
FGL	8 %	68 %	2 %	70 %
QBE	-90 %	-69 %	-90 %	-14 %
SGB	-12 %	-46 %	-88 %	-46 %
SUN	-28 %	22 %	24 %	31 %
TLS	23 %	-34 %	27 %	-34 %
WBC	-32 %	6 %	-32 %	28 %
WES	-58 %	27 %	-90 %	-13 %
WOW	11 %	-3 %	33 %	17 %
WPL	28 %	37 %	77 %	28 %
Mean	-18.9 %	3.25 %	-15.7 %	6.7 %

be noted that there is a relatively large variance in the accuracy value obtained using Hurst exponent based models.

To further explore the effectiveness of configuring the input window size of the EANN model using H values, a simple trading strategy was investigated. The strategy consisted of the following simple rules:

1. Set starting portfolio to $1000
2. Current Day: ANN predict Gain (Loss) in price
3. Current Day: Buy (short Sell) by investing 10% of portfolio
4. Next Day: Sell (Buy back) stock
5. Next Day: Calculate profit

The trading profits were recorded and annualized for each time series. Note, that in these experiments, trading commissions and costs were not factored into trades as they can vary based on size, quantity and context. Table 2 lists the average returns (as a percentage) when stocks in the "random range" ($0.48 < H < 0.52$) were removed from the stocks listed in Table 1. A comparison between the standard architectures and the Hurst exponent models (ERNN *vs* H-ERNN and MLP *vs* H-MLP) clearly indicates that by using the H value as mechanism for determining the input window size, significantly higher returns (p-value $>$ 0.05) can be found on average. Here, the estimated H value provides useful information about the upward/downward tendency of the time series, which in turn impact on the average return.

Fig 3 plots the average return *vs* H for the two different Hurst exponent configured models. The plots show that the time series that have H values closer to 0.5 have low or negative returns. Interestingly, there was no significant difference between the base EANN architectures (across the given stocks). However, these results are not conclusive and further simulations are required using other trading strategies before definitive statements are made.

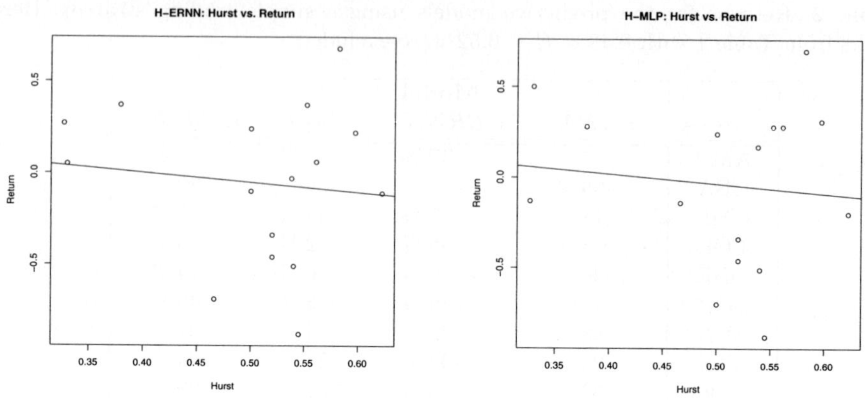

Fig. 3. Mean Returns vs Hurst exponent. *Left:* H-ERNN and *Right:* H-MLP.

We were also interested in identifying whether there was a strong relationship between the value of H and the most effective input window size of an EANN model. Therefore, we have varied the size of the input window and recorded the corresponding RMSE values for the two base EANNs. The input window was systematically increased by multiples of 3 (arbitrarily chosen value) over 17 cycles until the window size reached 51 time instants. Three different stocks were chosen for the analysis: AMP, ANZ and WPL with H values 0.55, 0.50, and 0.38 respectively.

Fig 4 plots the RMSE *vs* input window size for each of the stocks considered. An inspection of the plots across the spectrum of input window sizes does not reveal strong correlations (in fact, r^2 values were low). The results varied based on the time series, with the persistent and anti-persistent stocks suggesting trends across window spans and no apparent trends in ANZ (where $H = 0.50$). Further work is required before clear conclusions are drawn.

5   Summary and Future Work

This paper has examined a novel extension for EANNs used to predict financial time series. Here, we have incorporated specific domain knowledge into the network model by tailoring the input window size of the EANN for a given time series, using the Hurst exponent. The simulation results presented are very encouraging. When the alternative EANN models were compared in terms of RMSE, the Hurst-based models did not outperform the base EANN. However, it should be noted that the variance in the RMSE values across all stocks was larger for the Hurst-based models. When a simple trading strategy was used, there was a significant differences between models – both the H-ERNN and H-MLP outperformed the corresponding base models in terms of average return. For different time series, as the underlying value of H moved away from the random region ($H \approx 0.5$), there was a corresponding improvement in the average return using a simple trading strategy.

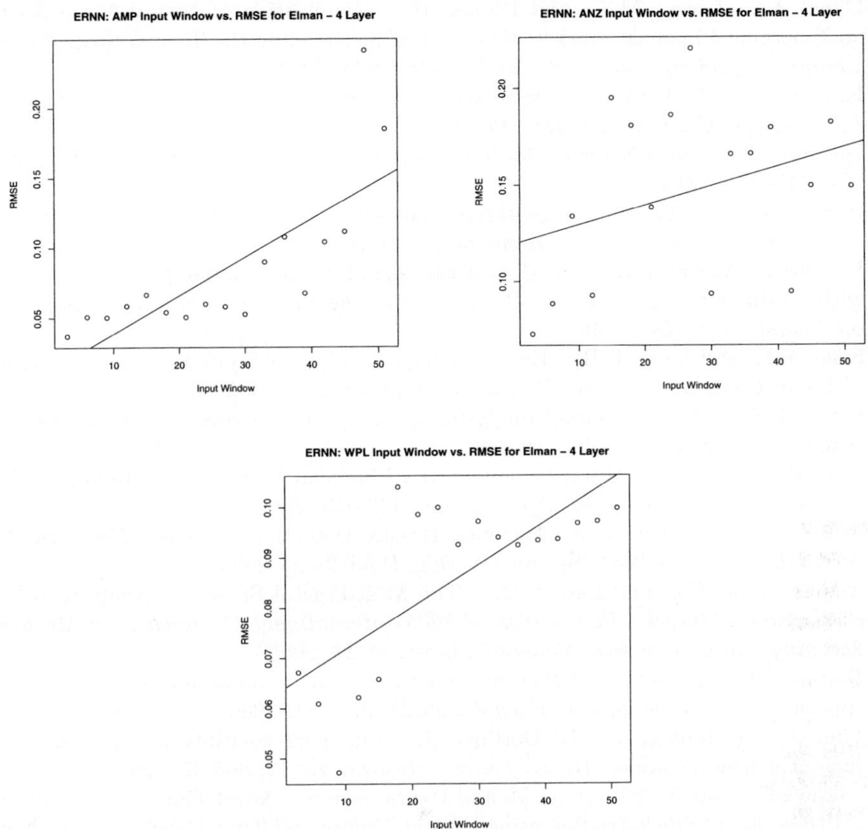

Fig. 4. Input time window sensitivity analysis. *Top Left:* AMP (*H*=0.55), *Top Right:* ANZ (*H*=0.50) and *Bottom:* WPL (*H*=0.38).

The estimate of the value of H provides insights into how predictable a given time series might be. An inherent assumption built into the R/S – Hurst exponent technique, is that the underlying process remains the same throughout the time series sample. In future work, we will investigate how estimates of H over sub samples of the original time-series and multi-fractal analysis techniques could be used as dynamic parameters for determining the size of the input window used in EANNs.

References

1. Abraham A., Philip N.S., and Saratchandran P.: Modeling Chaotic Behavior of Stock Indices Using Intelligent Paradigms. *International Journal of Neural, Parallel & Scientific Computations*, 11(1-2):143-160. 2003.
2. Cajuerio D.O and Tabak B.M. : The Hurst exponent over time: testing the assertion that emerging markets are becoming more efficient. *Physica A*, 336:521. 2004.

3. Chen, A.S., Leung, M.T., and Daouk, H. : Application of Neural Networks to an Emerging Financial Market: Forecasting and Trading the Taiwan Stock Index. *Computers and Operations Research,* 30:901-923. 2003.
4. Kim, D. et al. : Forecasting time series with genetic fuzzy predictor ensembles. *IEEE Trans. Fuzzy Syst.* 5:523-535. 1997.
5. Dorffner G. : Neural Networks for Time Series Processing. Neural Network World, 6(4) 447-468, 1996.
6. Hurst H.E. : Long-term storage of reservoirs: an experimental study. *Transactions of the American society of civil engineers.* 116:770-799. 1951.
7. Kimoto T., Asakawa K., Yoda M. and Takeoka M. : Stock market prediction system with modular neural network. *Proceedings of the International Joint Conference on Neural Networks.* 1990.
8. Kong Y.K. and Moon B.R. : Evolutionary ensemble for Stock Prediction. *Lecture Notes in Computer Science, Volume 3103* pp 1102 -1113. 2004.
9. Peters E.E. : *Fractal market analysis: applying chaos theory to investment and economics* New York: Wiley. 1994.
10. Qian B, Rasheed K. : Hurst Exponent and Financial Market Predictability. *FEA - Financial Engineering and Applications,* 437-043, 2004.
11. Resta M.: R/S Approach to Trends Breaks Detection. *Lecture Notes in AI, LNCS/LNAI series 3681.* Springer Verlag, Heidelberg. 2005.
12. Ruan J, Pang S.L. and Luo W.Q. : The MultiFractal Structure Analysis in the China Stock Market. *Proceedings of 2005 International Conference on Machine Learning and Cybernetics* Volume 5, Issue, 18-21. 2005.
13. Tekbas O.H. : Modelling of chaotic time series using a variable length windowing approach. *Chaos, Solitons & Fractals,* 29(2): 277-281. 2006.
14. Tino P., Schittenkopf C. and Dorffner G. : Financial volatility trading using recurrent neural networks. *IEEE-Neural Networks,* vol 12, 865-874. 2001.
15. Yakuwa F., Dote Y, Yoneyama M, and Uzurabashi S. : Novel Time Series Analysis & Prediction of Stock Trading using Fractal Theory and Time Delayed Neural Network. *IEEE International Conference on Systems, Man and Cybernetics* Volume 1, 5-8 Oct. vol.1 p.134-141. 2003.
16. Yao J.T., Tan C.L. and Poh H.L. : Neural Networks for Technical Analysis: A Study on KLCI. *International Journal of Theoretical and Applied Finance,* Vol. 2, No.2, pp221-241. 1999
17. Yao X, : Evolving artificial neural networks. *Proceedings of the IEEE Neural Networks* 8(9):1423–1447. 1999.
18. Yao X. and Yong Lim. : New Evolutionary System for Evolving Artificial Neural Networks. *IEEE Transactions on Neural Networks,* 8(3):694713, 1997.

BDDRPA*: An Efficient BDD-Based Incremental Heuristic Search Algorithm for Replanning*

Weiya Yue[3], Yanyan Xu[1,2], and Kaile Su[3,4],[**]

[1] Laboratory of Computer Science,
Institute of Software, Chinese Academy of Sciences, Beijing, China
[2] School of Information Science and Engineering,
Graduate University of the Chinese Academy of Sciences, Beijing, China
[3] Department of Computer Science, Sun Yat-sen University, Guangzhou, China
[4] Integrated and Intelligent Systems, Griffith University, Brisbane, Australia

Abstract. We introduce a new algorithm, BDDRPA*, which is an efficient BDD-based incremental heuristic search algorithm for replanning. BDDRPA* combines the incremental heuristic search with BDD-based search to efficiently solve replanning search problems in artificial intelligence. We do a lot of experiments and our experiment evaluation proves BDDRPA* to be a powerful incremental search algorithm. BDDRPA* outperforms breadth-first search by several orders of magnitude for huge size search problems. When the changes to the search problems are small, BDDRPA* needs less runtime by reusing previous information, and even when the changes reach to 20 percent of the size of the problems, BDDRPA* still works more efficiently.

1 Introduction

Many artificial intelligence systems have to adapt their plans continuously to changes in the world or changes of their models of the world, so the original plan might no longer apply or no longer be good and we thus need to replan for the new situation [JDOW99]. In these cases, most search algorithms replan from scratch, that is, solve the new search problem independently of the old ones. However, this can be inefficient in large domains with frequent changes and thus limit the responsiveness of artificial intelligence systems. Fortunately, the changes to the search problems are usually small, so it suggests that a complete recomputation of the best plan for the new search problem is unnecessary since some of the previous search results can be reused and we call this reusing search incremental search, and incremental heuristic search [KFB02] is adding heuristic method to incremental search when calculating the priorities for the vertices in the priority queue.

* Supported by the Australian Research Council grant DP0452628, National Basic Research 973 Program of China under grant 2005CB321902, National Natural Science Foundation of China grants 60496327, 10410638 and 60473004,and Guangdong Provincial Natural Science Foundation grants 04205407 and 06023195.
** Corresponding author.

A. Sattar and B.H. Kang (Eds.): AI 2006, LNAI 4304, pp. 627–636, 2006.
© Springer-Verlag Berlin Heidelberg 2006

During the last decade, powerful search techniques using an implicit state representation based on the reduced ordered binary decision diagram(BDD) [B86] have been developed in the area of symbolic model checking [M93]. Using blind exploration strategies BDD-based search methods have been successfully applied to verify systems with more than 10^{120} states. In [ER98, JBV02], BDD-based search algorithms have been developed that can be used for path finding problem in artificial intelligence.

One way of speeding up searches is incremental heuristic search, and a different way of speeding up searches is BDD-based search. The question is whether incremental heuristic search and BDD-based search can be combined. The answer is yes and in this paper, we develop such a new search algorithm called BDD-based incremental heuristic search algorithm for replanning(BDDRPA*), and our experiment results prove it to be an efficient incremental search algorithm.

The rest of our paper is organized as follows. In section 2, we briefly describe incremental heuristic search and BDD-based search methods. Then, in section 3 we illustrate the BDDRPA* algorithm and we evaluate it experimentally in a range of search and replanning domains in section 4. We draw conclusions in Section 5.

2   Incremental Heuristic Search and BDD-Based Search

2.1   Incremental Heuristic Search

Incremental search is a search method for replanning that reuses information from previous searches to find solutions to a series of similar search problems, which is potentially faster than solving each search problem from scratch. Incremental search solves dynamic shortest path problems, in which shortest paths have to be found repeatedly as the topology of the graph or its edge costs change [RR96a] and it can guarantee to find shortest paths. Incremental search methods have been developed for a long time and we can find them in [DP84, LC90, FMN98, FMN00]. Incremental heuristic search [KLLF04, KFB02] differs from incremental search only in the calculation of the priorities for vertices in the priority queue, that is, using the heuristic knowledge in the form of approximations of the goal distances to focus the search.

2.2   BDD-Based Search

An OBDD is an ordered binary decision diagram representing a Boolean function on a set of ordered variables. The BDD is a canonical and compact representation according to two reduction rules [B86]. The BDD's another advantage is that a large set of BDDs can share structure in a multi-rooted BDD and be efficiently manipulated by the *apply* function which can compute all sixteen logical operations.

Definition 1. *A search problem can be represented by a tuple (S, T, i, G):*

- *S is the set of states,*
- *$T \subseteq S \times S$ is a transition relation defining the search space and $(s, s') \in T$ iff there exists a transition from s to s',*

- *i is the initial state, and*
- *G is the set of goal states.*

A solution to a search problem is a path $\pi = s_0, ..., s_n$ where $s_0 = i$ and $s_n \in G$ and $\bigwedge_{i=0}^{n-1} (s_i, s_{i+1}) \in T$.

If we encode states as boolean vectors, then BDDs can be used to represent the characteristic function $\Phi(S)$ of a set of states and the transition relation $T(S, S')$. Let's consider an example shown in the middle graph of Figure 1. This is a simple four-connected gridworld problem with cells whose traversability changes over time. The cells in the gridworld problem are either traversable or untraversable. The gridworld problem is to repeatedly find a shortest path between two given cells, knowing both the topology of the graph and which cells are currently traversable.

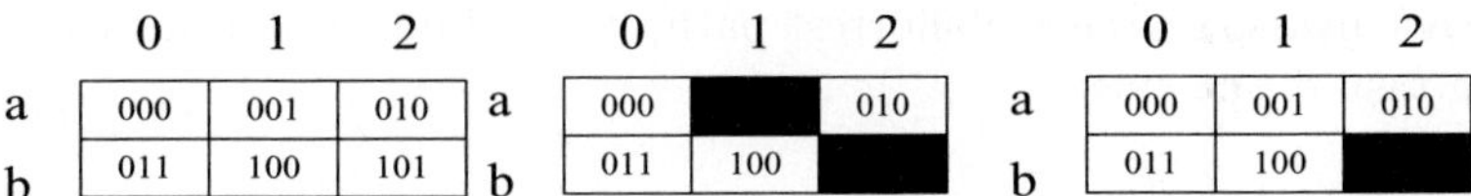

Fig. 1. The four-connected gridworld example

In the middle graph of Figure 1, a state s is represented by a boolean vector with three variables $(s_2 s_1 s_0)$. For example, the initial state a_0 is represented by a BDD for the expression $\neg s_0 \wedge \neg s_1 \wedge \neg s_2$ and the goal state b_2 is represented by $G = s_0 \wedge \neg s_1 \wedge s_2$ (In the example, b_2 is untraversable). In order to encode the transition relation, we need to refer to current state variables and next state variables, and

$$T(s_0, s_1, s_2, s_0', s_1', s_2')$$
$$= \neg s_0 \wedge \neg s_1 \wedge \neg s_2 \wedge s_0' \wedge s_1' \wedge \neg s_2'$$
$$\vee\ s_0 \wedge s_1 \wedge \neg s_2 \wedge \neg s_0' \wedge \neg s_1' \wedge \neg s_2'$$
$$\vee\ s_0 \wedge s_1 \wedge \neg s_2 \wedge \neg s_0' \wedge \neg s_1' \wedge s_2'$$
$$\vee\ \neg s_0 \wedge \neg s_1 \wedge s_2 \wedge s_0' \wedge s_1' \wedge \neg s_2'.$$

The main idea in BDD-based search is using BDDs when finding the next states of a set of states. This can be done by computing $\phi_{S^{i+1}} = \exists \overrightarrow{s}\,(T(\overrightarrow{s}, \overrightarrow{s'}) \wedge \phi_{S^i}(\overrightarrow{s}))$, and starting with ϕ_{S^0}, we can iterate the procedure to get the functions $\phi_{S^1}, \phi_{S^2}, \phi_{S^3}...$ and so on, until one of the characteristic functions has a non-empty intersection with the goal situation, and in this case a solution has been found.

3   BDDRPA*

BDDRPA* is a search algorithm which repeatedly determines shortest paths between two given nodes as the graph's nodes' traversability changes over time.

3.1   Notation and Definitions

BDDRPA* applies to finite graph search problem, on known graphs whose nodes' traversability changes over time. Let G denotes the finite graph and S the finite set of nodes of the graph, and a node $s \in S$ is blocked means that s can't reach any other nodes and any other nodes can't reach s, neither, and a blocked node s is unblocked has the opposite meaning. $succ(s) \subseteq S$ denotes the set of successors of $s \in S$, and similarly, $pred(s) \subseteq S$ denotes the set of predecessors of $s \in S$.

In our paper, a BDD which has only one path to the leaf node 1 will be called a cube, and a path p, which means a construction of several variables, can surely be transferred to a cube, vice versa. Because our work is focused on adding incremental search methods to BDDA* [ER98] so that it can be used for replanning problems, for convenience, we assume for every state, the heuristic function h is set to 0. BDDRPA* always determines a shortest path from the start node s_{start} to the goal node s_{goal}, knowing the topology of the graph. Since BDDRPA* always searches a shortest path, it also finds a plan for the changed planning task if one exists.

3.2   The Algorithm

Given a graph G, BDDRPA* searches the shortest path just like BDDA*, but if the topology of the graph changes, the transition function will change according to the topology and in this case, BDDA* needs a totally new search, but BDDRPA* doesn't need to do that and it will reuse the previous OBDDs to get the OBDD for the new transition function. Lemmas and theorems below will help to alter the previous OBDDs to get the new OBDD.

Lemma 1. *Given an OBDD F, if we delete a path p, which is represented by a cube c, from F, then we have the new OBDD $F' = F \land \neg c$.*

Proof. Trivial.

Lemma 2. *Given an OBDD F, if we add a path p, which is represented by a cube c, to F, then we have the new OBDD $F' = F \lor c$.*

Proof. Trivial.

Theorem 1. *Let the OBDD for the transition function of the graph G be T, and let V represent the OBDD for the node v, then if a node v in G is blocked, we can get the OBDD for the new transition function T' of the new graph G' through the two steps below:*

- *Let OBDD $O = T \land \neg V$;*
- *let S represent the OBDD of $succ(v)$, then $T' = O \land \neg S$.*

Proof. Since v is blocked, it can't reach any other node and on the other hand none of its adjacent nodes can reach it, either. The transition relation among other nodes would not be affected and this is the new property of T'.

- Since $O=T \wedge \neg V$, by lemma1 we know that in the new OBDD O the node v can't reach any other node;
- Since $T'=O \wedge \neg S$, by lemma1 we know that any of the nodes in original $succ(v)$ can't reach the node v, either.

By combing the two steps, we know that the transition relation among other nodes won't be affected by the two steps, and so the new OBDD is T'.

Theorem 2. *Let the OBDD for the transition function of the graph G be T, and if a blocked node v in G is unblocked, then we can get the OBDD for the new transition function T' of the new graph G' through the two steps below:*

- *Let the OBDD for the transition function of v to its adjacent unblocked nodes be T_v, and $O=T \vee T_v$;*
- *Let the OBDD for the transition function of v's adjacent unblocked nodes to v be T_s, and then $T'=O \vee T_s$.*

Proof. Since v is unblocked, it can reach its adjacent unblocked nodes and on the other hand its adjacent unblocked nodes can reach it, too. The transition relation among other nodes would not be affected and this is the new property of T'.

- Since $O=T \vee T_v$, by lemma2 we know that in the new OBDD O we can trace from node v to its adjacent unblocked nodes;
- Since $T'=O \vee T_s$, by lemma2 we know that in T' the node v's adjacent unblocked nodes can reach it.

By combing the two steps, we know that the transition relation among other nodes won't be affected by the two steps, and so the new OBDD is T'.

The pseudo code of BDDRPA* is displayed below.

```
01      BDD_initial, BDD_goal;
02      BDD_transition; //the transition function
03      BDD_open; //BDD to store the current nodes to be extended
04      BDD_trace; //BDD used for memorize the path
05   while(update_data){
06      BDD_open ← BDD_initial;
07      Reconstruct_transition();
08      while(BDD_open ∧ BDD_goal == BDD_0){
09         BDD_temp = Choose_shortest_cost(BDD_open);
10         update_BDD_open();//add the temp's successor nodes into BDD_open
11         if(BDD_open' == BDD_open){
12            The fix-point had been reached, so there is no path;
13            return;
14         }
15         update_trace();
16      }
17   }
```

In the pseudo code, the function $update\_BDD_{open}$ is used to update the *open* table, and here we differ from BDDA* in the two main aspects:

- In the function $update\_BDD_{open}$, when we choose a node to extend and then add its successor nodes to the *open*, there is no need to maintain a BDD to achieve the *arithematic add(+)*. We only need to count the successor nodes and calculate their current cost-values, and then use Theorem 1 and 2 to add them into *open*.
- When outputting the path, we must restore the related information. During the course, we know that every node only has one parent node, so just like as the transition function, we can make a BDD as transitions from the successor nodes to their parent node. Then when we reach a node, we can get its parent node through these two steps:
  $BDD_{temp} = trace \wedge successor\_node;$
  $parent\_node = ExistAbstract(BDD_{temp}, successor\_node);$
  The function *ExistAbstract* can *exist abstract* the variables from the BDD sucessor_node on the BDD_{temp}.

3.3   An Example

To describe the idea behind BDDRPA*, we use a path-planning example in a known four-connected gridworld with cells whose traversability changes over time, see Figure 1. For illustration, the cells in the left graph of Figure 1 are all unblocked, and then we block a_1 and b_2 to get the middle graph, and at last we unblock a_1 to get the right graph.

The transition function for the left graph of Figure 1 is in Figure 2. The initial state is a_0 and the goal state is b_2. Our aim is to find a shortest path from a_0 to b_2 and using the BDD-based search methods described in the last section we can get the characteristic functions ϕ_{S^0}, ϕ_{S^1}, ϕ_{S^2} and ϕ_{S^3} step by step in Figure 5. ϕ_{S^3} includes the goal state, so we have found a shortest path from a_0 to b_2.

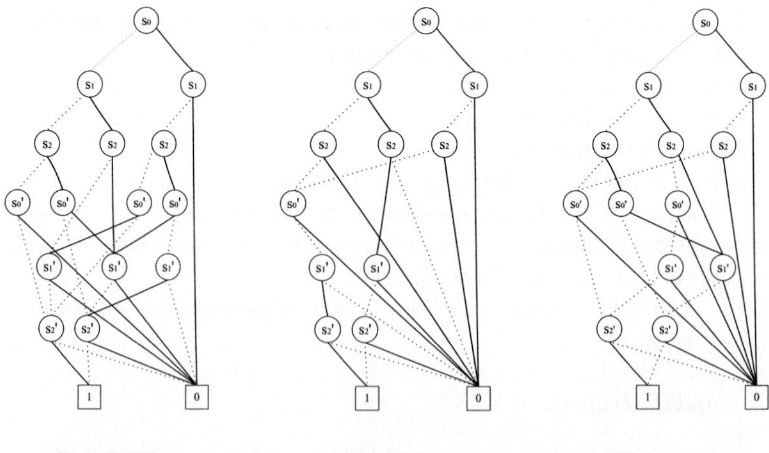

Fig. 2. T_1 Fig. 3. T_2 Fig. 4. T_3

Then the topology of the graph changes, and in the middle graph of Figure 1, a_1 and b_2 are both blocked and the goal state is changed to a_2. In order to get the new transition function, theorem 1 will be applied for two times and the OBDD for the new transition function is in Figure 3. At last we get the characteristic functions ϕ_{S^0}, ϕ_{S^1}, ϕ_{S^2} and ϕ_{S^3} step by step in Figure 6. ϕ_{S^3} equals to ϕ_{S^0}, and this means the algorithm enters a fix-point, so there is no path from a_0 to a_2.

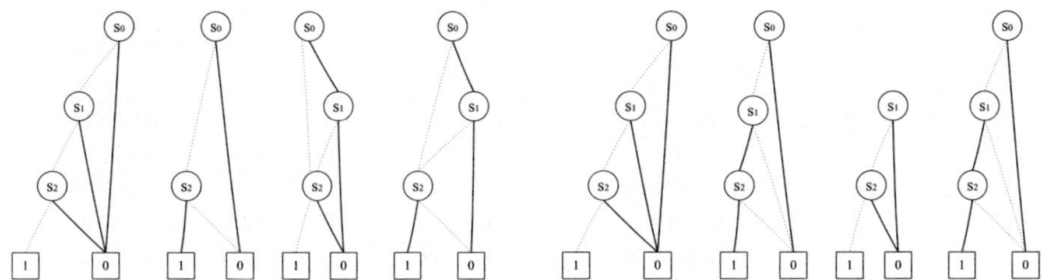

Fig. 5. $\phi_{S_{0-3}}$ of the left graph   **Fig. 6.** $\phi_{S_{0-3}}$ of the middle graph

Then the topology of the graph changes again, and in the right graph of Figure 1, a_1 is unblocked and the goal state is still a_2. Using theorem 2, we can get the new transition function in Figure 4. At last we get the characteristic functions ϕ_{S^0}, ϕ_{S^1} and ϕ_{S^2} step by step and we omit the corresponding figure due to the space limitations of this paper. ϕ_{S^2} includes the goal state a_2, so we have found a shortest path from a_0 to a_2.

4   Experimental Evaluation

In our experiment, we use the simple four-connected gridworld example. We first construct an $n * n$ grid-world with every node unblocked, and then we block a certain percent of all nodes randomly until there is no path from the initial node to the goal node. Assuming we need x times iteration in the above process, then we unblock the same percent nodes each time with x times until every node is unblocked. For each size of the gridworld, we test it for at least 100 times.

Our experiment results are shown in Table 1. The first column gives the maze size and the second column shows the percent of the number of the nodes that changes. Reconstruct_time means the average time to construct the transition BDDs with reusing the previous information and it is shown in the third column. Construct_time means the average time to construct the transition BDDs without reusing the previous information and it is shown in the fourth column. Search_time shows the average time of calculating the path and it is in the fifth column. BFS_time means the time needed by the breadth first search algorithm and it is in the sixth column. We use BFS_time to compare BDDRPA* with BFS. We choose BFS because the heuristic function in BDDRPA* is set to 0, and A* draws back to BFS. In Table 1, − means that the problem can't be solved in 2 hours.

All experiments are carried out on a PC with AMD Opteron242, 2GB DDR memory, under Redhat Linux 7.0 with CUDD 2.4.1. The time unit is second.

Table 1. Experiment results

maze size	percent	reconstruct_time	construct_time	search_time	BFS_time
20 * 20	0.01	0.000451328	0.0119292	0.0541062	0.0131416
20 * 20	0.02	0.000543478	0.0132609	0.0552989	0.0138043
20 * 20	0.05	0.00241071	0.0141072	0.0630357	0.0150892
20 * 20	0.1	0.00430232	0.0120931	0.0543023	0.0126744
20 * 20	0.2	0.00902175	0.0130435	0.0579348	0.0126087
50 * 50	0.01	0.0075	0.18047	0.5732	0.5233
50 * 50	0.02	0.0142424	0.185606	0.632576	0.623182
50 * 50	0.05	0.0319231	0.166346	0.586923	0.518269
50 * 50	0.1	0.0664474	0.163158	0.540526	0.472368
50 * 50	0.2	0.137045	0.160455	0.494773	0.453864
100 * 100	0.01	0.0383553	0.965592	2.9973	9.06257
100 * 100	0.02	0.0761224	0.996327	2.57235	8.31724
100 * 100	0.05	0.209286	0.954143	2.54743	8.04743
100 * 100	0.1	0.374444	0.860556	2.25333	6.5875
100 * 100	0.2	0.742353	0.864412	2.32735	6.79971
200 * 200	0.01	0.2175	5.04824	15.2285	168.593
200 * 200	0.02	0.440385	5.50692	14.8331	174.602
200 * 200	0.05	0.983077	4.48404	13.3331	137.487
200 * 200	0.1	2.00969	4.54688	11.1844	118.087
200 * 200	0.2	3.76875	4.05833	9.80333	100.242
500 * 500	0.01	1.98459	44.0953	166.445	-
500 * 500	0.02	4.02275	42.1457	172.706	-
500 * 500	0.05	8.59484	34.4082	118.162	-
500 * 500	0.1	21.175	38.4082	111.282	-
500 * 500	0.2	42.4877	43.9354	86.7131	-
1000 * 1000	0.01	17.0411	314.238	1270.92	-
1000 * 1000	0.02	26.6093	268.953	587.534	-
1000 * 1000	0.05	80.7475	352.554	885.439	-
1000 * 1000	0.1	145.647	288.021	776.43	-
1000 * 1000	0.2	245.64	310.311	528.666	-

From Table 1, we find that when the change percent is small, the reconstruct_time is trivial to construct_time, which also means that the time needed by BDDRPA* is mainly the search_time, and even when the change percent has reached to 20 percent, the reconstruct_time is still little. The construct_time is increasing directly according to the maze size, so when the maze size increases, the time saved by BDDRPA* will increase in direct ratio. We also find that if there is no path, the search_time is even trivial to the construct_time, so we will get more benefit from BDDRPA* in such conditions. The comparison of BDDRPA* and BFS shows that when the maze size is small, BFS is faster than

BDDRPA*, but with the increasing of the maze size, BFS is much more slower than BDDRPA*, so BDDPRPA* can solve huge size search problems.

Figure 7 gives a comparison of the experiment results when the percent equals to 0.01.

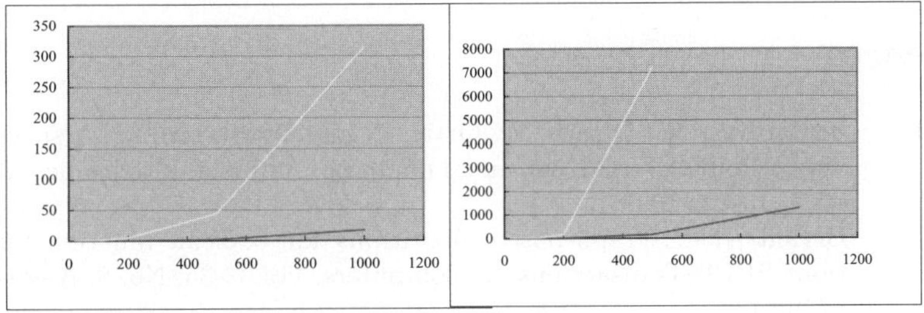

Fig. 7. The comparison

In Figure 7, the x-axis represents the size of the gridworld(For example, 200 means 200*200) and the y-axis represents the time-consuming(The time unit is second). In the left graph of Figure 7, we compare the reconstruct_time(the lower line) with construct_time(the upper line) explicitly, and obviously with reusing the previous information BDDRPA* works much better. In the right graph, we compare the search_time(the lower line) and BFS_time(the upper line) and it shows that BDDRPA* is more constant and efficient when the maze size becomes larger and larger.

Now we analyze the time and space complexity of BDDRPA*. In BDDRPA*, given a maze which is $n * n$, we need to construct its $BDD_{transition}$, BDD_{open} and BDD_{trace}. Since the node number is n^2, we need $log(n^2)$ BDDs to represent all the nodes. Because we use the breadth first algorithm, the cost values will also need $log(n^2)$ BDDs. At last, we need $6 * log(n)$ BDDs. $BDD_{transition}$, BDD_{open} and BDD_{trace} can share them. Because of the compact presentation of BDDs, BDDRPA*'s memory consumption is no doubt less than BFS, but since the memory an OBDD needs is not deterministic, it is impossible to count the memory consumption exactly. In BDDRPA*, when we update the data structures, for example, adding or deleting a node, we don't need to do comparison, we only need to calculate the related BDDs, so when the maze size increases, BDDRPA* needs less time to operate the data, and then needs less runtime. From the above analysis we know that BDDRPA* needs less runtime by reusing previous information, and so will speed up searches for huge size search problems.

5   Conclusion

In this paper, we develop a new search algorithm called BDDRPA*, which combines the incremental heuristic search with BDD-based search to efficiently solve replanning search problems in artificial intelligence. Our experiment evaluation proves BDDRPA* to be a powerful incremental search algorithm. BDDRPA* outperforms breadth-first search by several orders of magnitude for huge size search problems. When the changes to the search problems are small, BDDRPA* needs less runtime by reusing previous information, and even when the changes reach to 20 percent of the size of the problems, BDDRPA* still works efficiently.

References

[JDOW99] desJardins, M., Durfee, E., Ortiz, C, and Wolverton, M. A survey of research in distributed, continual planning. Artificial Intelligence Magazine 20(4):13-22. 1999.

[B86] Bryant, R.E. Graph-based algorithms for boolean function manipulations. IEEE Transactions on Computers, Vol. C-35, No. 8, August, 1986: 677-691.

[M93] McMillan, K.L. Symbolic Model Checking. Kluwer Academic Publ. 1993.

[ER98] Edelkamp, S., and Reffel, F. OBDDs in heuristic search. In Proceedings of the 22nd Annual German Conference on Advances in Artificial Intelligence(KI-98), 81-92. Springer, 1998.

[JBV02] Jensen, R.M., Bryant, R.E., and Veloso M.M. SetA*: An efficient BDD-Based Heuristic Search Algorithm. In Proceddings of AAAI-02, pages 668-673. AAAI Press, 2002.

[KLLF04] Koenig, S., Likhachev, M., Liu, Y., and Furcy, D. Incremental Heuristic Search in Artificial Intelligence. Artificial Intelligence Magazine, Vol 25(2), pages 99-112. Summer, 2004.

[KFB02] Koenig, S., Furcy, D., and Bauer, C. Heuristic search-based replanning. In Proceedings of the International Conference on Aritficial Intelligence Planning and Scheduling. 294-301. 2002.

[RR96a] Ramalingam, G. and Reps, T. 1996a. An incremental algorithm for a generalization of the shortest-path problem. Journal of Algorithms 21: 267-305.

[DP84] Deo, N. and Pang, C. Shortest-path algorithms: Taxonomy and annotation. Networks 14:275-323. 1984.

[LC90] Lin, C. and Chang, R. On the dynamic shortest path problem. Journal of Information Processing 13(4):470-476. 1990.

[FMN98] Frigioni, D., Marchetti-Spaccamela, A., and Nanni, U. Semidynamic algorithms for maintaining single source shortest path trees. Algorithmica 22(3):250-274. 1998.

[FMN00] Frigioni, D., Marchetti-Spaccamela, A., and Nanni, U. Fully dynamic algorithms for maintaining shortest pahts trees. Journal of Algorithms 34(2):251-281. 2000.

A Move Generating Algorithm for Hex Solvers

Rune Rasmussen, Frederic Maire, and Ross Hayward

Faculty of Information Technology,
Queensland University of Technology,
Gardens Point Campus,
GPO Box 2434,
Brisbane QLD 4001,
Australia
r.rasmussen@qut.edu.au,
f.maire@qut.edu.au,
r.hayward@qut.edu.au

Abstract. Generating good move orderings when searching for solutions to games can greatly increase the efficiency of game solving searches. This paper proposes a move generating algorithm for the board game called *Hex*, which in contrast to many other approaches, determines move orderings from knowledge gained during the search. This *move generator* has been used in Hex searches solving the 6x6 Hex board with comparative results indicating a significant improvement in performance. One anticipates this move generator will be advantageous in searches for complete solutions of Hex boards, equal to, and larger than, the 7x7 Hex board.

1   Introduction

Programs that can solve two player games explore a hierarchy of board positions. This hierarchy, known as a *game tree*, is a rooted tree whose nodes are all valid board positions and whose edges are legal moves [1]. Programs that analyze or solve games, such as Chess, Go or Hex, use minimax search algorithms with pruning to search their respective game trees [2,3][4][5]. Pruning is maximized if moves that trigger the pruning are searched first. The benefit of good move orderings is a smaller search. Move generating algorithms are known as *move generators* and they place moves in an order which maximize pruning. Move generators which generate moves as a function of a single board position are called *static*. A *dynamic* move generator generates moves as a function of a game tree [6]. Such move generators sample some or all of the board positions in a game tree to evaluate moves at the root board position.

This paper presents a dynamic move generator applied to the problem of solving the game of Hex. We demonstrate this move generator in a minimax search which van Rijswijck defines as the *Pattern Search* algorithm [7]. Hayward et al. apply a refinement of the pattern search algorithm in their *Solver* program, which can solve small Hex boards up to and including the 7x7 Hex board [2,3]. In Section 2, we introduce the game of Hex. Section 3, gives an overview of

A. Sattar and B.H. Kang (Eds.): AI 2006, LNAI 4304, pp. 637–646, 2006.

© Springer-Verlag Berlin Heidelberg 2006

the pattern search algorithm. In Section 4, we describe the details of our move generator. Section 5 gives experimental results for our move generator in solving small Hex boards.

2   The Game of Hex

The game of Hex is a two-player game played on a tessellation of hexagonal cells which covers a rhombic board (see Figure 1)[8]. Each player has a cache of coloured stones. The goal for the player with the black stones is to connect the black sides of the board. Similarly, the goal for the player with the white stones is to connect the white sides of the board. The initial board position is empty. Players take turns and place a single stone, from their respective cache, on an empty cell. The first player to connect their sides of the board with an unbroken chain of their stones is the winner. The game of Hex never ends in a draw [8].

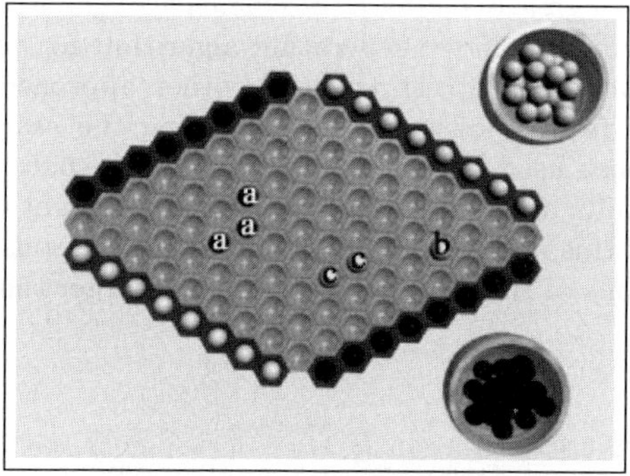

Fig. 1. A 9x9 Hex board

If a player forms an unbroken chain of stones, not necessarily between that player's sides, then any two stones in that chain are said to be *connected*. A *group* is a maximal connected component of stones [9,10]. Figure 1 shows seven groups where three of the seven groups have the labels *a*, *b* and *c*, respectively. The four sides of the board also constitute four distinct groups. A player wins a game, when the opposite sides for that player connect.

For the analysis of board positions, Anshelevich introduces the concept of a *sub-game* [9,10]. In a sub-game, the players are called *Cut* and *Connect*. Both *Cut* and *Connect* play on a subset of the empty cells between two disjoint targets, where a *target* is either an empty cell or one of *Connect's* groups. The player's roles are not symmetric, as *Connect* moves to form a chain of stones connecting the two targets, while *Cut* moves to prevent *Connect* from forming any such chain of stones. In this paper, we generalize sub-games to also include a subset of *Connect's* stones.

Definition 1 (sub-game). *A sub-game is a four-tuple (x, S, C, y) where x and y are targets. The set S, is a set of cells with Connect's stones and the set C is a set of empty cells. Finally, x, y, S and C are all disjoint.*

The set S is called the *support* and the set C is called the *carrier*. Anshelevich's sub-game is the case where S is the empty set [9,10]. A sub-game is a *virtual connection* if *Connect* can win this sub-game against a perfect *Cut* player. A virtual connection is *weak* if *Connect* must play first to win the sub-game. In addition, a virtual connection is *strong* if *Connect* can play second and still win the sub-game. Yang et al. define a *threat pattern* as a virtual connection, where the targets are two opposite sides of the board [11]. There are a number of rules to deduce virtual connections (see [12]), however, the most relevant deduction rule for this paper is the *OR* deduction rule [9,10].

Theorem 1 (OR Deduction Rule). *Let (x, S, C_i, y) be a set of n weak virtual connections with common targets x and y and common support S. If $\bigcap_{i=0}^{n} C_i = \emptyset$, then the sub-game $(x, S, \bigcup_{i=0}^{n} C_i, y)$ is a strong virtual connection.*

Let $M = \bigcap_{i=0}^{n} C_i$. If $M \neq \emptyset$ then *Cut* must move on a cell in M, otherwise, *Connect* has a move which can form a strong virtual connection. Hayward et al. call the set M, the *must-play region* [2,3].

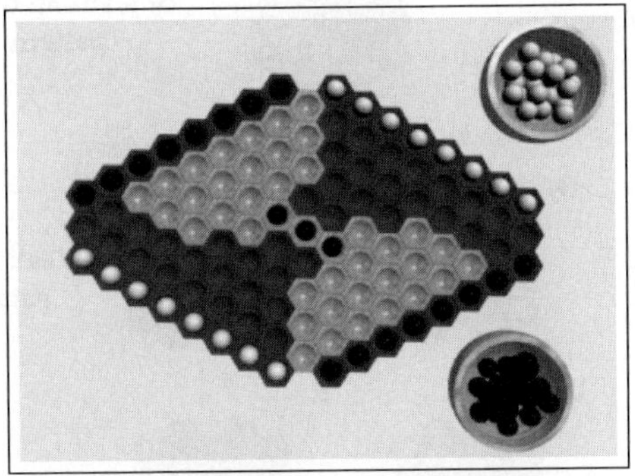

Fig. 2. An example of a threat pattern for player Black

This paper will restrict its discussion to threat patterns. Figure 2, gives a view of a threat pattern where Black is *Connect*. The lightly colored empty cells form the carrier and the black stones at the centre of the board form the support.

3   Pattern Search

The *Pattern Search* algorithm, by van Rijswijck, is a game tree search which deduces threat patterns [7]. Hayward et al. apply the *Pattern Search* algorithm in their Solver program to solve small Hex boards [2,3]. The pattern search algorithm is a depth-first traversal of the game tree, which deduces threat patterns as it backtracks from terminal board positions. The search switches between two modes, *Black mode* or *White mode*. In Black mode, the search tries to prove threat patterns for player Black and in White mode the search tries to prove threat patterns for player White.

Figure 3 gives an example of a search on a 3x3 Hex board in White mode. Player White is *Connect* and player Black is *Cut*. The diagram displays the carrier and the support on lightly coloured cells. The numbers on the stones indicate the order of the player's moves. On the left branch, the search visits terminal board position D where *Connect* is the winner. As the search backtracks from this terminal board position it removes *Connect's* move. At board position B, *Connect* has a weak threat pattern. The search backtracks once more and removes *Cut's* move. The search does a similar traversal of the right branch and deduces a weak threat pattern at board position C. At board position A, the search applies the OR deduction rule on the weak threat patterns found at B and C to deduce a strong threat pattern. On deducing that strong threat pattern, the search can backtrack and deal with board position A as it did with terminal board positions D and E.

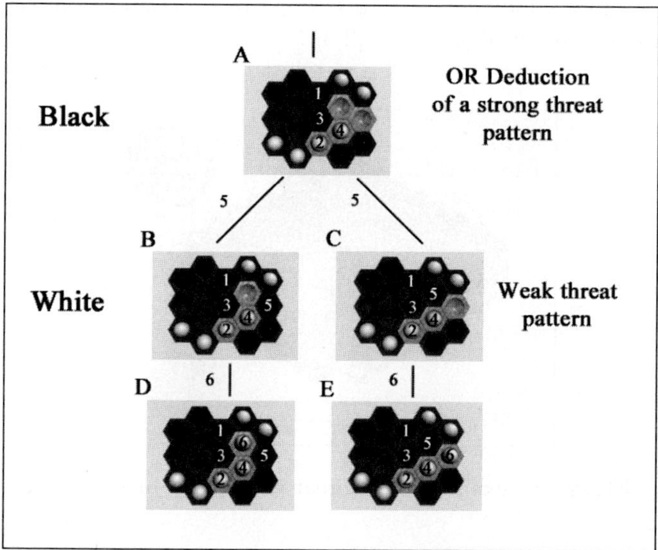

Fig. 3. The pattern search process for constructing threat patterns

A cut-off condition and a switch of mode colour can occur if Black has a winning move elsewhere in this search. Figure 4 follows on from the previous pattern search example where the search is in White mode. The search backtracks from board position A, where it removes *Connect's* move. Board position X has a weak threat pattern. The search backtracks to board position Z and removes Black's move. At Z, the search tries another move which is a winning move for player Black. This winning move results from the search deducing a strong threat pattern for Black at board position Y. The existence of a strong threat pattern for Black at Y indicates a search cut-off condition and a switch of mode colour. The search abandons White's threat pattern at Z and switches to Black mode. The search deals with Y as it did with terminal board positions D and E. The switch of mode colour means player Black is now *Connect* and player White is now *Cut*.

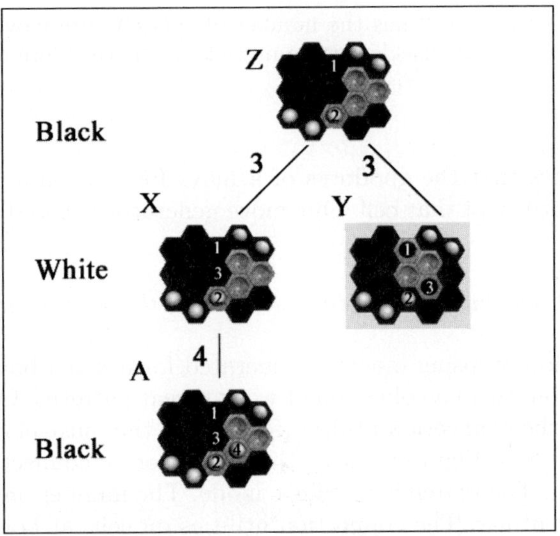

Fig. 4. The search is in White mode and switches to Black mode because there is a strong threat pattern for Black at Y

4   A Dynamic Move Generator for Hex Solvers

Given the example of Figure 4, the aim of our move generator at board position Z is to derive, from the threat patterns found in the subtree under board position X, a good move order. Since Black is the player who moves at board position Z and Black is *Cut*, the problem for our move generator is to order moves for player *Cut*. Given board position Z, our move generator recommends a move on the cell which has appeared most often with a *Connect* stone on terminal board positions where *Connect* is the winner.

Left of Figure 5 shows board position Z (from Figure 4) with *Connect*'s weak threat pattern. Each empty cell in the carrier is labeled with the number of terminal board positions where that cell had player *Connect*'s stone and *Connect* was the winner. Right of Figure 5 gives board position Y (also from Figure 4), where *Cut* has moved on that empty cell with the largest number, successfully cutting *Connect*'s weak threat pattern.

In the search for a weak threat pattern, the *Connection Utility* of a cell in the carrier of that weak threat pattern, is the number of terminal board positions where that cell has *Connect*'s stone and *Connect* is the winner. Our move

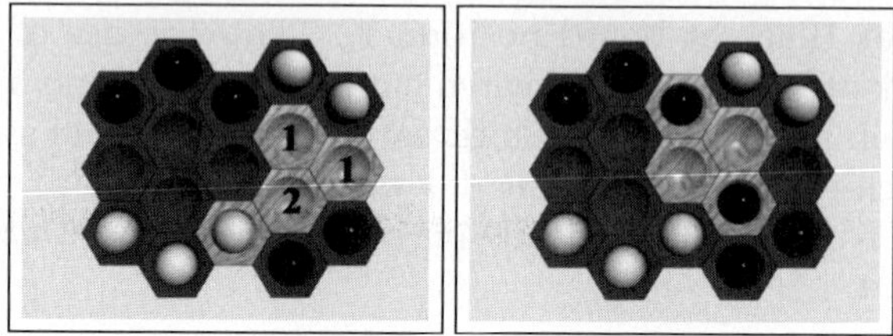

Fig. 5. Left: Each carrier cell has the number of times White moved on that cell to connect the targets. Right: Black's cutting move is the move White used most often to connect.

generator assumes that the goodness of a move for *Cut* on a cell is related to the connection utility of that cell. Our move generator is based on the following hypothesis.

Hypothesis 1. *The higher the connection utility the better is the move for Cut.*

When a succession of losing moves is generated for a given board position, our search derives from them a collection of weak threat patterns. Our move generator must derive the connection utilities for cells in the must-play region of these weak threat patterns. Figure 6 shows the derivation of connection utilities in a must-play region. The search is in White mode. The number on each carrier cell is its connection utility. The connection utilities on cells at board position A is the sum of connection utilities found at both board positions B and C and is restricted to the must-play region. From board position A, the search makes a move for Black on the cell with the largest connection utility. Board position D gives Black's winning threat pattern which causes a switch from White mode to Black mode. Board position D is treated as a terminal board position.

Algorithm 1, the *Connection Utility Search* algorithm (see Appendix), is an extension of van Rijswijck's pattern search algorithm with our move generator approach. Lines 6, 7 and 8 initializes connection utility vectors from terminal board positions. For every cell with a *Connect* stone on a given terminal board position, the value one is assigned to the corresponding element in a vector of connection utilities. Line 30 is a vector sum of connection utilities. This sum occurs for each new weak threat pattern found at a given board position. Line 32 reorders the moves according to the new vector of connection utilities. Moves are restricted to the must-play region at line 31. Algorithm 1 is a recursive procedure which takes an initial board position, performs a search and returns a carrier, a vector of connection utilities and a mode colour. The returning mode colour is the colour of the winning player. If the winning player is the first player then the returning carrier belongs to a weak threat pattern, otherwise, it belongs to a strong threat pattern. Algorithm 1 works explicitly on the carriers of threat patterns because their respective targets and support are implicit in the search. In execution, the mode colour changes if the mode colour prior to line 18 is different from the mode colour after line 18.

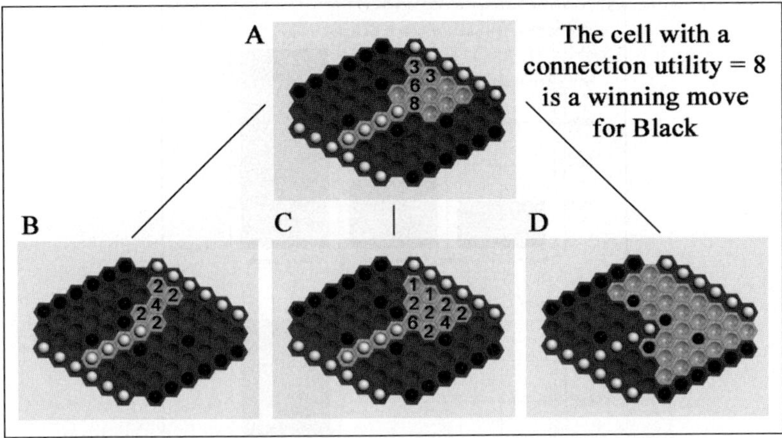

Fig. 6. The accumulation of connection utilities in a must-play region reveals a good move for Black

5   Demonstration of the Connection Utility Search

The aim of this experiment is to demonstrate the pruning performance of our move generator. Given the connection utility search extends the pattern search with our move generator, we can demonstrate the performance of our move generator by comparing implementations of these two search algorithms. The pattern search implementation will use a good static move generator called *Queen-bee* [1]. Thus, this experiment is comparing our move generator against a static move generator. In addition, an alpha-beta search with a Queen-bee move generator is included as a control and as a benchmark. The search with the greatest move generator performance is the search which visits the least number of nodes (board positions) in the game tree, to completely solve an empty Hex board of a given size. Each search solves completely the 3x3 and 4x4 Hex board. The 4x4 Hex board is a practical limit in this experiment. However, additions such as a transposition table can push this limit to larger board sizes.

5.1   Results

Table 1 and Figure 7 show that our move generator maximizes pruning to a greater extent than a Queen-bee move generator in a pattern search. It also

Table 1. The number of nodes visited for each search in the order of best-to-worst

Search	Nodes Visited 3x3	Nodes Visited 4x4
Connection Utility Search	2223	3765784
Alpha Beta	3305	9871596
Pattern Search	9613	790811318

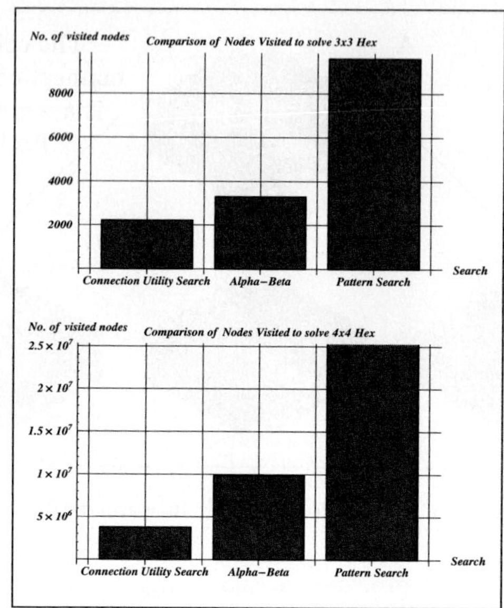

Fig. 7. A comparison of nodes visited by searches solving 3x3 and 4x4 Hex

shows that a pattern search can perform better than an alpha-beta search given an appropriate move generator.

6   Conclusion

This paper presents a move generator which is based on connection utilities for solving Hex boards. Our experiments show, through the performance of our move generator, that connection utilities provide a good heuristic for ordering moves in a game of Hex. In recent developments, the connection utility search was modified so that threat patterns were reused via a specialized transposition table. With this development, our search completely solved the 5x5 Hex board in 165 nodes and the 6x6 Hex board in 77000 nodes. On a 3GHz Intel P4 machine, search times were about ten seconds and about ninety minutes, respectively. The same search without our move generator failed to solve the 6x6 Hex board after running for several days. We are currently improving threat pattern reuse and indexing in order to solve the larger sized boards.

References

1. van Rijswijck, J.: Computer Hex: Are Bees Better than Fruitflies? Master of science, University of Alberta (2000)
2. Hayward, R., Bjrnsson, Y., M.Johanson, Kan, M., Po, N., Rijswijck, J.V.: Advances in Computer Games: Solving 7x7 HEX: Virtual Connections and Game-State Reduction. Volume 263 of IFIP International Federation of Information Processing. Kluwer Achedemic Publishers, Boston (2003)

3. Hayward, R., Björnsson, Y., Johanson, M., Po, N., Rijswijck, J.v.: Solving 7x7 hex with domination, fill-in, and virtual connections. In: Theoretical Computer Science. Elsevier Science (2005)
4. Reinefeld, A.: A minimax algorithm faster than alpha-beta. In van den Herik, H., Herschberg, I., Uiterwijk, J., eds.: Advances in Computer Chess 7, University of Limburg (1994) 237–250
5. Müller, M.: Not like other games - why tree search in go is different. In: Fifth Joint Conference on Information Sciences. (2000) 974–977
6. Abramson, B.: Control strategies for two player games. ACM Computing Survey **Volume 21**(2) (1989) 137–161
7. van Rijswijck, J.: Search and Evaluation in Hex. Master of science, University of Alberta (2002)
8. Gardener, M.: The game of hex. In: The Scientific American Book of Mathematical Puzzles and Diversions. Simon and Schuster, New York (1959)
9. Anshelevich, V.V.: An automatic theorem proving approach to game programming. In: Proceedings of the Seventh National Conference of Artificial Intelligence, Menlo Park, California, AAAI Press (2000) 198–194
10. Anshelevich, V.V.: A hierarchical approach to computer hex. Artificial Intelligence **134** (2002) 101–120
11. Yang, J., Liao, S., Pawlak, M.: On a Decomposition Method for Finding Winning Strategies in Hex game. Technical, University of Manitoba (2001)
12. Rasmussen, R., Maire, F.: An extension of the h-search algorithm for artificial hex players. In: Australian Conference on Artificial Intelligence, Springer (2004) 646–657

3.2.2 Selection Method

Superior animals are mostly used as seed animals to bring forth the young at domestic animal breeding farms. Seed selection, a method of individual selection used in the propagation of cattle and preservation of an individual, has been introduced to the evolution of GA [13]. If the random value generated between 0 and 1 for an individual belonging to the father is smaller than 0.9, the good chromosome will be selected within a seed size involving superior chromosomes in the ranking population. Otherwise, the chromosome will be randomly selected from the entire population. The mother will be selected randomly from the entire population. Those selected will be used as parents, and then returned to the population so that they can be used again later.

Table 2. Operation sequences and machines selected by chromosome

Job		Operations Sequence	#	First operation	Second operation	Third operation
1	1	①→②→③	1	M1	M2	M3
			2	(M2)	(M1)	
			3		(M3)	
	2	①→③→②	1	M1	M3	M2
			2	(M2)		(M1)
			3			(M3)
2	1	④→⑤→⑥	1	M1	M2	M3
			2	(M3)		(M1)
			3			(M2)
	2	④→⑥→⑤	1	M1	M3	M2
			2	(M3)	(M1)	
			3		(M2)	
	3	⑥→④→⑤	1	M3	M1	M2
			2	(M1)	(M3)	
			3	(M2)		
3	1	⑦→⑧→⑨	1	M1	M3	M2
			2	(M3)	(M1)	
			3		(M2)	
	2	⑦→⑨→⑧	1	M1	M2	M3
			2	(M3)		(M1)
			3			(M2)

3.2.3 Crossover and Mutation

The crossover operator should maintain and evolve a good order relationship of chromosomes. In this research the crossover operator first produces a random section and then inserts all the genes inside the section into parent 2. The position of insertion is just before the gene where the random section starts. For example, in parent 1 if the random section starts in the fourth place, then the position of insertion will be before the fourth gene in parent 2. Then all genes with the same index as the genes in the random section will be deleted in parent 2. To make the alternative operations

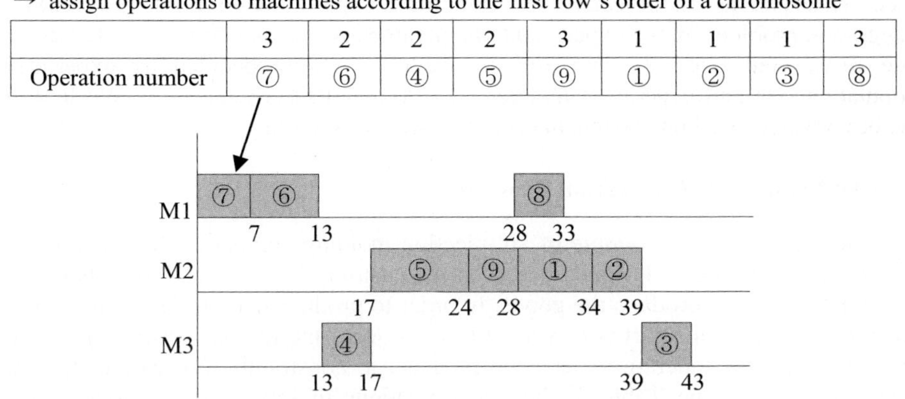

Fig. 2. A Gantt chart obtained from the representation

sequences coincide with the same job number, it will be corrected according to the alternative operations sequence of the initial job number. These processes will be performed by alternating parent 1 and 2, thus producing two children. After two offspring are evaluated, the better one will be sent as the next generation. The crossover operator has shown good performance in a previous study [13].

The mutation operator brings a change to the chromosome, thus maintaining diversity within the group. This research uses the two types of mutation operator [13] based on the neighborhood searching method [5].

3.2.4 Objective Function and Replacement

The minimum makespan in scheduling often means the highest efficiency of a machine. When a chromosome is represented as a permutation type, the makespan is produced by the process that assigns operations to the machines according to the sequence of genes from left to right. This operation is performed while maintaining the technological order of jobs and considering alternative operations sequences and alternative machines.

The next generation will be formed by selection among the current generation with a help of the genetic operator. The new individuals will be produced as many as the number of initial population and they form the next generation. By using elitism, bad individuals will be replaced with good individuals.

4   Evaluation of Performance

In order to evaluate the performance of the GA developed in this study, we used an example of a molding company, a typical job shop that has multiple process plans. Through this real case, we have compared two kinds of scheduling results, that is, the integrated scheduling considering the alternative machines and operations sequences and the traditional scheduling considering them sequentially. We used the representative job shop problem with alternative machines and integration problem of process planning and scheduling for the performance evaluation of the GA. We also used makespan as an objective function for comparing with prior researches of integration problem. In the experiment, the parameters - crossover rate (0.8), mutation rate (0.1), seed size (20) and elitism size (10) - were decided by experiment. Population size and generation number are 200 and 1000 respectively. We look for the best results based on 50 runs in each benchmark problem.

4.1 An Example of Integration Problem

First, we deal with the example of an injection molding company which is a typical job shop. A molding is designed based on the customer's order, and then its process planning is made to produce the goods. In order to produce one product, a lot of parts are necessary, and one part is considered as one job. One job has several operations, and some of these operations have no precedence constraints. For example, in the Elbow product of the Table 3, the two operations in job 2 can be done without precedence constraints. The shaded operations in Table 3, Table 4 and Table 5 have no precedence constraints. In particular, in job 5 of Table 3, MM operation is to be done prior to RD operation, but NCL operation can be done before MM operation or RD operation. That is, there are three alternative operations sequences. Some operations have alternative machines. There are three units of MM machine and two units of E machine in this case. Table 3, Table 4 and Table 5 show the process plans where their alternative machines and alternative operations sequence were decided by several tests. It is process plans producing the best makespan among the several process plans in considering process planning and scheduling separately.

Table 3. Process planning of "Elbow" product

Job	Operation sequence (processing time)					
1	MM2(20)	RD(15)				
2	MM1(20)	RD(10)				
3	LM(16)	MM3(30)	RD(12)			
4	LM(30)	MM2(30)	RD(12)			
5	MM1(18)	RD(3)	NCL(4)	E1(20)	SF1(10)	L(10)
6	MM2(30)	RD(12)	L(10)	E2(20)	SF1(15)	
7	MM3(5)	E1(4)	SF1(6)	L(10)	RD(5)	

Table 4. Process planning of "Picnic Case" product

Job	Operation sequence (processing time)				
1	LM(14)	RD(17)			
2	LM(14)	RD(5)			
3	MM1(4)	RD(8)			
4	RD(2)	MM1(1)			
5	MM2(6)	RD(2)			
6	MM3(4)	RD(6)	NCM(15)	E1(8)	SF1(10)
7	MM2(5)	RD(6)	NCM(17)	E2(12)	SF1(10)
8	MM3(3)	RD(4)	NCM(8)	E1(3)	SF1(5)
9	MM1(10)	AS(2)	SF1(3)		

Table 5. Process planning of "Cake Box" product

Job	Operation sequence (processing time)				
1	LM(18)	RD(20)			
2	LM(17)	RD(10)			
3	MM1(6)	RD(2)			
4	MM1(6)	RD(2)			
5	MM2(4)	RD(7)	E1(4)	NCM(19)	SF1(10)
6	MM2(4)	RD(8)	E2(4)	NCM(25)	SF1(10)
7	MM3(22)	RD(4)	E1(3)	SF2(5)	

Table 6. Comparison of integrated scheduling results

Product	Makespan	
	Non-Integration	Integration
Elbow	92 hours	88 hours
Picnic Case	65	62
Cake Box	69	61

We look for the best results based on 10 runs in this case. Table 6 shows the makespan obtained by separated scheduling and integrated scheduling of process planning and scheduling. Here, one can clearly see the improved makespan in the integrated scheduling. Since there are many solutions with the same value in the final population of GA, we have not suggested the final process plan.

4.2 Chambers' Problem

In an effort to examine the performance of an algorithm used in a job shop with alternative machines, Chambers [4] has added the specific machine according to the simple criterion in the benchmark problem of a classic job shop. He has revised an MT10 problem for making an alternative machine problem. The criterion is the sum of required processing times for each machine. As shown in Table 7, the newly suggested GA is showing a better solution.

Table 7. Results of Alternative Machine Problem through Revised MT 10 Problem

Problem	Dispatching	Chambers' Tabu search	Proposed GA
P1	1023	929	929
P1, P1	1023	929	929
P1,P1,P1	1023	936	929
P1, P2	982	913	909

4.3 Tan's Integration Problem

Tan [14] developed an LMIPM (Linear Mixed-Integer Programming Model). To test the performance of the model, he has suggested the problem that has five jobs with twenty operations. In this problem, most jobs and operations have more than four operations sequences and alternative machines. Also, each job has a different release time. In this problem, Tan has introduced a specially designed dispatching rule for creating a feasible initial solution. As shown in the Table 8, the newly suggested GA has obtained the optimal solution acquired by LMIPM. The optimal schedule was obtained four times in 50 runs of the proposed GA.

Table 8. Results from Tan's problem example of integrated scheduling

Algorithms	ECTF	OS	FS	SA	LMIPM	Proposed GA
Makespan	43.33	44.91	55.51	43.40	34.15*	34.15*

5 Conclusion

This study tried to develop a GA for integration of process planning and scheduling, and to show the possibility of improving makespan. Also, we have produced the optimal or improved solution for the job shop problem that has alternative operations sequences and alternative machines, thus proving the performance of the GA. This means that we can more effectively respond to the due date demanded by customers.

Hereafter, more research should be made on the problem of integration because of its complexity and difficulty. Although the combination of a mathematical programming approach and a meta-heuristic approach is believed to be a better solution for enhancing search capacity and efficiency, it still has problem of limitation of problem size which it is able to deal with and long searching time. Accordingly, more research on an effective local search method should be explored.

Acknowledgement

This work was supported by the Regional Research Centers Program(Research Center for Logistics Information Technology), granted by the Korean Ministry of Education & Human Resources Development.

References

1. Bierwirth, C., Mattfeld, D., Kopfer, H.: On Permutation Representations for Scheduling Problems. In Voigt, H M., et al. editors, Proceedings of Parallel Problem Solving from Nature IV, Springer Verlag, Berlin, Germany, (1996) 310-318
2. Brandimarte P.: Routing and scheduling in a flexible job shop by tabu search. Annals of Operations Research, 41 (1993) 157-183
3. Brandimarte P, Calderini M.: A heuristic bicriterion approach to integrated process plan selection and job shop scheduling. International Journal of Production Research, 33 (1995) 161-181
4. Chambers J.B.: Classical and flexible job shop scheduling by tabu search. Ph.D. Dissertation, University of Texas at Austin, (1996)
5. Cheng, R.: A study on Genetic Algorithms-based Optimal Scheduling Techniques. Ph.D. thesis, Tokyo Institute of Technology, (1997)

6. Huang S.H, Zhang H.C, Smith M.L.: A progressive approach for the integration of process planning and scheduling. IIE Transactions, 27 (1995) 456-464
7. Khoshnevis B, Chen Q.: Integration of process planning and scheduling functions. Journal of Intelligent Manufacturing, 1 (1990) 165-176
8. Kim K.H, Egbelu P.J.: Scheduling in a production environment with multiple process plans per job. International Journal of Production Research, 37 (1999) 2725-2753
9. Kim Y.K, Park K.T, Ko J.S.: A symbiotic evolutionary algorithm for the integration of process planning and job shop scheduling. Computer and Operations Research, 30 (2003) 1151-1171
10. Lee Y.H, Jeong C.S, Moon C.U.: Advanced planning and scheduling with outsourcing in manufacturing supply chain. Computers and Industrial Engineering, 43 (2002) 351-374
11. Nasr A, Elsayed E.A.: Job Shop Scheduling with Alternative Machines. International Journal of Production Research, 28 (1990) 1595-1609
12. Palmer G.J.: A simulated annealing approach to integrated production scheduling. Journal of Intelligent Manufacturing, 7 (1996) 163-176
13. Park B.J, Choi H.R, Kim H.S.: A hybrid genetic algorithms for job shop scheduling problems. Computers and Industrial Engineering, 45 (2003) 597-613
14. Tan W.: Integration of process planning and scheduling - a mathematical programming approach. Ph.D. Dissertation, University of Southern California, (1998) 97-98
15. Tan W, Khoshnevis B.: Integration of process planning and scheduling - a review. Journal of Intelligent Manufacturing, 11 (2000) 51-63
16. Weintraub A, Cormier D, Hodgson T, King R, Wilson J, Zozom A.: Scheduling With Alternatives: A Link Between Process Planning and Scheduling. IIE Transactions, 31 (1999) 1093-1102

A Robust Shape Retrieval Method Based on Hough-Radii

Xu Yang and Xin Yang

Institute of Image Processing & Pattern Recognition, Shanghai Jiaotong University,
800 Dong Chuan Road, Shanghai, 200240, P.R. China
feilang@sjtu.edu.cn, yangxin@sjtu.edu.cn

Abstract. A novel shape similarity retrieval algorithm (Hough-Radii) for 2-D objects is presented. The method uses a polar transformation of the contour points to get the shape descriptor that is invariant to translation, rotation and scaling. We take the maximum point in the generalized Hough transform (GHT) mapping array as the reference point for polar transform that is different from the traditional Centroid-Radii method where the geometric centre was taken as the origin. The effectiveness of our algorithm is illustrated in the retrieval of two databases of 99 and 216 shapes provided by Sebastian et al. The experimental results show the competitiveness of our approach to some others especially in the retrieval of partially occluded and missing images.

1   Introduction

More and more images have been generated in digital form around the world. There is an urgent demand for effective tools to facilitate the searching of images. Shape is an important visual feature and it is one of the basic features used to describe image content. There are many shape representation and analysis techniques in the literature [1, 2]. However, to find images in large databases efficiently by shape description is still a difficult task. This is because when a 3-D real world object is projected onto a 2-D image plane, one dimension of object information is lost. As a result, the shape extracted from the image only partially represents the projected object. Furthermore, shape is often corrupted with noise, defects, arbitrary distortion and occlusion. MPEG-7 has set several principles to measure a shape descriptor, that is good retrieval accuracy, compact features, general application, low computation complexity, robust retrieval performance and hierarchical coarse to fine representation [3]. A practical shape description method usually makes tradeoffs among the above features.

Chang et al. [4] described shapes by computing the distance of points of high curvature to the centroid. The radial distances were all normalized to the minimum to provide for scale invariance. Two contours were then assessed for shape similarity by a cyclic comparison of feature point radial distance. Ozugur et al. [5] again used the polar form but computed the distance of feature points to the centre of the smallest circle enclosing the shape. The polar representation was then broken into segments and each segment described by its Fourier coefficients and shape similarity is assessed using the Euclidean distance of the Fourier coefficients.

A. Sattar and B.H. Kang (Eds.): AI 2006, LNAI 4304, pp. 658–667, 2006.
© Springer-Verlag Berlin Heidelberg 2006

Tan and Thiang proposed a novel method for shape representation in their paper [6, 7]. The basic idea of their proposal is to use the centroid-radii model to represent shapes. In this method, lengths of the shape's radii from centroid to boundary are used to represent the shape. The interval between radii, measured in degrees, is fixed. To make this approach invariant to scale, all radius lengths can be normalized by dividing by the longest radius among them. In Tan and Thiang's method, two shapes are considered similar to each other if and only if the lengths of their radii at the respective angles differ by a trivial value. Thomas Bernier et al. [8] specially discussed the correspondence points problem of complicated shape objects. In the method, multiple distinctive vertices of contour boundary are extracted and used as comparative parameters to get the corresponding starting points.

Similarly, Shuang Fan [9] proposed the centroid-radii histogram model. Since the space information was omitted, this model was robust to noise and got features of rotationally invariant intrinsically. While Dalong Li et al. [10] proposed a shape-matching algorithm based on distance ratio distribution, and the standard deviation was used to measure the shape similarity, they also discussed the relationship between the number of sample boundary points and the shape retrieval accuracy.

All the above methods represent a shape by one dimensional function, which was got by polar transformation of the boundary points. So they are called Shape Signature [3].This method can be normalized easily to achieve the translation and scaling invariance. Furthermore, it can also be invariant to rotation by shifting the signature plot to the maximum radii as the starting point.

While in either case, the matching cost is too high for online retrieval. In addition to the high matching cost, shape signatures are sensitive to noise, and slight changes in the boundary will cause large errors in matching. This is because the above shape description methods all make use of the entire information of the contour, which means the centroid of the contour is calculated from the mean of all points lying on the boundary. Therefore, the position of the centroid will shift, when only a few boundary points have slight changes in position. Consequently, the relative distance of other points on the boundary to the centroid will change. As a result, the effect of the shape description is affected and this will directly results in the failure of shape matching.

To overcome the above two limitations of the Centroid-Radii method, we proposed the Hough-Radii method. The generalized Hough transform (GHT) is utilized to give the reference point as the origin of polar transformation for the query image. This can solve the reference point shifting problem greatly. And because the GHT can be used as filters to abandon obvious false matching images, so the retrieval efficiency is improved at the same time.

2 General Description of the Approach

In this paper a method is described which consists of two integrally linked aspects. The first is shape representation and the second is shape matching, through which two representations can be compared. While segmentation is required prior to representation, it is not the topic of the paper. Section 2.1 describes the method of representation in detail and Section 2.2 describes the comparison of representations.

2.1 Shape Description

2.1.1 Polar Transform Description [8]

First, the vector $\vec{R}$ from each contour point to the reference point is calculated. Then we get the signature plot by its angle and amplitude, so the most important thing to do is to get the reference point location for the polar transform. The centroid of the image is taken for granted as the origin of the polar transform on all the above method. It's calculated by

$$(x_c, y_c) = \left(\frac{\sum_{i=1}^{n} x_i}{n}, \frac{\sum_{i=1}^{n} y_i}{n} \right) \tag{1}$$

Where (x_i, y_i) is the i-th contour point and n is the total number of sample points on the boundary.

The amplitude R and angle θ of the signature plot is calculated by the following formula:

$$R = \sqrt{(x_i - x_c)^2 + (y_i - y_c)^2} \tag{2}$$

θ is the angle taken counter-clockwise, starting from the X axis direction and it's calculated by:

$$Angle(\theta) = \begin{cases} \dfrac{\pi}{2} & \textit{if } \Delta x = 0 \textit{ and } \Delta y \geq 0 \\[2ex] \dfrac{3\pi}{2} & \text{if } \Delta x = 0 \text{ and } \Delta y < 0 \\[2ex] arctan(\dfrac{\Delta y}{\Delta x}) + \pi & \text{if } \Delta x < 0 \\[2ex] arctan(\dfrac{\Delta y}{\Delta x}) + 2\pi & \text{if } \Delta x > 0 \text{ and } \Delta y < 0 \\[2ex] acrtan(\dfrac{\Delta y}{\Delta x}) & \text{if } \Delta x > 0 \text{ and } \Delta y \geq 0 \end{cases} \tag{3}$$

Where $\Delta y = y_i - y_c$, $\Delta x = x_i - x_c$, $(x_c\ y_c)$ is the location of centroid.

Once all the R and θ is got, plot the $R - \theta$(signature plot) curve in X-Y plane withθ as the x axis and R as the y axis. Since the signature plot takes only the relative location information between boundary points and centroid point, this description method is invariant to translation.

2.1.2 Scale Normalization

A shape based image retrieval system requires the algorithm has feature of scale invariance. In order to provide for scale invariance, we take the area based normalization method. This method is insensitive to the local boundary deformation, because area contains the most stable and effective information of the image size. First, define an area constant C_{area}, then every template image and the query image is compared to this constant to get the scaling factor α as follows:

$$\alpha = \sqrt{\frac{S}{C_{area}}} \tag{4}$$

Where S is the area of the shape enclosed.

After the scaling factor α is calculated, all radii lengths can be normalized by dividing with α. Thus the signature plots acquired in this way are invariant under scale.

Although the polar transform based Centroid−Radii model has features of translation and scale invariant, this shape descriptor is very sensitive to the boundary local deformation. From equation (1) we can see that the centroid is calculated by the mean position of all boundary points. Thus, even several points on the boundary have small changes in position; the coordinate of the centroid will change. And this lead to the centroid radii of all boundary points change, so the shape signature plot will change also. Then the matching failure rate is increased.

2.1.3 Hough-Radii Model

Hough transform was originally used as a method to detect line and circle in images. Ballard [11] developed it and proposed the generalized Hough transform which could detect arbitrary shape objects. We take its parameter mapping theory, and analyze the mapping point distribution to give the origin point for the polar transform shape description. The origin point got by this method is more robust to local deformations than centroid. Fig. 1. shows the signature plots got by two different methods, the fish template image is on the left and the query image with tail part missing is on the right.

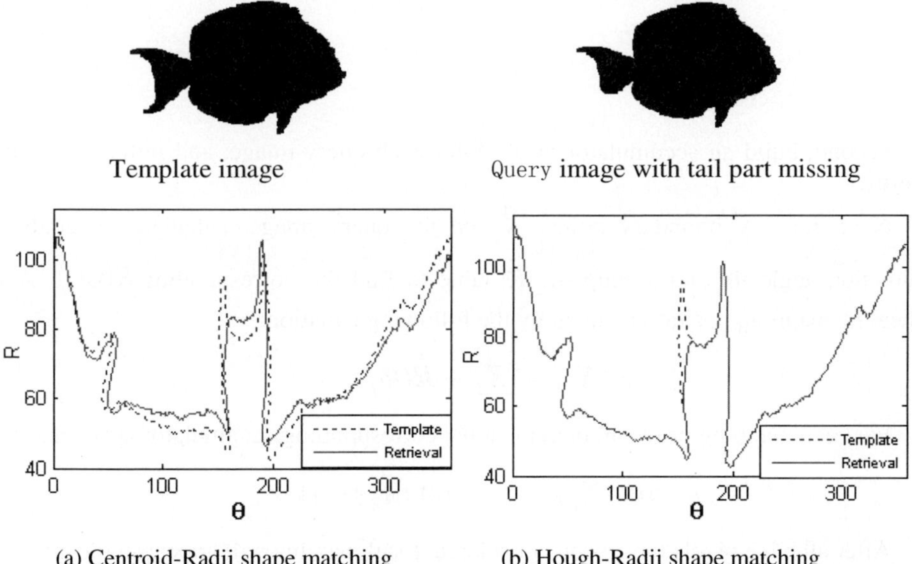

Fig. 1. Comparison of Shape signature plots of two fish images got by different methods

From the above figure we can see that there are much more cumulate error areas for the Centroid-Radii method. While error areas only happen at the tail for the Hough-Radii method, which means the error doesn't diffuse.

The procedures of the Hough-Radii are as follows: First, construct the $\vec{R}$-table from the template image. Choose the centroid (c) as reference point, for each boundary point (B), calculate the gradient direction Φ and the distance vector $\vec{R}$ to centroid c. Store $\vec{R}$ as a function of Φ. Table 1 shows the form of the $\vec{R}$-table diagrammatically.

Table 1. $\vec{R}$-table format

i	Φ_i	$\vec{R}(\Phi_i)$
0	0	$\{r \mid c - r = x, x \ in \ B, \Phi(x) = 0\}$
1	$\Delta\Phi$	$\{r \mid c - r = x, x \ in \ B, \Phi(x) = \Delta\Phi\}$
2	$2\Delta\Phi$	$\{r \mid c - r = x, x \ in \ B, \Phi(x) = 2\Delta\Phi\}$
...	...	...

Second, build an accumulator array A for each query image, and initialize it with zeros.

Next, for any boundary points $\vec{X}_i$ on the query image, calculate its gradient direction angle Φ_i and lookup the $\vec{R}$-table to find the corresponding $\vec{R}(\Phi_i)$. And then the mapping position is given by the following equation:

$$\text{ref}(\vec{X}_i) = \vec{X}_i + \vec{R}(\Phi_i) \tag{5}$$

For each mapping position, increment the corresponding accumulator array entries:

$$A(\text{ref}(\vec{X}_i)) = A(\text{ref}(\vec{X}_i)) + 1 \tag{6}$$

After all edge pixels have been considered, local maxima will appear in the array A if the query image is similar enough to the template image. Otherwise the value in array A is dispersed and no obvious maxima can be found, as is shown in Fig.2. As

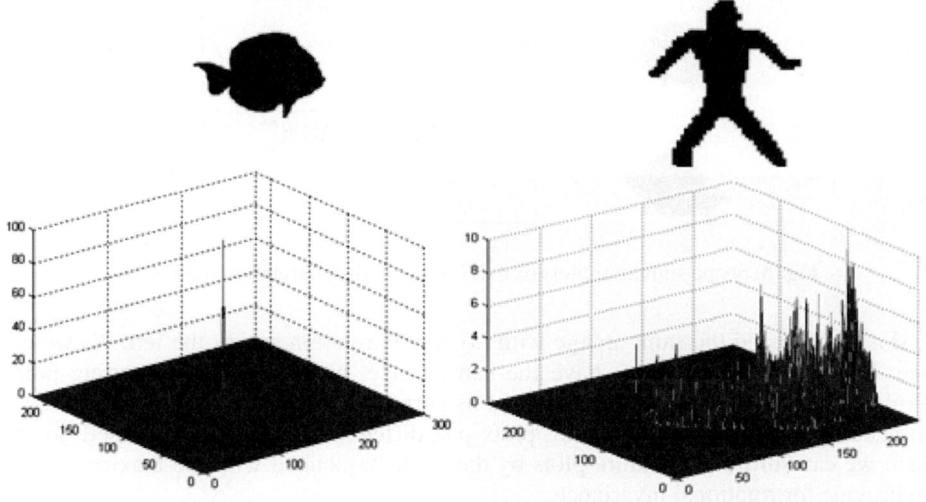

(a) Image of the same category   (b) Image of different category

Fig. 2. Accumulated array got by GHT for query images of different shapes

the $\vec{R}$-table of fish image template is utilized to human query image (b) and fish query image (a) with tail part missing, the accumulator A is concentrated for the fish image, however, is dispersive for the human image.

Summing up the number of mapping points falling into the 3×3 neighboring region of the maximum, and store it as V_m. If $V_m/N > \beta$ (where N is the total number of points on the boundary; β is threshold which depends on the shape deformation extent of the same class.), then take this maximum as origin of the polar transform and give the corresponding signature plot. Else, it can be concluded that the query image and template image do not belong to the same category. So there is no need to get its signature plot. As a result, with this method we can improve the retrieval accuracy and at the same time quicken the retrieval speed.

2.2 Shape Matching

The similarity of two shapes is inversely proportional to the area (accumulate error area) lying between their respective signature plot. So the similarity is given by the following equation:

$$S_{im} = \left(\int_{\theta=0}^{2\pi} \left| R_\theta - R'_\theta \right| d\theta \right)^{-1} \tag{7}$$

Where θ is the angle and R is the radii.

When the image rotates by arbitrary degree, its signature plot is just shifting by the corresponding interval. As is shown in Fig.3. .

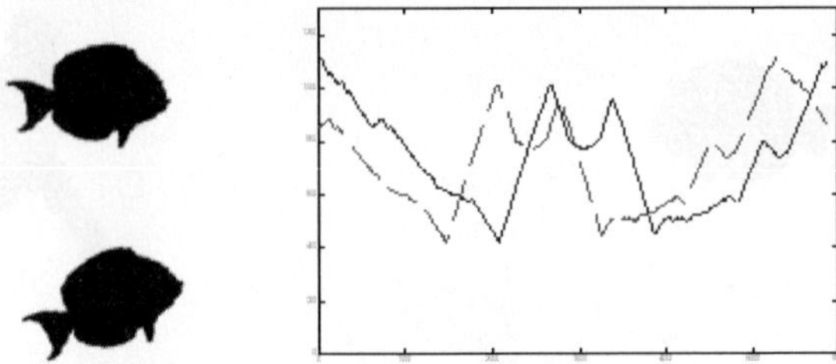

Fig. 3. Shape signature plots of fish images with different lying direction

Fish image and the same image with 60 degree rotation are on the left. As we can see that their signature plots have the same shape, but just differ in starting point. Which means for the template image and its corresponding query image with different direction positioned their signature plots just differ in starting points. Based on this fact, we can shift the signature plots by the angle associated with the maximum radii to provide for rotational invariance.

However, this assumes that the maximum radii will always occur at the same point while practical objects hold many exceptions to this assumption. To solve this situation, we take every maximum point as the starting point to match with the template image signature plot.

3   Comparative Performance Analysis

In order to compare our method to the traditional Centroid-Radii method, we test on the same image retrieval database which was build by Sebastian [12]. This database contains two sub-databases A and B. Sub-database A has 99 shapes which are classified into 9 categories with 11 shapes in each category. Sub-database B has 216 shapes which are classified into 18 categories with 12 shapes in each category.

Each shape in sub-database A was used as a template to which the whole database was retrieved. The retrieval result gives 10 most similar shapes (excluding the model itself) to the template. Ideal results would be that 10 other shapes of the same category were all retrieved. Some of the results obtained by T. Bernier et al. [8] are shown in Fig. 4. 15 nearest retrieved results and their match values are ranked by their similarity to the template image. It's clear that the first 10 images do not always belong to the correct category. So there exit some error matching.

To show the superiority of our approach to the Centroid-Radii in the presence of partial occlusion and part-based changes like articulation, we specially selected a small dataset of 10 shapes from sub-database A ((Fig. 5(a)). This dataset contains 5 Rabbit images and 5 Greebles images, which were mostly false retrieved by the Centroid-Radii method.

Each of the shapes was used as a template to which the other 9 were retrieved. 10 most similar shapes (excluding the model itself) to the model were ranked by their similarity to the template image (Fig. 5(b), (c)). Centroid-Radii description method utilizes the entire information of the contour, which makes it very sensitive to partial

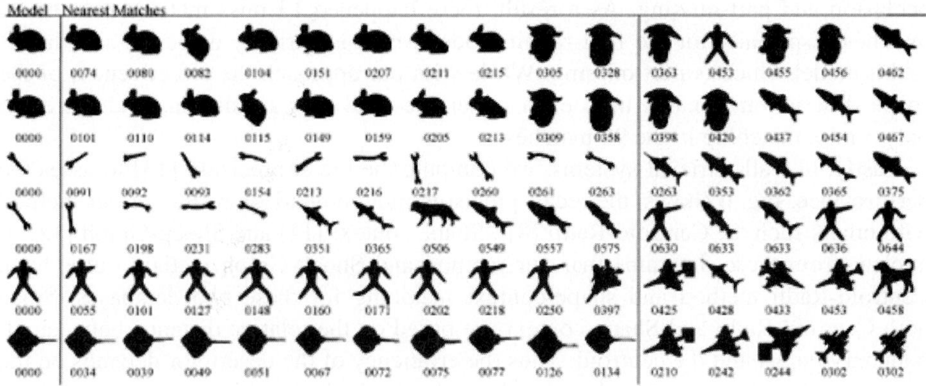

Fig. 4. Selected results obtained by T. Bernier et al [8]

(a) Test database specially selected of Rabbit and Greebles

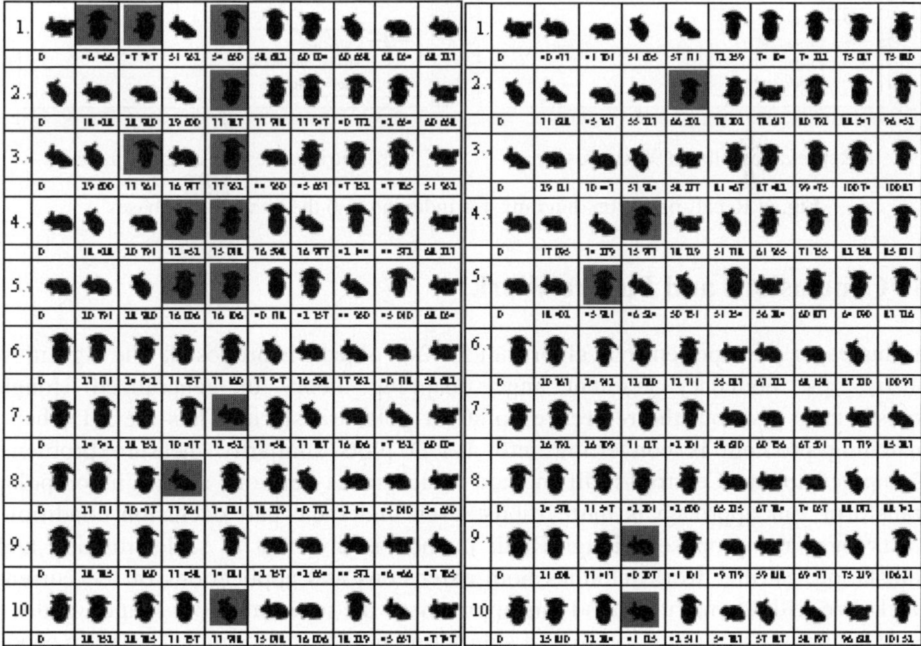

b) Retrieval results of Centroid-Radii   c) Retrieval results of Hough-Radii

Fig. 5. Comparison of the retrieval results on partially occluded/missing database

occlusion and part-missing. As a result, there happened 13 miss matching in the 90 matches. Especially for the first rabbit model which is partially occluded and the 3rd rabbit model which is part missing. While with our approach the effectiveness of the shape description remains high even under these adverse conditions, and there are only 5 miss matching in the 90 matches.

Lastly, like all retrieval systems, we compute the recall-precision [13] to asses the performance. Fig. 6 shows the recall-precision plot comparison results of our method with others such as: Centroid-Radii [8] , Shape context [14] and Shock-Graph [12]. It is clear from these diagrams that our method and Shock-Graph method outperform Centroid-Radii method and shape-context matching for these two databases. Since both Centroid-Radii and Shape context are based on the relative distance between all boundary points and the centroid. Thus the efficiency of the descriptor depends on the entirely information of the shape and consequently is sensitive to noise, such as partially occlusion. However, the plots remain high for our approach and Shock-Graph, whose description efficiency was not affected by local disturbance.

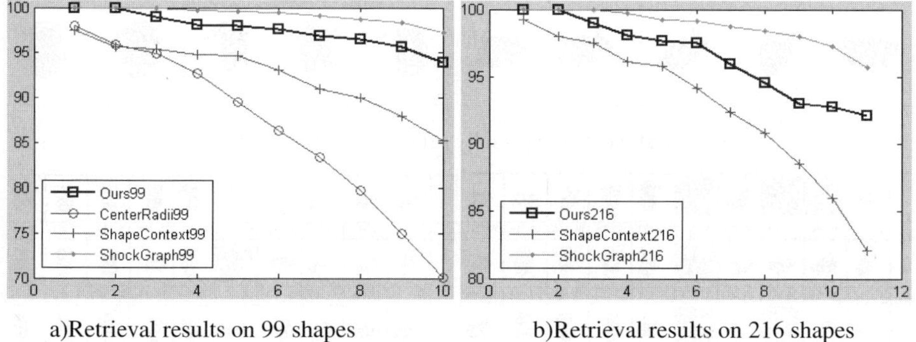

a)Retrieval results on 99 shapes   b)Retrieval results on 216 shapes

Fig. 6. Recall-precision diagrams for indexing into the two sub-databases

4   Conclusions and Future Work

In this paper, a shape retrieval algorithm for 2-D objects is proposed. The algorithm utilizes the boundary gradient angel mapping theory of generalized Hough transform to give out the reference point for polar transform. Thus the 2-D shape matching problem is converting into a 1-D plots similarity measure problem. Compared to the Centroid-radii method which takes centroid as origin for polar transform, our approach is more robust even under the part occlusion condition. Tests show the proposed Hough-Radii method is invariant to scale, rotation and translation. Further work includes both extensions and applications of the current work. We will attempt to extend this approach to the retrieval of 3D shapes, and we have begun to use our algorithm as a subroutine in indexing of mitral valve in echo-cardio graphic images.

References

1. S. Loncarin, "A survey of shape analysis techniques," Pattern Recognition, vol. 31, no. 5, pp. 983–1001, 1998
2. Zhang, D., Lu, G., "Review of shape representation and description techniques. Pattern Recognition," vol. 37, no. 1, pp. 1-19, 2004
3. H. Kim, J. Kim, "Region-based shape descriptor invariant to rotation, scale and translation," Signal Processing: Image Communication, vol. 16, pp. 87–93, 2000
4. C.C. Chang, S.M. Hwang, D.J. Buehrer, "A shape recognition scheme based on relative distances of feature points from the centroid," Pattern Recognition, vol. 24, pp. 1053–1063, 1991
5. T. Ozugur, Y. Denizhan, E. Panayirci, "Feature extraction in shape recognition using segmentation of the boundary curve," Pattern Recognition Letters, vol. 18, pp. 1049–1056, 1997
6. KL Tan, BC Ooi and LF Thiang, "Indexing Shapes in Image Databases Using the Centriod-Radii Model," Data and Knowledge Engineering, vol. 32, pp. 271-289, 2000
7. KL Tan, BC Ooi and LF Thiang, "Retrieving similar shapes effectively and efficiently, Multimedia Tools and Applications," vol. 19, no. 2, pp. 111–134, 2003
8. T. Bernier, JA Landry, "A new method for representing and matching shapes of natural objects," Pattern Recognition, vol. 36, pp. 1711-1723, 2003.
9. Shuang Fan, "Shape Representation and Retrieval Using Distance Histograms," Technical Report TR 01-14,department of computer science, University of Alberta, Edmonton, Alberta, Canada, October 2001.
10. Dalong Li, Steven Simske, "Shape Retrieval Based on Distance Ratio Distribution," Technical Report. HPL-2002-251
11. Ballard Dana H, "Generalizing the Hough transform to detect arbitrary shapes," Pattern Recognition, vol. 13, no. 2, pp. 11—122, 1981
12. T. Sebastian, P. N. Klein, and B. Kimia, "Recognition of shapes by editing their shock graphs," IEEE Transactions on Pattern Analysis and Machine Intelligence, vol. 26, no. 5, pp. 550—571, 2004
13. E. Milios, E. Petrakis, "Shape Retrieval Based on Dynamic Programming," IEEE Transactions on Image Processing, vol. 9, no. 1, pp. 141- 146, 2000.
14. S. Belongie, J. Malik, and J. Puzicha, "Shape Matching and Object Recognition Using Shape Contexts," IEEE Transactions on Pattern Analysis and Machine Intelligence, vol. 24, no. 4, pp. 509-522, 2002.

Intelligent Control of Mobile Agent Based on Fuzzy Neural Network in Intelligent Robotic Space

TaeSeok Jin[1], HongChul Kim[2], and JangMyung Lee[2]

[1] Dept. of Mechatronics Eng., DongSeo University, Busan, 617-716, Korea
jints@dongseo.ac.kr
[2] Dept. of Electronics Engineering, Pusan National University
{hit_pnu, jmlee}@pusan.ac.kr

Abstract. This paper introduces Fuzzy Neural Network controller to increase the ability of a mobile robot in reacting to the dynamic environments. States of robot and environment, for examples, the distance between the mobile robot and obstacles and the velocity of mobile robot, are used as the inputs of fuzzy logic controller. The navigation strategy is based on the combination of fuzzy rules tuned for both goal-approach and obstacle-avoidance. To identify the environments, a sensor fusion technique is introduced, where the sensory data of ultrasonic sensors and a vision sensor are fused into the identification process. Preliminary experiment and results are shown to demonstrate the merit of the introduced navigation control algorithm.

1 Introduction

The Intelligent Robotic Space(IRS) propagates mobile robots in the space, which act in the space in order to change the state of the space. These mobile robots are called mobile agents. Mobile Agents cooperating with each other and with the core of the IRS to realize intelligent services to inhabitants. Mobile robots become more intelligent through interaction with the IRS. Moreover, robots can understand the requests (e.g. gestures) from people, so that the robots and the space can support people effectively. The Intelligent Robotic Space can physically and mentally support people using robot and VR technologies; thereby providing satisfaction for people. These functions will be an indispensable technology in the coming intelligence consumption society (Fig. 1). An autonomous mobile robot is intelligent robot that performs a given work with sensors by identifying the surrounded environment and reacts on the state of condition by itself instead of human. Unlike general manipulator in a fixed working environment [1],[2] it is required intelligent processing in a flexible and variable working environment. Recently studies on a fuzzy-neural-network control are attractive in the field of autonomous mobile robot. The fuzzy-neural-network control is suitable for adopting sensor fusion techniques in order to increasing the ability of the mobile robot to react to dynamic environment as well as ensuring collision avoidance with both moving and non-moving obstacles [3],[4]. Generally, fuzzy inference approaches tend to de-emphasize the goal-directed navigation and focus

A. Sattar and B.H. Kang (Eds.): AI 2006, LNAI 4304, pp. 668–677, 2006.

© Springer-Verlag Berlin Heidelberg 2006

more upon handling reactive and reflexive situations. The results of the fuzzy inference controller generally do not tend towards the optimal path [5]. However, unexpected and rapidly moving obstacles are avoided safely than other methods [6],[7].

This paper is organized as follows. This section introduces the concept of Intelligent Robotic Space. Section 2 introduces the mobile agent and human behavior model, which is used to understand human behavior directly from observation. Section 3 shows experiment for observation of human behavior using the distributed sensory intelligence of the Intelligent Robotic Space and control the mobile agent using the acquired behavior.

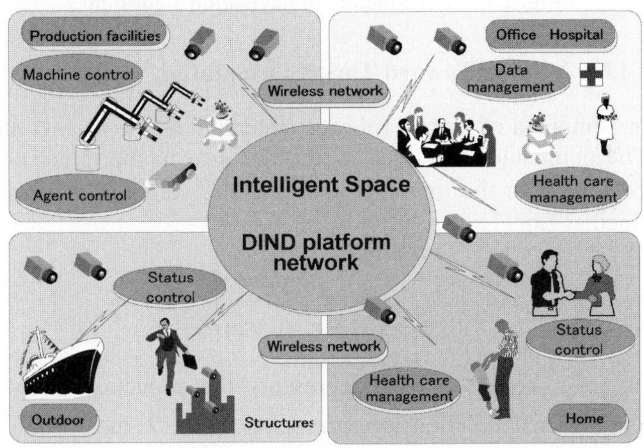

Fig. 1. Vision of IRS, as a human support system for more comfortable life

2   Fuzzy Neural Network Controller Design

2.1   Structure of Navigation Algorithm

The proposed fuzzy neural network controller is shown as Fig. 1. We define three major navigation goals, i.e., target orientation, obstacle avoidance and rotation movement; represent each goal as a cost function. Note that the fusion process has a structure of forming a cost function by combining several cost functions using weights.

In this fusion process, we infer each weight of command by the fuzzy algorithm that is a typical artificial intelligent scheme. With the proposed method, the mobile robot navigates intelligently by varying the weights depending on the environment, and selects a final command to keep the minimum variation of orientation and velocity according to the cost function. In the type of fuzzy neural networks based on BP, neurons are organized into a number of layers and the signals flow in one direction. There are no interactions and feedback loops among the neurons of same layer.

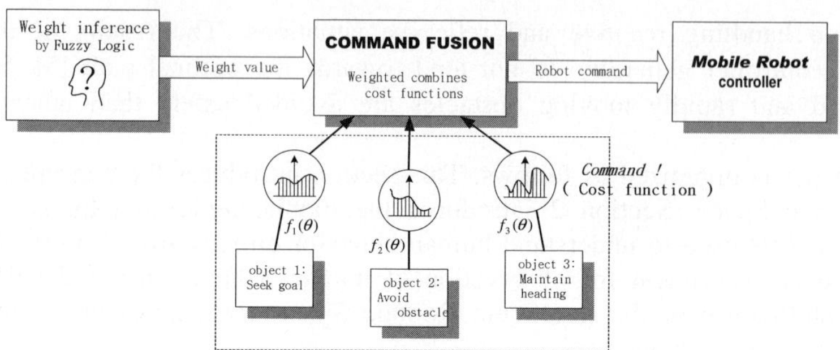

Fig. 2. Overall structure of navigation algorithm

2.2 Command for Moving Toward Target Orientation

The orientation command of mobile robot is generated as the nearest direction to the target point. The command is defined as the distance to the target point when the robot moves present with the orientation, θ and the velocity, v. Therefore, a cost function is defined as Eq. (1).

$$
\begin{aligned}
E_d(\theta) = &\{x_d - x_c + v \cdot \Delta t \cdot \cos\theta)\}^2 \\
&+ \{y_d - (y_c + v \cdot \Delta t \cdot \sin\theta)\}^2
\end{aligned}
\tag{1}
$$

where, v is $v_{\max} - k \cdot |\theta_c - \theta|$ and k represents the reduction ratio of rotational movement. Also the cost function is represented as the values in the most surface when the robot moves with velocity, and orientation, θ, as shown in Fig. 3.

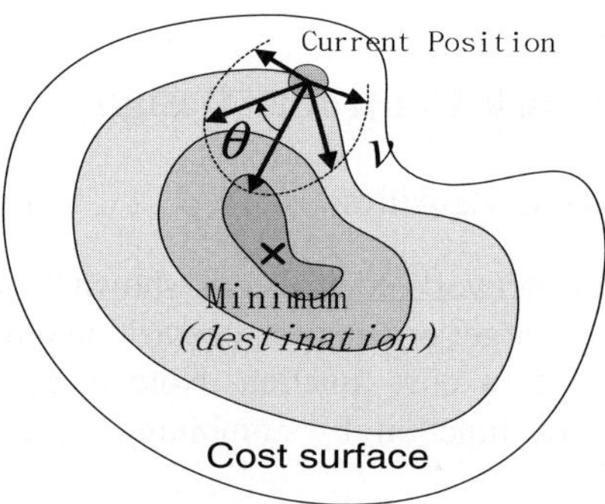

Fig. 3. Function of θ, v in cost surface

2.3 Command for Avoiding Obstacle

We use the method of representing the cost function for obstacle-avoidance as the shortest distance to an obstacle based upon the sensor data in the form of histogram. The distance information is represented as a form of second order energy, and represented as a cost function by inspecting it about all θ as shown in Eq. (2).

$$E_0(\theta) = d_{sensor}^2(\theta) \tag{2}$$

To navigate in a dynamic environment to the goal, the mobile robot should recognize the dynamic variation and react to it. For this, the mobile robot extracts the variation of the surrounded environment by comparing the past and the present. For continuous movements of the robot, the transformation matrix of a past frame *w.r.t* the present frame should be defined clearly.

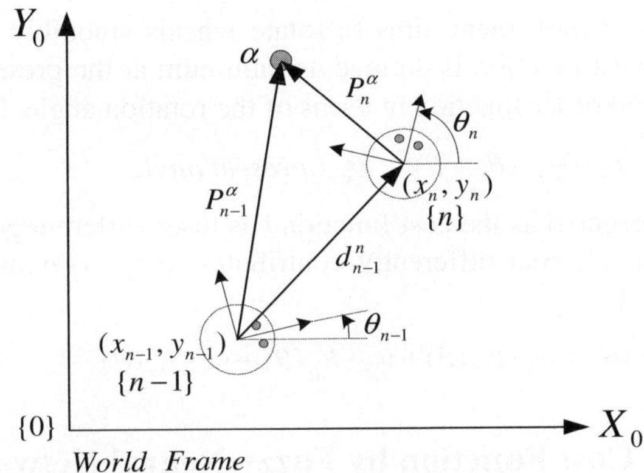

Fig. 4. Transformation of frame

In Fig. 4, a vector, P_{n-1}^α is defined as a position vector of the mobile robot *w.r.t* the $\{n-1\}$ frame and P_n^α is defined as a vector *w.r.t* the $\{n\}$ frame. Then, we obtain the relation between P_{n-1}^α and P_n^α as follow.

$$P_n^\alpha = R_{n-1}^n(P_{n-1}^\alpha - d_{n-1}^n) \tag{3}$$

Here, R_{n-1}^n is a rotation matrix from $\{n-1\}$ to $\{n\}$ frame defined as Eq. (4), and d_{n-1}^n is a translation matrix from $\{n-1\}$ frame to $\{n\}$ frame as shown in Eq. (5).

$$R_{n-1}^n = \begin{bmatrix} \cos(\Delta\theta) & \sin(\Delta\theta) \\ -\sin(\Delta\theta) & \cos(\Delta\theta) \end{bmatrix} \tag{4}$$
$$where \ \ \Delta\theta = \theta_n - \theta_{n-1}$$

$$d_{n-1}^n = \begin{bmatrix} \cos\theta_{n-1} & \sin\theta_{n-1} \\ -\sin\theta_{n-1} & \cos\theta_{n-1} \end{bmatrix} \begin{bmatrix} x_n - x_{n-1} \\ y_n - y_{n-1} \end{bmatrix} \tag{5}$$

According to the Eq. (3), the environment information measured in the $\{n-1\}$ frame can be represented *w.r.t* the $\{n\}$ frame. Thus, if W_{n-1}, and W_n are the environment information in the polar coordinates measured in $\{n-1\}$ and $\{n\}$ frames, respectively, we can represent W_{n-1} *w.r.t* the {n} frame, and extract the moving object by the Eq. (6) in the {n} frame.

$$movement = {}^nW_{n-1} \cdot ({}^nW_{n-1} - W_n) \tag{6}$$

where ${}^nW_{n-1}$ represents W_{n-1} transformed into the {n} frame.

2.4 Command for Minimizing Rotational Movements

Minimizing rotational movement aims to rotate wheels smoothly by restraining the rapid motion. The cost function is defined as minimum at the present orientation and is defined as a second order function in terms of the rotation angle, θ as Eq. (7).

$$E_r(\theta) = (\theta_c - \theta)^2 \quad \theta_c : present\ angle \tag{7}$$

The command represented as the cost function has three different goals to be satisfied at the same time. Each goal differently contributes to the command by a different weight, as shown in Eq. (8).

$$E(\theta) = w_1 \cdot E_d(\theta) + w_2 \cdot E_o(\theta) + w_3 \cdot E_r(\theta) \tag{8}$$

3 Inference of Cost Function by Fuzzy Neural Network

3.1 Structure of Control System

An interesting architecture for a fuzzy neural controller has been proposed by Jang [10] and our controller is based on this idea. Fuzzy neural controllers can be regarded as fuzzy controllers implemented using a neural network, often in the form of a modified radial basis function network. By this means the learning algorithms developed for neural networks become available for the training of the fuzzy controller.

We infer the weights of Eq. (8) by means of fuzzy algorithm. The main reason of using fuzzy neural network algorithm is that it is easy to reflect the human's intelligence into the robot control. Fuzzy inference system is developed through the process

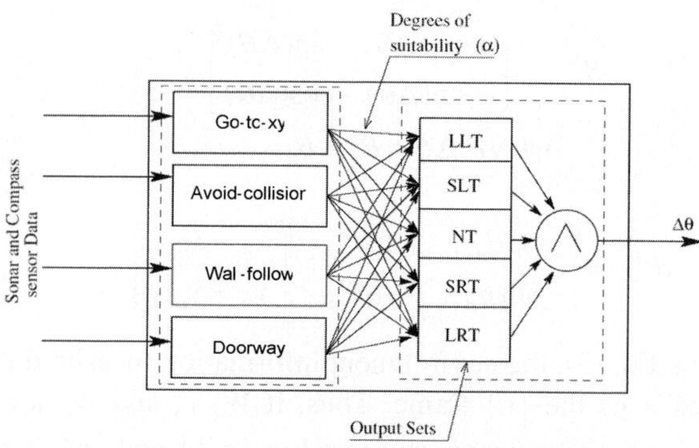

Fig. 5. Structure of fuzzy neural network control system

of setting each situation, developing fuzzy-neural with proper weights, and calculating weights for the commands. Fig. 5 shows the structure of a fuzzy-neural inference system. We define the circumstance and state of a mobile robot as the inputs of fuzzy-neural inference system, and infer the weights of cost functions. The inferred weights determine a cost function to direct the robot and decide the velocity of rotation. For the control of the mobile robot, the results are transformed into the joint angular velocities by the inverse kinematics of the robot.

3.2 Behavior Approximation with Fuzzy Neural Network

Fuzzy-Neural Netwok(FNN) is applied in the behavior approximation framework in order to handle the non-linear mapping of target tracking, obstacle avoidance. The FNN is class of adaptive network that is functionally equivalent to fuzzy inference systems. Takagi-Sugeno fuzzy inference system (TS-FIS) is an effort to develop a systematic approach to generating fuzzy rules from a given input-output data set [7]. A typical fuzzy rule of the TS-FIS model:

$$R = IF(x_1 \text{ is } A_{i1}) AND \ldots AND(x_j \text{ is } A_{ij}) AND \ldots AND(x_n \text{ is } A_{in})$$
$$THEN(y_i = w_{i0} + w_{i1}x_1 + \cdots + w_{ij}x_j + \cdots + w_{in}x_n) \tag{9}$$

where R_i denotes the ith fuzzy rule, $(i = 1 \ldots r)$, r is the number of fuzzy rules, $\vec{x}$ is the input vector, $\vec{x} = [x_1 \ldots, x_j \ldots, x_n]^T$, $A_{i,j}$ denotes the antecedent fuzzy sets, $(j = 1 \ldots n)$, y_i is the output of the ith linear subsystem, and w_{ij} are its parameters, $(l = 0 \ldots n)$. The nonlinear system of FNN forms a collection of loosely coupled multiple linear models. The degree of firing of each rule is proportional to the level of contribution of the corresponding linear model to the overall output of the TS-FIS model. For Gaussian-like antecedent fuzzy set, the degree of membership is

$$\vec{\mu}_{ij} = \vec{e}^{-\left\| x_j - x_{ij}^* \right\|^2}, \tag{10}$$

where x_j is the jth input, x_{ij}^* denotes the center of A_{ij} membership function, $\alpha = 4/r^2$ and r is positive constant, which defines the spread of the antecedent and the zone of the influence of the ith model (radius of the neighborhood of a data point); too large value of r leads to averaging. The firing level of rules are defined as Cartesian product or conjunction of respective fuzzy sets for this rule, The output of the TS-FIS model is calculated by the weighted averaging of individual rules' contributions,

$$y = \sum_{i=1}^{r} \lambda_i y_i = \sum_{i=1}^{r} \lambda_i x_e^T \pi_i, \tag{11}$$

where $\lambda_i = (\tau_i / \sum_{j=1}^{r} \tau_j)$ is the normalized firing level of the ith value, y_i represents the output of the i_{th} linear model, $\pi_i = [w_{i0}, w_{i1}, ..., w_{ij}, ..., w_{in}]^T$ is the vector of parameters of the ith linear model, and $x_e = [1 \quad x^T]^T$ is the expanded data vector.

4 Experimentation

4.1 Mobile Agent

This proposed navigation method is applied for a mobile robot named as Pioneer-DX that has been developed in Laboratory for Intelligent Robot, DSU as shown in Fig. 6. We use a DC motor for each wheel, and use a ball-caster for an assistant wheel. Two encoders, a gyro-sensor (ENV-05D), an ultrasonic sensor and a vision sensor are used for the navigation control. The gyro sensor is used for recognizing the orientation of robot by measuring the rotational velocity; the ultrasonic sensor (Polaroid 6500) is used for recognizing environment, which is rotated by a step motor within 180 degrees; the CCD camera (Samsung SFA-410ED) is used for detecting obstacles. A Pentium 4, 2.45Ghz processor is used as a main controller and an 80C196KC microprocessor is used as a joint controller.

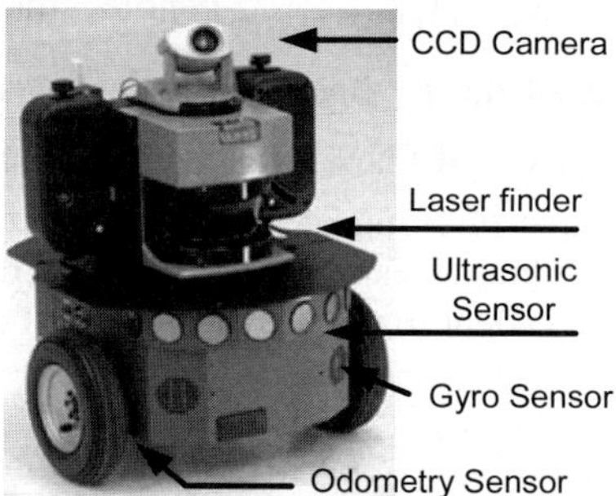

Fig. 6. Mobile agent, Pioneer-DX

4.2 Experiment Results

Ultrasonic sensor is good in distance measurement of the obstacles, but it also suffers from specular reflection and insufficient directional resolution due to its wide beam–opening-angle. So, we use a sensor fusion method to decide the distance and width of obstacles and avoid them during the navigation. Mobile agent examines whether measured value is data of distance to real obstacle or distance to its shadow. If difference of measured data by vision and ultrasonic sensor is within the error tolerance, mobile agent uses measured data by vision sensor as distance to obstacle. Otherwise, mobile agent uses measured data by vision sensor as distance to obstacle.

Fig. 7(a) depicts sensing coverage of vision and ultrasonic sensor used this experiment. Ultrasonic sensor can detect obstacles within 7m and Vision system can detect obstacles within the range of between 130cm and 870cm. Fig. 7(b) shows the mapped relation between CCD image and real obstacle. Eq. 9 and 10 are relation equations between distance to obstacle from mobile robot on real environment and pixel coordination about each direction x and y on CCD image.

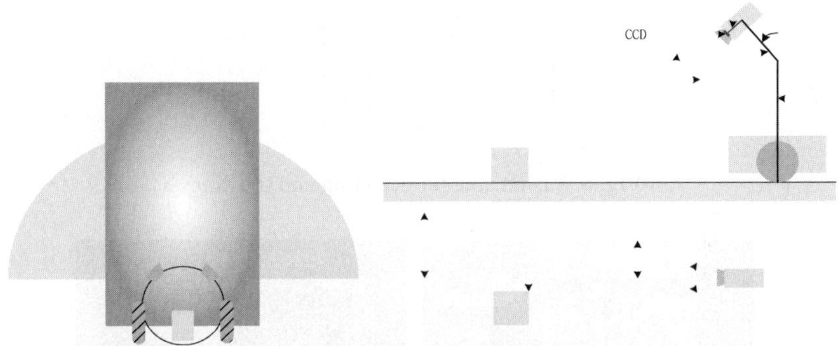

Fig. 7. Sensing coverage and mapping between CCD image and real obstacle

where, p_x, p_y are maximum values of each x and y coordination on CCD image frame. Parameter values used for experiment are shown in Table 1.

Table 1. Parameter values used for experiment

h_1	39cm	h_2	7.5cm
h_3	4cm	α	15°
θ_{rx}	27°	θ_{ry}	20°
p_x	320 pixel	p_y	240 pixel

We experiment the recognition of obstacle using above equations and parameter values as shown in Fig. 8. Fig. 9(a) is the image used on the experiment; Fig. 9(b) is the values resulted from matching after image processing. Fig. 9. shows that maximum matching error is within 4%. Therefore, it can be seen that above vision system is proper to apply to navigation. The mobile robot navigates along a corridor with 2m width and with some obstacles as shown in Fig. 10(a). The real trace of the mobile robot is shown in Fig. 10(b). It demonstrates that the mobile robot avoids the obstacles intelligently and follows the corridor to the goal.

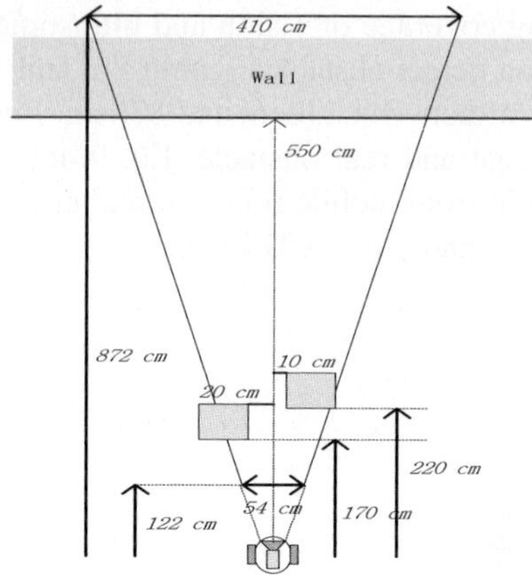

Fig. 8. Vision area for detecting and tracking

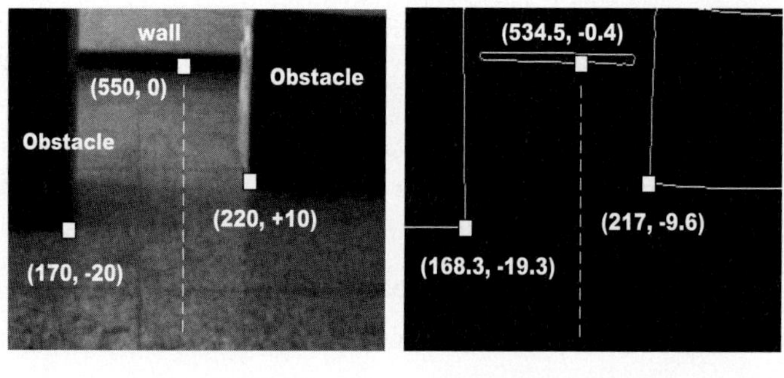

(a) Input image   (b) Result of matching

Fig. 9. Experimental result of the vision system

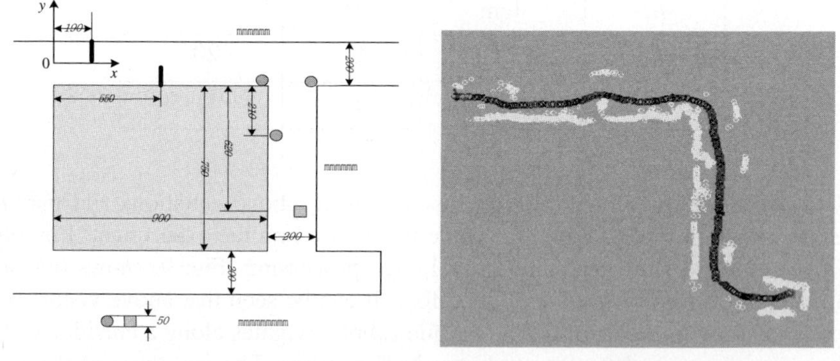

(a) Diagram of robot in a corridor   (b) Composed world map

Fig. 10. Navigation of the mobile robot in a corridor

5   Conclusions

A fuzzy-neural-network control algorithm for both obstacle avoidance and path planning is proposed so that it enables the mobile robot to reach to target point in the Intelligent Robotic Space safely and autonomously. To show the efficiency of proposed method, real experiments are performed. The experimental results show that the mobile agent, Pioneer-DX can navigate to the target point safely under unknown environments and also can avoid moving obstacles autonomously. Obstacle avoidance and target tracing is approximated with Fuzzy-Neural Networks. Note that mobile agent is not able to detect moving obstacles that are faster than the mobile agent. Also, it is difficult to estimate motion-vectors of obstacles that are navigating fast and irregularly.

Further researches on the prediction algorithm of the obstacles and on the robustness of performance are required. Also, we involve improving the detecting accuracy for the mobile robot and applying this system to complex environments where many people, mobile robots and obstacles coexist. Moreover, it is necessary to survey the influence of the mobile agent which maintains a flexible distance between the robot and the human, and introduces the knowledge of cognitive science and social science.

Acknowledgements

This work was supported by the 2006 Basic Research Promotion Grant funded by the Dongseo University.

References

1. Morioka, K., Lee, J.H., Hashimoto, H.: Human-Following Mobile Robot in a Distributed Intelligent Sensor Network. IEEE Trans. on Industrial Electronics, Vol.51, No.1 (2004) 229-237.
2. Appenzeller, G., Lee, J.H., Hashimoto, H.: Building Topological Maps by Looking at People: An Example of Cooperation between Intelligent Spaces and Robots. IEEE/RSJ Int. Conf. on Intelligent Robots and Systems, (1997) 1326-1333.
3. Home Page of Inter Process Communication Package at: http://www.ri.cmu.edu/ projects/project_394.html
4. Home Page of Player/Stage Project at: http://playerstage.sourceforge.net/
5. McFarland, D.: What it means for a robot behavior to be adaptive. In. J.-A. Meyer and S.W. Wilson (Eds.), From animals to animates: Proc. of the First Int. Conf. on Simulation of Adaptive Behavior, Cambridge, MA, MIT Press (1990) 22-28.
6. Kanayama, Y.: Two Dimensional Wheeled Vehicle Kinematics. Proc. of IEEE Conf. on Robotics and Automation, San-Diego, California (1994) 3079-3084.
7. Takagi, T., Sugeno, M.: Fuzzy Identification System and its Application to Modeling and Control. IEEE Trans. on Systems, Man and Cybernetics, 15, (1985) 116-132.
8. Jang, J.-S. R.: ANFIS: Adaptive-Network-based Fuzzy Interference Systems. IEEE Trans. on Systems, Man, and Cybernetics, 23(03): (1993) 665-685.

The rule table is encoded in a chromosome as shown in Fig. 3. The output bits are listed in lexicographical order of neighborhood. Interestingly, even though the radius is minimum possible for cell-cell interaction to take place, the rather large number of states per cell has made the search space of possible transition rules phenomenal (2^{1536}). For this reason we did not consider increasing the radius.

Table 1. A sample rule table for $(k,r) = (8,1)$

Index	Neighborhood			Output
0	000	000	000	010
1	000	000	001	100
2	000	000	010	000
⋮				
511	111	111	111	101

O	1	O	1	O	O	O	O	O	…	1	O	1
0		1		2							511	

Fig. 3. Representation of the rule table in a chromosome

2.2 Fitness Function

The fittest of each chromosome is evaluated by first decoding the rule table it represents. The rule is then applied to a simulated model of the brittle star robot whose initial position, (x_i,y_i) and final position, (x_f,y_f) in the desired direction of motion is recorded. Since the focus of this paper is on forward locomotion of the robot, the fitness of the chromosome is thus proportional to the straight-line distance covered by the robot.

$$F = \sqrt{(x_f - x_i)^2 + (y_f - y_i)^2} \tag{3}$$

2.3 Genetic Operators

Based on the fitness of the chromosomes in the population, GA operations; namely selection, crossover and mutation are applied to whole population. Firstly selection was performed using roulette wheel selection method, which offered fitter individuals better chance of mating. The selected pairs of chromosomes are crossed over using one-point crossover at randomly chosen locus with a crossover probability of P_C. Finally, mutation operation involving bit flips is applied with a probability of P_M to individual genes in chromosomes after the selection and crossover operations.

2.4 Simulation Results

The simulation was carried out over numerous trials using parameters shown in Table 2. In each trial, initial population of chromosomes of size (POP_SIZE) was randomly generated and GA was executed for a number of generations (MAX_GEN). For each generation, the fitness of all the chromosomes in the population is evaluated after which genetic operators are applied to the population to create the next generation.

In the fitness evaluation of each chromosome, the CA lattice structure representing all legs and its constituent modules in robot is initialized randomly with a fixed seed. The transition rule encoded in the chromosome being evaluated is applied for a number of successive iterations (SIM_STEPS) to transform the cell lattice. In each iteration the object model in the ODE environment is updated accordingly. The final position of the robot at the end of SIM_STEPS iteration is used in fitness calculation above (3).

It is worth mentioning that for comparability of results, the same seed for the random number generator is used for the initialization of all the lattice configurations in all the models discussed in this paper.

Table 2. Summary of simulation parameters

Parameter	Value
P_C	0.85
P_M	0.15
POP_SIZE	25
SIM_STEPS	1500
MAX_GEN	250

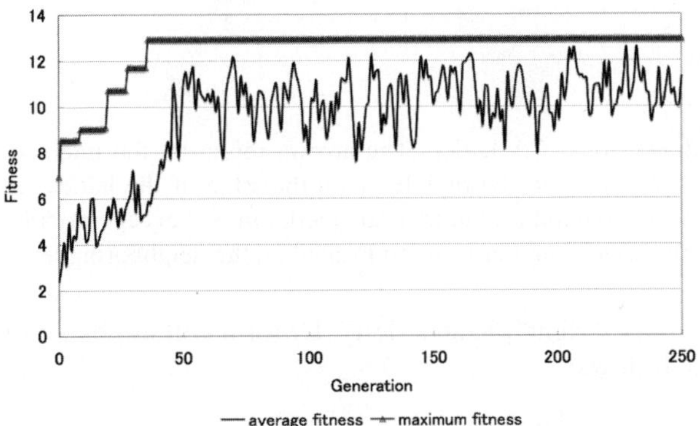

Fig. 4. Graph of average and maximum fitness of population against generation for the disconnected one-dimensional CA model of the robot

The evolution of transition rules across generations is captured in Fig. 4. Notably, in most of the trials there was premature convergence of GA. From the observations of the motion of simulated robot in ODE it became apparent that there was lack of coordination amongst the legs and thus the overall mobility of the robot in terms of maximizing the fitness function is affected. In summary the results of modeling the robot as set of disconnected one-dimensional cellular lattice have provided a valuable insight about the need for greater coordination between these one-dimensional structures. Thus our next step would be to connect the lattice into a two-dimensional structure.

3 Singular Transition Rule for Connected Leg Set

In this section the shortcomings of the model described in the previous section is addressed. In particular the set of one-dimensional disconnected CA lattice representing the legs are connected via modules closest to the central disc of the robot (Fig. 5), thus forming a two-dimensional structure. The notion of single transition rule for all the modules is adopted for reasons described in §2. Furthermore the radius describing the neighborhood of a cell is set to 1 (same as §2).

Fig. 5. The two-dimensional CA lattice. The arrows represent the interaction between a cell (possibly located somewhere in the middle or on the edge of the lattice) and its neighboring cells. Essentially the interaction is similar to the model in § 2 except modules closest to central disc interact with modules in similar position located on the neighboring leg.

The two neighbors ("right", R and "left", L) for a cell at position (i,j) in the n x m lattice is given as follows:

$$R = \begin{cases} S^t_{(i+1)j} & \text{if } j=0 \text{ and } i+1 < n \\ S^t_{(i+1-n)j} & \text{if } j=0 \text{ and } i+1 \geq n \\ S^t_{i(j+1-m)} & \text{if } j>0 \text{ and } j+1 \geq m \\ S^t_{i(j+1)} & \text{otherwise} \end{cases} \tag{4}$$

$$L = \begin{cases} S^t_{(i-1)j} & \text{if } j=0 \text{ and } i-1 \geq 0 \\ S^t_{(i-1+n)j} & \text{if } j=0 \text{ and } i-1 < 0 \\ S^t_{i(j-1)} & \text{otherwise} \end{cases} \tag{5}$$

3.1 Simulation Results

Using the same procedures and parameters described in §2, the simulations were carried out using the revised model. The results obtained (Fig. 6) indicated that performance of the robot worsened compared to the previous model, which used disconnected one-dimensional lattice. Closer examination of the motion pattern of the simulated robot revealed that the current two-dimensional CA lattice does introduce a degree of coordination between the legs. However since all the legs are following the same rule, eventually all the legs converge to the same formation (Fig. 7) thus being counterproductive. While the motivation behind using a singular transition rule for all the modules was to reinforce modularity of the robot, the results on the other hand provide strong justification for using differential transition rule.

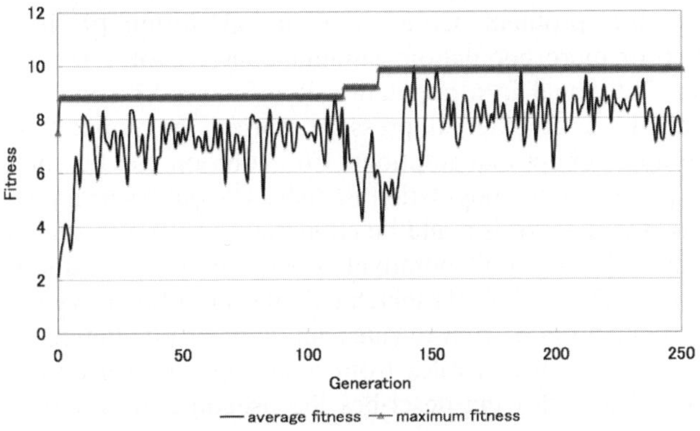

Fig. 6. Graph of average and maximum fitness of population against generation for the two-dimensional CA model of the robot

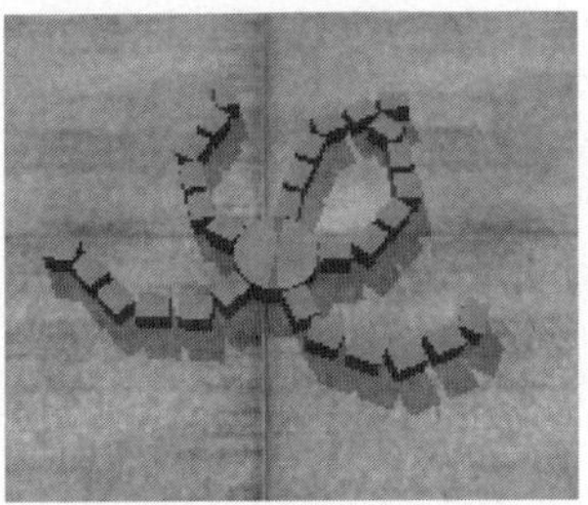

Fig. 7. Snapshot of the simulation for the robot using single transition rule for all the modules represented by a two-dimensional CA lattice

4 Differential Transition Rule for Connected Leg Set

Building on the earlier models, in this section we present the final model, which represents the robot using a two-dimensional CA lattice as in § 3, however there are two distinct transition rules for updating the cell states in the lattice. These rules are termed as control rule, *CR* and leg rule, *LR*. The position of a cell (i,j) in the lattice determines which of the two rules would be used to update its state and is given as follows.

$$Rule = \begin{cases} CR & if\ j = 0 \\ LR & otherwise \end{cases} \tag{6}$$

Essentially modules closest to the central disc are covered by control rule and modules located on other parts of the leg are taken care of by the leg rule.

4.1 Co-evolving the Rules

In the previous sections using GA to search for good transition rules often resulted in premature convergence to local optima. In this section we use co-evolutionary algorithm to remedy this problem. Co-evolutionary algorithm [9] is a computational model where two or more populations simultaneously evolve in a manner such that the fitness of an individual in a population is influenced by individuals from other populations. This kind of external pressure, forces the populations to adaptively evolve thus avoiding suboptimal solutions. Co-evolution can be competitive in a nature such as in [10] where the population includes opponents trying to beat each other in a game of Tic-Tac-Toe, or it could be cooperative in nature such as in [11] where populations of match strings collaboratively contribute individual string to form a set of binary vectors which is to closely match a target set of binary vectors.

In our case a population of control rules and another population of leg rules is co-evolved cooperatively such that rules from both population need to be combined to control the robot. The following describes the pseudo code for the co-evolutionary algorithm.

Create random population of control rules $\mathbf{P}_{CR}$ and leg rules $\mathbf{P}_{LR}$
LOOP UNTIL MAX_GENERATION
    Get best control rule *//Rule with the highest fitness; for initial population*
             //best rule is randomly chosen
    LOOP UNTIL MAX_COEVOL_GENERATION
        //Evaluate fitness of individual leg rules
        LOOP FOR all individuals in $\mathbf{P}_{LR}$
            • Transform lattice model using best control rule and current leg rule for SIM_STEPS
            • Get final robot position
            • Calculate fitness, $F = \sqrt{(x_f - x_i)^2 + (y_f - y_i)^2}$
        END LOOP
        Perform GA operations on $\mathbf{P}_{LR}$
    END LOOP

    Get best leg rule
    LOOP UNTIL MAX_COEVOL_GENERATION
        //Evaluate fitness of individual control rules
        LOOP FOR all individuals in $\mathbf{P}_{CR}$

- Transform lattice model using best leg rule and current control rule for SIM_STEPS
- Get final robot position
- Calculate fitness, $F = \sqrt{(x_f - x_i)^2 + (y_f - y_i)^2}$

        END LOOP

        Perform GA operations on $\mathbf{P}_{CR}$

    END LOOP

END LOOP

4.2 Simulation Results

The simulations were carried out again using the same parameters as in §2.4. The results show marked improvement in the fitness of the best rules discovered (Fig. 8). Noticeably the average fitness of the populations is less compared to simulation results from other models (Fig. 4&6). The reason for this is that the interaction between the populations within the scope of the co-evolutionary algorithm creates greater diversity thus requiring greater adaptation on the part of individuals in the populations. Consequently the stagnation of the search algorithm at local optima that was prevalent in the previous simulations was not encountered. Finally observations using the best rule discovered in controlling the movement of the simulated robot in ODE environment showed far greater coordination and fluidity of motion than any of the other models.

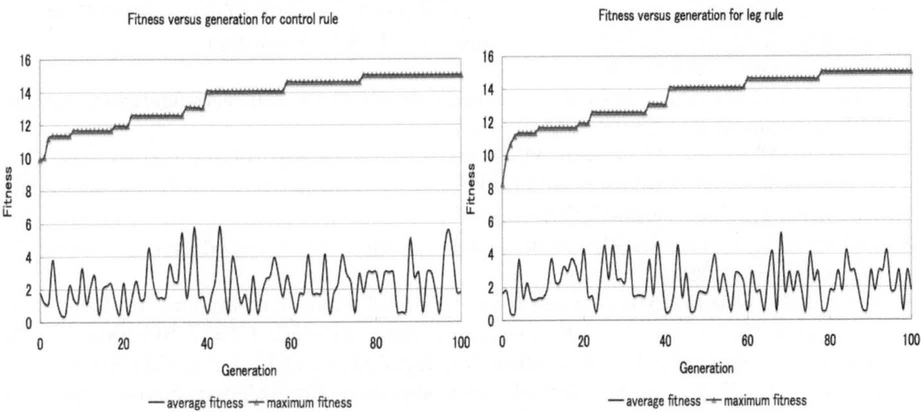

Fig. 8. Graph of fitness against generation for the control and leg rule

5 Conclusion

In this paper we demonstrated the feasibility of using cellular automata based architecture for control of a modular robotic system. Out of the three control models considered the model, which used differential transition rule on two-dimensional CA lattice representation of the robot, proved to be the most effective in controlling the motion of robot in a straight path.

Needless to say there are still much room for improvements. One of the crucial areas that need further research to be carried out deals with the fact that the rules obtained were tightly coupled with the initial state of the lattice configuration. Naturally the performance of the rules and the robot as a whole would get affected should the initial configuration be changed. To address this, in our future work we plan on using reinforcement learning as a means to train the robot to appropriately select rules based on any given class of initial configuration so as to accomplish some user-specified high-level goal.

References

1. Takashi, M.: Studies on Forward Motion of Modular Robot. MSc Dissertation, University of Ryukyu (2005)
2. Futenma, N., Yamada, K., Endo, S., Miyamoto, T.: Acquisition of Forward Locomotion in Modular Robot. Proceeding of the Artificial Neural Network in Engineering Conference, St. Louis, U.S.A (2005) 91-95
3. Crutchfield, J. P., Mitchell, M., Das, R.: The Evolutionary Design of Collective Computation in Cellular Automata. In: Crutchfield J. P., Schuster P.K. (eds): Evolutionary Dynamics-Exploring the Interplay of Selection, Neutrality, Accident, and Function. Oxford University Press, New York (2003) 361-411
4. Weisstein, E. W.: Life. Mathworld-A Wolfarm Web Resource. http://mathworld.wolfram.com/ Life.html
5. Butler, Z., Kotay, K., Rus, D., Tomita, K.: Cellular Automata for Decentralized Control of Self-Reconfigurable Robots. Proceedings of ICRA Workshop on Modular Self-Reconfigurable Robots (2001)
6. Stoy, K.: Using Cellular Automata and Gradients to Control Self-reconfiguration. Robotics and Autonomous Systems, Vol. 54 (2006) 135-141
7. Behring, C., Bracho, M., Castro, M., Moreno, J. A.: An Algorithm for Robot Path Planning with Cellular Automata. ACRI-2000: Theoretical and Practical Issues on Cellular Automata. Springer-Verlag (2000) 11-19
8. Goldberg, D. E.: Genetic Algorithms in Search, Optimization and Machine Learning. Addison-Wesley (1989)
9. Koza, J. R.: Genetic Evolution and Co-evolution of Computer Programs. In: Langton, C. G., Taylor, C., Farmer, J. D., Rasmussen, S. (eds): Artificial Life II, SFI Studies in the Science of Complexity, Vol. X. Addison-Wesley, Redwood City, CA (1991) 603-629
10. Rosin, C. D., Belew, R. K.: Methods for Competitive Co-evolution: Finding Opponents Worth Beating. In: Proc. 6th International conference on Genetic Algorithms. San Francisco (1995) 373-380
11. Potter, M. A., Jong K. A. D.: Cooperative Coevolution: An Architecture for Evolving Co-adapted Subcomponents. Evolutionary Computation, Vol. 8(1). MIT Press (2000) 1-29

A Novel Evolutionary Algorithm for Multi-constrained Path Selection

Qi Xiaogang[1], Liu Lifang[2], and Liu Sanyang[1]

[1] Department of Mathematics Science, Xidian University, Xi'an 710071, P.R. China
[2] School of computer, Xidian University, Xi'an 710071, P.R. China

Abstract. For the problem of multi-constrained path selection, a novel evolutionary algorithm named MCP_EA is proposed. Firstly, a novel coding technology named PNNC (Preceding Natural Number Coding, PNNC) is designed, and no circle exists on the path coded by PNNC. Secondly, a novel crossover operator called DCC operator (Dispersing Connection Crossover operator, DCC operator) is designed to guarantee the validity of the crossed paths and the diversity of the population. Thirdly, a novel mutation operator named selective mutation operator is proposed. Finally, the theoretical analysis proves that the algorithm converges to the satisfactory solution with probability 1.0. Extensive simulations show that the novel evolutionary algorithm outperforms the H_MCOP in performance for the problem, and is a promising algorithm for the problem with high performance.

1 Introduction

With the rapid development of the multimedia communication services, the problem of providing different quality-of-services (QoS) guarantees for different application is a challenging issue, in which QoS Routing (QoSR) is one of the most pivotal problems[1,2]. The objective of QoSR is to find a feasible path that satisfies multiple constraints for QoS applications, so we focus on the problem of multi-constrained path selection, as an NPC problem,which is also a big challenge for the upcoming next generation networks.In this paper,we propose MCP_EA, a novel evolutionary algorithm for multi-constrained path selection.

To omit the circle checking process,we propose a novel Preceding Natural Number Coding (PNNC) technology, and no circle does exist on the path coded by it. For the efficiency of crossover operation and the diversity of the population, we design a novel Dispersing Connection Crossover (DCC) operator, which can avoid producing a wrong path. Furthermore, we design a selective mutation operator based on the local section modification, which can increase the fitness of the optimal individual of each generation population, and finally meet the path satisfying the multiple constraints.

2 Problem Formulation

A network is usually represented by a directed graph $G = (V, E)$, where V is the set of nodes and E is the edges. Each a node $v_i \in V$ represents a router,

A. Sattar and B.H. Kang (Eds.): AI 2006, LNAI 4304, pp. 699–708, 2006.

© Springer-Verlag Berlin Heidelberg 2006

and a edge $e_{ij} \in E$ represents a link from v_i to v_j, and $e_{ij} = (v_i, v_j)$. In the graph G, each edge has k independent weights $(w_1(e), w_2(e), \cdots, w_k(e))$, which is called QoS metric $w(e)$, and $w_l(e) > 0$ for any $1 \leq l \leq k$. For a path $P_j = (v_j^0, v_j^1, \cdots, v_j^L)$, the weight $w_l(e)$ satisfies the additive property if $w_l(p) = \sum_{i=1}^{L} w_l(e_j^i)$, and $e_j^i = (v_j^{i-1}, v_j^i)$; the weight $w_l(e)$ satisfies the concave property if $w_l(p) = Min_{i=1}^{L} w_l(e_j^i)$; the weight $w_l(e)$ satisfies the multiplicative property if $w_l(p) = \prod_{i=1}^{L} w_l(e_j^i)$. Thus the k independent weights $(w_1(p_j), w_2(p_j), \cdots, w_k(p_j))$ can be computed according to the property of the each constraint, and let $w(p_j) = (w_1(p_j), w_2(p_j), \cdots, w_k(p_j))$.

So the problem of multi-constrained path selection seeks to find a path p from s to t if

$$w_l(p) \leq c_l(s, t)(1 \leq l \leq k). \tag{1}$$

where $c_l(s, t)$ is the lth element of the k-dimension constraint vector $c(s, t)$. we write $w(p) \leq c(s, t)$ in brief.

3   MCP_EA

3.1   Coding

The coding is one of important problems of solving the problem of multi-cons-trained path selection using evolutionary algorithm. In MCP_EA, we first pro-posed a novel PNNC, by which the length of each path coded is $|V|$, and no circle exists on it. For the path $p_j = (v_j^0, v_j^1, \cdots, v_j^L)$, we can describe the p_j as the preceding connection relation

$$v_j^0 \leftarrow v_j^1 \leftarrow \cdots \leftarrow v_j^L \tag{2}$$

where the start node is v_j^L, the end one is v_j^0, v_j^{i-1} is the preceding node of v_j^i for any $1 \leq i \leq L$, and $v_j^L = t$, $v_j^0 = s$.

The PNNC can be implemented by the following steps:

STEP1 For each path, we first construct an sequence whose length equals to $|V|$, and on which the value of each node equals to the serial number of the corresponding node.

STEP2 According to the preceding connection relation, the value of each node is replaced by the serial number of its preceding node.

For example, there exists a network and there is a path $p_1 = (1, 4, 8, 5, 9, 2)$ whose preceding connection relation is $1 \leftarrow 4 \leftarrow 8 \leftarrow 5 \leftarrow 9 \leftarrow 2$. The p_1 coded by PNNC is 0193186745 as shown in Fig. 1.

3.2   Selection of Initial Population

In the MCP_EA, let the number of population be N, the N/k optimal paths can be computed by the K-shortest loopless paths algorithm [3,4] for each con-straint weight respectively, then initial population $pop(0)$ can be constructed,

The serial number of node	0	1	2	3	4	5	6	7	8	9
The preceding node	pre[0]= 0	pre[1]= 1	pre[2]= 9	pre[3]= 3	pre[4]= 1	pre[5]= 8	pre[6]= 6	pre[7]= 7	pre[8]= 4	pre[9]= 5

Source node   Detination node

Fig. 1. A path coded by PNNC

and accordingly any individual in $pop(0)$ has some kind of native advantage in the later evolution.

3.3 Fitness Function

"Bucket principle" tells us that the capacity of a bucket is determined by the shortest wood block. In the problem, the feasibility of a path depends on the worst one among the k constraints. we design a fitness function

$$f(p) = Min_{l=1}^{k} r_l(p) \tag{3}$$

where

$$r_l(p) = \begin{cases} [c_l(s,t) - w_l(p)]/c_l(s,t) & \text{if } c_l(s,t) \text{ is constraint's upper limit,} \\ [w_l(p) - c_l(s,t)]/w_l(p) & \text{if } c_l(s,t) \text{ is constraint's lower limit.} \end{cases} \tag{4}$$

where $w_l(p)$ and the $c_l(s,t)$ are the the lth element of path weight vector $w(p)$ and constraint vector $c(s,t)$ respectively, and $r_l(p) \leq 1$.

For the $c_l(s,t)$, if it is the constraint's upper limit, the path p satisfies the relation

$$\begin{aligned} w_l(p) \leq c_l(s,t) &\Longleftrightarrow w_l(p) \geq 0; \\ w_l(p) > c_l(s,t) &\Longleftrightarrow w_l(p) < 0. \end{aligned} \tag{5}$$

For the $c_l(s,t)$, if it is the constraint's lower limit, the path p satisfies the relation

$$\begin{aligned} w_l(p) \geq c_l(s,t) &\Longleftrightarrow w_l(p) \geq 0; \\ w_l(p) < c_l(s,t) &\Longleftrightarrow w_l(p) < 0. \end{aligned} \tag{6}$$

Based on (5) and (6), if there exists a path p satisfying

$$r_l(p) \geq 0(1 \leq l \leq k), \tag{7}$$

the p is a satisfying solution for the problem, and we can deduce

$$f(p) \geq 0. \tag{8}$$

In our algorithm, we aim to improve the fitness of the optimal individual in the population of each iteration step by step, if there exist a path p satisfying the inequation (8), the algorithm will stop, and the path is what we required.

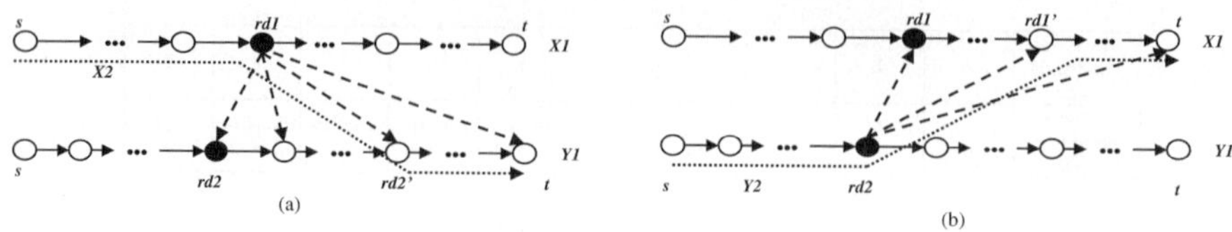

Fig. 2. DCC operator

3.4 Crossover

Considering the efficiency of the crossover operation, we design a DCC operator as shown in Fig.2.

The DCC operator is described as follows:

STEP1 Select a node on the parent $X1$ and $Y1$ excluding s and t, which is denoted by $rd1$ and $rd2$ respectively.

STEP2 Set the label of $rd1$ and nodes after $rd1$ on $X1$ be 1, the label of the others on $X1$ be 0;Set the label of $rd2$ and nodes after $rd2$ on $Y1$ be 1, the label of the others on $Y1$ be 0.

STEP3 Let $RD1$ denote the set of nodes whose label is 1 on $X1$;Let $RD2$ denote the set of nodes whose label is 1 on $Y1$.

STEP4 Find a section from $rd1$ to any node in $RD2$ with the minimal hops by the shortest path algorithm[5], if the added node is an element in $RD2$,the added one is denoted by $rd2'$, the section is denoted by p_{xy}, and then stop the calculating. In this way, another section p_{yx} can be calculated, the added node in $RD1$ is denoted by $rd1'$. The detailed operation is described in Fig.2.

STEP5 The offsprings are denoted by $X2$ and $Y2$ respectively, and

$$X2 = X1(s, rd1)//p_{xy}//Y1(rd2', t). \tag{9}$$

$$Y2 = Y1(s, rd2)//p_{yx}//X1(rd1', t). \tag{10}$$

STEP6 End.

The DDC operator has the following advantages:

(1) The DDC operator can rise the diversity of the population, and avoid converging to the local optimal solution.

(2) The DDC operator can shorten the running time of the MCP_EA.

As shown in Fig.3, there exist 2 parent individuals $X1$ and $Y1$,if we insist on connecting $rd1$ and $rd2$ but there is no edge $(rd1,rd2)$, a path p from $rd1$ to $rd2$ should be selected with the minimal hops, thus some nodes in $RD2$ may be selected. Suppose a node $rd2'$ in $RD2$ has been selected, a circle $Y1(rd2, rd2')//p(rd2', rd2)$ will result. However the circle has to be cut, so there are some unnecessary calculations that should be reduced. In our DDC operator, we have omit the redundant calculations by adopting the dispersing connection.

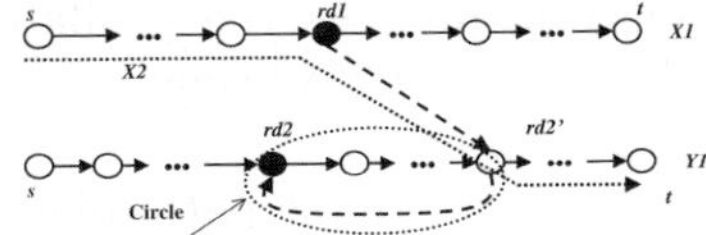

Fig. 3. The DDC can omit the unnecessary calculations

3.5   Mutation

Considering the fitness function, we design a selective mutation operator, which can improve the fitness of the path by selecting and enhancing the worst element of path's weight vector.The selective mutation operator is shown in Fig.4.

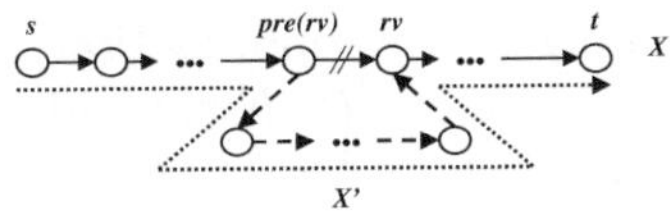

Fig. 4. The selective mutation can omit the unnecessary calculations

The selective mutation operator is described as follows:

STEP1 Randomly select a mutation node rv on the individual X.
STEP2 Delete the edge $(pre(rv), rv)$ in the network.
STEP3 According to the $w_i(X)$ and $c_l(s,t)$, the $r_l(X)$ can be calculated.
STEP4 Find out the minimal one among the $r_l(X)$ for any $1 \leq l \leq k$.
STEP5 Select the path from the $pre(rv)$ to rv based on the $w_l(e)$ by the SPT algorithm[5].
STEP6 Connect the sections $X(s, pre(rv))$,$p(pre(rv), rv)$, thus $X(rv, t)$, and there will result a new path X'.
STEP7 End.

3.6   Reproduction

Reproduction is mainly the operation to increase the number of good solution in the population in every iteration of the algorithm. In this paper, we adopt the tournament scheme, and let $k = 2$.In this way, the better individual replace the inferior one, and the best individual will be hold in the next generation.

3.7   The Proposed Algorithm

Based on the above operators, we describe the novel MCP_EA as follows.

(1) Set the parameters of the MCP_EA. The parameters include N, pr_c, pr_m, s, t and D, where N denotes the number of the population, pr_c denotes the

probability of the crossover, pr_m denotes the probability of the mutation, s denotes the source node, t denotes the destination node and D denotes the maximum of the iteration generation.

(2) Initialization. Based on the K-shortest paths algorithm[3,4], N individuals are selected from the initial population $pop(0)$. At the same time, let the current generation $t = 0$.

(3) Crossover. Provided that the crossover probability is pr_c, $N \cdot pr_c$ parent individuals can be randomly selected from current population $pop(t)$, and $N \cdot pr_c$ offsprings will be produced according to the DDC operator. The set of the offsprings is denoted by $sx(t)$.

(4) Mutation. Provided that the mutation probability is pr_m, $pr_m(1 + pr_c) \cdot N$ individuals will be randomly selected from the set $pop(t) \cup sx(t)$, and $pr_m(1 + pr_c) \cdot N$ new individuals will produced. The set of the new individuals is denoted by $sm(t)$.

(5) Reproduction. According to the reproduction operator and the fitness of each individual in $pop(t) \cup sx(t) \cup sm(t)$, N individuals can be selected as the next generation population $pop(t + 1)$. Let $t = t + 1$.

(6) If the $t = D$ or $f(p) \geq 0$, goto (7), otherwise goto (3), where D is the maximum of iteration generation, p is a required individual(path).

(7) End.

Theorem 1. *No circle exists on the path coded by the PNNC.*

Proof. Assume that $P_j = (v_j^0, v_j^1, \cdots v_j^{L-1}, v_j^L)$ is a path whose length is L, where $v_j^i (1 \leq i \leq L)$ is the ith node on P_j, s, t is the source and the destination node respectively, and $v_j^0 = s$, $v_j^L = t$. The preceding connection relation of P_j is $v_j^0 \leftarrow v_j^1 \leftarrow \cdots \leftarrow v_j^{L-1} \leftarrow v_j^L$. If there exists a circle on P_j, one of nodes will appear $m(m \geq 2)$ times, its preceding connection relation has to be set m times according to the PNNC, and its preceding connection relation is determined at the last time, which can be expressed as Figure 5.

Fig. 5. The preceding connection relation of P_j

In Figure 5, provided the node v_j^u, v_j^i, v_j^d is one identical node, which appears m times on P_j. The node v_j^d appears firstly, v_j^u appears lastly, and $pre(v_j^d) = v_j^{d-1}$, $pre(v_j^i) = v_j^{i-1}$, $pre(v_j^u) = v_j^{u-1}$, $1 \leq \cdots \leq u \leq \cdots \leq i \leq \cdots \leq d \leq \cdots \leq L$. Finally, its preceding connection relation is determined as v_j^{u-1}, and the sector $v_j^u \cdots \leftarrow v_j^{i-1} \leftarrow v_j^i \cdots \leftarrow v_j^{d-1}$ will disabled. The preceding connection relation of each node is uniquely determined, and each node on the path appears one and only one time, so there no circle exists on the path coded by the PNNC. ∎

Theorem 2. *No circle exists on the offsprings(path) after the crossover and mutation operations.*

Proof. According to Fig.2, $X2 = X1(s, rd1)//p_{xy}//Y1(rd2', d)$,
$Y2 = Y1(s, rd2//p_{yx}//X1(rd1', t))$, and $X2$ and $Y2$ are offsprings of $X1$ and $Y1$. According to Fig.3, path $X' = X(s, rv)//p(rv, pre(rv))//X(pre(rv), t)$ is mutated from X.

For $X2$, the sections $X1(s, rd1)$, p_{xy}, and $Y1(rd2', d)$ can be connected as $s \leftarrow \cdots \leftarrow rd1 \leftarrow \cdots rd2' \leftarrow \cdots \leftarrow t$. we suppose that there exists a circle on it, the circle will be deleted according to **Theorem 1**.

Similarly, no circle exists on $Y2$ and X'.   ¶

Theorem 3. *If there exists a path satisfying the multiple constraints, the proposed algorithm will convergence to it with probability 1.0.*

Proof. For the problem of multi-constrained path selection, if there exists a path p satisfying all constraints, the fitness of p satisfying $f(p) \geq 0$, and the optimal solution $f(p^*) = Max f(p)$, and we can deduce $f(p^*) \geq 0$.

According to **theorem 2.6** in reference [7], if the crossover probability $pr_c \in [0, 1]$, the mutation probability $pr_m \in (0, 1)$, and the optimal solution can be selected as an individual in the next population, the algorithm will convergence to the optimal solution. In the proposed algorithm, the crossover probability and the mutation probability satisfy the assumption, the $K = 2$ tournament selection is employed to select the excellent individuals as the members in the next generation population, so the algorithm will convergence to the global optimal solution p_* with probability 1.0^1.   ¶

We analyze the computation complexity of the MCP_EA as follows.

(1) For **Initialization**, its computation complexity is $N \cdot |V|^3/2$;
(2) For **Crossover**, its computation complexity is $3|V|^2 - |V|$;
(3) For **Mutation**, its computation complexity is $(3|V|^2 - |V|)/2$;
(4) For **Fitness computation**, its computation complexity is $k \cdot (1+pr_c)(1+ pr_m) \cdot |V| \cdot N$;
(5) For **Reproduction**, its computation complexity is $(1 + pr_c)(1+pr_m) \cdot N$;
So the computation complexity of the MCP_EA is $N \cdot |V|^3/2+\{[(1 + pr_c) \cdot pr_m + pr_c] \cdot N \cdot (3|V|^2 - |V|)/2+ (k \cdot |V| + 1)(1 + pr_c)(1 + pr_m) \cdot N\} * D$.

4   Performance Evaluation and Simulation

For the MCP_EA, we evaluate it in two ways, one is the success ratio(SR) of path selection, another is its CPU time for solving the problem under the different scale networks.

---

[1] Note that the proposed algorithm is with high computational complexity for the problem of multi-constrained path selection, so we aim to find a path with the positive fitness within the limited iteration times.

4.1   Under Accurate State Information

Similarly to the simulations by Turgay Korkmaz[8]and by Yong Cui[9], we first generate graphs[10],then associate two randomly generated weights with each link, as shown in Table 1. These weights are selected from uniform distributions under three types of correlation between them. The source and destination of a request are randomly generated such that the minimum hop-count between them is at least two.The constraints are also randomly generated, but their ranges are determined based on the best paths with respect to w_1 and w_2 as follows. Let p_1 and p_2 be two optimal paths with respect to w_1 and w_2 respectively. We take $c_1 \sim uniform[0.8*w_1(p_2), 1.2*w_1(p_2)]$ and $c_2 \sim uniform[0.8*w_2(p_1), 1.2*w_2(p_1)]$.

Table 1. Ranges of link weights and the correlation between them

Correlation	$w_1(e)$	$w_2(e)$
Positive correlation	$uniform[1, 50]$	$uniform[1, 100]$
No correlation	$uniform[1, 100]$	$uniform[1, 200]$
Negative correlation	$uniform[1, 50]$	$uniform[100, 200]$

We contrast the performance of MCP_EA and the other algorithms using the success ratio, which refers to the fraction of connection requests for which feasible paths are found by the given algorithm.In each run, 10 random graphs are generated. For each instance of a random graph with given link weights, 1000 connection requests are generated for graphs with 50,100, and 200 nodes respectively. The Success ratio of each algorithm is listed in table 2.

Table 2. The Success ratio of each algorithm

Correlation	SR	H_MCOP	MEFPA	MCP_EA
Positive correlation	$N = 50$	85.7%	83.9%	87.0%
	$N = 100$	91.6%	89.4%	92.2%
	$N = 200$	92.2%	90.1%	93.1%
No correlation	$N = 50$	86.2%	84.1%	87.3%
	$N = 100$	91.2%	89.0%	92.1%
	$N = 200$	91.7%	89.7%	92.4%
Negative correlation	$N = 50$	73.6%	70.2%	74.9%
	$N = 100$	79.9%	75.2%	80.8%
	$N = 200$	77.9%	73.8%	79.0%

4.2   Under Inaccurate State Information

In this experiment, we adopt the Relative Threshold Trigger (RTT). For some different threshold values, the performance of H_MCOP,MEFPA, and MCP_EA is compared. Suppose that $w_{l\_new}(e)$ is the updated value, and the $w_{l\_cur}w(e)$ is the practical value at the moment, if the $|w_{l\_new}(e) - w_{l\_cur}w(e)|/w_{l\_new}w(e) > th$, the $w_{l\_new}(e)$ will be updated, and th is the value of the threshold. At the same time, we assume that the $w_{l\_cur}w(e)$ is uniformly distributed over the range $w_{l\_new}(e)(1 - th), w_{l\_new}(e)(1 + th)$. The experiment results show that the performance of MCP_EA is higher than that of H_MCOP and MEFPA, and their performance under inaccurate state information is shown in Figure 10-18.

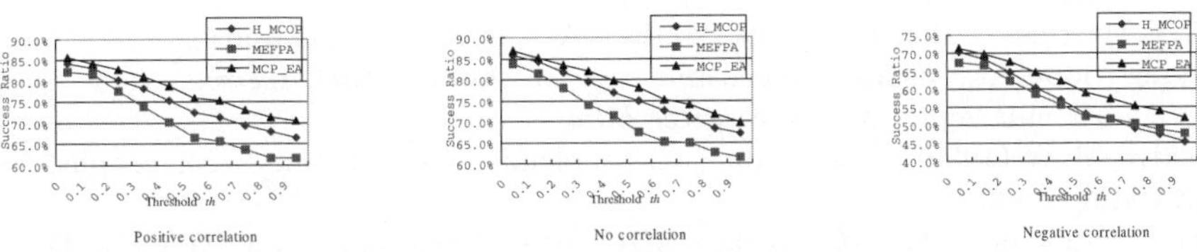

Fig. 6. Performance comparison under inaccurate state information, $N = 50$

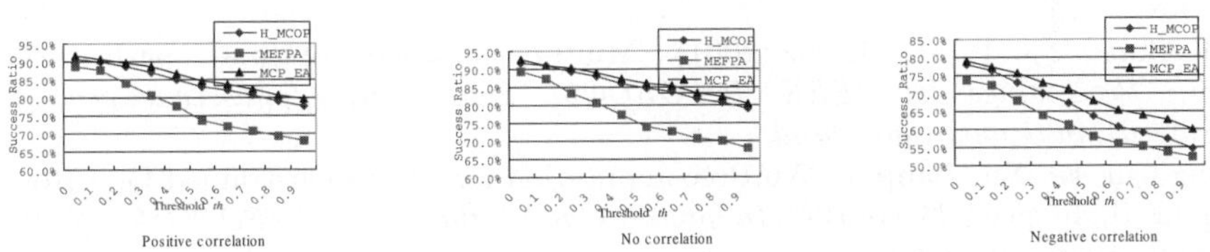

Fig. 7. Performance comparison under inaccurate state information, $N = 100$

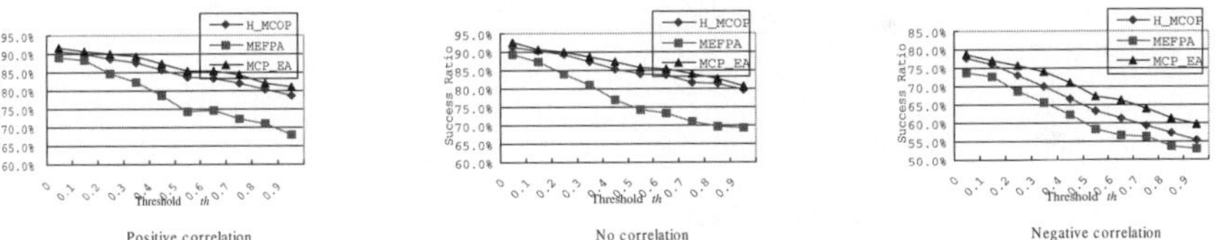

Fig. 8. Performance comparison under inaccurate state information, $N = 200$

5   Conclusion

For the problem of multi-constrained path selection, a novel evolutionary algorithm named MCP_EA is proposed.In the algorithm, a preceding coding technology is employed to avoid existing a circle on a path, and thus the circle check is omitted in the crossover and mutation operators. Based on the bucket principle, a novel fitness function is designed, if there is a path whose fitness is greater than zero, the path is a solution of the problem. In addition, a new crossover and mutation operators are presented respectively.Finally, theoretical analysis proves that the algorithm can convergence to the global optima with probability 1.0, computer simulation results also show that the algorithm performs better than the other algorithms for the problem.

References

1. CUI Yong, WU Jianping.(2002) Research on Internetwork QoS Routing Algorithms: a Survey. *Journal of Software*, 13(11),2065-2075.
2. ZHU Huiling, Kang Daming.(1994) Quality of service routing:problems and solutions. *ACTA ELECTRONICA SINICA*, 31(1), 1-7.
3. J.Y.Yen.(1971) Finding the k-th shortest paths in a network. *Management Science*, 17, 712-716.
4. E.Q.V.Martins, M.M.B.pascoal.(2001) A new implementation of Yen's ranking loopless paths algorithm. *Optimization 2001,Aveiro, July 2001. http://www.mat.uc.pt/ marta/research.html.*
5. Dijsktra E W.(1959) A note on two problems in connection with graphs *Numer.Math*, 1, 269-271.
6. CHEN Guoliang, WANG Xufa, ZHUANG Zhenquan, and WANG Dongsheng.(1996) Genetical Algorithm and its application. *People's Post and Telecommunications Press, 1996. (in Chinese).*
7. J.A.Bandy and U.S.R. Murty.(1976) Graph theory with applications. The macmillan press Ltd.
8. T.Korkmaz and M. krunz.(2001) Multi-constrained optimal path selection. *Proceedings of IEEE INFORCOM 2001. http://citeseer.ist.psu.edu/ kormaz01multiconstrained.html.*
9. Yong Cui, Ke Xu, Jianping Wu.Precomputation for Multi-constrained QoS Routing in High-speed Networks. *In proceedings of the IEEE INFOCOM 2003[C]. http://netlab.cs.tsinghua.edu.cn/ cuiy/.*
10. B. Waxman.(1988) Routing of multipoint connections. *IEEE JSAC*, 6(9), 1617-1622.

ODE: A Fast and Robust Differential Evolution Based on Orthogonal Design

Wenyin Gong[1], Zhihua Cai[1,2], and Charles X. Ling[2]

[1] School of Computer Science
China University of Geosciences, Wuhan 430074, P.R. China
`cug11100304@yahoo.com.cn, zcai6@uwo.ca`
[2] Dept. of computer science
The University of Western Ontario London, Ontario, N6A 5B7, Canada

Abstract. In searching for optimal solutions, Differential Evolution (DE), a type of genetic algorithms can find an optimal solution satisfying all the constraints. However, DE has been shown to have certain weaknesses, such as slow convergence,the accuracy of solutions are not high. In this paper, we propose an improved differential evolution based on orthogonal design, and we call it ODE (Orthogonal Differential Evolution). ODE makes DE faster and more robust. It uses a novel and robust crossover based on orthogonal design and generates an optimal offspring by a statistical optimal method. A new selection strategy is applied to decrease the number of generations and make the algorithm converge faster. We evaluate ODE to solve twelve benchmark function optimization problems with a large number of local minimal. Simulations results show that ODE is able to find the near-optimal solutions in all cases. Compared to other state-of-the-art evolutionary algorithms, ODE performs significantly better in terms of the quality, speed, and stability of the final solutions.

1 Introduction

In the last two decades, evolutionary algorithms (including genetic algorithms, evolution strategies, evolutionary programming, and genetic programming) have received much attention regarding their potential as global optimization techniques [1]. Inspired from the mechanisms of natural evolution and survival of the fittest, evolutionary algorithms (EAs) utilize a collective learning process of a population of individuals. Descendants of individuals are generated using randomized operations such as mutation and recombination. Mutation corresponds to an erroneous self-replication of individuals, while recombination exchanges information between two or more existing individuals. According to a fitness measure, the selection process favors better individuals to reproduce more often than those that are relatively worse.

Differential evolution (DE) [2] is an improved version of Genetic Algorithm (GA) for faster optimization, which has won the third place at the 1st International Contest on Evolutionary Computation on a real-valued function test-suite. DE is the best genetic algorithm approach. Unlike simple GA that uses binary coding for representing problem parameters, DE is a simple yet powerful population based, direct search algorithm with the generation-and-test feature for globally optimizing functions using real valued parameters. The DE's advantages are its simple structure, ease of use, speed and robustness.

A. Sattar and B.H. Kang (Eds.): AI 2006, LNAI 4304, pp. 709–718, 2006.

© Springer-Verlag Berlin Heidelberg 2006

Price & Storn [2] gave the working principle of DE with single strategy. Later on, they suggested ten different strategies of DE [10]. A strategy that works out to be the best for a given problem may not work well when applied for a different problem. What's more, the strategy and key parameters to be adopted for a problem are to be determined by trial & error. The main difficulty with the DE technique, however, appears to lie in the slowing down of convergence as the region of global minimum is approaching.

Recently, some researchers applied the orthogonal design method incorporated with EAs to solve optimization problems. Leung and Wang [5] incorporated orthogonal design in genetic algorithm for numerical optimization problems found such method was more robust and statistically sound. This method was also adopted by other researchers [6], [7] and [8] to solve optimization problems. Numerical results demonstrated that these techniques had a significantly better performance than the traditional EAs on the problems studied, and the resulting algorithm can be more robust and statistically sound.to find an optimal solution satisfying all the comstraints. However, DE has been shown to have certain weaknesses, such as slow convergence,the accuracy of solutions are not high.

In this paper, an improved version of the differential evolution based on the orthogonal design (ODE) is presented to make the DE faster and more robust. Orthogonal design method is nested in crossover operator with the statistical optimal method to select better genes as offspring, and consequently, enhances the performance of the CDE(conventional DE). By combining with two crossover operators, faster convergence and better solutions are obtained. Moreover, the ODE can remedy the main defect of the DE technique mentioned above with the orthogonal crossover operator. The advantages of ODE are its simplicity, efficiency, and flexibility. It is shown empirically that ODE has high performance in solving benchmark functions comprising many parameters, as compared with some existing EAs.

The rest of this paper is organized as follows. Section 2 briefly describes function optimization problem and some properties of the orthogonal design method. Section 3 presents the proposed ODE. In Section 4, we test our algorithm use twelve benchmark functions . This is followed by discussions and analysis of the optimization experiments for the ODE in Section 5. The last section, Section 6, is devoted to conclusions and future studies.

2   Preliminary

2.1   Problem Definition

A global minimization problem can be formalized as a pair (S, f) , where $S \subseteq R^n$ is a bounded set on R^n and $f : S \rightarrow R$ is an n-dimensional real-valued function. The problem is to find a point $X_{min} \in S$ such that $f(X_{min})$ is a global minimum on S. More specifically, it is required to find an $X_{min} \in S$ such that

$$\forall X \in S : f(X_{min}) \leq f(X) \tag{1}$$

where f does not need to be continuous but it must be bounded

$$l_i \leq x_i \leq u_i, i = 1, 2, \cdots, n \tag{2}$$

2.2 Orthogonal Design

Orthogonal design method [9] with both orthogonal array (OA) and factor analysis (such as the statistical optimal method) is developed to sample a small, but representative set of combinations for experimentation to obtain good combinations. OA is a fractional factorial array of numbers arranged in rows and columns, where each row represents the levels of factors in each combination, and each column represents a specific factor that can be changed from each combination. It can assure a balanced comparison of levels of any factor. The term "main effect" designates the effect on response variables that one can trace to a design parameter. The array is called orthogonal because all columns can be evaluated independently of one another, and the main effect of one factor does not bother the estimation of the main effect of another factor.

In a discrete single objective optimization problem, when there are N factors (variables) and each factor has Q levels, the search space consists of Q^N combinations of levels. When N and Q are large, it may not be possible to do all Q^N experiments to obtain optimal solutions. Therefore, it is desirable to sample a small, but representative set of combinations for experimentation, and based on the sample, the optimal may be estimated. The orthogonal design was developed for the purpose [9]. The selected combinations are scattered uniformly over the space of all possible combinations Q^N. And the orthogonal design is an important tool for robust design. Robust design is an engineering methodology for optimizing the product and process conditions which are minimally sensitive to the causes of variation, and which produce high-quality products with low development and manufacturing costs.

In general, the orthogonal array $L_M(Q^N)$ has the following properties.

1. For the factor in any column, every level occurs equal times.
2. For the two factors in any two columns, every combination of two levels occurs equal times to represent the experiments.
3. Any two experiments are not the same, so their results cannot be compared directly.
4. If any two columns of an orthogonal array are swapped, the resulting array is still an orthogonal array.
5. If some columns are taken away from an orthogonal array, the resulting array is still an orthogonal array with a smaller number of factors.

3 ODE: Orthogonal DE

In order to get rapid convergence as the region of global minimum, and make DE easy to manipulate and get better solutions, we modify the DE. The enhancements of the ODE are as follows:

3.1 Orthogonal Crossover Operator

3.1.1 Design of the Orthogonal Array

To design a minimal orthogonal array, in this research, we use the two level orthogonal array $L_{2^J}(2^{2^J-1})$, $R = 2^J$ denotes the number of the rows of orthogonal array and $C = 2^J - 1$ denotes the number of the columns. The orthogonal array needs to find a proper J to satisfy

Halting precision: $\varepsilon = 1 \times 10^{-30}$. This halting criterion used is to make the algorithm stop earlier when the results satisfy the precision of the problems.

To evaluate the efficiency of the proposed ODE approach, we also test the functions in CDE, which does not use the orthogonal design, with DE/rand/1/exp strategy. And for all experiments in CDE, we used the parameters mentioned above only expect for the number of fitness function evaluations, which is different for different problems.

4.2 Benchmark Functions

In order to evaluate the performance of the proposed method we used twelve benchmark functions ($f_{01} - f_{12}$) . All of the functions are **minimization** problems. Where functions $f_{01} - f_{07}$ are $f_{01} - f_{07}$ from Ref. [5]; functions $f_{08} - f_{10}$ are f_{09}, f_{12}, f_{13} from Ref. [5]; and functions f_{11}, f_{12} are f_{22}, f_{23} from Ref. [4]. For functions $f_{01} - f_{06}$ and f_{09}, f_{10}, the dimension of variables is 30; for functions f_{07}, f_{08}, the dimension of variables is 100; and the dimension of variables for functions f_{11}, f_{12} is 4. Functions $f_{01} - f_{08}$ are high-dimensional and multimodal problems with many local minima. Functions f_{09}, f_{10} are high-dimensional and unimodal problems. Also function f_{09} is a noisy quartic function, where *random* [0,1) is a uniformly distributed random variable in [0,1). Functions f_{11}, f_{12} are low-dimensional and multimodal problems with few local minima; and they are difficult to find the global minimal on each run for many algorithms [4]. The test functions are described as follows.

$$f_{01} = \sum_{i=1}^{N} (-x_i \sin(\sqrt{|x_i|})) \tag{6}$$

$$f_{02} = \sum_{i=1}^{N} (x_i^2 - 10\cos(2\pi x_i) + 10) \tag{7}$$

$$f_{03} = -20\exp(-0.2\sqrt{\frac{1}{N}\sum_{i=1}^{N} x_i^2}) - \exp(\frac{1}{N}\sum_{i=1}^{N}\cos(2\pi x_i)) + 20 + \exp(1) \tag{8}$$

$$f_{04} = \frac{1}{4000}\sum_{i=1}^{N} x_i^2 - \prod_{i=1}^{N}\cos(\frac{x_i}{\sqrt{i}}) + 1 \tag{9}$$

$$f_{05} = \frac{\pi}{N}\{10\sin^2(\pi y_i) + \sum_{i=1}^{N-1} (y_i - 1)^2 \cdot [1 + 10\sin^2(\pi y_{i+1})] + (y_N - 1)^2\}$$
$$+ \sum_{i=1}^{N} u(x_i, 10, 100, 4) \tag{10}$$

where

$$y_i = 1 + \frac{1}{4}(x_i + 1) \tag{11}$$

and

$$u(x_i, a, k, m) = \begin{cases} k(x_i - a)^m, & x_i > a \\ 0, & -a \le x_i \le a \\ k(-x_i - a)^m, & x_i < a \end{cases} \tag{12}$$

$$f_{06} = \frac{1}{10}\{\sin^2(3\pi x_1) + \sum_{i=1}^{N-1} (x_i - 1)^2 \cdot [1 + 10\sin^2(3\pi x_{i+1})]$$
$$+(x_N - 1)^2 \cdot [1 + \sin^2(2\pi x_N)]\} + \sum_{i=1}^{N} u(x_i, 5, 100, 4) \tag{13}$$

$$f_{07} = -\sum_{i=1}^{N} \sin(x_i) \sin^{20}\left(\frac{i \times x_i^2}{\pi}\right) \tag{14}$$

$$f_{08} = \frac{1}{N} \sum_{i=1}^{N} (x_i^4 - 16x_t^2 + 5x_i) \tag{15}$$

$$f_{09} = \sum_{i=1}^{N} x_i^4 + random[0, 1) \tag{16}$$

$$f_{10} = \sum_{i=1}^{N} |x_i| + \prod_{i=1}^{N} |x_i| \tag{17}$$

$$f_{11,12} = -\sum_{i=1}^{m} \left[(x - a_i)(x - a_i)^T + c_i\right]^{-1} \tag{18}$$

where for function f_{11}, $m = 7$; for function f_{12}, $m = 10$. The coefficients are not described here.

4.3 Experimental Results

Table 1 shows the results compared with OGA/Q [5] and CDE. Table 2 shows the results compared with FEP [4] and CDE. Where all results have been averaged over 50 independent trails on each test function in standard C++ and recorded: 1) the mean number of fitness function evaluations (MFFE), 2) the mean function value (Mean Best), and 3) the standard deviation of the functions (Std. Dev.). Some of the convergence results of the test functions compared with ODE and CDE are shown in Fig. 1.

Table 1. Comparison with ODE, CDE and OGA/Q on functions $f_{01} - f_{10}$ over 50 independent runs. A result in **Boldface** indicates that a better result or the global optimum (or best known solution) was reached. Where "Optimal" in column 11 indicates the global optimum (or best known solution), similarly hereinafter.

F	MFFE			Mean Best			Std. Dev.			Optimal
	ODE	CDE	OGA/Q	ODE	CDE	OGA/Q	ODE	CDE	OGA/Q	
f_{01}	**47,980**	72,100	302,166	**-12569.5**	**-12569.5**	-12569.4537	**0**	**0**	$6.4*10^{-4}$	-12569.5
f_{02}	**36,430**	159,600	224,710	**0**	**0**	**0**	**0**	**0**	**0**	0
f_{03}	**57,850**	206,300	112,421	$5.9*10^{-15}$	$4.9*10^{-15}$	$\mathbf{4.4*10^{-16}}$	**0**	$1.4*10^{-15}$	$4.0*10^{-17}$	0
f_{04}	**51,970**	119,600	134,000	**0**	**0**	**0**	**0**	**0**	**0**	0
f_{05}	**100,000**	**100,000**	134,556	$\mathbf{1.4*10^{-16}}$	$5.2*10^{-15}$	$6.0*10^{-6}$	$\mathbf{7.3*10^{-17}}$	$1.6*10^{-15}$	$1.2*10^{-6}$	0
f_{06}	**99,850**	100,000	134,143	$\mathbf{4.1*10^{-19}}$	$6.6*10^{-13}$	$1.9*10^{-4}$	$\mathbf{2.8*10^{-19}}$	$3.5*10^{-13}$	$2.6*10^{-5}$	0
f_{07}	**98,570**	100,000	302,773	**-97.211**	-59.30467	-92.83	0.3	0.830304	$\mathbf{2.6*10^{-2}}$	-99.2784
f_{08}	**98,570**	100,000	245,930	**-78.33233**	-78.3254	-78.30	$\mathbf{3.0*10^{-6}}$	0.001458	$6.3*10^{-3}$	-78.3324
f_{09}	**100,000**	**100,000**	112,652	$\mathbf{6.6*10^{-4}}$	0.01647	$6.3*10^{-3}$	$\mathbf{3.8*10^{-4}}$	0.004207	$4.1*10^{-4}$	0
f_{10}	**100,000**	**100,000**	112,612	**0**	$1.7*10^{-8}$	**0**	**0**	$3.1*10^{-9}$	**0**	0

Table 2. Comparison with ODE, CDE and FEP on functions f_{11}, f_{12} over 50 independent runs. A result in **Boldface** indicates that a better result or the global optimum (or best known solution) was reached.

F	MFFE			Mean Best			Std. Dev.			Optimal
	ODE	CDE	FEP	ODE	CDE	FEP	ODE	CDE	FEP	
f_{11}	**6,640**	7,100	10,000	**-10.4029**	-10.40285	-5.52	**0**	$1.246*10^{-7}$	2.12	-10.4029
f_{12}	**7,030**	7,900	10,000	**-10.5364**	-10.53638	-6.57	**0**	$6.576*10^{-9}$	3.14	-10.5364

5   Discussions and Analysis of Results

From table 1, table 2 and Fig. 1, we can get following conclusions.

- The proposed ODE can find optimal or close-to-optimal solutions and gave the smallest MFFE for all functions compared with the OGA/Q, CDE and FEP.
- For functions $f_{01} - f_{04}$ and $f_{10} - f_{12}$, the proposed ODE can find the global optimal solutions on each run. Hence, the "Std. Dev." of these functions are zero.
- For six functions (f_{01} and $f_{05} - f_{09}$), the ODE can obtain a better "Mean Best" solutions than the OGA/Q.
- Both the ODE and OGA/Q can give the same optimal or close-to-optimal solutions in three functions (f_{02}, f_{04} and f_{10}).
- Only except for the function f_{07}, the ODE gave a worse "Std. Dev." than the OGA/Q. However, it can give smaller standard deviations of function values in six functions (f_{01}, f_{03}, f_{05}, f_{06}, f_{08} and f_{09}) than the OGA/Q and in the remaining three functions (f_{02}, f_{04} and f_{10}), both of them are 0. Hence, the proposed ODE has a more stable solution quality.
- The ODE found similar standard deviations of function values as the CDE for three functions (f_{01}, f_{02} and f_{04}), both of them are 0. And for the remaining nine functions, ODE was able to obtain smaller standard deviations of function values than the CDE. Therefore, ODE has a more stable solution quality than the CDE.
- The proposed ODE requires fewer mean numbers of function evaluations than the OGA/Q, CDE and FEP and, hence, the proposed ODE has a lower computational time requirement.
- From Fig. 1, we can see that the ODE can get faster convergence than that of the CDE. Especially, for some functions (such as f_{05}, f_{09} and f_{10}), the proposed ODE can find the close-to-optimal solutions at the beginning of the iteration. The reason is that the two level orthogonal crossover operator with the statistical optimal method can exploit the optimum offspring.
- Compared with the FEP for functions f_{11} and f_{12}, the performance of the ODE is significantly better than the FEP in terms of the MFFE, Mean Best and Std. Dev.

These results in table 1, table 2 and Fig. 1 indicate that the proposed ODE can, in general, give better mean solution quality, faster convergence speed and more stable solution quality than the OGA/Q, CDE and FEP. Thereore, ODE can be more robust and statistically sounder.

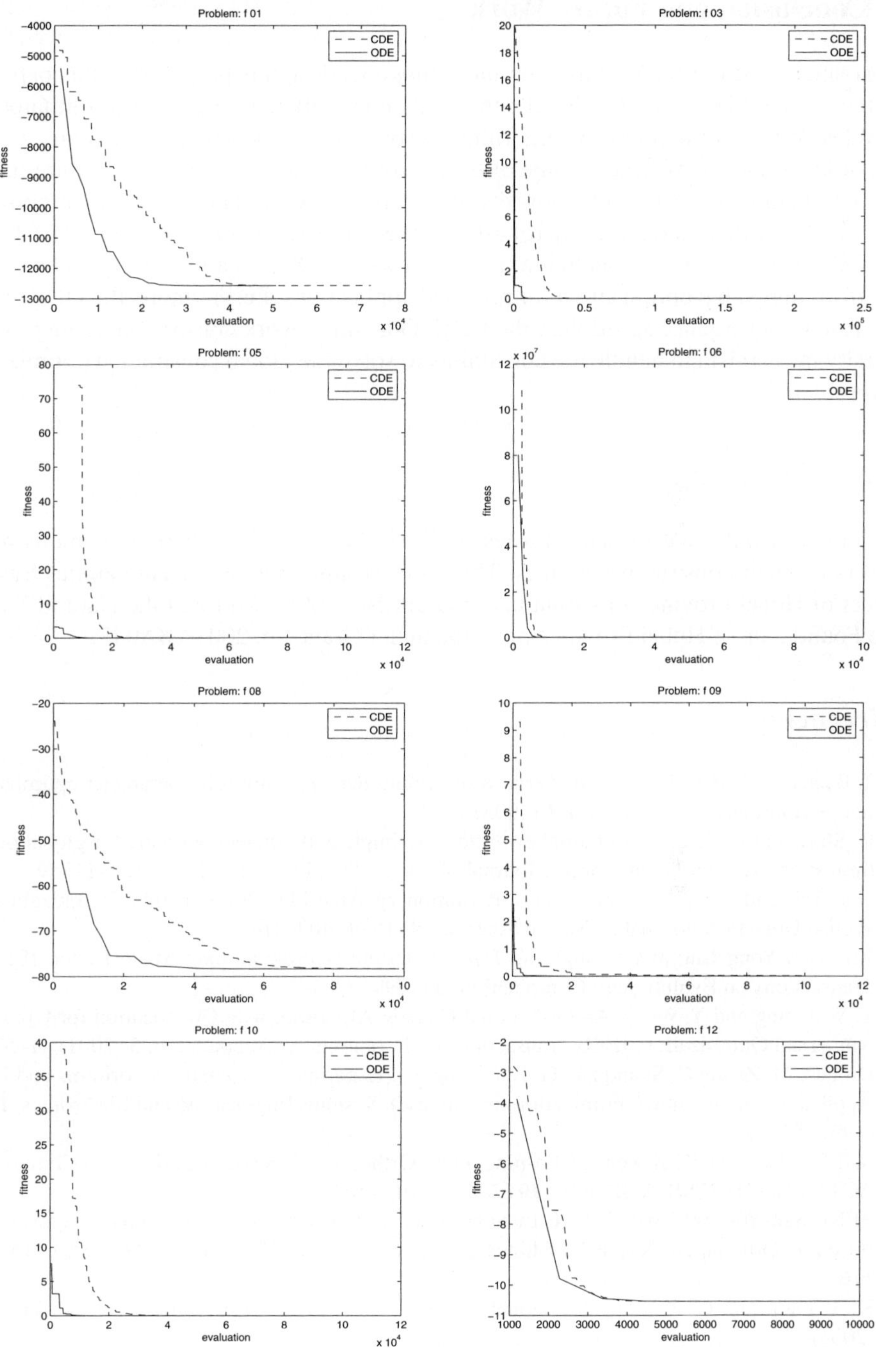

Fig. 1. Comparison with ODE and CDE on some convergence results of the test functions

6   Conclusion and Future Work

A fast and robust DE (ODE) based on the orthogonal design is proposed in this paper, and then it is used to solve the global numerical optimization problems with continuous variables. We used the proposed algorithm to solve twelve benchmark problems with high or low dimensions, where some of these problems have numerous local minimal. The computational experiments show that the proposed ODE can find optimal or close-to-optimal solutions, and it is more robust than the compared techniques (OGA/Q, CDE and FEP). It is, therefore, concluded that the proposed ODE is a promising technique in performing global optimization for practical applications. Furthermore, the ODE can have faster convergence speed than the CDE. Our future work consists on adding to a diversity mechanism to handle the constraints to solve the global constrained optimization problems.

Acknowledgments

The authors would like to thank Professor Y. W. Leung and the two anonymous reviewers for their constructive advices. This work is supported by the Humanities Base Project of Hubei Province of China under Grant No. 2004B0011 and the Natural Science Foundation of Hubei Province of China under Grant No. 2003ABA043.

References

1. T. Bäck and H.-P. Schwefel: An overview of evolutionary algorithms for parameter optimization. Evolutionary Computation. **1** (1993) 1–23
2. R. Storn and K. Price: Differential evolution–A simple and efficient heuristic for global optimization over continuous spaces. Journal of Global Optimization. **11** (1997) 341–359
3. Guo Tao and Kang Li-shan: A new Evolutionary Algorithm for Function Optimization. Wuhan University Journal of Nature Sciences. **4** (1999) 409–414
4. Xin Yao, Yong Liu, and Guangming Lin: Evolutionary Programming Made Faster. IEEE Transactions on Evolutionary Computation. **3** (1999) 82–102
5. Y. W. Leung and Y. Wang: An Orthogonal Genetic Algorithm with Quantization for Global Numerical Optimization. IEEE Transactions on Evolutionary Computation, **5** (2001) 41–53
6. Ding C. M, Zhang C. S. and Liu G. Z: Orthogonal Experimental Genetic Algorithms and Its Application in Function Optimization (in Chinese). Systems Engineering and Electronics, **10** (1997) 57–60
7. SHI Kui-fan, DONG Ji-wen, LI Jin-ping, et al: Orthogonal Genetic Algorithm (in Chinese). ACTA ELECTRONICA SINICA, **10** (2002) 1501–1504
8. ZENG San-You, WEI Wei, KANG Li-Shan, et al: A Multi-Objective Evolutionary Algorithm Based on Orthogonal Design (in Chinese). Chinese Journal of Computers, **28** (2005) 1153–1162
9. K. T. Fang and C. X. Ma: Orthogonal and Uniform Design (in Chinese). Science Press, (2001)
10. K. Price and R. Storn: Home Page of Differential Evolution. Available: http://www.ICSI.Berkeley.edu/~storn/code.html. (2003)

Cancer Classification by Kernel Principal Component Self-regression

Bai-ling Zhang

School of Computer Science and Mathematics
Victoria University, VIC 3011 Australia
bailing.zhang@vu.edu.au

Abstract. The classification of cancer based on gene expression data is one of the most important tasks in bioinformatics, and is essential for future clinical implementations of microarray based cancer diagnosis. In this paper, a novel procedure for classifying cancer using the gene expression data is proposed based on a Kernel Principal Component Self-regression (KPCSR) model. Developed from Kernel Principal Component Analysis (KPCA), the KPCSR model selects a subset of the principal components from the kernel space for the input variables to regress in order to accurately characterize each type of cancer. A modular scheme with class-specific KPCSR structure proves very efficient, from which each cancer class is assigned an independent KPCSR model for coding the corresponding gene expression information. The performance was measured on several public gene expression datasets involving human tumor samples, using 5-fold cross-validation and leave-one-out cross-validation (LOOCV) respectively. Experimental results has shown that the classification accuracies are better or comparable to the maximum accuracies based on the Support Vector Machine and k-Nearest Neighbor classifications combined with various gene selection schemes reported previously in the literature. These results suggest that our proposed method is useful for microarray based cancer classification.

1 Introduction

Accurate and objective diagnosis of human cancer has been greatly improved by the high-throughput microarray technologies, which can provide a global view of molecular changes involved in a cancer development by measuring simultaneously the activities and interactions of thousands of genes. The genome-wide expression patterns can be used to classify different type of cancers, to study varying levels of drug responses to facilitate drug discovery and development, and to predict clinical outcomes. In the past several years, extensive research has been carried out in the realm of cancerous tumor classification based on gene expression profiles, which are believed to be precise and objective molecular fingerprints for cancers.

The problem of cancer classification can be formulated as predicting binary or multi-category outcomes from the gene expression data. Many classical classifiers and machine-learning techniques have been attempted, with examples including

A. Sattar and B.H. Kang (Eds.): AI 2006, LNAI 4304, pp. 719–728, 2006.

© Springer-Verlag Berlin Heidelberg 2006

the k-nearest neighbor (kNN) methods, artificial neural networks, Bayesian approaches, support vector machines, and decision trees. Dudoit *et al.* have given an in-depth comparison of cancer classification performance from a number of classification algorithms and machine learning methods [Dudoit 2002].

Though much progress has been made, cancer classification based on gene expression data is still considered to be very challenging. Cancer microarray data is often characterised by very high dimensionality and very small number of available samples. Such a "small-sample-size (SSS)" problem with high dimensionality makes many classical classification methods not directly applicable. Means of effectively reducing the dimensionality and selecting relevant genes involved in different types of cancer remains a major research topic in which much effort has been made [Shen 2005].

Most of the prior reserach has been concentrated on binary classification, which aims to assign a label y (normal or cancerous), to a new gene expression pattern $\mathbf{x}$, based on the knowledge of M samples $\{(\mathbf{x}_1, y_1), (\mathbf{x}_2, y_2), ..., (\mathbf{x}_M, y_M)\}$. The problem of multiclass cancer classification is to determine which class a new tissue sample should be labelled given a collection of gene expression data belonging to different cancer types. In most situations, binary classification methodologies cannot be directly extended to produce comparable accuracies for simultaneous multiclass classification. Because of the large number of cancer types and subtypes, it is necessary to develop multiclass cancer classification procedures. Some methods have been developed utilizing one versus all (OVA) or all paired (AP) binary classifiers such as the support vector machine (SVM) [Vapnik 1998, Statnikov 2005]. For example, Ramaswamy *et al.* have employed SVM, weighted voting, and k-nearest-neighbors methods with OVA and AP schemes to distinguish 14 different tumor types [Ramaswamy 2001, Rifkin 2003].

Feature selection issue in cancer classification has been extensively studied. Gene selection, as it is often called, chooses a subset of many fewer genes which are more prominent in their contributions and removes features that are irrelevant. One earlier straightforward approach to select relevant genes is the application of standard parametric tests such as the t-test and a nonparametric test such as the Wilcoxon score test [Antoniadis et al., 2003, Tang 2006, Yang 2006, Wang 2005]. Another tool to find the underlying gene components and reduce the input dimensions is Principal component analysis (PCA) [Wang 2005], which can be efficient if the data points are linearly correlated. In order to extract informative genes from cancer microarray data and reduce dimensionality, a better alternative is the kernel principal component analysis (KPCA) [Rosipal 2001], which is a method of non-linear feature extraction closely related to methods applied in the Support Vector Machines.

This paper introduces a simple and efficient approach that can be applied to multi-class cancer classification. The technique is based on first performing Kernel Principal Component Analysis in kernel space and then applying self-regression from the extracted features (kernel principal components). The proposed model is a generalization of the Kernel Principal Component Regression (KPCR) [Rosipal 2001], a technique of Principal Component Regression (PCR) from the kernel space. The model is directly applied to describe each category of cancer in study and the corresponding attributes can be characterized by the features extracted from the kernel space. Computational results show that our algorithm is effective in classifying gene expression data.

2 Kernel Principal Component Self-Regression (KPCR)

2.1 Kernel Principal Component Analysis

Since the objective of PCA is to seek a lower dimensional subspace from which the projected reconstruction of the data is minimized, it does not perform well if the data points lie on a nonlinear manifold. Furthermore, PCA encodes the data based on second order dependencies and ignores higher-order statistics, which can capture important information for recognition.

The shortcomings of PCA can be overcome through the use of kernel functions with a method called kernel principal component analysis (KPCA). This performs the same procedure as PCA in a high dimensional space, F, related to the input by the (nonlinear) map $\boldsymbol{\Phi}(\mathbf{x}) : R \to F$. The map $\boldsymbol{\Phi}$ and the space F are determined implicitly by the choice of a kernel function k, which computes the dot product between two input examples x and y mapped into F via

$$k(\mathbf{x}, \mathbf{y}) = \boldsymbol{\Phi}(\mathbf{x}) \cdot \boldsymbol{\Phi}(\mathbf{y}) \tag{1}$$

where $(\cdot)$ is the vector dot product in the infinite-dimensional feature space F. The most commonly used kernel $k(\mathbf{x}, \mathbf{y})$ is:

$$k(\mathbf{x}, \mathbf{y}) = \exp\left(-\frac{||\mathbf{x} - \mathbf{y}||}{2\sigma^2}\right) \tag{2}$$

where σ is the width of the kernel.

In Kernel PCA, the kernel matrix k is first centralized; the result can be considered as the estimate of the covariance matrix of the new feature vector in F. Then the linear PCA is simply performed on the centralized k by finding a set of principal components in the span of vectors $\{\Phi(\mathbf{x})_r\}$ which represents the principal axes in the kernel space. Obviously, such a linear PCA on k may be viewed as a nonlinear PCA on the original data.

From the above discussion, KPCA can be formulated entirely in terms of inner products of the mapped examples. Thus, all inner products can be replaced by evaluations of the kernel function, which has two important consequences. First, inner products in F can be evaluated without computing $\boldsymbol{\Phi}$ explicitly, which allows us to work with a very high-dimensional, possibly infinite dimensional RKHS F. Second, if a positive definite kernel function is specified, explicit forms of $\boldsymbol{\Phi}$ or F are not needed to perform KPCA since only inner products are used in the computations.

2.2 Kernel Principal Component Self-Regression

In statistics, ordinary least squares (LS) regression is well-known. But one of main problems from LS regression is that its performance becomes poor when there are near multicollinearities in the matrix of explanatory variables [Rosipal 2001]. To solve the problem, principal components regression (PCR) has been proposed and successfully applied in some areas, particularly in chemometrics.

Traditional PCR technique attempts to estimate the values of a response variable at the basis of selected principal components of the explanatory variables. The significances of regressing the response variable on the PCs rather than directly on the explanatory variables include: (1) the explanatory variables are often highly correlated which may cause inaccurate estimations of the least squares (LS) regression coefficients, which can be avoided by using the PCs in place of the original variables since the PCs are uncorrelated; (2) the dimensionality of the regressors is reduced by taking only a subset of PCs for prediction [Rosipal 2001].

Let $\mathbf{y}$ be a vector of n observations of the dependent variable; Then PCR can be expressed as

$$\mathbf{y} = \mathbf{\Psi}\mathbf{w} + \epsilon \tag{3}$$

where $\mathbf{\Psi}$ represents an $(n \times M)$ matrix of zero-mean regressors $\{\mathbf{\Psi}_i\}_{i=1}^{n}$, $\mathbf{w}$ is a vector of regression coefficients and ϵ is the vector of error terms whose elements have equal variance, and are independent of each other. $\mathbf{\Psi}^T\mathbf{\Psi}$ is proportional to the sample covariance matrix.

Based on Equation 3, kernel PCA can be performed to extract its M eigenvalues $\left\{\tilde{\lambda}_i\right\}_{i=1}^{M}$ and corresponding eigenvectors $\{\mathbf{u}_i\}_{i=1}^{M}$. Then using the first p nonlinear principal components to create a linear model based on orthogonal regressors in a feature space F, the kernel PCR model (KPCR for short) can be formulated as

$$\begin{aligned}
\mathbf{y} &= f(\mathbf{x}, \mathbf{a}) \\
&= \sum_{k=1}^{p} v_k \beta_k(\mathbf{x}) + \mathrm{b} \\
&= \sum_{i=1}^{n} \alpha_i K(\mathbf{x}_i, \mathbf{x}) + \mathrm{b}
\end{aligned} \tag{4}$$

where $a_i = \sum_{k=1}^{p} v_k \tilde{\lambda}_k^{-1/2} \tilde{u}_i^k$, for $i = 1, 2, \cdots, n$, and b is a bias term.

An auto-associative data description can be implemented based on a self-regression variant of Equation 4, by simply replacing the response variables with explanatory variables:

$$\mathbf{x} = \sum_{i=1}^{n} \alpha_i K(\mathbf{x}_i, \mathbf{x}) + \mathrm{b} \tag{5}$$

which can be abbreviated as Kernel Principal Component Self-regression (KPCSR).

In summary, the KPCSR model provides a description of the nonlinear relationships between input and features from the kernel space. The model building involves two operations. The first is the kernel operation which transforms an input pattern to a high-dimensional feature space. The second is the mapping of the feature space back to the input space.

3   Experiments

3.1   Description of the Classification Procedure

We build a modular system for multiclass cancer classification, consisting of a set of separate KPCSR models, each offering the categorization of gene expression data for a specific cancer type. For the kth type, let the set of training data be $\mathbf{x}_1^{(m)}, \mathbf{x}_2^{(m)}, \cdots, \mathbf{x}_N^{(m)}$, $m = 1, \cdots, M$, where N is the number of training data for the mth cancer type and M the number of total types. When a gene expression data $\mathbf{x}$ with unknown cancer type is presented for classification, it is applied to all the KPCSR models to yield respective estimations. Then, a similarity score between the query data and each of the reconstructions is calculated to determine which reconstructed data best matches the query. Given the query gene expression $\mathbf{a}$ and a reconstructed gene expression $\mathbf{b}$, the similarity $\rho(a, b)$ can be defined by the cosine of the angle between them, formally

$$\rho(\mathbf{a}, \mathbf{b}) = \frac{\mathbf{a}^{\mathrm{T}}\mathbf{b}}{||\mathbf{a}|| \cdot ||\mathbf{b}||} \tag{6}$$

3.2   Data Sets Description

The following datasets were downloaded from the website given in [Yang 2006].

(1) The ALL/AML dataset (LEU). This dataset includes 72 samples from the two classes of *acute lymphoblastic leukemia (ALL)* and *acute myeloid leukemia (AML)*, which contain 47 and 25 samples, respectively. Every sample contains $7,129$ gene expression values. The original dataset had been preprocessed using the method in [Dudoit 2002] to remove about half the genes and subsequently every sample contains only $3,571$ gene expression values.

(2) The small round blue cell tumor (SRBCT) dataset. This dataset was first analyzed by Khan and his collaborators [Khan 2001]. Originally, gene expression levels of 6567 genes were measured using cDNA microarray for each sample, 2308 of which passed the filter that requires the red intensity of a gene to be greater than 20 and were kept for analyses. After an initial filtering, $2,308$ genes were included. The tumors were classified as the *Ewing family of tumors (EWS), Burkitt lymphoma (BL), neuroblastoma (NB)* and *rhabdomyosarcoma (RMS)* [Khan 2001]. Among the 83 samples, 29, 11, 18, and 25 samples belong to classes EWS, BL, NB and RMS, respectively.

(3) The brain tumor (GLIOMA) dataset. The dataset [Nutt 2003] contains in total 50 samples in four classes, *cancer glioblastomas (CG), non-cancer*

Table 4. On the GLIOMA dataset, comparison of the leave-one-out and 5-fold cross validation accuracies (KPCSR versus kNN and SVM with different gene selection methods)

Methods	kNN			SVM			KPCSR
	CS1	Cho's	F-test	CS1	Cho's	F-test	
5-fold CV	0.684	0.664	0.674	0.722	0.670	0.685	**0.665**
LOOCV	0.780	0.820	0.780	0.760	0.720	0.740	**0.768**

Table 5. On the LUNG dataset, comparison of the leave-one-out and 5-fold cross validation accuracies (KPCSR versus kNN and SVM with different gene selection methods)

Methods	kNN			SVM			KPCSR
	CS1	Cho's	F-test	CS1	Cho's	F-test	
5-fold CV	0.937	0.924	0.918	0.938	0.924	0.930	**0.946**
LOOCV	0.951	0.926	0.921	0.951	0.941	0.936	**0.955**

Table 6. On the DLBCL dataset, comparison of the leave-one-out and 5-fold cross validation accuracies (KPCSR versus kNN and SVM with different gene selection methods)

Methods	kNN			SVM			KPCSR
	CS1	Cho's	F-test	CS1	Cho's	F-test	
5-fold CV	0.903	0.920	0.902	0.918	0.930	0.926	**0.957**
LOOCV	0.922	0.935	0.922	0.961	0.961	0.961	**0.976**

Table 7. On the MLL dataset, comparison of the leave-one-out and 5-fold cross validation accuracies (KPCSR versus kNN and SVM with different gene selection methods)

Methods	kNN			SVM			KPCSR
	CS1	Cho's	F-test	CS1	Cho's	F-test	
5-fold CV	0.948	0.960	0.954	0.952	0.955	0.948	**0.967**
LOOCV	0.972	0.972	0.958	0.972	0.958	0.958	**0.969**

The above results show that the KPCSR based classification of the tissue samples can give very high accuracies. On the datasets of *SRBCT, CAR, LEU, GLIOMA* and *MLL*, both of the LOOCV accuracies and 5-fold CV accuracies are compariable to the SVM with gene selection methods from Yang and Cho [Cho 2003, Yang 2006]. On the *LUNG* and *DLBCL* datasets, both of the LOOCV accuracies and 5-fold CV accuracies are the best comparing with other methods reported previously in the literature.

4   Conclusion

We have introduced a data description model for classifying cancer types from the gene expression data. The model is built from Kernel Principal Component Regression (KPCR), which performs a least square reconstruction using the selected features from kernel space based on kPCA. Our proposed algorithm can be applied to both binary classification and multi-class classification and the efficient dimension reduction from the kPCA avoids the explicit gene selection step in most of previous cancer classification schemes. We have illustrated the effectiveness of the proposed algorithm in several public datasets involving human tumor gene expression data, and the results show that the procedure is able to distinguish different classes with high accuracy.

References

[Dudoit 2002] Dudoit, S., Fridlyand J, Speed T P (2002). Comparison of Discrimination Methods for the Classification of Tumors Using Gene Expression Data. Journal of the American Statistical Association, 97:7787.

[Golub 1999] Golub T R., Slonim D K., Tamayo P., Huard C., Gaasenbeek M., Mesirov J P., Coller H., Loh M L., Downing J R., Caligiuri M A., Bloomfield C D., Lander E S. (1999). Molecular Classification of Cancer: Class Discovery and Class Prediction by Gene Expression Monitoring. *Science*, 286:531537.

[Khan 2001] Khan J., Wei J S., Ringner M., Saal L H., Ladanyi M., Westermann F., Berthold F., Schwab M., Antonescu C R., Peterson C., Meltzer P S. (2001). Classification and diagnostic prediction of cancers using gene expression profiling and artificial neural networks. *Nature Medicine*, 7:673679.

[Nutt 2003] Nutt C L., Mani D R., Betensky R A., Tamayo P., Cairncross J G., Ladd C., Pohl U., Hartmann C., McLaughlin M E., Batchelor T T., Black P M., von Deimling A, Pomeroy S L., Golub T R., Louis D N. (2003). Gene Expression-based Classification of Malignant Gliomas Correlates Better with Survival than Histological Classification. Cancer Research, 63:16021607.

[Bhattacharjee, 2001] Bhattacharjee, A., Richards, W G., Staunton, J., Li, C., Monti, S., Vasa, P., Ladd, C., Beheshti, J., Bueno, R., Gillette, M., Loda, M., Weber, G., Mark, E J., Lander, E S., Wong, W., Johnson, B E., Golub, T R., Sugarbaker, D J., Meyerson, M. (2001) Classification of human lung carcinomas by mRNA expression profiling reveals distinct adenocarcinoma subclasses. *Proceedings of the National Academy of Sciences of USA*, 98:1379013795.

[Su 2001] Su, A I., Welsh, J B., Sapinoso, L M., Kern S G., Dimitrov, P., Lapp, H., Schultz, P G., Powell, S M., Moskaluk, C A., Frierson, H FJr, Hampton, G M. (2001). Molecular Classification of Human Carcinomas by Use of Gene Expression Signatures. *Cancer Research*, 61:73887393.

[Shipp, 2002] Shipp, M A., Ross, K N., Tamayo, P., Weng, A P., Kutok, J L., Aguiar, R C., Gaasenbeek, M., Amgel, M., Reich, M., Pinkus, G S., Ray, T S., Kovall, M A., Last, K W., Norton, A., Lister, T A., Mesirov, J., Neuberg, D S., Lander, E S., Aster, J C., Golub, T R. (2002). Diffuse large B-cell lymphoma outcome prediction by gene expression profiling and supervised machine learning. *Nature Medicine*, 8:6874.

computes the texture information of the image at multiple points, resulting in a collection of local features. However, images are often high dimensional and the number of feature vectors representing each image is significantly large. For a large-scale database, using local features directly is undesirable and generally intractable. Therefore, a method to represent the collection of all feature vectors adopted from the image by a single feature vector is required. In this paper, the Gaussian Probability Distribution Function (GPDF), in which the mean vector and covariance matrix are estimated from all local features vectors obtained from the image, is utilized to describe image content.

With regard to the feature vectors extracted from images, a classification model is employed for classifying the extracted feature vectors. Gaussian Mixture Models (GMMs) and maximum or minimum-likelihood classifier is used for modeling and classification, respectively. To cluster the GPDF data in GMMs, the convention k-mean and SOM clustering algorithms are widely used [3]. However, because of the selection of parameters such as learning rate and total number of iterations and the initialized conditions, the k-mean and SOM algorithms often give unstable results. Park proposed a competitive clustering algorithm called the Centroid Neural Network (CNN) [4]. Compared with the conventional k-mean and SOM, the CNN converges stably to suboptimal solutions [4]. However, the CNN was originally designed to handle deterministic data because of its quadric (Euclidean) distance measure.

In this paper, we propose a new classification model that can efficiently classify images by using the DCNN for clustering of GPDF data. The proposed classification model is designed for utilizing advantages of the localized image representation method. According to the localized representation method, each image is represented by a GPDF. In order to deal with GPDF data, the DCNN, which is based on the conventional CNN to exploit its advantageous features, employs a divergence measure as a distance measure for clustering of probability data [5]. By applying the DCNN to the clustering of probability data, the robust advantages of the localized representation method are incorporated and mixtures of GMMs can be efficiently formed.

The remainder of this paper is organized as follows: Section 2 presents image representation methods using the Discrete Cosine Transform (DCT). A short introduction of the DCNN is presented in Section 3. Section 4 presents experiments and results on several image data sets including comparisons with other conventional models. Conclusion are presented in Section 5.

2   Image Representation Using Discrete Cosine Transform

In general, texture descriptors of images can be extracted using the globalized representation and localized representation methods. Globalized representation describes the texture information of the whole image in a single vector. Although the globalized representation has been successfully applied for image recognition, it still suffers from some problems. In particular, keeping the dimensions of feature vectors such that they are tractable requires an image to be in low resolution or down-sampled images to keep the dimension in tractable. This, however, leads to a loss of valuable information in the image.

Another approach is to use the localized representation, which describes the content of the image by a collection of local features extracted from the image. These features are computed at different points of interest in the image. The localized representation does not place constraints on resolution (a larger image simply returns more feature vectors). Localized representation maintains the dimensions of the feature vectors in tractable by adjusting the sizes of blocks. Furthermore, this method is consequently more robust to occlusions and clutter.

In order to obtain the texture information from the image, conventional texture descriptors based on frequency domain analysis such as Gabor filters [6] and wavelet filters [7] are often used. However, these algorithms often require a high computational load for description extraction and are not suitable for real-time applications. In this paper, we use a standard method in image processing called the Discrete Cosine Transform (DCT) for extracting the texture information from each block of the image [8]. The DCT transforms the image from the spatial domain into the frequency domain. The 2-D DCT of a $M \times N$ block can be described as follows:

$$F(u,v) = \Lambda(u)\Lambda(v) \sum_{m=0}^{M-1} \sum_{n=0}^{N-1} f(m,n) \left(\cos \frac{\pi(2m+1)u}{2M} \right) \left(\cos \frac{\pi(2n+1)v}{2N} \right) \quad (1)$$

where $F(u,v)$ is the DCT coefficient in row u and column v of the DCT matrix, $f(m,n)$ is the intensity of the pixel in row m and column n in the block, and $\Lambda(u)$ and $\Lambda(v)$ are defined as:

$$\Lambda(u) = \begin{cases} \frac{1}{\sqrt{M}}, & u = 0 \\ \sqrt{\frac{2}{M}}, & 1 \leq u \leq M-1 \end{cases} \quad (2)$$

$$\Lambda(v) = \begin{cases} \frac{1}{\sqrt{M}}, & v = 0 \\ \sqrt{\frac{2}{M}}, & 1 \leq v \leq M-1 \end{cases} \quad (3)$$

The number of DCT coefficients in the block $M \times N$ is $M.N$, where M is the width of the block and N is the height of the block. It is noted that the high frequency component is not visible to human vision and is not efficient for texture retrieval. Therefore, to speed up the process of estimation of densities associated with each image, unnecessarily high frequency components are ignored.

As the texture information is extracted from images, each image is represented by a collection of feature vectors in which each feature vector stores the DCT coefficents obtained from each block in the image. One of the problems associated with the localized representation method is a rapidly increase in the number of feature vectors. For example, an image of 256×256 pixels with a block size of 8×8 pixels and 2 pixels overlapping window can return in 15,376 blocks. Directly using feature vectors obtained from blocks for the classification task induces a very high computational load. Therefore, instead of using a collection of individual feature vectors to represent the image, the image is modeled as a GPDF, where the mean vector and covariance matrix are estimated from a collection of feature vectors obtained from the image. The mean of the GPDF data can be estimated by:

$$\mu = \frac{1}{N} \sum_{i=1}^{N} x_i \tag{4}$$

while the covariance matrix can be estimated by

$$\Sigma = \frac{1}{N} \sum_{i=1}^{N} (x_i - \mu)(x_i - \mu)^T \tag{5}$$

where N is the number of blocks obtained from the image and x_i is an individual feature vector obtained from the block i.

3   Centroid Neural Networks with Divergence Measure

3.1   CNN Algorithm

The CNN [4] is an unsupervised competitive learning algorithm based on the classical k-means clustering algorithm [3]. It finds the centroids of clusters at each presentation of the data vector. The CNN first introduces definitions of the winner neuron and the loser neuron. When a data x_i is given to the network at the epoch (k), the winner neuron at the epoch (k) is the neuron with the minimum distance to x_i. The loser neuron at the epoch (k) to x_i is the neuron that was the winner of x_i at the epoch (k-1) but is not the winner of x_i at the epoch (k). The CNN updates its weights only when the status of the output neuron for the presenting data has changed when compared to the status from the previous epoch.

When an input vector $\mathbf{x}$ is presented to the network at epoch n, the weight update equations for winner neuron j and loser neuron i in CNN can be summarized as follows:

$$\mathbf{w}_j(n+1) = \mathbf{w}_j(n) + \frac{1}{N_j + 1}[\mathbf{x}(n) - \mathbf{w}_j(n)] \tag{6}$$

$$\mathbf{w}_i(n+1) = \mathbf{w}_i(n) - \frac{1}{N_i - 1}[\mathbf{x}(n) - \mathbf{w}_i(n)] \tag{7}$$

where $\mathbf{w}_j(n)$ and $\mathbf{w}_i(n)$ represent the weight vectors of the winner neuron and the loser neuron, iteration, respectively.

The CNN has several advantages over conventional algorithms such as SOM or k-means algorithm when used for clustering and unsupervised competitive learning. The CNN requires neither a predetermined schedule for learning gain nor the total number of iterations for clustering. It always converges to sub-optimal solutions while conventional algorithms such as SOM may give unstable results depending on the initial learning gains and the total number of iterations. More detailed description on the CNN can be found in [4,9].

Note that the CNN was designed for deterministic data because the distance measure used in the CNN is the quadric (Euclidean) distance.

3.2   Divergence-Based CNN (DCNN)

Though CNN has been applied successfully to deterministic data, it can not be used for clustering of probabilistic data such as GPDF data because it uses the Euclidean for its distance measure. In order to use for clustering of the probabilistic data such as GPDF data, the divergence distance measure is employed for its distance measure. After evaluating several divergence distance measures, the most popular Bhattacharyya distance measure is employed. The similarity measure between two distributions using Bhattacharyya distance measure is defined as follow:

$$D(G_i, G_j) = \frac{1}{8}(\boldsymbol{\mu}_i - \boldsymbol{\mu}_j)^T \left[\frac{\boldsymbol{\Sigma}_i + \boldsymbol{\Sigma}_j}{2} \right]^{-1} (\boldsymbol{\mu}_i - \boldsymbol{\mu}_j) + \frac{1}{2} \ln \frac{\left| \frac{\boldsymbol{\Sigma}_i + \boldsymbol{\Sigma}_j}{2} \right|}{\sqrt{|\boldsymbol{\Sigma}_i| |\boldsymbol{\Sigma}_j|}} \quad (8)$$

where $\boldsymbol{\mu}_i$ and $\boldsymbol{\Sigma}_i$ denote the mean vector and covariance matrix of Gaussian distribution G_i, respectively. $\boldsymbol{\mu}_j$ and $\boldsymbol{\Sigma}_j$, denote the mean vector and covariance matrix of Gaussian distribution G_j, respectively. T denotes the transpose matrix.

The concept of winner and loser in the CNN can be adopted for the DCNN without any change except for application of the divergence measure for the distance calculation. In this case, however GPDFs have two parameters to consider: mean, μ, and diagonal covariance, Σ. The weight update for mean is the same as the CNN weight update shown in Eq. (6) and Eq. (7). Like the CNN, the μ (mean) of the winner neuron moves toward the μ of input data vectors while the μ of the loser neuron moves away from the μ of input data vectors. One epoch of training the DCNN consists of two parts: the first concerns the weight updates for μ and the second concerns the weight updates for Σ. Weights updates for Σ are performed once at the end of each epoch while weights for μ are updated at every iteration. That is, after presentation of all the data vectors for weight updates for μ at each epoch, weight updates for μ are suspended until weight updates for Σ are performed with the grouping information of data vectors obtained for μ during the epoch. With this scenario, the diagonal covariance for each group is calculated by the following equation:

$$\boldsymbol{\Sigma}_{w_j}(n+1) = \frac{\sum_{k=1}^{N_j} [\boldsymbol{\Sigma}_{x_k}(n) + (\boldsymbol{\mu}_{x_k}(n) - \boldsymbol{\mu}_{w_j}(n))^2]}{N_j} \quad (9)$$

where $\boldsymbol{\mu}_{x_k}$ and $\boldsymbol{\Sigma}_{x_k}$ is the mean vector and the covariance matrix, respectively, of the vector data $\boldsymbol{x}_k$. $\boldsymbol{\mu}_{w_j}$ and $\boldsymbol{\Sigma}_{w_j}$ is the mean vector and covariance matrix, respectively, of the cluster j. N_j is the number of data vectors in the cluster j.

Table 1. Divergence Centroid Neural Network Algorithm

Algorithm DCNN(C,D)   [*C: number of clusters, D: number of data vectors*]
   [*Initialize weights* $\mathbf{w}_1 = (\mu_{w_1}, \Sigma_{w_1})$ *and* $\mathbf{w}_2 = (\mu_{w_2}, \Sigma_{w_2})$]
   Find the centroid, $\mathbf{c}$, of all data vector
   Initialize $\mathbf{w}_1$ and $\mathbf{w}_2$ around $\mathbf{c}$ with small ε :
   $\mu_{w_1} := \mu_c + \varepsilon,\ \mu_{w_2} := \mu_c - \varepsilon,\ \Sigma_{w_1} := \Sigma_{w_2} := \Sigma_c$
   $k := 2,\ epoch := 0$
   for $(k \leq C)$
      do
         $loser := 0$
         for $(n = 1)$ **to** D
            Apply a data vector $\mathbf{x}(n)$ to the network
            Find the winner neuron, j, using Divergence distance for $1 \leq j \leq k$.
            if $(epoch \neq 0)$ **then** Set i is winner neuron, i, for $\mathbf{x}(n)$ in previous epoch.
            if $(i \neq j)$, **then** neuron, i, is loser neuron.
            if $(epoch = 0$ or $i \neq j)$
               Run UpdateDCNNWeightMean(μ_{w_j} , μ_{w_i} , $epoch$)
               $loser := loser + 1$
            endif
         endfor   [*check for all data*]
         $\Sigma_{w_j}(epoch + 1) = \sum_q^{N_j}(\Sigma_{x_q j} + (\mu_{x_q j} - \mu_{w_j}(epoch))^2)/N_j$
         $j = 1, 2, \ldots, k$
         $epoch := epoch + 1$
      while $loser \neq 0$
      if $k \neq M$
         split the most erroneous group, j, by adding a small vector, ϵ, nearby group j
         $\mu_{w_{k+1}} = \mu_{w_j} + \epsilon,\quad \Sigma_{w_{k+1}} = \Sigma_{w_j}$
      endif
      $k := k + 1$
     endfor
end

Procedure UpdateDCNNWeightMean(μ_{w_j}, μ_{w_i}, $epoch$)
   [*Update the winner neuron,* $\mathbf{w}_j$, *and loser neuron,* $\mathbf{w}_i$]
   Update winner neuron : $\mu_{w_j}(n + 1) = \mu_{w_j}(n) + (\mu_x(n) - \mu_{w_j}(n))/(N_j + 1)$
   if $epoch \neq 0$ [*loser neuron is occurred only when epoch $\neq 0$*]
      Update losing neuron : $\mu_{w_i}(n + 1) = \mu_{w_i}(n) - (\mu_x(n) - \mu_{w_i}(n))/(N_i - 1)$
   endif
end

By using the divergence distance as its distance measure, the DCNN have abilities in clustering the probability data while it still keep advantageous features of the CNN. Because the CNN have been proven to outperform other conventional clustering algorithms such as k-means and SOM, the DCNN should show improvements over the k-means and SOM algorithms in probabilistic data. More details on the DCNN can be found in [5].

Table 1 shows a pseudocode of the DCNN algorithm.

4   Experiments and Results

The performance of the proposed classification model was evaluated on the Caltech image data set. The Caltech image data set consists of different image classes (categories) in which each class contains different views of an object. The Caltech image data were collected by the Computational Vision Group and can be downloaded at http://www.vision.caltech.edu/html-files/archive.html

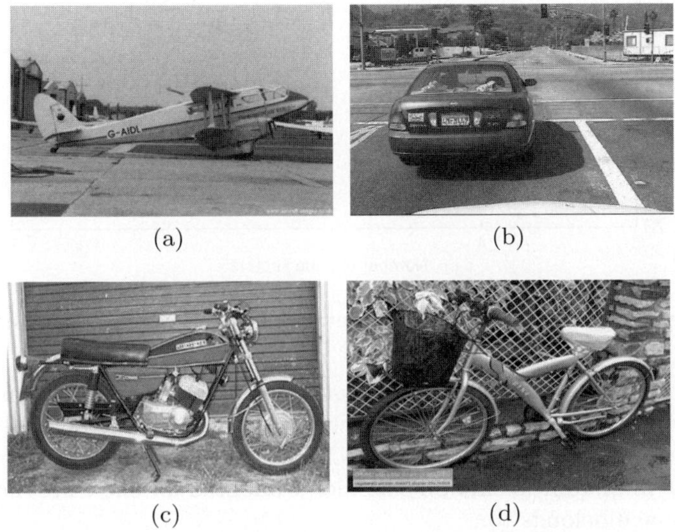

(a)   (b)

(c)   (d)

Fig. 1. (a) Airplane (b) Car (c) Motorbike (d) Bike

From these classes, we selected the 4 most easily confused classes: airplane, car, bike and motorbike for experiments. Each class consists of 200 images with different views resulting in a total of 800 images in the data set. From this data set, 100 images were randomly chosen for training while the remaining images were used for testing. The entire image are converted to grey scale and the same resolution. Fig. 1 shows an example of 4 image categories used in the experiments. Figs. 1(a), 1(b), 1(c), and 1(d) are examples of car, airplane, motorbike, and bike, respectively. For the localized representation, the images are transformed into a collection of 8×8 blocks. The block is then shifted by an increment of 2 pixels horizontally and vertically. The DCT coefficients of each block are then computed and return in 64 dimensional coefficients. Only the 32 lowest frequency DCT coefficients that are visible to the human eye were kept. Therefore, the feature vectors that are obtained from each block have 32 dimensions. In order to calculate the GPDF for the image, the mean vector and the covariance matrix are estimated from all blocks obtained from the image. Finally, a GPDF with 32-dimensional mean vectors and 32×32 covariance matrix is used to represent the content of the image.

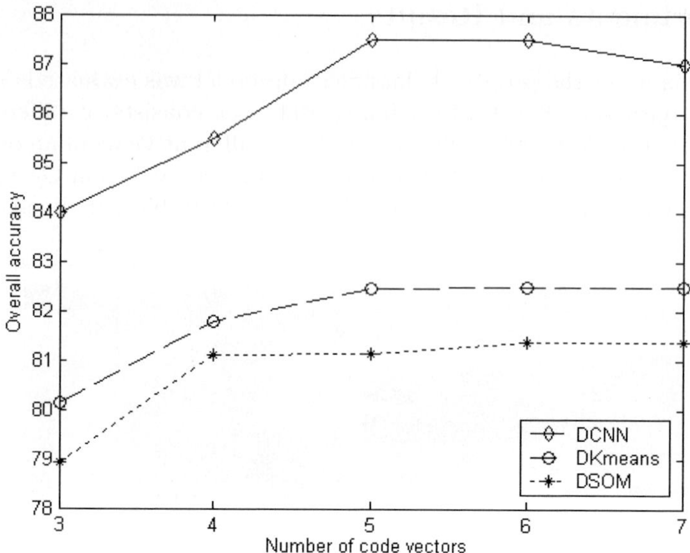

Fig. 2. Overall classification accuracies using different algorithms

After the GPDF data of all image classes are estimated, a minimum-likelihood (ML) classifier is obtained. The ML classifier is based on the GMM in which each mixture is estimated by using the DCNN clustering algorithm. After mixtures are built, the minimum-likelihood classifier is adopted for choosing the class that the tested image belongs to:

$$Class(x) = \arg\min_i D(\boldsymbol{x}, C_i) \tag{10}$$

$$D(G(\boldsymbol{x}; \mu, \Sigma), C_i) = \sum_{k=1}^{N_i} w_{ik} D(G(\boldsymbol{x}; \mu, \Sigma), G(\boldsymbol{x}; \boldsymbol{\mu}_{ik}, \boldsymbol{\Sigma}_{ik})) \tag{11}$$

where $\boldsymbol{x}$ is the tested image represented by a Gaussian distribution feature vector with mean vector, $\boldsymbol{\mu}$, and covariance matrix, $\boldsymbol{\Sigma}$. $\boldsymbol{\mu}_{ik}$ and $\boldsymbol{\Sigma}_{ik}$ represent for the mean vector and covariance matrix of cluster k in class C_i, respectively. w_{ik} is the weight component of cluster k in class C_i. N_i is the number of clusters in the class C_i.

Fig. 2 shows the classification accuracy of the classification model using the Divergence-based SOM (DSOM), the Divergence-based k-means (Dk-means), and the DCNN. In this figure, the number of code vector is varied from 3 code vectors to 7 code vectors in order to determine a sufficient number of code vectors to represent for the number of mixtures in GMMs. As can be seen from Fig. 2, the accuracies of all classification models increase when we increase the number of code vectors from 3 to 5 while they tend to saturate when the number of code vectors is increased from 5 to 7. This implies that using 5 code vectors for representing the mixtures of GPDF data is sufficient.

Table 2 shows the accuracy of the classification models using the DSOM, the Dk-means, and the DCNN. As can be seen from Table 2, the classification model using the DCNN always outperforms those using the DSOM and Dk-means. Accuracy improvement of 5.77% and 6.97% is achieved over the Dk-means and DSOM algorithms, respectively.

Table 2. Classification accuracy (%) of different algorithms using 5 code vectors

	Airplane	Car	Bike	Motorbike	Accuracy
DSOM	80	100	51.4	94.2	**81.4%**
Dk-means	82.8	100	52.8	94.2	**82.45%**
DCNN	84	100	68	98	**87.5%**

Table 3. Confusion matrix of image categories, using 5 code vectors

	Airplane	Car	Bike	Motorbike	Accuracy
Airplane	**84**	2.0	10.0	4.0	**84.0%**
Car	0.0	**100**	0.0	0.0	**100%**
Bike	12.0	0.0	**68.0**	20.0	**68.0%**
Motorbike	0.0	0.0	2.0	**98.0**	**98.0%**

Table 3 shows the confusion matrix that describes the classification results of the proposed classification model in detail. As can be inferred from Table 3, cars can be well discriminated from the others while bikes and motorbikes are easily confused. These are logical results because motorbikes and bikes are quite similar even to the human eye while the cars are significantly different.

5   Conclusion

A new image classification model using Divergence-based Centroid Neural Networks (DCNN) for clustering of Gaussian Distribution Probability Function (GPDF) data is proposed in this paper. The proposed classification model is designed for utilizing advantages of the localized image representation method in which each image is described by a Gaussian distribution feature vector. The localized representation not only overcomes some drawbacks of the globalized representation method but also describes the texture information of the image more efficiently. By using the DCNN for clustering of probability data, the proposed classification model outperforms its counterparts that use the conventional Dk-mean and DSOM algorithms. The proposed classification model provides an efficient tool for image classification systems.

Acknowledgment

This work was supported by the Korea Research Foundation Grant funded by the Korea Government (MOEHRD-KRF-2005-042-D00265).

References

1. Vailaya, A., Figueiredo, M.A.T., Jain, A.K., Zhang, H.J.: Image Classification for Content-Based Indexing. IEEE Trans. on Image Processing 10(1) (2001) 117-130
2. Smeulders, A.W.M., Worring, M., Santini, S., Gupta, A., Jain, R.: Content-Based Image Retrieval at The End of The Early Years . IEEE Trans. on Pattern Analysis and Machine Intelligence 22(12) (2000) 1349-1380
3. Hartigan, J.: Clustering Algorithms. New York, Wiley (1975)
4. Park, D.: Centroid Neural Network for Unsupervised Competitive Learning. IEEE Trans. on Neural Networks 11(2)(2000)520-528
5. Park, D.C., Kwon, O.H.: Centroid Neural Network with the Divergence Measure for GPDF Data Clustering. IEEE Trans. on Neural Networks (in review).
6. Daugman, J.G.: Complete Discrete 2D Gabor Transform by Neural Networks for Image Analysis and Compression. IEEE Trans. on Acoust., Speech, and Signal Processing 36 (1988) 1169-11179
7. Pun, C.M., Lee, M.C.: Extraction of Shift Invariant Wavelet Features for Classification of Images with Different Sizes. IEEE Trans. on Pattern Analysis and Machine Intelligence 26(9) (2004) 1228-1233
8. Huang, Y.L., Chang, R.F.: Texture Features for DCT-Coded Image Retrieval and Classification. ICASSP Proc. IEEE Int. Conf. on Acoustics, Speech, and Signal Processing, Vol. 6 (1999) 3013-3016
9. Park, D.C., Woo, Y.: Weighted Centroid Neural Network for Edge Reserving Image Compression, IEEE Trans. on Neural Networks 12 (2001) 1134-1146

Solving a Constraint Satisfaction Problem for Frequency Assignment in Low Power FM Broadcasting

Surgwon Sohn[1] and Geun-Sik Jo[2]

[1] Division of Computer Engineering, Hoseo University, Asan, Korea
`sohn@hoseo.ac.kr`
[2] School of Computer Science & Engineering, Inha University, Incheon, Korea
`gsjo@inha.ac.kr`

Abstract. We present an algorithm to solve the frequency assignment problem for low power FM broadcasting. To get a good suboptimal solution, some heuristics such as k-coloring variable ordering and mostly-used value ordering rule are provided. They enforce the backtracking process in a constraint satisfaction problem, so both the search space and computing time are greatly reduced. A lot of outstanding work on graph coloring problems has been achieved, and the theoretical lower bound of the chromatic number of random graph is one of them. Comparison between the theoretical lower bound and our computed approximate solution has been made for evaluation of proposed algorithm.

1 Introduction

Wireless communication is used in many different applications such as radio/TV broadcasting, mobile telecommunication, satellite communication, and military communication. In each of these specific area, frequency assignment problems (FAPs), also called channel assignment problems, have grown widely over the past years. Since 1970s, there has been a lot of interest in automatic frequency assignment for cellular telephones. In 1980, Hale[1] presented an excellent overview and references of the frequency planning for broadcast services of early years. Hale also introduced the relation of the FAP with graph coloring.

It is necessary for transmitters should be scattered around the region to cover an area for radio broadcasting. FM signals are transmitted on VHF frequencies in the range of 88 - 108 MHz and this band is usually partitioned into a set of channels, all with the same bandwidth 200KHz of frequencies. For this reason the channels are usually numbered from 1 to 100.

In mobile telephony and radio/TV broadcasting, the receivers are spread within a certain area. As the standard approach of determining signal strength at all locations in the area is in the literature[2,3], a grid of squares of predetermined size or the test points are designed to overlap the area. Radio/TV broadcasting is essentially same with the mobile communication instances except that all traffic for radio/TV broadcasting is unidirectional. Instances provided by

A. Sattar and B.H. Kang (Eds.): AI 2006, LNAI 4304, pp. 739–748, 2006.
© Springer-Verlag Berlin Heidelberg 2006

a major Italian radio broadcasting company were made available at the website, and results for these instances are presented by the literature[4].

Low Power FM(LPFM) radio service in USA was created by the Federal Communications Commission (FCC) in January 2000[5]. LPFM stations are not protected from interference that may be received from other classes of FM stations. LPFM stations are available to noncommercial educational entities and public safety and transportation organizations, but are not available to individuals or for commercial operations. These radio stations operate with an effective radiated power (ERP) of maximum 100 watts with facilities of 30 meters antenna height above average terrain (HAAT). The approximate service range of an 100-watt LPFM station is 5.6 kilometers radius. LPFM is also called a community radio nowadays in many countries[6] including Australia and Korea. Community radio stations tend to use transmitter powers varying from about a few watts right up to 10 or 100W depending on their circumstances.

Due to the vast increasing number of the radio stations and very limited availability of frequencies, it is very hard to find vacant frequencies to be assigned for the radio operators. Consequently, a major challenge to improve the frequency utilization efficiency is needed. Frequency planning deals with efficient assignment of frequencies to FM radio stations such that interference is avoided as far as possible.

Our research objective is to find a frequency assignment that satisfies all the constraints using minimum number of frequencies while maximizing the number of radio stations served for a given area. A given area is partitioned into cells, which are served by FM radio transmitters. The basic frequency assignment problem can be viewed as the constraint satisfaction problem (CSP) over a finite domain of frequencies while satisfying some kinds of constraints.

In this paper, we show the related work on FAP in section 2, the formal definition of the problem and the algorithms that utilize constraint satisfaction techniques in section 3. Furthermore, we show experimental results and evaluation in section 4, and provide conclusion in section 5.

2   Related Work

It has been noted by many authors[7,8,9] that the frequency assignment is a graph-coloring problem if only co-channel constraint is involved. In the FAP case, colors represent channels, vertices represent radio transmitters that are located at the cells, and links connect pairs of radio transmitters that cannot be assigned to the same channel. The objective is to find a solution with the minimum number of colors required to color all the vertices. Graph coloring is perhaps one of the most intensively investigated problems. In the language of complexity, the graph coloring problem is known to be NP-complete, so the time needed to solve this type of optimization problems grows exponentially with the increasing size of the problem.

To find an optimal solution, Giortzis tried branch and bound method as an integer linear programming. Exact method is too time consuming to be practical for most problems. However, a vast number of heuristic algorithms have been suggested over the years. Therefore meta-heuristics such as simulated annealing, tabu search and genetic algorithms are often used to provide sub-optimal solutions in a reasonable amount of time. Dunkin[10] and Thiel[11] also used hill climbing method in addition to these heuristics. Recently, Wang[12] has proposed a neural network system to solve a channel assignment problem for cellular communication. Some researchers such as Carlsson, Walsher, and Yokoo[13] have tried to handle FAP using CSP techniques in the applications of mobile phone networks.

These most of techniques produce approximate solutions, which will be close to the optimum. In order to assess the quality of these approximate solutions, some lower bounds are needed. The lower bound of the chromatic number of graph coloring has been studied for a few decades[14,15,16]. The lower bounds of Luczak and Achlioptas are used for comparison with our approximate solutions obtained in the experiments. On the other side, some lower bounds for frequency assignment problem in the applications of mobile phone networks have been treated by many authors[17,18].

3   Problem Description and Algorithms

Many combinatorial problems can be expressed as constraint satisfaction problems. As with many problems, the frequency assignment problem can be viewed as a CSP[8,9,13]. A CSP consists of a set of variables, each variable has associated with a finite set of possible values, and finally there is a set of constraints that restrict the values that can be simultaneously assigned to the variables.

3.1   Basic Concepts

Definition 1. *A CSP is a tuple $< V, D, C >$*

- *V is a finite set of variables, $\{V_1, V_2, \ldots, V_n\}$*
- *D is a finite set of possible values, $\{D_1, D_2, \ldots, D_n\}$, where D_i is domain of V_i*
- *C is a finite set of constraints, $\{C_1, C_2, \ldots, C_k\}$, where C_i restricts the values that V may take simultaneously.*

The arity of a constraint is the number of variables it constrains. A CSP is called binary if each constraint is unary or binary. This means that at most two variables are involved in any constraint. A binary CSP can be depicted as a constraint graph $G = (V, E)$. A vertex $v \in V$ is introduced for each given variable and an edge $e \in E$ is introduced for each given constraint. A solution to a CSP is an assignment of every variable to a value in its domain in such a way that all the constraints are satisfied. We can also find an optimal or sub-optimal solution for a given some objective. The total number of frequencies assigned to the transmitters can be considered as an objective. The problem is solved by a combination of systematic search through the set of all possible assignments to the decision variables, and constraint propagation that uses the constraints to prune the search space that cannot lead to a solution.

The traditional approach to minimizing the number of assigned frequencies is to perform graph coloring on the channel graph. Because each color can be identified with one channel, the minimal number of colors, which is called the chromatic number in graph theory, equals the minimal number of channels needed. Coloring this channel graph is non-trivial and NP-hard.

A graph G is a collection of points and lines connecting some subset of them. The points are most commonly known as a finite set $V(G)$ of vertices and lines are also known as a finite set $E(G)$ of edges which is unordered pairs (u, v) of distinct vertices. Two vertices u and v are adjacent if they are joined by a graph edge. The neighborhood of a vertex u in a graph is the set of all the vertices adjacent to v. The degree or valency of a vertex is the number of edge ends at that vertex. A k-coloring of a graph is an assignment of one of k possible colors to each vertex of G such that no two adjacent vertices receive the same color. The chromatic number $\chi(G)$ is the minimum k for which G has a k-coloring. The determination of $\chi(G)$ for arbitrary G is known to be NP-complete. A constraint graph G is a finite graph with each edge labeled with an integer in the set $\{1, 2, \ldots, k\}$. Let $V = V(G)$ denote the vertex set of G. Let $T_0, T_1, \ldots, T_k$ be sets of non-negative integers with $0 \in T_0$ and $T_0 \subseteq T_1 \subseteq \ldots \subseteq T_k$.

Definition 2. *A frequency assignment in G is a mapping $f : V \to F$ (where F is a set of non-negative integers $1, 2, \ldots, f$) such that if edge (u, v) is labeled i then*

$$|f(u) - f(v)| \notin T_i.$$

3.2   The CSP Model of Frequency Assignment for LPFM

In frequency assignment problem, each radio transmitter can be viewed as a variable, so V is the set of all transmitters. The set of possible frequencies make the domain for each transmitter variable. D_i is the domain of frequencies that the transmitter i can take. There exist following types of frequency separation constraints.

• Co-channel constraints (CCC) : A pair of transmitters located at different cells must not be assigned the same channel, unless they are sufficiently geographically separated. If f_i and f_j are the frequencies assigned to transmitters i and j , $d(i, j)$ is a distance between transmitter i and j, k is a sufficiently separated distance, then this gives rise to constraints of the form:

$$f_i \neq f_j \text{ ,if } d(i, j) \leq k$$

• Adjacent channel constraints (ACC) : any channels assigned to adjacent cells must have a specified frequency separation because there is still the potential for interference normally within one to three channel-distance. Therefore a number of constraints arise of the the following form:

$$|f_i - f_j| \geq c_{ij}$$

where $i \neq j$, and c_{ij} is the required number of channel separation.

The ACC can be represented as an NxN symmetric compatibility matrix corresponding to a constraint graph with N vertices. A convenient representation of interferences is by means of interference graph or constraint graph. Each transmitter is represented by a vertex $v \in V$. Two vertices u and v are connected by an edge $(u, v) \in E$. For each pair of transmitters (u, v), there is a value $c_{uv} \geq 0$ which measures the minimal frequency separation required for transmitters u and v.

• Co-site constraints (CSC) : any pair of channels in the same cell must have a specified frequency separation, typically 4 channels in FM broadcasting.

• IF constraint : any channels assigned to the same cell or adjacent cells must have an IF frequency separation. An intermediate frequency (IF) is a frequency to which a carrier frequency is shifted as an intermediate step in transmission or reception. It is the beat frequency between the signal and the local oscillator in a superheterodyne radio receiver. So, we should add IF constraint to a set of constraints.

• Blocked frequency constraint : some frequencies cannot be assigned to the cells because some commercial broadcasting operators already use the frequencies. These frequencies can be represented by k-element vector B.

• Preassigned frequency constraint : some frequencies are already occupied by local LPFM broadcasting within the service area, so these channels cannot be assigned.

As is clear from the problem statement, transmitters are contained within cells. For simplicity and maybe real case, all cells may have one radio transmitter at most. In cellular phone system, CSC and CCC must be exist in the constraints. But in our case, those constraints can be removed because only one channel is assumed being assigned to the cell. Some cells cannot have transmitter due to the geographic reason. As noted above, a FAP can be represented using a graph and this graph representation is also called a macro-graph and is shown in the paper[13].

3.3   CSP Algorithms for Frequency Assignment Problem

The CSP can be solved using the generate-and-test paradigm, but more efficient method uses the backtracking in which variables are instantiated sequentially. The backtracking method essentially performs a depth-first search of the search space. The first frequency d_1 from the domain D will be assigned to transmitter i, i.e. $f_i = d_k, k = 1$. The backtracking algorithm then checks for constraint violations with all already assigned frequencies $f_j, \forall j < i$. If there is no violation, the algorithm checks the next frequency in the domain. This means $f_i = d_{k+1}$. If all frequencies from the domain have been unsuccessfully assigned, the forward move fails and the algorithm backtracks from the previous transmitter. The backtracking algorithm is defined by a variable ordering and a value ordering rule. Experiments show that good variable and value ordering move a solution of CSP to the left of the search tree, so the solution can be found quickly by the simple or heuristic backtracking algorithm.

At level i of the backtracking algorithm, i variables have been assigned a value from their domain, $i = 0, 1, \ldots, n$. The variable ordering is usually based on some heuristic measure of the difficulty of a variable decision. A variable can be selected by static, i.e. determined in advance before the problem is solved, or dynamic. The "smallest domain first" (SDF) variable ordering heuristic is usually used because it minimizes the branch depth. It also has been called the $fail-first$ heuristic because it picks a variable that is most likely to cause a failure soon, thereby pruning the search tree. The idea is that if it is difficult to find a value for a given variable, then it is better to assign a value to this variable as early as possible. Unfortunately, in our case, the quality of the obtained solutions is not satisfactory because the domain of the variables has uniform size. Instead, we implemented a variant of the heuristic called $k-coloring$ variable ordering which is proved to obtain good results in a reasonable amount of time. Since the degree of the vertex information is not provided, we have to add a new data $degree$ to the model. The transmitters are ordered by decreasing of their degree, i.e. the number of incident edges in the constraint graph. The used frequency numbers is more minimized using this k-coloring heuristic.

Once a transmitter has been selected, the algorithm generates a frequency for it in a nondeterministic manner. Once again, the model specifies a heuristic for the ordering in which the frequencies must be tried. To reduce the number of frequencies, the model says to try first those values that were used most often in previous assignments. This is called a $mostly-used$ value ordering rule.

Besides the variable and value ordering rule, another important aspect of CSP is consistency enforcing techniques. Consistency enforcing techniques are applied in every node of the search tree and they eliminate values from the domain of a variable, which are known not to occur in any feasible solutions. Usually the more powerful the consistency enforcing technique is, the more computing time it takes. We describe two consistency enforcing techniques for binary constraints, namely forward checking and arc consistency which have proved to be the most useful in practice.

Whenever a variable u is assigned, the forward checking process looks at each unassigned variable v that is connected to u by a constraint and delete from v's domain any value that is inconsistent with the value chosen for u. Forward checking has detected that the partial assignment is inconsistent with the constraint of the problem, and the algorithm will therefore backtrack immediately. In comparison, forward checking is more work intensive than backtracking, however it fails earlier and thus greatly reduces the number of checked assignments compared with the simpler backtracking algorithm.

Although forward checking detects many inconsistencies, it does not detect all of them because it does not look far enough ahead. Constraint propagation is the general term for propagating the implications of a constraint on one variable onto other variables. The idea of arc consistency provides a fast method of constraint propagation that is substantially stronger than forward checking.

Arc (u, v) refers to a directed arc in the constraint graph, such as the arc from vertex or transmitter u to v. Given the current domains of u and v, the arc is consistent if, for every value x of u, there is some value y of v that is consistent with x. On the other hand, the arc (u, v) being consistent does not tell anything about the consistency of the reverse arc (v, u). The arc can be made consistent by deleting the value y from the domain of v.

Applying arc consistency has resulted in early detection of an inconsistency that is not detected by pure forward checking. Arc consistency checking can be applied either as a preprocessing step before the beginning of the search process, or as a propagation step after every assignment during search like forward checking. In either case, the process must be applied repeatedly until no more inconsistencies remain.

3.4   Some Lower Bounds on FAP

Lower bounds for the FAP can be used to assess the performance of the approximate solutions, and have been studied by several authors [11,17,18]. A clique of a graph is its maximal complete subgraph, and the problem of finding the size of a clique for a given graph is an NP-complete problem. The number of graph vertices in the largest clique of G denoted $\omega(G)$. For an arbitrary graph,

$$\omega(G) \geq \sum_{i=1}^{n} \frac{1}{n - d_i} \tag{1}$$

where d_i is the degree of graph vertex i. One of the most important invariants of a graph G is its chromatic number $\chi(G)$, namely the minimum number of colors required to color its vertices so that no pair of adjacent vertices has the same color. The chromatic number of a graph is equal to or greater than its clique number. In the case of complete graph K_n, $\omega(G)$ is n.

$$\chi(G) \geq \omega(G) \tag{2}$$

A random graph is a graph in which properties such as the number of graph vertices, graph edges, and connections between them are determined in some random way. The classical model of random graphs, in which each edge on n vertices is chosen independently with probability p, is denoted by $G(n, p)$. This model has been studied intensively in the past four decades. Since 1970s, work on $\chi(G(n, p))$ has been in the forefront of random graph theory, motivating some of the field's most significant developments. Indeed, one of the most fascinating facts about random graph is known[16]. $G(n, p)$ has some property almost surely if the probability that it has the property tends to 1 as $n \to \infty$.

Theorem 1. [15] *There exists a constant d_0 such that if $d = d(n) = np(n) > d_0$ and $p(n) \to 0$ then almost surely*

$$\frac{d}{2 \log d} \left(1 + \frac{\log \log d}{\log d} \right) < \chi(G(n, p)) < \frac{d}{2 \log d} \left(1 + \frac{30 \log \log d}{\log d} \right) \tag{3}$$

It is also known that for every $d \in (0, \infty)$ there exists an integer k_d such that almost surely $\chi(G(n, d/n))$ is either k_d or $k_d + 1$.

Theorem 2. *[16] Given $d \in (0, \infty)$, let k_d be the smallest integer k such that $d < 2k \log k$. With probability that tends to 1 as $n \to \infty$,*

$$\chi(G(n, d/n)) \in \{k_d, k_d + 1\}$$

Indeed, we determine $\chi(G(n, d/n))$ exactly for roughly half of all $d \in (0, \infty)$.

4   Computational Results and Evaluation

Modeling a geographical situation very close to the real world one would be too computationally expensive. For the simplicity, we choose a grid of square cells for the topology of the region. We consider a 5,000m by 5,000m square service area for each instance. The each cell is 500m x 500m size and we have all 100 cells for the service area. As is clear from the problem statement, transmitters are contained within cells, and all cells may have one radio transmitter at most.

Some cells may not have transmitter due to the geographic reason. Some frequencies cannot be assigned to the transmitter because commercial broadcasting operators already use the frequencies. The VHF frequency band for FM radio is in the range of 88 - 108 MHz, and is usually partitioned into a set of channels, all with the same bandwidth of 200KHz. For this reason the channels are usually numbered from 1 to 100. In Seoul metro area we choose 75 channels as a blocked frequency that may not be assigned. Of course, commercial FM broadcasting has different output powers and transmitter locations. We choose an experimental data performed by Korean radio station management agency[19].

Widely used intermediate frequencies (IF) are 10.7MHz in FM superheterodyne radio receiver. This makes IF constraint, so 10,700KHz/200KHz=53.5 (53 and 54 for real channels) should not be apart between adjacent transmitters within the service area.

We assume all LPFM transmitters are at 25m height above average terrain (HAAT) with the power of 1 watt effective radiated power (ERP). The radio service area is within about 2 km radius. All the experiments are done in Pentium IV/2.4GHz CPU, 2.6GB RAM with ILOG OPL3.7.

In general, a non-intelligent exhaustive search algorithm for graph coloring would be required to check total of 100^{100} assignments in case of using 100 cells and 100 channels. In our FAP case, 75 channels are removed from the domain of each transmitter due to blocked frequency constraint. Then the whole search space can be greatly reduced to 25^{100} while the total constraints are increased because we added adjacent constraint, IF constraint, preassigned frequency constraint in addition to co-channel constraint.

In order to assess how close the given assignment to the optimal solution, it is necessary to use lower bounds for the problem. Several papers give lower bounding techniques for the frequecy assignment problem[15,16,17,18]. Upper bound of this chromatic number can be obtained from Brooks' Theorem. If G is neither a complete graph nor a circuit of odd order, and if the maximal degree of any vertex in G is d, then $\chi(G) \leq d$.

The theoretical lower bound of $\chi(G)$ from Theorem 2, where d is less than n, shows 15 or 16 in case of using 25 channels and 75 cells. The computed clique size is 10, and the computed $\chi(G)$ is 16 when we applied both k-coloring variable and mostly-used value ordering rule. By comparison with theoretical lower bound, we can assume the computed approximate solution is close to optimum value. Table 1 shows the computed $\chi(G)$ when each heuristic is applied for backtracking algorithm. The result also shows that both k-coloring variable and mostly-used value ordering rule made a great performance over the other methods. It also shows that inappropriate combination of variable ordering and value ordering rule may increase the number of choice points or computing time. Choice points are the number of nodes internal to the search tree.

Table 1. Computed approximate solutions

	KVMV [a]	KVO [b]	MVO [c]	SB [d]
Computed $\chi(G)$	16	18	17	17
Choice points	41,526	41,466,336	3,444,675	7,869,970
Time (sec.)	8.17	1,807,02	394.88	788.02

[a] KVMV: both k-coloring variable and mostly-used value ordering
[b] KVO : k-coloring variable ordering only
[c] MVO : mostly-used value ordering only
[d] SB : simple backtracking

5   Conclusion and Future Work

Solving a frequency assignment problem is a NP-hard problem and many researchers have tried to get good approximate solutions using heuristic search. We present a CSP model and algorithms that are appropriate to the FAP to reduce the search space and computing time greatly. In the field of graph coloring problem, great progress has been done on the lower bound of chromatic number of the random graph, so the performance of algorithms presented in this paper can be assessed. By comparing the result that are obtained from the analytical calculation with the computer experiments, the solution can be regarded very close to or as good as an optimum value. The experiment also shows that applying value ordering rule without a good variable ordering rule, at least in FAP case, may cause the problem more complicated.

Currently output power of LPFM in Korea is 1 watt, but other countries use more powers and several kinds of powers within the same service area such as 10 or 100 watts. Then the constraint matrix for the adjacent channel constraints should be more complicated. In that case, we may need to use weight for the constraint edges.

References

1. W.K. Hale, Frequency Assignment:Theory and Applications, Proceedings of the IEEE, vol. 68, pp. 1497-1514, Dec. 1980.
2. J.F. Arnaud, Frequency Planning for Broadcast Services in Europe, Proceedings of the IEEE, vol. 68, pp. 1515-1522, Dec. 1980.
3. K. Aardal, et al, Models and Solution Techniques for Frequency Assignment Problems, Technical Report ZIB-Report01-40, 2001.
4. C. Mannino, A. Sassano, An Enumerative Algorithm for the Frequency Assignment Problem, Discrete Applied Mathematics, vol. 129, pp. 155-169, 2003.
5. Federal Communications Commission, http://www.fcc.gov/mb/audio/lpfm.
6. World Association of Community Radio Broadcasters, http://www.amarc.org.
7. A.I. Giortzis, L.F. Turner, Application of Mathematical Programming to the Fixed Channel Assignment Problem in Mobile Radio Networks, IEE Proceedings of Communication, vol. 144, pp.257-264, Aug. 1997.
8. M. Carlsson, M. Grindal, Automatic Frequency Assignment for Cellular Telephone Using Constraint Satisfaction Techniques, Proceedings of the Tenth International Conference on Logic Programming, pp. 647-665, 1993.
9. J.P. Walsher, Feasible Cellular Frequency Assignment Using Constraint Programming Abstractions, Proceedings of the Workshop on Constraint Programming Applications, 1996.
10. N. Dunkin, S. Allen, Frequency Assignment Problems: Representation and Solutions, Technical Report CSD-TR-97-14, Royal Holloway, University of London, 1997.
11. S.U. Thiel, S. Hurley, D.H. Smith, Frequency Assignment Algorithms, Technical Report Year 2, University of Wales Cardiff, 1997.
12. L. Wang, S. Li, F. Tian, X. Fu, A Noisy Chaotic Neural Network for Solving Combinatorial Optimization Problems: Stochastic Chaotic Simulated Annealing, IEEE Trans. on Systems, Man, and Cybernetics, vol. 34., no. 5, pp. 2119-2125, 2004.
13. M. Yokoo, K. Hirayama, Frequency Assignment for Cellular Mobile Systems Using Constraint Satisfaction Techniques, IEEE Int'l Proceedings of Vehicular Technology Conferences, 2000.
14. B. Bollobas, The Chromatic Number of Random Graphs, Combinatorica 8, pp.49-55, 1988.
15. T. Luczak, The Chromatic Number of Random Graphs, Combinatorica 11, pp.45-54, 1991.
16. D. Achlioptas, A. Naor, The Two Possible Values of the Chromatic Number of a Random Graph, Annals of Mathematics, 2005.
17. A. Gamst, Some Lower Bounds for a Class of Frequency Assignment Problems, IEEE Proceedings on Vehicular Technology, vol. VT-35, pp. 8-14, 1986.
18. D.H. Smith, S. Hurley, Bounds for the Frequency Assignment Problem, Discrete Mathematics 167/168, pp. 571-582, 1997.
19. D. Kim, E. Han, J. Park, Research on Active Use of Low Power Broadcasting, Technical Report KORA-2001-17, Korea Radio Station Management Agency, 2002.

An Intelligent Mobile Learning System for On-the-Job Training of Luxury Brand Firms*

Chien-Chang Hsu and Yao-Wen Tang

Department of Computer Science and Information Engineering,
Fu-Jen Catholic University
510 Chung Cheng Rd., Hsinchuang, Taipei, Taiwan 242

Abstract. Service quality is one of the key successful factors for the luxury brand firms. The shop employee is the direct person to provide the quality service to the customers. The employee must possess sufficient knowledge to introduce distinguishing features of company's merchandises. How to offer enough merchandise knowledge is a hard work for the firms every season. The mobile computing realizes the dreams of people to access information anytime and anywhere through wireless transmission channel. This paper proposes an intelligent system for on-the-job training of luxury brand firms to conduct mobile learning. The system uses WLAN and RFID to locate the learner and learning target. It uses collaborative filtering to estimate possible learning time of the learner from the other learners with similar preference ranking. It then uses estimated learning time to schedule the best learning objects for the learner by genetic algorithm. The system not only saves a large amount of expense of the firms but also enhances the service quality to retain their important customers.

Keywords: On-the-job training, RFID, Collaborative filtering, Genetic algorithm.

1 Introduction

Luxury brand firms usually sell high price or valuable merchandises to wealthy people in every country. Most firms are the well-known international firms, such as garment and ornamental companies. The important customers of these companies usually buy merchandises above ten thousand dollars every season or in one single purchase. These companies are eager to provide the best service to retain their customers. These firms are devoted to attract the customers by presenting new arrivals and fashion products to them every season. The store employees are the representatives of the firms to interact with the customers. They are the direct contacts that introduce new products to the customers. The employees are required have specific characters and communication skill to interact with their customers. She/He needs sufficient knowledge of all merchandise including design concept, material

---

* This work is partly supported by National Science Council of ROC under grants NSC 95-2745-E-030-004-URD.

A. Sattar and B.H. Kang (Eds.): AI 2006, LNAI 4304, pp. 749–759, 2006.

© Springer-Verlag Berlin Heidelberg 2006

name, style characteristics, and other pattern distinguishing features. It is the minimum requirements of the employees to possess the knowledge in order to introduce or explain all merchandise features. It is not an easy task to educate them in a few hours, not even to mention to update new product information in details. However, training cost is very high for these companies. For most companies, training courses are only available in one location such as the company headquarter. Providing trainings to employees worldwide is a very time consuming and troublesome task especially for the multi-national companies. These companies must spend extra costs in hiring relay persons and traveling expense of the employees. It is even worse for those companies with high turnover rate. They must train new employees or novices to be familiar with the products.

Mobile learning provides the characteristics of pervasive teaching anywhere and anytime [13]. It can overcome space and time limitations of the learner to acquire new knowledge. The learners can use cell phone or personal digital assistant (PDA) to find the appropriate courses. Context-awareness capability is also an important feature to support the ubiquitous learning. Many systems have been proposed to support pervasive teaching. For example, Bird-Watching Learning system (BWLs) and Butterfly-Watching Learning system (BWLs) integrate PDA, CCD camera, and Wi-Fi technique to watch birds and butterflies outdoors [2], [3], [4]. They use CCD camera of PDA to capture picture of the butterfly. The system then sends the pictures back to the server to search related information of the animal and insect. The server recognizes this kind of animal and insect and helps the learner to know the knowledge. eSchoolBag systems use PDA, wireless network, and notebook to construct ad hoc classroom for outdoor teaching environment. The environment can be used to deliver any useful information in museum and outdoor teaching course [1]. MOOsburg and MOOsburg++ construct a mobile virtual community. They provide virtual community map for the learner to conduct collaborative learning [12], [7].

However, most systems focus on mobile techniques to support ubiquitous learning. Both of them are static and passive systems, because time and task limitations of learner context are not considered. It is the necessary condition to carry the right learning course at the right time according to the learner profile and context model. The learner profile should contain dimension of identity, spatial-temporal, facility, activity, learner, and community [14]. The context model uses detectors to perceive the environment objects of the learner for conducting the context awareness. H. Ogata uses the RFID and PDA to provide a context-awareness language learning system [9], [10], [11]. The system uses RFID to detect the environment information to support different learning courses. But none of these systems have been proposed for on-the-job mobile learning system to improve the job skill of employees. Moreover, the cost of RFID tag is another problem to be considered in the return-of-investment (ROI) of a company. The price of a RFID tag depends on the tag size and the active or passive type. For the luxury brand companies, it is a very small investment compared to the high price merchandise. Therefore, it is a very appropriate place to use RFID in the luxury brand firms.

This paper proposes an intelligent mobile learning system for on-the-job training of luxury brand firms. The system contains two modules, context manager and course selector. The context manager is responsible for acquiring and analyzing the context information of the learner. It then updates the learner model to provide correct learner information. It uses collaborative filtering to predict available learning time of the learner. The course selector uses case-based reasoning to select the appropriate course for the learner in the specific context.

The rest of this paper is organized as follows. Section 2 introduces the architecture of intelligent mobile learning system for on-the-job training of luxury brand. Moreover, section 3 and 4 explore the context manager and course selector. Subsequently, section 5 demonstrates the application of the intelligent mobile learning system in the garment firm. Finally, section 6 concludes the work.

2   System Architecture

Figure 1 shows the architecture of intelligent mobile learning system for on-the-job training of luxury brand firms. The architecture contains two modules, namely, context manager and course selector. *Context manager* is responsible for acquiring and maintaining the learner information and predicting possible learning time. Context manager uses wireless local area network (WLAN) to communicate with the PDA of the learner to acquire his or her location and surrounding information. It then updates the learner model according to the new information of the learner. The learner model contains four types of learner information, which is, personal, spatial-temporal, environmental, and activity context. Personal context stores basic identification information and past learning history about the learner. Spatial-temporal context represents the starting time, ending time, and location of the learning activity. Environmental context contains facilities and other learners around the learner. Activity context is used to illustrate the operations and behavior of the learner during learning process. Moreover, context manager uses collaborative filtering to predict possible learning time of the learner. The possible learning time is estimated by comparing preference ranking between learners to find the learner with similar interesting and learning behavior. Pearson correlation is used to evaluate the strength of preference ranking between two learners. Context manager then uses context model to predict the learning plan of the learner.

Course selector uses case-based reasoning to select the best courses from the curriculum case library for the learner and schedule personalized learning objects. Curriculum case library contains precondition, learning path, learning time, location, and past learning cases. Each curriculum contains course name, location, learning time, learning goal, job skill, course level, learning tool, and learning target. The precondition identifies necessary condition of the learner to enter the course, such as, job position, pre-learning course, working task, learner location, and season of the year. Course selector then uses the predicted learning time and learner model to select similar cases from the course learning case library. It ranks similar courses of the learner according to similarity measurement. Finally, course selector uses learning conceptual diagram to arrange the learning objects according to the learning history of the learner.

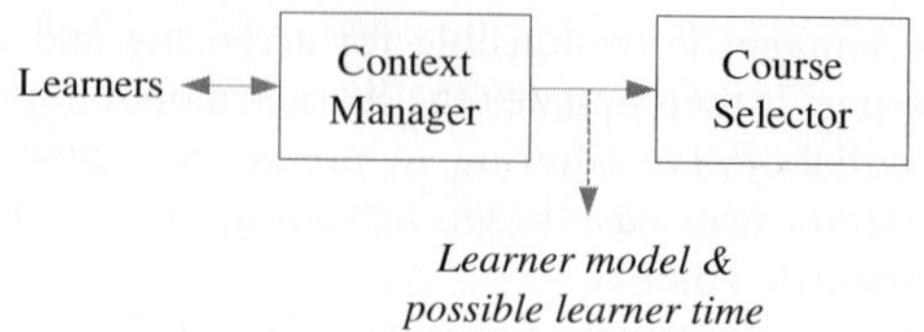

Fig. 1. Intelligent mobile learning system architecture

3   Context Manager

3.1   Learner Information Acquisition and Maintenance

Context manager is responsible for context awareness and information acquisition of the learner. Specifically, context awareness identifies the learner by login registration through WLAN. RFID reader in the learner side retrieves the related tag number from RFID tag around the learner and uploads it to the context manager by WLAN. Each tag number represents specific zone and product code of the merchandise. Context manager then uses product code to locate the learner location. Context manager also sends new joined learner, surrounding merchandises, and the nearest learners to the learner.

Learner model contains personal, spatial-temporal, environmental, and activity context of the leaner. Basically, *personal context* represents the basic learner information. It contains the learner's static and dynamic information. The former represents personal invariant information of the firm. It includes name, employee number, belonged department, position, professional career, and educational background of the learner. The latter contains personal knowledge map and past learning history of the leaner. Personal knowledge map represents the learner's professional literacy of the merchandises. It uses personal ontology to represent the learned knowledge. Table 1 displays the example knowledge of fashion shoes. Feature of knowledge includes design concept, pattern, sock, stamp, thread, eyelet,

Table 1. Knowledge of fashion informal footwear

Item no: 42-0139-595-01		
Pattern	Driving shoe	Design concept:
Type	Sport	The shoes are designed for driving. The
Stamp	Flat	exclusive sole pattern provides more safety
Thread	12 needle/inch	and comfort shoes to the customers. The
Eyelet	10	vamp covered with microporous which can
Sole	Rubber	absorb and expel sweat and keep water from coming in.
Material	Leather 90% and Suede 10%	Diagram:
Waterproof	Yes	
Sundries	Shoe brush and polish	
Size	#6 to #9	
Shoelace	Dark brown	

sole, material, waterproof, sundries, size, and shoelace of the shoes [8]. Learning history includes completed learning objects and courses by the learner. It usually includes course number, finished date, sequence, learning time, latest update time, examination score, and examination paper number. *Spatial-temporal context* is the occurring location and the time of the learning activity. This information is collected from the context collector by using WLAN and RFID devices. *Environmental context* records duty, surrounding merchandises, and the nearest learners of the employee. *Activity context* records tasks done by the learner. It contains routine tasks, customer arriving rate, and idle time of the employee. Figure 2 illustrates one example of learner model. Finally, context manager updates the learner model of above information.

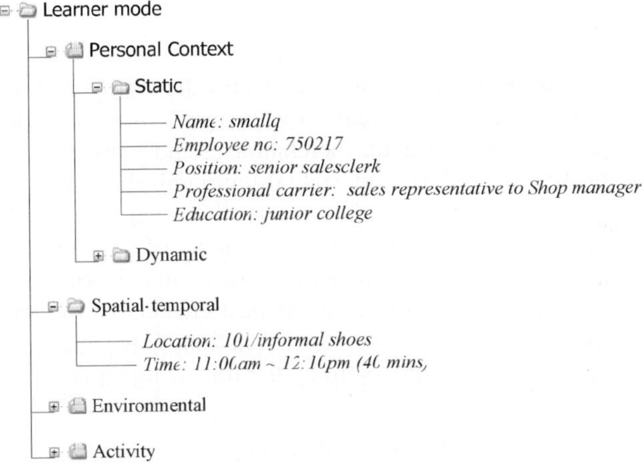

Fig. 2. Example of learner model

3.2 Learning Time Prediction

Context manager uses collaborative filtering to predicate possible learning time of the learner. Collaborative filtering uses the learner's preference ranking to find the learning cases with similar learning behaviors. Possible learning time, PS, is computed from similar cases of the learner with maximum preference ranking similarity.

$$PS = AT(u) \times \frac{T(v)}{AT(v)} \tag{1}$$

where u, v, AT, and T are the leaner, similar learner with maximum preference ranking, average learning time, and learning time of the predicate case. The learner with maximum preference ranking similarity is computed from Pearson correlation. It evaluates the correlation strength (ρ) of the learning behavior between learner u and v to search for the similar learners. The learner with maximum preference ranking similarity is the learner with maximum correlation strength.

$$\rho(u,v) = \frac{\sum_{j \in J_u \cap J_v}(r_{u,j} - \bar{r}_u) \quad \sum_{j \in J_u \cap J_v}(r_{v,j} - \bar{r}_v)}{\sqrt{\left(\sum_{j \in J_u \cap J_v}(r_{u,j} - \bar{r}_u)^2\right)\left(\sum_{j \in J_u \cap J_v}(r_{v,j} - \bar{r}_v)^2\right)}} \tag{2}$$

where r, $\bar{r}$, J_u, and J_v, are the learning case ranking, average learning case ranking, past learning cases of learner u, and past learning cases of learner v. Notably, similar learning cases are selected by using fuzzy c-means algorithm of the case selector in the next section to cluster past learning cases based on the learner model.

4 Course Selector

4.1 Case Selection

Case selection finds the most appropriate courses for the learners from the curriculum case library. Each learning case contains course name, user identification, location, learning time, sequence, finished date, merchandise, and learner preference ranking. It uses personal profile clustering and spatial-temporal similarity to select similar cases according to the estimated learning time and learner model. Personal profile clustering uses personal context and activity context of the learner to cluster the learner into the nearest cluster for finding the most similar learners. It uses fuzzy c-means clustering algorithm [5], [6] to find the best clusters of the learner. Fuzzy c-means algorithm minimizes the standard lost function, l, to assign the learners into cluster with varying degree of membership by the following equations.

$$l = \sum_{k=1}^{p}\sum_{i=1}^{n}[\mu_k(x_i)]^m \|x_i - c_k\|^2 \tag{3}$$

where p, n, μ_k, x_i, m, and c_k are the specified clusters, number of cases, membership of cluster k, i^{th} case, fuzzification parameter, and center of the kth cluster.

Spatial-temporal similarity selects the most similar cases of the learner in the same clusters. It uses location, learning time, merchandise, and learner preference ranking to compute the case similarity (Sim).

$$Sim = \sum_{i=1}^{n}(P_i - C_i) \times w_i \tag{4}$$

where n, P, C, and w are the number of cases, feature value of learners, feature value of case in the same clusters, and feature weight. It selects the cases with higher similarity value, above the threshold, as the candidate cases for arrangement of the learning objects.

4.2 Learning Object Scheduling

Learning object scheduling uses learning conceptual diagram to arrange learning objects from the similar cases. The learning conceptual diagram is represented by five tuples, G=(S, F, C, E, P), where S, F, C, E, P represent the start node, final node, learning concept, essential condition, and pass score (Fig. 3). The start node and final node represent the initial and last course of the diagram. The learning concept represents the knowledge chunk of the course. The essential condition contains specific concepts that must be learned prior the concept. Past score represent the minimum score to get the learning concept. A learning path means the path from the start node to the end node through a number of learning objects which must satisfy the essential condition of each learning object.

The course selector uses genetic algorithm to schedule the most appropriate learning path for the employee. Each bit of the chromosome was represented by one binary bit for one learning object. Genetic algorithm searches the best learning path by computing the maximum learning benefit (LB) as the fitness function under the estimated learning time constraint.

$$LB = \max imun \left(\sum_{i=1}^{n} Benefit(LO_i) \right) \tag{5}$$

$$Staisfy \ Total \ learning \ time \leq predicted \ learning \ time$$

where n, Benefit, and LO are the number of learning in the conceptual diagram, the benefit after the learning object learned, and the learning object. The benefit of each learning object is the inverse value of the node number of learning path. Genetic algorithm uses selection, crossover, and mutation operators to find the best offspring.

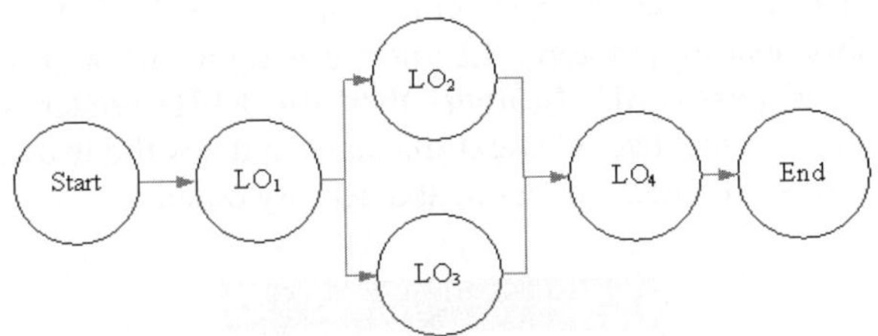

Fig. 3. Example of learning conceptual diagram

5   Application in the Garment Firms

Figure 4 shows the context diagram of the intelligent mobile learning system and the example merchandise with RFID tag in the garment firms. The merchandises of the garment firms contain man's suits, neckties, wallet, leather shoes, and other man's clothes. Figure 5 shows the example driving shoe as described in Table 1. The employee of the luxury brand firm uses handheld device, which is PDA with WLAN capability, to detect the RFID tag of the product. RFID reader is inserted to the CF slot of the PDA to read the zone code and product code of the clothing. Zone code contains merchandise information within the shop can be identified by the system to locate the employee. It uses 8 bits to represent the zone code. Product code is the item number of the product. It uses 32 bits to represent the product code. PDA sends the detected data to the backend mobile learning server. We use passive RFID tag and 13.25MHZ reader as the data collector (Fig. 6). As shown in Figure 6, gray box

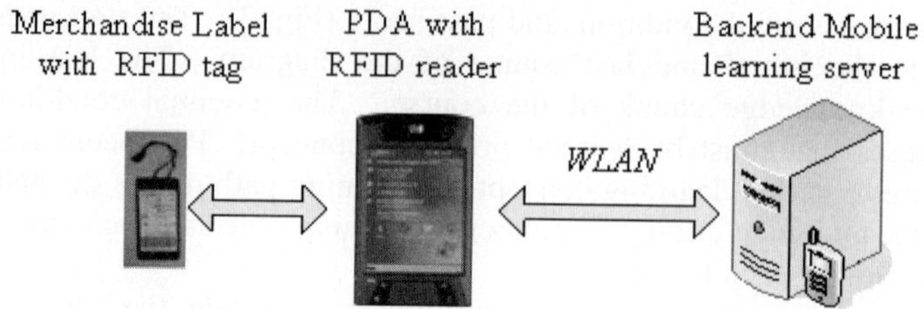

Fig. 4. Context diagram of the intelligent mobile learning system

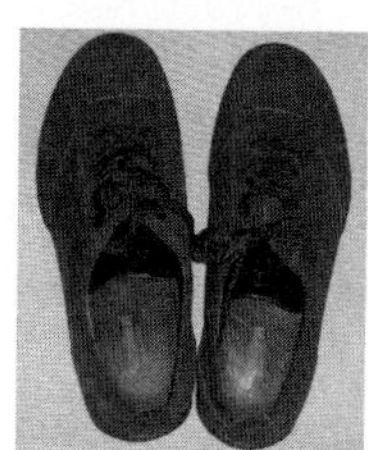

Fig. 5. Example merchandise of the garment firm

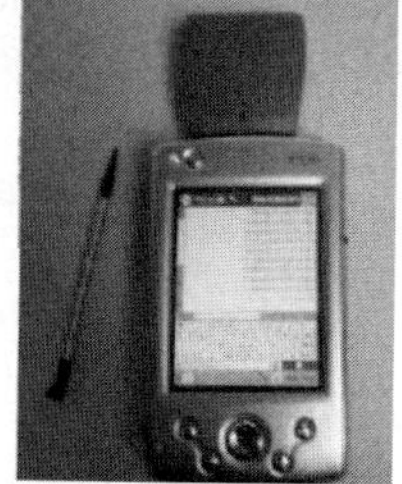

Fig. 6. Passive RFID tag and reader

in the CF slot of the PDA is the passive RFID reader. Figure 7 shows login frame of the intelligent mobile learning system. Learners use keyboard or point device to input the learner name and password. Learners then use RFID reader to read zone and product code of the merchandise. Context manager updates the learner model, that is, personal, spatial-temporal, environmental, and activity context.

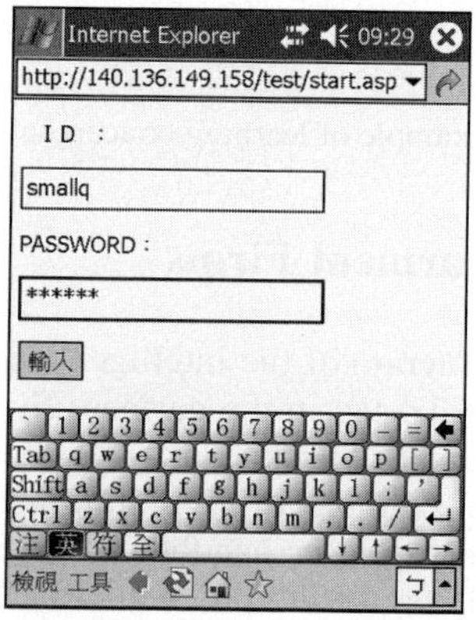

Fig. 7. Login frame of the intelligent mobile learning system

Context manager then estimates the possible learning time by collaborative filtering. Table 2 and 3 show the learning cases and the preference ranking with learning time of the similar learners. Table 4 shows the computed values by collaborative filtering. The most similar learner is Susan and the maximal correlation strength value was 1. The possible learning of Smallq is 54.74 minutes, that is,

$$PS = AT(u) \times \frac{T(v)}{AT(v)} = 43.8 \times \frac{55}{44} = 54.75 \cdot$$

Course selector uses fuzzy c-means algorithm to cluster the similar learners in the same cluster and selects the most similar cases as the shown examples in Table 4. Course selector then arranges the learning objects of the similar cases into conceptual diagram. Fig. 8 shows the recommended learning object after scheduling of genetic algorithm under the time constraint of 54.75 minutes.

Table 2. Example learning cases

	Case 1	Case 2	Case 3	Case 4	Case 5
Location	A	B	C	D	E
Position	Senior employee	Junior employee	Junior employee	Junior employee	Junior employee
Duty hours	09:00~17:00	24:00~08:00	12:00~20:00	13:30~18:30	09:00~13:00
Start learning time	13:05	14:20	15:50	10:30	11:00

Table 3. Example similar learners

		Case1	Case2	Case3	Case4	Case5	Predicting case
Susan	Ranking	6		8	9	5	8
Susan	Intervals	50		40	30	45	55
Angela	Ranking		3	6	6		7
Angela	Intervals		30	30	40		45
Candy	Ranking	7	6	8		7	7
Candy	Intervals	45	40	30		30	50
smallq	Ranking	7		9	10	6	
smallq	Intervals	50		40	45	40	?

Table 4. Collaborative filtering computation value

	$\overline{r}$	ρ	AT
Susan	7	1	44
Angela	5	-0.63	
Candy	7	0.41	
smallq	8		43.8

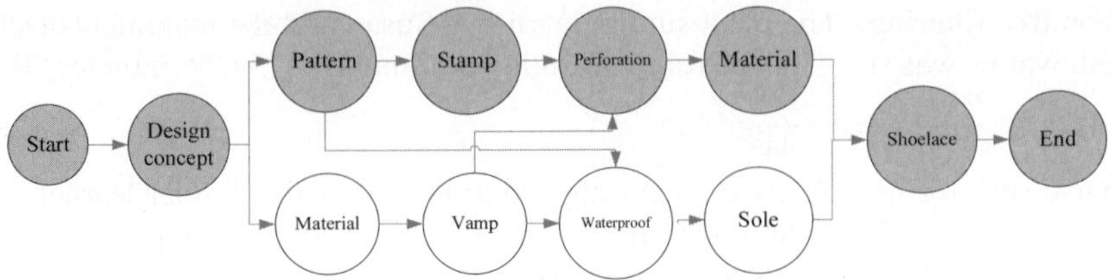

Fig. 8. Recommended learning objects

6   Conclusions

This paper proposed an intelligent mobile learning system for on-the-job training of luxury brand firms. The system uses context manager to collect learner information through WLAN. It uses RFID reader attached in the PDA as the front sensor to read location information of the learner by detecting the zone code from merchandise's RFID tag. Context manager maintains the learner model for estimating possible learning time. It uses collaborative filtering approach to find similar learner from the course preference ranking. The course selector uses fuzzy c-means algorithm to cluster learners with similar learner model in the same cluster. It then uses genetic algorithm to schedule the best learning object sequence under the possible constrained learning time.

References

1. Chang, C. -Y., Sheu, J. -P.: Design and Implementation of Ad Hoc Classroom and eSchoolbag Systems for Ubiquitous Learning, Proceedings of the IEEE International Workshop on Wireless and Mobile Technologies in Education (WMTE'02) (2002) 8-14
2. Chen, Y. -S., Kao, T. -C., Sheu, J. -P., Chiang, C. -Y.: A Mobile Scaffolding-Aid-Based Bird-Watching Learning System, Proceedings of the IEEE International Workshop on Wireless and Mobile Technologies in Education (WMTE'02) (2002) 15-22
3. Chen, Y. -S., Kao, T. -C., Sheu, J. -P.: A Mobile Learning System for Scaffolding Bird Watching Learning, Journal of Computer Assisted Learning, 19 (2003) 347-359
4. Chen, Y. -S., Kao, T. -C., Yu, G. -J., Sheu, J. -P,: A Mobile Butterfly-Watching Learning System for Supporting Independent Learning, Proceedings of the 2nd IEEE International Workshop on Wireless and Mobile Technologies in Education (WMTE'04) (2004) 11-18
5. Cox, E.: Fuzzy Modeling and Genetic Algorithms for Data Mining and Exploration, Morgan Kaufmann, San Francisco (2005)
6. Hoppner, F., Klawona, F.: Fuzzy Clustering Analysis: Methods for Classification, Data Analysis, and Image Recognition, Wiley and Sons, Chicester UK (1999)
7. Carroll, J. M., Rosson, M. B., Isenhour, P., Ganoe, C., Dunlap, D., Fogarty, J., Schafer, W., Mtre, C. V.: Designing Our Town: MOOsburg, International Journal of Human-Computer Studies 54(5) (2001) 725-751
8. Main, J., Dillon, T. S., Khosla, R.: Use of Neural Network for Case-Retrieval in a System for Fashion Shoe Design, Proceedings of the Industrial and Engineering Applications of Artificial Intelligence and Expert Systems (IEA/AIE'95) (1995) 151-158

9. Ogata, H., Yano, Y.: Combining Knowledge Awareness and Information Filtering in an Open-ended Collaborative Learning Environment, Proceedings of International Journal of Artificial Intelligence in Education 11(1) (2000) 33-46
10. Ogata, H., Yano, Y.: Context-Aware Support for Computer-Supported Ubiquitous Learning, Proceedings of the 2nd IEEE International Workshop on Wireless and Mobile Technologies in Education (WMTE'04) (2004) 27-34
11. Ogata, H., Yano, Y.: Knowledge Awareness Map for Computer-Supported Ubiquitous Language Learning, Proceedings of the 2nd IEEE International Workshop on Wireless and Mobile Technologies in Education (WMTE'04) (2004) 19-26
12. Farooq, U., Isenhour, P., Carroll, J. M., Rosson, M. B.: MOOsburg++: Moving Towards a Wireless Virtual Community, Proceedings of the International Conference on Wireless Networks (2002) Available at: http://cscl.ist.psu.edu/public/projects/mobile/docs/MOO.pdf
13. Wang, C. -Y., Liu, B. -J., Chang, K. -E., Horng, J. -T., Chen, G. -D.: Using Mobile Techniques in Improving Information Awareness to Promote Learning Performance, Proceedings of the 3rd IEEE International Conference on Advanced Learning Technologies (ICALT'03) (2003) 106-109
14. Wang, Y. -K.: Context Awareness and Adaptation in Mobile Learning, Proceedings of the 2nd IEEE International Workshop on Wireless and Mobile Technologies in Education (WMTE'04) (2004) 154-158

Studying the Performance of Unified Particle Swarm Optimization on the Single Machine Total Weighted Tardiness Problem

K.E. Parsopoulos[1,2] and M.N. Vrahatis[1,2]

[1] Computational Intelligence Laboratory (CI Lab), Department of Mathematics,
University of Patras, GR–26110 Patras, Greece
{kostasp, vrahatis}@math.upatras.gr
[2] University of Patras Artificial Intelligence Research Center (UPAIRC),
University of Patras, GR–26110 Patras, Greece

Abstract. Swarm Intelligence algorithms have proved to be very effective in solving problems on many aspects of Artificial Intelligence. This paper constitutes a first study of the recently proposed Unified Particle Swarm Optimization algorithm on scheduling problems. More specifically, the Single Machine Total Weighted Tardiness problem is considered, and tackled through a scheme that combines Unified Particle Swarm Optimization and the Smallest Position Value technique for the derivation of job sequences from real–valued particles. Experiments on well–known benchmark problems are conducted with promising results, which are reported and discussed.

1 Introduction

The allocation of resources to tasks is a problem that arises very often in real–world applications. In general, problems of this type are characterized as *scheduling* problems and they are NP–hard [1, 2, 3]. The main goal in scheduling problems is the assignment of jobs (tasks) to a single or many machines such that some criteria that involve the minimization of a single or many objective functions are met.

The *Single Machine Total Weighted Tardiness* (SMTWT) problem is an NP–hard scheduling problem [1]. In SMTWT, a number, n, of jobs have to be sequentially processed on a single machine. Each job, $j = 1, 2, \ldots, n$, has a processing time, p_j, a due date, d_j, by which it should be completed, and a weight, w_j. All jobs are assumed to be available for processing at time zero. If C_j denotes the completion time of job j in a job sequence, then the tardiness of job j is defined as:

$$T_j = \max\{0, C_j - d_j\}.$$

The main goal in SMTWT problems is to find a job sequence that minimizes the sum of the weighted tardiness:

$$T = \sum_{j=1}^{n} w_j T_j. \tag{1}$$

A. Sattar and B.H. Kang (Eds.): AI 2006, LNAI 4304, pp. 760–769, 2006.
© Springer-Verlag Berlin Heidelberg 2006

Instances of the SMTWT problem with large number of jobs cannot be solved to optimality with traditional branch–and–bound algorithms [4]. To this end, different heuristics such as Earliest Due Date and Apparent Urgency, as well as optimization algorithms such as Simulated Annealing, Tabu Search, Genetic Algorithms and Ant Colony Optimization have been successfully applied for tackling the SMTWT problem [5,6,7].

Recently, Particle Swarm Optimization (PSO) was applied on task assignment problems [8], as well as on the SMTWT problem [9] with promising results. In the latter case, the Smallest Position Value (SPV) representation technique was developed to transform a real–valued point to a sequence of jobs. Also, the Variable Neighborhood Search (VNS) technique [10] was employed and proved to enhance significantly the performance of PSO [9]. Unified Particle Swarm Optimization (UPSO) was recently introduced as a unified PSO scheme that combines the exploration and exploitation properties of different PSO variants [11]. Preliminary results on different problems indicate the superiority of UPSO against standard PSO variants [12,13,14,15].

This paper constitutes a first investigation of UPSO on the SMTWT problem. The SPV representation scheme is adopted in our study for the derivation of job sequences from real–valued vectors. Our primary intention in this preliminary study was to assess the performance of UPSO itself and compare it with standard PSO schemes. Therefore, in order to avoid possible affection of the algorithms' dynamic, techniques like VNS that proved to significantly enhance the performance of the algorithms are not considered.

The rest of the paper is organized as follows. PSO and UPSO are briefly described in Section 2 along with the SPV representation scheme. Experimental results are reported and discussed in Section 3, and the paper concludes in Section 4.

2   Background Material

The emergent (collective) behaviors observed in natural systems have attracted a lot of attention the late years by computer scientists. Swarms of insects and animal flocks that consist of members with very limited space of actions can produce more complex behaviors as a collective, providing inspiration for the development of algorithms that can tackle NP-hard problems effectively. *Swarm Intelligence* is a subject of Artificial Intelligence that investigates the collective behavior in decentralized, self–organized systems, and promotes the development of population–based, adaptive optimization algorithms that are characterized by stochasticity, noise–tolerance and minimum requirements regarding the form of the objective function (differentiability, continuity etc.) [16].

PSO is a Swarm Intelligence algorithm introduced in 1995 by Eberhart and Kennedy [17]. The inspiration behind its development lies on the emergent behavior and information exchange in socially organized colonies of simple agents [16]. PSO was primarily used in numerical optimization tasks. However, a plethora of applications have been developed and reported in the relative literature

up–to–date [16, 18, 19, 20]. For completeness purposes, the following subsections
are devoted to the description of PSO, UPSO and SPV.

2.1 Particle Swarm Optimization

PSO is a population–based algorithm. It employs a population, called a *swarm*,
of search points, called *particles*, to probe the search space. Assuming an n–
dimensional optimization problem,

$$\min_{x \in \mathcal{S}} f(x), \quad \mathcal{S} \subset \mathbb{R}^n,$$

then the particles are n–dimensional vectors,

$$x_i = (x_{i1}, x_{i2}, \ldots, x_{in})^\top, \quad i = 1, \ldots, N,$$

that constitute a swarm, $\mathbb{S} = \{x_1, \ldots, x_N\}$. The numbering of particles in $\mathbb{S}$ is
arbitrary. Each particle moves in the search space with an adaptable velocity,

$$v_i = (v_{i1}, v_{i2}, \ldots, v_{in})^\top,$$

and stores the best position,

$$p_i = (p_{i1}, p_{i2}, \ldots, p_{in})^\top \in \mathcal{S},$$

it has ever visited, i.e., the position with the lowest function value so far.

The computational strength of Swarm Intelligence algorithms originates from
the interaction of agents that constitute the swarm, either with their environment
(through stigmergy) or directly among them through information exchange. This
attribute gives rise to the concept of *neighborhood*, which determines the immediate "social" environment of the agent. In PSO's framework, each particle is
considered to have a neighborhood consisting of a number of other particles.
These particles influence its movement with their best experience (i.e., the best
positions they have discovered). However, the neighborhood is not defined directly in the search space by using a distance measure among particles, but rather
in the space of the particles' indices, in order to promote the algorithm's ability
for global search and avoid the computational burden of computing distances
among all particles at each iteration of the algorithm.

Different neighborhood topologies have been proposed and applied in the literature with promising results [21, 22, 23]. The most common neighborhood topology is the *ring* topology, where the immediate neighbors of the particle x_i are
the particles x_{i-1} and x_{i+1}, while x_1 is considered to be the particle that follows
immediately after x_N. Thus, a neighborhood of radius ρ of x_i consists of the
particles $x_{i-\rho}, \ldots, x_i, \ldots, x_{i+\rho}$. The ring topology is the neighborhood scheme
that we adopted in our study. There are two main variants of PSO with respect
to the size of neighborhood. In the *global* variant, the whole swarm is considered
as the neighborhood of each particle, while, in the *local* variant, strictly smaller
neighborhoods are used.

Let g_i be the index of the best particle in the neighborhood of x_i, i.e., the index of the particle that attained the best position among all the particles of the neighborhood. Then, the position of x_i is updated according to the equations [24]

$$v_i^{(k+1)} = \chi \left[v_i^{(k)} + \varphi_1 \left(p_i^{(k)} - x_i^{(k)} \right) + \varphi_2 \left(p_{g_i}^{(k)} - x_i^{(k)} \right) \right], \tag{2}$$

$$x_i^{(k+1)} = x_i^{(k)} + v_i^{(k+1)}, \tag{3}$$

where $i = 1, \ldots, N$; k is the iteration counter; χ is a parameter called *constriction coefficient* that controls the velocity's magnitude; $\varphi_1 = c_1 r_1$ and $\varphi_2 = c_2 r_2$, where c_1 and c_2 are positive acceleration parameters, called *cognitive* and *social* parameter, respectively, and r_1, r_2 are random vectors that consist of random values uniformly distributed in $[0, 1]$. All vector operations in Eqs. (2) and (3) are performed componentwise. A stability analysis of PSO, as well as recommendations regarding the selection of its parameters are provided in [24, 25].

The performance of an optimization algorithm depends heavily on the balance between its exploration and exploitation ability. In the global variant of PSO, all particles are attracted by the same best position, converging faster towards specific locations in the search space. Thus, it has better exploitation abilities, in contrast to the local variant where the information of the best position of each neighborhood is communicated slowly to the other particles of the swarm through their neighbors in the ring topology, thereby promoting exploration.

2.2   Unified Particle Swarm Optimization

UPSO has been recently proposed as a unified scheme that harnesses the local and global PSO variants, combining their exploration and exploitation capabilities [12, 13, 14, 15]. Let $\mathcal{G}_i^{(k+1)}$ denote the velocity update of the particle x_i in the global PSO variant and let $\mathcal{L}_i^{(k+1)}$ denote the corresponding velocity update for the local variant. Then, according to Eq. (2), we obtain:

$$\mathcal{G}_i^{(k+1)} = \chi \left[v_i^{(k)} + \varphi_1 \left(p_i^{(k)} - x_i^{(k)} \right) + \varphi_2 \left(p_g^{(k)} - x_i^{(k)} \right) \right], \tag{4}$$

$$\mathcal{L}_i^{(k+1)} = \chi \left[v_i^{(k)} + \varphi_1' \left(p_i^{(k)} - x_i^{(k)} \right) + \varphi_2' \left(p_{g_i}^{(k)} - x_i^{(k)} \right) \right], \tag{5}$$

where k denotes the iteration number; g is the index of the best particle of the whole swarm (global variant); and g_i is the index of the best particle in the neighborhood of x_i (local variant). The search directions $\mathcal{G}_i^{(k+1)}$ and $\mathcal{L}_i^{(k+1)}$ are combined in a single equation, resulting in the main UPSO scheme [11]:

$$\mathcal{U}_i^{(k+1)} = u\,\mathcal{G}_i^{(k+1)} + (1 - u)\,\mathcal{L}_i^{(k+1)}, \tag{6}$$

$$x_i^{(k+1)} = x_i^{(k)} + \mathcal{U}_i^{(k+1)}, \tag{7}$$

where $u \in [0, 1]$ is called the *unification factor* and it determines the influence of the global and local search direction in Eq. (6). The standard local and global PSO variant is obtained for $u = 0$ and $u = 1$, respectively. All intermediate

Table 1. An example of the Smallest Position Value representation scheme

Jobs	j	1	2	3	4	5
Particle	x_{ij}	1.45	-3.54	2.67	-2.29	-4.02
Sequence	s_{ij}	5	2	4	1	3

values of $u \in (0, 1)$ correspond to composite variants of PSO that combine the exploration and exploitation characteristics of the global and local variant.

UPSO can be further enhanced by incorporating a stochastic parameter in Eq. (6). This parameter imitates mutation in evolutionary algorithms, although, it is directed towards a direction that is consistent with the PSO dynamic. Thus, Eq. (6) can be written either as:

$$\mathcal{U}_i^{(k+1)} = r_3 \, u \, \mathcal{G}_i^{(k+1)} + (1 - u) \, \mathcal{L}_i^{(k+1)}, \tag{8}$$

which is mostly based on the local variant or, alternatively,

$$\mathcal{U}_i^{(k+1)} = u \, \mathcal{G}_i^{(k+1)} + r_3 \, (1 - u) \, \mathcal{L}_i^{(k+1)}, \tag{9}$$

which is mostly based on the global variant. The parameter $r_3 \sim \mathcal{N}(M, \Sigma)$ is a normally distributed parameter with mean vector M and variance matrix Σ. Based on the analysis of Matyas [26] for stochastic optimization algorithms, convergence in probability was proved for the schemes of Eqs. (8) and (9) [11].

2.3   The Smallest Position Value Representation Scheme

PSO and UPSO were designed to work primarily on real–valued search spaces. Thus, in different problems, proper representation schemes may be required. For example, a rounding scheme was used for the transformation of real to integer values in discrete search spaces [27, 12].

In SMTWT, each real–valued particle must correspond to a permutation of jobs. For this purpose, the SPV scheme [9] was used. More specifically, let n be the number of jobs. Then, the ith particle, $x_i = (x_{i1}, x_{i2}, \ldots, x_{in})^\top$, is n–dimensional and each component corresponds to one job, i.e., x_{ij} corresponds to the jth job. The sequence of jobs that corresponds to x_i is an integer vector

$$s_i = (s_{i1}, s_{i2}, \ldots, s_{in})^\top,$$

where s_{ij}, $j = 1, 2, \ldots, n$, is the assignment of job j in the processing order. The determination of s_{ij} is based on x_{ij} such that jobs with smaller values of x_{ij} are scheduled first. An example is provided in Table 1 for the particle $x_i = (1.45, -3.54, 2.67, -2.29, -4.02)^\top$. The smallest component of x_i is -4.02, which corresponds to the fifth job. Thus, job 5 is scheduled first and the first component of the sequence s_i takes the value $s_{i1} = 5$. The second smallest component of x_i is -3.54, which corresponds to job 2. Therefore, job 2 is the second job in the ordering, i.e., $s_{i2} = 2$, etc.

Thus, each particle of the swarm corresponds to a sequence of jobs that is used for the computation of the total tardiness using Eq. (1). Obviously, the same sequence vector corresponds to all particles with the same ordering in their components. This property could lead the algorithm to search stagnation if some assumptions that are usually made in static optimization problems and concern the bounding of particles and velocities within specific bounds are not abandoned in the case of SMTWT. The SPV representation scheme was applied successfully with the inertia weight version of PSO [28] on the SMTWT problem [9]. We adopted SPV also in our approach, with the constriction coefficient version of PSO and UPSO.

3   Results and Discussion

We investigated the performance of three UPSO variants that proved to be very efficient in static and dynamic optimization problems [12, 13, 14, 15]. More specifically, we used the main UPSO scheme of Eq. (6) with $u = 0.2$ and $u = 0.5$, as well as the scheme with mutation defined by Eq. (8) with $u = 0.1$; $r_3 \sim \mathcal{N}(M, \Sigma)$; $M = (0, \ldots, 0)^\top$; $\Sigma = \sigma^2 I$ with $\sigma = 0.01$, and I being the $n \times n$ identity matrix. We denote these variants as UPSO_1, UPSO_2, and UPSO_m, respectively. Their performance was compared with the performance of the global and local PSO variant. These variants are denoted as PSO_g and PSO_ℓ and they are obtained from the UPSO scheme for $u = 1$ and $u = 0$, respectively. For all algorithms, the default PSO parameter set, i.e., $\chi = 0.729$ and $c_1 = c_2 = 2.05$, was used [24], while the neighborhood radius was $\rho = 1$.

The established sets of randomly generated benchmark problems of 40 and 50 jobs, each containing 125 instances, that are provided via ORLIB [7, 9] were employed in our experiments. The swarm size was equal to $N = 10 \times n$, where n is the number of jobs (also equal to the dimension of the problem using the SPV representation scheme). For each problem instance, 25 independent experiments were conducted. At each experiment, the algorithm was applied until the optimal solution was detected or a maximum number of 2000 iterations was reached. The particles and velocities were initialized randomly in $[-1, 1]^n$, although no constraints were posed on them during the execution of the algorithm.

For each algorithm, we recorded the percentage, n_{opt}, of successful experiments, i.e., experiments where the optimal solution was found, as well as the expected number of iterations, which is defined as the mean number of iterations required in the successful experiments. Also, we computed the *average relative percent deviation* from the optimal solution, which is defined as:

$$\Delta_{\text{avg}} = \sum_{i=1}^{R} \left[\frac{1}{R} \left(\frac{100(s_b^i - s^*)}{s^*} \right) \right],$$

where s^* is the optimal solution; s_b^i is the best solution obtained in the ith experiment; and R is the total number of experiments for all instances per problem,

Table 2. Results for the SMTWT problems

# Jobs	Alg.	n_{opt}	Exp. It.	Δ_{avg}	Δ_{std}
	PSO_g	56.3%	229.6	1.692	8.931
	PSO_ℓ	47.0%	613.0	0.754	4.415
40	UPSO_1	62.5%	158.1	1.765	10.865
	UPSO_2	54.1%	131.1	2.147	11.053
	UPSO_m	65.8%	525.6	0.478	3.206
	PSO_g	42.2%	275.6	1.778	7.566
	PSO_ℓ	27.0%	338.4	1.292	3.898
50	UPSO_1	43.7%	182.3	1.483	6.954
	UPSO_2	37.7%	178.4	2.329	9.352
	UPSO_m	39.6%	472.3	0.720	3.029

i.e., $R = 25 \times 125 = 3125$. Furthermore, the standard deviation, Δ_{std}, of the relative percent deviation from the optimal solution was recorded. The values Δ_{avg} and Δ_{std} provide also an intuition regarding the behavior of the algorithm in cases where it failed to detect the optimal solution, since smaller values reveal a tendency of the algorithm to converge towards sub–optimal solutions that lie closer to the optimal one. All results are reported in Table 2. The values of Δ_{avg} and scaled n_{opt} in $[0, 1]$, per algorithm for the problems of 40 and 50 jobs are depicted also in the bar graphs of Figs. 1 and 2, respectively.

In the case of 40 jobs, UPSO_m had the best performance with respect to the number of successes, followed closely by UPSO_1 ($u = 0.2$). Also, UPSO_m exhibited the best values of Δ_{avg} and Δ_{std}. This is an indication of the good quality of sub–optimal solutions it detected in the unsuccessful experiments. However, this comes with a higher computational cost, since UPSO_m required a large number of iterations. Also, we notice that PSO_g performs better than PSO_ℓ in terms of the number of successful experiments, although, it has higher values of Δ_{avg} and Δ_{std}, i.e., it is less robust. This is due to the higher exploitation ability of PSO_g compared to PSO_ℓ, which results in faster convergence but with the risk of premature convergence. UPSO_1 increases the exploitation ability of PSO_ℓ, with an immediate impact on its performance and success rates, justifying the usefulness of the unified scheme.

Similar comments can be made for 50 jobs problem. In this case, UPSO_1 outperformed all other methods with respect to the number of successes, followed by PSO_g. UPSO_m has a slightly worst performance, although with significantly smaller values of Δ_{avg}, but higher expected number of iterations. The most balanced scheme, UPSO_2 ($u = 0.5$), had better performance than PSO_ℓ, but it was always outperformed by the rest variants, similarly to the case of 40 jobs.

Summarizing the results, UPSO_1, which constitutes a modified version of PSO_ℓ with increased exploitation ability, is the most promising scheme since it

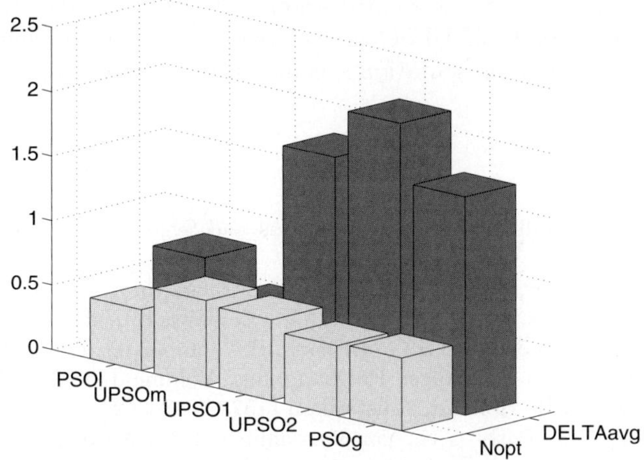

Fig. 1. Bar graph of n_{opt} normalized in $[0, 1]$, and Δ_{avg} for the 40 jobs problems

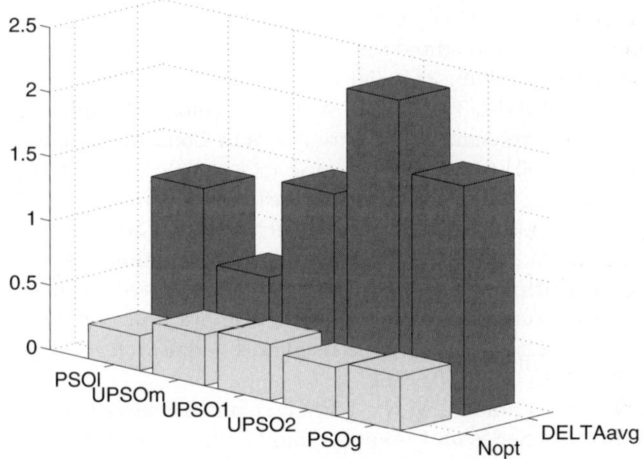

Fig. 2. Bar graph of n_{opt} normalized in $[0, 1]$, and Δ_{avg} for the 50 jobs problems

always exhibited robust behavior and required significantly smaller number of iterations than its competitive variants, UPSO_m and PSO_g. This justifies that the unified scheme can produce more efficient PSO variants.

4 Conclusions

A first study of UPSO on the Single Machine Total Weighted Tardiness problem was provided. Widely used benchmark problems were employed and results were reported and compared with that of established PSO variants. UPSO had the best performance and robust behavior, enhancing the standard PSO variants and justifying the usefulness of the unified scheme. Further investigation is needed to fully reveal the ability of UPSO in tackling scheduling problems, as well as possible impact of the representation scheme on the algorithm's performance.

References

1. Pinedo, M.: Scheduling: Theory, Algorithms and Systems. Prentice Hall, Englewood Cliffs (1995)
2. Johnson, D.S., Garey, M.R.: Computers and Intractability: A Guide to the Theory of NP–Completeness. W.H. Freeman & Co., Englewood Cliffs (1979)
3. Lenstra, J.K., Rinnooy Kan, H.G., Brucker, P.: Complexity of machine scheduling problem. In Studies in Integer Programming. Volume 1 of Annals of Discrete Mathematics. North–Holland, Amsterdam (1977) 343–362
4. Abdul-Razaq, T.S., Potts, C.N., Van Wassenhove, L.N.: A survey of algorithms for the single machine total weighted tardiness scheduling problem. Discrete Applied Mathematics **26** (1990) 235–253
5. Potts, C.N., Van Wassenhove, L.N.: Single machine tardiness sequencing heuristics. IIE Transactions **23** (1991) 346–354
6. Crauwels, H.A.J., Potts, C.N., Van Wassenhove, L.N.: Local search heuristics for the single machine total weighted tardiness scheduling problem. INFORMS Journal on Computing **10**(3) (1998) 341–350
7. den Besten, M., Stützle, T., Dorigo, M.: Ant colony optimization for the total weighted tardiness problem. In: Lecture Notes in Computer Science. Volume 1917. Springer-Verlag (2000) 611–620
8. Saldam, A., Ahmad, I., Al-Madani, S.: Particle swarm optimization for task assignment problem. Microprocessors and Microsystems **26** (2002) 363–371
9. Fatih Tasgetiren, M., Sevkli, M., Liang, Y.C., Gencyilmaz, G.: Particle swarm optimization algorithm for single machine total weighted tardiness problem. In: Proc. 2004 IEEE Congress on Evolutionary Computation. (2004) 1412–1419
10. Mladenovic, N., Hansen, P.: Variable neighborhood search. Computers and Operations Research **24** (1997) 563–571
11. Parsopoulos, K.E., Vrahatis, M.N.: UPSO: A unified particle swarm optimization scheme. In: Lecture Series on Computer and Computational Sciences, Vol. 1, Proc. Int. Conf. Comput. Meth. Sci. Engin. (ICCMSE 2004), VSP International Science Publishers, Zeist, The Netherlands (2004) 868–873
12. Parsopoulos, K.E., Vrahatis, M.N.: Unified particle swarm optimization for tackling operations research problems. In: Proc. IEEE 2005 Swarm Intelligence Symposium, Pasadena (CA), USA (2005) 53–59
13. Parsopoulos, K.E., Vrahatis, M.N.: Unified particle swarm optimization in dynamic environments. In Lecture Notes in Computer Science (LNCS). Volume 3449. Springer Verlag (2005) 590–599
14. Parsopoulos, K.E., Vrahatis, M.N.: Unified particle swarm optimization for solving constrained engineering optimization problems. In Lecture Notes in Computer Science (LNCS). Volume 3612. Springer Verlag (2005) 582–591
15. Parsopoulos, K.E., Vrahatis, M.N.: Parameter selection and adaptation in unified particle swarm optimization. (Mathematical and Computer Modelling) to appear.
16. Kennedy, J., Eberhart, R.C.: Swarm Intelligence. Morgan Kaufmann Publishers (2001)

17. Kennedy, J., Eberhart, R.C.: Particle swarm optimization. In: Proc. IEEE Int. Conf. Neural Networks. Volume IV., IEEE Service Center (1995) 1942–1948
18. Bonabeau, E., Dorigo, M., Théraulaz, G.: Swarm Intelligence: From Natural to Artificial Swarm Intelligence. Oxford University Press, New York (1999)
19. Bonabeau, E., Meyer, C.: Swarm intelligence: A whole new way to think about business. Harvard Business Review **79**(5) (2001) 106–114
20. Engelbrecht, A.P.: Fundamentals of Computational Swarm Intelligence. Wiley (2006)
21. Kennedy, J.: Bare bones particle swarms. In: Proc. IEEE Swarm Intelligence Symposium, Indianapolis, USA, IEEE Press (2003) 80–87
22. Mendes, R., Kennedy, J., Neves, J.: The fully informed particle swarm: Simpler, maybe better. IEEE Trans. Evol. Comput. **8**(3) (2004) 204–210
23. Li, X.: Adaptively choosing neighborhood bests using species in a particle swarm optimizer for multimodal function optimization. In LNCS, Vol. 3102, Springer (2004) 105–116
24. Clerc, M., Kennedy, J.: The particle swarm–explosion, stability, and convergence in a multidimensional complex space. IEEE Trans. Evol. Comput. **6**(1) (2002) 58–73
25. Trelea, I.C.: The particle swarm optimization algorithm: Convergence analysis and parameter selection. Information Processing Letters **85** (2003) 317–325
26. Matyas, J.: Random optimization. Automatization and Remote Control **26** (1965) 244–251
27. Parsopoulos, K.E., Vrahatis, M.N.: Recent approaches to global optimization problems through particle swarm optimization. Natural Computing **1**(2–3) (2002) 235–306
28. Shi, Y., Eberhart, R.C.: A modified particle swarm optimizer. In: Proc. IEEE CEC 1998, IEEE Service Center (1998) 69–73

Classification of Bio-data with Small Data Set Using Additive Factor Model and SVM

Hyeyoung Park and Minkook Cho

School of Electrical Engineering and Computer Science
Kyungpook National University, Deagu, Korea
hypark@knu.ac.kr, ucaresoft@paran.com

Abstract. Bio-data, which are obtained from human individuals, have been one of main applications of pattern classification these days. A critical property of bio-data classification is the small number of data in each class due to high cost of obtaining data from each individuals. Since most classification methods are based on the distribution of data in each class, the lack of data can be a main cause of low classification performance of conventional classifiers. To solve this problem, we propose a modified additive factor model for bio-data which has two factors; the individual factor and the environment factor. Under the proposed model, we estimate the distribution of environment factor which gives robust information even in case of small data set. We then define new similarity measures using the information. The similarity measure is applied to nearest neighbor method for classification. We also use the support vector machines (SVM) to find a sophisticated similarity measure. Through computational experiments, we confirm that the proposed model and similarity measure is appropriate enough to show better classification performance compared to conventional similarity measure as well as conventional SVM classifier.

Keywords: Bio-data classification, data generation model, additive factor model, similarity function, class factor, environment factor.

1 Introduction

Bio-data, which are some kinds of bio-signals obtained from human individuals, have useful information about each individual. Recently, bio-data have been widely applied to various applications such as security and medical systems. To get some meaningful information from bio-data, we common need pattern classification methods. In this paper, we mainly consider biometric systems, which classify given bio-data into several classes which of each corresponds to each individual.

Many well known pattern classification methods, which are related to machine learning, artificial intelligence, or signal processing area have been used for bio-data classification[1,2,5,8,10]. However, most previous studies have mainly considered a specific type of data, and have focused on the problem-dependent part such as sophisticated feature extraction appropriate for the given bio-data type.

A. Sattar and B.H. Kang (Eds.): AI 2006, LNAI 4304, pp. 770–779, 2006.

© Springer-Verlag Berlin Heidelberg 2006

Though the problem-dependent approaches can give some optimized results for the specific data, we also need to consider the common and core characteristics of bio-data, through which we expect to improve the classification performance.

There are three important characteristics of bio-data. First, in many bio-data classification problems such as biometrics, the number of classes is large. To find correct decision boundary for the large number of classes, complex classifiers with high capacity are preferred. However, this makes another problem due to the second characteristics of bio-data. That is, it is common that the number of data in each class is quite small because of high cost of data acquisition from individuals. The lack of data in a class makes it difficult to extract robust distribution property of the class. These characteristics of bio-data result in bad classification performances. Third, high dimensional input data are common. The high dimensionality emphasizes the sparsity of data in input space and makes it difficult to find statistically stable decision boundary.

As an approach to solve these problems, we exploit a data generation model that is appropriate for bio-data. Once the generation model is defined well to express the characteristics of bio-data, we expect to find a method for treating the problems mentioned above. Several generation models have recently been proposed, which describe underlying factors of observation. These include essentially additive factor models as used in principal component analysis[9], independent component analysis[3], and bilinear models[11]. However, the original additive factor model and independent component analysis do not consider class information when finding factors. Also these previous methods including bilinear models basically assume that sufficiently large number of data are given, which is common unsatisfied in bio-data classifications.

In this paper, we modify the additive factor model by adding class factor and environment factor. Under the modified model, we extract a statistically robust information and propose a method for defining new similarity measures using the statistical information and SVM. Through this approach, we expect to develop an efficient classification system for small number of data.

2   Data Generation Model for Bio-data

2.1   Additive Factor Model with Class and Environment Factor

We first discuss a generation model for bio-data. We assume that an observed bio data x can be decomposed into two mutually independent factors: individual-dependent factor and environment-dependent factor. Thus a random variable x for observed data can be written as a function of two distinct random variable ξ and δ, which is the form,

$$x = f(\xi, \delta). \tag{1}$$

Here, ξ represents an individual variable, which keeps some unique information in each individual (i.e. class). Thus, when the individual source of observed data is known, ξ is fixed and can be regarded as a constant value. In this case we denote the unique information vector for class C_i as $\boldsymbol{\xi}_i$. Under this assumption for an arbitrary data $\boldsymbol{x}_i$ in class C_i, we have

$$\boldsymbol{x}_{i\cdot} = f(\boldsymbol{\xi}_i, \boldsymbol{\delta}). \tag{2}$$

From this, we can see that any variation within a class is originated from $\boldsymbol{\delta}$, which represents environment variable.

We should argue that the above assumptions on the data generation model are plausible for bio-data. Considering image data of human face, it is natural to assume that a prototype of a human face is uniquely defined, and the diverse variations in different images are due to some environmental factors, such as illumination and other equipment noise. In our model, the prototype of face image corresponds to the individual variable $\boldsymbol{\xi}_i$ for each class, and the within-class variation corresponds to the environment variable $\boldsymbol{\delta}$ with class-independent randomness.

In order to explicitly define the generation model, we need to determine the shape of the function f. Though very diverse function shapes can be exploited, as a preliminary study, we propose a linear additive factor model,

$$\boldsymbol{x}_i = W\boldsymbol{\xi}_i + \boldsymbol{\delta}, \tag{3}$$

which means that a random sample $\boldsymbol{x}_i$ in class C_i is generated by the summation of a linearly transformed class prototype $\boldsymbol{\xi}_i$ and a random class-independent variation $\boldsymbol{\delta}$.

From this model, we can see that the distribution within a class can be explained by the distribution of the environment factor $\boldsymbol{\delta}$. Since the model assumes that the environment factor $\boldsymbol{\delta}$ is common for all classes, its distribution can be assumed to be identical for every class. We call this distribution property of $\boldsymbol{\delta}$ as the within-class variation, and try to estimate its probability density. Note that the estimation of within-class variation can be done not using small data set of each class but using the whole data set of all classes. Thus and we can expect that the estimation is statistically more stable, which will be discussed in the next subsection.

2.2   Extraction of Robust Statistical Information

As mentioned in the previous section, the estimation of the probability density of $\boldsymbol{\delta}$ can be done over all classes, because $\boldsymbol{\delta}$ is common for all classes. This is a main characteristics of the proposed model. Since it is common for bio-data that the number of data in each class is very small, we try to utilize this characteristics of $\boldsymbol{\delta}$ to get some statistically stable information from small data set.

Based on the class-independent property of $\boldsymbol{\delta}$, we estimate its distribution using all sample data $\boldsymbol{x}$. In order to omit the class-dependent information ($\boldsymbol{\xi}$) from observed data, we select two random samples $\boldsymbol{x}_i$ and $\boldsymbol{x}_i'$ from a class C_i, and take their substraction so as to obtain

$$\boldsymbol{x}_i - \boldsymbol{x}_i' = W\boldsymbol{\xi}_i + \boldsymbol{\delta} - (W\boldsymbol{\xi}_i + \boldsymbol{\delta}') = \boldsymbol{\delta} - \boldsymbol{\delta}'. \tag{4}$$

One can notify that the class information is excluded and it is possible to get some information related only environment variable δ.

We now define a new random variable y of the form,

$$y = x_i - x'_i, \qquad (i = 1, 2, \ldots, P), \tag{5}$$

where P is the number of classes. We then estimate the distribution of $y = \delta - \delta'$ instead of estimating that of δ directly. We should mention about the number of samples that can be used for the estimation. When the number of class is P and the number of data in each class is n, we can get $n(n-1)P$ samples for the random variable y. Thus, the obtained estimator using the samples is has smaller variance and we can expect to get statistically more stable information than the conventional class-based approaches.

Once we get the probability distribution $p(y)$, we can define a similarity measure between two input data x and x'. We first define the similarity between x and x' by the probability that the two data is given from a same class, which is shown as

$$S(x, x') = P\{x \in C_i \text{ and } x' \in C_i \text{ for some } i\}. \tag{6}$$

When the two data are given from a same class, the variable $x - x'$ subjects to the density function of y. Therefore, we can redefine the similarity measure as the probability that $x - x'$ subjects to the density function $p(y)$.

Under this definition, the shape of $p(y)$ determines the similarity measure. In this paper, we give two methods for determining $p(y)$. If we assume δ is subject to Gaussian distribution, we can get an explicit form of $p(y)$ and use it as similarity measure, which will be discussed in Section 3.2. On the other hands, we can use more sophisticated machine learning approaches to represent various type of $p(y)$. In Section 3.3, we exploit support vector machines to implicitly estimate $p(y)$ and find an optimized similarity measure. Once a similarity measure is determined, we can conduct classification task, which will be discussed in section 3.1.

3 Classification

3.1 Overall Structure of Classification System

Before discussing how to define similarity measure using the within-class variance δ, we give the overall structure of classification system used in this paper. Fig.1 shows the overall structure. In the stage of learning, we find a similarity measure using given data set. In the stage of classification, for a new observation x_{new}, we assign it to one of the classes. The conventional nearest neighbor method calculates the similarity between x_{new} and all data x_i, $S(x_{new}, x_i)$, in the training set to find the nearest neighbor, which can be written as

$$x_{nn} = \text{argmax}\{S(x_{new}, x_i) | x_i \in X\}. \tag{7}$$

As shown in the equation (7), the classification performance is determined by the similarity measure. In conventional nearest neighbor method, the similarity measure, which we note as S_E, is given from the Euclidean distance such as

$$S_E(\boldsymbol{x}, \boldsymbol{x}') = -\|\boldsymbol{x} - \boldsymbol{x}'\|^2 \tag{8}$$

By exploiting more appropriate similarity measure, we can expect to get better performance.

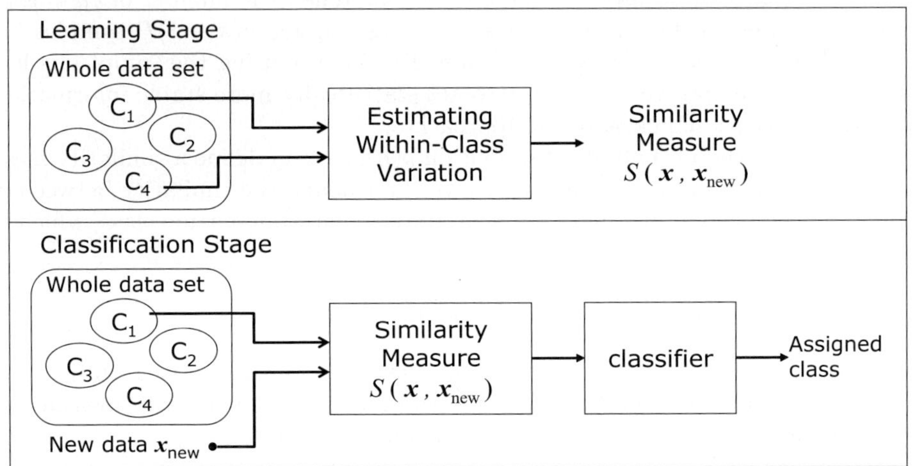

Fig. 1. Overall structure of classification system

3.2 Similarity Measure Under Gaussian Assumption

We proposed a new similarity measure based on the distribution of $\boldsymbol{y}$. In order to to estimate the probability density of $\boldsymbol{y}$, we first use the basic parametric density estimation method[7]. Under the fact that the environment factor can be considered as a implicit summation of multiple variation sources and by using central limit theorem, we can assume that the probability density function $\boldsymbol{\delta}$ is Gaussian. Then the probability density of $\boldsymbol{y}$ is also Gaussian, and parameters to be estimated are the mean $\boldsymbol{\mu}_y$ and the covariance matrix $\boldsymbol{\Sigma}_y$.

By the maximum likelihood estimation, we can easily obtain the estimate $\boldsymbol{\mu}_y$ and $\boldsymbol{\Sigma}_y$ as follows:

$$\hat{\boldsymbol{\mu}}_y = \frac{1}{n(n-1)P} \sum_{i=1}^{n(n-1)P} \boldsymbol{y}_i, \tag{9}$$

$$\hat{\boldsymbol{\Sigma}}_y = \frac{1}{n(n-1)P} \sum_{i=1}^{n(n-1)P} (\boldsymbol{y} - \hat{\boldsymbol{\mu}}_y)(\boldsymbol{y} - \hat{\boldsymbol{\mu}}_y)^t. \tag{10}$$

In the computational experiment of Section 5, we further assume that the covariance matrix is diagonal for the computational simplicity.

Using the estimates, we obtain the probability density $p(\boldsymbol{y})$ as,

$$p(\boldsymbol{y}) = p(\boldsymbol{x} - \boldsymbol{x}') \tag{11}$$

$$= \frac{1}{\sqrt{2\pi|\Sigma|}} \exp\left\{-\frac{1}{2}(\boldsymbol{x} - \boldsymbol{x}' - \hat{\boldsymbol{\mu}}_y)^t \hat{\boldsymbol{\Sigma}}_y^{-1}(\boldsymbol{x} - \boldsymbol{x}' - \hat{\boldsymbol{\mu}}_y)\right\} \tag{12}$$

Then a new similarity measure S_G can be determined as

$$S_G(\boldsymbol{x}, \boldsymbol{x}') \propto \log p(y) \tag{13}$$

$$\propto -\left\{(\boldsymbol{x} - \boldsymbol{x}' - \hat{\boldsymbol{\mu}}_y)^t \hat{\boldsymbol{\Sigma}}_y^{-1}(\boldsymbol{x} - \boldsymbol{x}' - \hat{\boldsymbol{\mu}}_y)\right\}. \tag{14}$$

When $\hat{\boldsymbol{\mu}}_y = 0$ and $\hat{\boldsymbol{\Sigma}}_y$ is the identity matrix, the similarity measure is equivalent to the Euclidean measure S_E. Therefore, the proposed similarity measure is said to be a kind of generalization of the classic similarity measure. We also should remark that the proposed similarity measure is different from that of Mahalanobis distance[4] because the covariance matrix $\boldsymbol{y}$ is different from that of $\boldsymbol{x}$. Using the proposed similarity measure and nearest neighbor method, we can assign the new data $\boldsymbol{x}_{new}$ to the class that $\boldsymbol{x}_{nn}$ with maximum similarity. In section 5, we will show some computational experiments how the proposed similarity measure affect the classification performance.

3.3   Similarity Measure Using SVM

In the previous section, we discussed the estimation of $\boldsymbol{y}$ and classification under the assumption that $\boldsymbol{\delta}$ is subject to Gaussian distribution. However, this assumption may not be appropriate in many cases. To solve this problem, we use the SVM and estimate the similarity measure directly without estimating $p(\boldsymbol{y})$, since what is necessary for classification is not the probability density but the similarity measure.

For a given two pair of data $\boldsymbol{x}$, and $\boldsymbol{x}'$, we made input data set as the difference vector $\boldsymbol{x} - \boldsymbol{x}'$ and define a desired output z for SVM as

$$z = 1 \quad \text{if } \boldsymbol{x} \text{ and } \boldsymbol{x}' \text{ are from a same class,} \tag{15}$$

$$z = -1 \quad \text{otherwise.} \tag{16}$$

The SVM is trained for all possible pairs from the training data set X. After training, we expect that the output of SVM, O_{SVM}, become positive for the data pairs from a same class. Otherwise, O_{SVM} is expected to become negative. Therefore, the output of SVM can be considered as a new similarity measure such as

$$S_V(\boldsymbol{x}, \boldsymbol{x}') = O_{SVM}(\boldsymbol{x} - \boldsymbol{x}') \tag{17}$$

Using the similarity measure and nearest neighbor method, we can classify the data $\boldsymbol{x}_{new}$.

Note that the proposed SVM method can also be applied to multi-class classification without any other process, while conventional SVM classifier need some extension of binary classifier to multiclass classification such as one-against-all (OVA) method[6]. In addition, since the conventional SVM design method are based on the distribution of each class, its performance may deteriorate when the number of data in each class is small.

4   Experimental Results

4.1   Data Description

In order to check the possibility of the proposed method, we conducted a number of computer experiments. First, we made two simple toy data sets that can explicitly show the data distribution. The test data are shown in Fig.2. The left one, TOY I data set, is composed of three classes. Each class is subject to a Gaussian distribution. For training, we used five data for each class. For testing, we used 1000 data for each class. The TOY I data set exactly matches to our Gaussian assumption on δ. The right one, TOY II data set, is also composed of three classes. However, each class is not subject to simple Gaussian but a mixture of two Gaussian clusters. This data set violates the Gaussian assumption on δ. For training set, we used only five data for each class. For testing, we used 1000 data for each class.

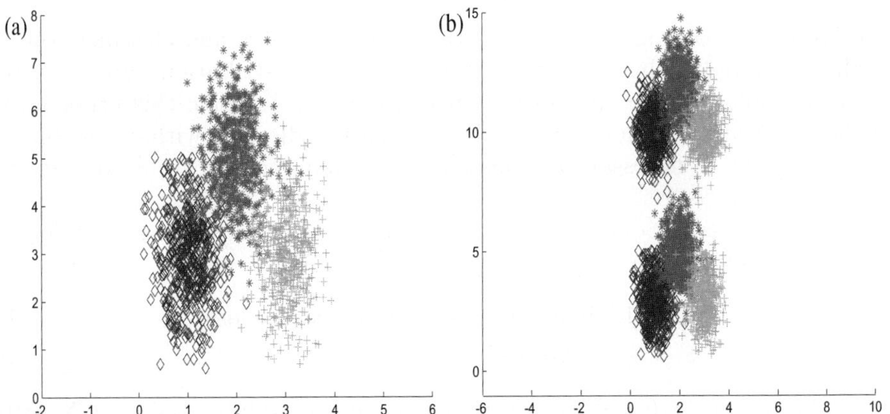

Fig. 2. Test data sets of Toy problem. (a) TOY I: one cluster data set (b) TOY II: two cluster data set.

We also conducted experimental comparisons using two real bio-data set, which are shown in Fig.3. The left one (a) is human iris images, which is one of the representative physiological feature for biometrics[8]. Primarily, we obtained 260 number of iris image from 14 different individuals as experiment data. We used 70 training data composed of 5 data randomly selected from data of each person, and used the rest 190 as test data. We applied a normalization process on the localized iris area to compensate for size variations due to the possible changes in the camera-to-face distance, and to facilitate the feature extraction process by converting the iris area represented by a polar coordinate system into a cartesian coordinate system. Since the iris images have 7200 pixels(225x32), we first applied the principal component analysis (PCA) to reduce the dimensions and obtained 70 dimensional feature vectors for each item of raw data.

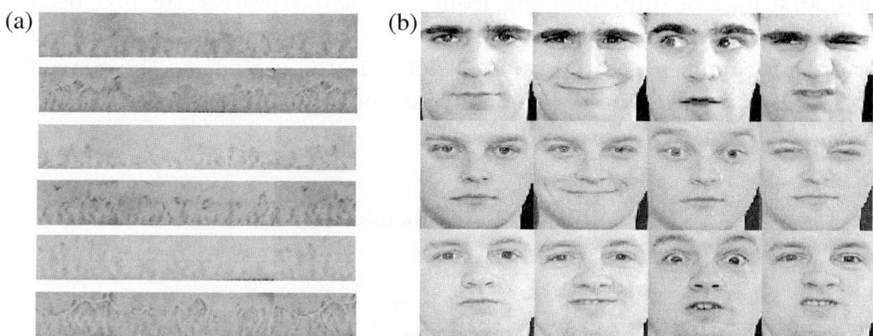

Fig. 3. Examples of (a)human iris image and (b)face image

The right one (b) is human face images that have different facial expressions. The data were obtained from the homepage of the Psychological Image Collection at Stirling (PICS, http://pics.psych.stir.ac.uk/). We use the facial images from the 69 persons, and there are four different images for each persons. This data set has the typical characteristic of bio-data, that is, the number of class is large and the number of data in each class is small. To get some average performance, we conducted 4-fold cross validation. For each trial, we used three images for each person (class) as training set, and the rest one for each person was used for testing. The size of image is 80×90, thus the dimension of the raw input is 7200. From each image data, we extracted 50 principal components by using the PCA.

4.2   Results

The experimental result is shown in Table 1. We compared four different methods; the naive nearest neighbor method with Euclidean similarity measure S_E, the proposed method with Gaussian similarity measure S_G, the proposed method with SVM similarity measure S_V, and the conventional SVM method (OVA method) for multiclass classification. When training SVM, we used RBF kernel function and its parameter was optimized through experiments.

From the results, we can see that the proposed method with S_V gives best performance in all data set. In the case of TOY I set, the performance of S_G and S_V are same, which support that the Gaussian assumption on δ is appropriate.

In the case of TOY II set in which the Gaussian assumption is destroyed, the performance of the method with S_G is not so good, though it is better than the naive method with S_E. By using the SVM measure S_V, we can improve the performance. In the case of human iris data, we can see remarkable improvement of performance comparing to the method with S_E. From the good performance of the method with S_G, we guess that the distribution of δ in this problems is similar to Gaussian. In the case of face data, the performance of the methods with S_E and S_G are bad. Since the variation within a class is caused by some specific expression, it may be quite different from Gaussian. However, the proposed SVM method still shows remarkably improved performance. In addition, we can see that the performances of the proposed method are better than the conventional SVM. The low performance of the conventional SVM may be due to the small number of training set.

Table 1. Experimental results

	Toy I	Toy II	IRIS	FACE
Euclidean (S_E)	90.9	90.9	90.5	70
Gaussian (S_G)	**92.5**	91.1	**98.9**	69.4
SVM (S_V)	**92.5**	**93.8**	**98.9**	**82.5**
SVM-OVA	91.5	91.8	97.4	79.2

5   Conclusions and Discussions

In this paper, we propose a data generation model for bio-data and a classification method with new similarity measure based on the model. Through the utilization of within-class variation δ, we could developed a classification method that can show stable performances when the number of data in each class is small. Especially, we assume a linear additive model and try to estimate δ using the difference of two samples $x - x'$. Though the linear additive model may be quite strict limitation, the experimental results shows the efficiency of the difference vectors, $x - x'$. As a further study, it is interesting to develop more general model. One approach is to exploit the pair vector (x, x') for SVM input to create a similarity measure, which is discussed in [6]. Through this, we are free from the assumption on the linear additive model, but computational complexity of SVM increases. Some comparative analysis on the two methods will be given in our previous works.

Acknowledgements

This work was supported by the Korea Research Foundation Grant funded by the Korean Government(MOEHRD) (KRF-2005-003-D00330)

References

1. Bartlett, M., and Sejnowsky, T., Viewpoint Invariant Face Recognition using Independent Component Analysis and Attractor Networks, *Naural Information Processing Systems-Natural and Synthetic,* 9, 817-823, 1997.
2. Belhumeur, P., Hespanha, J., and Kriegman, D., Eigenfaces vs. Fisherfaces: Recognition Using Class Specific Linear Projection, *IEEE trans. on Pattern Recogntion and Machine Intelligence,* 19(7), 711-720, 1997.
3. Bell, A. and Sejnowski, T., An information maximization approach to blind separation and bllind deconvolution, *Neural Compuation,* 7(6), 1129-1159, 1995.
4. C. Bishop, *Neural Networks for Pattern Recognition,* Oxford University Press, 1995.
5. Campbell, W., A Sequence Kernel and its Applications to Speaker Recognition, *Advances in Neural Information Processing Systems,* In Press, 2001.
6. M. Cho and H. Park, A Robist SVM Design for Multi-class Classification, *Lecture Notes in Artificial Intelligence,* 3809, 1335-1338, 2005.
7. K. Fukunaga, *Introduction to Statistical Pattern Recogntion,* 2ed, Academic Press, INC. 1990.
8. John G. Daugman, High Confidence Visual Recognition of Persons by a Test of Statistical Independence, *IEEE Trans. on Pattern Analysis and Machine Intelligence,* 15(11), 1148-1161, 1993.
9. Lattin, J., Analyzing Multivariate data, Thomson Learning, Inc., 2003.
10. Lee, O., Park, H., and Choi, S., PCA vs. ICA for Face Recogntion, *The 2000 International Technical Conference on Circuits/Systems, Computers,and Communications.* 873-876, 2000.
11. Tenenbaum, J. B. and Freeman, W.T., Separating Style and content with bilinear models, *Neural Computaion,* 12, 1247-1283, 2000.

Study of Dynamic Decoupling Method for Multi-axis Sensor Based on Niche Genetic Algorithm

Ding Mingli, Dai Dongxue, and Wang Qi

Dept. of Automatic Test and Control, Harbin Institute of Technology,
150001 Harbin, China
dingml@hit.edu.cn

Abstract. The dynamic coupling of a multi-axis sensor is defined as a situation that the output signal of one direction includes the additional value, which affected by the input in other direction. Obviously, the dynamic coupling error will decrease the sensor measurement precision seriously. In this paper, a new multi-axis sensor dynamic decoupling method is proposed based on a niche genetic algorithm (NGA). The method first gets the calibration data of the multi-sensor, and then uses the system identification method and the niche genetic algorithm to calculate and optimize the decoupling network based on the analysis method of transform function matrix. In addition, it also can avoid the high precision requirement to the sensor model in the traditional dynamic decoupling method. Finally, the simulation results show the correctness and the affectivity of the proposed method.

1 Introduction

The main function of the multi-axis sensor is to measure the parameters of the multi directions at the same time, such as the six-axis wrist force sensor used in the robot, the measurement unit in the helicopter which measures the changes of the pull force of the rotor, the pitching torque and the rolling torque, the strain or the piezoelectric multi-direction force measurement instrument used in the machine tool, and so on. But in the application of the multi-axis sensor, we can not ignore the cross coupling, which defines that the output signal of one direction is affect by the input signals from the other directions. Obviously, the performance of the multi-axis sensor is influenced seriously by the cross coupling. Generally, the cross coupling is mainly arisen from the reasons of the location deviation of the sensing element in the sensor, the location deviation of the sensor when installing it, and the low manufacture level of the sensor.

To solve the problem of the dynamic coupling, two kinds of ways are mentioned: One is to design the new structure sensor, improve the manufacture level and the sensor installation accuracy, but the result is not so satisfied due to the realization difficulty of this way; the other way is to use some decoupling method to compensate or modify the dynamic output signal. Some methods such as the stationary dynamic decoupling method[1], the iterative dynamic decoupling method[2] and the static and dynamic united decoupling method[3] are proposed in succession by the Prof. Xu Kejun, and another dynamic decoupling method based on diagonal predominance compensation[4] is presented by Prof. Zhang Weigong. The methods mentioned above

A. Sattar and B.H. Kang (Eds.): AI 2006, LNAI 4304, pp. 780–789, 2006.
© Springer-Verlag Berlin Heidelberg 2006

are all based on the analysis method of transform function matrix, namely when designing the decoupling network, these methods require the non-diagonal parameters of the system transform matrix to be zero. The idea has the advantages of simple decoupling network and easy realization. But it also has a disadvantage of the high precision requirement to the multi-sensor model. In fact, due to the existence of the measurement error and the disturbance in the system, and the limitation of the identification method, the identified system model is always not so correct that the decoupling precision drops down seriously. So a niche genetic algorithm (NGA)[5, 6] is applied to solve the problem in this paper. The NGA has the merits of fast and global convergence, and also can search multiple peak value of the object function. In this paper, based on the idea of the analysis method of transform function matrix, the decoupling network is built using the NGA. That is to say, we can easily build and optimize the decoupling network according to the calibrating data, and avoid setting up the transform function of the multi-sensor. In addition, the simulation case and discussion are investigated as well.

2   Analysis Method of Transform Function Matrix

Looking the multi-sensor model as a linear multi-input and multi-output (MIMO) system, we can treat the decoupling of the multi-sensor as a special decoupling problem of MIMO system, which is called Morgan problem with the character of $m = r$ (m is the number of input, r is the number of output). So using the matrix analysis method, the two-dimension sensor model in Fig. 1 can be express as

$$G = \begin{bmatrix} G_{11} & G_{12} \\ G_{21} & G_{22} \end{bmatrix}. \tag{1}$$

which can easily show the relationship between the input channel and the output channel, also their dynamic coupling situations. When $m = r$, the solution of Morgan problem is always exist. So the thought of designing a decoupling network is to make the system transform function matrix, which is constructed by the sensor transform function and the decoupling network, be a certain matrix with zero non-diagonal parameters or a diagonal predominance matrix.

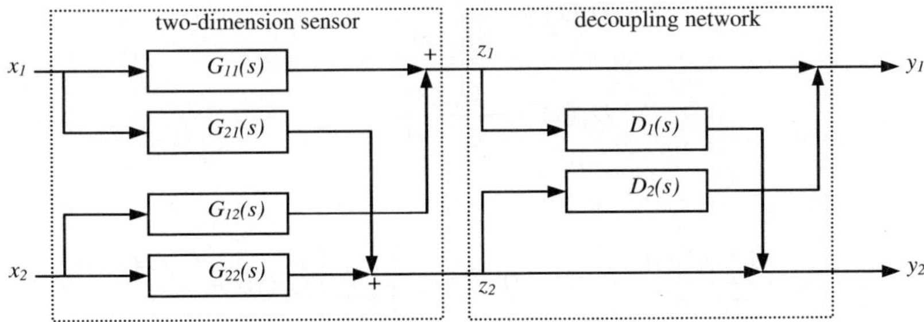

Fig. 1. Structure of the stationary decoupling method

Now, a stationary decoupling method is introduced as follows. In this method, the new decoupling parts are not added on the positive channel of every sensor output. However, the relatively simple decoupling parts are put between the channels of every sensor output to counteract the dynamic coupling of every dimension. The network structure is shown in Fig. 1. x_1 and x_2 are the two-dimension sensor input signals including the dynamic coupling, z_1 and z_2 are the corresponding coupling output signals. $G_{11}(s)$, $G_{21}(s)$, $G_{22}(s)$ and $G_{12}(s)$ are the transform functions of each sensor channel; $D_1(s)$ and $D_2(s)$ are the transform functions of the decoupling network; y_1 and y_2 are the corresponding decoupling output signals according to x_1 and x_2.

From Fig. 1, it is easy to get the coupling output expression of each channel as

$$z_1 = x_1 G_{11}(s) + x_2 G_{12}(s), \tag{2a}$$

$$z_2 = x_2 G_{22}(s) + x_1 G_{21}(s). \tag{2b}$$

Assume $Z = \begin{bmatrix} z_1 \\ z_2 \end{bmatrix}$, $G = \begin{bmatrix} G_{11}(s) & G_{12}(s) \\ G_{21}(s) & G_{22}(s) \end{bmatrix}$ and $X = \begin{bmatrix} x_1 \\ x_2 \end{bmatrix}$, so Eq. (2) can be rewritten like

$$Z = GX, \tag{3}$$

We also acquire the sensor output after decoupling from Fig. 1 as here

$$y_1 = z_1 + z_2 D_2(s). \tag{4a}$$

$$y_2 = z_2 + z_1 D_1(s) \tag{4b}$$

Let $Y = \begin{bmatrix} y_1 \\ y_2 \end{bmatrix}$ and $D = \begin{bmatrix} 1 & D_2(s) \\ D_1(s) & 1 \end{bmatrix}$, Eq. (4) can be rewritten as

$$Y = DZ. \tag{5}$$

and let

$$D_1(s) = -\frac{G_{21}(s)}{G_{11}(s)}. \tag{6a}$$

$$D_2(s) = -\frac{G_{12}(s)}{G_{22}(s)} \tag{6b}$$

so

$$Y = DGX = \begin{bmatrix} G_{11}(s) - \dfrac{G_{12}(s)G_{21}(s)}{G_{22}(s)} & 0 \\ 0 & G_{22}(s) - \dfrac{G_{21}(s)G_{12}(s)}{G_{11}(s)} \end{bmatrix} \begin{bmatrix} x_1 \\ x_2 \end{bmatrix} = \begin{bmatrix} G_{11}(s)x_1 - \dfrac{G_{12}(s)G_{21}(s)}{G_{22}(s)}x_1 \\ G_{22}(s)x_2 - \dfrac{G_{21}(s)G_{12}(s)}{G_{11}(s)}x_2 \end{bmatrix}, \tag{7}$$

It is obvious that the order of the decoupling part is low and the singularity of the coupling matrix can not influence the decoupling part. In addition, this method also can extend to the situation of three dimensions. From Eq. (7), the output of each channel after decoupling is only relative to the corresponding input of each channel. That is to say, if the sensor transform function is exactly correct, the stationary decoupling method can wholly remove the coupling error of each channel.

3   Niche Genetic Algorithm

The method mentioned above is set up based on the condition of exactly knowing the multi-axis sensor model. When the sensor model is not so precise, the decoupling effect will not be good, even can not fulfill the purpose of decoupling. To solve this problem, a dynamic decoupling method based on niche genetic algorithm (NGA) is proposed in this paper. The method uses the input and output calibrating data of the multi-axis sensor, in the condition of making the coupling part to be zero when dynamic decoupling, to calculate every part of the decoupling network directly in the Fig. 1 based on the method of system identification method. In addition, considering the measurement error of the input and the output signal and the modeling error of the system identification method, the method designs the corresponding NGA on the basis of the recursive least square algorithm (RLS) and optimizes the decoupling network.

NGA has been successfully applied to find a global optimum. In natural ecosystems, a niche can be viewed as an organism's task, which permits species to survive in their environment. Species are defined as a collection of similar organisms with similar features. For each niche, the physical resources are finite and must be shared among the population of that niche. By analogy, niching methods tend to achieve a natural emergence of niches and species in the search space. NGA keep the variety of the species and make the population evolve to the distributing direction of the individual with high quality through preserving the survival of the small-scale and low fitness species in population. Furthermore, NGA also can find more extremums of the object function in one search process. So the optimizing points distribute on many peak values of the object function, not only one peak value. Through this way, NGA not only increase the search efficiency, but also ensure the global convergence.

The mathematic model of the RLS which is used to acquire the decoupling model is

$$A(z^{-1})y(k) = B(z^{-1})u(k) + v(k), \tag{8}$$

where $u(k)$ and $y(k)$ are the system input and the system output respectively, $v(k)$ is noise, $A(z^{-1})$ and $B(z^{-1})$ are polynomial defined as

$$\begin{cases} A(z^{-1}) = 1 + a_1 z^{-1} + a_2 z^{-2} + \cdots + a_n z^{-n} \\ B(z^{-1}) = b_1 z^{-1} + b_2 z^{-2} + \cdots + b_m z^{-m} \end{cases}. \tag{9}$$

Let Eq. (8) be the initial population, use the NGA to optimize the population, and then get the satisfied system decoupling network. The algorithm steps can be expressed as follows.

(1) Use RLS get the initial population, and decide the encoding method;
(2) Does the population fulfill the termination condition? If it is, go to step (10);
(3) If it is not, calculate the fitness of each individual;
(4) Generate the middle population according to the fitness of each individual of the initial population;
(5) Intercross two individuals of the middle population according to cross probability P_c, and generate the new individuals to form the temporary new population;
(6) Mutate the individuals according to mutation probability P_m, and generate the new individuals to form the temporary new population;
(7) Calculate the fitness of each individual in the temporary new population;
(8) Let the initial population compete with the temporary new population, and generate the new population;
(9) Go to step (2);
(10) Decoding an optimum chromosome from the new population and acquire the decoupling part.

Now we explain some important steps.

3.1 Encoding Method

The advantages of the binary encoding method are simple and universal. But the parameters to be calculated in this paper are more and have the requirement of high precision. The binary encoding often makes the built chromosome length too long, and this situation causes the solution space acquired by the GA training process so large. So the results often overflow or it should take so much time to get the perfect results[7]. In fact, the process of the decoupling network optimization will need some mathematical operations, and the sign encoding method can not fulfill the requirement. The float encoding method has the advantages of high precision and large research space. So the float encoding method is applied in this paper. We encode the parameters in Eq. (8) into the chromosome $L = [a_1, a_2 \cdots, a_n, b_1, b_2, \cdots, b_m]$ according to the system. $L = [a_1, a_2 \cdots, a_n, b_1, b_2, \cdots, b_m]$. By the expression, we can design the corresponding useful crossover operator and mutation operator.

3.2 Fitness Function

Fitness function is one of the most important parts in the build and the optimization of the decoupling network. In this paper, the object is to make the dynamic coupling between each input channel smallest as possible. At the same time, with the time going on, the coupling signal should be near zero as possible. So we construct $f(x)$ as fitness function as follows.

$$f(x) = 1 \Big/ \sum_{i=0}^{n} ix_i^2 \,. \tag{10}$$

where x_i is decoupling signal at the i th time, n is the maximal sample number. The sum of x_i^2 is to express the value of the coupling signal, the multiply of i is to reflect the characteristic that the coupling signal is near to zero with the time going on. When i is becoming larger, the sum of $\sum_{i=0}^{n} ix_i^2$ is becoming larger, and the fitness function $f(x)$ is near to zero. This characteristic is just the one we want. If the decoupling algorithm goes well, the value of $\sum_{i=0}^{n} ix_i^2$ will be smaller, so we calculate the reciprocal of $\sum_{i=0}^{n} ix_i^2$ to fulfill the requirement of the fitness function.

3.3 Reproduction, Crossover and Mutation

Reproduction operator uses the remainder stochastic sampling (RSS) method. Reproduction is implemented by choosing the individual into the middle population randomly according to the different fitness value of each individual. This method has small selection error for that it can ensure that the individual with large fitness value relative to the average fitness value is chosen into the middle population.

In this paper, the simple two-point crossover can not fulfill the system requirement due to the large changeable encoding range of the float encoding method. So, the multi-point crossover is employed in the optimizing of the decoupling network. Define the following crossover operator when the alleles L_1 and L_2 of two chromosome is crossing

$$\begin{cases} L = L_1 & k \geq 0.75 \\ L = L_2 & 0.5 < k < 0.75 \\ L = (L_1 + L_2)/2 & k \leq 0.5 \end{cases} \tag{11}$$

Where k is a random number of uniform distribution, L is an allele of the new individual generated by the crossover process.

When the crossover process begins, the system produces a random number k to compare with the cross probability P_c. If k is smaller than P_c, the crossover process will work by Eq. (11), otherwise the crossover process will stop.

In the GA application using the float encoding method, mutation operator is a very important genetic operator. It will directly influence the research performance of GA. In the optimizing of the decoupling network, select the following equation as the mutation operator

$$L' = L + 0.5(k - 0.5)L \tag{12}$$

Where L is the gene of the chromosome before mutating, L' is the gene of the chromosome after mutating, k is a random number of uniform distribution in $[0, 1]$.

3.4 Realization of NGA

Through comparing the similarity of one individual and the other individual in the evolution process, the NGA restrains the growth of the similar individual, avoids the most individuals run to the same property and keeps the variety of the population,. The realization of the NGA can be expressed as follows.

Assume that the i th individual is L_i, the j th individual is L_j, $\|L_i - L_j\|$ is Euclidean distance between L_i and L_j, l is a small positive number, $f(\cdot)$ is the fitness function. First, compare the similarity of the two arbitrary individuals of the population (the similarity is expressed by the Euclidean distance). If $\|L_i - L_j\| < l$, it means that L_i and L_j have larger similarity, so the algorithm will push aside the individual with small fitness by putting a strong penalty function on it and make its fitness minimal. For example, if $f(L_i) > f(L_j)$, we let $f(L_j) = pena \cdot f(L_j)$ ($pena$ is a small positive number). In the later evolution, L_j will be eliminated by large probability. According to De Jong crowding strategy, the niche evolution environment can be setup.

In this paper, the NGA is applied to optimize the identified decoupling network acquired by using the RLS. The operations include encoding, calculating the fitness value, reproduction, crossover, mutation and the competition of the niche. After some generations' evolution, the satisfied dynamic decoupling network can be acquired.

4   Simulations and Results

In the simulations, if the sensor model is not so correct, it is obvious that the decoupling results acquired by the analysis method of transform function matrix will not be satisfied. In this paper, the simulations which take the two-dimension sensor model as an example, are performed to verify the affectivity of the algorithm. Let the two-dimension sensor model including the coupling parts as

$$\left\{ \begin{aligned} G_{11}(s) &= \frac{600s + 25 \times 10^6}{s^2 + 1000s + 25 \times 10^6} \\ G_{22}(s) &= \frac{700s + 2401 \times 10^4}{s^2 + 1176s + 2401 \times 10^4} \\ G_{12}(s) &= \frac{500s + 4 \times 10^6}{s^2 + 864s + 2304 \times 10^4} \\ G_{21}(s) &= \frac{400s + 6 \times 10^4}{s^2 + 960s + 2304 \times 10^4} \end{aligned} \right. \tag{13}$$

First, input the unit step signal into channel one, no signal into channel two, so a set of output signal is obtained; Then add some large Gauss noise in the output signal, to get a new output signal as the real output signal. According to the decoupling theory mentioned above, the input and the output data which is used to identify D_1 of the decoupling network is obtained. Use the RLS to identify D_1, the corresponding second order model is

$$z(k) + a_1 z(k-1) + a_2 z(k-2) = b_0 u(k) + b_1 u(k-1) \,, \tag{14}$$

Encode the coefficient of D_1 using the float encoding method and get the chromosomes of the initial population, let the data structure as

fitness value of the chromosome	a_1	a_2	b_1	b_2

Using the NGA, after some generations' evolution to the initial population, the satisfied decoupling part D_1 can be obtained. In the same way, the decoupling part D_2 also can be obtained. At last, let the optimized transfer function using the NGA as the real sensor model to verify the performance of the decoupling network. The results are shown in Fig. 2 ~ Fig. 5 respectively. In the simulations, the population size is 100, the number of generations is 50, $p_c = 0.5$ and $p_m = 0.01$.

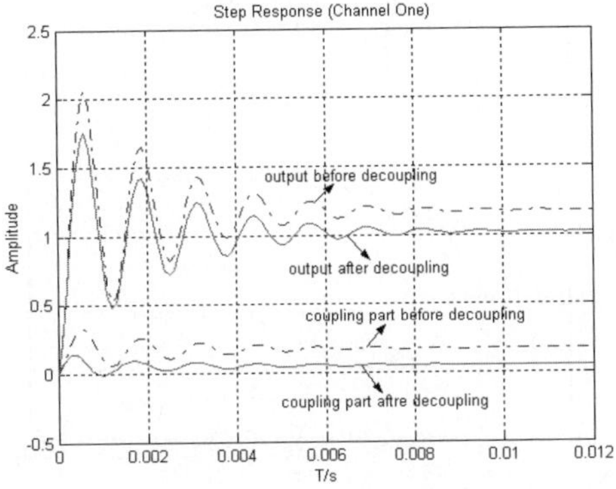

Fig. 2. Contrast of the decoupling effect of the initial population (Channel One)

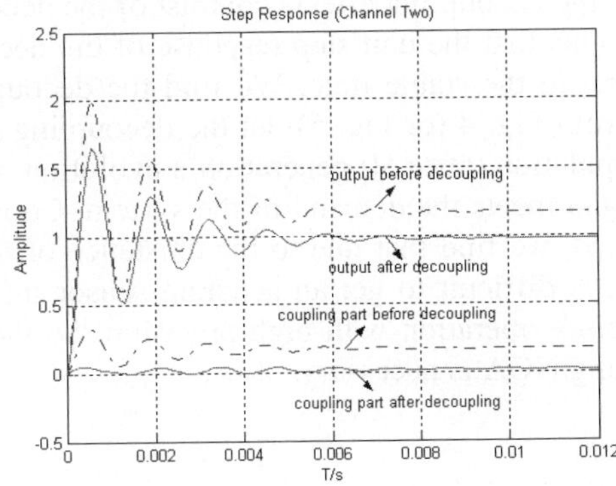

Fig. 3. Contrast of the decoupling effect of the initial population (Channel Two)

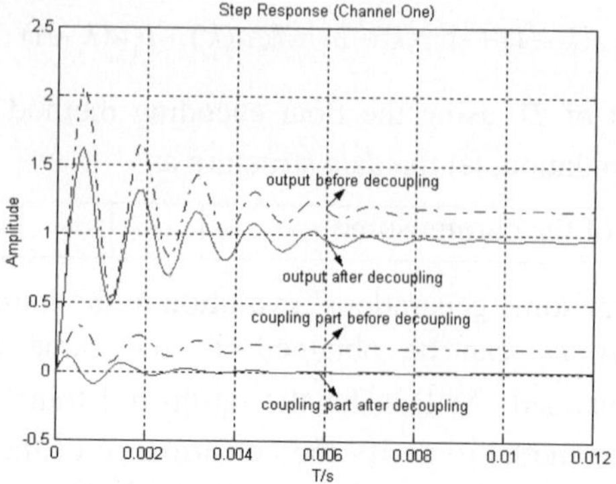

Fig. 4. Decoupling effect of the population after 50 generation's evolution (Channel One)

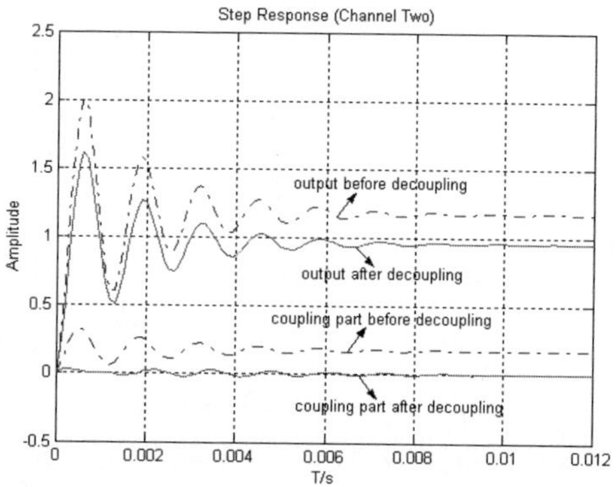

Fig. 5. Decoupling effect of the population after 50 generation's evolution (Channel Two)

Fig. 2 and Fig. 3 are the decoupling effects contrast of the decoupling network only built by RLS. It is obvious that the unit step response of the decoupling network still has some coupling error in the stable time. We find the decoupling effect is not so satisfied. In contrast, from Fig. 4 (or Fig. 5), let the decoupling network in Fig .2 (or Fig. 3)as the initial population, after 50 generation's evolution, its unit step response has no coupling error and meets the demand of the system. Comparing Fig. 2(or Fig. 3) with Fig. 4 (or Fig. 5), we find that due to the existence of the disturbance in the sensor output signal, it is difficult to get an accurate sensor model only using RLS, also realize the decoupling operation with high precision. So the optimization of the decoupling network using NGA is necessary.

5 Conclusions

In this paper, a new multi-axis sensor dynamic decoupling method is proposed based on a NGA. The method directly builds and optimizes the decoupling network according to the calibrating data, and avoids the high precision requirement to the transform function of the multi-sensor using the traditional method. In addition to the satisfied decoupling effect, the method has the advantage that the order of the acquired decoupling network can be changeable according to the change of the practical requirement. Also, the decoupling network has a considerable robustness because of the using of the NGA.

References

1. Xu Kejun, Yin Ming and Zhang Ying. A Kind of Dynamic Decoupling Method for Wrist Force Sensor[J]. Journal of Applied Science. 1999, 17(1): 39-44.
2. Xu Kejun, Li Cheng. Dynamic Decoupling and Compensation Methods of Multi-axis Force Sensors[J]. IEEE Transactions on Instrumentation and Measurement. 2000, 49(5): 935-941.
3. Xu Kejun and Li Cheng. The Stationary and Dynamic United Decoupling and Compensating Method in a Multi-axis Force Sensor[J]. Journal of Applied Science. 2000, 18(3): 189-191.
4. Song Guomin, Zhang Weigong, and Zhai Yujian. Study of Dynamic Decoupling of Sensor Based on Diagonal Predominance Compensation[J]. Chinese Journal of Scientific Instrument. 2001, 22(4): 165-167.
5. Deb K, Goldberg D E. An Investigation of Niche and Species Formation in Genetic Function Optimization[A]. Proceedings of the Third ICGA[C]. San Mateo. 1989, 42-50.
6. Jae-Kwang Kim, Dong-Hyeok Cho, Hyun-Kyo Jung, and Cheol-Gyun Lee. Niching Genetic Algorithm Adopting Restricted Competition Selection Combined With Pattern Search Method[J]. IEEE Transaction on Magnetic. 38(2): 1001-1004.
7. Xue Fuzhen, Tang Yan and Li Meng. The Design of a Multivariable Dynamic Precompensation Matrix Based on GA[J]. Industry Instrumentation and Automation. 2002, 5: 68-71.

User Action Based Adaptive Learning with Weighted Bayesian Classification for Filtering Spam Mail

Hyun-Jun Kim[1,2], Jenu Shrestha[2], Heung-Nam Kim[2], and Geun-Sik Jo[2]

[1] Samsung Electronics, Corporate Technology Operations, R&D IT Infra Group,
416, Maetan-3Dong, Yeongtong-Gu, Suwon-City, Gyeonggi-Do, Korea 443-742
`dannis@eslab.inha.ac.kr`
[2] Intelligent E-Commerce Systems Lab., School of Computer Science & Engineering,
Inha University, 253 Younghyun-dong, Nam-Gu, Incheon, Korea 402-751
`{jenustha, nami}@eslab.inha.ac.kr, gsjo@inha.ac.kr`

Abstract. Nowadays, e-mail is considered one of the most important communication methods, but most users suffer from Spam mail. To solve this problem, there has been much research. The previous research showed comparatively high performance, but for adaptation of real world, it requires several improvements. First, it needs personalized learning for better performance. We cannot make a strict definition of Spam, because the definition of any context depends on each user. Second, the concept drift or interest drift problem, that is, users' interest or any context's concept, may change over time. Therefore, many Spam filtering systems are using continuous learning schemes such as adaptive learning or incremental learning. However, these systems require user feedback or rating results manually, and this inconvenience causes slow learning and performance enhancement. In this research, we developed an adaptive learning system based on an automatic weighting environment. For the automatic weight, we categorized 6 user patterns (actions) on the mailing system whose weights are automatically adapted to the learning phase. From the experiment, we will demonstrate the Bayesian classification with an adaptive learning environment. By using suggesting ideas, we will analyze the comparison result with adaptive learning. Finally, from the experiment using real world data sets, we will prove its possibility for tracking the concept and interest drift problems.

Keywords: Adaptive Learning, Concept Drift, Weighted Bayesian, Spam Filtering.

1 Introduction

Nowadays, there is no doubt that e-mail is one of the most important communication methods in the world. Meanwhile, most people who send e-mails are suffering from unwanted (Spam) mail in their mailbox. According to the recent report from major internet service companies, 84.4% of total mail were Spam [1]. Considering this, much research has been done to prevent Spam mail using machine learning or statistical technologies such as Bayesian Classification [2], Support Vector Machine [3], Memory-based Approach [4] and so on. In particular, the Bayesian Classifier-based system is frequently used for filtering Spam mail because of its comparative ease of

A. Sattar and B.H. Kang (Eds.): AI 2006, LNAI 4304, pp. 790–798, 2006.
© Springer-Verlag Berlin Heidelberg 2006

use and high performance [5]. Therefore, many popular mail filtering systems are using Bayesian Classification [6][7][8]. In fact, while several featured filtering algorithms such as SVM, CBR, LSI show higher performance (recall and precision) than the Bayesian when we analyze the algorithms in the laboratory [3][9][10], in the real world, people pay relatively little attention to the other points except for the algorithm. We found that for an effective Spam mail filtering system, there are several factors that we have to consider besides just algorithms: personalized and continuous learning. In this research, we will show how a personalized Spam filtering system can be developed by using an automatic user action recognition scheme. Also, by using an adaptive learning scheme, we will show how the system performance increases. For the research, we use *LingSpam* [12] mail dataset and implemented a Bayesian classification-based Spam mail filtering system with a continuous learning environment.

The outline of the paper is as follows: section 2 concentrates on the related work - a history of Spam fighting and concept drift problems; section 3 refers to the suggested Spam filtering system using weighted Bayesian classification with adaptive learning; and we explain the experimental results in section 4. We conclude this study with prospects on future work in section 5.

2   Related Work

In the past several years, there have been many algorithms and applications developed to solve Spam problems. Although, some problems and defects still exist, there has been much improvement and some of them are being used as commercial filtering systems. In this section, we will discuss the efforts made to fight Spam mail and present background knowledge about concept drift and adaptive learning, which we will use to enhance the Spam filtering performance.

2.1   History of Spam Mail Filtering

The first approach to filtering Spam was based on rule or message [11]. Simply identifying Spam features from the email content, we could decide whether the mail were Spam or not. This approach worked comparatively well until statistics and machine learning approaches appeared. Because of rule-based system's simplicity of use and implementation, many mail filtering systems are using that approach. Bayesian classification was one of the better solutions using a statistical approach to overcome rule-based systems' limitations [5]. Although it was still not perfect, it showed comparatively good performance in the aspects of recall and precision [12]. Later, improved Bayesian Classification-based systems such as the Bayesian network and weighted Bayesian classification enhanced the performance of filtering Spam mails. Meanwhile, a number of methodologies such as SVM, and LSI suggested by many researchers [3][13][14], but different from the Bayesian approach, were difficult to use in real world because of its complexity of calculation and difficulties of implementation and management. Researchers are now starting to see that in the real world environment, we cannot just consider algorithms and methodologies. Even if we use currently perfect methodology for filtering Spam, it will not take long time to see it become defective because of the concept drift problem and intelligent Spammers [15].

Recently, CBR (Case-based Reasoning) has arisen as a good choice for filtering Spam [16][17], because CBR's life cycle, consisting of retrieve, reuse, revise and retain, supports tracking the concept drift problem. Hence, it is appropriate for a disjoint concept domain such as Spam [16]. Similar to CBR, adaptive learning and incremental learning schemes that support tracking concepts and features can be another useful way to track continuously changing Spam features and concepts. This paper analyzes their usefulness as Spam mail-filtering systems.

2.2 Concept Drift and Adaptive Learning

Frequently, user interests and concerns change with time. Therefore, machine learning-based systems such as automatic document classification and mail-filtering, need to consider this problem. A Spam-filtering system must track concept or user interest because Spam domain is one of the most rapidly changing fields in terms of its features and concepts. For example, any word can become a very strong Spam feature in the future, and user interest can change. 1994 marked the first attempt to track concept drift on a meeting-scheduling software assistant [18]. This system used a time window with some of the latest examples to reflect the shifts in a user's scheduling preferences. Similarly, STAGGER was developed to learn user interests to track changing concepts by adding new attributes or adjusting the connection weights for concept connections [19]. In addition, there was research done to learn user interests by monitoring web and e-mail habits [20]. This approach was based on content clustering and user rating, but it required user manual feedback. In this research, we propose a more enhanced and user-friendly feedback system for adaptive learning. In section 3, we will discuss the system architecture more in detail.

3 Spam Mail Filtering System with Adaptive Learning

Fig.1. shows a suggested Spam mail-filtering system architecture based on a weighted Bayesian classification and an adaptive learning environment. For the learning and classification, it uses the Naïve Bayesian classifier, and it uses a weighting scheme in the adaptive learning phase for faster improvement of performance.

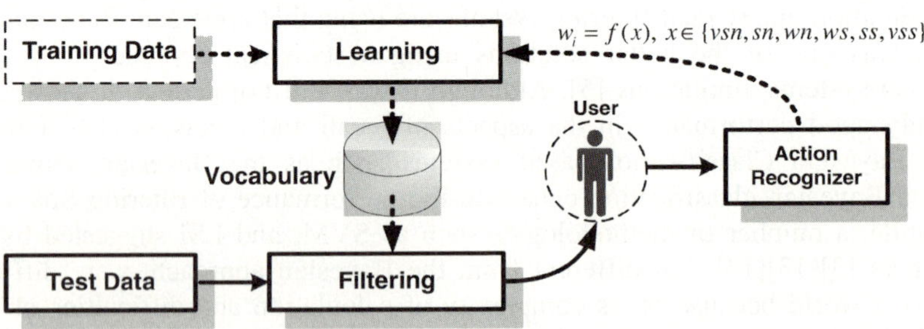

Fig. 1. Spam-Filtering Architecture based on Weighted Bayesian Classifier with Adaptive Learning Environment

Fig.1. shows the system architecture which has two phases, learning and testing. Once users receive mail in their mailboxes, they check the list of new mail, and react differently according to the mail's purpose. These actions are recognized by the Action Recognizer, and some of the mails are re-trained with automatically given weight. Compared to former systems using the Bayesian Classifier, the suggestion system runs without training data, and its performance improves over time. We will discuss this experimental result in section 4.

3.1 Categorization of User Actions on E-mail Service

As we mentioned earlier, there are already many systems using adaptive learning or incremental schemes, but they need feedback from users. For example, if a user considers some of the mail Spam, and wants to identify the Spam mail to the system, then he has to check the mail using a radio button to execute learning or blocking action. However, it is not easy to give some additional actions to the system for users, and also, this way only offers a simple re-learning. As a result, this kind of adaptive learning requires a lot of time for performance enhancement in the real world (Fig.4. in Section 4.3).

Table 1. Definition of 6 different User Activities

User Action Category (UAG)			
Cat. No.	*Symbol*	*Category*	*User Action*
1	*(VSN)*	*Very Strong NonSpam*	*Open + Reply + Save*
2	*(SN)*	*Strong NonSpam*	*Open + Reply, Open + Save*
3	*(WN)*	*Weak NonSpam*	*Open*
4	*(WS)*	*Weak Spam*	*Nothing*
5	*(SS)*	*Strong Spam*	*Delete*
6	*(VSS)*	*Very Strong Spam*	*Block*

In Table 1, we categorized 6 different user actions. The system can weight each action during the learning phase automatically. In section 3.2 we will discuss the optimal weighting function.

3.2 Performance Enhancement Using Adaptive Learning and WBC

For performance and user satisfaction, we need to consider that personalization issue. Since e-mail represents a disjoint concept and Spam's definition differs from user to user, it is not feasible to train data sets for all users. Instead, by using an adaptive learning scheme, training is done by users personally and continuously. To get affordable performance fast for filtering, we used a Weighted Bayesian Classifier (WBC). The Naïve Bayesian Classifier considers each attribute $< d_1, d_2, ..., d_n >$ as having the same importance. Most WBC systems just consider features of contents [11], but, as we mentioned in section 3.1, we will focus on user action for adding weight to the Bayesian classifier, and thereby getting faster performance enhancement.

$$c_w = \frac{\arg\max}{c_i \in C} P(c_i) \prod_{k=1}^{n} \left(\frac{w_{d_k}}{\max_j w_{d_j}} \right) P(d_k \mid c_i)$$

(1)

$$w_i = f(x), \quad x \in \{vsn, sn, wn, ws, ss, vss\}$$

Eq. (1) shows a weighted Bayesian classification. Let's assume that there is a document d_k, and we give a weight, w_{d_k}, to d_k resulting from its weight function $f(x)$ according to user action for d_k. For the normalization of a weighted value, we divided w_{d_k} by the denominator, $\max_j w_{d_j}$ which means the maximum weighted value according to a user action. For the weight function, we use Ivan Kychev's linear gradual forgetting function [21] as shown in Eq. (2). In his research, the function adapted to forget old features over time, but in this paper, we use that function to give weight to 6 different user actions.

$$w_i = -\frac{2k}{l-1}(i-1)+1+k, \quad k \in [0,1]$$

(2)

Where i is a counter of user actions from *VSS* to *VSN*, $i =\{1,...,l\}$. l is the length of the user actions. Also, k is a parameter that represents the percent of increasing the weight of user actions.

4   Experiments

The proposed system is implemented using Java, My-SQL. All the experiments are performed on a 3.4GHz Pentium PC with 1GB RAM, running on a Microsoft Window 2003 Server. We used a *LingSpam* mail set introduced by Androutsopoulos [12]. This mail set contains 4 types of mail – 'bare', 'stop-list', 'lemmatizer', 'lemmatizer+stop-list'. Each type of mail is created by a pre-processing procedure. In this experiment, we used the last type of mail, lemmatized and stop-list deleted mail, for training and testing the system.

Table 2. Data Set

	Training Data	Test Data
Spam Mail	*309*	*221*
NonSpam Mail	*213*	*83*
Total	*522*	*304*

4.1   Performance Measure

To evaluate the system's performance, we used the most commonly used evaluation measures - precision, recall and F1-measure [22] as defined below in Eq. (3).

$$Precision \ = \ \frac{Spam\ mails\ which\ are\ classified\ as\ Spam}{Mails\ which\ are\ classified\ as\ Spam}$$

$$Recall \ = \ \frac{Spam\ mails\ which\ are\ classified\ as\ Spam}{All\ spam\ mails}$$

$$F1 - measure \ = \ \frac{2*Precision*Recall}{Precision\ +\ Recall}$$

(3)

4.2 Definition of Weight for User Actions

To give different weight to 6 user actions, we used linear function as shown in Eq.(2). Although, this paper suggests only 6 user actions and its weights for a Spam mail filtering system, we can use this function for other domains which have more complex actions and situations. Fig.2. shows linear curbs according to weight function Eq.(2). For example, *wn* means a user action in which he opened a mail from the list, but even if the user opens the mail, the system cannot be sure whether the mail is Spam or not, because it could be a fraud Spam. Hence, we consider that it is non-Spam but don't give it weight. Also, in case of *ws*, it means that the user didn't open the mail from the list, so the system can consider that mail is not of interest to the user. However, there may also be an exception. The user may recognize the mail from the title, but sometimes he doesn't need to open it. In this case the system considers the mail as Spam, but doesn't give it any weight. Similarly, other actions - *vss, ss, sn, vsn* have their own weight according to their strength as Spam or non-Spam. We used constant k as 80% according to Ivan Kychev's linear gradual forgetting function [21].

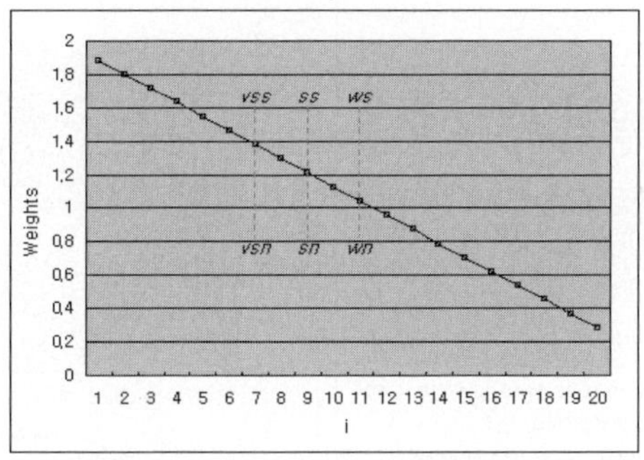

Fig. 2. A Linear Weight Function for Different User Action. (k=80%)

4.3 Result of the Experiment

For the verification of the proposed system, we will show two different results: First, performance enhancement when we use Bayesian classification under an adaptive learning environment (Fig 3); Second, a comparison analysis of two different environments, adaptive learning based on user patterns and user feedback. In the second experiment, we will show how much learning time takes to arrive at the targeted performance level for the proposed system when we start the system without any learning (Fig.4).

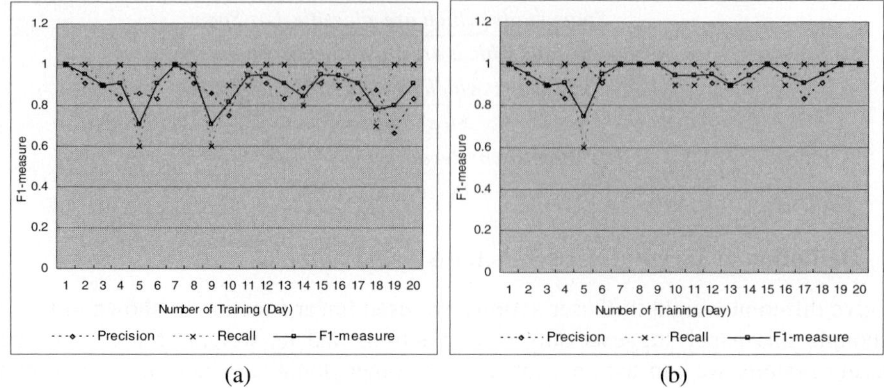

(a)   (b)

Fig. 3. Comparison Result between Bayesian Classification and Bayesian Classification with Adaptive Learning Environment

From Fig.3 (a) we can see that, by training the system only once, performance is unpredictable. Performance depends on the training and the test data set we are using. However, from Fig.3 (b), we can see that continuous learning with new mails leads to better performance, though not the best. Comparing these two graphs we can conclude that, the later is more stable. The graph shows that as time goes on from expt. no. 6 onwards; adaptive learning shows stable performance compared to normal learning, whose performance goes on fluctuating.

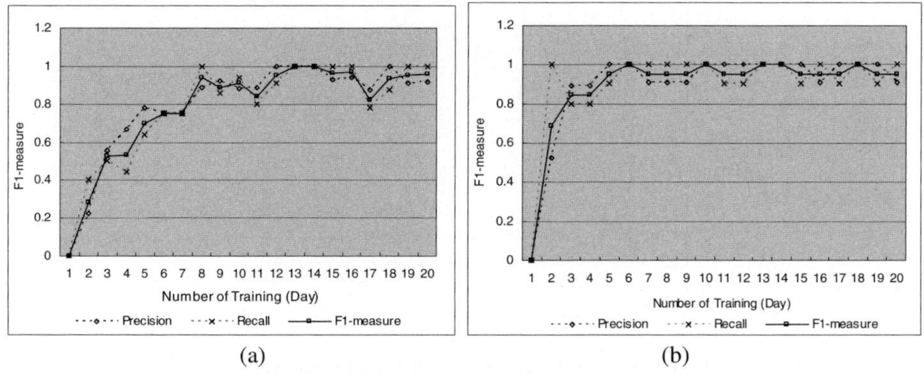

(a)   (b)

Fig. 4. Comparison Result of zero to goal performance between adaptive learning and the proposed system

Fig.4 mainly deals with personalized issues and adaptive learning. In adaptive learning, initially there is no learning, but once we start giving feedback to the system, performance gradually increases. This requires manual feedback from the user to the system, so the usability is not good and this system's performance is strongly depends on a user action. From the graph fig.4 (a), we can see that from expt. 12 onwards, it starts showing a better and more stable performance, though not as good as the user action based adaptive learning with weighted Bayesian classifier, Fig.4 (b). Meanwhile Fig.4 (b) shows that it takes little time to get stable performance. And also, it shows very stable performance from expt. 5 onwards from which we can conclude that in a very short period of time, we can achieve stable and better performance. Moreover by recognizing the user actions and giving weight on learning phase automatically, its usability is much better.

5   Conclusions and Future Work

In this paper, we proposed a performance enhancement system of Spam mail filtering using an adaptive learning scheme. The main advantage of this system is that we can achieve a better and more stable performance of the system in short period of training time. Furthermore, it deals with personalized issues and concept drift problems as well. The experimental results showed the proposed method has better performance in terms of learning time, filtering of Spam mail and usability compared to that of Bayesian classifier. For the future work, we will increase the performance by extending the number of category of user action considering contents reading time and other factors. Increasing the number of user categories from the 6 that we have implemented will definitely help to shorten the time necessary for stable performance. And also, we can prove the system's performance by testing on another domain such as document classification and collaborative filtering.

References

1. Korea Telecom.: www.kt.co.kr (2004)
2. Sahami, M., Dumais, S., Heckerman, D. and Horvitz, E.: A Bayesian Approach to Filtering Junk E-Mail. In Learning for Text Categorization, Proc. Of the AAAI Workshop, Madison Wisconsin. AAAI Technical Report WS-98-05, (1998) 55-62.
3. Thomas, G. and Peter, A. F.: Weighted Bayesian Classification based on Support Vector Machine, Proc. of the 18th International Conference on Machine Learning, (2001) 207-209.
4. Sakkis G., Androutsopoulos I., Paliouras G., Karkaletsis V., Spyropoulos C.D., Stamatopoulos P.: A Memory-Based Approach to Anti-Spam Filtering for Mailing Lists. Information Retrieval, 6 (1), Kluwer Academic Publishers (2000) 49-73.
5. Androutsopoulos, I., Koutsias, J., Paliouras, G., Karkaletsis, V., Sakkis, G., Spyropoulos, C., Stamatopoulos, P.: Learning to Filter Spam E-mail: A Comparison of a Naïve Bayesian and a Memory-Based Approach. 4th PKDD Workshop on Machine Learning and Textual Information Access, (2000).
6. The Apache SpamAssassin Project, http://Spamassassin.apache.org/
7. The SpamBayes Project, http://Spambayes.sourceforge.net/
8. H. J. Kim, H. N. Kim, J. J. Jung, G. S. Jo, : Spam mail Filtering System using Semantic Enrichment, Proc. of the 5th International Conference on Web Information Systems Engineering, (2004).
9. Cunningham P., Nowlan N., Delany S.J., Haahr M.: A Case-Based Approach to Spam Filtering that Can Track Concept Drift, The ICCBR'03 Workshop on Long-Lived CBR Systems, Trondheim, Norway, (2003).

10. Kevin R. G.: Using Latent Semantic Indexing to Filter Spam, ACM Symposium on Applied Computing, Data Mining Track, (2003).
11. Cohen, W. W.: Learning Rules that Classify E-Mail, Proc. Of the AAAI Spring Symposium on Machine Learning in Information Access, (1996).
12. Androutsopoulos, I., Koutsias, J., Chandrinos, K.V., Paliouras, G., Spyropoulos, C.D., : An Evaluation of Naive Bayesian Anti-Spam Filtering, Proc. of the ECML 2000 Workshop on Machine Learning in the New Information, (2000) 9-17.
13. Ferreira, J. T. A. S., Denison, D. G. T., Hand, D. J.: Weighted Naïve Bayes modeling for data mining, Technical report, Dept. of mathematics at Imperial College, (2001).
14. H. J. Kim, H. N. Kim, J. J. Jung, G. S. Jo, : On Enhancing The Performance of Spam mail Filtering System using Semantic Enrichment, Proc. of the 17th Australian Joint Conference on Artificial Intelligence (AI-04), (2004).
15. Koychev, I., Schwab I.: Adaption to Drifting User's Interests, Proc. Of the ECML200/MLnet Workshop "ML in the New Information Age", (2000).
16. Pádraig C., Niamh N., Sarah J. D., Mads H.: A Case-Based Approach to Spam Filtering that Can Track Concept Drift, Proc. of the ICCBR03 Workshop on Long-Lived CBR System, (2003).
17. Delany S. J., Cunningham P., Coyle L.: An Assessment of Case-Based Reasoning for Spam Filtering, Artificial Intelligence Review Journal, 24(3-4), Springer, (2005) 359-378.
18. Mitchell T., Caruana R., Freitag D., McDermott J. and Zabowski D.: Experience with a Learning Personal Assistant, Communications of the ACM 37.7, (1994) 81-91.
19. Schlimmer J., Granger R.: Incremental Learning from Noisy Data, Machine Learning1 (3), (1986) 317-357.
20. Grabtree I., Soltysiak S.: Identifying and Tracking Changing Interests, International Journal of Digital Libraries, Springer Verlag, vol. 2, (1998) 38-53.
21. Koychev, I.: Gradual Forgetting for Adaptation to Concept Drift, Proc. Of ECAI 2000 Workshop "Current Issues in Spatio-Temporal Reasoning", (2000) 101-106.
22. Y. Yang, and X. Liu.: A Re-examination of Text Categorization Methods, Proc. of the ACM SIGIR'99 Conference, (1999).

A Novel Clonal Selection Algorithm for Face Detection

Wenping Ma, Ronghua Shang, and Licheng Jiao

Institute of Intelligent Information Processing, P.O. Box 224, Xidian University,
Xi'an, 710071, P.R. China
wpma@mail.xidian.edu.cn

Abstract. Based on the Antibody Clonal Selection Theory of immunology, we put forward a novel clonal selection algorithm and apply it into face detection problem, named by CSAFD. The new detector is fast and reliable, which can detect faces with different sizes and various poses from both indoor and outdoor scenes. The goal of this research is to detect all regions that may contain faces while remaining a low false positive output rate. The new algorithm firstly abstracts the face template and then realizes the precise location of the face using clonal selection algorithm and template matching. When compared with immune genetic algorithm, standard genetic algorithm and evolutional genetic algorithm, the experimental results on images with moderately complex background scene showed, the new algorithm had a good performance and was much more accurate and robust.

1 Introduction

Face detection is a fundamental step before the recognition or identification procedure[1]. Its reliability and time response have a major influence on the performance and usability of a face recognition system. It is also a key step in almost any computational task related with the analysis of faces in digital images (access control, identification for law enforcement, borders control, identification and verification for credit cards and ATM, human-machine interfaces, passive recognition of criminals in public places or buildings, etc.). Given an arbitrary image, the goal of a face detection system is to find all contained faces and to determine the exact position and size of the regions containing these faces. When analyzing real-world scenes, face detection is a challenging task, which should be performed robustly and efficiently, regardless variability in scale, location, orientation, pose, illumination, artifacts (beard, eye-glasses, etc.), and facial expressions[1].

A representative face detection method with two steps: firstly, locating the face region[3~5], or assuming that the location of the face part is known[2], secondly, detecting the facial features in the face region based on edge detection, image segmentation, and template matching or active contour techniques. Another method is the visual learning or neural network approach[6]. Although the performance reported is quite well, and some of them can detect nonfrontal faces, approaches in this method are extremely computationally expensive. A relatively traditional approach of face detection is template matching and its derivations[7]. This approach uses a small image or a simple pattern that represents the average face as the face model. Face detection based on deformable shape models was also reported[8]. Although this method is

A. Sattar and B.H. Kang (Eds.): AI 2006, LNAI 4304, pp. 799–807, 2006.
© Springer-Verlag Berlin Heidelberg 2006

designed to cope with the variation of face poses, it is not suitable for generic face detection due to the high expense of computation.

A novel clonal selection algorithm for face detection based on Antibody Clonal Selection Theory, named by CSAFD, is presented in this paper, which can detect faces with different sizes and various poses from both indoor and outdoor scenes. The experimental results show that this method of face detection has a good performance in moderately complex background scene and is much more accurate and robust.

2   A Template Extraction of Facial Feature

Human face image belongs to a special class. The facial feature position of different person only varies in a local area. If we know the position and orientation of a face, we can use the spatial constraint among facial features to obtain the approximate position of each facial feature.

Eigenface is based on the global feature of face images, i.e. it uses the whole face image as training data. When there are gross variations in the input images that greatly affect the global feature, eigenface may fail[9]. However, in this situation, local features such as eyes, mouth and nose are often less affected. So these local features can help recognize faces. Brunelli and Poggio[11] proposed a template-matching method that used both global features and local features. Pentland et al. [10] extended eigenface to local features, getting eigeneyes, eigennoses and eigenmouths, all of which are called eigenfeatures. Each feature, no matter global or local, is in fact a region in the face image. Moreover, there are no reliable criterions to determine which facial feature and how many of them should be used. Thus the performance of those local-feature -based methods greatly depends on the experience of the operators.

We extract the face template from the images in the physics-based face database[11]. This database contains front view face images taken under four different illumination conditions. The database contains 111 persons and a minimum of 16 images per person. The general face template using the average outer edges[12] is shown in Fig.1.

Fig. 1. The general face template

3 The Novel Clonal Selection Algorithm for Face Detection

3.1 Clonal Selection Theory

The immune system's ability to adapt its B-cells to new types of antigen is powered by processes known as clonal selection and affinity maturation by hypermutation. The majority immune system inspired optimization algorithms are based on the applications of the clonal selection and hypermutation. The first immune optimization algorithm may be Fukuda, Mori and Tsukiyama's algorithm that included an abstraction of clonal selection to solve computational problems. But the clonal selection algorithm for optimization has been popularized mainly by de Castro and Von Zuben's CLONALG[13].

The clonal selection theory (F. M. Burnet, 1959) is used by the immune system to describe the basic features of an immune response to an antigenic stimulus[14]. The references [15] to [17] simulate this clonal selection mechanism above from different viewpoints and put forward the Clonal Selection Algorithms one after another. Just as the same as the Evolutionary Algorithms (EAs) , the Artificial Immune System Algorithms depend on the encoding of the parameter set rather than the parameter set itself. It establishes the idea that the cells are selected when they recognize the antigens and proliferate. When exposed to antigens, the immune cells which may recognize and eliminate the antigens can be selected in the body and mount an effective response against them. Its main ideas lie in that the antigen can selectively react to the antibody, which is a native production and spreads on the cell surface in the form of peptide. The reaction leads to cell proliferating clonally and the colony has the same antibody. Some clonal cells divide into antibody producing cells, and others become immune memory cells to boost the second immune response.

The clonal selection is a dynamic process of the immune system stimulated by the self-adapting antigen. From the viewpoint of the Artificial Intelligence, some biologic features such as learning, memory and antibody diversity can be used in artificial immune system[18].

3.2 Basic Definition

The new algorithm composes of two main processes: firstly, abstracts the face template based on the method in section 2, Secondly, realizes the precise detection of the face using clonal selection algorithm and template matching.

The major elements of CSAFD is established below:

Antigen: In the immunology, antigen is a kind of substance which can induce organism to immune response and can take place special responses with the corresponding antibodies or the T cells. In AIS, antigen usually means the problem and its constraints. Especially, for the face detection problem, Given a digital image $f(x, y)$ with $M \times N$, to confirm the image $f(x, y)$ whether contains the face region, which is similar to the face template obtained before. Suppose the template is a region $\omega(x, y)$ with $J \times K$, where $J < M$, $K < N$. Then the correlative function between $\omega(x, y)$ and $f(x, y)$ is defined as formula (1) [19]:

$$R(m,n) = \frac{\sum_x \sum_y f(x,y)\omega(x-m,y-n)}{\left[\sum_x \sum_y f^2(x,y)\right]^{1/2}}$$

(1)

Where $m = 0,1,2\,;\cdots,N-1$.

It can be taken a value of R by calculating formula (1) with arbitrary value of (m,n) in $f(x,y)$. when $\omega(x,y)$ removes along the image with the variation of m, n, We can take the value of $R(m,n)$. The maximum of $R(m,n)$ illuminates $\omega(x,y)$ is most similar to $f(x,y)$ in this position. The formula(1) adopts a unitary factor which is calculated in the whole area plotted out by $\omega(x,y)$, thus it varies as a displacement function. The antigen is defined as objective function $R(m,n)$, Similar to the function of antigen in immunology, it is the active factor for the artificial immune system algorithm.

Antibody: Antibodies represent candidates of the problem. The limited-length character string $a = a_1 a_2 \cdots a_l$ is the antibody coding of variable m、n. In our algorithm, we use binary coding. m is represented by 8 bits binary cluster as same as n, i.e. $1 = 16$.

Set I is called antibody space, namely $a \in I$. The antibody population $A = \{a_1, a_2, \cdots, a_n\} \in I^n$ is an n-dimension group of antibody a, namely,

$$I^n = \left\{A : A = (a_1, a_2, \cdots, a_n),\ a_k \in I,\ 1 \le k \le n\right\}$$

(2)

where the positive integer n is the antibody population size.

Antibody-Antigen Affinity: Antibody-Antigen Affinity is the reflection of the total combination power locates between antigen and antibodies. In AIS, it generally indicates values of objective functions or fitness measurement of the problem.

Cloning T_c^C **:** In the immunology, clone is the process of antibody proliferation. In AIS, the clonal operation to the antibody population is defined as:

$$Y(k) = T_c^C(A(k)) = [T_c^C(a_1(k))\quad T_c^C(a_2(k)),\quad \cdots \quad, T_c^C(a_n(k))]^T$$

(3)

where $T_c^C(a_{ci}(k)) = I_{ci} \times a_{ci}(k)$ $i = 1,2\cdots n$, I_{ci} is a q_{ci}-dimensional identity row vector. The process is called the q_{ci} clone of antibody a_i, namely $q_{ci}(k) = \hbar(n_c, \Theta_i)$, where Θ_i stands for the affinity function of antibody a_i and other antibodies, and n_c is the clonal scale.

Clonal Mutation T_m^C **:** According to the mutation probability p_m, the cloned antibody populations are mutated as follows:

$$Z(k) = T_m^C(Y(k)) = (-1)^{random \le p_m} Y(k)$$

(4)

Clonal Selection T_s^C : $\forall i = 1, 2, \cdots n$, if there are mutated antibodies

$$a_i'(k) = \max\{Z_i(k)\} = \{Z_{ij}(k) \mid \max R(Z_{ij}) \quad j = 1, 2, \cdots, q_i\} \quad , \quad \text{the probability of}$$

$a_i'(k)$ taking place of $a_i(k) \in A(k)$ is:

$$T_s^k\left(a_i(k) = a_i'(k)\right) = \begin{cases} 1 & \text{when } R(a_i(k)) < R(a_i'(k)) \\ 0 & \text{when } R(a_i(k)) \geq R(a_i'(k)) \end{cases} \tag{5}$$

After the clonal selection, the new population is:

$$A(k+1) = \{a_1(k+1), a_2(k+1), \cdots, a_i'(k+1), \cdots, a_n(k+1)\} \tag{6}$$

where $a_i'(k+1) = a_j(k+1) \in A(k+1) \, i \neq j$ and $R(a_i'(k+1)) = R(a_j(k+1))$, in which $a_j(k+1)$ is one of the best antibodies in $A(k+1)$.

3.3 Description of the Algorithm

Inspired by the Antibody Clonal Selection Theory, we proposed a novel clonal selection algorithm for face detection. In detail, the algorithm firstly abstracts the face template and then realizes the precise location of the face using clonal selection algorithm and template matching. The novel clonal selection algorithm can be implemented as Fig. 2.

The Novel Clonal Selection Algorithm for Face Detection (CSAFD)
Begin
   $k = 0$;
   Initialize $A(k)$ and algorithm parameters;
   Calculate the affinity of $A(k)$: $\{R(A(k))\} = \{R(a_1(k)), R(a_2(k)), \cdots R(a_n(k))\}$;
   While not finished do
      $k = k + 1$;
      Bring $A(k)$ from $A(k-1)$ by the Clonal Selection Operator including
         Clonal Operating T_c^C , Clonal Mutation Operating T_m^C , Clonal Selection
         Operating T_s^C ;
      Calculate the affinity of $A(k)$, namely, evaluate $A(k)$;
  End
End

Fig. 2. The Novel Clonal Selection Algorithm for Face Detection

4 Experimental Results

We run our algorithm on the test set of 60 gray-level images of size 500x400 pixels, the face size varied from 20×24 pixels to 200×240 pixels. These images contain faces at different scale but mainly in the front-parallel view with vertical orientation. 54 are successfully detected. giving a 90% detection rate. In order to validate the new algorithm, we have compared CSAFD with that of immune genetic algorithm(IGA) [20]、 the standard genetic algorithm(SGA) and evolutional genetic algorithm (EGA) [21].

The parameters setting of CSAFD is as follows: the size of antibody population $n_b = 10$, the clonal scale $n_c = 5$, the mutation probability $p_m = 2/l$, the max generation is 100; The parameters of IGA is set as: the size of population is 60, the crossover probability pc=0.85, the mutation probability pm=0.20, the probability of vaccination is 0.5, The renewed probability of vaccination is 0.5~0.8 varied with evolution process, the max generation is 100; The parameters of EGA is set as: the size of population is 60, the crossover probability pc=0.85, the mutation probability pm=0.15; The parameters of SGA is set as that of EGA. Some of the successful results are shown in Fig.3. The white rectangles in the figure indicate the detected face candidates. For every algorithm, the face detection is performed thirty times per test image. Each time we only try to detect faces of a specified size. There may be multiple partially overlapped rectangular regions that are considered as face candidates.

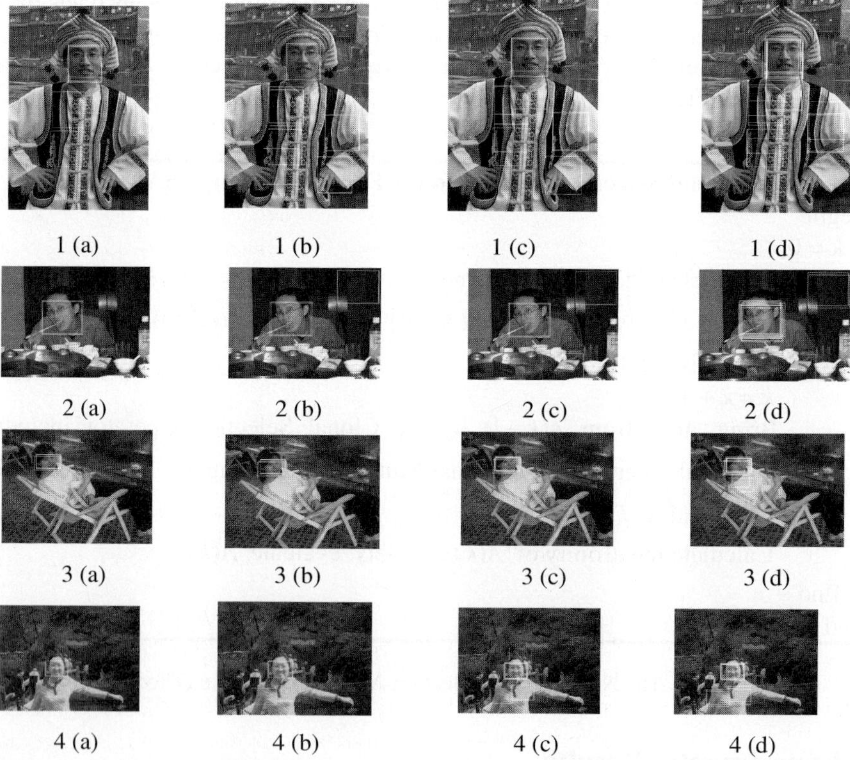

Fig. 3. Results of face detection on various test images (1(a) ~ 4(a) represent the results of CSA, 1(b) ~ 4(b) represent the results of IGA,1(c) ~ 4(c) represent the results of EGA, 1(d) ~ 4(d) represent the results of SGA)

From Fig.3 we can see that the new algorithm is able to cope with variations in orientation and viewpoint (to a small extent). The algorithm also seemed to be robust to distractions such as glasses. The time taken to run the new algorithm on images with a user-specified scale are about 4 seconds each.

For a more detailed verification of the results, we compared the accurate times with the test times and assess the accuracy of every algorithm.

Table 1. The test results among CSAFD, IGA, EGA, and SGA

test image	statistical data	CSAFD	IGA	EGA	SGA
1	accurate times / test times	25/30	15/30	12/30	6/30
2	accurate times / test times	30/30	25/30	23/30	17/30
3	accurate times / test times	28/30	17/30	17/30	21/30
4	accurate times / test times	30/30	28/30	21/30	21/30

Compared with IGA、 EGA and SGA, CSAFD is much more accurate and efficient. From Tables 1 it is apparent that the new method has the highest accurate rate. For all the images, the accurate times of CSAFD is higher than that of IGA、 EGA and SGA, especially for image 2 and image 4, CSAFD detects the face accurately in all 30 times which is impossible for IGA、 EGA and SGA. Therefore, the proposed algorithm is not sensitive to image noise and head-pose and is very robust. The experimental results showed that new algorithm could detect faces successfully in an uncontrolled environment with complex background.

5 Conclusion

In this paper, a novel clonal selection algorithm for face detection(CSAFD) is put forward, which is inspired by Antibody Clonal Selection Theory. The performance of CSAFD is evaluated via computer simulations. When compared with IGA、 EGA and SGA, the experimental results on images with moderately complex background scene showed, the new algorithm had a good performance and was much more accurate and robust.

Face detection is now an important research topic of computer vision and image understanding. It is more difficult to detect human face in gray-level images. A point of view is brought forward that we should manage to accelerate the velocity of face detection when the detection precision is assured to fit the practical application. Then our future works include increasing the size of the test set, improvement of the feature detection accuracy and the detection speed.

References

1. R. Chellappa, C. L Wilson, and C. S Barnes. Human and machine recognition of faces: A survey. Technical Report CAR-TR-731, University of Maryland, USA,1994
2. S.Y. Lee, Y.K. Ham, and R.-H. Park, "Recognition of Human Front Faces Using Knowledge-Based Feature Extraction and Neuro-Fuzzy Algorithm," *Pattern Recognition*, vol. 29, no. 11, pp. 1,863-1,876, 1996
3. C.H. Lee, J.S. Kim, and K.H. Park, "Automatic Human Face Location in a Complex Background Using Motion and Color Information," *Pattern Recognition*, vol. 29, no. 11, pp. 1,877-1,889, Nov. 1996
4. E. Saber and A. Tekalp, "Face Detection and Facial Feature Extraction Using Color, Shape and Symmetry Based Cost Functions," *ICPR'96*, pp. 654-658, 1996
5. Kin Choong Yow. Automatic human face detectionand localization [D]. London, U K: Department of Engineering, University of Cambridge, 1998
6. P. Juell and R. Marsh, "A Hierachical Neural Network for Human Face Detection," *Pattern Recognition*, vol. 29, no. 5, pp. 781-787, 1996
7. M. Lew and N. Huijsmanns, "Information Theory and Face Detection,"*ICPR'96*, pp. 601-605, 1996
8. A. Lanitis, C.J. Taylor, T.F. Cootes, and T. Ahmed, "Automatic Interpretation of Human Faces and Hand Gestures Using Flexible Models," *Proc. Int'l Workshop Automatic Face-and Gesture-Recognition*, pp. 98-103, 1995
9. Pentland A, Moghaddam B, Starner T. View-based and modular eigenspaces for face recognition. In: Proceedings of the IEEE Conference on Computer Vision and Pattern Recognition, Seattle, WA, 1994, pp. 84-91
10. Brunelli R, Poggio T. Face recognition: feature versus templates. IEEE Transactions on Pattern Analysis and Machine Intelligence, 1993, 15(10): 1042-1052
11. Soriano, E. Marszalec, M, Pietikainen. Color correction of face images under different illuminants by RGB eigenfaces[A], Proc. 2nd Audio and Video-Based Biometric Person Authentication Conference(AVBPA99)[C], PP.148-153, 1999
12. Stephen Karungaru, Minoru Fukumi and Norio Akamatsu. Feature Extraction for Face Detection and Recognition[A]. Proceedings of the 2004 IEEE International Workshop on Robot and Human Interactive Communication Kurashiki[C], Okayama Japan September 20-22, 2004
13. Leandro N. de Castro: Larning and Optimization Using the Clonal Slection Principle. IEEE Transactions on Evolutionary Computation, Vol. 6, No. 3. (2002) 239-251
14. Zhou Guangyan. Principles of Immunology. Shang Hai Technology Literature Publishing Company, 2000. (In Chinese)
15. L. N.De Castro, F. J.Von Zuben.: The Clonal Selection Algorithm with Engineering Applications. Proc. of GECCO'00, Workshop on Artificial Immune Systems and Their Applications, 2000:36-37
16. J. Kim and P. Bentley, "Toward an artificial immune system for network intrusion detection: An investigation of clonal selection with a negative selection operator," in Proc. Congress on Evolutionary Computation (CEC), Vol. 2, Seoul, Korea, (2001) 1244–1252
17. Haifeng DU, Licheng JIAO, Sun'an Wang. Clonal Operator and Antibody Clone Algorithms. Proceedings of the First International Conference on Machine Learning and Cybernetics, Beijing, (2002) 506-510
18. Dasgupta D, Forrest S Artificial immune systems in industrial applications. In: Proceedings of the Second International Conference on Intelligent Processing and Manufacturing of Materials. IEEE press, (1999) 257 ~ 267

19. Nanning ZHENG. Computer vision and pattern recognition [M]. Beijing: National Defence Industry Press, 1998
20. L. C. Jiao, L. Wang. A Novel Genetic Algorithm Based on Immunity[J]. IEEE Trans. on Systems, Man, And Cybernetics-Part A: Systems and Humans, September 2000, Vol.30, No.5, pp.552-561
21. Guoliang CHEN. Genetic algorithm and application[M]. Beijing: Posts and Telecommunications Press, 1996

Hardware Implementation of Temporal Nonmonotonic Logics

Insu Song and Guido Governatori

School of Information Technology & Electrical Engineering
The University of Queensland, Brisbane, QLD, 4072, Australia
{insu, guido}@itee.uq.edu.au

Abstract. In order to apply nonmonotonic logics for specifying industrial automation controllers, we define (1) a method to extend atemporal nonmonotonic logics with temporal operators and (2) a mapping of these new temporal nonmonotonic logics into a Metric Temporal Logic. This mapping provides a formal specification method for real-time temporal reasoning digital circuits for the temporal nonmonotonic logics. We present our method in the context of synthesizing custom digital hardware (called *agent chip*) automatically from high level agent specifications.

Keywords: agent, chip design, nonmonotonic logic, knowledge representation, temporal logic.

1 Introduction

Previously, Song and Governatori [12] described a method for synthesizing *agent chips* from high-level agent specifications. An agent chip is a custom hardware implementation of an agent specification, which performs several million times faster than conventional CPU-based systems. In the previous work, however, temporal expressions are not allowed in the specification language and a formal specification method for implementing temporal aspects of agent chips is missing. Time is a central factor in any automation controllers. Any industrial controllers must reason with timing issues of input and output control signals.

In this paper, we present a method for producing real-time temporal reasoning digital circuits for temporal nonmonotonic logics. This method extends the previous method for synthesizing *agent chips* [12]. Particularly, we provide nonmonotonic temporal logics for specifying real-time reasoning processors, which are suitable for reactive agent systems that are targeted for industrial automation controllers. The need for such a method is becoming acute since custom chips (e.g., FPGAs[1] and ASICs[2]) are now more and more affordable and vital applications such as sensor networks, smart sensors, autonomous mobile robots, and even small consumer electronic devices, require (a) low profile features, such as smaller size and power efficiency, and (b) high performance real-time features.

---

[1] FPGAs (Field Programmable Gate Arrays) are configurable digital circuits.
[2] ASICs (Application Specific Integrated Circuits).

A. Sattar and B.H. Kang (Eds.): AI 2006, LNAI 4304, pp. 808–817, 2006.
© Springer-Verlag Berlin Heidelberg 2006

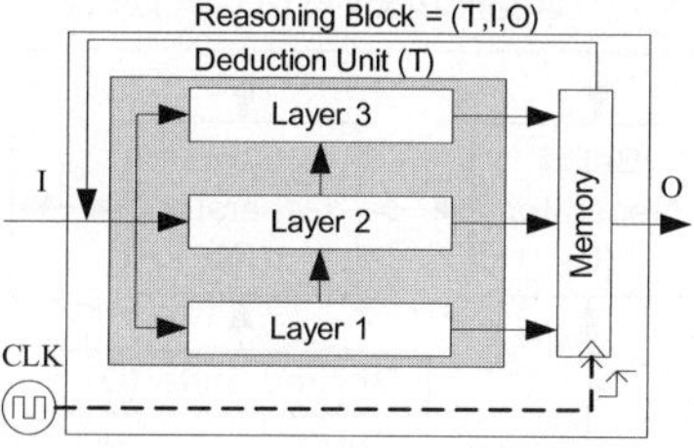

Fig. 1. A reasoning block with three layers. A reasoning block is specified by a knowledge base T, an input specification I, and an output specification O.

Extending any logical languages, however, particularly nonmonotonic logics, with temporal operators maintaining real-time reasoning property is not easy, because reasoning under temporal logic is generally undecidable or EXPSPACE-complete [1]. In this paper, however, we propose a simple solution for extending propositional languages with temporal operators for automation controllers while maintaining real-time reasoning capability of the agent chip described in [12].

The basic idea is that we separate the evaluation of temporal operators from the main reasoning process in order to convert temporal languages into atemporal languages. We, then, build a separate reasoning circuit especially for the evaluation of temporal formulas. We can, then, combine this temporal reasoning circuit and the agent chip circuit produced by the method presented in [12] to produce a complete agent chip from a high-level agent specification that allows temporal expressions.

Particularly, we extend propositional languages with bounded temporal operators of Metric-Temporal Logic [6,7] which is an extension of Linear Temporal Logic. Since many nonmonotonic logics can be mapped into classical logic (e.g., Default Logic [2], Defeasible Logic [12], Layered Argumentation System [12,11]), we can formulate various temporal nonmonotonic logics and temporal argumentation systems maintaining real-time reasoning property based on this idea.

This paper is organized as follows. In Section 2, we overview an agent-based chip specification method described in [12]. In Section 3, we then show how to separate temporal reasoning from the main reasoning process. In Section 4, we define a Metric temporal logic (MTL), a method to extend propositional languages, and a mapping of these languages into the MTL. In Section 5, we conclude this paper with some remarks and related works.

2 Agent Based Chip Specification

In order to provide the context of this paper, we briefly describe the agent chip architecture [12]. An agent chip is a hardware implementation of a high-level agent specification. It is basically an electronic circuit that produces outputs in reaction to its inputs, which are connected to the agent's environment. The basic building blocks of an agent chip are the *reasoning blocks* shown in Figure 1. Each reasoning block is specified by a knowledge base T, an input specification I, and an output specification O. The deduction unit of a reasoning block is implemented on FPGAs or ASICs as *combinational*

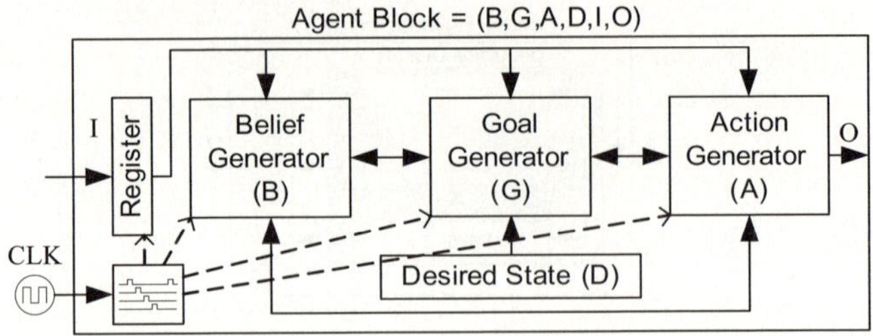

Fig. 2. Agent Block architecture: an example agent architecture consisting of three control blocks: belief generator (B), goal generator (G), and action generator (A)

logic circuits[3], which is organized in layers corresponding to the layers of the knowledge base T. The reasoning block also contains a *memory unit,* which provides temporal reasoning capabilities and other utility functions (e.g., counters, timers, arithmetic) that cannot be expressed in propositional languages. The circuit generation method of the deduction unit is described in [12] and the details of the language used in this method is described in [11]. The memory unit provides evaluation of temporal formulas separating temporal evaluation from the main reasoning process of the reasoning block. Section 4 defines a formal specification method for the implementation of the memory unit for supporting evaluation of temporal formulas based on a Metric-Temporal Logic (MTL). MTL[6] is an extension of Linear Temporal Logic to provide more convenient temporal operators for expressing bounded responses in formal specifications of real-time systems.

Figure 2 shows an example agent architecture called *Agent Block* which consists of three reasoning blocks: a belief generator (B), a goal generator (G), and an action generator (A). Each agent block is an autonomous system that performs actions in order to achieve a certain desired state D. An agent block operates in the following four-stage cycles:

1. At the beginning of the first stage of a cycle, the input signals in I are captured in the register. Then, the agent reasons about the current state of the world in B with its background knowledge, the memory contents of B, G, and A.
2. At the second stage of the cycle, the (belief) conclusions in B are captured in the memory unit of B. Then, the goal generator (G) decides what goals are needed to achieve the desired state with its goal description knowledge, the memory contents of B, G, and A.
3. At the third stage of the cycle, the (goal) conclusions in G are captured in the memory unit of G. Then, the set of goals instantiates a particular behavioral specification from a set of plans in the action generator (A) and produces appropriate actions for the output (O) with its action description knowledge base, the memory contents of B, G, and A.
4. At the last stage of the cycle, the (action) conclusions in A are captured in the memory unit of A which is connected to the output port of the agent block.

---

[3] A digital circuit that does not have internal states, i.e., the output signals are solely defined by the input signals of the circuit.

The four stage cycle is driven by the clock signal (CLK) shown in the figure. The register and the memory units of the three reasoning blocks keep the interconnecting signals steady while the input signals or the output signals of a reasoning block are used by other reasoning blocks. Consequently, the frequency of the clock is determined by the time required to draw all of the conclusions in each deduction unit. That is, the computation bottle-neck is in the combinational logic layers of reasoning blocks because the layers in a reasoning block are responsible for drawing all conclusions.

However, in agent chips, the combinational logic layers are implemented as pure combinational logic circuits, and thus each reasoning block can compute all of the conclusions almost instantaneously. For instance, our implementation of an agent chip consisting of 16,000 rules on a $Xilinx^{TM}$ Spartan-3 FPGA chip (currently each chip costs less than US\$9) can operates in 140 Mhz clock frequency (i.e., less than 8×10^{-9} seconds for each cycle) whereas conventional approaches based on more expensive CPUs (e.g., 3Ghz Pentium 4 CPU) require more than a minute for each cycle.

3   Separation of Temporal Evaluation

One important feature of the reasoning block is that its memory unit provides very efficient temporal reasoning capability by forming a closed-loop controller similarly to industrial automation controllers such as PLCs (Programmable Logic Controllers).

Example 1. As an example, suppose that when a cleaning robot has detected a signal on sensor sD, it needs to memorize the state that the area is dirty (denoted as d) even if sD is not true anymore until it detects another signal on sensor sC, which tells that the area is clean. If the agent concludes that the area is dirty, it turns on the vacuuming unit (vc). In most logical formalisms, this cannot be modelled without introducing the notion of time in the language of the knowledge base resulting in high computational complexity as shown below:

$$\{sD \rightarrow d,\ (P_1 d) \rightarrow d,\ sC \rightarrow \neg d,\ d \rightarrow vc\}$$

where all of the rules are defeasible rules (e.g., defeasible rules in Defeasible Logic [8]) and P_n is a bounded temporal operator denoting "Previous n time moments" for $n \geq 0$. Thus, $P_1 d$ denotes that d was true just one time moment ago. Since the rules are defeasible rules, when both sD and sC are true, it does not introduce inconsistency, but rather it just means that neither d nor $\neg d$ can be proved. We should note that, in general, reasoning under temporal logic is EXPSPACE-complete [1]. In a reasoning block, however, we can remove the notion of time in the object language by converting the formulas containing temporal operators into simple propositions referring to the evaluation results of the formulas in the memory unit of the reasoning block as shown below:

$$\{sD \rightarrow d,\ d('P',1) \rightarrow d,\ sC \rightarrow \neg d,\ d \rightarrow vc\}$$

where $d('P',1)$ is just a *proposition* denoting the previous conclusion of d: it denotes that the agent knows that d was true in the previous round of reasoning. Here, $d('P',1)$ refers to a state in the memory of a reasoning block: it is an input to the knowledge base of the deduction unit. That is, when the agent concludes that the area is dirty, the truth value of d is captured in the memory as $d('P',1)$.

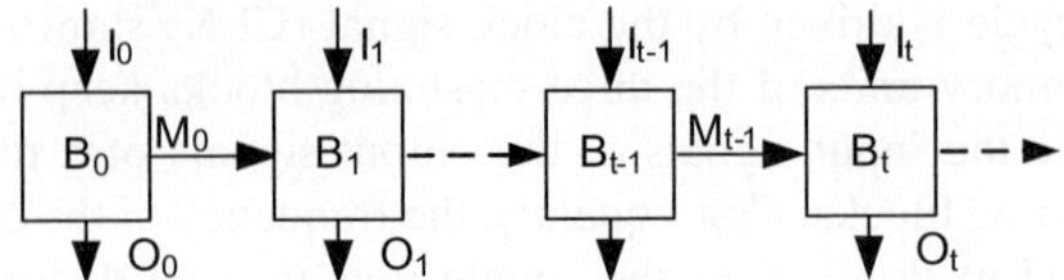

Fig. 3. Memory captures the previous state with which an agent can reason about time in a very simple way

Figure 3 illustrates how an agent's belief state (the state of the reasoning block representing the belief generator of an agent block) changes over time based on the current input state and the immediate previous state captured in memory M_{t-1}. This example demonstrates that the temporal operator (P_n) can be efficiently implemented in reasoning blocks. Similarly to PLCs, other bounded temporal operators of Metric-Temporal Logic [6] can also be efficiently implemented with the memory unit.

Therefore, symbolically, the object language for the reasoning blocks can be considered purely propositional and atemporal because the evaluation of the formulas with temporal operators is separated from the main reasoning process. Consequently, we can now study the object language of agents' knowledge bases and evaluation of temporal formulas separately.

4   Formal Temporal Specification Method of Agent Chips

In this section, we discuss a formal specification method of the memory units of reasoning blocks defined in Section 2. We also briefly discuss formal verification issues of agent chips. First, we give a brief introduction to Metric-Temporal Logic (MTL) which provides convenient operators for specifying quantitative requirements, such as bounded response. We refer the reader to [6,7] for more thorough treatments of MTL. Linear Temporal Logic (LTL) in comparison provides only the temporal operators for specifying qualitative requirements on execution sequences.

4.1   Introduction of Metric-Temporal Logic (MTL)

MTL extends Linear Temporal Logic (LTL) with the following bounded temporal operators $\Box_{\#d}q$, $\Diamond_{\#d}q$, and $pU_{\#d}q$ where $\# \in \{<, \leq\}$ and $d \in Nat$. For example, $\Diamond_{\leq d}$ expresses the notion "eventually within time d." Given a finite set P of propositions, the set of *MTL formulas* is inductively defined as follows:

$$\phi := true|false|p|\neg\phi|\phi_1 \wedge \phi_2|\phi_1 \vee \phi_2|\Box\phi|\Diamond\phi|X\phi|P\phi|$$
$$X_d\phi|P_d\phi|\Box_{\#d}\phi|\Diamond_{\#d}\phi|\phi_1 U_{\#d}\phi_2$$

where $p \in P$, $\Box$ denotes 'always', $\Diamond$ denotes 'possibly', X denotes 'next', P denotes 'previous', and U denotes 'until'. The time domain T of MTL is the set of natural numbers Nat. An interpretation of an MTL formula is a *trace* $h : Nat \rightarrow 2^P$ that maps to each time position $i \in T$ the set of propositions that hold at that position. We define a distance function to denote the time elapsed between time positions i and j in a trace:

$$dist(i,j) = |i - j| \times \delta$$

where δ is a time unit. In the case of the reasoning block with clock frequency f, δ is $1/f$ seconds.

The semantics of the temporal operators is defined over system traces and the distance function as follows:

1. $(h,i) \models p$ iff $p \in h(i)$.
2. $(h,i) \models Xp$ iff $(h,i+1) \models p$.
3. $(h,i) \models X_d p$ iff $(h,i+d) \models p$.
4. $(h,i) \models Pp$ iff $(h,i-1) \models p$.
5. $(h,i) \models P_d p$ iff $(h,i-d) \models p$.
6. $(h,i) \models \Box p$ iff $\forall j \geq i.\ (h,j) \models p$.
7. $(h,i) \models \Diamond p$ iff $\exists j \geq i.\ (h,j) \models p$.
8. $(h,i) \models \Box_{\#d} p$ iff $\forall j \geq i,\ (h,j) \models p$ and $dist(i,j)\#d$.
9. $(h,i) \models \Diamond_{\#d} p$ iff $\exists j \geq i,\ (h,j) \models p$ and $dist(i,j)\#d$.
10. $(h,i) \models pU_{\#d}q$ iff $\exists j \geq i,\ (h,j) \models q$ and $dist(i,j)\#d$ and $\forall i \leq k < j,\ (h,k) \models p$.

An *MTL theory* is a set of MTL formulas.

4.2   Mapping Knowledge Bases into MTL

Let L denote an atemporal logical language for a reasoning block. We now define special propositions for L to allow for the specification of temporal operations. We consider the following temporal operators for propositions q and ϕ in L in the antecedents of rules:

1. $q('P',d)$ meaning $P_d q$;
2. $\phi('X',d)$ meaning $X_d \phi$;
3. $\phi('G','\#',d)$ meaning $\Box_{\#d}\phi$;
4. $\phi('F','\#',d)$ meaning $\Diamond_{\#d}\phi$;
5. $\phi('U','p','\#',d)$ meaning $pU_{\#d}\phi$.

For example, $x('P',1) \rightarrow y$ means that if x was true just before, y is usually true.

We can also consider the following temporal operators in the consequents of rules for future-time events, but they will be interpreted as actions setting future-time events:

1. $q(Set,'X',d)$ meaning that the system must reset a timer so that $X_d q$ is true, i.e., q is true in the next d-th time units;
2. $q(Set,'G','\#',d)$ meaning that the system must reset a timer so that $\Box_{\#d}q$ is true;
3. $q(Set,'F','\#',d)$ meaning that the system must reset the a timer so that $\Diamond_{\#d}q$ is true;
4. $q(Set,'U','p','\#',d)$ meaning that the system must reset a timer so that $pU_{\#d}q$ is true.

For example, $x \rightarrow y(Set,'X',1)$ means that if x is true now, we usually set a timer so that y is (usually) true in the next time moment.

Let us now suppose L is a logical language that can be mapped into an equivalent classical propositional language: there exists a mapping Π of L into a propositional

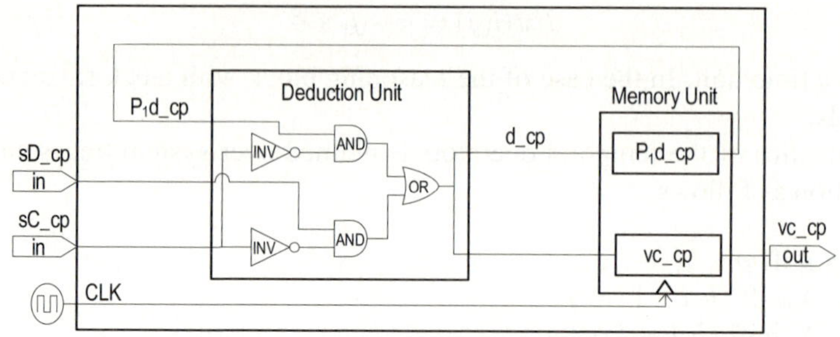

Fig. 4. An implementation of the temporal operator P_1 for the cleaning robot in Example 1. The signal lables sD_cp, sC_cp, vc_cp, d_cp, and P_1d_cp represent the positive literals sD, sC, vc, d, and P_1d, respectively. The value 1 of these signals means that the corresponding positive literals are provable. The value 0 of the signals means that the provability of these positive literals is unknown.

language. Defeasible Logic [12], Layered Argumentation System ([12] and [11]), and Default Logic ([2]) are such languages.

Then, mapping a theory T of L into an MTL theory is straightforward. We obtain the corresponding MTL theory $MT(\Pi(T))$ of $\Pi(T)$ by replacing the temporal information of each proposition p in $\Pi(T)$ with the temporal operators as follows :

1. Replace all propositions $a('X',d)$ in $\Pi(T)$ with $X_d a$;
2. Replace all propositions $a('P',d)$ in $\Pi(T)$ with $P_d a$;
3. Replace all propositions $a('G','\#',d)$ in $\Pi(T)$ with $\Box_{\#d} a$;
4. Replace all propositions $a('F','\#',d)$ in $\Pi(T)$ with $\Diamond_{\#d} a$;
5. Replace all propositions $a('U','b','\#',d)$ in $\Pi(T)$ with $bU_{\#d} a$.

The resulting MTL theory formally specifies how the memory unit should be implemented in hardware. As shown in Section 3 and Figure 4, the temporal operator P_1 can be directly implemented in reasoning blocks by just memorizing immediate previous conclusions. Figure 4 shows an implementation of operator P_1 for $d('P',1)$ (denoted as 'P₁d' in the figure) as a single-bit register in the memory unit of the cleaning robot. The digital circuit description in the deduction unit is automatically generated from the defeasible rules in Example 1 using a compiler that we have developed for Layered Argumentation System [11]. It is easy to verify that the combinational logic circuit performs the required nonmonotonic reasoning as described in the example.

Similarly, for $0 < n \leq N$, P_n can be implemented using an N-bit shift register: $(h, i - n) \vdash P_n q$ iff the n-th bit of the shift register is value 1. A shift register shifts its bits in one direction (e.g., from the least significant bit (LSB) to the most significant bit (MSB)) at every clock cycle. Figure 5 illustrates an implementation of operator P_n. It is straightforward to show that this implementation is sound and complete for P_n.

However, implementation of temporal formulas referring to the future time events is more complex. In this paper, however, we restrict our language to provide simple, yet practical, implementations of some future-time temporal operators as follows: we restrict our language so that the propositions ϕ in the future-time temporal formulas ($X_d \phi$,

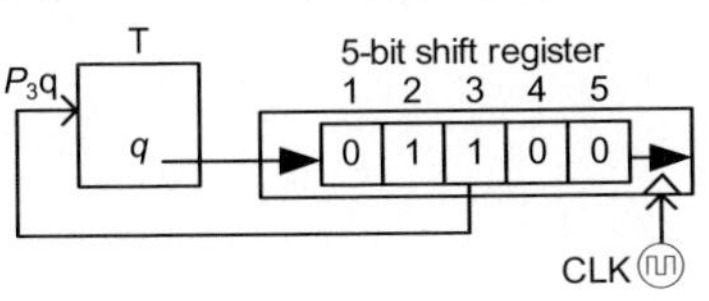
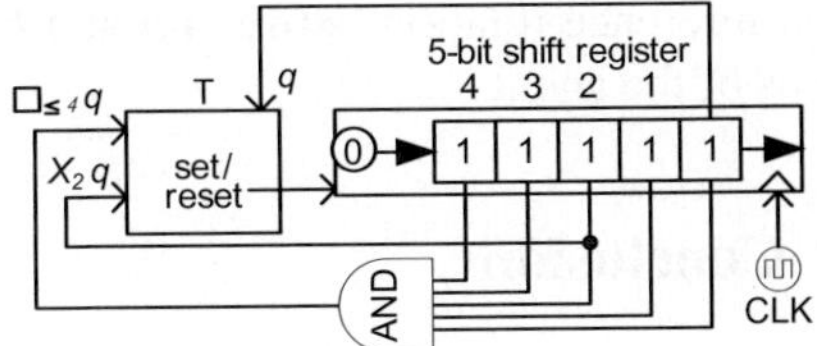

(a) An implementation of temporal operator P_n
via a 5-bit shift register where n ≤ 5.

(b) An implementation of temporal operator X_n
and $\Box_{\leq n}$ via a 5-bit shift register where n ≤ 4.

Fig. 5. Implementations of temporal operators

$\Box_{\#d}\phi$, $\Diamond_{\#d}\phi$, and $pU_{\#d}\phi$) in the antecedents of rules, are not allowed in the consequents of rules. That is, let us consider the special case that future events are determined by the input to the system or timers in the memory unit.

Then, similarly to P_n, for $0 < n < N$, $X_n q$ can be implemented using an $(N+1)$-bit shift register which always receive value 0 in its most significant bit. This can be used to test an event q that will occur within n time units. In other words, the event can be considered as an internal alarm clock that was set by the agent through the agent's action. Setting the n-th bit of the shift register with $q(Set,'X',n)$ will make $X_n q$ true. $\Box_{\leq n} q$ can be represented just as a conjunction of all of the bits in the shift register. Consequently, setting all of the bits in the shift register with $q('G','\leq',n)$ will make $\Box_{\leq n} q$ true. Figure 5 illustrates implementations of these temporal operators.

Other useful temporal operators can also be similarly implemented. This approach (using internal timers) is very similar to the implementations used in PLCs (Programmable Logic Controllers), the popular industrial controllers. Alternatively, an MTL theory or a subset of it can be directly used to synthesize real-time systems [9]. In this case, then, the temporal operators in the consequents of the rules should be interpreted as normal temporal formulas and mapped into proper MTL temporal formulas exactly like the temporal operators in the antecedents of rules. For example, instead of using $q(Set,'X',d)$, we should use $q('X',d)$ and map it to $X_d q'$.

4.3   Formal Verification Issues of Agent Chip Properties

Once we have the corresponding MTL theory, $MT(\Pi(T))$, of a knowledge base, T, we can formally test some system properties over $MT(\Pi(T))$ with existing proof theories of MTL. One interesting system property is testing for conditionals. For example, given an agent specification A, the following system properties might be interesting:

1. $A \vdash \Box(a \Rightarrow b)$ which says that, for all of the entire traces of agent A, b follows from a.
2. $A \vdash a \Rightarrow \Box(b \Rightarrow c)$ which says that, for all of the entire traces of agent A in which a is true, the agent has the property of $A \vdash \Box(b \Rightarrow c)$.

We have used a different arrow '$\Rightarrow$' to denote that the formula is not an MTL formula, but a conditional in some conditional logic. For instance, $A \vdash \Box(a \Rightarrow b)$ is not true for an agent that concludes b always even if a is false. The property test is asking whether there is an inference relation between a and b such that b follows from a in every possible traces of the agent.

5   Conclusion

The main result of this paper has been the solution to the problem of a real-time temporal reasoning that can be realized in small electronic chips. The key idea has been the separation of temporal evaluation from the main reasoning process and the mapping for generating formal specifications for temporal reasoning circuits. This mapping is a simple direct mapping and yet produces compact and scalable hardware specification. It is scalable, since it is mapped into MTL which is widely used in the formal specification of digital circuits (e.g., [9]).

In this paper, we have not prescribed any particular logical language for our method. The reason is because our method can be applied to most of nonmonotonic logical languages extended with *bounded temporal operators* provided that they can be mapped into a classical propositional logic.

Some important issues, which we have not discussed in detail in this paper, because of the limited space, are the evaluation of future temporal formulas ($X_d q$, $\Box_{\#d} q$, $\Diamond_{\#d} a$, and $pU_{\#d} a$) and the interpretation of temporal operators in the consequents of defeasible rules. For the former we have shown that a subset of language can be implemented as hardware straightforwardly. For the latter, we considered the temporal formulas in the consequents as actions for setting or resetting timers in the memory unit. PLCs (Programmable Logic Controllers), which are popular industrial automation controllers, have adopted this interpretation. In this case, they cannot refer to the past state, i.e., we cannot change what has happened and the facts of what we have concluded in the past. Of course conflicting actions must be resolved. Unlike PLCs, however, many nonmonotonic logics (e.g., Defeasible Logic [8], Defeasible Argumentation System [4], Layered Argumentation System [11]) can resolve conflicts between competing rules.

The most closely related work to this paper is "situated automata" [10], in which Kaelbling and Rosenchein [10] have developed a compiler that directly synthesizes digital circuits from a knowledge based model [3] of an agent's environment. Their language is based on a weak temporal Horn-clause language with the addition of *init* and *next* operators. While this work clearly highlighted the realization of epistemic theories of agents, they provide no formal semantics of the language. Their approach is also a model based top-down approach that the compiler generates a target system from a model of the world. Thus, the synthesis is not always guaranteed and resulting systems can be sub-optimal because in most cases the required solution is usually a small subset of the solutions of the model.

The results are important, since we can combine our method and atemporal logical languages to obtain temporal languages, which can be realized in hardware for real-time reasoning, such as temporal argumentation systems from [11] and [4], temporal Default Logics from [2], and temporal Defeasible Logics from [8]. Of course the reasoning blocks for these temporal languages can also be implemented in software. In fact, our method can be used to provide a very efficient implementation of the temporal Defeasible Logic detailed in [5].

References

1. Rajeev Alur and Thomas A. Henzinger. Real-time logics: complexity and expressiveness. Technical report, Stanford, CA, USA, 1990.
2. Rachel Ben-Eliyahu and Rina Dechter. Default reasoning using classical logic. *Artif. Intell.*, 84(1-2):113–150, 1996.
3. R. Fagin, J. Halpern, Y. Moses, and M. Vardi. *Reasoning about Knowledge.* MIT Press, Cambridge, MA, 1995.
4. Guido Governatori, Michael J. Maher, Grigoris Antoniou, and David Billington. Argumentation semantics for defeasible logics. *Journal of Logic and Computation*, 14(5):675–702, 2004.
5. Guido Governatori, Antonino Rotolo, and Giovanni Sartor. Temporalised normative positions in defeasible logic. In *Procedings of the 10th International Conference on Artificial Intelligence and Law*, pages 25–34. ACM Press, 2005.
6. Ron Koymans. Specifying real-time properties with metric temporal logic. *Real-Time Syst.*, 2(4):255–299, 1990.
7. Ron Koymans. *Specifying Message Passing and Time-Critical Systems with Temporal Logic.* Springer-Verlag New York, Inc., Secaucus, NJ, USA, 1992.
8. M. J. Maher and G. Governatori. A semantic decomposition of defeasible logics. In *AAAI '99*, pages 299–305, 1999.
9. A. Pnueli and R. Rosner. On the synthesis of a reactive module. In *Proc. POPL '89*, pages 179–190, New York, NY, USA, 1989. ACM Press.
10. Stanley J. Rosenschein and Leslie Pack Kaelbling. A situated view of representation and control. *Artif. Intell.*, 73(1-2):149–173, 1995.
11. Insu Song and Guido Goverantori. A compact argumentation system for agent system specification. In *Proc. STAIRS'06. In print: available at http://eprint.uq.edu.au/archive/00004138/01/InsuGuidoSTAIRS.pdf.* IOS Press, 2006.
12. Insu Song and Guido Governatori. Designing agent chips. In Peter Stone and Gerhard Weiss, editors, *5th International Conference on Autonomous Agents and Multi-Agent Systems*, pages 1311–1313. ACM Press, 10–12 May 2006.

Improvement of HSOFPNN Using Evolutionary Algorithm

Ho-Sung Park[1], Sung-Kwun Oh[2], and Tae-Chon Ahn[1]

[1] School of Electrical Electronic and Information Engineering, Wonkwang University,
344-2, Shinyong-Dong, Iksan, Chon-Buk, 570-749, South Korea
{neuron, tcahn}@wonkwang.ac.kr
[2] Department of Electrical Engineering, The University of Suwon, San 2-2 Wau-ri,
Bongdam-eup, Hwaseong-si, Gyeonggi-do, 445-743, South Korea
ohsk@suwon.ac.kr

Abstract. This paper presents genetically optimized Hybrid Self-Organizing Fuzzy Polynomial Neural Networks (gHSOFPNN). The architecture of the resulting gHSOFPNN results from a synergistic usage of the hybrid system generated by combining fuzzy polynomial neurons (FPNs)-based Self-Organizing Fuzzy Polynomial Neural Networks(SOFPNN) with polynomial neurons (PNs)-based Self-Organizing Polynomial Neural Networks(SOPNN). The augmented gHSOFPNN results in a structurally optimized structure and comes with a higher level of flexibility in comparison to the one we encounter in the conventional HSOFPNN. The GA-based design procedure being applied at each layer of gHSOFPNN leads to the selection of preferred nodes (FPNs or PNs) available within the HSOFPNN. The obtained results demonstrate superiority of the proposed networks over the existing fuzzy and neural models.

1 Introduction

While neural networks, fuzzy sets and evolutionary computing as the technologies of Computational Intelligence (CI) have expanded and enriched a field of modeling quite immensely, they have also given rise to a number of new methodological issues and increased our awareness about tradeoffs one has to make in system modeling. GMDH-type algorithms have been extensively used since the mid-1970's for prediction and modeling complex nonlinear processes [1]. While providing with a systematic design procedure, GMDH comes with some drawbacks. To alleviate the problems associated with the GMDH, Self-Organizing Polynomial Neural Networks (SOPNN)[2] Self-Organizing Fuzzy Polynomial Neural Networks (SOFPNN)[3], and Hybrid Self-Organizing Fuzzy Polynomial Neural Networks (HSOFPNN)[4] were introduced. In this paper, to address the above design problems coming with the development of conventional self-organizing neural networks, in particular, HSOFPNN, we introduce the gHSOFPNN with the aid of the genetic algorithm [5]. The determination of the optimal values of the parameters available within an individual PN(viz. the number of input variables, the order of the polynomial and the collection of preferred nodes) and FPN(viz. the number of input variables, the order of the polynomial, the collection of preferred nodes, the number of MFs for each input variable, and the selection of MFs) leads to a structurally and parametrically optimized network.

A. Sattar and B.H. Kang (Eds.): AI 2006, LNAI 4304, pp. 818–825, 2006.

© Springer-Verlag Berlin Heidelberg 2006

2   The Architecture and Development of HSOFPNN

2.1   Fuzzy Polynomial Neuron: FPN

This neuron, regarded as a generic type of the processing unit, dwells on the concept of fuzzy sets. When arranged together, FPNs build the first layer of the HSOFPNN. As visualized in Fig. 1, the FPN consists of two basic functional modules. The first one, labeled by **F**, is a collection of fuzzy sets that form an interface between the input numeric variables and the processing part realized by the neuron. The second module (denoted here by **P**) concentrates on the function – based nonlinear (polynomial) processing.

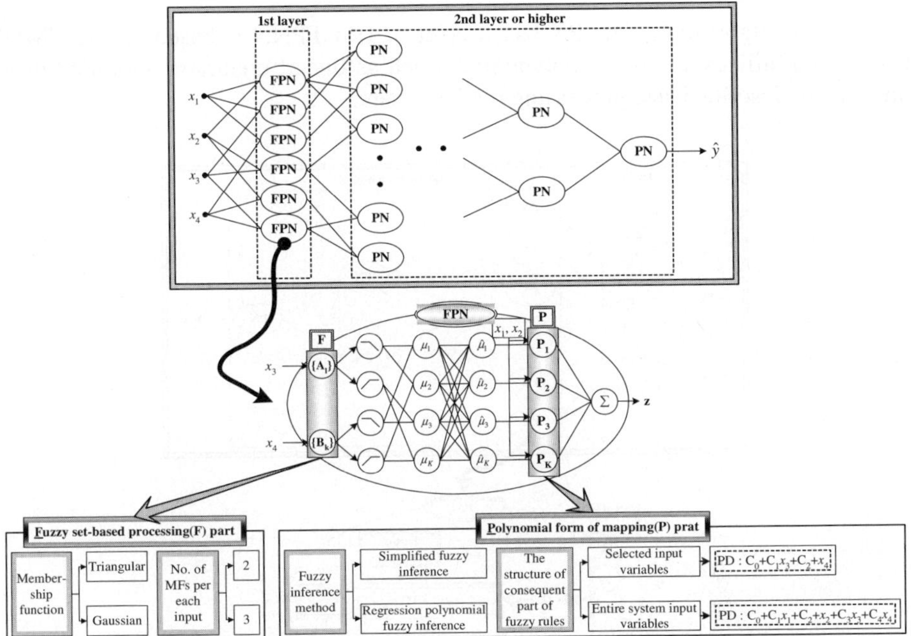

Fig. 1. A general topology of the FPN based layer of HSOFPNN

In other words, FPN realizes a family of multiple-input single-output rules. Each rule, refer again to Fig. 1, reads in the form

$$\text{If } x_p \text{ is } \boldsymbol{A}_l \text{ and } x_q \text{ is } \boldsymbol{B}_k \text{ then } z \text{ is } P_{lk}(x_i, x_j, \boldsymbol{a}_{lk}) \tag{1}$$

The activation levels of the rules contribute to the output of the FPN being computed as a weighted average of the individual condition parts (functional transformations) P_K.

$$z = \frac{\sum_{K=1}^{all\ rules} \mu_K P_K(x_i, x_j, \boldsymbol{a}_K)}{\sum_{K=1}^{all\ rules} \mu_K} = \sum_{K=1}^{all\ rules} \tilde{\mu}_K P_K(x_i, x_j, \boldsymbol{a}_K), \quad \tilde{\mu}_K = \mu_K \Bigg/ \sum_{L=1}^{all\ rules} \mu_L \tag{2}$$

Table 1. Different forms of the regression polynomials building a FPN and PN

Order of the polynomial			No. of inputs		
Order	Type		1	2	3
	FPN	PN			
0	Type 1		Constant	Constant	Constant
1	Type 2	Type 1	Linear	Bilinear	Trilinear
2	Type 3	Type 2	Quadratic	Biquadratic-1	Triquadratic-1
	Type 4	Type 3		Biquadratic-2	Triquadratic-2

1: Basic type, 2: Modified type

2.2 Polynomial Neuron : PN

The SOPNN algorithm in the PN based layer of HSOFPNN is based on the GMDH method and utilizes a class of polynomials such as linear, quadratic, modified quadratic, etc. to describe basic processing realized there.

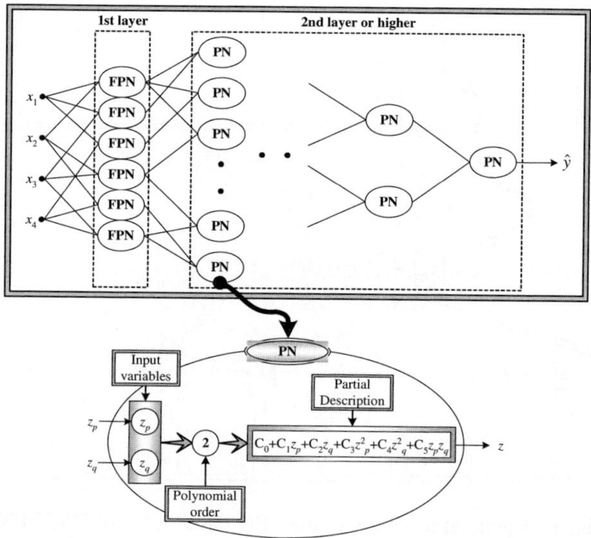

Fig. 2. A general topology of the PN based layer of HSOFPNN

The input-output relationship for the above data realized by the SOPNN algorithm can be described in the following manner

$$y = f(x_1, x_2, \ldots, x_N) \tag{3}$$

Where, $x_1, x_2, \cdots, x_N$ denote the outputs of the 1st layer of FPN nodes(the inputs of the 2nd layer(PN nodes)).

The estimated output $\hat{y}$ reads as

$$\hat{y} = c_0 + \sum_{i=1}^{N} c_i x_i + \sum_{i=1}^{N}\sum_{j=1}^{N} c_{ij} x_i x_j + \sum_{i=1}^{N}\sum_{j=1}^{N}\sum_{k=1}^{N} c_{ijk} x_i x_j x_k + \cdots \qquad (4)$$

The detailed PN involving a certain regression polynomial is shown in Table 1. The architecture of the PN based layer of HSOFPNN is visualized in Fig. 2.

3 Optimization of HSOFPNN by Genetic Algorithms

GAs has been theoretically and empirically demonstrated to provide robust search capabilities in complex spaces thus offering a valid solution strategy to problems requiring efficient and effective searching. It is eventually instructive to highlight the main features that tell GA apart from some other optimization methods: (1) GA operates on the codes of the variables, but not the variables themselves. (2) GA searches optimal points starting from a group (population) of points in the search space (potential solutions), rather than a single point. (3) GA's search is directed only by some fitness function whose form could be quite complex; we do not require it need to be differentiable.

In this study, for the optimization of the HSOFPNN model, GA uses the serial method of binary type, roulette-wheel used in the selection process, one-point crossover in the crossover operation, and a binary inversion (complementation) operation in the mutation operator. To retain the best individual and carry it over to the next generation, we use elitist strategy [5].

4 The Algorithm and Design Procedure of gHSOFPNN

Overall, the framework of the design procedure of the gHSOFPNN architecture comprises the following steps.

[Step 1] *Determine system's input variables.*
Define system's input variables $x_i(i=1, 2, \ldots, n)$ related to the output variable y.
[Step 2] *Form training and testing data.*
The input-output data set $(x_i, y_i)=(x_{1i}, x_{2i}, \ldots, x_{ni}, y_i)$, $i=1, 2, \ldots, N$ is divided into two parts, that is, a training and testing dataset.
[Step 3] *Decide initial information for constructing the HSOFPNN structure.*
Here we decide upon the essential design parameters of the HSOFPNN structure.
[Step 4] *Decide a structure of the PN and FPN based layer of HSOFPNN using genetic design.*
This concerns the selection of the number of input variables, the polynomial order, the input variables, the number of membership functions, and the selection of membership functions to be assigned at each node of the corresponding layer.
[Step 5] *Estimate the coefficient parameters of the polynomial in the selected node (PN or FPN).*
[Step 5-1] *In case of a PN (PN-based layer)*
The vector of coefficients $\mathbf{C}_i$ is derived by minimizing the root mean squared error between y_i and z_{mi}.

$$E = \sqrt{\frac{1}{N_{tr}} \sum_{i=0}^{N_{tr}} (y_i - z_{mi})^2} \tag{5}$$

Using the training data subset, this gives rise to the set of linear equations

$$\mathbf{Y=X}_i\mathbf{C}_i \tag{6}$$

Evidently, the coefficients of the PN of nodes in each layer are expressed in the form

$$y = f(x_1, x_2, \ldots, x_N) \quad \mathbf{C}_i = (\mathbf{X}_i^T \mathbf{X}_i)^{-1} \mathbf{X}_i^T \mathbf{Y} \tag{7}$$

[Step 5-2] *In case of a FPN (FPN-based layer)*
i) Simplified inference
The consequence part of the simplified inference mechanism is a constant. Using information granulation, the new rules read in the form

$$R^n : If \ x_1 \ is \ A_{n1} \ and \ \cdots \ and \ x_k \ is \ A_{nk} \ then \ y_n - M_n = a_{n0} \tag{8}$$

$$\hat{y} = \frac{\sum_{j=1}^{n} \mu_{ji} y_i}{\sum_{i=1}^{n} \mu_{ji}} = \frac{\sum_{j=1}^{n} \mu_{ji}(a_{j0} + M)}{\sum_{i=1}^{n} \mu_{ji}} = \sum_{j=1}^{n} \hat{\mu}_{ji}(a_{j0} + M_j) \tag{9}$$

$$\mu_{ji} = A_{j1}(x_{1i}) \wedge \cdots \wedge A_{jk}(x_{ki}) \tag{10}$$

The consequence parameters (a_{j0}) are produced by the standard least squares method.
ii) Regression polynomial inference
The use of the regression polynomial inference method gives rise to the expression.

$$R^n : If \ x_1 \ is \ A_{n1} \ and \ \cdots \ and \ x_k \ is \ A_{nk} \ then$$
$$y_n - M_n = f_n\{(x_1 - v_{n1}), (x_2 - v_{n2}), \cdots, (x_k - v_{nk})\} \tag{11}$$

$$\hat{y}_i = \frac{\sum_{j=1}^{n} \mu_{ji} y_i}{\sum_{j=1}^{n} \mu_{ji}} = \frac{\sum_{j=1}^{n} \mu_{ji}\{a_{j0} + a_{j1}(x_{1i} - v_{j1}) + \cdots + a_{jk}(x_{ki} - v_{jk}) + M_j\}}{\sum_{i=1}^{n} \mu_{ji}}$$
$$= \sum_{j=1}^{n} \hat{\mu}_{ji}\{a_{j0} + a_{j1}(x_{1i} - v_{j1}) + \cdots + a_{jk}(x_{ki} - v_{jk}) + M_j\} \tag{12}$$

The coefficients of consequence part of fuzzy rules obtained by least square method(LSE) as like a simplified inference.
[Step 6] *Select nodes (PNs or FPNs) with the best predictive capability and construct their corresponding layer.*
All nodes of this layer of the gHSOFPNN are constructed genetically. To evaluate the performance of PNs or FPNs constructed using the training dataset, the testing dataset is used. Based on this performance index, we calculate the fitness function. The fitness function reads as

$$F(fitness \ Function) = \frac{1}{1 + EPI} \tag{13}$$

where *EPI* denotes the performance index for the testing data (or validation data).
[Step 7] *Check the termination criterion.*
As far as the depth of the network is concerned, the generation process is stopped at a depth of less than three layers. This size of the network has been experimentally found to build a sound compromise between the high accuracy of the resulting model and its complexity as well as generalization abilities.
[Step 8] *Determine new input variables for the next layer.*
The outputs of the preserved nodes (z_{1i}, z_{2i}, ..., z_{Wi}) serves as new inputs to the next layer (x_{1j}, x_{2j}, ..., x_{Wj})($j=i+1$). This is captured by the expression

$$x_{1j} = z_{1i}, \, x_{2j} = z_{2i}, \, ..., \, x_{Wj} = z_{Wi} \tag{14}$$

The gHSOFPNN algorithm is carried out by repeating steps 3-8 of the algorithm.

5   Simulation Study

We demonstrate how gHSOFPNN can be utilized to predict future values of a chaotic time series. The performance of the network is also contrasted with some other models existing in the literature. The time series is generated by the chaotic Mackey-Glass differential delay equation [6] of the form:

$$\dot{x}(t) = \frac{0.2x(t-\tau)}{1+x^{10}(t-\tau)} - 0.1x(t) \tag{15}$$

To come up with a quantitative evaluation of the network, we use the standard RMSE performance index.

Table 2. Computational overhead and a list of parameters of the GAs and the HSOFPNN

	Parameters	1st layer	2nd layer	3rd layer
GAs	Maximum generation	100	100	100
	Total population size	150	150	150
	Selected population size	30	30	30
	Crossover rate	0.65	0.65	0.65
	Mutation rate	0.1	0.1	0.1
	String length	3+3+30+5+1	3+3+30+5	3+3+30+5
	Maximal no. of inputs to be selected(Max)	$1 \leq l \leq Max(2\sim3)$	$1 \leq l \leq Max(2\sim3)$	$1 \leq l \leq Max(2\sim3)$
HSOF PNN	Polynomial type (Type T) of the consequent part of fuzzy rules	$1 \leq T\_F \leq 4$	$1 \leq T\_P \leq 3$	$1 \leq T\_P \leq 3$
	Membership Function (MF) type	Triangular Gaussian		
	No. of MFs per each input(M)	2 or 3		

l, T_F, T_P : integer, T_F : Type of SOFPNN, T_P : Type of SOPNN.

Table 3 summarizes the results: According to the information of Table 2, the selected input variables (Node), the selected polynomial type (T), the selected no. of MFs (M), and its corresponding performance index (PI) was shown when the genetic optimization for each layer was carried out.

Table 3. Performance index of gHSOFPNN for nonlinear function process

Max	1st layer					2nd layer				3rd layer			
	Node(M)	T	MF	PI	EPI	Node	T	PI	EPI	Node	T	PI	EPI
	(a) In case of selected input												
2	4(3) 6(3)	3	G	0.0223	0.0220	4 6	2	0.0157	0.0158	3 16	2	0.0153	0.0153
3	1(3) 4(3) 6(3)	3	G	0.0021	0.0022	8 13 21	2	0.0015	0.0015	3 16 22	2	0.0014	0.0014
	(b) In case of entire system input												
2	3(3) 4(3)	3	G	4.1e-5	6.1e-5	5 26	2	2.7e-5	4.9e-5	7 13	3	2.7e-5	4.2e-5
3	1(3) 3(2) 4(3)	4	G	1.5e-5	4.7e-5	6 7 8	2	4.5e-6	4.0e-5	6 13 22	2	5.7e-6	2.9e-5

In case of entire system input, the result for network in the 3rd layer is obtained when using Max=2 with Type 3 polynomials (modified quadratic functions) and 2 node at input (node numbers are 7, 13); this network comes with the value of PI=2.7e-5, EPI=4.2e-5.

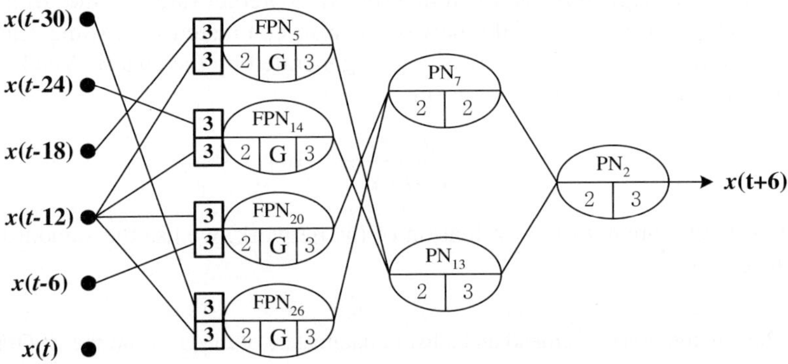

Fig. 3. Optimal gHSOFPNN architecture

Table 4 includes a comparative analysis of the performance of the proposed network with other models.

Table 4. Comparative analysis of the performance of the network; considered are models reported in the literature

	Model		PI	EPI
	Maguire et al.'s model [7]		0.014	0.009
	ANFIS [8]		0.0016	0.0015
HFPNN[9]	Triangular MF (5th layer)		3.4e-4	3.8e-4
	Gaussian-like MF (5th layer)		3.3e-5	5.7e-5
SONN [10]	Basic SONN	Case 1	1.5e-6	9.3e-5
	(5th layer)	Case 2	4.0e-5	7.4e5
Our model	Selected input	3rd layer (Max=2)	0.0153	0.0153
		3rd layer (Max=3)	0.0014	0.0014
	Entire system input	3rd layer (Max=2)	2.7e-5	4.2e-5
		3rd layer (Max=3)	5.7e-6	2.9e-5

6 Concluding Remarks

In this paper, we have introduced and investigated a class of genetically optimized Hybrid Self-Organizing Fuzzy Polynomial Neural Networks (gHSOFPNN) driven to genetic optimization regarded as a modeling vehicle for nonlinear and complex systems.

The GA-based design procedure applied at each stage (layer) of the HSOFPNN driven leads to the selection of the preferred nodes (or FPNs and PNs) with optimal local. These options contribute to the flexibility of the resulting architecture of the network.

Through the proposed framework of genetic optimization we can efficiently search for the optimal network architecture (being both structurally and parametrically optimized) and this design facet becomes crucial in improving the overall performance of the resulting model.

Acknowledgement. This paper was supported by Wonkwang University in 2006.

References

1. Ivakhnenko, A.G.: Polynomial theory of complex systems. IEEE Trans. on Systems, Man and Cybernetics. SMC-**1** (1971) 364-378
2. Oh, S.K., Pedrycz, W.: The design of self-organizing Polynomial Neural Networks. Information Science. **141** (2002) 237-258
3. Park, H.S., Park, K.J., Lee, D.Y, Oh, S.K.: Advanced Self-Organizing Neural Networks Based on Competitive Fuzzy Polynomial Neurons. Transactions of The Korean Institute of Electrical Engineers. **53D** (2004) 135-144
4. Oh, S.K., Pedrycz, W., Park, H.S.: Multi-layer hybrid fuzzy polynomial neural networks: a design in the framework of computational intelligence. Neurocomputing. **64** (2005) 397-431
5. Jong, D.K.A.: Are Genetic Algorithms Function Optimizers?. Parallel Problem Solving from Nature 2, Manner, R. and Manderick, B. eds., North-Holland, Amsterdam (1992)
6. Mackey, M.C., Glass, L.: Oscillation and chaos in physiological control system. Science. **197** (1977) 287-289
7. Maguire, L.P., Roche, B., McGinnity, T.M., McDaid, L.J.: Predicting a chaotic time series using a fuzzy neural network. Information Science. **112** (1992) 125-136
8. Jang, J.R.: ANFIS adaptive-network-based fuzzy inference systems. IEEE Trans. Systems, Man, Cybernetics. **23** (1993) 665-685
9. Oh, S.K., Kim, D.W.: Hybrid Fuzzy Polynomial Neural Networks. International Jouranl of Uncertainty, Fuzziness and Knwoledge-Based Systems. **10** (2002) 257-280
10. Oh, S.K., Pedrycz, W., Ahn, T.C.: Self-organizing neural networks with fuzzy polynomial neurons. Applied Soft Computing. **2** (2002) 1-10

Data Clustering and Visualization Using Cellular Automata Ants

Andrew Vande Moere, Justin J. Clayden, and Andy Dong

Key Centre of Design Computing and Cognition
The University of Sydney, Australia
{andrew, justin, adong}@arch.usyd.edu.au

Abstract. This paper presents two novel features of an emergent data visualization method coined "cellular ants": unsupervised data class labeling and shape negotiation. This method merges characteristics of ant-based data clustering and cellular automata to represent complex datasets in meaningful visual clusters. Cellular ants demonstrates how a decentralized multi-agent system can autonomously detect data similarity patterns in multi-dimensional datasets and then determine the according visual cues, such as position, color and shape size, of the visual objects accordingly. Data objects are represented as individual ants placed within a fixed grid, which decide their visual attributes through a continuous iterative process of pair-wise localized negotiations with neighboring ants. The characteristics of this method are demonstrated by evaluating its performance for various benchmarking datasets.

1 Introduction

This paper proposes a simple approach towards unsupervised data visualization. It uses principles of self-organization to determine the visual representation of complex, high-dimensional datasets. Self-organizing systems generally consist of a number of similar elements that perform numerous internal interactions, which can spontaneously generate an inherently complex pattern on a global level. The rules that govern this process are informed by local information only, without any reference to the global pattern. The proposed method, coined *cellular ants*, uses self-organization to determine the visual attributes of data items, including position, shape, color and size. By self-adapting the visual representation to data attributes, this approach goes beyond the traditional notion of using fixed and predefined data mapping rules.

The cellular ant method combines insights from ant-based clustering in the field of data mining and cellular automata in the field of artificial life with data mapping principles from the data visualization domain. It can be considered as a simple data clustering technique that is capable of creating visual representations similar to those of multidimensional scaling. As a non-optimized prototype, it demonstrates how simple behavior rules are capable of clustering complex, high-dimensional and large datasets. This work is built upon the methodology defined in [1], which introduced the data scaling in a toroidal grid. In this paper, two novel features are introduced: *color negotiation* (similar to data labeling or data clustering) and *shape negotiation*.

A. Sattar and B.H. Kang (Eds.): AI 2006, LNAI 4304, pp. 826–836, 2006.

© Springer-Verlag Berlin Heidelberg 2006

2 Related Work

Ant-based sorting was introduced by Deneubourg et al. [2] to describe different types of emergent phenomena in nature. Ants are represented as simple agents that are capable of roaming around in a toroidal grid, on which objects, representing data items, are randomly scattered. Ant actions are biased by probabilistic functions, so that ants are more likely to pick up objects that are isolated, and more likely to drop them in the vicinity of similar ones. A predefined *object distance measure variable* α determines this degree of similarity between pairs of data objects. Ant-based clustering has been used for data mining purposes, and has been combined with fuzzy-set theory [3], topographic maps [4], or bio-inspired genetic algorithms [5]. The cellular ant method differs from the standard method by mapping data items directly onto the ants themselves. Recent examples of this approach exist: Labroche et al. [6] associates data objects to ants and simulates meetings between them to dynamically build partitions, according to the data labels that best fit their genome.

Multi-dimensional scaling (MDS) displays the structure of distance-like datasets as geometrical pictures [7]. MDS representations are arranged in 2D space, in which the distance between pairs of data items denotes the degree of data similarity. Several similar data visualization techniques exist, for instance in combination with animation [8] or recursive pattern arrangements [9]. Multi-dimensional scaling differs from *clustering* in that clustering partitions data into classes, while MDS computes positions, without providing an explicit decomposition into groups. Self Organizing Maps (SOM) is an unsupervised clustering technique capable of detecting and spatially grouping similar data objects in topologically distinct classes [10]. This visualization method orders an initially random distribution of high-dimensional data objects as the emergent outcome of an iterative training process. In this paper, we describe how the cellular ant method is capable of unsupervised clustering, as it is capable of coloring ants in classes depending on an emergent data scaling topography.

Because the cellular ant methodology governs ants by principles of stigmergy and state density principles, it resembles that of *cellular automata*. Cellular automata is a computational method originally proposed by Ulam and Von Neumann [11]. It consists of a number of cells that each represents a discrete state (e.g. alive or dead). Cells are governed by behavior rules that are iteratively applied, and generally only consider the states of the neighboring cells. The cellular ant approach combines ant-based clustering and cellular automata, as the ants' reasoning takes into account grid cell states, rather than probabilistic functions. While ants can 'act' upon the environment and even change it to some degree, cellular automata 'make up' the environment itself. A recent clustering method [12] also maps data objects onto ants, that resemble cellular automata elements. It differs from our methodology as it does not order clusters, and is based on probability functions and internal ant states.

Agent-based visualizations have typically been used to display intrinsic relations (e.g. messages, shared interests) between agents for monitoring and engineering purposes [13], to represent complex fuzzy systems [14], or to support the choice of the most effective visualization method [15]. Other systems organize the visualization data flow, for instance by determining visualization pipeline parameters [16] or regulating rendering variables in distributed environments [17]. To our knowledge, agents have not yet been used to generate visualizations based on detected data correlations. A few simple prototype applications of agent-based data visualization have been developed that are capable to represent complex data properties through an emergent, decentralized process: for instance, the *infoticle* (information-particle) metaphor is capable of representing time-varying data properties as recognizable motion typologies of dynamic particle or flocking patterns [18].

3 Approach

3.1 Cellular Ant Concept

Each single normalized data item (e.g. database tuple, row, object) corresponds to a single agent, coined *cellular ant*. Each single ant (and thus data item) is represented as a single colored square cell within a toroidal, rectangular grid. Each ant is governed by a set of simple behavior rules. These behavior rules are applied simultaneously to all ants, in a discrete and iterative way. Each ant can only communicate with ants in its immediate vicinity, limited to its eight neighboring cells. The dynamic behavior of an ant only depends only on the data values it represents, and the data values of its immediate neighbors. A cellular ant is capable of determining its visual cues autonomously, as it can move around or stay put, swap its position with a neighbor, and adapt a color or shape size, by a process of pair-wise negotiating. Each cellular ant is determined by four different negotiation processes: data scaling, position swapping, color determination and shape size adaptation. A detailed description of the data scaling and the ant swapping methodologies can be found in [1]. This paper will instead focus on the recent additions that determine the other visual cues of an agent.

At initialization, ants are randomly positioned within a grid. Similar to classical MDS (CMDS) method, each ant calculates the Euclidian distance between its own normalized data item and that of each of its eight neighbors. This data distance measure represents an approximation of the similarity between pairs of data items, even when they contain multidimensional data values. Next, an ant will only consider and summate those ants of which the pair-wise similarity distance is below a specific data *similarity tolerance threshold value t*. Value t is conceptually similar to the *object distance measure* α in common ant-based clustering approaches. However, t originates from a cellular automata approach in that it is a fixed and discrete value, which generates a Boolean result (either a pair of data objects is "similar enough" or not) instead of a continuous similarity value (e.g. representing a numerical degree of similarity between pairs of data objects). Depending on the amount of ants in its neighborhood it considers as 'similar', an ant will then decide either to stay put, or to move. For instance, an ant decides to stay put when it has more than four similar neighbors. The value four was chosen from the experience of cellular automata simulations, which tend to generate interesting cell constellations for this number.

As a result, ants with similar data items group together emergently. However, these clusters have little visualization value as they only convey the relative amounts of data objects. Therefore, a positional *swapping* rule was introduced that orders clusters internally as well as globally in respect to data similarity. As a result, diagrams are generated that look conceptually similar to those of basic CMDS approaches.

3.2 Color Negotiation

Conceptually, the color of an ant can be considered as the representation of its assumed *data class* or *data label*, so that the resulting diagrams resemble that of (ant-based) data clustering in the domain of data mining, but inherit the additional capability of being spatially and visually ordered. At initialization, all ants are assigned an unspecific color (white). At each iteration, ants execute the following behavior rules. Each ant that has not been swapped (and thus is probably well placed within its neighborhood) and is fully surrounded by eight similar neighbors, considers the degree of data similarity with all of its neighbors. If this degree is below a predefined, discrete *color seed similarity threshold* c, it will request the system to be assigned a unique color. As a result, such ants will act as initial 'color seeds'. All other ants will consider whether their neighborhood contains four or more data objects that are smaller than t but larger than c. If so, such ant is 'satisfied' with its current position and will adopt the color of the most similar ant in its neighborhood. In practice, once colors are introduced within the grid, they will spread gradually over the ant population. Once the collection of ants is sufficiently ordered, several color seeds become introduced. Because of the multitude of pair-wise interactions, any surplus of colors (in respect to data clusters) will disappear, while any shortfall of colors will reemerge once a potential seed is surrounded by eight neighbors.

However, data clusters that contain less than nine members in a dataset cannot be recognized. In some cases, ants continuously 'swap' from one color to another, within a single visual cluster. This dynamic phenomenon generally indicates that the positional rules could not accurately spatially group two different data types, which nonetheless were recognized by the color clustering rule. A future research direction could consist of inversely informing the positional clustering of the color label values.

3.3 Size Negotiation

Instead of mapping a data value to a specific shape size, each ant can map one of its data attributes onto its size by negotiating with its neighbors. Conceptually, the size of an agent does not necessarily correspond to the 'exact' value of that data attribute, but rather how a data value locally relates to its neighborhood, and therefore whether clusters are *homogeneous* in respect of a specific data attribute. Because no direct, predefined mapping rule between value and visual cue exists, the shape size scale can automatically adapt to any data scale, in an autonomous and self-organizing way.

For each iteration step, the visual shape size of an ant is determined by following inducements. First, an ant A chooses a random neighboring ant B with whom it compares its one-dimensional data value DA and circular radius size SA, measured in screen pixels. Step size P is a predefined amount of pixels. Ant A evaluates whether its radius versus data value ratio is similar to that of ant B, and adapts its own as well as its neighbor's shape size accordingly. If, in comparison to ant B, its size SA is too large in relation to its data value DA, it will decide to 'shrink' by decreasing its amount of available pixels with P pixels, and then provides these P pixels to ant B.

$$\begin{cases} S_A > \dfrac{S_B}{D_B}.D_A \Rightarrow \begin{cases} S_A = S_A - P \\ S_B = S_B + P \end{cases} \\ S_A < \dfrac{S_B}{D_B}.D_A \Rightarrow \begin{cases} S_A = S_A + P \\ S_B = S_B - P \end{cases} \end{cases} \qquad \begin{cases} S_A > S_{max} \Rightarrow \begin{cases} S_A = S_A - P \\ S_B = S_B - P \end{cases} \\ S_A < S_{min} \Rightarrow \begin{cases} S_A = S_A + P \\ S_B = S_B + P \end{cases} \end{cases} \qquad (1)$$

These rules assure that no visual overlapping of ant shapes can occur. An additional rule checks whether ants do not grow too large or too small: when an ant becomes too large, it will 'punish' and shrink its neighbor, so that in the future, this 'action' will not longer be required. This constraint will emergently 'detect' the upper and lower shape size boundaries according to the data scale, and spreads throughout all ants. Because all ants are complying with these rules in random directions and over multiple iterations, a stable constellation of shape sizes appears in an emergent way.

3.4 Performance Measurements

A simple performance graph informs users of the actual visualization state. The number of similar ants in each ant neighborhood is squared for each ant, and summated over all ants. The visualization efficiency over time corresponds to the slope of the according graph: once a plateau value has been reached over a number of iterations, the visualization has reached a stable state and can be halted. Figure 1 captures the clustering performance of the 'Thyroid' dataset depending on varying variables similarity tolerance threshold value t (vertical) and color seed similarity treshold c, for the different amounts of iterations and different initial seeds. These 'solution space' diagrams enable users to pick the most appropriate variable values. The diagrams illustrate the 'hotspots' of effective clustering values, and the limited influence of the amount of iterations and the random initialization seeds on the quality of the results. Each initialization seed will result in different constellations and thus clustering error rates (see Table 1 for standard deviation).

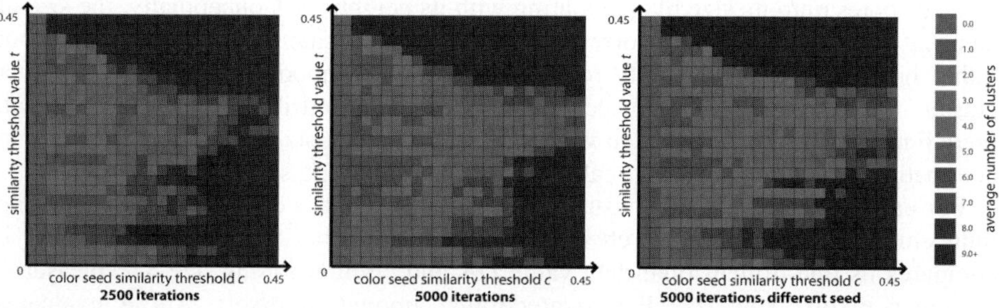

Fig. 1. A dot plot diagram of the Cellular Ant's method performance by varying the similarity tolerance threshold value t (vertical) and the color seed similarity threshold c (horizontal) for the Thyroid dataset (see Table 1). Color denotes the amount of data labels/clusters detected. These diagrams.

4   Application

4.1   Case Studies

The synthetic dataset visualized in Figure 2 consists of 500 data items with two data dimensions. Data objects and classes are distributed using a Gaussian distribution function to demonstrate the data scaling, color negotiation and size negotiation capabilities. The color negotiation successfully resulted in four distinct clusters or data class labels. As shown by the highlighted ants, the clusters are *internally* ordered: data items that are similar in data space, are positioned nearby each other in visualization space. Also, the clusters are *globally* ordered: clusters that are dissimilar in data space, have no 'common' borders in visualization space. For instance, the purple and yellow clusters (or blue and green) have no common orthogonally directed borders and have empty cells between their borders in visualization space, as they are diagonally positioned in data space, and thus have a larger 'global' data distance.

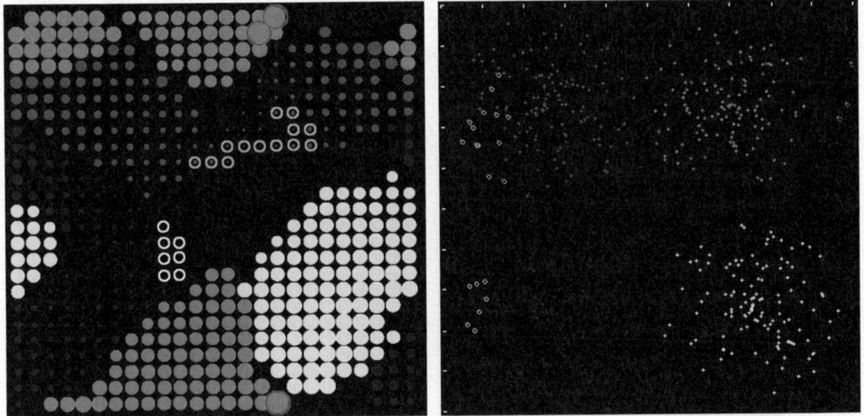

Fig. 2. Visualization with color and shape negotiation, size representing the 1st data attribute (left), and data scatterplot (right), on which the 1st data attribute is mapped on the X-axis. Corresponding ants (left) and data items (right) are highlighted in red, cyan and white.

The display of different circular shape sizes enables the user to understand how a single data attribute is distributed over the clustering representation. For instance, the three largest ants (highlighted in red) are positioned within an outlying green sub-cluster (see Fig. 2). The ants highlighted in cyan and white show that the smallest ants in shape size correspond to those ants with the smallest data value for that attribute.

Figure 3 shows two different clustering techniques of the car dataset, containing 38 items and 7 data dimensions, as taken from [19]. On the left, the multidimensional scaling technique positioned the cars in three apparent clusters (the color coding was artificially added by the authors for visual clarification). The cellular ant method, on the right, positioned the cars in a single visual cluster, but recognized 3 separate class labels that roughly correspond to those apparent clusters.

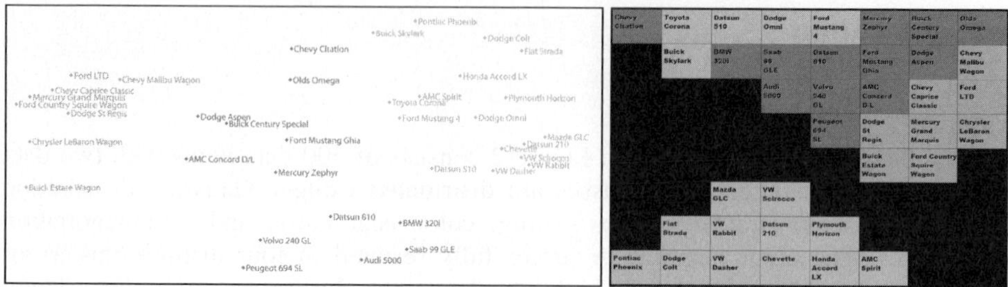

Fig. 3. The car dataset represented by MDS (left, based on [19], color coding artificially added by the authors) and the Cellular Ants representation with color negotiation (right)

Figure 4 illustrates how shape size negotiation is used to clarify data dependencies for high-dimensional datasets, without prior knowledge of the data scale and without using any predefined data mapping rules. Figure 4 uses the same representation as Figure 3, and maps a single data attribute to the decentralized shape size negotiation. As a result, one can investigate how the clusters are internally ordered for different attributes. Here, it shows the relative dominance of the cylinder count and MPG within specific clusters, and some cars visually stand out within the formed clusters.

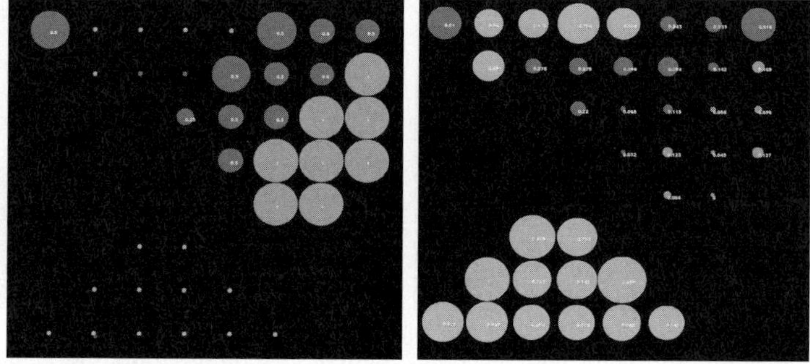

Fig. 4. A Cellular Ant representation in a toroidal grid using color and shape negotiation. Data attribute represented by the shape size: cylinders (left) and Miles per Gallon (MPG) (right).

The cellular ant method has been evaluated with typical benchmarking data, such as the IRIS dataset. The iteration timeline in Figure 5 (left) shows how several colors were introduced, but only three remained. In effect, the IRIS dataset is clustered in two distinct visual clusters, but the color negotiation recognizes that three different data classes exist, of which two are very similar. This interplay between visual and spatial clustering contains a high visualization value. The figure shows a momentary snapshot only: during the simulation the orange and yellow colors take over ants from each other. Using shape negotiation, one can investigate how a data attribute is relatively distributed over a cluster. As shown in Figure 5 (right), subclusters of high or low data values are made apparent, demonstrating the ordering power of the swapping rule. For instance, one can perceive that for attribute 4, the yellow type

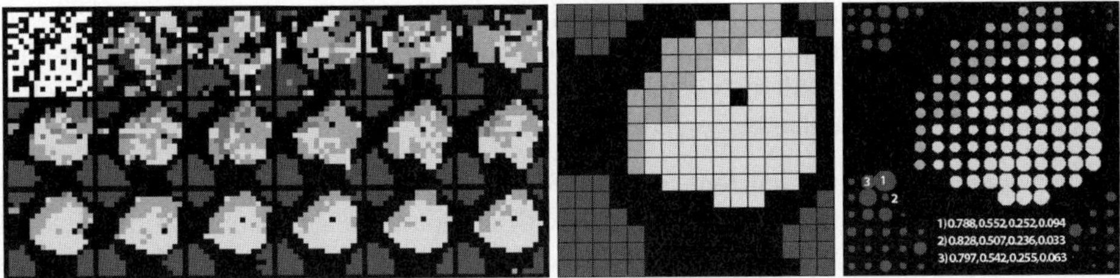

Fig. 5. Clustered IRIS dataset (150 data items, 4 attributes, 3 clusters, 1821 iterations) in a toroidal grid. Left: iteration timeline. Middle: resulting spatial clustering with color negotiation. Right: same result, with shape size negotiation for attribute 4.

(Virginia) has larger data values than the orange one (Versacolor), and that this attribute is very volatile for the red type (Setosa) (varying between values 0.1 and 0.3) when considering their relative numerical proportion to one another within the cluster.

Table 1 lists the performance of the color negotiation (or data classification) for various standard benchmarking datasets, after executing the cellular ant algorithm over 50 runs, each with a different, random initialization seed. The clustering error rate is calculated by counting the ants with correct colors over the whole population, and dividing this summation by the total amount of ants. In general, these results are worse but relatively similar to comparable clustering methods, such as reported in [6].

Table 1. Performance measurements of the color negotiation method for different benchmarking datasets. Averages are taken over 50 runs, each with a different random seed.

				#Clusters		Clustering Error	
Datasets	#Objects	#Attributes	#Clusters	Average	[std]	Average	[std]
Iris	150	4	3	2.68	[0.65]	0.37	[0.11]
Pima	768	8	2	1.14	[0.35]	0.36	[0.04]
Thyroid	215	5	3	3.45	[0.80]	0.41	[0.14]

5   Discussion

The performance of the current implementation depends on two variables: the data similarity tolerance threshold t and color seed similarity threshold c. The ant density (or the grid size determined by dividing the available cells in the grid by the dataset size) has been kept constant at about 75%. Similarly to the object distance measure α in common ant-based clustering approaches, the optimal value of t and c cannot be determined without prior knowledge of the dataset, unless the value is adaptable [20].

We consider the current implementation as a simple proof-of-concept prototype, and kept its implementation as simple as possible. Therefore, the scaling and clustering performance of the cellular ant method is not that effective as existing MDS methods. Its first aim is not to compete with alternative approaches, but rather to be considered as an early prototype towards more powerful cellular automata clustering algorithms, or towards data visualizations that are emergent and self-adaptive. As shown in the diagrams, the combination of spatial clustering with data class clustering can result in visual representations that are meaningful and useful. Following aspects can also be considered.

- **Performance.** In its current simple form of implementation, the amount of required iterations seems to be similar with comparable approaches in the field of ant-based data mining. However, the 'data-to-ant' model always requires less iteration steps because all data objects are able to move to increasingly ideal positions simultaneously. Similarly to existing ant-based data mining optimizations, the clustering performance could be addressed with increasing the data similarity cap value over time, so that clusters grow more rapidly and steadily. The method requires a considerable amount of calculations, as each ant is required to calculate many pair-wise dependencies for each iteration step.
- **Clustering Quality.** Grid density influences the clustering quality in two ways. Small grid densities do not assure that ants with equal colors (data labels) will be represented in a single spatial cluster, because two or more clusters might emerge without ever 'touching' and 'merging'. Too dense grids generate single, large groups with diverse labels and thus little visualization value.
- **Simplicity**. The current behavior rules have been kept as simple as possible, to demonstrate the potential value of the cellular automata-like decentralized negotiation for data mining and data visualization purposes. Further calculation optimization or solutions towards data size scalability can be accomplished by considering a combination of following three approaches: 1) real-time data optimization, including data approximation and gradual data streaming, 2) agent adaptation, which includes the distribution or balancing of loads between multiple agents, and 3) agent cooperation, by generating adaptive coalition formations of 'super agents' that have similar objectives, experience or goals.

6 Conclusion

This paper presented two new features of the cellular ant method: color (data clustering) and shape size negotiation. It combines ant-based data mining algorithms with cellular automata insights, or data scaling with data clustering to derive an approach that is capable of representing multidimensional datasets. The resulting diagrams are visually similar to those of ant-based data mining clustering approaches. However, the clusters are also similar to multi-dimensional scaling images, as they are ordered internally as well as globally over multiple data dimensions. As a simple prototype towards self-organizing visualization, inter-agent negotiations determine typical visual cues, such as position, color and size, depending on multidimensional data properties. Color negotiation can recognize data clusters of similar type. Shape size negotiation displays the relative distribution of a single data attribute and the internal structure of clusters. Conceptually, the self-adaptive, unsupervised data mapping process of the cellular ants proposes a conceptual alternative to the common fixed data mapping rules that are based on preconceived dataset assumptions.

Some of the limitations of method are caused by the simplicity of the rule-based approach, and its dependency on fixed, discrete cellular automata characteristics, instead of more continuous probability functions. Several optimizations can be accomplished, for instance by altering the data similarity tolerance threshold over time, or by informing agents of the global effectiveness. As a simple prototype, it demonstrates a potential future in which data visualization agents are capable of autonomously detecting complex data patterns and proactively acting upon them to make underlying data phenomena more visually apparent and the perceptual and cognitive understanding by humans more effective.

References

1. Vande Moere, A., Clayden, J.J.: Cellular Ants: Combining Ant-Based Clustering with Cellular Automata. International Conference on Tools with Artificial Intelligence (ICTAI'05). IEEE (2005) 177-184
2. Deneubourg, J., Goss, S., Franks, N., A., S.F., Detrain, C., Chretian, L.: The Dynamics of Collective Sorting: Robot-Like Ants and Ant-Like Robots. From Animals to Animats: 1st International Conference on Simulation of Adaptative Behaviour (1990) 356-363
3. Schockaert, S., De Cock, M., Cornelis, C., Kerre, E.E.: Fuzzy Ant Based Clustering. Lecture Notes in Computer Science 3172 (2004) 342-349
4. Handl, J., Knowles, J., Dorigo, M.: Ant-Based Clustering and Topographic Mapping. Artificial Life 12 (2005) 35-61
5. Ramos, V., Abraham, A.: Evolving a Stigmergic Self-Organized Data-Mining. International Conference on Intelligent Systems, Design and Applications, Budapest, (2004) 725-730
6. Labroche, N., Monmarché, N., Venturini, G.: A New Clustering Algorithm Based on the Chemical Recognition System of Ants. European Conference on Artificial Intelligence, Lyon, France (2003) 345-349
7. Torgerson, W.S.: Multidimensional Scaling. Psychometrika 17 (1952) 401-419
8. Bentley, C.L., Ward, M.O.: Animating Multidimensional Scaling to Visualize N-Dimensional Data Sets. Symposium on Information Visualization. IEEE (1996) 72 -73
9. Ankerst, M., Berchtold, S., Keim, D.A.: Similarity Clustering of Dimensions for an Enhanced Visualization of Multidimensional Data. Symposium on Information Visualization (Infovis'98). IEEE (1998) 52-60
10. Kohonen, T.: The Self-Organizing Map. Proceedings of the IEEE 78 (1990) 1464-1480
11. Von Neumann, J.: Theory of Self-Reproducing Automata. University of Illinois Press, Illinois (1966)
12. Chen, L., Xu, X., Chen, Y., He, P.: A Novel Ant Clustering Algorithm Based on Cellular Automata. Conference of the Intelligent Agent Technology (IAT'04). IEEE (2004) 148-154
13. Schroeder, M., Noy, P.: Multi-Agent Visualisation based on Multivariate Data. International Conference on Autonomous Agents. ACM Press, Montreal, Quebec, Canada (2001) 85-91
14. Pham, B., Brown, R.: Multi-Agent Approach for Visualisation of Fuzzy Systems. Lecture Notes in Computer Science 2659 (2003) 995-1004
15. Healey, C.G., Amant, R.S., Chang, J.: Assisted Visualization of E-Commerce Auction Agents. Graphics Interface 2001. Canadian Information Processing, Ottawa, (2001) 201-208
16. Ebert, A., Bender, M., Barthel, H., Divivier, A.: Tuning a Component-based Visualization System Architecture by Agents. International Symposium on Smart Graphics. Hawthorne, IBM T.J. Watson Research Center (2001)

17. Roard, N., Jones, M.W.: Agent Based Visualization and Strategies. Conference in Central Europe on Computer Graphics, Visualization and Computer Vision (WSCG), Pilzen (2006)
18. Vande Moere, A.: Time-Varying Data Visualization using Information Flocking Boids. Symposium on Information Visualization (Infovis'04). IEEE, Austin, USA (2004) 97-104
19. Wojciech, B.: Multivariate Visualization Techniques. Vol. 2006 (2001)
20. Handl, J., Meyer, B.: Improved Ant-based Clustering and Sorting in a Document Retrieval Interface. International Conference on Parallel Problem Solving from Nature (PPSN VII) LNCS 2439 (2002) 913-923

A Neuro-fuzzy Inference System for the Evaluation of New Product Development Projects

Orhan Feyzioğlu* and Gülçin Büyüközkan

Department of Industrial Engineering, Galatasaray University,
Çırağan Caddesi No: 36, Ortaköy 34357 İstanbul-Turkey
{ofeyzioglu, gbuyukozkan}@gsu.edu.tr

Abstract. As a vital activity for companies, new product development is also a very risky process due to the high uncertainty degree encountered at every development stage and the inevitable dependence on how previous steps are successfully accomplished. Hence, there is an apparent need to evaluate new product initiatives systematically and make accurate decisions under uncertainty. Another major concern is the time pressure to launch a significant number of new products to preserve and increase the competitive power of the company. In this work, we propose an integrated decision-making framework based on neural networks and fuzzy logic to make appropriate decisions and accelerate the evaluation process. We are especially interested in the two initial stages where new product ideas are selected and the implementation order of the corresponding projects are determined. We show that this two-staged intelligent approach allows practitioners to roughly and quickly separate good and bad product ideas by making use of previous experiences, and then, analyze a more shortened list rigorously.

1 Introduction

New product development (NPD) is the process by which an organization uses its resources and capabilities to create a new product or improve an existing one. Product development is seen as "among the essential processes for success, survival, and renewal of organizations, particularly for firms in either fast-paced or competitive markets" [2]. Markets are generally perceived to be demanding higher quality and higher performing products, in shorter and more predictable development cycle-times and at lower cost [18]. Along this line, new product idea selection and project launch are important cornerstones. In practice, it is observed that it is difficult to end NPD projects once they are begun [4, 25] and firms can make two types of erroneous decisions when evaluating their new product ideas: pursuing an unsuccessful idea and not developing a potentially successful new product. In either case, firms accrue big losses, and while the former leads to investment loses, the latter leads to missed investment opportunities [28]. NPD decisions, especially at the early stages of the development, are affected by a considerable amount of uncertainty causing elements. The uncertainty arises from multiple sources including technical, management and

---

* Corresponding author. Tel.: +90 212 227 4480 (ext.311); Fax: +90 212 259 5557

A. Sattar and B.H. Kang (Eds.): AI 2006, LNAI 4304, pp. 837–846, 2006.

© Springer-Verlag Berlin Heidelberg 2006

commercial issues, both internal and external to the project. Hence, it is critical to use a structured approach that is able to lower the risks caused by the uncertainty for NPD projects. In this work, we propose an integrated approach based on fuzzy logic and neural networks to make more rational selection decisions in a shortened time.

The rest of the paper is organized as follows. In the next section, we briefly expose the uncertainty factors affecting the NPD process. Then, we present a new intelligent decision making method that aims to accelerate the evaluation process while taking into account the uncertainty factors that affect it. The application of the approach is given as a case study in section 4 and the last section includes some concluding remarks and perspectives.

2   Risks Under Uncertainty in NPD Process

For projects such as encountered in NPD, it is necessary to define one or a number of objective functions, and then measure the likelihood of achieving certain target values. Examples of such functions include capital expenditure, completion time and so on. Project uncertainty is the probability that the objective function will not reach its planned target value [11, 13]. Meanwhile, risk is defined as the exposure to loss/gain, or the probability of occurrence of loss/gain multiplied by its respective magnitude. Risk management involves modeling the project's objective functions against project variables, such as cost and quantities of input resources, external factors, etc. Since the project variables are often stochastic in nature and dynamic (i.e., exhibiting varying degrees of uncertainty over time) it is quite natural that the objective functions will also exhibit uncertainty.

It can be observed that different approaches exist in the literature to define and analyze the uncertainty in NPD projects. Fox et al. [7] combined three dimensions of uncertainty as technical, market and process. They rated and categorized uncertainty along each dimension as being either low or high. When technical uncertainty is high, technologies used in the development of the project are neither existent nor proven at the start of the project, and/or are rapidly changing over time. When market uncertainty is high, the organization has little information regarding who the customer is, how the market is segmented and what are the needed channels of distribution. Finally, when process uncertainty is high, a significant portion of any or all of the engineering, marketing, and communications processes are relatively new, unstable, or evolving.

Similarly, Mullins and Sutherland [22] identified three levels of uncertainty that confront companies operating in rapidly changing markets. First, potential customers cannot easily articulate needs that a new technology may fulfill. Consequently, NPD managers are uncertain about the market opportunities that a new technology offers. Second, NPD managers are uncertain about how to turn the new technologies into new products that meet customer needs. Finally, senior management faces uncertainty about how much capital to invest in pursuit of rapidly changing markets as well as when to invest.

Miller and Lessard [20] identified three risk categories for engineering projects: ''completion risks'' which include technical, construction and operational risks, ''market related risks'' which include demand, financial and supply risks and finally, ''institutional risks'' which include social acceptability and sovereign risks.

We also refer to the work of Riek [24] where risks related to NPD are categorized as technical risks, commercial risks and human resource risks. If we analyze NPD from different perspectives, we can precise risk structure in a more detailed manner. As an example, we can allocate product positioning, pricing and customer uncertainties to marketing; organizational alignment and team characteristics uncertainties to organizations; concept, configuration and performance uncertainties to engineering design; supplier, material, design of production sequence and project management uncertainties to operations management. Efficient and effective NPD requires the appropriate management of all these uncertainty sources.

3   An Evaluation Approach for NPD Projects

The research in the intersection area of artificial intelligence and NPD is comparatively new. For a comprehensive overview of the application of the related techniques in NPD, we refer the interested readers to [23, 30]. We note that, Zaremba and Morel [30] identified neural networks and genetic search as the predominant techniques for the initial phases of NPD process. Here, we propose to use a two-stage decision making approach to accelerate the NP selection process and to improve its efficiency. The approach is especially relevant when there are numerous new ideas generating sources and it is difficult to rate all related products in a very detailed way and in a reasonable amount of time. It allows practitioners to roughly and quickly separate good and bad product ideas by making use of previous experiences, and then to analyze in details a more shortened list.

3.1   The Rough Evaluation Phase

This stage consists of a technique that merges neural networks and fuzzy logic. Artificial Neural Networks (ANN) [10, 17] make use of the way that the human brain learns and functions, possess the ability to learn from examples, have the ability to manage systems from their observed behavior, have the capacity to treat large amount of data and capturing complex interactions among the input variables. Meanwhile fuzzy logic [16, 29, 31] is used to deal with imprecise linguistic concepts or fuzzy terms, allows making rational decisions in an uncertain environment without loosing the richness of verbal judgment, and is highly suitable for approximate reasoning by incorporating fuzzy rules. Hence substantial improvements on NPD project selection can be made by merging the ANN and fuzzy set theory.

In this study, new product ideas generated individually or by groups of individuals have been collected by a formal system. Then, the preprocessing of ideas' database is left to an intelligent neuro-fuzzy inference system. Regarding to NPD, evaluations are mostly based on a scoring system with determined evaluation criteria. Therefore, translating if necessary these scores to eligibility percentages, one can build an input database. Our fuzzy inference system (FIS) maps the input space consisting of the information provided by past evaluations to the output space formed by the status of the ideas (i.e., "good" or "bad"). The system posses an internal mechanism that can learn the viewpoint of the company management towards products by making use of the extracted rules. It also reduces the decision-making effort when the number of applications is large. The details of the FIS are given in [1].

Neural network techniques aid the fuzzy modeling procedure to learn information about a historical data set, and compute the membership function parameters that best allow the associated FIS to track the given input/output data. ANFIS (adaptive network-based fuzzy inference system) is a class of adaptive networks that are functionally equivalent to FIS [14]. Using a given input/output data set, ANFIS constructs a FIS whose membership function parameters are adjusted using either a back propagation algorithm or a hybrid-learning algorithm. Therefore, using ANFIS, fuzzy systems can learn from the modeling data. The architecture of ANFIS is a feed-forward network that consists of five layers [14]. Fig. 1 shows the equivalent ANFIS architecture for a two-input Sugeno-type fuzzy inference system.

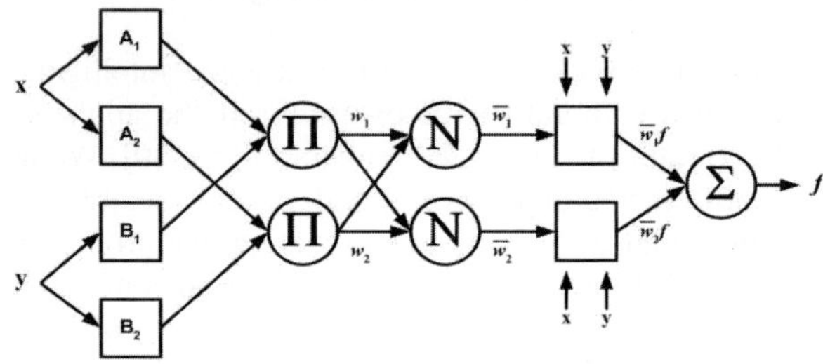

Fig. 1. ANFIS architecture for a two inputs, two rules Sugeno FIS

A rule in the first order Sugeno FIS has the form:

$$\text{If } x \text{ is } A_i \text{ and } y \text{ is } B_i \text{ then } f_i = p_i x + q_i y + r_i.$$

The output of a node in the first layer specifies to which degree a given input, x, satisfies a quantifier, A, i.e., the function of the node i in this layer is a membership function for the quantifier, A_i , of the form $\mu_{A_i}(x)$. Each membership function has a set of parameters that can be used to control that membership function. For example, a Gaussian membership function that has the form $\mu_{A_i}(x) = e^{-(x-c_i/\sigma_i)^2}$ and has two parameters, c_i and σ_i. Tuning the values of these parameters will vary the membership function, which means a change in the behavior of the FIS. In the second layer, a node output represents a firing strength of a rule. The node generates the output (firing strength) by multiplying the signals that come on its input, $w_i = \mu_{A_i}(x) \times \mu_{B_i}(y)$. The function of a node in the third layer is to compute the ratio between the ith rule's firing strength to the sum of all rules' firing strengths, $\bar{w}_i = w_i / w_1 + w_2$, where $\bar{w}_i$ is referred to as the normalized firing strength [14]. In the fourth layer, each node has a function of the form: $\bar{w}_i f_i = \bar{w}_i (p_i x + q_i y + r_i)$ where $\{p_i, q_i, r_i\}$ is the parameter set. The overall output is computed in the fifth layer by summing all the incoming signals, i.e., $f = \sum_i \bar{w}_i f_i = (w_1 f_1 + w_2 f_2)/w_1 + w_2$. During the learning process, the parameters are tuned until the desired response of the FIS is achieved [14].

3.2   The Exact Evaluation Phase

The exact evaluation phase consists of rating of project alternatives versus different and various criteria and the aim is to find a trade-off solution for these criteria. Therefore, a compromise operator should naturally be selected. Moreover, criteria satisfaction of NPD projects is usually expressed by means of unipolar difference scale (ie [0-100]). According to these considerations, the operators of the Choquet integral family are interesting because they cover many generalized mean operators (i.e. those comprised between the min and the max operators). They can be written under the form of a conventional weighted mean modified by effects coming from interactions between elementary performance evaluations and are coherent with the traditional performance expressions. We thus select this type of combination for decisional aggregation in performance measurement systems, especially the two-additive Choquet integral that considers only interactions by pair.

The two parameter types of the two-additive Choquet integral are [9] the weight v_i of each evaluation criterion (Shapley parameters), and the interaction parameters I_{ij} of any pair of evaluation criteria. Criteria weights sum up to 1 and interaction parameters range in [-1, 1].

Noting P_i as the satisfaction degree of a project for criterion $i = 1, \ldots, n$ and the overall score of that project as $Sc(P_1, \ldots, P_n)$, the associated combination function is

$$Sc\left(P_1,P_2,\ldots,P_n\right) = \sum_{i=1}^{n} P_i\left(v_i - \frac{1}{2}\sum_{j\neq i}\left|I_{ij}\right|\right) + \sum_{I_{ij}>0} \min\left(P_i,P_j\right)I_{ij} + \sum_{I_{ij}<0} \max\left(P_i,P_j\right)\left|I_{ij}\right|, \qquad (1)$$

with the property that

$$\left(v_i - 1/2\sum_{i\neq j}\left|I_{ij}\right|\right) \geq 0 \quad \text{for all } i = 1,\ldots,n. \qquad (2)$$

Positive I_{ij} implies that the simultaneous satisfaction of criteria i and j is significant for the aggregated evaluation, but a unilateral satisfaction has no effect. Negative I_{ij} implies that the satisfaction of either i and j is sufficient to have a significant effect on the aggregated evaluation. Null I_{ij} implies that no interaction exists; thus v_i acts as the weights in a common weighted mean.

The coefficients of importance v_m and $I_{mm'}$ being more natural to the decision maker, their determination by asking the decision maker is possible, but it must be verified that the monotonicity conditions in Eq.(2) is satisfied to use the transformation relation with the conventional fuzzy measure representation. Other methods based on the identification of these coefficients from experimental data exists but it is another problematic (see [8]), that is not the object of this study.

4   An Industrial Case Study

The subject company of our work is the local branch of an international toy-manufacturing firm. The new product database contains ideas generated from company designers, product managers, and also from employees and customers. Ideas are collected through a web based proposal system externally and also through an internal system where product managers introduce their proposals based on competitor products, benchmarking reports and marketing analysis reports. Therefore, large amount of data tend to accumulate over time. The marketing management team evaluates these ideas based on different tangible and intangible points. Note that evaluating only the risks and revenues of the investment is not sufficient for NPD projects since they have special characteristics which may establish intangible profits depending on their creativity and innovative features. We therefore summarized the available data into three indicators, namely *risk ratio*, *benefit ratio* and *strategic impact index*. These indicators form the input of our FIS and are expressed in terms of percentage. Their severity increases with the allocated value.

Since the objective is to classify ideas as "good" or "bad", two ANFIS models are built to recognize the corresponding idea status. The resulting architecture is sometimes called many ANFIS (MANFIS). One ANFIS model will be trained to provide a value close to 1 if the idea is good, the other model will perform the same for a bad idea. The discrimination process is done by presenting the features of the idea to be classified to each of the two ANFIS models. The result is two different responses and a voting scheme is applied to determine the class to which the idea belongs. The class (idea status) that is associated with the ANFIS model with the response closest to the value of 1 is chosen as the class to which the idea under investigation belongs. Obviously, only good ideas are kept for further analysis.

We have taken into consideration 1326 new product ideas examined by the company in the last five years. 468 of them were found to be acceptable at some extent, while the remaining 858 ideas were found not satisfactory at all or not correlating with the company goals and policies. We have divided this data into two training and test sets so as to train the neuro-fuzzy inference system. As there is no common understanding on how to separate the data in the literature, we have used the following rule of thumb. First, we considered only accepted ideas one by one. A uniform number is generated between 0 and 1, and if it shows to be less than 3/4, the chosen instance is added to the training set; otherwise it is added to the test set. Then, we applied the same rule for rejected ideas to complete the sets. The composition of the data set is given in Table 1.

To construct a fuzzy inference process [19] for the idea evaluation, the selection of a method to partition the input space to reflect the premise part of the fuzzy inference is an important consideration. As there are no preferable membership functions, we created an initial set of membership functions using grid-partitioning method [15]. Grid partitioning covers the whole input space with membership functions that have uniform distribution. We considered different number and type of membership functions. However, to reduce space, we give the summarized results in Table 1 concerning three Gaussian membership functions for each input variable where we obtained the least training and test errors.

The results were convincing with small (<%10) and close training and test errors. This implicitly implies that the past evaluations are very representative and robust, since usually test error is expected to be much higher than the training error. Then, it can be argued that the constructed FIS imitated properly the company attitude towards new ideas. The final point is that there are also some misaccepted and misrejected ideas. A close investigation shows that these ideas correspond to ones with almost equal membership function values for a given index. Since these cases are the hardest to classify, it is quite natural to expect this type of error.

Table 1. Summarized MANFIS evaluation results of new product ideas

	Real Value				MANFIS Results			
	Tot. Inst.	Accept.	Reject.	Accept.	Reject.	Misaccept.	Misreject.	Misclass.	Misclass.
Training Set	982	352	630	343	639	43	35	78	7.9%
Test Set	344	116	228	112	232	18	13	31	9.0%

The obtained FIS together with its adjusted parameters were implemented as a decision support software by the authors to integrate the approach with the actual company's system. Then, eighteen actual new product ideas covering one month were presented to the trained FIS. The outcome shows that only five of them were acceptable for further analysis. As the company's managers agreed on the results, we proceed to the second phase of our approach.

To determine project prioritization for resource allocation, our approach calls for a detailed analysis on the projects corresponding to the previously selected product ideas by means of Choquet integral aggregation method. The factors for new product project evaluation considered here were identified from the review of the related literature [3, 5, 6, 12, 21, 26] and interviews of the company managers: (1) R&D Resources (RDR): Human Resources (RHR), Development time (DTM), Technical Feasibility (TFE); (2) Production Resources (PRE): Human Resources (PHR), Available Equipment & Facilities (AEF), Available Suppliers (ASP), Available Production Process (APP); (3) Marketing Resources (MRE): Human Resources (MHR), Available Advertorial & Promotional Strategies (AAS), Available Distribution Channels (ADS), Available Service Channels (ASC); (4) Financial: Internal Rate of Return (IRR), Payback Period (PPD), Margin Rate (MRA), Sales Volume (SVL); (5) Management (MNG): Top Management Support (TMS), Production Management Competence (PMC), Organization (ORG); (6) Market (MRT): Competition Strength (CSR), Market Size (MSZ), Market Growth (MGR); (7) Product (PRD): Customer Satisfaction (CST), Product Quality (PQT), Alignment with Company Strategies (ACS), Windows of Opportunities (WOP).

The weight of the criteria and their possible interactions were determined by an interdisciplinary board of the company with a Delphi type method. We assumed that there is no direct interaction between two second level criteria belonging to different primary criteria. However, synergy may exist either between primary criteria or between secondary criteria that belong to the same primary criterion (see also [27]).

Finally, concurring five projects were rated for secondary criteria and the score related to each primary criterion is computed using Eq.(1) and given in Table 2 under the column "Choquet Integral". Then, the final scores of five projects are computed such that 71.54, 76.48, 77.06, 69.13, 71.67. As the highest scored projects are preferable, the implementation order is determined as 3 > 2 > 5 > 1 > 4 for this case.

It is quite natural to ask what could be if another evaluation method was used instead of the proposed one. While it is difficult to compare different methods, a general expectation is to get more robust results if more information is incorporated in the decision making process. Here, we will compare the previous results with the ones obtained through the simple "weighted mean". In this well-known technique, only criteria weights and project ratings are used to obtain the weighted average. Results are shown in Table 2 under the column "Weighted Mean". It can be remarked that there are minor but not insignificant modifications in overall project scores and the implementation order is changed as 2 > 3 > 1 > 5 > 4. As for example, a close examination of the 2nd project reveals that scores obtained with the Choquet Integral method tend to be lower and makes it second in the ordering. This is mainly due to the attributed lower scores for criteria pairs with significant positive interaction. This is an expected outcome, since positive interaction implies simultaneous satisfaction of two criteria important for the overall score, and being low in one of them leads to worsen results. Note that most of the identified interactions were of positive type in this study.

Table 2. Scores and ranking of projects based on the Choquet integral and weighted mean

	Choquet Integral					Weighted Mean				
	1	2	3	4	5	1	2	3	4	5
RDR	71.25	80.75	81.25	62.50	65.00	72.50	82.25	82.50	62.75	65.50
PRE	76.75	87.88	72.50	83.00	60.25	77.50	87.75	72.50	84.00	57.25
MRE	78.25	82.00	75.50	76.88	76.75	83.00	83.50	76.00	77.75	80.50
FIN	68.00	68.63	83.50	63.00	75.50	74.50	70.25	83.75	65.25	76.25
MNG	68.63	85.88	81.50	74.50	87.50	67.25	85.75	81.00	75.00	86.50
MRT	66.63	76.75	73.13	66.63	75.13	66.75	77.00	74.00	67.25	75.75
PRD	90.63	70.75	76.25	73.38	67.88	91.50	73.00	77.50	76.50	68.00
Score	71.54	76.48	77.06	69.13	71.67	75.96	79.30	78.59	72.11	73.56
Rank	4	2	1	5	3	3	1	2	5	4

5 Final Remarks and Perspectives

In this study, we aim to improve the quality of the decision-making and to increase the probability of success in NPD under uncertainty by introducing a new iterative methodology. First, we describe the risks that may be encountered during the NPD process. Then, we give the motivation behind our approach which incorporates fuzzy database mining and decision making techniques for the new product project evaluation. An industrial case study is given to elucidate the details of the method.

The approach is general in a sense that although in different sectors, companies exercising similar vast new product idea selection process and having a scoring system can adopt it quite easily. However, we have to also underline two limitations of this study:

- Without a reliable historical database, the neural network can not be trained and the FIS can only be equipped with theoretical understanding. This can lead to inconsistent results.

- There is a need for intellectual capital evaluation for highly innovative and creative, very few new product developing or highly R&D oriented companies.

We can easily articulate that our approach has offered significant savings by shortening the time and decreasing the steps necessary for the evaluation process while deviated in the totality from the traditional system results with almost 15%. Moreover, the company authorities were very pleased with the results and strongly supported the project. However, a long period of time is always necessary to observe the results of such a strategic level decision. Additionally, the product success is not only depending on catching the best idea but also on how to manage subsequent development stages. We keep trying to understand the sources of conflict and possible improvements on the approach.

Based on this work, our future extension can be to investigate other decision phases in NPD and to provide similar approaches to enrich the available literature. We aim to evaluate in a more detailed way, the influence of other methods on the final quality and accuracy of decisions. We also want to enhance our decision support system with new techniques to enable managers comparing different solutions and making more rigorous decisions.

Acknowledgements

The authors acknowledge also the financial support of the Galatasaray University Research Fund.

References

1. Bezdek, J.C., Dubois, D., Prade, H. (eds.): Fuzzy sets approximate reasoning and information systems. Kluwer Academic Publishers Group, Netherlands (1999)
2. Brown, S.L., Eisenhardt, K.M.: Product development: past research, present findings and future directions. Academy of Management Review. 20 (1995) 343–378
3. Carbonell-Foulquie, P., Munuera-Aleman, J.L., Rodriguez-Escudero, A.I.: Criteria employed fo go/no-go decisions when developing successful highly innovative products. Industrial Marketing Management. 33 (2004) 307–316
4. Cooper, R.G.: Perspective: third generation new product processes. Journal of Product Innovation Management. 11 (1994) 3–14
5. Danneels, E., Kleinschmidt, E.J.: Product innovativeness from the firm's perspective: its dimensions and their relation with project selection and performance. Journal of Product Innovation Management. 18 (2001) 357–373
6. Ernst, H.: Success factors of new product development: a review of the empirical literature. International Journal of Management Reviews. 4 (2002) 1–40
7. Fox, J., Gann, R., Shur, A., Glahn, L., Zaas, B.: Process uncertainty: a new dimension for new product development. Engineering Management Journal. 10 (1998) 19–27
8. Grabisch, M.:.k-order additive discrete fuzzy measures and their representation. Fuzzy Sets and Systems, 92 (1997)167–189

9. Grabisch, M., Roubens, M.: Application of the Choquet integral in multi-criteria decision making. Grabisch, M., Murofushi, T., Sugeno, M. (eds): Fuzzy Measures and Integrals: Theory and Applications. Physica, New York (2000) 348–374

10. Hammerstrom, D.: Neural networks at work. IEEE Spectrum Computer Applications. 30 (1993) 26–32

11. Hillson, D.: Extending the risk process to manage opportunities. International Journal of Project Management. 20 (2002) 235–240

12. Hultink, E.J., Hart, S., Robben, H.S.J., Griffin, A.: Launch decisions and new product success: an empirical comparison of consumer and industrial products. Journal of Product Innovation Management. 17 (2000) 5–23

13. Jaafari, A.: Management of risks, uncertainties and opportunities on projects: time for a fundamental shift. International Journal of Project Management. 19 (2001) 89–101

14. Jang, J.S.R.: ANFIS: Adaptive-Network-Based Fuzzy Inference System. IEEE Transaction on Systems, Man and Cybernetics. 23 (1993) 665–685

15. Jang, J.S.R., Sun, C.T., Mizutani, E.: Neuro-Fuzzy and Soft Computing: A Computational Approach to Learning and Machine Intelligence. Prentice Hall, USA (1997)

16. Klir, G.J., Yuan, B.: Fuzzy Sets and Fuzzy Logic: Theory and Applications. Prentice-Hall, Englewood Cliffs, NJ (1995)

17. Lin, C., Lee, C.: Neural Fuzzy Systems. Prentice Hall, Englewood Cliffs, NJ (1996)

18. Maffin, D., Braiden, P.: Manufacturing and supplier roles in product development. International Journal of Production Economics. 69 (2001) 205–213

19. Matlab Fuzzy Logic Toolbox. The Math Works Inc., Natick, MA, 1997

20. Miller, R., Lessard, D.: Understanding and managing risks in large engineering projects. International Journal of Project Management. 19 (2001) 437–443

21. Montoya, M.M., Calantone, R.: Determinants of new product performance: a review and meta-analysis. Journal of Product Innovation Management. 11 (1994) 397–417

22. Mullins, J.W., Sutherland, D.J.: New product development in rapidly changing markets: an exploratory study. Journal of Product Innovation Management. 15 (1998) 224–236

23. Rao, S. S., Nahm, A., Shi, Z., Deng, X., Syamil, A.: Artificial intelligence and expert systems applications in new product development–a survey. Journal of Intelligent Manufacturing. 10 (1999) 231–244

24. Riek, R.F.: From experience: Capturing hard-won NPD lessons in checklists. Journal of Product Innovation Management. 18 (2001) 301–313

25. Schmidt, J.B., Calantone, R.J.: Are really new product development projects harder to shut down? Journal of Product Innovation Management. 15 (1998) 111–123

26. Sun, H., Wing W.C.: Critical success factors for new product development in the Hong Kong toy industry. Technovation. 25 (2005) 293–303

27. Tzeng, G.H., Yang, Y.P.O., Lin, C.T., Chen, C.B.: Hierarchical MADM with fuzzy integral for evaluating enterprise intranet web sites. Information Sciences 169 (2005) 409 – 426

28. Özer, M.: Factors which influence decision making in new product evaluation. European Journal of Operational Research. 163 (2005) 784–801

29. Zadeh, L.A.: Fuzzy sets, Information and control. 8 (1965) 338–353

30. Zaremba, M.B., Morel, G: Integration and control of intelligence in distributed manufacturing. Journal of Intelligent Manufacturing. 14 (2003) 25–42

31. Zimmermann, H.J.: Practical Applications of Fuzzy Technologies. Kluwer Academic Publishers, Massachusetts (1999).

Extracting Minimum Unsatisfiable Cores
with a Greedy Genetic Algorithm⋆

Jianmin Zhang, Sikun Li, and Shengyu Shen

School of Computer Science, National University of Defense Technology
410073 ChangSha, China
{jmzhang, skli, syshen}@nudt.edu.cn

Abstract. Explaining the causes of infeasibility of Boolean formulas
has practical applications in various fields. We are generally interested
in a minimum explanation of infeasibility that excludes irrelevant information. A smallest-cardinality unsatisfiable subset, called a minimum
unsatisfiable core, can provide a succinct explanation of infeasibility and
is valuable for applications. However little attention has been concentrated on extraction of minimum unsatisfiable cores. In this paper, we
propose an efficient greedy genetic algorithm to derive an exact or nearly
exact minimum unsatisfiable core. It takes advantage of the relationship between maximal satisfiability and minimum unsatisfiability. We
report experimental results on practical benchmarks, as compared with
the branch-and-bound algorithm and the ant colony optimization.

1   Introduction

Explaining the causes of unsatisfiability of Boolean formulas is a key requirement
in many practical applications, such as artificial intelligence, formal verification
and electronic design. A paradigmatic example is repairing inconsistent knowledge from a knowledge base [1]. Additional examples include SAT-based model
checking on predicate abstraction, where analysis of unsatisfiability is an essential step for ensuring completeness of bounded model checking [2], and fixing
wire routing in FPGAs [3]. Many real-world problems can be formulated as constraint satisfaction problems, which can be translated into Boolean formulas in
conjunctive normal form (CNF). Modern Boolean satisfiability (SAT) solvers are
able to determine whether a formula is satisfiable or not. When a formula is unsatisfiable, it is often required to find an unsatisfiable core (UC), that is, a small
unsatisfiable subformula (US) of the original formula. Localizing an unsatisfiable
core is necessary to determine the underlying reasons for the failure. It is desirable to find a minimal unsatisfiable core, but checking minimal unsatisfiability
is a very hard problem known to be D^P-complete [4]. We are usually interested
in a minimal or minimum explanation of infeasibility. A minimum unsatisfiable
core provides a more succinct explanation of infeasibility and is more valuable

---

⋆ This work is supported by the National Science Foundation of China (NSFC) under
grant No. 60603088.

A. Sattar and B.H. Kang (Eds.): AI 2006, LNAI 4304, pp. 847–856, 2006.

© Springer-Verlag Berlin Heidelberg 2006

than a minimal unsatisfiable core. However, existing works have little concern regarding extraction of minimum unsatisfiable cores.

In this paper, we tackle the problem of finding a minimum unsatisfiable core from considerable real-world formulas by a compounded greedy and genetic algorithm. We develop the relationship between maximal satisfiability and minimal unsatisfiability [10,11], and present the relationship between maximal satisfiability and minimum unsatisfiability: a minimum UC is the minimum hitting set of the complements of maximal satisfiable subformulas. Our algorithm is able to compute an exact or nearly exact minimum unsatisfiable core from a Boolean formula. First, we derive the complements of maximal satisfiable subsets. The complement sets consist of the clauses not included in some maximal satisfiable subformula. Then, we extract a minimum unsatisfiable core by computing the minimum hitting set of those complement sets.

The paper is organized as follows. The next section surveys the related work on computing unsatisfiable cores. Section 3 introduces the basic definitions and notations used throughout the paper. Section 4 proposes the relationship between maximal satisfiability and minimum unsatisfiability. Section 5 presents our algorithm for extracting a minimum unsatisfiable core from CNF formulas. Section 6 shows and analyzes results for the well-known DaimlerChrysler's problem instances. Finally, Section 7 concludes this paper.

2   Related Work

The problem of finding an unsatisfiable core has been addressed rather frequently in the last few years, owing to its increasing importance in numerous fields. In [5], a method of adaptive core search is employed to find a small UC. This technique succeeds in finding small, but not minimal unsatisfiable cores for the problem instances. Additional algorithm is based on the ability of modern SAT solvers to produce resolution refutations [6], and it is able to obtain a small unsatisfiable subformula, with no guarantee that the size of the subformula is minimal. Another approach that allows for finding small, but not necessarily minimal unsatisfiable cores is called AMUSE [7]. A different algorithm that guarantees minimality is MUP [8], which is a prover of minimal unsatisfiability, as opposed to an unsatisfiable core extractor.

Experimental work on computing the unsatisfiable cores can be grouped into algorithms that find small or minimal UCs and those that find minimum UCs. Most of above methods mainly extract the small or minimal unsatisfiable cores from Boolean formulas. In the recent past, some research has considered the harder problem of finding the minimum unsatisfiable core. In [9], an approach is proposed to derive a minimum unsatisfiable core. This algorithm enumerates all possible subsets of a formula and checks which are unsatisfiable but all of whose subsets are satisfiable. In this way, it can find the solution exactly, but the technique is unlikely to scale well, due to the extreme size of its search space. Furthermore, its experimental results are obtained on small benchmarks.

Recently, the strong relationship between maximal satisfiability and minimal unsatisfiability has been exploited to develop a technique for finding all minimal unsatisfiable subformulas [10,11] and for finding a minimum unsatisfiable core [12]. In [12], the authors present a branch-and-bound algorithm that utilizes iterative maximal satisfiability solutions to generate lower and upper bounds on the size of the smallest-cardinality minimal UC, and branch on specific subformulas to find it. It can tackle considerable practical problem instances. This algorithm is the best previous approach we are aware of for extracting a minimum unsatisfiable core from a constraint set. In section 6, our approach will be compared with this branch-and-bound algorithm on the same benchmarks.

3   Preliminaries

Resolution is a proof system for CNF formulas with the following rule:

$$\frac{(A \vee x)(B \vee \neg x)}{(A \vee B)} . \tag{1}$$

Where A, B denote the disjunctions of literals, and $(A \vee x)$, $(B \vee \neg x)$ are the resolving clauses, and $(A \vee B)$ is the resolvent. The resolvent of the clauses x and $\neg x$ is the empty clause $(\bot)$. Every application of the resolution rule is called a resolution step. A sequence of resolution steps is that each one uses the result of the previous step as one of the resolving clauses of the current step. It is well-known that a CNF formula is unsatisfiable if it is possible to generate an empty clause by a finite sequence of resolution steps from the original clauses. The set of original clauses involved in the derivation of the empty clause is referred to as the unsatisfiable core.

Definition 1. (Unsatisfiable Core). *Given a formula φ, ψ is an unsatisfiable core for φ iff ψ is an unsatisfiable formula and $\psi \subseteq \varphi$.*

That is, an unsatisfiable subformula can be defined as any subset of the original formula which is unsatisfiable. It is obvious that there may exist many different unsatisfiable cores, with different numbers of clauses, for the same problem instance, such that some of these cores are the subsets of others.

Definition 2. (Minimal Unsatisfiable Core). *Given an unsatisfiable core ψ for a formula φ, ψ is a minimal unsatisfiable core iff removing any clause $\omega \in \psi$ from ψ implies that $\psi - \{\omega\}$ is satisfiable.*

For Boolean formulas in CNF, an unsatisfiable core is minimal if it becomes satisfiable whenever any of its clauses is removed. A minimal unsatisfiable core is irreducible, in other words, all of its proper subsets are satisfiable. There have been many different contributions to research on minimal UCs extraction.

Definition 3. (Minimum Unsatisfiable Core). *Consider the set of all unsatisfiable cores for a formula $\varphi : \{\psi_1, \ldots, \psi_j\}$. Then, $\psi_k \in \{\psi_1, \ldots, \psi_j\}$ is a minimum unsatisfiable core iff $\forall \psi_i \in \{\psi_1, \ldots, \psi_j\}, 1 \leq i \leq j : |\psi_k| \leq |\psi_i|$.*

That is to say, a minimum unsatisfiable core has the smallest cardinality of all unsatisfiable subsets of a formula. According to the definition, one may conclude that any unsatisfiable formula has at least one minimum unsatisfiable core. Whereas, existing works have paid little attention to the extraction of minimum unsatisfiable cores. We should note that in many cases the size of a minimal unsatisfiable core may be much larger than the size of a minimum unsatisfiable core, because the minimal UCs probably contain many redundant clauses which cannot be found by simple resolution rules.

4   Maximal Satisfiability and Minimum Unsatisfiability

Our approach is on the basis of the relationship between maximal satisfiability and minimum unsatisfiability. Firstly, we review the relationship between maximal satisfiability and minimal unsatisfiability. The definitions of a satisfiable subformula and a maximal satisfiable subformula are shown as follows:

Definition 4. (Satisfiable Subformula). *Consider a formula φ, ϕ is a satisfiable subformula for φ iff ϕ is a satisfiable formula and $\phi \subseteq \varphi$.*

Definition 5. (Maximal Satisfiable Subformula). *Given a satisfiable subformula ϕ for a formula φ, ϕ is a maximal satisfiable subformula iff adding any clause $\omega \in \{\varphi - \phi\}$ into ϕ implies that $\phi \cup \{\omega\}$ is unsatisfiable.*

We should note the relationship between maximal satisfiability and minimal unsatisfiability. From the above definition, every maximal satisfiable subformula (MSS) is a subset of the original formula in that it is satisfiable and is not augment any more. In general, given any MSS of an unsatisfiable formula, those clauses which are not in the MSS represent a correction to the formula in the sense that removing them will repair the infeasibility. Therefore, we define a "CoMSS" as the complement set of an MSS. This complementary view of MSSes provides the essential link to minimal UCs. The existence of any minimal UC causes the formula unsatisfiable. Therefore, to correct the infeasibility by removing clauses, at least one clause from each minimal UC must be removed so as to neutralize all of the minimal UCs. A CoMSS of a CNF formula is a set of clauses whose removal makes the formula satisfiable, thus each CoMSS must contain at least one clause from every minimal unsatisfiable core.

Definition 6. (Minimal Hitting Set). *Suppose $C = \{S_1, \ldots, S_n\}$ is a collection of sets. Then H is a hitting set of C iff $H \subseteq \bigcup_{1 \leq i \leq n} S_i$, and $H \cap S_i \neq \emptyset$ for each i with $1 \leq i \leq n$. H is a minimal hitting set iff $\forall\, R \subset H : R$ is not a hitting set.*

From the definition, a hitting set is a set that contains at least one element from each set in the collection. Then we can draw a conclusion that a CoMSS of a formula is a hitting set of the collection of all minimal UCs. Further, a CoMSS is a hitting set of the minimal UCs with the restriction that it is irreducible, that is, it is a minimal hitting set. According to the commutative relationship between the minimal UC and the CoMSS, one can infer that each minimal UC is a minimal hitting set of the CoMSSes. However, what is the relationship between the minimum unsatisfiable core and the CoMSS? First, we should clarify the relationship between the minimum UC and the minimal UC. It is easy to conclude that a minimum unsatisfiable core must be a minimal unsatisfiable core. The property of smallest cardinality of a minimum UC entails the irreducibility required to be a minimal UC.

Lemma 1. *A minimal unsatisfiable core for a formula φ is a minimal hitting set of the complements of maximal satisfiable subformulas for φ.*

Definition 7. (Minimum Hitting Set). *Given a collection C of sets and the set of all hitting sets of $C : \{H_1, \ldots, H_j\}$, then $H_k \in \{H_1, \ldots, H_j\}$ is a minimum hitting set iff $\forall\, H_i \in \{H_1, \ldots, H_j\}$, $1 \leq i \leq j : |H_k| \leq |H_i|$.*

We have already arrived at the conclusion that each minimal UC of a CNF formula is a minimal hitting set of the collection of CoMSSes. Thus a minimum UC is also a minimal hitting set of CoMSSes. According to the above definition, it is obvious that a minimum hitting set of a collection of sets must be a minimal hitting set of those sets. The smallest cardinality of a minimum hitting set ensures the irreducibility, in other words, its proper subsets are not hitting sets. However, a minimum hitting set is the smallest minimal hitting set, and is must not be reduced any more. Furthermore, a minimum unsatisfiable core is the smallest-cardinality minimal UC with the constraint that it cannot be any smaller. Consequently, we can infer the following conclusion:

Lemma 2. *A minimum unsatisfiable core for a formula φ is a minimum hitting set of the complements of maximal satisfiable subformulas for φ.*

We illustrate the relationship. For example, the CNF formula is

$$\varphi = (x_1) \wedge (\neg x_2) \wedge (\neg x_1 \vee x_2) \wedge (\neg x_2 \vee x_3) \wedge (\neg x_3) \ . \qquad (2)$$

The solutions of satisfiability and unsatisfiability are listed in Table 1. Each clause is encoded by its place in the sequence. For instance, the third clause $(\neg x_1 \vee x_2)$ is named "3". There are four maximal satisfiability subsets and four corresponding complements of the MSSes. The two minimal unsatisfiable subformulas are the minimal hitting set of the CoMSSes. Obviously, the minimum unsatisfiable core is the minimum hitting set of the CoMSSes.

Table 1. The SAT and UNSAT Solutions of φ

Maximal satisfiable subformulas	$\{2, 3, 4, 5\}\ \{1, 2, 4, 5\}\ \{1, 3, 5\}\ \{1, 3, 4\}$
Complements of MSSes	$\{1\}\ \{3\}\ \{2, 4\}\ \{2, 5\}$
Minimal unsatisfiable cores	$\{1, 2, 3\}\ \{1, 3, 4, 5\}$
Minimum unsatisfiable core	$\{1, 2, 3\}$

5   Extracting the Minimum Unsatisfiable Core

One naive method to compute a minimum UC is to generate all minimal UCs and then select the smallest one. This approach cannot be extend in that the number of minimal UCs of a formula can be exponential in the number of its variables. To solve this problem efficiently, we present a compounded greedy and genetic algorithm to compute a minimum unsatisfiable core, utilizing the relationship between maximal satisfiability and minimum unsatisfiability.

The genetic algorithm, originally introduced in [13], is a heuristic optimization method inspired by the mechanics of natural evolution, including survival of the fittest, reproduction and mutation. A population of potential solutions are refined iteratively by employing a strategy of Darwinian evolution and natural selection. This type of heuristic approach has been applied in many different fields, such as combinational optimization, self-adaptive controlling, machine learning, etc. The principle of an evolutionary mechanism has been proven to be a good framework for combinatorial optimization problems. We combine this approach with a greedy algorithm to solve the problem of the minimum UC extraction from CNF formulas. First of all, some definitions used in this algorithm, including the genetic operators, are introduced as follows:

Genetic encoding. The binary vector describes the chromosome code of CoMSSes. The sets are encoded as $\{x_1, x_2, \ldots, x_m\}$, where $x_i = 0$ or 1, $1 \leq i \leq m$. The binary vectors are called chromosomes, and each bit is called a gene, and all of the chromosomes are called the population. For example, if one set of CoMSSes is $S = \{1, 3, 4, 7\}$, then the corresponding binary genes are 1011001.

Fitness function. The fitness function is defined as $f(S) = H_S/|S|$, where H_S denotes the number of sets hit by all of the elements in S, and $|S|$ represents the cardinality of the set S.

Crossover operator. We use the single point crossover strategy. Suppose that $S_1 = \{x_1, x_2, \ldots, x_m\}$ and $S_2 = \{y_1, y_2, \ldots, y_m\}$ are two chromosomes, and we select a random integer number k, $1 \leq k \leq m$, then S_3 and S_4 are the crossover offspring of S_1 and S_2, such that $S_3 = \{s_i \mid \text{if } i \leq k, s_i \in S_1, \text{else } s_i \in S_2\}$, $S_4 = \{s_i \mid \text{if } i \leq k, s_i \in S_2, \text{else } s_i \in S_1\}$.

Mutation operator. Suppose a chromosome $S_1 = \{x_1, x_2, \ldots, x_m\}$, and we select a random integer number k, $1 \leq k \leq m$, then S_2 is the mutation offspring of S_1, such that $S_2 = \{s_i \mid \text{if } i = k, s_i = \neg x_i, \text{else } s_i = x_i\}$.

Inversion operator. Suppose a chromosome $S_1 = \{x_1, \ldots, x_k, x_{k+1}, \ldots, x_{l-1}, x_l, \ldots, x_m\}$, where k and l are two random numbers, $1 \leq k < l \leq m$, then S_2 is the inversion offspring of S_1, such that $S_2 = \{x_1, \ldots, x_l, x_{l-1}, \ldots, x_{k+1}, x_k, \ldots, x_m\}$.

Selection operator. Roulette wheel selection strategy are adopted. All chromosomes are evaluated by fitness function. We choose those that hit more sets and have smaller cardinality than others.

The proposed algorithm proceeds as follows:

1. Find all complements of maximal satisfiable subformulas for a formula φ;
2. Recode the clause number in CoMSSes to make the sets more compressed;
3. Adopt a greedy algorithm to derive two good near minimum unsatisfiable cores as the initial parents. We incrementally solve this problem, removing solutions as they are found to continue the search until all have been found;
4. Encode the sets of CoMSSes as binary vector and obtain the binary genes;
5. Apply the genetic operators, including crossover, mutation and inversion, to the selected parents, in order to give birth to the offspring;
6. Recombine the offspring and current parents to form a new population, and employ the selection operator to choose the best solutions as the new parents;
7. If the upper limit of iterations is reached, or the minimum unsatisfiable cores remain the same size for 10 successive offspring, stop and report the best solution found. Otherwise go to the step 5.

We note that the greedy algorithm has fast convergence, but often leads to local optimum. The genetic algorithm is a global evolutionary searching method, but its local convergent speed is slow. We combine the greedy algorithm and the genetic algorithm to make up for the shortage of each other. Firstly, we adopt the greedy algorithm to quickly obtain a few initial approximate optimal solutions. Then the genetic algorithm is used to compute the better near minimum unsatisfiable cores, taking advantage of its convergence to the global optimum.

To evaluate the efficiency of the greedy genetic algorithm, we implement an ant colony optimization in comparison with our approach. The ant colony algorithm, initially introduced in [14], is a cooperative heuristic searching algorithm inspired by the ethological study on the behavior of ants. The ant colony algorithm for deriving a minimum UC proceeds as follows:

1. The first two steps are the same as those of the greedy genetic algorithm;
2. Associate each clause i in the CoMSSes with a pheromone τ_i, which is a global heuristic indicating the favorableness for ants to select the next element; the pheromone on each clause is initially set by the value τ_0;
3. Each of m ants selects the first clause at random and crawls over the sets to incrementally find a minimum UC denoted by M_k, where $k = 1, \ldots, m$. Every ant decides its successor element based on the following probability:

$$
p_i(t) = \begin{cases} \dfrac{[\tau_i(t)]^\alpha \bullet [\eta_i(t)]^\beta}{\sum_{n \in N}[\tau_i(t)]^\alpha \bullet [\eta_i(t)]^\beta} & if \ i \in N, \\ 0 & otherwise. \end{cases}
\tag{3}
$$

Where N denotes the set of clauses which are not yet present in the solution and hit at least one remainder subset, and η_i is a local heuristic indicating the number of sets hit by the clause i. Extensive experimental studies show that the best values for three parameters are $\alpha = 1$, $\beta = 3$ and $\rho = 0.5$;
4. Evaluate the unsatisfiable core M_k generated by each ant. The quality of UCs is measured by its size $|M_k|$, that is, the number of clauses it contains;
5. Update the pheromone trails on the clauses using the following equation:

$$\tau_i(t+1) = \begin{cases} (1-\rho)\tau_i(t) + \sum_{k=1}^{m}(C - |M_k|) & \text{if } M_k \text{ is a UC,} \\ (1-\rho)\tau_i(t) & \text{otherwise,} \end{cases} \qquad (4)$$

where $0 < \rho < 1$ is the evaporation coefficient of pheromone, and C is the total number of clauses.

6. If the limit of iterations is reached, stop and report the best solution found. Otherwise go to step 3.

6 Experimental Results

To experimentally evaluate the effectiveness of our algorithm, we have selected 22 problem instances from the well-known DC family [15]. These instances are obtained from the validation and verification of automotive product configuration data and encode different consistency properties of the configuration data base which is used to configure DaimlerChrysler's Mercedes car lines. The size of 22 instances ranges from 4496 to 8686 clauses and 1608 to 1909 variables.

Our algorithm to extract a minimum unsatisfiable core is implemented in C++ . For generating all of the minimal UCs we use the algorithms in [10]. All experiments were conducted on a 1.6 GHz Athlon machine having 1 GB memory and running the Linux operating system. The limit time was 600 seconds. The experimental results are listed in Table 2. Table 2 gives the total number of minimal unsatisfiable subsets contained in each formula (MUSes), the number of clauses in the minimum unsatisfiable core (min), the number of clauses in the largest minimal unsatisfiable core (max), and the average size of all minimal unsatisfiable cores (ave). Table 2 also shows the runtime in seconds of the branch-and-bound algorithm (BaB time) on a 2 GHz Pentium 4 Linux machine with 1GB RAM [12]. Furthermore, Table 2 provides the runtime in seconds (ACO time) of the ant colony optimization and the number of clauses (ACO size) in the derived minimum unsatisfiable core. The last two columns present the runtime in seconds (GGA time) and the size of the minimum unsatisfiable core (GGA size) extracted by the greedy genetic algorithm. There are four instances where we failed to find all of the minimal UCs within the timeout. Therefore, their average sizes cannot be calculated and left blank in Table 2.

Table 2 illustrates that the greedy genetic algorithm strongly outperforms the branch-and-bound algorithm which is the best known approach, usually by about one order of magnitude. Our algorithm can find the exact minimum UCs for most formulas. A few minimum UCs extracted by the algorithm have one or two clauses more than the smallest-cardinality unsatisfiable core. Although two algorithms run on different processors, the performance of 1.6 GHz Athlon is generally inferior to the performance of 2 GHz Pentium. Moreover, this table also shows that the greedy genetic algorithm works better than the ant colony optimization, and both algorithms conduct on the same machine.

From this table, we may observe the following. For all problem instances, the percentage of clauses in the minimum UCs is quite small, in most cases less than 1%. Consequently, localizing a minimum unsatisfiable core is necessary to determine the underlying reasons for the failure. Further, the size of minimum UCs is much smaller than the size of the largest minimal UCs and the average size of all minimal UCs for many formulas. In some cases, the number of clauses in the minimum UC is half or one third of the number of clauses in the largest minimal UCs. Therefore, as compared with minimal UCs, a minimum UC provides a more succinct explanation of infeasibility and is more valuable for practical applications. Moreover, all minimal UCs of some problem instances cannot be found within a reasonable time limit, and thus it is unadvisable to find all of the minimal UCs for some formulas. In many cases, extracting all minimal unsatisfiable cores is intractable for complicated problem instances.

Table 2. Performance Results on Benchmarks

Benchmarks	MUSes	min	max	ave	BaB time	ACO time	ACO size	GGA time	GGA size
C220_FV_SZ_65	103442	23	40	35.1	18.85	2.83	23	1.21	23
C220_FV_SZ_46	>566210	17	≥64		78.44	16.45	17	12.20	17
C220_FV_RZ_14	80	11	18	14.5	33.89	1.51	11	0.15	11
C220_FV_RZ_13	6772	10	27	21.5	33.35	1.20	10	0.43	10
C220_FV_RZ_12	80272	11	35	26.5	50.58	2.75	11	0.60	11
C210_FW_RZ_59	15	140	173	156.8	56.69	7.92	140	0.89	140
C210_FS_RZ_40	15	140	173	156.8	36.24	7.79	140	0.69	140
C208_FA_UT_3255	52736	40	74	60.3	94.59	16.12	41	1.02	40
C208_FA_UT_3254	17408	40	74	59.7	95.27	13.80	41	0.90	40
C208_FA_SZ_87	12884	18	27	22.9	15.10	3.81	19	0.70	19
C208_FA_SZ_120	2	34	34	34.0	3.80	1.08	34	0.15	34
C202_FW_SZ_123	3	36	38	37.0	14.74	1.14	36	0.31	36
C202_FW_RZ_57	1	213	213	213.0	58.34	3.09	213	1.08	213
C202_FS_SZ_122	1	33	33	33.0	3.84	1.72	33	0.16	33
C202_FS_SZ_121	4	22	24	23.0	2.50	1.58	22	0.19	22
C202_FS_RZ_44	>125207	18	≥53		131.04	48.70	20	16.70	20
C208_FA_RZ_43	>20940	8	≥28		76.88	59.30	8	22.30	8
C170_FR_SZ_96	>99155	53	≥80		322.76	20.15	55	14.00	54
C170_FR_SZ_92	1	131	131	131.0	15.12	1.37	131	0.39	131
C170_FR_SZ_58	218692	46	63	56.4	15.329	3.90	47	0.79	47
C170_FR_RZ_32	32768	227	228	227.9	121.33	6.77	228	0.67	227
C168_FW_UT_851	102	8	16	14.1	59.91	1.75	8	0.57	8

7 Conclusions

Explaining the causes of infeasibility of Boolean formulas has practical applications in many fields, such as artificial intelligence, formal verification and electronic design. A smallest-cardinality unsatisfiable subformula, that is a minimum unsatisfiable core, provides a succinct explanation of infeasibility. In this paper, we present the relationship between maximal satisfiability and minimum unsatisfiability. A greedy genetic algorithm is proposed to extract the exact or nearly exact minimum unsatisfiable core. It utilizes the relationship between maximal satisfiability and minimum unsatisfiability. The experimental results indicate that the greedy genetic algorithm outperforms the branch-and-bound algorithm and the ant colony optimization. The results show that our algorithm efficiently finds the minimum unsatisfiable cores at the cost of occasional augmented size.

References

1. Mazure, L. Sais, and E. Gregoire. Boosting complete techniques thanks to local search methods. Annals of Mathematics and Artificial Intelligence, 22(3-4):319-331, 1998.
2. K. L. McMillan and N. Amla. Automatic abstraction without counterexamples. In Proceedings of the Ninth International Conference on Tools and Algorithms for the Construction and Analysis of Systems (TACAS'03), 2003.
3. G. J. Nam, K. A. Sakallah, and R. A. Rutenbar. Satisfiability-based layout revisited: Detailed routing of complex FPGAs via search-based Boolean SAT. In International Symposium on Field Programmable Gate Arrays (FPGA'99), 1999.
4. C. H. Papadimitriou, and D. Wolfe. The Complexity of Facets Resolved, In Journal of Computer and System Sciences, vol. 37, pp. 2-13, 1988.
5. R. Bruni and A. Sassano. Restoring Satisfiability or Maintaining Unsatisfiability by Finding Small Unsatisfiable Subformulae. In Proceedings of the Workshop on Theory and Application of Satisfiability Testing (SAT'01), 2001.
6. L. Zhang and S. Malik. Extracting small unsatisfiable cores from unsatisfiable Boolean formula. In Proceedings of the Sixth International Conference on Theory and Applications of Satisfiability Testing (SAT'03), 2003.
7. Y. Oh, M. N. Mneimneh, Z. S. Andraus, K. A. Sakallah, and I. L. Markov. AMUSE: a minimally-unsatisfiable subformula extractor. In Proceedings of the 41st Design Automation Conference (DAC'04), 2004.
8. J. Huang. MUP: A minimal unsatisfiability prover. In Proceedings of the Tenth Asia and South Pacific Design Automation Conference (ASP-DAC'05), 2005.
9. I. Lynce and J. P. Marques-Silva. On computing minimum unsatisfiable cores. In Proceedings of the Seventh International Conference on Theory and Applications of Satisfiability Testing (SAT'04), 2004.
10. M. H. Liffiton and K. A. Sakallah. On finding all minimally unsatisfiable subformulas. In Proceedings of the Eighth International Conference on Theory and Applications of Satisfiability Testing (SAT'05), 2005.
11. J. Bailey and P. J. Stuckey. Discovery of Minimal Unsatisfiable Subsets of Constraints Using Hitting Set Dualization. In Proceedings of the Seventh International Symposium on Practical Aspects of Declarative Languages (PADL'05), 2005.
12. M. N. Mneimneh, I. Lynce, Z. S. Andraus, J. P. Marques-Silva, and K. A. Sakallah. A branch and bound algorithm for extracting smallest minimal unsatisfiable formulas. In Proceedings of the Eighth International Conference on Theory and Applications of Satisfiability Testing (SAT'05), 2005.
13. J. H. Holland. Adaptation in Natural and Artificial Systems. University of Michigan Press, Ann Arbor, 1975; MIT Press, Cambridge, second edition, 1992.
14. A. Colorni, M. Dorigo, and V. Maniezzo. Distributed Optimization by Ant Colonies. In Proceedings of the First European Conference on Artificial Life, 1991.
15. SAT benchmarks from Automotive Product Configuration, http://www-sr. informatik.uni-tuebingen.de/~sinz/DC/.

A Partitioned Portfolio Insurance Strategy by Relational Genetic Algorithm

Jiah-Shing Chen and Yao-Tang Lin

Department of Information Management
National Central University
Jhongli, Taiwan 320
{jschen, s0443003}@cc.ncu.edu.tw

Abstract. This paper proposes a new portfolio insurance strategy called partitioned portfolio insurance (PPI) strategy and a relational genetic algorithm (RGA) based on relational encoding to optimize the new partitioned portfolio insurance strategy. Experimental results show that our PPI strategy is significantly better than the traditional PI strategy and our RGA works well for solving the portfolio partitioning problem.

Keywords: Partitioned Portfolio Insurance (PPI) Strategy, Relational Genetic Algorithm (RGA), Grouping Genetic Algorithm (GGA), relational encoding.

1 Introduction

In practice, portfolio insurance (PI) strategies are the most popular way to protect the investment assets by dynamically and continuously varying the risky asset position. However, traditional PI strategies are defensive and passive as the downside risk-aversion is their only concern. Conceptually, they avoid the downside risk with passive insurance mechanism, and are highly dependent on the underlying portfolio and the rebalancing frequency [2, 12]. Besides, they may incur a large amount of trading costs, including capital loss and transaction fee, form repeated buying-high and selling-low in flat markets. Several rebalancing disciplines, such as time discipline and market move discipline, have been proposed to address this problem, but no single discipline can effectively reduce costs in all market situations and the trade-off between precision and cost has to be considered [8]. For these reasons, more aggressive PI strategies are needed.

To create a more aggressive and systematic PI strategy, this paper develops a partitioned portfolio insurance (PPI) strategy by partitioning the portfolio into several similar sub-portfolios and then insuring the sub-portfolios individually. Since the securities in the portfolio often have different price rising/falling levels in the same period, if we can correctly partition the portfolio into several sub-portfolios with similar price rising/falling level and then insure the sub-portfolios individually, we have a better chance to capture the upside potential of rising sub-portfolios and to avoid the downside risk of falling sub-portfolios simultaneously. As a result, the PPI strategy has a better performance than traditional PI strategies do.

A. Sattar and B.H. Kang (Eds.): AI 2006, LNAI 4304, pp. 857–866, 2006.
© Springer-Verlag Berlin Heidelberg 2006

To analyze the PI and PPI strategies, a simplified model is built and the portfolio partitioning problem is then induced. The portfolio partitioning problem, which has a variable number of subsets and has no natural heuristic, is a special set partitioning problem. The set partitioning problem belongs to the class of NP-hard problems [15], many of which have been solved by genetic algorithms (GA) due to their comprehensive search capability. However, previous works of applying GA to solving the partitioning problem have non-unique encoding, require a fixed number of subsets and rely heavily on some problem-specific heuristics [4, 6, 9, 10, 13]. Hence, a new relational genetic algorithm (RGA), is designed in this paper to solve the induced portfolio partitioning problem. Since the RGA has a relation-based encoding, it is a unique encoding for the partitioning problems. In addition, the genetic operators of RGA are designed to work without heuristics and to support a variable number of subsets.

Experimental results show that our PPI strategy is significantly better than the traditional PI strategy and our RGA works well for solving the portfolio partitioning problem.

The rest of this paper is organized as follows. Section 2 reviews traditional PI strategies and previous works for solving the partitioning problem. Section 3 develops our partitioned portfolio insurance strategy. Section 4 describes our relational genetic algorithm. Section 5 presents some experimental results which demonstrate the feasibility of the PPI strategy and the RGA. Finally, conclusions and future works are summarized in section 6.

2   Background

2.1   Portfolio Insurance

Portfolio insurance is a specialized form of hedging which not only avoids the downside risk but also keeps the upside potential for the insured portfolio [1]. Two types of approaches for portfolio insurance have been proposed in the past, i.e., static and dynamic approaches.

Static approaches initiate active security positions and corresponding derivatives from the start, and then hold them until maturity. The most typical strategy of static approaches is protective put. Static approaches are very simple, but they have some limitations in practice which include limited specifications of derivatives (particular underlying assets and exercise prices), short maturities, and lack of early exercise rights [1]. Owing to the limitations of static approaches, dynamic approaches are usually adopted in practice.

Dynamic approaches distribute capital into two types of assets, active securities and risk-free securities, and then vary the active security position dynamically and continuously. In the process of varying positions, a trade-off between precision and cost has to be considered, and rebalancing disciplines are required [8]. As a result, a complete dynamic PI strategy should be composed of an effective dynamic insurance mechanism and an appropriate rebalancing discipline.

Dynamic insurance mechanisms focus on the insurance capability. Two types of mechanisms have been proposed and adopted in practice. They are model-driven mechanisms and equation-driven mechanisms. A model-driven and dynamic option-based portfolio insurance (OBPI) approach, named synthetic put option (SPO) strategy [11], is induced by the famous Black-Scholes option pricing model. In addition, some simple equation-driven approaches, such as constant proportion portfolio insurance (CPPI) strategy [3], time invariant portfolio insurance (TIPP) strategy [7], and modified stop loss (MSL) strategy [2], can be easily implemented by rebalancing according to some simple equations. On the other hand, rebalancing disciplines (exposure-adjustment strategies) describe the timing and ratios of exposure adjustments. Etzioni proposed three traditional rebalancing disciplines which include time discipline, market move discipline, and lag discipline [8].

2.2 Partitioning Problems

The partitioning (or set-partitioning) problem, also known as the grouping problem [9], is defined as partitioning a set into a collection of mutually disjoint subsets. In addition, some hard constraints and cost functions are pre-defined for individual partitioning problem. A solution to the problem must comply with various hard constraints, and the objective of partitioning is to optimize a cost function defined over the set of all valid partitions [9].

Well-known classic partitioning problems include bin packing [13], equal piles [10], graph/map coloring [6, 10], line balancing, and workshop layout [5]. They have been solved by using various approaches, such as greedy algorithms, tabu search [5], simulated annealing [14], and genetic algorithms [9, 15].

Genetic algorithms are suitable for solving partitioning problems due to their capability to search a large solution space. Previous works of applying GA on partitioning problems can be divided into two categories. One is based on object-oriented representations, which include permutation encoding and membership encoding. Another is based on a group-oriented representation proposed by Falkenauer [8]. The object-oriented representations result in a large degree of redundancy and the group-oriented representation requires some problem-specific heuristics. Neither can be used to solve the portfolio partitioning problem effectively.

3 Partitioned Portfolio Insurance Strategy

The partitioned portfolio insurance (PPI) strategy works by partitioning the portfolio into several sub-portfolios and then insuring the sub-portfolios individually. However, not all partitions result in better performance. To understand which kinds of partitions can produce good performance, a simplified analysis of PI and PPI strategies is developed below.

To simplify our analysis, a simplified dynamic PI strategy is adopted. This simple strategy works as follows:

1. Initiate an active security position P_0 at the beginning of the period ($t = 0$).
2. Rebalance using the market move rebalancing discipline when market move reaches a constant threshold $\delta - 1$.
3. Each rebalancing takes an equal amount of positive or negative position $|P_i| = P$.

This simple strategy can be viewed as being composed of two components, a buy-and-hold strategy with initial position and a rebalancing (adjustment) strategy. The buy-and-hold strategy initiates the first active security position P_0 and then holds it until the end of the period. And the rebalancing strategy takes the follow-up positions P_t ($t > 0$) and then also holds them until the end.

The terminal wealth W_T^{PI} of the insured portfolio can then be calculated by $W_T^{PI} = W_T^{BH} + RE_T$, where W_T^{BH} is the final wealth of initial active security position P_0 taken by buy-and-hold strategy plus the remaining cash C_0, and RE_T is the total final rebalancing effect of all follow-up rebalancing positions P_t . Thus, W_T^{BH} can also be denoted as $W_T^{BH} = P_0\left(1 + R_{0,T}\right) + C_0$, where $R_{0,T}$ is the rate of return of active security form rebalancing point 0 to rebalancing point T . RE_T can also be denoted as $RE_T = \sum_{t=1}^{T} P_t R_{t,T}$. In short, we have

$$W_T^{PI} = W_T^{BH} + RE_T = \left[P_0\left(1 + R_{0,T}\right) + C_0 \right] + \sum_{t=1}^{T} P_t R_{t,T} \tag{1}$$

RE_T can be further divided into two parts: the effect of net upward or downward movements/positions and the effect of offset movements/positions. The net upward or downward movements/positions contribute positive effect, and their effect becomes $P \cdot \sum_{i=0}^{|M|-1} \left(\delta^i - 1\right)$, where $M = \log_\delta\left(R_{0,T} + 1\right)$ is the number of ticks of security price movement. The offset movements/positions stand for repeated buying-high and selling-low, and their effect becomes $-P \cdot (\delta - 1)\left(\dfrac{T - |M|}{2}\right)$. As a result, RE_T becomes

$$RE_T = \sum_{t=1}^{T} P_t R_{t,T} \approx \left[P \cdot \sum_{i=0}^{|M|-1} \left(\delta^i - 1\right) \right] - \left[P \cdot (\delta - 1)\left(\frac{T - |M|}{2}\right) \right] \tag{2}$$

By the approximation $\delta^i - 1 \approx i\left(\delta - 1\right)$, RE_T can be simplified as

$$RE_T = \sum_{i=1}^{T} P_i R_{i,T} \approx P \cdot (\delta - 1)\left(\frac{M^2 - T}{2}\right) \tag{3}$$

Summing them all up, we have

$$W_T^{PI} \approx \left[P_0 \cdot \delta^M + C_0 \right] + \left[P \cdot (\delta - 1)\left(\frac{M^2 - T}{2}\right) \right] \tag{4}$$

From the equation of W_T^{PI}, we can see that the factor M determines the performance of the buy-and-hold strategy, and factors M and T influence rebalancing effect of the rebalancing strategy.

By a similar derivation, we can have equation for W_T^{PPI} as follows

$$W_T^{PPI} \approx \left[P_0 \cdot \delta^M + C_0 \right] + \sum_{j=1}^{K} \left[w_j \cdot P \cdot (\delta - 1) \left(\frac{M_j^{\,2} - T_j}{2} \right) \right]$$

(5)

where K is number of subsets, w_j is weight of subsets, and $\sum_{j=1}^{K} w_j = 1$.

The difference between W_T^{PPI} and W_T^{PI} is

$$W_T^{PPI} - W_T^{PI} = \frac{P \cdot (\delta - 1)}{2} \left[\left(\sum_{j=1}^{K} w_j M_j^{\,2} - M^2 \right) - \left(\sum_{j=1}^{K} w_j T_j - T \right) \right]$$

(6)

$$W_T^{PPI} - W_T^{PI} = \begin{cases} \frac{P \cdot (\delta - 1)}{2} \cdot \sum_{j=1}^{K} w_j \left(M_j - M \right)^2 \geq 0, & \text{if } \sum_{j=1}^{K} w_j T_j - T = 0 \\ \frac{P \cdot (\delta - 1)}{2} \cdot \left(T - \sum_{j=1}^{K} w_j T_j \right), & \text{if } \sum_{j=1}^{K} w_j M_j^{\,2} - M^2 = 0 \end{cases}$$

(7)

According to equation (7), if we can correctly partition the portfolio to keep similar numbers of rebalancing ($\sum_{j=1}^{K} w_j T_j - T = 0$) under PPI strategy, we have a better chance to enhance the performance ($W_T^{PPI} - W_T^{PI} \geq 0$).

4   Relational Genetic Algorithm

Since previous genetic algorithms can not be used to solve the portfolio partitioning problem effectively, a new relational genetic algorithm is proposed to solve the portfolio partitioning problem. The relational genetic algorithm uses a relation-oriented encoding with corresponding genetic operators.

4.1   Relation-Oriented Representation

It is well-known that an equivalence relation can be induced by a partition and vice versa. Thus, a partition can be uniquely represented by its induced equivalence relation matrix.

In this relational encoding, each chromosome is an equivalence relation matrix $M_R = \left[m_{ij} \right]$ which represents an equivalence relation R such that $m_{ij} = 1$ if objects i and j are in the same subset, and $m_{ij} = 0$ otherwise. An example is shown in figure 1(a).

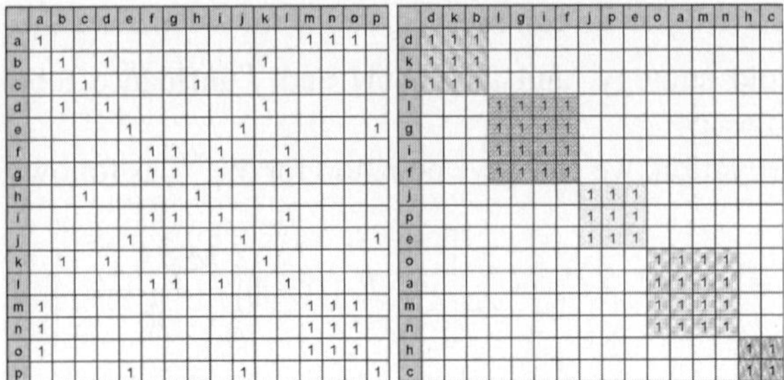

Fig. 1. (a) A partition encoded by an equivalence relation matrix. (b) Reordered matrix according to clusters.

Each equivalence relation matrix must be reflexive, symmetric, and transitive. These three properties must also be maintained in progress of evolution. Furthermore, relationships in same subset can be clustered together by reordering objects so that the groups can be identified more easily as shown in figure 1(b).

4.2 Genetic Operators

The two genetic operators of RGA, crossover and mutation, are redesigned to maintain the properties of equivalence relation matrix.

Crossover

Crossover operator for RGA consists of three traditional procedures and one additional repairing procedure. They are explained as fallows, and an example for one-point crossover is demonstrated.

1. **Select crossing points:** randomly select crossing points on diagonal in each parent matrix. One-point, two-point, and uniform crossovers are supported.

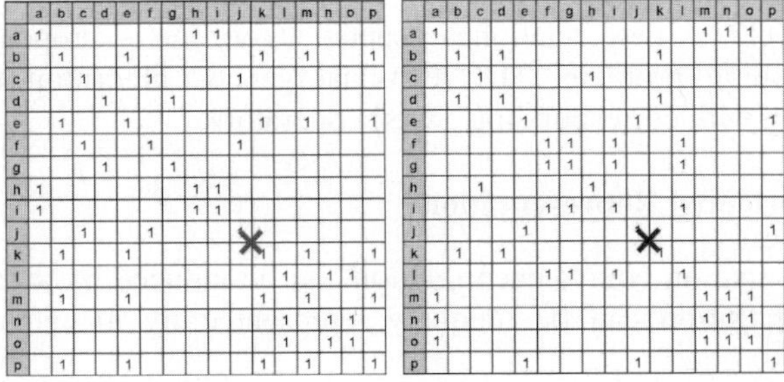

Fig. 2. Crossing points on diagonal are selected for one-point crossover

2. **Determine crossing and repairing areas:** further divide each matrix into several areas according to the selected crossing points. The genes in crossing areas will be

reserved until next generation and the genes in repairing areas will be disrupted. The disrupted genes are replaced by question marks temporarily and then rebuilt by repairing latter.

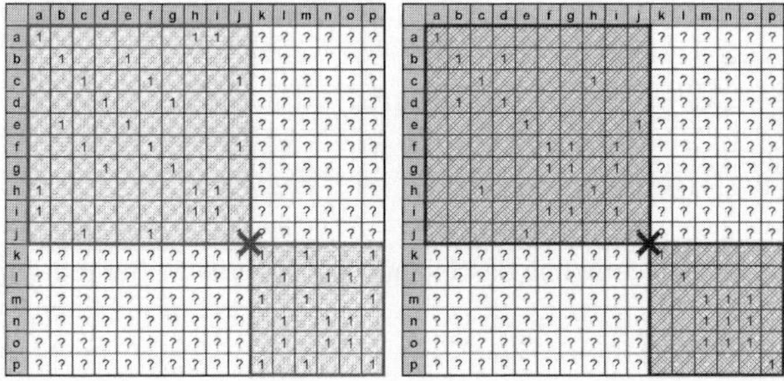

Fig. 3. Two crossing areas (colored) and two repairing areas (uncolored) are determined in each parent for one-point crossover. The genes in repairing areas are replaced by question marks.

3. **Swap crossing areas:** Swap crossing areas to generate descendants.

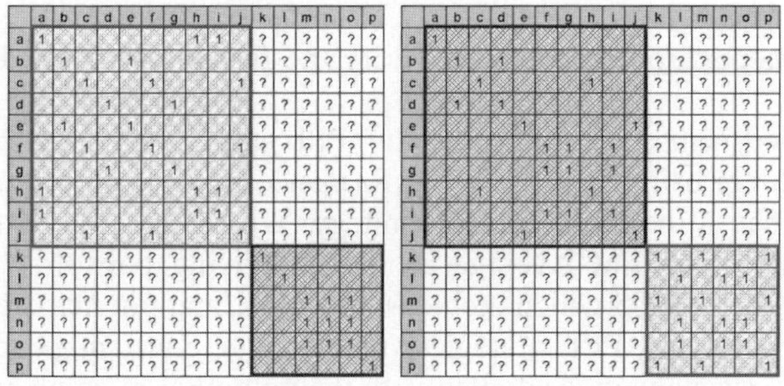

Fig. 4. Descendants inherit genes form parents by swapping crossing areas

4. **Repairing:** there are three objectives for repairing. They are controlling the number of subsets, maintaining symmetry, and maintaining transitivity. A simple recursive naïve (non-heuristic) repairing algorithm is illustrated as fallows.

```
Target = random(LowerBound,UpperBound)
//decide the new number of subsets

while (Subset > Target)
   calculate weights of each cell in repairing areas
   insert an 1 into repairing areas by roulette method
   determine symmetric and transitive closure
   determine new repairing areas
   recalculate the number of subset
end while
```

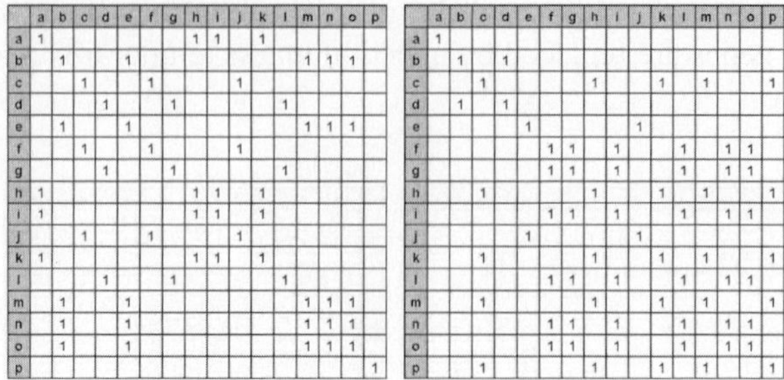

Fig. 5. Two complete descendants are generated after repairing

Mutation

The swap mutation operator in classical genetic algorithms is generalized to work on the equivalence relation matrix as follows: two randomly selected columns are swapped and their corresponding rows are also swapped to maintain the properties of the equivalence relation matrix.

5 Experiments

A set of experiments are performed to verify the feasibility of our PPI strategy and our RGA for PPI.

In these experiments, we optimize the PPI strategy by solving the portfolio partitioning problem using RGA. The performance of the optimized PPI strategy is compared with that of the traditional PI strategy. We use the 30 constituent securities in the Dow-Jones Industrial Average as the underlying portfolio. The experiment period is from 2002 to 2005 which includes bull, bear, and flat markets.

For traditional PI and PPI strategies, the popular constant proportion portfolio insurance (CPPI) mechanism is adopted with the multiplier m being set to 5, and floor F is set to be 80% of the initial wealth.

For RGA, two-point crossover is adopted because it outperforms others (one-point and uniform crossovers) in our pretests of the portfolio partitioning problem and classic partitioning problems. The population size of RGA is set to 200, generation is 500, selection method is stochastic remainder with replacement and elitism, crossover rate is 1, mutation rate is 0.01, and the fitness function is

$$Fitness = \sum_{i=1}^{K} w_i \left(A_i - F_i \right) \tag{8}$$

Finally, the sliding window method is adopted. The ratios of training period to testing period are 3:1, 2:1, or 1:1, and the lengths of testing periods are 60, 120, 240, or 480 days.

There are 51 different experiments in total. The average ROIs of the buy-and-hold strategy, the traditional PI strategy, and our PPI strategy in training and testing periods are shown in table 1. The statistical significances of three paired-samples t-tests for training and testing periods are shown in table 2.

Table 1. Average ROIs of three strategies in training and testing periods

	Training Periods	Testing Periods
Buy and Hold	3.0838	3.194
TPI	2.806	3.03
PPI	6.3109	3.6036

Table 2. Statistical significances of three paired-samples t-tests

	Training Periods	Testing Periods
ROI(PPI)-ROI(TPI)	3.93E-20	2.91E-08
ROI(PPI)-ROI(B&H)	4.32E-13	4.07E-04
ROI(TPI)-ROI(B&H)	0.231547	0.052472

According to the outcomes of experiments and statistical-tests shown in these two tables, our PPI strategy significantly outperforms the traditional PI strategy and the buy-and-hold strategy. The performances of the traditional PI strategy and the buy-and-hold strategy are not significant. In other words, our PPI strategy exploits the potential profit effectively, and our RGA works well for solving the portfolio partitioning problem.

6 Conclusion

This paper proposed a new partitioned portfolio insurance strategy which extends the traditional portfolio insurance strategy and contributes a more aggressive concept to portfolio insurance. We also analyzed the factors affecting the performance of portfolio insurance strategies by building a simplified analysis model. Simulation experiments and statistical tests show that our PPI strategy significantly outperforms the traditional PI strategy in bull, bear, and flat markets. In the other words, it not only avoids the downside risk, but also further explores the potential of profit successfully.

In addition, we designed a new relation-based genetic algorithm named relational genetic algorithm for the induced portfolio partitioning problem. The equivalence relation matrix encoding of the RGA has a 1-1 and onto correspondence with the class of all possible partitions. The encoding eliminates redundancy of previous GA representations and improves the performance of genetic search. The encoding can be used without requiring a fixed number of subsets in advance. We also designed a set of problem-independent operators for our RGA to solve such a partitioning problem which has no natural heuristics.

Our future works include incorporating other aggressive ideas for portfolio insurance, and applying our RGA to general partitioning problems.

References

[1] Abken, P.A., "An introduction to portfolio insurance," *Economic Review*, Federal Reserve Bank of Atlanta, issue Nov, pages 2-25, 1987.

[2] Bird, R., D. Dennis, and M. Tippett, "A Stop Loss Approach to Portfolio Insurance," *Journal of Portfolio Management*, Vol. 15, Issue 1, p35-40, Fall 1988.

[3] Black, F. and R. Jones, "Simplifying Portfolio Insurance," *Journal of Portfolio Management*, Vol. 14, Issue 1, p48-51, Fall 1987.

[4] Brown, E. C. and R. T. Sumichrast, "Evaluating performance advantages of grouping genetic algorithms," *Engineering Applications of Artificial Intelligence*, 18, p1-p12, 2005.

[5] Chiang, W-C., and P. Kouvelis, "An improved tabu search heuristic for solving facility layout design problem," *International Journal of Production Research*, 34, 9, pp.2565-2585, 1996.

[6] Eiben, A. E., J. K. van der Hauw, and J. I. van Hemert, "Graph coloring with adaptive evolutionary algorithms," *J. Heuristics*, vol. 4, no. 1, pp. 25-46, 1998.

[7] Estep, T. and M. Kritzman, "TIPP: Insurance without Complexity," *Journal of Portfolio Management*, Vol. 14, Issue 4, p38-42, Summer 1988.

[8] Etzioni, E.S., "Rebalance Disciplines for Portfolio Insurance," *Journal of Portfolio Management*, Vol. 13, Issue 1, p59-62, Fall 1986.

[9] Falkenauer, E., "The Grouping Genetic Algorithms — Widening The Scope of The GAs," *JORBEL — Belgian Journal of Operations Research, Statistics and Computer Science*, 33 (1,2), p79-102, 1992.

[10] Jones, D. R. and M. A. Beltramo, "Solving partitioning problems with genetic algorithms," *Proc. 4th Intnl. Conf. on Genetic Algorithms*, eds. K. R. Belew and L. B. Booker,Morgan Kaufmann, San Francisco, p442-449, 1991.

[11] Rubinstein, M. and H. E. Leland, "Replicating Options with Positions in Stock and Cash," *Financial Analysts Journal*, 37, p63-71, Jul/Aug 1981.

[12] Rubinstein, M., "Alternative Paths to Portfolio Insurance," *Financial Analysts Journal*, 41(4), p42-52, 1985.

[13] Smith, D., "Bin Packing with Adaptive Search," *In Proceeding of an International Conference on Genetic Algorithms and Their Application*, pp. 202-206, 1985.

[14] Johnson, D., C. Aragon, L. McGeoch, and C. Schevon, "Optimization by Simulated Annealing: An Experimental Evaluation; Part II, Graph Coloring and Number Partitioning," *Operations Research*, 39(3), 378–406, 1991.

[15] Holland, J., *Adaptation in Natural and Artificial System*, University of Michigan Press, 1975.

A Hybrid Genetic Algorithm for 2D FCC Hydrophobic-Hydrophilic Lattice Model to Predict Protein Folding

Md Tamjidul Hoque, Madhu Chetty, and Laurence S. Dooley

Gippsland School of Information Technology
Monash University, Churchill VIC 3842, Australia
{Tamjidul.Hoque, Madhu.Chetty,
Laurence.Dooley}@infotech.monash.edu.au

Abstract. This paper presents a Hybrid Genetic Algorithm (HGA) for the protein folding prediction (PFP) applications using the 2D face-centred-cube (FCC) Hydrophobic-Hydrophilic (HP) lattice model. This approach enhances the optimal core formation concept and develops effective and efficient strategies to implement generalized short pull moves to embed highly probable short motifs or building blocks and hence forms the hybridized GA for FCC model. Building blocks containing Hydrophobic (H) – Hydrophilic (P or Polar) covalent bonds are utilized such a way as to help form a core that maximizes the |fitness|. The HGA helps overcome the ineffective crossover and mutation operations that traditionally lead to the stuck condition, especially when the core becomes compact. PFP has been strategically translated into a multi-objective optimization problem and implemented using a swing function, with the HGA providing improved performance in the 2D FCC model compared with the Simple GA.

1   Introduction

Protein is a three dimensionally folded molecule composed of amino acids [1] linked together (called primary structure) in a particular order specified by DNA sequence of a gene. They are essential for functioning of the living cells as well as for providing structure. *Protein folding prediction* (PFP) is a problem of determining the native state of a protein from its primary structure and is of great importance [2] because three dimensionally folded structures determine the biological function and hence prove extremely useful in down streaming applications like drug design [3].

To investigate the underlying principles of protein folding, lattice protein models introduced by Dill [4] are highly regarded tools [5]. Protein conformation as a *self-avoiding walk* in the lattice model has been proven to be *NP-complete* [6] [7] so therefore a deterministic technique to folding prediction is not possible. Hence, a nondeterministic approach including robust strategies that can extract minimal energy conformations efficiently out of these models is of great importance. This is a very challenging task as there exist an inordinate number of possible conformations even for short amino acid sequences [8] [9].

So far, the most successful approach to the hard optimization problem like *PFP* is based on hybrid evolutionary approach [10]. A lattice model avoids the continuous conformational space that simplifies many of the required calculations. Among HP lattice models, 2D square and 3D cube models have been used mostly by the research

A. Sattar and B.H. Kang (Eds.): AI 2006, LNAI 4304, pp. 867–876, 2006.
© Springer-Verlag Berlin Heidelberg 2006

community [10-13] for the sake of simplicity but the parity problem within this model complicated the design approach without much benefit [12-13]. It was shown [17] that triangular model is parity problem free, that is an odd indexed amino acid or residue in the sequence position can be the neighbor of both odd and even indexed residues in the sequence and vice versa. Later, a full proof of the famous *Kepler Conjecture* [15] [16] problem was completed which implies that the *FCC* is the densest sphere-packing model, where a residue can have 12 neighbors in a 3D space and 6 neighbors forming a hexagon in 2D. Clearly, the most compact hydrophobic core (*H-Core*) [17] can be represented by *FCC* model [5].

Although triangular and *FCC* models have similar properties, it is shown Section 3 that the optimal *H-Core* is hexagonal rather than triangular. Therefore, while search is carried out using GA, the on going sub-optimal conformation (i.e. chromosome of the population) is guided towards the formation of a hexagonal *H-Core*. The highly likely motifs are remapped within the conformation based on a dynamic nuclei defined as *H-Core Centre* (HCC) [12-13, 18].

The pull move has been shown to be a very effective operator in [12-13, 19] for square and cube HP lattice. Here, in 2D *FCC*, we show that the pull move is easily implemented and that makes the deevloping conformation less destructive while highly likely motifs are mapped. The only difficulty is the handling of floating point operation in the *FCC* model, which can be overcome by an effective programming technique. As our main focus is to show the effectiveness of our strategies, the focus is confined within the 2D *FCC* model rather than the 3D model which in turn permits easy explanation. Further, the properties of *FCC* model relate well [20-21] to protein conformation amongst the known lattice models.

Several other outstanding nondeterministic approaches such as a number of versions *of Monte Carlo* (MC), *Simulated Annealing* (SA), *Tabu Search with GA* (GTB) and *Ant Colony Optimization* [23] are available for square and cube models. However, for the *FCC* model the GA is preferred as it consistently outperforms other approaches [10-13, 22], though it is noted that any approach including GA faces difficulties in the hard optimization problem like *PFP*. These will be investigated and overcome by using hybridization technqiues.

2   Two Dimensional FCC HP Lattice Model

Based on the observation that the hydrophobic forces dominate during protein folding, the HP model has been introduced by Dill [4] with amino acids being represented as a reduced set of *H* (Hydrophobic or Non-Polar) and *P* (Hydrophilic or Polar) only. The protein conformations of the sequence are placed as a *self-avoiding walk* (SAW) on a 2D hexagonal pattern in the *FCC* [20] model. The energy of a given conformation is defined as a number of topological neighboring (TN) contacts between those Hs, which are not sequential with respect to the sequence. The *PFP* can be formally defined as follows.

Given amino-acid sequence, $s = s_1, s_2, s_s, \cdots, s_m$, a conformation c needs to be formed where, $c^* \in C(s)$, energy $E^* = E(C) = \min\{E(c) \mid c \in C\}$ [23].

Here, m = total amino acids in the sequence and $C(s)$ is the set of all valid (i.e. SAW) conformations of s. If the number of TNs in a conformation c is q then the value of $E(c)$ is defined as $E(c) = -q$. In a 2D *FCC* HP model (Fig. 1, (a)), a non-terminal and a terminal residue both having 6 neighbors can have a maximum of 4 TNs and 5 TNs respectively.

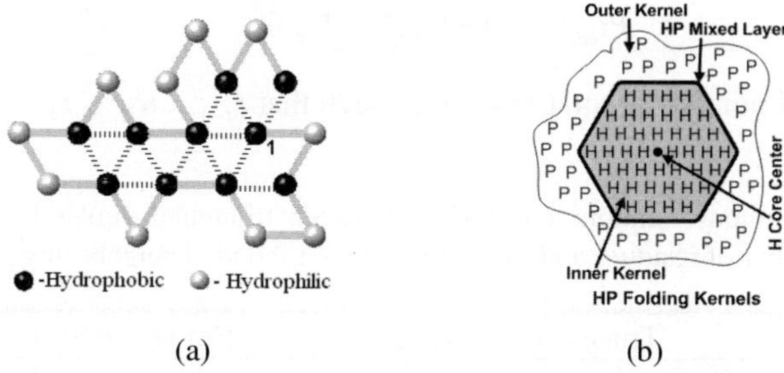

(a)   (b)

Fig. 1. (a) Conformation in 2D FCC HP model shown by solid line. Dotted line indicates TN. Fitness = -(TN Count) = -15. (b) 2D metaphoric HP folding kernels for the FCC model.

It is well known [17] that the *H*s form the protein core so freeing up energy while the *P*s, have an affinity with the solvent and so tend to remain in the outer surface. This paper visualizes the folded protein through the 2D *FCC* HP model as a three-layered kernel (Fig. 1(b)). The inner kernel called the *H-Core* [17] [24], is compact and mainly formed of *H*s while the outer kernel consists mostly of *P*s. The *H-Core Centre* (*HCC*) is defined in Section 4.2 as the average of the coordinates of *H*s. The composite thin layer between the two kernels consists of those *H*s that are covalent bonded with *P*s, which for the purpose of this paper is referred to as the *HP mixed layer*.

3   Optimal Shape of 2D H-Core in FCC Model

In this section, a proof is developed induction basis, for the optimum shape of the *H-Core* for FCC model compared with triangular model, rejecting other possible shapes for obvious reasons [12-13]. The sequence for the sake of this proof is assumed to be a segment of Hs only and it is a variation of that presented in [17, 12-13]. Table-1 shows that the *H-Core* tends to form a hexagonal rather than triangular shape and has maximal |fitness| whereas both approaches can have the same number of neighbors so that an optimal *H-Core* shape is hexagonal. The positioning of the *H*s inside the core (assuming a hexagonal boundary) can be categorized as *H* at the corner, *H* on the edge and *H* inside the interior which will respectively have 3, 4 and 6 neighbouring sides each.

Further, our concern is to compute the probability of an *H* to be appearing at a corner and on an edge. It is to be noted that for a hexagon, the number of residues at corner remain fixed but the number of residues on edge increase with the increasing size of the hexagon or the increasing number of *H*s. The total residues (T_t) within a hexagon in relation to 6 residues at corners and $6t$ residues on the edges can be expressed based on the recurrent equation (1), where $t = 0, 1, 2, 3, \ldots$ and $T_{-1} = 1$.

$$T_t = 6 + 6t + T_{t-1} \tag{1}$$

Therefore, the probability of an H being at corner is given by equation (2),

$$\Pr_{corner(t)} = \left(6/\left(T_t - T_{t-1}\right)\right) \tag{2}$$

and the probability of an H being on the edge is given by equation (3),

$$\Pr_{edge(t)} = 1 - \left(6/\left(T_t - T_{t-1}\right)\right) \tag{3}$$

where the actual number of total Hs is n_H, such that $T_{t-1} < n_H \leq T_t$.

Table 1. *H-Core* conformations comparison between triangular versus hexagonal, while m indicates the number of residues and losses indicate non-bonded neighboring positions

m	Triangular Shape		Hexagonal Shape	
	Conformation	Losses	Conformation	Losses
4		16		14
5		20		16
6		18		18
7		22		18

4 Highly Likely Sub-conformation for HP Mixed Layer

To form the cavity of *H-Core*, it is intuitive to think of placing the P of a -*HP*- segment on the opposite side of H with respect to the current *HCC*, while searching for the desired conformation. However, with such a straightforward placement, the cavity would tend to form a circular shape, which is not the desired hexagonal form. To address these problems, motif or sub-conformation that is highly probable to a sub-sequence (defined in Fig. 2) is forced to remap. The main idea is to form immediate TN and place P as far away as possible from *HCC* while concomitantly placing H as near as possible to *HCC*. To implement the same strategies in *FCC* HP model, various *moves* [12-13] are further simplified and merged which is the benefit of the *FCC* model over the HP-square or cube model.

Furthermore, the enforced placement of sub-conformations is not easy because their location in the lattice model is discrete and it can destruct already achieved sub-optimal sub-conformation. To address these problems, two broad categories of sub-sequences are defined; gS_H and gS_P, where $g \in$ N (N is natural number). These two categories completely cover the *HP mixed layer* including outer kernel. Let S_H

and S_p represent segments of H and P respectively. A segment refers to a contiguous string of length g, e.g. $2S_H$ means -*PHHP*-, i.e. $g = 2$ with the two boundary residues being of the opposite type. g is divided into even g_e and odd g_o numbers. For $1S_p$, $1S_H$, $2S_p$ and $2S_H$, there are few possible sub-conformations, so only highly potential sub-conformations (shown in Fig. 2) are chosen, based on embedded TN and core formation concepts. Collectively they are called *H-Core Boundary Builder Segments* (HBBS) and are mapped to potential sub-conformations which are referred to as the *H-Core Boundary Builder sub-Conformation* (HBBC). *HBBC* forms part of a corner (especially when $g = 1$ and through the composition with other group having $g = 2$) and an edge (especially when $g = 2$ and with the composition of the former group) of the *H-Core* boundary. The selection for mapping *HBBC* into *HBBS* can be probabilistically applied using (2) and (3). Due to the absence of the parity problem in the FCC model, we were able to simplify the desired sub-conformations and reduce their numbers compared to [12-13].

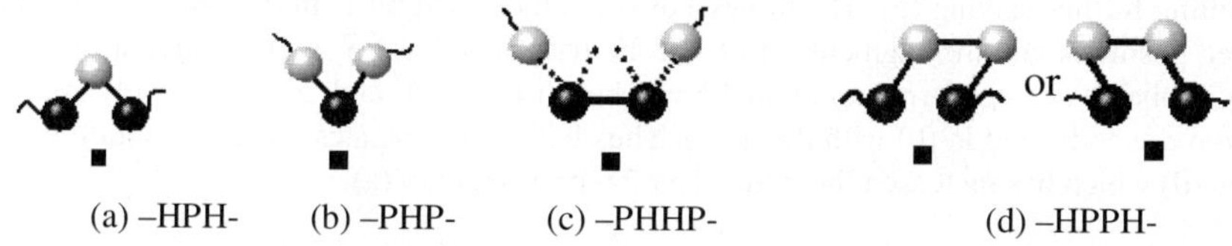

(a) –HPH–   (b) –PHP–   (c) –PHHP–   (d) –HPPH–

Fig. 2. Potential motifs sub-conformation for (a) $1S_p$ (b) $1S_H$ (c) $2S_p$ (d) $2S_H$. ●, ○ and ■ indicate an H, a P and the approximate position of HCC, respectively. Dotted line indicates alternate connection.

4.1 Probabilistic Constrained Fitness

While searching for an optimum conformation, if a sub-conformation corresponding to a particular sub-sequence exists in the *HP mixed layer* for a developing conformation, it is rewarded, otherwise penalized. This measure of fitness is referred to as the *Probabilistic Constrained Fitness* (PCF), so if any member of a *HBBC* corresponds to the related sub-sequence and the Hs are nearer to HCC than the Ps, then *PCF* will be decreased by 2 as reward, otherwise it will be penalized by an increase of 1 for a non-desired sub-conformation and 2 for a proper shape but having opposite of the desired directions (i.e. the position of Ps are closer to *HCC* than Hs).

4.2 Definitions for the Implementation of the Sub-conformations

To implement or remap *HBBC*, a number of terms need to be defined.

(i) *HCC* is the *H-Core centre*, calculated as the arithmetic mean of the coordinates of all Hs as in (4). Before enforcing to map a sub-conformation, the *HCC* (i.e. x_{HCC}, y_{HCC}) is updated to place H near *HCC* and P as far as possible from *HCC*.

$$x_{HCC} = \frac{1}{n_H} \sum_{i=1}^{n_H} x_i \text{ and } y_{HCC} = \frac{1}{n_H} \sum_{i=1}^{n_H} y_i \tag{4}$$

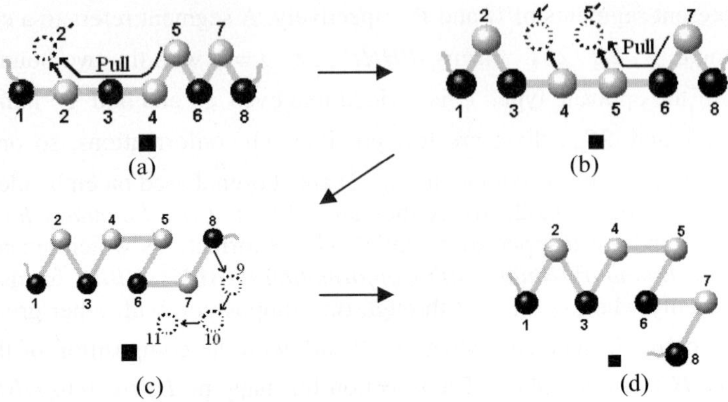

Fig. 3. The subsequence -123- in (a) need to remap to sub-conformation $2S_P$. If the position $2'$ is free then 2 can be placed at $2'$ and a pull (indicated in (a)) applied towards the higher indexed end. The pull moves 3 to 2, 4 to 3 and 5 to 4 and then finds a valid conformation without pulling further leaving (b). The |fitness| of (b) is increased by 1. In (b) assume, $4'$ and $5'$ are free positions and the segment 3 to 6 can be recognized as $2S_H$. To enforce a mapping to highly probable sub-conformation, 4 and 5 can be shifted to $4'$ and $5'$ respectively applying a pull move as indicated in (b) with the in (c). Thus 8 in (c) propagates through position 9, 10, 11 to give (d) which has increased the |fitness| by 2 with respect to (a).

(ii) *Pull move*: This has been shown to be very effective in [19] especially for enforcing any sub-conformation [12-13]. There are basically two-folded benefit: a) implementing the sub-conformations or motifs b) less distortion of other parts due to pulling required which may be in an optimal position as demonstrated in Fig. 3. For the *FCC* model we used the redefined pull move for the same purpose. As the parity problem is absent in *FCC* model, the pull move does not need to be moved diagonally to start as an ordinary pull to the next neighbor performs the same. This is because in *FCC* without the parity problem and with more neighbors, it is very likely to get a valid conformation, without need to propagating the pull often up to the terminal residue.

5　Implementation and Experiments

Although the additional constraint (*PCF*) formulates the multi-objectivities, the implementation is such that it ultimately maximizes the goal of original fitness $|F|$. The search process is mainly divided into two alternative phases namely, Phase 1 in which F dominates *PCF* and starts building the core. In the alternate Phase 2, *PCF* dominates which takes care of the proper formation of the *HP* mixed layer. Further, the enforcement to *HBBC* is performed in phase 2 since *PCF* helps the change sustain and stabilize. The *HBBC* implantation is done only if they are not found according to the highly likely sub-conformations for the corresponding sub-sequences. This action may reduce the already achieved fitness F, but it is expected that it will help reformulate a proper cavity that will maximize the H bonding inside core, which while shifting to the favorable phase would maximize the fitness $|F|$. As the phases alternate throughout the search process, the impact becomes such that F and PCF come up with common goal that is highly likely to be optimal. The total or combined fitness is defined as,

$$TF = \alpha(t) * F + \beta(t) * PCF \tag{5}$$

where t is t^{th} generation while search is carried out by GA. To alter the weight of α and β to dominate F and PCF over each other, a swing function (equation (6)) is used.

$$\delta(t) = A\,(1 + \cos \omega_m t)\cos \omega_0 t \tag{6}$$

where $\omega_m << \omega_0$, t = number of generation. The assignment of α and β is as,

$$\text{Phase 1: } \alpha(t) = \delta(t), \beta(t) = 1, \text{ when } \delta(t) > 0 \tag{7}$$

$$\text{Phase 2: } \alpha(t) = 1, \beta(t) = -\delta(t), \text{ when } \delta(t) < 0 \tag{8}$$

$$\text{Transient Phase: } \alpha(t) = 1, \beta(t) = 1, \text{ when } \delta(t) = 0 \tag{9}$$

For the typical value of $\delta(t)$ parameters are set as follows: amplitude $A=30$, $\omega_m =$ 0.004 and $\omega_0 =0.05$. The value of A is selected as, $2A \geq \max\left(|F|, |PCF|\right)$ where the upper limit of F is set using (10), which has been extended from [14].

$$F = -\left(2n_H + n_T\right) \tag{10}$$

Here, n_H is the total number of hydrophobic residues in a sequence and n_T is the number of hydrophobic residues at the terminal positions and $0 \leq n_T \leq 2$. Note, the minimum value of both $|\alpha(t)|$ and $|\beta(t)|$ equal 1 are maintained and never set to zero in (7), (8) and (9), so preserving the sub-conformation or schema developed in the alternate phase, possessing good features.

The search procedure is given in Algorithm-I. A simple GA which is hybridized with population size [27] of 200 is chosen for all sequences. The elite rate = 0.10, $p_c = 0.85$, $p_m = 0.5$ and a single point mutation by pivot rotation [10-11] is applied.

Table 2. Predictability of SGA versus HGA maximum |fitness| from 10 runs is shown

Sequence	SGA	HGA	Conformation (HGA)
HPHPPHHPHPPHPHHPPHPH	-11	-15	Fig.4 (a)
HHPPHPPHPPHPPHPPHPPHPPHH	-10	-13	Fig.4 (b)
PPHPPHHPPPPHHPPPPHHPPPPHH	-10	-10	Fig.4 (c)
P3(H2P2)2P3H7P2H2P4H2P2HP2	-16	-19	Fig.4 (d)
P2(HP2H)2HP510HP5(H2P2)2HP2H5	-26	-32	Fig.4 (e)
H(HP)4H4PHP3HP3HP3HP4HP3HP3HPH4(PH)4H	-21	-23	Fig.4 (f)
P2H3PH8P3H10PHP3H12P4H6PH(HP)2	-40	-46	Fig.4 (g)
H12(PH)2((P2H2)2P2H)3PHPH12	-33	-46	Fig.4 (h)
HHHPPHPHPHPPHPHPHPPH	-11	-14	Fig.4 (i)

The implementation of crossover and mutation is same as in [10-13] but without any special treatment (e.g. cooling). Roulette wheel is used for selection procedure.

Simulations are carried out for benchmark 2D problems [25]. For each of the sequences, standard Simple Genetic Algorithm (SGA) [26] and HGA run parallelly together, and stop when all of them become non-progressive. Therefore they run for equal amount of time. As the parameters of the developed HGA were not tuned, we avoid comparison based on the number of iterations required. The goal is to compare the predictability of the developed HGA approach. Results are shown in Table 2 with the conformations corresponding to the maximum |fitness| achieved, for future comparison. HGA outperformed predictabilities of SGA significantly.

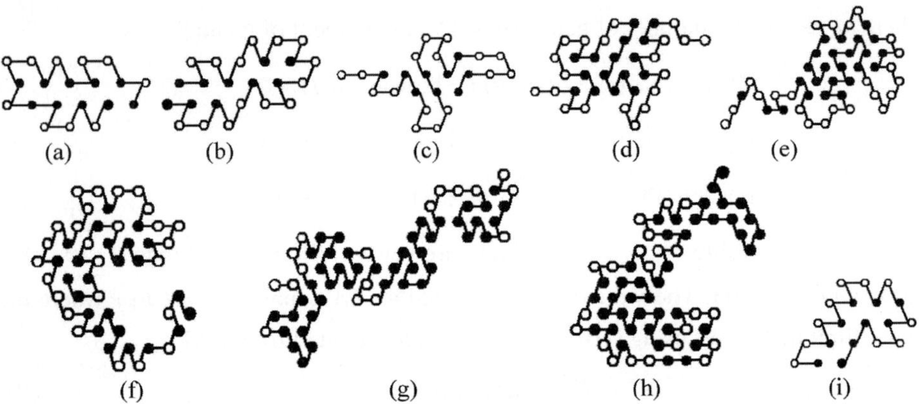

Fig. 4. (a) to (i) correspond to the conformation with maximum |fitness| achieved using HGA as indicated in Table-2

Algorithm-I. HGA for PFP Using 2D FCC Model

Input: Sequence S,
Output: Fitness of the optimum 2D FCC conformation.
  COMPUTE ***PCF;*** COMPUTE ***A*** (amplitude)
    t=0, F=0   /* Generation count and fitness initialization */
    Fillup the population with random (valid) conformation possible for S.
  While $F \neq Higher\_Target\_Value\_of\_F$ THEN
  { t = t + 1, COMPUTE $\delta(t)$, $\alpha(t)$, $\beta(t)$, TF
     CROSSOVER and then MUTATION
     IF $\delta(t) < 0$ THEN
         { FOR i =1 to *population_size* DO
           Check *chromosome_i* for any miss mapping of HBBC
             IF miss-mapping true then
           { Re-map the sub-sequence to corresponding HBBC using move-sets. }}
     COMPUTE *TF*
         Sort, Keep Elite, Reduce Twins.
       F $\leftarrow$ Best fitness found from the population.
  } END.

6 Discussion and Conclusions

There are two major drawbacks with SGA for *PFP*. GA computation is based on schema theorem, which states [10] that short, flexible schemata with above average performance will receive fast survival chance in the subsequent generations and schemata with below-average performance will decay nonlinearly fast too. So, one obstacle using SGA is that similarity within population grows which leads to stall or stuck condition, since crossover will most likely occur between twins. Those which are mutated are likely to be heavily dissimilar and would therefore be rejected by the selection process. To address this problem, twin removal was applied and elitism was used only to keep the found best. Secondly, as the optimum conformation is relatively compact, crossover and mutation confront more increasing collision and produce invalid conformation. Our specific implantation procedure of *HBBCs* moves the compact conformation without collision and the introduced move operator causes less destruction to the already gained fitness. The move creates probable reformation of the *H-Core* cavity to maximize the H-sides inside the *H-Core*. Hence, this approach makes change of the non-progressive situation in such a way, that it enhances the chance of gaining highly fitted conformations.

The novel strategies using HGA has also been extended for 2D *FCC* HP. It is shown that beyond removing the parity problem, *FCC* further simplifies the pull moves and reduces the need of higher number sub-conformations for remapping. As there is lack of previous works on 2D *FCC*, we compared the proposed HGA with the standard SGA, which shows significant improvement over predictability. Regarding future scope, parameter of HGA and the swing function are to be investigated further for optimization. It can also be extended for 3D *FCC* and real *PFP* as well. The overall framework we developed is robust enough and removes the causes of failure.

References

1. Allen, et al.: Blue Gene: A vision for protein science using a petaflop supercomputer. IBM System Journal (2001), Vol 40, No 2
2. Pietzsch,, J.: The importance of protein folding, http://www.nature.com/horizon/proteinfolding/background/importance.html.
3. Petit-Zeman, S.: Treating protein folding diseases, http://www.nature.com/horizon/proteinfolding/background/treating.html.
4. Dill, K.A.: Theory for the Folding and Stability of Globular Proteins. Biochem.,1985, 24: 1501
5. Backofen, R. and Will, S.: A constraint-based approach to fast and exact structure prediction in three-dimensional protein models. Journal of Constraints, 11 (1) pp. 5-30, Jan. 2006.
6. Crescenzi, P. and et al.: On the complexity of protein folding (extended abstract). ACM, International conference on Computational molecular biology, 1998, pp.597-603.
7. Berger, B. and Leighton, T.: Protein Folding in the Hydrophobic-Hydrophilic (HP) model is NP-Complete. ACM, International conference on Computational molecular biology (1998)
8. Schiemann, R., Bachmann, M. and Janke, W.: Exact Enumeration of Three – Dimensional Lattice Proteins. Elsevier Science, 2005.

9. Guttmann, A.J.: Self-avoiding walks in constrained and random geometries. In "Statistics of Linear Polymers in Disordered Media" ed. B. K. Chakrabarti, Elsevier, 2005, pp:59-101.
10. Unger, R. and Moult, J.: On the Applicability of Genetic Algorithms to Protein Folding. IEEE, 1993, pp. 715-725.
11. Unger, R. and Moult J.: Genetic Algorithm for Protein Folding Simulations, JMB, 1993.
12. Hoque, M. T., Chetty, M. and Dooley, L. S.:A New Guided Genetic Algorithm for 2D Hydrophobic-Hydrophilic Model to Predict Protein Folding. 2005 IEEE CEC, pp. 259-266.
13. Hoque, M. T., Chetty, M. and Dooley, L. S.:A Guided Genetic Algorithm for Protein Folding Prediction Using 3D Hydrophobic-Hydrophilic Model, 2006 IEEE WCCI.
14. Agarwala, R and et. al.: Local Rules for Protein Folding on a Triangular Lattice and Generalized Hydrophobicity in the HP Model. J. of Computational Biology, 4(2) pp. 275 – 296, 1997.
15. Sloane, N. J. A.: Kepler's Conjecture Confirmed. Nature 395, 435-436, 1998.
16. Hales, T. C.: A proof of the Kepler conjecture. Annals of Mathematics, 162, no. 3 (2005), 1065–1185.
17. Dill, K.A., Fiebig, K.M., Chan H.S.: Cooperativity in Protein-Folding Kinetics. Proceedings of the National Academy of Sciences USA, Biophysics (1993) Vol 90, pp.1942-1946
18. Shmygelska, A.:Search for Folding Nuclei in Native Protein Structures. Bioinformatics, 2005, 21: i394 - i402.
19. Lesh, N., Mitzenmacher, M. and Whitesides, S.: A Complete and Effective Move Set for Simplified Protein Folding. RECOMB, 2003, Berlin.
20. Backofen, R. and Will, S.: Optimally Compact Finite Sphere Packings -- Hydrophobic Cores in the FCC. CPM2001, LNCS, pp. 257-272.
21. Koehl, P. and Levitt, M.:A brighter future for protein structure prediction, Nature Structural Biology 6, 108 – 111, 1999.
22. Unger R. and Moult, J.: Genetic Algorithm for 3D Protein Folding Simulations. 5th International Conference on Genetic Algorithms, 1993, pp. 581-588.
23. Shmygelska, A. and Hoos, H.H.:An ant colony optimization algorithm for the 2D and 3D hydrophobic polar protein folding problem. BMC Bioinformatics 2005, 6(30).
24. Flebig and Dill.: Protein Core Assembly Processes. J. Chem. Phys., 1993, 98(4), 3475-87.
25. Hart, W. and Istrail, S.: HP Benchmarks, http://www.cs.sandia.gov/tech_reports/compbio/tortilla-hp-benchmarks.html
26. Vose, M. D.:The Simple Genetic Algorithm. 1999, The MIT Press.
27. Digalakis, J. G. and Margaritis, K. G.: An experimental Study of Benchmarking Functions for Genetic Algorithms, Intern. J. Computer Math., 2002, 79(4), 403-416.

Locality Preserving Projection on Source Code Metrics for Improved Software Maintainability

Xin Jin[1], Yi Liu[1], Jie Ren[1], Anbang Xu[2], and Rongfang Bie[1,*]

[1] College of Information Science and Technology,
Beijing Normal University, Beijing 100875, P.R. China
[2] Image Processing & Pattern Recognition Laboratory,
Beijing Normal University, Beijing 100875, P.R. China
xinjins@yahoo.com,
liuyi58@163.com,
henry_331166@163.com,
anbangxu@mail.bnu.edu.cn

Abstract. Software project managers commonly use various metrics to assist in the design, maintaining and implementation of large software systems. The ability to predict the quality of a software object can be viewed as a classification problem, where software metrics are the features and expert quality rankings the class labels. In this paper we propose a Gaussian Mixture Model (GMM) based method for software quality classification and use Locality Preserving Projection (LPP) to improve the classification performance. GMM is a generative model which defines the overall data set as a combination of several different Gaussian distributions. LPP is a dimensionality deduction algorithm which can preserve the distance between samples while projecting data to lower dimension. Empirical results on benchmark dataset show that the two methods are effective.

1 Introduction

Software project managers commonly use various metrics to assist in the design, maintaining and implementation of large software systems. Software quality analysis, which is an important aspect of software maintain, is an ongoing comparison of the actual quality of a product with its expected quality [16]. Better software quality analysis may result in improved software project maintainability. Software metrics are the key tool in software quality analysis. Many researchers have used data mining techniques to analyze the connection between software metrics and code quality [1, 8, 10, 12, 13, 14, 15, 17, 20]. These methods include: association analysis (association rules), clustering analysis (k-means, fuzzy c-means), classification analysis (Support Vector Machines (SVM), Decision Trees (DT), OneR, Nearest Neighbor, Naïve Bayes, Artificial Immune System (AIS), layered neural networks, Holographic networks, logistic regression, genetic granular classification, etc) and regression analysis. For classification, software metrics are the features and expert quality rankings the class labels.

---

* Corresponding author. rfbie@bnu.edu.cn

A. Sattar and B.H. Kang (Eds.): AI 2006, LNAI 4304, pp. 877–886, 2006.

© Springer-Verlag Berlin Heidelberg 2006

In this paper we propose to build Gaussian mixture model for software quality classification. Gaussian Mixture Model (GMM) technique has attracted many researchers in data mining and machine learning areas [5, 6, 7, 8]. We also propose to use Locality Preserving Projection (LPP) for extracting source code metrics to improve the performance of software quality classifiers.

The remainder of this paper is organized as follows. Section 2 describes Gaussian mixture model for software metrics. Section 3 presents two baseline classification algorithms for comparison. Section 4 presents the software metrics and the benchmark dataset. Section 5 presents the Locality Preserving Projection method. Section 6 presents performance measures and results. Conclusions are covered in Section 7.

2 Gaussian Mixture Model for Metrics Vector

The Gaussian Mixture Model (GMM) is a generative model. It defines the overall data set as a combination of several different Gaussian distributions. The parameters of the model are determined by the training data. Once the parameters are selected, the model can be used to predict the test data [4, 22, 24].

Given an ensemble of training source code metrics vectors, $X = \{x_1,...,x_m\}$, where $x_i \in R^d$ and assuming that the m vectors are statistically independent and identically distributed, the likelihood that the entire ensemble has been produced by class C_1 is,

$$P(X = \{x_1,...,x_m\} \mid C_1\}) = \prod_{i=m,1} P(x_i \mid C_1) \qquad (1)$$

GMM assumes that the likelihood of a vector can be expressed with a mixture of r Gaussian distributions/components,

$$P(x_i \mid C_1) = \sum_{j=1}^{r} P(j \mid C_1)P(x_i \mid j, C_1) \qquad (2)$$

where,

$$P(x_i \mid j, C_1) = \frac{1}{\sqrt{(2\pi)^d \mid \Sigma_{j,1} \mid}} \exp(-\tfrac{1}{2}(x_i - \mu_{j,1})^T \sum\nolimits_{j,1}^{-1}(x_i - \mu_{j,1})) \qquad (3)$$

where $P(j|C_1)$ is the prior probability of Gaussian j for class C_1 , and $P(x_i |j, C_1)$ is the likelihood of vector x_i being produced by Gaussian j within class C_1. The parameters of this Gaussian distribution are the mean vector $\mu_{j,1}$ and the diagonal covariance matrix $\Sigma_{j,1}$.

During training, basing on all the vectors for a given class, the task is to learn the parameters of the Gaussian mixture, i.e. the mixing weights, the mean vectors and the diagonal covariance matrices. The most widely used method to achieve this goal using the well-known Expectation-Maximization (EM) algorithm.

EM is an iterative algorithm that computes maximum likelihood estimates [23]. The EM algorithm works by using the log likelihood expression of the complete data set. The maximization is repeatedly performed over the modified likelihood [22]. Basically, the EM algorithm iteratively modifies the GMM parameters to decrease the negative log likelihood of the data set [21]. The class of each data vector x is known during the training process. For more details on the maximization steps please refer to [23].

The initialization of the model is performed by the k-means algorithm [11]. First, a rough clustering is done; and the number of clusters is determined by the number of printer models in the training set. Then each data point is assumed to belong to the closest cluster center. Initial prior probability, mean, and variances are calculated from these clusters. After that, the iterative part of the algorithm starts. The iteration will stop until there is no (or little) change to the parameters, then the mixture parameters are found.

Once the Gaussian mixture parameters for each class (that is, C_1 = fault-free, C_2 = fault-low, C_3 = fault-high) have been found, determining a test software module's source code metrics vector's class is straightforward. A test vector x is assigned to the class that maximizes $P(C_k|x)$, which is equivalent to maximizing $P(x|C_k)P(C_k)$ using Bayes rule,

$$P(C_k \mid x) = \frac{P(x \mid C_k)P(C_k)}{P(x)} \tag{4}$$

Eventually, the test vector x is classified into the class C_k that maximizes $P(x|C_k)$. Then the quality class of the test software module is obtained.

3 Traditional Methods

SVM: Support Vector Machine (SVM) is well founded theoretically because it is based on well developed statistical learning theory [6]. A SVM selects a small number of critical boundary samples from each class and builds a linear discriminant function (also called maximum margin hyperplane) that separates them as widely as possible. In the case that linear separation is impossible, the technique of kernel will be used to automatically inject the training samples into a higher-dimensional space, and to learn a separator in that space. In our experiments, we use the RBF kernel function: $K(\mathbf{x}, \mathbf{y}) = \exp(-\gamma \|\mathbf{x} - \mathbf{y}\|^2)$.

DT: Decision Tree (DT) is a famous inductive learning algorithm [7]. The nodes of the tree correspond to attribute (in our case software metrics) test, the links (to attribute values and the leaves) to the classes. To induce a DT, the most important attribute is selected according to attribute selection criteria and placed at the root; one branch is made for each possible attribute value. This divides the training samples into subsets. The process is repeated recursively for each subset until all instances at a node have the same classification, in which case a leaf is created. To classify a test sample we start at the root of the tree and follow the path corresponding to the example's values until a leaf node is reached and the classification is obtained.

4 MIS Source Code Metrics Data

The Medical Imaging System (MIS) is a commercial software system consisting of approximately 4500 routines written in about 400,000 lines of Pascal, FORTRAN and

PL/M assembly code. The MIS data were collected as the number of changes made to each module due to faults discovered during system testing and maintenance over an observation period of three years. Along with the above parameter, eleven software complexity measures (metrics) were provided [9, 16]. In this study, MIS is represented by a subset of the whole MIS data with 390 modules written in Pascal and FORTRAN. These modules consist of approximately 40,000 lines of code.

There are 12 source code metrics used in the MIS data [13]. Simple counting metrics such as the number of lines of source code or Halstead's number of operators and operands describe how many "things" there are in a program. More complex metrics such as McCabe's cyclomatic complexity or Bandwidth attempt to describe the "complexity" of a program, by measuring the number of decisions in a module or the average level of nesting in the module, respectively.

Software modules, which have no changes, could be deemed to be *fault-free*, and software modules with the number of changes being between 1 and 10 to be *fault-low*, while software modules with the number of changes being over 10 to be *fault-high*, which can be sought as potentially highly faulty modules where most of our testing and maintenance effort should be focused. Thus, we have three classes: fault-free, fault-low and fault-high. Our task is to develop a prediction model of software quality class on the basis of the values of the source code metrics.

5 Locality Preserving Projection

We believe feature extraction based dimensionality reduction can find a good low-dimension representation of the original source code metrics data by finding a good linear or nonlinear projection. To project to lower dimension, the distance between data should be preserved and thus it can be calculated to reflect the similarity of the original data in original space. Locality Preserving Projection (LPP) is the algorithm which is equipped with this property [2].

LPP is a linear approximation of the nonlinear Laplacian Eigenmap [3]. The algorithm procedure is formally stated below same as [5]:

1. Constructing the adjacency graph: Let G denote a graph with m nodes. We put an edge between node i and j if x_i and x_j are "close". In this paper, we use a "Supervised" method for constructing the adjacency graph: Nodes i and j are connected by an edge if they are in the same class.

2. Choosing the weights: Here, as well, we have two variations for weighting the edges. W is sparse symmetric $m{\times}m$ matrix with W_{ij} having the weight of the edge joining vertices i and j, and 0 if there is no such edge.

    (a) Heat kernel. [parameter $t \in \mathbf{R}$]. If nodes i and j are connected, put $W_{ij} = exp(-\|x_i - x_j\|^2/t)$. We call this weight method as "Kernel."

    (b) Simple-minded. [No parameter]. $W_{ij} = 1$ if and only if vertices i and j are connected by an edge. We call this weighting method as "Binary."

According to the variations of choosing the weights, we divide LPP into two sub-methods: LPP-SK and LPP-SB. In this paper, both sub-methods are investigated.

3. Eigenmaps: Compute the eigenvectors and eigenvalues for the generalized eigendecomposition problem:

$$XLX^{T} a = \lambda XDX^{T} a \tag{5}$$

where D is a diagonal matrix whose entries are column (or row, since W is symmetric) sums of W, $D_{ii} = \Sigma_j W_{ji}$. $L = D - W$ is the Laplacian matrix of the graph represented by W. The *ith* column of the matrix X is x_i. Let the column vectors $a_1, \dots , a_l$ be the solutions of Eq. (5), ordered according to their eigenvalues, $\lambda_1 < \cdots < \lambda_l$. Thus, the embedding is as follows:

$$x_i \rightarrow y_i = A^T x_i, A = (a_1, a_2, \cdots , a_l) \tag{6}$$

where y_i is a l-dimensional vector, and A is a $n \times l$ matrix.

6   Experiment Results

We estimated the prediction performance of the methods on the benchmark MIS dataset, available on [9]. We tested both of the two sub-types of supervised LPP: LPP-SK, LPP-SB, as described in Section 5. GMM model is compared with two traditional methods: Support Vector Machines (SVM) and Decision Tree (DT).

We performed 10-fold cross validation, i.e., each run takes one of the 10-folds as the test set and the remaining 9 folds as the training set. To ensure fair comparisons, all the methods are evaluated using the same folding.

The performance is evaluated by two measures, Weighted Average Precision (WAP) and Recall (REC).

Weighted Average Precision (WAP): Precision is the percentage of the predicted libraries for a given category that are classified correctly. The WAP is the average of all the precisions:

$$WAP = \frac{n_i \cdot \sum_{i=1}^{n} Precision_i}{n} \tag{7}$$

where $Precision_i$ is the sensitivity for class i, n is number of overall samples, n_i is the number of samples for class i. A perfect classifier should give a WAP of 1. The higher the WAP, the better the classifier.

Recall (REC): Recall of a class is defined by the ratio of the number of correct predictions of the class and the number of all samples of the class. A perfect classifier should give a REC of 1 for each class. The higher the REC, the better the classifier. We use REC of class *fault-high* to evaluate the ability of learning methods to find all the potentially highly faulty modules where most of our testing and maintenance effort should be focused.

Fig. 1 shows the best WAP for LPP based classifiers and the original classifiers. Supervised LPP, that is, LPP-SK and LPP-SB, can improves the performance of all three classifiers: GMM, SVM, and DT. LPP-SB based GMM achieves the highest WAP of 70.6%. GMM is better than the other two classifiers in achieving higher WAP.

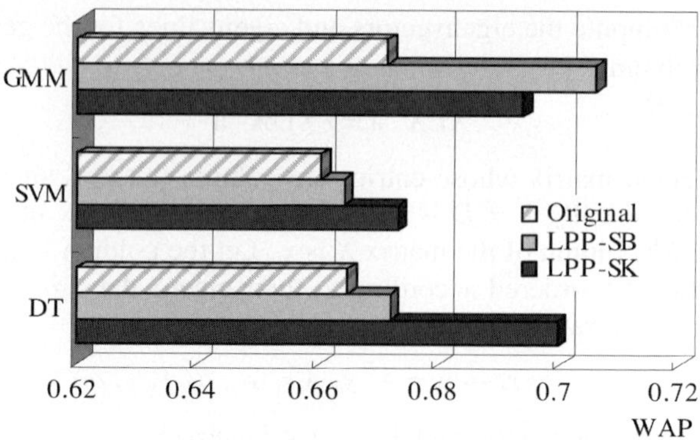

Fig. 1. Best WAP for LPP based classifiers and the original classifiers (Original)

Fig. 2 shows the best REC of class *fault-high* for LPP based classifiers and the original classifiers. Both LPP-SK and LPP-SB can improve the performance of the three classifiers: GMM, SVM, and DT. GMM is better than the other two classifiers in finding out most fault-high software modules. Supervised LPP based GMM achieves the highest REC (for *fault-high*) of 64.7%.

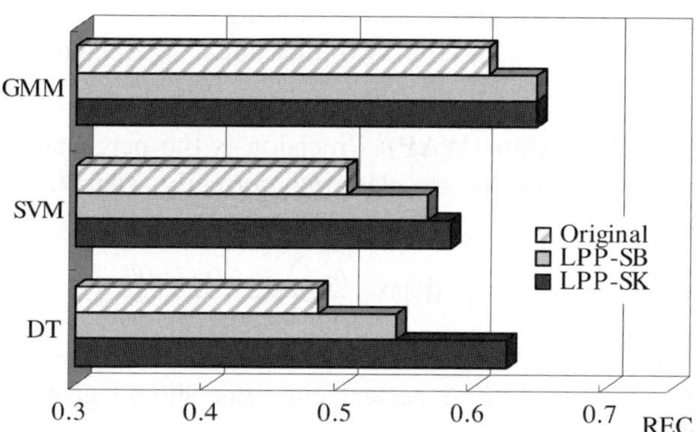

Fig. 2. Best REC of class *fault-high* for LPP based classifiers and the original classifiers (Original)

Fig. 3 and 4 show the WAP and REC of *fault-high* curves with different extracted metrics by LPP. As we can see from the curves, the three classifiers' performances are improved when we use LPP to transform the original 11-dimensional software metrics data to lower dimensions. LPP-SK is better than LPP-SB. In many cases, with only a few LPP extracted features, we can still achieve relatively high performance.

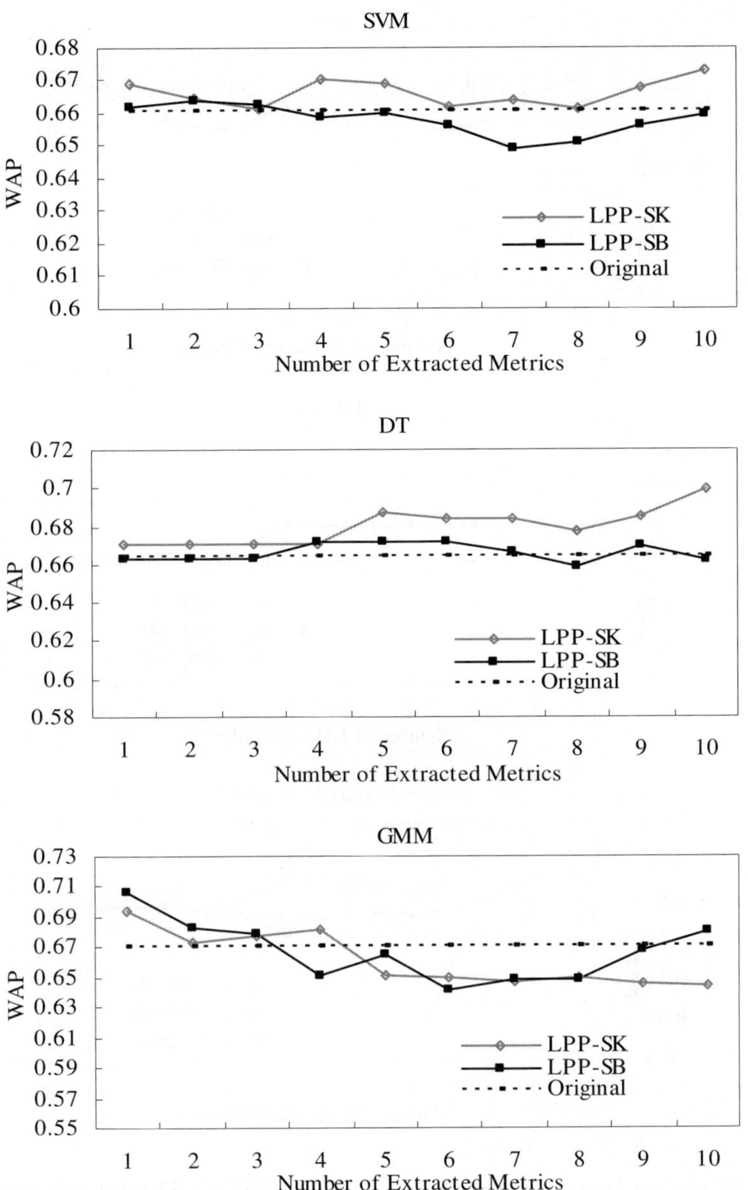

Fig. 3. WAP curves of GMM, SVM and DT with different number of extracted metrics by LPP. "Original" is without dimensionality reduction.

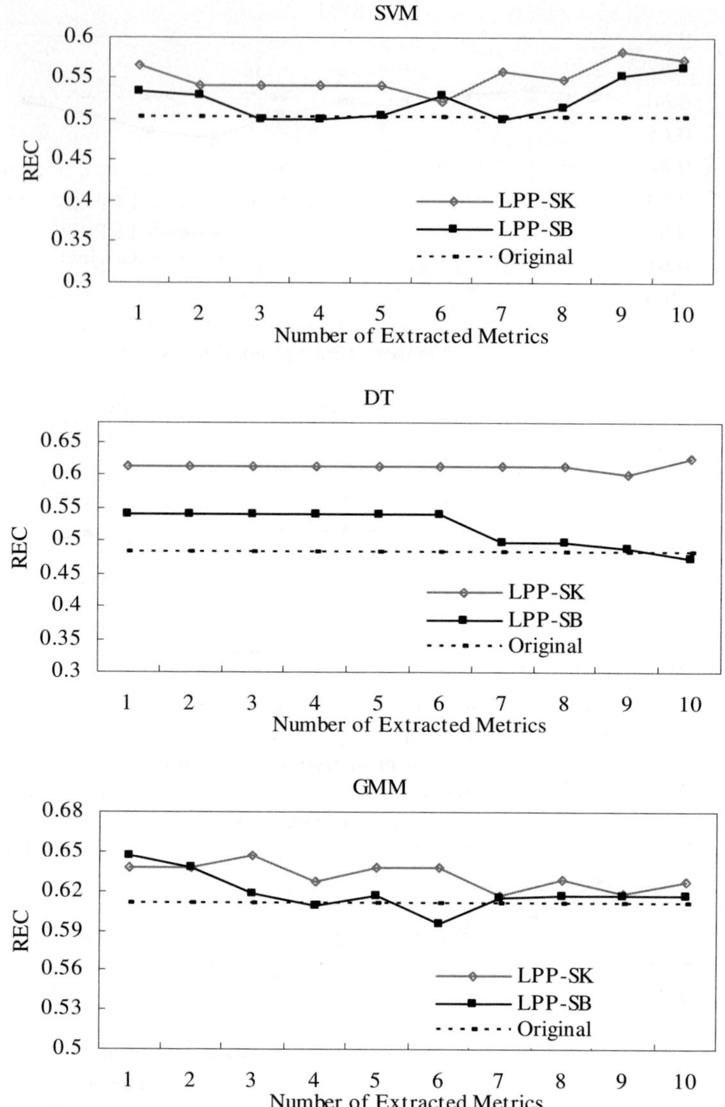

Fig. 4. REC of class *fault-high* curves for GMM, SVM and DT under different extracted metrics by LPP. "Original" is without dimensionality reduction.

7 Conclusions

One important aspect of efficient software project maintaining is source code based software quality classification, where software metrics are the features and expert quality rankings the class labels. In this study we introduce the Gaussian Mixture

Model (GMM) method for this task and use Locality Preserving Projection (LPP) to improve the classification performance. GMM is a generative model which defines the overall data set as a combination of several different Gaussian distributions. LPP is a dimensionality deduction algorithm which can preserve the distance between samples while projecting data to lower dimension. Experiment results indicate that LPP can improve the classification performance of all three learners: GMM, SVM, and DT. Basing on LPP, GMM is the best in finding fault-high software modules and in achieving high prediction precision.

Future work will consider additional learning techniques, and compare the performances with those obtained in this study. In addition, as suggested by one of the reviewers of this paper, whether the metric order when transforming high dimension to low dimension affects the classification performance deserves investigation.

Acknowledgments

This work was supported by the National Science Foundation of China under the Grant No. 10001006 and No. 60273015.

References

1. F. Xing, P. Guo, and M.R. Lyu: A novel method for early software quality prediction based on support vector machine. In Proceedings 16th International Symposium on Software Reliability Engineering, ISSRE'2005, Chicago, Illinois, November 8-11 (2005)
2. X. He, D. Cai, and W. Min: Statistical and computational analysis of locality preserving projection. In International Conference on Machine Learning (ICML), Bonn, Germany (2005)
3. M. Belkin and P. Niyogi. Laplacian eigenmaps and spectral techniques for embedding and clustering. In T. G. Dietterich, S. Becker, and Z. Ghahramani, editors, Advances in Neural Information Processing Systems 14, Cambridge, MA, MIT Press. (2002)
4. Marques Janet, Moreno Pedro J.: A Study of Musical Instrument Classification using Gaussian Mixture Models and Support Vector Machines. Compaq & DEC Technical Reports, CRL-99-4 (1999)
5. X. He and P. Niyogi: Locality preserving projection. In Advances in Neural Information Processing Systems 16 (NIPS 2003), Vancouver, Canada, MIT Press. (2003)
6. Corinna Cortes and Vladimir Vapnik: Support-vector Networks. Machine Learning, 20(3):273-297 (1995)
7. Ross Quinlan: C4.5: Programs for Machine Learning. Morgan Kaufmann Publishers, San Mateo, CA. (1993)
8. D. Garmus, D. Herron: Measuring the Software Process, Prentice Hall, Upper Saddle River, NJ (1996)
9. MIS: http://www.win.tue.nl/~jromijn/2IW30/2IW30_statistics/LYU/DATA/CH12 (Access 2006)
10. Xin Jin, Rongfang Bie, Xiaozhi Gao: An Artificial Immune Recognition System-based Approach to Software Engineering Management with Software Metrics Selection. Sixth International Conference on Intelligent System Design and Application (ISDA06) (2006)
11. Duda, R.O. and Hart, P.E.: Pattern Classification and Scene Analysis, John Wiley & Sons, New York. (1973)

12. Witold Pedrycz, Giancarlo Succi: Genetic Granular Classifiers in Modeling Software Quality. Journal of Systems and Software 76(3): 277-285 (2005)

13. W. Pedrycz, G. Succi, M. G. Chun: Association Analysis of Software Measures, Int. J of Software Engineering and Knowledge Engineering, 12(3): 291-316 (2002)

14. K.H. Muller, D.J. Paulish: Software Metrics. IEEE Press/Chapman & Hall, London (1993)

15. J. C. Munson, T. M. Khoshgoftaar: Software Metrics for Reliability Assessment, in Handbook of Software Reliability and System Reliability, McGraw-Hill, Hightstown, NJ (1996)

16. Scott Dick, Aleksandra Meeks, Mark Last, Horst Bunke, Abraham Kandel: Data Mining in Software Metrics Databases. Fuzzy Sets and Systems 145(1): 81-110 (2004)

17. W. Pedrycz, G. Succi, P. Musilek, X. Bai: Using Self-Organizing Maps to Analyze Object Oriented Software Measures. J. of Systems and Software, 59, 65-82 (2001)

18. P. K. Simpson: Fuzzy Min-Max Neural Networks. Part 1: Classification, IEEE Trans. Neural Networks, Vol. 3, pp. 776-786 (1992)

19. M.S. Bazaraa, H.D. Sherali, and C.M. Shetty: Nonlinear Programming: Theory and Algorithms, John Wiley &Sons Inc., New York (1993)

20. R. Subramanyan and M.S. Krishnan: Empirical Analysis of CK Metrics for Object-Oriented Design Complexity: Implications for Software Defects, IEEE Trans. Software Eng., Vol. 29, pp. 297-310, Apr (2003)

21. I. T. Nabney: Netlab: Algorithms for Pattern Recognition. Springer, pp. 79-113 (2001)

22. Gazi N. Ali, Pei-Ju Chiang, Aravind K. Mikkilineni, George T.-C. Chiu, Jan P. Allebach, Edward J. Delp: Application of principal components analysis and Gaussian mixture models to printer identification. In Proceedings of the IS&T's NIP20: International Conference on Digital Printing Technologies, pp. 301-305 (2004)

23. R. A. Render and H. F. Walker: Mixture Densities, maximum likelihood, and EM algorithm. SIAM review, Vol. 26, no. 2, pp. 195-239. April (1984)

24. H. Permuter and J.M. Francos and I.H. Jermyn: Gaussian Mixture Models of Texture and Colour for Image Database Retrieval. In Proc. IEEE International Conference on Acoustics, Speech and Signal Processing (ICASSP), Hong Kong, April (2003)

Ensemble Prediction of Commercial Bank Failure Through Diversification of Input Features

Sung Woo Shin[1,*], Kun Chang Lee[1], and Suleyman Bilgin Kilic[2]

[1] School of Business Administration, Sungkyunkwan University,
110-745, Seoul, Korea
`shinswoo@{skku.edu, kaist.ac.kr}`,
`kunchanglee@naver.com`
[2] Faculty of Economic and Administrative Science, Cukurova University
01330 Balcali, Adana, Turkey
`sbilgin@cu.edu.tr`

Abstract. As primary focus of banking regulation and supervision is being shifted toward internal risk management for all commercial banks, financial data mining task such as an early warning of bank failure becomes more critical than ever. In this study, we examine the effect of variable selection methods for intelligent bankruptcy prediction models. Moreover, an augmented stacked generalizer that utilizes diversified feature subsets during its learning phase is suggested as an effective ensemble method for promoting independencies among base prediction models. Empirical results show that the augmented stacked generalizer significantly improves overall predictability by reducing the more costly type-I error rate compared against both popular bagging and standard stacking procedures.

Keywords: Bankruptcy prediction; feature selection; stacked generalization; bagging; ensemble prediction; financial data mining.

1 Introduction

Since the mid-1980s bank failures have brought large adverse impact on real economies, especially causing enormous loss of output, economic recession, and severe currency crashes. The estimated international average loss of output is 7.3% of Gross Domestic Product (GDP), and the average duration of economic recession is 3.3 years [12]. Furthermore, bank failures bring additional restructuring costs: estimated average restructuring cost is 21.37% of GDP [8].

In this context, bank failure becomes very critical in terms of international financial stability, and therefore early warning capability in bank surveillance has come to the forefront of international banking regulators since accurate early warning could remarkably improve banks' risk management scheme by protecting them from any adverse economic conditions.

---

* Corresponding author.

A. Sattar and B.H. Kang (Eds.): AI 2006, LNAI 4304, pp. 887–896, 2006.
© Springer-Verlag Berlin Heidelberg 2006

Many academicians and practitioners have challenged to develop sophisticated and efficient analytical techniques that could help decision makers to capture risk signals in the early stage before failure. The study for predicting corporate failure has a long history over four decades. From the seminal work by Beaver [4], many efforts have been devoted to develop accurate prediction models. Traditionally, conventional statistical models including multivariate discriminant analysis [1], logit [15] and probit [20] have been employed for this domain. On the other hand, the artificial neural network (ANN) [17], [5] was introduced as an alternative, high-performance prediction model [14], [18]. Most recent studies reveal that the machine learning models such as ANN outperform the conventional statistical models in predicting corporate failure (see [2] for a detailed review).

This superiority might due to the nonlinear nature of real-world bankruptcy phenomenon: the class boundaries are not usually linearly separable. Until very recently, however, in many studies the methods of selecting salient variables (features) for these powerful approaches have been relying on the linear dependency tests that have been mainly employed by the traditional linear models. Therefore, the predictive power of the machine learning models can be underestimated and/or comparative evaluations using those models might not be correct.

In this study, we evaluate a performance of the conventional feature selection method that have been widely employed in this domain, termed as "ANOVA-Stepwise"—a combined form of a univariate t-test (ANOVA) and multivariate stepwise forward logit model—against a simple "information-theoretic" feature ranking method, by using three machine learning models on Turkish bank failure data set. In addition, we empirically show that an ensemble model composed of the two prediction approaches (ie., traditional logit and machine learning models) using these two different feature selection procedures can achieve enhanced prediction accuracy due to its weak interdependencies.

The rest of the paper is organized as follows. In Section 2, background knowledge for ensemble prediction and feature selection issues are discussed. The data and evaluation procedure used for this study are described in Section 3 and the next section summarizes the experimental results. Finally, the major findings and future research directions are summarized in the final section.

2 Background

Recently, researchers have started to develop general algorithms for improving classification performance by combining multiple classifiers under the name of committee or ensemble machine [10], [3], [11], [13]. It is well known that the ensemble approach can yield an improved predictability by aggregating partially independent individual members.

In this paper, we first examine the effect of feature selection method for individual members, and then evaluate an overall performance of ensemble prediction models that consist of heterogeneous individuals. A hypothesis behind this lies in that two different kinds of prediction models including the traditional and machine learning model can synergistically interact with each other when different feature selection procedures are appropriately used be the models.

Next, brief descriptions for ensemble prediction and feature selection issues are discussed.

2.1 Ensemble Prediction

The methods of combining independent estimator to improve their performance have a long history in statistics and econometrics. In literature, several terms—classifier ensemble, classifier committee, classifier fusion, mixture of experts, and consensus aggregation—are interchangeably used.

Representative Bagging (bootstrap aggregating) technique developed by Breiman [6] combines individual predictions by majority voting based on randomly redistributed sampling technique known as bootstrap procedure [9]. On the other hand, Metalearning—more complex approach beyond the scope of simple aggregation—can be viewed as an additional learning scheme. The metalearning framework consists of a variety of base classifiers and one meta-classifier that generalizes diverse behaviors of the base classifiers by learning how they interact with each other.

It is widely accepted that Stacking [19] is one of the effective metalearning techniques. However, it has a conflict problem when discrete class labels are used; that is, same input vectors with different desired output can be observed. To avoid this, *Bayes* posteriori probability or confidence value for each predicted class of base classifiers can be used. Unfortunately, however, this method is ineffective when the predicted behaviors among base classifiers are significantly inter-related.

In this study, we employ an augmented stacked generalizer (from herein the term *StackF* will be used) as the notion of "class-attribute-combiner" [7] using additional feature information in order to avoid the conflict problem. In "class-attribute-combiner", any random feature subsets can be added into a training vector with a crisp class label. In contrast, our *StackF* consists of following two steps: 1) directly injects salient features enabling superior performance of base classifiers and 2) indirectly injects another salient ones promoting independent predictions among base classifiers.

2.2 Feature Selection

A preceding generic task in predictive classification is to extract salient features for decision making. When the input vector contains too many features, probability of being extraneous noises that can cause classification errors will be increased. Furthermore, it is difficult to train a generalized model due to high computational complexity and/or high possibility of convergence to a local minimum of error surface. Thus, the selection of relevant attributes is vital for both constructing competitive classifiers and reducing data management costs.

The feature selection algorithms fall into two categories based on whether or not they perform selection independently of the learning algorithm. Independent selection is known as a *filter* approach, whereas the dependent method is called a *wrapper*.

Despite the computational efficiency, the major drawback of the filter approach is that the resulting feature subset may not be optimal for a particular prediction model. On the other hand, despite of a heavy computational load the wrapper can select an optimized feature subset for a specific classifier.

Traditionally, and until very recently, variable (ie., financial ratios) selection in the bankruptcy prediction literature has been primarily relied on ANOVA [4] followed by multivariate statistics such as stepwise multiple discriminant analysis (MDA) (eg., [1]) or stepwise logit. However, in case of using machine learning models rather than MDA and logit, there is no guarantee to obtain maximized class seperability using those selected features; that is, stepwise MDA and logit procedures can be viewed as the wrapper for MDA and logit, not for the machine learning models. Moreover, those criteria do not reflect nonlinear dependencies among variables.

In contrast, the *gain ratio* [16]—modified information gain which reduces its bias using split information—is one of the simplest feature ranking that can handle nonlinearity and reduce shortcomings from the classifier-specific wrapper approach. With this research motivation, ANOVA-Stepwise procedure will be validated by comparing its predictive performance against the gain ratio.

3 Experimental Setup

3.1 Data

In this study we tested publicly available Turkish bank failure data set that can be reached at www.tbb.org.tr/english/default.htm. The data set consists of total 57 privately owned commercial banks where 21 banks failed during the period of 1997-2003, and each bank is represented by five sets of 49 financial ratios over five years prior to failure. Consequently, the resulting data set is composed of 285 bank-years observations, 105 of which are labeled as failed.

3.2 Evaluation Procedure

To investigate robust performance of prediction models, we employed a 10-fold cross validation scheme. In each of the ten folds, stratified random sampling was used to avoid a selection bias due to the concentration of specific bank-year samples in training set. More specifically, for failed banks 90 bank-year samples (5 years observations of 18 banks) were included in each fold to predict 15 bank-years (5 years observations of 3 banks).

In each of the ten-fold cross validation evaluations, we used the following parameters for classifiers. The ANN employed was the multilayer perception with an error back-propagation learning algorithm [17]. Each continuous input feature was normalized while binary values (−1 and 1) were used for the class label. To avoid an overfitting problem, the maximum iteration to terminate was set to 500, and 10% of the training samples were reserved for a validation set. Relating to the ANN architecture, a single hidden layer consisting of 4 neurons was considered based upon preliminary evaluations. The learning and acceleration rates were initially set to 0.6 and 0.2, respectively, and gradually decreased using a weight-decay procedure.

For the kNN classifier, ten-fold cross validation tests on the training dataset were performed to determine the most promising neighborhood size k, by varying k over all the odd integers from 1 to 11. Doing so, we set the best performing k as 5.

For ensemble models, bagging consists of 30 ANNs based on preliminary evaluations while stacking consists of one ANN meta-classifier that makes a final prediction plus three base classifiers including kNN, C4.5, and ANN. To train a meta-classifier, we performed a stratified ten-fold cross validation scheme for each of the ten folds [7].

4 Results

In this section, we report the results of the two feature selection methods and their effect on the stacking with a focus on early warning capability and overall correctness.

4.1 Selecting Predictor Variables

Table 1 presents features selected by ANOVA-Stepwise procedure and gain ratio measure. After excluding 17 features that are not statistically significant at $p=0.1$ level by ANOVA test, stepwise forward logit selected seven features over 5 categories. On the other hand, six features were determined to be salient by the gain ratio because of the following reasons. First, there are many tied variables with the value just smaller than 0.15 (see Figure 1). Second, to achieve a fair comparison with ANOVA-Stepwise approximately equal number of features were considered. It is interesting to note that there is only one common variable between the two criteria.

Table 1. Variables selected by two criteria

Category	ANOVA-Stepwise	Gain ratio
Stability	(shareholders' equity + total income) / total assets	Standard capital ratio
	(shareholders' equity + total income) / (deposits + non-deposit funds)	
Assets Quality	total loans / total assets	
Liquidity	liquid assets / total assets	liquid assets / (deposits + non-deposit funds)
Income Structure		income before tax / average total assets
		total income / total expenditure
Branch Performance	total loans / # of branches	net income / # of branches
Activity Ratios	provisions for income tax / total income	provisions for income tax / total income
	reserve for seniority pay / # of employees	

In Figure 1, a scree plot displays the impact of 49 variables estimated by gain ratio sorted from large to small, as a function of the degree of relevance.

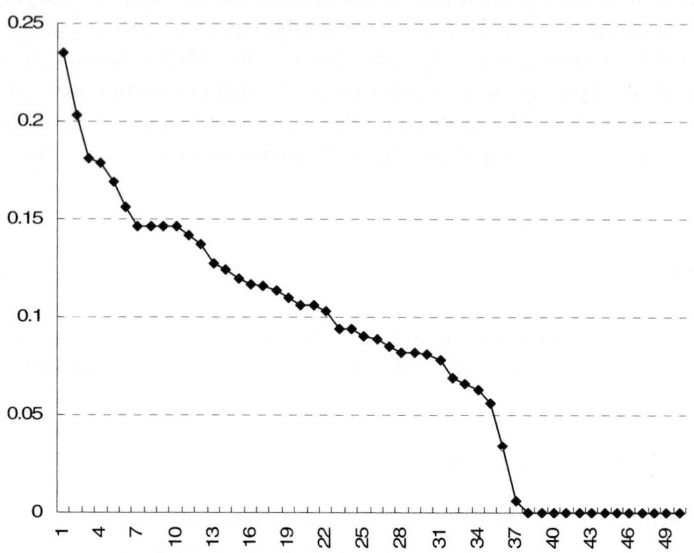

Fig. 1. Scree plot of 49 variables estimated by gain ratio

4.2   Comparing Feature Selection Methods

Table 2 presents comparative evaluations of the four individual prediction models using seven features selected by ANOVA-Stepwise procedure. In terms of an overall accuracy, ANN was slightly superior to both logit and kNN, while C4.5 decision tree showed a worst performance.

Interestingly, logit outperformed the others when prediction lag lies in less than T-4 (ie., short-term forecasting). Meanwhile, regarding an early warning capability, all the three machine learning models was superior to logit: ANN was outstanding in T-5, and both C4.5 and kNN similarly outperformed logit in T-4 and T-5.

Overall, in terms of the costly type-I error rate logit was the best while ANN showed lowest type-II error rate. It must be stressed that, however, this series of benchmarks was clearly beneficial to logit since seven features used by all four models were optimized by the stepwise logit procedure.

Figure 2 displays average relative strengths of gain ratio (labeled with "Gr") feature selection over ANOVA-Stepwise (labeled with "AS") method. Compared with the previous results of Table 2, overall accuracy of C4.5 was remarkably improved from 76.84% to 81.75% while type-I error of ANN decreased from 0.31 to the lowest 0.25 maintaining similar type-II error rate.

Table 2. Performance comparison of four individual models using ANOVA-Stepwise feature selection

Prior to Failure	Measure	Individual Models			
		Logit	**C4.5**	**kNN**	**ANN**
T-1	Overall	82.46%	68.42%	75.44%	77.19%
	Type-I	0.29	0.52	0.43	0.29
	Type-II	0.11	0.19	0.14	0.19
T-2	Overall	87.72%	75.44%	80.70%	77.19%
	Type-I	0.29	0.38	0.14	0.52
	Type-II	0.03	0.17	0.22	0.06
T-3	Overall	85.96%	82.46%	84.21%	84.21%
	Type-I	0.19	0.24	0.29	0.29
	Type-II	0.11	0.14	0.08	0.08
T-4	Overall	78.95%	84.21%	84.21%	85.96%
	Type-I	0.24	0.33	0.19	0.24
	Type-II	0.19	0.06	0.14	0.08
T-5	Overall	68.42%	73.68%	80.70%	82.46%
	Type-I	0.33	0.52	0.33	0.24
	Type-II	0.31	0.11	0.11	0.14
Average	Overall	**80.70%**	**76.84%**	**81.05%**	**81.40%**
	Type-I	0.27	0.40	0.28	0.31
	Type-II	0.15	0.13	0.14	0.11

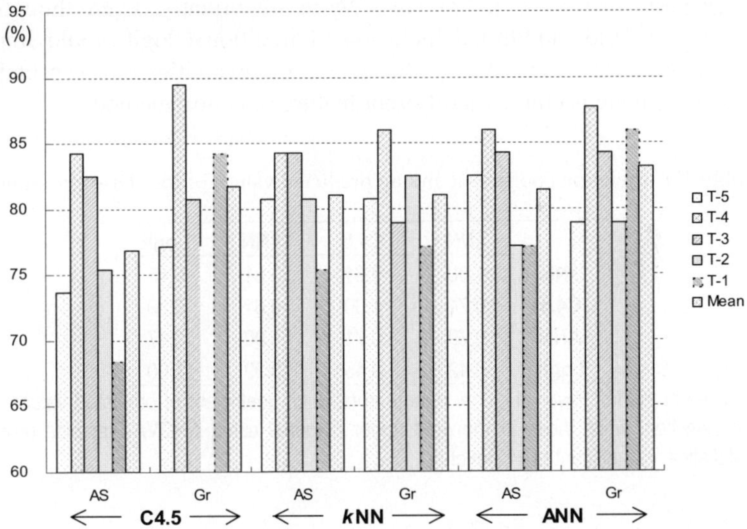

Fig. 2. Effect of feature selection using gain ratio criterion

Note that Gr outperformed AS in 10 cases among 15 comparisons (three models over 5 years period). These improvements suggest that the traditional AS is not an optimal solution for machine learning models and, instead the other simple feature selection criterion can significantly enhance its predictability.

Therefore, in many prior studies that have followed the traditional approach there might be a great possibility of underestimated predictability of machine learning models.

4.3 Promoting Ensemble Diversification

Next, we conduct series of experiments relating to ensemble prediction models in order to verify whether combining different approaches can improve overall prediction accuracies due to its independent prediction behavior partially resulted from independent feature subsets.

Table 3 shows the degree of dependencies among predicted values of four individual classifiers, each of which uses two different feature selection methods. The dependencies among three machine learning models measure by Pearson coefficient are raging from 0.61 to 0.81 for both selection methods while dependencies between those models and logit are below than 0.5. In particular, dependencies among machine learning models are higher when gain ratio criterion was used. Moreover, as expected, ANOVA-Stepwise procedure makes machine learning models highly correlated with logit.

Theses results of Figure 2 and Table 3 strongly suggest followings. Firstly, ensemble of machine learning models that utilize gain ratio would be beneficial due to the superior performance of individuals. More importantly, even though overall predictability is inferior additional inclusion of traditional logit would improve an ensemble performance by promoting the degree of diversification, especially when machine learning models employ a different feature selection method.

Table 3. Correlation coefficient among predicted values of four base classifiers

	*k*NN	C4.5	ANN	Logit
*k*NN	1.00	0.61	0.81	0.45
C4.5	0.71	1.00	0.61	0.39
ANN	0.79	0.71	1.00	0.47
Logit	0.42	0.46	0.39	1.00

Note: The values in the left lower diagonal denote correlation coefficient of predicted values using gain ratio feature selection, while those in the right upper diagonal using ANOVA-Stepwise procedure. All values are statistically significant at 1% level.

Table 4 contrasts the effectiveness of an inclusion of diversified models based on different feature subsets. First, we found that the standard stacking based on pure machine learning models (Stack) did not show any improvements against the simple bagging model. Second, stacking after inclusion of logit (Stack$^+$) slightly outperformed both bagging and standard stacking, indicating different approaches that use different feature selection methods can help stacking to achieve higher performance by learning diversified prediction behaviors. Third, most impressively, Stack$^+$ which utilizes six key features (ie., StackF$^+$) correctly predicts 86% for T-5 and shows 89% overall accuracy through10-fold cross validated evaluations over 5 year period. These results suggest that a classifier fusion of different approaches using different feature selection methods can be particularly beneficial when superior feature subsets are explicitly incorporated into an ensemble model.

Table 4. Performance comparison of five ensemble prediction models using gain ratio feature selection procedure

Prior to Failure	Measure	Ensemble Models				
		Bagging	**Stack**	**Stack$^+$**	**StackF**	**StackF$^+$**
T-1	Overall	85.96%	82.46%	85.96%	84.21%	91.23%
	Type-I	0.33	0.19	0.19	0.14	0.14
	Type-II	0.03	0.17	0.11	0.17	0.06
T-2	Overall	77.19%	82.46%	84.21%	85.96%	87.72%
	Type-I	0.38	0.19	0.19	0.24	0.24
	Type-II	0.14	0.17	0.14	0.08	0.06
T-3	Overall	82.46%	85.96%	87.72%	89.47%	92.98%
	Type-I	0.38	0.19	0.19	0.24	0.19
	Type-II	0.06	0.11	0.08	0.03	0.00
T-4	Overall	96.49%	91.23%	92.98%	84.21%	89.47%
	Type-I	0.10	0.14	0.10	0.29	0.10
	Type-II	0.00	0.06	0.06	0.08	0.11
T-5	Overall	80.70%	80.70%	78.95%	82.46%	85.96%
	Type-I	0.52	0.38	0.38	0.24	0.19
	Type-II	0.00	0.08	0.11	0.14	0.11
Average	Overall	**84.56%**	**84.56%**	**85.96%**	**85.26%**	**89.47%**
	Type-I	0.34	0.22	0.21	0.23	0.17
	Type-II	0.04	0.12	0.10	0.10	0.07

Note: The symbol "+" denotes an inclusion of logit to form the stacking model.

5 Conclusion

In this study, we have examined the effect of feature selection method for bank failure prediction, especially focusing on the performance of an ensemble model.

Our findings reveal that: 1) the traditional feature selection method such as AVOVA or ANOVA-Stepwise can lead wrong comparative evaluations of state-of-the-art models; and 2) proposed augmented stacked generalizer which utilizes an advantage of diversified feature subsets has a potential to improve both type-I error rate and early warning capability.

To sum up, future direction of greater scope lies in generalizing the proposed approach to other bankruptcy cases. In addition, work is necessary to apply more sophisticated feature selection procedures. An example would be a technique that can detect only probabilistically relevant financial ratios in a non-redundant fashion.

References

1. Altman E.I. 1968. Financial Ratios, Discriminant analysis and the prediction of corporate bankruptcy. *Journal of Finance* 589-609.
2. Atiya, A. 2001. Bankruptcy prediction for credit risk using neural networks: A survey and new results, *IEEE Trans. Neural Networks* **12** (4) 929-935.
3. Battiti, R., A. M. Colla. 1994. Democracy in neural nets: voting schemes for classification. *Neural Networks* **7** (4) 691-707.
4. Beaver W. 1968. Market prices, financial ratios, and the prediction of failure, *Journal of Accounting Research* 170-192.
5. Bishop, C. M. 1999. *Neural Networks for Pattern Recognition.* Oxford University Press.
6. Breiman. 1996. Bagging predictors. *Machine Learning* **24** (2) 123-140.
7. Chan, P. K., S. Stolfo. 1993. Meta-learning for multistrategy and parallel learning. *In Proc. Second International Workshop on Multistrategy Learning*, 150-165.
8. Claessens, S., Klingebiel, D., Laeven, L. 2001. Financial restructuring in banking and corporate sector crises: What policies to pursue. NBER Working Paper No. 8386.
9. Efron, B., R. Tibshirani. 1993. *An Introduction to the Bootstrap.* Chapman & Hall.
10. Hansen, L. K., P. Salamon. 1990. Neural network ensembles. *IEEE Trans. Systems, Man, and Cybernetics* **12** (10) 993-1001.
11. Ho, T. K., J. J. Hull, S. N., Srihari. 1994. Decision combination in multiple classifier systems. *IEEE Trans. Pattern Analysis and Machine Intelligence* **16** (1) 66-75.
12. Hutchison, M. M., McDill, K. 1999. Are all banking crises alike? The Japanese experience in international comparison. NBER Working Paper No. 7253.
13. Lam L., C. Y. Suen. 1997. Application of majority voting to pattern recognition: An analysis of its behavior and performance. *IEEE Trans. System, Man, and Cybernetics* **27** 553-568.
14. Odom, M. D., R., Sharda. 1990. A Neural Network Model for Bankruptcy Prediction, *International Joint Conference on Neural Networks* **2** 163-168.
15. Ohlson, J. A. 1980. Financial ratios and the probabilistic prediction of bankruptcy. *Journal of Accounting Research* **18** (1): 109 - 131.
16. Quinlan, J. R. 1993. *C4.5: programs for machine learning*, Morgan Kaufmann, San Mateo, California.
17. Rumelhart, D. E., G.E. Hinton, R.J. Williams. 1986. Learning internal representations by error propagation. in *Parallel Data Processing*, I, Chapter 8, the M.I.T. Press, Cambridge, MA, 318-362.
18. Tam, K., M. Kiang. 1992. Managerial application of neural networks: The case of bank failure predictions. *Management Science* **38** (7) 926-947.
19. Wolpert, D. H. 1992. Stacked generalization, *Neural Networks* **5** 241-259.
20. Zmijewski, M. E. 1984. Methodological Issues Related to the Estimation of Financial Distress Prediction Models, *Journal of Accounting Research* **22** 59-82.

LRCA: Enhanced Energy-Aware Routing
Protocol in MANETs

Kwan-Woong Kim[1], Jeong-Soo Lee[2], Kyoung-Jun Hwang[1], Yong-Kab Kim[1],
Mike M.O. Lee[3], Kyung-Taek Chung[4], and Byoung-Sil Chon[5]

[1] Div. of Electrical Electronic & Information Eng., Wonkwang Univ., Iksan, 570-749,
South Korea
`{watchbear, ykim}@wonkwang.ac.kr`
[2] Samsung electronics, 416 Maetan-3dong Yeongtong-Gu, Suwon, South Korea
`mrinteger@naver.com`
[3] Murdoch University, South Street, Murdoch, Western Australia 6150, Australia
`Mike.Lee@murdoch.edu.au`
[4] Kunsan Nat. Univ. San-69 Miryong-Dong, Kunsan, Chonbuk, 573-701, South Korea
`eoe604@kunsan.ac.kr`
[5] Chonbuk Nat. Univ. 664-14 Duckjin-Dong, Chonju, Chonbuk, 561-156, South Korea
`bschon@moak.chonbuk.ac.kr`

Abstract. Mobile Ad hoc Networks (MANETs) is power constrained since
mobile nodes operate with limited battery energy. So battery life is also
important research issue in the routing protocol design. During packet
transmission over multi-hop nodes, if a node failed, it should be caused link
failure and source node will perform route recovery. It results that increases
time of route recovery and packet loss rate. The objective of this study is to
reduce route failure caused by dead nodes which consume all the battery life.
We propose a new routing protocol based on AODV which provides an ability
of changing routes to neighbor nodes before some of intermediate nodes be
shutting down. From extensive simulations, results show that possibility of cut-
offs and time-delay caused by packet-loss have been decreased and also
improve overall performance by comparison with original AODV.

1 Introduction

A Mobile Ad hoc Network (MANET) [1,2] is an autonomous distributed system that
consists of a set of mobile nodes that move arbitrarily and use wireless links to
communicate with other nodes that reside within its transmission range. Because of
limited radio propagation range, mostly routes are multi-hop.

Routing protocols without consideration of energy consumption tend to use the
same paths for given traffic demands, which results in a quick exhaustion of energy of
the nodes along the paths if those traffic demands are long-lasting and concentrated.
This problem can become more serious where multimedia applications are being
supported since the traffic is long lasting and the amount of data being transferred is
quite large. As the use of it becomes increasingly widespread in ad hoc network
environments, energy conservation issues are becoming more significant.

A. Sattar and B.H. Kang (Eds.): AI 2006, LNAI 4304, pp. 897–901, 2006.

© Springer-Verlag Berlin Heidelberg 2006

Since the participating nodes equipped with battery, power conservation is critical to extending the lifetime of a functioning network. Actually, a number of routing algorithms have been proposed to provide multi-hop communication in wireless ad hoc networks taking into account energy.

Minimum Total Transmission Power Routing (MTPR) [7], Minimum Battery Cost Routing (MBCR) [8] and Min-Max Battery Cost Routing (MMBCR) [9] are examples of existing energy awared routing protocol. These protocols focused on finding the best path to maximize the system lifetime which is defined as duration from the beginning of the service to the first time of some node's energy depletion.

In this paper, we propose an enhanced routing protocol based on AODV named LRCA (Local Route Change Scheme in AODV) considering the battery life of mobile nodes. The rest of this paper is organized as follow. In the section 2, we describe the proposed approach. In the section 3, we evaluate performance of the proposed scheme through a set of simulations and compare it to the original AODV [3] [4]. Finally, we describe future work and conclusion.

2 Local Route Change Algorithm

As mentioned in previous section, the wireless mobile nodes have finite battery supply and in many cases the nodes are installed in an environment where it may be hard (or undesirable) to retrieve them to change or recharge the batteries such as wireless sensor networks. Because of limited power of mobile devices and difficultness/high costs of recharging battery, nodes could be failed by consuming all its energy and will not be available then all routes that traverse the failed node will be broken. When a link-breakage caused by node failure occurred, transmitting nodes will try to re-establish path to reach its destination. It results that increase packet latency and overhead of routing packets and also packet loss rate.

In this study, we will focus on the routing protocol design to reduce link-failure which caused by node failure and improve efficiency of routing protocol. To achieve this goal, Routing protocol should have ability of local route change from node which has low battery life to alternative node.

Figure 1 illustrates a scenario of the proposed scheme. When a node receives DATA packet, if this node has low battery life, it broadcasts HELP message to its neighbors with sender address 'A' and next-hop address 'C' as shown in figure 1 (a). When a node receive a HELP message, if a node has sufficient energy and two addresses 'A' and 'C' are belong to my neighbors than send an OK packet to the node that sent HELP as shown in figure 1 (b). The node 'B' choose the node address 'E' and send RCRQ(Route Change ReQuest) message with address 'D' to sender 'A' (figure 1 (c)). After receiving a RCRQ message, the node 'A' changes the route to the node 'E' from 'B' as like as figure 1 (d).

For providing local route change ability in AODV, we have designed new messages as depicted in figure 2: *HELP*, *OK* and *RCRQ*. Where 'energy-status' is the battery life of the transmitting node. The size of energy-status field is 2 bit and each value means that; '0' is full, '1' is high, '2' is medium and '3' is low.

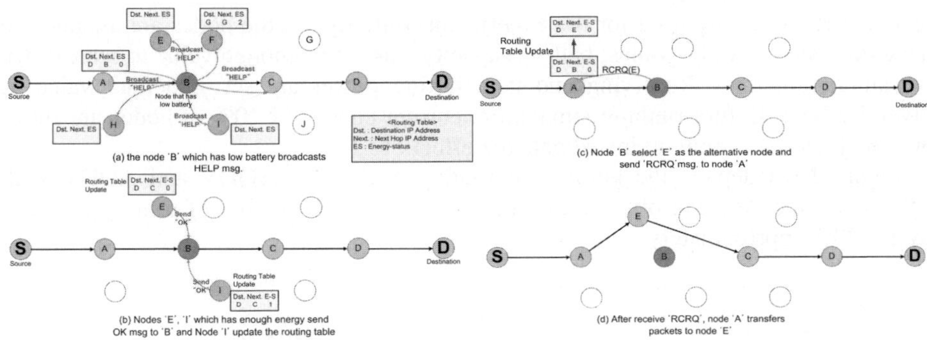

Fig. 1. An example scenario of LRCA

Determine energy status

In LRCA algorithm, we determine 'energy-status' with current battery life as bellows. Where E_{low} = low threshold of battery, E_{med} = medium threshold of battery, E_{high} = high threshold of battery, cap : the capacity of battery and en : current energy.

Initial Conditions: $E_{low} = 0.2 * cap;\ E_{med} = 0.4 * cap;\ E_{high} = 0.6 * cap;$
If $en < E_{low}$ than $energy\text{-}status = 3$;
Else if $E_{low} < en < E_{med}$ than $energy\text{-}status = 2$;
Else if $E_{med} < en < E_{high}$ than $energy\text{-}status = 1$;
Else $energy\text{-}status = 0$;

For gathering energy information of neighbor nodes, we modify HELLO message and neighbor list in AODV protocol by adding 2 bits of ES field.

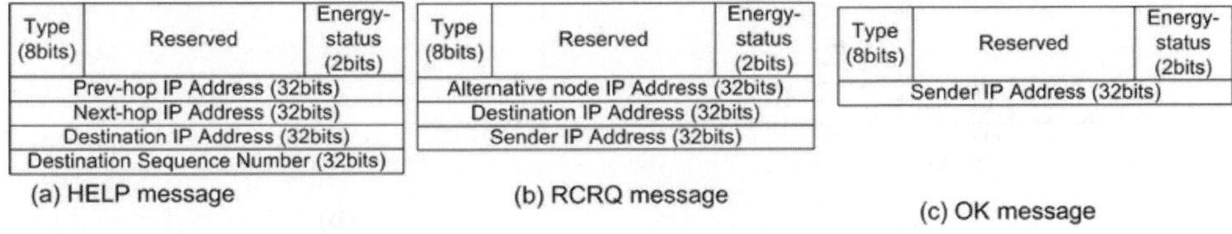

Fig. 2. New messages for LRCA

3 Numerical Results

In this section, we evaluate performance of our LRCA scheme using extensive simulations and compare its performance with traditional AODV by using NS2 simulator [10]. An ad hoc network used for simulations is that consists of 60 mobile nodes in 1.0 km × 1.0 km area. Each node uses IEEE 802.11 MAC protocol and the used channel model is Wireless channel/Wireless Physical propagation model. Traffic sources are CBR (Constant Bit Rate) and generate UDP packet in every 0.1 sec. The size of UDP packet is 1024 bytes. The simulation time is set to 200 seconds. From 5 to 15 CBR source are used for the experiment. Initially, all the nodes are assumed to have full battery of 200 joules; battery capacity was set to enough value to scale down the simulation time. Receiving and transmitting power are set to default value as given by NS2. Before running simulation, current energy of 20% of nodes are set to 30 – 60 joules randomly to investigate the effect of LRCA.

Figure 3-(a) depicts the number of routing packets of AODV and AODV with LRCA scheme. We can observe that LRCA reduces the number of routing message by 1 – 20% approximately.

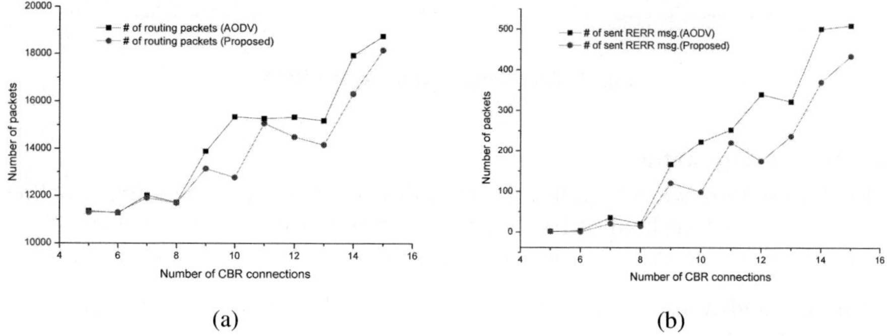

(a)   (b)

Fig. 3. (a) Numbers of routing messages (b) Numbers of RERR messages

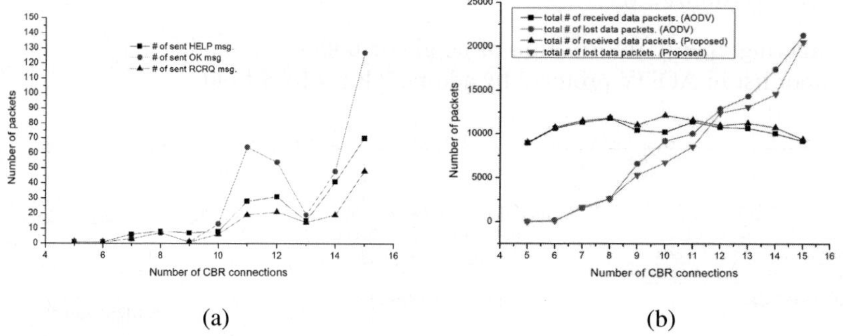

(a)   (b)

Fig. 4. (a) Number of HELP, OK, RCRQ msg (b) Numbers of received and lost packets

Numbers of RERR are depicted in figure 3 (b). Obviously, the proposed LRCA shows that reduce the number of RERR messages compared to original AODV. It means that the number of link failure of LRCA is less than original AODV and LRCA reduce re-route discovery by initiating local route change from a node which has low battery life to its neighbor node. Figure 4(a) depicts numbers of HELP, OK and RCRQ. As the number of CBR source increases, more control messages of LRCA are generated.

Figure 5 (b) shows the number of received and lost packets using both the original AODV and the AODV with LRCA scheme. As the number of CBR connections increase, more packets are injected to network and more packets are lost by collision. AODV with LRCA scheme improve the number of received packets and reduce packet loss as shown in figure 5(b).

4 Conclusion

Due to limited battery life of mobile node, energy is a main research issue in routing protocol design as well as mobility. If a node consumes all battery life then the node will not be available and all routes that traverse this node will be broken. RERR messages will be transmitted to source node and source nodes initiate route discovery.

This is quite negative effect to performance in MANETs in terms of routing overheads and packet transmission latency. To solve this problem, we proposed new algorithm named LRCA for route maintenance based on AODV. Our algorithm quite differ with other energy based routing protocols. Related works are focus on route discovery to minimize power consumption or effective power distribution. To minimize link breakage caused by power depletion of node, LRCA try to change path from a node which has low battery to its neighbor. To evaluate the validity of LRCA scheme, we compared its performance with the original AODV protocol by simulation. Results show that the proposed scheme can reduce link breakage, routing overhead and also improve end-to-end packet transmission and reduces packet loss.

Acknowledgement. This paper was supported by Research Institute of Engineering Technology Development, WonKwang University.

References

[1] C. E. Perkins. "Ad Hoc Networking". Addison-Wesley, Upper Saddle River, NJ, USA, Jan 2001.

[2] http://www.ietf.org/, IETF MANET Working Group.

[3] C. Perkins, E. Belding-Royer, and S. Das, "Ad hoc On-Demand Distance Vector (AODV) Routing," drift-ietf-manet-aodv-12.txt, Work in Progress, 2002.

[4] C. E. Perkins, E. M. Royer and S. R. Das, "Ad-hoc on demand distance vector routing," IETF RFC3561, http://www.ietf.org/rfc/rfc 3561.txt, 2003.

[5] Elizageth M. Royer & Chai-Keong Toh, "A Review of Current Routing Protocols for Ad hoc Mobile Wireless Networks," IEEE personal Communications, Apr. 1999.

[6] C. K. Toh, "Maximum Battery Life Routing to Support Ubiquitous Mobile Computing in Wireless Ad-hoc Networks," IEEE Communication Magazine, pp. 26-37, June 2001.

[7] J. Chang and L. Tassiulas, "Energy Conserving Routing in wireless Ad-hoc Networks," IEEE INFOCOM 2000, pp. 22-31, 2000.

[8] S. Singh, M. Woo, and C. S. Raghavendra, "Power-Aware Routing in Mobile Ad hoc Networks, " proc. MobiCom '98. Dallas, TX, Oct. 1998.

[9] C. K. Toh, "Associativity Based Routing for Ad hoc Mobile Networks, "Wireless Pers. Commun. J., Special Issue on Mobile Networking and Computing Systems, vol. 4, no. 2, Mar. 1997.

[10] S. McCanne, S. Floyd. NS. http://www.isi.edu/nsnam.ns/.

Enabling Agent Oriented Programming Using CORBA-Based Object Interconnection Technology for Ubiquitous Computing*

Hyongeun Choi and Tae-Hyung Kim

Department of Computer Science and Engineering, Hanyang University
1271 Sa 1-Dong, Ansan, Kyunggi-Do, 426-791, South Korea
{hechoi, tkim}@cse.hanyang.ac.kr

Abstract. This paper presents a real-time agent service layer (RT-ASL) that is built on CORBA, in order to use a high-level abstraction of software agents for ubiquitous computing. There are four important characteristics in designing RT-ASL: message passing communication mechanism for agent communications, agent service discovery mechanism, real-time agent communication language, and real-time generic scheduling interfaces. This paper clarifies the importance of such design decisions and their meanings in ubiquitous programming.

Keywords: Agent oriented programming, Distributed multi-agent systems, Ubiquitous computing, Distributed objects, CORBA.

1 Introduction

Since Yoav Shoham introduced the concept of agent oriented programming, it has been one of the highly studied research issues especially in the realm of distributed artificial intelligence [1]. According to his sentinel paper, an agent is an extended concept of objects to higher abstraction. The two notable features are the anthropomorphic abstraction and the use of common language called an agent communication language (ACL). Meanwhile, ubiquitous computing has emerged on the information technology scene just a few years ago but became one of the most heavily used buzzwords in these days. Many promising scenarios envisioned by ubiquitous computing would be only possible when programmers are conveniently able to write interoperable distributed programs under friendly programming environments. Since we have been experienced enough how much difficult and complicated the task of distributed programming is, it would be a far greater nightmare for application programmers to write intelligent programs, which are context-aware, autonomous, mobile, distributed and collaborative with each others, working in constantly changing computing environments. It is very hard to define the standard programming interfaces in object-oriented design style due to its evolving property. A new programming abstraction called software agent [2] can be regarded as a viable way to deal with such an add-on diversity and complexity due to the ubiquity of computing components. We believe the agent programming abstraction

---

* This research has been supported by the Ministry of Education and Human Resources Development, S. Korea, under the grant of the second stage of BK21 project.

A. Sattar and B.H. Kang (Eds.): AI 2006, LNAI 4304, pp. 902–906, 2006.
© Springer-Verlag Berlin Heidelberg 2006

beyond object orientation is an effective way of expressing the dynamic and continuously evolving computing environments since it provides a higher abstraction than the conventional object interconnection technology.

In this paper, we present a real-time agent service layer (RT-ASL) architecture on CORBA. Using this top-level service layer, programmers can write a multi-agent based distributed program, and the agent communication will be implemented transparently via the traditional CORBA services [3].

2 Related Work

Many software architectures have been studied to support distributed multi-agent systems. The DECAF (Distributed Environment-Centered Agent Framework) [4] and JADE (Java Agent Development Framework) [5] are early developed Java-based multi-agent system. They accept KQML (Knowledge Query Manipulation Language) or FIFA-ACL for agent communication, and provide system agents called matchmaker or directory facilitator for dynamic service registration and discovery. They intended to provide the full-fledged support for distributed multi-agent system, so they decided to rely on their own proprietary communication mechanism without using an existing software bus. Since it is not practical for an evolving technology like ubiquitous computing to use a non-standard mechanism, several other agent systems have tried to develop using CORBA as their communication infrastructure. KCobalt [6] used CORBA communication mechanism in order to implement KQML descriptions among distributed multi-agents. KCobalt provided an automated translation from KQML to CORBA IDL definitions. RTMAS [7] adopted the same approach with KCobalt but extended KCobalt implementation to include real-time agent communications. They extended the expressiveness power of the agent communication language KQML with real-time QoS requirements and capabilities through new performative parameters. In our RT-ASL implementation, we adopt a notification service to implement agent communications in an asynchronous loosely-coupled communication model for dynamic binding. We also adopt the real-time ACL as proposed in RTMAS [7], and a generic service discovery and real-time scheduling interface architecture.

3 Architecture for Agent Service Layer

Our RT-ASL architecture can be characterized as follows. First, an asynchronous loosely-coupled communication model is adopted for an agent to be implemented in object interconnection technology using CORBA Notification Service. RT-ACL messages are delivered in terms of structured CORBA events. Second, RT-ASL provides three system agents, which are facilitator, broker, and marketplace agents, respectively, so that programmers can construct their own agent-based ubiquitous applications. Third, the underlying agent communication language basically follows the standard specification in FIPA-ACL [8] with an extension of real-time capabilities. Last, the real-time scheduling architecture is incorporated in the layer. In this section, we present the internal design for the RT-ASL from these perspectives.

3.1 Overall RT-ASL Architecture

RT-ASL is built on top of the CORBA middleware. Fig. 1 shows the overall architecture of distributed multi-agent systems using RT-ASL. An agent working on the RT-ASL is thus a CORBA application in itself that uses RT-ASL APIs. An agent consists of `AgentMain()`, message queue, incoming message handler CORBA object, and other standard RT-ASL core APIs like `SendACLMsg()`, `GetACLMsg()`, and so on. `AgentMain()` is a main body of an agent program that should be written by an agent programmer. Incoming message handler object creates an event channel that is used for CORBA Notification Service. If an event is occurred, the message handler object translates the event object into an RT-ACL message and stores it in a message queue. An agent application continuously accesses the message queue and takes an RT-ACL message to execute a proper action according to the direction specified in the RT-ACL message.

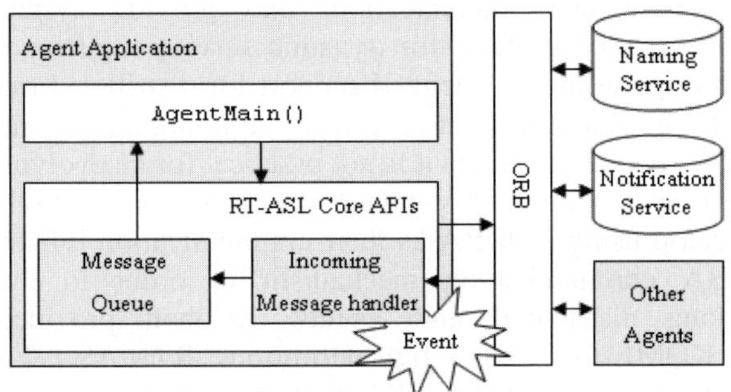

Fig. 1. RT-ASL Architecture

3.2 Message Transport for Real-Time Agent Communications

In RT-ASL, CORBA "Notification Service" is used as an underlying message transport for delivering agent communication languages among distributed agents. The CORBA Notification Service is built on top of the CORBA event service for backward compatibility with existing applications using the CORBA event service. Many important services that are useful for agent message transport are provided by the CORBA Notification Service like untyped events contained within an "Any" data type, typed events, structured events, event filtering, different QoS parameter settings, and so forth.

3.3 Agent Service Discovery

In distributed multi-agent systems, the service discovery is an essential mechanism for multiple agents to cooperate with each other in harmony. In our work, we provide three different kinds of system agents that are used for an agent to discover a service that it needs to ask, which are a facilitator agent, a broker agent, and a marketplace agent. The facilitator agent maintains its own catalog of registered services. If a client agent asks a particular service, it directly asks such a request to the facilitator agent, and the facilitator takes in charge of replying to the request. The broker agent simply mediates a service request between a client and a server. When a particular service is discovered, the broker informs it to the client agent, and the client directly sends a request to the server agent. A service agent does not have to register its service to the marketplace agent. Rather, a client agent registers its request to the marketplace agent, in advance. Whenever a service agent gets no message to process, it may voluntarily browse the marketplace agent to see if there is any request that can be served. In other words, a service agent actively tries to discover a request, in this case.

3.4 Real-Time ACL

For simple applications the messages communicated to and from agents may be quite simple. In this case, the agent communication language is nothing more different than an argument passing in a procedural way. However, the high-level abstraction of an agent enables us to express more complicated beliefs and intentions which should be regarded as knowledge. Our RT-ACL shares the same extensions with RT-KQML that is used in RTMAS [7], although we use the FIPA-ACL instead of KQML [9] as a base ACL for real-time extension as shown in Fig. 2.

```
(inform
      :sender   PatientAgent
      :receiver   PatientManagerAgent
      :content   (PatientID, Risklevel, Biosignal)
      :QoSInfo   (time, quality)
      :ontology   COPD
      ......
)
```

Fig. 2. RT-ACL with a real-time extension

3.5 Real-Time Scheduling Interface

RT-ASL provides a set of generic interfaces for the design-to-time real-time scheduling [10] architecture. In design-to-time scheduling uses the analyzed scheduling result by TAEMS (Task Analysis, Environment Modeling, and Simulation) in which there are three scheduling components: initial alternative generation, scheduling, and scheduling analysis. Thus, RT-ASL provides three scheduling interfaces that correspond to the TAEMS scheduling components, i.e. the interfaces for task tree generation, scheduling list generation, and scheduling analysis. We assume a service is organized hierarchically and consists of multi-agents. A service agent that responds to a client agent is a top-level task in a task tree, which manages the lower-level sub-functional agents with a proper scheduling algorithm.

4 Conclusion

We have presented the RT-ASL, a CORBA-based real-time agent service layer, which helps to write an agent-based ubiquitous application without having to wrestle with the complexities due to distributed object interconnection. We discussed why a higher abstraction than object is needed for ubiquitous computing environments, and why conventional static method invocation-based implementation of agents cannot solve the problem of object interconnection. In summary, we chose to use the CORBA Notification Service as an underlying communication infrastructure for agent communications. We also briefly presented the three different types of service discovery mechanism by which agent programmers have a freedom to choose for their own application, and a set of generic scheduling interfaces based on the design-to-time real-time scheduling architecture. We have implemented the prototype of RT-ASL on Redhat linux 9.0 with kernel version 2.4.20-8, using ACE+ TAO-1.3a [11] with latest patches version as a CORBA Ver. 2.6 compliant ORB.

References

1. Shoham Y.: Agent-oriented Programming, Artificial Intelligence, Vol. 60, No. 1 (1993)
2. Gensereth M.: Software Agents, Communications of ACM, Vol. 37, No. 7 (1994) 48-53
3. Object Management Group: The Common Object Request Broker Architecture: Core Specification, Version 3.0.3 (2004)
4. Graham J., Decker K.: Towards a Distributed, Environment-Centered Agent Framework, Proceedings of International Workshop on Agent Theories, Architectures and Languages (1999)
5. Bellifemine F., Poggi A., Rimassa G.: JADE – A FIPA-compliant Agent Framework, Proceedings of PAAM'99, London (1999) 97-108
6. Benech D., Desprats T.: A KQML-CORBA based Architecture for Intelligent Agents Communication in Cooperative Service and Network Management, Proceedings of IFIP/IEEE MMNS'97 (1997) 357-364
7. Dipippo L., Fay-Wolfe V., Nair L., Hodys E., Uvarov O.: A Real-Time Multi-Agent System Architecture for E-Commerce Applications, Proceedings of International Symposium on Autonomous Decentralized Systems (2001)
8. FIPA: The Foundations for Intelligent Physical Agents, http://www.fipa.org
9. Finin T., Fritzson R., McKay D., McEntire R.: KQML as an Agent Communication Language, Proceedings of the third International Conference on Information and Knowledge Management, Nov. (1994)
10. Garvey A., Lesser V.: Design-to-time Real-Time Scheduling, IEEE Transactions on Systems, Man and Cybernetics, Vol. 23, No. 6 (1993) 1491-1502
11. Schmidt D. C.: TAO Developer's Guide: Building a standard in performance, Version 1.3a, Object Computing Inc. (2003)

An Improved E-Learner Communities Self-organizing Algorithm Based on Hebbian Learning Law*

LingNing Li[1], Peng Han[2], and Fan Yang[2]

[1] Shanghai JiaoTong University, Shanghai 200030, China
[2] FernUniversitaet in Hagen, Hagen 58084, Germany

Abstract. In this paper we propose an improved E-Learner communities self-organizing algorithm based on Hebbian Learning Law, which can automatically group distributed e-learners with similar interests and make proper recommendations. Through similarity discovery, trust weights update and potential neighbors adjustment, the algorithm implements an automatic-adapted trust relationship with gradually enhanced satisfactions. It avoids difficult design work required for user preference representation or user similarity calculation. Hence it is suitable for open and distributed e-learning environments. Experimental results have shown that the algorithm has preferable prediction accuracy and user satisfaction. In addition, we achieve an improvement on both satisfaction and scalability.

1   Introduction

E-Learning which breaks the traditional classroom-based learning mode enables distributed e-learners to access various learning resources much more convenient and flexible. However, it also brings disadvantages due to distributed learning environment. Thus, how to provide personalized learning content is of high priority for e-learning applications.

An effective way is to group learners with similar interests into the same community[1]. Through strengthening connections and inspiring communications among the learners, learning of the whole community will get promotion. To achieve a better performance and a higher scalability, the organizational structure of the community would better be both self-organizing and adaptive[2].

We have investigated the behavior of students in the Network Education College of Shanghai Jiaotong University and found out that learners have strengthened trusts if they always share common evaluations or needs of learning resources, which is very similar with the Hebbian Learning theory proposed by Hebb D.O. based on his observation on bio-systems [3].

In this paper, we present an improved E-Learner communities self-organizing algorithm relying on the earlier work by F. Yang [4]. The algorithm uses corresponding feedback to adjust relationships between learners, aiming to find

---

* Supported by National Natural Science Foundation of China. Grant No. 60372078.

A. Sattar and B.H. Kang (Eds.): AI 2006, LNAI 4304, pp. 907–911, 2006.
© Springer-Verlag Berlin Heidelberg 2006

similar learners and provide facilities in their collaboration. Experimental results have shown that the algorithm has preferable prediction accuracy and user satisfaction. In addition, we achieve an improvement on both satisfaction and scalability.

2   Framework of Collaborative Agent

2.1   Agent Feature

Each agent in the community holds a set of resources, which are rated by the user. A bigger rating number means a better evaluation on one resource. Each agent can only be aware of its neighbors, while other agents are out of sight and can not be communicated with directly. No central agent or server exists, so it does not matter any agent breaks down or quits from the community. Hence, it forms a pure P2P environment.

We define TNL^i to be the trusted neighbor list of user u_i, storing the information of u_i's trusted neighbors and the related trust weights.

2.2   Query Behavior

During the operation of the system, a particular learner could be given lots of candidate resources for reading. He could then send out a request to his trusted neighbors asking for their advices. Such kind of request will be referred to "query".

If the neighbor finds that the queried resource has not been evaluated, it will then forward this query to its neighboring agents. The depth that a recommendation query will be forwarded is controlled by TTL (time-to-live), to avoid unlimited recurrence. Every time a query has been forwarded, the TTL will be decreased by one. Once the TTL reaches zero or the evaluation on the queried resource has been found, the forward process will cease and the corresponding result will be sent back along the query path.

3   Self-organizing Weight Update Algorithm

3.1   Trust Weight Update Algorithm

When u_i sends a query on r_k to t neighbors, ne_1, ne_2, $\cdots$, ne_t, the query spreads along t pathways and will get t ratings v_k^1, v_k^2, $\cdots$, v_k^t back before TTL decreases to 0. We compute the recommendation rating on r_k as:

$$*v_k^i = \sum_{j=1}^{t} \frac{w_{i,j} \cdot v_k^j}{\sum_{j=1}^{t} w_{i,j}} = \frac{\sum_{j=1}^{t} w_{i,j} \cdot v_k^j}{\sum_{j=1}^{t} w_{i,j}} \tag{1}$$

The rating v_k^j denotes an individual view from the neighbor ne_j. After reading the resource r_k, user u_i makes an rating v_k^i himself. The difference between v_k^j and v_k^i is the key to update trust weights between users.

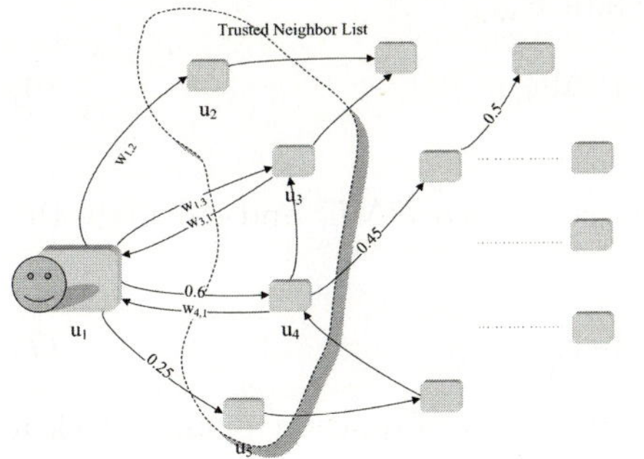

Fig. 1. Structure of Trusted Neighbor List

Fig. 2. Updating of Trust Weights Among Neighbors

Based on the Hebbian learning law, we define the trust weight increment strategy in the recommendation network as Equation 2

$$w_{i,j} = w_{i,j} + \Delta w_{i,j}, \quad \Delta w_{i,j} = \eta \cdot v_k^j v_k^i \tag{2}$$

where η is the learning rate.

We hope a great difference between v_k^j and v_k^i makes $w_{i,j}$ decrease while a little difference makes it increase. So a threshold T1 is set to determine the update direction. Besides, if two users have similar ratings in the higher interval or lower interval at [0,1], that means they both specially "like" or "dislike" this resource. It makes sense to strengthen the connection between these two users.

$$\kappa = \begin{cases} 1 + |v_k^j + v_k^i - 1| & \text{if } |v_k^j - v_k^i| \leq T1 \\ -1 & \text{if } |v_k^j - v_k^i| > T1 \end{cases} \tag{3}$$

If $\kappa > 0$, we say u_i get a positive response. At the same time, greater difference should lead to more decrease, and less difference should lead to more increase. So h is defined to emphasize these differences.

$$h = 1 + \left|2|v_k^j - v_k^i| - 1\right| \tag{4}$$

The learning rate should include the two factors above, besides, a pre-defined parameter β is need to control the overall rate. Hence,

$$\eta = \beta \cdot \kappa \cdot h \tag{5}$$

3.2   Potential Neighbor Structure Adaption

In order to speed the community organization process, we add a Potential Neighbor List (PNL) to store the neighbors with similar interests but without direct connections.

If a long query pathway (length>1) returns a similar rating with the current user, we can deduce the end user u_e in the pathway is qualified for a potential neighbor. The strategy of management for the PNL is as follows:

- If $u_e \in PNL^i$, then update its trust weight $w_{i,e}$:

$$w_{i,e} = w_{i,e} + \Delta w_{i,e}, \tag{6}$$

where $\Delta w_{i,e}$ is calculated by Equation 2.
- If $u_e \notin PNL^i$ and PNL^i is not full, insert u_e into PNL^i and calculate the trust weight as shown in Equation 7.

$$w_{i,e} = \Delta w_{i,e} \tag{7}$$

- If $u_e \notin PNL^i$ and PNL^i is full, calculate $w_{i,e}$ by Equation 7, and check if there exists any potential neighbor ne_l with a lower trust weight than $w_{i,e}$. If there is, insert u_e into PNL^i and delete ne_l.

After each update of the trust weight (in TNL as well as in PNL), check the user u_k with the highest trust weight w_k^{max} in PNL^i . Insert it into TNL^i if the list is not full. Otherwise, check the user with lowest trust weight w_k^{min} in TNL^i and whether it is lower than w_k^{max} . If it is, switch it with user u_k.

The update strategy is shown in Fig. 2, in contrast with Fig. 1.

4   Experiments and Evaluations

We define the term "satisfaction" as a main measurement of performances achieved by agents' behaviors. Each positive recommendation should lead to increasing of the receiver's satisfaction, otherwise, to decreasing.

So the satisfaction is related to the trust weight increment. In order to limit the satisfaction from -1 to 1 and create a nonlinear convex curve due to the principle of diminishing marginal utility, the satisfaction is calculated as:

$$\Delta^* w_{i,j} = \Delta^* w_{i,j} + \eta \cdot {}^* v_k^i v_k^i, \quad S = 2arctan(\rho \cdot \Delta^* w_{i,j})/\pi \tag{8}$$

where ρ controls the rate of the increment and $\Delta^* w_{i,j}$ is initialized to be 0.

The Hebbian Learning algorithm implemented by Yang[4] was used as a benchmark in contrast with the improved algorithm specified in this paper. The former was referred to "Traditional H. L." while the latter to "Improved H. L.".

As the total behaviors of the learners increase, the average satisfaction of the community climbs as illustrated in Fig. 3 with 500 users involved. After 5000 behaviors performed, the performance of the Improved H. L. will exceed the Traditional H. L., which proves its effectiveness in satisfying the recommendation need of learners.

Since each user only sends the query to the users selected, Hebbian Learning algorithm has a much lower calculation complexity than the memory based recommendation algorithm such as the Item-based Collaborative Filtering [5], which always has a time complexity of $\Omega(n^2)$. The Fig. 4 shows that the behaviors required by the Hebbian Learning algorithm increased almost linearly in relation to the number of users. In addition, the Improved H. L. achieves a little better performance than the Traditional H. L.

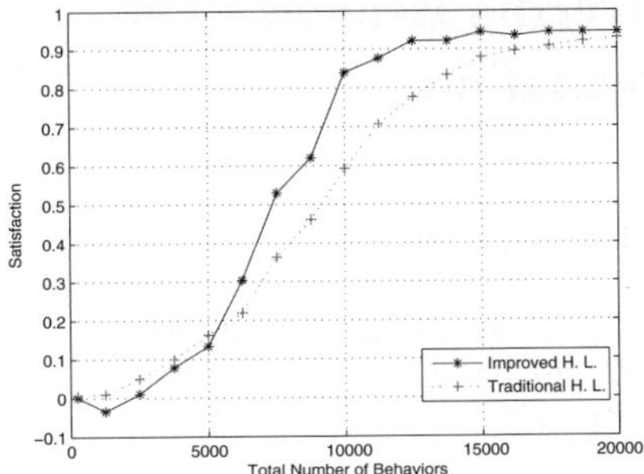

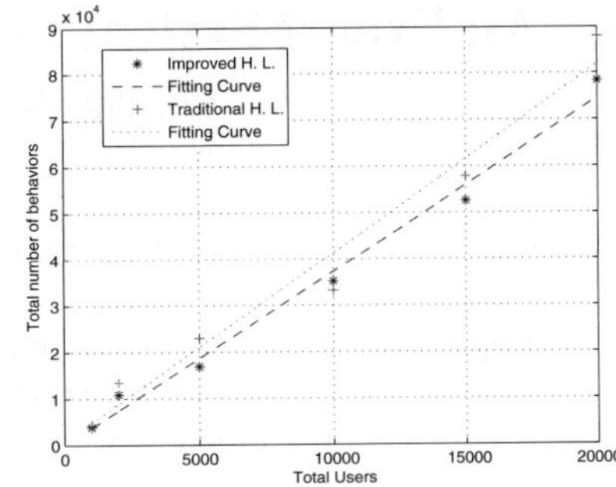

Fig. 3. Comparison of Satisfaction between Improved and Traditional Hebbian Learning Laws

Fig. 4. Scalability of the Self-organizing Algorithm

5   Conclusion

This paper has introduced a novel method to find and organize similar learners in an E-Learning community. This method is based on the Hebbian Learning law, and takes improvements relying on the earlier work by Dr. Yang[4].

The algorithm presented in this paper avoids difficult design work required for user preference representation or user similarity calculation, while reflect user preferences accurately. In addition, because the community is organized based on P2P communication and local interaction, it is suitable to work in open and distributed environments.

References

1. R. Shen, F. Yang, and P. Han. A dynamic self-organizing e-learner communities with improved multi-agent matchmaking algorithm. In *16th Australian Joint Conference on Artificial Intelligence (AI2003)*, pages 590–600, 2003.
2. P. Turner and N. Jennings. Improving the scalability of multi-agent systems. In *Proceedings of the First International Workshop on Infrastructure for Scalable Multi-Agent Systems*, pages 246–262, 2000.
3. D. O. Hebb. *The Organization of Behaviour.* 1949.
4. F. Yang. *Analysis, Design and Implementation of Personalized Recommendation Algorithms Supporting Self-organized Communities.* PhD thesis, Faculty of Mathematics and Information Technology FernUniversitaet at Hagen, 2005.
5. B. Sarwar, G. Karypis, J. Konstan, and J. Reidl. Item-based collaborative filtering recommendation algorithms. In *Proceedings of the 10th international conference on World Wide Web*, pages 285–295, May 2001.

An Expert System for Recovering Broken Relics Using 3-D Information

Ho Seok Moon* and Myoungho Oh

Department of Electronics and Information Science, Korea Military Acacdemy
Gongneung-Dong, Nowon-Ku, Seoul 139-799, South Korea
bawooi@korea.ac.kr, mhoh@kma.ac.kr

Abstract. An expert system is newly proposed to recover the broken fragments of relics into an original form. The system automatically assembles the fragments of tombstones by placing them in the right position. The system consists of three processes: aligning the front and letters of an object, identifying the matching directions, and determining the detailed matching positions. Least squares fitting, geometric and RGB error, and vector similarity methods are applied to the matching process. The system's algorithm has been simplified in contrast to the previous methods relying on relatively complex methods. 2-D translations via fragments-alignment save the computational load significantly. The benefit of proposed method is proven in the project of Ministry of Culture and Tourism and Ministry of Information and Communication of the Republic of Korea.

1 Introduction

The technology of retrieving digital images can have wider applications than ever before in today. One of the interesting applications is the digitalization of cultural assets. If cultural assets can be digitized as contents, they can be very useful to people in society. Due to their fragile and aging natures, special attention should be paid to cultural assets. It is very likely that computer vision and graphic technique can be very useful in the restoration task of excavated relics.

In the area of the reconstruction of fragments, there have been considerable studies. Most of them are on 2-D jigsaw puzzles. Among them, Wolfson [1] solved the puzzle by representing the boundary curve in polygonal approximation. Goldberg, Malon, and Bern [2] solved puzzles where fragments have border within more than four neighbors. Their ideas can be applicable to a variety class of puzzles. Other algorithms in 2-D problem can be found in Radack [3] and Wolfson et al. [4]. Most previous works on solving jigsaw puzzles mainly have been conducted on 2-D objects. However, 3-D solid modeling can be more reasonable in real-world problems.

Our research is distinct from previous researches in that the object of recovery is the *Hwaeomseokgyeong* that is the stone sutra in the *Hwaeom* temple in Korea. Since the stone sutra was designated as the Treasures No. 1040 by Cultural Properties Administration in 1991 [5], its recovery has been of special importance to the

---

* Corresponding author.

A. Sattar and B.H. Kang (Eds.): AI 2006, LNAI 4304, pp. 912–916, 2006.
© Springer-Verlag Berlin Heidelberg 2006

preservation of Korean cultural heritage. There have been few prior studies associated with the reconstruction of stone monuments like the *Hwaeomseokgyeong*. From both of economic and cultural standpoints, the theme is sure to be of great value.

We propose newly an expert system for recovering the broken fragments of relics into an original form by tracing the right matching positions. The proposed system works in the following sequences: (1) the raw fragments of relics are transformed into the converted 3-D fragment models through image-scanning with a 3-D range scanner: (2) they are preprocessed through least squares fitting to the front faces and letters thereon to obtain the 3-D vector data: and (3) the right directions and detailed matching positions for assembling are determined through the calculation of the vector similarity and the geometric and RGB error between the fragments of relics.

2 System Architecture and Procedure

The system for fragments-matching uses as an input the scanned data of broken fragments (see Fig. 1). They consist of vertexes and are stored into a database through triangulation. In order to ensure consistent data format, the scanning of all fragments should be made using the same device under the same conditions. Since the fragment information stored in the 3-D image is entered in arbitrary directions, it is difficult to match fragments without searching the front of the object and the vertical alignment of letters.

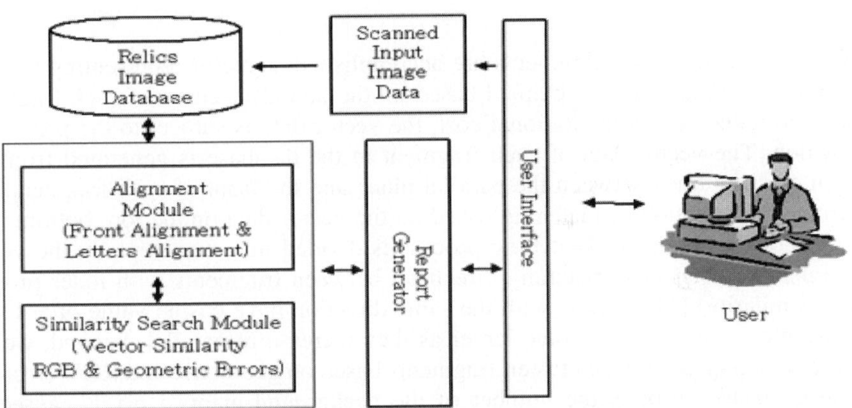

Fig. 1. Flow of Matching System

The front of the object corresponds to the face with letters engraved in intaglio(see Fig. 2.). By using this fact in matching fragments, we can align the front of a fragment parallel to X-Y plane. We make a plane by choosing three points in the front of the fragment and obtain normal vector of the plane. We calculate the angle of rotation of the plane to X-Y plane and rotate the fragment. We call this the front alignment.

Most sculptors, when engraving letters on the tombstone, predetermine the direction of arranging letters. The data on the letters appears, as shown in Fig. 2. Thus, it is possible to align the column of letters in line with the Y axis. As shown in Fig. 2, a bounding box is constructed using the minimum and maximum points among the vertexes representing the individual letter. The center of the bounding box is determined as an average of the coordinates of X and Y axes for each letter. We obtain the letters data and calculate the center of each letter. We find a line of the column of letters by using least squares fitting (LSF) technique and calculate the angle between the line and Z axis. We rotate the fragment. We call this letters alignment in this paper. LSF is used for the letters alignment. LSF is a method of finding a line (or a curve) such that the summation of vertical distances between data points and the line (or the curve) is minimized [6]. Once the fragments are in the front and letters alignments, the translation into X and Y axes only is required without the rotation onto X, Y, and Z axes in order to determine the matching positions.

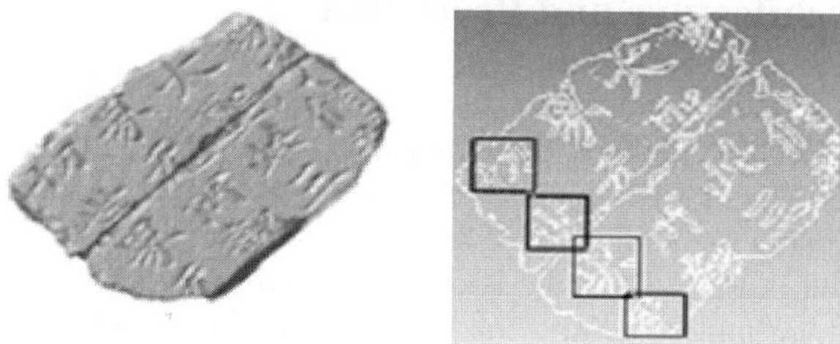

Fig. 2. Extraction of Letter Data Using the Cross (Overlapped) Lines between the Front of the Fragments and the Plane

After front and letters alignments are accomplished, a vector representing the feature of aligned fragments is obtained based on the boundary curve of each fragment. In order to reduce the computational cost, the vector data is subject to the process of decimation. The vector data of each fragment in the database is generated from the data on crossing lines between the parallel plane and the front of each fragment. We determine the direction of matching based on the vector data on the top, bottom, left, and right of the fragment. Matching process is divided into two steps by the vector data. First, we determine matching directions between fragments with inner product (vector similarity) [7]. Vectors with the same direction have cosine value of ± 1, and the absolute cosine value become larger as they increase in parallel. Second, we determine matching positions between fragments based on geometric and RGB errors.

The geometric error is the number of the background-mapped pixels caused by matching each captured image [8]. The error is calculated as a result of traversing the composite image of temporarily combined fragments. That is, it is the area size of the crack which appears when two fragments are matched. The background color-mapped pixel indicates the area size of the crack. Because this traversal method only finds the background color-mapped pixels with no other operations, the required time is independent of the complexity of the input models. It is only dependent on the resolution of the captured image. In order to match each captured image more effective, we considered RGB information of adjacent faces as well as geometric error.

The lengthwise RGB error can be obtained by comparing the RGB difference between the pixels of the same X coordinate. Likewise, the crosswise RGB error can be calculated with the RGB difference between the pixels of the same Y coordinate along the matching boundary.

The comparative advantage of our method lies in the new scheme designed for eliminating false matching positions by calculating the RGB difference as well as the simple geometric between the corresponding fragments. Equation (1) describes combined errors where the weight parameter (α) is chosen by the users. Where E_{geo} is the geometric error and E_{rgb} is RGB error, respectively,

$$E = \alpha E_{geo} + (1 - \alpha) E_{rgb} . \tag{1}$$

After the optimal matching position is found, each fragment is aligned at the corresponding place by performing a reverse operation on the transformation and scaling.

3   Experimental Results

We illustrate the recovery system using the *Hwaeomseokgyeong*, which is the stone sutra in the *Hwaeom* temple in Korea. Some 9 fragments were used as experimental fragments. Fig. 3 shows the pictures after final matching according to different values of α. The equal weighting between geometric and RGB information ($\alpha = 0.5$) performs the best, while $\alpha = 0.0$ or $\alpha = 1.0$ does not yield favorable results. Particularly, $\alpha = 0.2$ produced the poorest performance. The circled parts show false matching in Fig. 3. The performance is a result from the judgment by cultural arts experts.

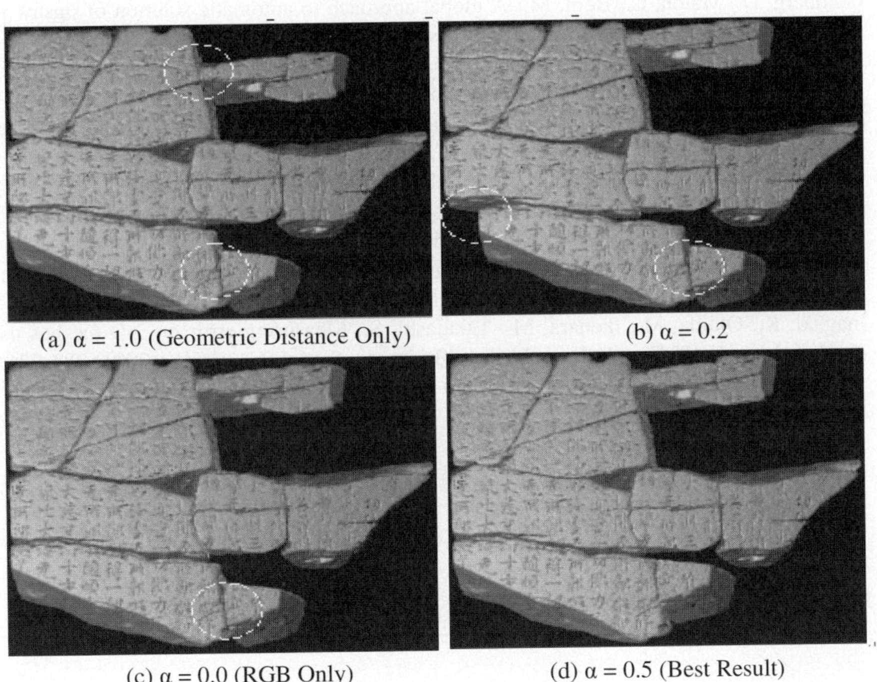

(a) $\alpha = 1.0$ (Geometric Distance Only)   (b) $\alpha = 0.2$

(c) $\alpha = 0.0$ (RGB Only)   (d) $\alpha = 0.5$ (Best Result)

Fig. 3. Matching Results According to α

4 Conclusion

In this paper, an expert system is developed for recovering damaged fragments through a series of systematic processes: arranging in the front and letter; identifying the matching directions; and determining the matching positions. The least squares fitting (LSF) was used for letters alignment. The identification of matching directions was made using the similarity technique based on inner product. The geometric and RGB errors were employed to find the matching positions. The system was applied to recovery projects of the *Hwaeomseokgyeong* that is the stone sutra in the *Hwaeom* temple in Korea.

The recovery system can be a pioneering expert system for recovering cultural assets such as monuments. This attempt can work as a stepping-stone for further studies in this area. In addition, the system can deal with both damaged and undamaged fragments. This can yield wider and more realistic applications. The system's algorithm was simplified in contrast to the previous methods relying on relatively complex methods. It turned out that the technique of 2-D translations via fragments alignment enables us to save the computational load significantly.

References

1. Wolfson, H.: On curve matching. IEEE Transactions on Pattern Analysis and Machine Intelligence, Vol. 2. (1990) 483-489
2. Goldberg, D., Malon, C., Bern, M.: A global approach to automatic solution of jigsaw puzzles. Proceedings of the 18th Annual Symposium on Computational Geometry, (2002) 82-87
3. Radack, G.M., Badler, N.I.: Jigsaw puzzles matching using a boundary-centered polar encoding. Computer Graphics Image Processing, Vol. 19. (1982) 1-17
4. Wolfson. H., Schonberg, E., Kalvin, A., Lambdan, Y.: Solving jigsaw puzzle using computer vision. Annals of Operations Research, Vol. 12. (1988) 51-64
5. Cultural Properties Administration.: Cultural properties yearbook, Daejeon, Korea (1991)
6. Dyer, S.A., He, X.: Least squares fitting of data by polynomials. IEEE Instrumentation and Measurement Magazine, Vol. 4. (2001) 46-51
7. Lee, W. P., Tsai, T.C.: An interactive agent-based system for concept-based web search. Expert Systems with Applications, Vol. 24. (2003) 365-373
8. Inagaki, K., Okuda, M., Ikehara, M., Takajashi, S.: Measuring error on 3-D meshes using pixel division. IEEE Fourth Workshop on Mutimedia Signal Processing, (2001) 281-286

Permutation Flow-Shop Scheduling
Based on Multiagent Evolutionary Algorithm

Kang Hu, Jinshu Li, Jing Liu, and Licheng Jiao

Institute of Intelligent Information Processing, Xidian University
710071 Xi'an, China
xdhukang@gmail.com

Abstract. Generated by integrating the multiagent systems and evolutionary algorithms, Multiagent Evolutionary Algorithm for PFSPs (MAEA-PFSPs) enables the agents to interact with their environment. With three designed behaviors, each agent increases energy as much as possible, so that MAEA-PFSPs find the optima. In the experiments, 29 benchmark PFSPs are used to compare with a GA-based algorithm, and good performance is obtained.

1   Introduction

In this paper, we integrate the multiagent systems and evolutionary algorithms (EAs) to form a new algorithm, Multiagent Evolutionary Algorithm for Permutation Flow-shop Scheduling Problems (MAEA-PFSPs). In MAEA-PFSPs, all agents live in a lattice-like environment. Using three designed behaviors, MAEA-PFSPs enables the agents to sense and act in the lattice environment, thus each agent increases its energy to find the optima. Experimental results show good performance of MAEA-PFSPs.

2   Multiagent Evolutionary Algorithm for Permutation Flow-Shop Scheduling Problems

2.1   Agent for Permutation Flow-Shop Scheduling Problems

One encoding method in common use for PFSPs is permutation of n jobs:

$$\forall P \in S, \ P = \langle P_1, P_2, \cdots, P_n \rangle \text{ and}$$
$$\forall i \neq j, \ P_i \neq P_j, \ i = 1, 2, \cdots, n, \ j = 1, 2, \cdots, n \tag{1}$$

Definition 1: Suppose a represents an agent for PFSPs, its energy is defined as

$$\forall a \in S, \ Energy(a) = -makespan(a) \tag{2}$$

A. Sattar and B.H. Kang (Eds.): AI 2006, LNAI 4304, pp. 917–921, 2006.

© Springer-Verlag Berlin Heidelberg 2006

Each agent is represented by the following structure:

```
Agent = record
   P:  P∈S;  E:  The energy of the Agent,  E = Energy(P);
   SL:  The flag for the self-learning behavior.
   S:   The stress the Agent suffers, which will be
        defined later.
End.
```

2.2 Environment of Agents

Definition 2: All agents live in a lattice-like environment, L. The size of L is $L_{size} \times L_{size}$. The neighbors of $L_{i,j}$, $Neighbors_{i,j}$, are defined as follows:

$$Neighbors_{i,j} = \left\{ L_{i',j}, L_{i,j'}, L_{i'',j}, L_{i,j''} \right\} \tag{3}$$

where $i' = \begin{cases} i-1 & i \neq 1 \\ L_{size} & i=1 \end{cases}$, $j' = \begin{cases} j-1 & j \neq 1 \\ L_{size} & j=1 \end{cases}$,

$i'' = \begin{cases} i+1 & i \neq L_{size} \\ 1 & i = L_{size} \end{cases}$, $j'' = \begin{cases} j+1 & j \neq L_{size} \\ 1 & j = L_{size} \end{cases}$.

The agent lattice can be represented as the one in Fig.1. Each circle represents an agent, and the paired numbers represent its position in the lattice. Two agents can interact with each other if and only if there is a line connecting them.

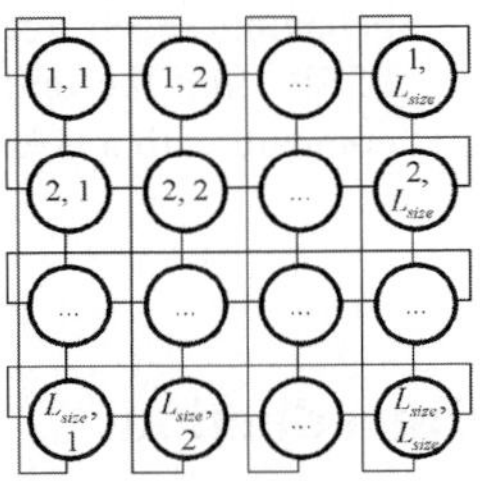

Fig. 1. The model of agent lattice

2.3 Behaviors of Agents

Competitive behavior: Suppose that $Max_{i,j}$ is the agent with maximum energy among the neighbors of $L_{i,j}$. If $L_{i,j}(E) \leq Max_{i,j}(E)$, $Max_{i,j}$ generates a child agent, $Child_{i,j}$, to replace $L_{i,j}$, and the method is shown in Algorithm 1.

Algorithm 1 Competitive behavior

```
Input:  Max_{i,j} : Max_{i,j}(P) = ⟨m₁,m₂,⋯,mₙ⟩ ;
Output: Child_{i,j} : Child_{i,j}(P) = ⟨c₁,c₂,⋯,cₙ⟩ ;
Random(n,i) is a random integer in [0,n] (not equal to i)
```

```
1  begin                                8          Swap(c_i, c_l);
2     Child_{i,j}(P) = Max_{i,j}(P);     9       end;
3     i := 1;                            10       i := i+1;
4     repeat                            11    until (i > n)
5        if (U(0,1) < P_c) then         12    Child_{i,j}(SL) := True;
6        begin                          13 end.
7           l := Random(n,i);
```

Self-learning behavior: We design the self-learning behavior for agents to improve the performance by making use of local search techniques.

Suppose that this behavior is performed on $L_{i,j}$. The details are shown below.

Algorithm 2 Self-learning behavior

Input: $L_{i,j} : L_{i,j}(P) = \langle p_1, p_2, \cdots, p_n \rangle$;

Output: $L_{i,j}$;

```
1   begin
2     repeat
3         Repeat := False ;
4         k := 1 ; Iteration := 1 ;
5         while  ( k ≤ n )  do
6         begin
7             Energy_old := L_{i,j}(E) ;
8             l := Random(n,k) ;
9             Swap(P_k, P_l) ;
10            Energy_new := L_{i,j}(E) ;
11            if ( Energy_new < Energy_old )
12                then  Swap(P_k, P_l) ;
13                else begin
14                if ( Energy_new > Energy_old )
15                    then  Repeat := True ;
16                    k := k + 1 ;
17                end;
18                if ( Iteration < n − 1 ) then
19                    Iteration := Iteration + 1 ;
20                else begin
21                    Iteration := 1 ;
22                    k := k + 1 ;
23                    end;
24            end;
25      until  ( Repeat = true )
26      L_{i,j}(SL) := False ;
27  end.
```

Mutation behavior: The agents with relatively low energy are under great stress of competition, and the stress is defined as follows:

Definition 3: The stress of $L_{i,j}$, $Agent_{i,j}(S)$, is equal to the number of agents in $Neighbors_{i,j}$ whose energy are higher than or equal to $Agent_{i,j}(E)$.

Algorithm 3 Mutation behavior

Input: $L_{i,j} : L_{i,j}(P) = \langle p_1, p_2, \cdots, p_n \rangle$

         P_m : A predefined real number in the range of 0~1

Output: $L_{i,j}$

```
1   begin
2       Compute Agent_{i,j}(S) ;
3       if ( U(0,1) < P_m × Agent_{i,j}(S) / 4 )
4       then begin
5           i := Random(n − Len_max) ;
6           l := Random(Len_max) + i ;
7           Inverse(P_i, P_l) ;
8           L_{i,j}(SL) := True ;
9           end;
10  end.
```

2.4 Implementation of MAEA-JSPs

At each generation, the competitive behavior is performed on each agent first. As a result, the agents with low energy are cleaned out from the agent lattice. Then, the self-learning behavior is performed and at the end is the mutation behavior.

Algorithm 4 Multiagent evolutionary algorithm for permutation flow-shop scheduling problems

Input: $Evaluation_{Max}$: The maximum number of evaluations

Output: A solution or an approximate solution;

L^t is the agent lattice at the tth generation. $Agent_B^t$ is the best agent in $L^0, L^1, \cdots L^t$, and $Agent_{tB}^t$ is the best agent in L^t . $Evaluations$ increases when any energy is computed.

```
1   begin
2     for i := 1 to L_size do
3       for j := 1 to L_size do
4       begin
5         Assign a random
6         permutation to L^0_{i,j}(P) ;
7         Compute L^0_{i,j}(E) ;
8         L^0_{i,j}(SL) := True ;
9       end;
10    Update Agent^0_B ;   t := 0 ;
11    repeat
12      for i := 1 to L_size do
13        for j := 1 to L_size do
14        begin
15          if ( L^t_{i,j} wins in the
16          competitive behavior)
17            then L^{t+1/2}_{i,j} := L^t_{i,j} ;
18            else L^{t+1/2}_{i,j} := Child_{i,j} ;
19          Compute L^{t+1/2}_{i,j}(E) ;
20        end;
21      Update Agent^{t+1/2}_{(t+1/2)B} ;
22      if ( Agent^{t+1/2}_{(t+1/2)B}(SL) = True )
23        then Perform the
24      self-learning behavior
25            on Agent^{t+1/2}_{(t+1/2)B}
26      for i := 1 to L_size do
27        for j := 1 to L_size do
28        Perform the mutation
29          behavior on L^{t+1/2}_{i,j}
30      Update Agent^{t+1}_{(t+1)B} ;
31      if ( Agent^{t+1}_{(t+1)B}(E) < Agent^t_B(E) )
32        then begin
33          Agent^{t+1}_B := Agent^t_B ;
34          Agent^{t+1}_{Random} := Agent^t_B ;
35        end;
36        else Agent^{t+1}_B := Agent^{t+1}_{(t+1)B} ;
37      t := t + 1 ;
38    until ( Makespan = C^* or
39      ( Evaluations >= Evaluation_Max ) ;
40  end.
```

3 Experimental Results

Car[6], Hel[7], and Rec[8] benchmark problems[1] are used to investigate the performance of MAEA-PFSPs. Comparing our results with the ones generated by Improved Genetic Algorithm (IGA) [9], we set the conditions as: $Evaluation_{Max} = 10000000$, $L_{size} = 10$, $P_c = 0.2$ and $P_m = 0.8$. BG and AG represent the best and the average of the percentage gaps between $Makespan$ and C^* in 100 independent runs. We set the result in **bold** when it is better than IGA and in *italic* when it is worse.

---

[1] http://people.brunel.ac.uk/~mastjjb/jeb/orlib/files/flowshop1.txt

Table 1. The experimental results of MAEA-PFSPs

Name	$n \times m$	C^*	IGA		MAEA-PFSPs		
			BG	AG	BG	AG	Standard Deviation
Car1	11×5	7038	0	0	0	0	0
Car2	13×4	7166	0	0	0	0	0
Car3	12×5	7312	0	0	0	0	0
Car4	14×4	8003	0	0	0	0	0
Car5	10×6	7720	0	0.61	0	**0**	0
Car6	8×9	8505	0	0.76	0	**0**	0
Car7	7×7	6590	0	0	0	0	0
Car8	8×8	8366	0	0	0	0	0
Rec01	20×5	1247	0	0.14	0	0	0
Rec03	20×5	1109	0	0.14	0	0	0
Rec05	20×5	1242	0	0.31	0	**0.20**	0.15
Rec07	20×10	1566	0	0.59	0	0	0
Rec09	20×10	1537	0	0.79	0	0	0
Rec11	20×10	1431	0	1.48	0	0	0
Rec13	20×15	1930	0.62	1.52	**0**	**0.18**	2.01
Rec15	20×15	1950	0.46	1.28	**0**	**0.15**	3.92
Rec17	20×15	1902	1.73	2.69	**0**	**0.11**	3.20
Rec19	30×10	2093	1.09	1.58	**0.28**	**0.61**	4.83
Rec21	30×10	2017	1.44	1.52	**0.44**	**1.12**	7.36
Rec23	30×10	2011	0.45	0.99	**0.44**	**0.50**	0.70
Rec25	30×15	2513	1.63	2.74	**0.43**	**0.93**	5.27
Rec27	30×15	2373	0.80	2.11	**0.56**	**0.98**	3.97
Rec29	30×15	2287	1.53	2.59	**0.78**	**1.19**	6.28
Rec31	50×10	3045	0.49	1.62	**0.10**	**1.52**	8.34
Rec33	50×10	3114	0.13	0.75	**0**	**0.12**	1.32
Rec35	50×10	3277	0	0	0	0	0
Rec37	75×20	4951	2.26	3.49	2.72	**3.13**	12.44
Rec39	75×20	5087	1.14	1.93	*1.61*	**1.92**	8.93
Rec41	75×20	4960	3.27	3.78	**2.70**	**3.01**	11.44
Hel01	100×10	513	-	-	0.38	0.38	0
Hel02	20×10	135	-	-	0	0.22	0.45

As can be seen, MAEA-PFSPs outperforms IGA at *BG* columns on 12 out of 29 problems, and on 17 out of 29 at *AG* columns. It finds the optimal solutions at least for one time on 20 problems with the results of the rest very close to the optima. What's more, the small values at the last column indicate the steadiness.

References

1. Carlier J.: Ordonnancements a Contraintes Disjonctives. RAIRO. Operations Research 12. (1978) 333-351
2. Heller J.: Some Numerical Experiments for an M × J Flow Shop and its Decision-Theoretical aspects. Operations Research 8. (1960) 178-184
3. Colin R. Reeves.: A Genetic Algorithm for Flowshop Sequencing. Computers and Operations Research. 22(1) (1995) 5-13
4. Wang, L.: Shop Scheduling with Genetic algorithms. Tsinghua University Press, Beijing, China (2003)

A Novel Mobile Epilepsy Warning System

Ahmet Alkan[1], Yasar Guneri Sahin[1], and Bekir Karlik[2]

[1] Yasar University, Department of Computer Engineering
Bornova Campus, 35500, Izmir, Turkey
{ahmet.alkan, yasar.sahin}@yasar.edu.tr
[2] Fatih University, Department of Computer Engineering
Buyukcekmece, 34500, Istanbul, Turkey
bkarlik@fatih.edu.tr

Abstract. This paper presents a new design of mobile epilepsy warning system for medical application in telemedical environment. Mobile Epilepsy Warning System (MEWS) consists of a wig with a cap equipped with sensors to get Electroencephalogram (EEG) signals, a collector which is used for converting signals to data, Global Positioning System (GPS), a Personal Digital Assistant (PDA) which has Global System for Mobile (GSM) module and execute Artificial Neural Network (ANN) software to test current patient EEG data with pre-learned data, and a calling center for patient assistance or support. The system works as individual sensors obtain EEG signals from patient who has epilepsy and establishes a communication between the patient and Calling Center (CC) in case of an epileptic attack. MEWS learning process has artificial neural network classifier, which consists of Multi Layered Perceptron (MLP) neural networks structure and back-propagation training algorithm.

1 Introduction

Epilepsy can present unique challenges to anyone with the disease. This already difficult situation can be exacerbated by poverty, lack of seizure control, transportation issues, and the stigma that still surrounds epilepsy [1]. For this reason, epileptic patients must be kept under control and observed and controlled with life-long care in order to prevent and/or minimize complications from unexpected seizures. By way of control some of medical pursuit systems have been developed to observe the current status of a patient.

While controlling the epileptic patients, their social life should not be affected because of any equipment that is used for medical control. In the other word, if any patient is wanted to be observed by medical center or his family by using some medical devices connected to him, these devices should not obstruct him to walk move freely within society, or to create discomfort. Of course, these devices must be controllable and give correct results as much as possible.

In this paper, a new mobile warning system have been investigated and presented that it can be used for epilepsy patients to be under controlled and observed for their epilepsy attacks. Briefly, the system is a combination of Global Positioning System (GPS) to get exact location of patient, Mobile Phone System to send data and current status of patients, a wig with a cap equipped with EEG sensors to get epileptic status of patients, a collector and a little software run on PDA or GSM that is used for

A. Sattar and B.H. Kang (Eds.): AI 2006, LNAI 4304, pp. 922–928, 2006.
© Springer-Verlag Berlin Heidelberg 2006

converting analog signals to digital and testing EEG signals with Artificial Neural Network algorithm, a cancel alarm button which is used in case of signaling errors, and an emergency call-center to detect emergencies and evaluate patient status.

About 1% of the world population suffers from epilepsy and 30% of epileptics are not helped by medication [2]. EEG signals involve a great deal of information about the function of the brain. Careful analyses of the EEG records can provide valuable insight into this extensive brain disorder. Therefore EEG is an important component in the diagnosis of epilepsy. On the other hand, the classification and evaluation of these signals are limited, there is no definite criterion evaluated by the experts for the visual analysis of EEG signals. Since time domain analysis is insufficient, generally frequency domain analysis, different signal processing and soft-computing techniques of these signals are used.

Last two decades, the recognition of the EEG signal characteristics can be carried out using a number of soft-computing approaches, such as Artificial Neural Networks (ANN) and Fuzzy Logic. For example; Alkan et al., have used logistic regression and artificial neural network for epileptic seizure detection [3]. In addition there are some other studies for detecting epilepsy using telemedicine facilities, those can be seen in [4],[5],[6].

Subasi has used dynamic fuzzy neural networks [7], Subasi et al., have used wavelet neural networks [8], parametric and subspace methods [9] for epileptic seizure detection. Ceylan and Ozbay [10], Alkan et al. [3], Kiymik et al. [11], have used different structures of MLP neural networks. Most of the research that has been carried out involves utilization of MLP neural networks with one hidden layer using the back-propagation algorithm. Very few researchers have used MLP neural networks with two hidden layers.

Alkan et al. [3] have reported a solution, based on classical, modern and subspace spectral analysis methods for EEG signal processing using two hidden layers. In their study, the spectral analysis methods are used as features for the feature sets. These feature sets are also classified for EEG signals using an MLP neural network structure, with a back-propagation supervisor training algorithm.

A MLP in association with the training algorithm is developed on detailed classification between set A (healthy volunteer, eyes open) and set E (epileptic seizure segments). With usage of the data obtained from Andrzejak et al. studies [12], each of the ANN classifier was trained so that they are likely to be more accurate for one class of EEG signals than the other classes. The correct classification rates and convergence rates of the proposed ANN model were examined and then performance of the ANN model was reported. Finally, some conclusions were drawn concerning the MLP on classification of the EEG signals. The classification problem may be divided into feature extraction, dimensionality reduction, and pattern recognition. The results of classification accuracy of this proposal study are found to be successful.

2 Mobile Epilepsy Warning System

Figure 1 describes the general diagram for mobile epilepsy warning system. The key point of this system is creating a warning signal in case of any emergency (epilepsy attack) and calling the nearest ambulance service and patient's related persons. The additional innovation of this proposed system is the apparatus is physically (innovative) and comfortable, enhancing the patient's sense of esteem, because the patient's apparatus is disguised as a wig or hat wired with EEG sensors, and nominally invisible to the inexperienced eye.

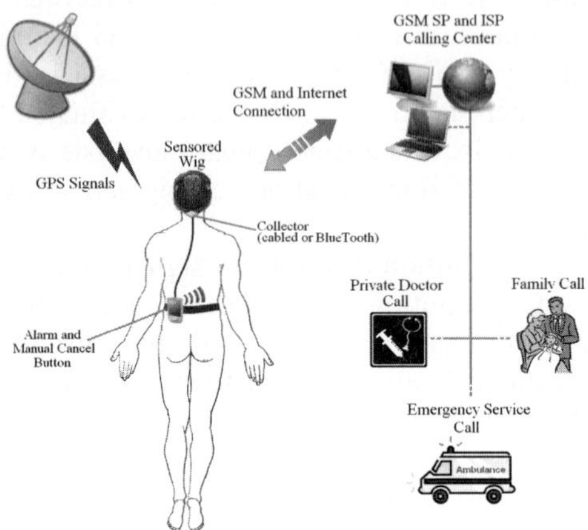

Fig. 1. Mobile warning system for epilepsy patients

System works in a simple way. First, wig equipped with sensors gets the EEG signal every 1 second and then sends these signals to a collector located the wig tail. This collector has a microchip that collects the signals from sensors and converts analog signals to digital data, than this data transferred to PDA fastened waist belt. ANN test software installed on PDA compares and tests these data with pre-learned data using a neural network algorithm. If there is any epileptic emergency condition then PDA gets current location of patient using GPS function, then waits 10 second (wait for cancellation and duration can be adjustable individually by patient or doctor), if there is no cancellation signal, next PDA creates a vibration and sends the emergency signal and little pocket of necessary data to Calling Center using SMS and ringing concurrently. Cancel button placed on PDA, which can be used for cancellation of emergency alert. If patient feels himself/herself good and thinks that this is a wrong alarm then this button can be used to cancel the alarm. This button is necessary because the wrong alert can be sent from software due to signal noises and false recognition. It is known that there is always a possibility to get wrong decision while using neural network algorithm for decision making.

If the cancel button is not pressed that means the patient is loosing himself/herself and the emergency alert is correct. That time, when the calling center gets the message then urgently transfers the data to private doctor's computer and calls the nearest ambulance service with exact position of patient using an ambulance system database and his/her parent. Up to this point this system can be called as a classical usage of telemedicine system on mobile environment. But we have to emphasis that this system supplies an excellent comfort to patients. In addition, this is not only a comfort; there also is a psychological and sociological aspect of using this system for a patient.

A hat can of course be used instead of wig. This system helps the patient to walk around and behave freely in society.

Figure 2 shows the sensors' settlement on cap. Top view of cap demonstrates the all possible sensors' location and example individual sensors' placement, which were determined by earlier measurement of epileptic points of patient. In order to select correct sensors location on the cap, these points must be designated in a laboratory environment by using pre-approved EEG device. Next, these points are pointed on the cap in their exact places and, then re-measurement must be done using the cap. If there is no problem using these points, then new epileptic and normal signal generated by the cap are stored in collector' memory. Thus, the cap is ready to be used in outside.

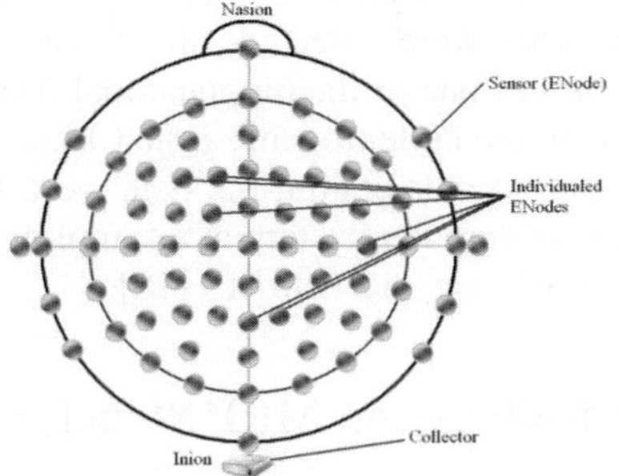

Fig. 2. Cap with sensors and Collector

Appearance and placement of cap and wig is shown Figure 3. The main objective of creating this type of a wig is that a patient can trip freely in society. In addition using this wig may support to prevent the patient from unexpected accident and emergency aid can be supplied by nearest ambulance and doctors.

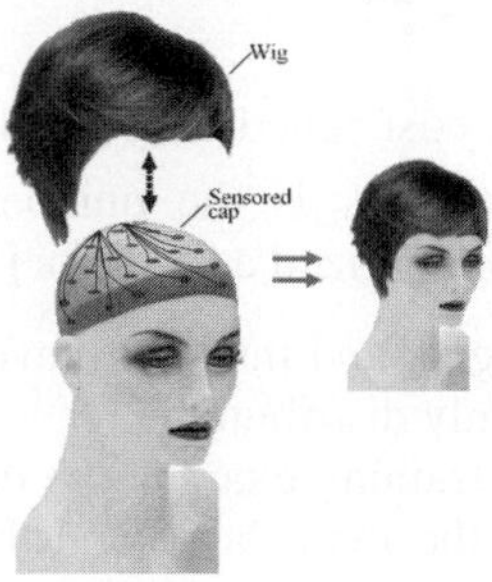

Fig. 3. Placement of cap and wig to head

In our applications, a detailed classification between set A (healthy volunteer, eyes open) and set E (epileptic seizure segments) was performed. Each of the ANN classifier was trained so that they are likely to be more accurate for one class of EEG signals than the other classes. The correct classification rates and convergence rates of the proposed ANN model were examined and then performance of the ANN model was reported. Finally, some conclusions were drawn concerning the MLP on classification of the EEG signals.

The used EEG sets (A and E) each containing 100 single channel EEG segments of 23.6 seconds duration, were composed for the study. These segments were selected and cut out from continuous multi-channel EEG recordings after visual inspection for artifacts, e.g., due to muscle activity or eye movements. In addition, the segments had to fulfill a stationary criterion Set A consisted of segments taken from surface EEG recordings that were carried out on five healthy volunteers using a standardized electrode placement scheme (Figure 2). Volunteers were relaxed in a wake state with eyes open [10].

Set E originated from the EEG archive of pre-surgical diagnosis. For the present study EEGs from five patients were selected, all of whom had achieved complete seizure control after resection of one of the hippocampal formations, which was therefore correctly diagnosed to be the epileptogenic zone (Figure 2).

After 12 bit analog-to-digital conversion, the data were written continuously onto the disk of a data acquisition computer system at a sampling rate of 173.61 Hz. Band-pass filter settings were 0.53–40 Hz (12 dB/oct.) [10].

3   Determination of Epilepsy on MLP Structure

In this study a three-layered feed-forward neural network is used and trained with the error back-propagation algorithm. The back propagation training through generalized delta rule of learning is an iterative gradient algorithm designed to minimize the root mean square error between the actual output of the multi-layer feed-forward neural network and the desired output. Each layer is fully connected to the previous layer, and has no other connection. The cost function of the MLP is given as

$$\varepsilon(n) = \frac{1}{2} \sum_{k=1}^{N_o} e_k^2(n) \tag{1}$$

where $\varepsilon(n)$ is the instantaneous cost function at iteration n, $e_k(n)$ is the error from output node k at iteration n and N_o is the number of output nodes. The back-propagation algorithm can be summarized as follows [11]:

1. *Initialization.* Set all the weights and threshold levels of the network to small random numbers that are uniformly distributed.
2. *Forward computation.* Let a training example be denoted by [*x(n)*, *d(n)*], with the input vector *x(n)* applied to the input layer and the desired response vector **d**(n) presented to the output layer.
3. *Backward computation.* Compute the local gradients (δ) of the network by progressing backward, layer-by-layer. For neuron j in output layer L, the local gradient is given by

Necessary formulas can be found at the study of Karlik B. et al. [13].

In this study, the learning rate and momentum coefficient are chosen to be 0.7 and 0.9 respectively. Simulation results show that the highest classification training accuracy is 97.78 % for 1000 iterations. In addition, simulation results showed that there is a decreasing in error level when the iteration count increases. Error levels according to iterations have been demonstrated in Figure 4. But this classification accuracy has been achieved in perfect laboratory environment. In practical aspect, test accuracy can reduce to approximately 50% because of many noises created by EEG sensors, movement of patient (even blood pressure), electromagnetic field created by antennas of PDA and so on. This means that almost half of attack warnings are false.

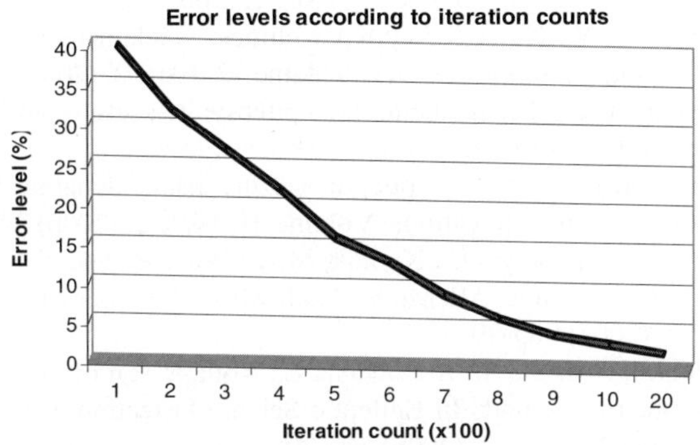

Fig. 4. Error levels according to iteration counts created by ANN in learning stage

4   Results and Conclusion

At the beginning, objective of this proposal study is to gather more reliable data to prevent epileptics from unexpected accidents, and to interfere as soon as possible in case of epileptic attacks. Data gathered so far has been achieved partly by getting EEG signals from a moving patient, which has made us aware the importance of more data for detection and decision making, particularly in relation to the usage of the cancellation button placed on PDA.

Using experimental data was obtained from a patient who has epileptic attack (simulation data was used on some of them) and average error level in detecting epileptic attack (testing) has been observed approximately 50%. There is an important point that there were no missed attacks, it means every attack has been detected, but some warnings were fake, and false warnings have been cancelled by manually. Therefore, all epileptic seizures were correctly identified. In addition, thanks to wig usage, this system supports the epileptics to feel themselves as healthy people, and they move relatively freely within society.

The system presented in this paper may be advanced and future function can be added. For example, new signal filtering techniques can be found and applied to reduce errors during converting signals to data. Thus the usage of this system can be widened and many people having epilepsy get the chance to use warning system.

References

1. Langfitt, J., Meador, K.: Want to Improve Epilepsy Care? Ask the Patient. Neurology, 62 (2004) 6-7
2. Adeli, H., Zhou, Z., Dadmehr, N.: Analysis of Eeg Records In An Epileptic Patient Using Wavelet Transform. J. Neuroscience Math. 123 (2003) 69–87.
3. Alkan, A., Koklukaya, E., Subasi, A.: Automatic Seizure Detection In EEG Using Logistic Regression and Artificial Neural Network. Journal of Neuroscience Methods, Volume 148, Issue 2, (2005) 167-176
4. Misra, U.K., Kalita, J., Mishra, S.K., Yadav, R.K.: Telemedicine in neurology: Underutilized potential, Neurology India, Volume 53, Issue 1, (2005) 27-31
5. Bingham, E., Patterson, V.: Nurse led epilepsy clinics: A telemedicine approach, Journal of Neurology Neurosurgery and Psychiatry, Volume 72, Issue 2, (2002) 216-216
6. Patterson, V., Bingham, E.: Telemedicine for epilepsy: A useful contribution, Epilepsia, Volume 46, Issue 5, (2005) 614-615
7. Subasi, A.: Automatic detection of epileptic seizure using dynamic fuzzy neural networks. Expert Systems with Applications, Volume 31, Issue 2, (2006) 320-328
8. Subasi, A., Alkan, A., Koklukaya, E., Kiymik, M.K.: Wavelet Neural Network Classification Of EEG Signals By Using AR Model With MLE. Preprocessing Neural Networks, Volume 18, Issue 7, (2005) 985-997
9. Subasi, A., Erçelebi, E., Alkan, A., Koklukaya, E.: Comparison of Subspace-Based Methods With AR Parametric Methods In Epileptic Seizure Detection. Computers in Biology and Medicine, Volume 36, Issue 2, (2006) 195-208
10. Ceylan, R., Ozbay, Y.: Comparison of FCM, PCA and WT Techniques for Classification ECG Arrhythmias Using Artificial Neural Network. Expert Systems with Applications, In Press, Corrected Proof, Available online (2006)
11. Kiymik, M.K., Akin, M., Subasi, A.: Automatic Recognition Of Alertness Level By Using Wavelet Transform And Artificial Neural Network. Journal of Neuroscience Methods, Volume 139, Issue 2, (2004) 231-240
12. Andrzejak, R.G., Lehnertz, K., Rieke, C., Mormann, F., David, P., Elger, C.E. : Indications of Nonlinear Deterministic And Finite Dimensional Structures In Time Series Of Brain Electrical Activity: Dependence On Recording Region And Brain State. Physics Rev. E. 64, 061907 (2001)
13. Karlik, B., Tokhi, O., Alci, M.: A Novel Technique for Classification of Myoelectric Signals for Prosthesis. CD-ROM Proceeding of IFAC'02, Barcelona, July (2002)

Modular Bayesian Networks
for Inferring Landmarks on Mobile Daily Life

Keum-Sung Hwang and Sung-Bae Cho

Dept. of Computer Science, Yonsei University
{yellowg, sbcho}@sclab.yonsei.ac.kr

Abstract. Mobile devices get to handle much information thanks to the convergence of diverse functionalities. Their environment has great potential of supporting customized services to the users because it can observe the meaningful and private information continually for a long time. However, most of the information has been generally ignored because of the limitations of mobile devices. In this paper, we propose a novel method that infers landmarks efficiently in order to overcome the problems. It uses an effective probabilistic model of Bayesian networks for analyzing various log data on the mobile environment, which is modularized to decrease the complexity. The proposed methods are evaluated with synthetic mobile log data generated.

1   Introduction

The mobile environment has very different characteristics from the ordinary personal computer environment. First of all, a mobile device can collect and manage various data about user. Besides, a mobile device can be adapted to fit user's preference or character since it is very private apparatus. Furthermore, a mobile device can collect everyday information effectively because a user brings it with him nearly every time, so it has a great potential to figure out and help its user. Such features of mobile device open the possibility of diverse and convenient services for user. However, there are some limitations of mobile devices. It has relatively insufficient memory capacity, lower CPU power (data-processing speed), and limited battery hours to desktop PC.

In this paper, we propose a novel method for effective analysis of mobile log data and extraction of semantic information and memory landmarks which are used as a special event for helping recall it [1]. The proposed method adopts Bayesian probabilistic model to efficiently manage various uncertainties occurring on mobile environment. We also propose a cooperative reasoning method for the modular model of Bayesian networks to work competently in mobile environments.

There already exist some attempts for analyzing log data and supporting expanded services. A. Krause, *et al.* clustered sensor and log data collected on mobile devices, learned a context classifier that reflected user's preference, and estimated user's situation to provide smart services to the user [2]. E. Horvitz, *et al.* proposed a method that detected and estimated landmarks by learning human's cognitive activity model from PC log data based on Bayesian approach [1]. However the models were based on the general learning method only for the small domain or for the PC domain.

A. Sattar and B.H. Kang (Eds.): AI 2006, LNAI 4304, pp. 929–933, 2006.

© Springer-Verlag Berlin Heidelberg 2006

2 Cooperative Reasoning of Modular Bayesian Networks

The overall process of landmark extraction from log data of mobile environments used in this paper is shown in Fig. 1. Various log data are preprocessed in advance, and then the landmark reasoning module detects the landmarks. The preprocessing module is operated by the techniques of simple rule reasoning and pattern recognition. The BN reasoning module performs complicated and probabilistic inference.

Bayesian network is a model that can express large probability distributions with relatively small cost to statistical mechanics. It has a structure of directed acyclic graph (DAG) that represents link relations of node, and has conditional probability tables (CPTs) that is constrained by the DAG structure [3].

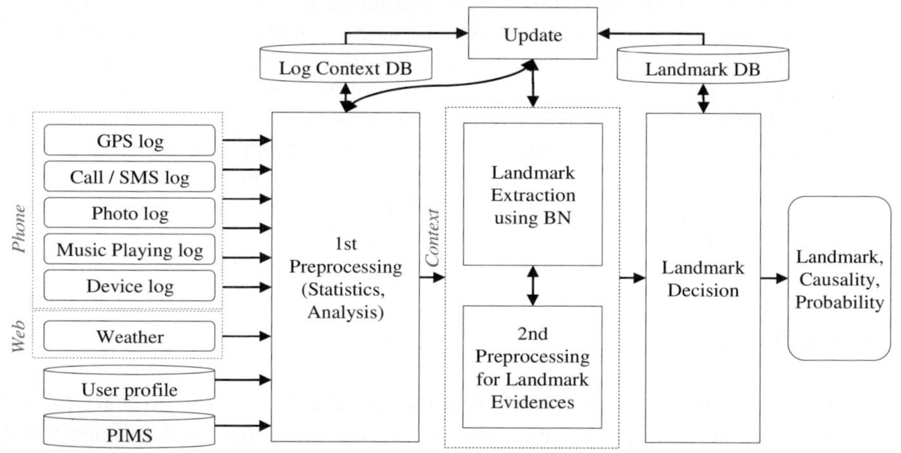

Fig. 1. The process of the landmark extraction from mobile log data

There are two general differences between the proposed Bayesian network and the conventional Bayesian network. Firstly, we modularize the Bayesian inference models according to their separated sub-domains (Fig. 2). Bayesian network model essentially requires more computing powers depending on the number of nodes and links. Especially, since the computational complexity of Bayesian inference is approximately proportional to $O(k^N)$, where k is the number of states and N is the number of causal nodes, the modularized BN is more efficient.

Secondly, in order to consider the co-causality of modularized BN, the proposed method operates 2-pass inference stages as shown in Fig. 2. A virtual linking technique is utilized in order to reflect the co-causal evidences more correctly. The technique is to add virtual node and regulate its conditional probability values (CPVs) to apply the probability of evidence [3]. In this paper, we have proposed and used the new virtual linking method that uses the prior probability of evidence node as virtual evidence parameter instead of adding virtual nodes. This method guarantees the maintenance of BN structure. For example, the structure of BN_1 is {A→B→C}, BN_2 is its modularized version; {A→B, B→C}, and given the evidence A, we can calculate the beliefs using the virtual link assumption and chain rule as follows [3]:

$$\text{BN}_1\text{'s } Bel(C) = P(A,C) = P(C|B)P(B|A)P(A) \qquad (1)$$

$$\text{BN}_2\text{'s } Bel(B) = P(A,B) = P(B|A)P(A) \qquad (2)$$

$$\text{BN}_2\text{'s } Bel(C) = P(B,C) = P(C|B) P(B) = P(C|B) P(A,B) = P(C|B) P(B|A)P(A) \qquad (3)$$

$$\therefore \ \ \text{BN}_2\text{'s } Bel(C) = \ \text{BN}_1\text{'s } Bel(C) \qquad (4)$$

where the proposed virtual link assumption is $P(B) = Bel(B)$.

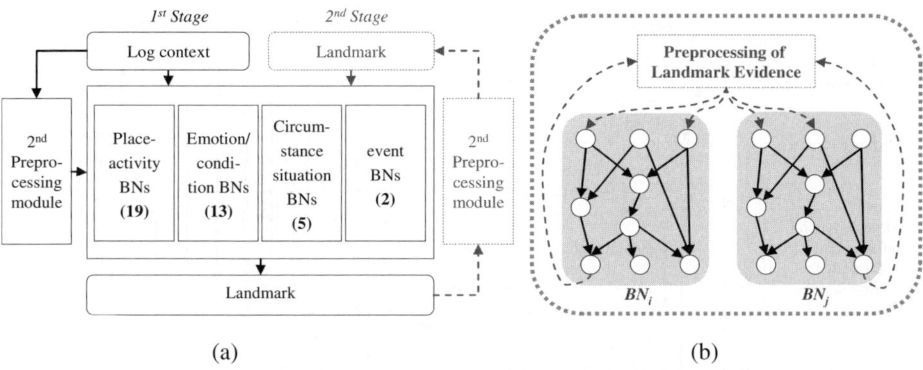

Fig. 2. (a) The 2-pass inference process for the cooperation of modular BNs. It uses the result from inference of BN as the evidences at the 2nd inference stage. The dotted line indicates the stream of the 2nd stage of inference processing. The parenthesis indicates the number of BNs contained. There are four kinds of BNs, and totally 39 BNs are used. (b) The modularized BNs for efficiency. The dotted line indicates a virtual link.

In this paper totally 39 BNs are designed with 638 nodes, 623 links and 4,205 CPV, which are summarized in Table 1. In average 16.6 nodes and 107.8 CPVs are used at the same time on the BN reasoning process since they are modularized and the computations are distributed. On the other hand, the monolithic BN for them has 462 nodes decreased from 638 because of the removal of duplicated nodes, but the number of parents and CPVs is increased. This means that it has much larger complexity.

Table 1. The information about the 39 modular BNs and the monolithic BN for them. Abbreviations: MonoBN – the monolithic BN corresponding to the modular BNs, NN – no. of nodes, NNR - no. of root nodes, NNI - no. of intermediate nodes, NNL - no. of leaf nodes, NL - no. of links, NP_{avg} - avg. no. of parents, NS - no. of states, NS_{avg} - avg. no. of states, NCPV - no. of CPVs, CPT_{max} - maximum size of CPT.

	NN	NNR	NNI	NNL	NL	NP_{avg}	NS	NS_{avg}	NCPV	CPT_{max}
39 BNs	638	375	135	128	623	0.98	1,279	2.00	4,205	64
MonoBN	462	235	111	116	588	1.27	927	2.01	4.869	512

3 Experimental Results

We have tested the proposed landmark reasoning model with a scenario in order to confirm the performance. The left side of Fig. 3 shows a scenario used. The BN set strongly related to the scenario is {food, photo, movement, nature, joy, home}. The probabilities are occurred when the related evidences are given. The right side of Fig. 3 shows the inference results. We can see the increment of the probabilities of the related landmarks at the corresponding time. For example, there are 'going-out-preparation' and 'shower' landmarks at 7~9 o'clock, 'eating' at 12~13 and 17~19 o'clock, 'walking-for' at 13~14 and 20~21 o'clock, 'joyful-photo' at 14~15 o'clock, and 'eating-out' and 'eating (western style)' landmarks at 17~19 o'clock.

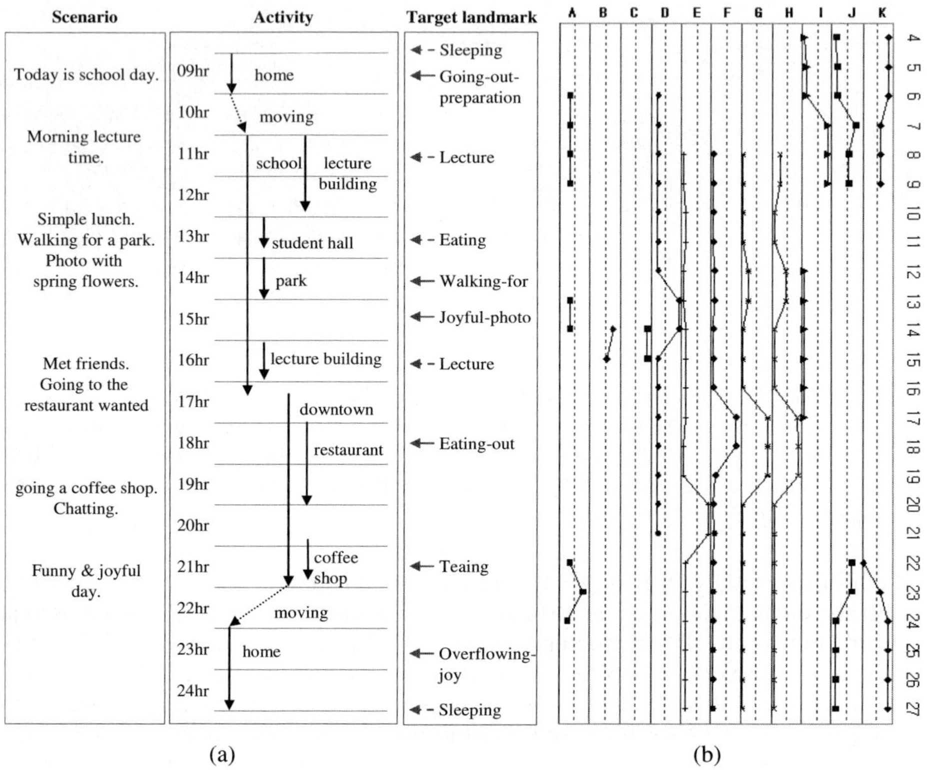

(a)   (b)

Fig. 3. (a) A scenario of an everyday life with mobile device of an undergraduate student for experiments. (b) The probability values of 11 target landmarks. The denoted time is from 4 o'clock to 27 o'clock (equal to 3 o'clock tomorrow). Abbreviations used for landmarks: A - Overflowing-joy, B - Photo (scenery), C - Joyful-Photo, D - Walking-for, E - Tea, F - Eating-out, G - Eating (western style), H - Eating, I - Going-out-preparation, J - shower, K - sleeping.

We grouped situations of everyday life on the two bases: usual/unusual and idle/busy. We made contexts for evidence because it is difficult to make the log data directly. For example, we made a context 'a lot of phone calls' instead of its phone call log data. The target landmarks are categorized 4 classes as shown in Table 2. A day of data set contains two landmarks, which are selected randomly. Each landmark has some candidates of evidence set and one of them is selected randomly. For example, the landmark 'annoying SMS' has candidates, {'a lot of phone calls', 'a lot of moving'}, {'a lot of moving', 'a lot of SMS'}, {'a lot of phone calls', 'a lot of SMS'}.

Table 2 shows the statistics of experimental results. We have excluded the landmarks related to the default place 'home' and the low-weight landmarks from main landmarks set. The false-positive error of 'usual/idle' class is high and the precision is low, because 'usual' class includes many places and landmarks that have evidence duplication. For example, since the landmark 'boarding ship' is caused by the evidence 'sea' or 'river', a landmark 'swim' can also be extracted. The false-positive error of 'usual/busy' class is low because the class includes relatively many landmarks that have distinct evidence. In the experiment, the overall recall rate was low as 75%. It results from the lack of tuning or the landmarks hard to detect.

Table 2. The experimental results with synthetic data. Two target objects (with few redundancies) are selected in each data set. 'unusual/busy' class data are composed of one 'unusual' landmark and one 'busy' landmark. Abbreviations: LM-Landmark, TP-true positive error rate (%), FP-false positive rate (%), FN-false negative rate (%), #-number.

Class	Time	Target LMs #	TP	FP	FN	Precision	Recall
usual/idle	30 days	60	46	14	14	0.767	0.767
unusual/idle	30 days	58	43	10	15	0.811	0.741
usual/busy	30 days	55	41	2	14	0.953	0.745
unusual/busy	30 days	60	46	8	14	0.852	0.767
Total	120 days	233	176	34	57	0.838	0.755

4 Concluding Remarks

In this paper, we have proposed a landmark inference model for it to be more suitable to mobile device environment. We have modularized BN structures for more efficient operation in mobile environment and proposed 2-step inference method of applying virtual node concept for cooperation of modular BNs. In some experimental results with artificial data, the intended landmarks are well extracted. Currently, this research is ongoing, and we will validate the performance and utility of the proposed method on the expanded real-world domain in the future.

Acknowledgement. This research was supported by the Ministry of Information and Communication, Korea under the Information Technology Research Center support program supervised by the Institute of Information Technology Assessment, IITA-2005-(C1090-0501-0019).

References

[1] E. Horvitz, S. Dumais, and P. Koch. "Learning predictive models of memory landmarks," *CogSci 2004: 26th Annual Meeting of the Cognitive Science Society*, pp. 1-6, 2004.

[2] A. Krause, A. Smailagic, and D.P. Siewiorek, "Context-aware mobile computing: Learning context-dependent personal preferences from a wearable sensor array," *IEEE Trans. on Mobile Computing*, vol. 5, no. 2, pp. 113-127, 2006.

[3] K.B. Korb and A.E. Nicholson, *Bayesian Artificial Intelligence*, Chapman & Hall/CRC, 2003.

[4] M. Raento, A. Oulasvirta, R. Petit, and H. Toivonen, "ContextPhone: A prototyping platform for context-aware mobile applications," *IEEE Pervasive Computing*, pp. 51-59, 2005.

A Bi-level Genetic Algorithm for Multi-objective Scheduling of Multi- and Mixed-Model Apparel Assembly Lines

Z.X. Guo, W.K. Wong, S.Y.S. Leung, J.T. Fan, and S.F. Chan

Institute of Textiles and Clothing, The Hong Kong Polytechnic University, Hunghom,
Kowloon, Hong Kong
zxguo@polyu.edu.hk,
{tcwongca, tcleungs, tcfanjt, tcchansf}@inet.polyu.edu.hk

Abstract. In this paper, a multi-objective scheduling problem of the multi- and mixed-model apparel assembly line (MMAAL) is investigated. A bi-level genetic algorithm is developed to solve the scheduling problem, in which a new chromosome representation is proposed to represent the flexible operation assignment including assigning one operation to multiple workstations as well as assigning multiple operations to one workstation. The proposed algorithm is validated using real-world production data and the experimental results show that the proposed algorithm can solve the proposed scheduling problem effectively.

1 Introduction

Scheduling is the allocation of available production resources over time to meet some performance criteria. A huge number of papers in this field have been published [1-5]. However, the assembly line scheduling problem, especially the multi- and mixed-model assembly line scheduling problem, has received little attention so far.

Today's apparel industry is characterized by unpredictable demand fluctuations, tremendous product variety, and long supply process. The multi- and mixed-model apparel assembly line (MMAAL) is most commonly adopted in the apparel industry, which is a flexible assembly line and has the features of both multi-model assembly line and mixed-model assembly line. However, as a traditional labor-intensive assembly form, the MMAAL scheduling mainly relies on managers' experience. The research in this field has not been reported yet. In this paper, we solve a MMAAL scheduling problem by minimizing the total earliness/tardiness (E/T) penalty costs and maximizing the smoothness of the production flow. To solve the scheduling problem, a bi-level genetic algorithm (BiGA) is developed.

The rest of this paper is organized as follows. In Section 2, the MMAAL scheduling problem is formulated. The BiGA is presented detailedly in Section 3 and experiments are conducted to test the performance of the BiGA in Section 4. Finally, the paper is summarized and the further research is suggested.

A. Sattar and B.H. Kang (Eds.): AI 2006, LNAI 4304, pp. 934–941, 2006.
© Springer-Verlag Berlin Heidelberg 2006

2 Problem Formulation

The MMAAL is composed of a certain number of sewing workstations. Each workstation is a physical location that accommodates an operator and a sewing machine. Multiple sewing machine types are involved and each type includes one or more machines. Multiple orders representing different product types can be produced simultaneously in the MMAAL. The assembly process of each garment comprises a series of manual sewing operations. According to a pre-determined sewing sequence, each operation can be processed on any of the workstations with the corresponding machine type. Moreover, each operator must only operate one machine at any instance of time.

In this paper, we let P_i denotes the i th order ($1 \leq i \leq p$), O_{il} denotes the l th operation of order P_i, M_{kj} denotes the j th machine of the k th machine type, S_i denotes the starting time of order P_i, and A_{ilkj} indicates if operation O_{il} is assigned to machine M_{kj} (if so, A_{ilkj} is equal to 1, otherwise A_{ilkj} is 0).

2.1 Objective Function

In the real-world MMAAL, the whole scheduling process of a scheduling problem comprises one or more scheduling statuses each of which represents an operation assignment status in the assembly line. For example, the scheduling shown in Fig.1(e) involves two scheduling status processing orders P_i and P_j respectively. The purpose of the scheduling problem is to determine the assignment A_{ilkj} of sewing operation O_{il} and the starting time S_i of order P_i with the consideration of following objectives:

Objective 1: to minimize the weighted sum Z of E/T penalties.

$$Z(S_i, A_{ilkj}) = \sum_{i=1}^{p} (\alpha_i \cdot (C_i - D_i) \cdot \lambda_i + \beta_i \cdot (D_i - C_i) \cdot (1 - \lambda_i)) \tag{1}$$

where α_i is the tardiness weight (the penalty cost per time unit of the delay) of order P_i while β_i is the earliness weight (the storage cost per time unit if order P_i is completed before the due date), D_i is the due date of order P_i which is pre-determined by the customer, C_i is the actual completion time for order P_i, and λ_i denotes if the tardiness of order P_i is greater than 0 (if so, λ_i is equal to 1, otherwise it is 0).

Objective 2: to maximize the smoothness of the production flow of the MMAAL, which is denoted by the balance index B,

$$B(S_i, A_{ilkj}) = \frac{\sum_r (SPT_r \cdot SB_r)}{\sum_r SPT_r} \tag{2}$$

where SPT_r is the processing time of the r th scheduling status, and SB_r is the balance index of the r th scheduling status, which is computed as follows,

$$SB_r = \sum_i PB_i / n \tag{3}$$

$$PB_i = \frac{\sum_l OPT_{il}}{o_i \cdot \max(OPT_{il})} \tag{4}$$

where n $(n \geq 1)$ is the number of orders in the r th scheduling status, PB_i is the balance index of order P_i, o_i is the number of operations of order P_i, and OPT_{il} is the average processing time of operation O_{il}.

2.2 Assumptions

The production in the MMAAL satisfied the following assumptions in this paper.

- Each workstation must process at least one operation.
- Once an operation is started, it cannot be interrupted.
- There is no materials shortage, machine breakdown and operator absenteeism.
- The MMAAL is empty initially (there is no WIP in each workstation).

3　Bi-level Genetic Algorithm for MMAAL Scheduling

If a machine is assigned to process different operations frequently in the real-world MMAAL, more additional setup times will be needed and the efficiencies of the operator will fluctuate and decrease inevitably. On the basis of switching operations as infrequently as possible in each machine, we optimize the scheduling problem proposed in Section 2. As two orders (orders P_i and P_j) are scheduled, there are five possible scheduling modes shown in Fig. 1 based on different production tasks and delivery dates. In these modes, 3 different scheduling statuses have been involved, including the status of processing order P_i solely, the status of processing order P_j solely, and the status of processing both orders P_i and P_j simultaneously.

In apparel production, it is commonly practiced that no more than two production orders are processed simultaneously in a MMAAL. In this paper, we consider the scheduling problem of two orders (2-order scheduling problem) by developing a BiGA which is composed of two GAs in different levels where the second-level GA (GA-2) is nested in the first-level GA (GA-1). Fig. 2 describes the steps involved in BiGA.

The following sub-sections describe the GA-1 and the GA-2 in detail.

3.1　First-Level Genetic Algorithm

On the basis of the scheduling modes and scheduling statuses shown in Fig. 1, the GA-1 can generate the optimal chromosome assigning operations to workstations.

3.1.1 Representation

In this paper, each chromosome in GA-1 represents a feasible solution including operation assignments of all scheduling statuses. A novel chromosome representation is developed in GA-1. Each chromosome is composed of 3 sub-chromosomes which represent the operation assignments of 3 different scheduling statuses. Each sub-chromosome is a sequence of genes whose length equals the number of workstations on which operations can be assigned. Each gene represents a workstation and the value of each gene represents the operation number(s) processed by the corresponding workstation. If the number of the machine type is t ($t \geq 1$), the genes in each sub-chromosome will be divided into t parts in turn. Each part represents a particular machine type. Each operation can only be assigned to the workstations which can handle it.

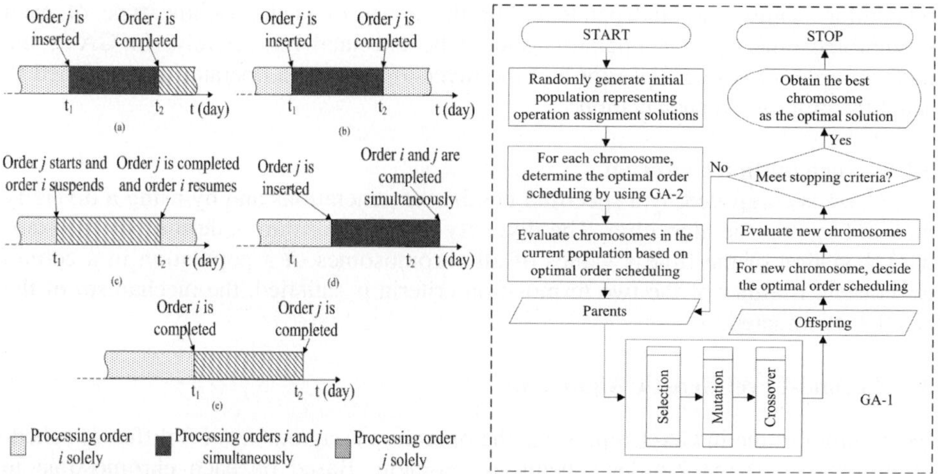

Fig. 1. Scheduling modes of processing two orders in a MMAAL

Fig. 2. Bi-level genetic algorithm

3.1.2 Initialization

The GA operates on a population of chromosomes. In GA-1, each chromosome is composed of 3 sub-chromosomes. Each sub-chromosome represents the operation assignment of one scheduling status, which is generated randomly.

3.1.3 Fitness

The GA-1 searches for the operation assignment with the highest fitness. In this paper, two objective functions should be optimized. For the second objective, ideally, the maximal balance index B (smoothness) of an assembly system is 100%. Balance index less than 100% implies extra production cost involved, i.e. imbalance penalty. We set the imbalance penalty as γ when the balance index decreases 1%. Consequently, the two objectives can be combined as $OBJ(S_i, A_{ilkj})$,

$$OBJ(S_i, A_{ilkj}) = Z(S_i, A_{ilkj}) + \gamma \cdot (1 - B(S_i, A_{ilkj})) \tag{5}$$

Thus, the fitness function *fitness* can be defined as

$$fitness = \frac{1}{OBJ(S_i, A_{ilkj}) + 1} \tag{6}$$

3.1.4 Selection

The tournament-selection technique [6] is used for the selection of chromosomes to survive to the next generation.

3.1.5 Genetic Operators

In this research, both genes of different machine types and genes of different scheduling statuses are independent. For the genes of each machine type of each sub-chromosome, genetic operators should be executed respectively. In GA-1, the uniform-order crossover operator and the inversion mutation operator are modified to adapt the proposed representation.

3.1.6 Termination Criterion

The GA-1 is controlled by a specified number of generations and by using a diversity measure to stop the algorithm. The diversity of the algorithm is defined by the standard deviation of the fitness values of all chromosomes of a population in a certain generation. If either of the two termination criteria is satisfied, the mechanism of the GA-1 is terminated.

3.2 Second-Level Genetic Algorithm

Each chromosome in GA-1 represents the operation assignments of 3 different scheduling statuses of the 2-order scheduling problem. Based on each chromosome in GA-1, GA-2 will determine the starting time of each scheduling status in each cycle, which can be considered as a simple unary first order function optimization problem.

The problem is easy to be optimized by a canonical real-coded GA. In this research, the BLX-α crossover operator [7], the non-uniform mutation operator [8] and the tournament-selection strategy [6] are used in GA-2. The fitness function and the termination criterion of GA-2 are the same with those of GA-1.

4 Experimental Results and Discussion

An experiment is conducted to demonstrate the performance of the proposed methodology using the industrial data from a MMAAL. In this experiment, the transportation time of semi-finished products and the setup time of each operation are known in advanced and included in the processing time.

In this experiment, 5 different scheduling scenarios (cases) are conducted respectively. In each scheduling scenario, two production orders with different production tasks are scheduled to meet the objectives (1) and (2) in a MMAAL with 7 "lock-stitch" sewing machines and 7 "overlock" sewing machines. The two types of machines are numbered as 1 to 7 and 8 to 14 respectively. Table 1 shows the production quantity of garments and the due date of each order in 5 scheduling cases. Furthermore, the earliness weights (dollars/day) of orders 1 and 2 are both 100, while the tardiness weights are 5000 and 3000 respectively.

Table 1. Data for orders of 5 scheduling cases

Cases	Case 1		Case 2		Case 3		Case 4		Case 5	
Order No.	1	2	1	2	1	2	1	2	1	2
Order size	2000	2500	2000	1500	2000	600	2000	600	2000	2500
Due date	20	25	22	22	15	10	15	4	20	28

The optimized operation assignments and scheduling results of the scheduling cases generated by the proposed BiGA are shown in Tables 2 and 3. In Table 2, the first row shows the workstation number, and each row of each case describes the optimized operation assignment of a scheduling status. The rows 3 to 7 of Table 3 describe the starting time to process the order(s) of each scheduling status, the processing time for processing the order(s) of one scheduling status, the completion time of each order, the penalty cost of each order and the balance index of each schedule respectively.

Table 2. Optimized operation assignments of 5 scheduling cases

Workstation No.		1	2	3	4	5	6	7	8	9	10	11	12	13	14
Case 1	Order 1	3,6	3,6	3,4	3	4,6	4	4,6	2,5	5	1	1,5	1	2	1,2
	Two orders	8	4	3	9	10	6	13	7	2	12	5	11	7	1
	Order 2	8,13	9,10	8	9,13	9,10	8	10	11,12	7,11	12	7,11	7	11,12	7
Case 2	Order 1	4	3	4,6	4,6	3	6	3	1,5	2	2,5	1	1	2,5	1
	Two orders	4	6	8	13	3,9	10	3,13	1	11,12	1,2	5,12	7,11	2,5	7
Case 3	Order 1	6	3,4	4	4,6	4,6	3	3	2,5	5	1	1,5	1	1,2	2
	Order 2	8	4	3	9	10	6	13	7	2	12	5	7	11	1
Case 4	Order 1	6	3,4	4	4,6	3,4	3	3,6	1	1	1,2	5	2	1,5	2,5
	Order 2	9,10	8,13	8	8	10	9	10,13	7,12	11,12	11,12	7,12	7,11	11	7
Case 5	Order 1	3	4	6	6	4	3	4	1	1	1	2	2	5	5
	Two orders	6,9	6	10,13	8	4	3,10	3	7,12	11	1	5	2	1,12	2,7
	Order 2	9	8	13	9	8	10	10	12	11	12	7	7	11	12

Table 3. Optimized scheduling results of 5 scheduling cases

Cases	Case 1			Case 2			Case 3			Case 4		Case 5		
Order No.	1	1&2	2	1	1&2	2	1	1&2	2	1	2	1	1&2	2
Starting time	0	5.72	20	0	2.6	/	0	4.17	/	0	1.17	0	6.12	20
Processing time	5.72	14.28	4.99	22	19.4	/	9.13	5.83	/	12.1	2.83	20	13.88	8
Completion time	20	/	24.99	22	/	22	14.96	/	10	14.93	4	20	/	28
Penalty cost	0	/	1	0	/	0	4	/	0	7	0	0	/	0
Balance index	90.61%			86.17%			94.45%			94.25%		84.17%		

As shown in Table 2, the optimized operation assignment is very flexible. For example, in the first scheduling status of Case 1, operation 3 is processed on workstations 1 to 4, and workstation 1 processes operations 3 and 6. Moreover, parallel workstations include workstations 1 and 2, workstations 5 and 7, etc.

As shown in Table 3, the optimized scheduling schemes of the 5 scheduling cases correspond to scheduling modes (a), (b), (c), (d) and (a) described in Section 3 respectively. Taking Case 1 for example, the MMAAL processes only order 1 from time 0 with processing time 5.72 units, then two orders are processed simultaneously with 14.28 time units, and finally order 2 is processed solely until time 24.99. Based on this schedule, order 1 is completed punctually and order 2 is completed by 0.01 time units in advance and its earliness penalty is only 1. The balance index is 90.61%.

In this section, the results of the experiment show that the proposed algorithm can schedule not only the processing of one operation on multiple workstations but also the processing of multiple operations on one workstation. The experiment also demonstrates the capability of the algorithm proposed for scheduling the real-world MMAAL. Each scheduling case is scheduled effectively and the penalty cost of each case is very low and negligible. The experiment covers all scheduling modes of two orders being processed in a MMAAL. These results show that the proposed BiGA can solve effectively the 2-order scheduling problem in the MMAAL.

5 Conclusions

This paper deals with a multi-objective scheduling problem in the MMAAL with parallel workstations and flexible operation assignment. A genetic optimization process called BiGA, has been developed to tackle the MMAAL scheduling problem. In the BiGA, a novel chromosome representation is proposed to tackle flexible operation assignments in the MMAAL including assigning one operation to multiple workstations and assigning multiple operations to one workstation. Experiments have been conducted to validate the proposed algorithm and the results have shown that the algorithm can solve the problem effectively. Further research will focus on the effects of various uncertainties in the MMAAL, including unpredictable customer orders, machine breakdown and operator absenteeism, etc.

Acknowledgements

The authors would like to thank Innovation and Technology Commission of The Government of the Hong Kong SAR and Genexy Company Limited for the financial support in this research project (Project No. UIT/62).

References

1. Yen, B., Wan, G.: Single Machine Bicriteria Scheduling: A Survey. Int. J. Ind. Eng-Theory. 3 (2003) 222-231
2. Mokotoff, E.: Parallel Machine Scheduling Problems: A Survey. Asia-Pac. J. of Oper. Res. 2 (2001) 193-242
3. Hejazi, S., Saghafian, S.: Flowshop-scheduling Problems with Makespan Criterion: A Review. Int. J. Prod. Res. 14 (2005) 2895-2929

4. Chan, F., Chan, H.: A Comprehensive Survey and Future Trend of Simulation Study on FMS Scheduling. J. Intell. Manuf. 1 (2004) 87-102
5. Blazewicz, J., Domschke, W., Pesch, E.: The Job Shop Scheduling Problem: Conventional and New Solution Techniques. Eur. J. Oper. Res. 1 (1996) 1-33
6. Goldberg, D. E.: Genetic Algorithms in Search, Optimization and Machine Learning. Massachusetts: Addison-Wesley (1989)
7. Eshelman, L.J., Schaffer, J.D.: Real-coded Genetic Algorithms and Interval Schemata. In: Whitley, L.D. (eds.): Foundations of Genetic Algorithms. Morgan Kaufmann: San Mateo, CA. (1993) 187-202
8. Michalewicz Z.: Genetic Algorithm + Data Structures = Evolution Programs. New York, USA: Springer-Verlag; 1992.

Membership Functions for Spatial Proximity

Jane Brennan[1] and Eric Martin[2]

[1] University of Technology, Sydney, Faculty of Information Technology, PO Box 123,
Broadway NSW 2007, Australia
`janeb@it.uts.edu.au`
[2] University of New South Wales, Kensington NSW 2002, Australia

Abstract. Formalising nearness has been the subject of extensive work, resulting in many membership functions based on absolute distance metrics, relative distance metrics, and combinations of those. The possible strengths and weaknesses of these functions have been discussed and argued at length, but strangely enough, no experiment seems to have been conducted to assess the merits and shortcomings of competing approaches. Conducting such experiments can be expected not only to provide an objective evaluation of the various measures that have been proposed, but also to suggest new measures that outperform all those being analysed. This paper fulfills these expectations, and gives further evidence that fuzzy logic provides fruitful and powerful methods to formalise qualitative reasoning and capture fundamental qualitative notions. The proposed fuzzy membership functions can be directly used in qualitative reasoning about spatial proximity in Geographic Information Systems, which are becoming more and more important in software development for diverse purposes such as Tourist Information Systems or property development.

1   Introduction

Love it or loathe it, fuzzy logic is, albeit its unfortunate choice of name, a very practical Artificial Intelligence technique that is used in applications such as control systems in washing machines, elevators, cars, etc. It is quite a valuable method that can generate precise solutions from approximate data, and can be of great use for developing computational models for vague concepts such as those described by natural language expressions. In this paper, we propose and evaluate fuzzy membership functions suitable to implement the notion of spatial proximity, generally represented by linguistic terms such as *near* or *close*, within Geographic Information Systems (GIS).

After a brief review of previous work, this paper will report on studies conducted to evaluate proximity membership functions against a data set of a distance network and predefined nearness information. We will conclude with recommendations for appropriate functions in the context of reachability across a distance network.

2   Factors That Influence Spatial Proximity Perception

Several fuzzy approaches have been devised to represent spatial proximity based on distance. In order to determine what exactly influences the perception of spatial nearness

A. Sattar and B.H. Kang (Eds.): AI 2006, LNAI 4304, pp. 942–949, 2006.
© Springer-Verlag Berlin Heidelberg 2006

within GIS data, Gahegan [3] conducted psycho-metric experiments. The objective of his experiments was to examine when or how people decide whether objects are near a chosen reference object on a GIS map. The pseudo-metric tests were conducted on a group of 50 subjects, who have all had some practical exposure to Geographic Information Systems. The subjects were asked to rate objects in diagrams representing geographic features on a map, according to how close they were to a given reference object. While Gahegan [3] pointed out that his tests were not necessarily conclusive, he could obtain some interesting results and make several observations that could be helpful in modelling spatial proximity. A first observation is that if a scene is devoid of additional objects, namely if only the reference object and the object to be located inhabit the scene, geometrical reasoning is applied. However, in the presence of other objects of the same type, proximity is partially defined by relative distance. Another observation is that proximity perception is impacted by the scale of the scene, which directly depends on the size of the area being considered. This paper is set out to define membership functions that address all of these points.

On the basis of his observations, Gahegan [3] suggested a contextual model of nearness relations in order to account for different influencing factors. Three kinds of metrics are considered in Gahegan's model: an absolute distance metric, a relative distance metric, and a combination of both. Using the absolute distance metric amounts to assuming that proximity is directly proportional to the Euclidian distance between the reference object and the object to be localised. As the scale of the area viewed by the user of a GIS map also seems to have an impact on the perception of proximity, Gahegan [3] further suggested that the bounding boxes of the area in the GIS can be used as scale indicator, the opposite corners of the bounding boxes providing the maximum distance. If the maximum possible distance between objects in the scene is used to normalise the distance between two objects A and B, it is then possible to represent the proximity between A and B by a fuzzy value. Similarly, but not exclusively, we will use the maximal distance between objects on the map of a country or state. This will enable proximity evaluations across several GIS maps if needed. In this paper, the maximal distance will be the largest distance between any two places in the country or state being considered. Absolute distance metrics result in continuous proximity with, for example, *very close* $>$ *close* $>$ *far*, but relations such as *closest* or *farthest* cannot be represented. For this and other reasons, Gahegan [3] proposed to treat proximity in terms of a relative distant metrics, in addition to the absolute distance metrics. More precisely, he suggested an ordinal approach to represent relative distance, by ranking the objects in the scene with respect to their distance to the reference object and the total number of objects. He pointed out that this approach could cause objects to be considered close to a given reference object A, even though such objects might be separated from A by a large distance. As the objects are ranked on the basis of their distance to A, this approach seems to be more absolute than relative in nature. Worboys [6] dealt with this problem in a more efficient way by calculating for each place the mean distance to all other places in the scene.

Gahegan [3] suggested to combine both the absolute and the relative distance. As both metrics offer fuzzy representations, he defined the membership function for absolute distance metrics and assumed a distribution function for nearness based on relative distance, and then combined these functions with the fuzzy union operator. This resulted in an object being considered *close* just in case it is geometrically OR relatively close.

Motivated by Gahegan's work, Guesgen and Albrecht [4] suggested to associate spatial binary relations such as *far from* or *close to*, or unary relations such as *downtown*, with fuzzy membership grades that could be calculated from the Euclidian distance between objects on a map. They did not test their suggestions against any data and did not provide any membership function for relative distance metrics. Guesgen [5] proposed to define proximity without any measure of distance by using the notion of fuzzy sets previously defined in Guesgen and Albrecht [4]. These fuzzy sets were used to reason about the relationship between proximity notions by means of transitive closure on ternary proximity relations such as "if B is closer to A than C is to A, and if C is closer to A than D is to A, then B is closer to A than D is to A." This is very similar to van Benthem's [1] approach in his logic of space.

There is no experimental data to give evidence that Gahegan's [3] or Guesgen and Albrecht's [4] fuzyy membership functions can be of practical use. None of their approaches bases fuzzy membership functions on truly relative distance. We therefore find it essential to evaluate Gahegan's [3] absolute distance metrics and Worboys' [6] relative distance metrics before considering whether and how to combine them using fuzzy logic operators.

Worboys [6] did some interesting studies on the qualitative location of cities and the relative distances between them. His definition of relative distance is not based on the comparative concept without Euclidean distance, but it does nonetheless incorporate the context of all places under consideration. He used the road distances between 48 cities in Great Britain, which he called objective distances, and determined their relative distances to each other by first calculating for each centre the mean of the distances to all remaining centres. The relative distance between a centre A and a centre B was then determined by dividing the objective distance between A and B by A's mean. This notion of relative distance is actually asymmetric: this method will most likely produce a different relative distance between A and B than between B and A. The relative distance can then be used to calculate fuzzy nearness values using the following definition: $nearness(x, y) = (relative\_distance(x, y) + 1)^{-1}$.

Places having high nearness values are therefore very close and low ones are not close. The greatest nearness is between a place and itself, with a value of 1. This approach does not suffer from the same restriction as Gahegan's approach. The objects do not need to be fairly evenly distributed. In his more recent work on environmental space, Worboys [7] used the number of subjects and the number of *yes* or *no* votes to calculate fuzzy membership values for nearness. This is a very interesting approach given his experimental data. However it is not practically applicable to do such a kind of data collection for every geographic area that GIS-users might need to work with.

A serious shortcoming of all the approaches that have been described is that none of them does actually evaluate the membership functions against any real data, in order to see how useful these functions are. As has been discussed in this section, we have been conducting experiments with several membership functions and evaluated them against proximity data. The following section introduces the functions we used.

3 Various Distance Metrics

As previously mentioned, in order to address all of the observations that Gahegan [3] made in the context of GIS users perceiving proximity, we will evaluate several spatial proximity functions based on absolute distance, relative distance, and combinations of both. Table 1 shows the fuzzy membership functions we evaluated in terms of their usefulness within GIS settings.

Table 1. Fuzzy Membership Functions

Absolute Distance Metrics:	$\mu_{abs}(A, B) = 1 - \frac{Dist(A,B)}{Max}$
Relative Distance Metrics:	$\mu_{rel}(A, B) = \frac{1}{(reldis(A,B)+1)}$
Fuzzy Union:	$\mu_{comb\_u}(A, B) = \mathrm{MAX}(\mu_{abs}(A, B), \mu_{rel}(A, B))$
Fuzzy Intersection:	$\mu_{comb\_i}(A, B) = \mathrm{MIN}(\mu_{abs}(A, B), \mu_{rel}(A, B))$

The fuzzy membership function based on absolute distance metrics is a derivation of Gahegan's [3] function with the maximum value Max being the maximum distance between all of the places in our data set; and $Dist$ being the distance between places A and B. For the fuzzy membership function based on relative distance metrics, we borrowed Worboy's [6] membership function, as we found that Gahegan's ordinal ranking approach is insufficient. Relative distance is calculated using the mean of each place A in the data set, calculated from the n places OP_i, $1 \leq i \leq n$, distinct from A and available in the set: $mean(A) = \frac{1}{n}\Sigma_{i=1}^{n} Dist(A, OP_i)$. The result of this is then used to calculate the relative distance between each two places: $reldis(A, B) = Dist(A, B) * mean(A)^{-1}$. While Gahegan [3] only suggested to combine the membership functions based on absolute and relative distance by applying the fuzzy union, we also investigated the application of the fuzzy intersection operator, which yielded interesting results. The fuzzy union operator will by definition always return the maximum membership function value for each data entry. While the fuzzy intersection operator will by definition always return the minimum membership function value for each data entry. We applied these functions to the data set described in the following section.

4 Data Set and Experiments

We encoded 34 places in the Australian state of New South Wales and the distances between them[1]. For each of the given places, we define the tourist region, the region and the regional area they are located in. For Sydney, a list of regions that are easily accessible for short trips is also supplied, thereby giving some indication of what is perceived and generally accepted as near to Sydney. This data set was collected from

---

[1] As given by "The Official Road Directory of New South Wales" by the Land Information Centre in Bathurst, The New South Wales Government.

the *Tourism New South Wales* site[2]. We will be able to use this information to evaluate our membership functions to see how well they suit the data and "nearness" information for the given places.

5   Results and Evaluation

The membership function based on absolute distance as illustrated in the left graph in Figure 1 shows a linear distribution, as expected from the function used. The maximum distance in the dataset is 1710 km. However, the relative membership function as illustrated in the right graph in Figure 1 shows quite a varied distribution, which is very different to Gahegan's more or less linear proposal of ordinal ranking. The issue arising from his kind of approach, that objects could possibly still be considered close to one another even when they are separated by a very large distance, is not a problem for the membership function we used, because ours is a function of the distance between the two places being considered.

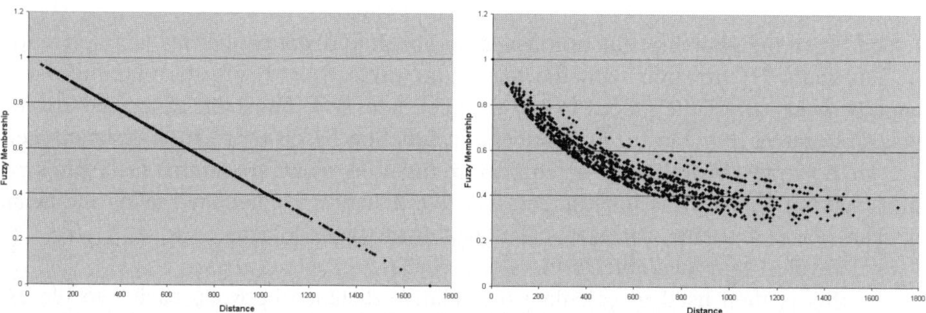

Fig. 1. Fuzzy Distribution Functions for Absolute and Relative Distance Metrics

The two combined membership functions give quite interesting results. On one hand they do support Gahegan's [3] suggestion that absolute measures are more appropriate for non-clustered objects, and relative measures for objects within clusters of same "kinded" objects. On the other hand, the results do contradict Gahegan's suggestion to use the union fuzzy operator for an efficient combined function. Because, when combining absolute and relative distance metrics functions by union we do get a linear distribution for distances until about 800km, where it changes into a more clustered distribution (see left graph in Figure 3).

This is even contradicting Gahegan's own terms that absolute distance metrics i.e., linear distributions in his case, are more suited for proximity assignments between objects that are located in virtually devoid areas. Figure 2 shows that the greater the distances between the places, the fewer places are within the area; which can be explained by the fairly isolated character of the Australian non-metropolitan areas. In order to comply with Gahegan's suggestion to use absolute distance metrics for only lightly and

---

[2] www.visitnsw.com.au

relative distance metrics for heavily populated areas, the membership function distribution should be the reverse of the result we obtained for the combined function using the fuzzy union operator.

When we applied the fuzzy intersection operator to our data set, we attained precisely such a distribution. The fuzzy intersection changes from a clustered to a linear distribution between 1000 and 1200 km. The right graph in Figure 3 shows this clearly. This is even contradicting Gahegan's own terms that absolute distance metrics i.e., linear distributions in his case, are more suited for proximity assignments between objects that are located in virtually devoid areas. Figure 2 shows that the greater the distances

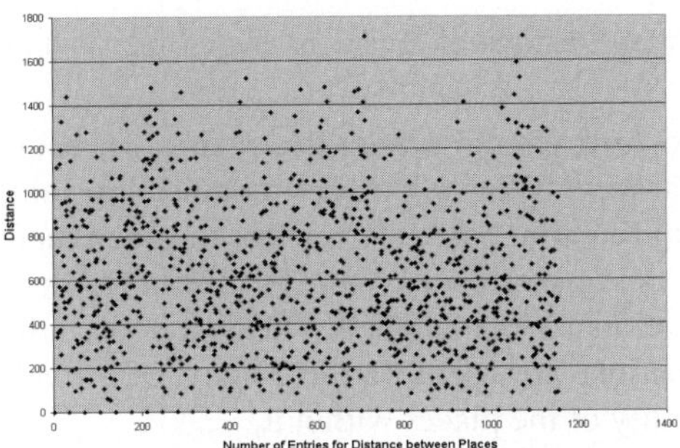

Fig. 2. Distance Distribution in the Data Set

between the places, the fewer places are within the area; which can be explained by the fairly isolated character of the Australian non-metropolitan areas. In order to comply with Gahegan's suggestion to use absolute distance metrics for only lightly and relative distance metrics for heavily populated areas, the membership function distribution should be the reverse of the result we obtained for the combined function using the fuzzy union operator.

When we applied the fuzzy intersection operator to our data set, we attained precisely such a distribution. The fuzzy intersection changes from a clustered to a linear distribution between 1000 and 1200 km. The right graph in Figure 3 shows this clearly. The clustered distribution is more appropriate for smaller distances, as there are more places within a smaller area, and the linear distribution will suit areas with fewer objects, which are to be found at greater distances in the given data set. This is perfectly consistent with Gahegan's [3] observations, although it is a different combination operator that gives the desired result.

We evaluated our membership distribution function values against the proximity information, namely *Sydney Surrounds* and regions, to appraise the usefulness of the functions and their combinations in the context of reachability within a road network.

For all the places that are within regions which are generally accepted to be in the Sydney surrounding area, all membership function values were not only well above the usual crossover point of 0.5, but also above 0.7. We tested the distances between all places in our dataset that are in the same region using this value as the crossover point. As geographic regions are generally defined with the perception of "everything" within

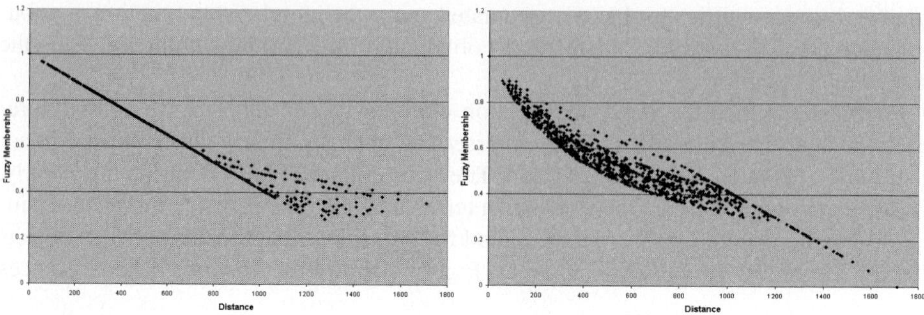

Fig. 3. Combined Fuzzy Distribution Functions using Union and Intersection Operator

a region being close to "everything" within this region. 16 out of 94 matches had two membership values that were below 0.7. These membership values were always the result of the membership function based on relative distance metrics and the combined function that returned the former one. When the threshold was lowered to the normal crossover point (0.5), all matches complied. This is a good indication that the investigated membership functions are useful in the context of a road distance network and the associated reachability of the places within it.

6   Conclusions and Future Work

In this paper, we have shown that while the membership functions based on absolute distance metrics and relative distance metrics, as proposed by Gahegan [3] and Worboys [6] respectively, do evaluate well against distance data, the combined membership function that proves useful in this context is the fuzzy intersection of the two former, and not their union as Gahegan [3] suggested. We will implement these proposed membership functions as part of our already existing qualitative reasoning extension to the Geographic Information System *AccuGlobe*, to further assess the benefits of fuzzy logic as a descriptor for spatial proximity notions.

We also plan to extend our data by including more information about regions and places within the regions that surround major places in the data set. This will allow to analyse further membership functions and to get more conclusive results for asymmetric relative distance relations. Moreover, we will use the expanded data set to automatically learn appropriate membership functions and then evaluate them against a data collection of another geographic area, such as another Australian state or a state within another country of a similar spatial distribution of places.

References

1. van Benthem, J.F.: The logic of time – a model theoretic investigation into the varieties of temporal ontology and temporal discourse. Kluwer Academic Publishing, Dordrecht, 1991.
2. Gahegan, M.: Support for the Contextual Interpretation of Data within an Object-Oriented GIS. Proceedings of Spatial Data Handling '94, T.C. Waugh and R.G. Healey (eds), Edinburgh, Scotland, pp. 988-1001, 1994.

3. Gahegan, M.: Proximity Operators for Qualitative Spatial Reasoning. Spatial Information Theory - A Theoretical Basis for GIS (COSIT'95). A. U. Frank and W. Kuhn (eds), Springer-Verlag, Berlin, Heidelberg, pp. 31-44, 1995.
4. Guesgen, H.-W. and Albrecht, J.: Imprecise reasoning in geographic information systems, In *Fuzzy Sets and Systems*, vol. 113(1), pp. 121-131, 2002.
5. Guesgen, H.-W.: Reasoning About Distance Based on Fuzzy Sets. In *Applied Intelligence*, vol. 17, pp. 265-270, 2002.
6. Worboys, M.F.: Metrics and topologies for geographic space. Advances in GIS Research II. Taylor and Francis, pp. 365-376, 1996.
7. Worboys M.F.: Nearness relations in environmental space. In *International Journal of Geographical Information Science*, vol. 12(7), pp. 633-651, 2001.

Engineering Evolutionary Algorithm to Solve Multi-objective OSPF Weight Setting Problem

Sadiq M. Sait, Mohammed H. Sqalli, and Mohammed Aijaz Mohiuddin

Computer Engineering Department
King Fahd University of Petroleum & Minerals
Dhahran-31261, Saudi Arabia
{sadiq, sqalli, aijazm}@ccse.kfupm.edu.sa

Abstract. Setting weights for Open Shortest Path First (OSPF) routing protocol is an NP-hard problem. Optimizing these weights leads to less congestion in the network while utilizing link capacities efficiently. In this paper, Simulated Evolution (SimE), a non-deterministic iterative heuristic, is engineered to solve this problem. A cost function that depends on the utilization and the extra load caused by congested links in the network is used. A *goodness* measure which is a prerequisite of SimE is designed to solve this problem. The proposed SimE algorithm is compared with Simulated Annealing. Results show that SimE explores search space intelligently due to its goodness function feature and reaches near optimal solutions very quickly.

1   Introduction

Non-deterministic iterative heuristics such as Simulated Annealing, Simulated Evolution, Genetic Algorithms etc., are stochastic algorithms that have found applications in myriad complex problems in science and engineering. They are extensively used for solving combinatorial optimization problems involving large, multi-modal search spaces, where regular constructive algorithms often fall short. One such domain that involves such non-convex problems is Routing - a fundamental engineering mechanism in computer communication networks. The optimization objective here is to determine the least expensive or best possible path between a source and destination. Routing can become increasingly complex in large networks because of the many potential intermediate nodes a packet traverses before reaching its destination [1]. To address this, the Internet is divided into smaller domains i.e., Autonomous Systems (AS). Each AS is a group of networks and routers under the authority of a single administration. An Interior Gateway Protocol (IGP) is used within an AS, while Exterior Gateway Protocols (EGP) are used to route traffic between them [2].

The TCP/IP suite has many routing protocols, one of them is the Open Shortest Path First (OSPF) Routing Protocol, used in today's Internet [1,3]. A computationally complex component related to the OSPF routing protocol is addressed - the OSPF Weight Setting (OSPFWS) problem. Proven to be NP-hard [4], it involves setting the OSPF weights on the network links such that the network is utilized efficiently.

A. Sattar and B.H. Kang (Eds.): AI 2006, LNAI 4304, pp. 950–955, 2006.
© Springer-Verlag Berlin Heidelberg 2006

2   Related Work

The use of non-deterministic iterative algorithms for solving the OSPFWS problem has been previously reported. In [4], a cost function based on the utilization ranges was formulated and Tabu search [5] was used. Dynamic shortest path algorithm was applied to find multiple equidistance shortest paths between source and destination nodes [6]. Ericsson et al applied genetic algorithm to the same problem [7]. Other papers on optimizing OSPF weights [8] have either chosen weights so as to avoid multiple shortest paths from source to destination, or applied a protocol for breaking ties, thus selecting a unique shortest path for each source-destination pair. Rodrigues and Ramakrishnan [8] presented a local search procedure similar to that of Fortz and Thorup [4]. The input to the algorithm for this problem is a network topology, capacity of links and a demand matrix. The demand matrix represents the traffic between each pair of nodes present in the topology. The methodology for deriving traffic demands from operational networks is described in [9]. For other related work the reader is referred to the contributions in [10,11].

In this paper SimE algorithm is engineered to solve the OSPFWS. An enhanced cost function proposed in our previous work is used for this problem [12].

3   Problem Statement

The OSPF Weight Setting (OSPFWS) problem can be stated as follows: Given a network topology and predicted traffic demands, find a set of OSPF weights that optimize network performance. More precisely, given a directed network $G = (N, A)$, a demand matrix D, and capacity C_a for each arc $a \in A$, it is required to find a positive integer $w_a \in [1, w_{max}]$ such that the cost function Φ is minimized; w_{max} is a user-defined upper limit. The chosen arc weights determine the shortest paths, which in turn completely determine the routing of traffic flow, the loads on the arcs, and the value of the cost function Φ. The quality of OSPF routing depends highly on the choice of weights [12]. Figure 1 depicts a topology with assigned weights in the range $[1, 20]$. A solution for this topology can be represented as $(18, 1, 7, 15, 3, 17, 5, 14, 19, 13, 18, 4, 16, 16)$. These elements (i.e., weights) are arranged in a specific order for simplicity. They are

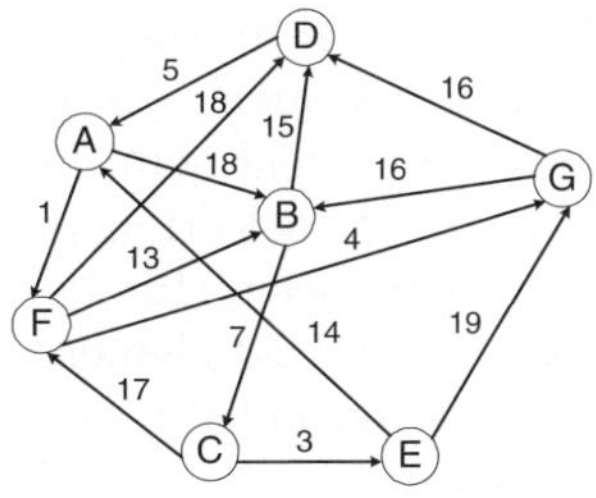

Fig. 1. Representation of a topology with assigned weights

ordered in the following manner: the outgoing links from node A listed first (i.e., AB, AF), followed by the outgoing links from node B (i.e., BC, BD), and so on.

Mathematical Model and Cost Function

Fortz and Thorup discussed a cost function based on the utilization ranges in their paper [4], which here is denoted as FortzCF. Through experimentation it was found that this cost function does not address the optimization of the number of congested links. In our previous work, the following new cost function was proposed [12].

$$\Phi = MU + \frac{\sum_{a \in SetCA} (l_a - c_a)}{E} \tag{1}$$

This new function, denoted as NewCF, contains two terms. The first is the maximum utilization (MU) in the network. The second term represents the extra load on the network divided by the number of edges present in the network to normalize the entire cost function. The motivation behind using such a cost function is to reduce the number of congested links, if any. Consequently, the network designer will need to upgrade fewer links in the network to avoid congestion, which in turn means a less expensive upgrade.

The steps to compute the cost function Φ for a given weight setting $\{w_a\}_{a \in A}$ and a given graph $G = (N, A)$ with capacities $\{c_a\}_{a \in A}$ and demands $d_{st} \in D$ are enumerated in [12]. This procedure is also described in [13].

4   Engineering SimE for OSPFWS

SimE is a general iterative heuristic proposed by Ralph Kling [14]. It falls in the category of algorithms which emphasize the behavioral link between parents and offspring, or between reproductive populations, rather than the genetic link. This scheme combines iterative improvement and constructive perturbation and saves itself from getting trapped in local minima by following a stochastic perturbation approach. It iteratively operates a sequence of **evaluation, selection, and allocation** steps on one solution.

The selection and allocation steps constitute a compound move from the current solution to another solution of the state space. SimE proceeds as follows - It starts with a randomly or constructively generated valid initial solution. A solution is seen as a set of movable elements. Each element e_i has an associated goodness measure g_i in the interval $[0, 1]$. In the evaluation step, the goodness of each element is estimated. In the consequent selection step, a subset of elements are selected and removed from the current solution. The lower the goodness of a particular element, the higher is its selection probability. A bias parameter B is used to compensate for inaccuracies of the goodness measure. Finally, the allocation step tries to assign the selected elements to better locations. Other than these three steps, some input parameters for the algorithm are set in an earlier step known as **initialization.**

Evaluation: SimE operates on a single solution termed as the population, which consists of elements. For the OSPFWS problem, each individual or element is a weight on a link. The *Evaluation* step consists of evaluating the goodness of each individual of the current solution. The goodness measure must be a number in the range $[0, 1]$. The goodness function is a key factor of Simulated Evolution and should be carefully formulated to handle the target objective of the given problem. As two cost functions are present two corresponding goodness functions are defined as follows. For FortzCF, the goodness function is taken as:

$$g_{ij} = 1 - \Phi_{i,j}/MC \tag{2}$$

For NewCF, the goodness function proposed is:

$$g_{ij} = \begin{cases} 1 - u_{ij} & \text{for } MU \le 1 \\ 1 - u_{ij}/MU + u_{ij}/MU^2 & \text{for } MU > 1 \end{cases} \tag{3}$$

Selection: In this stage of the algorithm, for each link i of the network, a random number $RANDOM \in [0, 1]$ is generated and compared with $g_i + B$, where B is the selection bias. If $RANDOM > g_i + B$, then weight w_i is selected for allocation. Bias B is used to control the size of the set of weights selected. For FortzCF, a bias value of -0.03 is found to be suitable through experimentation.

For NewCF, when maximum utilization is less than 1, a variable bias methodology [15] is used. The *variable bias* is a function of the quality of the current solution. When the overall solution quality is poor, a high value of bias is used, otherwise a low value is employed. The average weight goodness is a measure of how many "good" weights are present in a solution.

Allocation: During this stage of the algorithm, the selected weights are removed from the solution one at a time. For each removed weight, new weights are tried to obtain an overall better solution. For OSPFWS problem, the weight on a link lies in the range $[1,20]$. Different allocation schemes were tried for this problem, but we use the following scheme which provided good results both in quality and time. For all iterations, a weight window of value 4 is maintained. For example, if weight 6 is selected to change, then the values $\{4, 5, 7, 8\}$ are tried. This is done to moderate the perturbation of the solution state per iteration.

5   Results and Discussion

All the test cases used here are taken from the work by Fortz and Thorup [4]. Details regarding these test cases and their characteristics can be found in [13].

Figures 2 and 3 illustrate the behavior of the proposed SimE algorithm using NewCF. These figures respectively plot the average goodness and cardinality of the selection set for test case h50N148a, i.e., 50 nodes and 148 arcs. It is observed in Figure 2 that the average goodness increases with time. In the initial stages of the search, this rate of increase is significant and tends to slow down in later iterations. This slow stage suggests that on average, the weights have been assigned their values and the point of convergence has been reached. In Figure 3, the cardinality of the selection set is shown. Here it is observed that the number of

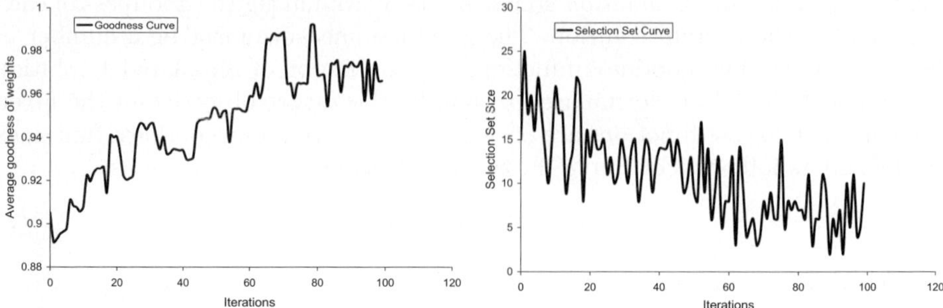

Fig. 2. Average goodness of weights for test case h50N148a NewCF

Fig. 3. Selection set size for test case h50N148a NewCF

selected elements (i.e., weights) decreases as the iterations increase. This suggests that the algorithm is reaching a level of convergence and therefore fewer number of weights are selected for perturbation. If the trends in Figures 2 and 3 are compared, the observation is that the convergence of the SimE algorithm towards a better solution is correlated with the average goodness of links.

Table 1 show the maximum utilization (MU) and number of congested link (NOC) values for all test cases with the highest demand values using Simulated Annealing (SA) [12] and SimE.

Table 1. MU and NOC corresponding to highest demand for all test cases (SA & SimE)

Test case	Demand value	SA				SimE			
		FortzCF		NewCF		FortzCF		NewCF	
		MU	NOC	MU	NOC	MU	NOC	MU	NOC
h100N280a	4605	1.341	8	1.341	7	1.341	23	1.341	5
h100N360a	12407	1.383	16	1.740	11	1.409	17	1.424	14
h50N148a	4928	1.271	9	1.411	9	1.532	10	1.386	9
h50N212a	3363	1.151	10	1.209	6	1.365	8	1.349	6
r100N403a	70000	1.441	95	1.826	58	1.407	81	1.402	45
r100N503a	100594	1.445	106	1.983	86	1.272	66	1.384	38
r50N228a	42281	1.218	38	1.394	22	1.316	33	1.339	20
r50N245a	53562	1.856	54	2.617	40	2.553	45	2.339	36
w100N391a	48474	1.401	1	1.424	1	1.495	1	1.284	1
w100N476a	63493	1.314	16	1.374	11	1.315	7	1.314	6
w50N169a	25411	1.252	5	1.249	3	1.252	11	1.260	5
w100N230a	39447	1.222	3	1.222	3	1.230	3	1.224	3
Average		1.358	30.083	1.566	21.417	1.457	25.417	1.421	15.667

6   Conclusion

In this paper, Simulated Evolution (SimE) is engineered to solve the the OSPFWS problem. A new proposed cost function [12] is employed which addresses the optimization of the number of congested links, and the required goodness functions are designed. The results obtained show that the engineered SimE heuristic is always able to find near-optimal solutions. Comparison with Simulated Annealing (SA) showed that the search performed by SimE is more intelligent, i.e., the solutions generated by SimE are of superior quality than that of SA and are obtained in better time.

Acknowledgement

Authors acknowledge King Fahd University of Petroleum & Minerals, Dhahran, Saudi Arabia, for all support of this research under Project # SAB-2006-10.

References

1. T. M. Thomas II. *OSPF Network Design Solutions.* Cisco Press, 1998.
2. U. Black. *IP Routing Protocols.* Prentice Hall Series, 2000.
3. J. T. Moy. *OSPF: Anatomy of an Internet Routing Protocol.* Addison-Wesley., 99.
4. Bernard Fortz and Mikkel Thorup. Internet traffic engineering by optimizing OSPF weights. *IEEE Conference on Computer Communications(INFOCOM)*, pages 519–528, 2000.
5. Sadiq M. Sait and Habib Youssef. *Iterative Computer Algorithms and their Application to Engineering.* IEEE Computer Society Press, December 1999.
6. Daniele Frigioni, Mario Loffreda, Umberto Nanni, and Giulio Pasqualone. Experimental analysis of dynamic algorithms for the single source shortest paths problem. *ACM Journal of Experimental Algorithms*, 1998.
7. Ericsson, M. Resende, and P. Pardalos. A genetic algorithm for the weight setting problem in OSPF routing. *J.Combinatorial Optimisation Conference*, 2002.
8. M. Rodrigues and K. G. Ramakrishnan. Optimal routing in data networks. *Presentation at International Telecommunication Symposium (ITS)*, 1994.
9. Anja Feldmann, Albert Greenberg, Carsten Lund, Nick Reigold, Jennifer Rexford, and Fred True. Deriving traffic demands for operational ip networks: Methodology and experience. *IEEE/ACM Transactions on Networking,*, 9(3), 2001.
10. Shekhar Srivastava, Gaurav Agrawal, Micha Pioro, and Deep Medhi. Determining link weight system under various objectives for OSPF networks using a lagrangian relaxation-based approach. *IEEE Transactions on Network and Service Management*, 2(1):9–18, Third quarter 2005.
11. Ashwin Sridharan, Roch Guerin, and Christophe Diot. Achieving near-optimal traffic engineering solutions for current OSPF/IS-IS networks. *IEEE INFOCOM*, 2003.
12. M. H. Sqalli, Sadiq M. Sait, and M. Aijaz Mohiuddin. An enhanced estimator to multi-objective OSPF weight setting problem. *Proceedings of 10th IEEE/IFIP Network Operations & Management Symposium (NOMS2006)*, April 2006.
13. Bernard Fortz and Mikkel Thorup. Increasing internet capacity using local search. *Technical Report IS-MG*, 2000.
14. R. M. Kling and P. Banerjee. ESP: Placement by Simulated Evolution. *IEEE Transactions on CAD*, Vol. 8(3):pp 245–256, March 1989.
15. Sadiq M. Sait, Habib Youssef, and Ali Hussain. Fuzzy simulated evolution algorithm for multiobjective optimization of VLSI placement. *IEEE Congress on Evolutionary Computation, Washington, D.C., U.S.A.*, pages 91–97, July 1999.

Intelligent Face Recognition: Local Versus Global Pattern Averaging

Adnan Khashman

Department of Electrical & Electronic Engineering
Near East University, Lefkosa, North Cyprus, Turkey
amk@neu.edu.tr

Abstract. Face recognition has lately attracted more research aimed at developing intelligent machine recognition which uses information within the encoded facial patterns to learn and recognize the objects. This paper investigates the efficiency of using Global and Local pattern averaging for facial data encoding prior to training a neural network using the averaged patterns. Averaging is a simple but efficient method that creates "fuzzy" patterns as compared to multiple "crisp" patterns, which provide the neural network with meaningful learning while reducing computational expense. A real-life application will be presented throughout recognizing the faces of 60 persons using our database and the ORL face database. Experimental results suggest that using pattern averaging; globally or locally, performs well as part of a fast and efficient intelligent face recognition system.

1 Introduction

Intelligent systems are being increasingly developed aiming to simulate our perception of various inputs (patterns) such as images, sounds…etc. Biometrics is an example of popular applications for artificial intelligent systems. Face recognition by machines can be invaluable and has various important applications in real life, such as, electronic and physical access control or security and defense applications. The development of an intelligent face recognition system requires providing sufficient information and meaningful data during machine learning of a face.

The use of neural networks for face recognition has been recently addressed [1], [2], [3], [4]. Other successful works suggested different methods for face recognition, such as Principal Component Analysis (PCA) [5], Linear Discriminant Analysis (LDA) [6] and Locality Preserving Projections (LPP) [7]. However, these methods are appearance-based or feature-based methods that search for certain global or local representation of a face.

This paper proposes the use of pattern averaging, globally or locally, as an efficient method for meaningful facial data preparation prior to training a neural network classifier. Pattern averaging creates a "fuzzy" feature vector as compared to multiple "crisp" feature vectors, which are used for training and generalizing the neural network. Two methods, namely, Global pattern averaging and Local pattern averaging, are presented in this work as a first phase of an intelligent face recognition system that uses the back propagation algorithm in its second phase.

A. Sattar and B.H. Kang (Eds.): AI 2006, LNAI 4304, pp. 956–961, 2006.
© Springer-Verlag Berlin Heidelberg 2006

The first method uses *global* pattern averaging of a face and its background and aims at simulating the way we see and recognize faces. This is based on the suggestion that a human "glance" of a face can be approximated in machines using pattern averaging, whereas, the "familiarity" of a face can be simulated by a trained neural network [3]. The second method uses *local* pattern averaging of essential facial features (eyes, nose and mouth). Here, multiple face images of a person with different facial expressions are used, where only eyes, nose and mouth patterns are considered. These essential features from different facial expressions are averaged and then used to train the supervised neural network [4]. A real-life application will be presented using both methods throughout recognizing the faces of 60 persons.

2 Face Image Databases

The databases used in this work comprise 270 face images of 60 persons, and are organized into two sets. The first set; from our face database, is used for implementing the Global averaging method and contains face and background images of 30 persons looking left, straight and right, thus providing 90 images; examples of which are shown in Fig. 1. The second set is used for implementing the Local averaging method and contains face images of another 30 persons with six different facial impressions (natural, smiley, sad, surprised and two random impressions), thus providing 180 images (90 from our databases and 90 from ORL face database [8]) as shown in Fig. 2.

Fig. 1. Examples of Global pattern averaging faces looking: a- left, b- straight, c- right

Fig. 2. Examples of Local pattern averaging faces: a- Own database, b- ORL database [8]

Each image set is divided into two groups: training and testing images. For Global averaging set each face has three different projections, which were captured while looking *Left*, *Straight* and *Right*. The neural network training images will be *Left* and *Right* looking faces (60 images) whereas the testing images will be *Straight* looking faces (30 images) as shown in Fig. 3a. For Local averaging set each person has six different face expressions. Four out of the six expressions for each person (natural, smiley, sad and surprised) are locally averaged and then used for training the neural network (120 feature images averaged to 30 images). The remaining two face expressions for each person are random expressions and will be used together with the *non-averaged* four training face expressions to test the trained neural network (180 images) as shown in Fig. 3b.

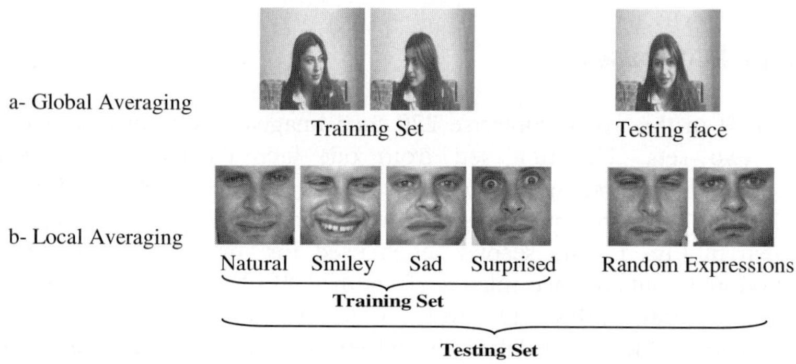

Fig. 3. Examples of training and testing face images

3 Pattern Averaging

Pattern averaging creates a "fuzzy" feature vector as compared to multiple "crisp" feature vectors, which are used for training the neural network. This section describes the two averaging methods that are applied prior to training a neural network as part of the intelligent face recognition system.

In *Global* averaging, a face image of size 100x100 pixels is segmented and the values of the pixels within each segment are averaged. The result average values are then used as input data for the neural network. Global averaging can be defined as:

$$PatAv_i = \frac{1}{S_k S_l} \sum_{l=1}^{S_l} \sum_{k=1}^{S_k} p_i(k,l), \qquad (1)$$

where k and l are segment coordinates in the x and y directions respectively, i is the segment number, S_k and S_l are segment width and height respectively, $P_i(k,l)$ is pixel value at coordinates k and l in segment i, $PatAv_i$ is the average value of pattern in segment I, that is presented to neural network input layer neuron i. Segment size of 10x10 pixels ($S_k = S_l = 10$) has been used and average values representing the image were obtained, thus resulting in 100 average values that were used as the input to the neural network for both training and testing.

In *Local* averaging, the essential features (eyes, nose and mouth) from four expressions (natural, smiley, sad and surprised) are approximated via averaging into one single vector that represents the person; this is *only* applied prior to training. The features undergo manual extraction, dimension reduction and then averaging which reduces the 120 training images to 30 averaged images. Finally, the averaged features ((272x1) pixel vectors) are presented to the neural network. Local averaging for each feature can be implemented using the following equation:

$$f_{avg} = \frac{1}{4} \sum_{i=1}^{4} f_i \,,$$
(2)

where f_{avg} is the feature average vector and f_i is feature in expression i of one person.

4 Neural Network Implementation

The back propagation algorithm is used for the implementation of the intelligent face recognition system due to its simplicity and efficiency in solving pattern recognition problems. Two different 3-layer neural networks are used for the arbitration of the two averaging methods. The outputs of the averaging methods are used for training and/or testing the neural network as its input data.

The neural arbitration of globally averaged faces uses a network with 100 neurons in the input layer, each receiving an averaged value of the face image segments. The hidden layer consists of 99 neurons, whereas the output layer has 30 neurons according to the number of persons. Fig. 4a shows the topology of this neural network and global data presentation to the input layer.

The neural arbitration of locally averaged faces uses a network with 272 neurons in the input layer which carry the values of the averaged features, a hidden layer with 65 neurons and an output layer with 30 neurons which is the number of persons. Fig. 4b shows this neural network design.

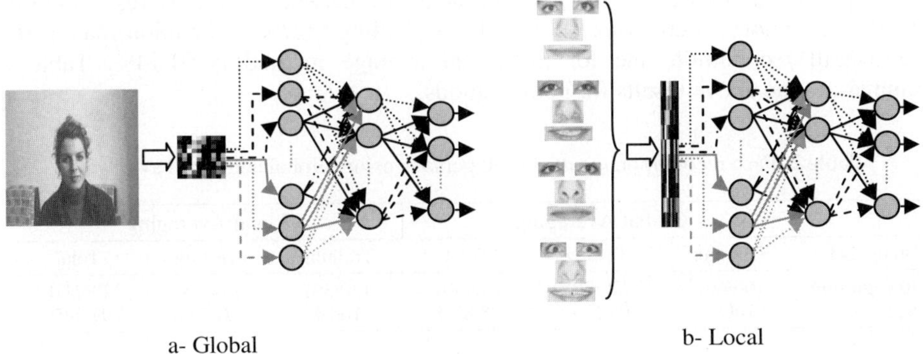

a- Global   b- Local

Fig. 4. Global and Local averaging neural networks

5 Results and Analysis

The implementation of Global or Local pattern averaging of face images as part of an intelligent face recognition system are carried out prior to training a neural network that uses the averaged representations of the face. Different neural networks were designed for each averaging method. The following results were obtained using a 2.4 GHz PC with 256 MB of RAM, Windows XP OS and Borland C^{++} compiler.

The Global averaging neural network learnt and converged after 4314 iterations and within 390 seconds, whereas the running time for the generalized neural network after training and using one forward pass was 0.21 seconds. The Local averaging neural network learnt and converged after 3188 iterations and within 265 seconds, whereas the running time for the generalized neural network after training and using one forward pass was 0.032 seconds. Table 1 shows the final parameters of the successfully trained neural networks.

Table 1. Global and Local averaging trained neural networks final parameters

	Input nodes	Hidden nodes	Output nodes	Learning rate	Momentum rate	Error	Iterations	Training time (seconds)	Run time (seconds)
Global	100	99	30	0.008	0.32	0.002	4314	390	0.21
Local	272	65	30	0.0495	0.41	0.001	3188	265	0.032

The Global average neural network implementation yielded 100% recognition rate of all training images (60 face images- looking left and right) as would be expected. Testing this neural network using test images (30 face images – looking straight) yielded a successful 96.67% recognition rate. Thus, the overall recognition rate for the Global average method is 98.89%. The Local average neural network implementation also yielded 100% recognition rate when using the 30 locally averaged face images in the training set. Testing was carried out using 180 face images which contain different face expressions that were not exposed to the neural network before. Here, 174 out of the 180 test images were correctly identified yielding 96.7% recognition rate. Thus, the overall recognition rate for the Local average method is 97.14%. Table 2 summarizes the above results for both methods.

Table 2. Face recognition results for 30 persons using Global and Local averaging

Image Set	Global Averaging			Local Averaging		
	Training	Testing	Total	Training	Testing	Total
Recognition Rate	(60/60) 100%	(29/30) 96.67%	(89/90) 98.89%	(30/30) 100%	(174/180) 96.70%	(204/210) 97.14%

6 Conclusions

This paper investigated the use of pattern averaging as a pre-processing phase in an intelligent face recognition system. The system uses a back propagation neural network classifier which receives at its input an averaged representation of a face image. Pattern averaging provides the neural network with a small fuzzy pattern reflecting the original larger crisp pattern, thus achieving two aims: reduction of computational and time cost and maintaining meaningful machine learning of faces that may have different expressions, orientations and backgrounds.

Two pattern averaging methods were investigated: Global averaging and Local averaging. Both methods have shown successful results which suggests they can be used as part of a robust intelligent face recognition system.

Global averaging can be successfully applied in real life where the faces are at different orientations and the image contains non-facial features such as hats, hair, eyeglasses and background.

Local averaging can also be applied successfully to identify faces with different expressions. Here, the image databases contain only the faces as the method uses essential facial features such as eyes, nose and mouth; therefore the existence of background and occlusions is irrelevant. Features from different face images of a person with different facial expression are averaged prior to training the neural network. Although the feature pattern values (pixel values) may change with the change of facial expression, the use of Local averaging of a face provides the neural network with an approximated understanding of the identity and is found to be sufficient for training a neural network to recognize that face with any expression.

Future work includes using larger face databases and investigating the use of averaging with images that contain more than one face. Additionally, work will be carried out on developing a single neural network that can be used with both Global and Local averaging for the purpose of face recognition.

References

1 Zhang, B., Zhang, H., Ge, S.: Face recognition by applying wavelet subband representation and kernel associative memory, IEEE Transactions on Neural Networks, Vol. 15 (2004) 166-177

2 Fan, X., Verma, B.: A Comparative Experimental Analysis of Separate and Combined Facial Features for GA-ANN based Technique, Proceedings of International Conference on Computational Intelligence and Multimedia Applications (2005) 279-284

3 Khashman, A.: Face Recognition Using Neural Networks and Pattern Averaging. Lecture Notes in Computer Science, Vol. 3972. Springer-Verlag, Berlin Heidelberg New York (2006) 98–103

4 Khashman, A., Garad, A.: Intelligent Face Recognition Using Feature Averaging. Lecture Notes in Computer Science, Vol. 3975. Springer-Verlag, Berlin Heidelberg New York (2006) 432-439

5 Turk, M., Pentland, A.P.: Face Recognition Using Eigenfaces, Proceedings of the IEEE Conference on Computer Vision and Pattern Recognition (1991) 586-591

6 Belhumeur, P.N., Hespanha, J.P., Kriegman, D.J.: Eigenfaces vs. Fisherfaces: Recognition Using Class Specific Linear Projection, IEEE Trans. PAMI, 19(7) (1997) 711-720

7 He, X., Yan, S., Hu, Y., Niyogi, P., Zhang, H.J.: Face Recognition Using Laplacianfaces, IEEE Trans. PAMI, 27(3) (2005) 328-340

8 Cambridge University, Olivetti Research Laboratory face database. http://www.uk.research. att.com/facedatabase.html

Economic Optimisation of an Ore Processing Plant with a Constrained Multi-objective Evolutionary Algorithm

Simon Huband[1], Lyndon While[2], David Tuppurainen[3],
Philip Hingston[1], Luigi Barone[2], and Ted Bearman[3]

[1] Edith Cowan University, Mt Lawley, Western Australia
[2] The University of Western Australia, Crawley, Western Australia
[3] Rio Tinto OTX, Perth, Western Australia

Abstract. Existing ore processing plant designs are often conservative and so the opportunity to achieve full value is lost. Even for well-designed plants, the usage and profitability of mineral processing circuits can change over time, due to a variety of factors from geological variation through processing characteristics to changing market forces.

Consequently, existing plant designs often require optimisation in relation to numerous objectives. To facilitate this task, a multi-objective evolutionary algorithm has been developed to optimise existing plants, as evaluated by simulation, against multiple competing process drivers. A case study involving primary through to quaternary crushing is presented, in which the evolutionary algorithm explores a selection of flow-sheet configurations, in addition to local machine setting optimisations. Results suggest that significant improvements can be achieved over the existing design, promising substantial financial benefits.

1 Introduction

Design in any sphere means the specification of a system that satisfies a given set of requirements whilst optimising performance parameters. In most industries, the principal parameter is profit, usually by some combination of minimising capital investment and operating costs, and maximising return on investment, throughput, and efficiency.

Usually, good designs cannot be derived analytically, and the design process will involve at least some aspect of trial-and-error or expert judgement. This has led many people to suggest that the process should be automated. Automation has often taken the form of a search, and in recent years this search has often utilised various artificial intelligence techniques.

The major contribution of this paper is the description of a case study that applies multi-objective evolutionary algorithms (EAs) to optimise the design of a comminution plant. The flexibility of the multi-objective evolutionary approach is illustrated by the ease with we are able to deal with two real-world complications: risk management, and complex feasibility conditions. The initial results from this study are promising.

A. Sattar and B.H. Kang (Eds.): AI 2006, LNAI 4304, pp. 962–969, 2006.
© Springer-Verlag Berlin Heidelberg 2006

2   Previous Applications of EAs to Plant Design

Ventner *et al.* [1] proposed an approach for plant design using learning classifier systems. Intelligent objects bid against each other for a position in the circuit. The simple flowsheets examined in the study demonstrated the concept, but most of the intelligence was in the form of subjective heuristics or empirical data. The strength of the work was that it showed that circuits could be assembled with the evolutionary approach, however full process optimisation of the assembled circuit remained elusive.

While *et al.* [2] used an evolution strategy to optimise the performance of a simple comminution circuit with a single cone crusher and a recirculating load. They allowed the algorithm to vary the shape of the crusher's liners and various operating parameters, trying to simultaneously maximise the quality of the product and the capacity of the circuit. Relative to an existing design, they reported an improvement of up to 12% in P80 (a product quality measure), or up to 224% in capacity, or simultaneous improvements of 7% in P80 and 192% in capacity. In [3], a multi-objective algorithm was used to optimise product quality and circuit cost, examining options for make, size, and number of processing units at various points in the circuit, as well as their operational settings.

Subsequently, other researchers have begun to report similar work in the minerals engineering literature: [4], [5], [6].

3   Case Study Problem

The processing circuit used in this study is a comminution circuit, the default version of which is shown in Fig. 1.

The term *comminution* is used to describe a collection of physical processes that can be applied to a stream of ore to reduce the sizes of the particles in the stream. The purpose of comminution is to transform raw ore into a more usable or more saleable product or to prepare it for further processing. A comminution circuit consists of a collection of processing units connected together (typically by conveyor belts), and may contain loops, typically re-cycling large particles through crushers until they reach the desired size. One or more streams of ore form the feed stream, entering the circuit typically from some preprocessing stage. One or more streams of transformed material exit the circuit as the product stream of the comminution process. More detailed information about comminution is available from [7].

In the circuit under study here, run of mine material is fed into a primary crusher, and the output is screened. Screen undersize goes directly into the fine ore bins, and the oversize is stacked in a coarse ore stockpile.

Reclaimed coarse ore is fed into a crushing plant consisting of an open circuit secondary crusher and a tertiary crusher in closed circuit with two single-deck screens. The screens are fed with the combined output of the secondary and

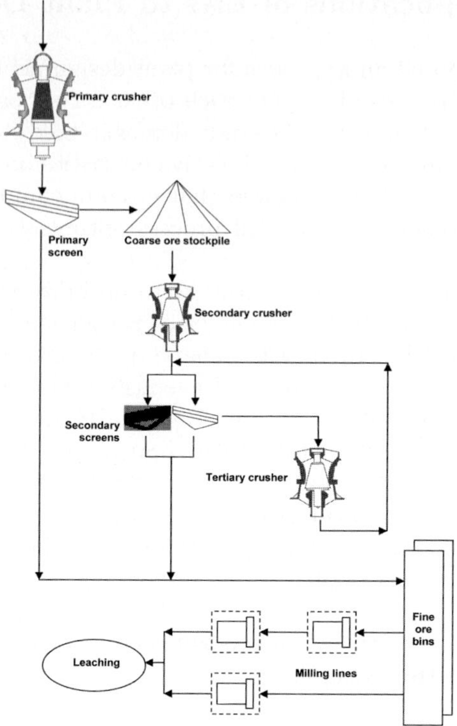

Fig. 1. The basic comminution circuit used in this study

tertiary crushers. Screen oversize is (re-)processed by the tertiary crusher and the undersize forms the crushing plant product and is taken to the fine ore bins.

The fine ore bins feed two parallel milling circuits and their combined product is treated in a leaching process. Each milling line has a selected target P80 value (the 80th percentile of particle size) which determines the capacity of that line. The milling lines have a pre-defined order in which they are utilised: material is fed into the preferred line until that line reaches capacity, then excess material is fed into the subsequent line(s).

The fundamental design goal of the circuit is to maximise the profit of the operating company. Circuit performance depends on the prevailing operating conditions, which are typically unknown or uncertain in advance. As such, a circuit that can perform well in response to changing circumstances is desirable; i.e., some risk management is required. This makes the problem well-suited to a multi-objective approach, which can return a population of designs offering a range of trade-offs between multiple objectives. In this study we use one objective that represents the profit of a circuit given the expected feed, and another objective that represents the profit given a harder feed that is more difficult to process. This enables us to evolve more robust circuits.

4   Design of the EA

In this study, we derived our initial population randomly and ran the algorithm for a fixed number of generations. This section examines the remaining algorithm design issues, with particular reference to the problem under discussion.

4.1   Representation of a Design

We vary the following design parameters:

- The percentage of product mineral in the feed.
- Independently, the closed-side settings of the secondary and tertiary crushers, i.e. the gap between the liners at their closest approach.
- Whether the design re-circulates the product of the tertiary crusher, or whether this crusher processes the material only once.
- The apertures of the decks in the primary screen, the type of this screen (either conventional inclined, or multi-slope), and its custom load factor.
- The aperture of each unit in the secondary screen-pair, and the type of these screens (either conventional inclined, or multi-slope).
- Independently, the charge level of each mill, i.e. the proportion of the interior volume that is occupied by balls or rods as appropriate.
- Independently, the target P80 of each milling line.
- The placement and types of the mills: either two milling lines with one ball mill each; or three lines with one ball mill each; or one line with one ball mill and another line with one rod mill followed by a ball mill (as in Figure 1).
- The feed order of the milling lines.

4.2   Evaluation of a Design

Designs are evaluated independently according to four objectives.

1. Maximise plant value when applied to "hard" material.
   This value takes into account set-up costs for the design, interest on the set-up costs, and summed daily net income over the design's lifetime.
2. Maximise plant value when applied to "soft" material.
   This objective considers the performance of the design when applied to soft material, characterised by both a lower P80 and lowered ore hardness values.
3. Minimise overflow in crusher feeds.
   A poor design might require a crushing machine to process material at a greater rate than its capacity. Excess material then overflows the mouth of the crusher and is lost. Obviously we want to minimise the amount of overflowing material: in fact we consider a design to be successful only if there is no overflowing material.
4. Minimise oversize material in crusher feeds.
   A poor design might present material to a crusher that is larger than it can accept. The amount of this material is to be minimised, and we consider a design to be successful only if there is no oversize material.

4.3   Treatment of Infeasible Designs

Some designs perform so poorly that they must be considered "infeasible". But such designs might have some useful features, so we do not discard them immediately. Instead, we quantify their "degree of infeasibility" and then select against them for future generations. For this we use the third and fourth objectives listed in Section 4.2, in a selection process (similar to the "constraint-domination" principle introduced in [8]) that is discussed in Section 4.4.

4.4   Selection of Designs

To select which designs become the parents of the next generation, we rank them according to their fitness values. We sort the population according to the fourth objective; then break ties according to the third objective; then break remaining ties by a Pareto ranking on the first and second objectives.

4.5   Mutation and Crossover

We used a uniform crossover variant with probability 0.8. Each design parameter was mutated with probability 0.5. Mutation varies depending on the nature of the parameter:

- Crusher identifiers: switch to one of the "neighbouring " crusher variants (a crusher variant which is different in only one attribute).
- Real-valued machine settings: NSGA-II's polynomial mutation variant [9] with distribution index 50, constrained to the machine's operating range.
- Integer unit counts: NSGA-II's polynomial mutation variant with distribution index 10, constrained to the range 1–20.

5   Experiments, Results, and Discussion

We ran the algorithm with a population of 100 for 500 generations. We performed five runs, each taking 12-18 hours on a modern PC.

Fig. 2 shows attainment surfaces that plot the increase through time of the performance of the population from each run. A 50% attainment surface shows the level of performance attained by half of the runs after a specified number of generations. The 0% attainment surface shows the level of performance attained by the best designs derived at the end of the runs. The plant value objectives are normalised relative to the existing design, therefore the performance of the existing design is (1, 1) by definition. We plot only feasible designs, so the two error objectives are 0 for all designs and need not be plotted.

Table 1 shows selected designs produced by the five runs, along with performance figures. The algorithm returns a *set* of designs, so we have chosen four interesting ones to compare with the existing design. We list the designs with

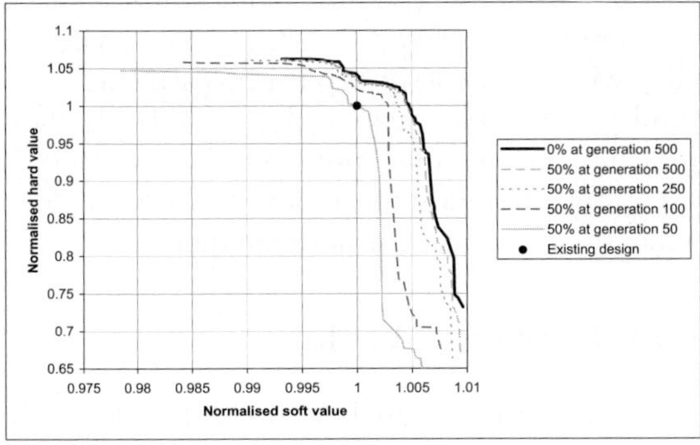

Fig. 2. Increase in normalised performance through time from five runs of the EA

Table 1. Details of the best designs from five runs of the evolutionary algorithm, and their normalised performance values

	Existing	Best soft value	Matched hard value	Matched soft value	Best hard value
Normalised soft value	1.0000	1.0096	1.0047	1.0002	0.9932
Normalised hard value	1.0000	0.7319	1.0021	1.0387	1.0624
Normalised % mineral in feed	100.00%	99.83%	99.91%	100.44%	101.01%
Screen-Prim-Config	CI	CI	CI	CI	CI
Screen-Prim-AppTop (mm)	65.0	50.7	66.0	67.7	47.7
Screen-Prim-AppBottom (mm)	17.0	28.5	26.8	29.0	20.7
Screen-Prim-CustomLoadFactor	160%	220%	201%	169%	164%
Crusher-Second-CSS (mm)	24.0	20.4	19.7	32.9	20.0
Screen-Second-App (mm)	19.0	14.7	18.1	18.4	22.0
Screen-Second-Config	CI	CI	CI	CI	CI
Closed Circuit	True	True	True	True	True
Crusher-Tert-CSS (mm)	12.0	12.0	12.0	12.0	12.0
RodMill	Retained	Retained	Retained	Discarded	Converted
FeedOrder	1,2	1,2	2,1	1,2	2,1,3
RodMill-Charge	40%	30%	32%		
BallMill-1-Charge	45%	27%	29%	38%	36%
BallMill-2-Charge	36%	30%	35%	38%	35%
BallMill-3-Charge					44%
Line-1-TargetP80 (μm)	298	351	315	370	324
Line-2-TargetP80 (μm)	306	267	284	327	332
Line-3-TargetP80 (μm)					480

the highest soft value, and the highest hard value. Both of these sacrifice one objective to maximise the other, so we also list two interesting designs that dominate the existing design: the design with the highest soft value that also beats the existing design in hard, and the design with the highest hard value that also beats the existing design in soft.

The algorithm has identified a number of realistic operational changes that could add significantly to the total earnings over the life of the project without any additional investment. Perhaps more importantly, examination of the optimised solutions shows great potential as a process diagnostic:

- All solutions fully utilise available crushing plant hours. This clearly identifies a bottle-neck that could be removed by simply improving the current low levels of availability without a costly capital investment;
- The gains in overall project value achieved are relatively modest. This suggests that the existing process may not contain any significant "hidden" values, and that capital upgrades must now be considered to provide the required improvements in operational performance.

6 Conclusions and Future Work

We have described the application of an EA to the problem of optimising the performance of a comminution plant, by fine-tuning the settings of individual machines, by tuning machines in the circuit to work well in tandem, and by making (minor) structural changes to the circuit. The performance of the plant is optimised independently for two different types of feed, and the algorithm returns a range of designs offering different trade-offs between them.

The results indicate that the algorithm is able to find designs that offer useful improvements in performance, particularly when applied to hard material, and that it is able to find designs that beat the existing design in both objectives. A full investigation promises significant financial benefits. We will allow the algorithm more scope to vary aspects of the design in the search for improvements, in particular we will allow a wider range of structural changes. We also plan to modify the definitions of our value objectives to more-closely mirror the usual metric of NPV (net present value).

Acknowledgments. This work was supported by Rio Tinto OTX and Australian Research Council Linkage Grant LP0347915.

References

1. Venter, J.J., Bearman, R.A., Everson, R.C.: A novel approach to circuit synthesis in mineral processing. Minerals Engineering 10 (3), 1997, pp. 287–299.
2. While, L., Barone, L., Hingston, P., Huband, S., Tuppurainen, D., Bearman, R.: A multi-objective evolutionary algorithm approach for crusher optimisation and flowsheet design. Minerals Engineering 17 (11/12), 2004, pp. 1063–1074.
3. Huband, S., Barone, L., Hingston, P., While, L., Tuppurainen, D. and Bearman, T.: Designing Comminution Circuits with a Multi-Objective EA, CEC 2005.
4. Guria, C., Verma, M., Mehrotra S.P., Gupta, S.K.: Multi-objective optimal synthesis and design of froth flotation circuits for mineral processing, using the jumping gene adaptation of genetic algorithm, Industrial & Engineering Chemistry Research 44 (8), 2005, pp. 2621–2633.
5. Guria, C., Verma, M., Mehrotra S.P., Gupta, S.K.: Simultaneous optimization of the performance of flotation circuits and their simplification using the jumping gene adaptations of genetic algorithm, International Journal of Mineral Processing 77 (3), pp. 165–185, 2005.
6. Svedensten, P., Evertsson, C., M.: Crushing plant optimisation by means of genetic evolutionary algorithm, Minerals Engineering 18 (5), 2005, pp. 473–479.

7. Napier-Munn, T., Morrell, S., Morrison, R.D., Kojovic, T.: Mineral Comminution Circuits. Julius Kruttschnitt Mineral Research Centre, 1996.
8. Deb, K., Pratap, A., Meyarivan, T.: Constrained test problems for multi-objective evolutionary optimization, EMO 2001.
9. Deb, K, Pratap, A., Agarwal, S., Meyarivan, T.: A fast and elitist multiobjective genetic algorithm: NSGA-II. IEEE Trans. on EC 6, 2002, pp. 182–197.

White and Color Noise Cancellation of Speech Signal by Adaptive Filtering and Soft Computing Algorithms

Ersoy Kelebekler and Melih İnal

Kocaeli University, Technical Education Faculty, Electronics and Computer Department,
Umuttepe Campus,
41380 İzmit, Kocaeli, Turkey
{ersoy, minal}@kou.edu.tr

Abstract. In this study, Gaussian white noise and color noise of speech signal are reduced by using adaptive filter and soft computing algorithms. Since the main target is noise reduction of speech signal in a car, ambient noise recorded in a BMW750i is used as color noise in the applications. Signal Noise Ratios (SNR) are selected as +5, 0 and -5 dB for white and color noise. Normalized Least Mean Square (NLMS), Recursive Least Square (RLS) and Genetic Algorithms (GA), Multilayer Perceptron Artificial Neural Network (MLP ANN), Adaptive Neuro-Fuzzy Inference System (ANFIS) are used as adaptive filter and soft computing algorithms, respectively. 5 female and 5 male speakers have been chosen as Speech data from database of Center for Spoken Language Understanding (CSLU) Speaker Verification version 1.1. Noise cancellation performances of the algorithms have been compared by means of Mean Squared Error (MSE). Also processing durations (second) of the algorithms are determined for evaluating possibility of real time implementation. While, the best result is obtained by GA for noise cancellation performance, RLS is the fastest algorithm for real time implementation.

1 Introduction

The most important effect which has influence on speech and speaker recognition systems is noise. The more noise causes the less intelligibility and recognition performance. When comparing with human ear, the modern speech recognition devices performance is comparatively low in noisy ambient [1].

Many methods have been improved to cancel ambient noise from speech signal. One of the best fundamental methods is adaptive filtering. The system has two inputs; one of them is speech signal and the other is noise signal used as reference signal. In this study, two kinds: white or color noise has been applied to the system as reference noise. White noise is Gaussian noise and color noise is a car ambient noise recorded in a BMW750i car. +5, 0 and -5 dB signal noise ratios (SNRs) have been chosen for both white and color noise.

5 female and 5 male speakers have been chosen as Speech data from database of Center for Spoken Language Understanding (CSLU) Speaker Verification version

A. Sattar and B.H. Kang (Eds.): AI 2006, LNAI 4304, pp. 970–975, 2006.

© Springer-Verlag Berlin Heidelberg 2006

Table 2. Features of ANFIS

Input number	1
Output number	1
Rule number	10
Type	Sugeno
andMethod	Prod
orMethod	Max
defuzzMethod	Wtaver
impMethod	Prod
aggMethod	Max

MSE values of ten speakers to compare the algorithm's performance of noise cancellation have been computed for each SNR value (+5, 0 and -5 dB) in each algorithm. Then ten speaker's MSE values have been averaged for each algorithm. Averaged MSE values of NLMS, RLS, GA, MLP ANN and ANFIS algorithms to compare the algorithm's performance of white and color noise cancellation are given in Fig. 1 and in Fig. 2, respectively.

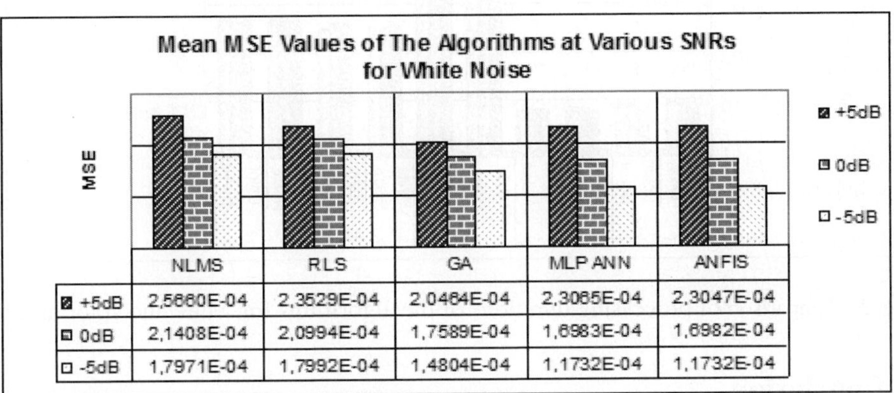

	NLMS	RLS	GA	MLP ANN	ANFIS
+5dB	2,5660E-04	2,3529E-04	2,0464E-04	2,3065E-04	2,3047E-04
0dB	2,1408E-04	2,0994E-04	1,7589E-04	1,6983E-04	1,6982E-04
-5dB	1,7971E-04	1,7992E-04	1,4804E-04	1,1732E-04	1,1732E-04

Fig. 1. Comparison of the algorithms for white noise

As seen in Fig. 1, GA gives the least mean square error value at +5 dB SNR. At 0 dB and -5 dB SNR values, the least mean square error values are taken by ANFIS and MLP ANN for white noise.

As seen in Fig. 2, GA is the algorithm which gives the least mean square error values at all SNRs for color noise.

Averaged processing durations of NLMS, RLS, GA, MLP ANN and ANFIS algorithms are given in Fig. 3 to compare processing time of the algorithms for white and color noise cancellation at various SNRs. The time axis of Fig. 3 is given logarithmic due to big differences between processing durations. RLS algorithm is the fastest algorithm for both two kinds of noise and all level SNRs as shown in Fig. 3.

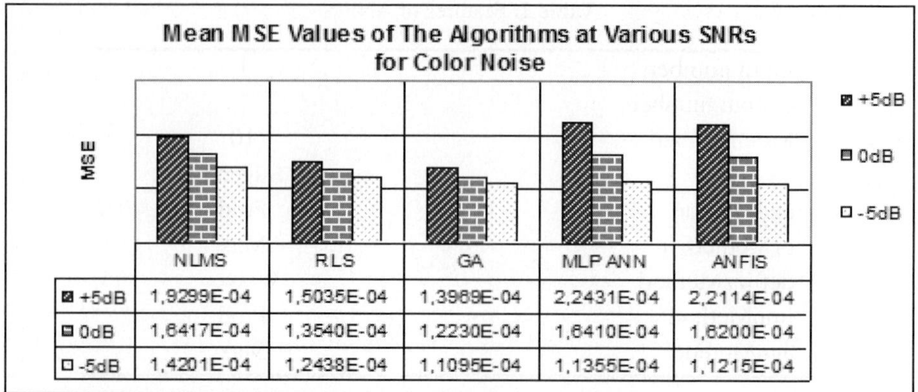

Fig. 2. Comparison of the algorithms for color noise

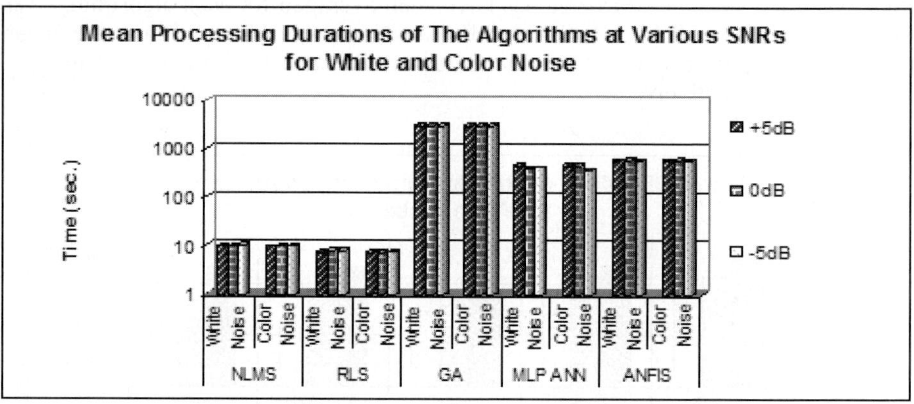

Fig. 3. Comparison of processing durations of the algorithms for white and color noise

3   Conclusion

Gaussian white noise and color noise which is recorded as ambient noise in BMW750i of speech signal are filtered by using adaptive filters and soft computing algorithms. Signal Noise Ratios (SNR) are selected as +5, 0 and -5 dB for both white and color noise. Clean speech data are constituted by 5 female and 5 male speakers from CSLU. Various experiments are performed for investigating the optimum result of the best algorithm by means of MSEs and processing durations. Although GA gives the best performance to filter color noise with the best minimum error values, it is the slowest algorithm by considering processing durations for our main target which is noise reduction of speech signal in car. So it does not seem possible to use GA in real time implementations. The best optimum noise reduction algorithm which can be used in real time implementations is determined by RLS algorithm considering its processing durations and performance of noise reduction.

Acknowledgment

Authors would like to thank the Center for Spoken Language Understanding, Oregon Graduate Institute, for supplying the Speaker Recognition V1.1 speech database.

References

1. Hussain, A., Campbell, D.J.: Binaural Sub-Band Adaptive Speech Enhancement Using Artificial Neural Networks. Speech Communication, Vol.25. Elsevier Science B.V., Amsterdam, The Netherlands (1998) 177-186
2. Liberti, J.C., Rappaport, T.S., Proakis, J.: Evaluation of Several Adaptive Algorithms for Canceller Acoustic Noise in Mobile Radio Environments. Vehicular Technology Conference (1991) 126-132
3. Shozakai, M., Nakamura, S., Shikano, K.: Robust Speech Recognition in Car Environments. Acoustic, Speech, and Signal Processing, Vol. 1. (1998) 269-272
4. Ortega, A., Lleida, E., Masgrau, E.: Speech Reinforcement System for Car Cabin Communications. IEEE Transaction on Speech and Audio Processing, Vol. 13. No. 5 (2005) 917-929
5. Goubran, R.A., Hebert, R., Hafez, H.M.: Acoustic Noise Suppression Using Regressive Adaptive Filtering. Vehicular Technology Conference, (1990) 48-53
6. Dhanjal, S.: Artificial Neural Networks in Speech Processing: Problems and Challenges. Communications, Computers and Signal Processing, Vol. 2. (2001) 510-513
7. Juang, C.-F., Lin, C.-T.: Noisy Speech Processing by Recurrently Adaptive Fuzzy Filters. IEEE Transaction on Fuzzy Systems, Vol. 9. No. 1 (2001) 139-152
8. Thevaril, J., and Kwan H.K.: Speech Enhancement using Adaptive Neuro-Fuzzy Filtering. Proceedings of 2005 International Symposium on Intelligent Signal Processing and Communication Systems, Hong Kong (2005) 753-756
9. Haykin, S.: Adaptive Filter Theory. 4th Edition, Prentice Hall Inc., New Jersey (2002)
10. Man, K.F., Tang, K.S., and Kwong, S.: Genetic Algorithms: Concepts and Applications. IEEE Trans. Industrial Electronics, Vol. 43. (1996) 519-533
11. Vasconcelos, J.A., Ramirez, J.A., Takahashi, R.H.C., and Saldanha, R.R.: Improvements in Genetic Algorithms. IEEE Trans. Magnetics, Vol. 37. No. 5 (2001) 3414-3417
12. Haykin, S.: Neural Networks a Comprehensive Foundation. Second Edition, Prentice-Hall, Inc. New Jersey (1999)
13. Jang, J.R.: ANFIS: Adaptive-Network-Based Fuzzy Inference System. IEEE Transaction on System, Man, and Cybernetics, Vol. 23. No. 3 (1993)
14. Jang, R., Gulley, N.: Fuzzy Logic Toolbox User's Guide. The Mathworks Inc. (1995)
15. Hagan, M.T, and Menhaj M.B.: Training Feedforward Networks with The Marquardt Algorithm. IEEE Trans. Neural Networks, Vol. 5. No. 6, (1994) 989-993
16. Lagaros, N.D., and Papadrakakis, M.: Learning Improvement of Neural Networks Used in Structural Optimization. Advances in Engineering Software, Vol. 35. (2004) 9-25
17. Nguyen, D. and Widrow, B.: Improving The Learning Speed of 2-Layer Neural Networks by Choosing Initial Values of The Adaptive Weights. International Joint Conference of Neural Networks, Vol. 3. (1990) 21-26

A Study on Time-of-Day Patterns for Internet User Using Recursive Partitioning Methods

Seong-Keon Lee[1], Seohoon Jin[2],
Hyun-Cheol Kang[3], and Sang-Tae Han[3]

[1] Department of Statistics, Sungshin Women's University,
249-1, Dongseon-Dong, Seongbuk-Ku, Seoul, 136-742, Korea
[2] Hyundai Capital,15-21 Youido-Dong, Youngdungpo-Gu,
Seoul, 150-706, Korea
[3] Department of Informational Statistics, Hoseo University,
29-1, Asan, 336-795, Korea

Abstract. As of the remarkable increasing of internet users, there have been some demands of analyzing the users web accessing patterns. Internet related companies want to know the internet users web accessing patterns to promote their own products to the users. For analyzing customer's time-of-day pattern for using internet as response vector that can be thought of as a discretized function, fitting ordinary decision trees may be unsuccessful because of their dimensionality. In this paper, we shall propose factor tree which would be more interpretable and competitive. Furthermore, using Korean internet company data, we will analyze time-of-day patterns for internet user.

1 Introduction

The landmark work of decision tree is the methodology of Breiman, Friedman, Olshen, and Stone (1984), who introduced classification tree for a univariate discrete/continuous response. There are various competing approaches to the work of Breiman et al. (1984), such as that of Hawkins and Kass (1982) and Quinlan (1992). These approaches are focused on the single response.

Recently, some decision trees for multiple responses have been constructed by Segal (1992) and Zhang (1998). Segal (1992) suggested a tree that can analyze continuous longitudinal response using Mahalanobis distance for within node homogeneity measures. Zhang (1998) suggested a tree that can analyze multiple binary response using generalized entropy criterion which is proportional to maximum likelihood of joint distribution of multiple binary responses. Furthermore, in real world application, responses which have many variables can be often observed such as functional data. However, naively applying multivariate decision trees to "long vector responses" is not successful.

The examples of multivariate decision trees that give unreasonable results to analyst are shown in the research of Yu and Lambert (1999). Yu and Lambert proposed new tree methodologies, spline tree and principal component tree for analyzing high dimensional response, applying two step procedures that reduce

A. Sattar and B.H. Kang (Eds.): AI 2006, LNAI 4304, pp. 976–980, 2006.

© Springer-Verlag Berlin Heidelberg 2006

the dimension of the responses and then constructing a tree to lower dimensional responses. Spline tree represents each response vector as a linear combination of spline basis functions and then fits a multivariate tree to the estimated coefficient vectors. Principal component tree uses the first several principal component scores as the response vector.

In this paper, factor analysis will be used to reduce the high dimension responses to low dimensions that have several independent explainable factors by using factor rotation. So, we shall propose a factor tree that has advantages in the view of interpretation. Finally, using a Korean internet company data set which consists of internet site member's demographic profiles and internet using pattern, we will investigate and compare the performance and results of the two step tree procedures, i.e., spline tree, principal component tree, factor tree.

2   Time-of-Day Patterns for Internet User

2.1   Data

The problem we face is to predict a customer's time-of-day internet usage pattern. For example, business customers who use internet in daytime are more frequent, while students use internet more frequently at night time. Therefore classification of the time of day patterns would give useful information, such as the optimal time bands of shaver/cosmetic banner advertisements in a web page. So, it is a good strategy to advertise goods in the internet by considering customer's profiles and their surfing time patterns.

The data which is used in the application consist of internet usage records of some internet site composed of 771 members. Table 1 and Table 2 show the data profile that have high dimension response vectors. Response variables are partitioned by 30 minutes, so they consist of 48 time intervals as responses. Explanatory variables consist of 5 categorical and 6 continuous variables, such as gender, age, job, etc. As we can see, naively applying multivariate decision trees to "long vector responses" like Table 1 may not be successful. So, dimension reduction techniques mentioned above should be used at first and then, we will construct and compare the tree results.

Table 1. Response variables Variable

Variable	Descriptions
y_1	monthly ♯ of visit to the Internet site between 0:00 a.m $\sim$ 0:30 a.m.
y_2	monthly ♯ of visit to the Internet site between 0:30 a.m $\sim$ 1:00 a.m.
$\vdots$	$\vdots$
y_{48}	monthly ♯ of visit to the Internet site between 11:30 a.m $\sim$ 0:00 a.m.

2.2   Factor Tree

We use iterative principal factor analysis to reduce the dimension of responses. As a result, we get eight factors that easily interpret the meaning by using

Table 2. Explanatory variables

Variable	Descriptions
♯ of connecting internet/month	♯ of count of using internet per month
Gender	Male, Female
Birth Year	Year of birth
Job	Student, General office worker, Financial worker, government officer,..., etc
Marital status	Married, Unmarried
Salary/month	0 Won ~ 10 million Won
Education	Under middle school, Under high school, Under college, Graduate school
Connecting Place	Usual place of using internet : Home, Company, Internet cafe, School, etc
Period of using internet	under 1 year, 1~2 years, 2~3years, 3~4 years, 4~5 years, over 5 years
Surfing time of internet/week	Under 0.5 hour, 0.5~1 hour, 1~2 hours, 2~3 hours, ..., over 50 hours
City	Living province : Seoul, Incheon, Busan, Daegu, ..., Jeju

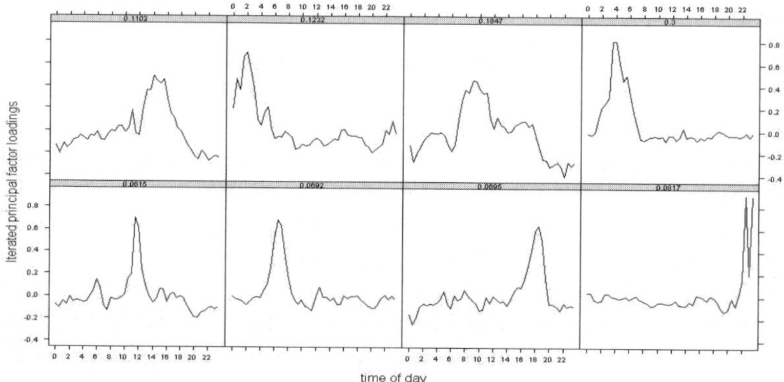

Fig. 1. Iterative principal factor loading (first eight factors)

varimax rotation method. Full tree has 45 terminal nodes and after the pruning procedure, the final tree has 9 terminal nodes, like Fig.2.

Seeing the factor loading, in Fig.1, factor profiles are more easily interpreted than principal component. Fig.1 represents that each factor has from 4 hour to 6 hour intervals which have higher factor loading. For example, the first factor, right upper side in Fig.1, has an interval which has higher loading between 2 a.m. and 7 a.m. So, the first factor could distinguish the customers who usually use internet during midnight.

Factor tree has more similar tree results to that of principal component tree than that of spline tree. The first split variable at the root node is also "connecting place", i.e., same as that of principal component tree. Many split variables which are chosen in principal tree are also chosen in factor tree, but they appear in different level of tree depth. Fig.3 represents the factor mean score at each terminal node of the factor tree. It is difficult to explain the meaning of principal component and spline coefficients, but factor can be easily explained by factor

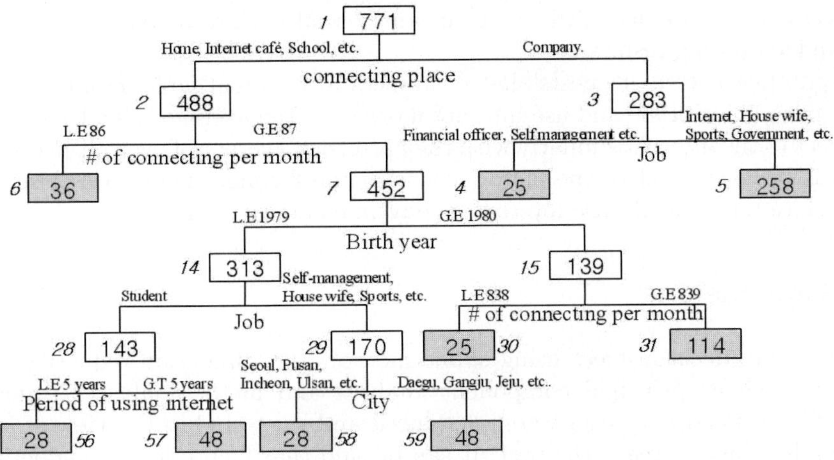

Fig. 2. Factor Tree

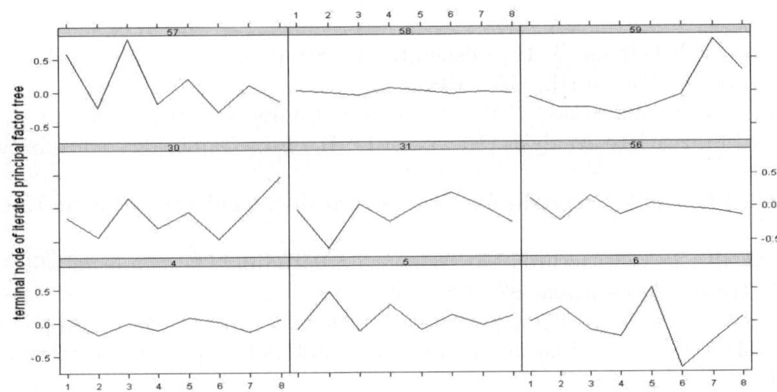

Fig. 3. mean factor score at terminal node

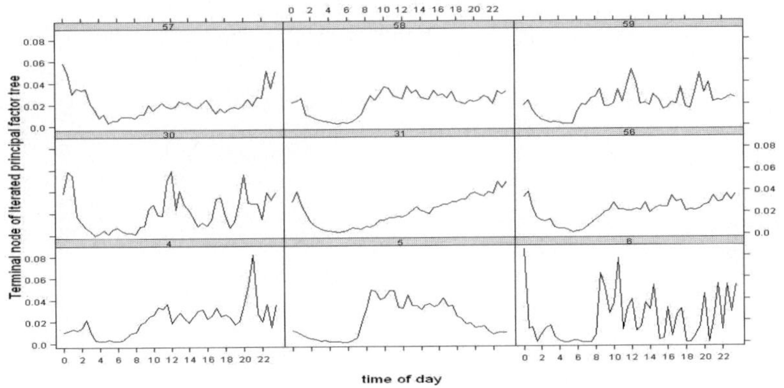

Fig. 4. Average internet using-time profiles at terminal node

rotation. Therefore, using the factor mean score at each terminal node is helpful for understanding results.

Fig.2, factor tree, suggests that customers in terminal node 56 and 58, who were born before 1980 and use internet more than the average using time, would connect to the internet similarly with the pattern of average of over-all customers. This is because most of the factor means in the terminal nodes are close to 0. Therefore they usually use internet at daytime and evening.

3   Conclusions

Multivariate decision trees using spline method and dimension reduction techniques, such as principal component analysis and factor analysis to analyze high dimensional responses were introduced and suggested.In the view of interpretability, factor tree is the best. It can be said that factor tree is competitive.

References

1. Breiman, L., Friedman, J. H., Olshen,R. A., Stone, C. J. : Classification and Regression Trees. Wadsworth, CA. (1984)
2. Hawkins, D.M. and Kass, G.V.: Automatic Interaction Detection. In Topics in Applied Multivariate Analysis. Hawkins, D. H., Ed.; Cambridge University Press. (1982)
3. Quinlan, J.R.: C4.5: Programs for machine learning. California: Morgan Kaufmann. (1992)
4. Segal, M. R. : Tree-Structured Methods for Longitudinal Data. Journal of the American Statistical Association. **87** (1992) 407–418.
5. Yu Y., Lambert, D. : Fitting Trees to Functional Data, With an Application to Time-of-Day Patterns. Journal of Computational and graphical Statistics. **8** (1999) 749–762.
6. Zhang, H. : Classification Trees for Multiple Binary Responses. Journal of the American Statistical Association. **93** (1998) 180–193.

Development of Data Miner for the Ship Design Based on Polynomial Genetic Programming

Kyung Ho Lee[1], June Oh[2], and Jong Hoon Park[2]

[1] Inha University, Department of Naval Architecture & Ocean Engineering, 253 Yonghyun-dong Nam-gu Incheon 402-751 South Korea
kyungho@inha.ac.kr
[2] Inha University, Graduate School, Department of Naval Architecture & Ocean Engineering, 253 Yonghyun-dong Nam-gu Incheon 402-751 South Korea

Abstract. Engineering data contains the experiences and know-how of experts. Data mining technique is useful to extract knowledge or information from the accumulated existing data. This paper deals with generating optimal polynomials using genetic programming (GP) as the module of Data Miner. The Data Miner for the ship design based on polynomial genetic programming is presented.

1 Introduction

Although Korean shipyards have accumulated a great amount of data, they do not have appropriate tools to utilize the data in practical works. Engineering data contains the experiences and know-how of experts. Data mining technique is useful to extract knowledge or information from the accumulated existing data. This paper presents a machine learning method based on genetic programming (GP), which can be one of the components for the realization of data mining [7]. The paper deals with polynomial GP for regression or approximation problems when the given learning samples are not sufficient. Polynomials are widely used in many engineering applications such as response surface modeling [1-4,6]. The purpose of this paper is to develop the data utilization tool to generate an empirical formula from the accumulated existing data for ship design with GP.

2 Polynomial Genetic Programming for Data Mining

Function Set for PGP
Our idea for easily generating a high order polynomial is to use Taylor series of mathematic functions in the function set. If high order series are taken, the GP tree produces a very complex polynomial. So, we determine to take only two or third order polynomials from whole Taylor series. We use the following function and terminal set.

$$F = \{+,-,*, g_i (i = 1,...,18)\} \quad T = \{one, rand, x_i (i = 1,...,n)\}$$

g_i is a low order Taylor series. Its description will not be given in this paper due to the space limitation. "one" return 1, and "rand" a random number, whose size is less than 1. x_i represents a variable. All functions and terminals have their weights.

A. Sattar and B.H. Kang (Eds.): AI 2006, LNAI 4304, pp. 981–985, 2006.
© Springer-Verlag Berlin Heidelberg 2006

Overfitting Avoidance
Without considering overfitting, the fitness function can be defined by (1).

$$\vartheta = \vartheta_{MSE} \tag{1}$$

where $\vartheta_{MSE} = 1/m \sum_{i=1}^{m} [f_{GP}(\overline{X}_i) - y_i]^2$.

The learning set takes the form of $L\{(\overline{X}_i, y_i), 0 \le y_i \le 1\}_{i=1,\ldots,n}$.

Here, $\overline{X}_i(x_{i,1},\ldots, x_{i,n}, 0 \le x_{i,j} \le 1)$ is an n-dimensional vector, and y_i is a desired output of the GP tree(f_{GP}) at $\overline{X}_i$. Since m is very small, the GP tree is overfitted if only $\vartheta = \vartheta_{MSE}$ is used. The EDS method[8] can be included in the fitness function for smooth fitting.

$$\vartheta_E = \vartheta_{MSE} + \lambda \hat{\vartheta}_{MSE} \tag{2}$$

where $\hat{\vartheta}_{MSE} = 1/p \sum_{i=1}^{p} [f_{GP}(\overline{X}_i^E) - y_i^E]^2$, and λ is a constant that determines the contribution of $\hat{\vartheta}_{MSE}$.

The extended data set $L^E\{(\overline{X}_i^E, y_i^E)\}_{i=1,\ldots,p}$ can be constructed by simple linear interpolations of closest learning samples. For the detailed description, see the literature[8]. To alleviate overfitting, we introduced the FNS method. If the GP tree contains several function nodes $(g_i T)$, where T is the subtree, and if the value of $(g_i T)$ is very large compared with others, then a slight change of T's value might cause a very large change of the GP tree's output. As shown in (3), the FNS method penalizes the GP tree if the tree contains such nodes.

$$\vartheta_{EF} = \vartheta_{MSE} + 0.1\hat{\vartheta}_{MSE} + \vartheta_{FNS} \tag{3}$$

where $\vartheta_{FNS} = 0$ if for all g_i in the GP tree are

$|(g_i T) - f_i(T)| \le \delta$, otherwise $\vartheta_{FNS} = \alpha$.

f_i is a mathematic function corresponding to g_i , α is very large positive number such as 1.0E10, and δ (0.1) is a tolerance. If $(g_i T)$ is large, then $(g_i T)$ is very different from f_i , and ϑ_{FNS} becomes α .

3 Validation for PGP with Small Learning Data

The function below shows the Goldstein-price function typically used as a benchmark problem for testing the performance of the optimization algorithms.

$$f(x_1, x_2) = (1 + (x_1 + x_2 + 1)^2 \cdot (19 - 14x_1 + 3x_1^{\,2} - 14x_2 + 6x_1 x_2 + 3x_2^{\,2}))$$
$$\cdot (30 + (2x_1 - 3x_2)^2 \cdot (18 - 32x_1 + 12x_1^{\,2} + 48x_2 - 36x_1 x_2 + 27x_2^{\,2}))$$
$$-2 \leq x_i \leq 2 \qquad i = 1,2$$

The learning set is prepared as only 5x5 grid type, and in the same manner the test set is made by 200x200 grid type. Fig.1.(b) shows the results of GP when previous (1) is used for the fitness function. The GP tree is severely overfitted. On the other hand, Fig.1.(c), when (3) used as the fitness function, the results give the smooth surface. Fig.2 shows the transformed polynomial form.

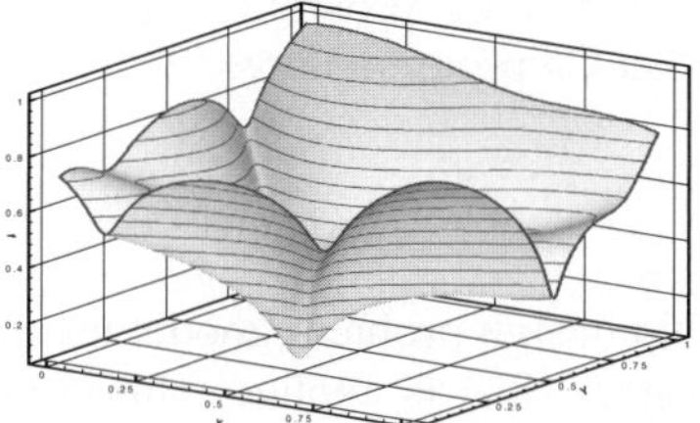

(a) Original function.

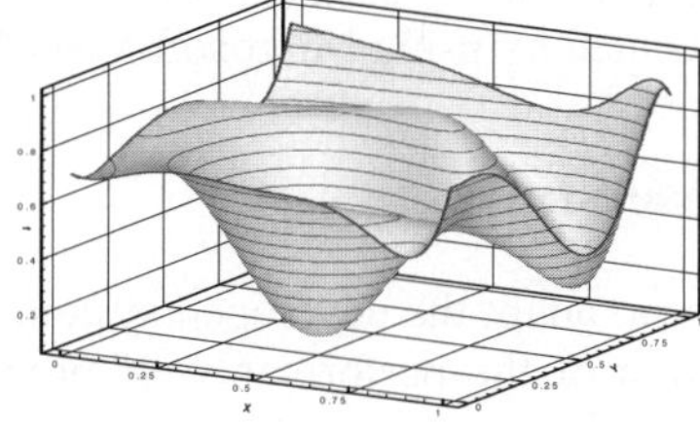

(b) Results of GP using (1) as a fitness function. The MSE of learning and test set are 3.386E-6 and 0.0712, respectively.

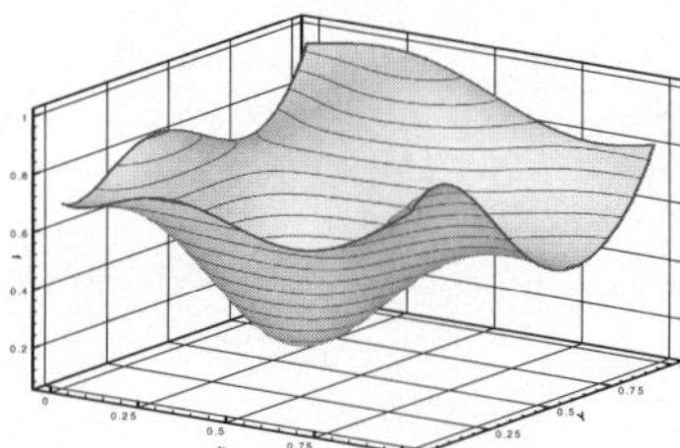

(c) Results of GP using (3) as a fitness function. The MSE of learning and test set are 3.645E-3 and 4.365E-3, respectively.

Fig. 1. The logarithm-scaled Goldstein-price function

```
0.693707 (-1.94697 (-2.56609 + 4.154 x2)
      (1. - 0.253758 (1.76982 - 3.07988 x1 - 1.45075 x2)²
          (1. - 0.520818 x2²)²)
      (0.598926 (1. + 0.556292 (1.66548 x1 + 0.751999
                  (-1.0079 x1 + 0.892572 x2)³ - 1.47491 x2) ^2) -
          0.97981 (1. - 0.439762 (-1.20691 x1³ + x1² (3.11693 -
                  2.61313 x2) - 1.88593 x1 (-2.20564 + x2)
                  (-0.179953 + x2) - 0.4537 (-2.94709 + x2)
                  (-1.1928 + x2) (0.561499 + x2)) ^2)) +
      1.542 (1.14388 (-1.08873 + x2) (0.0949399 + x2) +
          1.28987 (-0.327209
                  (1. - 0.0880898 (1.66792 - 4.38709 x1 - 0.462422 x2)²
                      (1. - 0.0417391 (1.10438 x1 - 0.286908 x2)²
                          (-2.54243 + x2)² (1.16337 + x2)²) ^2) +
              0.846979 (1. + 0.341689 (1. - 0.292192
                      (1.64987 - 1.24615 x1 - 3.537 x2)²
                      (1. - 0.473873 x2²)²) ^2)))))
```

Fig. 2. The polynomial obtained by GP

4 Development of Data Miner for Ship Design

In this section, Data Miner for data utilization by using polynomial genetic programming (PGP) and other modules is presented [9]. That is, the tool is contrived to apply to ship design under the case that the accumulated data is not enough to make learning process. The system is applied to real ship design problem in a Korean shipyard. Fig.3 shows the developed Data Miner by using GP. The data miner can make fitting functions with 3 types of GP such as GP with high order polynomial, linear model GP with polynomial (PLM-GP), and linear model GP with math functions (LM-GP). Users can make the process of function approximation by selecting arbitrary functions that they want to use. And the generated function tree can be converted to C code in order to integrate with other program. The system is implemented by using Microsoft Visual Studio .Net C# programming.

5 Case Study

In order to adapt the developed system to real ship design problem, model test for the estimation of the performance of propulsion system (K_T) is applied. Unfortunately,

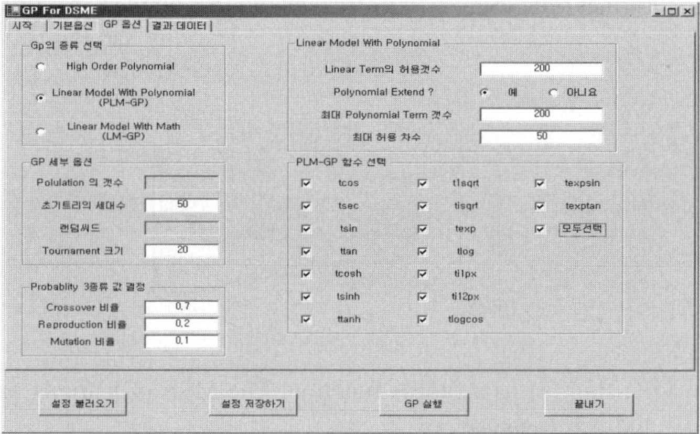

Fig. 3. Data Miner for Ship Design by using GP

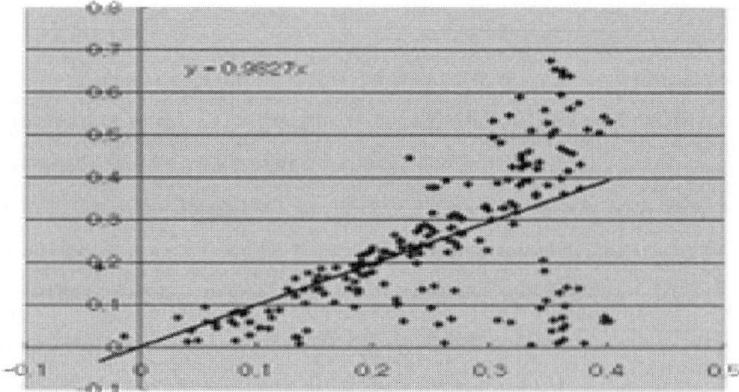

Fig. 4. Estimated value for K_T

the experiment data for the K_T is secret. So we generate 1000 data by using empirical formula. It means that the data contains some noise. The system automatically uses 800 data for training, and 200 data for test among 1000 learning data. Fig.4 shows the test result for 200 test data. The solid line means the expected location of target values. The dotted points are estimated results. The reasonable range of K_T is between 0.15 and 0.35, in which this system can estimate the K_T value very well.

6   Conclusions

In this paper, Data Miner as a data approximation/prediction tool to assist the ship designing process with insufficient learning samples is developed. First of all, Polynomial genetic programming technique is presented to make approximated function from the accumulated existing data. And also the Data Miner as a data miming tool based on the PGP is presented. The system can give consistent results with limited amount of learning samples, regardless of whether or not samples contain noise. The Data Miner is used in real preliminary ship design process to generate new empirical formula for new concept ships, such as LNG carrier, FPSO, and so on.

Acknowledgments. This work is supported by Advanced Ship Engineering Research Center (R11-2002-104-08002-0).

References

1. Simpson, T.W., Allen, J.K., and Mistree, F. : Spatial Correlation and Metamodels for Global Approximation in Structural Design Optimization, Proc. of DETC98, ASME (1998).
2. Malik, Z., Su, H., and Nelder, J. : Informative Experimental Design for Electronic Circuits, Quality and Reliability Engineering, vol.14 (1986) 177-188
3. Alotto, P., Gaggero, M., Molinari, G., and Nervi, M. : A Design of Experiment and Statistical Approach to Enhance the Generalized Response Surface Method in the Optimization of Multi-Minimas, IEEE Transactions on Magnetics, Vol.33, No.2 (1997) 1896-1899
4. Ishikawa, T. and Matsunami, M. : An Optimization Method Based on Radial Basis Function, IEEE Transactions on Magnetics, Vol.33, No. 2/II (1997) 1868-1871
5. Ott, R.L. : An Introduction to Statistical Methods and Data Analysis, Wadsworth Inc. (1993)
6. Myers, R.H. and Montgomery, D.C. : Response Surface Methodology: Process and Product Optimization Using Designed Experiments, John Wiley & Sons, Inc. (1995)
7. Koza, J.R. : Genetic Programming: On the Programming of Computers by Means of Natural Selection, The MIT Press (1992)
8. Yeun, Y.S., Lee, K.H., and Yang, Y.S. : Function Approximations by Coupling Neural Networks and Genetic Programming Trees with Oblique Design Trees, AI in Engineering, Vol.13, No3 (1999)
9. Lee, K.H., et al.: Data Analysis and Utilization Method Based on Genetic Programming in Ship Design, Lecture Notes in Computer Science, Vol.3981. Springer-Verlag (2006)

Detecting Giant Solar Flares Based on Sunspot Parameters Using Bayesian Networks

Tatiana Raffaelli, Adriana V.R. Silva, and Maurício Marengoni

Universidade Presbiteriana Mackenzie – São Paulo – Brazil
tati.raffaelli@gmail.com, asilva@craam.mackenzie.br,
mmarengoni@mackenzie.br

Abstract. This paper presents the use of Bayesian Networks (BN) in a new area, the detection of solar flares. The paper describes how to learn a Bayesian Network (BN) using a set of variables representing sunspots parameters such that the BN can detect and classify solar flares. Giant solar flares happen in the Sun's atmosphere quite frequently and as a consequence they can affect Earth. The work described here shows the relationship between the learned networks and the causality expected by solar physicists. The data used for learning and cross validation experiments show that the network substructures are easy to learn and robust enough to predict solar flares. The systems presented here are capable of detecting the flares within 72 hours, while the current method used today does the same work within 24 hours in advance only. It is also shown that sunspot parameters change over time, so different networks can be learned and perhaps combined in order to build a robust forecast system.

Keywords: Bayesian networks, fusion of information, forecast systems, sunspot, solar flares.

1 Introduction

Sunspots are the basic manifestation of solar activity since the Sun is not a static star [1]. The total number of sunspots and the related activities vary periodically [2] from a maximum to a minimum and so forth in a cycle of 11 years.

Sunspots are observed as dark regions on the Sun's surface and each sunspot group evolves in time. After a certain number of days, which is not fixed, the sunspot can either disappear or host an explosion. This explosion, or flare, is classified according to its X-ray flux as C, M or X, according to the magnitude of the peak burst intensity measured close to Earth in the 0,1 to 0,8 nm wavelength band [3]. A solar flare is the result of a sudden and intense release of the magnetic energy concentrated on the magnetic field located above sunspots [1]. During a solar flare radiation is emitted throughout all the electromagnetic spectrum (from gamma to radio wavelengths).

The material ejected from an X-class flare can reach the Earth, where it can affect its magnetic field. When this happens it can cause many problems such as damages to satellites, electrical networks, errors in navigation systems such as GPS systems, among others. Thus, the prediction of when a large flare will occur in the Sun with a few days notice is desired in order to take some precautions to avoid or minimize possible harmful consequences here on Earth.

A. Sattar and B.H. Kang (Eds.): AI 2006, LNAI 4304, pp. 986–991, 2006.

© Springer-Verlag Berlin Heidelberg 2006

Fusion of information is an important reasoning process. The Bayesian Network approach was chosen because it has been proved to be a very successful technique for fusion of information [4, 5, 6]. However, the design of the structure and probability tables of BNs is a time consuming process [4]. This task can be reduced using existing tools to learn the network automatically.

The work presented here is a first step to build a robust forecast system based on a set of Bayesian Network. The data used in this work was obtained from the Solar Monitor [7] from 1996 to 2006 including all the X-class flares that have occurred during this period. This work shows that Bayesian Network (BN) can be used to detect when X-class explosions will occur a few days in advance, based on the sunspot configuration and will answer the questions listed below:

1) Is the set of variables normally used by solar physicists good enough to model electromagnetic explosions in the Sun's atmosphere?

2) Is it possible to use BN as a model for the phenomenon described in question 1?

3) Can the BNs be learned from this set of variables?

2 Modeling

Bayesian Network is a technique used to search for a solution in situations where the information related to the problem is uncertain. A BN is a directed acyclic graph (DAG) [8] representing causality, where each variable in the problem is represented by a node and the arcs represent causality. Each node has an associated probability table. For more details in Bayesian Networks see Jensen [9].

One of the main criticisms to the Bayesian Network approach is the definition of the Bayesian Network structure and its probability tables. Fortunately it is possible to use training data and learn both the network structure and the probability tables. There are several tools (BN PowerConstructor, Hugin, BayesiaLab [10], MSBNx [11] among others) that can be used to learn the networks, in this work, because of previous experience, it was used the Bayesian Network PowerConstructor [12] to learn both the structure and the probability tables. The Hugin system [13] was used for Bayesian inference when testing the networks learned by BN PowerConstructor.

The BN PowerConstructor tool assumes that the variables have discrete and mutually exclusive values; also, the training data has to have a value for each variable. Thus, it is required to discretize the range for each variable used in this work. The BN PowerConstructor implements two different algorithms: one that requires node ordering and one that does not. For this work no ordering was provided. The intention was to obtain a BN structure completely based on the data collected.

The data set from the Solar Monitor site [7] presents sunspots parameters for each day. These parameters are listed bellow together with a brief description:

Number – denomination of a particular sunspot group
Date – date when the sunspot was observed
Location –sunspot's location on the Sun (heliocentric – latitude and longitude)
Hale – Magnetic classification of the sunspot
McIntosh – it classifies the sunspot group as a three letter code. The first letter describes the group configuration. The second letter describes the penumbra type of

the largest spot of the group, and the last letter describes the compactness (distribution) of the spots in the group

Area – total area of the sunspot (measured in millionth of solar disc area)

Nspots – number of spots inside a particular sunspot

Events – a solar flare is classified according to the magnitude of the peak burst intensity measured close to the Earth in the 0,1 to 0,8 nm wavelength band [3]. The classifications of the more intense flares are: C-class, M-class, and X-class.

The data collected from the Solar Monitor web site were used to create a database for sunspots. This database contains all sunspots that ended up producing a flare classified as larger than M5, which includes all the X-class flares, some sunspots from the M-class (index less than 5.0), some from the C-class, and some with no flares. The records were selected such that there were data for at least 5 days prior to the explosion. The data was divided into six groups, according to the day it was collected (D0, D-1, D-2, D-3, D-4, and D-5), where D0 is the data group at the day when the flare happened, and D-5 is the data group 5 days before the burst happened.

In order to determine the class of the flares within 72 hours of their explosion, the D-2 group of data was selected, that is, data from two days before the explosions. Each variable range in the data was discretized in groups based on physical meaning. For instance, the discretization criteria used for the Date, LocA, LocB, Area1 and Nspots1 variables considered ranges such that the data was split uniformly in each bucket. The date was grouped based on periods of activity's intensity: one group for the 2001 and 2002 period, one group for before 2001 and another for after 2002.

The algorithm used in the BN PowerConstructor software is able to construct a BN structure without any node ordering, however it is not always able to return all the arcs orientation (to learn how the algorithm treats these problems see [8]). In the work presented here, as expected, the arcs were not always oriented in the resulting BNs. Thus, it was necessary to do an extra study to detect the correct orientation for the arcs in the BN. Therefore to conclude the modeling step it was required to perform around fifty tests to check the correct arc orientation. The BN given by the software offered some possibilities for arcs directions. The tests performed respected the arcs created (showing the relation between the variables) and simulated all other combinations for the arcs direction. After the first network was created it was tested and the resultant possibilities were written down. The next step was to change an arc direction and then perform new tests also making notes of the possibilities. This step was repeated for every BN created when an arc direction was changed. After the completion of these tests a comparison between the results was performed and the arcs directions that resulted in the higher possibilities were selected.

At last some solar physicists were asked to define how the variables used in this work related to each other and, using this information a Bayesian Network showing the possible causality among the variables was constructed, as presented in Figure 1. This network will be used for comparisons with the networks learned from data.

3 Experiments

For the experiments, a database with 75 sunspots was created. These sunspots were divided into two groups: one with 57 sunspots used for learning the BN and one group with 18 sunspots used to test the BN. To validate the robustness of the network learned, three crosses-over were performed exchanging sunspots as detailed bellow:

- 18 sunspots from the test group with 18 sunspots from the training group.
- 9 sunspots from the test group with 9 sunspots from the training group.
- 9 sunspots from the test group with 9 sunspots from the training group. Three sunspots were maintained from the previous cross-over (2).

These four networks were learned and oriented, when required, as defined on the modeling section, and then they were tested with the sunspots on the test group. The first two networks are presented in Figure 2 and Figure 3. Figure 2 shows the first network learned and Figure 3 shows the network learned after the first cross-over.

The classification results of the testing data for each network and the average result are presented in Table 1. Table 1 shows three possible outcomes for each network: correct (C), undefined (U), and wrong (W). The correct label was assigned when the probability for the correct flare class was at least 10.01% larger than the second higher, the wrong label was assigned when the flare class classification was wrong at the end, and the undefined label was assigned when the system gives the correct classification, however the probability does not reach the 10% threshold.

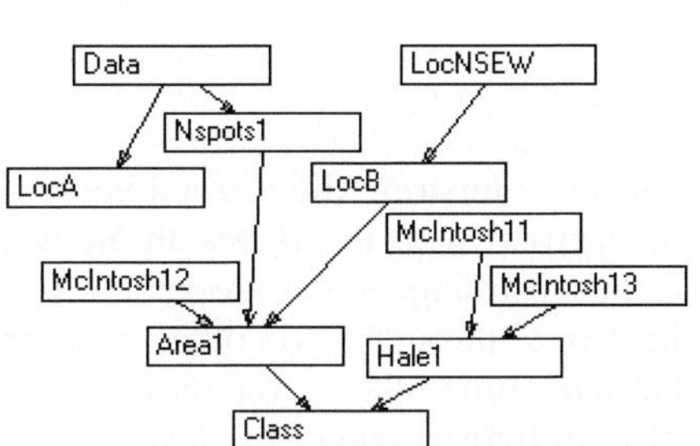

Fig. 1. BN manually designed

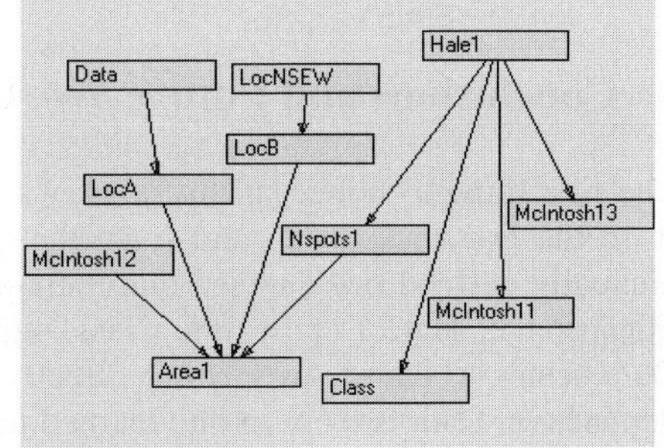

Fig. 2. Network one – the first D-2 learned network

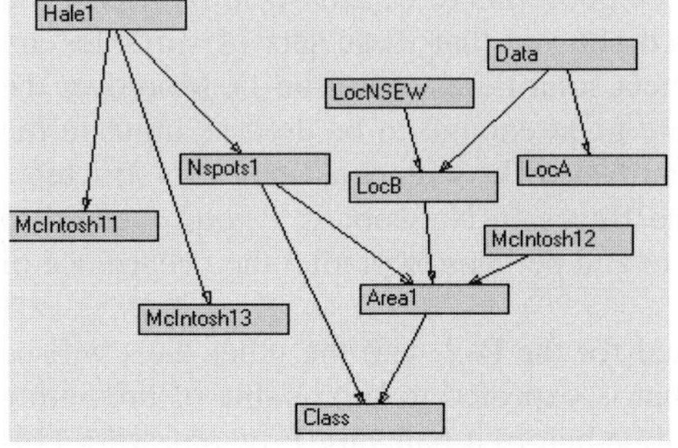

Fig. 3. Network two – the D-2 network learned after the first cross-over

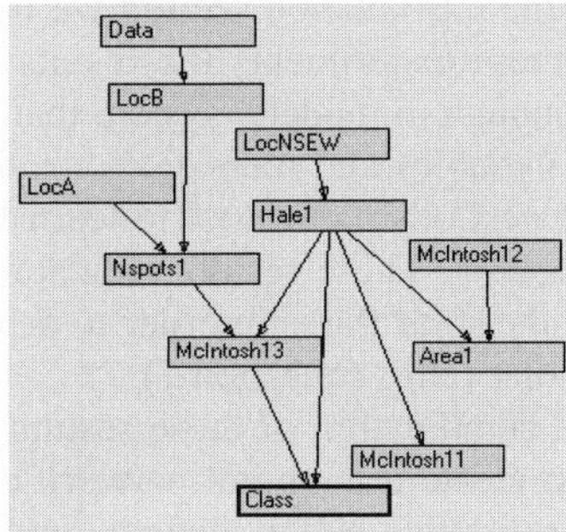

Fig. 4. Network five – the first D-4 network

A second experiment was performed to understand the difference between the networks using data from different days. For this experiment data from the D-4 group were used. The 75 records for the same sunspots were used and, again, they were divided into a training group of 57 sunspots and a testing group with 18 sunspots. This time only one cross-over was performed, changing 9 sunspots from the testing group with other 9 from the training group. Figure 4 shows the first network learned using the D-4 data. The results are also shown on table 1.

Table 1. The percentage of flares classification using both, training data and testing data

		Training data			Testing data		
		C	U	W	C	U	W
D-2	Network 1	77%	0%	23%	83%	0%	17%
	Network 2	82%	11%	7%	78%	0%	22%
	Network 3	77%	12%	11%	61%	17%	22%
	Network 4	79%	0%	21%	78%	0%	22%
	Average	79%	6%	15%	75%	4%	21%
D-4	Network 5	88%	0%	12%	83%	0%	17%
	Network 6	88%	0%	12%	78%	22%	0%
	Average	88%	0%	12%	81%	11%	8%

4 Conclusions and Future Work

The first thing to notice in this work was that when comparing the learned networks from the D-2 data group and from the D-4 data group with the Bayesian Network manually defined one can see that there are some substructures that are very similar. For instance, Hale 1 causes the Class, either directly or through some other variables, Data seems to cause Area either directly or through some other variable, LocA that depends on Data (except in one learned network), McIntosh11 and McIntosh13 which are always related to Hale1 (although in a different direction).

Table 1 shows that the Bayesian Network learned from D-2 data can determine the correct class for a flare about 80% of the time, which is a good indication that this approach can be used to predict giant flares using a sequence of data groups.

On the introduction section we listed three questions to be answered by this work. After these experiments the results obtained showed that these sets of variables are good enough to model a system that can detect solar flares (question 1). Moreover the set of variables and their discretization were good enough to be used as input in the BN PowerConstructor tool to learn Bayesian Networks with consistent sub-structures (question 3). These results indicate that the Bayesian Network is a good method to model the solar flares in order to detect them and perhaps to predict the occurrence of large solar flares (question 2).

The combination of these results obtained for the D-2 with the other days will be used to create a temporal network chain that is expected to be capable of predicting the flares with a better accuracy and more time before it happens. Previous work [14] have already pointed out the association between δ spots and large flares, however no quantitative estimate has been done so far. Moreover, the desired prediction is *when* a solar flare will occur, as opposed to where it happens. For solar physics standards, 80% probability of occurrence is a very good estimate.

References

1. Zirin, H.: Astrophysics of the Sun. Cambridge University Press (1988)
2. Lang, K. R.: Sun, Earth and Sky. Springer (1997) 75-93.
3. Solar Flares Classification, URL http://solar-center.stanford.edu/SID/docs/understanding-sid-data.pdf. Stanford Solar Center
4. Marengoni, M., Hanson, A. R., Zilberstein, S., Riseman, E.: Decision Making and Uncertainty Management in a 3D Reconstruction System. IEEE PAMI, Vol. 25, no 7 (2003) 852-858
5. Berler, A. and Shimony, S.E.: Bayes networks for sensor fusion in occupancy grids. Uncertainty in Artificial Intelligence (1997)
6. Pinz, A. and Prantl, M.: Active fusion for remote sensing image understanding, In: Desachy J. editor, Image and Signal Processing for Remote Sensing II, vol.2579. SPIE Proceedings (1995) 67-77
7. Solar Monitor, URL http://www.solarmonitor.org. Solar Data Analysis Center at NASA Goddard Space Flight Center in Greenbelt
8. Cheng, J., Bell, D., Liu, W.: Learning bayesian networks from data: An efficient approach based on information theory. Tech. Rep. – Department of Computer Science – University of Alberta (1998)
9. Jensen, F.V., An Introduction to Bayesian Networks. Springer –Verlag (1996)
10. Bayesia Ltd, Software. URL http://www.bayesia.com
11. Kadie, C.M., Hovel, D., Horvitz, E.: MSBNx: A Component-Centric Toolkit for Modeling and Inference with Bayesian Networks. Microsoft Research Technical Report MSR-TR-2001-67, (2001)
12. Marengoni, M.: Graphical Models for Computer Vision and Image Processing. RITA, Vol. VIII, no 1 (2001)
13. Andersen, S.K., Olesen, K.G., Jensen, F.V., and Jensen, F. HUGIN – a shell for building bayesian belief universes for expert systems. In: Proceedings of the 11th International Congress on Uncertain Artificial Intelligence (1989) 1080-1085
14. Zirin, H., Liggett: Association of delta spots and great flares, (1987), 113-267

The Virtual Reality Brain-Computer Interface System for Ubiquitous Home Control

Hyun-sang Cho, Jayoung Goo, Dongjun Suh, Kyoung Shin Park, and Minsoo Hahn

Digital Media Laboratory, Information and Communications University,
517-10 Dogok-dong, Gangnam-gu, Seoul 135-854, S. Korea
{haemosu, bucsu, linuxer, park, mshahn}@icu.ac.kr

Abstract. This paper presents a virtual reality brain computer interface (BCI) system which allows the user to interact physical objects in the ubiquitous home. The system is designed for motion disabled people to control real home facilities by a simple motor imagery and minimal physical body movements. While BCI research has mostly focused on improving classification algorithm, our BCI system uses a simple BCI classification method and the additional locking mechanism to compensate the BCI classification error. This paper describes the design, implementation, and user evaluation of our virtual reality BCI system for ubiquitous home control. The user study results showed the locking mechanism helped to improve user controllability which made the system more feasible and reduced user undesired BCI decision error rates.

1 Introduction

In the ubiquitous computing environment, the objects are intelligent and can be controlled remotely via a network. With the advent of ubiquitous computing technologies, it is possible to realize an affordable environment that allows motion disabled people to control the surrounding objects. Motion disabled people usually have a lot of psychological disorders such as depression due to their uncontrollable body. They also feel charging a burden on their family at home. Our research aim is to enable motion disabled people to feel as if they enjoy a real life. We believe that virtual reality (VR) can give them psychological compensates since it provides the sense of actually being there in the physical environment where the spatiality of the virtual objects matched with the position of real objects. However, motion disabled people need a specially designed VR system that will not require a lot of physical body movements for the system operation.

Recent research efforts show the feasibility of brain computer interface (BCI) used for motion disabled people [1, 2, 3]. They have achieved a mean classification rate of above 80%. However, BCI is not easily used for application domains due to 10~20% error rate. The random BCI classification errors will create a cyber sickness problem in the virtual environment when user navigation is mismatched with their will. Eventually, this problem will result in loosing user control. It is one of the serious obstacles of using virtual reality BCI system.

A. Sattar and B.H. Kang (Eds.): AI 2006, LNAI 4304, pp. 992–996, 2006.
© Springer-Verlag Berlin Heidelberg 2006

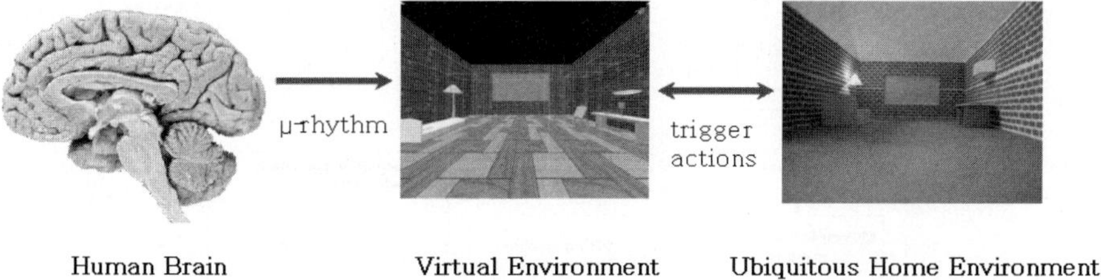

Fig. 1. The overview of the virtual reality brain-computer interface system for controlling a ubiquitous home environment

In this paper, we present a virtual reality BCI system for ubiquitous home network control. The system allows the user to control the objects in the virtual environment using user's motor-imagery, and this event triggers the control of objects in the ubiquitous computing environment for real executions. It uses motor-imagery Event-Related De-synchronization (ERD) brain signals of the EEG device for moving left or right direction and an auxiliary control device for moving forward in the virtual environment. We also provide the locking mechanism for enhancing system robustness and usability under high BCI classification error rates. The feasibility of this approach is confirmed through a user study. The result showed that the locking mechanism improved user controllability thereby enhancing system usability.

2   BCI-Based VR Interface for Ubiquitous Home Control

Fig. 1 shows the overview of our virtual reality BCI system for ubiquitous home control. The middle and the right image of Fig. 1 show a virtual world and a physical miniature mockup model replicated a real home. The system detects human motor-imagery brain signals (i.e., one's imagination about left or right hand moving) by measuring μ-rhythm from the EEG device. When a user imagines left or right hand moving, he/she can incrementally change the orientation viewpoint to left or right direction in the virtual environment. In this system, a user can freely navigate and trigger actions in the virtual environment using the additional direction device. In addition, the action events (in the virtual environment) trigger the events in the real ubiquitous home environment.

From our experiences with the early prototype of this system, users seemed to feel fatigue or cyber sickness induced by their unwilling wander in the virtual environment. This happened in association with a wrong BCI direction decision. To solve this cyber sickness problem, we provided a direction locking mechanism. That is, a user could lock the direction when he/she decided the correct direction which way he/she wanted to go. Then, the user could unlock (i.e., release the direction locking) so that he/she could change left or right direction (i.e., rotation in the virtual environment) using this BCI. We have developed two direction locking devices: a pedal-like device and a simple eye-blink detector (shown in Fig 3).

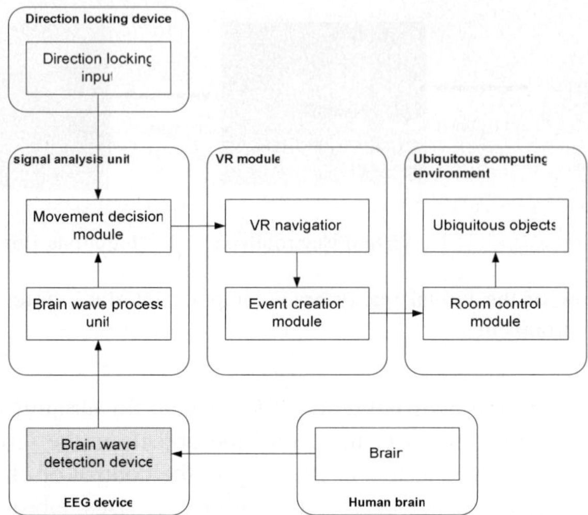

Fig. 2. The system architecture, describing the events triggering among modules

Fig. 2 shows the system architecture of our virtual reality BCI system for the ubiquitous computing environment controls. The system consists of three main modules: the signal analysis unit, the virtual environment, and the ubiquitous computing environment control unit. The signal analysis unit contains of three functional modules: the physical control signal input module, the EEG measurement equipment, and the brain signal analysis unit. We sampled brain wave with 256 Hz to make the classification every second. The virtual environment was implemented using YGdrasil (YG) [4].

3 User Evaluations

We conducted an experiment of participants using our virtual reality BCI system to evaluate the effect of the direction locking mechanism (i.e., with or without the direction locking for BCI uses in the virtual environment). Eight graduate and undergraduate male students volunteered as participants in this study. The range of the participants' age was 20 to 35 years old. They had high level computer experience and their typical computer usage time was more than three hours a day. However, none of them had prior experience with EEG equipments, BCI, and VR system. That is, all participants were novice users to our BCI and the virtual environment.

Fig. 3 shows a participant of our BCI system performing the VR navigation task. In this task, each participant was asked to navigate in the virtual environment to visit three object control regions (i.e., bed lamp, a curtain, and a table lamp) and trigger the operation (i.e., turning on/off the light and moving the curtain up/down). We examined participants with or without the direction locking mechanism that was provided as a foot pedal controller.

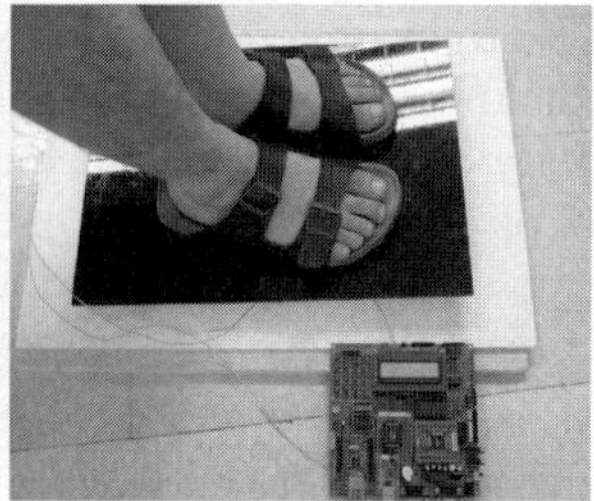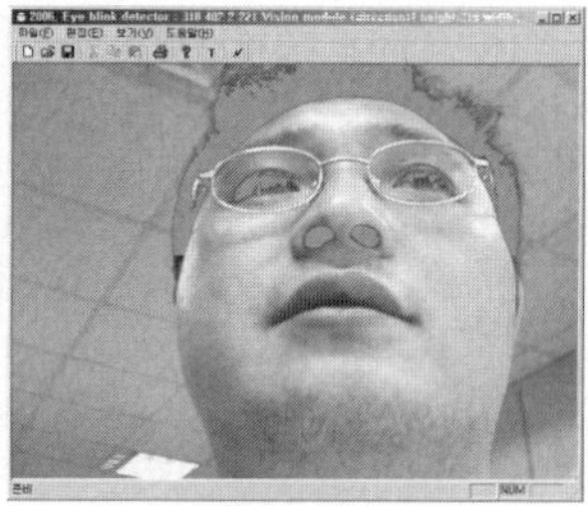

Fig. 3. A participant performed the VR navigation task visiting three object control regions such as a bed lamp, a curtain, and a table lamp(on the left). The participant performed the VR navigation task with or without the direction locking mechanism given as a foot pedal controller(on the center) or eye blink detector(on the right).

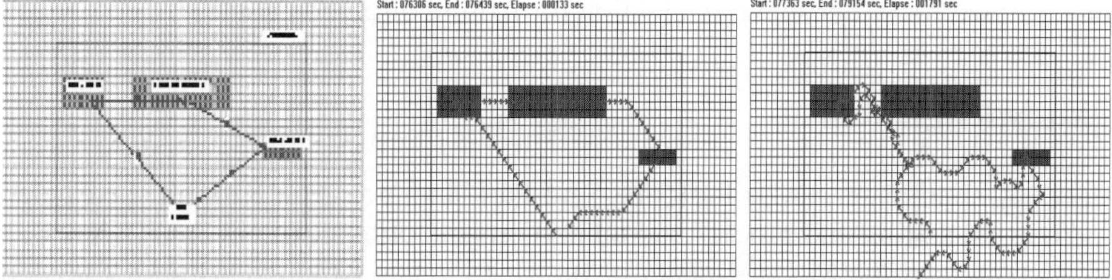

Fig. 4. The left image is the navigation path given to the participants in the virtual environment. The middle and the right image show one participant's actual navigation trails with and without the direction locking mechanism respectively.

The experiment result showed that our system had low BCI classification rate with novice users. In the past, an informal user evaluation with trained participants showed 70% on average and up to 90% correct classification rate. We suspect that there might have been a problem with the brain signal gathering or an inappropriate EEG electrode placement on the participant's head because the experiment was done by novice helpers to check the robustness of our system.

With this low BCI classification rate, we however, could confirm the effectiveness of this direction locking mechanism for VR navigation and interaction. The results showed it helped our participants were easily navigating in the virtual environment. The average completion time with direction locking was 147 seconds (35% faster) while the average completion time without locking was 414 seconds. In particular, two participants lost their controls in the virtual environment without locking which was never happened with locking. Fig. 4 shows the participant 5's actual navigation traces in the virtual environment with and without the direction locking mechanism. In Fig. 4, the colored areas indicate the object control regions.

4   Conclusions and Future Work

This paper presents a virtual reality brain computer interface (BCI) system designed for motion disabled people to control the ubiquitous home environment. In this system, users can change left or right direction turns by motor imagery (i.e., human imagination of a left or right hand movement) to navigate in the virtual environment and move forward by an auxiliary control device. Such user interactions in the virtual environment trigger the control of ubiquitous objects in the real home environment. In this paper, we also presented the additional direction locking mechanism to compensate the random BCI decision errors. The goal of this research was to develop a cheap and simple practical BCI system. Through the user evaluation, we found that this locking mechanism was imperative to user navigation and interaction in the virtual environment which also helped reduce the cyber sickness problem. The results also showed that user's navigation trails in the virtual environment was much shorter with the locking mechanism than without locking. We believe the locking mechanism can be used for the dynamic complex brain signal processing. We will also continue to develop other types of direction locking and forward input device, such as other brain signals, sound or jaw movements. We will envision expanding this system to support home or town for wider social interaction for motion disabled people.

Acknowledgement

I sincerely appreciate Ji-Min Park, Ye-Seul Park and You-Min Kim for their help on experimental setting and process and all participants of this experiment. This research was supported by the MIC (Ministry of Information and Communication), Korea, under the Digital Media Lab. support program supervised by the IITA (Institute of Information Technology Assessment).

References

1. Gert Pfurtscheller and Christa Newper: Motor Imagery and Direct Brain-Computer Communication. Proceedings of the IEEE, Vol.89 No.7, 2001.
2. Robert Leeb, Reinhold Scherer, Felix Lee, Horst Bischof, Gert Pfurtscheller: Navigation in Virtual Environments through Motor Imagery. Proc. of 9th Computer Vision Winter Workshop, pp. 99-108.
3. Baharan Kamousi, Zhongming Liu, and Bin He: Classification of Motor Imagery Tasks for Brain-Computer Interface Applications by Means of Two Equivalent Dipoles Analysis. IEEE Transactions on Neural System and Rehabilitation Engineering, Vol. 13, No. 2, June 2005.
4. Dave Pape, Josephine Anstey, Margaret Dolinsky, Edward J. Dambik: Ygdrasil--a framework for composing shared virtual worlds. Future Generation Computer Systems, volume 19, No. 6, p141-149, 2003.

Verb Prediction and the Use of Particles for Generating Sentences in AAC System

Eunsil Lee[1], Dohyun Nam[1], Youngseo Kim[2], Taisung Hur[3],
Yoseob Woo[1], and Hongki Min[1]

[1] Dept. of Information and Telecommunication Engineering, University of Incheon
177 Dowha-Dong, Nam-Gu, Incheon 402-749 Korea
{eslee, dhnam, yswooo, hkmin}@incheon.ac.kr
[2] Dept. of Biomedical Engineering, Seoul Health College, Kyoungki, Korea
yskorea@shjc.ac.kr
[3] School of Computing & Info. System, Inha Technical College, Incheon, Korea
tshur@inhatc.ac.kr

Abstract. This study proposes and develops a novel Augmentative & Alternative Communication (AAC) system which improves the verbal communication of speech disordered people. The success of AAC systems is largely determined by two factors: the selection of vocabularies appropriate to the occasions and the efficient utilization of limited display space in AAC systems. To meet the two vital requirements, vocabulary databases are built and embedded into our AAC system, thus the context-aware predicate prediction and the particle use in generated sentences even in a limited display space are feasible. We construct the databases by collecting utterances in everyday life dialogues by recording, sorting frequently used vocabularies, and applying semantic and lexical interpretation.

1 Introduction

Augmentative & Alternative Communication (AAC) systems, which help the communication of speech disordered people, are recently drawing more attention because it becomes more necessary that the speech disordered people should be provided with more chances to participate in education, social activities, religion, leisure, occupation, etc.

The core function of the AAC systems is about how to select the first vocabulary [1]. It is because, if wrong or improper vocabularies are selected by the AAC systems, it will dismay or frustrate the AAC system users [2]. Various prediction methods are used to enhance the user convenience of the AAC systems. In this study, we construct a vocabulary database through the following steps. Firstly we collect and analyze actual utterances, and then select frequently used vocabularies, finally decide the vocabularies to be included by referring to a lexical thesaurus and a sub-categorization dictionary. It was intended to develop a system to predict and recover the grammatical morphemes such as auxiliary words or endings of words, complements, annexed words and conjunctions to present a relevant sentence by assuming that they will be omitted at the time entry.

A. Sattar and B.H. Kang (Eds.): AI 2006, LNAI 4304, pp. 997 – 1001, 2006.
© Springer-Verlag Berlin Heidelberg 2006

2 Vocabulary Selection and Symbols

The vocabulary database construction is the first step when developing an AAC system. We target spontaneous utterances during break times in schools and in daily life environments to collect frequently used vocabularies from ordinary people. The collection places are home, restaurants, various transportation stations, schools, hospitals and shopping centers. In this study, we collected vocabularies by recording and analyzed a total of 9,559 vocabularies from the situational utterances of 24 schoolchildren and 20 adults at schools, and others. We also measured the frequency of each vocabulary, and calculated the ratio of the high frequency vocabularies to the entire vocabularies.

We analyze the relationship between vocabularies and symbols, design proper symbols representing the corresponding vocabulary database, and efficiently arrange the symbols in each area. We dealt also in actual sentence taking the mother change of the predicate part into account to the prototype, and an example to become a sentence generation in the symbol user interface is as follows.

3 Sentence Generation Through Predicate Prediction

The objective of our study is to help the speech disordered people to communicate more easily by providing them with visual symbols (each symbol has its corresponding Korean meaning), and having them choose appropriate symbols when they want to convey their intention to others. However, the visual symbolization of many words has to overcome the spatial limit when the symbols are shown in users' display panels.

We propose a predicate prediction method to resolve the spatial restriction of visual symbols. The predicate prediction is feasible in Korean language because the predicates locate at the end of sentences, which is quite different from the English sentence structure. Therefore, if a noun is given, the predicate can be predicted by particularizing the category of the noun by referring to a noun thesaurus and the sub-categorization of verbs. For this, we build a new thesaurus which is suitable for our AAC system rather than using established thesaurus for natural language processing. Our thesaurus has a hierarchical structure in which there are several upper domains and each upper domain contains multiple lower domains in it. In this structure, we firstly divide the words into upper domains such as transportation, shopping, home, school, hospital and restaurant, and then further sub-divide them into lower domains such as place, things, vehicle, event, action, quantity, food, person, time and others. To represent the upper and the lower domains of each word, we use logical 'right' and 'wrong,' namely, bit '1' and '0', respectively. In a sentence, the words are separated into an input part and a predicted predicate, and one-to-one meaning symbolization is performed for the nouns. In addition, a meaning marker is assigned to each noun by using the thesaurus. Through this process, a concept-based database is constructed: e.g. shoes belong to upper domains 'shopping,' 'home' and 'school' and a lower domain 'things', and a new product belongs to an upper domain 'shopping' and a lower domain 'things'. We use another database, which is a predicate database containing verb variations (declarative, interrogative and propositive) as well as upper-lower domains like the noun database. Because the predicate decides the particle, frequently used predicates are marked with '1' in upper domains, and instead of '1' particles used are put in lower domains.

With the constructed vocabulary database, we are able to predict the predicate by deriving meanings through combining the noun thesaurus specifying dependent relations with the meanings of vocabulary and a pattern dictionary containing situation information according to predicate with the results of dependent relation analysis. For this process, we use the sub-categorization dictionary. It is called selection limitation to derive meanings by combining the noun thesaurus specifying dependent relations with the meanings of vocabulary and a pattern dictionary containing situation information according to predicate with the results of dependent relation analysis and, through selection limitation, we can predict the predicate. Figure 3.1 shows the predicate prediction process by the selection limitation in which the thesaurus and the sub-categorization dictionary are used.

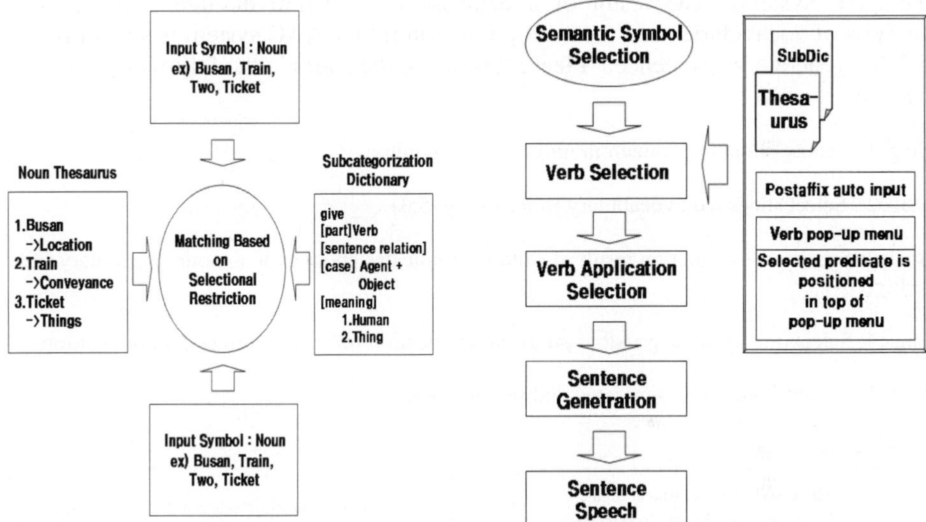

Fig. 3.1. The process of predicate prediction by the selection limitation

Fig. 4.1. The sentence generation process of the proposed system

4 Use of Particles

Our system also processes the particle after the predicate prediction is finished in order to endow constructed sentences with more syntactical completeness. Since the particle is determined after the predicate selection, frequent verbs are marked with '1' in upper domains and the particles are put in lower domains instead of '1'. The particle takes a form of either '은/는' or '을/를' depending on whether the preceding noun has a final consonant or not, thus only the basic form of the particle is put in. We can tell from Unicode whether the final consonant exists or not. Depending on whether the final consonant is 'ㄹ' or not, it is possible to transform the basic form of the particle into one of its variations. Figure 4.1 shows how our system generates a sentence when a user presses a symbol button. It firstly retrieves relevant meanings from the noun database and predicts the predicate in the predicate database. Then it transforms the selected predicate into a desired variation and also assigns an appropriate particle.

5   Results

The analysis result about the ratio of high frequency vocabularies to the entire ones is shown in Fig.5.1. It shows that top 10 high frequency vocabularies occupy 20% of the entire vocabularies; top 25 ones have 34%, and top 50 ones have 58%, respectively. Fig.5.2. shows the analysis of the utterance frequencies per part of speech. Verbs are used 1,378 times, nouns 790 times, pronouns 676 times, adverbs 582 times and adjectives 289 times; thus it is observed that verbs are most frequently used. It is a clear proof that prediction of verbs is more useful than that of other parts of speech in the AAC systems. The result of a database construction through such semantic analysis of vocabularies and the verb prediction in our AAC system is shown in Fig. 5.2. To generate the sentence, "Please, Show me the shoes.", the following stages are needed

Stage1: Choose shopping domain in initialization monitor.

Stage2: Select shoes icon vocabulary (refer to Fig. 5.3)

Stage3: Specify the prediction result of suitable predicates about an icon noun vocabulary (refer to Fig. 5.4)

Stage4: Select one from the possible predicate application forms for correct communication.

Stage5: A final sentence is generated and sent to TTS.

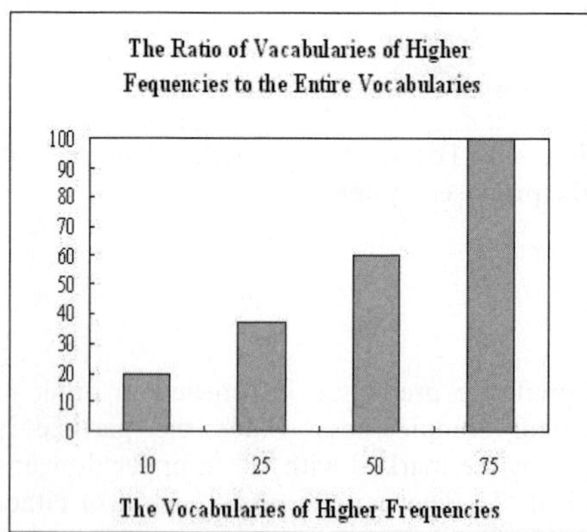

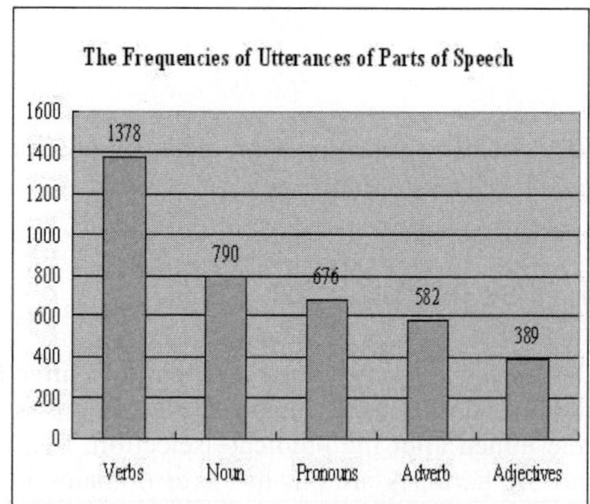

Fig. 5.1. Ratio of vocabularies of higher Frequencies to the entire vocabularies

Fig. 5.2. The freguencies of utterances parts of speech

Fig. 5.3. Vocabulary icon selection

Fig. 5.4. Application forms selection from predicted predicates

6   Conclusions

We visited schools and other sites to record spontaneous dialogues for the study. As one of the analysis results of our study, we presented a list of high frequency vocabularies and each frequency of respective parts of speech. It is reportedly known from other researches that a small number of vocabularies with high frequencies can cover most domains of the communication even if the entire vocabulary size is large. Our analysis results also showed that the verbs are more frequently used than other parts of speech. It provides a clear proof to our suggestion; the AAC systems equipped with the verb prediction method are more effective in the sentence generation. Our verb prediction required us to build a machine-readable lexical database and various electric dictionaries like a sub-categorization dictionary. Our system also processes the particle in order to improve the syntactical completeness of sentences. However, our AAC system may not be effective for all the speech disordered people. Also, we would like to note that various types of AAC systems might be needed depending on the level of disorders.

References

1   Sharon L. Glennen, DeCote, Ed.D, *"The Handbook of Augmentative and Alternative Communication."* Singular Publishing Group, Inc. 1996

2   Gittins, D. Icon-based Human-Computer Interaction, International Journal of Man-Machine Studies, 24, 519, 543, 1986

3   Kathleen F. McCoy and Patrick Demasco, "Some Applications of Natural Language Processing to the Field of Augmentative and Alternative Communication" In Proceedings of

4   the IJCAI-95 Workshop on Developing AI Applications for People with Disabilities, Montreal, Canada, August, 1995.

5   M. O'Donnell,""Input specification in the WAG sentence generation system", In INLG 8, 41-50,.1996

Abnormal Behavior Detection for Early Warning of Terrorist Attack

Xin Geng[1], Gang Li[1,*], Yangdong Ye[2], Yiqing Tu[1], and Honghua Dai[1]

[1] School of Engineering and Information Technology,
Deakin University, VIC 3125, Australia
{xge, gang.li, ytu, hdai}@deakin.edu.au
[2] School of Information Engineering,
Zhengzhou University, Zhengzhou , China
yeyd@zzu.edu.cn

Abstract. Many terrorist attacks are accomplished by bringing explosive devices hidden in ordinary-looking objects to public places. In such case, it is almost impossible to distinguish a terrorist from ordinary people just from the isolated appearance. However, valuable clues might be discovered through analyzing a series of actions of the same person. Abnormal behaviors of object fetching, deposit, or exchange in public places might indicate potential attacks. Based on the widely equipped CCTV surveillance systems at the entrance of many public places, this paper proposes an algorithm to detect such abnormal behaviors for early warning of terrorist attack.

1 Introduction

After September 11, the global "War on Terrorism" has become one of the main challenges of our time. Although great efforts have been taken all over the world to protect innocent people from terrorist attacks, it still appears to be, as the US President George W. Bush appropriately labeled, "a long war against a determined enemy". Besides the political and military actions, new technologies in various areas are urgently demanded to ensure the victory in this war.

One of the most common terrorist attack patterns is to bring explosive devices hidden in ordinary-looking objects to public places. Recent examples are the London bombings (7 July 2005) and the Bali bombings (1 October 2005), both resulting in massive casualty. The early warning of such attacks is extremely difficult since a terrorist carrying a camouflaged bomb is not evidently different in appearance from an ordinary people. Thus it is almost impossible to detect in advance the potential danger of such attacks through conventional CCTV surveillance systems mounted in public places.

Although the isolated appearance is insufficient to distinguish terrorists from ordinary people, clues of abnormal behaviors could be discovered by considering the combination of a sequence of actions. During the procedure of explosive device preparation, delivery and the final attack, the members of terrorist group

---

* Corresponding Author.

A. Sattar and B.H. Kang (Eds.): AI 2006, LNAI 4304, pp. 1002–1009, 2006.
© Springer-Verlag Berlin Heidelberg 2006

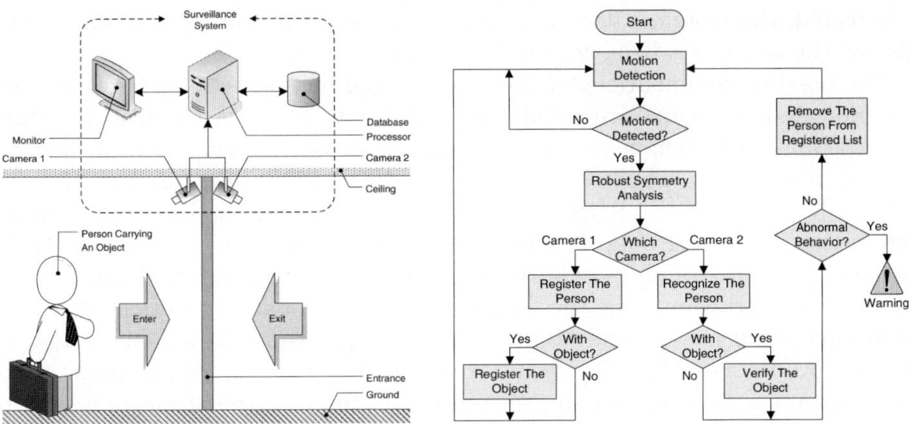

Fig. 1. Application Environment   **Fig. 2.** Flow Chart of ROSY

Table 1. Combination of the Enter/Exit Actions and the Indicated Behaviors

Enter	Exit	Behavior	Code
Without Object	Without Object	Normal	N1
With Object	With the Same Object	Normal	N2
Without Object	With Object	Abnormal (Fetch)	A1
With Object	Without Object	Abnormal (Deposit)	A2
With Object	With the Different Object	Abnormal (Exchange)	A3

need to fetch, exchange and deposit objects of certain size. These behaviors are very rare in many public places, but are not easy to be discovered by security officers because it is almost impossible for human officers to remember every passing individual and the object he/she is carrying. In this paper, a surveillance algorithm named ROSY (RObust SYmmetry) is proposed to automatically detect such abnormal behaviors and warn people before potential terrorist attacks.

The rest of this paper is organized as follows. In Section 2, the ROSY algorithm is explained in detail. Then the experimental results are reported and analyzed in Section 3. Finally conclusions are drawn in Section 4.

2   The ROSY Algorithm

The application environment of ROSY is illustrated in Fig. 1. The algorithm is designed to work with monochromatic stationary video sources, either visible or infrared. Camera 1 is used to image the entering person, and Camera 2 is used to image the exiting person. Images from both cameras are sent to the processor installed with ROSY. Each entering person is registered in a database, and each exiting person is checked with the database. If any abnormal behavior is detected, the system will send a warning message to the security officer and display the suspicious behavior on the monitor.

The possible combinations of the entering and exiting actions, together with the indicated behaviors are tabulated in Table 1. Three of the five situations indicate abnormal behaviors, corresponding to object fetching, deposit and exchange, respectively. The behaviors are coded as N1, N2, A1, A2, and A3.

The flow chart of the ROSY algorithm is show in Fig. 2. The algorithm assumes that the background is relatively steady. Thus motion detection can be achieved through training a Gaussian model for each background pixel over a short period and comparing the background pixel probability to that of an uniform foreground model. The result of motion detection is the silhouette map of the foreground, see Fig. 3(b) as an example. If significant motion is detected, which means a person is passing the entrance, the foreground image is sent to Robust Symmetry Analysis (RSA) to determine whether the person is carrying an object and segment the human body and the object. The details of RSA will be described in Section 2.1. If the person is entering (captured by Camera 1 in Fig. 1), then the person and the object (if any) are registered into the database. If the person is exiting (captured by Camera 2), then the person is recognized and the object (if any) is verified. If any abnormal behavior (A1, A2, A3) is detected, the algorithm will raise a warning message. Otherwise, it will remove the record of the person from the database. The registration and recognition/verification of persons and objects will be further discussed in Section 2.2.

2.1   Robust Symmetry Analysis

The basic idea of ROSY is from the observation that the frontal view of the human body is approximately symmetric, and such symmetry will be violated if people carries an object. The problem is that, in many cases, body motion will also cause some body parts to violate the symmetry. Take Fig. 3(a) as an example, except for the object, the arms and legs are also not symmetric about the middle line of the body. So the algorithm must be able to distinguish whether the asymmetry is caused by an object or body motion. Backpack [1] did this through shape periodicity analysis. Unfortunately there are no sufficient periodic images in the scenarios of this paper. On the one hand, while people pass an entrance, they often do something, such as slow down walking speed, stop to identify themselves, and open the door, all of which will violate their usual walking periodicity. On the other hand, the entrance is often located at relatively narrow places, such like the entrance of a subway train, or a door by the corridor, where no periodicity can be detected since the person appears in the scene for only a short time. Thus Backpack is not suitable for the application here. Instead, ROSY achieves this by Robust Symmetry Analysis (RSA), based on as few as one image. There are mainly two steps in RSA. The first step is to analyze the symmetry of the silhouette, which narrows down the interested region for the second step, appearance symmetry analysis.

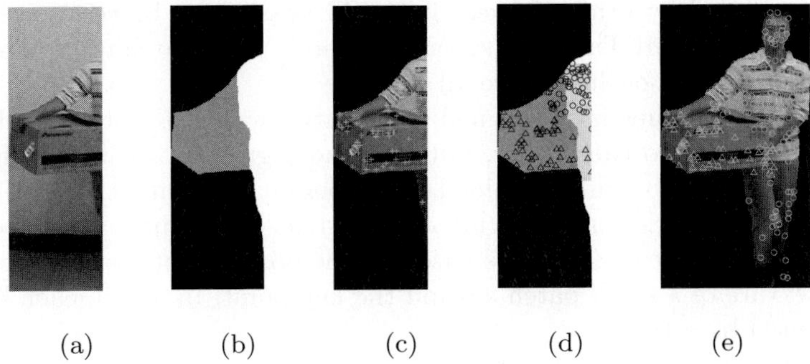

(a)   (b)   (c)   (d)   (e)

Fig. 3. Robust Symmetry Analysis. (a) Original image;(b) Symmetric axis and non-symmetric region;(c) SIFT key points;(d) Key points classification;(e) Segmentation.

Silhouette Symmetry Analysis. Fist of all, the symmetric axis of the human body is estimated. Suppose the height of the silhouette is h, the algorithm only consider the highest $h/10$ and the lowest $h/10$. The horizontal middle point of the upper $h/10$ silhouette m_1 is regarded as one end of the symmetric axis. The lower $h/10$ silhouette might consist of several disconnected regions. The horizontal distance from m_1 to the horizontal middle point of each region is calculated, and the mean of those middle points within a certain distance to m_1 is regarded as the other end of the symmetric axis m_2. The estimated symmetric axis of the silhouette is shown as a red line in Fig. 3(b).

Suppose the estimated symmetric axis is $[m_1, m_2]$, p_l and p_r are a pair of pixels on the silhouette boundary such that the line segment between them perpendicularly intersects with $[m_1, m_2]$ at p_s. Let $d(p_1, p_2)$ denote the distance between p_1 and p_2. Then the symmetry of a pixel p_x on the line segment $[p_l, p_r]$ is determined by

$$sym(p_x) = \begin{cases} false & \text{if } d(p_x, p_s) > min(d(p_l, p_s), d(p_s, p_r)) + \varepsilon \\ true & \text{otherwise,} \end{cases} \quad (1)$$

where $sym(p_x) = false$ means p_x is a nonsymmetric pixel, and ε is a predefined small number. As an example, the nonsymmetric region in Fig. 3(b) is denoted by green. In order to make the algorithm more robust against image noise and the slightly asymmetry of human body, only those nonsymmetric region larger than a certain size (predefined minimum width, height and area) is considered.

Appearance Symmetry Analysis. In order to segment the object and the human body, the appearance of both the symmetric and nonsymmetric regions are compared with the other side of the symmetric axis. ROSY extracts the SIFT features [2] for image matching purpose. The SIFT feature positions of the foreground in Fig. 3(a) are shown in Fig. 3(c). Each SIFT key point is modeled by its position, scale and orientation so that scale and rotation invariance can be achieved. Experimental results [2] have shown that reliable recognition is possible with as few as 3 SIFT features. Thus even when a large portion of the object is occluded by other objects, it is still possible to be recognized. All of these advantages of SIFT make it suitable for body part matching, where scaling, rotation, and partial occlusion are all possible.

The SIFT matching is performed in a batch way, i.e. not only the single feature is matched to each other, but also those groups of features that agree to the same object pose are examined. Thus before matching, the SIFT features in both the symmetric and nonsymmetric regions should first be clustered into groups. In ROSY, the clustering is based on the positions of the SIFT key points and the texture of a small patch around the key point. In detail, each key point is represented by a triplet

$$\mathbf{t} = <x, y, E> \tag{2}$$

where (x, y) is the 2D coordinate of the key point and E is the entropy of the 9-by-9 neighborhood around the key point.

$$E = -\sum_{i=0}^{L-1} p(z_i) \log_2 p(z_i), \tag{3}$$

where z_i is a random variable indicating intensity, $p(z)$ is the histogram of the intensity levels in the 9-by-9 neighborhood, L is the number of possible intensity levels. Then the elements of $\mathbf{t}$ are normalized so that they have zero mean and unity standard deviation. Clustering is done by applying a graph based method called Normalized Cuts [3] on the normalized $\mathbf{t}$. The clusters in both the symmetric and nonsymmetric regions are matched to the other side of the symmetric axis. For each cluster, if at least 3 key points match with the other side, and they agree on the same pose, then it is regarded as part of the human body, otherwise of the object. An example is shown in Fig. 3(d), where the nonsymmetric region is green, the examined symmetric region is yellow, the positions of the key points classified as from body part are marked by circles, and those classified as from object are marked by triangles. It can be seen that most key points are correctly classified. Finally all SIFT key points outside the colored region is classified as from the human body. The result of RSA is shown in Fig. 3(e). If no key point is classified as from the object, then the person is not carrying an object.

2.2   Abnormal Behavior Detection

ROSY maintains a database to remember whoever entered the entrance, and whatever they are carrying with them. Each entry consists of four fields: *img*, *withObject*, *bodySIFT*, and *objectSIFT*, which respectively store the image, whether the person is carrying an object, the SIFT key points from the body, and the SIFT key points from the object.

When Camera 2 in Fig. 1 images somebody, i.e. somebody is exiting the entrance, ROSY first recognizes the person from the records in the database. This is achieved through matching the *bodySIFT* of the current image to those stored in the database. The record with the maximum number of matching key

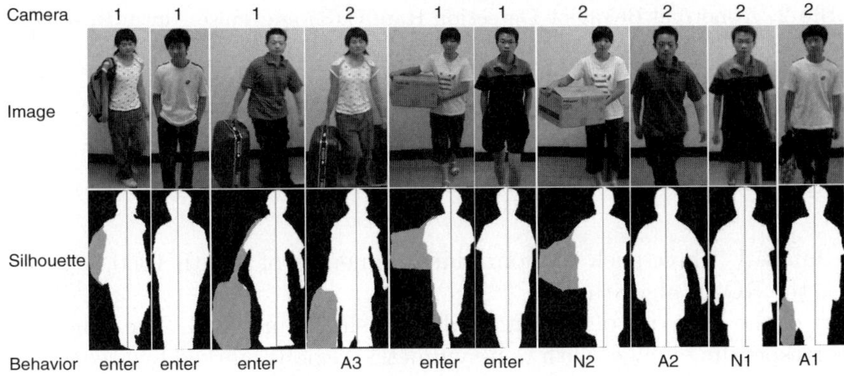

Fig. 4. Behavior detection in part of the random sequence

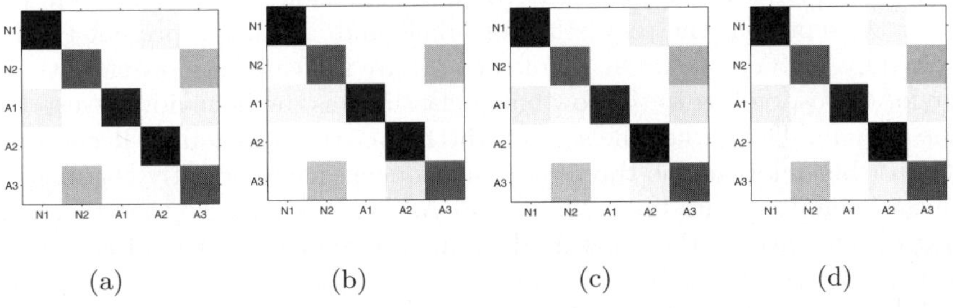

Fig. 5. Confusion matrices. (a) Trial 1; (b) Trial 2; (c) Trial 3; (d) Average

points that agree on the same pose is regarded as from the same person. Then the entering and exiting profiles of the same person are compared with each other. Based on the field *withObject*, three behaviors (N1, A1, A2) in Table 1 is decidable. If the person entered and is exiting both with an object, then the object is further verified to see whether it is the same one that the person brought in. This is achieved by matching the *objectSIFT* of the two images, if there are more than 3 matching key points agree on the same pose, then the objects are the same (behavior N2), otherwise they are different (behavior A3).

3   Experiment

We have collected video clips from 20 different persons. Each person entered and exited a room in 6 different states respectively, among which 4 states are with an object (each time a different one), and 2 states are without object (one is walking normally, the other is walking with some variation). There are totally 4 different objects: a backpack, a handbag, a luggage, and a box. The real scenario at the entrance of public place is simulated by creating a random sequence of the video clips. Part of the sequence is shown in the second line of Fig. 4 (labeled

Table 2. Abnormal Behavior Detection Rate (DR) and False Alarm Rate (FAR)

Trial	1	2	3	avg.	std.
DR (%)	88.16	90.06	89.26	89.16	0.95
FAR (%)	5.30	3.65	4.90	4.62	0.86

with 'image'). Three such random video sequences of length 1000 is generated to test the ROSY algorithm.

Part of the behavior detection results are shown in Fig. 4. The silhouettes of the corresponding images with non-symmetric regions marked by green are show in the line labeled 'silhouette'. Note that only when the image is captured by camera 2 does the algorithm detect a behavior listed in Table 1. Otherwise the behavior is 'enter'.

The performance of ROSY is illustrated as confusion matrices shown in Fig. 5, where rows represent the real behavior labels and columns represent the labels classified by ROSY. The intensity of each square indicates the probability of the behavior corresponding to the row being classified as the behavior corresponding to the column. The darker the square, the higher the probability. It can be seen that most behaviors in the three random sequences are correctly classified. The confusion mainly occurs between N2 and A3. This is because that there might be large difference in the view angle from the camera to the object and the object might be remarkably deformed. In such cases, automatic verification of the object is extremely hard. One way to solve this problem is to display any amphibolous behaviors between N2 and A3 on the monitor and let the security officer make the final decision.

As for abnormal behavior detection (A1, A2, A3), the detection rate (DR) and false alarm rate (FAR) of the 3 trials are tabulated in Table 2. It reveals that high detection rate and low false alarm rate can be achieved by ROSY. The average detection rate over all three trials is 89.16%, while the average false alarm rate is 4.62%. Moreover, the small standard deviation indicates the relatively steady performance of ROSY in different situations.

4   Conclusions

This paper proposes an abnormal behavior detection algorithm named ROSY for early warning of potential terrorist attack. Unlike previous work on abnormal behavior detection, ROSY aims to discover the abnormality concealed in a series of sub-behaviors, each of which alone is normal. This endows ROSY with the ability to detect behaviors like object fetching, deposit, and exchange, which is very rare in many public places and indicates potential danger of terrorist attack. Moreover, by using a novel technique called Robust Symmetry Analysis, ROSY can work on as few as one image, which makes it suitable to apply to the widely equipped CCTV systems at the entrance of many public places.

References

1. Haritaoglu, I., Cutler, R., Harwood, D., Davis, L.S.: Backpack: Detection of people carrying objects using silhouettes. In: ICCV, Kerkyra, Greece (1999) 102–107
2. Lowe, D.G.: Distinctive image features from scale-invariant keypoints. International Journal of Computer Vision **60**(2) (2004) 91–110
3. Shi, J., Malik, J.: Normalized cuts and image segmentation. IEEE Transactions on Pattern Analysis and Machine Intelligence. **22**(8) (2000) 888–905

A Two-Stage Region-Based Image Retrieval Approach Using Combined Color and Texture Features[*]

Yinghua Lu[1], Qiushi Zhao[1,2], Jun Kong[1,2],
Changhua Tang[1], and Yanwen Li[1,**]

[1]Computer School, Northeast Normal University, Changchun, Jilin Province, China
[2]Key Laboratory for Applied Statistics of MOE, China
{luyh, zhaoqs522, kongjun, tangch564, liyw085}@nenu.edu.cn

Abstract. An effective Content-Based Image Retrieval (CBIR) approach is proposed in this paper. In contrast with existing systems, the retrieval process is divided into two stages. Images are firstly classified into categories based on their semi-global features automatically using the Fuzzy C-Means (FCM) clustering algorithm, and the K Nearest Neighbor (KNN) algorithm is used to assign the query image into a proper category to get a candidate image set. As a consequence, most irrelevant images are pruned. For the second stage, a novel segmentation algorithm is applied to segment both the query image and the candidate images into regions approximately according to objects. Color and texture features are extracted from each region for finer level retrieval. The region-based features utilize local properties of objects in image, and it is suitable for complicated scenes. Finally, distance measure is applied to evaluate the image-level similarity. This coarse-to-fine mechanism provides an effective and efficient performance for our system, which is demonstrated in the experiments on the image database from COREL.

1 Introduction

With the rapid advancement of image acquisition and processing technology, a huge amount of digital images are generated at every moment. Traditional text-based management is not adequate to manipulate this enormous number of digital data. It is inefficient to index, annotate and retrieve images based on keywords, because human language cannot describe image contents accurately. Hence, it is necessary to develop an effective image retrieval methodology to solve this problem.

We propose a region-based image retrieval method which split the whole query process into two stages, classification and finer query (as shown in Fig. 1). This mechanism provides a high performance of accuracy

---

[*] This work is supported by science foundation for young teachers of Northeast Normal University, No. 20061002, China.
[**] Corresponding author.

A. Sattar and B.H. Kang (Eds.): AI 2006, LNAI 4304, pp. 1010–1014, 2006.
© Springer-Verlag Berlin Heidelberg 2006

The rest of this paper is organized as follows: Section 2 describes semi-global feature extraction and clustering-based classification. Section 3 presents an image segmentation method and region-based feature extraction. Similarity measure between images is illustrated in Section 4. Section 5 provides experimental results of the proposed method. Finally, conclusions are drawn in Section 6.

2   Semi-global Features and Image Classification

CBIR is a time-consuming task. We set the images in the database into classes according to their features beforehand. When users propose a query by an example image, it is assigned into a proper class, and within that class finer query is to be carried out. We employ semi-global features for classification to achieve high performance of efficiency without losing spatial information.

2.1   Feature Extraction

For extracting semi-global features, we split the image into five connected blocks (sub-images)[9]. Color and texture features are extracted from each sub-image:

Color: The mean values of the three coordinates of the L*a*b color space.
Texture: Based on the edge histogram descriptor (EHD) for MPEG-7, we propose a modified EHD (MEHD) which can capture more texture information by employing the four direction Sobel edge detectors [9] instead of the five filters used in EHD.

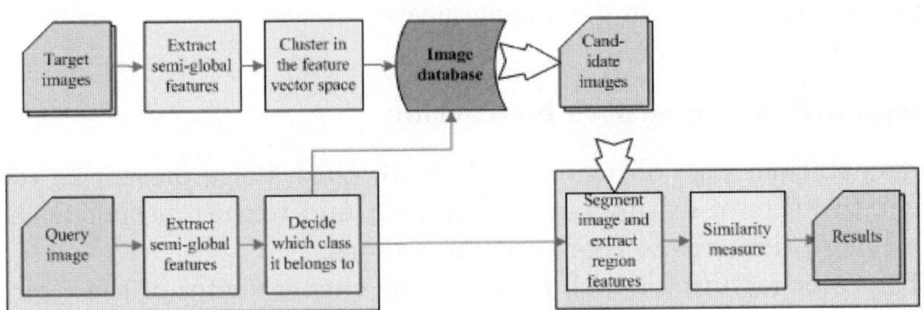

Fig. 1. Block diagram of proposed system

2.2   Image Classification

We apply the Fuzzy C-Means (FCM) and the the K Nearest Neighbor (KNN) algorithm in our method to address the image classification problem.

When users post a query sample, the system should decide which class the image belongs to, i.e. classify the test sample into the training samples it is closest to in the feature space.

3   Image Segmentation and Finer Retrieval

We propose a clustering-based segmentation method suitable for CBIR that uses both color and texture features, and it's fully automatic. This method divides an image into several homogenous regions coarsely according to objects. It is good enough for the retrieval task since an accurate segmentation is not necessary.

3.1   Image Segmentation Using FCM Clustering

The FCM algorithm is used to cluster the combined color and texture features extracted from an image into several groups, where each group in the feature space coarsely corresponds to one spatial region in the image space. Fig. 2 shows the segmentation results of two sample images from our test database.

Fig. 2. Results of image segmentation based on combined color and texture features. Left column: original images. Right column: segmentation results where each color represents a coherent region in the original image.

3.2   Region-Based Features Extraction

After segmentation, a set of regions will be obtained. For representing regions in the image, two sets of features including color and texture properties are used as under-lying primitives to represent the regions.

Color: The color feature for each region is the cluster center of the region.

Texture: The texture features for each region is a set of statistic properties of that region, namely, mean, standard deviation and uniformity (energy).

4   Similarity Measure on Image Level

In this paper, we propose a new distance measure suitable for region matching, called Region-based quadratic distance (R-quadratic distance), which is an extension of quadratic distance [10]. It achieves the effect of evaluating the image-level similarity by comparing regions of two images. The R-quadratic distance is defined as following:

$$d(r_1, r_2) = (r_1 - r_2)^T A(r_1 - r_2), \tag{1}$$

where r_1 and r_2 are the distributions of region properties of two images, respectively, in the form of feature vectors. $A = [a_{ij}]$ is a matrix and the weight a_{ij} denotes the similarity between regions i and j.

5   Experimental Results

We have tested our retrieval algorithm on a general-purpose image data-base with 1000 images from COREL[11].

Table 1 numerically lists the average retrieval precision of 7 image categories by applying the proposed approach and Qi's approach[6]. It shows that the proposed method obtains a higher average retrieval precision.

Table 1. Comparison of the average precision of 7 image categories by using our proposed approach and Qi's approach

Image category	Proposed	Qi's
Beach	0.4386	0.2800
Building	0.4259	0.5467
Dinosaur	0.9705	0.7567
Flower	0.5804	0.8233
Food	0.5944	0.5067
Horse	0.8055	0.8900
Vehicle	0.7335	0.5600
Average	**0.6498**	**0.6259**

Fig.3. compares the overall average retrieval precision and recall of the chosen 10 image categories from top 10 to 100 returned images. It shows that our proposed method obtains high overall retrieval effectiveness.

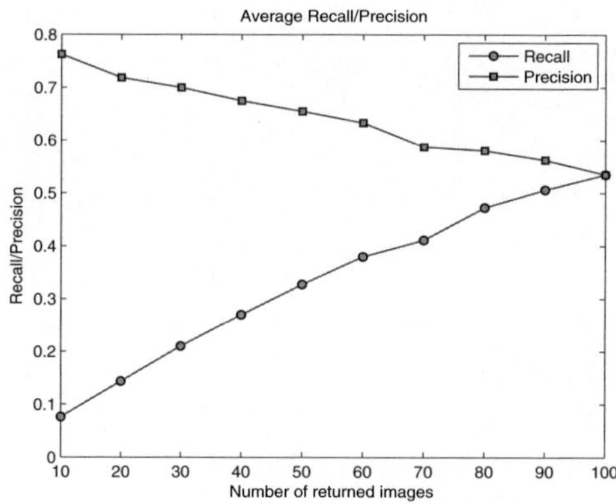

Fig. 3. Overall retrieval precision and recall of different retrieved image numbers

6   Conclusions

An effective content-based image retrieval system is proposed in this paper. In this approach, the retrieval process is divided into two stages. Images are firstly classified into categories based on their semi-global features automatically. For the second stage, the query image is assigned into a proper category before the finer query is carried out. Experimental results show that the mechanism above provides an effective and efficient performance for our system.

References

1. Swain M.J, Ballard D.H: Color indexing. International Journal of Computer Vision 7(1) :11-32 1991
2. Mehtre B. M, et al.: Color matching for image retrieval. Pattern Recognition Letters 16:325-331 1995
3. Yung-Kuan Chan, Chih-Ya Chen: Image retrieval system based on color-complexity and color-spatial features. The Journal of Systems and Software 71 65-70 2004
4. Guoping Qiu, Kin-man Lam: Frequency Layered Color Indexing for Content-Based Image Retrieval. IEEE transactions on image processing Vol.12 No.1 2003
5. Seong-O Shim, Tae-Sun Choi: Image Indexing By Modified Color Co-occurrence Matrix. ICASSP 2003
6. Xiaojun Qi, Yutao Han: A novel fusion approach to content-based image retrieval. Pattern Recognition 38 (2005) 2449-2465 2005
7. Jing, F., Mingjing, L., Zhang, H. J., Zhang, B: An effective region-based image retrieval framework. In Proc. ACM multimedia 2002
8. Carson, M. Thomas, S. Belongie, et al.: Blobworld: A system for region-based image indexing and retrieval. In Proc. Int. Conf. Visual Inf. Sys, 1999
9. Chin-Chuan Han, Hsu-Liang Cheng, Chih-Lung Lin et al.: Personal authentication using palm-print features. Pattern Recognition 36 (2003) 371-381
10. James Hafner, Harpreet S. Sawhney, Will Equitz et al.: Efficient Color Histogram Indexing for Quadratic Form Distance Functions. IEEE transactions on pattern analysis and machine intelligence, Vol. 17, No. 7, 1995
11. http://www.corel.com

Beyond Accuracy, F-Score and ROC: A Family of Discriminant Measures for Performance Evaluation[*]

Marina Sokolova[1], Nathalie Japkowicz[2], and Stan Szpakowicz[3]

[1] DIRO, Université de Montréal, Montreal, Canada
sokolovm@iro.umontreal.ca
[2] SITE, University of Ottawa, Ottawa, Canada
nat@site.uottawa.ca
[3] SITE, University of Ottawa, Ottawa, Canada
ICS, Polish Academy of Sciences, Warsaw, Poland
szpak@site.uottawa.ca

Abstract. Different evaluation measures assess different characteristics of machine learning algorithms. The empirical evaluation of algorithms and classifiers is a matter of on-going debate among researchers. Most measures in use today focus on a classifier's ability to identify classes correctly. We note other useful properties, such as failure avoidance or class discrimination, and we suggest measures to evaluate such properties. These measures – Youden's index, likelihood, Discriminant power – are used in medical diagnosis. We show that they are interrelated, and we apply them to a case study from the field of electronic negotiations. We also list other learning problems which may benefit from the application of these measures.

1   Introduction

Supervised Machine Learning (ML) has several ways of evaluating the performance of learning algorithms and the classifiers they produce. Measures of the quality of classification are built from a confusion matrix which records correctly and incorrectly recognized examples for each class. Table 1 presents a confusion matrix for binary classification, where tp are true positive, fp – false positive, fn – false negative, and tn – true negative counts.

Table 1. A confusion matrix for binary classification

Class \ Recognized	as Positive	as Negative
Positive	tp	fn
Negative	fp	tn

This paper argues that the measures commonly used now (accuracy, precision, recall, F-Score and ROC Analysis) do not fully meet the needs of learning problems in which

---

[*] We did this work while the first author was at the University of Ottawa. Partial support came from the Natural Sciences and Engineering Research Council of Canada.

A. Sattar and B.H. Kang (Eds.): AI 2006, LNAI 4304, pp. 1015–1021, 2006.
© Springer-Verlag Berlin Heidelberg 2006

the classes are *equally important* and where *several algorithms are compared*. Our findings agree with those of [1] who surveys the comparison of algorithms on multiple data sets. His survey, based on the papers published at the International Conferences on ML 2003–2004, notes that algorithms are mainly compared on accuracy.

2 Commonly-accepted Performance Evaluation Measures

The vast majority of ML research focus on the settings where the examples are assumed to be identically and independently distributed (IID). This is the case we focus on in this study. Classification performance without focussing on a class is the most general way of comparing algorithms. It does not favour any particular application. The introduction of a new learning problem inevitably concentrates on its domain but omits a detailed analysis. Thus, the most used empirical measure, *accuracy*, does not distinguish between the number of correct labels of different classes:

$$accuracy = \frac{tp + tn}{tp + fp + fn + tn} \tag{1}$$

Conversely, two measures that separately estimate a classifier's performance on different classes are sensitivity and specificity (often employed in biomedical and medical applications, and in studies which involve image and visual data):

$$sensitivity = \frac{tp}{tp + fn} ; specificity = \frac{tn}{fp + tn} \tag{2}$$

Focus on one class prevails in text classification, information extraction, natural language processing and bioinformatics, where the number of examples belonging to one class is often substantially lower than the overall number of examples. The experimental setting is as follows: within a set of classes there is a class of special interest (usually *positive*). Other classes are either left as is – multi-class classification – or combined into one – binary classification. The measures of choice calculated on the positive class[1] are:

$$precision = \frac{tp}{tp + fp} ; recall = \frac{tp}{tp + fn} = sensitivity \tag{3}$$

$$F - measure = \frac{(\beta^2 + 1) * precision * recall}{\beta^2 * precision + recall} \tag{4}$$

All three measures distinguish the correct classification of labels within different classes. They concentrate on one class (positive examples). Recall is a function of its correctly classified examples (true positives) and its misclassified examples (false negatives). Precision is a function of true positives and examples misclassified as positives (false positives). The F-score is evenly balanced when $\beta = 1$. It favours precision when $\beta > 1$, and recall otherwise.

A comprehensive evaluation of classifier performance can be obtained by the ROC:

$$ROC = \frac{P(x|positive)}{P(x|negative)} \tag{5}$$

---

[1] The same measures can be calculated for a negative class, but they are not reported.

$P(x|C)$ denotes the conditional probability that a data entry has the class label C. An ROC curve plots the classification results from the most positive to the most negative classification. Due to the wide use in cost/benefit decision analysis, ROC and the Area under the Curve (AUC) apply to learning with asymmetric cost functions and imbalanced data sets [2]. To get the full range of true positives and false negatives, we want easy access to data with different class balances. That is why ROC is used in experimental sciences, where it is feasible to generate much data. The study of radio signals, biomedical and medical science are a steady source of learning problems. Another possibility of building the ROC is to change the decision threshold of an algorithm. The AUC defined by one run is widely known as *balanced accuracy*:

$$AUC_b = (sensitivity + specificity)/2. \tag{6}$$

3   Critique of the Traditional ML Measures

We argue that performance measures other than accuracy, F-score, precision, recall or ROC do apply and can be beneficial. As a preamble, let us remind the reader that ML borrowed those measures from the assessment of medical trials [3] and from behavioural research [4], where they are intensively used. Our argument focusses on the fact that the last four measures are suitable for applications where one data class is of more interest than others, for example, search engines, information extraction, medical diagnoses. They may be not suitable if all classes are of interest and yet must be distinguished. For example, consider negotiations (success and failure of a negotiation are equally important) or opinion identification (markets need to know what exactly triggers positive and negative opinions).

In such applications, complications arise when a researcher must choose between two or more algorithms [5,6]. It is easy to ask but rather difficult to answer what algorithm we should choose if one performs better on one class and the other – on the other class. Here we present the empirical evidence of a case where such a choice is necessary. We studied the data gathered during *person-to-person* electronic negotiations (e-negotiations); for an overview of machine learning results refer to [6]. E-negotiations occur in various domains (for example, labour or business) and involve various users (for example, negotiators or facilitators).

The *Inspire* data [7] is the largest collection gathered through e-negotiations (held between people who learn to negotiate and may exchange short free-form messages). Negotiation between a buyer and a seller is successful if the virtual purchase has occurred within the designated time, and is unsuccessful otherwise. The system registers the outcome. The overall task was to find methods better suited to automatic learning of the negotiation outcomes – success and failure. Both classes were equally important for training and research in negotiations: the results on the positive class can reinforce positive traits in new negotiations; the results on the negative class can improve (or prevent) potentially weak negotiations. The amount of data was limited to 2557 entries, each of them a record of one bilateral e-negotiation. Successful negotiations were labelled as positive, unsuccessful – as negative. The data are almost balanced, 55% positive and 45% negative examples. The ML experiments ran Weka's Support Vector Machine (SVM) and Naive Bayes (NB) [8] with tenfold cross-validation; see Table 2.

Table 2. Traditional classification results

Measure	SVM	NB
Accuracy	77.4	76.8
F-score	81.2	78.9
Sensitivity	86.8	77.5
Specificity	65.4	75.9
AUC	76.1	76.7

Table 3. Classifier capabilities shown by traditional measures

Classifier	Overall effectiveness	Predictive power	Class effectiveness
SVM	superior	superior on positive examples	superior on positive examples
NB	inferior	superior on negative examples	superior on negative examples

4   Search for Measures

In this study, we concentrate on the choice of comparison measures. In particular, we suggest evaluating the performance of classifiers using measures other than accuracy, F-score and ROC. As suggested by Bayes's theory [9,10], the measures listed in the survey section have the following effect:

accuracy approximates how effective the algorithm is by showing the probability of the true value of the class label (assesses the overall effectiveness of the algorithm);
precision estimates the predictive value of a label, either positive or negative, depending on the class for which it is calculated (assesses the predictive power of the algorithm);
sensitivity/specificity approximates the probability of the positive/negative label being true (assesses the effectiveness of the algorithm on a single class);
ROC shows a relation between the sensitivity and the specificity of the algorithm;
F-score is a composite measure which favours algorithms with higher sensitivity and challenges those with higher specificity. See Table 3 for a summary.

Based on these considerations, we can conclude that SVM is preferable to NB. But will it always be the case? We will now show that the superiority of an algorithm (such as SVM) with respect to another algorithm largely depends on the applied evaluation measures. Our main requirement for possible measures is to bring in new characteristics for the algorithm's performance. We also want the measures to be easily comparable. We are interested in two characteristics of an algorithm:

- the confirmation capability with respect to classes, that is, the estimation of the probability of the correct predictions of positive and negative labels;
- the ability to avoid failure, namely, the estimation of the complement of the probability of failure.

Three measures that caught our attention have been used in medical diagnosis to analyze tests [3]. The measures are *Youden's index* [11], *likelihood* [12], and *Discriminant power* [13]. They combine sensitivity and specificity and their complements.

Youden's index. The avoidance of failure complements accuracy, or the ability to correctly label examples. Youden's index γ [11] evaluates the algorithm's ability to avoid failure – equally weights its performance on positive and negative examples:

$$\gamma = sensitivity - (1 - specificity) \tag{7}$$

Youden's index has been traditionally used to compare diagnostic abilities of two tests [12]. It summarizes sensitivity and specificity and has linear correspondence balanced accuracy (a higher value of γ means better ability to avoid failure):

$$\gamma = 2AUC_b - 1. \tag{8}$$

Likelihoods. If a measure accommodates both sensitivity and specificity, but treats them separately, then we can evaluate the classifier's performance to finer degree with respect to both classes. The following measure combining positive and negative likelihoods allows us to do just that:

$$\rho_+ = \frac{sensitivity}{1 - specificity}; \rho_- = \frac{1 - sensitivity}{specificity} \tag{9}$$

A higher positive and a lower negative likelihood mean better performance on positive and negative classes respectively. The relation between the likelihood of two algorithms A and B establishes which algorithm is preferable and in which situation [12]. Figure 1 lists the relations for algorithms with $\rho_+ \geq 1$. If an algorithm does not satisfy this condition, then "positive" and "negative" likelihood values should be swapped.

- $\rho_+^A > \rho_+^B$ and $\rho_-^A < \rho_-^B$ implies A is superior overall;
- $\rho_+^A < \rho_+^B$ and $\rho_-^A < \rho_-^B$ implies A is superior for confirmation of negative examples;
- $\rho_+^A > \rho_+^B$ and $\rho_-^A > \rho_-^B$ implies A is superior for confirmation of positive examples;
- $\rho_+^A < \rho_+^B$ and $\rho_-^A > \rho_-^B$ implies A is inferior overall;

Fig. 1. Likelihoods and algorithm performance

Relations depicted in Figure 1 show that the likelihoods are an easy-to-understand measure that gives a comprehensive evaluation of the algorithm's performance.

Discriminant power. Another measure that summarizes sensitivity and specificity is discriminant power (DP) [13]:

$$DP = \frac{\sqrt{3}}{\pi}(\log X + \log Y), \tag{10}$$

$$X = sensitivity/(1 - sensitivity), Y = specificity/(1 - specificity). \tag{11}$$

DP does exactly what its name implies: it evaluates how well an algorithm distinguishes between positive and negative examples. To the best of our knowledge, until now DP has been mostly used in ML for feature selection. The algorithm is a poor discriminant if $DP < 1$, limited if $DP < 2$, fair if $DP < 3$, good – in other cases.

Experiments. We applied Youden's index, positive and negative likelihood and DP to the results of learning from the data of e-negotiations. We calculate the proposed measures to evaluate the algorithms' performance – see Table 4.

Youden's index and likelihood values favour NB's performance. NB's γ is always higher than SVM's. This differs from the accuracy values (higher for SVM in two

Table 4. New classification results

Measure	SVM	NB
γ	0.522	0.534
ρ_+	2.51	3.22
ρ_-	0.20	0.30
DP	1.39	1.31

Table 5. Comparison of the classifiers' abilities by new measures

Classifier	Avoidance of failure	Confirmation of classes	Discrimination of classes
SVM	inferior	superior for negatives	limited
NB	superior	superior for positives	limited

experimental settings) and the F-score (higher in all the three settings). The higher values of γ indicate that NB is better at avoiding failure. Further observation shows that the positive and negative likelihood favour classifiers with more balanced performance over classifiers with high achievement on one class and poor results on the other. When a classifier performs poorly, the likelihood values support the class labels of the under-performing classifier. DP' values are rather low[2]. Its results are similar to those of accuracy, of F-score which prefer SVM to NB. We attribute this correspondence to a) positive correlation of all the three measures with sensitivity; b) close values of specificity and sensitivity in our case study. As expected, Youden's index values correspond to those of the AUC. Youden's index is the most complex measure and its results (NB consistently superior to SVM) do not correlate with the standard measures. This confirms that the ability to avoid failure differs from the ability of successful identification of the classification labels. Based on these results and relations given by the scheme from Figure 1, we summarize SVM's and NB's abilities in Table 5.

NB is marginally superior to SVM: we confirm our hypothesis that the superiority of an algorithm is related to how evaluation is measured.

The results reported in Tables 2, 3, 4 and 5 show that higher accuracy does not guarantee overall better performance of an algorithm. The same conclusion applies to every performance measure if it is considered separately from others. On the other hand, a combination of measures gives a balanced evaluation of the algorithm's performance.

5   Conclusion

We have proposed a new approach to the evaluation of learning algorithms. It is based on measuring the algorithm's ability to distinguish classes and thus to avoid failure in classification. We have argued this has not yet been done in ML. The measures which originate in medical diagnosis are Youden's index γ, the likelihood values ρ_-, ρ_+ , and Discriminant Power DP. Our case study of the classification of electronic negotiations has shown that there exist ML applications which benefit from the use of these measures. We also gave a general description of the learning problems which may employ γ, ρ_-, ρ_+ and DP. These problems are characterized by a restricted access to data, the need to compare several classifiers, and equally-weighted classes.

Such learning problems arise when researchers work with data gathered during social activities of certain group. We have presented some results at conferences with a focus

---

[2] Discriminant power is strong when it is close to 3.

more general than ML. We note that the particularly apt problems include knowledge-based non-topic text classification (mood classification [14], classification of the outcomes of person-to-person e-negotiations [6], opinion and sentiment analysis [15], and so on) and classification of email conversations [16]. All these studies involve data sets gathered with specific purposes in a well-defined environment: researchers discussing the time and venue of a meeting [16], bloggers labelling with their moods blogs posted on a Web site [14], participants of e-negotiations held by a negotiation support system [6]. All cited papers employ commonly used ML measures. For example, [6] and [15] report accuracy, precision, recall and F-score; other papers, especially on sentiment analysis, report only accuracy, for example [5].

Our future work will follow several interconnected avenues: find new characteristics of the algorithms which must be evaluated, consider new measures of algorithm performance, and search for ML applications which require measures other than standard accuracy, F-score and ROC.

References

1. Demsar, J.: Statistical comparisons of classifiers over multiple data sets. Journal of Machine Learning Research **7** (2006) 1–30
2. Chawla, N., Japkowicz, N., Kolcz, A., eds.: Special Issue on Learning from Imbalanced Data Sets. Volume 6(1). ACM SIGKDD Explorations (2004)
3. Isselbacher, K., Braunwald, E.: Harrison's Principles of Internal Medicine. McGraw-Hill (1994)
4. Cohen, J.: Statistical Power Analysis for the Behavioral Sciences. Lawrence Erlbaum Associates, Hillsdale, NJ (1988)
5. Pang, B., Lee, L., Vaithyanathan, S.: Thumbs up? sentiment classification using machine learning techniques. In: Proc Empirical Methods of Natural Language Processing EMNLP'02. (2002) 79–86
6. Sokolova, M., Nastase, V., Shah, M., Szpakowicz, S.: Feature selection for electronic negotiation texts. In: Proc Recent Advances in Natural Language Processing RANLP'05. (2005) 518–524
7. Kersten, G., et al.: Electronic negotiations, media and transactions for socio-economic interactions (2006) http://interneg.org/enegotiation/ (2002-2006).
8. Witten, I., Frank, E.: Data Mining. Morgan Kaufmann (2005)
9. Cherkassky, V., Muller, F.: Learning from Data. Wiley (1998)
10. Duda, R., Hart, P., Stork, D.: Pattern Classification. Wiley (2000)
11. Youden, W.: Index for rating diagnostic tests. Cancer **3** (1950) 32–35
12. Biggerstaff, B.: Comparing diagnostic tests: a simple graphic using likelihood ratios. Statistics in Medicine **19**(5) (2000) 649–663
13. Blakeley, D., Oddone, E.: Noninvasive carotid artery testing. Ann Intern Med **122** (1995) 360–367
14. Mishne, G.: Experiments with mood classification in blog posts. In: Proc 1st Workshop on Stylistic Analysis of Text for Information Access (Style2005). (2005) staff.science.uva.nl/gilad/pubs/style2005-blogmoods.pdf.
15. Hu, M., Liu, B.: Mining and summarizing customer reviews. In: Proc 10th ACM SIGKDD International Conf on Knowledge Discovery and Data Mining KDD'04. (2004) 168–177
16. Boparai, J., Kay, J.: Supporting user task based conversations via email. In: Proc 7th Australasian Document Computing Symposium. (2002)

Hybrid $O(n\sqrt{n})$ Clustering for Sequential Web Usage Mining

Jianhua Yang[1] and Ickjai Lee[2]

[1] School of Computing and Maths,
University of Western Sydney, Campbelltown, NSW2560, Australia
[2] School of Information Technology,
James Cook University, Townsville, QLD4811, Australia

Abstract. We propose a natural neighbor inspired $O(n\sqrt{n})$ hybrid clustering algorithm that combines medoid-based partitioning and agglomerative hierarchial clustering. This algorithm works efficiently by inheriting partitioning clustering strategy and operates effectively by following hierarchial clustering. More importantly, the algorithm is designed by taking into account the specific features of sequential data modeled in metric space. Experimental results demonstrate the virtue of our approach.

1   Introduction

This paper presents a medoid-based hybrid agglomerative clustering approach that combines partitioning clustering and hierarchical clustering. The proposed method overcomes the ineffectiveness of the former and the inefficiency of the latter while inheriting the efficiency of the former and the effectiveness of the latter. It initially partitions a set of data sequences into a number of temporary micro-clusters and then merges them into final clusters based on natural neighbour topological information derived from the Voronoi diagram. Our proposed method is able to identify quality clusters in $O(n\sqrt{n})$ time. Experimental results confirm the effectiveness and efficiency of our approach.

2   Clustering of Sequential Data

Partitioning clustering and hierarchical clustering are two distinctive types of methods used to group sequential data in arbitrary metric spaces. Clustering algorithms should be fairly designed to address the characteristics of sequences and specific requirements in the context of sequential data clustering. For instance, the discrete nature and similarity measures of transaction sequences remove vector-based operations, like averages (means). Inability of treating numerically attribute vectors rules out centroid-based partitioning clustering algorithms like k-MEANS. Moreover, the continuous center of mass of a cluster may not represent a prototypical object in that cluster. With sequential data, the vectorial mean is less meaningful to be a statistical indicator as used in STING.

Sequential data favor medoid-based partitioning approach. Medoids remove the concern of discrete nature of sequences and make clustering better suited

A. Sattar and B.H. Kang (Eds.): AI 2006, LNAI 4304, pp. 1022–1026, 2006.
© Springer-Verlag Berlin Heidelberg 2006

for metric space. While medoids are more robust estimators of location than means [1], they are computationally less tractable. The best interchange hill-climber is the Teitz and Bart heuristic [2] that requires quadratic time. One other example is CLARANS that is based on a sampling interchange hill-climbing. It requires $\Omega(n^2)$ time complexity and does not guarantee local optimality. Recently, a variant of medoid-based clustering for optimizing clustering criteria in absolute error has been proposed and applied to Web usage mining [3]. Different from the sampling approach, this randomized algorithm uses all the data available. The randomized interchange approach achieves $O(n\sqrt{n})$ time.

Hierarchical clustering is inefficient for large data sets since it typically requires quadratic time in the size of the input [4]. An algorithm from this family for sequential clustering tends to require CPU-time three orders of magnitude more than any other [5]. Matrix-based clustering approach [5,6] maintains a similarity matrix of each pair of sequences, and faces the problem of infeasible computational resources. The computation of the similarity matrix makes this method far from scalable. Recently, a scalable agglomerative hierarchical clustering method for visitation paths has been proposed [7]. This pattern-oriented algorithm introduces strong assumptions that restrict the similarity measure between sequences and groups of sequences. Its time complexity suffers from the cube of the number m of frequent patterns and the clustering quality degrades as m becomes smaller.

3   Hybrid Agglomerative Clustering

3.1   Observation on Partitioning Clustering Result

Let $V = \{v_1, \cdots, v_t\}$ be a set of literals, called *events*. An *eventset* $S = \{s_1, s_2, \cdots, s_r\}$ is a multiset of events, where $s_j \in V$ is an event. A *sequence* $u =< S_1 S_2 \cdots S_m >$ is an ordered list of eventsets, where S_i is an eventset. Clustering of sequential data is to partition a set $U = \{u_1, \cdots, u_n\}$ of sequences into groups by some natural similarity measure. While several similarity measures between sequences are widely adopted in the literature [8], partitioning clustering using those similarity measures suffers from a difficulty to group 'orphan' sequences.

Let us consider an artificially generated dataset with two clusters shown in Fig. 1(a). This dataset consists of 15 sequences: $u1 :< \{1\}\ \{2\}\ \{3\} >$, $u2 :< \{3\}\ \{4\}\ \{5\} >$, $u3 :< \{6\}\ \{7\}\ \{8\} >$, $u4 :< \{8\}\ \{9\}\ \{10\} >$, $u5 :< \{11\}\ \{12\}\ \{13\} >$, $u6 :< \{13\}\ \{14\}\ \{15\} >$, $u7 :< \{1\}\ \{10\}\ \{11\} >$, $u8 :< \{2\}\ \{9\}\ \{12\} >$, $u9 :< \{3\}\ \{8\}\ \{13\} >$, $u10 :< \{4\}\ \{7\}\ \{14\} >$, $u11 :< \{5\}\ \{6\}\ \{15\} >$, $u12 :< \{19\}\ \{24\}\ \{23\} >$, $u13 :< \{18\}\ \{24\}\ \{22\} >$, $u14 :< \{17\}\ \{24\}\ \{21\} >$, and $u15 :< \{16\}\ \{24\}\ \{20\} >$. Fig. 1(b) shows a clustering result from a medoid-based partitioning clustering approach [3]. Note that the clustering does not reveal the true pattern in the dataset by missing out sequences $u7$, $u8$, $u10$ and $u11$. This is because the similarity between two sequences will be 0 if they have no common events at all. In partitioning clustering, this indicates a data sequence without a direct link to any cluster representative will be classified as an 'orphan' item. Treating such orphan sequences as 'noise'

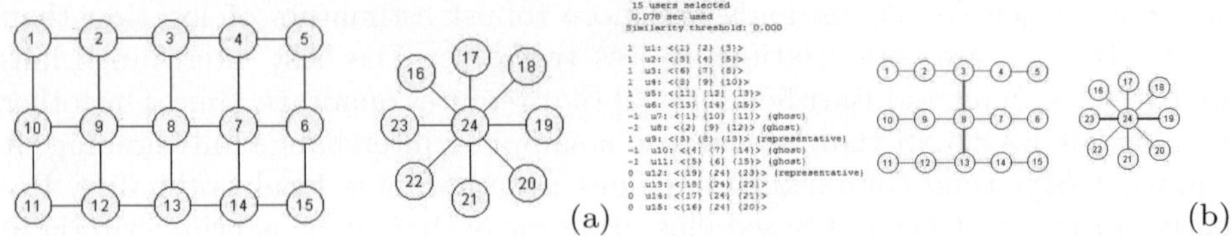

Fig. 1. Traditional partitioning clustering for sequential data: (a) A dataset with 15 sequences; (b) A partitioning clustering result ($k = 2$)

is one of the main causes to the ineffectiveness of partitioning clustering in sequence data mining. It should be noted that these orphan sequences do possess links to other non-representative sequences even though they do not have direct links to cluster representatives. They are not outliers. Yet the drawback of partitioning clustering can be mitigated when a large number k of partitions is used. This motivates us to start partitioning clustering with a larger number $k' > k$ of initial micro-clusters, and then merge neighbored similar micro-clusters together to form final k clusters.

3.2   Interactivity for Hierarchical Clustering

The merge process is a major computational bottleneck in agglomerative hierarchical clustering. The merge phase in our approach utilizes the "natural neighbor" concept from the Voronoi diagram to reduce the space and time requirement. The Voronoi diagram defines topologically meaningful near-by natural neighbours and contain many robust mathematical properties. These natural neighbours exhibit geometrically and topologically strong associations, they are indicative of the next desirable merge in hierarchical clustering. The merge in our approach is based on the total interactivity between natural neighbours as defined below.

Definition 1. *Total interactivity between clusters C_i and C_j, denoted by $S_{tiact}(C_i, C_j)$, is measured by*

$$Sim_{tiact}(C_i, C_j) = \frac{n_{ij}}{|C_i|} \cdot \frac{n_{ji}}{|C_j|},\qquad(1)$$

where n_{ij} is the number of sequences in C_i satisfying $Sim(u, C_j) - Sim(u, C_k) \geq \epsilon$ for $\forall C_k \in C \backslash C_i$, and n_{ji} is the number of sequences in C_j satisfying $Sim(u, C_i) - Sim(u, C_k) \geq \epsilon$ for $\forall C_k \in C \backslash C_j$, ϵ is a similarity threshold introduced.

4   Experimental Results of Web Usage Mining

In our experiments, we adopt the conventional assumption for converting entries in an access log to visitation paths [5,6,9]. To make the experimental results more comparable, we recreate benchmark test datasets based on the generation method suggested by [6] and also used by [3]. We generate k random walks of length $\rho = 8$. These nucleus paths are the representatives. The generation is also controlled by another parameter *branching factor br* $\in (0,1)$. This encodes the probability that a user represented by a given nucleus path will choose a different link than the prototypical nucleus path. The datasets generated exhibit various error rates, different levels of noise, and diverse degree of compactness.

4.1   On the Scalability of the Methods

The randomized medoid-based implementation [3] requires $O(n\sqrt{n})$ time. The construction of matrix for merging needs $O(\hat{k}k')$ time where $\hat{k}$ is the average number of neighboring clusters. Note that, $\hat{k} \ll k' \ll n$ in data-rich environments. The cluster merging process takes the same time as matrix construction, thus **HCS** requires $O(n\sqrt{n})$ time in total. Our experiments show HCS has only a tiny overhead on the top of the randomized medoid-based clustering. This confirms the merge phase using the natural neighbor concept is fast and scalable.

4.2   On the Quality of Clustering

From the cluster validity point of view, the error rate in a cluster structure detected can be an indication of the quality of a clustering method. Two quality indices, Jaccard Coefficient (JC) and Fowlkes-Mallows (FM) indices are widely used for external quality evaluation where the ground fact is known [10]. The two quality indices range from 0 to 1; methods with higher value of those indices are considered higher quality methods.

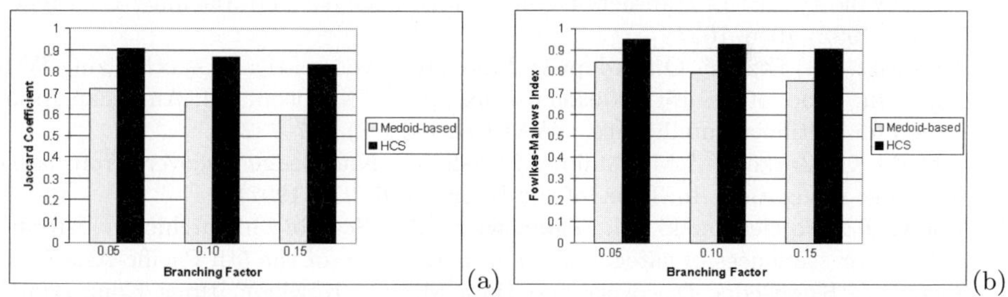

Fig. 2. Jaccard Coefficient and Fowlkes-Mallows Index: (a) Jaccard Coefficient; (b) Fowlkes-Mallows Index

Fig. 2 shows Jaccard Coefficient and Fowlkes-Mallows indices of HCS and the crisp version of randomized medoid-based algorithm with respect to various levels of branching factors and different sizes of data sequences generated around nucleus paths. Again we ran each of them 5 times and report confidence intervals with 95% accuracy. For both clustering algorithms, quality was much more affected by the level of branching factor than by the size of the dataset. This figure confirms that our hybrid approach not only outperforms the randomized medoid-based approach, but slowly decays as br grows. The partitioning-only method drops 17% of accuracy in Jaccard Coefficient as br grows from 0.05 to 0.15, while our approach declines 8%. Similar effects are observed with Fowlkes-Mallows Index. The partitioning-only approach drops 10% while our hybrid method declines only 4%.

5   Final Remarks

We have presented a hybrid clustering methods, **HCS**, which combines a medoid-based partitioning process with a hierarchical merging process. The proposed hybrid approach employs the topological neighborhood information among the pre-partitioned clusters for efficient merging process. Such topological information is naturally modeled by an underlying hypergraph corresponding to the Voronoi tessellation. Its $O(n\sqrt{n})$ time makes it more scalable to spotting discrete sequential patterns in high-dimensional pseudo-metric spaces.

References

1. Rousseeuw, P.J., Leroy, A.M.: Robust regression and outlier detection. John Wiley, New York (1987)
2. Teitz, M.B., Bart, P.: Heuristic methods for estimating the generalized vertex median of a weighted graph. Operations Research **16** (1968) 955–961
3. Estivill-Castro, V., Yang, J.: Clustering web visitors by fast, robust and convergent algorithms. Int. J. of Fundations of Computer Science **13**(4) (2002) 497–520
4. Murtagh, F.: Comments on parallel algorithms for hierarchical clustering and cluster validity. IEEE Transactions on Pattern Analysis and Machine Intelligence **14**(10) (1992) 1056–1057
5. Perkowitz, M., Etzioni, O.: Adaptive Web sites: Automatically synthesizing Web pages. In: Proc. of the 15th National Conf. on AI, Madison, WI, American Association for Artificial Intelligence, AAAI Press (1998) 727–732
6. Shahabi, C., Zarkesh, A.M., Adibi, J., Shah, V.: Knowledge discovery from users Web page navigation. In: Proc. of the IEEE RIDE'97. (1997)
7. Morzy, T., Wojciechowski, M., Zakrzewicz, M.: Scalable hierarchical clustering method for sequences of categorical values. In: Proc. of the 5th Pacific-Asia Conference on Knowledge Discovery and Data Mining, Kowloon, Hong Kong (2001) 282–293
8. Guralnik, V., Karypis, G.: A scalable algorithm for clustering sequential data. In: Proc. of the 1st IEEE Int. Conf. on Data Mining, San Jose, California, USA (2001) 179–186
9. Kato, H., Nakayama, T., Yamane, Y.: Navigation analysis tool based on the correlation between contents distribution and access patterns. In: Proc. of the Web Mining for E-Commerce Workshop, Boston, MA, USA (2000)
10. Theodoridis, S., Koutroumbas, K.: Pattern Recognition. Academic Press, San Diego (1998)

Clustering Algorithms for ITS Sequence Data with Alignment Metrics

Andrei Kelarev[1], Byeong Kang[1], and Dorothy Steane[2]

[1] School of Computing, University of Tasmania
Private Bag 100, Hobart, Tasmania 7001, Australia
{andrei.kelarev, bhkang}@utas.edu.au
http://www.comp.utas.edu.au/users/kelarev/
http://www.comp.utas.edu.au/users/bhkang/
[2] CRC Forestry and School of Plant Science
Private Bag 55, University of Tasmania, Hobart, Tasmania 7001, Australia
dorothy.steane@utas.edu.au
http://www.crcforestry.com.au
http://fcms.its.utas.edu.au/scieng/plantsci/

Abstract. The article describes two new clustering algorithms for DNA nucleotide sequences, summarizes the results of experimental analysis of performance of these algorithms for an ITS-sequence data set, and compares the results with known biologically significant clusters of this data set. It is shown that both algorithms are efficient and can be used in practice.

1 Introduction

The investigation of DNA data sets has broad applications in medical and health informatics and many branches of biology. Data mining and machine learning have been crucial in the development of these research areas (let us refer, for example, to Gedeon and Fung [3], Kang, Hoffman, Yamaguchi and Yeap [6], Li, Yang and Tan [10], Webb and Yu [13], Zhang, Guesgen and Yeap [17], Zhang and Jarvis [18]).

The present paper describes and investigates two clustering algorithms that can be used to group a data set of DNA sequences into clusters. In order to achieve significant correlation between clusterings produced by these machine learning algorithms and biological classifications, we have relied on measures of strong similarity between sequences. Such measures have not been considered for these algorithms before. Here we present the results of an experimental analysis of the performance of our new algorithms using a data set derived from the internal transcribed spacer (ITS) regions of the nuclear ribosomal DNA in *Eucalyptus*, and compare the results with clusterings published by Steane *et al.* [12].

For preliminaries on DNA molecules we refer to the monographs Baldi and Brunak [1], Durbin, Eddy, Krogh and Mitchison [2], Jones and Pevzner [5] and Mount [11]. Background information on clustering algorithms can be found in Witten and Frank [15].

A. Sattar and B.H. Kang (Eds.): AI 2006, LNAI 4304, pp. 1027 – 1031, 2006.

© Springer-Verlag Berlin Heidelberg 2006

2 Clusters of the k-Means Algorithm with Alignment Metrics

Both our algorithms use highly biologically significant alignment scores as a metric and obtain significant results. The novel character of our method is in using a sophisticated and highly informative distance metric based on alignment scores well known in bioinformatics. Every alignment produces an alignment score that measures the similarity of the nucleotide or amino acid sequences.

The alignment scores in our algorithms provide an accurate measure of similarity that is more significant biologically. Alignment scores have properties that differ from those of the Euclidean metrics and their simple modifications discussed, for example, by Witten and Frank [15] (Section 6.4). Hence our algorithms have to be designed differently. First, it is impossible to do simple calculations for the alignment score metric involved in various Euclidean clustering algorithms. For example, it is impossible to find a DNA sequence that is the "midpoint" or "mean" of two DNA sequences. Secondly, it is important to minimize the number of distance calculations, because each of them is time consuming.

Our first algorithm is a new version of the k-means clustering algorithm. The traditional *k-means* algorithm implemented in WEKA environment uses standard Euclidean distances (Witten and Frank [15], Section 4.8). Our algorithm uses the metric of alignment scores to establish similarity between sequences. The unusual character of this metric prohibits direct computation of the mean of a set of sequences in a cluster.

In order to make the algorithm faster, it is desirable to minimize the number of times the alignment scores have to be found. This is why the algorithm operates on the set of given sequences only and does not create any new sequences as means of the given ones. Every alignment score between each pair of the given sequences is found once during a pre-processing stage of the algorithms, and then these scores are looked up in a table during the process of looking for clusters.

The initialization stage proceeds as usual: k sequences are randomly chosen as centroids of clusters, and every other sequence is assigned to the cluster of its nearest centroid. Every iteration of the algorithm looks at each current cluster in turn. It then analyses all sequences of the cluster trying to determine which of them would be the best centroid for this particular cluster. Suppose that the algorithm is considering a cluster C. As a new centroid for this cluster our algorithm is going to choose the sequence s in C with the property that the sum of all distances from s to all other sequences in C is minimal. This is in fact precisely the property of the standard Euclidean means that is essential for the operation of the traditional k-means algorithm.

The average success rates of this method in comparison with five clusters obtained and published by Steane *et al.* [12] are represented in Figure 1. The diagram demonstrates how the success rates depend on the choice of the local alignment metric.

3 Nearest Neighbour Clustering Algorithm with Alignment Metrics

The second algorithm we implemented is an analog of the *nearest neighbour* clustering algorithm (Witten and Frank [15], Section 4.7). The standard nearest neighbour clustering algorithm implemented in WEKA could not be applied directly to the ITS dataset, because it handles data represented as points in an n-dimensional Euclidean space. Thus we had to encode a new version of the nearest neighbour algorithm based on optimal local alignments of the given sequences.

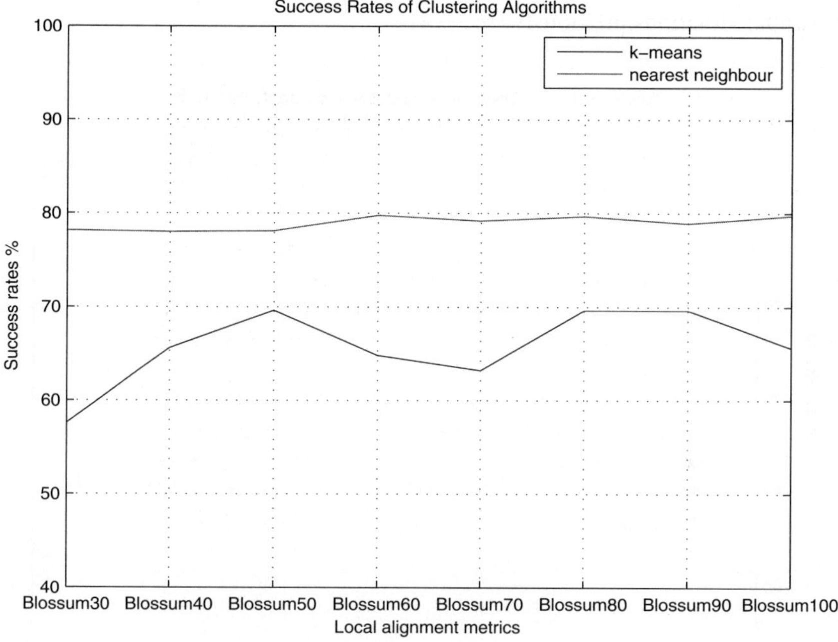

Fig. 1. Experimental data on success rates of clustering algorithms for several distance metrics

Given the number k of clusters, the algorithm computes all local alignment scores between all pairs of DNA sequences in the data set. It chooses random k sequences as representatives of the k clusters. For every other sequence s in the data set, it looks at all sequences which have been considered and allocated to the clusters and finds the nearest neighbour of s among these sequences. The algorithm then allocates s to the cluster of its nearest neighbour. This is repeated until all sequences have been assigned to clusters.

The average success rates of this method using various alignment metrics compared with the clusters obtained and published by Steane et al. [12] are represented in Figure 1.

4   Experimental Results

We investigated the groupings of an ITS dataset, displayed in Figures 2, 3 and 5 of Steane *et al.* [12]: The dataset includes many of different species from all subgenera and sections of *Eucalyptus*, as well as some other genera that are closely related to *Eucalyptus*. For a detailed description of the dataset we refer to [12].

Figure 1 summarizes the results of comparison between classifications obtained by our two algorithms and known classifications considered in the biological literature (see Steane *et al.* [12]). Our clustering algorithms use the strong similarity measures and have achieved high accuracy rates compared to other algorithms considered in other similar situations previously (see [14]).

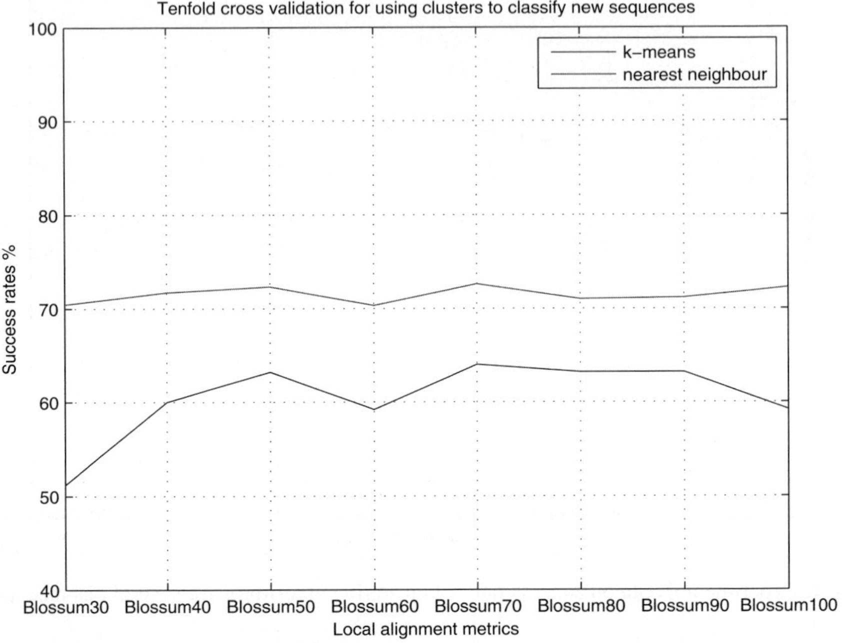

Fig. 2. Tenfold cross validation for the accuracy of allocating new sequences

We investigated the efficiency of the clusters produced by our algorithms for classifying new sequences (Figure 2), and established that both algorithms are accurate. The choice of algorithm is a trade-off between speed and accuracy. The nearest neighbor clustering algorithm is more accurate but is slower.

This research has been supported by the IRGS grant K14313 of the University of Tasmania.

References

1. Baldi, P. and Brunak, S.: Bioinformatics : The Machine Learning Approach. Cambridge, Mass, MIT Press, (2001).
2. Durbin, R., Eddy, S.R., Krogh, A. and Mitchison, G.: Biological Sequence Analysis. Cambridge University Press (1999).
3. Gedeon, T.D. and Fung, L.C.C.: AI 2003: Advances in Artificial Intelligence. Proc. 16th Australian Conference on AI, Perth, Australia, December 3-5, 2003. Lecture Notes in Artificial Intelligence 2903 (2003).

4. Gusfield, D.: Algorithms on Strings, Trees, and Sequences. Computer Science and Computational Biology, Cambridge University Press, Cambridge (1997).

5. Jones, N.C. and Pevzner, P.A.: An Introduction to Bioinformatics Algorithms. Cambridge, Mass, MIT Press, (2004). http://www.bioalgorithms.info/

6. Kang, B.H.: Pacific Knowledge Acquisition Workshop. Part of the 8th Pacific Rim Internat. Conf. on Artificial Intelligence, Auckland, New Zealand, (2004).

7. Kang, B.H., Kelarev, A.V., Sale, A.H.J. and Williams, R.N.: A model for classifying DNA code embracing neural networks and FSA. Pacific Knowledge Acquisition Workshop, PKAW2006, Guilin, China, 7-8 August 2006, 201-212 (2006).

8. Kelarev, A.V.: Ring Constructions and Applications. World Scientific, River Edge, NJ, (2002).

9. Kang, B.H., Kelarev, A.V., Sale, A.H.J. and Williams, R.N.: Labeled directed graphs and FSA as classifiers of strings. 17th Australasian Workshop on Combinatorial Algorithms, AWOCA 2006, 93-109 (2006).

10. Li, J., Yang, Q. and Tan, A.-H.: Data Mining for Biomedical Applications. Proc. PAKDD 2006, BioDM 2006, Singapore, April 9, 2006. Lecture Notes in in Artificial Intelligence 3916 (2006).

11. Mount, D.: Bioinformatics: Sequence and Genome Analysis. Cold Spring Harbor Laboratory, (2001). http://www.bioinformaticsonline.org/

12. Steane, D.A., Nicolle, D., Mckinnon, G.E., Vaillancourt, R.E. and Potts, B.M.: High-level relationships among the eucalypts are resolved by ITS-sequence data. Australian Systematic Botany 15, 49-62 (2002).

13. Webb, G.I. and Yu, X.: Advances in Artificial Intelligence: Proc. 17th Australian Joint Conference on Artificial Intelligence AI 2004, Cairns, Australia, December 4-6, 2004. Lecture Notes in Artificial Intelligence 3339 (2004).

14. WEKA, Waikato Environment for Knowledge Analysis, http://www.cs.waikato.ac.nz/ml/weka, viewed 20.06.2006.

15. Witten, I.H. and Frank, E.: Data Mining: Practical Machine Learning Tools and Techniques. Elsevier/Morgan Kaufman, Amsterdam (2005).

16. Yang, J. Y. and Ersoy, O.K.: Combined supervised and unsupervised learning in genomic data mining, School of Electrical and Computer Engineering, Purdue University, (2003).

17. Zhang, C., Guesgen, H.W. and Yeap, W.K.: Trends in Artificial Intelligence. 8th Pacific Rim Internat. Conf. on Artificial Intelligence PRICAI 2004, Auckland, New Zealand, August 9-13, 2004. Lecture Notes in Artificial Intelligence 3157 (2004).

18. Zhang, S. and Jarvis, R.: Advances in Artificial Intelligence. Proc. 18th Australian Joint Conference on Artificial Intelligence AI 2005, Sydney, Australia, December 5-9, 2005. Lecture Notes in Artificial Intelligence 3809 (2005).

A Pruning Technique to Discover Correlated Sequential Patterns in Retail Databases

Unil Yun

Electronics and Telecommunications Research Institute,
Telematics & USN Research Division, Telematics Service Convergence Research Team
161 Gajeong-dong, Yuseong-gu, Daejeon, 305-700, Korea
yununil@gmail.com

Abstract. In this paper, we suggest a pruning technique to discover weighted support affinity patterns in which an objective measure, sequential ws-confidence is developed to detect correlated sequential patterns with weighted support affinity patterns. Based on the pruning technique, we develop a weighted support affinity pattern mining algorithm (WSMiner). Our performance study shows that WSMiner is efficient and scalable for mining weighted support affinity patterns.

Keywords: data mining, sequential pattern mining, weighted support affinity.

1 Introduction

Large growth of data gives the motivation to find useful patterns among the huge data according to the requirement of real applications. Most sequential pattern mining algorithms use a support threshold to prune the search space. The previous sequential pattern mining algorithms may reduce the number of patterns but the approaches can not handle the issue to detect correlated sequential patterns containing items with similar levels of weighted support (total profits or total expenses of patterns) in retail databases. In this paper, we propose a pruning method to discover correlated sequential patterns with the weighted support affinity called WSMiner. We define the concept of a sequential weighted support affinity pattern that uses an objective measure, called sequential ws-confidence. The sequential ws-confidence measure considers both support and weight affinity simultaneously and prevents the generation of sequential patterns with substantially different weighted support levels, so more meaningful sequential patterns can be detected. The sequential weighted support affinity patterns can be applied in analysis of sequential patterns in retail databases.

The remainder of the paper is organized as follows. In section 2, we describe the problem statement and related work. In Section 3, we develop WSMiner (Weighted Support affinity Sequential pattern mining). Section 4 shows extensive experimental results. Finally, our conclusion is presented in section 5.

2 Problem Definition and Related Work

Let $I = \{i_1, i_2... i_n\}$ be a unique set of items. A sequence S is an ordered list of itemsets, denoted as $\langle s_1, s_2, .., s_m \rangle$, where s_j is an itemset which is also called an element of the

A. Sattar and B.H. Kang (Eds.): AI 2006, LNAI 4304, pp. 1032–1036, 2006.
© Springer-Verlag Berlin Heidelberg 2006

sequence, and $s_j \subseteq I$. That is, $S = \langle s_1, s_2, \ldots, s_m \rangle$ and s_j is $(x_{j1}x_{j2}\ldots x_{jk})$, where x_{it} is an item in the itemset s_i. An item can occur at most one time in an itemset of a sequence but it can occur multiple times in different itemsets of a sequence. The size $|S|$ of a sequence is the number of itemsets in the sequence. The length, $l(s)$, is the total number of items in the sequence. A sequence with length l is called an l-sequence. A sequence database, $D = \{S_1, S_2, \ldots, S_n\}$, is a set of tuples $\langle sid, s \rangle$, where sid is a sequence identifier and S_k is an input sequence. The support of a sequence α in a sequence database SDB $(support_S(\alpha) = | \{<sid, s>| (<sid, s> \in S) \wedge (\alpha \sqsubseteq s)\} |$ is the number of sequences in SDB that contain the sequence α. Given a support threshold, min_sup, a sequence α is called a frequent sequential pattern in the sequence database if the support of the sequence α is no less than the minimum support threshold $(Support_s(\alpha) \geq$ min_sup$)$.

The problem of sequential pattern mining is to find the complete set of frequent sequential patterns satisfying a minimum support in the sequence database. As related works, efficient sequential pattern mining algorithms [2, 3, 6, 8, 12] have been recently developed such as constraint-based sequential pattern mining [8], sequential pattern mining without using support thresholds [10] and closed sequential pattern mining [11]. These approaches may reduce the number of patterns but weak weighted support affinity patterns are still mined because they did not consider weights or priorities of the patterns. Recently, weighted sequential pattern mining [12] has been suggested to mine more important sequential patterns. This approach may reduce the number of patterns but weak affinity patterns are still found in the result sets. Before SPAM [2], SPADE [13] and PrefixSpan [7, 9] were two of the fastest algorithms in general sequential pattern mining. According to performance evaluations [2], SPAM outperforms SPADE on most datasets and PrefixSpan outperforms SPAM only slightly on very small datasets. Except for this case, SPAM outperforms PrefixSpan.

3   A Pruning Technique to Discover Correlated Sequential Patterns with the Weighted Support Affinity

In this section, we suggest a pruning technique to discover sequential weighted support affinity patterns. The sequential ws-confidence measure and the concept of the weighted support affinity pattern by the ws-confidence are suggested to answer the comparative analysis queries in market basket analysis which are not handled by the previous sequential pattern mining approaches. Specifically, the weak affinity patterns are efficiently eliminated with the sequential ws-confidence measure.

Table 1. Example of a sequence database

Sequence ID	Sequence
10	$\langle$a (abc) (ac) d (cf)$\rangle$
20	$\langle$(ad) abc (bcd) (ae) bcde$\rangle$
30	$\langle$a(ef) b (ab) c (df) ac$\rangle$
40	$\langle$ac (bc) eg (af) acb (ch) (ef)$\rangle$
50	$\langle$ba (ab) (cd) eg (hf)$\rangle$
60	$\langle$a (abd) bc (he)$\rangle$

3.1 The Sequential ws-Confidence

In retail databases, attribute values such as prices (profits) of items are used as a weight factor and the prices of items can be normalized within a specific weight range.

Definition 3.1 *Sequential ws-confidence*
A weighted support confidence of a sequential pattern $S = \{s_1, s_2, ..., s_m\}$, s_i is $(x_{i1}x_{i2}...x_{ik})$, where x_{it} is an item in the itemset s_i, denoted as sequential ws-confidence, is a measure that reflects the overall weighted support affinity among items within the sequence. It is defined as the ratio of the minimum weighted support of the items within the sequential pattern to the maximum weighted support of the items within the pattern. That is, this measure is defined as

$$\text{ws-conf}(S) = \frac{\text{Min}_{1 \leq m' \leq m,\, 1 \leq k' \leq legnth(s_{m'})} \{\text{support}(\{x_{m'k'} \subseteq s_{m'}\}) * \text{weight}(\{x_{m'k'} \subseteq s_{m'}\})\}}{\text{Max}_{1 \leq m'' \leq m,\, 1 \leq k'' \leq legnth(s_{m''})} \{\text{support}(\{x_{m''k''} \subseteq s_{m''}\}) * \text{weight}(\{x_{m''k''} \subseteq s_{m''}\})\}}$$

It may be other ways to examine the weighted support affinity levels. More complex definitions may detect more exact affinity levels. However, based on the definition of the ws-confidence, the ws-confidence satisfies the anti-monotone property for identifying strong weighted support affinity patterns.

Definition 3.2 *Weighted support affinity sequential patterns*
A sequential pattern S is defined as a Weighted support affinity sequential patterns if and if only (1) the support of the pattern is no less than a minimum support (Support $(S) \geq$ min_sup), and (2) the sequential ws-confidence of the pattern is no less than a minimum ws-confidence (wsconf $(S) \geq$ min_wsconf).

3.2 WSMiner Algorithm

Procedure WSMiner (WSP, α, L, S|α)

Parameter:
(1) α is a sequential pattern.
(2) L is the length of α,
(3) $S \mid_\alpha$ is the sequence database, SDB if α is null, otherwise, it is the α-projected database.
1. Scan $S \mid_\alpha$ once, count the support of items, and find frequent items, β in sequences: β is a weighted support affinity patterns if the following pruning conditions are satisfied.
   Condition 1: (support $\geq$ min_sup)
   Condition 2: (ws-confidence $\geq$ min_wsconf)
   (a) β can be assembled to the last itemset of α to form a weighted support sequential affinity pattern or (b) <β> can be appended to α to form a weighted support sequential affinity pattern.
2. For each frequent item β,
   Add it to α to form a sequential pattern $\alpha^{\grave{}}$, and output $\alpha^{\grave{}}$.
3. For each $\alpha^{\grave{}}$,
   Construct $\alpha^{\grave{}}$-projected database $S| \alpha^{\grave{}}$;
   Call WSMiner ($\alpha^{\grave{}}$, L+1, $S \mid \alpha^{\grave{}}$)
   End for

---

4 Performance Evaluation

In this section, we present our performance study over real and synthetic datasets. We report our experimental results on the performance of WSMiner in comparison with a recently developed algorithm, SPAM [2] and WSpan [12].

The IBM dataset generator is used to generate sequence datasets. It accepts essential parameters such as the number (D) of sequences (customers), the average number (C) of itemsets (transactions) in each sequence, the average number (T) of items (products) in each itemset, the average length (S) of maximal sequences, the average length (I) of transactions within the maximal sequences, and the number (N) of different items in the dataset. More detail information can be found in [1]. WSMiner was written in C++ and experiments were performed on a sparcv9 processor operating at 1062 MHz, with 2048MB of memory on a Unix machine. In our experiments, a random generation function was used to generate weights of items.

4.1 Performance Results

From Fig. 1 to Fig. 2, we report the evaluation results for D7C7T7S7I7 dataset with weights of items from 0.1 to 0.3. The main performance difference between WSMiner and other algorithms such as SPAM and WSpan results from using sequential ws-confidence. By increasing the sequential ws-confidence threshold, fewer patterns with a higher weighted support affinity can be found. SPAM mines a huge number of sequential patterns in this test so we could not show the number of patterns of SPAM in Fig. 1. For instance, the numbers of patterns of SPAM are 170,965 with a minimum support of 4%, 439,953 with a minimum support of 3%, and 1, 646,818 with a minimum support of 2%.

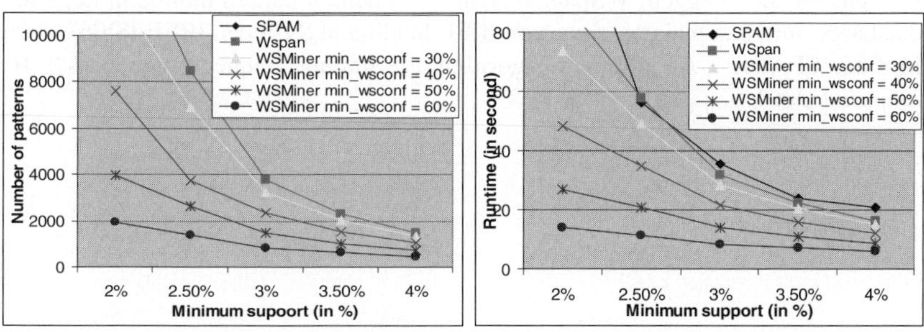

Fig. 1. Effectiveness of ws-confidence   Fig. 2. Processing time

5 Conclusion

In this paper, we presented the problem of mining correlated sequential patterns with the weighted support affinity. To detect the sequential patterns, we defined the sequential ws-confidence measure and the concept of weighted sequential affinity patterns. Using the measure, WSMiner algorithm is developed to detect correlated patterns with the weighted support affinity by pushing the sequential ws-confidence into sequential pattern mining. The performance analysis shows that WSMiner is efficient and scalable for mining weighted support affinity patterns.

References

[1] R. Agrawal, and R. Srikant, *Mining Sequential Patterns*, ICDE, 1995.

[2] J. Ayres, J. Gehrke, T. Yiu, and J. Flannick, Sequential Pattern Mining using A Bitmap Representation, SIGKDD'02, 2002.

[3] D. Y. Chiu, Y. h. Wu, and A. L. Chen, An Efficient Algorithm for Mining Frequent Sequences by a New Strategy without Support Counting, ICDE'04, 2004.

[4] Hong Chung, Xifeng Yan, and Jiawei Han, "SeqIndex: Indexing Sequences by Sequential Pattern Analysis", *SDM'05*, 2005.

[5] M. Ester, A Top-Down Method for Mining Most Specific Frequent Patterns in Biological Sequence Data, SDM'04, 2004.

[6] J. Han, J. Pei, B. Mortazavi-Asi, and et al, *FreeSpan: Frequent Pattern-Projected Sequential Pattern Mining*, SIGKDD'00, 2000.

[7] J. Pei, J. Han, B. Mortazavi-Asi, and H. Pino, PrefixSpan: Mining Sequential Patterns Efficiently by Prefix-Projected Pattern Growth, ICDE'01, 2001.

[8] J. Pei, J. Han, and W. Wang, Mining Sequential Patterns with Constraints in Large Databases, CIKM'02, 2002.

[9] J. Pei, J. Han, J. Wang, H. Pinto, Q. Chen, U. Dayal, and M. C. Hsu, *Mining Sequential Patterns by Pattern-Growth: The PrefixSpan Approach*, IEEE Transactions on Knowledge and Data Engineering, Oct, 2004.

[10] Petre Tzvetkov, Xifeng Yan, and Jiawei Han, "TSP: Mining Top-K Closed Sequential Patterns", *ICDM'03*, 2003.

[11] Xifeng Yan, Jiawei Han, and Ramin Afshar, "CloSpan: Mining Closed Sequential Patterns in Large Datasets", *SDM'03*, 2003.

[12] U. Yun, and J. J. Leggett, WSpan: Weighted Sequential pattern mining in large sequence databases, International conference on IEEE Intelligent Systems (IEEE IS`06), 2006.

[13] M.Zaki, SPADE: "An efficient algorithm for mining frequent sequences". *Machine Learning*, 2001.

Clustering Categorical Data Using Qualified Nearest Neighbors Selection Model*

Yang Jin and Wanli Zuo

College of Computer Science & Technology, Jilin University, Changchun, P.R. China
`mail_jinyang@yahoo.com.cn`

Abstract. ROCK is a robust, categorical attribute oriented clustering algorithm. The main contribution of ROCK is the introduction of a novel concept called links as a measure of similarity between a pair of data points. Compared with traditional distance-based approaches, links capture global information over the whole data set rather than local information between two data points. Despite its success in clustering some categorical databases, there are still some underlying weaknesses. This paper investigates the problems deeply and proposes a novel algorithm QNNS using Qualified Nearest Neighbors Selection model, which improves clustering quality with an appropriate selection of nearest neighbors. We also discuss a cohesion measure to control the clustering process. Our methods reduce the dependence of the clustering quality on the pre-specified parameters and enhance the convenience for end users. Experiment results demonstrate that QNNS outperforms ROCK and VBACC.

1 Introduction

Clustering has become one of the major areas of investigation in data mining as a useful technique for discovery of data distribution and patterns in the underlying data. Many traditional clustering algorithms, such as k-means [1] and centraloid-based [2] are designed mostly for numeric data, using some distance matrix to measure the similarity between two data points. While the use of the distance-based measurement could yield satisfactory results for numeric attributes, it is not appropriate for data sets with categorical attributes.

ROCK [3] proposes the notion of *link* and measured the similarity between pairs of data objects based on the number of links in common. Compared with distance-based approaches, links capture global information over the whole data set rather than local information between two data points. Despite ROCK's good performance in clustering standard databases such as Mushroom, there are still some shortcomings (See section 2). Some papers have proposed algorithms based on ROCK's idea such as VBACC [4] and QROCK [7]. But the improvements of these two algorithms are conducted under certain given conditions. For example, VBACC assumes the clustering algorithm can omit θ when the target dataset has high dimensionality. QROCK believes that θ and k

---

* This work was sponsored by Natural Science Foundation of China (NSFC) under Grant No. 60373099.

A. Sattar and B.H. Kang (Eds.): AI 2006, LNAI 4304, pp. 1037–1041, 2006.
© Springer-Verlag Berlin Heidelberg 2006

are relevant and an appropriate selection of θ can make the two parameters equivalent. Because QROCK aims to reduce the time complexity other than improve the clustering quality, we compare our proposed algorithm with ROCK and VBACC in section 4.

In this paper we propose a new categorical clustering algorithm called QNNS using Qualified Nearest Neighbors Selection model. QNNS can select appropriate neighbors automatically without pre-specifying θ. We also manage to control the clustering process by introducing a cohesion measure, which eliminates the need for the parameter k. Our study shows that QNNS generates better quality clusters than ROCK and VBACC on standard datasets.

2 The ROCK Algorithm

In this section we will have an in-depth discussion on ROCK's weaknesses. For the length limitation, we do not list the basic notions of ROCK such as *neighbor*, *link* and *goodness measure*. You can refer to [3] for more details.

The main shortcomings of ROCK lie in three aspects: (1) The similarity threshold θ requires to be given by the user and the clustering quality is sensitive to θ. If θ is overestimated then the number of neighbors considered will be reduced, which will produce more isolated objects. If θ is underestimated then some irrelevant points are included in the neighbor lists, making the purity of the clusters poorer. Therefore it is important for ROCK to initiate with a proper θ. However, it is hard to do so without sufficient prior-knowledge. (2) The degree of similarity is only used to decide whether two objects are neighbors and the information is actually valuable to guide the iterative clustering process. ROCK makes a polarization of the similarity information with θ. Two objects p_i and p_j are considered to be similar when $sim(p_i,p_j) \geq \theta$ and $sim(p_i,p_j)$ is set to be 1. Otherwise $sim(p_i,p_j)$ is set to be 0. This polarization makes the clustering results more sensitive to θ. (3) The number of desired clusters k need to be designated in advance and it acts as a critical condition to terminate the clustering process. However, it is also difficult to decide k reasonably.

3 The QNNS Algorithm

3.1 Qualified Nearest Neighbors Selection Model

Definition 1. *Unqualified nearest neighbors*
Suppose $p_1,\ldots,p_k$ are all the neighbors of i and they are arranged by $sim(p_j, i)$ in descending order, given a separator r and a percent ρ, if they satisfy the following condition,

$$\frac{\sum_{j=r+1}^{k} sim(p_j,i)}{\sum_{j=1}^{k} sim(p_j,i)} \leq \rho$$

then $p_{r+1},\ldots,p_k$ are called *unqualified nearest neighbors*.

Definition 2. *Qualified Nearest Neighbor Selection model*
Qualified Nearest Neighbor Selection model is to choose r and ρ dynamically and reasonably select all qualified nearest neighbors and exclude any unqualified neighbor. To make the algorithm parameter-free, we choose the average similarity of neighbors as the criterion. Let *ave_sim(i)* such as

$$ave\_sim(i) = \frac{\sum_{j=1}^{k} sim(p_j, i)}{N_i}$$

be the average similarity between i and its nearest neighbors, then all the neighbors which meet $sim(p_j,i) \geq ave\_sim(i)$ are qualified nearest neighbors. Therefore an efficient computation of connectivity matrix can be achieved as follows.

Algorithm 1. `Compute_connectivity(S)`

```
   1.    Compute nbrlist[i](list of neighbors of i) for every
point   i   in   S   and   the   corresponding   vallist[i](list  of
sim(pj,pi) where pj∈nbrlist[i]);
   2.    select all qualified neighbors for each i and adjust
nbrlist[i] and vallist[i];
   3.        For i=1 To |S| Do{
   4.           A=nbrlist[i];V=vallist[i];
   5.        For j=1 To |A|-1 Do
   6.           For l=j+1 To |A| Do{
   7.               y=v[j]×v[l]  ×(sj+1-posji)  ×(sl+1-posli)
×(max_neighbor_number²)/(sj×sl);
   8.        Mc[A[j],A[l]]=Mc[A[j],A[l]]+y; }
```

We can see from Algorithm 1 that it is not necessary to pre-specify θ in the computation of connectivity matrix. The goodness measure in our algorithm is:

$$g(C_i, C_j) = \frac{Mc[i, j]}{n_i \cdot n_j}$$

3.2 Cohesion Measure

We here introduce a criterion $E'(C)$ which is often used in the clustering algorithms for documents [5] as the cohesion measure.

$$E(C) = \sum_{r=1}^{k} \frac{\sum_{p_i, p_j \in C_r} sim(p_i, p_j)}{n_r} \qquad\qquad E'(C) = (1 - \frac{2k}{n})E(C)$$

This criterion tries to maximize the sum of the average similarity among all objects in each cluster and make a penalty against a big value of k. Experimental results show that $E'(C)$ is effective and the quality of clustering results at its extremum is also satisfactory. The whole clustering algorithm is as follows.

```
Algorithm 2. QNNS(S)
1.      Mc=Compute_connectivity(S); max_quality=0;
2.      For each s∈S Do
3.        q[s]=build_local_heap(Mc,s);
4.          Q=build_global_heap(s,q);
5.      Do { u=extract_max(Q); v=extract_max(q[u]);
6.            delete(Q,v);        w=merge(u,v);
7.          cur_quality= E'(C); Δ=cur_quality-max_quality;
8.            If (Δ<0 && |Δ|>(max_quality/4)) {  //If the
merging makes E'(C) drop drastically
9.              unmerge(u,v); break; }
10.          If (Δ>0) {
11.            update(m_clusters);
12.            max_quality=cur_quality; }
13.        For each x∈(q[u]∪q[v]) Do {
14.            Mc[x,w]=Mc[x,u]+Mc[x,v];
15.            delete(q[x],u); delete(q[x],v);
16.            insert(q[x],w,g(x,w));
17.            insert(q[w],x,g(x,w));
18.            update(Q,x,q[x]); }
19.          insert(Q,w,q[w]);
20.          deallocate(q[u]); deallocate(q[v]);
21.      }While(∃i, q[i]≠∅)
```

4 Experimental Results

In this section, we compare the performance of QNNS with ROCK and VBACC.
Table 1 describes the results of running on the standard VOTE and ZOO datasets.

Table 1. Experimental results on the VOTE and ZOO datasets

Algorithm	Pre-specified Parameter	VOTE		ZOO	
		clusters#	$f_{0.5}$	clusters#	$f_{0.5}$
ROCK	θ, k	63	0.85925	9	0.93000
VBACC	k	6	0.87297	8	0.87437
QNNS	null	7	0.90762	9	0.93000

Here we use $f_{0.5}$ [6], which is a useful evaluation metrics to measure the clustering
quality. In general, the bigger the value of $f_{0.5}$, the better is the clustering quality. In
some papers, $f_{0.5}$ is also delineated as f_1. We can draw the following conclusions: (1) On
the ZOO dataset, the values of $f_{0.5}$ are the same for QNNS and ROCK. They are much
better than those of VBACC. (2) On the VOTE dataset, the decreasing order of $f_{0.5}$ is
QNNS, VBACC and ROCK. Therefore the quality of the clustering results obtained by
QNNS is the best and ROCK is the poorest.

Because the main difference among ROCK, VBACC and QNNS lies in the selection of nearest neighbors, we hereby give the uniform representation of the time and space complexity. The worst-case time complexity is $O(n^2+nm_mm_a+n^2\log n)$ and the space complexity is $O(\min\{n^2,nm_mm_a\})$, where m_m is the maximum number of neighbors, m_a is the average number of neighbors. We conclude that in general, VBACC is the slowest one for it takes almost all neighbors into consideration. If θ and k can be pre-specified properly, ROCK would be the fastest one. However, it is difficult to accomplish this without sufficient prior-knowledge.

5 Conclusion

In this paper we conduct a thorough study of the ROCK algorithm and point out its shortcomings. Accordingly, we propose a new clustering algorithm QNNS using Qualified Nearest Neighbors model, which improves the quality of neighbors and incorporate similarity as weight into the iterative clustering process. We also introduce a cohesion measure to guide the clustering process and make QNNS able to find the optimal clustering results automatically without pre-specifying parameter k.

Our future work is to apply QNNS to the clustering tasks for high dimensional datasets such as web documents with proper dimension-reducing methods.

References

[1] Richard C Dubes, Anil K Jain. Algorithms for Clustering Data. Prentice Hall, 1998

[2] Tian Zhang, Raghu Ramakrishnan, Miron Livny: Birch: An Efficient Data Clustering Method for Very Large Databases. In: Proceedings of the ACM SIGMOD Conference on Management of Data, Montreal, Canada, 1996, 103-114

[3] S. Guha, R. Rastogi, K. Shim: ROCK: A Robust Clustering Algorithm for Categorical Attributes. In: Proceedings of the 15th international Conference on Data Engineering, 1999, Sydney, Australia. 1-11

[4] Gunjan K Gupta, Joydeep Ghosh. Value Balanced Agglomerative Connectivity Clustering. In: SPIE Conference on Data Mining and Knowledge Discovery III, April 2001.

[5] Michael Steinbach, George Karypis, Vipin Kumar. A Comparison of Document Clustering Techniques. Technical Report #00-034, University of Minnesota. 0-34

[6] Sebastiani,F. A Tutorial on Automatic Text Categorization. In: Proceedings of ASAI-99, 1st Argentinean Symposium on Artificial Intelligence, Buenos Aires, AR, 1999. 7-35

[7] M. Dutta, A. Kakoti Mahanta, Arun K. Pujari. QROCK: A Quick Version of the ROCK Algorithm for Clustering of Categorical Data. www.uohyd.ernet.in/smcis/dcis/./akpcs/qrock.pdf.1-12.

Product Recommendations for Cross-Selling in Electronic Business

Bharat Bhasker[1], Ho-Hyun Park[2,*], Jaehwa Park[2], and Hyong-Soon Kim[3]

[1] Indian Institute of Management, Lucknow, India
[2] School of Electrical and Electronics Engineering, Chung-Ang University, Korea
[3] Next Generation Internet Team, National Computerization Agency, Korea
bhasker@iiml.ac.in, {hohyun, jaehwa}@cau.ac.kr, khs@nca.or.kr

Abstract. A system applicable in electronic commerce environments that combines the strengths of both collaborative filtering and data mining for providing better recommendations is presented. It captures the item-to-item relationship through association rule mining and then uses purchase behaviour of collaborative users for generating the recommendations. It was implemented and evaluated on a set of real datasets. Our methodology results in improved quality of recommendations measured in terms of recall and coverage metrics.

1 Introduction

Several recommendation systems for electronic commerce have been deployed for overcoming the information overload problem to identify and recommend products in order to assist the customer in making the selection [1]. Various techniques such as Collaborative Filtering, Data Mining and Information Retrieval for processing the input information are used. Pure CF approach suffers from the two major problems; sparse availability of customers' rating information for products compared to the total number of items and scalability as the similarity of target user has to be computed with all other users [1]. Data mining techniques include association rule mining and web usage mining [2]. The methods based on data mining produce a small number of poor quality rules, especially for sparse datasets and it suffers from performance bottlenecks when there are a large number of products.

In order to overcome some of the limitations, combining methods of Collaborative Filtering and Data Mining approaches are proposed [4-5]. A faster association rule mining algorithm was used and reducing the size of database transactions to be mined by generating the rules using the transactions done by the collaborative users only. The approach suffers from a lower coverage as it loses the item-to-item relationships present in the large pool of transactions. The problem is more to be pronounced in the basket-to-purchase phase [3] where purchase transaction data rather than the click history are used.

In this paper, we suggest a methodology that offers better quality recommendations in the online retailing environment. The primary contributions of this paper include i) a methodology that combine the strengths of both data mining and collaborative

---

* Corresponding author.

A. Sattar and B.H. Kang (Eds.): AI 2006, LNAI 4304, pp. 1042–1047, 2006.

© Springer-Verlag Berlin Heidelberg 2006

filtering, ii) Better-personalized recommendations using buying behaviour of collaborative consumers for product rankings, and iii) evaluation of the proposed recommender system on three real-life datasets to show that the proposed system offers better quality of recommendation.

2 Cross-Sell Recommendation Methodology

In the final checkout phase, the recommendations should be strongly associated with the content of the shopping cart and it can be achieved by capturing the Item-to-Item association in the purchase history database. The entire purchase history provides us the rich data for mining the associations amongst items. The mining is likely to result in a large number of rules based on the consumer preferences of a large pool of users. Thus, it becomes important to rank the rules, as we would like to limit the number of recommendations to the top N items. Ranking the recommendations based solely on the support and confidence, as done quite often in association rule mining may not serve the purpose as it does not yield personalized product rankings.

In the proposed methodology, we handle the issue of top-N recommendation by identifying the set of users that have preferences similar to the target user and then using their purchase history for evaluating the interestingness. In order to present the recommendation procedure the following notation is used.

Let $C=\{C_1, C_2...C_m\}$, $P=\{P_1, P_2...P_n\}$ be the sets of customers and products offered respectively and **TgtC, TgtP** denote the customers for whom recommendations are to be generated and the products that have been placed in the shopping cart respectively. Customer Database, denoted as **CustomerDB**, consists of the details of customer purchase for all the products in **P**. More specifically, the database has data of the form $<C_k, P>$ for each customer C_k in **C**.

The recommendation strategy adopted by the proposed system is outlined as follows. First, Check to see if the shopping cart is not empty and is submitted for checkout, then we determine that the customer is in Basket-to-purchase phase of shopping cycle. Let $TgtP_c = (p_{c1}, p_{c2}, .. p_{ck})$ be the items that are in shopping cart submitted by the target customer (**TgtC**). Then mine association rules from **CustomerDB** with following constraints. i) Items in the rules antecedent are subset of $TgtP_c$. ii) Rule consequent has single item. iii) The minimum support, **DynSup**, used for the association rule mining is determined dynamically. Count the support for 1-itemsets in $TgtP_c$. The **DynSup** is taken as **x%** of the lowest frequent 1-item support in **TgtP**. This approach of dynamic determination of support ensures that customer's implicit interest is captured effectively and the need for ad hoc support specification is avoided. The value of **x** chosen in this paper is 50%, although the choice can be made flexibly, iv) To avoid performance bottlenecks limit the maximum number of rules to be mined to a predefined number **MaxRules**. v) Then select the rules with minimum confidence of 50% by default.

Next, examine all the consequents and make a list of potential recommendations. Identify collaborative users based on target customer's history, i.e.; use purchases stored in **CustomerDB**, to identify users with the similar buying preferences using Cosine/ Pearson similarity measure. Then extract purchase transactions of collaborative users. These transactions are used for calculating the interestingness of a rule.

For each rule, $\mathbf{R_i}$ (A→B), Compute the interest, often referred to as the lift (I_{Ri}), using the extracted purchase transaction history where, I_{Ri}=(P(A,B)/P(A)P(B)), Then, remove the rule from the set if the value of $I_{Ri} < 1$, as it implies negative correlation between the antecedent and consequent items. The candidates for recommendations are consequents of the discovered association rules. Some candidates may be consequent of several association rules. The recommendation score (rank) of the all such candidates is computed by combining the interestingness of all the rules. Remove all those products that have a negative feedback given by the target customer and Recommend top-N products to the target customer.

Table 1. The Purchase Transactions Dataset

tid	cid	Items	tid	cid	Items	tid	cid	Items
t1	c1	i1,i3,i5,i7,i9	t11	c4	i1,i3,i6,i8,i10	t21	c2	i9,i8
t2	c2	i1,i2,i3,i5,i7,i9	t12	c6	i1,i3,i5,i7,i9	t22	c3	i1,i3,i5
t3	c3	i1,i3,i4.i6,i7,i9	t13	c2	i2,i5,i7,i9	t23	c4	i1,i4,i6
t4	c4	i1,i2,i4,i5,i6,i7,i10	t14	c3	i1,i3,i5,i7,i10	t24	c5	i9,i10
t5	c5	i1,i2,i4,i5,i6,i8	t15	c4	i4,i5,i8,i10	t25	c5	i1,i2,i5,i7
t6	c6	i1,i3,i4,i5,i7,i10	t16	c5	i2,i5,i8	t26	c2	i4,i5,i6,i7
t7	c7	i1,i2,i4,i6	t17	c6	i1,i4,i7,i9	t27	c6	i2,i3,i5,i6
t8	c1	i2,i4,i5,i7,i10	t18	c3	i2,i5	t28	c4	i1,i3,i5,i7
t9	c4	i4,i6,i7,i9	t19	c4	i1,i3,i5,i8	t29	c2	i1,i3,i7
t10	c3	i1,i5,i8,i10	t20	c1	i1,i4	t30	c7	i1,i3,i4,i6

In order to illustrate the procedure, the customer purchase database shown in Table 1 is used. Let us assume the customer C3 with items i1, i3 in shopping cart is ready to checkout. The algorithm in that case will mine the association rules having combination of i1 and i3 as the antecedents. We assume the minimum support of 8 for illustration.

Using the Maximum Frequent Set Mining algorithm (MFS) [5], the algorithm mines (i1,i3,i5): 8, (i1,i3,i7):9, (i1,i5,i7):8, (i1,i4): 9, (i2,i5): 9, and (i4,i6): 8 as MFS. Consequently, the following association rules based on the components of the shopping cart are generated. i1, i3→i5, i1, i3→i7, i1→i4, i1→i5, i1→i7, i3→i5, i3→i7 We identify the collaborative users of C3. Using 0.49 as cut-off for similarity value shown in Table 2, C1, C4 and C6 are the collaborative users for C3.

Table 2. Similarity with C3 (Pearson's correlation coefficient)

Customer	Similarity	Customer	Similarity	Customer	Similarity
c1	0.6681	c2	0.3368	c5	0.0909
c3	1	c4	0.4975	c6	0.7662

We extract the purchase history for C3 and all the collaborative users of C3 and determine the interestingness as per the procedure a sample computation is as, I(i1, i3 →i5)=P(i1,i3,i5)/P(i1,i3).P(i5)=(6/19)/(8/19).(12/19)=1.18. Similarly, the interestingness of all other rules is computed and is as, $I(i1\rightarrow i5)=.974$, $I((i3\rightarrow i5)=1.23$, $I(i1,i3\rightarrow i7) = 1.425$, $I(i1\rightarrow i7) = 1.169$, $I(i3\rightarrow i7)=1.266$, $I(i1\rightarrow i4) = .974$.

As per our recommendation strategy, all the rules whose interestingness value is 1 or lower are removed. In the final set of rules, the item i5 is recommended by rules with interestingness of $I(i1,i3\rightarrow i5)=1.18$, $I((i3\rightarrow i5)=1.23$ and item i7 is recommended by rules with interestingness of $I(i1,i3\rightarrow i7)=1.425$, $I(i1\rightarrow i7)=1.169$, $I(i3\rightarrow i7) = 1.266$. The score for items i5 and i7 are computed as summation of the interestingness and are 2.41 and 3.86 respectively. The recommendation procedure here will recommend the items i7 followed by i5.

Table 3. Purchase Transactions of Collaborative Users of C3

Tid	cid	Items	tid	cid	Items	tid	cid	items
t1	C1	i1,i3,i5,i7,i9	t10	c3	i1,i5,i8,i10	t18	c3	i2,i5
t3	C3	i1,i3,i4.i6,i7,i9	t11	c4	i1,i3,i6,i8,i10	t20	c1	i1,i4
t4	C4	i1,i2,i4,i5,i6,i7,i10	t12	c6	i1,i3,i5,i7,i9	t22	c3	i1,i3,i5
t6	C6	i1,i3,i4,i5,i7,i10	t14	c3	i1,i3,i5,i7,i10	t23	c4	i1,i4,i6
t8	C1	i2,i4,i5,i7,i10	t15	c4	i4,i5,i8,i10	t27	c6	i2,i3,i5,i6
t9	C4	i4,i6,i7,i9	t17	c6	i1,i4,i7,i9	t28	c4	i1,i3,i5,i7

As seen from the presentation and illustration of the algorithm, our approach combines Association Rule Mining and collaborative filtering for personalizing the product rankings. Thus, our approach is quite distinct compared to existing methods on the following aspects: i) We offer recommendation for the final, i.e., Checkout phase and use complete purchase history for preserving the Item-to-Item relationship for better coverage. ii) In order to speed-up the process, this MFS are computed in advance. At the time of recommendations, the computed list of MFS is filtered to find the ones with minimum support and contain the products in the shopping cart. For association rule mining, we use MFS-Miner. iii) Recommendations generated from the entire purchase history are then ranked using the purchase transactions of collaborative users in our methodology.

3   Performance Results

For evaluation of the quality of recommendations, Recall and Coverage are used and three different real-life datasets and the characteristics of the datasets are used (Table 4). Two of the datasets (DS-1 and DS-2) are customer purchase data gathered from two major online retailers in India. The Msweb dataset corresponds to click-stream data of Microsoft's web site.

The dataset is divided into training (80%) and test sets (20%) through a random selection process. The test data is vertically partitioned in two parts in such a way that the first part as items in shopping cart/input and second part contains the one item called hidden item for checking the recommendation. The training data is used for computing association rules and similarity computations in our experiments. Using the test data portion of the data set, we obtain the target user information and input items and compute the recommendations using the training data. The system generates a set of products for recommendation, the Top-10 recommendations are used for evaluation.

Table 4. Dataset Characteristics

Dataset	Num. of Customers	Num. of Products	Density (%) (Non-zero entries/Total entries)	Avg. Basket
DS-1	8054	96	1.31	1.26
DS-2	359	105	7.13	7.51
Msweb	32711	294	1.02	3.02

We evaluated the metrics using proposed methodology (CF), a pure collaborative filtering approach (I) and the procedure described in [5] (II) for comparison purposes. The quality comparisons are shown using two similarity measures, the Pearson's correlation and the cosine. The results reveal that quality of recommendation is better for cosine similarity measure. The comparison of the results shown in Tables 5 reveals that our recommendation procedure yields better quality results.

Table 5. Quality of Recommendations on Real-life Datasets: Exp I,II and CF

Data set	Correlation						Cosine					
	Recall			Coverage			Recall			Coverage		
	I	II	CF	I	II	CF	I	II	CF	I	II	CF
DS-1	.329	.233	.085	.625	.567	1.00	.456	.433	.095	.714	.677	1.00
DS-2	.276	.200	.127	.872	.800	1.00	.357	.250	.085	.980	.925	1.00
MSWeb	.412	.348	.010	.643	.473	1.00	429	.388	.010	.618	.535	1.00

4 Summary

In this paper, we proposed a technique that was designed with the above objectives. The methodology was implemented and evaluated using the simulated datasets as well as the real-world datasets collected from e-commerce sites.

References

1. Deshpande, M., and Karypis, G. (2004) 'Item-Based Top-N Recommendation Algorithms', ACM Transaction on Information Systems, Vol. 22, No. 1, pp. 143-177.
2. Mobasher, B., Dai, H., Luo, T. and Nakagawa, M. (2001) 'Effective personalization based on association rule discovery from web usage data', In Proc. of the third ACM Workshop on Web Information and Data Management, Georgia, USA.
3. Lee, J., Podlaseck, M., Schonberg, E., and Hoch, R. (2001) 'Visualization and analysis of clickstream data of online stores for understanding of web merchandising', Data mining and knowledge discovery, Vol. 5, No. 1-2, pp. 59-84.

4. Prassas, G., Pramataris, K. and Papaemmanouil, O. (2001) 'Dynamic Recommendations in Internet Retailing', In Proc. of European Conf. on IS, Bled, Slovenia.
5. Srikumar, K. and Bhasker, B. (2005) 'Personalized Recommendations in E-Commerce', International Journal of Electronic Business, Vol 3, No. 1, pp. 4-27.

Enhancing DWT for Recent-Biased Dimension Reduction of Time Series Data

Yanchang Zhao, Chengqi Zhang, and Shichao Zhang

Faculty of Information Technology, University of Technology, Sydney, Australia
{yczhao, chengqi, zhangsc}@it.uts.edu.au

Abstract. In many applications, old data in time series become less important as time elapses, which is a big challenge to traditional techniques for dimension reduction. To improve Discrete Wavelet Transform (DWT) for effective dimension reduction in this kind of applications, a new method, *largest-latest-DWT*, is designed by keeping the largest k coefficients out of the latest w coefficients at each level of DWT transform. Its efficiency and effectiveness is demonstrated by our experiments.

Keywords: Time series, dimension reduction.

1   Introduction

The "curse of dimensionality" is a notorious problem in the field of data mining, because it not only causes the explosion of search space, but also questions the meaningfulness of looking for the nearest neighbor. Since time series data are usually of very high dimensionality, techniques of dimension reduction are often applied to time series before storing them. Popular techniques are SVD (Singular Value Decomposition), DFT (Discrete Fourier Transform) and DWT (Discrete Wavelet Transform) [5], and some methods proposed recently are landmarks [7], major minima and maxima [3], PIP [2], PAA [4], SWAT [1] and online amnesic approximation [6]. DWT is widely used for compression and dimension reduction, and some researchers proposed to use the largest coefficients to preserve the optimal amount of energy, or to choose the same subset of the coefficients for all time series [5]. Bulut and Singh designed SWAT algorithm to process queries over data streams that are biased towards the more recent values [1]. Their method uses Haar wavelets and maintains a single coefficient at each level.

In many applications, such as stock market and weather forecast, old data become less important as time elapses. Recent data are more important to judge the similarity between time series and predict the future trend than the details of old data. In such applications, a mechanism which favors the recent is called for. Nevertheless, most of the techniques for time series, except SWAT [1] and online amnesic approximation [6], treats every part equally.

In this paper, we will present a technique named *largest-latest-DWT* to tackle the above problem. By decomposing time series with DWT, the latest w coefficients at each level are put together to build a candidate set, which keeps the latest details. Then the largest k coefficients out of the candidates are used to

A. Sattar and B.H. Kang (Eds.): AI 2006, LNAI 4304, pp. 1048–1053, 2006.

© Springer-Verlag Berlin Heidelberg 2006

represent the original time series, which captures the largest fluctuations. Our experiments show that the proposed method is more accurate than both traditional DWT and SWAT [1]. This work is different from our previous paper [9] in that the paper [9] focuses on preserving recent-biased energy and the same subset of coefficients are kept for all time series, because the set of coefficients to keep are computed from a given recent-biased function only . However, this paper will present a method designed for recent-biased dimension reduction for specific time series, which is more effective than using the same subset of coefficients for all time series. What makes it different from our another paper [8] in that this paper achieves recent-biased dimension reduction by keeping the largest ones out of the latest coefficients, while the paper [8] tackles the problem with segmentation. Moreover, this work focused on improving DWT for recent-biased representation with various decay functions, while the paper [8] focuses on a generalized framework which can make most existing dimension-reduction methods ready for online recent-biased dimension reduction.

2   Enhancing DWT for Recent-Biased Dimension Reduction

Our idea comes from the observation that recent data are usually more important than old data in many applications. For example, there are three time series as follows: **A**: 7 -1 8 10 12 2 -5 -1; **B**: 4 2 5 13 <u>12 2 -5 -1</u>; **C**: <u>7 -1 8 10</u> 11 3 -2 -4 (Figure 1). For time series **A** and **B**, the second halves of them are the same, while the first halves different. On the contrary, the first halves of **A** and **C** are the same, while the second halves different. Out of **B** and **C**, which one is more similar to **A**? If we care more about recent data than old data, then **B** is more similar to **A** than **C** to **A**. However, the Euclidean distance between **A** and **B** is 6 , while that between **A** and **C** is 4.5, which indicates that **C** is more similar to **A** than **B** to **A**! For the above problem, bias should be given to recent data, so a technique of *largest-latest-DWT* will be designed in the following.

There are two traditional DWTs: one is DWT with the first k coefficients and the other is DWT with the largest k coefficients. The latter is more effective for preserving the energy of time series. To apply DWT to those scenarios where old data become less important as time goes by, a new method of selecting coefficients is designed as follows. The latest w coefficients at each level of DWT transform are selected and put together to build a candidate set. All coefficients are selected if there are less than w coefficients at a certain level. Then the largest k coefficients from the candidate set are used to represent the whole time series. The first selection of the latest w coefficients at each level helps to preserve more details of recent data and less details of old data, which reduces dimension in a recent-biased way. Then the second selection of the largest k coefficients out of the candidate set keeps the optimal energy of the time series. The DWT coefficients is organized in a hierarchy and the coefficients at level l are labeled as $C_{l1}, C_{l2}, C_{l3}, ...$ from present to past. Then, the largest k coefficients will be selected from the candidate set $\bigcup_{l=1..L}\{C_{l1}, C_{l2}, ..., C_{lm}\}$, where $m = \min\{w, 2^l\}$

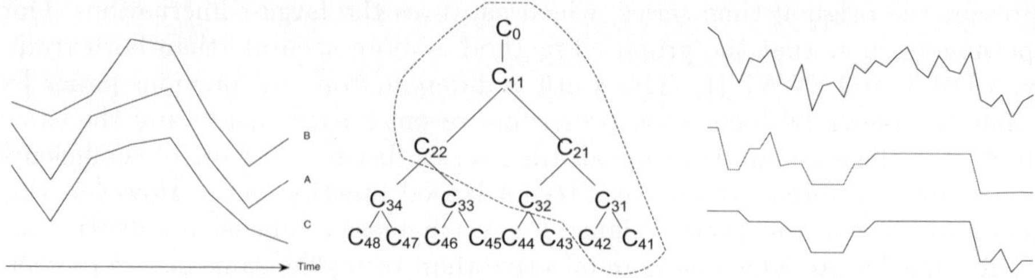

Fig. 1. **B** is more similar to **A** than **C** to **A** in recent-biased view

Fig. 2. The candidate set of coefficients when the latest two coefficients are used at each level

Fig. 3. A time series and the reconstructed ones with traditional DWT and our method (from top to bottom)

and L is the number of levels. Figure 2 shows an example of a 4-level hierarchy of coefficients where w is set to two. The coefficients circumscribed by the dotted line are candidates from which the largest k coefficients will be selected. For DWT transformation, Haar wavelet transform is used in this paper because it is very simple, widely used and of linear time complexity.

An example is shown in Figure 3. The curves from top to bottom shows respectively the original time series, the reconstructed ones with traditional DWT and the proposed method. The largest 10 coefficients are kept for both transforms. For our method, the candidate set is composed of at most 6 coefficients from each level, and then the largest 10 coefficients are selected from those candidates. The figure clearly shows that our method captures more recent details.

3   Experimental Results

The proposed method is implemented with Matlab 7 and compared with DWT and SWAT [1]. The accuracy is computed as $1 - \frac{\|(S'-S)\bullet E\|}{\|S\bullet E\|}$, where S and S' are respectively the original/reconstructed time series and E is a decay function [8].

The first dataset used is the control chart time series from UCI KDD Archive (http://kdd.ics.uci.edu/). It contains 600 examples of control charts and each example is composed of 60 values. We run the proposed method and traditional DWT on the dataset and the average accuracy is shown in Figure 4. The horizontal axis stands for w, the number of candidate coefficients at each level, while the vertical axis denotes accuracy. The horizontal dotted line shows the accuracy of traditional DWT by keeping the largest k coefficients. Because the accuracies of traditional DWT with different decay functions are nearly the same on this dataset, only one line in each subfigure is used to show its accuracy. The other lines show the accuracy of our method with linear (dash-dot line), exponential (solid line) and window (dash line) decay functions. Since SWAT uses only one coefficient (the latest one) at each level [1], it is the same as our method with $w=1$. As we can see from the figure, SWAT ($w=1$) is always of the lowest accuracy and the accuracy is much improved with our method. However, the accuracy

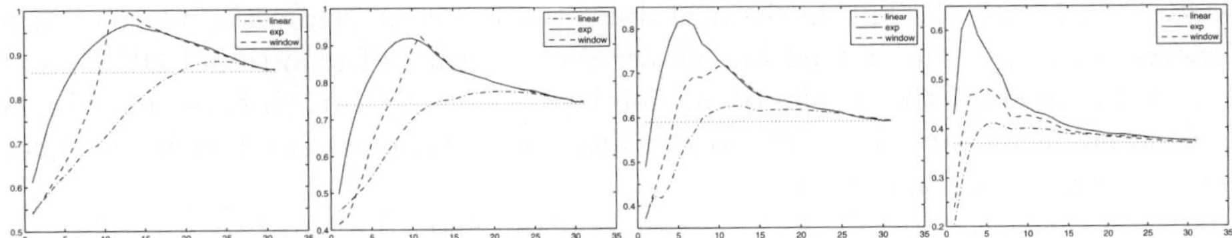

Fig. 4. Experimental results on control chart time series. The figures from left to right are the results with the largest (a) 40, (b) 30, (c) 20 and (d) 10 coefficients, respectively.

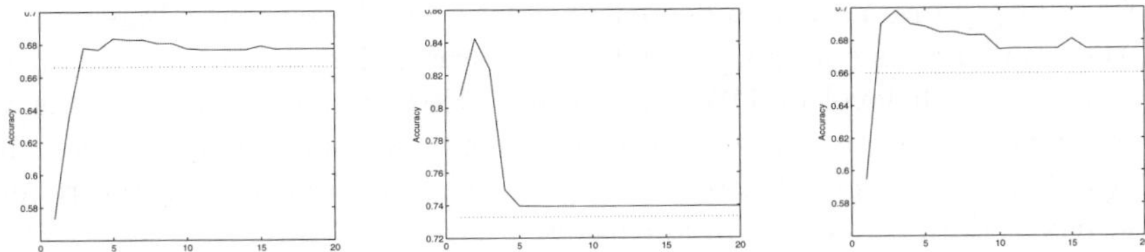

Fig. 5. Experimental results on petrol time series. The figures from left to right are respectively the results with (a) linear, (b) exponential and (c) window decay functions.

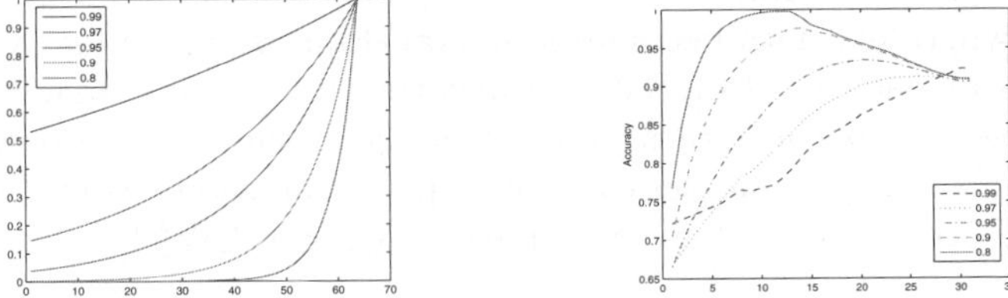

Fig. 6. Experimental results on various exponential decay functions. (a) Exponential decay functions, (b) Relationship between bias and w.

is lower than that of traditional DWT when w is small. The reason lies in that the number of candidate coefficients is less than k when w is small. For example, when $w=2$, there are only 11 candidate coefficients. Therefore, only 11 coefficients are used with the proposed method even when k is set to 40, so it is not surprising that traditional DWT is of higher accuracy than our method when $w=2$, as shown in Figure 4a. When the same number of coefficients are used for both methods, the proposed method is of higher accuracy than traditional DWT in most cases. When w is too large (e.g., when $w > 25$ in Figure 4a), the proposed method degrades to traditional DWT, because almost all coefficients become candidates.

The other dataset used is US monthly sales of petroleum and related products from Jan 1971 to Dec 1991 (http://www.robhyndman.info/forecasting/goto data.htm). There are 4 time series in the dataset which show the monthly sales of chemicals, coal, petrol and vehicles, and each time series is of 252 values. The largest 10 coefficients are kept for each time series and the average accuracy is given in Figure 5. The horizontal axis stands for w, while the vertical axis denotes accuracy. The dotted line shows the accuracy of traditional DWT, and the solid lines show the results of our method with (a)linear, (b)exponential and (c)window decay functions. The result also shows that our method is of higher accuracy than traditional DWT.

Another experiment is conducted on control chart data to test the relationship between the bias of decay function and w. The decay functions used are exponential function with bases as 0.99, 0.97, 0.95, 0.9 and 0.8 (Figure 6a). The smaller is the base, the function is more biased on recent data. The experimental result is shown in Figure 6b. In most cases, with the increase of w, the accuracy increases sharply first and then decreases slowly. The larger is the bias on recent data, the smaller w is enough to get high accuracy, that is, the less coefficients are needed at each level of DWT transform. However, when the bias is very little (eg. when the base is 0.99), more coefficients are necessary to achieve high accuracy. Therefore, the value of w should be chosen according to the required bias on recent data in specific applications.

4   Conclusions

We have designed a technique for efficiently representing time series data in applications where old data becomes less important as time goes by. The latest w coefficients at each level of DWT transformation are put together to build a candidate set, then the largest k coefficients in the candidate set are used to represent the whole time series. Our experiment shows that the proposed technique is more accurate than traditional DWT and SWAT.

Acknowledgements. This work was supported partly by Australian Research Council Discovery Projects (DP0449535,DP0559536,DP0667060), National Science Foundation of China (60463003,60496321,60496327) and Overseas Outstanding Talent Research Program of Chinese Academy of Sciences (06S3011S01).

References

1. A. Bulut, A.K. Singh: SWAT: hierarchical stream summarization in large networks. Proc. of ICDE'03.
2. T.-C. Fu, F.-L. Chung, V. Ng, R. Luk: Pattern discovery from stock time series using self-organizing maps. KDD'01 Workshop on Temporal Data Mining.
3. E. Fink, K. B. Pratt, H. S. Gandhi: Indexing of time series by major minima and maxima. Proc. of IEEE Int. Conf. on Systems, Man and Cybernetics, 2003.
4. E. Keogh, K. Chakrabarti, et al.: Dimensionality reduction for fast similarity search in large time series databases. Knowl. and Info. Systems, 3(3):263-286, 2000.
5. F. Mörchen: Time series feature extraction for data mining using DWT and DFT. Tech. Report No. 33, Dept. of Maths and CS, Philipps-University Marburg, 2003.
6. T. Palpanas, M. Vlachos, E. Keogh, D. Gunopulos, W. Truppel: Online amnesic approximation of streaming time series. Proc. of ICDE'04.

7. C.-S. Perng, H. Wang, S. R. Zhang, D. S. Parker: Landmarks: a new model for similarity-based pattern querying in time series databases. Proc. of ICDE'00.
8. Y. Zhao, S. Zhang: Generalized dimension-reduction framework for recent-biased time series analysis. IEEE Trans. on Knowl. and Data Eng., 18(2):231-244, 2006.
9. Y. Zhao, C. Zhang, S. Zhang: A recent-biased dimension reduction technique for time series data. Proc. of PAKDD'05.

Mining Large Patterns with Profit-Based Support in e-Commerce

Jin-Guk Jung, Supratip Ghose[1], and Geun-Sik Jo[2]

[1] Intelligent e-Commerce Systems Laboratory,
School of Computer Engineering, Inha University,
253 Yonghyun-Dong, Nam-Gu, Incheon, Korea 402-751
{gj4024, SGresearch}@gmail.com
[2] School of Computer Engineering, Inha University,
253 Yonghyun-Dong, Nam-Gu, Incheon, Korea 402-751
gsjo@inha.ac.kr

Abstract. In this paper, we propose an unique profit criterion as a new minimum support threshold for each item and exploit the criterion as multiple minimum supports in our algorithm. We then apply our profit-based association rule mining algorithm to generate large itemsets and show the result of our experiment. Experiment results carried on synthetic data set show that the proposed approach is efficient and effective in terms of reducing candidate itemsets and generating more profitable itemsets respectively.

1 Introduction

Traditional association rule algorithms [2,3,5,6] usually are amount-based approach which is difficult to mine large itemsets with high profit but low frequency. On the other hand, for users, these algorithms face the tasks of mining candidate association rules to cover the huge amount of item, and sometimes waste a lot of time to identify invalid rules. In this regard, invariably the association rules in themselves do not serve the ultimate purpose of the business people. Ultimately, most of algorithms for the problem of online shop basket analysis are based on the occurrence frequency of customer purchase in e-commerce environments [4].

In this paper, we propose a new approach to find efficiently large patterns incorporating the profit values of items for the subsequent generation of steps during the association rule mining process. This new model enables us to achieve our goal of producing useful patterns and rare item patterns without causing frequent items to generate too many meaningless patterns. We also present the Profit-based Association Rule mining Algorithm (PARA) that is inspired by MSapriori [6], which uses profit to decide the minimum support for every item.

The rest of the paper is organized as follows. In section 2, we introduce related works on mining large patterns. Section 3 describes our model to decide *support* and an exploration method with profit-based support. In Section 4, we present a case study in which we tested our method on synthetic data. Our conclusion is formulated in the final section.

A. Sattar and B.H. Kang (Eds.): AI 2006, LNAI 4304, pp. 1054–1058, 2006.
© Springer-Verlag Berlin Heidelberg 2006

2   Related Works

Since the Apriori algorithm [2] is introduced to mine association rules, much works recognize items as having equal weights, that is, they consider only the occurrence frequency of items in transactions when they find large itemsets. However, some other literatures address the importance of assigning weights to each item in a particular domain [7]. It just focuses on how weighted association rules can be generated by examining the weighting factors of the items included in generated large itemsets.

In [6], *minimum item supports* (MIS) is the value, used as a support threshold specified by the user, to reflect different natures and/or frequencies of items and the key element used to sort items in ascending order to avoid the problem. Therefore, the MSapriori algorithm can discover rare item rules without causing frequent items to generate too many unnecessary rules. This algorithm is based on level-wise search [3] and motivated by the work [5].

3   Profit-Based Association Rule Mining Algorithm (PARA)

We propose a new support threshold to make its function effective as involving the another property, *profit*, of an item. The support of an item is indicated in terms of profit support (PS). Let $PS(is)$ denote the PS value of an itemset *is*. The PS and $profit(is)$ are calculated by the following with the total number, n, determined by the subset of the itemset *is*.

$$PS(is) = \frac{frequency(is)}{\sum_{k=1}^{n} frequency(is_k)} + \frac{profit(is)}{\sum_{k=1}^{n} profit(is_k)}. \tag{1}$$

$$profit(is) = \sum_{k=1}^{m} profit(i_k). \tag{2}$$

In our model, the minimum support to prune meaningless itemsets is changed. We call this threshold *minimum profit support* (*MPS*) of items that appear in the pattern. The initial *MPS* is specified by the user. However, since the next iteration, *MPS* is determined by the following equation 3 where *1-itemset* is the set, *I*, of items.

$$MPS(k + 1) = MPS(k) * (1 - \frac{|k - itemset|}{|(k - 1) - itemset|})(k \geq 2). \tag{3}$$

The following pseudocode sketches out *PARA* and its related procedures. Note that we only give procedures which are different from the [6]. Step 1 of *PARA* decides the PS according to the equation 1. At this time we need to pass over the data, as we consider not only the profit but also the frequent count whenever we calculate the PS of each item.

For each subsequent pass, say iteration k, L_{k-1} is used to generate candidates C_k in order to find L_k. The *getnerateCandidate* procedure at step 5 generates the candidates which are all satisfied with the *MPS* value and then uses the sorted closure property to remove those having a subset that is not frequent and profitable from the candidates, C_k. It then scans the database. For each transaction, the *subset* function is used to find all subsets of the transaction t that are candidates and accumulates the count of these candidates in C_k (step 6-10). Then, we have to update the *PS* of all candidates for the next iteration, step 12. Finally, all those candidates satisfying the *MPS* value of the first item in each candidate form the set of large itemsets (step 12).

Algorithm: PARA
   Input: Database, D, of transactions; minimum support threshold, *minsup*;
        profit of each item, P.
   Output: L, large itemsets in D.

```
1:     PS = initPass(I, D);
2:     MPS = sort(I, PS);
3:     L₁ = findLargeItemsets(D, MPS);
4:     For(k = 2; L_{k-1} ≠ φ; k++){
5:         C_k = generateCandidate(L_{k-1}, MPS);
6:         For each transaction t∈D {
7:             C_t = subset(C_t, t);
8:             For each candidate c ∈ C_k
9:                 c.count++;
10:        }
11:        update(C_k, MPS);
12:        L_k = {c ∈ C_k|c.ps ≥ MPS(c[1])}
13:    }
14:    return L = ∪_k L_k;
```

generateCandidate procedure: This procedure conducts a similar task as *apriori-gen* in Apriori algorithm [2] and *candidate-gen* in MSapriori, and in turn, consists of two kinds of actions, namely, *join* and *prune*. In the join step, L_{k-1} is joined with L_{k-1}, whose last items only are different, to generate potential candidates as in the apriori-gen function. After the join step, the prune step takes the *sorted closure property* as prior knowledge to remove candidates that have a subset that is not frequent and profitable.

update procedure: We need to update the *PS* using the equation 1 and the *MPS* using the equation 3. In next step, we must identify whether the *PS* of each candidate is greater than or equal to the *MPS*. As we mentioned above, our strategy related to the *PS* and *MPS* ignores the different scale between two attributes when we weight each itemset according to profit with the count of the itemset.

4   Experimental Results

In this section, we carried out experiments to evaluate the performance and efficiency using our algorithm (*PARA*) in e-commerce environment. We generated the synthetic dataset with the profit of each item by using the IBM synthetic data generator [1] with the following parameters: 1,000 items, 10,000 transactions, 10 items per transaction on average, and 4 items per frequent itemset on average. We generated the profit of items for a single quantity as follows: 80% of items have medium profit ranging from \$1 to \$5, 10% of items have high profit ranging from \$5 to \$10, 10% of items have a low profit ranging from \$ 0.1 to \$1.

In our experiments, we evaluated the algorithm compared with the weighted association rule mining algorithm which uses the profit in order to calculate the weight and another amount-based association rules mining algorithm, MSapriori, with multiple minimum support as benchmark. These three algorithms have multiple minimum supports. The parameter α that controls *MIS* values in MSapriori is set by 4. The weighted algorithm used the weighted multiple minimum supports that were assigned in the same way as MSapriori. That is, the weight replaced the parameter β which is calculated as $1/\alpha$. We used five low *minsup* thresholds, 0.1%, 0.2%, 0.3%, 0.4%, 0.5% respectively.

In the left side of Figure 1, we can see that the number of candidate itemsets generated by our method is somewhat higher than other methods, because we need candidate itemsets that have some profitable items, though they have frequencies less than *minsup*. The weighted algorithm pruned the most of generated itemsets. However, regarding performance this is the worst method among algorithms used in our experiments, because the method could not generate candidate itemsets that have more than 5 items. That is why the rate of decreasing the number of candidate itemsets is steeper than others.

The center of Figure 1 shows the number of large itemsets that are all satisfied with *MPS*. We can see that the number of large itemsets found by our method is somewhat less than the MSapriori from 0.2% to 0.5%. While concerning *minsup*=0.1%, the number of large itemsets is relatively bigger than other algorithms, because the minimum support of each item puts profit and frequency together. Therefore, if we see our *minsup* with the point of view of frequency, the frequency of the *MPS* values is less than those of the *MIS* values or the minimum support used by traditional algorithms.

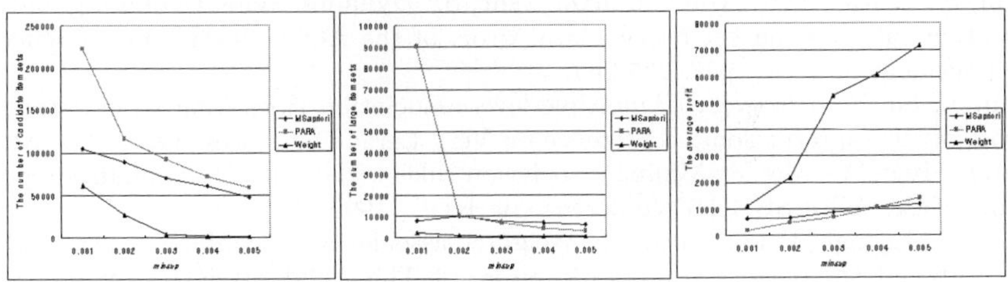

Fig. 1. The results of our experiment

The right side of Figure 1 shows the average profit of large itemsets. The weighted algorithm has the most profit because the frequency of large itemsets is relatively high and the number of large itemsets is too small. Therefore, this result does not give the merit of this algorithm. In contrast to that, our approach is less profitable than MSapriori before 0.4% of *minsup*, because large itemsets discovered by our method have lower frequencies than MSapriori, even if they have more profitable items. At $minsup = 0.5\%$, our method makes more profitable itemsets than MSapriori.

5   Conclusion

In this study, we first introduced the unique criterion, which integrated profit with occurrence frequency, to replace *minsup* for each item and exploited this criterion as multiple minimum supports. Secondly, we presented *PARA* which used the criterion to discover large itemsets. Finally, we evaluated *PARA* with the synthetic data. Experiment results carried on synthetic data set show that the proposed approach is efficient and effective in terms of reducing candidate itemsets and finding large itemsets respectively. In addition, as from $minsup=0.4\%$ our method makes more profitable itemsets.

Acknowledgments

This research was supported by the MIC(Ministry of Information and Communication), Korea, under the ITRC(Information Technology Research Center) support program supervised by the IITA(Institute of Information Technology Assessment)

References

1. Agrawal, R.: IBM Synthetic data generator. http://www.almaden.ibm.com/-software/projects/iis/hdb/Projects/data_mining/datasets/syndata.html
2. Agrawal, R., Imielinski, T., Swami, A.: Mining association rules between sets of items in large databases. Proc. of the ACM SIGMOD conference on Management of data, 1993, 207-216.
3. Agrawal, R., Srikant, R.: Fast algorithms for mining association rules. Proc. of the conference on Very Large Databases, 1994, 487-499.
4. Brin, S., Motwani, R., Ullman, J.D., Tsur, S.: Dynamic itemset counting and implication rules for market basket data. Proc. of the ACM SIGMOD conference on Management of data, 1997, 255-264.
5. Han, J., Fu, Y.: Discovery of multiple-level association rules from large databases. Proc. of 1995 international conference on Very Large Databases, 1995, 420-431.
6. Liu, B., Hsu, W., Ma, Y.: Mining association rules with multiple minimum supports. Proc. of the ACM SIGKDD conference on KDD, 1999, 337-341.
7. Tao, F., Murtagh, F. and Farid, M.: Weighted association rule mining using weighted support and significance framework. Proc. of The ACM SIGKDD conference on KDD, 2003, 661-666.

DynamicWEB: Profile Correlation Using COBWEB

Joel Scanlan, Jacky Hartnett, and Ray Williams

School of Computing, University of Tasmania
Hobart, Australia
`{jdscanla, j.hartnett, r.williams}@utas.edu.au`

Abstract. Establishing relationships within a dataset is one of the core objectives of data mining. In this paper a method of correlating behaviour profiles in a continuous dataset is presented. The profiling problem which motivated the research is intrusion detection. The profiles are dynamic in nature, changing frequently, and are made up of many attributes. The paper describes a modified version of the COBWEB hierarchical conceptual clustering algorithm called DynamicWEB. DynamicWEB operates at runtime, keeping the profiles up to date, and in the correct location within the clustering tree. Further, as there are a number of attributes within the domain of interest, the tree also extends multi-dimensionally. This allows for multiple correlations to occur simultaneously, focusing on different attributes within the one profile.

Keywords: Data Mining, Clustering Algorithms, Intrusion Detection.

1 Introduction

Fundamentally, data mining is the identification of patterns and relationships within large datasets. It has been implemented in a wide variety of domains, including biomedical research, economics, and security. The security domain, including systems such as those designed for intrusion detection, is the focus of the current research discussed in this paper.

Data mining algorithms and techniques are central to the computational power of many security systems. Automated intrusion detection systems have implemented a long list of data mining techniques with great success. The vast bulk of these systems have focused on the detection of malicious activity through matching a user's activity to a signature, or through detecting a pattern of activity that is inconsistent with a normal behaviour profile. Some of the data mining methods include: Instance-Based Learning, Neural Networks, Bayesian, Genetic Algorithms and Clustering.

The work that is outlined in this paper aims to employ clustering to generate profiles of malicious activity upon a network. This is not solely for the detection of unknown malicious users: it also functions to link existing profiles that may represent the same user operating under different IP addresses. This analysis is carried out using a substantially modified version of the COBWEB hierarchical incremental clustering algorithm [1] named DynamicWEB. The modifications allow for the clustering to occur using a dataset that is continuous in nature, with ever changing profiles, requiring the hierarchical structure to be dynamic at runtime.

A. Sattar and B.H. Kang (Eds.): AI 2006, LNAI 4304, pp. 1059–1063, 2006.

© Springer-Verlag Berlin Heidelberg 2006

The paper will explain the initial target dataset, before describing the Cobweb algorithm, and the modification being made to transform the program into DynamicWEB, will be examined.

1.1   The Data

The system is designed to operate on continuous data, with incoming events being used to update developing profiles. Thus, as the attributes that define each profile change with each update, there is a high likelihood of profiles being assigned and reassigned to different clusters. The mechanism for this updating process will be discussed later in the paper; however it is worth noting that the process occurs at the time the data is reviewed.

The most obvious attributes that is examined in the clustering, and one which is also the most readily available, is the time at which events occur. Indeed, similarities in the timing of network events have acted as a catalyst for this work. There are several aspects of time stamps to examine, such as session length, time between events or sessions, and time of session.

Other attributes that contribute to the profile are less variable, and relate to what the event recorded was about. These include things such as source and destination ports (representing the applications), source and destination IP address, and other alert information such as the kind of attack as labelled by the network device (such as an intrusion detection system).

2   Clustering with DynamicWEB

Clustering algorithms aim to divide data into its natural groups. Often these groups are not known beforehand: their discovery is the objective of the algorithm. There are several kinds of clustering algorithms. Some kinds allow an instance to be in only a single group, while others allow a given instance to be in multiple groups [2]. The clustering technique discussed in this paper is known as Incremental Clustering. It uses a hierarchical tree to sort the instances in a much easier way to visualise and understand. As each instance is added the knowledge of the tree grows which then allow information about the data set to be applied. The research being conducted by the authors uses a modified version of the COBWEB [1] which is an Incremental Clustering algorithm. COBWEB was originally designed to function using nominal attributes but was extended to use numerical in CLASSIT [3]. The method in which the clustering occurs within the tree itself was not modified.

2.1   DynamicWEB

The problem being addressed is one that involves the comparison and possible matching of activity profiles. The profiles, however, are not of a static nature as is the case, for example, when an instance is added to the COBWEB tree. Instead, they are dynamic, and require multiple updates at runtime. COBWEB is not designed to be searched efficiently for a previously clustered instance. The COBWEB tree sorts its contents based from the category utility. This in turn is calculated from the attribute values within an instance, not from an identifier relating to the instance. Therefore, any search of the COBWEB tree would result in a search length of n.

As the COBWEB tree is not designed to be searched for a particular instance or cluster, it also lacks a correct procedure to modify or delete an instance. In the original COBWEB, even if one could have searched the tree, the concept of modifying an instance in its current location is counter-productive. To modify an instance in its current location would in effect change the category utility of the cluster it resides in. This, without correction, would reduce the accuracy of any future category utility calculations, thus degrading the knowledge stored in the tree. Our extensions to COBWEB are designed to overcome these limitations.

The first extension made to COBWEB in its transformation into DynamicWEB was to create an indexing feature to the tree. Several methods were tested with an AVL tree was the most efficient and was adopted (with a hash table as second). The AVL tree is sorted using an identifier assigned to each profile, with a pointer to its location in the COBWEB tree being stored with it; effectively overlaying the COBWEB structure with a second data structure acting as an index.

Now that an instance can be quickly located within the tree, it can then be updated with new information. Updating an instance in its current location without checking for the resulting change in category utility damages the knowledge currently in the tree. While the change may not be dramatic, the variation may be sufficient for the

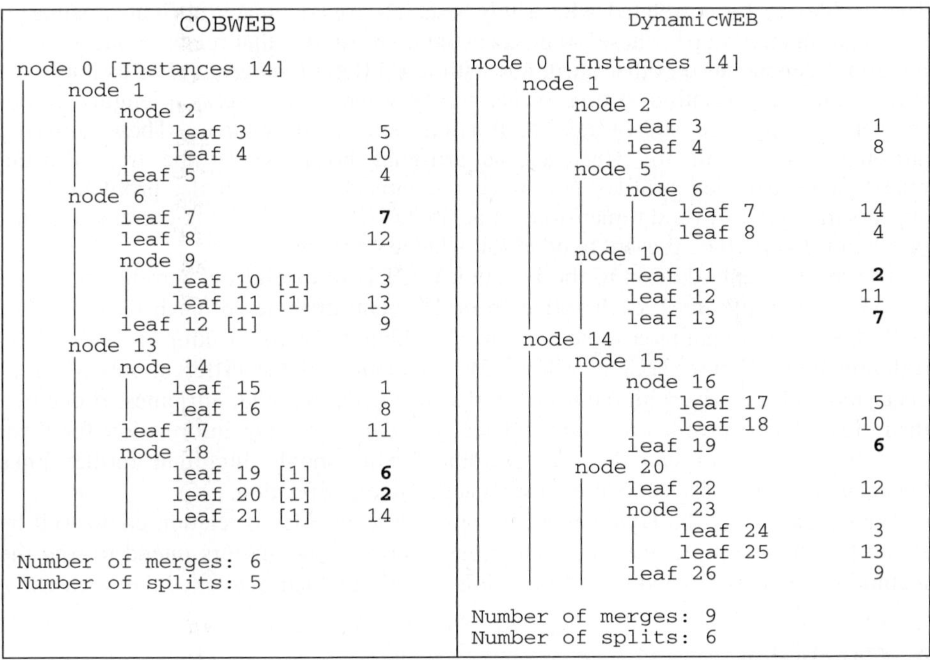

Fig. 1. The two trees above demonstrate how COBWEB and DynamicWEB have clustered the Weather [4] dataset. To illustrate the change 3 items (2, 6, and 7) were updated within the DynamicWEB tree.

instance to become better suited to another cluster. As such, when an instance is updated, it is removed from the tree and then re-added, thus covering the cases where an instance will be clustered to a different class. However, as this changes the content of the cluster. It has also changed the category utility describing the instances that are stored within the cluster. Several options that allow for the preservation of knowledge within the tree when this happens are now considered, with the best result being adopted. These options include removing the cluster (if it was the only profile in the cluster), merging it with a nearby cluster, or splitting the result into multiple clusters. Further operations are then carried out to confirm that no clusters were left parentless, out of place, or empty. In these cases again merging and splitting is considered, or elevating the profiles within the tree to better fitting clusters. The integrity of the tree, and the knowledge that it retains through these stages is the highest priority.

Figure 1 illustrates the change within a tree when 3 items are updated, being removed from the tree and re-clustered. The left-hand tree is the normal tree that is created using COBWEB. When items 2, 6 and 7 are updated using DynamicWEB the right hand tree results. The updating process did not reduce the clustering accuracy, improving it slightly by 7%. This is a result of the additional 3 merges and split that occurred during the 3 updates that resulted in the slight restructuring of the tree.

3   Multi-dimensional DynamicWEB

The profiles to be correlated within this research contain many attributes, some of which are in subgroups. These subgroups contain attributes that relate to one another, but aren't dependant on other attributes within a different subgroup. For example the various attributes relating to time (time, gap between events, session length) are not frequently related to source and destination port or IP address. These differing attribute groups will in effect act as artificial noise within the tree, causing relationships between profiles not to be recognised. As such the problem space appears to contain several dimensions of attributes which are open to correlation. This problem is likely to be present in other knowledge domains.

The most recent addition to the DynamicWEB is to extend the correlation engine to contain multiple trees, each sorted from different attributes of each instance. The AVL tree index contains a reference for each identity to its location in each of the different trees (Figure 3). This allows for correlation of the different subgroups to occur more efficiently, with the least interference from the other attributes. If one tree then clusters two profiles as being related, the corresponding information for them from the other trees can then be examined. This should highlight further links between the two profiles and allow for an accurate classification.

We envisage that the DynamicWEB approach to profile correlation could well be suited to other multiple attribute continuous datasets. The authors intend to trial the method in domains beyond that of intrusion detection which motivated the research.

4   Conclusion

In this paper a problem space has been outlined containing profiles that change over time. The research goal is to correlate these multi-attribute profiles, to the end of identifying any relationships present between them.

A clustering approach is explored, with the intention of grouping similar profiles together. The method described in this paper is called DynamicWEB, and is built upon the COBWEB incremental clustering algorithm described by Fisher [1]. DynamicWEB allows for instances within the tree to be changed, and re-clustered at runtime allowing it work with live data within the context of intrusion detection.

As the problem space contains a broad range of attributes, DynamicWEB also has the facility to allow for multiple clustering trees to function simultaneously upon the same profile, each focusing on different attributes within the profile.

The method described is not context specific and could be used in other domains requiring similar correlation to occur. It is this that the authors are currently exploring, to enable them to benchmark the system in a comparative qualitative manner.

References

1. Fisher, D.H., *Knowledge Acquisition Via Incremental Conceptual Clustering* Mach. Learn. , 1987 **2** (2): p. 139-172
2. Witten, I. and E. Frank, *Data Mining: Practical Machine Learning Tools and Techniques with Java implementations`*. 2000, San Francisco: Morgan Kaufmann Publishers. 371.
3. Gennari, J.H., P. Langley, and D. Fisher, *Models of incremental concept formation* Artif. Intell. , 1989 **40** (1-3): p. 11-61
4. Newman, D.J., et al., *{UCI} Repository of machine learning databases*. 1998, University of California, Irvine, Dept. of Information and Computer Sciences.

Finding Robust Models Using a Stratified Design

Rohan A. Baxter

Analytics Project,
Office of the Chief Knowledge Officer,
Australian Taxation Office
P.O. Box 900, Civic Square, ACT 2608
firstname.lastname@ato.gov.au

Abstract. Predictive performance in model selection is often estimated using out-of-sample validation and test datasets. The assumption is that the test and validation datasets are from the same population as the training dataset. This assumption may not apply in the common application context where the model is applied to scoring of future data. This paper proposes a sample design which can lead to better model performance and robust estimates of model generalization error. The sample design is shown applied to a collection scoring application.

1 Introduction

Out-of-sample validation datasets are used for model selection in data mining applications. Out-of-sample test datasets are also used for estimating the model generalization error[1]. The standard required assumption is that the three datasets, training, validation and test, are independent and identically-distributed (i.i.d.) samples from a population.

In most published data mining work on new modelling methods, this is all that is required for designing experiments to compare and evaluate the models. However, in data mining applications a further consideration is required: the purpose of the prediction model. In many practical scenarios, the modelling purpose is to score new data to make a prediction of the dependent variable being modelled. In other words, the application is one of prediction or forecasting. In these applications, we show in this paper that there can be an advantage to designing the train/validation/test datasets to be stratified by time. Where there is significant variation across time, one advantage of stratifying by time is better model performance where the model selection method uses a validation dataset. For example, a decision tree method can use the validation dataset to determine optimum tree size.

The second advantage is a more accurate estimate of generalization performance of the model. In data mining practice, I have observed many case studies where the modellers involved did not stratify by time, so the models did not account for time variation and so made over optimisitic predictions of model performance when deployed. The stratify-by-time sampling design proposal made here is relatively simple and we expect that some experienced data modellers will consider it obvious. However we have not found the sampling design proposed here considered in the standard texts [2,3,4] or in the research literature. This paper's contribution is therefore to specify the design and preconditions for its use, and to give an application example.

A. Sattar and B.H. Kang (Eds.): AI 2006, LNAI 4304, pp. 1064–1068, 2006.
© Springer-Verlag Berlin Heidelberg 2006

2   Stratify-by-Time Sampling Design

We assume that there is sufficient data for using the training/validation/test approach. For smaller datasets, where cross-validation is used instead, a stratify-by-time design may still help, but we do not consider it here further.

We assume that the model purpose is to make predictions by being applied to new data and is being built with past/current data. We consider that the past/current data can be grouped (or stratified) by time, as well as other possible attributes. Common stratification attributes include geographical, customer demographic and product type.

There will be sampling variation in the data between the time groups in the available data. We will have *a priori* expectations about how the available data will differ from the new data. Both sampling variation over time and expectations of differences between the new data and available data determine the optimum sampling design. The size and nature of the sampling variation across time will determine what benefits, if any, can arise by use of a stratify-by-time design. The candidate designs are shown in Figure 1.

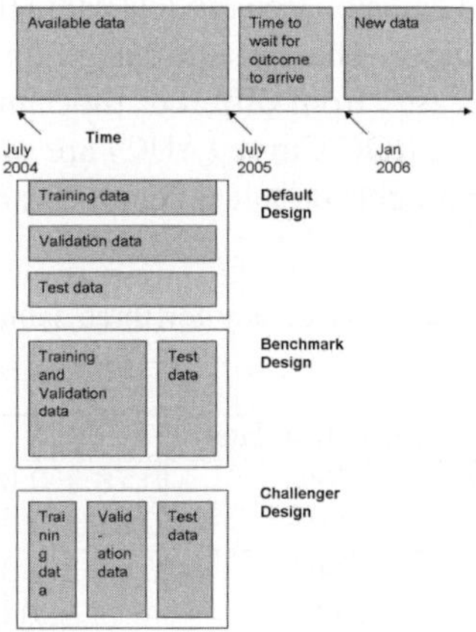

Fig. 1. Schematic diagram of candidate sampling designs across time

3   Results

We use an example from a collection scoring application, where a debtors likelihood of payment in the next 180 days is estimated.

63 variables based on debtors demographics, debt collection actions to date and prior payment history were obtained for each debt case. 800K debt cases (a subset of total debt case population) for July 1 2004 to June 30 2005 made up the available modelling data. Note that for deployment in January 2006, the most recent available labelled test data was June 2005, because 6 months is needed to obtain the case target outcome of debt payment within 180 days.

3.1 Sampling Design

We consider the three designs discussed in Section 2. The first is the *default design* scenario which partitions the available data randomly into training, validation and test datasets in 40%/30%/30% split. Therefore variation across time is ignored. This default design is commonly used, partly because the modelling tools available default to this design.

The second design is the *benchmark* design. We select training and validation cases from a random sample between July 2004 and December 2004, leaving the Jan 2005-June 2005 time period for the test sample. The training and validation samples are then obtained by a random 66% / 33% split.

The third design is our *challenger* design. It differs from the benchmark design in that we partition the training and validation datasets using the creation date for the case. Debt cases created before November 1 2004 are assigned to the training dataset, while debt cases created after November 1 2004 are assigned to the validation dataset.

3.2 Models

We fitted logistic regression, and decision tree models using the three candidate designs.

The results are shown in Table 1. The Default design has out-of-sample validation and test datasets but mixes up cases from different time intervals. The predictive accuracy results and Area under the ROC Curve (AUC) are the highest with this design. However the resulting classifier models do not reflect the variation due to time.

Table 1. Comparing AUC and Predictive Accuracy for three sampling designs and two model classes

Design	Classifier Test dataset	% Error Test dataset	AUC	Uses Validation data?
Default	Logistic	20%	0.78 ± 0.002	N
	Tree	18%	0.81 ± 0.002	Y
Benchmark	Logistic	34%	0.65 ± 0.002	N
	Tree	38%	0.77 ± 0.02	Y
Challenger	Logistic	34%	0.65 ± 0.002	N
	Tree	37%	0.79 ± 0.002	Y

The Benchmark design leads to models with reduced performance on the test dataset relative to the default design. This is because of the variance across time between training and test datasets. For a low-bias, high-variance classifier like the decision tree, the test AUC result has significantly higher variance than than for other designs and models. This is explained by noting that the size of the tree is sensitive to variation in training and validation data samples.

The Challenger design leads to more complex results. For the logistic regression classifier, its performance is similar to the Benchmark, because it doesnt use the validation dataset. For the tree classifier, the perfomance on the test dataset has improved. The out-of-time validation dataset has allowed model selection to select the tree size appropriate to the problem.

The ROC curves for the test dataset for test data for models using different sample designs are shown in Figure 2. This figure shows the over-optimism of the models developed using the default design, while those using the benchmark and challenger design demonstrate lesser but more realistic performance for new data.

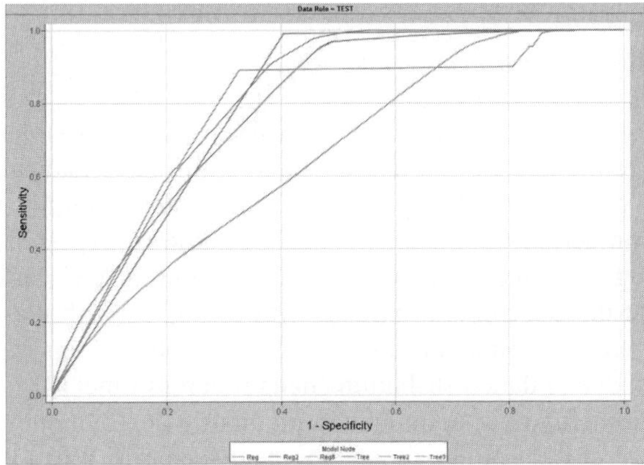

Fig. 2. ROC curves for Test data: Reg and Reg6 (overlaid): Benchmark/Challenger design LR, Reg2: Default Design, Tree: Challenger design, Tree2: Default design, Tree3: Benchmark Design

The logistic regression model was deployed in February 2006 and actual model performance has found to meet estimated model performance very well during first half of 2006. The concordance of expected performance with actual performance has been as important in establishing model credibility as its predictive accuracy. The empirical evidence in favour of the Challenger design here is suggestive rather than conclusive, since it is based on a single application. The observed improvement is minor here. We trust this is due to the nature of the application rather than the utility of stratified designs more generally.

4   Related Work

There are a number of examples where sample designs along the lines of those proposed here have been mentioned in the literature. Witten and Frank[4](p.145) give a credit risk example where data is available from New York and Florida and the purpose is to predict credit for customers in Nebraska. They recommend stratifying by state, with training dataset using New York and test dataset using Florida. Our sampling design provides the framework for this example and our empirical results support the example's intuition. Maindonald[5] gives a housing price prediction example, where purpose is to predict house prices for new suburbs.

Dhar and Stein[6] have proposed a Test Strategy Matrix, similar to the stratify-by-time design, proposed in this paper. Their focus was financial applications in credit risk and marketing. Their motivation is similar but with a focus on robust performance estimates. They discuss the issues but do not report on results in an actual deployment as we do here. We add to their approach by also recommending the use of an across-time validation dataset to improve model selection as well. Zadrozny and Elkan[7,8] have described a taxonomy of categories where distributions change between training and test datasets.

5   Conclusion

We note that the standard UCI[9] datasets don't have time variables in the available clean-up datasets, although the original datasets would have had time variables. The credit application datasets and the web visit datasets are examples where this is the case. With variation-across-time and new data with known expected differences from available data being near ubiquitous in financial data mining applications, this suggests a need to augment the benchmark datasets with examples with these characteristics. Hand has made a similar comment in the context of noting that simple classifiers are often good enough due to data distribution changes across time[10].

In order to avoid being over optimistic in predicting deployment results, the design should be chosen to closely mimic the deployment scenario. When this is done, using training/validation/test dataset stratify-by-time design can give better models (by using time-stratified validation dataset) and better generalization performance estimates (by using time-stratified test datasets).

References

1. Breiman, L.: Statistical modeling: The two cultures. Statistical Sciences **16** (2001) 199–231
2. Hand, D., Mannila, H., Smyth, P.: Principles of data mining. MIT Press, Cambridge, MA (2001)
3. Han, J., Kimber, M.: Data Mining: Concepts and techniques. Morgan Kaufmann, San Francisco (2001)
4. Witten, I., Frank, E.: Data mining: Machine Learning Tools and Techniques. Morgan Kaufmann (2005)
5. Maindonald, J.: The role of models in predictive validation (statistics for budding data miners). ISIS (2003)
6. Dhar, V., Stein, R.: Finding robust and usable models from data mining. PCAI (1998)
7. Zadrozny, B. In: Learning and Evaluating Classifiers under Sample Selection Bias. (2004)
8. Elkan, C.: Foundations of cost-sensitive learning (2001)
9. Blake, C., Keogh, E., Merz, C.: Uci repository of machine learning databases. UCI website (2001)
10. Hand, D.: Classifier technology and the illusion of progress. Statistical Science **21** (2006) 1–15

Clustering Transactional Data Streams

Yanrong Li and Raj P. Gopalan

Department of Computing, Curtin University of Technology
Kent Street, Bentley
Western Australia 6102
{liyl, raj}@computing.edu.au

Abstract. The challenge of mining data streams is three fold. Firstly, an algorithm for a particular data mining task is subject to the sequential one-pass constraint; secondly, it must work under bounded resources such as memory and disk space; thirdly, it should have capabilities to answer time-sensitive queries. Dealing with transactional data streams is even more challenging due to their high dimensionality and sparseness. In this paper, algorithms for clustering transactional data streams are proposed by incorporating the incremental clustering algorithm INCLUS into the *equal-width time window* model and the *elastic time window* model. These algorithms can efficiently cluster a transactional data stream in one pass and answer time sensitive queries at different granularities with limited resources.

1 Introduction

The challenge of designing algorithms for data stream mining is three fold: (1) The algorithm is subject to the sequential one-pass constraint over the data; (2) It must work under limited resources with respect to the unlimited data stream; (3) It should be able to reveal changes in the data stream over time. In the last few years, there has been active research into mining data streams for clustering, classification and frequent pattern discovery.

In data stream settings, the set of data points to be studied can be the whole data stream or a part of it depending on the purpose of mining. There exist four prominent models for filtering data points to be studied in data stream environments. The *landmark* model [1] assumes that data points to be computed are the entire data stream while the *sliding window* model [2, 3] assumes that only the most recent data in a data stream are of interest. In the *tilted time window* model [4], at any moment, a data stream is partitioned into windows of different granularities with respect to the time of their arrival. The most recent data has the finest granularity while the most remote has the coarsest. In the *pyramidal time window* model [5], a data stream is partitioned into windows based on various granularities, but only a certain number of windows is kept at any given time for each granularity. For approximation of changes in the data stream, both the *tilted time window* model and the *pyramidal time window* model can be employed but not the others.

Guha et al [1, 6] analytically studied the clustering problem for numeric data streams. They proposed a constant-factor approximation algorithm for the k-median problem in the *landmark* model. Babcock et al [3] presented a novel technique for

A. Sattar and B.H. Kang (Eds.): AI 2006, LNAI 4304, pp. 1069–1073, 2006.
© Springer-Verlag Berlin Heidelberg 2006

maintaining variance and k-medians in the *sliding window* model. Aggarwal et al [5] proposed the CluStream framework for clustering numeric data streams in the *pyramid time window* model. Based on the CluStream framework, SCLOPE [7] extended the CLOPE algorithm for clustering categorical data streams.

In this paper, we propose an *equal-width time window* model where the width of the window is the minimum granularity of interest for a particular application and an *elastic window* model where the size of windows is adaptively resized based on the changes in clustering. We have designed algorithms specific to transactional data stream clustering which incorporate INCLUS [8] into these models so that changes over the data stream can be estimated.

The rest of the paper is organized as follows. The proposed models and corresponding algorithms are described in Section 2 and the experimental results are given in Section 3. Section 4 concludes the paper.

2 Algorithm for Clustering Transactional Data Streams

A *transactional data stream D* consists of transactions T_1, T_2, ... over a set I of d distinct items (attributes) arriving at times t_1, t_2, ...

Definition 1 (Cluster snapshot). A cluster snapshot for a set of transactional data points in a time window is $C(H, t_s, t_f, N)$, where H is the histogram of the cluster C, t_s and t_f are the start and finish times of the window, and N is the total number of transactions in the cluster.

Definition 2 (Clustering snapshot). The clustering snapshot is the set of cluster snapshots for a time window.

Based on these definitions, the following properties can be obtained immediately.

Additive Property 1. If C_1 $(H_1, t_{s1}, t_{f1}, N_1)$ and $C_2(H_2, t_{s2}, t_{f2}, N_2)$ are two cluster snapshots in different clustering snapshots where $t_{f1} = t_{s2}$, then the cluster snapshot for $C=C_1 \cup C_2$, is $C(H_1+H_2, t_{s1}, t_{f2}, N_1+N_2)$.

Additive Property 2. If C_1 (H_1, t_s, t_f, N_1) and $C_2(H_2, t_s, t_f, N_2)$ are two cluster snapshots in the same clustering snapshot, then the cluster snapshot for $C=C_1 \cup C_2$ is $C(H_1+H_2, t_s, t_f, N_1+N_2)$.

Subtractive Property. If C_1 (H_1, t_s, t_{f1}, N_1) and $C_2(H_2, t_s, t_{f2}, N_2)$ are two cluster snapshots in different clustering snapshots where $t_{f2} > t_{f1}$, then the cluster snapshot for $C=C_2-C_1$, is $C(H_2-H_1, t_{f1}, t_{f2}, N_2-N_1)$.

Additive Property 1 can be used to merge cluster snapshots in consecutive windows while Additive Property 2 can be used to merge cluster snapshots in the same window.

Definition 3 (Clustering granularity). Clustering granularity is the time period upon which clustering is performed.

Definition 4 (Minimum clustering granularity). For a given application, the minimum granularity is the time period upon which the clustering information is to be maintained to enable the time sensitive queries at the same or coarser granularities.

We propose an *equal-width time window* model where the width of each window is equal to the minimum clustering granularity. The additive properties of cluster snapshots ensure that clustering for coarser time granularity can be obtained from the results of the minimum granularity. These properties also indicate that it is not necessary to store snapshots for every window of the finest granularity; consecutive windows with same clustering features can be merged to save disk space, thus making the window size stretchable, and so we call it the *elastic window* model. In the best case, only two clustering snapshots (including the snapshot for the first window) are stored when clustering features do not change over time.

We propose clustering algorithms by incorporating the INCLUS algorithm [8] into the *elastic window* model and the *equal-width time window* model, respectively. Due to space limitations, we refer readers to [8] for INCLUS and also describe only the algorithm for the *elastic window* model here. The algorithm is called CluTrans and has an online component (see *Algorithm 1*) and an offline component(see *Algorithm 2*).

Algorithm 1: Online component of CluTrans

Input: data stream D, minimum clustering granularity, minimum support, minimum similarity
Ouput: clustering snapshots
  1. create a cluster with the first transaction;
  2. while not end of the first widow
  3.   read the next transaction T;
  4.   allocate T to an existing cluster or a new cluster to maximize total similarity $Sim(P)$[8];
  5.   update the cluster representative of the modified cluster;
  6.   merge clusters according to the available memory if necessary;
  7: write clustering snapshots to disk;
  8. repeat for the rest of the data stream
  9.   read next transaction T;
  10. if T cannot be assigned to an existing cluster
  11.   write clustering snapshots to disk;
  12.   create a new cluster for T;
  13.   merge clusters according to the available memory if needed;
  14. else
  15.   allocate T to an existing cluster;

Algorithm 2: Offline component of CluTrans

Input: cluster snapshots, period1, period2.
Ouput: clusterings for period1 and period2.
  1. for each query period i
  2.   get all the clustering snapshots for the period;
  3.   compute clustering for the period using properties in Section 3;
  4. end for
  5. output clustering results;

Line 1-5 of *Algorithm 1* are the main part of INCLUS. Except for the first window, clustering snapshots will be stored to disk when changes occur in the data stream. When the number of clusters exceeds the available memory, some clusters will be merged and the cluster snapshots for the newly derived clusters can be obtained using Additive Property 2. When a user wants to compare clusterings at any two different time periods for a given granularity, *Algorithm 2* will retrieve all the cluster snapshots within the query period, merge clusters if necessary within the same clustering snapshots and/or different clustering snapshots by applying the additive and/or subtractive properties listed in Section 3, and then present the results to the user.

The offline component of the algorithm for the *equal width window* model consists of repeating lines 1-7 in *Algorithm 1* for each window while the offline component is the same as that for the *elastic window* model.

3 Evaluation of CluTrans Algorithm

First, we test the accuracy and performance of the algorithms using the Mushroom data set downloaded from UCI machine learning repository. It has 8124 records. We assume that the minimum clustering granularity is 812 transactions and test the algorithm under the equal-width window model. The other input parameters are minsup = 100% and minsim=60%. We use *purity*, the percentage of correctly clustered records, as the measure of accuracy. The result is shown in Table 1.

Table 1. Testing result for the Mushroom data set

	W_1	W_2	W_3	W_4	W_5	W_6	W_7	W_8	W_9	W_{10}
k	22	20	16	6	7	14	15	19	11	15
p(%)	92.1	99.3	99.3	100	99.9	99.1	97.4	99.3	100	100
t(s)	0.58	0.59	0.44	0	0.13	0.41	0.41	0.48	0.29	0.6

Note: W_i-ith window , k-number of clusters, p-purity, t-runtime in seconds

If we treat the Mushroom data set as streaming data, then we can see that the clustering is changing over time. In the first window, there are 22 clusters while in the fourth window there are only 6 clusters. The processing time for each window varies with an average of 2000 transactions per second. The average purity we obtained is 98.6% which is higher than that for any combination of micro-clusters and window sizes reported in SCLOPE [7].

Second, we perform scalability tests. Fig. 1(a) shows the results for the data set with 1000 attributes but varying number of transactions from 100K to 500K. Fig. 1(b) shows the results for data sets with 10K transactions but varying number of attributes from 500 to 4000. It can be seen that CluTrans is scalable as the number of transactions and the number of attributes increase.

Third, we compare the disk space usage by equal width window model and elastic window model. We obtain a new database by appending mushroom to itself 7 times to model a data stream where underlying data generation mechanism does not change. We run the algorithm for the elastic window model and equal-width model, respectively. The input parameters minsup=60% and minsim=45%, the window size is 8124, the size of mushroom data set. As expected, only two clustering snapshots are stored for the *elastic window* model while eight clustering snapshots are stored for the *equal-width window* model. The disk usage is largely reduced by using the *elastic window* model in this case.

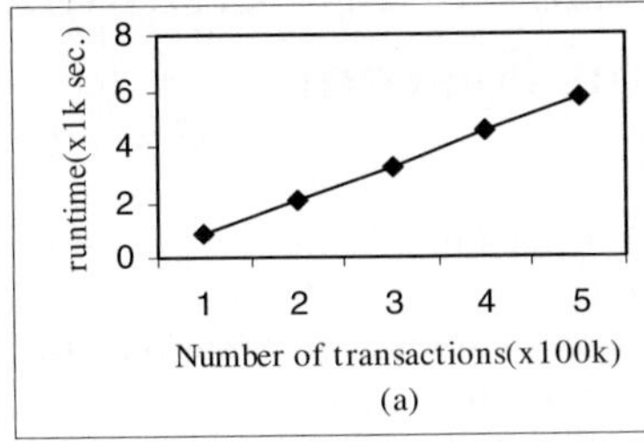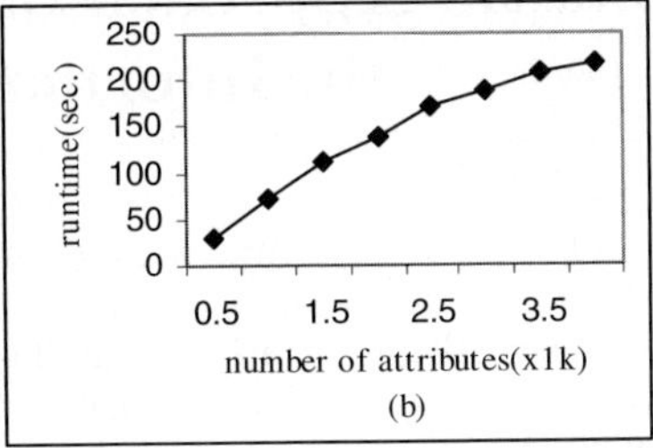

Fig. 1. Scalability test using synthetic data sets

4 Conclusions

In this paper, we proposed an *equal-width window* model for cluster analysis of data streams, where the width of the window is the minimum granularity of interest for a particular application. We also proposed an *elastic window* model where the size of windows is adaptively modified based on the changes in clustering, reducing the disk space requirements when clustering in the data streams does not change over time. INCLUS algorithm is incorporated into these models to detect clustering changes at different granularities. The empirical results show that the algorithms are accurate, efficient and scalable. The proposed future work is to address the systematic measurement of clustering changes in the data stream.

References

1. Guha, S., et al.: Clustering Data Streams: Theory and Practice, TKDE Special Issue on Clustering **15**(3): (2003) 515-528
2. O'Callaghan, L., et al.: Streaming-Data Algorithms for High-Quality Clustering. In: Proc. of ICDE (2002)
3. Babcock, B., et al: Maintaining Variance and k-Medians over Data Stream Windows. In: Proc. of PODS (2003)
4. Giannella, C., et al.: Mining Frequent Patterns in Data Streams at Multiple Time Granularities. In: H. Kargupta, et al (eds): Next Generation Data Mining (2003)
5. Aggarwal, C.C., et al: *A* framework for clustering evolving data streams. In: Proc. Of VLDB (2003)
6. Guha, S., et al: Clustering Data Streams. In: Proc. of FOCS (2000)
7. Ong, K.-L., et al.: SCLOPE: An Algorithm for Clustering Data Streams of Categorical Attributes. In: Proc. Of DaWaK (2004)
8. Li, Y. Gopalan, R.P. Clustering High Dimensional Sparse Transactional Data with Constraints. In Proc. of IEEE-GrC (2006). Atlanta, USA.

Axiomatic Approach of Knowledge Granulation in Information System

Jiye Liang[1] and Yuhua Qian[2]

Key Laboratory of Computational Intelligence and Chinese Information Processing of
Ministry of Education
School of Computer and Information Technology, Shanxi University
Taiyuan, 030006, People's Republic of China
[1]ljy@sxu.edu.cn,
[2]jinchengqyh@126.com

Abstract. Granular computing is potentially in knowledge discovery and data mining etc. In this paper, by introducing a partial relation $\preceq'$ with set size character to information system, an axiom definition of knowledge granulation for information system is presented, some existing the definitions of knowledge granulation become special forms. These results will be very helpful for understanding the essence of knowledge granulation and uncertainty measurement in information system.

Keywords: Information system, partial relation, knowledge granulation.

1 Introduction

The notion of information system (sometimes called data tables, attribute-value system, knowledge representation system, etc.), provides a convenient tool for the representation of objects in terms of their attribute values. Rough set theory has been introduced to deal with inexact, uncertain, or vague knowledge in information system. The use of the indiscernibility relation results in knowledge granulation.

According to whether or not there are missing data (null values), the information system can be classified into two categories: complete and incomplete. The knowledge granulation of an information system gives a measure of uncertainty about its actual structure[1-6]. In general, knowledge granulation can represent discernibility ability of knowledge in rough set theory, the smaller knowledge granulation is, the stronger its discernibility ability is [7-9]. Especially, several measures in information system closely associated with granular computing such as measure, information entropy, rough entropy and knowledge granulation and their relationships are discussed by Liang et al. in [6, 9]. In [7], combination granulation and combination entropy in information system are proposed, their gain function possesses intuitionistic knowledge content characteristic.

In this paper, we devote to the axiomatic approach of knowledge granulation in information system. In section 2, we review some basic concepts of information system, and proposed a new partial relation $\preceq'$ with set size character to an information system. In section 3, we give the axiom definition of knowledge

A. Sattar and B.H. Kang (Eds.): AI 2006, LNAI 4304, pp. 1074–1078, 2006.
© Springer-Verlag Berlin Heidelberg 2006

granulation to measure uncertainty of knowledge in formation system, prove that several existing knowledge granulations are all special instances of the definition. Finally, section 4 concludes the whole paper.

2   Preliminaries

An information system is a pair $S = (U, A)$, where,

(1) U is a non-empty finite set of objects;
(2) A is a non-empty finite set of attributes;
(3) for every $a \in A$, there is a mapping a, $a : U \to V_a$, where V_a is called the value set of a.

For an information system $S = (U, A)$, if $\forall a \in A$, every element in V_a is a definite value, then S is called a complete information system. If V_a contains a null value for at least one attribute $a \in A$, then S is called an incomplete information system, otherwise it is complete.

Let $S = (U, A)$ be an information system, $P, Q \subseteq A$. $K(P) = \{S_P(x_i) \mid x_i \in U\}$, $K(Q) = \{S_Q(x_i) \mid x_i \in U\}$. We define a partial relation $\preceq'$ with set size character as follows: $P \preceq' Q$ $(P, Q \in A)$, if and only if, for $K(P) = \{S_P(x_1), S_P(x_2), \cdots, S_P(x_{|U|})\}$, there exists a sequence $K'(Q)$ of $K(Q)$, where $K'(Q) = \{S_Q(x_1'), S_Q(x_2'), \cdots, S_Q(x_{|U|}')\}$, such that $|S_P(x_i)| \leq |S_Q(x_i')|$. If there exists a sequence $K'(Q)$ of $K(Q)$ such that $|S_P(x_i)| < |S_Q(x_i')|$, then we will call that P is strict granulation finer than Q, and denote it by $P \prec' Q$.

3   Axiomatic Construction of Knowledge Granulation

In 1979, the problem of fuzzy information granule was introduced by Zadeh in [10]. Especially, several measures in an information system closely associated with granular computing such as granulation measure, information entropy, rough entropy and knowledge granulation and their relationships were discussed in [6,9]. However, there exists no unified description for knowledge granulation. In the following, an axiom definition of knowledge granulation is given.

Definition 1. *Let $S = (U, A)$ be an information system, G be a mapping from the power set of A to the set of real numbers. We say that G is a knowledge granulation in an information system if G satisfies the following conditions:*

(1) $G(P) \geq 0$ for any $P \subseteq A$ (Non-negativity);

(2) $G(P) = G(Q)$ for any $P, Q \in A$ if there is a bijective mapping function $f : K(P) \to K(Q)$ such that $|S_P(u_i)| = |f(S_P(u_i))|$ ($\forall i \in \{1, 2, \cdots, |U|\}$), where $K(P) = \{S_P(x_i) \mid x_i \in U\}$ and $K(Q) = \{S_Q(x_i) \mid x_i \in U\}$ (Invariability);

(3) $G(P) < G(Q)$ for any $P, Q \in A$ with $P \preceq' Q$ (Monotonicity).

Theorem 1 (Extremum). *Let $S = (U, A)$ be an information system, $\forall P \in A$, then if $U/P = \omega = \{\{u\} \mid u \in U\}$, $G(P)$ achieves minimum value; if $U/P = \delta = \{U \mid u \in U\}$, $G(P)$ achieves maximum value; where ω denotes identity relation, δ denotes universal relation.*

In [14,15,17,18], some different kinds of knowledge granulations were given, we can prove that these knowledge granulations are all special forms under definition 1.

Definition 2. [17] *Let $S = (U, A)$ be a complete information system, $U/IND(A) = \{P_1, P_2, \cdots, P_m\}$. Knowledge granulation of A is defined by*

$$GK(A) = \frac{1}{|U|^2} \sum_{i=1}^{m} |P_i|^2. \tag{1}$$

Theorem 2. *GK in definition 2 is a knowledge granulation under definition 1.*

Proof. (1) Obviously, it is non-negative;

(2) Let $P, Q \subseteq A$, then $U/IND(P) = \{P_1, P_2, \cdots, P_m\}$ and $U/IND(Q) = \{Q_1, Q_2, \cdots, Q_n\}$ in complete information system can be uniformly denoted by $U/SIM(P) = \{S_P(u_1), S_P(u_2), \cdots, S_P(u_{|U|})\}$ and $U/SIM(Q) = \{S_Q(u_1), S_Q(u_2), \cdots, S_Q(u_{|U|})\}$.

Suppose that there be a bijective mapping function $f\colon U/SIM(P) \to U/SIM(Q)$ such that $|S_P(u_i)| = |f(S_P(u_i))|$ ($i \in \{1, 2, \cdots, |U|\}$) and $f(S_P(u_i)) = S_Q(u_{j_i})$ ($j_i \in \{1, 2, \cdots, |U|\}$), then we have that

$$GK(P) = \frac{1}{|U|^2} \sum_{i=1}^{m} |P_i|^2 = \frac{1}{|U|^2} \sum_{i=1}^{|U|} |S_P(u_i)|$$

$$= \frac{1}{|U|^2} \sum_{i=1}^{|U|} |S_Q(u_{j_i})| = \frac{1}{|U|^2} \sum_{i=1}^{|U|} |S_Q(u_i)|$$

$$= \frac{1}{|U|^2} \sum_{j=1}^{n} |Q_j|^2 = GK(Q);$$

(3) Let $P, Q \subseteq A$, $U/SIM(P) = \{S_P(u_1), S_P(u_2), \cdots, S_P(u_{|U|})\}$, $U/SIM(Q) = S_Q(u_1), S_Q(u_2), \cdots, S_Q(u_{|U|})$ and $P \prec' Q$, then for arbitrary $S_P(u_i)(i \leq |U|)$, there exists a sequence $\{S_Q(u_1'), S_Q(u_2'), \cdots, S_Q(u_{|U|}')\}$ such that $|S_P(u_i)| < |S_Q(u_i')|(i = 1, 2, \cdots, |U|)$. Hence, we obtain that

$$GK(P) = \frac{1}{|U|^2} \sum_{i=1}^{m} |P_i|^2 = \frac{1}{|U|^2} \sum_{i=1}^{|U|} |S_P(u_i)|$$

$$< \frac{1}{|U|^2} \sum_{i=1}^{|U|} |S_Q(u_i')| = \frac{1}{|U|^2} \sum_{i=1}^{|U|} |S_Q(u_i)|$$

$$= \frac{1}{|U|^2} \sum_{j=1}^{n} |Q_j|^2 = GK(Q).$$

Thus, GK in definition 2 is the knowledge granulation under definition 1.

Definition 3. [14] *Let $S = (U, A)$ be an incomplete information system, $U/SIM(A) = \{S_A(u_1), S_A(u_2), \cdots, S_A(u_{|U|})\}$. Knowledge granulation of A is defined by*

$$GK(A) = \frac{1}{|U|^2} \sum_{i=1}^{m} |S_A(u_i)|. \tag{2}$$

Theorem 3. *GK in definition 3 is a knowledge granulation under definition 1.*

Proof. Similar to theorem 2, it can be proved.

Definition 4. [18] *Let $S=(U, A)$ be a complete information system, $U/IND(A) = \{P_1, P_2, \cdots, P_m\}$. Combination granulation of A is defined by*

$$CG(A) = \sum_{i=1}^{m} \frac{|P_i|}{|U|} \frac{C^2_{|P_i|}}{C^2_{|U|}}. \tag{3}$$

Theorem 4. *CG in definition 4 is a knowledge granulation under definition 1.*

Proof. (1) Obviously, it is non-negative;

(2) Let $P, Q \subseteq A$, then $U/IND(P) = \{P_1, P_2, \cdots, P_m\}$ and $U/IND(Q) = \{Q_1, Q_2, \cdots, Q_n\}$ in complete information system can be uniformly denoted by $U/SIM(P) = \{S_P(u_1), S_P(u_2), \cdots, S_P(u_{|U|})\}$ and $U/SIM(Q) = \{S_Q(u_1), S_Q(u_2), \cdots, S_Q(u_{|U|})\}$.

Suppose that there be a bijective mapping function $f: U/SIM(P) \to U/SIM(Q)$ such that $|S_P(u_i)| = |f(S_P(u_i))|$ ($i \in \{1, 2, \cdots, |U|\}$) and $f(S_P(u_i)) = S_Q(u_{j_i})$ ($j_i \in \{1, 2, \cdots, |U|\}$), then we have that

$$CG(P) = \sum_{i=1}^{m} \frac{|P_i|}{|U|} \frac{C^2_{|P_i|}}{C^2_{|U|}} = \sum_{i=1}^{|U|} \frac{|S_P(u_i)|}{|U|} \frac{C^2_{|S_P(u_i)|}}{C^2_{|U|}}$$

$$= \sum_{i=1}^{|U|} \frac{|S_Q(u_{j_i})|}{|U|} \frac{C^2_{|S_Q(u_{j_i})|}}{C^2_{|U|}} = \sum_{i=1}^{|U|} \frac{|S_Q(u_i)|}{|U|} \frac{C^2_{|S_Q(u_i)|}}{C^2_{|U|}}$$

$$= \sum_{j=1}^{n} \frac{|Q_j|}{|U|} \frac{C^2_{|Q_j|}}{C^2_{|U|}} = CG(Q);$$

(3) Let $P, Q \subseteq A$, $U/SIM(P) = \{S_P(u_1), S_P(u_2), \cdots, S_P(u_{|U|})\}$, $U/SIM(Q) = S_Q(u_1), S_Q(u_2), \cdots, S_Q(u_{|U|})$ and $P \prec' Q$, then for arbitrary $S_P(u_i)(i \leq |U|)$, there exists a sequence $\{S_Q(u'_1), S_Q(u'_2), \cdots, S_Q(u'_{|U|})\}$ such that $|S_P(u_i)| < |S_Q(u'_i)|(i = 1, 2, \cdots, |U|)$. Hence, we obtain that

$$CG(P) = \sum_{i=1}^{m} \frac{|P_i|}{|U|} \frac{C^2_{|P_i|}}{C^2_{|U|}} = \sum_{i=1}^{|U|} \frac{|S_P(u_i)|}{|U|} \frac{C^2_{|S_P(u_i)|}}{C^2_{|U|}}$$

$$< \sum_{i=1}^{|U|} \frac{S_Q(u'_i)}{|U|} \frac{C^2_{|S_Q(u'_i)|}}{C^2_{|U|}} = \sum_{i=1}^{|U|} \frac{S_Q(u_i)}{|U|} \frac{C^2_{|S_Q(u_i)|}}{C^2_{|U|}}$$

$$= \sum_{j=1}^{n} \frac{|Q_j|}{|U|} \frac{C^2_{|Q_j|}}{C^2_{|U|}} = CG(Q).$$

Thus, CG in definition 4 is the knowledge granulation under definition 1.

Definition 5. [15] *Let $S = (U, A)$ be an incomplete information system, $U/SIM(A) = \{S_A(u_1), S_A(u_2), \cdots, S_A(u_{|U|})\}$. Combination granulation of A is defined by*

$$CG(A) = \frac{1}{|U|} \sum_{i=1}^{|U|} \frac{C^2_{|S_A(u_i)|}}{C^2_{|U|}}. \tag{4}$$

Theorem 5. *CG in definition 5 is a knowledge granulation under definition 1.*

Proof. Similar to theorem 4, it can be proved.

4   Conclusions

In the present research, the concept of partial relation $\preceq'$ with set size character in information system is introduced. Furthermore, we give the axiom definition of knowledge granulation to measure uncertainty of knowledge in formation system, prove that several existing knowledge granulations are all special instances of the definition. Present conclusions appears to be well suited for measuring uncertainty of knowledge in information system.

Acknowledgements. This work was supported by the national natural science foundation of China (No. 70471003, No. 60573074, No. 60275019), the foundation of doctoral program research of the ministry of education of China (No. 20050108004), the top scholar foundation of Shanxi, China, and key project of science and technology research of the ministry of education of China.

References

1. Chakik, F.E., Shahine, A., Jaam, J. and Hasnah, A.: An approach for constructing complex discriminating surfaces based on bayesian interference of the maximum entropy. *Information Sciences.* 163 (2004) 275-291
2. Düntsch, I. and Gediga, G.: Uncertainty measures of rough set prediction. *Artificial Intelligence.* 106 (1998) 109-137
3. Liang, J.Y., Chin, K.S., Dang, C.Y. and Richard C.M.Yam.: A new method for measuring uncertainty and fuzziness in rough set theory. *International Journal of General Systems.* 31 (4) (2002) 331-342
4. Kryszkiewicz, M.: Rules in incomplete information systems. *Information systems.* 113 (1999) 271-292
5. Liang, J.Y., Xu, Z.B.: The algorithm on knowledge reduction in incomplete information systems. *International Journal of Uncertainty, Fuzziness and Knowledge-Based Systems.* 24 (1) (2002) 95-103
6. Liang, J.Y., Shi, Z.Z., Li, D.Y. and Wierman, M.J.: The information entropy, rough entropy and knowledge granulation in incomplete information system. *International Journal of General Systems.* (to appear)
7. Qian, Y.H., Liang, J.Y.: Combination entropy and combination granulation in incomplete information system. *Lecture Notes in Artificial Intelligence.* 4062 (2006) 184-190
8. Leung, Y., Li, D.Y.: Maximal consistent block technique for rule acquisition in incomplete information systems. *Information Sciences.* 153 (2003) 85-106
9. Liang, J.Y., Shi, Z.Z.: The information entropy, rough entropy and knowledge granulation in rough set theory. *International Journal of Uncertainty, Fuzziness and Knowledge-Based Systems.* 12 (1) (2004) 37-46
10. Zadeh, L.A.: Fuzzy sets and information granularity. In: Gupta, M. and Yager, R. (Eds): Advances in Fuzzy Set Theory and Application. North-Holland, Amsterdam. (1979) 3-18

Spreading Design Patterns with Semantic Web Technologies

Susana Montero, Paloma Díaz, Ignacio Aedo, and Laura Montells

Laboratorio DEI. Dpto. de Informática
Universidad Carlos III de Madrid
Avda. de la Universidad 30. 28911 Leganés, Spain
`smontero@inf.uc3m.es, pdp@inf.uc3m.es, aedo@ia.uc3m.es,`
`lmontell@inf.uc3m.es`
`http://www.dei.inf.uc3m.es`

Abstract. There are a huge number of design patterns that can be used in specific software domains, not only in object-oriented software design. Several of these collections are disseminated as structured but informal documents, at best supported with hypertext tools on the web. These passive representations compromise usability of design patterns. We proposed to enrich the textual pattern description with semantic annotations that enable a machine and a human understanding, like in the Semantic Web. In this paper, we present an annotation tool to support the development of semantic pattern repositories. Moreover, in order to show the utility of this approach two software tools with different purposes have been developed: a semantic web repository to explore patterns and a pattern wizard to apply patterns into a design toolkit.

1   Introduction and Motivation

The oft-cited definition of a design pattern, *"a solution to a problem in a context"*, is so general that allows patterns to be formed from knowledge and expertise in any domain. Indeed, there are a huge number of patterns that can be used in specific domains such as object orientation, hci, security, hypermedia, e-commerce or intelligent systems. In contrast with the most commonly recognized pattern catalog [1], in most areas do not exist well-established pattern collections. In such cases, patterns are expressed in a very informal way and they are usually disseminated through publications and web sites where finding the appropriate pattern for a specific problem becomes a complicated if not impossible task. There have been proposed several approaches to formalize patterns to ease their reuse, as discussed in [4]. The use of a formal representation allows for a rigorous reasoning process but this advantage can turn into a strong disadvantage if the user cannot easily understand the patterns and their utility. Thus there is a need for a formalism that facilitates the description, localization and effective reuse of patterns, taking into account that the users of such patterns can be experts in the domain of application (e.g. e-commerce) who are not familiar with formal representations. Such representation should not only focus on the problem and solution gathered in the pattern but also in other parts of the pattern that can

A. Sattar and B.H. Kang (Eds.): AI 2006, LNAI 4304, pp. 1079–1083, 2006.
© Springer-Verlag Berlin Heidelberg 2006

provide valuable information to achieve a successful solution. For example, the intent of a pattern would allow for indexing patterns according to the problem addressed. Taking into account these issues, the Semantic Web technologies can provide all the infrastructure that allows to support different pattern usages [2].

In this context we have applied Semantic Web technologies to support different hypermedia/web pattern usages, combining an ontology-based framework for the formal representation of patterns with semantic annotations. In this way we provide a dual pattern representation that can be shared and reused among users and software systems [4]. This paper is focused on the presentation of the tool implemented to support this approach as well as two other semantic tools developed to spread the application of hypermedia/web patterns: a semantic repository of patterns and a pattern wizard for the integration of patterns into a hypermedia development method.

2 Building Repositories of Annotated Patterns

In order to built applications for patterns using the Semantic Web technologies, we need to have a pattern repository marked up with semantic information.

We have developed AnnotPat (Annotation Pattern) to allow writers and users patterns to annotate their design patterns according to a formal representation based on ontologies. The figure 1 shows the process to produce a pattern repository using the tool. As starting point, the tool needs the ontology that describes both the structure or armature of a pattern (format ontology) and domain-specific knowledge used in the patten description (domain ontology)in order to provide the user terms to make-up annotations. This ontology has to be compliant with our representation framework [4] and expressed in OWL, that is, the format ontology and the domain ontology are combined to describe the pattern

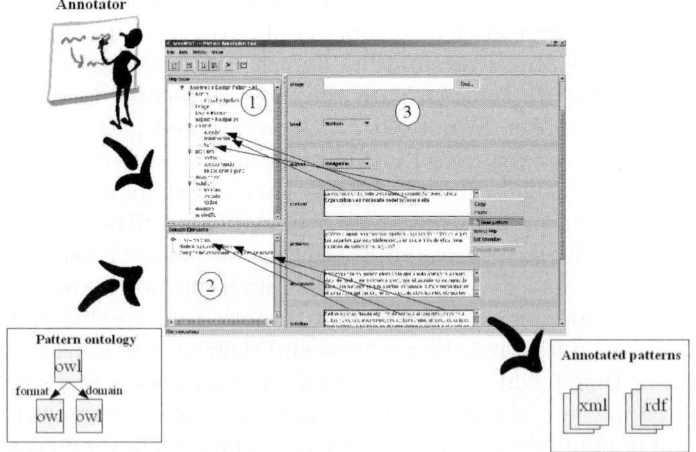

Fig. 1. The AnnotPat tool

ontology for a specific domain. This separation of domains allows our framework to be applied to patterns of any application domain. The annotation tool interface is divided in two sides. The left side shows the status of annotations made by the user separating the pattern format ones (1) and the domain knowledge ones (2). The right side shows the template pattern (3) where the textual descriptions are typed and annotations are made. After typing the pattern, the user can make annotations in order to link words with terms. For that, first she needs to highlight a relevant piece of text and does click with the right mouse button. A pop-up menu is displayed with the possible terms according to the pattern element selected, for easier use.

Regarding technical questions, the tool has been developed on Java version 1.4. As it has been mentioned above, ontologies are expressed on OWL language, whereas annotations are stored in XML format and formalized patterns as RDF files. Jena [3] tool is used to manage these file types and to infer knowledge.

3   Building Semantic Tools for Patterns

In order to show the utility of our approach, we have developed two semantic tools with different purposes, but sharing a pattern repository of hypermedia and web design patterns. These patterns have been annotated using AnnotPat according to the terms and the relationships contained in a domain-specific ontology for hypermedia patterns[1].

3.1   A Semantic Web Repository

The aim of this first tool is to make pattern repositories available for browsing, searching, and visualization through the Web, but in a more expressive way.

The figure 2 shows the interface of the semantic web repository [2]. It is divided into two areas. The left side (1) presents two navigation tools, an index of all patterns organized as a tree folder structure and a search box. The structured searches and links between information are based on annotations made with AnnotPat. The right side of the interface (2) shows the content of a pattern selected or the result of a query. Regarding technical questions, requests from client browsers are received by an Apache Tomcat web server. The semantic web application is supported by the Jena toolkit [3] that allows us to manipulate data and to make queries using the languages for the Semantic Web. Moreover, the results are formatted in HTML by means of XSLT transforms so that patterns can be displayed in conventional web browsers.

3.2   A Pattern Wizard to Apply Patterns into a Design Toolkit

The second tool has as aim the pattern application in a design environment. In particular, this wizard guides designers from requirements to a first modelling

---

[1] http://hypatterns.no-ip.info/ontologies/hypermediaPattern.owl
[2] http://hypatterns.no-ip.info:8080/HyPatterns/

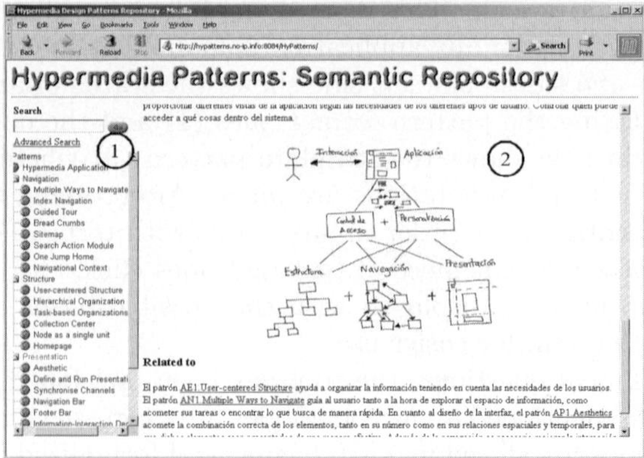

Fig. 2. The semantic web pattern repository

approach of a hypermedia application through the pattern solutions contained in the semantic repository. For each requirement a particular pattern is selected. After, the annotations made in the solution section of the selected patterns are used to compose different design models. The wizard comprises three phases: retrieve all pattern from the semantic repository so that the user can select a pattern; adapt the selected pattern solution to the context, that is, a template with pattern solution is shown according to the annotations to create an instance of the pattern; and compose all the solutions to generate different design models, since these laters are expressed in an OWL ontology as well. The figure 3 shows a screenshot of the wizard when a particular pattern is being instanciated. This wizard has been implemented with the same technologies aforementioned.

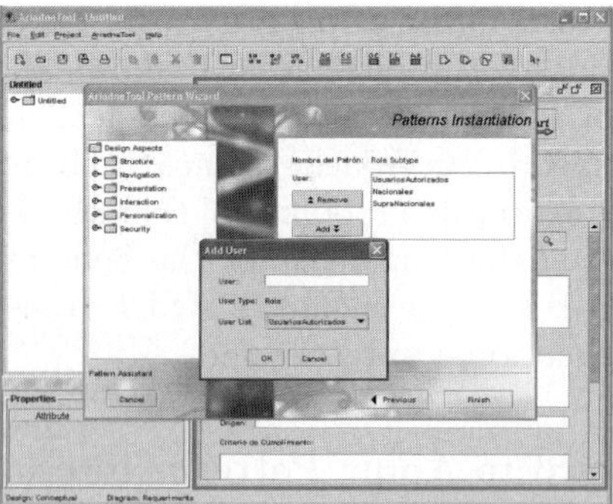

Fig. 3. The wizad pattern

4   Utility Evaluation

We have evaluated the utility of our approach, in particular the use of the AnnotPat tool and annotations as mechanism to formalize patterns. Ten users participated in the evaluation of our prototype. Five of them were familiarized themselves with the use of patterns for design activities, whereas the other five ones were their first contact with patterns. However, none of them had knowledge of ontologies. The evaluation consisted of three tasks. In the first task, the users performed a repository with several patterns using AnnotPat. The second task was to check the repository made from the user annotations could be used by other software tools, in particular, by the web semantic repository presented in the section 3.1. The third task was to compare AnnoPat with an ontology editor like Protégé to make a pattern instance from the same ontologies used by AnnotPat. Moreover, some questions about several parameters of usability such as utility, speed, reliability, interface design, interface adequacy, interface quality and readability were asked.

The overall results indicated that the third part of the participants were 'very satisfied' with respect to the tool usability. All the respondents strongly agreed that Annopat was more suitable to formalize patterns than Protégé. Despite of being the later a complete editor, the participants thought it was less intuitive and more complicated the formalization of a pattern.

5   Conclusions

In this paper, we presented a tool, AnnotPat, that allows designers and domain experts to enrich their domain-specific patterns with semantic annotations using domain concepts and terms. The most valuable contribution of this approach is to be both human and machine readable, since the text pattern description is linked to its formal representation, maintaining the essence of design patterns: express and communicate successful solutions to common problems between practitioners. Finally, we are working on exploring new usages of this approach as an explanation module to support designers decisions during the pattern application.

References

1. E. Gamma, R. Helm, R. Johnson, and J. Vlissides. *Design Patterns, Elements of Reusable Object-Oriented Software*. Addison-Wesley, 1994.
2. S. Henninger, M. Keshk, and R. Kinworthy. Capturing and disseminating usability patterns with semantic web technology. In *CHI Workshop on Perspectives on HCI Patterns: Concepts and Tools*, 2003.
3. http://jena.sourceforge.net/.
4. S. Montero, P. Díaz, and I. Aedo. A semantic representation for domain-specific patterns. In *Metainformatics International Symposium, MIS 2004*, volume 3511 of *LNCS*. Springer, 2004.

Detecting Anomalies and Intruders

Akara Prayote and Paul Compton

School of Computer Science and Engineering*
University of New South Wales, Sydney, NSW, 2052, Australia
{akarap, compton}@cse.unsw.edu.au

Abstract. Brittleness is a well-known problem in expert systems where a conclusion can be made, which human common sense would recognise as impossible e.g. that a male is pregnant. We have extended previous work on prudent expert systems to enable an expert system to recognise when a case is outside its range of experience. We have also used the same technique to detect new patterns of network traffic, suggesting a possible attack. In essence we use Ripple Down Rules to partition a domain, and add new partitions as new situations are identified. Within each supposedly homogeneous partition we use fairly simple statistical techniques to identify anomalous data. The special feature of these statistics is that they are reasonably robust with small amounts of data. This critical situation occurs whenever a new partition is added.

1   Introduction

Brittleness occurs when expert systems do not realise the limits of their own knowledge. The CYC project [4] is an attempt at a solution to this problem by building a knowledge base of common sense as a foundation on which other expert systems could be built on. A variety of applications have used CYC knowledge base, for example, in directed marketing and database cleansing[5].

Brittleness can also be characterised as a failure of the expert system to recognise when a case is outside its range of experience. To build a complete knowledge base that contains all possible knowledge is not easy as some data patterns may never occur in practice and expert justification is quite speculative when judging data patterns outside the expert's experience [1,2].

One attempt to address the brittleness of expert systems is a technique called "prudence" in the RDR paradigm [3,2]. In this work, for every rule the upper and lower bounds of each numerical variable in the data seen by the rule were kept, as well as a list of values seen for enumerated variables. A warning was raised when a new value or a value outside the range seen occurred. The idea was that the system would warn of new types of cases for which a new rule may have to be added. This approach worked well, but the false positive rate was about 15%, because of the simple way in which cases were compared to profiles. This paper extends this previous work using a probabilistic technique to decide if a value is

---

* Part of this work has been submitted elsewhere [6].

A. Sattar and B.H. Kang (Eds.): AI 2006, LNAI 4304, pp. 1084–1088, 2006.
© Springer-Verlag Berlin Heidelberg 2006

an outlier and allowing the expected range for a variable to decrease as well as increase over time. This is critical in dynamic domains where the type of cases seen may change.

The rest of paper is organized as follows. Section 2 discusses the algorithm to detect anomalies. In Section 3, the algorithm was applied to a medical domain as in [3,2]. It is important to note that the proposed algorithm is only for continuous attributes. A simple list of seen values is still preserved for categorical variables. Section 4 is a case study of the system in a dynamic domain. Here we chose an intrusion detection system as a test bed. We conclude the paper in Section 5.

2   Anomaly Detector

In developing a model representation for continuous attributes in dynamic domains, we have made some assumptions as follows: 1) provided a proper segmentation of the domain, an attribute's values should behave similarly; hence, forming a cluster of homogeneous data, 2) a homogeneous cluster of values follows a uniform distribution on an interval $[a, b]$ that is, $P(x < a) = 0$; $P(x > b) = 0$; $P(a \leq x \leq b) = \frac{1}{b-a}$ and the probability of a region $[a', b']$ inside $[a, b]$, i.e., $a' \geq a$ and $b' \leq b$, is $P([a', b']) = \frac{b'-a'}{b-a}$, 3) each variable is independent.

From the above assumptions, the probability that *all n objects* will fall inside a sub-region $[a', b']$ of the interval $[a, b]$, where $a' \geq a$ and $b' \leq b$, is $(\frac{b'-a'}{b-a})^n$. We use this probability to assess whether an object x seen after n objects have been observed, should be included n the model. If a is the minimum, b is the maximum and the object x is outside the range of $[a, b]$, e.g, $x > b$, the object x would only included in the model if the $(\frac{b-a}{x-a})^n > T$, where T is a confidence threshold that the interval should be extended to $[a, x]$. As well, if the range is extended with a new maximum or minimum, we apply the same calculation to the sub-range of the subsequent observed maximum and minimum. This is important as it is possible for the range to have been incorrectly extended by including an outlier, especially when little data has been seen. If T is less than the confidence threshold, the previous maximum or minimum is deleted and replaced by the observed maximum or minimum.

A key feature controlling the algorithm behaviour is the threshold T. Simulations were carried out to find the optimal range of T. The algorithm performed satisfactorily when the threshold was $1.0E - 44 < T < 1.0E - 2$. In the following studies, we used $T = 1.0E - 20$.

3   Anomaly Detection in a Medical Expert System

Following the previous approach [2], we built a knowledge-based system using machine learning (in this case Weka's J48). This KBS is used as a simulated expert in building an RDR KBS. That is, an RDR KBS is built by running cases through the RDR KBS and every time a conclusion is given which differs from the simulated expert's conclusion for that case a new rule is added with the conditions in this rule taken from the inference trace of the simulated expert. We also record whether a warning was generated and whether this was an appropriate warning or not (i.e. a false positive) and also whether a case was misclassified but no warning was raised (i.e. a false negative).

Table 1. Comparison between the original and model-based prudence. There were 20278 cases in the experiment. The metrics of interest are the number of false negative, false positive, true negative and true positive cases.

	False Negatives	False Positives	True Negatives	True Positives
Original prudence	0	3134	16843	301
Model-based prudence	0	2105	17842	301

The experiment was run with two prudence techniques, i.e., the original simple range prudence and model-based probabilistic prudence, on the Garvan data set[1]. The result, shown in Table 1, reveals that the model-based prudence significantly improves the anomaly detection by reducing false positives from 15% to 10% a significant improvement. It should be noted that both techniques had zero false negatives; i.e., prudence detected all the cases where the KBS had made a mistake.

4 Network Traffic Anomaly Detection

Traffic anomaly detection is now a standard task for network administrators, who with experience can generally differentiate anomalous from normal traffic. Many approaches have been proposed to automate this task. Most of them attempt to develop a sufficiently sophisticated model to represent the full range of normal traffic behaviour. The disadvantage with these approaches are 1) a large amount of training data for all acceptable traffic patterns is required to train the model, 2) sophisticated modelling techniques are required to cover the rang of traffic behaviour - the more coverage, the more sophisticated the model.

In contrast, RDR can be used to partition the problem space into smaller subspaces of more homogeneous traffic[2], each of which can be represented with

---

[1] In [2], three data sets from the UC Irvine Machine Learning Repository, i.e., Garvan, Chess, and Tic Tac Toe, were used. Only the Garvan data set contains continuous attributes. We also used a larger Garvan data set than that available through UC Irvine.

[2] While most RDR-based systems are used to capture knowledge from human experts, some RDR work can be characterised as segmenting a domain so that rules have local application. The segmentation can be carried out by anyone who can segment the domain in a reasonable way and does not necessarily need to be done by an expert. Using RDR's refinement structure it does not matter how many segments there are, or whether the best segmentation is initially chosen; the developer can keep adding segments until the domain is appropriately partitioned.

a separate model. The partitioning can be carried out very simply by adding an RDR rule whenever a new situation is encountered. The rule does not provide a conclusion, but simply partitions the space. With the learning algorithm mentioned in Section 2, the model should work reasonably well for new subspaces when little data has been observed.

The data used here are from RRDtool IP flow archives, collected by the network administrator of the School of Computer Science and Engineering, UNSW. Each archive contains seven days data with anomalies marked by hand. We used five consecutive sets of this data, i.e., 5 weeks of data. The system was run from a blind state on the first set of data. With the knowledge learned from the first series, and RDR partitioning, it was run again on the second series. This process was repeated through the five sets of data.

The results are as follows: With no pre-training, the system produced a false positive rate of 6%, with no false negatives on the first series. After the system had learnt some traffic behaviour, the false positive rate produced dropped to 2% on the second set, to1% on the third, increase to 2% on the fourth series. On the fifth series, the false positive rate climbed to 7%. The explanation for this increase (on the fourth and fifth series) is simply that the normal traffic is quite different from previous weeks. The first three weeks are during holidays, the forth series is the first week of the semester, where the pattern is starting to change, and the the fifth series is the second week of semester, and the semester pattern is more established. The profiles learned during recess did not cover these new behaviours; however, the RDR approach allows new partitions to be added at any time, and in the changeover the false positive rate is only 7%. It seems that during semester there is significant variability during the day and the week, but we have not gone far enough to reduce the false positives to zero.

5   Conclusion

Prudence is an attempt to address the brittleness of expert systems by attempting to flag when a case may be misclassified by the expert system. The major challenge in this is to reduce false positives, i.e., unnecessary warnings that a case is misclassified. As new rules may be introduced at any time, starting data collection afresh for each rule, the major challenge is that the technique be robust when there is little data. In this paper, prudence was implemented with the Outlier Estimation with Backward Adaptation algorithm (OEBA) described, which improves performance when little data has been seen. The probability of a new value being a member of the population is assessed, rather than simply raising a warning because the value is new. This gave a significant improvement by reducing the false positive rate, from 15% to 10%. We believe that we can further reduce the false positive rate by combining warnings and ranking cases according to the overall probability of an anomaly derived from all attributes. Again it should be noted that the false negatives are zero - no anomalous cases are missed.

With the current interest in a range of security problems, this type of technique has application beyond prudent expert systems, to detect anomalies in a range of situations. We have extended the approach to network traffic intrusion, an example of a dynamic domain. RDR is used to arbitrarily segment the problem space into sub-spaces of homogeneous traffic; each of which was maintained by a separate model again with the OEBA learning algorithm to enable anomaly detection to function reasonably when little data has been observed in a new partition. The system successfully detected traffic anomalies, with low false positive and false negative rate. The false negative rate was zero after one weeks training. It also yielded a better F-measure than the classic Holt-Winters algorithm.

In summary, model-based anomaly detection requires a deep understanding of the functionality and structure of the domain to construct models. In our framework, models are not needed to be established before problems are encountered; a series of simple sub-models can be constructed on the fly, incrementally creating what may be a very complex overall model. Because each sub-model is simple, we can use simple but robust techniques to detect anomalies and outliers. We believe there is a wide range of application for this approach beyond network intrusion detection and prudent expert systems. We also believe this ad-hoc approach is likely to find wider use than a pure model-based approach, because of the ad hoc nature of many domains.

Acknowledgements

We are grateful to the Thai government for funding an RTG scholarship and the University of New South Wales for a UIPA scholarship. We also thank Peter Linich, the network administrator at the school of CSE, for his support in providing audit data.

References

1. P. Compton and R. Jansen. A philosophical basis for knowledge acquisition. *Knowledge Acquisition*, 2:241–257, 1990.
2. P. Compton, P. Preston, G. Edwards, and B. Kang. Knowledge based system that have some idea of their limits. In *Proceedings of the 10th AAAI-Sponsored Banff Knowledge Acquisition for Knowledge-Based Systems Workshop*, Banff, Canada, 1996.
3. G Edwards, B Kang, P Preston, and P Compton. Prudent expert systems with credentials: Managing the expertise of decision support systems. *Int. J. Biomed. Comput.*, 40:125–132, 1995.
4. R.V. Guha and D. Lenat. Cyc: A midterm report. AI Magazine, 1990.
5. D. Lenat. A brief list of the applications. http://www.cyc.com/cyc/technology/cycandd/brieflist, 1994.
6. A. Prayote and P. Compton. Knowledge acquisition for anomaly detection. submitted to Internet Measurement Conference(IMC) 2006, 2006.

User Behavior Analysis of the Open-Ended Document Classification System*

Yang Sok Kim[1], Byeong Ho Kang[1], Young Ju Choi[1], SungSik Park[1], Gil Cheol Park[2], and Seok Soo Kim[2]

[1] School of Computing, University of Tasmania
Sandy Bay, Tasmania 7001, Australia
{yangsokk, bhkang, Y.J.Choi, sspark}@utas.edu.au
[2] School of Information & Multimedia, Hannam University
133 Ojung-Dong, Daeduk-Gu, Daejeon 306-791, Korea
{gcpark, sskim}@hannam.ac.kr

Abstract. Real-world document classification is an open-ended problem, rather than a close-ended problem, because the document classification domain continually evolves as the time passes. Unlike the close-ended document classification, the participants in the open-ended problem actively take part in the problem solving process. For this reason, it is important to understand the problem solver's behavioral characteristics. This paper proposes a thorough analysis of them. We found that the problem solving strategies are significantly different among participants because of individual differences in cognition among participants.

1 Introduction

Automated document classification has been a significant research topic. In the 1980's a common approach to document classification was rule-based, which involved a human in the construction of classifier. Though such a method provides accurate rules and has the additional benefit of being human understandable, the construction of such rules requires significant human input and the human needs some knowledge concerning the details of rule construction as well as domain knowledge, which become a bottleneck of this approach [1]. As an alternative, the machine learning (ML) approach has become the dominant one since 1990's. The ML system requires a set of pre-classified training documents and automatically produces a classifier from documents and the domain expert is needed only to classify a set of existing documents.

Although ML method produced accurate classifiers, there are a number of drawbacks as compared to a rule-based one [2]. Some limitations of the ML approach come from its assumptions about the document classification problem. Generally problems can be classified either close-ended or open-ended. Whereas the former usually has specified problem solving goals, correct answers and clearly defined criteria for success at problem solving, the latter typically has a lack of clearly defined

---

* This work is supported by the Asian Office of Aerospace Research and Development (AOARD) (Contract Number:FA5209-05-P-0253).

A. Sattar and B.H. Kang (Eds.): AI 2006, LNAI 4304, pp. 1089–1095, 2006.
© Springer-Verlag Berlin Heidelberg 2006

goals, no single correct solution to a problem, and no immediately obvious criteria for making a judgment as to what constitutes a correct solution [3, 4].

The ML approach is basically based on the close-ended assumption, because it uses finite and known data set and its significant goal is to find a method that successfully classifies documents into the predefined classes. In addition, there are lots of clearly defined success criteria of the classification method. Though this assumption has made considerable contributions, it is beneficial to look into the document classification problem from the open-ended viewpoint. First of all, sometimes the document classification problem, especially in the real-world document classification, becomes a kind of the open ended problem. For example, let's assume that we want to classify news articles from the web-based news provider. In this case, documents that should be classified are not known until they are presented and classes continually evolve over time, not pre-defined. Furthermore, most people share the intuition that it is not *a priori* clear that the results from the close-ended research will be generalized in the open-ended domains [3].

2 System Requirements and Implementation

We reviewed the research results of the open-ended education or intelligent tutoring systems and elicited the following three requirements for the open-ended document classification system:

Firstly, the system user should actively take part in the problem solving process. We select the rule-base approach because whereas the rule-based approach assumes the **active role of the problem solvers**, the ML approach excludes the system users from the learning process [5]. The problem solver can be either experts or novices in the domain. This approach is philosophically similar to that of the constructivist approach of knowledge acquisition [6]. Secondly, the open-ended system can provide multiple solutions for the same classification problem, because each user has individual differences in cognition [7]. Therefore, supporting **multiple classifications** is an essential requirement of the classification system. In the classification problem, this has a two-fold meaning. On one hand, it means that *(1) a case can be classified into multiple classes* without causing dissatisfactions among the problem solvers. On the other hand, *(2) multiple cases can be classified into one class* because of different reasons. Lastly, the system should support **negotiating interactions** between the system and the users. In the education, the more the problem solver knows about the domain and the learners, the more he/she can transfer more of his/her knowledge to the learner [4, 5, 8]. Likewise, the system and its users can interact to improve the system knowledge base. The system can provide current classification recommendation based on the current knowledge base and the user can either accept them or not. If the user does not want to accept the current recommendation, he/she can create another rule(s) to fix the current error. In this way, the system incrementally constructs its knowledge base over the time. Especially this is an essential requirement when the system user is a novice in a domain because he/she needs to refine the system knowledge base as he/she learns the domain knowledge.

An open-ended document classification system was developed with the MCRDR (Multiple Classification Ripple-Down Rules) algorithm, C++ program language, and

MySQL database to fulfill these requirements. MCRDR is based on the traditional rule-based system, but it overcomes traditional knowledge acquisition bottleneck problem by employing the exception-based knowledge representation and case-based validation [9, 10]. In this paper, we will focus not on the system itself, but on the user's behaviors because the system users have different roles in the problem solving process and it is important to understand how they act while constructing knowledge base. More detailed explanation about the classifier in [11, 12].

3 Inference and Knowledge Acquisition

A case is defined by attributes as follows:

$$CASE = T \cup B$$

where T is a distinct word set of hyperlink text and B is a distinct word set of the main content of the linked document. T and B are respectively defined as $T = \{t_1, t_2, ..., t_N\}$ and $B = \{b_1, b_2, ..., b_M\}$, where N and M are the number of distinct word and are greater that 0 (N, M $\geq$ 0). t_i and b_j are a word in the hyperlink text, which is text between <a> tags, and the hyperlinked document.

A rule structure is defined as follows:

IF
$$(TC \subset T) \; AND \; (BC \subset B) \; AND \; (AC \subset T \; OR \; AC \subset B)$$
THEN
$$Classify \; into \; folder \; F_i$$

where TC is a condition set for the hyperlink text, BC is a condition set for the hyperlinked document, and AC is a condition set for the hyperlink text or the hyperlinked document. Each set is defined as $TC = \{tc_1, tc_2, ..., tc_X\}$, $BC = \{bc_1, bc_2, ..., bc_Y\}$, and $AC = \{ac_1, ac_2, ..., ac_Z\}$, where tc_i is the word in the hyperlink text, bc_j is the word in the hyperlinked document, and ac_k is the word either in the hyperlink text or in the hyperlinked document. The number of words in each condition is greater than 0 (X, Y, Z $\geq$ 0).

In the inference process, the MCRDR-Classifier evaluates each rule node of the knowledge base (KB). If a case is selected from the case list (CL), the system evaluates rules from the root node and the inference result is provided by the last rule satisfied in a pathway. All children of the satisfied parent rule are evaluated, allowing for multiple conclusions. The conclusion of the parent rule is only given if none of the children are satisfied [10, 13].

The knowledge acquisition process begins when a case has been classified incorrectly or is missing a classification. Fig.1 illustrates knowledge acquisition algorithm. In the MCRDR-Classifier, the knowledge base (KB) is automatically maintained by the system. If a new knowledge acquisition process begins, the MCRDR-Classifier decides the location of a new rule according to the rule type. If there is no classification recommendation, the new rule is located under the root rule. If there is some inference result, but the user thinks it is wrong, a refining rule is located under the current firing rule. The stopping rule is a specific refining rule that has NULL conclusion.

```
begin
1.   User selects New Conclusion (NC)
2.   System generates Attribute List (AL) of Current Case
     If the new rule (NR) is refining rule,
              AL is attributes of current case that are not in the current firing rule's corner-
              stone case.
     Else
              AL is all attributes of current case
3.   User selects condition(s) from the AL
4.   System generates Tentative Firing Cases (TFC)
     If user agrees all TFC are also fired by NR
              End the KA process
     Else
              User selects Exclusive Case (EC) from TFC
              Recursively do again step 2 ~ 4.
end
```

Fig. 1. Knowledge Algorithm in the MCRDR-Classifier

4   Experiment and Results

The experiment is designed to analyze knowledge acquisition behaviors of the open-ended document classification users. This experiment was conducted by 20 Master and Hounours course students at the University of Tasmania for four months from August, 2005 ~ November, 2005. The Web monitoring system, called WebMon, continually collected newly updated documents from 9 well known information technology news Websites. Each participant could read the collected documents in real-time and train their own MCRDR-Classifiers. The classification structure (folder structure) of 86 folders was predefined for the experimental purpose, but each participant might use any number of folders for the classification, which totally depended on each participant's intention. In addition, the participants used a different number of documents based on his/her document filtering level.

4.1   Overall Classification Results

In total 12,784 articles were collected during the experiment period. In overall 95.6% documents (12,304) were used by the participants. The lowest ratio was 87.6% (BBC) and the highest ratio was 99.8% (ITNews). 13.0% documents were commonly used by all participants, which varied from 5.1% to 22.4%.

4.2   Classification Results by the Participants

Used Documents. Each participant classified documents into multiple folders by using the MCRDR-Classifier. Though the same monitored documents were provided to all the participants, the document usage results are very different. The smallest number is 1,775 and the largest number 21,045. The mean number is 11,693. The differences caused by two factors. Firstly, *the document filtering levels* are different among the participants. That is, the documents that each participant felt sufficiently important to classify were different among participants. Secondly, the *multiple classifications of each participant* were also different among the participants because some participants tended to classify document multiple classifications whereas others did not.

Used Folders. A total 86 hierarchical folders were used for this experiment: level one (8), level two (40), level three (31), and level four (8). The numbers of folders that were used for classification were different among participants. They varied from 14 to 73, with the mean number of used folders being 51.35. The most used folders were in level 2, and the least used folders in level 4. There is no evidence of the symmetric relationship between the number of used folder and the number of classified documents. For example, though P8, P9, and P10 used a very similar number of folders their correspondent document use is very different.

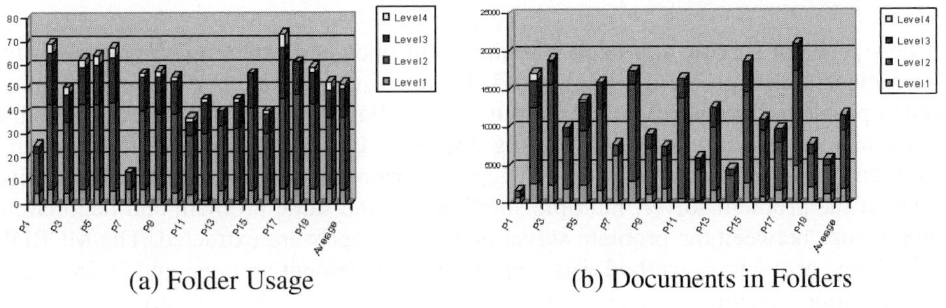

(a) Folder Usage   (b) Documents in Folders

Fig. 2. Folder Usage by Participants

4.3 Knowledge Acquisition

Rules. On average, the participants created 254 rules for 51 folders with 579 conditions to classify 11,693 documents. The minimum number of rules created was 59 (P13) and the maximum (P18, P19) 597. Documents per rule were 62, rules per folder numbered 5.3, with conditions per rule being 2.3. To examine relationships between document classification and rule creation, and between folder creation and rule creation, correlation values were calculated. The correlation between document classification and rule creation ($CR_{d,r}$) was 0.27 and folder (class) creation and rule creation ($CR_{f,r}$) 0.49.

Conditions. Participants were able to use three different types of condition words, which were seen in title (Type 1), seen in body (Type 2), and seen in both title and body (Type 3). Fig. 3. illustrates each participant's condition usage ratio of the three types of rules. Condition selecting depends on each participant's rule construction strategy. Whereas some participants mainly used title condition words (P5, P20), others frequently employed all conditions words (P7, P10, P17).

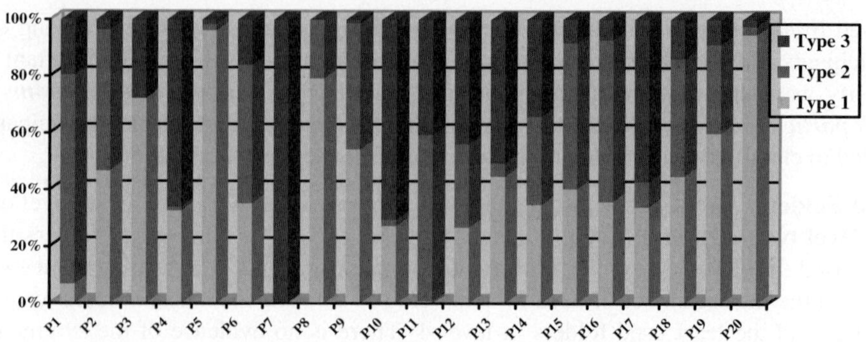

Fig. 3. Participants' Condition Usage Comparison

5 Conclusions

A new problem solving approach is required for open-end problems since they differ from close-ended problems. Real-world document classification is a kind of open-ended problem, because there are no pre-defined classes and cases, and it is possible to classify cases with various coexisting document classifications. We firstly examined the open-ended education experiences to obtain insights into this matter. Active roles of the problem solver, multiple solutions for the same problem and negotiating interactions between the problem solver and the learner were extracted. The MCRDR knowledge acquisition method was employed to implement an open-ended document classification system.

We conducted experiments to analyze knowledge acquisition behaviors. Twenty participants used the MCRDR-Classifier to classify real-world documents. The experiment results show that the participants have different problem solving behaviors while using the open-ended document classification problem.

References

1. Apte, C., F. Damerau, and S.M. Weiss, *Automated learning of decision rules for text categorization.* ACM Transactions on Information Systems (TOIS), 1994. **12**(3): p. 233 - 251.
2. Hirsch, L., M. Saeedi, and R. Hirsch. *Evolving Rules for Document Classification.* in *8th European Conference, EuroGP 2005.* 2005. Lausanne, Switzerland.
3. Goel, V. *Comparison of Well-Structured & Ill-Structured Task Environments and Problem Spaces.* in *Fourteenth Annual Conference of the Cognitive Science Society.* 1992. Hillsdale, NJ: Erlbaum.
4. Cook, J., *Bridging the Gap Between Empirical Data on Open-Ended Tutorial Interactions and Computational Models.* International Journal of Artificial Intelligence in Education, 2001. **12**: p. 85-99.
5. Andriessen, J. and J. Sandberg, *Where is Education Heading and How About AI?* International Journal of Artificial Intelligence in Education, 1999. **10**: p. 130-150.
6. Shaw, M.L.G. and J.B. Woodward, *Modeling expert knowledge.* Knowledge Acquisition, 1990. **2**(3): p. 179 - 206.

7. Dillon, R.F. and R.R. Schmeck, *Individual Differences in Cognition*. Vol. 1. 1983, New York, USA: Academic Press, Inc.

8. Hong, N.S., *The Relationship Between Well-Structured and Ill-Structured Problem Solving in Multimedia Simulation*, in *The Graduate School, College of Education*. 1998, The Pennsylvania State University.

9. Byeong Ho, K., *Validating Knowledge Acquisition: Multiple Classification Ripple Down Rules*, in *School of Computer Science and Engineering*. 1995, University of New South Wales.

10. Kang, B., P. Compton, and P. Preston. *Multiple Classification Ripple Down Rules : Evaluation and Possibilities*. in *9th AAAI-Sponsored Banff Knowledge Acquisition for Knowledge-Based Systems Workshop*. 1995. Banff, Canada, University of Calgary.

11. Park, S.S., Y.S. Kim, and B.H. Kang. *Web Document Classification: Managing Context Change*. in *IADIS International Conference WWW/Internet 2004*. 2004. Madrid, Spain.

12. Kim, Y.S., et al. *Adaptive Web Document Classification with MCRDR*. in *International Conference on Information Technology: Coding and Computing ITCC 2004*. 2004. Orleans, Las Vegas, Nevada, USA.

13. Compton, P. and D. Richards, *Generalising ripple-down rules*. Knowledge Engineering and Knowledge Management Methods, Models, and Tools. 12th International Conference, EKAW 2000. Proceedings (Lecture Notes in Artificial Intelligence Vol.1937), 2000: p. 380-386.

An Agent-Based Ontological Approach for Information Reorganisation

Li Li and Yun Yang

Faculty of Information and Communication Technologies
Swinburne University of Technology
PO Box 218, Hawthorn, Melbourne, VIC 3122, Australia
{lli, yyang}@ict.swin.edu.au

Abstract. This paper presents a novel approach for information reorganisation which is particularly suitable for distributed computing, such as Web services. An agent-based architecture is developed to cope with service agents' on and off and provide semantic support for implementation of complex e-business applications that may span diverse enterprises. A frame-based ontology representation is discussed. The representation is well suited for representing complex relationships that are unavailable in current ontology related work. Our approach aims to support dynamic information reorganisation in the Internet based e-business applications. The obtained information can serve for the knowledge acquisition purpose.

1   Introduction

An ontology is defined as an explicit specification of conceptualisation [2]. Ontologies facilitate the interoperability between heterogeneous systems by providing a formalisation that makes information machine-processable. Ontologies are seen as key enablers for the next generation Web, the Semantic Web [1]. On one hand, numerous ontologies have been developed and applied in two recent developments of the Semantic Web and Web services to bring about intelligent services, greater functionality and interoperability than current stand-alone services. It is known that any information system uses ontologies either implicitly or explicitly. On the other hand, the proliferation of Internet technology and globalisation of business environments give rise to the advent of a variety of ontologies. According to Papazoglou [5], e-business applications are based on the existence of standard ontologies for a vertical domain that establish a common terminology for sharing and resue. Obviously, it is hardly conceivable that engaged organisations will conform to a single ontology.

From this viewpoint it is necessary to develop relevant techniques to cope with various ontologies in e-business to facilitate system integration across organisations. Additionally, we also need to take ontology evolution into consideration as far as the ontology life cycle is concerned in e-business.

With the presence of agent technologies [6], flexibility and credibility, which are hard to achieve in other computational models, can be resolved in agent-based

A. Sattar and B.H. Kang (Eds.): AI 2006, LNAI 4304, pp. 1096–1100, 2006.
© Springer-Verlag Berlin Heidelberg 2006

applications where agents are designed to be able to perceive their environment, and respond in a timely fashion to changes occurred.

In this paper, we propose a novel approach for information reorganisation and ultimately for inter-organisational integration by taking advantage of agents' flexibility. Moreover, we present a clear view of ontological approach for information reorganisation in e-business. An important feature of our approach is its flexibility in coping with dynamic ontology changing in e-business environments.

This paper is organised as follows. Section 2 discusses a frame-based representation and its relation with ontology languages. Section 3 presents an architecture to deal with ontology changes in e-business where service agents may join and leave the system freely. Section 4 is about information reorganisation under the proposed architecture. Finally, Section 5 is conclusions.

2   Frame-Based Representation

The use of ontologies requires a well-designed, well-defined, and Web-compatible ontology language with supporting reasoning tools. The syntax of this language should be both intuitive to human users and compatible with existing Web standards. The frame-based language is a more suitable ontology description language with formally defined semantics and sufficient expressive power. Moreover, frame-based representations are compatible with current Web-standards. Additionally, there is a powerful supporting tool Protégé (*http://protege.stanford.edu/*) which can export frame-based Protégé ontologies into a variety of web-compatible formats such as XML Schema, RDF(S) and OWL. Another advantage is that Protégé users can rely on specific plug-ins to automatically generate Java source files from Protégé ontologies. This becomes extremely important when large ontologies are involved.

In this paper, an ontology O is defined as a triple (C, R, A) with C representing concepts, R for relations and A for axioms. In a frame-based representation, a frame associates with other frames by slots, where these relationships must meet the constraints specified by the axioms. A frame can be defined by

$$\frac{C_i :}{(R_{i1} : C_{j1}), (R_{i2} : C_{j2}), (R_{i3} : C_{j3})}$$

This indicates that the concept C associates with concepts C_{j1}, C_{j2}, C_{j2} with relationships R_{i1}, R_{i2}, R_{i3}, respectively. The domains and ranges of the values are defined, which will be specified in instances. It is known that existing work in ontology itself and ontology engineering rarely deal with relationships other than "is-a", which is neither exhaustive nor practical in real world applications. By using a frame-based representation, it comes true to represent constraints and applicability rules by calculus. With the presence of calculus, a concept definition can thus be represented as $R_i(C_i, C_j)$ which is intuitive to users. Ideally, systems such as CLisp (*http://www.ghg.net/clips/CLIPS.html*) and Jess (Java Expert System Shell, *http://www.jessrules.com/*) widely support the inference in predicate calculus.

3   System Architecture of Ontological Approach

The architecture must be flexible as agents join and leave freely. The flexibility of the system becomes evident when applications become more and more complex. We focus strictly on the problem of providing decentralised mediation of engaged service agents. In the following, we highlight the flexibility of the system in the architecture followed by description of functionalities of each component. In this architecture, an intelligent agent is defined as an agent which possesses characteristics defined in [6] and acts on behalf of an ontology as well. In order to accomplish complex tasks such as providing semantic support for business objects and business processes at a high-level, the most important quality of agents is that agents are able to learn and adapt to their environment. An agent can not achieve this without the support of underlying ontologies.

We have at least four types of agents. Figure 1 illustrates the system architecture and shows how agents collaborate in the Internet-based e-business environment, and how ontologies support enterprise system applications semantically.

- service agent (SA): An SA might be any service agent such as a supplier agent, a retailer agent, a procurement agent or a customer agent.
- scheduling agent (SCA): An SCA links a user agent with existing available service agents to provide necessary coordination and mediation. It is worth noting that this agent needs to reorganise information and acquire basic knowledge over time. Particularly in this paper, we treat this basic knowledge the same as an ontology. Information reorganisation process is defined as an ontology generation process. To this end, this dynamically developed ontology can be regarded as a high-level view for it to be (re)used in the system and serve for the knowledge acquisition purpose.
- facilitator agent (FA): An FA acts like a mediator. Though the architecture here is platform independent, we still use yellow-page service provided by most FIPA-compliant platforms. With the FA, agents may join and leave the system freely. Moreover, any newly created agents will be available once they register in the system. This is extremely important to ensure that agents only partially understanding other agents in a "conversation" can take the advantage of ontology mapping [3,4] to virtually expand their ontologies. In

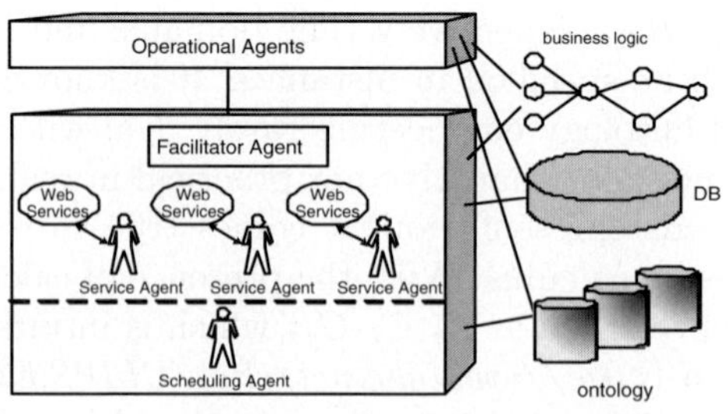

Fig. 1. System Architecture

our experiment this *FA* interacts with JADE Directory Facilitator to act as an interface agent of the system.

- operational agent (*OPA*): An *OPA* embeds processing logic associated with a business process. It also provides ontology engineering relevant operations. Ontology mapping is an example.

Ontology mapping aims to find corresponding concepts of similar ontologies. In some cases, mapping is just like a one-to-one mapping function. It is an equivalent relation in [4]. Additionally, other mappings such as inclusive relation may also exist. For a particular concept C_i, because $R_i(C_i, C_j)$ holds, a corresponding concept C_i' in another ontology can be found if it matches all/part slots as well as values of slots defined in C_i.

4   Information Reorganisation

Information reorganisation in e-business aims to provide a high-level view about the existing information for the system integration purpose. It is worth noting that this paper does not attempt ambitiously to solve all system integration issues, rather, it proposes an innovative ontological approach to allow agents to infer semantically in e-business with underlying ontologies. With the frame-based ontology representations, concepts and relationships can be represented in predicate calculus. Logic assertion and implication are described as follows:

- logic assertion: $\forall x, y R_i(?x, ?y) \longrightarrow R_i(?y, ?x)$ for symmetry relation.
  Example 1: Assume relationship R_i is semantic similar, then the formula $\forall x, y R_i(?x, ?y) \longrightarrow R_i(?y, ?x)$ exists.
- logic implication: $\forall x, y(\exists z) R_i(?x, ?y) \wedge R_j(?y, z) \longrightarrow R_k(?x, z)$ for transitivity, reflectivity and antisymmetry relations.
  Example 2: Assume relationships R_i, R_j, R_k are *"part-of"*, then the formula $\forall x, y(\exists z) R_i(?x, ?y) \wedge R_j(?y, z) \longrightarrow R_k(?x, z)$ exists.

With these in mind, information reorganisation in this paper will follow the steps below to respond to ontology changes in a timely manner:

(1) Finding (C_i, R_i, C_j) alike triples in available ontologies and putting them in a list. It is interesting that the form (C_i, R_i, C_j) is in accordance with RDF statement (*subject, predicate, object*);
(2) Counting appearance of triples in the above list if there exists equivalent relationship from other engaged ontologies, noting down the occurrence k in accordance with the number appearance of a triple. Then we get $((C_i, R_i, C_j), k)$, where k represents the strength of the triple in existing ontologies;
(3) Maintaining other relationships for tracking purpose in step (5);
(4) Sorting the list in step (2) in a descending order;
(5) Reorganising information to generate a new ontology by keeping the most general concepts and relationships and expand it to include other necessary conceptulisation specifications when needed.

ontology called "PetOwner", in which some pet owners, pets and their relations are specified. We searched the Web and found an ontology "Animal". However, this ontology is large. In particular, it contains a lot of animals that are not considered as pets normally such as tigers and lions. Thus, it is not practical to reason with or process the whole ontology. An interesting approach is to use the technique of forgetting and thus eliminate those animals that we do not want to keep from "Animal". As a result, a (smaller) sub-ontology of "Animal" is obtained, which keeps only the useful and desirable information in "Animal". Our approach is based on hex-programs whose external atoms are ontologies in OWL (see [2] for details).

2   Forgetting in Answer Set Programming

In this section we briefly review some basics of *answer set programming* [4] and *semantic forgetting* introduced in [3].

2.1   Answer Sets of Disjunctive Programs

A *disjunctive program* is a finite set of rules of the form

$$a_1 \vee \cdots \vee a_s \leftarrow b_1, \ldots, b_m, not\ c_1, \ldots, not\ c_n, \tag{1}$$

$s, m, n \geq 0$, where a, b's and c's are classical literals in a propositional language. A *literal* is a *positive literal* p or a *negative literal* $\neg p$ for some atom p. An *NAF-literal* is of the form *not* l where *not* is for the negation as failure and l is a (ordinary) literal. For an atom p, p and $\neg p$ are called *complementary*. For any literal l, its complementary literal is denoted $\tilde{l}$. To guarantee the termination of some program transformations, the body of a rule is a set of literals rather than a multiset.

Given a rule r of form (1), $head(r) = a_1 \vee \cdots \vee a_s$ and $body(r) = body^+(r) \cup not\ body^-(r)$ where $body^+(r) = \{b_1, \ldots, b_m\}$, $body^-(r) = \{c_1, \ldots, c_n\}$, and $not\ body^-(r) = \{not\ q \mid q \in body^-(r)\}$.

A rule r of the form (1) is *normal* or *non-disjunctive*, if $s \leq 1$; *positive*, if $n = 0$; *negative*, if $m = 0$; *constraint*, if $s = 0$; *fact*, if $m = 0$ and $n = 0$, in particular, a rule with $s = n = m = 0$ is the constant *false*.

A disjunctive program P is called *normal program* (resp. *positive program*, *negative program*), if every rule in P is normal (resp. positive, negative).

An *interpretation* X is a set of literals that contains no pair of complementary literals.

The answer set semantics. The *reduct* of P on X is defined as $P^X = \{head(r) \leftarrow body^+(r) \mid r \in P, body^-(r) \cap X = \emptyset\}$. An interpretation X is an *answer set* of P if X is a minimal model of P^X (by treating each literal as a new atom). $\mathcal{AS}(P)$ denotes the collection of all answer sets of P. P is *consistent* if it has at least one answer set.

Two disjunctive programs P and P' are *equivalent*, denoted $P \equiv P'$, if $\mathcal{AS}(P) = \mathcal{AS}(P')$.

As usual, B_P is the *Herbrand base* of logic program P, that is, the set of all (ground) literals in P.

2.2   Semantic Forgetting

A set X' is an *l-subset* of another set X, denoted $X' \subseteq_l X$, if $X' \setminus \{l\} \subseteq X \setminus \{l\}$. Similarly, a set X' is a strict *l*-subset of X, denoted $X' \subset_l X$, if $X' \setminus \{l\} \subset X \setminus \{l\}$. Two sets X and X' of literals are *l*-equivalent, denoted $X \sim_l X'$, if $(X \setminus X') \cup (X' \setminus X) \subseteq \{l\}$.

Definition 1. *Let P be a consistent disjunctive program, let l be a literal in P and let X be a set of literals.*

1. *For a collection $\mathcal{S}$ of sets of literals, $X \in \mathcal{S}$ is l-minimal if there is no $X' \in \mathcal{S}$ such that $X' \subset_l X$. $\min_l(\mathcal{S})$ denotes the collection of all l-minimal elements in $\mathcal{S}$.*
2. *An answer set X of disjunctive program P is an l-answer set if X is l-minimal in $\mathcal{AS}(P)$.*

Having the notion of minimality about forgetting a literal, we are now in a position to define the result of forgetting about a literal in a disjunctive program.

Definition 2. *Let P be a consistent disjunctive program and l be a literal. A disjunctive program P' is a result of forgetting about l in P, if P' represents l-answer sets of P, i.e., the following conditions are satisfied:*

1. *$B_{P'} \subseteq B_P \setminus \{l\}$ where B_Q denotes the set of (classical) literals occurring in Q.*
2. *For any set X' of literals with $l \notin X'$, X' is an answer set of P' iff there is an l-answer set X of P such that $X' \sim_l X$.*

Notice that the first condition implies that l does not appear in P'. An important difference of the notion of forgetting here from existing approaches to updating and merging logic programs is that only l and possibly some other literals are removed. In particular, no new symbol is introduced in P'.

For a consistent extended program P and a literal l, some program P' as in the above definition always exists. However, different such programs P' might exist. It follows from the above definition that they are all equivalent under the answer set semantics. Thus, we use $\mathsf{forget}(P, l)$ to denote a possible result of forgetting about l in P.

For further discussions about forgetting in disjunctive logic programming, the reader is referred to [3].

3   System Structure and Applying LPForget

LPForget has implemented two basic algorithms (Algorithm 1 and 2) for computing the result of forgetting introduced in [3]. Two variants of these algorithms

(Algorithm 3 and 4) [1] are also implemented in the system. Algorithm 3 is aimed for efficient implementation for forgetting while Algorithm 4 is designed for dealing with open-world reasoning tasks [1]. Besides these algorithms for computing forgetting, LPForget contains an application of forgetting for conflict resolution in the setting of multi-agents.

The system LPForget is implemented in Java and its structure is shown in the above diagram. It currently works on MS Windows since the version of DLV we adopted is of the Windows version. Obviously, the system can be easily adapted to Linux or Unix. The system can be downloaded from: http://www.cit.gu. edu.au/ kewen/LPForget/.

The main system can be executed by running the program LPForget.jar. Then one can choose to use the core module Forgetting or the application CRS. If the module Forgetting is chosen, then the users will be prompted to enter a disjunctive program, the literal to be forgotten, and the choice of algorithms. The result of forgetting can be obtained by clicking on the button "Compute". Notice that different algorithms may output syntactically different results of forgetting. However, the results are semantically equivalent.

The syntax of the input logic programs of the system coincides with that of DLV [6], as shown by the following example program P:

$$red : - pool, not\ blue.$$
$$blue : - pool, not\ red.$$
$$pool : - not\ tennisCourt.$$

This program represents some resident's requirements for a proposal of building a swimming pool and/or tennis court in a community complex: (1) if a swimming pool is built, then its colour can be either red or blue; (2) if a tennis court is not built, then a swimming pool should be built.

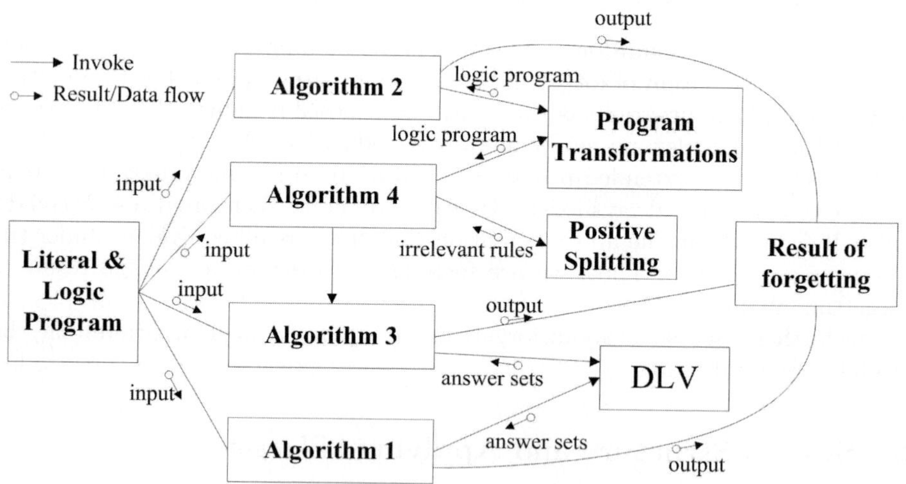

Fig. 1. System Structure of Forgetting

Some other residents may also have their requirements on the building proposal. Due to preferences conflict between different residents, the resident may have to give up some of his preferences and, for instance, say "it does not matter for me if the swimming pool is blue or not". Then we can "forget" about the colour "blue" from the above program and then get a new program $forget(P, blue) = \{pool : -.\}$, which means that only a swimming pool is built.

More examples can be found in the system website (Examples folder).

Acknowledgments. The authors would like to thank Rodney Topor for his helpful comments and contribution to this paper. This work was partially supported by the Australia Research Council (ARC) Discovery Projects 0666107, 0666540, the Austrian Science Funds (FWF) projects 17212, 18019, the European Commission project REWERSE (IST-2003-506779) and NICTA Queensland.

References

1. F. Cheng, T. Eiter, N. Robinson, A. Sattar, and K. Wang. LPForget: a system of forgetting in answer set programming. Technical report, 2006, available at `http://www.cit.gu.edu.au/~kewen/Papers/LPForget_long.pdf`
2. T. Eiter, G. Ianni, R. Schindlauer, H. Tompits, and K. Wang. Forgetting in managing rules and ontologies. In *Proceedings of the IEEE/WIC/ACM International Conference on Web Intelligence (WI-06)*, 2006 (accepted for publication).
3. T. Eiter, and K. Wang. Forgetting and conflict resolving in disjunctive logic programming. In *Proceedings of the 21st National Conference on Artificial Intelligence (AAAI-2006)*, pages 238-243, 2006.
4. M. Gelfond and V. Lifschitz. Logic programs with classical negation. In *Proceedings of the International Conference on Logic Programming*, pages 579-597, 1990.
5. J.Lang, P.Liberatore, and P.Marquis. Propositional independence: Formula-variable independence and forgetting. *Journal of Artificial Intelligence Research*, 18:391-443, 2003.
6. N. Leone, G. Pfeifer, W. Faber, T. Eiter, G. Gottlob, S. Perri, and F. Scarcello. The DLV system for knowledge representation and reasoning. *ACM Transactions on Computational Logic*, 7(3):499-562, 2006.
7. F. Lin and R. Reiter. Forget it. In *Proceedings of the AAAI Fall Symposium on Relevance*, pages 154-159. New Orleans (LA), 1994.
8. K.Wang, A.Sattar and K.Su. A theory of forgetting in logic programming. In *Proceedings of the 20th National Conference on Artificial Intelligence (AAAI-2005)*, pages 682-687, 2006.

Formalization of Ontological Relations of Korean Numeral Classifiers

Youngim Jung[1], Soonhee Hwang[2], Aesun Yoon[3], and Hyuk-Chul Kwon[1]

[1] Pusan National University, Department of Computer Science and Engineering,
[2] Pusan National University, Center for U-Port IT Research and Education,
[3] Pusan National University,
Department of French/Interdisciplinary Program of Cognitive Science,
Jangjeon-dong Geumjeong-gu, 609-735 Busan, S. Korea
{acorn, soonheehwang, asyoon, hckwon}@pusan.ac.kr

Abstract. Though many studies emphasized on the characteristics of numerals classifiers and the construction of database, or ontology of classifiers, few studies attempted to formalize them. This paper briefly presents preprocessing for the extraction of ontological relations from language resources, and formalizing those extracted relations with OWL DL. Then the overall result of building Korean numeral classifier ontology is illustrated.

1 Introduction

Developed and refined numeral classifiers are found in the languages of Asia, Central America, and South America. Chinese, Japanese and Korean can be more appropriately called 'classifier languages' than others, such as English or French [1, 5]. Numeral classifiers are more grammatically restricted than nominal classifiers, co-occurring next to numerals and quantifying only specific referents. The Korean classifiers are one of the most difficult atom words to handle in machine translation (MT), the fact reflected in the low accuracy of its translation into English or French, each of which lacks a classifier system [13]. Thus, classifier ontology and its applicability are highly required as language resources for MT and NLP, but relevant research is still insufficient. This study focuses on the building of Korean classifier ontology together with its formalization.

The present paper is structured as follows. In Section 2, we examine related studies on the characteristics of classifiers, including Korean classifiers, in more detail. In Section 3, preprocessing for extraction of semantic relations and the formalization of ontological relations with OWL are illustrated, and then the result will be discussed. Conclusions and future work follow in Section 4.

2 Related Studies

Related previous work has concentrated on (1) typological surveys and classification of classifiers [1]; (2) description of classifiers according to their meaning [5]; (3) semantic analysis for various classifiers [7]. However, only a few semantic features such as [±living], [±animacy], [±human], [±shape], [±function] were used for categorization, and thus event classifiers or generic classifiers were not included.

A. Sattar and B.H. Kang (Eds.): AI 2006, LNAI 4304, pp. 1106–1110, 2006.
© Springer-Verlag Berlin Heidelberg 2006

[3, 4] proposed a method for the generation of Korean and Japanese numeral classifiers using semantic categories from a thesaurus and dictionaries. However, the studies dealt with the limited number of Japanese and Korean classifiers and did not attempted to formalize the semantic relations. [13] built a database containing 952 Korean numeral classifiers; however, more than 500 classifiers fell into dummy categories without any semantic criteria, and neither the correlation of classifiers nor the salient semantic properties of co-occurring nouns were sufficiently described; thus, the relations between classifiers and co-occurring nouns were not constructed.

3   Formalization of Korean Classifiers

For building Korean classifier ontology (KCLOnto), Standard Korean Dictionary, the list of high-frequency Korean classifiers, large-scale corpora composed up of 7,778,848 words and 450,000 examples containing target classifiers and their co-occurring words, and WordNet and KorLex[1] are used as language resources.

3.1   Preprocessing for Building Korean Classifier Ontology

Since most available resources are not structured as natural language texts, natural language processing is prerequisite to efficiently and accurately extracting semantic relations from semi-structured dictionaries or raw corpus [2]. For example, a classifier, '*doe*' and its semantic relations are generated from the dictionary definition as shown in Figure 1[2].

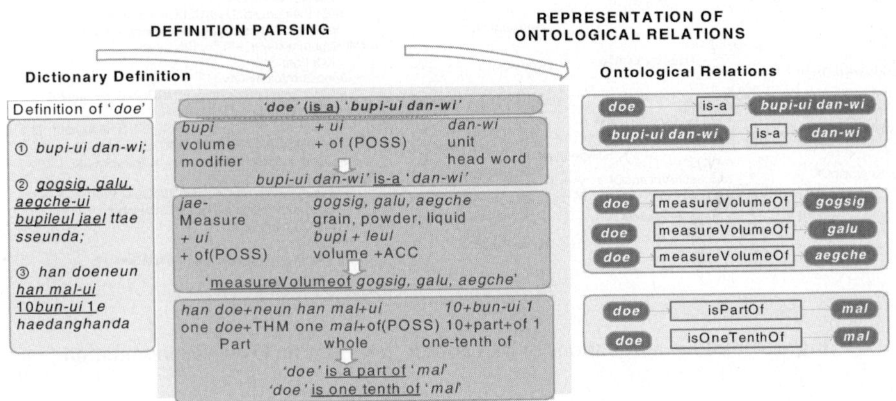

Fig. 1. Processing for Extracting Ontological Relations of a Classifier '*doe*'

3.2   Formalizing Ontological Relations

Based on the semantic relations extracted from language resources and the semantic analysis, major categories of Korean classifiers are suggested as in Table 1[10].

---

[1] KorLex is a Korean wordnet based on WordNet2.0 [10] and it covers nouns, verbs, adjectives and adverbs. KorLex Noun1.5 including 58,656 synsets was used in this study.

[2] The detailed description on the semi-automatic extraction of ontological relations from language resources is presented in [11].

Table 1. Main Categories of Korean Classifiers

Types	Description	Examples
Mensural-CL	measuring the amount of some entity	*doe* (CL of measuring volume of grain, powder)
Sortal-CL	classifying the kinds of quantified noun-referents	*songi* (CL of counting flowers)
Generic-CL	restricting quantified nouns to generic kinds	*gae* (CL of counting entities)
Event-CL	quantifying abstract events	*bal* (CL of counting shots)

The various semantic aspects of a classifier instance should be machine-interpretable. For that, we implemented this ontology in Protégé (http://protege.standford.edu) with OWL DL, which supports sufficient semantics using classes, instances, properties, and restrictions for semantic web. Ontological relations implemented in KCLOnto are categorized into four main sections: (1) the relation between a classifier and its lexical information, (2) the relation between synsets, (3) the relation between classifiers and nouns or adjectives, and (4) the relation between Korean classifiers and English units of measure or containers in WordNet nouns. For example, the ontological relations of *doe* are paraphrased as OWL, as shown in Figure 2.

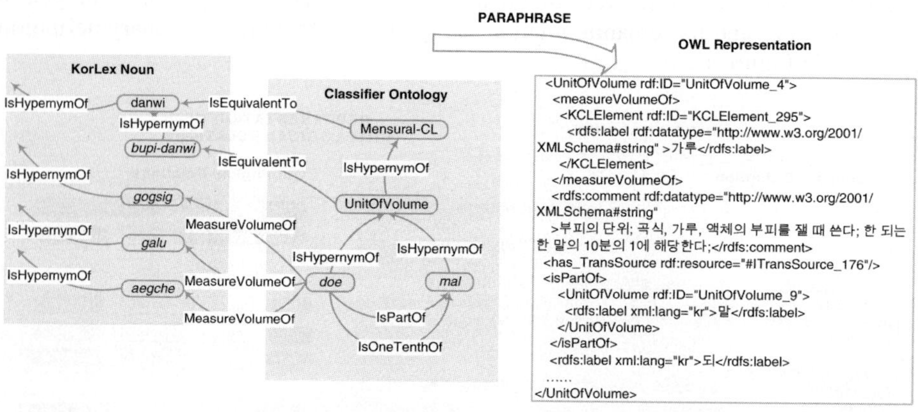


```
<UnitOfVolume rdf:ID="UnitOfVolume_4">
  <measureVolumeOf>
   <KCLElement rdf:ID="KCLElement_295">
    <rdfs:label rdf:datatype="http://www.w3.org/2001/
XMLSchema#string" >가루</rdfs:label>
   </KCLElement>
  </measureVolumeOf>
  <rdfs:comment rdf:datatype="http://www.w3.org/2001/
XMLSchema#string"
  >부피의 단위; 곡식, 가루, 액체의 부피를 잴 때 쓴다; 한 되는
한 말의 10분의 1에 해당한다;</rdfs:comment>
  <has_TransSource rdf:resource="#ITransSource_176"/>
  <isPartOf>
   <UnitOfVolume rdf:ID="UnitOfVolume_9">
    <rdfs:label xml:lang="kr">말</rdfs:label>
   </UnitOfVolume>
  </isPartOf>
  <rdfs:label xml:lang="kr">되</rdfs:label>
  ......
</UnitOfVolume>
```

Fig. 2. Ontological Relations of a Classifier '*doe*' and Its OWL Representation

While formalizing and constructing lexical information and various semantic relations, the 'Is-A' relation in the ontology (which is 'subclassOf') as well as the linguistic 'Is-A' relation (which is 'ishypernymOf') are required to be distinguished, because 'WordNet synsetoffset (WNSynsetOffset), KorLexSynsetOffset (KLSynsetOffset), and Standard Korean Dictionary ID (StdIdx)' are not hyponyms of 'ID' but subclasses of 'ID'. Thus, both 'isHypernymOf' and 'subclassOf' relations are generated to represent the linguistic 'Is-A' relations in KCLOnto.

Relations between a classifier and the noun that the classifier refers to and quantifies vary according to the kind of numeric attribute that is measured by the classifier: 'measureVolumeOf', 'measureWeightOf', 'measureLengthOf', 'count

BundleOf', 'countNumberOf', and others. All of these properties are bound under their super-properties such as 'quantifyOf' and 'quantifyClassOf'. If a classifier quantifies a specific noun (a synset having no 'isHypernymOf' relation), that classifier has a 'quantifyOf' relation with the noun, and if it quantifies a semantic class of nouns (a synset having a 'isHypernymOf' relation), it has a 'quantifyClassOf' relation with the noun class.

3.3 Results and Discussion

In total, 1,138 numeral classifiers were collected, and their taxonomy was constructed according to the 'Is-A' relations extracted from the dictionary, and a semantic-feature analysis by linguists. The size of the ontology is applicable to NLP applications, considering the size of classifier inventories created in a previous study.

We found that the sense granularity of noun classes quantified by classifiers differs depending on the types of the classifiers. For instance, mensural-CL and generic-CL can quantify a wide range of noun classes. By contrast, each sortal-CL and event-CL can combine only with a few specific noun classes. The selectional preferences between nouns and classifiers are represented by 'quantifyOf' and 'quantifyClassOf' relations. Table 2 shows the size of the classifiers and their representative ontological relations constructed in KCLOnto.

Table 2. Results of Korean Classifier Ontology

Types	Size		Relations	Size		Relations	Size
Mensural	772		isHypernymOf	1,350		hasDomain	696
Sortal	270		isHolonymOf	258		hasOrigin	657
Generic	7		isSynonymOf	142		hasStdIdx	442
Event	89		quantifyOf	2,973		isEquivalntToKL	696
Total	1,138		quantifyClassOf	287		isEquivalntToWN	734

All the procedures were performed semi-automatically for the improvement of efficiency and consistency. Compared with the previous studies noted in Section 2, this result shows that the comprehensive approach for building a Korean classifier ontology based on refined semantic analysis. With the formalization of various ontological relations of classifiers described in OWL, more accurate and wide application to NLP is expected. In addition, the constructed linkages to WordNet guarantee easier and prompter connections to ontologies in other languages, and can be applied to MT as fundamental multilingual resources.

4 Conclusions and Further Studies

As explained in this paper, the ontological relations of Korean numeral classifiers were extracted from natural language texts, and those various relations were formalized with OWL, which supports strong expressiveness in knowledge representation and expandability and linkage to other ontologies. The results show that the constructed ontology is sufficiently large to be applied to practical MT applications. As future work, for the purpose of evaluating the applicability of the suggested Korean classifier ontology, we will study and implement the sub-module for automatic matching nouns and numeral classifiers depending on their selectional preferences. In addition, based on the syntactic patterns and semantic features of the combination of nouns, numerals, and numeral classifiers in Korean texts, and their corresponding combination of numerals and nouns in English texts, we will apply KCLOnto to English-Korean machine translations system.

Acknowledgement

This work was supported by the National Research Laboratory (NRL) Program under Grant M10400000332-06J0000-33210.

References

1. Allan, K.: Classifiers. Language 53(2) (1977) 285-311
2. Alani, H., Kim, S., Millard, D.E., Weal, M.J., Hall, W., Lewis, P.H., Shadbolt, N.R.: Automatic Ontology-Based Knowledge Extraction from Web Documents, IEEE Intelligent Systems (2003) 14-21
3. Bond, F., Paik, K.: Classifying Correspondence in Japanese and Korean. In : 3rd Pacific Association for Computational Linguistics Conference: PACLING-97 (1997) 58-67
4. Bond, F., Paik, K.: Reusing an Ontology to Generate Numeral Classifiers. In : Proceedings of the 18th Conference on Computational Linguistics (2000) 90-96
5. Downing, P.: Pragmatic and Semantic Constraints on Numeral Quantifier Position in Japanese. Linguistics 29 (1993) 65-93
6. Fellbaum, C. (ed.): WordNet - An Electronic Lexical Database. MIT Press, Cambridge (1998)
7. Goddard, C.: Semantic Analysis, Oxford University Press, Oxford (1998)
8. Gruber, T.R.: A Translation Approach to Portable Ontologies, Knowledge Acquisition 5(2), (1993) 199-220
9. Hovy, E.H.: Methodologies for the Reliable Construction of Ontological Knowledge. In : Proceedings of ICCS2005, (2005) 91-106
10. Hwang, S., Jung. Y., Yoon, A., Kwon, H.: Building Korean Classifier Ontology Based on Korean WordNet. TSD-LNAI (2006) Accepted, to be published
11. Jung, Y., Hwang, S., Yoon, A., Kwon, H.: Extracting Ontological Relations of Korean Numerals Classifiers from Semi-structured Resources Using NLP Techniques. OTM Workshops 2006, LNCS 4278, pp.1038-1043
12. Matsumoto, Y.: Japanese Numeral Classifiers: a Study of Semantic Categories and Lexical Organization. Linguistics 31(4) (1993) 667-713
13. Nam, J.: A Study of Korean Numeral Classifiers for the Automatic French-Korean Translation of the Expression with Quantifiers. Review of French Studies (54) (2006) 1-28
14. Philpot, A.G., Fleischman, M., Hovy, E.H.: Semi-automatic Construction of a General Purpose Ontology, Proceedings of the International Lisp Conference, New York (2003), 1-8

Adaptive $\mathcal{ALE}$-TBox for Extending Terminological Knowledge

Ekaterina Ovchinnikova[1] and Kai-Uwe Kühnberger[2]

[1] University of Tübingen, Seminar für Sprachwissenschaft
e.ovchinnikova@gmail.com
[2] University of Osnabrück, Institute of Cognitive Science
kkuehnbe@uos.de

Abstract. Ontologies are usually considered as static data structures representing conceptual knowledge of humans. For certain types of applications it would be desirable to develop an algorithmic adaptation process that allows dynamic modifications of the ontology in the case new information is available. Dynamic updates can generate conflicts between old and new information resulting in inconsistencies. We propose an algorithm that can model the adaptation processes for conflicting and non-conflicting updates defined on $\mathcal{ALE}$-TBoxes.

1 Introduction

In artificial intelligence, there is an increasing interest in examining ontologies for a variety of applications in knowledge-based systems. A subsumption relation defined on concepts together with additional (optional) relations is usually considered as the core of every ontology. Several formalisms were proposed to represent ontologies. Probably the currently most important markup language for ontology design is the *web ontology language*[1] (*OWL*) in its three different versions: *OWL light*, *OWL DL*, and *OWL full*. Besides the markup specifications of representation languages, a related branch of research is concerned with logical characterizations of ontology representation formalisms using certain subsystems of predicate logic, so-called Description Logics (DL) [3]. It is well-known that *OWL* standards can be characterized by description logics.

The acquisition of ontological knowledge is one of the most important steps in order to develop new and intelligent web applications [4]. Because hand-coded ontologies of a certain size are usually tedious, expensive, and time-consuming to develop, there is the need for a (semi-)automatic extraction of ontological knowledge from given data. Furthermore rapid changes concerning the information theoretic background and the dramatic increase of available information motivates (semi-)automatic and consistent adaptation processes changing existing ontologies, if new data requires a modification. The aim of the present paper is to develop a theory of adapting terminological knowledge by extending the terminology with additional concepts and relations for conflicting and non-conflicting updates.

---

[1] See the documentation at http://www.w3.org/TR/owl-features/

A. Sattar and B.H. Kang (Eds.): AI 2006, LNAI 4304, pp. 1111–1115, 2006.
© Springer-Verlag Berlin Heidelberg 2006

2   Related Work

There is a variety of logical approaches in order to model inconsistencies. Examples are default logic, paraconsistent logic, or many-valued logic. Similarly several ontology extension mechanisms on the basis of classical description logics were proposed. This is quite often realized by the introduction of new construction methods like planning systems [2] or belief-revision processes [8]. The disadvantage is that standard DL-reasoners cannot be used easily for these extensions. Another tradition tries to model inconsistencies without extending the underlying DL: For example, in [9] a witness concept is demonstrating an occurring inconsistency, in order to characterize non-conservative extensions. Unfortunately this does not solve the problem, but just shows where the problem is. Another example of approaches not extending DL is based on learning techniques. Inductive logic programming is used in [7], where a new definition for a problematic concept is generated on the basis of positive examples. In this case as well as in cases where statistical learning techniques are used [6], information is lost due to the fact that the original definition is fully ignored.

We propose a solution in which the ontology remains formalized in DL and new information is adapted to the terminological knowledge automatically keeping the changes as minimal as possible.

3   Description Logics

DL provide a logical basis for representing ontologies [3]. In this paper, we consider mainly $\mathcal{ALE}$ and subsidiary $\mathcal{ALC}$, two relatively weak DLs.

A *TBox*[2] (terminological box) is a finite set of axioms of the form $C \equiv D$ (equalities) or $C \sqsubseteq D$ (inclusions) where C, D (called *concept descriptions*) are specified as follows in the $\mathcal{ALE}$-DL (A stands for atomic concept, R stands for role name): $C \rightarrow \top \mid \bot \mid A \mid \neg A \mid \forall R.C \mid \exists R.C \mid C_1 \sqcap C_2$. $\mathcal{ALC}$-DL additionally introduces the concept descriptions $\neg C$ and $C_1 \sqcup C_2$.

Concept descriptions are interpreted model-theoretically (see [3] for details). An interpretation $\mathcal{I}$ is called a *model* of a TBox $\mathcal{T}$ iff for all $C \sqsubseteq D \in \mathcal{T}$: $C^{\mathcal{I}} \subseteq D^{\mathcal{I}}$ and for all $C \equiv D \in \mathcal{T}$: $C^{\mathcal{I}} = D^{\mathcal{I}}$. D *subsumes* C *towards* $\mathcal{T}$ ($\mathcal{T} \models C \sqsubseteq D$) iff $C^{\mathcal{I}} \subseteq D^{\mathcal{I}}$ for every model $\mathcal{I}$ of $\mathcal{T}$. C is *satisfiable towards* $\mathcal{T}$ if there is a model $\mathcal{I}$ of $\mathcal{T}$ such that $C^{\mathcal{I}}$ is nonempty; otherwise C is *unsatisfiable* and $\mathcal{T} \models C \sqsubseteq \bot$. Subsumption algorithms (see [3]) are implemented in several reasoning systems[3].

The symbol $\doteq$ denotes the syntactical equality of concept descriptions. Atomic concepts occurring on the left side of an equality are called *defined*. The set $ax(\mathcal{T})$ collects *axiomatized* concepts occurring on the left side of an axiom in $\mathcal{T}$. *Unfolding* an acyclic $\mathcal{T}$ or a concept description towards $\mathcal{T}$ consists in a syntactical replacement of the defined concepts by their definitions. In order to make $\mathcal{T}$ *unfoldable* all inclusions must be replaced with equalities as follows: $C \sqsubseteq D \rightarrow C \equiv D \sqcap C^*$. A TBox $\mathcal{T}$ is *inconsistent* iff $\exists A \in ax(\mathcal{T}) : \mathcal{T} \models A \sqsubseteq \bot$.

---

[2] We restrict out attention to the TBox leaving the ABox for further investigations.

[3] Some DL reasoners are listed at `www.cs.man.ac.uk/~sattler/reasoners.html`

4   Automatic Adaptation to the New Axiom: Solutions

Informally, the *adaptation* of a TBox $\mathcal{T}$ to a new axiom for a concept X is a minimal modification of $\mathcal{T}$ such that X becomes satisfiable towards $\mathcal{T}$. By modification we mean the sequence of adding and deleting axioms in $\mathcal{T}$. Since the notion of minimal modification in general is rather vague, we suggest a constructive definition of the adaptation procedure introduced in Sec. 5.

A conflict between a new axiom $X \sqsubseteq Y$ and the original TBox $\mathcal{T}$ occurs if there exist a concept $C \in ax(\mathcal{T})$ such that $\mathcal{T} \models Y \sqsubseteq C$ and definition D of C conflicts with Y: $\mathcal{T} \models D \sqcap Y \sqsubseteq \bot$. The example below illustrates the modifications that can be performed to achieve a consistent TBox.

TBox: $\{$`Vehicle` $\sqsubseteq \forall$`Energy.Fuel` $\sqcap$ `Moves`, `Car` $\sqsubseteq$ `Vehicle` $\sqcap \exists$`Driver.`$\top\}$
New information: `ElectroCar` $\sqsubseteq$ `Vehicle` $\sqcap \exists$`Energy.`$\neg$`Fuel`
Adapted TBox: $\{$`Vehicle` $\sqsubseteq$ `Moves`, `FuelVehicle` $\sqsubseteq$ `Vehicle` $\sqcap \forall$`Energy.Fuel`,
   `ElectroCar` $\sqsubseteq$ `Vehicle` $\sqcap \exists$`Energy.`$\neg$`Fuel`, `Car` $\sqsubseteq$ `FuelVehicle` $\sqcap \exists$`Driver.`$\top\}$

In this example, $\exists$`Energy.`$\neg$`Fuel` in the definition of `ElectroCar` conflicts with $\forall$`Energy.Fuel` subsuming `Vehicle` while `Vehicle` subsumes `ElectroCar`. In the adapted TBox the conflicting information is deleted from the definition of `Vehicle` and moved to the definition of the new concept `FuelVehicle` which is supposed to capture the original meaning of `Vehicle`. `Vehicle` needs to be replaced with `FuelVehicle` in all axioms of the TBox except for the definition of `ElectroCar`. Although we assume that the new axiom has higher priority than the original TBox, we want to keep in the definition of `Vehicle` as much information not conflicting with the definition `ElectroCar` as possible. The conflicting information is moved to the definition of the new concept `FuelVehicle`.

The presented approach works for an $\mathcal{ALE}$-TBox. Disjunction may provide a problem even on the common sense level. If an axiom $X \sqsubseteq Y$ conflicts with a disjunctive definition D of a concept in $\mathcal{T}$, then there are two possibilities to make $\mathcal{T} \cup \{X \sqsubseteq Y\}$ consistent: to add X as a new disjunct to D or to adapt one or more disjuncts in D to Y. It seems to be impossible to find a general solution for this adaptation in logics with disjunction, but the information about the conflicts may give hints to the engineer of how to modify the original TBox.

5   Adaptation Algorithm for $\mathcal{ALE}$ TBoxes

We propose a procedure adapting an $\mathcal{ALE}$-TBox $\mathcal{T}$ to a new axiom. For the sake of simplicity, assume that the set of role names contains only one name R. For a concept description C in prenex conjunctive normal form the $\mathcal{ALE}$-normal form (compare [1]) $\mathcal{ALE}$-*NF* is the conjunction of the following concepts:

- Atomic and negated atomic concepts on the top-level[4] of C (the set *prim*)
- Value restriction $\forall R.val_R(C)$, where $val_R(C) := C_1 \sqcap ... \sqcap C_n$ if
  $\forall i \in \{1, ..., n\} : \forall R.C_i$ occurs on the top-level of C; otherwise $val_R(C) := \top$;

---

[4] The concepts occurring on the top-level of the concept description C are the concepts not embedded in any relational restriction in C.

- Existential restrictions of the top-level of C;
  $exr_R(C) := \{C' \mid \text{there exists } \exists R.C' \text{ on the top-level of } C\}$.

Definition 1. *The pair* $\mathfrak{C} = \langle C_c, C_n \rangle$ *is called conflict between the* $\mathcal{ALE}$*-concept descriptions* C_1 *and* C_2, *if* C_c *and* C_n *are specified as follows*

If $\exists i, j \in \{1,2\} : C_i \equiv \top$ *and* $C_{j \neq i} \not\equiv \bot$ *then* $C_c = \{\}, C_n = \{C_1\}$;
else if $\exists i, j \in \{1,2\} : C_i \equiv \bot$ *and* $C_{j \neq i} \not\equiv \bot$ *then* $C_c = \{C_1\}, C_n = \{\}$;
else if $C_1 \equiv C_2 \equiv \bot$ *then* $C_c = \{\}, C_n = \{\bot\}$; *else*

1. $C_c = \{C \in prim(C_1) \mid \exists C' : C' \dot{=} C \wedge \neg C' \in prim(C_2)\} \cup$
   $\{\forall R.X \mid D \dot{=} val_R(C_2) \sqcap \sqcap_{D_i \in exr_R(C_2)} D_i \wedge$
   $\mathfrak{C}(val_R(C_1), D) = \langle S_1, S_2 \rangle \wedge S_1 \neq \varnothing \wedge X \dot{=} \sqcap_{cc_i \in S_1} cc_i\} \cup$
   $\{\exists R.X \mid \exists D \in exr_R(C_1) : \mathfrak{C}(D, val_R(C_2)) = \langle S_1, S_2 \rangle \wedge S_1 \neq \varnothing \wedge X \dot{=} \sqcap_{cc_i \in S_1} cc_i\}$
2. $C_n = \{C \in prim(C_1) \mid C \notin C_c\} \cup \{\forall R.Y \mid D \dot{=} val_R(C_2) \sqcap \sqcap_{D_i \in exr_R(C_2)} D_i \wedge$
   $\mathfrak{C}(val_R(C_1), D) = \langle S_1, S_2 \rangle \wedge S_2 \neq \varnothing \wedge Y \dot{=} \sqcap_{cn_i \in S_2} cn_i\} \cup$
   $\{\exists R.Y \mid \exists D \in exr_R(C_1) : \mathfrak{C}(D, val_R(C_2)) = \langle S_1, S_2 \rangle \wedge$
   $((S_2 \neq \varnothing \wedge Y \dot{=} \sqcap_{cc_i \in S_2} cc_i) \vee Y \dot{=} \top))\}$

The *conflict* in Def. 1 is a tuple of two sets. The conflicting set C_c collects the concept descriptions subsuming C_1 which conflict with C_2. The non-contradicting set C_n collects all the other concept descriptions explicitly subsuming C_1.

Algorithm *Adapt* for adaptation of a TBox to an axiom
Input: an $\mathcal{ALE}$-TBox $\mathcal{T}$, an $\mathcal{ALE}$-axiom $X \sqsubseteq Y$
Output: adapted TBox $Adapt_{X \sqsubseteq Y}(\mathcal{T}) = \mathcal{T}'$

$\mathcal{T}' := \mathcal{T} \cup \{X \sqsubseteq Y\}$
$OC := \langle C_1, ..., C_n \rangle : \forall i, j \in \{1, ..., n\} : C_i \in prim(Y)$ **and** $(i < j \rightarrow \mathcal{T} \not\models C_i \sqsubseteq C_j)$
$i := 0$
FOR $i < n$
   $i := i + 1;\ C := C_i :\ C_i \in OC$
  IF $C \in ax(\mathcal{T}')$ THEN
     $Y \dot{=} C \sqcap Y', C \sqsubseteq D \in \mathcal{T}'$
     IF $unfolded_{\mathcal{T}'}\ D$ and $unfolded_{\mathcal{T}'}\ Y'$ are $\mathcal{ALE}$-concept descriptions THEN
       $\mathfrak{C}(unfolded_{\mathcal{T}'}\ D\ in\ \mathcal{ALE}\text{-}NF, unfolded_{\mathcal{T}'}\ Y'\ in\ \mathcal{ALE}\text{-}NF) = \langle C_c, C_n \rangle$
       IF $C_c \neq \varnothing$ THEN
         $\mathcal{T}' := (\mathcal{T}' \setminus (\{C \sqsubseteq D\} \cup \{Z \sqsubseteq C \sqcap Z' \in \mathcal{T}'\})) \cup$
         $\{C \sqsubseteq CN \mid CN \dot{=} \sqcap_{cn_i \in C_n} cn_i \wedge \mathcal{T}' \not\models C \sqsubseteq CN\} \cup$
         $\{C' \sqsubseteq C \sqcap CC \mid CC \dot{=} \sqcap_{cc_i \in C_c} cc_i\} \cup \{Z \sqsubseteq C' \sqcap Z' \mid Z \sqsubseteq C \sqcap Z' \in \mathcal{T}'\}$
END FOR

Algorithm *Adapt* defines an adaptation of a TBox $\mathcal{T}$ to a new axiom $X \sqsubseteq Y$ such that X becomes satisfiable towards $\mathcal{T}$. For the sake of simplicity we imply that X has not been axiomatized in $\mathcal{T}$ before, but it is easy to develop a more general procedure. For every axiomatized concept C occurring on the top-level of Y the conflict is computed for the rest of Y and the definition D of C. The conflicting concepts occurring on the top-level of D are deleted from the definition of C and moved to the definition of the new concept C' which is declared to be subsumed by C and subsume all subconcepts of C.

6 Conclusion and Future Work

We presented an approach for dynamically updating ontologies. The overall motivation is to provide a theoretical and algorithmic basis for a framework that allows to update ontological knowledge automatically handling the possible conflicts between the original ontology and the new incoming information. The main contribution of this paper is the algorithm *Adapt* adapting an $\mathcal{ALE}$-TBox to a new axiom and the notion of *conflict* (Def. 1).

We were mainly concerned with $\mathcal{ALE}$-DL, but the given approach can easily be extended to treat more expressive DLs. As sketched in Sec. 4 disjunctive definitions cannot be adapted fully automatically. Developing the prototype implementation of the *Adapt* algorithm and testing it on real data we suppose to find a solution for semi-automatic updates in the DL-logics allowing disjunction. There are two important theoretical issues we plan to examine. We will characterize the complexity of *Adapt* and the changes in the model theoretic semantics of an adapted ontology.

Acknowledgment. This line of research was partially supported by the grant MO 386/3-4 of the German Research Foundation (DFG).

References

[1] Baader, F., Küsters, R.: Non-Standard Inferences in Description Logics: The Story So Far. *International Mathematical Series*, volume 4, *Mathematical Problems from Applied Logic. New Logics for the XXIst Century* (2006) 1–75

[2] Baader, F., Lutz, C., Miličić, M. Sattler, U., Wolter, F.: Integrating Description Logics and Action Formalisms: First Results. In *Proc. of the 20th National Conference on Artificial Intelligence (AAAI'05)*, AAAI Press (2005) 572–577

[3] Baader, F., Calvanese, D., McGuinness, D., Nardi, D. Patel-Schneider (eds.), P.: *Description Logic Handbook: Theory, Implementation and Applications.* Cambridge University Press (2003)

[4] Berners-Lee, T., Hendler, J., Lassila, O.: The Semantic Web – A new form of Web content that is meaningful to computers will unleash a revolution of new possibilities. Scientific American (2001)

[5] Biemann, C., Shin, S., Choi, K.-S.: Semiautomatic Extension of CoreNet using a Bootstrapping Mechanism on Corpus-based Co-occurrences. *Proc. of the 20th International Conference on Comp. Ling. (Coling 2004)* (2004) 1227–1232

[6] Cohen, W., Hirsh, H.: Learning the CLASSIC Description Logic: Theoretical and Experimental Results. In *Proc. of the Fourth International Conference on Principles of Knowledge Representation and Reasoning (KR94)* (1994) 121–133

[7] Fanizzi, N., Ferilli, S., Iannone, L., Palmisano, I., Semeraro, G.: Downward Refinement in the ALN Description Logic. In *Proc. of the Fourth International Conference on Hybrid Intelligent Systems (HIS'04)* (2005) 68–73

[8] Flouris, G., Plexousakis, D., Antoniou, G.: Updating Description Logics using the AGM Theory. In Proc. *The 7th International Symposium on Logical Formalizations of Commonsense Reasoning* (2005)

[9] Ghilardi, S., Lutz, C., Wolter, F.: Did I damage my ontology: A Case for Conservative Extensions of Description Logics. In *Proc. of Principles of Knowledge Representation and Reasoning 2006 (KR06)* (to appear).

Preference Reasoning in Advanced Question Answering Systems

Farah Benamara[1] and Souhila Kaci[2]

[1] IRIT 118 route de Narbonne 31062 Toulouse France
benamara@irit.fr
[2] CRIL Rue de l'Université SP 16, 62307 Lens Cedex, France
kaci@cril.univ-artois.fr

Abstract. In this paper, we present an approach for the elicitation of bipolar preferences based on NL expressions and show how bipolarity can be used within an advanced question answering system, namely, the WEBCOOP system.

1   Introduction

It is known that preferences are crucial in decision making. Preferences are naturally bipolar. In fact, a user may express *positive preferences* or wishes, that indicate what he is looking for, what he considers as more or less satisfactory. On the other hand, he may also express *negative preferences*, or rejections, representing what he refuses, what he considers as more or less unacceptable. The distinction between positive and negative preferences is supported by cognitive psychology studies showing that this distinction makes sense, and the two types of preferences are processed separately in the brain, and are felt as different dimensions by people [4]. Bipolarity has been widely investigated in AI: preference and knowledge representations [3], decision theory [5], multi-agent systems [1], etc. However all these areas are based on a strong assumption: *negative preferences and positive preferences are distinguished and provided.* This is because preference elicitation is in general a serious bottleneck in AI and especially in the case of bipolar preferences since the borders between what is satisfactory and what is acceptable are not always clear.

Our aim in this paper is to integrate bipolar preferences in preference-based applications namely question answering systems (QAS for short). For that purpose, we present, in section 2, an approach for the elicitation of qualitative bipolar preferences form natural language (NL) expressions based on the analysis of specific linguistic markers, namely adverbs. We propose a general classification of preferences based on expression modalities called the illocutory force, that allow for the identification of positive preferences (wishes) and negative preferences (rejections) in an intuitive way. We then show, in section 3, how bipolarity can be used in an extented version of the WEBCOOP system [2], an advanced question answering system without dialogue and user model which goes beyond the extraction of paragraphs answering directly the question.

A. Sattar and B.H. Kang (Eds.): AI 2006, LNAI 4304, pp. 1116–1121, 2006.
© Springer-Verlag Berlin Heidelberg 2006

2   Bipolar Preferences

When both positive preferences and negative preferences are provided, they are called *bipolar preferences*. In this framework, we distinguish : (1) positive preferences, or wishes, that describe what is really satisfactory for the user. They delimit the set of *satisfactory answers*, denoted $\mathcal{S}_A$ and, (2) negative preferences, or rejections, that describe what is excluded by the user. They delimit the set of *acceptable answers*, denoted $\mathcal{A}_A$. Let us consider the following query: *"a hotel in Paris, not in the 18th district and preferably in the 10th district"*. In this query, we distinguish (1) the focus: *hotel*, (2) the set $\mathcal{S}_A$ of satisfactory answers: *"hotels in the 10th district"*, (3) the set $\mathcal{A}_A$ of acceptable answers: *"hotels in Paris but not in the 18th district"* and lastly, (4) the set of neutral answers: $\mathcal{N}_A = answers - (\mathcal{S}_A \cup \mathcal{A}_A)$. For e.g. *"hotels in Paris in the 8th district"*, since they are neither wished nor rejected. Bipolar preferences are meaningful only if the set of neutral answers $\mathcal{N}_A$ is not empty otherwise positive preferences and negative preferences are dual.

2.1   Elicitation of Bipolar Preferences from NL Expressions

Bipolar preferences are expressed in a sentence by means of adverbs that reflect the illocutory force of the modified word, such as: *seulement (only), absolument (absolutely), possiblement (possibly), etc.* The semantic interpretation of an adverb generally depends on the semantic category that it belong to. The following table describes how different categories of adverbs can identify the nature of a given preference i.e a wish or a rejection. The given classification is for French adverbs [6], since our application is dedicated to NL French queries.

Adverbs category / Preferences	Without modality (WITHOUT)	Affirmation adverbs (AFFIR)	Adverbs of doubt (DOUBT)	Adverbs of negation	Strong intensifier (STRONG)	Weak intensifier (WEAK)
Wishes	X		X	X	X	X
Rejections	X	X		X		

Fig. 1. Nature of preferences according to adverbs semantic categories

For example, the affirmation adverbs, denoted **AFFIR**, such as: absolument (absolutely), certainement (certainly), etc., induce rejections which correspond by complementation to integrity constraints (IC) that should be fulfilled. The adverbs of doubt, denoted **DOUBT**, such as preferably, possibly, etc. induces wishes which have to be satisfied as far as possible. We distinguish two subcategories here: POSS (possibly) and PREF (preferably). Adverbs of degree induce intensity and make preferences strong *(very, really)* or weak *(few, enough)*. Intensity adverbs (denoted STRONG and WEAK))can be applied only to the operators WITHOUT and DOUBT, while negation adverbs can be applied to all the categories. It is worth noticing that negation does not necessarily imply rejections. For e.g. in the question: *"I want a chalet absolutely not in Corsica"*,

"Corsica" is an imperative rejection and the preference *"not in Corsica"* becomes a constraint, whereas in the question *"I want a chalet not in Corsica if possible"*, the rejection *"not in Corsica"* is attenuated by the modality DOUBT. It is then considered as a wish.

More formally, a user's preference UP is represented under the form:

$MODALITY(INTENSITY(P))$, where, P is a first order formula,
$MODALITY \in \{WITHOUT, AFFIR, DOUBT\}$ and
$INTENSITY \in \{WITHOUT, STRONG, WEAK\}$, such that:
- If $MODALITY \in \{WITHOUT, AFFIR\}$ then UP is a constraint.
- If $MODALITY = DOUBT$ then UP is a wish.

Indeed, it is the modality that determines whether a preference is a constraint or a wish. For the sake of simplicity, the expression $MODALITY(WITHOUT(P))$ is written $MODALITY(P)$. Now we need to order expression modalities. Let $\succ$ be an order on user's preferences. Then:

- $\forall\ INTENSITY \in \{WITHOUT,\ STRONG,\ WEAK\}$,
$AFFIR(INTENSITY(P)) \succ WITHOUT(INTENSITY(P))$ and
$PREF(INTENSITY(P)) \succ POSS(INTENSITY(P))$.
- $\forall\ MODALITY \in \{WITHOUT,\ AFFIR,\ DOUBT\}$,
$MODALITY(STRONG(P)) \succ MODALITY(WITHOUT(P))$ and,
$MODALITY(WITHOUT(P)) \succ MODALITY(WEAK(P))$.

3   Bipolar Preferences in Advanced QAS

Our general framework is the WEBCOOP system [2]. WEBCOOP is a logic based question answering system (in the tourism domain – transportation and accommodations) that integrates knowledge representation and advanced reasoning procedures (query relaxation, intensional responses, detection of false presuppositions and miconceptions) to generate cooperative responses to french NL queries on the web. Like the majority of current QA systems, WEBCOOP answers factoïd questions, where the focus is an entity that should satisfies *positive preferences having the same priorities*. Our aim is to extend the WEBCOOP system to bipolar questions. Ideally, answers should satisfy all the wishes and falsify all the rejections. However this is not so systematic and *relaxation procedures* are then used to relax the wish and/or the rejection at the origin of the query failure.

Bipolar questions are represented in WEBCOOP as follows. Constraints and wishes are respectively represented by two sets of the form:

$$\mathcal{C} = \{\alpha_i(\lambda_i(c_i)) : i = 1, \cdots, n\}, \text{ and } \mathcal{W} = \{\beta_j(\lambda_j(w_j)) : j = 1, \cdots, m\},$$

where c_i (resp. w_j) are first order logic formulas, $\alpha_i \in \{WITHOUT, AFFIR\}$, $\beta_i \in \{POSS, PREF\}$ and $\lambda_i \in \{WITHOUT, WEAK, STRONG\}$. In addition, $\alpha_i(\lambda_i(c_i)) \succ \alpha_{i'}(\lambda'_i(c_{i'}))$ (resp. $\beta_j(\lambda_j(w_j)) \succ \beta_{j'}(\lambda'_j(w_{j'}))$) iff:

- $\alpha_i \succ \alpha_{i'}$, or $\alpha_i = \alpha_{i'}$ and $\lambda_i \succ \lambda_{i'}$, or
- $\alpha_i = \alpha_{i'}$ and $\lambda_i = \lambda_{i'}$, i.e the two constraints c_i and c_i' have the same priority, then c_i is prioritized over $c_{i'}'$ iff $i < i'$. That is, the first preference stated by the user in his question is the most important.

In the following, $\mathcal{C}^*$ (resp. $\mathcal{W}^*$) denotes the set of formulas of $\mathcal{C}$ (resp. $\mathcal{W}$) obtained after ignoring modality operators. The semantic representation of a bipolar question in WEBCOOP is a triple of the form $Q = (Focus, \mathcal{C}, \mathcal{W})$, where $Focus$ is a formula representing the expected answer type. Evaluating the question consists in finding answers which satisfy all the formulas appearing in the body of the query namely $Focus \wedge \mathcal{C}^* \wedge \mathcal{W}^*$, with the knowledge model of WEBCOOP. During the evaluation process, an answer should satisfy the wishes and not violate any constraint. In order to determine the reason of the query failure, the sets of acceptable answers and satisfactory answers are computed separately. Let $\mathcal{A}_A$ be the set of acceptable answers and $\mathcal{S}_A$ be the set of satisfactory answers associated to $Focus \wedge \mathcal{C}^*$ and $Focus \wedge \mathcal{W}^*$ respectively. The aim of evaluation process is to compute the set of direct answers, denoted $\mathcal{D}_A$, s.t. $\mathcal{D}_A = \mathcal{A}_A \cap \mathcal{S}_A$. However this set may be empty. In the case of bipolar queries, a false presupposition is identified when:

- **Case 1:** *the wishes cannot be satisfied together* i.e. $\mathcal{S}_A = \emptyset$ and $\mathcal{A}_A \neq \emptyset$.
- **Case 2:** *constraints are violated* i.e. $\mathcal{S}_A \neq \emptyset$ and $\mathcal{A}_A = \emptyset$.
- **Case 3:** *no satisfactory answers and no acceptable answers* i.e. $\mathcal{A}_A = \emptyset$ and $\mathcal{S}_A = \emptyset$.
- **Case 4:** *satisfied wishes and not violated constraints but satisfactory answers and acceptable answers do not overlap* i.e. $\mathcal{A}_A \neq \emptyset$ and $\mathcal{S}_A \neq \emptyset$ but $\mathcal{A}_A \cap \mathcal{S}_A = \emptyset$. When the reasons of the question failure are identified, the system has to find alternative answers. This implies to relax the appropriate wish and/or constraint in order to guarantee the satisfaction of the query.

3.1   Relaxation of Bipolar Questions in WEBCOOP

The relaxation mechanisms of WEBCOOP proceed gradually on each of the sub-queries responsible of the failure. They are modeled as a rewriting system, managed by a supervisor which consists in an iterative process running till a flexible solution is found that leads to a non-empty and coherent solution. The relaxation procedure returns the most immediate relaxations once the original query is exhausted. In the following we extend the relaxation mechanism of WEBCOOP to bipolar questions. These relaxations are based on the different ways to express preferences en NL. We show that the choice of a particular level of relaxation does not only depend on the nature of the query failure (namely, a wish or a constraint) but also on their associated enunciative operators.

A user preference UP in WEBCOOP, represented by $\mathrm{MODALITY(INTENSITY(P))}$, may be: (a) *a geographical localization* i.e a constant. The relaxation process consists in retrieving the nearest localization that satisfies the query. (b) *a spatial preference*, introduced by a localization preposition. The relaxation is based on the semantic interpretations of prepositions [2] and allows to replace the failed preference by another preposition that takes into account a larger (down) or a narrower (up) area. (c) *qualitative adjectives* (expensive, quite, etc.). In that case, adjectives are associated to values scale in our knowledge base (e.g. not expensive, not really expensive, moderately expensive, expensive) and the relaxation process consists in moving one step up or one step down in this scale.

The relaxation process depends on both modality and intensity expressions. Modalities guide the relaxation level whereas the intensity guides the sense of the relaxation i.e. going up or down. We define the relaxation process as shown in the table below.

Preferences	AFFIR(P)	WITHOUT(P)	WITHOUT(STRONG(P))	WITHOUT(WEAK(P))
Associated relaxation process	P is never relaxed	One level down relaxation	P is never relaxed	One level up relaxation
Preferences	POSS(P)	PREF(P)	PREF(STRONG(P))	PREF(WEAK(P))
Associated relaxation process	Three levels down of relaxation	Two levels down of relaxation	One level down relaxation	One level up relaxation

Fig. 2. Relaxation strategies of bipolar preferences

For example, the wishes introduced by the operator *PREF* are relaxed in case of failure. If there is *no intensity* then we authorize an additional relaxation level compared to the modality WITHOUT, namely *a "down" second relaxation level*. For example, in the query, *a hotel in Paris (preferably in the 14th district)*$_1$. The wish (1) can have two levels relaxation down:

–level 1: hotels in Paris near the 14th district.
–level 2: hotels in Paris not far from the 14th district. Moving from level i to level $i+1$ is allowed only when the set of answers obtained at the level i is empty. In this example, the geographic distance determines the depth of each relaxation level.

3.2   Relaxation Procedures of Bipolar Questions

Due to lake of space, we present in this section only the relaxation procedure associated to the case 1 introduced in section 3 where *wishes cannot be satisfied* independently of constraints. In this case, we have $\mathcal{S}_A = \emptyset$ while $\mathcal{A}_A \neq \emptyset$. At first sight we may relax wishes responsible of failure until $\mathcal{S}_A \neq \emptyset$. However this way is not efficient since we may have $\mathcal{D} = \mathcal{A}_A \cap \mathcal{S}_A = \emptyset$ even if $\mathcal{A}_A \neq \emptyset$ and $\mathcal{S}_A \neq \emptyset$. Indeed we should consider $\mathcal{C}$ when looking for the wish responsible of the failure. For the sake of simplicity we suppose that there is only one wish that is responsibe of the failure and the relaxation of that wish is sufficient to get a non-empty set of answers. This is formalized as follows. Suppose that the wish w_j is responsible of the failure. Then $\mathcal{A}(F \wedge C^* \wedge w_1 \wedge \cdots \wedge w_{j-1} \wedge w_{j+1} \wedge \cdots \wedge w_m) \neq \emptyset$ while $\mathcal{A}(F \wedge C^* \wedge w_1 \wedge \cdots \wedge w_{j-1} \wedge w_j \wedge w_{j+1} \wedge \cdots \wedge w_m) \neq \emptyset$. The relaxation of w_j in $\mathcal{W}^*$ gives a new set of wishes $\mathcal{W}'^* = \{w_1, \cdots, w_{j-1}, w'_j, w_{j+1}, \cdots, w_n\}$, where w'_j is the result of relaxing $\beta_j(\lambda_j(w_j))$. Following our hypotehsis we have $\mathcal{A}(Focus \wedge C^* \wedge \mathcal{W}'^*) \neq \emptyset$.

4   Conclusion

We present in this paper an approach for the elicitation of qualitative bipolar preferences based on NL expressions, namely adverbs. We then show how this representation is used within WEBCOOP, a cooperative QAS. This work has several future directions among which we plan to evaluate the linguistic and the cognitive adequacy of the relaxed responses w.r.t end users using psycholinguistics protocols as well as to study the interaction between bipolarity and fuzzy representation of knowledge.

References

1. L. Amgoud and S. Kaci, *Strategical considerations for negotiating agents*, AAMAS'05.
2. F. Benamara, *WEBCOOP, un système de question réponse coopératif pour le Web*, PHD thesis, Paul Sabatier University, Toulouse, 2004.
3. S. Benferhat and D. Dubois and S. Kaci and H. Prade, *Bipolar representation and fusion of preferences in the possibilistic logic framework*, KR02, pp. 421-432.
4. J.T. Cacioppo and G.G. Bernston, *The affect system: Architecture and operating characteristics*, Current Directions in Psychological Science, Vol. 8, pp. 133-137, 1999.
5. M. Ozturk and A. Tsoukias, *Positive and negative reasons in interval comparisons: Valued PQI interval orders*, IPMU04, pp. 983-989.
6. M. Riegel, J.C, Pellat, R. Rioul, *Grammaire méthodique du français*, PUF, 1994.

Clustering Data Manipulation Method for Ensembles

Matthew Spencer, John McCullagh, and Tim Whitfort

La Trobe University, Department of Computer Science & Computer Engineering
Bendigo, Victoria, Australia
mattspencer@dodo.com.au, {j.mccullagh, t.whitfort}@latrobe.edu.au

Abstract. Traditional data manipulation models such as Bagging and Boosting select training cases from throughout the problem space to generate diversity and improve performance. A new data manipulation model is proposed that dynamically assigns specialists to train on difficult clusters of training data. The model allows the expertise of specialists to overlap for difficult regions of the problem. It has been coupled with a dynamic combination model to exploit the diversity of specialist members. The model has been applied to an environmental problem and has demonstrated that dynamic modelling can enhance both the diversity of members and the accuracy of the ensemble.

Keywords: Ensemble, Data Manipulation, Clustering.

1   Introduction

An ensemble is a set of members whose estimates are combined in some way to effectively estimate unseen cases. The two most important characteristics of an ensemble are *accuracy* and *diversity*. If there is no disagreement between each of the ensemble members, the combined performance will be the same as any single member. Members that are too diverse may also become inaccurate, which can lead to poor ensemble accuracy. The need for a balance between diverse and accurate members to produce the optimal ensemble has been demonstrated [1].

Clustering models have been used in both unsupervised and supervised learning. The k-means clustering [2] approach is one such model that breaks up a set of data into smaller groups based on similar features. Another popular model that clusters data is the k-Nearest Neighbour [3]. This has been applied in various studies to label previously unseen cases according to its knowledge of similar training data. These models have been able to group cases with similar characteristics to reach a particular outcome.

Clustering has the potential of generating specialist members that are trained on a group of similar training patterns to both improve local performance and generate additional diversity. This paper presents a model that searches the problem space for difficult regions. Specialists are then trained on cases from these regions which can overlap with others to improve local performance in difficult areas.

A. Sattar and B.H. Kang (Eds.): AI 2006, LNAI 4304, pp. 1122–1126, 2006.

© Springer-Verlag Berlin Heidelberg 2006

2   Data Manipulation Research

Traditional data manipulation methods used to generate member diversity are Bagging [4] and Boosting [5]. Bagging randomly samples cases from a training set to encourage ensemble members to learn the entire problem in a different way. Boosting applies a weight to each case according to the last ensemble member's errors. Even though the average of each of the member estimates can reduce generalisation error, the ensemble's performance could be increased by improving local performance by specialisation.

The Mixture of Experts is an approach that divides a complex problem into smaller sub-problems [6]. Strict boundaries are assigned to each specialist for training and evaluation. Each ensemble member effectively becomes a specialist in a local region of the problem. As unseen cases are presented for evaluation, the member responsible for the case's local space is used to represent the ensemble estimate. Unfortunately, if a specialist is unable to effectively represent parts of its space, it may be difficult to allow the contribution of other appropriate specialists.

A novel data manipulation technique is proposed that dynamically generates data sets from difficult regions in the problem space. These sets are then used to train specialist members for the ensemble.

3   Dynamic Clustering

The Dynamic Clustering model is composed of one generalist and a set of specialists. The generalist member is trained on the whole training set. The specialists are trained on clusters of difficult training cases in the problem space. The two major features of Dynamic Clustering are cluster evaluation and cluster selection.

3.1   Cluster Evaluation

Dynamic Clustering identifies difficult regions by evaluating clusters of training cases for accuracy. To find the most appropriate cluster in the problem space for specialist training, n randomly selected cases are drawn from the training set, where n is the number of clusters to evaluate during each iteration. For each of these n cases, the nearest c training cases are evaluated for local accuracy, where c is the cluster size.

The Hamming distance is a method that has traditionally been used to determine the distance between cases, but it considers each of the inputs as being equal in contribution to distance. Instead of using this measure, the formula in Equation 1 is used to give larger weights to the inputs that have higher correlations to the target:

$$dist_j = \sum_{i=1}^{n} \left(|x_i - y_i| * r(i,t) \right) \tag{1}$$

where n is the number of inputs provided to predict the outputs, x_i is the value of input i for case x, y_i is the value of input i for case y, and $r(i,t)$ is the correlation between the factor i and target t for all training cases.

The ensemble's error is used for each training case to evaluate its difficulty. This research applies the Dynamic Fusion [7] combination model. This model assigns weights to each ensemble member according to its performance on local training data. Ensemble error has been chosen in preference to previous member error to search the current solution for potential areas of improvement. The cluster error is equal to the total error of all cases in the cluster divided by each case's weight, as shown in Equation 2:

$$err_{cluster} = \sum_{i=0}^{C}(err_i/w_i) \tag{2}$$

where $err_{cluster}$ is the cluster error, C is the cluster size, err_i is the ensemble error for case i, and w_i is the weight assigned to case i. Each training case is assigned an initial weight of 1.

3.2   Cluster Selection

Once the cluster error is calculated for each of the n clusters, the cluster with the largest error is selected. That cluster's nearest c cases are assigned to the next member as its training set. Each case that belongs to the next member's training set also receives an additional weight of one to discourage the cluster from being selected by future specialists.

4   Data

Environmental concerns have been the motivation for achieving a greater understanding of carbon transfer between the ocean, land and the atmosphere. The monitoring of carbon storage and transfer, as well as the accurate forecasting of carbon can assist in reducing the effects of environmental conditions such as global warming [8]. This research focuses on the estimation of organic carbon in the topsoil.

A nation-wide database has been used in this research, known as the ASRIS database [9]. Automated data, such as climate and landform maps from satellite images, as well as land use and lithology maps have been provided to assist in mapping organic carbon. Previous research [10] has shown a relative error of 0.68 using Cubist (also known as M5 [11]).

The majority of observations were taken from agricultural parts of Australia. Forty-one variables were used to model organic carbon. Of the 8,410 cases used in this study, 75% were randomly assigned to the training set, and 25% to the testing set.

5   Experimentation

5.1   Method

The Dynamic Clustering approach was compared to Bagging and Boosting. Each ensemble was trained with 10 Multi-Layer Perceptrons (MLP) using a 41-20-1 architecture for 50,000 training passes. Various parameters were trialled for Dynamic Clustering to optimise the performance of the ensemble. The number of random cases for cluster evaluation (n) was set to 50 to cover the majority of the problem space during each iteration. After a process of trial and error, a cluster size (c) of 500 was considered for this problem. Each ensemble's performance is evaluated using the relative error and the correlation between the ensemble's outputs to their target. The member correlation is calculated using the average correlation between all of the ensemble members' estimates.

5.2   Results and Discussion

The results in Table 1 show that the best performing ensemble was Dynamic Clustering with a cluster size of 500. Dynamic Clustering produced greater diversity than Bagging and Boosting for this problem, and outperformed previous experimental results [10]. This may be due to the improved local performance from the specialists or that the selected clusters complemented well with a dynamic combination model.

Table 1. Experimental results using Bagging, Boosting and Dynamic Clustering

Model	Correlation	Relative Error	Member Correlation
Cubist [10]	0.640	0.68	-
Bagging	0.663	0.68	0.99
Boosting	0.663	0.69	0.90
Dynamic Clustering	0.687	0.65	0.82

Static combination models such as simple averaging enforce the contribution of all members. If the member correlation is too low, these combination approaches can lead to poor ensemble performance. The diversity of each specialist has two effects. Firstly, each specialist is able to improve local performance in regions that the generalist was unable to model effectively. Secondly, the specialist is less likely to correctly estimate cases that are outside its local region, which increases estimation error. This enables the combination model to select members that are better suited to the input case and disregard members that are less suitable.

The Dynamic Clustering model shows encouraging improvements in performance when compared to other techniques. Results indicate that the Dynamic Fusion and clustering approaches complemented each other.

6   Conclusion

This research demonstrates potential in assigning local clusters of data to increase ensemble diversity and accuracy. The model demonstrates that local performance can be improved by the assignment of specialists that may overlap, coupled with a well-performing generalist. Dynamic Fusion has also demonstrated that it can differentiate between the appropriate and inappropriate ensemble members. The best model reached a relative error of 0.65, which is below previously published results [10].

Dynamic Clustering may also provide benefits outside the application used in this study. The identification of difficult regions may indicate the need for further data collection in a local area. The mapping of ensemble error may also assist in identifying where future samples could be taken to improve the model.

References

1. Kuncheva, L.: That elusive diversity in classifier ensembles. In: IbPRIA. First Iberian Conference on Pattern Recognition and Image Analysis (2003) 1126–1138
2. McQueen, J.: Some methods of classification and analysis of multivariate observations. In: 5th Berkeley Symp. on Mathematical Statistics and Probability. (1967) 281–297
3. Cover, T., Hart, P.: Nearest neighbour pattern classification. IEEE Transactions on Information Theory **13**(1) (1967) 21–27
4. Breiman, L.: Bagging predictors. Machine Learning **24**(2) (1996) 123–140
5. Freund, Y., Schapire, R.: Experiments with a new boosting algorithm. In: Machine Learning. Proceedings of the Thirteenth International Conference on Machine Learning, San Francisco, CA, Morgan Kaufmann (1996) 148–156
6. Jacobs, R., Jordan, M., Nowlan, S., Hinton, G.: Adaptive mixtures of local experts. Neural Computation **3** (1991) 79–87
7. Spencer, M., Whitfort, T., McCullagh, J., Bui, E.: Dynamic ensemble approach for estimating organic carbon using computational intelligence. In: ACST 2006. (2006)
8. Climate Change Science Program: Strategic plan for the U.S. climate change science program. Technical report (2003)
9. Johnston, R., Barry, S., Bleys, E., Bui, E., Moran, C., Simon, D., Carlile, P., McKenzie, N., Henderson, B., Chapman, G., Imhoff, M., Maschmedt, D., Howe, D., Grose, C., Schoknecht, N., Powell, B., Grundy, M.: ASRIS: The Database. Australian Journal of Soil Research **41** (2003) 1021–1036
10. Henderson, B., Bui, E., Moran, C., Simon, D.: Australia-wide predictions of soil properties using decision trees. Geoderma **124** (2004) 383–398
11. Quinlan, J.: Learning with continuous classes. 5th Australian Joint Conference on Artificial Intelligence, World Scientific (1992) 343–348

Improving the Performance of Multi-objective Genetic Algorithm for Function Approximation Through Parallel Islands Specialisation

A. Guillén[1], I. Rojas[1], J. González[1], H. Pomares[1], L.J. Herrera[1], and B. Paechter[2]

[1] Department of Computer Architecture and Computer Technology
Universidad de Granada. Spain
[2] School of Computing
Napier University. Scotland

Abstract. Nature shows many examples where the specialisation of elements aimed to solve different problems is successful. There are explorer ants, worker bees, etc., where a group of individuals is assigned a specific task. This paper will extrapolate this philosophy, applying it to a multiobjective genetic algorithm. The problem to be solved is the design of Radial Basis Function Neural Networks (RBFNNs) that approximate a function. A non distributed multiobjective algorithm will be compared against a parallel approach that emerges as a straight forward specialisation of the crossover and mutation operators in different islands. The experiments will show how, as in the real world, if the different islands evolve specific aspects of the RBFNNs, the results are improved.

1 Introduction

The problem to be tackled consists in designing an Radial Basis Function Neural Network (RBFNN) that approximates a set of given values. The use of this kind of neural networks is a common solution since they are able to approximate any function [8]. Formally, a function approximation problem can be formulated as, given a set of observations $\{(\boldsymbol{x}_k; y_k); k = 1, ..., n\}$ with $y_k = F(\boldsymbol{x}_k) \in \mathbb{R}$ and $\boldsymbol{x}_k \in \mathbb{R}^d$, it is desired to obtain a function $\mathcal{G}$ so $\sum_{k=1}^{n} ||y_k - \mathcal{G}(\boldsymbol{x}_k)||^2$ is minimum. The purpose of the design is to be able to obtain outputs from input vectors that were not specified in the original training data set.

An RBFNN $\mathcal{F}$ with d entries and one output has a set of parameters that have to be optimized:

$$\mathcal{F}(\boldsymbol{x}_k; C, R, \Omega) = \sum_{j=1}^{m} \phi(\boldsymbol{x}_k; \boldsymbol{c}_j, r_j) \cdot \Omega_j \tag{1}$$

where m is the number of RBFs, $C = \{\boldsymbol{c}_1, ..., \boldsymbol{c}_m\}$ is the set of RBF centers, $R = \{r_1, ..., r_m\}$ is the set of values for each RBF radius, $\Omega = \{\Omega_1, ..., \Omega_m\}$ is

A. Sattar and B.H. Kang (Eds.): AI 2006, LNAI 4304, pp. 1127–1132, 2006.
© Springer-Verlag Berlin Heidelberg 2006

the set of weights and $\phi(\boldsymbol{x}_k; \boldsymbol{c}_j, r_j)$ represents an RBF. The activation function most commonly used for classification and regression problems is the Gaussian function because it is continuous, differentiable, it provides a softer output and it improves the interpolation capabilities [1]. The procedure to design an RBFNN starts by setting the number of RBFs in the hidden layer, then the RBF centers $\boldsymbol{c}_j$ must be placed and a radius r_j has to be set for each of them. Finally, the weights Ω_j can be calculated optimally by solving a linear equation system [3].

We want to find both a network with the smallest number of neurons and one with the smallest error. This is a multiobjective optimisation problem. For some pairs of networks it is impossible to say which is better (one is better on one objective, one on the other) making the set of possible solutions partially sorted [7].

2   Multiobjective Algorithm for Function Approximation: MOFA

Within the many evolutionary algorithms that have been designed to solve multi-objective optimization problems, the Non-Dominated Sorting Genetic Algorithm II (NSGA-II) [2] has been shown to have an exceptional performance. This section will describe the adaptation of this genetic algorithm to solve the problem of designing RBFNN for function approximation.

2.1   Encoding RBFNN in the Individuals

As it was shown in Section 1, to design an RBFNN it is needed to specify: the number of RBFs, the position of the centers of the RBFs, the length of the radii, and the weights for the output layer.

The individuals in the population of the algorithm will contain those elements in a vector of real numbers, but the weights. The storage of the weights in the chromosome is useless since they can be obtained optimally by solving a linear equation system, and for each change in a radius or a center they have to be updated.

2.2   Crossover Operators

Standard crossover operators cannot be applied to our representation. Two specific crossover operators have been designed considering complete RBFs as genes to be exchanged by the chromosomes representing RBFNNs.

Crossover Operator 1: Neurons Exchange. This crossover operator, conceptually, would be the most similar one to a standard crossover because the individuals represent an RBFNN with several neurons and a crossover between two individuals would result in a neuron exchange. The operator will exchange only one neuron, selected randomly, between the two individuals.

This crossover operator exploits the genetic material of each individual in a simple and efficient way without modifying the structure of the network.

Crossover Operator 2: Addition of the Neuron with the Smallest Local Error. This operator consists in the addition into one parent of the neuron with the smallest local error belonging to the other parent. The local error of a neuron is defined as the sum of the errors between the real output and the output generated by the RBFNN for the input vectors that activate that neuron.

To make sure that the offsprings are different of their ancestors, before adding the neuron with the smallest local error, the crossover will make sure that the neuron to be added does not exist in the individual, otherwise the second neuron with the smallest local error will be considered and so on. This restriction combined with the way the NSGA-II proceeds helps to avoid the "competing convention" problem [4].

Once the offspring are obtained, they go through a refinement process that consists in the prune of the RBFs which doesn't influence the output of the RBFNN, to do this, all the weights that connect the processing units to the output layer are calculated and the neurons that don't have a significant weight will be removed.

2.3   Mutation Operators

The mutation operators add randomness into the process of finding good solutions. The mutation operators are desired to be as much simple as they can so it can be shown clearly, without the interference of introducing expert knowledge, the effects of the parallelization at a very basic level. For an RBFNN, two kind of modifications can be performed:

- Changes in the structure of the RBFNN (Increase and decrease operators): Addition and Deletion of an RBF. The first one adds an RBF in one random position over the input vectors space, setting its radius with a random value. All the random values are taken from an uniform distribution in the interval [0,1] since the input vectors and their output are normalized. The second operator is the opposite to the previous one, deleting a random existing RBF from the network. This mutation is constrained so that is not applied when the individual has less than two neurons.
- Changes in the elements of the structure (Movement operators): The third and the fourth operators refer to real coded genetic algorithms as presented in [5]. The third operator adds to all the coordinates of a center a random distance chosen in the interval [-0.5,0.5] from a uniform distribution. The fourth one has exactly the same behavior but changing the value of the radius of the selected RBF.

3   Island Specialisation

In order to obtain results of maximum quality and take advantage of each type of operator, a parallel implementation was studied. This implementation is based on the island paradigm where there are several islands (in our parallel implementation these map to processors) that evolve independent populations and, exchange information or individuals.

From the point of view of evolving the topology and structure of the RBFNN, we propose the following specialization: combine crossover 1 and the movement mutation operators in one island and the crossover 2 and the movement and increasing mutation operators. This combination of operators aims to boost the exploration capabilities of the crossover 2 and the exploitation ones of crossover 1. Since the crossover 2 explore more topologies by increasing the size of the NNs, the specialisation can be done by letting this island to produce only bigger networks. The other island with the crossover 1, will take care of exploitation of the chromosomes and will decrease the size of the networks.

4   Experiments

The experiments were performed using a two dimensional function that was generated using a Gaussian RBFNN with randomly chosen parameters over a grid of 25x25 points.

Although some metrics have been applied to compare nondominated sets [6], in [9] it is suggested that: *"as long as a derived/known Pareto front can be visualized, pMOEA[1] effectiveness may still best be initially observed and estimated by the human eye"* . Therefore the experimental analysis of the algorithms will be based on the plots of the resulting Pareto fronts.

The algorithms were executed using the same initial population of the same size and the same values for all the parameters. The crossover probability and mutation probability were 0.5, the size of the population 100 and the number of generations was fixed to 100. The probabilities might seem high but it's necessary to have a high mutation probability to allow the network to increase its size, and the centers and radii to move. The high crossover probability makes it possible to show the effects of the crossover operators. Several executions were made changing the migration rate, allowing the algorithms to have 2 (each 40 generations), 4 (each 20 generations), 9 (each 10 generations), and 19 (each 5 generations) migrations through the 100 generations.

The results are shown in Figure 1 where the Pareto fronts obtained from each algorithm are represented according to the objectives that have to be minimized, this is, the approximation error and the number of neurons in the network.

The parallel implementation will be compared with non distributed algorithms that are considered as those that have no communication. Within the non distributed algorithms, three different possibilities are determined by the crossover operators and all of these algorithms will use the four mutation operators at the same time. The first algorithm uses only the crossover 1, the second algorithm uses the crossover 2 and the third non distributed algorithm chooses randomly between the two crossover operators each time it has to be applied.

For the non distributed algorithms, as it is shown in Figure 1, the crossover operator 1 obtains better results for small networks but it is not able to grow them as much as the crossover operator 2 does. When the two operators are applied together the results improve significantly, obtaining a better Pareto front

---

[1] *pMOEA* stands for parallel MultiObjective Evolutionary Algorithms.

although the crossover operator 2 is able to design networks with a smaller approximation error when the size of them is big.

Figure 1 shows that the parallel model outperforms the three previous possibilities for all the migration rates. This means that the parallelization is positive independently of the migration rate. If the results are analyzed in detail, it is possible to see how, for low migration rates, the exploitation capabilities of the algorithm increases obtaining smaller networks with better approximation errors meanwhile for high migration rates, the exploration capabilities are boosted, obtaining a more complete Pareto.

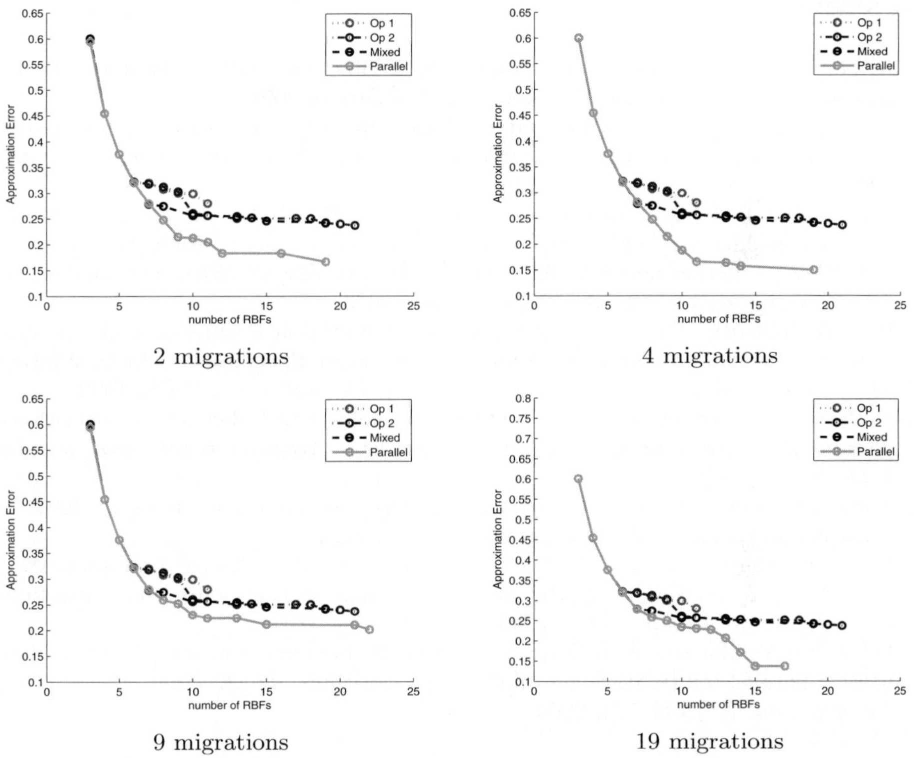

Fig. 1. Pareto front obtained performing several migration steps for the parallel algorithm and the non distributed approaches

5   Conclusions

The design of an RBFNN is a complex task that can be solved using evolutionary approaches that provide satisfactory results. This paper has presented a multiobjective GA that designs RBFNNs to approximate functions. This algorithm has been analyzed considering a non distributed approach that has been specialised by defining separate islands that applied different combinations of mutation and crossover operators. The results confirm that the specialisation on the different aspects of the design of RBFNNs through the parallel approach could lead to obtain better results.

Acknowledgements. This work has been partially supported by the Spanish CICYT Project TIN2004-01419 and the HPC-Europa programme, funded under the European Commission's Research Infrastructures activity of the Structuring the European Research Area programme, contract number RII3-CT-2003-506079.

References

1. A. G. Bors. Introduction of the Radial Basis Function (RBF) networks. *OnLine Symposium for Electronics Engineers*, 1:1–7, February 2001.
2. K. Deb, S. Agrawal, A. Pratap, and T. Meyarivan. A fast and elitist multiobjective genetic algorithm: NSGA-II. *IEEE Trans. Evolutionary Computation*, 6(2):182–197, 2002.
3. J. González, I. Rojas, J. Ortega, H. Pomares, F.J. Fernández, and A. Díaz. Multi-objective evolutionary optimization of the size, shape, and position parameters of radial basis function networks for function approximation. *IEEE Transactions on Neural Networks*, 14(6):1478–1495, November 2003.
4. P. J. B. Hancock. Genetic Algorithms and Permutation Problems: a Comparison of Recombination Operators for Neural Net Structure Specification. In D. Whitley, editor, *Proceedings of COGANN workshop, IJCNN, Baltimore*. IEEE, 1992.
5. F. Herrera, M. Lozano, and J. L. Verdegay. Tackling real-coded genetic algorithms: operators and tools for the behavioural analysis . *Artificial Intelligence Reviews*, 12(4):265–319, 1998.
6. J. Knowles and D. Corne. On metrics for comparing non-dominated sets, 2002. In Congress on Evolutionary Computation (CEC 2002).
7. V. Pareto. *Cours D'Economie Politique*, volume I and II. F. Rouge, Lausanne, 1896.
8. J. Park and J. W. Sandberg. Universal approximation using radial basis functions network. *Neural Computation*, 3:246–257, 1991.
9. D. A. van Veldhuizen, J. B. Zydallis, and G. B. Lamont. Considerations in engineering parallel multiobjective evolutionary algorithms. *IEEE Trans. Evolutionary Computation*, 7(2):144–173, 2003.

Generalized Unified Decomposition
of Ensemble Loss

Remco R. Bouckaert[1,2], Michael Goebel[3], and Pat Riddle[3]

[1] Xtal Mountain Information Technology, Auckland
`rrb@xm.co.nz`
[2] Computer Science Department, University of Waikato
`remco@cs.waikato.ac.nz`
[3] Department of Computer Science, University of Auckland, New Zealand
`{mgoebel, pat}@cs.auckland.ac.nz`

Abstract. Goebel et al. [4] presented a unified decomposition of ensemble loss for explaining ensemble performance. They considered democratic voting schemes with uniform weights, where the various base classifiers each can vote for a single class once only. In this article, we generalize their decomposition to cover weighted, probabilistic voting schemes and non-uniform (progressive) voting schemes. Empirical results suggest that democratic voting schemes can be outperformed by probabilistic and progressive voting schemes. This makes the generalization worth exploring and we show how to use the generalization to analyze ensemble loss.

This article is inspired by Goebel et al. [4], hereafter referred to as UDEL to reflect the title **U**nified **D**ecomposition of **E**nsemble **L**oss. They decomposed the loss of a classifier ensemble into the mean loss of the individual ensemble members and a loss term D which is a measure of the diversity of the ensemble members. However, they considered only ensembles where the various base classifiers each can vote for a single class once only. Here a GUDEL, a generalized UDEL is presented. This generalized decomposition additionally applies to ensembles with weighted votes, as well as to ensembles where the member classifiers output a probability distribution instead of a single vote. We also investigate several other voting schemes which are covered neither by UDEL or by GUDEL. Experiments show that simple democratic voting schemes can sometimes be outperformed by other voting schemes [1,5].

Preliminaries: Given input space $\mathbf{X} = \mathbf{X}_1 \times ... \times \mathbf{X}_v$, a set of class labels $Y = \{Y_1, ..., Y_n\}$, and an unknown but stationary probability distribution P over $\mathbf{X} \times Y$, a learner is usually given a finite sample $D = \{\langle \mathbf{x}_1, y_1 \rangle, ..., \langle \mathbf{x}_m, y_m \rangle\}$ drawn from $\mathbf{X} \times Y$ according to P and is required to produce a classifier c. Given a (previously unseen) test example $\mathbf{x} = \langle x_1, ..., x_v \rangle$, this classifier produces a prediction $\widehat{y}_c(\mathbf{x})$. The performance of models (and therefore implicitly the learners that produced those models) is measured using loss functions: A loss function $l : Y \times Y \to \Re$ measures the cost of making the prediction $\widehat{y}$ when the true value is y. For the case where Y is a set of class labels $Y = \{Y_1, ..., Y_n\}$, the most commonly used loss function is the zero-one-loss ($l_{01}(\widehat{y}, y) := 0$ iff $\widehat{y} = y$;

A. Sattar and B.H. Kang (Eds.): AI 2006, LNAI 4304, pp. 1133–1139, 2006.
© Springer-Verlag Berlin Heidelberg 2006

$l_{01}(\widehat{y}, y) := 1$ otherwise). The goal of a learner should be to produce a model with the smallest possible expected loss; i.e., a model which minimizes the average loss over examples drawn independently from $\mathbf{X} \times Y$ according to P. Some classifiers, rather than producing a prediction $\widehat{y}_c(\mathbf{x})$ directly, output instead a probability distribution $\widehat{P}_c(Y|\mathbf{x})$ over the set of class labels Y, such that, for any $y \in Y$, $\widehat{P}_c(y|\mathbf{x})$ reflects the degree of the classifier's internal belief that the test example $\mathbf{x}$ belongs to class y. In those cases, the ensemble prediction $\widehat{y}(\mathbf{x})$ is taken to be $\widehat{y}(\mathbf{x}) = \arg\min_{y' \in Y} \sum_{c \in C} w_c \int_{y'' \in Y} \widehat{P}_c(y''|\mathbf{x}) l(y'', y') dy''$.where C is an ensemble, and w_c the weight of the classifier c in the ensemble. This is the minimum average distance in terms of the loss function of the member predictions weighted by the members belief and their weigths.

Voting Schemes: A typical ensemble algorithm under 0-1 loss takes the predictions of the base classifiers giving the prediction $\widehat{y}(\mathbf{x}) := \arg\max_{y \in Y} \sum_{c \in C} w_c I(y = \widehat{y}_c(\mathbf{x}))$where $I(\cdot)$ is the indicator function ($I(y = \widehat{y}_c) = 1$ iff $\widehat{y}_c$ equals y, and 0 otherwise). When two or more classes get the same vote, one is chosen at random. This voting scheme is called *democratic voting* also known as *majority vote*. The base classifiers predictions are possibly weighted with a weight w_c. Throughout this article, we assume the weights are normalized to 1, i.e., $\sum_{c \in C} w_c = 1$.

A lot of commonly used base classifiers calculate a probability distribution $\widehat{P}_c(y|\mathbf{x})$ over y and pick the class with the highest probability as their prediction, for instance, decision trees, naive Bayes, and Bayesian network variants. So, instead of taking the class with the highest probability as a vote, one can take the class receiving the maximum average probabililty, where the average is taken over the weighted member classifiers. Formally, this gives $\widehat{y}(\mathbf{x}) := \arg\max_{y \in Y} \sum_{c \in C} w_c \widehat{P}_c(y|\mathbf{x})$where $\widehat{P}_c(y|\mathbf{x})$ is the probability that classifier c assigns to class y given observation $\mathbf{x}$. This voting scheme will be called *probabilistic voting*. Alternatively, it can be argued that the model with the highest confidence in its ability to classify should be making the decision. This *aristocratic voting* scheme returns the class $\widehat{y}(\mathbf{x}) := \arg\max_{y \in Y} (\max_{c \in C} w_c \widehat{P}_c(y|\mathbf{x}))$.

Balancing aristocratic and democratic arguments, the vote can be weighed according to a convex function of $\widehat{P}_c$, for example quadratic or exponential. The *progressive quadratic voting* scheme results in the following classification: $\widehat{y}(\mathbf{x}) := \arg\max_{y \in Y} \sum_{c \in C} w_c (\widehat{P}_c(y|\mathbf{x}))^2$ and *progressive exponential voting* scheme: $\widehat{y}(\mathbf{x}) := \arg\max_{y \in Y} \sum_{c \in C} w_c e^{\widehat{P}_c(y|\mathbf{x})}$. The standard bagging algorithm [3] uses the democratic voting scheme. It can be easily adapted to apply each of the above voting schemes by plugging in the appropriate voting scheme.

Decomposition: First, we need to define some terms. The *ensemble prediction* of an ensemble C for input $\mathbf{x}$ is denoted as $\widehat{y}(\mathbf{x})$ or $\widehat{y}$ if the input $\mathbf{x}$ is clear from the context. The way the ensemble prediction is calculated depends on the voting scheme applied.

Note that UDEL defines the ensemble prediction in terms of a loss function, while our definition does not take this into account. This specializes to the definition used by [4], which was $\widehat{y}(x) := \arg\min_{y \in Y} E_{c \in C}[l(y, \widehat{y}_c(x))]$, with $w_c := 1/|C|$ and $\widehat{P}_c(y'|\mathbf{x}) := 1$ iff $y' = \widehat{y}_c(\mathbf{x})$; $\widehat{P}_c(y'|\mathbf{x}) := 0$ otherwise for all $c \in C$ and $y' \in Y$.

Definition 1. *The* loss *of ensemble C on instance $\langle \mathbf{x}, y \rangle$ under loss function l is given by $L(\langle \mathbf{x}, y \rangle) := l(\widehat{y}(\mathbf{x}), y)$. The* expected loss *of ensemble C on the domain $\langle \mathbf{X}, Y, P \rangle$ is given by $L := E_P[L\langle \mathbf{x}, y \rangle] = \int_{X \times Y} L(\langle \mathbf{x}, y \rangle)p(\langle \mathbf{x}, y \rangle) \, d\langle \mathbf{x}, y \rangle$.*

Typically, we are interested in learning ensembles that have a low expected loss, where the expectation is taken over the instance distribution P.

Definition 2. *The* mean member loss *of ensemble C on instance $\langle \mathbf{x}, y \rangle$ under loss function l is given by $\overline{L}(\langle \mathbf{x}, y \rangle) := \sum_{c \in C} w_c E_{\widehat{P}_c}[l(\widehat{y}_c(\mathbf{x}), y)]$ which equals $\sum_{c \in C} \int_{y' \in Y} w_c l(y', y)\widehat{P}_c(y'|\mathbf{x})dy'$.*
The expected mean member loss *of ensemble C on the domain $\langle \mathbf{X}, Y, P \rangle$ is given by $\overline{L} := E_P[\overline{L}\langle \mathbf{x}, y \rangle] = \int_{X \times Y} \overline{L}(\langle \mathbf{x}, y \rangle)p(\langle \mathbf{x}, y \rangle)d\langle \mathbf{x}, y \rangle$.*

The mean member loss indicates how much the predictions of the individual base classifiers in the ensemble C differ from the true value of the class y. This definition differs from UDEL in that the classifier weights w_c are taken into account when taking the expectation over the set of classifiers C. This specializes to the definition in UDEL when $w_c = 1/|C|$ for all classifiers $c \in C$. Also, the expectation is taken over both C and Y while in UDEL only the expectation over C is taken. This will take into account their class probabilities instead of just their final predictions in the case where the base classifier is a distribution classifier. If the base classifier is not a distribution classifier, the definition still applies with $\widehat{P}_c(y'|\mathbf{x}) = I(y' = \widehat{y}(\mathbf{x}))$, where $I(\cdot)$ is the indicator function, i.e, $I(y' = \widehat{y}(\mathbf{x})) := 1$ iff y' equals $\widehat{y}(\mathbf{x})$, and 0 otherwise. In this case, the definition of mean member loss coincides with the one from UDEL, which was $\overline{L}(\langle x, y \rangle) := E_{c \in C}[l(\widehat{y}_c(x), y)]$.

Definition 3. *The* diversity *of ensemble C on input $\mathbf{x}$ under loss function l is $\overline{D}(\mathbf{x}) := \sum_{c \in C} w_c E_{\widehat{P}_c}[l(\widehat{y}_c(\mathbf{x}), \widehat{y}(\mathbf{x}))] = \sum_{c \in C} \int_{y' \in Y} w_c l(y', \widehat{y}(\mathbf{x}))\widehat{P}_c(y'|\mathbf{x})dy'$.*
The expected diversity *of ensemble C on the domain $\langle \mathbf{X}, Y, P \rangle$ is given by $\overline{D} := E_P[\overline{D}(\mathbf{x})] = \int_{X \times Y} \overline{D}(\mathbf{x})p(\langle \mathbf{x}, y \rangle)d\langle \mathbf{x}, y \rangle$.*

The diversity indicates how much the predictions of the individual base classifiers in the ensemble C differ from the ensemble prediction $\widehat{y}(\mathbf{x})$. Again, this definition differs from UDEL in that here the classifier weights w_c are taken into account, and in that the expectation is taken over both C and Y while in UDEL only the expectation over C is taken, such as $\overline{D}(x) := E_{c \in C}[l(\widehat{y}_c(x), \widehat{y}(x))]$.

Figure 1 shows the relations between the various terms just defined. It shows that the loss L is a function of the real class y and the ensemble prediction $\widehat{y}$, the expected mean member loss $\overline{L}$ is a function of y and the various base classifiers predictions $\widehat{y}_c$, and the expected diversity $\overline{D}$ is a function of the various base classifiers predictions $\widehat{y}_c$ and the ensemble prediction $\widehat{y}$.

Following UDEL, a decomposition of L is proposed as a function of $\overline{L}$ and $\overline{D}$, that is,

$$L\langle (\mathbf{x}, y) \rangle = f(\overline{L}(\langle \mathbf{x}, y \rangle), \overline{D}(\mathbf{x})) \tag{1}$$

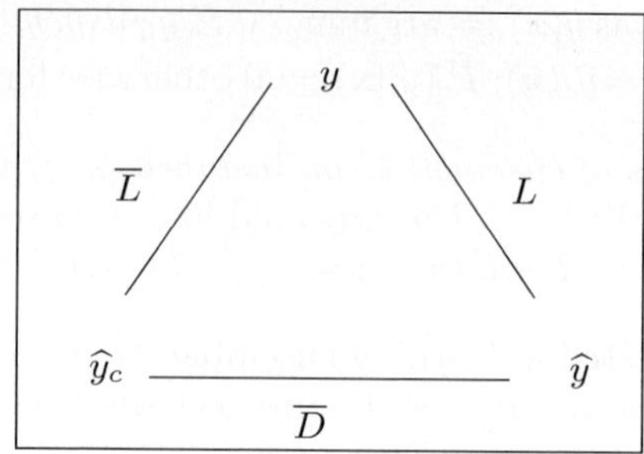

Fig. 1. Intuitive relations between L, $\overline{L}$, $\overline{D}$ and y, $\widehat{y}$ and $\widehat{y}_c$

for ensembles of weighted, probabilistic classifiers. This covers ensembles voting democratically or probabilistically with or without individual classifier weights. Though it does not cover the aristocratic or progressive voting schemes the theory can be easily extended by normalizing the votes in the schemes so that the votes add to unity for each particular instance. The extensions to the theory which cover these cases are discussed in the full paper. In the following section, details for 0-1 loss will be given.

Instantiating the decomposition for 0-1 loss: We first instantiate Equation 1 for the case where the class is binary, that is $Y = \{0, 1\}$, and the loss function l is the 0-1 loss, i.e., $l(y_1, y_2) = I(y_1 \neq y_2)$.

Lemma 1. *For 0-1 loss in two-class problems, the ensemble loss $L(\langle \mathbf{x}, y \rangle)$ can be written as $L(\langle \mathbf{x}, y \rangle) = f(\overline{L}(\langle \mathbf{x}, y \rangle), \overline{D}(\mathbf{x})) = \overline{L}(\langle \mathbf{x}, y \rangle) + k(x, y)\overline{D}(\mathbf{x})$ where $k(x, y) = -1$ iff $\widehat{y}(\mathbf{x}) = y$ and $k(x, y) = 1$ iff $\widehat{y}(\mathbf{x}) \neq y$.*

Refer to [2] for proofs of this and following lemmas.

For a given ensemble C, let $T := \{\langle \mathbf{x}, y \rangle \, | \widehat{y}(\mathbf{x}) = y\}$ be the set of instances which the ensemble classifies correctly, and let $F := \{\langle \mathbf{x}, y \rangle \, | \widehat{y}(\mathbf{x}) \neq y\}$ be the set of instances which the ensemble classifies incorrectly. Then, we can define $\overline{D}_T$ as the diversity over T and $\overline{D}_F$ as the diversity over F. More formally, $\overline{D}_T := \int_{\langle \mathbf{x}, y \rangle \in T} \overline{D}(\mathbf{x}) p(\langle \mathbf{x}, y \rangle \, | \, \langle \mathbf{x}, y \rangle \in T) d \langle \mathbf{x}, y \rangle$ and $\overline{D}_F := \int_{\langle \mathbf{x}, y \rangle \in F} \overline{D}(\mathbf{x}) p(\langle \mathbf{x}, y \rangle \, | \, \langle \mathbf{x}, y \rangle \in F) d \langle \mathbf{x}, y \rangle$. $\overline{D}_T$ denotes the expected ensemble diversity on correctly predicted examples, and $\overline{D}_F$ denotes the expected ensemble diversity on incorrectly predicted examples. $\overline{D}_T$, $\overline{D}_F$, and $\overline{D}$ will always be between 0 and 0.5. This can be easily seen for a two class dataset. If $\overline{D}$ became more than .5 than the ensemble would predict a new class thereby making $\overline{D}$ less than .5 again. The same argument holds for $\overline{D}_F$ and $\overline{D}_T$ and becomes even more obvious when there are more than two classes. The following lemmas hold:

Lemma 2. *Under 0-1 loss, the expected diversity $\overline{D}$ can be written as $\overline{D} = (1 - L)\overline{D}_T + L\overline{D}_F$.*

Lemma 3. *For 0-1 loss in two-class problems, the expected ensemble loss L can be written as*

$$L = \frac{\overline{L} - \overline{D}_T}{1 - \overline{D}_T - \overline{D}_F}.$$

Lemma 3 can be rewritten equivalently as

$$L = \overline{L} - (1 - L)\overline{D}_T + L\overline{D}_F. \tag{2}$$

Lemma 4. *For 0-1 loss in two-class problems, the expected ensemble loss L can be written as $L = \frac{\overline{L} + \overline{D} - 2\overline{D}_T}{1 - 2\overline{D}_T}$.*

Lemma 5. *For 0-1 loss in two-class problems, the expected ensemble loss L can be written as $L = \frac{\overline{L} - \overline{D}}{1 - 2\overline{D}_F}$.*

As discussed in the following section, Lemmas 3, 4 and 5 provide better insight in the behavior of ensemble classifiers.

Relating L, $\overline{L}$, $\overline{D}$, $\overline{D}_T$ and $\overline{D}_F$: There are two ways to manipulate $\overline{D}_T$ and $\overline{D}_F$:

1. by keeping T and F constant. In this case, the ensemble decisions do not change, hence the 0-1 loss does not change, but the diversities $\overline{D}_T$ and $\overline{D}_F$ can be manipulated independently.
2. by changing the boundaries of T and F. In this case, the 0-1 loss is affected, and so are $\overline{D}_T$ and $\overline{D}_F$.

Case 1: When keeping T and F constant and increasing $\overline{D}_T$ by α, $\overline{D}$ increases by $\alpha(1 - L)$ (by Lemma 2) and $\overline{L}$ increases by $\alpha(1 - L)$ (because of Equation 2 and L remains constant). Lemma 4 shows that L remains constant under such change. If $\overline{D}_F$ is increased by α, $\overline{D}$ increases by αL (again by Lemma 2), and likewise $\overline{L}$ decreases by αL (because of Equation 2). Lemma 5 confirms that L remains constant under such change.

Case 2: Now consider manipulating T and F by moving instances from F to T such that L decreases by α. Let M be the set of instances $\{\langle \mathbf{x}, y \rangle \in X \times Y\}$ that moved from F to T. For those instances holds $\overline{L} - \overline{D} = 0$ by Lemma 2. So, by Lemma 4, we have $L - \alpha = (\overline{L} - \overline{D} - \alpha)/(1 - 2\overline{D}_F)$, which after some manipulations gives $\overline{D}_F = L - \overline{L} + \overline{D}/2(L - \alpha)$, that is, $\overline{D}_F$ increases. This also works the other way around; increasing $\overline{D}_F$ helps decrease the expected ensemble loss L.

Let us look at a case study to make this a bit clearer. In Figure 2 we have three instances in the set T and two instances in the set F. We have 3 member classifiers for classifying each of these instances; their votes are given beside each instance. In case 1 we increase $\overline{D}_T$ while keeping T and F constant; for example changing the votes at instance 1 from TTT to TTF. This will cause an increase in $\overline{D}$ and $\overline{L}$ but L remains constant. Likewise a change in instance 4, from FFF to FTF will cause an increase in $\overline{D}_F$, $\overline{D}$, but a decrease in $\overline{L}$ while L remains constant. This highlights that just increasing or decreasing diversity can have no effect on the benefits derived from an ensemble.

Now let us examine case 2. If we move instance 3 from the T set to the F set by changing its votes from FTT to FTF, we will cause L to increase while $\overline{D}$ hasn't changed. $\overline{L}$ will also increase and $\overline{D}_F$ could either have increased or decreased. If we move instance 5 from the F set to the T set by changing its votes from FFT to FTT, we will cause L to decrease while $\overline{D}$ still hasn't changed. $\overline{L}$ will also decrease and $\overline{D}_T$ could either have increased or decreased. From this we can see, that increasing the diversity over the F set is the only way to lower the ensemble loss.

Why weighting may work: Instead of letting the classifiers in the ensemble vote uniformly ($\forall c \in C : w_c = 1/|C|$), the weights can be made proportional to the classification accuracy on the training data. This can be achieved by setting $\forall c \in C : w_c \propto 1 - L_c$, where $L_c = \int_{X \times Y} l(y, \hat{y}_c(\mathbf{x})) p(\langle \mathbf{x}, y \rangle) d\langle \mathbf{x}, y \rangle$ empirically estimates the expected loss of member classifier c on the training data, with m being the number of instances in the training set. We assume that when the classification accuracy on the training data is smaller, then the classification accuracy in general will be smaller, that is, performance on the training data can be seen as a predictor on the performance in general. This seems a reasonable assumption since it underlies all classifier learning algorithms (as long as overfitting is taken into account).

So, let's look at what happens when moving weight to classifiers in the ensemble that perform well on the training data. From the definition for expected loss $\overline{L}$ (definition 2) we see that it decreases. Also, the diversity $\overline{D}_F$ will decrease, hence from Lemma 5 we have that the expected loss will decrease (since the numerator decreases and the denominator increases).

Obviously the effect of increased performance by weighting is not the only mechanism at work. If this were the case, just taking the classifier in the ensemble with the highest accuracy on the training data and discarding the others would give a better classifier than using the whole ensemble in uniform voting. Experimental results suggest that this is generally not true (see [2] for details).

What diversity does for binary classes: Let's look at an example with a binary class, a set of four instances $\mathbf{X} = \{\mathbf{x}_1, \mathbf{x}_2, \mathbf{x}_3, \mathbf{x}_4\}$, and a set of three classifiers $C = \{c_1, c_2, c_3\}$, each classifier being able to classify two instances

<table>
<tr><td>T</td><td>F</td></tr>
<tr><td>1. TTT</td><td>4. FFF</td></tr>
<tr><td>2. TFT</td><td></td></tr>
<tr><td>3. FTT</td><td>5. FFT</td></tr>
</table>

Fig. 2. Case study of relations between $L, \overline{L}, \overline{D}$ and $y, \hat{y}$ and $\hat{y}_c$

correctly and two incorrectly. For simplicity, a democratic voting scheme is used, so that it is easy to illustrate when classifiers are correct or incorrect. For the purpose of this example, it is also assumed that the weights are uniform ($w_c = 1/3$), that there is no class noise ($\forall i \in \{1,..,4\} : y_i = f(\mathbf{x}_i)$), and that the examples are uniformly distributed ($\forall \langle \mathbf{x}_i, y_i \rangle : P(\langle \mathbf{x}_i, y_i \rangle) = 1/4$).

The same effect appears when the set of classifiers is larger than 3. So, in general, we want to have a high as possible $\overline{D}_T$ with an as low as possible $\overline{D}_F$.

Probabilistic voting can be interpreted as democratic voting where, instead of a single vote by a single classifier c_i, a set of classifiers $c_{i,1}, \ldots, c_{i,|Y|}$ vote democratically, and a portion $\widehat{P}_c(y = 0)$ of those votes for $y = 0$ while the rest votes for $y = 1$. Therefore, the same effect can be observed in probabilistically voting ensembles as well.

Conclusions: A generalized unified decomposition of ensemble loss for predicting ensemble performance was presented that generalizes the unified decomposition presented in UDEL by allowing for classifier weights and probabilistic voting schemes. While UDEL was limited to uniform weights and democratic voting schemes, where the various base classifiers each can vote for a single class once only, this article extends the result to more general voting schemes. Experimental results suggest (see [2] for details) that democratic voting can be outperformed by probabilistic and progressive voting schemes, hence a GUDEL is worth exploring. Specifically probabilistic voting schemes are shown to provide a real advantage over democratic voting schemes in many datasets. So they should be tried more often in real world problem solving situations. In addition the formalization of ensembles presented gave insight into why the voting schemes provided such differing results.

Another class of classifiers for multinomial classes can be considered which calculates confidence on the probabilities of the classes through standard Bayesian technique of representing it as Dirichlet distribution. A voting scheme based on the confidence in the probabilities could be devised giving more voting rights to base classifiers with high confidence in their probabilities. The GUDEL can be extended to such voting schemes (although preliminary experiments suggest this may not be a fruitful route to pursue).

References

1. E. Bauer and R. Kohavi. An empirical comparison of voting classification algorithms: bagging boosting and variants. Machine learning, 36(1-2): 105-139, 1999.
2. R. Bouckaert, M. Goebel, P Riddle. Generalized Unified Decomposition of Ensemble Loss (full version of this paper). Available from http://www.cs.waikato.ac.nz/~remco/gudelfull.ps.
3. L. Breiman. Bagging predictors. Machine Learning 24(2): 123-140, 1996.
4. M. Goebel, P. Riddle, M. Barley. A unified decomposition of ensemble loss for predicting ensemble performance. ICML 2002.
5. Y. Singer and R. Schapire. Improved boosting algorithms using confidence-rated predictions. Machine Learning, 1999.

Feature Selection with RVM and Its Application to Prediction Modeling

Dingfang Li and Wenchao Hu

School of Mathematics and Statistics, Wuhan University, Wuhan, 430072, P.R. China
dfli@whu.edu.cn, wchu80@sina.com

Abstract. We describe here a method named FSRVM-PLS for model construction using relevance vector machine (RVM). The most compelling feature of FSRVM-PLS is that it's not necessary to estimate parameters in the feature selection phase benefiting from a fully probabilistic framework. After evaluating the effectiveness of FSRVM on a synthetic data set, our method is applied successfully to the prediction of aqueous solubility and permeability.

Keywords: RVM, feature selection, solubility, permeability, prediction.

1 Introduction

Theoretical calculation of aqueous solubility and permeability is not feasible at present. Consequently, medicinal chemists have begun to pay closer attention to constructing models that are able to predict aqueous solubility and permeability of drug-like compounds[1-3]. Therefore, feature selection becomes very necessary for model construction when facing high dimensional data nowadays. The goal of feature selection is to minimize the number of features needed for an effective prediction of a targeted property[4-5]. However, exhaustive evaluation of all feature subsets is usually intractable.

To overcome the foregoing problems, a novel method that guarantees a sparse solution in the feature space has been studied presently, namely feature selection with relevance vector machine (FSRVM). The results show that FSRVM can select all the desired features. Then, an prediction modeling approach named FSRVM-PLS is proposed through combining FSRVM and PLS. Comparing the results on aqueous solubility data and Caco-2 permeability data as previous used in [5-6], FSRVM-PLS not only produces comparable performance as genetic algorithm-based partial least squares (GA-PLS), but also has many remarkable traits.

2 Feature Selection with Relevance Vector Machine

RVM is reported in several benchmark studies that RVM can yield nearly identical performance to, if not better than, that of SVM while using far fewer relevance vectors than the number of support vectors for SVM[7-10]. Compared with SVM, RVM does not need the tuning of any regularization parameter during the training phase. The kernel function does not need to satisfy Mercer's condition. Further more, the predictions are probabilistic.

A. Sattar and B.H. Kang (Eds.): AI 2006, LNAI 4304, pp. 1140–1144, 2006.
© Springer-Verlag Berlin Heidelberg 2006

The RVM makes predictions based on the following function:

$$y(x;\beta) = \sum_{i=1}^{N} \beta_i K(x,x_i) + \beta_0 \tag{1}$$

where $K(x,x_i)$ is a kernel function, and $\beta = (\beta_0,\beta_1,...\beta_N)^T$ are adjustable parameters (or 'weights'). Inferring weights procedures is under a fully probabilistic framework.

The solution of RVM is highly sparse, which means many weights associated with samples tend to become zero. It needs to stress that the solution is sparse in the instance space, but not in feature space where data is originally represented. In order to utilize RVM for feature selection, we employ linear kernel RVM. The linear kernel RVM makes predictions based on the following function:

$$y(x;\beta) = \sum_{p=1}^{n} \beta_p x_p + \beta_0 \tag{2}$$

Specifically, A Gaussian prior distribution of zero mean and variance $\sigma_{\beta_j}^2 \equiv \alpha_j^{-1}$ is defined over each weight:

$$p(\beta \mid \alpha) = \prod_{i=0}^{n} N(\beta_i \mid 0,\alpha_i^{-1}) \tag{3}$$

where the key to obtain sparsity is the use of $n+1$ independent hyperparameters $\alpha = (\alpha_0,\alpha_1,\alpha_2,...,\alpha_n)^T$, one per weight, which moderate the strength of the prior information.

Given a data set of input-target pairs $G = \{(x_i,t_i)\}_{i=1}^{N}$ (where x_i is the input vector, t_i is the desired real-valued labeling, and N is the number of the input records). Suppose the targets are independent and noise is assumed to be mean-zeros Gaussian with variance σ^2. Thus the likelihood of the complete data set can be written as:

$$p(t \mid \beta,\sigma^2) = (2\pi\sigma^2)^{-N/2} \exp\{-\frac{1}{2\sigma^2}\|t - \Phi\beta\|^2\} \tag{4}$$

where $t = (t_1,t_2,...,t_N)^T$, $\Phi = [1,[x_1,x_2,...,x_N]^T]$

Having defined the prior distribution and likelihood function, from Bayes' rule, the posterior over weights is thus given by:

$$p(\beta \mid t,\alpha,\sigma^2) = \frac{p(t \mid \beta,\sigma^2)p(\beta \mid \alpha)}{p(t \mid \alpha,\sigma^2)} \sim N(\beta \mid \mu,\Sigma) \tag{5}$$

where the posterior covariance and mean are respectively:

$$\Sigma = (\sigma^{-2}\Phi^T\Phi + A)^{-1} \tag{6}$$

$$\mu = \sigma^{-2}\Sigma\Phi^T t \tag{7}$$

with $A = diag(\alpha) = diag(\alpha_0,\alpha_1,...,\alpha_n)$.

The likelihood distribution over the training targets can be "marginalized" by integrating out the weights to obtain the marginal likelihood for the hyperparameters:

$$p(t \mid \alpha, \sigma^2) = \int p(t \mid \beta, \sigma^2) p(\beta \mid \alpha) d\beta \sim N(0, C) \tag{8}$$

where the covariance is given by $C = \sigma^2 I + \Phi A^{-1} \Phi^T$.

The estimated value of the model weights is given by the mean of the posterior distribution, which is also the maximum a posteriori (MAP) estimate of the weights, and depends on the value of the hyperparameter α and of the noise σ^2 whose estimated value is obtained by maximizing (8).

The updating procedure for α is given as follows:

$$\alpha_i^{new} = \frac{\gamma_i}{\mu_i^2} \tag{9}$$

where μ_i is the i-th posterior mean weight from (7) and γ_i is defined by:

$$\gamma_i = 1 - \alpha_i \Sigma_{ii} \tag{10}$$

wherein Σ_{ii} is the i-th diagonal element of the posterior weight covariance from (6) computed with the current α and σ^2 values.

For the noise variance σ^2, the update is as follows:

$$(\sigma^2)^{new} = \frac{\|t - \Phi\mu\|^2}{N - \sum_i \gamma_i} \tag{11}$$

The finally model obtained by linear kernel RVM can be regarded as minimize:

$$L(\beta) = \sum_i (y_i - \sum_p \beta_p x_{ip})^2 + \lambda \sum_p \log|\beta_p| \tag{12}$$

This model is built on the few features whose associated weights do not go to zero during the training process. These remaining features are obtained for prediction modeling. This method is named as FSRVM.

3 Effectiveness of FSRVM

The effectiveness of FSRVM on synthetic data is evaluated in this section. Our goal is to examine whether FSRVM can choose all features that are most predictive of a given outcome.

The synthetic data set is randomly generated with the similar rule used in [11] as follows:

$$y = x_1 + 2x_2 + 3x_3 + 2\sin x_4 + e^{x_5} + 0x_6 + \ldots + 0x_{12}$$

Here feature $x_1, x_2, \ldots, x_5$ are independently and identically distributed random variables following standard normal distribution. The 6th variable is $x_6 = x_1 + 1$ which is correlated to x_1. The 7th variable is $x_7 = x_2 x_3$ which relates to both x_2 and

x_3. Five additional standards normally distributed variables are also generated, and have nothing to do with the dependent y. We generate 200 examples.

As shown in table 1, FSRVM selects all desired features, and the coefficients of irrelevant features selected by FSRVM (x_8, x_{11}) are very small compared with that of relevant features (x_2, x_3, x_4, x_5, x_6). FSRVM does not select x_1 since either x_1 or x_6 could be represented by each other. From the results above, we can conclude that FSRVM is effective in selecting features needed for an effectively prediction of a targeted property.

Table 1. Regression coefficients of the feature subset selected by FSRVM

Symbols	Regression Coefficients
x2	2.0342
x3	2.9059
x4	1.2653
x5	1.3042
x6	1.064
x8	-0.050252
x11	-0.023229

4   Application to Modeling of Aqueous Solubility and Permeability

In this section, we constructed models using two datasets as previously used in GA-PLS[5-6]. The first dataset contains 211 drug-like compounds (Solubility) and 88 descriptors, while the second dataset involves 110 compounds (Caco-2) and 72 descriptors.

The performance of the model can be assessed using several metrics. The correlation coefficient r or squared correlation coefficient r^2 can be considered as a measurement of "self-consistency" of a model, or be used to assess the approximated generalization of a model using cross-validated prediction.

The results demonstrate that FSRVM is effective in reducing the dimensionality. The number of features selected by FSRVM for solubility and Caco-2 are 27 and 19 respectively, is less than features selected by GA-PLS (29 and 24 respectively).

5   Conclusion

In this study, a novel feature selection approach named FSRVM is established. Then a prediction modeling approach named FSRVM-PLS is proposed through combining FSRVM with PLS, and applied successfully in predicting aqueous solubility and permeability. Experiment results show that FSRVM-PLS is capable of providing comparable generalization performance to GA-PLS, while utilizing fewer features. Furthermore, there is no parameter needed to be estimated in our feature selection phase and is therefore faster than GA-PLS.

References

1. Johnson SR, Zheng W. Recent Progress in the Computational Prediction of Aqueous Solubility and Absorption. *The AAPS Journal* 2006; 8(1): E27-E40.
2. Hoffman, B., S. J. Cho, W. Zheng, S. Wyrick, D. E. Nichols, R. B. Mailman, and A. Tropsha. Quantitative structure-activity relationship modeling of dopamine $D1$ antagonists using comparative molecular field analysis, genetic algorithms-partial least squares, and k nearest neighbor methods. *J. Med. Chem.* 1999; 42: 3217-3226
3. Tropsha, A., and W. Zheng. Identification of the descriptor pharmacophores using variable selection QSAR: applications to database mining. *Curr. Pharm. Des.* 2001; 7: 599-612
4. Guyon and A. Elisseeff, An introduction to variable and feature selection. *Journal of Machine Learning Research* 2003; 3: 1157-1182
5. Wanchana S, Yamashita F, Hashida M. Quantitative structure/property relationship analysis on aqueous solubility using a genetic algorithm-combined partial least squares method. *Pharmazie* 2002; 57: 127-129
6. Yamashita F, Wanchana S, Hashida M. Quantitative structure/property relationship analysis of Caco-2 permeability using a genetic algorithm-based partial least squares method. *J Pharm Sci.* 2002; 91: 2230-2239
7. M.E.Tipping. Sparse Bayesian learning and the relevance vector machine. *J Mach Learn Res.* 2001; 1: 211-244
8. Schueler-Furman, Y Altuvia, A Sette, H Margalit. Applying Logistic Regression and RVM to Achieve Accurate Probablistic Cancer Diagnosis From Gene Expression Profiles. Protein Sci 2000; 9: 1838-1946
9. M.E. Tipping. The Relevance Vector Machine, *Advances in Neural Information Processing Systems*, Cambridge, MA: The MIT Press 2001; 12: 652-658
10. C.M. Bishop, Michael E. Tipping. Variational relevance vector machine. *Proceedings of the 16th Conference on Uncertainty in Artificial Intelligence,* Morgan Kaufmann 2000; 46-53
11. Jinbo Bi, Kristin Bennett, Mark Embrechts, Curt Breneman and Minghu Song. Dimensionality Reduction via Sparse Support Vector Machines. *Journal of Machine Learning Research* 2003; 3: 1229-1243

Hybrid Data Clustering Based on Dependency Structure and Gibbs Sampling

Shuang-Cheng Wang[1,2], Xiao-Lin Li[3], and Hai-Yan Tang[2]

[1] Department of Information Science
Shanghai Lixin University of Commerce, Shanghai, China
[2] China Lixin Risk Management Research Institute
Shanghai Lixin University of Commerce, Shanghai, China
{wangsc, tlx}@lixin.edu.cn
[3] National Laboratory for Novel Software Technology
Nanjing University, Nanjing 210093, China
lixl_126@126.com

Abstract. A new method for data clustering is presented in this paper. It can cluster data set with both continuous and discrete data effectively. By using this method, the values of cluster variable are viewed as missing data. At first, the missing data are initialized randomly. All those data are revised through the iteration by combining Gibbs sampling with the dependency structure that is built according to prior knowledge or built as star-shaped structure alternatively. A penalty coefficient is introduced to extend the MDL scoring function and the optimal cluster number is determined by using the extended MDL scoring function and the statistical methods.

1 Introduction

Clustering is a process that divides data into clusters according to a certain rule so as to represent the underlying structures compactly in the data set. Such a representation will lose some detail information, but it makes the problem more concise and apprehensible. As clustering plays an important role in data analysis and prediction, it has been extensively applied to information indexing, information abstraction, web page analysis, medical diagnosis, market analysis and so on.

Current clustering algorithms mainly focus on discrete or continuous data respectively. For hybrid data, the common approaches are to discrete continuous variables or numberalize discrete variables. As a result, some useful information for clustering is lost. Researches on direct clustering hybrid data are scarce. AutoClass algorithm [1-3] is one of the most representative hybrid data clustering algorithms. In the algorithm, given the upper bound for cluster number according to prior knowledge, it uses star-shaped structure and EM (expectation-maximization) algorithm to fix values for the cluster variable, and the cluster number is regulated dynamically in the clustering process by merging a smaller cluster into a larger one. There are some problems with AutoClass. On one hand, as EM algorithm is a greedy searching method, it is sensitive to the original value and is liable to be trapped in the local maximum. Furthermore, the iteration process may converge to the parameter space boundary that is always not the likelihood function maximum to cause deceptive convergence. On the other hand, there is no reliable criterion to determine whether to keep a cluster or not.

A. Sattar and B.H. Kang (Eds.): AI 2006, LNAI 4304, pp. 1145–1151, 2006.

© Springer-Verlag Berlin Heidelberg 2006

In this paper, a new algorithm named HDC-DSGS (hybrid data clustering based on dependency structure and Gibbs sampling) is presented to avoid the problems we stated above. HDC-DSGS takes the values of cluster variable as missing data. At the beginning, the missing data are initialized at random. Then Gibbs sampling [4, 5] and dependency structures are utilized to revise those values at each iteration. As the standard Gibbs sampling is a sampling process based on full conditional distribution and the sampling complexity is exponential in the number of variables, dependency structures built by prior knowledge or as star-shaped when prior knowledge is unavailable can be exploited to simplify this problem. Therefore, the sampling efficiency is improved. In theory, Gibbs sampling can iteratively converge to the global stationary distribution, the result is irrelevant to the original value, and the local optima can be escaped as well. To determine the optimal cluster number, a penalty coefficient is introduced to extend MDL [6] (minimal description length) metric and HDC-DSGS uses the extended MDL criterion and statistical methods to find the optimal cluster number. Experimental results show the validity of the algorithm.

Throughout the paper, we use $X_1,...,X_n$ to denote attribute variables, including both discrete and continuous variables, X_{n+1} is the cluster variable, $x_1,...,x_n x_{n+1}$ is their instantiation. D represents a data set with N cases, all data being generated from some hybrid probability distribution P randomly.

The paper is organized as follows: In section 1, the clustering method with fixed cluster number is presented. Section 2 studies method to determine the optimal cluster number. In Section 3 experiments are conducted from the aspects of accuracy and validity of the determination of optimal cluster number respectively.

2   Clustering with Fixed Cluster Number

Cluster number bounds M_{min} (generally is 2) and M_{max} are set according to prior knowledge. For selected number of cluster between M_{min} and M_{max} , all the missing values are initialized and get a complete data set denoted by $D^{(0)}$. Then all those values we padded in will be revised in the course of iteration. For each iteration, Gibbs sampling based on star-shaped structure assumption will be used to revise each missing value in the data set sequentially. When one missing value is revised, parameters for the star-shaped structure is regulated and reused to revise the next data.

Standard Gibbs sampling conducts sampling from full conditional distribution, that is, it samples values for variable X_i from distribution $p(x_i \mid x_1,...,x_{i-1},x_{i+1},...,x_{n+1})$. As the conditional set is of high dimensionality, the complexity is exponential in the number of variables. There is not yet any sound sampling approach for high dimension hybrid distribution. This problem can be solved by taking the dependency structure into consideration. When prior knowledge for structure is unavailable, the star-shaped structure can be used, denoted by S , as an alternative. All the following analyses are based on this star-shaped structure assumption.

According to the star-shaped structure, the joint probability distribution can be factored as:

$$p(x_1,...,x_n,x_{n+1} \mid S) = p(x_{n+1})\prod_{i=1}^{n} p(x_i \mid x_{n+1}) \qquad (1)$$

where $p(x_i \mid x_{n+1})$ denotes conditional probability for discrete variables, and conditional density for continuous variables.

Let $x_{(n+1)m}$ be the data to be revised for X_{n+1} in case m, $\hat{x}_{(n+1)m}$ be the value revised. X_{n+1} can take value only from $x_{n+1}^1,\ldots,x_{n+1}^{r_{n+1}}$. For iteration $k+1$, $D^{(k)} = D_{(n+1)1}^{(k)}$ is the data set before revision, $D_{(n+1)(m-1)}^{(k)}$ is the data set before $x_{(n+1)m}$ is revised, and $D^{(k+1)} = D_{(n+1)N}^{(k)}$ is the data set after revision and is to be used for the next iteration. The time complexity for one revision is $O(N)$.

For discrete variables, if $p^{(k)}\left(x_{im} \mid x_{(n+1)m}, D_{i(m-1)}^{(k)}, S\right) = 0$, the value will be Laplace-corrected [7] as

$$p^{(k)}\left(x_{jm} \mid x_{(n+1)m}, D_{(n+1)(m-1)}^{(k)}, S\right)$$
$$= (1/N)/\left(N\left(x_{(n+1)m} \mid D_{(n+1)(m-1)}^{(k)}\right) + N\left(x_{im} \mid D_{(n+1)(m-1)}^{(k)}\right)(1/N)\right) \tag{2}$$

where $N\left(x_{(n+1)m} \mid D_{(n+1)(m-1)}^{(k)}\right)$ and $N\left(x_{im} \mid D_{(n+1)(m-1)}^{(k)}\right)$ are case numbers for $X_{n+1} = x_{(n+1)m}$ and $X_i = x_{im}$ in $D_{(n+1)(m-1)}^{(k)}$ respectively.

For continuous variables, it is reasonable to assume a normal distribution, or assume kernel density or hybrid density as an alternative, and we get

$$p^{(k)}\left(x_{jm} \mid x_{(n+1)m}, D_{(n+1)(m-1)}^{(k)}, S\right) = g^{(k)}\left(x_{jm}; \mu_{x_{(n+1)m}}, \sigma_{x_{(n+1)m}} \mid D_{(n+1)(m-1)}^{(k)}, S\right) \tag{3}$$

where $g^{(k)}\left(x_{jm}; \mu_{x_{(n+1)m}}, \sigma_{x_{(n+1)m}} \mid D_{(n+1)(m-1)}^{(k)}, S\right) = \dfrac{1}{\sqrt{2\pi}\sigma_{x_{(n+1)m}}} e^{-\dfrac{\left(x_{jm} - \mu_{x_{(n+1)m}}\right)^2}{2\sigma_{x_{(n+1)m}}}}$, $\mu_{x_{(n+1)m}}$ is

the mean for attribute X_j in cluster $x_{(n+1)m}$ and $\sigma_{x_{(n+1)m}}^2$ is the variance.

According to the star-shaped structure,

$$p\left(x_{n+1} \mid x_{1m},\ldots,x_{nm}, D_{(n+1)(m-1)}^{(k)}, S\right) = \alpha p(x_{n+1}) \prod_{i=1}^{n} p\left(x_{im} \mid x_{n+1}, D_{(n+1)(m-1)}^{(k)}, S\right),$$

where α is a number irrelevant to X_{n+1}. After normalizing

$p(x_{n+1}) \prod_{i=1}^{n} p\left(x_{im} \mid x_{n+1}, D_{(n+1)(m-1)}^{(k)}, S\right)$, denote

$$w_{n+1}(h) = \frac{p\left(x_{n+1}^h \mid D_{(n+1)(m-1)}^{(k)}, S\right) \prod_{i=1}^{n} p\left(x_{im} \mid x_{n+1}^h, D_{(n+1)(m-1)}^{(k)}, S\right)}{\sum_{j=1}^{r_{n+1}} p\left(x_{n+1}^j \mid D_{(n+1)(m-1)}^{(k)}, S\right) \prod_{i=1}^{n} p\left(x_{im} \mid x_{n+1}^j, D_{(n+1)(m-1)}^{(k)}, S\right)}, h \in \{1,\ldots,r_{n+1}\}.$$

For random number λ, value for X_{n+1} is:

$$\hat{x}_{(n+1)m} = \begin{cases} x_{n+1}^{1}, & 0 < \lambda \leq w_{n+1}(1) \\ \cdots\cdots \\ x_{n+1}^{h}, \sum_{j=1}^{h-1} w_{n+1}(j) < \lambda \leq \sum_{j=1}^{h} w_{n+1}(j) \\ \cdots\cdots \\ x_{n+1}^{r_{n+1}}, & \lambda > \sum_{j=1}^{r_{n+1}-1} w_{n+1}(j) \end{cases} \tag{4}$$

3   Determining the Optimal Cluster Number

If the cluster number is unknown, the following approach should be used to determine the optimal cluster number in the algorithm.

MDL is a criterion for discrete variables. When there are continuous variables in the domain, the continuous variables should not be treated in the same way as a discrete one. In this paper, a penalty coefficient is introduced to settle this problem. Let $D_k(M_{\min} \leq k \leq M_{\max})$ be the final data set for cluster number k after iteration. The monotonic penalty coefficient bounds are obtained by calculating the extended MDL scoring criterion. Meanwhile the optimal cluster number can be found through statistical methods in response to the criteria.

3.1   Extended MDL Scoring Metric

Let $EMDL(S \mid D_k)$ be the extended MDL scoring metric,

$$EMDL(S \mid D_k) = \frac{\log N}{2}\left[|S_1| + \rho|S_2|\right] - L(S \mid D_k) \tag{5}$$

Where $|S_1| = r_{n+1} + \sum_{k} r_{n+1}(r_k - 1) = r_{n+1}\left[1 + \sum_{k}(r_k - 1)\right]$, r_{n+1} is the dimension of the cluster variable, r_k is the dimension of discrete attribute variable $_x$. $|S_2| = 2n_2 r_{n+1}$, n_2 is the number of continuous attributes. $\rho > 0$ is the penalty coefficient.

$$L(S \mid D_k) = \sum_{i=1}^{N} \log(p(u_i \mid D_k)) = N \sum_{i=1}^{n} \sum_{x_i, x_{n+1}} p(x_i, x_{n+1} \mid D_k)\log(p(x_i \mid x_{n+1}, D_k)). \text{ if }$$

X_i is a continuous attribute, the equation above can be calculated through conversion $p(x_i, x_{n+1} \mid D_k) = p(x_{n+1} \mid D_k)p(x_i \mid x_{n+1} \mid D_k)$.

3.2 Determining the Optimal Cluster Number

Let $L_i(\rho) = \dfrac{\log N}{2}\left(\|S_1\| + \rho\|S_2\|\right) - L(S_v \mid D_i), i = 2,...,M_v, \rho = \rho_{\min},...,\rho_{\max}$. Notice that ρ can take a different starting value and step according to domain knowledge. Find thresholds $\rho_{\min}$ which makes $\{L_i(\rho)\}$ monotonically non-increasing for $\rho = \rho_{\min} - 1$, that is $L_2(\rho_{\min} - 1) \geq L_3(\rho_{\min} - 1) \geq ... \geq L_{M_v}(\rho_{\min} - 1)$, but this monotonic property is destroyed for $\rho = \rho_{\min}$, and $\rho_{\max}$ which makes $\{L_i(\rho)\}$ monotonically non-decreasing for $\rho = \rho_{\max} + 1$, that is $L_2(\rho_{\max} + 1) \leq L_3(\rho_{\max} + 1) \leq ... \leq L_{M_v}(\rho_{\max} + 1)$, but this monotonic property is destroyed for $\rho = \rho_{\max}$. Let $I_\rho^* = \arg\min_{2 \leq i \leq M_v}\{L_i(\rho)\}$, $\rho_{\min} \leq \rho \leq \rho_{\max}$. Choose the value appearing most frequently in $\{I_\rho^*\}$ as the optimal cluster number. $\rho_{\min}$ is the boundary where awarding likelihood is dominating, while $\rho_{\max}$ is the boundary where penalizing likelihood is dominating. Those two kinds of likelihood both work between $\rho_{\min}$ and $\rho_{\max}$. The optimal cluster number is often the one that appears most frequently in $\{I_\rho^*\}$. If two cluster numbers appear with the same or almost the same frequency, it is difficult to figure out the better one. In this situation, domain knowledge or Ocams razor can be used to choose the better one.

4 Experiments

To verify the applicability of HDC-DSGS, experiments are conducted from aspects of validity of the determination of optimal cluster number and clustering accuracy on the data sets from UCI [8] data house.

Let $M_{\max} = 10$. The situation of finding optimal cluster numbers of 10 data sets is as follows:

Table 1. The situation of finding optimal cluster number

Data set	$\rho_{\min}$	$\rho_{\max}$	Finding optimal cluster number	Classes
Heart_disease	1	30	lcl=2(41.83%), lcl=3(46.27%), lcl=8(11.9%)	2
Hepatitis	Null	32	lcl=2(56.25%), lcl=4(43.75%)	2
Soybean	2	19	lcl=4(74.82%), lcl=7(25.18%)	4
Tic_tac_toe	4	28	lcl=2(33.85%), lcl=3(30.86%), lcl=4(35.29%)	2
Voting_records	Null	24	lcl=2(69.35%), lcl=4(11.23%), lcl=5(19.42%)	2
New_thyroid	Null	22	lcl=3(100%)	3
Iris	Null	8	lcl=3(62.5%), lcl=7(12.5%), lcl=9(25%)	3
Wdbc	3	16	lcl=2(41.66%), lcl=4(41.66%), lcl=8(16.68%)	2
Wine	1	9	lcl=3(85.72%), lcl=4(14.28%)	3
Glass	Null	12	lcl=3(12.35%), lcl=4(67.86%), lcl=6(19.79%)	4

From the above Table 1 we can see that HDC-DSGS can accurately find true cluster number in most cases. Even if acquired optimal cluster number is not real class, they also are very near. It is sure that HDC-DSGS can reliably find optimal cluster number.

To the data sets (hepatitis, soybean, voting_records, new_thyroid, iris, wdbc, wine and glass) of accurately finding optimal cluster number, the percentages of accurately finding respective cluster members are 86.73%, 97.3%, 88.27%, 96.26%, 93.33%, 91.39% , 96.07% and 64.35%. It is sure that HDC-DSGS can reliably find cluster members.

Table 2 compares the optimal cluster number and likelihood between AutoClass and HDC-DSGS, where $M_{max} = 10$. As AutoClass and HDC-DSGS are both based on star-shaped dependency structure, they can be evaluated by average negative log-likelihood (the lower is the better).

Table 2. Comparison of AutoClass and HDC-DSGS

Dataset	Classes	AutoClass		HDC-DSGS	
		Cluster Num.	Likelihood score	Cluster Num.	Likelihood score
Heart_disease	2	7	6.45	3	6.83
Hepatitis	2	4	8.43	2	8.22
Soybean	4	4	8.68	4	8.12
Tic_tac_toe	2	7	8.31	4	8.43
Voting_records	2	5	7.17	2	7.03
New_thyroid	3	3	2.43	3	2.28
Iris	3	7	2.24	3	2.13
Wdbc	2	7	42.76	2	25.87
Wine	3	5	13.81	3	13.1
Glass	4	5	5.71	4	2.65

Table 2 shows, in most cases, HDC-DSGS can determine the cluster number correctly and has a superior performance than AutoClass does. The clustering accuracy for HDC-DSGS is high and the clustering results are reliable. With respect to average negative log-likelihood, HDC-DSGS is better than AutoClass in most cases, except for a few datasets, where the parameter space has only one optimum and EM can find the optimum more accurately due to its local search property.

5 Conclusion

A hybrid data clustering algorithm based on dependency structure and Gibbs sampling, HDC-DSGS, is proposed in this paper. The algorithm can avoid the problems coming along with EM and provide a flexible framework for hybrid data clustering. The dependency structure can be constructed using prior knowledge, the continuous variable can take different form of density function, and the hierarchical clustering can also be incorporated into the framework easily. Experimental results show its superiority to AutoClass on clustering accuracy and determination of optimal cluster number.

Acknowledgments: This research is supported by the National Science Foundation of China under Grant No. 60675036, the Shanghai Academic Discipline Project under Grant No.P1601, the Key Research Project from Shanghai Municipal Education Commission under Grant No. 05zz66 and the Jiangsu Postdoctoral Foundation under Grant No.0601017B.

References

1. Chen S. M., Hsiao H. R.: A New Method to Estimate Null Values in Relational Database Systems Based on Automatic Clustering Techniques. Information Sciences: an International Journal 1-2 (2005) 47-69
2. Cheeseman P., Kelly J., Self M., Stutz J., Taylor W., Freeman D.: AutoClass: A Bayesian Classification System. In: Proceedings of the 15th International Conference on Machine Learning, edited by Laird J, San Mateo, CA: Morgan Kaufmann (1988) 54-64
3. Cheeseman P., Stutz J.: Bayesian Classification (AutoClass): Theory and Results. U. M. Fayyad, G. Piatetsky-Shapiro, P. Smyth, and R. Uthurusamy (Eds.), AAAI/MIT Press (1996) 153-180
4. Geman S., Geman D.: Stochastic Relaxation, Gibbs Distributions and the Bayesian Restoration of Images. IEEE Transactions on Pattern Analysis and Machine Intelligence 6 (1984) 721-742
5. Mao S. S., Wang J. L., Pu, X. L.: Advanced Mathematical Statistics. 1th ed., Beijing: China Higher Education Press, Berlin: Springer-Verlag (1998) 401-459
6. Lam W., Bacchus F.: Learning Bayesian Belief Networks: An Approach Based on the MDL Principle. Computational Intelligence 4 (1994) 269-293
7. Domingos P., Pazzani M.: On the Optimality of the Simple Bayesian Classifier Under Zero-one Loss. Machine Learning 2-3 (1997) 103-130
8. Murphy S. L., Aha D. W.: UCI Repository of Machine Learning Databases. http://www.ics.uci.edu/~mlearn/MLRepository.html

Linear Methods for Reduction from Ranking to Multilabel Classification

Mikhail Petrovskiy[*] and Valentina Glazkova

Computer Science Department of Lomonosov Moscow State University,
MSU, Vorobjovy Gory, Moscow, 119992, Russia
Phone: +7 495 9391789; Fax: +7 495 9391988
michael@cs.msu.su, glazv@mail.ru

Abstract. This paper is devoted to the multi-label classification problem. We propose two new methods for reduction from ranking to multi-label case. In existing methods complex threshold functions are being defined on the input feature space, while in our approach we propose to construct simple linear multi-label classification functions on the class relevance space using the output of ranking algorithms as an input. In our first method we estimate the linear threshold function defined on the class relevance space. In the second method we define the linear operator mapping class ranks into the set of values of binary decision functions. In comparison to existing methods our methods are less computationally expensive and in the same time they demonstrate competitive and in some cases significantly better accuracy results in experiments on well-known multi-label benchmarks datasets.

1 Introduction

Multi-label classification is an extension of traditional multi-class classification problem, where each sample may belong to several classes simultaneously. These problems arise in many important applications, such as text categorization [1,3,5], image recognition [4], bioinformatics [7,8], etc. A *multi-label classifier* must predict *all relevant* classes for a given sample. Existing multi-label classification methods can be divided into three groups: methods based on *one-against-rest binary decomposition* of multi-label problem into multiple independent binary classification subproblems [1,4] (for each class the individual classifier is built to separate samples, which belong to the class from samples, which do not belong to it); methods based on direct *posteriori class set probability estimation* [3,8] (they search the most expected combination of relevant classes); methods based on *reduction from ranking* [1,2,7]. Solving multi-label problem with ranking approach involves two stages. The first stage is learning a predictor that ranks classes according to their relevance for a given sample. The second stage is construction of the multi-label classification function to separate top ranked relevant classes from irrelevant. As a rule, *threshold methods* [1,2,7] are used. The problem is that usually a threshold depends on the sample and *fixed thresholds* [1,2] lead to poor precision. Thus, in addition to building ranking predictor one needs to estimate a threshold function. Elisseeff and Weston proposed

---

[*] Research is supported by grants: RFBR #05-01-00744, #06-01-00691 and MK-2111.2005.9.

A. Sattar and B.H. Kang (Eds.): AI 2006, LNAI 4304, pp. 1152–1156, 2006.
© Springer-Verlag Berlin Heidelberg 2006

to construct *Threshold function defined on the Input Feature Space* (TIFS) [7]. However, in many practical applications this approach does not work well, because the input space dimensionality is too high, and, as a result, time and space complexity are unacceptable. Besides, usually such threshold functions are imprecise.

To avoid these drawbacks we formulate two new methods for reduction from ranking to multi-label. They are denoted as TCRS (*Threshold function on the Class Relevance Space*) and LORM (*Linear Operator from class Ranks to Multi-label decision functions*). The key idea of these methods consists in *constructing multi-label classification functions* not on the input feature space, but on the *class relevance vector space*, i.e. the space of vectors, which coordinates are values of class ranks. In other words, we consider predicted ranks as new features of the sample. Usually the *dimensionality of class relevance space is much smaller* than dimensionality of the input space. Moreover, since predicted ranks are already the results of some nonlinear transformation of input features we can assume that dependence between class ranks and multi-label decision functions may be very simple. In this paper we consider the family of *linear functions*. All these allow greatly reduce the amount of calculations keeping the same or even better accuracy, as it was confirmed by experiments on well-know multi-label benchmark datasets.

2 Our Approach to Reduction from Ranking to Multi-label

Let us start with *ranking learning problem* and *multi-label learning problem* formal definitions. Assume that the *input feature space* is n-dimensional real space $X = \Re^n$ and the set of all possible classes is denoted by Y, where the number of different classes is $q = |Y|$. Given a *multi-label training set* $S = \{(\vec{x}_i, y_i) \mid 1 \le i \le m, \vec{x}_i \in \Re^n, y_i \subset Y\}$, where each *sample* $\vec{x} \in \Re^n$ is associated with *subset of relevant classes* $y_i \subset Y$. The goal of *ranking learning algorithm* is to analyze the training set S and produce a ranking function $r_S : X \times Y \to \Re$, such that for any given sample $\vec{x}$ it orders classes from Y according to their *relevance* to the sample. Class l is more relevant than class k iff $r(\vec{x}, l) > r(\vec{x}, k)$. The ideal ranking algorithm produces the ranking function that always predicts the *perfect order*, i.e. the order where all relevant classes are positioned higher than any irrelevant, but the order among relevant ones (and among irrelevant) is insufficient. On the other hand, the goal of *multi-label learning algorithm* is to analyze S and construct the decision function $F : X \to 2^q$ that for any $\vec{x}$ determines *all its relevant classes*. Unlike in ranking problem, the order of classes is insufficient. In fact, most multi-label methods are based on Hamming Loss minimization [5,8], while ranking methods re based on "confusion" (disorder) error minimization, such as *Ranking Loss, Average Precision* or *Coverage* [5,8].

Our first method TCRS is an improvement of the input space threshold method TIFS [7]. In our version we propose to find threshold function defined on q-dimensional class relevance vector space, instead of n-dimensional input feature space as in [7] ($q<<n$). Our threshold function is being found in a family of linear functions:

$$t(\vec{r}(x)) = \sum_{j=1}^{q} a_j r_j(x) + a_0 , \tag{1}$$

Here we assume that ranking algorithm has produced the perfect (or nearly perfect) order of class ranks $\vec{r}(\vec{x}_i) = (r_1(\vec{x}_i),...,r_q(\vec{x}_i)) \in \Re^q$ and any class l is relevant for $\vec{x}$ iff its rank $r_l(\vec{x})$ is among top $s(\vec{x})$ sorted classes. In that way, the task is to estimate the number of top relevant classes $s(\vec{x})$. For each training sample $\vec{x}_i$ we estimate the best threshold value t_i that minimizes the number of incorrectly classified relevant and irrelevant classes for $\vec{x}_i$:

$$t_i = \arg\min_t \left(\left| \{l \in y_i \mid r_l(\vec{x}_i) \leq t\} \right| + \left| \{l \in \overline{y}_i \mid r_l(\vec{x}_i) > t\} \right| \right). \tag{2}$$

Then we assume t_i to be the observation of real value of function (1) in $\vec{x}_i$ and we apply least squares method to estimate the coefficients of (1) over all training samples:

$$\min_{\vec{a}} \sum_{i=1}^{m} \left[t_i - \left(\sum_{j=1}^{q} a_j r_j(\vec{x}_i) + a_0 \right) \right]^2. \tag{3}$$

To predict the set of relevant classes for new given sample $\vec{x}_{test}$ we need to calculate ranks $r_{test} = \vec{r}(\vec{x}_{test}) = (r_1(\vec{x}_{test}),...,r_q(\vec{x}_{test}))$, substitute them into (1) and obtain threshold $t_{test} = a_0 + a_1 r_1(\vec{x}_{test}) + ... + a_q r_q(\vec{x}_{test})$. Then the size of the relevant classes set is estimated to the number of classes, for which the value of ranking function is above found threshold $s(\vec{x}_{test}) = \left| \{l \mid r_l(\vec{x}_{test}) > t_{test}, l \in Y\} \right|$.

Our second method LORM is based on the notation of *multi-label decision functions*, defined for each class separately. Let consider vector function $\vec{f}(\vec{x}) = \left(f_1(\vec{x}),...,f_q(\vec{x}) \right)$, where $f_l(\vec{x})$ are q individual binary decision functions defined for each class l. The sign of $f_l(\vec{x})$ gives a prediction whether class l is relevant for a given $\vec{x}$. We propose to find decision functions $\vec{f}(\vec{x})$ depending on ranking functions $\vec{r}(\vec{x})$ in a familiy of linear functions. It means to find a linear operator A mapping from $\vec{r}(\vec{x})$ to $\vec{f}(\vec{x})$: $\vec{f}(\vec{x}) = \tilde{A}\vec{r}(\vec{x}) + \tilde{b}$ Adding to a ranking vector a fictitious component $r_0(\vec{x}) \equiv 1$ we can reformulate the problem in homogeneous form $\vec{f}(\vec{x}) = A\vec{r}(\vec{x})$, where A is unknown linear operator, represented by $q \times (q+1)$ matrix. The advantage of such approach is that, unlike in threshold methods, we do not assume any more that the class order produced by ranking algorithm is perfect or really close to perfect. To find operator A we set $f_l(\vec{x}_i)$ to 1 if class l is relevant for the training sample $\vec{x}_i$ and -1 otherwise. Using learned ranking function $\vec{r}(\vec{x})$ we calculate $\vec{r}(\vec{x}_i) = (r_0(\vec{x}_i), r_1(\vec{x}_i),...,r_q(\vec{x}_i))$ for each $\vec{x}_i \in S$ and obtain m linear systems $\vec{f}(\vec{x}_i) = A\vec{r}(x_i)$ with the same $q \times (q+1)$ unknown matrix A and known $r_i(\vec{x}_i)$ and $f_l(\vec{x}_i)$. Now for each class l we can find l-*th* row of matrix A minimizing the mean-square error:

$$\min_{a_{lj}, j=0,q} \sum_{i=1}^{m} \left(f_l(\vec{x}_i) - \sum_{j=0}^{q} a_{lj} r_j(\vec{x}_i) \right)^2. \tag{4}$$

Solving (4) for all q classes with least squares method we obtain the matrix of linear operator A, which defines the mapping between class ranks and our multi-label decision functions $\vec{f}(\vec{x})$. To classify new sample $\vec{x}_{test}$, we need first to predict class ranks $\vec{r}(\vec{x}_{test}) = (1, r_1(\vec{x}_{test}), ..., r_q(\vec{x}_{test}))$ and then find values of q decision functions according to the formulas: $f_l(\vec{x}_{test}) = \vec{a}_l \vec{r}(\vec{x}_{test}) = a_{l0} + a_{l1} r_1(\vec{x}_{test}) + ... + a_{lq} r_q(\vec{x}_{test})$. If $f_l(\vec{x}_{test}) \geq 0$, we consider that class l is relevant for $\vec{x}_{test}$ and irrelevant otherwise. Though, the decision about relevance of each class is madden separately the correlation between classes is considered in our method, because the decision functions depend on ranks, which are correlated.

3 Experiments

We evaluated the performance of our methods TCRS ($t=ar$) and LORM ($f=Ar$) on Yeast (functional gene data) and Reuters-2000 (multi-topic text articles) multi-label benchmark datasets [9]. Two subsets of Reuters-2000 were used (case 1: 3000 training to 3000 test documents; case 2: 15000 to 15000). Our methods have been tested with following popular ranking methods: Multiclass-Multilabel Perceptron (MMP) [5]; ensemble of binary Perceptrons (bPerc) [5]; ranking k-Nearest Neighbors (kNN) [6]. For each selected ranking method we run series of experiments with different reduction settings: ">0" constant threshold (it is not applicable to kNN ranking); "TIFS" threshold function on the input space [7]; "TCRS" - our threshold method; LORM - our linear operator method. Each group of experiments corresponds to the same ranking learning method, but different reduction methods. Additionally, we have included two multi-label methods that are not based on ranking, but on one-against-rest binary decomposition (AdaBoost.MH [1]) and posteriori class set probability estimation (ML-KNN [8]). Experimental results are presented in the table.

Table 1. Experimental Results on Yeast and Reuters-2000 datasets

Method	Yeast (1500/917)		Reuters-2000 (3000/3000)		Reuters-2000 (15000/15000)	
	Rank. Loss	Hamming Loss	Rank. Loss	Hamming Loss	Rank. Loss	Hamming Loss
AdaBoost.MH [7]	0.298	0.237	0.0268	0.0160	0.0169	0.0146
ML-kNN [8]	0.168	**0.197**	0.0194	**0.0144**	0.0167	0.0135
MMP+(>0)	0.3779	0.4349	0.0241	0.2942	0.0137	0.3031
MMP+TIFS		0.2981		0.0157		**0.0126**
MMP+TCRS (t=ar)		0.2977		0.0155		**0.0126**
MMP+LORM (f=Ar)		**0.2034**		**0.0149**		0.0129
bPerc+(>0)	0.4195	0.4689	0.0210	0.0705	0.0099	0.0337
bPerc+TIFS		0.3131		0.0162		0.0128
bPerc+TCRS (t=ar)		0.3110		0.0158		**0.0127**
bPers+LORM (f=Ar)		**0.2131**		**0.0146**		0.0130
kNN+TIFS	0.1676	0.2093	0.0202	0.0146	0.0167	0.0131
kNN+TCRS (t=ar)		**0.1985**		**0.0141**		**0.0130**
kNN+LORM (f=Ar)		0.2064		0.0144		0.0137

Ranking Loss does not depend on the reduction method and estimates how good the class order was before reduction. Only Hamming Loss shows the quality of multi-label prediction. Experimental results demonstrate that our linear methods outperform their main competitors in each group. Besides, they demonstrate the accuracy comparable to one of the most precise probabilistic method ML-KNN and completely outperform popular AdaBoost.ML. Another interesting thing is that our linear operator method can significantly improve the accuracy of multi-label solution in the case when the produced ranking order is not perfect.

4   Conclusions

In this paper we propose the improvement of traditional approach to solving multi-label classification problem based on reduction from ranking. The key idea of the proposed improvement consists in defining multi-label decision functions on the class relevance vector space instead of input feature space. It allows greatly reduce the amount of calculations and construct multi-label decision functions in the family of linear functions. Following this idea we formulate two new methods. In our first method we have extended standard threshold method and proposed to define a threshold function on the class relevance vector space. Our second method is based on construction the linear operator, which maps ranking algorithm outputs directly to the values of multi-label vector decision function. Both methods lead to the mean squares error minimization and we use least squares method to find decision functions coefficients. Experiments on well-known multi-label benchmarks datasets have demonstrated high accuracy of developed methods.

References

1. Schapire R. E., Singer. Y.: Improved boosting algorithms using confidence-rated predictions. Machine Learning, 37(3), 1999, pp. 297-336.
2. Comite F. D., Gilleron R., Tommasi M.: Learning multi-label alternating decision tree from texts and data. Proceedings of MLDM 2003, LNCS 2734, Berlin, 2003, pp. 35–49.
3. McCallum A.: Multi-label text classification with a mixture model trained by EM. Working Notes of the AAAI'99 Workshop on Text Learning, Orlando, FL, 1999.
4. Boutell M. R., Luo J., Shen X., Brown C.M.: Learning multi-label scene classification. Pattern Recognition, 37, 2004, pp. 1757-1771.
5. Crammer C., Singer Y.: A family of additive online algorithms for category ranking. Journal of Machine Learning Research, 3, 2003, pp. 1025–1058.
6. Minh Duc Cao, Xiaoying Gao.: Combining Content and Citation for Scientific Document Classification. Proceedings of AI2005, LNAI 3809, 2005, pp. 143-152.
7. Elisseeff A., Weston J.: A kernel method for multi-labelled classification. Proceedings NIPS'14, Cambridge, 2002.
8. Zhang M.-L., Zhou Z.-H.: A k-nearest neighbor based algorithm for multi-label classification. Proceedings of IEEE International Conference GrC'05, 2005, pp. 718-721.
9. Chang, C.-C., Lin, C.-J. LIBSVM: a library for support vector machines, 2001. Software available at http://www.csie.ntu.edu.tw/~cjlin/libsvm.

Wavelet Kernel Matching Pursuit Machine

Qing Li, Licheng Jiao, and Shuiping Gou

Institute of Intelligent Information Processing, Xidian university, 224#, Xi'an,
710071, P.R. China
kingdomyangfan@hotmail.com

Abstract. Kernel Matching Pursuit Machine is a relatively new learning algorithm utilizing Mercer kernels to produce non-linear version of conventional supervised and unsupervised learning algorithm. But the commonly used Mercer kernels can't expand a set of complete bases in the feature space (subspace of the square and integrable space). Hence the decision-function found by the machine can't approximate arbitrary objective function in feature space as precise as possible. Wavelet technique shows promise for both nonstationary signal approximation and classification, so we combine KMPM with wavelet technique to improve the performance of the machine, and put forward a wavelet translation invariant kernel, which is a Mercer admissive kernel by theoretical analysis. The wavelet kernel matching pursuit machine is constructed in this paper by a translation-invariant wavelet kernel. It is shown that WKMPM is much more effective in the problems of regression and pattern recognition by the number of comparable experiments.

1 Introduction

The Kernel Matching Pursuit Machine (KMPM) is a new and promising classification technique proposed by Pascal Vincent, and can be seen as a form of boosting [1]. The performance of the KMPM shows comparable to that of Support Vector Machine (SVM), while typically requiring far fewer support points [2]. Many kinds of kernels can be used in KMPM, and preferably satisfy the Mercer condition for computation purposes [3], such as the RBF and polynomial kernels. However, the commonly used Mercer kernels can't expand a set of complete bases in the feature space (subspace of the square and integrable space). Hence the decision-function found by the machine can't approximate arbitrary objective function precisely in feature space, which means the decision function learned by KMPM is only the approximation to the objective function but not its reconstruction.

The application of wavelet technique to matching pursuit is originated by Mallat [4,5]. As we know, the wavelet functions are a set of complete bases in the subspace of the square and integrable space, which shows the great flexibility and validity in signal coding and information processing. Nowadays, the application field of wavelet technique includes many domains such as selective modifications of signals, detection and classification [6]. Since the success of combination of wavelet and matching pursuit, it is valuable for us to study the problem of whether a better performance could be obtained if we apply wavelet technique to KMPM. In this paper, we proposed a wavelet translation invariant kernel, Bubble wavelet kernel, which is also an admissible

A. Sattar and B.H. Kang (Eds.): AI 2006, LNAI 4304, pp. 1157–1161, 2006.
© Springer-Verlag Berlin Heidelberg 2006

Mercer kernel by theoretical analysis. The number of experiments shows the validity for improving the performance of KMPC.

2   Wavelet Mercer Kernel

In order to make use of the advantage of wavelet technique in KMPM, we propose wavelet Mercer kernel. First, some lemma must be cited.

Lemma 1 (Mercer dot-product wavelet kernel [2]). Let $\phi(x)$ be a mother wavelet, and let a and b denote the dilation and translation factors, respectively. If $\mathbf{x}, \mathbf{x}' \in R^n$, then dot-product wavelet kernel is

$$K(\mathbf{x}, \mathbf{x}') = \prod_{i=1}^{n} \phi\left(\frac{x_i - b_i}{a}\right) \phi\left(\frac{x_i' - b_i'}{a}\right). \tag{1}$$

Lemma 2 (Mercer translation invariant kernel [7]). A translation invariant kernel $K(\mathbf{x}, \mathbf{x}') = K(\mathbf{x} - \mathbf{x}')$ is an admissive Mercer kernel if and only if the Fourier transform of $K(\mathbf{x})$ satisfy

$$F[K](\omega) = (2\pi)^{-n/2} \int_{R^n} \exp(-i(\omega \cdot \mathbf{x}) K(\mathbf{x}) d\mathbf{x} \geq 0, \tag{2}$$

where $\mathbf{x} \in R^n$.

From lemma 1, the Mercer dot-product wavelet kernel of the mother wavelet $\phi(x)$ can be constructed directly, but the dilation and translation factors must be predefined. Such wavelet kernel is not convenient to practical use because the choice of the parameters in the kernel functions is quite difficult. From lemma 2, we only need to predefine dilation factor if we could design a wavelet Mercer translation invariant kernel, so the difficulty of the wavelet kernel used in KMPM will be greatly decreased. In the following, a practical wavelet translation invariant kernel is designed by the mother wavelet function chosen as Bubble wavelet, which is a widely used mother wavelet in practice. The form of the Bubble wavelet can be expressed as

$$\phi_B(t) = \frac{2}{\pi^{1/4}\sqrt{3\alpha}}\left(1 - \frac{t^2}{\alpha^2}\right)\exp\left(\frac{-t^2}{2\alpha^2}\right). \tag{3}$$

Theorem 3. Given the mother wavelet (3) and the dilation factor $\alpha \in R$. If $\mathbf{x}, \mathbf{x}' \in R^n$, the wavelet translation invariant kernel is

$$K(\mathbf{x}, \mathbf{x}') = \prod_{i=1}^{n} \phi_B(x - x') = \prod_{i=1}^{n} \frac{2}{\pi^{1/4}\sqrt{3\alpha}}\left(1 - \frac{(x_i - x_i')^2}{\alpha^2}\right)\exp\left(-\frac{(x_i - x_i')^2}{2\alpha^2}\right), \tag{4}$$

which is an admissive Mercer kernel.

3   Validation of WKMPM's Effectiveness

In the paper, we compare the wavelet function with the most commonly used kernel function, RBF kernel, and in each experiment of this paper, the parameters p for the RBF kernel and α for the Bubble wavelet kernel are selected by using cross validation that is in wide use. We adopt the parameters' notation of KMPM as follows: N—maximum of the basis functions; stops—KMPC stopping criterion (predefined accuracy); fitN—KMPC doing a back-fitting in every fitN-step. In the test of regression, we adopt the approximation error as $e_{ss} = \sqrt{\left(\sum_{i=1}^{l}(y_i - f_i)^2\right)/l}$, and both in the regression and in the classification tests, the loss function of the KMPM adopts squared loss. For avoiding the weak problem, each experiment has been performed 30 independent runs, and all experiments were carried on a Pentium Ⅳ 2.6Ghz with 512 MB RAM using Matlab6.01 compiler.

3.1   Regression Experiments

A well-known dataset, Boston housing dataset, choosing from the UCI machine learning database[1], has been tested in this experiment. The input space of Boston housing dataset is 13 dimensions. It contains 506 samples, and some of them are taken as the training examples while others as testing examples. Test Parameters: RBF kernel $K(\mathbf{x},\mathbf{y}) = \exp(-\|\mathbf{x}-\mathbf{y}\|^2/2p^2)$ with $p = 3.1e+2$, wavelet kernel $\alpha = 3.8e+2$ and parameters of KMPM $N=150,\, stops=0.01,\, fitN=5$. Table 1 lists the approximation errors using the two kernels.

Table 1. Approximation Results of Boston Housing Data

#.tr[2]	#.te	Kernel	# sp	Error
400	106	Wavelet	17	7.9801
		RBF	7	8.1536
450	56	Wavelet	18	7.9519
		RBF	7	8.1307

3.2   Experiments of Pattern Recognition

A. Two Spirals' Problem

Learning to tell two spirals apart is important both for purely academic reasons and for industrial application [8]. In the research of pattern recognition, it is a well-known problem for its difficulty. The parametric equation of the two spirals can be presented as follows:

---

[1] URL:http://www.ics.uci.edu/mlearn

[2] We adopt note '#' to present 'the number of' and 'tr', 'sp', 'te' means the training samples, support points of the KMPM and testing samples.

$$\text{spiral-1:} \quad \begin{aligned} x_1 &= (k_1\theta + e_1)\cos(\theta) \\ y_1 &= (k_1\theta + e_1)\sin(\theta) \end{aligned}$$
$$\text{spiral-2:} \quad \begin{aligned} x_2 &= (k_2\theta + e_2)\cos(\theta) \\ y_2 &= (k_2\theta + e_2)\sin(\theta) \end{aligned} \tag{5}$$

where k_1, k_2, e_1 and e_2 are parameters. In our experiment, we choose $k_1 = k_2 = 4$, $e_1 = 1$, $e_2 = 10$. We select RBF kernel function with $p = 8$ and bubble wavelet kernel with $\sigma = 18$. We uniformly generate 2000 samples, and randomly choose some of them as training data, others as test data.

KMPM Parameters: $N = 100$, $stops = 0.02$, $fitN = 5$. Table 2 lists the classification result by wavelet kernel and RBF kernel, respectively.

Table 2. Recognition Results of Two Spirals

Kernel	#.tr	#.sp	#.te	Accuracy
Wavelet	200	14	1800	95.93%
RBF	200	12	1800	95.80%
Wavelet	600	18	1400	98.04%
RBF	600	14	1400	96.48%

B. Classification on the Realistic Dataset

Waveform is one dataset of the UCI repository. It has 21 characteristic attributes with all including noise and one class attribute. It contains three classes, waveform 0, waveform 1 and waveform 3, whose number of data is 5000. When training KMPM, we choose one class as positive samples and others two classes as negative samples. The selection of the training samples is shown in table 3 and the left samples as test data. In the table 3, "Team i" means the classification on waveform i ($i = 1 \sim 3$) and "+, -" represent positive and negative samples, respectively.

Table 3. Selection of the Training

Group	# Training examples					
	Team 1		Team 2		Team 3	
	+	-	+	-	+	-
1	200	400	200	400	200	400
2	250	500	250	500	250	500

We select RBF kernel function $p = 7.68$ and bubble wavelet kernel $\sigma = 3.36$. KMPM Parameters: $N = 120$, $stops = 0.05$, $fitN = 4$. Table 4 lists the classification result by wavelet kernel and RBF kernel, respectively.

Table 4. Recognition Results of Waveform Dataset

Group	Kernel	Recognition rates		
		Team 1	Team 2	Team 3
1	Wavelet	91.34%	91.68	91.28%
	RBF	89.50%	89.87%	90.01%
2	Wavelet	92.00%	92.13%	91.89%
	RBF	90.01%	89.98%	90.12%

4 Concluding

KMPM is a class of learning algorithm to map the input space of the problems to a feature space and acquire a well control between the sparsity and precision. But the commonly used kernel in the machine can't expand a set of complete bases in feature space. Hence the decision-function found by the machine can't approximate arbitrary objective function as precise as possible. In this paper, the wavelet technique has been combined with the kernel matching pursuit machine to improve the performance of the machine. We construct the wavelet KMP machine through the Bubble wavelet translation invariant kernel. The Bubble wavelet kernel shows much higher performance than the commonly used Mercer kernel not only in regression but also in the problems of classification, which mainly because the wavelet kernel forms an complete basis of $L^2(R)$.

References

1. Pascal Vincent, Yoshua Bengio. Kernel matching pursuit. Machine Learning, 48:165--187, 2002.
2. Vapnik, V.N. Statistical Learning Theory. New York: Wiley, 1998.
3. Engel, Y., Mannor, S., Meir, R. The kernel recursive least-squares algorithm. IEEE Trans. Signal Processing, vol. 52, Issue: 8, pp. 2275-2285, August 2004.
4. Mallat S., Zhang Z. Matching pursuit with time-frequency dictionaries. IEEE Trans. Signal Proc. 41 (12), 3397-3415, 1993.
5. Mallat S. A theory for nuliresolution signal decomposition: The wavelet representation. IEEE Trans. Pattern Anal. Machine Intell., vol. 11, pp. 674-693, July 1989.
6. Mallat S. A wavelet tour of signal processing, Second Edition. Beijing: China Machine Press, 2003.
7. Zhang L., Zhou W. D., Jiao L. C.. Wavelet support vector machine. IEEE Trans. On Systems, Man, and Cybernetics. Part B: Cybernetics. Vol. 34, no. 1, February 2004.
8. Lang K.J., Witbrock M.J. Learning to tell two spirals apart. In Proc. 1989 Connectionist Models Summer School, 1989, pp.52-61.

Causal Discovery with Prior Information

R.T. O'Donnell, A.E. Nicholson, B. Han, K.B. Korb,
M.J. Alam, and L.R. Hope

Faculty of Information Technology, Monash University
{rodo, annn, bhan, korb, alam, lhope}@csse.monash.edu.au

Abstract. Bayesian networks (BNs) are rapidly becoming a leading tool in applied Artificial Intelligence (AI). BNs may be built by eliciting expert knowledge or learned via causal discovery programs. A hybrid approach is to incorporate prior information elicited from experts into the causal discovery process. We present several ways of using expert information as prior probabilities in the CaMML causal discovery program.

1 Introduction

Bayesian Networks (BNs) are popular graphical models for probabilistic reasoning. BNs can be hand-crafted, using expert elicited domain knowledge, or machine learned. Expert elicitation is time-consuming, expensive and heavily dependent upon human expertise. Automated learning methods are necessary to break through this "knowledge bottleneck", but these also suffer from limitations. Structural learning usually needs large and clean datasets. Also, the learning algorithms do not have common sense domain knowledge and hence the learned networks often contain rather simple errors. A hybrid approach is to introduce expert elicited knowledge into the learning process in the form of a constraint or prior probability. The hybrid approach both reduces the search space and biases the search, potentially improving the learning efficiency.

In this study, we present several ways in which experts may specify structural information about the domain along with a confidence in the information. We show how this can be incorporated as soft priors into the CaMML (Causal discovery via MML) program [1,2]. Elsewhere [3] we compare CaMML to a variety of BN learners and show (a) CaMML achieves comparable results without prior information and superior performance with it and (b) that CaMML is well calibrated to variations in the expert's skill and confidence.

2 BN Structural Learners

A Bayesian network is a directed acyclic graph (DAG) whose nodes represent random variables and arcs represent direct dependencies (e.g., causal relationships). Each node has a conditional probability table (CPT), quantifying the relationship between connected variables. An important concept for causal learning is the *statistical equivalence class* (SEC) [4]. Two BNs in the same equivalence

A. Sattar and B.H. Kang (Eds.): AI 2006, LNAI 4304, pp. 1162–1167, 2006.
© Springer-Verlag Berlin Heidelberg 2006

class can be parameterised to give an identical joint probability distribution. There is no way to distinguish between the two using only observational data, although they may be distinguished given experimental data.

BN structural learning algorithms can be classified into constraint-based (e.g. PC [5], implemented in Tetrad [6]) and metric-based. Metric-based methods such as K2 [7] and CaMML [1] search for a BN to minimise or maximise a metric; many different metrics have been used (see [8, Ch.8]). These learners also vary in the search method used, and in what is returned from the search; some (e.g., K2) return a DAG, others (e.g., GES [9]) learn only the SEC.

Here, we introduce expert prior information about the BN structure into CaMML. CaMML attempts to learn the best causal structure to account for the data, using a minimum message length (MML) metric with an MCMC search over the model space. MML [10] provides a Bayesian information-theoretic metric, making a tradeoff between prior probability (model complexity) and goodness of fit. The MML code for BNs describes: (1) network structure, (2) the parameters given this structure (CPTs), and (3) the data given these parameters. We have modified the encoding of the BN structure to incorporate expert priors. CaMML's structure code requires a probability for the existence of an arc (in either direction), called P_{arc}, which by default it estimates from the data.

In contrast to other metric learners that use a uniform prior over DAGs or SECs for their search, CaMML uses a uniform prior over Totally Ordered Models (TOMs). A TOM consists of (1) a DAG and (2) a full temporal ordering for that DAG. A single DAG may include several TOMs, just as an SEC several DAGs. Figures 1(a) and (b) show two DAGs that belong to the same SEC. However, Figure 1(a) represents a single TOM with a total ordering ABC while the DAG in Figure 1(b) represents two TOMs with orderings ABC and ACB. So under CaMML's default prior, the DAG in Figure 1(b) is twice as likely, reflecting the larger number of ways this DAG can be realized (see [8] for further discussion). CaMML also differs from other learners in using Metropolis sampling to estimate a distribution over the model space. For a fair comparison with other learners, here we restrict CaMML to using its single "best" BN, the DAG which best represents all DAGs in the highest posterior MML equivalence class.

Finally, existing BN learners do support the use of structural knowledge: K2 [7] *requires* a total ordering; Heckerman and Geiger [11] proposed the use of a

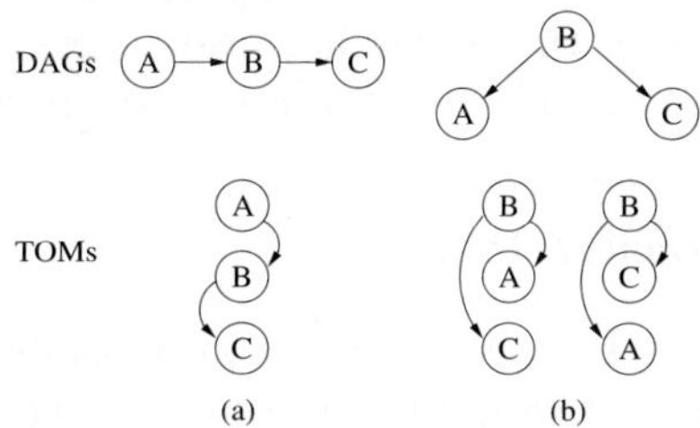

Fig. 1. These DAGs belong to the same SEC, but represent different TOMs: (a) one TOM with total ordering ABC; (b) two TOMS, with total orderings BAC and BCA

Minimum Weighted Spanning Tree (MWST) algorithm to learn a tree-like BN structure, which can then be used to initialise K2. Tetrad's implementations of Greedy Equivalence Search (GES) and PC allow the specification of 'temporal tiers', a partial ordering of variables; earlier versions of CaMML (e.g., [1]) allowed the specification of temporal orderings. However, these are all hard constraints. Heckerman and Geiger [12] proposed soft constraints by computing a prior distribution from the edit distance between an expert specified structure and the candidate structure, while Castelo and Siebes [13] use a prior over directed arcs. Neither method is supported in any BN structural learning package.

3   Structural Priors in CaMML

When eliciting structural information from experts, we want to give them a number of ways to describe relationships between variables. While specific, accurate information is ideal, we generally prefer accuracy over specificity. If the information we require is too specific, we may fail to get anything useful. Hence, we introduce several levels of structural information, each of which can be accompanied by a confidence level. The levels are presented below in (arguably) most specific to most general order.

Full structure. An expert may supply a fully specified network.

Direct causal connections between variables may be indicated. This requires a high level of knowledge of the causal process between the variables.

Direct relation. It may be known that two variables are related directly, but the direction of causality is unknown.

Causal dependency. This allows an expert to indicate that one variable is an ancestor of the other, when the mechanism between them remains unknown. For example, it is generally accepted that smoking causes lung cancer, however little is known about the detailed process.

Temporal order. In many domains it is clear that some variables come before others; we allow that to be indicated independent of other information.

Correlation. The most general sort of information we use is correlation. This implies that there is some connection between the nodes. It may be a causal dependency in either direction, or via a common ancestor.

During the elicitation process, the expert may respond with either a full structure, or any combination of the remaining prior types above. The system described below synthesizes the information into a coherent whole.

3.1   Pairwise Relationship Priors

CaMML allows an expert to specify priors on a combination of five types of pairwise relationships. Direct causal connections, direct relations and temporal order are considered "local" as only the variables involved are required to determine the status of the relationship. Indirect causal relation and correlation are "global" as the full network may be required to determine this.

Local priors are converted into four distinct relationships. $A \to B$, $B \to A$, $A \nrightarrow B$ and $B \nrightarrow A$ where $A \nrightarrow B$ represents $(A \prec B$ and not $A \to B)$. The distinction between $A \nrightarrow B$ and $B \nrightarrow A$ is required as the TOM prior treats them as distinct states. Expert specified "local" priors are mapped into this space by:

 - Fix $A \to B$ and $B \to A$ if they are specified by an expert.
 - Fix all other implied values. For example, if $P(A \to B) = 0.2$ and $P(A \prec B) = 0.3$ specified, then $P(A \nrightarrow B) = 0.1$ is implied.
 - Set remaining values proportional to $P(A \prec B) \times P(A - B)$ where $P(A \prec B)$ and $P(A - B)$ are expert priors if specified, or default priors if not specified.
 - If the generated prior does not match the expert priors, reject expert priors as inconsistent.

The message length of a TOM, t, based on local priors, is:

$$C_{local}(t) = \sum_{i,j} \begin{cases} -\log P_{i \to j} & \text{if } i \to j \\ -\log P_{j \to i} & \text{if } j \to i \\ -\log P_{i \nrightarrow j} & \text{if } i \nrightarrow j \\ -\log P_{j \nrightarrow i} & \text{if } j \nrightarrow i \end{cases}$$

where the sum is over all unique pairs (i, j).

Global Priors are mapped from expert specified priors on indirect causal and correlation relationships. Experts can specify $A \Rightarrow B$, A is an ancestor of B, and $A \sim B$, A and B are correlated. Internally these are transformed into $A \Rightarrow B$, $B \Rightarrow A$, $A \Leftrightarrow B$ and $A \nsim B$, where $A \Leftrightarrow B$ represents A and B have a common cause (direct or indirect), but one is not an ancestor of the other. $A \nsim B$ represents A and B are uncorrelated, there is no causal chain between them and they do not share a common ancestor.

We translate these global priors into CaMML's internal format in a similar way to our local priors. The default prior is calculated by sampling TOMs using our default value for P_{arc} and k, the number of nodes in the network.

$$C_{global}(t) = \sum_{i,j}^{g} \begin{cases} -\log P_{i \Rightarrow j} & \text{if } i \Rightarrow j \\ -\log P_{j \Rightarrow i} & \text{if } j \Rightarrow i \\ -\log P_{i \Leftrightarrow j} & \text{if } i \Leftrightarrow j \\ -\log P_{i \nsim j} & \text{if } i \nsim j \end{cases}$$

where g is the set of (i, j) pairs which have an expert specified prior on $i \Rightarrow j$, $j \Rightarrow i$ or $i \sim j$.

Encoding the TOM Structure. Summing these partial costs gives the total cost of the TOM's structure: $C(t) = C_{local}(t) + C_{global}(t) + \gamma$. The constant γ is used to normalise the probability distribution; i.e., γ enforces the efficiency requirement that $\sum_t^{TOMs} e^{-C(t)} = 1$. In practice, γ need never be calculated as all operations in the CaMML search use relative MML costs, so we omit it from further discussion.

Example Consider the example of Figures 2. We have an expert who is 70% sure that A causes B, 20% sure that B causes A and 60% sure of a link between A and C, $P_{arc} = 0.5$. CaMML converts these priors to the table shown in 2(b). During the sampling process, suppose the TOM of Figure 2(c) is sampled. Using Figure 2(b), the cost of Figure 2(c) would be $C(t) = -\ln(.05) - \ln(.3) - \ln(.25)$.

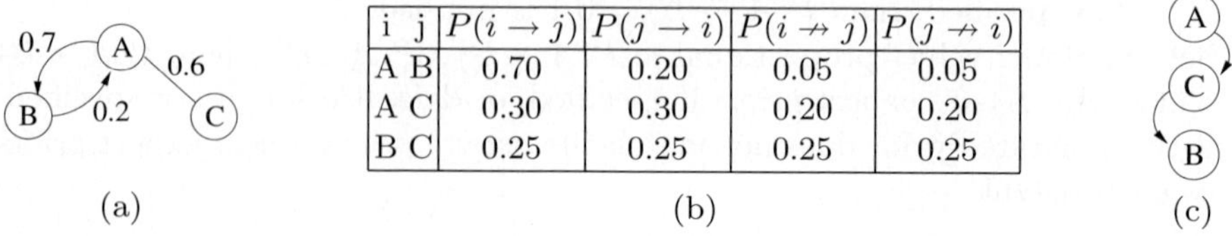

(a)   (b)   (c)

Fig. 2. (a) Expert specified network with local priors. (b) CaMML's interpretation of priors from (a). (c) Candidate TOM with ordering ACB.

3.2   Using an Expert Specified BN

Prior information can also be coded as a specific network structure, say DAG d, if that network (or subnet) is known to be near the truth. Priors over other networks can be computed via their edit distance from the given network. We also allow the expert to specify a confidence, expressed as a probability P_d. We apply P_d as probability of each arc in d. We have defined two edit distance functions: $ED_d(d_1, d_2)$, the edit distance between two DAGs d_1 and d_2; $ED_t(d, t)$, the edit distance from a DAG d to the TOM t itself. See [3] for details.

Message Length. Given these edit distances, we can generate priors over TOM space. The partial costs for a candidate TOM t using these edit distances are:

$$C_d(t) = -ED_d(d, d_t)(\log(P_d) - \log(1 - P_d))$$
$$C_t(t) = -ED_t(d, t)(\log(P_d) - \log(1 - P_d))$$

4   Conclusions

The aim of this research was to improve the performance of CaMML by incorporating prior structural knowledge – that is, knowledge about the relationships between the domain variables under consideration – elicited from experts. Our method can handle numerous types of structural information, providing much greater flexibility in the elicitation process. It also incorporates the expert's confidence, allowing both hard constraints (as supported by other BN learners) and soft constraints, so that experts are not forced into over- or underconfidence. In [3] we present experimental results showing that CaMML achieves results without prior information comparable to competitive algorithms, and superior performance with prior information. Furthermore, we show CaMML is well calibrated to variations in expert skill and confidence.

Acknowledgements. This research was supported by ARC Discovery Grant DP0450096.

References

1. Wallace, C., Korb, K.: Learning linear causal models by MML sampling. In Gammerman, ed.: Causal Models and Intelligent Data Management. Springer (1999)
2. O'Donnell, R.T., Allison, L., Korb, K.B.: Learning hybrid Bayesian networks by MML. Proc. 19th Australian Joint Conf. on AI, LNAI (2006)
3. O'Donnell, R.T., Nicholson, A.E., Han, B., Korb, K.B., Alam, M.J., Hope, L.R.: Incorporating expert elicited structural information in the CaMML causal discovery program. Technical Report TR 2006/194, Clayton School of IT, Monash University (2006)
4. Chickering, D.M.: A transformational characterization of equivalent Bayesian network structures. In: 11th UAI, San Francisco (1995) 87–98
5. Spirtes, P., Glymour, C., Scheines, R.: Causation, Prediction and Search. Second edn. MIT Press (2000)
6. Hayduk, L.A. In: Equivalent Models: TETRAD and Model Modification. Johns Hopkins University Press, Baltimore (1996) 121–154
7. Cooper, Gregory F.and Herskovits, E.: A Bayesian method for the induction of probabilistic networks from data. Machine Leaning **9** (1992) 309–347
8. Korb, K.B., Nicholson, A.E.: Bayesian Artificial Intelligence. Chapman & Hall/CRC, Boca Raton (2004)
9. Chickering, D.M.: Optimal structure identification with greedy search. Journal of Machine Learning Research **3** (2003) 507–554
10. Wallace, C.S.: Statistical and Inductive Inference by Minimum Message Length. Springer, Berlin, Germany (2005)
11. Heckerman, D., Geiger, D., Chickering, D.M.: Learning Bayesian networks: The combination of knowledge and statistical data. Machine learning **20** (1995) 197–243
12. Heckerman, D., Geiger, D.: Learning Bayesian networks: A unification for discrete and Gaussian domains. In: 11th UAI, San Fransisco, Morgan Kaufmann (1995) 274–84
13. Castelo, R., Siebes, A.: Priors on network structures. Int Jrn of Approximate Reasoning **24**(1) (2000) 39–57

Simulation of Human Motion for Learning and Recognition

Gang Zheng[1], Wanqing Li[1], Philip Ogunbona[1],
Liju Dong[1,2], and Igor Kharitonenko[1]

[1] School of Information Technology and Computer Science,
University of Wollongong, Australia
[2] College of Information Engineering, Shenyang University, P.R. of China
{gz207, wanqing, philipo, liju, igor}@uow.edu.au

Abstract. Acquisition of good quality training samples is becoming a major issue in machine learning based human motion analysis. This paper presents a method to simulate human body motion with the intention to establish a human motion corpus for learning and recognition. The simulation is achieved by a unique temporal-spatial-temporal decomposition of human body motion into actions, joint actions and actionlets based on the human kinematic model. The actionlet models the primitive moving phase of a joint and can be simulated based on the kinesiological study. A joint action is formed by proper concatenation of actionlets and an action is a group of synchronized joint actions. Methods for concatenation and synchronization are proposed in this paper for realistic simulation of human motion. Results on simulating "running" verifies the feasibility of the proposed method.

1   Introduction

Action recognition has been one of the most active research areas in Human Motion Analysis for the past few decades. Action recognition may be simply considered as a classification problem of time varying feature data. Recent research [1,2,3] has demonstrated that machine learning approach has the potential to lead to a generic solution. However, the success of machine learning approach hinges on the quality and amount of training samples which is difficult to be established because of the complication of human activities. This paper proposes the concept of building the motion corpus through simulation and generating good quality human motion data by decomposing the human activity into primitive motion elements that can be easily simulated.

Simulation of human motion is usually based on either human kinematic models or motion capture devices. In kinematic model, physical constraints and dynamic constraints are employed to solve the inverse kinematics for positioning the body [4,5,6]. Although the kinematics model is more accurate to simulate human motion, it is usually applied to actions generated by one and two joints due to its high computational load and has hardly been extended to slightly more complicated actions. In addition, the simulated motion sometimes looks unnatural. Motion capture devices usually acquire and represent the human action

A. Sattar and B.H. Kang (Eds.): AI 2006, LNAI 4304, pp. 1168–1172, 2006.
© Springer-Verlag Berlin Heidelberg 2006

data as discrete spatial and temporal points recording the position of the body and the joint angles with respect to a hierarchical skeleton model. Simulation of human motion using the acquired data then becomes interpolation and extrapolation [7,8,9]. Motion capture method has the advantage of make the simulation more smoothly and realistically than other method. Both of them are hardly applicable to the generation of motion data for learning and recognition since the former is limited to simple actions and the latter is infeasible to acquire large amount of data.

In this paper, we propose an approach to simulate human motion by decomposing it into primitive actions, referred to as *actionlet*. The actionlets can be modelled analytically based on the kinesiological study of human muscles or using captured motion data and appropriately derived variations. One of the obvious advantages of the proposed method is that virtually unlimited simple and complicated action data can be generated with a limited number of actionlets.

2   Kinematic Model and Motion Decomposition

Human body is often viewed as a system of rigid links or segments connected by joints. This paper concerns only the posture and its temporal changes. Only main joints are calculated in the 15-joint, 22-DOF kinematic model showed in Fig. 1(a). The DOFs of joints are indicated in the figure by the numbers after the joint names. Relative Euler angles [10] coordinate system is defined to represent the rotation of each segment.

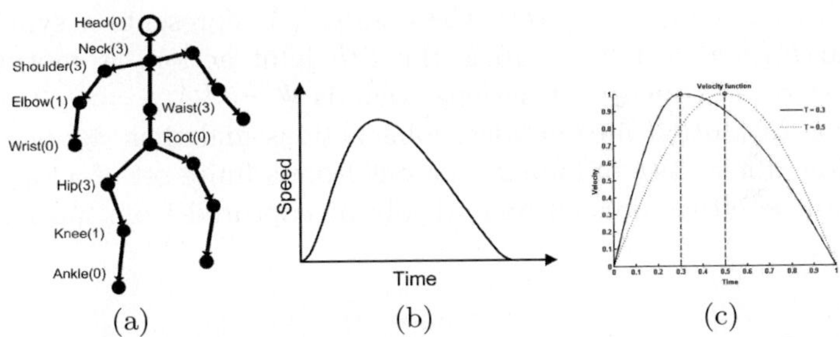

Fig. 1. The adopted kinematic model and velocity curves. (a) Kinematic model and main joints, (b) A typical velocity curve based on kinesiological study [11], (c) Two simulated velocity curves using Eq. 2.

Human motion has both spatial and temporal nature. Our approach uses both of them. We consider *human motion* as a concatenation of actions (temporal decomposition). Each action is formed by a set of coordinated *joint actions* (spatial decomposition). A joint action is further decomposed into a sequence of *actionlets* based on the kinesiological study [11,12]. (temporal decomposition), where each actionlet represents a cycle of acceleration and decceleration of a joint. Fig. 1(b) shows a typical velocity curve [11] of an actionlet.

Let $v(t)$ be a velocity function that satisfies $v(t) \in [T_b, T_e]$, $t \in [T_b, T_e]$, and $v(T_b) = v(T_e) = 0$. An actionlet in the period $[T_b, T_e]$ can be represented as Eq. 1. Eq. 2 is a demo of $v(t)$.

$$f(t) = \int_{T_b}^{T_e} v(t)dt; \quad T_b \leq t \leq T_e \tag{1}$$

$$v(t) = \begin{cases} h * \sin(\frac{\pi}{T} * t); & T_b \leq t \leq T \\ h * \sin(\frac{\pi}{1-T} * (t - (T - \frac{1-T}{2}))); & T \leq t \leq T_e, \end{cases} \tag{2}$$

Assuming a joint action $F(t)$ consists of N actionlets, then

$$F(t) = \{f_i(t) \,|\, \{1 \leq i \leq N, \, , \, T_{begin} \leq t \leq T_{end}\}, \tag{3}$$

where the $i'th$ actionlet happens in the time $[T_{bi} \leq t \leq T_{ei}]$, $1 \leq i \leq N$. The actionlets are subject to,

$$T_{ei} = T_{b(i+1)} \tag{4}$$

$$f_i(T_{ei}) = f_{i+1}(T_{b(i+1)}) \tag{5}$$

$$f_i'(t)\,|_{t=T_{bi}} = f_i'(t)\,|_{t=T_{ei}} = 0 \tag{6}$$

$$\sum_{i=1}^{N}(T_{ei} - T_{bi}) = T_{end} - T_{begin} \tag{7}$$

An action, Λ, is composed of a set of coordinated or synchronized actions, $\Lambda = synch(F_1(t), F_2(t), \cdots, F_{10}(t))$ where $synch(\cdot)$ represents a synchronizing function and $F_i(t), i = 1, 2, \cdots, 10$ is the $i'th$ joint action. A motion Ψ is a concatenation of a sequence of actions, that is $\Psi = \{\Lambda_i, i = 1, 2, \cdots, \}$ The decomposition of motion into actions, joint actions and then actionlets allows the simulation of a variety of human motions from a finite set of actionlets. The actionlets can be either defined analytically or captured from motion capture devices.

3   Simulation

With the proposed decomposition, simulation amounts to the composition of a specified sequence of actions using the actionlets. Any actionlet has to satisfy all possible physical constraints to which the corresponding joint may be subjected. There are two main composition operations required to build a specific type of motion (e.g. running) from actionlets: concatenation and synchronization. The realism of the simulated motion depends on how well these operations perform.

Concatenation is employed to build a joint action from actionlets or to form a motion from a sequence of actions. We apply the continuity of the Euler angles and their first order derivatives (speed) to all of the Euler angles that describe

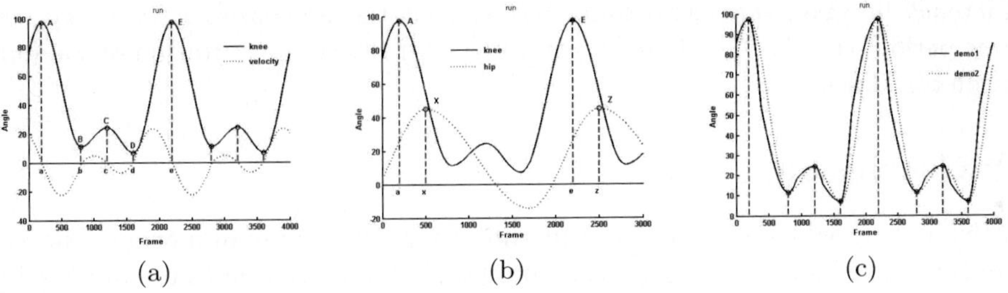

(a)   (b)   (c)

Fig. 2. Simulation of action "running". (a) Two types of actionlets for joint "knee" (AB and BC). (b) Synchronization of "knee"' and "hip" (c) Mutation result of "knee", actionlets were modelled by Eq. 2 with $T = 0.3$ (solid line) and $T = 0.5$ (dotted line).

the body posture (as expressed in Eqs. 4- 7). When two actionlets or actions do not meet the concatenation criteria, the two actionlets or actions will be either blended or a transition actionlet or action is inserted between them.

Synchronization is required when an action involves a number of joints. The joints have to act in a coordinated manner in order for the action to be realistic and meaningful. A simple action, such as "waving a hand", may involve one or two joints and complicated actions like "running" usually involve many joints. A typical example is "walking" where the arms and legs and the two joints (keen and hip) on the legs have to be synchronised properly. To simplify the synchronization problem, we introduce *joint action group* or *group of joint actions*, which is composed of a number of joint actions. The joints in a *joint action group* are usually connected together. For instance, joint actions of the left shoulder and elbow may be grouped together to become a group joint action.

Let $\Im_g(t)$ be a joint action group, then $\Im_g(t) \equiv \{F_k(t) \mid k \in \wp\}$, where $\wp$ is a set of valid joints. Assume $\Im_p(t)$ and $\Im_q(t)$ are two groups of joint actions, we consider both $\Im_p(t)$ and $\Im_q$ as periodic functions, the synchronization between them can be represented as $\Im_q(t) = \Im_p(sh * t + \phi)$, where sh denotes the temporal scale coefficient and ϕ is a phase variable.

(**Mutation**) Due to the demand for large number of samples of a particular action for learning, we introduce a mutation process in the simulation to generate variations of a particular action, instead of specifying an action for each sample. There are two ways to mutate the samples. One is to choose same type of actionlets but with different velocity functions from the actionlet database. Another is to perturb the parameters of the velocity functions as shown in Fig 1(c).

4   Implementation and Results

Fig. 2(a) shows one *joint action* (knee) while running at the temporal resolution of 3000 frames per second. The action was composed from two types of actionlets AB (or CD) and BC (or DE). Fig. 2(b) shows the synchronisation between joint "knee" and "hip" for running. *Actionlets AB, BC, CD, DE construct the joint action AE for "knee" and joint action XZ for hip. There is a phase $|A - X|$*

difference between these two *joint actions*, but the two *joint actions* have the same period, that is $|E - A| = |Z - X|$. Fig. 2(c) shows the mutation of the joint action for "knee".

5 Conclusion

In this paper, we have presented a method to simulate human motion. The system is designed for generating large samples of human motion that can be fed to machine learning algorithms for automatic identification and analysis of human motion. The system is based on novel scheme to decompose human motion into actions, joint actions and actionlets. This decomposition makes feasible a large scale simulation of human body motion in which not only a virtually unlimited types of actions can be simulated from a finite set of actionlets, but also an unlimited variations of one action can be generated.

References

1. W. Hu, T. Tan, L. Wang, and S. Maybank, "A survey on visual surveillance of object motion and behaviors," *IEEE Transactions on Systems, Man and Cybernetics, Part C*, vol. 34, no. 3, pp. 334–352, 2004.
2. X. Zhang, Y. Liu, and T. S. Huang, "Motion analysis of articulated objects from monocular images," *IEEE Transactions on Pattern Analysis and Machine Intelligence*, vol. 28, no. 4, pp. 625–636, April 2006.
3. R. D. Green and L. Guan, "Quantifying and recognizing human movement patterns from monocular video images Part I: A new framework for modelling human motion," *IEEE Trans Circuits and Systems for Video Technology*, vol. 14, no. 2, pp. 179–190, 2004.
4. R. Boulic, N. M. Thalmann, and D. Thalmann, "A global human walking model with real-time kinematic personification," *The Visual Computer*, vol. 6, no. 6, pp. 344–358, 1990.
5. Z. Xiao, X. Yang, and J. J. Zhang, "Motion data correction and extrapolation using physical constraints," in *IEEE Proceedings of the 9th International Conference on Information Visualisation (IV05)*, vol. 00, 2005, pp. 571 – 576.
6. A. C. Fang and N. S. Pollard, "Efficient synthesis of physically valid human motion," *Proceedings of ACM Transactions on Graphics*, vol. 22, issue 3, pp. 417–426, July 2003.
7. J. Faraway, "Human animation using nonparametric regression," Department of Statistics University of Michigan, Tech. Rep. 382, 2002.
8. A. Ijspeert, J. Nakanishi, and S. Schaal, "Movement imitation with nonlinear dynamical systems in humanoid robots," in *Proceedings of the IEEE International Conference on Robotics and Automation*, vol. 2, Washington, DC, May 2002, pp. 1398–1403.
9. J. Pettre, T. Simeon, and J. Laumond, "Planning human walk in virtual environments," vol. 3, Lausanne, Switzerland, Sept 2002, pp. 3048–3053.
10. V. M. Zatsiorsky, *Kinematics of Human Motion*, 1st ed. Human Kinetics, 1998.
11. G. L. Soderberg, *Kinesiology application to pathological motion*, 1st ed. Williams & Wilkins, 1986.
12. A. Tözeren, *Human Body Dynamics*. Springer, 1999.

Lazy Learning for Improving Ranking of Decision Trees

Han Liang[1,*] and Yuhong Yan[2]

[1] Faculty of Computer Science, University of New Brunswick
Fredericton, NB, Canada E3B 5A3
han.liang@unb.ca
[2] National Research Council of Canada
Fredericton, NB, Canada E3B 5X9
yuhong.yan@nrc.gc.ca

Abstract. Decision tree-based probability estimation has received great attention because accurate probability estimation can possibly improve classification accuracy and probability-based ranking. In this paper, we aim to improve probability-based ranking under decision tree paradigms using AUC as the evaluation metric. We deploy a lazy probability estimator at each leaf to avoid uniform probability assignment. More importantly, the lazy probability estimator gives higher weights to the leaf samples closer to an unlabeled sample so that the probability estimation of this unlabeled sample is based on its similarities to those leaf samples. The motivation behind it is that ranking is a relative evaluation measurement among a set of samples, therefore, it is reasonable to yield the probability for an unlabeled sample with reference to its extent of similarities to its neighbors. The proposed new decision tree model, *LazyTree*, outperforms C4.5, its recent improvement C4.4 and their *state-of-the-art* variants in AUC on a large suite of benchmark sample sets.

1 Introduction

A learning model is induced from a set of labeled samples represented by the vector of an attribute set $\mathbf{A} = \{A_1, A_2, \ldots, A_n\}$ and a class label C. Classic decision trees are typical decision boundary-based models. When computing class probabilities, decision trees use the observed frequencies at leaves for estimation. For instance, if a leaf contains 100 samples, 60 of which belong to the positive class and the others are in the negative class, then for any unlabeled sample that falls into the leaf, decision trees will assign the same positive probability of $\hat{p}(+|\mathbf{A_p} = \mathbf{a_p})$=60%, where $\mathbf{A_p}$ is the set of attributes from the leaf to the root. However, this incurs two problems: high bias (traditional tree inductive algorithm tries to make leaves homogeneous, therefore, the class probabilities are systematically shifted toward zero or one) and high variance (if the number of samples at a leaf is small, the class probabilities are unreliable) [9].

---

* This work was done when the author was a visiting worker at Institute for Information Technology, National Research Council of Canada.

A. Sattar and B.H. Kang (Eds.): AI 2006, LNAI 4304, pp. 1173–1178, 2006.

© Springer-Verlag Berlin Heidelberg 2006

Accurate probability estimation is important for problems like classification and probability-based ranking. Provost and Domingos' *Probability Estimation Trees* (PETs) [7] turn off pruning and collapsing in C4.5 to keep some branches that may not be useful for classification but are crucial for accurate probability estimation. The final version is called C4.4. PETs also use *Laplace* smoothing to deal with the pure nodes that contain samples from the same class. Instead of assigning a probability of 1 or 0, smoothing methods try to give a more modest estimation. Other smoothing approaches, such as m-Branch [2] and *Ling&Yan's* algorithm [6], are also developed. Using a probability density estimator at each leaf is another improvement to tackle the "uniform probability distribution" problem of decision trees. Kohavi [4] proposed an Naïve Bayes Tree (NBTree), in which a naïve Bayes is deployed at each leaf to produce probabilities. The intuition behind it is to take advantage of leaf attributes $\mathbf{A_l}(l)$ for probability estimation. Therefore, $p(c|\mathbf{e_t}) \approx p(c|\mathbf{A_p}(l), \mathbf{A_l}(l))$.

The main objective of this paper is to improve the ranking performance of decision trees. The improvement comes from two aspects. Firstly, decision trees work better when the sample set is large. After several splits of attributes, the samples at the subspaces can be too few on which to base the probability. There-fore, although employing a traditional tree inductive process, we stop the splits once the samples are reduced to some extent and deploy probability estimators at leaves. The probability estimators assign distinct probabilities to different samples. Thus, the probability generated by such a tree is more accurate than assigning a uniform probability for the samples falling into the same leaf. Sec-ondly, and more importantly, we observe that probability-based ranking is indeed a relative evaluation measurement where the correctness of ranking depends on the relative position of a sample among a set of other samples. In our paper, a lazy probability estimator that calculates the probability of an unlabeled sample based on its neighbors is designed for better ranking quality. The lazy probability estimator finds m closest neighbors at a leaf for the unlabeled sample and calcu-late a weight for each neighbor using a newly proposed similarity score function. We generate the probability estimates for this unlabeled sample by normalizing all weights of the neighbors at the leaf in terms of their class values. The new model is called *LazyTree*. AUC [3] is used to evaluate our method. On a large suite of 36 standard sample sets, empirical results indicate that *LazyTree* per-forms substantially better than C4.5, C4.4 and their variants with other methods designed for optimal ranking, such as m-Branch, *Ling&Yan's* algorithm and a voting strategy–*bagging*, in yielding accurate ranking.

2   Using Lazy Learner to Improve Tree-Based Ranking

2.1   The Lazy Learner

Our work aims to calibrate probability-based ranking of decision trees. Improve-ment comes from two aspects. Firstly, we want to trade-off between *bias* and *variance* of decision trees by deploying a probability estimator at each leaf. *Probability Estimator* is defined as:

Definition 1. *Given a set of unlabeled samples E and a set of class labels $C = \{c_i\}$, a **Probability Estimator** is a set of functions $p_i : E \mapsto [0,1]$, such that $\forall \mathbf{e} \in E, \sum p_i(\mathbf{e}) = 1$.*

The probability estimators give distinct probabilities to different samples. As a result, the probability estimation generated by such trees is more precise than the uniform probability assignment for the samples falling into the same leaf.

Secondly, and more essentially, we observe that probability-based ranking is indeed a ***relative evaluation measurement*** where the correctness of ranking depends on the relative position of a sample among a set of other samples. For instance, for a binary-class problem, if assigned class probabilities of a positive sample $\mathbf{e}_+$ and a negative sample $\mathbf{e}_-$ satisfy $p(+|\mathbf{e}_+) > p(+|\mathbf{e}_-)$, it is a correct ranking. For multi-classes, the right ranking means $p_i(\mathbf{e} \in c_i) > p_i(\mathbf{e}' \notin c_i)$.

These inspire us to use a ***lazy probability estimator*** which calculates the probability of a sample based on its neighbors. The lazy probability estimator finds m closest neighbors at a leaf (here m means all the samples at this leaf) for an unlabeled sample and calculate a weight for each neighbor using a newly proposed similarity metric.

Assume that sample $\mathbf{e}$ can be represented by an attribute vector as $< a_1(\mathbf{e}), a_2(\mathbf{e}), ..., a_n(\mathbf{e}) >$, where $a_i(\mathbf{e})$ denotes the value of ith attribute. The distance between two samples $\mathbf{e}_1$ and $\mathbf{e}_2$ is calculated in (1):

$$d(\mathbf{e_1}, \mathbf{e_2}) = \sqrt{\sum_{i=1}^{n} \delta(a_i(\mathbf{e_1}), a_i(\mathbf{e_2}))}, \tag{1}$$

$\delta(a_i(\mathbf{e_1}), a_i(\mathbf{e_2}))$ outputs zero if $a_i(\mathbf{e_1})$ is equivalent to $a_i(\mathbf{e_2})$, otherwise it outputs one.

For an unlabeled sample $\mathbf{e_t}$ and a set of labeled samples $\{\mathbf{e_i}|i = 1, ..., n\}$ falling in a leaf, we assign a weight to each of the labeled sample $\mathbf{e_i}$ based on its distance to $\mathbf{e_t}$ (as in 2):

$$w_i = 1 - \frac{d_i}{\sum_{i=1}^{n} d_i}. \tag{2}$$

In (2), $d_i = d(\mathbf{e_t}, \mathbf{e_i})$ is the distance of an unlabeled sample $\mathbf{e_i}$ to $\mathbf{e_t}$. Notice that for any two labeled samples $\mathbf{e_i}$ and $\mathbf{e_j}$, the shorter the distance to $\mathbf{e_t}$ (assuming that $d_i \leq d_j$), the larger weight the sample ($w_i \geq w_j$). That implies a labeled sample nearest to $\mathbf{e_t}$ contribute most when calculating the probability for $\mathbf{e_t}$.

We generate the probability estimates of the unlabeled sample by normalizing all weights of the labeled samples at a leaf in terms of their class values, as in (3):

$$p(c_j|\mathbf{e_t}) = \frac{\sum_{k=1}^{m} w_k^j + \frac{1}{|C|}}{\sum_{k=1}^{n} w_k + 1}, \tag{3}$$

where n represents the number of samples at a leaf and m is the number of samples that belong to c_j.

2.2 *LazyTree* Induction Algorithm

We deploy a lazy probability estimator at each leaf of a decision tree and call this model *LazyTree*. To induce the model, we adopt a heuristic search process, in which we exhaustively build all possible trees in each step and keep only the best one for the next level expansion. Suppose that finite k attributes are available. When expanding the tree at level q, there are k-q+1 attributes to be chosen. On each iteration, each candidate attribute is chosen as the root of the (sub) tree, the generated tree is evaluated, and we select the attribute that achieves the highest *gain ratio* as the next level node to grow the tree. We consider two criteria for halting the search process. We could stop splitting when none of the alternative attributes can statistically significantly upgrade the classification accuracy. Or, to avoid the "fragmentation" problem, there are at least 30 samples at the current node. Besides, we still permit splitting if the relative increment in accuracy is not a negative value, which is greedier than C4.5. The tree model is represented as T. An unlabeled sample $\mathbf{e_t}$ is assigned a set of class probabilities as in Algorithm 1.

Algorithm 1. $LazyTrees(T, \mathbf{e_t})$ return $\{p(c_j|\mathbf{e_t})\}$

T: a model with a set of leaves L
S_l: a set of labeled samples at a leaf l
$\mathbf{e_t}$: an unlabeled sample
$\{p(c_j|\mathbf{e_t})|c_j \in C\}$: a set of probability estimates of $\mathbf{e_t}$

> Dispatch e_t into one leaf l according to its attributes
> **for** each labeled sample $\mathbf{e_{train}} \in S_l$ **do**
> > Calculate the distance d_{train} between $\mathbf{e_{train}}$ and $\mathbf{e_t}$, by utilizing Equation 1
> > Calculate the sample weight w_{train}, in terms of its similarity to $\mathbf{e_t}$, by utilizing Equation 2
>
> **for** each class value $c_j \in C$ **do**
> > Use Equation 3 to compute $p(c_j|\mathbf{e_t})$
>
> Return a set of probability estimates $\{p(c_j|\mathbf{e_t})\}$ for the unlabeled sample $\mathbf{e_t}$

---

3   Empirical Study

More details of empirical study can be found in [5]. We used 36 standard sample sets from the UCI repository [1] and conducted three groups of experiments in terms of ranking within the *Weka* [8] Platform. Table 1 lists the properties of the sample sets. Numeric attributes were handled by decision trees themselves. Missing values were processed using the mechanism in *Weka*, which replaced all missing values with the modes and means from the training set. Besides, due to the relatively high time complexity of *LazyTree*, we made a re-sampling within *Weka* in sample set *Letter* and generated a new sample set named *Letter-2000*. The selection rate is 10%.

Table 1. Brief description of sample sets used in our experiments

Data Set	Size	Classes	Missing	Numeric	Sample Set	Size	Classes	Missing	Numeric
anneal	898	6	Y	Y	ionosphere	351	2	N	Y
anneal.ORIG	898	6	Y	Y	iris	150	3	N	Y
audiology	226	24	Y	N	kr-vs-kp	3196	2	N	N
autos	205	7	Y	Y	labor	57	2	Y	Y
balance	625	3	N	Y	letter-2000	2000	26	N	Y
breast	286	2	Y	N	lymph	148	4	N	Y
breast-w	699	2	Y	N	mushroom	8124	2	Y	N
colic	368	2	Y	Y	p.-tumor	339	21	Y	N
colic.ORIG	368	2	Y	Y	segment	2310	7	N	Y
credit-a	690	2	Y	Y	sick	3772	2	Y	Y
credit-g	1000	2	N	Y	sonar	208	2	N	Y
diabetes	768	2	N	Y	soybean	683	19	Y	N
glass	214	7	N	Y	splice	3190	3	N	N
heart-c	303	5	Y	Y	vehicle	846	4	N	Y
heart-h	294	5	Y	Y	vote	435	2	Y	N
heart-s	270	2	N	Y	vowel	990	11	N	Y
hepatitis	155	2	Y	Y	waveform-5000	5000	3	N	Y
hypoth.	3772	4	Y	Y	zoo	101	7	N	Y

In the first group of our experiments, *LazyTree* was compared to C4.5 and its PET variants including C4.5-L (C4.5 with *Laplace* estimation), C4.5-M (C4.5 with *m*-Branch) and C4.5-LY (C4.5 with *Ling&Yan*'s algorithm). In the second group, we compared *LazyTree* with C4.4 and its PET variants, which contain C4.4-nLa (C4.4 without *Laplace* estimation), C4.4-M (C4.4 with *m*-Branch) and C4.4-LY (C4.4 with *Ling&Yan*'s algorithm). In the last group, we made a comparison between LazyTree-B (*LazyTree* with *bagging*) and C4.5-B (C4.5 with *bagging*) and C4.4-B (C4.4 with *bagging*). Multi-class AUC was calculated by M-measure[3]. The AUC value on each sample set was measured via a ten-fold cross validation ten times, and we performed two-tailed t-tests with a significantly different probability of 0.95 to compare our model with others. Now, our observations are highlighted as follows.

1. *LazyTree* achieves remarkably good performance in AUC among C4.5 and its variants. *LazyTree* performs significantly better than C4.5 (31 wins and 0 loss). As the results show, C4.5 variants can improve the AUC values of C4.5. However, *LazyTree* considerably outperforms these models in AUC. In addition, decision trees with *m*-Branch or with *Ling&Yan*'s algorithm can not generate multiple probabilities for the samples falling into the same leaf.
2. *LazyTree* is the best model among C4.4 and its variants in AUC. *LazyTree* significantly outperforms C4.4 (10 wins and 1 loss). Since C4.4 is the *state-of-art* decision tree model designed for yielding accurate ranking, this comparison provides strong evidence to the ranking performance of *LazyTree*. *LazyTree* also outperforms most of C4.4 variants in AUC.
3. LazyTree-B performs greatly better than C4.5-B and C4.4-B. C4.5-B has a good ranking performance compared with other typical models. However, LazyTree-B produces better AUC values than C4.5-B (15 wins and 0 loss) and also performs better than C4.4-B (7 wins and 2 losses).

4. Besides having good performances on ranking, *LazyTree* also has better robustness and stability than other models. The average standard deviation of *LazyTree* in AUC is 4.37, the lowest among all models.

Generally speaking, *LazyTree* is a trade-off between the quality of probability-based ranking and the comprehensibility of results when selecting the best model.

4   Conclusion

In this paper, we analyzed that traditional decision trees have inherent defects in achieving precise ranking, and proposed to resolve those issues by representing the similarity between each sample at a leaf and an unlabeled sample. One key observation is that for a leaf, deploying a lazy probability estimator is an optimal alternative to produce a unique probability estimate for a specific sample, compared with directly using frequency-based probability estimation based on the leaf samples. Experiment results prove our expectation that *LazyTree* outperforms typical decision tree models in ranking quality. In our future work, other parameter-learning methods could be used to tune the probability estimation at a leaf. Additionally, we can find the right tree size for our model, i.e. based on C4.5 we use the *leave-one-out* technique to learn a weight for each sample at leaves, then continue fully splitting the tree, and at each leaf we normalize the weights and produce probability estimation of that leaf.

References

1. C. Blake and C.J. Merz. Uci repository of machine learning database.
2. P. A. Flach C. Ferri and J. Hernandez-Orallo. Improving the auc of probabilistic estimation trees. In *Proceedings of the 14th European Conference on Machine Learning (ECML2003)*. Springer, 2003.
3. D. J. Hand and R. J. Till. A simple generalisation of the area under the roc curve for multiple class classification problems. *Machine Learning*, 45, 2001.
4. Ron Kohavi. Scaling up the accuracy of naive-bayes classifiers: a decision-tree hybrid. In *Proceedings of the Second International Conference on Knowledge Discovery and Data Mining*, 1996.
5. H. Liang and Y. Yan. Lazy learning for improving ranking of decision trees. www.flydragontech.com/publications/2006/LazyLeaveTree_long.pdf, 2006.
6. C. X. Ling and R. J. Yan. Decision tree with better ranking. In *Proceedings of the 20th International Conference on Machine Learning (ICML2003)*. Morgan Kaufmann, 2003.
7. F. J. Provost and P. Domingos. Tree induction for probability-based ranking. *Machine Learning*, 52(30), 2003.
8. I. H. Witten and E. Frank. *Data Mining –Practical Machine Learning Tools and Techniques with Java Implementation*. Morgan Kaufmann, 2000.
9. B. Zadrozny and C. Elkan. Obtaining calibrated probability estimates from decision trees and naive bayesian classifiers. In *Proceedings of the 18th International Conference on Machine Learning (ICML2001)*. Springer, 2001.

Kernel Laplacian Eigenmaps for Visualization of Non-vectorial Data

Yi Guo[1], Junbin Gao[2,*], and Paul W.H. Kwan[1]

[1] School of Math, Stat. & Computer Science,
University of New England, Armidale, NSW 2351, Australia
{yguo4, kwan}@turing.une.edu.au
[2] School of Information Technology,
Charles Sturt University, Bathurst, NSW 2795, Australia
jbgao@csu.edu.au

Abstract. In this paper, we propose the Kernel Laplacian Eigenmaps for nonlinear dimensionality reduction. This method can be extended to any structured input beyond the usual vectorial data, enabling the visualization of a wider range of data in low dimension once suitable kernels are defined. Comparison with related methods based on MNIST handwritten digits data set supported the claim of our approach. In addition to nonlinear dimensionality reduction, this approach makes visualization and related applications on non-vectorial data possible.

1 Introduction

Recently, many approaches on dimensionality reduction and manifold learning have been proposed. Most of them attempt to map objects to a lower dimensional Euclidean space while preserving their proximity relations in the original high dimensional space,i.e, if two objects are close to each other in the original space, their images in the embedding space will still be close. Among these approaches, linear methods like MDS and PCA, as well as graph-based methods like LLE [1], Laplacian Eigenmaps (LE) [2], and LPP [3] are typical representatives.

One common characteristic shared by the above approaches is that they operate only on vectorial data. Moreover, they are spectral methods in the sense that the lower dimensional representations are derived from either the top or bottom eigenvectors of specially constructed matrices. However, in today's demanding applications like those in the field of multimedia and bioinformatics, input in the form of images, texts, or even gene expressions might not be readily represented as vectorial data. As such, they present unique challenges to approaches that assume vectorial input such as those mentioned above. Through feature mapping, KPCA [4] projects the mapped data on principal kernel "vectors" in feature space and avoids the vectorization of objects by the "kernel trick".

Recently, Ham et al. [5] explain several graph-based methods including LLE and LE by KPCA. But it doesn't imply that these methods are readily applied

---

* The author to whom all the correspondences should be addressed.

A. Sattar and B.H. Kang (Eds.): AI 2006, LNAI 4304, pp. 1179–1183, 2006.
© Springer-Verlag Berlin Heidelberg 2006

to non-vectorial objects since the graph construction still requires the vectorization of the data input. In this work, by means of kernel, we can obtain the pairwise similarity between objects in the feature space that can be converted to a distance measure, thus bypassing the hard vectorization problem. Through this, the visualization task can be restated as "if two objects are similar, their distance should be small in the lower dimensional Euclidean space." Note that we have used the term 'similar' rather than 'close' in order to emphasize similarity in the kernel sense instead of the usual Euclidean sense. Here, by providing a solution for optimizing a new objective function involving kernel within the framework of Laplacian eigenmaps, we are able to introduce a criterion for visualizing non-vectorial data that preserves their proximity relations in a lower dimensional Euclidean space. We call our method kernel Laplacian eigenmaps or KLE in short.

2   Kernel Laplacian Eigenmaps

Here, we will first give an overview of LE. In the following discussion, let $X = \{\mathbf{x}_1, \mathbf{x}_2, \ldots, \mathbf{x}_N\}$ be a set of input objects and $Y = \{\mathbf{y}_1, \mathbf{y}_2, \ldots, \mathbf{y}_N\}$ its projection on a lower dimensional manifold. The cost function to be minimized by the original LE algorithm was defined as

$$\sum_{ij} \|\mathbf{y}_i - \mathbf{y}_j\|^2 W_{ij} \ , \tag{1}$$

where W_{ij} is the weight assigned to the edge connecting $\mathbf{x}_i$ and $\mathbf{x}_j$ on an undirected neighborhood graph constructed according to the Euclidean distance between their vectorized representations. This neighborhood graph can be constructed by n nearest neighbors, that is, $\mathbf{x}_i$ and $\mathbf{x}_j$ are connected by an edge if $\mathbf{x}_i$ is among the n nearest neighbors of $\mathbf{x}_j$ or vice versa. n is a parameter whose value affects the final structure of the neighborhood graph being constructed. Once the neighborhood graph constructed, weights W_{ij} are assigned to the edges with 1/0 (1 for connected and 0 otherwise) or an exponentially decaying function, $W_{ij} = e^{-\frac{\|\mathbf{x}_i - \mathbf{x}_j\|^2}{t}}$, if there is an edge between $\mathbf{x}_i$ and $\mathbf{x}_j$ and 0 if not. By assigning weights to the set of edges on the neighborhood graph, a weight matrix whose entries capture the relations among objects in the original space can be obtained. Because the entries of the weight matrix reflect the structure of the neighborhood graph, their values are restricted to be the computed Euclidean distances between vectorized representation of the objects in LE.

In addition, the importance of the weight matrix to the overall performance of the original LE algorithm becomes clearer if we analyze equation (1). Because the LE preserves local distances, objects that are close to each other in the original space will be mapped close to each other in the embedding space. As the weight W_{ij} captures the proximity relation in the neighborhood graph, when its value increases, the Euclidean distance between $\mathbf{y}_i$ and $\mathbf{y}_j$ must decrease in order to minimize the summation. In order to avoid a trivial solution for example all $\mathbf{y}_i$

being some constant, LE imposes specific constraints in the optimization of the cost function.

From the above discussion, it is clear that the neighborhood graph from which the actual similarity matrix is derived enables enables the proximity relations among objects in high dimensional space to be preserved in lower dimensional embedding. It suggests that we could use kernel to evaluate their similarity, bypassing the neighborhood graph construction. In the following discussion, we assume a kernel function, denoted as $k(\mathbf{x}_i, \mathbf{x}_j)$ and corresponding kernel gram matrix $K_{ij} = k(\mathbf{x}_i, \mathbf{x}_j)$, has been appropriately chosen to encode the similarity between each pair of objects in X. We propose a new objective function

$$\sum_{ij} \|\mathbf{y}_i - \mathbf{y}_j\|^2 k(\mathbf{x}_i, \mathbf{x}_j) \tag{2}$$

to reflect this idea. It can be considered as the sum of products of the inner product between similarity "vectors" of non-vectorial objects and their distances in the lower dimensional embedding. While the former is in the kernel sense, the latter is in the usual Euclidean sense. So we minimize it for proximity preserving. Observe that equation (2) and (1) are similar with W_{ij} replaced by $k(\mathbf{x}_i, \mathbf{x}_j)$. Furthermore, we can interpret the kernel gram matrix as a fully connected graph whose weights are the values of the kernel which satisfy the condition of Laplacian in spectral graph theory [6]. This implies that the minimization of (2) can be solved under the framework of Laplacian Eigenmaps. The Laplacian will be computed as $\mathcal{L} = \mathcal{D} - \mathcal{K}$, where $\mathcal{D}$ is a diagonal matrix with $\mathcal{D}_{ii} = \sum_j K_{ij}$. The lower dimensional representation of the non-vectorial objects can thus be obtained by solving this generalized eigenvector system:

$$\mathcal{L}\mathbf{y} = \lambda \mathcal{D}\mathbf{y} \ , \tag{3}$$

and then select the m eigenvectors corresponding to the m smallest eigenvalues excluding those that are 0. We can also apply n-nearest neighbors method to preprocess K to eliminate too small entries thus endowing the algorithm the ability to handle locality information, that is, for every object $\mathbf{x}_i$, we let $K_{ij} = 0$ if $\mathbf{x}_j$ is not among the n most similar objects to $\mathbf{x}_i$ where similarity is measured by the kernel function $k(\cdot, \cdot)$. Since after this procedure, the entries of K still satisfy the condition of Laplacian mentioned above, and the algorithm is still applicable. Because this algorithm involves kernel and Laplacian Eigenmaps, we call it Kernel Laplacian Eigenmaps (KLE).

3   Experimental Evaluation

Experiments were conducted on a subset of the MNIST handwritten digits image database[1]. The test data consisted of 500 images with 50 images per digit. All images are in grayscale and have a uniform size of 28×28 pixels. The SCIGV kernel [7] was chosen to measure the similarity between the input objects. The lower dimensional embedding is in the form of a 2-D Euclidean space.

---

[1] The dataset can be found at http://yann.lecun.com/exdb/mnist/

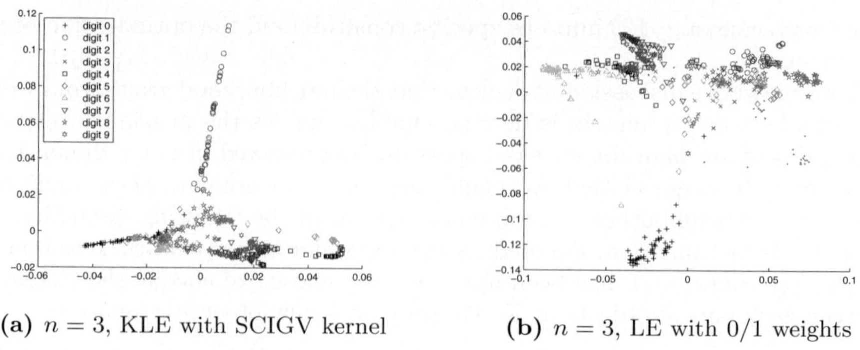

(a) $n = 3$, KLE with SCIGV kernel   **(b)** $n = 3$, LE with 0/1 weights

Fig. 1. The comparison of the performance of the KLE and LE

Figure 1 compares the result of our KLE algorithm with that of the LE. Visually, the result of our KLE algorithm reveals clusters for the different classes of digits that are more compact and have less overlappings than those of the LE, whereas the LE exhibits wider dispersion and overlappings of the objects in the 2-D embedding space. Furthermore, by analyzing the distribution of these clusters in relation to each other in our results, a commonly accepted belief regarding the structure of these digit classes was again verified. For example, the clusters of digits 0 and 1 were far apart from each other as well as from the other digits. This confirmed that the digits 0 and 1 have the best discriminatory accuracy which is a widely accepted belief. In addition, groups such as the one involving digits 9 and 7, and the one involving digits 6 and 8 both exhibit visually common features as can be inferred from the 2-D embedding space by their overlappings.

In our experiments, we also want to compare the effects of varying the value of n (a parameter used in constructing the weighted neighborhood graph by the nearest neighbor method) on the result of the embedding. If we observe the results[2] , it is clear that as the value of n increases, the result of the embedding begins to deteriorate. when n increases, KLE is exhibiting global rather than local features. As a result, KLE might fail to provide useful features when the value of n gets larger and is comparable to the number of samples in a single class.

4   Conclusions

In this paper, we proposed a novel algorithm, KLE, to embed non-vectorial objects in a lower dimensional space thus providing an effective technique for visualizing the structure of their proximity relations. Because vectorization of input objects is not required, visualization of a much wider range of data is possible given a proper kernel defined over the set of input objects is selected. In dimensionality reduction, it is important that a lower dimensional embedding of the objects faithfully reflects their proximity relations in the original space. A

---

[2] The result is available at http://turing.une.edu.au/~yguo4/kle

data dependent and specially designed kernel will definitely improve the result of KLE since the feature mapping associated with the kernel reveals the true proximity relations among the objects thus the lower dimensional representation will certainly be more meaningful.

In KLE, the value of n in the n nearest neighbors method used in preprocessing K is an adjustable parameter and should be carefully tuned in practice. As shown in our experiments, a larger n caused the result of the embedding to deteriorate. Whereas too small an n will incur degeneracy in the embedding. A general guidance on how to select an optimal value for n will be one of the foci in our future research.

Finally, possible applications of the proposed KLE algorithm include dimensionality reduction, exploratory data analysis, clustering, etc. for non-vectorial objects. In our future work, experiments conducted on text, voice, and biological data will be performed.

Acknowledgements

This work is supported by the National Natural Science Foundation of China (NSFC 60373090), the ARC DP Development Grant from Charles Sturt University and the University Research Grant from the University of New England.

References

1. Roweis, S.T., Saul, L.K.: Nonlinear dimensionality reduction by locally linear embedding. Science **290**(22) (2000) 2323–2326
2. Belkin, M., Niyogi, P.: Laplacian eigenmaps for dimensionality reduction and data representation. Neural Computation **15**(6) (2003) 1373–1396
3. He, X., Niyogi, P.: Locality preserving projections. In Thrun, S., Saul, L., Schölkopf, B., eds.: Advances in Neural Information Processing Systems 16. MIT Press, Cambridge, MA (2004)
4. Schölkopf, B., Smola, A.J., Müller, K.R.: Nonlinear component analysis as a kernel eigenvalue problem. Neural Computation **10** (1998) 1299–1319
5. Ham, J., Lee, D.D., Mika, S., Schölkopf, B.: A kernel view of the dimensionality reduction of manifolds. In: ICML '04: Proceedings of the twenty-first international conference on Machine learning, New York, NY, USA, ACM Press (2004) 47–54
6. Chung, F.R.K.: Spectral Graph Theory. Number 92 in Regional Conference Series in Mathmatics. AMS (1997)
7. Guo, Y., Gao, J.: An integration of shape context and semigroup kernel in image classification. submitted to Pattern Recognition Letters (2006)

An Evolutionary Approach for Clustering User Access Patterns from Web Logs

Rui Wu

School of Mathematics and Computer Science, Shanxi Normal University, Shanxi,
041004, China
wurui1971@eyou.com

Abstract. In this paper rough c-means is applied to cluster user access patterns, in which PSO(Particle Swarm Optimization) algorithm is employed to tune the threshold and relative importance of upper and lower approximations. The Davies-Bouldin clustering validity index is used as the fitness function that is minimized while arriving at an optimal clustering. The effectiveness of the algorithm is demonstrated by an experiment.

Keywords: web clustering, fuzzy variable, rough set, web user access pattern, PSO.

1 Introduction

Three important aspects of web mining, namely clustering, association and sequential analysis are often used to study important characteristics of web users. Clustering in web mining involves finding natural groupings of web user access patterns, web users or other web resources.

Due to a variety of reasons inherent in web browsing and web logging, the likelihood of bad or incomplete data is higher than conventional applications. The clusters of web data tend to have vague or imprecise boundaries. Therefore, the role of soft computing technique such as fuzzy theory and rough theory is highlighted. Krishnapuram and Joshi [4], Hathaway and Beadek [5] grouped web users into several disjoint clusters by using fuzzy set theory. Some other researchers tried to explore clustering method with rough theory. The upper and lower approximations of the rough set were used to model the clusters in [3,6]. De [3] applied rough approximation method to cluster web access patterns from web logs. Lingras [6] adopted the unsupervised rough set clustering method with genetic algorithms to group web visitors.

This paper proposes an evolutionary rough c-means clustering method to group user access patterns from web logs. The user access pattern denoting the unique surfing behavior of ith user can be represented as $s_i = \{(Url_{i_1}, t_{i_1}), (Url_{i_2}, t_{i_2}), \cdots, (Url_{i_l}, t_{i_l})\}(1 \leq i \leq n)$, where Url_{i_k} denotes kth visited web page, t_{i_k} denotes the time duration on Url_{i_k}, l is the total number of visited web pages. To easily gain the distance between any two user access patterns, each pattern is converted into a fuzzy vector with the same length, in which each element

A. Sattar and B.H. Kang (Eds.): AI 2006, LNAI 4304, pp. 1184–1188, 2006.
© Springer-Verlag Berlin Heidelberg 2006

represents the visited web page and time duration on it. Because these clusters tend to have ambiguous boundaries, a cluster is represented by a rough set denoted by its upper approximation and lower approximation in this paper. Then a rough c-means method based on the properties of rough set is adopted to cluster user access patterns denoted by fuzzy vectors. In order to optimize clustering results, PSO algorithm is employed to tune the threshold and the relative importance of the lower approximation and upper approximation parameters of the rough sets to minimize the fitness function.

The rest of the paper is organized as follows. The evolutionary algorithm of clustering user access patterns is proposed in section 2. The effectiveness of the method is demonstrated in section 3. Finally we conclude in section 4.

2 Evolutionary Rough c-Means Method for Clustering Web Access Patterns

2.1 Characterizing User Access Pattern as a Fuzzy Vector

Suppose there exist user access patterns $S = \{s_1, s_2, \cdots, s_n\}$, where $s_i(1 \leq i \leq n)$ represents a unique surfing behavior of ith user.

Let W be the universe of distinct m web pages visited by all users, U be the universe of s_i $(1 \leq i \leq n)$.

Each pattern $s_i \in S$ is a non-empty subset of U. Here, the temporal order of visited web pages has not been taken into account.

The user access pattern $s_i \in S(1 \leq i \leq n)$ can be denoted by a vector

$$V_i = < v_{i1}^t, v_{i2}^t, \cdots, v_{im}^t >, \tag{1}$$

where $v_{ik}^t = \begin{cases} t_{i_k}, & (Url_k, t_{i_k}) \in s_i \\ 0, & otherwise, \end{cases}$ $(1 \leq k \leq m)$.

Thus, each pattern $s_i(1 \leq i \leq n)$ is transformed as a real numerical vector with the same length. Furthermore, $v_{ik}^t(1 \leq k \leq m)$ is depicted by a fuzzy linguistic variable.

Membership functions of time durations on all web pages can be determined by expert opinion or by statistical methods. Each membership function corresponds to a fuzzy variable $\xi_i(1 \leq i \leq r)$. Then each numerical v_{ik}^t in vector V_i can be transformed as a corresponding fuzzy linguistic variable $\xi_j(1 \leq j \leq r)$ by the method introduced in [8]. Thus the web access pattern of ith user can be denoted as follows

$$f_{vi} = < \lambda_{i1}, \lambda_{i2}, \cdots, \lambda_{im} >, \tag{2}$$

where $\lambda_{ik} \in \{0, \xi_1, \xi_2, \cdots, \xi_r\}$ $(1 \leq k \leq m)$.

2.2 Algorithm for Clustering Fuzzy User Access Patterns Based on Rough c-Means

Assume there exist n user access patterns in web transaction S, $S = \{s_1, s_2, \cdots, s_n\}$. Given any two patterns s_i and s_j $(1 \leq i, j \leq n)$, they can be denoted as follows $f_{vi} = <\lambda_{i1}, \lambda_{i2}, \cdots, \lambda_{im}>$ and $f_{vj} = <\lambda_{j1}, \lambda_{j2}, \cdots, \lambda_{jm}>$, where $\lambda_{ik} \in \{0, \xi_1, \xi_2, \cdots, \xi_r\}$ and $\lambda_{jk} \in \{0, \xi_1, \xi_2, \cdots, \xi_r\}$ $(1 \leq k \leq m)$.

Their sum can be defined as

$$sum(s_i, s_j) \simeq sum(f_{vi}, f_{vj}) = <E[\lambda_{i1} + \lambda_{j1}], \cdots, E[\lambda_{im} + \lambda_{jm}]> . \tag{3}$$

Let $f(\xi) = \lambda_{ik} + \lambda_{jk}(1 \leq k \leq m)$, then $E[f(\xi)]$ can be gained by fuzzy simulation method seen in [7]. The distance between s_i and s_j can be defined as

$$d(s_i, s_j) \simeq d(f_{vi}, f_{vj}) = \sqrt{\frac{\sum_{k=1}^{n} (E[\lambda_{ik} - \lambda_{jk}])^2}{m}}. \tag{4}$$

similarly, $E[\lambda_{ik} - \lambda_{jk}]$ can also be computed by fuzzy simulation.

This paper adopts a rough c-means algorithm for clustering user access patterns. Here, ith cluster $U_i(1 \leq i \leq k)$ is characterized as a rough set. Then the centroid $\mathbf{m}_i$ of U_i is computed as

$$\mathbf{m}_i = \begin{cases} w_{low}\dfrac{\sum_{f_{vj} \in \underline{R}U_i} f_{vj}}{|\underline{R}U_i|} + w_{up}\dfrac{\sum_{f_{vj} \in (\overline{R}U_i - \underline{R}U_i)} f_{vj}}{|\overline{R}U_i - \underline{R}U_i|} & \overline{R}U_i - \underline{R}U_i \neq Null; \\[3mm] w_{low}\dfrac{\sum_{f_{vj} \in \underline{R}U_i} f_{vj}}{|\underline{R}U_i|} & otherwise, \end{cases} \tag{5}$$

where the parameter w_{low}/w_{up} controls the importance of the patterns lying within the lower/up approximation of a cluster in determining its centroid. $0.5 < w_{low} < 1$ and $w_{up} = 1 - w_{low}$. $|\underline{U}_i|$ indicates the number of patterns in the lower approximation of cluster U_i, while $|\overline{U}_i - \underline{U}_i|$ is the number of patterns in the rough boundary lying between the two approximations $\overline{U}_i$ and $\underline{U}_i$. Thus, the centroid $\mathbf{m}_i$ of ith cluster is a real-valued vector denoted by $\mathbf{m}_i = <c_{i1}, c_{i2}, \cdots, c_{im}>$.

The distance between pattern s_l $(1 \leq l \leq n)$ and the centroid $\mathbf{m}_i$ is defined as follows

$$d(s_l, \mathbf{m}_i) \simeq d(f_{vl}, \mathbf{m}_i) = \sqrt{\frac{\sum_{k=1}^{m} (E[\lambda_{lk} - c_{ik}])^2}{m}}. \tag{6}$$

The main steps of the clustering algorithm is described as follows.

step1: For each $s_i \in S$, denote s_i by a corresponding fuzzy vector f_{vi}.
step2: Assign initial means $\mathbf{m}_i$ for the c clusters.
       Here, choose randomly c user access patterns as the initial means.
step3: For each fuzzy vector f_{vl} $(1 \leq l \leq n)$ do
       For i=1 to c do
       Compute $d(f_{vl}, \mathbf{m}_i)$ according to Eq.(6).
step4: For each pattern s_l $(1 \leq l \leq n)$ do
       If $d(f_{vl}, \mathbf{m}_i) - d(f_{vl}, \mathbf{m}_j)$ $(i \neq j)$ is less than threshold δ
       then $s_l \in \overline{R}U_i$ and $s_l \in \overline{R}U_j$,

else if the distance $d(f_{vl}, \mathbf{m}_i)$ is minimum over the c clusters
then $s_l \in \underline{R}U_i$.

step5: $C_1 = \{\underline{R}U_1, \overline{R}U_1\}, \cdots, C_c = \{\underline{R}U_c, \overline{R}U_c\}$.

step6: Compute new mean for each cluster using Eq.(5).

step7: Repeat steps 3-6 until convergence.

step8: output C.

2.3 Evolutionary Optimization

It is observed that the performance of the algorithm is dependent on the choice of w_{low}, w_{up} and the threshold δ. In this paper we employed an evolutionary PSO algorithm, which was originally designed by Eberhat and Kennedy [2] to compute the optimal values of the parameters involved. In this paper we compute the optimal number of clusters by the Davies-Bouldin cluster validity index [1].

The optimal model is defined as follows

$$\begin{cases} \min \dfrac{1}{c} \sum_{k=1}^{c} \max_{l \neq k} \left\{ \dfrac{S(C_k) + S(C_l)}{d(C_k, C_l)} \right\} \\ \\ s.t. \quad 0 < w_{up} < 1, 0 < \delta < 0.5. \end{cases} \tag{7}$$

The main steps of optimal algorithm are provided below.

step1. $K = 1$, randomly choose N particles $(w_{up_i}, \delta_i)(i = 1, 2, \cdots, N$, and
$V_{iw_{up}}^K = 0, V_{i\delta}^K = 0$.

step2. Run algorithm 1, and compute Davies-Bouldin cluster validity index
$F(P_{iw_{up}}^K, P_{i\delta}^K)$.

step3. If $F(P_{iw_{up}}^s, P_{i\delta}^s) = \max\limits_{1 \leq l \leq K} F(P_{iw_{up}}^l, P_{i\delta}^l)$ then $pbest_i = (P_{iw_{up}}^s, P_{i\delta}^s)$.

If $F(P_{iw_{up}}^t, P_{i\delta}^t) = \max\limits_{1 \leq l \leq K, 1 \leq i \leq N} F(P_{iw_{up}}^l, P_{i\delta}^l)$ then $gbest = (P_{iw_{up}}^t, P_{i\delta}^t)$.

step4. $V_{iw_{up}}^{K+1} = wV_{iw_{up}}^K + c_1 r_1 (P_{iw_{up}}^s - w_{up_i}^K) + c_2 r_2 (P_{iw_{up}}^t - w_{up_i}^K)$.
$V_{i\delta}^{K+1} = wV_{i\delta}^K + c_1 r_1 (P_{i\delta}^s - \delta_i^K) + c_2 r_2 (P_{i\delta}^t - \delta_i^K)$.
w is the weight, c_1, c_2 are constants, and r_1, r_2 are random numbers in the interval $[0, 1]$.

step5. $w_{up_i}^{K+1} = w_{up_i}^K + V_{iw_{up}}^{K+1}$. $\delta_i^{K+1} = \delta_i^K + V_{i\delta}^{K+1}$.

step6. repeat steps 2-6 until convergence.

3 An Experiment

Assume 5,650 user access patterns are extracted from the information resource after a web log is downloaded. And these web access patterns are to be grouped into 5 clusters. Time durations on web pages correspond to trapezoidal fuzzy variables $\xi_1(5, 5, 30, 40), \xi_2(30, 60, 90, 120), \xi_3(90, 120, 150, 150)$.

The clustering results are shown in Table 1.

The same data is clustered into groups by other algorithms. The comparison with other algorithms are shown in Table 2.

By comparison, the evolutionary rough c-means gets better clustering results.

Table 1. Clustering result

cluster number	the included web access patterns
cluster 1	272
cluster 2	965
cluster 3	656
cluster 4	153
cluster 5	74

Table 2. Comparison of clustering results

clustering algorithm	Davies-Bouldin index
rough approximation method	0.780
fuzzy c-means	0.712
evolutionary rough c-means	0.656

4   Conclusion

In this paper, an evolutionary approach based on rough c-means in fuzzy environment is proposed to cluster user access patterns. This approach is useful to cluster web access patterns from web logs so that similar surfing behavior can be grouped into one class. Thus people can build up adaptive web server and design personalized service according to users' surfing behaviors.

References

1. Bezdek, J., Pal, N.: Some new indexes for cluster validity. IEEE Transactions on Systems, Man, and Cybernetics. Part-B 28(1998) 301-315.
2. Eberhat, R., Kennedy, J.: A new Optimizer Using Particles Swarm Theory, Proceedings of Sixth International Symposium on Micro Machine and Human Science, Japan, 1995, pp. 39-43.
3. De, S., Krishna, P.: Clustering web transactions using rough approximation. Fuzzy Sets and Systems. 148(2004) 131-138.
4. Krishnapram, R., Joshi, A., and etc: low compexity fuzzy relational clustering algorithms for web mining. IEEE Transactions on Fuzzy Systems. 9(2001) 595-607.
5. Hathaway, R., Beadek, J.: Switching regression models and fuzzy clustering. IEEE Transactions on Fuzzy Systems. 1(3)(1993) 195-204.
6. Lingras, P.: Rough set clustering for web mining, Proceedings of the 2002 IEEE International Conference on Fuzzy Systems (FUZZ-IEEE'02), Honolulu, HI, United States, Vol: 2, 2002, pp.1039-1044.
7. Liu, B., Liu, Y.: Expected value of fuzzy variable and fuzzy expected value models. IEEE Transactions on Fuzzy Systems. 10(2002) 445-450.
8. Ning, Y., Tang, W., Zhang, Y.: Analytic hierarchy process based on fuzzy simulation, Proceedings of the IEEE International Conference on Fuzzy Systems, Reno, NV, USA, 2005, pp.546-550.

Simplified Support Vector Machines Via Kernel-Based Clustering

Zhi-Qiang Zeng[1], Ji Gao[2], and Hang Guo[3]

Department of Computer Science and Engineering, Zhejiang University,
310027 Hangzhou, China
[1] lbxzzq@hotmail.com
[2] gaoji@mail.hz.zj.cn
[3] hangguo@zju.edu.cn

Abstract. Reduced set method is an important approach to speed up classification process of support vector machine (SVM) by compressing the number of support vectors included in the machine's solution. Existing works find the reduced set vectors based on solving an unconstrained optimization problem with multivariables, which may suffer from numerical instability or get trapped in a local minimum. In this paper, a novel reduced set method relying on kernel-based clustering is presented to simplify SVM solution. This approach is conceptually simpler, involves only linear algebra and overcomes the difficulties existing in former reduced set methods. Experiments on real data sets indicate that the proposed method is effective in simplifying SVM solution while preserving machine's generalization performance.

Keywords: Approximate solution; Reduced set vector; Kernel-based clustering.

1 Introduction

Recently, there has been considerable research on speeding up the classification process of Support Vector Machine (SVM) [1] by approximating the solution in terms of a reduced set of vectors. In [2,3,4], authors try to approximate the decision function with a much smaller number of reduced set vectors that are constructed incrementally. These approaches need to solve an unconstrained optimization problem with multivariables. Therefore, as in any nonlinear optimization problem, one may suffer from numerical instability or get trapped in a local minimum and the result obtained is thus sensitive to the initial guess [5].

This paper presents a novel reduced set approach to simplify SVM solution. The proposed method is conceptually simpler, involves only linear algebra and does not suffer from numerical instability or local minimum problems. Experiments on real data sets show the effectiveness of the proposed method in simplifying the SVM solution while preserving machine's generalization performance.

The paper is organized as follows. In section 2, the proposed method is introduced in detail. Experimental results are demonstrated in Section 3 to illustrate the effectiveness of the proposed method. Conclusions are included in Section 4.

A. Sattar and B.H. Kang (Eds.): AI 2006, LNAI 4304, pp. 1189 – 1195, 2006.
© Springer-Verlag Berlin Heidelberg 2006

2 Proposed Reduced Set Method

In [3], authors give an interpretation of reduced set vector related with cluster centroid. Motivated by that, we propose to organize positive and negative support vectors in clusters, respectively, in F, and replace images of support vectors within the cluster by the cluster centroid to simplify SVM decision function.

2.1 Kernel-Based Clustering

A simply unsupervised clustering (UC) algorithm proposed in [6] is applied to organize positive and negative support vectors in clusters, respectively. However, the simplification of SVM solution is processed in feature space. The cluster centroids generated by the UC algorithm in input space may not be suit for as a reduced vector set to approximate the decision function of SVM. To make use of the UC algorithm for the simplification, we have to extend the original UC algorithm into feature space by using kernel methods and develop the kernel-based unsupervised clustering (KUC) algorithm to generate cluster centroids in feature space instead of in input space. The KUC algorithm is described as following:

Suppose that the sample set for clustering is $X = \{x_1, x_2, ..., x_m\}$, $x_i \in R^d$, $i = 1, ...m$, the cluster radius is r, and φ is a nonlinear mapping function to project x_i into feature space F.

1) $C_1 = \{\varphi(x_1)\}, O_1 = \varphi(x_1), Cluster\_num = 1, Z = \{x_2, ...x_m\}$.

2) If $Z = \varnothing$, then STOP.

3) For a sample $x_i \in Z$, choose the cluster centroid O_j closest to $\varphi(x_i)$ from the existing cluster centroids, i.e,

$$O_j = \arg \min_{j} {}_{k=1}^{Cluster\_num} d(\varphi(x_i), O_k) . \tag{1}$$

4) If the distance $d(\varphi(x_i), O_j) \leq r$, add $\varphi(x_i)$ into cluster C_j, i.e, $C_j = C_j \cup \{\varphi(x_i)\}$, the centroid of this cluster is adjusted to the mean value of all the samples within the cluster in F:

$$O_j = \frac{n_j \times O_j + \varphi(x_i)}{n_j + 1} , \tag{2}$$

where n_j is the number of samples within the cluster. Adjust $n_j = n_j + 1$, and go to step 6.

5) If the distance $d(\varphi(x_i), O_j) > r$, add a new class $Cluster\_num=Cluster\_num+1$, $C_{Cluster\_num} = \{\varphi(x_i)\}$, $O_{Cluster\_num} = \varphi(x_i)$.

6) $Z = Z - \{x_i\}$, go to step 2.

In step 3, the distance between $\varphi(x_i)$ and k^{th} cluster centroid $O_k = (1/n_k)\sum_{p=1}^{n_k} \varphi(x_{k_p})$ can be calculated as:

$$d(\varphi(x_i), O_k) = \sqrt{k(x_i, x_i) - \frac{2}{n_k}\sum_{p=1}^{n_k} k(x_i, x_{k_p}) + \frac{1}{n_k^2}\sum_{p,q=1}^{n_k} k(x_{k_p}, x_{k_q})} , \tag{3}$$

where n_k denotes $|C_k|$ and $\varphi(x_{k_i}), i = 1, ..., n_k$, are the samples within the cluster C_k.

Finally, the cluster centroid obtained by KUC algorithm is the mean value of all the images of support vectors within the cluster, which assumes that each support vector exerts same force on cluster centroid and does not take into account the corresponding weights of support vectors. To amend the problem, we adjust the expression of cluster centroid as following:

$$O_k = \sum_{i=1}^{n_k} b_{k_i} \varphi(x_{k_i}) , \ b_{k_i} = \alpha_{k_i} / \sum_{i=1}^{n_k} \alpha_{k_i} , \ k = 1,\ldots,Cluster\_num, \qquad (4)$$

where $\alpha_{k_i}, i = 1,\ldots,n_k$, are corresponding weights of support vectors within the cluster C_k in F.

2.2 Solution of Pre-image Problem

It is clear that the cluster centroids in F derived from KUC algorithm cannot be used directly, thus, we must use their pre-images in input space. In [7], Kwok and Tsang present a method to find the approximate pre-images of patterns that are denoised in feature space via kernel principal component analysis (KPCA). We follow their strategy to seek the pre-images of cluster centroids generated by KUC algorithm.

The feature-space distance between cluster centroid O_k and an arbitrary point x_i can be calculated as following:

$$\tilde{d}_i^2(O_k,\varphi(x_i)) = k(x_i,x_i) - 2\sum_{p=1}^{n_k} b_{k_p} k(x_{k_p},x_i) + \sum_{p,q=1}^{n_k} b_{k_p} b_{k_q} k(x_{k_p},x_{k_q}) . \qquad (5)$$

Suppose the pre-image of O_k is z_k in input space, for the Gaussion kernel, the following simple relation holds true between $\tilde{d}_i^2(\varphi(z_k),\varphi(x_i))$ and $d_i^2(z_k,x_i)$ [8]:

$$\tilde{d}_i^2(\varphi(z_k),\varphi(x_i)) = \| \varphi(z_k) - \varphi(x_i) \|^2 = k(z_k,z_k) - 2k(z_k,x_i) + k(x_i,x_i)$$

$$\Rightarrow d_i^2(z_k,x_i) = -2\sigma^2 \ln(1 - \frac{1}{2}\tilde{d}_i^2(\varphi(z_k),\varphi(x_i))) . \qquad (6)$$

Because the feature-space distance $\tilde{d}_i^2(O_k,\varphi(x_i))$ is available from (5), the corresponding input-space distance between the desired approximate pre-image z_k of O_k and x_i can be calculated based on (6). Generally, the distances with neighbors are the most important in determining the location of any point. Hence, we will only consider the (squared) input-space distances between cluster centroid O_k and its n_k nearest neighbors $\{\varphi(x_{k_1}),\varphi(x_{k_2}),\ldots\varphi(x_{k_{n_k}})\}$ which are all the support vectors within the cluster C_k in F. Define a vector

$$d^2 = [d_1^2,d_2^2,\ldots d_{n_k}^2]^T , \qquad (7)$$

where d_i, $i=1,\ldots,n_k$, are the input-space distance between the desired pre-image of O_k and x_{k_i} . In [9], one attempts to find a representation of a further point with known distance from each of other points. Thus, we can use the idea to transform O_k back to the input space. For the n_k neighbors $\{x_{k_1},x_{k_2},\ldots,x_{k_{n_k}}\}\in R^d$ obtained by KUC algorithm, we will first center them at their centroid $\overline{x} = (1/n_k)\sum_{i=1}^{n_k} x_{k_i}$ and define a coordinate system in their span. First, we construct the $d \times n_k$ matrix $X = [x_{k_1}, x_{k_2}, ..., x_{k_{n_k}}]$ and a $n_k \times n_k$ centering matrix

$$H = I - \frac{1}{n_k}11^T,$$ (8)

where I is a $n_k \times n_k$ identity matrix and $1=[1,1,...,1]^T$ is a $n_k \times 1$ vector. The XH will center the x_{k_i}'s at their centroid

$$XH = [x_{k_1} - \overline{x}, x_{k_2} - \overline{x}, ..., x_{k_{n_k}} - \overline{x}].$$ (9)

Suppose that XH is of rank q, we can obtain the singular value decomposition (SVD) of the $d \times n_k$ matrix XH as:

$$XH = [E_1, E_2]\begin{bmatrix} \Lambda_1 & 0 \\ 0 & 0 \end{bmatrix}\begin{bmatrix} V_1^T \\ V_2^T \end{bmatrix} = E_1\Lambda_1 V_1^T = E_1\Gamma,$$ (10)

where $E_1 = [e_1, e_2, ...e_q]$ is a $d \times q$ matrix with orthonormal columns e_i, and $\Gamma = \Lambda_1 V_1^T = [c_1, c_2, ...c_{n_k}]$ is a $q \times n_k$ matrix with columns c_i being the projection of x_{k_i} onto the e_j's. Note that, $\| c_i \|^2 = \| x_{k_i} - \overline{x} \|^2, i = 1,...,n_k$, and collect this into a n_k-dimensional vector, as $d_0^2 = \left[\| c_1 \|^2, \| c_2 \|^2, ..., \| c_{n_k} \|^2 \right]^T$. It is clear that the location of the pre-image z_k is obtained by requiring $d^2(z_k, x_{k_i}), i = 1,...,n_k$ to be as close to those values in (7) as possible, i.e,

$$d^2(z_k, x_{k_i}) \simeq d_i^2, i = 1,...,n_k.$$ (11)

Define $\tilde{c} \in R^{q \times 1}$ via $E_1\tilde{c} = z_k - \overline{x}$, then

$$d_i^2 = \| z_k - x_{k_i} \|^2 = \| (z_k - \overline{x}) - (x_{k_i} - \overline{x}) \|^2 = \| \tilde{c} \|^2 + \| c_i \|^2 - 2(z_k - \overline{x})^T(x_{k_i} - \overline{x}), i = 1,...,n_k.$$ (12)

Following the steps in [9] and [7], we obtain that

$$\tilde{c} = \frac{1}{2}(\Gamma\Gamma^T)^{-1}\Gamma(d_0^2 - d^2) = \frac{1}{2}\Lambda_1^{-1}V_1^T(d_0^2 - d^2).$$ (13)

Finally, by transforming $\tilde{c}$ back to the original coordinated system in input space, the approximate pre-image of cluster centroid in F is

$$z_k = E_1\tilde{c} + \overline{x}.$$ (14)

2.3 Determine Optimal Coefficients of Reduced Set Vectors

With the derived pre-images of cluster centroids, the goal is now to find the optimal coefficients β_k to approximate $\sum_{i=1}^{n_k} \alpha_{k_i}\varphi(x_{k_i})$ by $\beta_k\varphi(z_k)$:

$$d(\beta_k) = \| \beta_k \varphi(z_k) - \sum_{i=1}^{n_k} \alpha_{k_i} \varphi(x_{k_i}) \|^2, k = 1, ..., Cluster\_num, \tag{15}$$

where z_k is pre-image of centroid of cluster C_k, x_{k_i}, $i=1,...,n_k$ are support vectors within the cluster C_k in F and α_{k_i} are corresponding weights. For the extreme, we have $\nabla_{\beta_k}(d(\beta_k)) = 0$ and obtain that

$$\beta_k = \sum_{i=1}^{n_k} \alpha_{k_i} k(z_k, x_{k_i}) / k(z_k, z_k) , k = 1, ..., Cluster\_num. \tag{16}$$

For Gaussion kernel, $k(z_k, z_k) = 1$, thus, $\beta_k = \sum_{i=1}^{n_k} \alpha_{k_i} k(z_k, x_{k_i})$.

In [3], the optimal coefficients $\beta = (\beta_1, \beta_2, ..., \beta_{N_z})$ are computed by

$$\beta = (K^z)^{-1} K^{zx} \alpha , \tag{17}$$

where $K_{ij}^z = \varphi(z_i)^T \varphi(z_j)$ and $K_{ij}^{zx} = \varphi(z_i)^T \varphi(x_j)$. The (17) take into consideration collective property of all reduced set vectors, thus, it always generates a solution that is better than original one. In our method, we adopt (17) to recompute corresponding optimal coefficients of desired reduced set vectors after iterative simplification step finished.

2.4 The Main Algorithm

The sketch of our reduced set method can be summarized as following:

1) Initializes cluster radius R, step length λ, stopping threshold τ
2) Organize the positive and negative support vectors in clusters in F, respectively, by KUC algorithm with cluster radius R.
3) Find the pre-images of cluster centroids in F following the approach described in section 2.2.
4) Regard these pre-images of cluster centroid as reduced set vectors and determine the optimal coefficients using (16) to construct a simplified SVM.
5) If the difference between simplified SVM solution and original one is bigger than a given stopping threshold τ
   go to step 6
  Else
    $R = R + \lambda$, go to step 2
6) Adopt the reduced set vectors obtained in previous iteration as desired reduced set vectors and recompute their optimal coefficients using (17) to generate resulting simplified SVM.

Generally, the simplified solution is different from the original one, which may lead to the degradation in generalization performance of simplified SVM. To control the situation, we monitor the difference between simplified SVM solution and the original one. If the generated simplified SVM makes the difference exceed a given threshold τ, the simplification process will stop. The difference between them can be measured as following:

$$\frac{\|\psi-\psi'\|^2}{\|\psi\|^2}=\frac{\|\sum_{i=1}^{N_S}\alpha_i\varphi(x_i)-\sum_{p=1}^{N_Z}\beta_p\varphi(z_p)\|^2}{\|\sum_{i=1}^{N_S}\alpha_i\varphi(x_i)\|^2}=1+\frac{\sum_{p,q=1}^{N_Z}\beta_p\beta_q k(z_p,z_q)-2\sum_{i=1}^{N_S}\sum_{p=1}^{N_Z}\alpha_i\beta_p k(x_i,z_p)}{\sum_{i,j=1}^{N_S}\alpha_i\alpha_j k(x_i,x_j)}. \tag{18}$$

3 Experiments and Discussions

The proposed method is implemented in Matlab 7.0 and VC++. The LIBSVM [10] is used for SVM implementation. USPS handwritten digit data set from UCI machine learning repository [11] is used in experiments.

In the experiment on the USPS data set of 7291 training patterns and 2007 test patterns, we simplify the SVM solution of ten binary classifiers, each trained to separate one digit from the rest. The Gaussion kernel is used with $g=2^{-8}$, and C is set to 10, which are obtained by doing 10-fold cross validation on a small subset.

Table 1 shows a different reduction in number of support vectors as well as different test errors with different values of stopping threshold τ. The last column displays number of reduced set vectors and test errors for 10-classification, and the numbers included in bracket denote corresponding reduction rate and test error rate, respectively.

From Table 1, it can be seen that the proposed method is able to reduce the support vectors even up to 90% on USPS data set with only 0.44% loss in generalization performance, which is better than results reported in [3] (90% reduction rate with 0.7% loss). Moreover, the results can be further improved by performing a subsequent global gradient descent in the space of all (z_i,β_i) to minimize $\|\varphi-\varphi'\|$ [2]. The corresponding results are displayed in the last row of Table 1, where RSV-26 stands for 26 reduced set vectors for 10-classification. Though the computation is expensive, the improved version leads to 90% reduction rate on the data set with only 0.1% loss in generalization performance, which is also better than corresponding results reported in [3] (90% reduction rate with 0.3% loss).

Table 1. Performance of simplified SVM on USPS data set

Stopping threshold τ		0	1	2	3	4	5	6	7	8	9	10-class
Original	# SVs	211	85	310	313	329	354	215	197	366	337	272 (0)
SVM	# Errors	17	15	33	24	32	23	11	14	25	17	21 (1.05%)
0.15	# RSVs	150	58	210	274	238	296	175	150	287	274	211 (22%)
	# Errors	15	15	28	21	32	23	12	16	25	19	21 (1.05%)
0.25	# RSVs	101	41	154	132	190	200	107	110	147	165	134 (51%)
	# Errors	17	15	13	31	34	23	20	16	30	20	22 (1.1%)
0.3	# RSVs	74	29	118	91	132	147	69	62	107	123	95 (65%)
	# Errors	18	16	15	34	34	25	21	16	31	22	23 (1.15%)
0.35	# RSVs	47	21	80	73	96	83	50	44	93	73	66 (76%)
	# Errors	22	18	19	35	37	31	24	19	35	24	26 (1.3%)
0.47	# RSVs	21	16	32	27	28	30	22	23	34	25	26 (90%)
	# Errors	22	21	30	43	41	33	27	23	38	26	30 (1.49%)
RSV-26	# Errors	17	16	24	21	35	28	19	18	29	24	23 (1.15%)

4 Conclusions

A reduced set method based on kernel-based clustering to simplify SVM solution is investigated in this paper. Compared with former reduced set methods, the proposed method possesses attractive advantages: (i) It is conceptually simpler, involves only linear algebra and does not suffer from numerical instability or local minimum problems. (ii) Each reduced set vector obtained by the method has an explicit meaning, which can be regarded as a representative of several closed original support vectors belonging to the same class in F. (iii) The proposed method is more effective in simplifying SVM solution while preserving machine's generalization performance.

References

1. V. Vapnik. "Statistical Learning Theory." Wiley, New York, NY, 1998.
2. C. J. C. Burges, "Simplified Support Vector Decision Rules", *13th International Conference on Machine Learning*, 1996, pp. 71-77.
3. B. Schoelkopf, S. Mika, C. J. C. Burges, P. Knirsch, K. Muller, G. Ratsch, and A. J. Smola, "Input space versus feature space in kernel-based methods," *IEEE Trans. Neural Netw.*, vol. 10, no. 5, pp.1000–1017, 1999.
4. Schoelkopf, B., & Smola, A. (2002). *"Learning with kernels."* Cambridge, MA: MIT Press.
5. Mika, S., Scholkopf, B., Smola, A., Muller, K., Scholz, M., & Ratsch, G. (1998). "Kernel PCA and denoising in feature spaces." *Advances in Neural Information Processing Systems* 11. San Mateo, CA:Morgan Kaufmann
6. LI Xiao-Li, LIU Ji-Min and SHI Zhong-Zhi. "A Chinese Web Page Classifier Based on Support Vector Machine and Unsupervised Clustering". *CHINESE J. COMPUTERS*, Jan. 2001,62-68
7. J. T. Kwok and I. W. Tsang, "The pre-image problem in kernel methods," *IEEE Transactions on Neural Networks*, vol. 15, pp. 1517-1525, 2004.
8. C. K. 1. Williams, "On a connection between kernel PCA and metric multidimensional scaling," *Machine Learning*, vol. 46, pp. 11-19, 2002.
9. J. C. Gower, "Adding a point to vector diagrams in multivariate analysis," *Biometrika*, vol. 55, pp. 582-585. 1968.
10. C.-C. Chang and C.-J. Lin.: "LIBSVM: a library for support vector machines", 2001.Software available at http://www.csie.ntu.edu.tw/~cjlin/libsvm
11. Murphy, P. M., &Aha, D.W. (1994). UCI repository of machine learning databases. Irvine, CA (Available at http://www.ics.uci.edu/~mlearn/MLRepository.html)

Protein Folding Prediction Using an Improved Genetic-Annealing Algorithm

Xiaolong Zhang and Xiaoli Lin

School of Computer Science and Technology,
Wuhan University of Science and Technology, Wuhan 430081 P.R. China
{xiaolong.zhang, xiaoli.lin}@mail.wust.edu.cn

Abstract. Based on the off-lattice *AB* model consisting of hydrophobic and hydrophilic residues, a novel hybrid algorithm is presented for searching the ground-state conformation of the protein. This algorithm combines genetic algorithm and simulated annealing. A kind of optimization of the crossover operators in the genetic algorithm is implemented, where a local adjustment mechanism is used to enhance the searching ability for optimal solutions of the off-lattice *AB* model. Experimental results demonstrate that the proposed algorithm is feasible and can insure the solution quality when used to search for native states with off-lattice *AB* model.

Keywords: Off-lattice *AB* Model, Genetic-Annealing Algorithm, Crossover Strategy, Local Adjustment.

1 Introduction

Predicting the native states of a protein from its given amino acid sequence is one of the most central problems in the field of bioinformatics. Based on the minimum free energy hypothesis that the native structure of protein is the one in which the free energy of the whole system is lowest [1], much work has been devoted to introduce theoretical computing methods to search for the structures of proteins, such as Monte Carlo [2], genetic algorithm [3] and dynamical approach [4] etc.

Since solving such a problem is too difficult to be approached with realistic potentials for the real protein, the highly simplified models are studied instead. One of the most prominent and widely used models is the off-lattice *AB* protein model in two dimensions proposed by Stillinger, etc.[5], where the hydrophobic monomers are labeled by *A* and the hydrophilic ones labeled by *B*. The energy function for the N-mer chain is defined as [5]:

$$\Phi = \sum_{i=2}^{n-1} V_1(\theta_i) + \sum_{i=1}^{n-2}\sum_{j=i+2}^{n} V_2(r_{ij}, \xi_i, \xi_j) \tag{1}$$

Where $V_1(\theta_i) = 1/4(1 - \cos\theta_i)$ and $V_2(r_{ij}, \xi_i, \xi_j) = 4(r_{ij}^{-12} - C(\xi_i, \xi_j)r_{ij}^{-6})$. r_{ij} denotes the distance between monomer i and j of the chain. V_1 is the bending potential and independent of the *A*, *B* sequence. The V_2 is Lennard-Jones potential

A. Sattar and B.H. Kang (Eds.): AI 2006, LNAI 4304, pp. 1196–1200, 2006.
© Springer-Verlag Berlin Heidelberg 2006

with a species-dependent coefficient $C\left(\xi_i,\xi_j\right)$ which is +1, +1/2 and -1/2 for *AA*, *BB* and *AB* pairs respectively, thus there is attractive for pairs of like monomers and repulsive for *AB* pairs of monomers.

Even for this simple model, it is still nontrivial to predict the native state for the protein folding problem. Therefore, the global optimization problem is a sticking point of the protein structure prediction. In this paper, we developed a new efficient method of conformational search called the local adjustment genetic-annealing algorithm (LAGAA). One of the greatest advantages of LAGAA is that it can improve the global searching ability, and compute the lowest energy of a protein in a more efficient way.

2 Improved Genetic-Annealing Algorithm

A novel hybrid algorithm that merges Genetic algorithm (GA) with Simulated annealing (SA) is proposed in this paper. This algorithm takes the advantage of GA by using its stochastic linearity combination crossover strategy, and the advantage of SA. Moreover, local adjustment mechanism is involved to enhance the searching ability for optimal solutions of the off-lattice *AB* model.

1)   Local adjustment mechanism

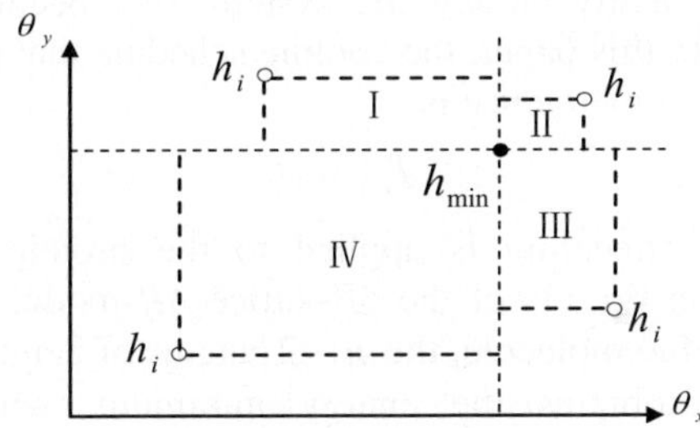

Fig. 1. A schematic diagram of local adjustment. The cycles indicate the individuals to be optimized, and the dots indicate the current individual with minimum energy.

Fig. 1 presents the schematic diagram of local adjustment for the tetramers. According to off-lattice *AB* model, the $\Phi(\theta)$ energy of each tetramer is determined by two angles of bend θ_x and θ_y. If the individuals h_i distribute in areas I , II , III , □, then the distance between each parameter of each individual h_i and the corresponding parameter of h_{min} can be described as:

$$\begin{cases} \Delta\theta_x = \theta_x^i - \theta_x^{min} \\ \Delta\theta_y = \theta_y^i - \theta_y^{min} \end{cases} \qquad (2)$$

Since searching smaller areas in which the global minimum lies will lead to a faster convergence to the desired solution, every parameter of h_i is adjusted by Eq. (3) to keep the searching in smaller areas.

$$\begin{cases} \theta^i_x = \theta^i_x + Rand\ (0..1)\Delta\theta_x \\ \theta^i_y = \theta^i_y + Rand\ (0..1)\Delta\theta_y \end{cases} \tag{3}$$

2)   Crossover strategy

For the main genetic evolution, LAGAA is the same as standard genetic algorithm in the reproduction process. But there is difference in the crossover processe. If the current solution with minimum energy is described as $h_{\min} = \{\theta^{\min}_1, \theta^{\min}_2, \cdots, \theta^{\min}_n\}$, and each solution selected is defined as $h_i = \{\theta^i_1, \theta^i_2, \cdots, \theta^i_n\}$, which is to do the crossover process with $h_{\min}$, and then the new solution is $h'_i = \{\theta^{i'}_1, \theta^{i'}_2, \cdots, \theta^{i'}_n\}$, where

$$\theta^{i'}_m = \begin{cases} r\theta^i_m + (1-r)\theta^{\min}_m, & \alpha < 0.8 \\ (x^i_m + x^{\min}_m)/2, & \alpha \ge 0.8 \end{cases}, \quad (1 \le m \le n) \tag{4}$$

The novel crossover mode is proposed here to bring the better characters of the parent into the next generation for improving the performance.

3)   Simulated annealing

In the annealing process, a melt, initially at high temperature and disordered, is slowly cooled so that the system at any time is approximately in thermodynamic equilibrium. As cooling proceeds, the system becomes more ordered and approaches a frozen ground state at $T = 0$. If the initial temperature of the system is too low or cooling is done insufficiently slowly the system may become trapped in a local minimum energy state. In this paper, the cooling schedule that provides necessary and sufficient conditions for convergence is

$$T_{i+1} = \sigma T_i$$

When the simulated annealing is applied to the protein folding problem, the potential energy function Eq. (1) of the off-lattice AB model becomes the objective function. For any n-residue molecule, the $n - 2$ angles of bend should be found when the objective function obtains the energy minimum, which is based on the thermodynamic hypothesis formulated by Anfinsen [1]: the natural structure of a protein in its physiological environment is the one of lowest free energy.

3   Experimental Results

LAGAA based on off-lattice AB model has implemented by C++ in windows XP. In our experiments, the parameters are set as following: the initial temperature is 100, the terminal temperature 10^{-7}, the drop rate of temperature 0.96, the loop times under same temperature 100, the mutation rate 0.8, the crossover rate 0.6, and the population scale 100.

Our experiment uses the same Fibonacci sequences in [6]. The purpose of the experiment is to see whether LAGAA algorithm is feasible and efficient for searching the low energy conformations in the off-lattice AB model. Table 1 lists the lowest energies obtained about the off-lattice AB model for all Fibonacci sequences with $13 \le N \le 55$, along with the values obtained by different methods for comparison. For $N < 13$, our energies are identical extremely to those of Frank H.

Stillinger [6] and our algorithm can find the ground states. Here E_{min} is the minimum energy obtained by Stillinger, while E_{PERM} is the lowest energy obtained by the pruned-enriched-Rosenbluth method (PERM) presented in Ref. [7]. E_{LAGAA} is the optimal energy obtained by LAGAA.

Table 1. The minimal energy of test sequences

N	SEQUENCE	E_{min}	E_{PERM}	E_{LAGAA}
13	ABBABBABABBAB	-3.2235	-3.2167	-3.2940
21	BABABBABABBABBABABBAB	-5.2881	-5.7501	-6.1896
34	ABBABBABABBABBABABBABABBABBABABBAB	-8.9749	-9.2195	-10.7068
55	BABABBABABBABBABABBABABBABBABABBAB BABABBABABBABBABABBAB	-14.4089	-14.9050	-18.4615

Table 1 shows that our results for length 13, 21, 34 are slightly better than the results obtained by other two methods, while for length 54, our result of the lowest energy is obviously improved. That is, LAGAA is the best among the given three methods for all the four sequences.

Table 2. Potential energies and radians of angles θ_i at the global minima for test sequences

N	E_{LAGAA}	θ_i / π
13	-3.2940	0.47479, 0.47741, -0.27642, 0.47093, 0.48360, -0.46668, 0.62175, -0.46118, 0.48274, 0.47198, -0.33155
21	-6.1896	0.16358, -0.61686, 0.14949, -0.11481, 0.06093, 0.52453, -0.61960, 0.16149, -0.13183, -0.09260, -0.57751, -0.29349, 0.02896, 0.52495, -0.61886, 0.14284, -0.11223, 0.07111, 0.40653
34	-10.7068	-0.09772, 0.33904, 0.56449, 0.09170, 0.14540, -0.16349, 0.62107, -0.52645, -0.04390, 0.30323, 0.57495, 0.09415, 0.13507, -0.50433, 0.62122, -0.19441, -0.12072, -0.33411, -0.55866, 0.61784, -0.14333, 0.10918, 0.10479, 0.57515, 0.29846, -0.03399, -0.52983, 0.61925, -0.13770, 0.16489, -0.30205, -0.57564
55	-18.4615	-0.13178, 0.61671, -0.50615, 0.06790, -0.02486, -0.14901, 0.61880, -0.51957, 0.17089, 0.09566, 0.54638, 0.35748, -0.11441, -0.15688, 0.62172, -0.51987, 0.07735, -0.02729, -0.15179, 0.61996, -0.54520, -0.36638, 0.05201, -0.59061, -0.03153, 0.00999, -0.13696, 0.62001, -0.52548, 0.14548, -0.04684, -0.35851, 0.47495, 0.47441, -0.11061, 0.61705, -0.16891, 0.14216, -0.12201, -0.14319, 0.61801, -0.52185, 0.03918, 0.05469, -0.35503, 0.47158, 0.48566, -0.48229, 0.62187, -0.20130, -0.05457, -0.30557, 0.53524

Table 2 is the computed results of the potential energies and radians of angles θ_i at the global minima for test sequences. Fig. 2 depicts the lowest energy conformations in two-dimensions obtained by LAGAA, corresponding to the energies and radians shown in Table 2, where the conformation of $n = 13$ has the single hydrophobic core, which is analogous to the real protein structure. But in other three conformations, the hydrophobic monomers form the clusters of particles. This is that hydrophobic residues are always flanked by hydrophilic residues along the chain. This shows that off-lattice AB model reflects the native characters of the real proteins in two-dimensions but it still is not perfect.

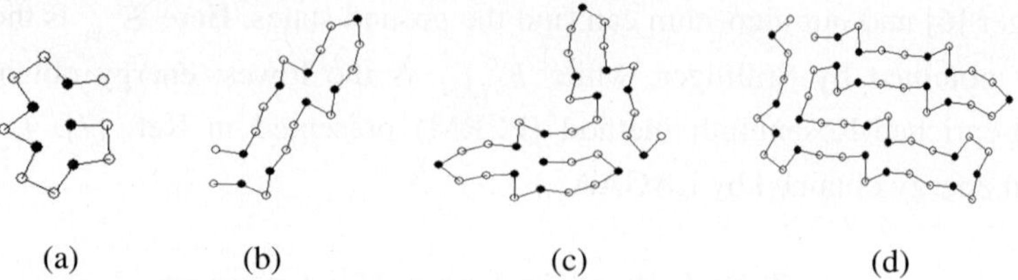

(a)   (b)   (c)   (d)

Fig. 2. The lowest energy conformations for the four Fibonacci sequences obtained by LAGAA algorithm. (a) $n=13$; (b) $n=21$; (c) $n=34$; (d) $n=55$. The black dots represent hydrophobic A monomers, white circles represent hydrophilic B monomers.

4 Conclusion

This paper proposes a hybrid genetic-annealing algorithm with novel crossover operator based on an off-lattice AB model that is used to locate the native structure of protein by the given potential function. To insure the solution quality, local adjustment mechanism is also added in the algorithm. Consequently, the algorithm appears to be superior performance and higher efficiency. As a future work, we will extend the proposed algorithm to three-dimensional off-lattice AB model to predict protein folding problem.

Acknowledgements

This work was supported in part by the Scientific Research Foundation for the Returned Overseas Chinese Scholars, State Education Ministry, and the Project (No.2004D006) from Hubei Provincial Department of Education, P. R. China.

References

1. C. B. Anfinsen. Principles that govern the folding of protein chains. Science, 181, 223-227, 1973.
2. A. Irback, C. Peterson, F. Potthast. Identification of amino acid sequences with good folding properties in an off-lattice model. 55: 860-867, 1997.
3. R. Kong, T. Dandekar. Improving genetic algorithms for protein folding simulations by systematic crossover. BioSystems, 50:17-25,1999.
4. A. Torcini, R. Livi, A. Politi. A dynamical approach to protein folding. J. Biol. Phys., 27: 181-186, 2001.
5. F. H. Stillinger, T.H. Head-Gordon, C. Hirshfel. Toy model for protein folding. Physical review, E48:1469-1477, 1993.
6. F. H. Stillinger. Collective aspects of protein folding illustrated by a toy model. Physical Review, E52:2872-2877, 1995.
7. H. P. Hsu, V. Mehra, W. Nadler, and P. Grassberger, J. Chem. Growth algorithms for lattice heteropolymers at low temperatures. Phys. 118, 2003.

Evolution of Characteristic Tree Structured Patterns from Semistructured Documents

Katsushi Inata, Tetsuhiro Miyahara, Hiroaki Ueda, and Kenichi Takahashi

Faculty of Information Sciences,
Hiroshima City University, Hiroshima 731-3194, Japan
{miyahara, ueda, takahasi}@its.hiroshima-cu.ac.jp

Abstract. Due to the rapid growth of Internet usage, semistructured documents such as XML/HTML files have been rapidly increasing. Genetic Programming is widely used as a method for evolving solutions from structured data and is shown to be useful for evolving highly structured knowledge. We apply genetic programming to the evolution of characteristic tree structured patterns from semistructured documents.

1   Introduction

We consider evolution of characteristic tree structured patterns from semistructured documents by using tag tree patterns [4], which are tree structured patterns and considered to be extended first order terms.

Due to the rapid growth of Internet usage, semistructured documents such as XML/HTML files have been rapidly increasing. Finding characteristic patterns from given data is a basic task in data mining and machine learning. But it is difficult to apply conventional methods for relational databases to finding characteristic tree structured patterns from semistructured documents, since semistructured documents such as Web documents have no rigid structure which relational databases have. So a new method for evolving characteristic patterns from semistructured documents is needed.

On the other hand, Genetic Programming (GP) [1,3] is Evolutionary Computation [2,7] and widely used as a search method for evolving solutions from structured data. GP is shown to be useful for evolving highly structured knowledge [5,8]. Then, in this paper, we report a new application of GP to the evolution of characteristic tree structured patterns from semistructured documents.

Many semistructured documents such as XML/HTML files are represented by rooted trees which have ordered children, and tags or text data as edge labels, and no vertex labels. To formulate a schema on such tree structured data, we use a tag tree pattern as a rooted tree pattern with ordered children and structured variables. A structured variable in a tag tree pattern can be substituted by an arbitrary tree.

This paper is organized as follows. In Section 2, we introduce tag tree patterns as tree structured patterns. In Section 3, we define our machine learning problem of evolving characteristic tag tree patterns and report an application of GP to the problem. In Section 4, we report experimental results on XML documents.

A. Sattar and B.H. Kang (Eds.): AI 2006, LNAI 4304, pp. 1201–1207, 2006.
© Springer-Verlag Berlin Heidelberg 2006

2   Tag Tree Patterns as Tree Structured Patterns

In this section, we briefly review tag tree patterns [4] which are used for representing characteristic tree structured patterns.

Let $T = (V_T, E_T)$ be a rooted tree with ordered children (or simply a **tree**) which has a set V_T of vertices and a set E_T of edges. Let E_g and H_g be a partition of E_T, i.e., $E_g \cup H_g = E_T$ and $E_g \cap H_g = \emptyset$. And let $V_g = V_T$. A triplet $g = (V_g, E_g, H_g)$ is called a **term tree**, and elements in V_g, E_g and H_g are called a *vertex*, an *edge* and a *variable*, respectively. A variable in a term tree can be substituted by an arbitrary tree. We assume that every edge and variable of a term tree is labeled with some words from specified languages. Λ and X denote a set of edge labels and a set of variable labels, respectively, where $\Lambda \cap X = \phi$. By regarding a variable as a special type of an edge, we can introduce the parent-child and the sibling relations in the vertices of a term tree. We use a notation $[v, v']$ to represent a variable $\{v, v'\} \in H_g$ such that v is the parent of v'. We assume that all variables in H_g have mutually distinct variable labels in X. A term tree with no variable is considered to be a tree.

A *substitution* θ on a term tree t is an operation which identifies the vertices of a variable x_i with the vertices of a substituted tree g_i, and replaces the variables x_i with the trees g_i, simultaneously. We assume that the parent of a variable is identified with the root of a substituted tree, and the child of a variable is identified with a leaf of a substituted tree. By $t\theta$ we denote the term tree which is obtained by applying a substitution θ to a term tree t.

For example, Fig. 1 shows a term tree t and trees t_1, t_2. In figures, a variable is represented by a box with lines to its parent and child vertices. The label inside a box is the variable label of a variable. The term tree t is defined as follows. $V_t = \{u_1, u_2, u_3, u_4, u_5, u_6\}$, $E_t = \{\{u_1, u_5\}, \{u_2, u_3\}, \{u_2, u_4\}\}$, $H_t = \{h_1, h_2\}$ with variables $h_1 = [u_1, u_2]$, $h_2 = [u_5, u_6]$, $\Lambda = \{a, b\}$, $X = \{x, y\}$. Consider the substitution θ which identifies the vertices u_1, u_2, u_5, u_6 with the vertices v_1, v_2, v_3, v_4 respectively and replaces the variables h_1 and h_2 with the trees t_1 and t_2 respectively. The term tree $t\theta$ is obtained by applying the substitution θ to the term tree t.

There are two kinds of words: *tags* and *keywords*. For example, in Fig. 2, "⟨Fruits⟩" and "⟨Name⟩" are tags, and "melon" and "watermelon" are keywords. A **tag tree pattern** is a term tree such that each edge label on it is any of a tag, a keyword, and a special symbol "?", which is a wildcard for any edge label. A tag tree pattern with no variable is called a *ground tag tree pattern*. Consider an edge e with an edge label L of a tag tree pattern and an edge e' with an edge label L' of a tree. We say that the edge e *matches* the edge e' if (1) L and L' are the same tag, (2) L and L' are keywords and L is a substring of L', or (3) L is "?" and L' is any tag or keyword. A ground tag tree pattern π *matches* a tree T if π and T have the same tree structure and every edge of π matches its corresponding edge of T. A tag tree pattern π **matches** a tree T if there exists a substitution θ such that $\pi\theta$ is a ground tag tree pattern and $\pi\theta$ matches T. For example, in Fig. 2, the tag tree pattern π matches the tree t_1 but does not match the trees t_2 and t_3.

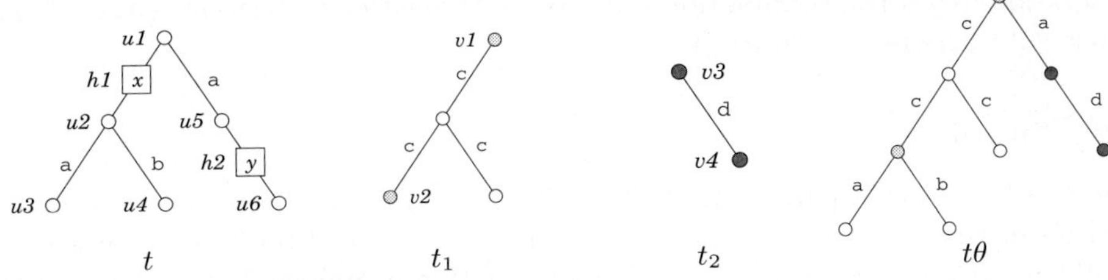

Fig. 1. A term tree t, trees t_1, t_2, and a term tree $t\theta$ obtained by a substitution θ

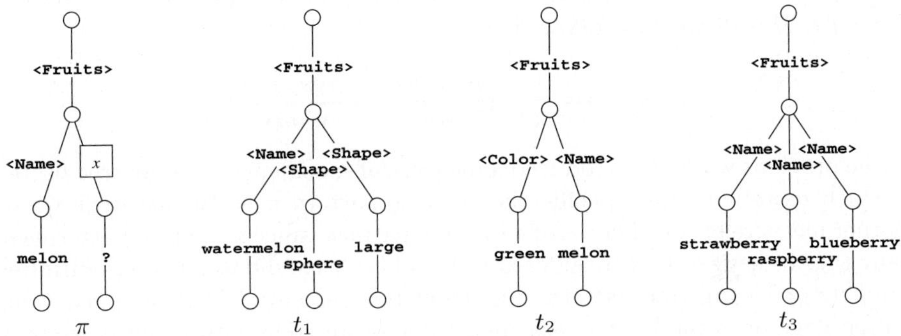

Fig. 2. A tag tree pattern π and trees t_1, t_2, t_3

3   Evolution of Characteristic Tree Structured Patterns Using Genetic Programming

3.1   Our Machine Learning Problem

We consider applying GP to the following machine learning problem.

Problem of Finding Characteristic Tag Tree Patterns
Inputs: A finite set of positive and negative tree structured data (or trees) $\mathcal{D} = \{T_1, T_2, \cdots, T_m\}$.
Problem: Find a tag tree pattern π such that π matches all positive tree examples in $\mathcal{D}$ and π matches no negative tree examples in $\mathcal{D}$. Such a tag tree pattern π is called a **characteristic tag tree pattern** w.r.t $\mathcal{D}$.

We apply GP method to the Problem of Finding Characteristic Tag Tree Patterns as follows. (1) Determine a finite set of tags Tag and a finite set of keywords KW, which are used in tag tree patterns, based on the set $\mathcal{D}$. (2) Generate the initial population of tag tree patterns in a random way by using Tag and KW. (3) Evaluate the fitness $fitness_{\mathcal{D}}(\pi)$ for each tag tree pattern π in the population. (4) Select individuals with probability in proportion to fitness. (5) Perform genetic operations such as crossover, inversion, mutation and reproduction on selected individuals, and generates individuals of the next generation. (6) If the termination criterion is fulfilled, then terminate the whole process. Otherwise return to (3).

3.2   Fitness

Let π be a tag tree pattern and $\mathcal{D} = \{T_1, T_2, \cdots, T_m\}$ a finite set of positive and negative tree examples. num_P denotes the number of positive examples T_i in $\mathcal{D}$, and num_N denotes the number of negative examples T_i in $\mathcal{D}$. $hits_P(\pi)$ denotes the number of positive examples T_i in $\mathcal{D}$ which π matches, and $hits_N(\pi)$ denotes the number of negative examples T_i in $\mathcal{D}$ which π does not match. The **frequency**, which is the ratio of explained examples over all examples, $freq_{\mathcal{D}}(\pi)$ of π w.r.t. $\mathcal{D}$ is defined as follows [8]:

$$freq_{\mathcal{D}}(\pi) = \frac{1}{2} \times \left(\frac{hits_P(\pi)}{num_P} + \frac{hits_N(\pi)}{num_N} \right).$$

The previous work [8] introduced a measure of specificness of a tag tree pattern π, which represents how specific the tag tree pattern π is. In this work we use another measure of specificness of a tag tree pattern, since a tag tree pattern with many tags or keywords is considered to be a better candidate. The **specificness** $specific(\pi)$ of a tag tree pattern π is defined as the sum of the assigned values of each edge or variable. The assigned value of an edge with a tag or keyword is 1.0, the assigned value of an edge with "?" is 0.2, and the assigned value of a variable is 0.1.

A desirable GP process first finds a tag tree pattern π with $freq_{\mathcal{D}}(\pi) = 1.0$ and then modifies π and finds a more specific tag tree pattern than π. So we define the **fitness** $fitness_{\mathcal{D}}(\pi)$ of π w.r.t. $\mathcal{D}$ as follows:

$$fitness_{\mathcal{D}}(\pi) = \begin{cases} freq_{\mathcal{D}}(\pi) & (freq_{\mathcal{D}}(\pi) < 1.0) \\ freq_{\mathcal{D}}(\pi) + \max\{1.0 - 1.0/specific(\pi), 0\} & (freq_{\mathcal{D}}(\pi) = 1.0) \end{cases}$$

For example, consider the tag tree pattern π and the set of trees $\mathcal{D} = \{t_1, t_2, t_3\}$ in Fig. 2. $specific(\pi)$ is 3.3. In the setting that t_1 and t_2 are positive examples, and t_3 is a negative example, we have $freq_{\mathcal{D}}(\pi) = 0.75$ then we have $fitness_{\mathcal{D}}(\pi) = 0.75$. In the setting that t_1 is a positive example, and t_2 and t_3 are negative examples, we have $freq_{\mathcal{D}}(\pi) = 1.0$ then we have $fitness_{\mathcal{D}}(\pi) = 1.7$.

3.3   Generation of Tag Tree Patterns

In order to avoid generating inappropriate tag tree patterns, we use the following generation algorithm [8]. (1) Extract a document structure consisting of elements such as tags or keywords from positive tree examples in $\mathcal{D}$. (2) Generate only tag tree patterns conforming with the extracted document structure in the processes of generating initial individuals and applying genetic operators.

We probabilistically apply the following 7 genetic operators: (1) crossover (Fig. 3), (2) inversion, (3) change-subtree, (4) add-subtree, (5) del-subtree, (6) replace-question-mark (replace a question mark with a tag or a keyword in an

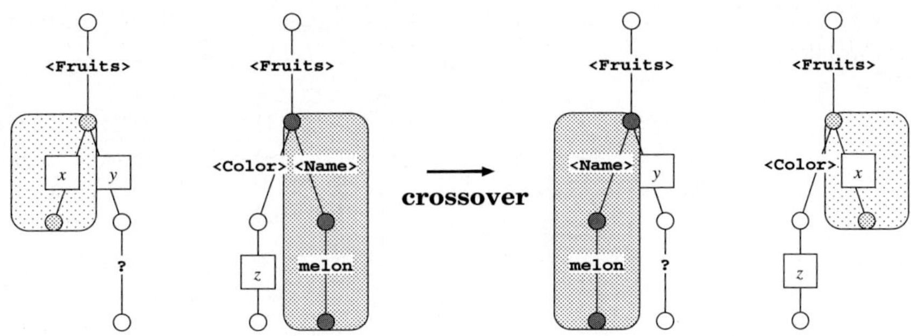

Fig. 3. A genetic operator crossover on tag tree patterns

Table 1. Parameters of GP

population size	50	probability of mutation		0.1
probability of reproduction	0.1	selection		elite tournament
probability of crossover	0.7			with size 5
probability of inversion	0.1	maximum number of generations		100

individual), (7) replace-tag-or-keyword (replace a tag or a keyword with a question mark in an individual). If we regard a variable as a special type of an edge, the operators (1)–(5) are considered usual tree-based GP operators.

4   Experiments

We have implemented our GP-based method for evolving characteristic tag tree patterns from semistructured documents. The implementation is by GCL (GNU Common Lisp) Version 2.2.2 on a SUN workstation Blade 2000 with clock 900MHz CPU. Our program uses the algorithm [6] for determining whether a tag tree pattern matches a tree or not. The first sample file (garment) is converted from a sample XML file about garment sales data. The file consists of 24 positive tree examples and 104 negative tree examples, and the average number of vertices of trees in the file is 23. The second sample file (dblp) is an extraction from the DBLP bibliographic database (http://dblp.uni-trier.de/xml/dblp.xml). Table 1 shows the parameters of our GP setting. The maximum number of generations in a GP run is 100. We have an experiment of 5 GP runs for each sample file.

First we report the experimental result on the first sample file (garment). In all 5 runs, our method finds tag tree patterns with frequency 1.0. Table 2(A) shows the data of the tag tree patterns with frequency 1.0 which our method generates first in each run. "best" ("worst") means the value of the tag tree pattern with best (worst) fitness in the set of tag tree patterns with frequency 1.0 which our method generates first in each run. "average" means the average value of the tag tree patterns with frequency 1.0 which our method generates first in each run. "generation of obtaining" means the generation when tag tree

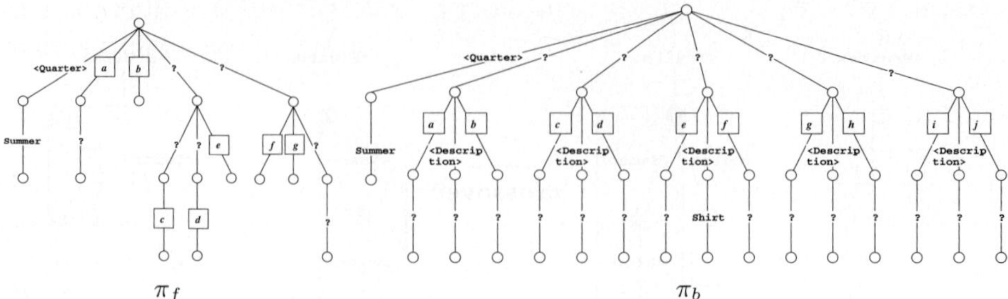

Fig. 4. A tag tree pattern π_f with frequency 1.0 obtained first and a tag tree pattern π_b with best fitness

Table 2. The data of (A) tag tree patterns with frequency 1.0 obtained first and (B) tag tree patterns with best fitness

	Table(A)			Table(B)		
	best	average	worst	best	average	worst
frequency	1.0	1.00	1.0	1.0	1.00	1.0
specificness	4.1	2.40	1.7	12.8	9.94	8.5
fitness	1.756	1.5414	1.412	1.922	1.8968	1.882
generation of obtaining	·2	3.6	2	45	61.0	40
time (sec) of obtaining	871	1501	924	26704	25194	17813

patterns with frequency 1.0 are generated first in each run. "time of obtaining" means the run time until tag tree patterns with frequency 1.0 are generated first in each run. The Table 2(A) shows that tag tree patterns with frequency 1.0 are generated in a relatively early generation. Fig. 4(left) shows the tag tree pattern π_f with best fitness in the set of tag tree patterns with frequency 1.0 which our method generates first in each run.

Table 2(B) shows the data of the tag tree patterns with best fitness in each run. "best" ("worst") means the value of the tag tree pattern with best (worst) fitness in the set of the tag tree patterns with best fitness in each run. "average" means the value of the tag tree patterns with best fitness in each run. "generation of obtaining" means the generation when tag tree patterns with best fitness are generated first in each run. "time of obtaining" means the run time until tag tree patterns with best fitness are generated first in each run. Table 2 shows that the specificness of the tag tree patterns with best fitness is greater than that of the tag tree patterns with frequency 1.0. This is because the fitness of many individuals is determined by specificness and such individuals influence the evolution of individuals after some generation. But the generation of obtaining tag tree patterns with best fitness is in middle stage and the run time of obtaining tag tree patterns with best fitness is very large.

Fig. 4(right) shows the tag tree pattern π_b with best fitness in the set of tag tree patterns with best fitness in each run. The tag tree pattern π_b shows structural feature and has larger size and more tags or keywords than tag tree patterns with frequency 1.0 obtained first. In order to evaluates the obtained tag tree pattern, we manually generates a target tag tree pattern which is characteristic of the sample file (garment). The frequency, specificness and fitness of this target tag tree pattern is 1.0, 13.6, 1.926. These values are almost equal to those of the best tag tree pattern π_b in Fig. 4.

In the second experimental results on the sample file (dblp), our method finds tag tree patterns with frequency 1.0 and the tag tree patterns with best fitness have structural feature characteristic to the sample file. These two experiments show that our GP-based method can successfully evolve characteristic tag tree patterns from semistructured documents.

5   Conclusions

In this paper, we have reported a new application of Genetic Programming to the evolution of characteristic tree structured patterns from semistructured documents. This work is partly supported by Grant-in-Aid for Scientific Research (C) No.16500084 from Japan Society for the Promotion of Science.

References

1. W. Banzhaf, P. Nordin, R.E. Keller, and F.D. Francone. *Genetic Programming: An Introduction : On the Automatic Evolution of Computer Programs and Its Applications.* Morgan Kaufmann, 1998.
2. X. Yao (ed.). *Evolutionary Computation: Theory and Applications.* World Scientific, 1999.
3. J. R. Koza. *Genetic Programming: On the Programming of Computers by Means of Natural Selection.* MIT Press, 1992.
4. T. Miyahara, Y. Suzuki, T. Shoudai, T. Uchida, K. Takahashi, and H. Ueda. Discovery of frequent tag tree patterns in semistructured web documents. *Proc. PAKDD-2002, Springer-Verlag, LNAI 2336*, pages 341–355, 2002.
5. A. Niimi and E. Tazaki. Genetic programming combined with association rule algorithm for decision tree construction. *Proc. 4th International Conference on Knowledge-Based Intelligent Engineering Systems & Allied Technologies (KES 2000)*, pages 746–749, 2000.
6. Y. Suzuki, R. Akanuma, T. Shoudai, T. Miyahara, and T. Uchida. Polynomial time inductive inference of ordered tree patterns with internal structured variables from positive data. *Proc. COLT-2002, Springer-Verlag, LNAI 2375*, pages 169–184, 2002.
7. K.C. Tan, M.H. Lim, X. Yao, and L. (eds.) Wang. *Recent Advances in Simulated Evolution and Learning.* World Scientific, 2004.
8. A. Watanabe, T. Miyahara, K. Takahashi, and H. Ueda. Application of genetic programming to discovery of characteristic tree structured patterns (in Japanese). *IEICE Technical Report*, 101(502):41–48, 2001.

Unsupervised Measurement of Translation Quality Using Multi-engine, Bi-directional Translation

Menno van Zaanen and Simon Zwarts

Division of Information and Communication Sciences
Department of Computing
Macquarie University
2109 Sydney, NSW, Australia
{menno, szwarts}@ics.mq.edu.au

Abstract. Lay people discussing machine translation systems often perform a round trip translation, that is translating a text into a foreign language and back, to measure the quality of the system. The idea behind this is that a good system will produce a round trip translation that is exactly (or perhaps very close to) the original text. However, people working with machine translation systems intuitively know that round trip translation is not a good evaluation method. In this article we will show empirically that round trip translation cannot be used as a measure of the quality of a machine translation system. Even when using translations of multiple machine translation systems into account, to reduce the impact of errors of a single system, round trip translation cannot be used to measure machine translation quality.

1   Introduction

Research in MT system evaluation concentrates on finding a fair way to measure the quality of MT systems. If a system is perceived to be better than another system by a user, an evaluation metric should also reflect this. However, evaluating MT systems is non-trivial. For instance, when several human translators translate a text from one language to another, there is a high likelihood that the resulting translations differ.

Automatic evaluation (where the computer compares the translation against one or more human translations) has advantages over human evaluation (where people directly analyse the outcome of MT systems). Not only is automatic evaluation much cheaper and less expertise-intensive (human translations only need to be produced once and can be reused afterwards), it is also faster.

There are several metrics that allow a computer to compare text translated by an MT system against pre-made human translations of the same text. Examples of such metrics are BLEU [1], F-Measure [2], and Meteor [3][1].

---

[1] In this article, we used the implementations of the BLEU score available from
http://www.ics.mq.edu.au/~szwarts/Downloads.php.

A. Sattar and B.H. Kang (Eds.): AI 2006, LNAI 4304, pp. 1208–1214, 2006.

© Springer-Verlag Berlin Heidelberg 2006

In this article, we investigate how far fully unsupervised, automatic MT evaluation can be taken. Unsupervised evaluation means that we evaluate using no human-made translation of any text. This requirement is much stricter than that of existing measures, that need one or more human translations of the text under investigation.

This article is structured as follows. We introduce new, unsupervised evaluation methods in section 2. Next we measure the usefulness of these approaches in section 3. Finally, we will mention possible future work in this direction and draw a conclusion.

2   Unsupervised MT System Evaluation

There are several ways of evaluating MT systems in an unsupervised way. Here we will discuss three different approaches. The first translates a text from the source to the target language and tries to evaluate the system based on that translation. The other two approaches rely on round-trip translations, where the source text is translated into the target language and then translated back to the source language. The evaluation takes place on the source language. All of these approaches will be treated in more detail.

2.1   One Way Translation

In the one-way quality assessment approach, we evaluate based on the outputs of different MT systems in the target language. The idea is that we can use the different outputs as reference translations with respect to the system that we are trying to measure. This means that we do not require any pre-translated text at all. In this approach, we use $c + 1$ different MT systems, which leads to one text under investigation and c reference texts.

The translation flow is similar to normal automatic translation (this means that for this new approach the traditional automatic evaluation tools can be used): the source text that will be used to evaluate is translated into the target language using both the MT system to be evaluated and several "reference" MT systems. The outputs of the reference MT systems are used as reference translations.

Evaluation using "round trip translation" (RTT) is very popular among lay people. The idea here is that a text in the source language is translated into the target language and that text is then translated back into the source language again. The evaluation is performed by comparing the original text with the back-translated text, all in the source language.

The advantage of RTT compared to other unsupervised evaluation methods is that we can use the original text as a reference text, which we know is a correct "translation". This is different from the other approaches, where the evaluation is performed on texts translated by other (unreliable[2]) MT systems.

The assumption behind RTT is that the closer the original sentences are to sentences translated to another language and back, the better the MT system

---

[2] The systems are unreliable in that their quality has not been measured on the text that is being evaluated.

is. The idea is that measuring the quality of the back-translation against the original text will give a measure of the forward translation.

There are, however, reasons to believe that RTT does not work. MT systems often make the same mistakes in the forward translation as in the back translation. For example, if the MT system completely failed in putting the words in the right order in the target language and again failed to reorder it during the back translation, the words in the round trip translation are still in the right order, even though the translation in the target language is incorrect.

RTT also gives incorrect results when the two translation steps are asymmetrical. For example, the forward translation may be perfect, but only the back translation is incorrect. This leads to a low RTT evaluation score, even though the forward translation is correct.

The effectiveness of RTT has been investigated earlier [4] and the results presented there are not promising. It seems that RTT evaluation has little correlation to evaluation based on human translations. Here we investigate this further. We aim to reduce the impact of errors of the back translation step by using more MT systems. This is done in the multi-engine RTT evaluation as described in the next section.

2.2   Multi-engine Round Trip Translation

To reduce the impact of the errors made by the MT system in RTT, we also investigated approaches using multiple MT systems. We identified two of these approaches, which we call *single RTT* and *multi RTT* (in contrast to the "standard" RTT, which we will call *default RTT* here).

The dashed box in Figure 1 illustrates single RTT evaluation. The translated text is back translated using several MT systems. S_{RTT} is now evaluated with the output of the reference systems ($S_{reference[0][x]}$) as reference translations.

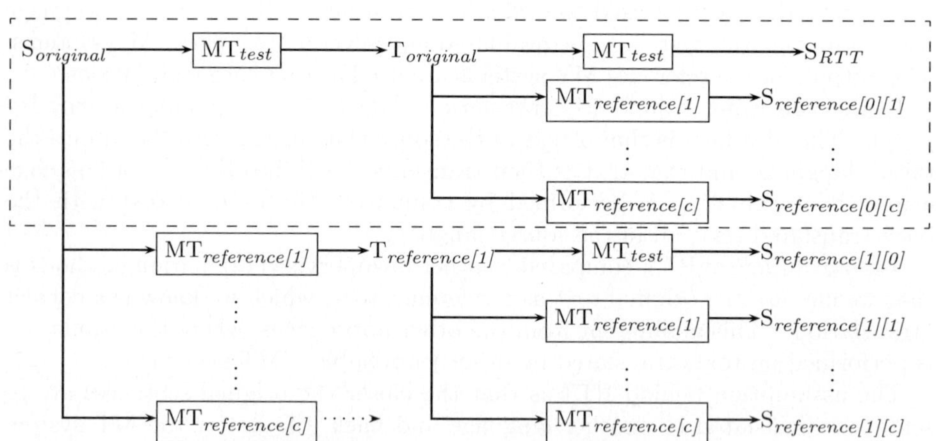

Fig. 1. Translation flow, round trip translation

Multi RTT (see complete Figure 1) is quite similar to single RTT. The idea is to generate even more reference texts by performing forward translations as well as back translations using multiple systems. The added amount of data allows us to average the translations to reduce the impact of the translation mistakes of the individual systems (in both the forward and back translation steps). Using $c + 1$ translation engines (including the one that is being tested) we generate $(c + 1)^2 - 1$ reference texts and compare these against one candidate text.

3 Experiments

We are interested in how effective the unsupervised evaluation methods (one-way translation, default RTT, single RTT, and multi RTT) are at measuring the quality of a single MT system without having to rely on human translated texts. To measure this, we have conducted several experiments that measure the performance of the unsupervised methods and we have compared these against a control experiment on the same texts. The control experiment is simply comparing the output of the MT system against human translated reference texts.

A method is successful if the result of the unsupervised evaluation method correlates highly with the results of the control experiment. We assume that the results of the control experiment have a high correlation with human translations.

3.1 Setting

Several parameters need to be set before the actual experiments can take place. First of all, MT systems behave differently on different types of texts. To reduce this impact, we have conducted the experiments on texts from different genres.

The choice of languages may also influence translation quality. We have used different language pairs to restrict the influence of language choice. Texts are translated from English to German or from French to English. Each of the texts consist of 100 sentences.

euEG, euFE English to German and French to English translations of the Europarl corpus [5]. These, quite formal texts, are extracts from sessions of the European Parliament over a couple of years;

maFE parts of a tourism website of Marseilles available in French and English translations are available. This text is less formal than the Europarl corpus;

miFE parts from the novel *20.000 Leagues Under the Sea* by Jules Verne. The original French version and an English translation are used.

3.2 Empirical Results

To find out whether unsupervised automatic MT evaluation is effective, we correlate the output of the different unsupervised methods with the standard evaluation method. Finding a high correlation between the unsupervised evaluation and the standard evaluation indicates that the unsupervised method provides similar results and is as usable.

We compute the BLEU scores of the different MT systems by comparing them against the human translated reference texts, which serves as the control data. We use the BLEU metric with n-gram sizes up to 4. The results can be found in Table 1.

Table 1. BLEU scores for the different MT-systems and corpora

MT system	euEG	euFE	maFE	miFE
system 1	0.110304	0.186572	0.186828	0.140024
system 2	0.106163	0.178278	0.230331	0.180803
system 3	0.105912	0.181388	0.205101	0.139894
system 4	0.100855	0.141064	0.147441	0.145592
system 5	0.095173	0.188239	0.179050	0.160152

Table 2. Formulas used to compute the measures

A.1 $\text{BLEU}(\text{T}_{original}, \text{T}_{reference[1]}, \ldots, \text{T}_{reference[c]})$

A.2 $\frac{1}{c} \sum_{i=1}^{c} \text{BLEU}(\text{T}_{original}, \text{T}_{reference[i]})$

B $\quad \text{BLEU}(\text{S}_{RTT}, \text{S}_{original})$

C.1 $\frac{1}{c} \sum_{i=1}^{c} \text{BLEU}(\text{S}_{reference[0][i]}, \text{S}_{original})$

C.2 $\text{BLEU}(\text{S}_{RTT}, \text{S}_{reference[0][1]}, \ldots, \text{S}_{reference[0][c]})$

C.3 $\frac{1}{c} \sum_{i=1}^{c} \text{BLEU}(\text{S}_{RTT}, \text{S}_{reference[0][i]})$

D $\quad \frac{1}{(c+1)^2-1} \sum_{i=0,j=0}^{c,c} \text{BLEU}(\text{S}_{RTT}, \text{S}_{reference[i][j]})$, where $i \neq 0 \vee j \neq 0$

As is seen in Table 1 there is no one MT system that consistently performs best (on different language pairs and genres). The unsupervised evaluation metrics should reflect these differences as well.

The first experiment of unsupervised evaluation measures the effectiveness of one-way translation as described in section 2.1. We calculate the BLEU score of the translation of the original text ($\text{T}_{original}$) with respect to the reference translations generated by the five MT systems ($\text{T}_{reference[x]}$ with $x = 1 \ldots 4$). This is computed using formula A.1 of Table 2.[3]

In experiment A.2 we compute the BLEU score of the translation with respect to each of the single reference translations, which results in four BLEU scores. BLEU expects all reference texts to be correct, which is certainly not the case here. Hence we propose averaging individual BLEU scores here.

Experiment B evaluates the default round trip translation. Only one MT system is used in the evaluation; this is essentially a reproduction of Somers's experiments [4]. Although he does not publish exact figures, we suspect that he used n-grams with $n=1$ only, whereas we use $n=4$.

In experiment C we measure the effectiveness of the single RTT approach (dashed box in Figure 1). We first compute the average BLEU score of the reference back

---

[3] Note that the names of the texts in the formulas refer to th names in the figure 1.

Table 3. Correlations of all experiments (formulas given in Table 2)

Exp.	Corr.	Exp.	Corr.	Exp.	Corr.	Exp.	Corr.
A.1	0.295	B	0.193	C.1	-0.063	D	0.341
A.2	0.441			C.2	0.532		
				C.3	0.401		
one-way		*default RTT*		*single RTT*		*multi RTT*	

translations with respect to the original text (C.1). We also compute the BLEU score of the round trip translation of the test system and use the round trip translations of the reference systems as reference texts (C.2). C.3 (like A.2) computes the average BLEU score using each of the reference texts separately.

Experiment D is similar to experiment C, only in this case multi RTT is evaluated (Figure 1). This means that more data points are available, because the forward translation step is also performed by several MT systems (in contrast to single RTT, where only one MT system is used in the forward translation). Due to space restrictions, we only show the results of the average BLEU score here. The other results are similar to those in experiment C.

Looking at the results presented in Table 3, we see that none of the experiments show a correlation that is high enough to use for a final MT quality assessment. The highest correlation is found in experiment C.2, where the round trip translation of the tested system is compared against all other round trip translations based on the forward translation of the tested system. Even the correlation of experiment D, which uses more data points (and as such may have given us more precise results) is lower.

4   Conclusion

In this article, we investigated unsupervised evaluation methods in the context of machine translation. Unsupervised evaluation is an automatic evaluation that removes the requirement of human translated reference texts by evaluating on texts that are translated from one language to another and back.

Unfortunately, the experiments show that unsupervised evaluation (including round trip translation which is often used by lay people) is unsuitable for the selection of MT systems. The research described here covers most possible combinations of one way and round trip translation using one or more MT systems. The highest correlation we found was 0.532. We think that this is not enough to base a clear verdict on. Essentially, RTT should never be used to illustrate MT system quality.

Bibliography

[1] Kishore Papineni, Salim Roukos, Todd Ward, and Wei-Jing Zhu. Bleu: a method for automatic evaluation of machine translation. In *Proceedings of the 40th Annual Meeting of the Association for Computational Linguistics (ACL)*, pages 311–318, Jul 2002. URL http://www1.cs.columbia.edu/nlp/sgd/bleu.pdf.

[2] Joseph P. Turian, Luke Shen, and I. Dan Melamed. Evaluation of machine translation and its evaluation. In *Proceedings of MT Summit IX*, 2003.

[3] Santanjeev Banerjee and Alon Lavie. Meteor: An automatic metric for MT Evaluation with improved correlation with human judgments. In *Proceedings of Workshop on Intrinsic and Extrinsic Evaluation Measures for MT and/or Summarization at the 43th Annual Meeting of the Association of Computational Linguistics (ACL-2005)*, June 2005.

[4] Harold Somers. Round-trip translation: What is it good for? In *Proceedings of the Australasian Language Technology Workshop*, 2005.

[5] Philipp Koehn. Europarl: A multilingual corpus for evaluation of machine translation, 2002. URL `http://people.csail.mit.edu/~koehn/publications/europarl.ps`.

Probabilistic, Multi-staged Interpretation of Spoken Utterances

Ingrid Zukerman, Michael Niemann, Sarah George, and Yuval Marom

Faculty of Information Technology, Monash University
Clayton, VICTORIA 3800, Australia
{ingrid, niemann, sarahg, yuvalm}@csse.monash.edu.au

Abstract. We describe *Scusi?*, a multi-stage, spoken language interpretation mechanism designed to be part of a robot-mounted dialogue system. *Scusi?*'s interpretation process maps spoken utterances to conceptual graphs, and the nodes in these graphs to concepts in the world. Maximum posterior probability is used to rank the (partial) interpretations produced at each stage of this process.

1   Introduction

The DORIS project aims to develop a spoken dialogue module for a robotic agent. Eventually, this module will be able to engage in a dialogue with users and plan physical actions (by interfacing with a planner). In this paper, we describe *Scusi?*, the speech interpretation module that is being developed within the DORIS framework. *Scusi?* was designed to achieve the following objectives: (1) provide a variety of possible interpretations of the input, ranked according to merit; (2) change its mind regarding the best interpretation if warranted by new information; and (3) assist the dialogue module in handling partial or faulty interpretations and recovering from them.

To achieve these objectives we use a multi-stage, probabilistic interpretation process based on that introduced in [1]. Each interpretation is an understanding of spoken input in terms of the system's information structures and domain knowledge. Our interpretation process comprises three main stages: (a) speech recognition, (b) parsing, and (c) semantic interpretation (Section 2). Each stage produces multiple candidate options, which are ranked according to their probability of being intended by the speaker. The probability of a candidate depends on the probability of its parents (generated in the previous stage of the interpretation process) and that of its components (Section 4).

Each of the final interpretations and the intermediate *sub-interpretations* are maintained by *Scusi?*, enabling it to return to previously "unpreferred" options, update previous sub-interpretations in light of new information, and possibly change its mind regarding the preferred interpretation. Additionally, by keeping track of the probability of different components of an interpretation, *Scusi?* can determine which parts of an interpretation can be used for further processing, and which should be re-examined in order to get a better interpretation.

2   Multi-stage Processing

Figure 1 illustrates the stages involved in processing spoken input. The first stage activates ViaVoice – an *Automatic Speech Recognizer (ASR)* – to generate candidate

A. Sattar and B.H. Kang (Eds.): AI 2006, LNAI 4304, pp. 1215–1220, 2006.

© Springer-Verlag Berlin Heidelberg 2006

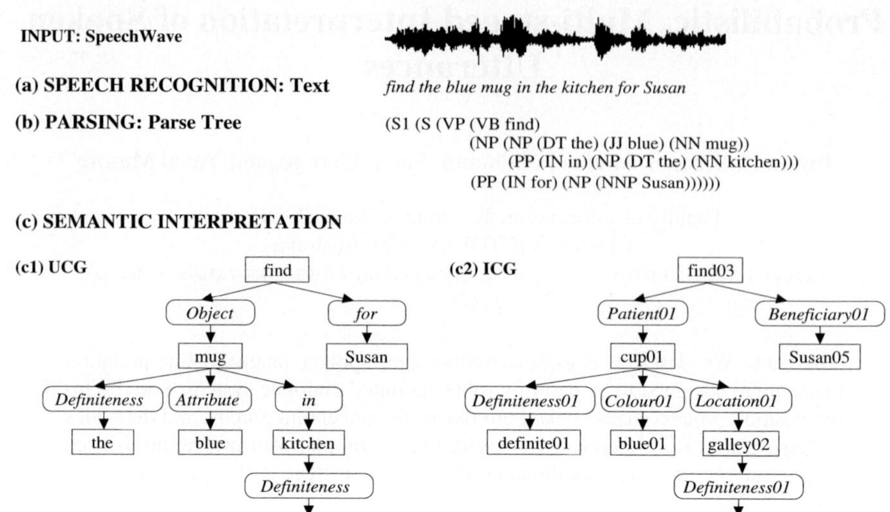

Fig. 1. Structures for the interpretation stages

sequences of words (*Text*) from a *Speech Wave*. Each Text has a score that represents how well its words fit the speech wave. This score is converted into a probability. The word sequences are then parsed using Charniak's probabilistic parser, which generates a set of *Parse Trees*. The last stage uses *Concept Graphs (CGs)* to perform semantic interpretation. Two types of CGs are generated (Section 3). First, the Parse Trees are mapped into a set of *Uninstantiated Concept Graphs (UCGs)*, which represent mainly syntactic information. Next, *Scusi?* proposes candidate *Instantiated Concept Graphs (ICGs)* from selected UCGs. This is done by nominating *Instantiated Concepts* from DORIS's knowledge base as a potential realization for every *Uninstantiated Concept* in a UCG.

2.1 Anytime Processing

The consideration of all possible options for each stage of the interpretation process is computationally intractable. To address this problem, we have adapted the *anytime* algorithm described in [1], which applies a selection-expansion cycle to build a search graph as follows. The selection step nominates a single sub-interpretation (Speech, Text, Parse Tree or UCG) to expand, and the expansion step generates only one child for that sub-interpretation. The selection step then nominates the next sub-interpretation to expand, which may be the one that was just expanded, its new child, or any other sub-interpretation in the search graph.

The nomination of a sub-interpretation is made on the basis of two factors: (1) the level in the search tree, and (2) the probability of its previously generated children. Preference is given to later stages in the search (e.g., UCGs rather than Texts) to encourage the early generation of complete interpretations, and within a stage, to sub-interpretations that have a "proven track record", i.e., that have previously produced high-probability children.

The selection-expansion cycle is repeated until one of the following happens: all the options are fully expanded, a time limit is reached, or the system runs out of memory. The results of the interpretation process are available to the dialogue module continuously. This enables the dialogue module to decide on an action on the basis of the progress of the interpretation process.

2.2   Features of the Interpretation Process

Our interpretation process implements an *iterative optimization method*, which retains all sub-interpretations, and allows even low-ranking sub-interpretations to generate children. This safeguards against getting stuck in local maxima, and supports a flexible and dynamic behaviour, allowing DORIS to improve its understanding and management of the dialogue in various key ways.

- *Reviewing the overall picture:* Each sub-interpretation is stored with its individual probability. As a result, the dialogue module is not restricted to choosing its actions solely on the basis of the high ranking ICGs. It can use other information obtained during the interpretation process to determine whether special attention needs to be paid to the results from other stages in the interpretation process. For instance, if the Text for the current top ICG has a low-probability due to the speech recognizer "mishearing" an unclear speech wave, the dialogue module can initiate a clarification sub-dialogue regarding the original input.
- *Recovery from partial success:* Interpretations that do not match all grammatical sub-categorization requirements or domain expectations have a low probability, but are still retained. This enables the dialogue module to initiate a focused recovery using elements from incomplete or flawed interpretations.
- *Updating probabilities:* New information can be added to concept graphs (e.g., as a result of user feedback or visual input) or to the knowledge base (e.g., after an action performed by a planner in the physical world). Our process allows the probabilities of existing sub-interpretations and final interpretations (ICGs) to be efficiently re-calculated and re-ranked in light of updates.
- *Background exploration:* Provided memory is available, *Scusi?* continues its search in the background, even after the time limit is reached. If a significantly better ICG is found, then a new interpretation can be proposed to the dialogue module.

3   Conceptual Graphs

Conceptual graphs represent entities and the relationships between them [2]. For instance, the CG in Figure 1(c2) indicates that there are two concepts find03 and cup01 that have a Patient01 Relationship.

Uninstantiated Conceptual Graphs.   A UCG represents the syntactic relationships found in a Parse Tree. Most phrases map to a Concept Node representing their headword. Slightly different Parse Trees may yield the same UCG. For instance, the blue Concept Node in Figure 1(c1) could be generated from an Adjectival Phrase instead of a stand-alone adjective (JJ) adjunct to the NP. The mapping from Parse Tree to UCG works the same way for declarative, imperative and interrogative sentences and for single words.

Instantiated Conceptual Graphs and the Knowledge Base. The generation of an ICG requires the selection of an *Instantiated Concept* from the knowledge base for each *Uninstantiated Concept* in a UCG. The knowledge base contains entries for specific real-world objects (e.g., cup03), general objects (e.g., CupClass01), domain actions (e.g., find02), abstract concepts (e.g., blue01) and relationships (e.g., Patient01). To postulate Instantiated Concepts for Uninstantiated ones, each Uninstantiated Concept in a UCG is associated with a list of Instantiated Concepts. Each entry in the list is assigned a probability on the basis of how well it matches the Uninstantiated Concept. To generate an ICG, one Instantiated Concept is iteratively selected from the list of each Uninstantiated Concept in the parent UCG, starting with the most probable candidates.

4   Probabilities of Interpretations

ICGs are ranked according to their posterior probability in light of the given speech wave and the conversational context, which is obtained from concepts recently seen or mentioned. The interpretation stages shown in Figure 1 lead to the following equation for the probability of an ICG.

$$\Pr(ICG|Speech, Context) = \alpha \Pr(ICG|Context) \times$$
$$\sum_{txt,prsTr,ucg} \left\{ \begin{array}{l} \Pr(UCG|ICG) \times \Pr(ParseTr|UCG) \times \\ \Pr(Text|ParseTr) \times \Pr(Speech|Text) \end{array} \right\}$$

where α is a normalizing constant. The summation is required since a sub-interpretation may have multiple parents.

At present, the probabilities required to perform this calculation are not available. Instead of providing probabilities in the required direction, the ASR provides probabilities from Speech Wave to Text (its scores are directly translated to probabilities), and the probabilistic parser from Text to Parse Tree. Hence, we approximate the calculation of the posterior probability of an ICG as follows.

$$\Pr(ICG|Speech, Context) \cong \alpha \Pr(ICG|Context) \times$$
$$\sum_{txt,prsTr,ucg} \left\{ \begin{array}{l} \Pr(ICG|UCG) \times \Pr(UCG|ParseTr) \times \\ \Pr(ParseTr|Text) \times \Pr(Text|Speech) \end{array} \right\}$$

We assume that $\Pr(UCG|ParseTr)=1$ since a UCG is directly built from the Parse Tree. $\Pr(ICG|UCG)$ depends only on the match between each ICG concept and its corresponding UCG concept, given the lexical items associated with the ICG concept. $\Pr(ICG|Context)$ represents whether the concepts in the ICG make sense in the current context, given their salience and the relationships they expect to be in.

5   Related Research

This research extends our previous work [1] in its use of conceptual graphs as a representation formalism and its expansion of the probabilistic interpretation framework to incorporate these knowledge structures.

Conceptual graphs were also used by Sowa and Way [3] and Shankaranarayanan and Cyre [4] for discourse interpretation. However, these researchers did not employ a probabilistic approach, did not retain multiple interpretations, and their usage of conceptual graphs resembled semantic grammars, which take advantage of domain related information early in the interpretation process. Miller *et al.* [5] and He and Young [6] also used semantic grammars for the interpretation of utterances from the ATIS corpus, but they applied a probabilistic approach similar to ours. Instead of using semantic grammars, *Scusi?*'s interpretation process initially uses generic, syntactic tools, and incorporates semantic- and domain-related information only in the final stage.

Our work resembles most that of Gorniak and Roy [7] in its use of a probabilistic parser and its integration of context-based expectations with alternatives obtained from spoken utterances. But Gorniak and Roy do not retain multiple interpretations, and they restrict the search space by training the parser on a domain-relevant corpus and providing tightly constrained domain expectations.

6   Conclusion

We have presented a multi-stage, probabilistic interpretation process that derives the meaning of spoken utterances in light of contextual information. Our interpretation process relies on an approximation of Bayesian propagation for the incremental calculation of the probability of interpretations; an iterative optimization method, which allows the examination of suboptimal sub-interpretations; and a unified knowledge representation formalism that can represent information from various sources.

Additionally, our interpretation process brings to bear semantic and domain knowledge only in the final stage of this process, retains all sub-interpretations and information about them, and can take new information into account. These features enable our system to make some sense of flawed interpretations, change its mind about the best interpretation, and provide sufficient information to a dialogue module for handling partial or flawed interpretations. Preliminary tests indicate that improved results are gained by examining sub-interpretations that are not ranked highly.

Acknowledgments

This research was supported in part by the ARC Centre for Perceptive and Intelligent Machines in Complex Environments. The authors thank Eugene Charniak for his modifications to his probabilistic parser, and Charles Prosser for his assistance in extracting multiple texts from ViaVoice.

References

1. Niemann, M., George, S., Zukerman, I.: Towards a probabilistic, multi-layered spoken language interpretation system. In: Proceedings of the Fourth IJCAI Workshop on Knowledge and Reasoning in Practical Dialogue Systems, Edinburgh, Scotland (2005) 8–15
2. Sowa, J.: Conceptual Structures: Information Processing in Mind and Machine. Addison-Wesley (1984)
3. Sowa, J., Way, E.: Implementing a semantic interpreter using conceptual graphs. IBM Journal of Research and Development **30**(1) (1986) 57–69

4. Shankaranarayanan, S., Cyre, W.: Identification of coreferences with conceptual graphs. In: ICCS'94 – Proceedings of the Second International Conference on Conceptual Structures, College Park, Maryland (1994)
5. Miller, S., Stallard, D., Bobrow, R., Schwartz, R.: A fully statistical approach to natural language interfaces. In: ACL96 – Proceedings of the 34th Conference of the Association for Computational Linguistics, Santa Cruz, California (1996) 55–61
6. He, Y., Young, S.: A data-driven spoken language understanding system. In: ASRU'03 – Proceedings of the IEEE Automatic Speech Recognition and Understanding Workshop, St. Thomas, US Virgin Islands (2003)
7. Gorniak, P., Roy, D.: Probabilistic grounding of situated speech using plan recognition and reference resolution. In: ICMI'05 – Proceedings of the Seventh International Conference on Multimodal Interfaces, Trento, Italy (2005)

New Hybrid Real-Coded Genetic Algorithm

Zhonglai Wang, Jingqi Xiong, Qiang Miao, Bo Yang, and Dan Ling

School of Mechatronics Eng., University of Electronic Science and Technology of China,
Chengdu, Sichuan, 610054, P.R. China
wangyang713@163.com

Abstract. Real-coded Genetic Algorithms (RGAs) usually meet the demand of continuous and continuous/discreet mixed space problems and have been widely applied in many fields. The paper proposed a new Hybrid Real-coded Genetic Algorithm (NHRGA), in which the idea of Particle Swarm Optimization (PSO) is introduced into mutation operator and physics field theory is also employed in algorithm operators. The NHRGA reduces the possibility of trapping into the local optimal solutions and improves the computation efficiency. A practical engineering example is given to demonstrate computation efficiency and robustness of the proposed method.

Keywords: Genetic Algorithms, Particle Swarm Optimization, operator, computation efficiency.

1 Introduction

Genetic Algorithms (GAs), first proposed by Holland in 1975, are a kind of highly parallel, random and adaptive searching algorithms. Now GAs have been applied in many fields, such as biology, combination optimization, engineering design and optimization and so on [1, 2, 3, 4].

Binary-coded and real-coded forms are usually found in genetic algorithms. RGAs can directly describe continuous parameter optimization problems without coding and decoding. Compared with binary-coded genetic algorithms, real-coded genetic algorithms have advantages as follows:

1) Solution space continuity. Because RGAs can express any solution in the continuous space, computation precision will be higher.

2) Uniform standard. Coding and decoding is not necessary, which can improve computation efficiency.

3) Without Hamming cliffs. RGAs can avoid Hamming cliffs.

In practical engineering application, many continuous or continuous/discrete mixed space problems are referred to. RGAs meet the demands of the practical engineering application and are widely focused on.

To improve the computation efficiency and reduce the possibility of trapping into the local optimum, many different crossover and mutation operators have been proposed. In this paper, both mutation operator and crossover operator are improved to reduce the possibility of trapping into the local optimal solution and increase the computation efficiency. The organization of this paper is as follows. Section 2 presents the improvement of mutation operator and crossover operator in more

A. Sattar and B.H. Kang (Eds.): AI 2006, LNAI 4304, pp. 1221–1225, 2006.
© Springer-Verlag Berlin Heidelberg 2006

details. In Section 3, a practical engineering example is used to demonstrate its computation efficiency and robustness. Conclusions are given in Section 4.

2 Analysis on Improved Operator

2.1 Mutation Operator Improvement

In PSO algorithm, every particle can ceaselessly update location in each step of iteration, which also means closer to the optimal location by iteration. In NHRGA, the aim of mutation is to let part of individuals closer to the optimal location. To reach the aim and make the algorithm simple, PSO algorithm can be rewritten as [5]:

$$X_i^{(t+1)} = X_i^{(t)} + e^{-d}\frac{\omega}{\log(r+1)}(X^{(t)}{}_{igbest} - X_i^{(t)}) \tag{1}$$

Where

$$d = \sqrt{(X_1^{(t)} - X_{1gbest}^{(t)})^2 + ... + (X_i^{(t)} - X_{igbest}^{(t)})^2 + ...(X_n^{(t)} - X_{ngbest}^{(t)})^2} \tag{2}$$

X_{igbest} denotes the optimal solution of the entire colony of the ith variable.

In order to avoid trapping into the local optimal location, field theory is introduced into the mutation operator, which can be expressed by the following equation:

$$K = m \times e^{(-\beta\frac{\|x-y\|}{\sigma})^k} \tag{3}$$

Let $d=x-y$ and Eq. (3) can be rewritten as:

$$K = m \times e^{(-\beta\frac{d}{\sigma})^k} \tag{4}$$

With the above two factors, mutation operator can be expressed as follow:

$$X_i^{(t+1)\,\prime} = K \times X_i^{(t+1)} \tag{5}$$

There are many parameters ω, β, σ, m and k, which need to be given. ω is between (0.4,0.9), while $\beta>0$, $\sigma>0$, $m>0$, k is a positive integer. However, the enactment of these parameters causes a little influence to the algorithm.

2.2 Crossover Operator Improvement

According to literature review, only when relatively best elitists appear in the next generation, does the convergence to the global optimal solution become possible [6]. The best five elitists and the fifteen mutated elitists get into the crossover pool. The detailed steps will be explained as follows.

Suppose that $X_m^{(t)}$ and $X_n^{(t)}$ are the two individuals in the t generation, $m \neq n$, i.e.

$$X_m^{(t)} = \left[X_{m1}^{(t)}, X_{m2}^{(t)}, ... X_{mi}^{(t)}, ... X_{mn}^{(t)}\right]$$
$$X_n^{(t)} = \left[X_{n1}^{(t)}, X_{n2}^{(t)}, ... X_{ni}^{(t)}, ... X_{nn}^{(t)}\right]$$

$$d = \left| X_{m}^{(t)} X_{n}^{(t)} \right| \tag{6}$$

If $F(X_{m}^{(t)}) > F(X_{m}^{(t)'})$, we can obtain

$$\begin{cases} X_{mi}^{(t)'} = X_{mi}^{(t)} + K \frac{1}{\lambda} \cdot rand \cdot (X_{mi}^{(t)} - X_{ni}^{(t)}) \\ X_{ni}^{(t)'} = X_{mi}^{(t)} - K \frac{1}{\lambda} \cdot rand \cdot (X_{mi}^{(t)} - X_{ni}^{(t)}) \end{cases} \tag{7}$$

If $F(X_{m}^{(t)}) < F(X_{m}^{(t)'})$, we have

$$\begin{cases} X_{mi}^{(t)'} = X_{ni}^{(t)} + K \frac{1}{\lambda} \cdot rand \cdot (X_{ni}^{(t)} - X_{mi}^{(t)}) \\ X_{ni}^{(t)'} = X_{ni}^{(t)} - K \frac{1}{\lambda} \cdot rand \cdot (X_{ni}^{(t)} - X_{mi}^{(t)}) \end{cases} \tag{8}$$

Where $K = m \times e^{(-\beta \frac{d}{\sigma})^{k}}$. Field theory is also applied in crossover operator to form an adaptive niche. According to the equations above, we know that one of filial generation lies between the two parents and the other lies beside the two parents but closer to the better one. However, it may cause some variables to go out of the constraint range. Thus, we make some rules as follows:

$$\text{if } X_{mi}^{(t)'} < a_i, \text{ then } X_{mi}^{(t)'} = a_i$$
$$\text{if } X_{ni}^{(t)'} < a_i, \text{ then } X_{ni}^{(t)'} = a_i$$
$$\text{if } X_{mi}^{(t)'} > b_i, \text{ then } X_{mi}^{(t)'} = b_i$$
$$\text{if } X_{ni}^{(t)'} > b_i, \text{ then } X_{ni}^{(t)'} = b_i.$$

3   Case Study

An example of flow transmission series system with four basic units, shown in Fig. 1, is given here. The system is a power station coal transportation system that can continuously supply coal to boilers [7].

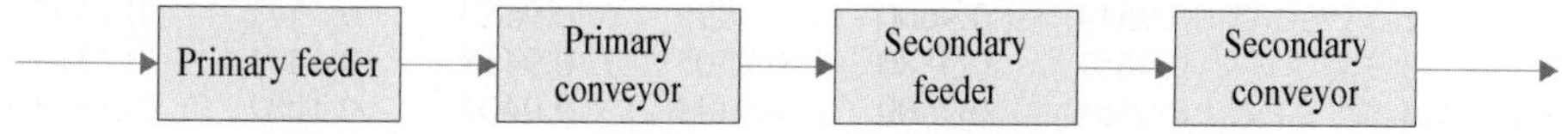

Fig. 1. The flowchart of a flow transmission series system

Each unit has different range of reliability according to its function demand and they are

$$0.85 \le R_1 \le 0.999,\ 0.90 \le R_2 \le 0.999,\ 0.88 \le R_3 \le 0.999,\ 0.83 \le R_4 \le 0.999$$

in sequence. Further more, the system reliability R should be greater than or equal to 0.8. To make sure every element satisfies the requirement of reliability, maintenance cost of the elements must be allocated according to the difficulty and time of maintenance activities. It can be expressed as [8]:

$$C = \sum_{i=1}^{4} \alpha_i \frac{-t_m}{\ln R_i} \tag{9}$$

Here t_m is the operation time, α_i is the parameter of maintenance cost. The question is how to decide R_1, R_2, R_3 and R_4 so as to

$$\min \quad C = \sum_{i=1}^{4} \alpha_i \frac{-t_m}{\ln R_i} \tag{10}$$

Let $[x_1, x_2, x_3, x_4] = [R_1, R_2, R_3, R_4]$, $t_m=20$ and $[\alpha_1, \alpha_2, \alpha_3, \alpha_4] = [0.8, 0.9, 0.9, 0.8]$. The mathematic model of this system can be expressed as:

$$\min \ C = (0.8 \times \frac{-20}{\ln x_1} + 0.9 \times \frac{-20}{\ln x_2} + 0.9 \times \frac{-20}{\ln x_3} + 0.8 \times \frac{-20}{\ln x_4})/10000$$

$$\text{s.t.} \quad 0.85 \le x_1 \le 0.999$$
$$0.90 \le x_2 \le 0.999$$
$$0.88 \le x_3 \le 0.999$$
$$0.83 \le x_4 \le 0.999$$
$$0.8 \le x_1 \times x_2 \times x_3 \times x_4$$

NHRGA is used to demonstrate its advantages in this problem. When $t=10$, the algorithm reaches the global optimum. Table 1, which shows the searching process of the NHRGA when solving the problem, is obtained.

From Table 1, we can see that when $t=10$, variables x_1, x_2, x_3 and x_4 can minimize cost C. That is, when $R_1=0.9532$, $R_2=0.9493$, $R_3=0.9428$ and $R_4=0.9417$, the system cost reaches the minimum 0.1219 ten thousands dollars.

Table 1. Variables x_1, x_2, x_3, x_4 and cost C

Generation(t)	mean x_1	mean x_2	mean x_3	mean x_4	mean C	min C
1	0.9474	0.9651	0.9587	0.9610	0.4926	0.1450
2	0.9514	0.9600	0.9495	0.9452	0.1547	0.1407
3	0.9551	0.9580	0.9505	0.9418	0.1401	0.1313
4	0.9564	0.9500	0.9443	0.9401	0.1287	0.1248
5	0.9555	0.9524	0.9404	0.9387	0.1272	0.1240
6	0.9561	0.9506	0.9414	0.9386	0.1267	0.1229
7	0.9555	0.9495	0.9423	0.9395	0.1262	0.1221
8	0.9546	0.9493	0.9426	0.9403	0.1260	0.1220
9	0.9531	0.9493	0.9427	0.9416	0.1259	0.1219
10	0.9532	0.9493	0.9428	0.9417	0.1257	0.1219

4   Conclusions and Discussions

Due to the similarity between PSO and the mutation operator of GA, PSO is applied to improve the mutation operator. If the distance between two individuals is bigger, the interaction between them will decrease sharply, which is similar to the field theory. Therefore, the algorithm forms a not-so-strict niche, which reduce the possibility of trapping into the local optimal solution. Meanwhile, NHRGA increases the computation precision and the computation efficiency. However, there are two issues to answer in our future work: How to define the index of diversity in every generation and control the degree of the diversity, and how to explain the diversity with pattern theory.

References

1. Huang, H.Z., Bo, R., Chen, W.: An integrated computational intelligence approach to product concept generation and evaluation. Mechanism and Machine Theory, 41(5) (2006) 567-583
2. Huang, H.Z., Zuo, M.J., Sun, Z.: Bayesian reliability analysis for fuzzy lifetime data. Fuzzy Sets and Systems, 157(12) (2006) 1674-1686
3. Huang, H.Z., Tian, Z.H., Zuo, M.J.: Intelligent interactive multiobjective optimization method and its application to reliability optimization. IIE Transactions, 37(11) (2005) 983-993
4. Zhang, X., Huang, H.Z., Yu, L.: Fuzzy preference based interactive fuzzy physical programming and its application in multi-objective optimization. Journal of Mechanical Science and Technology, 20(6) (2006) 731-737
5. Zhang, X. P., Du, Y.P., Qin, G.Q., Tan, Z.: Adaptive Particle Swarm Glgorithm with dynamically changing inertia weight. Journey of Xi'an JiaoTong University. 39 (2005) 1039-1042
6. Wang, L.: Convergence uniform criterion of Genetic Algorithm. Technique of Automation and Application., 26 (2004) 16-19
7. Lianianski, A., Levitin, G.: Multi-state system reliability. World Scientific, Singapore (2003)
8. Mettas, A.: Reliability allocation and optimization for complex systems. 2000 Proceedings Annal Reliability and Maintainability Symposium. (2000) 216-221

Comparison of Numeral Strings Interpretation: Rule-Based and Feature-Based N-Gram Methods

Kyongho Min[1] and William H. Wilson[2]

[1] School of Computer and Information Sciences,
Auckland University of Technology, New Zealand
kyongho.min@aut.ac.nz
[2] School of Computer Science and Engineering,
University of New South Wales, Sydney, Australia
billw@cse.unsw.edu.au

Abstract. This paper describes a performance comparison for two approaches to numeral string interpretation: manually generated rule-based interpretation of numerals and strings including numerals [8] vs automatically generated feature-based interpretation. The system employs three interpretation processes: word trigram construction with a tokeniser, rule-based processing of number strings, and *n*-gram based classification. We extracted numeral strings from 378 online newspaper articles, finding that, on average, they comprised about 2.2% of the words in the articles. For feature-based interpretation, we tested on 11 datasets, with random selection of sample data to extract tabular feature-based constraints. The rule-based approach resulted in 86.8% precision and 77.1% recall ratio. The feature-based interpretation resulted in 83.1% precision and 74.5% recall ratio.

1 Introduction

The domain of natural language processing focuses on sequences of alphabetical character strings, in a text. The text includes different types of data such as numeric (e.g. "801 voters") or alpha-numeric (e.g. "9:30pm") with/without special symbols (e.g. $ in "$2.5 million") [5]. Current NLP systems treat such strings as either a numeral (e.g. "801 voters") or as a named entity (e.g. Money for "$2.5 million") by using basic categories like PERSON, ORGANISATION, MONEY, etc. [2], [4].

In this paper, there are two numeral string types: *separate numeral strings* (e.g. quantity in "*5,000* peacekeepers") and *affixed numeral strings* (e.g. LENGTH in "a *10m* yacht"). The latter has meaningful semantic units attached (e.g. "m" in "10m") that result in less interpretation ambiguity. The interpretation of separate numeral strings uses their syntactic functional pattern (e.g. a modifier or a head of noun phrase) or structural pattern information rather than meaningful semantic clues.

FACILE [2] in MUC used rule-based named entity recognition system incorporating a chart-parsing technique. Semantic categories and semantic tags were used for a Chinese classification system in [7], [14] and for Japanese documents in [12]. The ICE-GB grammar [9] treated numerals as one of *cardinal, ordinal, fraction, hyphenated*, or *multiplier*, with two number features. Zhou and Su [15] classified numeral strings with surface and semantic features like FourDigitNum (e.g. 1990) as a year

A. Sattar and B.H. Kang (Eds.): AI 2006, LNAI 4304, pp. 1226–1230, 2006.
© Springer-Verlag Berlin Heidelberg 2006

form (see also [3] and [11] for time phrases in weather forecasts). Polanyi and van den Berg [10] studied anaphoric resolution of quantifiers with a quantifier logic framework.

We implemented two numeral interpretation systems: a manually generated rule-based interpretation method [8] and an automatically generated tabular feature-based method. Both systems are composed of word trigram construction with a tokeniser, rule-based processing of number strings, and n-gram based disambiguation of classification (e.g. a word trigram - left and right strings of a numeral string). In this paper, we will focus on an automatically generated tabular feature-based method.

In the next section, we will describe the disambiguation process for numeral strings based on two methods, in more detail, and focus on the disambiguation process based on the automatically generated tabular feature-based method. Section 3 will describe preliminary experimental results obtained with both methods, and discussion and conclusions follow.

2   Disambiguation of Numeral String Categories

There are two types of numeral strings: separate and affixed. A chart parsing algorithm [6] based on context-free rules was applied to the interpretation of affixed numeral strings using 40 syntactic and semantic categories. To interpret affixed numeral strings (e.g. "10m" in "a 10m yacht"), the system uses 64 context-free rules (e.g. LENGTH $\rightarrow$ number + length-unit) to represent their structural forms (see [8] for more detail).

2.1   Disambiguation Based on Manually Generated Rules

Word trigrams are used to disambiguate the syntactic/semantic categories of numeral strings after interpreting numeral strings. Disambiguation of categories is based on rules manually encoded by using sample data and each rule covers morpho-syntactic/semantic features of the word trigrams. Features used are based on POS, NUMBER (singular, plural) punctuation marks (e.g. comma), character case (e.g. capital letter), semantic category based on pattern matching (e.g. January $\Rightarrow$ MONTH), and verification of categories. With these manually generated rules extracted from sample data, called a rule-based method, the contextual information based on word trigrams is applied to disambiguate the multiple categories resulting from the numeral string interpretation process (see [8] for more detail).

2.2   Disambiguation Based on Automatically Generated Tabular Features

The other method to disambiguate the category of a numeral string is based on automatically generated tabular features from sample data. The features, based on a tabular format (set of attributes and values), are composed of the following attributes: direction (:D – e.g. left or right), offset (distance from the numeral string), Part-of-Speech (:POS), syntactic features (:SYNF – e.g. number, tense), prefix, suffix (e.g. prefixed/suffixed symbols or punctuation marks), and spec (:SPEC – e.g. specific semantic information like MONTH for "January").

Table 1. Examples of tabular features and their category indexes

ID	:D	:OFFSET	:POS	:SYNF	:PREFIX	:SUFFIX	:SPEC
1	L	1	NOUN	PLUR	NIL	NIL	NIL
2	R	1	PREP	"by"	NIL	NIL	CAPITAL
3	R	1	ADV	NIL	NIL	NIL	CAPITAL

Table 1, above, shows examples of feature tables extracted from sample data. The sample data's annotated category information is indexed using a field called ID (see Table 2). The tabular feature construction process is as follows: First, the tabulated features of an annotated numeral string with its word trigrams are automatically generated. Second, each feature table obtained by Step 1 is compared with an existing tabular database (Table 1). If a feature tabular is new, then add the feature table to tabular database (Table 1) with new indexed category information. Otherwise, modify its category indexing information in Table 2 (increment of its annotated category frequency by 1).

Table 2. Examples of category frequency of tabular features

ID	Category Frequency of Feature Tables
1	((MONEY 2) (RATE 1) (AGE 1) (QUANT 1) (NUMBER 1) (DATE 17))
2	((SCORES 3) (YEAR 1) (AGE 2) (NUMBER 2) (PLURAL 1) (DAY 5) (QUANT 4))
3	((RANGE 1) (FLOATNUMBER 32) (QUANT 67))

When disambiguating the category of a numeral string, the most frequent category based on retrieved tabular features from word trigrams of the numeral string is selected. This is called a tabular feature-based method. If a numeral string's feature vectors retrieved from its word trigrams were ID-1 and ID-3 in Table 1, then all category frequency information related to the feature tables are retrieved from Table 2 and summed up to select the category QUANT (e.g. (QUANT 68)).

3 Experimental Results

We collected online newspaper articles for nine days (378 articles). The articles cover a range of topics (e.g. domestic, international, sports, and economy). The articles included 4173 (2.3%) numeral strings among a total of 192528 strings. To test this system, we produced 11 sets of sample and test data. For each set, we randomly selected 2 days' articles (83 articles (22.0%) with 915 numeral strings (22.1%) on average) to extract tabular features and category frequency information (3799 category features (22.5%) on average) and the remaining 7 days' articles were used for testing this system.

Figure 1 shows the results of tests with the 11 datasets. The precision ratio for the 11 datasets, based on the tabular feature-based method is 74.5%, and the recall ratio is 68.1%, so the balanced F-measurement is 71.2% on average. The best result (S-4 in Fig. 1) shows 83.1% for precision, 74.5% for recall, and 78.6% for F-measurement. The worst result (S-1 in Fig. 1) shows 57.2% for precision, 61.1% for recall, and 59.1% for F-measurement.

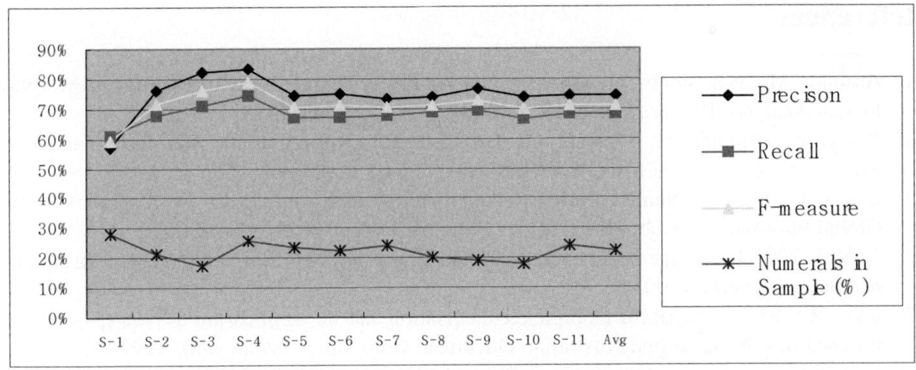

Fig. 1. Results for 11 test sets based on the tabular feature-based method

Table 3. Recall/Precision/F-measurement ratios for our two methods

Test Methods	Precision ratio (%)	Recall ratio (%)	F-measure (%)
Rule-based Method (Base)	86.8	77.1	81.6
Feature-based Method (average)	74.5	68.1	71.2
Feature-based Method (best)	83.1	74.5	78.6
Feature-based Method (worst)	57.2	61.1	59.1

Table 3 shows the manually generated rule-based method, based on conjunctive constraints relating to word trigrams, performs better than the automatically generated tabular feature-based method, based on disjunctive constraints relating to word trigrams. The performance difference between the rule-based method and the feature-based method (average) is 12.3% in precision ratio, 9.0% in recall ratio, and 10.4% in F-measurement. If the rule-based method is compared with the best feature-based method, then the performance difference is 3.7% in precision ratio, 2.6% in recall ratio, and 3.0% in F-measurement.

4 Discussion and Conclusions

For further improvement, the feature extraction techniques from sample data need to be improved, including the use of a POS tagger to get disambiguated lexical information for word trigrams. Second, the system may improve its performance with conjunctive tabular features of contextual information from word trigrams. In addition, the extension of this system to other data like biomedical corpora [13] is required to test the effectiveness of our approach.

In conclusion, separate and affixed numeral strings are frequently used in real text. However, there is no system that interprets numeral strings systematically; they are frequently treated as either numerals or nominal entities. In this paper, we discussed two numeral disambiguation methods of a numeral interpretation system. The manually generated rule-based disambiguation method performs better than the automatically generated tabular feature-based disambiguation method. The current disjunctive constraints on word trigrams successfully interpreted 83.1% of test data.

References

1. Asahara, M., Matsumoto Y.: Japanese Named Entity Extraction with Redundant Morphological Analysis. Proceedings of HLT-NAACL 2003. (2003) 8-15
2. Black, W., Rinaldi, F., Mowatt, D.: FACILE: Description of the NE system used for MUC-7. Proceedings of MUC-7. (1998)
3. Chieu, L., Ng, T.: Named Entity Recognition: A Maximum Entropy Approach Using Global Information. Proceedings of the 19th COLING. (2002) 190-196
4. CoNLL-2003 Language-Independent Named Entity Recognition. http://www.cnts.uia.ac.be/conll2003/ner/2. (2003)
5. Dale, R.: A Framework for Complex Tokenisation and its Application to Newspaper Text. Proceedings of the second Australian Document Computing Symposium. (1997)
6. Earley, J.: An Efficient Context-Free Parsing Algorithm. CACM. 13(2) (1970) 94-102
7. Maynard, D., Tablan, V., Ursu, C., Cunningham, H., Wilks, Y.: Named Entity Recognition from Diverse Text Types. Proceedings of Recent Advances in NLP. (2001)
8. Min, K., Wilson, W.H., Moon, Y.: Syntactic and Semantic Disambiguation of Numeral Strings Using an N-gram Method. in Shichao Zhang and Ray Jarvis (eds) AI 2005: Advances in Artificial Intelligence Proceedings of 18th Australian Joint Conference on Artificial Intelligence, Sydney, Australia, December 2005. Springer Lecture Notes in Artificial Intelligence 3809. (2005) 82-91
9. Nelson, G., Wallis, S., Aarts, B.: Exploring Natural Language - working with the British Component of the International Corpus of English, John Benjamins, The Netherlands. (2002)
10. Polanyi, L., van den Berg, M.: Logical Structure and Discourse Anaphora Resolution. Proceedings of ACL99 Workshop on The Relation of Discourse/Dialogue Structure and Reference. (1999) 110-117
11. Reiter E., Sripada, S.: Learning the Meaning and Usage of Time Phrases from a parallel Text-Data Corpus. Proceedings of HLT-NAACL2003 Workshop on Learning Word Meaning from Non-Linguistic Data. (2003) 78-8511.
12. Siegel, M., Bender, E. M.: Efficient Deep Processing of Japanese. Proceedings of the 3rd Workshop on Asian Language Resources and International Standardization. (2002)
13. Torii, M., Kamboj, S., Vijay-Shanker, K.: An investigation of Various Information Sources for Classifying Biological Names. Proceedings of ACL2003 Workshop on Natural Language Processing in Biomedicine. (2003) 113-120
14. Wang, H., Yu, S.: The Semantic Knowledge-base of Contemporary Chinese and its Application in WSD. Proceedings of the Second SIGHAN Workshop on Chinese Language Processing. (2003) 112-118
15. Zhou, G., and Su, J. Named Entity Recognition using an HMM-based Chunk Tagger. Proceedings of ACL2002. (2002) 473-480

Simulation Analysis on Effective Operation of Handling Equipments in Automated Container Terminal

Byung Joo Park, Hyung Rim Choi, Hae Kyoung Kwon, and Moo Hong Kang

Dong-A University, Department of Management Information System,
840 Saha-gu, Hadan-dong, Busan, 604-714, Korea
{a967500, hrchoi, hgkwon, mongy}@dau.ac.kr

Abstract. This research aims to discover effective integrated operation method of an automated container terminal, which includes the integration of ATC re-stacking rule, AGV dispatching rule, and container crane in order to minimize the makespan in time of loading. This study tried to find out the optimal way of reducing delay time, consequently minimizing the makespan. Among the many situations – the re-marshaling degree of blocks, the number of container and AGV – an empirical study should identify which combination of operation methods are most likely to be robust and effective. To this end, simulation models of equipment operation have been tested under various environments.

Keywords: Automated Container Terminal, Integrated Operation.

1 Introduction

An automated container terminal means that it makes use of automated handling equipment such as AGV (Automated Guided Vehicles) and ATC (Automated Transfer Crane). The effective operation of this automated handling equipment plays critical role for the productivity enhancement of a container terminal. The optimal operation of handling equipment reduces not only the staying time and waiting time of containership, thus lowering the transportation cost per unit, but also increases the terminal productivity, consequently reducing the expenses of the container terminal. Of course, the optimal operation of automated equipment cannot be done easily.

The many preceding studies on the optimal operation of automated handling equipment can be divided in two. One is to optimize the operation scheduling by individual equipment, and the other is to optimize the integrated operation scheduling for equipment of more than two. To optimize the routing of the single transfer crane, Kim and Kim [4] suggested the MIP (mixed integer programming) model based on dynamic programming, and again in 1999 they introduced a beam search algorithm for this model [5]. And Chen et al. [1] and Kim and Bae [3] made researches on the operation plan of AGV. Chen et al. also presented a network-based Greedy heuristics in dealing with AGV dispatching problems. In this research they assumed that the transfer cranes were always ready to load and unload into an AGV in order to simplify the matters. While suggesting a MIP model for the solution of AGV dispatching problems, Kim and Bae [3] aimed to minimize the loading/unloading time

A. Sattar and B.H. Kang (Eds.): AI 2006, LNAI 4304, pp. 1231–1238, 2006.
© Springer-Verlag Berlin Heidelberg 2006

of the container cranes through AGV dispatching. However, in the automated container terminals the container cranes, ATC, and AGV must be operated under mutual cooperation, the operation plan of individual equipment can cause a deadlock or a low performance. Because of this, an integrated scheduling of handling equipment is necessary. In order to minimize the total makespan of the whole automated equipments in an automated container terminal, Meersmans and Wagelmans [7] suggested, as an integrated scheduling, the Branch and Bound algorithm and Beam search method.

Although Meersmans and Wagelmans [7, 8] presented a new search method based on an integrated scheduling, there is a doubt whether it can be used practically on the spot. The reason is that all the automated equipment cannot be operated as planned at a given time. In real cases since unexpected situations happen sometimes during the operation hours of automated equipment, the scheduling should be revised, consequently requiring rescheduling and causing inefficiency. Therefore, this paper would like to suggest a new approach. When individual equipment finishes its job, it moves to the next transfer point and waits another job order. Let's call this process an "event". We believe that an event-oriented operation method of automated equipment based on dispatching rule is much more practical. Accordingly, capitalizing on AGV dispatching rule and ATC re-stacking rule, we assumed various situations of integrated operation – the degree of yard re-marshaling, the number of container and the number of AGV – thus seeking the most robust combination of the above rules. To this end this paper has made simulation models of automated equipment operation in the container terminal, then testing each model under diverse environments.

2 Operation Method of Handling Equipment in an Automated Container Terminal

2.1 Handling Equipments of an Automated Container Terminal

An automated container terminal uses automated handling equipment such as container cranes, ATC, and AGV. It also makes good use of information technologies in handling containers and automated equipment in accordance with job scheduling on a real time basis, consequently contributing to enhance the terminal's productivity and saving its manpower. Most automated container terminals are using ATC and AGV as main handling equipment. ATC, an automated crane, moves containers within the yard, or moves containers to the AGV or external trucks. AGV, an unmanned automated vehicle, loads containers from the ATC or container cranes, and travels between ATC and container cranes. Therefore, container handling can be done by mutual cooperation of this handling equipment. To load the containers in the yard ATC is used, and AGV moves these containers to the container cranes and waits for shipment. Manual container terminals use shuttle trucks to move containers in the yard, and so their personnel prepare job scheduling on a manual basis, and give a job order to the related workers. However, an automated container terminal demands a new operation policy.

2.2 Dispatching Rule of Automated Equipment

The dispatching rule can be easily used without applying a complicated mathematical model. In particular, under the circumstances of uncertainty and lack of information, the dispatching rule is very attractive. Of course, this cannot match the complicated model seeking an optimal solution. Nevertheless, thanks to its merits of speediness and simplicity, it is widely used under the dynamic circumstances.

2.2.1 Re-stacking Rule of ATC

In general the containers are stacked in multiple layers at the yard. And the productivity of yard operation is closely related to the minimization of re-stacking. This means that the storage location of containers should be rationally determined, and also at the time of transferring containers the storage location and the order of moving should be decided systematically. The container re-stacking can be divided in two. One is the proactive re-stacking that is done during the idle hours of container cranes in anticipation of next container movement (re-marshaling), and the other is the reactive re-stacking that is done when there are other containers placed on the containers to be moved right now [2].

The first one has to be dealt with in the yard planning. Therefore, this paper handles the reactive re-stacking policy. Practically speaking, the perfect proactive re-stacking for shipment is nearly impossible owing to a time limit. Because of this, the reactive re-stacking takes place. Therefore, the proper location of transferred containers, which leads to the minimization of the container re-stacking, is very important in reducing the shipping time and in enhancing the efficiency of handling equipment. The re-stacking rule of ATC, which decides the location of the containers to be transferred, is as follows.

(1) Random
The containers to be transferred should not be higher than the highest storage height, but can be stacked randomly. In other words, the moving containers should not have the highest stacking height, but can be freely placed in the nearest place.

(2) Closest Position
This is the way to select the nearest place in transferring containers as far as the place doesn't have the highest stacking height, and also the containers to be placed under the moving container should not be sooner in its order of shipment than that of the moving container.

(3) RSC (Remaining Stack Capacity)
Concerning the storage location (i, j), i = row, j = number of stacking, first calculate the value of RSC_{ij} before and after the re-stacking of the containers, and then the least difference in their value of RSC_{ij} is to be the location of the moving container. The RSC_{ij} of each container's storage location can be calculated in the following formula (1) [2].

$$RSC_{ij} = (H - h_{ij}) \times ((C_{ij} - C_n) + C_n) \tag{1}$$

H = the highest storage height
h_{ij} = the height of the present point (i, j)
C_{ij} = the handling sequence (value) of the highest container at the present point (i, j).
   If there is no container at the present point (i, j), $C_{ij} = C$
C = the handling sequence (value) of the last container.
C_n = the handling sequence (value) of containers $(n=1,2,......,C)$

2.2.2 AGV Dispatching Rule

The application of AGV dispatching rule can cause problems. As AGV is not able to load and unload the containers by itself, dependent scheduling between ATC and AGV, and between container cranes and AGV is to be carried out. This scheduling can cause blocking and deadlock. To solve the problems of blocking and deadlock, this paper has adopted an AGV-centered dispatching rule instead of a job (container)-centered rule. The idle AGV moves to the transfer point to have the container to be transferred. The AGV at the transfer point is to be assigned the container according to the order of arrival. The following is how the containers are assigned to AGV.

(1) Random
When the idle AGV arrives at the transfer point, any container among the containers, which are waiting for shipment by a container crane according to the order of shipment but not yet be assigned, can be assigned randomly.

(2) CCB (Container Crane Balancing)
When the idle AGV arrives at the transfer point, any container belonging to a container crane that has the largest number of containers to be handled according to the order of shipment but not yet be assigned can be assigned.

(3) CBB (Container Block Balancing)
When the idle AGV arrives at the transfer point, any container belonging to a block that has the largest number of containers to be handled according to the order of shipment but not yet be assigned can be assigned.

Table 1. Operation Method of Automated Handling Equipments

container crane	ATC	AGV
according to the order of shipment	random	random
	closest position	CCB
	RSC	CBB

2.3 Integrated Scheduling of Automated Equipments

This research aims to discover the integrated operation method of an automated container terminal through integration of ATC re-stacking, AGV dispatching rule, and container crane operation. First, in order to calculate the total makespan of container loading, the waiting time of individual equipment must be considered. For example, AGV has to wait until the ATC lifts up a container, and put it on the AGV. In addition, AGV ought to wait at the place of the container crane until the preceding AGV will finish moving its container. This means that the operation of a container crane, ATC, and AGV carries a waiting time. We tried to reduce this waiting time through diverse combination of all this equipment operation rules, consequently finding out the way to minimize the makespan. All the handling equipments are integrated at the time of container loading. At first, AGV will be assigned to the container block according to the order of container loading to be handled by a container crane. Then this AGV moves to the allotted block and get the container from ATC. And AGV travels to the loading point of a container crane. Here the container crane loads containers in accordance with its loading order. All these moving and loading are to be done by AGV dispatching rule and ATC re-stacking rule.

3 Simulation Model and Test

In our efforts to find out the best method in the integrated operation of automated equipment, we have made simulation models. The simulations achieved by the program developed on visual basic language, and been made on the basis of the "basic plan for Gwangyang automated container terminal" [6]. As mentioned above, the makespan of loading that is done by mutual cooperation of handling equipment includes the waiting time required during the operation of individual equipment.

3.1 Simulation Environment

The simulation environment of automated container terminal is as follows:

- Number of berth: 1
- Length of berth: 300 m
- Number of container crane: 4 units
- Number of block (vertical layout): 7 blocks (10 rows and 4 stacking per block)
- Speed of AGV: 6m/s on average
- Moving method of AGV: counterclockwise closed
- Number of lane for AGV: 5
- Speed of ATC: breadthwise moving speed 2m/s, lengthwise moving speed 3m/s, hoisting speed 1m/s
- ATC buffer: 5 units
- Speed of container crane: 1 lift/min on average

Each container crane has its own block, and the loading order of the block or by the container crane is also fixed. Therefore, the ultimate purpose is to minimize the makespan, that is, it means the earliest completion of the final container loading.

3.2 Simulation Parameters

To seek the most practical and robust operation rule, each operation method has been tested under the circumstances of following parameters.

- The degree of container re-marshaling within one block: 80%, 90%, and 100%
- Unit number of AGV: 14, 18, 22, and 26
- Number of containers to be handled: 300, 400, and 500

3.3 Test Results and Analysis

The makespan obtained in terms of each combination of AGV and ATC rules under the above parameters is shown in the Table 2, Table 3, and Table 4. We tried to seek the best combination of AGV rule and ATC rule through many simulations under the diverse environments. As a result, we can, in particular, find out that in case of 100% re-marshaling – i.e. under the condition that ATC re-stacking rule has no influence – the CCB rule of AGV has shown the best performance, and the combination with the RSC rule of ATC also has shown the best performance under any environment.

In particular, the lower the degree of re-marshaling, the better its result has turned out. Therefore, considering the low degree of re-marshaling in the real case of container terminals, this combination is more practical. These results have made much improvement in its performance compared with many other combinations including the random rule. Meanwhile, there has been the case that 90% re-marshaling has a shorter makespan than the 100% re-marshalling. This means that the shortened travel distance by effective ATC re-stacking can reduce the whole makespan. Also, the combination of the CCB rule resulted in poorer performance than the combination of the random rule. This means that if the order of shipment was fixed, the performance is more influenced by the container crane than the ATC. And input of AGV more than the proper level has not been of help in reducing the makespan.

Table 2. Makespan According to ATC-AGV Operation Rules (number of container = 300)

ATC Rule	AGV Rule	The degree of container re-marshaling											
		80%				90%				100%			
		Number of AGV				Number of AGV				Number of AGV			
		14	18	22	26	14	18	22	26	14	18	22	26
ran	ran	11760	11705	11468	11661	9688	9583	9606	9395	6038	5685	5176	5030
	ccb	11673	11685	11689	11684	10409	9898	9411	9403	5423	5238	5078	5118
	cbb	15361	12647	14553	14064	12470	11525	11434	10976	6900	6256	6137	5452
clos	ran	6148	5788	5351	5152	6076	5556	5336	5536	6038	5685	5176	5030
	ccb	5488	5280	5351	5215	5692	5336	5336	5185	5423	5238	5078	5118
	cbb	7165	6402	6360	5951	7702	6630	6644	6237	6900	6256	6137	5452
RSC	ran	6196	5554	5235	5037	6155	5592	5137	5068	6038	5685	5176	5030
	ccb	5414	5193	5042	5037	5440	5127	5085	4981	5423	5238	5078	5030
	cbb	7204	6542	6459	5775	7049	6807	6312	6314	6900	6256	6137	5452

Table 3. Makespan According to ATC-AGV Operation Rules (number of container = 400)

ATC Rule	AGV Rule	The degree of container re-marshaling											
		80%				90%				100%			
		Number of AGV				Number of AGV				Number of AGV			
		14	18	22	26	14	18	22	26	14	18	22	26
ran	ran	18382	18484	18473	18344	15897	15167	15116	14764	8295	7351	7144	6885
	ccb	18420	18411	18344	18346	15454	15490	15175	15011	7350	7045	6955	6695
	cbb	24054	21750	21897	20768	20154	17926	18341	18764	9229	8332	8383	7642
clos	ran	8373	7489	7348	6957	8168	7406	7235	6847	8295	7351	7144	6885
	ccb	7645	7202	7000	6955	7478	7100	6824	6801	7350	7045	6955	6695
	cbb	9730	8733	8649	8433	9818	9042	8893	8048	9229	8332	8383	7642
RSC	ran	8492	7577	7163	6992	8358	7403	6987	6876	8295	7351	7144	6885
	ccb	7500	7091	7042	6955	7249	7021	6983	6801	7350	7045	6955	6695
	cbb	10389	9230	8919	7949	9781	8873	6983	6801	9229	8332	8383	7642

Table 4. Makespan According to ATC-AGV Operation Rules (number of container = 500)

ATC Rule	AGV Rule	The degree of container re-marshaling											
		80%				90%				100%			
		Number of AGV				Number of AGV				Number of AGV			
		14	18	22	26	14	18	22	26	14	18	22	26
ran	ran	24396	23157	22725	22661	19522	18536	18387	17950	10369	9249	8826	8558
	ccb	24301	23125	22677	22655	19032	18520	18498	17888	9324	8868	8742	8405
	cbb	31484	28779	28695	28649	24819	24668	21317	21360	12038	10691	10714	9778
clos	ran	10632	9591	9151	9051	10453	9564	9121	8713	10369	9249	8826	8558
	ccb	9832	9388	9147	8824	9672	9166	8945	8712	9324	8868	8742	8405
	cbb	13233	11317	11317	11001	12413	11535	10894	10948	12038	10691	10714	9778
RSC	ran	10159	9643	9218	8793	10375	9590	9141	8773	10369	9249	8826	8558
	ccb	9655	9123	8924	8764	9760	9176	8864	8675	9324	8868	8742	8405
	cbb	13018	11092	10561	10481	12484	11002	10923	10738	12038	10691	10714	9778

4 Conclusion

At present, the equipment-related technologies are rapidly making progress, but the operation-related studies are relatively left behind. Our studies have been made on the operation method of automated equipment in the automated container terminal, and the operation of automated equipment is mainly performed by mutual cooperation of container crane, ATC, and AGV. In particular, since these mutual activities are performed under diverse dynamic environments, integrated operation of all the equipment is badly needed. Because of this, this paper is based on both AGV rule and ATC rule, while trying to suggest an event-oriented integrated operation method. We expect that the simulation models presented in this study will make practical contribution to the solution of diverse problems of the real automated container terminal.

Acknowledgement

This work was supported by the Regional Research Centers Program(Research Center for Logistics Information Technology), granted by the Korean Ministry of Education & Human Resources Development.

References

1. Chen, Y., Leong., T.Y., Ng, J.W.C., Demir, E.K., Nelso, B.L., Simchilevi, D.: Dispatching automated guided vehicles in a mega container terminal. Proceedings of the 1998 INFORMS, Montreal, Canada (1998)
2. Duinkerken, M.D., Evers, J.J.M., Ottjes, J.A.: A simulation Model for Integrating Quay Transport and Stacking Policies on Automated Container Terminals. Proceeding of the 15th European Simulation Multiconference, June (2001)
3. Kim, K.H., Bae, J.W.: A dispatching method for automated guided vehicles to minimize delays of containership operation. International Journal of Management Science, 5 (1999) 1-25
4. Kim, K.H., Kim, K.Y.: A Routing Algorithm for a Single Transfer Crane to Load Export Containers onto a Containership. Computers and industrial Engineering, 33 (1997) 673-679
5. Kim, K.H., Kim, K.Y.: An Optimal Routing Algorithm for a Transfer Crane in Port Container Terminals. Transportation Science, 33 (1999) 17- 33
6. Korea Maritime Institute: The Basic Development Plan for Gwangyang Port Automated Container Terminal. (2001)
7. Meersmans, P.J.M., Wagelmans, A.P.M.: Effective algorithms for integrated scheduling of handling equipment at automated container terminals. Erasmus Research Institute of Management, (2001) 1-20
8. Meersmans, P.J.M., Wagelmans, A.P.M.: Dynamic scheduling of handling equipment at automated container terminals. Report series research in management, ERASMUS Research Institute of Management, (2001) 1-26

A Study of Factors Affecting Due Date Predictibility for On-Time Delivery in a Dynamic Job Shop

Şerafettin Alpay and Nihat Yüzügüllü

Eskişehir Osmangazi University, Bademlik, 26030, Eskişehir, Turkiye
`salpay@ogu.edu.tr, nyuzugul@ogu.edu.tr`

Abstract. In this research, the factors have significant effect on due date predictability for on time delivery in a dynamic job shop were studied. For this aim a new due date assignment model was proposed and an experimental study in a simulated dynamic job shop was performed. The results indicate that the proposed due date assignment model is adequate to meet the assigned due dates for on time delivery and the number of jobs in the queues of the machines on a job's route is the most important factor affects the due date predictability.

Keywords: Dynamic Job Shop Scheduling, Simulation, On-time Delivery, Due Date Assignment, Dispatching.

1   Introduction

Due date based scheduling in the dynamic job shops has been examined in the related literature [1-7]. Actually, there are several job and shop characteristics or factors affect the due date predictability. Most used factors are processing time, number of jobs, jobs in route, time in route and jobs in queue [5, 7, 8, 9, 10].

In this study, a due date assignment model was defined and an experimental study in a dynamic job shop was performed to find out the most effective factor(s) according to the performance measures; Mean Absolute Lateness (MAL) and Mean squared Lateness (MSL). All simulation results were analyzed with ANOVA.

2   The Proposed Due Date Assignment Model

The factors classified as job and shop related characteristics are given below:

- *Job related factors :* Total processing time of a job (p), Number of operations on a job (n)
- *Shop related factors :* Total number of jobs on the machines and in their queues on a job's route when the job is released to the system (*JR),* Total processing time of the jobs on the machines and in their queues on a job's route when the job is released to the system (*TR*), Total number of jobs on the machines and in their queues out of a job's route when the job is released to the system (*JO*), Total processing time of the jobs on the machines and in their queues out of a job's route when the job is released to the system (*TO*), Total number of jobs in the queues of the machines on a job's route when the job is released to the system (*JQ*),

A. Sattar and B.H. Kang (Eds.): AI 2006, LNAI 4304, pp. 1239–1245, 2006.
© Springer-Verlag Berlin Heidelberg 2006

Predicted mean of flow times of the last completed three jobs (*Pred. F.*). Based on these factors and their pair-wise relationships, the proposed model is given as follows [11]:

$$
\begin{aligned}
d_i = {} & r_i + k_p\, p_i + k_n\, n_i + k_{JR}\, JR_i + k_{TR}\, TR_i + k_{JO}\, JO_i + k_{TO}\, TO_i + k_{JQ}\, JQ_i + k_{p.n} \\
& p_i.n_i + k_{p.JR}\, p_i.JR_i + k_{p.TR}\, p_i.TR_i + k_{p.JO}\, p_i.JO_i + k_{p.TO}\, p_i.TO_i + k_{p.JQ}\, p_i.JQ_i + \\
& k_{n.JR}\, n_i.JR_i + k_{n.TR}\, n_i.TR_i + k_{n.JO}\, n_i.JO_i + k_{n.TO}\, n_i.TO_i + k_{n.JQ}\, n_i.JQ_i + k_{JR.TR} \\
& JR_i\, TR_i + k_{JR.JO}\, JR_i\, JO_i + k_{JR.TO}\, JR_i\, TO_i + k_{JR.JQ}\, JR_i\, JQ_i + k_{TR.JO}\, TR_i\, JO_i + \\
& k_{TR.TO}\, TR_i\, TO_i + k_{TR.JQ}\, TR_i\, JQ_i + k_{JO.TO}\, JO_i\, TO_i + k_{JO.JQ}\, JO_i\, JQ_i + k_{TO.JQ}\, TO_i \\
& JQ_i + k_{pred\,f}\, Pred\,f_i
\end{aligned}
\tag{1}
$$

Where d_i denotes the due date of job i , r_i denotes the ready time of job i and, k_p , k_n , k_{JR} , k_{TR} , k_{JO} , k_{TO} , k_{JQ} , $k_{p.n}$, $k_{p.JR}$, $k_{p.TR}$, $k_{p.JO}$, $k_{p.TO}$, $k_{p.JQ}$, $k_{n.JR}$, $k_{n.TR}$, $k_{n.JO}$, $k_{n.TO}$, $k_{n.JQ}$, $k_{JR.TR}$, $k_{JR.JO}$, $k_{JR.TO}$, $k_{JR.JQ}$, $k_{TR.JO}$, $k_{TR.TO}$, $k_{TR.JQ}$, $k_{JO.TO}$, $k_{JO.JQ}$, $k_{TO.JQ}$, k_{predf} denote the allowance values of the corresponding factors in the model.

3 A New Dispatching Rule (CR+OPNSLK)

According to CR+OPNSLK rule, the modified due date $d_i^{\,*}$ of job i waiting for service at a machine at time t is defined as [11]:

$$
d_i^{\,*} = min\left[\, t + CR.p_{ik},\; d_i - \sum_{j=k+1}^{n_i} p_{ij}\, \right]
\tag{2}
$$

Where $d,\ p,\ CR$ denote original due date, processing time and Critical Ratio respectively for job i with n_i operations and its kth operation ($k \le n_i$) is imminent at time t.

4 Research Problem

Performance Measures : Two shop performance measures, Mean Absolute Lateness (MAL) and Mean Squared Lateness (MSL) [6, 11] were considered.
Dispatching Rules: Four commonly used dispatching rules were selected [6, 11]:
1. FIFO, 2. SPT, 3. CR+SPT, 4. All+CR+SPT
Due Date Assignment Models: The following due date assignment models were used [6, 11]: 1. TWK, 2. PPW, 3. DTWK, 4. DPPW
Shop Utilization: Two shop utilization rates were considered in the study: 90% and 80% [6, 11].

Experimental Study
Simulation model: The simulated dynamic job shop used in this study consists of 5 unique machines. Job arrivals at the shop form a Poisson process and the shop arrival rate is determined by the shop utilization setting. Each job requires one to nine operations, determined randomly according to a uniform probability distribution. Each operation is assigned randomly to one of the nine machines according to a uniform probability distribution, under the constraint that no two successive operations require the same machines. Operation processing times are exponentially

distributed with a mean of 1 unit of time. Other assumptions made in this system are; set-up times are included in the processing times, transportation times are excluded, the resources are available continuously, pre-emption of a job is not allowed and processing times of the jobs are known after their arrivals at the shop.

Experimental procedure: A three-factor full factorial design was used by this experiment to study the performance of the five due date assignment models. The first factor is due date assignment model that has five levels (based on factors 1.TWK, 2.PPW, 3.DTWK, 4.DPPW and 5.Proposed model). The second factor is the dispatching rule that has five levels (1.FIFO, 2.SPT, 3.CR+SPT, 4.All+CR+SPT, 5.CR+OPNSLK) and the third factor is the shop utilization rate that has two levels (1. 0,9, 2. 0,8). For each treatment 6 replications were performed. For each simulation run after the completion of first 1.000 jobs to reach steady state, information is collected on MAL and MSL for the next 10.000 jobs. The proposed due date assignment models correspond to each dispatching rule are given in Table 1 [11].

Table 1. Proposed due date assignment models for each dispatching rule

	For 90% shop utilization
FIFO	d= r − 1,38 + 2 n + 0,0358 TR + 0,839 JQ + 0,130 p.n -0,0539 p.JO + 0,0107 n.TO -0,130 n.JR + 0,0259 n.TR − 0, 0167 n.JQ − 0,000583 JR.TR + 0,000197 TR.JQ
SPT	d= r + 1,84 p + 0,733 JQ + 0,0321 p.TR + 0,0870 p.JO + 0,103 p.JQ - 0,0234 n.TR - 0,107 n.JQ - 0,00905 JR.JQ - 0,00376 TR.JO
CR+SPT	d= r + 0,517 p + 0,0603 TR + 1,38 JQ + 0,217 p.n - 0,0140 p.JR + 0,0241 p.JO - 0,124 n.JR + 0,0108 n.TR - 0,0484 n.JO - 0,0245 n.JQ -0,000582 JR.TR - 0,00632 JR.JO - 0,0131 JR.JQ + 0,00235 TR.JQ - 0,0213 Pred.F
ALL+CR +SPT	d= r + 0,879 + 0,679 n − 0,769 JR + 0,147 TR + 1,42 JQ + 0,213 p.n + 0,0311 p.JO − 0,00616 p.JQ − 0,0456 n.JR − 0,0593 n.JO − 0,0354 n.JQ − 0,0062 JR.JQ − 0,000089 TR.TO + 0,00115 TR.JQ − 0,0207 Pred.F
CR+ OPNSLK	d= r + 1,63 + 0,571 n -0,785 JR + 0,179 TR + 0,154 JQ +0,219 p.n +0,0402 p.JO − 0,0456 n.JR − 0,00947 n.TR − 0,0892 n.JO − 0,0376 n.JQ − 0,0124 JR.JQ − 0,000151 TR.TO + 0,00218 TR.JQ − 0,0119 Pred.F
	For 80% shop utilization
FIFO	d= r - 1,05 + 0,565 p + 1,21 n + 0,0798 TR + 0,565 JQ + 0,0657 p.n + 0,00597 n.TR + 0,00741 n.TO - 0,00144 JR.TR - 0,0156 JR.JQ + 0,00319 TR.JQ - 0,0507 Pred.F
SPT	d= r + 0,783 + 0,976 p + 0,521 n + 0,0987 p.JR + 0,193 p.JO + 0,0873 p.JQ - 0,149 n.JO - 0,0703 n.JQ + 0,0229 JO.JQ
CR+SPT	d= r + 0,747 p + 0,356 n + 0,0792 TR + 0,891 JQ + 0,0858 p.n - 0,0376 p.JO + 0,00758 p.TO - 0,0214 n.JQ - 0,00138 JR.TR - 0,0106 JR.JO - 0,0286 JR.JQ + 0,00577 TR.JQ + 0,0103 JO.JQ - 0,0289 Pred.F
ALL+CR +SPT	d= r + 0,837 p + 0,590 n - 0,262 JR + 0,151 TR + 0,917 JQ + 0,0542 p.n + 0,0167 p.JR + 0,00327 p.TO - 0,0135 n.TR - 0,0288 n.JO -0,000807 JR.TR - 0,0350 JR.JQ + 0,00669 TR.JQ - 0,0405 Pred.F
CR+ OPNSLK	d= r - 1,09 + 1,03 p + 0,680 n - 0,442 JR + 0,117 TR + 0,0224 TO + 1,02 JQ + 0,0515 p.n - 0,0289 n.JQ - 0,0112 JR.JQ + 0,00212 TR.JQ - 0,000778 JO.TO - 0,0441 Pred.F

5   Results and Discussions

All MAL and MSL results averaged over 6 runs of the experiments are summarized in Table 2.

Table 2. Experimental results for MAL and MSL (MSL results are given in parenthesis)

UTIL.	D.D.Mod.	FIFO	SPT	CR+SPT	ALL+CR+SP	CR+OPNSLK
%90	TWK	23.37 (1004.3)	13.30 (1216.6)	14.56 (457.1)	14.46 (454.9)	13.67 (338.2)
	PPW	19.73 (766.7)	14.44 (1401.3)	14.64 (458.2)	14.48 (446.9)	13.90 (354.2)
	DTWK	20.68 (891.9)	13.03 (1191.6)	7.62 (131.3)	7.54 (126.6)	7.42 (120.9)
	DPPW	14.77 (413.9)	18.09 (1442.0)	9.81 (184.1)	9.76 (182.1)	9.74 (180.8)
	Proposed	8.85 (166.8)	12.63 (1122.6)	5.27 (57.1)	5.31 (58.5)	5.12 (53.2)
%80	TWK	11.12 (232.7)	5.99 (165.6)	7.32 (121.5)	7.14 (113.2)	7.00 (92.7)
	PPW	9.65 (186.6)	13.15 (310.2)	7.26 (121.2)	7.14 (114.5)	6.93 (95.2)
	DTWK	10.23 (231.3)	6.65 (166.2)	4.59 (50.2)	4.58 (49.7)	4.52 (48.3)
	DPPW	9.20 (160.1)	10.22 (247.7)	7.22 (99.9)	7.22 (99.8)	7.22 (99.8)
	Proposed	5.65 (65.0)	5.62 (147.5)	3.42 (23.8)	3.39 (23.4)	3.41 (23.4)

The results in the tables show that CR+OPNSLK rule with almost all due date assignment models and the proposed due date assignment model with almost all dispatching rules and shop utilizations demonstrate the best MAL and MSL performances. The combination of the proposed due date assignment model with CR+OPNSLK is superior to almost all other combinations for both MAL and MSL performance. A three-factor ANOVA method was then applied to test the effects of due date assignment model, dispatching rule and shop utilization rate on the performance measures. All MAL and MSL statistics are shown in Table 3 and Table 4 respectively.

Table 3. F values for Three-Factor Analysis of Variance of MAL performance measure

Source	D.F.	S.S.	M.S.	F	P
Due Date Assignment Model	4	1582.15	395.54	191.82	0.00
Disp. Rule	4	1432.02	358.00	173.62	0.00
Shop utilization rate	1	2230.96	2230.96	1081.92	0.00
DDAM.*Disp.Rule	16	593.20	37.07	17.98	0.00
DDAM.*Shop U.R.	4	234.27	58.57	28.40	0.00
Disp.Rule*Shop U.R.	4	183.81	45.95	22.28	0.00
DDAM.*Disp.Rule*ShopU.R.	16	267.15	16.70	8.10	0.00
Error	250	515.51	2.06		
Total	299	7039.08			

Table 4. F values for Three-Factor Analysis of Variance of MSL performance measure

Source	D.F.	S.S	M.S.	F	P
Due Date Assignment Model	4	2555726	638931	17.93	0.00
Disp. Rule	4	15813098	3953274	110.97	0.00
Shop utilization rate	1	12319813	12319813	345.97	0.00
DDAM.*Disp.Rule	16	1549482	96843	2.72	0.00
DDAM.*Shop U.R.	4	1015234	253808	7.12	0.00
Disp.Rule*Shop U.R.	4	9325449	2331362	65.44	0.00
DDAM.*Disp.Rule*ShopU.R.	16	706584	44162	1.24	0.238
Error	250	8906329	35625		
Total	299	52191715			

As indicated by the statistics shown in Table 3 and Table 4, all three main factors and their pair-wise interactions have significant effects (at the 95 confidence level) on MAL and MSL. In Figure 1, all main factor effects are shown.

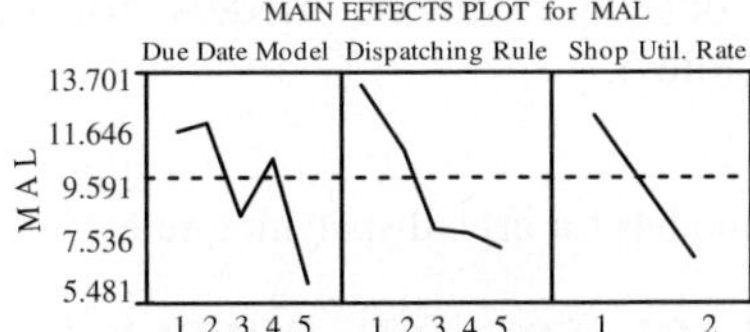

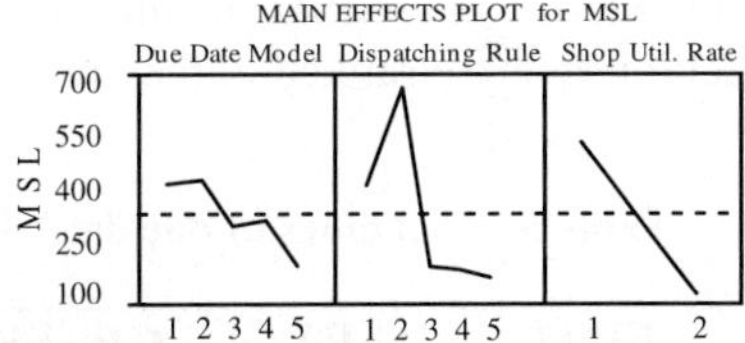

Fig. 1. Main effects plot for MAL and MSL

The plots of due date assignment model according to dispatching rule at 0.8 and 0.9 of shop utilization rate for MAL and MSL are given in Fig. 2 and Fig.3.

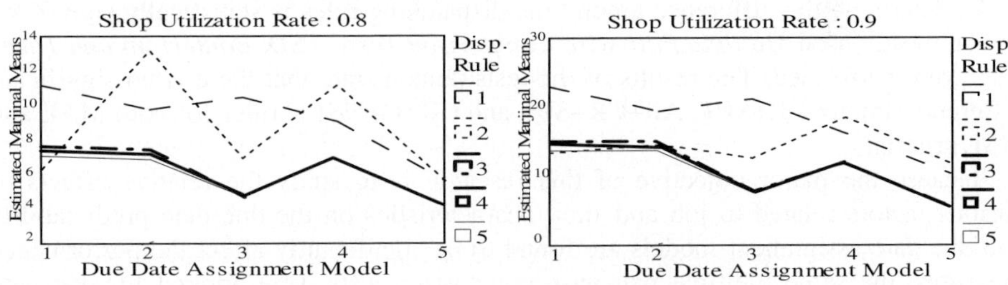

Fig. 2. Due date model & Dispatching rule plot at 0.8 and 0.9 utilization rates for MAL

At low shop utilization rate, the proposed due date assignment model with all dispatching rules demonstrates the best MAL performance. The next good assignment model is DTWK (numbered as 3). On the other hand, it is not easy to select the best dispatching rule for MAL performance. CR+SPT, All+CR+SPT and CR+OPNSLK rules show very similar MAL performances. These results do not vary so much as shop utilization rate increases although the MAL performances of the models deteriorate.

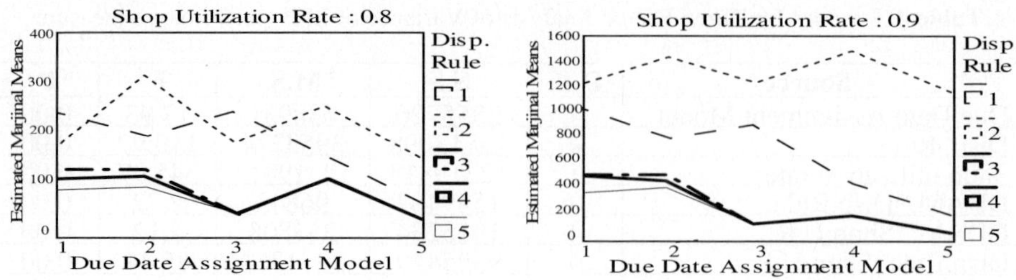

Fig. 3. Due date model & Dispatching Rule plot at 0.8 and 0.9 utilization rates for MSL

For the MSL performance, Fig. 3 shows that the proposed model is also best due date assignment model and its performance decreases as shop utilization rate increases. Other results are very similar to the results of MAL: DTWK model is the second good model and CR+SPT, All+CR+SPT and CR+OPNSLK rules have close MSL performance.

Both MAL and MSL results of all due date assignment models for each dispatching rule are ranked and summarized in Table 5.

Table 5. Final ranks of due date assignment models for each dispatching rule

FIFO	SPT	CR+SPT	ALL+CR+SPT	CR+OPNSLK
Proposed	Proposed	Proposed	Proposed	Proposed
DPPW	DTWK	DTWK	DTWK	DTWK
PPW	TWK	DPPW	DPPW	DPPW
DTWK	PPW	TWK	TWK	TWK
TWK	DPPW	PPW	PPW	PPW

To determine the difference among the dispatching rules is statistically significant or not, mostly used *Multiple Pair-wise Comparison Tests: LSD, Bonferroni and Tukey* tests were performed. The results of the tests demonstrate that there is no significant difference among CR+SPT, All+CR+SPT and CR+OPNSLK rules for both MAL and MSL criteria.

Because the major objective of this research is to study the relative effects of various factors related to job and shop characteristics on the due date predictability and due date assignment models are found to be significantly affect the performance measures, the same multiple pair-wise comparison tests were applied for due date assignment models. According to the results of the tests, the proposed due date assignment model is superior to the other models and the differences on MAL and MSL performances of the proposed model and other models are statistically significant. Although average MAL and MSL performances of CR+OPNSLK rule are very close to the performances of CR+SPT and All+CR+SPT rules, it is obvious that the combinations of CR+OPNSLK rule and TWK/PPW models will give better MAL and MSL performances than other two models can give. When we analyze the proposed models related to the dynamic dispatching rules at different shop utilization rates, we can see that the most effective factor on due date predictability in terms of selected performance measures is *the number of jobs in the queues of the machines on a job's route* (JQ). This result is consistent with the result reported by Chang [10]. The next factors can be count as the number of operations on a job (n) and total processing time (p) of a job.

6 Conclusions

In this research, a due date assignment model includes most used factors and a dynamic dispatching rule, CR+OPNSLK, were proposed. Experimental results show that the proposed model is very successful to predict the exact due dates and CR+OPNSLK rule is as adequate to meet the due dates as (or better on average than) CR+SPT or All+CR+SPT rules. The result also show that *the number of jobs in the queues of the machines on a job's route* (JQ) is the most important factor affects the due date predictability. The number of operations on a job and the total processing time information are the next important factors. On the other hand, the completion time of a job is affected by many different factors and to increase the quality of prediction of the exact due dates in a dynamic job shop, more factors should be considered in due date assignment.

The future works on this area may be developing better due date assignment models based on these three important factors and developing more intelligent systems (like agent based systems) capable of updating the allowance values of the factors dynamically to meet the assigned due dates.

References

1. A. Lengyel, I. Hatono, K. Ueda, Scheduling for on-time completion in job shops using feasibility function, Computers & Industrial Engineering, 45, (2003), 215-229.
2. E.A. Veral, Computer simulation of due-date setting in multi-machine job shops, Computers & Industrial Engineering, 41, (2001), 77-94
3. H.P.G. van Ooijen and J.W.M. Bertrand, Economic due-date setting in job-shops based on routing and workload dependent flow time distribution functions, Int. J. Production Economics 74 (2001) 261-268
4. I. Sabuncuoglu and A. Comlekci, Operation-based flowtime estimation in a dynamic job shop, Omega 30 (2002) 423–442
5. E.S. Gee and C.H. Smith, Selecting allowance policies for improved job shop performance, International Journal of Production Research, (1993), 1839-2852
6. Cheng, T.C.E. and Jiang, J., 1998, Job shop scheduling for missed due dates performance, Computers and Industrial Engineering, vol 34, No 2, 297-307.
7. E.J. Liao and H.T. Lin, A case study in a dual resource constrained job shop, International Journal of Production Research, 36(11), (1998), 3095-3111.
8. K.R. Baker, Sequencing rules and due-date assignment in a job shop . Management Science, 30, (1984), 1093-1104.
9. T.C.E. Cheng and M.C. Gupta, Survey of scheduling research involving due date determination decisions, European. Journal of Operations Research, 38, (1989), 156-166
10. F.C.R. Chang, A study of factors affecting due date predictability in a simulated dynamic job shop, Journal of Manufacturing Systems, 13(6), (1997),393-400.
11. S. Alpay, Dynamic Stochastic Multi-Machine Job Shop Scheduling, Eskisehir Osmangazi University, Ph.D. Dissertation, Turkey (2003)

A New Learning Method for S-GCM

Hamed Rahimov, Mohammad-Reza Jahedmotlagh, and Nasser Mozayani

Iran University of Science and Technology, Computer Engineering Faculty, Tehran, Iran
Hrahimov@comp.iust.ac.ir, {Jahedmr, Mozayani}@iust.ac.ir

Abstract. One of artificial neural network models with non-equilibrium dynamics is S-GCM. This model in comparison to Hopfield method has more capacity storage and more success rate, but yet, as an associative memory system has some weakness such as small storage rate and low speed of convergence. In this paper, a new learning method for S-GCM is proposed. In the proposed method, we use modified sparse matrix for learning method. Both the theory analyses and computer simulation results show that the performance of S-GCM can be improved greatly by using our learning method.

Keywords: Chaotic Associative Memory, Symmetric Globally Coupled Map (S-GCM), More Iterate More Store (MIMS) Learning.

1 Introduction

People are certain that chaotic dynamic behavior plays an important role in real neurons and neural networks [1,2]. Many researchers have attempted to model artificial neural networks with chaotic dynamics on the basis of deterministic differential equations or stochastic models. For example, Aihara et al. [3] have proposed deterministic difference equations, which describe an artificial neural network model, composed of chaotic neurons. This model has advantages in terms of computational time and memory for numerical analyses due to the spatiotemporal complex dynamics of the neurons. Adachi et al. [4] proposed a system based on model proposed by Aihara et al. A model based on coupled chaotic elements, called globally coupled map model, is proposed by Kaneko [5], and an improvement version of this model, called globally coupled map using the symmetric map, is proposed by Ishii et al. [6]. Ishii et al.'s studies show that in S-GCM both memory capacity and the basin volume for each memory are larger than those in the Hopfield model applying the same learning rule. Moreover, Inoue et al. [7,8] have presented a chaotic neuro-computer in which a neuron is composed of a pair of coupled oscillators. The neuro-computer runs on a deterministic rule, but it is capable of stochastic searching and solving difficult optimization problems. Ishii's modified global coupled chaotic system and Inoue's chaotic neuro-computer are two main chaotic neural networks used for pattern recognition and associative memory. Zhang et al. [9,10] proposed a one-dimensional, two-way coupled map network and a modified definition of an auto-associative matrix. Ishii et al. [11] proposed an associative memory system based on parametrically coupled chaotic elements. The proposed system was obtained by adding a new parameter control to Ishii's previously proposed system. A chaotic activity in an early association stage makes an efficient association over the memories

A. Sattar and B.H. Kang (Eds.): AI 2006, LNAI 4304, pp. 1246–1251, 2006.

© Springer-Verlag Berlin Heidelberg 2006

that are stored by means of autocorrelation learning. When the system successfully recalls the target memory, the system's motion is dominated by a spatially coherent oscillation, while unstable motions remain when the system fails to make the association. This system had a large memory capacity. In addition, Zheng et al. [12] proposed a new parameter control method for S-GCM. With the proposed method, the changes of the parameters are decided not only by the value of system partial energy, but also by the difference value of the partial energy. Results showed that the performance of S-GCM could be improved greatly by using the new parameter control method.

In the other hand, Menhaj et al. [13] proposed a learning method for Hopfield network, named sparse learning. They showed that the method learning is compatible to energy function of Hopfield model, and it cans storage target patterns in attractors with minimum energy. Additionally, they proved this learning method has more storage rate and more speed convergence contrast with modified Heb learning method. Based on an analysis of above-mentioned chaotic neural network models and their applications in information processing, to increase the recall speed and capacity of S-GCM, we proposed a new learning method named MIMS learning. Analyses show that the performance of S-GCM can be improved by using the MIMS learning method. Computer simulation results show that the recall speed is increased greatly by using the new method, too.

The rest of this paper is as following. In Section 2, the MIMS learning method is clarified. In Section 3, computer simulation results of solving associative memory problem, by using the conventional learning method and new method, respectively, are presented. Some conclusions are given in Section 4.

2   The MIMS Learning Method

Studies show that one of the disadvantages of S-GCM is that the recall speed is slow [9,10,12]. Additionally, another of the disadvantages of S-GCM is that the memory capacity is low [11]. To increase the recall speed and memory capacity of S-GCM, we propose the following learning method instead of modified Heb learning.

Let $\left\{\chi^1, \chi^2, ..., \chi^m \middle| \chi^k \in \{1,-1\}^N \right\}$ be a set of N-dimensional binary patterns to be stored. χ_i^k denotes the i th unit value in the k th binary pattern, and M is the number of stored patterns.

$$t_{ij} = \sum_{k=m}^{1} \sum_{s=k}^{1} \frac{1}{N} \prod_{k=1}^{m} (\chi_i^{(s)} + \chi_j^{(s)})$$

$$T = [t_{ij}], \qquad W = [w_{ij}], \qquad W+ = T^T \cdot T$$

(1)

Matrix $[w_{ij}]$ is called MIMS matrix. The method is like sparse method, with this difference that in sparse method, T matrix in firstly built according through stored patterns and then $T^T \cdot T$ product is considered as weight matrix. But in the innovative method, by adding every new pattern, the previous T matrix is also considered and $T^T \cdot T$ product is calculated and is added to the previous weight matrix. This process will continue till training all patterns. This way of creating the weight matrix will result this method to recur the stored patterns and in result it will be dependent on the sequence of storing patterns. In other words, primary patterns have more effect in creating the weight matrix in comparison to the patterns will be stored finally, it means they are recurred more and consequently they are stored and stick better in the memory. It seems this method of learning is more similar to the man's way of learning, as the patterns which we repetition during time we will keep them in our long-term memory better [14]. It cans storage target patterns in attractors with minimum energy. For the system, we proposed the following energy function:

$$E = -\sum_{k=m}^{1}\sum_{s=k}^{1}\sum_{i}\sum_{j} w_{ij} \cdot \left(\chi_i^{(s)} + \chi_j^{(s)}\right)^2 . \tag{2}$$

The S-GCM with this method learning searches for a local minimum of the energy function by making each partial energy, E_i , small and negative as follows:

$$\alpha_i(t+2) = \alpha_i(t) + \left[\alpha_i(t) - \alpha_{\min}\right] \cdot \tanh\left(\beta \cdot E_i(t)\right)$$

$$E_i(t) = -x_i(t) \cdot \sum_{j=1}^{N} w_{ij} \cdot x_j(t) \tag{3}$$

Where $\alpha_{\min}$ and β are constant parameters. x_i and $E_i(t)$ are the i th system state and the i th partial energy, respectively. ε is a constant parameter, and α is bifurcation parameter.

3 Experimental Results and Analysis

We use the conventional learning method and the new method to realize associative memory. The value of the parameters for the both methods is:

$$\beta = 2, \qquad N = 100, \qquad \alpha_{\min} = 3.4, \qquad \alpha_{\max} = 4, \qquad \varepsilon = 0.1$$

The training set contains five binary patterns (Fig. 1), and the size of each pattern is 10 by 10.

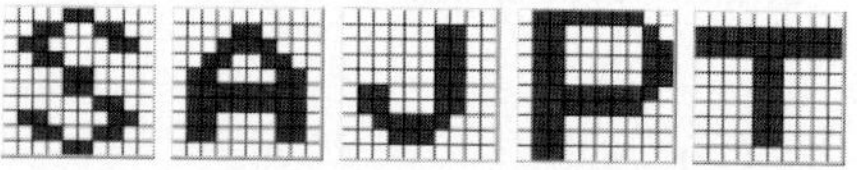

Fig. 1. Five stored patterns

To test the effectiveness of the proposed method, for each character we built 3 different patterns by adding 0%, 5%, and 10%, random pepper and salt noise to each pattern. Fig. 2 shows time-series of the overlap in S-GCM with these two methods when the initial overlap is set at various values. All of test patterns can be retrieved correctly and systems can recall the target pattern from a fairly distant initial state. Similar associations have been obtained for other stored patterns.

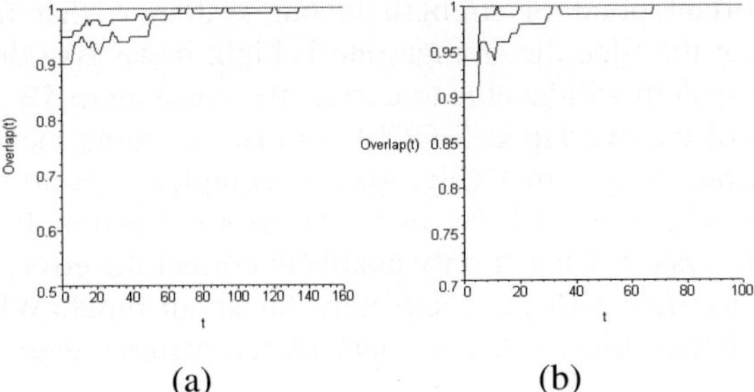

Fig. 2. Time-series of the overlap in S-GCM with two learning methods, P = 5 = 0.05N. (a) Heb learning method. (b) MIMS learning method. The abscissa denotes continuous-time t.

Let us show an example association process. When memorizes five 100-bit binary patterns shown in Fig. 1, and the input binary vector I is a 15% reversed pattern of "A", it associates "A" after scores of transitions. Figures 3(a) and 3(b) show this association process. In Fig.3(b), the abscissa denotes the association time t and the ordinate denotes the overlap values. In this figure, highly chaotic motions are observed at the early association stage. As time elapses, these motions become quiet, and the association is completed when the system falls into a 4-cluster frozen attractor. At this time, its binary representation O is successfully equivalent to "A" as Fig. 3(a) shows.

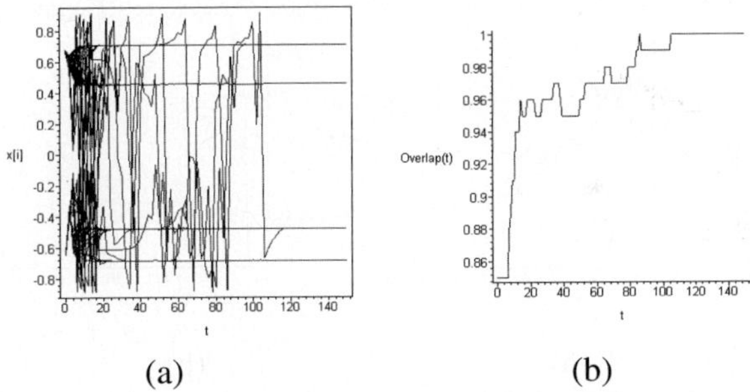

Fig. 3. Time-series in S-GCM with MIMS learning methods for A, (a) output values of all neurons (b) overlap association. The abscissa denotes continuous-time t.

In this system, there will be similar results for the three patterns, which are stored at the beginning, but for the two last patterns, there would not be an appropriate result. If we alter the sequence of storing the patterns, the result would be proper for the primary patterns and inappropriate for the final patterns. We offer a solution to solve this problem. Additionally, to contrast memory storage of these two learning methods, we store 100 Random patterns of 130 patterns and the size of each pattern is 10 by 10 for both methods. Then, in retrieval stage, we use stored patterns as input patterns.

The most important point of strength in our system is that for the first-stored patterns, i.e. even at the time the storage rate is high, if we give them as inputs with random noise, our system will be able to correct the noise up to 5%. In the Fig. 4 have shown time-series of the overlap in S-GCM with two learning methods. As shown in this figure, the output of system-1 (Fig. 4(a)) is completely gone far from the target patterns while system-2 (Fig. 4(b)) has done 3% error correction. In general, when the storage rate is high, system-1 is not only unable to correct the error, but also by giving the target pattern itself, it will go completely far as an input. While system-2 will perform the error correction for the primary-stored patterns even when the storage rate is so high.

The weakest point in the method of MIMS learning is in the final patterns or on the other hand, in the case of the patterns which has less influence in creating weight matrix. In order for the system to have suitable success rate for the final patterns, we will perform the method of learning as follows: for this purpose, we will store the patterns from the beginning to the end once and another time we store the same patterns from the end to the beginning. The result of our implementation indicates that applying this method will increases the final patterns of success rate, especially when the storage rate is low is more conspicuous. The reason for the final patterns of the success rate to be less than the primary patterns is that applying this method will not result in removing the sensitivity in sequence of the pattern storage. The strength of the final patterns in creating the weight matrix will increase more only to some extent, but yet, the primary patterns have more influence in creating the weight matrix.

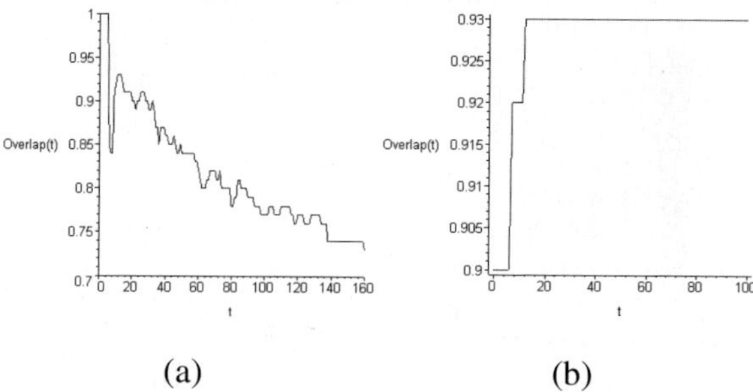

(a)   (b)

Fig. 4. Time-series of the overlap in S-GCM with two learning methods, (a) Heb learning, (b) MIMS learning

4 Conclusion

In this paper, MIMS learning method for S-GCM is proposed. The core of this method is that using MIMS matrix instead of covariance matrix for learning S-GCM. This method to recur the stored patterns and in result it will be dependent on the sequence of storing the patterns, on the other hand, primary patterns have more effect in creating the weight matrix in comparison to the patterns will be stored finally, it means they are recurred more and consequently they are stored and stick better in the memory.

Acknowledgments. Iran Telecommunication Research Center (ITRC) supported this work.

References

1. Yong, Y., Freeman, W.J.: Model of biological pattern recognition with spatially chaotic dynamics. Neural Networks. 3 (1992) 153-170
2. Skarda, C.A., Freeman, W.J.: How brains make chaos in order to make sense of the world. Behavior Brain Science. 10 (1987) 161-195
3. Chen, L., Aihara, K.: Chaotic simulated annealing by a neural network model with transient chaos. Neural Networks. 8 (1995) 915–930
4. Adachi, M., Aihara, K., Kotani, M.: An analysis of associative memory dynamics with a chaotic neural network. Proceedings of the International Symposium on Nonlinear Theory and its Applications. (1993) 1169-1172
5. Kaneko, K.: Clustering, coding, switching, hierarchical ordering, and control in a network of chaotic elements. Physica D. 41 (1990) 137-172
6. Ishii, S., Fukumizu, K., Watanabe, S.: A network of chaotic elements for information processing. Neural Networks. 9 (1996), 25–40
7. Inoue, M., Nagayoshi, A.: A chaos neuro-computer. Phys. Lett. A. 158 (1991) 373–376
8. Inoue, M., Fukushima, S.: A neural network of chaotic oscillators. Progress Letters. 87 (1992)
9. Zhang, Y., Yang, L., He, Z.: Chaotic neural network for associative memory. J. Appl. Sci. 17 (1999) 259–266
10. Zhang, Y., Yang, L., He, Z.: Study of chaotic neural network and its applications in associative memory. Neural Processing Letters. 9 (1999) 163-175
11. Ishii, S., Sato, M.: Associative memory based on parametrically coupled chaotic elements. Physica D. 121 (1998) 344-366
12. Zheng, L., Tang, X.: A new parameter control method for S-GCM. Pattern Recognition Letters. 26 (2005) 939-942
13. Menhaj, M.B. (ed.): Fundamentals neural networks. Computational intelligence, 2nd edn. Vol. 1, Amirkabir University, AKU press (2002) 222-229
14. Calvin, W.H., Ojemann, G.A. (eds.): Conversations with Neil's Brain. The Neural Nature of Thought & Language. Addison-Wesley (1994)

Empirical Verification of a Strategy for Unbounded Resolution in Finite Player Goore Games*

B. John Oommen[1], Ole-Christoffer Granmo[2], and Asle Pedersen[3]

[1] School of Computer Science, Carleton University, Ottawa, Canada**
oommen@scs.carleton.ca
[2] Department of ICT, Agder University College, Grooseveien 36, Grimstad, Norway
ole.granmo@hia.no
[3] InterMedium, Televeien 3, Grimstad, Norway

Abstract. This paper presents an *experimental* verification of a novel, fast and arbitrarily accurate solution to the Goore Game (GG). The latter game, introduced in [6], has the fascinating property that it can be resolved in a completely distributed manner with no inter-communication between the players. The game has recently found applications in many domains, including the field of sensor networks and Quality-of-Service (QoS) routing. In actual implementations of the solution, the players are typically replaced by Learning Automata (LA). The problem with the existing reported approaches is that the accuracy of the solution achieved is intricately related to the *number* of players participating in the game – which, in turn, determines the resolution, implying that arbitrary accuracy can be obtained only if the game has an *infinite* number of players. In this paper, we experimental demonstrate how we can attain an unbounded accuracy for the GG by utilizing no more than *three* stochastic learning machines, and by a recursive pruning of the solution space.

1 Introduction

This paper presents an *experimental* verification of a novel, fast and arbitrarily accurate solution to the Goore Game (GG) which is one of the most fascinating games studied in the field of artificial games. The GG can be best described using the informal formulation of [2]: *"Imagine a large room containing N cubicles and a raised platform. One person (voter) sits in each cubicle and a Referee stands on the platform. The Referee conducts a series of voting rounds as follows. On each round the voters vote "Yes" or "No" (the issue is unimportant) simultaneously and independently (they do not see each other) and the Referee counts the fraction, λ, of "Yes" votes. The Referee has a uni-modal performance criterion*

---

* This work was partially supported by NSERC, the Natural Sciences and Engineering Research Council of Canada.
** *Chancellor's Professor*; *Fellow : IEEE* and *Fellow : IAPR*. The Author also holds an *Adjunct Professorship* with the Dept. of ICT, Agder University College, Norway.

A. Sattar and B.H. Kang (Eds.): AI 2006, LNAI 4304, pp. 1252–1258, 2006.

© Springer-Verlag Berlin Heidelberg 2006

$G(\lambda)$, *which is optimized when the fraction of "Yes" votes is exactly* λ^*. *The current voting round ends with the Referee awarding a dollar with probability* $G(\lambda)$ *and assessing a dollar with probability* $1 - G(\lambda)$ *to every voter independently. On the basis of their individual gains and losses, the voters then decide, again independently, how to cast their votes on the next round."*

The game has many interesting and fascinating features which render it both non-trivial and "intriguing". These are listed below:

1. The game is a non-trivial *non-zero-sum* game.
2. Unlike the games traditionally studied in the AI literature (like Chess, Checkers, Lights-Out etc.) the game is essentially a *distributed* game.
3. The players of the game are ignorant of all of the game's "parameters". All they know is that they have to make a choice, for which they are either rewarded or penalized. They have no clue as to how many other players there are, how they are playing, or even of how/why they are rewarded/penalized.
4. The stochastic function used to reward or penalize the players can be completely arbitrary, as long as it is uni-modal.
5. The most "intriguing feature" of this game [2] is that if each voter updates its action based on either a Tsetlin automaton with large memory, or an absolutely expedient[1] algorithm, then the entire group will asymptotically optimize the Referee's performance criterion.

1.1 Applications to the Goore Game

The literature concerning the GG is scanty. It was initially studied in the general learning domain, and, as far as we know, was for a long time merely considered as an interesting pathological game. Recently, however, the GG has found important applications within two main areas, namely, QoS (Quality of Service) support in wireless sensor networks [1] and within cooperative mobile robotics as summarized in [7]. A description of the first application area follows.

In order to preserve energy in a battery driven network, Iyer *et al.* [1] proposed a scheme where a base station provided broadcasted QoS feedback to the sensors of the network. Using the GG perspective, a sensor is seen as a voter that chooses between transmitting data or remaining idle in order to preserve energy. Correspondingly, the base station takes the role of the Referee and rewards/punishes the sensors using a unimodal QoS performance criterion function with the goal of attaining an optimal resolution/energy-usage tradeoff.

1.2 Problems with Reported LA Solutions to the GG

The reported LA solution [2,5] to the GG is indeed both intriguing and actually, almost "mystical". Without a knowledge of the function $G(\cdot)$, of how their partners decide, or even a perception of how the Referee "manufactured" their responses, the LA converge to the optimal solution *without* any communication.

---

[1] We assume that the reader has a fairly fundamental knowledge of the field of Learning Automata. Excellent reviews of this material can be found in e.g. [2,5].

However, the main handicap associated with using it in real-life applications concerns the *accuracy* of the solution obtained, which is intricately linked to the number of LA used. If the number of LA involved in the game is d, the precision of the solution is bounded by $\frac{1}{d}$, and thus the solution can be arbitrarily accurate only as d is increased indefinitely - leading to very slow convergence.

1.3   Salient Aspects of the Paper

The contributions of the unabridged paper [8] are the following:

1. In [8], we report the first solution to the GG which needs only a finite number of LA. Indeed, the number of LA can be as small as 3.
2. In [8], we report the first GG solution which is arbitrarily accurate.
3. The solution we propose in [8] is recursive. To the best of our knowledge, there has been no other reported recursive solution.
4. The solution that we propose in [8] is "fast".

This present paper demonstrates experimentally that the theoretical results of [8] lead to a pragmatic solution. Indeed, usually, each epoch of the recursion converges within a few hundred iterations, and the accuracy of the solution increases exponentially with the number of recursive calls.

2   Continuous Goore Game with Adaptive d-Ary Search

The solution presented in [8] and sketched here is based on a strategy, the so-called Continuous Goore Game with Adaptive d-ary Search (CGG–AdS). The basic idea behind the CGG–AdS solution is to use d LA to play the GG, and then to use the results of *their* solution to systematically explore an arbitrarily accurate *sub*-interval of the current interval for the overall solution.

2.1   Notations and Definitions

Let $\Delta(t) = [\sigma, \gamma)$ s.t. $\sigma \leq \lambda^\star < \gamma$ be the current search interval at epoch t, containing the solution point $\lambda^\star$ whose left and right (smaller and greater) boundaries on the real line are σ and γ respectively. $\Delta(0)$ is initialized to be the unit interval. We partition $\Delta(t)$ into d equi-sized disjoint partitions Δ^j, $j \in \{1, 2, \ldots d\}$, such that, $\Delta^j = [\sigma^j, \gamma^j)$. To formally describe the relative locations of intervals we define an interval relational operator $\prec$ such that, $\Delta^j \prec \Delta^k$ iff $\gamma^j < \sigma^k$. Since points on the real interval are monotonic, $\Delta^1 \prec \Delta^2 \ldots \prec \Delta^d$.

2.2   Construction of the Learning Automata

In the interest of simplicity, we shall assume that we are dealing with a "team" of the well-known L_{RI} LA, $\{A^1, A^2, \ldots A^d\}$. In terms of notation, we assume that the actions offered to each LA, A^j, from the Environment in question are

$\{\alpha_0^j, \alpha_1^j\}$, and that the corresponding penalty probabilities are $\{c_0^j, c_1^j\}$ respectively. Similarly, we let $P_k^j(n)$ represent the component of the action probability vector of A^j for action α_k^j, where n represents the discretized time index.

Since the actions chosen by each LA can lead to one of the two possible decisions, namely Yes or No, the number of "Yes" votes can be any integer in the set $\{0, 1, \ldots, d\}$. Once the team of automata have made a decision regarding where they reckon $\lambda^\star$ to be (by virtue of their votes), the CGG–AdS reduces the size of the search interval by a factor of at least $\frac{2}{d}$. If k^+ is the number of "Yes" votes received, the new pruned search interval, Δ^{new}, for the subsequent learning phase (epoch) is generated according to the following pruning rule:

$$\Delta(t + 1) = \begin{cases} \Delta^1 & \text{If } k^+ = 0. \\ \Delta^m \cup \Delta^{m+1} & \text{If } k^+ = m; m \neq 0, d. \\ \Delta^d & \text{If } k^+ = d. \end{cases} \tag{1}$$

The above pruning rule leads to a pruning table which clearly, "prunes" the size of the interval, because $\Delta(t + 1)$ at the next epoch is, at most, of length $\frac{2}{d}$. Various examples of pruning tables for specific values of d are includes in [8].

2.3   Implementation of CGG–AdS Scheme

The CGG–AdS strategy presented in [8] and sketched above is fairly simple to implement - it uses a straightforward partitioning of the search interval, a simple decision table for pruning, and the well known L_{RI} scheme for playing the GG.

The CGG–AdS strategy has been implemented and tested with a wide range of inputs. The pseudo-code for the algorithms (in Fig. 3) is included in the Appendix, and sample traces are presented in [8] (omitted here in the interest of brevity) to illustrate the workings of the CGG–AdS strategy.

The pseudo-code in Fig. 3 shows the recursive organization of the search, including the systematic pruning of the search interval. Each pruning decision is obtained by consulting the above Pruning Decision Table, after observing the outcome of an L_{RI} GG that has been projected into the *current* search interval. The algorithm is then recursively invoked. The recursion is terminated when the width of the interval is less than twice the desired accuracy. Indeed, it is the projection of the L_{RI} solution to the GG into increasingly smaller search intervals that allows unbounded solution precision.

3   Empirical Results

Although we have done numerous experiments, we present here two specific examples, to highlight two crucial issues. In the experiments which we report, we used a Gaussian criterion function $G(\lambda) = a e^{-(\frac{\lambda^\star - \lambda}{b})^2}$, allowing the magnitude and peakedness of the function to be controlled by the parameters a and b respectively. This allowed us to simulate a variety of environments, shown in [8].

In the first experiment which we report, we considered the case when $G(\lambda)$ attains it maximum at 0.9123. The trace given in [8] shows the execution of the

CGG–AdS algorithm for the case when $d = 3$. In this example run, the initial search interval is $[0, 1)$ and $\lambda^\star$ is 0.9123, and the parameters a and b were set to 0.7 and 0.035 respectively - which means that the optimal value of $G(\lambda)$ is 0.7. The search terminated when the width (i.e., the resolution) of the interval was ≤ 0.0002. The reward factor θ of the automata was 0.9999 and $\epsilon = 0.05$. In every invocation of CGG–AdS, the results of the automata are given as the optimal number of "Yes" votes, k^+. We remark that at Step 18 in the recursion, the width of the interval $[0.9121, 0.9123]$ is 0.0002, at which point the estimated value for $\lambda^\star$ is the mid-point of the interval $[0.9121, 0.9123]$, which is 0.9122. We note that at this resolution, our scheme is very close to optimizing the performance criterion function because $G(0.9122) \approx 0.69999$. It should be mentioned that the traditional LA solution to the GG would require 10,000 LA to attain this level of precision. Hence, the power of our strategy !! Additional examples and traces for other executions of the solution are given in [8] (omitted here in the interest of space), and illustrated in Fig. 1 and Fig. 2.

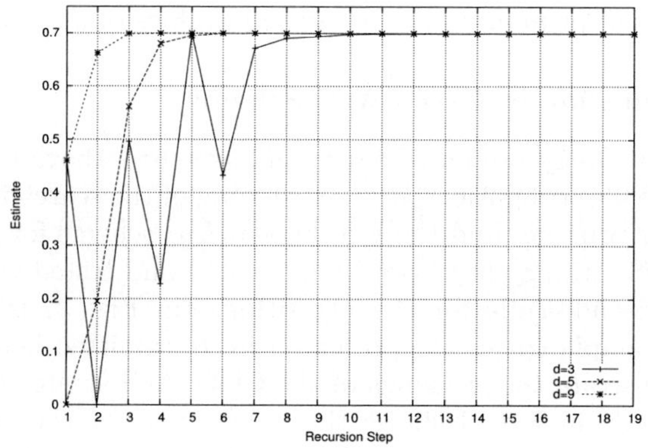

Fig. 1. Convergence of estimates for d=3,5,9. The unknown parameter $\lambda^\star$ =0.0.3139, and the optimal value of the criterion function is 0.7.

We first note that as the resolution increases at each recursion step, the accuracy of the $\lambda^\star$-estimates does not increase monotonically, as, perhaps, could have been expected. Instead, the estimates fluctuate, however, with decreasing variance. As argued in [8], this fluctuation is not a result of the random properties of our algorithm. Also, for larger number of LA, the positioning of the sub-partitions seems to become less significant, as seen in Fig. 2 for $d = 9$.

Also note that at any given recursion step, the speed of convergence seems to be determined by the magnitude that the best available estimate λ^+ differs from the inferior estimates. E.g., a function G with a $G(\lambda)$ that is flat around the optimal value λ^* may slow down convergence when the search interval falls within the flat area. However, such a flat search interval means that the candidate estimates are close to the optimal value of the criterion function, and the search could be terminated without significant loss of accuracy.

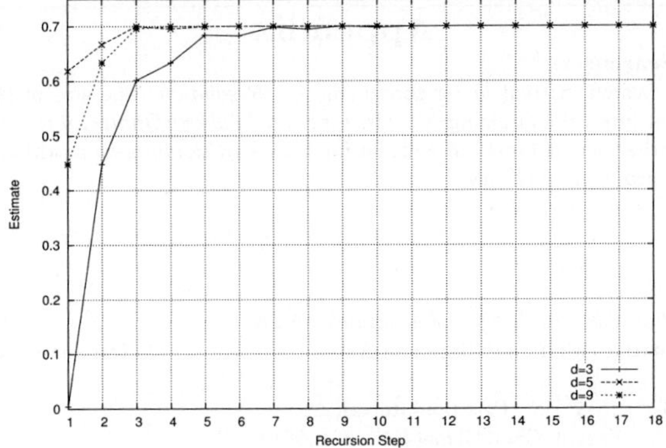

Fig. 2. Convergence of estimates for $d = 3,5,9$. The unknown parameter $\lambda^\star = 0.9123$, and the optimal value of the criterion function is 0.7.

4   Conclusions

In the unabridged version of this paper [8], we showed how we can attain an unbounded accuracy for the GG by using at most d LA, and by recursively pruning the solution space to guarantee that the retained domain contains the solution to the game with a probability as close to unity as desired. Indeed, d can be made as small as *three*. While the paper [8] contains the formal proofs and algorithms, this present paper experimentally demonstrates that the solution of [8] is feasible. Indeed, we believe that we have presented here the first *practical* implementation of the GG. We are currently investigating the application of these results to a variety of areas, including sensor networks and QoS routing.

References

1. R. Iyer and L. Kleinrock. Qos control for sensor networks. In *IEEE International Conference on Communications*, volume 1, pages 517–521, 2003.
2. K.S. Narendra and M.A.L. Thathachar. *Learning Automata*. Prentice-Hall, 1989.
3. B. J. Oommen, G. Raghunath, and B. Kuipers. Parameter learning from stochastic teachers and stochastic compulsive liars. *IEEE Transactions on Systems Man and Cybernetics*, To Appear, 2006.
4. M. A. L. Thathachar and M. T. Arvind. Solution of Goore game using models of stochastic learning automata. *J. Indian Institute of Science*, (76):47–61, 1997.
5. M. A. L. T. Thathachar and P. S. Sastry. *Networks of Learning Automata : Techniques for Online Stochastic Optimization*. Kluwer Academic, Boston, 2003.
6. M.L. Tsetlin. *Automaton Theory and the Modelling of Biological Systems*. Academic Press, New York and London, 1973.
7. B. Tung and L. Kleinrock. Using Finite State Automata to Produce Self-Optimization and Self-Control. *IEEE Transactions on Parallel and Distributed Systems*, 7(4):47–61, 1996.
8. B.J. Oommen, O.C. Granmo, and A. Pedersen. Achieving Unbounded Resolution in *Finite* Player Goore Games using Stochastic Automata, and its Applications. *Unabridged version of this paper*. Submitted for publication, 2006.

Appendix

Program Search(Δ)

Input : Δ: Search interval $[\sigma, \gamma]$ containing λ^*. *Resolution:* The size of the smallest significant interval containing λ^*. The function *MidPointOfInterval* returns the midpoint of the specified interval and the function *PartitionInterval* partitions the given interval into d sub-intervals.

Output : The estimate of λ^*.

Method :

Begin

    If (WidthOfInterval(Δ) $\leq$ Resolution) **Then**

        Return (MidPointOfInterval(Δ))   /* Terminate Recursion */

    Else

        $\{\Delta^0, \ldots, \Delta^d\} :=$ PartitionInterval(Δ)

        $k^+ :=$ ExecuteGooreGame($\{\Delta^0, \ldots, \Delta^d\}$)

        $\Delta^{new} :=$ ChooseNewSearchInterval($\{\Delta^0, \ldots, \Delta^d\}$, k^+, Decision-Table)

        Search(Δ^{new})   /* Tail Recursion */

    EndIf

END Program Search

Procedure ExecuteGooreGame($\{\Delta^0, \ldots, \Delta^d\}$)

Input: The partitioned search interval $\Delta = [\sigma, \gamma]$; the parameters θ and ϵ of the L_{RI} scheme; the performance criterion function $G(\lambda)$ of the Environment.

Output: A decision k^+ from the set $D = \{0, \ldots, d\}$. The decision represents the optimal number of "Yes" votes among D .

Method :

Begin

    For $i := 1$ **To** d **Do**

        $P_0^i := P_1^i := 0.5$

    While $\epsilon < P_0^i < 1 - \epsilon$ **For Any** $i \in \{1, \ldots, d\}$ **Do**

        $k := 0$

        For $i := 1$ **To** d **Do**

            $y_i :=$ ChooseAction(Δ^i) ; $k := k + y_i$

        EndFor

        For $i := 1$ **To** d **Do**

            If $(y_i = 0)$ **Then**

                $\beta :=$ GetFeedBack(σ^k)

                If $(\beta = 0)$ **Then**

                    $P_1^i := \theta.P_1^i;\ P_0^i := 1 - P_1^i$

                EndIf

            Else

                $\beta :=$ GetFeedBack(σ^k)

                If $(\beta = 0)$ **Then**

                    $P_0^i := \theta.P_0^i;\ P_1^i := 1 - P_0^i$

                EndIf

            EndIf

        EndFor

    EndWhile

    $k^+ := 0$

    For $i := 1$ **To** d **Do**

        If $(P_1^i \geq 1 - \epsilon)$ **Then**

            $k^+ := k^+ + 1$

        EndIf

    EndFor

    Return (k^+)

End Procedure ExecuteGooreGame

Fig. 3. Algorithm CGG–AdS: Overall search strategy

Robust Character Recognition Using a Hierarchical Bayesian Network

John Thornton, Torbjorn Gustafsson, Michael Blumenstein, and Trevor Hine

Institute for Integrated and Intelligent Systems, Griffith University, QLD, Australia
{j.thornton, t.gustafsson, m.blumenstein, t.hine}@griffith.edu.au

Abstract. There is increasing evidence to suggest that the neocortex of the mammalian brain does not consist of a collection of specialised and dedicated cortical architectures, but instead possesses a fairly uniform, hierarchically organised structure. As Mountcastle has observed [1], this uniformity implies that the same general computational processes are performed across the entire neocortex, even though different regions are known to play different functional roles. Building on this evidence, Hawkins has proposed a top-down model of neocortical operation [2], taking it to be a kind of pattern recognition machine, storing invariant representations of neural input sequences in hierarchical memory structures that both predict sensory input and control behaviour. The first partial proof of concept of Hawkins' model was recently developed using a hierarchically organised Bayesian network that was tested on a simple pattern recognition problem [3]. In the current study we extend Hawkins' work by comparing the performance of a backpropagation neural network with our own implementation of a hierarchical Bayesian network in the well-studied domain of character recognition. The results show that even a simplistic implementation of Hawkins' model can produce recognition rates that exceed a standard neural network approach. Such results create a strong case for the further investigation and development of Hawkins' neocortically-inspired approach to building intelligent systems.

Keywords: Bayesian network, cerebral cortex, character recognition.

1   The Hierarchical Bayesian Network

Figure 1 presents a simplified example of the hierarchical Bayesian network used in the current study. This network is based on George and Hawkins' original implementation [3] and consequently does not follow standard Bayesian update procedures, as the paper will explain. The network input consists of a feature vector of 100 real-valued elements, with a value range from 0 to 1. The particular form of the input was set by the feature extraction method used in the neural network implementation (described in Section 2). We retained this feature vector for the Bayesian network to ensure that our results reflect differences between the two computational architectures and are not confounded by effects from differing representations of the problem domain.

A. Sattar and B.H. Kang (Eds.): AI 2006, LNAI 4304, pp. 1259–1264, 2006.
© Springer-Verlag Berlin Heidelberg 2006

The Learning Phase: In Figure 1, each level one node receives a real number input from the feature vector, which is then classified into one of eight bit patterns or as an empty pattern. In the example, a bit set in position one classifies values > 0 and < 0.125, in position two ≥ 0.125 and < 0.25 and so on, with at most one bit set in any pattern. Bit patterns from level one are then sent to level two, where they are classified again, as follows: the level one eight bit patterns are projected into level two four bit patterns, such that if there is any bit in the first or second position of a child bit pattern, then position one in the level two bit pattern is set, and so on. The network undergoes supervised learning, so when the level two bit patterns are sent to level three they are then correctly classified against the actual character input.

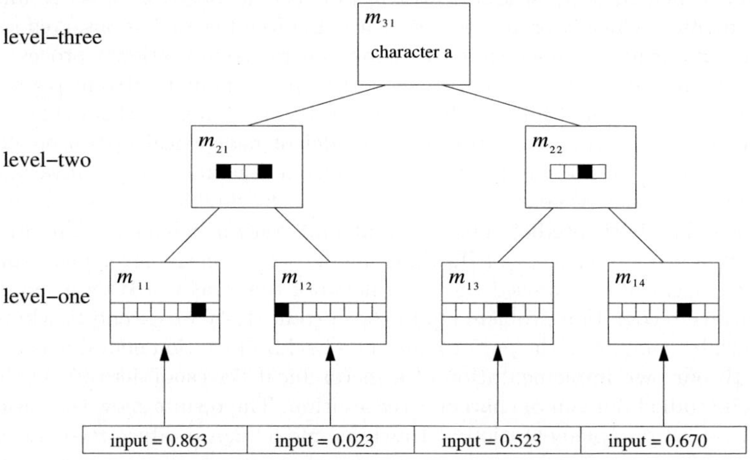

Fig. 1. A simple hierarchical network in the process of supervised learning

In the full implementation, a similar procedure is followed, except there are 100 level one nodes, each classifying the feature vector inputs into a bit pattern with up to 100 positions. These level one nodes then connect in groups of twenty-five to one of four level two nodes. The twenty-five associated level one bit patterns are then projected into single level two bit patterns of the same length to create level two language elements. Each time a level two module receives a new set of level one bit patterns, it checks to see if these patterns are classified under an existing level two language element (i.e. that they project onto an existing level two bit pattern). If not, a new level two language element is created and given an index. In either case, the index value of the level two language element is passed back to the level one modules. Each level one module then stores how many times a particular level one bit pattern has been classified against each level two language element. This is shown in left hand matrix in Figure 2. When learning is complete, this matrix is normalised by dividing each row element by the row sum to produce the *conditional probability distribution* (CPD). CPDs are generated in the same way for each level one and level two module (the level three root node does not generate a CPD because it does not communicate with a higher node).

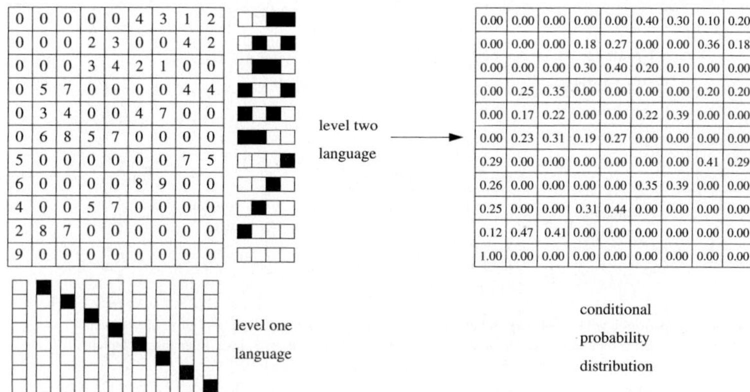

Fig. 2. Calculating a level one conditional probability distribution (CPD)

The Recognition Phase: The details of the recognition phase calculations are shown in Figure 3 for an internal node in the hierarchy. Both the level one nodes and the root node are special cases in that they only receive input either from above or below. In the root node case, it initially sends down a π vector containing m normalised equal probabilities, where m is the number of character types the network has been trained to recognise. In addition, a level one node needs to generate a λ vector containing the probabilities for each of the n possible language elements in the current input.

If we consider node m_{11} in relation to Figure 3, m_{11} will initially receive a π vector from level two (in the top left of the diagram) with all elements set to one (this means the level two node has no belief about which of its language elements are active). Then each column of the CPD matrix is multiplied by the corresponding λ vector element and each row by the corresponding π vector element (as all π elements equal one, only the λ elements will have an effect in the first iteration). This produces the array on the right of Figure 3. We then take the maximum likelihood value for each row to form the λ^{k+1} vector that is then sent to level two. We also take the maximum likelihood for each column to form the π^{k-1} vector. As we are at level one, this vector cannot be sent down, although it can be used to condition the probabilities of the next λ input vector (assuming we are seeing a meaningful temporal sequences of inputs).

Following the path of the λ^{k+1} vector upwards, the level two node will receive one such vector from each of its children. These vectors are combined by multiplying together all elements in corresponding positions to produce a new vector with the same number of elements. This process is repeated at level two, producing a new λ vector at level three, containing the combined beliefs of all

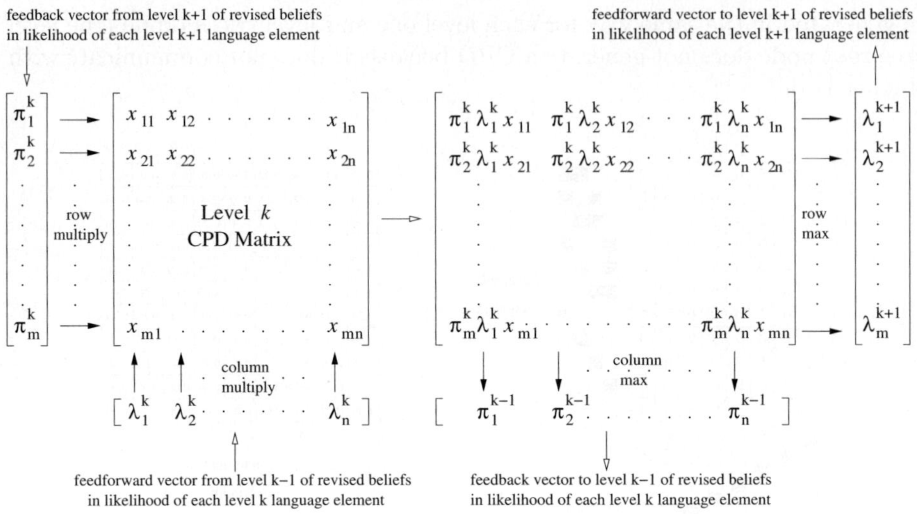

Fig. 3. Belief revision calculations for recognition

level two nodes about which level three language element is active. As the level three language elements are the actual characters we are trying to recognise, the solution to the character recognition problem is the λ element at this level with the highest probability.

2   Experimental Study

For the neural network implementation we used a modification to the back-propagation algorithm known as the Resilient Version (RPROP) [4]. RPROP is a locally adaptive learning scheme for batch learning in feed-forward neural networks. Because it only examines the direction of the gradient and uses a local adaptive approach to determine the size of the weight change, it generally converges faster than a standard backpropagation approach and has the added advantage that it does not require the constant modification of parameters to achieve near-optimal performance and convergence. The transition feature vector used to represent the input was calculated using the location of transitions from background to foreground pixels in the vertical and horizontal directions of a given character image (see [5] for more details).

The neural network and Bayesian implementations were tested on the well-known CEDAR database using the Buffalo Designed (BD) dataset of handwritten digits. [1] From the training set, six subsets were created containing 10, 25, 50, 100, 200 and 300 training digits per category and for testing, all (544) of the digits from the test set were used.

RPROP: Table 1 shows the best RPROP recognition rate was 92.28% achieved using 32 hidden layer neurons and 500 iterations. The results are for the best

---

[1] See http://www.cedar.buffalo.edu/Databases/CDROM1/

Table 1. RPROP character recognition results as percentages

Number of	Number of Iterations				
Hidden Units	100	200	300	400	500
18	88.97	90.99	91.17	91.17	90.99
24	91.36	90.44	90.99	91.54	89.34
32	90.63	91.91	90.44	91.91	**92.28**

of five runs and show the effects of varying the number of iterations and the number hidden units.

Bayesian Network: As previously described, we constructed a hierarchical Bayesian network in three levels with 100 level one nodes. During the initial experimentation we found that a tree structure with four level two nodes consistently produced better results, so we retained this structure for all the following experiments. However, it was unclear what size training set produced the best result, so we tried five different values: $[25, 50, 100, 200, 300]$. We also varied the number of level one elements between 26, 51, 76 and 101 elements and experimented with putting an upper limit on the size of the level two language. Once this limit is reached, any new language elements are classified under an existing element according to the minimum Hamming distance.

Table 2 shows that the best Bayesian recognition rate of 96.32% was achieved for a training set category size of 300, a level one language size of 76 and a level two language limit of $12,000$. However, recognition rates that exceeded the best RPROP results were achievable with a smaller level two language limit of $2,600$ and a faster mean recognition time of 4.39 seconds per character.

Table 2. Hierarchical Bayesian network character recognition results as percentages. Results that exceeded the best RPROP recognition rate are in bold, of these the 92.65% result had the best mean recognition time of 4.39 seconds per character and an overall training time of 12.19 seconds and the best result of 96.32% had a mean recognition time of 7.83 seconds per character and an overall training time of 15.99 seconds. All experiments were performed on an AMD Athlon 64 3000+ processor with 1 GB RAM.

Level One	Training Set Size / Level Two Alphabet Limit					
Alphabet Size	25/1000	50/2000	100/2600	200/2600	300/2600	300/12000
26	87.13	89.34	91.36	89.70	90.80	n/a
51	87.50	90.80	**92.65**	90.63	89.15	n/a
76	88.24	90.07	89.89	89.89	87.32	**96.32**
101	88.24	90.44	91.36	87.50	83.46	n/a

3   Conclusions

The final best recognition rate of 96.32% for the hierarchical Bayesian network compares very favourably with the 92.28% rate for RPROP. If we bear in mind that character recognition has been the province of neural network research for many years, and that the Bayesian network was doing recognition on a feature vector structure specifically developed for a neural network approach, then our results strongly recommend the new hierarchical Bayesian approach. In addition the results show that the Bayesian network can still achieve good recognition rates on quite small training sets. This is a distinct advantage in real world situations where only a small subset of the possible input is likely to be available for training. The main disadvantage of the hierarchical Bayesian approach is that it involves considerable computational overhead. While in the neocortex this overhead is counteracted by the massively parallel nature of neocortical processing, the matrix arithmetic required to propagate belief revision throughout the Bayesian network can result in recognition times lasting several seconds, in contrast to the virtually instantaneous recognition times for RPROP (this is partly balanced by the Bayesian training times being at least ten times faster than for RPROP). There are several ways of approaching this speed problem. One is to construct hardware that allows the matrix operations to occur in parallel (George and Hawkins are already exploring this option). A second more immediate approach is to develop a selective method for updating beliefs, such that strong beliefs tend to suppress similar, weaker beliefs. In this way the flow of information in the network could be considerably reduced. It would also be worth investigating problem representations that are more "natural" for the hierarchy, such as going back to original character images rather than using feature vectors. It should also be noted that this Bayesian network is only a partial implementation of Hawkins' model of the neocortex. Further improvements can be expected over a neural network approach in the area of detecting *moving* objects that allow the feedback properties of the network to influence recognition.

References

1. Mountcastle, V.: An organizing principle for cerebral function: the unit model and the distributed system. In Edelman, G., Mountcastle, V., eds.: The Mindful Brain. MIT Press, Cambridge, Mass (1978)
2. Hawkins, J., Blakeslee, S.: On intelligence. Henry Holt, New York (2004)
3. George, D., Hawkins, J.: A hierarchical Bayesian model of invariant pattern recognition in the visual cortex. In: Proceedings of the International Joint Conference on Neural Networks (IJCNN-05). (2005)
4. Riedmiller, M., Braun, H.: A direct adaptive method for faster backpropagation learning: The RPROP algorithm. In: Proceedings of the IEEE International Conference on Neural Networks, San Francisco, CA (1992) 586–591
5. Gader, P.D., Mohamed, M., Chiang, J.H.: Handwritten word recognition with character and inter-character neural networks. IEEE Transactions on Systems, Man, and Cybernetics-Part B: Cybernetics **27** (1997) 158–164

On the New Application of Call Patterns to CPM Testing of Prolog Programs*

Lingzhong Zhao[1,2], Tianlong Gu[1], Junyan Qian[1], and Guoyong Cai[1]

[1] Department of Computer Science, Guilin University of Electronic Technology, Guilin, China 541004
[2] Electronic Engineering School, Xidian University, Xi'an, China 710071
`gllzzhao@gliet.edu.cn`

Abstract. Information on call patterns is known to be useful for analysis and optimization of Prolog programs. Several call patterns semantics exists for Prolog or for the subset of Prolog. In this paper we propose a method to apply the call patterns semantics to CPM testing of Prolog programs. The method can be viewed as an attempt to improve the testing of Prolog programs by the result of program analysis. By analyzing the way in which procedures are used in a program, we can reduce the number of test frames and therefore the number of test cases generated in CPM testing of a procedure.

1 Introduction

Prolog is a well-known logic programming language. Call patterns are the atoms that are selected during the SLD-derivation of a goal. Information on call patterns is known to be useful for optimization of Prolog programs [2]. In this paper we argue that this information is also useful in Category Partition Method (CPM) testing of Prolog programs for controlling the number of generated test cases [5].

In traditional CPM testing, the control on the number of test cases is based on the specification of the program to be tested. We show, however, that if the code of a program is available, the number of generated test cases can be further reduced based on the implementation of the program; and this reduction will not much decrease our confidence in the quality of the tested programs.

The main idea can be illustrated by the following example. Suppose we want to test a program P: {main(x,y):-x=5,sub(x,y). sub(x,y):-...} with two procedures main and sub, where sub is an internal procedure. By *internal procedures* we mean the procedures in a Prolog program that are designed only for "internal" use, i.e. the procedures that are not directly used by the users of the program but by programmers to construct other procedures. Since every time sub is called its first argument x will be bounded to 5, we only have to test sub for the input data consistent with the property that x is ground or more specifically $x=5$, rather than for all possible input data. The information needed in this process can be obtained from a call patterns

---

* This work is supported by National Natural Science Foundation of China (No.60563005) and Guangxi Natural Science Foundation of China (No.0542036).

A. Sattar and B.H. Kang (Eds.): AI 2006, LNAI 4304, pp. 1265–1270, 2006.
© Springer-Verlag Berlin Heidelberg 2006

analysis of program P, which indicates that sub(5, y) is the only call pattern for sub that arises in the execution of goal main(x, y).

Several methods have been proposed in literature to improve the testing of Prolog programs, such as the PROTest II system [1] and the IDTS (Interactive Diagnosis and Testing) system [4]. Our method is different from theirs and can be viewed as an attempt to use the program analysis to improve CPM testing of Prolog programs.

2   CPM Testing and a Goal-Independent Call Patterns Semantics

Category Partition Method. CPM testing, a general method for functional testing of programs, is first proposed by Ostrand and Balker [5] and later formalized by Horváth *et al.* CPM testing can be outlined as follows: During the functional testing a program (or a procedure) cannot be tested for all possible properties of their parameters. Therefore the tester's first task is to identify the critical properties (*categories*) of the input parameters that will be investigated in the testing process. The categories are divided into disjoint classes called *choices* which presume that all elements within a choice have an identical observable behavior from the point of testing. Once the categories and choices for a program have been defined, they are written into a test specification. Based on the specification all possible *test frames* can be generated, each test frame covering exactly one choice from each category. In general there are many superfluous test frames among the generated ones, namely property combinations that have no real relevance. Those frames can be eliminated by associating a property or selector expression to a choice. Each test frame will give rise to a test case that satisfies the properties specified by the frame. Based on those test cases, a test script is produced with which the program (procedure) can be tested.

Example 1. Given a program P: $\{\texttt{p(x,y):-x}{\leqslant}\texttt{5,!.\ p(x,y):-q(2x),r(y).}$ $\texttt{q(x):-(x=10).\ q(x):-q(x).\ r(x):-x=3.}\}$, we assume that q is an internal procedure; the input domains of q and r are both N, the set of non-negative integers, and that of p is N^2. A simple test specification and the generated test frames for the procedure p and q are given below. For simplicity, we do not contain selector expressions in the test specification.

Test specification: p
Category one: first_argument
    Choice: c_0: $\{x|x{\leqslant}5, x{\in}N\}$
    Choice: c_1: $\{x|x{>}5, x{\in}N\}$
Category two: second_argument
    Choice: d_0: $\{x|x{<}3, x{\in}N\}$
    Choice: d_1: $\{x|x{=}3\}$
    Choice: d_2: $\{x|x{>}3, x{\in}N\}$
End of specification for p
The generated test frames are: $\{(c_0,d_0),$ $(c_0,d_1), (c_0, d_2), (c_1,d_0), (c_1,d_1), (c_1,d_2)\}$

Test specification: q
Category one: the_only_argument
    Choice: a_0: $\{x|x{<}10, x{\in}N\}$
    Choice: a_1: $\{x|x{=}10\}$
    Choice: a_2: $\{x|x{>}10, x{\in}N\}$
End of specification for q
The generated test frames are:
$\{(a_0),(a_1),(a_2)\}$

An example of the set of test cases produced for procedure q is $\{(3), (5), (8)\}$. Here we note that a test frame can be viewed as a set of constraints on the input arguments. For example, (c_0, d_0) can be regarded as the constraints store $\{x{\leqslant}5, y{<}3\}$. This observation is useful for relating CPM testing and constraint-based call patterns semantics in the following.

A Constraint-Based Call Patterns Semantics. Several call patterns semantics for logic programs have been proposed in literature [2,6]. In [6] Spoto and Levi proposed a goal-independent denotational semantics for Prolog which computes in a bottom-up manner the call patterns of the goals of the form $p(x_1, ..., x_n)$ with $x_1, ..., x_n$ being distinct free variables. In the following we say a procedure call to p is a most general goal if p is called with all its arguments as distinct variables. One feature of Spoto's semantics is that every possible call pattern (an atom) of a goal is associated with an observability constraint which describes the conditions for this atom to be a call pattern. Moreover it uses constraints to describe the substitutions associated with a call pattern. In this paper we use this semantics to illustrate our method.

In the following a triple <p, b, o> denotes a call pattern for p that can actually arise from the execution of a particular goal when b and o are both satisfiable in the constraint store obtained from the substitution associated with the goal, where b and o are both observability constraints defined in [6]. If a goal consists of most general sub-goals, then the associated constraint store is empty; and if it has an input substitution which assigns a to x, then $x=a$ is an element of the store. Intuitively b describes the substitutions at the moment p is called, and o puts further constraints on the substitutions in order to make the call to p be observable or really happen.

An observability constraint o is satisfiable in a constraint store S if and only if it is consistent with it, i.e. $S \land o = \{s \land o | s \in S\}$ is satisfiable. Note that o is not required to be entailed by S. The negation $-o$ of constraint o is satisfiable in S if and only if o is not satisfiable in S. We also allow the third element o of a triple to be a set of observability constraints; in this case o is satisfiable in S if and only if every element of o is consistent with it.

Obviously a triple <p, b, o> describes (a class of) input data for p which satisfy the properties described by b and o.

For each procedure p in a program P, the call patterns semantics of P associates to p a denotation which describes all possible call patterns of the most general goal $p(x_1, ..., x_n)$. For example, given a program P:

```
{p(x):-q(x),r(x).q(X):-x=4,!.q(x):-x=5.r(x):-x=5,q(x).},
```

the denotation of p will suggest that goal p(x) has 6 possible call patterns, i.e. <p, *true*, *true$_O$*>, <q, *true*, *true$_O$*>, <r, $x=4 \land x_r=4$, *true$_O$*>, <q, $x=4 \land x_q=4 \land x_q=5$, *true$_O$*>, <r, $x=5 \land x_r=5$, $-(x=4)$>, <q, $x=5 \land x_q=5$, $-(x=4)$>.

The notions are explained as follows. The variable x_r denotes the first argument of a procedure call to r; and if r has an arity no less than 2, y_r will be used to denote the second argument. In this way the input arguments of a goal can be distinguished from the local arguments of a procedure call. The second element b of a triple <p, b, o> is *true* if p is called with an empty substitution, and by $(x_p=m)$ we mean p is called with a substitution where x is bounded to m; *true$_O$* is a special observability constraint which is satisfiable in a constraint store S if and only if S is satisfiable. Given an empty constraints store corresponding to goal p(x), constraint $-(x=4)$, the negation of

($x{=}4$), is not satisfiable since ($x{=}4$) is satisfiable. Among the six possible call patterns of p(x), only the former three can really arise, and $<q, x{=}4{\wedge}x_q{=}4{\wedge}x_q{=}5, true_O>$ can not arise because ($x_q{=}4{\wedge}x_q{=}5$) is not consistent.

Given a finite analysis domain D, the call patterns semantics can be abstracted into a finitely computable semantics. The analysis w.r.t the domain D is in fact the process of computing an approximation of the concrete call patterns semantics. The analysis based on this semantics can be the type analysis, groundness or sharing analysis etc.

3   Using Call Patterns Semantics in CPM Testing

By the discussion in section 2, we know that generally there are many superfluous test frames generated in CPM testing of Prolog programs. Ostrand and Balker propose to eliminate those frames by associating a property or a selector expression to the choices of a category. The property and the selector expression are usually formulated based on the specification of a program; in other words, their reduction method is a specification-based method, which can be used both in white-box testing and black-box testing. This paper is concerned with the former, and in this case we argue that the number of test frames can be further reduced by the information on call patterns of a program. For our method to work we have to know the way procedures are used in a program, so this is in fact an implementation-based reduction method. The main idea can be described as follows.

In practice there often exist a number of *internal procedures* in a program. We can reduce the number of test frames generated in CPM testing of internal procedures using the information on call patterns.

Suppose there is an internal procedure p in a program such that every time p is called its input data satisfies certain properties described by condition c, and that we are sure that p could not be used in other situations. Then in the CPM testing we only have to test p for those input data that have properties consistent with c. In this way the number of generated test frames could be reduced. We'll show that, compared with the original testing this reduction will not much decrease our confidence in the quality of the tested programs. By the definition of call patterns, the needed information c can be obtained from the call patterns analysis of the program. Moreover we prefer a goal-independent call patterns analysis since it allows us to get the analysis result for any goal from the denotation of the most general goals [2,6]. In this paper we'll base the analysis on the goal-independent denotational call patterns semantics discussed in section 2.

In this section Prolog programs are assumed to have some *internal* procedures. For simplicity, we assume that users use non-internal procedures only in the most general form, i.e. when being called their arguments are distinct variables. This assumption forces us to perform the testing of non-internal procedures w.r.t. all the generated test frames for them. Next we show this assumption will not lead to the loss of generality. Let p(x_1, ..., x_n) be a non-internal procedure in a program and c be a constraint, if we first define a new non-internal procedure p' as "p'(x_1, ..., x_n):- c, p(x_1, ..., x_n)." and then change p to an internal procedure, then each goal p(x_1, ..., x_n) satisfying c can be substituted by the most general goal p'(x_1, ..., x_n), which has the same set of call patterns for internal procedures in the program as the original goal.

Suppose we want to test a program P that contains an internal procedure $p(x_1, ..., x_n)$ and a non-internal procedure $q(x_1, ..., x_m)$. The set of test frames for p is assumed to be F. If the call patterns analysis for goal $q(x_1, ..., x_m)$ produces a set C of call patterns for p such that $C=\{<p, b_1, o_1>,...,<p, b_i, o_i>\}$, then p will be called only on those input data described by the triples in C. So we only have to test those frames in F that are at least consistent with one of the triple in C, where a frame f is *consistent* with a triple $<p, b, o>$ if both b and o are satisfiable in f viewed as a constraint store.

The method discussed above can be extended to general case. What we need is to compute the set c_r of call patterns for any internal procedure r that arises in the executions of all non-internal procedures. Assuming program P has defined $(n+1)$ procedures $\{p_1, ..., p_n, r\}$, where $\{p_i \mid i=1,...,k\}$ is the set of non-internal procedures, and $\{p_i, r \mid i=k+1, ..., n\}$ the set of internal procedures. Let D be the call patterns semantics of P, and D[p] be the denotation of procedure p. The function `Call-Pat(r)` computing c_r for r can be described as follows.

```
Call-Pat(r)                          Add <r,bj,oj>∈D[pi] to
{   C=empty set;                     Cr;
    For each i=1,…,k                 Return(Cr);  }
```

Take example 1 again. According to the traditional CPM testing, procedure q should be tested w.r.t. all the test frames $\{(a_0),(a_1),(a_2)\}$. The call patterns analysis reveals that $<q, x=a \wedge y=b \wedge x_q=2a, a \not\leq 5>$ is the unique call pattern for q that arises in the derivation of goal $p(x, y)$ with substitution $(x/a, y/b)$, from which we know that $q(2a)$ is a call pattern only if $a \not\leq 5$; then constraint $(x_q=2a)$ allows us to conclude $x_q \not\leq 10$, so the frame (a_0) and (a_1) need not to be tested.

Next we take the example in [6] to show that the result of groundness and freeness analysis can also be used in our method. The adopted analysis domain is $\{\perp, \top\} \cup \{g(x), nf(x) \mid x$ is a variable$\}$. By $g(x)$ and $nf(x)$ it is meant that variable x is ground and not free, respectively.

Example 2. Given program P: $\{$`p(x):-x=4,!. p(x):-q(x). q(x):-(x=5). q(x):-p(x).`$\}$ with q as an internal procedure, its test specifications and generated test frames are as follows.

Test specification: p
Category one: the_only_argument
  Choice: c_0: $\{x \mid x$ is ground and $x \in$ N.$\}$
  Choice: c_1: $\{x \mid x$ is a free variable.$\}$
  Choice: c_2: $\{x \mid x$ is neither ground nor free variable.$\}$
End of specification for p
The generated test frames are:
  $\{(c_0), (c_1), (c_2)\}$

Test specification: q
Category one: the_only_ argument
  Choice: a_0: $\{x \mid x$ is ground and $x \in$ N.$\}$
  Choice: a_1: $\{x \mid x$ is a free variable.$\}$
  Choice: a_2: $\{x \mid x$ is neither ground nor free variable.$\}$
End of specification for q
The generated test frames are:
  $\{(a_0), (a_1), (a_2)\}$

The call patterns analysis of the program shows that goal $p(x)$ has a unique call pattern for q, i.e. $<q, \top, nf(x)>$, which means that the call pattern can arise only when

x is not a free variable. This information allows us to remove (a_1) from the set of test frames for procedure q.

Next we discuss the limitation of our method. Since the program to be tested may contain bugs in a procedure p, the following situation may occur. According to the intended semantics of the program, an internal procedure q may be called by p with substitutions consistent with a choice c. The call patterns analysis, however, draws an opposite conclusion that the call pattern for q is actually impossible. By our method q will be tested without covering choice c, which may have a consequence of decreasing the confidence in the quality of the program. Fortunately, in most cases the testing of p will find the bugs in p. Because in the testing of p the lack of the procedure call to q most likely leads to unexpected behavior. In this case the p will be debugged and corrected, a call patterns analysis will be performed w.r.t. the updated program, and our method can still be applied and this time covers choice c in the testing of q. The limitation of our method is that if the testing of p still produces results that are correct w.r.t. the intended semantics in the absence of the procedure call to q, which is a case that seldom happens in practice, the generated test frames will fail to cover choice c.

4 Conclusions

This paper presents a method for improving CPM testing of Prolog programs by the call patterns information derived from analyzing the target program based on a suitable call patterns semantics. In designing the method we can make use of many existing program analysis techniques [2,3,6,7]. We illustrate the method by a constraint-based call patterns semantics. Although our method has some limitation, it is useful when we want to reduce the number of generated test cases without much decreasing the confidence in the quality of a program.

References

[1] F. Belli, O. Jack. PROTest II, Testing Logic Programs, Technical report, 1992/13, ADT, October 1992.

[2] M. Gabbrielli and R. Giacobazzi. Goal Independency and Call Patterns in the Analysis of Logic Programs, In: Proceedings of the Ninth ACM Symposium on Applied Computing, 1994, pp. 394-399.

[3] J. M. Howe, A. King. Efficient Groundness Analysis in Prolog, Theory and Practice of Logic Programming, 2003, 3(1): 95-124.

[4] G. Kókai, L. Harmath, T. Gyimóthy. IDTS: a Tool for Debugging and Testing of Prolog Programs. In: Proceedings of LIRA'97, The 8th Conference on Logic and Computer Science, Novi Sad, Yugoslavia 1-4 September 1997, pp. 103-110.

[5] T.J. Ostrand, M.J. Balker. The Category-Partition Method for Specifying and Generating Functional Tests. Communications of ACM, 1988, 31(6): 676-686.

[6] F. Spoto, G. Levi, Abstract Interpretation of Prolog Programs, Lecture Notes in Computer Science, 1999, 1548: 455-470.

[7] P. Volpe: A first-order language for expressing sharing and type properties of logic programs. Science of Computer Programming, 2001, 39(1): 125-148.

Correlation Error Reduction of Images in Stereo Vision with Fuzzy Method and Its Application on Cartesian Robot

M. Ghayoumi[1,4], P. Porkar Rezayeyeh[2], and M.H. Korayem[3]

[1] Young Researchers Club
[2] Islamic Azad University, Damavand Branch, Tehran, Iran
[3] Robotic Research Laboratories, Mechanical Engineering Department
Science & Technology University, Narmak, Tehran, Iran
[4] Science and Research branch of IAU, Faculty of Engineering, Tehran, Iran
Ghayoumi_me@sr.iau.ac.ir, Payam_porkar@yahoo.com,
Hkorayem@iust.ac.ir

Abstract. Stereo vision is one of the most active research topics in machine vision. The most difficult task in stereo vision to getting depth is to find corresponding points in different images of the same scene. There are some approaches and correlation is one of them. This method has some errors and up to now has presented some methods to reduce these errors. In this paper a heuristic and fuzzy approach has been demonstrated for this purpose. Also the experimental test is presented on 3p robot. Simulation results have demonstrated improvement in compare with neural network method.

Keywords: Correlation Error- Stereo Vision- Fuzzy Model.

1 Introduction

Vision method at first was used for estimating robot errors more than one decade ago. So far, different companies and research centers have used for robot positioning, calibration, error estimation and error compensation with genetic algorithm, neural networks and fuzzy control algorithms. In general, recognition of 3D object requires two or more appropriately defined 2D images. With this approximation many methods have been proposed such as structure from motion [1-2], stereo lenses correspondence and shape [3-4]. Achour and Benkhelif presented a new approach for 3D scene reconstruction based on projective geometry without camera calibration. The contribution is to reduce the number of reference points to four points by exploiting some geometrical shapes contained in the scene [5]. In online application, these methods have some problems. There is a difficulty in finding the correspond-ence between one image and the others. One of these methods is stereo vision. The most important step in stereo vision is to find two points of two or more images. A general approach to be correlation that has some errors as discussed in [6].In this paper we applied a fuzzy approach to reduce these errors. Since fuzzy logic has applied in some domains such as process control, decision support system, optimization and a large class of robotic manipulators and other mechanical systems [7].

A. Sattar and B.H. Kang (Eds.): AI 2006, LNAI 4304, pp. 1271 – 1275, 2006.

© Springer-Verlag Berlin Heidelberg 2006

This paper is concerned with the aspect of improving correlation based stereo vision by reducing errors with a fuzzy system on a set of points. The experiments will be on Cartesian robot. So far a neural network approach has been used to get the optimum point in world coordinate for 3p robot [8]. Clearly there is no magic panacea for selecting a neural network for the best generalization, and also because of structure and foundation of neural networks, it has some errors. A fuzzy approach can be used to reduce these errors.

This paper is organized as follows: In Section 2, fuzzy model is described. Experimental test on 3p robot is presented in Section 3 and finally conclusions are given in section 4.

2   Fuzzy Model

Fuzzy logic controller utilizes fuzzy to convert the linguistic control strategy based on expert knowledge into an automatic control strategy. This section describes a fuzzy model for 3p robot. It also discusses a heuristic that have been applied to determine fuzzy input and output set. In first step the boarder points is obtained by exploiting some geometrical relations. Then the fuzzy model is applied to points of correlation area.

2.1   Getting the Border Points

The goal is to finding the best point of correlation area. For this propose, a heuristic method is used to find four points. These points are maximum and minimum coordinate of area's correlation in each of axes. Eq.1 gives the distance of correlation from these four points as follows:

$$\|\mathbf{X}_1 - \mathbf{X}_2\| = \sqrt{(x_1 - x_2)^2 + (y_1 - y_2)^2 + (z_1 - z_2)^2} \tag{1}$$

Where $\mathbf{X}_1, \mathbf{X}_2$ are the coordinate of two points [9].

2.2   Fuzzy Method

Recently fuzzy systems approaches have achieved superior performance. The identification of fuzzy models from input-output data of the process normally lead to representations which are difficult to understand fuzzy logic has had great success in running machinery that is computer operated. For instance, fuzzy systems used to formulate the human's knowledge. Fuzzy set theory and fuzzy logic have evolved into powerful tools for managing uncertainties inherent in complex systems [10-12]. In general, building a fuzzy system consists of three basic steps: structure identification (variable selection, partitioning input and output spaces and choosing membership functions), parameter estimation, and model validation. Fuzzy systems create a systematic process for replacing one knowledge base with a nonlinear mapping. Because of this, we will be able to use systems according to knowledge fuzzy system in engineering applications [13]. The area of correlation points is divided to four parts as shown in Fig.1.

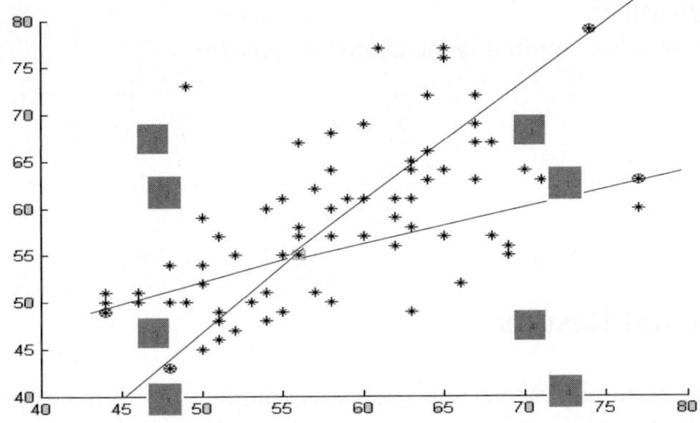

Fig. 1. The parts of correlation area

2.2.1 Fuzzification

The computational technique of inputs of fuzzy model is demonstrated in Fig.2. The distances of centers of images from best point are d_1 and d_2, respectively. The triangular membership functions have been used. Four inputs in fuzzy model are the distances of center of each area from the images (Fig. 1).

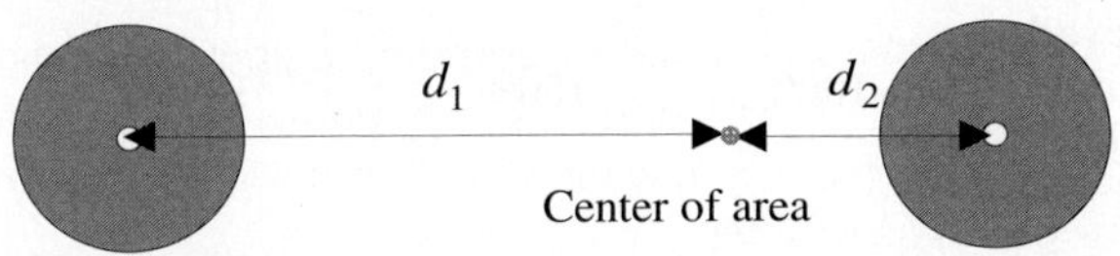

Fig. 2. Distances of centers of images from best point in heuristic method

The fuzzy controller employs four inputs: The relation of distances of points in each area in correlation image from centers of images. This fuzzy controller has only one control output.

2.2.2 Fuzzy Rule Base

The IF part of rules base includes the ratio of the distances of central point in each area from centers of images. The THEN part of these rules suggests to the center of correlation area:

IF input $r = \dfrac{d_1}{d_2} \succ 1$ THEN output is center of area (1) or center of area (2).

IF input $r = \dfrac{d_1}{d_2} = 1$ THEN output is center of correlation area.

IF input $r = \dfrac{d_1}{d_2} \prec 1$ THEN output is center of area (3) or center of area (4).

The above rules can be dedicated for any of areas.

2.2.3 Defuzzification

The defuzzifier which is applied is the center of gravity:

$$y^* = \frac{\sum_{l=1}^{M} \bar{y}^l w_l}{\sum_{l=1}^{M} w_l} \qquad (2)$$

3 Experimental Results

In this study, the accuracy of our approach was compared with neural network for the 3P robot [14-15]. In previous work, a neural network approach has been used [6], but because of structure and foundation of neural networks, it has some errors. Table.1 compares neural network method and fuzzy system on stereo vision method for 3P robot. It demonstrates in comparison the fuzzy method is more reliable.

Table 1. Comparison of neural network with fuzzy model in 3p robot

Type of Object	Circle	Cylinder	Cubic Rectangle	Cubic Square
Neural Network	98.3%	98%	97%	97.2%
Fuzzy Model	100%	100%	100%	100%

4 Conclusions

In this paper, a fuzzy model in stereo vision has been presented and applied to 3p robot. According to the simulation results, correlation errors have been reduced. Previously the best result in 3p robot by neural networks has been about 97% correctness for some objects in simulation, but with fuzzy approach we have achieved up to 100% correctness. Of course these results are in simulation and in real environment we have some errors. The errors of 3p robot have been discussed in [16]. The fuzzy model that is defined can be applied to a large class of robotic manipulators.

References

1. Seitz, S. M., and Dyer, C. R.: Complete Scene Structure from Four Point Correspondences,   5th Int. Conf on Computer Vision, Cambridge MA, (1995) 330–337.
2. Taylor, C. J., and Kriegman, D. J.: Structure and Motion from Line Segments in Multiple Images, IEEE Trans. on Pattern Analysis Machine Intelligence, Vol.17, (1995) 1021–1033.
3. Haralick, R. M., and Shapiro, L. G.: Computer and Robot Vision, Addison-Wesley Pub, (1992).

4. Grosso, E., Metta, G., Oddera, A., and Sandini, G.: Robust Visual Serving in 3-D Reaching Tasks, IEEE Trans on Robotics Automation, Vol.12, (1996) 732–742.
5. Achour, K., and Benkhelif, M.: A New Approach to 3D Reconstruction without Calibration, Pattern Recognition, Vol.34, (2001) 2467–2476.
6. Lopez, A., and Plat, F.: Dealing with Segmentation Errors in Region-Based Stereo Matching, Pattern Recognition ,Vol.33, (2000) 1325–1338.
7. Hsu, Y., Chen, G., Li, H.: A Fuzzy Adaptive Variable Structure Controller with Applications to Robot Manipulators, IEEE Transaction Systems man and Cybernetics part, Vol. 31,(2001) 331–340.
8. Korayem, M. H., Khoshhal, K., Aliakbarpour, H.: Vision Based Robot Simulation and Experiment for Performance Tests of Robot, International Journal of AMT, Vol.25, (2005) 1218–1231.
9. Hirschuller, H., Innocent, P. R., and Garibaldi, J.: Real Time Correlation Based Stereo Vision with Reduce Border Errors, International Journal of Computer Vision, (2000).
10. Bender, E. A.: Mathematical Methods in Artificial Intelligence, IEEE Computer Society Press, (1996).
11. Zhang, J., Knoll, A., and Schwert, V.: Situated Neuro-Fuzzy Control for Vision-Based Robot Localization, Robotics and Autonomous System, (1999).
12. Alexander.: Distributed Fuzzy Control of Multivariable System, Kulwer Academic Publishers, (1996).
13. Wang, L.: A Course in Fuzzy Systems and Control, Prentice Hall, (1997) 151–156.
14. Wang, L., Jin, Y.: Fuzzy Systems and Knowledge Discovery, Springer Lecture Notes in Artificial Intelligence, Berlin, (2005) 3613–3614.
15. Hahnel, D., Burgard, W., and Thrun, S.: Learning Compact 3D Models of Indoor and Outdoor Environments with a Mobile Robot, Robotics and Autonomous System, Vol.44, (2003) 15–27.
16. Korayem, M. H., Shiehbeiki, N., and Khanali, T.: Design, Manufacturing and Experimental Tests of Prismatic Robot for Assembly Line, International Journal of AMT, Vol. 29, No.3-4, (2006) 379–388.

A Hybrid Question Answering Schema Using Encapsulated Semantics in Lexical Resources

Bahadorreza Ofoghi, John Yearwood, and Ranadhir Ghosh

Centre for Informatics and Applied Optimization,
School of Information Technology and Mathematical Sciences, University of Ballarat
PO Box 663 Ballarat, Victoria 3353, Australia
bofoghi@students.ballarat.edu.au,
{j.yearwood, r.ghosh}@ballarat.edu.au

Abstract. Question Answering systems, as an extreme of the Information Retrieval field, could save lots of time and effort in satisfying a specific information need. In this regard, there are still many challenges to be resolved by current state-of-the-art systems as they cope with free texts. We propose a new hybrid question answering schema capable of answering questions with respect to different semantically related syntactically mismatched situations either in a structured or unstructured semantic format. We have exploited FrameNet and WordNet lexical resources and implemented the prototype system in a TREC-friendly fashion to obtain results comparable with outstanding participant systems in TREC 2004.

Keywords: Question Answering, Lexical Resources, FrameNet.

1 Introduction

Question Answering systems as Textual Information Retrieval systems have attracted lots of interest recently. From Information Retrieval point of view, there are two major approaches to specifying information needs: i) keyword-based queries, and ii) natural language questions. While the former is subject to resolution by systems known as Search Engines, the latter has been the impetus for Question Answering systems evolving in a way that allows direct communication with users in a more convenient and comprehensive fashion through a natural language. This is in contrast with Search Engines which require users skim and extract their information needs from the content of a number of documents with relevant and irrelevant parts mixed together.

Due to the time-consuming and tedious manner of textual information skimming inside documents, most users may prefer to have direct answers to their questions such as *"How old is the President? Who was the second person on the moon? When was the storming of the Bastille?"* [3].

We have been analyzing the results of participant systems in TREC 2003 and TREC 2004 to examine what failure situations may occur when answering the questions. From this, we have developed a new schema to cope with such cases based on a hybrid answer extraction module using two lexical resources FrameNet [1] and WordNet [5]. Results of the empirical implementation have been promising compared to the best performances in TREC 2004.

A. Sattar and B.H. Kang (Eds.): AI 2006, LNAI 4304, pp. 1276–1280, 2006.
© Springer-Verlag Berlin Heidelberg 2006

2 Lexical Resources

There are two major well-known lexical resources exploited in our question answering schema: We use WordNet which is a lexical reference system whose design is inspired by psycholinguistic theories of human lexical memory [5]. This system includes all English verbs, nouns, and adjectives organized into synonym sets. There are different relations between the synonym sets, also known as synsets. Each of the sets represents an underlying concept and from this point of view, this lexical system forms a concept hierarchy with different abstraction levels for concepts. However, the main lexical resource exploited in this work is FrameNet which is a lexical resource for English [1] that relies on an infrastructure based on Frame Semantics [2] [4]. Frame Semantics tries to emphasize the continuities between language and human experience, and FrameNet as a result of these efforts, is a framework to encapsulate the semantic information gained via Frame Semantics [7].

Generally, the main entity in FrameNet is a Frame as a kind of generalization over concepts semantically related to each other. The semantic relation between concepts in a Frame is realized with regard to the scenario of a real situation which may happen and cover the participant concepts rather than synonymy or other dictionary-oriented peer-to-peer relations. Since there are semantic relations between the circumstances covered by different Frames, a limited set of Frame-to-Frame relations has been defined in FrameNet which connect Frames to constitute a network of concepts and their pictures [8].

3 Hybrid Question Answering

Before explaining the schema of our question answering system, the failure situations where outstanding baseline question answering systems (as state-of-the-art systems) may fail to correctly reply to different types of questions are discussed. Further to our survey on the results obtained by systems participating in TREC 2004, it seems that the main reasons of reduced capability in answering more questions by such systems include: i) not accurate enough Passage Retrieval due to lack of the attention to the question answering specific challenges in passage retrieval, ii) ignorance or unawareness of scenario-based relations between language items, iii) far or wrong pronominal anaphora referencing, iv) inability in accessing indirectly available objects hidden behind a chain of lexical relations, v) failing in resolution of deep unstructured semantic relatedness between terms when there is only background semantic knowledge to resolve similarity issues, and vi) incapability of targeting adjectival answers where an answer could be realized only as a noun modifier. The major top level reason to explain why these failure situations, in the answer extraction phase, happen seems to be the nature of Named Entity Oriented approaches used by most question answering systems. Whilst in many cases, the correct answer to a question is either not of a solid Named Entity type or contains more than a unique Named Entity term.

A new question answering schema has been developed to address these complexities. The proposed schema has been specially designed to overcome these inadequacies to the extent that the exploited model and resources can contribute to the knowledge of the system. The high level illustration of this schema contains three main sections: i) the User Interface to communicate with end-users, ii) the QA Engine to formulate a query on top of the original question and then retrieve the most related documents and passages from which the candidate and actual answers are to be extracted, and iii) the Semantic Interface to play the role of the bridge between the system and the lexical resources already introduced in section 2. The Answer Extraction module, as the main part under our consideration, operates while processing the text of the top most passages to extract candidate answers to a given question. This module in our schema includes two models: i) Structured Semantic Model (SSM) and ii) Utility-Based Model (UBM). The first and most reliable answers may be results of the SSM which directly deals with semantic structures. In cases that such a model is not able to identify the candidate and/or actual answers, the other model would be exploited in order to find some other candidate answers.

SSM-The task of answer extraction in the Structured Semantic Model starts with annotating the question and the anaphorically resolved passages which potentially bear the candidate and/or actual answers. If the texts of the passages evoke the same Frame as the target Frame evoked by the question, then the following steps are followed: i) find the vacant Frame Element inside the annotated question string via question stems processing (e.g. where, when, what, how, …), ii) find the matching Frame inside the annotated passages, and iii) extract the value of the Frame Element inside the passage which matches with the vacant Frame Element of the same Frame inside the question.

In such cases, the information inside a single Frame is of importance. We refer to this term-level related analysis as *Intra-Frame Analysis* to underpin the scenario-based relations between syntactically different situations inside texts. For example, if a question asks *"When was it discovered?"* and the passage contains *"…it was first spotted in 1989…"* then because of the fact that both verbs "spot" and "discover" come from the same Frame, the Frame and Frame Element matching with regard to this Intra-Frame knowledge could result in the candidate answer "1989". The main advantage of using encapsulated semantics in FrameNet is to make the question answering system able to consider a more general scenario around the focused information need asked by the question.

If there is no match between the target Frame of a question and those Frames evoked in the passages, the answer extraction process continues in the second model known as UBM.

UBM-Despite the SSM, in the UBM's approach, there is no strict guarantee that the extracted answer is a good potential answer for the question. The process of answer extraction in this model starts with merging all top passages into one passage which consequently is broken down into its sentences. The semantic relatedness of each sentence to the question is measured based on an algorithm which has been already developed by Troy Simpson and Thanh Dao[1]. The sentences are ranked based on their scores and all Named Entities from these sentences are extracted. The set of Named Entities is further processed to be filtered against the category of the question already identified by the Question Analysis module of the system. After extracting and

---

[1] http://www.codeproject.com/csharp/semanticsimilaritywordnet.asp

filtering all Named Entities any remaining Named Entities are scored and ranked and finally, the top most Named Entities in the ranked list are reported as candidate answers.

4   Experiments

The experimental issues presented relate to the dataset, procedure, and results obtained. The dataset used in the experiment is the TREC 2004 question list and its corresponding text collection - AQUAINT[2]. We have used a subset of the factoid questions to test the new question answering schema. This subset contains 5 factoid questions correctly answered by the best-performing TREC 2004 participant, LCC [6] and 10 factoid questions not correctly answered by this system. Only 15 questions were used because manual annotation had to be performed in the absence of an automated annotator.

Table 1. Answer Extraction Results

Answer Extraction Model	Total No. of Questions	Questions Answered by our System		
		Number	Percent	MRR
SSM	15	7	46.66%	0.3889
UBM	15	6	40.00%	0.2244
		13	86.66%	0.6133
Questions Answered by LCC	5	5	100%	0.4733
Questions not Answered by LCC	10	8	80%	0.6833

Our experiments started with indexing the necessary documents of AQUAINT with respect to the related document set of each target in the TREC 2004 QA Track reported by TREC. Lemur's text indexing engine has been used to this end[3]. Some software components have been either implemented or wrapped and integrated in our system. In addition, support tools for Named Entity tagging, Anaphora Reference resolution[4], and Part-of-Speech tagging[5] were used.

Results summarized in Table 1 show that the SSM was able to correctly answer 7 questions and 6 questions were correctly answered by the UBM after the SSM failed. Totally 13 questions out of the 15 questions were answered correctly by our system which translates to 86.66% of the questions under consideration. The second part of the table illustrates that all 5 questions correctly answered by the baseline system,

---

[2]   http://www.ldc.upenn.edu/Catalog/docs/LDC2002T31/
[3]   http://www.lemurproject.org/lemur/overview.html
[4]   LingPipe: http://alias-i.com/lingpipe/download.html
[5]   Zamora: http://www.scientificpsychic.com/linguistics.html

LCC, have been correctly answered by our system. In addition, 8 out of 10 questions not answered by LCC have been correctly answered by our answering models.

The SSM and UBM models have 53.85% and 46.25% performance respectively over the 13 questions correctly answered given that they have been applied in order. The SSM has been the first model used and only in the cases that it has failed, has the UBM model been activated and used in the answer extraction task.

5 Conclusion

To enhance the ability of question answering systems with respect to answering a wider range of questions, a survey on failure situations of best-performance question answering systems participating in TREC 2004 has been performed and a novel hybrid question answering schema has been proposed and implemented. This system has been implemented and benefits from the lexical semantic information encapsulated in the two well-known English resources, FrameNet and WordNet. The results of this system have been promising in the limited testing of the failure situations of current outstanding question answering systems.

References

1. Baker, C. F., C. J. Fillmore and J. B. Lowe (1998). The Berkeley FrameNet Project. *International Conference on Computational Linguistics* **1**: 86 - 90.
2. Fillmore, C. J. (1976). Frame Semantics and the Nature of Language. *In Annals of the New York Academy of Sciences: Conference on the Origin and Development of Language and Speech* **280**: 20-32.
3. Hovy, E., L. Gerber, U. Hermjakob, M. Junk and C.-Y. Lin (2001). Question Answering in Webclopedia. *Proceedings of the TREC-9 Conference*: 655.
4. Lowe, J. B., C. F. Baker and C. J. Fillmore (1997). A Frame-Semantic Approach to Semantic Annotation. *SIGLEX Workshop on Tagging Text with Lexical Semantics: Why, What, and How?*
5. Miller, G., R. Beckwith, C. Fellbaum, D. Gross and K. J. Miller (1990). Introduction to WordNet: an On-Line Lexical Database. *International Journal of Lexicography*: 235-244.
6. Moldovan, D., S. Harabagiu, R. Girju, P. Morarescu, F. Lacatusu, A. Novischi, A. Badulescu and O. Bolohan (2002). LCC Tools for Question Answering. *Proceedings of the eleventh Text REtrieval Conference (TREC 2002)*.
7. Petruck, M. R. L. (1996). Frame Semantics. *Handbook of Pragmatics Online*.
8. Ruppenhofer, J., M. Ellsworth, M. R. L. Petruck and C. R. Johnson (2005). FrameNet: Theory and Practice. *http://framenet.icsi.berkeley.edu/*.

Structure Detection System from Web Documents Through Backpropagation Network Learning

Bok Keun Sun, Je Ryu, and Kwang Rok Han

Department of Computer Engineering, Hoseo University,
165 Baebang, Asan City, ChungNam, Korea
{bksun, ryuje, krhan}@office.hoseo.ac.kr

Abstract. This paper discusses a system that learns the structure of Web documents through a backpropagation network and infers the structure of new Web documents. The system first converts Web documents into the input of the backpropagation network through assigning ID to XPath. The learning system of the backpropagation network repeats learning until the error rate goes down below the level specified in the system. After learning, a new Web document is passed through the network, the system infers the structure of the document and extracts information suitable for the structure. The biggest advantages of this system are that there is no human intervention in the learning process and the network is designed to derive the optimal learning result by changing the internal factors and parameters in various ways. When the implemented system was evaluated, the average recall rate was 99.5% and the precision rate was 96.6%, suggesting the satisfactory performance of the system.

Keywords: Structure Detection, Backpropagation Network Learning.

1 Introduction

Standard templates are applied to Web documents such as news articles and book information at book stores, and such templates contain main data, header, advertisement, trailer, image, etc. However, with the explosive increase of Web documents and data, there are increasing demands for the automation of information retrieval including documents classification, search and retrieval. In this situation, various types of information contained in template-based documents are great obstacles to the automation of information retrieval. To cope with this problem, various researches are being made to use Web documents as data sources for data-based information retrieval, computation and application.

In our case, the system was designed based on the fact that each site has its unique document structure like a template. In a HTML document of each site, XPath containing pcdata (printable character data) can express the unique structure of the document. Thus, we developed a system that learns XPath data through backpropagation network and, based on the learned structure, infers the unique structure of new documents.

2 Related Works

Various researches are being made to use Web documents as data sources for information retrieval, computation and application. These researches on Web document

A. Sattar and B.H. Kang (Eds.): AI 2006, LNAI 4304, pp. 1281–1287, 2006.
© Springer-Verlag Berlin Heidelberg 2006

processing are based on diverse technologies including query language, natural language processing, grammatical rules, HTML tree structure processing and HTML table processing. Most of these works concern over how the structure of HTML documents should be recognized by their systems. Ashish[1], MDR[2], Omini[3], Kushmerick[4], STAVIES[5] works are making efforts to read Web page structure to extract data.

Studies on Web document structure itself often use the method of finding the table tag of HTML. The system of Wang[6] proposed a table detection algorithm based on machine learning in order to use it in information retrieval systems. The system was designed, taking note that HTML tables are commonly used to express information, and it carried out machine learning using 1,393 HTML documents collected from hundreds of sites. Chen[7] also used data mining with HTML table tags, and Hurst[8] and Yoshida[9] used table structure to understand document structure and process information.

In this study, we built a system that trains backpropagation network using XPath not of tables but of pcdata and recognizes document structure. Although most of table tags contain pcdata, many of them do not, so we chose the XPath of pcdata for learning data.

3 Backpropagation Network Learning

In this study, we built a backpropagation network learning system to examine the structure of Web documents. The backpropagation network learning model can learn a continuous function model if a sufficient number of units are provided to the hidden layer. As in Figure 1, if *Input(i)* is repeatedly multiplied by and added to the weights of the neural network, *Calculation(y)* is produced. Here, *Calculation(y)* is different from *Output(o)*, which is the expected output for given learning data. In the neural network, error e as large as *(y-o)* takes place. Then, in proportion to the error, the weight of the output layer is updated and then the weight of the hidden layer is updated. The order of updating weights is opposite to that of neural network processing. This learning process is repeated until error e reaches an appropriate level. The number of inputs in the input and output layers varies according to application area, and control constants used inside the network should also be optimized through experiment [10].

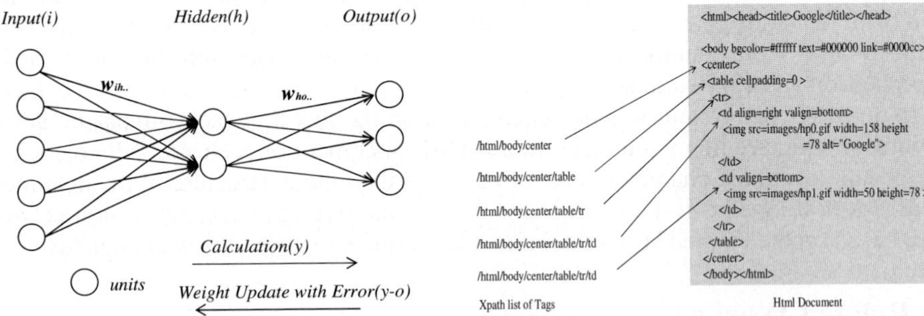

Fig. 1. Backpropagation Network Learning Model

Fig. 2. An Example of XPath in HTML Document

3.1 Network Input

For the input of the backpropagation network, we used vector created by processing XPath containing pcdata in Web documents. The XPath is an expression indicating the path of a specific element in an XML document. Because HTML is the subset of XML, XPath can be also applied to HTML as in figure 2. System structure for creating input vector is as in Figure 3.

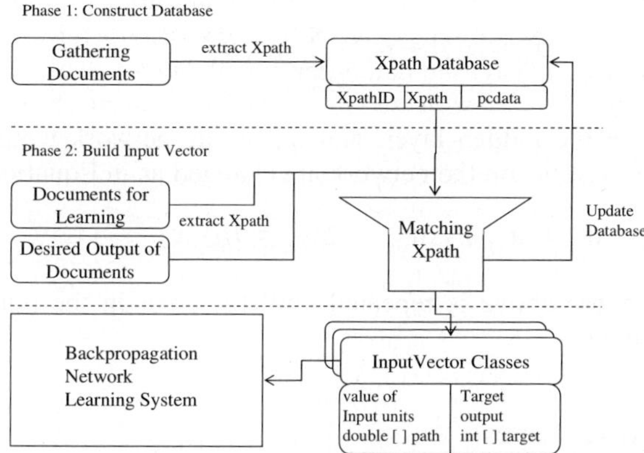

Fig. 3. Structure of Input Vector Building System

Phase 1 collects various kinds of documents, extracts XPath from the documents, and creates a database. All XPaths containing pcdata are extracted and stored in the database. To use as the input of the backpropagation network, Phase 2 creates Input-Vector class based on data in the database from Phase 1. It extracts the XPath of a learning document corresponding to the desired output and checks if it exists in the database. After XPath inspection is finished for a learning document, an InputVector class is generated and the class contains an array of XPathIDs and an array of outputs. In this paper, the total number of outputs was 20, and for each output 200 learning documents were collected and their InputVector classes were created.

3.2 Backpropagation Network Learning Algorithm

The backpropagation network proposed in this paper performs learning using Input-Vector classes as its input, and the backpropagation algorithm is multi-layer perceptron. The backpropagation algorithm is a least mean square algorithm that minimizes the mean square error between the actual output of the multi-layer perceptron and desired output, and error E_p for input vector P is calculated by Equation 1. Error E for all inputs is the sum of $E_p's$.

$$E_p = \frac{1}{2}\sum_j \left(t_{pj} - O_{pj}\right)^2 \tag{1}$$

Error in the output layer is calculated by Equation 2, which is obtained by differentiating Equation 1.

$$\delta_o = o_o(1 - o_o)(t_o - o_o) \tag{2}$$

o_o and t_o are the result and the target value of the output layer, respectively. Then the error in Equation 2 is back-propagated to the hidden layer. Receiving the back-propagated error, the hidden layer calculates an error reflecting weights. This process is expressed by Equation 3.

$$\delta_h = o_h(1 - o_h) \sum_{o \in outputs} w_{oh} \delta_o \tag{3}$$

Here, o_h is output in the hidden layer, and w_{oh} is the network weight in the output layer. By the value, weights on the network are changed as in Equation 4.

$$w_{ji} = w_{ji} + \Delta w_{ji}, \quad \Delta w_{ji} = \eta \delta_j x_{ji} \tag{4}$$

The learning process above is repeated until the error in the output layer comes within a specific range.

4 System Architecture

Figure 4 shows the entire structure of the system implemented in this study. The input vector building system in Chapter III is used in all of pre-processing, learning, and deduction & detection. Pre-processing converts HTML documents to the input of the network, and learning process trains the network using backpropagation algorithm. Lastly, the deduction & detection system recognizes the structure of a new HTML document using the trained network. The details of each process are as follows.

Preprocessing: Through the input vector building system, a HTML document is utilized in recording XPath into the database, training the network, and deduction & detection. For this, the HTML document is converted through a parser to an element tree, in which a HTML tag is the root. Then, XPath extraction process is performed to extract XPath beginning from the HTML tag of the element, in which the lowest node of the tree is pcdata. In the input vector building system, XPath is used in the three ways mentioned above.

Learning: A preprocessed InputVector class is used as input for learning. The learning algorithm of the network is described in Section 2 of Chapter III.

As the number of inputs for the learning network is 300 and the number of output is 10, an InputVector class has XPaths and outputs as many as the numbers of network inputs and outputs, respectively.

Deduction & Detection: After learning is completed in the learning process, the network is copied to a network for deduction & detection. For document classification, a new document is converted to InputVector through preprocessing and the value is used as the input of the deduction network. The InputVector passes through the network and gets an output. The output value is used to infer the structure of the input document. Using the inferred structure information and the XPath of the corresponding main content, the body of the main content is extracted. Here, the XPath of the main content corresponding to each structure is applied heuristics registered in advance by the user.

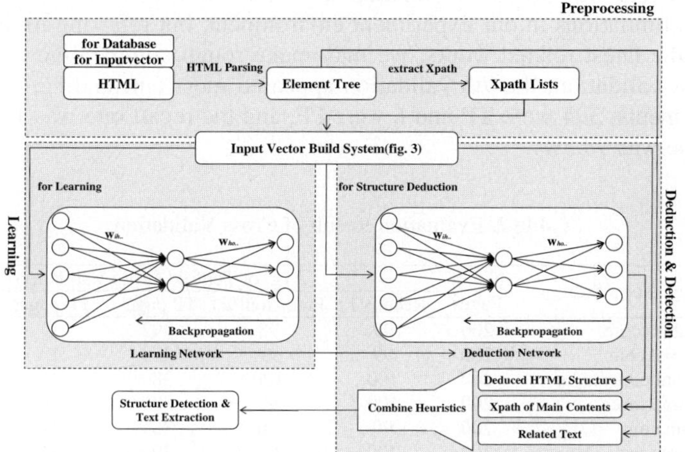

Fig. 4. Learning & Detection System Architecture

5   Experimental Results

Limitations: Because the system implemented in this study was designed to extract articles in news sites, Web documents used in the experiment for evaluation were limited to news site documents that have a template with fixed rules. For the same reason, each site's portal pages by category were excluded from the experiment.

Implementation: To implement the prototype of the system, we used Java 1.4 JDK. To store XPath into database and create InputVector, we made a simple browser with HTML parsing and rendering functions and used it as a basic program for HTML document processing.

To train the backpropagation network, we selected 20 news sites and collected 200 documents per site. A HTML document is converted to an InputVector, and each InputVector contains 300 input nodes, each of which goes to one of 20 output nodes. As in Table 1, the optimal values were set experientially for the numbers of hidden layers and internal parameters of the learning network based on experiment.

Table 1. Parameter values of Backpropagation Network

No. of Hidden Layer	Weight Cange Step	Iteration Count of Training	Momentum	Threshold
30	0.1	5,000	0.1	0.005

Evaluations: The results of learning was evaluated as in Figure 6. Recall and precision were measured using the tenfold cross validation method. According to the contents explained in [11], TP (True Positive), FP (False Positive) and FN (False Negative) pages were used in measuring, and the results of measuring are presented in Table 2. According to the result of evaluation, the average recall rate was 99.5% and the precision rate was 96.6%. Comparative evaluation with other systems was not made due to limitations in our experiment environment, but referring to the results of [5], one of the latest related works, we may make rough comparative evaluation. In tenfold cross validation, the 7th validation appeared most outstanding, in which, out of 400 documents, 394 were TP and 6 were FP, and the recall rate was 100% and the precision rate was 98.5%.

Table 2. Evaluation Result of Cross Validation

#	Web News Site	No. of Pages	Recall(%)	Precision(%)	TP pages	FP pages	FN pages
1	news.yahoo.com	200	99	98	392	8	2
2	news.naver.com	200	99	97	378	13	3
3	www.msnbc.msn.com	200	100	100	380	0	0
4	news.empas.com	200	100	99	380	4	0
5	www.cnn.com	200	99	96	384	14	2
6	news.daum.net	200	100	97	396	11	1
7	www.hankooki.com	200	99	92	400	33	3
8	news.bbc.co.uk	200	100	99	388	4	0
9	www.chosun.com	200	100	95	400	20	0
10	www.joins.com	200	100	94	400	24	0
11	www.donga.com	200	100	99	388	2	0
12	www.ytn.co.kr	200	99	97	368	10	2
13	www.ohmynews.com	200	99	98	370	7	3
14	www.khan.co.kr	200	99	97	390	13	3
15	isplus.joins.com	200	100	97	392	14	0
16	www.edaily.co.kr	200	99	95	370	20	2
17	www.cbs.co.kr	200	99	95	376	19	3
18	www.ktv.go.kr	200	99	92	396	35	3
19	www.latimes.com	200	100	97	386	12	0
20	www.hani.co.kr	200	100	99	372	3	1

According to results in [Table 2], FP of site #7 and #18 is slightly higher than that of other sites. Analyzing the cause, we found that part of pages in site #17 and #20 were similar in structure to those in site #7 and #18.

6   Conclusions

The present study proposed a method of recognizing Web document structure. For this, we designed a system that trains a backpropagation network using Web document structure and infers the structure of new Web documents, and implemented a prototype of the system that extracts the main content of the corresponding structure.

Evaluating the system, we obtained satisfactory results with the average recall rate of 99.5% and precision rate of 96.6%, but also found the problem that the FP level is somewhat higher in some pages. To solve this problem, we will apply heuristic rules by sites. What is more, we need to find more optimized internal parameters through more experiments.

We used the method of recognizing the structure of Web documents in order to extract the body of Web pages in news sites, and because Web pages in news sites were different in nature from those used in related works such MDR [2], OMINI [3] and STAVIES [5] as mentioned above, we could not make cross evaluation among the systems. More objective results are expected from cross evaluation among the systems in related works in the future.

References

1. N. Ashish and C.A. Knoblock, "Semi-Automatic Wrapper Generation for Internet Information Sources," Proc. Int'l Conf. Cooperative Information System, pp.160-169, 1997.
2. B. Liu, R.Grossman, and Y.Zhai, "Mining Data Records in Web Pages," Proc. Ninth ACM SIGKDD Int'l Conf. Knowledge Discovery and Data Mining, pp.601-606, 2003.
3. D. Buttler, L. Liu, and C. Pu, "A Fully Automated Object Extraction System for the World Wide Web," Proc. 2001 Int'l Conf. Distributed Computing Systems(ICDCS '01),pp.361-370,2001
4. N. Kushmerick, D.Weld, and R.Doorenbos, "Wrapper Induction for Information Extraction," Proc. Int'l Joint Conf. Artificial Intelligence(IJCAI-97),1997
5. K. Papadakis, D Skoutas, K Raftopoulos and A. Varvarigou, "STAVIES: A System for Information Extraction from Unknown Web Data Sources through Automatic Web Wrapper Generation Using Clustering Techniques," IEEE Transactions on Knowledge and Data Engineering, Vol 17, No 12, pp1638-1652, 2005.12
6. Y. Wang and J Hu, "A Machine Learning Based Approach for Table Detection on The Web," Proc. 11th Int'l Conf. World Wide Web, pp. 242-250 ACM Press, 2002.
7. HH. Chen, SC. Tsai and JH. Tsai, "Mining Tables from Large Scale HTML Texts", Proc. 18th Int'l Conf. Computational Linguistics(COLING), pp.166-172, 2000
8. M. Hurst "Challenges for Table Understanding on the Web," In Web Document Analysis, Proceedings of the 1st Int'l Workshop on Web Document Analysis, Seattle, WA, USA, September 2001.
9. M. Yoshida,K. Torisawa and J. Tsujii, "A Method to Integrated Tables of the World Wide Web," In Web Document Analysis, Proceedings of the 1st Int'l Workshop on Web Document Analysis, Seattle, WA, USA, September 2001.
10. S. Y. Kung, "Digital Neural Networks," Prentice Hall, pp.184-187, 1993
11. H. Witten and E Framk, "Data Mining:Practical Machine Learning Tools and Techniques with Java Implementations", Morgan Kaufmann Publishers, 1999.

An Effective Recommendation Algorithm for Improving Prediction Quality

Taek-Hun Kim and Sung-Bong Yang

Dept. of Computer Science, Yonsei University
Seoul, 120-749, Korea
{kimthun, yang}@cs.yonsei.ac.kr

Abstract. A recommender system utilizes in general an information filtering technique called collaborative filtering. To improve prediction quality, collaborative filtering needs reinforcements such as utilizing useful attributes of the items as well as a more refined neighbor selection. In this paper we present that the recommender systems that utilizing the attributes of the items in collaborative filtering improves prediction quality. The experimental results show that the recommender systems using the attributes provide better prediction qualities than other methods that do not utilize the attributes.

1 Introduction

A recommender system using collaborative filtering which we call it CF, calculates the similarity between the test customer and each of other customers who have rated the items that are already rated by the test customer. To improve prediction quality, collaborative filtering needs reinforcements such as utilizing useful attributes of the items as well as a more refined neighbor selection.

In this paper we show that the recommender systems that utilizing the attributes of each item for improving prediction quality in collaborative filtering. The recommender system exploits the attribute information from the item-attribute matrix and applies it to the prediction process. The experimental results show that the prediction quality is affected by using the attribute information for the test customer or the neighbors. Therefore the recommender systems using the attributes provide better prediction qualities than other methods that do not utilize the attributes.

2 Collaborative Filtering-Based Recommender Systems

Collaborative filtering creates a set of customer's preferences for each item, compares it to other customers' preferences. Collaborative filtering filters the proper information through it. That is, collaborative filtering calculates the degrees of similarity between the preference of the test customer and each of those of other customers using the correlation of preferences for the same item. Collaborative filtering then filters suitable items based on their preferences and the

A. Sattar and B.H. Kang (Eds.): AI 2006, LNAI 4304, pp. 1288–1292, 2006.

© Springer-Verlag Berlin Heidelberg 2006

similarities, and recommends the items to the test customer. Collaborative filtering works well for the recommender systems in general. Collaborative filtering is computed by (1) in [4].

There have been many investigations for selecting proper neighbors based on neighbor selection methods such as the k-nearest neighbor selection, the threshold-based neighbor selection, and the clustering-based neighbor selection. They are quite popular techniques for recommender systems[1][4].

3   Improving Prediction Quality in CF

Item-Attribute Matrix. In order to utilize the attributes of the items in the prediction process, we consider a matrix of item-attribute for a customer as in Fig. 1. This figure shows the information of the item-attribute for customer a. In this figure $r_{a,i}$ denotes customer a's rating for item i. Note that $r_{a,i} \in \{nil, 1, 2, 3, 4, 5\}$, where '$nil$' indicates "no rating". An item may have an arbitrary combination among j attribute values as its attribute values; $Attribute(i) \subset \{1, 2, 3, \ldots, j\}$. Note that $Attribute(i) \neq \emptyset$. And $A_{i,j}^a \in \{0, 1\}$.

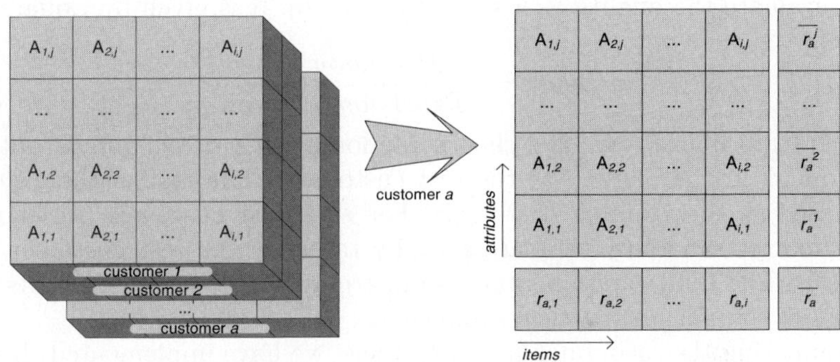

Fig. 1. The item-attribute matrix for customer a

Prediction Process. We propose Equation (1) as a new prediction formula in order to predict customer's preferences more accurately. In this equation, $A(\overline{r_{a,i}})$ and $A(\overline{r_{k,i}})$ are the averages of customer a and k's attribute values, respectively. In the equation, $N_{a,i}$ and $N_{k,i}$ is the number of 'valid' attributes for item i and $M_{a,j}$ and $M_{k,j}$ is the number of 'valid' items that has attribute j for the customer a and k, respectively. The valid attributes means $A_{i,j}^a = 1$ or $A_{i,j}^k = 1$ and the valid item means $r_{a,i}$ or $r_{k,i}$ is not 'nil', respectively. All other terms in the equation are similar terms defined for the Equation (1) in [4].

There are a lot of items which have different attributes for the item for new prediction. Therefore, if we retrieve the attributes of the items more accurately and use them to the prediction process with Equation (1), then we can get more accurate prediction quality than the case without considering the attributes because customers are in general very sensitive to the attributes of the items.

$$P_{a,i} = A(\overline{r_{a,i}}) + \frac{\sum_k \{w_{a,k} \times (r_{k,i} - A(\overline{r_{k,i}}))\}}{\sum_k |w_{a,k}|} \;, where$$

$$A(\overline{r_{a,i}}) = \frac{\sum_j \{A_{i,j}^a \times \overline{r_a^j}\}}{N_{a,i}} \;, \quad \overline{r_a^j} = \frac{\sum_i \{A_{i,j}^a \times r_{a,i}\}}{M_{a,j}} \;,$$

$$A(\overline{r_{k,i}}) = \frac{\sum_j \{A_{i,j}^k \times \overline{r_k^j}\}}{N_{k,i}} \;, \quad \overline{r_k^j} = \frac{\sum_i \{A_{i,j}^k \times r_{k,i}\}}{M_{k,j}} \;. \tag{1}$$

4   Experimental Results

In the experiments we used the MovieLens dataset of the GroupLens Research Center[5]. We used two types of evaluation metrics which are prediction accuracy metrics and recommendation list accuracy metrics. One of the statistical prediction accuracy metrics is MAE. MAE is the mean of the errors of the actual customer ratings against the predicted ratings in an individual prediction [3]. *Precision* and *recall* are also used for evaluating recommendation list in the *Information Retrieval Community*[2]. And the standard F-measure is used in order to evaluate the quality as a single measure[6]. It is given by Equation (2).

$$F-measure = \frac{2 \cdot Precision \cdot Recall}{Precision + Recall} \;. \tag{2}$$

For the experiment, we have chosen randomly 10% of customers out of all the customers in the dataset as the test customers. The rest of the customers are regarded as the training customers. For each test customer, we chose ten different movies randomly that are actually rated by the test customer as the test movies. The final experimental results are averaged over the results of ten different test sets for a statistical significance.

For comparing the recommendation systems, we have implemented three recommender systems each of which uses different neighbor selection method. The first system is CF with the k-nearest neighbor selection ($KNCF$). The second system is CF with the threshold-based neighbor selection ($TNCF$). The third one is CF with the clustering-based neighbor selection ($KMCF$).

For TNCF, three different systems have been implemented. The first system is TNCF with positive neighbors ($TNCF\_P$). The second one is TNCF with negative neighbors ($TNCF\_N$). And the last one is TNCF with both positive and negative neighbors ($TNCF\_A$).

For KMCF, we also have two different settings. One is KMCF with the Euclidean distance as a distance ($KMCF\_E$). And the other is KMCF with the Pearson correlation coefficient as a distance ($KMCF\_C$).

There are four cases of combination according to whether the attribute of the test customer or the attribute of the neighbor is applied to the prediction process or not. That is, case A doesn't use any attributes, case B uses only the attributes of the neighbors, case C uses only the attributes of the test customer, and case D uses both of them. Fig. 2 shows the experimental results and the improvement ratio for each case to case 'A' is shown in Table 1.

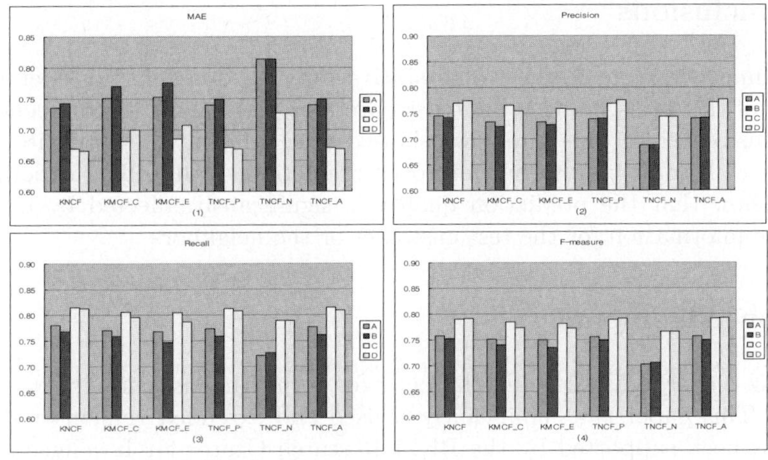

Fig. 2. Comparisons of various methods for each case

Table 1. The improvement ratio(%) to case A

		KNCF	KMCF_C	KMCF_E	TNCF_P	TNCF_N	TNCF_A
	B	-1.06	-2.56	-3.09	-1.25	-0.00	-1.24
MAE	C	8.97	9.31	9.00	9.32	10.75	9.35
	D	9.42	6.77	6.06	9.60	10.75	9.65
	B	-0.45	-1.25	-0.68	0.16	0.00	0.07
Precision	C	3.33	4.33	3.58	4.09	8.03	4.10
	D	3.94	2.83	3.40	4.90	8.03	4.87
	B	-1.44	-1.44	-2.72	-1.87	0.67	-2.12
Recall	C	4.52	4.63	4.77	4.99	9.34	4.75
	D	4.21	3.35	2.51	4.48	9.34	4.05
	B	-0.74	-1.41	-1.98	-0.88	0.49	-1.00
F-measure	C	4.19	4.51	4.16	4.51	8.94	4.49
	D	4.33	3.03	2.97	4.67	8.94	4.64

As shown in Fig. 2, each system has been tested both with and without utilizing the attributes. The results in the figure show that the systems using the attributes outperform those without considering them.

The experimental results show us that the prediction with case D that the attribute is applied to both the test customer and the neighbors, provides the best prediction quality in KNCF and TNCF. And in these methods case C shows also as good prediction qualities as case D. On the other hand KMCF provides the best results when we use case C in both KMCF_C and KMCF_E.

In the results we found that the neighbor selection methods in which the attribute information is applied to both the test customer and the neighbors or to only the test customer gets more accurate prediction quality than the system in which the attribute information is applied to only the neighbors or to neither the test customer nor the neighbors.

5   Conclusions

A recommender system with collaborative filtering does not consider the attributes of the items. So it is limited to produce high quality recommendations to the customers. In this paper we showed that utilizing the attributes of the items in collaborative filtering improves prediction quality. The experimental results show that the prediction quality is significantly affected by using the attribute information for the test customer or the neighbors.

Acknowledgements

We thank the GroupLens Research Center for permitting us to use the MovieLens dataset. This paper is a significantly revised and extended version of [4]. This work has been supported by the BK21 Research Center for Intelligent Mobile Software at Yonsei University in Korea.

References

1. B.M. Sarwar, G. Karypis, J.A. Konstan, J.T. Riedle: Recommender Systems for Large-Scale E-Commerce: Scalable Neighborhood Formation Using Clustering. Proceedings of the Fifth International Conference on Computer and Information Technology. (2002)
2. H. Nguyen, P. Haddawy: The Decision-Theoretic Interactive Video Advisor. Proceedings of the 15th Conference on UAI. (1999)
3. J.S. Breese, D. Heckerman, and C. Kadie: Empirical Analysis of Predictive Algorithms for Collaborative Filtering. Proceedings of the Conference on UAI. (1998)
4. T.H. Kim, S.B. Yang: Using Attributes to Improve Prediction Quality in Collaborative Filtering. LNCS, Vol. 3182. (2004)
5. MovieLens dataset, GroupLens Research Center, url: http://www.grouplens.org/.
6. R.J. Mooney, L. Roy: Content-Based Book Recommending Using Learning for Text Categorization. Proceedings of the ACM Conference on Digital Libraries. (2000)

Customer Online Shopping Behaviours Analysis Using Bayesian Networks

Zi Lu[1], Jie Lu[2], Chenggang Bai[3], and Guangquan Zhang[2]

[1] Faculty of Resources & Environment Sciences, Hebei Normal University, China
luzi@mail.hebtu.edu.cn
[2] Faculty of Information Technology, University of Technology, Sydney, Australia
{jielu, zhangg}@it.uts.edu.au
[3] Department of Automatic Control, Beijing University of Aeronautics and Astronautics, China
bcg@buaa.edu.cn

Abstract. This study applies Bayesian network technique to analyse the relationships among customer online shopping behaviours and customer requirements. This study first proposes an initial behaviour-requirement relationship model as domain knowledge. Through conducting a survey customer data is collected as evidences for inference of the relationships among the factors described in the model. After creating a graphical structure, this study calculates conditional probability distribution among these factors, and then conducts inference by using the Junction-tree algorithm. A set of useful findings has been obtained for customer online shopping behaviours and their requirements with motivations. These findings have potential to help businesses adopting more suitable online system development.

Keywords: E-services, Bayesian networks, Customer behaviours, Inference.

1 Introduction

Related research has also been conducted in evaluating the quality of web-based e-service [8]. Typical approaches used in the research field are testing, inspection and inquiry [3]. These approaches are often used together in conducting and analysing a web search or a desk survey. For example, Lu et al. [7] showed their assessment results for e-commerce development in businesses of New Zealand. Customer satisfactory evaluation has been conducted for obtaining customers' feedback and measure the degree of their satisfaction for current e-services provided. Questionnaire-based survey and multi-criteria evaluation systems are mainly used to conduct this kind of research. For example, Lin [6] examined customer satisfaction for e-commerce and proposed three main scales which play a significant role in influencing customer satisfaction. In general, the common research methods (e.g. surveys and focus groups) used in these researches more often reveal what customers *think* their motivations are, rather than what their motivations truly are. When respondents do not comprehend their true motivations, they tend to state how they think they *ought* to be motivated. As a business, it would more like to know if its customers want using e-service applications, their motivation, their requirement to the

A. Sattar and B.H. Kang (Eds.): AI 2006, LNAI 4304, pp. 1293–1297, 2006.
© Springer-Verlag Berlin Heidelberg 2006

features provided in their e-service websites, and which features are more important than others to attract new customers. These results will directly or indirectly support making business strategies in e-service application development. It is therefore very necessary to explore the possible relationships among customer online shopping behaviours and their requirements.

2 Customer Online Shopping Behaviours Model

A model to describe the relationships among customer online shopping behaviours and their requirements on the features of e-service is established as shown in Fig. 1. A1-A3 are regarding to customer online shopping behaviours. S1-S5 are regarding to customer requirements on the features provided in e-services.

A1	I like to experiment with new information technologies.
A2	I like to purchase goods/services online.
A3	I intend or have done purchasing goods/services online.

S1	I feel it is Attractive
S2	I feel it is Dependable
S3	I feel it is Reliable
S4	I feel it is Trustworthy
S5	It can meet my Demand

Fig. 1. Model of customer online shopping behaviours and their requirements

This study collected data from a sample with two groups of people: an Australian customer group, a Chinese customer group. A questionnaire is designed to have questions with five-point Likert scales: '1'—strong disagree, '5'—strong agree. The survey results are firstly used to identify which one, out of S1, S2, S3, S4, and S5, is as the most important factor for users to decide purchasing service online, and how customers assess the quality of e-service websites. Based on the results, this paper focuses on the analysis of relationships among the two sets of factors.

3 Bayesian Network Based Analysis

Bayesian network is a powerful knowledge representation and reasoning tool under conditions of uncertainty. A Bayesian network $B =< N, A, \Theta >$ is a directed acyclic graph (DAG) $< N, A >$ with a conditional probability distribution (CPD) for each node, collectively represented by Θ, each node $n \in N$ represents a variable, and each arc $a \in A$ between nodes represents a probabilistic dependency [1]. Based on Fig. 1 a graphical Bayesian network structure about the relationships among the factors of the two categories can be created.

3.1 Creating a Graphical Structure

These notes and relationships shown in Fig. 2 are considered as a result obtained from domain knowledge. In order to improve the Bayesian network, structural learning is needed by using the data collected, described in Section 2 to test the established relationships. A suitable structural learning algorithm, Greedy Hill-Climbing [4], is selected for the structural learning. The algorithm starts at a specific point in a space, checks all nearest neighbours, and then moves to the neighbour that has the highest score. If all neighbours' scores are less than the current point, a local maximum is thus reached. The algorithm will stop and/or restart in another point of the space. By running the Greedy Hill-climbing algorithm for structure learning from the collected data, an improved Bayesian network is obtained where three links between S_i ($i = 1$, 2, 3) and A3 are removed.

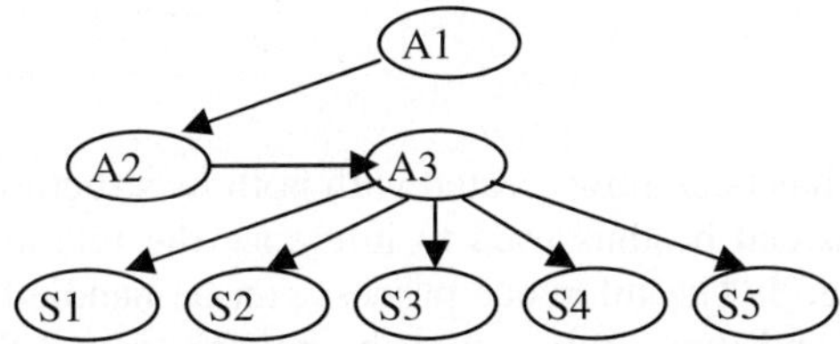

Fig. 2. Factors relation Bayesian network

3.2 Calculating the Conditional Probability Distributions

Once the improved Bayesian network is defined, the next step is to determine the conditional probabilities distribution for each node. Let $X = (X_0, \cdots, X_m)$ be a note set, X_i ($i=0, 1,\ldots m$) is a discrete node (variable), in the improved Bayesian network. The CPD of the node X_i is defined as $\theta^B_{x_i|Pa_i} = P(X_i = x_i \mid Pa_i = pa_i)$ [7], where Pa_i is the parent set of node X_i, pa_i is a configuration (a set of values) for the parent set Pa_i of X_i, and x_i is a value that X_i takes. Based on the data collected, the CPDs of all nodes shown in Fig. 3 are calculated.

Before using a Bayesian network to conduct inference, learning and establishing the parameters $\theta^B_{x_i|Pa_i}$ from the data collected should be completed. Based on Heckerman [7]'s suggestions, the Dirichlet distribution is choose as the prior distribution $\theta^B_{x_i|Pa_i}$ for using the Bayes method. The probability density of the Dirichlet distribution for variable $\theta = (\theta_1, \cdots, \theta_n)$ with parameter $\alpha = (\alpha_1, \cdots, \alpha_n)$ is defined by

$$\mathrm{Dir}(\theta \mid \alpha) = \begin{cases} \dfrac{\Gamma(\alpha)}{\prod\limits_{i=1}^{n} \Gamma(\alpha_i)} \prod\limits_{i=1}^{n} \theta_i^{\alpha_i - 1} & \theta_1, \cdots, \theta_n \geq 0, \sum\limits_{i=1}^{n} \theta_i = 1 \\ 0 & \text{others} \end{cases}$$

1296   Z. Lu et al.

where $\theta_1, \cdots, \theta_n \geq 0$, $\sum_{i=1}^{n} \theta_i = 1$, and $\alpha_1, \cdots, \alpha_n > 0$. The parameter α_i is interpreted as prior observation count for events governed by θ_i.

Let $\alpha_0 = \sum_{i=1}^{n} \alpha_i$. The mean value and variance of the distribution for θ_i can be calculated by [2]

$$\mathrm{E}\,\theta_i = \frac{\alpha_i}{\alpha_0}, \text{ and } \mathrm{Var}\,(\theta_i) = \frac{\alpha_i(\alpha_0 - \alpha_i)}{\alpha_0^2(\alpha_0 + 1)}.$$

After the prior distributions are determined, the Bayes method requires to calculate the posterior distributions of $\theta^B_{x_i|Pa_i}$ and then complete the Bayes estimations of θ_i. This study assumes that the state of each node can be one of the five values: 1 (very low), 2 (low), 3 (medium), 4 (high), and 5 (very high). Through running the approach, the CPDs of all nodes are obtained.

3.3 Inference

The Bayesian network has been now created with both its structure and all conditional probabilities defined. It can be thus used to inference the relationships among these factors identified in Fig. 3. The inference process can be handled by fixing the states of observed variables, and then propagating the beliefs around the network until all the beliefs (in the form of conditional probabilities) are consistent. Finally, the desired probability distributions can be read directly from the network. There are a number of algorithms for conducting inference in Bayesian networks, which make different tradeoffs between speed, complexity, generality, and accuracy. Junction-tree algorithm [5] is suitable for middle and small size of samples. The Junction-tree algorithm computes the joint distribution for each maximal clique in a decomposable graph. It contains three steps: construction, initialization, and message passing or propagation. The construction step is to convert a Bayesian network to a junction tree. The junction tree is then initialized so that to provide a localized representation for the overall distributions. After the initialization, the junction tree can receive evidences, which consists of asserting variables to specific states.

4 Result Analysis

Over all inference results obtained through running the Junction-tree algorithm, a set of significant results about the relationships among customer online shopping behaviours and their requirements are obtained. Just one result is particularly discussed in the paper as the limitation of number of pages. Assuming A1=4 (high), we can get the probabilities of the other nodes under the evidence. The result is shown in Table 1. The probability of a high A2 has increased from 0.3701 to 0.4276, which suggests that A1 and A2 are correlated to some extent, so that a high A1 tends to "cause" a high A2. This result means that when a customer likes to experiment with new information technology, he/she will like to purchase goods/services online.

Table 1. Probabilities of the nodes when A1=4 (high)

state Node	1	2	3	4	5
A_1	0	0	0	1	0
A_2	0.0211	0.0618	0.4683	0.4276	0.0211
A_3	0.0427	0.1810	0.2706	0.3917	0.1140
S_1	0.0023	0.1632	0.4046	0.3471	0.0828
S_2	0.0138	0.1172	0.4391	0.3701	0.0598
S_3	0.0138	0.0598	0.2667	0.4736	0.1862
S_4	0.0145	0.0528	0.2040	0.4591	0.2696
S_5	0.0021	0.0667	0.2503	0.5371	0.1438

5 Conclusions and Further Study

By applying Bayesian network technique this study analyses the relationships among customer online shopping behaviours and their requirements for the features of e-service websites. As a further study, we will address on the verification of the relationships between customer online shopping behaviour, customer requirement for websites' features and the quality of e-services provided by websites.

References

1. Cooper, G. F., Herskovits, E.: A Bayesian method for the induction of probabilistic networks from data. Machine Learning, 9 (1992) 309-347.
2. Gelman, A., Carlin J., Stern, H.., Rubin, D.: Bayesian Data Analysis. Chapman & Hall/CRC, Boca Raton. (1995)
3. Hahn J., Kauffman R. J.: Evaluating selling web site performance from a business value perspective, Proceedings of international conference on e-Business May 23-26, 2002, Beijing, China, 435-443.
4. Heckerman, D.: A tutorial on learning Bayesian networks. Technical Report MSRTR-95-06, Microsoft Research. (1996)
5. Lauritzen S., Spiegelhalter D.: Local Computations with Probabilities on Graphical Structures and their Application to Expert Systems, J. R. Statis. Soc. B, 50 (1988) 157-224.
6. Lin, C.: A critical appraisal of customer satisfaction and e-commerce, Managerial Auditing Journal 18 (2003) 202-212.
7. Lu, J., Tang S., McCullough, G.: An assessment for internet-based electronic commerce development in businesses of New Zealand, Electronic Markets: International Journal of Electronic Commerce and Business Media 11 (2001) 107-115.
8. Wade R. M., Nevo S.: Development and Validation of a Perceptual Instrument to Measure E-Commerce Performance, International Journal of Electronic Commerce 10 (2005) 123.

Author Index

Printing: Mercedes-Druck, Berlin
Binding: Stein+Lehmann, Berlin

Lecture Notes in Artificial Intelligence (LNAI)

Vol. 4304: A. Sattar, B.-H. Kang (Eds.), AI 2006: Advances in Artificial Intelligence. XXVII, 1303 pages. 2006.

Vol. 4293: A. Gelbukh, C.A. Reyes-Garcia (Eds.), MICAI 2006: Advances in Artificial Intelligence. XXVIII, 1232 pages. 2006.

Vol. 4285: Y. Matsumoto, R. Sproat, K.-F. Wong, M. Zhang (Eds.), Computer Processing of Oriental Languages. XXXX, 55000 pages. 2006.

Vol. 4274: Q. Huo, B. Ma, E.-S. Chng, H. Li (Eds.), Chinese Spoken Language Processing. XXXX, 8000 pages. 2006.

Vol. 4265: L. Todorovski, N. Lavrač, K.P. Jantke (Eds.), Discovery Science. XIV, 384 pages. 2006.

Vol. 4264: J.L. Balcázar, P.M. Long, F. Stephan (Eds.), Algorithmic Learning Theory. XIII, 393 pages. 2006.

Vol. 4259: S. Greco, Y. Hata, S. Hirano, M. Inuiguchi, S. Miyamoto, H.S. Nguyen, R. Słowiński (Eds.), Rough Sets and Current Trends in Computing. XXII, 951 pages. 2006.

Vol. 4253: B. Gabrys, R.J. Howlett, L.C. Jain (Eds.), Knowledge-Based Intelligent Information and Engineering Systems, Part III. XXXII, 1301 pages. 2006.

Vol. 4252: B. Gabrys, R.J. Howlett, L.C. Jain (Eds.), Knowledge-Based Intelligent Information and Engineering Systems, Part II. XXXIII, 1335 pages. 2006.

Vol. 4251: B. Gabrys, R.J. Howlett, L.C. Jain (Eds.), Knowledge-Based Intelligent Information and Engineering Systems, Part I. LXVI, 1297 pages. 2006.

Vol. 4248: S. Staab, V. Svátek (Eds.), Managing Knowledge in a World of Networks. XIV, 400 pages. 2006.

Vol. 4246: M. Hermann, A. Voronkov (Eds.), Logic for Programming, Artificial Intelligence, and Reasoning. XIII, 588 pages. 2006.

Vol. 4223: L. Wang, L. Jiao, G. Shi, X. Li, J. Liu (Eds.), Fuzzy Systems and Knowledge Discovery. XXVIII, 1335 pages. 2006.

Vol. 4213: J. Fürnkranz, T. Scheffer, M. Spiliopoulou (Eds.), Knowledge Discovery in Databases: PKDD 2006. XXII, 660 pages. 2006.

Vol. 4212: J. Fürnkranz, T. Scheffer, M. Spiliopoulou (Eds.), Machine Learning: ECML 2006. XXIII, 851 pages. 2006.

Vol. 4211: P. Vogt, Y. Sugita, E. Tuci, C.L. Nehaniv (Eds.), Symbol Grounding and Beyond. VIII, 237 pages. 2006.

Vol. 4203: F. Esposito, Z.W. Raś, D. Malerba, G. Semeraro (Eds.), Foundations of Intelligent Systems. XVIII, 767 pages. 2006.

Vol. 4201: Y. Sakakibara, S. Kobayashi, K. Sato, T. Nishino, E. Tomita (Eds.), Grammatical Inference: Algorithms and Applications. XII, 359 pages. 2006.

Vol. 4200: I.F.C. Smith (Ed.), Intelligent Computing in Engineering and Architecture. XIII, 692 pages. 2006.

Vol. 4198: O. Nasraoui, O. Zaïane, M. Spiliopoulou, B. Mobasher, B. Masand, P.S. Yu (Eds.), Advances in Web Mining and Web Usage Analysis. IX, 177 pages. 2006.

Vol. 4196: K. Fischer, I.J. Timm, E. André, N. Zhong (Eds.), Multiagent System Technologies. X, 185 pages. 2006.

Vol. 4188: P. Sojka, I. Kopeček, K. Pala (Eds.), Text, Speech and Dialogue. XV, 721 pages. 2006.

Vol. 4183: J. Euzenat, J. Domingue (Eds.), Artificial Intelligence: Methodology, Systems, and Applications. XIII, 291 pages. 2006.

Vol. 4180: M. Kohlhase, OMDoc – An Open Markup Format for Mathematical Documents [version 1.2]. XIX, 428 pages. 2006.

Vol. 4177: R. Marín, E. Onaindía, A. Bugarín, J. Santos (Eds.), Current Topics in Artificial Intelligence. XV, 482 pages. 2006.

Vol. 4160: M. Fisher, W. van der Hoek, B. Konev, A. Lisitsa (Eds.), Logics in Artificial Intelligence. XII, 516 pages. 2006.

Vol. 4155: O. Stock, M. Schaerf (Eds.), Reasoning, Action and Interaction in AI Theories and Systems. XVIII, 343 pages. 2006.

Vol. 4149: M. Klusch, M. Rovatsos, T.R. Payne (Eds.), Cooperative Information Agents X. XII, 477 pages. 2006.

Vol. 4140: J.S. Sichman, H. Coelho, S.O. Rezende (Eds.), Advances in Artificial Intelligence - IBERAMIA-SBIA 2006. XXIII, 635 pages. 2006.

Vol. 4139: T. Salakoski, F. Ginter, S. Pyysalo, T. Pahikkala (Eds.), Advances in Natural Language Processing. XVI, 771 pages. 2006.

Vol. 4133: J. Gratch, M. Young, R. Aylett, D. Ballin, P. Olivier (Eds.), Intelligent Virtual Agents. XIV, 472 pages. 2006.

Vol. 4130: U. Furbach, N. Shankar (Eds.), Automated Reasoning. XV, 680 pages. 2006.

Vol. 4120: J. Calmet, T. Ida, D. Wang (Eds.), Artificial Intelligence and Symbolic Computation. XIII, 269 pages. 2006.

Vol. 4114: D.-S. Huang, K. Li, G.W. Irwin (Eds.), Computational Intelligence, Part II. XXVII, 1337 pages. 2006.

Vol. 4108: J.M. Borwein, W.M. Farmer (Eds.), Mathematical Knowledge Management. VIII, 295 pages. 2006.

Vol. 4106: T.R. Roth-Berghofer, M.H. Göker, H.A. Güvenir (Eds.), Advances in Case-Based Reasoning. XIV, 566 pages. 2006.

Vol. 4099: Q. Yang, G. Webb (Eds.), PRICAI 2006: Trends in Artificial Intelligence. XXVIII, 1263 pages. 2006.

Vol. 4095: S. Nolfi, G. Baldassarre, R. Calabretta, J.C.T. Hallam, D. Marocco, J.-A. Meyer, O. Miglino, D. Parisi (Eds.), From Animals to Animats 9. XV, 869 pages. 2006.

Vol. 4093: X. Li, O.R. Zaïane, Z. Li (Eds.), Advanced Data Mining and Applications. XXI, 1110 pages. 2006.

Vol. 4092: J. Lang, F. Lin, J. Wang (Eds.), Knowledge Science, Engineering and Management. XV, 664 pages. 2006.

Vol. 4088: Z.-Z. Shi, R. Sadananda (Eds.), Agent Computing and Multi-Agent Systems. XVII, 827 pages. 2006.

Vol. 4087: F. Schwenker, S. Marinai (Eds.), Artificial Neural Networks in Pattern Recognition. IX, 299 pages. 2006.

Vol. 4068: H. Schärfe, P. Hitzler, P. Øhrstrøm (Eds.), Conceptual Structures: Inspiration and Application. XI, 455 pages. 2006.

Vol. 4065: P. Perner (Ed.), Advances in Data Mining. XI, 592 pages. 2006.

Vol. 4062: G. Wang, J.F. Peters, A. Skowron, Y. Yao (Eds.), Rough Sets and Knowledge Technology. XX, 810 pages. 2006.

Vol. 4049: S. Parsons, N. Maudet, P. Moraitis, I. Rahwan (Eds.), Argumentation in Multi-Agent Systems. XIV, 313 pages. 2006.

Vol. 4048: L. Goble, J.-J.C.. Meyer (Eds.), Deontic Logic and Artificial Normative Systems. X, 273 pages. 2006.

Vol. 4045: D. Barker-Plummer, R. Cox, N. Swoboda (Eds.), Diagrammatic Representation and Inference. XII, 301 pages. 2006.

Vol. 4031: M. Ali, R. Dapoigny (Eds.), Advances in Applied Artificial Intelligence. XXIII, 1353 pages. 2006.

Vol. 4029: L. Rutkowski, R. Tadeusiewicz, L.A. Zadeh, J.M. Zurada (Eds.), Artificial Intelligence and Soft Computing – ICAISC 2006. XXI, 1235 pages. 2006.

Vol. 4027: H.L. Larsen, G. Pasi, D. Ortiz-Arroyo, T. Andreasen, H. Christiansen (Eds.), Flexible Query Answering Systems. XVIII, 714 pages. 2006.

Vol. 4021: E. André, L. Dybkjær, W. Minker, H. Neumann, M. Weber (Eds.), Perception and Interactive Technologies. XI, 217 pages. 2006.

Vol. 4020: A. Bredenfeld, A. Jacoff, I. Noda, Y. Takahashi (Eds.), RoboCup 2005: Robot Soccer World Cup IX. XVII, 727 pages. 2006.

Vol. 4013: L. Lamontagne, M. Marchand (Eds.), Advances in Artificial Intelligence. XIII, 564 pages. 2006.

Vol. 4012: T. Washio, A. Sakurai, K. Nakajima, H. Takeda, S. Tojo, M. Yokoo (Eds.), New Frontiers in Artificial Intelligence. XIII, 484 pages. 2006.

Vol. 4008: J.C. Augusto, C.D. Nugent (Eds.), Designing Smart Homes. XI, 183 pages. 2006.

Vol. 4005: G. Lugosi, H.U. Simon (Eds.), Learning Theory. XI, 656 pages. 2006.

Vol. 3978: B. Hnich, M. Carlsson, F. Fages, F. Rossi (Eds.), Recent Advances in Constraints. VIII, 179 pages. 2006.

Vol. 3963: O. Dikenelli, M.-P. Gleizes, A. Ricci (Eds.), Engineering Societies in the Agents World VI. XII, 303 pages. 2006.

Vol. 3960: R. Vieira, P. Quaresma, M.d.G.V. Nunes, N.J. Mamede, C. Oliveira, M.C. Dias (Eds.), Computational Processing of the Portuguese Language. XII, 274 pages. 2006.

Vol. 3955: G. Antoniou, G. Potamias, C. Spyropoulos, D. Plexousakis (Eds.), Advances in Artificial Intelligence. XVII, 611 pages. 2006.

Vol. 3949: F.A. Savacı (Ed.), Artificial Intelligence and Neural Networks. IX, 227 pages. 2006.

Vol. 3946: T.R. Roth-Berghofer, S. Schulz, D.B. Leake (Eds.), Modeling and Retrieval of Context. XI, 149 pages. 2006.

Vol. 3944: J. Quiñonero-Candela, I. Dagan, B. Magnini, F. d'Alché-Buc (Eds.), Machine Learning Challenges. XIII, 462 pages. 2006.

Vol. 3937: H. La Poutré, N.M. Sadeh, S. Janson (Eds.), Agent-Mediated Electronic Commerce. X, 227 pages. 2006.

Vol. 3932: B. Mobasher, O. Nasraoui, B. Liu, B. Masand (Eds.), Advances in Web Mining and Web Usage Analysis. X, 189 pages. 2006.

Vol. 3930: D.S. Yeung, Z.-Q. Liu, X.-Z. Wang, H. Yan (Eds.), Advances in Machine Learning and Cybernetics. XXI, 1110 pages. 2006.

Vol. 3918: W.-K. Ng, M. Kitsuregawa, J. Li, K. Chang (Eds.), Advances in Knowledge Discovery and Data Mining. XXIV, 879 pages. 2006.

Vol. 3913: O. Boissier, J. Padget, V. Dignum, G. Lindemann, E. Matson, S. Ossowski, J.S. Sichman, J. Vázquez-Salceda (Eds.), Coordination, Organizations, Institutions, and Norms in Multi-Agent Systems. XII, 259 pages. 2006.

Vol. 3910: S.A. Brueckner, G.D.M. Serugendo, D. Hales, F. Zambonelli (Eds.), Engineering Self-Organising Systems. XII, 245 pages. 2006.

Vol. 3904: M. Baldoni, U. Endriss, A. Omicini, P. Torroni (Eds.), Declarative Agent Languages and Technologies III. XII, 245 pages. 2006.

Vol. 3900: F. Toni, P. Torroni (Eds.), Computational Logic in Multi-Agent Systems. XVII, 427 pages. 2006.

Vol. 3899: S. Frintrop, VOCUS: A Visual Attention System for Object Detection and Goal-Directed Search. XIV, 216 pages. 2006.

Vol. 3898: K. Tuyls, P.J. 't Hoen, K. Verbeeck, S. Sen (Eds.), Learning and Adaption in Multi-Agent Systems. X, 217 pages. 2006.

Vol. 3891: J.S. Sichman, L. Antunes (Eds.), Multi-Agent-Based Simulation VI. X, 191 pages. 2006.